Edible Medicinal and Non-Medicinal Plants

T.K. Lim

Edible Medicinal and Non-Medicinal Plants

Volume 4, Fruits

ISBN 978-94-007-4052-5   ISBN 978-94-007-4053-2 (eBook)
DOI 10.1007/978-94-007-4053-2
Springer Dordrecht Heidelberg London New York

Library of Congress Control Number: 2011932982

© Springer Science+Business Media B.V. 2012
This work is subject to copyright. All rights are reserved by the Publisher, whether the whole or part of the material is concerned, specifically the rights of translation, reprinting, reuse of illustrations, recitation, broadcasting, reproduction on microfilms or in any other physical way, and transmission or information storage and retrieval, electronic adaptation, computer software, or by similar or dissimilar methodology now known or hereafter developed. Exempted from this legal reservation are brief excerpts in connection with reviews or scholarly analysis or material supplied specifically for the purpose of being entered and executed on a computer system, for exclusive use by the purchaser of the work. Duplication of this publication or parts thereof is permitted only under the provisions of the Copyright Law of the Publisher's location, in its current version, and permission for use must always be obtained from Springer. Permissions for use may be obtained through RightsLink at the Copyright Clearance Center. Violations are liable to prosecution under the respective Copyright Law.
The use of general descriptive names, registered names, trademarks, service marks, etc. in this publication does not imply, even in the absence of a specific statement, that such names are exempt from the relevant protective laws and regulations and therefore free for general use.
While the advice and information in this book are believed to be true and accurate at the date of publication, neither the authors nor the editors nor the publisher can accept any legal responsibility for any errors or omissions that may be made. The publisher makes no warranty, express or implied, with respect to the material contained herein.

Printed on acid-free paper

Springer is part of Springer Science+Business Media (www.springer.com)

Acknowledgment

Special thanks to Malcolm Smith, Agri-Science Queensland, Department of Employment, Economic Development and Innovation, and Prof. Andrew Beattie, University of Western Sydney for some photo credits.

Disclaimer

The author and publisher of this work have checked with sources believed to be reliable in their efforts to confirm the accuracy and completeness of the information presented herein and that the information is in accordance with the standard practices accepted at the time of publication. However, neither the author nor publishers warrant that information is in every aspect accurate and complete and they are not responsible for errors or omissions or for consequences from the application of the information in this work. This book is a work of reference and is not intended to supply nutritive or medical advice to any individual. The information contained in the notes on edibility, uses, nutritive values, medicinal attributes and medicinal uses and suchlike included here are recorded information and do not constitute recommendations. No responsibility will be taken for readers' own actions.

Contents

Introduction 1

Fagaceae

Castanea sativa 6
Quercus infectoria 16

Grossulariaceae

Ribes nigrum 27
Ribes rubrum 43
Ribes uva-crispa 51
Ribes x nidigrolaria 56

Hypoxidaceae

Molineria latifolia 59

Myrsinaceae

Ardisia crenata 65
Ardisia elliptica 72

Olacaceae

Scorodocarpus borneensis 77

Oleaceae

Olea europaea 82

Orchidaceae

Vanilla planifolia 106

Oxalidaceae

Sarcotheca diversifolia 115

Pandanaceae

Pandanus conoideus 117
Pandanus dubius 124
Pandanus julianettii 128
Pandanus leram 131
Pandanus spiralis 134
Pandanus tectorius 136

Passifloraceae

Passiflora edulis 147
Passiflora foetida 166
Passiflora ligularis 174
Passiflora miniata 178
Passiflora quadrangularis 181

Pedaliaceae

Sesamum indicum 187

Phyllanthaceae

Antidesma bunius 220
Baccaurea angulata 225
Baccaurea dulcis 228
Baccaurea edulis 230
Baccaurea lanceolata 232
Baccaurea macrocarpa 236
Baccaurea motleyana 239
Baccaurea racemosa 243
Baccaurea ramiflora 248
Phyllanthus acidus 252
Phyllanthus emblica 258

Pinaceae

Pinus koraiensis 297

Pinus pinea 304

Piperaceae

Piper cubeba 311

Piper nigrum 322

Piper retrofractum 351

Rosaceae

Amelanchier alnifolia 358

Chaenomeles speciosa 364

Cydonia oblonga 371

Eriobotrya japonica 381

Fragaria x ananassa 395

Malus spectabilis 410

Malus x domestica 413

Mespilus germanica 437

Prunus armeniaca 442

Prunus avium 451

Prunus domestica 463

Prunus domestica subsp. insititia 476

Prunus dulcis 480

Prunus persica var. nucipersica 492

Prunus persica var. persica 498

Prunus salicina 509

Pseudocydonia sinensis 515

Pyrus bretschneideri 523

Pyrus communis 527

Pyrus pyrifolia 535

Pyrus ussuriensis 540

Rubus fruticosus aggr. 544

Rubus idaeus 555

Rubus occidentalis	570
Rubus ursinus x idaeus 'Boysenberry'	581
Rubus x loganobaccus	587
Sorbus domestica	590

Rutaceae

Aegle marmelos	594
Citrus 'Meyer'	619
Citrus amblycarpa	623
Citrus australasica	625
Citrus australis	629
Citrus garrawayi	631
Citrus hystrix	634
Citrus inodora	644
Citrus japonica 'Marumi'	647
Citrus japonica 'Meiwa'	651
Citrus japonica 'Nagami'	654
Citrus japonica 'Polyandra'	659
Citrus latifolia	662
Citrus maxima	667
Citrus medica	682
Citrus medica var. sarcodactylis	690
Citrus reticulata	695
Citrus reticulata Satsuma Group	721
Citrus reticulata 'Shiranui'	732
Citrus wintersii	739
Citrus x aurantiifolia	742
Citrus x aurantium Grapefruit Group	755
Citrus x aurantium Sour Orange Group	786
Citrus x aurantium Sweet Orange Group	806
Citrus x aurantium Tangelo Group	832
Citrus x aurantium Tangor Group	837
Citrus x floridana	843

Citrus x taitensis	846
Citrus x limon	849
Citrus x microcarpa	865
Clausena lansium	871
Limonia acidissima	884
Merrillia caloxylon	890
Poncirus trifoliata	893
Triphasia trifolia	900
Zanthoxylum simulans	904
Medical Glossary	912
Scientific Glossary	975
Common Name Index	999
Scientific Name Index	1009

Introduction

This book continues as volume 4 of a multi-compendium on *Edible Medicinal and Non-Medicinal Plants*. It covers edible fruits/seeds used fresh, cooked or processed into other by-products, or as vegetables, cereals, spices, stimulant, edible oils and beverages. It covers selected species from the following families: Fagaceae, Grossulariaceae, Hypoxidaxeae, Myrsinaceae Olacaceae, Oleaceae, Orchidaceae, Oxalidaceae, Pandanaceae, Passifloraceae, Pedaliaceae, Phyllanthaceae, Pinaceae, Piperaceae, Rosaceae and Rutaceae. However, not all the edible species in these families are included. The edible species dealt with in this work include to a larger extent lesser-known, wild and underutilized crops and also common and widely grown crops.

As in the preceding three volumes, topics covered include: taxonomy (botanical name and synonyms); common English and vernacular names; origin and distribution; agro-ecological requirements; edible plant part and uses; plant botany; nutritive and medicinal/ pharmacological properties with up-to-date research findings, traditional medicinal uses; other non-edible uses; and selected/cited references for further reading.

Fagaceae or more commonly known as the beech family comprises about 900 species of evergreen and deciduous trees and shrubs. It includes economically important species like the oak (*Quercus*), beech (*Fagus*) and the chestnut (*Castanea*) – all provide invaluable timber. The cork oak (*Quercus suber*) provides cork for bottles and other uses while *Castanea* also provides the delectable tasty and nutritive chestnut. Wood chips from the genus, *Fagus* are often used in flavoring beers. *Castanea sativa* and another lesser known species *Quercus infectoria* are covered in this volume. The latter has edible seeds that is used in food and nut galls used in herbal drinks and tea. Both species also possess medicinal properties.

Several edible species of *Ribes* of the family Grossulariaceae, namely gooseberry, black currant and red currant are covered in this volume. *Ribes* is a genus of about 150 species found in the temperate areas in the northern hemisphere. Besides being rich in nutrients, currants and gooseberries contain many phenolic compounds that include phenolic acids, flavonoids, anthocyanins and carotenoids with good antioxidant activity.

Hypoxidaceae is a family of flowering plants belonging to the Monocots, under the order Asparagales. Members are small to medium herbs, with grass-like leaves and an invisible stem, modified into a corm or a rhizome. Many are ornamental genera such as *Curculigo*, *Hypoxis*, and *Rhodohypoxis*. *Molineria latifolia* (*Curculigo latifolia*) produces an edible fruit which contains a high intensity sweetening protein called neoculin which is 430–2,070 times sweeter than sucrose on a weight basis (Yamashita et al. 1995; Kurihara 1992).

Myrsinaceae or Myrsine family comprises about 35 genera and about 1,000 species of trees, shrubs and lianas. The members occur in both temperate and tropical climates. Some economically important genera include *Ardisia* (medicine, oil, edible, wild vegetables), *Maesa* (edible, tea, dye), *Aegiceras* (tannin, fine fuel), *Embelia* (vermifuge, edible) and *Myrsine* (medicine, fine wood, tannin, fuel). Two *Ardisia* species, *Ardisia crenata* and *A. elliptica* are covered in this volume. Both species produce edible fruit and leaves. Both plants contain bioactive phytochemicals which impart numerous pharmacological properties that include antimicrobial, anticancer, antiplatelet and antimalarial properties.

Members of the family Olacaceae comprising 180–250 species from 23 to 27 genera, are found in tropical and warm-temperate regions worldwide. They comprised scandent shrubs, trees, or lianas, sometimes hemiparasitic. Edible fruit species of this family include *Ximenia* (false sandalwood, hog-plum) and *Scorodocarpus borneensis* (wood garlic, forest onion). The latter treated herein has fruit (nut), bark and leaves which are edible and medicinal. Many parts of wood garlic have medicinal properties, the nuts contain alkaloids and sesquiterpenes (Wiart et al. 2001), the leaves have megastigmanes and flavonoids (Abe and Yamauchi 1993).

The Oleaceae or olive family, comprises 30 genera and about 600 species of mesophytic shrubs, trees and occasionally vines. Many species have economic significance such as the olive (*Olea europaea*) valued for its fruit and oil extracted from it, the ashes (*Fraxinus*) important for their timber, and the ornamental plants forsythia, lilacs, jasmines, osmanthuses, privets and fringetrees. Olive leaves, fruit, pomace and oil have a host of pharmacological properties that include anticancer, antiathrogenic, cardioprotective, antiinflammatory, antihyperglycemic, antihyperlipidemic, antihypertensive, antiplatelet, antinociceptive antimicrobial and wound healing activities. Olive wood is hard and durable and olive branch is an ancient symbol for peace.

Orchidaceae, the orchid family, is a morphologically diverse and widespread family of monocots in the order Asparagales. This family is regarded to be the largest family of flowering plants having between 22,000 and 26,000 currently accepted species, found in 880 genera. The family includes *Orchis* (type genus) and many commonly cultivated orchids, such as *Phalaenopsis*, *Dendrobium*, *Epidendrum* and *Cattleya* and *Vanilla* (*Vanilla planifolia* which is covered in this volume). The dried seed pods of *Vanilla* are commercially important as flavouring in confectionery, dairy products, for perfume manufacture and aromatherapy.

The Oxalidaceae, or wood sorrel family, is a small family of eight genera of herbaceous plants, shrubs and small trees, with vast majority of the 900 species in the genus *Oxalis* (wood sorrels). The family is represented in this volume by an under-utilised, lesser known tropical fruit *Sarcotheca diversifolia*. The genus *Averrhoa* of which starfruit is a member, is often included in this family, but some botanists placed it in a separate family Averrhoaceae. Several currently recognised *Averrhoa* species with edible fruits have been dealt with in Volume 1 of the multicompendium under Averrhoaceae.

Pandanaceae is a large family of flowering plants found in the tropical and subtropical regions of the Old World from West Africa through to the Pacific. Pandanaceae is a highly variable genus complex of trees, climbing or scrambling shrubs that are adapts well from sea level in salted beaches to montane cloud- forest, and riverine forest habitat. One distinctive feature is that the stems have aerial prop roots to provide support and display sympodial branching. Seven species producing edible drupes are covered in this volume.

Passifloraceae, the passion fruit family, comprises about 530 species of flowering plant in about 18 genera of trees, shrubs, lianas and climbing plants, mostly found in tropical America and Asia. Seven species with edible fruits and medicinal properties are covered in this volume.

Pedaliaceae, the sesame family, is a small family of 13 genera and 70 species. Its native distribution is exclusively the Old World, in tropical and dry habitats, and its best-known member is *Sesamum indicum* (sesame). Sesame seeds are rich in lignans like sesamin, sesamolin and other bioactive phytochemicals responsible for its culinary and pharmacological attributes.

Phyllanthaceae comprises about 2,000 species grouped into 54–60 genera of mostly trees, shrubs and herbs. A few are climbers or succulents and one *Phyllanthus fluitans,* an aquatic plant. Several genera produce edible fruits such as *Phyllanthus, Upaca, Antidesma* and *Baccaurea.* Some species of the latter two genera are covered in this volume. *Baccaurea* was previously classified under the family Euphorbiaceae. Unlike many of the Euphorbiaceae, no member of Phyllanthaceae has latex and only a very few produce a resinous exudate.

Pinaceae or pine family comprises shrubs or trees and include many of the familiar conifers of commercial importance such as cedars, firs, hemlocks, larches, pines and spruces. The family is the largest extant conifer family with between 220 and 250 species in 11 genera found mostly in temperate regions but also in sub-arctic to tropical areas. Two *Pinus* species providing edible pine nuts are discussed in this volume.

Piperaceae, better known as the pepper family, is a large family of flowering herbs, shrubs and small trees. It has been reported to have 3,610 currently accepted species in five genera distributed pantropically. The vast majority of peppers can be found within the two main genera: *Piper* (2,000 species) and *Peperomia* (1,600 species). Three *Piper* species are covered including *Piper nigrum* which provide the important spice, peppercorns. *Piper nigrum* (black pepper) is used not only in human dietaries but also for a variety of other purposes such as medicinal, as a preservative, and in perfumery.

Rosaceae or the rose family is a medium-sized family of about 2,830 species in 95 genera of flowering herbs, shrubs, climbers and trees. Among the largest genera are *Prunus* (430) *Alchemilla* (270), *Sorbus* (260), *Crataegus* (260), *Cotoneaster* (260), and *Rubus*. *Rubus* consists of about 750 species (Daubeny 1996) which have been separated into blackberry (subgenus *Rubus* including *R. armeniacus, R. laciniatus* and *Rubus* hybrids) and raspberry (subgenus *idaeobatus* including, *Rubus idaeus* and *Rubus occidentalis*) types according to the abscission of the fruit; in raspberry this comes off a woody or fleshy receptacle which remains on the plant, and in blackberry the fruit separates from the plant with the soft, edible receptacle included (Clark et al. 2007). Also included in the genus are the hybridberries (including boysenberries, loganberries and other hybrid types). Most raspberry species are diploid ($2x=14$) as are a few blackberries, but the bulk of blackberry species and all hybridberries are polyploids, ranging from $3x=21$ to $18x=126$.

Rosaceae provides many economically important products which include edible fruits *Malus* spp. (apples), *Prunus* spp. (apricot, cherry, nectarine, peach, plums, prune, damson, sloe), *Cydonia* (quince), *Pyrus* (pears), *Eriobotrya* (loquat), *Rubus* (blackberry, boysenberry, loganberry, black and red raspberry), *Fragaria* (strawberry), *Mespilus* (medlar), *Amelanchier* spp. (serviceberry, Juneberry); *Prunus dulcis* (almond nuts); many ornamental trees and shrubs or hedge-plants, e.g. *Spiraea, Photinia, Cotoneaster, Kerria, Filipendula* (meadowsweets) *Pyracantha* (firethorns), *Crataegus* (hawthorns), *Rhodotypos, Rosa* (roses), *Sorbus* (mountain ash, rowan) and *Potentilla (*cinquefoils). Roses can be herbs, climbers, shrubs or small trees. Many of the edible fruit species in the genera *Malus, Pyrus, Prunus, Rubus, Cydonia, Mespilus, Eriobotrya, Fragaria, Chaenomeles* and *Sorbus* also have important nutrients and bioactive secondary phytochemicals with a diverse range of pharmacological activities.

Rutaceae, the rue or citrus family is a large, morphologically diverse, cosmopolitan family of flowering plants of 160 genera and 1,900 species with great economic importance in warm temperate and sub-tropical climate areas. The most economically important genus is *Citrus*. Several taxonomical studies employing molecular DNA techniques have supported a wider polyphyletic classification of Citreae and *Citrus*. Studies by Araújo et al. (2003) and Bayer et al. (2009) supported the broader definition of *Citrus* to include *Clymenia, Eremocitrus, Fortunella, Microcitrus, Oxanthera* and *Poncirus*. Bayer et al. (2009) also supported the monophyly of the subfamily Aurantioideae and the transfer of *Murraya* sensu stricto and *Merrillia* from Clauseneae to Citreae. Likewise, data from Guerra et al. (2000) supported segregation of *Bergera* from *Murraya*, and movement of *Murraya* sensu stricto and *Merrillia* from

Clauseneae to Citreae. The results of studies by Groppo et al. (2008) supported monophyly of Spathelioideae and Aurantioideae, but not of other subfamilies and tribes. Thus, the genus *Citrus* boast of many economic important edible fruits such as oranges, mandarins, citrons, grapefruits, pumello, lemons, limes, kumquats and a diverse host of *Citrus* hybrids. More taxonomical work is required to classify the many hybrids. Other edible non-Citrus fruits include white sapote (*Casimiroa edulis*), orangeberry (*Glycosmis pentaphylla*), clymenia (*Clymenia polyandra*), limeberry (*Triphasia trifolia*), elephant apple (*Limonia acidissima*), wampee (*Clausena lansium*) and the bael (*Aegle marmelos*). *Bergera kongii* although has edible fruit is cultivated mainly for its aromatic spicy leaves (curry leaf) and is treated in a later volume. Other important genera include *Ruta* (treated in later volume), *Zanthoxylum* and *Boronia,* a large Australian genus, some members of which are plants with highly fragrant flowers and are used in commercial ornamental, cutflower and oil production. *Zanthoxylum* is represented in this volume by *Z. simulans*, an important source of Szechuan pepper. Rutaceous species in general possess extraordinary array of secondary chemical metabolites, many have medicinal, antimicrobial, insecticidal, or herbicidal properties. *Citrus* species in particular are important sources of bioactive polyphenolic flavonoid compounds such as the flavones, flavonols and flavonones with many important pharmacological properties (antioxidant, anticancer, antiviral, antidiabetic, antilipidemic, antihypercholesterolemic, antiinflammatory, etc.).

Selected References

Abe F, Yamauchi T (1993) Megastigmanes and flavonoids from the leaves of *Scorodocarpus borneensis*. Phytochem 33(6):1499–1501

Araújo EF, de Queiroz LP, Machado MA (2003) What is *Citrus*? Taxonomic implications from a study of cpDNA evolution in tribe Citreae (Rutaceae subfamily Aurantioideae). Org Divers Evol 3:55–62

Bayer RJ, Mabberley DJ, Morton C, Miller CH, Sharma IK, Pfeil BE, Rich S, Hitchcock R, Sykes S (2009) A molecular phylogeny of the orange subfamily (Rutaceae: Aurantioideae) using nine cpDNA sequences. Am J Bot 96:668–685

Chen J, Pipoly JJ III (1996) Myrsinaceae R. Brown. In: Wu ZY, Raven PH (eds) Flora of China, vol 15, Myrsinaceae through Loganiaceae. Science Press/Missouri Botanical Garden Press, Beijing/St. Louis

Clark JR, Stafne ET, Hall HK, Finn CE (2007) Blackberry breeding and genetics. Plant Breed Rev 29:19–144

Cronquist A (1981) An integrated system of classification of flowering plants. Columbia University Press, New York

Daubeny HA (1996) Brambles. In: Janick J, Moore JN (eds) Fruit breeding, vol II, Vine and small fruit crops. Wiley, New York, pp 109–190

Elias TS (1971) The genera of Fagaceae in the southeastern United States. J Arnold Arbor 52:159–195

Groppo M, Pirani JR, Salantino MLF, Blanco SR, Kallunki JA (2008) Phylogeny of Rutaceae based on two non-coding regions from cpDNA. Am J Bot 95:985–1005

Gu C, Li CL, Lu LD, Jiang SY, Alexander C, Bartholomew B, Brach AR, Boufford DE, Ikeda H, Ohba H, Robertson KR, Spongberg SA (2003) Rosaceae A.L. Jussieu. In: Wu ZY, Raven PH, Hong DY (eds) Flora of China, vol 9, Pittosporaceae through Connaraceae. Science Press/Missouri Botanical Garden Press, Beijing/St Louis

Guerra M, dos Santos KG, Barros E, Silva AE, Ehrendorfer F (2000) Heterochromatin banding patterns in Rutaceae-Aurantioideae–a case of parallel chromosomal evolution. Am J Bot 87:735–747

Hoffmann P, Kathriarachchi HS, Wurdack KJ (2006) A phylogenetic classification of Phyllanthaceae. Kew Bull 61(1):37–53

Huxley AJ, Griffiths M, Levy M (eds) (1992) The new RHS dictionary of gardening (4 vols). Macmillan, New York

Jennings DL (1988) Raspberries and blackberries: their breeding, diseases and growth. Academic, London, 230 pp

Kurihara Y (1992) Characteristics of antisweet substances, sweet proteins, and sweetness-inducing proteins. Crit Rev Food Sci Nutr 32(3):231–252

Mabberley DJ (1997) A classification for edible citrus. Telopea 7(2):167–172

Mangion CP (2011) Piperaceae. In: Short PS, Cowie ID (eds) Flora of the Darwin region, vol 1. Department of Natural Resources, Environment, The Arts and Sport, Northern Territory Herbarium, Palmerston/Alice Springs, pp 1–3

Potter D, Eriksson T, Evans RC, Oh S, Smedmark JEE, Morgan DR, Kerr M, Robertson KR, Arsenault M, Dickson TA, Campbell CS (2007) Phylogeny and classification of Rosaceae. Plant Syst Evol 266(1–2):5–43

Qiu HX, Chiu HH, Kiu HX, Gilbert MG (2003) Olacaceae. In: Wu ZY, Raven PH, Hong DY (eds) Flora of China, vol 5, Ulmaceae through Basellaceae. Science Press/Missouri Botanical Garden Press, Beijing/St. Louis

Ravindran PN (2000) Black pepper, *Piper nigrum*. Harwood Acadiic, Amsterdam, 553 pp

Soepadmo E (1972) Fagaceae. In: Van Steenis CGGJ (ed) 1950+ Flora Malesiana. Series I. Spermatophyta, 11+ vols in parts. Djakarta and Leiden, vol 7, part 2, pp 265–403

Stevens PF (2001 onwards) Angiosperm phylogeny website. Version 9, June 2008 [and more or less continuously updated since]

Watson L, Dallwitz MJ (1992 onwards) The families of flowering plants: descriptions, illustrations, identification, and information retrieval. Version, 4 March 2011. http://delta-intkey.com

Wiart C, Martin MT, Awang K, Hue N, Serani L, Laprévote O, Païs M, Rhamani M (2001) Sesquiterpenes and alkaloids from *Scorodocarpus borneensis*. Phytochemistry 58(4):653–656

Yamashita H, Akabane T, Kurihara Y (1995) Activity and stability of a new sweet protein with taste-modifying action, curculin. Chem Senses 20(2):239–243

Castanea sativa

Scientific Name

Castanea sativa Mill.

Synonyms

Castanea castanea (L.) H. Karst. nom inval., *Castanea prolifera* (K. Koch) Hickel, *Castanea sativa* f. *discolor* Vuk., *Castanea sativa* var. *hamulata* A. Camus, *Castanea sativa* var. *microcarpa* Lavialle, *Castanea sativa* var. *prolifera* K. Koch, *Castanea sativa* var. *spicata* Husn., *Castanea sativa* var. *typica* Seemen, nom. inval., *Castanea vesca* Gaertn., *Castanea vulgaris* Lam., *Fagus castanea* L. (basionym), *Fagus castanea* var. *variegata* Weston, *Fagus procera* Salisb.

Family

Fagaceae

Common/English Names

Chestnut, Chestnut Tree, Edible Chestnut, European Chestnut, Italian Chestnut, Marron, Portuguese Chestnut, Spanish Chestnut, Sweet Chestnut.

Vernacular Names

Afrikaans: Kastaiing;
Brazil: Castanha Européia, Castanha Portuguesa (Portuguese);
Bulgarian: Kecteh;
Chinese: Ou Zhou Li, Wu Shu Li;
Croatian: Kestenjaste Boje;
Czech: Kaštanovník Jedlý, Kaštanovník Setý;
Danish: Ægte Kastanie, Ægte Kastanje;
Dutch: Europese Kastanje, Kastanje, Kastanjeboom, Tamme Kastanje, Tamme Kastanjeboom, Tame Kastanjeboom Soort;
Eastonian: Harilik Kastanipuu;
Finnish: Aito Kastanja, Jalokastanja;
French: Châtaigne, Châtaigner Commun, Châtaignier, Châtaignier Commun Marron, Marron Comestible;
German: Cheschtene, Cheste, Echte Kastanie, Edelkastanie, Edel-Kastanie, Edelkastanienbaum, Eßkastanie, Eßkastanienbaum, Ess-Kastanie, Essbare Kastanie, Keschte, Marone, Maroni;
Greek: Kastania;
Hungarian: Édes Gesztenye, Szelídgesztenye;
Icelandic: Kastaníuhneta;
Indonesia: Berangan:
Italian: Castagno, Castagno Comune, Castagno Domestico, Marone;
Japanese: Yooroppa Guri;
Latvian: Kastanis;

Lithuanian: Kaštonas;
Malaysia: Buah Berangan;
Norwegian: Edelkastanje, Ekte Kastanje, Kastanje;
Polish: Kasztan Jadalny;
Portuguese: Castanheiro-Comum, Castanheiro, Reboleiro;
Romanian: Castan, Castană;
Russian: Kashtan Nastoiashchii (Kaštan Nastojaščij), Kashtan Posevnoi (Kaštan Posevnoj);
Serbian: Kesten;
Slovaščina: Evropski, Pravi, Pravi Kostanj, Žlahtni Kostanj;
Slovencina: Gaštan Jedlý;
Spanish: Castaña, Castaño, Castaño Común, Castaño Regoldo, Regoldo;
Swedish: Äkta Kastanj, Kastanje;
Turkish: Kestane Ağacı;
Vietnamese: Cây Hạt Dẻ.

Origin/Distribution

A species of chestnut originally native to the Mediterranean in south-eastern Europe to Caucasus in Asia Minor. Wild or naturalized populations occur throughout southern Europe, northern Africa and southwestern Asia. It is cultivated in mild temperate regions in Europe and in the southern hemisphere and some subtropical regions.

Agroecology

Chestnut requires a mild cool climate Mediterranean or sub-temperate climate with good annual rainfall of 750–1,200 mm. The tree is frost sensitive. Sub-zero temperatures are injurious to the tree. The tree thrives in full sun on deep well-drained, fertile, sandy or loamy soils as it has a deep root system. It is highly tolerant of acidic soils, gravelly or stony soils but intolerant of calcareous soils, heavy clays and impermeable soils. The soil pH range is from 4 to 6.5 with an optimum from 5.5 to 6.5. It will tolerate partial shade under forest conditions. Adult trees are drought tolerant but adequate moisture is required for good growth and a good nut harvest.

Edible Plant Parts and Uses

Chestnuts can be eaten raw or dried but are usually eaten roasted, fried or cooked (boiled or steamed). Roasting, frying or cooking brings out the delicious, sweet chestnut flavour and floury texture. The dried or cooked nuts are used in confectionery, pastries, chocolates, puddings, desserts and cakes. They are used for flour, bread making, as a cereal substitute, coffee substitute, a thickener in soups and other cookery uses, as well as for fattening pig stock. A sugar can be extracted from it. The Italian and Corsican polenta (type of porridge) is made with sweet chestnut flour. A local variety of Corsican beer also uses chestnuts.

Botany

A medium to large, deciduous tree growing to 15–35 m high with a spreading crown and trunk diameter reaching 2 m and deeply fissured bark (Plate 1). Branchlets are tomentose. Leaves are elliptic to ovate-lanceolate, 14–28 cm × 5–9 cm, with 11–14 pairs of nerves more prominent on the under surface, serrated margin, acuminate tip, oblique base, coriaceous, glabrous, pale green and puberulous on the under surfaces (Plate 2). Inflorescences or catkins are unisexual (male) or androgynous with female flowers (cupules) at the base of an otherwise male inflorescence. Staminate catkins are pendulous, pubescent, creamy yellow, consists of male flowers in dense cymules. Male flowers are apetalous, with 4–6 (–9), scale-like, connate or distinct sepals, filiform filaments with dorsifixed or versatile anthers opening by longitudinal slits; and with or without a rudimentary pistil. Female inflorescences of 1–7 or more flowers subtended individually or collectively by a spiny protective cupule formed from numerous fused bracts, arranged individually or in small groups along an axis or at base of an androgynous inflorescence. Female flower

Plate 1 Tree habit with spreading crown

Plate 2 Elliptic leaves with serrated margins

Plate 3 Chestnut catkins

Plate 4 Spiny chestnut cupules

Plate 5 Harvested chestnuts

with 1–7 perianth, 1 pistil, inferior ovary with 3–6(–9) locules, style and carpels as many as locules; placentation axile; ovules anatropous, 2 per locule. Fruit is 1.3–2.5 cm in diameter, with each spiny cupule or burr (Plate 3) consisting of 1–4 nuts. Seed (nut) is brown to reddish-brown, usually solitary by abortion and is non-endospermic with a large embryo (Plate 5).

Nutritive/Medicinal Properties

Food value of raw peeled, European chestnuts per 100 g edible portion was reported as follows (USDA 2011): water 52.00 g, energy 196 kcal (820 kJ), protein 1.63 g, total lipid (fat) 1.25 g, ash 0.96 g, carbohydrate 44.17 g; minerals – calcium 19 mg, iron, 0.94 mg, magnesium 30 mg, phosphorus 38 mg, potassium 484 mg, sodium 2 mg, zinc 0.49 mg, copper 0.418 mg, manganese 0.336 mg, vitamins – vitamin C (total ascorbic acid) 40.2 mg, thiamine 0.144 mg, riboflavin 0.016 mg, niacin 1.102 mg, pantothenic acid 0.476 mg, vitamin b-6 0.352 mg, folate (total) 58 mcg, vitamin A 26 IU; lipids – fatty acids (total saturated) 0.235 g, 14:0 (myristic acid) 0.005 g, 16:0 (palmitic acid) 0.212 g, 18:0 (stearic acid) 0.012 g; fatty acids (total monounsaturated) 0.430 g, 16:1 undifferentiated (palmitoleic acid) 0.012 g, 18:1 undifferentiated (oleic acid) 0.413 g, 20:1 (gadoleic acid) 0.005 g; fatty acids (total polyunsaturated) 0.493 g, 18:2 undifferentiated (linoleic acid) 0.440 g, 18:3 undifferentiated (linolenic acid) 0.053 g; amino acids – tryptophan 0.018 g, threonine 0.058 g, isoleucine 0.064 g, leucine 0.096 g, lysine 0.096 g, methionine 0.038 g, cystine 0.052 g, phenylalanine 0.069 g, tyrosine 0.045 g, valine 0.091 g, arginine 0.116 g, histidine 0.045 g, alanine 0.109 g, aspartic acid 0.281 g, glutamic acid 0.210 g, glycine 0.084 g, proline 0.086 g and serine 0.081 g.

In an analysis conducted in Turkey (Ylldiz et al. 2009), the following nutrient data was reported for wild chestnut (*C. sativa*): moisture 54.84%, crude lipid 2.24%, crude protein, 8.93%, crude fibre, 3.92%, crude energy 4,046 kcal/100 g, ash 2.078%, dry matter 45.16%, total carbohydrate 11.21%, and ether soluble extract 4.44%. Zinc 5099.4 mg/kg, manganese 3031.9 mg/kg, sodium 1058.6 mg/kg and calcium 308.5 mg/kg were established as major minerals of chestnut fruit. Aluminium 21.52 mg/kg, boron 15.87 mg/kg, arsenic 7.82 mg/kg, bismuth 1.54 mg/kg, copper 0.12 mg/kg, chromium 5.72 mg/kg, strontium 15.22 mg/kg and titanium 768.72 mg/kg, iron 38.15 mg/kg were found in minor amounts.

Chestnut with a low fat content, completely free of cholesterol, a low sodium and high potassium content, moderate but high quality protein content, and rich in energy, vitamin C and amino acids especially lysine, tryptophan and sulphur containing amino acids provide a balanced and quality food (Bounous et al. 2000). Chestnuts also contained folate and vitamin Bs and A and the macro minerals. Chestnut was found to be characterized by the presence of seven organic acids: oxalic, cis-aconitic, citric, ascorbic, malic, quinic and fumaric acids (Ribeiro et al. 2007). Roasting, boiling and frying procedures led to significant reduction of total organic acids contents. Moisture was the major component in four Portuguese chestnut cultivars followed by carbohydrates, protein and fat, resulting in an energetic value lower than 195 kcal/100 g of fresh fruit (Barreira et al. 2009). Fatty acids (FA) profiling revealed a clear prevalence of C18:1 and C18:2, two FA very well-known due to their beneficial effects on human health, e.g., in the prevention of cardiovascular diseases. Triacylglycerols (TAG) profiling revealed that OLL, PLL, OOL and POL were the major compounds (O = oleic acid, P = palmitic acid, L = linoleic acid).

Chestnuts are mostly consumed as processed forms, and the different types of processing clearly affect the nutrient and non-nutrient composition of the fruits. Chestnut fruit industrial processing could be divided into four major stages: harvest, post-harvest storage (during 3 months at ±0°C), industrial peeling (by flame or fire – "brûlage" – at temperatures of 800–1,000°C) and freezing (tunnel with a CO2 flow at −65°C during 15–20 minutes) (De Vasconcelos et al. 2010b). Starch and hemicelluloses were the predominant polysaccharides but free sugars, mainly sucrose (saccharose), were also present. Vitamin C and vitamin E (predominantly δ-tocopherol) were present in the fruits with lower levels of carotenoids (lutein and lutein esters). The fibre and free sugars contents were found to increase during industrial processing, while the interaction of cultivar × processing stage had the greatest influence on the starch content. Various phenolics, mainly gallic and ellagic acids and lower levels of ellagitannins, were also detected. De Vasconcelos et al. (2010c) also found that potassium and phosphorous were predominant in the fruits of all cultivars for both harvest years. Calcium, magnesium, iron, zinc and manganese were also present. Fruits from both harvest years had a significant content of free sugars, with sucrose predominating, and these sugars were more affected by the processing stage. Significant levels of lutein, lutein esters, γ-tocopherol and vitamin C were also found in the chestnut fruits. The contents of vitamin C and carotenoids were adversely affected during industrial processing.

The cooking processes was found to significantly affect primary and secondary metabolite composition of chestnuts (Gonçalves et al. 2010). Roasted chestnuts had higher protein contents, insoluble and total dietary fibre and lower fat contents whilst boiled chestnuts had lower protein, but higher fat contents. Cooking increased citric acid contents, especially in roasted chestnuts. In contrast, raw chestnuts had higher malic acid contents than cooked chestnuts. Moreover, roasted chestnuts had significantly higher gallic acid and total phenolic contents, and boiled chestnuts had higher gallic and ellagic acids contents, when compared to raw chestnuts. The data confirmed cooked chestnuts to be a good source of organic acids and phenolics and with low fat contents, properties that associated with positive health benefits.

Other Phytochemicals

All three Portuguese cultivars were found to have high moisture contents but were low in ash, crude fat, and crude protein contents, with high starch and low fiber contents (De Vasconcelos et al. 2007). The free amino acid contents, including various essential amino acids, varied depending on the cultivar. All three cultivars also had a significant content of polyphenolics with gallic acid; ellagic acid was predominant among hydrolyzable and condensed tannins. Chestnuts have become increasingly important with respect to human health, for example, as an alternative gluten-free flour source.

The chestnut fruit processing had been reported to generate large amounts of residues as pericarp (outer shell; 8.9–13.5%) and integument (inner shell; 6.3–10.1%) and studies showed that these materials clearly had the potential as sources of valuable co-products (de Vasconcelos et al. 2010a). The analyses of the pericarp and integument of four Portuguese chestnut cultivars revealed significant contents of total phenolics, low molecular weight phenolics (gallic and ellagic acid), condensed tannins and ellagitannins including castalagin, vescalagin, acutissimin A and acutissimin B. The integument tissues had the highest levels of total phenolics and condensed tannins. The most efficient extraction solvent for the total phenolics, total condensed tannins and low molecular weight phenolics was 70:30 acetone:water at 20°C.

Some low molecular weight phenolic compounds and hydrolyzable tannins were found in hardwood extracts of *Castanea sativa* before and after toasting (Sanz et al. 2010). The low molecular weight phenolic compounds were lignin constituents as the acids gallic, protocatechuic, vanillic, syringic, ferulic, and ellagic, the aldehydes protocatechuic, vanillic, syringic, coniferylic, and sinapic, and the coumarin scopoletin. Vescalagin and castalagin were the main ellagitannins, and acutissimin was also identified in the wood. Some gallotannins were tentatively identified, including different isomers of di, tri, tetra, and pentagalloyl glucopyranose, and di and trigalloyl-hexahydroxydiphenoyl glucopyranose, comprising 20 different compounds, as well as some ellagic derivatives such as ellagic acid deoxyhexose, ellagic acid dimer dehydrated, and valoneic acid dilactone. These ellagic derivatives as well as some galloyl and hexahydroxydiphenoyl derivatives were tentatively identified for the first time in chestnut wood. Seasoned and toasted chestnut wood showed a very different balance between lignin derivatives and tannins because toasting resulted in the degradation of tannins and the formation of low molecular weight phenolic compounds from lignin degradation. A chlorine-free environmentally-friendly process was developed for extraction of 4-O-methylglucuronoxylans (MGX) from *Castanea sativa* hardwood (Barbat et al. 2010). Chestnut sawdust was first delignified using metalled phthalocyanine or porphyrin in presence of hydrogen peroxide. Then, MGXs were easily extracted by hot water.

The parts of chestnut such as: seed, peeled seed, brown seed shell, red internal seed shell, leaves, catkin, spiny bur, as well as the new and old chestnut bark were found to have varying amounts of total phenolics, total flavonoids and tannins (Živković et al. 2009a, b). The highest content of total phenolic compounds (3.28)% GAE (Gallic acid equivalent) was found in dry extract of catkin, while the lowest content (0.42)% GAE was obtained for the dry extract of the seeds. The phenolic contents (% GAE) in other tissues were as follows: peeled chestnut 0.59%, brown seed shell 1.19%, red internal seed shell 2.82%, leaf 1.40%, new chestnut bark 3.00%, old chestnut bark 1.70%, spiny burs 0.49%. The total flavonoid content % CE (catechin equivalent) in various plant parts were reported as: seeds 0.17%, brown seed shell 0.65%, peeled chestnut 0.09%, red internal seed shell 1.44%, catkin 0.60%, leaf 1.40%, new chestnut bark 0%, old chestnut bark 0.69%, spiny burs 0.13%. The total condensed tannins (CT) content was highest in red internal seed shell 15.29% CE (vanillin assay) and 3.12% CT (acid butanol assay). The total condensed tannins content in other plant parts were estimated as follows: seeds 0.39% CE (vanillin assay) 0.88% CT (butanol assay); peeled chestnut 0% CE (vanillin assay), 0% CT (acid butanol assay); brown seed shell 2.78% CE (vanillin assay), 1.67% CT (acid butanol assay); catkin 0.49%CE (vanillin assay), 0.95% CT (acid butanol assay); leaf 0% CE (vanillin assay) 0.08% CT (acid butanol assay); new chestnut bark 3.91% CE (vanillin assay), 1.89% CT (acid butanol assay); old chestnut bark 0.76% CE (vanillin assay), 0.58% CT (acid butanol assay); and spiny burs 0% CE (vanillin assay), 0.08% (acid butanol assay).

Antioxidant Activity

Studies had shown that the leaves had antioxidant properties. The leaf extract and an ethyl acetate fraction, which contained a high level of total phenolic compounds (29.1 g/100 g) demonstrated high antioxidant potentials, equivalent to at least those of reference compounds quercetin and vitamin E and standard extracts (pycnogenol, from French *Pinus maritima* bark, and grape marc extract) (Calliste et al. 2005). In another study, five phenolic compounds were identified in the leaf ethanol:water (7:3) extract, namely chlorogenic acid, ellagic acid, rutin, isoquercitrin and hyperoside (Almeida et al. 2008a). The content of total phenolics for *C. sativa* was 284 of gallic acid equivalents (GAE)/g of lyophilized extract. The extract presented a high potency to scavenge the tested reactive species viz. reactive oxygen species (ROS) such as superoxide radical, peroxyl radical, hydrogen peroxide, singlet oxygen and reactive nitrogen species (RNS) namely nitric oxide and peroxynitrite; all the IC_{50}s being found at the µg/ml level. The IC_{50} found for the iron chelation and DPPH (1,1-diphenyl-2-picryl hydrazyl) scavenging assays were 132.94 and 12.58 µg/ml, respectively.

The water soluble extracts obtained from leaves, catkins, and outer brown peel of *Castanea sativa* showed high antioxidant activity in scavenging ·OH and DPPH radical (Živković et al. 2009b). The highest content of total phenolic compounds ((4.24)% of GAE) and flavonoids ((2.41)% of GAE) were determined in dry extract of outer brown peel of Lovran's Marrone cultivar. The TF/TP (total flavonoid/total phenolic) ratio was from 11.86 for the extract of peeled chestnut to 56.83% determined in the outer brown peel of Lovran's Marrone cultivar. All extracts, except for sweet chestnut catkins, showed the ability to protect liposomes from peroxidation. Phenolic compounds, as active antioxidants, have the ability to enter and protect cell membranes from lipid peroxidation, thus overcoming the body's refractory response to the antioxidant supplements in the diet. It was shown that phenolics were easily accessible natural antioxidants that could be used as food supplements or for the treatment of pathophysiological conditions related to oxidative stress. In a recent study, all the assayed by-products (almond green husks, chestnut skins and chestnut leaves) revealed good antioxidant properties, with very low EC_{50} values (lower than 380 µg/ml), particularly for lipid peroxidation inhibition (lower than 140 µg/ml) (Barreira et al. 2010). The correlation between the bioactive compounds (total phenols and flavonoids) and DPPH (2,2-diphenyl-1-picrylhydrazyl) radical scavenging activity, reducing power, inhibition of β-carotene bleaching and inhibition of lipid peroxidation in pig brain tissue through formation of thiobarbituric acid reactive substances, was also obtained. Although, all the assayed by-products proved to have a high potential of application in new antioxidants formulations, chestnut skins and leaves demonstrated better results. Studies proposed that glycerine with a pH around 5 was best vehicle for topical formulations incorporating *C. sativa* leaf extract with antioxidant activity for the prevention of photoaging and oxidative-stress-mediated diseases (Almeida et al. 2010).

Gamma irradiation is a possible feasible alternative to substitute the traditional quarantine chemical fumigation with methyl bromide to extend post-harvest shelf-life of food. Results of studies appeared to indicate that storage favoured chestnuts antioxidant potential as assessed by DPPH (2,2-diphenyl-1-picrylhydrazyl) radical-scavenging activity, reducing power and inhibition of β-carotene bleaching capacity (Antonio et al. 2011). In addition, the application of gamma irradiation also appeared to be advantageous for antioxidant activity, independently of the dose used (0.27 or 0.54 kGy).

Photoprotective Activity

The in-vivo patch test carried out on 20 volunteers showed that, with respect to irritant effects, chestnut leaf extract could be regarded as safe for topical application (Almeida et al. 2008b). Topical application of natural antioxidants had proven to be effective in protecting the skin against ultraviolet-mediated oxidative damage and to provide a straightforward way to strengthen the endogenous protection system. Further studies proposed that glycerine with a pH around 5 was the best vehicle for topical formulations incorporating *C. sativa* leaf extract with antioxidant activity for the prevention of photoaging of the skin and oxidative-stress-mediated diseases (Almeida et al. 2010).

Anticancer Activity

The concentration of ellagic acid, a naturally occurring inhibitor of carcinogenesis, in chestnut fruits and bark was generally increased after hydrolysis due to the presence of ellagitannins in the crude extract (Vekiari et al. 2008). The concentration varied between 0.71 and 21.6 mg/g (d.w.) in non-hydrolyzed samples, and between 2.83 and 18.4 mg/g (d.w.) in hydrolyzed samples. In chestnut fruits, traces of ellagic acid were present in the seed, with higher concentrations in the pellicle and pericarp. However, all fruit tissues had lower concentrations of ellagic acid than the bark.

Xylans, 4-O-methylglucuronoxylan (MGX) and a homoxylan (HX), were purified from delignified holocellulose alkaline extracts of *Castanea sativa* (Barbat et al. 2008). The xylan, MGX inhibited the proliferation of A431 human epidermoid carcinoma cells with an IC_{50} value of 50 µM (Moine et al. 2007; Barbat et al. 2008). In addition, this xylan inhibited A431 cell migration and invasion. Preliminary experiments showing that secretion of metalloproteinases MMP2 and MMP9 by A431 tumour cells was inhibited by the purified MGX strongly suggesting that this mechanism of action may play a role in its anti-migration and anti-invasive properties.

Antigastroenteritic Activity

Seeds of *C. sativa* were reported to be used in paediatrics for treatment of gastroenteritis and as a gluten-free diet in cases of coeliac disease (Živković et al. 2009a, b). A new pyrrole alkaloid, methyl-(5-formyl-1*H*-pyrrole-2-yl)-4-hydroxybutyrate, was isolated from sweet chestnut seeds (Hiermann et al. 2002).

Antibacterial Activity

An ethyl acetate extract of *C. sativa* was shown to have pronounced antibacterial effects against seven of the eight strains of Gram-positive and Gram-negative bacteria used (MIC in the range of 64–256 μg/ml and MBC in the range of 256–512 μg/ml) (Basile et al. 2000). The active fraction contained rutin, hesperidin, quercetin, apigenin, morin, naringin, galangin and kaempferol. The highest bactericidal activity was shown by quercetin, rutin and apigenin. was tested.

Staphylococcal quorum sensing is encoded by the AGR locus and is responsible for the production of δ-hemolysin (Quave et al. 2011). Studies found that the extracts from three medicinal plants (*Ballota nigra*, *Castanea sativa*, and *Sambucus ebulus*) exhibited a dose-dependent response in the production of δ-hemolysin, indicating anti-quorum sensing activity in a pathogenic methicillin-resistant *Staphylococcus aureus* (MRSA) isolate.

Antithrombin Activity

Of selected Slovak medicinal plants, the methanol extract of *C. sativa* exhibited the most distinct inhibition activity to thrombin with $IC_{50} = 73.2$ μg/ml (Jedinak et al. 2010). The methanol extract also had strong inhibition activity to protease trypsin with IC_{50} values below 10 μg/ml.

Antidiarrhoeal Activity

Stop Fitan®, a dietary supplement based on the bioactive purified natural extract of chestnut (*Castanea sativa*) wood and *Saccharomyces boulardii*, a nonpathogenic yeast strain has been proposed as a co-adjuvant in the therapy of diarrhea (Budriesi et al. 2010). Studies showed that natural extract of chestnut wood rich in hydrolyzable tannins, exerted spasmolytic effects in guinea pig ileum and proximal colon. The findings, coupled with the antibacterial, antiviral, and antispasmodic properties of tannins, suggest that the combination of chestnut tannins and *S. boulardii* may be relevant to treat diarrhea by Stop Fitan®.

Hypersensitivity Problem

Only occasionally does chestnut cause hypersensitivity. There were only a few reported cases, depending on cross-reactivity in previously latex-hypersensitive patients. The case of oral allergy syndrome to chestnut appeared to be a manifestation of immediate IgE-dependent hypersensitivity resulting from direct contact between food and the oral mucosa (Antico 1996).

Traditional Medicinal Uses

In the Middle Ages the raw seeds were found useful in the treatment of heart disorders. *C. sativa* leaves were used in folk medicine as a tea in France to treat whooping cough and diarrhoea. Beside the leaves, the bark is also a good source of tannins. They are antiinflammatory, expectorant, tonic and astringent. The astringent activity is useful in the treatment of bleeding and diarrhoea. Leaf infusions are employed in respiratory diseases and are a common therapy for whooping cough, fevers and ague. The leaves are also employed in the treatment of rheumatism, to ease lower back pains and to relieve stiff muscles and joints. A decoction is used as gargle for treating sore throats. A hair shampoo can be made from infusing leaves and fruit husks. Leaves are used in homeopathy for therapy of depression and fatigue.

Other Uses

The Spanish chestnut is a magnificent huge shade tree for parks, estates and avenues and sometimes planted for erosion control in Mediterranean countries. The flowers provide good forage for bees. A blackish-brown dye is extracted from the leaves and the bark and also an oil which is medicinal. Tannin is obtained from the bark and used in tanning. The wood, leaves and seed husks also contain tannin. The seed meal can be used as a source of starch, for fattening stock and also for whitening linen cloth. A hair shampoo is made from the leaves and the husks of the fruits. The wood is good for carpentry, turnery, furniture, barrels, roof beams, props, basketry and fence posts. It is also a very good fuel.

Comments

Chestnut tree responds well to coppicing, a traditional method of woodland management.

Selected References

Almeida IF, Fernandes E, Lima JL, Costa PC, Bahia MF (2008a) Protective effect of *Castanea sativa* and *Quercus robur* leaf extracts against oxygen and nitrogen reactive species. J Photochem Photobiol B 91(2–3):87–95

Almeida IF, Valentão P, Andrade PB, Seabra RM, Pereira TM, Amaral MH, Costa PC, Bahia MF (2008b) In vivo skin irritation potential of a *Castanea sativa* (Chestnut) leaf extract, a putative natural antioxidant for topical application. Basic Clin Pharmacol Toxicol 103(5):461–467

Almeida IF, Costa PC, Bahia MF (2010) Evaluation of functional stability and batch-to-batch reproducibility of a *Castanea sativa* leaf extract with antioxidant activity. AAPS PharmSciTech 11(1):120–125

Antico A (1996) Oral allergy syndrome induced by chestnut (*Castanea sativa*). Ann Allergy Asthma Immunol 76(1):37–40

Antonio AL, Fernandes A, Barreira JC, Bento A, Botelho ML, Ferreira IC (2011) Influence of gamma irradiation in the antioxidant potential of chestnuts (*Castanea sativa* Mill.) fruits and skins. Food Chem Toxicol 49(9):1918–1923

Bailey LH (1976) Hortus third. A concise dictionary of plants cultivated in the United States and Canada. Liberty Hyde Bailey Hortorium, Cornell University. Wiley, London, 1312 pp

Barbat A, Gloaguen V, Moine C, Sainte-Catherine O, Kraemer M, Rogniaux H, Ropartz D, Krausz P (2008) Structural characterization and cytotoxic properties of a 4-O-methylglucuronoxylan from *Castanea sativa*. 2. Evidence of a structure-activity relationship. J Nat Prod 71(8):1404–1409

Barbat A, Gloaguen V, Sol V, Krausz P (2010) Aqueous extraction of glucuronoxylans from chestnut wood: new strategy for lignin oxidation using phthalocyanine or porphyrin/H2O2 system. Bioresour Technol 101(16):6538–6544

Barreira JC, Casal S, Ferreira IC, Oliveira MB, Pereira JA (2009) Nutritional, fatty acid and triacylglycerol profiles of *Castanea sativa* Mill. cultivars: a compositional and chemometric approach. J Agric Food Chem 57(7):2836–2842

Barreira JCM, Ferreira ICFR, Oliveira MPP, Pereira JA (2010) Antioxidant potential of chestnut *Castanea sativa* L., and almond, *Prunus dulcis* L., by-products. Food Sci Technol Int 16:209–216

Basile A, Sorbo S, Giordano S, Ricciardi L, Ferrara S, Montesano D, Castaldo Cobianchi R, Vuotto ML, Ferrara L (2000) Antibacterial and allelopathic activity of extract from *Castanea sativa* leaves. Fitoterapia 71(Suppl 1):S110–S116

Bounous G, Botta R, Beccaro G (2000) The chestnut: the ultimate energy source. Nutritional value and alimentary benefits. Nucis – Information Bulletin of the Research Network on Nuts (FAO-CIHEAM) 9:44–50

Budriesi R, Ioan P, Micucci M, Micucci E, Limongelli V, Chiarini A (2010) Stop Fitan: antispasmodic effect of natural extract of chestnut wood in guinea pig ileum and proximal colon smooth muscle. J Med Food 13(5):1104–1110

Calliste CA, Trouillas P, Allais DP, Duroux JL (2005) *Castanea sativa* Mill. leaves as new sources of natural antioxidant: an electronic spin resonance. J Agric Food Chem 53(2):282–288

Chevallier A (1996) The encyclopedia of medicinal plants. Dorling Kindersley, London, 336 pp

De Vasconcelos MCBM, Bennett RN, Rosa EAS, Ferreira-Cardoso JV (2007) Primary and secondary metabolite composition of kernels from three cultivars of Portuguese chestnut (*Castanea sativa* Mill.) at different stages of industrial transformation. J Agric Food Chem 55(9):3508–3516

De Vasconcelos MCBM, Bennett RN, Quideau S, Jacquet R, Rosa EAS, Ferreira-Cardoso JV (2010a) Evaluating the potential of chestnut (*Castanea sativa* Mill.) fruit pericarp and integument as a source of tocopherols, pigments and polyphenols. Ind Crops Res 31(2):301–311

De Vasconcelos MCBM, Bennett R, Rosa EAS, Ferreira-Cardoso JV, Nunes F (2010b) Industrial processing of chestnut fruits (*Castanea sativa* Mill.) – effects on nutrients and phytochemicals. Acta Hortic (ISHS) 866:611–617

De Vasconcelos MCBM, Nunes F, Viguera CG, Bennett RN, Rosa EAS, Ferreira-Cardoso JV (2010c) Industrial processing effects on chestnut fruits (*Castanea sativa* Mill.) 3. Minerals, free sugars, carotenoids and antioxidant vitamins. Int J Food Sci Technol 45(3):496–505

Gonçalves B, Borges P, Costa HS, Bennett R, Santos M, Silva AP (2010) Metabolite composition of chestnut (Castanea sativa Mill.) upon cooking: Proximate analysis, fibre, organic acids and phenolics. Food Chem 122(1):154–160

Hiermann A, Kedwani S, Schramm HW, Seger C (2002) A new pyrrole alkaloid from seeds of *Castanea sativa*. Fitoterapia 73:22–27

Jedinak A, Valachova M, Maliar T, Sturdik E (2010) Antiprotease activity of selected Slovak medicinal plants. Pharmazie 65(2):137–140

Miller G, Miller DD, Jaynes RA (1996) Chestnuts. In: Janick J, Moore JN (eds) Fruit breeding, vol 3, Nuts. Wiley, New York, 278 pp

Mills SY (1985) The dictionary of modern herbalism. Thorsons, Wellingborough

Moinc C, Krausz P, Chaleix V, Sainte-Catherine O, Kraemer M, Gloaguen V (2007) Structural characterization and cytotoxic properties of a 4-O-methylglucuronoxylan from *Castanea sativa*. J Nat Prod 70(1):60–66

Quave CL, Plano LR, Bennett BC (2011) Quorum sensing inhibitors of *Staphylococcus aureus* from Italian medicinal plants. Planta Med 77(2):188–195

Ribeiro B, Rangel J, Valentão P, Andrade PB, Pereira JA, Bölke H, Seabra RM (2007) Organic acids in two Portuguese chestnut (*Castanea sativa* Miller) varieties. Food Chem 100(2):504–508

Rutter PA, Miller G, Payne JA (1990) Chestnuts (*Castanea*). Acta Hortic 290(2):761–788

Sanz M, Cadahía E, Esteruelas E, Muñoz AM, Fernández de Simón B, Hernández T, Estrella I (2010) Phenolic compounds in chestnut (*Castanea sativa* Mill.) heartwood. Effect of toasting at cooperage. J Agric Food Chem 58(17):9631–9640

U.S. Department of Agriculture, Agricultural Research Service (2011) USDA National Nutrient Database for standard reference, release 24. Nutrient Data Laboratory Home Page. http://www.ars.usda.gov/ba/bhnrc/ndl

Vekiari SA, Gordon MH, García-Macías P, Labrinea H (2008) Extraction and determination of ellagic acid content in chestnut bark and fruit. Food Chem 110(4):1007–1011

Ylldiz MU, Özcan MM, Çalipr S, Demir F, Er F (2009) Physico-chemical properties of wild chestnut (*Castanea sativa* Mill.) fruit grown in Turkey. World Appl Sci J 6(3):365–372

Živković J, Mujić I, Zeković Z, Nikolić G, Vidović S, Mujić A (2009a) Extraction and analysis of condensed tannins in *Castanea sativa* Mill. J Cent Eur Agric 10(3):283–288

Živković J, Zeković Z, Mujić I, Vesna Tumbas V, Cvetkovi D, Spasojevi I (2009b) Antioxidant properties of chestnut phenolics. Food Technol Biotechnol 47(4): 421–427

Quercus infectoria

Scientific Name

Quercus infectoria G. Olivier.

Synonyms

Quercus infectoria ssp. *euinfectoria* A.Camus, *Quercus lusitanica* ssp. *infectoria* (G.Olivier) Mouillef., *Quercus lusitanica* var. *infectoria* (G. Olivier) A DC.

Family

Fagaceae

Common/English Names

Aleppo Oak, Asian Holly-Oak, Cyprus Oak, Downy Oak, Dyer's Oak, Gall Oak, Nut-Galls

Vernacular Names

Arabic: Afas, Afss, Ballut Afssi, Mazu, Uffes;
Czech: Dub Hálkovec;
Dutch: Eik Soort;
Eastonian: Tinditamm;
French: Chêne À Galles, Chêne d'Alep, Chêne d'Israel;
German: Gall-Eiche, Gallapfel-Eiche;
Hungarian: Kurdisztáni Tölgy;
India: Majuphal, Majuphul, Mazu, Muphal (Hindu), Machikai, Macike, Machi Kaayi (Kannada), Masikka, Mayakku (Malayalam), Majuphala, Maayaphal (Marathi), Ambastha, Majjaphala, Majuphal, Majuphul, Manjuphal, Mayakku, Mayaphala, Mayuka (Sanskrit), Cakkirakacikakkay, Cakkirakacikam, Civatakitakkay, Maasikkai, Macakkai, Macakkay, Macikkai, Macikkay, Machakai, Machikai, Machikkai, Maci, Masikkai, Mayakkay, Mayakkay (Tamil), Mashi Kaaya, Mashikaya, Masikaya (Telugu), Baloot., Mazu, Mazu Sabz, Mazu Subz (Urdu);
Indonesia: Manjakani;
Malaysia: Manjakani;
Persian: Mazu, Mazu-E-Sabz;
Spanish: Encina De La Agalla;
Swedish: Aleppoek;
Thai: Ben Ka Nee;
Turkish: Mazı Meşesi.

Origin/Distribution

Q. infectoria is indigenous to Turkey, Iran, Iraq, Kurdistan, Cyprus, East Aegean Islands, Greece, Lebanon and Syria. The tree is occasionally cultivated for production of tanning bark and for dye production of the wood.

Agroecology

In its native range, it is usually found in semi-humid to semi-arid forests in areas with mean annual rainfall of 400–1,100 mm from 900 to 2,000 m altitude. It is intolerant of frost. It grows on a wide range of soil types from acidic to alkaline, in full to partial sun.

Edible Plant Parts and Uses

The seed can be thoroughly washed in running water to remove the bitter tannins and cooked. The seeds can be dried, ground into a powder and used as a thickening in stews etc. or mixed with cereals for making bread. When roasted the seeds can be used as a coffee substitute. The nut gall (Plate 1) extract or powder is used as a herbal drink or tea for health purposes.

Botany

A semi-evergreen, small tree or shrub, 1–4 (–10) m with grey, scaly, ridged bark. Juvenile shoots are pubescent, reddish or yellowish-brown; buds reddish-brown, about 3 mm and pubescent. Leaves are alternate, very variable in size and colour, 40–70(–100) by 10–45 mm, leathery, glabrescent, ovate to narrowly oblong, rounded or wedge-shaped at base, margins often wavy with 4–8 crenate to saw-toothed lobes, or entire (at base of twigs); primary veins 6–11; petiole 1–15(–25) mm. Inflorescences unisexual, in axils of leaves or bud scales, usually clustered at base of new growth; staminate inflorescences lax, spicate; pistillate inflorescences usually stiff, with terminal cupule and sometimes 1-several sessile, lateral cupules. Fruit is a smooth nut, called an acorn, mucronate, ovoid elongated, 2–3.5 cm long, 1.8 cm in diameter; glabrous and is more or less enclosed in a scaly involucre called the cup or cupule. Cupules solitary or in pairs, approximately hemispherical or cyathiform, 10–18 mm in diameter with lanceolate strongly adpressed, greyish pubescent scales.

The galls or nut-galls are hard, corky, resinous, greyish-brown, nearly round excrescences formed on the young branches. The excrescences vary from the size of a large pea to that of a small hickory-nut. They are the result of a puncture made in the bark by an insect (*Diplolepis gallae tinctoriae*, or *Cynips quercufolii*) for the purpose of depositing its egg. A small tumour soon follows the puncture, and forms a very dense mass about the egg. The egg hatches into the fly while in these tumours, eating its way by a small opening. The nut gall has an integument with rugae-like surface interspersed by protruding blunt horn-like lumps (Soon et al. 2007) imparting to it a tuberculate appearance (Plate 1). Cross section of the gall revealed a whitish core and concentric circles of resinous materials constituting the middle layer.

Plate 1 Nut galls of *Quercus infectoria*

Nutritive/Medicinal Properties

Nut-galls abound in tannins (36–60%). The constituents of galls was found to comprise a large amount of tannins: gallic acid, syringic acid, ellagic acid, β-sitosterol, amentoflavone, hexamethyl ether, isocryptomerin, methyl betulate, methyl oleanate, hexagalloyl glucose (Aroonrerk and Kamkaen 2009) and ellagic acid-4- O -[β-D-glucopyranosyl]-10- O -[β-D-glucopyranosyl]-(4→1)-β-D-rhamnopyranoside, and 2-methyl-3-hydroxymethylene-4,5,6,7,

8-pentahydroxynaphthalene (Hamid et al. 2005). The nut-galls also contained useful minerals of carbon, oxygen, silica, magnesium, aluminium, potassium and calcium (Soon et al. 2007).

Studies had shown that the nutgall extract had antioxidant, antiinflammatory, antiviral, analgesic, antitremorine, skin whitening, antidiabetic, antibacterial, larvicidal, molluscicidal, anti protozoal, antivenom, and anti-amoebic activities.

Antioxidant Activity

Two compounds isolated from the ethanol extract of the galls of *Quercus infectoria* exhibited nitric oxide (NO) and superoxide inhibiting activity (Hamid et al. 2005). Their structures were established as ellagic acid-4- O -[β-D-glucopyranosyl]-10- O -[β-D-glucopyranosyl]-(4→1)-β-D-rhamnopyranoside and 2-methyl-3-hydroxymethylene-4,5,6,7,8-pentahydroxynaphthalene. Umachigi et al. (2008) reported *Quercus infectoria* nutgalls to have potent antioxidant activity. Its extract strongly scavenged DPPH radical with the IC_{50} being 0.25 mg/ml in a dose dependent manner. The total antioxidant capacity of the extract was found to be 152.91 nmol/g ascorbic acid. *Quercus infectoria* extract also moderately inhibited nitric oxide in dose dependent fashion with the IC_{50} being 0.258 mg/ml. It was also determined that the 50% aqueous alcoholic extract of *Quercus infectoria* inhibited $FeSO_4$ induced lipid peroxidation in a dose dependent manner. IC_{50} value was found to be 0.124 mg/ml. The decrease in the MDA (malondialdehyde) level with increases in the concentration of the extracts indicated the role of the extract as an antioxidant. The extract also moderately scavenged superoxide radical with the IC_{50} value of 1.024 mg/ml. Gallic acid was found to be 0.68% w/w in the methanolic extract. The high percentage of the gallic acid in the extract underpinned the potent antioxidant activity exhibited. The results obtained indicated that *Quercus infectoria* extract had potent antioxidant activity, achieved by scavenging abilities observed against DPPH, and lipid peroxidation.

Studies showed that *Q. infectoria* galls possessed potent antioxidant activity, when tested both in chemical as well as biological models (Kaur et al. 2008) Ethanolic gall extract was found to contain a large amount of polyphenols and to possess potent reducing power. HPTLC analysis of the extract found it to contain 19.925% tannic acid (TA) and 8.75% gallic acid (GA). The extract strongly scavenged free radicals including DPPH (IC_{50} approximately 0.5 µg/ml), ABTS (IC_{50} approximately 1 µg/ml), hydrogen peroxide (H_2O_2) (IC_{50} approximately 2.6 µg/ml) and hydroxyl (OH) radicals (IC_{50} approximately 6 µg/ml). Gall extract also chelated metal ions and inhibited Fe^{3+}-ascorbate-induced oxidation of protein and lipid peroxidation. Exposure of rat peritoneal macrophages to tertiary butyl hydroperoxide (tBOOH) induced oxidative stress and modified their phagocytic functions. These macrophages exhibited increased secretion of lysosomal hydrolases, and mitigated phagocytosis and respiratory burst. Activity of macrophage mannose receptor (MR) also declined following oxidant exposure. Pretreatment of macrophages with gall extract maintained antioxidant protection close to control values and significantly protected against all the investigated functional mutilations. MTT (3-(4,5-Dimethylthiazol-2-yl)-2,5-diphenyltetrazolium bromide) assay revealed the gall extract to enhance percent survival of tBOOH exposed macrophages.

Antiinflammatory Activity

Galls of *Quercus infectoria* were found to possess manifold therapeutic activities, with particular efficacy against inflammatory diseases (Kaur et al. 2004). Oral administration of gall alcoholic extract significantly inhibited carrageenan, histamine, serotonin and prostaglandin E2 (PGE2) induced paw oedemas, while topical application of gall extract inhibited phorbol-12-myristate-13-acetate (PMA) induced ear inflammation. The extract also suppressed various functions of macrophages and neutrophils relevant to the inflammatory response. In-vitro exposure of rat peritoneal macrophages to gall extract dose dependently ameliorated lipopolysaccharide (LPS) stimulated PGE2 and nitric oxide (NO) production and PMA triggered superoxide (O_2^{*-}) production. Gall extract also scavenged NO and O_2^{*-}. Probing into mechanism of NO inhibition in macrophages revealed gall extract to ameliorate the stimulation of inducible NO synthase (iNOS), respectively without any inhibitory effect on its catalytic activities even at higher concentrations. Gall extract also significantly inhibited formyl-Met-Leu-Phe (fMLP) triggered degranulation in neutrophils. These results suggested that alcoholic extract of galls of *Q. infectoria* exerted in-vivo antiinflammatory activity after oral or topical administration and also had the ability to curb the production of some inflammatory mediators. In separate studies, *Quercus infectoria* (QI) nutgall extract was found to exhibit potent antioxidant and antiinflammatory activities (Pithayanukul et al. 2009). Treatment of rats with nutgall extract reversed oxidative damage in hepatic tissues induced by carbon tetrachloride. It was suggested that the *Quercus infectoria* extract which was rich in hydrolysable tannins and known for their potent antioxidant and antiinflammatory activities, may potentially confer hepatoprotective effect against oxidative stress-induced liver injury.

Recent studies reported that the Thai herbal recipe for aphthous ulcer comprising *Quercus infectoria, Glycyrrhiza uralensis, Kaempferia galanga* and *Coptis chinensis* and the individual plant powder components exhibited potent anti-inflammatory activity (Aroonrerk and Kamkaen 2009). The four plant powders, *K. galanga, C. chinensis, G. uralensis* and *Q. infectoria* inhibited interleukin IL-6 production with IC_{50} value of 0.04, 0.07, 0.08 and 0.31 µg/ml respectively. They also displayed anti-PGE2 activities but lower than the aphthous powder and aphthous gel. The antiinflammatory activities were significantly higher than prednisolone and the COX-2 inhibitor. The plant powders and the herbal recipe had no growth inhibitory effect on the human gingival fibroblast cells even at the highest dose. No sign of irritation was noted during the dermal irritation test. It was postulated that the antiinflammatory activity of *Q. infectoria* was due to the presence of high tannin compounds and was mediated by either inhibiting the synthesis, release or action of mediators such as interleukin, serotonin, histamine and PGE2.

Antimicrobial Activity

The methanol extract of *Quercus infectoria*, exhibited potent antibacterial activity against the cariogenic bacterium *Streptococcus mutans* (Hwang et al. 2004). Recent studies also showed that methanolic extract had maximum anti-bacterial activity against the following dental pathogens – *Streptococcus mutans, Streptococcus salivarius, Staphylococcus aureus, Lactobacillus acidophilus* and *Streptococcus sanguis* (Vermani and Navneet 2009). The most susceptible bacteria were *S. sanguis* followed by *S. aureus, S. mutans, S. salivarius* and *L. acidophilus*. The MIC values showed that methanolic extract was more effective than the water extract.

Numerous research studies conducted by Voravuthikunchai and co-workers in Thailand showed that the nut gall extracts of *Q. infectoria* had potent antibacterial activity against a broad spectrum of bacterial pathogens. Among the 38 Thai medicinal plants tested, only eight species (21.05%) exhibited antimicrobial activity against enterohaemorrhagic *Escherichia coli* (Voravuthikunchai et al. 2004). *Acacia catechu, Holarrhena antidysenterica, Peltophorum pterocarpum, Psidium guajava, Punica granatum, Quercus infectoria, Uncaria gambir,* and *Walsura robusta* demonstrated antibacterial activity with inhibition zones ranging from 7 to 17 mm. The greatest inhibition zone against *Escherichia coli* was produced from the ethanolic extract of *Quercus infectoria*. Both aqueous and ethanolic extracts of *Quercus infectoria* and aqueous extract of *Punica granatum* were highly effective against *Escherichia coli* with the best MIC and MBC values of 0.09, 0.78, and 0.19, 0.39 mg/ml, respectively. These plant species may provide alternative but bioactive medicines for the treatment of enterohaemorrhagic *Escherichia coli* infection. Of nine Thai traditional medicinal plants that displayed activity against all 35 hospital isolates of methicillin-resistant *Staphylococcus aureus* (MRSA) tested, the ethanolic extracts of *Garcinia mangostana*, *Punica granatum* and *Quercus infectoria* were most effective, with MICs for MRSA isolates of 0.05–0.4, 0.2–0.4 and 0.2–0.4 mg/ml, respectively, and for *Staphylococcus aureus* of 0.1, 0.2 and 0.1 mg/ml, respectively (Voravuthikunchai and Kitpipit 2005). MBCs for MRSA isolates were 0.1–0.4, 1.6–3.2 and 0.4–1.6 mg/ml, and for *Staphylococcus aureus* were 0.4, 3.2 and 1.6 mg/ml, respectively. In another study, of four medicinal plants used for traditional remedies for diarrhea, *Acacia catechu*, *Peltophorum pterocarpum*, *Punica granatum*, and *Quercus infectoria*, the ethanolic extract of *Q. infectoria* was the most effective against all strains of *E. coli*, with MICs of 0.12–0.98 mg/ml and MBCs of 0.98–3.91 mg/ml (Voravuthikunchai and Limsuwan 2006). Ethanolic extracts of *Q. infectoria*, *P. pterocarpum*, and *P. granatum* were among the most effective extracts against the two strains of *E. coli* O157:H7.

Acetone, ethyl acetate, 95% ethanol and aqueous extracts of *Quercus infectoria* nut-galls demonstrated significant antibacterial activities against all strains of methicillin-resistant *Staphylococcus aureus* (MRSA) and methicillin-susceptible *Staphylococcus aureus* (MSSA) (Chusri and Voravuthikunchai 2008). Inhibition zones were in the range 11.75–16.82 mm. Both MRSA and MSSA strains exhibited minimum inhibitory concentration (MIC) and minimum bactericidal concentration (MBC) values at 0.13 and 0.13–1.00 mg/mL, respectively. In another paper they reported the use of *Q. infectoria* nut galls, containing up to 70% tannin content, was effective in the treatment of methicillin-resistant *Staphylococcus aureus* infections (Chusri and Voravuthikunchai 2009). The appearance of pseudomulticellular bacteria in the treated cells and the synergistic effect of the *Q. infectoria* extract with β-lactamase-susceptible penicillins suggested that the extract may interfere with staphylococcal enzymes including autolysins and β-lactamase. Subsequent studies by Voravuthikunchaia et al. (2008) indicated *Quercus infectoria* galls to be potentially a good source of antibacterial substances with broad spectrum of activities against antibiotic-resistant bacteria. Ethanol extracts of *Quercus infectoria* galls demonstrated a broad spectrum of activity against *Acinetobacter baumannii*, *Bacillus cereus*, *Enterobacter faecalis*, *Escherichia coli*, *Helicobacter pylori*, *Klebsiella pneumoniae*, *Listeria monocytogenes*, *Pseudomonas aeruginosa*, *Salmonella spp.*, *Shigella flexneri*, *Staphylococcus aureus*, *Streptococcus mutans* and *Streptococcus pyogenes*. The extracts of *Quercus infectoria* displayed remarkable activity against methicillin-resistant *Staphylococcus aureus* (MRSA) with MICs ranging from 0.02 to 0.4 mg/ml, and MBCs ranging from 0.4 to 1.6 mg/ml. More importantly, *Quercus infectoria* could exhibit strong antibacterial activity against all Gram-negative organisms. Its significant activity was shown with enterohemorrhagic *Escherichia coli* (EHEC), with MICs of 0.05–0.1 mg/ml and MBCs of 0.8–1.6 mg/ml.

Chusri et al. (2011) reported the minimal inhibitory concentration (MIC)/minimal bactericidal concentration (MBC) values of ethyl acetate I, ethyl acetate II, 95% ethanol and 30% ethanol fractions of *Q. infectoria* nutgalls against MRSA (methicillin resistant *Staphylcoccus aureus*) to be 0·06/0·25, 0·13/0·25, 0·25/0·5 and 0·5/1·00 mg/ml, respectively. Among its purified major components: ellagic acid, gallic acid, syringic acid and tannic acid, good MIC/MBC values were obtained with gallic acid (0·06/0·06mg/ml) and tannin acid (0·13/0·25 mg/ml). Both MRSA and *Staphylcoccus aureus* ATCC 25923 treated with the ethanol extract, ethyl acetate fraction I, gallic acid and tannic acid displayed significant loss of tolerance to low osmotic pressure and high salt concentration.

Voravuthikunchai and Suwalak (2008) reported that the fractions Qi2, Qi3, and Qi4 of *Q. infectoria* galls demonstrated good antibacterial activity against enterohemorrhagic *E. coli* O157:H7, with MICs and MBCs ranging from 250 to 500 μg/ml. *Escherichia coli* O157:H7 is one of the most important food-borne pathogens, causing non-bloody and bloody diarrhea, hemorrhagic colitis, and hemolytic uremic syndrome. The results indicated that fraction Qi4 markedly inhibited the release of verocytotoxin, VT1 and VT2 from VT-producing enterohemorrhagic E. coli (VTEC) cells at both inhibitory and subinhibitory concentrations. Use of antibiotics had been demonstrated to result in increased levels of verocytotoxin (VT) production as well as antibiotic resistance. Further, verotoxicity assay demonstrated that bacterial cultures treated with fraction Qi4 exerted less toxic effect on Vero cells. These in-vitro results distinctly indicated that the fraction Qi4 may constitute a promising natural food additive for the control of food poisoning by E. coli O157:H7 as well as other VTEC strains. Recent studies showed that ethanolic extract of *Q. infectoria* demonstrated inhibitory and bactericidal effects on all of the strains of Shiga toxigenic *Escherichia coli* tested with minimal inhibition concentrations (MICs) at 0.78–1.56 mg/ml and minimal bactericidal concentrations (MBCs) at 1.56–3.12 mg/ml (Suwalak and Voravuthikunchai 2009). *E. coli* cell numbers treated with 4×MIC of the extract decreased at least two log-fold within 4 hours and were completely killed within 12 hours Scanning electron microscopy revealed a complete loss of surface appendages and pronounced morphological changes of *E. coli* cells at MIC and 2×MIC. At 4×MIC, the damage to *E. coli* cells was extensive, and there was loss of their cellular integrity followed by cell collapse. The ethanolic extract of *Q. infectoria* modified the bacterial cell surface hydrophobicity enabling the extract to partition the lipids of the bacterial cell membrane, rendering the membrane more permeable and allowing leakage of ions and other cell contents, leading to cell mortality (Voravuthikunchai and Suwalak 2009).

Studies in Malaysia showed that the aqueous and acetone *Quercus infectoria* gall extracts displayed similarities in their antimicrobial activity on tested bacterial species indicating the galls to be potentially good source of antimicrobial agents (Basri and Fan 2005). Out of the six bacterial species tested, *Staphylococcus aureus* was the most susceptible to the gall extract of *Q. infectoria*. In contrast, the extracts showed weak inhibitory effect against *Staphylococcus epidermidis, Bacillus subtilis, Salmonella typhimurium* and *Pseudomonas aeruginosa* while there was no inhibition zone observed for *Escherichia coli* O157. The MIC values of the extracts ranged from 0.0781 to 1.25 mg/ml whereas the MBC values ranged from 0.3125 to 2.50 mg/ml. The MBC values of aqueous extract against *S. aureus* and *S. typhimurium* were higher than their MIC values. The MBC value of acetone extract against *S. aureus* was also higher than its MIC value. However, the MIC and MBC values of acetone extract against *S. typhimurium* were the same (1.25 mg/ml).

The ethanol extracts of *Quercus infectoria, Linusm usitatissium* and *Cinnamomum zeylanicium* were found to be more potent against *Escherichia coli* E45 and E62 isolates, than aqueous extracts (Khder and Muhammed 2010). The extracts exhibited most of the antibiotic activity against these two isolates irrespective of their antibiotic resistance behaviour. A comparative evaluation of plasmid elimination from *E. coli* E62 clinical isolate by sub-MIC of plant extracts showed that these extracts could cure plasmids effectively at their respective sub-MIC concentration. Sub-MIC of aqueous and ethanol of *Q. infectoria* cured 15 kb plasmid from E62 isolate.

Both aqueous and methanolic extracts of *Quercus infectoria* galls exhibited antibacterial activity against the Gram positive *Cellulosimicrobium cellulans* with MIC value of 0.5 mg/ml and MBC value of 2 mg/ml (Muskhazli et al. 2008). *C. cellulans* previously identified as *Oerskovia xanthineolytica* or *Brevibacterium fermentans* or *Arthrobacter luteus* is a virulent bacterium found in immunocomprised patients and has been reported to cause meningitis in infants and children.

Skin Whitening Activity

Rohana et al. (2004) reported that the *Q. infectoria* galls aqueous extract showed high potential in skin whitening and antioxidant properties as the extract inhibited the superoxide and DPPH radical scavenging activities, and tyrosinase activities. This was also confirmed by Vimala et al. (2007). They found that *Quercus infectoria* galls had a high DPPH scavenging activity of 99.2% and 94.7% tyrosinase inhibition. The gall extract significantly reduced the activity of tyrosinase enzyme which catalyses the biosynthesis of melanin, the colour pigment. Thus, regular application of such plant extracts could reduce pigment formation in the skin which will lead to a lighter tone of skin color. The plant extracts could also be used to treat black spots, freckles and hyperpigmentation due to accident scars and pregnancy. The results indicated that aqueous extract of *Quercus* galls had potential to be used for modern medicinal products as well as cosmetics and skin care products.

Alpha-Glycosidase Inhibitory/ Antidiabetic Activity

Hexagalloylglucose (3-O-digalloyl-1,2,4,6-tetra-O-galloyl-β-D-glucose), which was isolated from the methanol extract of the galls of *Quercus infectoria*, significantly inhibited α-glycosidases such as sucrase, maltase and isomaltase (Hwang et al. 2000). Its inhibitory activity was comparable to acarbose, a hypoglycemic agent, while the inhibitory activity on α-amylase was approximately ten times lower than that of acarbose. The results indicated that, when compared to acarbose, hexagalloylglucose may reduce the side effects by reducing inhibition of α-amylase.

Antihyperlipidemic/ Antihypercholesterolemic Activity

Quercus infectoria, *Rosa damascena* and *Myrtus communis* methanol extracts showed more than 50% inhibitory effect on β-hydroxy-β-methylglutaryl coenzyme A reductase (HMG CoA reductase) activity (Gholamhoseinian et al. 2010a). HMG CoA reductase is the key enzyme in cholesterol biosynthesis and inhibition of this enzyme had been reported to reduce the synthesis of cholesterol and could be used in the management of coronary artery disease. Preliminary in-vitro studies showed that *Q. infectoria* extract showed more than 50% inhibition on pancreatic lipase activity (Gholamhoseinian et al. 2010b). Pancreatic lipase is the most important enzyme in digestion of triglycerides. One of the strategies in prevention or treatment of obesity is altering metabolism of lipids by inhibition of dietary fat absorption.

Wound Healing Activity

Ethanol extract of the shade-dried leaves of *Quercus infectoria* exhibited a positive effect on wound healing, with a significant increase in the levels of the antioxidant enzymes, superoxide dismutase and catalase, in the granuloma tissue when tested in rats, using incision, excision and deadspace wound models, at two different dose levels of 400 and 800 mg/kg (Umachigi et al. 2008). In studies using the excision wound model, animals treated with the ethanol extract of *Q. infectoria* showed a significant decrease in the epithelization period, as evidenced by the shorter period for the fall of eschar compared to control. The extract also facilitated the rate of wound contraction significantly at both dose levels.

Analgesic/Antitremorine Activities

Early studies reported that various solvent extracted fractions of *Q. infectoria* galls possessed neuropharmacological activities (Dar et al. 1976). Fraction A (dried acetone-treated methanol extract dissolved in water) was active as an analgesic in rats and significantly reduced blood sugar levels in rabbits. Fraction B (chloroform-methanol extraction) exhibited CNS depressant activity. It potentiated the barbiturate sleeping time significantly without changing the onset time or the loss of the righting reflex. Further, Fraction B showed a moderate antitremorine activity by causing a delay in the onset and a decrease in the severity of tremorine-induced tremors. The local anaesthetic action of Fraction B was evident due to the complete blockade of the isolated frog sciatic nerve conduction. Pure syringic acid was isolated from the methanolic fraction of the galls of *Quercus infectoria* and found to be a CNS active component with significant local anaesthetic and sedative activity (Dar and Ikram 1979).

Larvicidal Activity

Ethyl-acetate extract of *Quercus infectoria* was found to be the most effective of all the five extracts tested for larvicidal activity against the fourth instar larvae of *Anopheles stephensi,* with LC_{50} of 116.92 ppm followed by gallotannin, n-butanol, acetone, and methanol with LC_{50} values of 124.62, 174.76, 299.26, and 364.61 ppm, respectively (Aivazi and Vijayan 2009). In separate studies, extracts and fractions of *Quercus lusitania* var. *infectoria* galls were found to have larvicidal activity against *Culex pipiens,* the urban nuisance mosquito (Redwane et al. 2002). Fraction F(2) had an interesting, low LC_{50} (24 h) of 60 ppm while the LC_{50} values of gallotannins were 335 and 373 ppm, respectively for the second and fourth instar period.

Molluscicidal Activity

The acetonic extract and gallotanin of *Quercus infectoria* galls exhibited high molluscicidal activity against *Bulinus truncates,* a vector of schistosomiasis (Redwane et al. 1998).

Antiprotozoal Activity

Dichloromethane and methanol extracts from the *Brucea javanica* seed and a methanol extract from *Quercus infectoria* nut-gall showed the highest inhibitory activity against the intestinal protozoan parasite, *Blastocystis hominis* (Sawangjaroen and Sawangjaroen 2005) At a concentration of 2,000 μg/ml, the three extracts killed 82%, 75% and 67% of the *Blastocystis hominis* samples tested and inhibited 94%, 100% and 76% of them, respectively. Metronidazole, used as a reference antiprotozoan drug, at a concentration of 40 μg/mL, killed 97% of the *Blastocystis hominis* isolates and inhibited all samples tested at levels that ranged from 1.25 to 20 μg/ml.

Antiamoebic Activity

Crude methanol extract of *Q. infectoria* displayed anti-amoebic effects against *Entamoeba histolytica* infecting the caecum of mice (Sawangjaroen et al. 2004). Extract of *Q. infectoria* nut gall at a concentration of 500 and of 250 mg/kg per day healed 26% and 13% of mice from amoebiasis, respectively. The severity of caecal wall ulceration was reduced in mice which received the extract.

Antiviral Activity

Of 71 plants used in Sundanese traditional medicine, *Q. infectoria* and *Syzygium aromaticum,* were the most active (>/=90% inhibition at 100 μg/mL) in inhibitory effects on hepatitis C virus (HCV) protease (Hussein et al. 2000). *Q. lusitanica (Q. infectoria)* extract was found to have good inhibitory effect on the replication of dengue virus type 2, both in conventional cell culture and proteomics technique (Muliawan et al. 2006). The extract exhibited dose-dependent in-vitro antiviral inhibition in C6/36 cells (cloned cells of *Aedes albopictus* larvae, vector of dengue fever). The extract at its maximum non-toxic concentration of 0.25 mg/ml completely inhibited 10–1,000 $TCID_{50}$ (median tissue culture infective dose) of virus, as reflected by the absence of cytopathic effect. The low dose of the extract (0.032 mg/ml) showed 100% inhibition with 10 $TCID_{50}$ of virus, but only 50% and 25% inhibition with 100 and 1,000 $TCID_{50}$, respectively. Using proteomics technique, the extract showed down-regulation of NS1 protein expression of infected C6/36 cells after treatment with this extract. The NS1 is a glycoprotein present in all flaviviruses and appears essential for virus viability.

Antivenom Activity

The aqueous extract of *Q. infectoria* galls was reported to have high hydrolysable tannin content which inhibited the lethality of the *Naja kaouthia*

(Thai cobra) venom (Pithayanukul et al. 2005). *Quercus infectoria* aqueous extract exhibited in-vitro inhibitory activity against *Naja kaouthia* but at much higher LD_{50} concentration of tannin than the aqueous extracts of *Pentace burmanica, Pithecellobium dulce,* and *Areca catechu*. The anti-venom activities of these plant polyphenols were attributed to the selective blocking of the nicotinic acetylcholine receptor and non-selectively by precipitation of the venom proteins. In another study, polyphenols from the extracts of *Areca catechu* and *Quercus infectoria* inhibited phospholipase A(2), proteases, hyaluronidase and L-amino acid oxidase of *Naja naja kaouthia* (NK) and *Calloselasma rhodostoma* (CR) venoms by in-vitro tests (Leanpolchareanchai et al. 2009). Both extracts also inhibited the hemorrhagic activity of *Calloselasma rhodostoma* venom and the dermonecrotic activity of *Naja naja kaouthia* venom by in-vivo tests. The inhibitory activity of plant polyphenols against local tissue necrosis induced by snake venoms may be caused by inhibition of inflammatory reactions, hemorrhage, and necrosis.

Antiparkinsonian Activity

Q. infectoria was found to have weak antiparkinsonian activity; the methanolic extract showed a weak activity in mice at a dose of 500 mg/kg i.p. (Dar et al. 1976).

Traditional Medicinal Uses

Nut-galls were reported to have a great medicinal value and had pharmacologically been reported to be astringent, antidiabetic, anti-tremor, local anesthetic, antipyretic and anti-Parkinson and used in the treatment of intertrigo, impetigo and eczema haemorrhages, chronic diarrhoea, dysentery, etc.

In Asian countries, the galls have been used for centuries for treating inflammatory diseases (Aroonrerk and Kamkaen 2009). Gargle of hot water extract of galls was claimed to be very effective against aphthous sores and putrid sore throat, while direct application of boiled and bruised galls on skin was claimed to effectively cure any swelling or inflammation. The application of powdered galls in the form of ointment and suppository was also utilised to cure hemorrhoids caused by inflammation of the skin. *Majuphal,* as it is widely known in Indian traditional medicine have been used as dental powder and in the treatment of toothache and gingivitis. *Quercus infectoria* (Manjakani) was claimed to be highly beneficial for the Malay Kelantanese postpartum women preparation and was thought to help in revitalization and full recovery of the reproductive functions; hazardous effects were not reported so far (Soon et al. 2007). Grieve (1971) reported that early studies showed that as part of postpartum care, the Arabs, Persians, Indians, Malays and Chinese had traditionally used *Quercus infectoria* gall nuts after childbirth to treat vaginal discharge and related postpartum infections. Its astringent property was held to play a role in the restoring of health, to tone and increase vigour of the vagina (Muhamad and Mustafa 1994).

Other Uses

An extract from the tannin-rich gallnut is mixed with ferrous sulphate together with a gum and colouring in order to make an ink. The gallnuts are also used to make a black dye besides providing a rich source of tannin and gallic acids.

Comments

Most of the gall nuts sold in Malaysia and Indonesia are imported from Turkey.

Selected References

Aivazi AA, Vijayan VA (2009) Larvicidal activity of oak *Quercus infectoria* Oliv. (Fagaceae) gall extracts against *Anopheles stephensi* Liston. Parasitol Res 104(6):1289–1293

Aroonrerk N, Kamkaen N (2009) Anti-inflammatory activity of *Quercus infectoria, Glycyrrhiza uralensis, Kaempferia galanga* and *Coptis chinensis*, the main components of Thai herbal remedies for aphthous ulcer. J Health Res 23(1):17–22

Basri DF, Fan SH (2005) The potential of aqueous and acetone extracts of galls of *Quercus infectoria* as antibacterial agents. Indian J Pharm 37(1):26–29

Chopra RN, Nayar SL, Chopra IC (1986) Glossary of Indian medicinal plants (including the supplement). Council Scientific Industrial Research, New Delhi, 330 pp

Chusri S, Voravuthikunchai SP (2008) *Quercus infectoria*: a candidate for the control of methicillin-resistant *Staphylococcus aureus* infections. Phytother Res 22(4):560–562

Chusri S, Voravuthikunchai SP (2009) Detailed studies on *Quercus infectoria* Olivier (nutgalls) as an alternative treatment for methicillin-resistant *Staphylococcus aureus* infections. J Appl Microbiol 106(1):89–96

Chusri S, Voravuthikunchai S (2011) Damage of staphylococcal cytoplasmic membrane by *Quercus infectoria* G. Olivier and its components. Lett Appl Microbiol 52:565–572. doi: 10.1111/j.1472-765X.2011.03041.x

Dar MS, Ikram M (1979) Studies on *Quercus infectoria*; isolation of syringic acid and determination of its central depressive activity. Planta Med 35:156–161

Dar MS, Ikram M, Fakouhi T (1976) Pharmacology of *Quercus infectoria*. J Pharm Sci 65(12):1791–1794

Duke JA, Bogenschutz-Godwin MJ, DuCellier J, Duke PA (2002) CRC handbook of medicinal plants, 2nd edn. CRC Press, Boca Raton, 936 pp

Foundation for Revitalisation of Local Health Traditions (2010) FRLHT Database. http://envis.frlht.org

Gholamhoseinian A, Shahouzehi B, Sharifi-far F (2010a) Inhibitory activity of some plant methanol extracts on 3-hydroxy-3-methylglutaryl coenzyme a reductase. Int J Pharm 6:705–711

Gholamhoseinian A, Shahouzehi B, Sharifi-far F (2010b) Inhibitory effect of some plant extracts on pancreatic lipase. Int J Pharm 6:18–24

Grieve M (1971) A modern herbal. Penguin, 2 vols. Dover Publications, New York, 919 pp

Hamid H, Kaur G, Abdullah S, Ali M, Athar M, Alam M (2005) Two new compounds from the galls of *Quercus infectoria* with nitric oxide and superoxide inhibiting ability. Pharm Biol 43(4):317–323

Hedge IC, Yaltırık F (1982) *Quercus* L. In: Davis PH (ed) Flora of Turkey and the East Aegean Islands, vol 7. Edinburgh University Press, Edinburgh, pp 659–683

Hussein G, Miyashiro H, Nakamura N, Hattori M, Kakiuchi N, Shimotohno K (2000) Inhibitory effects of Sudanese medicinal plant extracts on hepatitis C virus (HCV) protease. Phytother Res 14(7):510–516

Huxley AJ, Griffiths M, Levy M (eds) (1992) The new RHS dictionary of hardening (4 vols). Macmillan, London

Hwang JK, Kong TW, Baek NI, Pyun YR (2000) Alpha-glycosidase inhibitory activity of hexagalloylglucose from the galls of *Quercus infectoria*. Planta Med 66(3):273–274

Hwang JK, Shim JS, Chung JY (2004) Anticariogenic activity of some tropical medicinal plants against *Streptococcus mutans*. Fitoterapia 75(6):596–598

Kaur G, Hamid H, Ali A, Alam MS, Athar M (2004) Antiinflammatory evaluation of alcoholic extract of galls of *Quercus infectoria*. J Ethnopharmacol 90(2–3):285–292

Kaur G, Athar M, Alam MS (2008) *Quercus infectoria* galls possess antioxidant activity and abrogates oxidative stress-induced functional alterations in murine macrophages. Chem Biol Interact 171(3):272–282

Khder AK, Muhammed SA (2010) Potential of aqueous and alcohol extracts of *Quercus infectoria, Linusm usitatissium* and *Cinnamomum zeylanicium* as antimicrobials and curing of antibiotic resistance in *E. coli*. Curr Res J Biol Sci 2(5):333–337

Leanpolchareanchai J, Pithayanukul P, Bavovada R (2009) Anti-necrosis potential of polyphenols against snake venoms. Immunopharmacol Immunotoxicol 31(4):556–562

Muhamad Z, Mustafa AM (1994) Traditional Malay Medicinal Plants. Penerbit Fajar Bakti Sdn. Bhd., Kuala Lumpur

Muliawan SY, Lam SK, Devi S, Hasim O, Yusof R (2006) Inhibitory potential of *Quercus lusitanica* extract on dengue virus type 2 replication. Southeast Asian J Trop Med Public Health 37(Suppl 3):132–135

Muskhazli M, Nurhafiz Y, Nor Azwady AA, Nor Dalilah E (2008) Comparative study on the in vitro antibacterial efficacy of aqueous and methanolic extracts of *Quercus infectoria* gall's against *Cellulosimicrobium cellulans*. J Biol Sci 8:634–638

Pithayanukul P, Ruenraroengsak P, Bavovada R, Pakmanee N, Suttisri R, Saen-oon S (2005) Inhibition of *Naja kaouthia* venom activities by plant polyphenols. J Ethnopharmacol 97(3):527–533

Pithayanukul P, Nithitanakool S, Bavovada R (2009) Hepatoprotective potential of extracts from seeds of *Areca catechu* and nutgalls of *Quercus infectoria*. Molecules 14(12):4987–5000

Redwane A, Markouk M, Lazrek HB, Amarouch H, Jana M (1998) Laboratory evaluation of molluscicidal activity of extracts from *Cotula cinerea* (L) and *Quercus lusitania* var. *infectoria* galls (Oliv.). Ann Pharm Fr 56(6):274–276

Redwane A, Lazrek HB, Bouallam S, Markouk M, Amarouch H, Jana M (2002) Larvicidal activity of extracts from *Quercus lusitania* var. *infectoria* galls (Oliv.). J Ethnopharmacol 79(2):261–263

Rohana S, Vimala S, Abdull Rashih A, Mohd. Ilham A (2004) Skin whitening and antioxidant properties of *Quercus infectoria* galls. In: Chang YS, Mastura M, Nurhanan MY (eds) Proceedings of the seminar on medicinal plants. Forest Research Institute Malaysia (FRIM), Selangor, Malaysia, pp 188–191

Sawangjaroen N, Sawangjaroen K (2005) The effects of extracts from anti-diarrheic Thai medicinal plants on the in vitro growth of the intestinal protozoa parasite: *Blastocystis hominis*. J Ethnopharmacol 98(1–2):67–72

Sawangjaroen N, Sawangjaroen K, Poonpanang P (2004) Effects of *Piper longum* fruit. *Piper sarmentosum* root and *Quercus infectoria* nut gall on caecal amoebiasis in mice. J Ethnopharmacol 91(2–3):357–360

Soon LK, Hasni E, Law KS, Waliullah SS, Farid CG, Syed Mohsin SSJ (2007) Ultrastructural findings and elemental analysis of *Quercus infectoria* Oliv. Ann Microsc 7:32–37

Suwalak S, Voravuthikunchai SP (2009) Morphological and ultrastructural changes in the cell structure of enterohaemorrhagic *Escherichia coli* O157:H7 following treatment with *Quercus infectoria* nut galls. J Electron Microsc (Tokyo) 58(5):315–320

Umachigi SP, Jayaveera KN, Ashok Kumar CK, Kumar GS, Swamy BMV, Kumar DVK (2008) Studies on wound healing properties of *Quercus infectoria*. Trop J Pharm Res 7(1):913–919

Vermani A, Navneet P (2009) Screening of *Quercus infectoria* gall extracts as anti-bacterial agents against dental pathogens. Indian J Dent Res 20(3):337–339

Vimala S, Ilham MA, Rashih AA, Rohana S, Juliza M (2007) Antioxidant and skin whitening standardized extracts: cosmeceutical and neutraceutical products development and commercialization in FRIM. Sustainable management and utilization of medicinal plant resources. In: Proceedings of the international conference on medicinal plants, UPM and JPSM, pp 224–230

Voravuthikunchaia SP, Chusrib S, Suwalak S (2008) *Quercus infectoria* Oliv. Pharm Biol 46(6):367–372

Voravuthikunchai SP, Kitpipit L (2005) Activity of medicinal plant extracts against hospital isolates of methicillin-resistant *Staphylococcus aureus*. Clin Microbiol Infect 11(6):510–512

Voravuthikunchai SP, Limsuwan S (2006) Medicinal plant extracts as anti-*Escherichia coli* O157:H7 agents and their effects on bacterial cell aggregation. J Food Prot 69(10):2336–2341

Voravuthikunchai SP, Lortheeranuwat A, Jeeju W, Sririrak T, Phongpaichit S, Supawita T (2004) Effective medicinal plants against enterohaemorrhagic *Escherichia coli* O157:H7. J Ethnopharmacol 94(1):49–54

Voravuthikunchai SP, Suwalak S (2008) Antibacterial activities of semipurified fractions of *Quercus infectoria* against enterohemorrhagic *Escherichia coli* O157:H7 and its verocytotoxin production. J Food Prot 71(6):1223–1227

Voravuthikunchai SP, Suwalak S (2009) Changes in cell surface properties of shiga toxigenic *Escherichia coli* by *Quercus infectoria* G. Olivier. J Food Prot 72(8):1699–1704

Ribes nigrum

Scientific Name

Ribes nigrum L.

Synonyms

Botrycarpum nigrum (L.) A. Rich., *Grossularia nigra* (L.) Rupr., *Ribes cyathiforme* Pojark, *Ribes nigrum* forma *chlorocarpum* (Späth) Rehder, *Ribes nigrum* var. *chlorocarpum*, *Ribes nigrum* var. *europaeum* Jancz, *Ribes nigrum* var. *pauciflorum* (Turcz. ex Ledeb.) Jancz, *Ribes nigrum* var. *sibiricum* W.Wolf, *Ribes olidum* Moench nom. illeg., *Ribes pauciflorum* Turcz. ex Ledeb.

Family

Grossulariaceae

Common/English Names

Blackcurrant, European Blackcurrant, Garden Blackcurrant

Vernacular Names

Arabic: Kishmish Aswad;
Brazil: Groselha (Portuguese);
Chinese: Hei Cha Bao Zi;
Czech: Rybíz Černý;
Danish: Solbær, Vild Solbær;
Dutch: Zwarte Aalbes, Zwarte Bes, Zwarte Trosbes;
Eastonian: Must Sõstar;
Finnish: Mustaherukka, Mustaviinimarja;
French: Cassis, Cassissier, Gadellier Noir, Groseillier À Fruits Noirs, Groseillier Noir;
German: Ahlbeere, Cassis, Schwarze Johannisbeere, Schwarze Johannisbeeren;
Greek: Fragostafyla Mavra;
Hungarian: Fekete Ribiszke;
Icelandic: Sólber, Sólberjarunni, Svört Hlaupber;
Irish: Cuirín Dubh;
Italian: Ribes Nero, Ribes Nigrum;
Japanese: Kashisu, Kuro Fusa Suguri, Kurorasasuguri;
Korean: Komunkkachibapnamu, Komunsongimulaengdunamu;
Lithuanian: Juodųjų Serbentų;
Morocco: Nnbaq Aswad (Arabic);
Norwegian: Solbær;
Pakistan: Karan;
Polish: Porzeczka Czarna;
Portuguese: Groselha Negra , Groselheira Negra, Groselheira-Preta;
Russian: Smorodina Černaja;
Serbian: Crna Ribizla;
Slovaščina: Črni Ribez, Črno Grozdičje, Grozdičje Črno;
Slovencina: Ríbezľa Čierna;
Spanish: Casis, Grosella Negra, Grosellero Negro;
Swedish: Svarta Vinbär, Solbär, Svart Vinbär, Tistron;

Origin/Distribution

Ribes nigrum is indigenous to central and northern Europe, Caucasus, Central Siberia and Himalaya.

Agroecology

Black current is a temperate species, winter hardy with optimal temperature for the growth of 18–20°C and optimal temperature for photosynthesis 15–25°C. Temperatures above 35°C are detrimental and cause leaf fall. The crop requires a period of winter chill to terminate dormancy and stimulate bud break the following spring. The optimum temperature for meeting the chill requirement is between 0°C and 10°C, with general agreement that 5 ± 1°C is satisfactory (Shirazi 2003). Studies in Tasmania showed that 2°C was more effective than the winter chill model of 7.2°C adopted by the Tasmanian Blackcurrant industry (Westmore 2004).

Blackcurrant prefers moisture-retentive, well-drained, fertile loams or deep sandy loamy soils. It prefers damp fertile soils with subsoil water not nearer than 1–1.5 m from surface. It is the least drought-resistant species among currants. Blackcurrant abhors heavy clay, chalky soils and thin dry soils. It is intolerant of acid soils and prefers soils in pH range of 6.7–7.

Edible Plant Parts and Uses

Blackcurrants can be eaten fresh, on its own, or with ice cream or in mixed fruit salad. They are more often processed into jam, fruit jelly, compote, syrup juice, preserves, wines and liqueurs and other beverages. In UK pubs blackcurrant cordial is often mixed with cider in a drink called "Cider and Black" or "snakebite". A dash of blackcurrant juice is used in Guinness stout to augment the taste. Blackcurrant syrup when added to white wine is called *Kir* or *Kir Royale* when mixed with champagne. In Belgium and the Netherlands, macerated blackcurrants are also the primary ingredient in the apéritif *crème de cassis*. A universally available popular drink, "Ribena", is a juice drink made from blackcurrants.

Blackcurrants are also cooked and used in pies, sauces, meat dishes, desserts and confectionery and as relish for dishes. Blackcurrant is a standard ingredient of "*Rødgrød*", a popular kissel-like dessert in North German and Danish cuisines. In UK and Europe blackcurrant is used to flavour some confectionery. Dried blackcurrant is used as a main ingredient of *pemmican* which is a mixture of dried meat and fat (tallow). Japan imports considerable volume of New Zealand blackcurrants for uses as dietary supplements, snacks, functional food products and as quick-frozen (IQF) produce for culinary food products such as jams, jellies or preserves. An essential "*Níribine*" oil can be distilled from the buds for use in the food and alcohol industry.

Studies showed that blackcurrant jam with low sugar content and without additives and manufactured at 92°C and stored at 8°C still have good sources of vitamins and antioxidant after a year (Viberg et al. 1997). After 13 months of storage, at 8°C, 60% of the amount of ascorbic acid and 29% of the quantity of anthocyanins were retained. In the jam stored at higher temperatures less of both were retained. The β-carotene in the jam was found to be stable throughout the whole shelf-life study. Another study showed that black currant could be used as fermentation substrates for producing alcoholic beverages obtained by distillation of the fruits previously fermented with *Sacchromyces cerevisiae* (Alonso González et al. 2010). The amount of volatile compounds in the black currant distillate (121.1 g/hL absolute alcohol) was lower than the minimum limit (200 g/hL absolute alcohol) fixed by the European Council (Regulation 110/2008) for fruit spirits and did not pose health hazards.

Leaves have been reported to be used in soups. In Russia, blackcurrant leaves are used to flavour tea or preserves. Sweetened vodka is often infused with blackcurrant leaves or berries, to give a deep yellowish-green beverage with a sharp flavour and astringent taste. Dried leaves are used in herbal teas or tea blends. The leaves are utilized for preservation of mushrooms and vegetables.

Botany

A small, erect shrub growing to 1–2 m tall with glabrous branchlets which are pubescent when young. Leaves are alternate, petiolate (petiole 1–4 cm long), suborbicular, palmately lobed with five lobes, 3–5 cm long and broad, and with a serrated margin, base cordate, abaxial surface pubescent and glandular, adaxial surface puberulent when young, glabrescent (Plates 1 and 3). Flowers in 4–12-flowered racemes with pubescent rachis and pedicels and pubescent, lanceolate or ovate bracts (Plate 2). Flower bisexual, 5–7 mm across. Calyx yellowish green to pinkish, pubescent and yellow glandular; tube subcampanulate, 1.5–2.5 mm with lobes spreading or reflexed, ligulate, lobes. Petals ovate to ovate-elliptic, 2–3 mm, white to reddish, nectary disc prominent, green or purplish, circular, covering ovary; stamens slightly longer than petals; filaments linear with white sagittate anthers. Fruit, globose, 0.8–1 (–1.4) cm, sparsely glandular, green (Plates 3 and 4) turning to glossy, black, but sometimes brown or green, with persistent calyx and containing several to numerous seeds.

Plate 1 Blackcurrant leaves

Nutritive/Medicinal Properties

Nutrient composition of raw ripe *Ribes nigrum* fruits per 100 g edible portion was reported as: water 81.96 g, energy 63 kcal (264 kJ), protein 1.40 g, total lipid 0.41 g, ash 0.86 g, carbohydrate 15.38 g, Ca 55 mg, Fe 1.54 mg, Mg 24 mg, P 59 mg, K 322 mg, Na 2 mg, Zn 0.27 mg, Cu 0.086 mg, Mn 0.256 mg, Vitamin C 181 mg, thiamine 0.05 mg, riboflavin 0.05 mg, niacin 0.3 mg, panthothenic acid 0.398 mg, vitamin B-6 0.066 mg, vitamin A 230 IU, vitamin E (α-tocopherol) 1 mg, total saturated fatty acids 0.034 g, 16:0 (palmitic) 0.02 g, 18:0 (stearic) 0.056 g, total monounsaturated fatty acids 0.058 g, 16:1 undifferentiated (palmitoleic) 0.001 g, 18:1 undifferentiated (oleic) 0.056 g, total polyunsaturated fatty acids 0.179 g, 18:2 undifferentiated (linoleic) 0.107 g, and 18:3 undifferentiated (linolenic) 0.072 g (USDA 2011).

Plate 2 Blackcurrant flowers and buds

The total lipid content by weight of fruit seeds of the *Ribes* family was found to range from 18.3% in goose-berries (*Ribes uva crispa*) to 30.5% in black currants (*Ribes nigrum*) (Traitler

Plate 3 Unripe blackcurrant fruits

Plate 4 Close-up of unripe blackcurrant fruits

et al. 1984). Black currant seed oil was found to contain up to 19% by weight of γ-linolenic acid (γ-LA, C18:3, n-6). Black currant species thus represents one of the richest natural sources in γ-LA yet described. These oils had been reported to be promising for critically ill patients unable to convert linoleic acid into subsequent essential fatty acid fractions. Study of the fatty acid profile of 29 black currant genotypes found that α-linolenic, stearidonic, and γ-linolenic acid (GLA) contents, varied between 11.1% and 18.7%, between 2.5% and 4.5%, and between 11.6% and 17.4%, respectively (Del Castillo et al. 2004). Although GLA content was not strongly correlated with juice parameters, some genotypes had both high GLA contents and desirable juice characteristics.

Goffman and Galletti (2001) found that the highest total tocopherol content was found in *R. nigrum* (mean, 1,716 mg/kg oil), followed by *R. rubrum* (mean, 1,442 mg/kg oil). *R. grossularia* showed the lowest tocopherol content (mean, 786 mg/kg oil). The three species also differed markedly in tocopherol composition. *R. rubrum* had the highest content of δ-tocopherol (mean, 20.2%); *R. grossularia* had the highest level of γ-tocopherol (mean, 70.0%), and *R. nigrum* the highest percentage of α-tocopherol (mean, 34.8%), the most biologically active among the four tocopherols. As for γ-linolenic acid, the highest concentration was found in *R. nigrum*, up to 15.8% while *R. grossularia* and *R. rubrum* showed mean γ-linolenic acid levels of 8% and 6.2%, respectively. The present study indicated that seeds of *Ribes* species, especially *R. nigrum*, could be used as sources of gamma-linolenic acid and natural vitamin E. γ-linolenic acid is an essential fatty acid for humans with δ-6-desaturase deficiency; it is a precursor of prostaglandins, prostacyclins, and tromboxanes; and has anti-inflammatory and antitumoral effects. Tocopherols are natural antioxidants with biological activity, heart/vascular, and cancer protective properties.

Oil contents of Canadian black currant seeds varied from 27% to 33% (Bakowska-Barczak et al. 2009). The γ-linolenic acid content varied significantly among the cultivars (from 11% for Ben Conan to 17% for Ben Tirran). Among the 44 triacylglycerol identified, LLαLn, αLnLγLn, and PLγLn (where L=linoleoyl, αLn=α-linolenoyl, γLn=γ-linolenoyl, and P=palmitoyl) were the predominant ones. Black currant seed oil was also a good source of tocopherols (mean 1,143 mg/100 g of oil) and phytosterols (6,453 mg/100 g of oil on average). Quercetin-3-glucoside and p-coumaric acid were the main phenolic components in the seed residues. The high concentration of flavonols and phenolic acids was correlated with a high antioxidant activity of seed residue (average ABTS value of 1.5 mM/100 g and DPPH value of 1.2 mM/100 g). The data indicated Canadian black currant seed oil to be a good source of essential fatty acids, tocopherols, and phytosterols.

Besides genotype, latitude and weather conditions were found to impact on sugars, fruit acids, and ascorbic acid levels in black currant juice (Zheng et al. 2009). In comparison to black currants grown in northern Finland (latitude 66° 34′N), the berries grown in southern Finland (latitude 60° 23′N) had higher contents of fructose, glucose, sucrose, and citric acid (by 8.8%, 6.1%, 10.0%, and 11.7%, respectively) and lower contents of malic acid, quinic acid, and vitamin C (by 31.1%, 23.9%, and 12.6%). Fructose, glucose, and citric acid in cultivar Melalahti were not influenced by the weather, whereas their concentrations in Mortti and Ola correlated positively with the average temperature in February and July and negatively with the percentage of the days with a relative humidity of 10–30% from the start of the growth season until the day of harvest. Similarly latitude and weather conditions were found to impact on the regioisomer compositions of triacylglycerols like α-linolenoyldilinoleoylglycerol and γ-linolenoyldilinoleoylglycerol in currant seed oils (Leskinen et al. 2009). In $R.$ $rubrum$ the proportion of the symmetric regioisomer LAlaL among Ala/L/L (18:3(n-3)/18:2(n-6)/18:2(n-6)) was higher (14.1%) than in $R.$ $nigrum$ (12.1%). Generally in currants, the portion of LAlaL was lower in northern Finland (12.1%) than in southern Finland (13.5%), where temperature and radiation sums were higher.

Other Phytochemicals

The major flavonol glycosides isolated from ripe blackcurrants ($Ribes$ $nigrum$ cv. Silvergieters Schwarze) were myricetin 3-β-D-glucopyranoside, rutin, and isoquercitrin (Koeppen and Herrmann 1977). The presence of the 3-rutinosides and 3-glucosides of cyanidin and delphinidin was also confirmed. No free flavonoid aglycones was detected in the fresh berries. The major constituent fluorescing blue under ultraviolet light was isolated and characterized as 1-O-caffeyl-β-D-glucopyranose. Also isolated were 1-O-ferulyl- and 1-O-p-coumaryl-β-D-glucopyranose and hydroxycinnamyl-D-glucoses. Matsumoto et al. (2001) isolated from black currant fruits four anthocyanin components: delphinidin 3-O-β-rutinoside, cyanidin 3-O-β-rutinoside, delphinidin 3-O-β-glucoside, and cyanidin 3-O-β-glucoside. Fifteen anthocyanin structures were reported from an extract of black currant berries ($Ribes$ $nigrum$) (Slimestad and Solheim 2002). These were the 3-O-glucosides and the 3-O-rutinosides of pelargonidin, cyanidin, peonidin, delphinidin, petunidin, and malvidin, cyanidin 3-O-arabinoside, and the 3-O-(6″-p-coumaroylglucoside)s of cyanidin and delphinidin. The four main pigments (the 3-O-glucosides and the 3-O-rutinosides of delphinidin and cyanidin) comprised >97% of the total anthocyanin content. The amounts of anthocyanin rutinosides were found to be higher than the amount of the corresponding glucosides for all detected pigments having the same aglycon moiety.

Eleven delphinidin-, cyanidin-, malvidin-, petunidin-, and peonidin-based anthocyanins were detected, with the main components being delphinidin-3-O-glucoside (839 nmol/g of fresh weight), delphinidin-3-O-rutinoside (2,233 nmol/g), and cyanidin-3-O-rutinose (1,693 nmol/g) (Borges et al. 2010). The other anthocyanins were cyanidin-3-O-glucoside (327 nmol/g), delphinidin-3-O-(600-p-coumaroyl)glucoside (77 nmol/g), delphinidin-3-O-galactoside (52 nmol/g), petunidin-3-O-rutinoside peonidin-3-O-galactoside (103 nmol/g), malvidin-3-O-galactoside peonidin-3-O-glucoside (71 nmol/g), peonidin-3-O-rutinoside peonidin-3-O-rutinoside (126 nmol/g). In addition to anthocyanins, the black currants contained vitamin C (2,328 nmol/g) and smaller quantities of a caffeic acid-O-glucoside (76 nmol/g). Several flavonols myricetin-3-O-rutinoside (135 nmol/g), myricetin-O-glucuronide

(138 nmol/g), and myricetin-3-O-(600-malonyl) glucoside (29 nmol/g) were also detected. Several kaempferol and quercetin conjugates were also found such as kaempferol-3-O-rutinoside (12 nmol/g), kaempferol-3-O-galactoside (23 nmol/g), quercetin-3-O-rutinoside (77 nmol/g), quercetin-3-O-glucoside (83 nmol/g), and quercetin-3-O-(600-malonyl)glucoside (17 nmol/g).

Of the berries of *Ribes nigrum* (black and green currants) and *Ribes x pallidum* (red and white currants), the highest contents of anthocyanins (3,011 mg/kg fresh weight, expressed as the aglycon) and flavonol glycosides (100 mg/kg) were found in black currant (Maatta et al. 2001). The lack of anthocyanins in the colourless (green, white) berries was associated with elevated levels of phenolic acids, especially p-coumaric acid (80 mg/kg in green currant versus 45 mg/kg in black currant) and 4-hydroxybenzoic acid (18 mg/kg in white currant versus 3 mg/kg in red currant). The study demonstrated that the amounts of extractable (22–41 mg/kg) and non-extractable proanthocyanidins (32–108 mg/kg) were comparable to those of other phenolics, with the exception of anthocyanins in black currant. The results suggested that anthocyanins dominated in black and red currants, whereas proanthocyanidins and phenolic acids were the predominant phenolic compounds in green and white currants. Rubinskiene et al. (2005) reported that the highest contents of pigments were found in overripe *Ribes nigrum* berries. Delphinidin-3-rutinoside was the major component in the reddish coloured berries (onset of ripening), and cyanidin-3-rutinoside was the dominant pigment in the black ones (ripe berries). Cyanidin-3-rutinoside was found to be the most thermally stable anthocyanin.

Scalzo et al. (2008) found that *Ribes* genotypes had greater total anthocyanin contents than *Vaccinium* which was, in turn, higher than *Rubus*. However, all genera afforded rich sources of anthocyanins and individual crop types within genera varied substantially. Five percent (i.e., 223) of samples comprising mostly blackcurrants, some black raspberries and three ornamental blueberries had total anthocyanin contents >5,000 μg/g. *Ribes nigrum* samples were predominated by cyanidin and delphinidin rutinosides which, comprised almost 80% of the total anthocyanin contents. Non-structural carbohydrates of blackcurrant fruits from 17 UK-grown black currant cultivars were found to range from 85.09 to 179.92 mg/g fresh weight basis, while organic acids contents ranged from 36.56 to 73.35 mg/g FW. Relative concentrations of cyanidin 3-glucoside, cyanidin 3-rutinoside, delphinidin 3-glucoside and delphinidin 3-rutinoside were 3.1–7.9%, 35.4–47.0%, 7.6–12.5% and 36.9–50.9%, respectively (Bordonaba and Terry 2008).

Raw cutin (i.e., extractive-free isolated cuticular membrane) fraction from Finnish berries including *Ribes nigrum* was found to contain <50% polyester polymer cutin (Kallio et al. 2006). The major cutin monomers were C(16) and C(18) omega-hydroxy acids with mid-chain functionalities, mainly epoxy and hydroxyl groups. Generally, the dominant compounds were 9,10-epoxy-18-hydroxyoctadecanoic acid, 10,16-dihydroxyhexadecanoic acid, 9,10,18-trihydroxyoctadecanoic acid, 9,10-epoxy-18-hydroxyoctadec-12-enoic acid, and 18-hydroxyoctadec-9-enoic acid. The black currant cutin differed from that of the other berries with a significant component of hydroxyoxohexadecanoic acid (about 12% of total monomers).

Two novel nitrile-containing compounds, nigrumin-5-p-coumarate and nigrumin-5-ferulate, together with six known flavonoids, were isolated from black currant (*Ribes nigrum*) seeds (Lu et al. 2002). The chemical structures of nigrumin-5-p-coumarate and 5-ferulate were elucidated as 2-trans-p-coumaroyloxymethyl-4-β-D-glucopyranosyloxy-2(E)-butenenitrile and 2-trans-feruloyloxymethyl-4-β-D-glucopyranosyloxy-2(E)-butenenitrile, respectively. The non-conjugated octadecatetraenoic acid found in the seed oil of *Ribes nigrum* was found to be identical to the C18-polyunsaturated fatty acid previously isolated in a number of fish oils and seed oils (Moine et al. 1992). Comparison with authentic material prepared by chemical synthesis provided further confirmation of the (all-cis)-6,9,12,15-octadecatetraenoic acid structure.

Flavonoids (kaempferol, quercetin, myricetin) and phenolic acids (p-coumaric, caffeic, ferulic, p-hydroxybenzoic, gallic and ellagic acids) were detected in the fruits of 19 berries (Hákkinen et al. 1999). In the genus *Ribes*, quercetin was the main compound in gooseberry, red currant and black currant. The data suggested berries to have potential as good dietary sources of quercetin or ellagic acid. The major phenolic compound in red currant was found to be chlorogenic acid (da Silva Pinto et al. 2010). Of red and black currants, and red and green gooseberries, red currants had the highest α-glucosidase, α-amylase and angiotensin I-converting enzyme (ACE) inhibitory activities. A complex profile of anthocyanins, flavonols, flavan-3-ols and hydroxycinnamic acid derivatives were assayed in acetone-acetic acid (99:1, v/v) extracts of blue berries, red and black currants (Gavrilova et al. 2011). Anthocyanins comprised the highest content of total phenolic compounds in currants (>85%) and lower and variety dependent in blueberries (35–74%). Hydroxycinnamic acid derivatives comprised 23–56% of total phenolics in blueberries and 1–6% in currants. From flavan-3-ols, epigallocatechin was detected in currants.

Fifty-three volatile compounds were identified from blackcurrants and quantified through calibration curves (Harb et al. 2008). Terpene compounds included terpene alcohols: terpineal-4-ol, eucalyptol, β-linalool, nerol; terpene esters: nerol acetate, linalool acetate, citronellyl butyrate, α-terpineol acetate; monoterpenes: 3-carene, α-pinene, β-pinene, β-myrcene, D-limonene, 4-carene, β-phellandrene; and sesquiterpenes: β-caryophlene, α-farnesene, (z)-β-farnesene. Nonterpene volatiles included non-terpene alcohols: tetradecane, 10-undecen-1-ol, 1-octanol, 1-hexanol, 1-heptanol, eugenol; aldehydes: hexanal, 2-nonenal (E), nonanal, benzaldehyde; and branched or straight-chain esters: methyl decanoate, hexyl acetate, methyl benzoate and hexyl hexanoate, ethyl butanoate, heptyl butanoate, hexyl 2-methylbutanoate, ethyl benzoate, ethyl acetate, butyl 2-methylbutanoate and ethyl 2-methylbutanoate, ethyl octanoate, ethyl 2-butenoate, hexyl formiate, ethyl hexanoate, heptyl acetate, methyl octanoate, octyl acetate, octyl formiate and β-phenyethyl acetate. Fruit that were stored in air, for either 3 or 6 weeks, did not differ significantly from freshly harvested fruit with respect to total terpene volatiles. However, decreasing O2 levels and increasing CO2 levels retarded the capacity of 3-week stored fruit to synthesize terpenes, although prolonged storage under these conditions led to a partial recovery. Terpene alcohols reached a peak in 6-week air-stored fruit, and storing berries under a high CO_2 level (18 kPa) and/or decreasing O_2 level (2 kPa) resulted, in most cases, in lower biosynthesis of these alcohols compared to 6-week air-stored fruit. Non-terpene compounds, mainly esters and alcohols, were also increased in airstored fruit. Non-terpene esters differed greatly in storage, in particular the ester ethyl butanoate. Air-stored fruit at both sampling dates synthesized significantly higher amounts of esters than freshly harvested fruit but a significant decline was observed for branched butyl substances (2-methylbutanoate) after 6 weeks storage. Forty-nine aromatic compounds were quantified in black currant juice, and the thermal treatment resulted in concentration increases of most terpenes, aldehydes, furans, and phenols, whereas the concentration of esters slightly decreased (Varming et al. 2004). Higher temperatures and longer exposure times had larger effects on the aroma compounds. It was found that a 90°C thermal treatment of black currant juice, the temperature range used for conventional evaporation of black currant juice, had an effect on the aroma and sensory properties.

The glycerolipid composition of *R. nigrum* leaves was found to be largely characteristic of 16:3 plants but there was a minor contribution typical of 18:4 plants (Dobson 2000). The total fatty acid composition was unusual in that α-linolenic acid (alpha-18:3) occurred together with cis-7,10,13-hexadecatrienoic acid (16:3) and lower amounts of stearidonic acid (18:4) and γ-linolenic acid (γ-18:3). Monogalactosyldiacylglycerol contained the highest proportion of 16:3 with less in digalactosyldiacylglycerol. All lipids had γ-18:3 and 18:4 with the latter always higher than the former. The highest percentages of γ-18:3 and 18:4 were in phosphatidylcholine, but phosphatidylglycerol was particularly low in these acids. A flavonoid, kaempferol-3-0-(6″-0-

malonyl)-6-D-glucopyranoside was isolated from the leaves of black-currant (Pieri et al. 2002).

Amongst nine different berry extracts investigated for their free radical scavenging activity, black currant (*Ribes nigrum*) extract was recently found to be the second most effective (Bishayee et al. 2010). Black currant is known to have high contents of anthocyanins (250 mg/100 g fresh fruit). Black currant fruits have been used in Asian and European traditional medicine for the treatment of a variety of diseases. Compounds present in black currant juice were found to exert a number of health-promoting effects, including immunomodulatory, antioxidant, antimicrobial and antiinflammatory actions, inhibition of low-density lipoprotein, and reduction of cardiovascular diseases.

Antioxidant Activity

In black currant, the three main anthocyanins, delphinidin-3-O-glucoside (14.2%), delphinidin-3-O-rutinoside (32.8%), and cyanidin-3-O-rutinose (18.5%) along with vitamin C (17.5%), were the major contributors to the antioxidant capacity (AOC) of the extract (Borges et al. 2010) The flavonols myricetin-3-O-rutinoside and myricetin-3-O-glucuronide were each responsible for 1.9% of the total AOC, with the quercetin and kaempferol and quercetin conjugates (each <1%) providing minimal contribution.

All the crude extracts of the berries of several cultivars of *Ribes, Rubus*, and *Vaccinium* genera examined showed a significantly high activity towards chemically-generated superoxide radicals (Costantino et al. 1992). The activities were higher than those expected on the basis of the quantities of anthocyanins and polyphenols present in the samples. In addition, the extracts exhibited inhibitory activity towards xanthine oxidase. *Ribes nigrum* extracts showed the highest activity, being the richest in both anthocyanins and polyphenols. In contrast, *Ribes rubrum* extracts seemed to contain more active substances than the other crude extracts. Berries such as raspberry (*Rubus idaeus*), bilberry (*Vaccinium myrtillus*), lingonberry (*Vaccinium vitis-idaea*), and black currant (*Ribes nigrum*) were found to be rich in monomeric and polymeric phenolic compounds providing protection toward both lipid and protein oxidation as assessed by a lactalbumin-liposome system (Viljanen et al. 2004). In bilberries and black currants, anthocyanins contributed the most to the antioxidant effect by inhibiting the formation of both hexanal and protein carbonyls. Studies showed that black currant anthocyanins provided good antioxidant protection toward oxidation of tryptophan (Salminen and Heinonen 2008).

After treatment with ornithine decarboxylase inhibitor, a polyamine inhibitor (O-phosphoethanolamine, KF), and a phenol biosynthesis stimulator (carboxymethyl chitin glucan, CCHG), the antioxidant activities compared using LDL in-vitro oxidation assay were increased more markedly after treatment with KF in both black currant and chokeberry, though the regulators had the lower effect on the phenolic accumulation in black currant (Hudec et al. 2009). There was a strong correlation between the total phenolics in the both crops and anthocyanins, hydroxybenzoic acids, and hydroxycinnamic acids contents, respectively. Both regulators significantly changed the ratio of conjugated (rutin) to free (quercetin) flavonol mainly in black chokeberry.

Treatment of black currant pomace with commercial pectinolytic enzyme preparations, Macer8 FJ, Macer8 R, Novozym 89 protease and Pectinex BE significantly enhanced the contents of phenols extracted from the pomace (Landbo and Meyer 2001). A decrease in pomace particle sizes from 500–1,000 μm to <125 μm elevated the phenol yields 1.6–5 fold. Black currant pomace without seeds gave significantly higher yields of phenols than pomace with seeds and seedless wine pomace. Four selected black currant pomace extracts all exerted a pronounced antioxidant activity against human LDL oxidation in vitro when tested at equimolar phenol concentrations of 7.5–10 μM.

Four black currant anthocyanins, cyanidin 3-O-β-glucoside, cyanidin 3-O-β-rutinoside, delphinidin 3-O-β-glucoside, and delphinidin 3-O-β-rutinoside were quantitatively determined in black current juices (Nielsen et al. 2003). The antioxidant capacity of all 13 black currant juices was determined by TEAC and FRAP. Less than 70% of the antioxidant capacity of the juices could be due to vitamin C or the anthocyanin indicating that other very potent antioxidants were present in commercial black currant juices. In another study, phenol content in the black currant press residue (BPR) extracts was found to be 8–9 times higher than in the black currant pomace extracts (Kapasakalidis et al. 2006). Acid hydrolysis liberated a much higher concentration of phenols from the pomace than from the black currant press residue. The main anthocyanins constituted the main class of phenols and the following anthocyanins were indentified delphinidin-3-O-glucoside, delphinidin-3-O-rutinoside, cyanidin-3-O-glucoside, and cyanidin-3-O-rutinoside. Anthocyanins were present in considerably lower amounts in the pomace than in the BPR. Using the ABTS(*)(+) assay, BPR extracts prepared by solvent extraction exhibited significantly higher (7–10 times) radical scavenging activity than the pomace extracts, and BPR anthocyanins contributed significantly (74% and 77%) to the observed high radical scavenging capacity of the corresponding extracts.

Black currant berries contained very high levels of natural phenolic compounds; the leaves and buds were also found to be a good source of natural antioxidants (Tabart et al. 2011). They contained high amount of phenolic acids, flavonoids and carotenoids. An acetone mixture extract gave good yield of several classes of phenolic compounds that included flavonols, flavan-3-ols and anthocyanins.

Stability and Bioavailability of Anthocyanins

Studies found that anthocyanins present in commercial black currant juice remained stable during in-vitro digestion in gastric fluid regardless of the addition or not of pepsin into the medium (Uzunović and Vranić 2008). The anthocyanins remained stable during in-vitro digestion in simulated intestinal fluid without pancreatin. With pancreatin in the intestinal fluid there was a reduction in stability, and this also contributed to slight reduction of total anthocyanins content (−1.83%) in commercial black currant juice. Steinert et al. (2008) studied the bioavailability of blackcurrant anthocyanins using Caco-2 monolayers as an in-vitro model of the absorptive intestinal epithelium. They found that cell metabolism and translocation across the basolateral membrane may be the key determinants of anthocyanin absorption and bioavailability. Apical transport might occur to a much larger extent than the further translocation across the basolateral membrane.

Antiinflammatory Activity

Total flavonoids extracted from *Ribes nigrum* leaves and their two major components, rutin and isoquercitrin did not exhibit spasmodic nor relaxing activity on rat stomach strip (Chanh et al. 1986). They were not capable of inducing any biosynthesis and release of prostaglandin-like substances and did not act on prostaglandin E2 receptor receptors. They inhibited both biosynthesis and release of prostaglandin-like substances. Total flavonoids were more active than rutin and isoquercitrin: ID30 was 1.03 mg/ml for total flavonoids compared to 3.75 and 2.31 mg/ml for rutin and isoquercitrin respectively. A hydroalcoholic extract of black currant leaves exhibited antiinflammatory activity when tested on carrageenan-induced rat paw oedema (Declume 1989) The black currant extract and lyophilisate revealed significant antiinflammatory activity comparable to that seen with the reference substances, indomethacin and niflumic acid, but without their ulcerogenic potential, even at high doses during chronic treatment.

Pretreatment of rats with proanthocyanidins (PACs), isolated from blackcurrant (10, 30, 60 and 100 mg/kg, i.p.) reduced paw oedema induced by carrageenin in a dose and time-dependent manner (Garbacki et al. 2004; 2005). PACs also inhibited dose-dependently carrageenin-induced pleurisy in rats. They decreased lung injury, pleural exudate formation, polymorphonuclear cell infiltration, pleural exudate levels of TNF-alpha, IL-1beta and CINC-1 but did not affect IL-6 and IL-10 levels. PACs inhibited in vivo nitric oxide release; they lowered exudate levels of nitrite/nitrate (NOx). In indomethacin treated rats, the volume of pleural exudate was low; leukocytes and TNF-alpha, IL-1beta, IL-6 and IL-10 contents were reduced but not NOx. The data suggested that the antiinflammatory properties of PACs were achieved through a different pattern from those of indomethacin, mainly in an interference with the migration of the leukocytes. The antiinflammatory activity of proanthocyanidins was found to be related to an inhibition of leukocyte infiltration partially caused by a down-regulation of endothelial adhesion molecules, ICAM-1 and VCAM-1 and these compounds were capable of modulating TNF-alpha-induced vascular endothelial growth factor (VEGF) transcription.

Antiviral Activity

The major constituents of the fraction D separated from *R. nigrum* "Kurokarin" fruit extract were determined as anthocyanins (Knox et al. 2001). Further fractionation of fraction D yielded fractions A' to G'. The fraction E' consisted of 3-O-α-L-rhamnopyranosyl-β-D-glucopyranosyl-cyanidin and 3-O-β-D-glucopyranosyl-cyanidin, and the fraction F' consisted of 3-O-α-L-rhamnopyranosyl-β-D-glucopyranosyl-delphinidin and 3-O-β-D-glucopyranosyl-delphinidin. The fractions D' to G' showed potent antiviral activity against influenza viruses A and B. The additive antiviral effect of a combination of the fractions E' and F' was assessed. Anthocyanins in the fraction F' did not directly inactivate influenza viruses A and B, but they inhibited virus adsorption to cells and also virus release from infected cells. In further studies, Knox et al. (2003) showed that the concentration of Kurokarin extract required to inhibit the plaque formation of both influenza virus types IVA and IVB by 50% (IC_{50}) was 3.2 μg/ml. Both IVA and IVB were directly inactivated up to 99% by 10 μg/ml of the extract at pH 2.8, and 95–98% at pH 7.2. The growth of IVA in cells treated with 10 and 100 μg/ml of the extract for 6 hours after infection was completely suppressed. Virus titres in culture fluids of the cells treated with 100 μg/ml of Kurokarin extract for 1 hour at 8–9 hours post infection, were completely suppressed, indicating that the extract inhibited the virus release from the infected cells. Suzutani et al. (2003) reported that found *R. nigrum* (Kurokarin) fruit extract inhibited completely herpes simplex virus type 1 attachment on the cell membrane at a 100-fold dilution, as well as the plaque formation of herpes simplex virus types 1 and 2, and varicella-zoster virus by 50% at a 400-fold dilution or lower concentrations. This latter activity of inhibition of virus replication in cells was attributed to the inhibition of protein synthesis in infected cells from the early stage of infection.

Anticancer Activity

Black currant fruit juice was found to contain a polysaccharide-rich substance, which was named cassis polysaccharide (CAPS), with macrophage-stimulating activity (Takata et al. 2005). Its interleukin (IL)-1beta-inducing activity was extraordinarily high, compared with other fruit juice preparations. CAPS was found to consist of rhamnose, mannose, arabinose, galactose, xylose, and glucose in a molar ratio of 11.3:0.9:54.1:29.8:2.0:1.9. CAPS was partitioned into a soluble component (CAPS-l.m.) and a precipitable component (CAPS-h.m.) with mean molecular weights of 80,000 and 600,000 respectively. CAPS-l.m. rather than CAPS-h.m. appeared to play an important role in macrophage activation in-vitro. Oral administration of black currant juice and CAPS to Ehrlich carcinoma-bearing mice inhibited the growth of the solid tumour by 45% and 51% respectively. CAPS administration had a stimulatory effect on the release of IL-2, IL-10, interferon-gamma, and IL-4 from splenocytes in comparison with phosphate buffered saline treatment in tumour-bearing mice. The IL-4 concentration was, however, still lower than that exhibited by a group of normal mice. Cassis polysaccharide exhibited cytotoxicity directly against Ehrlich ascites cells with an estimated IC_{50} of 760 µg/ml.

The aqueous extract of black currant fruit skin yielded an anthocyanin-rich fraction with cyanidin-3-O-rutinoside as one of the major anthocyanins (Bishayee et al. 2010). This fraction exhibited a potent cytotoxic effect on HepG2 liver cancer cells. This effect was more marked than that of delphinidin and cyanidin, two major aglycones of anthocyanins present in black currant. This action was possibly attributable additive as well as synergistic effects in the anthocyanin-rich fraction of black currant skin. Bishayee et al. (2011) reported that the anthocyanin-rich black currant skin extract (BCSE) dose-dependently decreased the incidence, total number, multiplicity, size and volume of preneoplastic hepatic nodules in model of rat liver hepatocarcinogenesis induced by of diethylnitrosamine (DENA) followed by promotion with phenobarbital. The antihepatocarcinogenic effect of BCSE was confirmed by histopathological examination of liver sections. Immunohistochemical analysis of proliferating cell nuclear antigen and DNA fragmentation revealed BCSE-mediated inhibition of abnormal cell proliferation and induction of apoptosis in DENA-induced rat liver tumorigenesis respectively. Mechanistic studies revealed that BCSE-mediated proapototic signal during experimental hepatocarcinogenesis may be evoked via the up-regulation of Bax and down-modulation of Bcl-2 expression at the translational level. The author maintained that the results together with a safety profile of BCSE supported the development of black currant bioactive constituents as chemopreventive agents for human liver cancer.

Skin Cell Stimulation Activity

An arabinogalactan protein (F2) isolated from *Ribes nigrum* was found to stimulate significantly cellular dehydrogenase activities (MTT and WST-1 tests) of human skin cells (fibroblasts, keratinocytes) as well as the proliferation rate of keratinocytes at 10 and 100 µg/ml (Zippel et al. 2009). F2 did not affect the differentiation status of keratinocytes and did not exert any cytotoxic potential using the lactate dehydrogenase test. The fluorescein isothiocyanate-labeled polysaccharide was incorporated in a time-dependent manner into human fibroblasts via endosomal transport. This internalization of the polysaccharide was inhibited by Cytochalasin B.

Antimicrobial Activity

Ribes nigrum bud essential oils exhibited strong antibacterial activity against *Acinetobacter baumanii, Escherichia coli, Pseudomonas aeruginosa* and *Staphylococcus aureus*, as evidenced by the very low MIC values observed for the respective strains (Oprea et al. 2008). Lengsfeld et al. (2004) found that acidic, high molecular weight 1,3-linked galactans with side chains possessing 1,4-galacturonic acid, galactose and arabinose residues were responsible for the anti-adhesive qualities of black currant seed extracts. These polymers were able to block *Helicobacter pylori* surface receptors, thus inhibiting their interaction with specific binding factors located on human gastric epithelia. The juice, water and methanol extracts of the polyphenol–rich pomace of *Ribes nigrum* exhibited efficient antibacterial activity against both Gram positive *Bacillus subtilis* and *B. cereus,* and Gram negative bacteria *Escherichia coli* and *Serratia marcescens* except for the water pomace extract which was very much less inhibitory to *B. cereus* and the methanol pomace extract to *Serratia marcescens* (Krisch et al. 2008). The juice, water and methanol extracts of the polyphenol–rich pomace of *Ribes nigrum* exhibited anti-fungal activity against 8 *Candida* spp., namely *Candida glabrata, C. guilliermondii, C. inconspicua, C. lipolytica, C. norwegica, C. parapsilosis, C. tropicalis* and *C. zeylanoides* (Krisch et al. 2002). The MIC (minimum inhibitory concentration) values of the juice and pomace extracts ranged from 1.83 to 10.38 mg/ml of dry matter. The total phenolic content of the juice, methanol and water extracts of the pomace was 68.03, 145.60 and 178.39 µg gallic acid equivalent/mg dry weight.

Angioprotective Activity

Extract from *Ribes nigrum* inhibited 50% of the activity of porcine pancreas elastase at concentrations of 0.56 mg/ml against a synthetic substrate (Jonadet et al. 1986). Inhibition was less effective on activity of trypsin and alpha-chymotrypsin. Marked in-vivo angioprotective properties were shown by the compounds studied. The results suggested a possible role by these inhibitors in the protection of conjunctive and elastic tissues adversely affected by proteolytic enzymes.

Antiatherogenic Activity

Currant oil from *Ribes nigrum* is one of the few plant oils containing PUFAn-3 (15.3 mol%) in addition to PUFAn-6 (60.5 mol%) (Vecera et al. 2003). Plant-based n-3 polyunsaturated fatty acids (PUFA) possess a prospective antiatherogenic potential. Studies in rats found that after 3 weeks of feeding, the currant oil caused a significant decrease in blood glutathione (GSH) and an increase in $Cu(2+)$ induced oxidizability of serum lipids, but did not affect liver GSH and t-butyl hydroperoxide-induced lipoperoxidation of liver microsomes. Although currant oil did not cause accumulation of liver triacylglycerols as lard fat, the lipoprotein profile (VLDL, LDL, HDL) was not significantly improved after currant oil feeding. The consumption of PUFAn-3 was reflected in LDL as an increase in eicosapentaenoic and docosahexaenoic acid. The results suggested that currant oil affected positively the lipid metabolism in the liver, and did not cause the development of a fatty liver. However, adverse effects of currant oil on the antioxidant status in the blood still remained of concern.

Antiosteoarthritic Activity

Studies by Garbacki et al. (2002) suggested that the prodelphinidins fractions, the major compounds isolated from *R. nigrum* leaves may be useful as an additive agent in the prevention of osteoarthritis. Gallocatechin trimer (GC-GC-GC) displayed the higher stimulation of proteoglycans and type II collagen production (1 µg/ml) and the synthesis of prostaglandin E(2) was significantly reduced by gallocatechin dimer (GC-GC), gallocatechin-epigallocatechin (GC-EGC) and GC-GC-GC at 10 and 100 µg/ml. The inhibition of prostaglandin E(2) synthesis was confirmed by the in-vitro test on purified COX enzymes, showing the selectivity of prodelphinidins on COX-2.

Immuno-Enhancing Activity

It had been shown that the age-associated elevation in prostaglandin E(2) production contributed to the decline in T cell-mediated immune function with age. Wu et al. (1999) conducted a randomized, double-blind, placebo-controlled (soybean oil) study to study the effect of 2 months supplementation with black current seed oil (BCSO) on the immune response of 40 healthy elderly subjects aged 65 years +. The BCSO group exhibited a significant increase in proliferative response of peripheral blood mononuclear (PBMC) cells to the T cell mitogen phytohemagglutinin that was not significantly different from that noted in the placebo group. BCSO did not affect concanavalin A-induced mitogenic response, interleukin 2 and -1beta production, and PBMC membrane fluidity. Prostaglandin E(2) production was significantly lowered in the BCSO-supplemented group, and this change was significantly different from that of the placebo group. The researchers concluded that black currant seed oil had a moderate immune-enhancing action attributable to its ability to reduce prostaglandin E(2) production.

Protein Kinase Inhibition Activity

Wang et al. (1996) isolated and characterised condensed tannins based on procyanidin and/or prodelphinidin and having a cis or trans stereochemistry at positions 2 and, from various plant sources, including *R. nigrum*,. All the condensed tannin preparations were found to be potent inhibitors of rat liver cyclic AMP-dependent protein kinase catalytic subunit (cAK) with IC_{50} values ranging from 0.009 to 0.2 μM. The tannin preparations were also found to be excellent inhibitors of rat brain Ca^{2+}- and phospholipid-dependent protein kinase C (PKC) (IC_{50} values in the range 0.3–7 μM), wheat embryo Ca^{2+}-dependent protein kinase (CDPK) (IC_{50} values in the range 0.8–7 μM) and of calmodulin (CaM)-dependent myosin light chain kinase (MLCK) (IC50 values in the range 7–24 μM). One of the most effective preparations, that from the leaves of *Ribes nigrum*, exhibited IC_{50} values with respect to cAK, PKC, CDPK and MLCK of 0.009, 0.6, 2.0 and 16 μM, respectively. The sequence with regard to inhibition sensitivity by these condensed tannins was cAK > PKC > CDPK > MLCK. The *Ribes nigrum* preparation was found to be a competitive inhibitor of cAK with respect to both ATP and synthetic peptide substrate. These condensed tannin preparations were deemed to be the most potent plant-derived inhibitors of cAK yet found.

Atopic Activity

Linnamaa et al. (2010) conducted randomized, double-blind, placebo-controlled trial wherein 313 pregnant mothers were randomly assigned to receive blackcurrant seed oil (151) or olive oil as placebo (162) to ascertain the effect of dietary supplementation with the blackcurrant seed oil on the prevalence of atopy at 12 months of age. There was a significantly lower incidence of atopic dermatitis in the seed oil group than in the olive oil group at the age of 12 months (33.0% versus 47.3%). Similarly the severity (SCORAD index) was also lower in the seed oil group than in the olive oil group at 12 months of age. Dietary supplementation with blackcurrant seed oil was well tolerated and it transitionally lowered the incidence of atopic dermatitis. It could therefore be one potential tool in the prevention of atopic symptoms when used at an early stage of life.

Tocopherol Enhancing Activity

Dietary anthocyanins from blackcurrant did not affect feed intake, body weight, and organ weights in Sprague-Dawley rats (Frank et al. 2002). Dietary cyanidin-3-O-glucoside (C3G) elevated the concentrations of tocopherols in the liver and lungs. Cholesterol levels in plasma and liver were not affected by any of the regimens. C3G and blackcurrant concentrate lowered the relative amount of saturated fatty acids in the liver. The results indicated that dietary cyanidin-3-O-glucoside, and blackcurrant concentrate appeared to have little impact on cholesterol levels and the fatty acid pattern in the liver but appeared to be capable of sparing vitamin E in healthy, growing rats.

Traditional Medicinal Uses

Black currants are used as a remedy for colds and flu and the juice is used to stop diarrhoea and stabilise digestion. The raw juice is diuretic and diaphoretic and an excellent beverage for febrile diseases. Boiled, sugar-added juice is used to treat inflamed sore-throats. Lozenges are also prepared from it.

An infusion of the leaves is cleansing and diuretic. An infusion of leaves is used in the treatment of dropsy, rheumatic pain and whooping cough, and as a gargle for sore throats and mouth ulcers. It has also been used externally on slow-healing cuts and abscesses.

An infusion of the young roots is useful in the treatment of eruptive fevers. A bark decoction is useful for treating calculus, dropsy and haemorrhoidal tumours.

Other Uses

The leaves provide a yellow dye and the fruits a violet dye. The seed oil is used in skin and cosmetic preparations.

Comments

Blackcurrant are readily propagated from hardwood stem cuttings. The best time to take the cuttings is when the foliage has stopped growing or are being shed.

Selected References

Alonso González E, Torrado Agrasar A, Pastrana Castro LM, Orriols Fernández I, Pérez Guerra N (2010) Production and characterization of distilled alcoholic beverages obtained by solid-state fermentation of black mulberry (*Morus nigra* L.) and black currant (*Ribes nigrum* L.). J Agric Food Chem 58(4): 2529–2535

Bakowska-Barczak AM, Schieber A, Kolodziejczyk P (2009) Characterization of Canadian black currant (*Ribes nigrum* L.) seed oils and residues. J Agric Food Chem 57(24):11528–11536

Bishayee A, Háznagy-Radnai E, Mbimba T, Sipos P, Morazzoni P, Darvesh AS, Bhatia D, Hohmann J (2010) Anthocyanin-rich black currant extract suppresses the growth of human hepatocellular carcinoma cells. Nat Prod Commun 5(10):1613–1618

Bishayee A, Mbimba T, Thoppil RJ, Háznagy-Radnai E, Sipos P, Darvesh AS, Folkesson HG, Hohmann J (2011) Anthocyanin-rich black currant (*Ribes nigrum* L.) extract affords chemoprevention against diethylnitrosamine-induced hepatocellular carcinogenesis in rats. J Nutr Biochem 22(11):1035–1046

Bordonaba JG, Terry LA (2008) Biochemical profiling and chemometric analysis of seventeen uk-grown black currant cultivars. J Agric Food Chem 56(16): 7422–7430

Borges G, Degeneve A, Mullen W, Crozier A (2010) Identification of flavonoid and phenolic antioxidants in black currants, blueberries, raspberries, red currants, and Cranberries. J Agric Food Chem 58:3901–3909

Bown D (1995) Encyclopaedia of herbs and their uses. Dorling Kindersley, London, 424 pp

Chanh PH, Ifansyah N, Chahine R, Mounayar-Chalfoun A, Gleye J, Moulis C (1986) Comparative effects of total flavonoids extracted from *Ribes nigrum* leaves, rutin and isoquercitrin on biosynthesis and release of prostaglandins in the ex vivo rabbit heart. Prostaglandins Leukot Med 22(3):295–300

Chevallier A (1996) The encyclopedia of medicinal plants. Dorling Kindersley, London, 336 pp

Costantino L, Albasini A, Rastelli G, Benvenuti S (1992) Activity of polyphenolic crude extracts as scavengers of superoxide radicals and inhibitors of xanthine oxidase. Planta Med 58(4):342–344

Da Silva Pinto M, Kwon Y-I, Apostolidis E, Lajolo FM, Genovese MI, Shetty K (2010) Evaluation of red currants (*Ribes rubrum* L.), black currants (*Ribes nigrum* L.), red and green gooseberries (*Ribes uva-crispa*) for potential management of type 2 diabetes and hypertension using in vitro models. J Food Biochem 34:639–660

Declume C (1989) Anti-inflammatory evaluation of a hydroalcoholic extract of black currant leaves (*Ribes nigrum*). J Ethnopharmacol 27(1–2):91–98

Del Castillo ML, Dobson G, Brennan R, Gordon S (2004) Fatty acid content and juice characteristics in black currant (*Ribes nigrum* L.) genotypes. J Agric Food Chem 52(4):948–952

Dobson G (2000) Leaf lipids of *Ribes nigrum*: a plant containing 16:3, alpha-18:3, gamma-18:3 and 18:4 fatty acids. Biochem Soc Trans 28(6):583–586

Frank J, Kamal-Eldin A, Lundh T, Määttä K, Törrönen R, Vessby B (2002) Effects of dietary anthocyanins on tocopherols and lipids in rats. J Agric Food Chem 50(25):7226–7230

Garbacki N, Angenot L, Bassleer C, Damas J, Tits M (2002) Effects of prodelphinidins isolated from *Ribes nigrum* on chondrocyte metabolism and COX activity. Naunyn Schmiedebergs Arch Pharmacol 365(6): 434–441

Garbacki N, Tits M, Angenot L, Damas J (2004) Inhibitory effects of proanthocyanidins from *Ribes nigrum* leaves on carrageenin acute inflammatory reactions induced in rats. BMC Pharmacol 4:25

Garbacki N, Kinet M, Nusgens B, Desmecht D, Damas J (2005) Proanthocyanidins, from *Ribes nigrum* leaves, reduce endothelial adhesion molecules ICAM-1 and VCAM-1. J Inflamm (Lond) 2:9

Gavrilova V, Kajdžanoska M, Gjamovski V, Stefova M (2011) Separation, Characterization and quantification of phenolic compounds in blueberries and red and black currants by HPLC-DAD-ESI-MS(n). J Agric Food Chem 59(8):4009–4018

Goffman FD, Galletti S (2001) Gamma-linolenic acid and tocopherol contents in the seed oil of 47 accessions from several *Ribes* species. J Agric Food Chem 49(1):349–354

Hákkinen S, Heinonen M, Kárenlampi S, Mykkánen H, Ruuskanen J, Törrönen R (1999) Screening of selected flavonoids and phenolic acids in 19 berries. Food Res Int 32:345–353

Harb J, Bisharat R, Strief J (2008) Changes in volatile constituents of blackcurrants (*Ribes nigrum* L. Cv. "Titania") following controlled atmosphere storage. Postharvest Biol Technol 47(3):271–279

Hudec J, Kochanov R, Burdov M, Kobida L, Kogan G, Turianica I, Chlebo P, Hanakovac E, Slamka P (2009) Regulation of the phenolic profile of berries can increase their antioxidant activity. J Agric Food Chem 57(5):2022–2029

Jonadet M, Meunier MT, Villie F, Bastide JP, Lamaison JL (1986) Flavonoids extracted from *Ribes nigrum* L. and *Alchemilla vulgaris* L.: 1. In vitro inhibitory activities on elastase, trypsin and chymotrypsin. 2. Angioprotective activities compared in-vivo. J Pharmacol 17(1):21–27 (in French)

Kallio H, Nieminen R, Tuomasjukka S, Hakala M (2006) Cutin composition of five Finnish berries. J Agric Food Chem 54(2):457–462

Kapasakalidis PG, Rastall RA, Gordon MH (2006) Extraction of polyphenols from processed black currant (*Ribes nigrum* L.) residues. J Agric Food Chem 54(11):4016–4021

Keep E (1975) Currants and gooseberries. In: Janick J, Moore JN (eds) Advances in fruit breeding. Purdue University Press, West Lafayette, pp 197–268, 623 pp

Knox YM, Hayashi K, Suzutani T, Ogasawara M, Yoshida I, Shiina R, Tsukui A, Terahara N, Azuma M (2001) Activity of anthocyanins from fruit extract of *Ribes nigrum* L. against influenza A and B viruses. Acta Virol 45(4):209–215

Knox YM, Suzutani T, Yosida I, Azuma M (2003) Anti-influenza virus activity of crude extract of *Ribes nigrum* L. Phytother Res 17(2):120–122

Koeppen BH, Herrmann K (1977) Flavonoid glycosides and hydroxycinnamic acid esters of blackcurrants (*Ribes nigrum*). Phenolics of fruits 9. Z Lebensm Unters Forsch 164(4):263–268

Krisch J, Galgóczy L, Tölgyesi M, Papp T, Vágvölgyi C (2008) Effect of fruit juices and pomace extracts on the growth of Gram-positive and Gram-negative bacteria. Acta Biol Szeged 52(2):267–270

Krisch J, Ördögh L, Galgóczy L, Papp T, Vágvölgyi C (2002) Anticandidal effect of berry juices and extracts from *Ribes* species. Cent Eur J Biol 4(1):86–89

Landbo AK, Meyer AS (2001) Enzyme-assisted extraction of antioxidative phenols from black currant juice press residues (*Ribes nigrum*). J Agric Food Chem 49(7):3169–3177

Lengsfeld C, Deters A, Faller G, Hensel A (2004) High molecular weight polysaccharides from black currant seeds inhibit adhesion of *Helicobacter pylori* to human gastric mucosa. Planta Med 70(7):620–626

Leskinen HM, Suomela JP, Kallio HP (2009) Effect of latitude and weather conditions on the regioisomer compositions of alpha- and gamma-linolenoyldilinoleoylglycerol in currant seed oils. J Agric Food Chem 57(9):3920–3926

Linnamaa P, Savolainen J, Koulu L, Tuomasjukka S, Kallio H, Yang B, Vahlberg T, Tahvonen R (2010) Blackcurrant seed oil for prevention of atopic dermatitis in newborns: a randomized, double-blind, placebo-controlled trial. Clin Exp Allergy 40:1247–1255

Lu LT, Alexander C (2001) *Ribes* Linnaeus. In: Wu ZY, Raven PH (eds) Flora of China, vol 8, Brassicaceae through Saxifragaceae. Science Press/Missouri Botanical Garden Press, Beijing/St. Louis, pp 428–452

Lu Y, Foo LY, Wong H (2002) Nigrumin-5-p-coumarate and nigrumin-5-ferulate, two unusual nitrile-containing metabolites from black currant (*Ribes nigrum*) seed. Phytochemistry 59(4):465–468

Maatta K, Kamal-Eldin A, Törrönen R (2001) Phenolic compounds in berries of black, red, green, and white currants (*Ribes* sp.). Antioxid Redox Signal 3(6):981–993

Matsumoto H, Hanamura S, Kawakami T, Sato Y, Hirayama M (2001) Preparative-scale isolation of four anthocyanin components of black currant (*Ribes nigrum* L.) fruits. J Agric Food Chem 49(3):1541–1545

Moine G, Forzy L, Oesterhelt G (1992) Identification of (all-cis)-6,9,12,15-octadecatetraenoic acid in *Ribes nigrum* and fish oils: chemical and physical characterization. Chem Phys Lipids 60(3):273–280

Morin NR (2008) *Ribes nigrum* Linnaeus. In: Flora of North America Editorial Committee (eds) 1993+ Flora of North America North of Mexico, 16+ vols. New York/Oxford, vol 8, p 18

Nielsen IL, Haren GR, Magnussen EL, Dragsted LO, Rasmussen SE (2003) Quantification of anthocyanins in commercial black currant juices by simple high-performance liquid chromatography. Investigation of their pH stability and antioxidative potency. J Agric Food Chem 51(20):5861–5866

Oprea E, Radulescu V, Balotescu C, Lazar V, Bucur M, Mladin P, Farcasanu IC (2008) Chemical and biological studies of *Ribes nigrum* L. buds essential oil. Biofactors 34(1):3–12

Pieri G, Baghdikian B, Elias R, Ollivier E, Mahiou V, Balansard G (2002) Flavonoids from the leaves of *Ribes nigrum* L. identification of a malonyl flavonol and HPLC analysis. Etudes Chim Pharmacol. In: Fleurentin J, Pelt J-M, Mazars G (eds) From the sources of knowledge to the medicines of the future, pp 422–425

Porcher MH et al (1995–2020) Searchable world wide web multilingual multiscript plant name database. Published by The University of Melbourne, Australia. http://www.plantnames.unimelb.edu.au/Sorting/Frontpage.html

Rubinskiene M, Jasutiene I, Venskutonis PR, Viskelis P (2005) HPLC determination of the composition and stability of blackcurrant anthocyanins. J Chromatogr Sci 43(9):478–482

Salminen H, Heinonen M (2008) Plant phenolics affect oxidation of tryptophan. J Agric Food Chem 56(16):7472–7481

Scalzo J, Currie A, Stephens J, McGhie T, Alspach P, Horticulture and Food Research Institute of New Zealand Limited Hortresearch (2008) The anthocyanin composition of different *Vaccinium*, *Ribes* and *Rubus* genotypes. Biofactors 34(1):13–21

Shirazi AM (2003) Standardizing methods for evaluating the chilling requirements to break dormancy in seeds and buds (including geophytes): introduction to the workshop. Hortscience 38(3):334–335

Slimestad R, Solheim H (2002) Anthocyanins from black currants (*Ribes nigrum* L.). J Agric Food Chem 50(11):3228–3231

Steinert RE, Ditscheid B, Netzel M, Jahreis G (2008) Absorption of black currant anthocyanins by monolayers of human intestinal epithelial Caco-2 cells mounted in using type chambers. J Agric Food Chem 56(13):4995–5001

Suzutani T, Ogasawara M, Yoshida I, Azuma M, Knox YM (2003) Anti-herpes virus activity of an extract of *Ribes nigrum* L. Phytother Res 17(6):609–613

Tabart J, Kevers C, Evers D, Dommes J (2011) Ascorbic acid, phenolic acid, flavonoid and carotenoid profiles of selected extracts from *Ribes nigrum*. J Agric Food Chem 59(9):4763–4770

Takata R, Yamamoto R, Yanai T, Konno T, Okubo T (2005) Immunostimulatory effects of a polysaccharide-rich substance with antitumor activity isolated from black currant (*Ribes nigrum* L.). Biosci Biotechnol Biochem 69(11):2042–2050

Traitler H, Winter H, Richli U, Ingenbleek Y (1984) Characterization of gamma-linolenic acid in *Ribes* seed. Lipids 19(12):923–928

U.S. Department of Agriculture, Agricultural Research Service (2011) USDA National Nutrient Database for standard reference, release 24. Nutrient Data Laboratory Home Page. http://www.ars.usda.gov/ba/bhnrc/ndl

Uzunović A, Vranić E (2008) Stability of anthocyanins from commercial black currant juice under simulated gastrointestinal digestion. Bosn J Basic Med Sci 8(3):254–258

Varming C, Andersen ML, Poll L (2004) Influence of thermal treatment on black currant (*Ribes nigrum* L.) juice aroma. J Agric Food Chem 52(25):7628–7636

Vecera R, Skottová N, Vána P, Kazdová L, Chmela Z, Svagera Z, Walterá D, Ulrichová J, Simánek V (2003) Antioxidant status, lipoprotein profile and liver lipids in rats fed on high-cholesterol diet containing currant oil rich in n-3 and n-6 polyunsaturated fatty acids. Physiol Res 52(2):177–187

Viberg U, Ekström G, Fredlund K, Oste RE, Sjöholm I (1997) A study of some important vitamins and antioxidants in a blackcurrant jam with low sugar content and without additives. Int J Food Sci Nutr 48(1):57–66

Viljanen K, Kylli P, Kivikari R, Heinonen M (2004) Inhibition of protein and lipid oxidation in liposomes by berry phenolics. J Agric Food Chem 52(24): 7419–7424

Wang BH, Foo LY, Polya GM (1996) Differential inhibition of eukaryote protein kinases by condensed tannins. Phytochemistry 43(2):359–365

Westmore G (2004) Is climate change a threat to the blackcurrant industry in Tasmania? Nexus J Undergrad Sci Eng Technol 1:14–21

Wu D, Meydani M, Leka LS, Nightingale Z, Handelman GJ, Blumberg JB, Meydani SN (1999) Effect of dietary supplementation with black currant seed oil on the immune response of healthy elderly subjects. Am J Clin Nutr 70(4):536–543

Zheng J, Yang B, Tuomasjukka S, Ou S, Kallio H (2009) Effects of latitude and weather conditions on contents of sugars, fruit acids, and ascorbic acid in black currant (*Ribes nigrum* L.) juice. J Agric Food Chem 57(7):2977–2987

Zippel J, Deters A, Pappai D, Hensel A (2009) A high molecular arabinogalactan from *Ribes nigrum* L.: influence on cell physiology of human skin fibroblasts and keratinocytes and internalization into cells via endosomal transport. Carbohydr Res 344(8): 1001–1008

Ribes rubrum

Scientific Name

Ribes rubrum L.

Synonyms

Ribes domesticum Jancz., *Ribes rubrum* subsp. *vulgare* Domin, *Ribes rubrum* L. var. *sativum* Rchb., *Ribes sativum* (Rchb.) Syme, *Ribes sylvestre* (Lam.) Mert. & W. D. J. Koch, *Ribes vulgare* Lam., nom. illeg., *Ribes vulgare* Lam. var. *macrocarpum* Jancz., *Ribes vulgare* Lam. var. *sylvestre* Lam.

Family

Grossulariaceae

Common/English Names

Cultivated Currant, Common Currant, Garden Currant, Red Currant, Reps, Ribs, Risp, White Currant.

Vernacular Names

Arabic: Kishmish Ahmar;
Chinese: Ru Hong Cu Li;
Czech: Meruzalka Červená, Rybíz Červený;
Danish: Have-Ribs, Ribs;
Dutch: Ribes Sort, Rode Aalbes, Rode Bessen, Rodetrosbes, Witte Aalbes;
Eastonian: Punane Sõstar;
Finnish: Herukka, Lännenpunaherukka, Punaherukka, Valkea Viinimarja, Valkoherukka, Viinimarja;
French: Groseillier À Grappes, Groseillier Commun, Groseillier Rouge, Raisin-De-Mars;
German: Garten-Johannisbeere, Gewöhnliche Johannisbeere, Johannisbeere Johannisbeerstrauch, Ribisel, Rote Johannisbeere, Weiße Johannisbeere;
Greek: Fragostafyla Kokkina, Fragostafylla, Fragostafylo;
Hungarian: Kerti Ribiszke, Piros Ribiszke, Ribizke, Ribizli, Termesztett Ribiszke, Vadegres, Vörös Ribiszke;
Icelandic: Rauð Hlaupber;
Italian: Ribes Rosso, Ribisi;
Japanese: Aka Fusa Suguri, Aka-Suguri, Shiro Fusa Suguri;
Korean: Pulgunkkachibapnamu, Pulgunsongimulaengdunamu;
Morocco: Nnbaq Hhmar (Arabic);
Norwegian: Hagerips;
Polish: Czerwona Porzeczka, Porzeczka Czerwona, Porzeczka Zwyczajna;
Portuguese: Groselheira Vermelha;
Russian: Smorodina;
Slovaščina: Grozdičje Rdeče;
Slovencina: Ríbezľa Červená;
Spanish: Grosella Colorada, Grosella Roja, Grosellero Común, Grosellero Rojo;
Swedish: Röda Vinbär, Trädgårdsvinbär;

Origin/Distribution

The species is native to parts of western Europe (Belgium, France, Germany, Netherlands, Northern Italy Northern Spain and Portugal).

Agroecology

Currant fruit are cold climate species and are well-adapted to northern areas. They are often grown where severe winter cold precludes other tree fruit production. They are extremely cold-hardy, have long chilling requirements, intolerant of summer heat and have short maturity. Plants need only 120–140 frost-free days to mature fruit and complete their vegetative period. Fruit ripens in 90 days.

Currant bushes favor partial to full sunlight. Best sited in locations protected from strong winter winds and frost pockets. They are not fastidious of soil types but thrives in organically rich, medium moisture, well-drained loamy soil. They are relatively low-maintenance plants but regular irrigation is beneficial to production, due to shallow rooting. Overhead watering should be avoided to keep away foliar diseases.

Edible Plant Parts and Uses

Redcurrant fruit is slightly more acidic than the blackcurrant, and is cultivated mainly for jams, preserves, jellies and cooked dishes, rather than for eating raw. It is often served raw or as a simple supplement in salads, garnishes, or drinks when in season. In Scandinavia it is commonly used in fruit soups and summer puddings (*Rødgrød, Rote Grütze* or *Rode Grütt*). In Germany it is used together with custard or meringue as a filling for tarts. In Linz, Austria, it is the most frequently used filling for *Linzer torte*. In German-speaking areas, syrup or nectar processed from the red currant is added to soda water and relished as a refreshing drink, *Johannisbeerenschorle*. In the United Kingdom, redcurrant jelly is used as a traditional condiment with lamb in a Sunday roast.

In France, the highly delicate and hand-made *Bar-le-duc* or Lorraine jelly is a spreadable preparation traditionally made from white currants (albino red currants) or alternatively red currants. White currant an albino cultivar of red currant is less acidic and smaller than red currant, it has a sweetly tar flavour and is usually eaten raw.

Botany

Redcurrant is a small, deciduous shrub normally growing to 1–1.5 m tall, occasionally 2 m with gray stems. Leaves are alternate, green, 3-lobed, leaves 6×7cm, with cordate bases (Plates 1, 2, 3 and 4). The flowers are inconspicuous, yellow-green, pentamerous in 10–20 flowered, 4–8 cm long pendulous racemes (Plates 1 and 2). Flower has a nearly flat receptacle with a raised ring between stamens and styles; 5 orbicular-spatulate, patent, greenish-yellow sepals; campanulate corolla with 5 tiny petals; 5 stamens with anther lobes separated by a connective as wide as the lobes and 2 styles. Fruits are globose, pleasantly acid, glabrous, pale green maturing into bright red translucent edible berries about 8–12 mm diameter, with 3–10 berries on each raceme (Plates 2, 3, 4 and 5).

Nutritive/Medicinal Properties

Proximate nutrient composition of *Ribes rubrum* red and white currant fruits (exclude 2% stem refuse) had been reported per 100 g edible portion as follows: water 83.95 g, energy 56 kcal (234 kJ), protein 1.40 g, total lipid 1.40 g, ash 0.66 g, carbohydrate 13.80 g, total dietary fibre 4.3 g, total sugars 7.37 g, sucrose 0.61 g, glucose 3.22 g, fructose 3.53 g, Ca 33 mg, Fe 1 mg, Mg 13 mg, P 44 mg, K 275 mg, Na 1 mg, Zn 0.23 mg, Cu 0.107 mg, Mn 0.186 mg, Se 0.6 µg, vitamin C 41 mg, thiamin 0.040 mg, riboflavin 0.05 mg, niacin 0.1 mg, pantothenic acid 0.064 mg, vitamin B-6 0.070 mg, total folate 8 µg, choline 7.6 mg, vitamin A 2 µg RAE, vitain A 42 IU, β carotene 25 µg, lutein+zeaxanthin 47 µg, vitamin E (α-tocopherol) 0.10 mg, vitamin K (phylloquinone) 11 μg, total saturated fatty acids 0.017 g, 16:0 (palmitic) 0.010 g, 18:0 (stearic) 0.003 g, total monounsaturated fatty acids 0.028 g, 16:1 undifferentiated (palmitoleic) 0.053 g, 18:1 undifferentiated (oleic) 0.028 g, total polyunsaturated fatty acids 0.088 g, 18:2 undifferentiated (linoleic) 0.053 g and 18:3 undifferentiated (linolenic) 0.035 g (USDA 2011). Hägg et al. (1995), reported that the vitamin C contents in redcurrant grown in Finland ranged from 17 to 21 mg/100 g.

Goffman and Galletti (2001) found that the highest total tocopherol content was found in *R. nigrum* (mean, 1,716 mg/kg oil), followed by *R. rubrum* (mean, 1,442 mg/kg oil). *R. grossularia* showed the lowest tocopherol content (mean, 786 mg/kg oil). The three species also differed markedly in tocopherol composition. *R. rubrum* had the highest content of δ-tocopherol (mean, 20.2%); *R. grossularia* had the highest level of γ-tocopherol (mean, 70.0%), and *R. nigrum* the highest percentage of α-tocopherol (mean, 34.8%), the most biologically active among the four tocopherols. As for γ-linolenic acid, the highest concentration was found in *R. nigrum*, up to 15.8% while *R. grossularia* and *R. rubrum* showed mean γ-linolenic acid levels of 8% and 6.2%, respectively. The present study indicated that seeds of *Ribes* species, especially *R. nigrum*, could be used as sources of γ-linolenic acid and natural vitamin E. γ-linolenic acid is an essential fatty acid for humans with Δ-6-desaturase deficiency; it is a precursor of prostaglandins, prostacyclins, and tromboxanes; and it has antiinflammatory and antitumoral effects.

Plate 1 Flowers and leaves

Plate 2 Flowers and young fruits

Plate 3 Ripening red currant fruit

Plate 5 Harvested red currants

Plate 4 Close-view of red currant fruits

Tocopherols are natural antioxidants with biological activity, heart/vascular, and cancer protective properties.

Latitude and weather conditions were found to influence the regioisomer triacylglycerols (TAG) profiles of Ala/L/L (18:3(n-3)/18:2(n-6)/18:2(n-6)) and Gla/L/L (18:3(n-6)/18:2(n-6)/18:2(n-6)) in the seed oil of *Ribes* species and varieties in Finland (Leskinen et al. 2009). In *Ribes rubrum* the portion of the symmetric regioisomer LAlaL among Ala/L/L was higher (14.1%) than in *R. nigrum* (12.1%). Generally in currants, the proportion of LAlaL was lower in northern Finland (12.1%) than in southern Finland (13.5%), where temperature and radiation sums were higher. In *R. rubrum* varieties grown in the south, the proportion of LGlaL among Gla/L/L was significantly higher in the years 2005 and 2007 (30.7–32.0%) than in 2006 (24.2–25.4%), when temperature and radiation sums were higher and the amount of precipitation was lower.

Djordjević et al. (2010) found that the average amount of ascorbic acid in 11 Serbian redcurrant varieties varied from 50.5 to 71.6 mg/100 g FW, while concentration of invert sugars ranged from 6.0% to 9.0%. The highest amounts of total phenolics and anthocyanins were detected in variety Redpoll (153.4 mg GAE/100 g FW and 19.3 mg/100 g, respectively). Red currants were processed to juice, and the phenolic and anthocyanin contents changed as a result of processing. In berries, storage at −18°C caused the decrease of ascorbic acid content up to 49%, and a general reduction of total phenolics was also observed. In juices, total phenolics content increased after 1 year of storage. In both berries and juices total anthocyanins increased during storage by up to 85% and 50%, respectively. This study demonstrated that certain varieties, namely Redpoll, Jonkheer and London Market were good sources of phytochemicals, retaining the nutritional value during processing and storage.

Other Phytochemicals

Studies showed that anthocyanins dominated in black and red currants, whereas proanthocyanidins and phenolic acids were the predominant phenolic compounds in green and white currants (Maatta et al. 2001). The lack of anthocyanins in the colorless (green, white) berries was associated with increased levels of phenolic acids, especially p-coumaric acid (80 mg/kg in green currant vs. 45 mg/kg in black currant) and 4-hydroxybenzoic acid (18 mg/kg in white currant vs. 3 mg/kg in red currant). The contents of extractable (22–41 mg/kg) and nonextractable proanthocyanidins (32–108 mg/kg) were comparable to those of other phenolics, with the exception of anthocyanins in black currant. Distinct similarities were found in the relative distribution of conjugated forms of phenolic compounds among black, green, red, and white currants (*Ribes* spp.) (Määttä et al. 2003). Phenolic acids were found mainly as hexose esters. Flavonol glycosides and anthocyanin pigments were mainly found as 3-O-rutinosides and second as 3-O-glucosides. However, cyanidin 3-O-sambubioside and quercetin hexoside-malonate were notable phenolic compounds in red currant. Flavonol hexoside-malonates were identified and quantified in the berries of currants for the first time.

Traces of the stilbene, resveratrol (15.72 µg/g) was found in red currant (Ehala et al. 2005). A branched quercetin triglycoside: quercetin 3-O-(2″-O-α-L-rhamnopyranosyl-6″-O-α-L-rhamnopyranosyl)-β-D-gluocpyranoside was isolated from red currant leaves (Siewek et al. 1984).

Flavonoids (kaempferol, quercetin, myricetin) and phenolic acids (p-coumaric, caffeic, ferulic, p-hydroxybenzoic, gallic and ellagic acids) were detected in the fruits of 19 berries (Hákkinen et al. 1999). In the genus *Ribes*, quercetin was the main compound in gooseberry, red currant and black currant. The data suggested *Ribes* berries to have potential as good dietary sources of quercetin or ellagic acid.

The main components in the red currant extract were vitamin C (313 nmol/g) and anthocyanins cyanidin-3-O-rutinoside and cyanidin-3-O-(2″-O-xylosyl)rutinoside (both totalling 247 nmol/g) and cyanidin-3-O-sambubioside (81 nmol/g (Borges et al. 2010). The extract also contained a number of myricetin, kaempferol, and quercetin conjugates such as myricetin-3-O-rutinoside (4.6 nmol/g), myricetin-O-rhamnoside (3.7 nmol/g), kaempferol-O-rutinoside (5.1 nmol/g), kaempferol-3-O-galactoside (3.2 nmol/g), kaempferol-3-O-glucoside (5.4 nmol/g), quercetin-3-O-rutinoside (23 nmol/g), quercetin-3-O-galactoside (8.6 nmol/g), quercetin-3-O-glucoside (11 nmol/g), quercetin-3-O-(6″-O-malonyl)glucoside (4.6 nmol/g), and 1 (4-hydroxybenzoic acid-O-hexoside) (73 nmol/g) and caffeic acid-O-glucoside (16 nmol/g).

Twenty-five key astringent compounds were identified in red currant juice (Schwarz and Hofmann 2007b). Besides several flavonol glycosides, in particular, 3-carboxymethyl-indole-1-N-β-D-glucopyranoside, 3-methylcarboxymethyl-indole-1-N-β-D-glucopyranoside, and a family of previously not identified compounds, namely, 2-(4-hydroxybenzoyloxymethyl)-4-β-D-glucopyranosyloxy-2(E)-butenenitrile, 2-(4-hydroxy-3-methoxybenzoyloxymethyl)-4-β-D-glucopyranosyloxy-2(E)-butenenitrile, (E)-6-[3-hydroxy-4-(O-β-D-glucopyranosyl)phenyl]-5-hexen-2-one named dehydrorubrumin, and (3E,5E)-6-[3-hydroxy-4-(O-β-D-glucopyranosyl)phenyl]-3,5-hexadien-2-one named rubrumin, were identified. Determination of the oral astringency thresholds by means of the half-tongue test revealed that the lowest thresholds of 0.3 and 1.0 nmol/L were found for the nitrogen-containing 3-carboxymethyl-indole-1-N-β-D-glucopyranoside and 3-methylcarboxymethyl-indole-1-N-β-D-glucopyranoside, which did not belong to the group of plant polyphenols. A series of flavon-3-ol glycosides and (E)/(Z)-aconitic acid, four nitrogen-containing phytochemicals were found as the key astringent and mouth-drying compounds in red currants (*Ribes rubrum*) (Schwarz and Hofmann 2007a). These included the astringent indoles 3-carboxymethyl-indole-1-N-β-D-glucopyranoside and 3-methyl-carboxymethyl-indole-1-N-β-D-glucopyranoside, as well as the astringent, noncyanogenic nitriles 2-(4-hydroxybenzoyloxymethyl)-4-β-D-glucopyranosyloxy-2(E)-butenenitrile and 2-(4-hydroxy-3-methoxybenzoyloxymethyl)-4-β-D-glucopyranosyloxy-2(E)-butenenitrile. Using the half-tongue test, human recognition thresholds for the astringent and mouth-drying nitrogen compounds were determined to be between 0.0003 and 5.9 μmol/L (water).

Antioxidant Activity

All the crude extracts of the berries of several cultivars of *Ribes, Rubus*, and *Vaccinium* genera examined showed a significantly high activity towards chemically-generated superoxide radicals (Costantino et al. 1992). The activities were higher than those expected on the basis of the quantities of anthocyanins and polyphenols present in the samples. In addition, the extracts exhibited inhibitory activity towards xanthine oxidase. *Ribes nigrum* extracts showed the highest activity, being the richest in both anthocyanins and polyphenols. In contrast, *Ribes rubrum* extracts seemed to contain more active substances than the other crude extracts. Ferric reducing antioxidant power (FRAP) values of raspberry (*Rubus idaeus*), blackberry (*Rubus fructicosus*), raspberry × blackberry hybrids, red currant (*Ribes sativum*), gooseberry (*Ribes glossularia*) and Cornelian cherry (*Cormus mas*) cultivars ranged from 41 to 149 μmol ascorbic acid/g dry weight and protection of deoxyribose ranged from 16.1% up to 98.9% (Pantelidis et al. 2007). Anthocyanin content ranged from 1.3 mg in yellow-coloured fruit, up to 223 mg cyanidin-3-glucoside equivalents 100/g fresh weight in Cornelian cherry, whereas phenol content ranged from 657 up to 2,611 mg gallic acid equivalents/100 g dry weight. Ascorbic acid content ranged from 14 up to 103 mg/100 g fresh weight.

Vitamin C was found to contribute 47.5% of the total antioxidant capacity (AOC) of the red currant extract; anthocyanins (cyanidin-3-O-rutinoside and cyanidin-3-O-(2′-O-xylosyl)rutinoside were responsible for 21% of AOC of the extract with a number of unidentified components accounting for a further 23.5% (Borges et al. 2010). The extract also contained a number of myricetin, kaempferol, and quercetin conjugates but their contribution to the overall AOC, like 4-hydroxybenzoic acid-O-hexoside and a caffeic acid-O-glucoside, was relatively minor.

Among the following small fruits, strawberries (*Fragaria ananassa*), raspberries (*Rubus idaeus*) and red currants (*Ribes rubrum*), as well as two drupes, cherries (*Prunus avium*), and sour cherries (*Prunus cerasus*), red currants and strawberries exhibited the highest initial total phenol (TP) contents (322.40 and 335.47 mg GAE/100 g FW, respectively) and maintained the highest TP contents throughout storage at 4°C and 25°C temperatures (Piljac-Žegarac and Šamec 2010). Storage of fruits at 25°C as opposed to 4°C, facilitated faster spoilage. In addition, most fruits stored at 4°C, exhibited slightly higher antioxidant activity values at the end of storage as evaluated by all three antioxidant activity assays (DPPH, ABTS and FRAP assays) as opposed to fruits stored at 25°C.

Antidiabetic and Antihypertension Activities

Of red and black currant, and red and green gooseberries, red currants had the highest α-glucosidase, α-amylase and angiotensin I-converting enzyme (ACE) inhibitory activities (da Silva Pinto et al. 2010). The major phenolic compound in red currant was found to be chlorogenic acid. The data indicated that red currants could be good dietary sources with potential antidiabetic and antihypertension functions to complement overall dietary management of early stages of type 2 diabetes.

Anticancer Activity

Studies showed that feeding Min mice with a diet containing 10% freeze dried white currant for 10 weeks was effective in preventing cancer initiation and progression (Rajakangas et al. 2008). The white currant diet reduced the number of adenomas from 81 to 51 in the total small intestine of Min mice. In the distal part of the small intestine, white currant reduced the adenomas number from 49 to 29.5 and also the size of the adenomas from 0.88 to 0.70 mm. White currant reduced nuclear β-catenin and NF-kappaB protein levels in the adenomas. The growth of various cancer cell lines, including those of stomach, prostate, intestine and breast, was strongly inhibited by raspberry, black currant, white currant, gooseberry, velvet leaf blueberry, low-bush blueberry, sea buckthorn and cranberry juice, but not (or only slightly) by strawberry, high-bush blueberry, serviceberry, red currant, or blackberry juice (Boivin et al. 2007). No correlation was found between the anti-proliferative activity of berry juices and their antioxidant capacity. The inhibition of cancer cell proliferation by berry juices did not involve caspase-dependent apoptosis, but appeared to involve cell-cycle arrest, as evidenced by down-regulation of the expression of cdk4, cdk6, cyclin D1 and cyclin D3.

Traditional Medicinal Uses

The fruit is antiscorbutic, aperient, depurative, digestive, diuretic, laxative, refrigerant and sialagogue (Grieve 1971; Chiej 1984). The leaves contain the toxin hydrogen cyanide. A concoction of them is used externally to relieve rheumatic symptoms. They are also used in poultices to relieve sprains or reduce the pain of dislocations. English and German language herbalist sources consider redcurrant berries to have fever-reducing, sweat-inducing, menstrual-flow inducing, mildly laxative, astringent, appetite increasing, diuretic and digestive properties. Some of these proposed effects are probably, due to the verified high levels of vitamin C, fruit acids, and fibre the berries contain. Tea made from dried redcurrant leaves is said to ease the symptoms of gout and rheumatism, be useful in compresses for poorly healing wounds, and as a gargling solution for mouth infections. The fruit is used cosmetically in face-masks for firming up tired and lifeless skin.

Other Uses

The fruit yields a black dye and the leaves a yellow dye (Polunin 1969). Red currant can be used as a ground cover.

Comments

Currants are easily propagated by hardwood cuttings of 1-year old wood.

Selected References

Boivin D, Blanchette M, Barrette S, Moghrabi A, Béliveau R (2007) Inhibition of cancer cell proliferation and suppression of TNF-induced activation of NFkappaB by edible berry juice. Anticancer Res 27(2):937–948

Borges G, Degeneve A, Mullen W, Crozier A (2010) Identification of flavonoid and phenolic antioxidants in black currants, blueberries, raspberries, red currants, and cranberries. J Agric Food Chem 58:3901–3909

Chiej R (1984) The Macdonald encyclopaedia of medicinal plants. Macdonald & Co, London, 447 pp

Costantino L, Albasini A, Rastelli G, Benvenuti S (1992) Activity of polyphenolic crude extracts as scavengers of superoxide radicals and inhibitors of xanthine oxidase. Planta Med 58(4):342–344

Da Silva Pinto M, Kwon Y-I, Apostolidis E, Lajolo FM, Genovese MI, Shetty K (2010) Evaluation of red currants (Ribes rubrum L.), black currants (Ribes nigrum L.), red and green gooseberries (Ribes uva-crispa) for potential management of type 2 diabetes and hypertension using in vitro models. J Food Biochem 34:639–660

Djordjević B, Savikin K, Zdunić G, Janković T, Vulić T, Oparnica C, Radivojević D (2010) Biochemical properties of red currant varieties in relation to storage. Plant Foods Hum Nutr 65(4):326–332

Ehala S, Vaher M, Kaljurand M (2005) Characterization of phenolic profiles of Northern European berries by capillary electrophoresis and determination of their antioxidant activity. J Agric Food Chem 53(16):6484–6490

Goffman FD, Galletti S (2001) Gamma-linolenic acid and tocopherol contents in the seed oil of 47 accessions from several Ribes species. J Agric Food Chem 49(1):349–354

Grieve M (1971) A modern herbal. Penguin, 2 vols. Dover publications, New York, 919 pp

Hägg M, Ylikoski S, Kumpulainen J (1995) Vitamin C content in fruits and berries consumed in Finland. J Food Compos Anal 8(1):12–20

Hákkinen S, Heinonen M, Kárenlampi S, Mykkánen H, Ruuskanen J, Törrönen R (1999) Screening of selected flavonoids and phenolic acids in 19 berries. Food Res Int 32:345–353

Leskinen HM, Suomela JP, Kallio HP (2009) Effect of latitude and weather conditions on the regioisomer compositions of alpha- and gamma-linolenoyldilinoleoylglycerol in currant seed oils. J Agric Food Chem 57(9):3920–3926

Maatta K, Kamal-Eldin A, Törrönen R (2001) Phenolic compounds in berries of black, red, green, and white currants (*Ribes* sp.). Antioxid Redox Signal 3(6):981–993

Määttä KR, Kamal-Eldin A, Törrönen AR (2003) High-performance liquid chromatography (HPLC) analysis of phenolic compounds in berries with diode array and electrospray ionization mass spectrometric (MS) detection: *Ribes* species. J Agric Food Chem 51(23):6736–6744

Morin NR (2008) *Ribes rubrum* Linnaeus. In: Flora of North America Editorial Committee (eds) 1993+ Flora of North America, North of Mexico, 16+vols, New York, vol 8, p 14

Pantelidis GE, Vasilakakis M, Manganaris GA, Diamantidis G (2007) Antioxidant capacity, phenol, anthocyanin and ascorbic acid contents in raspberries, blackberries, red currants, gooseberries and Cornelian cherries. Food Chem 102(3):777–783

Piljac-Žegarac J, Šamec D (2010) Antioxidant stability of small fruits in postharvest storage at room and refrigerator temperatures. Food Res Int 44(1):345–350

Polunin O (1969) Flowers of Europe – a field guide. Oxford University Press, Oxford, 864 pp

Porcher MH et al. (1995–2020) Searchable world wide web multilingual multiscript plant name database. Published by The University of Melbourne, Australia. http://www.plantnames.unimelb.edu.au/Sorting/Frontpage.html

Rajakangas J, Misikangas M, Päivärinta E, Mutanen M (2008) Chemoprevention by white currant is mediated by the reduction of nuclear beta-catenin and NF-kappaB levels in Min mice adenomas. Eur J Nutr 47(3):115–122

Roach FA (1985) Cultivated fruits of Britain: their origin and history. Blackwell, Oxford, 349 pp

Schwarz B, Hofmann T (2007a) Isolation, structure determination, and sensory activity of mouth-drying and astringent nitrogen-containing phytochemicals isolated from red currants (*Ribes rubrum*). J Agric Food Chem 55(4):1405–1410

Schwarz B, Hofmann T (2007b) Sensory-guided decomposition of red currant juice (*Ribes rubrum*) and structure determination of key astringent compounds. J Agric Food Chem 55(4):1394–1404

Siewek F, Galensa R, Herrmann K (1984) Isolation and identification of a branched quercetin triglycoside from *Ribes rubrum* (Saxifragaceae). J Agric Food Chem 32(6):1291–1293

Tutin TG, Burgess NA, Chater AO, Edmondson JR, Heywood VH, Moore DM, Valentine DH, Walters SM, Webb DA (1993) Flora Europaea, vol 1, 2nd edn, Psilotaceae to Plantanaceae. Cambridge University Press, Cambridge. 629 pp

U.S. Department of Agriculture, Agricultural Research Service (2011) USDA National Nutrient Database for standard reference, release 24. Nutrient Data Laboratory Home Page. http://www.ars.usda.gov/ba/bhnrc/ndl

Ribes uva-crispa

Scientific Name

Ribes uva-crispa L.

Synonyms

Grossularia reclinata (L.) Miller, *Grossularia uva* Scop., *Grossularia uva-crispa* Mill., *Grossularia vulgaris* Spach, *Oxyacanthus uva-crispa* Chev., *Ribes grossularia* L., *Ribes pubescens* Opiz, *Ribes reclinatum* L.

Family

Grossulariaceae

Common/English Names

Carberry, Deberries, English Gooseberry, European Gooseberry, Feaberry, Feabes, Feverberry, Gooseberry, Goosegogs, Groser, Grozet, Honeyblobs, Wild Gooseberry.

Vernacular Names

Czech: Meruzalka Srstka, Srstka Obecná;
Danish: Stikkelsbaer;
Dutch: Kruisbes;
Eastonian: Aed-Karusmari;
Finnish: Karviainen;
Flemish: Stekbes;
French: Groseillier Épineux, Groseillier Des Haies, Groseille A Maquereaux, Groseillier Vert;
German: Stachelbeere;
Iceland: Broddber, Stikilsber;
Italian: Racinedda, Uva Spina;
Japanese: Maru Suguri;
Korean: Almulaengdunamu, Mulkkachibapnamu;
Norway: Stikkelsbaer;
Polish: Agrest;
Russian: Kryžovnik;
Scotland: Grozet;
Spanish: Grosellero Espinoso;
Swedish: Krusbär.

Origin/Distribution

Gooseberry is native to Europe (West, Central and Eastern), north-western Africa and the Caucasus, south-western Asia; often escaping from cultivation.

Agroecology

Gooseberry grows naturally in alpine thickets and rocky woods from France eastwards well into the Himalayas and peninsular India. It thrives in cool regions with a humid summer and winter chilling. It grows best in full sun for best fruit production. It grows in any soil type as long as it is moist and well-drained. It succeeds best in rich loam or black alluvium. Sandy soils are less suitable for gooseberries because they dry out too rapidly. They will grow on well-drained heavy soils that retain moisture. A thick organic mulch also helps keep the soil cool and moist.

Edible Plant Parts and Uses

Ripe fruit can be eaten fresh out of hand or cooked. They can be used in tarts and pies or processed into sauces, chutneys, jams, and dessert. The fruit can also be preserved in bottles for winter use. Gooseberries contain citric acid, pectins, sugars and mineral and are excellent for jellies. Yellow gooseberries have usually the richest flavour for dessert, and the best wine made from them very closely resembles champagne. Red gooseberries are more acidic. Young and tender leaves can be eaten in salads.

The fruit does not appear to be highly valued in the South of Europe, but further North is very popular for tarts, pies, sauces, chutneys, jams, and dessert, also for preserving in bottles for winter use.

Gooseberries are best known for their use in desserts such as "Gooseberry Fool" and "Gooseberry Crumble". In Portugal, gooseberries are relished as a beverage, being mostly used mixed with soda, water or even milk.

Botany

Gooseberry is a scrambling, deciduous, armed shrub, 0.5–2 m high with sharp, axillary prickles on the stem and branches and a shallow root system. Leaves are alternate, single, deeply crenate, 3–5 lobed, glossy dark green or pale grey green, glabrous to finely pubescent (Plates 1 and 2). Flowers are bell-shaped, inconspicuous, small, 1 cm across, regular with 5 yellowish-green, green flush with pink or white tepals, 5 stamens and 2 fused carpels. Flowers are borne singularly or in groups of 2–3 in leaf axils of 1-year old wood and on short spurs of older wood (Plates 2 and 3). Fruit is a tough-skinned, subglobose to obovate berry, 1.5–2.7 cm across,

Plate 1 Gooseberry twigs and leaves

Plate 2 Gooseberry flowers

Plate 3 Close-view of gooseberry flower

Plate 4 Ripe gooseberries

Plate 5 Harvested gooseberry fruits

green, yellowish-green (Plates 4 and 5), pink or purplish-red, glabrous or pubescent, juicy, acid to subacid.

Nutritive/Medicinal Properties

Nutrient composition of raw gooseberry per 100 g edible portion was reported as follows (Saxholt et al. 2008): energy 179 kJ; protein (total) 0.900 g, total N 0.144 g, fat (total) 0.550 g, saturated fatty acids 0.078 g, C16:0 (palmitic acid) 0.070 g, C18:0 (stearic acid) 0.08 g; monounsaturated fatty acids 0.066 g, C 16:1 (palmitoleic acid) 0.005 g, C18:1, n-9 (oleic acid) 0.061 g; polyunsaturated 0.295 g, C18:2, n-6 (linoleic acid) 0.174 g, C18:3, n-3 (α-linolenic acid) 0.122 g; total carbohydrate 10.1 g, available carbohydrate 6.9 g, dietary fibre 3.2 g; ash, 0.400 g, moisture 8,709 g; vitamin A 6.83 RE, ß-carotene equivalent 82 μg, vitamin E (α-tocopherol) 1.0 mg, vitamin B1(thiamine) 0.011 mg, vitamin B2 (riboflavin) 0.025 g, niacin 0.25 mg, vitamin B-6 0.015 mg, pantothenic acid 0.286 mg, biotin 0.5 μg, folates 11.7 μg, vitamin C 33 mg, Na 2 mg, K 176 mg, Ca 26.5 mg, Mg 8 mg, P 23.6 mg, Fe 0.4 mg, Cu 0.056 mg, Zn 0.16 mg, I 0.3 μg, Mn 0.17 mg, Cr 0.3 μg, Se 0.068 μg, Ni 2.54 μg, and tryptophan 9 mg.

Gooseberry is rich vitamin C, A and also contain vitamin Bs, pantothenic acid, vitamin E, biotin, folate and α-linolenic acid, an omega-3 fatty acid.

Flavonoids (kaempferol, quercetin, myricetin) and phenolic acids (p-coumaric, caffeic, ferulic, p-hydroxybenzoic, gallic and ellagic acids) were detected in the fruits of 19 berries (Hákkinen et al. 1999). In the genus *Ribes*, quercetin was the main compound in gooseberry, red currant and black currant. The data suggested berries to have potential as good dietary sources of quercetin or ellagic acid.

Antioxidant Activity

Goffman and Galletti (2001) found that the highest total tocopherol content was found in *R. nigrum* (mean, 1,716 mg/kg oil), followed by *R. rubrum* (mean, 1,442 mg/kg oil). *R. grossularia* showed the lowest tocopherol content (mean, 786 mg/kg oil). The three species also differed markedly in tocopherol composition. *R. rubrum* had the highest content of δ-tocopherol (mean, 20.2%); *R. grossularia* had the highest level of γ-tocopherol (mean, 70.0%), and *R. nigrum* the highest percentage of α-tocopherol (mean, 34.8%), the most biologically active among the four tocopherols. As for γ-linolenic acid, the highest concentration was found in *R. nigrum*, up to 15.8% while *R. grossularia* and *R. rubrum* showed mean γ-linolenic acid levels of 8% and 6.2%, respectively. The present study indicated that seeds of *Ribes* species, especially *R. nigrum*, could be used as sources of gamma-linolenic acid and natural vitamin E. γ-linolenic acid is an essential fatty acid for humans with δ-6-desaturase deficiency; it is a precursor of prostaglandins, prostacyclins, and tromboxanes; and it has antiinflammatory and antitumoral effects. Tocopherols are natural antioxidants with biological activity, heart/vascular, and cancer protective properties. Ferric reducing antioxidant power (FRAP) values of raspberry (*Rubus idaeus*), blackberry (*Rubus fructicosus*), raspberry × blackberry hybrids, red currant (*Ribes sativum*), gooseberry (*Ribes glossularia*) and Cornelian cherry (*Cormus mas*) cultivars ranged from 41 to 149 μmol ascorbic acid/g dry weight and protection of deoxyribose ranged from 16.1% up to 98.9% (Pantelidis et al. 2007). Anthocyanin content ranged from 1.3 mg in yellow-coloured fruit, up to 223 mg cyanidin-3-glucoside equivalents 100/g fresh weight in Cornelian cherry, whereas phenol content ranged from 657 up to 2,611 mg gallic acid equivalents/100 g dry weight. Ascorbic acid content ranged from 14 up to 103 mg/100 g fresh weight.

The total phenolic content of green gooseberries (*R. uva-crispa*) was found to be 3.2 mg/g fruit fresh weight (Da Silva Pinto et al. 2010). No correlation was found between total phenolics and antioxidant activity. The major phenolic compounds were quercetin derivatives in green gooseberries.

Antimicrobial Activity

Recent studies reported that fruit juices and pomace (skin, pulp, seeds) extracts from blackcurrant (*Ribes nigrum*), gooseberry (*Ribes uva-crispa*) and their hybrid plant (jostaberry, *Ribes* × *nidigrolaria*) inhibited growth of the most common human *Candida* species, with the exception of *C. albicans, C. krusei, C. lusitaniae* and *C. pulcherrima* (Krisch et al. 2009). *R. nigrum,* with the highest phenol content, exhibited the highest anticandidal activity. *R. uva-crispa* juice was found to exhibit antibacterial activity against two acne-inducing bacteria, *Staphylococcus aureus* and *Staphylococcus epidermidis* (Ördögh et al. 2010). The growth inhibition effect of *Ribes uva-crispa* (gooseberry) juice was stronger at acidic pH (MIC 0.40 mg/ml) than at neutral pH (MIC 5.30 mg/ml) against *Staphylococcus epidermidis*. Gooseberry juice exhibited 91% radical scavenging activity.

Traditional Medicinal Uses

In traditional medicine, the fruit is laxative. Stewed unripe gooseberries are employed as a spring tonic to cleanse the system. The juice was formerly regarded as panacea for inflammations.

The jelly made from the red berries is valuable for sedentary, plethoric, and bilious subjects. The fruit pulp is used cosmetically in face-masks for its cleansing effect on greasy skins. The leaves have been employed to treat gravel. A leaf infusion taken before the monthly periods is said to be a useful tonic for adolescent girls. The leaves contain tannin and have been employed as an astringent to treat dysentery and wounds.

Other Uses

Gooseberry is grown as a backyard crop or hedge. In the Middle Ages, gooseberry was used for magic purposes to overcome witchcraft and was probably selected for this purpose in monasteries.

Comments

There is frequent out-crossing of *Ribes uva-crispa* with other wild *Ribes* species and forms resulting in an abundance of hybrid and cultivars.

Selected References

Bailey LH (1976) Hortus Third. A concise dictionary of plants cultivated in the United States and Canada. Liberty Hyde Bailey Hortorium/MacMillan, 1312 pp

Chiej R (1984) Encyclopaedia of medicinal plants. MacDonald, London, 447 pp

Da Silva Pinto M, Kwon Y-I, Apostolidis E, Lajolo FM, Genovese MI, Shetty K (2010) Evaluation of red currants (*Ribes rubrum* L.), black currants (*Ribes nigrum* L.), red and green gooseberries (*Ribes uva-crispa*) for potential management of type 2 diabetes and hypertension using in vitro models. J Food Biochem 34:639–660

Goffman FD, Galletti S (2001) Gamma-linolenic acid and tocopherol contents in the seed oil of 47 accessions from several *Ribes* species. J Agric Food Chem 49(1):349–354

Grieve M (1971) A modern herbal, 2 vols. Penguin/Dover Publications, Harmondsworth/New York, 919 p

Häkkinen S, Heinonen M, Kárenlampi S, Mykkánen H, Ruuskanen J, Törrönen R (1999) Screening of selected flavonoids and phenolic acids in 19 berries. Food Res Int 32:345–353

Keep E (1975) Currants and gooseberries. In: Janick J, Moore JN (eds) Advances in fruit breeding. Purdue University Press, West Lafayette, pp 197–268

Krisch J, Ördögh L, Galgóczy L, Papp T, Vágvölgyi C (2009) Anticandidal effect of berry juices and extracts from *Ribes* species. Cent Eur J Biol 4(1):86–89

Morin NR (2008) *Ribes uva-crispa* Linnaeus. In: Flora of North America Editorial Committee (eds) 1993+. Flora of North America North of Mexico, 16+ vols. New York/Oxford, vol 8. p 42

Ördögh L, Galgóczy L, Krisch J, Papp T, Vágvölgyi C (2010) Antioxidant and antimicrobial activities of fruit juices and pomace extracts against acne-inducing bacteria. Acta Biol Szeged 54(1):45–49

Pantelidis GE, Vasilakakis M, Manganaris GA, Diamantidis G (2007) Antioxidant capacity, phenol, anthocyanin and ascorbic acid contents in raspberries, blackberries, red currants, gooseberries and Cornelian cherries. Food Chem 102(3):777–783

Saxholt E, Christensen AT, Møller A, Hartkopp HB, Hess Ygil K, Hels OH (2008) Danish food composition databank, Revision 7. Department of Nutrition, National Food Institute, Technical University of Denmark, Copenhagen. http://www.foodcomp.dk/

Ribes x nidigrolaria

Scientific Name

Ribes x nidigrolaria Rud. Bauer & A. Bauer.

Synonyms

(*Ribes divaricatum* x *Ribes nigrum* x *Ribes uva-crispa*) (Hort.).

Family

Grossulariaceae

Common/English Name

Jostaberry

Vernacular Names

German: Jochelbeere, jostabeere.

Origin/Distribution

Ribes x *nidigrolaria*, best known as Jostaberry, is an artificial, fertile F-2 amphidiploid hybrid of complex parentage from crossbreeding of the F1-hybrid (*Ribes nigrum* 'Langtraubige Schwarze' x *R. divaricatum*) resistant to mildew and the F-1 hybrid (*R. nigrum* 'Silvergieters Schwarze' x *R. grossularia* 'Grune Hansa') resistant to white pine blister rust by Dr. Rudolf Bauer in Germany (Bauer 1973, 1986; Bauer and Weber 1989). The first cultivar of *R. x nidigrolaria* was introduced by Dr Baeur in 1977 under the name josta. The resulting double hybrids carries the characteristics of their four genetical grand parents with combined mildew and rust resistance, resistance to other leaf-fall diseases and to gall mite and have sufficient self-fertility (Baeur 1973). This hybrid is sometimes erroneously referred to as *Ribes* x *culverwellii* MacFarlane. The latter is a nearly sterile hybrid of *R. nigrum* and *R. uva-crispa*. Despite its complex parentage *Ribes* x *nidigrolaria* is a fully fertile taxon and relatively easily escapes (Mang 1992; Weber 1995). *R. x nidigrolaria* has been recorded as an escape on riverbanks (between granite blocks) in Flanders, Belgium, probably bird-dispersed or from washed-up rhizomes (Verloove 2011). It was first seen along river Leie near Wervik, subsequently also along river Schelde (at its junction with Ringvaart). Further, in 2010 it was discovered in grassland (former dump) along motorway near Drongen.

Agroecology

Ribes x nidigrolaria grows in a wide variety of soil of pH 5.5–6.5 but prefers loamy soils. In Belgium it was found growing in nitrophilous, wet habitats as an escape. The plant prefers full sun but will also grow in partial shade. *Ribes x nidigrolaria* can survive very cold sub-zero winters.

Edible Plant Parts and Uses

The flavour of ripe jostaberry is similar to black currant but slightly milder and the fruits are larger. It is excellent for eating fresh or for processing into cordials, beverages, wine, jellies, preserves, pies and jams.

Botany

The plant is a vigorous, multibranched, small, eglandular, deciduous shrub, 1 m high but can reach 2 m. Unlike *Ribes divaricatum* and *R. uva-crispa*, *R. x nidigrolaria* lacks thorns. Branches are long-living like *R. uva-crispa*. Leaves are alternate, sub-orbicular, palmately lobed with 3–5 irregular lobes, deeply veined and with serrated margins (Plate 1). Inflorescence in relatively few flowered racemes (Plate 2). Corollas are greenish-purple and distinctly hairy (a character of *R. divaricatum*). Berry is globose larger than black currant but smaller than gooseberry, glossy black, glabrous and eglandular containing few to numerous seeds.

Plate 2 Jostaberry inflorescence

Nutritive/Medicinal Properties

The major anthocyanins in two cultivars of jostaberries (*R. × nidigrolaria*) were identified as delphinidin 3-glucoside, delphinidin 3-rutinoside, cyanidin 3-glucoside, cyanidin 3-rutinoside and cyanidin 3-O-β-(6″-E-p-coumaroylglucopyranoside) reflecting that this hybrid contained the major anthocyanins of both parents, black currant and gooseberry (Jordheim et al. 2007). Another pigment cyanidin 3-(6″-Z-p-coumaroylglucoside) was found in trace amount in jostaberries.

Antioxidant Activity

Total anthocyanin content value of *Ribes* x *nidigrolaria* was higher than those for cv Captivator (*Ribes uva-crispa*), but were lower than *R. nigrum* cultivars (Moyer et al. 2002a). Mean total anthocyanin in *Ribes* x *nidigrolaria* ORUS 6–10 selections was 74 mg/100 g, mean total phenolics

Plate 1 Jostaberry leaves

309 mg/100 g, mean oxygen radical absorbing capacity (ORAC) 28.1 μmol TE/g, and mean ferric reducing antioxidant power (FRAP) 39 μmol/g. The ORUS 6–10 series of *Ribes x nidigrolaria* contained dark shin and dark flesh but had the lowest in total anthocyanin content, as compared to that of fruits with nonpigmented flesh. In another paper, they reported that *R. x nidigrolaria* 'ORUS 6,' 'ORUS 8,' and 'ORUS 10' selections had antioxidant values intermediary between gooseberries and blackcurrants (Moyer et al. 2002b). Mean total anthocyanin content of the three cultivars was 73.67 mg/100 g berries, mean total phenolics was 313.67 mg gallic acid/100 g berries and mean oxygen radical absorbing capacity (ORAC) was 29.87 μmoles TE/g.

Antimicrobial Activity

Ribes x nidigrolaria juice was found to be inhibitory to Gram positive bacteria: *Bacillus subtilis* and *Bacillus cereus* and *Escherichia coli* (Gram negative) but was ineffective against *Serratia marcescens* (another Gram negative bacterium) (Krisch et al. 2008). *R. x nidigrolaria* juice inhibited the growth of two acne causing bacteria *Staphylococcus aureus* and *S. epidermidis* (Örödgh et al. 2010). Total phenol content of the fruit juice was determined to be 18.87 mg/g and DPPH radical scavenging capacity 84.3%. The methanol extract of the polyphenol–rich pomace of *Ribes x nidigrolaria* exhibited anti-fungal activity against seven *Candida* spp, namely *Candida glabrata, C. inconspicua, C. lipolytica, C. norwegica, C. parapsilosis, C. tropicalis* and *C. zeylanoides* (Krisch et al. 2002). The MIC (minimum inhibitory concentration) values of the extract ranged from 2.91 to 10.38 mg/ml of dry matter. The total phenolic content of the juice, methanol and water extracts of the pomace was 21.50, 50.51 and 45.86 μg gallic acid equivalent/mg dry weight respectively.

Other Uses

Jostaberry is also valued as an ornamental.

Comments

Hardwood cuttings provide an easy, quick way to propagate Jostaberry bushes, as they afford a very high chance of success.

Selected References

Bauer R (1973) True breeding for combined resistance to leaf, bud, and shoot diseases by amphidiploidy in *Ribes*. J Yugosl Pomol 7:17–19

Bauer R (1978) Josta, eineneue Beerenobstart, aus des Kreuzung Scharze Johannisbeere x Stachelbeere. Erwerbs-onstbau 20:116–119

Bauer A (1986) New results of breeding *Ribes nidigrolaria*: amphidiploid species hybrids between blackcurrant and gooseberry. Acta Hortic (ISHS) 183:107–110

Bauer A, Weber HE (1989) *Ribes* × *nidigrolaria* R. & A. Bauer und *Fragaria* × vescana R. & A. Bauer – Beschreibung zweier Hybridarten. Osnabrück Naturwiss Mitt 15:49–58

Jordheim M, Måge F, Andersen ØM (2007) Anthocyanins in berries of *Ribes* including gooseberry cultivars with a high content of acylated pigments. J Agric Food Chem 55(14):5529–5535

Krisch J, Ördögh L, Galgóczy L, Papp T, Vágvölgyi C (2002) Anticandidal effect of berry juices and extracts from *Ribes* species. Cent Eur J Biol 4(1):86–89

Krisch J, Galgóczy L, Tölgyesi M, Papp T, Vágvölgyi C (2008) Effect of fruit juices and pomace extracts on the growth of Gram-positive and Gram-negative bacteria. Acta Biol Szeged 52(2):267–270

Mang FWC (1992) Zur Verwilderung der Jostabeere, *Ribes x nidigrolaria* R. et A. Bauer an zwei Standorten in Hamburg. Osnabrück Naturwiss Mitt 17:175–178

Moyer RA, Hummer KE, Finn CE, Frei B, Wrolstad RE (2002a) Anthocyanins, phenolics, and antioxidant capacity in diverse small fruits: *Vaccinium, Rubus*, and *Ribes*. J Agric Food Chem 50:519–525

Moyer RA, Hummer KE, Wrolstad RE, Finn C (2002b) Antioxidant compounds in diverse *Ribes and Rubus* germplasm. Acta Hortic (ISHS) 585:501–505

National Botanical Garden of Belgium (2011 onwards) Manual of the alien plants of Belgium. http://alienplantsbelgium.be/

Ördögh L, Galgóczy L, Krisch J, Papp T, Vágvölgyi C (2010) Antioxidant and antimicrobial activities of fruit juices and pomace extracts against acne-inducing bacteria. Acta Biol Szeged 54(1):45–49

Verloove F (2011) *Fraxinus pennsylvanica, Pterocarya fraxinifolia* en andere opmerkelijke uitheemse rivierbegeleiders in België en NW-Frankrijk. Dumortiera 99:1–10

Weber HE (1995) Grossulariaceae. In: Weber HE (ed) Hegi G, Illustrierte Flora van Mitteleuropa, vol 4(2A), 3rd edn. Blackwell Wissenschafts-Verlag, Berlin, pp 48–68

Molineria latifolia

Scientific Name

Molineria latifolia (Dryand. ex W. T. Aiton) Herb. ex Kurz.

Synonyms

Aurota latifolia (Dryand. ex W.T.Aiton) Raf., *Curculigo latifolia* Dryand. ex W.T.Aiton (basionym).

Family

Hypoxidaceae, also placed in Amaryllidaceae, Liliaceae

Common/English Names

Curculigo, Lemba, Lumbah, Weevil Lily.

Vernacular Names

Indonesia: Marasi (Sundanese), Keliangau (Bangka), Doyo, Lemba (Kalimantan), Helai, Kah Kuhue Manak, Kehoang, Kohoang, Kuhueng Lumpoa, Kuhuang Manak, Mangih Tana, Lekuhoe, Louku Huang, Rou' Kuhoa Manak, Rou' Kuhoe, Rou' Kuhue, Rou' Lakuhoeng, Rou' Lekuhoe, Rou' Lekuhue, Rou Lekuhueng, Rou' Parakuwang, Rowei Lekuhong, Teung Lekuhve, Merap, (Kalimantan),Bangit Tanuk, Lekuan, Lekuhan, Lumpa (Daun Panjang), Lumpa Manuk, Luva, Luva', Luva Manuk, Luva Uuk Pla, Luwah, Luwak, Tuban Tlop, Uru' Luva' (Punan, Kalimantan), Congkok, Ketari;
Malaysia: Kelapa Puyoh, Lamba, Lemba, Lemba Kilat, Lumbah, Lumbah Padi, Lumbah Rimba, Nyiur Lember, Pinang Puyuh;
Thailand: Chaa Laan, Ma Phraao Nok Khum (Northern), Phraa Nok (Peninsular);
Vietnamese: Cồ Nốc Lá Rộng, Sâm Cau Lá Rộng.

Origin/Distribution

The species is found from China (Guangdong) to Malesia – Malaysia (Perak, Pahang, Sarawak, Sabah), Indonesia (Sumatra, Bangka, Lingga, Java, Kalimantan) and the Philippines (Palawan, Balabac, Samar).

Agroecology

In its native range, the plant is found in wet areas near streams in primary and secondary forests, from near sea level to 1,100 m altitude. It is a shade-loving plant, thriving under partly shaded or sunless conditions, with abundant water supply. It prefers fertile, well-drained soils, rich in organic matter.

Edible Plant Parts and Uses

The fruit is edible and taste like sweetened cucumber and is believed to increase appetite. The fruit also has taste modifying properties due to the presence of the protein neoculin (previously called curculin). After consumption of neoculin, water, sour solutions and substances taste sweet, for example a lemon eaten after taking neoculin elicits a sweet taste lasting for about 10 minutes. Neoculin is heat labile, high temperatures of above 50°C degrades the protein and destroys its "sweet-tasting" and "taste-modifying" properties, Thus, it is not suitable for use as sweetening agent in hot or processed foods. However, below this temperature both properties of neoculin are unaffected in basic and acidic solutions, so it has potential for use in fresh foods, frozen foods and as a table-top sweetener.

Food and soft drinks high in sugar content are significantly contributing to the problem of diabetes and obesity and related cardiovascular diseases. The use of artificial, low calorie or zero calorie sweeteners has become a dietary option in many nutritional guides and diet plans. Additionally, there is a need of a natural, low calorie or zero calorie sweetener. A great potential exists with natural sweet proteins, such as neoculin, to fill this multi-million dollar niche.

Botany

A tufted, stemless, andromonoeciou herb with erect rhizome and creeping stolons and thick, fibrous roots. Leaf long-petioled (10–100 cm), lanceolate to oblong-lanceolate, 18–40 × 3–8 cm, usually glabrous, both ends tapering, lateral nerves parallel to mid-rib, penni-nerved (Plates 1 and 2). Flowers yellow, inconspicuous, in axils of a large lanceolate, hairy bract, in densely several flowered 1.5–3 cm racemes. Lower flowers hermaphrodite, sessile to subsessile with long styles, upper flowers staminate with shorter styles and long-pedicelled. Flower perianth yellow; segments suboblong, 8–12 mm with involute margin

Plate 1 Top view of leaves

Plate 2 Leaves with long petioles

(Plates 3, 4 and 5); stamens slightly shorter than perianth segments; ovary cylindric, 1.5 cm long; style slender, subequaling stamens with subcapitate stigma. Berry is ovoid to oblong-ovoid, 2.5 cm, white, slightly hairy; beak 6–7 mm, pulp is sweet and edible. Seeds small, ribbed and verruculose.

Plate 3 Flowers arising from the base

Plate 4 Flowers and dried inflorescences

Nutritive/Medicinal Properties

Fruits of *Molineria latifolia* (*Curculigo latifolia*) was found to contain neoculin previously called curculin (Shirasuka et al. 2004; Masuda and Kitabatake 2006), Neoculin is a sweet-tasting protein that also has taste-modifying activity to convert sourness to sweetness (Yamashita et al. 1995; Kurihara 1992; Suzuki et al. 2004). Neoculin is considered to be a high-intensity sweetener, with a reported relative sweetness of 430–2,070 times sweeter than sucrose on a weight basis (Yamashita et al. 1990, 1995; Kurihara 1992). After consuming curculin, water elicited a sweet taste, and sour substances induced a stronger sense of sweetness (Yamashita et al. 1990, 1995). The maximum sweetness induced by 0.02 M citric acid or deionized water after curculin dissolved in a buffer of pH 6.0 was held in mouth for 3 minutes was equivalent to that of 0.35 M sucrose (Yamashita et al. 1995). The taste-modifying activity of curculin was unchanged when curculin was incubated at 50°C for 1 hour between pH 3 and 11.

Plate 5 Flowers with yellow tepals

Neoculin was found to be a large heterodimeric protein composed of an N-glycosylated acidic subunit (NAS) and a monomeric basic subunit (NBS), conjugated by disulfide bonds (Nakajima et al. 2006a; Shimizu-Ibuka et al. 2006; Okubo et al. 2008). Okubo et al. (2008) using protein gel blot analysis revealed the presence of a non-glycosylated NAS species. This suggested the presence of multiple NAS–NBS heterodimers in one cultivar. The neoculin acidic subunit (NAS) was found to consist of 113 amino acid residues weighing 12.5 kDa, while the neoculin basic subunit (NBS) consisted of 114 amino acid residues weighing 12.7 kDa (Suzuki et al. 2004; Shimizu-ibuka et al. 2006). Although these residues were different, 77% of the amino acid sequence was identical. The overall crystal structure of neoculin was found to be quite similar to those of monocot mannose-binding lectins (Shimizu-Ibuka et al. 2006, 2008). However, crucial topological differences were observed in the C-terminal regions of both subunits. In both subunits of neoculin, the C-terminal tails bent up to form loops fixed by inter-subunit disulfide bonds that were not observed in the lectins. In addition, distribution of electrostatic potential on the surface of neoculin was found to be unique and significantly different from those of the lectins, particularly in the basic subunit (NBS). The scientists found a large cluster composed of six basic residues on the surface of NBS, and speculated that it might be involved in the elicitation of sweetness and/or taste-modifying activity of neoculin. Hemagglutination assay data demonstrated that neoculin had no detectable agglutinin activity (Shimizu-Ibuka et al. 2008). DNA microarray analysis indicated that neoculin had no significant influence on gene expression in Caco-2 cell, whereas kidney bean lectin (*Phaseolus vulgaris* agglutinin) greatly influenced various gene expressions. These data strongly suggested that neoculin had no lectin-like properties, promoting its practical use in the food industry (Shimizu-Ibuka et al. 2008). Immunoblot analysis indicated that antiserum to curculin was faintly reactive with miraculin, but not with thaumatin or monellin (Nakajo et al. 1992).

Nakajima et al. (2006b) found in sensory tests, when acetate buffers with different pH values were placed on the tongue after tasting neoculin, a higher intensity of sweetness was detected at lower pH. The sweetness was also suppressed with the addition of lactisole. These results suggested that both the sweetness and the taste-modifying activity were mediated via the human sweet taste receptor hT1R2/T1R3. Koizumi et al. (2007) demonstrated that hT1R3 was required for the reception of neoculin and that the extracellular amino terminal domain (ATD) of hT1R3 was essential for the reception of neoculin. Kurimoto et al. (2007) found that the curculin heterodimer exhibited sweet-tasting and taste-modifying activities through its partially overlapping but distinct molecular surfaces. These findings suggested that the two activities of the curculin heterodimer were expressed through its two different modes of interactions with the human T1R2-T1R3 heterodimeric sweet taste receptor. Nakajima et al. (2008) found that the acid-induced sweetness of neoculin was ascribed to its pH-dependent agonistic-antagonistic interaction with human sweet taste receptor. At acidic pH, neoculin acted as an hT1R2-hT1R3 agonist but functionally changed into its antagonist at neutral pH. Findings by Morita et al. (2009) suggested that the sweetness of neoculin depended on structural change accompanying the pH change, with histidine residues playing a key role.

Northern blot analysis showed that the mRNA for curculin was first detected in *Curculigo latifolia* fruits at 2 weeks after pollination and remained at a constant level for the following 4 weeks (Abe et al. 1992). The content of curculin in the fruit of *Curculigo latifolia* was found to increase gradually until 3 weeks after artificial pollination and dramatically at 4 weeks, to finally reach 1.3 mg per fruit (Nakajo et al. 1992) The neoculin content of the fruit was high for 10 weeks after flowering, following which the yield decreased gradually (Okubo et al. 2008). The optimal period for harvesting the fruits with sensory activity coincided with this 10-week peak period during which the amount of neoculin was 1–3 mg in the whole fruit and 1.3 mg/g of pulp. Immunohistochemical staining showed that neoculin occurred in the whole fruit, especially at the basal portion. Okubo et al. (2010) found that *C. latifolia* exhibited self-incompatibility. The rate of fruit setting was shown to be 45% by cross-pollination and 4% by self-pollination. To improve the rate of fruit setting of *C. latifolia*, it was found necessary to pollinate compatible pollen by around the 15th day after the first flowering.

A quinone-ester gentisylquinonyl-2,6-dimethoxybenzoate was isolated from *Molineria latifolia* (Piyakarnchana et al. 1995). *Curculigo latifolia* var. *latifolia* was found to contain arbutin (1.10 mg/g) which is used as an ingredient in skin whitening creams (Thongchai et al. 2007).

Traditional Medicinal Uses

In Peninsular Malaysia, decoction of the rhizome with *Areca catechu* is drunk for menorrhagia or with *Hibiscus rosa-sinensis* used as a lotion for ophthalmia. Roots are prescribed as internal medicine for fever and fever with delirium. The leaves, shoot tips and roots are used as infusions in water. Flowers and roots are used as stomachic and diuretic in genito-urinary disorders.

Other Uses

In Peninsular Malaysia and Borneo, the tough, light-weight leaf fibres are made into fishing nets. In Borneo, they are also used to make ropes, twines, sarongs, rice bags, garments and fabrics. The cloth made from the fibres is known as 'lemba cloth'. The leaves are also used for wrapping fruits, vegetables and food in Indonesia and Malaysia. The leaves are also rolled into strings for tying. In Borneo, the leaves are used in magical healing ceremonies. The plant is also grown as an ornamental in southeast Asia, India, Africa, Europe and the United States.

Comments

C. latifolia is best propagated using rhizomes and corms as the seeds have a low and slow germination rate (Abdullah et al. 2010).

Selected References

Abdullah NAP, Saleh GB, Shahari R, Lasimin V (2010) Shoot and root formation on corms and rhizomes *of Curculigo latifolia* Dryand. J Agro Crop Sci 1(1):1–5

Abe K, Yamashita H, Arai S, Kurihara Y (1992) Molecular cloning of curculin, a novel taste-modifying protein with a sweet taste. Biochim Biophys Acta 1130(2):232–234

Backer CA, Bakhuizen van den Brink RC Jr (1968) Flora of Java, (spermatophytes only), vol 3. Wolters-Noordhoff, Groningen, 761 pp

Brink M (2003) *Curculigo* Gaertn. In: Brink M, Escobin RP (eds) Plant resources of South-East Asia No. 17: fibre plants. Backhuys Publisher, Leiden, pp 118–120

Burkill IH (1966) A dictionary of the economic products of the Malay Peninsula. Revised reprint, 2 vols. Ministry of Agriculture and Co-operatives, Kuala Lumpur, Malaysia. Vol 1 (A–H) pp 1–1240, vol 2 (I–Z) pp 1241–2444

Ji Z, Meerow AW (2000) Amaryllidaceae Jaume Saint Hilaire. In: Wu ZY, Raven PH (eds) Flora of China, vol 24 (*Flagellariaceae* through *Marantaceae*). Science Press/Missouri Botanical Garden Press, Beijing/St. Louis, pp 264–273

Koizumi A, Nakajima K, Asakura T, Morita Y, Ito K, Shmizu-Ibuka A, Misaka T, Abe K (2007) Taste-modifying sweet protein, neoculin, is received at human T1R3 amino terminal domain. Biochem Biophys Res Commun 358(2):585–589

Kurihara Y (1992) Characteristics of antisweet substances, sweet proteins, and sweetness-inducing proteins. Crit Rev Food Sci Nutr 32(3):231–252

Kurimoto E, Suzuki M, Amemiya E, Yamaguchi Y, Nirasawa S, Shimba N, Xu N, Kashiwagi T, Kawai M, Suzuki E, Kato K (2007) Curculin exhibits sweet-tasting and taste-modifying activities through its distinct molecular surfaces. J Biol Chem 282(46):33252–33256

Masuda T, Kitabatake N (2006) Developments in biotechnological production of sweet proteins. J Biosci Bioeng 102(5):375–389

Morita Y, Nakajima K, Iizuka K, Terada T, Shimizu-Ibuka A, Ito K, Koizumi A, Asakura T, Misaka T, Abe K (2009) pH-Dependent structural change in neoculin with special reference to its taste-modifying activity. Biosci Biotechnol Biochem 73(11):2552–2555

Nakajima K, Asakura T, Maruyama J, Morita Y, Oike H, Shimizu-Ibuka A, Misaka T, Sormachi H, Arai S, Kitamoto K, Abe K (2006a) Extracellular production of neoculin, a sweet-tasting heterodimeric protein with taste-modifying activity, by *Aspergillus oryzae*. Appl Environ Microbiol 72:3716–3723

Nakajima K, Asakura T, Oike H, Morita Y, Shimizu-Ibuka A, Misaka T, Sorimachi H, Arai S, Abe K (2006b) Neoculin, a taste-modifying protein, is recognized by human sweet taste receptor. Neuroreport 17(12):1241–1244

Nakajima K, Morita Y, Koizumi A, Asakura T, Terada T, Ito K, Shimizu-Ibuk A, Maruyama J, Kitamoto K, Misaka T, Abe K (2008) Acid-induced sweetness of neoculin is ascribed to its pH-dependent agonistic-antagonistic interaction with human sweet taste receptor. FASEB J 22(7):2323–2330

Nakajo S, Akabane T, Nakaya K, Nakamura Y, Kurihara Y (1992) An enzyme immunoassay and immunoblot analysis for curculin, a new type of taste-modifying protein: cross-reactivity of curculin and miraculin to both antibodies. Biochim Biophys Acta 1118(3):293–297

Okubo S, Asakura T, Okubo K, Abe K, Misaka T, Akita T, Abe K (2008) Neoculin, a taste-modifying sweet protein, accumulates in ripening fruits of cultivated *Curculigo latifolia*. J Plant Physiol 165(18):1964–1969

Okubo S, Yamada M, Yamaura T, Akita T (2010) Effects of the pistil size and self-incompatibility on fruit production in *Curculigo latifolia* (Liliaceae). J Jpn Soc Hortic Sci 79(4):354–359

Piyakarnchana N, Rukachaisirikul V, Chantrapromma K (1995) Synthesis of a quinone-ester from *Molineria latifolia* herb. Songklanakarin J Sci Technol 17(1):69–72

Shaari N (2005) Lemba (*Curculigo latifolia*) leaf as a new materials for textiles. In: 4th international symposium on environmentally conscious design and inverse manufacturing, Tokyo, Japan, 12–14 Dec 2005, pp 109–111

Shimizu-Ibuka A, Morita Y, Terada T, Asakura T, Nakajima K, Iwata S, Misaka T, Sorimachi H, Arai S, Abe K (2006) Crystal structure of neoculin: insights into its sweetness and taste-modifying activity. J Mol Biol 359(1):148–158

Shimizu-Ibuka A, Nakai Y, Nakamori K, Morita Y, Nakajima K, Kadota K, Watanabe H, Okubo S, Terada T, Asakura T, Misaka T, Abe K (2008) Biochemical and genomic analysis of neoculin compared to monocot mannose-binding lectins. J Agric Food Chem 56(13):5338–5344

Shirasuka Y, Nakajima K, Asakura T, Yamashita H, Yamamoto A, Hata S, Nagata S, Abo M, Sorimachi H, Abe K (2004) Neoculin as a new taste-modifying protein occurring in the fruit of *Curculigo latifolia*. Biosci Biotechnol Biochem 68(6):1403–1407

Suzuki M, Kurimoto E, Nirasawa S, Masuda Y, Hori K, Kurihara Y, Shimba N, Kawai M, Suzuki E, Kato K (2004) Recombinant curculin heterodimer exhibits taste-modifying and sweet-tasting activities. FEBS Lett 573(1–3):135–138

Thongchai W, Liawruangrath B, Liawruangrath S (2007) High-performance liquid chromatographic determination of arbutin in skin-whitening creams and medicinal plant extracts. Int J Cosmet Sci 29(6):488

Uddin MZ, Khan MS, Hassan MA (1999) *Curculigo latifolia* Dryand (Hypoxidaceae): a new angiospermic record for Bangladesh. Bangladesh J Plant Taxon 6(2):105–107

Yamashita H, Akabane T, Kurihara Y (1995) Activity and stability of a new sweet protein with taste-modifying action, curculin. Chem Senses 20(2):239–243

Yamashita H, Theerasilp S, Aiuchi T, Nakaya K, Nakamura Y, Kurihara Y (1990) Purification and complete amino acid sequence of a new type of sweet protein taste-modifying activity, curculin. J Biol Chem 265(26):15770–15775

Ardisia crenata

Scientific Name

Ardisia crenata Sims (Plate 3).

Synonyms

Ardisia bicolor E. Walker; *Ardisia crenata* var. *bicolor* (E. Walker) C. Y. Wu & C. Chen, *Ardisia crenulata* Lodd. et al., nom. nud., *Ardisia crispa* (Thunberg ex Murray) A. de Candolle, *Ardisia crispa* (Thunberg) A. de Candolle var. *taquetii* H. Léveillé, *Ardisia henryi* Hemsl., *Ardisia konishii* Hayata, *Ardisia kusukusensis* Hayata, *Ardisia labordei* H. Léveillé, *Ardisia lentiginosa* Ker Gawler, *Ardisia linangensis* C. M. Hu, *Ardisia miaoliensis* S. Y. Lu, *Bladhia crenata* (Sims) H. Hara, *Bladhia crispa* Thunberg, *Bladhia crispa* Thunberg var. *taquetii* (H. Léveillé) Nakai, *Bladhia lentiginosa* (Ker Gawler) Nakai var. *lanceolata* Masamune.

Family

Myrsinaceae also placed in Primulaceae.

Common/English Names

Coral Ardisia, Coral Bush, Coralberry, Coralberry Tree, Hen's-Eyes, Hilo Holly, Japanese Holy, Spear Flower, Spiceberry, Village Ardisia.

Vernacular Names

Afrikaans: Koraalbessieboom;
Chinese: Yun Chi Zi Jin Niu, Zhu Sha Gen, Chu Sar Gun;
Czech: Klíman Vroubkovaný;
Eastonian: Täkiline Ardiisia;
French: Baie Corail;
German: Gekerbte Spitzblume, Gewürzbeere, Korallenbeere, Spitzenblume;
Indonesia: Mata Ayam (Bangka), Popinoh (Lampung);
Khmer: Ping Chap;
Korean: Baek-Ryang-Geum;
Malaysia: Mata Ayam, Mata Pelandok;
Philippines: Atarolon, Tagpo (Tagalog);
Polish: Ardizja Drzewiasta;
Thai: Chamkhruea, Tinchamkhok, Tappla;
Vietnamese: Trouu Dua, Com Ngor Raw.

Origin/Distribution

The species' native range stretches from Asia temperate – Japan (Honshu (south), Kyushu, Ryukyu Islands, Shikoku); South Korea; Taiwan, China (Anhui, Fujian, Guangdong, Guangxi, Hainan, Hubei, Hunan, Jiangsu, Jiangxi, SW Xizang, Yunnan, Zhejiang) to Bhutan and Asia tropical, India, Sri Lanka, Myanmar; Thailand; Vietnam, Malaysia and Philippines.

Agroecology

In its native habitat, it is found in the forests, hillsides, valleys, shrubby areas, dark damp places, lowlands forest woods, in low mountains of Central China and S. Japan from 100 to 2,400 m elevation. The plant grows on all soil types from acid to alkaline soils provided they are well-drained. It prefers partial shade but can withstand full sun. The plant performs poorly in cold climate and is killed by hard freeze.

Edible Plant Parts and Uses

Young leaves are used in salads; the small fruits are sweet and edible.

Botany

Evergreen shrub growing to 1.2 m by 2.5 m with glossy, elliptic to lanceolate or oblanceolate leaves, 7–15 cm by 2–4 cm, leathery or papery, prominently punctate with margin sub-revolute, crenate, or undulate, with large vascularized marginal glands, apex acute or acuminate; lateral veins 12–18 on a short, 6–10 mm, glabrous petiole (Plates 1 and 2). Inflorescence sub-umbellulate or corymbose, terminal on terete branchlets, Flowers small, bisexual, with oblong-ovate, 1–1.5(–2.5) mm sepals, Petals nearly free, ovate, punctate, glandular papillose, 4–6 mm long, white or pinkish, Stamens shorter than petals; filaments nearly obsolete; anthers triangular-lanceolate and yellow, ovary glabrous fruit 6–8 mm across, bright red. Fruit pale green becoming a bright red, globose, 1-seeded drupe, 6–8 mm in diameter produced in clusters (Plates 1 and 2).

Nutritive/Medicinal Properties

The genus *Ardisia* including *A. crenata* is a rich source of novel and biologically potent phytochemical compounds and has the potential as a source of therapeutic agents (Kobayashi and de

Plate 1 Ripe fruits and leaves of spiceberry

Plate 2 Close-up of fruit and leaves

Plate 3 Plant label

Mejía 2005). The roots of *Ardisia crenata* contain saponins, prosaponins, sapogenins and depsipeptide and thus represent a novel source of health-promoting compounds and potential phytopharmaceuticals and therapeutic agents.

Phytochemicals have been found in various parts of the plant:

Fruit: hydroxybenzoquinones 2-hydroxy-5-methoxy-3-pentadecenyl(tridecenyl- and tridecyl-) benzoquinone (Ogawa and Natori 1968).

Leaves: depsipeptide FR900359 (Fujioka et al. 1988; Miyamae et al. 1989).

Roots: saponins, prosaponins, sapogenins and depsipeptide (Kobayashi and de Mejía 2005); emebelin, rapanone (Ogawa and Natori 1968); a triterpene: cyclamiretin A (Guan et al. 1987); bergenin, friedelin, β-sitosterol and rapanone (Ni and Han 1988); bergenin, an isocoumarin and its derivatives; a bergenin derivative, 11-O-syringylbergenin, spinasterol, series of fatty acids, β-sitosterol-β-D-glucoside, norbergenin and sucrose (Han and Ni 1989b); demethylbergenin, 11-O-syrinylbergenin (Han and Ni 1989a); a triterpenoid saponin, ardicrenin elucidated as cyclamiretin A-3-O-[α-L-rhamnopyranosyl-(1→4)-β-D-glucopyranosyl-(1→4)][β-D-glucopyranosyl-(1→2)]-α-L-arabinopyranoside and another partially hydrolyzed saponin was characterized as cyclamiretin A-3-O-β-D-glucopyranosyl-(1→2)-α-L-arabinopyranoside (Wang et al. 1992); two triterpenoid pentasaccharides, ardisicrenosides E and F (Jia et al. 1994a);

two triterpenoid saponins, ardisicrenoside A [3-β-O-(α-L-rhamnopyranosyl-(1→2)-[β-D-glucopyranosyl- (1→4)-[β-D-glucopyranosyl-(1→2)]-α-L-arabinopyranosyl)-13 β,28-epoxy-16α,30-oleananediol] and ardisicrenoside B-[3 β-O-(β-D-xylopyranosyl-(1→2)-[β-D-glucopyranosyl-(1→4)-[β-D-glucopyranosyl-(1→2)]-α-L-arabinopyranosyl)-13 β,28-epoxy-16 α,30-oleananediol], ardisiacrispins A and B (Jia et al. 1994c); two triterpenoid saponins, ardisicrenoside C (1) [3 β-O-(α-L-rhamnopyranosyl-(1→2)-β-D-glucopyranosyl-(1→4)- [β-D-glucopyranosyl-(1→2)]-α-L-arabinopyranosyl)-16 α, 28-dihydroxy-olean-12-en-30-oic acid 30-O-β-D-glucopyranosyl ester] and ardisicrenoside D (2) [3 β-O-(β-D-xylopyranosyl-(1→2)-β-D-glucopyranosyl-(1→4)-[β-D-glucopyranosyl-(1→2)]-α-L-arabinopyranosyl)-16α, 28-dihydroxy-olean-12-en-30-oic acid 30-O-β-D-glucopyranosyl ester] (Jia et al. 1994b); bergenin and bergenins 11-O-galloylbergenin and 11-O-syringylbergenin along with two new bergenin derivatives, 11-O-vanilloyl- and 11-O-(3′,4′-dimethylgalloyl)-bergenins (Jia et al. 1995); ardisicrenoside K and ardisicrenoside L characterised as 3β-O-[α-L-rhamnopyranosyl-(1→2)-β-D-glucopyranosyl-(1→4)-[β-D-glucopyranosyl-(1→2)]-α-L-arabinopyranosyl]-13β,28-epoxy-16-oxo-30,30-dimethoxy-oleanane and 3β-O-[β-D-xylopyranosyl-(1→2)-β-D-glucopyranosyl-(1→4)-[β-D-glucopyranosyl-(1→2)]-α-L-arabinopyranosyl]-13β,28-epoxy-16α, 20-dihydroxyoleanane (Liu et al. 2007); two minor triterpenoid saponins, ardisicrenosides G and H (Koike et al. 1999); ardisiacrenoside I (Zheng et al. 2008); a new triterpenoid saponin, 3β-O-{β-d-glucopyranosyl-(1→4)-[β-d-glucopyranosyl-(1→2)]-α-l-arabinopyranosyl}-16α, 28-dihydroxyolean-12-en-30-oic acid 30-O-β-d-glucopyranosyl ester (ardisicrenoside N), together with two known saponins, ardisicrenoside C and D (Liu et al. 2011).

cAMP Phosphodiesterase Inhibition Activity

Two triterpenoid saponins, ardisicrenoside C and ardisicrenoside D along with their prosapogenins and sapogenins showed inhibitory activity on cAMP phosphodiesterase (Jia et al. 1994b). Two triterpenoid pentasaccharides, ardisicrenosides E and F, isolated from the roots of *Ardisia crenata* also exhibited moderate inhibitory activity on cAMP phosphodiesterase (Jia et al. 1994a).

Abortifacient Activity

Two of the *Ardisia* saponins, ardisiacrispins A and B were found to be utero-contracting (Jansakul et al. 1987). At a concentration of 8 μg/ml both saponins gave contractive responses of the isolated rat uterus corresponding to 84% of the contraction caused by a standard dose of acetylcholine (0.2 μg/ml). In-situ intra-uterine injections of ardisiacrispin B isolated from *A. crenata* root, caused dose-dependent contraction of uterine smooth muscle in a manner similar to that of prostaglandin E_2 derivatives with no changes in mean arterial blood pressure (Jansakul 1995). Intra-uterine injection of ardisiacrispin B or prostaglandin E_2 did not cause softening of the cervix as observed with intra-uterine injection of saline, suggesting that ardisiacrispin B may exert a prostaglandin E_2 effect which may act at the prostaglandin E_- receptor but not by stimulation or enhancement of prostaglandin E_2 synthesis. The results suggested ardisiacrispin B may have potential to be used as alternative abortifacient drug to oxytocin or prostaglandin E_2 to terminate pregnancy

Antiplatelet Aggregation and Hypotensive Activities

A bioactive cyclic depsipeptide FR900359, was isolated from leaves of *Ardisia crenata*, (Fujioka et al. 1988; Miyamae et al. 1989). The cyclic depsipeptide was found to inhibit platelet aggregation in rabbits, decreased blood pressure and induced hypotension in anesthetized, normotensive rats. Their studies revealed that the biological activity of FR900359 may be due to the vulnerability of the *N*-methyldehydro-l-alanine residue to nucleophilic attack. Methylene chloride and methanol extracts of *Ardisia crenata* demonstrated >80% antithrombin activity using a chromogenic bioassay (Chistokhodova et al. 2002). A benzoquinonoid compound, 2-methoxy-6-tridecyl-1,4-benzoquinone was characterized as the potent PAF (platelet-activating factor) receptor-binding antagonist with nonspecific antiplatelet effects on platelet aggregation induced by various agonists including PAF, ADP, thrombin and collagen (Kang et al. 2001).

Anticancer/Antimetastatic Activity

An antimetastatic and cytostatic substance, termed AC7-1, was isolated from *Ardisia crispa* and identified as a benzoquinonoid compound, 2-methoxy-6-tridecyl-1,4-benzoquinone (Kang et al. 2001). The antimetastatic activities of AC7-1 were confirmed using various in-vitro and in vivo metastasis assays. AC7-1 strongly blocked B16-F10 melanoma cell adhesion to extracellular matrix (ECM) and B16-F10 melanoma cell invasion. AC7-1 also remarkably inhibited pulmonary metastasis and tumour growth in vivo. AC7-1 inhibited B16-F10 melanoma cell adhesion to only specific synthetic peptides including RGDS. These findings suggested that antimetastatic activities of AC7-1 could be caused by blocking integrin-mediated adherence indicating AC7-1 to be a potential candidate for the development of a new antimetastatic drug. *Ardisia crenata* plant extract was found to have in-vitro photo-cytotoxic activity using a human leukaemia cell line HL-60 (Ong et al. 2009). It was able to able to reduce the in-vitro cell viability by more than 50% when exposed to 9.6 J/cm(2) of a broad spectrum light when tested at a concentration of 20 μg/ml.

Ardisiacrenoside I, a new triterpenoid pentasaccharide with an unusual glycosyl glycerol side chain, was isolated from *Ardisia crenata* together with five closely related triterpenoid saponins (Zheng et al. 2008). Their cytotoxic activities were determined against several different human tumour cell lines by the 3-[4,5-dimethylthiazol-2-yl]-2,5-diphenyl tetrazolium bromide (MTT) method. Ardipusilioside was found to have antitumour activity (Zheng et al. 2008). Ardisiacrispin

(A+B), a mixture of ardisiacrispins A and B, with a fixed proportion (2:1) inhibited proliferation of several human cancer cell lines with IC_{50} values in the range of 0.9–6.5 µg/ml by sulphorhodamine B-based colorimetric assay, in which human hepatoma Bel-7402 was the most sensitive cell line (Li et al. 2008). Ardisiacrispin (A+B) induced dose-dependent apoptosis in human hepatoma Bel-7402 cells at doses of 1–10 µg/ml and resulted in the changes of the mitochondrial membrane depolarization, membrane permeability enhancement, and nuclear condensation in a dose-dependent manner and could disassemble microtubule in human hepatoma Bel-7402 cells. The findings suggested that ardisiacrispin (A+B) could inhibit the proliferation of Bel-7402 cells by inducing apoptosis and disassembling microtubule.

Of six *Ardisia* species, *Ardisia compressa* showed the highest topoisomerase II catalytic inhibition against liver cancer HepG2 cells followed by *A. crenata* (Newell et al. 2010). Total polyphenols ranged from 21 to 72 mg equivalents of gallic acid (GA)/g solid extract (SE). The following chemicals were found gallic acid, quercetin derivatives, ardisenone, ardisiaquinone, ardisianone, bergenin, norbergenin, and embelin. However, neither total polyphenol concentration nor antioxidant capacity correlated with anticancer capacity. Significant HepG2 cytotoxicity was also achieved by bergenin (IC_{50} = 18 µM) and embelin (IC_{50} = 120 µM). AC, bergenin, embelin, and quercetin showed a tendency to accumulate cells in the G1 phase and reduced G2/M leading to apoptosis. *A. crenata* was one of four *Ardisia* species with the greatest anticancer potential against liver cancer cells in-vitro. The saponin ardisicrenoside N showed cytotoxicity against MCI-7 and NCI-H460 cancer cell lines at 11.0 and 22.1 µmol/L in-vitro (Liu et al. 2011).

Antibacterial Activity

The ethanol extract of A. crenata exerted the significant anti-bacteria activity against α-*Streptococcus hemolyticus*, β-*Streptococcus hemolyticus* and *Staphylococcus aureus* in-vitro (Tian et al. 1998).

Antiinflammatory Activity

Intraperitoneal administration of the ethanol extract of *A. crenata* inhibited capillary permeability in mice and plantar swelling in rats induced by acetic acid and albumen respectively (Tian et al. 1998).

Antiplasmodial Activity

Studies of Malaysian medicinal plants by Noor Rain et al. (2007), found that *Ardisia crispa* (leaf extract) had antiplasmodial activity. The leaf extract demonstrated antiplasmodial activity against *Plasmodium falciparum* D10 strain (sensitive strain) with an IC_{50} at 5.90 µg/ml.

Miscellaneous Pharmacological Activities

Bergenin an isocoumarin found in various plant species including *A. crenata* had been reported to exhibit a wide range of biological activities including hepatoprotective (Lim et al. 2000), antifungal (Prithivirai et al. 1997), anti-HIV (Piacente et al. 1996), antiarrhythmic (Pu et al. 2002) and hypolipidemic (Jahromi et al. 1992).

Ma et al. (2009) isolated ardicrenin from the roots of *A. crenata* and developed an effective and economical method to extract ardicrenin which was found suitable for industrial production. The final product was a white powder and its purity and yield were 98% and 1.65 respectively.

Traditional Medicinal Uses

The genus *Ardisia* is widely used as the traditional medicine to cure diseases, e.g. pulmonary tuberculosis, hepatitis, chronic bronchitis and irregular menstruation (Kobayashi and de Mejía 2005). *Ardisia crenata* root is anodyne, depurative, febrifuge, antidotal and diuretic. Its roots have been used in traditional Chinese medicine for the treatment of several kinds of diseases including tonsillitis, tooth- ache, trauma, arthralgia, respiratory tract infections, and menstrual disorders and also to stimulate blood circulation. The root also has anti-fertility effects. An infusion is pectoral. The leaves are crushed and applied to scurf; it is also applied to the ears in the treatment of earache. The juice is used internally against fever, cough, and diarrhoea and also used to treat infections of the respiratory tract and menstrual disorders. In Thailand, the roots are used in combination with other medicinal plants to wash-out dirty blood in women who suffer from menstrual pain.

Other Uses

Ardisia crenata is a popular ornamental garden and potted plant.

Comments

In some countries like USA, *Ardisia crenata* is deemed an invasive weed.

Selected References

Backer CA, Bakhuizen van den Brink RC Jr (1965) Flora of Java, vol 2. Wolters-Noordhoff, Groningen, p 179

Burkill IH (1966) A dictionary of the economic products of the Malay Peninsula. Revised reprint, 2 vols. Ministry of Agriculture and Co-operatives, Kuala Lumpur, Malaysia. Vol 1 (A–H) pp 1–1240, Vol 2 (I—Z) pp 1241–2444

Chistokhodova N, Nguyen C, Calvino T, Kachirskaia I, Cunningham G, Howard Miles D (2002) Antithrombin activity of medicinal plants from central Florida. J Ethnopharmacol 81(2):277–280

Duke JA, Ayensu ES (1985) Medicinal plants of China, vols 1 and 2. Reference Publications, Inc., Algonac, 705 pp

Fujioka M, Koda S, Morimoto Y, Biemann K (1988) Structure of FR-900359: a cyclic despeptide from *Ardisia crenata* Sims. J Org Chem 53:2820–2825

Guan XT, Wang MT, Gong YM, Zhao TZ, Hong SH (1987) Sapogenin and secondary glucosides of coral ardisia (*Ardisia crenata*). Chin Tradit Herb Drugs 18(8):338–341

Han L, Ni MY (1989a) Chemical constituents of *Ardisia crenata* Sims. China J Chin Materia Med 14(12): 33–35 (in Chinese)

Han L, Ni MY (1989b) Studies on the chemical constituents of *Ardisia crenata* Sims. Zhongguo Zhong Yao Za Zhi 14(12):737–739, 762–763 (in Chinese)

Jahromi MAF, Chansouria JPN, Ray AB (1992) Hypolipidaemic activity in rats of bergenin, the major constituent of *Flluegea microcarpa*. Phytother Res 6:180–183

Jansakul C (1995) Some pharmacological studies of ardisiacrispin B, an utero-contracting saponin, isolated from *Ardisia crispa*. J Sci Soc Thai 24:14–26

Jansakul C, Baumann H, Kenne L, Samuelsson G (1987) Ardisiacrispin A and B, two utero-contracting saponins from *Ardisia crispa*. Planta Med 53:405–409

Jia ZH, Koike K, Nikaido T, Ohmoto T (1994a) Two novel triterpenoid pentasaccharides with an unusual glycosyl glycerol side chain from *Ardisia crenata*. Tetrahedron 50(41):11853–11864

Jia ZH, Koike K, Nikaido T, Ohmoto T, Ni M (1994b) Triterpenoid saponins from *Ardisia crenata* and their inhibitory activity on cAMP phosphodiesterase. Chem Pharm Bull (Tokyo) 42(11):2309–2314

Jia ZH, Koike K, Ohmoto T, Ni M (1994c) Triterpenoid saponins from *Ardisia crenata*. Phytochemistry 37(5): 1389–1396

Jia ZH, Mitsunaga K, Koike K, Ohmoto T (1995) New bergenin derivatives from *Ardisia crenata*. Nat Med 49:187–189

Kang YH, Kim WH, Park MK, Han BH (2001) Antimetastatic and antitumor effects of benzoquinonoid AC7-1 from *Ardisia crispa*. Int J Cancer 93:736–740

Kobayashi H, de Mejía E (2005) The genus *Ardisia*: a novel source of health-promoting compounds and phytopharmaceuticals. J Ethnopharmacol 96(3): 347–354

Koike K, Jia Z, Ohura S, Mochida S, Nikaido T (1999) Minor triterpenoid saponins from *Ardisia crenata*. Chem Pharm Bull 47(3):434–435

Lemmens RHMJ (2003) *Ardisia* Sw. In: Lemmens RHMJ, Bunyapraphatsara N (eds) Plant resources of South East Asia No. 12(3). Medicinal and poisonous plants. Prosea Foundation, Bogor, pp 77–81

Li M, Wei SY, Xu B, Guo W, Liu DL, Cui JR, Yao XS (2008) Pro-apoptotic and microtubule-disassembly effects of ardisiacrispin (A+B), triterpenoid saponins from *Ardisia crenata* on human hepatoma Bel-7402 cells. J Asian Nat Prod Res 10(7–8):739–746

Lim H-K, Kim H-S, Choi H-S, Oh S, Choi J (2000) Hepatoprotective effects of bergenin, a major constituents of *Mallotus japonicus*, on carbon tetrachloride-intoxicated rats. J Ethnopharmacol 72:469–474

Liu DL, Wang NL, Zhang X, Gao H, Yao XS (2007) Two new triterpenoid saponins from *Ardisia crenata*. J Asian Nat Prod Res 9(2):119–127

Liu DL, Zhang X, Wang SP, Wang NL, Yao XS (2011) A new triterpenoid saponin from the roots of *Ardisia crenata*. Chin Chem Lett 22(8):957–960

Ma Y, Pu SR, Cheng QS, Ma MD (2009) Isolation and characterization of ardicrenin from *Ardisia crenata* Sims. Plant Soil Environ 55(7):305–310

Miyamae A, Fujioka M, Koda S, Morimoto Y (1989) Structural studies of FR900359, a novel cyclic depsipeptide from *Ardisia crenata* Sims (Myrsinaceae). J Chem Soc Perkin Trans 55:873–878

Newell AM, Yousef GG, Lila MA, Ramírez-Mares MV, de Mejia EG (2010) Comparative in vitro bioactivities of tea extracts from six species of *Ardisia* and their effect on growth inhibition of HepG2 cells. J Ethnopharmacol 130(3):536–544

Ni MY, Han L (1988) The chemical constituents of *Ardisia crenata* Sims. Zhong Yao Tong Bao 13(12):33–34, 59

Noor Rain A, Khozirah S, Mohd Ridzuan MA, Ong BK, Rohaya C, Rosilawati M, Hamdino I, Badrul A, Zakiah I (2007) Antiplasmodial properties of some Malaysian medicinal plants. Trop Biomed 24(1):29–35

Ogawa H, Natori S (1968) Hydroxybenzoquinones from Myrsinaceae plants. II. Distribution among Myrsinaceae plants in Japan. Phytochemistry 7:773–782

Ong CY, Ling SK, Ali RM, Chee CF, Samah ZA, Ho AS, Teo SH, Lee HB (2009) Systematic analysis of in vitro photo-cytotoxic activity in extracts from terrestrial plants in Peninsula Malaysia for photodynamic therapy. J Photochem Photobiol B 96(3):216–222

Piacente S, Pizza C, De Tommasi N, Mahmood N (1996) Constituents of *Ardisia japonica* and their in vitro anti-HIV activity. J Nat Prod 59:565–569

Prithivirai B, Singh UP, Manickam M, Srivastava JS, Ray AB (1997) Antifungal activity of bergenin, a constituent of *Flueggea microcarpa*. Plant Pathol 46:224–228

Pu HL, Huang X, Zhao JH, Hong A (2002) Bergenin is the antiarrhythmic principle of *Fluggea virosa*. Planta Med 68:372–374

Tian ZH, He Y, Luo HM, Huang YQ (1998) Antibacterial and anti-inflammatory effects of *Ardisia crenata* Sims. Northwest Pharm J 13(3):109–110 (in Chinese)

Wang MT, Guan XT, Han XW, Hong SH (1992) A new triterpenoid saponin from *Ardisia crenata*. Planta Med 58(2):205–207

Zheng ZF, Xu JF, Feng ZM, Zhang PC (2008) Cytotoxic triterpenoid saponins from the roots of *Ardisia crenata*. J Asian Nat Prod Res 10(9–10):833–839

Ardisia elliptica

Scientific Name

Ardisia elliptica Thunberg.

Synonyms

Anguillaria solanacea (Roxb.) Poir., *Ardisia hainanensis* Mez, *Ardisia humilis* (Vahl.) Kuntze, *Ardisia ketoensis* Hayata, *Ardisia littoralis* Andr., *Ardisia polycephala* Wall., *Ardisia pyrgina* Saint Lager, *Ardisia pyrgus* Roemer & Schultes, *Ardisia solanacea* Roxb., *Ardisia squamulosa* Presl, *Bladhia elliptica* (Thunb.) Nakai, *Bladhia kotoensis* (Hayata) Nakai, *Bladhia solanacea* (Roxb.) Nakai, *Icacorea solanacea* (Roxb.) Britton, *Tinus humilis* (Vahl) Kuntze, *Tinus squamulosa* (Presl) Kuntze.

Family

Myrsinaceae also placed in Primulaceae

Common/English Names

Duck's Eye, Elliptical-Leaf Ardisia, Inkberry, Seashore Ardisia, Shoebutton, Shoebutton Ardisia.

Vernacular Names

Burmese: Krak-Ma.Oak;
Chinese: Ai Zi Jin Niu, Dong Fang Zi Jin Niu, Suan Tai Cai;
Cook Islands: Venevene Tinitō, Vine Tinitō (Maori);
French: Ardisie Elliptique, Ati Popa'a;
India (Orissa): Nbong Thithi (Bonda), Kitti Gocho (Gadaba), Kutti, Lidi Kutti, Reedikki (Kondh), Goli (Poraja);
Indonesia: Buni Keraton;
Malaysia: Rempenai, Mempanai, Cempenai, Penai, Buah Letus, Kayu Lampilan, Duan Bisa Hati, Mata Ayam, Mata Itek, Mata Pelanduk;
Pakistan: Halad;
Philippines: Bahagion (Bisaya), Kolen (Iloko), Katagpo (Tagalog);
Samoan: Togo Vao;
Tahitian: Ati Popa'A, Atiu;
Thailand: Ramyai (Southern), Langphisa (Eastern), Thurang Kasah, Cham, Pak Cham.

Origin/Distribution

Native distribution of the species is uncertain although the original range has invariably included India, Sri Lanka, China, Taiwan, Southeast Asia – Thailand, Vietnam, Malaya, Indonesia, New Guinea and the Philippines.

Agroecology

In its native range, it occurs as understorey in tidal swamps, mangroves habitats in the warm humid coastal zones as an understorey bush as well moist ravines and forests up to altitude of 1,200 m. It also establishes well in natural areas such as riparian habitats, hammocks, marsh islands, cypress stands, wet forests, monsoonal forests and in disturbed systems such as altered wetlands and fallow fields. The species is shade tolerant and is frost sensitive.

Edible Plant Parts and Uses

The fruit (Plate 5) is edible and taste slightly sour and lacks flavour. Young leafy shoots (Plate 2) can be eaten raw or cooked. In Orissa, the fruit are eaten by the Kondh, Poraja, Gadaba and Bonda tribal communities and the leaves used as vegetable by the Gadaba tribal community (Franco and Narasimhan 2009).

Botany

Branched glabrous, evergreen shrub up to 4 m tall with terete almost perpendicular branchlets. Leaves alternate, ovate, obovate or oblanceolate, 15–18 × 5–7 cm, leathery, glabrous, inconspicuously pellucid punctate, scrobiculate, base cuneate and slightly decurrent on petiole (4–8 mm long), margin entire, apex broadly acute to obtuse; with 12 prominent lateral veins on each side of midrib, marginal vein absent, coriaceous, pink-bronze when young turning dark green with age (Plates 2 and 3). Inflorescences terminal, subumbellate or cymose in pyramidal panicles, 8–17(–20) cm. Flowers pink or purplish red, 5–6 mm (Plate 1). Sepals broadly ovate, 1–2 mm, glabrous, punctate, base subauriculate, margin entire, apex acute. Petals nearly free, broadly ovate, inconspicuously pellucid and glabrous. Stamens subequalling petals with oblong-lanceolate, apiculate anthers. Ovary punctate, glabrous with numerous ovules in three series. Fruit dull red or purplish black, globose, 5–6 mm in diameter, densely punctate, on 1.5 cm long pedicel, form in dense clusters (Plates 4 and 5), containing a single spherical seed.

Plate 1 Flowers of *Ardisia elliptica*

Plate 2 Juvenile leaves

Plate 3 Mature leaves

Plate 4 Young fruit cluster

Plate 5 Mature and ripe fruits

Nutritive/Medicinal Properties

The genus *Ardisia* including *A. elliptica* was found to be a rich source of novel and biologically potent phytochemical compounds and to have the potential as a source of therapeutic agents (Kobayashi and de Mejía 2005).

Brazilian scientists reported that the major anthocyanin pigment in fruits of *A. humilis* was found to be malvidin-3-galactoside (Baldini et al. 1995). Others were delphinidin-3-glucoside, petunidin-3-galactoside, malvidin-3-glucoside and peonidin-3-glucoside. The anthocyanin content ranged from 547 to 613 mg/100 g fresh fruit. The benzene extract of the defatted leaves of *Ardisia solanacea* contained triterpenoid alcohols which, isolated and characterized, were: bauerenol, α-amyrin, β-amyrin (Ahmad et al. 1977).

TLC-densitometry of extracts revealed that the bergenin content was the highest in *Ardisia elliptica*, of 17 *Ardisia* species found in China (Liu et al. 1993). Bergenin was reported as a potent phytochemical with pharmacological attributes (Kobayashi and de Mejía 2005; see notes on *A. crenata*).

Anticancer Activity

Ardisia elliptica was found to have anticancer and antiviral activity. *Ardisia elliptica* plant extract was 1 of 9 Thai medicinal plants that exhibited antiproliferative activity against SKBR3 human breast adenocarcinoma cell line using MTT assay (Moongkarndia et al. 2004).

Antiviral Activity

Hot water extract of *Ardisia squamulosa* was found more effective in inhibiting adenovirus ADV-8 replication than the other four viruses (ADV-8, ADV-11, herpes simplex virus – HSV-1, HSV-2) (Chiang et al. 2003). Cell cytotoxic assay demonstrated that the tested hot water extract had CC_{50} values higher than their EC_{50} values.

Antiplatelet Activity

In recent studies, β-amyrin isolated from *A. elliptica* was found to be more potent than aspirin in inhibiting collagen-induced platelet aggregation (Ching et al. 2010). The IC_{50} value of β-amyrin was found to be 4.5 μg/ml (10.5 μM) while that of aspirin was found to be 11 μg/ml (62.7 μM), indicating that β-amyrin was six times as active as aspirin in inhibiting platelet aggregation. The leaf extract was found to inhibit platelet aggregation with an IC_{50} value of 167 μg/ml. The study showed that *A. elliptica* leaves inhibited collagen-induced platelet aggregation and one of the bioactive components responsible for the observed effect was determined to be β-amyrin.

Antibacterial Activity

Ardisia elliptica was also found to have antimicrobial activity. A hexane extract of its leaves afforded fractions containing hydrocarbons, apolar and polar fatty esters, triterpenoid alcohols (bauerenol; α-amyrin and β-amyrin), sterols (β-sitosterol) and polar compounds (Khan et al. 1991). The polar fraction was the most effective, being active at 5 mg/ml against *Pseudomonas aeruginosa* and at 2.5 mg/ml against nine other bacteria. The triterpenoid fraction (the largest fraction, comprising 38.6% of the total) at 10 mg/ml was active against all ten bacteria (7 Gram-positive and 3 Gram-negative); the sterol fraction was the only fraction inactive against all the bacteria. In another study, dried fruit extracts of *Ardisia elliptica* exhibited antibacterial activity against veterinary *Salmonella* (Phadungkit and Luanratana 2006). Three active anti-*Salmonella* compounds were isolated, namely, syringic acid, isorhamnetin and quercetin. The minimal inhibitory concentrations (MICs) of the isolated compounds ranged between 15.6 and 125.0 µg/ml.

Antiplasmodial Activity

Leaf extract of *A. humilis* was reported to have antiplasmodial activity against *Plasmodium falciparum* D10 strain (sensitive strain) but had no cytotoxic activity towards Madin-Darby bovine kidney cells. (Noor Rain et al. 2007).

Traditional Medicinal Uses

The genus *Ardisia* is widely used as the traditional medicine to cure diseases, e.g. pulmonary tuberculosis, hepatitis, chronic bronchitis and irregular menstruation (Kobayashi and de Mejía 2005). Its roots are used in Pakistani traditional medicine against fever, diarrhoea and rheumatism (Khan et al. 1991). In folkloric medicine, the leaves or roots are boiled and a decoction drunk for pains at the heart. The leaves or the roots are used to treat fever, diarrhoea and liver poisoning (Burkill 1966). In traditional Thai medicine *A. elliptica* has antipyretic activity and is used in diarrhoea, gonorrhoea and venereal diseases (Moongkarndia et al. 2004). It a medicinal plant traditionally used for alleviating chest pains, treatment of fever, diarrhoea, liver poisoning and parturition complications in Malaysia (Ching et al. 2010). In Orissa, India, the fruit is used for fits by the Kondh tribal community and for eye pain by the Poraja tribal community (Franco and Narasimhan 2009).

Other Uses

The plant is a popular ornamental for growing in pots or in garden landscape. The plant is useful for fuel and for use as vegetable stakes.

Comments

Frugivorous birds attracted to the numerous red to blackish fruits are the principal dispersal agents.

Selected References

Ahmad S-A, Catalano S, Marsili A, Morelli I, Scartoni V (1977) Chemical examination of the leaves of *Ardisia solanacea*. Planta Med 32:162–164

Baldini VLS, Iaderoza M, Draetta I, Dos S (1995) Anthocyanins from *Ardisia humilis*. Trop Sci 35(2):130–134

Burkill IH (1966) A dictionary of the economic products of the Malay Peninsula. Revised reprint, 2 vols. Ministry of Agriculture and Co-operatives, Kuala Lumpur. Vol 1 (A–H) pp 1–1240, vol 2 (I–Z) pp 1241–2444

Chen J, Pipoly JJ III (1996) Myrsinaceae R. Brown. In: Wu ZY, Raven PH (eds) Flora of China, vol 15. (Myrsinaceae through Loganiaceae). Science Press/Missouri Botanical Garden Press, Beijing/St. Louis, 387 pp

Chiang LC, Cheng HY, Liu MC, Chiang W, Lin CC (2003) In vitro anti-herpes simplex viruses and anti-adenoviruses activity of twelve traditionally used medicinal plants in Taiwan. Biol Pharm Bull 11:1600–1604

Ching JH, Chua TK, Chin LC, Lau AJ, Pang YK, Jaya JM, Tan CH, Koh HL (2010) β-Amyrin from *Ardisia elliptica* Thunb. is more potent than aspirin in inhibiting collagen-induced platelet aggregation. Indian J Exp Biol 48(3):275–9

Corner EJH (1952) Wayside trees of Malaya. Government Printing Office, Singapore/Kuala Lumpur, 772 p

Franco MF, Narasimhan D (2009) Plant names and uses as indicators of knowledge patterns. Indian J Tradit Knowl 8(4):645–648

Henderson MR (1959) Malayan wild flowers: dicotyledons. The Malayan Nature Society, Kuala Lumpur, 478 pp

Khan MTJ, Ashraf M, Nazir M, Ahmad W, Bhatty MR (1991) Chemistry and antibacterial activity of the constituents of *Ardisia solanacea* leaves. Fitoterapia 62:65–68

Kobayashi H, de Mejía E (2005) The genus *Ardisia*: a novel source of health-promoting compounds and phytopharmaceuticals. J Ethnopharm 96(3):347–354

Langeland KA, Burks KC (eds) 1998 Identification and biology of non-native plants in Florida's natural areas. UF/IFAS, Gainesville, Florida, 165 pp

Liu N, Li Y, Gua JX, Qian DG (1993) Studies on the taxonomy of the genus *Ardisia* (Myrsinaceae) from China and the occurrence and quantity of bergenin in the genus. Acta Acad Med Shanghai 20:49–54

Moongkarndi P, Kosem N, Luanratana O, Jongsomboonkusol S, Pongpan N (2004) Antiproliferative activity of Thai medicinal plant extracts on human breast adenocarcinoma cell line. Fitoterapia 75:375–377

Noor Rain A, Khozirah S, Mohd Ridzuan MAR, Ong BK, Rohaya C, Rosilawati M, Hamdino I, Amin B, Zakiah I (2007) Antiplasmodial properties of some Malaysian medicinal plants. Trop Biomed 24(1):29–35

Pacific Island Ecosystems at Risk (PIER) (1999) Invasive plant species: *Ardisia elliptica* Thunberg, Myrsinaceae. http://www.hear.org/pier/arell.htm

Phadungkit M, Luanratana O (2006) Anti-*Salmonella* activity of constituents of *Ardisia elliptica* Thunb. Nat Prod Res 20(7):693–696

Van den Bergh MH (1994) Minor vegetables. In: Siemonsma JS, Piluek K (eds) Plant resources of South-East Asia No 8. Vegetables. Prosea, Bogor, pp 280–310

Scorodocarpus borneensis

Scientific Name

Scorodocarpus borneensis (Baill.) Becc.

Synonyms

Ximenia borneensis Baillon.

Family

Olacaceae, also placed in Strombosiaceae, Erythropalaceae.

Common/English Names

Bawang Hutan, Forest Onion, Garlic Nut, Nutwood Garlic, Sindu, Wood Garlic, Woodland Onion.

Vernacular Names

Borneo: Bawang Hutan, Ja'oi, Kayu Hindu, Kesidu, Kisinduh, Mencorug, Sagad-Berauh, Sindok;
Brunei: Bawang Hutan;
Indonesia: Kayu Bawang Utan (Kalimantan), Kulim (Sumatra);
Peninsular Malaysia: Bawang Hutan, Kulim (Malay), Dali-Dali, Kalip (Sakai);
Sabah: Bawang Hutan (Malay);
Sarawak: Kesindu (Iban), Kayo Kesindo (Kelabit), Bawang Hutan (Malay);
Thailand: Kuleng.

Origin/Distribution

The species occurs in Peninsular Thailand, Peninsular Malaysia, Sumatra, and throughout the island of Borneo.

Agroecology

S. borneensis is found in undisturbed to slightly disturbed (open) mixed dipterocarp forests up to 900 m altitude. It occurs scattered but may be locally common or even gregarious in primary rain forest on alluvial sites near rivers and streams and on hillsides. In secondary forests, it is usually present as a pre-disturbance remnant.

Edible Plant Parts and Uses

The bark and nuts (endocarp) are used to flavour food (as onion/garlic substitute). The nuts are roasted and eaten with salt or grated. Young leaves are used as vegetable.

Botany

An evergreen, spreading, medium-sized tree (Plate 1) or rarely large up to 40 m tall, with 210 cm girth, fissured bark with weakly elongate, adherent scales and small, dense crown; all parts smelling of garlic. Leaves simple, alternate, green, ex-stipulate, oblong-elliptic, 10–22 cm by 4–9 cm wide, acuminate tip, base rounded to cuneate, margin entire, glabrous, thinly coriaceous, 5–6 pairs of secondary veins, tertiary nerves scalariform (Plate 2). Petioles 15 mm weakly keeled distally. Flowers in an axillary, short raceme, 4-5-merous; calyx cup-shaped, margin wavy to toothed; petals reflexed, white and hairy inside, pinkish on the outside; stamens 8 or 10, inserted in pairs about halfway on the petal; ovary superior, imperfectly 3-4-locular with a single ovule in each cell, style with 3–4 minutely lobed stigmas (Plates 3 and 4). Fruit a thinly fleshy, sub-globose to ovoid, 1-seeded, green drupe, up to 4 cm long, on a stout 1 cm stalk (Plates 5 and 6); endocarp (stone) woody with longitudinal strands.

Nutritive/Medicinal Properties

The nutritive value of the leaves per 100 g edible portion had been reported as follows: energy 93 kcal, moisture 66.5%, protein 3.7%, fat 3.6%, carbohydrate 11.6%, crude fibre 13.7%, ash 0.9%, P 46 mg, K 405 mg, Mg 33 mg, Mn 20 ppm, Zn 10 ppm and vitamin C 3.5 mg (Voon et al. 1988; Voon and Kueh 1999).

A new sesquiterpene, scodopin, and a mixture of three tryptamine-type alkaloids, scorodocarpines A-C, were isolated from the fruits of *Scorodocarpus borneensis*, together with a known hemisynthetic sesquiterpene, cadalene-β-carboxylic acid, which was isolated from the bark (Wiart et al. 2001). An aliphatic sulfur compound, bis-(methylthiomethyl)disulfide, a new sesquiterpene, scopotin, and a new indole alkaloid, 13-docosenoyl serotonine were isolated from the seed (Wiart 2001). Two new natural amino acids, (Rs)-3-[(methylthio)methylsulfinyl]-l-alanine and S-[(methylthio)methyl]-l-cysteine, were isolated from the fruit of *Scorodocarpus borneensis* (Kubota et al. 1998). C–S lyase-mediated enzymatic conversion showed that both amino acids play an important role in developing the main odorous components of methyl methylthiomethyl disulfide and bis(methylthiomethyl) disulfide.

Thirteen compounds including four megastigmanes and five flavonoids were isolated from the leaves of *Scorodocarpus borneensis* (Abe and Yamauchi 1993). One of the megastigmanes, scorospiroside, was elucidated as 3,5-dihydroxy-6,9-epoxymegastigmane-3-O-β-d-glucoside.

Some pharmacological properties reported from wood garlic are elaborated below.

Plate 1 Spreading tree habit

Scorodocarpus borneensis

Plate 2 Leaves and inflorescences

Plate 4 Open flower: reflex petal with white inner surface, pinkish outer surface

Plate 5 Fasicle of young wood garlic fruits

Plate 3 Pinkish–white flowers and pink buds

Antiplatelet Activity

Three pure compounds isolated from wood garlic, 2,4,5-trithiahexane (I), 2,4,5,7-tetrathiaoctane (II), and 2,4,5,7-tetrathiaoctane 2,2-dioxide (III),

Plate 6 Close-up of young wood garlic fruit with stout stalk

were shown to inhibit rabbit platelet aggregation induced by collagen, arachidonic acid, U46619, ADP (adenosine 5'-diphosphate), PAF (platelet aggregating factor), and thrombin (Lim et al. 1999). Compounds I, II, and III exhibited a stronger inhibitory effect against the thrombin-induced aggregation of GFP (gel-filtered platelets) than against the aggregation induced by the other agonists. In inhibiting collagen-induced aggregation, II was as potent as methyl allyl trisulfide and aspirin, with a marked disaggregation effect on the secondary aggregation by arachidonic acid. I, II, and III also suppressed U46619-induced aggregation. These results suggested that sulfur-containing compounds in wood garlic not only inhibit arachidonic acid metabolism but also suppress aggregation in association with the function of the platelet plasma membrane.

Antimicrobial Activity

Three novel sulfur-containing compounds 2,4,5,7-tetrathiaoctane 4,4-dioxide ($CH_3SCH_2SO_2SCH_2SCH_3$) (1); 5-thioxo-2,4,6-trithiaheptane 2,2-dioxide ($CH_3SO_2CH_2SCSSCH_3$) (2), and O-ethyl S-methylthiomethyl thiosulfite ($CH_3SCH_2SS(O)OCH_2CH_3$) (3) were isolated from *Scorodocarpus borneensis* (Lim et al. 1998). Compounds 1 and 2 exhibited antimicrobial activity against some bacteria and fungi. The volatiles of wood garlic fruit contained a large amount of ethanol, most of the components were sulfur-containing (Kubota and Kobayashi 1994). Methyl methylthiomethyl disulfide (I) and bis(methylthiomethyl) disulfide (II), two polysulfides were determined to be potent odor compounds of *S. borneensis* by a sensory evaluation. Relatively strong antimicrobial activity was observed in the ethanol extract of the fruit; II and methylthiomethyl (methylsulfonyl)methyl disulfide (III) were isolated as the active components. II exhibited relatively strong antifungal activity, while III, a novel compound, exhibited broader activities than II against bacteria and fungi. These results showed that the fruit of *S. borneensis* possesses useful properties for use as a natural preservative. bis-(methyithiomethyl)-disulfide appeared as the most active compound and the major constituent of the seeds extract of *Scorodocarpus borneensis*; it acted significantly on a methicillin resistant strain of *Staphylococcus aureus* (Wiart 2001). The crude petroleum ether seed extract and bis-(methyithtomethyl)-disulfide strongly inhibited in-vitro, the growth of pathogenic fungi, *Candida albicans*, *Candida lipolytica*, *Saccharomyces lipolytica*, and *Aspergillus ochraceous* and was formulated in external preparation by using commercial paraffin as excipient.

Anticancer Activity

Bis-(methylthiomethyl)-disulfide from the seeds of *Scorodocarpus bomeensis* showed a strong cytotoxic activity against CEM-SS leukemia cell-line and KU812F chronic myelogenous leukemia cell lines (Wiart 2001). 13-docosenoyl serotonine exhibited moderate cytoxicity activity against CEM-SS leukemia cell line while scopotin showed strong cyotoxic effect against CEM-Sleukemia cell line and moderate antimicrobial activity.

Traditional Medicinal Uses

In Peninsular Malaysia, an infusion of the bark is sometimes used as an antidote for Ipoh poisoning "antiaris" (*Antiaris toxicaria* Lesch.). In Sarawak, the leaves and bark are boiled and drank to treat leprosy and diabetes.

Other Uses

It provides a first class timber, heavy, hard dark-purplish red-brown with a garlic odour when fresh and peppery odour when dry. The timber is used for medium to heavy construction of beams, joists, posts, door and window frames, parquet flooring and planks for houses; rafters, piling for bridges, saltwater piling (with bark on), planks for boat keels, mine props, transmission posts, agricultural implements and sleepers in temporary railway lines. The bark is used for tanning. The shell of the fruit may be made into tobacco boxes and humming-tops.

Comments

The species is widespread in its native range and is not at risk of genetic erosion.

Selected References

Abe F, Yamauchi T (1993) Megastigmanes and flavonoids from the leaves of *Scorodocarpus borneensis*. Phytochemistry 33(6):1499–1501

Burkill IH (1966) A dictionary of the economic products of the Malay Peninsula. Revised reprint. 2 vols. Ministry of Agriculture and Co-operatives, Kuala Lumpur, vol 1 (A–H), pp 1–1240, vol 2 (I–Z), pp 1241–2444

Chai PPK (2006) Medicinal plants of Sarawak. Lee Ming Press, Kuching, 212 pp

Chua LSL (1998) *Scorodocarpus* Becc. In: Sosef MSM, Hong LT, Prawirohatmodjo S (eds) Plant resources of South-East Asia No 5(3). Timber trees: lesser-known timbers. Prosea Foundation, Bogor, pp 514–516

Heriyanto NM, Garsetiasih R (2004) Potensi pohon kulim (*Scorodocarpus borneesis* Becc.) di kelompok hutan Gelawan Kampar, Riau. Bul Plasma Nutfah 10(1):37–42 (in Indonesian)

Kubota K, Hirayama H, Sato Y, Kobayashi A, Sugawara F (1998) Amino acid precursors of the garlic-like odour in *Scorodocarpus borneensis*. Phytochemistry 49(1):99–102

Kubota K, Kobayashi A (1994) Sulfur compounds in wood garlic (*Scorodocarpus borneensis* Becc.) as versatile food components. In: Mussinan CJ, Keelan ME (eds) Sulfur compounds in foods, vol ACS Symposium Series., American Chemical Society, pp 238–246, Chapter 19

Lim H, Kubota K, Kobayashi A, Seki T, Ariga T (1999) Inhibitory effect of sulfur-containing compounds in *Scorodocarpus borneensis* Becc. on the aggregation of rabbit platelets. Biosci Biotechnol Biochem 63(2):298–301

Lim H, Kubota K, Kobayashi A, Sugawara F (1998) Sulfur-containing compounds from *Scorodocarpus borneensis* and their antimicrobial activity. Phytochemistry 48(5):787–790

Slik JWF (2006) Trees of Sungai Wain. Nationaal Herbarium Nederland. http://www.nationaalherbarium.nl/sungaiwain/

Voon BH, Chin TH, Sim CY, Sabariah P (1988) Wild fruits and vegetables in Sarawak. Sarawak Department of Agriculture, Kuching, 114 pp

Voon BH, Kueh HS (1999) The nutritional value of indigenous fruits and vegetables in Sarawak. Asia Pacific J Clin Nutr 8(1):24–31

Whitmore TC (1972) Olacaceae. In: Whitmore TC (ed) Tree flora of Malaya, vol 2. Longman, Kuala Lumpur, pp 299–307

Wiart C (2001) Antimicrobial and cytotoxic compounds of *Scorodocarpus borneensis* (Olacaceae) and *Glycosmis calcicola* (Rutaceae). PhD thesis, Universiti Putra Malaysia

Wiart C, Martin MT, Awang K, Hue N, Serani L, Laprévote O, Païs M, Rhamani M (2001) Sesquiterpenes and alkaloids from *Scorodocarpus borneensis*. Phytochemistry 58(4):653–656

Olea europaea

Scientific Name

Olea europaea L.

Synonyms

Olea officinarum Crantz, *Olea pallida* Salisb.

Family

Oleaceae (Synonyms include: Bolivariaceae, Forestieraceae, Fraxinaceae, Jasminaceae, Lilacaceae nom. illeg., Nyctanthaceae, Syringaceae).

Common/English Names

Black Olive, Common Olive, Cultivated Olive, European Olive, Green Olive, Iberian Olive, Mediterranean Olive, Olive, Olive Tree, Pickling Olive, Pyrene Oil.

Vernacular Names

Albanian: Ullir;
Amharic: Oliva, Wayra, Zayt;
Arabic: Zeitoun, Zeytoon, Zeytun, Zitoon, Zitun;
Armenian: Jitabdoogh, Jitaptugh, Jiteni, Zeytoon, Zeytun;
Azeri: Zeytun;
Basque: Oliba, Oliondo;
Belarusian: Aliva;
Brazil: Azeitona, Olive, Oliveira (Portuguese);
Bulgarian: Maslina;
Chinese: Qi-Dun-Guo, Ch'i- Tun-Kuo, Yang-Gan-Lan, Yang-Kan-Lan, Gan Lan Shu, Mu Xi Lan, Yuan Ya Zhong;
Croatian: Maslina;
Cyprus: Elea;
Czech: Oliva Evropská, Olivovník;
Danish: Oliven, Olietrae, Oliventrć;
Dhivehi: Zaithooni;
Dutch: Olijf, Olijfboom;
Eastonian: Őlipuus, Euroopa Őlipuu, Oliiv;
Esperanto: Olivo;
Farsi: Zeitun;
Finnish: Öljypuu;
French: Olive, Olive Commune, Olivier, Olivier De Culture;
Frisian: Oliif;
Gaelic: Ola, Sgolag;
German: Ölbaum, Olive, Olivenbaum;
Greek: Elia, Oleia;
Hebrew: Zayit;
Hungarian: Európai Olajfa, Olajfa, Olajbogyó (Tree), Olíva;
Icelandic: Ólífa;
India: Olibh, Jolpai (Bengali), Oliv (Gujerati), Zaitun, Jaitun, Jalapai (Hindu), Aliv, Julipe, Julpai (Kannada), Oleevu, Oli (Malayalam), Jaitun (Punjabi), Alivceti, Caitun, Cimaikkalikacceti, Cimaikkalikam, Olivai, Olivu, Saidun (Tamil), Jaitun (Telugu), Zaitun (Urdu);

Indonesian: Zaitun;
Irish: Ológ;
Italian: Oliva, Olivo, Ulivo;
Japanese: Oriibu, Oriibu No Ki;
Kazakh: Zäytwn, Zäytün Ağaşı (Tree);
Korean: Ol Li Bu;
Latin: Olea, Oliva;
Latvian: Olīvas;
Lithuanian: Alyvos, Europinis Alyvmedis;
Macedonian: Maslinka, Maslinovo Drvo (Tree), Maslinov Zejtin (Oil);
Malaysia: Zaitun;
Maltese: Żebbuġ;
Norwegian: Oliven;
Persian: Zeitun;
Peru: Aceituna;
Philippines: Langis Ng Oliba (Olive Oil), Oliba (Tagalog);
Polish: Drzewko Oliwkowe, Oliwka, Oliwka Europejska;
Portuguese: Azeitona (Fruit), Olivo, Oliveira (Tree);
Romanian: Măslină, Măslin (Tree);
Russian: Oliva, Maslina;
Serbian: Maslina, Maslinov;
Slovak: Oliva, Oliva Európska, Olivovník Európsky;
Slovaščina: Oljka, Oljka Divja;
Spanish: Aceituna, Aceituno, Oliva, Olivo;
Swahili: Zeituni, Mzeituni, Mzaituni;
Swedish: Oliv, Olivträd;
Taiwan: Gan Lan Shu, You Gan Lan;
Tajik: Zrytun;
Thai: Makok;
Turkish: Zeytin;
Turkmen: Zeýtin;
Ukrainian: Oliva;
Uzbek: Zaytun;
Vietnamese: Quả Ôliu (Fruit), Cây Ôliu (Olive Tree), Dầu Oliu(Olive Oil);
Welsh: Olewydden;
Yiddish: Eylbert, Eylbirt, Masline, Olive.

Origin/Distribution

Olive is native to the eastern Mediterranean region from Syria and the maritime parts of Asia minor and northern Iraq at the south end of the Caspian Sea. The olive tree is the oldest known cultivated tree in history. Olives were first cultivated in Africa, and then spread to Morocco, Algiers, and Tunisia. It was first cultivated in Crete and Syria over 5,000 years ago. Around 600 BC olive tree cultivation spread to Greece, Italy and other Mediterranean countries. It was introduced into Australia in the mid nineteenth century and has now become naturalised.

Agroecology

Olive thrives in places with a Mediterranean climate – dry, warm in the summer with a mild winter chill, and plenty of sun. It will also grow in a dry subtropical and sub temperate climate. Olives are now cultivated in many regions of the world with Mediterranean climates, such as South Africa, Chile, Australia, the Mediterranean Basin, Israel, Palestinian Territories and California and in areas with temperate climates such as New Zealand, under irrigation in the Cuyo region in Argentina which has a desert climate. They are also cultivated in the Córdoba Province, Argentina, which has a temperate climate with rainy summers and dry winters or dry subtropical climate in Cuba. Worldwide cultivation is concentrated between 30° and 45° latitudes in the northern and southern hemispheres, from sea-level to 900 m altitude but olive is also found growing at 1,000–3,150 m altitude. In tropical Africa wild olive occurs in montane woodland, rainforest and wooded grassland at these altitudes.

Olive needs full sun for fruit production, but also needs a slight winter chill, a vernalization period of 6–11 weeks below 9°C for the fruit to set. It is frost sensitive, freezing temperatures below −10°C will kill the tree. Frost in spring can damage young shoots and flowers, and the ripening fruits in late autumn. Optimum temperatures for shoot growth and flowering are 18–22°C. Temperatures above 30°C in spring can damage flowers, but the tree can withstand much higher temperatures in summer.

Olives can grow and be productive on a wide range of soils and soil quality. Productive groves occur on hardpan soils and poor soils except when these are waterlogged, saline or too alkaline (higher than pH 8.5). Deep, well-drained, light-textured soils are preferred. Rich, fertile soils promote productive vegetative growth at the expense of flowering and fruit bearing. Olives will not produce commercial crops without irrigation although they are extremely drought tolerant. Annual precipitation of 500–800 mm per year during the critical growth stage is preferable.

Edible Plant Parts and Uses

Fruit and leaves are edible. Because of its distinct flavour it is often considered a condiment. An edible manna is obtained from the tree. Olive leaves are used in the human diet as an extract, an herbal tea, and a powder. Olives are harvested at the green stage or left to ripen to a rich purple-black color (black olive) and harvested. Olive fruits are widely used, especially in the Mediterranean, as a relish and flavouring for foods. Pickled or otherwise prepared fruits are eaten as relish or used in bread, soups, salads, etc. The fruit is usually pickled or cured with water, brine, oil, salt or lye (2% sodium hydroxide). The lye treatment is necessary to remove a bitter glucoside compound (oleuropein) from the outer tissues of the olive. Oleuropein is highly toxic to bacteria and therefore needs to be removed in order for a fermentation to take place. The optimum fermentation temperature is 24°C. The fermentation period usually takes between 2 and 3 months. Once fermentation is complete, the olives are packed in airtight jars and sterilised which produces a good quality product with a long storage life. The cured fruits are eaten as a relish, stuffed with pimentos or almonds, or used in breads, soups, salads etc. 'Olives schiacciate' are olives picked green, crushed, cured in oil and used as a salad. They can also be dried in the sun and eaten without curing when they are called 'fachouilles'. Olives can also be flavoured by soaking them in various marinades, or removing the pit and stuffing them. Popular flavourings are herbs, spices, olive oil, feta, capsicum (pimento), chili, lemon zest, lemon juice, garlic cloves, wine, vinegar, juniper berries and anchovies.

The fruit (fleshy mesocarp) is also processed into an edible non-drying oil, that is used in salads and cooking and, because of its distinct flavour, is considered a condiment. Olive oil is mono-unsaturated and regular consumption is thought to reduce the risk of circulatory diseases. Olive oil is classified into two main quality classes: cold-pressed unrefined or virgin oil and refined olive oil. Commercially there are several grades of olive oil:

1. Extra Virgin olive oil – top grade and best quality. The oil has less than 1% acidity, the ripe olives have been picked and cold-pressed the same day, and the oil has a strong, green colour with a perfect aroma. Essentially, extra virgin Olive oil has the fruitiest and most pronounced flavour.
2. Virgin olive oil is the next grade. It has less than 2% acidity with good colour and aroma. This may be the result of the second, next day olive pressing or from the second-best grade of olives by cold pressing.
3. Fino or fine olive oil is a blend of extra virgin and virgin olive oils, with an acid content not above 3%.
4. Pure olive oil. This is much lighter in colour with little or no aroma. Pure olive oil is the result of a blend of virgin olive oil and refined olive oil, which is generally extracted from olive pulp, skin and/or pits. Refining is carried out using combination of pressure, heat or chemical solvents.
5. Light olive oil. This oil type is not lower in calories, but has been so finely filtered that is has lost most of its colour, flavour and fragrance. It has a higher smoke point than the other types of olive oil and is well suited to high-temperature frying.
6. Extra light: More of a marketing term than a grade. Usually highly processed, may be mixed with other oils, or may be just pure olive oil grade. The "light" refers to flavour rather than caloric content.
7. Lampante or pomace or cake: not intended for human consumption, and generally used for industrial purposes, such as soap making or lamp oil.

Botany

An evergreen, small, densely branched tree, 6–10 m high with rough grey bark and extensive, moderately deep root system. Branchlets and leaf blades abaxially densely silvery-grey lepidote. Leaves opposite, simple, entire, lanceolate or oblanceolate or narrowly elliptic, 1.5–9×0.5–2 cm, apex acute to acuminate, mucronate, base cuneate, upper surface glabrous and grey green (Plates 1 and 2). Flowers white, small, bisexual, 4-merous, subsessile, in axillary cymose panicles, 2–6 cm long. Flower with four-cleft, cup-shaped calyx, white corolla with short tube and 4 elliptical lobes, two short stamens with large anthers, and bifid stigma on a short style. Drupe ellipsoid, 2–3 cm, mesocarp thick and fleshy, epicarp green when immature (Plate 1) becoming red to purplish–black (Plate 2) or ivory-white when ripe and containing one seed, enclosed by the oval and hard endocarp.

Plate 1 Unripe green olives and leaves

Plate 2 Ripe, purplish-red olives and leaves

Nutritive/Medicinal Properties

The food value of ripe, canned, jumbo-super colossal olive per 100 g edible portion is: water 84.34 g, energy 81 kcal (339 kJ), protein 0.97 g, total lipid 6.87 g, carbohydrate 5.61 g, total dietary fibre 2.5 g, Ca 94 mg, Fe 3.32 mg, Mg 4 mg, P 3 mg, K 9 mg, Na 898 mg, Zn 0.22 mg, Cu 0.226 mg, Mn 0.020 mg, Se 0.9 μg, vitamin C 1.5 mg, thiamine 0.003 mg, niacin 0.022 mg, pantothenic acid 0.015 mg, vitamin B-6 0.012 mg, total choline 6.6 mg, vitamin A (RAE) 17 μg, vitamin A, 346 IU, lutein+zeaxanthin 510 μg, vitamin E (α-tocopherol) 1.65 mg, vitamin K (phylloquinone) 1.4 μg, total saturated fatty acids 0.909 g, 16:0 (palmitic acid) 0.758 g, 18:0 (stearic acid) 0.152 g; total monounsaturated fatty acids 5.0711 g, 16:1 undifferentiated (palmitoleic acid) 0.055 g, 18:1 undifferentiated (oleic acid) 4.995 g, 20:1 (gadoleic acid) 0.021 g; total polyunsaturated fatty acids 0.586 g, 18:2 undifferentiated (linoleic acid) 0.544 g, 18:3 undifferentiated (linolenic acid) 0.041 g, threonine 0.031 g, isoleucine 0.036 g, leucine 0.058 g, lysine 0.038 g, methionine 0.014 g, phenylalanine 0.034 g, tyrosine 0.027 g, valine 0.044 g, arginine 0.078 g, histidine 0.027 g, alanine 0.050 g, aspartic acid 0.107 g, glutamic acid 0.108 g, glycine 0.057 g, proline 0.047 g and serine 0.036 g (USDA 2011).

The food value of olive oil used for cooking or salad per 100 g edible portion is: energy 884 kcal (3,699 kJ), total lipid 100 g, Ca 1 mg, Fe 0.56 mg, K 1 mg, Na 2 mg, total choline 0.3 mg, betaine 0.1 mg, vitamin E (α-tocopherol) 14.35 mg, β-tocopherol 0.11 mg, γ-tocopherol 0.83 mg, vitamin K (phylloquinone) 60.2 μg, total saturated fatty acids 13.808 g, 16:0 (palmitic acid) 11.20 g, 17:0 (margaric acid) 0.022 g, 18:0 (stearic acid) 1.953 g, 20:0 (arachidic acid) 0.0414 g, 22:0 (behenic acid) 0.129 g; total monounsaturated fatty acids 72.961 g, 16:1 undifferentiated (palmitoleic acid) 1.255 g, 17:1 0.125 g, 18:1 undifferentiated (oleic acid)

71.269 g, 20:1 (gadoleic acid) 0.311 g; total polyunsaturated fatty acids 10.523 g, 18:2 undifferentiated (linoleic acid) 9.762 g, 18:3 undifferentiated (linolenic acid) 0.761 g and phytosterols 221 mg (USDA 2011).

The following phenolic compounds were reported in virgin olive oil (Bendini et al. 2007):

Benzoic and derivatives acids: 3-hydroxybenzoic acid, p-hydroxybenzoic acid, 3,4-dihydroxybenzoic acid, gentisic acid, vanillic acid, gallic acid, syringic acid;

Cinnamic acids and derivatives: o-coumaric acid, p-coumaric acid, caffeic acid, ferulic acid, sinapinic acid;

Phenyl ethyl alcohols: tyrosol [(p-hydroxyphenyl)ethanol] or p-HPEA 4-OH, hydroxytyrosol [(3,4-dihydroxyphenyl)ethanol] or 3,4-DHPEA;

Other phenolic acids and derivatives: p-hydroxyphenylacetic acid, 3,4-dihydroxyphenylacetic acid, 4-hydroxy-3-methoxyphenylacetic acid; 3-(3,4-Dihydroxyphenyl) propanoic acid;

Dialdehydic forms of secoiridoids: decarboxymethyloleuropein aglycon (3,4-DHPEA-EDA), decarboxymethyl ligstroside aglycon (p-HPEA-EDA) [EDA- elenolic acid];

Secoiridoid aglycons: oleuropein aglycon or 3,4-DHPEA-EA, ligstroside aglycon or p-HPEA-EA, aldehydic form of oleuropein aglycon, aldehydic form ligstroside aglycon;

Flavonols: (+)-taxifolin;

Flavones: apigenin, luteolin;

Lignans: (+)-pinoresinol, (+)-1-acetoxypinoresinol, (+)-1-hydroxypinoresinol;

Hydroxyisochromans: 1-phenyl-6,7-dihydroxyisochroman, 1-(3'-methoxy-4'-hydroxy) phenyl-6,7-dihydroxyisochroman.

All 18 samples of Portuguese olive oil studied showed similar qualitative profiles of tocopherols and tocotrienols with six identified compounds: α-tocopherol, β-tocopherol, γ-tocopherol, δ-tocopherol, α-tocotrienol and γ-tocotrienol (Cunha et al. 2006). Alpha-tocopherol was the main vitamin E isomer in all samples ranging from 93 to 260 mg/kg. The total tocopherols and tocotrienols ranged from 100 to 270 mg/kg. Geographic origin did not seem to influence the tocopherol and tocotrienol composition of the olive oils under evaluation. Simple phenols (hydroxytyrosol, tyrosol, vanillic acid, vanillin, p-coumaric acid, hydroxytyrosol, and tyrosol acetates), lignans (pinoresinol and 1-acetoxypinoresinol), flavonoids (apigenin and luteolin), and a large number of secoiridoid derivatives were identified from polar part of olive oil (Christophoridou et al. 2005). Simple phenols, such as p-coumaric acid, vanillic acid, homovanillyl alcohol, vanillin, free tyrosol, and free hydroxytyrosol, the flavonols apigenin and luteolin, the lignans (+) pinoresinol, (+) 1-acetoxypinoresinol and syringaresinol, two isomers of the aldehydic form of oleuropein and ligstroside, the dialdehydic form of oleuropein and ligstroside lacking a carboxymethyl group, and finally total hydroxytyrosol and total tyrosol reflecting the total amounts of free and esterified hydroxytyrol and tyrosol, respectively were identified in the polar fraction of extra virgin olive oil (Christophoridou and Dais 2009). Simple phenols such as hydroxytyrosol, tyrosol, vanillic acid, p-coumaric acid, ferulic acid, and vanillin were found in most of the Spanish virgin olive oils (Brenes et al. 1999). The flavonoids apigenin and luteolin were also found in most of the oils. The dialdehydic form of elenolic acid linked to tyrosol and hydroxytyrosol was also detected, as were oleuropein and ligstroside aglycons. The new compound was elucidated as being that of 4-(acetoxyethyl)-1,2-dihydroxybenzene. Concentrations of hydroxytyrosol, tyrosol, and luteolin increased with maturation of fruits. In contrast, the levels of glucoside aglycons diminished with maturation.

The HPLC chromatograms of the oils of the Meski, Sayali, and Picholine Tunisian olive varieties showed the presence of 15 triacylglycerols (TAG) species, among which triolein (OOO) was the most abundant (21–48%) (Sakouhi et al. 2010). In the Sayali cultivar, OOO was the predominant TAG species followed by palmitydiolein (POO) and linoleoyl-dioleoylglycerol (LOO). However, the minor TAG molecules were represented by α-linolenoyl-linoleoyl-oleoylglycerols (LnLO) and linoleoyl-3(1)-palmitoyl glycerol (LnLP). Human plasma and lipoprotein triacylglycerol concentrations were found to increase rapidly over fasting values and peaked twice at 2 and 6 h during the 7-h postprandial period (Abia et al. 1999). The triacylglycerols in the lipoprotein fraction at 2 h generally reflected the composition of the olive oil, however, the proportions of the individual molecular species were altered by the processes leading to their formation. Among the major triacylglycerols, the proportion of triolein (OOO; 43.6%) decreased, palmitoyl-dioleoyl-glycerol (POO; 31.1%) and stearoyl-dioleoyl-glycerol (SOO; 2.1%) were maintained and linoleoyl-dioleoyl-glycerol (LOO; 11.4%) and palmitoyl-oleoyl-linoleoyl-glycerol (POL; 4.6%) significantly increased compared with the composition of the triacylglycerols in the olive oil. Smaller amounts of endogenous triacylglycerol (0.8%), mainly constituted of the saturated myristic (14:0) and palmitic (16:0) fatty acids, were also identified. Analysis of total fatty acids suggested the presence of molecular species composed of long-chain polyunsaturated fatty acids of the (n-3) family, docosapentaenoic acid, [22:5(n-3)] and docosahexaenoic acid (DHA), [22:6(n-3)] and of the (n-6) family arachidonic acid, [20:4(n-6)]. The fastest conversion of lipoproteins to remnants occurred from 2 to 4 h and was directly related to the concentration of the triacylglycerols in the lipoprotein particle. The rates of clearance were significantly different among the major triacylglycerols (OOO, POO, OOL (dioleoyl-linoleoyl-glycerol) and POL) and among the latter ones and PLL (palmitoyl-dilinoleoyl-glycerol), POS (palmitoyl-oleoyl-stearoyl-glycerol) and OLL (oleoyl-dilinoleoyl-glycerol). OOO was removed faster and was followed by POO, OOL, POL, PPO (dipalmitoyl-oleoyl-glycerol), SOO (stearoyl-dioleoyl-gylcerol), PLL, POS and OLL. The main phenols found in Cornicabra virgin oils were the dialdehydic form of elenolic acid linked to tyrosol (p-HPEA-EDA), oleuropein aglycon, and the dialdehydic form of elenolic acid linked to hydroxytyrosol (3,4-DHPEA-EDA) (Gómez-Alonso et al. 2002). The five variables that most satisfactorily characterised the principal commercial Spanish virgin olive oil varieties were 1-acetoxypinoresinol, 4-(acetoxyethyl)-1,2-dihydroxybenzene (3,4-DHPEA-AC), ligstroside aglycon, p-HPEA-EDA, and RT 43.3 contents.

Olive seed is the only seed known to contain albumen. The fruit pulp was found to be rich in oil and potassium and to contain 50–60% (by weight) water, 15–30% oil, 2.5%, N-matter, 3–75% sugars, 3–6% cellulose, 1–2% ash, 2–2.5% polyphenols (Duran 1990). Olive oil was reported to contain a high percentage of the monounsaturated oleic acid; and the unsaponifiable fraction of olive oil comprises phenol, tocopherols, chlorophyll, pheophytin, carotene, squalene and aroma components, all imparting a high nutritional and biological value, resulting in good human health (Kiritsakis 1998, 1999). Olive oil, as a highly monounsaturated oil, is resistant to oxidation. Also the presence of phenols, tocopherols and other natural antioxidants prevent lipid oxidation within the body eliminating the formation of free radicals which may cause cancer. The aroma and the flavour compounds of olive oil, as well as the chlorophyll and pheophytin pigments, facilitate the absorption of the natural antioxidants.

Results of studies by Ortega-García and Peragón (2009) suggested the existence of a coordinated response between phenylalanine ammonia-lyase, polyphenol oxidase, and the concentration of total phenols oleuropein, hydroxytyrosol, and tyrosol during ripening in the four Spanish olive varieties. The concentration of total and specific phenols differed between varieties and were altered during ripening. Twenty-two compounds belonging to monosaccharides, disaccharides, trisaccharides, sugar carboxylic acids and alcohols, cyclic polyols (cyclic sugar alcohols), and derived compounds were determined and characterised in the vegetal tissues from olive fruits, leaves, and stems (Gómez-González et al. 2010). The sugar fraction in *O. europaea* is of great relevance because of the role of sugars in the metabolism of lipids, proteins, and antioxidants. Analysis of olive tissue extracts revealed that mannitol and glucose, the primary photosynthetic products along with fructose and sucrose, were the predominant sugar compounds in the investigated samples of the leaves and roots (Cataldi et al. 2000). These sugars constituted more than 90% of the total soluble carbohydrates in olive tissues. This is not surprising, as mannitol and glucose represent the major transport sugars in olive trees and contribute significantly to osmotic adjustment. Leaf sugar content (g dw) were determined as follows: myo-inositol 14.3 g (1.9%), mannitol 309 g (41%), galactose 4.8 g (0.6%), glucose 370 g (49.2%), sucrose 21.8 g (2.9%), fructose 20.7 g (2.8%), raffinose 2.7 g (0.4%), stachyose 9.7 g (1.2%). Root sugar content were determined as: myo-inositol 2.5 g (0.9%), mannitol 77.9 g (29%), galactose 1.2 g (0.4%), glucose 168 g (62.6%), sucrose 8.5 g (3.2%), fructose 6.5 g (2.4%), raffinose 1.0 g (0.4%), and stachyose 2.7 g (1.0%).

Eight flavonoidic compounds were identified and quantified in leaves of 18 Portuguese olive cultivars: luteolin 7,4′-O-diglucoside, luteolin 7-O-glucoside, rutin, apigenin 7-O-rutinoside, luteolin 4′-O-glucoside, luteolin, apigenin and diosmetin (Meirinhos et al. 2005). Seven phenolic compounds were identified and quantified in olive leaves: caffeic acid, verbascoside, oleuropein, luteolin 7-O-glucoside, rutin, apigenin 7-O-glucoside and luteolin 4′-O-glucoside (Pereira et al. 2007). The principal polyphenols recovered from olive leaves were luteolin 7-O-glucoside, apigenin 7-O-rutinoside and oleuropein, with smaller amounts of luteolin 3′,7-O-diglucoside, quercetin 3-O-rutinoside (rutin), luteolin 7-O-rutinoside and luteolin 3′-O-glucoside (Mylonaki et al. 2008). Two new secoiridoid glycosides, oleuricines A and B, together with five known triterpenoids, β-amyrin, oleanolic acid, erythrodiol, urs-2β,3β-dihydroxy-12-en-28-oic acid, and β-maslinic acid, were isolated from the ethyl-acetate-soluble part of ethanol extract of olive leaf (Wang et al. 2009). Olive fruits, leaves and olive oil are rich source of bioactive phytochemicals that impart a diverse range of pharmacological activities as elaborated below. Olive leaves have been used in the human diet as an extract, an herbal tea, and a powder, and they contain numerous potentially bioactive compounds that may have antioxidant, antihypertensive, antiatherogenic, anti-inflammatory, hypoglycemic, and hypocholesterolemic properties (El and Karakaya 2009). One of these potentially bioactive compounds is the secoiridoid oleuropein, which can constitute up to 6–9% of dry matter in the leaves. Other bioactive components found in olive leaves include related secoiridoids, flavonoids, and triterpenes.

The fruit and compression-extracted oil have a wide range of therapeutic and culinary applications (Waterman and Lockwood 2007). Olive oil also constitutes a major component of the "Mediterranean diet." The chief active components of olive oil include oleic acid, phenolic constituents, and squalene. The main phenolics include hydroxytyrosol, tyrosol, and oleuropein, which occur in highest levels in virgin olive oil and have demonstrated antioxidant activity. Antioxidants are believed to be responsible for a number of olive oil's biological activities. Oleic acid, a monounsaturated fatty acid, has shown activity in cancer prevention, while squalene has also been identified as having anticancer effects. Olive oil consumption has benefit for colon and breast cancer prevention. The oil has been widely studied for its effects on coronary heart disease (CHD), specifically for its ability to reduce blood pressure and low-density lipoprotein (LDL) cholesterol. Antimicrobial activity of hydroxytyrosol, tyrosol, and oleuropein has been demonstrated against several strains of bacteria implicated in intestinal and respiratory infections. Paiva-Martins et al. (2011) studied the structural behaviour of olive oil phenolic compounds hydroxytyrosol, oleuropein and the oleuropein aglycones 3,4-DHPEA-EA and 3,4-DHPEA-EDA-as well as some of their metabolites and established Raman spectroscopy as a rapid, non-destructive and reliable analytical technique for identifying these bioactive components in dietary extracts.

Antioxidant Activity

Olive Fruit

Oleuropein is the abundant phenolic compound in Chemlali olive, and its concentration increases during maturation (Bouaziz et al. 2004). An indirect relationship between oleuropein content in olive fruit and hydroxytyrosol was observed.

Weak changes in the amounts of the other phenolic monomers and flavonoids were also observed. The total phenolic content varied from 6 to 16 g/kg expressed as pyrogallol equivalents with Its highest level at the last maturation period. The antioxidant capacity of olive extracts as measured by the radical scavenging effect on 1,1-diphenyl-2-picrylhydrazyl produced IC_{50} values ranging from 3.2 to 1.5 µg/mL. There was a correlation between antioxidant activity and total phenolic content of samples. The antioxidant activity increased with maturation. This could be attributed to the increase of the total phenol level with fruit development. In another study, eight phenolic monomers and 12 flavonoids were also identified in Chemlali olives (Bouaziz et al. 2005). Oleuropein, a secoiridoid glycoside esterified with a phenolic acid, was the major compound. Five flavonoids were isolated and purified. The antioxidant activity of the extract and the purified compounds was evaluated by measuring the radical scavenging effect on 1,1-diphenyl-2-picrylhydrazyl and by using the beta-carotene-linoleate model assay. Acid hydrolysis of the extract enhanced its antioxidant activity. Hydroxytyrosol and quercetin showed antioxidant activities similar to that of 2,6-di-tert-butyl-4-methylphenol. A hydroxyl group at the ortho position at 3′ on the B ring of the flavonoid nucleus could contribute to the antioxidant activity of the flavonoids.

In a recent study, oleuropein was again the major phenolic compound at all stages of ripeness in the olive Chétoui cultivar. Unexpectedly, both phenolic compounds hydroxytyrosol and oleuropein exhibited the same trends during maturation (Damak et al. 2008). Indeed, the oleuropein levels decreased during the ripening process and were not inversely correlated with the concentrations of hydroxytyrosol. The antioxidant capacity of olive extracts was evaluated by measuring the radical scavenging effect on 1,1-diphenyl-2-picrylhydrazyl and the beta-carotene linoleate model system. The IC_{50} and AAC (Antioxidant Activity Coefficient) values of the olive extracts decreased from 3.68 to 1.61 µg/mL and from 645 to 431, respectively. There was a correlation between the antioxidant activity and the oleuropein concentration. The fatty acid composition was quantified in olive fruit during maturation and showed that fatty acids were characterized by the highest level of oleic acid, which reached 65.2%.

In a more recent study, oleuropein, the major olive fruit biophenolic compound, decreased significantly during all the ripeness stages, and its level decreased from 3.29 g/kg fresh olive (July) to 0.16 g/kg (October) in Dhokar cv. and from 5.7 g/kg (July) to 3.75 g/kg (October) in Chemlali cv (Jemai et al. 2009). This decrease inversely correlated with hydroxytyrosol concentrations until September. DPPH and ABTS assays showed that the more important antioxidant capacity of olive extracts was found at the last stage of maturation. The data obtained during the ripening indicated that polyphenol content and composition, in particular the oleuropein concentration, were in correlation with the measured beta-Glucosidase and esterase enzymatic activities. Glucosidase and esterase showed their maximum values in September reaching 179.75 and 39.03 U/g of olive pulp, respectively. Glucose and mannitol were the main sugars; they reached their highest level at the last stage of ripening: 8.3 and 79.8 g/kg respectively.

Some common phenolic compounds in identified in the extracts of leaves, fruits and seeds of olive, were verbascoside, rutin, luteolin-7-glucoside, oleuropein and hydroxytyrosol (Silva et al. 2006). Nüzhenide was also found in olive seeds and an oleuropein glucoside was also detected in olive tree leaves. There was a correlation between total antioxidant activity and total phenolic content with the exception of the seed extracts analysed. The apparent high antioxidant activity of seed extracts may be due to nüzhenide, a secoiridoid the major phenolic component of olive seeds. These results suggested a possible application of olive seeds as sources of natural antioxidants.

Studies also reported that olive polyphenols were absorbed and metabolized within the body, occurring in plasma mainly in the conjugated form with glucuronic acid and reaching C(max) in 1–2 hours after consumption (Kountouri et al. 2007). Excretion rates were maximum at 0–4 hours.

Tyrosol and hydroxytyrosol increased in plasma after intervention and total antioxidant potential increased. The results indicated that olive polyphenols possessed good bioavailability, which was in accordance with their antioxidant efficacy.

Olive Oil

Olive oil is the principal source of fats in the Mediterranean diet, which has been associated with a lower incidence of coronary heart disease and certain cancers (Visioli et al. 2002). Phenolic compounds, e.g., hydroxytyrosol and oleuropein, in extra-virgin olive oil are responsible for its peculiar pungent taste and for its high stability. Recent findings demonstrated that olive oil phenolics are powerful antioxidants, both in-vitro and in-vivo, and possess other potent biological activities that could partially account for the observed healthful effects of the Mediterranean diet. Accordingly, the incidence of coronary heart disease and certain cancers is lower in the Mediterranean area, where olive oil is the dietary fat of choice. The low EC_{50} values indicated both hydroxytyrosol and oleuropein compounds to be potent scavengers of superoxide radicals and inhibitors of neutrophils respiratory burst: whenever demonstrated in-vivo, these properties may partially explain the observed lower incidence of CHD and cancer associated with the Mediterranean diet (Visioli et al. 1998) Studies showed that extra virgin olive oils (EVOOs) with a low degradation level had a higher content of dialdehydic form of elenolic acid linked to 3′,4′-DHPEA (3′,4′-DHPEA-EDA) and a lower content the oleuropein derivatives hydroxytyrosol (3′,4′-dihydroxyphenylethanol, 3′,4′-DHPEA) than oils having intermediate and advanced degradation levels (Lavelli 2002). EVOOs with a low degradation degree were 3–5 times more efficient as 2,2-diphenyl-1-picrylhydrazyl radical (DPPH) scavengers and 2 times more efficient as inhibitors of the xanthine oxidase – catalyzed reaction than oils with intermediate and advanced degradation levels. The diaphorase (DIA)/NADH/juglone – catalyzed reaction was inhibited by EC_{50} values having low or intermediate degradation levels but not by the most degraded oils. The secoiridoid compounds typical of extra virgin olive oils (EVOOs) were the oleuropein and ligstroside derivatives, hydroxytyrosol, tyrosol, and tocopherols (Lavelli and Bondesan 2005). Destoning lowered slightly the α-tocopherol content in EVOOs but increased the total secoiridoid content and the antioxidant activity of EVOOs (up to 3.5-fold) as evaluated by the xanthine oxidase/xanthine system, generating superoxide radical and hydrogen peroxide, and by the 2,2-diphenyl-1-(2,4,6-trinitrophenyl) hydrazyl test.

In a double-blind, randomized, crossover study, 3 olive oils with low (LPC), moderate (MPC), and high (HPC) phenolic content were given as raw doses (25 ml/day) for 4 consecutive days preceded by 10-day washout periods to 12 healthy men (Weinbrenner et al. 2004). Short-term consumption of olive oils was found to decrease plasma oxidized LDL (oxLDL), 8-oxo-dG in mitochondrial DNA and urine, malondialdehyde in urine and increased HDL cholesterol and glutathione peroxidase activity in a dose-dependent manner. At day 4, oxLDL after MPC and HPC, and 8-oxo-dG after HPC administration (25 ml, respectively), were reduced when the men were in the postprandial state. Phenolic compounds in plasma increased dose dependently during this stage with the phenolic content of the olive oils at 1, 2, 4, and 6 hours, respectively. Their concentrations increased in plasma and urine samples in a dose-dependent manner after short-term consumption of the olive oils. The authors concluded that the olive oil phenolic content modulated the oxidative/antioxidative status of healthy men who consumed a very low-antioxidant diet.

Olive Leaves

Studies showed that the relative abilities of the flavonoids from olive leaf to scavenge the 2,2′-azinobis (3-ethylbenzothizoline-6-sulfonic acid) (ABTS)$\sqrt{}$+ radical cation were influenced by the presence of functional groups in their structure, mainly the B-ring catechol, the 3-hydroxyl group and the 2,3-double bond conjugated with the 4-oxo function (Benavente-García et al. 2000).

For the other phenolic compounds present in olive leaves, their relative abilities to scavenge the ABTS√+ radical cation were mainly influenced by the number and position of free hydroxyl groups in their structure. Also, both groups of compounds showed synergic behaviour when mixed, as occurs in the olive leaf. Total flavonoid and phenolic contents were significantly higher in the 80% ethanol extract, butanol, and ethylacetate fractions than hexane, chloroform and water fractions of olive leaf (Lee et al. 2009). Oleuropein was identified as a major phenolic compound with considerable contents in these major three fractions and the extract that correlated with their higher antioxidant and radical scavenging. Olive leaf was found to be a robust source of bioactive flavonoids regardless of sampling parameters such as olive cultivar, leaf age or sampling date (Goulas et al. 2010). Total flavonoids accounted for the 13–27% of the total radical scavenging activity assessed using HPLC-DPPH protocol. Luteolin 7-O-glucoside was one of the dominant scavengers (8–25%). Based on frequency of appearance, the contribution of luteolin (3–13%) was also considered important. Both the individual and combined phenolics from olive leaves exhibited good radical scavenging abilities, and also revealed superoxide dismutase (SOD)-like activity (Lee and Lee 2010).

Olive Mill Wastewater

Ingestion of an olive mill wastewater (OMWW) preparation containing olive phenolics by 98 healthy individuals found no difference in plasma antioxidant capacity observed between baseline and 1 h after the ingestion of the extract (Visioli et al. 2009). Conversely, a significant increase in total plasma glutathione concentration was measured. This increase involved both the reduced and oxidized forms of glutathione; hence, their ratio was unaffected by the treatment. The observed effects of OMWW on glutathione levels were postulated to be governed by the antioxidant response element (ARE)-mediated increase in phase II enzyme expression, including that of gamma-glutamylcysteine ligase and glutathione synthetase.

Anticancer Activity

Olive Oil

Owen et al., in their reviews in 2000, 2004 asserted olive oil, along with fruits, vegetables, and fish, to be an important constituent of the diet in the Mediterranean basin, and a major factor in preserving a healthy and relatively disease-free population. They presented evidence that it was the unique profile of the phenolic fraction, along with high intakes of squalene and the monounsaturated fatty acid, oleic acid, which confer its health-promoting properties. The major phenolic compounds identified and quantified in olive oil belonged to three different classes: simple phenols (hydroxytyrosol, tyrosol); secoiridoids (oleuropein, the aglycone of ligstroside, and their respective decarboxylated dialdehyde derivatives); and the lignans [(+)-1-acetoxypinoresinol and pinoresinol]. All three classes had been reported to have potent antioxidant properties. Recent studies had shown that olives and olive oil contained antioxidants in abundance. Olives contained acteosides, hydroxytyrosol, tyrosol and phenyl propionic acids. The authors maintained that high consumption of extra-virgin olive oils, particularly rich in the above phenolic antioxidants (as well as squalene, terpenoids and peroxidation-resistant lipid oleic acid), should afford considerable protection against cancer (colon, breast, skin), coronary heart disease, and ageing by inhibiting oxidative stress.

Research findings revealed that oleic acid (18:1n-9), the main olive oil's monounsaturated fatty acid, could suppress the over-expression of HER2 (erbB-2), a well-characterized oncogene playing a key role in the etiology, invasive progression and metastasis in several human cancers through several mechanisms (Menendez and Lupu 2006). First, exogenous supplementation of oleic acid significantly down-regulated HER2-coded p185(Her-2/neu) oncoprotein in human cancer cells. Second, oleic acid exposure specifically repressed the transcriptional activity of the human HER2 gene promoter in tumour-derived cell lines. Third, oleic acid treatment induced the up-regulation of the Ets protein PEA3 (a transcriptional repressor of the HER2 gene promoter) solely in cancer cells. Fourth, HER2 gene promoter bearing a PEA3 site-mutated sequence cannot be negatively regulated by oleic acid, while treatment with oleic acid failed to repress the expression of a human full-length HER2 cDNA controlled by a SV40 viral promoter. Fifth, oleic acid-induced inhibition of HER2 promoter activity did not occur if HER2 gene-amplified cancer cells did not concomitantly exhibit high levels of fatty acid synthase (FASN; oncogenic antigen-519).

Extra-virgin olive oil (EVOO) polyphenols lignans (i.e. 1-[+]-pinoresinol, 1-[+]-acetoxypinoresinol), flavonoids (i.e. apigenin, luteolin), and secoiridoids (i.e. deacetoxyoleuropein aglycone, ligstroside aglycone, oleuropein glycoside, oleuropein aglycone) were found to drastically suppress fatty acid synthase (FASN) protein expression in HER2 gene-amplified SKBR3 breast cancer cells (Menendez et al. 2008b). Fatty acid synthase (FASN), is a key enzyme involved in the anabolic conversion of dietary carbohydrates to fat in mammals. Equivalent results were observed in MCF-7 cancer cells engineered to overexpress the HER2 tyrosine kinase receptor, a well-characterized up-regulator of FASN expression in aggressive sub-types of cancer cells. EVOO-derived lignans, flavonoids and secoiridoids were significantly more effective than the mono-HER2 inhibitor trastuzumab (approximately 50% reduction) and as effective as the dual HER1/HER2 tyrosine kinase inhibitor lapatinib ($\geq$95% reduction) at suppressing high-levels of FASN protein in HER2-overexpressing SKBR3 and MCF-7/HER2 cells. EVOO single (i.e. tyrosol, hydroxytyrosol, vanillin) and phenolic acids (i.e. caffeic acid, p-coumaric acid, vanillic acid, ferulic acid, elenolic acid) failed to modulate FASN expression in SKBR3 and MCF-7/HER2 cells. These findings revealed for the first time that phenolic fractions, directly extracted from EVOO, may induce anti-cancer effects by suppressing the expression of the lipogenic enzyme FASN in HER2-overexpressing breast carcinoma cells. Further studies showed that EVOO-derived single phenols tyrosol and hydroxytyrosol and the phenolic acid elenolic acid failed to significantly decrease HER2 tyrosine kinase activity (Menendez et al. 2008a, 2009). The anti-HER2 tyrosine kinase activity IC_{50} values were up to 5-times lower in the presence of EVOO-derived lignans and secoiridoids than in the presence of EVOO-derived single phenols and phenolic acids. EVOO polyphenols induced strong tumoricidal effects by selectively triggering high levels of apoptotic cell death in HER2-positive MCF10A/HER2 cells but not in MCF10A/pBABE matched control cells. The researchers asserted that their findings not only molecularly supported recent epidemiological evidence revealing that EVOO-related anti-breast cancer effects primarily affect the occurrence of breast tumours over-expressing the type I receptor tyrosine kinase HER2 but further suggested that the stereochemistry of EVOO-derived lignans and secoiridoids might provide an excellent and safe platform for the design of new HER2 targeted anti-breast cancer drugs.

Virgin olive oil phenols inhibited proliferation of human promyelocytic leukemia cells (HL60) by inducing apoptosis and differentiation (Fabiani et al. 2006). Virgin olive oil phenol extract (PE) inhibited HL60 cell proliferation in a time- and concentration-dependent manner. Cell growth was completely blocked at a PE concentration of 13.5 mg/L; apoptosis was also induced. Two compounds isolated from PE, the dialdehydic forms of elenoic acid linked to hydroxytyrosol (3,4-DHPEA-EDA) and to tyrosol (*p*HPEA-EDA), were shown to possess properties similar to those of PE; they accounted for a part of the potent effects exerted by the complex mixture of compounds present in PE. In a subsequent study, oxidative DNA damage was found to be prevented by extracts of olive oil, hydroxytyrosol, and other olive phenolic compounds in human blood mononuclear cells and HL60 (human promyelocytic leukemia cells) cells (Fabiani et al. 2008). Hydroxytyrosol [3,4-dyhydroxyphenyl-ethanol (3,4-DHPEA)] and a complex mixture of phenols extracted from both virgin olive oil (OO-PE) and olive mill wastewater (WW-PE) reduced the DNA damage at concentrations as low as 1 μmol/L when coincubated in the medium with H_2O_2 (40 μmol/L). At 10 μmol/L 3,4-DHPEA, the protection was 93% in HL60 and 89% in PBMC. A similar protective activity was also shown by the dialdehydic form of elenoic acid linked to hydroxytyrosol (3,4-DHPEA-EDA) on both kinds of cells. Other purified compounds such as isomer of oleuropein aglycon (3,4-DHPEA-EA), oleuropein, tyrosol, [p-hydroxyphenyl-ethanol (p-HPEA)] the dialdehydic form of elenoic acid linked to tyrosol, caffeic acid, and verbascoside also protected the cells against H_2O_2-induced DNA damage although with a lower efficacy (range of protection, 25–75%). Overall, the results suggested that virgin olive oil and olive mill water waste may efficiently prevent the initiation step of carcinogenesis invivo, and the concentrations effective against the oxidative DNA damage could be easily reached with normal intake of olive oil.

Hydroxytyrosol, one of the major polyphenolic constituents of extra virgin olive oil, exerted strong antiproliferative effects against human colon adenocarcinoma cells via its ability to induce a cell cycle block in G2/M by inhibition of ERK1/2 phosphorylation and cyclin D1 expression (Corona et al. 2009). These findings are of particular relevance due to the high colonic concentration of HT compared to the other olive oil polyphenols and may help explain the inverse link between colon cancer and olive oil consumption.

Olive Fruit and Pomace

Studies showed that olive fruit skin extract composed of pentacyclic triterpenes with the main components maslinic acid (73.25%) and oleanolic acid (25.75%) exhibited antiproliferative effect on HT-29 human colon cancer cells (Juan et al. 2006). The dose-dependent effects showed antiproliferative activity at an EC_{50} value of 73.96 μmol/L of maslinic acid and 26.56 μmol/L of oleanolic acid without displaying necrosis. Apoptosis was confirmed by the microscopic observation of changes in membrane permeability in and detection of DNA fragmentation HT-29 cells incubated for 24 hours with olive fruit extract containing 150 and 55.5 μmol/L of maslinic and oleanolic acids, respectively. The extract containing 200 μmol/L maslinic acid and 74 μmol/L oleanolic acid increased caspase-3-like activity to sixfold that of control cells. The results revealed that the inhibition of cell proliferation without cytotoxicity and the restoration of apoptosis in colon cancer cells by maslinic and oleanolic acids present in olive fruit extracts. The anticancer activity observed for olive fruit extracts appeared to originate from maslinic acid but not from oleanolic acid (Juan et al. 2008a). Oleanolic acid showed moderate antiproliferative activity, with an EC_{50} of 160.6 μmol/l, and moderate cytotoxicity at high concentrations. In contrast, maslinic acid inhibited cell growth with an EC_{50} of 101.2 μmol/l, without necrotic effects. Oleanolic acid, lacking a hydroxyl group at the carbon 2 position, failed to activate caspase-3 as a prime apoptosis protease. In contrast, maslinic acid increased caspase-3-like activity at 10, 25 and 50 μmol/l by 3-, 3.5- and 5-fold over control cells, respectively. Studies showed that erythrodiol the precursor of pentacyclic triterpenic acids in olive also exhibited antiproliferative and proapoptotic activity in human colorectal adenocarcinoma HT-29 cells (Juan et al. 2008b). Erythrodiol inhibited cell growth with an EC_{50} value of 48.8 μM without any cytotoxic effects in a concentration range up to 100 μM. However, exposure of cells for 24 h to 50, 100, and 150 μM erythrodiol increased caspase-3-like activity by 3.2-, 4.8-, and 5.2-fold over that in control cells. In another study, maslinic acid, a pentacyclic triterpene, present in high concentrations in olive pomace was found to have antiproliferative activity on HT29 and Caco-2 colon-cancer cell lines (Reyes et al. 2006). At concentrations inhibiting cell growth by 50–80% (IC_{50} HT29 = 61 μM, IC_{80} HT29 = 76 μM and IC_{50} Caco-2 = 85 μM, IC_{80} Caco-2 = 116 μM), maslinic acid induced strong G0/G1 cell-cycle arrest and DNA fragmentation, and increased caspase-3 activity. However, maslinic acid did not alter the cell cycle or induce apoptosis in the non-tumoural intestine cell lines IEC-6 and IEC-18. The data revealed that in tumoral cancer cells, maslinic acid exerted a significant anti-proliferation effect by inducing an apoptotic process characterized by caspase-3 activation by a p53-independent mechanism, which occurred via mitochondrial disturbances and cytochrome c release.

Olive Leaves

Dry olive leaf extract (DOLE) was found to possess strong antimelanoma potential (Mijatovic et al. 2011). DOLE significantly inhibited proliferation and subsequently restricted clonogenicity of the B16 mouse melanoma cell line in vitro. Moreover, late phase tumour treatment with DOLE significantly reduced tumour volume in a syngeneic strain of mice. DOLE-treated B16 cells were blocked in the G(0)/G(1) phase of the cell cycle, underwent early apoptosis and died by late necrosis. Despite molecular suppression of the proapoptotic process, DOLE successfully promoted cell death mainly through disruption of cell membrane integrity and late caspase-independent fragmentation of genetic material. Fu et al. (2010) demonstrated using tetrazolium salt (MTT)-based assays that olive leaf extracts exhibited dose-dependent inhibitory effects on the metabolic status (cell viability) of three breast cancer models in-vitro. They identified several important isomers of secoiridoids and flavonoids in the extract.

Antiatherosclerotic, Antiatherogenic, Cardioprotective Activities

Recently, several studies had demonstrated olive oil phenolics to be powerful antioxidants, both in-vitro and in-vivo, and to exert additional potent biologic activities that could partially account for the observed cardioprotective effects of the Mediterranean diet (Visioli and Galli 2001). The antioxidant effects associated with olive oil consumption could explain part of this 'Mediterranean Paradox' (Covas et al. 2001). Virgin olive oils processed by two centrifugation phases and with low fruit ripeness were found to have the highest levels of antioxidant content. The total content of phenolic compounds (PC) from virgin olive oil could delay LDL oxidation. The Mediterranean diet, abundant in antioxidants, was found to be associated with a relatively low incidence of coronary heart disease. Olive oil and olives, containing the antioxidants hydroxytyrosol, oleuropein, and tyrosol, were found to be important components of this diet (Rietjens et al. 2007). In the study, hydroxytyrosol (10 µM) efficiently protected the aorta against the cumene hydroperoxide (CHP) induced impairment of the nitric oxide (NO$^-$)-mediated relaxation of rat aorta. Oleuropein, tyrosol, and homovanillic alcohol, a major metabolite of hydroxytyrosol, did not show protection. Moreover, hydroxytyrosol was found to be a potent OH$^-$ scavenger, attributable to its catechol moiety. In one study, a mixed response was obtained with hydroxytyrosol with regards to its role in atherosclerosis. The study reported that 10 week administration of an aqueous solution of hydroxytyrosol, showed no significant changes in HDL cholesterol, paraoxonase, apolipoprotein B or triglyceride levels (Acin et al. 2006). However, hydroxytyrosol administration decreased apolipoprotein A-I and increased total cholesterol, atherosclerotic lesion areas and circulating monocytes expressing Mac-1. The results indicated that administration of hydroxytyrosol in low cholesterol diets increased atherosclerotic lesion associated with the degree of monocyte activation and remodelling of plasma lipoproteins. The data supported the concept that phenolic-enriched products, out of the original matrix, could be not only non-useful but also harmful.

In a randomized sequential crossover design involving 21 hypercholesterolemic volunteers, intake of the polyphenol-rich (virgin olive oil) breakfast was associated with an improvement in endothelial function, as well as a greater increase in concentrations of nitrates/nitrites (NO$_x$) and a lower increase in lipoperoxides LPO and 8-epi prostaglandin-F2alpha than the ones induced by the low polyphenol fat meal (Ruano et al. 2005). A positive correlation was found to exist between NOx and enhanced endothelial function at the second hour ($R^2 = 0.669$). Furthermore, a negative correlation was found between Ischemic reactive hyperemia (IRH) and LPO ($R^2 = -0.203$) and 8-epi prostaglandin-F2alpha levels ($R^2 = -0.440$). The results demonstrated that a meal containing high-phenolic virgin olive oil improved ischemic reactive hyperemia during the postprandial state. In another study, lower plasma oxidized LDL and lipid peroxide levels, together with higher activities of glutathione peroxidase, were observed after Virgin olive oil (VOO) intervention in a placebo controlled, crossover, randomized trial involving 40 males with stable coronary heart disease (Fitó et al. 2005). Systolic blood pressure decreased after intake of VOO in hypertensive patients. No changes were observed in diastolic blood pressure, glucose, lipids, and antibodies against oxidized LDL. The results showed that consumption of VOO, rich in PC, could provide beneficial effects in CHD patients as an additional and complementary intervention to the pharmacological treatment. In another recent placebo-controlled, crossover, randomized trial involving 28 stable coronary heart disease patients, interleukin-6 and C-reactive protein decreased after virgin olive oil intervention over two periods of 3-weeks, preceded by 2-week washout periods (Fitó et al. 2008). No changes were observed in soluble intercellular and vascular adhesion molecules, glucose and lipid profile. In a separate randomized, crossover, controlled study, a linear increase in high-density lipoprotein (HDL) cholesterol levels was observed for low-, medium-, and high-polyphenol olive oil: mean change, 0.025, 0.032, and 0.045 mmol/L, respectively in participants (Covas et al. 2006). Total cholesterol-HDL cholesterol ratio decreased linearly with the phenolic content of the olive oil. Triglyceride levels decreased by an average of 0.05 mmol/L for all olive oils. Oxidative stress markers decreased linearly with increasing phenolic content. Mean changes for oxidized low-density lipoprotein levels were 1.21, −1.48, and −3.21 U/L for the low-, medium-, and high-polyphenol olive oil, respectively. The study showed olive oil to be more than a monounsaturated fat. Its phenolic content could also provide benefits for plasma lipid levels and oxidative damage.

Antiinflammatory Activity

The topical application of the olive oil compounds (0.5 mg/ear) produced a variable degree of antiinflammatory effect with both arachidonic acid (AA) or 12-O-tetradecanoylphorbol acetate (TPA) (de la Puerta et al. 2000). In the auricular edema induced by TPA, β-sitosterol and erythrodiol from the unsaponifiable fraction of the oil showed a potent antiedematous effect with a 61.4% and 82.1% of inhibition respectively, values not very different to that of the reference indomethacin (85.6%) at 0.5 mg/ear. The four phenolics, oleuropein, tyrosol, hydroxytyrosol and caffeic acid exerted a similar range of inhibition (33–45%). All compounds strongly inhibited the enzyme myeloperoxidase, indicating a reduction of the neutrophil influx in the inflamed tissues. The strongest inhibitor of arachidonic acid (AA) edema was the total unsaponifiable fraction with 34% inhibition similar to that obtained by the reference drug dexamethasone at 0.05 mg/ear. Among the phenolics, oleuropein also produced an inhibition of about 30% with the same dose, but all the other components were found less active in this assay. The anti-inflammatory effects exerted by both unsaponifiable and polar compounds might contribute to the potential biological properties reported for virgin olive oil against different pathological processes.

Phenolic compounds from extra virgin olive oil exhibited differential antiinflammatory effects in human whole blood cultures (Miles et al. 2005). Oleuropein glycoside and caffeic acid decreased the concentration of interleukin-1beta. At a concentration of $10(-4)$ M, oleuropein glycoside inhibited interleukin-1beta production by 80%, whereas caffeic acid inhibited production by 40%. Kaempferol decreased the concentration of prostaglandin E2. At a concentration of $10(-4)$ M, kaempferol inhibited prostaglandin E2 production by 95%. No effects were seen on concentrations of interleukin-6 or tumour necrosis factor-alpha and there were no effects of the other phenolic compounds. The data showed that some, but not all, phenolic compounds derived from extra virgin olive oil decreased inflammatory mediator production by human whole blood cultures. This may contribute to the antiatherogenic properties ascribed to extra virgin olive oil.

Studies reported that hydrolyzed olive vegetation water had antiinflammatory activity (Bitler et al. 2005). Olive vegetation water (OVW) or waste water is the wash water used in olive oil extraction, in addition to that endogenously contained in the olives. OVW is a complex emulsion that includes many potentially valuable components, e.g., oil, sugars, polyphenolics, and antioxidants such as,4-dihydroxyphenyl ethanol or hydroxytyrosol (HT), oleuropein and its various derivatives. In lipopolysaccharide (LPS)-treated BALB/c mice, a model system of inflammation, OVW at a dose of 125 mg/mouse (500 mg/kg) reduced serum TNF-α levels by 95%. In the human monocyte cell line, THP-1, OVW reduced LPS-induced TNF-α production by 50% at a concentration of 0.5 g/L (equivalent to ~0.03 g/L simple and polyphenols). OVW had no toxic effects in-vitro or in-vivo. When OVW was combined with glucosamine, a component of proteoglycans and glycoproteins that was shown to decrease inducible nitric oxide synthase production in cultured macrophage cells, the 2 compounds acted synergistically to reduce serum TNF-α levels in LPS-treated mice. These findings suggested that a combination of OVW and glucosamine may be an effective therapy for a variety of inflammatory processes, including rheumatoid and osteoarthritis. Olive vegetation water, a waste product resulting from olive oil extraction, is a source of hydrophilic antioxidants, which are highly prized by the cosmetics and health food sectors.

Studies showed that the n-hexane extract of olive fruit displayed 12.7–27.8% inhibition on the carrageenan-induced hind paw edema model at the 400 mg/kg dose, without inducing any apparent acute toxicity as well as gastric damage (Süntar et al. 2010).

Antidiabetic, Antihyperglycemic Activity

The oral administration of the olive leaves extract (0.1, 0.25 and 0.5 g/kg body wt) for 14 days significantly decreased the serum glucose, total cholesterol, triglycerides, urea, uric acid, creatinine, aspartate amino transferase (AST) and alanine amino transferase (ALT) while it increased the serum insulin in streptozotocin-induced diabetic rats but not in normal rats (Eidi et al. 2009). The antidiabetic effect of the extract was more effective than that observed with glibenclamide (600 μg/kg), a known antidiabetic drug.

Oleanolic acid another constituent of olive plant was shown to be an agonist for TGR5, a member of G-protein coupled receptor activated by bile acids and which mediates some of their various cellular and physiological effect (Sato et al. 2007). Oleanolic acid lowered serum glucose and insulin levels in mice fed with a high fat diet and it enhanced glucose tolerance. The results further emphasized the potential role of TGR5 agonists to improve metabolic disorders. Studies showed that dry olive leaf extract improved pancreatic islet-directed autoimmunity in diabetic mice by down-regulating production of proinflammatory and cytotoxic mediators (Cvjetićanin et al. 2010). In-vivo administration of the extract significantly reduced clinical signs of diabetes (hyperglycaemia and body weight loss) and led to complete suppression of histopathological changes in pancreatic islets. Concurrently, insulin expression and release were restored in extract-treated mice. The results suggested the potential use of a olive leaf extract-enriched diet for prophylaxis/treatment of human autoimmune type 1 diabetes, and possibly other autoimmune diseases.

Antihyperlipidemic/ Hypocholesterolemic Activity

Studies demonstrated that antioxidants, possibly phenolic compounds present only in extra virgin olive oil, may contribute to the endogenous antioxidant capacity of low density lipoprotein (LDL), resulting in an increased resistance to copper-mediated oxidation as determined in vitro (Wiseman et al. 1996). The lag phase before demonstrable oxidation occurred was significantly increased in the high polyphenol, extra virgin olive oil group when compared with combined results from the low polyphenol group (refined olive oil and Trisun high oleic sunflower seed oil), even though the LDL vitamin E concentration in the high polyphenol group was significantly lower. The study demonstrated that antioxidants other than vitamin E may also function against oxidation of LDL in-vitro. In separate studies, extra virgin olive oil intake did not affect fatty acid composition of LDL but significantly reduced the copper-induced formation of LDL hydroperoxides and lipoperoxidation end products as well as the depletion of LDL linoleic and arachidonic acid in patients with combined hyperlipidemia (Masella et al. 2001). A significant increase in the lag phase of conjugated diene formation was observed after dietary treatment. These differences were statistically correlated with the increase in plasma phenolic content observed at the end of the treatment with extra virgin olive oil; they were not correlated with LDL fatty acid composition or vitamin E content, which both remained unmodified after the added fat change. This report suggested that the daily intake of extra virgin olive oil in hyperlipidemic patients could reduce the susceptibility of LDL to oxidation, not only because of its high monounsaturated fatty acid content but probably also because of the antioxidative activity of its phenolic compounds. In a 2003 study, plasma total cholesterol and LDL cholesterol decreased significantly after 6 weeks of extra virgin olive oil dietary intervention in elderly lipidemic patients (Nagyova et al. 2003). A significant increase in the lag time of conjugated diene formation and the decrease in the rate of lipid oxidation were observed after olive oil consumption. The changes in the fatty acid profile were characterized by an increase in oleic acid content as well as by a decline in the content of linoleic acid and arachidonic acid. The data showed that the daily consumption of extra virgin olive oil in elderly lipidemic patients favourably affected serum lipoprotein spectrum and fatty acid composition that probably contributed to the increased resistance of serum lipids to oxidation.

In one study, all three olive oils differing in their phenolic content, caused an increase in plasma and LDL oleic acid content in a placebo-controlled, double-blind, crossover, randomized supplementation trial during three periods of 3 weeks separated by a 2-week washout period (Gimeno et al. 2007). Olive oils rich in phenolic compounds led to an increase in phenolic compounds in LDL. The concentration of phenolic compounds in LDL was directly correlated with the phenolic concentration in the olive oils. The increase in the phenolic content of LDL could account for the increase of the resistance of LDL to oxidation, and the decrease of the in-vivo oxidized LDL, observed in the frame of this trial. The results supported the hypothesis that a daily intake of virgin olive oil promoted protective LDL changes ahead of its oxidation. In a double-blind, randomized, crossover trial of 33 participants, intervention of refined olive oil (devoid of phenolic content), common olive oil and virgin olive did not modify the concentrations of serum and low-density lipoprotein cholesterol and triacylglycerol; but they exerted changes in the cholesterol, triacylglycerol, and phospholipid content of VLDL (Perona et al. 2011). The virgin olive oil consumption led to increased oleic and palmitic acids, as well as decreased linoleic acid, in very low-density lipoprotein (VLDL) triacylglycerol concentration. The main outcome was the significant dose-dependent linear trend between the phenolic compounds in the olive oils and the palmitic (16:0) and linoleic (18:2 n-6) acid and their corresponding triacylglycerol molecular species in VLDL.

Studies showed that a net effect of oleuropein and hydroxytyrosol, phenols in olive, on Cu^{2+}−induced LDL peroxidation was determined by a balance of their pro- and antioxidant capacities (Briante et al. 2004). Cu^{2+}-Induced LDL oxidation was inhibited by oleuropein and hydroxytyrosol in the initiation phase of the reaction at concentrations of phenols higher than that of Cu^{2+} ions. At lower concentration, both phenols anticipated the initiation process of LDL oxidation, thus exerting prooxidant capacities. It was observed that during Cu^{2+}-induced LDL oxidation in the presence of bioreactor eluates, there was evidence of a synergistic effect among phenolic compounds that enhanced their antioxidant capacities so avoiding the prooxidant effects. A 2008 study showed that the cholesterol-rich diet induced hyperlipidemia resulting in the elevation of total cholesterol (TC), triglycerides (TG) and low-density lipoprotein cholesterol (LDL-C). Administration of polyphenol-rich olive leaf extracts significantly lowered the serum levels of TC, TG and LDL-C and increased the serum level of high-density lipoprotein cholesterol

(HDL-C) (Jemai et al. 2008). Further, the content of thiobarbituric acid reactive substances (TBARS) in liver, heart, kidneys and aorta decreased significantly after oral administration of polyphenol-rich olive leaf extracts compared with those of rats fed a cholesterol-rich diet. In addition, these extracts increased the serum antioxidant potential and the hepatic superoxide dismutase (SOD) and catalase (CAT) activities. These results suggested that the hypocholesterolemic effect of oleuropein, oleuropein aglycone and hydroxytyrosol-rich extracts might be due to their abilities to lower serum TC, TG and LDL-C levels as well as slowing the lipid peroxidation process and enhancing antioxidant enzyme activity.

Administration of a low-dose (2.5 mg/kg of body weight) of hydroxytyrosol and a high-dose (10 mg/kg of body weight) of olive mill wastewaters (OMW) extract to Wistar rats significantly lowered the serum levels of total cholesterol (TC) and low-density lipoprotein cholesterol (LDL-C) while increasing the serum levels of high-density lipoprotein cholesterol (HDL-C) (Fki et al. 2007). In addition, the TBARS contents in liver, heart, kidney, and aorta decreased significantly after oral administration of hydroxytyrosol and OMW extract as compared with those of rats fed a cholesterol-rich diet. Further, OMW phenolics increased CAT and SOD activities in liver. The results suggested that the hypocholesterolemic effect of hydroxytyrosol and OMW extract might be due to their abilities to lower serum TC and LDL-C levels as well as slowing the lipid peroxidation process and enhancing antioxidant enzyme activity.

Antiobesity Activity

Olive oil and its phenolic compounds improved myocardial oxidative stress in standard-fed conditions (Ebaid et al. 2010). Study in male Wistar rats demonstrated that olive-oil, oleuropein and cafeic-acid enhanced fat-oxidation and optimized cardiac energy metabolism in obesity conditions. Obese-Olive, Obese-Oleuropein and Obese-Cafeic groups had higher oxygen consumption, fat-oxidation, myocardial beta-hydroxyacyl coenzyme-A dehydrogenase and lower respiratory-quotient than obese rats. Citrate-synthase was highest in Obese-Olive group. Myocardial lipid-hydroperoxide (LH) and antioxidant enzymes were unaffected by olive-oil and its compounds in obesity condition, whereas LH was lower and total-antioxidant-substances were higher in standard chow-olive and standard-oleuropein than in standard group.

Results of another study strongly suggested that an olive leaf extract (OLE) containing polyphenols such as oleuropein and hydroxytyrosol reversed the chronic inflammation and oxidative stress that induces the cardiovascular, hepatic, and metabolic symptoms in a rat model of high fat diet-induced obesity and diabetes without changing blood pressure (Poudyal et al. 2010).

Antihypertensive Activity

Susalit et al. (2011) reported on the antihypertensive effect and the tolerability of olive leaf extract in comparison with Captopril in patients with stage-1 hypertension in a double-blind, randomized, parallel and active-controlled clinical study. After 8 weeks of treatment, both the olive and Captopril groups experienced a significant reduction of systolic blood pressure (SBP) as well as diastolic blood pressure (DBP) from baseline; while such reductions were not significantly different between groups. A significant reduction of triglyceride level was observed in Olive group, but not in Captopril group. They concluded that olive leaf extract, at the dosage regimen of 500 mg twice daily, was similarly effective in lowering systolic and diastolic blood pressures in subjects with stage-1 hypertension as Captopril, given at its effective dose of 12.5–25 mg twice daily.

Extracts of African wild olive leaves containing 0.27% 1:1 mixture of oleanolic acid and ursolic acid, named oleuafricein; Greek olive leaves containing 0.71% oleanolic acid and Cape Town cultivar containing 2.47% oleanolic acid in a common dose of 60 mg/kg b.w. for 6 weeks treatment, prevented the development of severe hypertension and atherosclerosis and improved the insulin resistance of the experimental animals (Somova et al. 2003).

Antiplatelet Activity

Preincubation of platelet rich plasma with 2-(3,4-di-hydroxyphenyl)-ethanol (DHPE), a phenol component of extra-virgin olive oil, for at least 10 minutes resulted in maximal inhibition (Petroni et al. 1995). The IC_{50} (concentration resulting in 50% inhibition) of DHPE for ADP or collagen-induced PRP aggregations were 23 and 67 µM, respectively. At 400 µM DHPE, a concentration which completely inhibited collagen-induced PRP aggregation, thromboxane B2 production by collagen- or thrombin-stimulated PRP was inhibited by over 80%. At the same DHPE concentration, the accumulation of thromboxane B2 and 12-hydroxyeicosatetraenoic acid in serum was reduced by over 90% and 50%, respectively. The effects of PRP aggregation of oleuropein, another typical olive oil phenol, and of selected flavonoids (luteolin, apigenin, quercetin) were found to be much less active. In contrast, a partially characterized phenol-enriched extract obtained from aqueous waste from olive oil showed rather potent activities. The results suggested that components of the phenolic fraction of olive oil could inhibit platelet function and eicosanoid formation in vitro, and that other, partially characterized, olive derivatives share these biological activities.

Olive leaf polyphenols derived from olive leaves inhibited in-vitro platelet activation in healthy, non-smoking males (Singh et al. 2008). The active phenolic compounds in this extract are part of the secoiridoid family, known for their capacity to scavenge H2O2. Blood analysis revealed a significant dose-dependant reduction in platelet activity with olive extract concentrations of 1.0% v/v. The phenol content of high phenol olive oil ranged between 250 and 500 mg/kg, whereas the low phenol olive oil was 46 mg/kg (Dell'Agli et al. 2008). The compounds identified were hydroxytyrosol (HT), tyrosol (TY), oleuropein aglycone (OleA) and the flavonoids quercetin (QU), luteolin (LU) and apigenin (AP). Oleuropein aglycone was the most abundant phenol (range 23.3–37.7%) and luteolin was the most abundant flavonoid in the extracts. Oil extracts inhibited platelet aggregation with an 50% inhibitory concentration interval of 1.23–11.2 µg/ml. The inhibitory effect of individual compounds including homovanillyl alcohol (HVA) followed this order: OleA>LU>HT=TY=QU=HVA, while AP was inactive. All the extracts inhibited cAMP-PDE enzyme, while no significant inhibition of recombinant PDE5A1 enzyme (50 µg/ml) was observed. All the flavonoids and OleA inhibited cAMP-PDE, whereas HT, TY, HVA (100 microm) were inactive. The results indicated that olive oil extracts and some of its phenolic constituents inhibited platelet aggregation partly via cAMP-PDE inhibition. In another study, hydroxytyrosol acetate (HT-AC), a polyphenol present in virgin olive oil, hydroxytyrosol and acetylsalicylic acid inhibited platelet thromboxane B2 and leucocyte 6-keto-prostaglandin F1 alpha (6-keto-PF1 alpha) production (Correa et al. 2009). In quantitative terms HT-AC showed a greater antiplatelet aggregating activity than acetylsalicylic acid and a similar activity to that of acetylsalicylic acid. This effect involved a decrease in platelet thromboxane synthesis and an increase in leucocyte nitric oxide production.

Photopreventive Activity

Oral administration of an olive leaf extract and its component oleuropein separately to mice for 14 days was found to inhibit the increases in skin thickness induced by UVB radiation (Sumiyoshi and Kimura 2010). They also inhibited increases in the Ki-67- and 8-hydroxy-2′-deoxyguanosine-positive cell numbers, melanin granule area and matrix metalloproteinase-13 (MMP-13) expression. These preventive effects on UVB-induced skin damage might be caused in part by inhibiting the degradation of extracellular matrixes in the corium, and by the proliferation of epidermal cells through the inhibition of increases in MMP-13 levels and reactive oxygen species induced by irradiation.

Analgesic/Antinociceptive Activity

Studies showed that olive leaf extract had analgesic property in several models of pain and had useful influence on morphine analgesia in rats (Esmaeili-Mahani et al. 2010). The data showed that the extract (50–200 mg/kg i.p.) produced dose-dependent analgesic effect on tail-flick and hot-plate tests. Administration of 200 mg/kg extract (i.p.) caused significant decrease in pain responses in the first and the second phases of formalin test. In addition, the extract potentiated the antinociceptive effect of 5 mg/kg morphine and blocked low-dose morphine-induced hyperalgesia.

Neuroprotective Activity

Preincubation of brain cells with hydroxytyrosol, from olive mill waste-water, significantly attenuated the cytotoxic effect of both stressors Fe^{2+} or SNP (a nitric oxide donor), although with different efficiencies (Schaffer et al. 2007). Subchronic, but not acute, administration of 100 mg of hydroxytyrosol per kilogram body weight to mice for 12 days enhanced resistance of dissociated brain cells to oxidative stress, as shown by reduced basal and stress-induced lipid peroxidation. Also, basal mitochondrial membrane potential was moderately hyperpolarized, an effect suggestive of cytoprotection. Overall, the ex-vivo data provided the first evidence of neuroprotective effects of oral hydroxytyrosol intake. Schaffer et al. (2010) confirmed their previous observation of promising cytoprotection of brain PC12 cells by ortho-diphenol hydroxytyrosol (HT)-rich olive mill waste-water extract in different stressor paradigms. Further, correlation analyses revealed that the observed cytoprotective effects in PC12 cells were likely due to HT present in the extract.

Wound Healing Activity

Studies showed that the aqueous extract of *O. europaea* leaves displayed wound healing activity (Koca et al. 2011). The group of animals treated with the aqueous extract demonstrated increased contraction (87.1%) on excision and a significant increase in wound tensile strength (34.8%) on incision models compared to the other groups. Moreover, the antioxidant activity assay showed that aqueous extract had higher scavenging ability than the n-hexane extract. Secoiridoid oleuropein (4.61%) was identified as the major active compound.

Antimicrobial Activity

At low concentrations olive leafs extracts showed an unusual combined antibacterial and antifungal action against gram positive (*Bacillus cereus, B. subtilis* and *Staphylococcus aureus*), gram negative bacteria (*Pseudomonas aeruginosa, Escherichia coli* and *Klebsiella pneumoniae*) and fungi (*Candida albicans* and *Cryptococcus neoformans*) (Pereira et al. 2007). Olive leaf extract was found to be most active against *Campylobacter jejuni, Helicobacter pylori* and *Staphylococcus aureus* [including methicillin-resistant *S. aureus* (MRSA)], with minimum inhibitory concentrations (MICs) as low as 0.31–0.78% (v/v) (Sudjana et al. 2009). In contrast, the extract showed little activity against all other test organisms (n=79), with MICs for most ranging from 6.25% to 50% (v/v). Given this specific activity, olive leaf extract may have a role in regulating the composition of the gastric flora by selectively reducing levels of *H. pylori* and *C. jejuni*. Oleuropein and caffeic acid from olive leaves showed inhibition effects against microorganisms (Lee and Lee 2010). The antimicrobial effect of the combined phenolics was significantly higher than those of the individual phenolics.

Traditional Medicinal Uses

The olive oil is considered a cholagogue, a nourishing demulcent, emollient and laxative. Consuming the oil was reported to reduce gastric secretions and to be beneficial to patients suffering from hyperacidity. The oil is also used internally as a laxative and to treat peptic ulcers. It is used externally to treat pruritis, the effects of stings or burns and also as a base for liniments and ointments. Used with alcohol, olive oil is a good hair tonic and used with rosemary oil provides a good treatment for dandruff. Leaves and fruits of olive are used for the treatment of various kinds of diseases, i.e., rheumatism and hemorrhoids, and as a vasodilator in vascular disorders in Turkish folk medicine (Süntar et al. 2010).

The leaves are antiseptic, astringent, febrifuge and sedative. A decoction of leaves have a tranquillising effect on nervous tension and hypertension and is employed for treating obstinate fevers. Externally, leaves can be applied to abrasions.

The bark is regarded ad astringent, bitter and febrifuge. It is said to be a substitute for quinine in the treatment of malaria. In warm countries, the bark exudes a gum-like substance that has been used as a vulnerary.

Other Uses

Low-grade olive oils are used mainly for making soaps, lighting and as lubricant. Maroon and purple dyes are obtained from the whole fresh ripe fruits and blue and black dyes are obtained from the fruit skin. A yellow/green dye is obtained from the leaves. Olive trees are planted to stabilize dry dusty hillsides. The hard and beautiful grained wood is used in turnery and cabinet making, and is much valued.

Olive oil production industry is characterized by relevant amounts of liquid and solid by-products [olive mill wastewater (OMW) and olive husk (OH)], and by economical, technical and organizational constraints that make difficult the adoption of environmentally sustainable waste disposal approaches (Caputo et al. 2003). Waste treatment technologies aimed at energy recovery represent an interesting alternative. Olives are now being looked at for use as a renewable energy source, using waste produced from the olive plants as an energy source that produces 2.5 times the energy generated by burning the same amount of wood. The smoke released has no negative impact on the environment, and the ash left in the stove can be used for fertilizing gardens and plants. The process has been patented in the Middle East and the US.

Comments

The cultivated form of olive probably arose as ancient natural hybrid between the wild *Olea ferruginea* Royle and *O. laperrinii* Battand. & Trab.

Selected References

Abia R, Perona JS, Pacheco YM, Montero E, Muriana FJG, Ruiz-Gutiérrez V (1999) Postprandial triacylglycerols from dietary virgin olive oil are selectively cleared in humans. J Nutr 129:2184–2191

Acin S, Navarro MA, Arbones-Mainar JM, Guillen N, Sarria AJ, Carnicer R, Surra JC, Orman I, Segovia JC, de la Torre R, Covas M-I, Fernandez-Bolanos J, Ruiz-Gutierrez V, Osada J (2006) Hydroxytyrosol administration enhances atherosclerotic lesion development in Apo E deficient mice. J Biochem 140(3):383–391

Bartolini G, Petrucelli R (2002) Classification, origin, diffusion and history of the olive. FAO, Rome, 74 pp

Benavente-García O, Castillo J, Lorente J, Ortuño A, Del Rio JA (2000) Antioxidant activity of phenolics extracted from *Olea europaea* L. leaves. Food Chem 68(4):457–462

Bendini A, Cerretani L, Carrasco-Pancorbo A, Gómez-Caravaca AM, Segura-Carretero A, Fernández-Gutiérrez A, Lercker G (2007) Phenolic molecules in virgin olive oils: a survey of their sensory properties, health effects, antioxidant activity and analytical methods. An overview of the last decade. Molecules 12:1679–1719

Bitler CM, Viale TM, Damaj B, Crea R (2005) Hydrolyzed olive vegetation water in mice has anti-inflammatory activity. J Nutr 135(6):1475–1479

Bouaziz M, Chamkha M, Sayadi S (2004) Comparative study on phenolic content and antioxidant activity during maturation of the olive cultivar Chemlali from Tunisia. J Agric Food Chem 52(17):5476–5481

Bouaziz M, Grayer RJ, Simmonds MS, Damak M, Sayadi S (2005) Identification and antioxidant potential of flavonoids and low molecular weight phenols in olive cultivar chemlali growing in Tunisia. J Agric Food Chem 53(2):236–241

Bown D (1995) Encyclopaedia of herbs and their uses. Dorling Kindersley, London, 424 pp

Brenes M, García A, García P, Rios JJ, Garrido A (1999) Phenolic compounds in Spanish olive oils. J Agric Food Chem 47(9):3535–3540

Briante R, Febbraio F, Nucci R (2004) Antioxidant/ prooxidant effects of dietary non-flavonoid phenols on the Cu2+-induced oxidation of human low-density lipoprotein (LDL). Chem Biodivers 1(11):1716–1729

California Rare Fruit Growers (1997) Olive. [Internet] http://www.crfg.org/pubs/ff/olive.html

Caputo AC, Scacchia F, Pelagagge PM (2003) Disposal of by-products in olive oil industry: waste-to-energy solutions. Appl Therm Eng 23(2):197–214

Cataldi TRI, Margiotta G, Iasi L, Chio BD, Xiloyannis C, Bufo SA (2000) Determination of sugar compounds in olive plant extracts by anion-exchange chromatography with pulsed amperometric detection. Anal Chem 72:3902–3907

Chopra RN, Nayar SL, Chopra IC (1956) Glossary of Indian medicinal plants. (Including the supplement). Council Scientific Industrial Research, New Delhi, 330 pp

Christophoridou S, Dais P (2009) Detection and quantification of phenolic compounds in olive oil by high resolution 1H nuclear magnetic resonance spectroscopy. Anal Chim Acta 633(2):283–292

Christophoridou S, Dais P, Tseng LH, Spraul M (2005) Separation and identification of phenolic compounds in olive oil by coupling high-performance liquid chromatography with postcolumn solid-phase extraction to nuclear magnetic resonance spectroscopy (LC-SPE-NMR). J Agric Food Chem 53(12):4667–4679

Corona G, Deiana M, Incani A, Vauzour D, Dessì MA, Spencer JP (2009) Hydroxytyrosol inhibits the proliferation of human colon adenocarcinoma cells through inhibition of ERK1/2 and cyclin D1. Mol Nutr Food Res 53(7):897–903

Correa JA, López-Villodres JA, Asensi R, Espartero JL, Rodríguez-Gutiérrez G, De La Cruz JP (2009) Virgin olive oil polyphenol hydroxytyrosol acetate inhibits in vitro platelet aggregation in human whole blood: comparison with hydroxytyrosol and acetylsalicylic acid. Br J Nutr 101(8):1157–1164

Covas MI, Fitó M, Marrugat J, Miró E, Farré M, de la Torre R, Gimeno E, López-Sabater MC, Lamuela-Raventós R, de la Torre-Boronat MC (2001) Coronary disease protective factors: antioxidant effect of olive oil. Therapie 56(5):607–611 (In French)

Covas MI, Nyyssönen K, Poulsen HE, Kaikkonen J, Zunft HJ, Kiesewetter H, Gaddi A, de la Torre R, Mursu J, Bäumler H, Nascetti S, Salonen JT, Fitó M, Virtanen J, Marrugat J, EUROLIVE Study Group. (2006) The effect of polyphenols in olive oil on heart disease risk factors: a randomized trial. Ann Intern Med 145(5):333–341

Cunha SC, Amaral JS, Fernandes JO, Oliveira MB (2006) Quantification of tocopherols and tocotrienols in Portuguese olive oils using HPLC with three different detection systems. J Agric Food Chem 54(9): 3351–3356

Cvjetićanin T, Miljković D, Stojanović I, Dekanski D, Stosić-Grujicić S (2010) Dried leaf extract of Olea europaea ameliorates islet-directed autoimmunity in mice. Br J Nutr 103(10):1413–1424

Damak N, Bouaziz M, Ayadi M, Sayadi S, Damak M (2008) Effect of the maturation process on the phenolic fractions, fatty acids, and antioxidant activity of the Chétoui olive fruit cultivar. J Agric Food Chem 56(5): 1560–1566

de la Puerta R, Martínez-Domínguez E, Ruíz-Gutiérrez V (2000) Effect of minor components of virgin olive oil on topical antiinflammatory assays. Z Naturforsch C 55(9–10):814–819

Dell'Agli M, Maschi O, Galli GV, Fagnani R, Dal Cero E, Caruso D, Bosisio E (2008) Inhibition of platelet aggregation by olive oil phenols via cAMP-phosphodiesterase. Br J Nutr 99(5):945–951

Duran RM (1990) Relationship between the composition and ripening of the olive and quality of the oil. Acta Hortic 286:441–450

Ebaid GM, Seiva FR, Rocha KK, Souza GA, Novelli EL (2010) Effects of olive oil and its minor phenolic constituents on obesity-induced cardiac metabolic changes. Nutr J 9:46

Eidi A, Eidi M, Darzi R (2009) Antidiabetic effect of Olea europaea L. in normal and diabetic rats. Phytother Res 23(3):347–350

El SN, Karakaya S (2009) Olive tree (Olea europaea) leaves: potential beneficial effects on human health. Nutr Rev 67(11):632–638

Esmaeili-Mahani S, Rezaeezadeh-Roukerd M, Esmaeilpour K, Abbasnejad M, Rasoulian B, Sheibani V, Kaeidi A, Hajializadeh Z (2010) Olive (Olea europaea L.) leaf extract elicits antinociceptive activity, potentiates morphine analgesia and suppresses morphine hyperalgesia in rats. J Ethnopharmacol 132(1): 200–205

Fabiani R, De Bartolomeo A, Rosignoli P, Servili M, Selvaggini R, Montedoro GF, Di Saverio C, Morozzi G (2006) Virgin olive oil phenols inhibit proliferation of human promyelocytic leukemia cells (hl60) by inducing apoptosis and differentiation. J Nutr 136(3): 614–619

Fabiani R, Rosignoli P, De Bartolomeo A, Fuccelli R, Servili M, Montedoro GF, Morozzi G (2008) Oxidative dna damage is prevented by extracts of olive oil, hydroxytyrosol, and other olive phenolic compounds in human blood mononuclear cells and HL60 cells. J Nutr 138(8):1411–1416

Fitó M, Cladellas M, de la Torre R, Martí J, Alcántara M, Pujadas-Bastardes M, Marrugat J, Bruguera J, López-Sabater MC, Vila J, Covas MI, The members of the SOLOS Investigators (2005) Antioxidant effect of virgin olive oil in patients with stable coronary heart disease: a randomized, crossover, controlled, clinical trial. Atherosclerosis 181(1): 149–158

Fitó M, Cladellas M, de la Torre R, Martí J, Muñoz D, Schröder H, Alcántara M, Pujadas-Bastardes M, Marrugat J, López-Sabater MC, Bruguera J, Covas MI, SOLOS Investigators (2008) Anti-inflammatory effect of virgin olive oil in stable coronary disease patients: a randomized, crossover, controlled trial. Eur J Clin Nutr 62(4):570–574

Fki I, Sahnoun Z, Sayadi S (2007) Hypocholesterolemic effects of phenolic extracts and purified hydroxytyrosol recovered from olive mill wastewater in rats fed a cholesterol-rich diet. J Agric Food Chem 55(3): 624–631

Fu S, Arráez-Roman D, Segura-Carretero A, Menéndez JA, Menéndez-Gutiérrez MP, Micol V, Fernández-Gutiérrez A (2010) Qualitative screening of phenolic compounds in olive leaf extracts by hyphenated liquid chromatography and preliminary evaluation of cytotoxic activity against human breast cancer cells. Anal Bioanal Chem 397(2):643–654

Garrido Fernandez A, Fernandez Diez MJ, Adams MR (1997) Table olives, production and processing. Chapman & Hall, London, 495 pp

Gimeno E, de la Torre-Carbot K, Lamuela-Raventós RM, Castellote AI, Fitó M, de la Torre R, Covas MI, López-Sabater MC (2007) Changes in the phenolic content of low density lipoprotein after olive oil consumption in men. A randomized crossover controlled trial. Br J Nutr 98(6):1243–1250

Gómez-Alonso S, Salvador MD, Fregapane G (2002) Phenolic compounds profile of Cornicabra virgin olive oil. J Agric Food Chem 50(23):6812–6817

Gómez-González S, Ruiz-Jiménez J, Priego-Capote F, Luque de Castro MD (2010) Qualitative and quantitative sugar profiling in olive fruits, leaves, and stems by gas chromatography-tandem mass spectrometry (GC-MS/MS) after ultrasound-assisted leaching. J Agric Food Chem [Epub ahead of print]

Goulas V, Papoti VT, Exarchou V, Tsimidou MZ, Gerothanassis IP (2010) Contribution of flavonoids to the overall radical scavenging activity of olive (*Olea europaea* L.) leaf polar extracts. J Agric Food Chem 58(6):3303–3308

Grieve M (1971) A modern herbal. 2 vols. Penguin/Dover publications, Harmondsworth/New York, 919 pp

Hu SY (2005) Food plants of China. The Chinese University Press, Hong Kong, 844 pp

Jemai H, Bouaziz M, Fki I, Feki AE, Sayadi S (2008) Hypolipidimic and antioxidant activities of oleuropein and its hydrolysis derivative-rich extracts from Chemlali olive leaves. Chem Biol Interact 176(2–3):88–89

Jemai H, Bouaziz M, Sayadi S (2009) Phenolic composition, sugar contents and antioxidant activity of Tunisian sweet olive cultivar with regard to fruit ripening. J Agric Food Chem 57(7):2961–2968

Juan ME, Wenzel U, Ruiz-Gutierrez V, Daniel H, Planas JM (2006) Olive fruit extracts inhibit proliferation and induce apoptosis in HT-29 human colon cancer cells. J Nutr 136(10):2553–2557

Juan ME, Planas JM, Ruiz-Gutierrez V, Daniel H, Wenzel U (2008a) Antiproliferative and apoptosis-inducing effects of maslinic and oleanolic acids, two pentacyclic triterpenes from olives, on HT-29 colon cancer cells. Br J Nutr 100(1):36–43

Juan ME, Wenzel U, Daniel H, Planas JM (2008b) Erythrodiol, a natural triterpenoid from olives, has antiproliferative and apoptotic activity in HT-29 human adenocarcinoma cells. Mol Nutr Food Res 52(5):595–599

Kiritsakis A (1998) Olive oil- from the tree to the table, 2nd edn. Food and Nutrition Press, Inc., Trumbull

Kiritsakis A (1999) Composition of olive oil and its nutritional and health effect. Paper presented at the 10th international rapeseed congress, Canberra

Koca U, Süntar I, Akkol EK, Yılmazer D, Alper M (2011) Wound repair potential of *Olea europaea* L. leaf extracts revealed by in vivo experimental models and comparative evaluation of the extracts' antioxidant activity. J Med Food 14(1–2):140–146

Kountouri AM, Mylona A, Kaliora AC, Andrikopoulos NK (2007) Bioavailability of the phenolic compounds of the fruits (drupes) of *Olea europaea* (olives): impact on plasma antioxidant status in humans. Phytomedicine 14(10):659–667

Lavelli V (2002) Comparison of the antioxidant activities of extra virgin olive oils. J Agric Food Chem 50(26): 7704–7708

Lavelli V, Bondesan L (2005) Secoiridoids, tocopherols, and antioxidant activity of monovarietal extra virgin olive oils extracted from destoned fruits. J Agric Food Chem 53(4):1102–1107

Lee OH, Lee BY (2010) Antioxidant and antimicrobial activities of individual and combined phenolics in *Olea europaea* leaf extract. Bioresour Technol 101(10):3751–3754

Lee OH, Lee BY, Lee J, Lee HB, Son JY, Park CS, Shetty K, Kim YC (2009) Assessment of phenolics-enriched extract and fractions of olive leaves and their antioxidant activities. Bioresour Technol 100(23):6107–6113

Masella R, Giovannini C, Varì R, Di Benedetto R, Coni E, Volpe R, Fraone N, Bucci A (2001) Effects of dietary virgin olive oil phenols on low density lipoprotein oxidation in hyperlipidemic patients. Lipids 36(11): 1195–1202

Meirinhos J, Silva BM, Valentão P, Seabra RM, Pereira JA, Dias A, Andrade PB, Ferreres F (2005) Analysis and quantification of flavonoidic compounds from Portuguese olive (*Olea europaea* L.) leaf cultivars. Nat Prod Res 19(2):189–195

Menendez JA, Lupu R (2006) Mediterranean dietary traditions for the molecular treatment of human cancer: anti-oncogenic actions of the main olive oil's monounsaturated fatty acid oleic acid (18:1n-9). Curr Pharm Biotechnol 7(6):495–502

Menendez JA, Vazquez-Martin A, Garcia-Villalba R, Carrasco-Pancorbo A, Oliveras-Ferraros C, Fernandez-Gutierrez A, Segura-Carretero A (2008a) tabAnti-HER2 (erbB-2) oncogene effects of phenolic compounds directly isolated from commercial Extra-Virgin Olive Oil (EVOO). BMC Cancer 8:377

Menendez JA, Vazquez-Martin A, Oliveras-Ferraros C, Garcia-Villalba R, Carrasco-Pancorbo A, Fernandez-Gutierrez AA, Segura-Carretero A (2008b) Analyzing effects of extra-virgin olive oil polyphenols on breast cancer-associated fatty acid synthase protein expression using reverse-phase protein microarrays. Int J Mol Med 22(4):433–439

Menendez JA, Vazquez-Martin A, Oliveras-Ferraros C, Garcia-Villalba R, Carrasco-Pancorbo A, Fernandez-

Gutierrez A, Segura-Carretero A (2009) Extra-virgin olive oil polyphenols inhibit HER2 (erbB-2)-induced malignant transformation in human breast epithelial cells: relationship between the chemical structures of extra-virgin olive oil secoiridoids and lignans and their inhibitory activities on the tyrosine kinase activity of HER2. Int J Oncol 34(1):43–51

Mijatovic SA, Timotijevic GS, Miljkovic DM, Radovic JM, Maksimovic-Ivanic DD, Dekanski DP, Stosic-Grujicic SD (2011) Multiple antimelanoma potential of dry olive leaf extract. Int J Cancer 128(8): 1955–1965

Miles EA, Zoubouli P, Calder PC (2005) Differential anti-inflammatory effects of phenolic compounds from extra virgin olive oil identified in human whole blood cultures. Nutrition 21(3):389–394

Moutier N, van der Vossen HAM (2001) *Olea europaea* L. In: van der Vossen HAM, Umali BE (eds) Plant resources of South-East Asia no 14. Vegetable oils and fats. Backhuys Publishers, Leiden, pp 107–112

Mylonaki S, Kiassos E, Makris DP, Kefalas P (2008) Optimisation of the extraction of olive (*Olea europaea*) leaf phenolics using water/ethanol-based solvent systems and response surface methodology. Anal Bioanal Chem 392(5):977–985

Nagyova A, Haban P, Klvanova J, Kadrabova J (2003) Effects of dietary extra virgin olive oil on serum lipid resistance to oxidation and fatty acid composition in elderly lipidemic patients. Bratisl Lek Listy 104(7–8): 218–221

Ortega-García F, Peragón J (2009) Phenylalanine ammonia-lyase, polyphenol oxidase, and phenol concentration in fruits of *Olea europaea* L. cv. Picual, Verdial, Arbequina, and Frantoio during ripening. J Agric Food Chem 57(21):10331–10340

Owen RW, Giacosa A, Hull WE, Haubner R, Würtele G, Spiegelhalder B, Bartsch H (2000) Olive-oil consumption and health: the possible role of antioxidants. Lancet Oncol 1:107–112

Owen RW, Haubner R, Würtele G, Hull E, Spiegelhalder B, Bartsch H (2004) Olives and olive oil in cancer prevention. Eur J Cancer Prev 13(4):319–326

Paiva-Martins F, Rodrigues V, Calheiros R, Marques MP (2011) Characterization of antioxidant olive oil biophenols by spectroscopic methods. J Sci Food Agric 91(2):309–314

Pereira AP, Ferreira IC, Marcelino F, Valentão P, Andrade PB, Seabra R, Estevinho L, Bento A, Pereira JA (2007) Phenolic compounds and antimicrobial activity of olive (*Olea europaea* L. Cv. Cobrançosa) leaves. Molecules 12(5):1153–1162

Perona JS, Fitó M, Covas MI, Garcia M, Ruiz-Gutierrez V (2011) Olive oil phenols modulate the triacylglycerol molecular species of human very low-density lipoprotein. A randomized, crossover, controlled trial. Metabolism 60(6):893–899

Petroni A, Blasevich M, Salami M, Papini N, Montedoro GF, Galli C (1995) Inhibition of platelet aggregation and eicosanoid production by phenolic components of olive oil. Thromb Res 78(2):151–160

Poudyal H, Campbell F, Brown L (2010) Olive leaf extract attenuates cardiac, hepatic, and metabolic changes in high carbohydrate-, high fat-fed rats. J Nutr 140(5): 946–953

Reyes FJ, Centelles JJ, Lupiáñez JA, Cascante M (2006) (2Alpha,3beta)-2,3-dihydroxyolean-12-en-28-oic acid, a new natural triterpene from *Olea europaea*, induces caspase dependent apoptosis selectively in colon adenocarcinoma cells. FEBS Lett 580(27): 6302–6310

Rietjens SJ, Bast AJ, de Vente J, Haenen GRMM (2007) The olive oil antioxidant hydroxytyrosol efficiently protects against the oxidative stress-induced impairment of the NObullet response of isolated rat aorta. Am J Physiol Heart Circ Physiol 292(4): H1931–H1936

Ruano J, Lopez-Miranda J, Fuentes F, Moreno JA, Bellido C, Perez-Martinez P, Lozano A, Gomez P, Jimenez Y, Perez Jimenez F (2005) Phenolic content of virgin olive oil improves ischemic reactive hyperemia in hypercholesterolemic patients. J Am Coll Cardiol 46(10):1864–1868

Sakouhi F, Absalon C, Kallel H, Boukhchina S (2010) Comparative analysis of triacylglycerols from *Olea europaea* L. fruits using HPLC and MALDI-TOFMS. Eur J Lipid Sci Technol 11(5):574–579

Sato H, Genet C, Strehle A, Thomas C, Lobstein A, Wagner A, Mioskowski C, Auwerx J, Saladin R (2007) Anti-hyperglycemic activity of a TGR5 agonist isolated from *Olea europaea*. Biochem Biophys Res Commun 362(4):793–798

Schaffer S, Müller WE, Eckert GP (2010) Cytoprotective effects of olive mill wastewater extract and its main constituent hydroxytyrosol in PC12 cells. Pharmacol Res 62(4):322–327

Schaffer S, Podstawa M, Visioli F, Bogani P, Müller WE, Eckert GP (2007) Hydroxytyrosol-rich olive mill wastewater extract protects brain cells in vitro and ex vivo. J Agric Food Chem 55(13):5043–5049

Silva S, Gomes L, Leitão F, Coelho AV, Boas LV (2006) Phenolic compounds and antioxidant activity of *Olea europaea* L. fruits and leaves. Food Sci Technol Int 12(5):385–395

Singh I, Mok M, Christensen A-M, Turner AH, Hawley JA (2008) The effects of polyphenols in olive leaves on platelet function. Nutr Metab Cardiovasc Dis 18(2):127–132

Somova LI, Shode FO, Ramnanan P, Nadar A (2003) Antihypertensive, antiatherosclerotic and antioxidant activity of triterpenoids isolated from *Olea europaea*, subspecies *africana* leaves. J Ethnopharmacol 84(2–3):299–305

Sudjana AN, D'Orazio C, Ryan V, Rasool N, Ng J, Islam N, Riley TV, Hammer KA (2009) Antimicrobial activity of commercial *Olea europaea* (olive) leaf extract. Int J Antimicrob Agents 33(5):461–463

Sumiyoshi M, Kimura Y (2010) Effects of olive leaf extract and its main component oleuroepin on acute ultraviolet B irradiation-induced skin changes in C57BL/6J mice. Phytother Res 24(7):995–1003

Süntar IP, Akkol EK, Baykal T (2010) Assessment of anti-inflammatory and antinociceptive activities of *Olea europaea* L. J Med Food 13(2):352–356

Susalit E, Agus N, Effendi I, Tjandrawinata RR, Nofiarny D, Perrinjaquet-Moccetti T, Verbruggen M (2011) Olive (*Olea europaea*) leaf extract effective in patients with stage-1 hypertension: comparison with Captopril. Phytomedicine 18(4):251–258

U.S. Department of Agriculture, Agricultural Research Service (2011) USDA National nutrient database for standard reference, release 24. Nutrient Data Laboratory Home Page. http://www.ars.usda.gov/ba/bhnrc/ndl

van der Vossen HAM, Mashungwa GN, Mmolotsi RM, (2007) *Olea europaea* L. [Internet] Record from Protabase. In: van der Vossen HAM, Mkamilo GS (eds) PROTA (Plant Resources of Tropical Africa/Ressources végétales de l'Afrique tropicale), Wageningen, Netherlands. http://database.prota.org/search.htm

Visioli F, Bellomo G, Galli C (1998) Free radical-scavenging properties of olive oil polyphenols. Biochem Biophys Res Commun 247(1):60–64

Visioli F, Galli C (2001) Antiatherogenic components of olive oil. Curr Atheroscler Rep 3(1):64–67

Visioli F, Poli A, Galli C (2002) Antioxidant and other biological activities of phenols from olives and olive oil. Med Res Rev 22(1):65–75

Visioli F, Wolfram R, Richard D, Abdullah MI, Crea R (2009) Olive phenolics increase glutathione levels in healthy volunteers. J Agric Food Chem 57(5): 1793–1796

Wang XF, Li C, Shi YP, Di DL (2009) Two new secoiridoid glycosides from the leaves of *Olea europaea* L. J Asian Nat Prod Res 11(11):940–944

Waterman E, Lockwood B (2007) Active components and clinical applications of olive oil. Altern Med Rev 12(4):331–342

Weinbrenner T, Fitó M, de la Torre R, Saez GT, Rijken P, Tormos C, Coolen S, Albaladejo MF, Abanades S, Schroder H, Marrugat J, Covas MI (2004) Olive oils high in phenolic compounds modulate oxidative/antioxidative status in men. J Nutr 134:2314–2321

Wiseman SA, Mathot JN, de Fouw NJ, Tijburg LB (1996) Dietary non-tocopherol antioxidants present in extra virgin olive oil increase the resistance of low density lipoproteins to oxidation in rabbits. Atherosclerosis 120(1–2):15–23

Vanilla planifolia

Scientific Name

Vanilla planifolia H.C. Andrews.

Synonyms

Epidendrum vanilla L., *Myrobroma fragrans* Salisbury nom. illeg., *Notylia planifolia* (Andrews) Conzatti, *Vanilla aromatica* Sw., *Vanilla bampsiana* Geerinck, *Vanilla carinata* Rolfe, *Vanilla domestica* (L.) Druce, *Vanilla duckei* Huber, *Vanilla epidendrum* Mirb., *Vanilla fragrans* (Salisbury) Ames, *Vanilla majaijensis* Blanco, *Vanilla mexicana* P. Miller, *Vanilla rubra* (Lam.) Urb., *Vanilla sativa* Schiede, *Vanilla sylvestris* Schiede, *Vanilla viridflora* Blume.

Family

Orchidaceae

Common/English Names

Bali Vanilla, Bourbon Vanilla, Commercial Vanilla, Common Vanilla, Flat-Leaved Vanilla, Indian Ocean Vanilla, Indonesian Vanilla, Java Vanilla, Mauritius Vanilla, Mexican Vanilla, Seychelles Vanilla, Tahitian Vanilla, West Indian Vanilla, Vanilla, Vanilla Vine.

Vernacular Names

Amharic: Vanila;
Arabic: Al-Fanilya;
Armenian: Vanil;
Basque: Bainila;
Belarusian: Vanil͡;
Brazil: Baunilha;
Bulgarian: Vanilia;
Chinese: Fan Ni Lan, Hsaing Ts'ao Lan, Xiang Cao, Xiang Cao Dou, Xiang Jia Lan, Xiang Lan, Xiang Zi Lan, Wahn Nei La;
Columbia: Bejuquillo, Bejuquillo, Vainilla;
Croatian: Vanilija;
Cuba: Cuyanquillo, Flor Negra, Lombricera;
Czech: Vanilka, Vanilka Pravá, Vanilovník Plocholistý;
Danish: Ægte Vanilje, Vanilje, Vanille;
Dutch: Vanielje, Vanieljesoort, Vanille;
Eastonian: Harilik Vanill;
Esperanto: Vanilo;
Farsi: Vanil;
Finnish: Vanilja;
French: Vanille Bourbon, Vanille D'indonésie, Vanille Des Seychelles, Vanille Du Méxique, Vanillier, Vanille, Vrai Vanillier;
Frisian: Fanylej, Fanille;
Gaelic: Faoineag;
Georgian: Vanili;
German: Echte Vanille, Gewürzvanille, Große Vanille, Vanille;
Greek: Vanilia, Vanillia;

Hebrew: Vanil;
Hungarian: Vanília;
Icelandic: Vanilla;
India: Bhenila (Bengali), Vanilla (Hindu), Vyanilla, Venila (Kannada), Bhenila (Maithili), Vanila (Malayalam), Vanile (Punjabi), Vanikkodi, Vanila (Tamil);
Indonesia: Panili (Javanese, Sundanese);
Irish: Fanaile;
Italian: Vaniglia;
Japanese: Banira;
Kazakh: Vanil;
Korean: Panilla;
Latvian: Smaržīgā Vaniļa, Vanilla;
Lithuanian: Vanilė, Kvapioji Vanilė;
Macedonian: Vanila;
Malaysia: Panili, Anggrek Panili;
Maltese: Vanilja;
Mexico: Tilisúchil, Tilisóchil, Tilijuche, Tiljuche, Tilsúchil, Tilsóchil, Tlilsóchil, Tlilxóchitl (Mayan), Segnexanté, Vainilla, Vainilla De Papantla, Semenquete, Vainilla Mansa (Spanish);
Nepal: Bhenila (Nepali), Bhenila (Nepalbhasa);
Norwegian: Vanilje;
Persia: Vanil (Farsi);
Philippines: Vanilia (Tagalog);
Polish: Wanilia, Wanilia Płaskolistna;
Portuguese: Baunilha, Baunilheira;
Romanian: Vanilie;
Russian: Vanil', Vanil'nyi, Vanil' Ploskolistnaia;
Serbian: Vanila;
Slovak: Vanilka;
Slovaščinan: Vanilija;
Spanish: Vainilla, Vainillero, Vainillero De Flores Aromáticas;
Sri Lanka: Vanila (Sinhala);
Swahili: Lavani;
Swedish: Vanilj;
Thai: Wanila, Waanilaa;
Turkish: Vanilya;
Ukrainian: Vanil';
Yiddish: Vanil.

Origin/Distribution

Vanilla planifolia originated from Mesoamerica – Mexico and Guatemala. The Totonac Indians of Papantla in north-central Vera Cruz, were the earliest to cultivate vanilla and the oldest use of vanilla use related to the pre-Columbian Maya of southeasten Mexico (Lubinsky et al. 2008). It has been cultivated and escaped or persisted in many areas of the tropics and the south Pacific. Today, the most important exporters are Madagascar and Réunion (formerly called Bourbon), even before México. In Asia, Indonesia is the most successful producer.

Agroecology

In its natural habitat, vanilla is found in the shade of humid, evergreen tropical forest and watershed areas climbing up trees. Vanilla performs best under hot humid tropical condition in areas with 1,500–3,000 mm annual rainfall uniformly distributed throughout the year and with optimum temperatures of 20–32°C. Most plantings are found within 20° north degrees and south of the equator from seal level to 1,500 m altitude. The vine is often cultivated under the shade under plantings of *Areca*, coconut and *Ficus* spp. It thrives in friable, well drained, loamy soil rich in organic matter in the pH range of 5.5–7. Mulching and regular fertilisation (especially with organic manures) are beneficial for good growth. Vanilla flowers although hermaphrodite requires cross-pollination for fruit set and in nature this is carried out by euglossine bees. Under commercial situation, artificial manual pollination is carried out and regular foliar fertilisation is also practised.

Edible Plant Parts and Uses

Vanilla is sold and used as dried fermented beans (Plate 4), vanilla extract, vanilla essence, vanilla powder and vanilla oleoresin. Most vanilla is used in the food industry in dairy products, followed by beverages, baked goods and confections etc. vanilla is often used as a background note or flavour enhancer to round out the flavour profiles of many products. The main application of natural vanilla is for flavouring ice creams and soft drinks. Vanilla is an important flavour component in colas and is also used in cream sodas, root beer, some fruit beverages, tea and coffee.

Plate 1 Vanilla vine

Plate 2 Vanilla flowers

Plate 3 Young vanilla pods

Plate 4 Dried vanilla pods

Vanillin or vanilla flavours are used in many alcoholic beverages, such as whiskeys, cordials and cocktails, to round out and smooth the harsh edges of the alcohol. Vanilla is also used as a kitchen spice for domestic use.

Vanilla is widely used in the flavouring of chocolate, ice-cream, cakes, biscuits, sweets, candies, dry pastries, puddings, gruels, sauces, syrups, milk-based sweet drinks and other confectionary. Another popular product is vanilla sugar which is used for butter puddings, bread, crème brulée, sugar glazed fruits and other desserts. Vanilla sugar is prepared by adding a pierced vanilla pod to sugar in an air-tight container. Vanilla pods are often used whole (being split and the tiny seeds scraped into the mixture) to infuse flavour into cream and custard based sauces. Vanilla is also used as flavouring in medicinal syrup and tobacco.

Botany

A succulent, herbaceous, perennial vine climbing trees or other support to a height of 12–15 m by means of long adventitious roots opposite the leaves (Plate 1). Stem is cylindrical long 1–2 cm diameter, dark green simple or branched. Leaves are alternate, fleshy and sub-sessile. Lamina flat, oblong-elliptical to lanceolate 8–25×5–8 cm, apex acute to acuminate. Inflorescence a short axillary raceme, 5–10 cm long, having 6–30 flowers (Plate 2). Flowers: sepals and petals erect-spreading, yellow-green, fleshy, rigid; sepals 3 – oblanceolate, 3.5–5.5×1.1–1.3 cm, margins straight, apex acute to obtuse; petals elliptic-oblanceolate, abaxially keeled, thinner than sepals, 3.5–5.5×1.1–1.3 cm, apex acute to obtuse; lip adnate to column for 1.5–2 cm, yellow-green, becoming dark yellow toward apex, lamina gullet-like, cuneate, rhomboid, with apical retuse lobule; disc with central tuft of retrorse scales, an hairy to apex; column white, slender, 3–3.5 cm, margins slightly sinuate, adaxially bearded; pollinia – 2, yellow; ovary pedicellate, 3–5 cm. Fruit pendulous, narrow cylindrical capsule, 10–25 cm by 0.8–1.5 cm, obscurely 3 angled (Plate 3) splitting longitudinally when ripe. Seeds numerous, globose, 0.4 mm diameter and black.

Nutritive/Medicinal Properties

The nutrient value of vanilla extract per 100 g edible portion was reported as: water 52.58 g, energy 288 kcal (1,205 kJ), protein 0.06 g, fat 0.06 g, ash 0.26 g, carbohydrate 12.65 g, sugars 12.65 g, Ca 11 mg, Fe 0.12 mg, Mg 12 mg, P 6 mg, K 148 mg, Na 9 mg, Zn 0.11 mg, Cu 0.072 mg, Mn 0.230 mg, thiamine 0.011 mg, riboflavin 0.095 mg, niacin 0.425 mg, pantothenic acid 0.035 mg, vitamin B-6 0.026 mg, total saturated fatty acids 0.010 g, 16:0 (palmitic acid) 0.005 g, 18:0 (stearic acid) 0.001 g, total monounsaturated fatty acids 0.010 g, 18:1 undifferentiated (oleic acid) 0.008 g, total polyunsaturated fatty acids 0.04 g, 18:2 undifferentiated (linoleic acid) 0.003 g, 18: undifferentiated 0.001 g and ethyl alcohol 34.4 g (USDA 2011).

Vanilla planifolia bean was found to contain more than a hundred compounds such as alkaloids, flavonoids, glycosides, phenols, phenol ether, alcohols, carbonyl compounds, acids, esters, vitamins, minerals, reducing sugars, lactones, aliphatic and aromatic carbohydrates, heterocyclic compounds and other phytochemicals. The principal constituent was found to be vanillin, a methylprotocatechuic aldehyde comprising about 85% of the total volatiles other important compounds are p-hydroxybenzaldehyde (up to 9%) and p-hydroxybenzyl methyl ether (1%),4-hydroxybenzyl alcohol, vanillyl alcohol, 3,4-dihydroxybenzaldehyde, 4-hydroxybenzoic acid, vanillic acid, p-coumaric acid, ferulic acid, and piperonal. Two stereoisomeric vitispiranes (2,10, 10-trimethyl-1,6- and methylidene-1-oxaspiro (4,5)dec-7-ene), although only occurring in traces, also influenced the aroma of vanilla (Schulte-Elte et al. 1978).

Older vanilla pods (8 months after pollination were found to have a higher content of glucovanillin, vanillin, p-hydroxybenzaldehyde glucoside, p-hydroxybenzaldehyde, and sucrose, while younger pods (3 months or more after pollination) had more bis[4-(β-d-glucopyranosyloxy)-benzyl]-2-isopropyltartrate (glucoside A), bis[4-(β-d-glucopyranosyloxy)-benzyl]-2-(2-butyl) tartrate (glucoside B), glucose, malic acid, and homocitric acid (Palama et al. 2009). Quantification of compounds showed that free vanillin could reach 24% of the total vanillin content after 8 months of development in the vanilla green pods. Ten phenolic compounds, namely vanillic acid, 4-hydroxybenzyl alcohol, vanillyl alcohol, 3,4-dihydroxybenzaldehyde, 4-hydroxybenzoic acid, 4-hydroxybenzaldehyde, vanillin, p-coumaric acid, ferulic acid, and piperonal were quantitatively determined in vanilla beans using RP-HPLC method (Sinha et al. 2007).

Glucosides of vanillin, vanillic acid, p-hydroxy benzaldehyde, vanillyl alcohol, p-cresol, creosol and bis[4-(β-D-glucopyranosyloxy)-benzyl]-2-isopropyltartrate and bis[4-(β-D-glucopyranosyloxy)-benzyl]-2-(2-butyl)tartrate were identified in vanilla green extract (Dignum et al. 2004). In the overall vanilla aroma, minor compounds p-cresol, creosol, guaiacol and 2-phenylethanol were found to have a high impact as shown by GC-olfactometry analysis of cured vanilla beans. They investigated the kinetics of the β-glucosidase activity from green vanilla beans towards eight glucosides naturally occurring in vanilla and towards p-nitrophenol. For glucosides of p-nitrophenol, vanillin and ferulic acid the enzyme had a Km value (the concentration of substrate required to produce 50% of the Vmax value) of about 5 mM. For other glucosides (vanillic acid, guaiacol and creosol) the Km-values were higher (>20 mM). The Vmax (the maximal speed of activity of the enzyme) was between 5 and 10 IU/mg protein for all glucosides tested. Glucosides of 2-phenylethanol and p-cresol were not hydrolysed. Beta-glucosidase was found not to have a high substrate specificity for the naturally occurring glucosides compared to the synthetic p-nitrophenol glucoside.

Beta-d-glucosidase, was reported to be the enzyme responsible for hydrolysis of glucovanillin into vanillin and glucose. Odoux et al. (2003b) showed that ß-glucosidase activity increased from the epicarp towards the placental zone, whereas glucovanillin was exclusively located in the placentae and papillae of vanilla bean. Activity of the enzyme was observed in the cytoplasm (and/or the periplasm) of mesocarp and endocarp cells, with a more diffuse pattern observed in the papillae. A possible mechanism for the hydrolysis of glucovanillin and release of the aromatic aglycon vanillin was postulated to involve the decompartmentation of cytoplasmic (and/or periplasmic) ß-glucosidase and vacuolar glucovanillin.

Vanilla bean β-d-glucosidase was purified and characterised by Odoux et al. (2003a) and found to be a tetramer (201 kDa) made up of four identical subunits (50 kDa). The optimum pH was 6.5, and the optimum temperature was 40°C at pH 7.0. Km values (the concentration of substrate required to produce 50% of the Vmax value) of vanilla bean β-d-glucosidase for p-nitrophenyl-β-d-glucopyranoside and glucovanillin were 1.1 and 20.0 mM, respectively; Vmax values (the maximal speed of activity of the enzyme) were 4.5 and 5.0 µkat/mg. The β-d-glucosidase was competitively inhibited by glucono-δ-lactone and 1-deoxynojirimycin, and not inhibited by 2 M glucose. The β-d-glucosidase was not inhibited by N-ethylmaleimide and DTNB and fully inhibited by 1.5–2 M 2-mercaptoethanol and 1,4-dithiothreitol. The enzyme showed decreasing activity on p-nitrophenyl-β-d-fucopyranoside, p-nitrophenyl-β-d-glucopyranoside, p-nitrophenyl-β-d-galactopyranoside, and p-nitrophenyl-β-d-xylopyranoside. The enzyme was also active on prunasin, esculin, and salicin and inactive on cellobiose, gentiobiose, amygdalin, phloridzin, indoxyl-β-d-glucopyranoside, and quercetin-3-β-d-glucopyranoside.

Other phytochemicals like nine 4-demethylsterols were identified in *V. fragrans* beans; the 4-demethylsterol fraction of *V. fragrans* was characterized by a high content of 24-methylene cholesterol (27–40%) and of β-sitosterol (35–46%) (Ramaroson-Raonizafinimanana et al. 1998). The beans' age modified the ratio 24-methylene cholesterol/β-sitosterol in *V. fragrans*. Four other demethylsterols in *V. fragrans* (brassicasterol, 0.02%; stigmasta-5,23-dien-3β-ol, 1.43%; stigmasten-22-ol, 0.1%; and fucosterol, 0.5%) were identified from the 4-demethylsterol fraction. 24-Methylene cholesterol and β-sitosterol were also isolated. Four triterpene alcohols were identified in *V. fragrans*, including cycloartenol (0.9–1.6%) from the triterpene alcohol fraction, 24-dihydrotirucallol (17–23%) from the triterpene alcohol fraction, tirucall-7-en-3β-ol (6–7.5%) from the triterpene alcohol fraction, and in a higher content cyclosadol (66–69%) from the triterpene alcohol fraction. Beta-dicarbonyl compounds representing approximately 28% of the neutral lipids, that is, 1.5%, in the epicuticular wax of immature vanilla beans, and approximately 10% of the neutral lipids, that is, 0.9%, in mature beans (Ramaroson-Raonizafinimanana et al. 2000). Five β-dicarbonyl compounds have been identified including 16-pentacosene-2,4-dione; 18-heptacosene-2,4-dione; 20-nonacosene-2,4-dione; 22-hentriacontene-2,4-dione; and 24-tritriacontene-2,4-dione. Among them (Z)-18-heptacosene-2,4-dione, or nervonoylacetone, was synthesized in two steps starting from nervonic acid. The major constituent, nervonoylacetone, represented 74.5% of the beta-dicarbonyl fraction. Long-chain gamma-pyrones representing 7–8% of the neutral lipids in epicuticular wax of mature vanilla beans were identified (Ramaroson-Raonizafinimanana et al. 1999). Three γ-pyrones were identified, including 2-(10-nonadecenyl)-2, 3-dihydro-6-methyl-4H-pyran-4-one, 2-(12-heneicosenyl)-2, 3-dihydro-6-methyl-4H-pyran-4-one, and 2-(14-tricosenyl)-2, 3-dihydro-6-methyl-4H-pyran-4-one. The major constituent was 2-(14-tricosenyl)-2,3-dihydro-6-methyl-4H-pyran-4-one, which represented 70.3% of the γ-pyrone fraction.

Sixty-five volatiles were identified in a pentane/ether extract from cured vanilla beans by GC–MS analysis (Pérez-Silva et al. 2006). Aromatic acids, aliphatic acids and phenolic compounds were the major volatiles. By GC–O analysis of the pentane/ether extract, 26 odour-active compounds were detected. The compounds guaiacol, 4-methylguaiacol, acetovanillone and vanillyl alcohol, found at much lower concentrations in vanilla beans than vanillin, proved to be as intense as vanillin. Seven velvety orosensory active molecules were identified in cured vanilla beans (*Vanilla planifolia*) (Schwarz and Hofmann 2009). Among these, 5-(4-hydroxybenzyl)vanillin, 4-(4-hydroxybenzyl)-2-methoxyphenol, 4-hydroxy-3-(4-hydroxy-3-methoxybenzyl)-5-methoxybenzaldehyde, (1-O-vanilloyl)-(6-O-feruloyl)-β-d-glucopyranoside, americanin A, and 4',6'-dihydroxy-3',5-dimethoxy-[1,1'-biphenyl]-3-carboxaldehyde were previously not reported in vanilla beans. In another study, vanillin was found in higher amounts in cured beans of *V. planifolia* (1.7–3.6% of dry matter) than in *V. tahitensis* (1.0–2.0%), and anisyl compounds were found in lower amounts in *V. planifolia* (0.05%) than in *V. tahitensis* (1.4–2.1%) (Brunschwig et al. 2009). Ten common and long chain monounsaturated fatty acids (LCFA) were also identified LCFA derived from secondary metabolites have discriminating compositions as they reach 5.9% and 15.8% of total fatty acids, respectively in *V. tahitensis* and *V. planifolia*.

Natural vanilla extract and artificial vanilla extract were found to contain the polar aromatic flavour compounds vanillin, ethyl vanillin, 4-hydroxybenzaldehyde, 4-hydroxybenzoic acid, 4-hydroxybenzyl alcohol, vanillic acid, coumarin, piperonal, anisic acid, and anisaldehyde (Belay and Poole 1993). The ratio of 4-hydroxybenzoic acid, 4-hydroxybenzaldehyde and vanillic acid to vanillin in natural vanilla extracts was used to confirm the authenticity of extracts purchased in the United States of America and the United Kingdom. Natural vanilla extracts purchased in Mexico and Puerto Rico were identified as counterfeit products based on changes in the above ratio and the presence of synthetic flavour compounds such as ethyl vanillin and coumarin.

Vanillin extract is produced from vanilla beans. For production of vanillin (4-hydroxy 3-methoxy benzaldehyde) extract from vanilla beans, Waliszewski et al. (2007) recommended a hydration process in 5% ethanol during 48 hours enzymatic pretreatment with stable cellulolytic preparations up to 12 hours. This combination of pretreatment was found to double vanillin content in the ethanolic extract, yielding a product of excellent sensory properties. They found as much as 12–15% of vanillin on a dry basis to be present in mature vanilla pods before harvesting, and if enzymatic hydrolysis of glucovanillin was completed during an adequate curing process, as much as 6–7% of vanillin content could be expected in cured beans. They also found that after 24 hours of bean hydration, the results of reducing sugar after enzymatic digestion were higher compared to non-hydrated beans. Total mean cellulose content in vanilla bean was 26.7 g per 100 g dry weight. Studies by Ranadive (1992) reported that vanillin, *p*-hydrobenzoic acid, *p*-hydroxybenzaldehyde and vanillic acid are present in green vanilla beans as glycosides and are released by B-glucosidase enzymatic action during bean ripening or upon curing (fermentation process).

Young vanilla leaves were found to have higher levels of glucose, bis[4-(β-d-glucopyranosyloxy)-benzyl]-2-isopropyltartrate (glucoside A) and bis[4-(β-d-glucopyranosyloxy)-benzyl]-2-(2-butyl)-tartrate (glucoside B), whereas older leaves had more sucrose, acetic acid, homocitric acid and malic acid (Palama et al. 2010).

Antioxidant Activity

Vanilla has antioxidant activity. Vanilla exhibited the highest antioxidant activity in the peroxidase-based assay (H_2O_2) among seven dessert spices (anise, cinnamon, ginger, licorice, mint, nutmeg, and vanilla) analysed (Murcia et al. 2004). When the Trolox equivalent antioxidant capacity (TEAC) assay was used to provide a ranking order of antioxidant activity, the result in decreasing order of antioxidant capacity was cinnamon≅propyl gallate (common food antioxidant)>mint>anise>BHA (butylated hydroxyanisole, common food antioxidant)>licorice≅vanilla>ginger>nutmeg>BHT (butylated hydroxytoluene, common food antioxidant). Vanilla has also been reported to increase catecholamines in the human body (including epinephrine, more commonly known as adrenaline), which makes it mildly addictive. In another study, at a concentration of 200 ppm, the vanilla extract showed 26% and 43% of antioxidant activity by β-carotene-linoleate and DPPH methods, respectively, in comparison to corresponding values of 93% and 92% for BHA (Shyamala et al. 2007). Major compound identified in the extract included vanillic acid, 4-hydroxybenzyl alcohol, 4-hydroxy-3-methoxybenzyl alcohol, 4-hydroxybenzaldehyde and vanillin. Interestingly, 4-hydroxy-3-methoxybenzyl alcohol and 4-hydroxybenzyl alcohol exhibited antioxidant activity of 65% and 45% by β-carotene-linoleate method and 90% and 50% by DPPH methods, respectively. In contrast, pure vanillin exhibited much lower antioxidant activity. The present study suggested the potential use of vanilla extract components as antioxidants for food preservation and in health supplements as nutraceuticals.

Anti-sickle Cell Activity

Vanillin has also been reported to have promising potential in treating sickle cell disease in humans. Unlike normal red blood cells, the bent sickle cells are too stiff to pass through capillaries and into smaller blood vessels. As a result, they clog and damage the blood vessels. The results can include severe pain, stroke, anemia, life-threatening infections, and damage to the lungs and other organs. Studies found that vanillin, a food additive, covalently bound with sickle haemoglobin (Hb S), and inhibited cell sickling development (Zhang et al. 2004). Studies using transgenic sickle mice, which nearly exclusively develop pulmonary sequestration upon exposure to hypoxia, showed that oral administration of a vanillin prodrug, MX-1520 prior to hypoxia exposure significantly reduced the percentage of sickled cells in the blood. The survival time under severe hypoxic conditions was prolonged from 6.6 minutes in untreated animals to 28.8 minutes and 31 minutes for doses of 137.5 and 275 mg/kg respectively. Intraperitoneal injection of MX-1520 to bypass possible degradation in the digestive tract showed that doses as low as 7 mg/kg prolonged the survival time and reduced the percentage of sickled cells during hypoxia exposure. These results demonstrated the potential for MX-1520 to be a new and safe anti-sickling agent for patients with sickle cell disease.

Antimicrobial Activity

Vanilla was found in recent studies to inhibit bacterial quorum sensing (Choo et al. 2006). Bacteria quorum sensing signals function as a switch for virulence. The microbes only become virulent when the signals indicate that they have the numbers to resist the host immune system response. Vanilla extract was found to significantly reduced violacein production in Tn-5 mutant, *Chromobacterium violaceum* in a concentration-dependent manner. The results suggested that the intake of vanilla-containing food materials might promote human health by inhibiting quorum sensing and preventing bacterial pathogenesis.

Studies revealed that both leaf and stem extracts of *V. planifolia* to be potent antimicrobials against all the pathogenic organisms studied: *Escherichia coli* and its mutant K12, *Proteus vulgaris, Enterobacter aerogens, Bacillus cereus, Streptococcus faecalis, Klebsiella Pneumoniae, Salmonella typhi, Serratia marcescens* and *Pseudomonas aeruginosa* (Shanmugavalli et al. 2009). Among various solvent extracts studied, ethanolic leaf extract showed higher degree of inhibition followed by ethyl acetate and chloroform. In the present study, the antibacterial sensitivity was maximum in ethanolic leaf and lowest inhibition was observed in petroleum ether. The aqueous extract did not show any antibacterial activity. 18 unidentified alkaloids and 11 flavonoids were found in the extract.

The ethanol extract of *Vanilla planifolia* leaves was found to effectively inhibit the growth of both the gram-positive and gram-negative bacteria. High inhibitory activity was exhibited in vitro against *Escherichia coli, E. coli* mutant K12 and *Proteus vulgaris*, moderate inhibitory activity against *Enterobacter aerogenes, Streptococcus feacalis* and *Pseudomonas aeruginosa;* and low degree of inhibition against *Klebsiella pneumonia* and *Salmonella typhi*.

Larvicidal Activity

Studies showed that the ethyl acetate and aqueous butanol fractions of the alcoholic extracts of leaves and stems of *Vanilla fragrans* exhibited toxic bioactivity against mosquito larvae (Sun et al. 2001). Bioactivity of the ethyl acetate fraction was found to be much greater than that from the butanol fraction in mosquito larvae toxicity. The water fraction appeared to contain no substances that impaired mosquito larval growth. The ethyl acetate fraction was found to contain 4-ethoxymethylphenol, 4-butoxymethylphenol, vanillin, 4-hydroxy-2-methoxycinnamaldehyde, and 3,4-dihydroxyphenylacetic acid. 4-Ethoxymethylphenol was the predominant compound, but 4-butoxymethylphenol showed the strongest toxicity to mosquito larvae.

Traditional Medicinal Uses

In traditional medicine, vanilla beans are used as aphrodisiac, carminative, emmenagogue and stimulant and has been claimed to reduce or cure fevers, spasms and caries. The vanilla pods have been used as drugs against nervous diseases and hypochondria.

Other Uses

Vanilla is also used in pharmaceutical products, perfumery, air-fresheners, incense, candles, house-hold, baby and personal care products, aromatherapy and as aroma for tobacco and alcoholic beverages. Apart from flavouring food products vanilla is widely used as an odour maskant for paints, industrial chemicals, rubber tires and plastics etc. It is also used as insect repellent.

Comments

Commercially, vanilla is always propagated by stem cuttings. Vanilla can also be established from tissue culture (micropropagation) of callus masses, protocorms, root tips and stem nodes (Ravishankar 2004).

Selected References

Ackerman JD (2003) *Vanilla* Miller. In: Flora of North America Editorial Committee (eds) 1993+. Flora of North America North of Mexico, 12+ vols, vol 26. Oxford University Press, New York/Oxford, p 510. Published on the internet. http://www.efloras.org. Accessed 2 Oct 2009

Belay MT, Poole CF (1993) Determination of vanillin and related flavor compounds in natural vanilla extracts and vanilla-flavored foods by thin layer chromatography and automated multiple development. Chromatographia 37(7–8):365–373

Brunschwig C, Collard FX, Bianchini JP, Raharivelomanana P (2009) Evaluation of chemical variability of cured vanilla beans (*Vanilla tahitensis* and *Vanilla planifolia*). Nat Prod Commun 4(10):1393–1400

Choo JH, Rukayadi Y, Hwang JK (2006) Inhibition of bacterial quorum sensing by vanilla extract. Lett Appl Microbiol 42(6):637–641

Correll DS (1953) Vanilla. Its botany, cultivation and economic importance. Econ Bot 7.291–358

Council of Scientific and Industrial Research (CSIR) (1976) The wealth of India. A dictionary of Indian raw materials and industrial products, raw materials 10. Publications and Information Directorate, New Delhi

Dignum MJW, van der Heijden R, Kerler J, Winkel C, Verpoorte R (2004) Identification of glucosides in green beans of *Vanilla planifolia* Andrews and kinetics of vanilla β-glucosidase. Food Chem 85(2):199–205

Huxley AJ, Griffiths M, Levy M (eds) (1992) The new RHS dictionary of gardening, 4 vols. Macmillan, London

Lubinsky P, Bory S, Hernández JH, Kim SC, Gómez-Pompa A (2008) Origins and dispersal of cultivated vanilla (*Vanilla planifolia* Jacks. [Orchidaceae]). Econ Bot 62(2):127–138

Murcia MA, Egea I, Romojaro F, Parras P, Jimenez AM, Martinez-Tome M (2004) Antioxidant evaluation in dessert spices compared with common food additives influence of irradiation procedure. J Agric Food Chem 52(7):1872–1881

Odink J, Korthals H, Knijff JH (1988) Simultaneous determination of the major acidic metabolites of catecholamines and serotonin in urine by liquid chromatography with electrochemical detection after a one-step sample clean-up on sephadex G-10; influence of vanilla and banana ingestion. J Chromatogr 424(2): 273–283

Odoux E, Chauwin A, Brillouet J-M (2003a) Purification and characterization of vanilla bean (*Vanilla planifolia* Andrews) β-d-Glucosidase. J Agric Food Chem 51(10):3168–3173

Odoux E, Escoute J, Verdeil J-L, Brillouet J-M (2003b) Localization of ß-D-glucosidase activity and glucovanillin in vanilla bean (*Vanilla planifolia* Andrews). Ann Bot 92:437–444

Palama TL, Fock I, Choi YH, Verpoorte R, Kodja H (2010) Biological variation of *Vanilla planifolia* leaf metabolome. Phytochemistry 71(5–6):567–573

Palama TL, Khatib A, Choi YH, Payet B, Fock I, Verpoorte R, Kodja H (2009) Metabolic changes in different developmental stages of *Vanilla planifolia* pods. J Agric Food Chem 57(17):7651–7658

Pérez-Silva A, Odoux E, Brat P, Ribeyre F, Rodriguez-Jimenes G, Robles-Olvera V, García-Alvarado MA, Günata Z (2006) GC–MS and GC–olfactometry analysis of aroma compounds in a representative organic aroma extract from cured vanilla (*Vanilla planifolia* G. Jackson) beans. Food Chem 99(4):728–735

Porcher MH et al (1995–2020) Searchable world wide web multilingual multiscript plant name database. Published by The University of Melbourne, Australia. http://www.plantnames.unimelb.edu.au/Sorting/Frontpage.html

Ramaroson-Raonizafinimanana B, Gaydou EM, Bombarda I (1998) 4-Demethylsterols and triterpene alcohols from two *Vanilla* bean species: *Vanilla fragrans* and *V. tahitensis* 4-Demethylsterols and triterpene alcohols from two *Vanilla* bean species: *Vanilla fragrans* and *V. tahitensis*. JAOCS 75(1): 51–55

Ramaroson-Raonizafinimanana B, Gaydou EM, Bombarda I (1999) Long-chain gamma-pyrones in epicuticular wax of two vanilla bean species: *V. fragrans* and *V. tahitensis*. J Agric Food Chem 47(8): 3202–3205

Ramaroson-Raonizafinimanana B, Gaydou EM, Bombarda I (2000) Long-chain aliphatic beta-diketones from epicuticular wax of vanilla bean species. Synthesis of nervonoylacetone. J Agric Food Chem 8(10):4739–4743

Ranadive AS (1992) Vanillin and related flavor compounds in vanilla extracts and from beans of various global origins. J Agric Food Chem 40(10):1992–1994

Ranadive AS (1994) Vanilla – cultivation, curing, chemistry, technology and commercial products. In: Charalambous G (ed) Spices, herbs and edible fungi: development in food science 34. Elsevier Science Publishers, Amsterdam, pp 517–577; 780 pp

Ravishankar GP (2004) Efficient micropropagation of *Vanilla planifolia* Andrews under influence of thidiazuron and coconut milk. Indian J Biotechnol 3(1): 113–118

Schulte-Elte KH, Gautschi F, Renold W, Hauser A, Fankhauser P, Limacher J, Ohloff G (1978) Vitispiranes, important constituents of vanilla aroma. Helv Chim Acta 61(3):1125–1133

Schwarz B, Hofmann T (2009) Identification of novel orosensory active molecules in cured vanilla Beans (*Vanilla planifolia*). J Agric Food Chem 57(9): 3729–3737

Shanmugavalli N, Umashankar V, Raheem (2009) Antimicrobial activity of *Vanilla planifolia*. Indian J Sci Technol 2(3):37–40

Shyamala BN, Naidu MM, Sulochanamma G, Srinivas P (2007) Studies on the antioxidant activities of natural vanilla extract and its constituent compounds through in vitro models. J Agric Food Chem 55(19): 7738–7743

Sinha AK, Verma SC, Sharma UK (2007) Development and validation of an RP-HPLC method for quantitative determination of vanillin and related phenolic compounds in *Vanilla planifolia*. J Sep Sci 30(1):15–20

Straver JTG (1999) *Vanilla planifolia* H.C. Andrews. In: de Guzman CC, Siemonsma JS (eds) Plant resources of South East Asia No 13. Spices. Backhuys Publishers, Leiden, pp 228–233

Sun RQ, Sacalis JN, Chin CK, Still CC (2001) Bioactive aromatic compounds from leaves and stems of *Vanilla fragrans*. J Agric Food Chem 49(11):5161–5164

U.S. Department of Agriculture, Agricultural Research Service (2011) USDA national nutrient database for standard reference, release 24. Nutrient Data Laboratory Home Page. http://www.ars.usda.gov/ba/bhnrc/ndl

Waliszewski KN, Ovando SL, Pardio VT (2007) Effect of hydration and enzymatic pretreatment of vanilla beans on the kinetics of vanillin extraction. J Food Eng 78(4):1267–1273

Zhang C, Li X, Lian L, Chen Q, Abdulmalik O, Vassilev V, Lai CS, Asakura T (2004) Anti-sickling effect of MX-1520, a prodrug of vanillin: an in vivo study using rodents. Br J Haematol 125(6):788–795

Sarcotheca diversifolia

Scientific Name

Sarcotheca diversifolia (Miq.) Hallier f.

Synonyms

Connaropsis acuminata Pearson, *Connaropsis diversifolia* (Miq.) Kurz, *Connaropsis grandiflora* Ridl., *Rourea diversifolia* Miq., *Santalodes diversifolium* (Miq.) O.Kuntze, *Sarcotheca acuminata* (Pearson) Hall. f., *Sarcotheca subtriplinervis* Hall. f.

Family

Oxalidaceae

Common/English Name

Jungle Belimbing

Vernacular Names

Brunei: Kerapa-Kerapa, Perapan Macas, Tebarus;
Borneo (Sabah, Sarawak, Kalimantan): Parapa (Dusun), Belimbing Bulat, Belimbing Hutan, Buah Piang, Buah Picing, Iba Jantan, Kadazan, Kandis, Kerapa-Kerapa, Jiwang, Perapan Macas, Piang, Tabaus, Tebarus, Tutong, Ubah Gandis;
Indonesia: Asam, Asam Kalimbawan (Kalimantan).

Origin/Distribution

This species is found in Sumatra, Borneo (Sarawak, Brunei, Sabah, West-, Central- and East-Kalimantan).

Agroecology

It is strictly a tropical species. In its native habitat, it occurs as middle-storey, sub- canopy tree in undisturbed to disturbed mixed dipterocarp and keranga forests, secondary forests, on slopes up to 500 m altitude. It grows on sandy soils, sandy clays and clayey soils.

Edible Plant Parts and Uses

The sour fruit can be eaten fresh with salt or made into pickles and preserves (Plate 2). The acid fruit is eaten as a vegetable or added to curries.

Botany

A sub-canopy tree, reaching a height of 30 m with a trunk diameter of 30 m. Leaves are alternate, exstipulate and trifoliate. Leaflets are oval, lateral leaflet 3–9.5 cm long by 1–4 cm wide, caducous,

Plate 1 Fresh, ripe, obovoid, greenish-yellow of *Sarcotheca diversifolia*

Plate 2 Savoury pickled fruit (*top*), fresh fruit and leaves of *Sarcotheca diversifolia*

middle leaflet 5.5–18 cm long by 2–7 cm wide, pale green, glabrous, acuminate apex, 1–5 pairs of secondary veins, glabrous. Petioles 4–7 mm long. Flowers heterostylous, with long- and short-style, 13 mm diameter, pink-violet-purple with 10 stamens and 2 ovules per locule, in loosely-branched axillary or pseudoterminal thyrso-panicles. Fruit subglobose to obovoid, greenish-yellow drupe, 1.8–3.1 cm by 1–2 cm wide with 5 shallow episeptal furrows and persistent calyx and stamens at the base (Plate 1).

Nutritive/Medicinal Properties

No information on its nutritive food value or medicinal value has been published.

Other Uses

The wood has been used for indoor construction, roofing furniture, plywood and agricultural implements.

Comments

Sarcotheca diversifolia is closely related to other timber species in the genus, namely *S. glauca, S. griffithii, S. macrophylla, S. ochracea* and *S. rubrinervis*.

Selected References

Argent G, Saridan A, Campbell E, Wilkie P (eds) (1997) Manual of the larger and more important non dipterocarp trees of Central Kalimantan. 2 vol. Forest Research Institute, Samarinda, 685 pp

Chung RCK (1998) *Sarcotheca* Blume. In: Sosef MSM, Hong LT, Prawirohatmodjo S (eds) Plant resources of South-East Asia. No. 5(3): timber trees: lesser-known timbers. Prosea Foundation, Bogor, pp 505–507

Cockburn PF (1972) Oxalidaceae. In: Whitmore TC (ed) Tree flora of Malaya, vol 1. Longman, Kuala Lumpur, pp 347–350

Slik JWF (2006) Trees of Sungai Wain. Nationaal Herbarium Nederland. http://www.nationaalherbarium.nl/sungaiwain/

Veldkamp JF (1967) A revision of *Sarcotheca* Bl. and *Dapania* Korth. (Oxalidaceae). Blumea 15:519–543

Wong TM (1982) A dictionary of Malaysian timbers. Revised by Lim SC, Chung RCK. Malayan forest records No. 30. Forest Research Institute Malaysia, Kuala Lumpur, 201 pp

Pandanus conoideus

Scientific Name

Pandanus conoideus Lamarck.

Synonyms

Bryantia butyrophora Webb, *Pandanus butyrophorus* (Webb) Kurz, *Pandanus ceramicus* Knuth nom. superfl., *Pandanus cominsii* Hemsl., *Pandanus cominsii* var. *augustus* B.C.Stone, *Pandanus cominsii* var. *micronesicus* B.C.Stone, *Pandanus englerianus* Martelli, *Pandanus erythros* H.St.John, *Pandanus hollrungii* Warb., *Pandanus hollrungii* f. *caroliniana* Martelli, *Pandanus latericius* B.C.Stone, *Pandanus magnificus* Martelli, *Pandanus minusculus* B.C.Stone, *Pandanus plicatus* H.St.John, *Pandanus ruber* H.St.John, *Pandanus subumbellatus* Becc. ex Solms, *Pandanus sylvestris* Kunth nom. illeg.

Family

Pandanaceae

Common/English Names

Buah Merah, Marita, Oil Pandan, Pandanus Nut, Red Fruit, Red Pandanus.

Vernacular Names

Indonesia: Buah Merah (Malay), Kuansu (Wamena), Kuansu, Sak (Papua Barat) Pandan Seran (Alf Seram, Maluku), Saun (Buru, Maluku), Kleba (North Halmahera, Maluku), Siho, Garoko Ma Ngauku (Maluku);
Papua New Guinea: Aran, Arang, Marita (Tok Pisin), Abare (Huli) Opar (Mendi), Dapu (Kewa), Pangu (Wira), Apare (Duna), Neka (Imbongu) in higher areas of Southern Highlands Province; Kayo (Etoro), Oka (Kaluli) Alakape (Onabasolo), Oga (Hawalisi), Abare (Foi), Sina (Podopa), Anga (Samberigi), Hase (Fasu), Anga (Pole) in lower areas of Southern Highlands Province; Simaho (Ankave) in the Gulf Province.

Origin/Distribution

Distribution of marita is limited to New Guinea and some of the islands to the west (Ceram, Buru and Ternate) in Indonesia to West Pacific. It grows in all Papua regions, especially in Jayawijaya mountain area (Wamena, Tolikara Kelila, Bokondini, Karubaga, Kobakma, Kenyam and Pasema), Jayapura, Manokwari, Nabire, Timika, and Ayamaru Sorong. Within Papua New Guinea, marita is found in all mainland provinces, particularly in the highlands and in the Momase Region, and sometimes on Manus and West New Britain. It is most common in the

Plate 1 Tree habit

following provinces: Eastern Highlands, Morobe, Western Highlands, Southern Highlands, East Sepik, Simbu, Madang and Sandaun. The species is also reported in the wild in Maluku.

Agroecology

Marita is a tropical species. Marita grows best in moist locations, often under shade, and tolerates water-logged soils. It is found from low (10–50 m) altitude in inland situations up to 1,700 m and occasionally as high as 1,980 m in Papua New Guinea. In Papua, red fruit plant is found growing in the area between 2 and 2,300 m altitude with temperature in the range of 23–33°C and relative humidity 73–98%. It thrives in loose, fertile soil rich in humus with pH 4.3–5.3.

Edible Plant Parts and Uses

The fruit is cut into pieces then boiled, roasted or cooked in a stone oven. The pulp and seeds are removed from the core, mashed with water and strained to produce a thick, rich red sauce. This pleasant-tasting, oily, vitamin-A-rich, ketchup-like sauce is used to flavour other foods such as sago, sweet potato, banana, pumpkin and green vegetables. Oil extracted from the fruit is used as valuable food flavouring because it contains high nutrients, such as β-carotene, also utilized as natural colourant that does not contain heavy metals and pathogenic microorganisms (Limbongan and Malik 2009). The fruit is also eaten by directly sucking off the edible mesocarp. Marita fruit is commonly sold in food markets on the New Guinea mainland, especially in the highlands.

Botany

A branching, dioecious evergreen aborescent shrub, 4-7-(16) m high with erect, grey brownish, white-speckled, woody stem (Plate 1) and prickly prop roots and suckering around the mother plant (Plate 2). Leaves are sessile, densely arranged in corkscrew spirals towards the terminal of the stem and branches, broadly linear, strap-shaped, 1–2 m long by 5–8 cm wide, bright green, glaucous beneath, thin to firm, apex acute, margins and mid rib prickly (Plates 1, 2 and 3). Male inflorescence unknown. Female inflorescence oblong-cylindrical head enclosed by bracts, stigma flat and broad. Fruit head short and oblong

Plate 2 Suckers at the base of the mother plant

Plate 4 Oblong maroon fruit

Plate 3 Spirally arranged leaves at the apex of branches

Plate 5 Brown fruit

to elongate and cylindrical, when ripe, 30–100 cm long and 5–25 cm diameter, weighing 3–8 kg, composed of densely packed, aggregation of individual small, connate, angled, fibrous drupes with red oily pericarp; endocarp small comparatively thin. Typically, the fruit head changes from green to bright orange, red, maroon and also to brown or yellow as it matures (Plates 4, 5, 6 and 7).

Nutritive/Medicinal Properties

Plate 6 Red elongated fruit

Pandanus conoideus fruit was found to contain per 100 g: moisture 0.7 g, energy 868 kcal, protein 0 g, lipid 94.2%, carbohydrate 5.1 g, Na 3 mg, α-carotene 130 μg, β-carotene 1,980 μg, β-cryptoxanthin 1,460 μg and vitamin E (α-tocopherol) 21.2 mg (Waspodo and Nishigaki 2007). The fruit was also found to be extremely

Plate 7 Orange elongated fruit

rich in α-carotene, β-carotene, β-cryptoxanthin and vitamin E. β-cryptoxanthin level in Buah Merah was relatively high, more than 1 mg per 100 g extract oil.

The fruit was found to contain 19.4% red fat (dry weight basis) a solid at room temperature. The major fatty acid components of the oil comprised oleic acid 69.0% of total fatty acids, palmitic acid 18.2% and linoleic acid 7.8%. The oil was also found to have an unusually high level of free fatty acids (90.0%) which may have resulted from the presence in the fruit of fat splitting enzymes similar to lipases (Southwell and Harris 1992). The fatty acid composition found in the fruit were saturated fatty acids 14:0 (myristic acid) 0.1%, 16:0 (palmitic acid) 19.7%, 18:0 (stearic acid) 1.8%, 20:0 (arachidic acid) 0.2%; monounsaturated fatty acids 16:1 (n-9) (palmitoleic acid) 0.9%, 17:1 (n-8) (heptadesenic acid) 0.2%, 18:1 (n-9) oleic acid (omega-9) 64.9%, 18:1 (n-7) (cis-vaccenic acid) 3.1%, 20:1 (n-9) (eicosenoic acid) (omega-9) 0.2%; polyunsaturated fatty acids 18:2 (n-6) (linoleic acid) (omega-6) 8.6%, 18:3 (n-3) (α-linolenic acid) (omega-3) 0.7% (Waspodo and Nishigaki 2007). Buah Merah fruit was also found to be rich in omega 3, 6 and 9 fatty acids, in α-carotene, β-carotene, β-cryptoxanthin and α-tocopherol, all powerful antioxidants that could help in preventing many diseases including cancers.

Recent studies identified nicotiflorin and shikimic acid in the leaves of *P. conoideus* from Irian Jaya (Papau) (Suciati 2008). Nicotiflorin from other plant sources had been reported to have neuroprotective activities (Li et al. 2006; Huang et al. 2007). Shikimic acid from the Chinese star anise is used as a base material for production of Tamiflu (oseltamivir) (Nie et al. 2009).

Some of red fruit reported pharmacological properties are elaborated below.

Antioxidant Activity

The total flavonoid content of red fruit was found to be 260.03 mg RE/g extract in the methanol extract, 31.21 mg in the chloroform extract and 697.12 mg in the ethyl acetate extract (Rohman et al. 2010). Among the three extracts the ethyl acetate extract exerted the highest antiradical activities. The ethyl acetate extract of red fruit and its fractions were found to strongly scavenge DPPH radical with IC_{50} value ranged from 5.25 to 53.47 μg/ml. Its antiradical scavenging activity was moderately correlated with the total phenolics content ($R^2=0.645$) and with flavonoid content ($R^2=0.709$). The ethyl acetate extract and its fractions were also able to reduce Fe^{3+} to Fe^{2+} with reduction activity ranging from 91.26 to 682.18 μg ascorbic acid equivalent/g extract or fraction. The extract and its fractions also had metal chelating activity lower than that of EDTA (IC_{50} 18.19 μg/ml). The results indicated that Red fruit can be used as natural antioxidant source to prevent diseases associated with free radicals.

Anticancer Activity

Several studies have been reported on the anticancerous activity of the fruit extract. However, clinical studies are still wanting. Buah Merah oil was found to inhibit the growth of lung cancer cell, A549 at 500 mg of fruit extract/ml dissolved in dimethylsulfoxide (DMSO) (Waspodo and Nishigaki 2007). The concentration of β-cryptoxanthin in Buah Merah oil was equivalent to 0.015 μg/ml. From this study, it was suggested that very small quantity of β-cryptoxanthin could cause the growth inhibition of lung cancer cells associated with smoking. The extract of *Pandanus conoideous* fruit at 0.21 ml/200 g body weight exhibited good lung carcinogenesis inhibition in 7,12-dimethylbenz[a]anthrasene (DMBA)-induced lung cancer in female Sprague–Dawley rats (Mun'im et al. 2006). Another study showed that Buah Merah extract was cytotoxic to myeloma cancer cells. The concentration of 5 mg/ml had been shown to kill 24.75% of myeloma cells (Kurnijasanti and I'tisom 2008). Buah Merah oil exhibited antitumour effects in either tumour volumes or tumour weights against mouse Sarcoma-180, Lewis lung cancer and human non-small lung cancer, A549 bearing mice (Nishigaki et al. 2010). Buah Merah had no oral toxicity in rats and was found to be safe and to have potential chemo-preventive functional food against the lung cancer. Animal studies showed that Buah Merah oil could reduce the serum level of proinflammatory cytokines, such as TNF-α (tumour necrosis factor- α), IFN-γ (interferon- γ) and IL-1 (interleukin-1) in experimental colorectal cancer model mice (Oeij and Khiong 2010). Moreover, Buah Merah also improved the clinical score of diarrhoea in those mice.

Fruit extract from the yellow-form of *P. conoideus* exhibited anti cancer activities against the breast cancer cell line T47D by inducing apoptosis of the cells with an LC_{50} of 0.25 µl/ml (Astirin et al. 2009) The percentage of apoptotic cells recorded at the various concentrations of the crude fruit extract 0.125 µl/ml, 0.0625 µl/ml, and 0.03125 µl/ml were 34.38, 30.03 and 21.07 respectively. The yellow-fruited form was found to contain higher levels of tocopherol and β-carotene than that of red fruited form. It also contained α-carotene (9,500 ppm), β-carotene (240 ppm), tocopherol (10,400 ppm) and omega-3, -6 and -9 fatty acids. Studies found that the ethyl acetate extract of *P. conoideus* fruit induced apoptosis of cervix cancer cells by upregulation of p53ser46 expression and downregulation of pAkt and mTOR (Achadiyani et al. 2010).

Antihyperchlolesteromic Activity

P. conoideus fruit extract was found to increase levels of small HDL particles in normal and diabetic rats (Winarto 2007). The combination of *P. conoideus* and glibenclamide was reported to produce significant increased levels of small HDL cholesterol particles compared with that of the standard drug, glibenclamide in control groups. Dyslipidemia contributed significantly to coronary heart disease in diabetic patients, in whom lipid abnormalities included hypertriglyceridemia, low HDL cholesterol, and increased levels of small dense LDL particles. Subjects with diabetes were at high risk for many long-term complications, including early mortality and coronary artery disease. There were unsubstantiated claims that Buah Merah could be useful as a in retro-viral drug against HIV AIDS.

Hypoglycaemic Activity

A combination of *Padanus conoideus* oil and glibenclamide produced significant reduction in blood glucose in diabetic rats (Winarto et al. 2009). Significant increase in histoscore, number and diameter of Langerhans islets were observed in *P. coinoideus* treated rats. The results showed that *P. conoideus* could increase the hypoglycaemic effect of glibenclamide in diabetic rats.

Longevity Effect

Using spontaneous hypertensive rat-stroke prone rats model, three Indonesian indigenous herbal medicines Noni juice and Extra Virgin Coconut Oil and Buah merah were found effective in prolonging life compared to control rats (Yoshitomi et al. 2010).

Antiinflammatory Activity

Red fruit oil at a dose of 0.9 ml/kg b.w. significantly inhibited inflammation in female wistar rats after carrageen induced inflammation with inhibition rates of 72.5, 69.2 and 85% at 3, 4, and 24 hours respectively, compared to control (Sukandar et al. 2005). At the higher rate of 1.8 ml/kg the inhibition rate was higher at 97% 24 hours after inflammation induction. Currently in Indonesia, red fruit (*Pandanus conoideus*) has been considered to be an alternative therapy of inflammatory bowel disease characterized by chronic inflammation, which can lead to colorectal cancer (Khiong et al. 2010).

They proposed that red fruit could reduce inflammation by inhibiting the activity of NF-κB. They asserted that previous studies proved that red fruit had the capacity to increase the proliferation of immune cells; increased lymphocyte proliferation might induce the production of anti-inflammatory cytokines, such as IL-10 and IL-22 and subsequently inhibit the activation of NF-κB, COX-2, and the production of pro-inflammatory cytokines.

Immunomodulatory Activity

In-vitro studies showed that exposure of Swiss Webster mice's macrophage cell to *P. conoideus* extract could increase the macrophage's phagocytotic activity, with the optimal dose at 0.25 μg/dl (Ratnawati et al. 2007).

Traditional Medicinal Uses

Red fruit has been known to local inhabitants in Papua for many generations as a natural food supplement containing medicinal qualities and as a dye. It is known in Papua that red fruit oil has been used in ethnic tribal communities for stamina, illnesses and because it is a natural product it does not carry the side effects associated with long term use of medicines for degenerative diseases (Anonymous 2006). It is traditionally believed to be a good supplement as a skin and eye medicine, and as a vermifuge. Local communities in Indonesia believed that fruit of *P. conoideus* can treat several degenerative diseases such as cancer, arteriosclerosis, rheumatoid arthritis, and stroke (Budi and Paimin 2004). The special usage of the oil is to cure some diseases, such as cancer, HIV, malaria, cholesterol and diabetes mellitus (Limbongan and Malik 2009).

Other Uses

Leaves are used in making *tikar* mats. Marita is one of the most important cultivated trees in orchards and around houses in highland Papua New Guinea. The large, conical syncarps weighing up to 8 kg bearing numerous, bright red (rarely yellow) drupes are often sold at roadside and urban markets.

Dregs of red fruit oil extraction can be used as feed supplement for poultry (Limbongan and Malik 2009). Oil from drupes is also used as hair and body oil, as polish for arrow shafts, and for paints and dyes. The leaves are sometimes used for thatching materials. Leaves, stem bark, and roots from red fruit plant are used to make ropes, yarn, seat cover, even as a bed for sleeping. The leaves, stem bark and root of the red fruit are used in making handicrafts by the indigenous Papuans. Young leaves are used as a substitute for cigarette wrap.

Comments

In Papua, there are around 30 cultivars but only four have high economic value: merah panjang (long red), merah pendek (short red), cokelat (brown), and kuning (yellow) (Anonymous 2006; Astirin et al. 2009). In Papua New Guinea, marita is grown by 1.5 million people, which is 59% of the rural population (Bourke 1996). This makes it the third-most commonly grown fruit in Papua New Guinea, behind banana and papaya.

Selected References

Achadiyani, Kurnia D, Septiani L, Faried A (2010) The role of buah merah (*Pandanus conoideus* Lam.): ethyl acetate fraction on apoptotic induction of cervix cancer cell lines. Majalah Kedokteran Bandung 42(4S):23–24 (in Indonesian)

Anonymous (2006) Buah Merah. http://www.buah-merah.info/bm-why.php

Anonymous (2010) Red fruit. http://www.buah-merah.info/RedFruit.pdf

Astirin OP, Harini M, Handajani NS (2009) The effect of crude extract of *Pandanus conoideus* Lamb. var. yellow fruit on apoptotic expression of the breast cancer cell line (T47D). Biodiversitas 10(1):44–48

Bourke RM (1996) Edible indigenous nuts in Papua New Guinea. In: Stevens ML, Bourke RM, Evans BR (eds) South Pacific Indigenous nuts. ACIAR proceedings no. 69. Australian Centre for International Agricultural Research, Canberra, pp 45–55

Budi M, Paimin FR (2004) Red fruit (*Pandanus conoideus*, L). Jakarta, Penebar Swadaya, pp 3–26, 47–56, 67–68

French BR (1982) Growing food in the Southern Highlands Province of Papua New Guinea. AFTSEMU

(Agricultural Field Trials, Surveys, Evaluation and Monitoring Unit), World Bank Papua New Guinea Project

French BR (1986) Food plants of Papua New Guinea – a compendium. Ashgrove, Australia and Pacific Science Foundation, 408 pp

Govaerts R, Radcliffe-Smith (2010) World checklist of Pandanaceae. The Board of Trustees of the Royal Botanic Gardens, Kew. Published on the Internet. http://www.kew.org/wcsp/. Retrieved 1 Feb 2010

Huang JL, Fu ST, Jiang YY, Cao YB, Guo ML, Wang Y, Xu Z (2007) Protective effects of nicotiflorin on reducing memory dysfunction, energy metabolism failure and oxidative stress in multi-infarct dementia model rats. Pharmacol Biochem Behav 86(4):741–748

Khiong K, Adhika OA, Chakravitha M (2010) Inhibition of NF-κB pathway as the therapeutic potential of red fruit (*Pandanus conoideus* Lam.) in the treatment of inflammatory bowel disease. Jurnal Kedokteran Maranatha 9(1):69–75 (in Indonesian)

Kurnijasanti R, I'tisom R (2008) Penggunaan antikanker sari buah merah (*Pandanus conoideus*) terhadap kultur sel myeloma (Use of anticancerous red fruit (*Panadus conoideus*) against myeloma cell culture). Majalah Ilmu Faal Indonesia 7(2) (in Indonesian)

Li RP, Guo ML, Zhang G, Xu XF, Li Q (2006) Neuroprotection of nicotiflorin in permanent focal cerebral ischemia and in neuronal cultures. Biol Pharm Bull 29(9):1868–1872

Limbongan J, Malik A (2009) Opportunity of red fruit crop (*Pandanus conoideus* Lamk.) development in Papua Province. Jurnal Penelitian dan Pengembangan Pertanian 28(4):134–141 (in Indonesian)

Mun'im A, Andrajati R, Susilowati H (2006) Tumorigenesis inhibition of water extract of red fruit (*Pandanus conoideus* Lam.) on Sprague–Dawley rat female induced by 7,12 dimethylbenz[a]anthracene (DMBA). Indones J Pharm Sci 3(3):153–161 (in Indonesian)

Nie LD, Shi XX, Ko KH, Lu WD (2009) A short and practical synthesis of oseltamivir phosphate (Tamiflu) from (−)-shikimic Acid. J Org Chem 74(10):3970–3973

Nishigaki T, Dewi F, Hirose K, Shigematsu H (2010) Safety and anti-tumour effects of *Pandanus conoideus* (Buah Merah) in animals. Paper presented at the international conference on nutraceutical and functional foods, Inna Grand Bali Beach, Sanur, Bali, Indonesia, 13–15 Oct 2010

Oeij AA, Khiong TK (2010) The effect of Buah merah oil (*Pandanus conoideus* Lam) towards proinflammatory cytokines profile and clinical score of experimental colorectal cancer in mice. Paper presented at the 14th international congress of immunology meeting, Kobe, Japan. Immunology Meeting Abstracts 22(Suppl 1 Pt 3):iii137–iii142

Purseglove JW (1972) Tropical crops. Monocotyledons. 1 & 2. Longman, London, 607 pp

Ratnawati H, Handoko Y, Purba LH (2007) Effect of administration of buah merah (*Pandanus conoideus* Lam.) extract on the phagocytotic activity of macrophage. Jurnal Kedokteran Marantha 7(1):1–14 (in Indonesian)

Rohman A, Riyanto S, Yuniarti N, Saputra WR, Utami R, Mulatsih W (2010) Antioxidant activity, total phenolic, and total flavaonoid of extracts and fractions of red fruit (*Pandanus conoideus* Lam). Int J Food Res 17:97–106

Southwell K, Harris R (1992) Chemical characteristics of *Pandanus conoideus* fruit lipid. J Sci Food Agric 58(4):593–594

St John H (1961) Revision of the genus Pandanus Stickman, part 7. New species from Borneo, Papua, and the Solomon Islands. Pac Sci 15(4):576–590

St John H (1968) Revision of the genus *Pandanus* Stickman, part 29. New Papuan species in the Section Microstigma collected by C. E. Carr. Pac Sci 22(4):514–519

Stone BC (1966) Further additions to the flora of Guam. Micronesica 2(1):47–50

Stone BC (1992) *Pandanus* Parkinson. In: Verheij EWM, Coronel RE (eds) Plant resources of South-East Asia, no. 2. Edible fruits and nuts. Prosea Foundation, Bogor, pp 240–243

Suciati (2008) Secondary metabolites and acetylcholinesterase inhibitors from *Fagraea* spp. and *Pandanus* spp. Master of philosophy thesis, school of molecular and microbial sciences, University of Queensland, Australia

Sukandar EY, Suwendar, Adnyana IK (2005) Uji aktivas antiinflamasi minyak bauh merah (*Pandanus conoideus* Lamk.) pada tikus wistar betina. Acta Pharm Indone 30(3):76–79 (in Indonesian)

Tarepe T, Bourke RM (1982) Fruit crops in the Papua New Guinea highlands. In: Bourke RM, Kesavan V (eds) Proceedings of the second Papua New Guinea food crops conference. Department of Primary Industry, Port Moresby, pp 86–100

Walter A, Sam C (2002) Fruits of Oceania. ACIAR monograph no 85. Australian Centre for International Agricultural Research, Canberra, 329 pp

Waspodo IS, Nishigaki T (2007) Novel chemopreventive herbal plant buah merah (*Pandanus conoideus*) for lung cancers. Paper presented at the PATPI conference, Bandung, Indonesia, 17–18 July 2007. http://www.buahmerah.jp/PATPIBM2007.7.pdf

Winarto (2007) Pengaruh minyak buah merah (*Pandanas conoideus* Lam.). (Influence of red fruit oil). Jurnal Ilmu Kedokteran (J Med Sci) 1(1):Sept 2007 (in Indonesian)

Winarto, Madiyan M, Anisah M (2009) The effect of *Pandanus conoideus* Lam. Oil on pancreatic B-cells and glibenclamide hypoglycaemic effect of diabetic Wistar rats. Berkala Ilmu Kedokteran 41(1):11–19

Yoshitomi H, Nishigaki T, Surono I, Gao M (2010) Longevity of spontaneous hypertensive rat-stroke prone rats (SHR-SP) by *Morinda citrifolia* (Noni) fruit juice, *Cocos nucifera* (Extra virgin coconut oil) and *Pandanus conoideus* (Buah Merah) oil. Paper presented at the international conference on nutraceutical and functional foods, Inna Grand Bali Beach, Sanur, Bali, Indonesia, 13–15 Oct 2010

Pandanus dubius

Scientific Name

Pandanus dubius K. Sprengler.

Synonyms

Barrotia gaudichaudii Brongn., *Barrotia macrocarpa* (Vieill.) Brongn., *Barrotia tetrodon* Gaudich., *Hombronia edulis* Gaudich, *Pandanus andamanensium* Kurz, *Pandanus bagea* Miq. nom. superfl., *Pandanus bidur* Jungh. ex Miq., *Pandanus bidoer* Jungh., *Pandanus compressus* Martelli, *Pandanus dubius* var. *compressus* (Martelli) B.C.Stone, *Pandanus edulis* (Gaudich) de Vriese nom. illeg., *Pandanus hombronia* F. Muell., *Pandanus kafu* var. *confluentus* Kaneh, *Pandanus latifolius* Perrot, *Pandanus latissimus* Blume ex Miq., *Pandanus leram* Kurz nom. illeg., *Pandanus macrocarpus* Vieill., *Pandanus odoratus* Thunb. nom. illeg., *Pandanus pacificus* J.H.Veitch, *Pandanus tetrodon* (Gaudich.) Balf.f., *Pandanus yamagutii* Kaneh.

Family

Pandanaceae

Common/English Names

Bakong, Knob-Fruited Screwpine, Pandanus, Pandanus Nut, Screw Pine.

Vernacular Names

Banaban: Te Kaina;
Chamorro: Pahong;
Federated States of Micronesia: Mweng kaki, Kipar-N-Ai;
Fiji: Fala, Vadra;
Guam: Pahong;
India: Vilayati Keora;
Indonesia: Pandan Bidur;
Japanese: Dyubiusu;
Kiribati: Tekaureiko;
Micronesian: Pahong;
Palauan: Pohuae, Pohu;
Papua New Guinea: Bakong (Tok Ples);
Polynesia: Hosoa;
Rotuman: Hosoa;
Solomon Islands: Meongehe, Meou;
Vanuatu: Wavax (Banks – Mosina), Na-Vak (Tongoa – Namakura);

Origin/Distribution

Wild distribution of the species are found in The Andaman and Nicobar Islands through southeast Asia in Java, Moluccas to the Western Pacific region – Papua New Guinea, Solomon Islands, Palau, Northern Marianas, Guam, Kiribati, Federated States of Micronesia, Rotuma and Vanuatu.

Agroecology

The species is found in coastal littoral environment on beaches, rocky shores and limestone outcrops in full sun. It tolerates saline conditions and strong winds. It is also cultivated in its native range.

Edible Plant Parts and Uses

The white seed kernel of ripe fruits is edible, its taste is similar to that of coconut. In Vanuatu, children eat the extracted kernels. The seed kernels are also eaten in Guam, the Philippines and Rota island in Northern Mariana Islands in Micronesia. The kernels are cooked and stored.

Botany

An evergreen, dioecious tree, 3–10 m high with male and female flowers produced on different plants with a loose crown and prominent, robust, thick prop roots and aerial roots (Plate 1). Leaves are strap-shaped (lanceolate), with shallowly denticulate margins, dark green, up to 2 m long by 11–16 cm wide (Plates 2 and 3). Female spadix consist of a globose head with densely crowded carpels, each containing a single cell with a single ovule and with a sessile stigma. Fruit is a stalked, solitary syncarp, globose 20–30 cm diameter consisting of numerous ellipsoid to elongate, 10×6×4 cm

Plate 1 Prop roots

Plate 2 Leaves and fruiting syncarps

Plate 3 Leaves and immature fruiting syncarp

Plate 4 Close-view of fruiting syncarp

drupes (carpels), pressed against one another, in the upper quarter, bluish-pale green, with a 5–6 angular top, the mid section shiny orangey brown, lower quarter of the drupe is fibrous and yellowish-white (Plates 2, 3 and 4). The mesocarp is fleshy and fibrous, and the edible seed is hard, yellowish-white, round, 1–1.5 cm diameter located at the base of the drupe.

Nutritive/Medicinal Properties

No information has been published on the nutrient value of the edible seed.

Two new alkaloids, dubiusamine-A and dubiusamine-B were isolated from the crude base of *Pandanus dubius* (Tan et al. 2010).

Young shoots have medicinal use in cases of food poisoning.

Other Uses

Leaves are dried and used for plaiting mats, basketry, coarse wicker-work, native bags and making umbrellas and roof thatching. The plant is grown in gardens for its fragrant flowers and as a nice landscape plant. The entire fruit head is used for fiestive decoration.

Comments

In Papua New Guinea, it is of occasional occurrence and of minor use.

Selected References

Backer CA, Bakhuizen van den Brink RC Jr (1968) Flora of Java (Spermatophytes only), vol 3. Wolters-Noordhoff, Groningen, 761 pp

Brown WH (1951–1957) Useful plants of the Philippines. Reprint of the 1941–1943 edition. 3 vols. Technical Bulletin 10. Department of Agriculture and Natural Resources, Bureau of Printing, Manila, vol 1 (1951), 590 pp, vol 2 (1954), 513 pp, vol 3 (1957), 507 pp

French BR (1986) Food plants of Papua New Guinea – a compendium. Australia and Pacific Science Foundation, Ashgrove, 408 pp

Govaerts R, Radcliffe-Smith (2010) World checklist of Pandanaceae. The Board of Trustees of the Royal Botanic Gardens, Kew. Published on the internet. http://www.kew.org/wcsp/. Retrieved 1 Feb 2010

Liberty Hyde Bailey Hortorium (1976) Hortus Third. A concise dictionary of plants cultivated in the United States and Canada. Liberty Hyde Bailey Hortorium, Cornell University. Wiley, 1312 pp

McClatchey W (1996) The ethnopharmacopoeia of Rotuma. J Ethnopharmacol 50:147–156

Rickard PP, Cox PA (1984) Custom umbrellas (Poro) from *Pandanus* in Solomon Islands. Econ Bot 38(3): 314–321

Stone BC (1976) The Pandanaceae of the New Hebrides, with an essay on intraspecific variation in *Pandanus tectorius*. Kew Bull 31:47–70

Stone BC (1978) Studies in Malesian Pandanaceae. XVII. On the taxonomy of 'Pandan Wangi': a *Pandanus* cultivar with scented leaves. Econ Bot 32:285–293

Tan MA, Kitajima M, Kogure N, Nonato MG, Takayama H (2010) Isolation and total syntheses of two new alkaloids, dubiusamines-A (I) and -B (II), from *Pandanus dubius*. Tetrahedron 66(18):3353–3359

Walter A, Sam C (2002) Fruits of Oceania. ACIAR monograph no. 85. Australian Centre for International Agricultural Research, Canberra, 329 pp

Wardah, Setyowati FM (2009) Ethnobotanical study on the genus *Pandanus* L.f. in certain areas in Java, Indonesia. Biodiversitas 10(3):146–150

Yen DE (1974) Arboriculture in the subsistence of Santa Cruz, Solomon Islands. Econ Bot 28:247–284

Pandanus julianettii

Scientific Name

Pandanus julianettii Martelli.

Synonyms

None.

Family

Pandanaceae

Common/English Names

Karuka, Karuka Nut, Pandanus Nut

Vernacular Names

Papua New Guinea: Xweebo (Ankave), Yase (Baruya), Anga (Huli) Ank (Mendi), Aga (Kewa), Ama (Wiru), Anga (Duna), Amo (Imbomgu), maisene (Pole).
  Vernacular names of different ethnic groups in PNG.

Origin/Distribution

Karuka is found both wild and cultivated at high elevations in New Guinea – Papau and Papua New Guinea.

Agroecology

Karuka is occurs naturally in a narrow altitudinal band 5–7°S in the highlands in the central cordillera of New Guinea and on the Huon Peninsula. Karuka grows between 1,300 and 3,300 m and in locations with a mean annual precipitation of 2,000–5,000 mm. It is found as individual trees and in large groves in primary forest and in woody regrowth and cane grass. It does poorly in open sites and short grasslands. It is adaptable to both well-drained and poorly drained sites but does best in fairly fertile soil along the banks of small creeks, in hollows around the edges of hills and edges of small clearings in the bush.

Edible Plant Parts and Uses

The white kernel (endosperm) of the fruit is eaten raw or eaten after roasting or smoking. Fruits after smoking or roasting are stored in rafters and taken to the local markets. The creamy-white kernel is delicious, sweetish and tastes like coconut and is an important snack food and seasonal staple food for the highland communities being one of the few high protein plant foods in the region. The core of the fruiting head is spongy and has a honeycomb-like appearance after the nuts are separated and can be cooked and eaten. The weight of edible kernels in a fruit is about 8% of the total fresh weight or about 0.5 kg.

The whole fruit bunch can be stored in damp waterlogged ground for a few months if there are too many fruit to use at any one time. These fruit are collected again and cooked and used as if they had just been harvested. Particularly for nuts harvested ripe there are two ways to use them. They can be eaten fresh after cooking or they can be dried and stored and eaten later without cooking.

Plate 1 Roasted halves of karuka fruit

Botany

Dioecious, sparsely branched, erect tree, 10–30 m high, trunks straight of 30 cm diameter with basal prop roots and with male and female flowers on separate trees. The leaves grow in pairs opposite each other and they are twisted to look like a spiral going up the trunk. Leaves are large, thick, leathery, broadly linear, 3–4 m long by 8–12 cm wide with attenuate apex, leaf apex twin-pleated with upwards directed prickles. Male inflorescence consists of a large branched dense spadix bearing 12 long cylindric spikes each spike consists of numerous staminate phalange each phalange with a 3 mm column and 6–9 sub-sessile anther at the top of the column, no perianth. Female inflorescence consists of a solitary ovoid or ellipsoid multiple fruiting heads (syncarps) (Plates 1 and 2), subtended by creamy-white bracts and composed of numerous (700–1,000) densely packed slender (finger-like) one celled carpels ripening as drupes each 8–10 cm × 1.5 cm. The endocarp is thin but hard (Plate 4) and the endosperm is white, copious, oily, sweetish with coconut-like taste and edible (Plate 3). Each mature head and stalk can weigh up to 16 kg. The central core of the matured head

Plate 2 Smaller sections of unroasted fruit

Plate 3 Separated drupes whole and peeled showing the white kernel

Plate 4 Roasted karuka nuts (endocarps) on sale in a local market in PNG

is spongy and has a honeycomb-like appearance when separated from the nuts. The spongy core can be cooked and eaten.

Nutritive/Medicinal Properties

The nutrient value of the edible kernel had been reported by French (1982) as follows: moisture 9%, energy 540–700 cals, protein 11.9–14.1 g, Fe 419 mg and the spongy core contained 8.5 g protein.

Other Uses

Trunks and prop roots of the tree are used as building materials and the leaves are used to make bush shelters.

Comments

B.C. Stone regards *Pandanus julianettii* as a possible cultigen of *Pandanus brosimos* Merr. & Perry.

Selected References

Bourke RM (1996) Edible indigenous nuts in Papua New Guinea. In Stevens ML, Bourke RM, Evans BR (eds) South Pacific Indigenous nuts. ACIAR proceedings No. 69. Australian Centre for International Agricultural Research, Canberra, pp 45–55

French BR (1982) Growing food in the Southern Highlands Province of Papua New Guinea. AFTSEMU (Agricultural Field Trials, Surveys, Evaluation and Monitoring Unit), World Bank Papua New Guinea Project

French BR (1986) Food plants of Papua New Guinea – a compendium. Australia and Pacific Science Foundation, Ashgrove, 408 pp

Govaerts R, Radcliffe-Smith (2010) World checklist of Pandanaceae. The Board of Trustees of the Royal Botanic Gardens, Kew. Published on the internet. http://www.kew.org/wcsp/. Retrieved 1 Feb 2010

Purseglove JW (1972) Tropical crops. Monocotyledons. 1 & 2. Longman, London, 607 pp

Rose CJ (1982) Preliminary observations on the Pandanus nut (*Pandanus julianettii* Martelli). Proceedings of the second Papua New Guinea food crops conference, Department of Primary Industry, Port Moresby, PNG, pp 160–167

Stone BC (1992) *Pandanus* Parkinson. In: Verheij EWM, Coronel RE (eds) Plant resources of South-East Asia. No. 2. Edible fruits and nuts. Prosea Foundation, Bogor, pp 240–243

Tarepe T, Bourke RM (1982) Fruit crops in the Papua New Guinea highlands. In: Bourke RM, Kesavan V (eds) Proceedings of the second Papua New Guinea food crops conference. Department of Primary Industry, Port Moresby, pp 86–100

Pandanus leram

Scientific Name

Pandanus leram Voigt.

Synonyms

Pandanus indicus (Gaudich.) Warb, *Pandanus leram* var. *macrocarpus* Kurz, *Pandanus mellori* Boden-Kloss, *Roussinia indica* Gaudich.

Family

Pandanaceae

Common/English Names

Nicobar-Breadfruit (Plate 3), Pandan Wong

Vernacular Names

Nicobar Islands: mukung (Shompen), Larohm;
Portuguese: Melori;

Origin/Distribution

The species is indigenous to the Nicobar and Andaman Islands and the south coasts of Sumatra and western Java.

Agroecology

A strictly tropical species found in low swampy areas away from the seashore in lowlands and along water courses where the soil is moist and heavy. *P. leram* is the dominant species in this low swampy vegetation flora. In its native habitat the mean annual maximum and minimum temperature is 25 and 30°C and with high annual rainfall up to 3,900 mm and high relative humidity of 77–88%.

Edible Plant Parts and Uses

Ripe fruit forms the staple food or daily bread of Shompen folks in Nicobar (Hedrick 1972; Sharief and Rao 2007; Sharief 2008). The ripe globose fruit is harvested and the wedge-shaped fruitlets are inedible when raw and are consumed after

Plate 1 Strap-shaped leaves clustered at the end of the shoots

Plate 3 Tree label

Plate 2 Aerial prop roots

cooking in water, the mealy mass is scooped off and prepared into various dishes. This is the *melori* of the Portuguese and the *larohm* of the natives. The flavor of the mass thus prepared strongly resembles that of apple marmalade. The wedge-shaped sections are also cooked by covering with thick layers of leaves of *Macaranga nicobarica* and cooked for 1–2 hours, the mealy pulp is scooped off and mixed with other food ingredients like pig fat and sugar to prepare different types of dishes.

Botany

An erect, evergreen tree, 12–16 m high, with robust, aerial basal prop roots (Plates 1 and 2). Leaves bright green, clustered towards the shoot apex, linear or strap-shaped, 2 m long, 6–8 cm at the proximal end and tapering to a sharp point at the distal apex, mid-rib and margin with short, fine prickles (Plates 1 and 2). Staminate inflorescence large, pendent, spicate-racemose. Flower bracts cream, stigma erect or oblique. Pistillate head, globose to subglobose, ripening orange or yellow-orange, made up of 90–110 carpellate phalange.

Nutritive/Medicinal Properties

The aggregate fruit pulp and seeds of *Pandanus leram* were found to contain 75.8% and 57.1% moisture, 0.6% and 0.9% total mineral content, 0.4% and 7.1% protein, 8.1% and 3.3% fibre, 0.5% and 23.7% total lipid, and 14.6% and 7.9% non-fibre carbohydrates, respectively (Katiyar et al. 1989). The seeds were found to be nutritionally rich in comparison with the fruit pulp, but constituted only a small fraction (3%) of the total fruit. Palmitic (56.4%), oleic (26.5%) and linoleic acids (16.4%) were the major fatty acids in the seed oil.

The tender young leaves are pounded with coconut oil and rubbed on the body to remove fatigue (Verma et al. 2010).

Other Uses

Dried fruit with fibres are used as tooth brush. The leaves are used for thatching. The dissected split leaves are made into brooms and weave into mats (Sharief 2008; Sharief and Rao 2007).

Comments

P. leram propagates readily from seed, but it is also widely propagated from branch cuttings by local people in the Nicobar Islands.

Selected References

Govaerts R, Radcliffe-Smith (2010) World checklist of Pandanaceae. The Board of Trustees of the Royal Botanic Gardens, Kew. Published on the Internet. http://www.kew.org/wcsp/. Retrieved 1 Feb 2010

Hedrick UP (1972) Sturtevant's edible plants of the world. Dover Publications, New York, 686 pp

Katiyar SK, Kumar N, Bhatia AK (1989) A chemical study of *Pandanus lerum* fruit grown in the Andaman and Nicobar Islands. Trop Sci 29(2):137–140

Negi SS (1996) Biosphere reserves in India, landuse, biodiversity and conservation. Indus Publishing Company, New Delhi, 221 pp

Sharief MU (2008) Tribal artifacts of Nicobari folk of Nicobar Archipelago. Indian J Tradit Knowl 7(1):42–49

Sharief MU, Rao RR (2007) Ethnobotanical studies of Shompens – a critically endangered and degenerating ethnic community in Great Nicobar Island. Curr Sci 93(10):1623–1628

Stone BC (1975) *Pandanus leram* var. *andamanensium* (Kurz) B.C.Stone. Ceylon J Sci Biol Sci 11(2):118

Verma C, Bhatia S, Srivastava S (2010) Traditional medicine of the Nicobarese. Indian J Tradit Knowl 9(4):779–785

Pandanus spiralis

Scientific Name

Pandanus spiralis R. Br.

Synonyms

Pandanus convexus H.St.John, *Pandanus spiralis* var. *convexus* (H.St.John) B.C.Stone, *Pandanus integer* H.St.John, *Pandanus spiralis* R.Br. var. *spiralis*, *Pandanus spiralis* var. *flammeus* B.C.Stone, *Pandanus spiralis* var. *multimammillatus* B.C.Stone, *Pandanus spiralis* var. *septemlocularis* B.C.Stone, *Pandanus semiarmatus* H.St.John, *Pandanus thermalis* H.St.John, *Pandanus spiralis* var. *thermalis* (H.St.John) B.C.Stone, *Pandanus yirrkalaensis* H.St.John, *Pandanus darwinensis* H.St.John, *Pandanus darwinensis* var. *darwinensis*, *Pandanus latifructus* H.St.John, *Pandanus darwinensis* var. *latifructus* (H.St.John) B.C.Stone.

Family

Pandanaceae

Common/English Names

Pandanus Palm, Screw Palm, Screwpine.

Vernacular Names

Australia: An-Yakngarra, Gunga (Aboriginal)

Origin/Distribution

The species is native to the Northern Territory, extreme north of Western Australia and Northern Queensland.

Agroecology

The plant is most commonly found growing in poorly drained areas such as along watercourses, swamp fringes, edges of floodplains, or coastal fringes, dune systems and in woodlands and open forests in the warm monsoonal tropics of northern Australia. It often forms dense stands in areas that are seasonally flooded.

Edible Plant Parts and Uses

The seeds from ripe fruits are eaten raw after extraction from the woody, fibrous fruit. The seed is very tasty, and is often likened to pine kernels or peanuts. The seeds can be extracted and ground into flour. The soft fleshy, white leaf bases are also eaten raw and contains carbohydrates.

Botany

A shrub to small tree, dioecious, much branched, erect growing to 10 m with grey-brown bark with distinct leaf scars. The old dead leaves stay attached to the trunk for some time looking like grass skirts. Leaves are spirally arranged and clustered at the tip of stem and branches. Leaves are broadly linear (strap-like) green, glaucous, 1.5 m long by 5–10 cm wide with many small spines at the margins and midrib (Plate 1). Flowers are small, white, scented in dense terminal spikes to 10 cm long and enclosed by leafy bracts. Fruit is subglobose to ovoid aggregate head, 30 cm diameter, composed of 10–25 wedge-shaped, woody carpels, 7–10 cm long, 5–8 cm wide, deep orange to red when ripe (Plate 1). Each carpel has 5–7 seeds.

Nutritive/Medicinal Properties

The edible kernel contains niacin and minerals in particular potassium, sodium, magnesium and calcium. Analysis carried out in Australia revealed that following food composition of the kernel per 100 g edible portion: energy 2,342 kJ, moisture 3 g, N 3 g, protein, 25.9 g, Fat 46–8 g, ash 1.7 g, total dietary fiber 21.3 g, Ca 62 mg, Copper 1.6 mg, iron 6.6 mg, magnesium 160 mg, potassium 267 mg, sodium 344 mg and zinc 0.8 mg (Brand Miller et al. 1993).

In traditional medicine, mashed leaves of *Pandanus spiralis* were reported used by the aborigines to cure headaches by applying it around the head and the white portion of the leaf bases are used as antiseptic packing for wounds. The core of the trunk was traditionally used for a number of complaints. Pounded and/or boiled, it was found useful for diarrhoea and stomach pain but also for mouth sores and toothache and to relieve headaches and flu. In some cases the pith from the prop roots was used. There are records from Groote Eylandt, in the Northern Territory, of pandanus seeds being consumed for contraceptive purposes but there is no evidence to support their effectiveness.

Other Uses

The leaves are used to weave neck-bands, armbands, baskets, mats and dillybags and also used to make fish traps. The white cabbage of the growing points is used to make a green dye. Trunks are used to make digderidoos and rafts.

Comments

P. spiralis is readily propagated from seeds or stem cuttings.

Selected References

Aboriginal Communities of the Northern Territory (1993) Traditional aboriginal medicines in the Northern Territory of Australia. Conservation Commission of the Northern Territory of Australia, Darwin, p 651

Australian Plant Census. IBIS database, Centre for Plant Biodiversity Research, Council of Heads of Australian Herbaria, viewed 12 Dec 2007. http://www.chah.gov.au/apc/index.html

Brand Miller J, James KW, Maggiore P (1993) Tables of composition of Australian aboriginal foods. Aboriginal Studies Press, Canberra

Govaerts R, Radcliffe-Smith (2010) World checklist of Pandanaceae. The Board of Trustees of the Royal Botanic Gardens, Kew. Published on the internet. http://www.kew.org/wcsp/. Retrieved 1 Feb 2010

Wightman G, Andrews M (1991) Bush tucker identikit. Conservation Commission of The Northern Territory, Darwin

Plate 1 Ripe fruiting syncarp

Pandanus tectorius

Scientific Name

Pandanus tectorius Parkinson ex Du Roi.

Synonyms

Corypha laevis (Lour.) A.Chev., *Pandanus absonus* H.St.John, *Pandanus adscendens* H.St.John, *Pandanus aequor* H.St.John, *Pandanus aitutakiensis* H.St.John, *Pandanus akiakiensis* H.St.John, *Pandanus alloios* H.St.John, *Pandanus amplexus* H.St.John, *Pandanus angulatus* H.St.John, *Pandanus angulosus* H.St.John, *Pandanus anisos* H.St.John, *Pandanus aoraiensis* H.St.John, *Pandanus apionops* H.St.John, *Pandanus arapepe* H.St.John, *Pandanus asauensis* H.St.John, *Pandanus ater* H.St.John, *Pandanus baptistii* Misonne, *Pandanus bassus* H.St.John, *Pandanus bathys* H.St.John, *Pandanus bergmanii* F.Br., *Pandanus bicurvatus* H.St.John, *Pandanus blakei* H.St.John, *Pandanus boraboraensis* H.St.John, *Pandanus bothreus* H.St.John, *Pandanus bowenensis* H.St.John, *Pandanus brachypodus* Kaneh., *Pandanus brownii* H.St.John, *Pandanus cacuminatus* H.St.John, *Pandanus carolinensis* Martelli, *Pandanus chamissonis* Gaudich., *Pandanus charancanus* Kaneh., *Pandanus chelyon* H.St.John, *Pandanus christophersenii* H.St.John, *Pandanus citraceus* H.St.John, *Pandanus collatus* H.St.John, *Pandanus complanatus* H.St.John, *Pandanus cooperi* (Martelli) H.St.John, *Pandanus coronatus* Martelli, *Pandanus coronatus* f. *minor* Martelli, *Pandanus crassiaculeatus* H.St.John, *Pandanus crassus* H.St.John, *Pandanus cylindricus* Kaneh., *Pandanus cylindricus* var. *sinnau* Kaneh., *Pandanus cymatilis* H.St.John, *Pandanus decorus* K.Koch, *Pandanus dicheres* H.St.John, *Pandanus dilatatus* Kaneh., *Pandanus discolor* auct., *Pandanus distinctus* Martelli, *Pandanus divaricatus* H.St.John, *Pandanus divergens* Kaneh., *Pandanus dotyi* H.St.John, *Pandanus douglasii* Gaudich., *Pandanus drakei* H.St.John, *Pandanus drolletianus* Martelli, *Pandanus duriocarpoides* Kaneh., *Pandanus duriocarpus* Martelli, *Pandanus edwinii* H.St.John, *Pandanus elevatus* H.St.John, *Pandanus enchabiensis* Kaneh., *Pandanus erythrophloeus* Kaneh., *Pandanus extralittoralis* H.St.John, *Pandanus eyesyes* Kaneh., *Pandanus fahina* H.St.John, *Pandanus faramaa* H.St.John, *Pandanus fatuhivaensis* H.St.John, *Pandanus fatyanion* (Kaneh.) Hosok., *Pandanus feruliferus* H.St.John, *Pandanus filiciatilis* H.St.John, *Pandanus fischerianus* Martelli, *Pandanus fischerianus* f. *bergmanii* (F.Br.) B.C.Stone, *Pandanus fischerianus* f. *bryanii* B.C.Stone, *Pandanus fischerianus* f. *compressus* B.C.Stone, *Pandanus fischerianus* var. *bryanii* B.C.Stone, *Pandanus fischerianus* var. *cooperi* (Martelli) B.C.Stone, *Pandanus fischerianus* var. *rockii* (Martelli) B.C.Stone, *Pandanus fragrans* Gaudich., *Pandanus futunaensis* H.St.John, *Pandanus gambierensis* H.St.John, *Pandanus glomerosus* H.St.John, *Pandanus grantii* H.St.John, *Pandanus guamensis* Martelli, *Pandanus haapaiensis* H.St.John, *Pandanus heronensis* H.St.John, *Pandanus hivaoaensis*

H.St.John, *Pandanus horneinsularum* H.St.John, *Pandanus hosinoi* Kaneh., *Pandanus hosokawae* Kaneh., *Pandanus houmaensis* H.St.John, *Pandanus hubbardii* H.St.John, *Pandanus impar* H.St.John, *Pandanus inarmatus* H.St.John, *Pandanus inermis* Roxb., *Pandanus inflexus* H.St.John, *Pandanus infundibuliformis* H.St.John, *Pandanus insularis* Kaneh., *Pandanus intraconicus* H.St.John, *Pandanus intralaevis* H.St.John, *Pandanus jaluitensis* Kaneh., *Pandanus jonesii* (F.Br.) H.St.John, *Pandanus kafu* Martelli, *Pandanus kamptos* H.St.John, *Pandanus koidzumii* Hosok., *Pandanus korrensis* Kaneh., *Pandanus kraussii* H.St.John, *Pandanus kusaiensis* Kaneh., *Pandanus laculatus* H.St.John, *Pandanus laevis* Kunth, *Pandanus laevis* Lour., *Pandanus lakatwa* Kaneh., *Pandanus lambasaensis* H.St.John, *Pandanus laticanaliculatus* Kaneh., *Pandanus laticanaliculatus* var. *edulis* Kaneh., *Pandanus lauensis* H.St.John, *Pandanus licinus* H.St.John, *Pandanus limitaris* H.St.John, *Pandanus longifolius* H.L.Wendl., *Pandanus macfarlanei* Martelli, *Pandanus macrocephalus* Kaneh., *Pandanus makateaensis* H.St.John, *Pandanus malatensis* Blanco, *Pandanus mangarevaensis* H.St.John, *Pandanus mariaensis* H.St.John, *Pandanus marquesasensis* H.St.John, *Pandanus matukuensis* H.St.John, *Pandanus mbalawa* H.St.John, *Pandanus meetiaensis* H.St.John, *Pandanus menne* Kaneh., *Pandanus menziesii* Gaudich., *Pandanus metius* H.St.John, *Pandanus minysocephalus* H.St.John, *Pandanus mooreaensis* H.St.John, *Pandanus moschatus* Miq., *Pandanus moschatus* Rumph. ex Voigt, *Pandanus motuensis* H.St.John, *Pandanus nandiensis* H.St.John, *Pandanus notialis* H.St.John, *Pandanus oblatiapicalis* H.St.John, *Pandanus oblaticonvexus* H.St.John, *Pandanus obliquus* Kaneh., *Pandanus odontoides* Hosok., *Pandanus odoratissimus* f. *major* Martelli, *Pandanus odoratissimus* var. *laevigatus* Martelli, *Pandanus odoratissimus* var. *laevis* (Warb.), *Pandanus odoratissimus* var. *oahuensis* Martelli, *Pandanus odoratissimus* var. *parksii* Martelli, *Pandanus odoratissimus* var. *pyriformis* Martelli, *Pandanus odoratissimus* var. *savaiensis* (Martelli) Martelli, *Pandanus odoratissimus* var. *setchellii* Martelli, *Pandanus odoratissimus* var. *spurius* Willd., *Pandanus odoratissimus* var. *suvaensis* Martelli, *Pandanus okamotoi* Kaneh., *Pandanus onoilauensis* H.St.John, *Pandanus orarius* H.St.John, *Pandanus otemanuensis* H.St.John, *Pandanus ovalauensis* H.St.John, *Pandanus pachys* H.St.John, *Pandanus palkilensis* Hosok., *Pandanus palmyraensis* H.St.John, *Pandanus pansus* H.St.John, *Pandanus paogo* H.St.John, *Pandanus papeariensis* Martelli, *Pandanus papenooensis* H.St.John, *Pandanus parhamii* H.St.John, *Pandanus parksii* H.St.John, *Pandanus patulior* H.St.John, *Pandanus pedunculatus* R.Br., *Pandanus pedunculatus* var. *insularis* B.C.Stone, *Pandanus pedunculatus* var. *malagunensis* B.C.Stone, *Pandanus pedunculatus* var. *rendovensis* B.C.Stone, *Pandanus planus* H.St.John, *Pandanus politus* Martelli, *Pandanus ponapensis* Martelli, *Pandanus prismaticus* Martelli, *Pandanus prolixus* H.St.John, *Pandanus pseudomenne* Hosok., *Pandanus pulposus* (Warb.) Martelli, *Pandanus pulposus* var. *cooperi* Martelli, *Pandanus pusillus* H.St.John, *Pandanus pyriformis* (Martelli) H.St.John, *Pandanus radiatus* H.St.John, *Pandanus raiateaensis* H.St.John, *Pandanus raivavaensis* Martelli, *Pandanus raroiaensis* H.St.John, *Pandanus rectangulatus* Kaneh., *Pandanus repens* Miq., *Pandanus rhizophorensis* H.St.John, *Pandanus rhombocarpus* Kaneh., *Pandanus rikiteaensis* H.St.John, *Pandanus rimataraensis* H.St.John, *Pandanus rockii* Martelli, *Pandanus rotensis* Hosok., *Pandanus rotundatus* Kaneh., *Pandanus rurutuensis* H.St.John, *Pandanus sabotan* Blanco, *Pandanus saipanensis* Kaneh., *Pandanus saltuarius* H.St.John, *Pandanus samak* Hassk., *Pandanus sanderi* Sander, *Pandanus savaiensis* (Martelli) H.St.John, *Pandanus seruaensis* H.St.John, *Pandanus sinuosus* H.St.John, *Pandanus sinuvadosus* H.St.John, *Pandanus smithii* H.St.John, *Pandanus spurius* (Willd.) Miq., *Pandanus spurius* var. *weteringii* Martelli, *Pandanus stradbrookeensis* H.St.John, *Pandanus subaequalis* H.St.John, *Pandanus subcubicus* H.St.John, *Pandanus subhumerosus* H.St.John, *Pandanus subradiatus* H.St.John, *Pandanus suvaensis* (Martelli) H.St.John, *Pandanus taepa* (F.Br.) H.St.John, *Pandanus tahaaensis* H.St.John, *Pandanus tahitensis* Martelli,

Pandanus tahitensis var. *exiguus* J.W.Moore, *Pandanus tahitensis* var. *niueana* B.C.Stone, *Pandanus takaroaensis* H.St.John, *Pandanus tamaruensis* J.W.Moore, *Pandanus tapeinos* H.St.John, *Pandanus taravaiensis* H.St.John, *Pandanus tectorius* f. *convexus* B.C.Stone, *Pandanus tectorius* f. *laevis* (Warb.) Masam., *Pandanus tectorius* f. *philippinensis* Martelli, *Pandanus tectorius* var. *acutus* Kaneh., *Pandanus tectorius* var. *angaurensis* Kaneh., *Pandanus tectorius* var. *australianus* Martelli, *Pandanus tectorius* var. *brongniartii* Martelli, *Pandanus tectorius* var. *chamissonis* (Gaudich.) Martelli, *Pandanus tectorius* var. *cocosensis* B.C.Stone, *Pandanus tectorius* var. *douglasii* (Gaudich.) Martelli, *Pandanus tectorius* var. *drolletianus* (Martelli) B.C.Stone, *Pandanus tectorius* var. *exiguus* (J.W.Moore) B.C.Stone, *Pandanus tectorius* var. *fatyanion* Kaneh., *Pandanus tectorius* var. *ferreus* Y.Kimura, *Pandanus tectorius* var. *fragrans* Martelli, *Pandanus tectorius* var. *heronensis* (H.St.John) B.C.Stone, *Pandanus tectorius* var. *incrassatus* B.C.Stone, *Pandanus tectorius* var. *javanicus* Martelli, *Pandanus tectorius* var. *jonesii* F.Br., *Pandanus tectorius* var. *laevigatus* (Martelli) B.C.Stone, *Pandanus tectorius* var. *laevis* Warb., *Pandanus tectorius* var. *menziesii* (Gaudich.) Martelli, *Pandanus tectorius* var. *microcephalus* Martelli, *Pandanus tectorius* var. *novocaledonicus* Martelli, *Pandanus tectorius* var. *novoguineensis* Martelli, *Pandanus tectorius* var. *oahuensis* (Martelli) B.C.Stone, *Pandanus tectorius* var. *ongor* Kaneh., *Pandanus tectorius* var. *parksii* (Martelli) J.W.Moore, *Pandanus tectorius* var. *pedunculatus* (R.Br.) Domin, *Pandanus tectorius* var. *pulposus* Warb., *Pandanus tectorius* var. *samak* (Hassk.) Warb., *Pandanus tectorius* var. *sanderi* (Sander) B.C.Stone, *Pandanus tectorius* var. *sandvicensis* Warb., *Pandanus tectorius* var. *savaiensis* Martelli, *Pandanus tectorius* var. *spiralis* Martelli, *Pandanus tectorius* var. *stradbrookensis* (H.St.John) B.C.Stone, *Pandanus tectorius* var. *sumbavensis* Martelli, *Pandanus tectorius* var. *suringaensis* Martelli, *Pandanus tectorius* var. *taepa* F.Br., *Pandanus tectorius* var. *timorensis* Martelli, *Pandanus tectorius* var. *tubuaiensis* (Martelli) B.C.Stone, *Pandanus tectorius* var. *uapensis* F.Br., *Pandanus tectorius* var. *yorkensis* (H.St.John) B.C.Stone, *Pandanus tectorius* var. *zollingeri* Martelli, *Pandanus temehaniensis* J.W.Moore, *Pandanus terrireginae* H St.John, *Pandanus tessellatus* Martelli, *Pandanus tikeiensis* H.St.John, *Pandanus tima* H.St.John, *Pandanus timoeensis* H.St.John, *Pandanus tolotomensis* Glassman, *Pandanus tomilensis* Kaneh., *Pandanus tongaensis* H.St.John, *Pandanus trapaneus* H.St.John, *Pandanus tritosphaericus* H.St.John, *Pandanus trukensis* Kaneh., *Pandanus tubuaiensis* Martelli, *Pandanus tupaiensis* H.St.John, *Pandanus uea* H.St.John, *Pandanus utiyamae* Kaneh., *Pandanus vahitahiensis* H.St.John, *Pandanus vandra* H.St.John, *Pandanus vangeertii* auct., *Pandanus variegatus* Miq., *Pandanus veitchii* Mast., *Pandanus virginalis* H.St.John, *Pandanus viri* H.St.John, *Pandanus viridinsularis* H.St.John, *Pandanus volkensii* Kaneh., *Pandanus yorkensis* H.St.John, *Pandanus yunckeri* H.St.John.

The species is very polymorphous. Numerous varieties and forms exist which have been described under a host of names.

Family

Pandanaceae

Common/English Names

Beach Pandan, Hala, Hala Tree, Pandan, Pandanas, Pandanas Palm, Screw Pine, Tahitian Screwpine, Textile Screw-Pine, Thatch Screw-Pine, Veitch Screw-Pine.

Vernacular Names

Andaman Islands: Oro;
Arabic: Kadi;
Burmese: Tsatthapu;
Chinese: Lu Dou Shu;
Chuuk: Deipw, Fach, Far;
Colombia: Palma De Tornillo;
Fiji: Vadra, Voivoi;
French: Pandanus, Baquois, Vacouet, Vacquois;

Danish: Skruepalme;
Dutch: Pandan, Schroefpalm;
Eastonian: Lõhnav Pandan;
Finnish: Kairapalmu;
German: Pandanuspalme, Schraubenbaum, Schraubenpalme;
Guam: Kafu;
Hawai'i: Hala, Pū Hala;
Hebrew: Ha-Pandanus;
Hungarian: Illatos Pandanusz, Pandánusz Víz, Panpung Víz;
India: Ketakiphul, Keteki (Assamese), Keori, Ketaki, Ketaky, Keya (Bengali),Kewoda (Gujerati), Keora, Kevda, Kewda, Kewara, Kewra (Hindu), Kedige, Kedigĕ, Ketake, Ketakĕ, Tāḷe Hū, Tale Hu (Kannada), Kaida, Kaitha, Thala (Malayalam), Kenda, Ketaki, Ketakī, Keura, Kevḍā, Kewda, Kewra (Marathi), Kia (Oriya), Ketaka, Ketaki (Sanskrit), Talai, Tāḷai, Tazhai (Tamil), Mogheli, Mogil Mogli Chettu, Mugali (Telugu), Keora (Urdu);
Indonesia: Pandan Pudak, Pandan Pudak Duri, Pandan Pudak Emprit (Java), Pandan Pudak, Pandan Samk Laut (Sundanese);
Italian: Ananasso Della China, Panda Odorosa, Pandano;
Japanese: Adan, Shima Tako No Ki;
Kiribati: Te Kaina
Korean: A-Dan, Adan;
Kosrae: Mweng;
Lithuanian: Kvapusis Pandanas;
Malaysia: Mengkuang Laut, Mengkuang Duri, Mengkuang Layer, Padan Berdahan, Pandan Laut, Padan Darat, Pandan Todak, Pandan Pudak;
Marshall Islands: Bōb;
Nauruan: Épo;
Norwegian: Skrupalme;
Palauan: Ongor;
Philippines: Baroi (Bikol), Panhakad (Bisaya), Pandin (Ibanag), Panglan, Pandin (Iloko), Laha, Padan, Uhañgo (Ivatan), Alasas (Pampangan), Panglan (Sambali), Laha (Sulu), Pandan-Dagat, Alasas, Dasa, Pandin (Tagalog);
Pohnpei: Binu (Kapingamarangi Atoll), Hala (Nukuoro Atoll), Ajbwirōk, Anewetāk, Kipar (Pingelap Atoll), Kipar, Deipw (Sapwuahfik Atoll), Kipar (Mokil Atoll);
Portuguese: Pândano;
Russian: Pandanus Aromatnejshi;
Samoan: Fala, Lau Fala;
Spanish: Bacua, Pandan, Pandano;
Sri Lanka: Mudukeyiya (Sinhalese),
Swedish: Skruvpalm;
Thai: Karaket, Lamchiek;
Tokelau Islands: Fala;
Tongan: Fa, Fafa, Laufala, Falahola, Kukuvalu, Lou'Akau;
Tuvalu: Fala, Lau Fala;
Vanuatu: Pandanas (Bislama), Xer (Banks, Mosina), Feveo (Maexo, Sungawadaxa), Butsu, Vip (Penteote, Apma), Na- Barau (Tongoa, Namakura), No-Xixo (Torres, Hiu);
Vietnamese: Cay-Jua, Dúa Trô, Dứa Gỗ;
Yapese: Choy, Fach, Far.

Origin/Distribution

Wild distribution of *Pandanus tectorius* occurs in the exposed coastal headlands beaches and near-coastal forests of south Asia (south India, Sri Lanka), southeast Asia (Myanmar, Thailand, Malaysia, Indonesia, Philippines), eastward through Papua New Guinea and tropical northern Australia (the Port Macquarie area to Cape York and Torres Strait islands in Queensland) and extending throughout the Pacific islands, including Melanesia (Solomon Islands, Vanuatu, New Caledonia, and Fiji), Micronesia (Palau, Northern Marianas, Guam, Federated States of Micronesia, Marshall Islands, Kiribati, Tuvalu, and Nauru), and Polynesia (Wallis and Futuna, Tokelau, Samoa, American Samoa, Tonga, Niue, Cook Islands, French Polynesia, and Hawai'i). It is cultivated in the native areas as well as in Central India and Saudi Arabia.

Agroecology

Pandanus tectorius is a robust, hardy plant for tropical, sub-tropical and warm temperate maritime areas where frost is not a problem. It occurs usually from sea level to 20 m elevation in the Pacific islands and up to 60 m in southeast Asia, but has also been cultivated up to 600 m in

Hawaii. Its native habitats are found in strandline and coastal vegetation, including grassy or swampy woodlands, secondary forests, and scrub thickets developed on makatea (raised fossilized coralline limestone terraces). It commonly occurs on the margins of mangroves and swamps. *Pandanu*s also occurs as an understory tree in plantations and forests on atolls and larger islands either planted or naturalized. It thrives in areas with mean annual temperature of 24–28°C with a mean maximum temperature of 28–36°C and mean minimum temperature of 17–25°C. It can tolerate a minimum cool temperature of 12°C. The species is adapted to localities with summer bimodal and uniform rainfall distribution with mean annual rainfall of 1,500–400 mm but is tolerant to short drought periods of a few months. The tree is extremely resistant to strong winds, even tropical cyclonic and salt-laden winds. *Pandanu*s prefers a free-draining soil but will tolerate seasonally waterlogged conditions. In its native range, it is found on various littoral soils, especially sandy and rocky beaches, including raised coralline terraces and recent basalt from lava flows. It is adaptable to a wide diversity of soil types from light to heavy-textured soil types, including brackish/saline soils, light-coloured, infertile coralline atoll sands, sodic soils, alkaline sands, thin soils over limestone, and peaty swamps from pH of 4–9. It occurs in open, exposed areas in full sun but will tolerate intermediate partial shading.

Edible Plant Parts and Uses

Pandanus tectorius fruits are edible and it is reported to form a major source of staple food in Micronesia including the Marshall Islands, Federated States of Micronesia, and Kiribati. *Pandanu*s are also widely consumed on Tokelau and Tuvalu. In parts of Micronesia, chewing the ripe fruit segments (keys) is a common, pleasurable, and highly social activity. The ripe red segments of the fruit are also roasted and lower fibrous parts are eaten. The seeds found in woody cavities in each segment can also be roasted and eaten. Juice and jam may also be prepared from the fruit. Juice pressed from the fruits is acid-sweet with a pungent flavour (Miller et al. 1956). It is being produced commercially in the Marshall Islands. *Pandanus* can also be made into flour that is consumed in different ways, usually prepared as a drink. Pandanus pulp is preserved in several different ways. A paste, which is compared to dates in taste, texture, and appearance, is made by boiling and baking the ripe segments, followed by extracting, processing, and drying the pulp. Preserved *Pandanus* pulp mixed with coconut cream makes a tasty, sweet food item. Cultivars with large amounts of pulp are preferred, and the taste and flavour differs among cultivars. In Marshall Islands male *Pandanus* flowers are believed to have aphrodisiac properties and are eaten as masticatory.

Botany

Pandanus tectorius is a robust, small, laxly and widely branched, dioecious tree usually 5–6 m high but may reach 18 m (Plates 1 and 2). The trunk, 12–25 cm across, is often divided near the base, with or without prop roots, often with aerial roots descending from the branches. The bark is greyish- or reddish-brown and smooth. The stem and branches are ringed with distinct undulating leaf scars (Plate 3) sometimes with rows of prickles. Leaves are glaucous, deep green, sessile, linear 1–3 m long by 11–16 cm wide, acuminate, tapering with sheathing bases and spiny midribs and spiny or smooth margins (Plate 4). Leaves are crowded at the top of stems and arranged in three spirals in a screw-like arrangement which gives rise to the common name (Plate 2). Male and female flowers occur at the shoot apex in separate plants in inflorescence head. Male flowers are fragrant, tiny, white, pendant, arranged in racemes or branched in clusters in cylindric spikes, with large white showy bracts and numerous stamens on short filaments (stemonophore) with basifixed anthers longer than filament. Female flowers occur crowded together in globose, ellipsoid to ovoid heads, each carpel with 5–18 stigmas. Fruit an aggregate fruit, globose, subglobose, ellipsoid to ovoid, pineapple-like,

Pandanus tectorius

Plate 1 Tree habit with aerial prop roots

Plate 2 Leaves crowded at the apices of branches

Plate 3 Stem and branches with characteristic leaf scars

Plate 4 Cluster of strap-shaped leaves

Plate 5 Immature green fruiting head

large, 10–30 cm long by 8–20 cm across, comprising of tightly crowded, wedge-shaped fleshy drupes or phalanges (keys) (Plates 5, 6, 7 and 8). There are 1–15 carpels per phalange, and these are arranged either radially or in parallel rows. The central apical sinuses range from 1 to 28 mm deep. When ripe, the color of the basal section of the phalanges varies from pale yellow to dark yellow, orange, and orange/red. For intact fruiting heads the visible apical portion of the phalange is typically green with brown markings at

Plate 6 Ripe, orange fruiting head

Plate 8 Detached drupes (phalanges or keys)

maturity, turning yellow with age, after falling. The endocarp or pyrene (internal tissue surrounding the seeds) is dark reddish-brown, hard and bony. The mesocarp comprises apical and basal sections. The apical section formed in the apex of each carpel comprises an elongated cavity of aerenchyma cells consisting of a few longitudinal fibres and white membranes. The basal section is fibrous and fleshy, about 10–30 mm long. The seeds are obovoid, ellipsoid, or oblong; 6–20 mm long; red-brown and whitish and jelly-like inside.

Nutritive/Medicinal Properties

Nutrient value of the *Pandanus* fruit per 100 g edible portion was reported as: water 69 g, energy 595 kJ (144 kcal), protein 4.9 g, total fat total fat 8.3 g, available carbohydrate 12 g, dietary fibre 5.4 g, Na traces, K 89 g, Ca 13 g, Fe 0.7 g, total vitamin A equivalent 119 µg, B-carotene equivalent 714 µg, thiamine 0.14 mg, riboflavin 0.03 mg, niacin 1 mg, vitamin C 8 mg (Dignan et al. 1994). Nutrient composition of *Pandanus* paste was reported as: energy 1,350 kJ (326 kcal), protein 2.2 g, total fat 1.4 g, available carbohydrate 76 g, Ca 134 mg, Fe 5.7, total vitamin A equivalent 1,080 µg, thiamine 0.04 mg, riboflavin 0.06 mg, niacin 2 mg, and vitamin C 2 mg (Dignan et al. 1994).

A 100 g portion of the raw edible *Pandanus* kernel was found to comprise mainly water (25 g), energy 1,550 kJ (376 kcal), protein 15 g, total fat 30 g, available carbohydrate 11 g, dietary fibre

Plate 7 Fruiting head with some phalanges (drupes) removed

4.6 g, Na 229 mg, Ca 10 mg, Mg 154 mg, Fe traces, Zn, 2.4 mg, total vitamin A equivalent 302 μg, thiamine 0.38 mg, riboflavin 0.1 mg, niacin 4 mg, and vitamin E 1 mg (Dignan et al. 1994).

The edible cultivars contained a range of carotenoid levels (21–902 μg ß-carotene/100 g), with higher levels in cultivars having deeper yellow-orange coloured fruit; ten cultivars had significant levels that met estimated vitamin A requirements within normal consumption patterns (Englberger et al. 2006); yellow fruited cultivars contained low levels of carotenoids, while the orange fruited ones, which were also well liked as a food in the community, contained higher levels at maxima of 190 μg/100 g and 393 μg/100 g for α- carotene and β-carotene, respectively (Englberger et al. 2003a). The *Pandanus* cultivars contained substantial concentrations of provitamin A carotenoids (β- and α-carotene, β-cryptoxanthin, lutein, zeaxanthin, and lycopene) (110–370 μg β-carotene/100 g) and total carotenoids (990–5,200 μg/100 g) (Englberger et al. 2009). Pandanus paste contained 1,400 μg β-carotene/100 g, 5,620 μg total carotenoids/100 g, and 10 vitamins (including 10.8 mg/100 g vitamin C).

The fruit of these varieties has considerable potential for alleviating vitamin A deficiency in Micronesia (Englberger et al. 2003b). Carotenoid-rich foods protect against vitamin A deficiency, anaemia, and chronic disease, including cancer, heart disease and diabetes, which are serious problems in Micronesia. As carotenoid-rich food may protect against diabetes, heart disease, and cancer, the consumption of *Pandanus* may also alleviate these serious emerging problems of the Pacific.

The essential oil obtained from the ripe fruit of *Pandanus tectorius* was found to have large amounts of isopentenyl and dimethylallyl acetates and cinnamates (Vahirua-lechat et al. 1996).

Pandanus tectorius contained triterpenes and phytosterols which were found to exhibit antituberuclar activity (Tan et al. 2008). The chloroform extract of *Pandanus tectorius* Soland. var. *laevis* leaves afforded a new tirucallane-type triterpene, 24,24-dimethyl-5β-tirucall-9(11), 25-dien-3-one (1), squalene and a mixture of the phytosterols stigmasterol and β-sitosterol. Compound (1) inhibited the growth of *Mycobacterium tuberculosis* H37Rv with a MIC of 64 μg/mL, while squalene and the sterol mixture have MICs of 100 and 128 μg/mL, respectively.

Methanol leaf extract of *P. tectorius* (0.5 mg/ml) exhibited tryrosinase inhibitory activity of 26.7%. Tyrosinase inhibitors are important in preventing the enzymatic oxidation (browning) of food and to prevent inhibition of melanisation in animals (Masuda et al. 2005). Over the years tyrosinase inhibitors have attracted strong interest in both the food and cosmetic industries.

All parts of *P. tectorius* tree has been used in traditional medicine in south and southeast Asia and the Pacific islands (Adam et al. 2003; Del Rosario and Esguerra 2003; Burkill 1966; Kapoor 1989; Miller et al. 1956; Nguyen 1993; Stuart 2010; Thomson et al. 2006). *Pandanus* is a very important medicinal plant, with certain varieties sometimes preferred for particular treatments. In Hawaii, the fruits, male flowers, and aerial are used individually or in combination with other ingredients to treat a wide range of illnesses, including digestive and respiratory disorders. In Kiribati, *Pandanus* leaves are used in treatments for cold, flu, hepatitis, dysuria, asthma, boils and cancer, while the roots are used in a decoction to treat hemorrhoids. In Palau, the leaves are used to alleviate vomiting and the root is used to make a drink that alleviates stomach cramps. In Ayurveda, leaves have been used for leprosy, smallpox, scabies, syphilis and leukoderma; and for filarial disease, leucorrhea and as emmenagogue in traditional Indian systems. In India, the male inflorescence is distilled to make an essence which is medicinal. Anthers of male flowers are used for treating earaches and headaches. Poultice of fresh leaves mixed with oil were used for headaches and a leaf decoction used for arthritis and stomach spasms. Pulverized dried leaves have been used to facilitate wound healing. Poultice of mash of cabbage of plant, mixed with salt and juice of *Citrus microcarpa*, for abscesses. In the Philippines, the leaves are reported to be useful against leprosy, small pox, rabies and heart and brain diseases. In Peninsular Malaysia, it was reported that an infusion of the pith of the plant was taken as antidote for poisoning. In Papua

New Guinea, the bark was scraped into a solution of wild ginger leaf and the solution drank to sedate mental patients. A decoction of fresh or dried prop roots was drank as tea as a diuretic. Decoction of roots was believed to have aphrodisiac and cardiotonic properties. Roots were chewed to strengthen gums. A root decoction was used for arthritis and to prevent spontaneous abortion. Decoction of roots combined with sap of banana plant for urethral injections was used for a variety of urinary complaints. In Vietnam, the roots are regarded as diuretic and used to treat oliguria and other urinary complaints.

Other Uses

Besides the edible and medicinal uses of *Pandanus tectorius*, various parts of the tree have a myriad of other important uses (Burkill 1966; Little and Skolmen 1989; Meilleur et al. 1997; Thomson et al. 2006; Panda et al. 2009; Quisumbing 1978).

Tree

Pandanus is widely planted as an ornamental in home gardens, especially as a boundary hedge along front fences in the Pacific islands. When established on the seaward slopes and crests of frontal dunes, *Pandanus* helps to stabilise soil by binding the sand and prevent wind and water erosion. All parts of the tree may be used for production of compost, as well as in mulching and improving fertility and organic matter levels in sandy, coralline soils. In Kiribati, *Pandanus* leaves are used for mulching in giant swamp taro pits.

Roots

Dried aerial prop roots are used in making slats used for walls of houses and food cupboards, cordage, skipping ropes, paint brushes and basket handles. In Kiribati, fish traps are made out of the aerial roots and A black dye from the roots used in weaving.

Trunk/Wood

The stems provide timber that are used in house construction, making ladders in the Marshall Islands, The trunk of one variety is used to make the masts of traditional canoes. Erstwhile, the wood was used to make lances and batons. The wood has many uses that include as headrests, pillows, vases and as aid for string making and implement for extracting coconut cream. The trunk and branches are occasionally used as fuelwood where other fuelwood is scarce. The trunk provides a source of glue or caulking for canoes. *Pandanus* charcoal was employed in various mixtures to dye and waterproof canoes. The trunk of female trees have been used to make water pipes after removal of the soft core.

Leaves

In Hawaii, *Pandanus* leaves were the traditionally employed as the main material for making canoe sails. *Pandanus* leaves are used to weave traditional floor mats, baskets (including for ladies and to keep valuables), sugar bags, hats, fans, pillows as well as other plaited wares in Mauritius, Peninsular Malaysia and the Pacific islands. In the Philippines, the very strong and durable "sabutan hats" are made from young leaves, mats are also made from leaves. They are commonly used as thatching materials for walls and roofing in traditional houses. Young leaves, are reportedly used as fodder for domestic animals such as pigs and horses. In Kiribati and the Marshall Islands, the leaves are formed into a ball for use in a kicking game. In Micronesia, the leaves are used as wrappings for tobacco and cigarettes. In many Pacific islands *Pandanus* leaves are used to weave traditional items of attire, including mats for wearing around the waist in Tonga, as well as hats, fans and various types of baskets. Leaves, often neatly cut are used for making leis and garlands.

Flowers

In South Asia, Southeast Asia and Polynesia, the male flowers and preparations derived from them are used to scent clothes, coconut oil and incorporated into cosmetics, soaps, hair oils, and incense sticks. Male flowers picked from uncultivated *Pandanus* are used alone or in combination with other flowers to perfume coconut oil in Polynesia. In Hawai'i, the male flowers were used to scent tapa. The highly fragrant male flowers are widely used for decoration, and used in making garlands or leis. *Pandanus* constitutes one of the major bioresources of Ganjam coast, Orissa in India; used mainly in small scale perfume industry for aromatic compound extracted from the male inflorescences. *P. tectorius* produce aroma of high quality and yield, composed of primarily phenyl ethyl methyl ether (66.8–83%) and terpinen-4-ol (5–12%) along with a number of other phyto-chemical compounds. In Indonesia, the male flowers are used in religious and social ceremonies.

Fruit

In the northern Pacific, the discarded, dried keys are much appreciated as fuelwood for cooking because they are slow burning and therefore preferred for barbecues. The dried, exposed fibrous bristles of a dried key are used as a brush for decorating tapa, with the hard, woody outer end acting as a handle. A high quality, uniquely Pacific perfume is made from the aromatic fruits of selected traditional cultivated varieties in the Cook Islands. Fragrant fruits are also used in garlands and leis.

Comments

Pandanas tectorius is usually grown from seeds or large stem cuttings. Plants grown from cuttings fruit in 4–6 years, earlier than those from seeds.

Selected References

Abbott IA (1992) *La'au Hawai'i*: traditional Hawaiian use of plants. Bishop Museum Press, Honolulu

Adam IE, Balick MJ, Lee RA (eds) (2003) Useful plants of Pohnpei: a literature survey and database: a report of the Micronesia Ethnobotany Project. Institute of Economic Botany/New York Botanical Garden, New York, 431 pp

Backer CA, Bakhuizen van den Brink RC Jr (1968) Flora of Java (*Spermatophytes* only), vol 3. Wolters-Noordhoff, Groningen, 761 pp

Burkill IH (1966) A dictionary of the economic products of the Malay Peninsula. Revised reprint, 2 vols. Ministry of Agriculture and Co-operatives, Kuala Lumpur, Malaysia. Vol 1 (A–H) pp 1–1240, vol 2 (I–Z) pp 1241–2444.

Del Rosario AG, Esguerra NM (2003) Medicinal plants in Palau, vol 1. PCC-CRE Publication 28/03. Palau Community College, Koror, 51 pp

Dignan CA, Burlingame BA, Arthur JM, Quigley RJ, Milligan GC (1994) The Pacific Islands food composition tables. South Pacific Commission, Palmerston North, New Zealand, 147 pp

Englberger L, Aalbersberg W, Fitzgerald MH, Marks GC, Chand K (2003a) Provitamin A carotenoid content of different cultivars of edible pandanus fruit. J Food Comp Anal 16(2):237–247

Englberger L, Aalbersberg W, Schierle J, Marks GC, Fitzgerald MH, Muller F, Jekkein A, Alfred J, van der Velde N (2006) Carotenoid content of different edible *Pandanus* fruit cultivars of the Republic of the Marshall Islands. J Food Compos Anal 19(6–7):484–494

Englberger L, Fitzgerald MH, Marks GC (2003b) Pacific pandanus fruit: an ethnographic approach to understanding an overlooked source of provitamin A. Asia Pac J Clin Nutr 12:138–144

Englberger L, Schierle J, Hofmann P, Lorens A, Albert K, Levendusky A, Paul Y, Lickaneth E, Elymore A, Maddison M, de Brum I, Nemra J, Alfred J, Velde NV, Kraemer K (2009) Carotenoid and vitamin content of Micronesian atoll foods: Pandanus (*Pandanus tectorius*) and garlic pear (*Crataeva speciosa*) fruit. J Food Compos Anal 22(1):1.8

Foundation for Revitalisation of Local Health Traditions (2008) FRLHT database. http://envis.frlht.org

Govaerts R, Radcliffe-Smith A (2010) World checklist of Pandanaceae. The Board of Trustees of the Royal Botanic Gardens, Kew. Published on the Internet; http://www.kew.org/wcsp/. Retrieved 1 Feb 2010

Holdsworth DK, Pilokos B, Lambes PE (1983) Traditional plants of New Ireland, Papua New Guinea. Int J Crude Drug Res 21(4):161–168

Isaacs J (2002) Bush food: aboriginal food and herbal medicine. New Holland, Sydney

Kapoor LD (1989) CRC handbook of ayurvedic medicinal plants. CRC Press, Boca Raton, 424 pp

Little EL Jr, Skolmen RG (1989) Common forest trees of Hawaii (native and introduced), Agricultural handbook no 679. USDA, Washington, DC

Low T (1989) Bush Tucker – Australia's wild food harvest. Angus & Robertson, Sydney, 233 pp

Masuda T, Yamashita D, Takeda Y, Yonemori S (2005) Screening for tyrosinase inhibitors among extracts of seashore plants and identification of potent inhibitors from *Garcinia subelliptica*. Biosci Biotechnol Biochem 69(1):197–201

Meilleur BA, Maigret MB, Manshardt R (1997) Hala and Wauke in Hawai'i. Bishop Mus Bull Anthropol 7:1–55

Miller CD, Murai M, Pen F (1956) The use of Pandanus fruit as food in Micronesia. Pac Sci 10(1):3–16

Murai M, Pen F, Miller CD (1958) Some tropical South Pacific Island foods. Description, history, use, composition, and nutritive value. University of Hawaii Press, Honolulu, 159 pp

Ng LT, Yap SF (2003) *Pandanus* Parkinson. In: Lemmens RHMJ, Bunyapraphatsara N (eds) Plant resources of South-East Asia No. 12(3): medicinal and poisonous plants 3. PROSEA Foundation, Bogor, pp 321–323

Nguyen VD (1993) Medicinal plants of Vietnam, Cambodia and Laos. Mekong Printing, Santa Ana, 528 pp

Panda KK, Das AB, Panda BB (2009) Use and variation of *Pandanus tectorius* Parkinson (*P. fascicularis* Lam.) along the coastline of Orissa, India. Genet Res Crop Evol 56(5):629–637

Porcher MH et al (1995–2020) Searchable World Wide Web multilingual multiscript plant name database. The University of Melbourne, Melbourne, Australia. http://www.plantnames.unimelb.edu.au/Sorting/Frontpage.html

Quisumbing E (1978) Medicinal plants of the Philippines. Katha Publishing Co., Quezon City, 1262 pp

St. John H (1979) Revision of the genus *Pandanus* Stickman. Part 42. *Pandanus tectorius* Parkins. ex Z and *Pandanus odoratissimus* L.f. Pac Sci 33(4):395–401

Stone BC (1976) The Pandanaceae of the New Hebrides, with an essay on intraspecific variation in *Pandanus tectorius*. Kew Bull 31(1):47–70

Stone BC (1982) *Pandanus tectorius* Parkins. in Australia: a conservative view. Bot J Linn Soc 85:133–146

Stone BC (1992) *Pandanus* Parkinson. In: Verheij EWM, Coronel RE (eds) Plant resources of South-East Asia No. 2. Edible fruits and nuts. Prosea, Bogor, pp 240–243

Stuart GU (2010) Philippine alternative medicine. Manual of some Philippine medicinal plants. http://www.stuartxchange.org/OtherHerbals.html

Tan MA, Takayama H, Aimi N, Kitajima M, Franzblau SG, Nonato MG (2008) Antitubercular triterpenes and phytosterols from *Pandanus tectorius* Soland. var. *laevis*. J Nat Med 62(2):232–235

Thomson LAJ, Englberger L, Guarino L, Thaman RR, Elevitch C (2006) *Pandanus tectorius* (pandanus), ver. 1.1. In: Elevitch CR (ed) Species profiles for Pacific Island agroforestry. Permanent Agriculture Resources (PAR), Hōlualoa, Hawaii. http://www.traditionaltree.org

Vahirua-lechat I, Menut C, Roig B, Bessiere JM, Lamaty G (1996) Isoprene related esters, significant components of *Pandanus tectorius*. Phytochemistry 43(6):1277–1279

Walter A, Sam C (2002) Fruits of Oceania, ACIAR monograph no. 85. Australian Centre for International Agricultural Research, Canberra, 329 pp

Passiflora edulis

Scientific Name

Passiflora edulis Sims.

Synonyms

Passiflora edulis var. *pomifera* (M. Roem.) Mast., *Passiflora edulis* var. *rubricaulis* (Jacq.) Mast., *Passiflora edulis* var. *verrucifera* (Lindl.) Mast., *Passiflora diaden* Vell., *Passiflora gratissima* A. St.-Hil., *Passiflora incarnata* L., *Passiflora iodocarpa* Barb. Rodr., *Passiflora middletoniana* Paxton, *Passiflora minima* Blanco, *Passiflora pallidiflora* Bertol., *Passiflora picroderma* Barb. Rodr., *Passiflora pomifera* M. Roem., *Passiflora rigidula* J. Jacq., *Passiflora rubricaulis* J. Jacq., *Passiflora vernicosa* Barb. Rodr., *Passiflora verrucifera* Lindl.

Family

Passifloraceae

Common/English Names

Black Passionfruit, Granadilla, Maracuya, Passion Fruit, Purple Granadilla, Purple Passion Fruit, Purple Water Lemon, Red Passionfruit, Sweet Cup, Yellow Passionfruit.

Vernacular Names

Passiflora edulis: Sims f. *flavicarpa*
Afrikaans: Grenadella;
Brazil: Flor-Da-Paixão, Granadilho, Maracuj, Maracujá, Maracujá-Comum, Maracujá-De-Comer, Maracujá-De-Ponche, Maracujá-Do-Mato, Maracujá-Doce, Maracujá-Mirim, Maracujá-Peroba, Maracujá-Preto, Maracujá-Redondo (Portuguese);
Chinese: Ji Dan Guo;
Colombia: Curuba, Curuba Redonda, Gulupa;
Cook Islands: Ka'Atene Papa'Ā, Katingapapa'Ā, Pārapōutini Papa'Ā (Maori);
Cuba: Ceibey;
Czech: Mučenka Jedlá;
Danish: Granatblomst, Gul Passionsblomst, Passionsfrugt;
Dutch: Eetbare Passiebloem, Paarse-Passievrucht, Passiebloem, Passievrucht, Passie Vrucht;
Eastonian: Purpur-Kannatuslill;
Fijian: Qarandila;
Finnish: Kärsimyshedelmä, Passiohedelmä;
French: Grenadille, Fruit De La Passion, Grenadille Pourpre, Maracaju Pourpre, Maracudja, Passiflore Comestible, Gouzou, Pomme-Liane Violette;
French Guiana: Couzou;
German: Maracuja, Purpurgranadille Purpur-Granadille, Granadilla, Passionsfrucht;
Hawaiian: Liliko'I;
Hungarian: Golgotavirág Gyümölcse;
India: Louki;

Indonesia: Buah Negeri, Markisa, Pasi, Konyal;
Italian: Granadiglia, Frutto Della Passione, Passiflora Commestibile, Granatiglia;
Japanese: Kudamonotokeiso;
Kenya: Matunda;
Laos: Linmangkon;
Malaysia: Buah Susu, Markisa;
Mexico: Granadita De China;
Palauan: Kudamono;
Philippines: Maraflora, Pasionaria;
Pohnpeian: Pompom;
Polish: Meczennica Jadalna;
Portuguese: Maracujá, Maracujá-Pequeno, Maracujá-Roxo, Maracujá-Suspiro;
Puerto Rico: Fruta De La Pasión, Parcha;
Samoan: Pasio;
Slovenian: Granadilja, Marakuja, Pasijonka;
Spanish: Granadilla Morada, Maracuyá, Maracuyá Púrpura, Granadilla China, Parchita Maracuyá;
Swedish: Passionsfrukt;
Taiwan: Xi Fan Lian;
Thailand: Lin Mang Kon Saowarot;
Tongan: Vaine Tonga;
Venezuela: Parchita, Fruta De La Pasión, Parcha;
Vietnam: Chum Bap;
West Indies: Couzou;
Passiflora edulis: Sims f. *edulis*
Dutch: Paarse-Passievrucht, Passievrucht;
English: Black Passionfruit, Passion Fruit, Purple Granadilla, Purple Passionfruit, Red Passionfruit;
Eastonian: Purpur-Kannatuslill;
French: Fruit De La Passion, Grenadille, Grenadille Pourpre, Maracaju Pourpre, Pomme-Liane Violette;
German: Eßbare, Passionsfrucht, Purpurgranadille, Purpur-Granadilla;
Indonesia: Markisa;
Malaysia: Markisah;
Portuguese Maracujá;
Spanish: Fruta De La Passion, Granada De Castilla, Granadilla, Maracu.
Passiflora edulis Sims f. flavicarpa O. Deg.
Dutch: Gele-Passievrucht;
English: Brown-Seeded Passionfruit, Clock Flower, Clock Plant, Golden Passionfruit, Malaysian Passionfruit, Philippines Passionfruit, Thai Passionfruit, Yellow Granadilla, Yellow Passionfruit;
French: Grenadille À Fruit Jaune, Pomme Liane Jaune;
German: Gelbe Grenadille;
Indonesia: Konyal;
Portuguese: Maracujá-Amarelo, Maracujá-Azedo;
Spanish: Maracuyá Amarillo, Parcha Amarilla;
Suriname: Maracuja.

Origin/Distribution

Passiflora edulis is native to south America. There are two distinct forms: f. *edulis* with purple fruits and f. *flavicarpa* Degener with larger yellow fruits; f. *flavicarpa* occurs in the tropical lowlands, f. *edulis* occurs in cooler regions at higher altitudes. The purple passion fruit is indigenous from southern Brazil through Paraguay to northern Argentina. It has been stated that the yellow form is of unknown origin, or perhaps native to the Amazon region of Brazil, or is a hybrid between *P. edulis* and *P. ligularis*. Cultivated and escaped worldwide in tropical and subtropical areas, purple passionfruit is widely grown in India, New Zealand, the Caribbean, Brazil, Ecuador, California, southern Florida, Hawaii, Australia, East Africa, Israel, South Africa, and in Fujian, Guangdong, Yunnan (China) and Taiwan. It is also grown in the cooler highlands in tropical areas of southeast Asia and Papua New Guinea. The yellow form is grown pan-tropically usually in the warmer lowlands.

Agroecology

The purple passionfruit is subtropical in its climatic requirement and thrive at altitudes above 1,100–2,000 m. It prefers cool, frost free climate but does tolerate mild frost but does not flower below 1,000 m and perform poorly in intense summer heat. It will grow in areas with 900 mm annual precipitation that is well distributed throughout the year.

The yellow passionfruit is near tropical in its requirement, is intolerant of frost and grows best in the warmer lowlands from near seal level to 800 m. It grows well in areas with more than 1,500–2,500 mm annual rainfall. Both types need protection from strong winds.

Both types are adaptable to a wide range of soils, including clayey and sandy soils, the former requiring drainage and the latter needs manuring. Both do best in a free-draining, friable, moist and fertile soil rich in organic matter.

Edible Plant Parts and Uses

The ripe aromatic fruit is eaten fresh and the best way is to cut the fruit in halves and scoop out the aril (pulp) with seeds intact with a spoon or eaten with ice-cream, yoghurt and in fruit salads. The arils are also processed into juice, nectar, syrup, cordial and carbonated drinks, sherbets, jams, jellies or used in pastries and cakes or to flavour ice-cream and yoghurt. Passion fruit juice or syrup is an essential ingredient of some cocktails, such as the hurricane and the Peruvian maracuya sour. In Brazil, passion fruit is made into a common dessert passion fruit "mousse", and the seeds are routinely used to decorate the tops of certain cakes. Passion fruit juice is also widely consumed. In the Dominican Republic, passionfruit is used to make juice, jams, and the chinola flavoured syrup which is used on shaved ice. The fruit is also commonly eaten raw sprinkled with sugar. In Australia, passionfruit is sold commercially fresh and canned. It is usually eaten fresh on its own or in fruit salads, and used as topping for desserts like pavlova (a meringue cake), cheesecake, and vanilla slice. The juice is also made into mixed fresh fruit juices, nectar, cordial and carbonated drinks. In South Africa, passion fruit is used to flavour yogurt. It is also used to flavour soft drinks such as Schweppes Sparkling Granadilla and numerous cordial drinks. In Hawaii, passion fruit flavoured syrup is a popular topping for shave ice. Ice cream and mochi are also flavoured with passionfruit as well as many other desserts such as cookies, cakes, and ice cream. Passionfruit is also processed into jam, jelly, as well as a butter. In Indonesia, passion fruit is consumed fresh or the juice is strained and cook with sugar to form a thick syrup which is diluted with water and served with ice-cubes. The yellow variety is used usually for juice processing, while the purple variety is sold in fresh fruit markets.

Botany

A robust, glabrous, herbaceous perennial climber, woody at base with slender, striate green stem armed with rameal, axillary tendrils. Leaves alternate, glossy green, membranous, 6–13 cm by 8–13 cm, deeply trilobed, middle lobe ovate, lateral lobes ovate-oblong, margin glandular-serrate, apex acute, acuminate, distinctly 3-nerved with prominent laterals, stipules small, linear-lanceolate to 10 mm long. Flowers solitary, axillary, fragrant, whitish or pale violet, 4–7 cm across, on 5–7 cm long trigonous pedicel subtending 3 large, foliaceous bracts at the top of the pedicel, hypanthium 1 × 1.2 cm. Calyx, tubular-campanulate, whitish green with 5 lobes patent or reflexed. Corolla 5, white inserted at the margin of the calyx tube. Corona at the fringe of the calyx tube, 5-seriate, two outer row with long radiating, curly white tipped with purple base threads and the 3 inner rows with shorter white-puple threads. Stamens 5 with large yellowish anther, ovary ovoid, yellowish with trifid style and reniform pale green stigma. Fruit globose (4–6 cm) or ovoid (6–10 cm by 5–7 cm), smooth, glabrous, glossy green and dotted when young turning to yellow (yellow form) or dark purple (purple form) when mature with a tough rind. Seeds numerous (up to 250) small, hard, dark brown or black, pitted, ovoid seeds in the fruit cavity. Each seed enclosed in a juicy, orange, sweet to acid-sweet, aromatic aril (pulp). There are also hybrid cultivars with pink and maroon-coloured fruits.

P. edulis **f.** *edulis* (Plates 1, 2, 3, 4 and 5) – purple passionfruit has purple or reddish fruit, fruits are usually globose to subglobose, 4–6 cm diameter, with green tendrils and leaves. Purple passion fruit is self-fertile, but pollination is best under humid conditions.

P. edulis **f.** *flavicarpa* **Degener** (Plates 6, 7, 8 and 9) – yellow passionfruit has canary yellow and larger globose to ovoid fruit, 6–10 cm by 5–7 cm, reddish purple tinged tendrils and leaves, more showy flowers with deeper violet corona and more vigorous growth. The flowers of the yellow form are perfect but self-sterile and require cross pollination.

Plate 1 Immature purple passionfruit and trilobed leaves

Plate 4 Red passionfruit

Plate 2 Purple passionfruit on sale in a local market

Plate 5 Red passionfruit halved showing the pulp and seeds

Plate 3 Ripe purple passionfruit

Plate 6 Immature yellow passionfruit and leaves

Plate 7 Close-view of immature yellow passionfruit

Plate 8 Ripening yellow passionfruit

Plate 9 Ripe yellow passionfruit

Nutritive/Medicinal Properties

Nutrient value per 100 g edible portion of the pulp and seeds of purple skinned passionfruit, *Passiflora edulis* f. *edulis* comprised the following (Wills et al. 1986): energy 304 KJ, moisture 74.4 g, nitrogen 0.48 g, protein 3.0 g, fat 0.3 g, ash 0.6 g, fructose 1.9 g, glucose 2.3 g, sucrose 1.5 g, total sugars 5.7 g, available carbohydrate 5.7 g, total dietary fibre 13.9 g; minerals – calcium 10 mg, iron 0.6 mg, magnesium 28 mg, potassium 200 mg, sodium 19 mg, zinc 0.8 mg; vitamins – thiamin 0.03 mg, riboflavin 0.14 mg, niacin 2.5 mg, niacin derived from tryptophan or protein 0.5 mg, niacin equivalents 3.0 mg, vitamin c 18 mg, alpha carotene 410 μg, beta carotene 360 μg, cryptoxanthin 370 μg, beta carotene equivalents 750 μg, retinol equivalents 125 μg; organic acids – citric acid 3.5 g and malic acid 0.5 g.

Another analysis conducted in USA reported the following nutrient value of purple passion fruit, *Passiflora edulis* f. *edulis*, per 100 g edible portion minus 48% shell refuse (USDA 2011): water 72.93 g, energy 97 kcal (406 kJ), protein 2.20 g, total lipid 0.70 g, ash 0.8 g, carbohydrate 23.38 g, total dietary fibre 10.4 g, total sugars 11.20 g, Ca 12 mg, Fe 1.60 mg, Mg 29 mg, P 68 mg, K 348 mg, Na 28 mg, Zn 0.10 mg, Cu 0.08 mg, Se 0.6 μg, vitamin C 30 mg, thiamine 0 mg, riboflavin 0.130 mg, niacin 1.5 mg, vitamin B-6 0.1 mg, total folate 14 μg, total chloline 7.6 mg, vitamin A, 64 μg RAE, vitamin A 1271 IU, β-carotene 743 μg, β-cryptoxanthin 41 μg, vitamin E (α-tocopherol) 0.02 mg, vitamin K (phylloquinone) 0.7 μg, total saturated fatty acids 0.059 g, 16:0 (palmitic) 0.045 g, 18:9 (stearic) 0.014 g; total monounsaturated fatty acids 0.086 g, 18:1 undifferentiated (oleic) 0.086 g; total polyunsaturated fatty acids 0.411 g, 18:2 undifferentiated (linoleic) 0.410 and 18:3 undifferentiated (linolenic) 0.001 g.

Fresh passion fruit is known to be high in provitamin A (α-, β-carotenes), potassium, and dietary fibre and also has moderate levels of vitamin C and niacin.

Yellow passion fruit, *Passiflora edulis* f. *flavicarpa*, juice was reported to have the following nutrient composition per 100 g (USDA 2011): water 84.21 g, energy 60 kcal (251 kJ), protein 0.67 g, total lipid 0.18 g, ash 0.49 g, carbohydrate 14.45 g, total dietary fibre 0.2 g, total sugars 14.25 g, Ca 4 mg, Fe 0.36 mg, Mg 17 mg, P 25 mg, K 278 mg, Na 6 mg, Zn 0.06 mg, Cu 0.050 mg, Se 0.1 μg, vitamin C 18.2 mg, thiamine 0 mg, riboflavin 0.101 mg, niacin 2.240 mg, vitamin

B-6 0.060 mg, total folate 8 μg, total choline 4 mg, vitamin A, 47 μg RAE, vitamin A 943 IU, β-carotene 525 μg, α-carotene 35 μg, β-cryptoxanthin 47 μg, vitamin E (α-tocopherol) 0.01 mg, vitamin K (phylloquinone) 0.4 μg, total saturated fatty acids 0.015 g, 16:0 (palmitic) 0.012 g, 18:9 (stearic) 0.004 g; total monounsaturated fatty acids 0.022 g, 18:1 undifferentiated (oleic) 0.022 g; total polyunsaturated fatty acids 0.106 g and 18:2 undifferentiated (linoleic) 0.105 g.

Other Fruit Phytochemicals

Yellow passion fruit rind was found to be a potentially rich source of naturally low-methoxyl pectin (Yapo and Koffi 2006). The extracted pectins were found to be rich in an hydrogalacturonic acid and had a low degree of methyl esterification, low contents of acetyl groups, neutral sugar and proteinaceous material. Their gelling ability and viscoelastic properties were comparable to those of a commercial citrus low-methoxyl pectin. The extraction of safe pectin products with good functional properties from yellow passion fruit by-product, was developed using two natural acid extractants, namely, pure lemon juice and citric acid solvent (Yapo 2009). The results showed that both of them solubilise, from cell wall material, pectins characterised by high galacturonic acid content (64–78% w/w), degree of esterification (52–73), viscosity-average molecular weight (70–95 kDa) and capable of forming gels in the presence of high soluble solids (sucrose) content and acid.

The total dietary fibre in alcohol-insoluble material from yellow passion fruit (YPF) rind was found to be >73% dry matter of which insoluble dietary fibre accounted for >60% (w/w) (Yapo and Koffi 2008). Non-starchy polysaccharides were the predominant components (approximately 70%, w/w), of which cellulose appeared to be the main fraction. The water holding and oil holding capacities of the fibre-rich material were >3 g of water/g of fibre and >4 g of oil/g of fibre, respectively.

A total of 51 volatile components were identified in yellow passion fruit essence including 20 alcohols, 20 esters, 2 aldehydes, 4 ketones, 1 acetal, 1 furan and 1 hydrocarbon (Jordan et al. 2000). Alcohols were the major components comprising 56.94% of the total components. The most abundant compounds identified were linalool (15.3%) (characterised by its floral, citrus and lemon flavour),1-octanol (11.51%) (fatty, citrus and green), 1-hexanol (9.03%) (herbaceous, fragrant and sweet) and α-terpineol (fragrant, floral and lilac). Other alcohols above 1% include 3-methyl-1-butanol (2.80%), geraniol (2.44%), benzyl alcohol (1.6%) undecanol (1.45%), cis-3-Hexen-1-ol (1.17%) and terpinene-4-ol (1.05%). A new alcohol identified was 3-methyl-2-buten-1-ol. The next most abundant compounds were the esters comprising 30.38% of the total volatile compounds; ethyl hexanoate (11.11%), ethyl butanoate (9.44%), ethyl benzoate (3%) and phenylmethyl acetate (1.44%) were the predominant esters. The two aldehydes made up 4.59% of the total volatile components and comprised benzene acetaldehyde (0.22%) and benzaldehyde (4.57%). Ketones comprised 3.3% of the total with cyclopentanone (1.5%) and 3-hydroxy-2-butanone (1%) predominating. The acids made up 1.2% of the total volatiles – hexanoic acid (0.9%) and octanoic acid (0.4%). Of the miscellaneous components, the acetal 1,1-Diethoxy ethane (1.06%) was important for its contribution to a strong, tart, fruity, refreshing green odour. The furan identified was cis-linalool oxide (0.62%) and the hydrocarbon methyl cyclopentane (1.77%).

(+)-cis-2-methyl-4-propyl-1,3-oxathiane, the main component responsible of the passion fruit aroma was enantioselective synthesised (Scafato et al. 2009). Together with 3-mercaptohexan-1-ol and 3-mercaptohexyl acetate, already known to contribute to the aroma of passion fruit (*Passiflora edulis*), 3-mercapto-3-methylbutan-1-ol and 3-mercapto-3-methylbutyl acetate were identified for the first time in passion fruit (Tominaga and Dubourdieu 2000). The precursor of 3-mercaptohexan-1-ol as S-(3-hexan-1-ol)-L-cysteine, in the form of trimethylsilylated derivatives from the juice of this fruit was also identified. The presence of free and combined forms of these volatile thiols in this fruit was also reported. A thioesterase activity towards volatile organosulfur compounds (VOSCs) was identified in ripening purple passion fruit (Tapp et al. 2008). VOSCs are high impact aroma chemicals characteristic of tropical fruits which are active as both free thiols and the respective thioesters. The major thioesterase in the fruit was found to be a wall-bound protein in the mesocarp. The results suggested that cell wall hydrolases in tropical fruit may have additional useful roles in biotransforming VOSCs.

About 150 volatiles were identified in purple passion fruit with esters and terpenoids being the most abundant classes of flavour compounds (Whitfield and Last 1986). Among the terpenoids numerous monoterpenes were found as degradation products of non-volatile precursor compounds such as hydroxylated linalool derivatives as well as monoterpene glycosides. The carotenoid-derived C13 nortepenoids have been suggested to be derived by similar pathways. Juice of purple passion fruit contains a wide range of terpenoids (Winterhalter 1990). In addition to the main component linalool and the other monoterpenoids, the following C_{13} nortepenoid aglycones were identified: 4-hydroxy-β-ionol; 4-oxo-β-ionol; 4-hydroxy-7,8-dihydro-β-ionol; 4-oxo-7,8-dihydro-β-ionol; 3-oxo-α-ionol; isomeric 3-oxo retro-α-ionols, 3-oxo-7,8-dihydro-α-ionol; 3-hydroxy-1,1,6 – trimethyl-1,2,3,4-tetrahydronaphthalene; vomifoliol and dehydrovomifoliol. Other strong odoriferous flavour compound in purple passion fruit included C13 norisoprenoids i.e. 4 isomers of megastigma-4,6,8-triene (Whitfield and Last 1986) and isomeric edulan I and II (Whitfield et al. 1973; Whitfield and Stanley 1977) key flavour components of purple passion fruit juice (Winterhalter 1990). The precursor to the isomeric edulans in purple passion fruit was identified as 3-hydroxyl-retro α-ionol (Herderich and Winterhalter 1991). The novel C_{13}-glucoside β-ionyl-β-D-glucopyranoside was identified in purple passionfruit as a genuine precursor of isomeric megastigma-4,6,8-trienes (Herderich et al. 1993).

The majority of the volatiles components in purple-skinned passionfruit identified were esters derived largely from combinations of the alkan-1-ols ($C_{1,2,4,6,8}$) the alkan-2-ols ($C_{5,7}$),(Z)-and (E)-hex-3-en-1-ol,(2)-hex-4-en-1-ol,and benzyl alcohol, with acids of even carbon number (acetic, butanoic, hexanoic, octanoic, hex-3-enoic, oct-3-enoic, and 3-hydroxyhexanoic) (Murray et al. 1972). The non-ester constituents included many of the above free alcohols, acetaldehyde, alkan-2-ones ($C_{3,5,7,9,11}$), monoterpenes ((E)-β-ocimene, 1,8-cineole, linalool, α-terpineol, citronellol, geraniol, citronellyl acetate, (Z) and(E) five-membered ring linalool oxides), and 1,1,6-trimethyl- 1,2-dihydronaphthalene, β-ionone, γ-hexanolac-tone, γ-octanolactone, and the lactone of 2-hydroxy-2,6,6-trimethylcyclohexylidene-acetic acid. Preliminary odour assessment of the constituents indicated volatile passionfruit flavour to be complex and made up of many components, especially certain esters.

Passiflora edulis was reported to be rich in glycosides and phenols. From the fruit of *P. edulis*, 6-O-α-L-arabinopyranosyl-β-D-glucopyranosides of linalool, benzyl alcohol and 3-methylbut-2-en-1-ol were identified (Chassagne et al. 1996b) Cyanogenic β-rutinoside ((R)-mandelonitrile α-L-rhamnopyranosyl-β-D-glucopyranoside) was isolated (Chassagne and Crouzet 1998); β-D-glucopyranoside and 6-O-α-L-rhamnopyranosyl-β-D-glucopyranoside of methyl salicylate and the β-D-glucopyranoside of eugenol were characterized in purple passion fruit (Chassagne et al. 1997). Prunasin was found to be the most important cyanogenic glycoside in peel (285 mg/kg for *P. edulis* f. *flavicarpa*), whereas amygdalin (31 mg/kg for *P. edulis*) and the two compounds tentatively identified as mandelonitrile rhamnopyranosyl β-D-glucopyranosides were mostly found in the juice (99 mg/kg for *P. edulis* f. *flavicarpa*). Different amounts of sambunigrin were found in the juice and the peel (from 0.4 mg/kg in *P. edulis* juice to 15.5 mg/kg in *P. edulis* f. *flavicarpa* peel) (Chassagne et al. 1996a). The glycoside 2,5-Dimethyl-4-hydroxy-3-(2H) furanone (furaneol) was identified for the first time in bound form in purple (*P. edulis*) and yellow passion fruit (*P. edulis* f. *flavicarpa*) (Chassagne et al. 1999). Several terpene diols: 2,6-dimethyl-1,8-octanediol, (E)- and (Z)-2,6-dimethylocta-2,7-diene-1,6-diol, 2,6-dimethylocta-3,7-dien-2,6-diol and 2,6-dimethylocta-1,7-dien-3,6-diol were identified in both purple and yellow passion fruit

(Chassagne et al. 1999). α-ionol derivatives oxygenated in position 3 seemed to be characteristic of purple passion fruit whereas β-ionol compounds oxygenated in position 3 were the major norisoprenoids identified as the aglycone in yellow passion fruit. Significant concentrations of bound aromatic alcohols were found in purple and yellow passion fruit whereas phenolics could be considered as characteristic of purple varieties. C-glycosylflavonoids identified as isoorientin, vicenin-2, spinosin, and 6,8-di-C-glycosylchrysin were also present in the yellow passion fruit pericarp (Sena et al. 2009). Two new ionones I and II were isolated for from purple passion fruit (Naf et al. 1977).

Anthocyanins identified in the passion fruit rind included cyanidin 3-glucoside (97%), small amounts of cyanidin 3-(6″-malonylglucoside) (2%) and pelargonidin 3-glucoside (1%) (Kidoy et al. 1997) and cyanidin-3-O-galactopyranoside (Chang and Su 1998). The following 13 carotenoids from yellow passion fruit (*Passiflora edulis*) were identified: phytoene, phytofluene, ζ-carotene (principal carotenoid), neurosporene, β-carotene, lycopene, prolycopene, monoepoxy-β-carotene, β-cryptoxanthin, β-citraurin, antheraxanthin, violaxanthin, and neoxanthin (Mercadente et al. 1998) and from passion fruit α-carotene, γ-carotene, β-cryptoxanthin, β-apo-10'-carotenol (Goday and Rodriguez 1994). Other volatile compounds identified in the fruit rind of *P. edulis* included 2-tridecanone, (62.9%), (9Z)-octadecenoic acid (16.6%), 2-pentadecanone, (6.2%), hexadecanoic acid (3.2%), 2-tridecanol (2.1%), octadecanoic acid (2%) and caryophyllene oxide (2%) (Arriaza et al. 1997).

Proximate analysis showed that 'Tainung No. 1' passion fruit seeds had a high amount of protein (10.8%) and were rich in oil (23.40%) (Liu et al. 2008). The seeds were found to be a good source of minerals. They contained considerable amounts of sodium (2.980 mg/g), magnesium (1.540 mg/g), potassium (0.850 mg/g), and calcium (0.540 mg/g). The passion fruit seeds contained the 17 amino acids found naturally in plant protein (except for tryptophan which was not analyzed). The essential amino acids accounted for 34% of the 17 amino acids. The amino acid score of passion fruit seeds protein was 74 and the limiting amino acids were methionine and cystine. The oil extracted by solvent and supercritical dioxide carbon was liquid at room temperature and the color was golden-orange. The specific gravity of the oil was about 0.917. Fatty acid composition of the seed oil indicated that the oil contained two essential fatty acids (linoleic acid and linolenic acid), but the content of linoleic acid (72.69%) was by far greater than that of linolenic acid (0.26%). The present analytical results indicated that passion fruit seed to be a potentially valuable non-conventional source for high-quality oil.

Phytochemicals in Leaf and Stem

Phytochemical analysis revealed the presence of carbohydrates, glycosides, flavonoids, resins and balsams, alkaloids, and phenolic compounds in all the plant parts of *P. edulis* (Akanbi et al. 2011). Tannins were present in the leaf and fruit extracts but absent in the stem whereas saponins were present in the leaf and stem but not detected in the fruit sample. Terpenes were not detected in any part of the plant. C-glycosyl flavonoids namely orientin, isoorientin, vitexin and isovitexin were found in the leaves and pericarp of *P. edulis* var. *flavicarpa*, and *P. edulis* var. *edulis* (Zucolotto et al. 2011).

Glucosides were also found in the leaves and stem. The new cyanogenic glycosides (2R)–β-D-allopyranosyloxy-2-phenylacetonitrile and (2S)–β-D-allopyranosyloxy-2-phenylacetonitrile plus (2R)–prunasin and (2S)-sambunigrin the major cyanogens of the fruits and the alloside of benzyl alcohol were found in leaf and stem material of *P. edulis* (Seigler et al. 2002). Leaves also contained benzylic β-D-allopyranosides 1 and 2, representatives of a rare class of natural glycosides with D-allose as the only sugar constituent (Christensen and Jaroszewski 2001). The glycoside 1 was the first known cyanogenic glycoside containing a sugar different from D-glucose attached directly to the cyanohydrin center. The methanol extract of air dried leaves, yielded a cyclopropane triterpine glycoside named Passiflorine (22R), (24S)-22,31-epoxy-24-methyl-1α,3β,24,31-tetrahydroxy-9,19-*cyclo*-9β-lanostan-28-oic acid β-d-glucosyl ester (Bombardelli et al. 1975). C-glucosylflavones isolated from the butanolic fraction of an aqueous extract of *P. edulis* var. *flavicarpa* leaves included isoorientin, vicenin-2 and spinosin (Zucolotto et al. 2009). *P. edulis* also contained flavonoid glycosides, viz., luteolin-6-C-chinovoside, luteolin-6-C-fucoside (Mareck et al. 1991); cyclopentenoid cyanohydrin glycosides passicapsin and passibiflorin (Olafsdottir et al. 1989); three cyanogenic glycosides, passicoriacin, epicoriacin and epitetraphyllin B, the structure of which were reassigned as epivalkenin, taraktophyllin and volkenin respectively, based on a reinterpretation of their spectral data (Seigler and Spencer 1989). Passionfruit leaves also contained C-deoxyhexosyl flavonoids which had antioxidant activity (Ferreres et al. 2007). Sixteen apigenin or luteolin derivatives were characterized, which included four mono-C-glycosyl, eight O-glycosyl-C-glycosyl, and four O-glycosyl derivatives. With the exceptions of C-hexosyl luteolin and C-hexosyl apigenin, all the compounds exhibited a deoxyhexose moiety. The flavanoids orientin and isoorientin were also detected in the leaves (Pereira et al. 2004).

Four cycloartane triterpenoids and six related saponins were reported from *Passiflora edulis* (Yoshikawa et al. 2000a). Four cycloartane triterpenes, cyclopassifloic acids A, B, C and D, and six related saponins, cyclopassiflosides I, II, III, IV, V and VI, were isolated from the leaves and stems of *Passiflora edulis*. Cyclopassifloic acids A-D were assigned as 22(R), 24(S)-1α,3β,22,24,31-pentahydroxy-24-methylcycloartan -28-oic acid; 24(S)-1α,3β,24, 31-tetrahydroxy-24-methylcycloartan-28-oic acid; 20(S),24(S)-1α, 3β,21, 24,31-pentahydroxy-24-methylcycloartan-28-oic acid; and 22(R)-1α,3β,22-trihydroxy-24-oxocycloartan-28-oic acid, respectively. Cyclopassiflosides I–VI, in turn, were established as the 28-O-βa-D-glucopyranosides of cyclopassifloic acids A–D. Finally, cyclopassiflosides III and V were demonstrated as the 28, 31-bis-O-β-D-glucopyranosides of cyclopassifloic acids B and C, respectively. Also obtained were the known compounds passiflorin and passifloric acid. Three new cycloartane triterpenes, cyclopassifloic acids E, F and G, and their saponins, cyclopassiflosides VII, VIII, IX, X and XI, were isolated from the leaf and stem parts of *Passiflora edulis* (Yoshikawa et al. 2000b).

P. edulis leaves were found to contain alkaloids. The alkaloid identified included harmine, harmalol, harmaline and harman, a β-carboline alkaloid, with highest concentration (0.12 mg%) in the leaves (Slaytor and McFarlane 1968; Lutomski and Malek 1975; Lutomski et al. 1975). *P. edulis* aqueous extracts, also contained coumarin, long-chain fatty acids and lactones (Khanh et al. 2006).

Six compounds identified as luteolin 6-C-β-D-glucopyranoside, luteolin 6-C-β-D-chinovoside, luteolin 6-C-β-L-fucoside, apigenin 8-C-β-D-glucopyranoside, apigenin-6-C-β-D-glucopyrano-4'-O-α-L-rhamnopyranoside and 5,8-epidioxyergosta-6, 22-dien-3ol were isolated from the stem of *P. edulis* f. *flavicarpa* (Zhou et al. 2009).

Scientific studies have shown *P. edulis* to have numerous pharmacological properties that supported all its traditional medicinal uses.

Antioxidant Activity

The leaves of *Passiflora alata* and *Passiflora edulis*, traditionally used in American countries to treat both anxiety and nervousness by folk medicine, were found to be rich in polyphenols, which had been reported as natural antioxidants (Rudnicki et al. 2007). Studies verified the antioxidant activities of *P. edulis* and *P. alata* hydroalcoholic leaf extracts in in-vitro and ex-vivo assays. *P. alata* showed a higher total reactive antioxidant potential than did *P. edulis*. The antioxidant activities of both extracts were significantly correlated with polyphenol contents. In addition, both extracts attenuated ex-vivo iron-induced cell death, quantified by lactate dehydrogenase leakage, and effectively protected against protein damage induced by iron and glucose. These findings demonstrated that the *P. alata* and *P. edulis* leaf extracts had potent in-vitro and ex-vivo antioxidant properties.

Passionfruit leaves also contained C-deoxyhexosyl flavonoids which exhibited antioxidant activity (Ferreres et al. 2007). Sixteen apigenin or luteolin derivatives were characterized, which included four mono-C-glycosyl, eight O-glycosyl-C-glycosyl, and four O-glycosyl derivatives. With the exceptions of C-hexosyl luteolin and C-hexosyl apigenin, all the compounds exhibited a deoxyhexose moiety. Moreover, the uncommon C-deoxyhexosyl derivatives of luteolin and apigenin were also identified for first time in *P. edulis* leaves. The antioxidative capacity of passion fruit leaves was determined against DPPH radical and several reactive oxygen species (superoxide radical, hydroxyl radical, and hypochlorous acid), revealing it to be concentration-dependent, although a pro-oxidant effect was observed for hydroxyl radical.

The petroleum ether and chloroform fractions of ethanol extract of leaf and stem from *Passiflora edulis* showed potent antioxidant activity, of which the chloroform and petroleum ether fraction of stem demonstrated the strongest antioxidant activity with the IC_{50} value of 51.28 and 54.01 μg/ml, respectively using DPPH free radical assay (Ripa et al. 2009). Ethanol extract of *Passiflora edulis* leaves showed strong potential antioxidant activity in 1,1-diphenyl-2-picryl hydrazyl radical reducing power method assay exhibiting an IC_{50} value of 875 μg/ml (Sunitha and Devaki 2009).

The polyphenolic compound piceatannol which occurs in passion fruit seeds in high amount and its dimer, scirpusin B exhibited potent antioxidant activity (1,1-diphenyl-2-picrylhydrazyl (DPPH)) and vasorelaxant effect when tested in rat thoracic aorta (Sano et al. 2011). Sscirpusin B exerted a greater antioxidant activity and vasorelaxant effect compared with piceatannol. Additionally, the vasorelaxation effects of the compounds were induced via the NO derived from the endothelium. The results suggested the potential of polyphenols in passion fruit seeds to be used against cardiovascular diseases (CVDs).

Antiinflammatory and Wound Healing Activities

In Brazilian countryside, cataplasm made from *Passiflora edulis* leaves was reported to be used by the population as a healing agent for infections and skin inflammations (Garros et al. 2006). No significant difference in the rate of wound healing was detected in rats with wounds treated with *Passiflora edulis* hydro-alcoholic extract or in control rats treated with distilled water (Garros et al. 2006). However, a significant increase in the number of fibroblastic cells was seen on the 7th post-operative day, and significantly greater collagen deposition was observed on the 14th day post-operative day in rats from the *Passiflora* group. In separate studies, *Passiflora edulis* leaf extract (250 mg/kg) administered by intraperitoneal route (i.p.) was found to inhibit the leukocyte, neutrophils, myeloperoxidase, nitric oxide, TNFalpha and IL-1beta levels in the pleurisy induced by carrageenan (Montanher et al. 2007). The extract (250–500 mg/kg, i.p.) also inhibited total and differential leukocytes in the pleurisy induced by bradykinin, histamine or substance P. Several mechanisms, including the inhibition of pro-inflammatory cytokines (TNFalpha, IL-1beta), enzyme (myeloperoxidase) and mediators (bradykinin, histamine, substance P, nitric oxide) release and/or action, appeared to account for *Passiflora edulis'* activities. Vargas et al. (2007) reported that the aqueous leaves extracts of *Passiflora alata* (100–300 mg/kg, i.p.) *and Passiflora edulis* (100–1,000 mg/kg, i.p.) possessed significant antiinflammatory activity on carrageenan-induced pleurisy in mice. Treatment with the extracts inhibited leukocyte migration and reduced the formation of exudate. Further, a significant inhibition of myeloperoxidase and adenosine-deaminase activities was observed at the doses tested (100 or 250 mg/kg, i.p.). At the same doses, a significant decrease of serum C-reactive protein was also observed.

In the inflammation assay induced by carrageenan, aqueous extract of yellow passion fruit leaves (100 mg/kg, i.p.), butanolic fraction (50 mg/kg, i.p.), aqueous residual fraction (100 mg/kg, i.p.) and dexamethasone (0.5 mg/kg, i.p.) were found to inhibit the leukocyte, neutrophil, myeloperoxidase, nitric oxide, and interleukin-1 beta (IL-1β) levels (Beninca et al. 2007). The aqueous extract and butanolic and aqueous residual fractions, but not dexamethasone, decreased macrophage inflammatory protein-2 (MIP-2) levels. Only dexamethasone inhibited mononuclear cells. In inflammation induced by histamine, the aqueous extract, butanolic and aqueous residual fractions, and dexamethasone inhibited total and differential leukocytes. In inflammation induced by substance P, the aqueous extract, butanolic and aqueous residual fractions, and dexamethasone also inhibited total leukocytes and mononuclears. Neutrophils were only inhibited by aqueous extract, butanolic fraction, and dexamethasone. The study indicated that the active principle(s) present in the *P. edulis* aqueous extract and its two fractions showed pronounced antiinflammatory properties, inhibiting cell migration, proinflammatory cytokines, enzymes and mediators.

Several other studies reported dry leaf extracts from *Passiflora edulis* to have an antiinflammatory effect on wound healing in rats. The intraperitoneal use of *Passiflora edulis* extract was shown to influence favorably the healing of gastric sutures in rats because of the increase in the fibroblastic proliferation on the 7th post operative day (Silva et al. 2006). All animals presented adequate healing of the abdominal wall with no clinical signs of infections or dehiscence. *Passiflora edulis* extract was also reported to enhance the healing of midline abdominal incisions in rats, especially the histological (collagenization and capillary neoformation) and tensiometric aspects i.e. maximal breaking and deformation strength (Gomes et al. 2006). In another study, the use of *Passiflora edulis* hydroalcoholic leaf extract administered by intraperitoneal injection in male wistar rats resulted in less acute inflammation, greater fibroblastic proliferation, collagenous formation and capillary neo-formation on rats' bladder wound healing (Gonçalves Filho et al. 2006). In a separate study, the peri-operative administration of the hydro-alcoholic extract of *Passiflora edulis* had a positive influence on the healing of colonic anastomosis in rats based on the following parameters evaluated macroscopic aspects of the wall and abdominal cavity, perianastomotic (adherences), bursting pressure, inflammatory tissue reaction on the anastomotic wound (Bezerra et al. 2006). The butanolic fraction obtained from an aqueous extract of *P. edulis* var. *flavicarpa* leaves (50 and 100 mg/kg, I.P.) showed antiinflammatory activity by inhibiting leukocytes and neutrophils (Zucolotto et al. 2009). Sub-fraction C showed itself to be more effective than the other sub-fractions. Isoorientin, vicenin-2 and were isolated from the active sub-fraction C derived from the butanolic fraction. The sub-fraction C (50 mg/kg, i.p.), as well as its major isolated compounds (25 mg/kg, i.p.), inhibited leukocytes and neutrophils. Additionally, the butanolic fraction and isoorientin also inhibited myeloperoxidase activity. The present study showed that the C-glucosylflavones isolated from *P. edulis* leaves could be responsible for the antiinflammatory effect of *P. edulis* on the mouse model of pleurisy induced by carrageenan. In a randomized, double-blind, placebo-controlled trial with parallel-group design of 33 patients with knee osteoarthritis, supplementation of passion fruit peel extract pills substantially alleviated osteoarthritis symptoms (Farid et al. 2010). This beneficial effect of the extract may be due to its antioxidant and antiinflammatory properties

Anticancer Activity

Fruit decoctions of *Passiflora edulis* and *P. foetida* var. *albiflora* exhibited inhibitory activity of gelatinase MMP-2 and MMP-9, two metalloproteases involved in the tumour invasion, metastasis and angiogenesis (Puricelli et al. 2003). Both water extracts, at different concentrations, inhibited the enzymes.

The phytochemical composition of passion-fruit juice (PFJ) was also shown to have valuable anti-cancer activity, when tested in a BALB/c 3T3 neoplastic transformation model (Rowe et al. 2004). A higher concentration of PFJ compared with a lower concentration was effective in reducing the number, size, and invasiveness of transformed foci. When incubated with another mammalian cell line, the MOLT-4, PFJ was unable to alter the cell cycle kinetics while at the same time was successful in inducing the activity of caspase-3, an enzyme that commits the cell to apoptosis. This suggested that phytochemicals found in PFJ were able to produce the changes in transformed foci due to apoptotic mechanisms rather than by a reduction in cell proliferation. The authors maintained that beneficial results were achieved at levels that could theoretically be attained in the plasma after consumption of the juice.

Passiflin, a novel dimeric antifungal protein from seeds of the passion fruit was found to potently inhibit proliferation of MCF-7 breast cancer cells with an IC_{50} of 15 μM (Lam and Ng 2009).

In the brine shrimp lethality bioassay, the crude chloroform and petroleum ether extracts of *P. edulis* leaf and stem possessed considerable cytotoxic activity. The chloroform and petroleum ether extracts of stem and leaf exhibited significant cytotoxic potentials with the LC_{50} value of 6.63 μg/ml, 6.89 μg/ml and 7.91 μg/ml, 11.17 μg/ml respectively (Ripa et al. 2009).

Antiviral Activity

Passiflora edulis root extract also showed antiviral activity (Müller et al. 2007). Test results were expressed as 50% cytotoxicity (CC_{50}) for MTT assay and 50% effective (EC_{50}) concentrations for viral cytopathic effect (CPE), and these were used to calculate the selectivity indices (SI = CC(50)/EC(50)) of each tested material. *Passiflora edulis* extract showed values of SI >7 against herpetic herpes simplex virus (HSV-1 KOS) and 29-R strains.

Anxiolytic Activity

Most of the pharmacological investigations of *Passiflora edulis* had been focused on the central nervous system (CNS) activities, such as anxiolytic, anticonvulsant and sedative actions. The anxiolytic activity of hydroethanol extracts of *P. alata* and *P. edulis* leaves was demonstrated using the elevated plus-maze test (Petry et al. 2001). The extracts presented anxiolytic activity in dosages around 50, 100 and 150 mg/kg.

In separate studies, the spray-dried powders of *P. alata* and *P. edulis* showed anxiolytic activity in doses of 400 and 800 mg/kg supporting their use in Brazilian folk medicine for its reputed sedative and anxiolytic properties (Reginatto et al. 2006). Anti-anxiety activity of *P. edulis* was evaluated on the performance of mice in the elevated plus maze, openfield, and horizontal-wire tests (Coleta et al. 2001). Coleta et al. (2006) reported that the aqueous extracts of *P. edulis* presented an anxiolytic-like activity without any significant effect upon the motor activity whilst the total flavonoid fraction (TFF) presented an anxiolytic-like activity but compromised motor activity. Through fractionation of TFF it was possible to isolate and characterize luteolin-7-O-[2-rhamnosylglucoside] which showed an anxiolytic-like activity without compromising motor activity. The results indicated that flavonoids in the leaves were partly involved in the neuropharmalogical activity.

Phytochemical analysis showed that the content of flavonoids of the aqueous extract of *P. edulis* was almost twice that of *P. alata* and that differences in contents of flavonoids could explain the lower active doses of the aqueous extract of *P. edulis* in inducing anxiolytic-like effects compared to *P. alata* (Barbosa et al. 2008). The research findings suggested that, distinct from diazepam, the aqueous extract of both species of *Passiflora* induced anxiolytic-like effects in rats without disrupting memory process. Aqueous extract of *Passiflora edulis* had been reported to exhibit non specific CNS depressant effects in mice, rats and healthy human volunteers, whereas, it was also noted that some samples of *Passiflora edulis* had a "non-specific" CNS-depressant effect (Maluf et al. 1991). In another report on CNS depressant effects of *Passiflora edulis*, it was reported that the aqueous extract of the plant prolonged barbiturate-induced as well as morphine-induced sleep time in mice and also "partially" blocked the amphetamine-induced stimulant effects (Do et al. 1983).

In a recent study, the aerial part of *Passiflora edulis* f. *flavicarpa* was reported to be anxiolytic at low dose but sedative at high dose (Deng et al. 2010). In the elevated plus-maze (EPM) test, single-dose oral administration of ethanolic extract (EE) (300 and 400 mg/kg), n-BuOH extract (BE)

(125 and 200 mg/kg), aqueous extract (AE) (200 and 300 mg/kg), subfractions of BE BEF-I (200 mg/kg), BEF-II (200 mg/kg), BEF-III (100 mg/kg), or isoorientin (20 mg/kg), a flavonoid component isolated from BEF-III resulted in anxiolytic-like effects, but a sedative-like activity was produced at higher doses, such as 300 mg/kg of BE, 200 mg/kg of BEF-III, or 40 and 80 mg/kg of isoorientin. The results of the SA (spontaneous activity) test manifested that treatment with 400 mg/kg of EE, 300 mg/kg of BE, or 40 and 80 mg/kg of isoorientin compromised motor activity in mice, which accorded with the results of the EPM test. Flavonoids are important active constituents. Since the aqueous extract contained little flavonoids, it was conjectured that there were other components responsible for the anxiolytic effect of *Passiflora edulis* f. *flavicarpa* besides flavonoids. In a recent study, the ethanol extracts of leaves of *Passiflora edulis* 'flavicarpa' displayed anxiolytic activity at 400 mg/kg, while those of *Passiflora edulis* 'edulis' exhibited sedative effect at 400 mg/kg (Li et al. 2011). The six major flavonoid compounds isolated from the leaves of *Passiflora edulis* 'flavicarpa', lucenin-2, vicenin-2, isoorientin, isovitexin, luteolin-6-C-chinovoside, and luteolin-6-C-fucoside, were not detected in *Passiflora edulis* 'edulis.

In another study, the aqueous extract (AE), the butanolic fraction (BF), and the aqueous residual fraction (ARF) obtained from the pericarp of *P. edulis flavicarpa* were found to be involved in the putative neuropharmacologic effects in mice (Sena et al. 2009). AE, BF, and ARF increased the total time spent in the light compartment of the light:dark box, an anxiolytic-like effect, and AE also potentiated the hypnotic effects of ethyl ether, a sedative effect. Analysis indicated the predominance of C-glycosylflavonoids in these extracts and fractions, which were identified as isoorientin, vicenin-2, spinosin, and 6,8-di-C-glycosylchrysin.

Antihypertensive Activity

Research showed that orally administered methanol extract of *Passiflora edulis* rind (10 or 50 mg/kg) or luteolin (50 mg/kg), one of constituent polyphenols of the extract, significantly lowered systolic blood pressure in spontaneously hypertensive rats (Ichimura et al. 2006). The extract was found to contain 20 µg/g dry weight of luteolin and 41 µg/g dry weight of luteolin-6-C-glucoside. It also contained gamma-aminobutyric acid (GABA, 2.4 mg/g dry weight by LC-MS/MS or 4.4 mg/g dry weight by amino acid analysis) which has been reported to be an antihypertensive material. Since the extract contained a relatively high concentration of GABA, the antihypertensive effect of the extract in the hypertensive rats might be due mostly to the GABA-induced antihypertensive effect and partially to the vasodilatory effect of polyphenols including luteolin.

The diet of spontaneously hypertensive rats supplemented with the purple passion fruit peel (PFP) extract, a mixture of bioflavonoids, phenolic acids, and anthocyanins at 50 mg/kg significantly lowered systolic blood pressure by 12.3 mm Hg and markedly decreased serum nitric oxide level by 65% compared with the control group (Zibadi et al. 2007). In a 4-week randomized, placebo-controlled, double-blind trial, the systolic and diastolic blood pressure of the PFP extract–treated group decreased significantly compared with the placebo group. No adverse effect was reported by the patients. In a rat liver toxicity assay, no hepatotoxicity was observed after 9 h incubation in the presence of PFP extract, (20 µg/ml). The PFP extract also revealed hepatoprotection against chloroform (1 mmol/L)-induced liver injury. The results suggested that the antihypertensive effect of the PFP extract may, in part, be mediated through nitric oxide modulation. The results also suggested that the PFP extract may be offered as a safe alternative treatment to hypertensive patients.

Hypocholesterolemic/ Antihyperlipidemic Activity

Studies showed that the consumption of insoluble fibre-rich fraction (FRF) diet prepared from defatted *Passiflora edulis* seed, relative to cellulose diet could effectively decrease the levels of serum triglyceride, serum total cholesterol, and liver cholesterol, and increase the levels of total lipids, cholesterol, and bile acids in faeces (Chau and Huang 2005). The consumption of insoluble FRF also increased the faecal bulk and moisture. The marked cholesterol- and lipid-lowering effects of insoluble FRF might be partly attributed to its ability to enhance the excretion of lipids and bile acids via faeces. The results suggested that insoluble FRF could be a potential hypocholesterolemic ingredient for fibre-rich functional foods.

The offsprings of passion fruit juice treated non-diabetic rats and passion fruit juice treated streptozotocin-diabetic rats showed significantly reduced total cholesterol, triglyceride, and low-density lipoprotein cholesterol levels and an increased high-density lipoprotein cholesterol level after 30 days (Barbalho et al. 2011). The use of passion fruit juice was found to improve lipid profiles, suggesting that it may have beneficial effects in the prevention and treatment of dyslipidemias and hyperglycemia.

Studies in diabetic rats showed that pectin from *P. edulis* fruit had antiinflammatory, hypoglycemic and hypotriglyceridemic properties (Silva et al. 2011). Pectin administration decreased blood glucose and triglyceride levels in diabetic male wistar rats. Pectin also decreased edema volume and release of myeloperoxidase. It also significantly decreased neutrophil infiltration and partially decreased immunostaining for tumour necrosis factor-α and inducible nitric oxide synthase.

Antiasthmatic Activity

Most clinical symptoms of asthma of the purple passion fruit peel (PFP) extract-treated group were moderated significantly compared to the baseline (Watson et al. 2008). Purple passion fruit peel (PFP) extract comprised a novel mixture of bioflavonoids. The prevalence of wheeze, cough, as well as shortness of breath was reduced significantly in group treated with PFP extract, whereas the placebo caused no significant improvement. Purple passion fruit peel extract supplementation resulted in a marked increase in forced vital capacity as placebo showed no effect. However, no significant improvement was observed in the forced expiratory volume at 1 s of those supplemented with PFP extract. No adverse effect was reported by any of study participants. Results suggested that PFP extract may be safely offered to asthmatic subjects as an alternative treatment option to reduce clinical symptoms.

Antimicrobial Activity

Extract of *Passiflora edulis* exhibited mild in-vitro, anti-fungal activity against three keratinophilic fungi: *Microsporum gypseum, Chrysosporium tropicum* and *Trichophyton terrestre* (Qureshi et al. 1997). All the extracts (hexane, water, ethylacetate and methanolic extract) of the leaf, stem and fruit showed antimicrobial activity against two gram positive bacteria, *Bacillus subtilis* and *Staphylococcus aureus*, and four gram negative bacteria *Pseudomonas aeruginosa, Salmonella paratyphi, Klebsiella pneumoniae* and *Escherichia coli* (Akanbi et al. 2011). Amongst the extracts examined, hexane extracts significantly exhibited the highest antimicrobial activity against all the bacteria tested. The antimicrobial activity was found to be dependent on the type of solvent used for extraction as well as the part of the plant used. The methanolic leaf extract showed high in-vitro inhibitory activity against *Bacillus subtilis* and *E.coli* and was also inhibitory *Staphylococcus aureus* and *Salmonella typhi* when compared with standard ciprofloxacin under similar conditions (Kannan et al. 2011).

A novel dimeric, 67-kDa, antifungal protein from the seeds of passion fruit, designated as passiflin, impeded mycelial growth in *Rhizotonia solani* with an IC_{50} of 16 μM and potently inhibited proliferation of MCF-7 breast cancer cells with an IC_{50} of 15 μM (Lam and Ng 2009). It exhibited an N-terminal amino acid sequence closely resembling that of bovine β-lactoglobulin. Its dimeric nature was rarely found in antifungal proteins. Passiflin was found to be distinct from β-lactoglobulin. There was no cross-reactivity of passiflin with anti-β-lactoglobulin antiserum. Intact β-lactoglobulin lacked antifungal and antiproliferative activities and was much smaller in molecular size than passiflin.

An antifungal peptide of 5.0 kDa, Pe-AFP1, purified from passion fruit (*Passiflora edulis*) seeds inhibited the development of the filamentous fungi *Trichoderma harzianum, Fusarium oxysporum*, and *Aspergillus fumigatus* with IC_{50} values of 32, 34, and 40 µg/ml, respectively, but not of *Rhizoctonia solani, Paracoccidioides brasiliensis* and *Candida albicans* (Franco 2006; Pelegrini et al. 2006). The pepetide had similarities to 2S albumin proteins.

The crude chloroform extract and petroleum ether extracts of passion fruit leaf and stem showed varying antibacterial activity ranged against twelve bacteria four Gram-positive (*Bacillus megaterium, Bacillus subtilis, Staphylococcus aureus* and *Sarcina lutea*) and eight Gram-negative (*Salmonella paratyphi, Salmonella typhi, Vibrio parahemolyticus, Vibrio mimicus, Escherichia coli, Shigella dysenteriae, Shigella boydii* and *Pseudomonas aeruginosa*) (Ripa et al. 2009). The crude chloroform extract of the leaf at a concentration of 500 µg/disc showed moderate activity but no activity was observed by the petroleum ether extract against most of the tested organisms, except *Bacillus megaterium* and *Pseudomonas aeruginosa* having positive effect. The crude chloroform and petroleum ether extracts of stem showed notable antibacterial activity at a concentration 500 µg/disc against twelve microorganisms. In case of the stem, the chloroform extract showed the highest activity against the growth of *Vibrio mimicus*. The extract also showed good activity against the growth of *Vibrio parahemolyticus, Shigella dysenteriae* and *Shigella boydii*. The petroleum ether stem extract showed moderate inhibitory activity.

Melanogenesis Inhibition Activity

The concentration of polyphenols was found to be higher in *P. edulis* seed (PF-S) than in the rind (PF-R) or pulp (PF-P) ethanol extracts (Matsui et al. 2010). Treatment of melanoma cells with PF-S led to inhibition of melanogenesis. In addition, the production of total soluble collagen was elevated in dermal fibroblast cells cultured in the presence of PF-S. PF-R and PF-P did not yield these effects. Further, the removal of polyphenols from PF-S led to the abolishment of the effects described above. Piceatannol (3,4,3′,5′-tetrahydroxy-trans-stilbene) was found to be present in passion fruit seeds in large amounts and was the major component responsible for the PF-S effects observed on melanogenesis and collagen synthesis.

Traditional Medicinal Uses

Passiflora edulis has been used as a sedative, diuretic, anthelmintic, anti-diarrheal, stimulant, tonic and also in the treatment of hypertension, menopausal symptoms, colic of infants in South America (Chopra et al. 1986; Kirtikar and Basu 1975). In Madeire, the fruit of *Passiflora edulis* is regarded as a digestive stimulant and is used as a remedy for gastric carcinoma (Watt and Breyer-Brandwijk 1962). In Nagaland (India), fresh leaves of *Passiflora edulis* are boiled in little amount of water and the extract is drunk for the treatment of dysentery and hypertension (Jamir et al. 1999).

In Brazilian folk medicine, *Passiflora edulis* has been commonly used as a sedative, tranquilizer, antiinflammatory drug, intermittent fever and also for the treatment of inflammatory cutaneous wounds, lesions and erysipelas. The pulp of the fruit is stimulant and tonic. In Suriname, the leaves of passion fruit are used to settle edgy nerves and are employed also for colic, diarrhoea, dysentery, and insomnia.

Other Uses

Studies showed that *Passilfora edulis* contained strong allelopathic potential (Khanh et al. 2006). Aqueous extracts of *P. edulis* strongly suppressed germination and growth of lettuce, radish and two major paddy rice weeds, *Echinochloa crusgalli* and *Monochoria vaginalis*. Ten newly identified substances in *P. edulis* extracts, including coumarin, long-chain fatty acids and lactones, may be responsible for the inhibitory activity of *P. edulis*. Coumarin and the lactones showed greater inhibition of germination and growth of

E. crusgalli than the fatty acids. The authors suggested that *P. edulis* may be used as a natural herbicide to reduce the dependency on synthetic herbicides.

Comments

There is a controversy over the synonym of *Passiflora incarnata* with *Passiflora edulis*. The designation by Sir William J. Hooker in 1843, followed by the citation of *P. edulis* as the synonym of *P. incarnata* in Index Kewensis of 1895, not only substantiated the controversial identity but also caused confusion to researchers. The prevailing confusion might have led to improper selection of the bioactive plant, thereby accounting for inconclusive and contradictory pharmacological reports on either of the two plants. Recently, researchers have reported that the two entities should remain as two separate species instead of being synonymous. Using a range of key identification parameters they differentiated *P. incarnata* from *P. edulis*. Various leaf constants such as vein-islet number, vein-termination number, stomatal number, and stomatal index are different for the two species. Physicochemical parameters such as ash values and extractive values and the thin layer chromatography profile of the petroleum ether extract of *P. incarnata* and *P. edulis* are also distinct and different.

Selected References

Akanbi BO, Bodunrin OD, Olayanju S (2011) Phytochemical screening and antibacterial activity of *Passiflora edulis*. Researcher 3(5):9–12

Arriaza AMC, Craveiro AA, Machado MIL, Pouliquen YBM (1997) Volatile constituents from fruit shells of *P. edulis* Sims. J Essent Oil Res 9:235–236

Backer CA, Bakhuizen van den Brink RC Jr (1963) Flora of Java, vol 1. Noordhoff, Groningen, 648 pp

Barbalho SM, Damasceno DC, Spada AP, Lima IE, Araújo AC, Guiguer EL, Martuchi KA, Oshiiwa M, Mendes CG (2011) Effects of *Passiflora edulis* on the metabolic profile of diabetic wistar rat offspring. J Med Food 14(12):1490–1495

Barbosa PR, Valvassori SS, Bordignon CL Jr, Kappel VD, Martins MR, Gavioli EC, Quevedo J, Reginatto FH (2008) The aqueous extracts of *Passiflora alata* and *Passiflora edulis* reduce anxiety-related behaviours without affecting memory process in rats. J Med Food 11(2):282–288

Beninca JP, Montanher AB, Zucolotto SM, Schenkel EP, Frode TS (2007) Evaluation of the anti-inflammatory efficacy of *Passiflora edulis*. Food Chem 104: 1097–1105

Bezerra JA, Campos AC, Vasconcelos PR, Nicareta JR, Ribeiro ER, Sebastião AP, Urdiales AI, Moreira M, Borges AM (2006) Extract of *Passiflora edulis* in the healing of colonic anastomosis in rats: a tensiometric and morphologic study. Acta Cir Bras 21(suppl 3): 16–25 (in Portuguese)

Bombardelli E, Bonati A, Gabetta B, Martinelli E, Mustich G (1975) Passiflorine, a new glycoside from *Passiflora edulis*. Phytochemistry 14:2661–2665

Chang YW, Su JD (1998) Antioxidant activity of major anthocyanins from skins of passion fruit. Shipin Kexue 25:651–656

Chassagne D, Boulanger R, Crouzet J (1999) Enzymatic hydrolysis of edible *Passiflora* fruit glycosides. Food Chem 66:281–288

Chassagne D, Crouzet J (1998) A cyanogenic glycoside from *Passiflora edulis* fruits. Phytochemistry 49: 757–759

Chassagne D, Crouzet J, Bayonove CL, Baumes RL (1996a) Identification and quantification of passion fruit cyanogenic glycosides. J Agric Food Chem 44:3817–3820

Chassagne D, Crouzet J, Bayonove CL, Brillout JN, Baumes RL (1996b) 6-O-l-Arabinopyranosyl-d-glucopyranosides as aroma precursors from passion fruit. Phytochemistry 41:1497–1500

Chassagne D, Crouzet J, Bayonove CL, Brillout JN, Baumes RL (1997) Glycosidically bound eugenol and methyl salicylate in the fruit of edible *Passiflora* species. J Agric Food Chem 45:2685–2689

Chau CF, Huang YL (2005) Effects of the insoluble fiber derived from *Passiflora edulis* seed on plasma and hepatic lipids and fecal output. Mol Nutr Food Res 49:786–790

Chopra RN, Nayar SL, Chopra IC (1986) Glossary of Indian medicinal plants. (Including the supplement). Council Scientific Industrial Research, New Delhi, 330 pp

Christensen J, Jaroszewski JW (2001) Natural glycosides containing allopyranose from the passion fruit plant and circular dichroism of benzaldehyde cyanohydrin glycosides. Org Lett 3(14):2193–2195

Coleta M, Campos MG, Cotrim MD, Cunha AP (2001) Comparative evaluation of *Melissa officinalis* L., *Tilia europaea* L., *Passiflora edulis* Sims. and *Hypericum perforatum* L. in the elevated plus maze anxiety test. Pharmacopsychiatry 34(1):S20–S21

Coleta M, Batista MT, Campos MG, Carvalho R, Cotrim MD, Lima TC, Cunha AP (2006) Neuropharmacological evaluation of the putative anxiolytic effects of *Passiflora edulis* Sims, its sub-fractions and flavonoid constituents. Phytother Res 20(12):1067–1073

Deng J, Zhou Y, Bai M, Li H, Li L (2010) Anxiolytic and sedative activities of *Passiflora edulis* f. *flavicarpa*. J Ethnopharmacol 128(1):148–153

Dhawan K, Dhawan S, Sharma A (2004) *Passiflora*: a review update. J Ethnopharmacol 94:1–23

Dhawan K, Kumark S, Sharma A (2001) Comparative biological activity study on *Passiflora incamata* and *P. edulis*. Fitoterapia 72:698–702

Do V, Nitton B, Leite JR (1983) Psychopharmacological effects of preparations of *Passiflora edulis* (Passion flower). Cienc Cult 35:11–24

Facciola S (1990) *Cornucopia*. A source book of edible plants. Kampong Publications, Vista, 677 pp

Farid R, Rezaieyazdi Z, Mirfeizi Z, Hatef MR, Mirheidari M, Mansouri H, Esmaelli H, Bentley G, Lu Y, Foo Y, Watson RR (2010) Oral intake of purple passion fruit peel extract reduces pain and stiffness and improves physical function in adult patients with knee osteoarthritis. Nutr Res 30(9):601–606

Ferreres F, Sousa C, Valentão P, Andrade PB, Seabra RM, Gil-Izquierdo A (2007) New C-deoxyhexosyl flavones and antioxidant properties of *Passiflora edulis* leaf extract. J Agric Food Chem 55(25): 10187–10193

Fouqué A (1972) Espèces fruitières d'Amérique tropicale. IV. Les Passiflorées. Fruits 27:368–382

Franco OL (2006) An antifungal peptide from passion fruit (*Passiflora edulis*) seeds with similarities to 2S albumin proteins. Biochim Biophys Acta 1764(6): 1141–1146

Garros IC, Campos AC, Tâmbara EM, Tenório SB, Torres OJ, Agulham MA, Araújo AC, Santis-Isolan PM, Oliveira RM, Arruda EC (2006) Extract from *Passiflora edulis* on the healing of open wounds in rats: morphometric and histological study. Acta Cir Bras 21(suppl 3):55–65 (in Portuguese)

Goday HT, Rodriguez ADB (1994) Occurrence of cis isomers of provitamin A in Brazilian fruits. J Agric Food Chem 42:1306–1313

Gomes CS, Campos AC, Torres OJ, Vasconcelos PR, Moreira AT, Tenório SB, Tâmbara EM, Sakata K, Moraes Júnior H, Ferrer AL (2006) *Passiflora edulis* extract and the healing of abdominal wall of rats: morphological and tensiometric study. Acta Cir Bras 21(suppl 2):9–16 (in Portuguese)

Gonçalves Filho A, Torres OJ, Campos AC, Tâmbara Filho R, Rocha LC, Thiede A, Lunedo SM, Barbosa RE, Bernhardt JA, Vasconcelos PR (2006) Effect of *Passiflora edulis* (passion fruit) extract on rats' bladder wound healing: morphological study. Acta Cir Bras 21(suppl 2):1–8 (in Portuguese)

Green PS (1972) *Passiflora* in Australasia and the Pacific. Kew Bull 26:539–558

Gurnah AM (1992) *Passiflora edulis* Sims. In: Verheij EWM, Coronel RE (eds) Plant resources of South-East Asia No. 2: Edible fruits and nuts. Prosea Foundation, Bogor, pp 244–248

Herderich M, Winterhalter P (1991) 3-hydroxyl-retro-a-ionol: a natural precursor of isomeric edulans in purple passion fruit (*Passiflora edulis* Sims). J Agric Food Chem 39:127–1274

Herderich M, Winterhalter P, Schreier P (1993) b-Ionyl-b-D-glucopyranoside: a natural precursor of isomeric megastigma-4,6,8-trienes in purple passionfruit (*Passiflora edulis* Sims). Nat Prod Res 2(3):227–230

Ichimura T, Yamanaka A, Ichiba T, Toyokawa T, Kamada Y, Tamamura T, Maruyama S (2006) Antihypertensive effect of an extract of *Passiflora edulis* rind in spontaneously hypertensive rats. Biosci Biotechnol Biochem 70(3):718–721

Jamir TT, Sharma HK, Dolui AK (1999) Folklore medicinal plants of Nagaland, India. Fitoterapia 70:395–401

Jordan MJ, Goodner KL, Shaw PE (2000) Volatile components in tropical fruit essences: yellow passion fruit (*Passiflora edulis* Sims. f. *flavicarpa* Degner) and banana (*Musa sapientum* L.). Proc Fla State Hortic Soc 113:284–286

Kannan S, Devi BP, Jayakar B (2011) Antibacterial evaluation of the methanolic extract of *Passiflora edulis*. Hygeia J D Med 3(1):46–49

Khanh TD, Chung IM, Tawata S, Xuan TD (2006) Weed suppression by *Passiflora edulis* and its potential allelochemicals. Weed Res 46(4):296–303

Kidoy L, Nygard AM, Andersen OM, Pedersen AT, Aksnes DW, Kiremire BT (1997) Anthocyanins in fruits of *Passiflora edulis* and *P. suberosa*. J Food Compos Anal 10(1):49–54

Kirtikar KR, Basu BD (1975) Indian medicinal plants, 4 vols, 2nd edn. Jayyed Press, New Delhi

Lam SK, Ng TB (2009) Passiflin, a novel dimeric antifungal protein from seeds of the passion fruit. Phytomedicine 16(2–3):172–180

Li H, Zhou P, Yang Q, Shen Y, Deng J, Li L, Zhao D (2011) Comparative studies on anxiolytic activities and flavonoid compositions of *Passiflora edulis* '*edulis*' and *Passiflora edulis* '*flavicarpa*'. J Ethnopharmacol 133(3):1085–1090

Liu S, Yang F, Li J, Zhang C, Ji H, Hong P (2008) Physical and chemical analysis of *Passiflora* seeds and seed oil from China. Int J Food Sci Nutr 59(7–8): 706–715

Lutomski J, Malek B (1975) Pharmacological investigations on raw materials of the genus *Passiflora*. 4. The comparison of contents of alkaloids in some harman raw materials. Planta Med 27:381–386 (in German)

Lutomski J, Malek B, Rybaika L (1975) Pharmacochemical investigation of the raw materials from *Passiflora* genus 2. Pharmacochemical estimation of juices from the fruits of *P. edulis* and *P. edulis* forma flavicarpa. Planta Med 27:112–121 (in German)

Maluf E, Barros HMT, Frochtengarten ML, Benti R, Leite JR (1991) Assessment of the hypnotic/sedative effects and toxicity of *Passiflora edulis* aqueous extract in rodents and humans. Phytother Res 5:262–266

Mareck U, Herrmann K, Galensa R, Wray V (1991) 6-C-chinovoside and 6-C-fucoside of luteolin from *Pasiflora edulis*. Phytochemistry 30(10):3486–3487

Martin FW, Nakasone HY (1970) The edible species of *Passiflora*. Econ Bot 24:333–343

Matsui Y, Sugiyama K, Kamei M, Takahashi T, Suzuki T, Katagata Y, Ito T (2010) Extract of passion fruit (*Passiflora edulis*) seed containing high amounts of piceatannol inhibits melanogenesis and promotes collagen synthesis. J Agric Food Chem 58(20): 11112–11118

Mercadente AZ, Britton G, Rodriguez ADB (1998) Carotenoids from yellow passion fruit (*Passiflora*). J Agric Food Chem 46:4102–4106

Montanher AB, Zucolotto SM, Schenkel EP, Fröde TS (2007) Evidence of anti-inflammatory effects of *Passiflora edulis* in an inflammation model. J Ethnopharmacol 109(2):281–288

Morton JF (1987) Passion fruit. Fruits of warm climates. Julia F. Morton, Miami, pp 320–328

Müller V, Chávez JH, Reginatto FH, Zucolotto SM, Niero R, Navarro D, Yunes RA, Schenkel EP, Barardi CR, Zanetti CR, Simões CM (2007) Evaluation of antiviral activity of South American plant extracts against herpes simplex virus type 1 and rabies virus. Phytother Res 21(10):970–974

Murray KE, Shipton J, Whitfield FB (1972) The chemistry of food flavour. I. Volatile constituents of passionfruit, *Passiflora edulis*. Aust J Chem 25(9):1921–1933

Naf F, Decorzant R, Willhalm B, Velluz A, Winter M (1977) Structure and synthesis of two novel ionones identified in the purple passionfruit (*Passiflora edulis* Sims). Tetrahedron Lett 16:1413–16

Ochse JJ, Bakhuizen van den Brink RC (1931) Fruits and fruitculture in the Dutch East Indies. G. Kolff & Co., Batavia-C, 180 pp

Olafsdottir ES, Cornett C, Jaroszewski JW (1989) Natural cyclopentenoid cyanohydrin glycosides. Part VIII. Cyclopentenoid cyanohydrins glycosides with unusual sugar residues. Acta Chem Scand 43:51–55

Pacific Island Ecosystems at Risk (PIER) (1999) *Passiflora edulis* Sims, Passifloraceae. http://www.hear.org/Pier/species/passiflora_edulis.htm

Pelegrini PB, Noronha EF, Muniz MA, Vasconcelos IM, Chiarello MD, Oliveira JT, Franco OL (2006) An antifungal peptide from passion fruit (*Passiflora edulis*) seeds with similarities to 2S albumin proteins. Biochim Biophys Acta 1764(6):1141–1146

Pereira CA, Yariwake JH, Lancas FM, Wauters JN, Tits M, Angenot L (2004) A HPTLC densitometric determination of flavonoids from *Passiflora alata*, *P. edulis*, *P. incarnata* and *P. caerulea* and comparison with HPLC method. Phytochem Anal 15(4):241–248

Petry RD, Reginatto F, de Paris F, Gosmann G, Salgueiro JB, Quevedo J, Kapczinski F, Ortega GG, Schenkel EP (2001) Comparative pharmacological study of hydroethanol extracts of *Passiflora alata* and *Passiflora edulis* leaves. Phytother Res 15(2):162–164

Popenoe W (1974) Manual of tropical and subtropical fruits. Hafner Press, New York, pp 241–245, Facsimile of the 1920 edition

Puricelli L, Dell'Aica I, Sartor L, Garbisa S, Caniato R (2003) Preliminary evaluation of inhibition of matrix-metalloprotease MMP-2 and MMP-9 by *Passiflora edulis* and *P. foetida* aqueous extracts. Fitoterapia 74(3):302–304

Qureshi S, Rai MK, Agrawal SC (1997) In-vitro evaluation of inhibitory nature of extracts of 18 plant species of Chhindwara against 3 keratinophilic fungi. Hindustan Antibiot Bull 39:56–60

Reginatto FH, De-Paris F, Petry RD, Quevedo J, Ortega GG, Gosmann G, Schenkel EP (2006) Evaluation of anxiolytic activity of spray dried powders of two South Brazilian *Passiflora* species. Phytother Res 20(5): 348–351

Ripa FA, Haque M, Nahar L, Islam MM (2009) Antibacterial, cytotoxic and antioxidant activity of *Passiflora edulis* Sims. Eur J Sci Res 31(4):592–598

Rowe CA, Nantz MP, Deniera C, Green K, Talcott ST, Percival SS (2004) Inhibition of neoplastic transformation of benzo[alpha]pyrene-treated BALB/c 3T3 murine cells by a phytochemical extract of passion-fruit juice. J Med Food 7(4):402–407

Rudnicki M, Oliveira MR, Pereira TV, Reginatto FH, Pizzol FD, Moreira JCF (2007) Antioxidant and antiglycation properties of *Passiflora alata* and *Passiflora edulis* extracts. Food Chem 100(2):719–724

Sano S, Sugiyama K, Ito T, Katano Y, Ishihata A (2011) Identification of the strong vasorelaxing substance scirpusin B, a dimer of piceatannol, from passion fruit (*Passiflora edulis*) seeds. J Agric Food Chem 59(11): 6209–6213

Scafato P, Colangelo A, Rosini C (2009) A new efficient enantioselective synthesis of (+)-cis-2-methyl-4-propyl-1,3-oxathiane, a valuable ingredient for the aroma of passion fruit. Chirality 21(1):176–182

Seigler DS, Pauli GF, Nahrstedt A, Leen R (2002) Cyanogenic allosides and glucosides from *Passiflora edulis* and *Carica papaya*. Phytochemistry 60:873–882

Seigler DS, Spencer KC (1989) Corrected structures of passicoriacin, epicoriacin and epitetraphyllin B and their distribution in the Flacourtiaceae and Passifloraceae. Phytochemistry 28(3):931–932

Sena LM, Zucolotto SM, Reginatto FH, Schenkel EP, De Lima TC (2009) Neuropharmacological activity of the pericarp of *Passiflora edulis flavicarpa* Degener: putative involvement of C-glycosylflavonoids. Exp Biol Med 234(8):967–975

Silva DC, Freitas AL, Pessoa CD, Paula RC, Mesquita JX, Leal LK, Brito GA, Gonçalves DO, Viana GS (2011) Pectin from *Passiflora edulis* shows anti-inflammatory action as well as hypoglycemic and hypotriglyceridemic properties in diabetic rats. J Med Food 14(10): 1118–1126

Silva JR, Campos AC, Ferreira LM, Aranha Júnior AA, Thiede A, Zago Filho LA, Bertoli LC, Ferreira M, Trubian PS, Freitas AC (2006) Extract of *Passiflora edulis* in the healing process of gastric sutures in rats: a morphological and tensiometric study. Acta Cir Bras 21(suppl 2):52–60 (in Portuguese)

Slaytor M, McFarlane IJ (1968) The biosynthesis and metabolism of harman in *Passiflora edulis*—I: the biosynthesis of harman. Phytochemistry 7(4):605–611

Sunitha M, Devaki K (2009) Antioxidant activity of *Passiflora edulis* Sims leaves. Indian J Pharm Sci 71(3):310–311

Tapp EJ, Cummins I, Brassington D, Edwards R (2008) Determination and isolation of a thioesterase from passion fruit (*Passiflora edulis* Sims) that hydrolyzes volatile thioesters. J Agric Food Chem 56(15): 6623–6630

Tominaga T, Dubourdieu D (2000) Identification of cysteinylated aroma precursors of certain volatile thiols in passion fruit juice. J Agric Food Chem 48(7):2874–2876

U.S. Department of Agriculture, Agricultural Research Service (2011) USDA national nutrient database for standard reference, release 24. Nutrient Data Laboratory Home Page. http://www.ars.usda.gov/ba/bhnrc/ndl

Vargas AJ, Geremias DS, Provensi G, Fornari PE, Reginatto FH, Gosmann G, Schenkel EP, Fröde TS (2007) *Passiflora alata* and *Passiflora edulis* spray-dried aqueous extracts inhibit inflammation in mouse model of pleurisy. Fitoterapia 78(2):112–119

Watson RR, Zibadi S, Rafatpanah H, Jabbari F, Ghasemi R, Ghafari J, Afrasiabi H, Foo LY, Faridhosseini R (2008) Oral administration of the purple passion fruit peel extract reduces wheeze and cough and improves shortness of breath in adults with asthma. Nutr Res 28(3):166–171

Watt JM, Breyer-Brandwijk MG (1962) The medicinal and poisonous plants of Southern and Eastern Africa, 2nd edn. E. and S. Livingstone, Edinburgh, 1457 pp

Whitfield FB, Stanley G (1977) The structure and stereochemistry of edulan I and II and the stereochemistry of the 2,5,5,8a-tetramethyl-3,4,4a,5,6,7,8, 8a-octahydro-2H-1-benzopyrans. Aust J Chem 30(5):1073–1091

Whitfield FB, Stanley G, Murray KE (1973) Concerning the structures of edulan I and II. Tetrahedron Lett 2:95–98

Whitfield JB, Last JH (1986) The flavour of the passion fruit. In: Brunke EJ (ed) Progress in essential oil research. De Gruyter, Berlin, pp 3–48

Wills RBH, Lim JSK, Greenfield H (1986) Composition of Australian foods. 31. Tropical and sub-tropical fruit. Food Technol Aust 38(3):118–123

Winterhalter P (1990) Bound terpenoids in the juice of the purple passion fruit (*Passiflora edulis* Sims). J Agric Food Chem 38(2):452–455

Yapo BM (2009) Lemon juice improves the extractability and quality characteristics of pectin from yellow passion fruit by-product as compared with commercial citric acid extractant. Bioresour Technol 100(12):3147–3151

Yapo BM, Koffi KL (2006) Yellow passion fruit rind – a potential source of low-methoxyl pectin. J Agric Food Chem 54(7):2738–2744

Yapo BM, Koffi KL (2008) Dietary fiber components in yellow passion fruit rind – a potential fiber source. J Agric Food Chem 56(14):5880–5883

Yoshikawa K, Katsuta S, Mizumori J, Arihara S (2000a) Four cycloartane triterpenoids and six related saponins from *Passiflora edulis*. J Nat Prod 63(9):1229–1234

Yoshikawa K, Katsuta S, Mizumori J, Arihara S (2000b) New cycloartane triterpenoids from *Passiflora edulis*. J Nat Prod 63(10):1377–1380

Zhou YJ, Li HW, Tan F, Deng J (2009) Studies on the chemical constituents of *Passiflora edulis* f. *flavicarpa*. Zhong Yao Cai 32(11):1686–1688 (in Chinese)

Zibadi S, Farid R, Moriguchi S, Lu Y, Foo L, Tehrani P, Ulreich J, Watson R (2007) Oral administration of purple passion fruit peel extract attenuates blood pressure in female spontaneously hypertensive rats and humans. Nutr Res 27(7):408–416

Zucolotto SM, Fagundes C, Reginatto FH, Ramos FA, Castellanos L, Duque C, Schenkel EP (2011) Analysis of c-glycosyl flavonoids from South American *Passiflora* species by HPLC-DAD and HPLC-MS. Phytochem Anal doi:10.1002/pca.1348

Zucolotto SM, Goulart S, Montanher AB, Reginatto FH, Schenkel EP, Fröde TS (2009) Bioassay-guided isolation of anti-inflammatory C-glucosylflavones from *Passiflora edulis*. Planta Med 75(11):1221–1226

Passiflora foetida

Scientific Name

Passiflora foetida L.

Synonyms

Dysosmia ciliata (Dryand.) M. Roem., *Dysosmia fluminensis* M. Roem., *Dysosmia foetida* (L.) M. Roem., *Dysosmia gossypifolia* (Desv. ex Ham.) M. Roem., *Dysosmia hastata* (Bertol.) M. Roem., *Dysosmia hibiscifolia* (Lam.) M. Roem., *Dysosmia hircina* Sweet ex M. Roem. Nom. illeg., D*ysosmia nigelliflora* (Hook.) M. Roem., *Dysosmia polyadena* (Vell.) M. Roem., *Granadilla foetida* (L.) Gaertn., *Passiflora baraquiniana* Lem., *Passiflora ciliata* Dryand., *Passiflora ciliata* Dryand. var. *polyadena* Griseb., *Passiflora ciliata* Dryand. var. *quinqueloba* Griseb., *Passiflora ciliata* Dryand. var. *riparia* C. Wright ex Griseb., *Passiflora foetida* L. forma *suberecta* Chodat & Hassl., *Passiflora foetida* L. forma *latifolia* Kuntze, *Passiflora foetida* L. forma *longifolia* Kuntze, *Passiflora foetida* L. forma *quinqueloba* (Griseb.) Mast., *Passiflora foetida* L. var. *acapulcensis* Killip, *Passiflora foetida* L. var. *arizonica* Killip, *Passiflora foetida* L. var. *balansae* Chodat, *Passiflora foetida* L. var. *ciliata* (Dryand.) Mast., *Passiflora foetida* L. var. *eliasii* Killip, *Passiflora foetida* L. var. *fluminensis* (M.Roem.) Killip, *Passiflora foetida* L. var. *galapagensis* Killip, *Passiflora foetida* L. var. *gardneri* Killip, *Passiflora foetida* L. var. *glaziovii* Killip, *Passiflora foetida* L. var. *gossypifolia* (Desv. ex Ham.) Mast., *Passiflora foetida* L. var. *hastata* (Bertol.) Mast., *Passiflora foetida* L. var. *hibiscifolia* (Lam.) Killip, *Passiflora foetida* L. var. *hirsuta* Mast., *Passiflora foetida* L. var. *hirsutissima* Killip, *Passiflora foetida* L. var. *hispida* (DC. ex Triana & Planch.) Killip, *Passiflora foetida* L. var. *isthmia* Killip, *Passiflora foetida* L. var. *lanuginosa* Killip, *Passiflora foetida* L. var. *maxonii* Killip, *Passiflora foetida* L. var. *mayarum* Killip, *Passiflora foetida* L. var. *salvadorensis* Killip, *Passiflora foetida* L. var. *sericea* Chodat & Hassl., *Passiflora foetida* L. var. *subpalmata* Killip, *Passiflora foetida* L. var. *variegata* G. Mey., *Passiflora gossypiifolia* Desv. ex Ham., *Passiflora hastata* Bertol., *Passiflora hibiscifolia* Lam., *Passiflora hibiscifolia* Lam. var. *velutina* Fenzl ex Jacq., *Passiflora hispida* DC. ex Triana & Planchon, *Passiflora hirsuta* Lodd. Nom. illeg., *Passiflora nigelliflora* Hook., *Passiflora polyadena* Vell., *Passiflora vesicaria* L., *Passiflora variegata* Mill., *Tripsilina foetida* (L.) Rafinesque.

Family

Passifloraceae

Common/English Names

Bush Passion Fruit, Foetid Passion Flower, Granadilla Cimarrona, Hispid granadilla, Love-In-A-Mist Passionflower, Mossy Passionflower,

Running Pop, Stinking Granadilla, Stinking Passionflower, Stinking Passion Flower, Tagua Passion Flower, Wild Maracuja, Wild Passionfruit, Wild Water Lemon.

Vernacular Names

Bolivia: Pedon;
Brazil: Maracujá-Da-Pedra; Maracujá De Cobra, Maracujá-Fedido (Portuguese);
Chamorro: Dulce, Kinahulo' Atdao;
Chinese: Long Zhu Guo;
Chuukese: Bombom;
Colombia: Bejuco;
Costa Rica: Bombillo;
Cuba: Caguajosa, Canizo;
Czech: Mučenka Smrdutá;
Dutch: Sneekie Markoesa;
Eastonian: Lehkav Kannatuslill;
Fijian: Loliloli Ni Kalavo, Qaranidila, Sou;
French: Passiflore Fétide, Pomme-Liane Collant, Marie-Gougeat, Toque Molle;
Galapagos Islands: Bedoca (Spanish);
German: Übelriechende Passionsblume;
Haiti: Mariegouya;
Hawaiian: Lani Wai, Pohāpohā;
I-Kiribati: Te Biku;
India: Junuka Phul (Assamese), Gharibel, Gudsar, Phophni-Ki-Bel (Hindu), Kukkiballi, Kukke Balli, Kukki Balli (Kannada), Cirrancantiya (Malayalam), Lam Radhikanachan (Manipuri), Mukkopeera (Sanskrit), Siruppunaikkali, Ciru Punai-K-Kali, Mupparisavalli, Sirupunaikkali, Cirupunaikkali (Tamil), Tellajumiki, Tella Jumiki, Gaju Tige, Adavi Motala, Thellajumiki (Telugu);
Indonesia: Buwah Tikus (Malay), Ceplukan Blungsun (Javanese), Ermut, Ermut Pacean, Kaceprak, Kaceprek, Mermut, Permot, Rajutan, Randa Bolong (Sundanese);
Japanese: Kusa Tokeisou;
Kwara'Ae: Kakalifaka, Kwalo Kakali;
La Réunion: Fetid Pa, Grenadier Marron, Passiflore Poc-Poc, Ti Grenadelle (French);
Malaysia: Timun Padang, Timun Dendang, Poko Lang Bulu;
Mexico: Clavellin Blanco (Spanish);
Nauruan: Oatamo, Watamo;
Nigeria: Ninge-Ninge;
Niuean: Vine Vao;
Palauan: Kudamono;
Papiamento: Koron'e La Birgui;
Peru: Bedoca, Puru-Puru;
Philippines: Lurunggut, Masaflora (Bikol), Tauñgon (Cebu-Bisaya), Masaplora (Iloko), Lupok-Lupok (Ilongo), Pasionaria Que Hiede (Spanish), Melon-Melonan, Pasionariang-Mabaho, Pasyonaryang-Mabaho, Prutas-Baguio (Tagalog);
Pohnpeian: Pompom, Pwomwpwomw;
Portuguese: Maracujá-Catinga, Maracujá De Cobra, Maracujá De Lagartinho, Maracujá De Sapo; Maracujá De Estalo, Maracujá De Raposa, Maracujá Fedorento, Maracujá De Cheiro, Maracujá Hirsuto Do Sul;
Saipan: Dulce, Ka Thoc Rock, Kinahulo' Atdao;
Samoan: Pāsio Vao, Pasio Vao, Pāsio Vao;
Senegal: Maribisab;
Spanish: Sandía De Culebra, Bombillo, Caguajasa, Canizo, Cuguazo, Pasionaria Hedionda, Pasionaria De La Candelaria, Tagua-Tagua, Parcha De Culebra, Bejuco Canastilla, Cinco-Llagas, Bedoca, Vedoca, Pedón, Taguatagua, Pasiflora Hedionda, Granadilla De Culebra;
Suriname: Markosea;
Thailand: Ka Thok Rok, Rok, Kra Prong Thong, Tam Leung Tong, Tan Lueng Farang, Thao Sing To, Phak Khee Hit, Phal Kheap Farang, Yieo Wua, Yaa Thalok Baat, Rok Chaang;
Tongan: Vaine 'Ae Kuma, Vaine 'Initia;
Venezuela: Parchita De Culebra;
Vietnamese: Mác Quánh Mon (Tày), Chùm Bao, Co Hồng Tiên, Dây Lưới, Dây Nhãn Lồng, Lạc Tiên, (Thái);
Yapese: Tomates.

Origin/Distribution

The species is native to tropical northern south America and West Indies. It is now naturalised in many tropical areas globally and is deemed a pantropical weed in many countries.

Agroecology

The species grows wild as woody climbers amongst trees at the edge of forests, secondary thickets, plantations, coastal dunes, river and creek banks and in gorges, rocky slopes, hill slopes, valley plains, disturbed sites, abandoned fields, roadsides, moist uplands from sea level to 1,000 m on all soil types – rocky or stony soil, gravelly soil, sand, loam, clay. It favours wet areas and thrives in disturbed natural vegetation forming a dense ground cover which prevents or delays the establishment of other species.

Plate 1 Flower of *Passiflora foetida*

Edible Plant Parts and Uses

Yellow ripe fruit can be eaten raw. Young leaves and tender leafy shoots are cooked as vegetables usually in soups and as lalab in Indonesia. The seeds are edible. Dry leaves are used in herbal tea in Vietnamese folk medicine to relieve sleeping problems.

Plate 2 Leaves and immature fruits

Botany

Annual or perennial fetid scrambling or climbing vine with hispid, green cylindrical stems 1.5–5 m long and axillary tendrils. Leaves are alternate, stipulate, hispid-hirsute (Plate 2), base cordate with three palmate lobes having acute apices, 2.5–10 cm long by 2.5–10 cm wide, margin entire or shallowly dentate on 2–5 cm long petioles. Flowers solitary in axils, 4–5 cm wide white with purple centre, 3 cm across, on pedicels 3–7 cm long (Plate 1). Flowers subtended by prominent involucre of three pale green bracts that are 1- to 3-pinnately divided into numerous segments, the ultimate segments glandular; sepals 5, white, linear, 1.5 cm long, pale; 5 linear petals and within corolla is a collar or corona of two rows of purple filaments with white tips, connate and adnate to gynophores, anthers versatile, oblong and dorsifixed; carpels 3, syncarpous; style 3, each with 2- or 3-lobed stigma; ovary intermediate, 1-locular with many ovules. Fruit an indehiscent

Plate 3 Close-view of immature fruit and involucre

berry or capsule, pale green turning yellow, orange, or reddish when ripe, globular, dry inflated, 2–3 cm across, with thin leathery-skinned, nearly 2 cm thick and enveloped by the shaggy involucral pinnatifid bracts (Plates 2, 3 and 4); seeds numerous, wedge-shaped to ovate, about 4 mm long, with coarse reticulate pattern

Plate 4 Ripe orangey yellow fruits

centrally each side, seeds covered with scanty white, sweet, fragrant, mucilaginous and pulp.

Nutritive/Medicinal Properties

The nutrient composition of the raw fruit per 100 g edible portion (Brand Miller et al. 1993) was reported as: moisture: 64.2 g. energy 262 kJ, N 0.78 g, protein 4.9 g, fat 2.7 g, ash 1.3 g, available carbohydrate 0 g, total dietary fibre 9.8 g, Ca 10 mg, Cu 0.7 mg, Fe 8.4 mg, K 491 mg, Na 2 mg, Zn 1.1 mg, riboflavin 0.04 mg, niacin (derived from protein or tryptophan) 0.8 mg, niacin equivalents 0.8 mg, vitamin C 5 mg.

Passiflora foetida seeds were found to contain fatty acids (g/100 g): total lipids 8.8 g, total saturated fatty acids 11.8 g, total MUFA 16.4 g, total PUFA 71.8, 18:1n-9 (oleic acid) (omega-9)16.2 g, 18:2n-6 (linoleic acid) (Omega-6) 71.4 g, and 18:3 n-3 (α-linolenic acid) (Omega-3) 0.4 g (Siriamornpun et al. 2005).

The major phytoconstituents of this plant were found to be alkaloids, phenols, glycosides, flavonoids and cyanogenic compounds (Dhawan et al. 2004); and passifloricins, polypeptides and α-pyrones (Echeverri et al. 2001).

Passiflora foetida was found to have flavonoids pachypodol, 7,4 -dimethoxyapigenin, ermanin), 4,7-O-dimethyl-naringenin, 3,5-dihydroxy-4,7-dimethoxy flavanone (Echeverri and Suarez 1985, 1989), and polyketides α -pyrones (Echeverri et al. 2001). Three polyketides α -pyrones, named passifloricins A, B and C, were isolated from the plant resin. *Passiflora foetida* was found to have harmaline (Krishnaveni and Thaakur 2009). Harmaline is a fluorescent psychoactive indole alkaloid from the group of harmala alkaloids and β-carbolines.

Leaves of *Passiflora foetida* were found to contain C-glycosylflavonoids (Ulubelen et al. 1982). *P. foetida* var. *hispida* was found to contain chrysoeriol, apgenin, isovitexin, vitexin, 2″xylosylvitexin, 2″xylosylisovitexin, luteolin 7-β-D-glucoside, kaempferol, luteolin, mixture of isoschaftoside and vicenin-2 and also glucose, galactose and saccharose. *P. foetida* var. *hisbiscifolia* was found to contain isovitexin, vitexin, 2″xylosylvitexin, 2″xylosylisovitexin, apigenin 7-β-D-glucoside, schaftoside, vicenin-2, a mixture of isoschaftoside and vicenin-2 and also glucose, galactose and saccharose. *P. foetida* was found to also contain cyanohydrin glycosides (Andersen et al. 1998). Five cyanohydrin glycosides with a cyclopentene ring, tetraphyllin A, tetraphyllin B, tetraphyllin B sulphate, deidaclin and volkenin, were isolated from *Passiflora foetida* grown from seeds collected on the Galápagos Islands.

Some pharmacological properties of the plant reported are elaborated below.

Antioxidant Activity

The plant parts of *P. foetida* had varying antioxidant potential as assessed by various assays (Sasikala et al. 2011). The total phenolic content determination showed that *P. foetida* ethanolic peel extract possessed higher (10.09%) phenolic content over the other extracts such as ethanol root extract (9.3%), petroleum ether leaf extract (7.80%), petroleum ether peel extract (6.95%), hot water seed extract (5.42%), and petroleum ether flower extract (5.26%). Among the different samples analysed, the tannin content was found to be highest (4.73%) in ethanol root extract, whereas the other sample extracts showed low tannin content. The ascorbic acid content of *P. foetida* was found to be higher in petroleum ether extract of peel (2.4%) and petroleum ether extract of leaf (2.3%), while the other extracts showed content ranging from 1.2% to 1.7%. The highest free radical scavenging activity was exerted by ethanol leaf extract (595.23 µg). The free radical scavenging activity was found to be least in petroleum ether peel extract (980.39 µg). The scavenging effects of various solvent extracts with the DPPH radical in the ascending order was shown by ethanol leaf extract > petroleum ether root extract > ethanol peel extract > hot water peel extract > ethanol seed extract > petroleum ether seed extract > petroleum ether flower extract > hot water seed extract > ethanol root extract > hot water root extract > ethanol flower extract > hot water flower extract > hot water leaf extract > petroleum ether leaf extract > petroleum ether peel extract.

Potent reducing power was noted for the samples of ethanol root extract and hot water peel extract (0.57 OD_{700}) (Sasikala et al. 2011). Much lower reducing power was found for the petroleum ether root extract (0.27 OD_{700}). All the sample extracts exhibited the ability to chelate metal irons. Among the different samples extracts, the ethanol flower extract showed higher metal chelating activity (6748.1 mg/g) than other sample extracts. In the hydroxyl radical scavenging activity the ethanol peel extract showed the highest level of scavenging activity (92.5%). The flower samples exhibited moderate scavenging activity (62.5%, 64.2%, and 65.5%) compared to other sample extracts. The nitric oxide radical scavenging activity of different solvent extracts also varied. The higher free radical scavenging activity was exerted by ethanol peel extract (120.48 µg) followed by ethanol flower extract > hot water leaf extract > hot water peel extract > ethanol root extract > ethanol leaf extract > hot water leaf extract > hot water flower extract > petroleum ether seed extract > ethanol seed extract > hot water seed extract > petroleum ether leaf extract > petroleum ether flower extract > ethanol leaf extract > petroleum ether peel extract.

In the β-carotene bleaching/linoleic acid peroxidation inhibition activity, the highest antioxidant activity was shown by ethanol leaf extract with 18.52% inhibition when compared to other extracts (Sasikala et al. 2011). RBC (red blood corpuscles) haemolysis was found to be a more sensitive system for evaluating the antioxidant properties of the phytoceuticals. At the concentration of 500 µg in the final reaction mixture, all the sample extracts (leaf, flower, fruit, seed, root) of *P. foetida* exhibited the antihaemolytic activity. Inhibition of haemolysis was found to be the highest in hot water root extract (89.09%) followed by hot water leaf extract (84.96%) and hot water flower extract (83.33%) compared to other extracts.

Anticancer Activity

Studies showed *Passiflora foetida* to have antitumorous activity. The water extracts of fruits of *Passiflora edulis* and *P. foetida* var. *albiflora,* at different concentrations, inhibited the activity of gelatinase MMP-2 and MMP-9 enzymes, two metallo-proteases involved in the tumour invasion, metastasis and angiogenesis (Puricelli et al. 2003). Ethanolic extract of *Passiflora foetida* aerial part exhibited anti-proliferative activity against SKBR3 human breast adenocarcinoma cell line using MTT assay (Moongkarndi et al. 2004).

Methanol extracts of *Passiflora foetida* stem showed DPPH radical-scavenging activity of more than 70% at 100 µg/ml, while the stem and fruit (300 µg/ml) strongly inhibited the melanin production of B16 melanoma cells without significant cytotoxicity (Arung et al. 2009).

Hepatoprotective Activity

Fruits of *P. foetida* were found to possess hepatoprotective activity (Anandan et al. 2009). Pretreatment of the fruit decoction reduced the biochemical markers of hepatic injury like serum glutamate pyruvate transaminase, serum glutamate oxaloacetate transaminase, alkaline phosphatase, total bilirubin and gamma glutamate transpeptidase. Histopathological observations also revealed that pre-treatment with the decoction protected the animals from CCl4 induced liver damage. This property was attributed to the flavonoids present in the fruits of *P. foetida*.

Gastroprotective/Activity

P. foetida ethanol extract treatment significantly reduced the ulcer index and significantly increased the gastric pH of both ethanol and aspirin-induced ulcer rats (Sathish et al. 2011). *P. foetida* showed significant reduction in lipid peroxidation and increase in reduced glutathione levels. The observations confirmed that *P. foetida* plant had antiulcer and antioxidant activities.

Antimicrobial/Insecticidal Activities

The methanolic extract of *P. foetida* roots showed significant activity against *Klebsiella pneumoniae, Pseudomonas aeruginosa* and *Escherichia coli* when compared with the standard antibiotics Levofloxacin, Amikacin and Sparfloxacin (Baby et al. 2010). Earlier studies demonstrated that leaf and fruit extract exhibited antibacterial activities. Studies showed the leaf and fruit extract exhibited varying antibacterial activity against four human pathogenic bacteria i.e. *Pseudomonas putida Vibrio cholerae Shigella flexneri* and *Streptococcus pyogenes* (Mohanasundari et al. 2007). Results indicated that the antibacterial properties of ethanol and acetone extracts. exhibited better antibacterial activity against *Pseudomonas putida, Vibrio cholerae* and *Shigella flexneri*. The acetone extract showed an excellent antibacterial activity against *Vibrio* cholerae followed by *Pseudomonas putida, Shigella flexneri* and *Streptococcus. pyogenes*. But in case of ethanol extract showed higher spectrum of antibacterial activity in *Pseudomonas putida, Vibrio cholerae, Shigella flexneri* and *Streptococcus pyogenes*. Among the two plant parts tested, the leaf extracts exhibited better antibacterial activity than the fruits. In another study, methanolic extract of *P. foetida* plant parts exhibited antimicrobial activity against *Escherichia coli* as tested by agar diffusion and turbidity assays and also antioxidant activity (Bendini et al. 2006). An unknown component, tentatively identified as structural isomer of isoschaftoside, appeared to correlate with antimicrobial activity. Isoschaftoside was also indentified in the leaves of *P. foetida* (Ulubelen et al. 1982).

Ten flavonoids were also isolated from its resin. One of them, ermanin, demonstrated high deterrent activity at 40 ppm against *Dione juno* larvae (Echeverri et al. 1991).

Toxicity Studies

The results of studies by Chivapat et al. (2011) suggested that the ethanolic leaf extract of *P. foetida* at the given doses of 16, 160, 800 and 1,600 mg/kg/day for 6 months did not induce any harmful effects in the rats. The ethanolic leaf extract of *P. foetida* did not affect the body weights, food intake and relative organ weights of Wistar rats, and nor did cause abnormal changes of hematological and biochemical values. Histopathological alterations in the various organs of all extract-treated group did not show any significance, except the adrenal glands of the highest dose male group showed the appearance of fatty infiltration in the cortex; however this phenomenon was regarded to be physiological rather than pathological.

Traditional Medicinal Uses

In fokloric medicine, *Passiflora foetida* leaf infusion has been used to treat hysteria and insomnia in Nigeria (Nwosu 1999). In India, *Passiflora foetida* is traditionally used by the tribes and native medical practitioners for the treatment of diarrhoea, throat and ear infections, liver disorders, tumours, itches, fever and skin diseases and for wound dressing (Chopra et al. 1986). The leaves are applied on the head for giddiness and headache; a decoction is given in biliousness and asthma. The fruit is used as an emetic. In La Reunion, the leaves are considered emmenagogue and are also prescribed in hysteria. In Brazil, the herb is used in the form of lotions or poultices for erysipelas, and skin diseases with inflammation (Chopra et al. 1944). In Malaysia, *Passiflora foetida* has been used for treatment of asthma and in Argentina, to treat epilepsy.

In Vietnam, the leafy shoots and leaves are used medicinally for treating neurasthenia (fatigue, anxiety and listlessness), palpitation, insomnia, early menstruation, oedema, itching and coughs. It is reported to tranquillize the nervous system, remove toxic heat and cool the liver. The decoction of leaves and fruits have also been reported to be used to treat asthma and biliousness, leaves and root decoction for emmenagogue, and for hysteria. The herbal extracts have been used in traditional medicine to cure many diseases like diarrhoea, intestinal tract, throat, ear infections, fever and skin diseases. The leaf paste is applied on the head for giddiness and headache in India. In Brazil, the herb is used in the form of lotions or poultices for erysipelas and skin diseases with inflammation.

Other Uses

Passiflora foetida is also used as cover plant and as hedge plant.

Comments

Unripe fruit is poisonous. *Passiflora foetida* is deemed a noxious weed in many countries.

Selected References

Anandan R, Jayakar B, Manavalan R (2009) Hepatoprotective activity of the decoction of the fruits of *Passiflora foetida* Linn. on CCl4 induced hepatic injury in rats. J Pharm Res 2(12):1857–1859

Andersen L, Adsersen A, Jerzy W, Jaroszewski JW (1998) Cyanogenesis of *Passiflora foetida*. Phytochemistry 47(6):1049–1050

Arung ET, Kusuma IW, Christy EO, Shimizu K, Kondo R (2009) Evaluation of medicinal plants from Central Kalimantan for antimelanogenesis. J Nat Med 63(4): 473–480

Baby E, Balasubramaniam A, Manivannan R, Jose J, Senthilkumar N (2010) Antibacterial activity of methanolic root extract of *Passiflora foetida* Linn. J Pharm Sci Res 2(1):38–40

Backer CA, Bakhuizen van den Brink RC Jr (1963) Flora of Java, vol 1. Noordhoff, Groningen, 648 pp

Bendini A, Cerretani L, Pizzolante L, Toschi TG, Guzzo F, Ceoldo S, Marconi AM, Andreetta F, Levi M (2006) Phenol content related to antioxidant and antimicrobial activities of *Passiflora* spp. Extr Eur Food Res Technol 223(1):102–109

Brand Miller J, James KW, Maggiore P (1993) Tables of composition of Australian aboriginal foods. Aboriginal Studies Press, Canberra

Burkill IH (1966) A dictionary of the economic products of the Malay Peninsula. Revised reprint, 2 vols. Ministry of Agriculture and Co-operatives, Kuala Lumpur, Malaysia. Vol 1 (A–H) pp 1–1240, vol 2 (I–Z) pp 1241–2444

Chivapat S, Bunjob M, Shuaoprom A, Bansidhi J, Chavalittumrong P, Rangsripipat A, Sincharoenpoka P (2011) Chronic toxicity of *Passiflora foetida* L. extract. Int J Appl Res Nat Prod 4(2):24–31

Chopra RN, Badhwar RL, Ghosh S (1944) Poisonous plants of India. Public Service Commission, Govt. of West Bengal, Calcutta, pp 469–472

Chopra RN, Nayar SL, Chopra IC (1986) Glossary of Indian medicinal plants. (Including the supplement). Council Scientific Industrial Research, New Delhi, 330 pp

Dhawan K, Dhawan S, Sharma A (2004) *Passiflora*: a review update. J Ethnopharmacol 94:1–23

Echeverri F, Suarez GE (1985) Flavonoids from the surface of *Passiflora foetida* L. (Passifloraceae). Actual Biol 14:58–60

Echeverri F, Suarez GE (1989) Flavonoids from *Passiflora foetida* and deterrant activity. Rev Latinoam Quim 20:6–7

Echeverri F, Cardona G, Torres F, Pelaez C, Quiñones W, Renteria E (1991) Ermanin: an insect deterrent flavonoid from *Passiflora foetida* resin. Phytochemistry 30(1):153–155

Echeverri F, Arango V, Quiñones W, Torres F, Escobar G, Rosero Y, Archbold R (2001) Passifloricins, polyketides alpha-pyrones from *Passiflora foetida* resin. Phytochemistry 56(8):881–885

Global Invasive Species Database (2005) *Passiflora foetida* (vine, climber). http://www.invasivespecies.net/database/species/ecology.asp?fr=1&si=341&sts=

Holm LG, Plucknett DL, Pancho JV, Herberger JP (1977) The world's worst weeds: distribution and biology. East-west Center/University Press of Hawaii, Honolulu, 609 pp

Jansen PCM, Jukema J, Oyen LPA, van Lingen RG (1992) Minor edible fruits and nuts. In: Verheij EWM, Coronel RE (eds) Plant resources of South-East Asia no. 2. Edible fruits and nuts. Prosea Foundation, Bogor, pp 313–370

Krishnaveni A, Thaakur SR (2009) Quantification of harmaline content in *Passiflora foetida* by HPTLC technique. J Pharm Res 2(5):789–791

Mohanasundari C, Natarajan D, Srinivasan K, Umamaheswari S, Ramachandran A (2007) Antibacterial properties of *Passiflora foetida* L. – a common exotic medicinal plant. Afr J Biotechnol 6(23):2650–2653

Moongkarndi P, Kosem N, Luanratana O, Jongsomboonkusol S, Pongpan N (2004) Antiproliferative activity of Thai medicinal plant extracts on human breast adenocarcinoma cell line. Fitoterapia 75:375–377

Nadkarni KM, Nadkarni AK (1982) Indian materia medica with ayurvedic, unani-tibbi, siddha, allopathic, homeopathic, naturopathic & home remedies, vol 2, 2nd edn. Sangam Books, Bombay

National Institute of Materia Medica (1999) Selected medicinal plants in Vietnam, vol 2. Science and Technology Publishing House, Hanoi, 460 pp

Nwosu MO (1999) Herbs for mental disorders. Fitoterapia 70(1):58–63

Ochse JJ, Bakhuizen van den Brink RC (1980) Vegetables of the Dutch Indies, 3rd edn. Ascher & Co., Amsterdam, 1016 pp

Pacific Island Ecosystems at Risk (PIER) (1999) *Passiflora foetida* L., Passifloraceae. http://www.hear.org/Pier/species/passiflora_foetida.htm

Pongpangan S, Poobrasert S (1985) Edible and poisonous plants in Thai forests. Science Society of Thailand, Science Teachers Section, Bangkok, 206 pp

Puricelli L, Dell'Aica I, Sartor L, Garbisa S, Caniato R (2003) Preliminary evaluation of inhibition of matrix-metalloprotease MMP-2 and MMP-9 by *Passiflora edulis* and *P. foetida* aqueous extracts. Fitoterapia 74(3):302–304

Sasikala V, Saravana S, Parimelazhagan T (2011) Evaluation of antioxidant potential of different parts of wild edible plant *Passiflora foetida* L. J Appl Pharm Sci 1(4):89–96

Sasikala et al. (2011) Nwosu inserted: Nwosu MO (1999) Herbs for mental disorders. Fitoterapia 70(1):58–63

Sathish R, Sahu A, Natarajan K (2011) Antiulcer and antioxidant activity of ethanolic extract of *Passiflora foetida* L. Indian J Pharmacol 43:336–339

Siriamornpun S, Yang LF, Li D (2005) Alpha linolenic acid content in edible wild seeds in Thailand. Asia Pac J Clin Nutr 14(Supplement):S100

Stuart GU (2010) Philippine alternative medicine. Manual of some Philippine medicinal plants. http://www.stuartxchange.org/OtherHerbals.html

Tanaka Y, Nguyen VK (2007) Edible wild plants of Vietnam: the Bountiful Garden. Orchid Press, Bangkok, 175 pp

Ulubelen A, Topcu G, Mabry TJ, Dellamonica G, Chopin J (1982) C-glycosylflavonoids from *Passiflora foetida* var. *hispida* and *P. foetida* var. *hibiscifolia*. J Nat Prod 45(1):103

Wang YZ, Krosnick SE, Jørgensen PM (2007) *Passiflora* Linnaeus. In: Wu ZY, Raven PH, Hong DY (eds) Flora of China, vol 13 (Clusiaceae through Araliaceae). Science Press/Missouri Botanical Garden Press, Beijing/St. Louis

Passiflora ligularis

Scientific Name

Passiflora ligularis Juss.

Synonyms

Passiflora lowei Heer ex Regel, *Passiflora serratistipula* Moc. & Sessé ex DC., *Passiflora tiliaefolia* Sessé & Moc. non L.

Family

Passifloraceae

Common/English Names

Granadilla, Sweet Granadilla, Grenadia, Sweet Passion Fruit, Water lemon.

Vernacular Names

Bolivia: Granadilla;
Brazil: Maracujá-Urucú (Portuguese);
Costa Rica: Granadilla;
Danish: Sød Granadil, Amaril;
Dutch: Zoete Markoesa;
Eastonian: Keeljas Kannatuslill;
Ecuador: Granadilla;
French: Barbadine, Granadille, Grenadelle, Grenadille Douce, Grenadille Des Montagnes;
German: Granadille, Süße Granadilla, Zungenförmige Passionsblume;
Guatemala: Granada China, Granadilla Común, Cranix;
Hawaiian: Lani Wai, Lemi Wai, Lemona;
Indonesia: Buah Belebar, Buah Selaseh, Buah Susu, Markusa Leutik;
Jamaica: Granaditta;
Mexico: Granadilla;
Papua New Guinea: Sugar Fruit;
Peru: Apicoya, Granadilla, Tintin;
Spanish: Cranix, Granadita, Parcha Dulce;
Venezuela: Granadita, Granadilla De China, Parchita Amarilla, Parcha Dulce, Parcha Importada.

Origin/Distribution

Sweet Granadilla is native to the Andes Mountains between Bolivia and Venezuela, with Peru as the main producer. It grows as far south as northern Argentina and as far north as Mexico. Outside of its native range it grows in Florida, New Zealand, China and in tropical highlands of East Africa, South Africa Sri Lanka, Jamaica, Indonesia, Hawaii, Papua New Guinea and Australia. The major producing countries are Peru, Venezuela, Colombia, Ecuador, Brazil, South Africa, and Kenya. The main importing countries are the United States, Canada and Europe (Belgium, Holland, Switzerland, and Spain).

Agroecology

Sweet granadilla thrives best in the cool sub-tropics in areas with temperatures ranging from 15°C to 18°C and between 600 and 1,000 mm of annual precipitation. It occurs wild and is cultivated in its natural range at 900–2,700 m elevations. It can grow in the cool highlands in the tropics like in Indonesia, Papua New Guinea, Jamaica, Sri Lanka and elsewhere in the tropics. It is grown at high elevations in the tropics, between 2,100 and 2,700 m in Ecuador, and between 800 and 3,000 m in Bolivia and Colombia, and at lower elevations in the subtropics. It is naturally adapted to high cool humid mesic forests and wet rainforests. The plant is intolerant of heat and can withstand short periods of light frost. It grows on loamy to well drained, light clayey soils and on moist volcanic soils like in Indonesia and Hawaii. It prefers a soil pH range of 6.1–7.5.

Edible Plant Parts and Uses

The fresh, juicy, aromatic arils surrounding the seeds are eaten together, scooped from halved fruit with a spoon. The fruits are also eaten served over ice-cream or processed into juice, nectars and used in cakes.

Botany

A robust, vigorous liana (vine) with terete or weakly angled, striate stems and a woody base, climbing by tendrils up trees. Leaves cordate, 8–20 cm long by 6–15 cm wide, apex pointed, margin entire, conspicuously veined, medium green adaxially and pale green abaxially, glabrous, borne on petioles with 4–8 elongate, usually paired filiform nectaries (glands) and with oblong-ovate, 20–40 mm long stipules (Plate 1). Flowers pendent, showy, large, campanulate, 8–12 cm in diameter, peduncles solitary or paired, bracts ovate, 3–5 cm long, 1–3 cm wide; hypanthium 0.5–0.9 cm long; sepals and petals greenish white or white tinged with violet; corona in 5–7 rows, the filaments with white and purple bands, 3 cm long (Plate 1). The fruit is broad-ellipsoid, ovoid to sub-globose, 6–8 cm long, green with purple blush on the sunny side and minutely white-dotted when unripe (Plate 2), orange-yellow with white specks when ripe (Plates 3 and 4). The rind is smooth, thin, firm and brittle externally, white and soft on the inside, enclosing numerous black, flat, pitted seeds. Each seed is enveloped by greyish to yellowish-white, mucilaginous, very juicy, aromatic aril (pulp).

Nutritive/Medicinal Properties

Proximate nutrient composition of the pulp and seeds together per 100 g edible portion based on analyses made in Ecuador, El Salvador, Costa Rica and Guatemala (Morton 1987) was reported as: moisture 69.9–79.1 g, protein 0.340–0.474 g, fat 1.50–3.18 g, crude fibre 3.2–5.6 g, ash 0.87–1.36 g, calcium 5.6–13.7 mg, phosphorus 44.0–78.0 mg, iron 0.58–1.56 mg, carotene 0.00–0.035 mg, thiamine 0.00–0.002 mg, riboflavin 0.063–0.125 mg, niacin 1.42–1.813 mg, ascorbic acid 10.8–28.1 mg.

A new polysaccharide with a high molecular weight (greater than 1×10^6 Da) consisting of six different sugar residues: xylose, glucose, galactose, galactosamine, an unknown component, and fucose in the relative ratio of 1:0.5:0.2:0.06:0.05:traces was extracted and characterized from the peels of *Passiflora ligularis* (granadilla) fruits (Tommonaro et al. 2007). The formation of a biodegradable film using this novel xyloglucan was reported, and the anticytotoxic activity of the polysaccharide was also observed using the brine shrimp bioassay. Considerable antioxidant activity (Trolox equivalent antioxidant capacity (TEAC) value of 0.32 µM/mg fresh product) of this xyloglucan was noted in the lipophilic extracts of *Passiflora ligularis* fruit, indicating, that the fruit could provide an alternative source of bioactive compounds.

Plate 1 Flowers and leaves of sweet granadilla

Plate 2 Young, immature sweet granadilla fruits

Plate 3 Ripe sweet granadilla fruits

Passiflora ligularis

Plate 4 (**a**, **b**) Ripe sweet granadilla fruits on sale in local markets

Other Uses

None, sweet granadilla is grown for its fruit and fragrant, ornamental flowers.

Comments

The plant is propagated from seeds or cuttings.

Selected References

Fouqué A (1972) Espèces fruitières d'Amérique tropicale. IV. Les Passiflorées. Fruits 27:368–382

Green PS (1972) *Passiflora* in Australasia and the Pacific. Kew Bull 26:539–558

Jansen PCM, Jukema J, Oyen LPA, van Lingen RG (1992) Minor edible fruits and nuts. In: Verheij EWM, Coronel RE (eds) Plant resources of South-East Asia no. 2: edible fruits and nuts. Prosea Foundation, Bogor, pp 313–370

Morton JF (1987) Sweet Granadilla. In: Fruits of warm climates. Julia F. Morton, Miami, pp 330–331

National Research Council (1989) Lost crops of the Incas: little-known plants of the Andes with promise for worldwide cultivation. BOSTID/National Research Council/National Academy Press, Washington, D.C., 428 pp

Tommonaro G, Rodríguez CS, Santillana M, Immirzi B, Prisco RD, Nicolaus B, Poli A (2007) Chemical composition and biotechnological properties of a polysaccharide from the peels and antioxidative content from the pulp of *Passiflora ligularis* fruits. J Agric Food Chem 55(18):7427–7433

Wagner WL, Herbst DR, Sohmer SH (1999) Manual of the flowering plants of Hawaii, Revised edn. Bernice P. Bishop Museum special publication/University of Hawai'i Press/Bishop Museum Press, Honolulu, 1919 pp (two volumes)

Passiflora miniata

Scientific Name

Passiflora miniata Vanderplank.

Synonyms

Passiflora coccinea hort.

Family

Passifloraceae

Common/English Names

Monkey-Guzzle, Passionflower, Red Granadilla, Red Passion Flower, Red Passion Vine, Red Passionflower, Scarlet Passion Flower, Scarlet Passionflower.

Vernacular Names

Bolivia: Pachio-Tutumillo;
Brazil: Maracujá-De-Flor-Vermelha, Maracujá-Poranga (Portuguese);
French: Grenadille À Fleurs Rouges, Liane Serpent, Passiflore Écarlate;
French Guiana: Liane Trèfle, Pomme Rose;
German: Rote Passionsblume;
Guyana: Monkey-Guzzle;
Portuguese: Maracuja Poranga, Maracujá-Tomé-Açú, Maridi-Oúra, Tomé-Açú;
Spanish: Granadilla Rojo, Pachio De Flor Roja, Passionaria Roja;
Suriname: Snekie Marcoesa (Dutch).

Origin/Distribution

The species is native to the Amazon region of Peru, Brazil, Columbia and Bolivia as well as Venezuela and the Guianas.

Agroecology

In its native range, it occurs in the tropical Amazonian lowlands, but in Bolivia it can be found at altitudes of about 2,000 m in the western foothills of the Andes. It occurs on non-inundated lateritic soil. Like all passion fruit, it prefers a well-drained soil enriched with organic matter, such as well-rotted manure or compost. It thrives in a warm sunny position – a north to north-westerly aspect against a masonry wall would be necessary in cooler zones. It requires deep watering at least once a week during warm weather and should be kept well-mulched.

Edible Plant Parts and Uses

The aril of ripe fruit is eaten fresh and made into juice.

Botany

A fast growing tendril-climbing vine, rufo-tomentose throughout with narrowly linear-lanceolate, shallowly serrulate stipules and with reddish-purplish stems. Leaves are oval-oblong, 6–14 cm long, 3–7 cm wide, subcordate, shallowly crenulate, medium-green, glabrous or sparingly pubescent above, ferruginous-tomentose beneath (Plates 1 and 2). Peduncles up to 8 cm long; bracts ovate and coriaceous. Flowers scarlet or red, bisexual; perianth 10 (5 sepals, 5 petals), strap-like, bright scarlet red, reflex backwards, with a central, short upright corona with three series of corona filaments, the outer series being purple in colour and the inner series being white (Plates 1 and 2). Stamens 5, reddish filaments free at apex, connate into tube around ovary, with five linear, yellowish-green dorsifixed, versatile, oblong anthers. Ovary yellowish-tomentose, stipitate on androgynophore, styles and stigmas 3, pinkish-red and stigma 3 capitate. Fruit subglobose or ovoid, about 5 cm across, orange or yellow in colour and are mottled green with edible aril covering the minutely reticulate seeds.

Plate 2 Close-view of flower

Plate 1 Flowers, buds and leaves of red passion flower

Nutritive/Medicinal Properties

No information on the nutrient composition of the fruit has been published.

A novel cyclopentenoid cyanogenic glycoside (1-(6-O-β-D-rhamnopyranosyl-β-D-glucopyranosyloxy)-cyclopent-2-en-1-nitrile-4-sulphate) named passicoccin was isolated from *Passiflora coccinea* (Spencer and Seigler 1985).

Other Uses

Red Passion flower is a popular ornamental for growing on a pergola, trellis, arch or fence and it is also used as a rootstock for other *Passiflora* species.

Comments

A well-documented *Passiflora* species from Bolivia, Brazil and Colombia, which has been extensively cultivated under the erroneous name *Passiflora coccinea* Aubl., is here described as *P. miniata* Vanderplank (Vanderplank 2006). The true *Passiflora coccinea* has two series of corona filaments with the outer series being white or pale pink, large floral bracts and upright pear-shaped fruits that are golden-brown in colour. The *Passiflora coccinea* of cultivation has three series of corona filaments, the outer series being purple in colour and the two inner series being white, small floral bracts and large, subspherical, pendulous and mottled fruits.

Selected References

Killip EP (1938) The American species of Passifloraceae [prim.]. Publ Field Mus Nat Hist Bot Ser 19(1): 1–331

Macbride JF (1941) Passifloraceae, flora of Peru. Publ Field Mus Nat Hist Bot Ser 13(4/1):90–132

Spencer K, Seigler DS (1985) Passicoccin: a sulphated cyanogenic glycoside from *Passiflora coccinea*. Phytochemistry 24(11):2615–2617

Ulmer T, MacDougal JM (2004) *Passiflora*: passion-flowers of the world. Timber Press, London, 430 pp

Vanderplank J (1996) Passion flowers, 2nd edn. MIT Press, Cambridge, 224 pp

Vanderplank J (2006) 562. *Passiflora miniata*. Passifloraceae. Curtis's Bot Mag 23(3):223–230

Passiflora quadrangularis

Scientific Name

Passiflora quadrangularis L.

Synonyms

Passiflora grandiflora Salisb., *Passiflora macrocarpa* M.T. Mast., *Passiflora quadrangularis* L. var. *variegata* hort., *Passiflora sulcata* Jacq., *Passiflora tetragona* M. Roem.

Family

Passifloraceae

Common/English Names

Baden, Barbadine, Giant Granadilla, Giant Grenadilla, Granadille True, Granadilla, Grenedilla, Grenadine, Square Stalked Passion Flower, Square-Stem Passion Flower.

Vernacular Names

Bolivia: Quijón, Granadilla Real, Parcha Granadina, Parcha De Guinea, Sandia De Pasión;
Brazil: Maracujá-Açu, Maracujá-Assú, Maracujá De Caiena, Maracujá-Grande, Maracujá-Mamao, Maracujá-Melão, Maracujá-Suspiro, Maracuya-Acu, Maracuja Silvestre;
Chinese: Da Guo Xi Fan Lian;
Columbia: Badea, Corvejo;
Cook Islands: Kūkuma, Maratini, Maratini, Papatini, Pāpatīni, Pāpatīni, Pārapōtini (Maori);
Czech: Mučenka Obrovská;
Danish: Kæmpegranadil, Barbadin;
Dutch: Djari Markoesa, Granadilla, Markiza, Markoesa, Vierhoekige Passiebloem;
Eastonian: Suureviljaline Kannatuslill;
Ecuador: Tumbo, Tambo, Tumbo Costero;
El Salvador: Granadilla Grande, Granadilla De Fresco, Granadilla Para Refrescos;
French: Barbadine, Grenadille Géante, Passiflore Quadrangulaire;
German: Granadilla, Granadillas, Melonengranadille, Königs-Grenadille, Riesen-Grenadille, Riesen-Königsgranadille;
Hungarian: Óriás Passiógyümölcs;
Indonesia: Gardanela (Alfurese, N Sulawesi), Ansimon Bolanda, Antjimon Eropa, Buah Eropa, Labu Europa (Batak), Sumangga (Boeol, Sulawesi), Labu Belanda (Jambi), Manisah, Markisa, Markisat (Java), Air Bis, Bis, Kerebis, Rebis (Lampong), Kerbis (Lingga), Manesa, Markesah (Madurese), Erbis, Kerbis, Belewa (Malay), Rubis (Palembang), Erbis (Singkep), Erbis, Herbis, Markisat (Sundanese);
Italian: Passiflora Quadrangolare;
Mangarevan: Para Patini;
Malaysia: Timun Belanda, Timun Hutan, Akar Mentimun, Telur Dewa, Gendola;
Mexico: Sandía De Le Pasión;
Niuean: Palasini, Palatini, Vinē Palasini;

Palauan: Kudamono; *Peru*: Tumbo, Tambo, Tumbo Costero;
Philippines: Kasaflora, Paróla (Illoko), Granadilla (Tagalog);
Portuguese: Guassú, Guassú, Maracujá Açú, Maracujá Assu, Maracujá Cascudo, Maracujá De Caiena, Maracujá De Quatro Quinhas, Maracujá Do Igapó, Maracujá Do Pará, Maracujá Grande, Maracujá Mamão, Maracuja Melão, Maracuja-Caiana, Uauaçu, Martírio Quadrangular;
Samoan: Pasio;
Spanish: Badea, Badera, Corvejo, Granadilla De Fresco, Granadilla Grande, Granadilla Para Refrescos, Granadilla Real, Parcha De Guinea, Parcha Granadilla, Parcha Granadina, Pasionaria, Quijón, Sandía De La Pasión;
Sri Lanka: Seemaisora Kai;
Surinam: Grote Markoesa, Groote Markoesa;
Swedish: Barbadin;
Tahitian: Para Pautini;
Taiwan: Da Guo Xi Fan Lian;
Thailand: Ma Thuarot (Lamphun), Sao Warot, Sukhon Tharot (Bangkok), Taeng Kalaa (Chiang Mai);
Tongan: Pasione;
Tongarevan: Pālapōtini, Pārapōtini;
Venezuela: *Badea,* Parcha Granadina;
Vietnam: Chum Bao Duá Duá Gang Táy.

Origin/Distribution

The species is reported to be indigenous to tropical America, but its exact origin is uncertain probably northern South America. It has become naturalized in moist habitats in tropical lowlands especially in central and south America. It is cultivated all over in South and Central America, Hawaii, Southeast Asia, India, Australia, West Africa, the Pacific islands and other tropical regions.

Agroecology

Giant Granadilla has become naturalised in many tropical countries particularly in wet forests and thickets. It is truly a topical species requiring warm diurnal temperatures and high humidity. It grows from near sea level to 1,000 m altitude but in Ecuador it also grows up to 2,000 m. It is cold sensitive and will be killed by cold winters. It grows on a wide range of soils from alluvial, sandy, loams, loamy clays, volcanic and granitic soils but does best in deep, fertile, humus-rich, moist and well drained soils.

Edible Plant Parts and Uses

Ripe fruit flesh is eaten fresh after removal of the inner skin. Since it is rather bland it is often added to papaya, banana or pineapple slices in fruit salad flavoured with lime or lemon juice. The flesh is also cooked with sugar or honey and eaten as dessert or is canned in syrup or candied. In Indonesia, the flesh and arils are eaten together with sugar and shaved ice or used in rujak. In Australia, the flesh is eaten by adding orange juice and cream. Flesh and arils are also stewed and used as pie fillings.

The aril (pulp) is the primary product enjoyed fresh eaten with seeds intact scooped with spoon, with ice cream, or processed into fruit juice, frozen sherbets, syrup or canned nectar. The fruit can be juiced to prepare refreshing chilled drinks. Granadilla juice or nectar is often blended with orange, pineapple, guava or papaya juice. It is also made into cordials and squashes and carbonated beverages. The juice and nectar is bottled in Indonesia and served in restaurants. In Australia, the whole ripe fruit is also processed into wine by mashing and adding brandy and allow the mixture to ferment for a few weeks. The young, unripe fruit may be prepared into a savoury *sayur* or *sambal goreng* or steamed or boiled and served as a vegetable in *sayur* or eaten as *lalab* in Indonesia. The young fruit may be cut up into finger-sized slices, breaded and cooked in butter with milk, pepper and nutmeg for the Europeans in Java. The roots of old vines is baked and eaten as a substitute for yam in Jamaica but the roots and also leaves have been stated to be poisonous (Burkill 1966).

Botany

Giant Granadilla is a robust, perennial tendrilled vine with thick, yellowish green, quadrangular (4-angled) stem, glabrous when young becoming fistular when old, 5–50 m long, woody at the base. Tendrils are axillary, to 30 cm long, coiled, flanked by ovate to ovate-lanceolate stipules, 2–3.5 cm long (Plate 3). Leaves are alternate, ovate to broadly elliptic-ovate (Plate 2), 7.8–13 cm long, 6–10.2 cm wide, glabrous, margins entire, base shallowly cordate, apex abruptly acuminate, glabrous on both surfaces, pinnatinerved with 10–12 pairs of nerves prominent abaxially, dark green adaxially and pale green abaxially, petioles long, trigonous, with 4–6 usually paired, globose nectaries 1–2 mm in diameter. Flowers are solitary, pendulous, shortly stalked, fragrant, showy, large, campanulate, 7–10 cm across, hypanthium 0.6–0.9 cm long; bracts 3 on top of peduncle, ovate and sessile, sepals 5 and petals 5, white or white tinged pink or purple; corona at the margin of the hypanthium in 4–5 rows, innermost rows consisting of short, unequal threads red with white blotches, outer 2 rows consisting of 3 cm long, radiating, erect threads, reddish-brown below and lilac blotches above (Plate 1). Stamens 5, connate with gynophore into a column (gynophore), marked with yellowish green and violet dots. Ovary yellowish –green, ellipsoid, 1.2–1.5 cm long, glabrous, longitudinally furrowed, 1 locule with numerous parietal ovules, stigma reniform or cordiform. Berry very large, ovoid-oblong or ellipsoid, 15–30 cm long by 10–20 cm across with rounded ends, yellowish green or pale green or tinged with pink (Plates 1, 4 and 5). Beneath the rind is the fleshy edible mesocarp, 2–3 m thick, spongy, juicy, sweetish, white or yellowish white. The central

Plate 1 Flowers and granadilla fruit

Plate 3 Terminal shoot with axillary tendrils and stipules

Plate 2 Leaves and angular stem

Plate 4 Giant granadilla fruit

Plate 5 Granadilla fruit halved

cavity contains some juice and masses of flattened-obovoid, 1.25 cm long, purplish-brown seeds. Each seed is enclosed by whitish, yellowish, partly yellow or purple-pink, sweet-acid, edible aril.

Nutritive/Medicinal Properties

Nutrient composition of the fruit flesh of barbadine in terms of 100 g edible portion had been reported as: energy 77 cal., moisture %, protein 2.6 g, fat 1.9 g, carbohydrate 15.5 g, fibre 4.9 g, ash 1.0 g, Ca 9 mg, P 36 mg, Fe 0.6 mg, and vitamin C 20 mg (Leung et al. 1968). Another analysis carried out in El Salvador (Morton 1987) provided the following nutrient composition of the flesh per 100 g edible portion: moisture 94.4 g, protein 0.112 g, fat 0.15 g, crude fibre 0.7 g, ash 0.41 g, calcium 13.8 mg, phosphorus 17.1 mg, iron 0.80 mg, carotene 0.004 mg, riboflavin 0.033 mg, niacin 0.378 mg, ascorbic acid 14.3 mg. Food value of the aril and seed is: moisture 78.4 g, protein 0.299 g, fat 1.29 g, crude fibre 3.6 g, ash 0.80 g, calcium 9.2 mg, phosphorus 39.3 mg, iron 2.93 mg, carotene 0.019 mg, thiamine 0.003 mg, riboflavin 0.120 mg, and niacin 15.3 mg.

Free volatile extract of barbadine fruit was characterized as a mixture of 57 components with (5E)-2,6-dimethyl-5,7-octadiene-2,3-diol, (2E)-2,6-dimethyl-2,5-heptadienoic acid, benzoic acid, furaneol, benzyl alcohol, 2,6-dimethyl-5-hepten-1-ol, dimethylheptadienoic acid, and 2,6-dimethylheptenol, as major components (Osorio et al. 2002). Three of those: (E)-2-pentenol, 2,6-dimethyl-5-hepten-1-ol, and (2E)-2,6-dimethyl-2,5-heptadienoic acid, displayed a strong resemblance to the aroma of the fresh fruit. Volatile compounds released by enzymatic hydrolysis of glycosidic and phosphate extracts were all present in the aroma of the fruit pulp. This fact strongly suggests an active role of glycosides and phosphates in the generation of fruit aroma. Furthermore, the new monoterpenoids trans-4-hydroxylinalool 3,6-oxide and cis-4-hydroxylinalool 3,6-oxide were characterized as major transformation products of (5E)-2,6-dimethyl-5,7-octadiene-2,3-diol under acidic conditions. In earlier studies, 3,7-Dimethyl-3(E)-octene-1,2,6,7-tetraol, a monoterpene isolated from *Passiflora quadrangularis* fruit pulp, was established to be a 12:42:14:32 mixture of (2R,6R)-, (2R,6S)-, (2S,6R)- and (2S,6S)-stereoisomers (Osorio et al. 1999). Monoterpenoids (2E)-2,6-dimethyl-2,5-heptadienoic acid, (2E)-2,6-dimethyl-2,5-heptadienoic acid β-D-glucopyranosyl ester, (5E)-2,6-dimethyl-5,7-octadiene-2,3-diol, and (3E)-3,7-dimethyl-3-octene-1,2,6,7-tetrol were isolated from the fruit pulp along with the known 2,5-dimethyl-4-hydroxy-3(2H)-furanone β-D-glucopyranoside (Osorio et al. 2000).

Cyclopropane triterpene glycosides isolated from the methanolic extract of the leaves of *Passiflora quadrangularis* included quadranguloside (9,19-cyclolanost-24Z-en-3β,21,26-triol-3,26-di-O-gentiobioside) (Orsini et al. 1985), 9,19-cyclolanosta-22,25-epoxy-3β-21,22(R)-triol-3β-O-gentiobioside and 9,19-cyclolanosta-21,24-epoxy-3β-25,26-triol-3β-O-gentiobioside, together with oleanolic acid-3-sophoroside (Orsini et al. 1987). C-glycosyl flavonoids namely orientin, isoorientin, vitexin and isovitexin were found in the leaves but not in the pericarp of *P. quandragularis* (Zucolotto et al. 2011). A set of two diastereomers of phenylcyano glycosides, (7S)-phenylcyanomethyl 1′-O-α-L-rhamnopyranosyl-(1→6)-β-D-glucopyranoside and (7R)-phenylcyanomethyl 1′-O-α-L-rhamnopyranosyl-(1→6)-β-D-glucopyranoside were isolated from the methanol extract of dried vines of *P. quadrangularis* (Saeki et al. 2011).

Four anthocyanin pigments were identified from Giant granadilla flowers: cyaniding 3 monoglucoside, malvidin 3:5 diglucoside, petunidin 3:5 diglucoside and delphinidin 3:5 diglucoside (Halim and Collins 1970).

Some pharmacological activities of Giant granadilla reported are presented below.

Flavonoids and Antioxidant Activity

In callus culture, *P. quadrangularis* turned out to have a faster growth rate and a more friable texture and was selected to investigate its capacity to produce four glycosyl flavonoids (orientin, isoorientin, vitexin, isovitexin). In callus cultures only small amounts of isoorientin were found, while the concentration of the other flavonoids was below the detection limit. UV-B irradiation of calluses was able to increase the production of all four glycosyl flavonoids. After a 7-day exposure of cultures to UV-B light, the production of isoorientin reached concentrations similar to those found in fresh leaves from glasshouse-grown plants. Elicitation with methyl jasmonate also enhanced orientin, vitexin and isovitexin concentrations, even though the stimulation was about 6-fold weaker for orientin and vitexin and about 40-fold for isovitexin, than that exerted by UV-B treatment. Callus cultures treated with the UV-B dose which most enhanced flavonoid production showed a higher antioxidant activity compared to untreated calluses, with an increase ranging from 28% to 76%. Results show that the secondary metabolite biosynthetic capacity of *Passiflora* tissue cultures could be enhanced by appropriate forms of elicitation.

Hemolytic Activity

The leaves were also found to contain haemolysin (Yuldasheva et al. 2005). Hemolytic activity was found in the in n-butanol factions. Hemolysins and cytolysins present in many tropical plants, are potential bactericidal and anticancer drugs. The data suggested that membrane cholesterol was the primary target for this hemolysin and that several haemolysin molecules form a large transmembrane water pore. The properties of the *Passiflora* hemolysin, such as its frothing ability, positive colour reaction with vanillin, selective extraction with n-butanol, HPLC profile, cholesterol-dependent membrane susceptibility, formation of a stable complex with cholesterol, and rapid erythrocyte lysis kinetics indicated it to be probably a saponin.

Anxiolytic Activity

The hydroalcohol extract of *Passiflora quadrangularis* leaves exhibited anxiolytic activity in dosages around 100, 250 and 500 mg/kg, as expressed by elevation of the time spent on the open arms in the plus-maze; a decrease of freezing and an increase of deambulation and rearing in the open field test (De Castro et al. 2007). The hydroalcohol extract showed results similar to diazepam on the holeboard. No positive results were found for the aqueous extract.

Antivenom Activity

Leaves and branches of Giant granadilla also exhibited moderate neutralizing ability (21–72%) against the haemorrhagic effect of *Bothrops atrox* venom at doses up to 4 mg/mouse (Otero et al. 2000).

Traditional Medicinal Uses

Passiflora quadrangularis plant has been used for hypertension in traditional ethno-medicine. The fruit is valued as antiscorbutic and stomachic in the tropics. In Brazilian folk medicine, the fruit flesh is used as a sedative to relieve nervous headache, asthma, diarrhoea, dysentery, neurasthenia and insomnia. The seeds contain a cardiotonic principle, are sedative, and, in large doses, narcotic. The leaf decoction is used as a vermifuge and for bathing skin afflictions. Leaf poultices are applied in liver disorders. The root is used as an emetic, diuretic, vermifuge and narcotic. It is applied as a soothing poultice when powdered and mixed with oil.

Fermented juice has been used for body cleansing. Giant granadilla is used throughout the Caribbean as a sedative and for headaches. Leaf tea is taken for high blood pressure and diabetes (Seaforth et al. 1983).

Other Uses

The plant is also used as hedge planting.

Comments

The plant is readily propagated by seeds.

Selected References

Backer CA, Bakhuizen van den Brink RC Jr (1963) Flora of Java (spermatophytes only), vol 1. Noordhoff, Groningen, 648 pp

Burkill IH (1966) A dictionary of the economic products of the Malay Peninsula. Revised reprint, 2 vols. Ministry of Agriculture and Co-operatives, Kuala Lumpur, Malaysia. Vol. 1 (A–H) pp 1–1240, vol 2 (I–Z) pp 1241–2444

de Castro PC, Hoshino A, da Silva JC, Mendes FR (2007) Possible anxiolytic effect of two extracts of *Passiflora quadrangularis* L. in experimental models. Phytother Res 21(5):481–484

Dhawan K, Dhawan S, Sharma A (2004) *Passiflora*: a review update. J Ethnopharmacol 94:1–23

Fouqué A (1972) Espèces fruitières d'Amérique tropicale. IV. Les Passiflorées. Fruits 27:368–382

Green PS (1972) Passiflora in Australasia and the Pacific. Kew Bull 26:539–558

Halim MM, Collins RP (1970) Anthocyanins of *Passiflora quadrangularis*. Bull Torrey Bot Club 97(5):27–28

Lans CA (2006) Ethnomedicines used in Trinidad and Tobago for urinary problems and diabetes mellitus. J Ethnobiol Ethnomed 2:45

Leung W-TW, Busson F, Jardin C (1968) Food composition table for use in Africa. FAO, Rome, 306 pp

Martin FW, Nakasone HY (1970) The edible species of *Passiflora*. Econ Bot 24:333–343

Morton JF (1987) Giant granadilla. In: Morton JF (ed) Fruits of warm climates. Julia F. Morton, Miami, pp 328–330

Notodimedjo S (1992) *Passiflora quadrangularis* L. In: Verheij EWM, Coronel RE (eds) Plant resources of South-East Asia no. 2: edible fruits and nuts. Prosea Foundation, Bogor, pp 248–249

Ochse JJ, Bakhuizen van den Brink RC (1931) Fruits and fruit culture in the Dutch East Indies. G. Kolff & Co, Batavia-C, 180 pp

Ochse JJ, Bakhuizen van den Brink RC (1980) Vegetables of the Dutch Indies, 3rd edn. Ascher & Co., Amsterdam, 1016 pp

Orsini F, Pelizzoni F, Verotta L (1985) Quadranguloside, a cycloartane triterpene glycoside from *Passiflora quadrangularis*. Phytochemistry 25(1):191–193

Orsini F, Pelizzoni F, Ricca G, Verotta L (1987) Triterpene glycosides related to quadranguloside from *Passiflora quadrangularis*. Phytochemistry 26(4):1101–1105

Osorio C, Duque C, Koami T, Fujimoto Y (1999) Stereochemistry of (3E)-3,7-dimethyl-3-octene-1,2,6,7-tetraol isolated from *Passiflora quadrangularis*. Tetrahedron Asym 10(22):4313–4319

Osorio C, Duque C, Fujimoto Y (2000) Oxygenated monoterpenoids from badea (*Passiflora quadrangularis*) fruit pulp. Phytochemistry 53(1):97–101

Osorio C, Duque C, Suárez M, Salamanca LE, Urueña F (2002) Free, glycosidically bound, and phosphate bound flavor constituents of badea (*Passiflora quadrangularis*) fruit pulp. J Sep Sci 25(3):147–154

Otero R, Núñez V, Barona J, Fonnegra R, Jiménez SL, Osorio RG, Saldarriaga M, Díaz A (2000) Snakebites and ethnobotany in the northwest region of Colombia. Part III: neutralization of the haemorrhagic effect of Bothrops atrox venom. J Ethnopharmacol 73(1–2):233–241

Pacific Island Ecosystems at Risk (PIER) (1999) *Passiflora quadrangularis* L., Passifloraceae http://www.hear.org/Pier/species/passiflora_quadrangularis.htm

Saeki D, Yamada T, Kajimoto T, Muraoka O, Tanaka R (2011) A set of two diastereomers of cyanogenic glycosides from *Passiflora quadrangularis*. Nat Prod Commun 6(8):1091–1094

Seaforth CE, Adams CD, Sylvester Y (1983) A guide for the medicinal plants of Trinidad & Tobago. Commonwealth Secretariat/Marlborough House, Pall Mall/London, 222 pp

Yuldasheva LN, Carvalho EB, Catanho MT, Krasilnikov OV (2005) Cholesterol-dependent hemolytic activity of *Passiflora quadrangularis* leaves. Braz J Med Biol Res 38(7):1061–1070

Zucolotto SM, Fagundes C, Reginatto FH, Ramos FA, Castellanos L, Duque C, Schenkel EP (2011) Analysis of c-glycosyl flavonoids from South American *Passiflora* species by HPLC-DAD and HPLC-MS. Phytochem Anal. doi:10.1002/pca.1348

Sesamum indicum

Scientific Name

Sesamum indicum L.

Synonyms

Anthadenia sesamoides Lem., *Capraria integerrima* Miq., *Dysosmon amoenum* Raf., *Sesamum africanum* Todaro, *Sesamum brasiliense* Vell., *Sesamum luteum* Retz., *Sesamum malabaricum* J. Burm., *Sesamum mulayanum* N. C. Nair, *Sesamum oleiferum* Moench, *Sesamum orientale* L., *Sesamum trifoliatum* Mill., *Volkameria orientalis* Kuntze, *Volkameria sesamoides* Kuntze.

Family

Pedaliaceae

Common/English Names

Beniseed, Benneseed, Gingelly Sesame, Sesame, Semsem.

Vernacular Names

Albanian: Suzami;
Amharic: Selit;
Arabic: Juljulan, Simsim, Sumsum, Zelzlane;
Armenian: Shooshma, Shooshmayi Good (Seeds), Shushma, Shushmayi Kut;
Azeri: Küncüt, Hint Küncütü;
Belarusian: Kunžut, Sezam;
Brazil: Gergilim;
Bulgarian: Susam;
Burmese: Hnan Zi;
Catalan: Sèsam;
Chamorro: Ahonholi;
Chinese: Chi Ma, Hak Chi Mah (Black Sesame), Hei zhi ma, Hú má, Hu ma ren, Wuh Ma, Zhi Ma, Zi Moa;
Croatian: Sezam;
Czech: Sezam, Sezam indický, Sezam východní, Sezamové Semínko;
Danish: Indisk Sesam, Sesam;
Dutch: Sesamkruid, Sesamzaad;
Esperanto: Sezamo;
Eastonian: Harilik Seesam, Kunžuut;
Farsi: Konjed;
Finnish: Seesami;
French: Sésame, Sésame Blanc, Teel, Till;
Galician: Sésamo;
German: Sesam, Vanglo;
Greek: Sesami, Sesamon, Sousami;
Hebrew: Sumsum, Shumshum;
Huasa: Ridi;
Hungarian: Szezám, Szézámfű, Szézámmag;
Icelandic: Sesamfrae;
India: Til (Assamese), Til (Bengali), Thileyo, Thileyokoli (Dhivehi), Spin (Garo), Tal (Gujarati), Gingli, Kali Til, Saphed Til, Til (Hindu), Acchellu, Ellu, Tila (Kannada), Til (Maithili), Chitelu, Ellu, Thilam (Malayalam),

Til, Ashadital, Bariktil (Marathi), Chhawchii (Mizoram), Rasi (Oriya), Til (Punjabi), Til, Tila (Sanskrit), Ellu, Yellu (Tamil), Nuvvulu, Tillu (Telugu), Enme (Tulu), Til, Konjed (Urdu);
Indonesian: Wijen;
Irish: Seasaman;
Italian: Sesame, Sesamo;
Japanese: Goma, Kuro Goma (Black Sesame), Koba, Shima;
Kazakh: Künjit;
Khasi: Neiong, Nei;
Korean: Chamggae, Cham-Kkae, Ggaessi, Ggae, Kkae, Ssisaem;
Laotian: Man Nga, Nga;
Latvian: Sēzama Sēklas;
Lithuanian: Indinis Sezamas, Sezamas;
Maltese: Ġulġlien;
Malaysia: Bene, Bijan;
Naga (Tankhul): Hāngsi;
Nepali: Til, Hamo, Til (Newari);
Pahlavi: Kunijd;
Papiamento: Zjozjolí;
Philippines: Langa (Bikol), Langa, Lunga (Bisaya), Langa (Ibanag), Longis (Ifugao), Lenga (Iloko), Langis (Pampangan), Linga (Sambali), Ajonjoli (Spanish), Lunga (Sulu), Langa, Linga, Lingo (Tagalog);
Polish: Sezam Indyjski;
Portuguese: Gergelim, Gimgelim, Sésamo;
Romanian: Susan;
Russian: Kunzhut, Sezam;
Serbian: Sezam, Susam, Suzam;
Slovak: Sezam Indický, Sezam;
Slovenian: Sezama;
Spanish: Ajonjoli, Alegría, Sesame, Sésamo;
Sri Lanka: Tala (Sinhala);
Swahili: Simsim, Ufuta, Wangila;
Swedish: Sesam;
Thai: Ngaa, Nga Dam, Nga Khao;
Tibetan: Telu, Til kara, Khyuma; *Turkish*
Turkish: Susam;
Turkmen: Künji;
Ukrainian: Sezam;
Uzbek: *Uzbek*: Kunjut;
Vietnamese: Cây Vừng, Mè, Hắc Chi Ma, Vừng;
Yiddish: Sumsum, Sezam, Kunzhit.

Origin/Distribution

Sesame is believed to have originated in Africa, and is regarded to be the oldest oilseed crop known to man. Evidence from interspecific hybridization and phytochemical analysis indicate that the progenitor of sesame occurred in the Indian subcontinent (Bedigian et al. 1985). From here, sesame was introduced to Mesopotamia in the Early Bronze Age and by 2000 BC where it became a crop of enormous importance. Mesopotamia became the hub of distribution of sesame into the Mediterranean. By the second century BC, sesame became a prominent oil crop in China. Today sesame is cultivated pantropically.

Agroecology

Sesame is a crop of the tropics and subtropics. With newer cultivars and summer plantings its range has extended into the temperate areas. Sesame is cultivated mainly between 25°S and 25°N, but extends further to 40°N in China, Russia and the United States, 30°S in Australia and 35°S in South America. It is grown from sea level to 1,800 m altitude. The crop has an optimal day temperature of 25–27°C, below 20°C growth is retarded and below 10°C germination is suppressed Sesame requires 90–120 frost free days.

Yield is optimal with well distributed annual rainfall of 500–650 mm during the growing season. Rainfall late in the season prolongs growth and increases shattering losses. Strong wind can cause shattering at harvest and result in yield losses. Sesame is very drought-tolerant, due in part to an extensively branched root system which also improves soil structure. Sesame is intolerant of wet conditions. Sesame is adaptable to many soil types, but it thrives best on well-drained, fertile soils of medium texture and with pH ranging from 5.5 to 8.0, but most cultivars are intolerant of salinity. Growth and subsequent yield will be reduced on gravelly or sandy soils due to their poor moisture retention capacity.

Edible Plant Parts and Uses

Sesame seeds, both pale and dark coloured, and sesame oil, are widely used in various cuisines all over the world. The small sesame seed is used whole in cooking for its rich nutty flavour, and also yields sesame oil. In general, the paler varieties of sesame appear to be more prized in the West and Middle East, while both the pale and black varieties are valued in the Far East. Sesame seeds are used as spice for flavouring in food dishes, pastries and cakes, and other food industries. Sesame oil from the seed is used in cooking salad oils and margarine. Sesame oil and foods fried in sesame oil have a long shelf life because the oil contains an antioxidant called sesamol. Black sesame seed are especially good on salmon and other fish dishes. The simplest and now commonest use of sesame is as whole seeds sprinkled over cakes, breads (bagels and hamburger buns), in cookies and wafers, sushi food, and steamed rice noodle rolls called '*chee cheong fun*' like poppy seeds. Sesame seeds may be baked into crackers, often in the form of sticks. Sesame seeds can be ground into a paste, *tahini* (Sesame butter) or powder and used as flour, added to bread, vegetables, and used to make sweetmeat, halva, and for the preparation of rolls, crackers, cakes and pastry products in commercial bakeries. The seeds can also be fermented into *tempeh*. The seeds can also be sprouted and used in salads. Leaves raw or cooked can be used in soups and as potherbs.

About a third of Mexico sesame seed crop is purchased by McDonald enterprise for their sesame seed buns. In Mexican cuisine, sesame is used in the popular sauce *mole rojo* or *mole poblano* to accompany baked turkey. Sesame seeds are often sprinkle over artisan breads and baked in traditional form to coat the smooth dough, especially on whole wheat flat breads or artisan nutrition bars, such as *alegrías*. Good acceptance was obtained with the bread prepared with 30% sesame flour + 70% wheat flour (Salgado and Gonçalves 1988). Its external and internal appearance, as well as its organoleptic characteristics were close to the bread with 100% wheat flour. Sesame flour at the 50% proportion gave a bread of medium quality.

Toasted sesame seeds are a common spice in Eastern Asia; it is often sprinkled over Chinese, Korean and Japanese dishes. Black sesame appears frequently in Chinese, Japanese and Korean dishes where meat or fish is rolled in the seeds before cooking for a crunchy coating. In Chinese cuisine, sesame seeds and oil are used in dishes like dim sum, sesame seed ball (*mátuǎn* or *jin deui*), and Chinese noodles. Chinese sesame paste (*zhi ma jiang*) is made from toasted sesame seeds and has a very strong flavour resembling Chinese sesame oil; it is used mainly for salad dressings and sauces for cold appetizers like Sichuanese *guai wei ji si,* a salad dish made from precooked chicken meat cut in fine slivers with a dressing of soy sauce, sugar, black vinegar (*hei cu*), sesame paste, sesame seeds, chilli oil and toasted Sichuan pepper. Sesame oil is used as a condiment to flavour hot and sour Szechuan soup *suanla tang* and noodles. Sesame paste is also made in a delicious dessert soup called *chi ma wu*. In Vietnam, sesame seeds are used in *bánh rán,* deep fried glutinous rice balls and as a condiment flavour in various noodles. Sesame oil and roasted or raw seeds is also very popular in Korean cuisine, used to marinate meat and vegetables. Chefs in tempura restaurants blend sesame and cottonseed oil for deep-frying. Japanese *tempura* is made by deep-frying battered vegetables in a mixture of one part sesame oil and ten parts vegetable oil. In Japanese cuisine, *goma-dofu* is made from sesame paste and starch. Dark sesame oil forms part of *shichimi togarashi*, an exotic spice blend of Japan Szechuan pepper. A simpler mixture from toasted black sesame seeds with about 10% salt called *gomashio* is a popular Japanese tabletop condiment, usually sprinkled over dry colourful rice dishes and noodle dishes.

In Maharashtra, southwest India and Myanmar, a hot-pressed sesame oil is the preferred cooking medium. In Manipur, black sesame is widely used in popular dishes in 'Thoiding' and in 'Singju' a kind of hot spicy salad comprising vegetables, gingers and chillies. In Assam, black sesame seeds are used to make *til pitha* (pancake) and *tilor laru* (sesame seed balls) during

Assamese festive occasions (*bihu*). In Punjab and Tamil Nadu, a sweet sesame ball is made from sesame seeds and sugar, called *pinni* (Urdu), *ell urundai* (Tamil), *ellundai* (Malayalam), *yellunde* (Kannada) and *tilgul* (Marathi). In Tamil Nadu, sesame oil used extensively in their cuisine, *milagai podi*, a ground powder made of sesame, lentils and dry chili with jaggery, is used to enhance flavor and consumed along with other traditional foods such as *dosa* (bread) and *idli* (savoury cake of rice and lentils).

In South Asian, Middle East, and East Asian cuisines, popular treats are made from sesame mixed with honey, jaggery or syrup and roasted into a sesame candy. In the Democratic Republic of Congo and Angola, ground sesame seeds or *wangila* is a favourite dish cooked usually with smoke fish or lobsters. In Togo, the seeds are used in soups.

In the Middle East, sesame seeds are popularly used for *tahini*, starch and as the key ingredient for the confection *halvah*. *Tahini* is used as a flavouring for hummus, a sauce for kebabs and bread dips. In Jordan, Syria and Lebanon, dried, untoasted sesame seeds are used with sumac and thyme in the spice mixture *zahtar,* and the Egyptian *dukka*.

Plate 1 Sesame plant habit

Botany

An erect annual herb growing to 1.2 m high branched with quadrangular, pubescent or glabrescent short branches or unbranched (Plates 1 and 2) depending on varieties. Leaves opposite or alternate (Plates 2 and 3); petiole 3–11 cm on lower leaves; leaf blade variable lanceolate to ovate, or 3-parted, 4–20 by 2–10 cm; upper stem leaves oblong to linear-lanceolate, 0.5–2.5 cm wide, base cuneate, margin entire. Flowers white (Plates 1, 2 and 3), pink, or mauve-pink with darker markings, with strong unpleasant odour, bracteate; pedicel up to 5 mm long with 2 basal glands. Calyx 2–6 mm long, lobes linear to narrowly lanceolate, hairy and persistent. Petals obtuse, 2–3 cm long, pubescent. Stamens 1 cm long. Ovary 1–2 mm long, pilose, oblong. Petals 2–3 cm long, pubescent, obtuse. Capsule narrowly oblong, loculicidal, acuminate at apex and rounded at base, 1.5–3 cm × 6–7 mm; beak broad and short. Seeds black, brown or white, oval, small, 2–3 mm long, 1–1.5 mm wide, smooth, arranged horizontally in capsule (Plates 4a, 4b, 5, 6a and 6b).

Nutritive/Medicinal Properties

Nutrient value of whole, dried sesame seeds per 100 g edible portion was reported as: water 4.69 g, energy 573 kcal (2,397 kJ), protein 17.73 g, total lipid (fat) 49.67 g, ash 4.45 g, carbohydrate 23.45 g, total dietary fibre 11.8 g, total sugars 0.30 g, Ca 975 mg, Fe 14.55 mg, Mg 351 mg, P 629 mg, K 468 mg, Na 11 mg, Zn 7.75 mg, Cu 4.082 mg, Mn 2.460 mg, Se 34.4 μg, thiamine 0.791 mg, riboflavin 0.247 mg, niacin 4.515 mg, pantothenic acid 0.050 mg, vitamin B-6 0.790 mg, total folate 97 μg, β-carotene 5 μg,

Plate 2 Flowers and fruits

Plate 3 Fruits and leaves

vitamin E (α-tocopherol) 0.25 mg, total saturated fatty acids 6.957 g, 14:0 (myristic) 0.124 g, 16:0 (palmitic) 4.441 g, 18:0 (stearic) 2.090 g, total monounsaturated fatty acids 18.759 g, 16:1 undifferentiated (palmitoleic) 0.149 g, 18:1 undifferentiated (oleic) 18.521 g, 20:1 (gadoleic) 0.070 g, total polyunsaturated fatty acids 21.773 g, 18:2 undifferentiated (linoleic) 21.375 g, 18:3 undifferentiated (linolenic) 0.376 g, phytosterols 714 mg, tryptophan 0.388 g, threonine 0.736 g, isoleucine 0.763 g, leucine 1.358 g, lysine 0.569 g, methionine 0.586 g, cystine 0.358 g, phenylalanine 0.940 g, tyrosine 0.743 g, valine 0.990 g, arginine 2.630 g, histidine 0.522 g, alanine 0.927 g, aspartic acid 1.646 g, glutamic acid 3.955 g, glycine 1.215 g, proline 0.810 g and serine 0.967 g (USDA 2011). Nutrient value of sesame cooking oil per 100 g edible portion was reported as: energy 884 kcal (3,699 kJ), total lipid 100 g, total choline 0.2 mg, vitamin E (α-tocopherol) 1.40 mg, vitamin K (phylloquinone) 13.6 μg, total saturated fatty acids 14.200 g, 16:0 (palmitic) 8.9 g, 18:0 (stearic) 4.8 g, total monounsaturated fatty acids 39.700 g, 16:1 undifferentiated (palmitoleic) 0.200 g, 18:1 undifferentiated (oleic) 39.3 g, 20:1 (gadoleic) 0.200 g, total polyunsaturated fatty acids 41.700 g, 18:2 undifferentiated (linoleic) 41.300 g, 18:3 undifferentiated (linolenic) 0.300 g and phytosterols 865 mg (USDA 2011). Sesame seed and wheat germ had the highest total phytosterol content (400–413 mg/100 g) while Brazil nuts the lowest (95 mg/100 g) (Phillips et al. 2005). White sesame (WS) seed contained 22.20% protein and 52.61% fat while black sesame (BS) seeds contained 20.82% protein and 48.40% fat (Kanu 2011). Moisture was higher in WS than BS but ash was higher in BS than WS and the amount was significantly different. Carbohydrate was higher in BS than WS. Vitamins and sugars varied in quantity for the two seed types. Oleic and linoleic, were the major unsaturated fatty acids while

Plate 4a White sesame seeds

Plate 6a Black sesame seeds

Plate 4b Close-up white sesame seeds

Plate 6b Close-up black sesame seeds

Plate 5 Brown sesame seeds

palmitic and stearic were the main saturated fatty acids significantly observed in both seed types. Both types were higher in essential amino acids with the exception of lysine.

Sesame seed was found to contain 5.7% moisture, 20% crude protein, 3.7% ash, 3.2% crude fiber, 54% fat and 13.4% carbohydrate (Nzikou et al. 2009). The seeds were found to be good sources of minerals. Potassium (851.35 mg/100 g) was the highest, followed in descending order by phosphorus (647.25 mg/100 g), magnesium (579.53 mg/100 g), calcium (415.38 mg/100 g) and sodium (122.50 mg/100 g). The physical properties of the oil extracts showed the state to be liquid at room temperature. Sesame oil was found to contain high levels of unsaturated fatty acids, especially oleic (up to 38.84%) and linoleic (up to 46.26%), classifying it in the oleic-linoleic acid group. The dominant saturated acids were palmitic (up to 8.58%) and stearic (up to 5.44%). No essential differences in the oil components were found among the three sesame varieties (Yoshida et al. 2007). γ-Tocopherol was present in highest concentration, and δ-, and α-tocopherols in very small amounts. Sesamin and sesamolin were the main lignan components. Separation of the complex mixture of total triacylglycerol, provided 12 different groups of triacylglycerol. With a few exceptions, the major TAG components were SM2 (6.5–6.7%), SMD (19.8–20.7%), M2D (15.0–26.3%), MD2 (23.6–35.0%), and D3 (7.7–10.7%) (where S=saturated fatty acid, M=monoene, D=diene, and T=triene).

Oleic (O) and linoleic (L) acids were the major fatty acids followed by palmitic (P) and stearic acids in *Sesamum indicum* and three wild species *Sesamum alatum*, Thonn., *S. radiatum*, Schum & Thonn. and *S. angustifolium*, (Oliv) Engl (Kamal-Eldin et al. 1992b). The major triacylglycerols were: LLO (20–25%), LLL (10–20%), LOO (15–19%), PLL (8–11%) and PLO (6–10%). *S. alatum* was also different from the other three species in having higher percentages of PLO (10.1%) and OOO (8.7%) compared to 6.3–8.1% of PLO and 3.4–4.9% of OOO in the other three species. Oils from the three wild species contained more unsaponifiable material (2.3–2.4%) compared with the cultivated *S. indicum* species (1.1–1.3%) (Kamal-Eldin et al. 1992a). Considerable differences were observed in the total sterol contents and the relative proportions of the three sterol fractions in the oils from the four species studied. Sitosterol, campesterol, stigmasterol and Δ5-avenasterol were the major desmethyl sterols in all four species. The monomethyl sterol fraction consisted primarily of obtusifoliol, gramisterol, cycloeucalenol and citrostadienol. Cycloartenol and 24-methylene cycloartanol were the predominant dimethyl sterols. The oils from wild seeds were characterized by higher percentages of unsaponifiables (4.9%, 2.6% and 3.7%, respectively) compared to *S. indicum* (1.4–1.8%), mainly due to their high contents of lignans (Kamal-Eldin and Appelqvist 1994). Total sterols accounted for circa 40%, 22%, 20% and 16% of the unsaponifiables of the four species, respectively. The four species were different in the relative percentages of the three sterol fractions (the desmethyl, monomethyl and dimethyl sterols) and in the percentage composition of each fraction. Campesterol, stigmasterol, sitosterol and $Δ^5$-avenasterol were the major desmethyl sterols, whereas obtusifoliol, gramisterol, cycloeucalenol and citrostandienol were the major monomethyl sterols, and α-amyrin, β-amyrin, cycloartenol and 24-methylene cycloartanol were the main dimethyl sterols in all species. Differences were also observed among the four species in sterol patterns of the free sterols compared to the sterol esters. *Sesamum alatum* contained less tocopherols (210–320 mg/kg oil), and *S. radiatum* and *S. angustifolium* contained more tocopherols (ca. 750 and 800 mg/kg oil, respectively) than did *S. indicum* (490–680 mg/kg oil). The four species were comparable in tocopherol composition, with γ-tocopherol representing 96–99% of the total tocopherols. The four species varied widely in the identity and levels of the different lignans. The percentages of these lignans in the oils of *S. indicum* were sesamin (0.55%) and sesamolin (0.50%). *Sesamum alatum* showed 1.37% of 2-episesalatin and minor amounts of sesamin and sesamolin (0.01% each). *Sesamum radiatum* was rich in sesamin (2.40%) and contained minor amounts of sesamolin (0.02%), where *S. angustifolium* was rich in sesangolin (3.15%) and also contained considerable amounts of sesamin (0.32%) and sesamolin (0.16%).

A survey of 11 diverse *Sesamum indicum* genotypes from eight countries revealed that sesamin, α-tocopherol, δ-tocopherol and γ-tocopherol levels were 0.67–6.35 mg/g, 0.034–0.175 μg/g, 0.44–3.05 μg/g and 56.9–99.3 μg/g respectively, indicating that the sesame seed accessions contained higher levels of sesamin and γ-tocopherol compared with α-tocopherol and δ-tocopherol (Williamson et al. 2008). Significant differences were observed among the 11 different sesame genotypes suggesting that genetic, environmental and geographical factors influence sesamin and desmethyl tocopherol content.

Sesame seeds germinated readily in dark chambers maintained near 100% relative humidity at 35°C without presoaking reaching >99% germination rate in 4 days with the final moisture content of 2% (w/w) (Hahm et al. 2009). With, germinated derooted sesame seeds (DSS) showed marked reduction in fat content (23%) and were found rich in linolenic acid, P, and Na, increasing from 0.38% (w/w), 445 mg/100 g, and 7.6 mg/100 g before germination to 0.81% (w/w), 472 mg/100 g, and 8.4 mg/100 g after germination, respectively. DSS after germination contained considerable amount of Ca (462 mg/100 g), higher than that of soybean. Germinated DSS presented an excellent source of sesamol (475 mg/100 g), a potent natural antioxidant, and α-tocopherol (32 mg/100 g), the most active form of vitamin E.

Sesame seeds were found to contain small amounts of lipid soluble lignans sesamolinol, sesaminol and pinoresinol which occurred as glucosides; sesaminol glucosides sesaminol 2′- O-β-d-glucopyranoside, sesaminol 2′-O-β-d-glucopyranosyl (1→2)-O-β-d-glucopyranoside and sesaminol 2′-O-β-d- glucopyranosyl (1»2)-O-[β-d-glucopyransyl (1»6)]-[β-d-glucopyranoside] (Osawa et al. 1985; Katsuzaki et al. 1992, 1993, 1994a). Three novel lignan glucosides were isolated as the water-soluble antioxidative components from the 80% ethanol extracts of sesame seed; their structures were determined as pinoresinol 4′-O-β-D-glucopyranosyl(1→6)-β-D-glucopyranoside (KP1), pinoresinol 4′-O-β-D-glucopyranosyl-(1→2)-β-D-glucopyranoside (KP2), and the lignan triglucoside pinoresinol 4′-O-β-D-glucopyranosyl(1→2)-O-[β-D-glucopyranosyl-(1→6)]-β-D-glucopyranoside (KP3) (Katsuzaki et al. 1993, 1994a). These lignan glucosides possessed unique glucosidic linkages, in particular KP3, with branched (1→2)- and (1→6)-linkages. The most abundant lignan glucosides in sesame seed were found to be sesaminol triglucoside (Ryu et al. 1998). The content of sesaminol triglucoside in 100 g seeds ranged from 14.1 to 91.3 mg with a mean value of 68.4 mg; that of sesaminol diglucoside from 8.2 to 18.3 mg with a mean value of 11.6 mg; and that of sesaminol monoglucoside from 5.4 to 19.5 mg with a mean value of 8.3 mg. The total content of sesaminol glucoside was 88.3 mg in 100 g of sesame seeds. Also, the sesamolinol content in 100 g sesame seed ranged from 17.6 to 28.5 mg, with a mean of 20.5 mg. A new lignan glucoside, sesamolinol diglucoside [2-(3-methoxy-4-(O-β-D-glucopyranosyl(1→6)-O-β-D-glucopyranoside)phenoxyl)-6-(3,4-methylenedioxyphenyl)-cis-3,7-dioxabicyclo-(3.3.0)-octane] was found in sesame seeds (both black and white) in levels ranging from <5 to 232 mg/100 g of seeds (Moazzami et al. 2006a). Analysis of 65 different samples of sesame seeds indicated that the content of sesaminol triglucoside ranged from 36 to 1,560 mg/100 g of seed (mean 637) and that of sesaminol diglucoside ranged from 0 to 493 mg/100 g of seed (mean 75) (Moazzami et al. 2006b). No significant difference was found between sesaminol glucoside contents in black and white sesame seeds.

A study found great variation in the types and amounts of lignans in sesame seeds, seed products and oils (Moazzami et al. 2007). The total lignan content of 14 sesame seeds ranged between 405 and 1,178 mg/100 g and the total lignan content in 14 different products, including tahini, ranged between 11 and 763 mg/100 g.

The biosynthesis of the sesame lignans, (+)-sesamin and (+)-sesamolin in sesame seeds, was found to result from stereoselective coupling of E-coniferyl alcohol to give (+)-pinoresinol which was then subsequently metabolized (Kato et al. 1998). This enantiomer (+)-pinoresinol is metabolized further in maturing seeds to afford (+)-piperitol, (+)-sesamin, and (+)-sesamolin. Among 16 lignans identified from sesame perisperm (coat), two new lignans, (+)-saminol and (+)-episesaminone- 9-O-β-D-sophoroside were isolated (Grougnet et al. 2006). Additionally, the relative stereochemistry of (-)-sesamolactol, previously reported as todolactol A epimer, was unequivocally defined using X-ray crystallography. A new anthraquinone derivative, Z)-6,7-dihydroxy-2-(6-hydroxy-4-methyl-3-pentenyl)anthraquinone, named anthrasesamone F was isolated from the seeds of *Sesamum indicum* (Kim and Park 2008).

Lipids from all-seeded microwaved roasted sesame strains (black, brown and white) were comparable in their total fatty acid composition, with linoleic, oleic, stearic and palmitic acids as the major acids (Yoshida et al. 1995). The total lipids were isolated into the following five fractions: triacylglycerols (TAG), diacylglycerols, free fatty acids, polar lipids and steryl esters. The TAG were slightly and randomly hydrolysed by microwaves, but was still representing 900 g/kg of the total lipids at 30 minutes of roasting. Although burning and bitter tastes occurred at the time, the tocopherols and lignans still amounted to over 80% of the original level. Major lipid components of sesame seeds were triacylglycerols and phospholipids, while steryl esters, free fatty acids and sn-1,3- and sn-1,2-diacylglycerols (DAGs) were minor ones (Yoshida et al. 2001). Following electric-oven roasting, a significant increase was observed in free fatty acids and in both forms of DAG (primarily sn-1,3-DAG). The greatest phospholipids losses were observed in phosphatidylethanolamine, followed by phosphatidylcholine and phosphatidylinositol. On the other hand, the amounts of γ-tocopherol and sesamin remained at over 80% and 90% respectively of the original levels after roasting at 220°C. The principal characteristics of the positional distribution of fatty acids were still retained after 25 minutes of roasting: unsaturated fatty acids, especially linoleic and/or oleic, were predominantly concentrated in the sn-2-position, and saturated fatty acids, especially stearic and/or palmitic, primarily occupied the sn-1- or sn-3-position. The results suggested that unsaturated fatty acids located in the sn-2-position were significantly protected from oxidation during roasting at elevated temperatures. The concentrations of radiation-induced hydrocarbons in sesame seeds increased almost linearly with the applied doses of 0.5–4 kGy (Lee et al. 2008a). The hydrocarbons, 1,7-hexadecadiene and 8-heptadecene, were detected only in the irradiated samples before and after three types of treatments (steaming, roasting, and oil extraction) at doses ³0.5 kGy, but they were not detected in non-irradiated samples before and after treatment. These two hydrocarbons could be used as markers to identify irradiated sesame seeds. The concentrations of the three detected 2-alkylcyclobutanones, 2-dodecylcyclobutanone (2-DCB), 2-tetradecylcyclobutanone (2-TCB), and 2- (5′-tetradecenyl)cyclobutanone (2-TeCB), also linearly increased with the irradiation dose. These compounds could be detected at irradiated doses above 0.5 kGy but not in non-irradiated samples. The three types of treatments had no significant effect on the levels of 2-alkylcyclobutanones.

Sesame Seed Oil

The percentages of individual acids of fatty acids of sesame seed oil were found to be: palmitic, 11%; stearic, 6%; arachidic, 1%; oleic, 43%; linoleic, 39% (Sengupta and Roychoudhury 1976). Triglyceride composition of the was composed of 8%, 41% and 51%, GS2U, GSU2 and GU3 respectively.

Besides abundant oleosin, three minor proteins, Sop 1, 2, and 3, were found in sesame (*Sesamum indicum*) oil bodies (Lin et al. 2002). The gene encoding Sop1, named caleosin for its calcium-binding capacity, had recently been cloned. Sop2, tentatively named steroleosin may represent a class of dehydrogenases/reductases involved in plant signal transduction regulated by various sterols. In subsequent study, the found that sesame seed oil bodies comprised a triacylglycerol matrix shielded by a monolayer of phospholipids and proteins (Lin et al. 2005). These surface proteins include an abundant structural protein, oleosin, and at least two minor protein classes termed caleosin and steroleosin. Two steroleosin isoforms (41 and 39 kDa), one caleosin (27 kDa), and two oleosin isoforms (17 and 15 kDa) were identified in oil bodies isolated from sesame seeds. Studies suggested that succinylation of α-globulin the major protein fraction from *Sesamum indicum* improved the functional characteristics of α-globulin such as emulsion activity and emulsion stability and increased values for bulk density, water absorption capacity, oil absorption capacity, foam capacity and foam stability (Zaghloul and Prakash 2002).

Sesame (*Sesamum indicum* L.) seed and oil were found to contain abundant lignans, including sesamin, sesamolin and lignan glycosides (Rangkadilok et al. 2010). Studies showed that there was a large variation of sesamin and sesamolin contents in sesame and sesame oil products in Thailand. The sesamin and sesamolin contents in seeds showed that the mean values of sesamin and sesamolin were 1.55 mg/g (range n.d.–7.23 mg/g) and 0.62 mg/g (range n.d.–2.25 mg/g), respectively. The range of total tocopherols of these sesame lines was 50.9–211 µg/g seed. In commercial sesame oils, the ranges of sesamin and sesamolin were 0.93–2.89 mg/g oil and 0.30–0.74 mg/g oil, respectively, and tocopherol contents were 304–647 µg/g oil. The study revealed the extensive variability in sesamin, sesamolin and tocopherol contents among sesame products. The content of sesamin and sesamolin in ten commercial virgin and roasted sesame oils was in the range of 444–1,601 mg/100 g oil (Moazzami et al. 2007). In five refined sesame oils, sesamin ranged between 118 and 401 mg/100 g seed, episesamin between 12 and 206 mg/100 g seed, and the total contents of sesaminol epimers between 5 and 35 mg/100 g seed, and no sesamolin was found.

The mean total lignin content in 14 commercial brands of sesame oils was found to be 11.5 mg/g; 82% and 15% of the lignans were sesamin, and sesamolin, respectively (Wu 2007). The level of sesamol increased after heating at 180°C for 20 minutes. Heating at 200°C for 20 minutes caused a significant loss of sesamolin and sesamol. Cooking at temperatures above 200°C will cause loss of some lignans, but sesamin, a source of phytoestrogen, was relatively heat-stable. Ingestion of 10 g of sesame oil was calculated to be adequate to provide the level of lignans that might benefit cardiovascular health, as found by other studies. The sesame oil prepared at a 200°C roasting temperature gave the best flavour score (Yen 1990). The highest value of sesamol and γ-tocopherol was in oils roasted at 200–220°C. The fatty acid content of the oil was reduced considerably, especially in oleic and linoleic acids, when the roasting temperature was over 220°. The amounts of chlorophyll and sesamolin decreased with increasing roasting temperature. The phospholipid content was reduced from 690 mg/kg in unroasted oil to 0 mg/kg in the oil prepared using a 260° roasting temperature. The oils from roasted and roasting plus steaming sesame seeds showed higher oxidative stability than other processed oils (Abou-Gharbia et al. 2000). Different lipid classes and subclasses present and their fatty acid composition influenced the oxidative status of sesame oil. Neutral lipids constituted about 91.0%, monoglycolipids 2.4% and diglycolipids 3.5%, while phospholipids constituted 3.0% of the total lipids. Moreover, different processing treatments show considerable effects on lipid fractions. The influence on all components after roasting plus steaming was more pronounced than microwaving treatment.

Several studies have been reported on the flavor components of sesame oil (Soliman et al. 1986; Shimoda et al. 1996, 1997; Park et al. 1995; Schieberle 1996; Schieberle et al. 1996; Cadwallader and Heo 2001; Tamura et al. 2010). The following volatile compounds : 1-(5-methyl-2-furanyl)-1-propanone; 3-formylthiophene; 2-propyl-4-methylthiazole; 2-ethyl-4-methyl-1 H-pyrrole; 2-ethyl-6-methylpyrazine; 2-ethyl-5-methylpyrazine; 4,5-dimethylisothiazole; 4,5-dimethylthiazole; 2,6-diethylpyrazine; 2-ethyl-2,5-dimethylpyrazine; 1-(2-pyridinyl)ethanone, and 1-(1-methyl-1 H-pyrrol-2-yl)ethanone were considered to be principal contributors of sesame seed oil flavor (Shimoda et al. 1996). Soliman et al. (1986) identified 32 volatile constituents from white sesame seeds, including 10 aldehydes, 5 ketones, 6 alcohols, 4 esters and 7 pyrazines were identified. Among them 4-(5-methyl-2-furyl)-3-buten-2-one was reported for the first time in the aroma of roasted white sesame seeds. Park et al. (1995) identified 66 volatiles in toasted sesame oil. Among these, dimethylsulfide, 4-methylthiazole, 2,4-dimethylthiazole, methylpyrazine, 2,5-dimethylpyrazine, dihydro-4,5-dimethyl-2(3 H)-furanone, dodecane, tetradecane, 2,6-bis(1,1-dimethylethyl)-4-methyl phenol, N,N-bis (2-hydroxyethyl)-dodecamide, cyclododecane, 2,3-dihydro-1,1,3-trimethyl-3-phenyl-1H-indene, tetradecanoic acid, pentadecanoic acid, hexa-decenal, hexadecanoic acid, 9-octadecenoic acid, ocladecanoic acid, [1,1′:3′,1′I-tephenylJ-1′-ol and bis (2-ethyl hexyl)phthalate were the major volatiles identified.

Sesame Volatiles and Flavour Components

Aroma extract dilution analysis (AEDA) of an extract prepared from moderately roasted sesame (180°C; 10 minutes) afforded 41 odour-active volatiles (Schieberle 1996). Of the 18 aroma compounds showing very high Flavour Dilution factors in the range of 128–2,048, ten compounds [2-furfurylthiol; 2-phenylethylthiol; 2-methoxyphenol; 4-hydroxy-2,5-dimethyl-3(2H)-furanone; 2-pentylpyridine; 2-ethyl-3,5-dimethylpyrazine; acetylpyrazine; (E,E)-2,4-decadienal; 2-acetyl-1-pyrroline and 4-vinyl-2-methoxy-phenol] were quantified and their odour activity values (OAV; ratio of concentration to odour threshold) were calculated. On the basis of high OAVs in oil, especially 2-acetyl-1-pyrroline (roasty), 2-furfurylthiol (coffee-like), 2-phenylethylthiol (rubbery) and 4-hydroxy-2,5-dimethyl-3(2H)-furanone (caramel-like) were elucidated as important contributors to the overall roasty, sulphury odour of the crushed sesame material. Among the 41 odorants, a garlic, carbide-like smelling compound with a comparatively high flavour dilution factor was identified as 4-methyl-3-thiazoline (Schieberle et al. 1996). The amount of volatile flavor compounds in sesame oil is greatly affected by the roasting process. Shimoda et al. (1997) reported that the ratio of the amount of volatile components in deep-roasted oils was increased by 2–7 times in deep-roasted oil as compared with that of light-roasted oils. The recoveries of total volatiles were 9,726 and 2,014 ppb from deep- and light-roasted oils, respectively. The relative amount of monoalkylpyrazines decreased in contrast to the increases of di alkylpyrazines and trialkylpyrazines with increase in the degree of roasting. 1H-pyrrole-2-carboxyaldehyde, the most abundant pyrrole, was the only volatile that decreased in deep-roasted oil. The concentration of 4,5-dimethylisothiazole, 4,5-dimethylthiazole, 2-propyl-4-methylthiazole, and 2-butyl-5-methylthiazole increased in deep-roasted oil. Hexanal, (E)-2-heptenal, and (E,E)-2,4-decadienal occurred in almost the same levels. Guaiacol and 2-furanmethanethiol increased from 32 to 321 ppb and from 6 to 40 ppb, respectively, in deep-roasted oil. Forty-nine odorants were detected from roasted sesame seed oils with detection volumes (DVs) from 8 to 1,000 nL (Cadwallader and Heo 2001). Those detected with DVs from 8 to 40 nL were 1-octen-3-one, 4,5-epoxy-(E)-2-decenal, 2-acetyl-3-methylpyrazine, 2-methoxyphenol, 2,3-diethyl-5-methylpyrazine, 3-methylbutanal, (E)-2-nonenal, 2-methoxy-4-vinylphenol, and an unidentified compound (plastic aroma note).

Aroma extract dilution analysis revealed 32 odorants from freshly pan-roasted sesame seeds in the Flavour Dilution (FD) factor range of 2 – 2,048, 29 of which could be identified (Tamura et al. 2010). The highest FD factors were found for the coffee-like smelling 2-furfurylthiol, the caramel-like smelling 4-hydroxy-2,5-dimethyl-3(2 H)-furanone, the coffee-like smelling 2-thenylthiol (thiophen-2-yl-methylthiol), and the clove-like smelling 2-methoxy-4-vinylphenol. In addition, 9 odor-active thiols with sulfurous, meaty, and/or catty, black-currant-like odors were identified for the first time in roasted sesame seeds. Among them, 2-methyl-1-propene-1-thiol, (Z)-3-methyl-1-butene-1-thiol, (E)-3-methyl-1-butene-1-thiol, (Z)-2-methyl-1-butene-1-thiol, (E)-2-methyl-1-butene-1-thiol, and 4-mercapto-3-hexanone were previously unknown as food constituents. The relatively unstable 1-alkene-1-thiols represented a new class of food odorants and were suggested as the key contributors to the characteristic, but quickly vanishing, aroma of freshly ground roasted sesame seeds. The most important compounds identified in the roasted sesame oils were 2-furfurylthiol and guaiacol (Ho and Shahidi 2005). 2-Furfurylthiol, with an intense coffee-like odor, increased from 16 ppb in roasted oil processed at 160°C for 30 minutes to 158 ppb in the oil processed at 200°C for 30 minutes. Guaiacolhas a burnt and smoky odor and the amount of guaiacol increased from 147 ppb in roasted oil processed at160°C for 30 minutes to 718 ppb in the oil processed at 200°C for 30 minutes. Other odor active compounds identified included acetylpyrazine, 2-ethyl-3,5-dimethylpyrazine, 2,3-diethyl-5-methylpyrazine, trans,trans-2,4-decadienal and 2-ethyl-5-methylpyrazine.

Sesame Seed Flour/ Debittered and Defatted Sesame Seed Flour/Meal

A study on the functional and nutritional properties of sesame flour, concentrate and enzymatic hydrolysates, demonstrated that nitrogen solubility of the hydrolysates was improved, in water (85%) and at different pHs (91–95%), by the action of neutrasa 0.5 L and alcalasa 0.6 L enzymes, yielding a product with good emulsifying and improved foaming properties (Saad and Pérez 1984). The flour and the concentrate had PER (protein efficiency ratio) values of 1.2. Supplementation of one of the hydrolysates with soya hydrolysate (1:1), improved the PER to a value similar to that of casein.

Defatted sesame meal (approximately 40–50% protein content) is very important as a protein source for human consumption due to the presence of sulfur-containing amino acids, mainly methionine (Bandyopadhyay and Ghosh 2002). Sesame protein isolate (SPI) was produced from dehulled, defatted sesame meal and used as a starting material to produce protein hydrolysate by papain. Protein hydrolysates were found to have better functional properties than the original SPI. Significant increase in protein solubility, emulsion activity index, and emulsion stability index were observed. The greatest increase in solubility was observed between pH 5.0 and 7.0. The molecular weight of the hydrolysates was also reduced significantly during hydrolysis. These improved functional properties of different protein hydrolysates would make them useful products, especially in the food, pharmaceutical, and related industries

There was slight increase (about 10%) in protein content of sprouted sesame seeds; sprouting was to remove the bitter taste (Badifu and Akpagher 1996). The foaming capacity of flours from untreated, sprouted and boiled seeds were 34.6%, 38.5% and 11.5%, respectively. The flour from the boiled seeds had the highest foam stability. The emulsion capacity of flours from the untreated or sprouted seeds was the same (27.6 g oil/g sample) while that from boiled seeds was 12.9 g oil/g sample. Emulsion stability with prolonged storage appeared to be more with flours from the sprouted or boiled seeds than that from the untreated ones. The water absorption properties of flours from the untreated, sprouted and boiled seeds were 8.0, 5.9 and 6.5 g water/g sample, respectively whereas the oil absorption capacity was the same (5.9 g oil/g sample). The bitter taste in flours from the untreated or sprouted seeds was high. The bitter taste was not detected in flour from boiled seeds and the functional properties of the flour were not deleteriously affected except foaming and emulsion capacity. Thus boiling debittered the seed and the quality of sesame flour obtained after boiling could still serve its role in traditional dishes and in the formulation of some other conventional food products. Defatting increased the crude protein, ash, crude fiber, carbohydrate and mineral contents of sesame flour (Egbekun and Ehieze 1997). Defatted flour showed comparatively better foam capacity and stability, water absorption and emulsion capacities than full fat flour but diminished bulk density and oil absorption capacity. Nitrogen solubility was pH dependent with a minimum at pH 4 and maximum at pH 8. Maximum nitrogen solubility (95%) was recorded for defatted flour while that for the full fat flour was 60%.

Phytochemicals in Other Plant Parts

From a water extract of *Sesamum indicum* plant, two new, and six known phenylethanoid glycosides, and three new triglycosides which had the same sugar sequence were isolated (Suzuki et al. 1993). Six flavones were isolated from sesame flowers: apigenin (1), ladanetin (2), ladanetin-6-O-β-D-glucoside (3), apigenin-7-O-glucuronic acid (4), pedalitin (5), and pedalitin-6-O-glucoside (6) (Hu et al. 2007a). Ten compounds were isolated from the 95% ethanol extract of the flower and elucidated as latifonin (1), momor-cerebroside (2), soya-cerebroside II (3), 1-O-β-D-glucopyranosyl- (2S, 3S, 4R, 5E,9Z)-2-N-(2′-hydroxytetracosanoyl) 1,3,4-trihydroxy-5,9-octadienine (4), 1-O-β-D-glucopyranosyl-(2S, 3S, 4R, 8Z)-2-N-(2′R) 2′-hydroxytetracosanoyl) 3,4-dihydroxy-8-octadene (5), (2S, 1″ S) -aurantiamide acetate (6), benzyl alcohol-O-(2′-O-β-D-xylopyranosyl, 3′-O-β-D-glucopyranoside)-β-D-glucopyranoside (7), β-sitosterol (8), daucosterol (9) and D-galacititol (10) (Hu et al. 2007b). Compounds 2–4 were cerebroside, being rare to be found in land plants and were reported to possess many bioactivities. Sesamin, sesamolin, stigmasterol, β-sitosterol and stigmasterol-3-O-β-D-glucoside were isolated from the petroleum ether fraction of the alcoholic extract of residual aerial parts of sesame after seed collection (Khaleel et al. 2007). Ferulic acid, rhamnetin, verbascoside, kaempferol-3-O-β-D-glucuronide and mequelianin (quercetin-3-O-β-D-glucuronide) were isolated from the butanol fraction. The content of the major constituents, namely sesamin and sesamolin, were also determined. A red naphthoquinone, named hydroxysesamone, (2,5,8-trihydroxy-3-(3-methyl-2-butenyl)-1,4-naphthoquinone) was isolated from the roots of Sesamum indicum together with a known yellow naphthoxirene derivative, 2,3-epoxy-2,3-dihydro-5,8-dihydroxy-2-(3-methyl-2-butenyl)-1,4-naphthoquinone, named 2,3-epoxysesamone (Hasan et al. 2001). Three anthraquinones, named anthrasesamones A, B and C, were isolated from the roots of Sesamum indicum, and their respective structures were characterised as 1-hydroxy-2-(4-methylpent-3-enyl)anthraquinone, 1,4-dihydroxy-2-(4-methylpent-3-enyl)anthraquinone and 2-chloro-1,4-dihydroxy-3-(4-methylpent-3-enyl) anthraquinone (Furumoto et al. 2003). Two known anthraquinones were also isolated from the roots and characterized as 2-(4-methylpent-3-enyl) anthraquinone and (E)-2-(4-methylpenta-1,3-dienyl)anthraquinone. Anthrasesamone C is a rare chlorinated anthraquinone in higher plants. Another two anthraquinone derivatives, named anthrasesamones D and E, were isolated from the roots and characterised as 1,2,4-trihydroxy-3-(4-methylpent-3-enyl)anthraquinone and 1,2-dihydroxy-3-(4-methylpent-3-enyl)anthraquinone (Furumoto et al. 2006).

Biological Activities of Sesame Lignans

Sesame is a rich source of lignin phytoestrogens. Sesame contains very high levels (up to 2.5%) of furofuran lignans mainly sesamin, sesamolin, and sesaminol glucosides, all with beneficial neutraceutical functions (Namiki 1995, 2007; Kamal-Eldin et al. 2011). Research had indicate the novel synergistic effect of sesame lignans with tocopherols for the anti-aging effects of sesame resulting from the inhibition of metabolic decomposition of tocopherols by sesame lignans (Namiki 2007). Sesame lignans modulate fatty acid metabolism, lowering fatty acid concentration in liver and serum due to acceleration of fatty acid oxidation and suppression of fatty acid synthesis, and the controlling influence on the ratio of n-6/n-3 polyunsaturated fatty acids under excess intake of either n-6 or n-3 fatty acids in the diet. Sesame lignans have been reported to lower the cholesterol concentration in serum, especially in combination with tocopherol, due to the inhibition of absorption from the intestine and suppression of biosynthesis in the liver. Studies also showed other useful activities of sesame lignans like acceleration of alcohol decomposition in the liver, antihypertensive, immunoregulatory, anticarcinogenic, antioxidant, cardiovascular activities and other functions (Namiki 1995, 2007; Kamal-Eldin et al. 2011) as elaborated below.

Antioxidant Activity

Unsaponifiables from the brown sesame variety were markedly different in their composition from those of the white variety (Mohamed and Awatif 1998). The brown variety contained higher amounts of total sterols and tocopherols but lower amounts of sesamin, sesamolin and total hydrocarbons than the white variety. Roasting the seeds at 180°C for 30 minutes increased some effective antioxidant compounds. These included relatively higher percentages of sesamol, $\Delta^{24,28}$ ethylidene sterols (Δ^5 and Δ^7-avenasterols), squalene, as well as tocopherols and some active browning substances. Additionally, unsaponifiable matter from unroasted (USM) and roasted white sesame seeds (RSM) was added individually to sunflower oil at levels of 0.02%, 0.05% and 0.1% and their effectiveness was compared with a control (no additives) at 63°C. Results indicated that both USM and RSM had antioxidant activity which increased with increasing concentration. Compared to USM, the RSM was a better antioxidant in most cases. Moreover, the addition of 0.1% RSM gave a strong antioxidative efficiency and this could be used as an alternative natural antioxidant for food applications.

Studies suggested that sesamolin and its metabolites the sesame lignans, sesaminol and sesamolinol may contribute to the antioxidative properties of sesame seeds and oil (Kang et al. 1998). Lipid peroxidation activity, measured as 2-thiobarbituric acid reactive substances, was significantly lower in the kidneys and liver of the sesamolin-fed rats than in the controls. Liver weight was significantly greater in the sesamolin-fed rats than in the controls. In addition, the amount of 8-hydroxy-2′-deoxyguanosine excreted in the urine was significantly lower in the sesamolin-fed rats. The results supported the hypothesis that sesame lignans reduced susceptibility to oxidative stress. In a subsequent study, the authors found that feeding defatted sesame flour to rabbits did not protect cholesterol-induced hypercholesterolemia, but may decrease susceptibility to oxidative stress in rabbits fed cholesterol, perhaps attributable to the antioxidative activity of sesaminol (Kang et al. 1998). Serum total cholesterol, phospholipid, triglyceride and HDL cholesterol concentrations were unaffected by the addition of the defatted sesame flour. Lipid peroxidation activity, measured as 2-thiobarbituric acid reactive substances (TBARS), was lower in the liver and serum of rabbits fed the deffated sesame flour plus cholesterol than in rabbits fed the cholesterol diet.

Dachtler et al. (2003) used the Rancimat assay to study the effect of the sesame oil extracts as well as pure sesame lignans and γ-tocopherol on the oxidative stability of sunflower oil (lignan-free). The Rancimat assay revealed the following oxidative stability order: sesame oil extract < sesame oil deodorizer distillate < sunflower oil (no added sesame oil extracts) < sesamol < sesaminol-enriched sesame oil extract. In addition, the Trolox® equivalent antioxidant capacity (TEAC) assay revealed a slightly different antioxidant activity order: sesamin < sesame oil extract < sesaminol-enriched sesame oil extract < sesamol. The authors concluded that the sesaminol-enriched extract exerted strong antioxidant activity and was therefore suitable to increase the oxidative stability of edible oils high in polyunsaturated fatty acids. Using a nanosecond pulse radiolysis technique, 5-hydroxy-1,3-benzodioxole sesamol efficiently scavenged hydroxyl, one-electron oxidizing, organo-haloperoxyl, lipid peroxyl, and tryptophanyl radicals (Joshi et al. 2005). In biochemical studies, it was found to inhibit lipid peroxidation, hydroxyl radical-induced deoxyribose degradation, and DNA cleavage.

Studies demonstrated that the graded-dose (25–1,000 μg/ml) of aqueous and ethanolic seed extracts from *S. indicum* markedly scavenged the nitric oxide, superoxide, hydroxyl, 1,1-diphenyl-2-picrylhydrazyl and 2,2′-azinobis-(3-ethylbenzothiazoline-6-sulfonic acid) radicals and, showed metal chelating ability as well as reducing capacity in Fe^{3+} ferricyanide complex and ferric reducing antioxidant power assays (Visavadiya et al. 2009). In biological models, both extracts were found to inhibit metal-induced lipid peroxidation in mitochondrial fractions, human serum and LDL oxidation models. In lipoprotein kinetics study, both extracts significantly increased lag phase time along with reduced oxidation rate and conjugated dienes production. Ethanolic extract of *S. indicum* showed higher amounts of total polyphenol and flavonoid content as compared to their counterpart. Overall, ethanolic extract of *S. indicum* possessed strong antioxidant capacity and offered effective protection against LDL oxidation susceptibility. Sesame ethanol seed extract showed 92.00% inhibition and 56.00% reduction ability in hydrogen donation and reducing power assays, respectively at maximum concentration of the extract tested (Nahar and Rokonuzzaman 2009). The antioxidant activity of the extract in all these in-vitro assays was compared with standard antioxidant (ascorbic acid). Results from the linoleic acid system showed that the antioxidant activity of black sesame seed extracted by supercritical carbon dioxide extraction $SC-CO_2$ followed the order: extract at 35°C, 20 MPa > BHT > extract at 55°C, 40 MPa > extract at °C, 30 MPa > Trolox > solvent extraction > α-tocopherol. The $SC-CO_2$ extracts exhibited significantly higher antioxidant activities comparable to that by n-hexane extraction. The extracts at 30 MPa presented the highest antioxidant activities assessed in the DPPH method. At 20 MPa, the EC_{50} increased with temperature, which indicated that the antioxidant activity was decreased in a temperature-dependent manner. The significant differences of antioxidant activities were found between the extracts by $SC-CO_2$ extraction and n-hexane. However, no significant differences were exhibited among the extracts by $SC-CO_2$ extraction. The vitamin E concentrations were also significantly higher in $SC-CO_2$ extracts than in n-hexane extracts, and its concentrations in extracts corresponded with the antioxidant activity of extracts.

The free radical scavenging capacity (RSC) of sesame antioxidants expressed by the second-order rate constant (k_2) was calculated for the quenching reaction with (DPPH) radical and compared with those of butylated hydroxytoluene and α-tocopherol (Suja et al. 2004). The k_2 values for sesamol, sesamol dimer, sesamin, sesamolin, sesaminol triglucoside, and sesaminol diglucoside were 4.00×10^{-5}, 0.50×10^{-5}, 0.36×10^{-5}, 0.13×10^{-5}, 0.33×10^{-5} and 0.08×10^{-5} μM^{-1} s^{-1}, respectively.

The methanol extract of sesame (*Sesamum indicum*) seeds afforded 29 compounds including seven furofuran lignans (Kuo et al. 2011). Among these isolates, (+)-samin (1) was obtained from the natural source for the first time. In addition, (-)-asarinin (30) and sesamol (31) were generated by oxidative derivation from (+)-sesamolin (2) and (1)-sesamin (3), two abundant lignans found in sesame seeds. The in-vitro antioxidant potential of the seven isolated lignans (1–7) and the two derivatives (30 and 31) were examined for the scavenging activities on DPPH free radicals and superoxide anions in addition to chelating capability of ferrous ions and reducing power. The results suggested that, besides the well-known sesamolin and sesamin, the minor sesame lignans (+)-(7 S,8′R,8R)-acuminatolide (5), (-)-piperitol (6), and (+)-pinoresinol (7) were also adequate active ingredients and may be potential sources for nutritional and pharmacological utilization based on their in-vitro antioxidant potential.

Total phenolic, flavonoid and flavonol contents in sesame cake extract were 1.94 (mg gallic acid equivalent (GAE)/g dry weight (DW)), 0.88 (mg quercetain equivalent (QE)/g DW), and 0.40 (mg QE/g DW), respectively (Mohdaly et al. 2011). Sesame cake extract exerted protective effect in stabilizing sunflower oil (SFO) and soybean oil (SBO). It exhibited stronger antioxidant activity in SFO and SBO than butylated hydroxytoluene (BHT) and butylated hydroxyanisole (BHA), while its antioxidant activity was less than that of tert-butyl hydroquinone (TBHQ) as evaluated by (DPPH) radical scavenging capacity, and β-carotene/linoleic acid test system.

Lee et al. (2008b) investigated the effects of lignan compounds sesamol, sesamin, and sesamolin, extracted from roasted sesame oil, on oxidation of methyl linoleate (ML) during heating. conjugated dienoic acid (CDA) contents, *p*-anisidine value (PAV), and methyl linoleate decreased with heating time at 180°C. The antioxidant activity of sesame oil lignan compounds, sesamol, sesamin, and sesamolin in methyl linoleate oxidation during heating tended to be higher than that of α-tocopherol. The contents of lignan compounds in samples decreased with heating time due to their degradation, but the degradation rates were lower than that of α-tocopherol. This study suggested sesame oil lignan compounds could be used as antioxidants in oil at high temperatures for deep-fat frying due to their higher effectiveness and stability than α-tocopherol.

Roasting of sesame seeds for longer time and at higher temperature, generated more sesamol in sesame oil (Lee et al. 2010). Sesame oil from sesame seeds roasted at 247°C for 28 min had the highest oxidative stability. Higher oxidative stability of sesame oil may be related to the continuous generation of sesamol from the degradation of sesamolin during thermal oxidation rather than the initial antioxidant content.

Studies showed that sesame oil had the highest FRAP (Ferric Reducing/Antioxidant Power) value (803 μM), followed by canola oil (400 μM), and sunflower, peanut, corn and olive oils (100–153 μM) (Cheung et al. 2007). Oils with higher intrinsic antioxidant content showed higher resistance to oxidation.

Lipid and Alcohol Metabolism Activity

Animal studies showed sesamin, a sesame lignan, to be a potent inducer of hepatic fatty acid oxidation (Ashakumary et al. 1999). Dietary sesamin dose-dependently increased both mitochondrial and peroxisomal palmitoyl-coenzyme A (CoA) oxidation rates. Mitochondrial activity almost doubled in rats on the 0.5% sesamin diet. Peroxisomal activity increased more than tenfold in rats fed a 0.5% sesamin diet in relation to rats on the sesamin-free diet. Dietary sesamin greatly increased the hepatic activity of fatty acid oxidation enzymes and induced increase in the gene expression of mitochondrial and peroxisomal fatty acid oxidation enzymes. In contrast, dietary sesamin decreased the hepatic activity and mRNA abundance of fatty acid synthase and pyruvate kinase, the lipogenic enzymes. However, this lignan increased the activity and gene expression of malic enzyme, another lipogenic enzyme. An alteration in hepatic fatty acid metabolism may therefore account for the serum lipid-lowering effect of sesamin in the rat. Studies showed that as dietary level of sesamin increased up to 0.4%, the activity and gene expression of enzymes involved in fatty acid synthesis including acetyl-CoA carboxylase, fatty acid synthase, ATP-citrate lyase and glucose-6-phosphate dehydrogenase decreased (Ide et al. 2001). Dietary sesamin dose-dependently decreased the sterol regulatory element binding protein-1 (SREBP-1) mRNA level, and the value in rats fed a 0.4% sesamin diet was approximately one-half that in those fed a sesamin-free diet. The findings suggested that dietary sesamin-dependent decrease in lipogenic enzyme gene expression was attributed to the suppression of the gene expression of SREBP-1 as well as the proteolysis of the membrane-bound precursor form of this transcriptional factor to generate the mature form. Dietary studies in rats showed that a diet containing sesamin and fish oil in combination synergistically increased hepatic fatty acid oxidation primarily through up-regulation of the gene expression of peroxisomal fatty acid oxidation enzymes (Ide et al. 2004). Dietary sesamin increased fatty acid oxidation enzyme activities in all groups of rats given different fats (palm, safflower, fish oil). A diet containing sesamin and fish oil in combination appeared to increase many of these parameters synergistically. In particular, the peroxisomal palmitoyl-CoA oxidation rate and acyl-CoA oxidase activity levels were much higher in rats fed sesamin and fish oil combination than in animals fed sesamin and palm or safflower oil combination.

Studies suggested that sesamin ingestion regulated the transcription levels of hepatic metabolizing enzymes for lipids and alcohol in rats (Kiso 2004). Twenty-four hours after sesamin ingestion, over 40% of the dose of sesamin was detected in bile as glucuronides of 2-(3, 4-methylenedioxyphenyl) -6-(3, 4-dihydroxyphenyl)-cis-dioxabicyclo[3.3.0] octane and 2-(3, 4-dihydroxyphenyl)-6-(3, 4-dihydroxyphenyl)-cis-dioxabicyclo[3.3.0] octane. Both metabolites showed strong radical scavenging activities against not only superoxide anion radical but also hydroxyl radical. Sesamin could be classified as a pro-antioxidant. Gene expression of hepatic lipid oxidation enzymes were increased but the transcription of the genes encoding the enzymes for fatty acid synthesis was decreased. Further in sesamin rats, the gene expression of aldehyde dehydrogenase was increased about three-fold, whereas alcohol dehydrogenase, liver catalase and CYP2E1 were not changed. Sesamin had been reported to have multiple functions such as stimulation effect of ethanol metabolism in mice and human, and prevention of ethanol-induced fatty liver in rats (Kiso et al. 2005). Results of a DNA microarray analysis in rats suggested that sesamin ingestion regulated the transcription levels of hepatic metabolizing enzymes for alcohol and lipids. The gene expression levels of the early stage enzymes of β-oxidation including long-chain acyl-CoA synthetase, very long-chain acyl-CoA synthetase and carnitine palmitoyltransferase were not changed, however, those of the late stage enzymes of β-oxidation including trifunctional enzyme in mitochondria, and acyl-CoA oxidase, bifunctional enzyme and 3-ketoacyl-CoA thiolase in peroxisomes, were significantly enhanced by sesamin ingestion. Also, in sesamin rats, the gene expression of aldehyde dehydrogenase was increased about three-fold, whereas alcohol dehydrogenase, liver catalase and CYP2E1 were not changed

Further studies suggested that sesamin regulated the metabolism of lipids, xenobiotics, and alcohol at the mRNA level (Tsuruoka et al. 2005). The ingestion of sesamin dissolved in olive oil up-regulated the expression of 38 genes, 16 of which encode proteins possessing a lipid-metabolizing function, and 16 of which encode proteins possessing a xenobiotic/endogenous substance metabolizing function. In particular, sesamin significantly increased the expression of β-oxidation-associated enzymes in peroxisomes and auxiliary enzymes required for degradation, via the β-oxidation pathway, of unsaturated fatty acids in mitochondria. Sesamin ingestion also resulted in an increase in the gene expression of acyl-CoA thioesterase involved in acyl-CoA hydrolase and very-long-chain acyl-CoA thioesterase and also induced the expression of the gene for aldehyde dehydrogenase, an alcohol-metabolizing enzyme. Dietary sesamin and docosahexaenoic and eicosapentaenoic acids were found to synergistically increase the gene expression of enzymes involved in hepatic peroxisomal fatty acid oxidation in rats (Arachchige et al. 2006). Sesamin and sesamolin dose-dependently increased the activity and mRNA abundance of various enzymes involved in hepatic fatty acid oxidation in rats (Lim et al. 2007). The increase was much greater with sesamolin than with sesamin. In contrast, they decreased the activity and mRNA abundance of hepatic lipogenic enzymes despite dose-dependent effects. Sesamin and sesamolin were equally effective in lowering parameters of lipogenesis. Sesamin compared to sesamolin was more effective in reducing serum and liver lipid levels despite sesamolin more strongly increasing hepatic fatty acid oxidation. Differences in bioavailability may contribute to the divergent effects of sesamin and sesamolin on hepatic fatty acid oxidation.

Ide et al. (2009) found that compared to a lignan-free diet, a diet containing sesamin, episesamin and sesamolin caused more than 1.5- and 2-fold changes in the expression of 128 and 40, 526 and 152, and 516 and 140 hepatic genes, respectively. The lignans modified the mRNA levels of not only many enzymes involved in hepatic fatty acid oxidation, but also proteins involved in the transportation of fatty acids into hepatocytes and their organelles, and in the regulation of hepatic concentrations of carnitine, CoA and malonyl-CoA. Sesame lignans stimulated hepatic fatty acid oxidation by affecting the gene expression of various proteins regulating hepatic fatty acid metabolism. The changes in the gene expression were generally greater with episesamin and sesamolin than with sesamin. In terms of amounts accumulated in serum and the liver, the lignans ranked in the order sesamolin, episesamin and sesamin. The differences in bioavailability among these lignans appeared to be important to their divergent physiological activities.

Supplementation of sesame seeds at levels of 200 g/kg to the experimental diets in rats increased both the hepatic mitochondrial and the peroxisomal fatty acid oxidation rate (Sirato-Yasumoto et al. 2001). Increases were greater with sesame cultivars rich in lignans than with the conventional cultivar Maskin. Noticeably, peroxisomal activity levels were >3 times higher in rats fed diets containing sesame seeds from lignan lines than in those fed a control diet without sesame. Diets containing seeds from lignin rich lines, compared to the control and Masekin diets, also significantly increased the activity of hepatic fatty acid oxidation enzymes including acyl-CoA oxidase, carnitine palmitoyltranferase, 3-hydroxyacyl-CoA dehydrogenase, and 3-ketoacyl-CoA thiolase. In contrast, diets containing sesame lowered the activity of enzymes involved in fatty acid synthesis including fatty acid synthase, glucose-6-phosphate dehydrogenase, ATP-citrate lyase, and pyruvate kinase. Serum triacylglycerol concentrations were lower in rats fed diets containing sesame from lignin rich lines than in those fed the control or Masekin diet. It was apparent that sesame rich in lignans more profoundly affected hepatic fatty acid oxidation and serum triacylglycerol levels. Thus, consumption of sesame rich in lignans resulted in physiological activity to alter lipid metabolism in a potentially beneficial manner.

In a randomized crossover study, 16 postmenopausal women were supplemented in their diets with food bars containing either 25 g unground flaxseed, sesame seed, or their combination (12.5 g each) (flaxseed + sesame seed bar, FSB) for 4 week each, separated by 4 week washout periods (Coulman et al. 2009). Total serum n-3 fatty acids increased with flaxseed and FSB while serum n-6 fatty acids increased with sesame seed. Urinary lignans increased similarly with all treatments. Plasma lipids and several antioxidant markers were unaffected by all treatments, except serum γ-tocopherol (GT), which was increased with both sesame seed and FSB. The findings indicated that fatty acids and lignans from unground seed in food bars were absorbed and metabolized; however, except for serum γ-tocopherol, the 25 g unground seed had minimal antioxidant and lipid-lowering effects in postmenopausal women.

Hypolipidemic and Hypocholesteremic Activity

(+)-sesamin was found to be a potent and specific inhibitor of delta 5 desaturase in polyunsaturated fatty acid biosynthesis in the arachidonic acid-producing fungus, *Mortierella alpine* and rat liver microsomes (Shimizu et al. 1991). (+)-Sesamolin, (+)-sesaminol and (+)-episesamin also inhibited only delta 5 desaturases of the fungus and liver. In normocholesterolaemic stroke-prone spontaneously hypertensive (SHRSP) rats fed a regular diet, both sesamin and episesamin significantly increased the concentration of serum total cholesterol, which was due to an increase of high density lipoprotein (HDL) subfraction rich in apoE (apoE-HDL) (Ogawa et al. 1995). In addition, both compounds effectively decreased serum very low density lipoprotein (VLDL). In the liver, only episesamin significantly decreased the activity of microsomal acyl-CoA:cholesterol acyltransferase. In hypercholesterolaemic SHRSP fed a high-fat and high-cholesterol diet (HFC diet), only episesamin improved serum lipoprotein metabolism with an increase in apoA-I and a decrease in apoB. In the liver, both sesamin and episesamin significantly suppressed cholesterol accumulation. Only episesamin significantly increased the activity of microsomal cholesterol 7α-hydroxylase. These results indicated that sesamin may be effective in preventing cholesterol accumulation in the liver. In comparison with sesamin, episesamin may be effective in the regulation of cholesterol metabolism in the serum and liver. A study on male patients with hypercholesterolemia suggested that sesamin could reduce serum cholesterol especially LDL-C, a risk factor for atherosclerosis (Hirata et al. 1996) Sesamin treatment significantly reduced total cholesterol, LDL-C, and apoprotein (apo) B compared to placebo group. Results of studies suggest that sesame ingestion benefited postmenopausal women by improving blood lipids, antioxidant status, and possibly sex hormone status (Wu et al. 2006). After sesame treatment, plasma total cholesterol (TC), LDL-C, the ratio of LDL-C to HDL-C, thiobarbituric acid reactive substances in oxidized LDL, and serum dehydroepiandrosterone sulfate decreased significantly. The ratio of α- and γ-tocopherol to TC increased significantly by 18% and 73%, respectively. Serum sex hormone-binding globulin and urinary 2-hydroxyestrone increased significantly by 15% and 72%, respectively, after sesame treatment.

Studies showed that LDL receptor-deficient mice fed an atherogenic diet had an almost three-fold increase in serum cholesterol levels but no effect was observed for triglyceride levels (Peñalvo et al. 2006). Stanol ester alone or together with sesamin significantly attenuated the elevation of the cholesterol levels. Sesamin alone did not affect the elevation of the diet-induced cholesterol level and it did not enhance the effect of stanol ester.

Administration of sesame seed powder to hypercholesteraemic rats resulted in a significant decrease in plasma, hepatic total lipid and cholesterol contents and, plasma LDL-cholesterol contents with an elevation in plasma HDL-cholesterol content (Visavadiya and Narasimhacharya 2008). Further, these animals also exhibited enhanced fecal excretion of cholesterol, neutral sterol and bile acid along with increases in hepatic HMG-CoA reductase activity and bile acid concentration. Additionally sesame seed feeding improved the hepatic antioxidant status (catalase and SOD enzyme activities) with a decrease in lipid peroxidation. No significant changes in lipid and antioxidant profiles occurred in the normocholesteraemic rats administered with sesame seed powder. These beneficial effects of sesame seed on hypercholesteraemic rats appeared to be due to its fibre, sterol, polyphenol and flavonoid content, enhancing the fecal cholesterol excretion and bile acid production and as well as increasing the antioxidant enzyme activities.

Antidiabetic Activity

Flavonoids from *Sesamum indicum* elicited hypolipidemic and hypoglycemic activities in rats and raised the hemoglobin levels (Anila and Vijayalakshmi 2000). Studies indicated that hot-water extract from defatted sesame seeds and the methanol eluent fraction in the diet had a reductive effect on the plasma glucose concentration of diabetic KK-Ay mice, and this effect was suggested to have been caused by the delayed glucose absorption (Takeuchi et al. 2001). The alcoholic, petroleum ether and butanol extracts of sesame residual aerial plant parts significantly restored the reduced levels of glutathione in the hyperglycaemic diabetic rats (Khaleel et al. 2007). The total alcoholic extract was the most potent exhibiting comparable activity to that of vitamin E at the tested doses. All tested extracts showed a reductive effect on blood glucose level of diabetic rats. The total alcoholic extract showed a more powerful effect than its fractions. The stronger activity of the total alcoholic extract over its fractions might be attributed to a synergistic effect of the lignans present mainly in the petroleum ether fraction, and the phenolic compounds present in the butanol fraction. Sesame lignans, rhamnetin, ferulic acid, verbascoside and mequelianin had been reported to possess antioxidant activity. The researchers concluded that the alcoholic extract may be useful in alleviating oxidative stress and attenuating the hyperglycaemic response associated with diabetes.

In an open label study of type 2 diabetes mellitus patients, combination therapy of sesame oil and glibenclamide showed an improved anti-hyperglycaemic effect with 36% reduction of glucose and 43% reduction of glycated haemoglobin, HbA1c compared to sesame oil or glibenclamide monotherapy (Sankar et al. 2011). Significant reductions in the plasma TC, LDL-C and TG levels were noted in sesame oil (20%, 33.8% and 14% respectively vs. before treatment) or combination therapies (22%, 38% and 15% respectively vs. before treatment). Plasma HDL-C was significantly improved in sesame oil (15.7% vs. before treatment) or combination therapies (17% before treatment). Significant improvement was observed in the activities of enzymatic and non-enzymatic antioxidants in patients treated with sesame oil and its combination with glibenclamide. The authors asserted that, sesame oil exhibited synergistic effect with glibenclamide and could provide a safe and effective option for the drug combination that may be very useful in clinical practice for the effective improvement of hyperglycemias.

Anticancer Activity

The alcohol extract from *Sesamum indicum* flower inhibited tumour growth in sarcoma 180 (S180) and Hep22 (H22) tumorigenic mice (Xu et al. 2003). Studies found sesamin and episesamin isolated from unroasted sesame seed oil induced apoptosis in human lymphoid leukemia Molt 4B cells (Miyahara et al. 2000). Exposure of human lymphoid leukemia Molt 4B cells to sesamin and episesamin led to both growth inhibition and the induction of programmed cell death (apoptosis). The results suggested that growth inhibitions by sesamin and episesamin of Molt 4B cells resulted from the induction of apoptosis in the cells. Harikumar et al. (2010) found that sesamin, a lipid-soluble lignin isolated from *S. indicum*, inhibited the proliferation of a wide variety of tumour cells including leukemia, multiple myeloma, and cancers of the colon, prostate, breast, pancreas, and lung. Sesamin also potentiated tumour necrosis factor-α-induced apoptosis and this correlated with the suppression of gene products linked to cell survival (e.g., Bcl-2 and survivin), proliferation (e.g., cyclin D1), inflammation (e.g., cyclooxygenase-2), invasion (e.g., matrix metalloproteinase-9, intercellular adhesion molecule 1), and angiogenesis (e.g., vascular endothelial growth factor). Sesamin down regulated constitutive and inducible NF-κB activation induced by various inflammatory stimuli and carcinogens, and inhibited the degradation of IκBα, the inhibitor of NF-κB, through the suppression of phosphorylation of IκBα and inhibition of activation of IκBα protein kinase. The inhibition of IκBα protein kinase activation was found to be mediated through the inhibition of TAK1 kinase. Their results showed that sesamin may have potential against cancer and other chronic diseases through the suppression of a pathway linked to the NF-κB signalling.

Studies in mice with MCF-tumours, lignan-rich sesame seed negated the tumour-inhibitory effect of tamoxifen by reducing apoptosis but beneficially interacted with tamoxifen on bone in ovariectomized athymic mice (Sacco et al. 2008). Sesame seed combined with tamoxifen induced higher bone mineral content, bone mineral density, and biomechanical strength in the femur and lumbar vertebrae than either treatment alone.

Vitamin E (α-Tocopherol) Enhancing Activity

Vitamin E has been recognised as an important dietary component with antiaging function (Meydani 1992). Studies in rats demonstrated that γ-tocopherol in sesame seed exerted vitamin E activity equal to that of α-tocopherol through a synergistic interaction with sesame seed lignans (Yamashita et al. 1992). Indices of vitamin E activity were ascertained as changes in red blood cell hemolysis, plasma pyruvate kinase activity, and peroxides in plasma and liver. The sesame seed diet had high vitamin E activity, whereas this activity was low in the γ-tocopherol diet. In an additional experiment, sesame lignin (sesaminol or sesamin)-fed groups exhibited vitamin E activity comparable to that observed in the sesame seed-fed group in the earlier experiment.

Yamashita et al. (1995) reported that sesame seed lignans enhanced vitamin E activity in rats fed a low α-tocopherol diet and caused a marked increase in α-tocopherol concentration in the blood and tissue of rats fed an α-tocopherol-containing diet with sesame seed or its lignans. Sesamin and α-tocopherol were found to synergistically suppress lipid-peroxide in rats fed a high docosahexaenoic acid diet (Yamashita et al. 2000). TBARS concentrations in plasma and liver were significantly increased by docosahexaenoic acid, but were completely suppressed by sesamin. α-tocopherol concentrations in plasma and liver decreased by addition of docosahexaenoic acid, but were restored to control level with sesamin. Both flaxseed and sesame seed were reported to contain more than 40% fat, about 20% protein, and vitamin E, mostly γ-tocopherol and considerable amounts of plant lignans (Yamashita et al. 2003). However, flaxseed contained 54% α-linolenic acid, but sesame seed only 0.6%, and the chemical structures of flaxseed and sesame lignans were different. Dietary studies in rats showed that that sesame seed and its lignans induced higher γ-tocopherol and lower TBARS concentrations; whereas flaxseed lignans had no such effects (Yamashita et al. 2003). Further, α-linolenic acid produced strong plasma cholesterol-lowering effects and higher TBARS concentrations. Further studies suggested that the sesame lignan seminol increased tocopherol concentrations in animals by suppressing the conversion of γ-tocopherol to γ-CEHC (2, 7, 8-trimethyl-2(2′-carboxyethyl)-6-hydroxychroman), a γ-tocopherol metabolite (Yamashita et al. 2007). HMR (7-hydroxy-matairesinol), a structurally different dibenzylbutyrolactone type lignin from sesame seed, did not have such properties.

Dietary docosahexaenoic acid elevated the thiobarbituric acid reactive substance (TBARS) concentration and also increased the red blood-cell hemolysis induced by the dialuric acid in rats while dietary sesamin and sesaminol lowered the TBARS concentrations and decreased the red blood hemolysis (Ikeda et al. 2003). Additionally, dietary sesamin and sesaminol elevated the α-tocopherol concentrations in the plasma, liver, and brain of the rats fed a diet with or without

DHA. The results suggested that dietary sesame lignans decreased lipid peroxidation as a result of elevating the α-tocopherol concentration in rats fed docosahexaenoic acid. Further studies by Ikeda et al. (2007) suggested that dietary sesame seed and its lignin, sesamin stimulated ascorbic acid synthesis as a result of the induction of UDP-glucuronosyltransferase 1A and 2B-mediated metabolism of sesame lignan in rats. Data from ODS (Osteogenic Disorder Shionogi) inherently scorbutic rats, also suggested that dietary sesame seed enhanced antioxidative activity in the tissues by elevating the levels of two antioxidative vitamins, vitamin C and E. Sesame seed elevated the γ-tocopherol concentration in the various ODS rat tissues and the ascorbic acid concentrations in the kidney, heart and lung, while reducing the thiobarbituric acid reactive substance concentration in the heart and kidney.

The tocopherols, the major vitamers of vitamin E, are believed to play a role in the prevention of human aging-related diseases such as cancer and heart disease (Cooney et al. 2001). Studies in humans found that consumption of as little as 5 mg of γ-tocopherol per day over a 3-day period from sesame seeds significantly elevated serum γ-tocopherol levels (19.1% increase) and depressed plasma β-tocopherol (34% decrease) (Cooney et al. 2001). No significant changes in baseline or post-intervention plasma levels of cholesterol, triglycerides, or carotenoids were seen for any of the intervention groups. All subjects consuming sesame seed-containing muffins had detectable levels of the sesame lignan sesamolin in their plasma. The results suggested that consumption of moderate amounts of sesame seeds appeared to significantly increase plasma γ-tocopherol and alter plasma tocopherol ratios in humans. The results were in accord with the effects of dietary sesame seeds observed in rats leading to elevated plasma γ-tocopherol and enhanced vitamin E bioactivity.

Antihypertensive Activity

Sesamin (lignin from sesame oil) feeding in rats ameliorated the development of deoxycorticosterone acetate (DOCA)-salt-induced vascular hypertrophy in both the aorta and mesenteric artery (Matsumura et al. 1995). The treatment with DOCA and salt for 5 weeks significantly increased the weight of the left ventricle plus the septum, however, this increase was significantly suppressed in the sesamin group. These findings strongly suggested sesamin to be useful as a prophylactic treatment in the development of hypertension and cardiovascular hypertrophy. Dietary feeding of rats with sesamin markedly reduced the two-kidney, one-clip (2K, 1C) renal hypertension and also ameliorated vascular hypertrophy (Kita et al. 1995). Sesamin feeding was much more effective as an antihypertensive regimen in salt-loaded SHRSP (stroke-prone spontaneously hypertensive) rats than in unloaded SHRSP rats, thereby suggesting that sesamin was more useful as a prophylactic treatment in the malignant status of hypertension and/or hypertension followed by water and salt retention (Matsumura et al. 1998). Dietary sesamin also efficiently improved the abnormal vasodilator and vasoconstrictor responses in DOCA-salt hypertensive animals (Matsumura et al. 2000). Sesamin ameliorated the DOCA suppressed acetylcholine (ACh)-induced endothelium-dependent relaxation of aortic rings. This improvement appeared to be related to a nitric oxide (NO)-dependent component of ACh-induced action, because sesamin feeding did not affect the responses to ACh in the presence of NO synthase inhibitor. Sesamin feeding and triple therapy also significantly improved the DOCA-salt-induced impairment of endothelium-dependent relaxation (Nakano et al. 2003). In addition, dietary sesamin prevented DOCA salt-induced increases in aortic NADPH oxidase activity and subunit mRNA expression (Nakano et al. 2008). All these effects may contribute to the anti-oxidant and antihypertensive activity of sesamin. In a double-blind, cross-over, placebo-controlled trial of 25 middle-aged subjects with mild hypertension, 4 week administration of 60 mg sesamin significantly decreased blood pressure by an average of 3.5 mmHg systolic BP and 1.9 mmHg diastolic BP compared with placebo (Miyawaki et al. 2009). The results suggested that sesamin had an antihypertensive effect in humans. Epidemiological studies suggested that a 2–3 mmHg decrease in blood pressure reduced the rate of cardiovascular diseases; therefore, it is considered that blood pressure reduction achieved by sesamin may be meaningful to prevent cardiovascular diseases.

The petroleum ether soluble fraction of sesame roots at concentrations up to 180 ug/ml concentration significantly inhibited phenylephrine-induced and KCl-induced contraction in isolated rat aorta in a concentration dependent manner (Suresh Kumar et al. 2008). The vasorelaxant activity was not blocked by propranolol (10 µM), atropine (1 µM) indomethacin (10 µM) and glibenclamide (10 µM). However, in absence of functional endothelium, the extract exhibited little relaxation, indicating that the vasorelaxant activity of sesame root extract was chiefly mediated through endothelium-dependent pathway.

Separate studies suggested that the enhancement of endothelium-dependent vasorelaxation induced by sesamin metabolites SC-1 m (piperitol), SC-1 (demethylpiperitol), SC-2 m [(1R, 2S,5R,6S)-6-(4-hydroxy-3-methoxyphenyl)-2-(3,4-dihydroxyphenyl)-3,7-dioxabicyclo[3,3,0]octane], and SC-2 [(1R,2S,5R, 6S)-2,6-bis(3,4-dihydroxyphenyl)-3,7-dioxabicyclo-[3,3,0]octane] was one of the important mechanisms of the in-vivo antihypertensive effect of sesamin (Nakano et al. 2006). SC-1, SC-2 m, and SC-2, but not SC-1 m, exhibited potent radical-scavenging activities against the xanthine/xanthine oxidase-induced superoxide production. On the other hand, SC-1 m, SC-1, and SC-2 m produced endothelium-dependent vasorelaxation in phenylephrine-pre-contracted rat aortic rings, whereas SC-2 had no effect. Neither SC-1 m nor SC-1 changed the expression level of endothelial nitric oxide synthase protein in aortic tissues. The antihypertensive effects of sesamin feeding were not observed in chronically NG-nitro-L-arginine -treated rats or in deoxycorticosterone acetate-salt-treated endothelial nitric oxide synthase-deficient mice.

The results of a randomized study involving 22 women and 9 men with prehypertension showed that 4-week administration of black sesame meal significantly decreased systolic blood pressure (129.3 vs. 121.0 mmHg,) and malondialdehyde (MDA) level (1.8 vs. 1.2 µmol/L,), and increased vitamin E level (29.4 vs. 38.2 µmol/L) (Wichitsranoi et al. 2011). In the black sesame meal group, the change in systolic blood pressure tended to be positively related to the change in MDA ($R = 0.50$), while the change in diastolic blood pressure was negatively related to the change in vitamin E ($R = -0.55$). The finding suggested the possible antihypertensive effects of black sesame meal on improving antioxidant status and decreasing oxidant stress and may have a beneficial effect on prevention of cardiovascular diseases.

Cardiovascular Activity

Sesamol was found to induced nitric oxide release from human umbilical vein endothelial cells in a dose-dependent manner, the expression of endothelial NO synthase (eNOS) at both transcription and translation levels; and NO synthase (NOS) activity in endothelial cells (Chen et al. 2005). The content of cGMP was also increased by sesamol through nitric oxide signalling. The transcription of eNOS induced by sesamol was confirmed through the activation of PI-3 kinase-Akt (protein kinase B) signalling. The results demonstrated that sesamol induced NOS signalling pathways in human umbilical vein endothelial cells and suggested a role for sesamol in cardiovascular reactivity in-vivo.

In a randomized, placebo-controlled crossover intervention trial of overweight or obese men and women, supplementation with 25 g/day of sesame could significantly increase the exposure to mammalian lignans (Wu et al. 2009). However, this did not cause any improvement in markers of cardiovascular disease risk in overweight or obese men and women. Urinary excretion of the mammalian lignans, enterolactone and enterodiol, increased by approximately eightfold. Blood lipids and blood pressure were not altered. In addition, markers of systemic inflammation (C-reactive protein, interleukin-6, tumour necrosis factor-α) and lipid peroxidation (F(2)-isoprostanes) were not affected.

Antiinflammatory Activity

The data from studies suggested that sesame seed oil containing sesamin and Quil A, a fat emulsifying saponin when present in the mice diet exerted cumulative effects that resulted in a decrease in the levels of dienoic eicosanoids with a reduction in interleukin IL-1β, prostaglandin-E2 and thromboxane-B2 and a concomitant elevation in the levels of IL-10 that were associated with a marked increase in survival in mice (Chavali et al. 1997). Chronic ethanol drinking, at the dietary level of 23% (w/w), significantly increased the plasma IgA and IgM concentrations in rats, irrespective of the presence of 0.1% and 0.2% sesaminol, but the effects disappeared with 0.2% sesamin (Nonaka et al. 1997). A significant IgG-elevating effect of these lignans was also found. Although ethanol drinking did not influence splenic leukotriene B4 production, sesaminol tended to decrease it dose dependently, while sesamin increased the plasma prostaglandin E2 concentration. These results suggested that sesaminol and sesamin appeared to have a diverse effect on the plasma levels of immunoglobulins and eicosanoids.

Following a lethal dose of lipopolysaccharide endotoxin injection, all control animals died, survival was 40% in the sesame seed oil group and 27% and 50%, respectively, in those fed Quil-A-supplemented control and sesame seed oil diets. Further studies showed that sesamin, sesamol and other lignans in sesame seed oil appeared to be responsible for an increase in mice survival after caecal ligation and puncture and also for an increase in the IL-10 levels in response to a nonlethal dose of endotoxin in mice (Chavali et al. 2001). The authors asserted that lignans present in the non-fat portion of sesame seed oil (SSO) could inhibit delta-5 desaturase activity, resulting in an increase in the accumulation of dihomo-γ-linolenic acid and, subsequently, decreased the production of proinflammatory dienoic eicosanoids with a concomitant increase in the secretion of less inflammatory monoenoic eicosanoids. Studies showed that sesamin significantly inhibited lipopolysaccharides-stimulated IL-6 mRNA and protein, and to a lesser degree TNF-α, in BV-2 microglia (Jeng et al. 2005). Sesamin and sesamolin also reduced LPS-activated p38 mitogen-activated protein kinase (MAPK) and nuclear factor (NF)-κB activations. The results suggested that sesamin inhibited LPS-induced IL-6 production by suppression of p38 MAPK signal pathway and NF-κB activation.

Animal studies showed that sesaminol triglucoside, the main sesame lignin, had antiinflammatory and estrogenic activities via metabolism of intestinal microflora (Jan et al. 2010). After oral administration of sesaminol triglucoside to Sprague-Dawley rats, the concentrations of major sesaminol triglucoside metabolites in rectum, cecum, colon, and small intestines were higher than those in liver, lung, kidney, and heart. The study demonstrated that sesaminol triglucoside may be metabolized to form the catechol metabolites first by intestinal microflora and then incorporated via intestine absorption into the cardiovascular system and transported to other tissues. Sesaminol triglucoside metabolites significantly reduced the production of IL-6 and TNF-α in RAW264.7 murine macrophages stimulated with lipopolysaccharide. The estrogenic activities of sesaminol triglucoside metabolites were also established by ligand-dependent transcriptional activation through estrogen receptors.

Oestrogenic Activity

After 8 weeks of a diet rich in sesame pericarp, the expression of oestrogen receptors (ERα and ERβ) in the prostate and uterus tissues of male and female Wistar rats were determined (Anagnostis and Papadopoulos 2009). Significant increase in the expression of ERβ in prostate and uterus was evident. No statistically significant change was observed in the expression of ERα in uterus but in prostate, the increase was more evident. In both tissues, a shift of the ratio of ERα: ERβ in favour of ERβ was evident, indicating, a beneficial effect of the diet provided upon the health status of the animals. It was suggested that this effect could be attributed to the lignans present in the pericarp which exerted phyto-oestrogenic activity.

Neuroprotective and Cognitive Activity

Dietary sesaminol glycosides (SG) from sesame seeds showed a protective effect against Abeta-induced learning and memory deficits in mice in the passive avoidance and the Morris water maze test (Um et al. 2009). Injection of β-amyloid protein (Abeta)(25–35) in mice caused significant neuronal loss in the CA1 and CA3 regions of the hippocampus, but SG supplement showed decrease of the Abeta(25–35) induced neuronal loss. The SG supplementation significantly decreased thiobarbituric acid reactive substance values and 8-hydroxy-2′-deoxyguanosine (8-OHdG) levels in brain tissue. SG also reversed the activity of glutathione peroxidase (GPx), which is decreased by Abeta. The results suggested that sesaminol glycosides protected against cognitive deficits induced by Abeta (25–35), in part through its antioxidant activity.

Defatted sesame seeds extract (DSE) (0.1–10 μg/ml) significantly reduced e neuronal cell death and inhibited lipid peroxidation induced by oxygen-glucose deprivation followed by reoxygenation ischemia in the rat brain (Jamarkattel-Pandit et al. 2010). DSE (30, 100 and 300 mg/kg, p.o.) given twice at 0 and 2 hours after onset of ischemia reduced brain infarct volume dose-dependently and improved sensory-motor function. The results showed that DSE may be effective in ischemia models by an antioxidative mechanism.

Studies showed that oral administration of sesamin (30 mg/kg) twice, 30 minutes before the onset of ischemia and 12 hours after reperfusion in rats reduced the neurological deficits in terms of behaviour and reduced the level of thiobarbituric acid reactive species (TBARS), and protein carbonyl (PC) in the different areas of the brain when compared with rats with brain injury after middle cerebral artery occlusion (MCAO) (Khan et al. 2010). A significantly depleted level of glutathione and its dependent enzymes (glutathione peroxidase [GPx] and glutathione reductase [GR]) in MCAO group were protected significantly in MCAO group treated with sesamin. The study suggested that sesamin may be able to attenuate the ischemic cell death and may play a crucial role as a neuroprotectant in regulating levels of reactive oxygen species in the rat brain. Thus, sesamin may be a potential compound in stroke therapy.

Hepatoprotective Activity

Studies suggested sesamin to be a prodrug and the enzymatic metabolites containing the catechol moieties in their structures, namely (1R,2S,5R,6S)-6-(3,4-dihydroxyphenyl)-2-(3,4-methylenedioxyphenyl)-3,7-dioxabicyclo[3,3,0]octane and (1R,2S,5R,6S)-2,6-bis(3,4-dihydroxyphenyl)-3,7-dioxabicyclo[3,3,0]octane, were responsible for the protective effects of sesamin against oxidative damage in the rat liver (Nakai et al. 2003). All the metabolites exhibited strong radical scavenging activities. The same metabolites were found as glucuronic acid and/or sulfic acid conjugates in substantial amounts in rat bile after oral administration of sesamin. Another study showed that *Sesamum indicum* had potent hepatoprotective activity against carbon tetrachloride induced hepatic damage in rats (Kumar et al. 2011b). The substantially elevated SGOT (serum glutamic oxaloacetic transaminase), SGPT (serum glutamic pyruvic transaminase), Alkaline phosphatase, acid phosphatase, total protein albumin and total bilirubin were restored to normal levels by the ethanol extract of *S. indicum* seeds. This was further supported by histological examination of the rat's liver sections.

Antityrosinase Activity

Sesamol a phenolic degradation product of sesamolin from sesame seed inhibited both diphenolase and monophenolase activities with midpoint concentrations of 1.9 and 3.2 μM, respectively (Kumar et al. 2011a). It was a competitive inhibitor of diphenolase activity and a non-competitive inhibitor of monophenolase activity. Sesamol inhibited melanin synthesis in mouse melanoma B16F10 cells in a concentration dependant manner with 63% decrease in cells exposed to 100 µg/ml sesamol. Apoptosis was induced by sesamol, limiting proliferation.

Wound Healing Activity

In the excision and burn wound models, the sesame seed and oil treated animals displayed significant reduction in period of epithelization and wound contraction (50%) (Kiran and Asad 2008). In the incision wound model, a significant increase in the breaking strength was observed. Seeds and oil treatment (250 mg and 500 mg/kg; po) in dead space wound model, produced a significant increase in the breaking strength, dry weight and hydroxyproline content of the granulation tissue. The results suggested that sesame seed and oil applied topically or administered orally possessed wound healing activity. Using incision, excision and dead space wounds inflicted on albino rats, the tensile strength significantly increased with sesamol at 471.40 g when compared to control at 300.60 g in normal and sesamol suppressed healing (Shenoy et al. 2011). No significant change was observed in duration of wound contraction and lysyl oxidase when compared to control at 2.98 mg. Sesamol treated rats showed a significant rise in hydroxyproline levels at 6.45 mg when compared to control at 1.75 mg. The results indicated that sesamol, the main anti-oxidative constituent contained mainly in the processed sesame seed oil, could be a promising drug in normal as well as delayed wound healing processes.

Analgesic Activity

Sesame ethanol seed extract showed a significant dose-dependent inhibition on the writhing response produced by induction of acetic acid (Nahar and Rokonuzzaman 2009). The extract produced about 48.19% and 75.46% writhing inhibition at the doses of 250 and 500 mg/kg, respectively, which was comparable to the standard drug ibuprofen where the inhibition was about 71.82% at the dose of 25 mg/kg.

Fertility Activity

The results obtained in studies of male Wistar rats showed that ethanolic extract of *Sesamum indicum*, vitamin C and their combination were capable of significantly increasing body weight gain, seminal parameters, testosterone level, and body antioxidant activities (Ashamu et al. 2010) The results suggested that they promoted fertility via their testosterone-increasing effects and their antioxidant effects.

Metabolism of Sesame lignans

Coulman et al. (2005) demonstrated in a randomized crossover study of healthy postmenopausal women that precursors from unground whole flaxseed and sesame seed were converted by the bacterial flora in the colon to mammalian lignans enterolactone and enterodiol and that sesame seed, alone and in combination with flaxseed, produced mammalian lignans equivalent to those obtained from flaxseed alone. Sesame seed was found to be a rich source of mammalian lignan precursors and Sesamin was one of the major precursors of mammalian lignans as observed in vitro and in rats (Liu et al. 2006). The total plant lignan concentration in sesame seed (2,180 µmol/100 g) was higher than that in flaxseed (820 µmol/100 g). In-vitro fermentation with human faecal inoculum showed conversion of sesamin to the mammalian lignans. when fed to female Sprague-Dawley rats for 10 days, sesamin (15 mg/kg body weight) and a 10% sesame seed diet resulted in greater urinary mammalian lignan excretion (3.2 and 11.2 µmol/day, respectively), than the control (< 0.05 µmol/day).

Sesame lignin, sesaminol triglucoside with methylenedioxyphenyl moieties in its structure, was metabolized, via intestinal microbiota, to a catechol moiety (Jan et al. 2009). The major sesaminol triglucoside metabolite was characterized as 4-[((3R,4R)-5-(6-hydroxybenzo[d][1,3]dioxol-5-yl)-4-(hydroxymethyl)tetrahydrofuran-3-yl)methyl]benzene-1,2-diol. Sesaminol triglucoside could be converted to enterolactone and enterodiol by rat intestinal microflora which may have protective effects against hormone-related diseases such as breast cancer. Studies in Sprague-Dawley rats indicated that sesame sesaminol and its epimer2-episesaminol were poorly absorbed prior to reaching the rectum and that substantial amounts pass from the small to the large intestine, where they were metabolized by the colonic microflora to tetrahydrofuranoid metabolites (Jan et al. 2011). Sesaminol in plasma was largely present as phase II conjugates, and the seven metabolites were detected as the 2-episesaminol, sesaminol-6-catechol, methylated sesaminol-catechol, R,R-hydroxymethylsesaminol-tetrahydrofuran, S,R-hydroxymethylsesaminol-tetrahydrofuran, enterolactone, and enterodiol. Excretions of sesaminol in urine and faeces within the 24 hour period were equivalent to 0.02% and 9.33% of the amount ingested, respectively.

Genotoxicity Studies

Studies found that sesamin did not damage DNA in-vivo and that sesamin and episesamin had no genotoxic activity (Hori et al. 2011). Episesamin showed negative results in the Ames test (bacterial reverse mutation assay) with and without S9 mix, in the in-vitro chromosomal aberration test in cultured Chinese hamster lung cells with and without S9 mix, and in the in-vivo comet assay using the liver of Sprague-Dawley rats. Sesamin showed negative results in the Ames test with and without S9 mix. In the in-vitro chromosomal aberration test, sesamin did not induce chromosomal aberrations in the absence of S9 mix, but induced structural abnormalities at cytotoxic concentrations in the presence of S9 mix. Oral administration of sesamin at doses up to 2.0 g/kg did not cause a significant increase in either the percentage of micronucleated polychromatic erythrocytes in the in-vivo bone marrow MN test or in the % DNA in the comet tails in the in-vivo comet assay.

Allergy Problem

All ten patients tested had positive IgE antibodies and skin prick tests (SPTs) to sesame (Pastorello et al. 2001). The major, clinically most important allergen of sesame seeds was a protein with molecular mass of about 9,000. It was not glycosylated, the amino acid sequence showed it was a 2S albumin with a pI of 7.3; the small and the large subunits, forming the whole protein, showed pI values of 6.5 and 6.0. In another study, 24 of the 28 subjects diagnosed as allergic to sesame had sesame-specific IgE (Wolff et al. 2003). A 14 kDa protein belonging to the 2S albumin family was recognised by 22 of the 24 sera used. The reactivity of the 14 kDa protein with most of the sera indicated that was the major sesame allergen, later identified as 2S albumin precursor; and its peptide which reacted positively in the dot blot test evidently contained an epitope(s). Some minor sesame allergens, of higher molecular weight, were also found. Four sesame seed allergens were identified paving the first step toward generating recombinant allergens for use in future immunotherapeutic approaches (Beyer et al. 2002). The IgE-binding protein at 45 kd, which was recognized by 75% of the 20 patients, was found to be a 7S vicilin-type globulin, a seed storage protein of sesame and named Ses i 3. The protein at 7 kd was found to be a 2S albumin, another seed storage protein of sesame and named Ses i 2. Further, the proteins at 78 and 34 kd were found to be homologous to the embryonic abundant protein and the seed maturation protein of soybeans, respectively. The authors maintained that the detection of conserved IgE binding epitopes in common food allergens might be a useful tool for predicting cross-reactivity to certain foods.

Traditional Medicinal Uses

Sesame leaves, flowers, roots, seeds, and seed oil have been employed for various ailments in traditional medicine especially in Asia and Africa (Dalziel 1955; Burkill 1966; Grieve 1971; CSIR 1972; Duke and Ayensu 1985; Chopra et al. 1986;

Bown 1995; Chevallier 1996). The leaves, seeds, and seed oil are official in various national pharmacopoeias.

Sesame seeds are considered astringent, emollient, nourishing, tonic, lenitive, diuretic, laxative, antiphlogistic and galactogogue. A poultice made of the seeds is applied to ulcers, blisters, head sores on children and venereal sores in women. A plaster made of the ground seeds is applied to burns and scalds. A powder made from the roasted and decorticated seed is used externally and internally as an emollient. The seeds are especially useful in haemorrhoids and constipation, taken in decoction or as sweetmeats. A compound decoction of the seeds with linseed is used in coughs and as an aphrodisiac. Ground to a paste with water, they are given with butter for bleeding piles; if taken in large quantities, they are capable of producing abortion. The seed is used as tonic for the liver and kidney. It is taken internally for convalescence, chronic dry constipation, dental caries, osteoporosis, stiff joints, and dry cough and to treat premature hair loss and greying. Seed is employed to increase milk production in nursing mothers. The seeds are given as a laxative for children.

Both the seeds and the oil are used as demulcents in dysentery and urinary diseases in combination with other similar medicine types. Sesame oil is mildly laxative, emollient and demulcent and also promotes menstruation. Sesame oil is used as an antirheumatic in massage treatment. Sesame oil is used to treat dry constipation in the elderly. Mixed with lime water, sesame oil is used externally for burns, boils and ulcers.

The leaves, which are rich in mucilage, mixed with water, are employed in infantile cholera, diarrhoea, dysentery, catarrh, bladder troubles, acute cystitis, and strangury. An infusion of the leaves is also used as demulcent. The leaves also used as emollient poultices. The leaves are also astringent. A lotion prepared from leaves and roots are used as a hair-wash, and are considered to stimulate hair growth. A decoction of the root is used to treat asthma and coughs.

In traditional Chinese medicine, the dried flowers have been used to cure alopecia, frostbite and constipation; the flowers have been employed as a cure for verruca vulgaris and verruca plana (Hu et al. 2007a, b).

Other Uses

The production of biodiesel from sesame seed oil was found to be a viable alternative to diesel fuel (Saydut et al. 2008). Transesterification showed improvement in fuel properties of sesame seed oil which was obtained in 58wt/wt.%, by traditional solvent extraction.

Sesame oil can be used in the manufacture of soaps, paints, perfumes, pharmaceuticals and insecticides. The oil is also used as illuminant and is used in barrier creams to protect the skin from harmful UV light radiation. When added to the insecticide pyrethrum it acts as a synergist, doubling its potency. Sesame meal, left after the oil is pressed from the seed, is an excellent high-protein (34–50%) feed for poultry and livestock. The addition of sesame to the high lysine meal of soybean produces a well balanced animal feed.

Chlorosesamone, hydroxysesamone and 2,3-epoxysesamone isolated from sesame roots all showed antifungal activities toward *Cladosporium fulvum* (Hasan et al. 2001).

Comments

Sesame oil is a high-quality edible oil and is one of the most expensive cooking oils.

Selected References

Abou-Gharbia HA, Shehata AAY, Shahidi F (2000) Effect of processing on oxidative stability and lipid classes of sesame oil. Food Res Int 33(5):331–340

Anagnostis A, Papadopoulos AI (2009) Effects of a diet rich in sesame (*Sesamum indicum*) pericarp on the expression of oestrogen receptor alpha and oestrogen receptor beta in rat prostate and uterus. Br J Nutr 102(5):703–708

Anila L, Vijayalakshmi NR (2000) Beneficial effects of flavonoids from *Sesamum indicum, Emblica officinalis* and *Momordica charantia*. Phytother Res 14(8): 592–595

Arachchige PG, Takahashi Y, Ide T (2006) Dietary sesamin and docosahexaenoic and eicosapentaenoic acids synergistically increase the gene expression of enzymes involved in hepatic peroxisomal fatty acid oxidation in rats. Metabolism 55(3):381–390

Ashakumary L, Rouyer I, Takahashi Y, Ide T, Fukuda N, Aoyama T, Hashimoto T, Mizugaki M, Sugano M (1999) Sesamin, a sesame lignan, is a potent inducer of hepatic fatty acid oxidation in the rat. Metabolism 48(10):1303–1313

Ashamu E, Salawu E, Oyewo O, Alhassan A, Alamu O, Adegoke A (2010) Efficacy of vitamin C and ethanolic extract of *Sesamum indicum* in promoting fertility in male Wistar rats. J Hum Reprod Sci 3(1):11–14

Badifu GI, Akpagher EM (1996) Effects of debittering methods on the proximate composition, organoleptic and functional properties of sesame (*Sesamum indicum* L.) seed flour. Plant Foods Hum Nutr 49(2):119–126

Bandyopadhyay K, Ghosh S (2002) Preparation and characterization of papain-modified sesame (*Sesamum indicum* L.) protein isolates. J Agric Food Chem 50(23):6854–6857

Bedigian D (2004) History and lore of sesame in Southwest Asia. Econ Bot 58(3):329–353

Bedigian D, Seigler DS, Harlan JR (1985) Sesamin, sesamolin and the origin of sesame. Biochem Syst Ecol 13:133–139

Beyer K, Bardina L, Grishina G, Sampson HA (2002) Identification of sesame seed allergens by 2-dimensional proteomics and Edman sequencing: seed storage proteins as common food allergens. J Allergy Clin Immunol 110(1):154–159

Bown D (1995) Encyclopaedia of herbs and their uses. Dorling Kindersley, London, 424 pp

Burkill IH (1966) A dictionary of the economic products of the Malay Peninsula. Revised reprint, 2 vols. Ministry of Agriculture and Co-operatives, Kuala Lumpur, Malaysia. Vol. 1 (A–H) pp 1–1240, Vol. 2 (I–Z) pp 1241–2444

Cadwallader KR, Heo J (2001) Aroma of roasted sesame oil: characterization by direct thermal desorption-gas chromatography-olfactometry and sample dilution analysis. In: Leland JV, Schieberle P, Buettner A, Acree TE (eds) Gas chromatography-olfactometry: the state of the art, chapter 16. American Chemical Society, Washington, DC, pp 187–202

Chavali SR, Utsunomiya T, Forse RA (2001) Increased survival after cecal ligation and puncture in mice consuming diets enriched with sesame seed oil. Crit Care Med 29(1):140–143

Chavali SR, Zhong WW, Utsunomiya T, Forse RA (1997) Decreased production of interleukin-1-beta, prostaglandin-E2 and thromboxane-B2, and elevated levels of interleukin-6 and -10 are associated with increased survival during endotoxic shock in mice consuming diets enriched with sesame seed oil supplemented with Quil-A saponin. Int Arch Allergy Immunol 114(2)153–160

Chen PR, Tsai CE, Chang H, Liu TL, Lee CC (2005) Sesamol induces nitric oxide release from human umbilical vein endothelial cells. Lipids 40(9):955–961

Cheung SC, Szeto YT, Benzie IF (2007) Antioxidant protection of edible oils. Plant Foods Hum Nutr 62(1):39–42

Chevallier A (1996) The encyclopedia of medicinal plants. Dorling Kindersley, London, 336 pp

Chopra RN, Nayar SL, Chopra IC (1986) Glossary of Indian medicinal plants. (Including the supplement). Council Scientific Industrial Research, New Delhi, 330 pp

Cooney RV, Custer LJ, Okinaka L, Franke AA (2001) Effects of dietary sesame seeds on plasma tocopherol levels. Nutr Cancer 39(1):66–71

Coulman KD, Liu Z, Hum WQ, Michaelides J, Thompson LU (2005) Whole sesame seed is as rich a source of mammalian lignan precursors as whole flaxseed. Nutr Cancer 52(2):156–165

Coulman KD, Liu Z, Michaelides J, Quan Hum W, Thompson LU (2009) Fatty acids and lignans in unground whole flaxseed and sesame seed are bioavailable but have minimal antioxidant and lipid-lowering effects in postmenopausal women. Mol Nutr Food Res 53(11):1366–1375

Council of Scientific and Industrial Research (CSIR) (1972) The wealth of India. A dictionary of Indian raw materials and industrial products (raw materials 9). Publications and Information Directorate, New Delhi

Dachtler M, van de Put FHM, van Stijn F, Beindorff CM, Fritsche J (2003) On-line LC-NMR-MS characterization of sesame oil extracts and assessment of their antioxidant activity. Eur J Lipid Sci Technol 105(9):488–496

Dalziel JM (1955) The useful plants of West Tropical Africa (reprint of 1937 edn) Crown Agents for Overseas Governments and Administrations, London, 612 pp

Duke JA, Ayensu ES (1985) Medicinal plants of China, vol 1 and 2. Reference Publications, Inc., Algonac, 705 pp

Egbekun MK, Ehieze MU (1997) Proximate composition and functional properties of fullfat and defatted beniseed (*Sesamum indicum* L.) flour. Plant Foods Hum Nutr 51(1):35–41

Furumoto T, Iwata M, Feroj Hasan AF, Fukui H (2003) Anthrasesamones from roots of *Sesamum indicum*. Phytochemistry 64(4):863–866

Furumoto T, Takeuchi A, Fukui H (2006) Anthrasesamones D and E from *Sesamum indicum* roots. Biosci Biotechnol Biochem 70(7):1784–1785

Grieve M (1971) A modern herbal. Penguin, 2 vols. Dover publications, New York, 919 pp

Grougnet R, Magiatis P, Mitaku S, Terzis A, Tillequin F, Skaltsounis AL (2006) New lignans from the perisperm of *Sesamum indicum*. J Agric Food Chem 54(20):7570–7574

Hahm TS, Park SJ, Martin Lo Y (2009) Effects of germination on chemical composition and functional properties of sesame (*Sesamum indicum* L.) seeds. Bioresour Technol 100(4):1643–1647

Harikumar KB, Sung B, Tharakan ST, Pandey MK, Joy B, Guha S, Krishnan S, Aggarwal BB (2010) Sesamin manifests chemopreventive effects through the suppression of NF-kappa B-regulated cell survival, proliferation, invasion, and angiogenic gene products. Mol Cancer Res 8(5):751–761

Hasan AF, Furumoto T, Begum S, Fukui H (2001) Hydroxysesamone and 2,3-epoxysesamone from roots of Sesamum indicum. Phytochemistry 58(8):1225–1228

Hirata F, Fujita K, Ishikura Y, Hosoda K, Ishikawa T, Nakamura H (1996) Hypocholesterolemic effect of sesame lignan in humans. Atherosclerosis 122(1): 135–136

Ho CT, Shahidi F (2005) Flavor components of fats and oils. In: Shahidi F (ed) Bailey's industrial oil and fat products, vol 1, 6th edn. Wiley, Hoboken, pp 387–411

Hori H, Takayanagi T, Kamada Y, Shimoyoshi S, Ono Y, Kitagawa Y, Shibata H, Nagao M, Fujii W, Sakakibara Y (2011) Genotoxicity evaluation of sesamin and episesamin. Mutat Res 719(1–2):21–28

Hu Q, Xu J, Chen S, Yang F (2004) Antioxidant activity of extracts of black sesame seed (Sesamum indicum L.) by supercritical carbon dioxide extraction. J Agric Food Chem 52(4):943–947

Hu YM, Wang H, Ye WC, Zhao SX (2007a) Flavones from flowers of Sesamum indicum. Zhongguo Zhong Yao Za Zhi 32(7):603–605 (in Chinese)

Hu YM, Ye WC, Yin ZQ, Zhao SX (2007b) Chemical constituents from flos Sesamum indicum L. Yao Xue Xue Bao 42(3):286–291 (in Chinese)

Ide T, Ashakumary L, Takahashi Y, Kushiro M, Fukuda N, Sugano M (2001) Sesamin, a sesame lignan, decreases fatty acid synthesis in rat liver accompanying the down-regulation of sterol regulatory element binding protein-1. Biochim Biophys Acta 1534(1):1–13

Ide T, Hong DD, Ranasinghe P, Takahashi Y, Kushiro M, Sugano M (2004) Interaction of dietary fat types and sesamin on hepatic fatty acid oxidation in rats. Biochim Biophys Acta 1682(1–3):80–91

Ide T, Nakashima Y, Iida H, Yasumoto S, Katsuta M (2009) Lipid metabolism and nutrigenomics – impact of sesame lignans on gene expression profiles and fatty acid oxidation in rat liver. Forum Nutr 61:10–24

Ikeda S, Abe C, Uchida T, Ichikawa T, Horio F, Yamashita K (2007) Dietary sesame seed and its lignan increase both ascorbic acid concentration in some tissues and urinary excretion by stimulating biosynthesis in rats. J Nutr Sci Vitaminol (Tokyo) 53(5):383–392

Ikeda S, Kagaya M, Kobayashi K, Tohyama T, Kiso Y, Higuchi N, Yamashita K (2003) Dietary sesame lignans decrease lipid peroxidation in rats fed docosahexaenoic acid. J Nutr Sci Vitaminol (Tokyo) 49(4):270–276

Jamarkattel-Pandit N, Pandit NR, Kim MY, Park SH, Kim KS, Choi H, Kim H, Bu Y (2010) Neuroprotective effect of defatted sesame seeds extract against in vitro and in vivo ischemic neuronal damage. Planta Med 76(1):20–26

Jan KC, Hwang LS, Ho CT (2009) Biotransformation of sesaminol triglucoside to mammalian lignans by intestinal microbiota. J Agric Food Chem 57(14): 6101–6106

Jan KC, Ku KL, Chu YH, Hwang LS, Ho CT (2010) Tissue distribution and elimination of estrogenic and anti-inflammatory catechol metabolites from sesaminol triglucoside in rats. J Agric Food Chem 58(13): 7693–7700

Jan KC, Ku KL, Chu YH, Hwang LS, Ho CT (2011) Intestinal distribution and excretion of sesaminol and its tetrahydrofuranoid metabolites in rats. J Agric Food Chem 59(7):3078–3086

Jeng KC, Hou RC, Wang JC, Ping LI (2005) Sesamin inhibits lipopolysaccharide-induced cytokine production by suppression of p38 mitogen-activated protein kinase and nuclear factor-kappaB. Immunol Lett 97(1):101–106

Joshi R, Kumar MS, Satyamoorthy K, Unnikrisnan MK, Mukherjee T (2005) Free radical reactions and antioxidant activities of sesamol: pulse radiolytic and biochemical studies. J Agric Food Chem 53(7):2696–2703

Kamal-Eldin A, Appelqvist LA (1994) Variations in the composition of sterols, tocopherols and lignans in seed oils from four Sesamum species. J Am Oil Chem Soc 71(2):149–156

Kamal-Eldin A, Appelqvist LA, Yousif G, Iskander GM (1992) Seed lipids of Sesamum indicum and related wild species in Sudan. Part 3. The sterols. J Sci Food Agric 59(3):327–334

Kamal-Eldin A, Moazzami A, Washi S (2011) Sesame seed lignans: potent physiological modulators and possible ingredients in functional foods & nutraceuticals. Recent Pat Food Nutr Agric 3(1):17–29

Kamal-Eldin A, Yousif G, Iskander GM, Appelqvist LA (l992) Seed lipids of Sesamum indicum, L. and related wild species in Sudan. I. Fatty acids and triacylglycerols. 1. Lipid/Fett 7:254–259

Kang MH, Kawai Y, Naito M, Osawa T (1999) Dietary defatted sesame flour decreases susceptibility to oxidative stress in hypercholesterolemic rabbits. J Nutr 129(10):1885–1890

Kang MH, Naito M, Tsujihara N, Osawa T (1998) Sesamolin inhibits lipid peroxidation in rat liver and kidney. J Nutr 128(6):1018–1022

Kanu PJ (2011) Biochemical analysis of black and white sesame seeds from China. Am J Biochem Mol Biol 1:145–157

Kato MJ, Chu A, Davin LB, Lewis NG (1998) Biosynthesis of antioxidant lignans in Sesamum indicum seeds. Phytochemistry 47(4):583–591

Katsuzaki H, Kawasumi M, Kawakishi S, Osawa T (1992) Structure of novel antioxidative lignin glucosides isolated from sesame seed. Biosci Biotechnol Biochem 56:2087–2088

Katsuzaki H, Kawakishi S, Osawa T (1993) Structure of novel antioxidative lignan triglucoside isolated from sesame seed. Heterocycles 36(5):933–936

Katsuzaki H, Kawakishi S, Osawa T (1994a) Sesaminol glucosides in sesame seed. Phytochemistry 35:773–776

Katsuzaki H, Osawa T, Kawakishi S (1994b) Chemistry and antioxidative activity of lignan glucosides in sesame seed. In: Food phytochemicals for cancer prevention, ACS symposium series vol 547, chapter 28, pp 275–280

Khaleel AES, Gonaid MH, El-Bagry RI, Sleem AA, Shabana M (2007) Chemical and biological study of the residual aerial parts of *Sesamum indicum* L. J Food Drug Anal 15(3):249–257

Khan MM, Ishrat T, Ahmad A, Hoda MN, Khan MB, Khuwaja G, Srivastava P, Raza SS, Islam F, Ahmad S (2010) Sesamin attenuates behavioral, biochemical and histological alterations induced by reversible middle cerebral artery occlusion in the rats. Chem Biol Interact 183(1):255–263

Kim KS, Park SH (2008) Anthrasesamone F from the seeds of black *Sesamum indicum*. Biosci Biotechnol Biochem 72(6):1626–1627

Kiran K, Asad M (2008) Wound healing activity of *Sesamum indicum* L seed and oil in rats. Indian J Exp Biol 46(11):777–782

Kiso Y (2004) Antioxidative roles of sesamin, a functional lignan in sesame seed, and it's effect on lipid- and alcohol-metabolism in the liver: a DNA microarray study. Biofactors 21(1–4):191–196

Kiso Y, Tsuruoka N, Kidokoro A, Matsumoto I, Abe K (2005) Sesamin ingestion regulates the transcription levels of hepatic metabolizing enzymes for alcohol and lipids in rats. Alcohol Clin Exp Res 29(11 Suppl):116S–120S

Kita S, Matsumura Y, Morimoto S, Akimoto K, Furuya M, Oka N, Tanaka T (1995) Antihypertensive effect of sesamin. II. Protection against two-kidney, one-clip renal hypertension and cardiovascular hypertrophy. Biol Pharm Bull 18(9):1283–1285

Kumar CM, Sathisha UV, Dharmesh S, Rao AG, Singh SA (2011a) Interaction of sesamol (3,4-methylenedioxyphenol) with tyrosinase and its effect on melanin synthesis. Biochimie 93(3):562–569

Kumar M, Kamboj A, Sisodis AA (2011b) Hepatoprotective activity of *Sesamum indicum* Linn. against CCl4-induced hepatic damage in rats. Int J Pharmaceut Biol Arch 1(2):710–715

Kuo PC, Lin MC, Chen GF, Yiu TJ, Tzen JT (2011) Identification of methanol-soluble compounds in sesame and evaluation of antioxidant potential of its lignans. J Agric Food Chem 59(7):3214–3219

Lee J, Kausar T, Kwon JH (2008a) Characteristic hydrocarbons and 2-alkylcyclobutanones for detecting gamma-irradiated sesame seeds after steaming, roasting, and oil extraction. J Agric Food Chem 56(21): 10391–10395

Lee JY, Lee YS, Choe EN (2008b) Effects of sesamol, sesamin, and sesamolin extracted from roasted sesame oil on the thermal oxidation of methyl linoleate. LWT-Food Sci Technol 41(10):1871–1875

Lee SW, Jeung MK, Park MH, Lee SY, Lee JH (2010) Effects of roasting conditions of sesame seeds on the oxidative stability of pressed oil during thermal oxidation. Food Chem 118(3):681–685

Lim JS, Adachi Y, Takahashi Y, Ide T (2007) Comparative analysis of sesame lignans (sesamin and sesamolin) in affecting hepatic fatty acid metabolism in rats. Br J Nutr 97(1):85–95

Lin LJ, Tai SS, Peng CC, Tzen JT (2002) Steroleosin, a sterol-binding dehydrogenase in seed oil bodies. Plant Physiol 128(4):1200–1211

Lin LJ, Liao PC, Yang HH, Tzen JT (2005) Determination and analyses of the N-termini of oil-body proteins, steroleosin, caleosin and oleosin. Plant Physiol Biochem 43(8):770–776

Liu Z, Saarinen NM, Thompson LU (2006) Sesamin is one of the major precursors of mammalian lignans in sesame seed (*Sesamum indicum*) as observed in vitro and in rats. J Nutr 136(4):906–912

Matsumura Y, Kita S, Morimoto S, Akimoto K, Furuya M, Oka N, Tanaka T (1995) Antihypertensive effect of sesamin. I. Protection against deoxycorticosterone acetate-salt-induced hypertension and cardiovascular hypertrophy. Biol Pharm Bull 18(7):1016–1019

Matsumura Y, Kita S, Ohgushi R, Okui T (2000) Effects of sesamin on altered vascular reactivity in aortic rings of deoxycorticosterone acetate-salt-induced hypertensive rat. Biol Pharm Bull 23(9):1041–1045

Matsumura Y, Kita S, Tanida Y, Taguchi Y, Morimoto S, Akimoto K, Tanaka T (1998) Antihypertensive effect of sesamin. III. Protection against development and maintenance of hypertension in stroke-prone spontaneously hypertensive rats. Biol Pharm Bull 21(5): 469–473

Meydani M (1992) Protective role of dietary vitamin E on oxidative stress in aging. Age (Omaha) 15:89–93

Miyahara Y, Komiya T, Katsuzaki H, Imai K, Nakagawa M, Ishi Y, Hibasami H (2000) Sesamin and episesamin induce apoptosis in human lymphoid leukemia Molt 4B cells. Int J Mol Med 6(1):43–46

Miyawaki T, Aono H, Toyoda-Ono Y, Maeda H, Kiso Y, Moriyama K (2009) Antihypertensive effects of sesamin in humans. J Nutr Sci Vitaminol (Tokyo) 55(1): 87–91

Mkamilo GS, Bedigian D (2007) *Sesamum indicum* L. [internet] record from protabase. In: van der Vossen HAM and Mkamilo GS (eds) PROTA (Plant Resources of Tropical Africa/Ressources Végétales de l'Afrique Tropicale), Wageningen. http://database.prota.org/search.htm

Moazzami AA, Andersson RE, Kamal-Eldin A (2006a) Characterization and analysis of sesamolinol diglucoside in sesame seeds. Biosci Biotechnol Biochem 70(6):1478–1481

Moazzami AA, Andersson RE, Kamal-Eldin A (2006b) HPLC analysis of sesaminol glucosides in sesame seeds. J Agric Food Chem 54(3):633–638

Moazzami AA, Haese SL, Kamal-Eldin A (2007) Lignan contents in sesame seeds and products. Eur J Lipid Sci Technol 109(10):1022–1027

Mohamed HMA, Awatif II (1998) The use of sesame oil unsaponifiable matter as a natural antioxidant. Food Chem 62(3):269–276

Mohdaly AAA, Smetanska I, Ramadan MF, Sarhan MA, Mahmoud A (2011) Antioxidant potential of sesame (*Sesamum indicum*) cake extract in stabilization of sunflower and soybean oils. Ind Crop Prod 34(1): 952–959

Nahar L, Rokonuzzaman (2009) Investigation of the analgesic and antioxidant activity from an ethanol extract of seeds of *Sesamum indicum*. Pak J Biol Sci 12(7): 595–598

Nakai M, Harada M, Nakahara K, Akimoto K, Shibata H, Miki W, Kiso Y (2003) Novel antioxidative metabolites in rat liver with ingested sesamin. J Agric Food Chem 51(6):1666–1670

Nakano D, Itoh C, Ishii F, Kawanishi H, Takaoka M, Kiso Y, Tsuruoka N, Tanaka T, Matsumura Y (2003) Effects of sesamin on aortic oxidative stress and endothelial dysfunction in deoxycorticosterone acetate-salt hypertensive rats. Biol Pharm Bull 26(12):1701–1705

Nakano D, Kwak CJ, Fujii K, Ikemura K, Satake A, Ohkita M, Takaoka M, Ono Y, Nakai M, Tomimori N, Kiso Y, Matsumura Y (2006) Sesamin metabolites induce an endothelial nitric oxide-dependent vasorelaxation through their antioxidative property-independent mechanisms: possible involvement of the metabolites in the antihypertensive effect of sesamin. J Pharmacol Exp Ther 318(1):328–335

Nakano D, Kurumazuka D, Nagai Y, Nishiyama A, Kiso Y, Matsumura Y (2008) Dietary sesamin suppresses aortic NADPH oxidase in DOCA salt hypertensive rats. Clin Exp Pharmacol Physiol 35(3):324–326

Namiki M (1995) The chemistry and physiological functions of sesame. Food Rev Int 11(2):281–329

Namiki M (2007) Nutraceutical functions of sesame: a review. Crit Rev Food Sci Nutr 47(7):651–673

Nonaka M, Yamashita K, Iizuka Y, Namiki M, Sugano M (1997) Effects of dietary sesaminol and sesamin on eicosanoid production and immunoglobulin level in rats given ethanol. Biosci Biotechnol Biochem 61(5):836–839

Nzikou JM, Matos L, Bouanga-Kalou G, Ndangui CB, Pambou-Tobi NPG, Kimbonguila A, Silou T, Linder M, Desobry S (2009) Chemical composition on the seeds and oil of sesame (*Sesamum indicum* L.). Adv J Food Sci Technol 1(1):6–11

Ochse JJ, Bakhuizen van den Brink RC (1980) Vegetables of the Dutch Indies, 3rd edn. Ascher & Co., Amsterdam, 1016 pp

Ogawa H, Sasagawa S, Murakami T, Yoshizumi H (1995) Sesame lignans modulate cholesterol metabolism in the stroke-prone spontaneously hypertensive rat. Clin Exp Pharmacol Physiol Suppl 22(1):S310–S312

Osawa T, Nagata M, Namiki M, Fukuda Y (1985) Sesamolinol, a novel antioxidant isolated from sesame seeds. Agric Biol Chem 49:3351–3352

Park D, Maga JA, Johnson DL, Morini G (1995) Major volatiles in toasted sesame seed oil. J Food Lipids 2:259–268

Pastorello EA, Varin E, Farioli L, Pravettoni V, Ortolani C, Trambaioli C, Fortunato D, Giuffrida MG, Rivolta F, Robino A, Calamari AM, Lacava L, Conti A (2001) The major allergen of sesame seeds (*Sesamum indicum*) is a 2S albumin. J Chromatogr B Biomed Sci Appl 756(1–2):85–93

Peñalvo JL, Hopia A, Adlercreutz H (2006) Effect of sesamin on serum cholesterol and triglycerides levels in LDL receptor-deficient mice. Eur J Nutr 45(8): 439–444

Phillips KM, Ruggio DM, Ashraf-Khorassani M (2005) Phytosterol composition of nuts and seeds commonly consumed in the United States. J Agric Food Chem 53(24):9436–9445

Porcher MH et al (1995–2020) Searchable world wide web multilingual multiscript plant name database. Published by The University of Melbourne, Australia. http://www.plantnames.unimelb.edu.au/Sorting/Frontpage.html

Rangkadilok N, Pholphana N, Mahidol C, Wongyai W, Saengsooksree K, Nookabkaew S, Satayavivad J (2010) Variation of sesamin, sesamolin and tocopherols in sesame (*Sesamum indicum* L.) seeds and oil products in Thailand. Food Chem 122(1):724–730

Ryu SN, Ho CT, Osawa T (1998) High performance liquid chromatographic determination of antioxidant lignan glycosides in some varieties of sesame. J Food Lipid 5(1):17–28

Saad R, Pérez C (1984) Functional and nutritional properties of modified proteins of sesame (*Sesamum indicum*, L.). Arch Latinoam Nutr 34(4):749–762 (in Spanish)

Sacco SM, Chen J, Power KA, Ward WE, Thompson LU (2008) Lignan-rich sesame seed negates the tumor-inhibitory effect of tamoxifen but maintains bone health in a postmenopausal athymic mouse model with estrogen-responsive breast tumors. Menopause 15(1):171–179

Salgado JM, Gonçalves CM (1988) Sesame seed (*Sesamum indicum*, L.). II. Use of sesame flour in protein mixtures. Arch Latinoam Nutr 38(2):312–322 (in Portuguese)

Sankar D, Ali A, Sambandam G, Rao R (2011) Sesame oil exhibits synergistic effect with anti-diabetic medication in patients with type 2 diabetes mellitus. Clin Nutr 30(3):351–358

Saydut A, Duz MZ, Kaya C, Kafadar AB, Hamamci C (2008) Transesterified sesame (*Sesamum indicum* L.) seed oil as a biodiesel fuel. Bioresour Technol 99(14):6656–6660

Schieberle P (1996) Odour-active compounds in moderately roasted sesame. Food Chem 55(2):145–152

Schieberle P, Güntert M, Sommer H, Werkhoff P (1996) Structure determination of 4-methyl-3-thiazoline in roasted sesame flavour. Food Chem 56(4):369–372

Sengupta A, Roychoudhury SK (1976) Triglyceride composition of *Sesamum indicum* seed oil. J Sci Food Agric 27(2):165–169

Shenoy RR, Sudheendra AT, Nayak PG, Paul P, Kutty NG, Rao CM (2011) Normal and delayed wound healing is improved by sesamol, an active constituent of *Sesamum indicum* (L.) in albino rats. J Ethnopharmacol 133(2):608–612

Shimizu S, Akimoto K, Shinmen Y, Kawashima H, Sugano M, Yamada H (1991) Sesamin is a potent and specific inhibitor of delta 5 desaturase in polyunsaturated fatty acid biosynthesis. Lipids 26(7):512–516

Shimoda M, Nakada Y, Nakashima M, Osajima Y (1997) Quantitative comparison of volatile flavor compounds in deep-roasted and light-roasted sesame seed oil. J Agric Food Chem 45:3193–3196

Shimoda M, Shiratsuchi H, Nakada Y, Wu Y, Osajima Y (1996) Identification and sensory characterization of volatile flavor compounds in sesame seed oil. J Agric Food Chem 44:3909–3912

Sirato-Yasumoto S, Katsuta M, Okuyama Y, Takahashi Y, Ide T (2001) Effect of sesame seeds rich in sesamin and sesamolin on fatty acid oxidation in rat liver. J Agric Food Chem 49(5):2647–2651

Soliman MM, El-Sawy AA, Fadel HM, Osman F (1986) Identification of volatile flavour components of roasted white sesame seed. Acta Alimentaria 15(4):251–263

Suja KP, Jayalekshmy A, Arumughan C (2004) Free radical scavenging behavior of antioxidant compounds of sesame (*Sesamum indicum* L.) in DPPH(*) system. J Agric Food Chem 52(4):912–915

Suresh Kumar P, Patel JS, Saraf MN (2008) Mechanism of vasorelaxant activity of a fraction of root extract of *Sesamum indicum* Linn. Indian J Exp Biol 46(6):457–464

Suzuki N, Miyase T, Ueno A (1993) Phenylethanoid glycosides of *Sesamum indicum*. Phytochemistry 34(3): 729–732

Takeuchi H, Mooi LY, Inagaki Y, He P (2001) Hypoglycemic effect of a hot-water extract from defatted sesame (*Sesamum indicum* L.) seed on the blood glucose level in genetically diabetic KK-Ay mice. Biosci Biotechnol Biochem 65(10):2318–2321

Tamura H, Fujita A, Steinhaus M, Takahisa E, Watanabe H, Schieberle P (2010) Identification of novel aroma-active thiols in pan-roasted white sesame seeds. J Agric Food Chem 58(12):7368–7375

Tsuruoka N, Kidokoro A, Matsumoto I, Abe K, Kiso Y (2005) Modulating effect of sesamin, a functional lignan in sesame seeds, on the transcription levels of lipid- and alcohol-metabolizing enzymes in rat liver: a DNA microarray study. Biosci Biotechnol Biochem 69(1):179–188

Um MY, Ahn JY, Kim S, Kim MK, Ha TY (2009) Sesaminol glucosides protect beta-amyloid peptide-induced cognitive deficits in mice. Biol Pharm Bull 32(9):1516–1520

U.S. Department of Agriculture, Agricultural Research Service (USDA) (2011) USDA National Nutrient Database for standard reference, release 24. Nutrient Data Laboratory Home Page. http://www.ars.usda.gov/ba/bhnrc/ndl

Visavadiya NP, Narasimhacharya AV (2008) Sesame as a hypocholesteraemic and antioxidant dietary component. Food Chem Toxicol 46(6):1889–1895

Visavadiya NP, Soni B, Dalwadi N (2009) Free radical scavenging and antiatherogenic activities of *Sesamum indicum* seed extracts in chemical and biological model systems. Food Chem Toxicol 47(10):2507–2515

Watt JM, Breyer-Brandwijk MG (1962) The medicinal and poisonous plants of southern and eastern Africa, 2nd edn. E. and S. Livingstone, Edinburgh/London, 1457 pp

Weiss EA (1983) Oilseed crops. Longman, London, 660 pp

Weiss EA, de la Cruz QD (2001) *Sesamum orientale* L. In: van der Vossen HAM, Umali BE (eds) Plant resources of South-East Asia no 14. Vegetable oils and fats. Backhuys, Leiden, pp 123–128

Wichitsranoi J, Weerapreeyakul N, Boonsiri P, Settasatian C, Settasatian N, Komanasin N, Sirijaichingkul S, Teerajetgul Y, Rangkadilok N, Leelayuwat N (2011) Antihypertensive and antioxidant effects of dietary black sesame meal in pre-hypertensive humans. Nutr J 10:Article 82

Williamson KS, Morris JB, Pye QN, Kamat CD, Hensley K (2008) A survey of sesamin and composition of tocopherol variability from seeds of eleven diverse sesame (*Sesamum indicum* L.) genotypes using HPLC-PAD-ECD. Phytochem Anal 19(4):311–322

Wolff N, Cogan U, Admon A, Dalal I, Katz Y, Hodos N, Karin N, Yannai S (2003) Allergy to sesame in humans is associated primarily with IgE antibody to a 14 kDa 2S albumin precursor. Food Chem Toxicol 41(8): 1165–1174

Wu JH, Hodgson JM, Puddey IB, Belski R, Burke V, Croft KD (2009) Sesame supplementation does not improve cardiovascular disease risk markers in overweight men and women. Nutr Metab Cardiovasc Dis 19(11):774–780

Wu WH (2007) The contents of lignans in commercial sesame oils of Taiwan and their changes during heating. Food Chem 104(1):341–344

Wu WH, Kang YP, Wang NH, Jou HJ, Wang TA (2006) Sesame ingestion affects sex hormones, antioxidant status, and blood lipids in postmenopausal women. J Nutr 136(5):1270–1275

Xu H, Yang X, Yang J, Qi W, Liu C, Yang Y (2003) Antitumor effect of alcohol extract from *Sesamum indicum* flower on S180 and H22 experimental tumor. Zhong Yao Cai 26(4):272–273 (in Chinese)

Yamashita K, Iizuka Y, Imai T, Namiki M (1995) Sesame seed and its lignans produce marked enhancement of vitamin E activity in rats fed a low alpha-tocopherol diet. Lipids 30(11):1019–1028

Yamashita K, Ikeda S, Obayashi M (2003) Comparative effects of flaxseed and sesame seed on vitamin E and cholesterol levels in rats. Lipids 38(12):1249–1255

Yamashita K, Kagaya M, Higuti N, Kiso Y (2000) Sesamin and alpha-tocopherol synergistically suppress lipid-peroxide in rats fed a high docosahexaenoic acid diet. Biofactors 11(1–2):11–13

Yamashita K, Nohara Y, Katayama K, Namiki M (1992) Sesame seed lignans and gamma-tocopherol act synergistically to produce vitamin E activity in rats. J Nutr 122(12):2440–2446

Yamashita K, Yamada Y, Kitou S, Ikeda S, Abe C, Saarinen NM, Santti R (2007) Hydroxymatairesinol and sesaminol act differently on tocopherol concentrations in rats. J Nutr Sci Vitaminol (Tokyo) 53(5):393–399

Yen GC (1990) Influence of seed roasting process on the changes in composition and quality of sesame oil. J Sci Food Agric 50(4):563–570

Yoshida H, Abe S, Hirakawa Y, Takagi S (2001) Roasting effects on fatty acid distributions of triacylglycerols and phospholipids in sesame (*Sesamum indicum*) seeds. J Sci Food Agric 81:620–626

Yoshida H, Shigezak J, Takagi S, Kajimoto G (1995) Variations in the composition of various acyl lipids, tocopherols and lignans in sesame seed oils roasted in a microwave oven. J Sci Food Agric 68:407–415

Yoshida H, Tanaka M, Tomiyama Y, Mizushina Y (2007) Antioxidant distributions and triacylglycerol molecular species of sesame seeds (*Sesamum indicum*). J Am Oil Chem Soc 84(2):165–172

Zaghloul M, Prakash V (2002) Effect of succinylation on the functional and physicochemical properties of alpha-globulin, the major protein fraction from *Sesamum indicum* L. Nahrung 46(5):364–369

Antidesma bunius

Scientific Name

Antidesma bunius (L) Sprengel.

Synonyms

Antidesma andamanicum Hook.f., *Antidesma bunius* Spreng. var. *cordifolium* (C. Presl) Müll. Arg., *Antidesma bunius* var. *floribundum* (Tul.) Müll.Arg. *Antidesma bunius* (L.) Spreng. var. *genuinum* Müll.Arg., *Antidesma bunius* var. *pubescens* Petra Hoffm., *Antidesma bunius* var. *sylvestre* (Lam.) Müll.Arg., *Antidesma bunius* var. *wallichii* Müll.Arg., *Antidesma ciliatum* Presl, *Antidesma colletii* Craib, *Antidesma cordifolium* Presl, *Antidesma crassifolium* (Elmer) Merr., *Antidesma floribundum* Tul., *Antidesma glabellum* K.D.Koenig ex Benn., *Antidesma glabrum* Tul., *Antidesma retusum* Zipp. ex Span., *Antidesma rumphii* Tul., *Antidesma stilago* Poir. nom. illeg., *Antidesma sylvstre* Lam., *Antidesma thorelianum* Gagnep., *Bunius sativus* Rumph., *Sapium crassifolium* Elmer, *Stilago bunius* L.

Family

Phyllanthaceae also placed in Euphorbiaceae.

Common/English Names

Bignay, Black Currant Tree, Chinese Laurel, Currant Tree, Currentwood, Nigger's Cord, Salamander Tree, Wild Cherry.

Vernacular Names

Australia: Moi-Kin, Chunka (Queensland Aboriginal);
Chinese: Wu Jue Cha, Wu-Yuer-Cha, Wu-Jueh-Ch'a;
Dutch: Salamanderboom, Woeni;
French: Antidesme;
German: Lorbeerblättriger Flachsbaum, Salamanderbaum;
India: Bor-Heloch, Bor Heloch (Assamese), Bol-Aborak, Bol Aborak (Garo), Himalcheri (Hindu), Kareekomme, Naayikomme, Naayikoote, Nayikute, Kari Komme, Naikuti, Naayi Kote (Kannada), Dieng Soh Silli (Khasi), Mail-Kombi, Karivetti, Cerutali, Cherutali, Cheruthali, Noelitali, Nulitali, Nulittali (Malayalam), Amati, Aamatee, Bhumy-Sadpay (Marathi), Nolaiali, Nolaidali, Nolaitali, Nolaittali, Noyilatali, Neralaitali (Tamil), Janupullari, Janu Polari, Anepu, Aanepoo, Jaanupolaari (Telugu);
Indonesia: Buah Monton (Batak, Sumatra), Bune (Bima, Timor), Bunih (Buginese, Sulawesi),

Attor (Flores), Takuti (Gorontalo, Sulawesi), Buneh, Boeni, Wuni, Buni (Java), Bernai, Bonia, Menerk, Njam (Lampung, Sumatra), Burneh (Madurese), Bune Tedong (Makassar, Sulawesi), Buni (Malay), Katakuti, Kutikata (Malay, Maluka), Kiti-Kata (Malay, Timor), Rambai Tiris (Singkep), Barune, Huni, Huni Gede, Huni Wera, Wuni (Sundanese), Kiti-Kata (Timor), Barune (Sumbawanese), Guna, Haju Wune, Wuler (Lesser Sunda Islands);
Japanese: Buni NoKi, Nanyou Gomishi, Saramando No Ki;
Laotian: Kho Lien Tu;
Malaysia: Buni, Berunai, Bras-Bras Hitam;
Nepal: Himalcheri;
Philippines: Dokodoko, Mutagtamanuk (Bagobo), Bignay, Bignai (Bikol), Bugney, Bugnay, Bugnei, Bungai (Bontok), Bignai, Bignay, Bugnay, Bungai (Cebu Bisaya), Vunnay, Bugnay, Bundei, Vunnai, Bungai, Paginga (Ibanag), Bugney, Bugnei (Ifugao), Bugnay, Bungai (Iloko), Bignay, Bignai, (Mangyan), Isip (Pampangan), Bungai, Bugnay (Panay Bisaya), Oyhip, Bignai, Bignay, (Sambali), Bignay, Bignai, Bignay-Kalabaw (Tagalog);
Portuguese: Candoeira;
Spanish: Bignai;
Thai: Ba Mao Ruesi, Maeng Mao Khwai, Mao Chang, Mao Luang, Ma Mao Dong, Ma Mao Luang;
Vietnamese: Choi Moi, Liên Tu.

Origin/Distribution

It is difficult to establish the natural geographic distribution of the species as it is impossible to distinguish truly wild occurrences. It seems to be absent in Peninsular Malaysia (except Singapore) and nearly absent in Borneo. It is common in the wild from the lower Himalayas in India, Sri Lanka, and southeast Asia (but not Malaya) to the Philippines, PNG, Solomon Islands and northern Australia. It is cultivated in India (incl. Andaman and Nicobar Islands), Sri Lanka, southern China (Hong Kong, Hainan and Guangdong province), Myanmar, Laos, Vietnam, Thailand, Sumatra, Singapore, Borneo, Java, Philippines, Sulawesi, Lesser Sunda Islands, Moluccas, New Guinea, Christmas Islands (Indian Ocean, Australia), Tahiti, Hawaiian Islands, Cuba, Honduras and Florida. It is an abundant and invasive species in the Philippines; rarely cultivated in Malaysia, and grown in every village in Indonesia where the fruits are marketed in bunches.

Agroecology

The tree is not strictly tropical for it has proved to be hardy up to central Florida. It thrives from sea-level up to 2,100 m altitude. In its natural area, it occurs in a wide range of habitats that include: wet evergreen forest, dipterocarp forest, teak forest, on river banks, at forest edges, along roadsides, in bamboo thickets, in semi-cultivated and cultivated areas, in shady or open habitats usually in secondary but also in primary vegetation. It thrives best in full sun and on a wide range of soils from sand, loam or clay and on (coral) limestone or granite bedrock. It is tolerant of infertile soil and occasional waterlogging.

Edible Plant Parts and Uses

Ripe fruits are eaten fresh or cooked. Acid green fruits are used as flavoring in fish soup dishes. The fruits are also made into jam, preserves or are used in combination with other fruits, because of their high pectin content to make jelly. The fruits are also utilised in the production of syrup, soft drinks, wine, liqueur and brandy or are used in sauces for fish dishes. The pulp can be used for desserts like cakes, bavarois or ice cream. The young, tender leaves are eaten with rice in Indonesia and the Philippines. The leaves are often combined with other vegetables as flavouring. The leafy shoots are used for tea in China.

Botany

An evergreen, dioecious, perennial tree, 15–30 m high with a straight trunk, bole diameter of 20–85 cm, yellow brown bark, terete branchlets glabrous to densely ferrugineous-pubescent, usually branching near the base with a dense and irregular crown (Plates 1 and 2). Leaves are distichous with petiole furrowed and short, glabrous to ferrugineous-pubescent. Stipules are linear-lanceolate, pubescent and caduceus. The leaf lamina is oblong to elliptic, more rarely obovate, base obtuse or rounded to shallowly cordate, apex acuminate, obtuse or acute, entire, coriaceous, glabrous, glossy green above, yellowish green adaxially with 5–10 pairs lateral veins. Staminate inflorescences are 6–15 cm long, axillary, consisting of 3–8(–14) branches with deltoid to elliptic, pubescent bracts. Staminate flowers measure 3–4 by 3 mm, sessile; calyx, cupular, sepals 3 or 4, green, glabrous to ferrugineous-pubescent on both sides with fimbriate margin; disc variable, stamens 3 or 4, 2–3 mm long, exserted; pistillode clavate to cylindrical and short. Pistillate inflorescences are 4–17 cm long, axillary, simple or 4-branched, axes glabrous to pubescent with deltoid, short, pubescent to pilose bracts. Pistillate flowers are 2.5–3 by 1.5 mm; pedicels 0.5–2 mm long, pubescent to glabrous; calyx 1–1.5 by 1.5 mm, cupular, sepals 3, green, glabrous to pilose on both sides, ovary ellipsoid, glabrous to very sparsely pilose, style terminal to subterminal with 3–6 stigmas. Infructescences are 10–17 cm long, robust and fruiting pedicels 2–9 mm long, pubescent to glabrous. Fruits are globose or ovoid, glabrous, 5–11 mm by 4–7 mm, green turning yellow to pink to red and bluish-violet when ripe (Plates 1, 2 and 3), juicy with a single, straw-coloured, compressed, oval, 6–8 mm by 4.5–5.5 mm, ridged or fluted, very hard kernel.

Plate 1 Bignay foliage and fruit at various stages of development

Plate 2 Clusters of immature bignay fruit near top of irregular canopy

Plate 3 Close-up of developing bignay fruit cluster

Nutritive/Medicinal Properties

Nutrient composition of bignay fruit per 100 g of edible portion based on analyses made in Florida and the Philippines (Morton 1987) was reported as – moisture 91.11–94.80 g, protein 0.75 g, ash 0.57–0.78 g, calcium 0.12 mg, phosphorus 0.04 mg, iron 0.001 mg, thiamine 0.031 mg, riboflavin 0.072 mg, niacin 0.53 mg. Another analysis made in Australia (Brand Miller et al. 1993) reported the following: energy 12 kJ, moisture 81.7 g, nitrogen 0.11 g, protein 0.7 g, fat 0 g, ash 0.5 g, available carbohydrate 0 g, calcium 29 g, potassium 137 g, sodium 29 g, niacin equivalent 0.1 mg and vitamin C 69 mg.

Studies reported three different kinds of flavonoids in bignay fruit i.e., catechin, procyanidin B1 and procyanidin B2 which occurred in varying levels in different cultivars (Butkup and Samappito 2008). These three chemical compounds were the major flavonoids in bignay fruit. The highest amount of procyanidin B1 was found with Lompat followed by Maeloogdog with values of 4,122.75 and 3,993.88 mg/100 g of fresh weight (FW), respectively and the highest amount of procyanidin B2 was found with Sangkrow 2 followed by Fapratan with values of 5,006.39 and 3,689.42 mg/100 g FW, respectively Catechin contents in fruits of the 15 cultivars varied from 73.39 to 316.22 mg/100 g of fresh weight. The major group of organic acids in Mao Luang berries were: tartaric acid (7.97–12.16 mg/g FW), ascorbic acid (10.01–16.55 mg/g FW), citric acid (4.44–11.73 mg/g FW) and benzoic acid (8.13–17.43 mg/g FW) and the minor group included malic acid (3.05–4.52 mg/g FW), lactic acid (1.12–4.09 mg/g FW), oxalic acid (1.00–1.45 mg/g FW) and acetic acid (0.19–0.69 mg/g FW) (Samappito and Butkhup 2008b).

Thai scientist also reported on the chemical contents of fresh ripe fruit and brewed red wines of both non-skin contact and skin contact fermentation techniques of Mao Luang ripe fruits (Samappito and Butkhup 2008a). Ripe Mao Luang fruits on fresh weight of 100 g berries had the following mean values: fresh weight 65.62 g, juice:solids 3.28, pH3.51, total soluble solid (TSS), 16.50^0 brix, total organic acids (TOA), 49.36 mg/L, TSS: TOA 28. 10%, total flavonoids contents (TFC) 397.90 mg/L, total phenolic acids viz. gallic, caffeic, vanillic, ellagic and ferulic acid (TPA) 76.04 mg/L, total procyanidin B1 and B2 contents (TPC) 156.21 mg/L, and reducing sugars (184.32 g/L). Skin contact Mao Luang red wine gave higher amounts of flavonoids, phenolic acids, anthocyanins of procyanidin B1 and procyanidin B2, organic acids than non-skin contact red wine. The differences were highly significant. Additionally, ethanol (%) and total acidity (g/L citric acid) were much higher for skin contact wine than non-skin contact wine but a reverse was found with total soluble solids (^{0}brix) and pH where non-skin contact wine gave higher mean values than skin contact wine. In a recent study, they found that total phenolic content (TP) of *A. bunius* fruit accounted for 19.60–8.66 mg GAE/g FW as assayed by the Folin-Ciocalteu method (Butkhup and Samappito 2011). TP gradually decreased from the immature to ripe fruit stages. Total anthocyanin, however, was highest at the over-ripe stage with mean value of 141.9 mg/100 g FW. Highest antioxidant activity (DPPH) was found at the immature stage accompanied by the highest content of gallic acid and TP. The main polyphenol compounds namely procyanidin B2, procyanidin B1, (+) – catechin, (-)-catechin, rutin and *trans*-resveratrol increased during fruit development and ripening. Other phenolics like gallic, caffeic and ellagic acids significantly decreased during fruit development and ripening. At the overripe stage the fruit possessed the highest antioxidants and would be the best time to harvest.

Scientists in the Philippine reported that methanolic crude extracts of the leaves and fruits of bignay contains unidentified compounds with potential cytotoxic activity using the *Artemia salina* assay (Micor et al. 2005).

In folkloric medicine, bignay leaves are regarded as sudorific and employed in treating snakebite in Asia. The leaves are acid and diaphoretic, and are boiled with pot-herbs, and employed by the natives in syphilitic affections. The ripe fruit is subacid, and is esteemed for its cooling qualities. Filipinos believe the fruit can cure various ailments from parched tongue, lack of appetite, indigestion, high blood pressure to diabetes.

Other Uses

The bark yields a strong fibre for rope and cordage. The timber is reddish and hard and is utilized for the production of beams, tresses, rafters, light constructions, firewood and charcoal. The wood has been experimentally pulped for making cardboard. It is also planted as an ornamental, windbreak, hedge, living fence, in agroforestry, in home gardens and roadside tree

Comments

Bignay is easily established from stem cuttings as seeds may not be viable due to inadequate pollination.

Selected References

Backer CA, van den Brink Bakhuizen RC Jr (1963) Flora of Java (spermatophytes only), vol 1. Noordhoff, Groningen, p 648

Brand Miller J, James KW, Maggiore P (1993) Tables of composition of Australian aboriginal foods. Aboriginal Studies Press, Canberra

Burkill IH (1966) A dictionary of the economic products of the Malay Peninsula. Revised reprint, 2 vols. Ministry of Agriculture and Co-operatives, Kuala Lumpur, vol 1 (A–H), pp 1–1240, vol 2 (I–Z), pp 1241–2444

Butkhup L, Samappito S (2008) An analysis on flavonoids contents in mao luang fruits of fifteen cultivars (*Antidesma bunius*), grown in northeast Thailand. Pak J Biol Sci 11(7):996–1002

Butkhup L, Samappito S (2011) Changes in physio-chemical properties, polyphenol compounds and anti-radical activity during development and ripening of Maoluang (*Antidesma bunius* L. Spreng) fruits. J Fruit Ornam Plant Res 19(1):85–99

Foundation for Revitalisation of Local Health Traditions (2008) *FRLHT Database* htttp://envis.frlht.org

Hoffmann P (2006) *Antidesma* in Malaysia and Thailand. Royal Botanic Gardens, *Flora Malesiana*. 292 pp. http://www.nationaalherbarium.nl/euphorbs/specA/Antidesma.htm

Hu SY (2005) Food plants of China. The Chinese University Press, Hong Kong, 844 pp

Li B, Hoffmann P (2008) *Antidesma* Burman ex Linnaeus. In: Wu ZY, Raven PH, Hong DY (eds) Flora of China, vol 11, *Oxalidaceae* through *Aceraceae*. Science Press/Missouri Botanical Garden Press, Beijing/St. Louis

Micor JRL, Deocaris CC, Mojica ERE (2005) Biological activity of bignay [*Antidesma bunius* (L.) Spreng] crude extract in *Artemia salina*. J Med Sci (Pakistan) 5(3):195–198

Molesworth Allen B (1967) Malayan fruits. An introduction to the cultivated species. Moore, Singapore, 245 pp

Morton J (1987) Bignay. In: Fruits of warm climates. Julia F. Morton, Miami, pp 210–212

Ochse JJ, van den Brink Bakhuizen RC (1931) Fruits and fruitculture in the Dutch East Indies. G. Kolff & Co, Batavia-C, 180 pp

Samappito S, Butkhup L (2008a) An analysis on flavonoids, phenolics and organic acids contents in brewed red wines of both non-skin contact and skin contact fermentation techniques of Mao Luang ripe fruits (*Antidesma bunius*) harvested from Phupan Valley in Northeast Thailand. Pak J Biol Sci 11(13):1654–1661

Samappito S, Butkhup L (2008b) An analysis on organic acids contents in ripe fruits of fifteen Mao Luang (*Antidesma bunius*) cultivars, harvested from dipterocarp forest of Phupan Valley in northeast Thailand. Pak J Biol Sci 11(7):974–981

Stuart GU (2010) Philippine alternative medicine. Manual of some Philippine medicinal plants http://www.stuartxchange.org/OtherHerbals.html

Baccaurea angulata

Scientific Name

Baccaurea angulata Merr.

Synonyms

None.

Family

Phyllanthaceae also placed in Euphorbiaceae.

Common/English Names

Red Angled Tampoi, Wild Carambola

Vernacular Names

Brunei: Embaling Bobou (Dusun), Belimbing Hutan;
Kalimantan (Indonesia): Asem Ketiak, Pidau, Umbing, Umbung;
Sabah: Embaling (Dusun), Belimbing Uchong, Pelawak, Popotong, Tampoi Hutan, Liposu;
Sarawak: Uchong, Ujung (Iban), Belimbing Bukit, Belimbing Dayak, Belimbing Hutan, Belimbing Merah, Tampoi Belimbing (Malay).

Origin/Distribution

The species is indigenous to Borneo.

Agroecology

Its natural habitat is in the tropical primary and secondary riverine and non-riverine rain forest in the island of Borneo. A shade loving tree that occurs in sandstone or lateritic soils from sea level to 800 m elevation.

Edible Plant Parts and Uses

The fruit is edible. The sour to acid sweet arillode and pericarp are eaten fresh or cooked as vegetable in dishes. Chilled juice from the fruit provides a refreshing, thirst-quenching drink.

Botany

A perennial, medium-sized tree 6–21 m high. Leaves are elliptic to obovate, 12–40 cm by 4–14 cm, thick, coriaceous, with cuneate to attenuate base, cuspidate to acuminate apex, dark green and borne on 2–12.5 cm long petioles with glabrous to hairy stipules. Staminate inflorescences are cauline, few clustered together. Staminate flowers, 2–2.6 mm across, pubescent, with 4–5 obovate sepals, 6 staminodes, 6 stamens, and a hollow obtriangular pistillode. Pistillate inflorescences are cauline to ramiflorous, solitary or up to 7 clustered together with 8–many flowered, red. Pistillate flower, 4–10 mm across, pale yellow to cream yellow to greenish, 5–6 pubescent elliptic, persistent to caducous sepals, 3-locular, urn-shaped ovary with 6 wings and persistent to caducous stigma, and no style. Fruit are borne on trunk and branches (Plate 1). Fruits are obovoid with tapeing apex, 5–6 angled, star-shaped in cross section, 1–3-seeded berries, 5 cm by 2.5 cm, with raised glands present, red to purple to pink to red-brown (Plates 1, 2, and 3); pericarp 1–2 mm thick, column 22–25 mm long, straight; pedicel 4–8 mm long. Seeds are globose to ellipsoid, laterally flattened, 16–23 by 7–16 by 4–9.5 mm, dicotyledonous and endospermic with cream to greenish testa. Arillode is white and edible (Plate 3).

Plate 2 Angled, glossy pinkish red, obovoid fruits with tapering apices

Plate 3 White arillode and thick pericarp

Nutritive/Medicinal Properties

The fruit was reported to contain protein, carbohydrate, fibre, minerals and vitamin C. The nutrient composition per 100 g edible portion of the raw fruit pulp is: energy 93 kcal, moisture 73.8%, protein 3.8%, fat 0.2%, carbohydrate 21.9%, crude fibre 2.1%, ash 1.0%, P 39 mg, K 352 mg, Ca 21 mg, Mg 21 mg, Fe 6u g, Mn 5ug, Cu 2.6ug, Zn 5.8ug and vitamin C 0.1 mg (Voon and Kueh 1999).

No published information is available on its medicinal value.

Other Uses

Like other *Baccaurea* species, the species provides 'tampoi' timber that is suitable for medium construction, under-cover, posts, beams, joists, rafters, furniture and plywood.

Plate 1 Clusters of angled, obovoid star-shaped, cauliflorous fruit

Comments

The tree is propagated from seeds.

Selected References

Airy Shaw HK (1975) The Euphorbiaceae of Borneo. Kew Bull Add Ser 4:46

Haegens RMAP (2000) Taxonomy, phylogeny, and biogeography of *Baccaurea*, *Distichirhops*, and *Nothobaccaurea* (*Euphorbiaceae*). Blumea Suppl 12:1–216

Voon BH, Chin TH, Sim CY, Sabariah P (1988) Wild fruits and vegetables in Sarawak. Sarawak Department of Agriculture, Kuching, 114pp

Voon BH, Kueh HS (1999) The nutritional value of indigenous fruits and vegetables in Sarawak. Asia Pac J Clin Nutr 8(1):24–31

Baccaurea dulcis

Scientific Name

Baccaurea dulcis (Jack) Müll.Arg.

Synonyms

Baccaurea suvrae Chakrab & M. Gangop., *Pierardia dulcis* Jack.

Family

Phyllanthaceae, also placed in Euphorbiaceae

Common/English Names

Chupak, Ketupa

Vernacular Names

Borneo: Apor-apor (Bassap), pendal nyumbo, tampoi paya (Iban), pas, tampoi paya (Malay), boenjan, kapul, kapul putih, kelawat'n petik, kepsoet awoet, kulibon, puak burong;
Indonesia: Kapundang (Sundanese), ketupa (Sumatra), Cupa, tupa, cupak (Java, Sulaswesi);
Peninsular Malaysia: Tampoi paya, Rambai, Chupa, Tupa.

Origin/Distribution

The species is found in Sumatra, Java, Borneo (Sabah, Central-, South- and East-Kalimantan).

Agroecology

It is found wild in undisturbed mixed dipterocarp to sub-montane forests up to 1,100 m altitude but mainly below 700 m elevation. In secondary rain forests it is usually present as a pre-disturbance remnant tree on sandy soils. It is also found in swamp forest on alluvial sites.

Edible Plant Parts and Uses

The arillode and seed-coat is acid-sweet and edible. The fruit is commonly sold in local markets during the harvesting season in Java and Sumatra. Fruit can be pickled, use in stews or fermented to make wine.

Botany

Sub-canopy, perennial tree 8–30 m tall, with a trunk diameter of 5–60 cm and brown to red-brown bark, glabrous twigs which are pubescent when young. Leaves spirally arranged on pubescent,

12–77 mm long petiole and triangular stipules, pubescent outside and caducous. Lamina is usually elliptic, 8–30 cm by 3–12 cm, papery, dark green, base rounded to cuneate, acuminate apex, entire margin and with secondary veins (7 or) 8–10 per side. Staminate inflorescences are axillary to just below the leaves, solitary to few clustered together, 1.5–12.5 cm long, flowers scattered along inflorescence usually with 1 spatulate bract per branchlet. Staminate flowers are 1–2.5 mm across, yellowish-green to yellow to white; with 4 or 5 ovate to elliptic sepals with apex recurved, 3–5 glabrous, 0.1–0.3 mm long stamens with straight filaments and pale yellow anthers. Pistillate inflorescences are ramiflorous to cauline, 2–6 clustered together, 2–18 cm long with 1–3 persistent bracts per branchlet. Pistillate flowers are 3–7 mm across, greenish-yellow, with 4 or 5, ovate sepals; globose, 2–4 by 1.5–4 mm, bilocular, densely hairy ovary, 1–2, densely hairy style and stigmas 1–2 mm long, cleft and caducous. Fruits are globose, 2–4-seeded indehiscent berries, 5–6 cm diameter, (orange-) brown to whitish to yellow (Plate 1); pericarp 4–10 mm thick; column 19–25 mm long, straight; pedicel 5–14 mm long. Seeds are globose to ellipsoid, laterally flattened, 13–20 mm by 8–21 mm by 2–8 mm. Arillode is white to yellow.

Plate 1 Cupak fruit – globose and thin-skinned

Nutritive/Medicinal Properties

No published information is available on its nutritive value. Like other *Baccaurea* species the bark can be applied medicinally to treat skin diseases and inflammation of the eyes.

Other Uses

Like other *Baccaurea* species, the tree is often cultivated as a support tree for rattan cultivation, as shade and avenue trees. Its timber is used for posts, planks, house beams and boat construction.

Comments

The tree is propagated from seeds.

Selected References

Backer CA, van den Brink RCB Jr (1963) Flora of java. (Spermatophytes only), 1. Noordhoff, Groningen, 648pp

Burkill IH (1966) A dictionary of the economic products of the Malay Peninsula. Revised reprint. 2 vols, Ministry of Agriculture and Co-operatives, Kuala Lumpur, vol 1 (A–H), pp. 1–1240, vol 2 (I–Z), pp 1241–2444

Haegens RMAP (2000) Taxonomy, phylogeny, and biogeography of *Baccaurea*, *Distichirhops*, and *Nothobaccaurea (Euphorbiaceae)*. Blumea Suppl 12, 216pp

Lestari R (2008) Characterisation and selection of underutilised tropical fruit *Baccaurea dulcis* (Jack) Mull. Arg. in West Java, Indonesia. Poster presentation. 4th international symposium on tropical and subtropical fruits, 3–7 November 2008, Bogor, West Java, Indonesia (Abstr. PA 31)

Sastrapradja S (1977) Buah-buahan. Bogor: Proyek Sumber Daya Ekonomi, Lembaga Biologi Nasional, Indonesia, 133pp (In Indonesian)

Slik JWF (2006) Trees of Sungai Wain. Nationaal Herbarium Nederland. http://www.nationaalherbarium.nl/sungaiwain/

Uij T (1992) *Baccaurea* Lour. In: Verheij EWM, Coronel RE (eds) Plant resources of South-East Asia No 2. Edible fruits and nuts. PROSEA, Bogor, pp 98–100

Baccaurea edulis

Scientific Name

Baccaurea edulis Merr.

Synonyms

None

Family

Phyllanthaceae, also placed in Euphorbiaceae

Common/English Names

Tampoi, Tampoi Merah

Vernacular Names

Borneo: Apor-Apor (Bassap), Pendal Nyumbo, Tampoi Paya (Iban), Pas, Tampoi Paya (Malay), Boenjan, Kapul, Kapul Putih, Kelawat'n Petik, Kepsoet Awoet, Kulibon, Pasin, Puak, Puak Burong, Pugi, Tampoi, Tampoi Hutan, Tampoi Merah.

Origin/Distribution

The species is native to Borneo (Sarawak, Brunei, Sabah, West- and East-Kalimantan).

Agroecology

A tropical species, thrives in a hot, wet and humid areas. It occurs wild in lowland primary and secondary rain forest, and swamp forest on alluvial and sandy soils from sea level to 700 m elevation.

Edible Plant Parts and Uses

Arillode and seed-coat are edible, sweet to acid sweet. The fruit is sold in local markets during the fruiting season.

Botany

A small to medium-sized tree, 8–33 m tall with a girth of 5–60 cm and red-brownish bark. Leaves are spirally arranged on pubescent, 1.2–7.7 cm long petioles, stipules, pubescent, deltoid to elliptic, 6–11 mm by 2.5–4.5 mm. Lamina is dark-green, elliptic, 8–29 cm by 12–73 cm, papery, base cuneate to rounded, apex acute to cuspidate, margin entire, glabrous above and subglabrous beneath with 7–10 secondary veins per side. Staminate inflorescences are axillary, solitary to a few clustered together, 15–125 mm long, branched, pubescent with many scattered flowers on 0.5–1.7 mm pedicel, along the branchlets and with one bract per branchlet. Male flowers are small, 1–2.5 mm, yellowish-green to yellowish-white, sepals 4–5, ovate to elliptic, 0.7–1.5 by

Plate 1 Ripe tampoi fruit

0.5–1 mm, apex recurved, pubescent; stamens 3–5, glabrous with pale yellow anthers. Pistillate inflorescences are ramiflorous to cauline, 2–6 clustered together, 2–18 cm long, 1–1.5 mm thick, densely hairy, 10–25-flowered. Female flower is 3–7 mm across, greenish-yellow; sepals 4 or 5, ovate, 2–3.5 by 1–2 mm, pubescent; ovary globose, 2–4 by 1.5–4 mm, 2-locular, densely hairy; style, densely hairy; stigmas 1–2 mm long, cleft. Fruits are globose, 2–4-seeded berries, 5–6 cm diameter, glabrous, densely hairy when young, orangey-brown to whitish to yellow (Plate 1), cauliflorous; pericarp 4–10 mm thick; column 19–25 mm long, straight; pedicel 5–14 mm long. Seeds are globose to ellipsoid, laterally flattened, 13–20.5 by 8–21 mm; arillode is white to yellow.

Nutritive/Medicinal Properties

No published information is available on its nutritive and medicinal values.

Other Uses

A minor and lesser-known timber tree.

Comments

The tree is established from seeds.

Selected References

Airy Shaw HK (1975) The Euphorbiaceae of Borneo. Kew Bull Add Ser 4:1–245

Haegens RMAP (2000) Taxonomy, phylogeny, and biogeography of *Baccaurea, Distichirhops, and Nothobaccaurea (Euphorbiaceae)*. Blumea Suppl 12:1–216

Slik JWF (2006) Trees of Sungai Wain. Nationaal Herbarium Nederland. http://www.nationaalherbarium.nl/sungaiwain/

Baccaurea lanceolata

Scientific Name

Baccaurea lanceolata (Miq.) Muell. Arg.

Synonyms

Adenocrepis lanceolatus (Miq.) Müll.Arg., *Baccaurea glabriflora* Pax & K.Hoffm., *Baccaurea pyrrhodasya* (Miq.) Müll.Arg., *Hedycarpus lanceolatus* Miq. (basionym), *Pierardia pyrrhodasya* Miq.

Family

Phyllanthaceae, also placed in Euphorbiaceae

Common/English Names

Lampaong

Vernacular Names

Borneo: Limpasu (Banjarese; Bundu Tuhan), Ampusu' (Bidayuh), Asam Pauh, Empaong, Lampaong, Lampawong, Lampong (Iban), Buah Lepasu, Lipasu, Nipassu (Dusun), Kalampesu, Lempahong (In Kalimantan), Buah Lipauh (Kelabit), Kelepesoh (Kenyah), Tampoy (Malay), Buah Lepesuh (Punan), Empawang, Lapahung, Lempawong, Paong;
Peninsular Malaysia: Asam Pahong, Rambai Hutan, Medang Kelawar, Asam Pahung, Asam Paung, Limpanong, Pahu Asam, Pahu Temuangi;
Indonesia: Tegeiluk (Mentawai), Kaloe Goegoer, Langsat Hutan, Lempaong, Lempaoe-Oeng, Peng (Sumatra), Lingsoe (Javanese);
Sarawak: Pisau (Bidayuh), Limpa'ong (Iban), Bua'pau (Kelabit), Lepesu (Penan).

Origin/Distribution

The species is found in Thailand, Peninsular Malaysia, Sumatra, Borneo (Sarawak, Brunei, Sabah, West-, Central-, South- and East-Kalimantan) and Philippines. Rarely in cultivation in fruit gardens.

Agroecology

It occurs on sandy, loamy or clayey soil in the primary and secondary tropical rain forest, on slopes and in riverine forest from sea level to 1,300 m in its native range. It is more common in the lowlands.

Edible Plant Parts and Uses

The pericarp and arillode are edible but sour and therefore are eaten with sugar or salt. In Sarawak, the fruit is often eaten with chicken rice and is used in cooking.

Plate 1 Much-branched tree with large leaves and cauliflorous and ramiflorous clusters of of fruit

Plate 2 Pendant clusters of developing subglobose fruits

Botany

A perennial medium-sized, much-branched tree, 3–30 m tall, with gnarled trunk of 50–60 cm diameter. Leaves are spirally arranged on 16–184 mm long glabrous petioles with caducous stipules. Lamina is papery, glossy dark green, ovate to elliptic, 9.2–45 by 3.7–26.5 cm, base attenuate, apex acute to cuspidate, upper surface glabrous with reticulate venation and raised glands. Staminate inflorescences are cauline, many clustered together, 3.5–18 cm long, 0.1–1 mm thick, glabrous to sparsely hairy, many-flowered, flowers scattered along inflorescence, yellow to pink to cream-white; bracts 1 per branchlet, 0.4–1.5 mm long, densely hairy outside, (sub)glabrous inside, margin ciliate, hyaline. Staminate flowers are 2–7 mm across, yellow to pink to purple to cream-white; pedicelled, with 3–5 obovate to spatulate sepals glabrous outside, sparsely to densely hairy inside; 3–5 rudimentary petals, 3–5 staminodes, 3–5 stamens and globose to cylindrical, velutinous, solid pistillode. Pistillate inflorescences are cauline, few clustered together, 20–25-flowered, yellowish brown to reddish with 3 bracts per branchlet, (sub)glabrous outside, glabrous inside with ciliate margin. Pistillate flowers are 3.5–10 mm across on 0.8–4.5 mm long, pink, pedicel; yellow to orange to purple to reddish cream to whitish; with 4 or 5, ovate to obovate, caducous sepals densely hairy outside, (sub)glabrous inside; 2–8, reduced petals, a globose, 3- or 4-locular, sparsely to densely hairy ovary measuring 1.5–3.2 by 1.5–3.2 mm, no style and persistent to caducous cleft stigmas. Fruits are cauliflorous to ramiflorous (Plates 1–2), are globose to ellipsoid, 1–4-seeded berries, 38 by 60 mm, with raised glands present, green when young, turning to creamy white or mauve and

Plate 3 Close-up of creamy-white fruits

Plate 4 More matured fruits

to pale yellowish-brown when mature (Plates 1, 2, 3, 4); pericarp 1–10 mm thick; pedicel 2.5–17 mm long. Seeds are ellipsoid, laterally flattened, 12.2–26 by 8–15 by 4.8–9 mm with yellow to whitish testa. Arillode is white to grey and translucent.

Nutritive/Medicinal Properties

The proximate nutrient composition of the fruit of *B. lanceolata* per 100 g edible portion based on analyses made in Sarawak (Voon and Kueh 1999) was reported as: water 92.4%, energy 18 kcal, protein 0.2%, fat 0.2%, carbohydrates 3.7%, crude fibre 2.2%, ash 0.8%, P 6 mg, K 126 mg, Ca 35 mg, Mg 11 mg, Fe 0.3 mg, Mn 2 ppm, Cu 1.5 ppm, Zn 6.3 ppm, vitamin C 0.6 mg.

In traditional medicine in Sarawak, the leaves are pounded in bamboo and mixed with water and the decoction is drank to treat stomach-ache in the Kelabit community. A poultice of the fruit is applied to swellings on the body by the Bidayuh. To prevent drunkenness, the Penan pound the bark and drink the sap before consuming alcohol.

Other Uses

Its timber is used for house construction. The Iban used the fruit as protection against charms. During the Gawai, harvest festival, the fruit is consumed before paying house visits.

Comments

The tree is readily propagated by seeds.

Selected References

Airy Shaw HK (1975) The Euphorbiaceae of Borneo. Kew Bull Add Ser 4:46

Burkill IH (1966) A dictionary of the economic products of the Malay Peninsula. Revised reprint, 2 vols. Ministry of Agriculture and Co-operatives, Kuala Lumpur, vol 1 (A–H), pp 1–1240, vol 2 (I–Z), pp 1241–2444

Chai PPK (2006) Medicinal plants of Sarawak. Lee Ming Press, Kuching, 212pp

Haegens RMAP (2000) Taxonomy, phylogeny, and biogeography of *Baccaurea*, *Distichirhops*, and *Nothobaccaurea* (*Euphorbiaceae*). Blumea Suppl 12:1–216

Slik JWF (2006) Trees of Sungai Wain. Nationaal Herbarium Nederland. http://www.nationaalherbarium.nl/sungaiwain/

Voon BH, Chin TH, Sim CY, Sabariah P (1988) Wild fruits and vegetables in Sarawak. Sarawak Department of Agriculture, Kuching, 114pp

Voon BH, Kueh HS (1999) The nutritional value of indigenous fruits and vegetables in Sarawak. Asia Pac J Clin Nutr 8(1):24–31

Whitmore TC (1972) Euphorbiaceae. In: Whitmore TC (ed) Tree flora of Malaya, vol 2. Longman, Kuala Lumpur, pp 34–135

Baccaurea macrocarpa

Scientific Name

Baccaurea macrocarpa (Miq.) Müll.Arg.

Synonyms

Baccaurea borneensis (Müll.Arg.) Müll.Arg., *Baccaurea griffithii* Hook. F., *Mappa borneensis* Müll.Arg., *Pierardia macrocarpa* Miq. (basionym).

Family

Phyllanthaceae, also placed in Euphorbiaceae

Common/English Names

Greater Tampoi, Tampui

Vernacular Names

Borneo: Pasin, Pasim Salai, (Bassap Dyak), Pegak, Pekang, (Dayak Tunjung), Puak, Tampoi (Iban); Setai (Kenyah, Kalimantan), Djentikan (Malay, Kutei, Kalimantan), Tampoi, Tampoi Laki, Tampoi Hutan, (Malay, Kedayan), Bua'Abu (Punam, Malinau, Kalimantan), Bua Lifoh (Lun Daye, Mentarang, Kalimantan), Teraie (Merap, Malinau, Kalimantan), Tetai (Kenyah Uma' Lung, Kalimantan), Buah Setei, Embah cerila, Empak Kapur, Embak kapur, Jantikan, Kapul, Pasim salai, Pasin, Pegak, Pekang, Puak tampoi, Setai, Tampoi, Tampoi hutan, Tampoi laki, Terai;
Indonesia: Medang, Tampui (Bangka), Tampoei Daoen, Tampoei Benez (Sumatra);
Peninsular Malaysia: Merkeh (Kelantan), Ngeke (Malay), Lara (Temuan), Rambai, Tampoi Batang, Tampoi, Tampoi Kuning, Tampoi Putih, Tampui, Medang Kelawar.

Origin/Distribution

The species is native to Peninsular Malaysia, Sumatra, Borneo (Sarawak, Brunei, Sabah, West-, Central-, South- and East-Kalimantan).

Agroecology

A tropical species that thrives in a hot, wet and humid areas. It occurs in the wild in primary rain forest (undisturbed mixed dipterocarp forests), riverine rain forest and peat swamp forest, submontane and keranga forests from sea level up to 1,600 m altitude. In secondary forests it is usually present as a pre-disturbance remnant tree. It is found on alluvial sites and hillsides on soils with red clay, sandy clay and granitic sand. It is also and cultivated in gardens.

Edible Plant Parts and Uses

The arillode is edible, sweet to acid sweet; eaten fresh or made into conserve. The fruit is commonly sold in local markets and roadside stalls.

Botany

Small to medium-sized, sub-canopy, evergreen tree (Plate 1), 5–18 (–27) m tall with a girth of 120 cm, buttressed with greyish brown bark. Leaves are spirally arranged on glabrous, 22–145 mm long petioles with glabrous, 2–9 mm by 1–5 mm stipules. Lamina is large, ovate to obovate, averaging 20 cm by 13 cm, leathery to papery; base attenuate to cuneate (to rounded); apex (obtuse to) acuminate to cuspidate, upper surface glabrous, lower subglabrous with 6–10 secondary veins per side. Staminate inflorescences are ramiflorous to cauline, axillary, solitary to few clustered together, 50–130 mm long, pubescent, branched, with flowers scattered along inflorescence and 1–3 bracts per branchlet. Staminate flowers are small, 0.7–2 mm across, greenish to yellowish-white, pedicel 1–2 mm, sepals 5 elliptic, stamens 5, glabrous, yellowish disc absent. Pistillate inflorescences are ramiflorous to cauline, solitary to 3 clustered together, 3.5–18 cm by 2–3 mm thick, subglabrous to densely hairy, 8–many-flowered on 3–7.5 mm long pedicel with 1–3 bracts per branchlet. Pistillate flowers are small 2–4.5 mm across; sepals 4–6, ovate, pubescent, persistent; ovary globose to cylindrical, 3- or 4-locular, tomentose; style sparsely hairy; stigmas cleft, persistent to caducous. Fruits are depressed globose to subglobose (Plates 2–3), (2- or) 3–6-seeded, fleshy capsules, 3–6.5 cm by 3.4–7.5 cm by 3.4–7.5 cm, glabrous, raised glands present, brown to yellow to orange to dull red to dark green; pericarp 4–11 mm thick (Plate 3); column 16–32 mm long, straight, pedicel 7–30 mm long. Seeds are brown, globose to ellipsoid, laterally flattened, 13–23 by 11–18.5 mm,

Plate 1 Vertical branches with inflorescence primordia and young leaves

Plate 2 Cluster of 2–3 oblate fruit of *Baccaurea macrocarpa*

Plate 3 Fruit of *Baccaurea macrocarpa* halved to show the edible white arillode and thick rind

arillode white (Plate 3) to yellow to sometimes orange.

Nutritive/Medicinal Properties

Proximate nutrient composition of the fruit arillode per 100 g edible portion was reported as: 127 kcal, moisture 66.6%, protein 1.5 g, fat 4.4 g, carbohydrates 27.9, dietary fibre 2.2 g, ash 0.9 g, P 54 mg, K 293 mg, Ca 10 mg, Mg 20 mg, Fe 20 µg, Mn 3 µg, Cu 7.3 µg, Zn 18.3 µg and vitamin C 0.1 mg. (Voon and Kueh, 1999).

In a recent study conducted in Malaysia by Khoo et al. (2008), the total carotene content (mg/100 g) of selected underutilized tropical fruit in decreasing order was jentik-jentik (*Baccaurea polyneura*) 19.83 mg>Cerapu 2 (*Garcinia prainiana*) 15.81 mg>durian nyekak 2 (*Durio kutejensis*) 14.97 mg>tampoi kuning (*Baccaurea reticulata*) 13.71 mg>durian nyekak (1) 11.16 mg >cerapu 1 6.89 mg>bacang 1 (*Mangifera foetida*) 4.81>kuini (*Mangifera odorata*) 3.95 mg>jambu susu (*Syzygium malaccense*) 3.35 mg>bacang (2) 3.25 mg>durian daun (*Durio lowianus*) 3.04 mg>bacang (3) 2.58 mg>tampoi putih (*Baccaurea macrocarpa*) 1.47 mg>jambu mawar (*Syzygium jambos*) 1.41 mg. β-carotene content was determined by HPLC to be the highest in jentik-jentik 17,46 mg followed by cerapu (2) 14.59 mg, durian nyekak (2) 10.99 mg, tampoi kuning 10.72 mg, durian nyekak (1) 7.57 mg and cerapu (1) 5.58 mg. These underutilized fruits were found to have acceptable amounts of carotenoids and to be potential antioxidant fruits.

Other Uses

It provides a strong timber.

Comments

The tree is easily established from seeds.

Selected References

Airy Shaw HK (1975) The Euphorbiaceae of Borneo. Kew Bull Addit Ser 4:1–245

Burkill IH (1966). A dictionary of the economic products of the Malay Peninsula. Revised reprint, 2 vols. Ministry of Agriculture and Co-operatives, Kuala Lumpur. vol 1 (A–H), pp 1–1240, vol 2 (I–Z), pp 1241–2444

Corner EJH (1952) Wayside trees of Malaya. GPO, Singapore, 772pp

Haegens RMAP (2000) Taxonomy, phylogeny, and biogeography of *Baccaurea*, *Distichirhops*, and *Nothobaccaurea* (*Euphorbiaceae*). Blumea Suppl 12:1–216

Khoo HE, Ismail A, Mohd.-Esa N, Idris S (2008) Carotenoid content of underutilized tropical fruits. Plant Foods Hum Nutr 63:170–175

Munawaroh E, Purwanto Y (2009). Studi hasil hutan non kayu di Kabupaten Malinau, Kalimantan Timur (In Indonesian). Paper presented at the 6th basic science national seminar, Universitas Brawijaya, Indonesia, 21 Feb 2009. http://fisika.brawijaya.ac.id/bss-ub/proceeding/PDF%20FILES/BSS_146_2.pdf

Slik JWF (2006). Trees of Sungai Wain. Nationaal Herbarium Nederland. http://www.nationaalherbarium.nl/sungaiwain/

Voon BH, Chin TH, Sim CY, Sabariah P (1988) Wild fruits and vegetables in Sarawak. Sarawak Department of Agriculture, Kuching, 114pp

Voon BH, Kueh HS (1999) The nutritional value of indigenous fruits and vegetables in Sarawak. Asia Pac J Clin Nutr 8(1):24–31

Whitmore TC (1972) Euphorbiaceae. In: Whitmore TC (ed) Tree flora of Malaya, vol 2. Longman, Kuala Lumpur, pp 34–136

Baccaurea motleyana

Scientific Name

Baccaurea motleyana (Müll.Arg.) Müll.Arg.

Synonyms

Baccaurea pubescens Pax & K.Hoffm., *Pierardia motleyana* Müller Argoviensis (basionym).

Family

Phyllanthaceae, also placed in Euphorbiaceae

Common/English Names

Common Rambai, Rambai

Vernacular Names

Borneo: Pekan, Rambai (Iban), Rambai (Malay), Ulup-Lavai (Punan) Bua Trai, Pekang, Rambai, Ramei;
Chinese: Duo Mai Mu Nai Guo;
Indonesia: Rambai (Sumatra), Menteng (Java), Sekoyun (Lun Daye, Mentarang, Kalimantan), Pahae (Merap, Malinau, Kalimantan);
Peninsular Malaysia: Rambai, Rambeh;
Philippines: Rambi (Tagalog);
Singapore: Rambai, Buah Jentik, Asam Lambun;
Thailand: Rambi, Mafai-Farang, Ramai, Lam-Khae (Pattani), Raa-Maa Tee-Ku (Narathiwat).

Origin/Distribution

The species is native to Thailand, Peninsular Malaysia, Sumatra, Java, Borneo (Sarawak, Brunei, Sabah, West-, South- and East-Kalimantan) and Moluccas. It is cultivated in tropical areas elsewhere in southeast Asia, Northern Australia and Yunnan in China.

Agroecology

In its native range, it occurs in lowland primary and secondary, tropical rain forest, or open scrub vegetation and rarely riparian forest. It is often cultivated in home gardens. It grows on alluvial soils, yellow clay, sand or limestone soils from 10 to 750 m altitude.

Edible Plant Parts and Uses

Rambai is primarily grown for its fruit. The arillode is sweet to acid sweet, eaten raw or made into jams and conserves. The juice of any variety may be used to make drinks by sweetening and diluting according to taste and served over ice. The fruit may be fermented and processed into wine.

Botany

Evergreen, sub-canopy, dioecious, much branched tree 6–12 (–20 m) high, with trunk diameter of 15–40 cm and short buttresses and yellowish-brown, scaly to flaky bark and large, spreading canopy (Plates 1–2). Branchlets are terete and densely hairy. Leaves are spirally arranged, on 5–10 cm pubescent petioles with lanceolate and caducous stipules. Lamina is large, elliptic, obovate-lanceolate, or elliptic-lanceolate, 20–35× 7.5–17 cm, papery, pubescent abaxially and on mid-vein adaxially, base rounded to shallowly cordate, apex acute to shortly acuminate; dark-green, glossy, with 12–16 pairs lateral veins (Plates 2–3). Panicles are raceme-like, axillary, ramiflorous or cauliflorous, 13–35 cm with lanceolate bracts. The small, fragrant male and female, greenish-yellow, apetalous flowers are borne on separate trees. Staminate flowers are 2–3 mm across on 1.5–2.5 mm long, pubescent pedicel; sepals oblong-ovate, 1–1.5 mm, acute at apex, grey papillose-puberulent; stamens 4–6; disc present, pistillode retuse at apex. Pistillate flowers on pedicels to 2 mm; sepals oblong, 4–5 mm, grey papillose-puberulent; ovary ovoid or globose, 3-locular, sericeous. The fruit, in showy strands dangle from the older branches and trunk (Plates 2, 3, 4). Fruits are globose to ellipsoid (Plate 4), 2.5–4 cm, 3-seeded indehiscent berries, greenish-yellow to whitish-yellow when ripe, pericarp 0.5–1.5 mm thick; column 18–23 mm long, straight. Seeds are ellipsoid, laterally flattened, brown, 13–20 by 9–15 mm. Arillode is translucent, white to rarely purple.

Nutritive/Medicinal Properties

Food nutrient composition of fresh arillode of *Baccaurea motleyana* per 100 g edible portion was reported by Tee et al. (1997) as follows: energy 64 kcal, water 83.7 g, protein 0.4 g, fat

Plate 1 Tree habit with large dome-shaped canopy

Plate 2 Flaky bark and much branched tree

Plate 3 Pendant strings of ripening fruits

Plate 4 Large leaves and pendant clusters of rambai fruit

0.4 g, carbohydrate 14.6, fibre 0.1 g, ash 0.2 g, Ca 5 mg, P 13 mg, Fe 0.2 mg, Na 2 mg, K 111 mg, vitamin B1 0.02 mg, vitamin B2 0.05 mg, niacin 0.2 mg and vitamin C 6.2 mg. Leung et al. (1972) reported the following nutrient value per 100 g edible portion: energy 65 kcal, moisture 79.0%, protein 0.2 g, fat 0.1 g, carbohydrates 16.1 g, dietary fibre 0 g, ash 3 g, P 20 mg, K 0 mg, Ca 13 mg, Fe 0.8 and vitamin C 5 mg.

(E)-Hex-2-enal was the major component of rambai fruit volatiles which also contained high levels of methyl 2-hydroxy-3-methylbutanoate, methyl 2-hydroxy-3-methylpentanoate and methyl 2-hydroxy-4-methylpentanoate (Wong et al. 1994).

The petroleum ether, chloroform and ethanol extracts of rambai fruit peel exhibited antibacterial activity against *Staphylococcus aureus, Bacillus cereus, Bacillus subtilis, Proteus vulgaricus* and *Escherichia coli* (Suhaila et al. 1994). Only the ethanol extract exhibited antibacterial activity against *Pseudomonas aeruginosa*.

A few uses of the plant have been reported in traditional folk medicine. Squeezed cambium and inner bark has been used as remedy for eye inflammation. The bark has been used as an ingredient of a concoction of many ingredients and administered internally after childbirth in protective medicaments.

Other Uses

Its timber is of low quality but is used for posts. The bark is rich in tannins and yields a mordant for dyes. A black dammar oozes from the bark.

Comments

The tree can be propagated from seeds or stem cuttings.

Selected References

Backer CA, van den Brink RCB Jr (1963) Flora of Java, vol 1. Noordhoff, Groningen, 648pp

Burkill IH (1966) A dictionary of the economic products of the Malay Peninsula. Revised reprint, 2 vols. Ministry of Agriculture and Co-operatives, Kuala Lumpur. vol 1 (A–H), pp 1–1240, vol 2 (I–Z), pp 1241–2444

Chai PPK (2006) Medicinal plants of Sarawak. Lee Ming Press, Kuching, 212pp

Haegens RMAP (2000) Taxonomy, phylogeny, and biogeography of *Baccaurea*, *Distichirhops*, and *Nothobaccaurea* (*Euphorbiaceae*). Blumea Suppl 12:1–216

Leung W-TW, Butrum RR, Huang Chang F, Narayana Rao M, Polacchi W (1972) Food composition table for use in East Asia. FAO, Rome, 347pp

Li B, Gilbert MG (2008) *Baccaurea* Loureiro. In: Wu ZY, Raven PH, Hong DY (eds) Flora of China, vol 11, Oxalidaceae through Aceraceae. Science Press/Missouri Botanical Garden Press, Beijing/St. Louis

Molesworth Allen B (1967) Malayan fruits. An introduction to the cultivated species. Moore, Singapore, 245pp

Morton JF (1987) Rambai. In: Fruits of warm climates. Julia F. Morton, Miami, p 220

Munawaroh E, Purwanto Y (2009) Studi hasil hutan non kayu di Kabupaten Malinau, Kalimantan Timur (In Indonesian). Paper presented at the 6th basic science national seminar, Universitas Brawijaya, Indonesia, 21 Feb 2009. http://fisika.brawijaya.ac.id/bss-ub/proceeding/PDF%20FILES/BSS_146_2.pdf

Purseglove JW (1968) Tropical crops: dicotyledons, vol 1 and 2. Longman, London, 719pp

Slik JWF (2006) Trees of Sungai Wain. Nationaal Herbarium Nederland. http://www.nationaalherbarium.nl/sungaiwain/

Suhaila M, Zahariah H, Norhashimah AH (1994) Antimicrobial activity of some tropical fruit wastes (guava, starfruit, banana, papaya, passionfruit, langsat, duku, rambutan and rambai). Pertanika J Trop Agric Sci 17(3):219–227

Tee ES, Noor MI, Azudin MN, Idris K (1997) Nutrient composition of Malaysian foods, 4th edn. Institute for Medical Research, Kuala Lumpur, p 299

Uij T (1992) *Baccaurea* Lour. In: Verheij EWM, Coronel RE (eds) Plant resources of South-East Asia No 2. Edible fruits and nuts. PROSEA, Bogor, pp 98–100

Wong KC, Wong SW, Siew SS, Tie DY (1994) Volatile constituents of the fruits of *Lansium domesticum* (duku and langsat) and *Baccaurea motleyana* (Muell. Arg.) Muell. Arg. (rambai). Flav Fragr J 9(6):319–324

Baccaurea racemosa

Scientific Name

Baccaurea racemosa (Reinw. ex Blume) Müll. Arg.

Synonyms

Baccaurea bhaswatii Chakrab. & M.Gangop., *Baccaurea wallichii* Hook.f., *Coccomelia racemosa* Reinw. ex Blume (basionym), *Pierandia racemosa* (Reinw. ex Blume) Blume, *Pierardia racemosa* (Reinw. ex Blume) Miq.

Family

Phyllanthaceae, also placed in Euphorbiaceae

Common/English Names

Bencoi (red fruited variety), Kapundung, Menteng (yellowish-white fruited variety), Rambi.

Vernacular Names

Borneo: Engkumi, Kayu Masam, Kokonau, Kunau, Kunyi, Longkumo, Moho Liox, Tunding Undang, Umbarian;
Dutch: Menteng;
French: Rambeh;
Indonesia: Tangkilang, Kapundung (Bali), Haoundung, Ninggih (Batak, Sumatra), Kisip (Bengkoelen, Sumatra), Roesip, Kisip (Sumatra), Menteng, Kapundung, Jerek, Jirek (Java), Menteng, Rambai, Tampui (Lampung, Sumatra), Modung (Madurese), Kapundueng (Minangkabau), Bowo (Nias), Bencoi (Red Variety, Sundanese), Menteng (White Variety Sundanese), Kapundung, Kepundung (Singkep);
Peninsular Malaysia: Asam Tambun, Kapunddung, Jinteh Merah, Rambi, Tamut, Tampoi;
Borneo – Brunei, Sarawak, Sabah, Kalimantan: Kokonau (Dusun), Engkumi, Kayu Masam, Kokonau, Kunau, Kunyi, Longkumo, Moho liox, Tunding undang, Umbarian.

Origin/Distribution

The species is indigenous to Thailand, Peninsular Malaysia, Sumatra, Java, Lesser Sunda Islands, Borneo (Sarawak, Brunei, Sabah, West-, Central- and East-Kalimantan), Celebes, Moluccas. Cultivated in Sumatra, Java and Bali.

Agroecology

Menteng is a tropical species and occurs wild in undisturbed mixed dipterocarp primary and secondary rain forest, riverine forest, or fresh water swamp forest to sub-montane forests up to

1,500 m altitude. In secondary forests usually present as a pre-disturbance remnant tree. It is common on alluvial and dry (hillsides and ridges) sites. It grows on a wide range of soils, from sandy to clayey soil (granite to yellow or red sandy clay) to peat swamps.

Edible Plant Parts and Uses

Its acid-sweet arillode is edible fresh stewed, pickled, fermented or made into drinks but is not suitable for making conserves.

Botany

Under-storey, evergreen, perennial tree reaching 15–20 m tall with pale to grey-brown bark, straight trunk, 25–70 cm bole, candelabriform branching pattern with strong, sub-glabrous branchlets (Plates 2 and 8) and dense, irregular crown (Plate 1). Leaves are alternate, on 12–77 mm long petiole with elliptic to triangular stipules, 3–7.5 by 1–1.5 mm, glabrous to sparsely hairy. Lamina is simple, ovate to oblong to obovate, 5.8–22 by 2.3–18.8 cm, chartaceous, base cuneate, apex obtusely acuminate, margin entire, lower surface without disc-like glands, nerves 4–10 per side (Plates 1, 3, 9, 10). Inflorescences racemous, pendulous usually on older branches

Plate 2 Trunk with candelabriform branches of menteng

in axils of fallen leaves or less often from the trunk. Staminate inflorescences are cauliflorous to just below the leaves, 1–5 clustered, up to 10 cm long; bracts 0.2–1.1 mm long, bracteoles minute. Flowers apetalous (Plates 3, 4, 9). Male flowers are 1–3 mm in diameter, creamy yellow to white; pedicel 0.4–2.5 mm long; sepals 4–5, ovate to obovate; stamens 4–8. Pistillate inflorescences are cauliflorous to axillary, single (to few together), up to 28 cm long. Female flowers are 3–9.2 mm in diameter, greenish yellow, or pinkish; pedicel 1–6.7 mm long; sepals 4–5, obovate to lanceolate, ovary 1-3-locular; stigmas short 0.3 mm long, not lobed. Fruits are sub-globose to globose berries, 14–30 by 16–25 mm, green (Plate 5) turning to yellowish-white (Plates 6–7) or reddish (Plate 10) when ripe, glabrous and indehiscent. Seeds are obovoid to ellipsoid, laterally flattened, 9–11.5 by 6.5–9.2 mm. Arillode is blue to purple to violet.

Plate 1 Dense, irregular canopy of large chartaceous menteng leaves

Plate 3 Pendulous inflorescences of menteng

Plate 4 Apetalous, creamy-white menteng flowers

Plate 5 Immature, green menteng fruits

Plate 6 Bunches of menteng fruit being sold in the local market in Java

Plate 7 Harvested menteng fruit bruises easily with short shelf life

Plate 8 Trunk with candelabriform branches of bencoi

Plate 9 Pink flower buds and leaves of bencoi

Plate 10 Red fruited bencoi fruits in pedenlous clusters

Nutritive/Medicinal Properties

No published information is available on its nutritive value and medicinal value and uses.

Other Uses

Menteng is also planted as a shade tree in villages and for support for rattan. The timber is used for furniture, house and boat construction. Menteng yields a fibre that can be used for paper manufacture, and the bark and leaves yield a dye.

Comments

Both the red fruited and yellowish-white fruited *B. racemosa* are commonly sold in local markets in Java, Indonesia.

Selected References

Backer CA, van den Brink Bakhuizen Jr (1963) Flora of Java. (Spermatophytes only), vol 1. Noordhoff, Groningen, 648pp

Chayamarit K, van Welzen PC (2005) Flora of Thailand, vol 8 Part 1: Euphorbiaceae (Genera A-F). Forest Herbarium, Royal Forest Department, Bangkok, 303pp

Haegens RMAP (2000) Taxonomy, phylogeny, and biogeography of *Baccaurea*, *Distichirhops*, and *Nothobaccaurea* (*Euphorbiaceae*). Blumea Suppl 12:1–216

Ochse JJ, van den Brink BRC (1931) Fruits and fruitculture in the Dutch East Indies. G. Kolff & Co, Batavia-C, 180pp

Slik JWF (2006) Trees of Sungai Wain. Nationaal Herbarium Nederland. http://www.nationaalherbarium.nl/sungaiwain/

Uij T (1992) *Baccaurea* Lour. In: Verheij EWM, Coronel RE (eds) Plant resources of South-East Asia No 2. Edible fruits and nuts. PROSEA, Bogor, pp 98–100

Baccaurea ramiflora

Scientific Name

Baccaurea ramiflora Lour.

Synonyms

Baccaurea cauliflora Loureiro, *Baccaurea flaccida* Müll.Arg., *Baccaurea oxycarpa* Gagnep., *Baccaurea pierardi* Wall., *Baccaurea propinqua* Müll.Arg., *Baccaurea sapida* (Roxb.) Müll. Arg., *Baccaurea* wrayi King ex Hook.f., *Gatnaia annamica* Gagnep., *Pierardia flaccida* Wall., *Pierardia sapida* Roxb.

Family

Phyllanthaceae, also placed in Euphorbiaceae

Common/English Names

Baccaurea, Burmese Grape, LanternTree, Mafai.

Vernacular Names

Chinese: Mu Nai Guo;
Bangladesh: Kusumtenga;
Burmese: Kanazo, Kanaso, Krak-Hsu-Ro:Ni;
India: Leteku (Assamese), Lotqua (Bengali), Dojuka (Garo), Lutka, Latka, Kataphal, Lutco (Hindu), Dieng Sohmyndong, Soh Ramdieng (Khasi), Moktok (Manipuri), Pangkai (Mizoram);
Indonesia: Tampoi Kuning, Tampoi Merah (Kalimantan), Mafai Setambun, Tajam Molek, Pupor;
Khmer: Phnhiew;
Laotian: F'ai;
Malaysia: Pupor, Tempui, Tampoi, Tampoi Kuning, Tampoi Merah, Rambai;
Nepal: Kala Bogoti;
Thailand: Mafai (General), Khi Mi (Northern), Sae-Khruea-Sae (Karen-Mae Hong Son), Ham Kang (Phetchabun), Pha-Yio (Khmer-Surin), Mafai Pa (Eastern, South-Eastern), Mafai Ka, Som Fai (Peninsular);
Vietnamese: Dâu Da Đất, Giau Gia Dat, Giau Tien, Dzau Mien Dzu O'i.

Origin/Distribution

Wild distribution of the species occurs in India (Assam), Burma, China (Yunnan, Hainan), Vietnam, Laos, Thailand, Andaman and Nicobar Islands, Peninsular Malaysia. It is commonly cultivated in home backyards in Peninsular Malaysia, Burma and Thailand.

Agroecology

A tropical species, occurring in primary rain forest from 50 to 1,700 m above sea level and cultivated. It grows in well-drained soils derived from sand, and granite. Fruiting throughout the year with flowering in December to June and September.

Edible Plant Parts and Uses

The sweet to sour pulpy arillode of ripe fruit is eaten fresh. To consume the fruit one is advised to break the fruit open with the fingers and/or peel the skin. The pulp is then eaten directly and usually the seeds are also swallowed. The rind of the fruits is occasionally used for making chutney. Squash-making has increased the value of the fruits as the fruit is rich in vitamin C. In Thailand, mafai drink is quite popular. The fruits are also are utilized for making wine. The young tender leaves and flowers are also eaten. The flower is eaten raw in northeast India.

Botany

Evergreen, medium sized tree 5–20 m high with bole diameter of 5–60 cm, gray-brown bark, buttressed trunk and hispid branchlets becoming glabrescent with age. Leaves with raised glands and borne on 3–8 cm, glabrous petiole with stipules 2.5–6 by 1–2.5 mm, caducous (to late caducous), glabrous to sparsely hairy outside. Lamina is obovate-oblong, oblanceolate, or oblong, 9–15 × 3–8 cm, papery, green adaxially, yellowish-green abaxially, glabrous on both surfaces, base cuneate, margin entire or shallowly repand, apex shortly acuminate to acute with 4–9 pairs lateral veins depressed above, raised below. Flowers are small, dioecious, apetalous, many flowered, compound into raceme-like panicles. Staminate inflorescences are densely papillose, often fascicled on branchlets as well as on trunk, to 15 cm with bracts ovate-lanceolate, 2–3 mm, chestnut-yellowish and puberulent outside. Male flowers are yellow with 4 or 5, oblong sepals 5–6 mm and puberulent outside, 4–8 stamens and terete, bipartite pistillode. Pistillate inflorescences to 35 cm long with bracts as in male inflorescence. Female flowers are yellow with 4–6, oblong-lanceolate, 6 mm sepals, puberulent outside, 3-celled, ovoid or globose ovary (Plate 2), densely ferruginous and hispid, very short, ca. 0.5 mm styles and depressed stigma bifid at apex. Fruit is a baccate berry, globose, ovoid to slightly pear-shaped, 2–3.7 × 1.4–3 cm, yellowish (Plate 3), pinkish,

Plate 1 Globose red-fruited variety of *Baccaurea ramiflora*

Plate 2 (**a**, **b**) Fruit with 3-loculed and translucent white, juicy arillode

Plate 3 (**a**, **b**) Pinkish-pale yellowish-fruited Mafai with translucent pinkish-white juicy arillode

purplish to red (Plates 1 and 2) when ripe, indehiscent, 3-celled. Arillode is white to pinkish white, translucent and seeds are flat-elliptic or rotund, 1–1.3 cm with purplish red testa.

Nutritive/Medicinal Properties

The nutrient composition of raw Mafai fruit per 100 g edible portion conducted in Thailand was reported as moisture 88.2 g, energy 48 kcal, protein 0.7 g, fat 0.3 g, carbohydrate 10.5 g, ash 0.3 g, Ca 2 mg, Fe 3.3 mg and vitamin C 55 mg (Puwastein et al. 2000).

Recent studies in China reported that the leaves contain phenols with antioxidant properties (Yang et al. 2007). Seven phenolic compounds including two new ones, 6′- O-vanilloylisotachioside and 6′- O-vanilloyltachioside, exhibited potent antioxidant activities against hydrogen peroxide-induced impairment in PC12 cells, and also exhibited significant DPPH radical-scavenging activities with IC_{50} values of 86.9, 142.9, 15.2, 37.6, 35.9, 30.2, and 79.8 μM, respectively.

Hill tribes in northern Thailand used the bark, root bark and wood in decoction of dried and grinded material as folkloric medicine. In Mizoram, India, the plant called *pangkai* is used for stomach ache, colic and stomach ulcer in traditional ethnomedicine.

Other Uses

The wood is used for furniture, house-posts and cabinet-work. The bark serves as a mordant in dyeing.

Comments

B. ramiflora occurs in both red- and white-fruited varieties.

Selected References

Burkill IH (1966) A dictionary of the economic products of the Malay Peninsula. Revised reprint, 2 vols. Ministry of Agriculture and Co-operatives, Kuala Lumpur. vol 1 (A–H), pp 1–1240, vol 2 (I–Z), pp 1241–2444

Chayamarit K, van Welzen PC (2005) Flora of Thailand, vol 8 Part 1: Euphorbiaceae (Genera A-F). Forest Herbarium, Royal Forest Department, Bangkok, 303pp

Environment and Forest Department Government of Mizoram, India (2009) Medicinal plants in use in Mizoram. http://www.forest.mizoram.gov.in/index.php?option=com_content&task=blogsection&id=21

Foundation for Revitalisation of Local Health Traditions (2008) FRLHT Database. http://envis.frlht.org

Haegens RMAP (2000) Taxonomy, phylogeny, and biogeography of *Baccaurea*, *Distichirhops*, and *Nothobaccaurea* (*Euphorbiaceae*). Blumea Suppl 12, 216

Hu SY (2005) Food plants of China. The Chinese University Press, Hong Kong, 844pp

Kayang H (2007) Tribal knowledge on wild edible plants of Meghalaya, northeast India. Indian J Trad Knowl 6(1):177–181

Li B, Gilbert MG (2008) *Baccaurea* Loureiro. In: Wu ZY, Raven PH, Hong DY (eds) Flora of China, vol 11, Oxalidaceae through Aceraceae. Science Press/Missouri Botanical Garden Press, Beijing/St. Louis

Puwastien P, Burlingame B, Raroengwichit M, Sungpuag P (2000) ASEAN food composition tables 2000. Institute of Nutrition, Mahidol University, Salaya, 157pp

Subhadrabanhdu S (2001) Under utilized tropical fruits of Thailand. FAO Rap publication 2001/26. 70pp

Yang XW, Wang JS, Ma YL, Xiao HT, Zuo YQ, Lin H, He HP, Li L, Hao XJ (2007) Bioactive phenols from the leaves of *Baccaurea ramiflora*. Planta Med 73(13): 1415–1417

Phyllanthus acidus

Scientific Name

Phyllanthus acidus (L.) Skeels

Synonyms

Averrhoa acida L. basionym, *Cicca acida* (L.) Merr., *Cicca acidissima* Blanco, *Cicca disticha* L., *Cicca nodiflora* Lam., *Cicca racemosa* Lour., *Diasperus acidissimus* (Blanco) Kuntze, *Phyllanthus cicca* Müll. Arg., *Phyllanthus cicca* var. bracteosa Müll. Arg., *Phyllanthus cochinchinensis* (Lour.) Müll. Arg., *Phyllanthus distichus* (L.) Merr., *Phyllanthus distichus* f. *nodiflorus* (Lam.) Müll. Arg., *Phyllanthus longifolius* Jacq., *Tricarium cochinchinense* Lour.

Family

Phyllanthaceae, also placed in Euphorbiaceae.

Common/English Names

Country Gooseberry, Indian Gooseberry, Malay Gooseberry, Otaheite Gooseberry, Star Gooseberry, Tahitian Gooseberry, West Indian Gooseberry.

Vernacular Names

Brazil: Groselha (Portuguese);
Burmese: Thinbozihyoo, Mak-Hkam-Sang-Paw;
Chinese: E Mei Shu;
Costa Rica: Grosella (Spanish);
Chamorro: Iba;
Cuba: Cerezo Occidental, Grosella (Spanish);
El Salvador: Ciruela Corteña, Guinda, Pimienta (Spanish);
French: Grosella, Cerisier De Tahiti, Cherimbelier, Phyllanthe Sour, Surelle, Surette De La Martinique;
French West Indies: Groseillier Des Antilles (French);
German: Sternstachelbeerbaum;
Ghana: Dunyan (Huasa);
Guatemala: Grosella (Spanish);
Guinea-Bissau: Azedinha (Crioulo);
India: Chalmeri, Chota Aonla, Harfauri, Harparauri, Harphanevadi, Harpharevadi (Hindi), Rajamvali (Konkani), Gihori (Manipuri), Harparrevdi, Harpharori, Roi-Avala (Marathi), Lavali, Laveni, Pandu, Skandhaphara (Sanskrit), Aranelli (Tamil), Harfarauri (Urdu); Jimbling, Chalmeri, Harpharori, Kirinelli, Hariphal, Rachayusirika;
Indonesia: Ceremoi (Aceh), Cermen (Bali), Careme (Madurese), Ceremai (Malay), Carameng

(Sulawesi), Careme, Cerme (Sundanese) Karsinta, Kemlaka, Kemloko;
Jamaica: Cheramina, Jimbling, Short Jimbelin;
Kampuchea: Kântûet, Kântouot Srôk, Sloek Morom
Laotian: Nhom Baan, Mak-Nhom,
Malaysia: Cermai, Ceremai, Cermela, Camin-camin, Kemangur;
Mexico: Manzana Estrella (Spanish);
Nepalese: Harii Phala, Kaathe Amalaa, Paate Amalaa;
Nicaragua: Grosella (Spanish);
Philippines: Layoan (Bikol), Bangkiling, Poras, Kagindi (Bisaya), Bagbagútut Karmay, Karmai, Karamai, Bagbagutut (Iloko), Iba (Pampangan), Iba, Bangkiling, Karmai (Tagalog);
Portuguese: Cerejeira-Do-Taiti;
Puerto Rico: Cereza Amarilla, Cerezo De La Tierra, Cerezo Comun (Spanish);
Samoan: Vine;
Senegal: Azedina (Crioulo);
Spanish: Grosellero, Guinda;
Tahiti: Surette, Mue;
Thailand: Mayom;
Venezuela: Cerezo Agrio (Spanish);
Vietnam: Chùm Ruôt, Tâm Ruôt.

Origin/Distribution

Phyllanthus acidus is probably native to the coastal region of north-eastern Brazil, and has been frequently wrongly ascribed to Madagascar, India or Polynesia. It is now naturalized and cultivated pan-tropically in India, Thailand, Myanmar, Indonesia, South Vietnam, Laos, Peninsular Malaysia, Polynesia and all the larger islands of the West Indies.

Agroecology

Phyllanthus acidus thrives well in the tropics and sub tropics at low and medium altitudes in places with a short or prolonged dry season. The tree prefers hot, humid tropical lowlands. In north-eastern Brazil, the tree has been found in coastal forest and in Southeast Asia and El Salvador it is cultivated on humid sites, up to 1,000 m elevation. It tolerates a wide range of soils including very sandy soils. It quite cold hardy surviving the cold winter in Tampa, Florida. In the Pacific islands it occurs in open disturbed places, quarries and in home gardens.

Edible Plant Parts and Uses

The mature sour fruits may be eaten fresh but usually they are sprinkled with salt to remove the acidity. The fruits are excellent for processing into pickles (Plate 2) and sweetened dried fruits. In the Philippines, the fruit juice is used in cold drinks and the fruit is also processed into vinegar. In Malaysia, ripe and unripe fruit are cooked and served as a relish, or made into a thick syrup or sweet preserve. Pickled fruits are available in bottles in local markets. The fruits, combined

Plate 1 Cluster of fruit and leaves of star gooseberry

Plate 2 Pickled star gooseberry fruits

with other fruits are used in chutney or jam, because of their gelling properties. In Indonesia, the sour fruits are used as a condiment in cooking to flavour dishes and served as a substitute for *assam* and used as an ingredient in *sambal* and *sayur* or used in *rojak* mixture. Young leaves are eaten as cooked vegetable in Indonesia, Thailand and India.

Studies in Thailand showed that star gooseberry wine can be processed from the fruits by feermenting with the yeast, *Saccharomyces cerevesiae* (Sibounnavong et al. 2010). All the different formulations (different ratios of fruit juice and sugar) gave significantly higher amount of ethyl acohol than all the formulations of carambola wine.

Botany

A small, deciduous, sparingly-branched tree, 4–9 m tall with rough, grey, lenticelled bark and terete, glabrous branchlets clustered at the apex of the branches imparting an open and spreading crown. Stipules are deltoid-ovate, acuminate and 1–1.3 mm long. Leaves are pinnate, 20–40 cm long. Leaflets are alternate, simple, entire, shortly petiolate, broadly ovate to ovate-lanceolate, 4–9 cm × 2–4.5 cm, with obtuse to rounded base, acute apex and 3–9 pairs nerves (Plate 1). The petiole is 2.5 mm long. Flowers are small, pink, axillary in fascicles, with 2–6 staminate flowers in proximal axils; pistillate flowers are cauliflorous on old branchlets, rarely in distal axils. Staminate flowers are 4-merous with free filaments and anthers that dehisce vertically. Pistillate flowers are 4-merous, borne on a stout pedicel, with deeply lobed or split disk, connate, deeply bifid styles, staminodes and a superior, glabrous, 6–8 lobed ovary. Fruit is drupaceous, oblate, 1.5–2.5 cm diameter, shallowly 6- or 8-lobed, greenish–yellow to pale yellow, waxy and glossy and borne on 2–4 mm long pedicels (Plates 1–2). The flesh is firm, thin, sour enclosing a hard, bony, grooved stone containing 6–8 smooth, globose, 5–8 mm diameter seeds. Fruits develop in dense clusters along the old branchlets (Plate1).

Nutritive/Medicinal Properties

The nutritive values of Otaheite gooseberry fruit (per 100 g edible portion) are 28 kcal of energy, 91.7 g moisture, 0.7 g protein, 6.4 g carbohydrate, 0.6 g crude fiber, 5 mg calcium, 23 mg phosphorous, 0.4 mg iron, 0.01 mg thiamin, 0.05 mg riboflavin and 8 mg vitamin C (Ministry of Public Health, Thailand 1970).

Seventy-seven compounds: 45 terpenes, 18 esters, 7 acids, 4 aldehydes 2 phenols, 1 alcohol were identified from the volatile components of star gooseberry fruit (Pino et al. 2008). The total concentration of volatiles was 109 mg/kg fresh fruit, terpenes 100.1 mg/kg and acids 6.7 mg/kg. Among the terpenes many monoterpenes and sesquiterpenes were identified, chief constituents were by epi-α-muurolol (32.9 mg/kg) and α-cardinol (22.1 mg/kg). Hexadecanoic acid (3.8 mg/kg) was the predominant acid.

From the bark of *Phyllanthus acidus* the pentacyclic triterpenoids, phyllanthol and olean-12en-3β-ol (β-amyrin) were isolated (Sengupta and Mukhopadhyay 1996).

Some of the reported pharmacological activities of *P. acidus* plant parts are presented below.

Hepatoprotective and Antioxidant Activities

Oral administration of *Phyllanthus acidus* methanolic extracts to rats in rats with acute liver damage induced by carbon tetrachloride CCl(4) attenuated CCl(4)-induced increase in serum glutamate-oxalate-transaminase (GOT) and CCl(4)-induced increase in serum glutamate-pyruvate-transaminase (GPT) (Lee et al. 2006). Concurrently, the extract elevated the activity of liver reduced glutathione peroxidase (GSH-Px). The protective effects of *P. acidus* extract correlated with a reduction in liver infiltration and focal necrosis. These data demonstrated that *P. acidus* had hepatoprotective and antioxidant activities. Recent studies showed that *P. acidus* extracts and silymarin exhibited significant hepatoprotective effect against CCl4-induced oxidative damage (Jain et al. 2011). This was evident from the decreases of serum aspartate transaminase (AST), alanine transaminase (ALT), alkaline phosphatase (ALP), levels and as lipid peroxidation and increases in the levels of total protein, reduced glutathione (GSH), superoxide dismutase (SOD), catalase (CAT) and glutathione peroxidase (GPx) compared with control group. The biochemical results were substantiated with results of histopathological sections of the liver tissues. *P. acidus* extracts considerably shortened the duration of hexobarbitone-induced sleeping time in mice compared with control group and displayed remarkable DPPH-scavenging activity. The findings suggested that the hepatoprotective effect of *P. acidus* against CCl4-induced oxidative damage may be related to its antioxidant and free radical-scavenging potentials. The results of another study in wistar rats suggested that the aqueous extract of *P. acidus* leaves had significant hepatoprotective activity on acetaminophen and thioacetamide induced hepatotoxicity, which may be related to its high phenolic and flavonoid content and antioxidant properties (Jain and Singhai 2011). Acetaminophen and thioacetamide administration caused severe hepatic damage in rats as evident from significant rise in serum AST, ALT, ALP, total bilirubin and concurrent depletion in total serum protein. *P. acidus* extracts and silymarin prevented the toxic effects of acetaminophen and thioacetamide on the above serum parameters. The aqueous extract was found to be more potent than the corresponding ethanolic extract against both toxicants. The phenolic and flavonoid content (175.02 and 74.68 µg/ml, respectively) and 2,2-diphenyl-1-picrylhydrazil (DPPH) [IC_{50} = (33.2) µg/ml] scavenging potential was found maximum with aqueous extract as compared to ethanolic extract.

Anticancer Activity

Two novel water-soluble norbisabolane glycosides, phyllanthusol A and phyllanthusol B, isolated from the methanol extract of the roots of *P. acidus* were found to exhibit cytotoxic activity (Vongvanich et al. 2000). Phyllanthusols A (1) and B (2) exhibited cytotoxicity against BC (Breast cancer line) with EC_{50} at 4.2 and 4.0 µg/ml for 1 and 2, respectively and human carcinoma cell line KB with EC_{50} at 14.6 and 8.9 µg/ml for 1 and 2, respectively. KB is the cell line derived from a human carcinoma of the nasopharynx.

Anticystic Fibrosis Activity

Studies reported that extracts of *Phyllanthus acidus* had promising potential in treating cystic fibrosis (CF) (Sousa et al. 2007). *P. acidus* extract and co-application of its isolated components, adenosine, kaempferol and DHBA (2,3-dihydroxybenzoic acid) had similar activating effects on ion transport in mouse trachea. The herbal extract corrected defective electrolyte transport in cystic fibrosis airways by various parallel mechanisms including (1) increasing the intracellular levels of second messengers cAMP and Ca^{2+}, thereby activating Ca^{2+}-dependent Cl^- channels and residual CFTR-Cl^- conductance; (2) stimulating basolateral K^+ channels; (3) redistributing cellular localization of CFTR; (4) directly activating CFTR; and (5) inhibiting ENaC through activation of CFTR. These combinatorial effects on epithelial transport may provide a novel complementary nutraceutical treatment for the cystic fibrosis lung disease.

Hypotensive Activity

Recent studies showed that *P. acidus* leaf extract had hypotensive activity (Leeya et al. 2010). The hypotensive activity was attributed to the direct action of adenosine, 4-hydroxybenzoic acid, caffeic acid, hypogallic acid, and kaempferol were isolated from the n-butanol leaf extract. These five compounds had a direct action on the blood vessels of anesthitized rats by stimulating release of nitric oxide from the vascular endothelium, in part through stimulation of soluble guanylate cyclase, and opening of K(ATP) and K(Ca) channels in the vascular smooth muscle.

Antimicrobial Activity

Studies showed that *Phyllanthus acidus* leaf extract exhibited antimicrobial activity against *Eschericia coli, Staphylococcus aureus* and *Candida albicans* (Jagessar et al. 2008). The antimicrobial activity was selective and solvent dependent with the ethanolic extract, the most potent and hexane the least. In general, the order of antimicrobial activity followed the sequence: CH_3CH_2OH extract > EtOAc extract > CH_2Cl_2 extract > hexane extract. In another study, methanolic extracts of *Phyllanthus acidus* were also reported to possess strong in-vitro antibacterial activity against *Escherichia coli* and *Staphylococcus aureus* (Meléndez and Capriles 2006).

Antiplasmodial and Anticancer Activities

Recent studies showed that the leaf ethyl acetate, acetone, and methanol extracts of *P. acidus* exhibited good antiplasmodial activity with IC_{50} values of 9.37, 14.65, 12.68 µg/mL respectively and with selectivity indices of 4.88, 3.35, 3.42 for human laryngeal cancer cell line (HEp-2) and >11.75, >3.41, >3.94 for vero cells respectively (Bagavan et al. 2011).

Traditional Medicinal Uses

Phyllanthus acidus has been used in traditional ethnomedicine in Asia. The fruit is acid and astringent. The acrid latex of various parts of the tree is emetic and purgative. In India, the fruits are taken as liver tonic, to enrich the blood. The syrup is prescribed as a stomachic; and the seeds are cathartic. The decoction of the leaves is good diaphoretic and is also used as a demulcent in cases of gonorrhoea. The leaves are mucilaginous and demulcent and are given in gonorrhoea and are also administered as a sudorific. In the Philippines, a decoction of the leaves is applied to urticaria, and the fruit, which is astringent, is given at the same time, to eat. The bark yields a decoction, which is employed in bronchial catarrh. The root is drastically purgative and regarded as toxic in Malaya but is boiled and the steam inhaled to relieve coughs and headache. In Borneo, leaves are used, with pepper, for poulticing for lumbago, or sciatica, and the root is used externally to treat psoriasis of the soles of feet. In Java, the root infusion is taken in very small doses to alleviate asthma and the bark is heated with coconut oil and spread on eruptions on feet and hands. A leaf decoction is employed for urticaria. In Thailand, the extract from the root to cure skin diseases especially relief from itching. Leaves are used as one of the ingredients in Thai medicine to control fever. The juice of the root bark, which contains saponin, gallic acid, tannin and a crystalline substance which may be lupeol, has been employed in criminal poisoning. It was reported to produce headaches, sleepiness, and deaths accompanied by severe abdominal pains.

Other Uses

Bark extracts have nematicidal activity against the pine wood nematode, *Bursaphelenchus xylophilus* (Mackeen et al. 1997). The bark is used in India as a tanning agent. The wood is light-brown, fine-grained, attractive, fairly hard, strong, tough, durable if seasoned and used for utensils and other small objects. The tree also provides fuel-wood.

Comments

Star gooseberry is usually grown from seed but may also be propagated by budding, greenwood cuttings, or air-layers.

Selected References

Bagavan A, Rahuman AA, Kamaraj C, Kaushik NK, Mohanakrishnan D, Sahal D (2011) Antiplasmodial activity of botanical extracts against *Plasmodium falciparum*. Parasitol Res 108(5):1099–1109

Brown WH (1951–1957) Useful plants of the Philippines. Reprint of the 1941–1943 edn. 3 vols. Technical Bulletin 10. Department of Agriculture and Natural

Resources. Bureau of Printing, Manila. vol 1 (1951), 590pp, vol 2 (1954), 513pp, vol 3 (1957), 507pp

Burkill IH (1966) A dictionary of the economic products of the Malay Peninsula. Revised reprint, 2 vols. Ministry of Agriculture and Co-operatives, Kuala Lumpur. vol 1 (A–H), pp 1–1240, vol 2 (I–Z), pp 1241–2444

Council of Scientific and Industrial Research (CSIR) (1950) The wealth of India: a dictionary of Indian raw materials and industrial products, vol 2, Raw materials. Publications and Information Directorate, New Delhi

Jagessar RC, Mars A, Gomes G (2008) Selective antimicrobial properties of *Phyllanthus acidus* leaf extract against *Candida albicans, Escherichia coli* and *Staphylococcus aureus* using Stokes disc diffusion, well diffusion, streak plate and a dilution method. Nat Sci 6(2):24–38

Jain NK, Lodhi S, Jain A, Nahata A, Singhai AK (2011) Effects of *Phyllanthus acidus* (L.) Skeels fruit on carbon tetrachloride-induced acute oxidative damage in livers of rats and mice. Zhong Xi Yi Jie He Xue Bao 9(1):49–56

Jain NK, Singhai AK (2011) Protective effects of *Phyllanthus acidus* (L.) Skeels leaf extracts on acetaminophen and thioacetamide induced hepatic injuries in Wistar rats. Asian Pac J Trop Med 4(6):470–474

Lee CY, Peng WH, Cheng HY, Chen FN, Lai MT, Chiu TH (2006) Hepatoprotective effect of *Phyllanthus* in Taiwan on acute liver damage induced by carbon tetrachloride. Am J Chin Med 34(3):471–482

Leeya Y, Mulvany MJ, Queiroz EF, Marston A, Hostettmann K, Jansakul C (2010) Hypotensive activity of an n-butanol extract and their purified compounds from leaves of *Phyllanthus acidus* (L.) Skeels in rats. Eur J Pharmacol 649(1–3):301–313

Mackeen MM, Ali AM, Abdullah MA, Nasir RM, Mat NB, Razak AR, Kawazu K (1997) Antinematodal activity of some Malaysian plant extracts against the pine wood nematode, *Bursaphelenchus xylophilus*. Pest Manag Sci 51(2):165–170

Meléndez PA, Capriles VA (2006) Antibacterial properties of tropical plants from Puerto Rico. Phytomedicine 13(4):272–276

Ministry of Public Health (1970) Tables of nutrition values in Thai food per 100 gm of edible portion. Office of the Prime Minister, Royal Thai Government, Bangkok

Molesworth Allen B (1967) Malayan fruits: an Introduction to the cultivated species. Moore, Singapore, 245pp

Morton JF (1987) Otaheite gosseberry. In: Fruits of warm climates. Julia F. Morton, Miami, pp 217–219

Nguyen VD (1993) Medicinal plants of Vietnam. Self Publ, Cambodia and Laos, 528pp

Ochse JJ (1927) Indische Vruchten. Volkslectuur, Weltevreden, 330pp

Ochse JJ, Soule MJ Jr, Dijkman MJ, Wehlburg C (1961) Tropical and subtropical agriculture. 2 vols. Macmillan, New York. 1446pp

Pacific Island Ecosystems at Risk (PIER) (2006) *Phyllanthus acidus* (L.) Skeels, Euphorbiaceae. http://www.hear.org/pier/species/phyllanthus_acidus.htm

Pino JA, Cuevas-Glory LF, Marbot R, Feuntes V (2008) Volatile compounds of grosella (*Phyllanthus acidus* (L.) Skeels) fruit. Rev Cenic Cienc Quim 39(1):3–5

Purseglove JW (1968) Tropical crops: dicotyledons, vol 1 and 2. Longman, London, 719pp

Sengupta P, Mukhopadhyay J (1996) Terpenoids and related compounds—VII: Triterpenoids of *Phyllanthus acidus* Skeels. Phytochemistry 5(3):531–534

Sibounnavong P, Daungpanya S, Sidtiphanthong S, Keoudone C, Sayavong M (2010) Application of *Saccharomyces cerevisiae* for wine production from star gooseberry and carambola. J Agric Tech 6(1):99–105

Sousa M, Ousingsawat J, Seitz R, Puntheeranurak S, Regalado A, Schmidt A, Grego T, Jansakul C, Amaral MD, Schreiber R, Kunzelmann K (2007) An extract from the medicinal plant *Phyllanthus acidus* and its isolated compounds induce airway chloride secretion: a potential treatment for cystic fibrosis. Mol Pharmacol 71(1):366–376

Subhadrabanhdu S (2001) Under utilized tropical fruits of Thailand. FAO Rap publication 2001/26. 70pp

Vongvanich N, Kittakoop P, Kramyu J, Tanticharoen M, Thebtaranonth Y (2000) Phyllanthusols A and B, cytotoxic norbisabolane glycosides from *Phyllanthus acidus* Skeels. J Org Chem 65(17):5420–5423

Phyllanthus emblica

Scientific Name

Phyllanthus emblica L.

Synonyms

Cicca emblica Kurz, *Diasperus emblica* (L.) Kuntze, *Dichelactina nodicaulis* Hance, *Emblica arborea* Raf., *Emblica officinalis* Gaertn., *Mirobalanus emblica* Burm., *Phyllanthus glomeratus* Roxb. ex Wall., *Phyllanthus mairei* H. Lév., *Phyllanthus mimosifolius* Salisb.

Family

Phyllanthaceae, also placed in Euphorbiaceae.

Common/English Names

Amla, Emblic, Indian Gooseberry, Malacca Tree, Myrobalan.

Vernacular Names

Arabic: Amlag, Amlaj, As Sanânir;
Burmese: Mai Kham, Ziphiyu-si, Shabju;
Chinese: An Mo Le, Yu Gan Zi;
Danish: Grå Myrobalan;
Dutch: Kembaka Mylobalanen Baum;
French: Emblique Officinale, Groseillier De Ceylan, Myrobalan Emblic;
German: Amalanbaum, Ambla-Baum, Graue Myrobalane, Myrobalanenbaum;
India: Amluki (Assam), Amalaki (Ayurvedic), Amia, Aamloki, Amloki (Bengali), Ambala, Amla (Gujarati), Aamla, Amalaci, Amalak, Amla, Amlika, Amvala, Anola, Anuli, Anvula, Anvurah, Anwerd, Anwiya, Aonla, Aoula, Auna, Aungra, Aunra, Anwala, Aola, Awala, Daula, (Hindu), Aamalaka, Aamalakee, Amalaka, Ambla, Betta Nelli, Bettada Nelli, Bettada Nelli Kaayi, Cattu, Chattu, Chattu Daadi Kaayi, Dadi, Dhanya, Dhatri, Dodda Nelli, Nelli, Nellikai, Nellka, Nelli Chattu, Nilika, Perunelli, Sudhe (Kannada), Amalakam, Nelli, Nellikka, Nellikai, Nellikkaya, Nellimaram, Nilica, Nilicamaram, Nilika (Malayalam), Amla (Manipuri), Aamlee, Amla, Anvala, Aonli, Avala, Arola, Bhuiawali (Marathi), Sinhlu (Mizoram), Aonla, Aula (Oriya), Ambli, Ambul, Ambal, Amla (Punjabi), Adiphala, Akara, Amalaka, Amalakam, Amalaki, Amamalakam, Amlika, Amraphala, Amrita, Amritaphala, Amruthaphala, Bahuphali, Dhatri, Dhatrika, Dhatriphala, Jatiphala, Karshaphala, Kayastha, Pancharasa, Parvakeeta, Rochani, Shadarasa, Shanta, Shiva, Shriphala, Shriphali, Tishya, Triphala, Vayastha, Vrishya, Vrittaphala (Sanskrit), Alakam, Alikam, Amalagam, Amalai, Amalakam,

Amalakamalam, Amalaki, Amalakki, Amalam, Amarikam, Amirtai, Amirtapalai, Amirtapalam, Ampal, Ampalkamaram, Amukam, Amutai, Amuttam, Anantai, Andakaram, Antakaram, Antakaramaram, Antakolam, Antakolam, Antakoram, Antor, Atikoram, Attakolam, Attakoram, Avutanci, Cilata, Cinkitakam, Cinkitakamaram, Ciripalam, Cirottam, Citupalam, Civai, Civam, Indul, Intul, Intulam, Intulapam, Intuli A, Intuli X, Intuli, Iratai, Kaiyirkani, Kanakam, Kantamulakam, Kantattiri, Kantumalalam, Kapi, Kapitattiri, Kapitattirimaram, Kattunelli, Kayastam, Kokkam, Konkal, Konkam, Korankam, Korankamikam, Korankamikamaram, Kotam, Kotimukakkini, Kotimukakkinimaram, Makanti, Makantika, Makantikam, Malainelli, Mamalakam, Mamalakkay, Mankantikamaram, Mintu, Miruntikam, Miruntikamaram, Miruntu, Mirutupala, Mirutupala, Mirutupalamaram, Mitintu, Mituntumaram, Nelli, Nelli-Kay, Nellie Kai, Nellikai, Nellikkay, Nellikkai, Nellie Poo, Pancatcaravaruni, Parainatavanci, Parainatavancimaram, Perunelli, Tamalaikkay, Tantiri, Tattari, Tattili, Tattiri, Tiripalati,, Tattinimaram, Tattirimaram, Tattiripalai, Tattiripalam, Tecomantaram, Tecomantiram, Tecomaram, Tiriciyam, Tiriciyapalam, Toppi, Toppunelli, Totti, Tottiki, Tottikimaram, Tuttarikam, Tuppunelli, Vanamalakam, Yankoram, Yantikoram (Tamil), Aamalakamu, Aamalaki, Amalakamu, Amalaki, Asereki, Assereki, Nelli, Nelli Kaaya, Oosree, Pedda, Peddavusirika, Pullayusirika, Triphalam, Usari, Usarika, Userakee, Useri, Userikai, Usiri, Usirika, Usirikaaya, Usirikaya, Usiriki, Usrikayi, Usiuka, Usri, Vusirika, Vusirikaya (Telugu), Amia (Unani), Aamla, Amla, Aonla (Urdu), Amla (Uriya);
Indonesia: Kemlaka (Java), Mlakah (Madurese), Kimalaka, Malaka (Sundanese), Balaka, Balangka (Sumatra);
Khmer: Kan Tot, Kam Lam, Kam Lam Ko, Kântouot Préi;
Laos: Mak Kham Pom;
Malaysia: Kik (Semang), Melaka, Asam Melaka, Amlaka, Kayu Laka, Laka-Laka (Malay);
Nepalese: Amalaa, Rikhiya;
Persian: Amla, Amelah, Ameleh, Amuleh;
Philippines: Nelli;
Russian: Fillantus Emblika;
Spanish: Mirobalano, Neli;
Sri Lanka: Ambula, Awusada-nelli, Nelli, Nellika (Sinhalese);
Thailand: Ma-Khaam Pom (General), Kan-Tot (Chanthaburi), Kam-Thuat (Ratchaburi), Mang-Luu And San-Yaa-Saa (Karen-Mae Hong Son);
Tibetan: Kyou-rhoo-rah;
Vietnamese: Chùm Ruôt Nui, Kam Lam, Me Rùng.

Origin/Distribution

Phyllanthus emblica is indigenous to Nepal, India and Sri Lanka, throughout South-East Asia to southern China. It is widely cultivated for its fruits throughout its natural area of distribution, particularly in India, and also in the Mascarene Islands (Réunion, Mauritius), the West Indies (Cuba, Trinidad), Central America (Honduras, Costa Rica), and the West Indies and Japan. Formerly it was cultivated in Madagascar. It is commonly cultivated in home gardens in India, Malaysia, Singapore and southern China.

Agroecology

Indian gooseberry grows in tropical and subtropical regions from near sea-level to 1,500 m altitude. It grows equally well in arid and wet or humid conditions. It has been reported to thrive in areas too dry and on soil too poor for most other fruit crops. It is light demanding plant, common in grassy areas, brush and village groves. It is photosensitive, only producing flowers at a day-length between 12 and 13.5 h. The species is not fastidious of soil type and grows on a wide range of soil types ranging from sandy loam to clay, light or heavy, and slightly acidic to slightly alkaline. In a highly alkaline soil (pH 8.0) nutritional deficiencies are common. It flourishes in deep, fertile soil. It is moderately drought resistant but some cultivars may be sensitive to drought and frost. It is fire-resistant and is one of the first trees to recover after a fire. The tree is rather slow-growing, bearing fruit after 5–8 years.

Edible Plant Parts and Uses

Emblic fruit is extremely rich in vitamin C and edible fresh, cooked, dried or pickled. In Andra Pradesh, the fruit is cooked in various cuisines in a dhal (lentil) preparation and in a sweet dish *amle ka murabbah* indigenous to the northern part of India (wherein the berries are soaked in sugar syrup for a long time) which is traditionally consumed after meals. Fresh fruits are also baked in tarts. It is a common practice for Indian housewives to cook the fruits with sugar and saffron and give one or two to a child every morning. In Indonesia and Malaysia, the fruit is used to as sour seasoning in sambal and sayor in place of tamarind pulp. Ripe or half-ripe fruits are used to make sweetmeats, jams, jellies, "emblic myrobalan" fruit preserves, candies, emblic powder, chutney, pickles and relishes. Emblic preserves are manufactured and marketed in large quantities in India. Preserved amblics are available in enamelled cans or crystallised as a confection. Emblic flesh can be dried and made into chips. The fruit juice is also used to flavour vinegar. In Thailand, the fruit is widely used by local Thais to quench the thirst when walking in the forest. In China, phyllanthus drink prepared from fruit extract is commonly known, and wine made from fruit extract is sold in the market.

Plate 1 Ripe emblic fruits

Plate 2 Unripe emblic fruits and leaves

Botany

A small to medium-sized, low, deciduous tree, 5–25 m tall; trunk often crooked and gnarled, up to 35 cm in diameter; bark thin, smooth, grey, peeling in patches, with numerous knobs, main branches angular and densely pubescent. Leaves on the main branches are deltoid-squamiform, resembling stipules. The branchlets are glabrous or finely pubescent, 10–20 cm long, usually deciduous bearing biseriate alternate, thin, simple, light green, subsessile, closely-set leaves, resembling pinnate leaves with narrowly oblong, 5–25 mm × 1–5 mm lamina, rounded and more or less oblique base, acute or obtuse and mucronate apex and pinnatinerved (Plates 1 and 2). Flowers are greenish-yellow, fascicled in axils of leaves or fallen leaves, unisexual. The male flowers are numerous at base of young twigs, the female flowers solitary and higher up the twig. Male flowers are shortly pedicellate, with six pale green perianth lobes 1.5–2.5 mm long and three entirely connate stamens, six disk glands alternating with the perianth lobes; female flowers sessile, with six somewhat larger perianth lobes, a cup-like six-angular disk, and a superior three-celled, glabrous ovary crowned by three styles, connate for more than half of their length and deeply bifid at apex. Fruit a depressed globose drupe 2–4 cm in diameter, pale green changing to yellow when mature with six vertical striations (Plates 1 and 2). The fruit skin is thin, translucent and adherent to the very crisp, juicy, concolorous flesh. A slightly hexagonal stone is tightly embedded in the centre of the flesh; stone has three slightly dehiscent compartments, each usually containing two trigonous, 4–5 mm × 2–3 mm seeds.

Nutritive/Medicinal Properties

Nutrient composition of raw fresh emblic fruit per 100 g edible portion was reported as: moisture 81.8 g, energy 58 kcal, protein 0.5 g, fat 0.1 g, carbohydrate 13.7 g, fibre 3.4 g, ash 0.5 g, total carotene 9 ug, vitamin C 600 mg, thiamin 0.03 mg, riboflavin 0.01 mg, niacin 0.2 mg, iron 1.2 mg, calcium 50 mg and phosphorus 20 mg (Gopalan et al. 2002). Another analysis of the fruit conducted in the Finlay Institute Laboratory, Havana (Morton 1960) per 100 g of edible portion reported the following: moisture 77.1 g, protein 0.07 g, fat 0.2 g, carbohydrates 21.8 g, fibre 1.9 g, ash 0.5 g, calcium 12.5 mg, phosphorous 26.0 mg, iron 0.48 mg, carotene 0.01 mg, thiamine 0.03 mg, riboflavin 0.05 mg, niacin 0.18 mg, tryptophan 3.0 mg, methionine 2.0 mg, lysine 17.0 mg and ascorbic acid 625 mg. Analysis conducted in Australia by Brand Miller et al. (1993) reported the following nutrient composition: energy 54 kJ, moisture 78.4 g, nitrogen 0.1 g, protein 0.6 g, fat 0.3 g, ash 0.4 g, total dietary fibre 4.1 g, Cu 0.1 mg, Fe 0.9 mg, Mg 8 mg, K 215 mg, Zn 0.5 mg, niacin equivalent 0.1 mg and vitamin C 316 mg.

The emblic is highly nutritious, and could be an important dietary source of vitamin C, minerals and amino acids (Barthakur and Arnold 1991). The edible fruit tissue of the emblic contained about three times as much protein and 160 times as much ascorbic acid as apples (*Malus pumila* Mill.). The fruits also contained considerably higher concentrations of most minerals and amino acids than apples. Glutamic acid, proline, aspartic acid, alanine and lysine constituted 29.6, 14.6, 8.1, 5.4 and 5.3%, respectively, of the total amino acids. The concentration of each amino acid, except cystine, was much higher than in apples. Six new phenolic constituents, L-malic acid 2-O-gallate (1), mucic acid 2-O-gallate (5), mucic acid 1,4-lactone 2-O-gallate (6), 5-O-gallate (8), 3-O-gallate (10), and 3,5-di-O-gallate (11), were isolated from the fruit juice of *Phyllanthus emblica* together with their methyl esters (2–4, 7, 9) (Zhang et al. 2001c). Compounds 5, 6, and 8 were the major phenolic constituents of the juice.

Six new ellagitannins, phyllanemblinins A–F (1–6), were isolated from *Phyllanthus emblica* (Zhang et al. 2001a). Phyllanemblinins A (1) and B (2) were confirmed to be ellagitannins having a tetrahydroxybenzofuran dicarboxyl group and a hexahydroxydiphenoyl group, respectively. Phyllanemblinin C (3) with a new acyl group at the glucose 2, 4-positions was found to be structurally related to chebulagic acid. Phyllanemblinins D (4), E (5), and F (6) were found to be positional isomers of neochebuloyl 1(β)-O-galloylglucose. Phytochemical investigations reported the isolation of the two new flavonoids, kaempferol-3-O-α-L-(6″-methyl)-rhamnopyranoside and kaempferol-3-O-α-L-(6″-ethyl)-rhamnopyranoside from *Phyllanthus emblica* (Rehman et al. 2007). Eleven compounds were isolated from *Phyllanthus emblica* and identified as gallic acid, ellagic acid, 1-O-galloyl-β D-glucose, 3, 6-di-O-galloyl-D-glucose, chebulinic acid, quercetin, chebulagic acid, corilagin, 3-ethylgollic acid (3-ethoxy-4, 5-dihydroxy-benzoic acid), isostrictiniin, and 1, 6-di-O-galloyl-β-D-glucose (Zhang et al. 2003). Two acylated flavanone glycosides,(S)-eriodictyol 7-O-(6″-O-trans-p-coumaroyl)-β-D-glucopyranoside (1) and (S)-eriodictyol 7-O-(6″-O-galloyl)-β-D-glucopyranoside (2) were isolated from the leaves and branches of *Phyllanthus emblica* together with a new phenolic glycoside, 2-(2-methylbutyryl) phloroglucinol 1-O-(6″-O-β-D-apiofuranosyl)-β-D-glucopyranoside (3), as well as 22 known compounds (Zhang et al. 2002).

From the roots of *Phyllanthus emblica* were isolated: three ester glycosides, named phyllaemblicins A, B, and C, and a methyl ester of a highly oxygenated norbisabolane, phyllaemblic acid, along with 15 tannins and related compounds (Zhang et al. 2000a); phyllaemblic acid, a highly oxygenated norbisabolane (Zhang et al. 2000b); three bisabolane-type sesquiterpenoids, phyllaemblic acids B and C and phyllaemblicin D, together with two new phenolic glycosides, 2-carboxylmethylphenol 1-O-β-D-glucopyranoside and 2,6-dimethoxy-4-(2-hydroxyethyl)phenol 1-O-β-D-glucopyranoside, (Zhang et al. 2001b).

The review by Calixto et al. (1998), supported the experience of traditional medicine that

P. emblica and other *Phyllanthus* species might have beneficial therapeutic actions in the management of certain disorders such as inflammatory reactions, intestinal problems, genitourinary, hepatic antiviral disorders, and as antinociceptive agents. Several compounds including alkaloids, flavonoids, lignans, phenols, and terpenes were isolated from these plants and some of them interacted with most key enzymes.

In a review on the pharmacological properties of *Emblica officinalis*, Baliga and Dsouza (2011) reported that numerous preclinical studies had shown that amla possessed antipyretic, analgesic, antitussive, antiatherogenic, adaptogenic, cardioprotective, gastroprotective, antianemia, antihypercholesterolemia, wound healing, antidiarrheal, antiatherosclerotic, hepatoprotective, nephroprotective, and neuroprotective properties. Further, experimental studies had shown that amla and some of its phytochemicals such as gallic acid, ellagic acid, pyrogallol, some norsesquiterpenoids, corilagin, geraniin, elaeocarpusin, and prodelphinidins B1 and B2 also possessed antineoplastic effects. Additionally, Amla had also been reported to possess radiomodulatory, chemomodulatory, chemopreventive effects, free radical scavenging, antioxidant, antiinflammatory, antimutagenic and immunomodulatory activities, properties efficacious for the treatment and prevention of cancer.

Antioxidant Activity

Phyllanthus emblica was one of several Nepalese fruits that exhibited potent antioxidant activity as compared to vitamin C using the 1,1-diphenyl-2-picryl hydrazyl radical assay (Chalise et al. 2010). Aqueous extract of *Emblica officinalis* was found to be a potent inhibitor of lipid peroxide production and scavenger of hydroxyl and superoxide radicals in vitro (Jose and Kuttan 1995). The amount of the fresh pulp needed to inhibit 50% of lipid peroxidation was 1,000 µg/ml in the Fe^{2+}/ascorbate system and 640 µg/ml in the Fe^{3+}/ADP-ascorbate assay. The concentration needed for 50% inhibition of superoxide-scavenging activity was 107 µg/ml in the photoreduction of riboflavin assay and 232 µg/ml in the xanthine-xanthine oxidase assay, and that for hydroxyl radical scavenging (deoxyribose degradation asssay) was 3,400 µg/ml. An indigenous drug preparation containing *Emblica officinalis*, used in India as a health tonic, was also found to have potent antioxidant activity.

The active tannoids of *Emblica officinalis* (EOT) consisting of emblicanin A (37%), emblicanin B (33%), punigluconin (12%) and pedunculagin (14%), administered in doses of 5 and 10 mg/kg, i.p., and deprenyl (2 mg/kg, i.p.), induced an increase in both frontal cortical and striatal oxidative free radical scavenging enzymes, superoxide dismutase (SOD), catalase (CAT) and glutathione peroxidase (GPX), activity, with concomitant decrease in lipid peroxidation in these brain areas when administered once daily for 7 days (Bhattacharya et al. 1999). Acute single administration of EOT and deprenyl had insignificant effects. The results also indicated that the antioxidant activity of *E. officinalis* may reside in the tannoids of the fruits of the plant, which had vitamin C-like properties, rather than vitamin C itself.

Studies showed that the amla extract acted as a very good antioxidant against γ-radiation induced lipid peroxidation (LPO) (Khopde et al. 2001). Similarly, it was found to inhibit the damage to antioxidant enzyme superoxide dismutase (SOD). The antioxidant activity of the amla extract was found to be both dose- and concentration-dependent. The amount of ascorbic acid in amla was found to be 3.25–4.5% w/w. However, no inhibition of LPO was observed in microsomes. Cyclic voltammetry of the amla extract was carried out to estimate the ascorbic acid equivalents, which was found to be 9.4% w/w of amla. This value was found to be in agreement when compared with the reactivity of both amla and ascorbic acid towards ABTS radical, a stable free-radical. Based on these results it was concluded that amla was a more potent antioxidant than vitamin C.

Total phenolic content in emblica fruit from six regions in China was found to range from 81.5 to 120.9 mg gallic acid equivalents (GAE)/g and the flavonoid content s varied from 20.3 to 38.7 mg quecetin equivalents (QE)/g, while proanthocyanidin content ranged from 3.7 to 18.7 mg catechin equivalents (CE)/g (Liu et al. 2008). Among all the methanolic extracts analyzed, the Huizhou sample exhibited a significantly higher phenolic content than other samples and was found to have the strongest antioxidant activities in scavenging DPPH radicals, superoxide anion radicals, and had the highest reducing power. The Chuxiong sample showed the best performance in chelating iron and in inhibiting lipid peroxidation and exhibited a stronger inhibition activity of the hydroxyl radicals compared with other samples. High correlation coefficient was found between the phenolic content and superoxide anion radical scavenging activity, but no significant correlation was found between the former and hydroxyl radical scavenging activity. Methanolic extracts of emblica fruit from some selected regions exhibited stronger antioxidant activities compared to those of the commercial antioxidant compounds (quercetin and BHA).

In a recent study, all four commercial *E. officinalis* fruit extracts produced positive responses in the total phenol, total flavonoid and total tannin assays (Poltanov et al. 2009). The presence of predominantly (poly) phenolic analytes, e.g. ellagic and gallic acids and corilagin, was confirmed. Despite ascorbic acid being a major constituent of *E. officinalis* fruits, the furanolactone could not be identified in one of the samples. The extracts demonstrated varying degrees of antioxidative efficacy. The extract designated IG-3 was consistently amongst the most effective extracts in the iron(III) reduction and 1,1-diphenyl-2-picrylhydrazyl and superoxide anion radical scavenging assays while the extract designated IG-1 demonstrated the best hydroxyl radical scavenging activity. All extracts appeared to be incapable of chelating iron(II) at realistic concentrations.

Among methanol, hexane, ethyl acetate, and water fractions of dried fruit rind of *P. emblica*, only the ethyl acetate phase showed strong nitric oxide (NO) scavenging activity in-vitro, when compared with the water and hexane phases (Kumaran and Karunakaran 2006). From the ethyl acetate fraction five compounds showing strong NO scavenging activity were identified to be gallic acid, methyl gallate, corilagin, furosin, and geraniin. Gallic acid was found to be a major compound in the ethyl acetate extract and geraniin showed highest NO scavenging activity among the isolated compounds. In another study, free gallic and ellagic acids and emblicanins A and B in the *E. officinalis* extract were separated by TLC and identified and were found capable of scavenging of DPPH radicals (Pozharitskaya et al. 2007). It was established that the DPPH scavenging activity of emblicanins A and B was 7.86 and 11.20 times more than that of ascorbic acid and 1.25 and 1.78 times more than gallic acid, respectively. From the estimated ID_{50} values, it was observed that the increasing order of activity was emblicanin B > emblicanin A > gallic acid > ellagic acid > ascorbic acid.

Amalki was found to be rich in total phenolic content, vitamin E like activity, reducing power and antioxidant activity (Shukla et al. 2009). Total antioxidant activity of aqueous extract of amalki, at 1 mg/ml concentration was 7.78. At similar concentrations the total antioxidant activity of alcoholic extract of amalaki, was 6.67 mmol/l respectively. Amalki was also found to be rich in phenolic compounds (241 mg/g gallic acid equivalent). Both aqueous and alcoholic extracts of amalaki dose-dependently inhibited activity of rat liver glutathione S-transferase (GST) in-vitro. The aqueous extracts of amalki also showed protection against t-BOOH induced cytotoxicity and production of ROS in cultured C6 glial cells.

In a study with the electron spin resonance (ESR) spin-trapping method on large-scale screening with ethanol extracts of approximately 1,000 kinds of herbs, four herbal extracts *Punica granatum* (peel), *Syzygium aromaticum* (bud), *Mangifera indica* (kernel) and *Phyllanthus emblica* (fruit) were found to have prominently potent ability to scavenge superoxide anions (Saito et al. 2008a; Niwano et al. 2011). Experiments with the Fenton reaction and photolysis of hydrogen peroxide induced by UV irradiation, both of which generate hydroxyl radicals, showed that all four extracts had potent ability to directly scavenge hydroxyl radicals. In addition, the scavenging activities against superoxide anions and hydroxyl radicals of the extracts of *P. granatum* (peel), *M. indica* (kernel) and *P. emblica* (fruit) proved to be heat-resistant. These three extracts would be for food processing in which thermal devices are used, because of their heat resistance. *Phyllanthus emblica* (fruit), 1 of 4 herbal extracts allowed to be used as foodstuff under Japanese legal regulations, was found to have potent ability to reduce the signal intensity of 5,5-dimethyl-1-pyrroline-N-oxide (DMPO)-OOH, a spin adduct formed by DMPO and superoxide anion (Saito et al. 2008b). The extract directly scavenged superoxide anions, and the superoxide scavenging potential was comparable to that of L-ascorbic acid. Further, polyphenol determination indicated that the activity was partially attributable to polyphenols.

Scartezzini et al. (2006) found that the emblica fruit contained ascorbic acid (0.40%, w/w), and that the Ayurvedic method of processing named "Svaras Bhavana", increased the healthy characteristics of the fruit by imparting a higher antioxidant activity and a higher content of ascorbic acid (1.28%, w/w). It was found that vitamin C accounted for approximately 45–70% of the antioxidant activity.

γ-radiation induced strand break formation in plasmid DNA (pBR322) was effectively inhibited by triphala and its constituents (*Emblica officinalis* (T1), *Terminalia chebula* (T2) and *Terminalia belerica* (T3)) in the concentration range 25–200 μg/ml with a percentage inhibition by T1 (30–83%), T2 (21–71%), T3 (8–58%) and triphala (17–63%) (Naik et al. 2005). They also inhibited radiation induced lipid peroxidation in rat liver microsomes effectively with IC_{50} values less than 15 μg/mL. The extracts were found to possess the ability to scavenge free radicals such as DPPH and superoxide. The total phenolic contents present in these extracts were determined and expressed in terms of gallic acid equivalents and were found to vary from 33–44%. These studies revealed that all three constituents of triphala were active and they exhibited slightly different activities under different conditions. T1 showed greater efficiency in lipid peroxidation and plasmid DNA assay, while T2 has greater radical scavenging activity.

Aller-7/NR-A2, a novel, safe polyherbal formulation was found to have antioxidant activity. Aller 7 was developed for the treatment of allergic rhinitis using a unique combination of extracts from seven medicinal plants that included *Phyllanthus emblica, Terminalia chebula, Terminalia bellerica, Albizia lebbeck, Piper nigrum, Zingiber officinale* and *Piper longum* (D'Souza et al. 2004). Aller-7 exhibited concentration-dependent scavenging activities toward biochemically generated hydroxyl radicals (IC_{50} 741.73 μg/ml); superoxide anion (IC_{50} 24.μg/ml by phenazine methosulfate-nicotinamide adenine dinucleotide [PMS-NADH] assay and IC_{50} 4.27 μg/ml by riboflavin/nitroblue tetrazolium [NBT] light assay), nitric oxide (IC_{50} 16.34 μg/ml); 1,1-diphenyl-2-picryl hydrazyl (DPPH) radical (IC_{50} 5.62 μg/ml); and 2,2-azino-bis-ethyl-benzothiozoline-sulphonic acid diammonium salt (ABTS) radical (IC_{50} 7.35 μg/ml). Aller-7 inhibited free radical-induced hemolysis in the concentration range of 20–80 μg/ml. Aller-7 also significantly inhibited nitric oxide release from lipopolysaccharide-stimulated murine macrophages. These results indicated Aller-7 to be a potent scavenger of free radicals.

Cardioprotective/Antiatherogenic Activity

An emblicanin-A (37%) and -B (33%) enriched fraction of fresh juice of emblica fruits exhibited antioxidant activity against ischemia-reperfusion (IRI)-induced oxidative stress in rat heart (Bhattacharya et al. 2002). IRI induced a significant decrease in the activities of cardiac superoxide dismutase, catalase and glutathione peroxidase, with a concomitant increase in lipid peroxidation. These IRI-induced effects were prevented by the administration of emblic fruit extract (50 and 100 mg/kg body wt.) and vitamin E (200 mg/kg body wt.) as standard given orally twice daily for 14 days prior to the sacrifice of the animals and initiation of the perfusion experiments. The study confirmed the antioxidant effect of *E. officinalis* and indicated that the fruits of the plant may have a cardioprotective effect.

Phyllanthus emblica, widely used to treat atherosclerosis-related diseases, was found to inhibit ox-LDL via its compounds of soluble tannin, corilagin (β-1-O-galloyl-3,6-(R)-hexahydroxydiphenoyl-d-glucose), and its analogue Dgg16 (1,6-di-O-galloyl-β-d-glucose) (Duan et al. 2005). Both corilagin and Dgg16 were able to decrease MDA (malonadhyde), prevented monocyte adherence to human umbilical vein endothelial cells, ECV-304 cells and also inhibited the rat vascular smooth muscular cells (VSMC) proliferation activated by oxidized low-density lipoprotein (ox-LDL). Ox-LDL was reported as the main etiologic factor in atherogenesis, and antioxidants were accepted as effective treatment of atherosclerosis. The results suggested that the two compounds were effective in inhibiting the progress of atherosclerosis by alleviating oxidation injury or by inhibiting ox-LDL-induced VSMC proliferation, which may be promising mechanisms for treating atherosclerosis.

Phyllanthus emblica-ethanol extract (100 µg/ml) showed the highest cardioprotective effect (approximately 12-fold doxorubicin IC_{50} increase) against doxorubicin toxicity among ethanol and water extracts of four Thai medicinal plants (Wattanapitayakul et al. 2005). The data demonstrated that antioxidants from natural sources may be useful in the protection of cardiotoxicity in patients who receive doxorubicin. Doxorubicin is an important and effective anticancer drug widely used for the treatment of various types of cancer but its clinical use is limited by dose-dependent cardiotoxicity.

Antihypercholesterolemic/ Hypolipidaemic Activity

Emblica officinalis was reported to significantly reduce serum cholesterol, aortic cholesterol and hepatic cholesterol in rabbits (Thakur 1985). It did not influence euglobulin clot lysis time, platelet adhesiveness or serum triglyceride levels. Studies showed that rabbits fed standard cholesterol-rich diet and given 1.0 g fresh Amla daily showed significantly lower mean serum cholesterol levels at the end of the second week than their counterparts fed standard cholesterol-rich diet without additional substance and those fed standard cholesterol-rich diet and given 10 mg vitamin C daily (Mishra et al. 1981) At the end of the third and fourth weeks the differences were even more pronounced. Results indicated that amla had a hypocholesterolemic effect. Another study reported that both normal and hypercholesterolaemic men aged 35–55 years showed a decrease in cholesterol levels after supplementation with amla diet for 28 days (Jacob et al. 1988). Two weeks after withdrawing the supplement, the total serum cholesterol levels of the hypercholesterolaemic subjects rose significantly almost to initial levels. A separate study reported that serum cholesterol, TG, phospholipid and LDL levels were lowered by 82%, 66%, 77% and 90%, respectively in cholesterol-fed rabbits (rendered hyperlipidaemic by atherogenic diet and cholesterol feeding) after administration of *E. officinalis* fresh juice at a dose of 5 ml/kg body weight per rabbit per day for 60 days (Mathur et al. 1996). Similarly, the tissue lipid levels showed a significant reduction following *E. officinalis* juice administration. Aortic plaques were regressed. *E. officinalis* juice treated rabbits excreted more cholesterol and phospholipids, suggesting that the mode of absorption was affected. Results indicated that *E. officinalis* juice was an effective hypolipidaemic agent and could be used as a pharmaceutical tool in hyperlipidaemic subjects

In a 2005 study, SunAmla and ethyl acetate (EtOAc) extract of amla significantly inhibited thiobarbituric acid (TBA)-reactive substance level in the Cu^{2+}-induced low-density lipoprotein (LDL) oxidation and the effects were stronger than those of probucol (Kim et al. 2005). In addition, the administration of SunAmla (at a dose of 20 or 40 mg/kg body weight/day) or ethyl acetate extract of amla (at a dose of 10 or 20 mg/kg body weight/day) for 20 days to rats fed 1% cholesterol diet significantly reduced total, free and LDL-cholesterol levels in a dose-dependent manner. The ethyl acetate extract of amla exhibited more potent serum cholesterol-lowering effect than SunAmla in the same amount. In addition, the serum TBA-reactive substance level was also significantly decreased after oral administration of SunAmla or ethyl acetate extract of amla. These results suggested that amla may be effective for hypercholesterolemia and prevention of atherosclerosis. The scientists also reported that the lipid levels, such as cholesterol and TAG (triacylglycerol), in serum and liver were markedly elevated in aged control rats, while they were significantly decreased by the administration of amla at a dose of 40 or 10 mg/kg body weight per d for 100 days (Yokozawa et al. 2007a). The PPARα (Peroxisome proliferator-activated receptor α) protein level in liver was reduced in aged control rats. The PPARα is a nuclear receptor protein known to regulate the transcription of genes involved in lipid and cholesterol metabolism. However, the oral administration of amla significantly increased the hepatic PPARα protein level. Further, oral administration of amla significantly inhibited the serum and hepatic mitochondrial thiobarbituric acid-reactive substance levels in aged rats and the elevated expression level of bax was significantly decreased, while the level of bcl-2 led to a significant increase. The expressions of hepatic NF-kappaB, inducible NO synthase (iNOS), and cyclo-oxygenase-2 (COX-2) protein levels were also increased with ageing. However, amla extract reduced the iNOS and COX-2 expression levels by inhibiting NF-kappaB activation in aged rats. These results indicated that amla may prevent age-related hyperlipidaemia through attenuating oxidative stress in the ageing process.

Administration of 20% alcohol (5 g/kg body weight) to rats significantly increased cholesterol/phospholipid (C/P) ratio, lipid peroxidation and the activities of Na+/K+ and Mg2+ ATPases in erythrocyte membranes as well as augmented nitric oxide (NO) levels (Reddy et al. 2009). Alcohol administration significantly increased erythrocyte membrane anisotropic values and altered membrane individual phospholipid content. Administration of Emblica fruit extract (250 mg/kg body weight) to alcoholic rats resulted in significant reduction of NO levels, erythrocyte membrane lipid peroxidation, C/P ratio, activities of Na+/K+ and Mg2+ ATPases and fluorescent anisotropic values. Further, Emblica fruit extract administration to alcoholic rats beneficially modulated membrane properties as evidenced from the contents of total phospholipids as well individual phospholipid classes. The tannoid principles present in emblica offered protection against alcohol induced adverse effects in rats.

Flavonoids from *Emblica officinalis* and *Mangifera indica* effectively reduced lipid levels in serum and tissues of rats induced hyperlipidemia (Anila and Vijayalakshmi 2002). Hepatic HMG CoA reductase activity was significantly inhibited in rats fed *E. officinalis* flavonoids. But increase of this enzyme was observed in rats administered *M. indica* flavonoids. The degradation and elimination of cholesterol was highly enhanced in both the groups. In *E. officinalis*, the mechanism of hypolipidemic action was by the concerted action of inhibition of synthesis and enhancement of degradation. In the other group (*M. indica*) inhibition of cholesterogenesis was not encountered but highly significant degradation of cholesterol was noted, which may be the pivotal factor for hypolipidemic activity in this case. Though the mechanisms differ in the two cases, the net effect was to lower lipid levels. Another study showed significant increase in the total cholesterol, low density lipoprotein (LDL), very low density lipoprotein (VLDL), and free fatty acid in hypercholesteremic rats were significantly reduced in Triphala treated hypercholesteremic rats (Saravanan et al. 2007). Triphala is an Ayurvedic herbal formulation comprising *Terminalia chebula, Terminalia belerica*, and *Emblica officinalis*. The data demonstrated that Triphala formulation was associated with hypolipidemic effects on the experimentally induced hypercholesteremic rats.

Administration of ethyl acetate extract of *Emblica officinalis* to wistar rats was found to ameliorate the high fructose-induced metabolic syndrome, including hypertriacylglycerolaemia and hypercholesterolaemia (Kim et al. 2010). Also, the elevated levels of hepatic TAG and total cholesterol in rats given the high-fructose diet were significantly reduced by 33.8 and 24.6%, respectively, on the administration of the ethyl acetate extract of amla at the dose of 20 mg/kg with the regulation of sterol regulatory element-binding protein (SREBP)-1 expression. In addition, oral administration of the extract at the dose of 20 mg/kg significantly inhibited the increased serum and hepatic mitochondrial thiobarbituric acid-reactive substance. The findings suggested that fructose-induced metabolic syndrome was attenuated by the polyphenol-rich fraction of *E. officinalis*.

Antidiabetic Activity

Studies provided evidence supporting the efficacy of amla for relieving the oxidative stress and improving glucose metabolism in diabetes (Rao et al. 2005). Amla extracts showed strong free radical scavenging activity. Amla also showed strong inhibition of the production of advanced glycosylated end products. The oral administration of amla extracts to the diabetic rats slightly improved body weight gain and also significantly alleviated various oxidative stress indices of the serum of the diabetic rats. The elevated serum levels of 5-hydroxymethylfurfural, a glycosylated protein and an indicator of oxidative stress, were significantly reduced dose-dependently in the diabetic rats fed amla. Similarly, the serum level of creatinine, yet another oxidative stress parameter, was also reduced. Further, thiobarbituric acid-reactive substances levels were significantly reduced with amla, indicating a reduction in lipid peroxidation. In addition, the decreased albumin levels in the diabetic rats were significantly improved with amla. Amla also significantly improved the serum adiponectin levels. Another study showed that aqueous *Phyllanthus emblica* fruit extract, in a dose of 200 mg/kg body weight, significantly decreased the blood glucose level after its intraperitoneal administration in alloxan-induced diabetic rats (Qureshi et al. 2009). The aqueous extract also induced hypotriglyceridemia by decreasing triglycerides levels at 0, 1, 2 and 4 hours in diabetic rats. In addition, the extract was also found to improve liver function by normalizing the activity of liver-specific enzyme alanine transaminase (ALT).

E. officinalis extract inhibited rat lens and recombinant human aldose reductase with IC_{50} values 0.72 and 0.88 mg/ml respectively (Suryanarayana et al. 2004). It was demonstrated that the hydrolysable tannoids of *E. officinalis* were responsible for aldose reductase inhibition in sugar cataract, as enriched tannoids of *E. officinalis* exhibited remarkable inhibition against both rat lens and human aldose reductase with IC_{50} of 6 and 10 μg/ml respectively. The inhibition of aldose reductase by *E. officinalis* tannoids is 100 times higher than its aqueous extract and comparable to or better than quercetin. *E. officinalis* is also a rich source of ascorbic acid, however, ascorbic acid did not inhibit aldose reductase even at 5 mM concentration. Furthermore, the isolated tannoids not only prevented the aldose reductase activation in rat lens organ culture but also sugar-induced osmotic changes. In another study, emblica and its enriched tannoids did not prevent streptozotocin-induced hyperglycemia in rats as assessed by blood glucose and insulin levels, however, these supplements delayed cataract progression (Suryanarayana et al. 2007). The results suggested that emblica and its tannoids supplementation inhibited aldose reductase activity as well as sorbitol formation in the lens. Emblica also prevented aggregation and insolubilization of lens proteins caused by hyperglycemia. The results provided evidence that emblica and an enriched fraction of emblica tannoids were effective in delaying development of diabetic cataract in rats.

In a recent study, standardized doses of 100, 200, 300, and 400 mg/kg body weight of the aqueous emblic seed extract were administered orally to normal and streptozotocin (STZ)-induced type 2 diabetic rats in order to define its glycemic potential (Mehta et al. 2009). The maximum decrease of 27.3% in the blood glucose level of normal rats was observed at 6 hours during fasting blood glucose studies, with the dose of 300 mg/kg identified as the most effective dose. The same dose produced a fall of 25.3% in the same models during the glucose tolerance test (GTT) at 3 hours after glucose administration. However, the dose of 300 mg/kg of aqueous seed extract in sub- and mild-diabetic animals produced a maximum fall of 34.1% and 41.6%, respectively, during the GTT at 3 hours after glucose administration. This evidence indicated that the aqueous extract of *E. officinalis* seeds had definite hypoglycemic potential as well as antidiabetic activity.

Oral administration of *Phyllanthus emblica* fruit ethanolic extract (PFEet) at doses of 200 mg/kg body weight for 45 days resulted in a significant reduction in blood glucose and a significant increase in plasma insulin in diabetic rats (Krishnaveni et al. 2010). Diabetic rats fed PFEet showed a significant reduction in total cholesterol (TC), very low density lipoprotein-cholesterol (VLDL-C), LDL-cholesterol, free fatty acids (FFA), phospholipids (PL), triglycerides (TG) and decreased HDL-cholesterol and an elevation in HDL-C. The observations showed that *Phyllanthus emblica* had antidiabetic and its beneficial effects on lipid profile, thus could be recommended for use as a natural supplementary herbal remedy in patients suffering from diabetes mellitus.

Kusirisin et al. (2009) found that five plants namely *Phyllanthus emblica, Terminalia chebula, Morinda citrifolia, Kaempferia parviflora* and *Houttuynia cordata,* traditionally used in diabetic patients, exhibited strong antioxidant activity. *P. emblica* showed stronger antioxidative activity assayed by the 2,2′-azinobis-(3-ethylbenzothiazoline-6-sulfonic acid) diammonium salt (ABTS) method as well as inhibition of TBARS and anti-glycation activity than the other plants. The investigation showed that total polyphenol and tannin content of *Phyllanthus emblica* and the flavonoid content of *Houttuynia cordata* were the highest. The results implied these plants to be potential sources of natural antioxidants with free radical scavenging activity and might be used for reducing oxidative stress in diabetes.

Anticancer Activity

Addition of 2 ml of the *Phyllanthus emblica* juice dialyzate (JD) to a system containing 22 μmol/L NaNO2 and 22 μmol/L morpholine was found to effectively block N-nitrosomorpholine (NMOR) formation, with the inhibition of 93%, while only 49.2% of inhibition was observed from the low concentration of vitamin C (20.5 μmol/L), equivalent to those present in the JD (Hu 1990). In rats gavaging with $NaNO_2$ and proline (40 μmol/L), co-administration of 1.3 ml of the undialyzed juice containing 36 μmol/L of vitamin C reduced the urinary excretion of N-nitrosoproline (NPRO) from 69.62 nmol/L to 4.24 nmol/L. Similarly, it was found in human studies that 13 ml of the juice dramatically diminished the NPRO excretion from 75.10 nmol/L to 22.79 nmol/L in the 24-hour urine of 12 healthy volunteers, who had ingested 300 mg $NaNO_3$ and 500 mg proline. Again, the juice showed a higher inhibition in both cases in inhibiting NPRO synthesis than the equivalent of vitamin C solution. N-nitrosomorpholine (NMOR) and N-nitrosoproline (NPRO) are human carcinogens. The results suggested that only a fraction of inhibition was accounted for by vitamin C in the juice.

Extract of *Emblica officinalis* significantly inhibited hepatocarcinogenesis induced by N-nitrosodiethylamine (NDEA) in a dose dependent manner (Jeena et al. 1999). Animals treated with NDEA alone showed 100% tumour incidence and significantly elevated tissue levels of drug metabolizing enzymes such as glutathione S-transferase (GST) and aniline hydroxylase (AH) and levels of γ-glutamyl transpeptidase (GGT). Serum levels of lipid peroxide (LPO), alkaline phosphatase (ALP) and glutamate pyruvate transaminase (OPT), which are markers of liver injury, were also elevated. Morphology of liver tissue and levels of marker enzymes indicated that the extract offered protection against chemical carcinogenesis. Separate studies indicated that *Emblica officinalis* polyphenol fraction (EOP) treatment could induce apoptosis in Dalton's Lymphoma Ascites (DLA) and CeHa cell lines (Rajeshkumar et al. 2003). At 200 μg/ml, EOP induced membrane blebbing, chromatin condensation and intenucleosomal breaks as evident from the morphology and DNA ladder pattern obtained in gel electrophoresis. The results also suggested that EOP treatment could decrease the liver tumour development induced by N-nitrosodiethylamine (NDEA). Animals administered (oral) with NDEA (0.02%, 2.5 ml/rat, 5 days a week, 20 weeks) developed visible liver tumours by the end of the 20th week and the liver weight raised to 5.2 g/100 g body weight. Only 11% of the animals treated with EOP (60 mg/kg, oral, 5 days a week for 20 weeks) developed visible liver tumours by this period and the liver weights were reduced to 3.2 g/100 g body weight. γ-glutamyl transpeptidase activity elevated to 88.4 U/l in serum of NDEA treated group was reduced to 48.4 U/l by EOP treatment. Elevated levels of serum alkaline phosphatase (ALP), glutamate pyruvate transaminase (GPT), bilirubin, liver glutathione S-transferase (GST) and glutathione (GSH) in the NDEA administered group were significantly reduced by EOP treatment. The EOP was found to scavenge superoxide and hydroxyl radicals and inhibit lipid peroxidation in-vitro.

Aqueous extract of *Emblica officinalis* (EO) was found to be cytotoxic to L 929 cells in culture in a dose dependent manner (Jose et al. 2001). L 929 is a murine aneuploid fibrosarcoma cell line used to assay TNF-α and TNF-β. The LC_{50} value of EO for inhibition was found to be 16.5 μg/ml. EO and Chyavanaprash (a nontoxic herbal preparation containing 50% EO) extracts were found to reduce ascites and solid tumours in mice induced by DLA cells. Animals treated with 1.25 g/kg body weight of EO extract increased life span of tumour bearing animals (20%) while animals treated with 2.5 g/kg b.wt. of chyavanaprash produced 60.9% increased in the life span. Both EO and Chyavanaprash significantly reduced the solid tumours. EO extract was found to inhibit cell cycle regulating enzymes cdc 25 phosphatase in a dose dependent manner. The results suggested that antitumour activity of EO extract may partially be due to its interaction with cell cycle regulation.

Extracts from *Emblica officinalis* were the most active in inhibiting in-vitro cell proliferation of human tumour cell lines, including human erythromyeloid K562, B-lymphoid Raji, T-lymphoid Jurkat, erythroleukemic HEL cell lines, after comparison to those from *Terminalia arjuna, Aphanamixis polystachya, Oroxylum indicum, Cuscuta reflexa, Aegle marmelos, Saraca asoka, Rumex maritimus, Lagerstroemia speciosa*, Red Sandalwood (Khan et al. 2002). Pyrogallol as the common compound present both in unfractionated and n-butanol fraction of *Emblica officinalis* extracts was identified as the active principle. In another study, 18 main compounds, including four norsesquiterpenoids (1–4) and 14 phenolic compounds (5–18) isolated previously from *Phyllanthus emblica*, together with a main constituent, proanthocyanidin polymers (19) identified at this time from the roots, exhibited antiproliferative activities against MK-1 (human gastric adenocarcinoma), HeLa (human uterine carcinoma), and B16F10 (murine melanoma) cells (Zhang et al. 2004). All of the phenolic compounds including the major components 5–8 from the fruit juice, 8, 9, and 12 from the branches and leaves, and 19 from the roots showed stronger inhibition against B16F10 cell growth than against HeLa and MK-1 cell growth. Norsesquiterpenoid glycosides 3 and 4 from the roots exhibited significant antiproliferative activities, although their aglycon 1 and monoglucoside 2 showed no inhibitory activity against these tumor cells.

In a recent study, *P. emblica* extract demonstrated growth inhibitory activity, with a certain degree of selectivity against the two cancer cell lines, human hepatocellular carcinoma (HepG2), and lung carcinoma (A549) cells tested (Pinmai et al. 2008). Synergistic effects (CI<1) for *P. emblica*/doxorubicin or cisplatin at different dose levels were demonstrated in A549 and HepG2 cells. In another study, pretreatment with defatted methanolic emblic fruit extract (100 and 200 mg/kg b.w.) showed significant partial recovery of pathological manifestations as compared to diethylnitrosoamine and 2-acetylaminoflourine – treated group animals and suppressed the tumour forming potential of 2-acetylaminoflourine at both the doses (Sultana et al. 2008). The study demonstrated that *Emblica officinalis* had the potential to suppress carcinogen-induced response in rat liver.

In separate studies, emblic fruit extract (EO) inhibited tumour incidences of DMBA (7, 12-dimethyabenz(a)anthrecene) induced skin tumorigenesis in Swiss albino mice (Sancheti et al. 2005). The tumour incidence, tumour yield, tumour burden and cumulative number of papillomas were found to be higher in the control (without EO treatment) as compared to experimental animals (EO treated). The study demonstrated the chemopreventive potential of *Emblica officinalis* fruit extract on skin carcinogenesis in mice.

Ngamkitidechakul et al. (2010) reported that *Phyllanthus emblica* extract at 50–100 µg/mL significantly inhibited cell growth of six human cancer cell lines, A549 (lung), HepG2 (liver), HeLa (cervical), MDA-MB-231 (breast), SK-OV3 (ovarian) and SW620 (colorectal). However, the extract was not toxic against MRC5 (normal lung fibroblast). Apoptosis in HeLa cells was also observed as the extract caused DNA fragmentation and increased activity of caspase-3/7 and caspase-8, but not caspase-9, and up-regulation of the Fas protein indicating a death receptor-mediated mechanism of apoptosis. Treatment of the extract on mouse skin resulted in over 50% reduction of tumour numbers and volumes in animals treated with DMBA/TPA. Additionally, 25 and 50 µg/mL of the extract inhibited invasiveness of MDA-MB-231 cells in the in vitro Matrigel invasion assay.

A new cytotoxic acylated apigenin glucoside (apigenin-7-O-(6″-butyryl-β-glucopyranoside)) was isolated from the methanolic extract of the leaves of *Phyllanthus emblica* together with the known compounds; gallic acid, methyl gallate, 1,2,3,4,6-penta-O-galloylglucose and luteolin-4′-O-neohesperiodoside (El-Desouky et al. 2008).

The results of studies showed that progallin A isolated from the acetic ether part of the leaves of *Phyllanthus emblica* had low immune toxicity and strongly inhibited the proliferation of human hepatocellular carcinoma BEL-7404 cells in a time- and dose-dependent manner (Zhong et al. 2011). It induced apoptosis of BEL-7404 cells and the apoptosis rates and the number of apoptotic cells significantly increased with prolongation of the action time. Flow cytometry coupled with immunofluorescence staining and western blot indicated that progallin induced apoptosis of human hepatocellular carcinoma BEL-7404 cells by up-regulation of Bax expression and down-regulation of Bcl-2 expression.

Anticancer Activity of Triphala

Experimental studies in the past decade have shown that Triphala (an Ayurvedic formulation consisting of fruit powder of three different plants namely *Terminalia chebula, Terminalia belerica* and *Emblica officinalis*) is useful in the prevention of cancer and that it also possesses antineoplastic, radioprotective and chemoprotective effects (Baliga 2010). Studies showed that Triphala in diet significantly reduced the benzo(a) pyrene [B(a)P] induced forestomach papillomagenesis in mice (Deep et al. 2005). In the short term treatment groups, the tumour incidences were lowered to 77.77% by both doses of Triphala mixed diet. In the case of long-term treatment the tumour incidences were reduced to 66.66% and 62.50% respectively by 2.5% and 5% triphala containing diet. Tumour burden was 7.27 in the B(a)P treated control group, whereas it reduced to 3.00 by 2.5% dose and 2.33 by 5% dose of Triphala. In long-term studies the tumour burden was reduced to 2.17 and 2.00 by 2.5% and 5% diet of Triphala, respectively. It was observed that Triphala was more effective in reducing tumour incidences compared to its individual constituents. Triphala also significantly increased the antioxidant status of animals which might have contributed to the chemoprevention.

Separate studies revealed that acetone extract of "Triphala" showed a significant cytotoxic effect on cancer cell-lines Shionogi 115 (S115), MCF-7 breast cancer cells and PC-3 and DU-145 prostate cancer cell-lines (Kaur et al. 2005). The major phenolic compounds in the most potent acetone extracts were isolated and purified and found to be gallic acid. A separate 2006 studies showed that MCF 7 with wild type p53 was more sensitive to Triphala than T 47 D, which was p53 negative (Sandhya and Mishra 2006). Triphala induced loss of cell viability was determined by MTT assay. Moreover, Triphala inhibited the clonogenic growth of MCF 7 cells, which was significantly recovered by pifithrin-α, the p53 inhibitor. Exogenous addition of antioxidants, glutathione (GSH) and N-Acetyl-Cysteine (NAC) inhibited the anti-proliferative ability of Triphala in both MCF 7 and T47 D and induced significant apoptosis in both the cell lines in a dose dependant manner but magnitude of apoptosis was significantly higher in MCF 7 than in T 47-D cells. Triphala was also found to induce dose and time dependent increase in intracellular reactive oxygen species in both the cell lines. The results demonstrated that MCF 7 and T 47 D cells exhibited differential sensitivity to Triphala, which seemed to be dependant on their p53 status. It was concluded that p53 status of cancer cells formed an important factor in predicting the response of cancer cells to prooxidant drugs.

In another study, the viability of treated human breast cancer cell line (MCF-7) and a transplantable mouse thymic lymphoma (barcl-95) was found to decrease with the increasing concentrations of Triphala (Sandhya et al. 2006a). On the other hand, treatment of normal breast epithelial cells, MCF-10, human peripheral blood mononuclear cells, mouse liver and spleen cells, with similar concentrations of Triphala did not affect their cytotoxicity significantly. The drug treatment was found to induce apoptosis in MCF-7 and barcl-95 cells in-vitro as determined by annexin-V fluorescence and proportion of apoptotic cells was found dependent on Triphala concentration. MCF-7 cells treated with Triphala when subjected to single cell gel electrophoresis, revealed a pattern of DNA damage, characteristic of apoptosis. Studies on Triphala treated MCF-7 and barcl-95 cells showed significant increase in intracellular reactive oxygen species (ROS) in a concentration dependent manner. In-vivo, direct oral feeding of Triphala to mice (40 mg/kg body weight) transplanted with barcl-95 produced significant reduction in tumour growth as evaluated by tumour volume measurement. It was also found that apoptosis was significantly higher in the excised tumour tissue of Triphala fed mice as compared to the control, suggesting the involvement of apoptosis in tumour growth reduction. These results suggest that Triphala possessed ability to induce cytotoxicity in tumour cells but spared the normal cells. The differential response of normal and tumour cells to Triphala in-vitro and the substantial regression of transplanted tumour in mice fed with Triphala indicated its potential use as an anticancer drug for clinical treatment.

Preclinical studies demonstrated that Triphala was effective in inhibiting the growth of human pancreatic cancer cells in both cellular and in vivo model (Shi et al. 2008). Exposure of Capan-2 cells to the aqueous extract of Triphala for 24 hours resulted in the significant decrease in the survival of cells in a dose-dependent manner with an IC_{50} of about 50 µg/ml. Triphala-mediated reduced cell survival correlated with induction of apoptosis, which was associated with reactive oxygen species (ROS) generation. The data also suggested that the growth inhibitory effects of Triphala was mediated by the activation of ERK and p53 and showed potential for the treatment and/or prevention of human pancreatic cancer.

Anticancer Activity of Kalpaamruthaa

Kalpaamruthaa (KA), a herbal preparation containing *Semecarpus anacardium*, *E. officinalis* and honey exhibited anticancer effect against mammary carcinoma in rats. Kalpaamruthaa exhibited therapeutic efficacy on lipid peroxidation and antioxidant status (Ramprasath et al. 2006), on oxidative damage in mammary gland mitochondrial fraction (Arulkumaran et al. 2006), on marker enzymes, body weight and volume (Veena et al. 2006a), on biochemical alteration (Veena et al. 2006c) and protective effects on altered glycoprotein component levels and membrane stability (Veena et al. 2006b).

A significant increase in the levels of lipid peroxides (LPO), of reactive oxygen species (ROS) and a decreased levels of antioxidants observed in mammary carcinoma bearing rats were found to be reverted back to near normal levels on treatment with Kalpaamruthaa, a modified Siddha preparation (Ramprasath et al. 2006). Simultaneous treatment with Kalpaamruthaa showed more effect than post treatment with Kalpaamruthaa. Drug control animals showed no significant changes in the levels of ROS when compared with control animals. On administration of Kalpaamruthaa, at the dosage level of 300 mg/kg body weight the levels of the marker enzymes and lysosomal enzymes and the changes in the body weights and volume were significantly normalized in a dose dependent manner in mammary carcinoma bearing rats (Veena et al. 2006a). Cancerous animals with mammary cancer showed a significant decrease in their body weights and significant increase in tumour weights (Veena et al. 2006b). The levels of Carcino Embryonic Antigen (CEA) in cancer rats were significantly higher, when compared to that of control rats. The levels of these glycoproteins in cancer animals were significantly increased when compared with control animals. The levels of erythrocyte membrane, liver and kidney ATPases were significantly decreased in cancer condition when compared with control animals. Treatment of Kalpaamruthaa reverted back the pathologic condition by decreasing the level of above enzymes more effectively than *Semecarpus anacardium* nut milk extract. No significant changes were observed in drug control animals when compared with control animals in all the above studies. These results suggested that the therapeutic efficacy of the Kalpaamruthaa was increased on amalgamation of *Semecarpus anacardium, Emblica officinalis* and honey.

The mitochondrial fraction of untreated dimethyl benz[a]anthracene (DMBA)-induced mammary gland showed 2.61-fold increase in lipid peroxidation level and abnormal changes in the activities/levels of mitochondrial enzymic (superoxide dismutase, glutathione peroxidase and glutathione reductase) and non-enzymic (glutathione, vitamin C and vitamin E) antioxidants were observed (Arulkumaran et al. 2006). DMBA treated rats also showed decline in the activities of mitochondrial enzymes such as succinate dehydrogenase, malate dehydrogenase, α-ketoglutarate dehydrogenase and isocitrate dehydrogenase. In contrast, rats treated with Kalpaamruthaa showed normal lipid peroxide level and antioxidant defenses. The results of the present study highlighted the improved antioxidant property of Kalpaamruthaa over sole treatment of *S. anacardium* nut milk extract. The increased levels of total cholesterol, free cholesterol, phospholipids, triglycerides and free fatty acids and decreased levels of ester cholesterol in plasma, liver and kidney found in cancer suffering animals were reverted back to near normal levels on treatment with Kalpaamruthaa and *S. anacardium* (Veena et al. 2006c). In mammary carcinoma bearing animals, the activities of total lipase, cholesterol ester synthase, and cholesterol ester hydrolase were significantly increased whereas lipoprotein lipase and lecithin cholesterol-acyl transferase were decreased. The levels of very low-density lipoprotein (VLDL) and low-density lipoprotein (LDL) were increased and the level of high-density lipoprotein (HDL) was decreased. These alterations were restored upon treatment with Kalpaamruthaa as well as *Semecarpus anacardium* when compared to cancer animals. The effects of Kalpaamruthaa were found to be more effective than *Semecarpus anacardium*. No significant alterations were observed in herbal preparation control animals when compared to control animals.

Anticlastogenic, Antigenotoxic, Antimutagenic Activity

Emblic fruit extract significantly reduced the frequency of chromosomal aberrations (CA)/cell, the percentage of aberrant cells and the frequency of micronuclei induced by all doses of nickel chloride (10, 20 and 40 mg/kg body wt.) in the bone marrow cells of mice (Dhir et al. 1991). Nickel, a major environmental pollutant is known for its clastogenic and carcinogenic potential. Dietary inhibitors of mutagenesis and carcinogenesis are of particular importance since they may have a role in cancer prevention. Ascorbic acid, a major constituent of the fruit, fed for 7 consecutive days in equivalent concentration as that present in the fruit, however, could only alleviate the cytotoxic effects induced by low doses of nickel; at the higher doses it was ineffective. The greater efficacy of emblica fruit extract could be due to the interaction of its various natural components rather than to any single constituent.

In another study, oral administration of *Phyllanthus emblica* fruit extract for 7 days before exposure to lead and aluminium through intraperitoneal injections reduced the frequency of micronuclei induced by all doses of both metals in bone marrow cells of *Mus musculus* (Roy et al. 1992). Priming with comparable doses of synthetic ascorbic acid reduced micronuclei formation induced by both doses of aluminium and only the lower dose of lead. With the higher dose of lead (20 mg/kg body wt.) priming with ascorbic acid increased the frequency of micronuclei when compared with mice administered lead alone. The greater efficacy of *Phyllanthus* fruit extract in alleviating metal-induced clastogenicity may be due to the combined action of all ingredients in the crude extract, rather than to ascorbic acid alone.

The protective action of crude extracts of *Phyllanthus emblica* fruits (PFE) against lead (Pb) and aluminium (Al)-induced sister chromatid exchanges (SCEs) was studied in bone marrow cells of *Mus musculus* (Dhir et al. 1993). The modifying effect of the crude extract was compared with that of comparable amounts of synthetic ascorbic acid (AA), a major component of the fruits. Oral administration of PFE or AA for 7 consecutive days before exposure of mice to the metals by intraperitoneal injections reduced the frequencies of SCEs induced by both metals. PFE afforded a more pronounced protective effect than synthetic ascorbic acid in counteracting the genotoxicity induced by both Al and Pb: This difference was significant with Pb. A 2007 study showed that at higher concentration of lead, a significant increase in the percentage of sperm head abnormalities was noted but when animals primed with *Phyllanthus* fruit extract (PFE), a reduction in the frequency of sperm head abnormalities was observed (Madhavi et al. 2007). The data suggested that *Phyllanthus emblica* played a key role in inhibition of heavy metal mutagenesis in mammals.

Water, acetone and chloroform extracts of *E. officinalis* fruit reduced sodium azide and NPD induced his+revertants significantly in TA100 and TA97 strains of *Salmonella typhimurium* respectively (Grover and Kaur 1989). The chloroform extract was less active as compared to water and acetone extracts. Autoclaving of water extract for 15 minutes did not reduce its activity. The enhanced inhibitory activity of the extracts on pre-incubation suggested the possibility of desmutagens in the extracts. Besides ascorbic acid, a constituent of the extract, the role of other antimutagenic factors in the extract could not be discounted.

Oral administration of *Emblica officinalis* (EO) fruit extract in various concentrations (100, 250, 500 mg/kg b.wt) for seven consecutive days prior to a single intraperitoneal injection of 7,12-dimethylbenz(a)anthracene (DMBA) decreased the frequency of bone marrow micronuclei induced in Swiss albino mice (Banu et al. 2004). Significant increases in the liver antioxidants, such as glutathione (GSH), glutathione peroxidase (GPx), glutathione reductase (GR) and detoxifying enzyme glutathione-S-transferase (GST), were found in the fruit extract treated group. The extract also reduced the hepatic levels of the activating enzymes cytochrome (Cyt) P450 and Cyt b5. These increases in the carcinogen treated group, which emphasized its protective effect against the carcinogen. There was a dose-dependent effect of the extract against the genotoxin with the maximum effect at 500 mg/kg b.wt. The protection against genotoxicity induced by the rodent carcinogen afforded by EO may be associated with its antioxidant capacity and through its modulatory effect on hepatic activation and detoxifying enzymes. A study made in an Ames histidine reversion assay using TA98 and TA100 tester strains of *Salmonella typhimurium* revealed that water extract of Triphala (a composite mixture of *Terminalia bellerica, T. chebula and Emblica officinalis*) was ineffective in reducing the revertants induced by the directacting mutagens, 4-nitro-o-phenylenediamine (NPD) and sodium azide, and the indirect-acting promutagen, 2-aminofluorene (2AF), but the chloroform and acetone extracts showed inhibition of mutagenicity induced by both direct and S9-dependent mutagens (Kaur et al. 2002). A significant inhibition of 98.7% was observed with acetone extract against the revertants induced by S9-dependent mutagen, 2AF, in co-incubation mode of treatment.

Dietary supplementation with extract of fruit of *Emblica officinalis* to mice in-vivo significantly reduced the cytotoxic effects of a known carcinogen, 3,4-benzo(a)pyrene in Swiss albino mice (Nandi et al. 1997). The cytotoxic effects were significantly lower in the mice given the fruit extract with the carcinogen than in those given the carcinogen alone. In a 1999 study, dietary administration of a crude aqueous extract of *Emblica officinalis* fruit (685 mg/kg bw) reduced significantly the cytotoxic effects of sodium arsenite administered orally in Swiss albino mice (Biswas et al. 1999). The crude extract reduced arsenic damage (chromosomal aberrations and damaged cells) bringing the cells almost to the normal level.

In a 1994 study, oral administration of *Phyllanthus emblica*, was found to enhance natural killer (NK) cell activity and antibody dependent cellular cytotoxicity (ADCC) in syngeneic BALB/c mice, bearing Dalton's lymphoma ascites (DLA) tumour. *P. emblica* elicited a 2-fold increase in splenic NK cell activity on day 3 post tumour inoculation (Suresh and Vasudevan 1994). Enhanced activity was highly significant on days 3, 5, 7 and 9 after tumour inoculation with respect to the untreated tumour bearing control. A significant enhancement in ADCC was documented on days 3, 7, 9, 11 and 13 in drug treated mice as compared to the control. An increase in life span (ILS) of 35% was recorded in tumour bearing mice treated with *P. emblica*. This increased survival was completely abrogated when NK cell and killer (K) cell activities were depleted either by cyclophosphamide or anti-asialo-GM1 antibody treatment. These results indicated: (a) an absolute requirement for a functional NK cell or K cell population in order that *P. emblica* could exert its effect on tumour bearing animals, and (b) the antitumour activity of *P. emblica* was mediated primarily through the ability of the drug to augment natural cell mediated cytotoxicity.

Amla fruit exhibited cytoprotective and immunomodulating activities. Studies showed that chromium (Cr) treatment resulted in enhanced cytotoxicity, free radical production, lipid peroxidation and decreased glutathione peroxidase (GPx) activity and diminished glutathione (GSH) levels (Sai Ram et al. 2002). There was a significant inhibition of both lipopolysaccharide and concanavalin-A-stimulated lymphocyte proliferation. Chromium also inhibited Con A stimulated interleukin-2 and γ-interferon production significantly and enhanced apoptosis and DNA fragmentation. Amla significantly inhibited Cr-induced free radical production, restored the anti-oxidant status back to control level and inhibited apoptosis and DNA fragmentation induced by Cr. Also, Amla relieved the immunosuppressive effects of Cr on lymphocyte proliferation and even restored the IL-2 and γ-IFN production considerably. Similar results were observed in murine macrophages (Sai Ram et al. 2003). Amla fruit extract inhibited chromium induced immunosuppression and restored both phagocytosis and γ-IFN production by macrophages significantly. Chromium (VI) at 1 μg/mL concentration was highly cytotoxic. It enhanced free radical production and decreased reduced glutathione (GSH) levels and glutathione peroxidase (GPx) activity in macrophages. The presence of Amla resulted in an enhanced cell survival, decreased free radical production and higher antioxidant levels similar to that of control cells. Further, chromium (VI) treatment resulted in decreased phagocytosis and γ-interferon (γ-IFN) production.

Adaptogenic Activity

Emblica officinalis is one of six *rasayana* (rejuvenating) herbs used in Ayurvedic medicine that have been reported to have adaptogenic properties (Rege et al. 1999). These plants were found to offer protection against a variety of biological, physical and chemical stressors, as judged by using markers of stress responses and objective parameters for stress manifestations. Using a model of cisplatin induced alterations in gastrointestinal motility, the ability of these plants to exert a normalizing effect, irrespective of direction of pathological change was tested. All the plants reversed the effects of cisplatin on gastric emptying. *Emblica officinalis* strengthened the defence mechanisms against free radical damage induced during stress. The effect of *Emblica officinalis* appeared to depend on the ability of target tissues to synthesize prostaglandins.

Tannoids of emblic was found to exhibit antistress activity using model of chronic unpredictable footshock-induced stress-induced perturbations in oxidative free radical scavenging enzymes in rat brain frontal cortex and striatum (Bhattacharya et al. 2000b). Chronic stress, administered over a period of 21 days, induced significant increase in rat brain frontal cortical and striatal superoxide dismutase (SOD) activity, concomitant with significant reduction in catalase (CAT) and glutathione peroxidase (GPX) activity. The changes in the enzyme activities was accompanied by an increase in lipid peroxidation, in terms of augmented thiobarbituric acid-reactive products. Administration of emblic tannoids (emblicanin A, emblicanin B, punigluconin, and pedunculagin of E) (10 and 20 mg, po) for 21 days, concomitant with the stress procedure, induced a dose-related alteration in the stress effects. Thus, a tendency towards normalization of the activities of SOD, CAT and GPX was noted in both the brain areas, together, with reduction in lipid peroxidation. The results indicate that the reported antistress rasayana activity of *E. officinalis* may be, at least partly due to its tendency to normalize stress-induced perturbations in oxidative free radical scavenging activity. In another study, using a rat model of tardive dyskinesia (TD) induced by haloperidol, the tannoid principles of *E. officinalis* (EOT) administered concomitantly with haloperidol inhibited all three tardive dyskinesia parameters (chewing movements, buccal tremors and tongue protusion) assessed (Bhattachary et al. 2000a) as did vitamin E. The effect of sodium valproate, a Gaba-mimetic agent, remained statistically insignificant. The results suggested that EOT exerted a prophylactive effect against neuroleptic-induced tardive dyskinesia which was likely to be due to its earlier reported antioxidant effects in rat brain areas, including striatum.

Phyllanthus emblica, *Rhodiola rosea* (golden root), *Eleutherococcus senticosis* (Siberian ginseng) are regarded as adaptogens, which are harmless herb having pharmaceutical benefits due to their balancing, regulative and tonic functions (Chen et al. 2008). Data generated suggested that the antioxidant potential of the adaptogens were proportional to their respective polyphenol content. The supplementation of adaptogen extracts containing high levels of polyphenols may not only have adaptogen properties, but may decrease the risk of complications induced by oxidative stress. Studies suggested that Triphala (*Terminalia chebula*, *Terminalia belerica* and *Emblica officinalis*) supplementation could be regarded as a protective drug against stress (Dhanalakshmi et al. 2007). Following cold-stress exposure (8°C for 16 hours/daily for 15 days), significant increase in lipid peroxidation and corticosterone levels were observed in Wistar strain albino rats. Upon exposure to the cold-stress, a significant increase in immobilization with decrease in rearing, grooming, and ambulation behavior was seen in open field. Oral administration of Triphala (1 g/kg/animal body weight) for 48 days significantly prevented these cold stress-induced behavioral and biochemical abnormalities in albino rats.

Antimicrobial Activity

Of 8 Euphorbiaceae plants tested for antibacterial activity, the methanol extract of *P. emblica* plant was the most potent with MIC (30 μg/ml) and MBC (40 μg/ml) (Goud et al. 2008). Basant, a polyherbal cream was found to have inhibitory effect on bacterial, fungal and viral genital pathogens (Talwar et al. 2008). Basant has been formulated using diferuloylmethane (curcumin), purified extracts of *Emblica officinalis* (Amla), purified saponins from *Sapindus mukorossi*, *Aloe vera* and rose water along with pharmacopoeially approved excipients and preservatives. Basant inhibited the growth of WHO strains and clinical isolates of *Neisseria gonorrhoeae*, including those resistant to penicillin, tetracycline, nalidixic acid and ciprofloxacin. It displayed pronounced inhibitory action against *Candida glabrata*, *Candida albicans* and *Candida tropicalis* isolated from women with vulvovaginal candidiasis, including three isolates resistant to azole drugs and amphotericin B.

A separate study revealed that both individual and combined aqueous and ethanol extracts of Triphala (*Terminalia chebula*, *Terminalia belerica* and *Emblica officinalis*) had antibacterial activity against the following bacterial isolates: *Pseudomonas aeruginosa*, *Klebsiella pneumoniae*, *Shigella sonnei*, *S. flexneri*, *Staphylococcus aureus*, *Vibrio cholerae*, *Salmonella paratyphi-B*, *Escherichia coli*, *Enterococcus faecalis* and *Salmonella typhi* obtained from HIV infected patients (Srikumar et al. 2007).

Aqueous infusion and decoction of *Emblica officinalis* exhibited potent antibacterial activity against *Escherichia coli, Klebsiella pneumoniae, Klebsiella ozaenae, Proteus mirabilis, Pseudomonas aeruginosa, Salmonella typhi, Salmonella paratyphi A, Salmonella paratyphi B* and *Serratia marcescens* but did not show any antibacterial activity against Gram negative urinary pathogens (Saeed and Tariq 2007).

Short-term feeding of amla in mice caused a decrease in bacterial colonization but it was not significant (Saini et al. 2008). On the contrary, the decrease in bacterial load was significant on long-term feeding. Maximum decrease in malondialdehyde (MDA) levels and increase in phagocytic activity and nitrite levels on long-term feeding was noted. The results suggested that dietary supplementation with amla protected against *Klebsiella pneumoniae* colonization of lungs on long-term feeding in experimental model. The chloroform soluble fraction of the methanolic extract of ripe amlaki fruit containing alkaloids, exhibited significant antimicrobial activity against some Gram positive and Gram negative pathogenic bacteria and strong cytotoxicity having a LC_{50} of 10.257 μg/ml in the brine shrimp lethality bioassay (Rahman et al. 2009). Both essential oils of *P. emblica* obtained by hydrodistillation (HD-EO) and supercritical fluid extraction (SFE-EO) showed a broad spectrum of antimicrobial activity against all the tested Gram-positive and Gram-negative bacteria and three pathogenic fungi (Liu et al. 2009b). Gram-positive bacteria were more sensitive to the investigated oils than Gram-negative bacteria. SFE-EO exhibited a higher antifungal activity compared to HD-EO. The main components of both essential oils were β-caryophyllene, β-bourbonene, 1-octen-3-ol, thymol, and methyleugenol.

Antiviral Activity

The methanol extract of fruits of *Phyllanthus emblica* showed significant inhibitory activity against human immunodeficiency virus-1 (HIV) reverse transcriptase with $IC_{50} \leq 50$ μg/ml (el-Mekkawy et al. 1995). Through a bioassay guided-fractionation of the methanol extract of the fruit of *P. emblica*, putranjivain A (1) was isolated as a potent inhibitory substance with $IC_{50}=3.9$ μM, together with 1,6-di-O-galloyl-β-D-glucose (2), 1-O-galloyl-β-D-glucose (3), kaempferol-3-O-β-D-glucoside (4), quercetin-3-O-β-D-glucoside (5), and digallic acid (6). The inhibitory mode of action by 1, 2, and 6 was non-competitive with respect to the substrate but competitive with respect to a template-primer.

Basant, a polyherbal cream formulated using diferuloylmethane (curcumin), purified extracts of *Emblica officinalis* (Amla), purified saponins from *Sapindus mukorossi, Aloe vera* and rose water along with pharmacopoeially approved excipients and preservatives exhibited antiviral activity (Talwar et al. 2008). Basant displayed a high virucidal action against human immunodeficiency virus HIV-1NL4.3 in CEM-GFP reporter T and P4 (Hela-CD4-LTR-βGal) cell lines with a 50% effective concentration (EC_{50}) of 1:20,000 dilution and nearly complete (98–99%) inhibition at 1:1,000 dilution. It also prevented the entry of HIV-1(IIIB) virus into P4-CCR5 cells (EC_{50} approximately 1:2,492). Two ingredients, Aloe and Amla, inhibited the transduction of human papillomavirus type 16 (HPV-16) pseudovirus in HeLa cells at concentrations far below those that were cytotoxic and those used in the formulation. Basant was found to be totally safe according to pre-clinical toxicology carried out on rabbit vagina after application for 7 consecutive days or twice daily for 3 weeks. Basant was found to have the potential of regressing vulvovaginal candidiasis and preventing *N. gonorrhoeae*, HIV and HPV infections.

Phyllaemblic acid, a novel highly oxygenated norbisabolane was isolated from the roots of *Phyllanthus emblica* (Zhang et al. 2000b). Liu et al. (2009a) isolated three new norsesquiterpenoid glycosides, 4′-hydroxyphyllaemblicin B (1) and phyllaemblicins E (2) and F (3), from the roots of *Phyllanthus emblica*, together with three known compounds, phyllaemblic acid (4), phyllaemblicin B (5), and phyllaemblicin C (6). The isolated compounds, together with two other known analogues, phyllaemblic acid methyl ester

(7) and phyllaemblicin A (8), were evaluated for their antiviral activity toward coxsackie virus B3 (CVB3) by an in vitro cytopathic effect inhibitory assay. Compounds 5–7 exhibited strong anti-CVB3 activity. The polyphenolic compound, 1,2,4,6-tetra-O-galloyl-β-D-glucose isolated from the traditional Chinese medicine *Phyllanthus emblica* was found to an anti-hepatitis B virus (HBV) activity (Xiang et al. 2010). It caused cytotoxicity of HepG2.2.15 as well as HepG2 cells and reduced both HBsAg and HBeAg levels in HepG2.2.15 culture supernatant.

A polyphenolic compound 1,2,4,6-tetra-O-galloyl-β-d-glucose isolated from *Phyllanthus emblica* was found to inhibit in-vitro herpes simplex virus type 1 (HSV-1) and type 2 (HSV-2) infection (Xiang et al. 2011). The compound directly inactivated HSV-1 particles, leading to the failure of early infection, including viral attachment and penetration and also suppressed the intracellular growth of HSV-1 within a long period post-infection. It inhibited HSV-1 E and L gene expressions as well as viral DNA replication but did not affect the RNA synthesis of IE gene. The results suggested that the compound exerted anti-HSV activity both by inactivating extracellular viral particles and by inhibiting viral biosynthesis in host cells.

Hepatoprotective Activity

Extracts of *Phyllanthus emblica* or as a constituents of several commercial polyherbal formulations available in the market were reported to show hepatoprotective effects. A 50% alcoholic extract of *P. emblica* and quercetin isolated from it exhibited significant hepatoprotective effect against country made liquor (CML) and paracetamol challenge in albino rats and mice at the dose of 100 mg/100 g, p.o. and quercetin at the dose of 15 mg/100 g, p.o., respectively (Gulati et al. 1995). *Emblica officinalis* (EO) and Chyavanaprash (CHY) extracts were found to inhibit the hepatotoxicity produced by acute and chronic CCl(4) administration as seen from the decreased levels of serum and liver lipid peroxides (LPO), glutamate-pyruvate transaminase (GPT), and alkaline phosphatase (ALP) (Jose and Kuttan, 200). Chronic CCl(4) administration was also found to produce liver fibrosis as seen from the increased levels of collagen-hydroxyproline and pathological analysis. EO and CHY extracts were found to reduce these elevated levels significantly, indicating that the extract could inhibit the induction of fibrosis in rats. In another study, pretreatment with *E. officinalis* at doses of 100 and 200 mg/kg body weight, prior to CCl4 intoxication showed significant reduction in the levels of SGOT, SGPT, LDH, glutathione-S-transferase, LPO and DNA synthesis (Sultana et al. 2005). There was also increases in reduced glutathione, glutathione peroxidase and glutathione reductase. The pretreatment of *E. officinalis* for 7 consecutive days showed a profound pathological protection to liver cell as depicted by univacuolated hepatocytes. The results suggested that *E. officinalis* inhibited hepatic toxicity in Wistar rats. Separate studies showed that oral administration of methanolic extract of *P. emblica* at a dose of 1.0 g/kg attenuated CCl(4)-induced increases in serum glutamate-oxalate-transaminase (GOT) and serum glutamate-pyruvate-transaminase (GPT) (Lee et al. 2006). Data indicated that the extract was hepatoprotective.

In another study, *E. officinalis* fruits inhibited thioacetamide-induced oxidative stress and hyper-proliferation in rat liver (Sultana et al. 2004). Prophylactic treatment with *E. officinalis* for 7 consecutive days before thioacetamide administration inhibited serum glutamic oxaloacetic transaminase (SGOT), serum glutamic pyruvic transaminase (SGPT) and γ-glutamyl transpeptidase (GGT) release in serum compared with treated control values. It also modulated the hepatic glutathione (GSH) content and malanodialdehyde (MDA) formation. The plant extract caused a marked reduction in levels of GSH content and simultaneous inhibition of MDA formation. *E. officinalis* also caused a reduction in the activity of glutathione-S-transferase (GST), glutathione reductase (GR), glucose 6-phosphate dehydrogenase (G6PD). Glutathione peroxidase (GPx) activity was increased after treatment with the plant extract at doses of 100 mg/kg and 200 mg/kg. Prophylactic treatment with the plant caused a significant down-regulation of ornithine decarboxylase activity and potent inhibition in the rate of DNA synthesis. It was concluded that the acute effects of thioacetamide in rat liver were prevented by pre-treatment with *E. officinalis* extract. This was also confirmed in other studies. A hydroalcoholic (50%) extract of *Emblica officinalis* (fruit) reduced the severity of hepatic fibrosis induced by carbon tetrachloride and thioacetamide (Tasduq et al. 2005a; Mir et al. 2007). Improved liver function was observed by measuring the levels of aspartate aminotransaminase (AST), alanine aminotransferase (ALT), alkaline phosphatase (ALP) and bilirubin in serum. Other hepatic parameters monitored were the levels of glutathione (GSH), lipid peroxidation (LPO) and hydroxyproline and the activities of catalase, glutathione peroxidase (GPx), Na+, K+-ATPase and cytochrome P450 (CYP 450 2E1) (aniline hydroxylation). The results suggested that the extract effectively reversed such alterations with significant regenerative changes suggestive of its preventive role in prefibrogenesis of liver possibly due to its promising antioxidative activity.

In a separate study, Panda and Kar (2003) found that oral administration of the ethanolic extract from emblic fruits at a dose of 250 mg/kg/d (p.o.) for 30 days in hyperthyroid mice reduced triiodothyronine and thyroxine concentrations by 64% and 70% respectively as compared to a standard antithyroid drug, propyl thiouracil that decreased the levels of the thyroid hormones by 59% and 40% respectively. The emblic extract also maintained nearly normal value of glu-6-pase activity in hyperthyroid mice and decreased hepatic lipid peroxidation and increased the superoxide dismutase (SOD) and catalase (CAT) activities in hyperthyroid mice, exhibiting its hepatoprotective nature. Studies conducted in Thailand reported that *Phyllanthus emblica* extract (0.5 and 1 mg/ml) increased cell viability of rat primary cultured hepatocytes being treated with ethanol (96 μl/m) by increasing % MTT (3-(4,5-dimethylthiazol-2-yl) -2,5-diphenyltetrazolium bromide), and decreasing the release of transaminase (Pramyothin et al. 2006). Pretreatment of rats with emblic extract at oral dose of 25, 50 and 75 mg/kg or silymarin, a reference hepatoprotective agent, at 5 mg/kg, 4 hours before ethanol, lowered the ethanol induced levels of aspartate aminotransferase (AST), alanine aminotransferase (ALT) and interleukin-1β (IL-1β). The 75 mg/kg extract dose gave the best result similar to silymarin. Treatment of rats with emblic extract (75 mg/kg/day) or silymarin (5 mg/kg/day) for 7 days after 21 days with ethanol (4 g/kg/day, p.o.) enhanced liver cell recovery by bringing the levels of AST, ALT, IL-1β back to normal. Histopathological studies confirmed the beneficial roles of emblic extract and silymarin against ethanol induced liver injury in rats.

Prefeeding of dehydrated *E. officinalis* powder at 5% and 10% levels was found to counteract the adverse effects of hexachlorocyclohexane (HCH)-induced changes in the rat liver (Anilakumar et al. 2007). HCH induced significant elevation in hepatic malondialdehyde, conjugated dienes and hydroperoxides in the rat liver. The prefeeding of amla at 10% level decreased the formation of these lipid peroxides significantly. The HCH-induced impairment in hepatic catalase, G-6-PDH and SOD activities were modulated by amla at the 10% level of intake. Prefeeding of amla at 5% and 10% levels appeared to reduce the HCH-induced rise in renal GGT activity. The results showed elevation of hepatic antioxidant system and reduction of cytotoxic products as a result of prefeeding of amla, which were otherwise affected by the hexachlorocyclohexane administration.

Studies reported a 50% hydroalcoholic extract *Emblica officinalis* fruit (EO-50) exhibited hepatoprotective effect against antituberculosis (anti-TB) drugs rifampicin (RIF), isoniazid (INH) and pyrazinamide (PZA)-induced hepatic injury (Tasduq et al. 2005a). The hepatoprotective activity of EO-50 was found to be due to its membrane stabilizing, antioxidative and CYP 2E1 inhibitory effects.

Phyllanthus emblica, Tinospora cordifolia and their combination protected against antitubercular drugs (isoniazid, rifampicin and pyrazinamide) induced hepatic damage (Panchabhai et al. 2008). The antituberculosis treatment (ATT), when given for 90 days, induced significant degeneration and necrosis (score: 7.5) associated with morphological changes. However, no change was found in the serum bilirubin and liver enzymes. Co-administration of silymarin (positive control, 50 mg/kg) with ATT protected against necrosis (score: 1.5). *Tinospora cordifolia* (100 mg/kg) showed a reduction in liver damage (score: 6.5), which was not statistically significant. On the other hand, *Phyllanthus emblica* (300 mg/kg) prevented the necrotic changes to a significant extent (grade 1.0). Combination of *Tinospora cordifolia* and *Phyllanthus emblica* in their therapeutic doses (1:3) significantly prevented the necrosis (score: 3.5). Similar effects were seen even when the doses were halved and were comparable to the silymarin group. Thus, this study proves the synergistic protective effects exerted by the combination of *Tinospora cordifolia* and *Phyllanthus emblica* when co-administered with ATT.

The protective role of the fruits of *Emblica officinalis* (500 mg/kg b.wt.) was studied in adult Swiss albino mice against arsenic induced hepatopathy (Sharma et al. 2009). Arsenic treated group (NaAsO2, 4 mg/kg b.wt.) had a significant increase in serum transaminases and lipid peroxidation (LPO) content in liver, whereas significant decrease was recorded in hepatic superoxide dismutase (SOD), catalase (CAT), glutathione-S-transferase (GST) and serum alkaline phosphatase activity. Combined treatment of emblic and arsenic (pre and post) lowered the serum transaminases and LPO content in liver whereas significant increase was noticed in SOD, CAT, GST and serum alkaline phosphatase activities. Liver histopathology showed that emblic fruit extract had reduced karyolysis, karyorrhexis, necrosis and cytoplasmic vacuolization induced by NaAsO2 intoxication. It was concluded that pre- and post-supplementation of *E. officinalis* fruit extract significantly reduced arsenic induced oxidative stress in liver. In another study, gallic acid extracted from leaves of *Phyllanthus emblica* was found to inhibit cell proliferation of human hepatocellular carcinoma BEL-7404 cells and induced apoptosis in a time dependent manner (Zhong et al. 2009).

Nephroprotective Activity

Studies demonstrated that *E. officinalis* or its medicinal preparations may prove to be useful as a component of combination therapy in cancer patients under cyclophosphamide treatment regimen (Haque et al. 2001). Cyclophosphamide (CP) is one of the most popular alkylating anti-cancer drugs in spite of its toxic side effects including immunotoxicity, hematotoxicity, mutagenicity and a host of others. Emblic extract treatment at a dose of 100 mg/kg body weight per os (p.o.) for 10 days resulted in the modulation of these parameters in normal as well as CP (50 mg/kg)-treated animals. Emblic extract in particular was very effective in reducing CP-induced suppression of humoral immunity. Emblic extract treatment in normal animals modulated certain antioxidants of kidney and liver. In CP-exposed animals, emblic pretreatment provided protection to antioxidants of kidney. Not only were the reduced glutathione levels significantly increased but emblic extract treatment resulted in restoration of antioxidant enzymes in CP-treated animals.

Studies reported that could can attenuate age-related renal dysfunction by oxidative stress (Yokozawa et al. 2007b). The administration of SunAmla or ethyl acetate extract of amla reduced the elevated levels of serum creatinine and urea nitrogen in the aged rats. In addition, the tail arterial blood pressure was markedly elevated in aged control rats as compared with young rats, while the systolic blood pressure was significantly decreased by the administration of SunAmla or ethyl acetate extract of amla. Furthermore, the oral administration of SunAmla or ethyl acetate extract of amla significantly reduced thiobarbituric acid-reactive substance levels of serum, renal homogenate, and mitochondria in aged rats, suggesting that amla would ameliorate oxidative stress under aging. The increases of inducible nitric oxide synthase (iNOS) and cyclooxygenase (COX)-2 expression in the aorta of aging rats were also significantly suppressed by SunAmla extract or ethyl acetate extract of amla, respectively. Moreover, the elevated expression level of bax, a proapoptotic protein, was significantly decreased after oral administration of SunAmla or ethyl acetate extract of amla. SunAmla or ethyl acetate extract of amla also reduced the iNOS and COX-2 expression levels by inhibiting NF-kappaB activation in the aged rats. The results indicated that amla would be a very useful antioxidant for the prevention of age-related renal disease.

Recent studies also showed that supplementation with Amla extract for 4 months reduced the plasma oxidative marker, 8-iso-prostaglandin and increased plasma total antioxidant status in uremic patients (Chen et al. 2009). However, Amla extract did not influence hepatic or renal function, or diabetic and atherogenic indices in uremic patients.

Gastroprotective/Antiulcerogenic Activity

Gastric ulcers induced by oral administration of absolute ethanol (5 ml/kg) to fasted rats were reduced dose-dependently by oral pretreatment of animals with either *E. officinalis* fruit juice or its methanol extract (25–100 mg/kg) (Rajeshkumar, et al. 2001). Ethanol administration caused severe gastric damage with an ulcer index of 4.6, a 44% reduction in glutathione (GSH) content of gastric mucosa, an increase in stomach weight due to inflammation (1.24 g/100 g body weight), intraluminal bleeding (100%), and increased mortality rate (44%). *E. officinalis* fresh fruit juice administration (50 mg/kg) 30 minutes before alcohol challenge, reduced the ulcer index to 1.8, limited the depletion of GSH to 15.2%, reduced the stomach weight to 0.75 g/100 g body weight, and afforded 100% protection against mortality and intraluminal bleeding. Administration of indomethacin (25 mg/kg) and histamine (10 mg/kg) increased the ulcer index to 2.2 and 1.8 respectively, and a 28.5% and 20.6% depletion in mucosal GSH, respectively, as compared to normal rats. *E. officinalis* administration showed a dose-dependent protective effect against gastric damage induced by indomethacin and histamine. The protection afforded by *E. officinalis* fruits was found to be better than that of ranitidine (50 mg/kg). The results of the present study suggested the novel cytoprotective activity of *E. officinalis* fruits on gastric mucosal cells.

The methanolic extract of *Emblica officinalis* (EOE) exhibited ulcer protective effect in different acute gastric ulcer models in rats induced by aspirin, ethanol, cold restraint stress and pyloric ligation and healing effect in chronic gastric ulcers induced by acetic acid in rats (Sairam et al. 2002). Further study on gastric mucosal factors showed that it significantly decreased the offensive factors like acid and pepsin outputs and increased the defensive factors like mucin secretion, cellular mucus and life span of mucosal cells. EOE showed significant antioxidant effect in stressed animals and did not have any effect on cell proliferation in terms of DNA µg/mg protein or glandular weight. The results showed that EOE had significant ulcer protective and healing effects and this might be due to its effects both on offensive and defensive mucosal factors. In another study, oral administration of ethanol extract Amla extract at doses 250 mg/kg and 500 mg/kg significantly inhibited the development of gastric lesions in all test models used including pylorus ligation (Al-Rehaily et al. 2002). It also caused significant decrease of the pyloric-ligation induced basal gastric secretion, titratable acidity and gastric mucosal injury. Besides, Amla extract offered protection against ethanol-induced depletion of stomach wall mucus and reduction in nonprotein sulfhydryl concentration. Histopathological analyses were in good agreement with pharmacological and biochemical findings. The results indicated that Amla extract possessed antisecretory, antiulcer, and cytoprotective properties.

Ethanol extracts of *Piper betel, Emblica officinalis, Terminalia bellerica*, and *Terminalia chebula* exhibited healing activity against the indomethacin-induced stomach ulceration (Bhattacharya et al. 2007). Compared to autohealing, all the drugs accelerated the healing process, albeit to different extents. The relative healing activities of the extracts was $P.\ betel > E.\ officinalis > T.\ bellerica \sim T.\ chebula$, that correlated well with their in vivo antioxidant and mucin augmenting activities. The excellent healing activity of the extracts of *P. betel* and *E. officinalis* indicated a major role of mucin protection and regeneration in the healing of nonsteroidal anti-inflammatory drugs mediated stomach ulceration.

A modified indigenous Siddha formulation Kalpaamruthaa (KA), containing *Semecarpus anacardium* nut milk extract (SA), dried powder of *Emblica officinalis* (EO) fruit and honey exhibited analgesic, antipyretic and ulcerogenic properties (Mythilypriya et al. 2007a). KA exhibited an enhanced effect on all properties compared with that found with sole SA treatment, due to synergistic and additive interactions within the complex mixture of phytochemicals present in KA. Pepticare, a herbomineral formulation of the Ayurveda medicine consisting of the herbal drugs: *Glycyrrhiza glabra, Emblica officinalis* and *Tinospora cordifolia*, exhibited anti-ulcer and antioxidant activity in rats (Bafna and Balaraman 2005). The reduction in ulcer index in both the models (pylorus-ligation and on ethanol-induced gastric mucosal injury) along with the reduction in volume and total acidity, and an increase in the pH of gastric fluid in pylorus-ligated rats proved the anti-ulcer activity of Pepticare. It was also found that Pepticare was more potent than *G. glabra* alone in protecting against pylorus-ligation and ethanol-induced ulcers. The increase in the levels of superoxide dismutase, catalase, reduced glutathione and membrane bound enzymes like Ca^{2+} ATPase, Mg^{2+} ATPase and $Na+K+$ ATPase and decrease in lipid peroxidation in both the models proved the anti-oxidant activity of the formulation. Thus it was concluded that Pepticare possessed antiulcer activity, which could be attributed to its antioxidant mechanism of action.

Antiinflammatory Activity

Studies reported that *Emblica officinalis* extract had antiinflammatory property and identified pyrogallol as an active compound responsible for the antiinflammatory effect (Nicolis et al. 2008). *Emblica officinalis* (EO) extract strongly inhibited the *Pseudomonas aeruginosa* dependent expression of the neutrophil chemokines IL-8, GRO-α, GRO-γ, of the adhesion molecule ICAM-1 and of the pro-inflammatory cytokine IL-6. *Pseudomonas aeruginosa* causes morbidity and mortality in patients with cystic fibrosis (CF) chronic lung inflammatory disease which is characterized by chronic infection and inflammation of the lungs. Pyrogallol, one of the compounds extracted from EO, inhibited the *P. aeruginosa*-dependent expression of these pro-inflammatory genes similarly to the whole EO extract, whereas a second compound purified from EO, namely 5-hydroxy-isoquinoline, had no effect. In another study, *Emblica officinalis* extract showed a marked reduction in inflammation and edema in adjuvant induced arthritic (AIA) rat (Ganju et al. 2003). At cellular level immunosuppression occurred during the early phase of the disease. There was mild synovial hyperplasia and infiltration of few mononuclear cells in amla treated animals. The induction of nitric oxide synthase (NOS) was significantly decreased in treated animals as compared to controls.

Antiinflammatory activity was found in the water fraction of methanol extract of emblic leaves using carrageenan- and dextran-induced rat hind paw oedema models (Asmawi et al. 1993). The effects of the same fraction were tested on the synthesis of mediators of inflammation such as leukotriene B4 (LTB4), platelet-activating factor (PAF) and thromboxane B2 (TXB2), and on LTB4- and N-formyl-L-methionyl-L-leucyl-L-phenylalanine (FMLP)-induced migration of human polymorphonuclear leucocytes (PMNs) in-vitro. The water fraction of the methanol extract inhibited migration of human polymorphonuclear leucocytes PMNs in relatively low concentrations. It did not inhibit leukotriene B4 (LTB4) or platelet-activating factor synthesis in human PMNs or thromboxane B2 (TXB2) synthesis in human platelets during clotting, suggesting that the mechanism of the antiinflammatory action found in the rat paw model does not involve inhibition of the synthesis of the measured lipid mediators. Subsequently, the scientists reported that methanol, tetrahydrofuran, and 1,4-dioxane extracts (50 µg/ml) of emblic leaves inhibited leukotriene B4-induced migration of human polymorphonuclear leukocyte (PMN) by 90% and N-formyl-L-methionyl-L-leucyl-L-phenylalanine (FMLP)-induced degranulation by 25–35% (Ihantola-Vormisto et al. 1997). Diethyl ether extract (50 µg/ml) inhibited calcium ionophore A23187-induced leukotriene B4 release from human PMNs by 40%, thromboxane B2 production in platelets during blood clotting by 40% and adrenaline-induced platelet aggregation by 36%. The results showed that the leaves of *Phyllanthus emblica* had inhibitory activity on PMNs and platelets, which confirmed the antiinflammatory and antipyretic properties of this plant as suggested by its use in traditional medicine in China, India, Indonesia, and the Malay Peninsula. The data suggested that the plant leaves contained as yet unidentified polar compound(s) with potent inhibitory activity on PMNs and chemically different apolar molecule(s) which inhibited both prostanoid and leukotriene synthesis.

Studies found that Kalpaamruthaa (KA) a modified indigenous Siddha formulation consisting of *Semecarpus anacardium* nut milk extract (SA), *Emblica officinalis* (EO) and honey had antiinflammatory effects (Mythilypriya et al. 2008). It was observed that the drug KA (dose of 150 mg/kg body weight) exhibited enhanced effect on antiinflammatory and antiarthritic properties than sole SA treatment and the collective effect of KA might be due to the combined interactions of the phytochemicals such as flavonoids, tannins and other compounds such as vitamin C present in KA.

A novel, safe polyherbal formulation (Aller-7/NR-A2) had been developed for the treatment of allergic rhinitis using a unique combination of extracts from seven medicinal plants including *Phyllanthus emblica, Terminalia chebula, Terminalia bellerica, Albizia lebbeck, Piper nigrum, Zingiber officinale* and *Piper longum* (Pratibha et al. 2004). At a dose of 250 mg/kg, Aller-7 demonstrated 62.55% inhibition against compound 48/80-induced paw edema in Balb/c mice, while under the same conditions prednisolone at an oral dose of 14 mg/kg exhibited 44.7% inhibition. Aller-7 significantly inhibited compound 48/80-induced paw edema at all three doses of 175, 225 or 275 mg/kg in Swiss Albino mice, while the most potent effect was observed at 225 mg/kg. Aller-7 (120 mg/kg, p.o.) demonstrated 31.3% inhibition against carrageenan-induced acute inflammation in Wistar Albino rats, while ibuprofen (50 mg/kg, p.o.) exerted 68.1% inhibition. Aller-7 also exhibited a dose-dependent (150–350 mg/kg) antiinflammatory effect against Freund's adjuvant-induced arthritis in Wistar Albino rats and an approximately 63% inhibitory effect was observed at a dose of 350 mg/kg. The trypsin inhibitory activity of Aller-7 was determined, using ovomucoid as a positive control. Ovomucoid and Aller-7 demonstrated IC_{50} concentrations at 1.5 and 9.0 µg/ml, respectively. The results indicated this novel polyherbal formulation to be a potent antiinflammatory agent to ameliorate the symptoms of allergic rhinitis.

Ethanol amla extract exhibited biphasic activity in non-steroidal antiinflammatory drug (NSAID)-induced gastro-ulcerated mice, with healing effect observed at 60 mg/kg and an adverse effect at 120 mg/kg (Chatterjee et al. 2011). The switch from anti-oxidant to pro-oxidant shift and immunomodulatory property could be the major cause for its biphasic effect, as evident from the total antioxidant status, thiol concentration, lipid peroxidation, protein carbonyl content followed by mucin content, PGE(2) synthesis and cytokine status. Further, buthionine sulfoxamine (BSO) pretreatment confirmed the potential impact of antioxidative property in the healing action of amla extract. However, amla extract efficiently reduced pro-inflammatory cytokine (TNF-α and IL-1β) levels and considerably upregulated antiinflammatory cytokine (IL-10) concentration. It was concluded that, gastric ulcer healing by the ethanol amla extract was driven in a dose-specific manner through the harmonization of the antioxidative property and modulation of antiinflammatory cytokine level.

Antiarthritic/Chondroprotective Activity

Extracts of *Emblica officinalis* were able to induce programmed cell death of mature osteoclasts (Penolazzi et al. 2008). Osteoclasts (OCs) are involved in rheumatoid arthritis and in several pathologies associated with bone loss. *Emblica officinalis* increased the expression levels of Fas, a gene that encodes one of several proteins important to apoptosis. Gel shift experiments demonstrated that *Emblica officinalis* extracts act by interfering with NF-kB activity, a transcription factor involved in osteoclast biology. The data obtained demonstrated that *Emblica officinalis* extracts selectively compete with the binding of transcription factor NF-kB to its specific target DNA sequences. The study showed that induction of apoptosis of osteoclasts could be an important strategy both in interfering with rheumatoid arthritis complications of the bone skeleton leading to joint destruction, and preventing and reducing osteoporosis. The study indicated the potential application of *Emblica officinalis* extracts as an alternative tool for therapy applied to bone diseases.

In another study, aqueous extracts of both *Phyllanthus emblica* fruit powders A and B significantly inhibited the activities of hyaluronidase and collagenase type 2 in vitro (Sumantran et al. 2008). In the explant model of cartilage matrix damage, extracts of glucosamine sulphate and emblic powder B (0.05 mg/ml) exhibited statistically significant, long-term chondroprotective activity in cartilage explants from 50% of the patients tested. This result is important since glucosamine sulphate is the leading nutraceutical for osteoarthritis. Emblic powder A induced a statistically significant, short-term chondroprotective activity in cartilage explants from all of the patients tested. The data provided pilot pre-clinical evidence for the use of *P. emblica* fruits as a chondroprotective agent in osteoarthritis therapy.

Separate studies showed that the levels/activities of reactive oxygen species (ROS)/reactive nitrogen species (RNS), myeloperoxidase and lipid peroxide were increased significantly and the activities of enzymic and non-enzymic antioxidants were in turn decreased in arthritic rats, whereas these changes were reverted to near normal levels upon *Semecarpus anacardium* (SA) and Kalpaamruthaa treatment (Mythilypriya et al. 2007b, 2008). Kalpaamruthaa is a modified indigenous Siddha preparation constituting *Semecarpus anacardium* nut milk extract (SA), *Emblica officinalis* and honey. Kalpaamruthaa showed an enhanced antioxidant potential than sole treatment of SA in adjuvant induced arthritic rats. Kalpaamruthaa via enhancing the antioxidant status in adjuvant induced arthritic rats than sole SA treatment proved to be an important therapeutic modality in the management of rheumatoid arthritis and thereby instituting the role of oxidative stress in the clinical manifestation of the disease rheumatoid arthritis (Mythilypriya et al. 2007b). The profound antiinflammatory, antiarthritic and antioxidant efficacy of Kalpaamruthaa compared to SA alone might be due to the synergistic action of the polyphenols such as flavonoids, tannins and other compounds such as vitamin C and hydroxycinnamates present in Kalpaamruthaa.

Antitussive Activity

Studies showed that at a higher dose (200 mg/kg body weight) of emblic perorally was more effective, than at a lower dose, especially in decreasing the number of cough efforts (NE), frequency of cough (NE/minute) and the intensity of cough attacks in inspirium (IA+) and expirium (IA-) was more pronounced (Nosál'ová et al. 2003). These results showed that the cough suppressive activity of *E. officinalis* was dose-dependent. The antitussive activity of emblic was less effective than the classical narcotic antitussive drug codeine, but more effective than the non-narcotic antitussive agent dropropizine. It was postulated that the antitussive activity of the dry extract of *Emblica officinalis* was due not only to antiphlogistic, antispasmolytic and antioxidant efficacy effects, but also to its effect on mucus secretion in the airways.

Anticataleptic Activity

Aqueous extract of the fruits of *Emblica officinalis* was found to have protective effect on haloperidol (1.0 mg/kg intraperitoneal administration)-induced catalepsy in mice (Pemminati et al. 2009). The effects of the test drug EO (0.8, 2.0 and 4.0 mg/kg doses) and the standard neuroleptic drugs scopolamine (1.0 mg/kg) and ondansetron (0.5 and 1.0 mg/kg doses) were assessed after single and repeated dose administration for 7 days, 30 min prior to the haloperidol. A significant reduction in the cataleptic scores was observed in all the test drug treated groups as compared to the control, with maximum reduction in the dose 4.0 mg/kg group. Similarly, the maximum reduction in super oxide dismutase (SOD) activity was observed in the dose 4.0 mg/kg group. The study suggested that EO had significantly reduced oxidative stress and the cataleptic score induced by haloperidol. It could be used to prevent drug-induced extrapyramidal side effects.

Antiepileptic Activity

Golechha et al. (2010) reported that the hydroalcoholic extract of *Emblica officinalis* had antiepileptic activity. They showed that ten days of 500 and 700 mg/kg intra-peritoneal doses of the extract completely abolished the generalized tonic seizures and also improved the retention latency in passive avoidance task. Further, the extract dose-dependently ameliorated the oxidative stress induced by pentylenetetrazole (PTZ).

Wound Healing Activity

Ascorbic acid and tannins of low molecular weight, namely emblicanin A (2,3-di-O-galloyl-4,6-(S)-hexahydroxydiphenoyl-2-keto-glucono-δ-lactone) and emblicanin B (2,3,4,6-bis-(S)-hexahydroxydiphenoyl-2-keto-glucono-δ-lactone) present in *Emblica officinalis*, had been shown to exhibit a very strong antioxidant action. Studies showed that these antioxidants would support the repair process in wounds (Sumitra et al. 2009). Emblic increased cellular proliferation and cross-linking of collagen at the wound site, as evidenced by an increase in the activity of extracellular signal-regulated kinase 1/2, along with an increase in DNA, type III collagen, acid-soluble collagen, aldehyde content, shrinkage temperature and tensile strength. Higher levels of tissue ascorbic acid, α-tocopherol, reduced glutathione, superoxide dismutase, catalase, and glutathione peroxidase supported the fact that emblic application promoted antioxidant activity at the wound site.

Analgesic and Antipyretic Activities

Treatment with flavonoid rich fruit extract (ethyl acetate:methanol fraction) of *E. officinalis* and quercetin in diabetic rats showed significant increase in tail flick latency in hot immersion test and pain threshold level in hot plate test compared to control rats (Kumar et al. 2009). Diabetic rats exhibited a significant hyperalgesia (nociception) as compared to control rats. The changes in lipid peroxidation status and antioxidant enzymes (superoxide dismutase and catalase) levels observed in diabetic rats were significantly restored by *E. officinalis* extract and quercetin treatment. Both, *E. officinalis* extract and quercetin attenuated diabetic induced axonal degeneration. The study provided experimental evidence of the preventive and curative effect of *E. officinalis* on nerve function and oxidative stress in animal model of diabetic neuropathy. Diabetic neuropathic pain is an important microvascular complication in diabetes mellitus and oxidative stress plays a vital role in associated neural and vascular complications.

Extracts of *Emblica officinalis* fruits was found to possess potent antipyretic and analgesic activity (Perianayagam et al. 2004). A single oral dose of ethanol (EEO) and aqueous (AEO) extracts of *Emblica officinalis* fruits (500 mg/kg, i.p.) showed significant reduction in brewer's yeast induced hyperthermia in rats. EEO and AEO also elicited pronounced inhibitory effect on acetic acid-induced writhing response in mice in the analgesic test. Both, EEO and AEO did not show any significant analgesic activity in the tail-immersion test.

Radioprotective Activity

A review study reported that *Emblica officinalis* was one of several botanicals that exhibited protection against radiation-induced lethality, lipid peroxidation and DNA damage (Jagetia 2007). The dose of *Emblica officinalis* fruit pulp extract found to be most effective against sublethal γ radiation (9 Gy) radiation was 100 mg/kg body weight (Singh et al. 2005). This dose increased the survival time and reduced the mortality rate of mice significantly. Furthermore, body weight loss in *Emblica officinalis* administered irradiated animals was significantly less in comparison with animals who were given radiation only. Oral administration of *Emblica officinalis* before exposure to γ radiation was found to be effective in protecting mice against the hematological and biochemical modulation in peripheral blood (Singh et al. 2006). A significant increase in the RBC, WBC, hemoglobin, and hematocrit values was observed in the animals pretreated with *E. officinalis* extract as compared to the hematological values observed in the irradiated group. Furthermore, radiation sickness was greatly inhibited in those mice that were irradiated with prior treatment of E. officinalis. A significant decrease in glutathione (GSH) content and increase in lipid peroxidation (LPx) level were also observed in irradiated animals; whereas *E. officinalis* pretreated irradiated animals exhibited a significant increase in GSH content and decrease in LPx level, but such remained below the normal. The results suggested that *E. officinalis* pretreatment provided protection against irradiation to Swiss albino mice.

Administration of *Emblica officinalis* fruit pulp significantly increased total leukocyte count, bone marrow viability and hemoglobin levels, which were lowered by irradiation (7 Gy) (Hari Kumar et al. 2004). Emblic significantly enhanced the activity of the various antioxidant enzymes catalase (CAT), superoxide dismutase (SOD), glutathione peroxidase (GPX), and glutathione-S-transferase (GST) as well as glutathione system in the blood. Treatment with emblic also lowered the elevated levels of lipid peroxides in the serum. The data clearly indicated that the extract significantly reduced the bioeffects of radiation.

Recent studies showed that mice receiving *Emblica officinalis* extract prior to γ-irradiation (5 Gy) had a higher number of crypt cells and mitotic figures when compared with non-extract-treated control at all the autopsy intervals (Jindal et al. 2009). Irradiation of animals resulted in a dose-dependent elevation in lipid peroxidation and a reduction in glutathione as well as catalase concentration in the intestine at 1 hour post-irradiation. In contrast, emblic extract treatment before irradiation caused a significant depletion in lipid peroxidation and elevation in glutathione and catalase levels.

Studies found that radiation induced mortality was reduced by 60% in mice fed with Triphala (TPL), an Ayurvedic formulation (1 g/kg body weight/day) orally for 7 days prior to γ-irradiation at 7.5 Gy followed by post-irradiation feeding for 7 days (Sandhya et al. 2006b). It was concluded that TPL protected whole body irradiated mice and TPL induced protection was mediated through inhibition of oxidative damage in cells and organs. TPL seems to have potential to develop into a novel herbal radio-protector for practical applications.

Studies suggested amla fruit effectively inhibits UVB-induced photo-aging in human skin fibroblast via its strong ROS scavenging ability (Adil et al. 2010). It stimulated UVB inhibited cellular proliferation and protected pro-collagen 1 against UVB-induced depletion via inhibition of UVB-induced MMP-1 in skin fibroblasts. It also prevented UVB disturbed cell cycle to normal phase.

Majeed et al. (2011) showed that *Phyllanthus emblica* fruit extract significantly protected against ultraviolet-B (UVB) irradiation-induced reactive oxygen species (ROS) and collagen damage in normal human dermal fibroblasts indicating promising cosmeceutical benefits against photoaging. At a concentration of 0.5 mg/ml, emblic extract showed a significant response of 9.5-fold protection from UVB induced-collagen damage as compared to untreated cells while a known active, ascorbic acid, at a concentration of 0.5 mg/ml, showed 3.7-fold protection untreated cells showed 84% induction in ROS on UVB irradiation as compared to the non-irradiated cells. Emblic extract treatment inhibited the induction of ROS to 15% at a concentration of 0.5 mg/ml while ascorbic acid inhibited the induction in ROS to 64% at the same concentration level.

Antiaging Activity

Amla extract stimulated proliferation of fibroblasts in a concentration-dependent manner, and also induced production of procollagen in a concentration- and time-dependent manner (Fujii et al. 2008). Conversely, matrix metalloproteinase, MMP-1 production from fibroblasts was dramatically decreased, but there was no evident effect on MMP-2. Tissue inhibitor of metalloproteinase-1 (TIMP-1) was significantly increased by amla extract. From these results, it appeared that amla extract worked effectively in mitigative, therapeutic and cosmetic applications through control of collagen metabolism. In a recent study, *P. emblica* extract, at a concentration of 0.1 mg/ml, significantly increased the type I pro-collagen level up to 1.65-fold, and 6.78-fold greater than that of an untreated control (Chanvorachote et al. 2010). Emblica extract caused an approximately 7.75-fold greater type I pro-collagen induction compared to the known herbal collagen enhancer asiaticoside at the same treatment concentration (0.1 mg/ml). Additionally, emblic extract inhibited collagenase activity in a dose-dependent manner. The results suggested that emblic extract had a promising pharmacological effect that benefited collagen synthesis and protected against its degradation and could be used as a natural anti-aging ingredient. In an earlier study, a standardized extract of *Phyllanthus emblica* (trade named Emblica) was found to have a long-lasting and broad-spectrum antioxidant activity (Chaudhuri 2002). The product had no pro-oxidation activity induced by iron and/or copper because of its iron and copper chelating ability. Emblic helped protect the skin from the damaging effects of free radicals, non-radicals and transition metal-induced oxidative stress. Emblic was found suitable for use in anti-aging, sunscreen and general purpose skin care products.

Antidiarrheal Activity

The methanol extract of emblic fruit showed a significant inhibitory effect on rats with diarrhea induced by castor oil and magnesium sulphate (Perianayagam et al. 2005). The methanol extract produced a significant reduction in gastrointestinal motility in charcoal meal tests in Wistar albino rats. It also significantly inhibited PGE2-induced enteropooling as compared to control animals. The results obtained established the efficacy and substantiated the use of this herbal remedy as a nonspecific treatment for diarrhoea in folk medicine. *Phyllanthus emblica* fruit extract was found to possess antidiarrheal and spasmolytic activities, mediated possibly through dual blockade of muscarinic receptors and Ca(2+) channels (Mehmood et al. 2011). The crude amla extract containing alkaloids, tannins, terpenes, flavonoids, sterols and coumarins, suppressed castor oil-induced diarrhea and intestinal fluid accumulation in mice at 500–700 mg/kg. In isolated rabbit jejunum, the extract relaxed carbachol and K(+) (80 mM)-induced contractions, in a pattern similar to that of dicyclomine. The preincubation of guinea pig-ileum with the extract (1 mg/ml), produced a non-parallel rightward shift with suppression of the maximum response, similar to that of dicyclomine, suggesting anticholinergic and Ca(2+) channel blocking (CCB)-like antispasmodic effect.

Pancreas-Protective Activity

Preliminary studies showed that dogs pretreated with *Emblica officinalis* (28 mg/kg daily for 15 days) before inducing pancreatitis had less acinar cell damage and significantly lower total inflammatory score than those induced with acute pancreatitis and the rise in serum amylase was also lower (Thorat et al. 1995). Data showed that *Emblica officinalis* had protective effect against acute necrotising pancreatitis in dogs.

Immunomodulatory Activity

Triphala, an ayurvedic rayana formula consisting of equal parts of *Terminalia chebula*, *Terminalia belerica* and *Emblica officinalis,* was found to have immunomodulatory activity (Srikumar et al. 2005). In Triphala administration (1 g/kg/day for 48 days), avidity index was found to be significantly enhanced in the Triphala immunized group of albino rats, while the remaining neutrophil functions and steroid levels were not altered significantly. However the neutrophil functions were significantly enhanced in the Triphala immunized group with a significant decrease in corticosterone level. Upon exposure to the noise-stress, the neutrophil functions were significantly suppressed and followed by a significant increase in the corticosterone levels were observed in both the noise-stress and the noise-stress immunized groups. These noise-stress-induced changes were significantly prevented by Triphala administration in both the Triphala noise-stress and the Triphala noise-stress immunized groups. The study revealed that oral administration of Triphala appeared to stimulate the neutrophil functions in the immunized rats and stress induced suppression in the neutrophil functions were significantly prevented by Triphala.

Antiamnesic Activity

Anwala churna (*Emblica officinalis*), an Ayurvedic preparation at, 50, 100, and 200 mg/kg, p.o in rats produced a dose-dependent improvement in memory scores of young and aged rats (Vasudevan and Parle 2007). The elevated plus-maze and Hebb-Williams maze was used as exteroceptive behavioral models for testing memory. Diazepam-, scopolamine-, and ageing induced amnesia served as the interoceptive behavioral models. Furthermore, Anwala churna reversed the amnesia induced by scopolamine (0.4 mg/kg, i.p.) and diazepam (1 mg/kg, i.p.). It was concluded that Anwala churna may prove to be a useful remedy for the management of Alzheimer's disease due to its multifarious beneficial effects such as memory improvement and reversal of memory deficits.

Anti-Ochratoxin Activity

Chakraborty and Verma (2009); Verma and Chakraborty (2008b) found that oral administration of ochratoxin to mice followed by *Emblica officinalis* aqueous extract treatment or oral administration of ochratoxin along with *Emblica officinalis* aqueous extract ameliorated ochratoxin-induced spermatotoxic effect and caused significant recovery in all the sperm parameters as well as in fertility rate. Oral administration of ochratoxin for 45 days caused, as compared to vehicle control, dose-dependent significant reduction in cauda epididymal sperm count, sperm motility, sperm viability and fertility rate caused, significant, amelioration in ochratoxin-induced lipid peroxidation by increasing the contents of non-enzymatic GSH (glutathione) and TAA (total ascorbic acid) and activities of enzymatic SOD (superoxide dismutase), CAT (catalase), GPX (glutathione peroxidase), GRX (glutathione reductase) and GST (glutathione transferase) antioxidants in the testis of mice as compared with those given ochratoxin alone animals. However oral administration of aqueous extract of *Emblica officinalis* alone did not cause any significant changes in above mentioned parameters. They found also that oral administration of *Emblica officinalis* aqueous extract (2 mg/animal/day) and ochratoxin for a period of 45 days caused a significant amelioration in the ochratoxin-induced lipid peroxidation and ochratoxin-induced reduction in DNA, RNA and protein contents in the liver and kidney of mice (Verma and Chakraborty 2008a; Chakraborty and Verma 2010).

Antivenom Activity

Methanolic root extract of *Emblica officinalis* was found to possess antivenom activity (Alam and Gomes 2003). The extract significantly antagonized the *Vipera russellii* and *Naja kaouthia* venom induced lethal activity both in in-vitro and in-vivo studies. *V. russellii* venom-induced haemorrhage, coagulant, defibrinogenating and inflammatory activity was significantly neutralized by the extract.

Antiplasmodial Activity

Bagavan et al. (2011) found promising antiplasmodial (*Plasmodium falciparum*) activity in the extracts from two plants, *Phyllanthus emblica* leaf 50% inhibitory concentration (IC_{50}:7.25 mug/ml (ethyl acetate extract), 3.125 mug/ml (methanol extract), and *Syzygium aromaticum* flower bud, IC_{50}:13 mug/ml, (ethyl acetate extract) and 6.25 mug/ml (methanol extract). The above mentioned plant extracts were also found to be active against chloroquine-resistant strains. Cytotoxicity study with *P. emblica* leaf and *S. aromaticum* flower bud, extracts showed good therapeutic indices. These results demonstrated that leaf ethyl acetate and methanol extracts of *P. emblica* and flower bud extract of *S. aromaticum* may serve as antimalarial agents even in their crude form. Water extracts of *Phyllanthus emblica, Terminalia chebula* and *Terminalia bellerica* exhibited in-vitro and in-vivo antiplasmodial activity with good selectivity (Pinmai et al. 2010). All plant extracts showed antimalarial activity (IC_{50} values ranging from 14.0 to 15.41 µg/ml). The water extract of *Terminalia bellerica* had the highest in vitro antiplasmodial activity followed by *Phyllanthus emblica* and *Terminalia chebula*. The cytotoxic activity was exhibited by all plant extracts on Vero cells with IC_{50} values of 157.86–238.70 mg/ml. All of the plant extracts showed selectivity with the selectivity index (SI) ranging from 11 to 17. The extracts at 250 mg/kg/day exerted in-vivo antiplasmodial activity with good suppression activity ranging from 53.40% to 69.46%. All plant extracts contained flavonoids, hydrolysable tannins, saponin and terpenes.

Antiparasitic Activity

All plant extracts including that of *P. emblica* showed moderate toxic effect on parasites after 24 hours of exposure; methanol extract of *P. emblica* had an LC_{50} of 225.57 ppm against the adult cattle tick *Haemaphysalis bispinosa*, LC_{50} of 60.60 ppm against sheep fluke *Paramphistomum cervi*, and LC_{50} of 54.82 ppm against Japanese encephalitis mosquito vector, *Culex tritaeniorhynchus* (Bagavan et al. 2009).

Traditional Medicinal Uses

According to Baliga and Dsouza (2011) *Emblica officinalis* is arguably the most important medicinal plant in the Indian traditional system of medicine, the Ayurveda. Various parts of the plant are used to treat a range of diseases, but the most important is the fruit. The fruit is used either alone or in combination with other plants to treat many ailments that include common cold and fever; as a diuretic, laxative, liver tonic, refrigerant, stomachic, restorative, alterative, antipyretic, anti-inflammatory, hair tonic; to prevent peptic ulcer and dyspepsia, and as a digestive. *Phyllanthus emblica*, is of great importance in Asian traditional medicine especially in South Asia in the Ayurvedic and Unani (Graceo – arab) traditional medicinal systems (Krishnaveni and Mirunalini 2010). *Phyllanthus emblica* is highly nutritious and could be an important dietary source of vitamin C, amino acids, and minerals. The plant also contains phenolic compounds, tannins, phyllembelic acid, phyllembelin, rutin, curcum-inoids, and emblicol. All parts of the plant are used for medicinal purposes, especially the fruit, which has been used in Ayurveda as a potent rasayana and in traditional medicine for the treatment of diarrhea, jaundice, and inflammation. Amla is used as a *rasayana* (rejuvenating tonic) to promote longevity, and traditionally to enhance digestion, to treat constipation, reduce cough, reduce fever, purify the blood, alleviate asthma, strengthen the heart, benefit the eyes stimulate hair growth, enliven the body, enhance intellect. Amla is the primary and most important constituent of an ancient Ayurvedic polyherbal formulation called *Chyavanaprash* that is used as a *rasayan*a, and in the treatment of chronic lung and heart diseases, infertility and mental disorders. This formula, which contains 43 herbal ingredients as well as clarified butter, sesame oil, sugar cane juice, and honey. For sexual rejuvenation, Chyavanprash is added into warm milk or spread on toast, and consume every day.

Triphala composed of the three medicinal fruits *Emblica officinalis, Terminalia chebula* and *Terminalia belerica* is an important herbal preparation in the traditional Indian system of medicine, Ayurveda (Baliga 2010). Triphala is an antioxidant-rich herbal formulation and possesses diverse beneficial properties. Triphala is used to cleanse the body tissues, pacify all three ayurvedic doshas, and act as a rasayana to promote good health and long life. Triphala is administered for chronic dysentery, biliousness, hemorrhoids, enlarged liver, and other disorders. Another valued rasayana that contains Amalaki as the primary constituent is *Brahma rasayana*, which endows the person that takes it the vigor resembling an elephant, intelligence, strength, wisdom and right attitude. Kalpaamruthaa is a modified indigenous Siddha preparation constituting *Semecarpus anacardium* nut milk extract, *Emblica officinalis* and honey.

The fruit is the most commonly used plant part, and the fresh fruit is preferred. Anti-ascorbutic benefits have been attributed to the fruits, which are known as the Emblic Myrobalans. Fresh fruit is purgative and used as poultices. Fruit juice is taken internally for dyspepsia and as a diuretic. Unripe fruit is cored and the exuding juice is used topically in conjunctivitis. The unripe fruits are also made into pickles and given as aperients before meals to stimulate the appetite in anorexia. The dried fruit is used as a decoction to treat ophthalmia when applied externally, and is used internally as a hemostatic and antidiarrheal. The boiled, reconstituted dried fruit, blended into a smooth liquid with a small quantity of honey or jaggery added, is useful in anorexia, anemia, biliousness dyspepsia, and jaundice. This is also an excellent restorative in chronic rhinitis and fever, with swollen and dry red lips and rashes about the mouth. The dried fruit prepared as a decoction and taken on a regular basis is useful in menorrhagia and leucorrhea, and is an excellent post-partum restorative. A powder prepared from the dried fruit is an effective expectorant as it stimulates the bronchial glands. An infusion made by steeping dried fruit overnight in water also serves as an eyewash, as does an infusion of the seeds. A liquor made from the fermented fruits is prescribed as a treatment for indigestion, anaemia, jaundice, some cardiac problems, nasal congestion and retention of urine.

Emblic leaves, too, are taken internally for indigestion and diarrhea or dysentery, especially in combination with buttermilk, sour milk or fenugreek. A decoction of the leaves is used as a mouthwash and as a lotion for sore eyes. The plant is considered an effective antiseptic in cleaning wounds, and it is also one of the many plant palliatives for snakebite and scorpion stings. The flowers, considered refrigerant and aperient, and the roots is said to be emetic.

The bark is strongly astringent and used in the treatment of diarrhea. The milky sap of the fresh bark is mixed with honey and turmeric and administered in cases of gonorrhea. The root bark, mixed with honey, is applied to inflammations of the mouth. The seeds are burnt, powdered and mixed in oil as a useful pruritis for scabies or itch. The seeds are used in treating asthma, bronchitis, diabetes and fevers. The seeds are fried in ghee and ground in congee is applied to the forehead to stop bleeding from the nose.

In Thailand ma-khaam pom fruits are traditionally used as an expectorant, antipyretic, diuretic, antidiarrhoeal and antiscurvy. Emblic has been used for anti-inflammatory and antipyretic treatments by rural populations in its growing areas. Malays use a decoction of its leaves to treat fever. In Indonesia, the dried fruit is used for dysentery and a poultice of it on the head the pulp of the fruit is smeared on the head to dispel headache and vertigo caused by excessive heat.

Other Uses

The hard but flexible red wood, that is susceptible to warping and splitting, is used for minor construction, furniture, implements, gunstocks, hookahs and ordinary pipes. The wood is also used to clarify water in crude aqueducts and inner braces for wells. Branches and chips of the wood are thrown into muddy streams for clarification and to impart a pleasant flavor. The wood serves also as fuel and a source of charcoal. The tannin-rich bark, as well as the immature fruit and leaves, are highly valued and widely employed for tanning in India and Thailand, often in combination with other tanning materials such as chebulic myrobalan (*Terminalia chebula*) and beleric myrobalan (*Terminalia bellirica*). The twig bark is particularly esteemed for tanning leather. Leaves and fruits are used for animal fodder, whereas leaves can also be applied as green manure. An essential oil is distilled from the leaves for use in perfumery. The leaves and immature fruits are employed for dyeing matting, bamboo wickerwork, silk and wool into light-brown or yellow-brown hues. Grey and black colours are obtained when iron sulphate salts are used as mordants. Matting can be dyed dark colours with a decoction of the bark. The fruits are used to prepare a black ink and a hair dye. In Hinduism, amla is regarded as a sacred tree worshipped as Mother Earth.

Comments

Emblic is usually propagated by seeds, cuttings, air-layering and various forms of budding.

Selected References

Adil MD, Kaiser P, Satti NK, Zargar AM, Vishwakarma RA, Tasduq SA (2010) Effect of *Emblica officinalis* (fruit) against UVB-induced photo-aging in human skin fibroblasts. J Ethnopharmacol 132(1):109–114

Alam MI, Gomes A (2003) Snake venom neutralization by Indian medicinal plants (*Vitex negundo* and *Emblica officinalis*) root extracts. J Ethnopharmacol 86(1):75–80

Al-Rehaily AJ, Al-Howiriny TA, Al-Sohaibani MO, Rafatullah S (2002) Gastroprotective effects of 'Amla' *Emblica officinalis* on in vivo test models in rats. Phytomedicine 9(6):515–522

Anila L, Vijayalakshmi NR (2002) Flavonoids from *Emblica officinalis* and *Mangifera indica* – effectiveness for dyslipidemia. J Ethnopharmacol 79(1):81–87

Anilakumar KR, Nagaraj NS, Santhanam K (2007) Reduction of hexachlorocyclohexane-induced oxidative stress and cytotoxicity in rat liver by *Emblica officinalis* Gaertn. Indian J Exp Biol 45(5):450–454

Arulkumaran S, Ramprasath VR, Shanthi P, Sachdanandam P (2006) Restorative effect of Kalpaamruthaa, an indigenous preparation, on oxidative damage in mammary gland mitochondrial fraction in experimental mammary carcinoma. Mol Cell Biochem 291(1-2):77–82

Asmawi MZ, Kankaanranta H, Moilanen E, Vapaatalo H (1993) Anti-inflammatory activities of *Emblica officinalis* Gaertn leaf extracts. J Pharm Pharmacol 45(6):581–584

Bafna PA, Balaraman R (2005) Anti-ulcer and anti-oxidant activity of pepticare, a herbomineral formulation. Phytomedicine 12(4):264–270

Bagavan A, Kamaraj C, Elango G, Abduz Zahir A, Abdul Rahuman A (2009) Adulticidal and larvicidal efficacy of some medicinal plant extracts against tick, fluke and mosquitoes. Vet Parasitol 166(3–4):286–292

Bagavan A, Rahuman AA, Kamaraj C, Kaushik NK, Mohanakrishnan D, Sahal D (2011) Antiplasmodial activity of botanical extracts against *Plasmodium falciparum*. Parasitol Res 108(5):1099–1109

Baliga MS (2010) Triphala, Ayurvedic formulation for treating and preventing cancer: a review. J Altern Complement Med 16(12):1301–1308

Baliga MS, Dsouza JJ (2011) Amla (*Emblica officinalis* Gaertn), a wonder berry in the treatment and prevention of cancer. Eur J Cancer Prev 20(3):225–239

Banu SM, Selvendiran K, Singh JP, Sakthisekaran D (2004) Protective effect of *Emblica officinalis* ethanolic extract against 7,12-dimethylbenz(a) anthracene (DMBA) induced genotoxicity in Swiss albino mice. Hum Exp Toxicol 23(11):527–531

Barthakur NN, Arnold NP (1991) Chemical analysis of the emblic (*Phyllanthus emblica* L.) and its potential as a food source. Sci Hort 47(1–2):99–105

Bhattachary SK, Bhattacharya D, Muruganandam AV (2000a) Effect of *Emblica officinalis* tannoids on a rat model of tardive dyskinesia. Indian J Exp Biol 38(9):945–947

Bhattacharya A, Chatterjee A, Ghosal S, Bhattacharya SK (1999) Antioxidant activity of active tannoid principles of *Emblica officinalis* (amla). Indian J Exp Biol 37(7):676–680

Bhattacharya A, Ghosal S, Bhattacharya SK (2000b) Antioxidant activity of tannoid principles of *Emblica officinalis* (amla) in chronic stress induced changes in rat brain. Indian J Exp Biol 38(9):877–880

Bhattacharya SK, Bhattacharya A, Sairam K, Ghosal S (2002) Effect of bioactive tannoid principles of *Emblica officinalis* on ischemia-reperfusion-induced oxidative stress in rat heart. Phytomedicine 9(2):171–174

Bhattacharya SK, Chaudhuri SR, Chattopadhyay S, Bandyopadhyay SK (2007) Healing properties of some Indian medicinal plants against indomethacin-induced gastric ulceration of rats. J Clin Biochem Nutr 41(2):106–114

Biswas S, Talukder G, Sharma A (1999) Protection against cytotoxic effects of arsenic by dietary supplementation with crude extract of *Emblica officinalis* fruit. Phytother Res 13(6):513–516

Brand Miller J, James KW, Maggiore P (1993) Tables of composition of Australian aboriginal foods. Aboriginal Studies Press, Canberra

Burkill IH (1966) A dictionary of the economic products of the Malay Peninsula. Revised reprint, 2 vols. Ministry of Agriculture and Co-operatives, Kuala Lumpur. vol 1 (A–H), pp 1–1240, vol 2 (I–Z), pp 1241–2444

Calixto JB, Santos AR, Cechinel-Filho V, Yunes RA (1998) A review of the plants of the genus *Phyllanthus*: their chemistry, pharmacology and therapeutic potential. Med Res Rev 18(4):225–258

Chakraborty D, Verma R (2009) Spermatotoxic effect of ochratoxin and its amelioration by *Emblica officinalis* aqueous extract. Acta Pol Pharm 66(6):689–695

Chakraborty D, Verma R (2010) Ameliorative effect of *Emblica officinalis* aqueous extract on ochratoxin-induced lipid peroxidation in the kidney and liver of mice. Int J Occup Med Environ Health 23(1): 63–73

Chalise JP, Acharya K, Gurung N, Bhusal RP, Gurung R, Skalko-Basnet N, Basnet P (2010) Antioxidant activity and polyphenol content in edible wild fruits from Nepal. Int J Food Sci Nutr 61(4):425–432

Chanvorachote P, Pongrakhananon V, Luanpitpong S, Chanvorachote B, Wannachaiyasit S, Nimmannit U (2010) Type I pro-collagen promoting and anti-collagenase activities of *Phyllanthus emblica* extract in mouse fibroblasts. J Cosmet Sci 60(4):395–403

Chatterjee A, Chattopadhyay S, Bandyopadhyay SK (2011) Biphasic effect of *Phyllanthus emblica* L. extract on NSAID-induced ulcer: an antioxidative trail weaved with immunomodulatory effect. Evid Based Compl Alt Med 2011:146808

Chaudhuri RK (2002) Emblica cascading antioxidant: a novel natural skin care ingredient. Skin Pharmacol Appl Skin Physiol 15(5):374–380

Chen TS, Liou SY, Chang YL (2008) Antioxidant evaluation of three adaptogen extracts. Am J Chin Med 36(6):1209–1217

Chen TS, Liou SY, Chang YL (2009) Supplementation of *Emblica officinalis* (Amla) extract reduces oxidative stress in uremic patients. Am J Chin Med 37(1):19–25

Chopra RN, Nayar SL, Chopra IC (1986) Glossary of Indian medicinal plants. (Including the supplement). Council Scientific Industrial Research, New Delhi, 330 pp

Council of Scientific and Industrial Research (CSIR) (1950) The wealth of India: a dictionary of Indian raw materials and industrial products, vol 2, Raw materials. Publications and Information Directorate, New Delhi

D'Souza P, Amit A, Saxena VS, Bagchi D, Bagchi M, Stohs S (2004) Antioxidant properties of Aller-7, a novel polyherbal formulation for allergic rhinitis. J Drugs Exp Clin Res 30(3):99–109

Deep G, Dhiman M, Rao AR, Kale RK (2005) Chemopreventive potential of Triphala (a composite Indian drug) on benzo(a)pyrene induced forestomach tumorigenesis in murine tumor model system. J Exp Clin Cancer Res 24(4):555–563

Dhanalakshmi S, Devi RS, Srikumar R, Manikandan S, Thangaraj R (2007) Protective effect of Triphala on cold stress-induced behavioral and biochemical abnormalities in rats. Yakugaku Zasshi 127(11): 1863–1867

Dhir H, Agarwal K, Sharma A, Talukder G (1991) Modifying role of *Phyllanthus emblica* and ascorbic acid against nickel clastogenicity in mice. Cancer Lett 59(1):9–18

Dhir H, Roy AK, Sharma A (1993) Relative efficiency of *Phyllanthus emblica* fruit extract and ascorbic acid in modifying lead and aluminium-induced sister-chromatid exchanges in mouse bone marrow. Environ Mol Mutagen 21(3):229–236

Duan W, Yu Y, Zhang L (2005) Antiatherogenic effects of *Phyllanthus emblica* associated with corilagin and its analogue. Yakugaku Zasshi 125(7):587–591

El-Desouky SK, Ryu SY, Kim YK (2008) A new cytotoxic acylated apigenin glucoside from *Phyllanthus emblica* L. Nat Prod Res 22(1):91–95

el-Mekkawy S, Meselhy MR, Kusumoto IT, Kadota S, Hattori M, Namba T (1995) Inhibitory effects of Egyptian folk medicines on human immunodeficiency virus (HIV) reverse transcriptase. Chem Pharm Bull (Tokyo) 43(4):641–648

Fujii T, Wakaizumi M, Ikami T, Saito M (2008) Amla (*Emblica officinalis* Gaertn.) extract promotes procollagen production and inhibits matrix metalloproteinase-1 in human skin fibroblasts. J Ethnopharmacol 119(1):53–57

Ganju L, Karan D, Chanda S, Srivastava KK, Sawhney RC, Selvamurthy W (2003) Immunomodulatory effects of agents of plant origin. Biomed Pharmacother 57(7):296–300

Golechha M, Bhatia J, Arya DS (2010) Hydroalcoholic extract of *Emblica officinalis* Gaertn. affords protection against PTZ-induced seizures, oxidative stress and cognitive impairment in rats. Indian J Exp Biol 48(5):474–478

Gopalan G, Rama Sastri BV, Balasubramanian SC (2002) Nutritive value of Indian foods. National Institute of Nutrition, Indian Council of Medical Research, Hyderabad

Goud MJ, Komraiah A, Rao KN, Ragan A, Raju VS, Charya MA (2008) Antibacterial activity of some folklore medicinal plants from South India. Afr J Tradit Complement Altern Med 5(4):421–426

Grover IS, Kaur S (1989) Effect of *Emblica officinalis* Gaertn. (Indian gooseberry) fruit extract on sodium azide and 4-nitro-o-phenylenediamine induced mutagenesis in *Salmonella typhimurium*. Indian J Exp Biol 27(3):207–209

Gulati RK, Agarwal S, Agrawal SS (1995) Hepatoprotective studies on *Phyllanthus emblica* Linn. and quercetin. Indian J Exp Biol 33(4):261–268

Haque R, Bin-Hafeez B, Ahmad I, Parvez S, Pandey S, Raisuddin S (2001) Protective effects of *Emblica officinalis* Gaertn. in cyclophosphamide-treated mice. Hum Exp Toxicol 20(12):643–650

Hari Kumar KB, Sabu MC, Lima PS, Kuttan R (2004) Modulation of haematopoetic system and antioxidant enzymes by *Emblica officinalis* Gaertn. and its protective role against gamma-radiation induced damages in mice. J Radiat Res (Tokyo) 45(4):549–555

Hu JF (1990) Inhibitory effects of *Phyllanthus emblica* juice on formation of N-nitrosomorpholine in vitro and N-nitrosoproline in rat and human. Zhonghua Yu Fang Yi Xue Za Zhi 24(3):132–135 (In Chinese)

Ihantola-Vormisto A, Summanen J, Kankaanranta H, Vuorela H, Asmawi ZM, Moilanen E (1997) Anti-inflammatory activity of extracts from leaves of *Phyllanthus emblica*. Planta Med 63(6):518–524

Jacob A, Pandey M, Kapoor S, Saroja R (1988) Effect of the Indian gooseberry (amla) on serum cholesterol levels in men aged 35–55 years. Eur J Clin Nutr 42(11):939–944

Jagetia GC (2007) Radioprotective potential of plants and herbs against the effects of ionizing radiation. J Clin Biochem Nutr 40(2):74–81

Jansen PCM (2005) *Phyllanthus emblica* L. [Internet] record from protabase. In: Jansen, PCM, Cardon D (eds) PROTA (Plant Resources of Tropical Africa/Ressources végétales de l'Afrique tropicale), Wageningen, Netherlands. http://database.prota.org/search.htm

Jeena KJ, Joy KL, Kuttan R (1999) Effect of *Emblica officinalis*. *Phyllanthus amarus* and *Picrorrhiza kurroa* on N-nitrosodiethylamine induced hepatocarcinogenesis. Cancer Lett 136(1):11–16

Jindal A, Soyal D, Sharma A, Goyal PK (2009) Protective effect of an extract of *Emblica officinalis* against radiation-induced damage in mice. Integr Cancer Ther 8(1):98–105

Jose JK, Kuttan R (1995) Antioxidant activity of *Emblica officinalis*. J Clin Biochem Nutr 19:63–70

Jose JK, Kuttan R (2000) Hepatoprotective activity of *Emblica officinalis* and Chyavanaprash. J Ethnopharmacol 72(1–2):135–140

Jose JK, Kuttan G, Kuttan R (2001) Antitumour activity of *Emblica officinalis*. J Ethnopharmacol 75(2–3): 65–69

Kaur S, Arora S, Kaur K, Kumar S (2002) The in vitro antimutagenic activity of Triphala – an Indian herbal drug. Food Chem Toxicol 40(4):527–534

Kaur S, Michael H, Arora S, Härkönen PL, Kumar S (2005) The in vitro cytotoxic and apoptotic activity of Triphala – an Indian herbal drug. J Ethnopharmacol 97(1):15–20

Kennard WC, Winters HF (1960) Some fruits and nuts for the tropics. USDA Agric Res Serv Misc Publ 801: 1–135

Khan MT, Lampronti I, Martello D, Bianchi N, Jabbar S, Choudhuri MS, Datta BK, Gambari R (2002) Identification of pyrogallol as an antiproliferative compound present in extracts from the medicinal plant *Emblica officinalis*: effects on in vitro cell growth of human tumor cell lines. Int J Oncol 21(1):187–192

Khopde SM, Priyadarsini KI, Mohan H, Gawandi VB, Satav JG, Yakhmi JV, Banavaliker MM, Biyani MK, Pittal JP (2001) Characterizing the antioxidant activity of Amla (*Phyllanthus emblica*) extract. Curr Sci 81: 185–190

Kim HJ, Yokozawa T, Kim HY, Tohda C, Rao TP, Juneja LR (2005) Influence of amla (*Emblica officinalis* Gaertn.) on hypercholesterolemia and lipid peroxidation in cholesterol-fed rats. J Nutr Sci Vitaminol (Tokyo) 51(6):413–418

Kim HY, Okubo T, Juneja LR, Yokozawa T (2010) The protective role of amla (*Emblica officinalis* Gaertn.) against fructose-induced metabolic syndrome in a rat model. Br J Nutr 103(4):502–512

Kirtikar KR, Basu BD (1975) Indian medicinal plants. 4 vols. 2nd edn. Jayyed Press, New Delhi

Krishnaveni M, Mirunalini S (2010) Therapeutic potential of *Phyllanthus emblica* (amla): the ayurvedic wonder. J Basic Clin Physiol Pharmacol 21(1):93–105

Krishnaveni M, Mirunalini S, Karthishwaran K, Dhamodharan G (2010) Antidiabetic and antihyperlipidemic properties of *Phyllanthus emblica* Linn. (Euphorbiaceae) on streptozotocin induced diabetic rats. Pak J Nutr 9(1):43–51

Kumar NP, Annamalai AR, Thakur RS (2009) Antinociceptive property of *Emblica officinalis* Gaertn. (Amla) in high fat diet-fed/low dose streptozotocin induced diabetic neuropathy in rats. Indian J Exp Biol 47(9):737–742

Kumaran A, Karunakaran R (2006) Nitric oxide radical scavenging active components from *Phyllanthus emblica* L. J Plant Foods Hum Nutr 61(1):1–5

Kusirisin W, Srichairatanakool S, Lerttrakarnnon P, Lailerd N, Suttajit M, Jaikang C, Chaiyasut C (2009) Antioxidative activity, polyphenolic content and antiglycation effect of some Thai medicinal plants traditionally used in diabetic patients. Med Chem 5(2):139–147

Lee CY, Peng WH, Cheng HY, Chen FN, Lai MT, Chiu TH (2006) Hepatoprotective effect of *Phyllanthus* in Taiwan on acute liver damage induced by carbon tetrachloride. Am J Chin Med 34(3):471–482

Liu X, Zhao M, Wang J, Yang B, Jiang Y (2008) Antioxidant activity of methanolic extract of emblica fruit (*Phyllanthus emblica* L.) from six regions in China. J Food Compost Anal 21(3):219–228

Liu Q, Wang YF, Chen RJ, Zhang MY, Wang YF, Yang CR, Zhang YJ (2009a) Anti-coxsackie virus B3 norsesquiterpenoids from the roots of *Phyllanthus emblica*. J Nat Prod 72(5):969–972

Liu XL, Zhao MM, Luo W, Yang B, Jiang YM (2009b) Identification of volatile components in *Phyllanthus emblica* L. and their antimicrobial activity. J Med Food 12(2):423–428

Madhavi D, Devil KR, Rao KK, Reddy PP (2007) Modulating effect of *Phyllanthus* fruit extract against lead genotoxicity in germ cells of mice. J Environ Biol 28(1):115–117

Majeed M, Bhat B, Anand S, Sivakumar A, Paliwal P, Geetha KG (2011) Inhibition of UV-induced ROS and collagen damage by *Phyllanthus emblica* extract in normal human dermal fibroblasts. J Cosmet Sci 62(1):49–56

Mathur R, Sharma A, Dixit VP, Varma M (1996) Hypolipidaemic effect of fruit juice of *Emblica officinalis* in cholesterol-fed rabbits. J Ethnopharmacol 50(2):61–68

Mehmood MH, Siddiqi HS, Gilani AH (2011) The antidiarrheal and spasmolytic activities of *Phyllanthus emblica* are mediated through dual blockade of muscarinic receptors and Ca2+ channels. J Ethnopharmacol 133(2):856–865

Mehta S, Singh RK, Jaiswal D, Rai PK, Watal G (2009) Anti-diabetic activity of *Emblica officinalis* in animal models. Pharm Biol 47(11):1050–1055

Ministry of Public Health (1970) Tables of nutrition values in Thai food per 100 gm of edible portion. Office of the Prime Minister, Royal Thai Government, Bangkok

Mir AI, Kumar B, Tasduq SA, Gupta DK, Bhardwaj S, Johri RK (2007) Reversal of hepatotoxin-induced prefibrogenic events by *Emblica officinalis*–a histological study. Indian J Exp Biol 45(7):626–629

Mishra M, Pathak UN, Khan AB (1981) *Emblica officinalis* Gaertn and serum cholesterol level in experimental rabbits. Br J Exp Pathol 62(5):526–528

Molesworth Allen B (1967) Malayan fruits: an introduction to the cultivated species. Moore, Singapore, 245 pp

Morton JF (1960) The emblic (*Phyllanthus emblica* L.). Econ Bot 14:119–128

Morton JF (1987) Emblic. In: Fruits of warm climates. Julia F. Morton, Miami, pp 213–217

Mythilypriya R, Shanthi P, Sachdanandam P (2007a) Analgesic, antipyretic and Ulcerogenic properties of an indigenous formulation – Kalpaamruthaa. Phytother Res 21(6):574–578

Mythilypriya R, Shanthi P, Sachdanandam P (2007b) Restorative and synergistic efficacy of Kalpaamruthaa, a modified Siddha preparation, on an altered antioxidant status in adjuvant induced arthritic rat model. Chem Biol Interact 168(3):193–202

Mythilypriya R, Shanthi P, Sachdanandam P (2008) Synergistic effect of Kalpaamruthaa on antiarthritic and antiinflammatory properties – its mechanism of action. Inflammation 31(6):391–398

Naik GH, Priyadarsini KI, Bhagirathi RG, Mishra B, Mishra KP, Banavalikar MM, Mohan H (2005) In vitro antioxidant studies and free radical reactions of triphala, an ayurvedic formulation and its constituents. Phytother Res 19(7):582–586

Nandi P, Talukder G, Sharma A (1997) Dietary chemoprevention of clastogenic effects of 3,4-benzo(a)pyrene by *Emblica officinalis* Gaertn. fruit extract. Br J Cancer 76(10):1279–1283

Ngamkitidechakul C, Jaijoy K, Hansakul P, Soonthornchareonnon N, Sireeratawong S (2010) Antitumour effects of *Phyllanthus emblica* L.: induction of cancer cell apoptosis and inhibition of in vivo tumour promotion and in vitro invasion of human cancer cells. Phytother Res 24:1405–1413

Nicolis E, Lampronti I, Dechecchi MC, Borgatti M, Tamanini A, Bianchi N, Bezzerri V, Mancini I, Grazia Giri M, Rizzotti P, Gambari R, Cabrini G (2008) Pyrogallol, an active compound from the medicinal plant *Emblica officinalis*, regulates expression of pro-inflammatory genes in bronchial epithelial cells. Int Immunopharmacol 8(12):1672–1680

Niwano Y, Saito K, Yoshizaki F, Kohno M, Ozawa T (2011) Extensive screening for herbal extracts with potent antioxidant properties. J Clin Biochem Nutr 48(1):78–84

Nosáľová G, Mokrý J, Hassan KM (2003) Antitussive activity of the fruit extract of *Emblica officinalis* Gaertn. (Euphorbiaceae). Phytomedicine 10(6–7):583–589

Ochse JJ, van den Brink RCB (1980) Vegetables of the Dutch Indies, 3rd edn. Ascher & Co., Amsterdam, 1016 pp

Panchabhai TS, Ambarkhane SV, Joshi AS, Samant BD, Rege NN (2008) Protective effect of *Tinospora cordifolia Phyllanthus emblica* and their combination against antitubercular drugs induced hepatic damage: an experimental study. Phytother Res 22(5):646–650

Panda S, Kar A (2003) Fruit extract of *Emblica officinalis* ameliorates hyperthyroidism and hepatic lipid peroxidation in mice. Pharmazie 58(10):753–755

Pemminati S, Nair V, Dorababu P, Gopalakrishna HN, Pai MRSM (2009) Effect of aqueous fruit extract of *Emblica officinalis* on haloperidol induced catalepsy in albino mice. J Clin Diagn Res 3:1657–1662

Penolazzi L, Lampronti I, Borgatti M, Khan MT, Zennaro M, Piva R, Gambari R (2008) Induction of apoptosis of human primary osteoclasts treated with extracts from the medicinal plant *Emblica officinalis*. BMC Complement Altern Med 8:59

Perianayagam J, Sharma S, Joseph A, Christina A (2004) Evaluation of anti-pyretic and analgesic activity of *Emblica officinalis* Gaertn. J Ethnopharm 95:83–85

Perianayagam JB, Narayanan S, Gnanasekar G, Pandurangan A, Raja S, Rajagopal K, Rajesh R, Vijayarajkumar P, Vijayakumar SG (2005) Evaluation of antidiarrheal potential of *Emblica officinalis*. Pharm Biol 43(4):373–377

Pinmai K, Chunlaratthanabhorn S, Ngamkitidechakul C, Soonthornchareon N, Hahnvajanawong C (2008) Synergistic growth inhibitory effects of *Phyllanthus emblica* and *Terminalia bellerica* extracts with conventional cytotoxic agents: doxorubicin and cisplatin against human hepatocellular carcinoma and lung cancer cells. World J Gastroenterol 14(10):1491–1497

Pinmai K, Hiriote W, Soonthornchareonnon N, Jongsakul K, Sireeratawong S, Tor-Udom S (2010) In vitro and in vivo antiplasmodial activity and cytotoxicity of water extracts of *Phyllanthus emblica, Terminalia chebula*, and *Terminalia bellerica*. J Med Assoc Thai 93(suppl7):S120–S126

Poltanov EA, Shikov AN, Dorman HJ, Pozharitskaya ON, Makarov VG, Tikhonov VP, Hiltunen R (2009) Chemical and antioxidant evaluation of Indian gooseberry (*Emblica officinalis* Gaertn., syn. *Phyllanthus emblica* L.) supplements. Phytother Res 23(9):1309–1315

Pozharitskaya ON, Ivanova SA, Shikov AN, Makarov VG (2007) Separation and evaluation of free radical-scavenging activity of phenol components of *Emblica officinalis* extract by using an HPTLC-DPPH method. J Sep Sci 30(9):1250–1254

Pramyothin P, Samosorn P, Poungshompoo S, Chaichantipyuth C (2006) The protective effects of *Phyllanthus emblica* Linn. extract on ethanol induced rat hepatic injury. J Ethnopharmacol 107(3):361–364

Pratibha N, Saxena VS, Amit A, D'Souza P, Bagchi M, Bagchi D (2004) Anti-inflammatory activities of Aller-7, a novel polyherbal formulation for allergic rhinitis. Int J Tissue React 26(1–2):43–51

Purseglove JW (1968) Tropical crops: dicotyledons, 1 and 2nd edn. Longman, London, 719 pp

Qureshi SA, Asad W, Sultana V (2009) The effect of *Phyllantus emblica* Linn. on Type - II Diabetes, triglycerides and liver – specific enzyme. Pak J Nutr 8(2):125–128

Rahman S, Akbor MM, Howlader A, Jabbar A (2009) Antimicrobial and cytotoxic activity of the alkaloids of Amlaki (*Emblica officinalis*). Pak J Biol Sci 12(16):1152–1155

Rajak S, Banerjee SK, Sood S, Dinda AK, Gupta YK, Gupta SK, Maulik SK (2004) *Emblica officinalis* causes myocardial adaptation and protects against oxidative stress in ischemic-reperfusion injury in rats. Phytother Res 18(1):54–60

Rajeshkumar NV, Pillai MR, Kuttan R (2003) Induction of apoptosis in mouse and human carcinoma cell lines by *Emblica officinalis* polyphenols and its effect on chemical carcinogenesis. J Exp Clin Cancer Res 22(2):201–212

Rajeshkumar NV, Therese M, Kuttan R (2001) *Emblica officinalis* fruits afford protection against experimental gastric ulcers in rats. Pharm Biol 39(5):375–380

Ramprasath VR, Kumar BS, Shanthi P, Sachdanandam P (2006) Therapeutic efficacy of Kalpaamruthaa on lipid peroxidation and antioxidant status in experimental mammary carcinoma in rats. J Health Sci 52(6):748–757

Rao TP, Sakaguchi N, Juneja LR, Wada E, Yokozawa T (2005) Amla (*Emblica officinalis* Gaertn.) extracts reduce oxidative stress in streptozotocin-induced diabetic rats. J Med Food 8(3):362–368

Reddy VD, Padmavathi P, Paramahamsa M, Varadacharyulu N (2009) Modulatory role of *Emblica officinalis* against alcohol induced biochemical and biophysical changes in rat erythrocyte membranes. Food Chem Toxicol 47(8):1958–1963

Rege NN, Thatte UM, Dahanukar SA (1999) Adaptogenic properties of six rasayana herbs used in Ayurvedic medicine. Phytother Res 13(4):275–291

Rehman HU, Yasin KA, Choudhary MA, Khaliq N, Choudhary MI, Malik S (2007) Studies on the chemical constituents of *Phyllanthus emblica*. Nat Prod Res 21(9):775–781

Roy AK, Dhir H, Sharma A (1992) Modification of metal-induced micronuclei formation in mouse bone marrow erythrocytes by *Phyllanthus* fruit extract and ascorbic acid. Toxicol Lett 62(1):9–17

Saeed S, Tariq P (2007) Antibacterial activities of *Emblica officinalis* and *Coriandrum sativum* against Gram negative urinary pathogens. Pak J Pharm Sci 20(1):32–35

Sai Ram M, Neetu D, Deepti P, Vandana M, Ilavazhagan G, Kumar D, Selvamurthy W (2003) Cytoprotective activity of Amla (*Emblica officinalis*) against chromium (VI) induced oxidative injury in murine macrophages. Phytother Res 17(4):430–433

Sai Ram M, Neetu D, Yogesh B, Anju B, Dipti P, Pauline T, Sharma SK, Sarada SK, Ilavazhagan G, Kumar D, Selvamurthy W (2002) Cyto-protective and immunomodulating properties of Amla (*Emblica officinalis*) on lymphocytes: an in-vitro study. J Ethnopharmacol 81(1):5–10

Saini A, Sharma S, Chhibber S (2008) Protective efficacy of *Emblica officinalis* against Klebsiella pneumoniae induced pneumonia in mice. Indian J Med Res 128(2):188–193

Sairam K, Rao ChV, Babu MD, Kumar KV, Agrawal VKK, Goel RK (2002) Antiulcerogenic effect of methanolic extract of *Emblica officinalis*: an experimental study. J Ethnopharmacol 82(1):1–9

Saito Y, Kohno M, Yoshizaki F, Niwano Y (2008a) Antioxidant properties of herbal extracts selected from screening for potent scavenging activity against superoxide anions. J Sci Food Agr 88(15):2707–2712

Saito K, Kohno M, Yoshizaki F, Niwano Y (2008b) Extensive screening for edible herbal extracts with potent scavenging activity against superoxide anions. Plant Foods Hum Nutr 63(2):65–70

Sancheti G, Jindal A, Kumari R, Goyal PK (2005) Chemopreventive action of *Emblica officinalis* on skin carcinogenesis in mice. Asian Pac J Cancer Prev 6(2):197–201

Sandhya T, Lathika KM, Pandey BN, Bhilwade HN, Chaubey RC, Priyadarsini KI, Mishra KP (2006b) Protection against radiation oxidative damage in mice by Triphala. Mutat Res 609(1):17–25

Sandhya T, Lathika KM, Pandey BN, Mishra KP (2006a) Potential of traditional ayurvedic formulation, Triphala, as a novel anticancer drug. Cancer Lett 231(2):206–214

Sandhya T, Mishra KP (2006) Cytotoxic response of breast cancer cell lines, MCF 7 and T 47 D to triphala and its modification by antioxidants. Cancer Lett 238(2):304–313

Saravanan S, Srikumar R, Manikandan S, Jeya Parthasarathy N, Sheela Devi R (2007) Hypolipidemic effect of triphala in experimentally induced hypercholesteremic rats. Yakugaku Zasshi 127(2):385–388

Saw LG, LaFrankie JV, Kochummen KM, Yap SK (1991) Fruit trees in a Malaysian rain forest. Econ Bot 45(1):120–136

Scartezzini P, Antognoni F, Raggi MA, Poli F, Sabbioni C (2006) Vitamin C content and antioxidant activity of the fruit and of the Ayurvedic preparation of *Emblica officinalis* Gaertn. J Ethnopharmacol 104(1–2):113–118

Sharma A, Sharma MK, Kumar M (2009) Modulatory role of *Emblica officinalis* fruit extract against arsenic induced oxidative stress in Swiss albino mice. Chem Biol Interac 180(1):20–30

Shi Y, Sahu RP, Srivastava SK (2008) Triphala inhibits both in vitro and in vivo xenograft growth of pancreatic tumor cells by inducing apoptosis. BMC Cancer 8:294

Shukla V, Vashistha M, Singh SN (2009) Evaluation of antioxidant profile and activity of amalaki (*Emblica officinalis*), spirulina and wheat grass. Indian J Clin Biochem 24(1):70–75

Singh I, Sharma A, Nunia V, Goyal PK (2005) Radioprotection of Swiss albino mice by *Emblica officinalis*. Phytother Res 19(5):444–446

Singh I, Soyal D, Goyal PK (2006) *Emblica officinalis* (Linn.) fruit extract provides protection against radiation-induced hematological and biochemical alterations in mice. J Environ Pathol Toxicol Oncol 25(4):643–654

Srikumar R, Parthasarathy NJ, Devi RS (2005) Immunomodulatory activity of Triphala on neutrophil functions. Biol Pharm Bull 28(8):1398–1403

Srikumar R, Parthasarathy NJ, Shankar EM, Manikandan S, Vijayakumar R, Thangaraj R, Vijayananth K, Sheeladevi R, Rao UA (2007) Evaluation of the growth inhibitory activities of Triphala against common bacterial isolates from HIV infected patients. Phytother Res 21(5):476–480

Subhadrabanhdu S (2001) Under utilized tropical fruits of Thailand. FAO Rap publication 2001/26. 70 pp

Sultana S, Ahmed S, Sharma S, Jahangir T (2004) Emblica officinalis reverses thioacetamide-induced oxidative stress and early promotional events of primary hepatocarcinogenesis. J Pharm Pharmacol 56(12):1573–1579

Sultana S, Ahmed S, Khan N, Jahangir T (2005) Effect of Emblica officinalis (Gaertn) on CCl4 induced hepatic toxicity and DNA synthesis in Wistar rats. Indian J Exp Biol 43(5):430–436

Sultana S, Ahmed S, Jahangir T (2008) Emblica officinalis and hepatocarcinogenesis: a chemopreventive study in Wistar rats. J Ethnopharmacol 118(1):1–6

Sumantran VN, Kulkarni A, Chandwaskar R, Harsulkar A, Patwardhan B, Chopra A, Wagh UV (2008) Chondroprotective potential of fruit extracts of Phyllanthus emblica in osteoarthritis. Evid Based Compl Alt Med 5(3):329–335

Sumitra M, Manikandan P, Gayathri VS, Mahendran P, Suguna L (2009) Emblica officinalis exerts wound healing action through up-regulation of collagen and extracellular signal-regulated kinases (ERK1/2). Wound Repair Regen 17(1):99–107

Suresh K, Vasudevan DM (1994) Augmentation of murine natural killer cell and antibody dependent cellular cytotoxicity activities by Phyllanthus emblica, a new immunomodulator. J Ethnopharmacol 44(1):55–60

Suryanarayana K, Kumar PA, Saraswat M, Petrash JM, Reddy GB (2004) Inhibition of aldose reductase by tannoid principles of Emblica officinalis: implications for the prevention of sugar cataract. Mol Vis 12(10): 148–154

Suryanarayana P, Saraswat M, Petrash JM, Reddy GB (2007) Emblica officinalis and its enriched tannoids delay streptozotocin-induced diabetic cataract in rats. Mol Vis 13:1291–1297

Talwar GP, Dar SA, Rai MK, Reddy KV, Mitra D, Kulkarni SV, Doncel GF, Buck CB, Schiller JT, Muralidhar S, Bala M, Agrawal SS, Bansal K, Verma JK (2008) A novel polyherbal microbicide with inhibitory effect on bacterial, fungal and viral genital pathogens. Int J Antimicrob Agents 32(2):180–185

Tasduq SA, Kaisar P, Gupta DK, Kapahi BK, Maheshwari HS, Jyotsna S, Johri RK (2005a) Protective effect of a 50% hydroalcoholic fruit extract of Emblica officinalis against anti-tuberculosis drugs induced liver toxicity. Phytother Res 19(3):193–197

Tasduq SA, Mondhe DM, Gupta DK, Baleshwar M, Johri RK (2005b) Reversal of fibrogenic events in liver by Emblica officinalis (fruit), an Indian natural drug. Biol Pharm Bull 28(7):1304–1306

Thakur CP (1985) Emblica officinalis reduces serum, aortic and hepatic cholesterol in rabbits. Experientia 41(3): 423–424

Thorat SP, Rege NN, Naik AS, Thatte UM, Joshi A, Panicker KN, Bapat RD, Dahanukar SA (1995) Emblica officinalis: a novel therapy for acute pancreatitis–an experimental study. HPB Surg 9(1):25–30

van Holthoon FL (1999) Phyllanthus L. In: de Padua LS, Bunyapraphatsara N, Lemmens RHMJ (eds) Plant resources of South-East Asia No 12(1). Medicinal and poisonous plants 1. Backhuys Publishers, Leiden, pp 381–392

van Schaik-van Banning AJJ (1991) Phyllanthus emblica L. In: Lemmens RHMJ, Wulijarni-Soetjipto N (eds) Plant resources of South-East Asia No 3. Dye and tannin producing plants. Pudoc, Wageningen, pp 105–108

Vasudevan M, Parle M (2007) Effect of Anwala churna (Emblica officinalis Gaertn.): an ayurvedic preparation on memory deficit rats. Yakugaku Zasshi 127(10): 1701–1707

Veena K, Shanthi P, Sachdanandam P (2006a) Anticancer effect of Kalpaamruthaa on mammary carcinoma in rats with reference to glycoprotein components, lysosomal and marker enzymes. Biol Pharm Bull 29(3): 565–569

Veena K, Shanthi P, Sachdanandam P (2006b) Protective effect of kalpaamruthaa on altered glycoprotein component levels and membrane stability in mammary carcinoma. Int J Cancer Res 2(4):315–329

Veena K, Shanthi P, Sachdanandam P (2006c) The biochemical alterations following administration of Kalpaamruthaa and Semecarpus anacardium in mammary carcinoma. Chem Biol Interac 161(1):69–78

Verma R, Chakraborty D (2008a) Alterations in DNA, RNA and protein contents in liver and kidney of mice treated with ochratoxin and their amelioration by Emblica officinalis aqueous extract. Acta Pol Pharm 65(1):3–9

Verma R, Chakraborty D (2008b) Emblica officinalis aqueous extract ameliorates ochratoxin-induced lipid peroxidation in the testis of mice. Acta Pol Pharm 65(2):187–194

Wattanapitayakul SK, Chularojmontri L, Herunsalee A, Charuchongkolwongse S, Niumsakul S, Bauer JA (2005) Screening of antioxidants from medicinal plants for cardioprotective effect against doxorubicin toxicity. Basic Clin Pharmacol Toxicol 96(1):80–87

Xiang YF, Ju HQ, Li S, Zhang YJ, Yang CR, Wang YF (2010) Effects of 1,2,4,6-tetra-O-galloyl-β-D-glucose from P. emblica on HBsAg and HBeAg secretion in HepG2.2.15 cell culture. Virol Sin 25(5):375–380

Xiang YF, Pei Y, Qu C, Lai Z, Ren Z, Yang K, Xiong S, Zhang Y, Yang C, Wang D, Liu Q, Kitazato K, Wang Y (2011) In vitro anti-herpes simplex virus activity of 1,2,4,6-tetra-O-galloyl-β-d-glucose from Phyllanthus emblica L. (Euphorbiaceae). Phytother Res doi: 10.1002/ptr.3368.

Yokozawa T, Kim HY, Kim HJ, Okubo T, Chu DC, Juneja LR (2007a) Amla (Emblica officinalis Gaertn.) prevents dyslipidaemia and oxidative stress in the ageing process. Br J Nutr 97(6):1187–1195

Yokozawa T, Kim HY, Kim HJ, Tanaka T, Sugino H, Okubo T, Chu DC, Juneja LR (2007b) Amla (Emblica officinalis Gaertn.) attenuates age-related renal dysfunction by oxidative stress. J Agric Food Chem 55(19):7744–7752

Zhang YJ, Abe T, Tanaka T, Yang CR, Kouno I (2001a) Phyllanemblinins A-F, new ellagitannins from *Phyllanthus emblica*. J Nat Prod 64(12):1527–1532

Zhang YJ, Abe T, Tanaka T, Yang CR, Kouno I (2002) Two new acylated flavanone glycosides from the leaves and branches of *Phyllanthus emblica*. Chem Pharm Bull(Tokyo) 50(6):841–843

Zhang L-Z, Zhao W-H, Guo Y-J, Tu G-Z, Lin S, Xin L-G (2003) Studies on chemical constituents in fruits of Tibetan medicine *Phyllanthus emblica*. Zhongguo Zhongyao Zazhi 28(10):940–943, In Chinese

Zhang YJ, Tanaka T, Iwamoto Y, Yang CR, Kouno I (2000a) Novel norsesquiterpenoids from the roots of *Phyllanthus emblica*. J Nat Prod 63(11):1507–1510

Zhang YJ, Tanaka T, Iwamoto Y, Yang CR, Kouno I (2000b) Phyllaemblic acid, a novel highly oxygenated norbisabolane from the roots of *Phyllanthus emblica*. Tetrahedron Lett 41(11):1781–1784

Zhang YJ, Tanaka T, Iwamoto Y, Yang CR, Kouno I (2001b) Novel sesquiterpenoids from the roots of *Phyllanthus emblica*. J Nat Prod 64(7):870–873

Zhang YJ, Tanaka T, Yang CR, Kouno I (2001c) New phenolic constituents from the fruit juice of *Phyllanthus emblica*. Chem Pharm Bull 49(5):537–540

Zhang YJ, Nagao T, Tanaka T, Yang CR, Okabe H, Kouno I (2004) Antiproliferative activity of the main constituents from *Phyllanthus emblica*. Biol Pharm Bull 27(2):251–255

Zhong ZG, Huang JL, Liang H, Zhong YN, Zhang WY, Wu DP, Zeng CL, Wang JS, Wei YH (2009) The effect of gallic acid extracted from leaves of *Phyllanthus emblica* on apoptosis of human hepatocellular carcinoma BEL-7404 cells. Zhong Yao Cai 32(7):1097–1101 (In Chinese)

Zhong ZG, Wu DP, Huang JL, Liang H, Pan ZH, Zhang WY, Lu HM (2011) Progallin A isolated from the acetic ether part of the leaves of *Phyllanthus emblica* L. induces apoptosis of human hepatocellular carcinoma BEL-7404 cells by up-regulation of Bax expression and down-regulation of Bcl-2 expression. J Ethnopharmacol 133(2):765–772

Pinus koraiensis

Scientific Name

Pinus koraiensis Siebold & Zuccarini

Synonyms

Apinus koraiensis (Sieb. et Zucc.) Moldenke, *Pinus cembra* β *excelsa* Maxim. ex Rupr., *Pinus strobus* Thunb. non L., *Pinus cembra* var. *manchurica* Mast., *Pinus cembra* var. *mandschurica* (Rupr.) Carr., *Pinus mandschurica* Ruprecht, *Pinus prokoraiensis* Y. T. Zhao & al., *Strobus koraiensis* (Sieb. & Zucc.) Moldenke.

Family

Pinaceae

Common/English Names

Cedar Pine, Chinese Pinenut, Hinggan Red Pine, Korean Cedar, Korean Pine, Korean Pinenut, White Pine.

Vernacular Names

Catalan: Pi De Corea;
Chinese: Chao Xian Song, Guo Song, Hai Song, Hai Sung, Han Song, Hong Guo Song, Hong Song, Hung Song;
Czech: Borovice Korejská;
Danish: Koreafyr;
Eastonian: Korea Seedermänd;
Finnish: Koreansembra;
French: Pin De Corée;
German: Korea-Kiefer;
Hungarian: Koreai Fenyõ;
Icelandic: Kóreufura;
Italian: Pignoli, Pinoli;
Japanese: Chosenmatsu, Chosen-Goya, Chousen Goyou Matsu, Minimatsu;
Korean: Channamu, Jatnamu;
Polish: Sosna Koreańska;
Slovakian: Borovica Kórejská;
Spanish: Pino De Corea, Piñón;
Swedish: Koreansk Tall;
Taiwan: Hai Song;
Turkish: Kore Çamı.

Origin/Distribution

It is native to eastern Asia: Manchuria (Ussuri River basin) in northeast China, Primorsky Krai and Khabarovsk Krai in the far east of Russia, Korea and Japan -Honshu southward from Tochigi Prefecture and northward from Gifu Prefecture, and Shikoku.

Agroecology

A cold temperate species. In its natural range in northern-eastern Asia, it grows at moderate altitudes, typically 600–900 m, whereas further south, it is a mountain tree, growing at 1,300–2,600 m altitude in Japan. It prefers sandy to sandy loam soil with adequate drainage.

Edible Plant Parts and Uses

Shelled pine nuts have been eaten in Europe and Asia for a long time. They are also a source of dietary fibre. Pine nuts are frequently added to meat, fish, and vegetable dishes. They constituted an essential ingredient of pesto sauce. Pine nuts are used as topping for the *pignoli cookie*, an Italian specialty confection, is made of dough formed from almond flour. They are widely used in the cuisine of southwestern France, in dishes such as the *salade landaise*. They are also used in chocolates and desserts such as *baklava*. In New Mexico in the southwest United States, pine nut coffee a typically dark roast coffee having a deep, nutty flavour is a speciality and is made from roasted and lightly salted pine nuts. Pine nuts are also a widely used ingredient in a diverse range of Middle Eastern cuisine, such as *kibbeh*, *sambusek*, ladies' fingers and many other dishes. In Korea, pine nuts are used for making rice-biscuits and sweets. The kernels can also be dried and ground into a powder then used as a flavouring and thickener in soups. The green cones are used for making wine.

Botany

A tall, large, branched, pyramidal, monoecious, evergreen tree, 40–50 m tall; trunk to 1 m diameter.; bark grey-brown or grey, fissured longitudinally into irregularly oblong plates; branchlets densely red-brown, occasionally yellow pubescent; winter buds reddish brown, oblong-ovoid, slightly resinous. Leaves (needles), 5 per fascicle (bundle), dark bluish-green, straight, almost triangular in cross section, 6–12 cm, base with a deciduous sheath (Plate 1). Flowers axillary. Pollen cones crowded at base of new shoots, ellipsoid. Female cones crowded near end of new shoots (1st-year branchlets) in groups of 1–5, conelets green, ovoid. Cones erect, on a short peduncle, ovoid or cylindric-ovoid, 9–11 × 5–6 cm, indehiscent or slightly dehiscent at maturity, with seeds exposed but not shed; scales woody, broadly rhomboid, about 2.5 cm long and wide, green on upper half, brown on lower half, apex with recurved spiny boss. Seeds are triangular-obovoid, large, 12–16 mm long by 10 mm wide and 7 mm thick, with a hard seedcoat covering a glossy yellowish kernel (Plate 2).

Plate 1 Fascicles of leaves (needles)

Plate 2 Shelled pine kernels (nuts)

Nutritive/Medicinal Properties

Nutrient composition of pine nuts (for *Pinus koraiensis* and *Pinus pinea*) exclude 23% shell refuse had been reported per 100 g edible portion as: water 2.28 g, energy 673 kcal (2,816 kJ), protein 13.69 g, fat 68.37 g, ash 2.59 g, carbohydrate

13.08 g, sugars 3.59 g, sucrose 3.45 g, glucose 0.07 g, fructose 0.07 g, starch 1.43 g, Ca 16 mg, Fe 5.53 mg, Mg 251 mg, P 576 mg, K 597 mg, Na 2 mg, Zn 6.45 mg, Cu 1.324 mg, Mn 8.802 mg, Se 0.7 μg, vitamin C 0.8 mg, thiamine 0.364 mg, riboflavin 0.227 mg, niacin 4.387 mg, pantothenic acid 0.313 mg, vitamin B-6 0.094 mg, total folate 34 μg, total choline 55.8 mg, betaine 0.4 mg, vitamin A, 1 μg RAE, vitamin A 29 IU, β-carotene 17ug, lutein+zeaxanthin 9 μg, vitamin E (α-tocopherol) 9.33 mg, γ-tocopherol 11.15 mg, vitamin K (phylloquinone) 53.0 μg, total saturated fatty acids 4.899 g, 16:0 (palmitic acid) 3.212 g, 18:0 (stearic acid) 1.390 g, 20:0 0.229 g, 22:0 0.068 g, total monounsaturated fatty acids 18.764 g, 16:1 undifferentiated (palmitoleic acid) 0.017 g, 18:1 undifferentiated (oleic acid) 17.947 g, 20:1 (gadoleic acid) 0.801 g, 20:2 n-6 c,c (Eicosadienoic acid) 0.404 g, total polyunsaturated fatty acids 34.071 g, 18:2 undifferentiated (linoleic acid) 33.150 g, 18:3 undifferentiated (linolenic acid) 0.164 g, 18:3 n-3 c,c,c (α linolenic acid) 0.112 g, 18:3 n-6 c,c,c (γ linolenic acid) 0.052 g, 20:3 undifferentiated (eicosatrienoic acid) 0.353 g, phytosterols 141 mg, tryptophan 0.107 g, threonine 0.370 g, isoleucine 0.542 g, leucine 0.991 g, lysine 0.540 g, methionine 0.259 g, cystine 0.289 g, phenylalanine 0.524 g, tyrosine 0.509 g, valine 0.687 g, arginine 2.413 g, histidine 0.341 g, alanine 0.684 g, aspartic acid 1.303 g, glutamic acid 2.926 g, glycine 0.691 g, proline 0.673 g, and serine 0.835 g (USDA 2011).

Pine nuts contain about 31 g of protein per 100 g of nuts, the highest of any nut or seed and is a rich source of dietary fibre. A vicilin-type 7S seed storage protein was isolated from defatted Korean pine-nut extract (Jin et al. 2007). The 7S vicilin-type globulin from Korean pine was purified and characterised (Jin et al. 2008). It was found to consist of 4 major bands and was stable up to 80°C.

Pinus koraiensis seed oil was found to consist of 21 fatty acids with chain lengths from 14:0 to 22:0 (Yoon et al. 1989). The seed oil contained three nonmethylene-interrupted polyenoic (NMIP) acids — 18:3Δ5, 9, 12, 20:2Δ5, 11, and 20:3Δ5, 11, 14 as omega 5 fatty acids. The total amounts of three omega-5 fatty acids was 12.38%; and among these 18:3Δ5, 9, 12 acid was greatest, 11.14%. The seed contained oleic acid (28.40%) and linoleic acid (47.92) as major fatty acids. The ratio of poly unsaturated to saturated fatty acids was 7.04.

The triacylglycerol (TG) composition of *Pinus koraiensis* seed oil, was found to contain Δ5 nonmethylene-interrupted (NMI) fatty acids (FA) with the main acid being pinolenic, 18:3 Δ5,9,12 (Imbs et al. 1998). Species of TG with unsaturation degrees of 1–7 and trace amounts of saturated and octaenoic TG species were found. Except for minor compounds, 26 TG molecular species of 32 main components were quantitatively determined. The main species were oleoyl dilinoleoylglycerol (14.7%), dilinoleoyl pinolenoylglycerol (10.7%), palmitoyl oleoyl linoleoylglycerol (8.3%), triolein (7.6%), and dioleoyl, linoleoylglycerol (7.4%). Seven TG species contained Δ5 NMI acyl groups. Of these, the major were dilinoleoyl pinolenoylglycerol (10.7%), stearoyl linoleoyl pinolenoylglycerol (6.5%) dioleoyl, pinolenoylglycerol (5.4%), and palmitoyl linoleoyl pinolenoylglycerol (5.5%). TG species with two or three NMI acyl groups were not detected.

Hydrodistillation of the *P. koraiensis* cones yielded 1.07% (v/w) of essential oil, which was almost three times the amount of essential oil extracted from the needles of the same plant (Lee et al. 2008). Eighty-seven components, comprising about 96.8% of the total oil, were identified. The most abundant oil components were limonene (27.90%), α-pinene (23.89%), β-pinene (12.02%), 3-carene (4.95%), β-myrcene (4.53%), isolongifolene (3.35%), (-)-bornyl acetate (2.02%), caryophyllene (1.71%), and camphene (1.54%).

Eight compounds isolated from *P. koraiensis* were identified as 8 (14)-podocarpen-13-on-18-oic acid (1), 15-hydroxydehydroabietic acid (2), 12-hydroxyabietic acid (3), lambertianic acid (4), dehydroabietic acid (5), sandaracopimaric acid (6), β-sitosterol (7), daucosterol (8) (Yang et al. 2008b). Two new diterpenoid acids were isolated from *Pinus koraiensis* and elucidated as 7-oxo-1 3β,15-dihydroxyabiet-8(14)-en-18-oic acid (1) and 7-oxo-12α, 13β,15-trihydroxyabiet-8(14)-en-18-oic acid (2) (Yang et al. 2008a).

Antioxidant Activity

The optimal conditions of supercritical carbon dioxide (SC-CO(2)) extraction with which to obtain highest yield of *Pinus koraiensis* nut were determined to be 5760.83 psi, 50°C and 3.0 hours with an extraction yield of 458.5 g/kg (Chen et al. 2011). Nine compounds, constituting about 99.98% of the total oil, were identified. The most abundant polyunsaturated fatty acids identified in the oil, linoleic acid and α-linolenic acid, constituted 41.79% and 15.62% of the oil, respectively. Moreover, the results on their antioxidant activities showed that the oil could improve the activities of superoxide dismutase (SOD) and glutathione peroxidase (GSH-Px), total antioxidant capacity (T-AOC), and reduce the content of malondialdehyde (MDA) significantly, in the rat's serum. These results indicated that *P. koraiensis* nut oil obtained by SC-CO(2) extraction had excellent in-vivo antioxidant activities.

Anticancer Activity

Procyanidins from *Pinus koraiensis* bark was found to exhibit antitumour activity on mice bearing U14 cervical cancer (Li et al. 2007). Treatment on mice bearing U14 cervical cancer with the extract (158 and 250 mg/kg body weight, p.o.) could inhibit U14 cervical carcinoma growth up to 47.68% and 58.94%. The extract enhanced the activity of SOD (superoxidate dismutase) and decreased MDA (malondialdehyde) content. Additionally, extract treatment significantly inhibited the expression of Ki-67, mutant p53 and Bcl-2 protein. The mechanism of the extract antitumour activity might be associated with free radical production inhibition and regulation of the expression of Ki-67, mutant p53 and Bcl-2 protein.

Antiplatelet/Hypotensive Activities

The effects of dietary Korean pine (*Pinus koraiensis*)-seed oil containing a peculiar trienoic acid (cis-5,cis-9,cis-12-18:3, pinolenic acid, approximately 18%) were found to be beneficial on various lipid variables compared in rats with those of flaxseed oil, safflower oil and evening primrose oil under experimental conditions where the effects of different polyunsaturated fatty acids could be estimated (Sugano et al. 1994). In Sprague-Dawley rats fed on diets containing 100 g fat and 5 g cholesterol/kg, the hypocholesterolaemic activity of pinolenic acid was intermediate between α-linolenic and linoleic acids. Analysis of the fatty acid composition of liver phosphatidylcholine indicated that, in contrast to α-linolenic acid, pinolenic acid does not interfere with the desaturation of linoleic acid to arachidonic acid. However, the effects on ADP-induced platelet aggregation and aortic prostacyclin production were comparable. When spontaneously hypertensive rats were fed on diets containing 100 g fat/kg but free of cholesterol, γ-linolenic and pinolenic acids, as compared with linoleic acid, increased prostacyclin production and tended to reduce platelet aggregation. Moreover, pinolenic acid attenuated the elevation of blood pressure after 5 weeks of feeding.

Hypolipidemic Activity

The fatty acids compositions of *Pinus koraiensis* seeds appeared ideal for hypolipidemic effects and had been claimed for curing and/or preventing degenerative chronic diseases, such as heart disease and diabetes (Yoon et al. 1989). *P. koraiensis* seed oil was found to contain two particular fatty acids of the delta5-unsaturated polymethylene-interrupted fatty acid (delta5-UPIFA) family: all-cis-5,9,12-1 8:3 (pinolenic) and/or all-cis-5,11,14-20:3 (sciadonic) acids (Asset et al. 1999). In a study rats fed for 28 days with *Pinus koraiensis* seed oil had their serum triglycerides decreased by 16% and very low density lipoprotein (VLDL)-triglycerides by 21%. There was a tendency of high density lipoprotein to shift toward larger particles in pine seed oil-supplemented rats.

Supplementation of mice with *Pinus koraiensis* nut oil caused a significant reduction in body weight gain and liver weight (37.4% and 13.7%, respectively) (Ferramosca et al. 2008). A marked decrease in plasma triglycerides and total cholesterol (31.8% and 28.5%, respectively) was also found in pine nut oil-fed animals. Liver lipids were also positively influenced by pine nut oil. The mitochondrial and cytosolic enzyme activities involved in hepatic fatty acid synthesis were strongly reduced both in pine nut oil and maize oil-fed animals, thus suggesting that the beneficial effects of pine nut oil were not due to an inhibition of hepatic lipogenesis.

Satiety Effect

The results of a randomized, placebo-controlled, double-blind cross-over trial including 18 overweight post-menopausal women suggested that Korean pine nut may work as an appetite suppressant through an increasing effect on satiety hormones and a reduced prospective food intake (Pasman et al. 2008). Cholecystokinin (CCK-8) was higher 30 minutes after pine nut FFA (free fatty acid) and 60 minutes after pine nut TG (triglycerides) when compared to placebo. GLP-1 (glucagon like peptide-1) was higher one hour after pine nut FFA compared to placebo. Over a period of 4 hours the total amount of plasma CCK-8 was 60% higher after pine nut FFA and 22% higher after pine nut TG than after placebo. The appetite sensation "prospective food intake" was 36% lower after pine nut FFA relative to placebo. The results of another study confirmed, that pine nut oil (PinnoThin™) with high content of pinolenic acid, was capable of increase of satiety hormones, cholecystokinin and glucagon like peptide 1, release and suppressed appetite (Frantisek and Meuselbach 2010). It did not have any effect on hunger inducing factor ghrelin. Toxicity study showed that pine nut oil was not toxic and to be safe for obesity treatment. The results of another study of a cross-over, double-blind placebo-controlled randomised counterbalanced design in 42 overweight female volunteers suggested that Korean pine nut oil, PinnoThin™ may exert satiating effects consistent with its known action on endogenous cholecystokinin (CCK) and glucagon like peptide-1 (GLP-1)release, and previously observed effects on self-reported appetite ratings (Hughes et al. 2008). Two g FFA PinnoThin™, given 30 minutes prior to an ad-libitum buffet test lunch, significantly reduced food intake (gram) by 9% compared to olive oil control. No significant effect of PinnoThin™ on macronutrient intake or ratings of appetite were observed. However, the latest scientific opinion by European Food Safety Authority (EFSA) Panel on Dietetic Products, Nutrition and Allergies (2011) considered that pine nut oil from *Pinus koraiensis* in the specific preparation was sufficiently characterised in relation to the claimed effect "satiety". The Panel considered that an increase in satiety leading to a reduction in energy intake, if sustained, might be a beneficial physiological effect. On the basis of the data presented, the Panel concluded that a cause and effect relationship had not been established between the consumption of "pine nut oil from *Pinus koraiensis*." and a sustained increase in satiety leading to a reduction in energy intake.

Immunomodulatory Activity

Studies with male Brown-Norway rats fed purified diets containing safflower oil (SFO, linoleic acid, 18:2 n-6), evening primrose oil (EPO, γ-linolenic acid, 6,9,12- 18:3 n-6) or Korean pine seed oil (PSO, 5,9,12- 18:3) revealed that the relative population of CD4+ T-lymphocytes in the spleen was significantly lower in rats fed SFO than in those fed EPO or PSO, while that of CD8+ subsets remained unchanged (Matsuo et al. 1996). There was a significant increase in the splenic production of IgG and IgE in the PSO group compared to the SFO group, while EPO significantly increased IgE. The periodical response patterns of the serum levels of IgG and IgE varied depending on the source of dietary fats, and the initial rise of total immunoglobulins tended to be higher in the EPO group. These observations not only indicate specific roles of γ-linolenic acid but also diverse influences of different octadecatrienoic acids in various immune measurements.

Antimicrobial Activity

The essential oil of *P. koraiensis* cones was confirmed to have significant antimicrobial activities, especially against pathogenic fungal strains such as *Candida glabrata* and *Cryptococcus neoformans* (Lee et al. 2008).

Toxicity Studies

A sub-chronic (13 weeks) oral toxicity study in rats and an in-vitro genotoxicity study with Korean pine nut oil (KPNO) concluded that KPNO could be considered to be non-genotoxic in the in-vitro reverse mutation test (Ames test) (Speijers et al. 2009). A No Observable Adverse Effect Level (NOAEL) of 15% was established for KPNO. This NOAEL corresponded to a mean of 8,866 and 10,242 mg KPNO/kg body weight/day for males and females, respectively. For both sexes, the NOAEL was achieved at the highest dose tested.

Destaillat et al. (2011) confirmed that consumption of *Pinus armandii* nut which is not reported as edible pine nuts by the Food and Agriculture Organization (FAO) may lead to dysgeusia. Based on the present study and previous work, the authors advised import companies to trade pine nuts from traditionally recognized species such as *P. pinea, P. sibirica, P. koraiensis*, or *P. gerardiana*.

Traditional Medicinal Uses

Pine nut seed is analgesic, antibacterial and antiinflammatory and is used in Korea in the treatment of earache, epistaxis and to promote milk flow in nursing mothers. The turpentine obtained from the resin of the trees is antiseptic, diuretic, rubefacient and vermifuge. Externally it is a very beneficial treatment for a variety of skin complaints, wounds, sores, burns, boils etc. and is used in the form of liniment plasters, poultices, herbal steam baths and inhalers. The stem bark is used for treating burns and skin ailments.

Other Uses

Besides being cultivated for pine seed harvest, it is widely employed in reforestation programs. In Europe and North America it is planted as ornamental. The timber is used for construction, carpentry, bridge building, vehicles, furniture, and wood pulp. The seeds besides being edible and medicinal, are used as a source of soap and lubricating oil. Turpentine is obtained from the timber and roots. Turpentine has a wide range of uses including as a solvent for waxes etc., for making varnish and rosin the substance left after turpentine is removed is used by violinists on their bows. The bark yields tannin; the trunk yields resin and gum. A tan or green dye is obtained from the needles (leaves).

Comments

Korean pine is the most widely traded pine nut in international commerce and is widely planted and harvested particularly in northeast China and Korea. Another important species for pine nuts in Asia is *Pinus gerardiana* in the western Himalaya.

Selected References

Asset G, Staels B, Wolff RL, Baugé E, Madj Z, Fruchart JC, Dallongeville J (1999) Effects of *Pinus pinaster* and *Pinus koraiensis* seed oil supplementation on lipoprotein metabolism in the rat. Lipids 34(1):39–44

Chen X, Zhang Y, Wang Z, Zu Y (2011) In vivo antioxidant activity of *Pinus koraiensis* nut oil obtained by optimised supercritical carbon dioxide extraction. Nat Prod Res 25(19):1807–1816

Destaillats F, Cruz-Hernandez C, Giuffrida F, Dionisi F, Mostin M, Verstegen G (2011) Identification of the botanical origin of commercial pine nuts responsible for dysgeusia by gas-liquid chromatography analysis of fatty acid profile. J Toxicol 2011:316789

Duke JA, Ayensu ES (1985) Medicinal plants of china, vol 1 & 2. Reference Publications, Inc, Algonac, 705 pp

Earle CE (2010) The gymnosperm database. http://www.conifers.org/pi/pin/koraiensis.htm

EFSA Panel on Dietetic Products, Nutrition and Allergies (2011) Scientific Opinion on the substantiation of a health claim related to "pine nut oil from *Pinus koraiensis* Siebold & Zucc" and an increase in satiety leading to a reduction in energy intake (ID 551) pursuant to Article 13(1) of Regulation (EC) No 1924/20061. EFSA J 9(4):2046

Ferramosca A, Savy V, Einerhand AWC, Zara V (2008) *Pinus koraiensis* seed oil (PinnoThinTM) supplementation reduces body weight gain and lipid concentration in liver and plasma of mice. J Anim Feed Sci 17:621–630

Frantisek Z, Meuselbach K (2010) Effect of *Pinus koraiensis* seed oil on satiety hormones CCK and GLP-1 and appetite suppression. Transl Biomed 1:3–5

Fu L, Li N, Elias TS, Mill RR (1999) Pinaceae Lindley. In: Wu ZY, Raven PH (eds) Flora of China, vol 4, Cycadaceae through Fagaceae. Science Press/Missouri Botanical Garden Press, Beijing/St Louis

Grieve M (1971) A modern herbal. 2 vols. Dover publications, Penguin, New York, 919pp

Hughes GM, Boyland EJ, Williams NJ, Monnen L, Scott C, Kirkham TC, Harrold JA, Keizer HG, Halford JC (2008) The effect of Korean pine nut oil (PinnoThin™) on food intake, feeding behaviour and appetite: a double-blind placebo-controlled trial. Lipids Health Dis 7:6

Imbs AB, Nevshupova NV, Pham LQ (1998) Triacylglycerol composition of *Pinus koraiensis* seed oil. J Am Oil Chem Soc 75(7):865–870

Iwatsuki K, Yamazaki T, Boufford DE, Ohba H (eds) (1995) Flora of Japan, vol 1, Pteridophyta and Gymnospermae. Kodansha, Tokyo, pp. xv and 263–288

Jin T, Fu TJ, Kothary MH, Howard A, Zhang YZ (2007) Crystallization and initial crystallographic characterization of a vicilin-type seed storage protein from *Pinus koraiensis*. Acta Crystallogr Sect F Struct Biol Cryst Commun 63(Pt 12):1041–3

Jin T, Albillos SM, Chen YW, Kothary MH, Fu TJ, Zhang YZ (2008) Purification and characterization of the 7S vicilin from Korean pine (*Pinus koraiensis*). J Agric Food Chem 56(17):8159–8165

Lee JH, Yang HY, Lee HS, Hong SK (2008) Chemical composition and antimicrobial activity of essential oil from cones of *Pinus koraiensis*. J Microbiol Biotechnol 18(3):497–502

Li K, Li Q, Li J, Zhang T, Han Z, Gao D, Zheng F (2007) Antitumor activity of the procyanidins from *Pinus koraiensis* bark on mice bearing U14 cervical cancer. Yakugaku Zasshi 127(7):1145–1151

Little EL, Critchfield WB (1969) Subdivision of the Genus *Pinus*. (Miscellanous Publ. 1144). U.S.D.A. Forest Serv, Washington, 51pp

Matsuo N, Osada K, Kodama T, Lim BO, Nakao A, Yamada K, Sugano M (1996) Effects of gamma-linolenic acid and its positional isomer pinolenic acid on immune parameters of brown-Norway rats. Prostaglandins Leukot Essent Fatty Acids 55(4):223–229

Pasman WJ, Heimerikx J, Rubingh CM, van den Berg R, O'Shea M, Gambelli L, Hendriks HF, Einerhand AW, Scott C, Keizer HG, Mennen LI (2008) The effect of Korean pine nut oil on in vitro CCK release, on appetite sensations and on gut hormones in post-menopausal overweight women. Lipids Health Dis 7:10

Pemberton RW, Lee NS (1996) Wild food plants in South Korea; market presence, new crops, and exports to the United States. Econ Bot 50:60–67

Porcher MH et al (1995–2020) Searchable World Wide Web Multilingual Multiscript Plant Name Database. Published by The University of Melbourne Australia. http://www.plantnames.unimelb.edu.au/Sorting/Frontpage.html

Rosengarten F Jr (1984) The book of edible nuts. Walker and Company, New York, 384pp

Speijers GJ, Dederen LH, Keizer H (2009) A sub-chronic (13 weeks) oral toxicity study in rats and an in vitro genotoxicity study with Korean pine nut oil (PinnoThin TG). Regul Toxicol Pharmacol 55(2):158–165

Sugano M, Ikeda I, Wakamatsu K, Oka T (1994) Influence of Korean pine (*Pinus koraiensis*)-seed oil containing cis-5, cis-9, cis-12-octadecatrienoic acid on polyunsaturated fatty acid metabolism, eicosanoid production and blood pressure of rats. Br J Nutr 72(5):775–783

U.S. Department of Agriculture, Agricultural Research Service (USDA) (2011) USDA National Nutrient Database for Standard Reference, Release 24. Nutrient Data Laboratory Home Page, http://www.ars.usda.gov/ba/bhnrc/ndl

Usher G (1974) A dictionary of plants used by man. Constable, London, 619pp

Yang X, Zhang H, Zhang Y, Ma Y, Wang J (2008a) Two new diterpenoid acids from *Pinus koraiensis*. Fitoterapia 79(3):179–181

Yang X, Zhang YC, Zhang H, Wang J (2008b) Isolation and identification of diterpenoids from *Pinus koraiensis*. Zhong Yao Cai 31(1):53–55 (In Chinese)

Yoon TH, Im KJ, Koh ET, Ju JS (1989) Fatty acid compositions of *Pinus koraiensis* seed. Nutr Res 9(3):357–361

Pinus pinea

Scientific Name

Pinus pinea L.

Synonyms

Apinus pinea (L.) Neck. ex Rydb., *Pinus domestica* Mathiol, *Pinus domestica* Matthews, *Pinus esculenta* Opiz, *Pinus fastuosa* Salisb., *Pinus maderiensis* Ten., *Pinus pinea* L. var. *maderiensis* (Ten.) Carr., *Pinus sativa* C. Bauch, *Pinus sativa* Garsault, nom. inval., *Pinus umbraculifera* Tournef.

Family

Pinaceae

Common Names

European Stone Pine, Italian Stone Pine, Mediterranean Stone Pine, Parasol Pine, Pignolia-Nut Pine, Stone Pine, Umbrella Pine.

Vernacular Names

Arabic: Sanawbar;
Argentina: Pino Domestic;
Basque: Belorita;
Catalan: Pi Campaner, Pi De Pinions, Pi Pinyer, Pi Pinyoner;
Chinese: Shi Song, Yi Da Li San Song, Yi Da Li Song Yi Da Li Wu Zhen Song;
Czech: Borovice Hustokvětá, Borovice Pinie, Pinie;
Danish: Pinie, Ponje;
Dutch: Parasolden, Parasolpijn;
Esperanto: Pinio;
Finnish: Hopeapinja, Pinja;
French: Pignet, Pignon, Pin Bon, Pin De Pierre, Pin D'italie, Pin Franc, Pin Parasol, Pin Pignon, Pin Pinier, Pinier;
Galician: Piñeiro Manso, Pino Manso;
German: Italienische Steinkiefer, Penea, Pinea, Pinie, Piniekiefer, Schirmkiefer, Schirmpinie;
Greek: Koukounaria;
Italian: Pino, Pino A Ombrello, Pino Da Pinocchi, Pino Da Pinoli, Pino S'italia, Pino Di Pietro, Pino Domestico, Pino Gentile, Pino Italico, Pigna Pinoler, Pignare, Pignolo, Pignu Manzu, Pinocchio;
Japanese: Itaria Kasa Matsu, Kasa Matsu;
Portuguese: Pinheiro-Manso;
Russian: Piniia, Sosna Ital'ianskaia;
Slovaščina: Bor Pinija, Pinija;
Slovencina: Borovica Pínia;
Spanish: Pino Albar, Pino De Comer, Pino De La Tierra, Pino De Piñón, Pino Doncel, Pino Manso, Pino Piñón, Pino Piñonero, Pino Piñonero Europeo, Pino Real;
Swedish: Pinja;
Taiwan: Li Song;
Turkish: Çam, Çam Fıstık Ağacı, Fıstık Çamı

Origin/Distribution

Stone pine is native to the Mediterranean region, from Portugal to Asia Minor. Stone pine nut, *Pinus pinea* was confirmed as an autochthonous species of the Iberian peninsula in Spain (Martinez and Monetero 2004). Pine nuts has been cultivated for its nuts for over 6,000 years in Europe and harvested from wild trees for far longer. Most of the seeds are collected or harvested in Spain (especially in Catalonia, Castilla, León), Portugal, Italy (Sicily), Turkey and China from both wild and cultivated stands.

The kernels were eaten preserved in honey in Pliny's era. Seeds of *P. pinea* were found in household deposits in the ruins of Pompeii and in the garbage heaps in Roman military camps in England probably part of the rations for the Roman legions.

Agroecology

The agroecology of stone pine has been well elucidated by (Mutke et al. 2003, 2005, 2011; Calama and Montero 2005; Calama et al. 2007, 2008; Varo and Tel 2010; Sánchez-Gómez et al. 2011).

In its native range, stone pine occurs at low to montane elevations from sea level to 1,500 m in Lebanon. It also grows up to 2,000 m in Chile. It thrives in areas with mean annual rainfall of 400–800 mm or higher and optimum temperature range of 10–18°C, with average temperature in the warmest month of 21–26°C and 3–11°C in the coldest month. It is regarded as thermophilous and xerophilous. Very cold winters and frost can damage its reproductive organs, stunt growth and may cause the pine to wither away. Stone pine prefers well-drained acidic soils but is tolerant of soil pH from 5 to 9. It prefers sandy, sandy loams or gravelly soils with water holding capacity of at lease 60 mm. Areas of too compact clays or silt is unsuitable as it interferes with root development. Most natural stonepine woodlands grow on incipient soil types such as arenosols, regosols, lithosols or mature cambisols or luvisols on better sites. Stone pine can pioneer poor habitats such as limestone because of its robust root systems. Stone pine shows variability in adaptive traits in response to drought. Amelioration with organic matter will improve soil structure and water and nutrient retention capacities, irrigation improves tree vigour and cone yield. Strong aerial salt tolerance makes it suitable for maritime plantings.

Edible Plant Parts and Uses

Pine nuts have been eaten in Europe and Asia since the Paleolithic period. The kernels can be eaten raw, roasted or prepared into sweet meats, cakes, puddings, stuffings, soups and sauces like the popular Italian gourmet sauce, *pesto*. Pine nuts are commonly added to fish, meat, salads and vegetable dishes or baked into bread. *biscotti ai pinoli*, an Italian specialty confection is made of almond flour formed into a dough similar to that of a macaroon and then topped with pine nuts. In Spain, a sweet is made of small marzipan balls covered with pine nuts, painted with egg and lightly cooked. Pine nuts are also featured in the *salade landaise* of southwestern France. Pine nuts are also widely used in Middle Eastern cuisine, such as *kibbeh* (meat, lamb with spices), *sambusek* (fried stuffed pastry with meat, cheese and spices) desserts such as *baklava* (rich, sweet pastry made of layers of filo pastry filled with chopped nuts and sweetened with syrup or honey), and many others.

Botany

An erect, slender monoecious tree 20–25 m high with a short, sometime sinuous trunk, 1 m in diameter, and a dome-shaped or umbrella shaped not conic crown because of its radiating branches. The trunk sometimes forked close to the ground giving rise to multiple trunks. Bark is thick, plated, deeply fissured, red-brown to orange with blackish edges to the plates. The leaves (needles) are 10–18(–28) cm long, about 1.5 mm thick, with serrulate margins and are borne in groups of two, held by a basal sheath. Juvenile needles are glaucous (bluish-grey and covered with wax)

(Plate 1), older needles are mid-green and glossy (Plate 2). The pollen cones (male inflorescences or microsporophylls) are yellow and are short and cylindrical in shape in the basal part of the current year's growth. The seed cones (female inflorescences or macrosporophylls or strobiles) are borne on short stout stalks, singly, or sometimes in groups of two or three, in the sub-terminal portion of the current year's growth and persist for several years, ripening in the third year (Plate 3). They are flat-based ovoid, 8–15 cm long and up to 13 cm wide when closed, sessile or briefly pedunculate, symmetrical, initially reddish-brown with violet-brown highlights, later a shiny brown, smooth and resin covered (Plates 4 and 5). Scales are large and convex with an obtuse umbo and bear two seeds. The egg-shaped seeds are pale brown covered by sooty powder, large, 15–20 mm long and 8–10 mm wide, hard-shelled with a vestigial, 3–8 mm yellow-buff wing and an edible kernel (Plate 6).

Plate 3 Young strobili

Plate 1 Juvenile glaucous needles (leaves)

Plate 4 Open female cone

Plate 2 Mature mid green needles

Plate 5 Open female cone side view

Plate 6 Pine-nut seed (kernels)

Nutritive/Medicinal Properties

Proximate nutrient value of *P. pinea* kernels per 100 g edible portion was reported as energy 2,925 g, moisture 1.8 g, protein 13 g, N 2.45 g, fat 70 g, ash 2.5 g, dietary fibre 5.1 g, total sugars 3.4 g, sucrose 3.4 g, starch 1.1 g, available carbohydrates 4.5 g, oxalic acid 0.1 g, Ca 11 mg, Cu 1.2 mg, Fe 4.1 mg, Mg 23 mg, Mn 6.9 mg, P 569 mg, K 600 mg, Na 3 mg, S 110 mg, Zn 5.3 mg, Se 1 μg, thiamin 0.57 mg, riboflavin 0.19 mg, niacin 4.3 mg, niacin equivalent 6.47 mg, pantothenic acid 0.48 mg, pyridoxine (B6) 0.02 mg, biotin (B7) 17 μg, total folate 34 μg, β-carotene equivalents 10 μg, retinol equivalents 2 μg, α-tocopherol 12 mg, γ-tocopherol 9.4 mg, vitamin E 12.94 mg, total SFA 4.2 g, 16 FD (palmitic) 3.61 g, C18FD 0.2 g (stearic), C22 FD (behenic) 0.07 g, C24 FD (lignoceric) 0.13 g; total MUFA 22.95 g, C18:1FD (oleic) 22.22 g, C20:1FD (gadoleic) 0.74 g; total PUFA 39.8 g, and C18:2w6FD (linoleic) 39.75 g (FSANZ 2010). Another proximate analysis of stone pine seeds showed the following composition: moisture 5%, ash 4.5%, fat 44.9%, crude protein 31.6%, total soluble sugars 5.15% and energy value 583 kcal/100 g (Nergiz and Dönmez 2004). Oleic and linoleic acids were the major unsaturated fatty acids, while palmitic, stearic and lignoceric acids were the main saturated ones. Potassium, phosphorus and magnesium were the predominant elements present in the seeds. Zinc, iron and manganese were also detected in appreciable amounts. The contents of ascorbic acid, thiamine and riboflavin were found to be 2.50, 1.50 and 0.28 mg/100 g, respectively. Triacylglycerols with ECNs (equivalent chain-lengths) of 44 and 48 were dominant in the oil.

Based on the analysis of 27 different Portuguese populations, pine nuts (*Pinus pinea*) were characterised by high contents of fat (47.7 g per 100 g dry matter (DM)), protein (33.8 g per 100 g DM) and phosphorus (1,130 mg per 100 g DM) and low contents of moisture (5.9 g per 100 g DM) and starch (3.5 g per 100 g DM) (Evaristo et al. 2010). They were also found to be a good source of zinc, iron and manganese. *Pinus pinea* seeds were found to contain 25% proteins on a dry-weight basis (Nasri and Triki 2007). *P. pinea* was found to accumulate globulins as major storage proteins in seeds (75% of total storage proteins), composed of several subunits of 10–150 kDa. The albumin fraction (15%) constituted three subunits of 14, 24 and 46 kDa. Glutelins, the least soluble fraction, constituted a small proportion (10%). Prolamins also constituted a very small percentage (1–2%).

In a study of metal elements Mn, Zn, Ni and Cu in pine nut (*Pinus pinea*) from different geographical origin in Spain and Portugal, the highest concentration of metals studies were found from the region Faro and at the lowest from Cataluña (Gómez-Ariza et al. 2006). The most abundant element was Mn at concentrations ranging from 26 μg/g (Cataluña) to 559 μg/g (Faro). Zn was also present at high concentration in samples, from 25 μg/g (Cataluña) to 113 μg/g (Faro). Mn allows a good discrimination between samples, Cu was the only one associated to fractions at MW >70 kDa in sample from Cádiz, and profiles of Ni and Zn are clearly different in terms of abundance of peaks.

The tocopherol (α-tocopherol, γ-tocopherol, and δ-tocopherol) contents of *P. pinea* seeds were, respectively, 15.34, 1,681.75 and 41.87 ppm (Nasri et al. 2009). Lipids (mainly triacylglycerols) averaged 48% on a dry weight basis. Triacylglycerols with an equivalent carbon number of 44 (32.27%) and of 46 (30.91%) were dominant. The major triacylglycerol was LLO (24.06%). Tocopherols and triacylglycerols were present at remarkably high levels, thus making *P. pinea* oil a valuable source of antioxidants and unsaturated fatty acids. The fatty acid profile of *Pinus pinea* comprised unsaturated oil with several unusual polymethylene-interrupted unsaturated fatty acids (Nasri et al. 2005a). Linoleic acid was the major fatty acid followed by oleic, palmitic and stearic acids. *Pinus pinea* L. and *Pinus halepensis* Mill seeds were found to be rich in lipids, 34.63–48.12 % on a dry weight basis (Nasri et al. 2005b). Qualitatively, fatty acid composition of both species was identical. For *P. halepensis,* linoleic acid was the major fatty acid (56.06 % of total fatty acids) followed by oleic (24.03 %) and palmitic (5.23 %) acids. For *P. pinea*, the same fatty acids were found with the proportions 47.28 %, 36.56 %, and 6.67 %, respectively. Extracted fatty acids from both species were mainly unsaturated, respectively, 89.87 % and 88.01 %. *Pinus halepensis* cis-5 olefinic acids were more abundant (7.84 % compared to 2.24 %). Results revealed the potential nutraceutical value of *Pinus* seeds as new sources of fruit oils rich in polyunsaturated fatty acids and cis-5 olefinic acids

P. pinea oil unsaponifiable matter was found to contain very high levels of phytosterols (> or =4,298 mg/kg of total extracted lipids), of which β-sitosterol was the most abundant (74 %) (Nasri et al. 2007). Aliphatic alcohol contents were 1,365 mg/kg of total extracted lipids, of which octacosanol was the most abundant (41 %). Two alcohols (hexacosanol and octacosanol), usually absent in common vegetable oils, were detected in *P. pinea* oils. There were almost no differences in the total unsaponifiable matter of the seven Mediterranean populations studied. However, sterol and aliphatic alcohol contents showed some variability, with Tunisian and Moroccan populations showing very different and higher contents.

A field study of two pine species common in the Iberian Peninsula (*Pinus pinaster* Ait. and *Pinus pinea* L.) with naturally contaminated samples from eight sites (four in Portugal and four in Catalonia, Spain) showed that pine needles were more suitable biomonitors for polycyclic aromatic hydrocarbons (PAHs), yielding concentrations from 2 to 17 times higher than those found in bark (Ratola et al. 2009). The levels varied according to the sampling site, with the sum of the individual PAH concentrations between 213 and 1,773 ng/g (dry weight). Phenanthrene was the most abundant PAH, followed by fluoranthene, naphthalene and pyrene.

Allergy and Pine Mouth

Pine nut allergy had been reported (Nielsen 1990) among atopic patients (boys) (MA et al. 2002), children (Rubira et al. 1998) and young girls (Ibáñez et al. 2003). An in-vivo open oral provocation with pine nuts (*Pinus pinea*) confirmed information about systemic reaction after ingestion of pine nuts (Nielsen 1990). Pine nuts are employed in sweets and cakes and, as in the present case, in green salads.

In-vitro tests suggested a systemic IgE allergic reaction. Ibáñez et al. (2003) demonstrated specific IgE to pine nut by skin prick test in young girl patients. The patients had IgE against a pine nut protein band with apparent molecular weights of approximately 17 kDa that could be considered as the main allergen. Rubira et al. (1998) reported pine nut allergy in children of ages ranging from 12 months to 6 years. All patients had a personal history of atopy. Symptoms on ingesting pine nut were severe systemic reactions in three cases. Two of the children had allergic reactions to other nuts. In all cases, both the skin test and the specific IgE serum test were positive.

A case of cacogeusia, specifically metallogeusia (a perceived metallic or bitter taste) was reported in a 36-year-old male following ingestion of 10–15 roasted pine nuts (Munk 2010). Patients occasionally described abdominal cramping and nausea after eating the nuts, raw, cooked, and processed nuts. In the past years USFDA had received a number of consumer complaints regarding a bitter metallic taste associated with pine nuts. 12–48 hours after consuming pine nuts, and lasting on average between a few days and 2 weeks. It was exacerbated by consumption of any other food during this period and significantly decreased appetite and enjoyment of food. The symptoms decreased over time with no apparent adverse clinical side effects. FDA was able to confirm "pine mouth" to be an adverse food reaction to pine nuts that was clearly distinct from a typical food allergy. Over the last 10 years, complaints were increasingly reported from consumers that experienced dysgeusia following the consumption of pine nuts (Destaillats et al. 2011). Their study confirmed that consumption of *Pinus armandii* nuts (not reported as edible by FAO) may lead to dysgeusia. *P. armandii* nuts were identified in all the samples pure or in mixture with *P. koraiensis* nuts. They advised that import companies to trade pine nuts from traditionally recognized species such as *P. pinea, P. sibirica, P. koraiensis,* or *P. gerardiana*.

Toxicology

Studies in rabbits showed that 49 days of *Pinus pinea* seed supplementation did not cause any negative effects on the blood or organ parameters tested (Yildiz-Gulay et al. 2010). No significant differences were detected in mean hemoglobin, hematocrit, red blood cell count, white blood cell count, plasma protein, mean corpuscular hemoglobin concentration, percent neutrophil, eosinophil, basophil, lymphocyte and monocyte of rabbits in control and treatment groups. Liver, kidney, lung, heart and body weights between control and treatment groups did not differ. Moreover, no apparent changes in liver, kidney, liver, testis and brain were detected by gross post mortem and histopathological examination to suggest toxic effect of oral use of *Pinus pinea* seeds for 49 days.

Traditional Medicinal Uses

The turpentine obtained from the resin of pine trees is antiseptic, diuretic, rubefacient and vermifuge (Grieve 1971). It is a valuable remedy used internally in the treatment of kidney and bladder complaints and is used both internally and as a rub and steam bath in the treatment of rheumatic affections. It is also very beneficial to the respiratory system and so is useful in treating diseases of the mucous membranes and respiratory complaints such as coughs, colds, influenza and tuberculosis. Externally it is a very beneficial treatment for a variety of skin complaints, wounds, sores, burns, scald and boils and is used in the form of liniment plasters, poultices, herbal steam baths and inhalers.

Other Uses

Stone pine is used as shelter belts, erosion control, for timber production and resin production beside for food purposes. Resin and turpentine are obtained from the stone pine. The resins are obtained by tapping the trunk, or by destructive distillation of the wood. Turpentine consists of an average of 20% of the oleo-resin and is separated by distillation. Turpentine has a wide range of uses including as a solvent for waxes etc., for making varnish, medicinal products. The resin is used by violinists on their bows and also in making sealing wax, varnish, etc. Pitch can also be obtained from the resin and is used for waterproofing, as a wood preservative. The wood is used for carpentry, furniture making, buildings, pine mulch and fire-wood.

Studies showed that medium density fibreboards (MDF) can be made from various mixtures of wood fibres and stone pine (*Pinus pinea*) cones (Ayrilmis et al. 2009). The addition of 10% cone flour also improved water resistance of the MDF panels made using urea–formaldehyde (UF) resin. The results showed that pine cone can be used as a renewable biological formaldehyde catcher as an alternative to the traditional formaldehyde catchers for E1 Class MDF manufacture

Comments

The tree is fairly easily propagated from seed, preferably fresh seed. Soaking the seeds for 48 hours will facilitate germination. The optimum temperature for seed germination is about 17–19°C. Temperature above 25°C can inhibit seedling establishment while at temperatures below about 10°C seeds become dormant. Cuttings can be used if taken from very young trees less than 10 years old. Single leaf fascicles with the base of the short shoot is used. Disbudding the shoots some weeks before taking the cuttings can help. However, cuttings are normally slow to grow away. Tip-cleft grafting of a leader shoot with a bud scion of selected clone can also be used.

Selected References

Abellanas B, Veroz R, Butler I (1997) The stone pine in Andalusia, Spain: current situation and prospects. FAO Nucis Newsl 6:29–32

Ayrilmis N, Buyuksari U, Avci E, Koc E (2009) Utilization of pine (*Pinus pinea* L.) cone in manufacture of wood based composite. For Ecol Manage 259(1):65–70

Calama R, Gordo FJ, Mutke S, Montero G (2008) An empirical ecological-type model for predicting stone pine (*Pinus pinea* L.) cone production in the Northern Plateau (Spain). For Ecol Manage 255(3–4):660–673

Calama R, Madrigal G, Candela JA, Montero G (2007) Effects of fertilization on the production of an edible forest fruit: stone pine (*Pinus pinea* L.) nuts in south-west Andalusia. Invest Agrar Sist Recur For 16(3):241–252

Calama R, Montero G (2005) Cone and seed production from stone pine (*Pinus pinea* L.) stands in Central Range (Spain). Eur J For Res 126(1):23–35

Destaillats F, Cruz-Hernandez C, Giuffrida F, Dionisi F, Mostin M, Verstegen G (2011) Identification of the botanical origin of commercial pine nuts responsible for dysgeusia by gas-liquid chromatography analysis of fatty acid profile. J Toxicol 2011:3167789

Evaristo I, Batista D, Correia I, Correia P, Costa R (2010) Chemical profiling of Portuguese *Pinus pinea* L. nuts. J Sci Food Agric 90(6):1041–1049

Farjon A (2010) A handbook of the world's conifers, vol I–II. E.J Brill, Leiden

Food Standards Australia New Zealand (FSANZ) (2010) NUTTAB 2006. Australian Food Composition Tables. http://www.foodstandards.gov.au/consumerinformation/nuttab2010/nuttab2010onlinesearchabledatabase/onlineversion.cfm

Gómez-Ariza JL, Arias-Borrego A, García-Barrera T (2006) Multielemental fractionation in pine nuts (*Pinus pinea*) from different geographic origins by size-exclusion chromatography with UV and inductively coupled plasma mass spectrometry detection. J Chromatogr A 1121(2):191–199

Grieve M (1971) A modern herbal. Penguin. 2 Vols. Dover publications, New York. 919pp.

Harrison SG (1951) Edible pine kernels. Kew Bull 3:371–375

Ibáñez MD, Lombardero M, San Ireneo MM, Muñoz MC (2003) Anaphylaxis induced by pine nuts in two young girls. Pediatr Allergy Immunol 14(4):317–319

MA A, Maselli JP, Sanz Mf ML, Fernández-Benítez M (2002) Allergy to pine nut. Allergol Immunopathol (Madr) 30(2):104–108

Martinez F, Montero G (2004) The *Pinus pinea* (L.) woodlands along the coast of south-western Spain: data for a new geobotanical interpretation. Plant Ecol 175:1–18

Munk MD (2010) Pine mouth syndrome: cacogeusia following ingestion of pine nuts (genus: *Pinus*). An emerging problem? J Med Toxicol 6(2):158–159

Mutke S, Gordo J, Climent J, Gil L (2003) Shoot growth and phenology modelling of grafted stone pine (*Pinus pinea* L.) in inner Spain. Ann For Sci 60(6):527–537

Mutke S, Gordo J, Gil L (2005) Variability of Mediterranean stone pine cone yield: yield loss as response to climate change. Agric For Meteorol 132(3/4):263–272

Mutke S, Calama R, Sc G-M, Montero G, Gordo J, Bono D, Gil L (2011) Mediterranean stone pine: botany and horticulture. In: Janick J (ed) Horticultural reviews, vol 39. Wiley, Hoboken, pp 153–201

Nasri N, Fady B, Triki S (2007) Quantification of sterols and aliphatic alcohols in Mediterranean stone pine (*Pinus pinea* L.) populations. J Agric Food Chem 55(6):2251–2255

Nasri N, Khaldi A, Fady B, Triki S (2005a) Fatty acids from seeds of *Pinus pinea* L.: composition and population profiling. Phytochemistry 66(14):1729–1735

Nasri N, Khaldi A, Hammami M, Triki S (2005b) Fatty acid composition of two Tunisian pine seed oils. Biotechnol Prog 21(3):998–1001

Nasri N, Tlili N, Ben Ammar K, Khaldi A, Fady B, Triki S (2009) High tocopherol and triacylglycerol contents in *Pinus pinea* L. seeds. Int J Food Sci Nutr 60(Suppl 1):161–169

Nasri N, Triki S (2007) Storage proteins from seeds of *Pinus pinea* L. C R Biol 330(5):402–409, In French

Nergiz C, Iclal Dönmez I (2004) Chemical composition and nutritive value of *Pinus pinea* L. seeds. Food Chem 86(3):365–368

Nielsen NH (1990) Systemic allergic reaction after ingestion of pine nuts, *Pinus pinea*. Ugeskr Laeger 152(48):3619–3620 (In Danish)

Porcher MH et al (1995–2020) Searchable World Wide Web Multilingual Multiscript Plant Name Database. Published by The University of Melbourne. Australia. http://www.plantnames.unimelb.edu.au/Sorting/Frontpage.html

Ratola N, Lacorte S, Barceló D, Alves A (2009) Microwave-assisted extraction and ultrasonic extraction to determine polycyclic aromatic hydrocarbons in needles and bark of *Pinus pinaster* Ait. and *Pinus pinea* L. by GC-MS. Talanta 77(3):1120–1128

Rubira N, Botey J, Eseverri JL, Marin A (1998) Allergy to pine nuts in children. Allerg Immunol (Paris) 30(7):212–216

Sánchez-Gómez D, Velasco-Conde T, Cano-Martín FJ, Guevara MA, Cervera MT, Aranda I (2011) Interclonal variation in functional traits in response to drought for a genetically homogeneous Mediterranean conifer. Environ Exp Bot 70(2–3):104–109

U S Food and Drug Administration (USFDA) (2011) 'Pine mouth' and consumption of pine nuts. http://www.fda.gov/Food/ResourcesForYou/Consumers/ucm247099.htm

Varo O, Tel AZ (2010) Ecological features of the *Pinus pinea* forests in the north-west region of Turkey (Yalova). Ekoloji 19(76):95–101

Wikipedia (2010) Pinenuts. http://en.wikipedia.org/wiki/Pine_nut

Yildiz-Gulay O, Gulay MS, Ata A, Balic A, Demirtas A (2010) The effects of feeding *Pinus pinea* seeds on some blood values in male New Zealand white rabbits. J Anim Vet Adv 9(20):2655–2658

Piper cubeba

Scientific Name

Piper cubeba L. f.

Synonyms

Cubeba officinalis Raf.

Family

Piperaceae

Common/English Names

Cubeb, Cubeb Pepper, False Pepper, Java Pepper, Javanese Pepper, Javanese Peppercorn, Tailed Pepper.

Vernacular Names

Arabic: Kabâbah, Hhabb El'arûs;
Armenian: Hendkapeghpegh;
Bangladesh: Kabab Chini;
Belarusian: Iavanski Perac;
Brazil: Pimenta-Cubeba (Portuguese);
Bulgarian: Kubeba;
Chinese: Bi Cheng Qie, Cheng Qie, Bi Cheng Qie;
Czech: Pepřovník Dlouhý, Pepřovník Kubeba;
Danish: Cubeberpeber, Kubeber, Kubeber-Peber;
Dhivehi: Kabbaabu;
Dutch: Cubebe, Cubebepeper, Staartpeper;
Eastonian: Kubeebapipar;
Farsi: Kubabah;
French: Cubèbe, Poivre À Queue, Poivre De Java, Poivrier Cubèbe, Quibebes;
German: Javanischer Pfeffer, Kubebenpfeffer, Kubeben-Pfeffer, Schwanzpfeffer, Stielpfeffer, Stiel-Pfeffer;
Greek: Koubeba;
India: Kabab Chini (Bengali), Tadamari (Gujerati), Cubab-Chinee, Kabab Chini, Sheetal Chini (Hindu), Kaba-Chini (Maithili), Vaalmilagu (Malayalam), Kankol (Marathi), Kabachin, Kabab Chini (Oriya), Chinamilagu, Sinamilagu, Valmilagu (Tamil), Chalavamiriyaalu, Tokamiriyalu (Telugu), Kabab Chini (Urdu);
Indonesia: Kemukus, Temukus (Java), Rinu Katencar, Rinu Caruluk (Sudanese);
Italian: Cubebe;
Hungarian: Jávai Bors, Hosszú Bors, Kubéba Bors;
Italian: Cubebe, Pepe A Coda;
Japanese: Kubeba, Kubebu;
Korean: Chaba Huchu, Jaba Huchu, Kubepu, Kyubebu, Philjunggadongul, Pilchingga;
Lithuanian: Kubebos Pipirai;
Macedonian: Crniot Piper;
Malaysia: Lada Berkor, Kemukus, Kemungkus, Cabai Berekor, Lada Ekor;
Nepal: Ghanda Maric, Kabaab Chiinii, Thulo Pipla;
Norwegian: Kubebapepper;
Pakistan: Dhumkimirch, Kavabchini;

Polish: Pieprz Kubeba;
Portuguese: Cubeba;
Romanian: Piper De Cubebe;
Russian: Dikij Perets, Kubeba, Yavanskij Perets, Perets Kubebe;
Serbian: Biber Krupan, Biber Krupni;
Slovak: Kubéba, Piepor Kubébový;
Slovenian: Poper Kubeba;
Spanish: Cubeba;
Swedish: Kubebapeppar;
Thai: Prik Hang;
Tibetan: Ka Ko La;
Turkish: Hind Biberi, Hind Biberi Tohomu, Kebabe, Kebebe, Kebabiye Biber, Kebebiye, Kuyruklu Biber;
Vietnamese: Tiêu Thất;
Yiddish: Kebebe.

Origin/Distribution

Piper cubeba is indigenous to Indonesia – Greater Sunda Islands. It is cultivated in Java, Sumatra, Malaysia and Sri Lanka.

Agroecology

It grows in full or partial shade in the moist deciduous tropical forests, edges of mangrove forest and is grown from sea level to 700 m. It is also found along streams, forest tracks and margins.

Edible Plant Parts and Uses

During the Middle Ages, cubeb was one of the most valuable spices in Europe and today its culinary importance has been surpassed by medicinal uses. In powdered form cubeb was used as a seasoning for meat and in sauces. One common medieval recipe was for *sauce sarcenes* which is composed of almond milk and spices including cubeb. In Poland during the fourteenth century, a vinegar referred to as *Ocet Kubebowy* was infused with cubeb, cumin and garlic and used for meat marinades. Cubeb was often candied and eaten whole as an aromatic confectionery.

Cubeb is still used to enhance the flavor of savory soups. In Moroccan cuisine, cubeb is relished in savoury dishes and pastries like *markouts* and the renowned spice mixture *Ras el hanout* a popular mixture of herbs and spices that is used across the Middle East and North Africa. *Ras el hanout* is used in *pastilla*, the Moroccan squab/young pigeon, and almond pastry, is sometimes rubbed on meats, and incorporated into couscous or rice. Today, cubeb is used as a spice to impart flavour to food in south and southeast Asia for example in *gulés* (curries) in Indonesia. It is very popular in West African cooking. They are sold whole and should be crushed or ground before use in cooking. Cubebol a patented compound from cubeb oil is traded by a Swiss company as a cooling and refreshing agent and used in various products like chewing gum, drinks, sorbets, gelatine-based confectionaries and even tooth paste. Cubeb is also used to flavour alcoholic and non-alcoholic beverages and drinks. Bombay Sapphire gin is flavored with cubeb and grains of paradise (*Aframomum melegueta).* Pertsovka, a dark-brown, Russian pepper vodka with a fiery taste, is prepared from infusion of cubeb and capsicum peppers.

Botany

A perennial, climbing woody shrub with glabrous, jointed, cylindrical and striate stem somewhat thickened at the nodes and rooting at the nodes. Stem perennial, smooth, climbing, jointed. Leaves alternate, ovate-oblong or lanceolate, acuminate apex, somewhat unequal base, entire, 10–15 cm by 4-cm wide, wavy, leathery, deep green, smooth, prominently nerved below, on short, stout petioles. Flowers unisexual, dioecious, minute, sessile, each with a bract at the base, without calyx or corolla, densely crowded in small, long, cylindrical, stalked, solid spikes coming off opposite the leaves, two stamens to each flower on male plants, and three pistils on the pistillate plant. Fruit globose, yellowish red berry, 0.6–08 cm across, smooth, wrinkled when dried, hard, one seeded (globose) and stalked (Plates 1 and 2).

Plate 1 Dried Cubeb pepper berries

Plate 2 Close-view of dried cubeb pepper berries

Nutritive/Medicinal Properties

Piper cubeba fruit was found to be a good source of the essential elements while toxic elements are found in trace amounts (Fatima et al. 2011). It was found to be a good source of K (2.10%) and iron. The phytochemical profile of *Piper* species was found to be characterized by the presence of typical classes of compounds such as alkaloids, amides, benzoic acids, chromenes, propenylphenols, lignans, neolignans, sesquiterpenes, terpenes, steroids, kawapyrones, piperolides, chalcones, dihydrochalcones, flavones and flavanones (Jensen et al. 1993; Parmar et al. 1997). Numerous studies had been conducted on the composition of the oil from the berries and one from the leaves.

Cubeba Oil

The main components of berry oil from Sri Lanka were found to be cubebol (31%), α-cubebene (5.1%) and αcopaene (8.1%) (Terhune et al. 1974). Shankaracharya et al. (1995) identified 53 components in the commercial berry oil of which sabinene (28%) and cubebol (16%) were the main components. The main components of the berry oil from India were cubebol (23.6%), a-pinene (18.2%), β-elemene (7.3%), β-cubebene (5.6%) and δ-cadinene (4.7%) (Sumathykutty et al. 1999). In a number of *P. cubeba* berry oil samples, Lawrence (1980) identified 71 components with α-cubenene (7–9%), α-copaene (10–14%), β-cubebene (7–11%), δ-cadinene (9–10%) and cubebol (9–10%) as the main components. Cubenol was also found. In a subsequent analysis of a commercial berry oil sample, Lawrence (2001) found sabinene (30%) to be the main component, whereas cubebol was only present at 5.7%. More recently, Singh et al. (2008) reported the main component of the essential berry oil of *Piper cubeba* to be ß-cubebene (18.94%) followed by cubebol (13.32%), sabinene (9.60%), α-copaene (7.41%) and ß-caryophyllene (5.28%) with many other components in minor amount. All the oleoresins showed the presence of 85 components. The major component in all the oleoresins was cubebol (stereoisomer). The percentage of cubebol in the diethyl ether extract was 32.38, in the ethanol extract 25.51, in the petroleum benzene extract 42.89, in the chloroform extract 28.00 and in methanol extract 19.03.

Hydrodistillation of the berries of *Piper cubeba* yielded 11.8% (w/w) and the leaves 0.9% (v/w) oil (Bos et al. 2007). In total 103 components were identified in the berries, representing 59.6% of the oil. In the leaves, 62 components could be identified, corresponding with 77.9% of the oil. Cubeb berry oil and leaf oils had no large qualitative differences in the composition, although the berries contained a considerable amount of constituents in traces (<0.05%) that were not found in the leaves. Sabinene (9.1%), β-elemene (9.4%), β-caryophyllene (3.1%), epi-cubebol (4.3%), and cubebol (5.6%) were the main components of the berry oil. trans-Sabinene hydrate (8.2%), β-caryophyllene (5.0%), epi-cubebol (4.2%), γ-cadinene (16.6%) and cubebol (4.8%) were the main components of the leaf oil. The main monoterpenes in the berry oil were α-thujene (2.5%), α-pinene (1.8%), sabinene (9.1%), and limonene (2.3%), were a-pinene (3.2%), sabinene (3.8%), β-pinene (3.8%) and limonene (3.4%) were the principal monoterpenes in the leaf oil. In the oxygenated monoterpene fractions (3.6% and 10.6%, respectively, for the berry and leaf oil), trans-sabinene hydrate was the main component (2.5% and 8.2%, respectively). α-Copaene (3.8%), β-elemene (9.4%), β-caryophyllene (2.5%), were the main sesquiterpenes (23.7%) in the berry oil, where β-caryophyllene (5.0%), and γ-cadinene (16.6%) were the main sesquiterpenes (30.9%) in the leaf oil. Remarkable is the high content of γ-cadinene in the leaf oil, whereas it was present in the berry oil in only small amounts (0.1%). From the oxygenated sesquiterpenes (15.5% and 18.6%, respectively in berries and leaves), epi-cubebol (4.6% and 4.2%) and cubebol (5.6% and 4.8%), were the main components in both oils. Other major components were guaiol (2.9%) in the berry oil; and γ-cadinol (2.7%) and α-cadinol (1.9%) in the leaf oil.

Other Phytochemicals

Alkaloids: The alkaloid, piperine was isolated from *P. cubeba* (Hadom and Jungkunz 1951). Piperine is responsible for the pungency of cubeb pepper.

Lignans: A bisepoxylignan, ashantin was isolated from *P. cubeba* (Haensel and Pelter 1969) and bisasarin (Yang et al. 1982). From the hot petroleum extract of *Piper cubeba* fruits, six lignans were isolated (Prabhu and Mulchandani 1985). Two of these, were characterized as (2R,3R)-2-(3″,4″,5″-trimethoxybenzyl)-3-(3′,4′-methylenedioxybenzyl)-1,4-butanediol [(−)-dihydroclusin] and (3R,4R)-3,4-bis-(3,4,5-trimethoxybenzyl)tetra-hydro-2-furanol [(−)-cubebinin]. (−)Cubebin, (−)-hinokinin, (−)-clusin and (−)-dihydrocubebin were also found. Six more lignans were isolated from the hot petroleum extract of *Piper cubeba* fruits (Badheka et al. 1986). Of these, three compounds were characterized as (2R,3R)-2-(5″-methoxy-3″, 4″-methylenedi oxybenzyl) butrylactone [(−)-cubebinone],(2R,3R)-2-(3″,4″-methylenedioxybenzyl)-3-(3′,4′,5′-trimethoxybenzyl)butyrolactone [(−)-isoyatein] and (2R,3R)-2-(3″,4″, 5″-trimethoxybenzyl)-3-(3′,4′-dimethoxybenzyl) butyrolactone [(−)-di-O-methyl thujaplicatin methyl ether, i.e. (−)-thujaplicatin trimethyl ether]. The other three compounds were identified as (−)-yatein, (−)-cubebininolide and (2R,3R)-2-(3″,4″-methylenedioxybenzyl)-3-(3′,4′-dimethoxybenzyl) butyrolactone. Seven additional compounds were isolated from *Piper cubeba* and characterized as heterotropan, magnosalin, 2,4,5-trimethoxybenzaldehyde, α-O-ethyl cubebin, β-O-ethyl cubebin, 5″-methoxyhinokinin and the monoacetate of dihydrocubebin (Badheka et al. 1987). The following lignans were isolated from *Piper cubeba* :(-)-clusin, (-)-dihydroclusin, (-)-yatein, (-)-hinokinin, and (-)-dihydrocubebin (Usiaetal.2005a),(8R,8′R)-4-hydroxycubebinone and (8R,8′R,9′S)-5-methoxyclusin, ethoxyclusin (15), and (-)-dihydroclusin (17) (Usia et al. 2005b).

Terpene compounds: The following terpene compounds were isolated from *P.cubeba*: (+)-4-carene, 1.4-cineol, 4-Isopropyl-1-methylcyclohex-l-en-ol (Rao et al. 1928). In the *P. cubeba* berry oil of Indian origin, the following terpenes: α-cadinene and α-copaene were identified (Razdan and Bhattacharvya 1954, 1955). Ikeda et al. (1962) examined the monoterpene hydrocarbon fraction of *P. cubeba* oil and reported the following as the main components α-pinene (12.1%), ce-thujene (13.2%), sabinene (47.1%), and β-phellandrene (12.7%), together with eight other monoterpene hydrocarbons. Opdyke (1976) reported the following terpenesfrom *P. cubeba*: 1,8-cineole, α-Cubebene, *p*-cymene, limonene, myrcene, β-ocimene, α-phellandrene, β-phellandrene, α-pinene, β-pinene, sabinene, α-terpinene, γ-terpinene, terpinolene and α-thujene. A sesquiterpene hydrocarbon, bicyclosesquiphellandrene was isolated from *Piper cubeba* oil (Terhune et al. 1974). The following terpenes were isolated from *P. cubeba*: cubebol, germacrene D and (-)-muurolene (Shankaracharya et al. 1995).Two new sesquiterpenes ($5\alpha,8\alpha$)-2-oxo-1(10),3,7(11)-guaiatrien-12,8-olide and ($1\alpha,2\beta,5\alpha,8\alpha,10\alpha$)-1,10-epoxy-2-hydroxy-3,7(11)-guaiadien-12,8-olide were isolated from *Piper cubeba* (Usia et al. 2005b).

Miscellaneous compounds: From the petrol extract of *Piper cubeba*, new oxygenated cyclohexanes were isolated and their structures determined, as (+)-(2S,3R,4R,5R)1-benzoyloxy methyl cyclohex-1(6)-ene-2,3,4,5 tetrol-3-benzoate (piperenol A) and (+)-(1S,2S,3S,4R) 1-benzoyloxy methylcyclohex-5-ene-1,2,3,4 tetrol-4-benzoate (piperenol B) besides the known oxygenated cyclohexanes, (+)-crotepoxide and (+)-zeylenol (Taneja et al. 1991). Further investigation of the petrol extract of *Piper cubeba* yielded two new minor oxygenated cyclohexanes, (–)-rel-(2S,3R,4R,5R)-2,3,4,5-tetraacetoxy-1-benzoyloxy methylcyclohex-1(6)-ene-2,3,4,5-tetrol[(–)-piperenol C] and (+)-(2S,3R,4R,5R)-2,4,5-triacetoxy-1-benzoyloxy methylcyclohex-1(6)-ene-2,3,4,5-tetrol-3-benzoate[(+)-piperenol A-triacetate] (Koul et al. 1996). In addition, two rare neolignans were isolated and identified as (–)-kadsurin A and (–)-piperenone.

Some pharmacological properties reported on *Piper cubeba* include:

Antioxidant Activity

The n-hexane, dichloromethane (DCM) and methanol (MeOH) extracts of the dried berries (fruit) of *Piper cubeba* showed antioxidant activity(2,2-diphenyl-1-picrylhydrazyl (DPPH)) in the qualitative assay, the most prominent antioxidant activity was observed with the MeOH extract in the quantitative assay with a RC_{50} value of $2.71 \times 10-1$ mg/ml (Chitnis et al. 2007). The antioxidant potency of the DCM extract was about 3 fold less ($RC_{50} = 6.50 \times 10-1$ mg/ml) than that of the MeOH extract. Like all the *Piper* species,

P. cubeba was found to have glutathione (GSH) content of around 1–2 µM/g tissue and to exhibit catalyse activity (Karthikeyan and Rani 2003). The antioxidant components of *Piper* species were known to constitute a very efficient system in scavenging a wide variety of reactive oxygen species. Antioxidant potential of *Piper* species was further confirmed by their ability to curtail in vitro lipid peroxidation by around 30–50% with concomitant increase in GSH content.

Using the Fenton-like reaction [Fe(II)+H_2O_2], 16 compounds from *Piper cubeba* (CNCs) were found to inhibit 5,5-dimethyl-1-pyrroline-N-oxide, DMPO –OH radical formation ranging from 5% to 57% at 1.25 mmol/L concentration (Aboul-Enein et al. 2011). The examined CNCs also showed a high DPPH (2,2-diphenyl-1-picrylhydrazyl) antiradical activity (ranging from 15% to 99% at 5 mmol/L concentration). Furthermore, the results indicated that seven of the 16 tested compounds may catalyse the conversion of superoxide radicals generated in the potassium superoxide/18-crown-6 ether system, thus showing superoxide dismutase-like activity. The data obtained suggested that radical scavenging properties of CNCs might have potential application in many plant medicines.

High antioxidant (DPPH scavenging) activity was found in *Piper cubeba* ethanol extract 77.61% with IC_{50} value of 10.54 µg/ml compared to *Piper nigrum* 74.61% and IC_{50} value of 14.5 µg.ml (Nahak and Sahu 2011). *P. cubeba* had the highest total phenolic content of 123.1 µg/g compared to *P. nigrum* with 62.3 µg/g. *P. cubeba* was found to contain alkaloid, glycosides, steroid, flavonoid, tannins and antraquinones while *P. nigrum* contained the same plus terpenoid and reducing sugars.

Anticancer Activity

A number of polyhydroxy cyclohexanes had been isolated from *Piper cubeba* and shown to display tumour inhibitory, antileukemic and antibiotic activities (Taneja et al. 1991). An ethanolic extract of *P. cubeba*, designated P9605 exhibited anti-estrogenic, anticancer and antiinflammatory properties (Yam et al. 2008b). The extract significantly inhibited growth induced by β-estradiol in MCF-7, a human breast cancer cell line. It inhibited aromatase activity, which was responsible for transforming androgens into estrogens. Additionally the extract inhibited the activities of cyclo-oxygenases (COX-1 and COX-2) and 5-lipo-oxygenase (5-LOX), and attenuated the induction of interleukin 6 (IL-6) in differentiated THP-1 cells stimulated by lipopolysaccharide (LPS). The results supported the potential use of P9605 in phytotherapy against benign prostatic hyperplasia (BPH). The scientists also found that the P9605 extract inhibited proliferation in androgen-dependent LNCaP human prostate cancer cells by reducing DNA synthesis and inducing apoptosis (Yam et al. 2008a). P9605 potently inhibited 5 α-reductase II activity, which was responsible for converting testosterone to its active form, dihydrotestosterone (DHT), in the prostate. It also acted as an antagonist at recombinant wild-type androgen receptors (AR). P9605 suppressed cell growth and prostate-specific antigen (PSA) secretion stimulated by physiological concentrations of DHT in LNCaP cells. Further, it down-regulated androgen receptors levels. The findings suggested that P9605 may potentially retard the growth of androgen-dependent prostate cancer via several mechanisms.

Antiinflammatory and Analgesic Activities

Studies also showed the antiinflammatory and analgesic effects of three dibenzylbutyrolactone lignans, (−)-hinokinin (2), (−)-6,6′-dinitrohinokinin (3), and (−)-6,6′-diaminohinokinin (4), obtained by partial synthesis from (−)-cubebin (1), in different animal models (da Silva et al. 2005). It was observed that compounds from (−)-cubebin and (−)-hinokinin inhibited the edema formation in the rat paw edema assay at the same level and that all responses were dose dependent. Also, at the dose of 30 mg/kg, compounds (−)-cubebin, (−)-hinokinin (2), (−)-6,6′-dinitrohinokinin (3), and (−)-6,6′-diaminohinokinin inhibited the edema formation by 53%, 63%, 54%, and 82%, respectively. In the acetic acid-induced writhing test in mice, compounds 2 and 4 produced inhibition levels of 97% and 92%, respectively, while 3 displayed lower effect (75%), which was still higher than 1.

The medicinal plant extract (*Piper cubeba* (fruit), *Physalis angulata* (flower), *Rosa hybrida* (flower) displayed antiinflammatory activities as determined by carrageenan-induced paw edema, arachidonic acid-induced ear edema and formaldehyde-induced arthritis in mice (Choi and Hwang 2003). These plant extracts clearly exhibited inhibitory effects against acute and subacute inflammation by oral administration (200 mg/kg). Also, administration (200 mg/kg, p.o.) of plant extracts for 1 week significantly inhibited type IV allergic reaction in mice as evaluated by using 2,4-dinitrofluorobenzene (DNFB)-induced contact hypersensitivity reaction (type IV) . in a subsequent study the intake of medicinal plant extract (*Piper cubeba* (fruit), *Physalis angulata* (flower), *Rosa hybrida* (flower)) in rats resulted in an increase in antioxidant enzyme activity and HDL-cholesterol, and a decrease in malondialdehyde, which may reduce the risk of inflammatory and heart disease (Choi and Hwang 2005). After 3 weeks, the superoxide dismutase (SOD) activity of the *Piper cubeba* group and the catalase activity of the *Piper cubeba* and *Rosa hybrida* groups were significantly increased compared with the control group, while the SOD and catalase activities of the *Physalis angulata* group were not significantly changed, thiobarbituric acid reactive substance (TBARS), a marker of lipid peroxidation, was significantly lower in all experimental groups compeered with the control group. No significant changes occurred in the triglyceride (TG, total- and LDL-cholesterol) of all groups, but the HDL-cholesterol of the *Physalis angulata* group was significantly increased. This study showed that the intake of medicinal plants in rats resulted in an increase in antioxidant enzyme activity and HDL-cholesterol, and a decrease in malondialdehyde, which may reduce the risk of inflammatory and heart disease.

Antimicrobial Activity

A crude ethanol extract from *Piper cubeba* seeds, (−)-cubebin and its semi-synthetic derivatives were found to be active against oral pathogens (Silva et al. 2007). The crude ethanol extract was more active against *Streptococcus salivarius* (MIC value of 80 μg/ml). (−)-Cubebin displayed MIC values ranging from 0.20 mm for *Streptococcus mitis* to 0.35 mm for *Enterococcus faecalis*. The natural product (−)-cubebin and its semi-synthetic derivative (−)-hinokinin displayed bacteriostatic activity at all evaluated concentrations, as well as fungicidal activity against *Candida albicans* at 0.28 mm. The O-benzyl cubebin derivative showed fungistatic and fungicidal effects against *C. albicans* at 0.28 and .35 mm, respectively. Also, the other dibenzylbutyrolactone derivatives [(−)-6, 6′-dinitrohinokinin and (−)-O-(N,N-dimethylaminoethyl)-cubebin] displayed bacteriostatic and fungistatic effects at the evaluated concentrations. Moreover, the semi-synthetic derivative (−)-6, 6′-dinitrohinokinin was the most active compound against all the evaluated microorganisms. Another earlier study showed that the essential oil of *P. cubeba* exhibited maximum activity against *Streptococcus faecalis, Bacillus pumilus* and *Pseudomonas solanacearum* (Kar and Jain 1971). The combinations of *Litsea chinensis, P. cubeba* and *Colubrina asiatica* displayed the maximum inhibitory response indicating synergistic or potentiating effect.

The essential oil and oleoresins of *Piper cubeba* exhibited moderate to strong antimicrobial and antioxidant activities (Singh et al. 2007, 2008). The radical scavenging capacity of both essential oil and oleoresin on 2, 2′-diphenyl-1-picrylhydrazyl (DPPH) radical were (71.2%) and (69.77%) respectively at 25 μL/ml. It was relatively lower in comparison with synthetic antioxidants (BHA – 96.41%; BHT – 95.91%). The results obtained from reducing power, chelating effect and hydroxyl radical scavenging effect also supported the antioxidant of essential oil and oleoresin. The essential oil and oleoresin showed 100% mycelial zone inhibition against *Penicillium viridicatum* at 3,000 and 2,000 ppm respectively in the poison food method. The essential oil revealed 100% clear zone inhibition against *Aspergillus flavus* at all tested concentrations. None of the extracts namely n-hexane, dichloromethane and methanol extracts of the dried berries (fruit) of *Piper cubeba* showed any antibacterial property against *Bacillus subtilis, Escherichia coli*, and ampicillin resistant *Escherichia coli* (Chitnis et al. 2007). While both the n-hexane and the dichloromethane extracts inhibited the growth of *Bacillus cereus, Pseudomonas aeruginosa* and *Staphylococcus aureus*, the methanol extract was active only against *B. cereus* and *P. aeruginosa*. The most potent antibacterial activity was displayed by the n-hexane extract against *B. cereus* with an MIC value of 1.56 mg/ml. All antibacterial activities of the extracts were found to be bacteriostatic rather than bactericidal.

Antiviral Activity

A water extract of *Piper cubeba*, was reported to be active (≥90% inhibition at 100 μg/ml) in inhibitory effects on hepatitis C virus (HCV) protease (Hussein et al. 2000).

Trypanocidal Activity

Five ( ) cubebin derivative compounds from *P. cubeba* namely, (−)-O-acetyl cubebin (3), (−)-O-benzyl cubebin (4), (−)-O-(N, N-dimethylaminoethyl)-cubebin (5), (−)-hinokinin (6) and (−)-6, 6′-dinitrohinokinin (7), exhibited trypanocidal activity against free amastigote forms of *Trypanosoma cruzi*, the asogic agent of Chagas' disease (de Souza et al. 2005). It was observed that 6 was the most active compound ($IC_{50} = 0.7$ μM), and that 4 and 5 displayed moderate activity against the parasite, giving IC_{50} values of 5.7 and 4.7 μM, respectively. In contrast, it was observed that compound 3 was inactive and that 7 displayed low activity with IC50 values of $\cong 1.5 \times 10^4$ and 95.3 μM, respectively. (−)-Hinokinin, a dibenzylbutyrolactone lignan, obtained by partial synthesis from (−)-cubebin isolated from the dry seeds of *Piper cubeba*, exhibited significant trypanocidal activity both in vitro and in vivo. Further studies showed that (−)-hinokinin not only has no genotoxic effect, but is also effective in reducing the chromosome damage induced by the chemotherapeutic agent doxorubicin (DXR). (−)-Hinokinin exerted a significant antioxidant effect on parasite mitochondria in the protocol used, which might be one possible mechanism by which this compound may exert a protective effect on the chromosome damage induced by the free radicals generated by DXR.

Antileishmanial Activity

Piper cubeba and *Piper retrofractum* was found to possess antileishmanial activity (Bodiwala et al. 2007). The n-hexane, ethyl acetate, methanol, and acetone extracts of *Piper cubeba* and *P. retrofractum* exhibited significant in vitro activity at 100 μg/ml against promastigotes of *Leishmania donovani*. Two lignans, cubebin and hinokinin, were isolated from the hexane extract of *P. cubeba*; and one bis-epoxy lignan, (−)-sesamin, and two amides, pellitorine and piplartine, were isolated from the hexane and methanol extracts of *P. retrofractum*. Cubebin and piplartine showed significant antileishmanial activity in vitro at 100 μM and were further tested in vivo in a hamster model of visceral leishmaniasis. Piplartine showed activity at 30 mg/kg dose.

Antiparasitic Activity

Magalhães et al. (2011) suggested that *Piper cubeba* essential oil was efficacious against cercariae, schistosomula, and adult worms of the *Schistosoma mansoni*. At concentrations of 100 and 200 μg/ml, it caused a total absence of mobility after 120 hours. At concentrations from 12.5 to 50 μg/ml, it caused a reduction in the viability of cercariae and schistosomula when compared with the negative control groups. At concentrations ranging from 50 to 200 μg/ml, separation of all the coupled adult worms was observed after 24 hours of incubation, resulting in a reduction in egg production. The main chemical constituents of the essential oil were identified as sabinene (19.99%), eucalyptol (11.87%), 4-terpineol (6.36%), β-pinene (5.81%), camphor (5.61%), and δ-3-carene (5.34%). The essential oil exerted significant cytotoxicity at the concentration of 200 μg/ml after 24 hours treatment.

Antiulcer Activity

The methanolic extract of the fruits of *Piper cubeba* (400 mg/kg) showed maximum inhibition of gastric acid, free acid and total acid to 23.61%, 66.94% and 56.71% respectively using model of gastric in rats which were induced by pyloric ligation (Parvez et al. 2010). The ulcer index in the *Piper cubeba* treated animals was found to be significantly less in all the models compared to control and standard drug, treated cases. The antiulcer activity of *Piper cubeba* was, however, less than that of Omeprazole. The results suggested that *Piper cubeba* possessed significant antiulcer property which could be due to cytoprotective action of the drug or strengthening of gastric mucosa with the enhancement of mucosal defence.

Cytochrome P450 Inhibition Activity

Five methylenedioxyphenyl lignans namely (-)-clusin (1), (-)-dihydroclusin (2), (-)-yatein (3), (-)-hinokinin (4), and (-)-dihydrocubebin (5), isolated from *Piper cubeba* were found to be potent and selective inhibitors against cytochrome P450 3A4 (CYP3A4) (Usia et al. 2005a). All lignans (1–5) inhibited CYP3A4 in a time-, concentration-, and NADPH-dependent manners and thus appeared to be the mechanism-based inhibitors of CYP3A4. Among them, (−)-clusin (1) and (−)-dihydroclusin (2) were found to be the most potent CYP3A4 inactivator. The scientists also tested two new lignans, (8R,8′R)-4-hydroxycubebinone (1) and (8R,8′R,9′S)-5-methoxyclusin (2), and two new sesquiterpenes,

(5 α,8 α)-2-oxo-1(10),3,7(11)-guaiatrien-12,8-olide (3) and (1 α,2 β,5 α,8 α, 10 α)-1,10-epoxy-2-hydroxy-3,7(11)-guaiadien-12,8-olide (4), along with 16 known compounds (5-20) for their inhibitory activity on the metabolism mediated by CYP3A4 or CYP2D6 using [N-methyl-(14)C] erythromycin or [O-methyl-(14)C] dextromethorphan as a substrate, respectively (Usia et al. 2005b). The compounds (8R,8′R,9′S)-5-methoxyclusin (2), (−)-clusin (10), (−)-yatein (13), ethoxyclusin (15), and (−)-dihydroclusin (17), having one methylenedioxyphenyl moiety in their structures, showed very potent and selective inhibitory activity against CYP3A4 with IC_{50} values (0.44–1.0 μM) identical to that of the positive control, ketoconazole (IC_{50}, 0.72 μM).

Genotoxic Activity

Studies by Junqueira et al. (2007) showed that *Piper cubeba* seed extract was genotoxic in-vivo when administered orally to mice and rats. At 1.5 g/kg, the highest dose tested, the extract induced a statistically significant increase in both the mean number of micronucleated polychromatic erythrocytes and the level of DNA damage in the rodent cell types analysed.

Molluscicidal Activity

The dried berries powder of *P. cubeba*, dried fruit powder of *P. longum* and *Tribulus terrestris* singly as well as in binary and tertiary combination exhibited molluscicidal activity against the snail *Indoplanorbis exustus* in a time and concentration-dependent (Pandey and Singh 2009).

Traditional Medicinal Uses

Cubeb pepper is a popular medicinal plant which has been extensively used in Europe since the Middle Ages, as well as in many other countries, including Arabia, India, Indonesia, Malaysia and Morocco in traditional medicine.

Cubeb berry is considered a carminative, diuretic, expectorant, stimulant, stomachic, antiasthmatic, irritant, sedative, anti-dysenteric and antiseptic. It acts particularly on mucous tissues, and arrests excessive discharges, especially from the urethra. Cubeb berry also has a local stimulating effect on the mucous membranes of the urinary and respiratory tracts. It exercises an influence over the urinary apparatus, rendering the urine of deeper colour. It has been employed in the treatment of gonorrhoea, gleet, leucorrhoea, chronic bladder diseases, acute prostatitis bronchial affections, dysentery and in spermatorrhea. In England, various preparations of cubeb including *oleum cubebae* (oil of cubeb), tinctures, fluid extracts, oleo-resin compounds, and vapors, were employed for throat complaints. Cubeb was commonly included in lozenges designed to alleviate bronchitis, exploiting the antiseptic and expectoral properties of the drug. The most important therapeutic application of cubeb, however, was in treating gonorrhea. Cubeb berry has been shown to be effective in easing the symptoms of chronic bronchitis. In India, a cubeb paste is used as a mouthwash, and dried cubebs is used internally for oral and dental diseases, loss of voice, halitosis, fevers, and cough. It is also used for digestive ailments and is effective in treating dysentery. The herb has often been associated with the reproductive system and has been used to treat cystitis, leucorrhea, urethritis, and prostate infections. In India, Unani physicians use a paste of cubeb berries externally of the male and female genitals to intensify sexual pleasure during coitus. Indian physicians and Arab physician during the middle ages employ cubeb berries as a main ingredient in an aphrodisiac remedy for infertility. In Malaysia, cubeb was used in many medicinal mixtures administered as tonics, indigestion mixtures and pick-me-ups after childbirth and for rheumatism. It is also prescribed for external application. In Indonesia, cubeb berries have also been used for the treatment of abdominal pain, asthma, diarrhoea, dysentery, gonorrhoea, enteritis and syphilis. In China, cubeb is used in traditional medicine for its warming properties. In Tibet cubeb is one of the six fin herbs in *bzang po* drug.

Other Uses

Cubeb is also used for its fragrance in soaps and perfumes, and can also be found as a flavouring in tooth paste and tobacco (cubeb cigarettes) besides food. In 2000, Shiseido cosmetics company in Japan, patented a line of anti-aging products containing formulas made from several herbs, including cubeb. Cubeb berries are used in love-drawing magic spells by practitioners of hoodoo, an African-American form of folk magic.

Comments

Taken in excessive doses, cubeb berry can cause nausea, vomiting, burning pain, griping and purging.

Selected References

Aboul-Enein HY, Kładna A, Kruk I (2011) Radical scavenging ability of some compounds isolated from *Piper cubeba* towards free radicals. Luminescence 26(3):202–207
Backer CA, van den Bakhuizen Brink RC Jr (1963) Flora of Java, vol 1. Wolter-Noordhoff, Groningen, 647pp
Badheka LP, Prabhu BT, Mulchandani NB (1986) Dibenzylbutyrolactone lignans from *Piper cubeba*. Phytochemistry 25(2):487–489
Badheka LP, Prabhu BT, Mulchandani NB (1987) Lignans of *Piper cubeba*. Phytochemistry 26(7):2033–2036
Bodiwala HS, Singh G, Ranvir Singh R, Dey CS, Sharma SS, Bhutani KK, Pal I (2007) Antileishmanial amides and lignans from *Piper cubeba* and *Piper retrofractum*. J Nat Med 61(4):418–421
Bos R, Woerdenbag HJ, Kayser O, Quax WJ, Ruslan K, Elfami (2007) Essential oil constituents of *Piper cubeba* L. fils. from Indonesia. J Essent Oil Res 19(1):14–17
Burkill IH (1966) A dictionary of the economic products of the Malay Peninsula. Revised reprint. 2 volumes. Ministry of Agriculture and Co-operatives, Kuala Lumpur, Malaysia. Vol. 1 (A–H) pp 1–1240, Vol. 2 (I–Z) pp 1241–2444
Chitnis R, Abichandani M, Nigam P, Nahar L, Sarker SD (2007) Actividad antibacteriana y antioxidante de los extractos de *Piper cubeba* (Piperaceae) Antioxidant and antibacterial activity of the extracts of *Piper cubeba* (Piperaceae). Ars Pharm 48(4):343–350
Choi EM, Hwang JK (2003) Investigations of anti-inflammatory and antinociceptive activities of *Piper cubeba, Physalis angulata and Rosa hybrida*. J Ethnopharmacol 89(1):171–175
Choi EM, Hwang JK (2005) Effect of some medicinal plants on plasma antioxidant system and lipid levels in rats. Phytother Res 19(5):382–386
da Silva R, de Souza GHB, da Silva AA, de Souza VA, Pereira AC, de Royo VA, E Silva MLA, Donate PM, de Matos Araújo ALS, Carvalho JCT, Bastos JK (2005) Synthesis and biological activity evaluation of lignan lactones derived from (−)-cubebin. Bioorg Med Chem Lett 15(4):1033–1037
de Souza VA, da Silva R, Pereira AC, de Royo VA, Saraiva J, Montanheiro M, de Souza GHB, da Silva Filho AA, Grando MD, Donate PM, Bastos JK, Albuquerque S, E Silva MLA (2005) Trypanocidal activity of (−)-cubebin derivatives against free amastigote forms of *Trypanosoma cruzi*. Bioorg Med Chem Lett 15(2):303–307
Fatima I, Waheed S, Zaidi JH (2011) Essential and toxic elements in three Pakistan's medicinal fruits (*Punica granatum, Ziziphus jujuba* and *Piper cubeba*) analysed by INAA. Int J Food Sci Nutr doi:10.3109/09637486.2011.627842
Hadorn H, Jungkunz R (1951) Pepper and cubeb. Pharm Acta Helvetia 26:25
Haensel R, Pelter A (1969) Relative und absolute konfiguration von yangambin und aschantin. Arch Pharm 302:940–942
Hussein G, Miyashiro H, Nakamura N, Hattori M, Kakiuchi N, Kunitada Shimotohno K (2000) Inhibitory effects of Sudanese medicinal plant extracts on hepatitis C virus (HCV) protease. Phytother Res 14(7):510–516
Ikeda RM, Stanley WL, Vannier SH, Spitier EM (1962) The monoterpene hydrocarbon composition of some essential oils. J Food Sci 27:455–458
Jensen S, Hansen J, Boll PM (1993) Lignans and neolignans from Piperaceae (Review). Phytochemistry 33:523–530
Junqueira APF, Perazzo FF, Souza GHB, Maistro EL (2007) Clastogenicity of *Piper cubeba* (Piperaceae) seed extract in an in vivo mammalian cell system. Genet Mol Biol 30(3):656–663
Kar A, Jain SR (1971) Antibacterial evaluation of some indigenous medicinal volatile oils. Plant Foods Hum Nutr 20(3):231–237
Karthikeyan J, Rani P (2003) Enzymatic and non-enzymatic antioxidants in selected *Piper* species. Indian J Exp Biol 41(2):135–140
Koul JL, Koul SK, Taneja SC, Dhar KL (1996) Oxygenated cyclohexanes from *Piper cubeb*. Phytochemistry 41(4):1097–1099
Lawrence BM (1980) Progress in essential oils. Perfum Flavor 5(5):27–32
Lawrence BM (2001) Progress in essential oils. Perfum Flavor 26(4):78–81
Magalhães LG, de Souza JM, Wakabayashi KA, da S Laurentiz R, Vinhólis AH, Rezende KC, Simaro GV, Bastos JK, Rodrigues V, Esperandim VR, Ferreira DS, Crotti AE, Cunha WR, E Silva ML (2011) In vitro efficacy of the essential oil of *Piper cubeba* L. (Piperaceae) against *Schistosoma mansoni*. Parasitol Res doi: 10.1007/s00436-011-2695-7

Medola JF, Cintra VP, Pesqueira E, Silva EP, de Andrade RV, da Silva R, Saraiva J, Albuquerque S, Bastos JK, E Silva MLA, Tavares DC (2007) (-)-Hinokinin causes antigenotoxicity but not genotoxicity in peripheral blood of Wistar rats. Food Chem Toxicol 45(4):638–642

Nahak G, Sahu RK (2011) Phytochemical evaluation and antioxidant activity of *Piper cubeba* and *Piper nigrum*. J Appl Pharm Sci 1(8):153–157

Opdyke DLJ (1976) Monographs on fragrance raw materials. Food Cosmet Toxicol 14(Suppl):729

Pandey JK, Singh DK (2009) Molluscicidal activity of *Piper cubeba* Linn., *Piper longum* Linn. and *Tribulus terrestris* Linn. and their combinations against snail *Indoplanorbis exustus* Desh. Indian J Exp Biol 47(8):643–648

Parmar VS, Jain SC, Bisht KS, Jain R, Taneja P, Jha A, Tyagi OD, Prasad AK, Wengel J, Olsen CE, Boll PM (1997) Phytochemistry of the genus *Piper*. Phytochemistry 46:597–673

Parvez M, Gayasuddin M, Basheer M, Janakiraman K (2010) Screening of *Piper cubeba* (Linn) fruits for anti-ulcer activity. Int J PharmTech Res 2(2):1128–1132

Prabhu BR, Mulchandani NB (1985) Lignans from *Piper cubeba*. Phytochemistry 24(2):329–331

Rao BS, Shintre VP, Simonsen JL (1928) Constituents of some Indian essential oils. XXIII. Essential oil from the fruits of *Piper cubeba* Linn. J Soc Chem Ind 47:92T

Razdan RK, Bhattacharvya SC (1954) Sesquiterpenes from *Piper cubeba* Linn. Part I. Perfum Essent Oil Rec 45:181–183

Razdan RK, Bhattacharvya SC (1955) Sesquiterpenes from *Piper cubeba* Linn. Part II. Perfum Essent Oil Rec 46:8–13

Sastroamidjojo S (1997) Obat Asli. Dian Rakyat, Jakarta, 149pp

Shankaracharya NB, Rao LJ, Nagalakshmi S, Puranaik J (1995) Studies on the chemical composition of cubeb (*Piper cubeba* Linn.). PAFAZ J 17(1):33–38

Silva ML, Coímbra HS, Pereira AC, Almeida VA, Lima TC, Costa ES, Vinhólis AH, Royo VA, Silva R, Filho AA, Cunha WR, Furtado NA, Martins CH, Carvalho TC, Bastos JK (2007) Evaluation of *Piper cubeba* extract, (-)-cubebin and its semi-synthetic derivatives against oral pathogens. Phytother Res 21(5):420–422

Singh G, Kiran S, Marimuthu P, de Lampasona MP, de Heluani CS, Catalán CAN (2008) Chemistry, biocidal and antioxidant activities of essential oil and oleoresins from *Piper cubeba* (seed). Int J Essent Oil Ther 2(2):50–59

Singh G, Marimuthu P, de Heluani CS, Catalan CAN (2007) Chemical constituents, antioxidative and antimicrobial activities of essential oil and oleoresin of tailed pepper (*Piper cubeba* L). Int J Food Eng 3(6):Article 11

Sumathykutty MA, Rao JM, Padmakumart KP, Narayanan CS (1999) Essential oil constituents of some *Piper* species. Flavor Fragr J 14:279–282

Taneja SC, Koul SK, Pushpangadan P, Dhar KL, Daniewski WM, Schilf W (1991) Oxygenated cyclohexanes from *Piper* species. Phytochemistry 30(3):871–874

Terhune SJ, Hogg JW, Lawrence BM (1974) Bicyclosesquiphellandrene and 1-epibicyclosesquiphellandrene: two new dienes based on the cadalene skeleton. Phytochemistry 13:1183–1185

Usia T, Watabe T, Kadota S, Tezuka Y (2005a) Metabolite-cytochrome P450 complex formation by methylenedioxyphenyl lignans of *Piper cubeba*: mechanism-based inhibition. Life Sci 76(20):2381–2391

Usia T, Watabe T, Kadota S, Tezuka Y (2005b) Potent CYP3A4 inhibitory constituents of *Piper cubeba*. J Nat Prod 68(1):64–68

Utami D, Jansen PCM (1999) *Piper* L. In: de Guzman CC, Siemonsma JS (eds) Plant resources of South-East Asia no 13 spices. Backhuys Publishers, Leiden, pp 183–188

Yam J, Kreuter M, Drewe J (2008a) *Piper cubeba* targets multiple aspects of the androgen-signalling pathway. A potential phytotherapy against prostate cancer growth? Planta Med 74(1):33–38

Yam J, Schaab A, Kreuter M, Drewe J (2008b) *Piper cubeba* demonstrates anti-estrogenic and anti-inflammatory properties. Planta Med 74(2):142–146

Yuan Y, Wang C, Zhou X (1982) Isolation of bisasarin from *Piper cubeba*. Zhongcaoyao 13:378, 392

Piper nigrum

Scientific Name

Piper nigrum L.

Synonyms

Muldera wightiana Miq., *Muldera multinervis* Miq., *Piper aromaticum* Lam., *Piper trioicum* Roxb.

Family

Piperaceae

Common/English Names

Black Pepper, Green Pepper, Indian Pepper, Pepper, White Pepper.

Vernacular Names

Albanian: Piper;
Amharic: Kundo Berbere;
Arabic: Filfil, Fulful, Fulful Aswad;
Armenian: Bghbegh, Pghpegh;
Azeri: Bibər, İstiot, Qara Istiot;
Basque: Piper;
Belarusian: Čorny Perac, Perac;
Brazil: Pimenta-Da-Índia, Pimenta-Do-Reino;
Burmese: Nayukon, Nga Youk Kuan, Ngayokkaung;
Bulgarian: Cheren Piper, Piper, Pipereni Zurna;
Catalan: Pebre, Pebre Negre;
Chinese: Bai Hu Jiao, Hak Wuh Jiu, Hei Hu Jiao, Hu Chiao, Hu Jiao, Woo Jiu;
Croatian: Biber, Crni Papar, Papar;
Czech: Černý Pepř, Pepř, Pepřovník Černý;
Danish: Peber, Sort Peber;
Dutch: Peper, Zwarte En Witte Peper, Zwarte Peper;
Eastonian: Must Pipar, Pipar;
Esperanto: Nigra Pipro, Pipro;
Finnish: Mustapippuri, Pippuri, Valkopippuri, Viherpippuri;
French: Poivre Blanc, Poivre Commun, Poivre Noir;
Frisian: Piper;
Gaelic: Piobar, Piobar Dubh;
Galician: Pementa;
Georgian: P'ilp'ili, Pilpili, P'eritsa, Pertisa, Shavi P'ilp'ili, Shavi P'eritsa;
German: Grüner Pfeffer, Pfeffer, Schwarzer Pfeffer; Weißer Pfeffer;
Greek: Peperi, Pipéri, Piperi Mauro;
Hebrew: Pilpel, Pilpel Shahor;
Hungarian: Bors, Borscserje, Fekete Bors, Fűszerbors;
Icelandic: Pipar, Svartur Pipar;
India: Jaluk, Gol Morich (Assamese), Golmarich, Gulmorich, Kaalaamorich, Kalomarich (Bengali), Aseymirus (Dhivehi), Kali Mirch (Dogri), Kaalaamirich, Kaalaamirii, Kalamiri, Klamirich,

Kalomirich, Mari (Gujarati), Golmirch, Gulki, Kaalii Mirch, Malimirch (Hindu), Menasinaballii, Mensinballi, Menasina, Menasina-Kallu, Menasin Kallu, Menasu, Menusu. (Kannada), Kurumilagu, Kuru Mulagu, Kurumulaku, Nallamulaku, Yavanapriyam (Malayalam), Kaaliimirii, Mire (Marathi), Gola Maricha, Kala Marichamaricha (Oriya), Kali Marich, Kali Mirch (Punjabi), Maricha, Vella, Krishnan, Krishnadi (Sanskrit), Milagoo, Milaagu, Yavanappiriyam (Tamil), Miryaalatiga, Miryalatige, Miriyaalu, Miriyamu, Savyamu (Telugu), Kalimirch (Urdu);
Indonesia: Marica, Marica Hitam, Micha (Javanese), Lada Pedes (Sudanese), Lada, Lado Ketek, Lada Kobon (Sumatra);
Irish: Piobar;
Italian: Pepe, Pepe Nero;
Japanese: Burakku Peppaa, Koshou, Pepaa, Peppaa;
Kashmiri: Marts;
Kazakh: Burış;
Khmer: Môrech, Mrech;
Korean: Huchunamu, Pullaek Pepo, Pepeo-Bullaek;
Latvian: Melnie Pipari;
Lithuanian: Juodieji Pipirai;
Laotian: Mak Phik Noi, Phi Noi, Phik Noy, Phik Thai;
Macedonian: Crn Piper;
Malaysia: Lada, Lada Hitam, Lada Putih, Lada Sula;
Maltese: Bżar;
Nepali: Kalo Marichmarich;
Newari: Haku Male, Male;
Norwegian: Pepper, Svart Pepper;
Ossetian: Byrts, Tsyvzy;
Persian: Felfel Siah;
Philippines: Paminta (Tagalog);
Polish: Czarny Pieprz, Pieprz;
Portuguese: Pimenta, Pimenta Negra;
Provençal: Pebre, Peure;
Romanian: Piper, Piper Negru;
Russian: Perets Bélyi, Perets Chërnyi, Pjerets, Zelyony Pjerets;
Serbian: Papar, Biber;
Spanish: Pimentero Común, Pimienta, Pimienta Negra, Pimineta Nigra;
Sri Lanka: Gammiris, Miris (Sinhalese);
Slovaščina: Črni Poper, Poper;
Slovencina: Peprovnik;
Swahili: Pilipili;
Swedish: Peppar, Svart Peppar;
Tajik: Murch;
Thai: Phrik Thai;
Tibetan: Nalesham; Pho ba ril bu;
Turkish: Kara Biber, Siah Biber;
Turkmen: Burç;
Ukranian: Peets;
Vietnamese: Cây Tiêu, Hạt-Tiêu, Hồ Tiêu, Tiêu;
Yiddish: Fefer.

Origin/Distribution

The species is a native of the dense evergreen forests of the Western Ghats in South West India, now widely cultivated pantropically. The major producing countries are India, Indonesia, Sarawak, Malaysia and Brazil. It is also cultivated also in Sri Lanka, Myanmar, Thailand, Cambodia, Laos, Vietnam, New Guinea, on many Pacific islands, the Antilles, Madagascar, Zanzibar and in West Africa (Ghana to Angola).

Agroecology

Pepper thrives in warm areas with 1,750–2,500 mm annual rainfall and temperatures around 25–30°C and humidity above 85%. It grows in full sun or partial shade on well-composted, moist, well-drained, fertile soils rich in organic matter. It is drought and frost sensitive and abhors waterlogged soil. Black pepper is a tropical vine and needs artificial support (wooden, timber, cement) or trunk of live tree as support.

Edible Plant Parts and Uses

Pepper was the first spice to be traded globally and is the most important spice traded in terms of quantity and value.

Ripe and unripe berries and dried berries are used as spice in many food dishes. Dried fruits are called peppercorns. Dried ground pepper or whole peppercorn represent one of the most common spices in European and Asian cuisine. Depending on harvest time and processing, peppercorns can be black, white, green and red (actually, reddish brown) (Plates 3, 4, 5 and 6). The traditional types are black and white – the pepper of commerce and are used universally for flavouring food. Black and white pepper is used in various dishes for seasoning – white pepper in sauces or porridge, with mash potatoes and black pepper with steak or red meat. White pepper is popularly used in Western cooking, commonly recommended for white (cream-based) sauces and in conjunction with fish dishes such as *gefilte* stuffed fish. Dried green peppercorns are a more recent innovation, but are now rather common in Western countries. Red peppercorns, however, are still a very rare commodity. Red peppercorns are prepared from reddish brown ripe berries by drying or freeze drying of ripe berries. Black pepper is produced from the still-green mature but unripe berries. The berries are cooked briefly in hot water, both to clean them and to prepare them for drying. Drying is usually done in the sun for a few days or by a drying machine – oven or microwave. Once dried, black pepper or black peppercorn is produced. White pepper is processed from ripe berries by removal of the fruit pulp by soaking

Plate 1 Pepper vine growing on a tree trunk with green pepper berries

Plate 2 Pepper berries and leaves

Plate 3 Green peppercorns (berries)

Plate 4 White peppercorns

Plate 5 Black peppercorns

Plate 6 Close-up of white, red and black peppercorns

the berries in water for several days to a week and drying the seed. Green pepper, like black, is made from the unripe berries. Dried or dehydrated green peppercorns are treated in a manner that retains the green colour, such as treatment with sulphur dioxide or freeze drying. Fresh, unpreserved green pepper berries are also used in many Asian cuisine particularly Thai cuisines and are excellent in Thai stir-fries, Thai curries and Thai curry paste. Their flavour is described as piquant, fresh and with a bright aroma. Green pepper is also used in Western cooking, often incorporated into mustard or bottled condiments. Pickled peppercorns, also green, are unripe berries preserved in brine or vinegar and are often used as a spicy accompaniment to cold foods.

Other value added products of pepper include lemon pepper, garlic pepper, sauces and marinades that have pepper as the vital and main ingredient, pepper paste, pepper mayonnaise, pepper tofu, pepper cookies, pepper sweets, pepper toffees and pepper candies, pepper-flavoured crackers, pepper-flavoured fish and prawn crackers. Pepper is also used to flavour certain beverages and liquors such as pepper tea, pepper coffee, pepper flavoured milk, pepper flavoured puddings and pepper flavoured ice cream. Pepper oil and or piperine are added to certain liquors to impart a pungent and exotic taste and aroma. Double encapsulated pepper products are also available. The contained flavour or aroma is released only at high temperatures such as during steam cooking or baking. Pepper oil and or oleoresin are blended in vegetable oil or hydrogenated fat called fat-based pepper for use in products such as mayonnaise. Pepper oleoresin is also used in the meat processing industry.

Botany

Perennial woody climber up to 4 m high. The pepper plant is normally grown with support, either on a living tree or a post (which can be made of cement or wood) or trellis. Robust glabrous, woody stem with distinct enlarged nodes and rooting at the nodes. It has a shallow root system. Two types of branches are produced, vegetative (climbing) and reproductive (fruiting). Leaves alternate, simple, entire, ovate to ovate-oblong, 10–15×5–9 cm, thick, more or less leathery, glabrous, base rounded, usually slightly oblique, apex acute; veins 5–7(–9), reticulate veins prominent (Plates 1 and 2). Petiole 1–2 cm Flowers polygamous, usually monoecious, small produce on pendulous Spikes which are leaf-opposed, peduncle bracts spatulate-oblong,

3–3.5×ca. 0.8 mm, adaxially adnate to rachis, only margin and broad, rounded apex free, shallowly cupular. Stamens 2, 1 on each side of ovary; filaments thick, short; anthers reniform. Ovary globose; stigmas 3 or 4, rarely 5. Drupe green borne in cluster (spike) 20–30 cm long (Plates 1, 2 and 3), becoming red when ripe, drying black when unripe (Plate 6), globose, 3–4 mm in diam., sessile containing a single seed.

Nutritive/Medicinal Properties

Analyses carried out in the United States (2011) reported black pepper (spice) to have the following proximate composition minus piperine (per 100 g value): water 12.46 g, energy 251 kcal (1,050 kJ); protein 10.39 g, total lipid 3.26 g, ash 4.49 g, carbohydrates 63.95 g, total dietary fibre 25.3 g, total sugars 0.64 g, sucrose 0.02 g, glucose 0.24 g, fructose 0.23 g, galactose 0.15 g, Ca 443 mg, Fe 9.17 mg, Mg 171 mg, P 158 mg, K 1,329 mg, Na 20 mg, Zn 1.19 mg, Cu 1.330 mg, Mn 12.753 mg, F 34.2 μg, Se 4.9 μg, vitamin C 0 mg, thiamine 0.108 mg, riboflavin 0.180 mg, niacin 1.143 mg, pantothenic acid 1.399 mg, vitamin B-6 0.291 mg, total folate 17 μg, choline 11.3 mg, betaine 8.9 mg, vitamin A 547 IU (27 μg RAE), α-carotene 12 μg, β-carotene 310 μg, β-cryptoxanthin 25 μg, lycopene 20 μg, lutein+zeaxanthin 454 μg, vitamin E (α-tocopherol) 1.04 mg, γ-tocopherol 6.56 mg, vitamin K (phylloquinone) 163.7 μg; total saturated fatty acids 1.392 g, 6:0 0.012 g, 8:0 0.102 g, 10:0 0.036 g, 12:0 0.093 g,14:0 0.030 g, 16:0 0.533 g, 18:0 0.327 g; total monounsaturated fatty acids 0.739 g, 14:1 0.016 g, 16:1 undifferentiated 0.077 g, 18:1 undifferentiated 0.647 g, 18:1 c 0.647 g; total polyunsaturated fatty acids 0.998 g, 18:2 undifferentiated 0.694 g, 18:3 undifferentiated 0.152 g, 18:3 n-3 c,c,c (ALA) 0.152 g, 20:3 undifferentiated 0.152 g; phytosterols 92 mg; tryptophan 0.058 g, threonine 0.244 g, isoleucine 0.366 g, leucine 1.014 g, lysine 0.244 g, methionine 0.096 g, cystine 0.138 g, phenylalanine 0.446 g, tyrosine 0.483 g, valine 0.547 g, arginine 0.308 g, histidine 0.159 g, alanine 0.616 g, aspartic acid 1.413 g, glutamic acid 1.413 g, glycine 0.441 g, proline 1.413 g, and serine 0.409 g.

Analyses carried out in the United States reported white pepper to have the following proximate composition minus piperine (per 100 g value): water 11.42 g, energy 296 kcal (1,238 kJ); protein 10.40 g, total lipid 2.12 g, ash 1.59 g, carbohydrates 68.61 g, total dietary fiber 26.2 g, Ca 265 mg, Fe 14.1, Mg 90 mg, P 176 mg, K 73 mg, Na 5 mg, Zn 1.13 mg, Cu 0.910 mg, Mn 4.3 mg, Se 3.1 μg, vitamin C 21 mg, thiamine 0.022 mg, riboflavin 0.126 mg, niacin 0.212 mg, vitamin B-6 0.1 mg, total folate 10 μg, total saturated fatty acids 0.626 g, total monounsaturated fatty acids 0.789 g, total polyunsaturated fatty acids 0.616 g, phytosterols 55 mg (USDA 2011).

There is a difference in the nutrient composition and levels between black and white pepper which reflect on the maturity and ripeness of the harvested berries and the way the pepper berries are processed to form black and white pepper. Black pepper is produced from the still-green unripe berries. The berries are cooked briefly in hot water, both to clean them and to prepare them for drying. Drying is usually done in the sun for a few days or by a drying machine – oven or microwave. Once dried, black pepper or black peppercorn is produced. White pepper is processed from ripe berries by removal of the fruit pulp by soaking the berries in water for several days to a week and drying the seed.

Black pepper contain higher levels of the minerals, vitamins than white pepper and also contain β-carotene, β-cryptoxanthin, lycopene, lutein and zeaxanthin.

Other Phytochemicals

Pepper Fruit

Some major sesquiterpene hydrocarbon constituents of black pepper were identified as δ-elemene, α-copaene, β-elemene, α-cis-bergamotene, α-santalene, α-trans-bergamotene, β-caryophyllene, α-humulene, β-selinene, α-selinene, β-bisabolene, δ-cadinene and calamine (Muller and Jennings 1967). Some minor volatile sesuqiterpene hydrocarbons identified from black pepper were: α-cubebene, isocaryophyllene and γ-muurolene and a major constituent was identified as β-farnesene (Muller et al. 1968). The following optically active monoterpenes: (±)-linalool, (+)-α-phellandrene, (−)-limonene, myrcene, (−)-α-pinene, 3-methylbutanal and methylpropanal were identified as the most potent odorants of black pepper (Jagella and Grosch 1999a). Additionally, 2-isopropyl-3-methoxypyrazine and 2,3-diethyl-5-methylpyrazine were detected as important odorants of a black pepper sample with a mouldy, musty off-flavour. Nineteen odorants were quantified in white pepper (Jagella and Grosch 1999b). These included limonene, linalool, α-pinene, 1,8-cineole, piperonal, butyric acid, 3-methylbutyric acid, methylpropanal, and 2- and 3-methylbutanal as key odorants of white pepper. The faecal off-flavour was caused by skatole and was enhanced by the presence of p-cresol. Oxygenated compounds isolated from black pepper included: linalool, 1-terpinen-4-ol, α-terpineol, cryptone, carvone, p-cymene-8-ol, trans-carveol, cis-carveol, safrole, ar-curcumene, methyl eugenol, nerolidol and myristicin (Russell and Jennings 1969). Piperine, a pungent component of black pepper oleoresin was isolated and its structure elucidated as trans-5-(3,4-methylenedioxyphenyl)-2-pentenoic acid piperidide (Traxler 1971).

Piper nigrum fruit had been reported to contain alkamides: N-trans-feruloyltyramine, coumaperine (Nakatani et al. 1980), guineensine (Nakatani and Inatani 1981), N-isobutyl amide of octadeca-trans-2-cis-4-dienoic acid (Siddiqui et al. 1997), dipiperamides A, B, C, D, E, retrofractamide A, brachyamide A, neopellitorine B, richolein, ilepcimide (Tsukamoto et al. 2002a, b), a pyrrolidine, isopiperolein B (Srinivas and Rao 1999), piperamide, guineensine (Singh et al. 2004), 2E, 4E, 8Z-N-isobutyleicosatrienamide, pellitorine, trachyone, pergumidiene, isopiperoleine B (Reddy et al. 2004), pipsaeedine, pipbinine, piptaline (Siddiqui et al. 2004b), kalecide, sarmentine, chingchengenamide, pipnoohine, pipyahyine (Siddiqui et al. 2004c), piperine, piperoleines A, B, piperyline, piperanine, pipercide, feruperine, pipernonaline, dehydropipernonaline, nigramide R pipercyclobutanamide A, (Subehan et al. 2006), retrofractamide A, pipercide, piperchabamide D, pellitorin, dehydroretrofractamide C and dehydropipernonaline (Rho et al. 2007), guineensine, retrofracamide C, (2E,4Z,8E)-N-[9-(3,4-methylenedioxyphenyl)-2,4,8-nonatrienoyl]piperidine, pipernonaline, piperrolein B, piperchabamide D, pellitorin and dehydropipernonaline (Lee et al. 2008); flavonoids: glycosides of kaempferol, rhamnetin and quercetin in considerable concentration, and a lesser amount of isorhamnetin (Voesgen and Herrmann 1980); alkaloids: pipercyclobutanamides A and B (Fujiwara et al. 2001); and lignans: hinokinin (Tsukamoto et al. 2002a, b).

The following major aroma compounds were found in black pepper: germacrene D (11.01%), limonene (10.26%), β-pinene (10.02%), α-phellandrene (8.56%), β-caryophyllene (7.29%), α-pinene (6.40%) and cis-β-ocimene (3.19%) (Jirovetz et al. 2002). The main compounds such as β-caryophyllene, germacrene D, limonene, β-pinene, α-phellandrene and α-humulene, as well as minor constituents such as δ-carene, β-phellandrene, isoborneol, α-guaiene, sarisan, elemicin, calamenene, caryophyllene alcohol, isoelemicin, T-muurolol, cubenol and bulnesol, were of greatest importance for the characteristic pepper odor notes. Further aroma impressions were attributed to mono- and sesquiterpenes, hexane, octane and nonane derivatives. Musenga et al. (2007) extracted 11 aromatic and terpenic compounds namely terpinen-4-ol, caryophyllene oxide, limonene, α-pinene, 3-carene, β-pinene, αa-humulene, β-caryophyllene, α-phellandrene, eugenol and piperine from pepper extracts using capillary electrochromatography.

Black pepper was found to contain about 5–9% of the alkaloids piperine and piperettine and about 1–2.5% of volatile oil, the major constituents of which are α and β-pinene, limonene, and phellandrene (Tewtrakul et al. 2000). The main pungent principle in the green berries of pepper was piperine. Generally the piperine content of black or white peppercorns varied within the range of 3–8 g/100 g, whereas the content of the minor alkaloids piperyline and piperettine was estimated as 0.2–0.3 and 0.2–1.6 g/100 g, respectively (Tainter and Grenis 1993). Other alkaloids present included chavicine, isopiperine and isochavicine and minor alkaloids piperalin A, pieralein B, piperanine (Wood et al. 1988). The acetone extract of pepper showed the presence of 18 components accounting for 75.59% of the total amount; piperine

(33.53%), piperolein B (13.73%), piperamide (3.43%) and guineensine (3.23%) were the major components (Singh et al. 2004). Piperine is known to be the pungent principle in black and white ground pepper and green whole pepper berries (Schulz et al. 2005). It was found that piperine occurs more or less in the whole perisperm of the green fruit. Recovery of piperine from dried fruit powder was in the high range of 98.57–97.25% from *P. nigrum* and 96.50–97.50% from *P. longum* (Hamrapurkar et al. 2011).

Piptigrine was isolated from the dried ground seeds of *Piper nigrum* along with the known amides piperine and wisanine (Siddiqui et al. 2004a). The petroleum ether and ethyl acetate fractions of dried ground seeds of *Piper nigrum* afforded 16 compounds (1–16) including one new insecticidal amide, pipwaqarine (1) and six constituents (3,4,6,7,11,15) previously unreported from this plant. (Siddiqui et al. 2005). Studies on the petroleum ether soluble and insoluble fraction of ethanol extract of dried ground seeds of *Piper nigrum* resulted in the isolation and structure elucidation of one new and 11 known compounds which included three hitherto unreported constituents, namely, cinnamylideneacetone, 3,4-methylenedioxyphenylpropiophenone and 2-hydroxy-4,5-methylenedioxypropiophenone (Siddiqui et al. 2008).

Using vibrational microscopy Schulz et al. (2005) found the following contents (in mg/100 g) for piperine 2.75–5.07, essential oil 0.80–3.60 and terpenoids in black and white peppers: α-pinene nd (not determined) -0.18, β-pinene 0.003–0.29, sabinene nd-0.75, myrcene nd-0.08, α-phellandrene nd-0.16, δ-3-carene 0.005–0.80, limonene 0.01–0.52 , β-caryophyllene 0.26–1.22, caryophyllene oxide 0.005–0.14. The following terpenoids were detected in hydrodistilled pepper oils: α-pinene 0.26–5.39, β-pinene 0.69–11.96, sabinene nd-20.9, myrcene 0.21–3.26, α-phellandrene nd-6.61, δ-3-carene 0.72–32.65, limonene 1.64–22.48, β-caryophyllene 12.84–65.68, caryophyllene oxide 0.22–8.35.

Pepper Essential Oil

Pepper oil contains a complex mixture of monoterpenes (70–80%), sesquiterpenes (20–30%), and small amounts of oxygenated compounds, with no pungent principles present. Concentration and composition vary, depending on sources, cultivars and methods of extraction.

Some physical and chemical properties of black pepper berries were reported as follows: oleoresin 10.6 wt%, piperine 5.8 wt%, essential oil 1.7(v/w)%, water content 11 wt%, real density 1,545 kg/m^3, bulk density 793 kg/m^3 and bed porosity 0.49 (Kumoro et al. 2010). Pepper essential oil was found to be a mixture consisting of 90% hydrocarbons and 10% oxygenated terpenes and aromatic compounds (Ferreira et al. 1999). The hydrocarbon fraction was composed of monoterpenes (70–80%) and sesquiterpenes (20–30%), which appeared to possess the desirable attributes of pepper flavor. Though oxygenated terpenes were relatively minor constituents, they contributed to the characteristic odor of pepper oil (Pino et al. 1990). They detected 46 components including (E)-β-ocimene, δ-guaiene, (Z) (E)-farnesol, δ-cadinol and guaiol, which were reported for the first time. Constituents of black pepper essential oil (% area) detected by hydrodistillation and supercritical fluid extraction using carbon dioxide (values in brackets) (Kumoro et al. 2010) included: monterpenes:- α-tujene 0.45 (0.43), α-pinene 2.15 (1.80), sabinene 4.80 (5.50), β-pinene 4.20 (3.50), δ-carene 4.30 (5.00), limonene 6.30 (6.80); oxygenated monoterpenes:- linalool 0.59 (1.30), myrtenol 0.27 – (nd), caryone 0.29 – (nd); sesqueterpenes:- δ-elemene 0.30 (0.23), α-cubebene 0.38 (0.52), α-copaene 5.40 (6.85), calarene 2.20 (2.70), δ-gurjunene 0.30 (2.65), β-caryophyllene 2.36 (12.40), α-humulene 0.48 (3.30), α-selinene 2.65 (3.70), δ-cadinene 2.15 (3.75); oxygenated sesqueterpenes:- elemol 7.96 (6.40), caryophyllene oxide 12.50 (7.94) and β-eudesmol 3.90 (1.80).

Wrolstad and Jennings (1965) identified the following monoterpenes in the volatiles of black pepper oil: α-pinene, β-pinene, sabinene, β-carene, limonene, ρ-cymene, α-thujene, α-phellandrene, myrcene, β-phellandrene, γ-terpinene, and terpinolene. Schulz et al. (2005) reported the volatile fraction of pepper oil to be dominated by monoterpene hydrocarbons (30–70%) such as α- and β-pinene, limonene, sabinene, myrcene, α-phellandrene, and δ-3-carene. Sesquiterpenes such as α- and β-caryophyllene and β-farnesene made up 25–45% of the total essential oil content, whereas oxygen-containing sesquiterpenes such as caryophyllene oxide occurred in amounts of 4–14%. The pinenes possessed a turpentine eucalyptus- like smell, whereas the odour of the other monoterpenes was described as fresh and pine-needle-like. Generally the so-called "top-peppery note" was assigned to the monoterpene group; the "pepper odor" to the sesquiterpenes, and the oxygen-containing sesquiterpenes were responsible for the "body" of the pepper aroma. Kapoor et al. (2009) found 54 volatile components that amounted to 96.6% of the total weight of pepper essential oil. β-caryophyllene (29.9%) was found as the major component along with limonene (13.2%), β-pinene (7.9%), sabinene (5.9%), and several other minor components. The major component of both ethanol and ethyl acetate oleoresins was found to contain piperine (63.9% and 39.0%), with many other components in lesser amounts.

Monoterpene hydrocarbons of black pepper oil were isolated and identified as α-pinene, β-pinene, sabinene, β-carene, limonene, and p-cymene. α-thujene, α-phellandrene, myrcene, β-phellandrene, γ-terpinene, and terpinolene were also found (Wrolstad and Jennings 1965). The monoterpenes isolated also contained small quantities of α-terpinene and greater quantities of γ-terpinene, p-cymene, and terpinolene. Cis-p-2-methen-1-ol, cis-p-2,8-menthadien-1-ol and transpinocarveol was detected in black pepper oil (Richard and Jennings 1971). m-mentha-3(8),6-diene (isosylveterpinolene) was found to the extent of 0.12–0.17% in black and green pepper oil (Nussbaumer et al. 1999).

Ninety four percent of the compounds were identified in fresh pepper berry oil (Sasidharan and Menon 2010). Limonene was the predominant compound (18%), followed by β-pinene (14.2%), and β-caryophyllene (13.2%). Monoterpene hydrocarbons accounted for 54.4% and some of the major monoterpene hydrocarbons were α-pinene (12.1%), α-thujene (1.5%), δ-3-carene(3.2%), and sabinene (3.3%). monoterpene oxygenated compounds amounted to 5.1% ,and some of the compounds identified were linalool, terpinen-4-ol, and α-terpineol. The total sesquiterpene hydrocarbon content was 20.3% of which the main compound was β-caryophyllene (13.2%), followed by α-humulene (1.6%), cuparene (2.7%), δ-cadinene (0.7%). The total sesquiterpene oxygenated compounds content was 14.8% of which the main (E)-nerolidol (4.1%) was the main compound followed by Farnesol (Z,E) (2.1%).

Ninety six percent of the compounds were found in dry pepper berry oil (Sasidharan and Menon 2010). The main compound was β-caryophyllene (20.4%), followed by limonene (16%). The monoterpene hydrocarbon content came down to 46.7%. α- and β- pinenes, δ-3-carene, p-cymene, camphene were some of the monoterpene hydrocarbons. Monoterpene oxygenated compounds amounted to 2.4%. The sesquiterpene hydrocarbon compounds content constituted 45.4% of which β-caryophyllene was the main compound, and followed by α-humulene (2.8%), α-copaene (1.9%), α-guaine (2%), δ-guaine (1.6%), β-bisabolene (2.1%), cuparene (3.5%). The sesquiterpene oxygenated compounds amounted to 1.9%. Liu et al. (2007) separated and identified 30 compounds from the essential oil of *P. nigrum*. The main components were β-caryophyllene (23.49%), δ-3-carene (22.20%), D-limonene (18.68%), β-pinene (8.92%) and α-pinene (4.03%).

Fifty-five compounds were identified in the oils of four major cultivars of black pepper from Kerala viz. Thevanmundi, Poonjaranmunda, Valiakaniakadan, and Subhakara (Jayalekshmy et al. 2003). The main components of Thevanmundi oil were sabinene (4.5–16.2%), β-pinene (3.7–8.7%), limonene (8.3–18.0%) and β-caryophyllene (20.3–34.7%). Poonjaranmunda oil contained as major compounds β-pinene (6.0–11.7%) limonene (14.9–15.8%), β-ocimene. Forty-nine components accounting for 99.39% of the total amount was identified in pepper oil, and the major components were β-caryophyllene (24.24%), limonene (16.88%), sabinene (13.01%), β-bisabolene (7.69%) and α-copaene (6.3%) (Singh et al. 2004). Fifty-five compounds were identified in the oils of two major cultivars of black pepper from Kerala viz. Vellanamban and Sreekara and one Indonesian cultivar grown in Kutching in Kerala (Menon and Padmakumari 2005). The main components of vellanamban oil were sabinene (3.9–18.8%), β-pinene (3.9–10.9%), limonene (8.3–19.8%) and β-caryophyllene (28.4–32.9%). Sreekara oil contained as major compounds β-pinene (0–11.2%), limonene (20.1–22.1%) and β-caryophyllene (16.8–23.1%). Kutching oil contained α-pinene (2.3–5.4%), sabinene (6.7–13.3%), limonene (14.5–17.5%) and β-caryophyllene (20.8–39.1%). Clery et al. (2006) analysed the N-containing compounds in black pepper oil by capillary GC, GC/MS and HPLC techniques. The most abundant nitrogenous compounds were 1-formylpiperidine (28 ppm), 1-acetylpiperidine (22 ppm) and 2,3,5,6-tetramethylpyrazine (7 ppm). In addition to previously reported compounds, 20 previously unreported pyrazines, pyridines and piperidines were identified. Aroma assessment revealed that 2-ethyl-3,5-dimethylpyrazine (1.5 ppm; roasted green vegetable, coffee and cocoa notes) and 3-ethyl-2,5-dimethylpyrazine (490 ppb; earthy and roasty notes) contributed most significantly to the overall aroma.

Orav et al. (2004) found that the most abundant compounds in pepper oils were (E)-β-caryophyllene (1.4–70.4%), limonene (2.9–38.4%), β-pinene (0.7–25.6%), δ-3-carene (1.7–19.0%), sabinene (0–12.2%), α-pinene (0.3–10.4%), eugenol (0.1–41.0%), terpinen-4-ol (0–13.2%), hedycaryol (0–9.1%), β-eudesmol (0–9.7%), and caryophyllene oxide (0.1–7.2%). Green pepper corn obtained by a sublimation drying method gave more oil (12.1 mg/g) and a much higher content of monoterpenes (84.2%) in the oil than air-dried green pepper corn (0.8 mg/g and 26.8%, respectively). The oil from ground black pepper contained more monoterpenes and less sesquiterpenes and oxygenated terpenoids as compared to green and white pepper oils. After 1 year of storage of pepper samples in a glass vessel at room temperature, the amount of the oils isolated and the content of terpenes decreased, and the amount of oxygenated terpenoids increased. Different from other pepper samples, 1 year storage of green peppercorn raised the oil content more than twice of both drying methods. Fifty-four components representing about 96.6% of the total weight were found in black pepper essential oil (Kapoor et al. 2009). β-caryophylline (29.9%) was found as the major component along with limonene (13.2%), β-pinene (7.9%), sabinene (5.9%), and several other minor components. The major component of both ethanol and ethyl acetate oleoresins was found to contain piperine (63.9% and 39.0%), with many other components in lesser amounts. The antioxidant activities of essential oil and oleoresins of black pepper were evaluated against mustard oil by peroxide, p-anisidine, and thiobarbituric acid (Kapoor et al. 2009).

Fifty-five compounds were identified in the essential oil samples of four major black pepper cultivars the chemical composition of which varied widely due to variations in agroclimatic conditions and growing seasons (Menon et al. 2000). For example sabinene varied from 0.2% in cv. Subhakara to 17.1% in cv. Valiakaniakadan. Delta-3-carene varied from 0% in cv. Valiakaniakadan in 1990 to 23.4% in cv. Subhakara in 1991. Limonene varied from 8.3% in cv. Thevanmundi oil to 22.7% in cv Subhakra oil. Alpha-cubebene varied from 0.1% in cv. Poonjaranmunda oil (in the season 1990) and in cv. Subhakara oil (in the seasons 1990 and 1992) to 6.8% in cv. Thevanmundi oil (in the season 1992). Beta-caryophyllene, the main sesquiterpene hydrocarbon of the pepper oil, varied from 7.6% in cv. Subhakara (season 1990) sample to 38.4% in cv. Valiakaniakadan oil (season 1992). Elemol content varied from 0.7% in cv. Subhakara oil (season 1990) to 9.6% in cv. Thevanmundi oil (season 1992). Another oxygenated compound, caryophyllene oxide varied from 0.4% in cv. Subhakara oil (season 1991) and in cv. Poonjaranmunda oil (season 1991) to 6% in cv. Subhakara oil (season 1990). Studies by Liang et al. (2010) indicated that the chemical compositions of pepper essential oils were greatly influenced by different production areas. The main constituents in three kinds of pepper essential oils from different regions were: α-pinene; cyclohexene, 1-methyl-4-(1-methylethylidene)-; 3-cyclohexen-1-ol, 4-methyl-1-(1-methylethyl)-; β-pinene; limonen-6-ol, pivalate; (E)-3(10)-caren-4-ol; and *trans*-caryophyllene. Bicyclo [3.1.0]hexane, 4-methylene-1-(1-methylethyl)-; isocaryophillene; and bufa-20,22-dienolide, 14-hydroxy-3-oxo-, (5á)- were detected only in essential oil of black pepper from Guang

Dong province. 2-cyclohexen-1-ol, 1-methyl-4-(1-methylethyl)-, trans-caryophyllene; copaene; eudesma-4(14),11-diene; and trans-9-octadecenoic acid, trimethylsilyl ester were detected only in essential oil of black pepper from YunNan province. 1,2-dihydropyridine, 1-(1-oxobutyl)-; and 1-chloroeicosane were detected only in essential oil of black pepper from FuJian province.

Pepper Leaves

The pepper leaf oil was found to contain 63.3% sesquiterpene hydrocarbons and 32.4% sesquiterpene oxygenated compounds oil (Sasidharan and Menon 2010). The main compound of the oil was α-bisabolol (25.3%), followed by α-cubebene (20.2%). Some of the other compounds were β-bisabolene (15%), β-elemene (5.4%) and β-caryophyllene (4%). Sesquiterpene alcohols like elemol and nerolidols were also present in the oil. Pepper leaf was found to contain lignans: (-)-cubebin, (-)-3,4-dimethoxy-3,4-desmethylenedioxycubebin, (-)-3-desmethoxycubebinin (Matsuda et al. 2004).

About 50 compounds were identified in the stem and leaf oils of *Piper nigrum* (Pino et al. 2003), accounting for more than 99.8% and 99.5% of the oils, respectively. The leaf oil was dominated by globulol (49.7%) and α-pinene (21.7%), while the major components of the stern oil were β-caryophyllene (19.5%), α-pinene (19.3%), α-terpinene (12.71%) and globulol (12.6%). Young shoots and leaves contained only traces of piperine but considerable amounts of piperine (ca 0.2%) were found, however, in mature, fully differentiated shoots, reaching the contents of commercially available white or black pepper of lower quality (Semler and Gross 1988). Roots had about 0.03% piperine.

Pepper Roots

Seven new amide alkaloids, named N-isobutyl-4-hexanoyl-4-hydroxypyrrolidin-1-one (1), (±)-erythro-1-(1-oxo-4,5-dihydroxy-2E-decaenyl)piperidine (2), (±)-threo-1-(1-oxo-4,5-dihydroxy-2E-decaenyl)piperidine (3), (±)-threo-N-isobutyl-4,5-dihydroxy-2E-octaenamide (4), 1-(1,6-dioxo-2E,4E-decadienyl)piperidine (5), 1-[1-oxo-3(3,4-methylenedioxy-5-methoxyphenyl)-2Z-propenyl]piperidine (6), and 1-[1-oxo-5(3,4-methylenedioxyphenyl)-2Z,4E-pentadienyl] pyrrolidine (7), were isolated from the roots of *Piper nigrum*, together with 32 known amides that included piperine, isopiperine, fagaramide, piperettine, pellitorin, piperanine, kalecide, piperyline, piperoleine B, isochavicine, neopellitorine B, cinnamonpyrrolidide, piperettyline, tricholein, sarmentineilepcimide, pipernonaline, sarmentosine, brachyamide B, pipercycliamide (Wei et al. 2004b). Two phytosterols elucidated as stigmastane-3,6-dione and stigmast-4-ene-3,6-dione, were isolated from the roots of *Piper nigrum* (Wei et al. 2004a). Fifteen novel dimeric amide alkaloids possessing a cyclohexene ring, nigramides A-O (1–15), as well as four novel dimeric amide alkaloids possessing a cyclobutane ring, nigramides P-S (17–20), were isolated from the roots of *Piper nigrum* (Wei et al. 2005).

Black pepper (*Piper nigrum*) is used not only in human dietaries but also for a variety of other purposes such as medicinal, as a preservative, and in perfumery (Srinivasan 2007). All these uses can be attributed to its important major bioactive alkaloid, piperine. Dietary piperine, by favorably stimulating the digestive enzymes of pancreas, enhances the digestive capacity and significantly reduces the gastrointestinal food transit time. Pepper and piperine have also a diverse array of pharmacological properties such as antioxidant, anticancer, hepatoprotective, antiinflammatory, antidepressant, antiatherogenic, antihypertensive, drug potentiating and other activities elaborated below.

Antioxidant Activity

Volatile pepper oil and acetone pepper extract were identified as a better antioxidant for linseed oil, in comparison with butylated hydroxyanisole (BHA) and butylated hydroxytoluene (BHT) (Singh et al. 2004). The hydrolyzed and nonhydrolyzed extracts of black pepper contained significantly more polyphenols compared with those of white pepper (Agbor et al. 2006). For either of these peppercorns, the hydrolyzed extract contained significantly more polyphenols compared with the nonhydrolyzed extract. A dose-dependent effect was observed in the free radical and reactive oxygen species scavenging activities of all the extracts, with the black pepper extracts being the most effective. Phenolic compounds from green pepper were found to have high antioxidant potential (Chatterjee et al. 2007). EC_{50} values of the major phenolic compounds of green pepper namely, 3,4-dihydroxyphenyl ethanol glucoside, 3,4-dihydroxy-6-(N-ethylamino) benzamide and phenolic acid glycosides were found to be 0.076, 0.27 and 0.12 mg/ml, respectively, suggesting a high radical scavenging (DPPH) activity of these phenolics. These phenolics also protected plasmid DNA damage upon exposure to gamma radiation as high as 5 kGy and inhibited lipid oxidation as revealed by β-carotene–linoleic acid assay.

P. nigrum could be considered as a potential source of natural antioxidant (Singh et al. 2008). The free radical scavenging activity of the different fractions of petroleum ether extract of *P. nigrum* (PEPN) increased in a concentration dependent manner. In DPPH free radical scavenging assay, the activity of fraction R3 and R2 were found to be almost similar. The R3 (100 μg/ml) fraction 55.68% nitric oxide radicals generated from sodium nitroprusside. PEPN scavenged the superoxide radical generated by the xanthine/xanthine oxidase system. The fraction R2 and R3 in the doses of 1,000 μg/ml inhibited 61.04% and 63.56%, respectively. The amounts of total phenolic compounds were determined and 56.98 μg pyrocatechol phenol equivalents were detected in 1 mg of R3. *P. nigrum* was richest in glutathione peroxidase and glucose-6-phosphate dehydrogenase, green pepper was richest in peroxidase and vitamin C while vitamin E was more in *P. longum* and *P. nigrum*. Catalase activity was high in *Piper longum*, followed by *Piper cubeba*, green pepper, *Piper brachystachyum* and *Piper nigrum*. All the *Piper* species had GSH content of around 1–2 nM/g tissue. The antioxidant components of *Piper* species constitute a very efficient system in scavenging a wide variety of reactive oxygen species. Antioxidant potential of *Piper* species was further confirmed by their ability to suppress in-vitro lipid peroxidation by around 30–50% with concomitant increase in GSH content.

Studies indicated that carcinogens (7,12, dimethyl benzanthracene, dimethyl aminomethyl azobenzene and 3-methyl cholenthrene) treatment induced glutathione (GSH) depletion with substantial increase in thiobarbituric reactive substances and enzyme activities in male rats but piperine treatment with carcinogens resulted in inhibition of thiobarbituric reactive substances (Khajuria et al. 1998). Piperine mediated a significant increase in the GSH levels and restoration in gamma-GT and Na+-K+-ATPase activity. The data indicated a protective role of piperine against the oxidative alterations by carcinogens. The data suggested that piperine modulated the oxidative changes by inhibiting lipid peroxidation and mediating enhanced synthesis or transport of GSH thereby replenishing thiol redox. They further showed that Lipid peroxidation content, measured as thiobarbituric reactive substances (TBARS), was increased with piperine treatment although conjugate diene levels were not altered in the rat intestinal mucosa and epithelial cells (Khajuria et al. 1999). A significant increase in glutathione levels was observed, whereas protein thiols and glutathione reductase activity were not altered. The study suggested that increased TBARS levels may not be a relevant index of cytotoxicity, since thiol redox was not altered, but increased synthesis transport of intracellular GSH pool may play an important role in cell hemostasis.

The following phenolic amides: piperine (1), N-trans-feryloyl tyramine (2a) and its acetyl and methyl derivatives, diacetyl N-trans-feryloyl tyramine (2b), trimethyl N-trans-feryloyl tyramine (2c), coumaperine [N-5-(4-hydroxyphenyl)-2E,4E-pentadienoyl piperidine] (3a) and its derivative, acetyl coumaperine (3b), N-trans-feruloyl piperidine (4), N-5-(4-hydroxy-3-methoxyphenyl)-2E,4E-pentadienoyl piperidine (5), and N-5-(4-hydroxy-3-methoxyphenyl)-2E-pentenoyl piperidine (6) were identified from *P. nigrum* (Nakatani et al. 1986). All the phenolic amides possessed significant antioxidant activities that were more effective than the naturally occurring antioxidant, alpha-tocopherol at the same concentration (0.01%). One amide, feruperine, had antioxidant activity as high as the synthetic antioxidants, butylated hydroxyanisole (BHA) and butylated hydroxytoluene (BHT). Naturally occurring antioxidants, therefore, may surpass BHA and BHT in their ability to inactivate mutagens in food.

Studies by Vijayakumar et al. (2004) indicated supplementation with black pepper or the active principle of black pepper, piperine, could reduce high-fat diet induced oxidative stress to the cells. Significantly elevated levels of thiobarbituric acid reactive substances (TBARS), conjugated dienes (CD) and significantly lowered activities of superoxide dismutase (SOD), catalase (CAT), glutathione peroxidase (GPx), glutathione-S-transferase (GST) and reduced glutathione (GSH) in the liver, heart, kidney, intestine and aorta were observed in rats fed the high fat diet as compared to the control rats. Simultaneous supplementation with black pepper or piperine lowered TBARS and CD levels and maintained SOD, CAT, GPx, GST, and GSH levels to near those of control rats.

Both water extract (WEBP) and ethanol extract (EEBP) of black pepper exhibited strong total antioxidant activity in six different assay, namely, total antioxidant activity, reducing power, 1,1-diphenyl-2-picryl-hydrazyl (DPPH) free radical scavenging, superoxide anion radical scavenging, hydrogen peroxide scavenging, and metal chelating activities (Gülçin 2005). The 75 µg/ml concentration of WEBP and EEBP showed 95.5% and 93.3% inhibition on peroxidation of linoleic acid emulsion, respectively. On the other hand, at the same concentration, standard antioxidants such as butylated hydroxyanisole (BHA), butylated hydroxytoluene (BHT) and alpha-tocopherol exhibited 92.1%, 95.0%, and 70.4% inhibition on peroxidation of linoleic acid emulsion, respectively. The total phenolics content of water and ethanol extracts were determined by the Folin-Ciocalteu procedure and 54.3 and 42.8 µg gallic acid equivalent of phenols was detected in 1 mg WEBP and EEBP.

Both the black pepper oil and oleoresins showed strong antioxidant activity in comparison with butylated hydroxyanisole (BHA) and butylated hydroxytoluene (BHT) but lower than that of propyl gallate (PG) (Kapoor et al. 2009). In addition, their inhibitory action by FTC method, scavenging capacity by DPPH (2,2′-diphenyl-1-picrylhydrazyl radical), and reducing power were also confirmed, proving the strong antioxidant capacity of both the essential oil and oleoresins of pepper.

The methanolic extracts of the leaves of three pepper spices (*Piper guineense, Piper nigrum* and *Piper umbellatum*) exhibited a marked polyphenolic concentration and dose dependent free radical scavenging activity (Agbor et al. 2007). The free polyphenolic concentration of the three spices was in the order *P. umbellatum* (15.9 mg/g) > *P. guineense* (12.6 mg/g) > *P. nigrum* (9.8 mg/g). The three *Piper* extracts exhibited a 79.8–89.9% scavenging effect on DPPH, an 85.1–97.9% scavenging effect on nitric oxide at a dose level of 10 mg/ml and a 47.1–51.6% scavenging effect on superoxide radical at a dose level of 8 mg/ml extraction. The *Piper* extracts also exhibited a 57–76.1% scavenging effect on hydroxyl radical at 5 mg/ml, a 0.4–0.6 reducing power and an 88.3–93.9% metal chelating activity at a dose level of 8 mg/ml of extract. Thus, these *Piper* species could play a role in the modulation of free radical induced disorders. High antioxidant (DPPH scavenging) activity was found in *Piper cubeba* ethanol extract 77.61% with IC_{50} value of 10.54 µg/ml compared to *Piper nigrum* 74.61% and IC_{50} value of 14.5 µg.ml (Nahak and Sahu 2011). *P. cubeba* had the highest total phenolic content of 123.1 µg/g compared to *P. nigrum* with 62.3 µg/g. *P. cubeba* was found to contain alkaloid, glycosides, steroid, flavonoid, tannins and antraquinones while *P. nigrum* contained the same plus terpenoid and reducing sugars.

Anticancer Activity

Administration of coumaperine from white pepper, to male F344 rats provided protection against the initiation stage of rat hepatocarcinogenesis induced by diethylnitrosamine (Kitano et al. 2000). This was attributed to inhibition of proliferation of proliferating cell nuclear antigen (PCNA)-positive cells. Studies demonstrated the antimetastatic activity of piperine, an alkaloid of

Piper nigrum and *Piper longum*. Simultaneous administration of piperine with tumour induction by B16F-10 melanoma cells produced a significant reduction (95.2%) in tumour nodule formation in C57BL/6 mice (Pradeep and Kuttan 2002). Increased lung collagen hydroxyproline (22.37 µg/mg protein) in the metastasized lungs of the control animals compared to normal animals (0.95 µg/mg protein) was significantly reduced (2.59 µg/mg protein) in the piperine-treated animals. The high amount of uronic acid (355.83 µg/100 mg tissue) in the metastasized control animals was significantly reduced (65 µg/100 mg tissue) in the animals treated with piperine. Lung hexosamine content was also significantly reduced in the piperine-treated animals (0.98 mg/100 mg lyophilized tissue) compared to the untreated tumour-bearing animals (4.2 mg/100 mg lyophilized tissue). The elevated levels of serum sialic acid and serum gamma glutamyl transpeptidase activity in the untreated control animals was significantly reduced in the animals treated with piperine. The piperine-treated animals even survived the experiment (90 days). Histopathology of the lung tissue also correlated with the lifespan of the drug-treated animals. In another study, piperine was found to inhibit production of nitric oxide (NO) and tumour necrosis factor-alpha (TNF-alpha) level using in-vitro as well as in-vivo systems (Pradeep and Kuttan 2003). The level of nitrite in the LPS stimulated Balb/C mice (95.3 µM) was reduced in the piperine treated animals (25 µM) significantly. Nitrite level in the Concanavalin-A (Con-A) treated control animals (83.1 µM) was also significantly reduced to 18 µM in the piperine treated mice. The drastically elevated levels of TNF-alpha in the lipopolysaccharide (LPS) stimulated animals (625.8 µg/ml) was lowered in the piperine treated animals (105.8 µg/mL). Piperine also inhibited the Con-A induced TNF-alpha production. Piperine could inhibit the nitrite production by in-vitro activated macrophages (116.25 µM) to the normal level (15.67 µM) at concentration of 5 µg/ml. In-vitro L929 bioassay also revealed the inhibition of TNF-alpha production by the piperine treatment.

Piplartine {5,6-dihydro-1-[1-oxo-3-(3,4,5-trimethoxyphenyl)-trans-2-propenyl]-2(1 H)pyridinone} and piperine {1-[5-(1,3-benzodioxol-5-yl)-1-oxo-2,4-pentadienyl]piperidine} were found to display cytotoxicity in the brine shrimp lethality assay, sea urchin eggs development, 3-(4,5-dimethyl-2-thiazolyl)-2,5-diphenyl-2 H-tetrazolium bromide (MTT) assay using tumour cell lines and lytic activity on mouse erythrocytes (Bezerra et al. 2005). Piperine showed higher toxicity in brine shrimp (DL_{50} = 2.8 µg/ml) than piplartine (DL_{50} = 32.3 µg/ml). Both piplartine and piperine inhibited the sea urchin eggs development during all phases examined, first and third cleavage and blastulae, but in this assay piplartine was more potent than piperine. In the MTT assay, piplartine was the most active with IC_{50} values in the range of 0.7–1.7 µg/ml. In subsequent study, Bezerra et al. (2006) found that piperine and piplartine inhibited solid tumour development in mice transplanted with Sarcoma 180 cells. The inhibition rates were 28.7% and 52.3% for piplartine and 55.1% and 56.8% for piperine, after 7 days of treatment, at the lower and higher doses, respectively. The antitumour activity of piplartine was related to inhibition of the tumour proliferation rate but piperine did not inhibit cell proliferation as observed in Ki67 immunohistochemical analysis. Histopathological analysis of liver and kidney showed that both organs were reversibly affected by piplartine and piperine treatment, but in a different way. Piperine was more toxic to the liver, leading to ballooning degeneration of hepatocytes, accompanied by microvesicular steatosis in some areas, than piplartine which, in turn, was more toxic to the kidney, leading to discrete hydropic changes of the proximal tubular and glomerular epithelium and tubular hemorrhage in treated animals. In further study, piplartine was found to enhance the therapeutic effectiveness of chemotherapeutic drugs and that the combination of 5-fluorouracil (5-FU) with piplartine or piperine could improve immunocompetence hampered by 5-FU (Bezerra et al. 2008). The results indicated that either piplartine- or 5-FU-treated animals showed a low inhibition rate when they were used individually at low doses of 28.67% and 47.71%, respectively, but when they were combined at the same dose, the inhibition rate increased significantly to 68.04%.

Piplartine was also found to suppress leukemia growth and reduce cell survival, triggering both apoptosis and/or necrosis, depending on the concentration used (Bezerra et al. 2007). The antiproliferative activity of piplartine appeared to be related to the inhibition of DNA synthesis. Piplartine-mediated reduction in cell number was associated with an increasing number of dead cells at a concentration of 10 μg/ml. These findings were corroborated by morphologic analysis. However, at the lowest concentration (2.5 μg/ml), piplartine-treated cells exhibited typical apoptotic morphological changes. The increase in caspase-3 activity was also observed in lysates of piplartine-treated cells. Rho et al. (2007) using a bioactivity-guided fractionation of MeOH extracts of the fruits of *Piper nigrum*, isolated the following alkamides: retrofractamide A(1), pipercide (2), piperchabamide D (3), pellitorin (4), dehydroretrofractamide C (5) and dehydropipernonaline (6). The IC_{50} values for acyl coenzyme A:cholesterol acyltransferase (ACAT), an intracellular enzyme involved in cholesterol accumulation, for the compounds were 24.5 (1), 3.7 (2), 13.5 (3), 40.5 (4), 60 (5) and 90 μM (6), according to the results of an ACAT enzyme assay system using rat liver microsomes. These compounds all inhibited cholesterol esterification in HepG2 (human hepatocellular liver carcinoma) cells.

Hwang et al. (2011) demonstrated that the anti-invasive effects of piperine may occur through inhibition of PKCα (protein kinase Cα) and ERK (extracellular signal-regulated kinase) phosphorylation and reduction of NF-κB (nuclear factor kappa-light-chain-enhancer of activated B cells) and AP-1 (activator protein-1) activation, leading to down-regulation of matrix metalloproteinase-9 (MMP-9) expression that was enhanced by phorbol-12-myristate-13-acetate (PMA). Thus, piperine had potential as a potent anticancer drug in therapeutic strategies for fibrosarcoma metastasis.

All compounds alkylamides and piperine derived from black pepper suppressed proinflammatory transcription factor TNF-induced NF-kappaB activation, but alkyl amides, compound 4 from black pepper and 5 from hot pepper, were most effective (Liu et al. 2010). The human cancer cell proliferation inhibitory activities of piperine and alklyl amides in Capsicum and black pepper were dose dependant. The inhibitory concentrations 50% (IC_{50}) of the alklylamides were in the range 13–200 μg/ml. The extracts of black pepper at 200 μg/ml and its compounds at 25 μg/ml inhibited lipid peroxidation by 45–85%, COX enzymes by 31–80% and cancer cells proliferation by 3.5–86.8%. Overall, these results suggested that black pepper and its constituents exhibited antiinflammatory, antioxidant and anticancer activities.

Piper nigrum roots afforded the following alkaloids: pellitorine, (E)-1-[3′,4′-(methylenedioxy) cinnamoyl]piperidine, 2,4-tetradecadienoic acid isobutyl amide, piperine, sylvamide, cepharadione A, piperolactam D and paprazine (Ee et al. 2009). Cytotoxic activity screening of the plant extracts indicated some activity. Pellitorine, which was isolated from the roots of *Piper nigrum*, showed strong cytotoxic activities against HL60 and MCT-7 cell lines (Ee et al. 2010). Microbial transformation of piperine gave a new compound 5-[3,4-(methylenedioxy)phenyl]-pent-2-ene piperidine. Two other alkaloids (E)-1-[3′,4′-(methylenedioxy) cinnamoyl]piperidine and 2,4-tetradecadienoic acid isobutyl amide were also found.

Antimutagenic Activity

Black pepper was found to be effective in reducing mutational events induced by and the promutagen agent ethyl carbamate using the wing Somatic Mutation And Recombination Test (SMART) in *Drosophila melanogaster* (El Hamss et al. 2003). Suppression of metabolic activation or interaction with the active groups of mutagens could be mechanisms by which the spice exerted its antimutagenic action.

Melanocyte Stimulatory Activity

The aqueous extract of black pepper at 0.1 mg/ml was observed to cause nearly 300% stimulation of the growth of a cultured mouse melanocyte line, melan-a, in 8 days (Lin et al. 1999). Piperine (1-piperoylpiperidine), the main alkaloid from

Piper nigrum fruit, also significantly stimulated melan-a cell growth. Both *Piper nigrum* extract and piperine induced morphological alterations in melan-a cells, with more and longer dendrites observed. The augmentation of growth by piperine was effectively inhibited by RO-31-8220, a selective protein kinase C (PKC) inhibitor, suggesting that PKC signalling is involved in its activity. In subsequent studies, amides such as piperine, guineensine and pipericide, from pepper fruits were found to have stimulatory effect on melanocyte proliferation in-vitro and supported the traditional use of *P. nigrum* extracts in the depigmentary skin disorder, vitiligo (Lin et al. 2007). A crude chloroform extract of *P. nigrum* containing piperine was more stimulatory than an equivalent concentration of the pure compound, suggesting the presence of other active components. Piperine (1), guineensine (2), pipericide (3), N-feruloyltyramine (4) and N-isobutyl-2E, 4E-dodecadienamide (5) were isolated from the chloroform extract. Their activity was compared with piperine and with commercial piperlongumine (6) and safrole (7), and synthetically prepared piperettine (8), piperlonguminine (9) and 1-(3, 4-methylenedioxyphenyl)-decane (10). Compounds 6–10 either occurred in *P. nigrum* or were structurally related. Compounds 1, 2, 3, 8 and 9 stimulated melanocyte proliferation, whereas 4, 5, 6, 7 and 10 did not.

A methanolic extract from the leaves of *Piper nigrum* showed a significant stimulatory effect on melanogenesis in cultured murine B16 melanoma cells (Matsuda et al. 2004). Activity-guided fractionation of the methanolic extract led to the isolation of two known lignans, (-)-cubebin (1) and (-)-3,4-dimethoxy-3,4-desmethylenedioxycubebin (2), together with a new lignan, (-)-3-desmethoxycubebinin (3). Among these lignans, 1 and 2 showed a significant stimulatory activity of melanogenesis without any significant effects on cell proliferation.

Gastrointestinal/Gastroprotective Activity

P. nigrum may protect the colon by decreasing the activity of b-glucuronidase and mucinase. Histopathological studies also showed lesser infiltration into the submucosa, fewer papillae and lesser changes in the cytoplasm of the cells in the rats colon in the black pepper groups (Nalini et al. 1998). Piperazine at doses of 25, 50, 100 mg/kg i.g. dose-dependently protected rats and mice from gastric ulceration induced by stress, indometacin, HCl, and pyloric ligation induced by stress, indometacin, HCl, and pyloric ligation (Bai and Xu 2000). The inhibitory rates were 16.9%, 36.0%, and 48.3% in stress ulcers; 4.4%, 51.1%, and 64.4% in indometacin ulcers; 19.2%, 41.5%, and 59.6% in HCl ulcers; 4.8%, 11.9%, and 26.2% in pyloric ligation ulcers, respectively. Piperazine also inhibited the volume of gastric juice, gastric acidity, and pepsin A activity.

Piperine, an alkaloid of black pepper was found to inhibit gastric emptying of solids/liquids in rats and gastrointestinal transit in mice in a dose and time dependent manner (Bajad et al. 2001). Piperine significantly inhibited gastric emptying of solids and gastrointestinal transit at the doses extrapolated from humans. One week oral treatment of 1 and 1.3 mg/kg in rats and mice, respectively, did not produce a significant change in activity as compared to single dose administration. Gastric emptying inhibitory activity of piperine was independent of gastric acid and pepsin secretion. Piperine was also found to increase gastric secretion in white albino rats (Ononiwu et al. 2002). Increasing the dose from 20 mg/kg weight to 142 mg/kg body weight produced dose dependent increases in gastric acid secretion. When compared with control basal acid secretion, these increases were significant. Piperine was however about 40 times less effective than histamine in increasing gastric acid secretion. In a separate study, piperine (2.5–20 mg/kg, i.p.) was found to dose-dependently reduce castor oil-induced intestinal fluid accumulation (Capasso et al. 2002). The inhibitory effect of piperine was strongly attenuated in capsaicin-treated mice but it was not modified by the vanilloid receptor antagonist capsazepine. The results suggested that piperine reduced castor oil-induced fluid secretion via a mechanism involving capsaicin-sensitive neurons, but not capsazepine-sensitive vanilloid receptors. Results of studies by McNamara et al. (2005) suggested that the effects of piperine at human vanilloid receptor TRPV1 were similar to those of capsaicin except for its propensity to induce greater receptor desensitisation and to exhibit a greater efficacy than capsaicin itself. Piperine also caused greater tachyphylaxis in response to repetitive agonist applications

The medicinal use of pepper and piperine in gastrointestinal motility disorders had been shown to be due to the presence of spasmodic (cholinergic) and antispasmodic (opioid agonist and Ca^{2+} antagonist) effects (Mehmood and Gilani 2010). When tested in isolated guinea pig ileum, the crude extract of pepper (1–10 mg/ml) and piperine (3–300 μM) caused a concentration-dependent and atropine-sensitive stimulant effect. In rabbit jejunum, pepper extract and piperine relaxed spontaneous contractions, similar to loperamide and nifedipine. In mice, the pepper extract and piperine exhibited a partially atropine-sensitive laxative effect at lower doses, whereas at higher doses it caused antisecretory and antidiarrheal activities that were partially inhibited in mice pretreated with naloxone, similar to loperamide.

Antiinflammatory Activity

Piperine (1-peperoyl piperidine) isolated from *Piper nigrum* was found to exhibit antiinflammatory activity in rats in different acute and chronic experimental models like carrageenin-induced rat paw edema, cotton pellet granuloma, and croton oil-induced granuloma pouch (Mujumdar et al. 1990a). Piperine acted significantly on early acute changes in inflammatory processes and chronic granulative changes. Exudative changes in both acute and chronic models, however, were insignificant. The methanolic extract of *P. nigrum* leaves exhibited antiinflammatory activity (Hirata et al. 2008). Oral administration of the methanolic leaf extract showed a potent dose-dependent inhibition of dinitrofluorobenzene (DNFB)-induced cutaneous reaction at 1 hour [immediate phase response (IPR)] after and 24 hours [late phase response (LPR)] after DNFB challenge in mice which were passively sensitized with anti-dinitrophenyl (DNP) IgE antibody. Ear swelling inhibitory effect of PN-ext on very late phase response (vLPR) in the model mice was significant but weaker than that on IPR. Oral administration of pepper leaf extract inhibited picryl chloride (PC)-induced ear swelling in PC sensitized mice. The extract exhibited in-vitro inhibitory effect on compound 48/80-induced histamine release from rat peritoneal mast cells. Two lignans of the extract, (-)-cubebin and (-)-3,4-dimethoxy-3,4-esmethylen edioxycubebin, were identified as major active principles having histamine release inhibitory activity. Bae et al. (2010) examined the effects of piperine on lipopolysaccharide (LPS)-induced inflammatory responses. Administration of piperine to mice inhibited LPS-induced endotoxin shock, leukocyte accumulation and the production of tumour necrosis factor-alpha (TNF-alpha), but not of interleukin (IL)-1beta and IL-6. In peritoneal macrophages, piperine inhibited LPS/poly (I:C)/CpG-ODN-induced TNF-alpha production. Piperine also inhibited LPS-induced endotoxin shock in TNF-alpha knockout (KO) mice. The results suggested that piperine inhibited LPS-induced endotoxin shock through inhibition of type 1 interferon production.

Antidiabetic Activity

Oral administration of aqueous extract of *Piper nigrum* seeds and *Vinca rosea* flowers to alloxan induced diabetic rats once a day for 4 weeks led to significant lowering of blood sugar level and reduction in serum lipids (Kaleem et al. 2005). The decreased levels of antioxidant enzymes, catalase and glutathione peroxidase in alloxan induced diabetic rats were returned to normal in insulin, *P. nigrum* and *V. rosea* treated rats. Lipid peroxidation levels were significantly higher in diabetic rats and it was slightly higher in insulin, *P. nigrum* and *V. rosea* treated rats as compared to control rat. These results suggested that oxidative stress plays a key role in diabetes, and treatment with *P. nigrum* and *V. rosea* were useful in controlling not only the glucose and lipid levels but these components may also be helpful in strengthening the antioxidants potential. Piperine treatment of normal rats enhanced hepatic GSSG (oxidized glutathione) level by

100% and decreased renal GSH (reduced glutathione) concentration by 35% and renal glutathione reductase activity by 25% when compared to normal controls (Rauscher et al. 2000). Treatment of streptozotocin-induced diabetic Sprague-Dawley rats with piperine reversed the diabetic effects on GSSG concentration in the brain, on renal glutathione peroxidase and superoxide dismutase activities, and on cardiac glutathione reductase activity and lipid peroxidation. Piperine treatment did not reverse the effects of diabetes on hepatic GSH concentrations, lipid peroxidation, or glutathione peroxidase or catalase activities; on renal superoxide dismutase activity; or on cardiac glutathione peroxidase or catalase activities. The data indicated that subacute treatment with piperine for 14 days was only partially effective as an antioxidant therapy in diabetes.

Bioassay-guided isolation of chloroform extracts of the fruits of *Piper longum* and *Piper nigrum*, using an in-vitro DGAT (acyl CoA:diacylglycerol acyltransferase) inhibitory assay, lead to isolation of a new alkamide named (2E,4Z,8E)-N-[9-(3,4-methylenedioxyphenyl)-2,4,8-nonatrienoyl]piperidine (2), together with four known alkamides: retrofractamide C (1), pipernonaline (3), piperrolein B (4), and dehydropipernonaline (5) (Lee et al. 2006). Compounds 2–5 inhibited DGAT with IC_{50} values of 29.8 (2), 37.2 (3), 20.1 (4), and 21.2 (5) µM, respectively, but the IC_{50} value for 1 was more than 900 µM. This finding indicated that compounds possessing piperidine groups (2–5) could be potential DGAT inhibitors. DGAT inhibitors have emerged as a potential therapy for the treatment of obesity and type 2 diabetes.

Hepatoprotective Activity

Piperine, an active alkaloidal constituent of *Piper longum* and *Piper nigrum* exerted a significant protection against tert-butyl hydroperoxide and carbon tetrachloride hepatotoxicity by reducing both in vitro and in vivo lipid peroxidation, enzymatic leakage of GPT and AP, and by preventing the depletion of GSH and total thiols in the intoxicated mice (Koul and Kapil 1993), Piperine showed a lower hepatoprotective potency than silymarin. Black pepper was found to have a modulatory effect on the hepatic biotransformation system in mice (Singh and Rao 1993). The results revealed a significant and dose-dependent increase in glutathione S-transferase (GST) and acid-soluble sulfhydryl content in the experimental mice except the one maintained on 0.5% black pepper diet for 10 days. The level of malondialdehyde (MDA) was lowered in the group fed on 2% black pepper diet for 20 days. Being a potential inducer of detoxication system, the possible chemopreventive role of black pepper in chemical carcinogenesis was suggested. The ethanol extract of *P. nigrum* root was found to be an efficient hepatoprotective and antioxidant agent against CCl_4-induced liver injury (Bai et al. 2011). The extract decreased activities of alanine transaminase (ALT) and aspartate transanimase (AST) in rat serum; decreased lipid peroxidation (MDA) and increased glutathione (GSH) % in the rats liver homogenate, as compared with those of the CCl_4 positive control rats. The hepatoprotective effect of ethanol extract was also supported by the histopathological observations.

Antiatherogenic/Antihyperlipidemic Activity

Administration of the *Piper* species (*Piper guineense, Piper nigrum* and *Piper umbellatum*) for 12 weeks in atherogenic diet fed hamsters prevented the collapse of the antioxidant system and the increase of plasma parameters restoring them towards normality (Agbor et al. 2010). The *Piper* species also prevented LDL oxidation by increasing the time (lag time) for its oxidation. The atherogenic diet of rodent chow supplemented with 0.2% cholesterol and 10% coconut oil induced a collapse of the erythrocyte antioxidant defense system (significant decrease in superoxide dismutase, catalase and glutathione peroxidase activities). The atherogenic diet also induced an increase in plasma total cholesterol, triglyceride, thiobarbituric acid reactive substances (TBARS), oxidation of low density lipoprotein cholesterol (LDL) and accumulation of foam cells in the aorta a hall mark for atherosclerosis. The results suggested that these *Piper* species had significant antioxidant and antiatherogenic effect against atherogenic diet intoxication.

Studies showed that piperine supplementation appreciably protected erythrocytes from oxidative stress by improving the antioxidant status in high fat diet fed antithyroid drug treated hyperlipidemic rats (Vijayakumar and Nalini 2006a). Concurrent piperine supplementation along with high fat diet and antithyroid drug administration normalized erythrocyte osmotic fragility, reduced lipid peroxidation, and improved the enzymic and non-enzymic antioxidant status compared to those rats that did not receive piperine. Studies also showed that piperine possessed thyrogenic activity, thus modulating apolipoprotein levels and insulin resistance in high fat diet-fed rats, opening a new view in the management of dyslipidemia by dietary supplementation with nutrients (Vijayakumar and Nalini 2006b). The simultaneous administration of piperine and high fat diet significantly reduced plasma lipids and lipoproteins levels, except for HDL, which was significantly elevated. Piperine supplementation also improved the plasma levels of apo A-I, T3, T4, testosterone, and I and significantly reduced apo B, TSH, and insulin to near normal levels.

Antihypertensive Activity

In-vitro studies with rabbit aortic rings showed that piperine possessed a blood pressure-lowering effect mediated possibly through Ca^{2+} channel blockade (CCB) while consistent fall in blood pressure was restricted by associated vasoconstrictor effect (Taqvi et al. 2008). Additionally, species selectivity existed in the CCB effect of piperine.

Oral administration of piperine in wistar rats was able to partially prevent the increase of blood pressure caused by chronic L-NAME (N(G)-nitro-L-arginine methyl ester) administration (Hlavackova et al. 2010). L-NAME increased the blood pressure, cross-sectional area of aorta, media thickness, elastin and smooth muscle cells actin (SMCA) synthesis and phosphotungstic acid hematoxylin (PTAH) positive myofibrils relative and absolute content in the aortic media, whereas it decreased percentual content of inducible NO synthase (iNOS), elastin and SMCA. Piperine decreased the blood pressure rise from the third week of treatment, synthesis of elastin and the percentual and absolute content of PTAH positive myofibrils, however, it did not affect other parameters. The authors asserted that the piperine effect was probably caused by the blockage of voltage-dependent calcium channels and supported by filamentous actin disassembly.

Spasmolytic Activity

Black pepper and its main ingredient piperine were found to relieve menorrhalgia in women and supported the use of black pepper by women to relief menorrhalgia in Iranian traditional medicine (Naseri and Yahyavi 2007). The extract reduced the uterus contractions in rats induced by KCl and oxytocin dose dependently. The spasmolytic effect of extract on the KCl-induced contractions was not reduced by L-NAME, phentolamine and naloxone but propranolol reduced the extract activity. In Ca^{2+}-free De Jalon solution with high potassium (60 mM), the extract reduced the contractions induced by cumulative concentrations of CaCl2 dose dependently. The results suggested that the spasmolytic effect of the extract on rat uterus was mediated via voltage dependent calcium channels and β-adrenoceptors could also be involved in this action.

Neuroprotective and CNS Activity

Studies showed that administration of piperine to adult male Wistar rats at a period of 2 weeks before and 1 week after the intracerebroventricular administration of ethylcholine aziridinium ion significantly improved memory impairment and neurodegeneration in hippocampus (Chonpathompikunlert et al. 2010). Piperine also demonstrated neurotrophic effect in hippocampus. The possible underlying mechanisms was postulated to be partly associated with the decrease lipid peroxidation and acetylcholinesterase enzyme. Results of studies by Fu et al. (2010) suggested that the neuroprotective effects of piperine might be associated with suppression of synchronization of neuronal networks, presynaptic glutamic acid release, and Ca^{2+} overloading. Piperine effectively inhibited the synchronized oscillation of intracellular calcium in rat hippocampal neuronal networks and represses spontaneous synaptic activities in terms of spontaneous synaptic currents (SSC) and spontaneous excitatory postsynaptic currents (sEPSC). Additionally, pretreatment with piperine exerted a protective effect on glutamate-induced decrease of cell viability and apoptosis of hippocampal neurons.

Anticonvulsant, Antidepressant, Sedative and Analgesic Activities

Mori et al. (1985) showed that convulsions of E1 mice were completely suppressed by 60 mg/kg of piperine injected intraperitoneally. The ED_{50} was 21.1 mg/kg. The brain 5-HT level was significantly higher in the cerebral cortex of piperine treated mice than in control mice. This increase may be related directly to the mechanism of inhibition of convulsions by piperine. In contrast, lower levels of 5-HT were observed in the hippocampus, midbrain and cerebellum. The dopamine level in the piperine treated mice was markedly higher only in the hypothalamus, while the norepinephrine levels were lower in every part of the brain. Amide alkaloids such as piperine, pellitorine, 3,4-dimethyinenedioxycinnamoyl-piperidine, and ß-sitesterol isolated from the ethanol extract of *Piper nigrum* roots showed anticonvulsive, sedative, and analgesic activities to mice (Hu et al. 1996). LD_{50} of the ethanol and the water extracts to mice were respectively 12.66 and 424.38 g/kg converted into crude material. The anticonvulsant actions of the principal component of pepper, piperine was postulated to underpin the effectiveness of a traditional Chinese medicine, a mixture of radish and pepper in treating epilepsy (D'Hooge et al. 1996). They found that piperine significantly block convulsions induced by intracerebroventricular injection of threshold doses of kainate, but to have no or only slight effects on convulsions induced by L-glutamate, N-methyl-D-aspartate or guanidinosuccinate.

Li et al. (2007) demonstrated the antidepressant-like effects of piperine and its derivative, antiepilepsirine. After 2 weeks of chronic administration, piperine and antiepilepsirine at doses of 10–20 mg/kg significantly reduced the duration of immobility in both forced swimming test and tail suspension test, without accompanying changes in locomotor activity in the open-field test. They elucidated that the antidepressant effect might depend on the augmentation of the neurotransmitter synthesis or the reduction of the neurotransmitter reuptake. Antidepressant properties of piperine were proposed to be mediated via the regulation of serotonergic system, whereas the mechanisms of antidepressant action of antiepilepsirine might be due to its dual regulation of both serotonergic and dopaminergic systems. Wattanathorn et al. (2008) showed that administration of piperine at doses ranging from 5, 10 and 20 mg/kg BW once daily for 4 weeks to male Wistar rats possessed antidepression like activity and cognitive enhancing effect at all treatment duration. The results suggested piperine may be used as potential functional food to improve brain function.

Liao et al. (2009) demonstrated that the antidepressant-like effects of piperine and its mechanisms might be involved by up-regulation of the progenitor cell proliferation of hippocampus and cytoprotective activity. After a week of administration, piperine significantly reduced the duration of immobility forced swimming test and tail suspension test. Piperine exerted protective effect on neuroblastoma cells and increased proliferation of hippocampus neural progenitor cells. Studies showed that piperine significantly reduced the immobility time in the forced swim test and tail suspension test in mice (Mao et al. 2011). Piperine treatment also significantly potentiated the number of head-twitches of mice induced by 5-HTP (a metabolic precursor to 5-HT). The results suggested that the antidepressant-like effect of piperine was mediated via the serotonergic system by enhancing 5-HT (5-hydroxytryptophan) content in mouse brain.

Antithyroid Activity

Daily oral administration (2.50 mg/kg) of piperine to mice for 15 days lowered the serum levels of both the thyroid hormones, thyroxin (T_4) and triiodothyronine (T_3) as well as glucose concentrations with a concomitant decrease in hepatic 5′D enzyme and glucose-6-phosptase (G-6-Pase) activity (Panda and Kar 2003). However, no significant alterations were observed in animals treated with 0.25 mg/kg of piperine in any of the activities studied except an inhibition in serum T(3) concentration. The decrease in T(4), T(3) concentrations and in G-6-Pase were comparable to that of a standard antithyroid drug, Proylthiouracil (PTU). The hepatic lipid-peroxidation (LPO) and the activity of endogenous antioxidants, superoxide dismutase (SOD), and catalase (CAT) were not significantly altered in either of the doses (0.25 and 25 mg/kg). It appeared that the action of *P. nigrum* on thyroid functions was mediated through its active alkaloid, piperine.

Hydrocholagoguic Effect

Administration of black pepper (250 mg/kg b.w.) to rats by gavage caused an increase in bile solids while with other treatments (500 mg black pepper and piperine at 12.5 or 25 mg/kg body weight) bile secretion or dry matter in bile was not changed (Ganesh Bhat and Chandrasekhara 1987). Dietary feeding of black pepper caused an increase in bile flow with a concomitant decrease in bile solids – a hydrocholagoguic effect. Cholesterol and bile acid output were not affected by black pepper or piperine at either level irrespective of the mode of administration; in contrast, the secretion of uronic acids in bile was enhanced by both levels of pepper as also of piperine indicating possible excretion of some of the components of black pepper or of piperine as glucuronides.

Drug Potentiation Activity

Piperine, a major alkaloid isolated from *Piper nigrum* was found to potentiate pentobarbitone sleeping time in dose dependant manner, with peak effect at 30 minutes (Mujumdar et al. 1990b). Blood and brain pentobarbitone levels were higher in piperine treated animals. Piperine treatment in rats, treated chronically with phenobarbitone, significantly potentiated pentobarbitone sleeping time, as compared to the controls. There was no alteration in barbital sodium sleeping time. It was postulated that piperine inhibited liver microsomal enzyme system and thereby potentiated the pentobarbitone sleeping time.

Piperine was found to be an inhibitor of multidrug resistance (Li et al. 2011). Piperine could potentiate the cytotoxicity of anti-cancer drugs in resistant sublines, such as MCF-7/DOX and A-549/DDP, which were derived from MCF-7 and A-549 cell lines. At a concentration of 50 µM piperine could reverse the resistance to doxorubicin 32.16 and 14.14 folds, respectively. It also re-sensitized cells to mitoxantrone 6.98 folds. Further, long-term treatment of cells by piperine inhibited transcription of the corresponding ABC transporter genes.

Studies showed that piperine inhibited both the drug transporter P-glycoprotein and the major drug-metabolizing enzyme CYP3A4 (Bhardwaj et al. 2002). Piperine inhibited digoxin and cyclosporine A transport in Caco-2 cells with IC_{50} values of 15.5 and 74.1 µM, respectively. CYP3A4-catalyzed formation of D-617 and norverapamil was inhibited in a mixed fashion. Because both proteins are expressed in enterocytes and hepatocytes and contribute to a major extent to first-pass elimination of many drugs, the data indicated that dietary piperine could affect plasma concentrations of P-glycoprotein and CYP3A4 substrates in humans, in particular if these drugs were administered orally. Two new bisalkaloids, dipiperamides D and E, were isolated as inhibitors of a drug metabolizing enzyme cytochrome P450 (CYP) 3A4 from the white pepper, *Piper nigrum* (Tsukamoto et al. 2002b). Three new bisalkaloids, dipiperamides A, B, and

C, isolated from the white pepper were found to inhibit cytochrome P450 (CYP) 3A4 activity (Tsukamoto et al. 2002a). Dipiperamides D and E showed potent CYP3A4 inhibition with IC_{50} values of 0.79 and 0.12 µM, respectively, and other metabolites from the pepper were moderately active or inactive. Of 19 alkamides isolated from *Piper nigrum* that inhibited human liver microsomal cytochrome P450 2D6 (CYP2D6), compounds 15 and 17 showed more than 50% decrease of the CYP2D6 residual activity after 20 minutes preincubation (Subehan et al. 2006). Both compounds 15 and 17 showed that the characteristic time- and concentration-dependent inhibition.

Piperine at non-cytotoxic concentrations ranging from 10 to 100 µM, was observed to inhibit P-gp mediated efflux transport of [(3)H]-digoxin across L-MDR1 and Caco-2 cell monolayers (Han et al. 2008). The acute inhibitory effect was dependent on piperine concentration, with abrogation of [(3)H]-digoxin polarized transport attained at 50 µM of piperine. In contrast, prolonged (48 and 72 h) co-incubation of Caco-2 cell monolayers with piperine (50 and 100 µM) increased P-gp activity through an up-regulation of cellular P-gp protein and MDR1 mRNA levels. Peroral administration of piperine at the dose of 112 µg/kg body weight/day to male Wistar rats for 14 consecutive days also increased intestinal P-gp levels. Their data suggested that caution should be exercised when piperine was to be co-administered with drugs that are P-gp substrates, particularly for patients whose diet relies heavily on pepper. In a more recent study, P-glycoprotein (P-gp) inhibitory activity of the aqueous pepper extract was found to be comparable with that of pure piperine and was significantly higher than the alcoholic extract (Aher et al. 2009). Pure piperine and the aqueous extract exhibited significant P-gp inhibitory activity compared with control, which was irrespective of oral pretreatment dose and duration levels. No significant effect of oral pretreatment duration of the aqueous extract was observed. Another recent study showed piperine significantly enhanced the oral exposure of fexofenadine in rats likely by the inhibition of P-glycoprotein-mediated cellular efflux during the intestinal absorption, suggesting that the combined use of piperine or piperine-containing diet with fexofenadine may require close monitoring for potential drug-diet interactions (Jin and Han 2010).

Otoprotective Activity

Piperine was found to protect House Ear Institute-Organ of Corti 1 (HEI-OC1) cells against cisplatin-induced apoptosis through the induction of heme oxygenase (HO)-1 expression (Choi et al. 2007). Piperine (10–100 µM) induced the expression of HO-1 in dose- and time-dependent manners. The results demonstrated that the expression of HO-1 by piperine was mediated by both c-Jun N-terminal kinase (JNK) pathway and translocated nuclear factor-E2-related factor-2 (Nrf2).

Antimicrobial Activity

The amide alkaloid constituents isolated from pepper berries namely 2E, 4E, 8Z-N-isobutyleicosatrienamide, pellitorine, trachyone, pergumidiene and isopiperolein B exhibited antimicrobial activity (Reddy et al. 2004). All the isolated compounds were active against *Bacillus subtilis, Bacillus sphaericus,* and *Staphylococcus aureus* amongst Gram positive bacteria, and *Klebsiella aerogenes* and *Chromobacterium violaceum* among Gram negative bacterial strains. Pepper oil was found to be 100% effective in controlling the mycelial growth of *Fusarium graminearum* (Singh et al. 2004). The acetone pepper extract retarded 100% mycelial growth of *Penicillium viridicatum* and *Aspergillus ochraceus*. The volatile oils from fresh and dried pepper berries and pepper leaves were found to have antimicrobial activity (Sasidharan and Menon 2010). Fresh berry oil (FBO) was found to be more active than standard, tetracycline against *Bacillus subtilis* and *Pseudomonas aeruginosa*, on par with reference compound, nystatin against *Aspergillus niger* but weaker towards *Penicillium* spp, *Candida albicans, Saccharomyces cerevisiae* and *Trichoderma* spp. Dried berry oil (DBO) was more active towards *Pseudomonas aeruginosa* and *Penicillium* spp, equal to reference compound against *Bacillus subtilis, Candida albicans* and *Saccharomyces cerevisiae* and weaker towards *Aspergillus niger* and *Trichoderma* spp. Pepper leaf oil (PLO) was weaker towards all the organisms studied than the references. FBO recorded MIC values of 1 µg/ml for *Bacillus subtilis* 2.5 µg/ml *for Pseudomonas aeruginosa* and 8.5 µg/ml for *Aspergillus niger*. DBO had MIC values of 5 µg/ml for *Penicillium* spp. and 4 µg/ml for *Candida albicans* and only 0.8 mug/ml for *Saccharomyces cerevisiae*. In the case of *Trichoderma* spp both FBO and DBO required 10 µg/ml. Piperine significantly enhanced accumulation and decreased the efflux of ethidium bromide in *Mycobacterium smegmatis*, suggesting that it had the ability to inhibit mycobacterial efflux pump (Jin et al. 2011).

ICAM Binding Inhibition Activity

Eight alkamides isolated from the ethanol extracts of the fruits of *Piper longum* and *Piper nigrum* were elucidated as follows: guineensine (1), retrofracamide C (2), (2E,4Z,8E)-N-[9-(3,4-methylenedioxyphenyl)-2,4,8-nonatrienoyl]piperidine (3), pipernonaline (4), piperrolein B (5), piperchabamide D (6), pellitorin (7), and dehydropipernonaline (8) (Lee et al. 2008). Compounds 3–5, 7, and 8 were found to inhibit potently the direct binding between ICAM-1 (intercellular cell adhesion molecules) and LFA-1 (lymphocyte function associated antigen-1) of THP-1 cells (human monocytic leukaemia cell line) in a dose-dependent manner, with IC_{50} values of 10.7, 8.8, 13.4, 13.5, and 6.0 µg/ml, respectively.

Immunomodulatory Activity

The purified polysaccharides from *Piper nigrum* seeds afforded two active fractions PN-Ib and PN-IIa, purified anti-complementary polysaccharides (Chun et al. 2002). PN-Ib had an average molecular mass of 21 kDa contained 88.5% glucose and other negligible minor monosaccharides, while PN-IIa showed a different monosaccharide composition, which contained a significant proportion of galactose, arabinose, galacturonic acid and rhamnose. None of the anti-complementary activity of any polysaccharide was changed by pronase digestion or polymyxin B treatment, but they were decreased by periodate oxidation. Based upon these results, the usefulness of purified anti-complementary polysaccharides from *Piper nigrum* was suggested as a supplement for immune enhancement.

Cadmium (Cd), a well known environmental carcinogen, is a potent immunotoxicant; it causes marked thymic atrophy and splenomegaly in rodents and induces apoptosis in murine lymphocytes and alters the immune functions (Pathak and Khandelwal 2007, 2008). Addition of piperine in various concentrations (1, 10 and 50 µg/ml) ameliorated the toxic effects of cadmium. The reported free radical scavenging property of piperine and its antioxidant potential could be responsible for the modulation of intracellular oxidative stress signals. These in turn appeared to mitigate the apoptotic pathway and other cellular responses altered by cadmium. The findings strongly indicated the antioxidative, antiapoptotic and chemo-protective ability of piperine in blastogenesis, cytokine release and restoration of splenic cell population indicating its therapeutic usefulness in immuno-compromised situations. In another study, of the three herbals examined for antiimmunotoxic activity, piperine displayed maximum efficacy. All the three doses of piperine (1, 10 and 50 µg/ml) increased cell viability in a dose dependent manner, whereas curcumin and picroliv were also effective, but to a lesser degree. Restoration of ROS and GSH was most prominent with piperine. The antiapoptotic potential was directly proportional to their antioxidant nature. In addition, Cd altered blastogenesis, T and B cell phenotypes and cytokine release were also attenuated best with piperine. The ameliorative potential was in order of piperine > curcumin > picroliv. Studies by Majdalawieh and Carr (2010) found that black pepper and cardamom exerted immunomodulatory roles and antitumour activities. The black pepper and cardamom aqueous extracts significantly enhanced splenocyte proliferation in a dose-dependent, synergistic fashion. The extracts significantly enhanced the cytotoxic activity of natural killer cells, indicating their potential anti-cancer effects. The findings suggested that black pepper and cardamom constituents could be used as potential therapeutic tools to regulate inflammatory responses and prevent/attenuate carcinogenesis.

Antiandrogenic Activity

The extracts of *P. nigrum* leaf, *P. nigrum* fruit and *P. cubeba* fruit showed potent testosterone 5-alpha reductase inhibitory activity (Hirata et al. 2007). Activity-guided fractionation of *P. nigrum* leaf extract led to the isolation of (-)-cubebin (1) and (-)-3,4-dimethoxy-3,4-desmethylenedioxycubebin (2). Additionally, it was found that piperine, a major alkaloid amide of *P. nigrum* fruit, showed potent inhibitory activity, thus a part of the inhibitory activity of *P. nigrum* fruit may depend on piperine. In addition, the *P. nigrum* leaf extract showed in-vivo antiandrogenic activity using the hair regrowth assay in testosterone sensitive male C57Black/6CrSlc strain mice.

Carcinogenic Effect

Painting and feeding of mice with 2 mg of an extract from black pepper on 3 days a week for 3 months resulted in a significant increase in tumour number of tumour-bearing mice (Shwaireb et al. 1990). Tumour incidence was reduced in those groups of experimental animals receiving 5 or 10 mg Vitamin A-palmitate twice weekly for 3 months by feeding or painting during and subsequent to application of pepper extract. However, feeding of mice with powder of black pepper in diet (50 g/3 kg food) had no impact on carcinogenesis. The researchers (Wrba et al. 1992) found that in mice, injection of safrole, tannic acid (constituents of black pepper) or methylcholanthrene (MCA) during the preweaning period induced tumours in different organs. Safrole and tannic acid were weak carcinogens when compared with MCA which was used as a carcinogenic control substance. Force feeding of d-limonene (one of the pepper terpenoids) for a long time to the mice which were injected with any of the above three substances reduced their carcinogenic activity, while force feeding of piperine (one of black pepper alkaloids) was ineffective.

Insecticidal Activity

Several insecticidal amides, such as pipericide, (E,E)-N-(2-methylpropyl)-2,4,12-tridecadienamide, and (E,E,E)-11-(1,3-benzodioxol-5-yl)-N-(2-methylpropyl)- 2,4,10-undecatrien-amide, had been isolated from *P. nigrum* (Miyakado et al. 1979; Su and Horvat 1981). The amide alkaloids from pepper fruit such as pellitorine, guineensine, pipercide, and retrofractamide exhibited insecticidal activity against third instar larvae of *Culex pipiens pallens, Aedes aegypti,* and *Aedes togoi* (Park et al. 2002). The compound most toxic to *C. pipiens pallens* larvae was pipercide (0.004 ppm) followed by retrofractamide A (0.028 ppm), guineensine (0.17 ppm), and pellitorine (0.86 ppm). Piperine (3.21 ppm) was least toxic. Against *A. aegypti* larvae, larvicidal activity was more pronounced in retrofractamide A (0.039 ppm) than in pipercide (0.1 ppm), guineensine (0.89 ppm), and pellitorine(0.92 ppm). Piperine (5.1 ppm) was relatively ineffective. Against *A. togoi* larvae, retrofractamide A (0.01 ppm) was much more effective, compared with pipercide (0.26 ppm), pellitorine (0.71 ppm), and guineensine (0.75 ppm). Again, very low activity was observed with piperine (4.6 ppm). A new insecticidal amide piptigrine possessing highly extended conjugation was isolated from the dried ground seeds of *Piper nigrum* along with the known amides piperine and wisanine (Siddiqui et al. 2004a). Piptigrine exhibited toxicity of 15.0 ppm against fourth instar larvae of *Aedes aegypti* Liston. Pipsaeedine and pipbinine from ground pepper fruit exhibited toxicities of 45.0 and 40.0 ppm, respectively, against fourth instar larvae of *Aedes aegypti* (Siddiqui et al. 2004b). Pipnoohine and pipyahyine from dried ground whole pepper fruits exhibited toxicity at 35.0 and

30.0 ppm respectively against fourth instar larvae of *Aedes aegypti* (Siddiqui et al. 2004c).

The petroleum ether and ethyl acetate fractions of dried ground seeds of *Piper nigrum* afforded 16 compounds (1–16) including one new insecticidal amide, pipwaqarine elucidated as 1-[13-(3′,4′-methylenedioxyphenyl)-2E,4E,12E-tridecatrienoyl]-N-isopentylamide, and six constituents (3,4,6,7,11,15) (Siddiqui et al. 2005). Pipwaqarine exhibited toxicity of 30 ppm against fourth instar larvae of *Aedes aegypti*. Fractionation of *Piper nigrum* ethanol extract, yielded the larvicidal amides piperolein-A and piperine (Simas et al. 2007). Comparing LC_{50} values, the ethanol extract (0.98 ppm) was the most toxic, followed by piperolein-A (1.46 ppm) and piperine (1.53 ppm). Ethanolic extracts of dried fruits of three species of peppercorns: Long pepper, *Piper longum*, black pepper, *Piper nigrum*, and white pepper, *Piper nigrum* exhibited the larvicidal activity against the different instars of field-collected Indian strain of dengue fever mosquito *Aedes aegypti* (Kumar et al. 2010, 2011). Against early fourth instar, the ethanolic extracts of Black and White *P. nigrum* proved to be 30–40% less toxic than the extracts of *P. longum*, whereas against third instars, white pepper extracts exhibited 7% more efficacy than that of black pepper and 47% more toxicity than that of long pepper. The results also revealed that the extracts of all the three pepper species were 11–25 times more toxic against the third instar larvae as compared to the early fourth instars. It was observed, however, that the larvae of *Ae. aegypti* were most susceptible against a mixture of the three extracts when taken in 1:1:1. The larvae treated with all the pepper species showed initial abnormal behavior in their motion followed by excitation, convulsions, and paralysis, leading to 100% kill indicating delayed larval toxicity and effects of the extracts on the neuromuscular system.

N-isobutylamide alkaloids (pellitorine, guineensine, pipercide and retrofractamide A) derived from *P. nigrum* were found to have insecticidal activity (Park 2011). On the basis of 24-hour LD_{50} values, the compound most toxic to female *Culex pipiens pallens* was pellitorine (0.4 μg/female) followed by guineensine (1.9 μg/female), retrofractamide A (2.4 μg/female) and pipercide (3.2 μg/female). LD_{50} value of chlorpyrifos was 0.03 μg/female. Against female *Aedes aegypti*, the insecticidal activity was more pronounced in pellitorine (0.17 μg/female) than in retrofractamide A (1.5 μg/female), guineensine (1.7 μg/female), and pipercide (2.0 μg/female). LD_{50} value of chlorpyrifos was 0.0014 μg/female.

Traditional Medicinal Uses

Like all eastern spices, pepper was historically both a seasoning and a medicine. The pungency in pepper is due to the piperine compound. Black peppercorns have featured in traditional remedies in Ayurvedic, Unani and Siddha medicine in India for centuries and in Jamu preparations in Indonesia. It has been prescribed for illnesses such as constipation, diarrhoea, earache, gangrene, heart disease hernia, indigestion, insect bites, insomnia, joint pains, lung diseases, liver problems, tooth decay and toothache. Various sources also recommend pepper to treat eye problems, often by applying salves or poultices made with pepper directly to the eye.

Other Uses

Pepper oil is a constituent in certain perfumes and some famous brands includes 'Charlie' of Revlon and 'Poison' of Christian Dior. In the past, Egyptians used it in embalming mixture and also as an air-purifier. Its insecticidal properties are also being exploited for household use and agriculture.

Piper nigrum has insecticidal potential. Toxicity (mortality) of *Piper nigrum* extracts against the larva of *Spodoptera litura* in decreasing order were: hexane (LD_{50}: 1.8 mg/g) > acetone (LD_{50}: 18.8 mg/g) > chloroform (LD_{50}: NA, the toxicity was very low) > essential oil (no mortality) (Fan et al. 2011). Insect development and growth index observations showed that the hexane extract had antifeedant properties resulting in severe growth inhibition of *Spodoptera litura*.

Comments

Vietnam is the world's leading producer and exporter of pepper, producing 34% of the world's *Piper nigrum*. Other main producers include India (19%), Brazil (13%), Indonesia (9%), Malaysia (8%), Sri Lanka (6%), China (6%), and Thailand (4%).

Selected References

Agbor GA, Vinson JA, Oben JE, Ngogang JY (2006) Comparative analysis of the in vitro antioxidant activity of white and black pepper. Nutr Res 26(12):659–663

Agbor GA, Vinson JA, Oben JE, Ngogang JY (2007) In vitro antioxidant activity of three *Piper* species. J Herb Pharmacother 7(2):49–64

Agbor GA, Vinson JA, Sortino J, Johnson R (2010) Antioxidant and anti-atherogenic activities of three *Piper* species on atherogenic diet fed hamsters. Exp Toxicol Pathol doi:10.1016/j.etp.2010.10.003

Aher S, Biradar S, Gopu CL, Paradkar A (2009) Novel pepper extract for enhanced P-glycoprotein inhibition. J Pharm Pharmacol 61(9):1179–1186

Backer CA, Bakhuizen van den Brink RC Jr (1963) Flora of Java, vol 1. Wolter-Noordhoff, Groningen, 647 pp

Bae GS, Kim MS, Jung WS, Seo SW, Yun SW, Kim SG, Park RK, Kim EC, Song HJ, Park SJ (2010) Inhibition of lipopolysaccharide-induced inflammatory responses by piperine. Eur J Pharmacol 642(1–3):154–162

Bai YF, Xu H (2000) Protective action of piperine against experimental gastric ulcer. Acta Pharmacol Sin 21(4):357–359

Bai X, Zhang W, Chen W, Zong W, Guo Z, Liu X (2011) Anti-hepatotoxic and anti-oxidant effects of extracts from *Piper nigrum* L. root. Afr J Biotechnol 10(2):267–272

Bajad S, Bedi KL, Singla AK, Johri RK (2001) Piperine inhibits gastric emptying and gastrointestinal transit in rats and mice. Planta Med 67:176–179

Bezerra DP, de Castro FO, Alves AP, Pessoa C, de Moraes MO, Silveira ER, Lima MA, Elmiro FJ, Costa-Lotufo LV (2006) In vivo growth-inhibition of Sarcoma 180 by piplartine and piperine, two alkaloid amides from *Piper*. Braz J Med Biol Res 39(6):801–807

Bezerra DP, de Castro FO, Alves AP, Pessoa C, de Moraes MO, Silveira ER, Lima MA, Elmiro FJ, de Alencar NM, Mesquita RO, Lima MW, Costa-Lotufo LV (2008) In vitro and in vivo antitumor effect of 5-FU combined with piplartine and piperine. J Appl Toxicol 28(2):156–163

Bezerra DP, Militão GC, de Castro FO, Pessoa C, de Moraes MO, Silveira ER, Lima MA, Elmiro FJ, Costa-Lotufo LV (2007) Piplartine induces inhibition of leukemia cell proliferation triggering both apoptosis and necrosis pathways. Toxicol In Vitro 21(1):1–8

Bezerra DP, Pessoa C, de Moraes MO, Silveira ER, Lima MA, Elmiro FJ, Costa-Lotufo LV (2005) Antiproliferative effects of two amides, piperine and piplartine, from *Piper* species. Z Naturforsch C 60(7–8):539–543

Bhardwaj RK, Glaeser H, Becquemont L, Klotz U, Gupta SK, Fromm MF (2002) Piperine, a major constituent of black pepper, inhibits human P-glycoprotein and CYP3A4. J Pharmacol Exp Ther 302(2):645–650

Burkill IH (1966) A dictionary of the economic products of the Malay Peninsula. Revised reprint, 2 volumes. Ministry of Agriculture and Co-operatives, Kuala Lumpur, Malaysia, vol 1 (A–H), pp 1–1240, vol 2 (I–Z), pp 1241–2444

Capasso R, Izzo AA, Borrelli F, Russo A, Sautebin L, Pinto A, Capasso F, Mascolo N (2002) Effect of piperine, the active ingredient of black pepper, on intestinal secretion in mice. Life Sci 71:2311–2317

Chatterjee S, Niaz Z, Gautam S, Adhikari S, Variyar PS, Sharma A (2007) Antioxidant activity of some phenolic constituents from green pepper (*Piper nigrum* L.) and fresh nutmeg mace (*Myristica fragrans*). Food Chem 101(2):515–523

Choi BM, Kim SM, Park TK, Li G, Hong SJ, Park R, Chung HT, Kim BR (2007) Piperine protects cisplatin-induced apoptosis via heme oxygenase-1 induction in auditory cells. J Nutr Biochem 18(9):615–622

Chonpathompikunlert P, Wattanathorn J, Muchimapura S (2010) Piperine, the main alkaloid of Thai black pepper, protects against neurodegeneration and cognitive impairment in animal model of cognitive deficit like condition of Alzheimer's disease. Food Chem Toxicol 48(3):798–802

Chun H, Shin DH, Hong BS, Cho WD, Cho HY, Yang HC (2002) Biochemical properties of polysaccharides from black pepper. Biol Pharm Bull 25(9):1203–1208

Clery RA, Hammond CJ, Wright AC (2006) Nitrogen-containing compounds in black pepper oil (*Piper nigrum* L.). J Essent Oil Res 18(1):1–3

Council of Scientific and Industrial Research (CSIR) (1969) The wealth of India. A dictionary of Indian raw materials and industrial products (Raw materials 8). Publications and Information Directorate, New Delhi

D'Hooge R, Pei YQ, Raes A, Lebrun P, van Bogaert PP, de Deyn PP (1996) Anticonvulsant activity of piperine on seizures induced by excitatory amino acid receptor agonists. Arzneimittelforschung 46(6):557–560

Ee GC, Lim CM, Lim CK, Rahmani M, Shaari K, Bong CF (2009) Alkaloids from *Piper sarmentosum* and *Piper nigrum*. Nat Prod Res 23(15):1416–1423

Ee GC, Lim CM, Lim CK, Rahmani M, Shaari K, Bong CF (2010) Pellitorine, a potential anti-cancer lead compound against HL6 and MCT-7 cell lines and microbial transformation of piperine from *Piper nigrum*. Molecules 15(4):2398–2404

El Hamss R, Idaomar M, Alonso-Moraga A, Muñoz SA (2003) Antimutagenic properties of bell and black peppers. Food Chem Toxicol 41(1):41–47

Fan LS, Muhamad R, Omar D, Rahmani M (2011) Insecticidal properties of *Piper nigrum* fruit extracts and essential oils against *Spodoptera litura*. Int J Agric Biol 13:517–522

Ferreira SRS, Nikolov ZL, Doraiswamy LK, Merireles MAA, Petenate AJ (1999) Supercritical fluid extraction of black pepper (*Piper nigrum* L.) essential oils. J Supercrit Fluid 14(3):235–245

Fu M, Sun ZH, Zuo HC (2010) Neuroprotective effect of piperine on primarily cultured hippocampal neurons. Biol Pharm Bull 33(4):598–603

Fujiwara Y, Naithou K, Miyazaki T, Hashimoto K, Mori K, Yamamoto Y (2001) Two new alkaloids, pipercyclobutanamides A and B, from *Piper nigrum*. Tetrahedron Lett 42(13):2497–2499

Ganesh Bhat B, Chandrasekhara N (1987) Effect of black pepper and piperine on bile secretion and composition in rats. Nahrung 31:913–916

George CK, Abdullah A, Chapman K (2005) Pepper production guide for Asia and the Pacific. International Pepper Community and Food and Agricultural Organization of United Nations, Bangkok/Rome, p 219

Gülçin I (2005) The antioxidant and radical scavenging activities of black pepper (*Piper nigrum*) seeds. Int J Food Sci Nutr 56:491–499

Hamrapurkar PD, Jadhav K, Zine S (2011) Quantitative estimation of piperine in *Piper nigrum* and *Piper longum* using high performance thin layer chromatography. J Appl Pharm Sci 1(3):117–120

Han Y, Chin Tan TM, Lim LY (2008) In vitro and in vivo evaluation of the effects of piperine on P-gp function and expression. Toxicol Appl Pharmacol 230(3):283–289

Hirata N, Naruto S, Inaba K, Itoh K, Tokunaga M, Iinuma M, Matsuda H (2008) Histamine release inhibitory activity of *Piper nigrum* leaf. Biol Pharm Bull 31(10): 1973–1976

Hirata N, Tokunaga M, Naruto S, Iinuma M, Matsuda H (2007) Testosterone 5 alpha-reductase inhibitory active constituents of *Piper nigrum* leaf. Biol Pharm Bull 30(12):2402–2405

Hlavackova L, Urbanova A, Ulicna O, Janega P, Cerna A, Babal P (2010) Piperine, active substance of black pepper, alleviates hypertension induced by NO synthase inhibition. Bratisl Lek Listy 111(8):426–431

Hu S, Ao P, Tan H (1996) Pharmacognostical studies on the roots of *Piper nigrum* l. ii: chemical and pharmacological studies. Acta Hortic (ISHS) 426:175–178

Hwang YP, Yun HJ, Kim HG, Han EH, Choi JH, Chung YC, Jeong HG (2011) Suppression of phorbol-12-myristate-13-acetate-induced tumor cell invasion by piperine via the inhibition of PKCα/ERK1/2-dependent matrix metalloproteinase-9 expression. Toxicol Lett 203(1):9–19

Jagella T, Grosch W (1999a) Flavour and off-flavour compounds of black and white pepper (*Piper nigrum* L.). I. Evaluation of potent odorants of black pepper by dilution and concentration techniques. Eur Food Res Technol 209:16–21

Jagella T, Grosch W (1999b) Flavour and off-flavour compounds of black and white pepper (*Piper nigrum* L.). III. Desirable and undesirable odorants of white pepper. Eur Food ResTechnol 209:27–31

Jayalekshmy A, Menon AN, Padmakumari KP (2003) Essential oil composition of four major cultivars of black pepper (*Piper nigrum* L.). J Essent Oil Res 15:155–157

Jin J, Zhang J, Guo N, Feng H, Li L, Liang J, Sun K, Wu X, Wang X, Liu M, Deng X, Yu L (2011) The plant alkaloid piperine as a potential inhibitor of ethidium bromide efflux in *Mycobacterium smegmatis*. J Med Microbiol 60(Pt 2):223–229

Jin MJ, Han HK (2010) Effect of piperine, a major component of black pepper, on the intestinal absorption of fexofenadine and its implication on food-drug interaction. J Food Sci 75(3):H93–H96

Jirovetz L, Buchbauer G, Ngassoum MB, Geissler M (2002) Aroma compound analysis of *Piper nigrum* and *Piper guineense* essential oils from Cameroon using solid-phase microextraction-gas chromatography, solid-phase microextraction-gas chromatography-mass spectrometry and olfactometry. J Chromatogr A 976(1–2):265–275

Kaleem M, Sheema SH, Bano B (2005) Protective effects of *Piper nigrum* and *Vinca rosea* in alloxan induced diabetic rats. Indian J Physiol Pharmacol 49(1):65–71

Kapoor IP, Singh B, Singh G, De Heluani CS, De Lampasona MP, Catalan CA (2009) Chemistry and in vitro antioxidant activity of volatile oil and oleoresins of black pepper (*Piper nigrum*). J Agric Food Chem 57(12):5358–5364

Karthikeyan J, Rani P (2003) Enzymatic and non-enzymatic antioxidants in selected *Piper* species. Indian J Exp Biol 41(2):135–140

Khajuria A, Johrn RK, Zutshi U (1999) Piperine mediated alterations in lipid peroxidation and cellular thiol status of rat intestinal mucosa and epithelial cells. Phytomedicine 6(5):351–355

Khajuria A, Thusu N, Zutshi U, Bedi KL (1998) Piperine modulation of carcinogen induced oxidative stress in intestinal mucosa. Mol Cell Biochem 189(1–2):113–118

Kitano M, Wanibuchi H, Kikuzaki H, Nakatani N, Imaoka S, Funae Y, Hayashi S, Fukushima S (2000) Chemopreventive effects of coumaperine from pepper on the initiation stage of chemical hepatocarcinogenesis in the rat. Cancer Sci 91:674–680

Koul IB, Kapil A (1993) Evaluation of the liver protective potential of piperine, an active principle of black and long peppers. Planta Med 59(5):413–417

Kumar S, Warikoo R, Wahab N (2010) Larvicidal potential of ethanolic extracts of dried fruits of three species of peppercorns against different instars of an Indian strain of dengue fever mosquito, *Aedes aegypti* L. (Diptera: Culicidae). Parasitol Res 107(4):901–907

Kumar S, Warikoo R, Wahab N (2011) Relative larvicidal efficacy of three species of peppercorns against dengue fever mosquito, *Aedes aegypti* L. J Entomol Res Soc 13(2):27–36

Kumoro AC, Hasan M, Singh H (2010) Extraction of Sarawak black pepper essential oil using supercritical carbon dioxide. Arab J Sci Eng 35(2B):7–16

Lee SW, Rho MC, Park HR, Choi JH, Kang JY, Lee JW, Kim K, Lee HS, Kim YK (2006) Inhibition of diacylglycerol acyltransferase by alkamides isolated from the fruits of *Piper longum* and *Piper nigrum*. J Agric Food Chem 54(26):9759–9763

Lee SW, Kim YK, Kim K, Lee HS, Choi JH, Lee WS, Jun CD, Park JH, Lee JM, Rho MC (2008) Alkamides from the fruits of *Piper longum* and *Piper nigrum* displaying potent cell adhesion inhibition. Bioorg Med Chem Lett 18(16):4544–4546

Li S, Lei Y, Jia Y, li N, Wink M, Ma Y (2011) Piperine, a piperidine alkaloid from *Piper nigrum* re-sensitizes P-gp, MRP1 and BCRP dependent multidrug resistant cancer cells. Phytomed 19(1):83–87

Li S, Wang C, Li W, Koike K, Nikaido T, Wang MW (2007) Antidepressant-like effects of piperine and its derivative, antiepilepsirine. J Asian Nat Prod Res 9(3–5):421–430

Liang R, Shi S, Ma Y (2010) Analysis of volatile oil composition of the peppers from different production areas. Med Chem Res 19(2):157–165

Liao H, Liu P, Hu Y, Wang D, Lin H (2009) Antidepressant-like effects of piperine and its neuroprotective mechanism. Zhongguo Zhong Yao Za Zhi 34(12):1562–1565 (in Chinese)

Lin Z, Hoult JR, Bennett DC, Raman A (1999) Stimulation of mouse melanocyte proliferation by *Piper nigrum* fruit extract and its main alkaloid, piperine. Planta Med 65(7):600–603

Lin Z, Liao Y, Venkatasamy R, Hider RC, Soumyanath A (2007) Amides from *Piper nigrum* L. with dissimilar effects on melanocyte proliferation in-vitro. J Pharm Pharmacol 59(4):529–536

Liu L, Song G, Hu Y (2007) GC–MS analysis of the essential oils of *Piper nigrum* L. and *Piper longum* L. Chromatographia 66(9–10):785–790

Liu Y, Yadev VR, Aggarwal BB, Nair MG (2010) Inhibitory effects of black pepper (*Piper nigrum*) extracts and compounds on human tumor cell proliferation, cyclooxygenase enzymes, lipid peroxidation and nuclear transcription factor-kappa-B. Nat Prod Commun 5(8):1253–1257

Majdalawieh AF, Carr RI (2010) In vitro investigation of the potential immunomodulatory and anti-cancer activities of black pepper (*Piper nigrum*) and cardamom (*Elettaria cardamomum*). J Med Food 13(2): 371–381

Mao QQ, Xian YF, Ip SP, Che CT (2011) Involvement of serotonergic system in the antidepressant-like effect of piperine. Prog Neuropsychopharmacol Biol Psychiatry 35(4):1144–1147

Matsuda H, Kawaguchi Y, Yamazaki M, Hirat N, Naruto S, Asanuma Y, Kaihatsu T, Kubo M (2004) Melanogenesis stimulation in murine B16 melanoma cells by *Piper nigrum* leaf extract and its lignan constituents. Biol Pharm Bull 27(10):1611–1616

McNamara FN, Randall A, Gunthorpe MJ (2005) Effects of piperine, the pungent component of black pepper, at the human vanilloid receptor (TRPV1). Br J Pharmacol 144(6):781–790

Mehmood MH, Gilani AH (2010) Pharmacological basis for the medicinal use of black pepper and piperine in gastrointestinal disorders. J Med Food 13(5): 1086–1096

Menon AM, Padmakumari KP (2005) Studies on essential oil composition of cultivars of black pepper (*Piper nigrum* L.) – V. J Essent Oil Res 17:153–155

Menon AN, Padmakumari KP, Jayalekshmi A, Gopalakrishnan M, Narayanan CS (2000) Essential oil composition of four major cultivars of black pepper (*Piper nigrum* L.). J Essent Oil Res 12:431–434

Miyakado M, Nakayama I, Yoshioka H, Nakatani N (1979) The Piperaceae amides I: structure of pipercide, a new insecticidal amide from *Piper nigrum* L. Agric Biol Chem 43:1609–1611

Mori A, Kabuto H, Pei YQ (1985) Effects of piperine on convulsions and on brain serotonin and catecholamine levels in E1 mice. Neurochem Res 10(9):1269–1275

Mujumdar AM, Dhuley JN, Deshmukh VK, Raman PH, Naik SR (1990a) Anti-inflammatory activity of piperine. Jpn J Med Sci Biol 43(3):95–100

Mujumdar AM, Dhuley JN, Deshmukh VK, Raman PH, Thorat SL, Naik SR (1990b) Effect of piperine on pentobarbitone induced hypnosis in rats. Indian J Exp Biol 28(5):486–487

Muller CJ, Creveling RK, Jennings WG (1968) Some minor sesquiterpene hydrocarbons of black pepper. J Agric Food Chem 16(1):113–117

Muller CJ, Jennings CG (1967) Constituents of black pepper. Some sesquiterpene hydrocarbons. J Agric Food Chem 15(5):762–766

Musenga A, Mandrioli R, Ferranti A, D'Orazio G, Fanali S, Raggi MA (2007) Analysis of aromatic and terpenic constituents of pepper extracts by capillary electrochromatography. J Sep Sci 30(4):612–619

Nahak G, Sahu RK (2011) Phytochemical evaluation and antioxidant activity of *Piper cubeba* and *Piper nigrum*. J Appl Pharm Sci 1(8):153–157

Nakatani N, Inatani R (1981) Isobutyl amides from pepper (*Piper nigrum* L.). Agric Biol Chem 45:1473–1476

Nakatani N, Inatani R, Fuwa H (1980) Structures and syntheses of two phenolic amides from *Piper nigrum* L. Agric Biol Chem 44:2831–2836

Nakatani N, Inatani R, Ohta H, Nishioka A (1986) Chemical constituents of peppers (*Piper* spp.) and application to food preservation: naturally occurring antioxidative compounds. Environ Health Perspect 67:135–142

Nalini N, Sabitha K, Viswanathanp MVP (1998) Spices and glycoprotein metabolism in experimental colon cancer rats. Med Sci Res 26(11):781

Naseri MKG, Yahyavi H (2007) Spasmolytic activity of *Piper nigrum* fruit aqueous extract on rat non-pregnant uterus. Iran J Pharmacol Ther 6(1):35–40

Nussbaumer C, Cadalbert R, Kraft P (1999) Identification of m-Mentha3(8),6-diene (Isosylveterpinolene) in black pepper oil. Helv Chim Acta 82(1):53–58

Ononiwu IM, Ibeneme CE, Ebong OO (2002) Effects of piperine on gastric acid secretion on albino rats. Afr J Med Sci 31:293–295

Orav A, Stulova I, Kailas T, Müürisepp M (2004) Effect of storage on the essential oil composition of *Piper* nigrum L. fruits of different ripening states. J Agric Food Chem 52(9):2582–2586

Panda S, Kar A (2003) Piperine lowers the serum concentrations of thyroid hormones, glucose and hepatic 5′D activity in adult male mice. Horm Metab Res 35(9): 523–526

Park IK (2011) Insecticidal activity of isobutylamides derived from *Piper nigrum* against adult of two mosquito species. *Culex pipiens pallens* and *Aedes aegypti*. Nat Prod Res. doi:10.1080/14786419.2011.628178

Park IK, Lee SG, Shin SC, Park JD, Ahn YJ (2002) Larvicidal activity of isobutylamides identified in *Piper nigrum* fruits against three mosquito species. J Agric Food Chem 50(7):1866–1870

Pathak N, Khandelwal S (2007) Cytoprotective and immunomodulating properties of piperine on murine splenocytes: an in vitro study. Eur J Pharmacol 576(1–3): 160–170

Pathak N, Khandelwal S (2008) Comparative efficacy of piperine, curcumin and picroliv against Cd immunotoxicity in mice. Biometals 21(6):649–661

Pino JA, Aguero J, Fuentes V (2003) Chemical composition of the aerial parts of *Piper nigrum* L. from Cuba. J Essent Oil Res 15:209–210

Pino JA, Rodriguez-Feo G, Borges P, Rosado A (1990) Chemical and sensory properties of black pepper oil. Nahrung 34(6):555–560

Pradeep CR, Kuttan G (2002) Effect of piperine on the inhibition of lung metastasis induced B16F-10 melanoma cells in mice. Clin Exp Metastasis 19(8):703–708

Pradeep CR, Kuttan G (2003) Effect of piperine on the inhibition of nitric oxide (NO) and TNF-alpha production. Immunopharmacol Immunotoxicol 25(3):337–346

Purseglove JW (1968) Tropical crops: dicotyledons, vol 1 and 2. Longman, London, 719 pp

Rauscher FM, Sanders RA, Watkins JB 3rd (2000) Effects of piperine on antioxidant pathways in tissues from normal and streptozotocin-induced diabetic rats. J Biochem Mol Toxicol 14(6):329–334

Reddy SV, Srinivas P, Praveen B, Kishore KH, Raju CU, Murthy S, Rao JM (2004) Antibacterial constituents from the berries of *Piper nigrum*. Phytomedicine 11(7–8):697–700

Rho MC, Lee SW, Park HR, Choi JH, Kang JY, Kim K, Lee HS, Kim YK (2007) ACAT inhibition of alkamides identified in the fruits of *Piper nigrum*. Phytochemistry 68(6):899–903

Richard HM, Jennings WG (1971) Volatile composition of black pepper. J Food Sci 36(4):584–589

Russell GF, Jennings WG (1969) Constituents of black pepper. Oxygenated compounds. J Agric Food Chem 17(5):1107–1112

Sasidharan I, Menon AN (2010) Comparative chemical composition and antimicrobial activity of berry and leaf essential oils of *Piper nigrum* L. Int J Biol Med Res 1(4):215–218

Schulz H, Baranska M, Quilitzsch R, Schütze W, Lösing G (2005) Characterization of peppercorn, pepper oil, and pepper oleoresin by vibrational spectroscopy methods J. Agric Food Chem 53(9):3358–3363

Semler U, Gross GG (1988) Distribution of piperine in vegetative parts of *Piper nigrum*. Phytochemistry 27(5):1566–1567

Shwaireb MH, Wrba H, el-Mofty MM, Dutter A (1990) Carcinogenesis induced by black pepper (*Piper nigrum*) and modulated by vitamin A. Exp Pathol 40(4):233–238

Siddiqui BS, Begum S, Gulzar T, Farhat FN (1997) An amide from fruits of *Piper nigrum*. Phytochemistry 45(8):1617–1619

Siddiqui BS, Gulzar T, Begum S, Afshan F (2004a) Piptigrine, a new insecticidal amide from *Piper nigrum* Linn. Nat Prod Res 18(5):473–477

Siddiqui BS, Gulzar T, Begum S, Afshan F, Sattar FA (2004b) Two new insecticidal amide dimers from fruits of *Piper nigrum* Linn. Helv Chim Acta 87(3):660–666

Siddiqui BS, Gulzar T, Begum S, Afshan F, Sattar FA (2005) Insecticidal amides from fruits of *Piper nigrum* Linn. Nat Prod Res 19(2):143–150

Siddiqui BS, Gulzar T, Begum S, Afshan F, Sultana R (2008) A new natural product and insecticidal amides from seeds of *Piper nigrum* Linn. Nat Prod Res 22(13): 1107–1111

Siddiqui BS, Gulzar T, Mahmood A, Begum S, Khan B, Afshan F (2004c) New insecticidal amides from petroleum ether extract of dried *Piper nigrum* L. whole fruits. Chem Pharm Bull 52(11):1349–1352

Simas NK, Lima Eda C, Kuster RM, Lage CL, de Oliveira Filho AM (2007) Potential use of *Piper nigrum* ethanol extract against pyrethroid-resistant *Aedes aegypti* larvae. Rev Soc Bras Med Trop 40(4):405–407

Singh A, Rao AR (1993) Evaluation of the modulatory influence of black pepper (*Piper nigrum* L.) on the hepatic detoxication system. Cancer Lett 72(1–2):5–9

Singh G, Marimuthu P, Catalan C, de Lampasona MP (2004) Chemical, antioxidant and antifungal activities of volatile oil of black pepper and its acetone extract. J Sci Food Agric 84(14):1878–1884

Singh R, Singh N, Saini BS, Rao HS (2008) In vitro antioxidant activity of pet ether extract of black pepper. Indian J Pharmacol 40(4):147–151

Srinivas PV, Rao JM (1999) Isopiperolein B: an alkamide from *Piper nigrum*. Phytochemistry 52(5):957–958

Srinivasan K (2007) Black pepper and its pungent principle-piperine: a review of diverse physiological effects. Crit Rev Food Sci Nutr 47(8):735–748

Su HCF, Horvat R (1981) Isolation, identification, and insecticidal properties of *Piper nigrum* amides. J Agric Food Chem 29:115–118

Subehan UT, Kadota S, Tezuka Y (2006) Mechanism-based inhibition of human liver microsomal cytochrome P450 2D6 (CYP2D6) by alkamides of *Piper nigrum*. Planta Med 72(6):527–532

Tainter DR, Grenis AT (1993) Spices and seasonings. A food technology handbook. Wiley, New York, 226 pp

Taqvi SI, Shah AJ, Gilani AH (2008) Blood pressure lowering and vasomodulator effects of piperine. J Cardiovasc Pharmacol 52(5):452–458

Tewtrakul S, Hase K, Kadota S, Namba T, Komatsu K, Tanaka K (2000) Fruit oil composition of *Piper chaba*

Hunt, *Piper longum* L. and *Piper nigrum* L. J Essent oil Res 12(5):603–608

Traxler JT (1971) Piperanine, a pungent component of black pepper. J Agric Food Chem 19(6):1135–1138

Tsukamoto S, Cha BC, Ohta T (2002a) Dipiperamides A, B, and C: bisalkaloids from the white pepper *Piper nigrum* inhibiting CYP3A4 activity. Tetrahedron 58(9):1667–1671

Tsukamoto S, Tomise K, Miyakawa K, Cha BC, Abe T, Hamada T, Hirota H, Ohta T (2002b) CYP3A4 inhibitory activity of new bisalkaloids, dipiperamides D and E, and cognates from white pepper. Bioorg Med Chem 10(9):2981–2985

U.S. Department of Agriculture, Agricultural Research Service (2011) USDA National Nutrient Database for standard reference, release 24. Nutrient Data Laboratory Home Page. http://www.ars.usda.gov/ba/bhnrc/ndl

Vijayakumar RS, Nalini N (2006a) Efficacy of piperine, an alkaloidal constituent from *Piper nigrum* on erythrocyte antioxidant status in high fat diet and antithyroid drug induced hyperlipidemic rats. Cell Biochem Funct 24(6):491–498

Vijayakumar RS, Nalini N (2006b) Piperine, an active principle from *Piper nigrum*, modulates hormonal and apo lipoprotein profiles in hyperlipidemic rats. J Basic Clin Physiol Pharmacol 17(2):71–86

Vijayakumar RS, Surya D, Nalini N (2004) Antioxidant efficacy of black pepper (*Piper nigrum* L.) and piperine in rats with high fat diet induced oxidative stress. Redox Rep 9:105–110

Voesgen B, Herrmann K (1980) Flavonol glycosides of pepper (*Piper nigrum* L.), clove (*Syzygium aromaticum* (L.) Merr. and Perry), and allspice (*Pimenta dioica* (L.) Merr.). 3. Spice phenols. Z Lebensm Unters Forsch 170(3):204–207 (in German)

Wattanathorn J, Chonpathompikunlert P, Muchimapura S, Priprem A, Tankamnerdthai O (2008) Piperine, the potential functional food for mood and cognitive disorders. Food Chem Toxicol 46(9):3106–3110

Wei K, Li W, Koike K, Pei Y, Chen Y, Nikaido T (2004a) Complete 1H and 13C NMR assignments of two phytosterols from roots of *Piper nigrum*. Magn Reson Chem 42(3):355–359

Wei K, Li W, Koike K, Pei Y, Chen Y, Nikaido T (2004b) New amide alkaloids from the roots of *Piper nigrum*. J Nat Prod 67(6):1005–1009

Wei K, Li W, Koike K, Pei Y, Chen Y, Nikaido T (2005) Nigramides A-S, dimeric amide alkaloids from the roots of *Piper nigrum*. J Org Chem 70(4):1164–1176

Wood AB, Barrow ML, James DJ (1988) Piperine determination in pepper (*Piper nigrum* L.) and its oleoresins – a reversed-phase high-performance liquid chromatographic method. Flavour Fragr J 3:55–64

Wrba H, el-Mofty MM, Schwaireb MH, Dutter A (1992) Carcinogenicity testing of some constituents of black pepper (*Piper nigrum*). Exp Toxicol Pathol 44(2):61–65

Wrolstad RE, Jennings WG (1965) Volatile constituents of black pepper. III. The monoterpene hydrocarbon fraction. J Food Sci 30:274–279

Piper retrofractum

Scientific Name

Piper retrofractum Vahl

Synonyms

Chavica maritima Miquel, *Chavica officinarum* Miquel, *Chavica peepuloides* Wight, *Chavica retrofracta* (Vahl) Miq., *Piper chaba* Hunter, *Piper longum* Blume non L., *Piper officinarum* (Miquel) C. DC.

Family

Piperaceae

Common Names

Balinese Pepper, Jaborandi Pepper, Java Pepper, Long Pepper, Java Long Pepper, Javanese Long Pepper.

Vernacular Names

Chinese: Jia Bi Ba, Bi Ba, Zhao Wa Chang Guo Hu Jiao;
Danish: Medicinsk Peber;
Dutch: Javaanse Lange Peper;
Eastonian: Jaava Pipar;
French: Piovre Long De Java, Poivre Long;
German: Balinesischer Pfeffer, Bengalischer Pfeffer, Jawanischer Langer Pfeffer, Jaborandi-Pfeffer, Langer Pfeffer, Stangenpfeffer;
Hungarian: Bali Bors;
India: Chabya, Chai (Bengali), Pippal (Gujerati), Chab, Chavi, Gajj Pippal, Gajphal, Gaj-Pipal, Pippal, Pippli (Hindu), Chavya (Kannada), Chavyam (Malayalam), Kankala, Chabchini, Chavala, Miravela (Marathi), Cavika, Cavya, Cavyaka, Chavaka, Chavana, Chavi, Chaviaka, Chavika, Chavya, Chawika, Chuve, Gajapippaleemoola, Gajapippali, Gandhanakuli, Hasti, Hastipippali, Katuka, Katukapini, Kola, Kolaka, Kolavalli, Kolavallika, Krikara, Kutilasaptaka, Nakuli, Purandara, Sainhali Sreyasi, Tikshnakarikanavali, Uchhishta, Ushana, Ushanah, Vashira (Sanskrit), Anai Tippili, Anaiddippili, Cavikai, Caviyam, Chavyam, Kantirai, Milaku Ver, Tippali, Vanapippili (Tamil), Chaikama, Saevamu, Saevasu (Telugu), Peepal Chab, Kabab Chini, Chab, Kankol Mirch (Urdu);
Indonesia: Lada Panjang (Sumatra) Cabia (Sulawesi) Cabe Jamu, Cabe Jawa, Cabe Panjang, Chabean, Chabe Alas, Chabe Sula (Javanese) Cabé Bali;
Malaysia: Ladang Panjang, Lada Sulah, Cabai Jawa, Bakek, Kedawak;
Philippines: Amaras, Kamara (Iloko), Boyo-Boyo (Tagbanua), Kayuñgo, Litlit, Sabia, Salimara, Soag-matsing, Subon-manok (Tagalog);
Thailand: Dipli, Dipli-Chuak;
Tibetan: Dbyi Mon Dkar Po, Lca Ba, Dbyi Mo I Lo Ma, Dbyi Moi Lo Ma, Tsa-Bya;

Swedish: Långpeppar;
Vietnam: Tiêu Dội, Tiêu Long.

Origin/Distribution

Java Long Pepper occurs wild in Indonesia, Malaysia, Philippines, Thailand and Vietnam. It is also found in the Ryuku Islands and Yunnan, and is cultivated in Guangdong, China and Kampuchea. Long pepper is cultivated only to a limited extent on Java, Bali and some neighbouring islands, as the species occurs in sufficient quantities in the wild.

Plate 1 Leaves and developing leaf-opposed inflorescences

Agroecology

Java Long pepper grows wild on trunks in the shade or partial shade in open humid deciduous tropical forest and thickets on poor soils and along beaches from sea level to 600 m elevation. Cultivated in backyards in villages in southeast Asia.

Edible Plant Parts and Uses

Dried unripe and ripe infructescences are used as a spice for curries, preserves and pickles. Both dry and unripe spikes are available in the local wet markets in Java.

Plate 2 Dried Java long pepper infructescences

Botany

A dioecious, perennial, soft woody climbers or creeper, glabrous except for rachis and stigmas and with clasping roots. Stems brownish when dry, 2 mm thick, cylindrical and striated. Petiole 5–11 mm, long; leaf blade narrowly elliptic, ovate-oblong, or elliptic, 8.5–16 × 3.2–7.5 cm, coriaceous, glabrous, densely glandular, base with both sides rounded or asymmetric and oblique (Plate 1). Spikes erect or patent and leaf-opposed (Plate 1). Male spikes 5–6.5 cm; peduncle slightly longer than petioles; bracts orbicular, 1–1.2 mm wide, peltate and sessile. Stamens 2 or 3; filaments very short to nearly absent; anthers broadly ellipsoid. Female spikes 3–4 cm × 7–10 mm; peduncle and bracts as in male spikes. Ovary immersed in rachis; stigmas 3, ovate-acute, recurved and persistent. Infructescences cylindrical, 2–4 cm by 4–8 mm on a 1 cm stalk; berries, spherical, arranged connate on rachis, apex rounded, hard green, pungent when unripe becoming reddish-brown and soft when ripe (Plates 2 and 3). Seed globose, small 2–2.5 cm across.

Nutritive/Medicinal Properties

Two piperidine alkaloids, piperoctadecalidine and pipereicosalidine were isolated from the fruits of *Piper retrofractum* along with known piperidine

Plate 3 Close-up of Java long pepper infructescences

alkaloids, guineensine, piperine and pipernonaline (Ahn et al. 1992). The aerial parts of *Piper retrofractum* was found to contain amides such as retrofractamide-D, retrofractamide A (*N*-isobutyl-9(3′,4′-methylenedioxyphenyl) 2*E*,4*E*,8*E*-nonatrienamide) and retrofractamide C.(Banerji et al. 1985, 2002). The plant also contained sesamin and 3,4,5-trimethoxydihydrocinnamic acid as well as two higher homologues of retrofractamide A, viz. pipericide (retrofractamide B) and retrofractamide D. A new lignan, epimers of [8R,8′R]-9-hydroxy,3,4-dimethoxy,3′,4′-methylene dioxy-9,9′ epoxy lignan, was isolated from the chloroform-soluble fraction of a crude alcoholic extract of *Piper chaba* (Bhandari et al. 1998). Two new dimeric alkaloids, chabamide F (1) and chabamide G (2), containing pyrrolidine rings, were isolated from the roots of *Piper chaba* (Rao et al. 2009). A new amide, piperchabamide F, and two new phenylpropanoid glycosides, piperchabaosides A and B, were isolated from 80% aqueous acetone extract from fruit of *Piper chaba* (Morikawa et al. 2009).

Other constituents from the fruit, stem bark and roots are discussed below in association with the pharmacological properties.

Antioxidant Activity

Studies reported seven constituents from *P. retrofractum* fruits: piperic acid, β-sitosterol and phenolic amides, guineensine, pellitorine, piperine, methyl piperate and N-siobutyl-2E, 4E, 8Z-eicosatrienamideguineensine (Nakatani et al. 1986). All the phenolic amides compounds were found to possess significant antioxidant activities that were more effective than the naturally occurring antioxidant, α-tocopherol. Naturally occurring antioxidants, therefore, may surpass food preservatives like BHA (butylated hydroxyanisole) and BHT (butylated hydroxytoluene) in their ability to inactivate mutagens in food.

Hepatoprotective Activity

Methanolic extract from the fruit was found to have a hepatoprotective effect on D-galactosamine (D-GalN)/lipopolysaccharide (LPS)-induced liver injury in mice (Matsuda et al. 2008). From the ethyl acetate-soluble fraction, a new amide constituent named piperchabamide E together with twenty known amide constituents (e.g., piperine, piperchabamides A-D, and piperanine) and two aromatic constituents were isolated as the hepatoprotective constituents. The 1,9-decadiene structure between the benzene ring and amide moiety were suggested to be important for strong inhibition of D-GalN/tumour necrosis factor-α (TNF-α)-induced death of hepatocytes. Further a principal amide constituent, piperine, dose-dependently inhibited increase in serum GPT (glutamic-pyruvic transaminase) and GOT (glutamic-oxaloacetic transaminase) levels at doses of 2.5–10 mg/kg (p.o.) in D-GalN/LPS-treated mice, and this inhibitory effect was suggested to depend on the reduced sensitivity of hepatocytes to TNF-α. The aqueous acetone (80%) extract from fruit of *Piper chaba* was found to have a hepatoprotective effect on D-GalN/LPS-induced liver injury in mice (Matsuda et al. 2009). Among the isolates, several amide constituents inhibited D-GalN/tumour necrosis factor-α (TNF-α)-induced death of hepatocytes, and the following structural requirements were suggested: (i) the amide moiety was essential for strong activity; (ii) the 1,9-decadiene structure between the benzene ring and the amide moiety tended to enhance the activity. Moreover, a principal constituent, piperine, exhibited strong in-vivo hepatoprotective effect at a dose of 5 mg/kg, p.o. and its mode of action was suggested to depend on the reduced sensitivity of hepatocytes to TNF-α.

Anticancer Activity

Studies demonstrated that E-piplartine isolated from the roots to be a promising candidate to use in combinatorial treatments to combat cancer (Jyothi et al. 2009). While Z-piplartine (cis-piplartine) failed to induce cytotoxicity (even at high concentrations, 50 µM), E-piplartine induced a dose-dependent cytotoxicity (2–24 µM) in different tumour cells: rat histiocytoma (BC-8), mouse embryonal carcinoma (PCC4), mouse macrophages (P388D1 and J774), and human neuroblastoma (IMR32) tumour cells. The combinatorial treatment of piplartine with diferuloylmethane (curcumin), an antiinflammatory and anticancer agent, significantly enhanced the piplartine induced cytotoxicity in tumour cells. Diferuloylmethane itself was not cytotoxic at 15 µM concentration; however, potentiated the piplartine induced cytotoxicity. The tumour cell killing with piplartine was preceded by G1 cell cycle arrest, and surpassed diferuloylmethane induced G2/M arrest when used in combination. Seven Thai medicinal plants including *P. chaba* exhibited promising in-vitro cytoxic activity against the human cholangiocarcinoma CL-6 cell line with survival of less than 50% at the concentration of 50 µg/ml (Mahavorasirikul et al. 2010). *P. chaba* had potent cytotoxic activity with mean IC_{50} value of 40.74 µg/ml. All possessed high activity against human laryngeal (Hep-2) cell with mean IC_{50} ranging from 18.93 to 32.40 µg/ml. The extract from *Piper chaba* (IC_{50} = 18.63 µg/ml, SI (selective index) = 9.8) and Pra-Sa-Prao-Yhai recipe (IC_{50} = 20.99 µg/ml, SI = 12.5.) exhibited the most promising and most selective cytotoxic activity against Hep-2 cell line.

Gastroprotective Activity

The fruit also possess gastroprotective activity. From the aqueous acetone extract of the fruit, four new amides named piperchabamides A (1), B (2), C (3), and D (4) were isolated (Morikawa et al. 2004) In addition, the gastroprotective effects of the principal constituents, piperine (5), piperanine (6), pipernonaline (7), dehydropipernonaline (8), piperlonguminine (9), retrofractamide B (10), guineensine (11), N-isobutyl-(2 E,4 E)-octadecadienamide (12), N-isobutyl-(2 E,4 E,14 Z)-eicosatrienamide (13), and methyl piperate (14). Of these, compounds 5–10 and 12–14 significantly inhibited ethanol-induced gastric lesions at a dose of 25 mg/kg, p. o., while 5, 7, 8, 10, 12, and 13 also significantly inhibited indomethacin-induced gastric lesions at the same dose.

Antiobesity Activity

Kim et al. (2011) isolated piperidine alkaloids from *P. retrofractum* fruits (PRPA), including piperine, pipernonaline, and dehydropipernonaline, as the antiobesity constituents through a peroxisome proliferator-activated receptor δ (PPARδ) transactivation assay. PRPA treatment activated AMP-activated protein kinase (AMPK) signalling and PPARδ protein and also regulated the expression of lipid metabolism-related proteins in 3 T3-L1 adipocytes and L6 myocytes. In the animal model, oral PRPA administration (50, 100, or 300 mg/kg/day for 8 weeks) significantly reduced high-fat diet – induced body weight gain without altering the amount of food intake. Fat pad mass was reduced in the PRPA treatment groups, as evidenced by reduced adipocyte size. In addition, elevated serum levels of total cholesterol, low-density lipoprotein cholesterol, total lipid, leptin, and lipase were suppressed by PRPA treatment. PRPA also protected against the development of nonalcoholic fatty liver by decreasing hepatic triglyceride accumulation. Consistent with the in-vitro results, PRPA activated AMPK signaling and altered the expression of lipid metabolism-related proteins in liver and skeletal muscle. The findings demonstrated that PRPA attenuated high-fat diet-induced obesity by activating AMPK and PPARδ, and regulated lipid metabolism, suggesting their potential antiobesity effects.

Antileishmanial Activity

Piper retrofractum also possess antileishmanial activity (Bodiwala et al. 2007). The n-hexane, ethyl acetate, methanol, and acetone extracts of *Piper cubeba* and *P. retrofractum* exhibited significant in vitro activity at 100 μg/ml against promastigotes of *Leishmania donovani*. Two lignans, cubebin and hinokinin, were isolated from the hexane extract of *P. cubeba*; and one bis-epoxy lignan, (−)-sesamin, and two amides, pellitorine and piplartine were isolated from the hexane and methanol extracts of *P. retrofractum*. Cubebin and piplartine showed significant antileishmanial activity in vitro at 100 μM and were further tested in-vivo in a hamster model of visceral leishmaniasis. Piplartine showed activity at 30 mg/kg dose.

Larvicidal Activity

Piper retrofractum showed significantly high level of in-vitro activity against 3rd and 4th instar larvae of the mosquito *Culex quinquefasciatus* and *Aedes aegypti* (Chansang et al. 2005). Extracts of unripe and ripe fruits showed different levels of activity against *Culex quinquefasciatus* larvae. Both extracts were equi-toxic to a *Bacillus sphaericus* resistant and susceptible strains, both from Thailand. The ripe fruit extract was somewhat more active against *Aedes aegypti* than *Culex quinquefasciatus*. Another ripe fruit extract was much more toxic to both mosquito species.

Other Pharmacological Activities

Chabamide, a novel piperine dimer, isolated from the stems showed antimalarial activity with an IC_{50} value of 2.7 μg/ml and antituberculosis activity with the minimum inhibitory concentration (MIC) of 12.5 μg/ml (Rukachaisirikul et al. 2002).

Another study reported that the methanol extract of the stem bark displayed analgesic, antiinflammatory, diuretic, antidiarrhoeal, effect on gastrointestinal motility and CNS depressant activity in mice and rat at 125, 250 and 500 mg/kg body weight doses (Taufiq-Ur-Rahman et al. 2005). The extract at given doses significantly and dose dependently reduced the frequency of acetic acid induced writhing in mice, prolonged the tail flicking latency in mice, reduced carrageenan-induced paw edema volume in rat, delayed the onset as well as reduced the frequency of castor oil induced diarrhoeal episodes in mice. It also decreased gastrointestinal motility and prolonged pentobarbitone induced sleeping time in mice. However at the same doses, the extract exhibited moderate diuretic activity only at the highest dose.

Toxicity Studies

Studies showed that the aqueous extract of *P. chaba* fruits did not produce acute or subchronic toxicity in either female or male rats at doses of 5,000 mg/kg body weight for acute toxicity and 300–1,200 for subchronic toxicity (Jaijoy et al. 2010). The results showed no abnormalities in treated groups as compared to the controls. Neither gross abnormalities nor histopathological changes were observed

Traditional Medicinal Uses

Java long pepper has various applications in traditional medicine in southeast Asia. Immature spikes and ripe fruits are used as spice and remedy. The latter contains also dried parts of the rhizome or roots. In Indonesia, the fruits are used for their anti-flatulent, expectorant, antitussive, antifungal, and appetizing properties in traditional medicine. In Malaysia, the aromatic spice is taken as tonic for languidness and after childbirth, for digestive and intestinal disorders and administered internally for degenerative organs, for cramps, for congestion of the liver, and ulceration of the bones. It is also prescribed for the skin irritation, numbness and in fevers as well as for nostril ulcerations. In Vietnam, the fruit is regarded to have warming properties and to be good for digestion. The plant is used to treat liver diseases, jaundice and oedema. The plant is crushed and used as poultice for pains and rheumatism. In the Philippines, the root is chewed and the saliva swallowed, or the root is brewed in decoction as a cure for colic. It is also good for dyspepsia and gastralgia. In Thailand, its fruits have a "hot" flavour that is used to aid food digestion, blood circulation, asthma, overall health, treat influenza and hypertension, and act as an antiflatulent. When combined with *Piper nigrum* and "khing haeng" (*Zingiber* sp.), it is an important ingredient of trikatu medical prescriptions within Thai traditional medicine. Trikatu in Sanskrit means "Three hot flavoured items". They are used to improve four body elements (earth, water, wind and fire), food digestion, and food nutrient absorption and to act as an antitussive and diaphoretic, and also to treat influenza and flatulence. In India, the fruit is considered stimulant, anti-catarrhal and carminative and are much used as adjuncts to medicines for cough, cold and hoarseness.

Other Uses

Java long pepper is an important ingredient in various formulations of *jamu* (Indonesian traditional medicine), thus is also called *cabe jamu*. It is equally widely used in *jamu* and as a culinary spice in Indonesia.

Comments

Java long pepper is more pungent than Indian long pepper and black pepper. The best time to harvest Java long pepper is when the green spike turns reddish at the top and the spike is then sun-dried for several days.

Selected References

Ahn JW, Ahn MJ, Zee OP, Kim EJ, Lee SG, Kim HJ, Kubo I (1992) Piperidine alkaloids from *Piper retrofractum* fruits. Phytochemistry 31(10):3609–3612

Atal CK, Ojha JN (1965) Studies on the genus *Piper*. Part IV. Long peppers of Indian commerce. Econ Bot 19:157–164

Backer CA, Bakhuizen van den Brink RC Jr (1963) Flora of Java, vol 1. Wolter-Noordhoff, Groningen, 647pp

Banerji A, Bondopadhyay D, Sarkar M, Siddhanta AK, Pal SC, Ghosh S, Abraham D, Shoolery JN (1985) Structural and synthetic studies on the retrofractamides - amide constituents of *Piper retrofractum*. Phytochemistry 24:279–284

Banerji A, Sarkar M, Datta R, Sengupta P, Abraham K (2002) Amides from *Piper brachystachyum* and *Piper retrofractum*. Phytochemistry 59(8):897–901

Bhandari SPS, Babu UV, Garg HS (1998) A lignan from *Piper chaba* stems. Phytochemistry 47(7):1435–1436

Bodiwala HS, Singh G, Ranvir Singh R, Dey CS, Sharma SS, Bhutani KK, Pal I (2007) Antileishmanial amides and lignans from *Piper cubeba* and *Piper retrofractum*. J Nat Med 61(4):418–421

Burkill IH (1966) A dictionary of the economic products of the Malay Peninsula. Revised reprint. 2 vols. Ministry of Agriculture and Co-operatives, Kuala Lumpur, vol 1 (A–H), pp 1–1240, vol 2 (I–Z), pp 1241–2444

Chansang U, Zahiri NS, Bansiddhi J, Boonruad T, Thongsrirak P, Mingmuang J, Benjapong N, Mulla MS (2005) Mosquito larvicidal activity of aqueous extracts of long pepper (*Piper retrofractum* Vahl) from Thailand. J Vector Ecol 30(2):195–200

Chaveerach A, Piya Mokkamul P, Sudmoon R, Tanee T (2006) Ethnobotany of the genus *Piper* (Piperaceae) in Thailand. Ethnobot Res Appl 4:223–231

ENVIS (2010) Medicinal plants of conservation concern. Environmental information System (ENVIS) Centre, Foundation for Revitalisation of Local Health Traditions (FRLHT), Ministry of Environment and Forests, India. http://envis.frlht.org/about_envis.htm

Jaijoy K, Vannasiri S, Piyabhan P, Lerdvuthisopon N, Boonraeng S, Khonsung P, Lertprasertsuke N, Sireeratawong S (2010) Acute and subchronic toxicity study of the water extract from the fruits of *Piper chaba* Hunter in rats. Int J Appl Res Nat Prod 3(4):29–35

Jyothi D, Vanathi P, Mangala Gowri P, Rama Subba Rao V, Madhusudana Rao J, Sreedhar AS (2009) Diferuloylmethane augments the cytotoxic effects of piplartine isolated from *Piper chaba*. Toxicol In Vitro 23(6):1085–1091

Kim KJ, Lee MS, Jo K, Hwang JK (2011) Piperidine alkaloids from *Piper retrofractum* Vahl. protect against high-fat diet-induced obesity by regulating lipid metabolism and activating AMP-activated protein kinase. Biochem Biophys Res Commun 411(1):219–225

Mahavorasirikul W, Viyanant V, Chaijaroenkul W, Itharat A, Na-Bangchang K (2010) Cytotoxic activity of Thai medicinal plants against human cholangiocarcinoma, laryngeal and hepatocarcinoma cells in vitro. BMC Complement Altern Med 10:55

Matsuda H, Ninomiya K, Morikawa T, Yasuda D, Yamaguchi I, Yoshikawa M (2008) Protective effects of amide constituents from the fruit of *Piper chaba* on D-galactosamine/TNF-α-induced cell death in mouse hepatocytes. Bioorg Med Chem Lett 18(6): 2038–2042

Matsuda H, Ninomiya K, Morikawa T, Yasuda D, Yamaguchi I, Yoshikawa M (2009) Hepatoprotective amide constituents from the fruit of *Piper chaba*: structural requirements, mode of action, and new amides. Bioorg Med Chem 17(20):7313–7323

Morikawa T, Matsuda H, Yamaguchi I, Pongpiriyadacha Y, Yoshikawa M (2004) New amides and gastroprotective constituents from the fruit of *Piper chaba*. Planta Med 70(2):152–159

Morikawa T, Yamaguchi I, Matsuda H, Yoshikawa M (2009) A new amide, piperchabamide F, and two new phenylpropanoid glycosides, piperchabaosides A and B, from the fruit of *Piper chaba*. Chem Pharm Bull(Tokyo) 57(11):1292–1295

Nakatani N, Inatani R, Ohta H, Nishioka A (1986) Chemical constituents of peppers (*Piper* spp.) and application to food preservation: naturally occurring antioxidative compounds. Environ Health Perspect 67:135–142

Purseglove JW (1968) Tropical crops: dicotyledons. 1 & 2. Longman, London, 719pp

Rao VRS, Kumar GS, Sarma VUM, Raju SS, Babu KH, Babu KS, Babu TH, Rekha K, Rao JM (2009) Chabamides F and G, two novel dimeric alkaloids from the roots of *Piper chaba* Hunter. Tetrahedron Lett 50(23):2774–2777

Rukachaisirikul T, Prabpai S, Champung P, Suksamrarn A (2002) Chabamide, a novel piperine dimer from stems of *Piper chaba*. Planta Med 68(9):853–855

Tanaka Y, Nguyen VK (2007) Edible wild plants of Vietnam: the bountiful garden. Orchid Press, Bangkok, 175pp

Taufiq-Ur-Rahman M, Shilpi JA, Ahmed M, Faiz Hossain C (2005) Preliminary pharmacological studies on *Piper chaba* stem bark. J Ethnopharmacol 99(2):203–209

Tseng YC, Xia N, Gilbert MG (1999) Piperaceae C. Agardh. In: Wu ZY, Raven PH (eds) Flora of China, vol 4 (Cycadaceae through Fagaceae). Science Press/Missouri Botanical Garden Press, Beijing/St. Louis

Uphof JC Th (1968) Dictionary of economic plants, 2nd edn. (1st edn. 1959) Cramer, Lehre, 591pp

Utami D, Jansen PCM (1999) *Piper* L. In: de Guzman CC, Siemonsma JS (eds) Plant resources of South-East Asia No 13 Spices. Backhuys Publishers, Leiden, pp 183–188

Amelanchier alnifolia

Scientific Name

Amelanchier alnifolia (Nutt.) Nutt. Ex M. Roem.

Synonyms

Amelanchier canadensis var. *alnifolia* (Nutt.) Torr. & A. Gray, *Amelanchier carrii* Rydb., *Amelanchier leptodendron* Lunell, *Amelanchier macrocarpa* Lunell, *Amelanchier sanguinea* var. *alnifolia* (Nutt.) P. Landry, *Aronia alnifolia* Nutt.

Family

Rosaceae

Common/English Names

Alder Leaf Shadbush, Mountain Juneberry, Pacific Serviceberry, Rocky Mountain Blueberry, Saskatoon, Saskatoon Berry, Saskatoon Serviceberry, Serviceberry, Sarvisberry, Shadbush, Juneberry, Western Serviceberry, Western Shadbush.

Vernacular Names

Canada: Hlighag, Saskatoon, Sgan Gam (Indigenous);
Czech: Muchovník Olšolistý;
Danish: Ellebladet Bærmispel;
Eastonian: Lepalehine Toompihlakas;
Finnish: Marjatuomipihlaja;
French: Amélanchier À Feuilles D'aulne;
German: Erlenblättrige Felsenbirne;
Icelandic: Hlíðaramall;
Norwegian: Taggblåhegg;
Swedish: Bärhäggmispel, Grovsågad Häggmispel, Sen Häggmispel, Västamerikansk Häggmispel.

Origin/Distribution

This species is native to North America from Alaska, Western Canada and western (southwards to North California, Utah and Colorado) and north central Unites States. In Canada, the species is found in British Columbia, Alberta, Saskatchewan, Manitoba, Ontario, Quebec, North West Territories and Nunavut.

Agroecology

In its native range, it is found in thickets, woodland margins, banks of streams, canyons and hill-sides from plains to subalpine, from sea level to 3,000 m altitude. It is frost resistant to −20°C. It prefers a rich, well-drained loamy soil in but will grow in any sandy or clayey soil that is not water-logged or too dry. It is quite drought tolerant and is also salt tolerant. It thrives in a sunny position or semi-shade.

Edible Plant Parts and Uses

Ripe fruit is edible and is sweet with a hint of apple. The ripe fruit is eaten fresh out of hand or baked into pies, pastries, processed into preserves, jams, jellies, spreads or dried and used like raisins in cereals and snack food. The fruits are also made into cider, wine, beer, or tea. The Canadian indigenous people used the fruit in soups, stews, meat dishes, dried cakes and in a dried meat preparation called *pemmican* to which serviceberries are incorporated to impart flavour and to act as preservative. Saskatoon berry juice was used to marinate other foods such as black tree lichen or roots to sweeten them (Turner 1997). The advent of new and innovative methods of processing, freezing and packaging have greatly increased fruit uses and growers in Canada are promoting it as a superfruit like the berry fruits.

The leaves are used as a substitute for tea.

Botany

A deciduous, multi-stemmed shrub or small tree, mostly 1–5 m high with a fastigiate crown and smooth ashy-gray bark. Twigs are slender, reddish-brown becoming glabrous during flowering. Leaves are alternate, simple, oval-obovate to nearly round, 2–5 cm long by 1.8–4 cm wide, pinnately-veined with a rounded to sub-acute apex, rounded bases, finely serrated margin, and born on 1–2 cm long petioles (Plate 1). Very young leaves are conduplicate in bud; less than half-expanded and unfolded, pubescent abaxially. Inflorescences in erect racemes with 4–20-flowers. Flower has a campanulate hypanthia, pubescent sepals erect or loosely spreading after flowering, 5 white broadly linear to oblong, 1–2 cm petals, 15–20 stamens, glabrous to pubescent ovary with 5 styles. Fruit a pome, globose to subglobose, 7–15 mm across, glabrous, wax-coated, maroon-purple, juicy and sweet (Plate 1).

Plate 1 Leaves and young developing fruits

Nutritive/Medicinal Properties

Saskatoon berries appear to be an excellent source of manganese, magnesium and iron and a relatively good source of calcium, potassium, copper, and carotene (Mazza 1982). Proximate nutrient value of raw Saskatoon berries per 100 g edible portion (dwb) was reported as: water 80 g, protein 9.7 g, fat 4.2 g, fibre 19 g, total sugars 11.4 g, Ca 0.44 g, P 0.16 g, Mg 0.2 mg, S 0.06 g, Fe 67.5 pp, NA 31.8 ppm, Mn 67.5 ppm, Cu 7.23 ppm, Zn 16.5 ppm, Ba 34.8 mg, Mo 0.38 mg, Al 74.5 mg, carotene 29.7 ppm (Mazza 1982). In a subsequent paper, Mazza (2005) reported the following nutrient composition in raw saskatoon berries per 100 g value, total dietary fibre 5.9 g, total sugars 11.4 g, Ca 42 mg, Fe 1 mg, Mn 1.4 mg, K 162 mg, Na 0.5 mg, vitamin C 3.6 mg, vitamin A 11 IU, vitamin E 1.1 mg, folate 4.6 μg, riboflavin 3.5 mg, pantothenic acid 0.3 mg, pyridoxine 0.03 mg and biotin 20 μg.

Total solids content in saskatoon berries ranged from 20% to 29.4% fresh weight with 15.9–23.4% sucrose and 8–12% reducing sugars (Mazza 1979, 1982). Wolfe and Wood (1972) found that the sugar content increased slowly as the fruit matured and then accelerated considerably before ripening. They found that fructose content decreased greatly (25%) after the fruit ripened while the glucose content remained unaltered. Anthocyanins, total phenolics, sugars, sugars-acids ratios and anthocyanins-phenolics ratios in Saskatoon berries (*Amelanchier alnifolia*) increased with fruit ripening (Green and Mazza 1986). Titratable acidity and pH differed among cultivars but showed little change with fruit development. Anthocyanin contents ranged from 25.1 to 178.7 mg/100 g fruit, while total phenolics and soluble solids ranged from 0.17% to 0.52% and 9.6–18.7%, respectively. Significant correlation with anthocyanins was observed for total phenolics content, titratable acidity, pH and sugar-acid ratio. The results suggested that high contents of anthocyanins in Saskatoon berries were associated with high total phenolics and acids and low pH and sugar-acid ratios. The pH values of fruit ranged from 4.2 to 4.4 and titratable acidity values (% malic acid) from 0.36% to 0.49% (Mazza 1979; Green and Mazza 1986). The major acid in saskatoon berries was malic acid (Wolfe and Wood 1972) and the predominant aroma component was found to be benzaldehyde (Mazza and Hodgins 1985). The total anthocyanin content in fresh saskatoon berries was found to be 86–125 mg/100 g of fresh berries (Mazza 1982). The main anthocyanins identified were cyanidin 3-galactoside, (61% of the total anthocyanins) and cyanidin 3-glucoside (21%). Also detected were cyanindin 3-xyloside, chlorogenic acid and rutin. Hellström et al. (2007) found that saskatoon berries contained proanthocyanidins from dimers through heptamers and higher polymers. Saskatoon proanthocyanidins were generally of the procyanidin type, consisting mainly of epicatechin units linked by B-type bonds. Bakowska-Barczak et al. (2007) found saskatoon berries to have the following anthocyanidins: cyanidin, delphinidin, pelargonidin, petunidin, peonidin, and malvidin.

Ozga et al. (2007) characterised the following phenolic compounds in Saskatoon berries: cyanidin 3-O-galactoside, cyanidin 3-O-glucoside, cyanidin 3-O-arabinoside, and cyanidin 3-O-xyloside identified as the four major anthocyanins in mature fruit. The quercetin-derived flavonols, quercetin 3-O-glucoside, quercetin 3-O-galactoside, quercetin 3-O-arabinoside, quercetin 3-O-xyloside, quercetin 3-O-arabinoglucoside, quercetin 3-O-robinobioside, and quercetin 3-O-rutinoside were also identified in mature fruit extracts. In addition, two chlorogenic acid isomers (hydroxycinnamates), 3-O-caffeoylquinic acid and 5-O-caffeoylquinic acid were detected. The total content of the anthocyanin-, flavonol-, and hydroxycinnamate-type phenolics detected in mature 'Smoky' saskatoon fruit was 140, 25, and 96 mg/100 g fresh weight, respectively.

Studies by Hosseinian and Beta (2007) showed Saskatoon berries and wild blueberries to have high potential value for fruit growers as well as the food and nutraceutical manufacturers because of their high anthocyanin contents. The total anthocyanin content of Manitoba fruits followed the order: Saskatoon berry and blueberry (high anthocyanin berries), raspberry and chokecherry (medium anthocyanin berries), strawberry (low anthocyanin berries), and seabuckthorn (negligible anthocyanin berries). Saskatoon berry and wild blueberry presented a high content of total anthocyanins (562.4 and 558.3 mg/100 g, respectively). Saskatoon berry and wild blueberry contained higher amounts of delphinidin 3-glucoside (Dp-3-glc), malvidin 3-glucoside (Mv-3-glc), and malvidin 3-galactoside (Mv-3-gal). Dp-3-glc was 263.8 (mg/100 g) in Saskatoon berry and 84.4 (mg/100 g) in wild blueberry, whereas the corresponding values for Mv-3-glc in these berries were 47.4 and 139.6 (mg/100 g), respectively.

Lutein was found to be the predominant carotenoid in mature (purple) berries of 5 saskatoon cultivars ranging from a low value of about 300 to a high of 1,000 µg/100 g of fresh berries (Mazza and Cottrell 2008). The corresponding values for zeaxanthin were 60–120 µg/100 g of fresh berries. Levels of lutein, zeaxanthin and β-carotene were much higher in the green berries (48,000–146,000 µg/100 g, fresh weight (FW)) than in the more mature fruit. In mature fruit, the Smoky berries from Alberta had in general higher levels of lutein, zeaxanthin and β-carotene than the other varieties. Two cyanogenic glucosides, prunasin and amygdalin, was also detected in all saskatoon berries analyzed. In all varieties, the levels of the two glucosides were very low in the green stage and then essentially the same for the following maturity stages. In mature (purple) berries the content of amygdalin ranged from 43 to 129 mg/kg of fresh berries, and the content of prunasin was 5–19 mg/kg FW. Prunasin a cyanogenic glycoside was also detected in twigs and leaves especially during flowering (Majak et al. 1981). Cooking or drying destroys the cyanogenic glycoside in the leaves (Kershaw 2000).

The seed oil content of 17 cultivars of Saskatoon berries was found to vary from 9.4% (cv. 'Pasture') to 18.7% (cv. 'Thiessen') (Bakowska-Barczak et al. 2009). The seed oils contained mainly linoleic acid in the range from 47.3% (cv. 'Success') to 60.1% (cv. 'Lee 3') and oleic acid in the range from 26.3% (cv. 'Lee 3') to 38.1% (cv. 'Success'). The total tocopherol content varied from 1,053 to 1,754 mg/kg of oil. α-tocopherol was the predominant vitamin E compound in all berry seed oils, accounting for 87% of total tocopherols. The major sterols were β-sitosterol, δ(5)-avenasterol, and campesterol. The sterols content in seed oil varied from 7,357 mg/kg of oil (cv. 'Success') to 15,771 mg/kg of oil (cv. 'Lee 3'). Thirteen triacylglycerols (TAG) were identified in the seed oils, among which LLL, LLO, LOO, LLP, LOP (L, linoleoyl; O, oleoyl; P, palmitoyl) represented 88% of the total TAG. TAG composition suggested good oxidative stability of the Saskatoon berry seed oil, which could be suitable for food and industrial applications. The authors also maintained that Saskatoon berry seed oil may serve as potential dietary source of tocopherols, sterols, and unsaturated fatty acids.

Antioxidant Activity

Two cultivars (Thiessen and Smoky) of Saskatoon berries were found to possess free radical scavenging activities in a concentration-dependent manner (Hu et al. 2005). Cultivar Thiessen exhibited higher activity compared to cv. Smoky due to its relatively abundant anthocyanin content. Total anthocyanin content significantly correlated to free radical scavenging activities. It was found that the free radical scavenging components, i.e., anthocyanin, occurred predominantly in the ethyl acetate and n-butanol extracted fractions, suggesting that active components were more likely to occur in glycoside forms. HPLC subsequently confirmed the existence of cyanidin-3-O-galactoside and cyanidin-3-O-glucoside as leading anthocyanins in the Saskatoon berries. In addition, Saskatoon berries extracts from both cultivars inhibited peroxy-radical induced intracellular oxidation in a concentration-dependent fashion without affecting cell viability.

The polyphenol contents and antioxidant activities were assessed for 17 Saskatoon berry cultivars grown in Canada in fresh and stored fruits at −20°C for 9 months. Of 17 Saskatoon berry cultivars, the Nelson cultivar was the richest in total polyphenol, anthocyanin, and procyanidin contents (801, 382, and 278 mg/100 g fresh weight, respectively) (Bakowska-Barczak and Kolodziejczyk 2008). This cultivar exhibited the highest antioxidant potential measured with DPPH and ABTS radicals (2.8 and 5.0 mM/100 g FW, respectively). Cultivar-dependent changes in polyphenol content were observed after freezer storage at −20°C for 9 months. In the Lee 2 cultivar, significant increases in anthocyanin and flavonol contents occurred, while in the Lee 3 and Martin cultivars considerable decreases were observed. During the freezer storage, the antioxidant activity remained unchanged except for the Smokey which showed to be the most sensitive cultivar during storage. The Nelson and Lee 2 were the most stable cultivars during storage.

In the lipid peroxidation inhibitory assay, the anthocyanin mixture at 10 ppm presented activity of 72% compared with 89, 87 and 98% for commercial anti-oxidants butylated hydoxyanisole, butylated hydroxytoluene, and tertbutylhydroxyquinone at 1.67, 2.2 and 1.67 ppm, respectively. At 10 ppm, compounds 1-3 inhibited lipid peroxidation by 70, 75 and 78%, respectively (Adhikari et al. 2005).

Antiinflammatory Activity

Bioactive anthocyanins found in the fruits of *Amelanchier alnifolia* included cyanidin 3-galactoside (1) 155 mg/100 g, cyanidin 3-glucoside

(2) 54 mg/100 g and cyanidin (3) (Adhikari et al. 2005). At 100 ppm, the anthocyanin mixture inhibited cyclo-oxygenase (COX)-1 and COX-2 enzymes at 66 and 67% respectively. Anthocyanins 1 and 2 and cyanidin (3) inhibited COX-1 enzyme 50.5, 45.62 and 96.36%, respectively, at 100 ppm, whereas COX-2 inhibition was the highest for cyanidin at 75%. Cyclo-oxygenase enzymes are involved in mechanisms of pain and inflammation.

Antiviral Activity

Methanolic extract *A. alnifolia* plant was found to be active at non-toxic concentration against enteric coronavirus (McCutcheon et al. 1995).

Antidiabetic Activity

Burns Kraft et al. (2008) found that nonpolar constituents including carotenoids, from 4 wild berry species including Saskatoon berries, were potent inhibitors of aldose reductase (an enzyme involved in the etiology of diabetic microvascular complications), whereas the polar constituents, mainly phenolic acids, anthocyanins, and proanthocyanidins, were hypoglycemic agents and strong inhibitors of IL-1β and COX-2 gene expression. Berry samples also exhibited the ability to regulate lipid metabolism and energy expenditure in a manner consistent with improving metabolic syndrome. The results demonstrated that berries like Saskatoon traditionally consumed by tribal cultures contained a rich array of phytochemicals with the capacity to promote health and protect against chronic diseases, such as diabetes.

Traditional Medicinal Uses

Indigenous people in Canada used to juice for treating stomach ailments and as a laxative. Eye and ear-drops were made from ripe berries. The boiled bark is used as a disinfectant. The root infusion was used to prevent miscarriage after an injury. Thompson people made a tea from the twigs and stem and administer it to women just after birth and as a bath. A potent tonic from the bark was given to women after delivery to hasten discharge of the placenta (Turner et al. 1990) Saskatoon berry juice was ingested to relieve stomach upset and boiled berry juice was used as ear drops (Kershaw 2000).

Other Uses

It can be used as a wind-break plant. Its wood is hard, strong, fine-grained and used for tool handles, canes, canoe crossbars, tipi stakes, tipi closure pins and small implements. The young stems are used to make rims, handles and used in basket making. The shoots and young stems were used to make basket rims and handles, arrows, combs, digging sticks, salmon spreaders and pipes. Saskatoon berries provide a purple dye.

Comments

Saskatoon berries can be propagated by seeds, layering or suckers.

Selected References

Adhikari DP, Francis JA, Schutzki RE, Chandra A, Nair MG (2005) Quantification and characterisation of cyclo-oxygenase and lipid peroxidation inhibitory anthocyanins in fruits of *Amelanchier*. Phytochem Anal 16:175–180

Bailey LH (1976) Hortus third. A concise dictionary of plants cultivated in the United States and Canada. Liberty Hyde Bailey Hortorium, Cornell University, Wiley, 1312pp

Bakowska-Barczak AM, Kolodziejczyk P (2008) Evaluation of Saskatoon berry (*Amelanchier alnifolia* Nutt.) cultivars for their polyphenol content, antioxidant properties, and storage stability. J Agric Food Chem 56(21):9933–9940

Bakowska-Barczak AM, Marianchuk M, Kolodziejczyk P (2007) Survey of bioactive components in Western Canadian berries. Can J Physiol Pharmacol 85(11): 1139–1152

Bakowska-Barczak AM, Schieber A, Kolodziejczyk P (2009) Characterization of Saskatoon berry

(*Amelanchier alnifolia* Nutt.) seed oil. J Agric Food Chem 57(12):5401–5406

Burns Kraft TF, Dey M, Rogers RB, Ribnicky DM, Gipp DM, Cefalu WT, Raskin I, Lila MA (2008) Phytochemical composition and metabolic performance-enhancing activity of dietary berries traditionally used by Native North Americans. J Agric Food Chem 56(3):654–660

Green RC, Mazza G (1986) Relationships between anthocyanins, total phenolics, carbohydrates, acidity and colour of saskatoon berries. Can Inst Food Sci Technol J 19:107–113

Harris RE (1976) The saskatoon - Canada's national fruit. Canada Agric 21:28–29

Hedrick UP (1972) Sturtevant's edible plants of the world. Dover Publications, New York, 686pp

Hellström J, Sinkkonen J, Karonen M, Mattila P (2007) Isolation and structure elucidation of procyanidin oligomers from saskatoon berries (*Amelanchier alnifolia*). J Agric Food Chem 55(1):157–164

Hosseinian FS, Beta T (2007) Saskatoon and wild blueberries have higher anthocyanin contents than other Manitoba berries. J Agric Food Chem 55(26):10832–10838

Hu C, Kwok BHL, Kitts DD (2005) Saskatoon berries (*Amelanchier alnifolia* Nutt.) scavenge free radicals and inhibit intracellular oxidation. Food Res Int 38(8-9):1079–1085

Kershaw L (2000) Edible and medicinal plants of the rockies. Lone Pine Publishing, Edmonton, p 69

Landry P (1975) Le concept d'espèce et la taxinomie du genre *Amelanchier* (Rosacées). Bull Soc Bot Fr 122:249

Majak W, McDiarmid RE, Hall JW (1981) The cyanide potential of saskatoon serviceberry and chokeberry. Can J Anim Sci 61:681–686

Mazza G (1979) Development and consumer evaluation of a native fruit product. Can Inst Food Sci Technol J 12(4):166–169

Mazza G (1982) Chemical composition of saskatoon berries (*Amelanchier alnifolia* Nutt.). J Food Sci 47:1730–1731

Mazza G (2005) Compositional and functional properties of saskatoon berry and blueberry. Int J Fruit Sci 5(3):99–118

Mazza G, Cottrell T (2008) Carotenoids and cyanogenic glucosides in saskatoon berries (*Amelanchier alnifolia* Nutt.). J Food Comp Anal 21(3):249–254

Mazza G, Davidson CG (1993) Saskatoon berry: a fruit crop for the prairies. In: Janick J, Simon JE (eds) New crops. Wiley, New York, pp 516–519

Mazza G, Hodgins MW (1985) Benzaldehyde, a major aroma component of saskatoon berries. HortSci 20:742–744

McCutcheon AR, Roberts TE, Gibbons E, Ellis SM, Babiuk LA, Hancock RE, Towers GH (1995) Antiviral screening of British Columbian medicinal plants. J Ethnopharmacol 49(2):101–110

Ozga JA, Saeed A, Wismer W, Reinecke DM (2007) Characterization of cyanidin- and quercetin-derived flavonoids and other phenolics in mature saskatoon fruits (*Amelanchier alnifolia* Nutt.). J Agric Food Chem 55(25):10414–10424

Turner NJ (1997) Food plants of interior first peoples. UBC Press, Vancouver, p 139

Turner NJ, Thompson LC, Thompson MT, York AZ (1990) Thompson ethnobotany. Royal British Columbia Museum, Victoria, pp 253–257

Wolfe FH, Wood FW (1972) Non-volatile organic acid and sugar composition of saskatoon berries during ripening. Can Inst Food Sci Technol J 4:29–30

Chaenomeles speciosa

Scientific Name

Chaenomeles speciosa (Sweet) Nakai

Synonyms

Chaenomeles lagenaria (Loiseleur-Deslongchamps) Koidzumi, *Cydonia japonica* (Thunberg) Persoon var. *lagenaria* (Loiseleur-Deslongchamps) Makino, *Cydonia lagenaria* Loiseleur-Deslongchamps ex Duhamel, *Cydonia speciosa* Sweet (basionym).

Family

Rosaceaae

Common/English Names

Chinese Flowering Quince, Chinese-Quince, Flowering-Quince, Ornamental Quince Japanese Quince.

Vernacular Names

Chinese: Zhou Pi Mu Gua, Tie Geng Hai Tang;
Czech: Kdoulovec Japonský, Kdoulovec Lahvicovitý, Kdoulovec Žlutoplodý;
Danish: Stor Japankvæde;
Eastonian: Sile Ebaküdoonia;
French: Cognassier Ornemental De Chine;
German: Chinesische Zierquitte;
Hungarian: Japánbirs;
Iceland: Stóri Eldrunni;
Japanese: Boke;
Korean: San-Dang-Hwa;
Polish: Pigwowiec Chinski;
Slovencina: Dulovec Nádherný;
Spanish: Membrillero Del Japón;
Swedish: Storrosenkvitten;
Tibetan: Bse-Yab.

Origin/Distribution

The species is native to Eastern Asia – China and Korea.

Agroecology

A cool temperate species. It thrives in full sun or partial sun on fertile, well-drained soil with a broad pH range of 5.6–7.8.

Edible Plant Parts and Uses

The fruit is very hard and astringent and very unpleasant to eat raw, and are eaten cooked. They are suitable for making liqueurs, jams, jellies, marmalade and preserves, as they contain abundant pectin.

Botany

Deciduous shrubs, to 2–3 m tall, with Light brown, smooth bark and thorns and arching, dense, tangled crown. Branchlets slender purplish brown or blackish brown, terete, glabrous, Stipules reniform or suborbicular, rarely ovate, large, 5–10 mm, herbaceous, dark green, glabrous, sharply doubly serrate at margin, apex acute; petiole ca. 1 cm, initially sparsely pubescent, glabrescent; Leaves alternate, ovate to elliptic, rarely narrowly elliptic, 3–9 × 1.5–5 cm, glabrous or pubescent abaxially along veins on leaves of shoots, base cuneate to broadly cuneate, margin shortly serrate, apex acute or obtuse (Plate 1). Pedicel absent or short, ca. 3 mm, subglabrous. Flowers precocious, 3–5-fascicled on second year branchlets, 3–5 cm in diameter. Hypanthium campanulate, glabrous. Sepals erect, suborbicular, rarely ovate, 3–4 mm, abaxially glabrous, adaxially pubescent and caducous. Petals scarlet, rarely pinkish or white, ovate or suborbicular, base shortly clawed, apex rounded. Stamens 40–50. Styles 5, as long as stamens, connate at base, glabrous or slightly pubescent. Pome fragrant, green with purplish blush (Plate 1) turning to yellow when ripe, globose or ovoid, 4–6 cm in diameter.

Plate 1 Leaves and fruit

Nutritive/Medicinal Properties

Average fruit weight of *Chaenomeles* fruit was 27–211 g and the juice yield was 42–50% based on fresh weight of fruits (Hellin et al. 2003). The juice was very acidic, with a pH of 2.5–2.8, and the titratable acidity was at most 4.2% calculated as anhydrous citric acid. The following nutrient composition was reported on *Chaenomeles* fruit juice per 100 ml: vitamin C 45–109 mg, phenols 210–592 mg, malic acid 5.1 g and also quinic acid and succinic acid; ten amino acids were detected – phosphoserine, aspartic acid, threonine, serine, asparagine, glutamic acid, alanine, phenylalanine, γ- aminobutyric acid and lysine. The most abundant amino acid was glutamic acid (up to 14 mg/100 ml), followed by phosphoserine and aspartic acid (6 mg/100 ml). Sodium, ammonium, potassium, magnesium and calcium were detected, with potassium present in the highest amount (up to 241 mg/100 ml), fluoride (up to 139 mg/100 ml) and chloride (9 mg/100 ml) were also found. Nine carbohydrates were detected: stachyose, raffinose, sucrose, glucose, xylose, rhamnose, fructose, inositol and sorbitol. Of the carbohydrates, fructose (up to 2.3 g/100 ml), glucose (up to 1.1 g/100 ml) and sorbitol (up to 0.5 g/100 ml) were present in the highest amounts.

The fruit was found to contain more vitamin C than lemons (up to 150 mg/100 g) and to be rich in pectin. Fragrances and associated volatile levels from peel and flesh of *Chaenomeles speciosa* fruit in green ripened stage and full ripen stage were quite different (Zheng et al. 2010). The content of ethyl butyrate (12–15%) was highest in the peel of green or ripe fruit. Butyric acid, 2-methyl-, ethyl ester and ethyl caproate were also high. Linalool was detected in the peel and flesh. Terpenes were low in all samples but edulans especially edulan 1 was high.

Seven compounds were isolated from the fruit and identified as oleanolic acid, betulinic acid, 3-O-acetyl pomolic acid, ethyl chlorogenate, protocatechuic acid, gallic acid, and kojic acid (Yin et al. 2006). Thirteen compounds identified as 3,4-dihydroxybenzoic acid, quercetin, methyl 3-hydroxybutanedioic ester, oleanolic acid, masilinic acid, 3-O-acetyl ursolic acid, speciosaperoxide, ursolic acid, tormentic acid, 3β-acetoxyurs-11-en-13β, 28-olide, roseoside, vomifoliol and (6S,7E,9R)-6,9-dihydroxy-4,7-megastigmadien-3-one 9-O-[β-D-xylopyranosyl (1→6)-glucopyranoside] were isolated from *C. speciosa* fruit (Zhang et al. 2010). Protocatechuic acid (Li et al. 2010), olenolic acid and urolic acid (Li and He 2005) were isolated from the fruits.

Forty compounds, constituting about 85.13% of the total oil, were identified from essential oil of *C. speciosa* fruit (Xie et al. 2007). The main constituents were β-caryophyllene (12.52%), α-terpineol (5.41%), terpinen-4-ol (4.56%) and 1,8-cineole (4.31%). The fruit also contained terpenoid compounds such as speciosaperoxide; 3β-acetoxyurs-11-en-13β,28-olide; 3-O-acetyl ursolic acid; oleanolic acid; ursolic acid; masilinic acid and tormentic acid, and three known norsesquiterpenoids, roseoside, vomifoliol and (6S, 7E,9R)-6,9-dihydroxy-4,7-megastigmadien-3-one 9-O-[β-d-xylopyranosyl (1→6)-glucopyranoside] (Song et al. 2008). None of these compounds exhibited inhibitory activity against T-and B-lymphocyte proliferation.

Seven compounds obtained from the ethyl acetate fraction of *C. speciosa* were identified as cinnamic acid, 2′-methoxyaucuparin, 2-hydroxylbutanedioic acid-4-methylester, esculetin, p-hydroxybenzoic acid, chlorogenic acid, caffeic acid (Yang et al. 2009). Three triterpenoid compounds were separated from *C. lagenaria* fruit and identified as 3-O-acetyl ursolic acid, 3-O-acetyl pomolic acid and betulinic acid (Guo et al. 1998).

Some of the reported pharmacological properties of the plant are:

Antioxidant Activity

Four compounds were isolated from ethyl acetate extract of the dried fruits of *Chaenomeles speciosa* and identified as hydroquinone (1), 3,4-dihydroxybenzoic acid (2), quercetin (3), and methyl 3-hydroxylbutanedioic ester (4) (Song et al. 2007). Compounds(1–3) exhibited antioxidant effects in 2,2′-diphenyl-1-picrylhydrazyl (DPPH) radical scavenging assay.

Chaenomeles speciosa powder (CSP) was found to be rich in rich in vitamin C and polyphenols (Tang et al. 2010). The ferric reducing antioxidant power of CSP was 173 μmol Fe^{2+}/g and the scavenging activity on 1,1-diphenyl-2-picrylhydrazyl free radical (DPPH•) and O^{2-} were 945 μg DPPH•/g and 700 U/ml, respectively. Both 5 and 10% CSP dietary supplement significantly reduced serum low-density lipoprotein cholesterol and total cholesterol levels in ApoE−/− mice which was fed a high-fat diet for 16 weeks. Comparing with normal group, there was a significant increase of glutathione peroxidase activity and total antioxidative capacity, and a reduction of relative atherosclerotic plaque area of aortic sinus and aortic arch in ApoE−/− mice supplemented with 5% and 10% CSP. The results suggested that CSP had potent antioxidative ability and potential antiatherosclerotic effects.

Antidiarrheal Activity

Chaenomeles speciosa fruit extract was found to inhibit the heat labile enterotoxin (LT)-induced diarrhea in mice by blocking the binding of the B subunit of LT (LTB) to ganglioside GM1 (Chen et al. 2007a). The ethyl acetate (EA) soluble fraction was the most active fraction that significantly abolished the LTB and GM1 interaction. Further, the oleanolic acid, ursolic acid, and betulinic acid from the EA fraction, blocked the toxin binding effects, resulting in the suppression of LT-induced diarrhoea. The findings suggested that oleanolic acid, ursolic acid, and betulinic acid were the active constituents from the fruit and might be considered as lead therapeutic agents in the treatment of enterotoxin-induced diarrhea.

Antiinflammatory/Antiarthritic Activity

Glucosides of *Chaenomeles speciosa* (GCS) was found to exhibit antiinflammatory and immune responses in collagen-induced arthritis (CIA) rats

(Chen and Wei 2003). The administration of GCS inhibited the inflammatory response and restored body weight and the weight of immune organs of CIA rats. The antiinflammatory and immunoregulatory actions and therapeutic effect on CIA rats were found to be due to G protein-AC-cAMP transmembrane signal transduction of synoviocytes, which played a crucial role in pathogenesis of arthritis. GCS was found to have therapeutic effect on adjuvant arthritis in rats, which may be related to modulation function of thymocyte T cells and inhibiting the productions of proinflammatory cytokines secreted by peritoneal macrophages (Dai et al. 2003a, b). Treatment with GCS (60, 120 mg/kg) and Actarit (60 mg/kg) for 5 days significantly diminished the secondary hind paw swelling, as well as relieved the pain response and the polyarthritic symptoms of the whole body as compared with untreated adjuvant arthritis group. Both treatments restored the diminished thymocytes proliferation ConA induced in adjuvant arthritis rats. GCS reduced the enhanced productions of interleukin1, TNF α and PGE 2 from peritoneal macrophages in adjuvant arthritis rats. Glucosides of *Chaenomeles speciosa* were found to have antinociceptive effects, which related to its inhibitory effects on peripheral inflammatory mediators (Wang et al. 2005). The glucosides at different doses inhibited mice's writhing response and second phase of formalin response. They also suppressed the increased arthritic flexion scores in adjuvant arthritis rats. On 28 days after inflammation induction, the glucosides (60, 120 mg/kg) decreased the concentration of prostaglandin E and tumour necrosis factor-α of synovial cells in the adjuvant arthritis rats. Further studies found that GCS down-regulated the levels of serum antibodies in rats with adjuvant arthritis, which may be related to its efficacy in the treatment of rats with adjuvant arthritis (Chen et al. 2007b).

The 10% ethanol fraction (C3) of *C. speciosa* was found to have stronger antiinflammatory effects compared with other fractions at the same dose using carrageenan-induced paw edema in rats (Li et al. 2009). Chlorogenic acid was found to be one of the active constituents responsible for the antiinflammatory effect. Compared with controls, fraction C3 demonstrated significant antiinflammatory activity in the xylene-induced ear edema test, acetic acid-induced peritoneal capillary permeability test, and the cotton pellet granuloma test in mice or rats; it also showed marked analgesic activity in the acetic acid-induced abdominal contraction test and formalin-induced paw licking test in mice and rats. However, fraction C3 showed no significant effect in the hot plate test in mice. The findings supported the use of the *C. speciosa* for treating pain and inflammation.

Three compounds from *C. speciosa* fruit exhibited antioxidant, antiinflammatory and antiviral activities (Zhang et al. 2010). 3,4-dihydroxybenzoic acid (1) displayed high inhibitory activities on DPPH and neuramindase with IC_{50} values of 1.02 μg/ml and 1.27 μg/ml respectively, and quercetin (2) also showed significant inhibitory action on DPPH and NA with IC_{50} values of 3.82 μg/mlL and 1.90 μg/ml. Compounds 1, 2 and methyl 3-hydroxybutanedioic ester inhibited the production of TNF-α by 22.73%, 33.14% and 37.19% at 5 μg/ml compared with the control. Quercetin was found to be active on the release of IL-6 in RAW264.7 macrophage cells, with an inhibitory rate of 39.79%. The compounds may have a cocktail-like role in the treatment of avian influenza, and *C. speciosa* components, especially quercetin, might be a potent source for antiviral and antiinflammatory agents.

Human β2-Adrenoceptor Activity

Chaenomeles speciosa was one of six herbal extracts studied that were suggested to be potential agonists of human β2-adrenoceptor (Wang et al. 2009). The active fraction from the ethanol extract showed significant effects on active reporter gene expression with an EC_{50} of 4.8 μg/ml. Human β2-adrenoceptors are involved in many physiological activities in the human body such as smooth muscle relaxation, lipolysis in adipose tissue, histamine-release inhibition from mast cells, increase renin secretion from kidney, insulin release from pancreatic β cells, etc.

Immunomodulatory Activity

Chaenomeles speciosa glycosides (GCS) (240 mg/kg) decreased the thymus index and spleen index in contact hypersensitive (CHS) mice (Zheng et al. 2004). GCS (60,120,240 mg/kg) and also inhibited the ear swelling of CHS mice and splenocyte proliferation induced by Concanavalin A. GCS (240 mg/kg) decreased CD4+ CD8+ T lymphocytes subsets ratio and restored the CD4+CD8 -subsets ratio in CHS mice. GCS also decreased the TGF-β 1 and IL-2 levels but increased the IL-4 levels in mice thymus with CHS. The findings showed that GCS could suppress mice contact hypersensitive reaction and restore the balance of T lymphocytes subsets in mice thymus with CHS, and also could modulate the cytokines production by CD4+T lymphocytes of CHS mice. Chaenomeles speciosa broth exhibited protective effects on the immunosuppressive mouse model induced by cyclophosphamide (Shi et al. 2009). Administration of the broth increased serum hemolysin and lymphocyte transformation rate and downregulated mRNA expression of foxp3, TGF-β, PD1, Fas, Bax compared with the cyclophosphamide mice.

Antiparkinsonian Activity

Chaenomeles fruit (FQ) was found to be a selective, potent dopamine transporter (DAT) inhibitor and had antiparkinsonian-like effects that were mediated possibly by DAT suppression (Zhao et al. 2008). FQ at concentrations of 1–1,000 μg/ml concentration-dependently inhibited dopamine uptake by Chinese hamster ovary (CHO) cells and by synaptosomes. FQ had a slight inhibitory action on norepinephrine uptake by CHO cells and no inhibitory effect on γ-aminobutyric acid (GABA) uptake by CHO cells or serotonin uptake. FQ mitigated 1-methyl-4-phenylpyridinium-induced toxicity in D8 cells. In behavioral studies, FQ alleviated rotational behavior in 6-hydroxydopamine-treated rats and improved deficits in endurance performance in 1-methyl-4-phenyl-1,2,3,6-tetrahydropyridine (MPTP)-treated mice. Immunohistochemistry revealed that FQ markedly reduced the loss of tyrosine hydroxylase-positive neurons in the substantia nigra in MPTP-treated mice.

Antimicrobial Activity

The essential oil C. speciosa fruit was reported to show a broad spectrum of antimicrobial activity against all the tested bacterial strains (Xie et al. 2007). The essential oil had higher sensitivity to Gram-positive than Gram-negative bacteria.

Teratogenic Study

Studies showed that glycosides of Chaenomeles speciosa (GCS) exerted no teratogenic effect on pregnant mice embryos (Lu et al. 2008).

Traditional Medicinal Uses

Chaenomeles speciosa has been cultivated widely in Chongqing for hundreds of years and is also a traditional herb for the Chinese because of its many health preventive potentials (Tang et al. 2010). The fruit is reported to be analgesic, antiinflammatory, antispasmodic, astringent and digestive in traditional medicine. A decoction is used internally in the treatment of nausea, arthritis, leg oedema, cholera and associated cramps. Fruit-containing cocktails have been applied to the treatment of neuralgia, migraine, and depression in traditional Chinese medicine (Zhao et al. 2008). The fruit is a traditional Chinese medicine used for the treatment of dyspepsia rheumatoid arthritis, prosopalgia, and hepatitis and various inflammatory diseases (Zhang et al. 2010).

Other Uses

The species is a popular ornamental hedge and a ground cover plant.

Comments

The plant is readily propagated from cuttings.

Selected References

Bailey LH (1949) Manual of cultivated plants most commonly grown in the continental United States and Canada (Revised edition). The Macmillan Co., New York, 1116pp

Bown D (1995) Encyclopaedia of herbs and their uses. Dorling Kindersley, London, 424pp

Chen Q, Wei W (2003) Effects and mechanisms of glucosides of *Chaenomeles speciosa* on collagen-induced arthritis in rats. Int Immunopharmacol 3(4):593–608

Chen JC, Chang YS, Wu SL, Chao DC, Chang CS, Li CC, Ho TY, Hsiang CY (2007a) Inhibition of *Escherichia coli* heat-labile enterotoxin-induced diarrhea by *Chaenomeles speciosa*. J Ethnopharmacol 113(2):233–239

Chen Y, Wei W, Wu H, Tang LQ, Wang XY, Yang YQ (2007b) Down-regulation effect of glucosides of *Chaenomeles speciosa* on the levels of serum antibodies in rats with adjuvant arthritis. Chin Pharmacol Bull 23(7):941–944 (In Chinese)

Dai M, Wei W, Shen YX, Zheng YQ (2003a) Glucosides of *Chaenomeles speciosa* remit rat adjuvant arthritis by inhibiting synoviocyte activities. Acta Pharmacol Sin 24(11):1161–1166 (In Chinese)

Dai M, Wei W, Wang NP, Chen Q (2003b) Therapeutic effect of glucosides of *Chaenomeles speciosa* on adjuvant arthritis in rats. Chin Pharmacol Bull 19(3):340–343 (In Chinese)

Duke JA, Ayensu ES (1985) Medicinal plants of China, vols 1 & 2. Reference Publications, Inc., Algonac, 705pp

Grieve M (1971) A modern herbal, vol 2. Penguin/Dover Publications, New York, 919pp

Guo X, Zhang L, Quan S, Hong Y, Sun L, Liu M (1998) Isolation and identification of triterpenoid compounds in the fruits of *Chaenomeles lagenaria* (Loisel.) Koidz. Zhongguo Zhong Yao Za Zhi 23(9):546–547 (In Chinese)

Hellin P, Vila R, Jordan MJ, Laencina J, Rumpunen K, Ros JM (2003) Characteristics and composition of *Chaenomeles* fruit juice. In: Japanese quince - potential fruit crop for Northern Europe, Department of Crop Science, Swedish University of Agricultural Sciences, pp 127–139

Ku TC, Spongberg SA (2003) *Chaenomeles* Lindley. In: Wu ZY, Raven PH, Hong DY (eds) Flora of China, vol 9 (*Pittosporaceae* through *Connaraceae*). Science Press/Missouri Botanical Gardern Press, Beijing/St. Louis

Kunkel G (1984) Plants for human consumption. An annotated checklist of the edible Phanerogams and Ferns. Koeltz Scientific Books, Koenigstein

Li H, He L (2005) Determination of olenolic acid and urolic acid in Fructus chaenomelis by HPLC-evaporative light-scattering detection method. Chin Hosp Pharm J 25:259–261 (In Chinese)

Li X, Yang YB, Yang Q, Sun LN, Chen WS (2009) Anti-inflammatory and analgesic activities of *Chaenomeles speciosa* fractions in laboratory animals. J Med Food 12(5):1016–1022

Li HR, Gao FX, Ren YJ (2010) Determination of protocatechuic acid in *Chaenomeles lagenaria* (Loisel) Koidz by RP-HPLC. China Pharmacist 2:179–180 (In Chinese)

Lu JT, Xu DX, Sun MF, Wei LZ, Shen T, Wei W (2008) Study on teratogenicity of glycosides of *Chaenomeles speciosa*. Carcinogen, Teratogen Mutagen 20(1):27–29. (In Chinese)

Shi JJ, Liu CQ, Li B, Qin XL, Chen J, Wang JJ, Yang F (2009) Protecting effect of *Chaenomeles speciosa* broth on immunosuppressive mice induced by cyclophosphamide. Zhong Yao Cai 32(9):1418–1421 (In Chinese)

Song YL, Feng ZB, Cheng YX, Gao JM (2007) Chemical components of *Chaenomeles speciosa* (Sweet) Nakai. Acta Bot Boreali Occident Sin 27:831–833 (In Chinese)

Song YL, Zhang L, Gao JM, Du GH, Cheng YX (2008) Speciosaperoxide, a new triterpene acid, and other terpenoids from *Chaenomeles speciosa*. J Asian Nat Prod Res 10(3–4):217–222

Tang Y, Yu X, Mi M, Zhao J, Wang J, Zhang T (2010) Antioxidative property and antiatherosclerotic effects of the powder processed from *Chaenomeles speciosa* in ApoE–/– mice. J Food Biochem 34:535–548

Wang H, Li SH, Zhao CK, Zeng X (2009) A system for screening agonists targeting β2-adrenoceptor from Chinese medicinal herbs. J Zhejiang Univ Sci B 10(4):1673–1681 (In Chinese)

Wang NP, Dai M, Wang H, Zhang LL, Wei W (2005) Antinociceptive effect of glucosides of *Chaenomeles speciosa*. Chin J Pharmacol Toxicol 19(3):169–174 (In Chinese)

Xie XF, Cai XQ, Zhu SY, Zou GL (2007) Chemical composition and antimicrobial activity of essential oils of *Chaenomeles speciosa* from China. Food Chem 100(4):1312–1315

Yang YB, Yang Y, Li X, Yang Z, Wu ZJ, Zheng YL, Sun LN (2009) Studies on the chemical constituents of *Chaenomeles speciosa*. Zhong Yao Cai 32(9):1388–1390 (In Chinese)

Yin K, Gao HY, Li XN, Wu LJ (2006) Chemical constituents *of Chaenomeles speciosa* (Sweet.) Nakai. J Shengyang Pharm Univ 23:760–763 (In Chinese)

Zhang L, Cheng YX, Liu AL, Wang HD, Wang YL, Du GH (2010) Antioxidant, anti-inflammatory and anti-influenza properties of components from *Chaenomeles speciosa*. Molecules 15:8507–8517

Zhao G, Jiang ZH, Zheng XW, Zang SY, Guo LH (2008) Dopamine transporter inhibitory and antiparkinsonian effect of common flowering quince extract. Pharmacol Biochem Behav 90(3):363–371

Zheng H, Kong YQ, Zhang RG, Yu LS, Zhang H, Gan J, Wang YQ (2010) Analysis of volatiles of *Chaenomeles speciosa* (Sweet) Nakai from Yunnan by TCT-GC/MS. J Yunnan Agric Univ 25(1):135–142 (In Chinese)

Zheng YQ, Wei W, Dai M, Wang NP (2004) Glucosides of *Chaenomeles speciosa* suppressed contact hypersensitivity response via modulating the thymus T lymphocytes subsets in mice. Chin Pharmacol Bull 20(9):1016–1019 (In Chinese)

Cydonia oblonga

Scientific Name

Cydonia oblonga Mill.

Synonyms

Cydonia communis Poiret, *Cydonia cydonia* Persoon, *Cydonia europaea* Savi, *Cydonia vulgaris* Pers., *Pyrus cydonia* L, *Sorbus cydonia* Crantz.

Family

Rosaceae

Common/English Names

Apple-shaped Quince, Common Quince, Quince, Quince Seeds, Quince Tree.

Vernacular Names

Albanian: Ftua;
Arabic: Habbus Safarjal (Seeds), Safarjal;
Brazil: Marmelo;
Chinese: Wen Po;
Czech: Kdouloň, Kdouloň Obecná;
Danish: Almindelig Kvæde, Kvæde, Kvædetræ, Pærekvæde;
Dutch: Japanse Kwee, Kwee, Kweeboom, Kweepeer, Kweepeer Sort, Lijsterbes Sort;
Eastonian: Harilik Küdoonia;
Finnish: Kvitteni;
French: Cognassier, Cognassier À Fruit Comestible, Coing, Graines De Coing;
Gaelic: Cainche;
German: Echte Quitte, Kittenbaum, Kötte, Köttenbaum, Kütte, Küttenbaum, Quittenbaum, Quitte, Quittenbaum, Schmeckbirne;
Greek: Kydoni, Kydonion;
Hebrew: Habush;
Hungarian: Birs, Birsalma, Birskörte;
India: Behidana (Seeds);
Iran: Beh;
Italian: Cotogno, Mela, Mela Cotogna, Melocotogno, Pomo Cotogno;
Japanese: Marumero;
Lebanon: Sfarjel;
Malta: Gamm Ta' L-Isfargel;
Norwegian: Kvede;
Persian: Safarjal, Tukhme Safarjal (Seeds);
Polish: Pigwa, Pigwa Pospolita;
Portuguese: Cidónia, Marmeleiro, Marmelo;
Russian: Aiva, Ajva Aiva Obyknovennaia;
Serbian: Dunja;
Slovaščina: Kutina, Kutina Navadna;
Spanish: Membrillero, Membrillo;
Swedish: Kvitten, Kvitten-Arter;
Turkish: Ayva, Ayva Agh.

Origin/Distribution

The species is indigenous to Western Asia; its primary wild area is probably limited to the Caucasus – Armenia, Azerbaijan, Turkmenistan, and the Russian Federation – Ciscaucasia, Dagestan. Partially connected with this core area are populations in Iran, Anatolia, Syria, Turkmenia and Afghanistan.

Agroecology

Quince grows in the warm temperate to cool temperate zone from 0°C to 25°C. Quince is frost hardy and requires a cold chilling period of 100–450 hours below 7°C to flower properly. Quince flowers later in the spring than pears, because some vegetative growth must occur before the flowers appear. The tree is self fertile, however, yield can be enhanced from cross fertilization.

Edible Plant Parts and Uses

Quince is primarily grown for its edible fruit, although its flowers are also edible. Only cultivars with a soft skin, matured in a warm climate can be eaten raw. Most varieties of quince are too hard, astringent and sour to eat raw. They are roasted, baked or stewed or processed into jams, marmalades, pastes, jelly and quince pudding. Quince paste is still widely made in France and Spain, while in Argentina, Chile and Uruguay quince is cooked into a reddish jello-like block or firm reddish paste known as *dulce de membrillo* which is then eaten as a spread in sandwiches and with cheese. Boiled quince is also popular in desserts such as the *murta con membrillo* that combines *Ugni molinae* with quince. In Spain, the fruit is cooked into a firm reddish paste and is eaten with *manchego* cheese. In Syria, quince is cooked in pomegranate paste (*dibs rouman*) with shank meat and *kibbeh* (a middle eastern meat pie with burghul and mince meat) and is called "*kibbeh safarjalieh*". Because of its strong and intensive fragrance and flavour, the fruit it is used in mixed products including juices pure and mixed, liqueurs, wine, cider, fruits in liqueur, jam, marmalade, jelly, and dried jelly as a special "quince-bread". Quinces have long been grown for flavouring apple pies, ices and confections. Quince juice mixes well with other fruit juices. In Bosnia, the quince is made into brandy. Quince wine was popular in Britain in the nineteenth century, and was reputed to be beneficial to asthma sufferers.

Quince scalding water, rich in phenolic compounds, flavonoids, organic acids and sugars could be use in low-fat yogurt production (Trigueros et al. 2011). Quince scalding water had inhibitory effect against lactic acid bacteria, probably due to its high content of polyphenols. As a consequence, quince scalding water enriched yogurts had higher pH and lower lactic acid content compared to control yogurts. Such changes were reflected in their rheological and textural properties: soft yogurts of higher deformability and lower elastic behavior and viscosity.

Botany

Quince is a deciduous, unarmed, perennial shrub or small tree, 4–8 m high with crowded gnarled branches and branchlets. Branchlets are purplish red when young, turning purplish brown with age, terete, initially densely tomentose becoming glabrous when old. Leaves are alternate, petiolate (0.8–1.5 cm long) with caducous, tomentose ovate stipules (Plate 1). Lamina is simple, ovate to oblong, 5–10×3–5 cm, lower surface pale green with conspicuous veins and densely villous, upper surface dark green glabrous or sparsely pubescent when young, base rounded or subcordate, margin entire, apex acute or emarginate. Flowers are 4–5 cm across on tomentose pedicel with caducous, ovate bracts with campanulate hypanthium which is densely tomentose abaxially; sepals 5, ovate or broadly lanceolate, 5–6 mm, longer than hypanthium, both surfaces tomentose, margin glandular serrate, apex acute; petals 5, white or pinkish, 1.8 cm long; stamens less than 1/2 as long as petals. styles nearly as long as stamens, densely villous basally (Plate 2).

Plate 1 Leaves and developing quince fruits

Plate 2 Flower buds

Plate 3 Ripe quince fruit

Fruit is fragrant, light green turning yellow (Plates 1 and 3) when ripe, pear-shaped, 3–5 cm across, densely tomentose with persistent reflexed sepals and stout, tomentose fruiting pedicel, 5 mm long. Pulp is firm, fleshy and aromatic.

Nutritive/Medicinal Properties

The food value per 100 g edible portion of raw ripe quince fruit (exclude 39% of core, seeds and parings) is: water 83.80 g, energy 57 kcal (238 kJ), protein 0.4 g, total lipid 0.10 g, ash 0.4 g, carbohydrate 15.3 g, total dietary fibre 1.9 g, Ca 11 mg, Fe 0.70 mg, Mg 8 mg, P 17 mg, K 197 mg, Na 4 mg, Zn 0.04 mg, Cu 0.130 mg, Se 0.69 µg, vitamin C (total ascorbic acid) 15 mg, thiamine 0.020 mg, riboflavin 0.030 mg, niacin 0.2 mg, pantothenic acid 0.081 mg, vitamin B-6 0.040 mg, total folate 3 µg, vitamin A 2 µg RAE, vitamin A 40 IU, total saturated fatty acids 0.010 g, 16:0 (palmitic acid) 0.007 g, 18:0 (stearic acid) 0.002 g, total monounsaturated fatty acids 0.036 g, total polyunsaturated fatty acids 0.050 g, 18:1 undifferentiated (oleic acid) 0.036 g and 18:2 undifferentiated (linoleic acid) 0.049 g (USDA 2011). Quince is not a nutrient rich fruit, it has moderate amount of vitamin A, and low amounts of other vitamins and minerals. However, it is rich in health beneficial phenolic compounds.

A GC/FID (gas chromatography/flame ionization detector) methodology was employed for the determination of 21 free amino acids in quince fruit (pulp and peel) and jam (Silva et al. 2003). The detection limit values for amino acids were low, between 0.004 and 0.115 µg/mL, and the method was precise. The GC/FID procedure was rapid, sensitive, reproducible, accurate and low cost and could be useful in the quality control of quince products. Twenty-one free amino acids were found in quince fruit (pulp and peel) and quince jam (homemade and industrially manufactured) (Silva et al. 2004b). Generally, the highest content in total free amino acids and in glycine was found in peels. The three major free amino acids detected in pulps were aspartic acid, asparagine, and hydroxyproline. For quince peels, usually, the three most abundant amino acids were glycine, aspartic acid, and asparagine. Similarly, for quince jams the most important free amino acids were aspartic acid, asparagine, and glycine or hydroxyproline. All samples of quince fruit (pulp and peel) and quince jam (homemade and industrially manufactured) presented a similar organic acid profile composed of at least six identified organic acids: citric, ascorbic, malic, quinic, shikimic, and fumaric acids (Silva et al. 2002b). Several samples also contained oxalic acid. A homo-monoterpenic compound (trans-9-amino-8-hydroxy-2,7-dimethylnona-2,4-dienoic acid glucopyranosyl ester) was isolated, identified and quantified in quince pulps, peels and jams (Sousa et al. 2007). The compound can be used as a tool for the characterization of quince and its jam.

Total soluble solids (TSS) of five Spanish quince clones were found to range from 11.5°Brix to 14.7°Brix, with fructose and glucose as predominant sugars; clone MEMB3 yielded the highest sugar content of all (17.93%) (Rodríguez-Guisado et al. 2009). Malic was the main organic acid (0.78%) followed by tartaric (0.22%), while quince juice yielded very low citric acid (0.009–0.014%). Quince generally showed high crude fibre contents (8.14% for MEMB1), low fat contents and can weigh up to 290 g.

Various parts of the quince fruit were found to be rich in phenolic compounds. The total phenolic content of the pulp and peel parts ranged from 37–47 to 105–157 mg/100 g of fresh weight, respectively (Fattouch et al. 2007). Chlorogenic acid (5-O-caffeoylquinic acid) was the most abundant phenolic compound in the pulp (37%), whereas rutin (quercetin 3-O-rutinoside) was the main one in the peel (36%). Another study reported that the pulp contained mainly caffeoylquinic acids (3-, 4-, and 5-O-caffeoylquinic acids and 3,5-dicaffeoylquinic acid) and one quercetin glycoside, rutin (in low amount) (Silva et al. 2002a). The peels presented the same caffeoylquinic acids and several flavonol glycosides: quercetin 3-galactoside, kaempferol 3-glucoside, kaempferol 3-rutinoside, and several unidentified compounds (probably kaempferol glycoside and quercetin and kaempferol glycosides acylated with p-coumaric acid). The highest content of phenolics was found in peels.

Fifty-nine secondary metabolites were isolated from quince fruit peels, among them, five metabolites, 3β-(18-hydroxylinoleoyl)-28-hydroxyurs-12-ene, 3β-linoleoylurs-12-en-28-oic acid, 3β-oleoyl-24-hydroxy-24-ethylcholesta-5, 28(29)-diene, tiglic acid 1-O-β-d-glucopyranoside and 6,9-dihydroxymegastigmasta-5,7-dien-3-one 9-O-β-d-gentiobioside were isolated for the first time (Alesiani et al. 2010). A cytosolic carotenoid cleavage enzyme with a molecular weight of 20 kD was isolated and partially purified from quince fruit (Fleischmann et al. 2002). Using β-carotene as substrate, the enzyme activity was detected spectrophotometrically at a wavelength of 505 nm.

The difference between quince pulp and peel phenolic profiles was more apparent during principal component (PC) analysis (Silva et al. 2005b). Two PCs accounted for 81.29% of the total variability, PC1 (74.14%) and PC2 (7.15%). PC1 described the difference between the contents of caffeoylquinic acids (3-O-, 4-O-, and 5-O-caffeoylquinic acids and 3,5-O-dicaffeoylquinic acid) and flavonoids (quercetin 3-galactoside, rutin, kaempferol glycoside, kaempferol 3-glucoside, kaempferol 3-rutinoside, quercetin glycosides acylated with p-coumaric acid, and kaempferol glycosides acylated with p-coumaric acid). PC2 related the content of 4-O-caffeoylquinic acid with the contents of 5-O-caffeoylquinic and 3,5-O-dicaffeoylquinic acids. Two main principal component characterised the quince jam phenolic composition (54.4% of all variance): PC1 (37.4%) and PC2 (17.0%) (Silva et al. 2006). The analyses of 17 Portuguese quince jam samples showed that all the samples presented a similar profile composed of at least eight identified phenolic compounds, several unidentified characteristic procyanidin polymers, and sodium benzoate as preservative of quince jams (Silva et al. 2002a). Several samples also contained arbutin, suggesting that these quince jam samples were fraudulently adulterated with pear puree. The use of non-polar sorbents technique was found useful in the evaluation of commercial quince jams genuineness (Silva et al. 2001). The detection limit values for phenolic compounds were between 0.1 and 1.6 mg/ml and the method was precise. Generally, the recovery values were high, except for arbutin. All samples of quince jellies samples presented a similar phenolic profile with at least eight identified phenolic compounds and also contained 5-HMF (5-(hydroxymethyl) furfural), a sugar derivative, as the major compound (Silva et al. 2000b).

An HPLC diode array detection methodology for separating 13 phenolic compounds from quince purees could also be applied to the detection of apple and/or pear fraudulently added to quince puree (Andrade et al. 1998). The presence of apple can be detected by phloretin 2′-xylosylglucoside and phloretin 2′-glucoside, while that of pear detected by the presence of arbutin. In addition, 3-O-caffeoylquinic acid was found to be present at an appreciable amount (~23.4%) in quince puree, while the sample of pear puree contained only 8.2% was absent in apple puree.

Quince seeds exhibited a phenolic profile composed of 3-O-caffeoylquinic, 4-O-caffeoylquinic, 5-O-caffeoylquinic and 3,5-dicaffeoylquinic acids, lucenin-2, vicenin-2, stellarin-2, isoschaftoside, schaftoside, 6-C-pentosyl-8-C-glucosyl chrysoeriol and 6-C-glucosyl-8-C-pentosyl chrysoeriol (Silva et al. 2005a). Six organic acids constituted the organic acid profile of quince seeds: citric, ascorbic, malic, quinic, shikimic and fumaric acids. The free amino acid profile was composed of 21 identified free amino acids and the three most abundant were glutamic acid, aspartic acid and asparagine. The following C-glycosyl flavones: vicenin-2 (6,8-di-C-glucosyl apigenin), lucenin-2 (6,8-di-C-glucosyl luteolin), stellarin-2 (6,8-di-C-glucosyl chrysoeriol), isoschaftoside (6-C-arabinosyl-8-C-glucosyl apigenin), schaftoside (6-C-glucosyl-8-C-arabinosyl apigenin), 6-C-pentosyl-8-C-glucosyl chrysoeriol and 6-C-glucosyl-8-C-pentosyl chrysoeriol were identified in quince seed (Ferreres et al. 2003).

Solvent partition of quince wax with n-hexane or acetone yielded an insoluble (crystalline) and a soluble (oily) fraction (Lorenz et al. 2008). The insoluble fraction consisted of saturated n-aldehydes, n-alcohols and free n-alkanoic acids of carbon chain lengths between 22 and 32, with carbon chain lengths of 26 and 28 dominating. Also odd-numbered unbranched hydrocarbons, mainly C27, C29 and C31, were detected particularly in the acetone-insoluble fraction (total, 15.8%). Triterpenoic acids were separated from the hexane-insoluble matter and identified as a mixture of ursolic, oleanolic and betulinic acids. The major constituents of the hexane-soluble fraction were glycerides of linoleic [$\Delta(9,12)$, 18:2] and oleic [$\Delta(9)$, 18:1] acids, accompanied by free linoleic, oleic and palmitic acids (C16). In addition, β-sitosterol, $\Delta(5)$-avenasterol as well as trace amounts of other sterols were detected. Finally the carotenoids phytoene and phytofluene were identified and quantified yielding 1.0% and 0.3% of the quince wax, respectively.

Quince leaves presented a common organic acid profile composed of six constituents: oxalic, citric, malic, quinic, shikimic and fumaric acids (Oliveira et al. 2008). Total organic acid content varied from 1.6 to 25.8 g/kg dry matter (mean value of 10.5 g/kg dry matter). Quinic acid was the major compound (72.2%), followed by citric acid (13.6%). Significant differences were found in malic and quinic acids relative abundances and total organic acid contents at various times of the year. *Cydonia oblonga* leaves also contained phenolic compounds. Quince leaves presented a common phenolic profile composed by nine compounds (Oliveira et al. 2007): 3- O-, 4- O- and 5-O-caffeoylquinic acids, 3,5-O-dicaffeoylquinic acid, quercetin-3- O-galactoside, quercetin-3-O-rutinoside, kaempferol-3- O-glycoside, kaempferol-3- O-glucoside, and kaempferol-3-O-rutinoside. 5- O-caffeoylquinic acid was the major phenolic compound (36.2%), followed by quercetin 3- O-rutinoside (21.1%). Quince leaves had higher relative contents of kaempferol derivatives than fruits (pulps, peels, and seeds), especially in regards to kaempferol-3- O-rutinoside (12.5%). *C. oblonga* leaf total phenolic content was very high, varying from 4.9 to 16.5 g/kg dry matter (mean value of 10.3 g/kg dry matter), indicating that these leaves can be used as a good and cheap source of bioactive constituents. Two new ionone glucosides 9-O-β-D-glucopyranosides of (6R)-3-oxo-4-hydroxy-7,8-dihydro-α-ionol 1 and 3-oxo-5,6-epoxy-β-ionol 2 were isolated from quince leaves (Lutz et al. 2002). The novel quince leaf constituents were characterized as peracetates 1a and 2a.

The chloroform-methanol extract of quince plant was shown to contain four new sesterterpene esters, namely 24,25-O-diacetylvulgaroside (1), 25-O-acetylvulgaroside (2), 24-O-acetyl-25-O-cinnamoylvulgaroside (3), and 25-O-cinnamoylvulgaroside (4) (De Tommasi et al.

1996a). Four new flavonol glycosides (1–4) and nine new α-ionol-derived glycosides (5–13) together with the known 3-oxo-α-ionol 9-O-β-d-apiofuranosyl-(1→6)-β-d-glucopyranoside (14), vomifoliol 9-O-β-d-glucopyranoside (roseoside) (15), and vomifoliol 9-O-β-d-apiofuranosyl-(1→6)-β-d-glucopyranoside (16) were isolated from the methanol extract of the aerial parts of quince plant (De Tommasi et al. 1996b).

Fibre-rich products (e.g. powder) could be obtained from quince wastes with useful functional and physiological properties (de Escalada Pla et al. 2010). The products obtained presented interesting hydration properties comparable to those reported for citrus and apple pulps. At the same time, all dried fractions showed high spontaneous water absorption rate in kinetics assay. Oil absorption seemed to essentially depend on the microstructural characteristics of the fibre powders whereas water absorption were determined by the material's hydrophilicity.

Quince leaves were found to constitute a promising natural source of bioactive phytochemicals such as phenolics and organic acids, with antioxidant and antiproliferative properties suitable for application in nutritional/pharmaceutical fields (Oliveira et al. 2010). Comparisons with green tea, considered by the scientific community as an effective natural antioxidant, were established. Phenolics from quince fruit and leaves were found to possess antimicrobial, antioxidant, anticancer, hypoglycaemic antiulcerative, antimicrobial, antiviral, antiallegic and immunomodulatory activities.

Antioxidant Activity

In one study, thirteen fatty acid esters of cinnamyl alcohols, three fatty acid esters of hydroxybenzoic acid, three fatty acid esters of hydroxybenzaldehyde, three glucosides of aromatic acids, four chlorogenic acids, two flavonols, and a benzylamine were isolated and identified from organic extract of quince peels (Fiorentino et al. 2008). The chlorogenic acids and the flavonols exhibited more antioxidant and radical scavenger capacity than the positive standards α-tocopherol and ascorbic acid. Another research showed that the strength of antioxidant activity and 2,2'-diphenyl-1-picrylhydrazyl (DPPH) radical scavenging activity of the fruit phenolics varied according to different in-vitro evaluation systems, whereas the antioxidative property of rat blood increased in all rats orally administered fruit phenolics (Hamauzu et al. 2005, 2006). Quince was found to have considerable amounts of hydroxycinnamic derivatives mainly composed of 3-caffeoylquinic acid and 5-caffeoylquinic acid and polymeric procyanidins. The antioxidant functions of quince and Chinese quince phenolic extracts were superior to that of chlorogenic acid standard or ascorbic acid evaluated in both the linoleic acid peroxidation system and the DPPH radical scavenging system.

The phenolic fraction of the methanol extract of quince fruit (pulp, peel, and seed) and jam exhibited a stronger antioxidant activity than the whole methanolic extract (Silva et al. 2004a). Organic acid fractions were always the weakest in terms of antiradical activity, implying that the phenolic fraction contributed more to the antioxidant potential of quince fruit and jam. The methanol peel extract showed the highest antioxidant capacity. Among the phenolic fractions, the seed fraction was the one that exhibited the strongest antioxidant activity. The IC_{50} values of quince pulp, peel, and jam phenolic fractions were strongly correlated with caffeoylquinic acids and phenolics total contents. For organic acid fractions, the peel was the one that had the strongest antiradical activity. The IC_{50} values of quince pulp, peel, and jam organic acid fractions were correlated with the ascorbic acid and citric acid contents.

Quince leaf was found to have comparable antioxidant activity as green tea (*Camellia sinensis*) (Costa et al. 2009). Quince leaf exhibited a significantly higher reducing power than green tea (mean value of 227.8 and 112.5 g/kg dry leaf, respectively). Quince leaf extracts showed similar DPPH (2,2'-diphenyl-1-picrylhydrazyl) radical-scavenging activities (EC_{50} mean value of 21.6 μg/ml) but significantly lower than that presented by green tea extract (EC_{50} mean value of 12.7 μg/ml). Caffeoylquinic acid was found to be the major phenolic compound in quince leaf extract. Under the oxidative action of AAPH (2,2′-azobis(2-amidinopropane) (dihydrochloride)), quince leaf methanolic extract significantly protected the erythrocyte membrane from hemolysis in a similar manner to that found for green tea (IC_{50} mean value of 30.7 and 24.3 µg/ml, respectively).

Anticancer Activity

Studies revealed that quince leaf and fruit extracts exhibited marked antiproliferative activities (Carvalho et al. 2010). The extracts from quince leaf showed concentration-dependent growth inhibitory activity toward human colon cancer cells ($IC_{50} = 239.7$ µg/ml), while no effect was observed in renal adenocarcinoma cells. Concerning the fruit, seed extracts exhibited no effect on colon cancer cell growth, whereas strong antiproliferative efficiency against renal cancer cells was observed for the highest concentration assayed (500 µg/ml). The antiproliferative activity of pulp and peel extracts was low or absent in the selected range of extract concentrations.

Antimicrobial Activity

Quince peel extract was found to be the most active in inhibiting bacterial growth with minimum inhibitory and bactericide concentrations in the range of $102^{-5} \times 10^3$ µg polyphenol/ml (Fattouch et al. 2007). It was postulated that chlorogenic acid acted in synergism with other components of the extracts to exhibit their total antimicrobial activities.

Antiviral Activity

Chloroform-methanolic extract of quince fruit yielded four new sesterterpene esters, namely 24, 25-O-diacetylvulgaroside, 25-O-acetylvulgaroside, 24-O-acetyl-25-O-cinnamoylvulgaroside, and 25-O-cinnamoylvulgaroside that exhibited antiviral (HIV) activity (De Tommasi et al. 1996a, b).

Hypoglycaemic Activity

Oral administration of *Cydonia oblonga* ethanol leaf extract (500 mg/kg) for 5 days in diabetic rats caused a decrease in blood glucose levels by 33.8% (Aslan et al. 2010). Moreover, the *Cydonia oblonga* extract induced significant alleviation on only heart tissue TBARS levels (44.6%) but did not restore reduced glutathione (GSH) levels in kidney, liver, and heart tissues of diabetic rats.

Antiallergic Activity

Gencydo(®), a combination of lemon (*Citrus limon*) juice and aqueous quince (*Cydonia oblonga*) extract had been used as a traditional topical treatment of allergic disorders. Studies by (Gründemann et al. 2011) revealed that Gencydo(®) downregulated soluble mediators, which were essential for the initiation and maintenance of allergic reactions. Gencydo(®) reduced the degranulation and histamine release of IgE-activated basophilic cells and mast cells and inhibited the IgE- and PMA/A23187-induced increases in IL-8, TNF-α and GM-CSF production in mast cells. The effects were comparable to that of the used concentration of azelastine and dexamethasone. Furthermore, Gencydo(®) partially blocked eotaxin release from human bronchial epithelial cells, but had no impact on the viability and activation of GM-CSF-activated eosinophil granulocytes.

Immunomodulatory Activity

Hot water quince extract reduced the induction of intracellular cyclooxygenase (COX)-2 expression but not COX-1 expression in mouse bone marrow-derived mast cells (Kawahara and Iizuka 2011). The extract reduced the elevation of interleukin-13 and tumour necrosis factor-α expression level. The extract also suppressed these cytokine expressions and leukotriene C(4) and prostaglandin D(2) production in mouse bone marrow-derived mast cell. The results suggested that quince hot water extract had an inhibitory effect on broad range of the late-phase immune reactions of mast cells.

Antiulcerative Activity

In the ethanol-induced gastric ulcer, pre-administration of Chinese quince and quince phenolics suppressed the occurrence of gastric lesions in rats, whereas apple phenolics seemed to promote ulceration (Hamauzu et al. 2006). The trend of myeloperoxidase activity was similar to that of the ulcer index. The results showed that Chinese quince and quince phenolics might have health benefits by acting both in blood vessels and on the gastrointestinal tract.

Traditional Medicinal Uses

Quince has been used as an herbal medicine since ancient times.

The fruit is antivinous, astringent, cardiac, carminative, digestive, diuretic, emollient, expectorant, pectoral, peptic, refrigerant, restorative, stimulant and tonic. An infusion has been employed for sore throat, diarrhoea and haemorrhage of the bowel. It is effective against inflammation of the mucous membranes, intestines and stomach. The fruit, and its juice, can be used as a mouthwash or gargle to treat mouth ulcers, gum problems and sore throats Quince is also used in the cosmetic industry and for medicinal cosmetics. The unripe fruit is very astringent, a syrup made from it is used in the treatment of diarrhoea and is particularly safe for children. The fruit is rich in pectin. Pectin is said to protect the body against radiation and has a beneficial effect on the circulatory system and helps to reduce blood pressure. In Malta, a teaspoon of quince jam dissolved in a cup of boiling water relieves intestinal discomfort. The leaves contain tannin and pectin. Tannin can be used as an astringent. The stem-bark is astringent and used for ulcers in Chinese herbal medicine.

The seed is a mild but reliable laxative, astringent and antiinflammatory. The seeds are used as a remedy for pneumonia and lung disease in Iran. In parts of Afghanistan, the quince seeds are collected and boiled and then ingested to combat pneumonia. The seeds are also rich in pectin, soaked or boiled in water, release the mucilage from the seed coat. The mucilage has a soothing and demulcent action when taken internally and is used in the treatment of sore throat and respiratory diseases, especially in children. This mucilage is also applied externally to minor burns and use for eye lotions. Studies in Iran reported that quince seed mucilage proved to hasten wounds more rapidly than a commercial wound healing cream (1% phenytoin) or eucerin cream without mucilage (Hemmati and Mohammadian 2000). Quince leaves have been used as decoction or infusion, in folk medicine for their sedative, antipyretic, anti-diarrheic and antitussive properties and for the treatment of various skin diseases (De Tommasi et al. 1996a; Oliveira et al. 2007). *Cydonia oblonga* leaves are used as a folk remedy for the treatment of diabetes in Turkey (Aslan et al. 2010). In Germany, Gencydo(®), a combination of lemon (*Citrus limon*) juice and aqueous quince (*Cydonia oblonga*) extract has been used traditionally in anthroposophical medicine for treating patients with allergic rhinitis or asthma (Gründemann et al. 2011).

Other Uses

Quinces are very widely used as rootstocks for pears and is also suitable for loquat. Quince has the property of dwarfing the growth of pears, of forcing them to produce more precociously, and to bear more fruits, instead of vegetative growth, and of accelerating the maturity of the fruit. Quince flower is also a rich nectar source for bees. Quince fruit because of their strong and rich fragrance was once popularly used as room deodorisers. Mucilage from the seed coat is used as a gum arabic substitute to add gloss to a material. Quince leaves contain 11% tannin and can be used for tanning.

Comments

Four other species previously included in the genus *Cydonia* have now been moved to separate genera. These are the Chinese Quince, *Pseudocydonia sinensis*, a native of China, and the three flowering quinces of eastern Asia in the genus *Chaenomeles*.

Selected References

Alesiani D, Canini A, D'Abrosca B, DellaGreca M, Fiorentino A, Mastellone C, Monaco P, Pacifico S (2010) Antioxidant and antiproliferative activities of phytochemicals from Quince (*Cydonia vulgaris*) peels. Food Chem 118(2):199–207

Andrade PB, Carvalho ARF, Seabra RM, Ferreira MA (1998) A previous study of phenolic profiles of quince, pear, and apple purees by HPLC diode array detection for the evaluation of quince puree genuineness. J Agric Food Chem 46(3):968–972

Aslan M, Orhan N, Orhan DD, Ergun F (2010) Hypoglycemic activity and antioxidant potential of some medicinal plants traditionally used in Turkey for diabetes. J Ethnopharmacol 128(2):384–389

Carvalho M, Silva BM, Silva R, Valentão P, Andrade PB, Bastos ML (2010) First report on *Cydonia oblonga* Miller anticancer potential: differential antiproliferative effect against human kidney and colon cancer cells. J Agric Food Chem 58(6):3366–3370

Chevallier A (1996) The encyclopedia of medicinal plants. Dorling Kindersley, London, 336pp

Chopra RN, Nayar SL, Chopra IC (1986) Glossary of Indian medicinal plants (Including the supplement). Council Scientific Industrial Research, New Delhi, 330pp

Costa RM, Magalhães AS, Pereira JA, Andrade PB, Valentão P, Carvalho M, Silva BM (2009) Evaluation of free radical-scavenging and antihemolytic activities of quince (*Cydonia oblonga*) leaf: a comparative study with green tea (*Camellia sinensis*). Food Chem Toxicol 47(4):860–865

de Escalada Pla MF, Uribe M, Fissore EL, Gerschenson LN, Rojas AM (2010) Influence of the isolation procedure on the characteristics of fiber-rich products obtained from quince wastes. J Food Eng 96(2):239–248

De Tommasi N, De Simone F, Pizza C, Mahmood N (1996a) New tetracyclic sesterterpenes from *Cydonia vulgaris*. J Nat Prod 59(3):267–270

De Tommasi N, Piacente S, De Simone F, Pizza C (1996b) Constituents of *Cydonia vulgaris*: isolation and structure elucidation of four new flavonol glycosides and nine new (a-ionol-derived glycosides). J Agric Food Chem 44(7):1676–1681

Duke JA, Ayensu ES (1985) Medicinal plants of China, vol 1&2. Reference Publications, Inc., Algonac, 705pp

Facciola S (1990) Cornucopia. A source book of edible plants. Kampong Publication, Vista, 677pp

Fattouch S, Caboni P, Coroneo V, Tuberoso CIG, Angioni A, Dessi S, Marzouki N, Cabras P (2007) Antimicrobial activity of Tunisian quince (*Cydonia oblonga* Miller) pulp and peel polyphenolic extracts. J Agric Food Chem 55(3):963–969

Ferreres F, Silva BM, Andrade PB, Seabra RM, Ferreira MA (2003) Approach to the study of C-glycosyl flavones by ion trap HPLC-PAD-ESI/MS/MS: application to seeds of quince (*Cydonia oblonga*). Phytochem Anal 14(6):352–359

Fiorentino A, D'Abrosca B, Pacifico S, Mastellone C, Piscopo V, Caputo R, Monaco P (2008) Isolation and structure elucidation of antioxidant polyphenols from quince (*Cydonia vulgaris*) peels. J Agric Food Chem 56(8):2660–2667

Fleischmann P, Studer K, Winterhalter P (2002) Partial purification and kinetic characterization of a carotenoid cleavage enzyme from quince fruit (*Cydonia oblonga*). J Agric Food Chem 50(6):1677–1680

Gründemann C, Papagiannopoulos M, Lamy E, Mersch-Sundermann V, Huber R (2011) Immunomodulatory properties of a lemon-quince preparation (Gencydo(®)) as an indicator of anti-allergic potency. Phytomedicine 18(8–9):760–768

Hamauzu Y, Yasui H, Inno T, Kume C, Omanyuda M (2005) Phenolic profile, antioxidant property, and anti-influenza viral activity of Chinese quince (*Pseudocydonia sinensis* Schneid.), quince (*Cydonia oblonga* Mill.), and apple (*Malus domestica* Mill.) fruits. J Agric Food Chem 53(4):928–934

Hamauzu Y, Inno T, Kume C, Irie M, Hiramatsu K (2006) Antioxidant and antiulcerative properties of phenolics from Chinese quince, quince, and apple fruits. J Agric Food Chem 54(3):765–772

Hemmati AA, Mohammadian F (2000) An investigation into the effects of mucilage of quince seeds on wound healing in rabbit. J Herbs Spices Med Plants 7(4):41–46

Huxley AJ, Griffiths M, Levy M (eds) (1992) The new RHS dictionary of gardening, 4 vols. Macmillan, London

Kawahara T, Iizuka T (2011) Inhibitory effect of hot-water extract of quince (*Cydonia oblonga*) on immunoglobulin E-dependent late-phase immune reactions of mast cells. Cytotechnology 63(2):143–152

Ku TC, Spongberg SA (2003) *Cydonia* Miller. In: Wu ZY, Raven PH, Hong DY (eds) Flora of China, vol 9 (*Pittosporaceae* through *Connaraceae*). Science Press/Missouri Botanical Garden Press, Beijing/St. Louis

Lorenz P, Berger M, Bertrams J, Wende K, Wenzel K, Lindequist U, Meyer U, Stintzing FC (2008) Natural wax constituents of a supercritical fluid CO(2) extract from quince (*Cydonia oblonga* Mill.) pomace. Anal Bioanal Chem 391(2):633–646

Lutz A, Schneider M, Winterhalter P (2002) Isolation of two new ionone glucosides from quince (*Cydonia oblonga* Mill) leaves. Nat Prod Lett 16:119–122

Magalhães AS, Silva BM, Pereira JA, Andrade PB, Valentão P, Carvalho M (2009) Protective effect of quince (*Cydonia oblonga* Miller) fruit against oxidative hemolysis of human erythrocytes. Food Chem Toxicol 47(6):1372–1377

Oliveira AP, Costa RM, Magalhães AS, Pereira JA, Carvalho M, Valentão P, Andrade PB, Silva BM (2010) Targeted metabolites and biological activities of *Cydonia oblonga* Miller leaves. Food Res Int doi: 10.1016/j.foodres.2010.10.021

Oliveira AP, Pereira JA, Andrade PB, Valentão P, Seabra RM, Silva BM (2007) Phenolic profile of *Cydonia oblonga* Miller leaves. J Agric Food Chem 55(19): 7926–7930

Oliveira AP, Pereira JA, Andrade PB, Valentão P, Seabra RM, Silva BM (2008) Organic acids composition of *Cydonia oblonga* Miller leaf. Food Chem 111(2): 393–399

Rodríguez-Guisado I, Hernández F, Melgarejo P, Legua P, Martínez R, Martínez JJ (2009) Chemical, morphological and organoleptical characterisation of five Spanish quince tree clones (*Cydonia oblonga* Miller). Sci Hortic 122(3):491–496

Silva BM, Andrade PB, Ferreres F, Domingues AL, Seabra RM, Ferreira MA (2002a) Phenolic profile of quince fruit (*Cydonia oblonga* Miller) (pulp and peel). J Agric Food Chem 50(16):4615–4618

Silva BM, Andrade PB, Ferreres F, Seabra RM, Oliveira MBPP, Ferreira MA (2005a) Composition of quince (*Cydonia oblonga* Miller) seeds: phenolics, organic acids and free amino acids. Nat Prod Res 19(3):275–281

Silva BM, Andrade PB, Martins RC, Seabra RM, Ferreira MA (2006) Principal component analysis as tool of characterization of quince (*Cydonia oblonga* Miller) jam. Food Chem 94(4):504–512

Silva BM, Andrade PB, Martins RC, Valentão P, Ferreres F, Seabra RM, Ferriera MA (2005b) Quince (*Cydonia oblonga* Miller) fruit characterization using principal component analysis. J Agric Food Chem 53:111–122

Silva BM, Andrade PB, Mendes GC, Seabra RM, Ferreira MA (2002b) Study of the organic acids composition of quince (*Cydonia oblonga* Miller) fruit and jam. J Agric Food Chem 50(8):2313–2317

Silva BM, Andrade PB, Mendes GC, Valentão P, Seabra RM, Ferreira MA (2000a) Analysis of phenolic compounds in the evaluation of commercial quince jam authenticity. J Agric Food Chem 48:2853–2857

Silva BM, Andrade PB, Seabra RM, Ferreira MA (2001) Determination of selected phenolic compounds in quince jams by solid-phase extraction and HPLC. J Liq Chromatogr Relat Technol 24(18):2861–2872

Silva BM, Andrade PB, Valentão P, Ferreres F, Seabra RM, Ferreira MA (2004a) Quince (*Cydonia oblonga* Miller) fruit (pulp, peel, and seed) and jam: antioxidant activity. J Agric Food Chem 52(5):4705–4712

Silva BM, Andrade PB, Valentão P, Mendes GC, Seabra RM, Ferreira MA (2000b) Phenolic profile in the evaluation of commercial quince jellies authenticity. Food Chem 71(2):281–285

Silva BM, Casal S, Andrade PB, Seabra RM, Oliveira MB, Ferreira MA (2003) Development and evaluation of a GC/FID method for the analysis of free amino acids in quince fruit and jam. Anal Sci 19: 1285–1290

Silva BM, Casal S, Andrade PB, Seabra RM, Oliveira MBPP, Ferreira MA (2004b) Free amino acid composition of quince (Cydonia oblonga Miller) fruit (pulp and peel) and jam. J Agric Food Chem 52(5): 1201–120

Sousa C, Silva BM, Andrade PB, Valentão P, Silva A, Ferreres F, Seabra RM, Ma F (2007) Homomonoterpenic compounds as chemical markers for *Cydonia oblonga* Miller. Food Chem 100(1):331–338

Trigueros L, Pérez-Alvarez JA, Viuda-Martos M, Sendra E (2011) Production of low-fat yogurt with quince (*Cydonia oblonga* Mill.) scalding water. LWT- Food Sci Technol 44(6):1388–1395

US Department of Agriculture, Agricultural Research Service (USDA) (2011) USDA national nutrient database for standard reference, release 24. Nutrient Data Laboratory Home Page. http://www.ars.usda.gov/ba/bhnrc/ndl

Eriobotrya japonica

Scientific Name

Eriobotrya japonica (Thunb.) Lindl.

Synonyms

Crataegus bibas Lour., *Mespilus japonica* Thunb., *Photinia japonica* Franch. & Sav.

Family

Rosaceae

Common/English Names

Chinese Loquat, Japanese Loquat, Japanese Medlar, Japanese Plum, Loquat, Naspli.

Vernacular Names

Afrikaans: Lukwart;
Brazil: Ameixa-Do-Japão, Nêspera, Nespereira-Do-Japão
Burmese: Tayok Hninthi;
Chinese: Pi Ba, Pi Ba Ye;
Czech: Mišpule, Mišpule Japonská;
Danish: Japanmispel, Japansk Mispel;
Dutch: Japanse Mispel, Lokwat;
Eastonian: Jaapani Villpööris, Vili: Nispero;
Finnish: Japaninmispeli, Lokvatti;
French: Bibasse (Fruit), Bibassier (Tree), Néfle Du Japon (Fruit), Néflier Du Japon (Tree);
German: Japanische Mispel, Japanische Wollmispel;
Greek: Iaponika Mousmoula, Mespilea E Iaponiki, Mousmoulia E Koini;
Hungarian: Japán Naspolya;
Italian: Nespola Del Giappone, Nespolo Del Giappone;
Japanese: Biwa;
Khmer: Tôn Leap;
Korean: Bipanamu;
Nepalese: Lokat, Lukaath, Suvana Aalu;
Polish: Niesplik Japonski;
Portuguese: Nêsperas Do Japão, Ameixa Amarella, Ameixa Do Japa, Nespereira (Tree);
Russian: Eriobotriia, Lokva, Mushmula Iaponskaia;
Slovenian: Japonska Nešplja, Lokvat;
Spanish: Nispero De España, Níspero Del Japón (Tree), Níspola Del Japón, Nectarina Nispero, Níspero De Japón Nespera, Nispolero, Nespora (Fruit), Nispolero;
Swedish: Japansk Mispel;
Thai: Lokhwoot, Pi Pae;
Vietnamese: Ti Ba Diêp, Nhót Tây;
Zambia: Muzhanje (Shona);
Zimbabwe: Muzhanje (Shona).

Origin/Distribution

Loquat is native to western China, wild ancestors of the cultivated loquat can still be found in the mountains of Sizhuan and Yunnan. It is widely cultivated along the Yangtze River and southwards especially in Suzhou. It was introduced to Japan by Buddhist monks in the Tang dynasty. Loquat is widely cultivated in suitable areas in subtropical to warm-temperate zones and in the tropics in the highlands. Besides China, other centres of production are in east Asia (Japan), south Asia (India), Australia, South Africa, Central and South America (Brazil) and around the Mediterranean Sea.

Agroecology

The loquat is adapted to a subtropical to mild-temperate climate. In China, it grows naturally at altitudes between 900 and 2,000 m. In India, it grows up to 1,500 m and in Guatemala at elevations of 900–1,200 m, yielding poorly at lower elevations. Well-established trees can tolerate a drop in temperature to −11.11°C. In Japan, the lower temperature limit 7°C, temperatures below this is detrimental to flowers and fruit resulting in flower drop and fruit abortion. Extreme summer heat is detrimental to the crop, and dry, hot winds cause leaf scorch.

Loquat will grow on a wide range of soils of moderate fertility, from light sandy loam to heavy clay and even oolitic limestone, but needs good drainage. It abhors water-logged conditions.

Edible Plant Parts and Uses

Loquat is eaten as a fresh fruit and sometimes mixed with sliced banana, orange sections and grated coconut in fresh fruit salads or fruit cups. Firm, slightly immature fruits are best for making pies or tarts. The fruits are also commonly used to make jam, marmalades, jelly, chutney, and are delicious poached or stewed in light syrup. Fruit can also be chopped and cooked as a sauce. Spiced loquats are also prepared with cloves, cinnamon, lemon and vinegar and bottled in glass jars. Loquat fruit are also made into beverages and makes an excellent light wine. Taiwan exports canned loquat in syrup. Seeds are also edible and have a pleasant flavour when cooked. Roasted seeds are used as a coffee substitute. In China, fresh roots are used as food, cooked slowly with pig's throttles or chicken and yellow rice wine.

Botany

A small, evergreen tree, to 10 m high, with a short trunk, rounded canopy and greyish-rusty tomentose stout twigs (Plate 1). Stipule is subulate and pubescent, petiole short to sub sessile and tomentose. Leaves are glossy dark green, margin entire, shallowly serrate apically, coriaceous, large, obovate-oblong or elliptic, 12–30 long, 3–9 cm wide, apex acute, base cuneate, abaxially rusty grey tomentose, 11–12 pairs of lateral veins (Plates 3 and 4). Flowers are fragrant, ivory- white, 1.2–2 cm across, numerous (30–100) borne in large terminal panicles 10–20 cm long, tomentose throughout (Plate 2). Sepals are triangular-ovate, apex obtuse; petals 5, white, oblong or ovate, apex obtuse or emarginate; stamens 20; ovary rusty pubescent, 5-loculed, with 2 ovules per locule; styles 5, free. Fruit occurs in clusters of 4–30, green (Plates 3 and 4) turning to yellow or orangey-yellow, globose or obovate, 3–4 cm by 2–5 cm across, skin leathery rusty tomentose later glabrescent, juicy pulp, sweet to sub acid (Plates 5, 6, 7). Seed is usually 1-3-(5), brown, angular-ellipsoid, 15 mm by 8 mm (Plate 7).

Plate 1 Young Loquat tree

Eriobotrya japonica

Plate 2 Loquat flowers

Plate 3 Loquat fruit clusters and leaves

Plate 5 Ripening loquat fruits

Plate 4 Loquat leaves with rusty grey tomentose abaxial surfaces

Plate 6 Harvested ripe loquat fruits

Plate 7 Whole and halved loquat fruit with seed

Nutritive/Medicinal Properties

Analyses carried out in the United States reported raw loquat fruit, *Eriobotrya japonica* (minus 35% seeds, and skin) to have the following nutrient composition (per 100 g edible portion): water 86.73 g, energy 47 kcal (197kj), protein 0.43 g, total lipid 0.20 g, ash 0.50 g, carbohydrates 12.14 g, total dietary fibre 1.7 g, Ca 16 mg, Fe 0.28 mg, Mg 13 mg, P 27 mg, K 266 mg, Na 1 mg, Zn 0.05 mg, Cu 0.04 mg, Mn 0.148 mg, vitamin C 1.0 mg, thiamin 0.019 mg, riboflavin 0.0.024 mg, Niacin 0.180 mg, vitamin B-6 0.10 mg, total folate 14mcg, vitamin A 1,528 IU, total saturated fatty acids 0.04og, total monounsaturated fatty acids 0.008 g, 16:1 undifferentiated fatty acids 0.008 g, total polyunsaturated fatty acids 0.091 g, 18:2 undifferentiated fatty acids 0.077 g, 18:3 undifferentiated fatty acids 0.013 g, phytosterols 2 mg, tryptophan 0.005 g, threonine 0.015 g, isoleucine 0.015 g, leucine 0.026 g, lysine 0.023 g, methionine 0.004 g, cystine 0.006 g, phenylalanine 0.014 g, tyrosine 0.013 g, valine 0.021 g, arginine 0.014 g, histidine 0.007 g, alanine 0.024 g, aspartic acid 0.058 g, glutamic acid 0.061 g, glycine 0.020 g, proline 0.027 g and serine 0.020 g (USDA 2011).

Loquat is extremely rich in vitamin A and is low in fats; it also has phytosterols, many essential amino acids and contains phytonutrient antioxidants beneficial to health.

Other Phytochemicals

Fruit: Phenolic compounds in loquat fruit were identified as 5-caffeoylquinic acid (chlorogenic acid), neochlorogenic acid, hydroxybenzoic acid, 5-p-feruloylquinic acid, protocatechuic acid, 4-caffeoylquinic acid, epicatechin, o-coumaric acid, ferulic acid, and p-coumaric acid (Ding et al. 2001). Neochlorogenic acid was found to be dominant in the early stages of loquat fruit development. Both the concentrations and types of phenolic compounds were high in young fruit but then decreased steadily during growth. Chlorogenic acid levels increased during ripening and became predominant in ripe fruit. In all of the cultivars, the types of phenolic compounds were similar but the total phenolic content varied from 81.8 to 173.8 mg/100 g of fresh pulp. In the biosynthetic pathway of chlorogenic acid, the enzyme activities of phenylalanine ammonia-lyase (PAL), 4-coumarate:CoA ligase (CL), and hydroxycinnamoyl CoA:quinate hydroxycinnamoyl transferase (CQT) were high at the early stage of growth, diminished to low levels about 3 weeks prior to harvest, but then rose to a peak at 1 week before harvest. The changes of these enzyme activities appeared to be associated with variations in chlorogenic acid concentration during development, maturation, and ripening of loquat fruit.

Twenty-five carotenoids were identified in the fruits of 5 Brazilian loquat cultivars (Ferreira de Faria et al. 2009). All-trans-β-carotene (19–55%), all-trans-β-cryptoxanthin (18–28%), 5,6:5′,6′-diepoxy-β-cryptoxanthin (9–18%) and 5,6-epoxy-β-cryptoxanthin (7–10%) were the main carotenoids. The total carotenoid content ranged from 196 μg/100 g (cv. Néctar de Cristal) to 3,020 μg/100 g (cv. Mizumo). Cultivars Mizauto, Mizuho, Mizumo and Centenária showed provitamin A values between 89 and 162 μg RAE/100 g, and could be considered good source of this provitamin.

Seed: A hydroxynitrile lyase, a flavoprotein enzyme containing FAD (flavin adenine dinucleotide) as a prosthetic group isolated from loquat seeds was found to be active toward aromatic and aliphatic aldehydes, and showed a preference for smaller substrates over bulky ones (Ueatrongchit et al. 2008). The enzyme showed excellent stability with regard to pH and temperature.

Flowers: oleanolic acid ursolic acid 2α, 3α,19α-trihydroxyurs-5,12-dien-28-acid, and 2β, 3β, 23α-trihydroxyolean-12-en-28-acid were isolated from loquat flowers (Cheng et al. 2001).

Leaves: Three new flavonoid glycosides, together with 15 known flavonoids, were isolated from loquat leaves, and characterized as (2S)- and (2R)-naringenin 8-C-α-L-rhamnopyranosyl-(1→2)-β-D-glucopyranosides, and cinchonain Id 7-O-β-D-glucopyranoside, respectively (Kobayashi and Takamatsu 2000). Two new megastigmane glycosides, eriojaposides A and B, and a new acylated triterpenoid were isolated along with nine known compounds from a loquat leaf extract (Ito et al. 2001). Nine triterpenes were isolated and identified as methyl betuliate, oleanolic acid, ursolic acid, methyl maslinate, methyl corosolate, (methyl 2α-3α-dihydroxyurs-12-en-28-oate), maslinic acid, corosolic acid, tormentic acid, euscaphic acid from loquat leaf (Lv et al. 2008).

Pharmacological studies have shown that various extracts of loquat possess hepatoprotective, gastroprotective, antiinflammatory, antioxidant, hypoglycemic or antihyperglycemic, antiallergic, antiviral, anticancer and antiplatelet properties.

Antioxidant Activity

Studies showed that the ethanol extracts of all three loquat fruit parts (peel, flesh and seed) and the water extract of the peels exhibited a strong ability to scavenge 1,1-diphenyl-2-picrylhydrazyl (DPPH) radical (Koba et al. 2007). Among extracts of different loquat parts, the ethanol extract of loquat seeds was the most potent. The ethanol extract of the seed was also effective in suppressing the oxidation of linoleic acid which was demonstrated by a slow discoloration of β-carotene/linoleic acid conjugation system. The ethanol extract of loquat seeds as compared to other extracts could also suppress significantly the 2,2'-azobis(4-methoxy-2,4-dimethylvaleronitrile) (MeO-AMVN)-induced LDL oxidation. When the content of total polyphenolic compounds in different loquat parts (peel, flesh and seed) was examined, a significantly higher level of total polyphenols was found in the seed than the peels and flesh. Significant levels of polyphenolic compounds such as chlorogenic acid, cyanidine glucoside, epicatechin, epigalocatechin gallate and procyanidin B2 in the ethanol extract of different loquat parts were identified and quantified. The latter two compounds were found mainly in the ethanol extract of loquat seeds, but not in peels and flesh. Therefore, it was suggested that the high ability to scavenge free radicals and suppress the LDL oxidation exerted by the ethanol extract of loquat seeds was partially due to the high content of polyphenolic compounds in the seeds.

Another study reported that the radical scavenging activities and inhibitory activities on lipid peroxidation differed among the solvent fractions and components of loquat seed extract (Yokota et al. 2006). In the n-butanol, methanol and water fractions, radical scavenging activity and inhibitory activity on lipid peroxidation were high. In addition, these fractions contained abundant polyphenols, and the radical scavenging activity increased with the polyphenol content. In the low-polar n-hexane and ethyl acetate fractions, the radical scavenging activity was low, but the lipid peroxidation inhibition activity was high. These fractions contained β-sitosterol, and the inhibitory activity on lipid peroxidation was high. Based on these findings, the antioxidative activity of *Eriobotrya japonica* seed extract may be derived from many components involved in a complex mechanism, resulting in high activity. *Eriobotrya japonica* seed extract was also reported useful for chemotherapy of (5-fluorouracil)-induced mucositis in hamsters (Takuma et al. 2008). After 10 days of treatment, epithelial injury and bacterial infection were noted in the tap water group but not in the seed extract treated group. The plasma lipid peroxide level was significantly lower in this group than that in the tap water group.

Flavonoids and phenolics were found to be abundant in loquat flowers (Zhou et al. 2011). The average content of flavonoids and phenolics of loquat flower of five cultivars were 1.59 and 7.86 mg/g DW, respectively, when using ethanol as extraction solvent. The antioxidant capacity values of ABTS method was highest, followed by DPPH, the lowest was FRAP, when using vitamin C equivalent antioxidant capacity (VCEAC) as unit.

The methanol extract and its fractions of *Eriobotrya japonica* leaves also showed strong antioxidant activity. The antioxidant activity of ethyl acetate and n-butanol soluble fractions were stronger than the others, antioxidant chlorogenic acid, quercetin-3-sambubioside from n-butanol fraction, and methyl chlorogenate, kaempferol- and quercetin-3-rhamnosides, together with the inactive ursolic acid and 2 α-hydroxyursolic acid from ethyl acetate fraction were isolated from the fractions (Jung et al. 1999). Antioxidant flavonoids and chlorogenic acid also showed prominent inhibitory activity against free radical generation in dichlorofluorescein (DCF) assay. Eighteen compounds (8 hydroxycinnamic acid derivatives and 10 flavonoid glycosides) were identified from the leaves and fruits (peel and flesh) of six improved loquat cultivars (Ferreres et al. 2009). Loquat leaves exhibited the lowest amounts of phenolics. 3- And 5-caffeoylquinic, and 5-feruloylquinic acids were the major compounds. Generally, 'Mizauto' cultivar presented the highest phenolic content. All loquat materials exhibited DDPH scavenging capacity, in a concentration-dependent manner, with the leaves being the most active and appeared to be related to the flavonoid content. Polat et al. (2010) found that total phenolic content varied between loquat cultivars, from 129 mg gallic acid equivalent/kg fresh fruit weight in cv 'Baduna 5' to 578 mg gallic acid equivalent/kg. 'Hafif Çukurgöbek' also had the highest FRAP mean (12.1 mmol Trolox Equivalent (TE)/kg fw) and had the highest pH and high soluble solids.

Antiinflammatory Activity

Studies by Huang et al. (2006) showed that triterpene acids of loquat leaves had antiinflammatory activity. Intra-gastric administration of loquat triterpene acids to lipoploysaccharide-induced chronic bronchitis rats inhibited heme oxygenase-1 (HO-1) expression and malondialdehyde (MDA) production and up-regulated superoxide dismutase which might be one of molecular mechanisms of the antiinflammatory effects. Loquat triterpene acids also inhibited nuclear factor κ B (NF-kB) activation in alveolar macrophages from chronic bronchitis rats and led to down-regulation of tumour necrosis factor-α (TNF-α), interleukin-1 (IL-1), prostaglandin E2 (PGE2) and leukotriene (LTB4) expression, which might be a mechanism for the antiinflammatory effects in chronic bronchitis rats (Huang et al. 2007). Similar antiinflammatory results and underlying mechanism with loquat leaf tritepenoic acids were reported in chronic bronchitis rats by (Ge et al. 2009). In chronic bronchitis rats treated with triterpenoic acids from *Eriobotrya japonica* (TAL), (50, 150 and 450 mg/kg) for 3 weeks, the total number of leukocyte, the differential counts of neutrophils and alveolar macrophage in bronchoalveolar lavage fluid, the levels of levels of cytokine tumour necrosis factor alpha (TNF-α), interleukin (IL)-8, in the supernatants of lung homogenate, and the expression of nuclear factor κB and intercellular adhesion molecule-1 on bronchial epithelium in chronic bronchitis rats were significantly increased, while the level of interleukin -10 (IL-10) was decreased compared with the normal and sham groups (Ge et al. 2009). The data indicated that TAL has antiinflammatory effect in the rats with chronic bronchitis. Administration of strawberry, loquat, mulberry and bitter melon fruit juices increased interleukin IL-10 production by lipopolysaccharid-stimulated murine peritoneal macrophages in dose-dependent manners (Lin and Tang 2008). Concurrently, the levels of IL-1β, IL-6 and/or TNF-α were decreased. The results suggested that strawberry, loquat, mulberry, and bitter melon juices exhibited a prophylactic effect on LPS-induced inflammation of peritoneal macrophages via increasing antiinflammatory cytokine and/or decreasing proinflammatory cytokines secretions.

In one study, leaves of *Eriobotrya japonica* (Pi-Pa-Ye, PPY), a traditional Chinese medicine for the treatment of pulmonary inflammatory diseases, was found capable of suppressing lipopolysaccharide -induced cytokine productions in inflamed lungs in a dose-dependent manner (Lee et al. 2008). The suppression of PPY on the cytokine productions resulted from the inhibition of inhibitory κB-α phosphorylation and nuclear factor-κB (NF-κB) activation. Analysis of the anti inflammatory effects of ursolic acid and oleanolic acid, the triterpene compounds present in PPY, showed that ursolic acid significantly inhibited LPS-induced IL-8 production, NF-κB activation, and inducible nitric oxide synthase (iNOS) mRNA expression, whereas oleanolic acid did not have these effects. The findings suggested the potential mechanisms of PPY and its active component, ursolic acid, in the treatment of pulmonary inflammation. Results of studies by Huang et al. (2009) indicated that loquat leaf extract significantly decreased inducible nitric oxide synthase (iNOS) and nitric oxide induction in alveolar macrophage of chronic bronchitis rats. This effect was related to its inhibition of p38 MAPK signal transduction.

Recent studies showed that epicatechin isolated from *Eriobotrya japonica* suppressed the production of pro-inflammatory cytokines, IL6 and IL-8, enhanced the production of antiinflammatory cytokine, IL-10, and stimulated NFκB p65 translocation to nucleus in phytohemagglutinin (PHA)+lipopolysaccharide (LPS), stimulated whole blood culture (Al-Hanbali et al. 2009) Epicatechin has the potential as an immunomodulating agent for treating several diseases such as autoimmune diseases.

Uto et al. (2010) also reported that loquat leaf extract had antiinflammatory activity. Their results indicated that the antiinflammatory effect might result from inhibition of iNOS and COX-2 expression through the downregulation of NF-κB activation and MAPK phosphorylation in lipopolysaccharide -stimulated RAW264 murine macrophage cells.

Recent studies showed that loquat leaf extract could be used to inhibit the activation of a wide variety of inflammation-related genes and inflammatory mediators (Choi et al. 2011). In addition, 78 of the 158 genes suppressed by the LPS when compared to the control were up-regulated by the leaf extract treatment. The authors suggested that loquat leaf extract may be applied to alleviate the inflammation of periodontal diseases.

The results of studies by Cha et al. (2011a) indicated that n-butanol extract of *Eriobotrya japonica* leaves had potent inhibitory effects on the inflammatory mediators including nitric oxide, iNOS, COX-2, TNF-α and IL-6 via the attenuation of NF-κB translocation to the nucleus in IFN-γ/LPS activated murine peritoneal macrophage model. The n-butanol leaf also showed excellent antinociceptive activity in both central and peripheral mechanism as a weak opioid agonist. Based on these results, loquat leaves may possibly be used as an antiinflammatory and an analgesic agent for the treatment of pains and inflammatory diseases.

Antihyperglycemic/Antihyperlipidemic Activity

Eriobotrya japonica seeds were found to have hypoglycemic activities. *Eriobotrya japonica* seeds (10% powered) suppressed the increment of blood glucose in rats and mice fed on a diet containing 10% powdered *Eriobotrya japonica* seeds for 4 months. The treatment also effectively improved the glucose tolerance in the KK-A(y) mice, these actions being mainly exerted by the ethanol extract of the seeds (Tanaka et al. 2008).

The total sesquiterpenes (EJA-1) 30 g (crude drug)/kg from loquat leaf had significant effect on lowering blood glucose level in normal or/and alloxan-diabetic mice (Li et al. 2007). The tests of maximum dosage and acute toxicity showed that the leaf extract was safe (MD=360 g/kg, LD_{50}=400.1 g/kg). The pharmacological tests on anti-hyperglycemic effects of EJA-0 and EJA-1 extracts showed folium eriobotryae to be a good candidate as medicine for treatment of diabetes mellitus. Loquat leaves also yielded euscaphic acid which significantly lowered plasma glucose levels in normoglycemic mice treated with euscaphic acid compared to mice treated with 0.5% CMC-Na solution only (Chen et al. 2008b). Moreover, the dosage of

50 mg/kg exerted a significant hypoglycemic effect in alloxan-diabetic mice after oral administration.

Sesquiterpenes glycoside and polyhydroxylated triterpenoids isolated by methanol extraction of *Eriobotrya japonica* also possessed hypoglycemic activity (De Tommasi et al. 1991). The sesquiterpene glycoside 3 and the polyhydroxylated triterpenoids 5 and 6 produced a marked inhibition of glycosuria in genetically diabetic mice (C57BL/KS-db/db/Ola). In addition, the polyhydroxylated triterpenoids were able to reduce blood glucose levels in normoglycemic rats.

Another sesquiterpene glycoside, nerolidol-3-O-α-l-rhamnopyranosyl($1\rightarrow 4$)-α-l-rhamnopyranosyl($1\rightarrow 2$)-[α-l-rhamnopyranosyl($1\rightarrow 6$)]-β-d-glucopyranoside isolated from dried loquat leaves elicited antidiabetic effect (Chen et al. 2008a). The extract exerted a significant hypoglycemic effect at doses of 25 and 75 mg/kg in alloxan-diabetic mice and had a slight effect in normal mice.

Studies showed that *Eriobotrya japonica* water extract (EJWE) and the compounds derived from it: cinchonain Ib, procyanidin B-2, chlorogenic acid and epicatechin, exhibited effects on insulin secretion from INS-1 cells following oral administration in rats (Qa'dan et al. 2009). The results showed that EJWE increased significantly insulin secretion from INS-1 cells in a dose-dependent manner. Oral administration of EJWE at 230 mg/kg to rats, however, decreased plasma insulin level for as long as 240 minutes post-administration and caused a transient drop of blood glucose at 15 and 30 minutes post-administration. In contrast, cinchonain Ib enhanced significantly insulin secretion from INS-1 cells, whereas epicatechin inhibited significantly insulin secretion from INS-1 cells. Also, cinchonain Ib enhanced significantly (150%) plasma insulin level in rats for as long as 240 minutes after 108 mg/kg oral administration but did not induce any change in blood glucose level. The study concluded that cinchonain Ib had an insulinotropic effect and suggested the possible use of cinchonain Ib for managing type 2 diabetes. The study also supported the traditional use of immersing the dried loquat leaves in a hot water drink for the treatment of diabetes mellitus in Jordan.

Studies showed that a dose of 300 mg/kg of the total triterpene acid (TTA) fraction extracted from loquat leaves was the most effective dose to cause significant hypoglycemic and/or hypolipidemic effects on normal, alloxan and streptozotocin-induced diabetic mice (Lü et al. 2009). This dose also significantly lowered the glycosylated serum protein (GSP), total cholesterol (TC) and triglyceride (TG) level in severely diabetic mice. Further, TTA increased superoxide dismutase activity (SOD) and the serum insulin level of diabetic mice. The results suggested the triterpene acid fraction from loquat leaves to have a high antidiabetic potential along with a good hypolipidemic profile.

Shih et al. (2010) demonstrated in mice that loquat was effective in ameliorating the high fat diet-induced hyperglycemia, hyperleptinemia, hyperinsulinemia and hypertriglyceridemia, as well as in decreasing the levels of free fatty acid (FFA). Loquat increased the adipose PPARγ (peroxisomal proliferator-activated receptor γ) and hepatic PPARα mRNA levels. Loquat significantly decreased the body weight gain, weights of white adipose tissue and visceral fat. Loquat not only suppressed the hepatic mRNA levels of enzymes involved in fatty acid and triacylglycerol synthesis and lowered the sterol regulatory element binding protein-1c (SREBP-1c) mRNA level, but also affected fatty acid oxidation enzyme levels. The authors concluded that the findings provided a nutritional basis for the use of loquat as a functional food factor that may have benefits for the prevention of hyperlipidemia and diabetes.

Anticancer/Antimetastatic Activity

Three new flavonoid glycosides, together with 15 known flavonoids, were isolated from loquat leaves, and characterized as (2S)- and (2R)-naringenin 8-C-α-L-rhamnopyranosyl-($1\rightarrow 2$)-β-D-glucopyranosides, and cinchonain Id 7-O-β-D-glucopyranoside, respectively (Kobayashi and Takamatsu 2000). These polyphenols were also assessed for cytotoxic activity against two human oral tumour (human squamous cell carcinoma and human salivary gland tumour) cell lines.

In an antiproliferative assay using five human cancer cell lines representing different tissues (breast, lung and ovary) *Eriobotrya japonica* showed strong cytotoxicity in ER-negative breast cancer (MDA-MB-231), cervix epitheloid (HeLa) and lung (A549) carcinoma cell lines (Kang et al. 2006). *E. japonica* also exhibited potent estrogenic activity using a recombinant yeast system with both a human estrogen receptor (ER) expression plasmid and a reporter plasmid.

Three new flavonoid glycosides (Polypehnols) isolated from the leaves of *Eriobotrya japonica*, and characterized as (2S)- and (2R)-naringenin 8-C-α-L-rhamnopyranosyl-(1→2)-β-D-glucopyranosides, and cinchonain Id 7-O-β-D-glucopyranoside, polyphenols also demonstrated cytotoxic activity against two human oral tumour (human squamous cell carcinoma and human salivary gland tumour) cell lines (Ito et al. 2000). Selective cytotoxicity of the procyanidin oligomer mixture mainly composed of undecameric procyanidin between tumour and normal gingival fibroblast cells was also observed.

Ethyl acetate leaf extract and the isolated active components, ursolic acid and 2α-hydroxyursolic acid significantly inhibited MMP-2 and MMP-9 activities and expressions via down-regulation of NF-κB translocation to the nucleus in B16F10 cell (Cha et al. 2011b). In addition, the cell migration and invasion were down-regulated and lung metastasis was significantly suppressed in-vivo by the leaf extract. The results demonstrated that loquat leaf extract may be used as valuable antimetastatic agent for the treatment of cancer metastasis.

Among 11 triterpene acids isolated from loquat leaves tested, four compounds, δ-oleanolic acid (4), ursolic acid (5), 3-O-(E)-p-coumaroyl tormentic acid (8), and betulinic acid (10), exhibited potent Topo I inhibitory activity (IC_{50} 20.3–36.5 μM) and cytotoxicity against human leukemia HL60 cancer cells (EC_{50} 5.0–8.1 μM) (Kikuchi et al. 2011). Compound 8 exhibited induction of apoptosis and markedly reduced the levels of procaspases-3 and 9. In contrast, compound 8 exerted almost no influence on the expression of caspase-8. Further, compound 8 increased significantly Bax/Bcl-2 ratio and activated caspase-2. These results suggested that compound 8 induced apoptotic cell death in HL60 via mainly mitochondrial pathway by, at least in part by Topo I inhibition. The results suggested compound 8 may be a promising lead compound for developing an effective drug for treatment of leukemia.

Targeting the baculoviral inhibitor of apoptosis proteins repeat (BIR) 3 of X-linked inhibitor of apoptosis proteins (XIAP) represents an innovative strategy for the design of chemosensitizers. From the methanol leaf extract of *E. japonica* an enriched mixture of acylated flavonol monorhamnosides (AFMR) was obtained showing chemosensitizing potential in combination with etoposide in X-linked inhibitor of apoptosis proteins (XIAP) -overexpressing Jurkat cells (Pfisterer et al. 2011). The main constituent kaempferol-3-O-α-l-(2″,4″-di-E-p-coumaroyl)-rhamnoside (3) was isolated from the enriched fraction and identified as XIAP baculoviral inhibitor of apoptosis proteins repeat (BIR) 3 ligand with a dose-dependent affinity (IC_{50} 10.4 μM). Further, 3 induced apoptosis in XIAP-overexpressing Jurkat cells and activated caspase-9 in combination with etoposide.

Two hydrophilic extracts of *E. japonica* leaves elicited in-vivo immunomodulatory effect on interferon-γ (IFN-γ), interleukin-17 (IL-17), and transforming growth factor-β 1 (TGF-β1) in the tissues of normal and Meth-A-fibrosarcoma bearing mice (Alshaker et al. 2011). Intraperitoneal (i.p.) administration of 10 μg of EJHE and EJHE-water residue (WR), prepared from butanol extraction, increased significantly IFN-γ production in the spleen and lung tissues at 6–48 h and suppressed significantly TGF-β1 production levels in the spleen in normal mice for as long as 48 h. Triple i.p. injections, of 10 μg EJHE increased significantly IFN-γ production in the spleen while only EJHE-WR increased significantly IFN-γ, TGF-β1 and IL-17 production within the tumour microenvironment of Meth-A fibrosarcoma. There was significant prolongation of survival time of mice inoculated i.p. with Meth-A cells followed by three times/week for 8 weeks of i.p. administration of EJHE-WR. The latter prolonged survival effect was not seen with EJHE. The results highlighted the therapeutic value of EJHE-WR as an anticancer agent.

Antiviral Activity

Among the triterpene esters from chloroform extracts of loquat, namely, 23-trans-p-coumaroyltormentic acid, 23-cis-p-coumaroyltormentic acid, 3-O-trans-caffeoyltormentic acid, and 3-O-trans-p-coumaroylrotundic acid, in addition to three common ursolic acid derivatives 5, 6, and 7, only 3-O-trans-caffeoyltormentic acid significantly reduced rhinovirus infection (De Tommasi et al. 1992). The compounds were ineffective towards human immunodeficiency virus type 1 (HIV-1) and Sindbis virus replication.

Antiplatelet Activity

Two sesquiterpene namely 3-O-α-L-rhamnopyranosyl-(1-->4)-α-L-rhamnopyranosyl-(1-->2)-[α-L-(4-trans-feruloyl)-rhamnopyranosyl-(1-->6)]-β-D-glucopyranosyl nerolidol (2) and ferulic acid (3) from loquat leaves 3 inhibited 50% of the TF (tissue factor) activity at concentrations of 2 and 369 μM/TF units, respectively (Lee et al. 2004). Tissue factor (tissue thromboplastin) is a membrane bound glycoprotein, which accelerates blood clotting.

Hepatoprotective Activity

Loquat seed extract containing unsaturated linolenic and linoleic acids and the sterol β-sitosterol was found to improve liver function (Nishioka et al. 2002). L-asparate aminotransferase (AST), L-alanine aminotransferase (ALT), and hydroxyproline levels and liver fibrosis rates were significantly lower, and retinoid levels were significantly higher in hepatopathic rats treated with 70% ethanolic and methanolic extracts of the seed than in water-treated control rats. This suggested that the positive effect on liver function of the extracts varied depending on the extracting solvent used. Both extracts of the seeds inhibited the development of liver fibrosis in hepatopathic rats, thus exhibiting potent improvement. Studies showed that pretreatment of HepG2 cells transfected with CYP2E1 with ethanol loquat leaf extract led to a decrease in intracellular reactive oxygen species formation and an increase in hepatic antioxidant activity (Bae et al. 2010). The results suggest loquat leaves to have hepatoprotective profile against ethanol-induced toxicity in-vitro.

Studies by Yoshioka et al. (2010) found that increases in hepatic alanine aminotransferase and aspartate aminotransferase levels, fatty droplets formation and lipid peroxidation were significantly inhibited in rats fed with loquat seed extract. Expression of 8-hydroxy-2′-deoxyguanosine and 4-hydroxy-2-nonenal was lower in the rats given the seed extract than in the diet-only group. Antioxidant enzyme activity in liver tissue was higher than in the diet-only group. The results showed that E. japonica seed extract inhibited fatty liver, inflammation and fibrosis, suggesting its usefulness in the treatment of non-alcoholic steatohepatitis.

Renal Protective Activity

Eriobotrya japonica seed extract was also found effective in reducing the adverse oxidative stress of adriamycin-induced renal disorder (Hamada et al. 2004). Studies showed that increases in indices of renal function, plasma urea nitrogen, were significantly inhibited in rats treated with the seed extract compared to rats treated with tap water. In addition, the renal tissue level of reduced glutathione was significantly higher in rats that ingested the extract, while the lipid peroxide levels in plasma and renal tissue were significantly low. However, no effect on renal tissue antioxidative enzymes was observed, suggesting that loquat seed extract exhibited direct antioxidative action.

Gastroprotective Activity

Eriobotrya japonica ethanol seed extract was also reported to exhibit gastroprotective activity (Yokota et al. 2008). The extract inhibited the formation of gastric mucosal injury in rats when administration was initiated 14 days before induction of gastric mucosal injury. Results of studies suggested that deep sea water (DSW), when combined with *Eriobotrya japonica* seed extract (ESE), inhibited antioxidative enzymes, and enhanced the gastric mucosal protecting effect of the seed extract (Yokota et al. 2011). The injury area and lipid peroxide levels in plasma and gastric tissue were significantly reduced in the ESE and ESE+DSW groups compared with the control and DSW group. The plasma and gastric tissue antioxidative enzyme levels were significantly higher in the ESE and ESE+DSW groups than in the control group.

Cognitive Activity

Kim et al. (2011) demonstrated that *E. japonica* extract scavenged approximately 40% of DPPH radicals and treatment of *E. japonica* extract inhibited Aβ(1-42)-mediated neuronal cell death. Further, treatment of *E. japonica* extract efficiently suppressed the increase in intracellular ROS triggered by the Aβ(1-42) peptide. Mice pre-treated with the *E. japonica* extract showed restoration of alternation behavior and reversal of Aβ(1-42)-induced memory impairment. Consequently, *E. japonica* extract substantially inhibited the increase in lipid peroxidation and restored superoxide dismutase activity. These results suggested that *E. japonica* protected from oxidative stress and cognitive deficits induced by the Aβ peptide.

Antiaging Activity

Results of studies suggested that treatment with an adequate concentration of *Eriobotrya japonica* seed extract rendered bradykinin-induced Ca(2+) dynamics in senescent cells similar to those in young cells (Muramoto et al. 2011). The application of bradykinin transiently elicited intracellular calcium ion (Ca(2+)) increased in most of the young fibroblasts, whereas these responses were scarcely observed or were significantly attenuated in senescent cells. Therefore, *Eriobotrya japonica* seed extract could retard and/or protect against cellular aging and may be useful for elucidating the antiaging processes.

Antiallergic Activity

Eriobotrya japonica seeds extract and leaves have anti-allergic activity. Oral administration of the extract dramatically inhibited ear swelling due to allergic contact dermatitis caused by repeated application of two antigens, 4-ethoxymethylene-2-phenyl-2-oxazolin-5-one (oxazolone) and dinitrofluorobenzene (DNFB), respectively (Sun et al. 2007). The increase of histamine content in inflamed ear tissue induced by oxazolone and DNFB was significantly antagonized by orally administered extract. Oral administration of the extract also suppressed eosinophil peroxidase and myeloperoxidase activity in both models. Tumour necrosis factor-α (TFN-α) in the inflamed region caused by repeated application of DNFB was also significantly suppressed. The findings suggested that the seed extract may be effective for treating allergic contact dermatitis. This was also confirmed by another study. The administration of a loquat seed extract inhibited histamine release from rat mast cells, suggesting its usefulness in allergic disease treatment (Onogawa et al. 2009). In an experiment using a guinea-pig allergic rhinitis model, this extract reduced the frequency of sneezing and nose-scratching. The results suggested that *Eriobotrya japonica* seed extract may contribute to the relief of allergic disease-related symptoms.

Leaves of *Eriobotrya japonica* inhibited compound 48/80-induced systemic anaphylactic reactions and serum histamine release in mice (Kim et al. 2009). LEJL dose-dependently decreased the IgE-mediated passive cutaneous anaphylaxis and histamine release from mast cells. Furthermore, LEJL decreased the production of tumour necrosis factor-α in phorbol 12-myristate 13-acetate and A23187-stimulated human mast cells. These findings provided evidence that LEJL could be a candidate as an antiallergic agent. Oral administration of loquat seed extract significantly inhibited development of allergic dermatitis in rats based on lower ear thickness and serum immunoglobulin E (IgE) levels (Sun et al. 2010). Oral administration of the extract significantly decreased interleukin-4 (IL-4) while significantly increasing IL-10 in lesional skin, and the lower levels of I interferon -γ and IL-2 were reversed by oral administration of ESE. The infiltration of eosinophils in the lesional skin was decreased by oral administration of the extract. These results suggested that the extract exerted anti-allergic actions by improving the balance of Th1/Th2 cytokines in allergic dermatitis.

Toxicity Study

The 28-day repeated-dose oral toxicity study demonstrated that BC-AF, an antiacne formulation, consisting of the extracts of danshen (*Salvia miltiorrhiza*), licorice (*Glycyrrhiza uralensis*), and loquat leaf (*Eriobotrya japonica*), produced no effects in either male or female rats following oral administration of up to 10 g/kg (Li et al. 2011).

Traditional Medicinal Uses

All parts of the plant are used in traditional Chinese medicine. Loquat is recommended for pharyngolaryngitis, cough, thirst and constipation. Dried leaves are deemed to have cool energy an bitter flavour and to relive cough, nosebleed and coughing up blood. They are also used in treatment for diarrhoea and depression. A type of loquat syrup is used for soothing the throat like a cough drop. Mixed with other ingredients and known as *pipa ga* (loquat paste) it acts as a demulcent and an expectorant as well as to soothe the digestive and respiratory systems. The fragrant flowers contain essential oils, extracted for cosmetics and are used as expectorants, for cough and common colds. The leaves have used to treat skin diseases, as well as to relieve inflammation, pain, coughing, chronic bronchitis and sputa. The leaves are known to have many physiological actions such as antiinflammatory, antitussive, and expectorant.

Other Uses

The fragrant flowers contain essential oils, that can be extracted for use in cosmetics. The seed oil can be used for making soap.

Comments

There are two main cultivar groups of loquat: Chinese group characterised by orange skin and flesh, not juicy but highly aromatic; Japanese group characterised by yellow skin, white flesh, juicy and mildly aromatic.

Selected References

Al-Hanbali M, Ali D, Bustami M, Abdel-Malek S, Al-Hanbali R, Alhussainy T, Qadan F, Matalka KZ (2009) Epicatechin suppresses IL-6, IL-8 and enhances IL-10 production with NF-kappaB nuclear translocation in whole blood stimulated system. Neuro Endocrinol Lett 30(1):131–138

Alshaker HA, Qinna NA, Qadan F, Bustami M, Matalka KZ (2011) *Eriobotrya japonica* hydrophilic extract modulates cytokines in normal tissues, in the tumor of Meth-A-fibrosarcoma bearing mice, and enhances their survival time. BMC Complement Altern Med 11:9

Bae D, You Y, Yoon HG, Kim K, Lee YH, Kim Y, Baek H, Kim S, Lee J, Jun W (2010) Protective effects of loquat (*Eriobotrya japonica*) leaves against ethanol-induced toxicity in HepG2 cells transfected with CYP2E1. Food Sci Biotechnol 19(4):1093–1096

Cha DS, Eun JS, Jeon H (2011a) Anti-inflammatory and antinociceptive properties of the leaves of *Eriobotrya japonica*. J Ethnopharmacol 134(2):305–312

Cha DS, Shin TY, Eun JS, Kim DK, Jeon H (2011b) Anti-metastatic properties of the leaves of *Eriobotrya japonica*. Arch Pharm Res 34(3):425–436

Chen J, Li WL, Wu JL, Ren BR, Zhang HQ (2008a) Euscaphic acid, a new hypoglycemic natural product from Folium Eriobotryae. Pharmazie 63(10):765–767

Chen J, Li WL, Wu JL, Ren BR, Zhang HQ (2008b) Hypoglycemic effects of a sesquiterpene glycoside isolated from leaves of loquat (*Eriobotrya japonica* (Thunb.) Lindl.). Phytomedicine 15:98–102

Cheng L, Liu Y, Chen L, Luo J (2001) Studies on the triterpenoidal saponins from flowers of *Eriobotrya japonica*. Hua Xi Yi Ke Da Xue Xue Bao 32(2):283–285

Choi YG, Seok YH, Yeo S, Jeong MY, Lim S (2011) Protective changes of inflammation-related gene expression by the leaves of *Eriobotrya japonica* in the

LPS-stimulated human gingival fibroblast: microarray analysis. J Ethnopharmacol 135(3):636–645

De Tommasi N, De Simone F, Cirino G, Cicala C, Pizza C (1991) Hypoglycemic effects of sesquiterpene glycosides and polyhydroxylated triterpenoids of *Eriobotrya japonica*. Planta Med 57(5):414–416

De Tommasi N, De Simone F, Pizza C, Mahmood N, Moore PS, Conti C, Orsi N, Stein ML (1992) Constituents of *Eriobotrya japonica*. A study of their antiviral properties. J Nat Prod 55(8):1067–1073

Ding CK, Chachin K, Ueda Y, Imahori Y, Wang CY (2001) Metabolism of phenolic compounds during loquat fruit development. J Agric Food Chem 49(6): 2883–2888

Duke JA, Ayensu ES (1985) Medicinal plants of China, vols 1 and 2. Reference Publications, Inc., Algonac, 705pp

Facciola S (1990) Cornucopia. A source book of edible plants. Kampong Publications, Vista, 677pp

Ferreira de Faria A, Hasegawa PN, Chagas EA, Pio R, Purgatto E, Mercadante AZ (2009) Cultivar influence on carotenoid composition of loquats from Brazil. J Food Comp Anal 22(3):196–203

Ferreres F, Gomes D, Valentão P, Gonçalves R, Pio R, Chagas EA, Seabra RM, Andrade PB (2009) Improved loquat (*Eriobotrya japonica* Lindl.) cultivars: variation of phenolics and antioxidative potential. Food Chem 114(3):1019–1027

Ge JF, Wang TY, Zhao B, Lv XW, Jin Y, Peng L, Yu SC, Li J (2009) Anti-inflammatory effect of triterpenoic acids of *Eriobotrya japonica* (Thunb.) Lindl. leaf on rat model of chronic bronchitis. Am J Chin Med 37(2):309–321

Hamada A, Yoshioka S, Takuma D, Yokota J, Cui T, Kusunose M, Miyamura M, Kyotani S, Nishioka Y (2004) The Effect of *Eriobotrya japonica* seed extract on oxidative stress in adriamycin-induced nephropathy in rats. Biol Pharm Bull 27(12):1961–1964

Hu SY (2005) Food plants of China. The Chinese University Press, Hong Kong, 844pp

Huang Y, Li J, Cao O, Yu SC, Lv XW, Jin Y, Zhang L, Zou YH, Ge JF (2006) Anti-oxidative effect of triterpene acids of *Eriobotrya japonica* (Thunb.) Lindl. leaf in chronic bronchitis rats. Life Sci 78(23): 2749–2757

Huang Y, Li J, Meng XM, Jiang GL, Li H, Cao Q, Yu SC, Lv XW, Cheng WM (2009) Effect of triterpene acids of *Eriobotrya japonica* (Thunb.) Lindl. leaf and MAPK signal transduction pathway on inducible nitric oxide synthase expression in alveolar macrophage of chronic bronchitis rats. Am J Chin Med 37(2):1099–1111

Huang Y, Li J, Wang R, Wu Q, Li Y-H, Yu S-C, Cheng W-M, Wang Y-Y (2007) Effect of triterpene acids of *Eriobotrya japonica* (Thunb.) Lindl. leaf on inflammatory cytokine and mediator induction from alveolar macrophages of chronic bronchitic rats. Inflamm Res 56(2):76–82

Ito H, Kobayashi E, Li SH, Hatano T, Sugita D, Kubo N, Shimura S, Itoh Y, Yoshida T (2001) Megastigmane glycosides and an acylated triterpenoid from *Eriobotrya japonica*. J Nat Prod 64(6):737–740

Ito H, Kobayashi E, Takamatsu Y, Li SH, Hatano T, Sakagami H, Kusama K, Satoh K, Sugita D, Shimura S, Itoh Y, Yoshida T (2000) Polyphenols from *Eriobotrya japonica* and their cytotoxicity against human oral tumor cell lines. Chem Pharm Bull(Tokyo) 48(5):687–693

Jung HA, Park JC, Chung HY, Kim J, Choi JS (1999) Antioxidant flavonoids and chlorogenic acid from the leaves of *Eriobotrya japonica*. Arch Pharm Res 22(2):213–218

Kang SC, Lee CM, Cho IH, Lee JH, Oh JS, Kwak JH, Zee OP (2006) Evaluation of oriental medicinal herbs for estrogenic and antiproliferative activities. Phytother Res 20(11):1017–1019

Kikuchi T, Akazawa H, Tabata K, Manosroi A, Manosroi J, Suzuki T, Akihisa T (2011) 3-O-(E)-p-coumaroyl tormentic acid from *Eriobotrya japonica* leaves induces caspase-dependent apoptotic cell death in human leukemia cell line. Chem Pharm Bull(Tokyo) 59(3):378–381

Kim MJ, Lee J, Seong AR, Lee YH, Kim YJ, Baek HY, Kim YJ, Jun WJ, Yoon HG (2011) Neuroprotective effects of *Eriobotrya japonica* against β-amyloid-induced oxidative stress and memory impairment. Food Chem Toxicol 49(4):780–784

Kim SH, Kwon YE, Park WH, Jeon H, Shin TY (2009) Effect of leaves of *Eriobotrya japonica* on anaphylactic allergic reaction and production of tumor necrosis factor-alpha. Immunopharmacol Immunotoxicol 31(2): 314–319

Koba K, Matsuoka A, Osada K, Huang YS (2007) Effect of loquat (*Eriobotrya japonica*) extracts on LDL oxidation. Food Chem 104(1):308–316

Kobayashi E, Takamatsu Y (2000) Polyphenols from *Eriobotrya japonica* and their cytotoxicity against human oral tumor cell lines. Chem Pharm Bull 48(5):687–693

Lee CH, Wu SL, Chen JC, Li CC, Lo HY, Cheng WY, Lin JG, Chang YH, Hsiang CY, Ho TY (2008) *Eriobotrya japonica* leaf and its triterpenes inhibited lipopolysaccharide-induced cytokines and inducible enzyme production via the nuclear factor-kappa B signaling pathway in lung epithelial cells. Am J Chin Med 36(6):1185–1198

Lee MH, Son YK, Han YN (2004) Tissue factor inhibitory sesquiterpene glycoside from *Eriobotrya japonica*. Arch Pharm Res 27(6):619–623

Li S, Zou Y, Jiao K, Qiao X, Jiao R, Wang J (2011) Repeated-dose (28 days) oral toxicity study in rats of an antiacne formula (BC-AF) derived from plants. Drug Chem Toxicol 34(1):77–84

Li WL, Wu JL, Ren BR, Chen J, Lu CG (2007) Pharmacological studies on anti-hyperglycemic effect of Folium Eriobotryae. Am J Chin Med 35(4): 705–711

Lin JY, Tang CY (2008) Strawberry, loquat, mulberry, and bitter melon juices exhibit prophylactic effects on

LPS-induced inflammation using murine peritoneal macrophages. Food Chem 107(4):1587–1596

Lu HC (2005) Chinese natural cures. Black Dog & Leventhal Publishers, New York, 512pp

Lü HC, Chen J, Li WL, Ren BR, Wu JL, Kang HY, Zhang HQ, Adams A, De Kimpe N (2009) Hypoglycemic and hypolipidemic effects of the total triterpene acid fraction from Folium Eriobotryae. J Ethnopharmacol 122(3):486–491

Lv H, Chen J, Li WL, Zhang HQ (2008) Studies on the triterpenes from loquat leaf (*Eriobotrya japonica*). Zhong Yao Cai 31(9):1351–1354 (In Chinese)

Morton JF (1987) Loquat. In: Fruits of warm climates. Julia F. Morton, Miami, pp 103–108

Muramoto K, Quan RD, Namba T, Kyotani S, Miyamura M, Nishioka Y, Tonosaki K, Doi YL, Kaba H (2011) Ameliorative effects of *Eriobotrya japonica* seed extract on cellular aging in cultured rat fibroblasts. J Nat Med 65(2):254–261

Nguyen TH, Verheij EWM (1992) *Eriobotrya japonica* (Thunb.) Lindley. In: Verheij EWM, Coronel RE (eds) Plant resources of South-East Asia No 2. Edible fruits and nuts. Prosea, Bogor, pp 161–164

Nishioka Y, Yoshioka S, Kusunose M, Cui T, Hamada A, Ono M, Miyamura M, Kyotani S (2002) Effects of extract derived from *Eriobotrya japonica* on liver function improvement in rats. Biol Pharm Bull 2(8):1053–1057

Onogawa M, Sun G, Takuma D, Hamada A, Yokota J, Yoshioka S, Kusunose M, Miyamura M, Kyotani S, Nishioka Y (2009) Animal studies supporting the inhibition of mast cell activation by *Eriobotrya japonica* seed extract. J Pharm Pharmacol 61(2):237–241

Pfisterer PH, Shen C, Nikolovska-Coleska Z, Schyschka L, Schuster D, Rudy A, Wolber G, Vollmar AM, Rollinger JM, Stuppner H (2011) In silico discovery of acylated flavonol monorhamnosides from *Eriobotrya japonica* as natural, small-molecular weight inhibitors of XIAP BIR3. Bioorg Med Chem 19(2):1002–1009

Polat AA, Çalişkan O, Serçe S, Saraçoğlu O, Kaya C, Özgen M (2010) Determining total phenolic content and total antioxidant capacity of loquat cultivars grown in Hatay. Pharmacogn Mag 6(21):5–8

Qa'dan F, Verspohl EJ, Nahrstedt A, Petereit F, Matalka KZ (2009) Cinchonain Ib isolated from *Eriobotrya japonica* induces insulin secretion in vitro and in vivo. J Ethnopharmacol 124(2):224–227

Shih CC, Lin CH, Wu JB (2010) *Eriobotrya japonica* improves hyperlipidemia and reverses insulin resistance in high-fat-fed mice. Phytother Res 24(12):1769–1780

Sun GC, Liu YQ, Zhu JL, Iguchi M, Yoshioka S, Miyamura M, Kyotani S (2010) Immunomodulatory effect of *Eriobotrya japonica* seed extract on allergic dermatitis Rats. J Nutr Sci Vitaminol 56(2):145–149

Sun GC, Zhang YX, Takuma D, Onogawa M, Yokota J, Hamada A, Yoshioka S, Kusunose M, Miyamura M, Kyotani S, Nishioka Y (2007) Effect of orally administered *Eriobotrya japonica* seed extract on allergic contact dermatitis in rats. J Pharm Pharmacol 59(10):1405–1412

Takuma D, Sun GC, Yokota J, Hamada A, Onogawa M, Yoshioka S, Kusunose M, Miyamura M, Kyotani S, Nishioka Y (2008) Effect of *Eriobotrya japonica* seed extract on 5-fluorouracil-induced mucositis in hamsters (pharmacognosy). Biol Pharm Bull 31(2):250–254

Tanaka K, Nishizono S, Makino N, Tamaru S, Terai O, Ikeda I (2008) Hypoglycemic activity of *Eriobotrya japonica* seeds in type 2 diabetic rats and mice. Biosci Biotechnol Biochem 72(3):686–693

U.S. Department of Agriculture, Agricultural Research Service (2011) USDA national nutrient database for standard reference, Release 24. Nutrient Data Laboratory Home Page, http://www.ars.usda.gov/ba/bhnrc/ndl

Ueatrongchit T, Kayo A, Komeda H, Asano Y, H-Kittikun A (2008) Purification and characterization of a novel (R)-hydroxynitrile lyase from *Eriobotrya japonica* (Loquat). Biosci Biotechnol Biochem 72(6):1513–1522

Uto T, Suangkaew N, Morinaga O, Kariyazono H, Oiso S, Shoyama Y (2010) Eriobotryae folium extract suppresses LPS-induced iNOS and COX-2 expression by inhibition of NF-kappaB and MAPK activation in murine macrophages. Am J Chin Med 38(5): 985–994

Wu ZY, Raven PH, Hong DY (eds) (2003) Flora of China, vol 9 (Pittosporaceae through Connaraceae). Science Press/Missouri Botanical Garden Press, Beijing/St. Louis

Yokota J, Kitaoka T, Jobu K, Takuma D, Hamada A, Onogawa M, Yoshioka S, Kyotani S, Miyamura M (2011) *Eriobotrya japonica* seed extract and deep sea water protect against indomethacin-induced gastric mucosal injury in rats. J Nat Med 65(1):9–17

Yokota J, Takuma D, Hamada A, Onogawa M, Yoshioka S, Kusunose M, Miyamura M, Kyotani S, Nishioka Y, Nishioka Y (2006) Scavenging of reactive oxygen species by *Eriobotrya japonica* seed extract. Biol Pharm Bull 29(3):467–471

Yokota J, Takuma D, Hamada A, Onogawa M, Yoshioka S, Kusunose M, Miyamura M, Kyotani S, Nishioka Y (2008) Gastroprotective activity of *Eriobotrya japonica* seed extract on experimentally induced gastric lesions in rats. Nat Med (Tokyo) 62(1):96–100

Yoshioka S, Hamada A, Jobu K, Yokota J, Onogawa M, Kyotani S, Miyamura M, Saibara T, Onishi S, Nishioka Y (2010) Effects of *Eriobotrya japonica* seed extract on oxidative stress in rats with non-alcoholic steatohepatitis. J Pharm Pharmacol 62(2):241–246

Zhang HZ, Peng SA, Cai LH, Fang DQ (1990) The germplasm resources of the genus *Eriobotrya* with special reference on the origin of *E. japonica* Lindl. Acta Hortic Sin 17:5–12

Zhou C, Sun C, Chen K, Li X (2011) Flavonoids, phenolics, and antioxidant capacity in the flower of *Eriobotrya japonica* Lindl. Int J Mol Sci 12(5):2935–2945

Fragaria x ananassa

Scientific Name

Fragaria x ananassa **Duchesne ex Rozier.**

Synonyms

Fragaria ananassa Duchesne, *Fragaria* x *ananassa* (Weston) Duchesne & Naudin, *Fragaria x ananassa* hort (illeg.), *Fragaria bonariensis* (Juss.) Persoon, *Fragaria calyculata* Duchesne, in Lam., *Fragaria caroliniensis* Duchesne, *Fragaria chilensis β ananassa* Seringe, *Fragaria chilensis δ tincta* Seringe, *Fragaria chilensis γ calyculata* Seringe, *Fragaria chiloensis* var. *ananassa* L.H. Bailey, *Fragaria chiloensis* (L.) Miller var. *ananassa* (Duchesne ex Rozier) Seringe, *Fragaria chiloensis* (Linnaeus) Miller var. *ananassa* Weston, *Fragaria chiloensis* (L.) Duchesne ex Weston var. *ananassa* Weston, *Fragaria chiloensis* (L.) Miller var. *ananassa* (Duchesne ex Rozier) Seringe, *Fragaria* x *cultorum* Thorsrud & Reisaeter, *Fragaria hybrida* Duchesne, *Fragaria grandiflora* Ehrhart, *Fragaria* x *grandiflora* Ehrhart nom. illeg., *Fragaria* x *magna* auct., *Fragaria tincta* Duchesne, *Fragaria vesca ananassa* Desf., *Fragaria vesca ε ananas* Aiton, *Fragaria vesca* Linnaeus var. *sativa* Linnaeus, *Potentilla* x *ananassa* (Duchesne ex Rozier) Mabbley.

Family

Rosaceae

Common/English Names

Garden Strawberry, Hybrid Strawberry, Large-Fruited Strawberry, Pine Strawberry, Pineapple strawberry, Strawberries, Strawberry.

Vernacular Names

Brazil: Morango, Morango-Grande (Portuguese);
Chinese: Cao Mei, Da Hua Cao Mei;
Czech: Jahoda Aromatická, Jahodník Velkoplodý, Květy Jahodníku, Zahrodna Truskalca;
Danish: Ananasjordbaer, Have-Jordbær;
Dutch: Aardbei, Ananasaardbei, Cultuuraardbei, Grootbloemaardbei;
Estonian: Aedmaasikas;
Finnish: Mansikka, Puutarhamansikka, Tarhamansikka;
French: Fraisier, Fraisier Ananas, Fraise De Culture, Fraisier À Gros Fruits, Fraisier Cultivé;
German: Ananaserdbeere, Erdbeere, Erdbeerstaude, Grosse Erdbeere, Gartenerdbeere, Kulturerdbeere;
Indonesia: Stroberi;

Italian: Fragola, Frangola Ananasso, Fragola Coltivata, Pianta Di Fragola;
Japanese: Ichigo, Oranda-Ichigo, Sutoroberii;
Malaysia: Stroberi;
Norwegian: Hage-Jordbær;
Polish: Poziomka;
Portuguese: Morangueiro, Morangueiro Agrícola, Morangueiro Comum, Morangueiro Cultivado;
Russian: Zemljanika;
Slovakian: Jahoda Ananásová;
Spanish: Fresa, Fresa Ananás Fresal, Frutilla, Frutilla Ananás, Fresón;
Swedish: Ananassmultron, Smultron, Jordgubbe, Smultronsläktet;
Thai: Satroboery;
Turkish: Ananas Çileği, Bahçe Çileği;
Vietnamese: Cây Dâu Tây, Dâu Tây, Quả Dâu Tây.

Origin/Distribution

*Fra*garia x *ananassa* originated spontaneously in mixed plots of *Fra*garia *virginiana* P. Miller and *Fra*garia *chiloensis* (L.) P. Miller in horticultural areas in north west France and in some botanical gardens of Europe between circa 1715 and 1760. The hybridogenic origin and the octoploid level are the basis for a high variability of all important characters. The five main producers of strawberries in the world are: USA (California), Spain, Japan, Poland and Italy.

Agroecology

Strawberry is cultivated worldwide in the temperate, Mediterranean, taiga and subtropical zones and in the tropical highlands. Most cultivation is between latitudes 28°N and 60°N and S, where summer temperatures hovers between 15°C and 40°C and winter temperatures hovers between 15°C and −40°C. It grows best in full sun, in moist, well-drained, fertile soil. It tolerates partial shade but fruit production is reduced under such conditions. It grows on a wide range of soils from acidic to basic soils and abhors water-logged conditions which cause rots and diseases.

Edible Plant Parts and Uses

The large, aromatic and attractive fruits are used for fresh consumption and are variously processed into preserves, jams and jellies. Strawberry is popularly used in ice cream, milkshakes, sherbets, smoothies, custards and yoghurts. It is also used in pies, tarts and biscuits. Strawberries are also dried for use in cereal bars. Young leaves are also edible raw.

Botany

A herbaceous perennial with a very short stem bearing a rosette of leaves, runners (stolons) and roots, 10–40 cm high (Plate 1). Leaves spirally arranged on 2–10 cm long, soft, pubescent petioles, trifoliolate. Leaflets shortly petiolulate to sessile, obovate or rhombic, pale green and sparsely hairy below, and dark green and subglabrous above, base broadly cuneate on central leaflet, oblique on lateral ones, margin acutely serrated, apex rounded (Plates 1 and 2). Inflorescence cymose, 5-15-flowered, proximally with a shortly petiolate, leaflet-like bract. Flowers bisexual but function as either male or female, 1.5–2 cm across, 5-6-merous; epicalyx segments elliptic-lanceolate with entire margin, sepals segments ovate, petals sub-orbicular, white, stamens 25–37, unequal, sterile in female functional flowers. Carpels numerous, not developed in

Plate 1 Strawberry plants grown in hydroponics

Plate 2 Strawberry fruit and leaves

Plate 3 Ripening strawberries

Plate 4 Harvested red strawberries

male functional flowers. Fruit an aggregate compound fruit or pseudocarp (false fruit), obovoid or ovoid, large up to 4.5 cm by 5.5 cm, composed of numerous ovaries each with a single ovule and crowned by persistent appressed sepals ripening red (Plates 1, 2, 3 and 4). Achenes (true fruit) acutely ovoid, sunken in the swollen torus, 1.25 mm by 1–1.25 mm, smooth.

Nutritive/Medicinal Properties

Nutrient composition of raw strawberry fruit per 100 g edible portion (Saxholt et al. 2008) was reported as: moisture 89.6 g, energy 170 kJ, total protein 0.675 g, total N 0.108 g, total fat 0.6 g, saturated fatty acids 0.05 g, monounsaturated fatty acids, 0.083 g, polyunsaturated fatty acids 0.348 g, available carbohydrate 7.3 g, total sugars 4.57 g, dietary fibre 1.49 g, ash 0.425 g; vitamin A 3.33 RE, β-carotene equivalent 40 μg, Vitamin E (α-tocopherol) 0.45 mg, vitamin K 20 μg, Vitamin B1 (thiamin) 0.021 mg, vitamin B2 (riboflavin) 0.018 mg; niacin equivalents 0.783 mg – niacin 0.6 mg, tryptophan 0.183 mg; vitamin B6 0.047 mg, pantothenic acid 0.34 mg, biotin 1.1 μg, folates 117 μg, vitamin C (L-ascorbic acid) 76 mg; Na 4 mg, K 187 mg, Ca 20.4 mg, Mg 12.4 mg, P 22.7 mg, Fe 0.27 mg, Mn 0.43 mg, Zn 0.1 mg, I 0.6 μg, Cr 0.3 μg, Se 0.2 μg, Ni 3.76 μg; amino acids – isoleucine 34 mg, leucine 54 mg, lysine 49 mg, methionine 14 mg, cystine 5 mg, phenylalanine 31 mg, tyrosine 15 mg, threonine 31 mg, tryptophan 11 mg, valine 39 mg, arginine 56 mg, histidine 16 mg, alanine 42 mg, aspartic acid 133 mg, glutamic acid 125 mg, glycine 35 mg, proline 33 mg, serine 42 mg; sugars – fructose 2.39 g, glucose 2.10 g, saccharose 0.08 g; fatty acids – C16:0 (palmitic acid) 0.05 g, C18:1, n-9(oleic acid) 0.083 g, C 18:2, n-6 (linoleic acid) 0.182 g, C18:3, n-3 (α-linolenic acid) 0.166 g, total n-3 fatty acids (ω 3 fatty acids) 0.166 g and total n-6 fatty acids (ω 6 fatty acids) 0.182 g.

Folate content in 13 different Swedish strawberry cultivars varied from 335 μg/100 g of dry matter (DM) for cv. Senga Sengana to 644 μg/100 g of DM for cv. Elsanta (Strålsjö et al. 2003). The study indicated high folate retention in intact berries during storage until 3 or 9 days at 4°C (71–99%) and also in most tested commercial products (79–103%). On the basis of these data, fresh strawberries as well as processed strawberry products were recommended to be good folate sources. For instance, 250 g (fresh weight) of strawberries (approximately 125 μg of folate) would supply approximately 50% of the recommended daily folate intake in various European countries (200–300 μg/day) or 30% of the U.S. recommendation (400 μg/day).

Vitamin C contents in different strawberry varieties ranged from 56 to 99 mg/100 g fresh weight (Hägg et al. 1995). Vitamin C contents in berries varied during different years. Frozen storage destroyed about 34% of the vitamin C contents in strawberries. The average concentration of vitamin C in strawberries ranged from 32.4 mg/100 g to 84.7 mg/100 g (Hakala et al. 2003). Strawberries were found to be a good source of potassium (1.55–2.53 g/kg), magnesium (0.11–0.23 g/kg) and calcium (0.16–0.29 g/kg). The lead content was in general below its detection limit (0.004 mg/kg). The cadmium level in the Finnish berries was lower than 0.016 mg/kg. Among the four strawberry cultivars studies, there was a 2–5-fold variation for ascorbic acid, chlorogenic acid, ellagic acid, and total antioxidative capacity, measured in both water-soluble and water-insoluble extracts (Olsson et al. 2004). Unripe berries contained lower concentrations of chlorogenic acid and p-coumaric acid and also quercetin and kaempferol compared with riper berries. During cold storage for up to 3 days, relatively few changes in the concentration of the different antioxidants occurred. The dominating sugars in strawberries were fructose and glucose, but considerable amounts of sucrose were also present, and their contents varied among cultivars, giving a predicted glycemic index of about 81. Verbascose, raffinose, and stachyose were found in only minor amounts. The study showed that the concentration of a number of bioactive compounds in strawberries varied according to cultivar, ripening stage, and storage.

Twenty-five defined anthocyanin pigments were detected in strawberries, most of them containing pelargonidin as aglycone; some cyanidin derivatives were also found (da Silva et al. 2007). Glucose and rutinose constituted the usual substituting sugars, although arabinose and rhamnose were also tentatively identified. Some minor anthocyanins showed acylation with aliphatic acids. Anthocyanin-derived pigments, namely 5-carboxypyranopelargonidin-3-glucoside and four condensed pigments containing C–C linked anthocyanin (pelargonidin) and flavanol (catechin and afzelechin) residues were also detected. Total anthocyanin content of strawberries ranged between 200 and 600 mg/kg, with pelargonidin 3-glucoside constituting 77–90% of the anthocyanins followed by pelargonidin 3-rutinoside (6–11%) and cyanidin 3-glucoside (3–10%).

Phenolic compounds detected in strawberries included: p-coumaric acid derivative, ellagic acid, p-coumaroylglucose, quercetin, quercetin 3-glucoside, quercetin 3-glucuronide, kaempferol 3-glucoside, and anthocyanins cyanidin 3-glucoside, pelargonidin 3-glucoside, pelargonidin 3-rutinoside (Gil et al. 1997). Strawberries were reported to contain flavonoids (kaempferol, quercetin, myricetin) and phenolic acids (p-coumaric, caffeic, ferulic, p-hydroxybenzoic, gallic and ellagic acids) with beneficial effects on health as antioxidants and anticarcinogens (Häkkinen et al. 1999; Häkkinen and Törrönen 2000). Total content of the phenolics detected in strawberry cultivars ranged from 42.1 to 54.4 mg/100 g. Ellagic acid was the dominant acid in strawberries where it accounted for 51% of all acids found. Studies found that ellagic acid contents after 3 months of storage at −20°C varied between 31.5 (strawberry 'Senga Sengana') and 68.6 mg/100 g FW (fresh weight) (arctic bramble) (Häkkinen et al. 2000). Ellagic acid content in strawberry jam (23.8 mg/100 g FW.) was 80% of that in unprocessed strawberries. The content of ellagic acid in strawberries and red raspberries was reduced by 40% and 30%, respectively, during the 9 months of storage at −20°C.

Seventeen structurally well-defined phenolic compounds including phenylpropanoids, flavonols, flavan-3-ols, and anthocyanins were found in the ripe fruits of two cultivars of the commercial strawberry (*Fragaria* x *ananassa*) as well as in accessions of *F. vesca*, *F. moschata*, and *F. chiloensis* (Muñoz et al. 2011). Metabolic analysis revealed that the majority of the compounds accumulated in a genotype-dependent manner. The presence of biosynthetic enzymes such as phenylalanine ammonia-lyase, cinnamic acid 4-hydroxylase, chalcone synthase, and flavonoid 3'-hydroxylase could partially explain the different levels of polyphenolics observed in the *Fragaria* species.

Both *cis*- and *trans*-resveratrol were detected in strawberry achenes (seeds) and pulp (receptacle tissue). Resveratrol was found to be higher in achenes than in fruit pulp (Wang et al. 2007). The contents of resveratrol in strawberries were affected by genotype variations, fruit maturation, cultural practices, and environmental conditions. High growing temperature (25°C and 30°C) or enriched CO_2 in the atmosphere significantly enhanced resveratrol content of strawberries. Advancing maturation also elevated resveratrol content. The mature pulp and achenes contained higher levels of resveratrol than the immature fruit. Adding compost as a soil supplement or preharvest application of methyl jasmonate (MJ) also significantly enhanced the level of resveratrol in strawberry fruit. Among the plants grown in hill plasticulture, fruits of 'Ovation (B28)', 'Mohawwk', 'Earliglow', and 'B35' had higher amounts of resveratrol than fruits of other genotypes. 'Ovation' contained the highest amount of resveratrol among strawberries grown in matted row, whereas 'Latestar' contained the least. Ten of 14 tested genotypes (all except 'Allstar', 'Delmarvel', 'Northeaster', and 'MEUS 8') had higher amounts of resveratrol when grown in hill plasticulture compared to matted row.

Newly identified sulfur volatiles in strawberries included methyl thiopropionate, ethyl thiobutanoate, methyl thiohexanoate, methyl (methylthio)acetate, ethyl (methylthio)acetate, methyl 2-(methylthio)butyrate, methyl 3-(methylthio) propionate, ethyl 3-(methylthio)propionate, and methyl thiooctanoate (Du et al. 2011). Most sulfur volatiles increased with increasing maturity, with only concentrations of hydrogen sulfide and methanethiol remaining relatively consistent at all five maturity stages. At the white and half red stages, most sulfur volatiles consisted of various alkyl sulfides. At three-quarter red (commercial ripe), full ripe, and overripe stages, the majority of sulfur volatiles consisted of sulfur esters. Most sulfur volatiles increased dramatically between the commercial ripe, full ripe, and overripe stages, increasing as much as 100% between full ripe and overripe. Principal component analysis indicated that sulfur volatiles could be used to distinguish overripe from full ripe and commercial ripe berries.

Strawberries and Health

Strawberries are rich in phenolic compounds that may help defend the body against several diseases and conditions, including cancer, cardiovascular disease, diabetes and neurological decline. Data from numerous scientific studies suggests the antioxidants in strawberries may help reduce levels of oxidized low density lipoprotein cholesterol, a risk factor for cardiovascular disease. Strawberries are rich in flavonoids that may confer cardioprotection by inhibiting platelet aggregation and thromboxane synthesis. Anthocyanins in strawberries may protect the neuronal cells from inflammation associated with declines in cognitive function.

Strawberry fruits have been reported to contain phenolic compounds that have antioxidant, anticancer, antiatherosclerotic and anti-neurodegenerative properties (Seeram et al. 2006a, b). The nature, size, solubility, degree and position of glycosylation and conjugation of food phenolics were found to influence their absorption, distribution, metabolism and excretion in humans. Phenolics in strawberries were identified as ellagic acid (EA), EA-glycosides, ellagitannins, gallotannins, anthocyanins, flavonols, flavanols and coumaroyl glycosides. The anthocyanidins were pelargonidin and cyanidin, found predominantly as their glucosides and rutinosides. The major flavonol aglycons were quercetin and kaempferol found as their glucuronides and glucosides.

In strawberries, the most abundant of the bioactive phytochemicals were reported to be ellagic acid, and certain flavonoids: anthocyanin, catechin, quercetin and kaempferol (Hannum 2004). These compounds in strawberries also possessed potent antioxidant power. Antioxidants help lower risk of cardiovascular events by inhibition of LDL-cholesterol oxidation, promotion of plaque stability, improved vascular endothelial function, and decreased tendency for thrombosis. In addition, strawberry extracts had been shown to inhibit COX enzymes in-vitro, which would modulate the inflammatory process. Individual compounds in strawberries had demonstrated anticancer activity in several different experimental systems, blocking initiation of carcinogenesis, and suppressing progression and proliferation of tumours. Preliminary animal studies had indicated that diets rich in strawberries may also have the potential to provide benefits to the aging brain.

In the Women's Health Study, Sesso et al. (2007) examined strawberry intake for both its prospective association with cardiovascular disease risk in 38,176 women and its cross-sectional association with lipids and C-reactive protein (CRP) in a subset of 26,966 women. They found that strawberry intake was unassociated with the risk of CVD incident, lipids, or C-reactive protein in middle-aged and older women, though higher strawberry intake may slightly reduce the likelihood of having elevated C-reactive protein levels.

Antioxidant Activity

Strawberries like other berries provide unique antioxidants, anthocyanins, which give berries their red and blue hues but also act as potent antioxidants. Specific antioxidants present in strawberries include quercetin, kaempferol, chlorogenic acid, p-coumaric acid, ellagic acid and vitamin C (Olsson et al. 2004).

The phenolic compound found in strawberry fruits were cyanidin-3-glucoside (1), pelargonidin (2), pelargonidin-3-glucoside (3), pelargonidin-3-rutinoside (4), kaempferol (5), quercetin (6), kaempferol-3-(6'-coumaroyl) glucoside (7), 3,4,5-trihydroxyphenyl-acrylic acid (8), glucose ester of (E)- p-coumaric acid (9), and ellagic acid (10) (Zhang et al. 2008). Among the pure compounds, the anthocyanins 1 (7,156 µM Trolox/mg), 2 (4,922 µM Trolox/mg), and 4 (5,514 µM Trolox/mg) were the most potent antioxidants.

The amount of total phenolics varied between 617 and 4,350 mg/kg in fresh berries (blackberries, red raspberries, blueberries, sweet cherries and strawberries), as gallic acid equivalents (GAE) (Heinonen et al. 1998). In the copper-catalyzed in-vitro human low-density lipoprotein oxidation assay at 10 µM gallic acid equivalents (GAE), berry extracts inhibited hexanal formation in the order: blackberries > red raspberries > sweet cherries > blueberries > strawberries. In the copper-catalyzed in-vitro lecithin liposome oxidation assay, the extracts inhibited hexanal formation in the order: sweet cherries > blueberries > red raspberries > blackberries > strawberries. Red raspberries were more efficient than blueberries in inhibiting hydroperoxide formation in lecithin liposomes. HPLC analyses showed high anthocyanin content in blackberries, hydroxycinnamic acid in blueberries and sweet cherries, flavonol in blueberries, and flavan-3-ol in red raspberries. The antioxidant activity for LDL was associated directly with anthocyanins and indirectly with flavonols, and for liposome it correlated with the hydroxycinnamate content. Berries thus contribute a significant source of phenolic antioxidants that may have potential health effects.

On the basis of the wet weight of the fruits (edible portion), strawberry had the highest ORAC activity (15.36 µmol of Trolox equivalents per gram) followed by plum, orange, red grape, kiwi fruit, pink grapefruit, white grape, banana, apple, tomato, pear, and honeydew melon (Wang et al. 1996). On the basis of the dry weight of the fruits, strawberry (153.6 µmol TE per g DM) again had the highest ORAC activity followed by plum, orange, pink grapefruit, tomato, kiwi fruit, red grape, white grape, apple, honeydew melon, pear, and banana. The ORAC of strawberry juice extract was 12.44 µmole TE per g fruit. Fruit and vegetables rich in anthocyanins (e.g. strawberry, raspberry and red plum) demonstrated the highest antioxidant activities, followed by those rich in flavanones (e.g. orange and grapefruit) and flavonols (e.g. onion, leek, spinach and green cabbage), while the hydroxycinnamate-rich fruit (e.g. apple, tomato, pear and peach) consistently elicited the lower antioxidant activities (Proteggente et al. 2002). The TEAC (Trolox Equivalent Antioxidant Capacity), the FRAP (Ferric Reducing Ability of Plasma) and ORAC (Oxygen Radical Absorbance Capacity) values for each extract were relatively similar and well-correlated with the total phenolic and vitamin C contents. The antioxidant activities (TEAC) in terms of 100 g fresh weight uncooked portion size were in the order: strawberry >> raspberry = red plum >> red cabbage >>> grapefruit = orange > spinach > broccoli > green grape approximately/= onion > green cabbage > pea > apple > cauliflower approximately/=tomato approximately/= peach = leek > banana approximately/= lettuce. Blackberries (*Rubu*s sp.) and strawberries (*Fragaria* × *ananassa*) had the highest ORAC (oxygen radical absorbance capacity) values during the green stages, whereas red raspberries (*Rubus idaeus*) had the highest ORAC activity at the ripe stage (Wang and Lin 2000). Total anthocyanin content increased with maturity for all species of fruits. Compared with fruits, leaves were found to have higher ORAC values. In fruits, ORAC values ranged from 7.8 to 33.7 µmol of Trolox equivalents (TE)/g of fresh berries (35.0–162.1 µmol of TE/g of dry matter), whereas in leaves, ORAC values ranged from 69.7 to 182.2 µmol of TE/g of fresh leaves (205.0–728.8 µmol of TE/g of dry matter). As the leaves become older, the ORAC values and total phenolic contents decreased. The results showed a linear correlation between total phenolic content and ORAC activity for fruits and leaves. For ripe berries, a linear relationship existed between ORAC values and anthocyanin content. Of the ripe fruits tested, on the basis of wet weight of fruit, cv. Jewel black raspberry and blackberries were the richest source for antioxidants. On the basis of the dry weight of fruit, strawberries had the highest ORAC activity followed by black raspberries (cv. Jewel), blackberries, and red raspberries.

Aaby et al. (2005) found that strawberries contained 1% achenes on a fresh weight basis; however, they contributed to about 11% of total phenolics and 14% of antioxidant activities in strawberries. Ellagic acid, ellagic acid glycosides, and ellagitannins were the major contributors to the antioxidant activities of achenes. The predominant anthocyanin in the flesh was pelargonidin-3-glucoside, whereas achenes consisted of nearly equal amounts of cyanidin-3-glucoside and pelargonidin-3-glucoside. Phenolic content and antioxidant activity of strawberry achenes were reduced by industrial processing. However, the levels were still high and strawberry waste byproduct could thus be a possible source of nutraceuticals or natural antioxidants. About 40 phenolic compounds including glycosides of quercetin, kaempferol, cyanidin, pelargonidin, and ellagic acid, together with flavanols, derivatives of p-coumaric acid, and ellagitannins, were identified in strawberry fruits (Aaby et al. 2007). Quercetin-3-malonylhexoside and a deoxyhexoside of ellagic acid were reported for the first time. Antioxidative properties of individual components in strawberries were estimated by their electrochemical responses. Ascorbic acid was the single most significant contributor to electrochemical response in strawberries (24%), whereas the ellagitannins and the anthocyanins were the groups of polyphenols with the highest contributions, 19% and 13% at 400 mV, respectively.

Free phenolic contents varied between strawberry cultivars, differing by 65% between the highest (Earliglow) and the lowest (Allstar) ranked strawberry cultivars (Meyers et al. 2003). The water soluble bound and ethyl acetate soluble bound phenolic contents averaged 5% of the total phenolic content of the cultivars. The total flavonoid content of Annapolis was 2-fold higher than that of Allstar, which had the lowest content. The anthocyanin content of the highest ranked cultivar, Evangeline, was more than double that of the lowest ranked cultivar, Allstar. Overall, free phenolic content was weakly correlated with total antioxidant activity, and flavonoid and anthocyanin content did not correlate with total antioxidant activity. Studies by Ozsahin et al. (2011) confirmed that flavonoid ingredients of three different varieties of strawberry (r Camarosa, Selva and Dorit) fruit that had a scavenging effect against the radicals (DPPH* and OH*). Strawberry also inhibited lipid peroxidation, in the group given strawberries, the level of malondialdehyde (MDA)-2-thiobarbituric acid was markedly reduced.

Studies indicated that high oxygen treatments exerted the most effects on fruit quality and antioxidant capacity of strawberry fruit in the first 7 days of storage (Zheng et al. 2007). While fruit quality parameters such as titratable acidity, total soluble solids and surface colour were only slightly affected by differing levels of O2, the higher oxygen concentration treatments significantly reduced decay. Oxygen concentrations higher than 60 kPa also promoted increases in ORAC values, total phenolics and total anthocyanins as well as individual phenolic compounds during the initial 7 days of storage.

The mean total antioxidant activity (TAA) (sum of hydrophilic and lipophilic antioxidant activities) for freeze-dried strawberries based on an 'as consumed' weight was significantly higher compared to fresh, frozen strawberries and jam (Marques et al. 2010). The mean TAA based on dry weight for fresh strawberries was significantly higher than for freeze-dried, frozen and jam. Results concurred with previous studies reporting strawberries to be a valuable source of antioxidants for consumers.

Anthocyanins were detected in two strawberry jams at very low content (Da Silva Pinto et al. 2007). Kaempferol glycosides were the main flavonoids present (from 0.38 to 1.05 mg/100 g fresh weight, FW), while quercetin glycosides were present in the range 0.14–1.20 mg/100 g FW. Free and total ellagic acid content ranged from 0.4 to 2.9 mg/100 g FW, and from 17.0 to 29.5 mg/100 g FW, respectively. Total phenolics varied from 58 to 136 mg/100 g FW, and the antioxidant capacity from 0.55 to 0.76 μmol BHT (butylhydroxytoluene) equivalents/g FW. Taken together, the results indicated that jams could also represent a good source of antioxidant compounds, although compared to the fruit important losses appeared to occur.

Postharvest studies of strawberries at 20°C storage for 3 days showed that pelargonidin-3-glucoside, the major anthocyanin, increased with the increase of shelf life period, while cyanidin-3-glucoside and pelargonidin-3-rutinoside were found at lower concentrations (Goulas and Manganaris 2011). The potent radical scavenging activity, evaluated with four in-vitro assays, showed a higher antioxidant capacity after 3 and 1 day of shelf life. Further, the antioxidant effect of strawberry fruit extracts on lipid substrates and on an emulsion system showed a significant inhibition in the formation of conjugated diene hyperoxides.

The results of studies by Cao et al. (1998) showed that the total antioxidant capacity of serum determined as ORAC, TEAC and FRAP, increased significantly by 7–25% during the 4-hour period following consumption of red wine, strawberries, vitamin C or spinach. The total antioxidant capacity of urine determined as ORAC increased by 9.6%, 27.5%, and 44.9% for strawberries, spinach, and vitamin C, respectively, during the 24-hour period following these treatments. The plasma vitamin C level after the strawberry drink, and the serum urate level after the strawberry and spinach treatments, also increased significantly. However, the increased vitamin C and urate levels could not fully account for the increased total antioxidant capacity in serum following the consumption of strawberries, spinach or red wine. The researchers concluded that the consumption of strawberries, spinach or red wine, rich in antioxidant phenolic compounds, could increase the serum antioxidant capacity in humans.

A recent report published in the American Journal of Clinical Nutrition analyzed over 1,000 foods and beverages for antioxidant capacity (Halvorsen et al. 2006). On the basis of typical serving sizes, blackberries, walnuts, strawberries, artichokes, cranberries, brewed coffee, raspberries, pecans, blueberries, ground cloves, grape juice, and unsweetened baking chocolate were at the top of the ranked list in total antioxidant capacity (AOX) per serving. Strawberries ranked third in total antioxidant capacity per serving, superseded only by blackberries and walnuts which showed higher antioxidant capacity. For comparison, the researchers found that a serving of strawberries provided 3.6 mmol antioxidants/serving while blueberries were 2.7 mmol AOX/serving, sour cherries were 2.2 mmol AOX/serving and oranges, 1.3 mmol/serving.

Anticancer Activity

Ethanol extracts from two strawberry cultivars, Sweet Charlie and Carlsbad, and two blueberry cultivars, Tifblue and Premier fruits strongly inhibited CaSki and SiHa cervical cancer cell lines and MCF-7 and T47-D breast cancer cell lines (Wedge et al. 2001). Strawberry extracts rich in antioxidant enzymes glutathione peroxidase, superoxidedismutase, guaiacol peroxidase, ascorbate peroxidase, and glutathione reductase, inhibited the proliferation of human lung epithelial cancer cell line A549 and decreased tetradecanoylphorbol-13-acetate (TPA) -induced neoplastic transformation of JB6 P+ mouse epidermal cells (Wang et al. 2005). Pretreatment of JB6 P+ mouse epidermal cells with strawberry extract resulted in the inhibition of both UVB- and TPA-induced AP-1 and NF-kappaB transactivation. The results suggested that the ability of strawberries to block UVB- and TPA-induced AP-1 and NF-kappaB activation may be due to their antioxidant properties and their ability to reduce oxidative stress. The strawberries may be highly effective as a chemopreventive agent that acts by targeting the down-regulation of AP-1 and NF-kappaB activities, blocking MAPK signaling, and suppressing cancer cell proliferation and transformation.

The berry extracts (blackberry, black raspberry, blueberry, cranberry, red raspberry and strawberry) rich in phenolics such as anthocyanins, flavonols, flavanols, ellagitannins, gallotannins, proanthocyanidins, and phenolic acids, inhibited the growth of human oral (KB, CAL-27), breast (MCF-7), colon (HT-29, HCT116), and prostate (LNCaP) tumour cell lines at concentrations ranging from 25 to 200 µg/ml (Seeram et al. 2006a). With increasing concentration of berry extract, increasing inhibition of cell proliferation in all of the cell lines were observed, with different degrees of potency between cell lines. Black raspberry and strawberry extracts showed the most significant pro-apoptotic effects against the COX-2 expressing colon cancer cell line, HT-29. Meyers et al. (2003) found that the proliferation of HepG(2) human liver cancer cells was significantly inhibited in a dose-dependent manner after exposure to all strawberry cultivar (Earliglow, Annapolis, Evangeline, Allstar, Sable, Sparkle, Jewel, and Mesabi) extracts, with Earliglow exhibiting the highest antiproliferative activity and Annapolis exhibiting the lowest. No relationship was found between antiproliferative activity and antioxidant content.

Purified ellagitannins from strawberries were found to have antiproliferative activity (Pinto Mda et al. 2010). It was observed that ellagic acid had the highest percentage inhibition of cell proliferation. The strawberry extract had lower efficacy in inhibiting the cell proliferation, indicating that in the case of this fruit there was no synergism. Polyphenol-rich strawberry extract was found to have antiproliferative effect against human cervical cancer (HeLa) cells (McDougall et al. 2008). The most effective extracts (strawberry > arctic bramble > cloudberry > lingonberry) gave EC_{50} values in the range of 25–40 µg/ml of phenols. These extracts were also effective against human colon cancer (CaCo-2) cells. The strawberry, cloudberry, arctic bramble, and the raspberry extracts shared common polyphenol constituents, especially the ellagitannins, which had been shown to be effective antiproliferative agents. Crude strawberry extracts (250 µg/ml) and pure phenolic compounds cyanidin-3-glucoside, pelargonidin, pelargonidin-3-glucoside, pelargonidin-3-rutinoside, kaempferol, quercetin, kaempferol-3-(6'-coumaroyl) glucoside, 3,4,5-trihydroxyphenyl-acrylic acid, glucose ester of (E)-p-coumaric acid (9), and ellagic acid (100 µg/ml) inhibited the growth of human oral (CAL-27, KB), colon (HT29, HCT-116), and prostate (LNCaP, DU145) cancer cells with different sensitivities observed between cell lines (Zhang et al. 2008). OptiBerry, a combination of wild blueberry, wild bilberry, cranberry, elderberry, raspberry seeds, and strawberry, exhibited high antioxidant efficacy as shown by its high oxygen radical absorbance capacity (ORAC) values, novel antiangiogenic and antiatherosclerotic activities, and potential cytotoxicity towards *Helicobacter pylori*, a noxious pathogen responsible for various gastrointestinal disorders including duodenal ulcer and gastric cancer, as compared to individual berry extracts (Zafra-Stone et al. 2007). OptiBerry also significantly inhibited basal MCP-1 and inducible NF-κβ transcriptions as well as the inflammatory biomarker IL-8, and significantly reduced the ability to form hemangioma and markedly decreased

EOMA cell-induced tumour growth in an in-vivo model. Overall, berry anthocyanins triggered genetic signalling in promoting human health and disease prevention.

Studies showed that dietary freeze-dried strawberries effectively inhibited N-nitrosomethylbenzylamine (NMBA)-induced tumorigenesis in the rat esophagus (Carlton et al. 2001). At 30 weeks, 5% and 10% freeze-dried strawberries in the diet caused significant reductions in esophageal tumour multiplicity of 24% and 56%, respectively. A significant decrease in O6-methylguanine levels was observed in the esophageal DNA of animals fed strawberries, suggesting that one or more components in strawberries influenced the metabolism of NMBA to DNA-damaging species.

Fisetin, found in high level in strawberry (Arai et al. 2000) had also been reported to have anti-cancer activity: against human colon cancer cells (Lu et al. 2005), pancreatic cancer (Murtaza et al. 2009) and prostate cancer (Khan et al. 2008).

The cytotoxic effects of strawberry polyphenol-rich extract were investigated on normal cells and tumour cells namely a human prostate epithelial cell line (P21) and two tumour cell lines (P21 tumour cell line 1 and 2) derived from the same patient, and a normal human breast epithelial cell line (B42) and a tumour line derived from it (B42 clone 16) (Weaver et al. 2009). The strawberry extract was cytotoxic with doses of approximately 5 μg/ml causing a 50% reduction in cell survival in both the normal and the tumour lines. The extracts were also cytotoxic to peripheral blood human lymphocytes stimulated with phytohaemagglutinin but higher levels (>20 μg/ml for 50% reduction in cell survival) were required. After fractionation of the strawberry sample, the cytotoxicity was retained in the tannin-rich fraction and this fraction was considerably more toxic to all cells (normal or tumour cell lines or lymphocytes) than the anthocyanin-rich fraction. Established prostate (LNCaP and PC-3) and breast (MCF-7) tumour cell lines were more resistant to the strawberry extract with concentrations of 50 μg/ml required for 50% reduction in cell survival. From these findings, the researchers concluded that there was little evidence to assume that polyphenols from strawberry had a differential cytotoxic effect on tumour cells relative to comparable normal cells from the same tissue derived from the same patient.

Antimutagenic Activity

Fresh juices and organic solvent extracts from the fruits of strawberry, blueberry, and raspberry inhibited the production of mutations by the direct-acting mutagen methyl methanesulfonate and the metabolically activated carcinogen benzo[a]pyrene (Hope Smith et al. 2004). Juice from strawberry, blueberry, and raspberry fruit significantly inhibited mutagenesis caused by both carcinogens. Ethanol extracts from freeze-dried fruits of strawberry cultivars (Sweet Charlie and Carlsbad) and blueberry cultivars (Tifblue and Premier) were also tested. Of these, the hydrolyzable tannin-containing fraction from Sweet Charlie strawberries was most effective at inhibiting mutations.

Antiangiogenic Activity

The oxygen radical absorbance capacity (OARC) antioxidant values of strawberry powder and grape seed proanthocyanidin extract (GSPE) were higher than cranberry, elderberry or raspberry seed but significantly lower than the other samples studied. Wild bilberry and blueberry extracts possessed the highest ORAC values (Roy et al. 2002). Each of the berry samples studied significantly inhibited both H_2O_2 as well as TNF α induced vascular endothelial growth factor expression by the human keratinocytes. This effect was not shared by other antioxidants such as α-tocopherol or GSPE but was commonly shared by pure flavonoids. Matrigel assay using human dermal microvascular endothelial cells showed that the edible berries hindered angiogenesis.

Antithrombotic Activity

Strawberry varieties KYSt-4 (Nohime), KYSt-11 (Kurume IH-1) and KYSt-17 (Kurume 58) showed significant antiplatelet activity both in-vitro and, after oral administration, in-vivo (Naemura et al. 2005). Both KYSt-11 and KYSt-17, but not KYSt-4, significantly reduced flow-mediated vasodilation; that is, caused endothelial dysfunction. Significant correlation was found between antiplatelet and antioxidant activities or total phenolic compounds. Of the tested strawberry varieties, KYSt-4, KYSt-11 and KYSt-17 showed significant antithrombotic effect. The dual mechanism of the effect may involve a direct inhibition of both platelet function and antioxidant activities. Among various strawberry varieties tested, a particular variety (KYSt-4, Nohime) showed a significant antithrombotic effects in humans while the experimentally inactive variety (KYSt-10) as well as the relevant control (water) were ineffective (Naemura et al. 2006) Daily intake of an antithrombotic diet may offer a convenient and effective way of prevention of arterial thrombotic diseases.

Antiatherosclerotic Activity

In an 8 weeks randomized controlled trial, short-term freeze-dried strawberry supplementation was found to improve selected atherosclerotic risk factors, including dyslipidemia and circulating adhesion molecules in 27 subjects with metabolic syndrome (Basu et al. 2010). Strawberry supplementation significantly decreased total and low-density lipoprotein cholesterol and small low-density lipoprotein particles. Strawberry supplementation further decreased circulating levels of vascular cell adhesion molecule-1. Serum glucose, triglycerides, high-density lipoprotein cholesterol, blood pressure, and waist circumference were not affected.

Antidiabetic Activity

Results of in-vitro studies suggested the ellagitannins and ellagic acid from strawberries to have good potential for the management of hyperglycemia and hypertension linked to type 2 diabetes (Pinto Mda et al. 2010). Purified ellagitannins had high α-amylase and angiotensin I-converting enzyme (ACE) inhibitory activities. However, these compounds had low α-glucosidase inhibitory activity. In-vitro studies showed that of polyphenols, phenolic acids and tannins (PPTs) from strawberry inhibited glucose transport from the intestinal lumen into cells and also the GLUT2 (glucose transporter −2)-facilitated exit on the basolateral side. Further, pelargonidin-3-O-glucoside (IC_{50} = 802 µM) contributed 26% to the total inhibition by the strawberry extract.

Among fruits and vegetable, strawberry is a rich source of the flavonol fisetin, a potent antioxidant. Strawberries contained (160 µg/g) of fisetin 5–10-fold more than apples (26.9 µg/g) and persimmon (10.6 µg/g), and 25–40 fold more than lotus roots (5.8 µg/g), onions (4.8 µg/g) and grapes (3.9 µg/g) (Arai et al. 2000). Studies showed that fisetin or a synthetic derivative may have potential therapeutic use for the treatment of diabetic complications such as kidney failure (Maher et al. 2011). Fisetin lowered the elevation of α-oxoaldehyde methylglyoxal (MG)-protein glycation that was associated with diabetes and ameliorated multiple complications of the disease.

Antiinflammatory Activity

Administration of strawberry, loquat, mulberry and bitter melon fruit juices increased IL-10 production by lipopolysaccharide-stimulated murine peritoneal macrophages in dose-dependent manners (Lin and Tang 2008). Concurrently, the levels of IL-1β, IL-6 and/or TNF-α were decreased. The results suggested that strawberry, loquat, mulberry, and bitter melon juices exhibited a prophylactic effect on LPS-induced inflammation of peritoneal macrophages via increasing anti-inflammatory cytokine and/or decreasing pro-inflammatory cytokines secretions.

In a cross-over placebo study of 24 overweight adults, concurrent consumption of strawberry beverage with the high-carbohydrate, moderate-fat meal (HCFM) significantly increased the postprandial concentrations of pelargonidin sulfate and pelargonidin-3-O-glucoside compared with the placebo beverage (Edirisinghe et al. 2011). The strawberry beverage significantly attenuated the postprandial inflammatory response as measured by high-sensitivity C-reactive protein and interleukin IL-6 induced by the HCFM. It was also associated with a reduction in postprandial insulin response. Overall, the data reflected the favourable effects of strawberry antioxidants on postprandial inflammation and insulin sensitivity. Strawberry ethanol fruit extract at 500 mg/kg showed significant amelioration of experimentally induced inflammatory bowel disease in albino rats, which may be attributed to its antioxidant and anti-inflammatory properties (Kanodia et al. 2011). The extract showed significant prevention of increase in colon weight and disease activity index along with decrease in macroscopic and microscopic lesion score as compared to control group. Significant improvement was observed in the levels of myeloperoxidase, catalase and superoxide dismutase, except glutathione. However, the effect of the extract was significantly less than 5-aminosalisylic acid.

Anti-allergic Activity

The flavonoids isolated from strawberry were found to suppress the degranulation from Ag (antigen)-stimulated rat basophilic leukemia RBL-2H3 cells to varying extent (Itoh et al. 2009). The intracellular free Ca(2+) concentration ([Ca(2+)]i) was elevated by Fc epsilonRI activation, but these flavonoid treatments reduced the elevation of [Ca(2+)]i by suppressing Ca(2+) influx. Kaempferol strongly suppressed the activation of spleen tyrosine kinase (Syk) and phospholipase Cγ (PLCγ). The findings thus suggested that suppression of Ag-stimulated degranulation by the flavonoids was mainly due to suppression of [Ca(2+)]i elevation and Syk activation. The results suggested that strawberry would be of some ameliorative benefit for the allergic symptoms. In another study, the scientists reported that among the eight isolated phenolic constituents of strawberry, linocinnamarin, 1-O-trans-cinnamoyl-β-d-glucopyranose, and cinnamic acid exhibited antigen (Ag)-stimulated degranulation in rat basophilic leukemia RBL-2H3 cells (Ninomiya et al. 2010). Treatment with both linocinnamarin and cinnamic acid markedly suppressed antigen-stimulated elevation of intracellular free Ca(2+) concentration and reactive oxygen species (ROS). Both linocinnamarin and cinnamic acid suppressed Ag-stimulated spleen tyrosine kinase (Syk) activation. These results indicated that inhibition of antigen-stimulated degranulation by linocinnamarin and cinnamic acid was primarily due to inactivation of Syk/phospholipase Cγ (PLCγ) pathways. The findings suggested that linocinnamarin and cinnamic acid isolated from strawberry could be beneficial agents for alleviating symptoms of type I allergy.

Neuroprotective Activity

The cell viability test using the 3-(4,5-dimethylthiazol-2-yl)-2,5-diphenyltetrazolium bromide (MTT) reduction assay showed that strawberry phenolics significantly reduced oxidative stress-induced neurotoxicity (Heo and Lee 2005). Strawberry showed the highest neuronal cell protective effects among the samples. The overall relative neuronal cell protective activity of three fruits by three tests followed the decreasing order strawberry > banana > orange. The protective effects appeared to be due to the higher phenolic contents including anthocyanins, and anthocyanins in strawberries.

Drug Interaction Activity

A new glycoside, 2-β-d-glucopyranosyloxy-4,6-dihydroxyisovalerophenone (3), was isolated from strawberry fruit along with kaempferol-3-β-D-(6-O-trans-p-coumaroyl) glucopyranoside (1) and kaempferol-3-β-D-(6-O-cis-p-coumaroyl) glucopyranoside (2) (Tsukamoto et al. 2004). Compounds 1 and 2 inhibited activity of a drug-metabolizing enzyme, CYP3A4.

Traditional Medicinal Uses

Strawberry has been used in traditional medicine; both leaves and fruit have appeared in early pharmacopoeias (Grieve 1971). They have been used as laxative, diuretic and astringent.

Leaves are used in tea to treat dysentery. Roots are also astringent and used in diarrhoea. Strawberry is also a useful dentifrice for removing teeth discoloration and cosmetic for skin conditioning.

Ancient Romans used the fruit to treat melancholy, fainting, throat infections, inflammations, fevers, kidney stones, halitosis, gout, and diseases of the blood, liver and spleen.

Other Uses

Strawberry pigment extract can be used as a natural acid/base indicator due to the different colour of the conjugate acid and conjugate base of the pigment.

Comments

Strawberries can be grown from seeds but commercially are propagated from runners (stolons).

Selected References

Aaby K, Skrede G, Wrolstad RE (2005) Phenolic composition and antioxidant activities in flesh and achenes of strawberries (*Fragaria ananassa*). J Agric Food Chem 53(10):4032–4040

Aaby K, Ekeberg D, Skrede G (2007) Characterization of phenolic compounds in strawberry (*Fragaria x ananassa*) fruits by different HPLC detectors and contribution of individual compounds to total antioxidant capacity. J Agric Food Chem 55(11):4395–4406

Arai Y, Watanabe S, Kimira M, Shimoi K, Mochizuki R, Kinae N (2000) Dietary intakes of flavonols, flavones and isoflavones by Japanese women and the inverse correlation between quercetin intake and plasma LDL cholesterol concentration. J Nutr 130:2243–2250

Bailey LH (1976) Hortus third. A concise dictionary of plants cultivated in the United States and Canada. Liberty Hyde Bailey Hortorium/Cornell University/Wiley, New York, 1312pp

Basu A, Fu DX, Wilkinson M, Simmons B, Wu M, Betts NM, Du M, Lyons TJ (2010) Strawberries decrease atherosclerotic markers in subjects with metabolic syndrome. Nutr Res 30(7):462–469

Cao G, Russell RM, Lischner N, Prior RL (1998) Serum antioxidant capacity is increased by consumption of strawberries, spinach, red wine, or vitamin C in elderly women. J Nutr 128:2383–2390

Carlton PS, Kresty LA, Siglin JC, Morse MA, Lu J, Morgan C, Stoner GD (2001) Inhibition of N-nitrosomethylbenzylamine-induced tumorigenesis in the rat esophagus by dietary freeze-dried strawberries. Carcinogenesis 22(3):441–446

da Silva FL, Escribano-Bailón MT, Alonso JJP, Rivas-Gonzalo JC, Santos-Buelga C (2007) Anthocyanin pigments in strawberry. LWT- Food Sci Technol 40(2):374–382

Da Silva Pinto M, Lajolo FM, Genovese MI (2007) Bioactive compounds and antioxidant capacity of strawberry jams. Plant Foods Hum Nutr 62(3):127–131

Du X, Song M, Rouseff R (2011) Identification of new strawberry sulfur volatiles and changes during maturation. J Agric Food Chem 59(4):1293–1300

Edirisinghe I, Banaszewski K, Cappozzo J, Sandhya K, Ellis CL, Tadapaneni R, Kappagoda CT, Burton-Freeman BM (2011) Strawberry anthocyanin and its association with postprandial inflammation and insulin. Br J Nutr 16:1–10

Gil MI, Holcroft DM, Kader AA (1997) Changes in strawberry anthocyanins and other polyphenols in response to carbon dioxide treatments. J Agric Food Chem 45:1662–1667

Goulas V, Manganaris GA (2011) The effect of postharvest ripening on strawberry bioactive composition and antioxidant potential. J Sci Food Agric 91(10):1907–1914

Grieve M (1971) A modern herbal, 2 vols. Penguin/Dover publications, New York, 919pp

Hägg M, Ylikoski S, Kumpulainen J (1995) Vitamin C content in fruits and berries consumed in Finland. J Food Comp Anal 8(1):12–20

Hakala M, Lapveteläinen A, Huopalahti R, Kallio H, Tahvonen R (2003) Effects of varieties and cultivation conditions on the composition of strawberries. J Food Comp Anal 16(1):67–80

Häkkinen SH, Törrönen AR (2000) Content of flavonols and selected phenolic acids in strawberries and *Vaccinium* species: influence of cultivar, cultivation site and technique. Food Res Int 33(6):517–524

Häkkinen SH, Heinonen IM, Kärenlampi SO, Mykkänen HM, Ruuskanen J, Törrönen AR (1999) Screening of selected flavonoids and phenolic acids in 19 berries. Food Res Int 32(5):345–353

Häkkinen SH, Kärenlampi SO, Mykkänen HM, Heinonen IM, Törrönen AR (2000) Ellagic acid content in berries: influence of domestic processing and storage. Eur Food Res Technol 212(1):75–80

Halvorsen BL, Carlsen MH, Phillips KM, Bøhn SK, Holte K, Jacobs DR Jr, Blomhoff R (2006) Content of redox-active compounds (ie, antioxidants) in foods consumed in the United States. Am J Clin Nutr 84:95–135

Hancock JF (1999) Strawberries. Crop production science in horticulture, Ser. 11. CAB International, Wallingford, 237pp

Hannum SM (2004) Potential impact of strawberries on human health: a review of the science. Crit Rev Food Sci Nutr 44(1):1–17

Heinonen IM, Meyer AS, Frankel EN (1998) Antioxidant activity of berry phenolics on human low-density lipoprotein and liposome oxidation. J Agric Food Chem 46(10):4107–4112

Heo HJ, Lee CY (2005) Strawberry and its anthocyanins reduce oxidative stress-induced apoptosis in PC12 cells. J Agric Food Chem 53(6):1984–1989

Hope Smith S, Tate PL, Huang G, Magee JB, Meepagala KM, Wedge DE, Larcom LL (2004) Antimutagenic activity of berry extracts. J Med Food 7(4):450–455

Itoh T, Ninomiya M, Yasuda M, Koshikawa K, Deyashiki Y, Nozawa Y, Akao Y, Koketsu M (2009) Inhibitory effects of flavonoids isolated from *Fragaria ananassa* Duch on IgE-mediated degranulation in rat basophilic leukemia RBL-2H3. Bioorg Med Chem 17(15):5374–5379

Kanodia L, Borgohain M, Das S (2011) Effect of fruit extract of *Fragaria vesca* L. on experimentally induced inflammatory bowel disease in albino rats. Indian J Pharmacol 43(1):18–21

Khan N, Afaq F, Syed DN, Mukhtar H (2008) Fisetin, a novel dietary flavonoid, causes apoptosis and cell cycle arrest in human prostate cancer LNCaP cells. Carcinogenesis 29(5):1049–1056

Lin JY, Tang CY (2008) Strawberry, loquat, mulberry, and bitter melon juices exhibit prophylactic effects on LPS-induced inflammation using murine peritoneal macrophages. Food Chem 107(4):1587–1596

Lu X, Jung J, Cho HJ, Lim DY, Lee HS, Chun HS, Kwon DY, Park JH (2005) Fisetin inhibits the activities of cyclin-dependent kinases leading to cell cycle arrest in HT-29 human colon cancer cells. J Nutr 135(12):2884–2890

Maher P, Dargusch R, Ehren JL, Okada S, Sharma K, Schubert D (2011) Fisetin lowers methylglyoxal dependent protein glycation and limits the complications of diabetes. PLoS ONE 6(6):e21226

Manzano S, Williamson G (2010) Polyphenols and phenolic acids from strawberry and apple decrease glucose uptake and transport by human intestinal Caco-2 cells. Mol Nutr Food Res 54(12):1773–1780

Marques KK, Renfroe MH, Brevard PBB, Lee RE, Gloeckner JW (2010) Differences in antioxidant levels of fresh, frozen and freeze-dried strawberries and strawberry jam. Int J Food Sci Nutr 61(8):759–769

McDougall GJ, Ross HA, Ikeji M, Stewart D (2008) Berry extracts exert different antiproliferative effects against cervical and colon cancer cells grown in vitro. J Agric Food Chem 56(9):3016–3023

Meyers KJ, Watkins CB, Pritts MP, Liu RH (2003) Antioxidant and antiproliferative activities of strawberries. J Agric Food Chem 51(23):6887–6892

Muñoz C, Sánchez-Sevilla JF, Botella MA, Hoffmann T, Schwab W, Valpuesta V (2011) Polyphenol composition in the ripe fruits of *Fragaria* species and transcriptional analyses of key genes in the pathway. J Agric Food Chem 59(23):12598–12604

Murtaza I, Adhami VM, Hafeez BB, Saleem M, Mukhtar H (2009) Fisetin, a natural flavonoid, targets chemoresistant human pancreatic cancer AsPC-1 cells through DR3-mediated inhibition of NF-kappaB. Int J Cancer 125(10):2465–2473

Naemura A, Mitani T, Ijiri Y, Tamura Y, Yamashita T, Okimura M, Yamamoto J (2005) Anti-thrombotic effect of strawberries. Blood Coagul Fibrinolysis 16(7):501–509

Naemura A, Ohira H, Ikeda M, Koshikawa K, Ishii H, Yamamoto J (2006) An experimentally antithrombotic strawberry variety is also effective in humans. Pathophysiol Haemost Thromb 35(5):398–404

Ninomiya M, Itoh T, Ishikawa S, Saiki M, Narumiya K, Yasuda M, Koshikawa K, Nozawa Y, Koketsu M (2010) Phenolic constituents isolated from *Fragaria ananassa* Duch. inhibit antigen-stimulated degranulation through direct inhibition of spleen tyrosine kinase activation. Bioorg Med Chem 18(16):5932–5937

Olsson ME, Ekvall J, Gustavsson KE, Nilsson J, Pillai D, Sjöholm I, Svensson U, Akesson B, Nyman Margareta GL (2004) Antioxidants, low molecular weight carbohydrates, and total antioxidant capacity in strawberries: effects of cultivar, ripening, and storage. J Agric Food Chem 52:2490–2498

Ozsahin AD, Gokce Z, Yilmaz O, Kirecci OA (2011) The fruit extract of three strawberry cultivars prevents lipid peroxidation and protects the unsaturated fatty acids in the Fenton reagent environment. Int J Food Sci Nutr doi:10.3109/09637486.2011.628646

Pinto Mda S, de Carvalho JE, Lajolo FM, Genovese MI, Shetty K (2010) Evaluation of antiproliferative, anti-type 2 diabetes, and antihypertension potentials of ellagitannins from strawberries (*Fragaria* × *ananassa* Duch.) using in vitro models. J Med Food 13(5):1027–1035

Porcher MH et al (1995–2020) Searchable World Wide Web multilingual multiscript plant name database. The University of Melbourne, Melbourne. http://www.plantnames.unimelb.edu.au/Sorting/Frontpage.html

Proteggente AR, Pannala AS, Paganga G, Van Buren L, Wagner E, Wiseman S, Van De Put F, Dacombe C, Rice-Evans CA (2002) The antioxidant activity of regularly consumed fruit and vegetables reflects their phenolic and vitamin C composition. Free Radic Res 36(2):217–233

Roy S, Khanna S, Alessio HM, Vider J, Bagchi D, Bagchi M, Sen CK (2002) Anti-angiogenic property of edible berries. Free Radic Res 36(9):1023–1031

Saxholt E, Christensen AT, Møller A, Hartkopp HB, Hess Ygil K, Hels OH (2008) Danish food composition databank, Revision 7. Department of Nutrition, National Food Institute, Technical University of Denmark. http://ww.food.comp.dk/

Seeram NP, Adams LS, Zhang Y, Lee R, Sand D, Scheuller HS, Heber D (2006a) Blackberry, black raspberry, blueberry, cranberry, red raspberry, and strawberry extracts inhibit growth and stimulate apoptosis of human cancer cells in vitro. J Agric Food Chem 54(25):9329–9339

Seeram NP, Lee R, Scheuller HS, Heber D (2006b) Identification of phenolic compounds in strawberries by liquid chromatography electrospray ionization mass spectroscopy. Food Chem 97(1):1–11

Sesso HD, Jenkins JM, Gaziano DJ, Buring JE (2007) Strawberry intake, lipids, C-reactive protein, and the risk of cardiovascular disease in women. J Am Coll Nutr 26(4):303–310

Strålsjö LM, Witthöft CM, Sjöholm IM, Jägerstad MI (2003) Folate content in strawberries (*Fragaria x ananassa*): effects of cultivar, ripeness, year of harvest, storage, and commercial processing. J Agric Food Chem 51(1):128–133

Sukumalanandana C, Verheij EWM (1992) *Fragaria x ananassa* (Duchesne) Guedes. In: Verheij EWM, Coronel RE (eds) Plant resources of South-East Asia, No. 2. Edible fruits and nuts. Prosea Foundation, Bogor, pp 171–175

Tsukamoto S, Tomise K, Aburatani M, Onuki H, Hirorta H, Ishiharajima E, Ohta T (2004) Isolation of cytochrome P450 inhibitors from strawberry fruit, *Fragaria ananassa*. J Nat Prod 67(11):1839–1841

Wang SY, Lin HS (2000) Antioxidant activity in fruits and leaves of blackberry, raspberry, and strawberry varies with cultivar and developmental stage. J Agric Food Chem 48:140–146

Wang H, Cao G, Prior RL (1996) Total antioxidant capacity of fruits. J Agric Food Chem 44(3):701–705

Wang SY, Chen CT, Wang CY, Chen P (2007) Resveratrol content in strawberry fruit is affected by preharvest conditions. J Agric Food Chem 55(20):8269–8274

Wang SY, Feng R, Lu Y, Bowman L, Ding M (2005) Inhibitory effect on activator protein-1, nuclear factor-kappa b, and cell transformation by extracts of strawberries (*Fragaria X ananassa* Duch). J Agric Food Chem 53(10):4187–4193

Weaver J, Briscoe T, Hou M, Goodman C, Kata S, Ross H, McDougall G, Stewart D, Riches A (2009) Strawberry polyphenols are equally cytotoxic to tumourigenic and normal human breast and prostate cell lines. J Oncol 34(3):777–786

Wedge DE, Meepagala KM, Magee JB, Smith SH, Huang G, Larcom LL (2001) Anticarcinogenic activity of strawberry, blueberry, and raspberry extracts to breast and cervical cancer cells. J Med Food 4(1):49–51

Wu ZY, Raven PH, Hong DY (eds) (2003) Flora of China, vol 9 (Pittosporaceae through Connaraceae). Science Press/Missouri Botanical Garden Press, Beijing/St. Louis

Zafra-Stone S, Yasmin T, Bagchi M, Chatterjee A, Vinson JA, Bagchi D (2007) Berry anthocyanins as novel antioxidants in human health and disease prevention. Mol Nutr Food Res 51(6):675–683

Zhang Y, Seeram NP, Lee R, Feng L, Heber D (2008) Isolation and identification of strawberry phenolics with antioxidant and human cancer cell antiproliferative properties. J Agric Food Chem 56(3):670–675

Zheng YH, Wang SY, Wang CY, Zheng W (2007) Changes in strawberry phenolics, anthocyanins, and antioxidant capacity in response to high oxygen treatments. LWT 40:49–57

Malus spectabilis

Scientific Name

Malus spectabilis (Aiton) Borkhausen

Synonyms

Malus domestica Borkhausen var. *spectabilis* (Aiton) Likhonos, *Malus microcarpa* A. Savatier var. *spectabilis* (Aiton) Carrière, *Pyrus spectabilis* Aiton, *Malus domestica* var. *spectabilis* (Aiton) Likhonos, *Malus microcarpa* var. *spectabilis* (Aiton) Carrière.

Family

Rosaceae

Common/English Names

Asiatic Apple, Chinese Crabapple, Chinese Flowering Apple, Crabapple, Doubleflower Chinese Crabapple, Riversii Chinese Crabapple.

Vernacular Names

Chinese: Hai Tang Hua;
German: Pracht-Apfel.

Origin/Distribution

Malus spectabilis is native to China.

Agroecology

In its native temperate habitat, it is found in plains and mountain regions at elevations of 500–2,000 m. It is frost hardy and thrives in full sun in well-drained mildly acidic to alkaline soils.

Edible Plant Parts and Uses

Crabapples can be eaten raw or cooked. It can be used in juicing to offset the sweeter apple or pear. Crabapples are an excellent source of pectin and their juice can be processed into a ruby coloured preserves with a spicy flavour and into jellies. Crabapples can be made into cider with an interesting flavour.

Botany

Small deciduous tree, 3.5–8 m high. Branchlets reddish brown, terete, puberulous when young, glabrous when old. Stipules lanceolate and caducous. Leaves elliptic or narrowly elliptic,

Plate 1 Immature fruit and leaves

5–8 × 2–3 cm, both surfaces sparsely pubescent when young becoming glabracent, base rounded to broadly cuneate, apex acuminate to obtuse, margin adpressed serrulate. Flowers, bisexual, 4–5 cm across, in 4–6 flowered sub-umbel corymb. Flower hypanthium campanulate; sepals deltoid-ovate; petals dark pink buds opening to single mid-pink, fading to white, ovate, 2–2.5 cm, base shortly clawed, apex rounded; stamens 20–25, unequal, yellow; ovary 4–5 locules with 2 ovules per locule, style 4–5. Pome, pale green with pinkish tinge becoming yellow when ripe, subglobose 3–5 cm diameter, not impressed at apex, convex at base, subglabrous with persistent sepals (Plate 1).

Nutritive/Medicinal Properties

The nutrient composition of raw crabapples per 100 g edible portion had been reported as: water 78.94 g, energy 76 kcal (318 kJ), protein 0.40 g, total lipid (fat) 0.30 g, ash 0.42 g, carbohydrate 19.95 g, Ca 19 mg, Fe 0.36 mg, Mg 7 mg, P 15 mg, K 194 mg, Na 1 mg, Cu 0.067 mg, Mn 0.115 mg, vitamin C 8 mg, thiamine 0.030 mg, riboflavin 0.020 mg, niacin 0.1 mg, vitamin A 2 μg RAE, vitamin A 40 IU, total saturated fatty acids 0.048 g, 12:0 (lauric acid) 0.001 g, 14:0 (myristic acid) 0.001 g, 16:0 (palmitic acid) 0.040 g, 18:0 (stearic acid) 0.006 g, total monounsaturated fatty acids 0.012 g, 16:1 undifferentiated (palmitoleic acid) 0.001 g, 18:1 undifferentiated (oleic acid) 0.011 g, total polyunsaturated fatty acids 0.088 g, 18:2 undifferentiated (linoleic acid) 0.073 g, 18:3 undifferentiated (linolenic acid) 0.015 g, tryptophan 0.004 g, threonine 0.014 g, isoleucine 0.016 g, leucine 0.025 g, lysine 0.025 g, methionine 0.004 g, cystine 0.005 g, phenylalanine 0.011 g, tyrosine 0.008 g, valine 0.019 g, arginine 0.013 g, histidine 0.006 g, alanine 0.014 g, aspartic acid 0.070 g, glutamic acid 0.042 g, glycine 0.016 g, proline 0.014 g and serine 0.016 g (USDA 2011).

A total of 37 compounds comprising aldehydes, esters and alcohols as the major compounds were identified from the ripe fruits of six crabapple varieties (Red Splendor, Strawberry Parfait, Pink Spire, Radiant, Sparkler, and Flame) (Li et al. 2008). The main aroma compound was 2-hexenal the content of which was 45.37%, 21.98%, 33.56%, 32.21%, 38.60%, and 45.88% in the respective varieties. Other major aroma volatiles were 3-hexenal, hexanal, 2,4-hexadienal, benzaldehyde, and diethyl phthalate. The relative content of aldehydes and esters decreased as alcohols increased in the Red Splendor and Strawberry Parfait fruit as it ripened. For Red Splendor, the main volatile was still 2-hexenal, but the relative content decreased to 42.89%, and the relative content of alcohols increased by 13.86% and aldehydes and esters declined by 12.16% and 7.18%, respectively. For Strawberry Parfait, the main volatile was changed to cyclohexanol, and the relative content increased to 46.43%, while the relative content of alcohols increased by 49.03% as aldehydes and esters declined by 23.74% and 9.34%, respectively.

Other Uses

Crabapples are widely grown as ornamental trees, grown for their beautiful flowers or fruit, with numerous cultivars selected for these qualities and for resistance to fire-blight disease. They are also popular in bonsai culture.

Some crabapples are used as rootstocks for domestic apples to add beneficial characteristics such as cold hardiness. They are also valued as pollinisers in apple orchards.

Comments

Malus spectabilis is one of the most popular ornamental trees in China, widely cultivated in the eastern and northern regions. The cultivated var. *riversii* (G. Kirchner) Rehder has double, pink flowers, and the cultivated f. *albiplena* Schelle, has double, white flowers.

Selected References

Hedrick UP (1972) Sturtevant's edible plants of the world. Dover Publications, New York, 686pp

Huxley AJ, Griffiths M, Levy M (eds) (1992) The new RHS dictionary of gardening, 4 vols. MacMillan, London

Ku TC, Spongberg SA (2003) *Malus* Miller. In: Wu ZY, Raven PH, Hong DY (eds) Flora of China, vol 9 (Pittosporaceae through Connaraceae). Science Press/Missouri Botanical Garden Press, Beijing/St. Louis

Li XL, Kang L, Hu JJ, Li XF, Shen X (2008) Aroma volatile compound analysis of SPME headspace and extract samples from crabapple (*Malus* sp.) fruit using GC-MS. Agric Sci China 7(12):1451–1457

Morgan J, Richards A (1993) The book of apples. Ebury Press, London, 304pp

Tanaka T (1976) Tanaka's cyclopaedia of edible plants of the world. Keigaku Publishing, Tokyo, 924pp

U.S. Department of Agriculture, Agricultural Research Service (USDA) (2011) USDA national nutrient database for standard reference, release 24. Nutrient Data Laboratory Home Page, http://www.ars.usda.gov/ba/bhnrc/ndl

Malus x domestica

Scientific Name

Malus x domestica Borkhausen.

Synonyms

Malus communis Poiret, *Malus dasyphylla* Borkh., *Malus dasyphylla* var. *domestica* Koidz., *Malus domestica* subsp. *pumila* (Miller) Likhonos, *Malus niedzwetzkyana* Dieck ex Koehne, *Malus malus* (L.) Britton, nom. inval., *Malus pumila* auct., *Malus pumila* Mill., *Malus pumila* auct. var. *domestica* (Borkh.) C. K. Schneid., *Malus sylvestris* auct., *Malus sylvestris* Mill. subsp. *mitis* (Wallr.) Mansf., *Malus sylvestris* var. *domestica* (Borkh.) Mansf., *Malus sylvestris* auct. var. *domestica* (Borkh.) Mansf., *Pyrus malus* L., *Pyrus malus* var. *pumila* Henry.

Family

Rosaceae

Common/English Names

Apple, Apple Tree

Vernacular Names

Arabic: Tuffahh;
Brazil: Maça, Maçanzeira, Maceira, Macieira;
China: Ping Guo;
Czech: Jabloň Domácí;
Danish: Abild, Æble, Æbletræ (Tree), Almindelig Æble;
Dutch: Appel, Vruchtappel;
Eastonian: Aed-Õunapuu;
Finnish: Omena, Paratiisiomena, Tarhaomenapuu
French: Pomme, Pommier, Pommier Commun;
Gaelic: Aabhail;
German: Apfel, Apfelbaum, Echter Apfelbaum, Kultur-Apfel;
Hungarian: Alma, Nemes Alma(Fa);
Icelandic: Eplatré, Garðepli;
India: Seb (Hindu), Badara, Mushtinanan, Seba, Seva (Sanskrit), Applepazham (Tamil);
Indonesia: Epal;
Italian: Mela, Melo, Melo Commune, Pomo;
Japanese: Ringo, Seiyou Ringo;
Korean: Sagwanamu;
Malaysia: Epal;
Norwegian: Apal;
Philippines: Mansanas;
Polish: Jabłoń Domowa;
Portuguese: Macieira;
Russian: Iabloko, Iablonia;
Serbian: Jabuka;

Slovaščina: Jablana, Žlahtna Jablana;
Slovencina: Jabloň Domáca;
Sorbic: Jablušina;
Spanish: Manzana, Manzano;
Swedish: Äpple, Äppel, Äppelträd;
Thai: Aeppen, Aoppoen;
Turkish: Elma;
Vietnamese: Pom.

Origin/Distribution

Recent genomic studies by Velasco et al. (2010) identified the progenitor of the cultivated apple as *M. sieversii*. This species was found in the Ili Valley, on the northern slopes of the Tien Shan mountains at the border of northwest China and Kazakhstan. Leaves taken from trees in this area were analyzed for DNA composition, which showed them all to belong to the species *M. sieversii*, with some genetic sequences common to *Malus domestica*. However, another chloroplast DNA analysis found that *Malus sylvestris* had also contributed to the genome of *M. domestica* (Coart et al. 2006). A closer relationship than presently accepted was found between *M. sylvestris* and *M. domestica* at the cytoplasmic level, with the detection of eight chloroplast haplotypes shared by both species. Hybridization between *M. sylvestris* and *M. domestica* was also apparent at the local level with sharing of rare haplotypes among local cultivars and sympatric wild trees. Indications of the use of wild *Malus* genotypes in the (local) cultivation process of *M. domestica* and cytoplasmic introgression of chloroplast haplotypes into *M. sylvestris* from the domesticated apple were found. Only one of the *M. sieversii* trees studied displayed one of the three main chloroplast haplotypes shared by *M. sylvestris* and *M. domestica*. Thus the origin of the domestic apple is still unresolved. There are more than 7,500 known cultivars of apples, resulting in a diverse range of desired characteristics.

Agroecology

Apples are best adapted to the cool temperate areas from about 35–50° latitude. Apples thrives best in cool temperate climates with high light intensity, warm (not hot) days and cool nights. Apples require about 1,000–1,600 hours of chilling (7.2°C) to break dormancy. Wood and buds are hardy to −40°C but open flowers and developing fruitlets are damage by brief exposure to −2°C or less.

In the tropics, apples can be grown in cool areas in the high elevation of 800–1,200 m where temperatures hovers from 16 to 27°C, rainfall 1,600–3,200 mm, relative humidity 75–85% and sunshine more than 50% duration.

Apples can be grown in a wide variety of soils but does best in deep, fertile, well-drained, loamy soils with pH 6–7.

Edible Plant Parts and Uses

Apples are widely eaten fresh and are often eaten baked or stewed. Apples are also dried and eaten or reconstituted by soaked in water, alcohol or some other liquid, for subsequent use. Apples can be canned or juice. Apple juice has surpassed orange juice consumption by children in the USA. Apples are milled to produce apple cider (non-alcoholic sweet beverage) and filtered for apple juice. Apple juice can be fermented to process cider, ciderkin (weak alcoholic cider) and vinegar, Distillation of apple juice affords various alcoholic beverages such as apfelwein (apple wine), applejack (strong alcoholic beverage) and Calvados (apple brandy). Apples are widely used in pies, pastries, cakes, jams, sauces, apple butter, apple jellies, meat dishes. In the United Kingdom, traditional toffee apples made by coating apples in hot toffee are relished, and candy apples and caramel apples in USA. Fruit pectins and apple seed oil are also produced.

Botany

The tree is small and deciduous, reaching 5–12 m tall, with a broad, often densely twiggy crown. Branchlets brown, tomentose when young, glabrous when old. Stipules caducous and lanceolate. Leaves, alternate, simple, ovate, or broadly elliptic 5–11 cm × 3–6 cm, both surfaces densely puberulous when young, adaxially glabrecent, pinnately veined, apex acute, base broadly cuneate or rounded, margin serrated on 2–5 cm petiole (Plates 1, 2 and 3). Flowers in 3–7 flowered corymb at the apices of branchlets, 3–4 cm across, pedicel tomentose; bracts linear-lanceolate, caducous. Flower – hypanthium tomentose; sepals lanceolate-deltoid, longer than hypanthium, tomentose; petals 5 white with tinge of pink, obovate, base shortly clawed, apex rounded; stamens 20, unequal; ovary 5-loculed with 2 ovules per locule, style 5 tomentose at the base (Plate 1). Pome subglobose, obovoid to ellipsod, variable size and colour, 5–10 cm diameter, pale green, yellow, pink to red, impressed at the base, sepals persistent and fruting pedicel short and thickened (Plates 3, 4, 5, 6, 7 and 8). Seeds small and black-brown.

Plate 3 Maturing apples and leaves

Plate 1 Apple blossoms

Plate 4 Red Delicious apple

Plate 2 Immature apples

Plate 5 Indonesian cultivar 'Anna'

Plate 6 Granny Smith

Plate 7 Indonesian cultivar 'Mana Lagi'

Nutritive/Medicinal Properties

Fruit Nutrients

The nutrient composition of raw apples (*Malus domestica*) with skin based on analytical data (per 100 g edible portion) for red delicious, golden delicious, gala, granny smith, and fuji varieties was reported as: water 85.56 g, energy 52 kcal (218 kJ), protein 0.26 g, total lipid (fat) 0.17 g, ash 0.19 g, carbohydrate 13.81 g, total dietary fibre 2.4 g, total sugars 10.39 g, sucrose 2.07 g glucose (dextrose) 2.43 g, fructose 5.90 g, starch 0.05 g, Ca 6 mg, Fe 0.12 mg, Mg 5 mg, P 11 mg, K 107 mg, Na 1 mg, Zn 0.04 mg, Cu 0.027 mg, Mn 0.035 mg, F 3.3 μg, vitamin C 4.6 mg, thiamin 0.017 mg, riboflavin 0.026 mg, niacin 0.091 mg, pantothenic acid 0.061 mg, vitamin B-6 0.041 mg, total folate 3 μg, total choline 3.4 mg, betaine 0.1 mg, vitamin A 54 IU, vitamin A 3 μg RAE, β-carotene 27 μg, β-cryptoxanthin 11 μg, lutein + zeaxanthin 29 μg, vitamin E (α-tocopherol) 0.18 mg, vitamin K (phylloquinone) 2.2 μg, total saturated fatty acids 0.028 g, 14:0 (myristic acid) 0.001 g, 16:0 (palmitic acid) 0.024 g, 18:0 (stearic acid) 0.003 g, total monounsaturated fatty acids 0.007 g, 18:1 undifferentiated (oleic) 0.007 g, total polyunsaturated fatty acids 0.051 g, 18:2 undifferentiated (linoleic acid) 0.043 g, 18:3 undifferentiated (linolenic acid) 0.009 g,

Plate 8 (a) Gala apple whole; (b) sliced

phytosterols 12 mg, tryptophan 0.001 g, threonine 0.006 g, isoleucine 0.006 g, leucine 0.013 g, lysine 0.012 g, methionine 0.001 g, cystine 0.001 g, phenylalanine 0.006 g, tyrosine 0.001 g, valine 0.012 g, arginine 0.006 g, histidine 0.005 g, alanine 0.011 g, aspartic acid 0.070 g, glutamic acid 0. 025 g, glycine 0.009 g, proline 0.006 g, and serine 0.010 g (USDA 2011).

Fruit Volatile Compounds

Over 300 volatile compounds had been reported associated with the aroma profile of apples (Dixon and Hewett 2000). These compounds included alcohols, aldehydes, carboxylic esters, ketones, ethers, acids, bases, acetals, and hydrocarbons (Dimick and Hoskin 1983). Compounds in the volatile profile of apples were reported to be dominated by esters (78–92%) and alcohols (6–16%) (Paillard 1990). The most abundant compounds were even numbered carbon chains including combinations of acetic, butanoic, and hexanoic acids with ethyl, butyl, and hexyl alcohols (Paillard 1990). The important aroma volatile compounds reported include: aldehydes:- acetaldehyde, trans-2-hexenal (Flath et al. 1967), hexanal (Paillard 1990); alcohols:- ethanol (Teranishi et al. 1987), propan-1-ol, butan-1-ol, heaxa-1-ol, (Flath et al. 1967), 2 methyl butan-1-ol (Buttery et al. 1973); esters:- ethyl-2-methyl butanoate (Flath et al. 1967), propyl acetate (Takeoka et al. 1996), ethyl acetate, butyl acetate, pentyl acetate, hexyl acetate, ethyl butanoate, ethyl-2-methyl butanoate, ethyl propionate, ethyl pentanoate, ethyl hexanoate (Teranishi et al. 1987; Takeoka et al. 1996); propyl butanoate, 2 methyl butyl acetate (Teranishi et al. 1987). Dixon and Hewett (2000) reported that hypoxic treatment of fresh fruit could induce significant increases in volatile concentrations that could be used in production of high quality essences from apple juice. In addition the use of hypoxia to enhance volatile concentrations may be a beneficial side effect when such treatments are used for disinfestations purposes. It is possible that given equal efficacy, hypoxia could be either preferred or used as an adjunct to heat treatments to eradicate insects.

Other Fruit Phytochemicals

Main structural classes of apple constituents include hydroxycinnamic acids, dihydrochalcones, flavonols (quercetin glycosides), catechins and oligomeric procyanidins, as well as triterpenoids in apple peel and anthocyanins in red apples (Gerhauser 2008). The phenolic compound profile in apples varies with varieties. The highest levels of phenolic compounds were found in the peel of four apple varieties (Golden and Red Delicious, Granny Smith and Green Reineta) (Escarpa and Gonzalez 1998). High levels of catechins and flavonol glycosides, especially rutin, were found in apple peels. Chlorogenic acid was the major compound in the pulp for all apple varieties studied except for Granny Smith. Significant quantitative differences between the apple varieties were also found, the Golden Delicious variety showing the lowest content of phenolic compounds and Green Reineta variety the highest. Studies in Poland found the Champion apple variety to be the best source of phenolic acids and epicatechin compared to the Jonica variety (Malik et al. 2009). Chlorogenic predominated in the Champion variety whereas in the Jonica variety, chlorogenic and homovanilic acids were dominant. The highest concentration of chlorogenic acid was detected in the pulp of Jonica variety around the cities of Puławy and Lublin, whereas homovanilic acid was the highest in the other samples collected from the vicinity of Stryjno and Góry Markuszowskie. Among the Jonica and Champion varieties of apples collected from various orchards in the vicinity of Lublin, the highest content of epicatechin (13,12 mg/kg) was found in the pulps of Champions variety collected in Puławy. Studies reported the following phenolic compound in 14 French apple varieties: monomeric catechins, proanthocyanidins, hydroxycinnamic acids, and dihydrochalcones (Sanoner et al. 1999). Depending on the variety, the total polyphenol concentration varied from 1 to 7 g/kg of fresh cortex. Cider varieties generally showed a higher polyphenol concentration than the dessert apple Golden Delicious, bitter varieties had the highest concentration. For all varieties, procyanidins were the domiant group, mainly constituted of (−)-epicatechin units with a small proportion of (+)-catechin as a terminal unit. Of the apple polyphenols, procyanidins were found to bind with cell wall material, leading to decreased levels of polyphenols found in apple juices (Renard et al. 2001). Hydroxycinnamic acids and (−)-epicatechin did not bind to cell walls.

The air-dried and freeze-dried apple (Rome Beauty) peels had the highest total phenolic, flavonoid, and anthocyanin contents (Wolfe and Liu 2003). On a fresh weight basis, the total phenolic and flavonoid contents of these samples were similar to those of the fresh apple peels. Freeze-dried peels had a lower water activity than air-dried peels on a fresh weight basis. The total phenolic content of apple peel powder was 3,342 mg gallic acid equivalents/100 g dried peels, the flavonoid content was 2,299 mg catechin equivalents/100 g dried peels, and the anthocyanin content was 169.7 mg cyanidin 3-glucoside equivalents/100 g dried peels. These phytochemical contents were a significantly higher than those of the fresh apple peels if calculated on a fresh weight basis. Apple peel extracts had higher total soluble phenolic content and related antioxidant capacity than pulp extracts (Barbosa et al. 2010). Quercetin derivatives, protocatechuic acid, chlorogenic acid, and p-coumaric acid were detected, and the amount varied significantly between aqueous and ethanolic extracts. Compared with apple pulps, peels were found to be richer in phenolics (Chinnici et al. 2004). Flavonols, flavanols, procyanidins, dihydrochalcones, and hydroxycinnamates were the identified phenolic classes in apple peel tissue, and the most abundant compounds were epicatechin, procyanidin B2, and phloridzin (Chinnici et al. 2004). Pulps were poorer in phytochemicals. Their major phenolics were procyanidins and hydroxycinnamates. Flavonols in amounts <20 mg/kg fresh weight (fw) were also found. In both peels and pulps, integrated production samples were richer in polyphenols. Among the 14 compounds identified, only phloridzin had a tendency to appear higher in organic peels. Apples also contain soluble polysaccharides. The yield of soluble polysaccharides in peeled apples was found to range from 0.43% to 0.88%, with molecular weight ranging 223–848 kDa (Ker et al. 2010). All belonged to peptidoglycans. Among the 14 amino acids found, seven were essential amino acids. Additionally, 50% of apple samples consisted of glucoarabinan, 37.5% comprising taloarabinan and the remaining 12.5% containing alloglucan. Further, the soluble polysaccharides consisted of a huge amount of myo-inositol (>5.61%) and uronic acid (>11.7%). Talose, allose and fucose were found for the first time in the soluble polysaccharides of apples. The most abundant phenolic compounds found in Annurca apple peel ethanol extract were rutin, epicatechin, dicaffeoylquinic acid, and caffeic acid; these compounds constituted 27.43%, 24.93%, 16.14%, and 15.3% of the total phenols, respectively (Fratianni et al. 2011).

Apple peels were found to contain hypophasic carotenoids mainly composed of violaxanthin, zeaxanthin and lutein and epiphasic carotenoids composed of a high content of β,β-carotene and β-cryptoxanthin (Molnár et al. 2005). Seven additional carotenoids were isolated from apple peels: (all-E)-luteoxanthin, (all-E)-neoxanthin, (9'Z)-neoxanthin, (all-E)-antheraxanthin, (all-E)-violaxanthin, (9Z)-violaxanthin and (all-E)-lutein (Molnár et al. 2010).

Organic acids found in apple peels included: glyoxylic, isocitric, malic, quinic, shikimic, glyceric, α-oxoglutaric, pyruvic; and in apple pulp: pyruvic, malic, quinic, shikimic, citramalic, glyceric, and α-oxoglutaric (Salunkhe and Kadam 1995).

Apple Juice Chemicals

Delage et al. (1991) found the following phenolic compounds chlorogenic acid, p-coumaric acid, protocatechuic acid, (+)-catechin, (−)-epicatechin, phloridzin and di-, tri- and tetrameric procyanidins in apple juice. Phenolic and furfural compounds found in apple juice comprised chlorogenic and coumaroylquinic acids and phloridzin as the major phenolic components and caffeic, p-coumaric, ferulic, gallic and protocatechuic acids, and catechin as the minor phenolics; and 5-hydroxymethyl-2-furaldehyde and 2-furaldehyde (Kermasha et al. 1995). Organic acids found in apple juice included: malic, quinic, succinic, lactic, glucoronic, citramalic, mucic acids (Salunkhe and Kadam 1995).

Apple Pomace Chemicals

The predominant phenolic compounds in apple pomace were phloridzin, chlorogenic acid and quercetin glycosides (Schieber et al. 2003). While the polyphenolics recovered from apple pomace may be used as natural antioxidants or as functional food ingredients, extended fields of application may be obtained for decolorised, refined apple pectins. Six high-purified polyphenols were identified in apple pomace: chlorogenic acid, quercetin-3-glucoside/quercetin-3-glacaside, quercetin-3-xyloside, phloridzin, quercetin-3-arabinoside and quercetin-3-rhamnoside (Cao et al. 2009). HPLC analysis indicated that the major polyphenols of apple pomace consisted of chlorogenic acid, caffeic acid, syrigin, procyanidin B2, (−)-epicatechin, cinnamic acid, coumaric acid, phlorizin and quercetin, of which procyanidin B2 had the highest content of 219.4 mg/kg (Bai et al. 2010). The actual yield of polyphenols was 62.68 mg gallic acid equivalent per 100 g dry apple pomace.

Apple Seed Oil

Oil content extracted from apple seeds ranged from 20.69 to 24.32 g/100 g (Tian et al. 2010). The protein, fibre and ash contents were found to be 38.85–49.55, 3.92–4.32 and 4.31–5.20 g/100 g, respectively. The extracted oils exhibited an iodine value of 94.14–101.15 g I/100 g oil; refractive index (40°C) was 1.465–1.466; density (25°C) was 0.902–0.903 mg/ml; saponification value was 179.01–197.25 mg KOH/g oil; and the acid value was 4.036–4.323 mg KOH/g oil. The apple seed oils consisted mainly of linoleic acid (50.7–51.4 g/100 g) and oleic acid (37.49–38.55 g/100 g). Other prominent fatty acids were palmitic acid (6.51–6.60 g/100 g), stearic acid (1.75–1.96 g/100 g) and arachidic acid (1.49–1.54 g/100 g).

Leaf Phytochemicals

In apple, the dihydrochalcone phloridzin (phloretin 2′-O-glucoside) was found to be dominant representing more than 90% of the soluble phenolics in the leaves (Gosch et al. 2009). Apple leaves were found to contain flavonoids dihydrochalcones such as phloridzin, sieboldin and trilobatin (Dugé de Bernonville et al. 2010).

Scientific studies suggested that apples and apple products possessed a wide range of biological activities that include antioxidant, antiproliferative, anti-diabetic, anti-inflammatory, lipid oxidation inhibition, and cholesterol lowering activities which may contribute to health beneficial effects against cardiovascular disease, asthma and pulmonary dysfunction, diabetes, obesity, and cancer (Boyer and Lui 2004).

Antioxidant Activity

Apples were reported to contain a variety of phytochemicals, including quercetin, catechin, phloridzin and chlorogenic acid, all potent antioxidants (Boyer and Lui 2004). Eberhardt et al. (2000) reported that 100 g of fresh apples had an antioxidant activity equivalent to 1,500 mg of vitamin C. All the apple polyphenols epicatechin, its dimer (procyanidin B2), trimer, tetramer and oligomer, quercetin glycosides, chlorogenic acid, phloridzin and 3-hydroxy-phloridzin displayed strong antioxidant activities (Lu and Foo 2000). Their DPPH-scavenging activities were 2–3 times and superoxide anion radical-scavenging activities were 10–30 times better than those of the antioxidant vitamins C and E. Sun et al. (2002) found that cranberry had the highest total antioxidant activity (177.0 μmol of vitamin C equiv/g of fruit), followed by apple, red grape, strawberry, peach, lemon, pear, banana, orange, grapefruit, and pineapple. Results of studies indicated that flavonoids such as quercetin, epicatechin, and procyanidin B(2) rather than vitamin C contributed significantly to the total antioxidant activity of apples (Lee et al. 2003). They found that the average concentrations of major phenolics and vitamin C in six apple cultivars were as follows (mg/100 g of fresh weight of apples): quercetin glycosides, 13.20; procyanidin B(2), 9.35; chlorogenic acid, 9.02; epicatechin, 8.65; phloretin glycosides, 5.59; vitamin C, 12.80. A highly linear relationship ($R^2 > 0.97$) was attained between concentrations and total antioxidant capacity of phenolics and vitamin C. Relative vitamin C equivalent antioxidant capacity (VCEAC) values of these compounds were in the order quercetin (3.06)>epicatechin (2.67)>procyanidin B(2) (2.36)>phloretin (1.63)>vitamin C (1.00)>chlorogenic acid (0.97). In another study, the estimated contribution of major phenolics and vitamin C to the total antioxidant capacity of 100 g of fresh apples was as follows: quercetin (40.39 VCEAC)>epicatechin (23.10)> procyanidin B(2) (22.07)>vitamin C (12.80)> phloretin (9.11)>chlorogenic acid (8.75). Processing was found to impact on the bioactivity of apple products (van der Sluis et al. 2002). Raw juice obtained by pulping and straight pressing or after pulp enzyming had an antioxidant activity that was only 10% and 3%, respectively, of the activity of the fresh apples. The levels of flavonoids and chlorogenic acid in the juice were reduced to between 50% (chlorogenic acid) and 3% (catechins). Most of the antioxidants were retained in the pomace rather than being transferred into the juice. In apple juice, 45% of the total measured antioxidant activity could be ascribed to the analyzed antioxidants. For three apple cultivars tested (Elstar, Golden Delicious, and Jonagold), the processing methods had similar effects.

Integrated apple peels gave the highest total antioxidant capacities (TAC) (18.56 mM/kg fw), followed by organic peels (TAC=14.96), integrated pulps (TAC=7.12), and organic pulps (TAC=6.28) (Chinnici et al. 2004). In peels, the top contributors to the antioxidant activity were found to be flavonols, flavanols, and procyanidins, which accounted for about 90% of the total calculated activity whereas in pulps, the TAC was primarily derived from flavanols (monomers and polymers) together with hydroxycinnamates. A good correlation between the sum of polyphenols and the radical scavenging activities was found. Among the single classes of compounds, procyanidins (in peels and pulps) and flavonols (in peels) were statistically correlated to the TAC. The apple peel (Rome Beauty) powder had a total antioxidant activity of 1,251 µmol vitamin C equivalent/g, similar to fresh Rome Beauty peels on a fresh weight basis (Wolfe and Liu 2003). One gram of powder had an antioxidant activity equivalent to 220 mg of vitamin C. Apple peel powder may be used in a various food products to add phytochemicals and promote good health. In four varieties of apples (Rome Beauty, Idared, Cortland, and Golden Delicious) commonly used in apple sauce production, the total phenolic and flavonoid contents were highest in the peels, followed by the flesh + peel and the flesh (Wolfe et al. 2003). Idared and Rome Beauty apple peels had the highest total phenolic contents (588.9 and 500.2 mg of gallic acid equivalent/100 g of peels, respectively). Rome Beauty and Idared peels had highest flavonoids (306.1 and 303.2 mg of catechin equivalent/100 g of peels, respectively). Idared apple peels had the most anthocyanins, with 26.8 mg of cyanidin 3-glucoside equivalent/100 g of peels. The peels all had significantly higher total antioxidant activities than the flesh + peel and flesh of the apple varieties examined. Idared peels had the greatest antioxidant activity (312.2 µmol of vitamin C equivalent/g of peels). In separate study, apple peel and pulp were found to have significantly higher antioxidant potentials than in pear peel and pulp as measured by 1,1-diphenyl-2-picrylhydrazyl (DPPH), β-carotene bleaching (β-carotene), and nitric oxide inhibition radical scavenging (NO) assays (Leontowicz et al. 2003). The ethanol extract of apple peels exhibited the strongest inhibition of lipid peroxidation as a function of its concentration and was comparable to the antioxidant activity of butylated hydroxyanisole. The polyphenols, phenolic acids, and flavonoids contributed to the antioxidant potential which correlated well with polyphenols and flavonoids. The correlation coefficients between polyphenols and antioxidant activities by DPPH, β-carotene, and NO were as follows: 0.9207, 0.9350, and 0.9453. In contrast, the correlation coefficient between the content of dietary fiber and the antioxidant activities test was low. The content of all studied indices in apple and pear peel was significantly higher than in peeled fruits. Diets supplemented with fruit peels exercised a significantly higher positive influence on plasma lipid levels and on plasma antioxidant capacity of rats than diets with fruit pulps.

In-vitro studies found apple skin extracts to be effective inhibitors of oxidation of polyunsaturated fatty acid in a model system (Huber and Rupasinghe 2009). The antioxidant capacity measured by Folin-Ciocalteu ranged from 16.2 to 34.1 mg GAE/100 g DW, ferric reducing antioxidant power varied from 1.3 to 3.3 g TE/100 g DW, oxygen radical absorbance capacity ranged from 5.2 to 14.2 g TE/100 g DW, and percent inhibition of oxidation of methyl linolenate from 73.8% to 97.2% among the apple genotypes. The total phenolic concentrations of methanolic extracts of skins of the apple genotypes varied from 150 to 700 mg/100 g DW. The test for 2,2-diphenyl-1-picryl-hydrazyl free radical-scavenging activity showed that the ethanol extract of annurca apple peel rich in polyphenols (rutin, epicatechin, dicaffeoylquinic acid, and caffeic acid) possessed an impressive antioxidant capacity (50% effective concentration of 2.50 µg/g of product) (Fratianni et al. 2011).

Honeycrisp and Red Delicious apple varieties had the highest total phenolic contents and a significant correlation with antioxidant capacity ($R^2 = 0.91$) (Barbosa et al. 2010). Quercetin-rich apple peel extract and apple pomace were found to effectively reduced ROS (reactive oxygen species)-DNA damage in Caco-2 cells (with the former being more potent), whereas apple juice was only moderately effective (Bellion et al. 2010). Glutathione peroxidise activity was reduced by all the extracts in the following order apple juice > apple pomace > apple peel. Direct antioxidant activity decreased in the order apple juice > apple peel > apple pomace. The data suggested that apple phenolics at low, nutritionally relevant concentrations may protect intestinal cells from ROS-induced DNA damage, mediated by cellular defense mechanisms rather than by antioxidant activity.

During long-term cold storage (120 days at 1°C) as well as during an additional 7 day storage of fruits at 16°C, total phenols, total antioxidant activity (TAA), and radical scavenging activity (RSA) in the peel of two apple cultivars (Jonagold and S'ampion) increased considerably, irrespective of the storage conditions (Leja et al. 2003). A slight decrease in anthocyanins was observed in apples stored in air, while the controlled atmosphere treatment (2% CO_2/2% O_2) did not cause any significant changes.

Results of a study suggested that the two farming systems (organic/conventional) did not result in differences in the bioavailability of apple polyphenols in humans (Stracke et al. 2010). In a randomized cross-over short term intervention study of six men after 1 kg intake of apples phloretin and coumaric acid plasma concentrations increased significantly in both intervention groups, without differences between the two farming systems. In a double-blind, randomized long term intervention study of 43 healthy individuals, consumption of organically or conventionally grown apples did not result in increasing polyphenol concentrations in plasma and urine compared to the control group suggesting no accumulation of apple polyphenols or degradation products in humans.

Apple polyphenols have beneficial bioactive effects such as antioxidant activity in-vivo, but can also exert prooxidative effects in-vitro (Bellion et al. 2009). From their findings they cautioned that the generation of hydrogen peroxide in-vitro by polyphenols had to be taken into consideration when interpreting results of such cell culture experiments. However they maintained that high polyphenol concentrations, favoring substantial H_2O_2 formation, were not expected to occur in-vivo, even under conditions of high end nutritional uptake.

The dihydrochalcone phloridzin from apple leaves exhibited high antioxidant activity in the oxygen radical absorbance capacity (ORAC) assay, and sieboldin in young leaves had high 1,1-diphenyl-2-picrylhydrazyl (DPPH) free radical scavenging activity (Dugé de Bernonville et al. 2010).

Inhibition of Lipid Oxidation/ Hypolipidemic Activities

Six commercial apple juices and extracts of the peel (RP), flesh (RF) and whole fresh Red Delicious apples (RW), tested at 5 µM gallic acid equivalents (GAE), all inhibited low density lipoprotein oxidation (Pearson et al. 1999). The inhibition by the juices ranged from 9% to 34%, and inhibition by RF, RW and RP was 21%, 34% and 38%, respectively. The phenolic composition of six commercial apple juices, and of the peel (RP), flesh (RF) and whole fresh Red Delicious apples (RW), was found to comprise several classes of phenolic compounds: cinnamates, anthocyanins, flavan-3-ols and flavonols. Phloridzin and hydroxy methyl furfural were also detected. The profile of total phenolic concentration in the apple juices was: hydroxy methyl furfural, 4–30%; phloridzin, 22–36%; cinnamates, 25–36%; anthocyanins, n.d.; flavan-3-ols, 8–27%; and flavonols, 2–10%. The phenolic profile of the Red Delicious apple extracts differed from those of the juices. The profile of phenolic classes in fresh apple extracts was: hydroxy methyl furfural, n.d.; phloridzin, 11–17%; cinnamates, 3–27%; anthocyanins, n.d.–42%; flavan-3-ols, 31–54%; and flavonols, 1–10%. Liver microsomal lipid peroxidation was decreased in rats pretreated with apple juice by 52–87% when compared to animals given toxicants N-nitrosodiethylamine (NDEA) or carbon tetrachloride alone (Kujawska et al. 2011). Pretreatment with juice protected antioxidant enzymes: catalase, glutathione peroxidase and glutathione reductase but not superoxide dismutase. The plasma activity of paraoxonase 1 was reduced by both toxicants and was increased by 23% in the apple/carbon tetrachloride group. A rise in plasma protein carbonyls caused by the xenobiotics was reduced by 20% only in apple/NDEA-treated rats. Also, in this group of animals, a 9% decrease in DNA damage in blood leukocytes was observed.

Soluble polysaccharides of apples comprising all peptidoglycans and also a huge amount of myo-inositol (>5.61%) and uronic acid (>11.7%), may play a synergistic role in the hypolipidemic effect (Ker et al. 2010). The biological value of soluble polysaccharides was attributable to the differential effect of soluble polysaccharides and the synergistic effect exerted by its unique soluble polysaccharides profile, high myo-inositol and uronic acid contents.

Cardiovascular/Antiatherosclerotic/ Antihypercholesterolemic Activities

Leontowicz et al. (2002) found that diets with apples and to a lesser extent with peaches and pears improved lipid metabolism and increased the plasma antioxidant potential especially in rats fed with added cholesterol. The total polyphenols of peeled fruits and fruit peels of apples were higher than in pears or peaches. Caffeic, p-coumaric and ferulic acids and the total radical-trapping antioxidative potential (TRAP) values in peeled apples and their peels were significantly higher than in peaches and pears. Dietary fibre were similar in all three fruits. They concluded that the higher content of biologically active compounds and the better results in rats makes apple preferable for dietary prevention of atherosclerosis and other diseases. The content of all studied indices in peels was significantly higher than peeled fruits. A good correlation between the total polyphenols and the TRAP values was found in all fruits. Aprikian et al. (2002) found that a moderate supply of dessert apples elicited interesting effects on lipid and peroxidation parameters. When cholesterol fed rats were supplemented with lyophilized apples (15%), there was a significant drop in plasma cholesterol and liver cholesterols and an increase in high-density lipoproteins (HDL) giving. Furthermore, they found that cholesterol excretion increased in the faeces of rats fed apples, suggesting reduced cholesterol absorption. Concomitantly apple supplementation gave higher FRAP (ferric reducing antioxidant power) plasma levels than controls together with a reduced malondialdehyde excretion in urine. In another study, Aprikian et al. (2003) found that combined apple pectin and high polyphenol freeze-dried apple lowered plasma and liver cholesterol, triglycerides, and apparent cholesterol absorption to a much greater extent than either apple pectin alone or apple phenolic fraction alone in rats. Their study suggested a beneficial interaction between fruit fibre and polyphenolic components on large intestine fermentations and lipid metabolism and also supports the benefits of eating whole fruits as opposed to dietary supplements.

Studies showed that consumption of apple, pear or orange juice increased total plasma antioxidant capacity, total cholesterol, high-density lipoprotein-cholesterol, and low-density lipoprotein-cholesterol in non-smokers (Alvarez-Parrilla et al. 2010). In smokers, fruit/juice supplementation decreased total cholesterol without inducing increase in total antioxidant capacity. Prospective epidemiological studies had reported that a higher fruit and vegetable intake is associated with a lower risk of coronary heart disease. Hansen et al. (2010) found in a median follow-up of 7.7 years of the Diet, Cancer and Health cohort study in 1993–1997, 1,075 incident acute coronary syndrome (ACS) cases were identified among 53 383 men and women, aged 50–64 years. They found an inverse association for apple intake with ACS. This association was also seen among women, albeit borderline significant. Their results supported previously observed inverse associations between fresh fruit intake, particularly apples, and ACS risk.

Apple juice (5 and 10 ml) significantly decreased total cholesterol, total glyceride, C-reactive protein, fibrinogen, factor VII levels, atherosclerotic lesions in the right and left coronary arteries and increased nitrite and nitrate in rabbits compared to those fed a cholesterolemic diet (Setorki et al. 2009). Ten ml but not 5 ml apple juice significantly reduced LDL-C and increased HDL-C. No significant difference was found between 5 and 10 ml apple juice groups with regard to C-reactive protein P, nitrite, nitrate, fibrinogen, factor VII, total glyceride, HDL-C and total cholesterol concentrations. The anti atherosclerotic effect of apple juice was attributed to its antioxidant and antiinflammatory properties.

Apple procyanidins exerted a potent vasorelaxation effect in 1.0 μM phenylephrine-contractive rat thoracic aorta (Matsui et al. 2009). The procyanidin-induced vasorelaxation was found to be associated with NO-cGMP pathway in combination with hyperpolarization due to multiple activation of Ca(2+)-dependent and -independent K(+) channels.

Anticancer Activity

Apple extracts and it phytochemicals, especially oligomeric procyanidins, had been shown to influence multiple mechanisms relevant for cancer prevention in in-vitro studies (Gerhauser 2008). These include antimutagenic activity, modulation of carcinogen metabolism, antioxidant activity, antiinflammatory mechanisms, modulation of signal transduction pathways, antiproliferative and apoptosis-inducing activity, as well as novel mechanisms on epigenetic events and innate immunity. Apple products had been shown to prevent skin, mammary and colon carcinogenesis in animal models. Epidemiological observations indicated that regular consumption of one or more apples a day may reduce the risk for lung and colon cancer (Boyer and Lui 2004; Gerhauser 2008).

Several in-vitro studies had shown apples to have antiproliferative activity. Cranberry showed the highest in-vitro inhibitory effect on proliferation of HepG(2) human liver-cancer cells followed by lemon, apple, strawberry, red grape, banana, grapefruit, and peach (Sun et al. 2002). Eberhardt et al. (2000) found that whole-apple extracts inhibited the growth of Caco-2 colon cancer cells and Hep G2 liver cancer cells in-vitro in a dose-dependent manner. They proposed that the unique combination of phytochemicals in the apples were responsible for inhibiting the growth of tumour cells. The antiproliferative activity of apples was found to vary with apple varieties (Liu et al. 2001). At a dose of 50 mg/ml, Fuji apple extracts inhibited Hep G2 human liver cancer cell proliferation by 39% and Red Delicious extracts inhibited cell proliferation by 57%. Northern Spy apples had no effect on cell proliferation. They found that total phenolic and flavonoid content was positively related to antioxidant activity and inhibition of cell proliferation. Wolfe et al. (2003) found that apple peels

(Rome Beauty, Idared, Cortland, and Golden Delicious varieties) were more effective in inhibiting the growth of HepG(2) human liver cancer cells than whole apple or apple flesh. Rome Beauty apple peels showed the most bioactivity, inhibiting cell proliferation by 50% at the low concentration of 12.4 mg of peels/ml. The freeze-dried apple peels of Rome Beauty exerted a strong antiproliferative effect on HepG(2) liver cancer cells with a median effective dose (EC_{50} of 1.88 mg/ml) (Wolfe and Liu 2003). This was lower than the EC_{50} exhibited by the fresh apple peels.

The results of studies by Ding et al. (2004) showed that an extract from fresh apple peel may inhibit tumour promoter-induced carcinogenesis and associated cell signalling and suggested that the chemopreventive effects of fresh apple may be through its antioxidant properties by blocking reactive oxygen species-mediated AP-1-MAPK activation. Oral administration of apple peel extracts decreased the number of nonmalignant and malignant skin tumours per mouse induced by 12-O-tetradecanolyphorbol −13-acetate (TPA) in 7,12-dimethylbenz(a)anthracene-initiated mouse skin. This inhibitory effect appeared to be mediated by the inhibition of ERKs and JNK activity. ESR analysis indicated that apple extract strongly scavenged hydroxyl (OH) and superoxide O_2^- radicals. Reagan-Shaw et al. (2010) found that apple (gala) peel extract exhibited potent antiproliferative effects. The peel extract elicited a significant decrease in growth and clonogenic survival of human prostate carcinoma CWR22Rnu1 and DU145 cells and breast carcinoma Mcf-7 and Mcf-7:Her18 cells. The peel extract treatment caused a marked concentration-dependent decrease in the protein levels of proliferative cell nuclear antigen, a marker for proliferation and generated a marked increase in maspin, a tumour suppressor protein that negatively regulates cell invasion, metastasis, and angiogenesis.

The study of Lapidot et al. (2002) suggested that apple antioxidants did not directly inhibit tumour cell proliferation, but instead they indirectly inhibit cell proliferation by generating H_2O_2 in reaction with the cell culture media. However, studies by Liu and Sun (2003) demonstrated that apple extracts did not generate H_2O_2 formation in WME, DMEM, or DMEM/Ham F12 media, and H_2O_2 addition to culture medium did not inhibit Hep G2 cell proliferation or Caco-2 colon cancer cell proliferation. Additionally, the addition of catalase did not obstruct the antiproliferative activity of apple extracts.

Hypophasic carotenoids of apple peel mainly composed of violaxanthin, zeaxanthin and lutein showed slightly higher cytotoxic activity against three human tumour cell lines (squamous cell carcinoma HSC-2, HSC-3, submandibular gland carcinoma HSG) and human promyelocytic leukemic HL-60 cells than against three normal human oral cells (gingival fibroblast HGF, pulp cell HPC, periodontal ligament fibroblast HPLF), suggesting a tumour-specific cytotoxic activity and displayed much higher multidrug resistance -reversing activity than (±)-verapamil (Molnár et al. 2005).

Liu et al. (2005) found whole apple extracts prevented mammary cancer in a rat model in a dose-dependent manner at doses comparable to human consumption of one, three, and six apples a day. In further studies they isolated 13 triterpenoids from apple peels; most of the triterpenoids showed high potential anticancer activities against human HepG2 liver cancer cells, MCF-7 breast cancer cells, and Caco-2 colon cancer cells (He and Liu 2007). Among the compounds isolated, 2α-hydroxyursolic acid, 2α-hydroxy-3β-{[(2E)-3-phenyl-1-oxo-2-propenyl]oxy}olean-12-en-28-oic acid, and 3β-trans-p-coumaroyloxy-2α-hydroxyolean-12-en-28-oic acid showed higher antiproliferative activity toward HepG2 cancer cells. Ursolic acid, 2α-hydroxyursolic acid, and 3β-trans-p-coumaroyloxy-2α-hydroxyolean-12-en-28-oic acid exhibited higher antiproliferative activity against MCF-7 cancer cells. All triterpenoids tested showed antiproliferative activity against Caco-2 cancer cells, especially 2α-hydroxyursolic acid, maslinic acid, 2α-hydroxy-3β-{[(2E)-3-phenyl-1-oxo-2-propenyl]oxy}olean-12-en-28-oic acid, and 3β-trans-p-coumaroyloxy-2α-hydroxyolean-12-en-28-oic acid, which displayed much higher antiproliferative activities. The results showed the triterpenoids isolated from apple peels may be partially responsible for the anticancer activities of whole apples. He and Liu (2008) isolated 29 compounds, including triterpenoids, flavonoids, organic acids and plant sterols which showed antiproliferative and antioxidant activities from Red Delicious apple peels. On the basis of the yields of isolated flavonoids, the major flavonoids in apple peels were quercetin-3-O-β-D-glucopyranoside (82.6%), then quercetin-3-O-β-D-galactopyranoside (17.1%), followed by trace amounts of quercetin (0.2%), (−)-catechin, (−)-epicatechin, and quercetin-3-O-α-L-arabinofuranoside. Among the compounds isolated, quercetin and quercetin-3-O-β-D-glucopyranoside showed potent antiproliferative activities against HepG2 and MCF-7 cells, with EC_{50} values of 40.9 and 49.2 µM to HepG2 cells and 137.5 and 23.9 µM to MCF-7 cells, respectively. Six flavonoids (18–23) and three phenolic compounds (10, 11, and 14) showed potent antioxidant activities. Caffeic acid (10), quercetin (18), and quercetin-3-O-β-D-arabinofuranoside (21) showed higher antioxidant activity, with EC_{50} values of <10 µM. Most tested flavonoids and phenolic compounds had high antioxidant activity when compared to ascorbic acid and might be responsible for the antioxidant activities of apples. Apple phytochemical extracts significantly inhibited human breast cancer MCF-7 and MDA-MB-231 cell proliferation at concentrations of 10–80 mg/ml (Sun and Liu 2008). The apple extracts were found to significantly induced G1 arrest in MCF-7 cells in a dose-dependent manner at concentrations. At concentrations of 15, 30, and 50 mg/ml, apple extracts caused a greater increase in the G1/S ratio in MDA-MB-231 cells when compared with MCF-7 cells. Cyclin D1 and Cdk4 proteins, the two major G1/S transit regulators, decreased in a dose-dependent manner after exposure to apple extracts. The results suggested that the antiproliferative activities of apple phytochemical extracts toward human breast cancer cells might be due to the modulation effects on cell cycle machinery. In further studies, Liu et al. (2009) showed that fresh apples potently and dose-dependently suppressed 7,12-dimethylbenz(a)anthracene (DMBA)-induced mammary cancers in rats. Tumour multiplicity and proportions of adenocarcinoma masses decreased with increasing apple extracts. The expression of proliferating cell nuclear antigen (PCNA), cyclin D1, and Bcl-2 decreased, and Bax expression and apoptosis increased with increasing apple extracts. The antiproliferative activity of MCF-7 human breast cancer exerted by apple extract and quercetin 3-β-d-glucoside (Q3G) was twofold to fourfold greater than apple extract and Q3G alone, indicating a synergistic effect in antiproliferative activity (Yang and Liu 2009).

Polyphenol-enriched apple juice extracts were fractionated to identify components with cancer chemopreventive potential (Zessner et al. 2008). Regression analyses indicated that 1,1-diphenyl-2-picrylhydrazyl (DPPH) radical scavenging potential correlated with the sum of low molecular weight (LMW) antioxidants (such as chlorogenic acid, flavan-3-ols, and flavonols) and procyanidins, whereas peroxyl radicals were more effectively scavenged by LMW compounds than by procyanidins. Quercetin aglycone was identified as a potent Cyp1A inhibitor, whereas phloretin and (−)-epicatechin were the most potent cyclooxygenase 1 (Cox-1) inhibitors. Aromatase and Cyp1A inhibitory potential and cytotoxicity toward HCT116 colon cancer cells increased with increasing content in procyanidins. Hypermethylation of the promoter of hMLH1 colon cancer gene and subsequent microsatellite instability had been reported to occur in approximately 12% of sporadic colorectal cancers (CRC) (Fini et al. 2007). Annurca apple polyphenol extract was found to have potent demethylating activity through the inhibition of DNA methyltransferase proteins. The authors maintained that the lack of toxicity in Annurca extracts makes them excellent candidates for the chemoprevention of colorectal cancers.

Pinova and Braeburn apple pomace extracts showed potent antiproliferative activity against cervix epithelioid carcinoma (HeLa) and colon adenocarcinoma (HT-29) human cancer cell lines (Cetkovic et al. 2011). HeLa cells were found more sensitive than HT-29 cells to the extracts. The relationship between radical scavenging activities and phenolic contents or flavonol glycosides ($R^2 \geq 0.80$) was high, but there were no significant correlations between the total phenolic contents or individual phenolic compounds and the antiproliferative activity.

An oligogalactan composed of five galacturonic acids from apple pectin exhibited protective efficacy against intestinal toxicities and carcinogenesis in a mouse model of colitis-associated colon cancer induced by 1,2-dimethylhydrazine and dextran sodium sulphate (Liu et al. 2010). The apple oligogalactan (AOG) decreased the elevated levels of TLR4 and tumour necrosis factor-α (TNF-α) induced by inflammation in-vivo in this model system. In in-vitro studies, AOG alone only slightly increased the levels of protein expression and messenger RNA of TLR4, phosphorylation of IκBα and production of TNF-α in HT-29 cells. However, AOG significantly decreased the elevation of all the biomarkers induced by lipopolysaccharide (LPS) when it was combined with LPS. AOG was active against inflammation and carcinogenesis through targeting LPS/TLR4/NF-κB pathway. The results indicated that the apple oligogalactan AOG may be useful for treatment of colitis and prevention of carcinogenesis.

Apple extract rich in flavonoids were found to inhibit proliferation of HT29 human colon cancer cells and modulate expression of genes involved in the biotransformation of xenobiotics (Veeriah et al. 2006). Treatment of preneoplastic cells derived from colon adenoma (LT97) with apple fruit extract induced expression of 30 and 46 genes expressed over cut-off in Superarray and custom array, respectively (Veeriah et al. 2008). Of 87 genes spotted on both arrays, 4 genes (CYP3A7, CYP4F3, CHST7, GSTT2) were regulated with similar directional changes. Expression of selected phase II genes (GSTP1, GSTT2, GSTA4, UGT1A1, UGT2B7), regulated on either array were also confirmed. The enzyme activities of glutathione S-transferases and UDP-glucuronosyltransferases were altered by treatment of LT97 cells with apple extract. The observed altered gene expression patterns in LT97 cells, resulting from apple extract treatment, suggested a possible protection of the cells against some toxicological insults. Studies demonstrated that induction of the phase II gene glutathione S-transferase T2 (GSTT2) by apple polyphenols protected colon epithelial cells against cumene hydroperoxide-induced genotoxic damage (Petermann et al. 2009; Miene et al. 2009). Apple extract was found to enhance expression of glutathione S-transferases (e.g., GSTT2) in human colon adenoma cells (LT97). Storage of apple extract caused changes in phenolic composition along with loss of activity regarding GSTT2 induction and amplified growth inhibition. Apple extract was also found to protect against oxidatively induced DNA damage.

Epidemiologic studies suggested an inverse correlation between apple consumption and colon cancer risk (Koch et al. 2009). In rat studies they found that under the cancer promoting condition of obesity, apple juice did not show cancer-preventive bioactivity.

A Polish case-control study of 592 incident cases of colorectal cancer and a comparison group of 765 patients without colorectal cancer found that the adjusted risk of colorectal cancer inversely correlated with daily number of apple servings (Jedrychowski et al. 2010). The reduced risk of colorectal cancer of border significance level was already observed at the consumption of at least one apple a day, but at the intake of more than one apple a day the risk was reduced by about 50%. Neither the consumption of vegetables nor other fruits displayed beneficial effects on the risk of colorectal cancer. The observed protective effect of apple consumption on colorectal risk may result from their rich content of flavonoid and other polyphenols, which can inhibit cancer onset and cell proliferation.

Studies by Lhoste et al. (2010) found that using a model of human microbiota-associated rats (HMA), fed a human-type diet and injected with 1–2,dimethylhydrazine (DMH), aberrant crypt foci numbers and multiplicity induced by DMH were not reduced by apple proanthocyanidin-rich extract at 0.001% and 0.01% in drinking water. They maintained that the modulating role of human gut microbiota should be taken into account in colon carcinogenesis models and in using proanthocyanidin extracts as dietary supplements for humans.

Antiviral Activity

Methanolic and acetonic extracts of apple pomace were able to inhibit both herpes simplex virus HSV-1 and HSV-2 replication in Vero cells by more than 50%, at non-cytotoxic concentrations (Suárez et al. 2010). Selectivity indexes (SI) ranged from 9.5 to 12.2. Acetone extraction yielded the higher amounts of phenolic compounds. Among the polyphenols analysed, quercetin glycosides were the most important family, followed by dihydrochalcones.

Antimicrobial Activity

Apple seed oil was found to be almost completely active against bacteria but not mildews, with inhibitory concentration (MIC) ranging from 0.3 to 0.6 mg/ml (Tian et al. 2010). The observed biological properties showed that the oil had a good potential for use in the food industry and pharmacy. Annurca apple peel ethanol extract exhibited antimicrobial activity against *Bacillus cereus* and *Escherichia coli* serotype O157:H7 (Fratianni et al. 2011). No activity was observed against the probiotic lactobacilli tested or against *Staphylococcus aureus*. The apple peel extracts also displayed antiquorum sensing activity tested by using the microorganism *Chromobacterium violaceum*. The results indicated the potential of apple peel extract for treating some microbial infections through cell growth inhibition or quorum sensing antagonism, thereby validating the health benefits of apples.

Antihyperglycaemic Activity

In a randomised study of hypercholesterolemic, overweight, non-smoking Brazilian women (aged 30–50 years), placed on a diet consisting of 55% of energy from carbohydrate, 15% from protein, and 30% from fat, dietary supplementation of apples or pear s resulted in a significantly greater decrease of blood glucose compared to those supplemented with oat cookies (De Oliviera et al. 2003). However, the glucose:insulin ratio was not statistically different from baseline to follow-up.

Aqueous apple pulp extracts exhibited high α-amylase and α-glucosidase inhibitory activities (Barbosa et al. 2010). However, the peel extracts had the highest α-glucosidase inhibitory activity along with low α-amylase inhibitory activity. No correlation between α-amylase inhibitory activity and total phenolic content was observed. However, positive correlations between α-glucosidase inhibitory activity and total phenolics in aqueous ($r = 0.50$) and ethanolic ($r = 0.70$) extracts were observed. Native fructose, FructiLight extracted from apple was found to improve glucose tolerance in mice (Dray et al. 2009). FructiLight, had a very low impact on glycemic and insulin response during acute treatment compared to other sugars based on glycemic index and exhibited beneficial properties when administrated for long term treatment. As with two other sugars extracted from apple (FructiSweetApple and FructiSweet67), FructiLight exposure during 21 weeks in beverage promoted an enhancement of glucose tolerance compared to glucose treatment without affecting food intake and weight. Nishigaki et al. (2010) showed that fresh apple extract treatment at 100 or 250 µg of human umbilical vein endothelial cells (HUVEC) exposed to glycated protein (GFBS) either alone or combined with iron chelate, significantly decreased the level of lipid peroxidation and returned the levels of antioxidants cytochrome c reductase and glutathione S-transferase to near normal in a dose-dependent manner. The extracts recovered viability of HUVEC damaged by GFBS-iron treatment in a concentration-dependent manner. The findings suggested a protective effect of apple extract on HUVEC against glycated protein/iron chelate-induced toxicity, which suggested that apple extract could exert a beneficial effect by preventing diabetic angiopathies. The dihydrochalcone sieboldin from apple leaves exhibited ability to prevent oxidative-dependent formation of advanced glycation end-products (AGEs) and phenylephrine-induced vasocontraction of isolated rat mesenteric arteries, provided interesting information concerning a potential use of sieboldin as a therapeutic (Dugé de Bernonville et al. 2010) The results also confirmed the bioactivity of dihydrochalcones as functional antioxidants in the resistance of apple leaves to oxidative stress.

Anti-asthmatic Activity

A study in United Kingdom involving a survey of 607 individuals with asthma and 864 individuals without asthma found that apple consumption and wine intake was inversely related to severity of asthma indicating a protective effect of flavonoids (Shaheen et al. 2001). Intake of dietary selenium was also negatively associated with asthma. In a study involving 1,601 adults in Australia, apple and pear intake was found to be associated with a decreased risk of asthma and a decrease in bronchial hyper-reactivity, but total fruit and vegetable intake was not associated with asthma risk or severity (Woods et al. 2003). Specific antioxidants, such as vitamin E, vitamin C, retinol, and β-carotene, were not associated with asthma or bronchial hypersensitivity.

A study of over 13,000 adults in the Netherlands found that apples had a beneficial effect on incidence of chronic obstructive pulmonary disease (Tabak et al. 2001). Apple, pear and catechin intake was positively associated with pulmonary function and negatively associated with chronic obstructive pulmonary disease but not tea (also rich in catechin). Smoking was strongly associated with chronic obstructive pulmonary disease, independent of dietary effects. A prospective cohort study of 2,512 Welshmen aged 45–59 also found a strong positive association between lung function and the number of apples eaten per week (Butland et al. 2001). Good lung function, indicated by high maximum FEV (forced expiratory volume in 1 s), was associated with high intakes of vitamin C, vitamin E, β-carotene, citrus fruit, apples, and the frequent consumption of fruit juices/squashes. However, the association with citrus fruit and fruit juice/squash lost significance after adjustment for smoking. Apple consumption remained positively correlated with lung function after taking into account possible confounders such as smoking, body mass index, social class, and exercise.

Anxiolytic Activity

Studies showed that aged rats fed with the annurca apple enriched diet showed a significant decrease in the anxiety level (Viggiano et al. 2006). The aged rats improved in the ability to sustain long-term potentiation P, reaching the level of the young rats superoxide dismutase (SOD) activity was increased in the aged rats fed with the standard diet whereas SOD activity in the hippocampus of the aged rats treated with annurca apple was at the level of the young animals. The results suggested that a diet rich in annurca apple could have an important role in health-care during aging.

Hepatoprotective Activity

Apple polyphenols were found to have significant protective effect against acute hepatotoxicity induced by CCl(4) in mice, which may be due to its free radical scavenging effect, inhibition of lipid peroxidation, and its ability to increase antioxidant activity (Yang et al. 2010). Apple polyphenols significantly prevented the increase in serum alanine aminotransferase and aspartate aminotransferase levels in acute liver injury induced by CCl(4) and produced a marked amelioration in the histopathological hepatic lesions coupled to weight loss. Apple polyphenols reduced malondialdehyde formation and enhanced superoxide dismutase activity and GSH (reduced glutathione) concentration in the hepatic homogenate in apple polyphenolic-treated groups compared with the CCl(4)-intoxicated group. Apple polyphenols also exhibited antioxidant effects on FeSO(4)-L-Cys-induced lipid peroxidation in rat liver homogenate and DPPH free radical scavenging activity in-vitro.

Gastroprotective Activity

Apple extracts prevented exogenous damage to human gastric epithelial cells in-vitro induced by xanthine-xanthine oxidase or indomethacin and to the rat gastric mucosa in-vivo (Graziani et al. 2005). The apple extracts caused a fourfold increase in intracellular antioxidant activity, prevented its decrease induced by xanthine-xanthine oxidase, counteracted xanthine-xanthine oxidase induced lipid peroxidation, and decreased indomethacin injury to the rat gastric mucosa by 40%. This effect appeared to be associated with the antioxidant activity of apple phenolic compounds such as catechin or chlorogenic acid (the main phenolic components of apple extracts) which were equally effective as apple extracts in preventing oxidative injury to gastric cells. The findings suggested that a diet rich in apple antioxidants might exert a beneficial effect in the prevention of gastric diseases related to generation of reactive oxygen species. Apple peel extract (60% of total polyphenols; 58% of flavonoids; 30% of flavan-3-ols and procyanidins) displayed an inhibiting effect on the multiplication of two *Helicobacter pylori* strains with a minimum inhibitory concentration (MIC) value of 112.5 μg gallic acid equivalent (GAE)/ml (Pastene et al. 2009). The apple peel extract inhibited the respiratory burst of neutrophils induced by *H. pylori*, phorbol myristate acetate (PMA), and formyl-methionyl-leucyl-phenylalanine (fMLP) in concentration-dependent manner. The result suggested that apple peel polyphenols had an attenuating effect on the damage to gastric mucosa caused by neutrophil generated reactive oxygen species and, particularly, when *H. pylori* displayed its evasion mechanisms. In-vitro studies showed that apple peel polyphenol-rich extract significantly prevented vacuolating *H. pylori* toxin (VacA) induced vacuolation in HeLa cells with an IC_{50} value of 390 μg of gallic acid equivalents (GAE)/ml (Pastene et al. 2010). The extract also displayed an in-vitro antiadhesive effect against *Helicobacter pylori*. In-vivo studies in mice found a significant inhibition with a 20–60% reduction of *H. pylori* attachment at concentrations between 0.250 and 5 mg of GAE/ml. Orally administered apple peel polyphenols also showed an antiinflammatory effect on *H. pylori*-associated gastritis, lowering malondialdehyde levels and gastritis scores. In Caco-2 cells, apple peel polyphenol extract prevented deleterious mitochondrial oxidative and cell viability alterations induced by indometacin possibly through its ability to scavenge reactive oxygen species (Carrasco-Pozo et al. 2010). The extract was found to actively scavenge O_2^-, hydroxyl and peroxyl radicals. Such free radical-scavenging activity of the extract suggested that its ability to protect mitochondria and prevent the oxidative and lytic damage induced by indometacin conventional antiinflammatory agent, arose from its potent antioxidant capacity.

Hypophasic carotenoids, violaxanthin, zeaxanthin and lutein of Golden delicious apple peel showed potent anti-*H. pylori* activity (MIC (50)=36 μg/ml), comparable to metronidazole (MIC_{50}=45 μg/ml) (Molnár et al. 2005). The MIC_{50} values of anti-*H. pylori* activity of (all-E)-luteoxanthin, (all-E)-neoxanthin and (9′Z)-neoxanthin were 7.9, 11 and 27 μg/mL, respectively (Molnár et al. 2010). Other carotenoids and, β-carotene did not exhibit potent anti-*H. pylori* activity (MIC_{50}>100 μg/ml).

Antiinflammatory Activity

Apple extract powders from three different manufacturers showed similar, but clearly different, antiinflammatory activities, and had substantially different total phenolic contents, and different chemical compositions (Lauren et al. 2009). The most active fractions were those that contained epicatechin, catechin with phloridzin and quercetin glycosides, or those that contained procyanidin polymers. In-vitro studies showed that apple juice extract significantly inhibited the expression of NF-κB regulated proinflammatory genes (TNF-α, IL-1β, CXCL9, CXCL10), inflammatory relevant enzymes (COX-2, CYP3A4), and transcription factors (STAT1, IRF1) in LPS/IFN-γ stimulated MonoMac6 cells without significant effects on the expression of house-keeping genes. (Jung et al. 2009). The bioactive constituents procyanidin B(1), procyanidin B(2), and phloretin were responsible for the antiinflammatory activity and may serve as transcription-based inhibitors of proinflammatory gene expression. Studies showed that administration of Marie Ménard apples, rich in polyphenols and used at present only in the manufacturing of cider, ameliorated colon inflammation in transgenic rats developing spontaneous intestinal inflammation, suggesting the possible use of these and other apple varieties to control inflammation in inflammatory bowel diseased patients (Castagnini et al. 2009). Rats fed Marie Ménard apples had reduced myeloperoxidase activity and reduced cyclooxygenase-2 and inducible NO synthase gene expression in the colon mucosa and significantly less diarrhoea, compared with control rats. A down-regulation of the pathways of prostaglandin synthesis, mitogen-activated protein kinase (MAPK) signalling and TNFα-NF-κB was observed in Marie Ménard-fed rats.

Cognitive/Behavioral Activity

Studies demonstrated that apple juice concentrate prevented the increase in oxidative damage to brain tissue and decline in cognitive (maze) performance observed when transgenic mice lacking apolipoprotein E (ApoE−/−) were maintained on a vitamin-deficient diet and challenged with excess iron (included in the diet as a pro-oxidant) (Tchantchou et al. 2005). Further they found dietary supplementation with apple juice concentrate alleviated the compensatory increase in glutathione synthase transcription and activity that accompanied dietary- and genetically-induced oxidative stress (Tchantchou et al. 2004). Their findings provided further evidence that the antioxidant potential of apple juice concentrate could compensate for dietary and genetic deficiencies that otherwise promoted neurodegeneration. They demonstrated that apple juice concentrate, administered ad libitum in drinking water, could compensate for the increased reactive oxygen species and decline in cognitive performance in maze trials observed when normal and transgenic mice lacking apolipoprotein E were deprived of folate and vitamin E (Rogers et al. 2004). Additionally, they demonstrated that this protective effect was not derived from the sugar content of the concentrate. They also demonstrated that apple juice concentrate administered in drinking water, maintained acetylcholine levels that otherwise declined when adult and aged mice were maintained on the vitamin-deficient, oxidative stress-promoting diet (Chan et al. 2006). The findings presented a likely mechanism by which consumption of antioxidant-rich foods such as apples could prevent the decline in cognitive performance that accompanied dietary and genetic deficiencies and aging.

Apple juice was found to alleviate the neurotoxic consequences of exposure of cultured neuronal cells to amyloid-β (Ortiz and Shea 2004). Apple juice prevented the increased generation of reactive oxygen species (ROS) normally induced by Abeta treatment and prevented Abeta-induced calcium influx and apoptosis. The results suggested that the antioxidant potential of apple products can prevent Abeta-induced oxidative damage that contributes to the decline in cognitive performance during normal aging and in neurodegenerative conditions such as Alzheimer's disease. Chan and Shea (2009) demonstrated that dietary deficiency in folate and vitamin E, coupled pro-oxidant stress induced by dietary iron, increased amyloid-β (Abeta) levels in normal adult mice. This increase was potentiated by apolipoprotein E (ApoE) deficiency as shown by treatment of transgenic mice homozygously lacking murine ApoE. Dietary supplementation with apple juice concentrate in drinking water alleviated the increase in Abeta for both mouse genotypes. More recent studies indicated that supplementation with apple juice concentrate can compensate for genetic as well as dietary insufficiency in folate in a murine model of genetic folate compromise (Chan et al. 2011). MTHFR+/− mice deficient in methylene tetrahydrofolate reductase activity exhibited significantly impaired cognitive performance in standard reward-based T maze and the non-reward-based Y maze tests as compared to MTHFR+/+ when maintained on the complete diet; supplementation with apple juice concentrate improved the performance of MTHFR+/− to the level observed for MTHFR+/+ mice. MTHFR+/+ and +/− demonstrated virtually identical neuromuscular performance in the standard paw grip endurance test when maintained on the complete diet, and displayed similar, non-significant declines in performance when maintained on the deficient diet. Supplementation of either diet with apple juice concentrate dramatically improved the performance of both genotypes.

In an open-label clinical study of 21 institutionalized individuals with moderate-to-severe Alzheimer's disease, consumption of apple juice daily for a month did not improve cognitive activities but attenuated behavioral and psychotic symptoms associated with dementia (Remington et al. 2010).

Anti-toxin Activity

Crude polyphenol extract from immature apples dose-dependently inhibited cholera toxin catalyzed ADP-ribosylation of agmatine (Saito et al. 2002). Additionally the extract reduced cholera toxin induced fluid accumulation two diarrhea models for in vivo mice. On fractionation the FAP3 and FAP4 fractions, which possessed highly polymerized catechin compounds, strongly inhibited the ADP-ribosylation, indicating that the polymerized structure of catechin was responsible for the inhibitory effect of the crude extract. FAP2, which contained compounds with monomeric, dimeric, and trimeric catechins, inhibited the ADP-ribosylation only partially, but significantly. FAP1, which contained non-catechin polyphenols, did not significantly inhibit the CT-catalyzed ADP-ribosylation of agmatine. In another study, dilutions of freshly prepared apple juices and Apple Poly (a commercial apple polyphenol preparation) inhibited the biological activity of staphylococcal enterotoxin A (SEA) produced by *Staphylococcus aureus*, without any significant cytotoxic effect on the spleen cells (Rasooly et al. 2010). Additional studies with antibody-coated immunomagnetic beads bearing specific antibodies against the toxin revealed that SEA added to apple juice appeared to be largely irreversibly bound to the juice constituents.

Probiotic Effect

Studies in rats showed that a 4-week consumption of apple pectin (7% in the diet) increased the population of butyrate-glucuronidase and β-glucuronidase producing Clostridiales, and decreased the population of specific species within the Bacteroidetes group in the rat gut (Licht et al. 2010). Similar changes were not found by consumption of whole apples, apple juice, purée or pomace. Shinohara et al. (2010) found in a study of eight healthy adult humans that apple consumption improved intestinal environment and apple pectin was one of the effective bioactive component. Ingestion of apples was found to increase the faecal population of *Bifidobacterium* and *Lactobacillus* and also *Streptococcus* and *Enterococcus*. The lecithinase-positive clostridia, including *Clostridium perfringens*, Enterobacteriaceae and *Pseudomonas* tended to decrease. Several isolates of *Bifidobacterium, Lactobacillus, Enterococcus*, and the *Bacteroides fragilis* group utilized apple pectin, most isolates of *Escherichia coli, Collinsela aerofaciense, Eubacterium limosum*, and *Clostridium perfringens* could not. Additionally, the concentrations of faecal acetic acid tended to increase on apple intake, while faecal ammonia and sulphide tended to decrease on apple intake.

Bioavailability of Apple Phytochemicals

Studies involving ten healthy ileostomy subjects showed that oligomeric procyanidins, D-(−)-quinic acid and 5-caffeoylquinic acid reaching the ileostomy bags were considerably higher after apple smoothie consumption than after the consumption of cloudy apple juice or cider (Hagl et al. 2011). The results suggested that the food matrix might affect the colonic availability of polyphenols, and apple smoothies could be more effective in the prevention of chronic colon diseases than both cloudy apple juice and apple cider.

Allergy Problem

Consumption of apples can provoke severe allergic reactions, in susceptible individuals, due to the presence of the allergen Mal d 3, a nonspecific lipid transfer protein, found largely in the fruit skin (Sancho et al. 2006). The scientists found that pre- and postharvest treatments (i.e., storage) could modify the allergen load in apple peel, the highest levels being found in overly mature and freshly harvested fruits. During storage, levels of Mal d 3 decreased in all cultivars (cvs. Cox, Jonagored, and Gala), the rate of overall decrease being greatest under controlled atmosphere conditions. A separate study of 22 Spanish patients with oral allergy syndrome after apple ingestion found that the ten apple varieties tested differed in the antigenic and allergenic profiles and the content of the allergen Mal d 3 (Carnés et al. 2006). Another allerge, Mal d 2 a thaumatin-like protein and important allergen of apple fruits was found to be associated with IgE-mediated symptoms in apple allergic individuals (Krebitz et al. 2003). Purified recombinant Mal d 2 displayed the ability to bind IgE from apple-allergic individuals equivalent to natural Mal d 2. In addition, the recombinant thaumatin-like Mal d 2 exhibited antifungal activity against *Fusarium oxysporum* and *Penicillium expansum*, implying a function in plant defence against fungal pathogens.

Traditional Medicinal Uses

Raw apples are eaten for indigestion and for constipation and is said to be good for gout. Apples have been found useful in acute and chronic dysentery among children.

Other Uses

Dried apple pomace was found to be a cost-effective and good feed adjunct for broiler chicks (Zaffar et al. 2005).

Comments

China is the leading producer of apples followed by the United States.

Selected References

Alvarez-Parrilla E, De La Rosa LA, Legarreta P, Saenz L, Rodrigo-García J, González-Aguilar GA (2010) Daily consumption of apple, pear and orange juice differently affects plasma lipids and antioxidant capacity of smoking and non-smoking adults. Int J Food Sci Nutr 61(4):369–380

Aprikian O, Duclos V, Guyot S, Besson C, Manach C, Bernalier A, Morand C, Remesy C, Demigne C (2003) Apple pectin and a polyphenol rich apple concentrate are more effective together than separately on cecal fermentations and plasma lipids in rats. J Nutr 133: 1860–1865

Aprikian O, Levrat-Verny M, Besson C, Busserolles J, Remesy C, Demigne C (2002) Apple favourably affects parameters of cholesterol metabolism and of anti-oxidative protection in cholesterol fed rats. Food Chem 75:445–452

Bai XL, Yue TL, Yuan YH, Zhang HW (2010) Optimization of microwave-assisted extraction of polyphenols from apple pomace using response surface methodology and HPLC analysis. J Sep Sci 33(23–24):3751–3758

Barbosa AC, Pinto Mda S, Sarkar D, Ankolekar C, Greene D, Shetty K (2010) Varietal influences on antihyperglycemia properties of freshly harvested apples using in vitro assay models. J Med Food 13(6):1313–1323

Bellion P, Digles J, Will F, Dietrich H, Baum M, Eisenbrand G, Janzowski C (2010) Polyphenolic apple extracts: effects of raw material and production method on antioxidant effectiveness and reduction of DNA damage in Caco-2 cells. J Agric Food Chem 58(11):6636–6642

Bellion P, Olk M, Will F, Dietrich H, Baum M, Eisenbrand G, Janzowski C (2009) Formation of hydrogen peroxide in cell culture media by apple polyphenols and its effect on antioxidant biomarkers in the colon cell line HT-29. Mol Nutr Food Res 53(10):1226–1236

Boyer J, Lui RH (2004) Apple phytochemicals and their health benefits. Nutr J 3:5

Butland B, Fehily A, Elwood P (2001) Diet, lung function, and lung decline in a cohort of 2512 middle aged men. Thorax 55(2):102–108

Buttery RG, Guadagni DG, Ling LC (1973) Flavor compounds: volatiles in vegetable oil and oilwater mixtures. Estimations of odour thresholds. J Agric Food Chem 21:198–201

Cao X, Wang C, Pei H, Sun B (2009) Separation and identification of polyphenols in apple pomace by high-speed counter-current chromatography and high-performance liquid chromatography coupled with mass spectrometry. J Chromatogr A 1216(19):4268–4274

Carnés J, Ferrer A, Fernández-Caldas E (2006) Allergenicity of 10 different apple varieties. Ann Allergy Asthma Immunol 96(4):564–570

Carrasco-Pozo C, Gotteland M, Speisky H (2010) Protection by apple peel polyphenols against indometacin-induced oxidative stress, mitochondrial damage and cytotoxicity in Caco-2 cells. J Pharm Pharmacol 62(7):943–950

Castagnini C, Luceri C, Toti S, Bigagli E, Caderni G, Femia AP, Giovannelli L, Lodovici M, Pitozzi V, Salvadori M, Messerini L, Martin R, Zoetendal EG, Gaj S, Eijssen L, Evelo CT, Renard CM, Baron A, Dolara P (2009) Reduction of colonic inflammation in HLA-B27 transgenic rats by feeding Marie Ménard apples, rich in polyphenols. Br J Nutr 102(11):1620–1628

Cetkovic GS, Savatovic SM, Canadanovic-Brunet JM, Cetojevic-Simin DD, Djilas SM, Tumbas VT, Skerget M (2011) Apple pomace: antiradical activity and antiproliferative action in HeLa and HT- 29 human tumor cell lines. J BUON 16(1):147–153

Chan A, Shea TB (2009) Dietary supplementation with apple juice decreases endogenous amyloid-beta levels in murine brain. J Alzheimers Dis 16(1):167–171

Chan A, Graves V, Shea TB (2006) Apple juice concentrate maintains acetylcholine levels following dietary compromise. J Alzheimers Dis 9(3):287–291

Chan A, Ortiz D, Rogers E, Shea TB (2011) Supplementation with apple juice can compensate for folate deficiency in a mouse model deficient in methylene tetrahydrofolate reductase activity. J Nutr Health Aging 15(3):221–225

Chinnici F, Bendini A, Gaiani A, Riponi C (2004) Radical scavenging activities of peels and pulps from cv. Golden Delicious apples as related to their phenolic composition. J Agric Food Chem 52(15):4684–4689

Coart E, Van Glabeke S, De Loose M, Larsen AS, Roldán-Ruiz I (2006) Chloroplast diversity in the genus *Malus*: new insights into the relationship between the European wild apple (*Malus sylvestris* (L.) Mill.) and the domesticated apple (*Malus domestica* Borkh.). Mol Ecol 15(8):2171–2182

de Oliviera M, Sichieri R, Moura A (2003) Weight loss associated with a daily intake of three apples or three pears among overweight women. Nutr 19:253–256

Delage E, Bohuon G, Baron A, Drilleau JF (1991) High-performance liquid chromatography of the phenolic compounds in the juice of some French cider apple varieties. J Chromatogr A 555(1–2):125–136

Dimick PS, Hoskin JC (1983) Review of apple flavor – state of the art. CRC Crit Rev Food Sci 18(4):387–409

Ding M, Lu Y, Bowman L, Huang C, Leonard S, Wang L, Vallyathan V, Castranova V, Shi X (2004) Inhibition of AP-1 and neoplastic transformation by fresh apple peel extract. J Biol Chem 279(11):10670–10676

Dixon J, Hewett EW (2000) Factors affecting apple aroma/flavour volatile concentration: a review. N Z J Crop Hortic Sci 28:155–173

Dray C, Colom A, Guigné C, Legonidec S, Guibert A, Ouarne F, Valet P (2009) Native fructose extracted from apple improves glucose tolerance in mice. J Physiol Biochem 65(4):361–368

Dugé de Bernonville T, Guyot S, Paulin JP, Gaucher M, Loufrani L, Henrion D, Derbré S, Guilet D, Richomme P, Dat JF, Brisset MN (2010) Dihydrochalcones: implication in resistance to oxidative stress and bioactivities against advanced glycation end-products and vasoconstriction. Phytochemistry 71(4):443–452

Eberhardt M, Lee C, Liu RH (2000) Antioxidant activity of fresh apples. Nature 405:903–904

Escarpa A, Gonzalez M (1998) High-performance liquid chromatography with diode-array detection for the performance of phenolic compounds in peel and pulp from different apple varieties. J Chromatogr A 823: 331–337

Fini L, Selgrad M, Fogliano V, Graziani G, Romano M, Hotchkiss E, Daoud YA, De Vol EB, Boland CR, Ricciardiello L (2007) Annurca apple polyphenols have potent demethylating activity and can reactivate silenced tumor suppressor genes in colorectal cancer cells. J Nutr 137(12):2622–2628

Flath RA, Black DR, Guadagni DG, McFadden WH, Schultz TH (1967) Identification and organoleptic evaluation of compounds in Delicious apple essence. J Agric Food Chem 15:29–35

Fratianni F, Coppola R, Nazzaro F (2011) Phenolic composition and antimicrobial and antiquorum sensing activity of an ethanolic extract of peels from the apple cultivar Annurca. J Med Food 14(9):957–963

Gerhauser C (2008) Cancer chemopreventive potential of apples, apple juice, and apple components. Planta Med 74(13):1608–1624

Gosch C, Halbwirth H, Kuhn J, Miosic S, Stich K (2009) Biosynthesis of phloridzin in apple (*Malus domestica* Borkh.). Plant Sci 176(2):223–231

Graziani G, D'Argenio G, Tuccillo C, Loguercio C, Ritieni A, Morisco F, Del Vecchio BC, Fogliano V, Romano M (2005) Apple polyphenol extracts prevent damage to human gastric epithelial cells in vitro and to rat gastric mucosa in vivo. Gut 54(2):193–200

Hagl S, Deusser H, Soyalan B, Janzowski C, Will F, Dietrich H, Albert FW, Rohner S, Richling E (2011) Colonic availability of polyphenols and D-(-)-quinic acid after apple smoothie consumption. Mol Nutr Food Res 55(3):368–377

Hansen L, Dragsted LO, Olsen A, Christensen J, Tjønneland A, Schmidt EB, Overvad K (2010) Fruit and vegetable intake and risk of acute coronary syndrome. Br J Nutr 104(2):248–255

He X, Liu RH (2007) Triterpenoids isolated from apple peels have potent antiproliferative activity and may be partially responsible for apple's anticancer activity. J Agric Food Chem 55(11):4366–4370

He X, Liu RH (2008) Phytochemicals of apple peels: isolation, structure elucidation, and their antiproliferative and antioxidant activities. J Agric Food Chem 56(21):9905–9910

Huber GM, Rupasinghe HP (2009) Phenolic profiles and antioxidant properties of apple skin extracts. J Food Sci 74(9):C693–C700

Jackson JE (2003) Biology of apples and pears. Cambridge University Press, Cambridge, 488 pp

Jedrychowski W, Maugeri U, Popiela T, Kulig J, Sochacka-Tatara E, Pac A, Sowa A, Musial A (2010) Case-control study on beneficial effect of regular consumption of apples on colorectal cancer risk in a population with relatively low intake of fruits and vegetables. Eur J Cancer Prev 19(1):42–47

Jung M, Triebel S, Anke T, Richling E, Erkel G (2009) Influence of apple polyphenols on inflammatory gene expression. Mol Nutr Food Res 53(10):1263–1280

Ker YB, Peng CH, Chyau CC, Peng RY (2010) Soluble polysaccharide composition and myo-inositol content help differentiate the antioxidative and hypolipidemic capacity of peeled apples. J Agric Food Chem 58(8):4660–4665

Kermasha S, Goetghebeur M, Dumont J, Couture R (1995) Analyses of phenolic and furfural compounds in concentrated and non-concentrated apple juices. Food Res Int 28(3):245–252

Koch TC, Briviba K, Watzl B, Fähndrich C, Bub A, Rechkemmer G, Barth SW (2009) Prevention of colon carcinogenesis by apple juice in vivo: impact of juice constituents and obesity. Mol Nutr Food Res 53(10):1289–1302

Krebitz M, Wagner B, Ferreira F, Peterbauer C, Campillo N, Witty M, Kolarich D, Steinkellner H, Scheiner O, Breiteneder H (2003) Plant-based heterologous expression of Mal d 2, a thaumatin-like protein and allergen of apple (*Malus domestica*), and its characterization as an antifungal protein. J Mol Biol 329(4):721–730

Ku TC, Spongberg SA (2003) *Malus* Miller. In: Wu ZY, Raven PH, Hong DY (eds) Flora of China, vol 9 (Pittosporaceae through Connaraceae). Science Press/Missouri Botanical Garden Press, Beijing/St. Louis

Kujawska M, Ignatowicz E, Ewertowska M, Markowski J, Jodynis-Liebert J (2011) Cloudy apple juice protects against chemical-induced oxidative stress in rat. Eur J Nutr 50(1):53–60

Kusumo S, Verheij EWM (1992) *Malus domestica* Borkh. In: Verheij EWM, Coronel RE (eds) Plant resources of South-East Asia No 2. Edible fruits and nuts. PROSEA, Bogor, pp 200–203

Lapidot T, Walker M, Kanner J (2002) Can apple antioxidants inhibit tumor cell proliferation? Generation of H2O2 during interaction of phenolic compounds with cell culture media. J Agric Food Chem 50:3156–3160

Lauren DR, Smith WA, Adaim A, Cooney JM, Wibisono R, Jensen DJ, Zhang J, Skinner MA (2009) Chemical composition and in vitro anti-inflammatory activity of apple phenolic extracts and of their sub-fractions. Int J Food Sci Nutr 7:188–205

Lee K, Kim Y, Kim D, Lee H, Lee C (2003) Major phenolics in apple and their contribution to the total antioxidant capacity. J Agric Food Chem 51:6516–6520

Leja M, Mareczek A, Ben J (2003) Antioxidant properties of two apple cultivars during long-term storage. Food Chem 80(3):303–307

Leontowicz H, Gorinstein S, Lojek A, Leontowicz M, Ciz M, Soliva-Fortuny R, Park Y, Jung S, Trakhtenberg S, Martin-Belloso O (2002) Comparative content of some bioactive compounds in apples, peaches, and pears and their influence on lipids and antioxidant capacity in rats. J Nutr Biochem 13:603–610

Leontowicz M, Gorinstein S, Leontowicz H, Krezeminski R, Lojek A, Katrich E, Ciz M, Martin-Belloso O, Soliva-Fortuny R, Haruenkit R, Trakhtenberg S (2003) Apple and pear peel and pulp and their influences on plasma lipids and antioxidant potential in rats fed cholesterol-containing diets. J Agric Food Chem 51(19):5780–5785

Lhoste EF, Bruneau A, Bensaada M, Cherbuy C, Philippe C, Bruel S, Sutren M, Rabot S, Guyot S, Duée PH, Latino-Martel P (2010) Apple proanthocyanidins do not reduce the induction of preneoplastic lesions in the colon of rats associated with human microbiota. J Agric Food Chem 58(7):4120–4125

Licht TR, Hansen M, Bergström A, Poulsen M, Krath BN, Markowski J, Dragsted LO, Wilcks A (2010) Effects of apples and specific apple components on the cecal environment of conventional rats: role of apple pectin. BMC Microbiol 10:13

Liu L, Li YH, Niu YB, Sun Y, Guo ZJ, Li Q, Li C, Feng J, Cao SS, Mei QB (2010) An apple oligogalactan prevents against inflammation and carcinogenesis by targeting LPS/TLR4/NF-κB pathway in a mouse model of colitis-associated colon cancer. Carcinogenesis 31(10):1822–1832

Liu RH, Eberhardt M, Lee C (2001) Antioxidant and antiproliferative activities of selected New York apple cultivars. NY Fruit Q 9:15–17

Liu RH, Liu J, Chen B (2005) Apples prevent mammary tumors in rats. J Agric Food Chem 53(6):2341–2343

Liu RH, Sun J (2003) Antiproliferative activity of apples is not due to phenolic-induced hydrogen peroxide formation. J Agric Food Chem 51:1718–1723

Liu JR, Dong HW, Chen BQ, Zhao P, Liu RH (2009) Fresh apples suppress mammary carcinogenesis and proliferative activity and induce apoptosis in mammary tumors of the Sprague-Dawley rat. J Agric Food Chem 57(1):297–304

Lu Y, Foo L (2000) Antioxidant and radical scavenging activities of polyphenols from apple pomace. Food Chem 68:81–85

Malik A, Kiczorowska B, Zdyb J (2009) The content of phenolic acids in the edible parts of selected varieties of apples. Rocz Panstw Zakl Hig 60(4):333–336 (in Polish)

Matsui T, Korematsu S, Byun EB, Nishizuka T, Ohshima S, Kanda T (2009) Apple procyanidins induced vascular relaxation in isolated rat aorta through NO/cGMP pathway in combination with hyperpolarization by multiple K+channel activations. Biosci Biotechnol Biochem 73(10):2246–2251

Miene C, Klenow S, Veeriah S, Richling E, Glei M (2009) Impact of apple polyphenols on GSTT2 gene expression, subsequent protection of DNA and modulation of proliferation using LT97 human colon adenoma cells. Mol Nutr Food Res 53(10):1254–1262

Molnár P, Deli J, Tanaka T, Kann Y, Tani S, Gyémánt N, Molnár J, Kawase M (2010) Carotenoids with anti-*Helicobacter pylori* activity from Golden Delicious apple. Phytother Res 24(5):644–648

Molnár P, Kawase M, Satoh K, Sohara Y, Tanaka T, Tani S, Sakagami H, Nakashima H, Motohashi N, Gyémánt N, Molnár J (2005) Biological activity of carotenoids in red paprika, Valencia orange and Golden delicious apple. Phytother Res 19:700–707

Morgan J, Richards A (1993) The book of apples. Ebury Press, London, 304 pp

Nishigaki I, Rajkapoor B, Rajendran P, Venugopal R, Ekambaram G, Sakthisekaran D, Nishigaki Y (2010) Effect of fresh apple extract on glycated protein/iron chelate-induced toxicity in human umbilical vein endothelial cells in vitro. Nat Prod Res 24(7):599–609

Ortiz D, Shea TB (2004) Apple juice prevents oxidative stress induced by amyloid-beta in culture. J Alzheimers Dis 6(1):27–30

Paillard NMM (1990) The flavour of apples, pears and quinces. In: Morton LD, MacLeod AJ (eds) Food flavours, Part C. The flavour of fruits. Elsevier Science Publishing Company Inc, Amsterdam, pp 1–41

Pastene E, Speisky H, García A, Moreno J, Troncoso M, Figueroa G (2010) In vitro and in vivo effects of apple peel polyphenols against *Helicobacter pylori*. J Agric Food Chem 58(12):7172–7179

Pastene E, Speisky H, Troncoso M, Alarco J, Guillermo Figueroa G (2009) In vitro inhibitory effect of apple peel extract on the growth of *Helicobacter pylori* and respiratory burst induced on human neutrophils. J Agric Food Chem 57(17):7743–7749

Pearson D, Tan C, German B, Davis P, Gershwin M (1999) Apple juice inhibits low density lipoprotein oxidation. Life Sci 64:1919–1920

Petermann A, Miene C, Schulz-Raffelt G, Palige K, Hölzer J, Glei M, Böhmer FD (2009) GSTT2, a phase II gene induced by apple polyphenols, protects colon epithelial cells against genotoxic damage. Mol Nutr Food Res 53(10):1245–1253

Qian GZ, Lie LF, Tang GG (2010) (1933) Proposal to conserve the name *Malus domestica* against *M. pumila*, *M. communis*, *M. frutescens*, and *Pyrus dioica* *(Rosaceae)*. Taxon 59(2):650–652

Rasooly R, Do PM, Friedman M (2010) Inhibition of biological activity of staphylococcal enterotoxin A (SEA) by apple juice and apple polyphenols. J Agric Food Chem 58(9):5421–5426

Reagan-Shaw S, Eggert D, Mukhtar H, Ahmad N (2010) Antiproliferative effects of apple peel extract against cancer cells. Nutr Cancer 62(4):517–524

Remington R, Chan A, Lepore A, Kotlya E, Shea TB (2010) Apple juice improved behavioral but not cognitive symptoms in moderate-to-late stage Alzheimer's disease in an open-label pilot study. Am J Alzheimers Dis Other Demen 25(4):367–371

Renard C, Baron A, Guyot S, Drilleau J (2001) Interactions between apple cell walls and native apple polyphenols' quantification and some consequences. Int J Biol Macromol 29:115–125

Rogers EJ, Milhalik S, Orthiz D, Shea TB (2004) Apple juice prevents oxidative stress and impaired cognitive performance caused by genetic and dietary deficiencies in mice. J Nutr Health Aging 8(2):92–97

Saito T, Miyake M, Toba M, Okamatsu H, Shimizu S, Noda M (2002) Inhibition by apple polyphenols of ADP-ribotransferase activity of cholera toxin and toxin-induced fluid accumulation in mice. Microbiol Immunol 46:249–255

Salunkhe DK, Kadam SS (1995) Handbook of fruit science and technology: production, composition, storage, and processing. CRC Press, Boca Raton, 632 pp

Sancho AI, Foxall R, Rigby NM, Browne T, Zuidmeer L, van Ree R, Waldron KW, Mills EN (2006) Maturity and storage influence on the apple (*Malus domestica*) allergen Mal d 3, a nonspecific lipid transfer protein. J Agric Food Chem 54(14):5098–5104

Sanoner P, Guyot S, Marnet N, Molle D, Drilleau JF (1999) Polyphenol profiles of French cider apple varieties (*Malus domestica* sp.). J Agric Food Chem 47(12):4847–4853

Schieber A, Hilt P, Streker P, Endreß HU, Rentschler C, Reinhold Carle R (2003) A new process for the combined recovery of pectin and phenolic compounds from apple pomace. Innov Food Sci Emerg Technol 4(1):99–107

Setorki M, Asgary S, Eidi A, Rohani AH, Esmaeil N (2009) Effects of apple juice on risk factors of lipid profile, inflammation and coagulation, endothelial markers and atherosclerotic lesions in high cholesterolemic rabbits. Lipids Health Dis 8:39

Shaheen S, Sterne J, Thompson R, Songhurst C, Margetts B, Buerney P (2001) Dietary antioxidants and asthma in adults- population based case-control study. Am J Respir Crit Care Med 164:1823–1828

Shinohara K, Ohashi Y, Kawasumi K, Terada A, Fujisawa T (2010) Effect of apple intake on fecal microbiota and metabolites in humans. Anaerobe 16(5):510–515

Stracke BA, Rüfer CE, Bub A, Seifert S, Weibel FP, Kunz C, Watzl B (2010) No effect of the farming system (organic/conventional) on the bioavailability of apple (*Malus domestica* Bork., cultivar Golden Delicious) polyphenols in healthy men: a comparative study. Eur J Nutr 49(5):301–310

Suárez B, Álvarez AL, García YD, del Barrio G, Lobo AP, Parra F (2010) Phenolic profiles, antioxidant activity and in vitro antiviral properties of apple pomace. Food Chem 120(1):339–342

Sun J, Chu Y, Wu X, Liu RH (2002) Antioxidant and antiproliferative activities of common fruits. J Agric Food Chem 50:7449–7454

Sun J, Liu RH (2008) Apple phytochemical extracts inhibit proliferation of estrogen-dependent and estrogen-independent human breast cancer cells through cell cycle modulation. J Agric Food Chem 56(24): 11661–11667

Tabak C, Arts I, Smit H, Heederik D, Kromhout D (2001) Chronic obstructive pulmonary disease and intake of catechins, flavonols, and flavones. Am J Respir Crit Care Med 164:61–64

Takeoka GR, Buttery RG, Ling L (1996) Odour thresholds of various branched and straight chain acetates. Lebensm Wiss Technol 29:677–680

Tchantchou F, Chan A, Kifle L, Ortiz D, Shea TB (2005) Apple juice concentrate prevents oxidative damage and impaired maze performance in aged mice. J Alzheimers Dis 8(3):283–287

Tchantchou F, Graves M, Ortiz D, Rogers E, Shea TB (2004) Dietary supplementation with apple juice concentrate alleviates the compensatory increase in glutathione synthase transcription and activity that accompanies dietary- and genetically-induced oxidative stress. J Nutr Health Aging 8(6):492–496

Teranishi R, Buttery RG, Schamp N (1987) The significance of low threshold odor compounds in aroma research. In: Martens M, Dalen GA, Russwurm H (eds) Flavour science and technology. Wiley, New York, pp 515–527

Tian HL, Zhan P, Li KX (2010) Analysis of components and study on antioxidant and antimicrobial activities of oil in apple seeds. Int J Food Sci Nutr 61(4):395–403

U.S. Department of Agriculture, Agricultural Research Service (USDA) (2011) USDA national nutrient database for standard reference, release 24. Nutrient Data Laboratory Home Page, http://www.ars.usda.gov/ba/bhnrc/ndl

van der Sluis A, Dekker M, Skrede G, Jongen W (2002) Activity and concentration of polyphenolic antioxidants in apple juice. 1. Effect of existing production methods. J Agric Food Chem 50:7211–7219

Veeriah S, Kautenburger T, Habermann N, Sauer J, Dietrich H, Will F, Pool-Zobel BL (2006) Apple flavonoids inhibit growth of HT29 human colon cancer cells and modulate expression of genes involved in the biotransformation of xenobiotics. Mol Carcinog 45(3):164–174

Veeriah S, Miene C, Habermann N, Hofmann T, Klenow S, Sauer J, Böhmer F, Wölfl S, Pool-Zobel BL (2008) Apple polyphenols modulate expression of selected genes related to toxicological defence and stress response in human colon adenoma cells. Int J Cancer 122(12):2647–2655

Velasco R, Zharkikh A, Affourtit J, Dhingra A, Cestaro A, Kalyanaraman A, Fontana P, Bhatnagar SK, Troggio M, Pruss D, Salvi S, Pindo M, Baldi P, Castelletti S, Cavaiuolo M et al (2010) The genome of the domesticated apple (*Malus* x *domestica* Borkh.). Nat Genet 42(10):833–839

Viggiano A, Viggiano A, Monda M, Turco I, Incarnato L, Vinno V, Viggiano E, Baccari ME, De Luca B (2006) Annurca apple-rich diet restores long-term potentiation and induces behavioral modifications in aged rats. Exp Neurol 199(2):354–361

Wolfe K, Liu RH (2003) Apple peels as a value-added food ingredient. J Agric Food Chem 51:1676–1683

Wolfe K, Wu X, Liu RH (2003) Antioxidant activity of apple peels. J Agric Food Chem 51(3):609–614

Woods R, Walters H, Raven J, Wolfe R, Ireland P, Thien F, Abramson M (2003) Food and nutrient intakes and asthma risk in young adults. Am J Clin Nutr 78:414–421

Yang J, Li Y, Wang F, Wu C (2010) Hepatoprotective effects of apple polyphenols on CCl4-induced acute liver damage in mice. J Agric Food Chem 58(10):6525–6531

Yang J, Liu RH (2009) Synergistic effect of apple extracts and quercetin 3-beta-d-glucoside combination on antiproliferative activity in MCF-7 human breast cancer cells in vitro. J Agric Food Chem 57(18):8581–8586

Zaffar F, Idrees M, Ahmed Z (2005) Use of apple by-products in poultry rations of broiler chicks in Karachi. Pak J Physiol 1(1–2)

Zessner H, Pan L, Will F, Klimo K, Knauft J, Niewöhner R, Hümmer W, Owen R, Richling E, Frank N, Schreier P, Becker H, Gerhauser C (2008) Fractionation of polyphenol-enriched apple juice extracts to identify constituents with cancer chemopreventive potential. Mol Nutr Food Res 52(Suppl 1):S28–S44

Mespilus germanica

Scientific Name

Mespilus germanica L.

Synonyms

Mespilus sylvestris Mill.

Family

Rosaceae

Common Names

Medlar, Medlar Tree

Vernacular Names

Chinese: Ou Cha;
Czech: Mišpule obecná;
Danish: Mispel;
Dutch: Mispel, Mispelboom;
Eastonian: Harilik astelpihlakas
Finnish: Mispeli;
French: Merlier, Néfle Commune (Fruit), Néflier, Néflier Commun (Tree);
German: Aschperln, Asperl, Deutsche Mispel, Dürgen, Dürrlitzen, Dörrlitzen, Echte Mispel, Hespelein, Hundsärsch, Mispel, Mispelbaum, Mispelche, Nespoli, Nispel;
Greek: Mespilea E Germaniki;
Hungarian: Naspolya;
Iran: Kondos;
Italian: Nespola, Nespolo, Nespolo volgare;
Japanese: Seiyou Karin;
Polish: Nieszpułka zwyczajna
Portuguese: Nêsperas, Nespereira, Nespereira (Tree), Nespereira-Da-Europa;
Russian: Mushmula, Mushmula Obyknovennaia;
Slovenian: Navadna nešplja
Spanish: Níspero (Tree), Níspero Común, Níspero Europeo, Nísperoeuropeo, Níspola (Fruit), Nispolero;
Swedish: Mispel, Tysk mispel;
Turkish: Mumula, Mušmula;
Ukranian: Mushmula.

Origin/Distribution

Medlar is indigenous to southwest Asia and possibly also southeastern Europe – from northern Turkey (some occurrence in Greece and on the Crimea) to the Caucasus and Transcaucasus and the north-eastern part of Iran.

Agroecology

It requires warm summers and mild winters and prefers sunny, dry locations and well-drained, slightly acidic soil from sea level to 2,000 m altitude.

Edible Plant Parts and Uses

Fruits are edible but is harvested and left to blet for a few weeks before the flesh is spooned out and eaten raw. Bletting is to allow the hard fruit to senesce, mellow and soften (become over ripe). Bletted fruit is used to make medlar jelly, marmalade, syrup, candied fruit, mixed jam or baked to make a thick, sauce that goes well with rich meats. Medlar is also consumed with cheese as a dessert. Another speciality is medlar cheese made with the fruit pulp, eggs and butter. Medlar is also used as condiment for making fruit wine. Unripe fruit can be pickled with vinegar and sugar. Medlar fruit is widely consumed in some countries such as Turkey.

Botany

A large shrub or small, deciduous tree growing to 8 m high. The leaves are dark green, large, simple, elliptic-oblong, 8–15 cm long and 3–4 cm wide, apex acute to obtuse, base cuneate, margin entire and pubescent (Plates 1, 2, 3). Flowers, solitary, white, 3–5 cm across excluding the 5 wide spreading persistent sepals, with 5 petals. Fruit reddish-brown, globose to sub-globose pome, 3–4 cm (to 6 cm) across crowned by 5 persistent erect to erecto-patent sepals imparting a hollow appearance to the fruit (Plates 1–2).

Plate 1 Medlar fruit and leaves

Plate 2 Close-up of the stylar fruit end

Plate 3 Side view of medlar fruit

Nutritive/Medicinal Properties

The food value of raw medlar fruit per 100 g edible portion was reported as energy 88 kcal (370 kJ), moisture 74.0 g, protein 0.5 g, total lipid 0.1 g, total carbohydrate 24 g, fibre 1.3 g, ash 0.6 g, Ca 41 mg, Fe 1.2 mg, thiamine 0.06 mg, riboflavin 0.03 mg, niacin 0.2 mg, and β-carotene equivalent 12 µg (Sabry and Rizek 1982).

Studies found ripe medlar fruit to be rich in potassium (7,370 µg/g dry wt), calcium (1,780 µg/g dry wt), phosphorus (1,080 µg/g dry wt), magnesium (661 µg/g dry wt) and sodium (183 µg/g dry wt) (Glew et al. 2003c). During fruit development, Al, Ba, Fe, Mn, P, Sr, and Zn were highest in unripe fruits while the concentrations of K, Ca, Mg and Cu gradually decreased throughout development. The ripe medlar fruit is an important source of nutritionally needed minerals and trace elements, in particular Ca, Cu, Fe, K, Mg, Mn, Na and Zn. Fructose, glucose and sucrose were identified as the principal sugars and their levels varied remarkably during development. The fructose level increased continually through development reaching its maximum of 1,200 mg/100 g fresh weight by 161 DAF (days after fruit drop) while the increase of sucrose reached maximum at 131 DAF and had decreased at 161 DAF (Glew et al. 2003a). After some fluctuations at 69 DAF, glucose level remained high (686 mg/100 g fresh weight) at 161 DAF, when compared with Stage IV (131 DAF). While the level of malic acid increased continually, the ascorbic acid level decreased dramatically through fruit development: both acids reached their maximum and minimum levels at 161 DAF, i.e. 428 and 8.4 mg/100 g fresh weight, respectively. The total amino acid composition also changed in decreasing trend throughout development and remained low at 161 DAF. In the ripe fruit, glutamate and aspartate were the major amino compounds identified. In 1964, two new antibiotic cyclopentoid monoterpenes were isolated and identified from the fruit. These were genipic acid and genipinic acid, its carbomethoxyl derivative (Haciseferogullari et al. 2005). Medlar fruit was found to have moisture content 72.15%, crude oil 4.09%, protein 11.4%, fibre 3.71%, energy 16.5 kcal/g, and ash 1.96%, acidity 0.28%, pH 4.26, water-soluble extract 68.89%, alcohol-soluble 53.35% and ether-soluble 2.41% (dry basis (Hacıseferogullari et al. 2005). Potassium (8,052.91 mg/kg) was present in the highest concentration followed by S, Ca, B and P and traces of Cr, Ti and V were also detected

Fructose, glucose, and sucrose were the major soluble sugars, and citric, malic, and ascorbic acids were the major organic acids determined in medlar fruits (Glew et al. 2003b). Among the fatty acids, palmitic acid (16:0) and stearic acid (18:0), oleic acid (18:1), linoleic acid (18:2n-6) and linolenic acid (18:3n-3) were the major fatty acids Sucrose was highest at 1 week after harvest (WAH) (228.4 mg/100 g fresh weight) and then decreased, remaining very low at 4 WAH (1.4 mg/100 g fresh weight). As for the levels of fructose and glucose, their levels shifted up to 2,230.8 and 845.2 mg/100 g fresh weight at 2 and 3 WAHs, then the levels fell to their lowest concentration. The levels of the three organic acids were high at the beginning, except malic acid level at 2 WAH, all acids levelled off through the latter weeks of post harvest period. In medlar fruit, the levels of saturated palmitic acid (16:0) and stearic acid (18:0), and unsaturated oleic acid (18:1), linoleic acid (18:2n-6) and linolenic acid (18:3n-3) were most abundantly detected throughout medlar ripening (pulp softening and darkening). The level of palmitic and stearic acids as well as the level of linoleic and linolenic acids were the highest at 1 WAH and then suddenly decreased as the medlar softened and the pulp turned slightly (2 WAH) and fully (3 WAH) brown through 2 and 4 WAHs. In addition to these prominent fatty acids, a remarkable decrease was also obtained in the content of some other fatty acids (C10-15, C16:1, C20-24).

During the early stages of medlar fruit development, polyphenol oxidase (PPO) activity and the level of ascorbic acid gradually decreased, whereas in the post-ripening stage PPO activity increased (Aydin and Kadioglu 2001). Ascorbic acid level increased in the pre-ripening stage followed by a decrease in the post-ripening period. Similarly, POD (polyphenol peroxidase) activity decreased during development of the fruit and increased in the pre-ripening and post-ripening stages. The level of glucose gradually increased during fruit development and ripening. Contents of pentoses, hexoses and soluble proteins decreased during fruit development, but increased in the stage of ripening. These observations suggested that the increase in PPO and POD activities as well as in sugar and protein contents has an important role in reducing the astringent taste of the medlar fruits.

Among the five mono- and di-phenolic substrates examined during ripening, p-hydroxyphenylpropionic acid, L-3,4-dihydroxyphenylalanine, catechol, 4-methylcatechol and tyrosine, 4-methylcatechol was selected as the best polyphenoloxidase substrate for all ripening stages (Ayaz et al. 2002, 2008). A range of pH 3.0–9.0 was also tested and the highest enzyme activity was at pH 7.0 throughout ripening. The studies concluded that as medlar fruit ripen there was no significant changes in the optimum values of polyphenoloxidases, although their kinetic parameters change. As the fruit ripening progressed through ripe to over-ripe, in contrary to polyphenoloxidase activity, there was an apparent gradual decrease in total fruit phenolic concentrations.

Antioxidant Activity

Of the various medlar plant parts, stem bark extract (aqueous and methanol) showed best activity in 1,1-diphenyl-2-picryl hydrazyl (DPPH) radical scavenging activity with $IC_{50}=10.7$ and 11.4 μg/ml, respectively (Nabavi et al. 2011). All extracts (leaf, fruit, stem bark) showed weak Fe^{2+} chelating ability. Methanol extract of fruit had better activity in nitric oxide scavenging model than others ($IC_{50}=247\ 2$ μg/ml). The leaves and bark extracts showed good reducing power than fruit extract. In reducing powers, there were no significant differences among the stem bark and leaves extracts that were comparable with vitamin C. Extracts exhibited good antioxidant activity in the ferric thiocyanate (FTC) method. They manifested almost the same pattern of activity as vitamin C and butylated hydroxyanisole (BHA) at different incubation times (until 72nd hour) but stem bark extract showed higher peroxidation inhibition than vitamin C and BHA at the 96th hour. The extracts were capable of scavenging H_2O_2 in a concentration-dependent manner. Leaves methanol extract showed good activity that was comparable with quercetin. Bark and leaf extracts had higher total phenolic and flavonoid contents than fruit.

Eight phenolic acids (protocatechuic, 4-hydroxybenzoic, syringic, 3-hydroxybenzoic, caffeic, salicylic, 4-coumaric and sinapic) were determined in medlar fruit (Gruz et al. 2011). The concentrations of phenolic acids were found to decrease as the fruit ripening progressed, except for insoluble ester-bound phenolics, which increased at the early stages of maturity and decreased only during the ripe to over-ripe stage of maturity. The DPPH scavenging activity also decreased during fruit maturation, suggesting a decrease of natural antioxidants in fruit. A strong correlation between TPC and antioxidant capacity was found as measured by the DPPH method.

The lyophilized medlar fruit extract was found to have antioxidant and radical scavenging activity as evaluated by the following assays all expressed in Trolox equivalent : 1, 1,-diphenyl-2-picryl-hydrazyl radicals (DPPH*) scavenging 0.62 TE; N,N-dimethyl-p-phenylenediamine (PMPD*) scavenging 0.81 TE ; superoxide anion radical (O2*) scavenging 1.41 TE; hydrogen peroxide, ferric ions reducing 0.69TE, cupric ions reducing ability 0.43TE, FRAP (Ferric_ reducing antioxidant power) reducing ability 0.36TE and ferrous ion chelating activity 2.76 TE. (Gülçin et al. 2011).Total phenolic content was found as 25.08 mg GAE and total flavonoids was found as 2.39 mg quercetin equivalent. Caffeic acid 4.9 mg/kg and p-courmaric 2.4 mg/kg were the predominant phenolic compounds. Catechol, p-hyrdoxybenzoic acid, vanillin, syringic acid and gallic acid were not detected. a-tocopherol, 13.4 mg and ascorbic acid 184.6 mg were detected. The following compounds were also detected (in mg/kg) ferulic acid 2.4 mg, quercetin 2.4 mg, pyrogallol 3.6 mg, and ellagic acid 0.2 mg.

Traditional Medicinal Uses

In traditional medicine, the pulp of the fruit is considered laxative, the leaves are astringent and the seed is lithontripic. In Iran, medlar has been used for treatment of diseases associated with high blood pressure, heart tonic, heart rate. The leaf extract was found useful for mouth and throat infection, fruit as a relaxant and purgative for diarrhoea and the seed for expelling bladder stone. The medlar fruit is also used as treatment for constipation, as diuretic, and to rid the kidney and bladder of stones (Haciseferogullari et al. 2005; Glew et al. 2003b).

Other Uses

Unripe fruit are used as astringent drug and for tanning.

Comments

Mespilus is closely related to the genera *Amelanchier* and *Crategus* within the subfamily Spiraeoideae.

Selected References

Ayaz FA, Demir O, Torun H, Kolcuoglu Y, Colak A (2008) Characterization of polyphenoloxidase (PPP) and total phenolic contents in medlar (*Mespilus germanica* L.) fruit during ripening and over ripening. Food Chem 106(1):291–298

Ayaz FA, Gle RH, Huang HS, Chuang LT, VanderJagt DJ, Strnad M (2002) Evolution of fatty acids in medlar (*Mespilus germanica* L.) mesocarp at different stages of ripening. Grasas y Aceites 53(3):352–356

Aydin N, Kadioglu A (2001) Changes in the chemical composition, polyphenol oxidase and peroxidase activities during fruit development and ripening of medlar fruits (*Mespilus germanica* L.). Bulg. J Plant Physiol 27(3–4):85–92

Bailey LH (1949) Manual of cultivated plants most commonly grown in the continental United States and Canada, (Revised Edition). The Macmillan Co., New York, 1116pp

Chiej R (1984) Encyclopaedia of medicinal plants. MacDonald, London, 447pp

Glew RH, Ayaz FA, Sanz C, Vanderjagt DJ, Huang HS, Chuang LT, Strnad M (2003a) Changes in sugars, organic acids and amino acids in medlar (*Mespilus germanica* L.) during fruit development and maturation. Food Chem 83(3):363–369

Glew RH, Ayaz FA, Sanz C, Vanderjagt DJ, Huang HS, Chuang LT, Strnad M (2003b) Effect of postharvest period on sugars, organic acids and fatty acids composition in commercially sold medlar (*Mespilus germanica* Dutch) fruit. Eur Food Res Technol 216(5): 390–394

Glew RH, Ayaz FA, Vanderjagt DJ, Millson M, Dris R, Niskanen R (2003c) Mineral composition of medlar (*Mespilus germanica*) fruit at different stages of maturity. J Food Qual 26(5):441–447

Gruz J, Ayaz FA, Torun H, Strnad M (2011) Phenolic acid content and radical scavenging activity of extracts from medlar (*Mespilus germanica* L.) fruit at different stages of ripening. Food Chem 124(1): 271–277

Gülçin I, Topal F, Sarikaya SBO, Bursal E, Bilsel G, Gören A (2011) Polyphenol contents and antioxidant properties of medlar (*Mespilus germanica* L.). Rec Nat Prod 5(3):158–175

Hacıseferogullari H, Ozcan M, Sonmete MH, Ozbek O (2005) Some physical and chemical parameters of wild medlar (*Mespilus germanica*) fruit grown in Turkey. J Food Eng 69(1):1–7

Nabavi SF, Nabavi SM, Ebrahimzadeh MA, Asgarirad H (2011) The antioxidant activity of wild medlar (*Mespilus germanica* L.) fruit, stem bark and leaf. Afr J Biotechnol 10(2):283–289

Sabry ZI, Rizek RL (1982) Food composition tables for the near east. FAO Food and Nutrition Paper. FAO Rome, 275pp

Tabatabaei NS, Mazandaranee M (2008) Autocology and ethnopharmacology of *Mespilus germanica* L. in the North of Iran. AIP Conf Proc 971:248–251

Whiteman K (1998) The new guide to fruit. Anness Publishing Limited, London, 128pp

Prunus armeniaca

Scientific Name

Prunus armeniaca L.

Synonyms

Amygdalus armeniaca (L.) Dumort., *Armeniaca armeniaca* Huth nom. illeg. tautonym, *Armeniaca bericoccia* Delarbre, nom. illeg., *Armeniaca communis* Besser, *Armeniaca epirotica* G. Gaertn., B. Mey., & Scherb., *Armeniaca macrocarpa* Poit. & Turp. ex Duhamel, *Armeniaca vulgaris* Lam., *Prunus armeniaca* var. *typica* Maxim, *Prunus armeniaca* L. var. *vulgaris* (Lam.) Zabel, *Prunus tiliaefolia* Salisb. nom. illeg. superfl.

Family

Rosaceae

Common/English Names

Apricot, Chinese Almond, Common Apricot, Siberian Apricot.

Vernacular Names

Albanian: Kajsi;
Argentina: Damasco;
Armenian: Tziran;
Bosnian: Kajsija;
Brazil: Abricó, Damasco, Damasqueiro (Portuguese);
Catalan: Albercoc;
Chile: Damasco;
Chinese: Ku Xing Ren, Xing, Xing Xin Shu;
Croatian: Kajsija;
Czech: Meruňka, Meruňka Obecná;
Danish: Abrikos, Almindelig Abrikos;
Dutch: Abrikoos;
Eastonian: Aprikoos, Harilik Aprikoosipuu;
Esperanto: Abrikoto;
Finnish: Aprikoosi;
French: Abricotier, Abricotier Commun;
German: Aprikose, Aprikosenbaum, Marille;
Greek: Βερίκοκο;
Haitian: Zabriko;
Hungarian: Kajszibarack, Sárgabarack;
Icelandic: Apríkósa;
Iran: Zard-ālū;
Irish: Aibreog;
Italian: Abricocco Comune, Albicocco, Armenillo;
Japanese: Anzu;
Korean: Sal-Goo, Hoeryngbaeksalkunamu;
Kurdish: Mijmij, Qeysî, Zerdelî, Hêrûg;
Latvian: Aprikoze;
Lithuanian: Abrikosas;
Maltese: Berquqa;
Norwegian: Aprikos;
Polish: Morela;
Portuguese: Damasco, Damasci Italbrac, Damasqueiro;

Roman: Kajsija, Zerdelica;
Romanian: Caisă;
Russian: Abrikos Obyknovennyj;
Slovenia: Marelica, Marhuľa Obyčajná;
Spanish: Albaricoque, Albaricoquero, Albercoquer, Chabacano, Chabacano Italbrac Mexico, Damasco, Damasquino, Damasquillo;
Swedish: Aprikos;
Turkish: Kayisi.

Origin/Distribution

The native range of apricot is somewhat unclear due to its extensive prehistoric cultivation, but has been regarded to be the northern, north-western and north-eastern provinces of China (Qinghai, Gansu, Shaanxi, Hebei, Liaoning) and possibly also Korea and Japan. Domestic cultivation in China dates back over 3,000 years ago. It spread to Asia Minor and was introduced to Europe through Greece and Italy by the Romans. Apricot was introduced into North America by English travellers and by Spanish missionaries into California. Apricot is extensively cultivated in Eurasia and America. Secondary centres of diversity with locally adapted races can be found in Middle Asia, Caucasus, Iran, and less so in southern Europe and southern USA.

Agroecology

In its native range, apricot is found in sparse forests on mountain slopes, slopes, gullies, from 700–3,000 m altitude. Apricot is a cool temperate climate species although it can grow in a Mediterranean climate. It is winter-hardy enough to survive temperatures down to −30°C, they prefer stable winter temperature and are sensitive to winter temperature fluctuations. Spring frost is the limiting factor. Apricots require chilling requirement to break dormancy of 600–1270 chill units and a heat requirement for flowering of 4078 and 5879 growing degree hours (GDH) (Ruiz et al. 2007). Apricots prefers deep, fertile, well-drained, well-aerated soils with a pH of 6.0–7.0 It grows on loessial, loamy and marl or sandy soils, and are intolerant of saline and waterlogged soils. During the dry months irrigation is required to sustain good yields. Watering is carried out before and after blossoming, 10–15 days before the beginning of fruit maturity and after harvest. Apricot is responsive to potash fertilisers.

Edible Plant Parts and Uses

Ripe apricots are delicious when eaten fresh or slightly chilled on their own or in fruit salads. Whole or halved apricots coated lightly with honey on skewers and grilled are fabulous. The fruits can be poached in water or fruit juice to which clove or cinnamon can be added to enhance the flavour. The fruit can be frozen, preserved and canned. Apricots can be made into excellent jams, jellies, puree, nectar, juice, drinks and sauces. Apricot jams are fabulous with toast. Apricot nectar or juice is an excellent and nutritious drink and is available in cans or bottles. Puree can be made into sauces which are excellent with cold cut or in meat sandwiches. Dried apricots are nutritious and delicious and are excellent snacks. Apricots fruits both fresh or dried or sauces can be used in wide array of desserts – pancakes, cakes, bread, muffins, croissant, pies, crumble, custard tart, strudel, flan, ice-cream, smoothes, milk-shakes, apricot fruit bars, apricot coconut candy balls and apricot cream. Apricots are excellent with low fat cottage cheese.

Apricot seeds especially those grown in central Asia and around the Mediterranean area are so sweet that they may be substituted for almonds. In Turkey, the sweet kernels of some cultivars are eaten as roasted and salted tidbits and in baked products. The Italian liqueur Amaretto and Amaretti Biscotti are flavoured with extract of apricot kernels rather than almonds. In India, apricot seed oil is used for cooking.

Botany

A small- to medium-sized tree with a dense, spreading, round or elongated oblong canopy 4–8(–12) m tall with grayish brown, longitudinally splitting bark. young branchlets are reddish

Plate 1 Apricot inflorescence

Plate 2 Close-up of apricot flower

Plate 3 Immature apricot fruits and leaves

brown and lenticellate. Winter buds are purplish red, ovoid, glabrous or puberulous. Leaves are alternate and borne on 2–3.5 cm petioles, glabrous or pubescent, basally usually with 1–6 nectaries (Plates 3, 4, 5). Lamina is broadly ovate to subcordate, 5–9 × 4–8 cm, both surfaces glabrous, green, margin crenate, apex acute to shortly acuminate and base cuneate. Flowers solitary or occasionally paired, opening before leaves, 2–4.5 cm across, bisexual; sepals 5 purplish green, ovate to ovate-oblong, reflexed after anthesis, petals 5 white, pink, or tinged with red, orbicular to obovate; stamens 20–100, slightly shorter than petals with white filaments and yellow anthers; ovary pubescent with style slightly longer to nearly as long as stamens (Plates 1–2). Fruit a drupe, globose, ovoid, or rarely obovoid, 2.5–4 cm in diameter, smooth at maturity, pubescent when young, pale green (Plate 4) turning yellow or orange often flushed with red with a fleshy, succulent, white, yellow or orange-coloured outer layer (mesocarp) surrounding a hard, globose, ovoid, or ellipsoid, compressed laterally stone (endocarp) ridged along suture containing the seed (Plates 3, 4, 5, 6). Seeds are flat, obovate with dense light painted skin, bitter or sweet.

Nutritive/Medicinal Properties

Food value of raw, apricot fruit (refuse 7% pit) per 100 g edible portion was reported as follows (USDA 2011): water 86.35 g, energy 48 kcal

Plate 4 Fruit laden apricot tree

Plate 5 Ripening apricot fruits

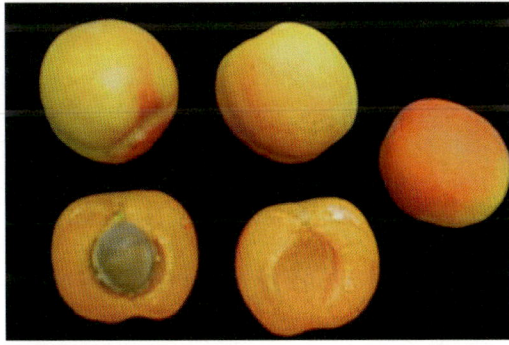

Plate 6 Ripe apricot fruit and seed

(201 kJ), protein 1.40 g, total lipid (fat) 0.39 g, ash 0.75 g, carbohydrate 11.12 g; fibre (total dietary) 2.0 g, total sugars 9.24 g, sucrose 5.87 g, glucose 2.37 g, fructose 0.94 g, maltose 0.06 g; minerals – calcium 13 mg, iron 0.39 mg, magnesium 10 mg, phosphorus 23 mg, potassium 259 mg, sodium 1 mg, zinc 0.20 mg, copper 0.078 mg, manganese 0.077 mg, selenium 0.1 μg; vitamins – vitamin C 10 mg, thiamine 0.030 mg, riboflavin 0.040 mg, niacin 0.600 mg, pantothenic acid 0.240 mg, vitamin B-6 0.054 mg, folate (total) 9 μg, total choline 2.8 mg, vitamin A 1926 IU (96 μg RAE), vitamin E (α-tocopherol) 0.89 mg, vitamin K (phylloquinone) 3.3 μg, lipids – fatty acids (total saturated) 0.017 g, 16:0 (palmitic acid) 0.014 g, 18:0 (stearic acid) 0.003 g; fatty acids (total monounsaturated) 0.170 g, 18:1 undifferentiated (oleic acid) 0.170 g; fatty acids (total polyunsaturated) 0.077 g, 18:2 undifferentiated (linoleic acid) 0.077 g, phytosterols 18 mg, β-carotene 1094 μg, α-carotene 19 μg, β-cryptoxanthin 104 μg, lutein + zeaxanthin 89 μg; amino acids – tryptophan 0.015 g, threonine 0.047 g, isoleucine 0.041 g, leucine 0.077 g, lysine 0.097 g, methionine 0.006 g, cystine 0.003 g, phenylalanine 0.052 g, tyrosine 0.029 g, valine 0.047 g, arginine 0.045 g, histidine 0.027 g, alanine 0.068 g, aspartic acid 0.314 g, glutamic acid 0.157 g, glycine 0.040 g, proline 0.101 g and serine 0.083 g.

Apricot is extremely rich in pro-vitamin A carotenes and vitamin A and phenolic compounds. Among the 37 apricot varieties, the total carotenoid content ranged from 1,512 to 16,500 μg/100 of edible portion, with β-carotene as the main pigment followed by β-cryptoxanthin and γ-carotene (Ruiz et al. 2005b). The carotenoid content was correlated with the color measurements, and the hue angle in both flesh and peel was the parameter with the best correlation ($R^2 = 0.92$ and 0.84, respectively). Four phenolic compound groups, procyanidins, hydroxycinnamic acid derivatives, flavonols, and anthocyanins, were identified in the fruit of 37 apricot varieties (Ruiz et al. 2005a). Chlorogenic and neochlorogenic acids, procyanidins B1, B2, and B4, and some procyanidin trimers, quercetin 3-rutinoside, kaempferol 3-rhamnosyl-hexoside and quercetin 3-acetyl-hexoside, cyanidin 3-rutinoside, and 3-glucoside, were detected and quantified in the skin and flesh of the different cultivars. The total phenolics content, ranged between 32.6 and 160.0 mg/100 g of edible tissue. No correlation between the flesh colour and the phenolic content of the different cultivars was observed.

The major fatty acids in apricot kernel oil were oleic, linoleic and palmitic (El-Aal et al. 1986). Chloroform-methanol extracts consisted mainly of neutral lipids in which triglycerides were predominant components. The triglycerides consisted of six types of glycerides. Glycolipids and phospholipids were the minor fractions of the total lipids and their major constituents were acylsterylglycosides (62·3%) and phosphatidyl choline (72·2%), respectively. Evaluation of the crude apricot kernel oil added to different types of biscuits and cake revealed that it has excellent properties and is comparable with corn oil at the same level. In another study, the total oil contents of apricot kernels in Turkish apricot cultivars ranged from 40.23% to 53.19% (Turan et al. 2007). Oleic acid contributed 70.83% to the total fatty acids, followed by linoleic (21.96%), palmitic (4.92%), and stearic (1.21%) acids. The sn-2 fatty acid position is mainly occupied with oleic acid (63.54%), linoleic acid (35.0%), and palmitic acid (0.96%). Eight triacyglycerol species were identified: LLL, OLL, PLL, OOL+POL, OOO+POO, and SOO (where P, palmitoyl; S, stearoyl; O, oleoyl; and L, linoleoyl), among which mainly OOO+POO contributed to 48.64% of the total, followed by OOL+POL at 32.63% and OLL at 14.33%. Four tocopherol and six phytosterol isomers were identified and quantified; among these, γ-tocopherol (475.11 mg/kg of oil) and β-sitosterol (273.67 mg/100 g of oil) were predominant.

Gezer et al. (2011) reported apricot kernel of 5 Turkish cultivars to have moisture content 2.18–3.25%, 28.26–42.48% crude oil, 96.75–97.82% dry matter, 15.7–18.3% crude protein, crude fibre 28.26–42.48% and ash 15.3–17.1%. The mineral content of the kernel comprised 11,090–9,190 ppm K, 1,344.5–2,909.6 ppm Ca, 1,323.7–1,960 ppm Mg, 4967.5–9387.2 P, 964.9–1347.7 ppm Na, 97.9–138.8 ppm B, 55.0–87.3 ppm Al, 51.0–60.7 Zn, 30.2–45.2 ppm Fe, 10.3–22.0 ppm Cu, 6.5–14.4 ppm Mn, 4.3–7.7 ppm Cr and 0.6–4.5 ppm Ni.

Three isomeric A-type proanthocyanidins were isolated from the root of *Prunus armeniaca*. They were identified as ent-epiafzelechin-($4\alpha \rightarrow 8; 2\alpha \rightarrow O \rightarrow 7$)-epiafzelechin (mahuannin A), ent-epiafzelechin-($4\alpha \rightarrow 8; 2\alpha \rightarrow O \rightarrow 7$)-(+)-afzelechin and ent-epiafzelechin-($4\alpha \rightarrow 8; 2\alpha \rightarrow O \rightarrow 7$)-(−)-afzelechin, (Prasad et al. 1998; Rawat et al. 1999) and a new aromatic glycoside, 4-O-glycosyloxy-2-hydroxy-6-methoxyacetophenone (Prasad 1999).

Antioxidant Activity

Considerable variation was observed in 29 apricot cultivars in the total phenol content (0.3–7.4 mg gallic acid equivalent/g FW(fresh weight)) and total antioxidant capacity (0.026–1.858 mg ascorbic acid equivalent/g FW), with the American origin cultivars Robada and NJA(2) and the new cultivar Nike exhibiting the greatest values (Drogoudi et al. 2008). The cultivar Tomcot and hybrid 467/99 had the highest content of total carotene (37.8 μg β-carotene equivalent/g FW), which was up to four times greater as compared with the rest of studied genotypes. The dominant sugar in fruit tissue was sucrose, followed second by glucose and third by sorbitol and fructose-inositol. The new cultivars Nike, Niobe, and Neraida contained relatively higher contents of sucrose and total sugars, while Ninfa and P. Tirynthos contained relatively higher contents of K, Ca, and Mg. Correlation analysis suggested that late-harvesting cultivars/hybrids had greater fruit developmental times ($R^2=0.817$) and contained higher sugar ($R^2=0.704$) and less Mg contents ($R^2=-0.742$) in fruit tissue. The total antioxidant capacity was better correlated with the total phenol content ($R^2=0.954$) as compared with the total carotenoid content ($R^2=0.482$). Weak correlations were found between the fruit skin colour and the antioxidant contents in flesh tissue.

Studies demonstrated that 1-methylcyclopropene (1-MCP) treated fruits exhibited higher superoxide dismutase (SOD) activity, whereas unspecific peroxidase (POX) activity was significantly higher only after 21 days at 2°C (Egea et al. 2010). Treated fruits also exhibited better retention of ascorbate and carotenoids and higher TEAC (trolox equivalent antioxidant capacity) during storage. In accordance with these observations, lower ion leakage values were detected in 1-MCP-treated apricots. The results suggested that 1-MCP conferred a greater resistance to oxidative stress. This, along with the reduction in ethylene production, could contribute to the increase in commercial life and nutritional value observed in 1-MCP-treated apricots.

In contrast to extracts of the bitter apricot kernels, both the water and methanol extracts of sweet kernels had antioxidant potential (Yiğit et al. 2009). The highest percent inhibition of lipid peroxidation (69%) and total phenolic content (7.9 μg/ml) were detected in the methanol extract of sweet kernels (Hasanbey) and in the water extract of the same cultivar, respectively. Roasting was found to affect he antioxidant properties of peeled, defatted and roasted apricot kernel flours (Durmaz and Alpaslan 2007). Contrary to browning degree, radical scavenging power (RSP), reducing power (RP), and total phenolic content (TPC) did not increase linearly but showed a maximum for 10 minutes of roasting. Roasting reduced the anti-lipid peroxidative activity (ALPA) values, thus unroasted sample showed the highest ALPA value. RSP, RP and TPC measurements of all samples, were in high correlation ($R^2=0.92$). Primary metabolites (sugars, organic acids) and secondary metabolites (phenolics) were quantified in the fruit of 13 apricot cultivars (Schmitzer et al. 2010). Total sugars ranged from 59.2 to 212.5 g/kg FW and total organic acids from 4.2 to 20.8 g/kg FW. Four hydroxycinnamic acids and three flavonols were quantified; their content was significantly higher in skin compared to pulp. Similarly, antioxidative potential was significantly higher in skin and ranged from 125.4 to 726.5 mg ascorbic acid equivalents/kg FW. A positive correlation between total phenolic content and antioxidant potential was determined.

Cardioprotective Activity

Studies demonstrated in-vivo cardio-protective activity of apricot-feeding related to its antioxidant phenolic contents in rats subjected to myocardial ischemia-reperfusion (Parlakpinar et al. 2009) Infarct sizes were found significantly decreased in 10% (55.0%) and 20% (57.0%) apricot-fed groups compared to control group (68.7%). Light and electron microscopic evaluations of the heart also demonstrated similar beneficial effects on ischemia-reperfusion injury in apricot-fed both groups. Total phenolic contents, DPPH radical scavenging and ferric-reducing power as in vitro antioxidant capacities of rat chows were significantly increased after supplementation with apricot for each ratio. Cu, Zn Superoxide dismutase (Cu, Zn SOD) and catalase (CAT) activities were increased, and lipid peroxidation was decreased significantly in the hearts of 20% apricot-fed group after I/R.

Gastroprotective Activity

Studies showed that apricot and β-carotene had potent protective effect against methotrexate-induced intestinal oxidative damage in rats and ameliorated MTX (methotrexate)-induced intestine damage at biochemical and histological levels

(Vardi et al. 2008). Single or combined application of apricot and β-carotene ameliorated all of these hazardous effects in antioxidant system in MTX-treated groups. The hazardous effects observed In the MTX group included fusion and shortening in the villus, epithelial desquamation, crypt loss, inflammatory cell infiltration in the lamina propria, goblet cell depletion and microvillar damage were observed in the small intestine. Parallel to histological results, malondialdehyde (MDA) content and myeloperoxidase (MPO) activity were found to be increased, whereas superoxide dismutase (SOD), catalase (CAT), glutathione peroxidase (GP-x) activities and glutathione (GSH) content were decreased in the MTX group.

Hepatoprotective Activity

Studies showed that apricot feeding had beneficial effects on CCl4-induced liver steatosis and damage probably due to its antioxidant nutrient (β-carotene and vitamin) contents and high radical-scavenging capacity (Ozturk et al. 2009). Dietary intake of apricot could reduce the risk of liver steatosis and damage caused by free radicals. In the CCl4 group, vacuolated hepatocytes and hepatic necrosis were seen, especially in the centrilobular area. Hepatocytes showed an oedematous cytoplasmic matrix, large lipid globules and degenerated organelles. The area of liver injury was found significantly decreased with apricot feeding. Malondialdehyde and total glutathione levels and catalase, superoxide dismutase and glutathione peroxidase activities were significantly changed in the carbon tetrachloride group and indicated increased oxidative stress. Apricot feeding decreased this oxidative stress and ameliorated histological damage.

Antinociceptive and Antiinflammatory Activities

The intramuscular injection of amygdalin significantly reduced the formalin-induced tonic pain in both early (the initial 10 minutes after formalin injection) and late phases (10–30 minutes following the initial formalin injection) (Hwang et al. 2008). During the late phase, amygdalin did reduce the formalin-induced pain in a dose-dependent manner in a dose range less than 1 mg/kg. Molecular analysis targeting c-Fos and inflammatory cytokines such as tumour necrosis factor-α (TNF-α) and interleukin-1 β (IL-1β) also showed a significant effect of amygdalin, which matched the results of the behavioural pain analysis. These results suggested amygdalin to be effective at alleviating inflammatory pain and that it can be used as an analgesic with antinociceptive and antiinflammatory activities.

Antimicrobial Activity

Flavonoid glycosides, 4′,5,7-trihydroxy flavone-7-O-[β-D-mannopyranosyl $(1''' \rightarrow 2'')$]-β-D-allopyranoside (1) and 3,4′,5,7-tetrahydroxy-3′,5′-di-methoxy flavone 3-O-[α-L-rhamnopyranosyl $(1''' \rightarrow 6'')$]-β-D-galactopyranoside (2), were isolated from the butanolic fraction of the fruits (Rashid et al. 2007). The butanolic extract exhibited antibacterial activity against both Gram positive and Gram negative bacteria. The butanolic extract was more effective against growth of Gram-positive bacteria (MIC values 31.25–250 μg/ml) with the highest activity exhibited against *Micrococcus luteus* (MIC 31.25 μg/ml) This was followed by *Bacillus subtilis* and *Corynebacterium diptheriae* both with MIC 62.5 μg/ml. The MICs for *Staphylococcus epidermis* and *Mycobacterium smegmatis* were 125 μg/ml. The MICs for *Staphylococcus aureus, Streptococcus faecalis, Streptococcus pyogenes* and *Mycobacterium fortuitum* were 250 μg/ml. The MIC for Methicillin resistant *Staphylococcus aureus* (MRSA) was 78.125 μg/ml. The MICs of the extract for the Gram negative bacteria range from 125 μg/ml for the *Salmonella typhi, Salmonella paratyphi* A, *Salmonella paratyphi* B, *Proteus mirabilis, Proteus vulgaris* and *Shigella dysenteriae* to 250 μg/ml for *Enterobacter aerogenes*. The lowest activity was against the enteropathogenic *Escherichia coli* with an MIC of 500 μg/ml.

In another study, the most effective antibacterial activity was observed in the methanol and water extracts of bitter apricot kernels and in the methanol extract of sweet kernels against the

Gram-positive bacteria *Staphylococcus aureus* (Yiğit et al. 2009). Additionally, the methanol extracts of the bitter kernels were very potent against the Gram-negative bacteria *Escherichia coli* (0.312 mg/ml MIC value). Significant anticandida activity was also observed with the methanol extract of bitter apricot kernels against *Candida albicans*, 0.625 mg/ml MIC value.

Apricot kernel extract (AKE) effectively suppressed the production of androstenone which was generated by the metabolism of a skin-resident microorganism, *Corynebacterium xerosis* (Someya et al. 2006). Androstenone (5α-androst-16-en-3-one) is a key compound in body malodors, and that female subjects were more sensitive than male subjects to androstenone. (R)-Prunasin and (S)-prunasin, which were nitrile compounds, were isolated and identified as the active constituents in AKE and they strongly suppressed the bacterial metabolism. Amygdalin was not included in the active fraction and it did not have any effect on suppressing androstenone generation.

Testicular Protective Activity

Apricot was found to ameliorate alcohol induced testicular damage in rats (Kurus et al. 2009). Rats given ethanol had severe histopathological changes in seminiferous tubules and germ cells as well as tubular degeneration and atrophy. Sertoli and Leydig cell counts in the interstitial tissue were decreased. Biochemical parameters revealed tissue oxidative stress. Rats given ethanol and apricot diet for 3 months produced similar alterations but to a lesser extent. Rats in the control group, and those fed with apricot diet for 3 months; 6 months and those fed apricot diet for 3 months, and then ethanol + apricot diet for 3 months had no histopathological alterations.

Traditional Medicinal Uses

Apricot fruit is antipyretic, antiseptic, emetic and ophthalmic. The salted fruit is antiinflammatory and antiseptic. It is used medicinally in Vietnam in the treatment of respiratory and digestive diseases. In India, apricot is used in Unani medicine as an anti-diarrhoeic, anti pyretic, emetic, anthelmintic, in liver diseases, piles (Prasad 1999). In Europe, apricots were long considered an aphrodisiac. The seed is analgesic, anthelmintic, antiasthmatic, antispasmodic, antitussive, demulcent, emollient, expectorant, pectoral, sedative and vulnerary In Korea, apricot seed is used for cough, phlegm and common cold. The seed oil lubricates the intestine and produces laxative action. The seed contains amydalin or 'laetrile', a substance that has also been called vitamin B17. This has been spuriously claimed to have a positive effect in the treatment of cancer, but there is refuted by scientific evidence to have no such benefits. The pure substance is almost harmless, but on hydrolysis it yields hydrocyanic acid, a very rapidly acting poison – it should thus be treated with caution. The flowers are tonic, promoting fecundity in women. The inner bark and/or the root are used for treating poisoning. A root decoction is also used to soothe inflamed and irritated skin conditions and also used in the treatment of asthma, coughs, acute or chronic bronchitis and constipation.

Other Uses

Apricot is also used as an ornamental plant. Its flowers provide good foraging for bees producing good honey. Apricots kernels are used in the production of oils, benzaldehyde, cosmetics, active carbon and aroma perfumes. Apricot seed provides an edible, semi-drying oil that has been used for lighting. The oil has a softening and moisturising effect and is used in perfumery and cosmetics, and also in pharmaceuticals. Green and dark grey-green dyes can be obtained from the leaves and fruit respectively. The hard, durable wood has been used for making agricultural implements.

Comments

Hybrids between plums and apricots have been produced recently which are said to be finer fruits than either parent. A "plumcot" is 50% plum, 50% apricot; an "aprium" is 75% apricot, 25% plum; and the most popular hybrid, the "pluot" is 75% plum, 25% apricot, "peachcot" between peach and apricot

On 4 November 2011, the Food Standards of Australia New Zealand warned consumers against eating raw apricot kernels following the discovery of high levels of a naturally occurring toxin in some available products that can release hydrocyanic acid in the gut.

Selected References

Chevallier A (1996) The encyclopedia of medicinal plants. Dorling Kindersley, London, 336pp

Drogoudi PD, Vemmos S, Pantelidis G, Petri E, Tzoutzoukou C, Karayiannis I (2008) Physical characters and antioxidant, sugar, and mineral nutrient contents in fruit from 29 apricot (*Prunus armeniaca* L.) cultivars and hybrids. J Agric Food Chem 56(22): 10754–10760

Durmaz G, Alpaslan M (2007) Antioxidant properties of roasted apricot (*Prunus armeniaca* L.) kernel. Food Chem 100(3):1177–1181

Egea I, Flores FB, Martínez-Madrid MC, Romojaro F, Sánchez-Bel P (2010) 1-Methylcyclopropene affects the antioxidant system of apricots (*Prunus armeniaca* L. cv. Búlida) during storage at low temperature. J Sci Food Agric 90(4):549–555

El-Aal MHA, Khalil MKM, Rahma EH (1986) Apricot kernel oil: characterization, chemical composition and utilization in some baked products. Food Chem 19(4):287–298

Gezer I, Haciseferogullari H, Arslan D, Özcan MM, Arslan D, Asma BM, Ünver A (2011) Physicochemical properties of apricot (*Prunus armeniaca* L.) kernels. Southwest J Hort Biol Environ 2(1):1–13

Gönülsen N (1996) *Prunus* germplasm in Turkey (pp 56–62). In: IPGRI Working Group on *Prunus* (ed), Report on the Working Group on *Prunus*: 5. Meeting. FAO, Rome 70pp

Hwang H-J, Kim P, Kim C-J, Lee H-J, Shim I, Yin CS, Yang Y, Hahm D-H (2008) Antinociceptive effect of amygdalin isolated from *Prunus armeniaca* on formalin-induced pain in rats. Biol Pharm Bull 31(8):1559–1564

Kurus M, Ugras M, Ates B, Otlu A (2009) Apricot ameliorates alcohol induced testicular damage in rat model. Food Chem Toxicol 47(10):2666–2672

Lu L-T, Bartholomew B (2003) Armeniaca Scopoli. In: Wu ZY, Raven PH, Hong DY (eds) Flora of China, vol 9, Pittosporaceae through Connaraceae. Science Press/Missouri Botanical Garden Press, St. Louis/Beijing

Natural Products Research Institute (1998) Medicinal plants in the Republic of Korea. Seoul National University, WHO Regional Publications, Western Pacific Series No 21. 316pp

Ozturk F, Gul M, Ates B, Ozturk IC, Cetin A, Vardi N, Otlu A, Yilmaz I (2009) Protective effect of apricot (*Prunus armeniaca* L.) on hepatic steatosis and damage induced by carbon tetrachloride in Wistar rats. Br J Nutr 102(12):1767–1775

Parlakpinar H, Olmez E, Acet A, Ozturk F, Tasdemir S, Ates B, Gul M, Otlu A (2009) Beneficial effects of apricot-feeding on myocardial ischemia-reperfusion injury in rats. Food Chem Toxicol 47(4):802–808

Porcher MH et al (1995–2020) Searchable World Wide Web Multilingual Multiscript Plant Name Database. Published by The University of Melbourne. Australia. http://www.plantnames.unimelb.edu.au/Sorting/Frontpage.html

Prasad D (1999) A new aromatic glycoside from the roots of *Prunus armeniaca*. Fitoterapia 70(1):266–268

Prasad D, Joshi RK, Pant G, Rawat MSM, Inoue K, Shingu T, He ZD (1998) An A-type proanthocyanidin from *Prunus armeniaca*. J Nat Prod 61(9):1123–1125. http://pubs.acs.org/doi/abs/10.1021/np970383n - np970383nAF2

Rashid F, Ahmed R, Mahmood A, Ahmad Z, Bibi N, Shahana Urooj Kazmi SU (2007) Flavonoid glycosides from *Prunus armeniaca* and the antibacterial activity of a crude extract. Arch Pharmacal Res 30(8):932–937

Rawat MSM, Prasad D, Joshi RK, Pant G (1999) Proanthocyanidins from *Prunus armeniaca* roots. Phytochemistry 50(2):321–324

Ruiz D, Campoy JA, José Egea J (2007) Chilling and heat requirements of apricot cultivars for flowering. Environ Exp Bot 61(3):254–263

Ruiz D, Egea J, Gil MI, Tomás-Barberán FA (2005a) Characterization and quantitation of phenolic compounds in new apricot (*Prunus armeniaca* L.) varieties. J Agric Food Chem 53(24):9544–9552

Ruiz D, Egea J, Tomás-Barberán FA, Gil MI (2005b) Carotenoids from new apricot (*Prunus armeniaca* L.) varieties and their relationship with flesh and skin color. J Agric Food Chem 53(16):6368–6374

Schmitzer V, Slatnar A, Mikulic-Petkovsek M, Veberic R, Krska B, Franci Stampar F (2010) Comparative study of primary and secondary metabolites in apricot (*Prunus armeniaca* L.) cultivars. J Sci Food Agric 91(5):860–866

Someya K, Mikoshiba S, Okumura T, Hiroki Takenaka H, Ohdera M, Shirota O, Kuroyanagi M (2006) Suppressive effect of constituents isolated from kernel of *Prunus armeniaca* on 5 α-Androst-16-en-3-one generated by microbial metabolism. J Oleo Sci 55(7):353–364

Turan S, Topcu A, Karabulut I, Vural H, Hayaloglu AA (2007) Fatty acid, triacylglycerol, phytosterol, and tocopherol variations in kernel oil of Malatya apricots from *Turkey*. J Agric Food Chem 55(26): 10787–10794

U.S. Department of Agriculture, Agricultural Research Service (2011) USDA National Nutrient Database for Standard Reference, Release 24. Nutrient Data Laboratory Home Page. http://www.ars.usda.gov/ba/bhnrc/ndl

Usher G (1974) A dictionary of plants used by man. Constable. 619pp

Yiğit D, Yiğit N, Mavi A (2009) Antioxidant and antimicrobial activities of bitter and sweet apricot (*Prunus armeniaca* L.) kernels. Braz J Med Biol Res 42(4): 346–352

Prunus avium

Scientific Name

Prunus avium (L.) L.

Synonyms

Cerasus avium (L.) Moench, *Cerasus avium* var. *aspleniifolia* G. Kirchn., *Cerasus dulcis* P. Gaertn., *Cerasus nigra* Miller, *Druparia avium* Clairv., *Prunus avium* var. *aspleniifolia* (G. Kirchn.) H. Jaeger, *Prunus avium* (L.) L. var. *sylvestris* Dierb., *Prunus cerasus* L. var. *avium* L. (basionym), *Prunus macrophylla* Poir.

Family

Rosaceae

Common/English Names

Bing Cherry, Bird Cherry, Crab Cherry, Gean, Mazzard, Sweet Cherry, Wild Cherry.

Vernacular Names

Arabic: Karaz Barrî, Kuryaz;
Brazil: Cereja, Cerejeira;
Bulgaria: Cheresha;
Chinese: Ou Zhou Tian Ying Tao, Tian Guo Ying Tao, Tian Ying Tao, Ye Ying Tao;
Croatian: Tresnja;
Czech: Toešeň Ptačí, Třešeň Obecná, Třešeň Ptačí;
Danish: Fugle-Kirsebær, Kirsebær, Sødkirsebær;
Dutch: Kers, Wilde Kersenboom, Zoete Kers, Zoete Kersen, Zoete Kerseboom;
Estonian: Magus Kirsipuu, Vili: Maguskirss;
Finnish: Imeläkirsikka, Kirsikka;
French: Amèrise, Cerise Douce, Cerisier Des Oiseaux, Merise, Merisier;
German: Bauernkirsche, Gemeine Vogelkirsche, Herzkirsche, Kirschbaum, Kirsche, Süßkirsche, Süßkirschenbaum, Vogel-Kirsche, Waldkirsche, Wildkirsche;
Greek: Kerasia;
Hebrew: Dudevan;
Hungarian: Cseresznye, Madárcseresznye, Vadcseresznye
Icelandic: Fuglakirsiber;
India: Gilaas (Hindu), Kerii (Urdu);
Italian: Ciliegia Dolce, Ciliegia Di Monte, Ciliegio, Ciliegio Montano, Ciliegio Selvatico, Ciliegio Visciolo, Ciregiolo;
Japanese: Kanka Outou, Outou, Sakuranbo, Seiyo-Mizakura;
Korean: Beo Jji, Beojinamu, Tanpotnamu, Yangaengdonamu, Yang Beot, Yangsalgu;
Macedonian: Creshna;
Malay: Buah Céri;
Norwegian: Morell, Søtkirsebær;
Persian: Gilas;
Polish: Czereśnia, Czereshnia Dzika, Czereshnia Ptasia, Trzeshnia Dzika;

Portuguese: Agriota, Cereja, Cerejeira, Cerejeira-Brava;
Romanian: Cireş;
Russian: Chereshnia, Vishnia Ptich'ia;
Serbian: Tresnja;
Slovenian: Čerešňa Vtáčia, Cheshnja;
Spanish: Cerasus Dulce, Cerecera, Ceresera, Cereza, Cereza Silvestre, Cerezo, Cerezo Borde, Cerezo Bravío, Cerezo De Aves, Cerezo De Monte, Cerezo Silvestre, Guereciga, Guindo Zorrero, Maroviña, Picota;
Swedish: Fågelbär, Fågelkörsbär, Körsbärsträd, Sötkörs, Sötkörsbär, Vildkörsbär;
Turkish: Kiraz, Yabani Kiraz.

Origin/Distribution

P. avium is indigenous to the area between the Black and Caspian seas of Asia Minor and spread to Europe prior to human civilization. Domestic cultivation probably began with Greeks, and later perpetuated by the Romans. Sweet cherries came to the USA with English Colonists in 1629. It is cultivated in other temperate areas in both hemispheres.

Agroecology

Sweet cherries thrive best in cool and somewhat dry temperate areas where the danger of late frost is restricted. They have a high chill requirement to break bud dormancy. Sweet cherries are adapted traditionally to high chill areas-usually from 800 to 1,700 h chilling (Richards et al. 1995). Medium chill cultivars are being used in crosses with *Prunus pleioceracus, Prunus campanulata* and other low chilling cherry species to develop low chill selections. Sweet cherries does best in full-sun and in areas where rain does not fall during harvest as rain causes fruit cracking. High humidity and rain during fruit development is also bad as they increase incidence of brown rot fruit disease. Rain and hail during ripening and harvesting season can cause yield losses up to 90% in sweet cherry cultivations (Mucha-Pelzer et al. 2006) Particularly, high yield losses after precipitation are due to the cracking and the following rotting process through bacteria and fungi. A fertile, well-drained, light sandy loams are preferred. Mulching is also important. Flooded or waterlogged soils should be avoided as it impedes growth and reduce productivity. Irrigation is required on sandy soils and during the dry seasons for optimum productivity.

Edible Plant Parts and Uses

Ripe sweet cherries are excellent when eaten fresh or lightly chilled. Cherries can be frozen, canned, dried or preserved in brine (maraschino). Canned and frozen sweet cherries are a tasty ingredients in many recipes. They're also ready to eat and delicious straight from the can or freezer. The most popular canned sweet cherries are reddish purple to deep purple Bing Cherries and golden yellow to rosy pink Royal Anne Cherries. Maraschino cherries are made mostly from sweet cherries, but a small proportion of sour cherries are brined for this purpose. Cherries with clear flesh are picked slightly early, steeped in Marasca, a liqueur distilled from the fermented juice of wild cherries. Maraschino cherries, sometimes called cocktail cherries, are sweet cherries that have been tinted red with food coloring, flavored with almond extract, sweetened with sugar and packed in a sugar syrup. maraschino cherries goes very well with ice cream sundae or a banana split. Dried cherries, a relatively new cherry product, are delicious as snacks, wonderful in salads, and increasingly popular in recipes for everything from appetizers to entrees and desserts. Sweet cherries fresh, frozen, canned or dried are used in a wide array of food, snacks and desserts. Some notable food recipes include pork with cherry salsa, cherry orange chicken, cabernet beef jubilee, cherry rice pilaf and *cherry pierogi* (dumplings). Snacks and desserts with cherries include cherry pineapple fruit salad, fresh fruit ambrosia, *cherry clafouti*, frozen cheese fruit salad, cherry pies, cherry custard pie, cherry bread pudding, cherry-banana-pecan bread, pineapple cherry upside-down cake, traditional cherry fruit cake, black forest cake, pancakes, waffles, cherry chocolate chip muffins, pastries, cherry cookies, brownies, cherry fruit bars, cherry crunch bars, cherry oatmeal powerbars, cherry milkshakes, cherry smoothies, sherberts and cherries with ice cream.

The gum from bark wounds is aromatic and can be chewed as a substitute for chewing gum.

Botany

A medium-sized to large, deciduous, perennial tree, 5–25 m high with erect-pyramidal shaped canopy and dark brownish-black bark. Young branchlets are green becoming greyish-brown with age (Plate 1). Winter buds are ovoid-ellipsoid and glabrous. Petioles are glabrous, 2–7 cm long with two nectaries at the base, stipule are linear with serrated margins. Leaves are alternate, obovate-elliptic to elliptic-ovate, 3–13 × 2–6 cm, green and glabrous above, pale green and sparsely pubescent below, base cuneate to rounded, apex shortly acuminate, margin obtusely incised and serrted (Plates 2, 3, and 4). Flowers in 2–5 flowered umbellate inflorescence on short spurs with multiple buds at tips opening at same time as leaves. Flowers bisexual, fragrant, 1.5–3 cm across, on 2–6 cm long pedicels with distinct, glabrous, cyathiform hypanthium; sepals 5 elliptic lobes, recurved after anthesis, margin entire, apex obtuse; petals 5s white, obovate, apex

Plate 2 Sweet cherries and leaves

Plate 1 A heavy fruiting well-trained sweet cherry tree

Plate 3 Ripening sweet cherries

Plate 4 Clusters of ripening sweet cherries

Plate 5 Harvested ripe sweet cherries

emarginate; stamens numerous 30–35; style as long as stamen, ovary perigynous, superior. Fruit a fleshy drupe bright red, maroon to purplish black (Plates 2, 3, 4, and 5) sometimes golden yellow, subglobose to ovoid, 1.5–2.5 cm in diameter, mesocarp red sometimes white, succulent, juicy and fleshy enclosing a smooth, hard endocarp containing one seed.

Nutritive/Medicinal Properties

Food value of raw, sweet cherries (refuse 8% pit, pedicels) per 100 g edible portion was reported as follows (USDA 2011): water 82.25 g, energy 63 kcal (263 kJ), protein 1.06 g, total lipid (fat) 0.20 g, ash 0.48 g, carbohydrate 16.01 g; fibre (total dietary) 2.1 g, total sugars 12.82 g, sucrose 0.15 g, glucose 6.59 g, fructose 5.37 g, maltose 0.12 g, galactose 0.59 g; minerals – calcium 13 mg, iron 0.36 mg, magnesium 11 mg, phosphorus 21 mg, potassium 222 mg, sodium 0 mg, zinc 0.07 mg, copper 0.060 mg, manganese 0.070 mg, fluoride 2 μg, selenium 0 μg; vitamins – vitamin C (total ascorbic acid) 7.0 mg, thiamin 0.027 mg, riboflavin 0.033 mg, niacin 0.154 mg, pantothenic acid 0.199 mg, vitamin B-6 0.049 mg, folate (total) 4 μg, total choline 6.1 mg, vitamin A 64 IU (3 μg RAE), vitamin E (α-tocopherol) 0.07 mg, β-tocopherol 0.01 mg, γ-tocopherol 0.04 mg, vitamin K (phylloquinone) 2.1 μg, lipids – fatty acids (total saturated) 0.038 g, 14:0 (myristic acid) 0.001 g, 16:0 (palmitic acid) 0.027 g, 18:0 (stearic acid) 0.009 g; fatty acids (total monounsaturated) 0.047 g, 16:1 undifferentiated (palmitoleic acid) 0.001 g, 18:1 undifferentiated (oleic acid) 0.047 g; fatty acids (total polyunsaturated) 0.052 g, 18:2 undifferentiated (linoleic acid) 0.027 g, 18:3 0.026 g; phytosterols 12 mg, β-carotene 38 μg, lutein+zeaxanthin 85 μg; amino acids – tryptophan 0.009 g, threonine 0.022 g, isoleucine 0.020 g, leucine 0.030 g, lysine 0.032 g, methionine 0.010 g, cystine 0.010 g, phenylalanine 0.024 g, tyrosine 0.014 g, valine 0.024 g, arginine 0.018 g, histidine 0.015 g, alanine 0.026 g, aspartic acid 0.569 g, glutamic acid 0.083 g, glycine 0.023 g, proline 0.039 g and serine 0.030 g.

The major non-volatile constituents of sweet cherry cultivars varied widely among cultivars: glucose (5.2–8.8 g/100 g of fresh weight (FW)), fructose (4.4–6.4 g/100 g of FW), sorbitol and mannitol (2.2–8.0 g/100 g of FW), and malic acid (502.7–948.3 mg/100 g of FW) (Girard and Kopp 1998). Three principal components accounted for 53.3% of the total variation among 50 volatile compounds: (E)-2-Hexenol, benzaldehyde, hexanal, and (E)-2-hexenal were predominant flavor volatiles. Fruit fresh weight ranged from 8.8 to 14.5 g per fruit, soluble solids concentration (SSC) from 13.5 to 24.5 degrees Brix, and SSC/TA (titratable acidity ratio) from 18.3 to 29.0.

Studies showed that ascorbic acid, total antioxidant activity (TAA), and total phenolic compounds decreased during the early stages of sweet cherry development but exponentially increased from ripening stage 8, which coincided with the anthocyanin accumulation and fruit darkening (Serrano et al. 2005). TAA showed positive correlations ($R^2=0.99$) with both ascorbic acid and total phenolic compounds and also with the anthocyanin concentration from stage 8. Studies showed that during fruit ripening, there were significant increases in weight, soluble solids content, fructose, total phenols, total anthocyanins and total antioxidant activity; a non-significant decrease in firmness; and significant decreases in the colour parameters of both skin and flesh and in glucose (Serradilla et al. 2011). Five anthocyanins were found, of which the most abundant was cyanidin-3-O-rutinoside; 3 hydroxycinnamic acids, of which was p-coumaroylquinic acid predominated; a flavonol (rutin); and a flavan-3-ol (epicatechin) in small amount. Because of the increased levels of bioactive compounds associated with it, ripening stage 5 was considered to represent the highest nutritional and functional quality.

Total phenolics in sweet and sour cherries per 100 g ranged from 92.1 to 146.8 and from 146.1 to 312.4 mg gallic acid equivalents, respectively (Kim et al. 2005). Total anthocyanins of sweet and sour cherries ranged from 30.2 to 76.6 and from 49.1 to 109.2 mg cyanidin 3-glucoside equivalents, respectively. Anthocyanins such as cyanidin and peonidin derivatives were dominant phenolics. Hydroxycinnamic acids consisted of neochlorogenic acid, chlorogenic acid, and p-coumaric acid derivatives. Glycosides of quercetin, kaempferol, and isorhamnetin were also found. In another study, *P. avium* fruit was found to contain 78.8 mg GAE/100 g fresh weight total phenolics, 19.6 mg CE/100 g fresh weight total flavonoids, flavonoids/phenolics ratio of 0.25 (Marinova et al. 2005). Sweet and sour cherries are rich in polyphenolic compounds particularly phenolic acids and anthocyanins; Anthocyanins contributed more to the antioxidant activity of all fruits (~90%) than flavonols, flavan-3-ols and phenolic acids (~10%) (Jakobek et al. 2009). A biphasic reaction was observed between DPPH radicals and phenols, with 'fast' and 'slow' scavenging rates which might be important in the biological activity of these cherries. Sweet cherries were found to have 2010.67 mg/kg total phenols (TP), 192.5 total anthocyanins (TA), TA/TP ratio of 0:1 and antioxidant activities in terms of DPPH 4.22 umol TE/g and, ABTS (2,2,-azino-bis(3-ethylbenzothiazoline-6-sulfonic acid) diammonium salt) 13.62 umol TE/g (Jakobek et al. 2007). Flavonols identified in sweet cherry were quercetin (9 mg/kg), kaempferol (6 mg/kg) and myricetin (0.17 mg /kg). Among five investigated phenolic acid, only caffeic acid (1 ng/kg) and p-coumaric acid (5 mg/kg) were identified in the study.

The phenolic compounds hydroxycinnamates, anthocyanins, flavonols, and flavan-3-ols of sweet cherry cultivars Burlat, Saco, Summit, and Van were quantified and analyzed at partially ripe and ripe stages and during storage at $15\pm5°C$ (room temperature) and $1–2°C$ (cool temperature) (Gonçalves et al. 2004). Neochlorogenic and p-coumaroylquinic acids were the main hydroxycinnamic acid derivatives, but chlorogenic acid was also identified in all cultivars. The 3-glucoside and 3-rutinoside of cyanidin were the major anthocyanins. Peonidin and pelargonidin 3-rutinosides were the minor anthocyanins, and peonidin 3-glucoside was also present in cvs. Burlat and Van. Epicatechin was the main monomeric flavan-3-ol with catechin present in smaller amounts in all cultivars. The flavonol rutin was also detected. Cultivar Saco contained the highest amounts of phenolics [227 mg/100 g of FW] and cv. Van the lowest (124 mg/100 g FW). Phenolic acid contents generally decreased with storage at $1–2°C$ and increased with storage at $15\pm5°C$. Anthocyanin levels increased at both storage temperatures. In cv. Van, the anthocyanins increased up to 5-fold during storage at $15\pm5°C$ (from 47 to 230 mg/100 g FW). Flavonol and flavan-3-ol contents remained quite constant. The major organic acid in sweet cherry cultivars (Van, Noir de Guben, Larian and 0–900 Ziraat) was found as malic acid (8.54–10.02 g/k FW); other organic acids found were citric, shikimic and fumaric acids (Kelebek and Selli 2011). With regard to sugars, glucose was present in the highest amounts (44.71–48.31 g/kg FW). Other sugars presented included sucrose, fructose and sorbitol. A total of 11 phenolic compounds were identified and quantified in the sweet cherry cultivars, including hydroxycinnamic acids (3), anthocyanins (5), flavan-3-ols (2) and flavonol

(1) compounds. Total phenolic contents ranged from 88.72 (Van) to 239.54 (Noir de Guben) mg/100 g FW, while antioxidant activities ranged from 2.02 to 7.75 μm Trolox equivalents/g FW.

Sweet cherries contained melatonin, a free radical scavenger and modulator of the sleep-wake cycle in mammals' and serotonin the main intermediate in melatonin biosynthesis (Rodríguez 2009). The limits of detection of the proposed analytical method were 4.3 ng/ml for melatonin and 3.0 ng/ml for serotonin. The highest melatonin amounts were found in 'Burlat' sweet cherries (22.4 ng/100 g of fresh fruit), while the highest serotonin contents were found in the cultivar 'Ambrunés' (37.6 ng/100 g of fresh fruit).

Non-flavonoid (16 compounds) and flavonoid (27 compounds) polyphenols were identified in cherry heartwood used for cooperage (Sanz et al. 2010). The non-flavonoids found were lignin constituents, and include compounds such as protocatechuic acid and aldehyde, p-coumaric acid, methyl vanillate, methyl syringate, and benzoic acid, but not ellagic acid, and only a small quantity of gallic acid. In seasoned wood, the researchers found a great variety of flavonoid compounds which had not been found in oak wood for cooperage, mainly, in addition to the flavan-3-ols (+)-catechin, a B-type procyanidin dimer, and a B-type procyanidin trimer, the flavanones naringenin, isosakuranetin, and eriodictyol and the flavanonols aromadendrin and taxifolin. Seasoned and toasted cherry wood showed different ratios of flavonoid to non-flavonoid compounds, since toasting results in the degradation of flavonoids, and the formation of non-flavonoids from lignin degradation. In contrast, cherry wood had no hydrolyzable tannins, which were very important in oak wood used in cooperage.

Cherries, and in particular sweet cherries, are a nutritionally dense food rich in anthocyanins, quercetin, hydroxycinnamates, potassium, fibre, vitamin C, carotenoids, and melatonin (McCune et al. 2011). UV concentration, degree of ripeness, postharvest storage conditions, and processing, each could significantly alter the amounts of nutrients and bioactive components. These constituent nutrients and bioactive food components support the potential preventive health benefits of cherry intake in relation to cancer, cardiovascular disease, diabetes, inflammatory diseases, and Alzheimer's disease (Ferretti et al. 2010; McCune et al. 2011). In-vitro and in-vivo animal studies had shown that cherries exhibited relatively high antioxidant activity, low glycemic response, COX 1 and 2 enzyme inhibition, and other anticarcinogenic effects. However, well-designed cherry feeding studies are needed to further substantiate any health benefits in humans (McCune et al. 2011). Other pharmacological properties of sweet cherries reported include antimicrobial, neuroprotective, photoprotective, diuretic, antigenotoxic activities.

Antioxidant Activity

Harvesting sweet cherry 4 days later than the commercial harvest date coupled with storage at 2°C for 16 days and a further period of 2 days at 20°C gave the highest antioxidant capacity for total antioxidant activity (TAA), in the hydrophilic (H-TAA) or lipophilic (L-TAA) fractions although important differences existed among cultivars (Serrano et al. 2009).

The amount of total phenolics varied between 617 and 4,350 mg/kg in fresh berries (blackberries, red raspberries, blueberries, sweet cherries and strawberries), as gallic acid equivalents (GAE) (Heinonen et al. 1998). In the copper-catalyzed in-vitro human low-density lipoprotein oxidation assay at 10 μM gallic acid equivalents (GAE), berry extracts inhibited hexanal formation in the order: blackberries>red raspberries>sweet cherries>blueberries>strawberries. In the copper-catalyzed in-vitro lecithin liposome oxidation assay, the extracts inhibited hexanal formation in the order: sweet cherries>blueberries>red raspberries>blackberries>strawberries. Red raspberries were more efficient than blueberries in inhibiting hydroperoxide formation in lecithin liposomes. HPLC analyses showed high anthocyanin content in blackberries, hydroxycinnamic acid in blueberries and sweet cherries, flavonol in blueberries, and flavan-3-ol in red raspberries. The antioxidant activity for LDL was associated directly with anthocyanins and indirectly with flavonols, and for liposome it correlated with the hydroxycinnamate content. Berries thus contribute a significant source of phenolic antioxidants that may have potential health effects.

The antioxidant activity of anthocyanins from cherries was comparable to the commercial antioxidants, tert-butylhydroquinone, butylated hydroxytoluene and butylated hydroxyanisole, and superior to vitamin E, at a test concentration of 125 µg/ml (Seeram et al. 2001). The presence and levels of cyanidin-3-glucosylrutinoside 1 and cyanidin-3-rutinoside 2 were determined in both fruits. The yields of pure anthocyanins 1 and 2 in 100 g Balaton and Montmorency tart cherries, sweet cherries and raspberries were 21, 16.5; 11, 5; 4.95, 21; and 4.65, 13.5 mg, respectively. Total sugars (glucose, fructose, sucrose and sorbitol) in 13 sweet cherries cultivars ranged from 125 to 265 g/kg fresh weight (FW) and total organic acids (malic, citric, shikimic and fumaric) ranged from 3.67 to 8.66 g/kg FW (Usenik et al. 2008). Total phenolic content ranged from 44.3 to 87.9 mg gallic acid equivalents/100 g FW and antioxidant activity ranged from 8.0 to 17.2 mg ascorbic acid equivalent antioxidant capacity mg/100 g FW. The correlation of antioxidant activity with total phenolics content and content of anthocyanins was cultivar dependent.

Total phenolics in the tart and sweet cherry juices and wines were 56.7–86.8 mg of gallic acid equivalents (GAE)/l and 79.4–149 mg GAE/l, respectively (Yoo et al. 2010). Total anthocyanins in the cherry juices and wines were 7.9–50.1 mg of cyanidin 3-glucoside equivalents (CGE)/l and 29.6–63.4 mg CGE/l, respectively. Both cherry juices and wines exhibited protective effects against oxidative stress induced by H_2O_2 on V79-4 lung fibroblast cells and also enhanced the activities of antioxidative enzymes, such as superoxide dismutase and catalase, in a dose-dependent manner. The protection of V79-4 cells from oxidative stress by phenolics was mainly attributable to anthocyanins. The positive correlation between the protective effects against oxidative stress in V79-4 cells and the antioxidant enzyme activities was stronger for cyanidin 3-glucoside than for cyanidin 3-rutinoside.

Cherries contain bioactive anthocyanins that are reported to possess antioxidant, anticancer, antiinflammatory, antidiabetic and antiobese properties. Mulabagal et al. (2009) reported that red sweet cherries contained cyanidin-3-O-rutinoside as major anthocyanin (>95%). The sweet cherry cultivar "Kordia" (aka "Attika") showed the highest cyanidin-3-O-rutinoside content, 185 mg/100 g fresh weight. The red sweet cherries "Regina" and "Skeena" were similar to "Kordia", yielding cyanidin-3-O-rutinoside at 159 and 134 mg/100 g fresh weight, respectively. The yields of cyanidin-3-O-glucosylrutinoside and cyanidin-3-O-rutinoside were 57 and 19 mg/100 g fresh weight in the sour cherry cultivars "Balaton" and 21 and 6.2 mg/100 g fresh weight in "Montmorency", respectively, in addition to minor quantities of cyanidin-3-O-glucoside. The water extracts of "Kordia", "Regina", "Glacier" and "Skeena" sweet cherries gave 89, 80, 80 and 70% of lipid peroxidation (LPO) inhibition, whereas extracts of "Balaton" and "Montmorency" were in the range of 38–58% at 250 µg/ml. Methanol and ethyl acetate extracts of the yellow sweet cherry "Rainier" containing β-carotene, ursolic, coumaric, ferulic and cafeic acids inhibited LPO by 78 and 79%, respectively, at 250 µg/ml. In the cyclooxygenase (COX) enzyme inhibitory assay, the red sweet cherry water extracts inhibited the enzymes by 80–95% at 250 µg/ml. However, the methanol and ethyl acetate extracts of "Rainier" and "Gold" were the most active against COX-1 and -2 enzymes. Water extracts of "Balaton" and "Montmorency" inhibited COX-1 and -2 enzymes by 84, and 91 and 77, and 87%, respectively, at 250 µg/ml. Cyclooxygenase (COX) is an enzyme that is responsible for the formation of prostanoids. The three main groups of prostanoids: prostaglandins, prostacyclins, and thromboxanes are involved in the inflammatory response.

Anticancer Activity

The sweet cherry (*P. avium*) extract obtained with a mixture of CO_2: ethanol (90:10, v/v) exhibited the highest antioxidant activity (181.4 µmol

TEAC/g) and was the most effective in inhibiting the growth of human colon cancer cells (ED_{50} = 0.20 mg/ml) (Serra et al. 2010). Perillyl alcohol was found to be one of the major compounds responsible for antiproliferative properties of cherry extracts and polyphenols, in particular sakuranetin and sakuranin, appeared to be the major contributors of the antioxidant capacity. Saco sweet cherry and two exotic cultivars (Ulster and Lapin) proved to have higher contents of phenolic compounds, highest antioxidant activity and were the most effective in inhibiting human cancer cells derived from colon (HT29) and stomach (MKN45) (Serra et al. 2010). Correlation of the data obtained showed that anthocyanins were the major contributors to the antioxidant capacity and antiproliferative effect of cherries. In addition, hydroxycinnamic acids (neochlorogenic acid, chlorogenic acid and p-coumaroylquinic acid), flavan-3-ols (catechin and epicatechin) and flavonols (rutin and quercetin-3-glucoside) also played important roles in protection against oxidative stress.

Antiinflammatory Activity

The cyanidin in cherries could protect against the paws swelling in Freund's adjuvant-induced arthritis (AIA) rats, and alleviate the inflammatory reaction in the joint (He et al. 2005, 2006). anthocyanins at 20 and 10 mg/kg had less effect on the inflammatory factors and antioxidative capacity of Freund's adjuvant-induced arthritis than the high dose of 40 mg/kg. Anthocyanins at 40 mg/kg significantly decreased the levels of TNFα in serum and prostaglandin E2 in paws, simultaneously improving the anti-oxidative status of AIA. At the high dosage total antioxidative capacity (T-AOC) was potentiated, the activity of glutathione (GSH), superoxide dismutase (SOD) increased and the level of malonaldehyde (MDA) in serum decreased. The results suggested that the Anthocyanins from cherries have potential antiinflammatory and anti-oxidative effects and could be one of the potential candidates for the alleviation of arthritis. The studies supported earlier findings by Blazso and Gabor (1994) that the extract of its fruit stalks may reduce inflammations. Cherry fruit stalk extracts had also been reported to exert a positive effect on the cardiovascular system and smooth muscles (Hetenyi and Valyi-Nagy 1969a, b).

The physiologic effects of cherry consumption with respect measured plasma urate, antioxidant and inflammatory markers was studied in 10 healthy women (22–40 years old) (Jacob et al. 2003). Five hours post-consumption, plasma urate declined to 183 μmole/l from a pre-consumption level of 214 μmole/l. Urinary urate increased significantly post-consumption, with peak excretion of 350 μmol/mmol creatinine 3 hours post-consumption compared with 202 at baseline. Plasma C-reactive protein (CRP) and nitric oxide (NO) concentrations had decreased marginally 3 hours post-consumption, whereas plasma albumin and tumour necrosis factor-α were unchanged. The vitamin C content of the cherries was solely as dehydroascorbic acid, but post-consumption increases in plasma ascorbic acid indicated that dehydroascorbic acid in fruits was bioavailable as vitamin C. The decrease in plasma urate after cherry consumption supported the reputed anti-gout efficacy of cherries. Cherries and cherry juice have been used since the 1950s by sufferers of gout and arthritis to ease their symptoms. The trend toward decreased inflammatory indices (CRP and NO) indicated that compounds in cherries may inhibit inflammatory pathways.

The physiologic effects of bing cherry consumption was studied in 18 healthy men and women. Cherry consumption did not affect the plasma concentrations of total-, HDL-, LDL-, and VLDL- cholesterol, triglycerides, subfractions of HDL, LDL, VLDL, and their particle sizes and numbers (Kelley et al. 2006). It also did not affect fasting blood glucose or insulin concentrations or a number of other chemical and hematological variables. However, results suggested a selective modulatory effect of sweet cherries on CRP, NO, and RANTES (regulated upon activation, normal T-cell expressed, and secreted). Such antiinflammatory effects may be beneficial for the management and prevention of inflammatory pathways in cardiovascular diseases.

Cherry extracts had also been reported to inhibit the action of cyclooxygenase-1 (COX-1) and COX-2 enzymes, important components of the inflammatory process and the sensation of pain (Seeram et al. 2001; McCune et al. 2011). Anthocyanins from raspberries *Rubus idaeus* and sweet cherries *Prunus avium* demonstrated 45% and 47% cyclooxygenase-I and cyclooxygenase-II inhibitory activities, respectively, when assayed at 125 µg/ml (Seeram et al. 2001). The cyclooxygenase inhibitory activities of anthocyanins from these fruits were comparable to those of ibuprofen and naproxen (antiinflammatory agents) at 10 µM concentrations. Anthocyanins cyanidin-3-glucosylrutinoside and cyanidin-3-rutinoside were present in both cherries and raspberry.

Antimicrobial Activity

Sweet cherry fruit extracts exhibited high antimicrobial and antioxidant activities (Ahn et al. 2009). The hot-water fruit extract contained about 40% sugars, and the solvent fraction yields for hexane, ethyl acetate (EA), butanol, and water residue were 0.01%, 3.45%, 16.30%, and 80.24%, respectively (Ahn et al. 2009). Contents of total polyphenol and total flavonoid of the fractions were 1.24–5.24%, and 0–3.76%, respectively. Among the fractions, the ethyl acetate fraction showed the highest total polyphenol and total flavonoid concentrations. The ethyl acetate fraction and butanol fraction exhibited strong antibacterial activity against *Listeria monocytogenes*, *Staphylococcus aureus*, and *Salmonella typhimurium* with minimal inhibitory concentration (MIC) of 0.5–1.0 mg/ml. But the extract and fractions tested were not active to *Pseudomonas aeruginosa*. The hexane fraction showed anti-Candida activity with 0.5–1.0 mg/ml of MIC. The fraction also showed strong activity against different multi-antibiotics resistant strains such as *C. albicans* CCARM 14,020. Antioxidative activity assay showed that ethyl acetate fraction had a strong DPPH scavenging activity and a reducing power. The IC_{50} values of vitamin E and ethyl acetate fraction were 15.5 and 195.5 µg/ml, respectively.

Sweet cherry wood contain the following flavonoid compounds: δ-catechin, naringenin, pruning, aromadendrin (traces), eriodictyol, taxifolin, chrysin, aequinoctin, genistein, prunetin (Hasegawa 1957) and the heartwood has eight known flavonoid aglycones, namely pinocembrin, pinostrobin, dihydrowogonin, naringenin, sakuranetin, aromadendrin-7-methylether, chrysin and tectochrysin (Vinciguerra et al. 2003). Some of the flavanones and flavones from the heartwood and resin of *Prunus avium* exhibited human cytochrome P450 1A1, 3A4 and 19 (aromatase) inhibition, and for antifungal activity against a panel of pathogenic fungi (McNulty et al. 2009). Two flavonol glycosides, quercetin 3-rutinosyl-4′-glucoside and kaempferol 3-rutinosyl-4′-glucoside, and a flavanone glucoside, dihydrowogonin 7-glucoside, were identified in sweet cherry leaves (Mo et al. 1995). Flavonoids are known to have a potent positive role in human health.

Antigenotoxic Activity

Among the fruit juices, sweet cherry juice exhibited the highest inhibitory effect on 2-amino-3-methylimidazo[4,5-f]quinoline (IQ) genotoxicity ($IC_{50} = 0.17\%$), followed by juices from kiwi fruit, plum and blueberry ($IC_{50} = 0.48$–0.71%) (Platt et al. 2010). The juices from watermelon, blackberry, strawberry, black currant, and Red delicious apple showed moderate suppression, whereas sour cherry, grapefruit, red currant, and pineapple juices were only weakly active. In most cases, fruits and vegetables inhibited 2-amino-1-methyl-6-phenylimidazo[4,5-b]pyridine (PhIP) genotoxicity less strongly than IQ genotoxicity.

Diuretic Activity

Powdered cherry fruit stalks had diuretic activity (Hooman et al. 2009). After administration of cherry stalk, the mean of urine calcium, sodium, chloride, and urine volume increased, but the amount of urine potassium and urine osmolality did not change. No adverse reaction was observed. Powdered cherry stalk increased mild urine volume confirming thus the claimed diuretic effect of the plant. Administration of cherry stalk caused urinary sodium and chloride rising less than loop diuretics but higher than the others. Because of rising calcium excretion, it should be used with cautioun in people with urolithiasis.

Photoprotective Activity

Oral administration of 450 mg/kg body weight per day of cherry fruit pulp methanol extract to Swiss Albino mice for 15 consecutive days before exposure to 10 Gy of γ-radiation was found to afford maximum protection in terms of body weight and survivability of the mice in comparison to other doses (Sisodia et al. 2009). Oral administration of a *P. avium* fruit extract significant halted significant elevation of lipid peroxidation and depletion in glutathione and protein levels in blood serum and spleen in irradiated mice (Sisodia et al. 2011). γ radiation-induced deficit in blood sugar, cholesterol, and hematological constituents could also be modulated by supplementation of cherry fruit extract before and after irradiation. This photoprotective effect was postulated to the possible synergistic action of various antioxidants, minerals, vitamins, etc., present in the cherry fruit.

Neuroprotective Activity

Cherry phenolics protected neuronal cells (PC 12) from cell-damaging oxidative stress in a dose-dependent manner mainly due to anthocyanins (Kim et al. 2005). The results showed cherries to be rich in phenolics, especially in anthocyanins, with a strong anti-neurodegenerative activity in diseases like dementia and that they could serve as a good source of biofunctional phytochemicals in our diet.

Bioavailability Studies

The bioavailability of phenolic compounds from frozen sweet cherries was investigated by a digestion process involving pepsin-HCl digestion (to simulate gastric digestion) and pancreatin digestion with bile salts (to simulate small intestine conditions) and dialyzed to assess serum- and colon-available fractions (Fazzari et al. 2008). Following pancreatic digestion and dialysis, the total phenolics in the IN (serum-available) fraction was about 26–30% and the OUT (colon-available) fraction was about 77–101%. The anthocyanin content in the IN fraction was 15–21%, and in the OUT fraction, it was 52–67%. Immature cherries had higher % total phenolics in the IN fraction than mature or over-mature cherries. However, immature cherries had the lowest concentrations of these compounds, making the actual bioavailable amounts of these compounds lower than for mature and over-mature fruit.

Traditional Medicinal Uses

In traditional medicine, the fruit stalks are astringent, antitussive, diuretic and tonic. The fruit stalks (cherry tails) are sold by herbal druggists in Iran and are used as decoction for relief of renal stones, oedema and hypertension (Kirtikar and Basu 2001). The decoction is also used in the treatment of cystitis, oedema, bronchial complaints, looseness of the bowels and anaemia. An aromatic resin from the trunk has been used as an inhalant in the treatment of persistent coughs. *P. avium* like all *Prunus* species contain the exceedingly poisonous compounds amygdalin and prunasin in the seeds which break down in water to form hydrocyanic acid (cyanide or prussic acid). In small amounts, they have been reported to stimulate respiration and improve digestion.

Other Uses

The leaves yield a green dye and the fruit a dark grey/dark green dye. The hard reddish cherry wood is valued for turnery, cooperage and used for making fine furniture, cabinets, musical instruments, and carvings.

Comments

Cherry trees are readily propagated from hardwood cuttings or graftings.

Selected References

Ahn L, Seon MI, Ryu H-Y, Kang D-K, Jung I-C, Sohn HY (2009) Antimicrobial and antioxidant activity of the fruit of *Prunus avium*. Han'gug Mi'saengmul Saengmyeong Gong Haghoeji 37(4):371–376

Blazso G, Gabor M (1994) Anti-inflammatory effects of cherry (*Prunus avium* L.) stalk extract. Pharmazie 49:540–541

Bown D (1995) Encyclopaedia of herbs and their uses. Dorling Kindersley, London, 424pp

Chiej R (1984) The Macdonald encyclopaedia of medicinal plants. Macdonald & Co., London, 447pp

Facciola S (1990) Cornucopia. A source book of edible plants. Kampong Publication, Vista, 677pp

Faust M, Suranyi D (1997) Origin and dissemination of cherry. Hort Rev 19:263–317

Fazzari M, Fukumoto L, Mazza G, Livrea MA, Tesoriere L, Marco LD (2008) In vitro bioavailability of phenolic compounds from five cultivars of frozen sweet cherries (*Prunus avium* L.). J Agric Food Chem 56(10):3561–3568

Ferretti G, Bacchetti T, Belleggia A, Neri D (2010) Cherry antioxidants: from farm to table. Molecules 15(10):6993–7005

Girard B, Kopp TG (1998) Physicochemical characteristics of selected sweet cherry cultivars. J Agric Food Chem 46(2):471–476

Gonçalves B, Landbo AK, Knudsen D, Silva AP, Moutinho-Pereira J, Rosa E, Meyer AS (2004) Effect of ripeness and postharvest storage on the phenolic profiles of Cherries (*Prunus avium* L.). J Agric Food Chem 52(3):523–530

Grae I (1974) Nature's colors – dyes from plants. MacMillan Publishing Co., New York

Grieve M (1971) A modern herbal, 2 vols. Penguin/Dover publications, New York, 919pp

Hasegawa M (1957) Flavonoids of various *Prunus* species. VI. The flavonoids in the wood of *Prunus aequinoctialis*, *P. nipponica*, *P. Maximowiczii* and *P. avium*. J Am Chem Soc 79(7):1738–1740

He YH, Xiao C, Wang YS, Zhao LH, Zhao HY, Tong Y, Zhou J, Jia HW, Lu C, Li XM, Lu AP (2005) Antioxidant and anti-inflammatory effects of cyanidin from cherries on rat adjuvant-induced arthritis. Zhongguo Zhong Yao Za Zhi 30(20):1602–1605 (In Chinese)

He YH, Zhou J, Wang YS, Xiao C, Tong Y, Tang JC, Chan AS, Lu AP (2006) Anti-inflammatory and anti-oxidative effects of cherries on Freund's adjuvant-induced arthritis in rats. Scand J Rheumatol 35(5):356–358

Heinonen IM, Meyer AS, Frankel EN (1998) Antioxidant activity of berry phenolics on human low-density lipoprotein and liposome oxidation. J Agric Food Chem 46(10):4107–4112

Hetenyi E, Valyi-Nagy T (1969a) Pharmacology of cherry (*Prunus avium*) stalk extract. I. Effect on smooth muscle. Acta Physiol Acad Sci Hung 35:183–188

Hetenyi E, Valyi-Nagy T (1969b) Pharmacology of cherry (*Prunus avium*) stalk extract. II. Cardiovascular effects. Acta Physiol Acad Sci Hung 35:189–197

Hooman N, Mojab F, Nickavar B, Pouryousefi-Kermani P (2009) Diuretic effect of powdered *Cerasus avium* (cherry) tails on healthy volunteers. Pak J Pharm Sci 22(4):381–383

Jacob RA, Spinozzi GM, Simon VA, Kelley DS, Prior RL, Hess-Pierce B, Kader AA (2003) Consumption of cherries lowers plasma urate in healthy women. J Nutr 133(6):1826–1829

Jakobek L, Šeruga M, Novak I, Medvidović-Kosanović M (2007) Flavonols, phenolic acids and antioxidant activity of some red fruits. Deutsche Lebensmittel-Rundschau 103(8):369–378

Jakobek L, Šeruga M, Šeruga B, Novak I, Medvidović-Kosanović M (2009) Phenolic compound composition and antioxidant activity of fruits of *Rubus* and *Prunus* species from Croatia. Int J Food Sci Technol 44(4):860–868

Kelebek H, Selli S (2011) Evaluation of chemical constituents and antioxidant activity of sweet cherry (*Prunus avium* L.) cultivars. Int J Food Sci Technol 46:2530–2537

Kelley DS, Rasooly R, Jacob RA, Kader AA, Mackey BE (2006) Consumption of Bing sweet cherries lowers circulating concentrations of inflammation markers in healthy men and women. J Nutr 136(4):981–986

Kim DO, Heo HJ, Kim YJ, Yang HS, Lee CY (2005) Sweet and sour cherry phenolics and their protective effects on neuronal cells. J Agric Food Chem 53(26):9921–9927

Kirtikar KR, Basu BD (2001) Indian medicinal plants, vol 4, 2nd edn. Oriental Enterprises, Uttranchal, pp 1330–1349

Marinova D, Ribarova F, Atanassova M (2005) Total phenolics and total flavonoids in Bulgarian fruits and vegetables. J Univ Chem Technol Metall 40(3):255–260

McCune LM, Kubota C, Stendell-Hollis NR, Thomson CA (2011) Cherries and health: a review. Crit Rev Food Sci Nutr 51(1):1–12

McNulty J, Nair JJ, Bollareddy E, Keskar K, Thorat A, Crankshaw DJ, Holloway AC, Khan G, Wright GD, Ejim L (2009) Isolation of flavonoids from the heartwood and resin of *Prunus avium* and some preliminary biological investigations. Phytochemistry 70(17–18):2040–2046

Mo YY, Geibel M, Bonsall RF, Gross DC (1995) Analysis of sweet cherry (*Prunus avium* L.) leaves for plant signal molecules that activate the syrB gene required for synthesis of the phytotoxin, syringomycin, by *Pseudomonas syringae* pv *syringae*. Plant Physiol 107(2):603–612

Mucha-Pelzer T, Müller S, Rohr F, Mewis I (2006) Cracking susceptibility of sweet cherries (*Prunus avium* L.) under different conditions. Commun Agric Appl Biol Sci 71(2 Pt A):215–223

Mulabagal V, Lang GA, DeWitt DL, Dalavoy SS, Nair MG (2009) Anthocyanin content, lipid peroxidation and cyclooxygenase enzyme inhibitory activities of sweet and sour cherries. J Agric Food Chem 57(4):1239–1246

Platt KL, Edenharder R, Aderhold S, Muckel E, Glatt H (2010) Fruits and vegetables protect against the genotoxicity of heterocyclic aromatic amines activated by human xenobiotic-metabolizing enzymes expressed in immortal mammalian cells. Mutat Res 703(2):90–98

Porcher MH et al (1995–2020) Searchable World Wide Web Multilingual Multiscript Plant Name Database. Published by The University of Melbourne, Australia. http://www.plantnames.unimelb.edu.au/Sorting/Frontpage.html

Richards GD, Kantharajah K, Sherman WB, Porter GW (1995) Progress with low-chill sweet cherry breeding in Australia and USA. Acta Hort (ISHS) 403:179–181

Rodríguez AB (2009) Detection and quantification of melatonin and serotonin in eight Sweet Cherry cultivars (*Prunus avium* L.). Eur Food Res Technol 229(2):223–229

Sanz M, Cadahía E, Esteruelas E, Muñoz AM, Fernández De Simón B, Hernández T, Estrella I (2010) Phenolic compounds in cherry (*Prunus avium*) heartwood with a view to their use in cooperage. J Agric Food Chem 58(8):4907–4914

Seeram NP, Momin RA, Nair MG, Bourquin LD (2001) Cyclooxygenase inhibitory and antioxidant cyanidin glycosides in cherries and berries. Phytomedicine 8(5):362–369

Serra AT, Seabra IJ, Braga MEM, Bronze MR, de Sousa HC, Duarte CMM (2010) Processing cherries (*Prunus avium*) using supercritical fluid technology. Part 1: Recovery of extract fractions rich in bioactive compounds. J Supercrit Fluid 55(1):184–191

Serradilla MJ, Lozano M, Bernalte MJ, Ayuso MC, López-Corrales M, González-Gómez D (2011) Physicochemical and bioactive properties evolution during ripening of 'Ambrunés' sweet cherry cultivar. LWT Food Sci Technol 44(1):199–205

Serrano M, Díaz-Mula HM, Zapata PJ, Castillo S, Guillén F, Martínez-Romero D, Valverde JM, Valero D (2009) Maturity stage at harvest determines the fruit quality and antioxidant potential after storage of sweet cherry cultivars. J Agric Food Chem 57(8):3240–3246

Serrano M, Guillén F, Martínez-Romero D, Castillo S, Valero D (2005) Chemical constituents and antioxidant activity of sweet cherry at different ripening stages. J Agric Food Chem 53(7):2741–2745

Sisodia R, Sharma K, Singh S (2009) Acute toxicity effects of *Prunus avium* fruit extract and selection of optimum dose against radiation exposure. J Environ Pathol Toxicol Oncol 28(4):303–309

Sisodia R, Singh S, Mundotiya C, Meghnani E, Srivastava P (2011) Radioprotection of Swiss albino mice by *Prunus avium* with special reference to hematopoietic system. J Environ Pathol Toxicol Oncol 30(1):55–70

U.S. Department of Agriculture, Agricultural Research Service (2011) USDA National Nutrient Database for Standard Reference, Release 24. Nutrient Data Laboratory Home Page. http://www.ars.usda.gov/ba/bhnrc/ndl

Usenik V, Fabčič J, Stampar F (2008) Sugars, organic acids, phenolic composition and antioxidant activity of sweet cherry (*Prunus avium* L.). Food Chem 107(1):185–192

Vinciguerra V, Luna M, Bistoni A, Zollo F (2003) Variation in the composition of the heartwood flavonoids of *Prunus avium* by on-column capillary gas chromatography. Phytochem Anal 14(6):371–377

Webster AD, Looney NE (1995) Cherries: crop physiology, production and uses. CAB International, Wallingford, 513pp

Yoo KM, Al-Farsi M, Lee HJ, Yoon HG, Lee CY (2010) Antiproliferative effects of cherry juice and wine in Chinese hamster lung fibroblast cells and their phenolic constituents and antioxidant activities. Food Chem 123(3):734–740

Prunus domestica

Scientific Name

Prunus domestica L.

Synonyms

Prunus communis Hudson, *Prunus domestica* Rouy & Camus, *Prunus domestica* subsp. *oeconomica* (Borkhausen) C. K. Schneider, *Prunus domestica* var. *damascena* Linnaeus, *Prunus sativa* L. subsp. *domestica* Rouy & Camus, *Prunus sativa* Rouy & Camus subsp. *domestica* (Linnaeus) Rouy & E. G. Camus.

Family

Rosaceae

Common/English Names

Common Plum, European Plum, Gage, Garden Plum, Plum, Prune, Prune Plum.

Vernacular Names

Afrikaans: Pruim;
Arabic: Barqûq, Iggâss;
Armenian: Salor;
Basque: Aran;
Brazil: Ameixa-Comum, Ameixa-Européia, Ameixa-Preta, Ameixa-Roxa, Ameixa-Vermela;
Bosnian: Šljiva;
Catalan: Pruna;
Chinese: Ou Zhou Li, Li Zi, Mei Zi;
Croatian: Šljiva;
Czech: Sliva;
Danish: Blomme;
Dutch: Pruim, Pruimenboom;
Eastonian: Aedploomipuu, Harilik Ploomipuu, Ploom, Ploomipuu;
Esperanto: Pruno;
Finnish: Luumu, Luumupuu;
French: Prunier, Prunier Commun, Prunier Domestique, Prunier Cultivé;
German: Bauernpflaume, Echte Pflaume, Hauspflaume, Kultur-Pflaume, Pflaume, Pflaumenbaum, Zwetsche, Zwetschge, Zwispeln;
Greek: Damáskino;
Hebrew: Shazif;
Hungarian: Kerti Szilva, Szilva;
Icelandic: Plóma;
Italian: Prugna, Pruno, Prugno, Susina, Susino;
India: Alu Bukhara (Hindu), Arukam (Malayalam), Heikha (Manipuri), Alpagodapandu (Telugu), Aalu Bukhara (Urdu);
Indonesia: Plum;
Japanese: Seiyou Sumomo, Seiyou Sumomo, Puramu, Yooroppa Sumomo;
Latvian: Plūme;
Malaysia: Plum;
Nepalese: Aalu Bakharaa, Alu Bakhara;
Persian: Aalu;

Polish: Shliwa Domowa, Šliwa Domowa, Śliwka;
Portuguese: Abrunheiro, Abrunheiro-Manso, Abrunho, Agruñeiro, Ameixa, Ameixeira, Ameixoeira;
Roman: Šljiva;
Russian: Sliva Domašnaja;
Scots: Ploum;
Serbian: Šljiva;
Slovakian: Bystrické Slivky, Slivka, Slivky;
Slovenian: Češplja, Cheshplja, Sliva;
Spanish: Ciruela, Ciruelo, Pruna, Prunero, Pruno;
Swedish: Plommon;
Thai:
Turkish: Erik;
Vietnamese: (Quả) Mận, (Trái) Mận;
West Frisian: Prom;
Zulu: Umplaamuzi.

Origin/Distribution

Prunus domestica is believed to have originated in the area of the Caucusus and Asia Minor. First findings of primitive cultivars were reported to occur in Central Europe about 500 BC, probably known already by the Celtic and Teutonic tribes. Subsequently the fruit was distributed to countries in central Europe by the Romans. It is probable that high quality cultivars originated from southeast Europe after the Middle Ages, and distributed throughout Europe by the seventeenth century. Plum is now cultivated globally in temperate to warm-temperate regions, predominantly in Central, S and SE Europe, further on in N Africa, W Asia, India and North America, Australia, New Zealand, South Africa and in south America.

Prunus domestica, the domestic European plum, is a hexaploid species. Its origin is still doubtful. It is thought to have originated from natural amphiploid crosses of two wild species, the sloe plum, *Prunus spinosa* (4n), and the cherry plum, *Prunus cerasifera* (2n). Another hypothesis is that the species arose from selections of hexaploid accessions from the gene-pool of *P. cerasifera*.

Agroecology

Plums are robust, vigorous trees that thrive in areas with cold winters, short springs and a long warm summer for optimum productivity. Plum like other *Prunus* species has a winter chill requirement or vernalization to break dormancy. "Chill hours" or "chill units" refers to the hours of temperature below 7.2°C and above 0°C that occur while the tree is dormant. The trees require a certain number of chill hours for buds to break in a timely manner and start the growing season that follows the winter cold period. Depending on the cultivars and localities the chill hour requirements ranges from 250 to 500 chill hours. Frosts do not generally harm a plum tree during the winter period but frosts over flowering during spring can devastate a crop. Both the flowers and young fruit are susceptible to this frost and can all be killed if temperatures dip below 0°C. Traditional frost control measures such as smoke pots have been surpassed by over-head sprinkler frost fighting, propylene tarp or other cover material and now in some cases wind machines. Electric heat sources such as light bulbs can also be used.

Plums prefer a sunny well-drained, fertile soil. A well-drained, sandy loam is ideal and establishment on hilltops or slopes with good air circulation for spring frost protection is most preferable. Planting on gentle slope can cause the frosts to be less severe during the sensitive flowering period in the spring. They are slightly more tolerant of damp conditions than apricots but abhor water-logged conditions. Clay loams are tolerated as long as drainage is good. Sandy soils are also suitable especially if they are grafted on nematode resistant rootstock.

Rainfall by itself does not tend to harm plum trees but it does raise humidity and increase the incidences of fungal and bacterial diseases such as spot and blast during flushing, flowering and fruit development. During periods of dry weather plums will benefit from irrigation, especially before harvest as it allows the fruit to fill up.

Plate 1 Plum flowers in fascicles

The recommended growing degree day (GDD) value for plums is similar to that for peaches, nectarines and apricots and is at least 800 GDD at a base of 10°C, although a few cultivars are able to successfully fruit in regions with only 600 GDDs. GDD is calculated by taking the average of the daily high and low temperature each day compared to a baseline winter low (usually 10°C). As an equation; GDD=((High+Low)/2)-Baseline. GDDs are typically measured from the winter low.

Edible Plant Parts and Uses

Ripe, plum fruit is sweet and juicy and it can be eaten fresh or made into jams, preserves, pastry, plum dumplings, and other foods. Plum juice can be fermented into plum wine; when distilled, this produces a brandy known in Serbia as *Slivovitz*, in Romania *as Tzuica*, in Hungary as *Palink*a and *rakia* or *rakija* in Albania, Bosnia, Croatia and Serbia. Acid plums are used for cooking. Dried plums are also known as prunes. Prunes are also sweet and juicy and are also rich antioxidant polyphenols. Prune juice is also a popular drink and is beneficial to health. Various flavours of dried plum are available at Chinese grocers and specialty stores worldwide such as salty plums, spicy, liquorice and ginseng flavoured dried plums. Such Asian store also sell pickled plums.

Prune kernel oil is made from the fleshy seed kernel inside the pit. Plum flowers are also edible. They are used as garnish for salads and ice-cream or brewed into a tea plum-jam and dried prunes.

Botany

A small, branched, deciduous tree, 4–15 m high with reddish-brown, glabrous branches, unarmed or armed with few spines and reddish-brown to greyish-green, pubescent branchlets. Winter buds are reddish-brown and glabrous. Stipules linear and glandular and petioles densely pubescent, 1–2 cm long. Leaves are alternate, deep green, elliptic to obovate, 4–10×2.5–5 cm, pubescent below, glabrous above, margins serrulate, apex acute to obtuse, base cuneate with a pair of nectaries, and with 5–7 pairs of lateral veins (Plates 3, 5, and 9). Flowers are solitary or in fascicles of 3 at the tip of branchlets, pedicellate (10–12 mm), 10–15 mm across (Plates 1 and 2). Hypanthium is pubescent; sepals 5, ovate with acute apex, imbricate; petals 5, white or greenish-white, obovate, apex rounded to obtuse, imbricate, on rim of hypanthium; stamens 20–30, in 2 whorls; filaments unequal; carpel 1; ovary superior, 1-loculed, glabrous or sometimes villous; style terminal, elongated. Drupe red, purple, purple-black, green, yellow or golden yellow, usually

Plate 2 Close-up of plum flowers

Plate 3 Developing common garden plum fruits and leaves

Plate 4 Ripening common garden plums with wax bloom

globose to oblong, rarely subglobose, 3–6 cm in diameter often glaucous with a whitish bloom (Plates 3, 4, 5, 6, 7, 8, and 9); mesocarp fleshy, not splitting when ripe, endocarp broadly ellipsoid, laterally compressed, pitted.

Nutritive/Medicinal Properties

Food value of raw, plum fruit (refuse 6% pit) per 100 g edible portion was reported as follows (USDA 2011): water 87.23 g, energy 46 kcal (192 kJ), protein 0.70 g, total lipid (fat) 0.28 g, ash 0.37 g, carbohydrate 11.42 g; fibre (total dietary) 1.4 g, total sugars 9.92 g, sucrose 1.57 g, glucose 5.07 g, fructose 3.07 g, maltose 0.08 g, galactose 0.14 g; calcium 6 mg, iron 0.17 mg, magnesium 7 mg, phosphorus 16 mg, potassium 157 mg, sodium 0 mg, zinc 0.10 mg, copper 0.057 mg, manganese 0.052 mg, fluoride 2 μg, vitamin C (total ascorbic acid) 9.5 mg, thiamine 0.028 mg, riboflavin 0.026 mg, niacin 0.417 mg, pantothenic acid 0.135 mg, vitamin B-6 0.029 mg, folate (total) 5 μg, total choline 1.9 mg, vitamin A 345 IU (17 μg RAE), vitamin E (α-tocopherol) 0.26 mg, γ-tocopherol 0.08 mg, vitamin K (phylloquinone) 6.4 μg, total saturated fatty acids 0.017 g, 16:0 (palmitic acid) 0.014 g, 18:0 (stearic

Plate 5 (**a**) Whole black plum (**b**) Halved black plum with sweet-yellow flesh

Plate 6 (**a**) Whole Stanley prune plum and (**b**) Halved Stanley Prune plum with sweet greenish-yellow flesh

Plate 7 Golden plum with sweet orangey-yellow flesh

Plate 8 Teagan blue plum with sweet, golden-yellow flesh

Plate 9 Green gage plum

acid) 0.003 g; total monounsaturated fatty acids 0.134 g, 16:1 undifferentiated (palmitoleic acid) 0.002 g, 18:1 undifferentiated (oleic acid) 0.132 g; total polyunsaturated fatty acids 0.044 g, 18:2 undifferentiated (linoleic acid) 0.044 g, phytosterols 7 mg, β-carotene 190 μg, β-cryptoxanthin 35 μg, lutein+zeaxanthin 73 μg; tryptophan 0.009 g, threonine 0.010 g, isoleucine 0.014 g, leucine 0.015 g, lysine 0.016 g, methionine 0.008 g, cystine 0.002 g, phenylalanine 0.014 g, tyrosine 0.008 g, valine 0.016 g, arginine 0.009 g, histidine 0.009 g, alanine 0.028 g, aspartic acid 0.352 g, glutamic acid 0.035 g, glycine 0.009 g, proline 0.027 g and serine 0.023 g.

In plums, the anthocyanin content increased with the red colour intensity (Vizzotto et al. 2006). Red/purple-flesh plums generally had higher phenolic content (400–500 mg chlorogenic acid/100 g fw) than the other plums. Carotenoid content in plums was similar for all varieties. AOA (antioxidant activity) tended to be higher in red/purple-flesh varieties as compared to light colored flesh plums. The best correlations were between the AOA and the total phenolics content of the fruit. The phenolic compounds hydroxycinnamates, procyanidins, flavonols, and anthocyanins were detected and quantified in red plum cultivars (Tomás-Barberán et al. 2001). As a general rule, the peel tissues contained higher amounts of phenolics, and anthocyanins and flavonols were almost exclusively located in this tissue. There was no clear trend in phenolic content with ripening of the different cultivars. The plum cultivars Black Beaut and Angeleno were especially rich in phenolics.

From the heartwood of *Prunus domestica* the following components have been isolated: (i) a new dihydroflavonol, 5,7-dihydroxy-4′-methoxy dihydroflavonol (dihydrokaempferide), (ii) a new flavonol prudomestin whose constitution is established as 5,7-dihydroxy-8,4′-dimethoxyflavonol, (iii) kaempferol (Nagarajan and Seshadri 1964). Besides these, a second dihydroflavonol and a leucoanthocyanidin were also isolated.

Prunes are good source of energy in the form of simple sugars, but do not mediate a rapid rise in blood sugar concentration, possibly because of high fibre, fructose, and sorbitol content (Stacewicz-Sapuntzakis et al. 2001). Prunes were found to contain large amounts of phenolic compounds (184 mg/100 g), mainly as neochlorogenic and chlorogenic acids, which may aid in the laxative action and delay glucose absorption. Phenolic compounds in prunes had been found to inhibit human LDL oxidation in-vitro, and thus might serve as preventive agents against chronic diseases, such as heart disease and cancer. Additionally, high potassium content of prunes (745 mg/100 g) might be beneficial for cardiovascular health.

Plums and prunes are rich in phenolic compounds and potent antioxidant property which endowed it with many pharmacological functions that include anticancer, laxative, anti-osteoporosis, antidiabetic, antihypercholesterolemic, anxiolytic, antimicrobial and mitigating cognitive deficit mitigating activities.

Antioxidant Activity

Fruit and vegetables rich in anthocyanins (e.g. strawberry, raspberry and red plum) demonstrated the highest antioxidant activities, followed by those rich in flavanones (e.g. orange and grapefruit) and flavonols (e.g. onion, leek, spinach and green cabbage), while the hydroxycinnamate-rich fruit (e.g. apple, tomato, pear and peach) consistently elicited the lower antioxidant activities (Proteggente et al. 2002). The TEAC (Trolox Equivalent Antioxidant Capacity), the FRAP (Ferric Reducing Ability of Plasma) and ORAC (Oxygen Radical Absorbance Capacity) values for each extract were relatively similar and well-correlated with the total phenolic and vitamin C contents. The TEAC antioxidant activities in terms of 100 g FW uncooked portion size were in the order: strawberry >> raspberry = red plum >> red cabbage >>> grapefruit = orange > spinach > broccoli > green grape approximately/= onion > green cabbage > pea > apple > cauliflower tomato approximately/= peach = leek > banana approximately/= lettuce.

The anthocyanin content of plums ranged from 44.1 to 231.29 mg cyanidin 3-glucoside/100 g fresh tissue (Cevallos-Casals et al. 2002). The total phenolic content ranged from 298 to 563 mg chlorogenic acid/100 g fresh tissue. The antioxidant activity ranged from 1,254 to 3,244 μg Trolox equivalent /g fresh tissue for the plums. Correlation analysis indicated that the anthocyanin content and phenolic content was well correlated with the anti-oxidant activity. Plum extracts showed good antimicrobial activity and some potential as a colourant.

The total phenolic contents of various plum cultivars widely varied from 125.0 to 372.6 mg/100 g expressed as gallic acid equivalents (Kim et al. 2003). The level of total flavonoids in fresh plums ranged between 64.8 and 257.5 mg/100 g expressed as catechin equivalents. Antioxidant capacity, expressed as vitamin C equivalent antioxidant capacity (VCEAC), ranged from 204.9 to 567.0 mg/100 g with an average of 290.9 mg/100 g of fresh weight. Cv. Beltsville Elite B70197 showed the highest amounts of total phenolics and total flavonoids and the highest VCEAC. A positive relationship (correlation coefficient $R^2=0.977$) was presented between total phenolics and VCEAC, suggesting polyphenolics would play an important role in free radical scavenging. The level of IC_{50} value of superoxide radical anion scavenging activity of the plum cultivars ranged from 13.4 to 45.7 mg of VCEAC/100 g. Neochlorogenic acid was the predominant polyphenolic among fresh plums tested. Flavonols found in plum were commonly quercetin derivatives. Rutin was the most predominant flavonol in plums. Various anthocyanins containing cyanidin aglycon and peonidin aglycon were commonly found in all plums except for cv. Mirabellier and NY 101. Further the scientists reported that the superoxide radical scavenging activity (SRSA) levels of the polyphenols were closely related to their chemical structures; cyanidin showed the lowest IC_{50} among the polyphenols examined, and aglycones are more effective than their glycosides (Chun et al. 2003a). BY 69–339 cultivar exhibited the lowest IC_{50} among the 11 plum cultivars, which means the highest antioxidant activity in scavenging superoxide radicals, followed by French Damson, Cacaks Best, Beltsville Elite B70197, Empress, Castleton, Stanley, NY 6, NY 101, Mirabellier, and NY 9. IC_{50} values showed a higher correlation with total flavonoids ($R^2=0.8699$) than total phenolics ($R^2=0.8355$), which indicated that flavonoids might contribute to the total SRSA more directly than other polyphenols. Anthocyanins in plums appeared to be the major contributors to the total SRSA, except for two yellow cultivars having no anthocyanins. Chlorogenic acid was the predominant phenolic acid, and it also exhibited SRSA significantly in the range of 1.0–94.9%. Quercetins were the major flavonols in plums. However, they showed relatively low contribution to the total SRSA. The scientists also found a good linear relationship between the amount of total phenolics and total antioxidant capacity ($R^2=0.9887$) (Chun et al. 2003b). The amount of total flavonoids and total antioxidant capacity also showed a good correlation ($R^2=0.9653$). Although the summation of individual antioxidant capacity was lower than the total antioxidant capacity of plum samples, there was a positive correlation ($R^2=0.9299$) of total antioxidant capacity of plum samples with the sum of the vitamin C equivalent antioxidant capacity (VCEAC) s calculated from individual phenolics. Chlorogenic acids and glycosides of cyanidin, peonidin, and quercetin were major phenolics among 11 plum cultivars. The antioxidant capacity of chlorogenic acids and anthocyanins showed higher correlation (R^2) of 0.7751 and 0.6616 to total VCEAC, respectively, than that of quercetin glycosides ($R^2=0.0279$). Chlorogenic acids were a major source of antioxidant activity in plums, and the consumption of one serving (100 g) of plums can provide antioxidants equivalent to 144.4–889.6 mg of vitamin C.

In recent studies, the 2,2′-azino-bis(3-ethylbenzothiazoline-6-sulfonic acid) (ABTS) and cellular tests revealed that the total antioxidant capacities, expressed as vitamin C equivalents, ranged from 691.2 to 2,164.64 mg and from 613.98 to 2,137.59 mg per 100 g of fresh weight, respectively, suggesting plums to be rich in natural antioxidants and appreciably protect granulocytes from oxidative stress (Bouayed et al. 2009). The results showed a linear correlation between either total phenolic or flavonoid contents and total antioxidant capacity, revealing that these compounds contributed significantly to the antioxidant activity of plums. The results also suggested that individual polyphenolics contributed directly to the total protective effect of plums.

Neochlorogenic acid was found to be the most important phenolic acid in all the Norwegian plum cultivars studies (Slimestad et al. 2009). Together with other phenolic acids, this compound varied significantly in amount among the cultivars. Cyanidin 3-rutinoside was found to account for >60% of the total anthocyanin content. Minor amounts of flavonols (rutin and quercetin 3-glucoside) were detected in all cultivars. Total antioxidant capacity varied from 814 to 290 μmol of Trolox 100/g of fresh weight.

Thiolysis of the phenolic compounds showed that the flesh and skin contained a large proportion of flavan-3-ols, which account, respectively, for 92% and 85% in 'Golden Japan' (GJ), 61% and 44% in 'Green Gage' (GG-V), 62% and 48% in 'Green Gage' (GG-C), 54% and 27% in 'Mirabelle' (M) and 45% and 37% in 'Green Gage' (GG-F) (Nunes et al. 2008). Terminal units of procyanidins observed in plums were mainly (+)-catechin (54–77% of all terminal units in flesh and 57–81% in skin). The GJ plums showed a phenolic composition different from all of the others, with a lower content of chlorogenic acid isomers and the presence of A-type procyanidins as dimers and terminal residues of polymerized forms. The average degree of polymerization (DPn) of plum procyanidins was higher in the flesh (5–9 units) than in the skin (4–6 units). Procyanidin B7 was observed in the flesh of all 'Green Gage' plums and in the skin of the Portuguese ones "Rainha Claudia Verde".

Studies show that prunes and prune juice may provide a source of dietary antioxidants (Donovan et al. 1998). The mean concentrations of phenolics were 1,840 mg/kg, 1,397 mg/kg, and 441 mg/L in pitted prunes, extra large prunes with pits, and prune juice, respectively Hydroxycinnamates, especially neochlorogenic acid, and chlorogenic acid predominated, and these compounds, as well as the prune and prune juice extracts, inhibited the oxidation of low-density lipoprotein (LDL). The pitted prune extract inhibited LDL oxidation by 24, 82, and 98% at 5, 10, and 20 μM gallic acid equivalents (GAE). The prune juice extract inhibited LDL oxidation by 3, 62, and 97% at 5, 10, and 20 μM GAE.

Prune were found to contain neochlorogenic acid, cryptochlorogenic acid and chlorogenic acid in the ratio 78.7:18.4:3.9, respectively (Nakatani et al. 2000). Each chlorogenic isomer showed antioxidative activities which were almost the same as evaluated by scavenging activity on superoxide anion radicals and inhibitory effect against oxidation of methyl linoleate. Furthermore, hydrolysis of EtOH extract residue led to higher levels of total phenolics and ORAC, and these results suggested the existence of conjugated antioxidant components in prunes.

The methanol eluate exhibited the strongest antioxidant activity among the separated fractions of prune extract evaluated by oxygen radical absorbance capacity (ORAC) (Kayano et al. 2002). Further purification of the MeOH eluate led to isolation of a novel compound, 4-amino-4-carboxychroman-2-one, together with four known compounds (p-coumaric acid, vanillic acid β-glucoside, protocatechuic acid, and caffeic acid), The ORAC values of these isolated compounds showed 0.15–1.43 units (μmol of Trolox equiv)/μmol, and the new compound showed a remarkable synergistic effect on caffeoylquinic acid isomers. The antioxidant activity of the MeOH eluate was highly dependent on the major prune components, caffeoylquinic acid isomers, with a contribution from the new synergist. Prunes showed high antioxidant activity on the basis of the oxygen radical absorbance capacity (ORAC), and their major antioxidant components were caffeoylquinic acid isomers (Kayano et al. 2003). A novel bipyrrole compound identified to be 2-(5-hydroxymethyl-2′,5′-dioxo-2′,3′,4′,5′-tetrahydro-1′H-1,3′-bipyrrole) carbaldehyde, and 7 phenolic compounds were isolated from prunes for the first time (Kayano et al. 2004a). The degree of contribution caffeoylquinic acid isomers to the ORAC was found to be 28.4%; the remaining ORAC is dependent on other antioxidant phenolic compounds. The contribution of caffeoylquinic acid (CQA) isomers to the antioxidant activity of prunes was revealed to be 28.4% on the basis of oxygen radical absorbance capacity (ORAC); hence, it was indicated that residual ORAC is dependent on unknown antioxidant components (Kayano et al. 2004b). Of the total 28 compounds isolated, four abscisic acid related compounds, a chromanon, and a bipyrrole were novel. Each CQA isomer in prunes showed high antioxidant activities when measured by the oil stability index (OSI) method, O^{2-} scavenging activity, and ORAC. Other isolated compounds such as hydroxycinnamic acids, benzoic acids, coumarins, lignans, and flavonoid also showed high ORAC values.

Four new abscisic acid related compounds (1–4), together with (+)-abscisic acid (5), (+)-β-D-glucopyranosyl abscisate (6), (6S,9R)-roseoside (7), and two lignan glucosides ((+)-pinoresinol mono-β-D-glucopyranoside (8) and 3-(β-D-glucopyranosyloxymethyl)-2- (4-hydroxy-3-methoxyphenyl) 5 (3-hydroxypropyl)-7-methoxy-(2R,3S)-dihydrobenzofuran (9)) were isolated from the antioxidative ethanol extract of prunes (*Prunus domestica*) (Kikuzaki et al. 2004). The structures of 1–4 were elucidated to be rel-5-(3S,8S-dihydroxy-1R,5S-dimethyl-7-oxa-6-oxobicyclo[3,2,1]oct-8-yl)-3-methyl-2Z,4E-pentadienoic acid (1), rel-5-(3S,8S-dihydroxy-1R,5S-dimethyl-7-oxa-6-oxobicyclo[3,2,1] oct-8-yl)-3-methyl-2Z,4E-pentadienoic acid 3′-O-β-d-glucopyranoside (2), rel-5-(1R,5S-dimethyl-3R,4R,8S-trihydroxy-7-oxa-6-oxobicyclo[3,2,1]oct-8-yl)-3-methyl-2Z,4E-pentadienoic acid (3), and rel-5-(1R,5S-dimethyl-3R,4R,8S-trihydroxy-7-oxabicyclo[3,2,1]- oct-8-yl)-3-methyl-2Z,4E-pentadienoic acid (4). The ORAC values of abscisic acid related compounds (1–7) were very low. Two lignans (8 and 9) were more effective antioxidants whose ORAC values were 1.09 and 2.33 μmol of Trolox equivalent/μmol, respectively. An antioxidative oligomeric proanthocyanidin from prunes composed of epicatechin and catechin units showed greater potency than chlorogenic acid another potent antioxidative component in prunes (Kimura et al. 2008).

Anxiolytic Activity

Chlorogenic acid, a polyphenol from *Prunus domestica* (Mirabelle), exhibited anxiolytic effects coupled with antioxidant activity (Bouayed et al. 2007). Chlorogenic acid (20 mg/kg) induced a decrease in anxiety-related behaviors suggesting an anxiolytic-like effect of this polyphenol. The anti-anxiety effect was blocked by flumazenil suggesting that anxiety was reduced by activation of the benzodiazepine receptor. In-vitro, chlorogenic acid protected granulocytes from oxidative stress.

Antiosteoporosis Activity

Osteoporosis is a common debilitating disorder that affects both female and male, albeit more so in women. Aside from existing drug therapies, certain lifestyle and nutritional factors are known to reduce the risk of osteoporosis. Dried prunes being an important source of boron, had been postulated to play a role in prevention of osteoporosis (Stacewicz-Sapuntzakis et al. 2001). A serving of prunes (100 g) would fulfill the daily requirement for boron (2–3 mg).

Among nutritional factors, recent studies suggested that dried plum, or prunes (*Prunus domestica*) was the most effective fruit in both preventing and reversing bone loss (Hooshmand and Arjmandi 2009). Animal studies and a 3-month clinical trial had shown that dried plum had positive effects on bone indices. The animal data indicated that dried plum not only protected against but more importantly reversed bone loss in two separate models of osteopenia. Initial animal study indicated that dried plum prevented the ovariectomy-induced reduction in bone mineral density (BMD) of the femur and lumbar vertebra (Arjmandi et al. 2002). In another study (Deyhim et al. 2005), found that dried plum as low as 5% (w/w) restored femoral and tibial BMD to the level of intact rats. Dried plum reversed the loss of trabecular architectural properties such as trabecular number and connectivity density, and trabecular separation in ovarietomized rats. Varying doses of dried plum were also able to significantly improve trabecular microarchitectural properties in comparison with ovariectomized controls. Analysis of BMD and trabecular bone structure by microcomputed tomography (microCT) revealed that dried plum enhanced bone recovery during reambulation following skeletal unloading and had comparable effects to parathyroid hormone. In addition to the animal studies, the 3-month clinical trial indicated that the consumption of dried plum daily by postmenopausal women significantly increased serum markers of bone formation, total alkaline phosphatase, bone-specific alkaline phosphatase and insulin-like growth factor-I by 12%, 6%, and 17%, respectively (Arjmandi et al. 2002). Higher levels of both serum IGF-I and BSAP were associated with greater rates of bone formation. Additional studies showed that dried plum prevented osteopenia in androgen deficient male rats, and these beneficial effects may be attributed in part to a decrease in osteoclastogenesis via down-regulation of receptor activator of NFκ-B ligand (RANKL) and osteoprotegerin and stimulation of bone formation mediated by serum insulin-like growth factor (IGF)-I (Franklin et al. 2006).

In a study involving 236 women with 1–10 years postmenopausal and not on hormone replacement therapy or any other prescribed medication, dried plum daily ingestion significantly increased bone mineral density of ulna and spine in comparison with dried apple when assessed at 12 months (Hooshmand et al. 2011). In comparison with corresponding baseline values, only dried plum significantly decreased serum levels of bone turnover markers including bone-specific alkaline phosphatase and tartrate-resistant acid phosphatase-5b. The findings of the study confirmed the ability of dried plum in improving bone mineral density in postmenopausal women in part due to suppressing the rate of bone turnover.

Anticancer Activity

Ethanol fraction from concentrated prune juice (PE) dose-dependently reduced the viable cell number of Caco-2, KATO III, but does not reduce the viable cell number of human normal colon fibroblast cells (CCD-18Co) used as a normal cell model (Takashi et al. 2006). PE treatment for 24 hours led to apoptotic changes in Caco-2 such as cell shrinkage and blebbed surfaces due to the convolutions of nuclear and plasma membranes and chromatin condensation, but this was not observed in CCD-18Co. PE induced nucleosomal DNA fragmentation typical of apoptosis in Caco-2 after 24 hours of treatment. These results show that PE induced apoptosis in Caco-2.

Laxative Effect

Because of their sweet flavor and well-known mild laxative effect, prunes are considered to be an epitome of functional foods. Dried prunes contain approximately 6.1 g of dietary fiber per 100 g, while prune juice is devoid of fiber due to filtration before bottling (Stacewicz-Sapuntzakis et al. 2001). The laxative action of both prune and prune juice could be explained by their high sorbitol content (14.7 and 6.1 g/100 g, respectively).

Antihypercholesterolemic Activity

Consumption of prunes as a source of dietary fiber in men with mild hypercholesterolemia was found to have beneficial effect (Tinker et al. 1991). Plasma low-density-lipoprotein cholesterol was significantly lower after the prune period (3.9 mmol/L) than after the grape-juice-control period (4.1 mmol/L). Faecal bile acid concentration of lithocholic acid was significantly lower after the prune period (0.95 mg bile acid/g dry weight stool) than after the grape-juice-control period (1.20 mg bile acid/g dry weight stool). Both fecal wet and dry weights were approximately 20% higher after the prune period than after the grape-juice-control period. Total bile acids (mg/72 h) did not significantly differ between experimental periods.

Antihypertensive Activity

In an 8-week placebo controlled clinical trial involving 259 pre-hypertensive volunteers, single dose of prune comprising prune juice and dried prune (*P. domestica*) daily caused a significant reduction in blood pressure (BP) (Ahmed et al. 2010a). With the double dose of prunes, only systolic BP was reduced significantly. Control group had significantly increased serum HDL whereas prune groups had significantly reduced serum cholesterol and LDL.

Hepatoprotective Activity

In an 8 week clinical trial involving 166 healthy volunteers, ingestion of prune juice and prune fruit (single or double dose) was found to have a beneficial effect in hepatic disease (Ahmed et al. 2010b) There was significant reduction of serum alanine transaminase and serum alkaline phosphatase by the lower dose of prunes. There was no change in serum aspartate transaminase and bilirubin.

Cognitive Activity

Studies showed that supplementation with *Prunus domestica* mitigated age-related deficits in cognitive function in rats (Shukitt-Hale et al. 2009). Rats that drank plum juice from 19 to 21 months of age had improved working memory in the Morris water maze, whereas rats fed dried plum powder were not different from the control group, possibly due to the smaller quantity of phenolics consumed in the powder group compared with the juice group.

Antimicrobial Activity

Prunusins A (1) and B (2), the new C-alkylated flavonoids, isolated from the seed kernels of *Prunus domestica* showed significant antifungal activity against pathogenic fungus *Trichophyton simmi* (Mahmood et al. 2010). 3, 5, 7, 4'-Tetrahydroxyflavone (3) and 3, 5, 7-trihydroxy-8, 4'-dimethoxyflavone (4) were also isolated.

Oil fractions, obtained from n-hexane extract of *Prunus domestica* shoots contained hentricontane (35.7%), ethyl hexadecanoate (21.7%) and linoleic acid (16.16%) (Mahmood et al. 2009). Bioassay screening of oil showed moderate antibacterial activity against *Salmonella* group (Gram +ve and –ve), moderate antifungal activity against *Microsporum canis* and good antioxidant activity by DPPH radical scavenging method.

Antidiabetic Activity

Purunusides A-C, new homoisoflavone glucosides together with the known compounds β-sitosterol and 6,7-methylenedioxy-8-methoxy coumarin were isolated from n-butanol and ethyl acetate soluble fractions of *Prunus domestica* (Kosar et al. 2009). The purunusides A-C showed potent inhibitory activity against the enzyme α-glucosidase. α-glucosidase inhibitors and acted as antidibaetic drugs, preventing digestion of carbohydrates in the digestive tract, thereby lowering the after-meal glucose levels.

Respiratory Toxin

Plum (*Prunus domestica*) seeds contain the cyanogenic diglucoside (R)-amygdalin and lesser amounts of the corresponding monoglucoside (R)-prunasin, which releases the respiratory toxin, hydrogen cyanide upon tissue disruption (Poulton and Li 1994).

Traditional Medicinal Uses

Dried plum fruit or prunes are used as laxative and also as stomachic. The seed contains amygdalin and prusin which break down in water to form hydrocyanic acid and prussic acid. Amydalin has been illegally used for treating cancer but has now been proven to be not effective for cancer (see *Prunus persica*). The bark is occasionally used as a febrifuge.

Other Uses

Green, dark grey and yellow dyes can be obtained from the leaves, fruit and bark respectively. The gum from the stem can be used as adhesive. Ground seeds are used in cosmetics in the production of face masks for dry skin. The wood can be used for making musical instruments.

Comments

Prunus domestica has a long history of cultivation. Three subspecies have been recognised, each with a host of horticultural varieties:

P. domestica ssp. *domestica* the common plums;

P. domestica ssp. *institia*, the damson, bullace or mirabelle

P. domestica ssp. *italica,* the greengage.

Selected References

Ahmed T, Sadia H, Batool S, Janjua A, Shuja F (2010a) Use of prunes as a control of hypertension. J Ayub Med Coll Abbottabad 22(1):28–31

Ahmed T, Sadia H, Khalid A, Batool S, Janjua A (2010b) Report: prunes and liver function: a clinical trial. Pak J Pharm Sci 23(4):463–466

Arjmandi BH, Khalil DA, Lucas EA, Georgis A, Stoecker BJ, Hardin C, Payton ME, Wild RA (2002) Dried plums improve indices of bone formation in postmenopausal women. J Womens Health Gend Based Med 11(1):61–68

Bouayed J, Rammal H, Dicko A, Younos C, Soulimani R (2007) Chlorogenic acid, a polyphenol from *Prunus domestica* (Mirabelle), with coupled anxiolytic and antioxidant effects. J Neurol Sci 262(1–2):77–84

Bouayed J, Rammal H, Dicko A, Younos C, Soulimani R (2009) The antioxidant effect of plums and polyphenolic compounds against H(2)O(2)-induced oxidative stress in mouse blood granulocytes. J Med Food 12(4):861–868

Bown D (1995) Encyclopaedia of herbs and their uses. Dorling Kindersley, London, 424pp

Byrne D, Vizzotto M, Cisneros-Zevallos L, Ramming DW, Okie WR (2004) Antioxidant content of peach and plum genotypes. HortScience 39(4):798

Cevallos-Casals BA, Byrne DH, Cisneros-Zevallos L, Okie WR (2002) Total phenolic and anthocyanin content in red-fleshed peaches and plums. Acta Hort (ISHS) 592:589–592

Chun OK, Kim DO, Lee CY (2003a) Superoxide radical scavenging activity of the major polyphenols in fresh plums. J Agric Food Chem 51(27):8067–8072

Chun OK, Kim DO, Moon HY, Kang HG, Lee CY (2003b) Contribution of individual polyphenolics to total antioxidant capacity of plums. J Agric Food Chem 51(25):7240–7245

Deyhim F, Stoecker BJ, Brusewitz GH, Devareddy L, Arjmandi BH (2005) Dried plum reverses bone loss in an osteopenic rat model of osteoporosis. Menopause 12(6):755–762

Donovan JL, Meyer AS, Waterhouse AL (1998) Phenolic composition and antioxidant activity of prunes and prune juice (*Prunus domestica*). J Agric Food Chem 46(4):1247–1252

Facciola S (1990) Cornucopia. A source book of edible plants. Kampong Publication, Vista, 677pp

Franklin M, Bu SY, Lerner MR, Lancaster EA, Bellmer D, Marlow D, Lightfoot SA, Arjmandi BH, Brackett DJ, Lucas EA, Smith BJ (2006) Dried plum prevents bone loss in a male osteoporosis model via IGF-I and the RANK pathway. Bone 39(6):1331–1342

Gil MI, Tomás-Barberán FA, Hess-Pierce B, Kader AA (2002) Antioxidant capacities, phenolic compounds, carotenoids, and vitamin C contents of nectarine, peach, and plum cultivars from California. J Agric Food Chem 50(17):4976–4982

Grieve M (1971) A modern herbal, 2 vols. Penguin/Dover publications, New York, 919pp

Gu C, Bartholomew B (2003) *Prunus* Linnaeus. In: Wu ZY, Raven PH, Hong DY (eds) Flora of China, vol 9 (*Pittosporaceae* through *Connaraceae*). Science Press/Missouri Botanical Garden Press, Beijing/St. Louis

Hedrick UP (1972) Sturtevant's edible plants of the world. Dover Publications, New York, 686pp

Hooshmand S, Arjmandi BH (2009) Viewpoint: dried plum, an emerging functional food that may effectively improve bone health. Ageing Res Rev 8(2):122–127

Hooshmand S, Chai SC, Saadat RL, Payton ME, Brummel-Smith K, Arjmandi BH (2011) Comparative effects of dried plum and dried apple on bone in postmenopausal women. Br J Nutr 106(6):923–930

Kayano S, Kikuzaki H, Ikami T, Suzuki T, Mitani T, Nakatani N (2004a) A new bipyrrole and some phenolic constituents in prunes (*Prunus domestica* L.) and their oxygen radical absorbance capacity (ORAC). Biosci Biotechnol Biochem 68(4):942–944

Kayano S, Kikuzaki H, Yamada NF, Aoiki A, Kasamatsu K, Yamasaki Y, Ikami T, Suzuki T, Mitani T, Nakatani N (2004b) Antioxidant properties of prunes (*Prunus domestica* L.) and their constituents. Biofactors 21(1–4):309–313

Kayano S, Kikuzaki H, Yamada NF, Mitani T, Nakatani N (2002) Antioxidant activity of prune (*Prunus domestica* L.) constituents and a new synergist. J Agric Food Chem 50:3708–3712

Kayano S, Yamada NF, Suzuki T, Ikami T, Shioaki K, Kikuzaki H, Mitani T, Nakatani N (2003) Quantitative evaluation of antioxidant components in prunes (*Prunus domestica* L.). J Agric Food Chem 51(5):1480–1485

Kikuzaki H, Kayano S, Fukutsuka N, Aoki A, Kasamatsu K, Yamasaki Y, Mitani T, Nakatani N (2004) Abscisic acid related compounds and lignans in prunes (*Prunus domestica* L.) and their oxygen radical absorbance capacity (ORAC). J Agric Food Chem 52(2):344–349

Kim DO, Chun OK, Kim YJ, Moon HY, Lee CY (2003) Quantification of polyphenolics and their antioxidant capacity in fresh plums. J Agric Food Chem 51(22):6509–6515

Kimura Y, Ito H, Kawaji M, Ikami T, Hatano T (2008) Characterization and antioxidative properties of oligomeric proanthocyanidin from prunes, dried fruit of *Prunus domestica* L. Biosci Biotechnol Biochem 72(6):1615–1618

Kosar S, Fatima I, Mahmood A, Ahmed R, Malik A, Talib S, Chouhdary MI (2009) Purunusides A-C, alpha-glucosidase inhibitory homoisoflavone glucosides from *Prunus domestica*. Arch Pharm Res 32(12):1705–1710

Mahmood A, Ahmed R, Kosar S (2009) Phytochemical screening and biological activities of the oil components of *Prunus domestica* Linn. J Saudi Chem Soc 13(3):273–277

Mahmood A, Fatima I, Kosar S, Ahmed R, Malik A (2010) Structural determination of prunusins A and B, new C alkylated flavonoids from *Prunus domestica*, by 1D and 2D NMR spectroscopy. Magn Reson Chem 48(2):151–154

McLaren GF, Glucina PG (1995) Prunes: prospects for a new fruit crop in New Zealand. Hortnet. http://www.hortnet.co.nz/publications/science/prune.htm

Nagarajan GR, Seshadri TR (1964) Flavonoid components of the heartwood of *Prunus domestica* Linn. Phytochemistry 3(4):477–484

Nakatani N, Kayano S, Kikuzaki H, Sumino K, Katagiri K, Mitani T (2000) Identification, quantitative determination, and antioxidative activities of chlorogenic acid isomers in prune (*Prunus domestica* L.). J Agric Food Chem 48:5512–5516

Nunes C, Guyot S, Marnet N, Barros AS, Saraiva JA, Renard CM, Coimbra MA (2008) Characterization of plum procyanidins by thiolytic depolymerization. J Agric Food Chem 56(13):5188–5196

Porcher MH et al (1995–2020) Searchable World Wide Web Multilingual Multiscript Plant Name Database. Published by The University of Melbourne, Australia. http://www.plantnames.unimelb.edu.au/Sorting/Frontpage.html

Poulton JE, Li CP (1994) Tissue level compartmentation of (R)-amygdalin and amygdalin hydrolase prevents large-scale cyanogenesis in undamaged *Prunus* seeds. Plant Physiol 104(1):29–35

Proteggente AR, Pannala AS, Paganga G, Van Buren L, Wagner E, Wiseman S, Van De Put F, Dacombe C, Rice-Evans CA (2002) The antioxidant activity of regularly consumed fruit and vegetables reflects their phenolic and vitamin C composition. Free Radic Res 36(2):217–233

Shukitt-Hale B, Kalt W, Carey AN, Vinqvist-Tymchuk M, McDonald J, Joseph JA (2009) Plum juice, but not dried plum powder, is effective in mitigating cognitive deficits in aged rats. Nutrition 25(5):567–573

Slimestad R, Vangdal E, Brede C (2009) Analysis of phenolic compounds in six Norwegian plum cultivars (*Prunus domestica* L.). J Agric Food Chem 57(23):11370–11375

Stacewicz-Sapuntzakis M, Bowen PE, Hussain EA, Damayanti-Wood BI, Farnsworth NR (2001) Chemical composition and potential health effects of prunes: a functional food? Crit Rev Food Sci Nutr 41(4):251–286

Takashi F, Takao I, Xu JW, Katsumi I (2006) Prune extract (*Prunus domestica* L.) suppresses the proliferation and induces the apoptosis of human colon carcinoma Caco-2. J Nutr Sci Vitaminol 52(5):389–391

Tinker LF, Schneeman BO, Davis PA, Gallaher DD, Waggoner CR (1991) Consumption of prunes as a source of dietary fiber in men with mild hypercholesterolemia. Am J Clin Nutr 53(5):1259–1265

Tomás-Barberán FA, Gil MI, Cremin P, Waterhouse AL, Hess-Pierce B, Kader AA (2001) HPLC-DAD-ESIMS analysis of phenolic compounds in nectarines, peaches, and plums. J Agric Food Chem 49(10):4748–4760

U.S. Department of Agriculture, Agricultural Research Service (2011) USDA National Nutrient Database for Standard Reference, Release 24. Nutrient Data Laboratory Home Page. http://www.ars.usda.gov/ba/bhnrc/ndl

Vizzotto M, Cisneros-Zevallos L, Byrne DH, Ramming DW, Okie WR (2006) Total phenolic, carotenoid, and anthocyanin content and antioxidant activity of peach and plum genotypes. Acta Hort 713:453–456

Prunus domestica subsp. *insititia*

Scientific Name

Prunus domestica L. **subsp.** **insititia** (**L.**) C. K. Schneider

Synonyms

Prunus domestica subsp. *insititia* (L.) Bonnier & Layens, *Prunus domestica* L. subsp. *insititia* (L.) C. K. Schneider (Bullace Group), *Prunus domestica* L. subsp. *insititia* (L.) Poiret, *Prunus domestica* L. var. *insititia* (L.) Fiori & Paoletti, *Prunus insititia* L.

Family

Rosaceae

Common/English Names

Bullace Plum Damask Plum, Damson, Damson Plum, Eurasian Wild Plum, Persian Gum.

Vernacular Names

Austria: Steinkrieche;
Chinese: Ou Ya Ye Li, Wu Jing Zi Li;
Czech: Slíva, Slivoň Obecná, Švestka Domácí;
Danish: Kræge;
Dutch: Kroos, Kroosjes;
Eastonian: Kreegipuu;
Finnish: Kriikuna, Kriikunapuu
French: Crèque, Crèquier, Pruneaulier, Prunier À Greffer, Prunier De Saint Julien, Prunier Sauvage;
German: Haferpflaume, Hafer-Pflaume, Krieche, Kriechenpflaume, Kriechenpflaume, Pflaumenbaum;
Hungarian: Kék Ringló, Kökényszilva;
India: Aaluucaa, Aaluuchaa, Alubukhara, Alucha (Hindu), Aluka (Urdu);
Italian: Prugnola Ciliegia, Prugnola Da Siepe, Susina, Susina Salvatica, Susino, Susino Di Macchia;
Nepalese: Aalucaa, Aluca;
Norwegian: Kreke;
Polish: Śliwa Lubaszka;
Portuguese: Abrunheiro, Ameixieira, Cabrunho;
Russian: Sliva Nenastoiashchaia, Ternosliva;
Slovaščina: Cibora, Trnasta Sliva;
Solvencina: Slivka Guľatoplodá;
Spanish: Ciruelo Silvestre, Endrino, Endrino De Injertar, Endrino Grande, Endrino Mayor, Endrino Prunero, Espino De Injertar, Niso, Niso Ciruelo De San Julián;
Swedish: Krikon;
Ukranian: Ternosliva.

Origin/Distribution

The species is native to south western Asia and Europe. It has escaped and become naturalised from Southeast Russia through Central Europe to

Western France. It is cultivated in Europe, Western Asia, India, North Africa and North America.

Agroecology

As described for plums, it thrives in areas with a temperate climate. It does best in full sun in moist, well-drained fertile, sandy loams to clayey loams with pH of 4.5–6. It has a low winter chill requirement. It is cold hardy tolerating frost to −20°C. Rain close to harvest is detrimental causing the fruits to crack.

Plate 1 Leaves and fruits

Edible Plant Parts and Uses

The ripe fruit is eaten fresh, made into pies and puddings, preserves or processed into jams and jellies. Fermented damson fruits is processed into wine, gin and in Slavic countries processed into the plum spirit *Slivovitz*.

Botany

A deciduous shrub or tree growing to 6 m tall. Branches grayish black, glabrous, sometimes spiny; branchlets brown, tomentose. Stipules lorate with acuminate apex. Petioles pubescent. Leaves obovate, elliptic, or rarely oblong, 3.5–8 cm by 2-4cm, pubescent becoming glabrescent, abaxially pale green, adaxially dark green, base cuneate to broadly cuneate and with a pair of nectaries, margin coarsely serrate, apex acute to obtuse; midvein and secondary veins prominent (Plate 1). Flowers solitary or in 2-3 flowered fascicles. Hypanthium glabrous. Sepals narrowly ovate to oblong; petals white and inconspicuously purplish veined, broadly obovate, base broadly cuneate and with a short claw, apex obtuse; stamens 20–25, ovary glabrous, superior; stigma disc-shaped. Drupe green becoming purplish-black, oval to ovoid pointed at one end (Plates 1 and 2), 2.5–4 cm diameter glabrous; endocarp small, more or less flattened, nearly smooth. Pulp smooth-textured, juicy and greenish-yellow.

Plate 2 Immature glaucous fruits

Nutritive/Medicinal Properties

For nutrient values of damsom plums see chapter on plums (*Prunus domestica*).

The contribution of the seed and pericarp to the content of malic, quinic, citric and fumaric acids, and sucrose, fructose and glucose was determined during development and ripening of damson plum fruits (Garcia-Marino et al. 2008). In whole fruit, malic and quinic acids were the predominant organic acids and their levels varied significantly, the highest being found at the beginning of the late-green stage of fruit development. The content of citric and fumaric acids was low but fluctuated remarkably towards development and ripening. In the seed, the levels of malic, quinic and fumaric acid were lower in ripening than at the beginning of maturation, and a significant synthesis of citric occurred from the middle of maturation onwards. In the mesocarp, however, malic, quinic, and citric acids peaked in the middle of maturation, whereas fumaric acid increased appreciably towards ripening. In the epicarp, quinic and malic peaked at the beginning of ripening and maturation. In the seed, all soluble sugars peaked at the middle of maturation, while fructose and glucose (the most abundant soluble sugars) tended to be stored during ripening, sucrose (the most abundant in the edible part of fruit) decreased. All the soluble sugars tend to increase in mesocarp and epicarp throughout maturation and ripening.

Other Phytochemicals

The flower opening of damson plum (*Prunus insititia*) was accompanied by an increase in the content of free-polyamines in the sepals, petals and sex organs, the ovary being most active in accumulating spermine (De Dios et al. 2006). The fertilization process and senescence brought on a decline in ovarian spermine, but stimulated putrescine and spermidine content in the sepals. The mesocarp contributed the most to the total content in free polyamines throughout damson fruit development. The control of S-adenosylmethionine (SAM) distribution towards ethylene and/or polyamine appeared to differ during the development of the endocarp, as the only peak of free-putrescine (detected in S2) coincided with the highest 1-aminocyclopropane-1-carboxylic acid (ACC) accumulation and ethylene production. On the contrary, in S3 it was probable that SAM was transformed preferentially into free-polyamines, given that free-spermidine and spermine were hardly detectable in S1 and S2 phases of fruit development. The endocarp of this climacteric fruit produced only ethylene at the end of the S1 phase and throughout S2, in which there was a great accumulation in ACC and its conjugate, 1-(malonyl-amino)-cyclopropane-1-carboxylic acid (MACC). The greatest amounts of ACC and MACC were observed in the ripening mesocarp and epicarp.

Traditional Medicinal Uses

The bark of the root and branches is regarded to be febrifuge and considerably styptic. An infusion of the flowers has been used as a mild purgative for children (Grieve 1971). Although no specific mention has been seen for this species, all members of the genus *Prunus* are known to contain amygdalin and prunasin, that hydrolyses to form hydrocyanic acid (cyanide or prussic acid). In small amounts this exceedingly poisonous compound stimulates respiration, improves digestion and gives a sense of well-being (Bown 1995).

Other Uses

Damson trees are fairly wind resistant and can be grown as a shelter-belt hedge.

A green dye can be obtained from the leaves and a grey to green dye from the fruit (Grae 1974).

Comments

There is general consensus that damson (*P. insititia*) is the ancestor of the large-fruited domestic plums (Woldring 1997–1998). Ramming and Cociu (1991) accepted *Prunus insititia* as the species name and commented that it hybridizes with *P. domestica*. According to Depypere et al. (2009) morphometric and AFLP-PCR (amplified fragment length polymorphism-polymerase chain reaction) studies revealed this taxon *Prunus insititia*; both genetically and morphologically to share similar characters with *P. domestica* supporting its inclusion in *P. domestica*.

Selected References

Bailey LH (1976) Hortus third. A concise dictionary of plants cultivated in the United States and Canada. Liberty Hyde Bailey Hortorium, Cornell University. Wiley, 1312pp

Bown D (1995) Encyclopaedia of herbs and their uses. Dorling Kindersley, London, 424pp

Childers NF (1973) Culture of plums. In: Modern fruit science orchard & small fruit culture, 5th edn. Horticultural Publications, New Brunswick

De Dios P, Matilla AJ, Gallardo M (2006) Flower fertilization and fruit development prompt changes in free polyamines and ethylene in damson plum (*Prunus insititia* L.). J Plant Physiol 163(1):86–97

Depypere L, Chaerle P, Breyne P, Mijnsbrugge KV, Goetghebeur P (2009) A combined morphometric and AFLP based diversity study challenges the taxonomy of the European members of the complex *Prunus* L. section *Prunus*. Plant Syst Evol 279:219–231

Garcia-Marino N, Torre FDL, Matilla AJ (2008) Organic acids and soluble sugars in edible and nonedible parts of damson plum (*Prunus domestica* L. subsp *insititia* cv. Syriaca) fruits during development and ripening. Food Sci Technol 14(2):187–193

Grae I (1974) Nature's colors - dyes from plants. MacMillan Publishing Co, New York

Grieve M (1971) A modern herbal, 2 vols. Penguin, Dover publications, New York, 919pp

Gu C, Bartholomew B (2003) *Prunus* Linnaeus. In: Wu ZY, Raven PH, Hong DY (eds) Flora of China, vol 9 (*Pittosporaceae* through *Connaraceae*). Science Press/Missouri Botanical Garden Press, Beijing/St. Louis

Hanelt P (ed) (2001) Mansfeld's encyclopedia of agricultural and horticultural crops, vol 1–6. Springer, Berlin/Heidelberg/New York

Hartmann W, Neumüller M (2009) Plum breeding. In: Jain SM, Priyadarshan PM (eds) Breeding plantation tree crops. Springer, New York, pp 161–231

Huxley AJ, Griffiths M, Levy M (eds) (1992) The new RHS dictionary of gardening, 4 vols. Macmillan, New York

Porcher MH et al (1995–2020) Searchable World Wide Web Multilingual Multiscript Plant Name Database. Published by The University of Melbourne, Australia. http://www.plantnames.unimelb.edu.au/Sorting/Frontpage.html

Ramming DW, Cociu V (1991) Plums (*Prunus*). Acta Hort 290:235–290

Woldring H (1997–1998) On the origin of plums: a study of sloe, damson, cherry plum, domestic plums and their intermediate forms. Palaeohistoria 39(40):535–562

Prunus dulcis

Scientific Name

Prunus dulcis (Mill.) D.A.Webb

Synonyms

Amygdalus amara Duhamel, *Amygdalus communis* L., *Amygdalus communis* var. *amara* (Duhamel) Candolle, *Amygdalus communis* var. *dulcis* (Miller) Candolle, *Amygdalus communis* var. *fragilis* (Borkhausen) Seringe, *Amygdalus dulcis* Mill., *Amygdalus fragilis* Borkhausen, *Amygdalus sativa* Miller, *Prunus amygdalus* (Linnaeus) Batsch, *Prunus amygdalus* var. *amara* (Duhamel) Focke, *Prunus amygdalus* var. *dulcis* (Miller) Koehne, *Prunus amygdalus* var. *fragilis* (Borkhausen) Focke, *Prunus amygdalus* var. *sativa* (Miller) Focke, *Prunus communis* (L.) Arcang., *Prunus communis* var. *dulcis* (Miller) Borkhausen, *Prunus communis* var. *fragilis* (Borkhausen) Focke, *Prunus communis* var. *sativa* (Miller) Focke, *Prunus dulcis* var. *amara* (Duhamel) H. L. Moore.

Family

Rosaceae

Common/English Names

Almond, Bitter Almond, Sweet Almond

Vernacular Names

Afrikaans: Amandel;
Albanian: Bajame, Bajamja;
Amharic: Lawz;
Arabic: Lawz, Lawzah, Luz;
Aramaic: Badam, Luz, Qataraq, Shegd, Shegda;
Azerbaijan: Badam;
Basque: Alemndra, Amanda;
Belarusian: Mihdaly, Mindai;
Brazil: Amêndoa-Amarga, Amêndoa-Doce, Amendoeira (Portuguese);
Breton: Alamandez, Alamandez Dous;
Bulgarian: Badem;
Catalan: Ametler;
Chinese: Hahng Yahn (Cantonese), Bian Tao, Ben Tao, Ba Tan Hsung, Xing Ren (Mandarin);
Coptic: Karia;
Croatian: Badem, Mendula;
Czech: Mandloň Obecná, Mandloň Obecná Hořká;
Danish: Bittermandel, Mandel;
Dutch: Amandel, Amandelboom;
Esperanto: Migdalo;
Eastonian: Harilik Mandlipuu;
Farsi: Badam;
Finnish: Manteli;
French: Amandier, Amandier Amer, Amandier Commun;
Frisian: Mangel;
Gaelic: Almon, Cno Ghreugach;
Garo: Badam Pol;
Georgian: Nushi;

German: Bittermandelbaum, Knackmandel, Mandel, Mandelbaum, Süsser Mandelbaum;
Greek: Amigdalia, Amygdalia;
Hebrew: Shaqed;
Hungarian: (Édes Vagy Keserű) Mandula, Keserű Mandula;
Iceland: Mandla;
India: Badam (<u>Assamese</u>), Badam, Katbadam (<u>Bengali</u>), Bada Pol (<u>Garo</u>), Badam (<u>Gujarati</u>), Badam (<u>Hindi</u>), Badamu, Badaami, Baadaami (<u>Kannada</u>), Badam, Badam Kayu, Badamkotta (<u>Malayalam</u>), Baadaam (<u>Marathi</u>), Badam (<u>Punjabi</u>), Alabukhara, Vatadah, Vatamah (<u>Sanskrit</u>), Padam, Paruppu, Vatamkottai, Vaadumai, Vatumai (<u>Tamil</u>), Badamvittulu, Baadaamamu, Baadaamu, Badamupappu, Paarsibaadami (<u>Telugu</u>), Badam (<u>Urdu</u>);
Italian: Mandorlo;
Japanese: Amendo;
Kazakhstan: Badam, Badamgül, Ïtbadam;
Khais: Budam;
Korean: Amondu;
Latvian: Mandele;
Lithuanian: Migdolai:
Macedonia: Badem;
Malay: Badam;
Maltese: Lewż;
Mongolian: Büjls;
Norwegian: Mandel;
Persian: Baadaam;
Polish: Migdal, Migdałowiec Pospolity, Migdał Zwyczajny;
Portuguese: Amêndoa (Fruto), Amendoeira, Amendoeira-Amarga, Amendoeira-Doce;
Romanian: Migdală, Migdal;
Russian: Mindal' Obyknovennyj;
Serbian: Badem;
Slovaščina: Mandljevec;
Slovencina: Mandľa Obyčajná;
Spanish: Almendra, Almendro, Migdalujo;
Swahili: Lozi;
Swedish: Mandel;
Tajikstan: Bodom;
Thai: Alomon, Aelmon;
Turkish: Baadaam, Badem;
Ukrainian: Mygdal;
Uzbekistan: Bodom;
Vietnamese: Hạnh;
Welsh: Almon;
Yiddish: Mandl.

Origin/Distribution

Almond and related species are indigenous to the Mediterranean climate region of the Middle East (Syria, Lebanon, Israel and Jordan and Turkey eastward to Pakistan). The almond and its close relative, the peach, probably evolved from the same ancestral species. Almonds were domesticated at least by 3000 BC, and perhaps much earlier since wild almonds have been unearthed in Greek archaeological sites dating to 8000 BC. Traditional regions of almond cultivation are located around the Mediterranean Sea and southwest Asia with extension to south western Russia and Ukraine, Caucasus, Middle Asia and Himalaya. Later introduced into North and South America, South Africa and Australia. California is the world leading almond producer. In 2002, there were over 500,000 acres of almonds in California, making it the most widely planted tree crop in the state.

Agroecology

Almond grows in the following climatic regimes ranging from Cool Temperate Moist to Wet through Subtropical Thorn to Moist Forest Life zones. It prefers areas with mild winters and long rain-less, hot summers with low humidity. It thrives in areas with 400–1,470 mm annual rainfall but requires supplemental irrigation in dry areas for good growth and yield of well-filled nuts. Although relatively cold hardy, it prefers frost free areas with annual temperature of 10.5–19.5°C. Critical temperatures below which damage occurs varies with the stage of flowering and fruit development: pink bud stage −4°C to −7°C, full bloom stage −1°C to −2°C and small nut stage 0.6°C. Almond has a low winter chilling requirement (or short rest period), and the relatively low amount of heat required to bring the trees into bloom, the almond is generally the earliest deciduous fruit or nut tree to flower, hence extremely subject to frost injury where moderately late-spring frosts prevail. Almond does well in the hot, dry interior valleys of California, where the nuts mature satisfactorily. The almond tree has been successfully grown on wide range of soils. Almond will tolerate poor soils but does best in deep, well-drained sandy loam soils with pH range of 5.3–8.3. Almond trees have high nitrogen and phosphorus requirements.

Edible Plant Parts and Uses

Almonds kernels are eaten raw, dried, cooked, roasted or dried and ground into a powder for use in bakery and confectionery. A standard one cup serving of almond flour contains 20 g of carbohydrates, of which 10 g is dietary fibre, for a net of 10 g of carbohydrate per cup. This makes almond flour very desirable for use in cake, pastries, cookies, candies and bread recipes by people on carbohydrate-restricted diets. Crushed almond pieces are often sprinkled over desserts, particularly sundaes and other ice cream based dishes, and used in cakes cookies and pastries. Almonds are used in desserts like baklava, nougat, macaroons and marzipan. In China, almonds are used in a popular dessert when they are mixed with milk and then served hot. In Indian cuisine, almonds are the base ingredient for *pasanda*-style curries. In Spain, almonds of the Marcona variety are traditionally served after being lightly fried in oil, and are also used by Spanish chefs to prepare a dessert called *turrón*. Almonds are also made into a spread called almond butter which is good for people with peanut allergy. Almond can be processed into almond milk for an analog to dairy, and soy-free choice and for lactose intolerant consumers and vegans. Almonds can be made into almond syrup – an emulsion of sweet and bitter almonds usually made with barley syrup or in a syrup of orange-flower water and sugar. *Oleum Amygdalae* or almond oil is used as a substitute for olive oil and is used as a flavouring agent in baked goods and also perfumery and medicines.

Botany

A small, deciduous tree, 3–8 m high with unarmed spreading horizontal branches and greyish brown bark and an open spreading crown. Leaves are alternate, medium green, lanceolate to elliptic lanceolate 3–9 cm by 1.2.5 cm, sparsely pilose when young becoming glabrescent with broadly cuneate to rounded base, shallowly densely serrate, margin and acute to shortly acuminate apex (Plates 1 and 2). Flowers are usually borne laterally on spurs or short lateral branchlets or laterally on long shoots. Flowers solitary, white to pink, actinomorphic, pentamerous, 2–5 cm across, on 3–4 mm pedicels, appearing before the foliage. Hypanthium cylindrical and glabrous

Plate 1 Dense almond foliage and branchlets

Plate 2 Almond leaves and immature fruit

Plate 3 Whole and halved almond fruit

Plate 5 Almond nuts (endocarp) with the kernel

Plate 4 Almond fruit showing the thick hull, the hard endocarp, testa and white kernel

Plate 6 Almond kernels with brown testa

outside. Sepals 5, broadly oblong to broadly lanceolate, margin pubescent, apex obtuse. Petals 5, white or pinkish, oblong to obovate-oblong, base tapering to a narrow claw, apex obtuse to emarginate. Stamens elongated, unequal in length. Ovary perigynous and densely tomentose. Style longer than stamens. Fruit an obliquely oblong to oblong-ovoid drupe 3–6 cm by 2–4 cm, pubescent, the tough mesocarp splitting at maturity to expose the endocarp (stone) (Plates 2, 3, 4, 5). Endocarp yellowish white to brown, ovoid, broadly ellipsoid, or shortly oblong, asymmetric on both sides, 2.5–4 cm, hard to fragile, ventral suture curved and acutely keeled, dorsal suture generally straight, surface smooth and pitted with or without shallow furrows (Plates 4 and 5) enclosing a flattened, long-ovoid, brown-coated seed containing a white kernel (endosperm) (Plates 3, 4, 6).

Nutritive/Medicinal Properties

Proximate food value of almond nut (*Prunus dulcis*) per 100 g edible portion (60% shells as refuse) (USDA 2011) was reported as: water 4.70 g, energy 575 kcal (2,408 kJ), protein 21.22 g, total lipid (fat) 49.42 g, ash 2.99 g, carbohydrate 21.67 g, fibre (total dietary) 12.2 g, sugars (total) 3.89 g, sucrose 3.60 g, glucose (dextrose) 0.12 g, fructose 0.09 g, maltose 0.04 g, galactose 0.05 g, starch 0.74 g; calcium 264 mg, iron 3.72 mg, magnesium 268 mg, phosphorus 484 mg, potassium 705 mg, sodium 1 mg, zinc 3.08 mg, copper 0.996 mg, manganese 2.285 mg, selenium 2.5 μg; thiamine 0.211 mg, riboflavin 1.014 mg, niacin 3.385 mg, pantothenic acid 0.469 mg, vitamin B-6 0.143 mg, total folate 50 μg, choline (total) 52.1 mg, betaine 0.5 mg, vitamin A 1 IU, vitamin E (α-tocopherol) 26.22 mg, β-tocopherol 0.29 mg, γ-tocopherol 0.65 mg, δ-tocopherol 0.05 mg; total saturated fatty acids 3.731 g, 14:0 (myristic acid) 0.006 g, 16:0 (palmitic acid) 3.044 g, 17:0 (margaric acid) 0.007 g, 18:0 (stearic acid) 0.658 g, 20:0 (arachidic acid) 0.013 g, 22:0 (behenic acid) 0.002 g; total monounsaturated fatty acids 30.889 g, 16:1 undifferentiated (palmitoleic acid) 0.243 g, 16:1 c (palmitoleic acid cis) 0.231 g, 16:1 t (palmitoleic acid trans) 0.012 g, 17:1 (heptadecenoic acid) 0.025 g, 18:1 undifferentiated (oleic acid) 30.611 g, 18:1 c (oleic acid cis) 30.611 g, 20:1 (gadoleic acid) 0.010 g; total polyunsaturated fatty acids 12.070 g, 18:2 undifferentiated (linoleic acid) 12.061 g, 18:2 n-6 c,c (linoleic acid n-6, cic, cis) 12.055 g, 18:2 CLAs 0.001 g, 18:2 t (trans linoleic acid) not further defined 0.005 g, 18:3 undifferentiated (linolenic acid) 0.006 g, 18:3 n-3 c,c,c (α-linolenic acid) 0.006 g, 20:2 n-6 c,c (eicosadienoic acid) 0.004 g; total trans fatty acids 0.017 g, total trans-monoenoic fatty acids 0.012 g; stigmasterol 4 mg, campesterol 5 mg, β-sitosterol 132 mg β-carotene 1 μg, lutein + zeaxanthin 1 μg; amino acids: tryptophan 0.214 g, threonine 0.598 g, isoleucine 0.702 g, leucine 1.488 g, lysine 0.580 g, methionine 0.151 g, cystine 0.189 g, phenylalanine 1.120 g, tyrosine 0.452 g, valine 0.817 g, arginine 2.446 g, histidine 0.557 g, alanine 1.027 g, aspartic acid 2.911 g, glutamic acid 6.810 g, glycine 1.469 g, proline 1.032 g, serine 0.948 g, phytosterols (δ-5-avenasterol, sitostanol, campestanol, and other minor phytosterols) 31 mg.

Almonds are a rich source of vitamin E containing 26 mg/100 g, monounsaturated fats (31 g/100 g) – mainly palmitoleic acid and oleic acid that are responsible for lowering LDL cholesterol. They are also rich in energy, proteins, essential amino acids, dietary fibre, niacin, and minerals like Ca, K, P, Fe and Zn. They also contains selenium, thiamine, riboflavin, pantothenic acid, vitamin B6, folate, betaine and phytosterols.

The food composition of three marketing varieties of almonds: Carmel, Mission, and Nonpareil were found to be as follows: moisture, lipids, protein, ash, sugars, and tannins ranges were 3.05–4.33%, 43.37–47.50%, 20.68–23.30%, 3.74–4.56%, 5.35–7.45%, and 0.12–0.18%, respectively (Ahrens et al. 2005). No detectable hemagglutinating and trypsin inhibitory activities were present in the three varieties of almonds tested. Amino acid analyses indicated the sulfur amino acids (methionine + cysteine), lysine, and threonine to be the first, second, and third limiting amino acids in almonds when compared to the recommended amino acid pattern for children 2–5-year old. However, compared to the recommended amino acid pattern for adults, sulfur amino acids were the only limiting amino acids in almonds tested. Analysis of fatty acid composition of soluble lipids of almonds grown in California indicated that palmitic (C16:0), oleic (C18:1), linoleic (C18:2), and α-linolenic (C18:3) acid, respectively, accounted for 5.07–6.78%, 57.54–73.94%, 19.32–35.18%, and 0.04–0.10%; of the total lipids (Sathe et al. 2008). Oleic and linoleic acid were inversely correlated ($R^2 = -0.99$) and together accounted for 91.16–94.29% of the total soluble lipids. In China, the major fatty acids in Taiyuan almond oil were found to be about 68% oleic acid (C18:1), 25% linoleic acid (C18:2), 4.6–4.8% palmitic acid (C16:0) and a little of palmitoleic acid (C16:1), stearic acid (C18:0) (Shi et al. 1999). A trace of arachidic acid (C20:0) was also found.

Almond fruit consists of four portions: kernel or meat (seed), middle shell (endocarp), outer green shell cover or almond hull and a thin leathery layer known as brown skin of meat or seedcoat (testa) (Esfahlan et al. 2010). The nutritional importance of almond fruit is attributable to its kernel. In the past decades, different phenolic compounds were characterised and identified in almond seed extract and its skin, shell and hull as almond by-products. Polyphenols are important micronutrients in the human diet, and evidence for their role in the prevention of degenerative diseases such as cancer and cardiovascular diseases is emerging. The health effects of polyphenols depend on the amount consumed and on their bioavailability. Esfahlan et al. (2010), had recently comprehensively reviewed the importance of almond and its by-products. They also reviewed the antioxidant properties and potential use as natural dietary antioxidant, as well as their other beneficial compounds and applications of various phenolic compounds present in almond and its by-products. Polyphenol metabolism may produce several classes of metabolites that could often be more biologically active than their dietary precursor and could also become a robust new biomarker of almond polyphenol intake. In a metabolomic study of human urinary metabolome modifications, ingestion of a dietary supplement of almond skin phenolic compounds (flavan-3-ols and flavonols) identified conjugates of hydroxyphenylvaleric, hydroxyphenylpropionic, and hydroxyphenylacetic acids in the human urinary samples (Llorach et al. 2010).

Total phenols ranged from 127 to 241 mg gallic acid equivalents/100 g of fresh weight of the major almond varieties in California (Milbury et al. 2006). The analyses produced a data set of 18 flavonoids and three phenolic acids in the skins and kernels. The predominant flavonoids were isorhamnetin-3-O-rutinoside and isorhamnetin-3-O-glucoside (in combination), catechin, kaempferol-3-O-rutinoside, epicatechin, quercetin-3-O-galactoside, and isorhamnetin-3-O-galactoside at 16.81, 1.93, 1.17, 0.85, 0.83, and 0.50 mg/100 g of fresh weight almonds, respectively Using the existing approach of calculating only the aglycone form of flavonoids for use in the U.S. Department of Agriculture nutrient database, whole almonds would provide the most prevalent aglycones of isorhamnetin at 11.70 (3.32), kaempferol at 0.60 (0.17), catechin at 1.93 (0.55), quercetin at 0.72 (0.20), and epicatechin at 0.85 (0.24) mg/100 g of fresh weight (mg/oz serving), respectively. These data can lead to a better understanding of the mechanisms of action underlying the relationship between almond consumption and health-related outcomes and provide values for whole and blanched almonds suitable for inclusion in nutrient databases. Four flavonol glycosides, isorhamnetin rutinoside, isorhamnetin glucoside, kaempferol rutinoside, and kaempferol glucoside, were detected and quantified in almond skin in analyses made in Canada (Frison and Sporns 2002). In all almond varieties, isorhamnetin rutinoside was the most abundant flavonol glycoside, and the total content ranged from 75 to 250 μg/g.

Antioxidant Activity

Portuguese regional and commercial almond cultivars exhibited dirreferential degrees of antioxidant activity as assayed by different biochemical models: DPPH (2,2-diphenyl-1-picrylhydrazyl) radical scavenging activity, reducing power, inhibition of β-carotene bleaching, inhibition of oxidative hemolysis in erythrocytes, induced by 2,2′-azobis(2-amidinopropane) dihydrochloride (AAPH), and inhibition of thiobarbituric acid reactive substances (TBARS) formation in brain cells (Barreira et al. 2008) Bioactive compounds such as phenols and flavonoids were also obtained and correlated to antioxidant activity. The results obtained were quite heterogeneous, revealing significant differences among the cultivars assayed. Duro Italiano cv. revealed better antioxidant properties, presenting lower EC_{50} values in all assays, and the highest antioxidants contents. The protective effect of this cultivar on erythrocyte biomembrane hemolysis was maintained during 4 hours.

Almond skin is a rich a source of bioactive polyphenols. Nine phenolic compounds were isolated from the ethyl acetate and n-butanol fractions of almond skins (Sang et al. 2002). These compounds were identified as 3′-O-methylquercetin 3-O-β-D-glucopyranoside (1); 3′-O-methylquercetin 3-O-β-D-galactopyranoside (2); 3′-O-methylquercetin 3-O-α-L-rhamnopyranosyl-(1→6)-β-D-glucopyranoside (3); kaempferol 3 O-α-L-rhamnopyranosyl-(1→6)-β-D-glucopyranoside (4); naringenin 7-O-β-D-glucopyranoside (5); catechin (6); protocatechuic acid (7); vanillic acid (8); and p-hydroxybenzoic acid (9). Compounds 6 and 7 showed very strong DPPH radical scavenging activity. Compounds 1–3, 5, 8, and 9 showed strong activity, whereas compound 4 exhibited very weak activity. Garrido et al. (2008) reported a total of 31 phenolic compounds corresponding to flavan-3-ols (33–56% of the total of phenolic compounds identified), flavonol glycosides (9–36%), hydroxybenzoic acids and aldehydes (6–26%), flavonol aglycones (1.7–18%), flavanone glycosides (3–7.7%), flavanone aglycones (0.69–5.4%), hydroxycinnamic acids (0.65–2.6%), and dihydroflavonol aglycones (0–2.8%) in the skins from three different varieties of almonds. The total contents of phenolic compounds identified were significantly higher (around two-fold) in the roasted samples than in the blanched, freeze-dried almonds. Industrial oven drying of the blanched almond skins produced a two-fold increase in the contents of phenolic compounds. The antioxidant activity (ORAC values) was higher for the roasted samples (0.803–1.08 mmol Trolox/g), followed by the samples subjected to blanching + drying (0.398–0.575 mmol Trolox/g) and then the blanched freeze-dried samples (0.331–0.451 mmol Trolox/g). Roasting was the most suitable type of industrial processing of almonds to obtain almond skin extracts with the greatest antioxidant capacity. Proanthocyanidins, including a series of A- and B-type procyanidins and propelargonidins up to heptamers, and A- and B-type prodelphinidins up to hexamers were also found in almond skins (Monagas et al. 2007). Flavanols and flavonol glycosides were the most abundant phenolic compounds in almond skins, constituting 38–57% and 14–35% of the total quantified phenolics, respectively.

Studies showed that almond skin flavonoids (ASF) possessed antioxidant capacity in-vitro, were bioavailable and acted in synergy with vitamins C and E to protect LDL against oxidation in hamsters (Chen et al. 2005). ASF from 0.18 to 1.44 μmol gallic acid equivalent (GAE)/L increased the lag time to LDL oxidation in a dose-dependent manner. Combining ASF with vitamin E or ascorbic acid extended the lag time >200% of the expected additive value. In subsequent studies, Chen et al. (2007) demonstrated the almonds skin polyphenolics (ASP) and quercetin reduced the oxidative modification of apo B-100 and stabilize LDL conformation in a dose-dependent manner, acting in an additive or synergistic fashion with vitamin C and E. The scientists also showed that ASP acted as antioxidants and induced quinine reductase activity, but these actions were dependent upon their dose, method of extraction, and interaction with antioxidant vitamins C and E (Chen and Blumberg 2008). Almond green husks (Cvs. Duro Italiano, Ferraduel, Ferranhês, Ferrastar and Orelha de Mula) were found to have good antioxidant properties, with very low EC_{50} values (<380 μg/mL), particularly for lipid peroxidation inhibition (<140 μg/mL) (Barreira et al. 2010). Correlation between total phenol – flavonoid contents and DPPH (2,2-diphenyl-1-picrylhydrazyl) radical scavenging activity, reducing power, inhibition of β-carotene bleaching and inhibition of lipid peroxidation in pig brain tissue through formation of thiobarbituric acid reactive substances, were also obtained.

Phenols levels in roasted peanut and hazelnut skins were higher than that of almond skins, but their flavan-3-ol profiles, differed considerably. Peanut skins were low in monomeric flavan-3-ols (19%) in comparison to hazelnut (90%) and almond (89%) skins (Monagas et al. 2009). In contrast, polymeric flavan-3-ols in peanut and almond skins occurred as both A- and B-type proanthocyanidins, but in peanuts the A forms (oligomers up to DP (degree of polymerization) 12) were predominant, whereas in almonds, the B forms (up to DP8) were more abundant. In contrast, hazelnuts were mainly constituted by B-type proanthocyanidins (up to DP9). The antioxidant capacity as determined by various methods (i.e., total antioxidant capacity, ORAC, DPPH test, and reducing power) was higher for whole extracts from roasted hazelnut and peanut skins than for almond skins; however, the antioxidant capacities of the high molecular weight fractions of the three types of nut skins were equivalent despite their different compositions and degrees of polymerization.

The methanol extract of almond hulls (Nonpareil variety) was found to contain 5-O-caffeoylquinic acid (chlorogenic acid), 4-O-caffeoylquinic acid (cryptochlorogenic acid), and 3-O-caffeoylquinic acid (neochlorogenic acid) in the ratio 79.5:14.8:5.7 (Takeoka and Dao 2003). The chlorogenic acid concentration of almond hulls was 42.52 mg/100 g of fresh weight. At an equivalent concentration (10 μg/1 g of methyl linoleate) almond hull extracts had higher antioxidant activity than α-tocopherol. At higher concentrations (50 μg/1 g of methyl linoleate) almond hull extracts showed increased antioxidant activity that was similar to chlorogenic acid and morin [2-(2,4-dihydroxyphenyl)-3,5,7-trihydroxy-4H-1-benzopyran-4-one] standards (at the same concentrations).

The data indicated almond hulls to be a potential source of these dietary antioxidants. The sterols (3β,22E)-stigmasta-5,22-dien-3-ol (stigmasterol) and (3β)-stigmast-5-en-3-ol (β-sitosterol) (18.9 mg and 16.0 mg/100 g) of almond hull, respectively.

Studies showed that almond had antioxidant property and retarded the aging process (Wang et al. 2004). D-gal-induced aging rats fed with almond at 65 g/kg feedstuff, 97.5 g/kg feedstuff after 30 days had their levels of malonaldehyde (MDA) significantly reduced, but the lowest dose group 32.5 g was not significantly different. At the same time, the activities of SOD (superoxide dismutase), GSH-Px (glutathione peroxidise) and the liquid fluidity of erythrocyte membrane increased significantly.

The total phenolic contents of ethanolic extracts of brown skin and green shell cover of almond were ten and nine times higher than that of the whole seed, respectively (Wijeratne et al. 2006). Brown skin extract at 50 ppm effectively inhibited copper-induced oxidation of human LDL cholesterol compared to whole seed and green shell cover extracts, which reached the same level of efficacy at 200 ppm. Green shell cover extract at 50 ppm level completely arrested peroxyl radical-induced DNA scission, whereas 100 ppm of brown skin and whole seed extracts was required for similar efficiencies. All three almond extracts exhibited excellent metal ion chelation efficacies. Tall the extract were found to contain quercetin, isorhamnetin, quercitrin, kaempferol 3-O-rutinoside, isorhamnetin 3-O-glucoside, and morin as the major flavonoids.

In a pilot study, almond consumption was found to have preventive effects on oxidative stress and DNA damage caused by smoking (Jia et al. 2006). After the almond intervention in a randomized, crossover clinical trial with 60 healthy male soldiers (18–25 years) who were habitual smokers (5–20 cigarettes/daily) (Li et al. 2007), serum α-tocopherol, plasma superoxide dismutase (SOD), glutathione peroxidase (GPX) increased significantly in smokers by 10%, 35%, and 16%, respectively and urinary 8-hydroxydeoxyguanosine (8-OHdG) and malondialdehyde (MDA) and peripheral lymphocyte DNA strand breaks decreased significantly by 28%, 34%, and 23%. In smokers, after almond supplementation, the concentration of 8-OHdG remained significantly greater than in nonsmokers by 98%. These results suggested almond intake could enhance antioxidant defences and diminish biomarkers of oxidative stress in smokers.

Anticancer Activity

Among the main components of almond hulls – oleanolic, ursolic, and betulinic acids, the 2-hydroxy analogues alphitolic, corosolic, maslinic acids, as well as the related aldehydes, namely, betulinic, oleanolic, and ursolic, and from a more polar fraction, the β-sitosterol 3-O-glucoside (Amico et al. 2006). Betulinic acid showed potent antiproliferative activity toward MCF-7 human breast cancer cells ($GI_{50} = 0.27$ μM), higher than the anticancer drug 5-fluorouracil. *Prunus dulcis* also contained the antitumour compound dihydroquercetin, taxifolin.

Almond kernel also contains amygdalin which is a major constituent in laetrile the abbreviated name for laevomandelonitrile. Amygdalin is often confounded with laetrile. Laetrile, has been erroneously dubbed as vitamin B17 and to have a positive effect in the treatment of cancer. Laetrile does not meet the requirements of a vitamin and its cancer curing claim is doubtful and bogus as reported by numerous studies. One study involving 178 cancer patients treated with amygdalin (Laetrile) plus a "metabolic therapy" program consisting of diet, enzymes, and vitamins generated no substantive benefits in terms of cure, improvement or stabilization of cancer, improvement of symptoms related to cancer, or extension of life span (Moertel et al. 1982). The hazards of amygdalin therapy were evidenced in several patients by symptoms of cyanide toxicity or by blood cyanide levels approaching the lethal range. This study confirmed amygdalin (Laetrile) to be a toxic drug not effective as a cancer treatment. Laetrile has not been approved for this use by the United States' Food and Drug Administration.

Antihyperlipidemic Activity

Almonds are known to have a number of nutritional benefits, including cholesterol-lowering effects and protection against diabetes. They are also a good source of minerals and vitamin E, associated with promoting health and reducing the risk for chronic disease. In one study, almonds used as snacks in the diets of hyperlipidemic subjects was found to significantly reduce coronary heart disease risk factors, probably in part because of the non-fat (protein and fibre) and monounsaturated fatty acid components of the nut (Jenkins et al. 2002). The full-dose almonds produced the greatest reduction in levels of blood lipids. Significant reductions from baseline were seen on both half- and full-dose almonds for LDL cholesterol (4.4% and 9.4% respectively) and LDL:HDL cholesterol (7.8% and 12.0% respectively) and on full-dose almonds alone for lipoprotein(a) (7.8%) and oxidized LDL concentrations (14.0%), with no significant reductions on the control diet.

In a randomised crossover study comparing the effects of whole almonds, taken as snacks, with the effects of low saturated fat (<5% energy) whole-wheat muffins (control) in the therapeutic diets of hyperlipidemic subjects, Mean body weights differed ≤300 g between treatments, although the weight loss on the half-dose almond treatment was greater than on the control (Jenkins et al. 2008a). At 4 weeks, the full-dose almonds reduced serum concentrations of malondialdehyde (MDA) and creatinine-adjusted urinary isoprostane output compared with the control. Serum concentrations of α-tocopherol or γ-tocopherol were not affected by the treatments. Almond antioxidant activity was demonstrated by their effect on 2 biomarkers of lipid peroxidation, serum MDA and urinary isoprostanes. Antioxidant activity provided an additional possible mechanism, in addition to lowering cholesterol, that may account for the reduction in CHD (coronary heart disease) risk with nut consumption. In another randomized crossover study, 27 hyperlipidemic men and women were provided 3 isoenergetic (mean, 423 kcal/day) supplements each for 1 month comprising 22.2% of energy and consisted of full-dose almonds (73±3 g/day), half-dose almonds plus half-dose muffins, and full-dose muffins (Jenkins et al. 2008b). At the end of 4 weeks mean body weights differed by less than 300 g between treatments. No differences were seen in baseline or treatment values for fasting glucose, insulin, C-peptide, or insulin resistance as measured by homeostasis model assessment of insulin resistance. However, 24-hour urinary C-peptide output as a marker of 24-hour insulin secretion was significantly reduced on the half-and full-dose almonds by comparison to the control after adjustment for urinary creatinine output. The reductions in 24-hour insulin secretion appeared to be a further metabolic advantage of nuts that in the longer term may help to explain the association of nut consumption with reduced CHD risk.

Antidiabetic Activity

A separate study reported that almond-enriched diets do not alter insulin sensitivity in healthy adults or glycemia in patients with diabetes (Lovejoy et al. 2002). Almond consumption did not change insulin sensitivity significantly, although body weight increased and total and LDL cholesterol decreased by 21% and 29%, respectively in study 1. In study 2, total cholesterol was lowest with the HFA (high fat, high almond) diet (4.46, 4.52, 4.63, and 4.63 mmol/L with the HFA, HFC (High fat control), LFA (low fat high almond), and LFC (low fat control) diets, respectively). HDL cholesterol was significantly lower with the almond diets; however, no significant effect of fat source on LDL:HDL was observed. Glycemia was unaffected. Almonds had beneficial effects on serum lipids in healthy adults and produced changes similar to high monounsaturated fat oils in diabetic patients.

Consumption of almonds was found to lower postprandial glucose excursions and to decrease the risk of oxidative damage to proteins (Jenkins et al. 2006). Glycemic indices for the rice (38) and almond meals (55) were less than for the potato meal (94), as were the postprandial areas under the insulin concentration time curve.

No postmeal treatment differences were seen in total antioxidant capacity. However, the serum protein thiol concentration increased following the almond meal (15 mmol/L), indicating less oxidative protein damage. Therefore, lowering postprandial glucose excursions may decrease the risk of oxidative damage to proteins. Almonds were postulated to lower this risk by decreasing the glycemic excursion and by providing antioxidants. These actions may relate to mechanisms by which nuts are associated with a decreased risk of coronary heart disease.

Tocopherol Improvement Activity

Almonds in the diet was found to simultaneously improve plasma α-tocopherol concentrations and reduce plasma lipids in healthy adults in a randomized, crossover feeding trial (Jambazian et al. 2005). Incorporating almonds into the diet helped meet the revised Recommended Dietary Allowance of 15 mg/day α-tocopherol and increased lipid-adjusted plasma and red blood cell α-tocopherol concentrations. A significant dose-response effect was observed between percent energy in the diet from almonds and plasma ratio of α-tocopherol to total cholesterol.

Prebiotic Activity

Almond also has prebiotic activity. Addition of finely ground almond altered the composition of gut bacteria by stimulating the growth of *Bifidobacteria* and *Eubacterium rectale* resulting in a higher prebiotic index (4.43) than was found for the commercial prebiotic fructo-oligosaccharides (4.08) at 24 hours of incubation (Mandalari et al. 2008). No significant differences in the proportions of gut bacteria groups were detected in response to deffated finely ground almonds.

Almond Allergy and Other Effects

Almond proteins can cause severe anaphylactic reactions in susceptible individuals. Two IgE-binding almond proteins were N-terminally sequenced and identified as almond 2S albumin and conglutin γ (Poltronieri et al. 2002).

Raw and roasted almonds were found to contain advanced glycation endproducts (AGEs) (Zhang et al. 2011). Carboxymethyl-lysine (CML) and carboxyethyl-lysine (CEL) were found in both raw and roasted almonds. Pyralline (Pyr) was identified for the first time in roasted almonds and accounted for 64.4% of free plus bound measured AGEs. Argpyrimidine (Arg-p), and pentosidine (Pento-s) were below the limit of detection in all almond samples tested. Free AGEs accounted for 1.3–26.8% of free plus bound measured AGEs, indicating that protein-bound forms predominated. The roasting process significantly increased CML, CEL, and Pyr formation in almonds. AGEs had been reported to play a role as proinflammatory mediators in gestational diabetes (Pertyńska-Marczewska et al. 2009) and also had been implicated in the progression of age related diseases such as diabetes and atheroscelrosis (Tan et al. 2006).

Traditional Medicinal Uses

Almond kernel and almond oil have been used in traditional folk medicine for cancer (especially bladder, breast, mouth, spleen, and uterus), carcinomata, condylomata, corns, indurations, kidney stones, gallstones and tumours. Almond is also used as folk remedy for asthma, constipation, cold, corns, cough, dyspnea, eruptions, gingivitis, heartburn, itch, lungs, prurigo, skin, sores, spasms, stomatitis, and peptic ulcers. It is reported to be alterative, astringent, carminative, cyanogenetic, demulcent, discutient, diuretic, emollient, laxative, pectoral, lithontryptic, nervine, sedative, stimulant and tonic. Almond oil is emollient and applied to dry skins and is also often used as a carrier oil in aromatherapy. The leaves have been used in the treatment of diabetes.

Other Uses

The seed contains amygdallin, under the influence of water and in the presence of emulsion it can be hydrolysed to produce benzaldehyde (the almond aroma, formula C6H5 CHO) and prussic acid. Almond oil extracted from the kernels is an excellent lubricant for watches. Sweet almond oil is widely used for cosmetic creams, perfumes, sopas and lotions because it has a softening and moisturising effect on the skin. Bitter almond oil in a crisis, might conceivably be used as an energy source. The burnt hulls is rich in potassium, it is used in soap making and a valuable absorbent for coal gas. After the almonds are harvested, the nuts are passed through a machine which removes the hulls. The hulls can be used in cattle and sheep rations. They have been mixed 1:1 with barley and fed together with alfalfa hay with excellent results. Varieties with soft hulls are superior to varieties with hard hulls. Other parts of fruit such as shells and hulls were burned as fuel. A green dye can be obtained from the leaves and fruits. A yellow dye is obtained from the roots and leaves. The gum exuded from the tree has been used as an adhesive and substitute for tragacanth.

Comments

The world's leading producer of almonds is the United States; other major producers include Greece, Iran, Italy, Morocco, Portugal, Spain, Syria and Turkey. In the United States, production is concentrated in California, with almonds being California's sixth leading agricultural product and its top agricultural export.

Selected References

Ahrens S, Venkatachalam M, Mistry AM, Lapsley K, Sathe SK (2005) Almond (*Prunus dulcis* L.) protein quality. Plant Foods Hum Nutr 60(3):123–128

Amico V, Barresi V, Condorelli D, Spatafora C, Tringali C (2006) Antiproliferative terpenoids from almond hulls (*Prunus dulcis*): identification and structure-activity relationships. J Agric Food Chem 54(3):810–814

Barreira JC, Ferreira IC, Oliveira MB, Pereira JA (2008) Antioxidant activity and bioactive compounds of ten Portuguese regional and commercial almond cultivars. Food Chem Toxicol 46(6):2230–2235

Barreira JC, Ferreira IC, Oliveira MB, Pereira JA (2010) Antioxidant potential of chestnut (*Castanea sativa* L.) and almond (*Prunus dulcis* L.) by-products. Food Sci Technol Int 16(3):209–216

Bown D (1995) Encyclopaedia of herbs and their uses. Dorling Kindersley, London, 424 pp

Chen CY, Blumberg JB (2008) In vitro activity of almond skin polyphenols for scavenging free radicals and inducing quinone reductase. J Agric Food Chem 56(12):4427–4434

Chen CY, Milbury PE, Chung SK, Blumberg J (2007) Effect of almond skin polyphenolics and quercetin on human LDL and apolipoprotein B-100 oxidation and conformation. J Nutr Biochem 18(12):785–794

Chen CY, Milbury PE, Lapsley K, Blumberg JB (2005) Flavonoids from almond skins are bioavailable and act synergistically with vitamins C and E to enhance hamster and human LDL resistance to oxidation. J Nutr 135(6):1366–1373

Duke JA, Ayensu ES (1985) Medicinal plants of China, vols 1 and 2. Reference Publications, Inc. Algonac, 705 pp

Esfahlan AJ, Jamei R, Esfahlan RJ (2010) The importance of almond (*Amygdalus* L.) and its by-products. Food Chem 120:349–360

Foundation for Revitalisation of Local Health Traditions (2008) FRLHT database. http://envis.frlht.org

Frison S, Sporns P (2002) Variation in the flavonol glycoside composition of almond seedcoats as determined by maldi-tof mass spectrometry. J Agric Food Chem 50(23):6818–6822

Garrido I, Monagas M, Gómez-Cordovés C, Bartolomé B (2008) Polyphenols and antioxidant properties of almond skins: influence of industrial processing. J Food Sci 73(2):C106–C115

Grae I (1974) Nature's colors – dyes from plants. MacMillan Publishing Co, New York

Grieve M (1971) A modern herbal, 2 vols. Penguin/Dover publications, New York, 919 pp

Jambazian PR, Haddad E, Rajaram S, Tanzman J, Sabaté J (2005) Almonds in the diet simultaneously improve plasma alpha-tocopherol concentrations and reduce plasma lipids. J Am Diet Assoc 105(3):449–454

Jenkins DJ, Kendall CW, Josse AR, Salvatore S, Brighenti F, Augustin LSA, Ellis PR, Vidgen E, Rao AV (2006) Almonds decrease postprandial glycemia, insulinemia, and oxidative damage in healthy individuals. J Nutr 136:2987–2992

Jenkins DJ, Kendall CW, Marchie A, Josse AR, Nguyen TH, Faulkner DA, Lapsley KG, Blumberg J (2008a) Almonds reduce biomarkers of lipid peroxidation in older hyperlipidemic subjects. J Nutr 138(5):908–913

Jenkins DJ, Kendall CW, Marchie A, Josse AR, Nguyen TH, Faulkner DA, Lapsley KG, Singer W (2008b) Effect of almonds on insulin secretion and insulin resistance in nondiabetic hyperlipidemic subjects: a randomized controlled crossover trial. Metabolism 57(7):882–887

Jenkins DJ, Kendall CW, Marchie A, Parker TL, Connelly PW, Qian W, Haight JS, Faulkner D, Vidgen E, Lapsley KG, Spiller GA (2002) Dose response of almonds on coronary heart disease risk factors: blood lipids, oxidized low-density lipoproteins, lipoprotein(a), homocysteine, and pulmonary nitric oxide: a randomized, controlled, crossover trial. Circulation 106(11): 1327–1332

Jia X, Li N, Zhang W, Zhang X, Lapsley K, Huang G, Blumberg J, Ma G, Chen J (2006) A pilot study on the effects of almond consumption on DNA damage and oxidative stress in smokers. Nutr Cancer 54(2):179–183

Li N, Jia X, Chen CY, Blumberg JB, Song Y, Zhang W, Zhang X, Ma G, Chen J (2007) Almond consumption reduces oxidative DNA damage and lipid peroxidation in male smokers. J Nutr 137(12):2717–2722

Llorach R, Garrido I, Monagas M, Urpi-Sarda M, Tulipani S, Bartolome B, Andres-Lacueva C (2010) Metabolomics study of human urinary metabolome modifications after intake of almond (*Prunus dulcis* (Mill.) D.A. Webb) skin polyphenols. J Proteome Res 9(11):5859–5867

Lovejoy JC, Most MM, Lefevre M, Greenway FL, Rood JC (2002) Effects of diets enriched in almonds on insulin action and serum lipids in adults with normal glucose tolerance or type 2 diabetes. Am J Clin Nutr 76(5):1000–1006

Lu LT, Bartholomew B (2003) *Amgydalus* Linnaeus. In: Wu ZY, Raven PH, Hong DY (eds) Flora of China, vol 9 (Pittosporaceae through Connaraceae). Science Press/Missouri Botanical Garden Press, Beijing/St. Louis

Mandalari G, Nueno-Palop C, Bisignano G, Wickham MS, Narbad A (2008) Potential prebiotic properties of almond (*Amygdalus communis* L.) seeds. Appl Environ Microbiol 74(14):4264–4270

Milbury PE, Chen CY, Dolnikowski GG, Blumberg JB (2006) Determination of flavonoids and phenolics and their distribution in almonds. J Agric Food Chem 54(14):5027–5033

Moertel CG, Fleming TR, Rubin J, Kvols LK, Sarna G, Koch R, Currie VE, Young CW, Jones SE, Davignon JP (1982) A clinical trial of amygdalin (Laetrile) in the treatment of human cancer. N Engl J Med 306(4): 201–206

Monagas M, Garrido I, Lebrón-Aguilar R, Bartolome B, Gómez-Cordovés C (2007) Almond (*Prunus dulcis* (Mill.) D.A. Webb) skins as a potential source of bioactive polyphenols. J Agric Food Chem 55(21): 8498–8507

Monagas M, Garrido I, Lebrón-Aguilar R, Gómez-Cordovés MC, Rybarczyk A, Amarowicz R, Bartolomé B (2009) Comparative flavan-3-ol profile and antioxidant capacity of roasted peanut, hazelnut, and almond skins. J Agric Food Chem 57(22):10590–10599

Pertyńska-Marczewska M, Głowacka E, Sobczak M, Cypryk K, Wilczyński J (2009) Glycation endproducts, soluble receptor for advanced glycation endproducts and cytokines in diabetic and non-diabetic pregnancies. Am J Reprod Immunol 61(2):175–182

Poltronieri P, Cappello MS, Dohmae N, Conti A, Fortunato D, Pastorello EA, Ortolani C, Zacheo G (2002) Identification and characterisation of the IgE-binding proteins 2S albumin and conglutin gamma in almond (*Prunus dulcis*) seeds. Int Arch Allergy Immunol 128(2):97–104

Sang S, Lapsley K, Jeong WS, Lachance PA, Ho CT, Rosen RT (2002) Antioxidative phenolic compounds isolated from almond skins (*Prunus amygdalus* Batsch). J Agric Food Chem 50(8):2459–2463

Sathe SK, Seeram NP, Kshirsagar HH, Heber D, Lapsley KA (2008) Fatty acid composition of California grown almonds. J Food Sci 73(9):C607–C614

Shi Z, Fu Q, Chen B, Xu S (1999) Analysis of physicochemical property and composition of fatty acid of almond oil. Se Pu 17(5):506–507 (In Chinese)

Takeoka GR, Dao LT (2003) Antioxidant constituents of almond [*Prunus dulcis* (Mill.) D.A. Webb] hulls. J Agric Food Chem 51(2):496–501

Tan KC, Chow WS, Lam JC, Lam B, Bucala R, Betteridge J, Ip MS (2006) Advanced glycation endproducts in nondiabetic patients with obstructive sleep apnea. Sleep 29(3):329–333

U.S. Department of Agriculture, Agricultural Research Service (2011) USDA national nutrient database for standard reference, Release 24. Nutrient Data Laboratory Home Page, http://www.ars.usda.gov/ba/bhnrc/ndl

Wang H, Zhang S, Guo A, Zhang J (2004) Effects of almond on D-gal-induced aging rats. Wei Sheng Yan Jiu 33(2):222–224 (In Chinese)

Wijeratne SS, Abou-Zaid MM, Shahidi F (2006) Antioxidant polyphenols in almond and its coproducts. J Agric Food Chem 54(2):312–318

Zhang G, Huang G, Xiao L, Mitchell AE (2011) Determination of advanced glycation endproducts by LC-MS/MS in raw and roasted almonds (*Prunus dulcis*). J Agric Food Chem 59(22):12037–12046

Prunus persica var. nucipersica

Scientific Name

Prunus persica var. *nucipersica* (Suckow) C.K. Schneid.

Synonyms

Amygdalus persica var. *nectarina* Aiton, *Amygdalus persica* var. *nucipersica* Suckow, *Prunus nucipersica* Borkh., *Prunus persica nucipersica* (Suckow.) C.K. Schneid, *Prunus persica* var. *nectarina* (Aiton.) Maxim., *Prunus persica* var. *nucipersica* (Borkh.) C.K. Schneid., *nom. Illeg*, *Persica vulgaris* Mill. var. *nectarina* (Aiton) Holub.

Family

Rosaceae

Common/English Names

Fresh Nectarine, Nectarine, Smooth-Skinned Peach, Table Nectarine.

Vernacular Names

Chinese: You Tao, Yu T'ao;
Czech: Broskvoň Obecná Nektarinka;
Danish: Nektarin;
Dutch: Nectarine, Naakte Perzik;
Eastonian: Nektariin;
Finnish: Nektariini;
French: Brugnon, Nectarine, Nectarine De Table, Pêche À Peau Lisse;
German: Nektarine, Nektarinenbaum;
Hungarian: Csupaszbarack, Kopaszbarack, Nektarin;
Italian: Nettarina;
Japanese: Yutou, Nekutarina, Nekutarin, Zubaimomo;
Korean: Poksunga Namu;
Maltese: Nuċiprisk;
Portuguese: Nectarina;
Russian: Nektarin;
Spanish: Nectarina;
Swedish: Nektarin.

Origin/Distribution

Nectarine is a smooth-skinned variety or mutation from the peach, *Prunus persica*. Several genetic studies have concluded in fact that nectarines are created due to a recessive gene in its peach parent. Nectarines have arisen many times from peach trees, often as bud sports. Its origin is unknown. Peaches probably originated from China, being one of the first fruit crop domesticated about 4,000 years ago.

Agroecology

Nectarines have similar agro-ecological requirements as peaches. Both are cool temperate species requiring a winter chill period for flowering and fruiting. On average, most cultivars have chill requirements of 600–900 hours, though some newer cultivars have lower winter chill requirement of 350–400 hours. Nectarines are sensitive to spring frost. They grow best in full sun for optimum yield and high quality coloured fruits. Nectarines like peaches thrives best in deep, well-drained, sandy loam to sandy clay loam with a preferred pH near to 6.5. Poorly drained soils and waterlogged soil should be avoided.

Edible Plant Parts and Uses

The ripe fruit and skin can be eaten fresh out of hand. The fresh fruit can be used in ice creams, pies, jams. The fruit can also be cooked and dried for later use. Flowers are edible, eaten in salad or used as a garnish. A tea can be brewed from the flowers. A white liquid can be distilled from the flowers, having a flavour similar to the seed. The seed kernel can be eaten raw or cooked. However it is best avoided raw especially if it is bitter because of hydrocyanic acid as cases of toxicity have been reported. The gum from the stem has been reported to be used for chewing.

Botany

A small, vigorous, deciduous tree 3–7 m high with a spreading crown of 3 m and dark brown scabrous bark. Petiole 10–15 mm, longitudinally grooved with reniform nectaries. Leaves alternate, simple, lanceolate, oblong-lanceolate to elliptic-lanceolate, 7–15 × 2–3.5 cm, apex acuminate, base cuneate, margin finely serrate or serrulate, green, both surfaces glabrous (Plates 2 and 3). Flowers solitary, opening before leaves, 2–3.5 cm in diam. Pedicel very short 4–4.5 mm. Hypanthium green with a red tinge, shortly campanulate, glabrous. Sepals 5 ovate to oblong, upper surface glabrous, lower pubescent; petals pink (Plate 1) or white, alternately arranged to sepals, orbicular, apex rounded, base narrows at point of attachment, both upper and lower surfaces glabrous; Stamens – numerous 30–(–44)–50, filament white, 14–17 mm, anther reddish; ovary glabrous, style slightly longer than stamens. Drupe globose to subglobose, (3–)5–7(–10) cm in diameter, glossy golden yellow with

Plate 1 Nectarine blossoms

Plate 2 Nectarine fruits and leaves

Plate 3 Ripening nectarines

Plate 4 Harvested nectarines

Plate 5 Mature but unripe nectarines on sale in a north Vietnam market

Plate 6 Ripe golden-yellow fleshed nectarine

Plate 7 Ripe white-fleshed nectarines

large blushes of red or completely red to maroon (Plates 2, 3, 4, 5, 6, 7, 8). The skin exocarp is thin, smooth, enclosing the firm, fleshy, juicy sweet to acid sweet mesocarp, white or yellow or orangey-yellow (Plates 6, 7, 8) free-stone or cling-stone. The bony endocarp (pit) large, ellipsoid to suborbicular, compressed on both sides, surface longitudinally and transversely furrowed and pitted, surrounding a single, large, ovate seed.

Plate 8 Whole and halved white-fleshed nectarine

Nutritive/Medicinal Value

Food value of raw, nectarine fruit (refuse 9% pit) per 100 g edible portion was reported as follows (USDA 2011): water 87.59 g, energy 44 kcal (185 kJ), protein 1.06 g, total lipid (fat) 0.32 g, ash 0.48 g, carbohydrate 10.55 g; fibre (total dietary) 1.7 g, total sugars 7.89 g, sucrose 4.87 g, glucose 1.57 g, fructose 1.37 g; minerals – calcium 6 mg, iron 0.28 mg, magnesium 9 mg, phosphorus 26 mg, potassium 201 mg, sodium 0 mg, zinc 0.17 mg, copper 0.086 mg, manganese 0.054 mg; vitamins – vitamin C (total ascorbic acid) 5.4 mg, thiamin 0.034 mg, riboflavin 0.027 mg, niacin 1.125 mg, pantothenic acid 0.185 mg, vitamin B-6 0.025 mg, folate (total) 5 μg, total choline 6.2 mg, betaine 0.2 mg, vitamin A 332 IU (17 μg RAE), vitamin E (α-tocopherol) 0.77 mg, β-tocopherol 0.01 mg, γ-tocopherol 0.01 mg, δ-tocopherol 0.01 mg, vitamin K (phylloquinone) 2.2 μg, lipids – fatty acids (total saturated) 0.025 g, 16:0 (palmitic acid) 0.023 g, 18:0 (stearic acid) 0.002 g; fatty acids (total monounsaturated) 0.088 g, 16:1 undifferentiated (palmitoleic acid) 0.002 g, 18:1 undifferentiated (oleic acid) 0.086 g; fatty acids (total polyunsaturated) 0.113 g, 18:2 undifferentiated (linoleic acid) 0.111 g, 18:3 undifferentiated 0.002 g, β-carotene 150 μg, β-cryptoxanthin 98 μg, lutein + zeaxanthin 130 μg; amino acids – tryptophan 0.005 g, threonine 0.009 g, isoleucine 0.009 g, leucine 0.014 g, lysine 0.016 g, methionine 0.006 g, cystine 0.005 g, phenylalanine 0.011 g, tyrosine 0.007 g, valine 0.013 g, arginine 0.009 g, histidine 0.008 g, alanine 0.017 g, aspartic acid 0.568 g, glutamic acid 0.034 g, glycine 0.011 g, proline 0.010 g, and serine 0.018 g.

Sugars (glucose, fructose, sucrose and sorbitol) and organic acids (citric, malic, shikimic and fumaric acid) in fruits were identified in all cultivars of peach and nectarine studied in Slovenia (Colaric et al. 2004). Sucrose was the major sugar and malic and citric acids were the predominant organic acids. The content of fructose ranged from 6.76 to 12.97 g/kg, glucose from 5.43 to 11.11 g/kg, sucrose from 46.14 to 70.7 g/kg and sorbitol from 0.40 to 2.80 g/kg of fruits. The content of citric acid ranged from 1.71 to 8.34 g/kg, malic acid from 3.2 to 8.05 g/kg, shikimic acid from 127 to 809 mg/kg and fumaric acid from 1.56 to 6.09 mg/kg of fruits. The content of total sugars ranged from 61.53 to 93.70 g/kg and the content of total organic acids ranged from 7.06 to 14.69 g/kg of fruits.

The volatile components in nectarine fruit were found to consist of 10 lactones, 8 C_6 aldehydes and alcohols, 8 terpenoids, 3 esters, and 4 other compounds. Besides C_6 components a series of saturated and unsaturated γ and δ lactones ranging from chain length C_6–C_{12} with concentration maxima for γ and δ decalactone formed the major class of constituents (Engel et al. 1988).

Glycosidically bound volatile constituents of yellow-fleshed clingstone nectarines (cv. Springbright) were identified and quantified at three stages of maturity (Aubert et al. 2003). Forty-five bound aglycons were identified in yellow-fleshed nectarine. Thirty were terpene derivatives, and the most abundant ones were (E)- and (Z)-furan linalool oxides, linalool, α-terpineol, (E)-pyran linalool oxide, 3,7-dimethylocta-1,5-diene-3,7-diol, linalool hydrate, 8-hydroxy-6,7-dihydrolinalool, (E)- and (Z)-8-hydroxylinalools, and (E)- and (Z)-8-hydroxygeraniols. The group of C_{13} norisoprenoids included 3-hydroxy-β-damascone, 3-hydroxy-7,8-dihydro-β-ionone, 3-oxo-α-ionol, 3-hydroxy-7,8-dihydro-β-ionol, 3-hydroxy-β-ionone, 3-oxo-7,8-dihydro-α-ionol, 3-hydroxy-5,6-epoxy-β-ionone, 3-oxo-retro-α-ionol (isomers I and II), 3-hydroxy-7,8-dehydro-β-ionol, 4,5-dihydrovomifoliol, and vomifoliol. Generally, levels of bound compounds, in particular monoterpenols and C13 norisoprenoids, increased significantly with maturation. δ-decalactone was the only lactone found in the enzymatic hydrolysate of yellow-fleshed nectarine, but its level was much lower than that of its free form.

The phenolic compounds hydroxycinnamates, procyanidins, flavonols, and anthocyanins were detected and quantified in nectarine cultivars (Tomás-Barberán et al. 2001). As a general rule, the peel tissues contained higher amounts of phenolics, and anthocyanins and flavonols were almost exclusively located in this tissue. No clear differences in the phenolic content of nectarines were detected between white flesh and yellow flesh cultivars. There was no clear trend in phenolic content with ripening of the different cultivars. Some cultivars, however, had a very high phenolic content. For example, the white flesh nectarine cultivar Brite Pearl (350–460 mg/kg hydroxycinnamates and 430–550 mg/kg procyanidins in flesh) and the yellow flesh cv. Red Jim (180–190 mg/kg hydroxycinnamates and 210–330 mg/kg procyanidins in flesh), contained 10 times more phenolics than cultivars such as Fire Pearl (38–50 mg/kg hydroxycinnamates and 23–30 mg/kg procyanidins in flesh).

The gum exudate polysaccharide from the trunk of nectarine (PPNEC) was found to compose of arabinose, xylose, mannose, galactose, and uronic acids in 37:13:2:42:6 Molar ratio and had molecular weight of 3.93×10^6 g/mol (Simas-Tosin et al. 2009). Methylation analysis of PPNEC indicated a highly branched structure with relatively high amounts of di- (16%) and tri-O-substituted (9%) Galp units and non-reducing end-units of Araf (26%) and Xylp (17%). Combination with 13C NMR data, showed the presence of α-l-Araf (non-reducing end, 3-O-, 5-O-, and 2,5-di-O-subst.), β-l-Arap (4-O- and 2,4-di-O-subst.), β-d-Galp (3-O-, 2,3-di-O-, 3,6-di-O-, and 3,4,6-tri-O-subst.), and α- and/or β-d-Xylp non-reducing end-units. PPNEC had structures similar to those of polysaccharide from peach tree gum, although in different proportions and with a lower molecular weight.

Antioxidant Activity

Gil et al. (2002) reported on the antioxidant capacities, phenolic compounds, carotenoids, and vitamin C contents of nectarine as follows. The ranges of total ascorbic acid (vitamin C) (in mg/100 g of fresh weight) were 5–14 mg (white-flesh nectarines), 6–8 mg (yellow-flesh nectarines); total carotenoids concentrations (in μg/100 g of fresh weight) were 7–14 μg (white-flesh nectarines), 80–186 μg (yellow-flesh nectarines), and total phenolics (in mg/100 g of fresh weight) were 14–102 mg (white-flesh nectarines), 18–54 mg (yellow-flesh nectarines). The contributions of phenolic compounds to antioxidant activity were much greater than those of vitamin C and carotenoids. There was a strong correlation (0.93–0.96) between total phenolics and antioxidant activity. Anthocyanin, carotenoids and vitamin C play important role in the antioxidant capacity of ripened nectarine fruits and their content was strongly dependent on harvest date (Ghasemi et al. 2011). Total phenol content and flavonoids decreased with nectarine fruit maturity and ripening. Sucrose levels in ripened nectarine decreased due to conversion to fructose or glucose. Antioxidant capacity and contents of total phenolics, anthocyanins, flavonoids, and vitamin C of peach and nectarine were found to be influenced by genotype and flesh colour traits (Cantín et al. 2009; Abidi et al. 2011).

Traditional Medicinal Uses

As described for peaches.

Other Uses

Similar to those described for peaches.

Comments

Nectarine is easily propagated from semi-hardwood or hardwood cuttings and by bud-grafting as they do not come true from seeds.

Selected References

Abidi W, Jiménez S, Moreno MÁ, Gogorcena Y (2011) Evaluation of antioxidant compounds and total sugar content in a nectarine [*Prunus persica* (L.) Batsch] progeny. Int J Mol Sci 12(10):6919–6935

Aubert C, Ambid C, Baumes R, Günata Z (2003) Investigation of bound aroma constituents of yellow-fleshed nectarines (*Prunus persica* L. cv. Springbright) changes in bound aroma profile during maturation. J Agric Food Chem 51(21):6280–6286

Bown D (1995) Encyclopaedia of herbs and their uses. Dorling Kindersley, London, 424 pp

Cantín CM, Moreno MA, Gogorcena Y (2009) Evaluation of the antioxidant capacity, phenolic compounds, and vitamin C content of different peach and nectarine [*Prunus persica* (L.) Batsch] breeding progenies. J Agric Food Chem 57(11):4586–4592

Colaric M, Stamar F, Hudina M (2004) Contents of sugars and organic acids in the cultivars of peach (*Prunus persica* L.) and nectarine (*Prunus persica* var. *nucipersica* Schneid.). Acta Agric Slovenica 83(1):53–61

Council of Scientific and Industrial Research (1969) The wealth of India. A dictionary of Indian raw materials and industrial products. (Raw materials 8). Publications and Information Directorate, New Delhi

Duke JA, Ayensu ES (1985) Medicinal plants of China, vols 1 and 2. Reference Publications, Inc., Algonac, 705 pp

Engel KH, Flath RA, Buttery RG, Mon TR, Ramming DW, Teranishi R (1988) Investigation of volatile constituents in nectarines. 1. Analytical and sensory characterization of aroma components in some nectarine cultivars. J Agric Food Chem 36(3):549–553

Facciola S (1990) *Cornucopia*. A source book of edible plants. Kampong Publications, Vista, 677 pp

Ghasemi Y, Ghasemnezhad A, Atashi S, Mashayekhi K, Ghorbani M (2011) Variations in antioxidant capacity of nectarine fruits (*Prunus persica* cv. red-gold) affected by harvest date. Int J Plant Prod 5(3):311–318

Gil MI, Tomás-Barberán FA, Hess-Pierce B, Kader AA (2002) Antioxidant capacities, phenolic compounds, carotenoids, and vitamin c contents of nectarine, peach, and plum cultivars from California. J Agric Food Chem 50(17):4976–4982

Grieve M (1971) A modern herbal, 2 vols. Penguin/Dover publications, New York, 919 pp

Huxley AJ, Griffiths M, Levy M (eds) (1992) The new RHS dictionary of gardening, 4 vols. MacMillan, New York

Porcher MH et al (1995–2020) Searchable World Wide Web multilingual multiscript plant name database. The University of Melbourne, Melbourne. http://www.plantnames.unimelb.edu.au/Sorting/Frontpage.html

Simas-Tosin FF, Wagner R, Santos EMR, Sassaki GL, Gorin PAJ, Iacomini M (2009) Polysaccharide of nectarine gum exudate: comparison with that of peach gum. Carbohydr Polym 76(3):485–487

Šoferistov EP, Kravcova TA (1990) Phylogenetic relationships of *Persica vulgaris* Mill. var. *nectarina* (Maxim.) Holub with various representatives of subfamily Prunoideae, family Rosaceae on the basis of an immunochemical analysis of the seed proteins. Rastiteln Resursy 26:11–22

Tomás-Barberán FA, Gil MI, Cremin P, Waterhouse AL, Hess-Pierce B, Kader AA (2001) HPLC-DAD-ESIMS analysis of phenolic compounds in nectarines, peaches, and plums. J Agric Food Chem 49(10):4748–4760

U.S. Department of Agriculture, Agricultural Research Service (2011) USDA national nutrient database for standard reference, Release 24. Nutrient Data Laboratory Home Page, http://www.ars.usda.gov/ba/bhnrc/ndl

Prunus persica var. *persica*

Scientific Name

Prunus persica (L) Batsch var. *persica*

Synonyms

Amygdalus persica L., *Amygdalus persica* [unranked] *aganonucipersica* Schübler & Martens, *Amygdalus persica* var. *aganonucipersica* (Schübler & Martens) T. T. Yü & L. T. Lu, *Amygdalus persica* [unranked] *aganopersica* Reichenbach, *Amygdalus persica* var. *compressa* (Loudon) T. T. Yü & L. T. Lu, *Amygdalus. persica* [unranked] *scleronucipersica* Schübler & Martens, *Amygdalus. persica* var. *scleronucipersica* (Schübler & Martens) T. T. Yü & L. T. Lu, *Amygdalus persica* [unranked] *scleropersica* Reichenbach, *Amygdalus persica* var. *scleropersica* (Reichenbach) T. T. Yü & L. T. Lu, *Persica ispahanensis* Thouin, *Persica platycarpa* Decaisne, *Persica vulgaris* Mill., *Persica vulgaris* var. *compressa* Loudon, *Prunus persica* (Linnaeus) Batsch, *Prunus persica* var. *compressa* (Loudon) Bean, *Prunus persica* subsp. *platycarpa* (Decaisne) D. Rivera et al., *Prunus persica* var. *platycarpa* (Decaisne) L. H. Bailey.

Family

Rosaceae

Common/English Names

Peach, Peach Tree, Peaches

Vernacular Names

Albanian: Bukuroshe;
Arabic: Khawkh, Khokh, Khoukh;
Brazil: Nectarina, Pêssego, Pessegueiro;
Bulgarian: Praskova;
Catalan: Préssec;
Chinese: Da Tao Ren, Hao Ren, Mao Tao, Shou Tao, Tao, Tao Ren, Tao Zi;
Croatian: Breskva;
Czech: Broskvoň Obecná;
Danish: Fersken;
Dutch: Perzik, Perzikboom;
Eastonian: Harilik Virsikupuu;
Finnish: Persikka;
French: Pêcher, Pêcher Commun;
German: Echter Pfirsich, Pfirsich, Pfirsichbaum;
Greek: Robakinon;
Hebrew: Afarseq;
Hungarian: Barrack, Kerti Őszibarack, Őszibarack;
Icelandic: Ferskja;
India: Adoo, Aru (Hindu), Pichesu (Kannada), Chumbhrei (Manipuri), Pishu (Oriya), Aaruu (Urdu);
Indonesia: Persik;
Italian: Persico, Pesco;

Korean: Boksanamu, Poksunga, Poksunganamu;
Japanese: Ke Momo, Momo, Piichi;
Laotian: Khai;
Latvian: Persiks;
Lithuanian: Persikas;
Malaysia: Persik;
Maltese: Ħawħ;
Persian: Hulu;
Philippines: Peras (Tagalog);
Polish: Brzoskwinia, Brzoskwinia Zwyczajna, Przerzedzanie Brzoskwin;
Portuguese: Pessego;
Romanian: Piersica;
Russian: Persik Obyknovennyj;
Serbian: Breskva;
Slovenia: Breskev, Broskyňa Obyčajná;
Spanish: Albérchigo, Durazno, Melocotonero, Pavía, Persico, Prescal, Melocotón;
Swedish: Persika, Persiketräd, Prunusväxter;
Thai: Hung Mon, Makmuan, Tho;
Turkey: Şeftali;
V*ietnamese*: Đào;
Zulu: Umumpetshisi.

Origin/Distribution

Peaches probably originated from China, being one of the first fruit crop domesticated about 4,000 years ago. Cultivars grown today are derived largely from ecotypes native to southern China, an area with climate similar to that of the southeastern USA, a major peach growing region. Peaches were introduced to Persia (Iran) along silk trading routes and was given the species epithet *persica* denoting Persia which was then believed to be its source of origin. Along the route of migration secondary centres of diversity originated (Middle Asia, Transcaucasus). Greeks and especially Romans distributed the peach throughout Europe and England around 300–400 BC. During the sixteenth to seventeenth centuries, Portuguese explorers brought the peach to south America and the Spaniard explorers introduced it to the northern Florida coast of North America. Native Americans and settlers distributed the peach across North America into southern Canada.

Agroecology

Peach is a cool climate species that adapts well to temperate or sub-temperate areas with cool winters and a warm summer. Peach tree requires a winter chilling period for flowering and fruiting. On average, most cultivars have chilling requirements of 600–900 h. It grows best in full sun. The tree is not frost hardy. Peach trees are extremely sensitive to poorly drained soils. In areas of poor drainage, roots will die, resulting in stunted growth and eventual death of the tree. Although peach trees will grow well in a wide range of soil types, a deep soil ranging in texture from a sandy loam to a sandy clay loam is preferred. Sites previously established with peaches should be avoided since they succumb the tree to peach decline disease referred to as "peach tree short live", which greatly affects its growth, development and productivity. A ring nematode (*Criconemella xenoplax*) has been implicated as the predisposing agent for PTSL, and they move fastest in sandy soils.

Edible Plant Parts and Uses

Cultivated peaches are divided into "freestone" and "clingstone" cultivars, depending on whether the flesh adheres to the stone (endocarp). Both kinds can have either white or yellow or orangey yellow flesh. Peaches with white flesh typically are very sweet, while yellow-fleshed peaches typically have an acidic tang coupled with sweetness, though this also varies greatly. The ripe fruit and skin can be eaten fresh out of hand. Most suitable for this and also for freezing are "freestone" cultivars. The "clingstone" types are suitable for processing to juices, jams, pies, pastries or used in ice-cream. The fruit can also be cooked and dried for later use. Special cultivars with high sugar content and intensive flavour can be used for distillery.

Flowers are edible, eaten in salad or used as a garnish. A tea can be brewed from the flowers. A white liquid can be distilled from the flowers, having a flavour similar to the seed. The seed kernel can be eaten raw or cooked. However it is best avoided raw especially if it is bitter because of hydrocyanic acid as cases of toxicity have been reported. The gum from the stem has been reported to be used for chewing.

Botany

A small deciduous, branched tree, 3–8 m tall with a broad and more or less horizontally spreading crown and dark reddish brown, scabrous bark. Branchlets are dark brownish-green, glabrous, and lenticellate. Winter buds are conical, pubescent, with obtuse apex and occur in fascicle of 2–3. Petiole robust, 1–2 cm, with or without 1 to several nectaries. Leaves are alternate, simple, green, lanceolate, oblong-lanceolate, elliptic-lanceolate, or obovate-oblanceolate, 7–15 × 2–3.5 cm, abaxially with or without a few hairs in vein axils, adaxially glabrous, base broadly cuneate, margin finely to coarsely serrate, apex acuminate (Plates 3, 4, 5). Flowers are bisexual, pentamerous, shortly pedicellate or subsessile with a reddish-green, campanulate, sparsely pubescent hypanthium; sepals ovate to oblong, pubscent, as long as hypanthium; petals pink, deep pink (Plates 1 and 2) or white, oblong-elliptic to broadly obovate; stamens 20–30; anthers purplish red; ovary pubescent with a style, nearly as long as stamens. Fruit a drupe, usually globose to oblate, also ovoid to broadly ellipsoid, 4–12 cm diameter, tomentulose (velvety), greenish white to orangey yellow, usually with red blushes on exposed side with a conspicuous ventral suture (Plates 3, 4, 5, 6). Mesocarp (flesh) is white, greenish-white, yellow, orangey yellow, or red, succulent, sweet to acid-sweet, fragrant; endocarp is large, hard, ellipsoid to suborbicular, compressed on both sides, surface longitudinally and transversely furrowed and pitted, free from mesocarp or compactly adhering to it, apex acuminate. Seed is red-brown, oval shaped and 1.5–2 cm long, bitter.

There is also a flattish, doughnut-shaped, sweet, white-fleshed and small-seeded peach cultivar (Plates 7, 8, 9) which is commercially grown in Australia.

Nutritive/Medicinal Properties

Food value of raw, peach fruit (refuse 4% pit) per 100 g edible portion was reported as follows (USDA 2011): water 88.87 g, energy 39 kcal (165 kJ), protein 0.91 g, total lipid (fat) 0.25 g, ash 0.43 g, carbohydrate 9.54 g; fibre (total dietary) 1.5 g, total sugars 8.39 g, sucrose 4.76 g, glucose 1.95 g, fructose 1.53 g, maltose 0.08 g, galactose 0.06 g; minerals – calcium 6 mg, iron 0.25 mg, magnesium 9 mg, phosphorus 20 mg, potassium 190 mg, sodium 0 mg, zinc 0.17 mg, copper 0.068 mg, manganese 0.061 mg, fluoride 4 μg, selenium 0.1 μg; vitamins – vitamin C (total ascorbic acid) 6.6 mg, thiamine 0.024 mg, riboflavin 0.031 mg, niacin 0.806 mg, pantothenic acid 0.153 mg, vitamin B-6 0.025 mg, folate (total) 4 μg, total choline 6.1 mg, betaine 0.3 mg, vitamin A 326 IU (16 ug RAE), β-carotene 162 μg, β-cryptoxanthin 67 μg, lutein + zeaxanthin 91 μg, vitamin E (α tocopherol) 0.73 mg, γ-tocopherol 0.02 mg, vitamin K (phylloquinone) 2.6 μg, phytosterols 10 mg, total saturated fatty acids 0.019 g, 16:0 (palmitic acid) 0.017 g, 18:0 (stearic acid) 0.002 g, total monounsaturated fatty acids 0.067 g, 16:1 undifferentiated (palmitoleic acid) 0.002 g, 18:1 undifferentiated (oleic acid) 0.065 g, total polyunsaturated fatty acids 0.086 g, 18:2 undifferentiated (linoleic acid) 0.084 g, 18:3 undifferentiated 0.002 g, tryptophan 0.010 g, threonine 0.016 g, isoleucine 0.017 g, leucine 0.027 g, lysine 0.030 g, methionine 0.010 g, cystine 0.012 g, phenylalanine 0.019 g, tyrosine 0.014 g, valine 0.022 g, arginine 0.018 g, histidine 0.008 g, alanine 0.028 g, aspartic acid 0.418 g, glutamic acid 0.056 g, glycine 0.021 g, proline 0.018 g and serine 0.032 g.

Sugars (glucose, fructose, sucrose and sorbitol) and organic acids (citric, malic, shikimic and fumaric acid) in fruits were identified in all peach and nectarine cultivars studied in Slovenia (Colaric et al. 2004). Sucrose was the major sugar and malic and citric acids were the predominant organic acids. The content of fructose ranged from 6.76 to 12.97 g/kg, glucose from 5.43 to 11.11 g/kg, sucrose from 46.14 to 70.7 g/kg and sorbitol from 0.40 to 2.80 g/kg of fruits. The content of

Plate 1 Peach blossoms

Plate 2 Close-up of peach blossoms

citric acid ranged from 1.71 to 8.34 g/kg, malic acid from 3.2 to 8.05 g/kg, shikimic acid from 127 to 809 mg/kg and fumaric acid from 1.56 to 6.09 mg/kg of fruits. The content of total sugars ranged from 61.53 to 93.70 g/kg and the content of total organic acids ranged from 7.06 to 14.69 g/kg of fruits.

The phenolic compounds hydroxycinnamates, procyanidins, flavonols, and anthocyanins were detected and quantified in peach cultivars (Tomás-Barberán et al. 2001). As a general rule, the peel tissues contained higher amounts of phenolics, and anthocyanins and flavonols were almost exclusively located in this tissue. No clear differences in the phenolic content of peaches were detected or between white flesh and yellow flesh cultivars. There was no clear trend in phenolic content with ripening of the different cultivars. Some cultivars, however, had a very high phenolic content. Among white flesh peaches, cultivars Snow King (300–320 mg/kg hydroxycinnamates and 660–695 mg/kg procyanidins in

Plate 3 Immature peaches and leaves

Plate 6 Harvested peaches

Plate 4 Near-ripe peaches and leaves

Plate 7 Doughnut peach cultivar (*top* and *bottom* views)

Plate 5 Ripe peaches

Plate 8 Doughnut peach (*top* and *lateral* views)

flesh) and Snow Giant (125–130 mg/kg hydroxycinnamates and 520–540 mg/kg procyanidins in flesh) showed the highest content. Chlorogenic acid, catechin, epicatechin, rutin and cyanidin-3-glucoside were detected as the main phenolic compounds of ripened peach fruits (Andreotti et al. 2008). The concentration was always higher in peel tissue, with average values ranging

Plate 9 Flat, sweet, white-fleshed and small seeded doughnut peach

from 1 to 8 mg/g dry weight (DW) depending on cultivar.

Peaches were found to contain the following organic acids – citric acid, malic acid, tartaric acid, quinic acid, mucic acid, galacturonic acid, chlorogenic acid and neo-chlorogenic acid (David et al. 1956) and the following tannins – catechin, catechol, tannic acid and chlorogenic acids (Johnson et al. 1951). The anthocyanin in freestone peach was identified as a J-mono-glucoside of cyanidin (Hsia et al. 1965) and a leucocyanidin was obtained from immature Elberta peaches (Hsia et al. 1964). The presence of (2R: 3S) (+)-catechin and certain chlorogenic acids with their isomers were confirmed.

The gum exudate polysaccharide from the trunk was found to compose of Ara, Xyl, Man, Gal, and uronic acids in 37:13:2:42:6 M ratio and had Mw 3 of 5.61×10^6 g/mol for peach gum polysaccharide (Simas-Tosin et al. 2009). Peach tree gum had structures similar to those of polysaccharide from nectarine tree gum, although in different proportions and with a lower molecular weight. From the heartwood of peach β-sitosterol and its D-glucoside, hentriacontane, hentricontanol and the flavonoids naringenin, dihy-drokaempferol, kaempferol and quercetin were isolated (Chandra and Sastry 1988).

Peaches are rich in antioxidant phenolic compounds which impart to it anticancer, photogenotoxic-protective and other pharmacological properties.

Antioxidant Activity

In selected clingstone peach cultivars, inhibition of low-density lipoprotein (LDL) oxidation was found to vary from 17.0% to 37.1% in peach flesh extract, from 15.2% to 49.8% in whole peach extract, and from 18.2% to 48.1% in peel extract (Chang et al. 2000). Total phenols were 432.8–768.1 mg/kg in flesh extract, 483.3–803.0 mg/kg in whole extract, and 910.9–1922.9 mg/kg in peel extract. The correlation coefficient between relative LDL antioxidant activity and concentration of total phenols was 0.76. The lowest polyphenol oxidase (PPO) and specific activities were found in the Walgant cultivar, followed by Kakamas and 18-8-23. These three cultivars were found to combine the desirable characteristics of strong antioxidant activity, low PPO activity, and lower susceptibility to browning reactions.

The ranges of total ascorbic acid (vitamin C) (in mg/100 g of fresh weight) were 6–9 mg (white-flesh peaches), and 4–13 mg (yellow-flesh peaches) (Gil et al. 2002). Total carotenoids concentrations (in μg/100 g of fresh weight) were 7–20 μg (white-flesh peaches) and 71–210 μg (yellow-flesh peaches). Total phenolics (in mg/100 g of fresh weight) were 28–111 mg (white-flesh peaches) and 21–61 mg (yellow-flesh peaches). The contributions of phenolic compounds to antioxidant activity were much greater than those of vitamin C and carotenoids. There was a strong correlation (0.93–0.96) between total phenolics and antioxidant activity.

Carotenoid content was higher in yellow-flesh (2–3 mg β-carotene/100 g fresh weight) than in white or red-flesh peaches (0.01–1.8 mg β-carotene/100 g fresh weight) (Vizzotto et al. 2006). Antioxidant activity (AOA) as evaluated by 2,2-diphenyl-1-picrylhydrazyl (DPPH) was about twofold higher in red-flesh varieties than in white/yellow-flesh peach varieties. Among the peaches, the AOA was best correlated with phenolic content. Studies in three different peach cultivars showed that Trolox equivalent antioxidant capacity (TEAC), measured from harvest to 7 days postharvest, was influenced mainly by vitamin C content, whereas polyphenols and carotenoids seemed to play a secondary role (Valle et al. 2007).

Although TEAC was similar in the three cultivars, only cv 'Luisa Berselli' significantly increased the total radical-trapping potential (TRAP) in human plasma at 1, 2 and 4 hours after ingestion of peaches. Sugar moiety, condensed and glycoside phenols were suggested to be involved in the higher effect on plasma TRAP of cv 'Luisa Berselli'. Polyphenolic compounds at harvest were high in the 3 cultivars rating 288 mg/kg for 'Maria Serena', 405 mg/kg for 'Luisa Berselli' and 549 mg/kg for 'Stark Earlyglo'. Vitamin C at 4 days after harvest rated 85 mg/kg for 'Luisa Berselli', 70 mg/kg for 'Stark Earlyglo' and 52 mg/kg for Maria Serena'. 'Luisa Berselli' had lower levels of carotenoids (lutein, zexanthin, β-cryptoxanthin, β-carotene, β-cis-carotene) than 'Maria Serena' and 'Stark Earlyglo' which had similar contents.

The anthocyanin content of the peaches ranged from 7.64 to 50.01 mg cyanidin 3-glucoside/100 g fresh tissue (Cevallos-Casals et al. 2002). The total phenolics content for peaches ranged from 99 to 449 mg chlorogenic acid/100 g fresh tissue. The antioxidant activity (AOA) ranged from 440 to 1,784 ug equivalent Trolox/g fresh tissue for peaches. Correlation analysis indicated that the anthocyanin content and phenolic content was well correlated with the antioxidant activity. Anthocyanin (red pigments) and phenolic contents were higher in red-flesh than in white/yellow-flesh peaches. Carotenoid (orange pigments) content was higher in yellow-flesh than in white or red-flesh peaches (Byrne et al. 2004). AOA was about twofold higher in red-flesh varieties than in white/yellow-flesh varieties. Among the peaches, the AOA was well correlated with both phenolic and anthocyanin content. These results suggested that red-flesh peach varieties have a greater potential health benefit based on antioxidant content and AOA as compared to the white/yellow-flesh varieties. Antioxidant capacity and contents of total phenolics, anthocyanins, flavonoids, and vitamin C of peach and nectarine were found to be influenced by genotype and flesh colour traits (Cantín et al. 2009).

Studies showed that the extracts of peels and flesh of Maciel, Leonense, and Eldorado peach cultivars presented free radical scavenging activity in all concentrations tested, with a concentration-dependent action (Rossato et al. 2009). The immediate inhibition of chemiluminescence and the duration of this inhibition were significantly higher with the extracts than with the major compound (chlorogenic acid) alone, and it can be due to a synergistic or additive effect of other antioxidants present in the extracts. The 50% inhibitory concentration (IC_{50}) values for peach extract and chlorogenic acid were 1.19 and 8.43 μg/ml, respectively, when total radical-trapping antioxidant potential was evaluated, whereas IC_{50} values of 0.41 and 1.89 μg/ml was found when total antioxidant reactivity was evaluated in peach extract and chlorogenic acid, respectively. Chlorogenic acid presented a good contribution to antioxidant reactivity and potential, but the fruit extracts provided better antioxidant action.

Anticancer Activity

Studies demonstrated that supplementation with PPFE (*Prunus persica* flesh extract) might protect against cisplatin-induced toxicity in cancer patients (Lee et al. 2008). In a xenograft model with the repeated administration of a low-dose cisplatin (5 mg/kg body weight) for 15 days, and in an acute toxicity model with a single administration of a high-dose cisplatin (45 mg/kg body weight) over a 16 hours period, the consecutive administration of PPFE in combination with and prior to the cisplatin injection reversed the cisplatin-induced decrease in the liver weight as a percentage of total body weight, and the cisplatin-induced increases in the serum alanine aminotransferase and aspartate aminotransferase levels caused by liver damage. Moreover, the oral administration of PPFE significantly recovered the reduced glutathione level and inhibited lipid peroxidation in the cisplatin-treated mice. The administration of PPFE alone significantly inhibited the growth of CT-26 colon carcinoma xenografted onto mice without adverse effects (Lee et al. 2009). The combination of PPFE and cisplatin enhanced the inhibitory effect of cisplatin against tumour growth. In a xenograft model involving the repeated administration of low-dose cisplatin for 15 days, and in an acute toxicity model involving a single administration of high-dose cisplatin over a 16 hours period, the administration of PPFE in combination with and prior to the cisplatin injection reversed the cisplatin-induced reduction in the kidney weight. PPFE blocked the increases in the serum blood urea nitrogen and creatinine levels associated with the kidney damage. Moreover, the administration of PPFE induced a significant reduction in cisplatin-induced oxidative stress. These results indicated that PPFE may promote the therapeutic efficacy of cisplatin therapy, while attenuating its inherent nephrotoxicity. Cisplatin (cis-diamminedichloroplatinum II) is one of the most effective chemotherapeutic agents used in the treatment of a variety of human solid tumours.

The ethanol extract of the flowers of *Prunus persica* (Ku-35) at 100–1,000 μg/mL inhibited the amount of 14C-arachidonic acid/metabolites release from UVB-irradiated keratinocytes (Kim et al. 2000). It was also demonstrated that Ku-35 possessed the protective activity against UV-induced cytotoxicity of keratinocytes and fibroblasts. In addition, Ku-35 was revealed to protect UVB-induced erythema formation using guinea pigs in preliminary in-vivo study. All these results indicate that the flowers of *P. persica* extract may be beneficial for protecting UV-induced skin damage when topically applied. Further the scientists found that Ku-35 (50–200 μg/mL) was found to inhibit UVB- as well as UVC-induced DNA damage as measured by the COMET assay in the skin fibroblast cell (NIH/3T3) (Heo et al. 2001). In addition, Ku-35 inhibited UVB- or UVC-induced lipid peroxidation, especially against UVB-induced peroxidation at higher than 10 μg/mL. Ku-35 also had a the protective effect against UVB-induced non-melanoma skin cancer in mice. The application of Ku-35 clearly resulted in a delay of tumour development compared to the control. In tumour incidence, 100% mice in the control group and the low dose treatment of Ku-35 had tumours, whereas 94.1% of the mice had tumours after the high dose treatment of Ku-35 at the end of experiment (28 weeks). In tumour multiplicity, low and high treatments of Ku-35 resulted in 25.9% and 53.9% reduction. The findings indicated that Ku-35 protected against photogenotoxicity in NIH/3T3 fibroblasts. The possible action mechanism of Ku-35 may be through its antioxidant activity without pro-oxidant effect. Ku-35 could also show a delay of tumour development against UVB-induced skin carcinogenesis. These results suggested that Ku-35 extract may be useful for protecting UV-induced DNA damage and carcinogenesis when topically applied. They also reported that *P. persica* flower extract clearly inhibited UVB-induced erythema formation dose dependently when topically applied (IC_{50} = 0.5 mg/cm^2) (Kim et al. 2002). It also inhibited UVB-induced ear oedema (49% inhibition at 3.0 mg/ear). From the ethanol extract of peach flowers, four kaempferol glycoside derivatives were successfully isolated. Among the derivatives isolated, the content of multiflorin B was highest (3.3%, w/w). Multiflorin B inhibited UVB-induced erythema formation (80% inhibition at 0.3 mg/cm^2), indicating that this compound was one of the active principles of the extract. All these results suggested that *P. persica* flower extract may be useful for protection against UVB-induced skin damage when topically applied.

Central Nervous system Activity

Prunus persica seed extract (PPE) at 2.5 g/kg and tacrine at 5 mg/kg showed significant effects on acetylcholine concentration in the hippocampus of rats for more than 6 hours (Kim et al. 2003). At these doses, the maximum increases were observed at about 1.5 hours after administration of PPE, and at about 2 hours with tacrine, and were 454% and 412% of the pre-level, respectively. Tacrine (9-amino-1,2,3,4-tetrahydroacridine hydrochloride), is a well-known and centrally acting acetylcholinesterase (AChE) inhibitor, which had been developed for the treatment of Alzheimer's disease. The results suggested that oral administration of PPE and tacrine increased acetylcholine concentration in the synaptic cleft of the hippocampus mostly through AChE inhibition, and that PPE had a potent and long-lasting effect on the central cholinergic system.

Antiaging Activity

Croteau et al. (2010) showed that repair of various oxidative DNA lesions was more efficient in liver extracts derived from mice fed fruit-enriched diets. There was a decrease in the levels of formamidopyrimidines in peach-fed mice compared with the controls. Additionally, microarray analysis revealed that NTH1 repair protein was upregulated in peach-fed mice. The results suggested that an increased intake of fruits might modulate the efficiency of DNA repair, resulting in altered levels of DNA damage and improved genome integrity. DNA damage and oxidative stress are some important contributing factors to aging.

Allergy Activity

The lipid transfer protein Pru p 3 and Pru p1 the putative peach member of the Bet v 1 family, were identified as major peach fruit allergens (Ahrazem et al. 2007). A differential distribution between peel and pulp and different solubility properties were found for Pru p 3, Pru p 1, and peach profilin. Mean Pru p 3 levels were 132.86, 0.61, and 16.92 μg/g of fresh weight of peels, pulps, and whole fruits, respectively. The corresponding mean Pru p 1 levels were 0.62, 0.26, and 0.09 μg/g of fresh weight. Most US cultivars showed higher levels of both allergens than Spanish cultivars.

Traditional Medicinal Uses

Nearly all parts of the tree are used locally in traditional folk medicine. The flowers are diuretic, sedative and vermifuge and have been used for treatment of constipation and oedema. The leaves are considered to be astringent, demulcent, diuretic, expectorant, febrifuge, laxative, vermicidal and mildly sedative. Leaf decoctions have been employed for gastritis, whooping cough, coughs and bronchitis, to help relive vomiting and morning sickness during pregnancy. Dried, pulverised leaves has been used to heal wounds and sores. Gum from the stem alterative, astringent, demulcent and sedative and bark demulcent, diuretic, expectorant and sedative. Bark has been used in the treatment of gastritis, whooping cough, coughs and bronchitis. The root bark has been used in the treatment of dropsy and jaundice. The seed is antiasthmatic, antitussive, emollient, haemolytic, laxative, anti-inflammatory, analgesic, and sedative. It has been used internally in the treatment of constipation in the elderly, coughs, asthma and menstrual disorders. In Korea seed extracts have been used for constipation, laryngitis, menostasis, dermatopathy and contusion.

Peach seed like other *Prunus* species seeds contain amygdalin, wrongly dubbed as vitamin B17, Nitriloside or commercially as Laetrile. Laetrile has been spuriously claimed to have anticancer properties and have been illegally used in the treatment of human cancer. Laetrile is not registered for cancer use in USA, Europe and Australia. Recent scientific reviews of research papers and clinical trials all refute the claim that laetrile has beneficial effects for cancer patients and is not supported by sound clinical data (Dorr and Paxinos 1978; Ellison et al. 1978; Moertel et al. 1981, 1982; Milazzo et al. 2007; Queensland Heath 2006). The United States National Cancer Institute (Ellison et al. 1978) and Queensland Health (2006) have issued warnings that read Amygdalin (Laetrile) is a toxic drug that is not effective in treating or controlling cancer. In a large clinical trial conducted in the United States, Moertel et al. (1982) reported that no substantive benefit was observed in terms of cure, improvement or stabilization of cancer, improvement of symptoms related to cancer, or extension of life span. The hazards of amygdalin therapy were evidenced in several patients by symptoms of cyanide toxicity or by blood cyanide levels approaching the lethal range. Toxicity of cyanide poisoning has also been reported by other researchers (Humbert et al. 1977; Sadoff et al. 1978; Morse et al. 1979; Lee et al. 1982).

Other Uses

Various coloured dyes are obtained from the fruit (dark-grey to green) and leaves (green). A semi-drying oil expressed from the seeds are used in cosmetics, often as a substitute for almond oil in skin creams. The oil is also used as fuel in India. The gum from the stem is used as adhesive.

Peaches are revered in China, Japan, Korea, and Vietnam not only for its edible fruit and medicinal attributes, its beautiful ornamental flowers but also for its association with folklores and social ethnotraditions. The peach often plays an important part in Chinese tradition and is symbolic of long life. In China, the peach was said to be consumed by the immortals due to its mystic virtue of conferring longevity on all who ate them. One of the Chinese Eight Immortals, is often depicted carrying a Peach of Immortality. Due to its delicious taste and soft texture, in ancient China "peach" was also a slang word for "young bride". Momotaro, or "Peach Boy" one of Japan's most noble and semi-historical heroes who fights evil, was born from within an enormous peach floating down a stream.

Comments

Commercially, peaches are propagated by asexual methods such as grafting, budding (e.g. T-budding) and stem cuttings as they do not come true to type from seeds.

Selected References

Ahrazem O, Jimeno L, López-Torrejón G, Herrero M, Espada JL, Sánchez-Monge R, Duffort O, Barber D, Salcedo G (2007) Assessing allergen levels in peach and nectarine cultivars. Ann Allergy Asthma Immunol 99(1):42–47

Andreotti C, Ravaglia D, Ragaini A, Costa G (2008) Phenolic compounds in peach (*Prunus persica*) cultivars at harvest and during fruit maturation. Ann Appl Biol 153(1):11–23

Bown D (1995) Encyclopaedia of herbs and their uses. Dorling Kindersley, London, 424 pp

Byrne D, Vizzotto M, Cisneros-Zevallos L, Ramming DW, Okie WR (2004) Antioxidant content of peach and plum genotypes. Hortscience 39(4):798

Cantín CM, Moreno MA, Gogorcena Y (2009) Evaluation of the antioxidant capacity, phenolic compounds, and vitamin C content of different peach and nectarine [*Prunus persica* (L.) Batsch] breeding progenies. J Agric Food Chem 57(11):4586–4592

Cevallos-Casals BA, Byrne DH, Cisneros-Zevallos L, Okie WR (2002) Total phenolic and anthocyanin content in red-fleshed peaches and plums. Acta Hortic (ISHS) 592:589–592

Chandra S, Sastry MS (1988) Phytochemical investigations on *Prunus persica* heart wood. Indian J Pharm Sci 50(6):321–322

Chang S, Tan C, Frankel EN, Barrett DM (2000) Low-density lipoprotein antioxidant activity of phenolic compounds and polyphenol oxidase activity in selected clingstone peach cultivars. J Agric Food Chem 48(2): 147–151

Colaric M, Stamar F, Hudina M (2004) Contents of sugars and organic acids in the cultivars of peach (*Prunus persica* L.) and nectarine (*Prunus persica* var. n*ucipersica* Schneid.). Acta Agric Slovenica 83(1):53–61

Council of Scientific and Industrial Research (CSIR) (1969) The wealth of India. A dictionary of Indian raw materials and industrial products. (Raw Materials 8). Publications and Information Directorate, New Delhi

Croteau DL, de Souza-Pinto NC, Harboe C, Keijzers G, Zhang Y, Becker K, Sheng S, Bohr VA (2010) DNA repair and the accumulation of oxidatively damaged DNA are affected by fruit intake in mice. J Gerontol A Biol Sci Med Sci 65(12):1300–1311

David JJ, Luh BS, Marsh GL (1956) Organic acids in peaches. Food Res 21(2):184–194

Dorr RT, Paxinos J (1978) The current status of laetrile. Ann Intern Med 89:389–397

Duke JA, Ayensu ES (1985) Medicinal plants of China, vols 1 and 2. Reference Publications, Inc., Algonac, 705 pp

Ellison NM, Byar DP, Newell GR (1978) Special report on laetrile: the NCI laetrile review. Results of the National Cancer Institute's retrospective Laetrile analysis. N Engl J Med 299:549–552

Facciola S (1990) Cornucopia. A source book of edible plants. Kampong Publications, Vista, 677 pp

Gil MI, Tomás-Barberán FA, Hess-Pierce B, Kader AA (2002) Antioxidant capacities, phenolic compounds, carotenoids, and vitamin C contents of nectarine, peach, and plum cultivars from California. J Agric Food Chem 50(17):4976–4982

Grieve M (1971) A modern herbal, 2 vols. Penguin/Dover publications, New York, 919 pp

Hedrick UP (1972) Sturtevant's edible plants of the world. Dover Publications, New York, 686 pp

Heo MY, Kim SH, Yang HE, Lee SH, Jo BK, Kim HP (2001) Protection against ultraviolet B- and C-induced DNA damage and skin carcinogenesis by the flowers of *Prunus persica* extract. Mutat Res 496(1–2):47–59

Hsia C, Claypool LL, Abernethy JL, Esau P (1964) Leucoanthocyan material from immature peaches. J Food Sci 29(6):723–729

Hsia CL, Luh BS, Chichester CO (1965) Anthocyanin in freestone peaches. Food Res 30(1):5–12

Humbert JR, Tress JH, Braico KT (1977) Fatal cyanide poisoning: accidental ingestion of amygdalin. JAMA 238(6):482

Huxley AJ, Griffiths M, Levy M (eds) (1992) The new RHS dictionary of gardening, 4 vols. MacMillan, New York

Johnson G, Mayer MM, Johnson DK (1951) Isolation and characterization of peach tannins. Food Res 16(3): 169–180

Kim YH, Yang HE, Kim JH, Heo MY, Kim HP (2000) Protection of the flowers of *Prunus persica* extract from ultraviolet B-induced damage of normal human keratinocytes. Arch Pharm Res 23(4):396–400

Kim YH, Yang HE, Park BK, Heo MY, Jo BK, Kim HP (2002) The extract of the flowers of *Prunus persica*, a new cosmetic ingredient, protects against solar ultraviolet-induced skin damage in vivo. J Cosmet Sci 53(1): 27–34

Kim YK, Koo BS, Gong DJ, Lee YC, Ko JH, Kim CH (2003) Comparative effect of *Prunus persica* L. Batsch-water extract and tacrine (9-amino-1,2,3,4-tetrahydroacridine hydrochloride) on concentration of extracellular acetylcholine in the rat hippocampus. J Ethnopharmacol 87(2–3):149–154

Lee M, Berger HW, Givre HL, Jayamanne DS (1982) Near fatal laetrile intoxication: complete recovery with supportive treatment. Mt Sinai J Med 49:305–307

Lee CK, Park KK, Hwang JK, Lee SK, Chung WY (2008) The extract of *Prunus persica* flesh (PPFE) attenuates chemotherapy-induced hepatotoxicity in mice. Phytother Res 22(2):223–227

Lee CK, Park KK, Hwang JK, Lee SK, Chung WY (2009) Extract of *Prunus persica* flesh (PPFE) improves chemotherapeutic efficacy and protects against nephrotoxicity in cisplatin-treated mice. Phytother Res 23(7):999–1005

Lu L-T, Bartholomew B (2003) *Amygdalus* Linnaeus. In: Wu ZY, Raven PH, Hong DY (eds) Flora of China, vol 9 (Pittosporaceae through Connaraceae). Science Press/Missouri Botanical Garden Press, Beijing/St. Louis

Milazzo S, Lejeune S, Ernst E (2007) Laetrile for cancer: a systematic review of the clinical evidence. Support Care Cancer 15(6):583–595

Moertel CG, Ames MM, Kovach JS, Moyer TP, Rubin JR, Tinker JH (1981) A pharmacologic and toxicological study of amygdalin. JAMA 245:591–594

Moertel CG, Fleming TR, Rubin J, Kvols LK, Sarna G, Koch R, Currie VE, Young CW, Jones SE, Davignon JP (1982) A clinical trial of amygdalin (Laetrile) in the treatment of human cancer. N Engl J Med 306(4): 201–206

Morse DL, Boros L, Findley PA (1979) More on cyanide poisoning from laetrile. N Engl J Med 301:892

Natural Products Research Institute (1998) Medicinal plants in the Republic of Korea, Western Pacific Series No 21. Seoul National University/WHO Regional Publications, Manila, 316 pp

Porcher MH et al (1995–2020) Searchable World Wide Web multilingual multiscript plant name database. The University of Melbourne, Australia. http://www.plantnames.unimelb.edu.au/Sorting/Frontpage.html

Queensland Health (2006) Drugs and poisons: amygdalin/laetrile. Patient information. Queensland Health, Queensland Government. http://www.health.qld.gov.au/ph/documents/ehu/32469.pdf

Rossato SB, Haas C, Raseira Mdo C, Moreira JC, Zuanazzi JA (2009) Antioxidant potential of peels and fleshes of peaches from different cultivars. J Med Food 12(5): 1119–1126

Sadoff L, Fuchs K, Hollander J (1978) Rapid death associated with laetrile ingestion. JAMA 239(15):1532

Simas-Tosin FF, Wagner R, Santos EMR, Sassaki GL, Gorin PAJ, Iacomini M (2009) Polysaccharide of nectarine gum exudate: comparison with that of peach gum. Carbohydr Polym 76(3):485–487

Tomás-Barberán FA, Gil MI, Cremin P, Waterhouse AL, Hess-Pierce B, Kader AA (2001) HPLC-DAD-ESIMS analysis of phenolic compounds in nectarines, peaches, and plums. J Agric Food Chem 49(10):4748–4760

U.S. Department of Agriculture, Agricultural Research Service (2011) USDA national nutrient database for standard reference, Release 24. Nutrient Data Laboratory Home Page, http://www.ars.usda.gov/ba/bhnrc/ndl

Valle AZD, Mignani I, Pinardi A, Galvano F, Ciappellano S (2007) The antioxidant profile of three different peaches cultivars (*Prunus persica*) and their short-term effect on antioxidant status in human. Eur Food Res Technol 225(7):167–172

Vizzotto M, Cisneros-Zevallos L, Byrne D, Okie WR, Ramming DW (2006) Total phenolic, carotenoids, and anthocyanin content and antioxidant activity of peach and plum genotypes. Acta Hort 713:453–455

Prunus salicina

Scientific Name

Prunus salicina Lindley

Synonyms

Prunus thibetica Franch., *Prunus triflora* Roxb.

Family

Rosaceae

Common/English Names

Asian Plum, Californian Plum, Chinese Plum, Japanese Blood Plum, Japanese Box Plum, Japanese Plum, Nai Plum, Ussurian Plum, Willow-Leaf Cherry.

Vernacular Names

Brazil: Ameixa-Japonêsa (Portuguese);
Chinese: Dong Bel Li, Li, Li Zi, Ri Ben Li,;
Czech: Slivoň Vrbová;
Danish: Japansk Blomme, Pilebladet Blomme;
Dutch: Japanse Pruim, Kaspruim;
Eastonian: Pajulehine Ploomipuu;
Finnish: Japaninluumu
French: Prune Japonaise, Prune Sanguine Du Japon, Prune Japonaise À Chair Rouge, Prunier Du Japon, Prunier Japonais;
German: Chinesische Pflaume, Chinesischer Pflaumenbaum, Dreibluetige Pflaume, Japanische Pflaume, Japanischer Pflaumenbaum;
Hungarian: Japán Szilva, Kínai Fűzringló, Kínai Szilva;
Icelandic: Víðiplóma;
Italian: Ciliegia Del Giappone, Prugna Del Giappone, Susino Giapponese;
Japanese: Batankyō, Sumomo;
Korean: Churinam, Jadu, jadunamu;
Laotian: 'Mân 'Luang, Tsi Keu;
Malay: Ijas Jepang;
Portuguese: Ameixeira-Japonêsa;
Russian: Sliva Iaponskaia;
Slovaščina: Kitajskojaponska Sliva;
Spanish: Cirolero Japonés, Ciruela Japonesa, Ciruelo Japonés;
Swedish: Kinesiskt Plommon;
Vietnamese: Prun.

Origin/Distribution

Prunus salicina is native to China and is commonly cultivated in China, Korea and Japan. Improved cultivars were developed mainly in Japan and were distributed and cultivated in other temperate localities including in Australia and USA.

Agroecology

In its native range, the tree is found in sparse forests, forest margins, thickets, scrub, along trails in mountains, stream sides in valleys from 200 to 2,600m altitude. It is widely cultivated in China – Anhui, Fujian, Gansu, Guangdong, Guangxi, Guizhou, Hebei, Heilongjiang, Henan, Hubei, Hunan, Jiangsu, Jiangxi, Jilin, Liaoning, Ningxia, Shaanxi, Shandong, Shanxi, Sichuan, Yunnan, Zhejiang and Taiwan.

Prunus salicina has winter chill requirement to overcome bud dormancy. Chill requirement for three groups of cultivars were 700–900 chill units (Rumayor-Rodriguez 1995) whereas low chill cultivars have lower chill requirements of 200–350 chill units (Rouse and Sherman 1997). Japanese plum thrives in areas with warm summers and mild springs. Late spring frosts may kill flowers and summer rains may lead to diseases. Humid conditions during the ripening of the fruit are undesirable because they favour the development of brown rot.

Edible Plant Parts and Uses

The ripe fruit is excellent when eaten fresh. The fruit is also used in canning and made into jam, jellies, juice wine and liquor. In China, a liquor is made from the fruits. In Japan, half-ripe fruits are used as a flavouring in a liqueur called *sumomo shu*. Fruits are also dried, preserved or candied, flavoured with sugar, salt, and liquorice.

Botany

A deciduous, small-medium-sized, branched, unarmed tree, 9–12 m high. Branches and branchlets are reddish-brown glabrous (Plates 3, 4, 5) or pubescent and winter buds are purplish-red usually glabrous. Stipules are linear, margin glandular, apex acuminate. Leaves are alternate, on 1–2 cm long petiole with 2 nectaries at the apex. Lamina is oblong-obovate, narrowly elliptic, or rarely oblong-ovate, 6–8(–12)×3–5 cm, dark green and glossy above, apex acute to shortly with 6–7 pairs of lateral veins. caudate, base cuneate, margin doubly crenate (Plates 3, 4, 5). Flowers usually 3 in a fascicle, 1.5–2.2 cm in diameter on 1–1.5 cm pedicels (Plates 1 and 2). Hypanthium campanulate; sepals, 5 oblong-ovate, outside glabrous, margin loosely serrate, apex acute to obtuse ,imbricate; petals 5, white, oblong-obovate, base cuneate; ovary glabrous, superior, 1-ovuled; style elongated; stigma disc-shaped, stamens numerous 20–30. Fruit a drupe yellow, red, or purple, globose (Plate 3, 4, 5, 6, 7), ovoid, or conical, 3.5–5 cm in diameter to 7 cm in diameter in horticultural forms, glaucous; mesocarp fleshy, reddish (Plate 7) or yellow-pink; endocarp ovoid to oblong, rugose.

Nutritive/Medicinal Properties

Japanese plum was reported to have the following mean nutritional composition range based on 6 cultivars of *Prunus salicina* (Lozano et al. 2009): moisture 80.65–89.40%, energy 183–331 kJ/100 g, protein 0.38–0.96%, ash 0.19–0.53%, carbohydrate 10.04–17.9%, dietary fibre 0.84–1.50%, glucose 4.41–10.27%, fructose 1.82–4.79%, sucrose 0.65–4.34%. Total soluble solids (TSS) ranged from 13.4 to 17.8 °Brix, TSS/total acidity 9.26–20.86, acidity (% malic acid) 0.85–1.85%, pH 3.14–3.42. Total phenolic content 95.54–202.46 mg/100 g, anthocyanin 10.93–28.93 mg/100 g and total antioxidant activity 258.6–946.52 mg Trolox/100 g. Of the cultivars, 'Black Amber' had the highest phenolic and anthocyanin contents and the greatest total antioxidant activity. Of the volatile compounds, those present in the greatest proportion are the esters 4-hexen-1-ol acetate and hexyl acetate, guanidine, and 3-hexen-1-ol. 'Fortune; stood out as the richest cultivar with esters. Other compounds identified in the volatile fraction include 2-butanone + heaxan, 2 methyl-3-buten-2-ol, ethyl acetate, trichloromethne, 3-methylbutanol, hexanal, ethyl butanoate, butyl acetate, (Z)-3-hexen-1-ol, (E)-2-hexen-1-ol, furan tetrahydro-2,4-dimethyl-trans-, *m*-xylene, styrene, nonane, isopropylben-

Plate 1 Fascicle of Japanese blood plum blooms

Plate 2 Close-up of Japanese blood plum flower

zene, α-methylstyrene, cyclotetrasiloxane octamethyl-, ethyl hexanoate, decane, (E)-2-hexen-1-ol-acetate, 2-ethyl-1-hexanol, 2-hydroxy-1-phenyl-ethanone, α α dimethyl benzenemethanol, N,4-dimethyl-benzeamine, 4-hexenyl propanoatr, butyl hexanoate, (E)-2-hexenyl butanoate, ethyl octanoate, ciclodecanol, 4-hexenyl butanoate, 4-hexenyl hexanoate, hexyl hexanoate, butyl caprylate, 2-hexenyl hexanoate, tetradecane and butylated hydroxytoluene. The C6 compounds presented aromatic notes of freshly cut grass and green apples. The esters had pleasant fruity aromas. In terms of their volatile fraction, Japanese plums were relatively poor in

Plate 3 Immature Japanese blood plums and leaves

Plate 6 Harvested ripe Japanese blood plums

Plate 4 Prolific bearing Japanese blood plum tree

Plate 7 Blood–red flesh of ripe Japanese blood plum

Plate 5 Near ripe Japanese blood plum with wax bloom

volatile compounds in comparison with other stone fruits. However, there were significant differences between the cultivars studied, with 'Fortune' emerging as the richest in volatile compounds and especially in esters.

P. salicina cultivars differed in their relative contens of fructose, glucose, sorbitol, and sucrose and of organic acids (malic, shikimic and succinic, citric, tartaric and fumaric acid) (Singh and Singh 2008). Fructose was the major sugar followed by glucose, sorbitol and sucrose. Sugar concentration in Japanese plum varied with cultivar, maturity and harvest season. Late-maturing plum cultivars accumulated more sugar. The changes in individual sugars during ripening of these cultivars were not significant suggesting a low possibility of sugar interconversion. The percentage of sucrose to the total sugar concentration varied greatly among these cultivars. In 'Blackamber', sucrose contributed only 8% whereas its contribution was 17% and 18% in 'Amber Jewel' and 'Angeleno', respectively. Malic acid accounted for the major share of the total organic acids accounting for 80–89% of the total acids concentration followed by succinic, shikimic (both 10–18%), and citric acids. There were also traces of tartaric and fumaric acids. Robertson et al. (1992) found that the average firmness of the maturity Japanese plums was 25 N. Mean chemical compositions for all cultivars were as follows: soluble solids – 127 g/kg; acidity – 174.4 g/kg; soluble solids to acidity ratio – 7.4, and total sugar content – 96 g/kg.

P. salicina aside from being a nutritious fruit with antioxidant polyphenolic compounds also has some pharmacological activities which include antioxidant, anticancer, immunostimulatory and glucosyltransferase inhibition activities.

Anticancer Activity

Studies showed that an acetone extract of immature plums possessed cytotoxic effects, which were related to the activity of the total polyphenols in the fruits (Yu et al. 2007). Apoptosis in MDA-MB-231 cells mediated by the immature plums was associated with an increase in Bax levels and a reduction in Bcl-2 levels and the cleavage of caspase 3, caspase 7, caspase 9 and poly-(ADP-ribose) polymerase. These results indicated that immature fruit of P. salicina Lindl. cv. Soldam can be regarded as a safe and promising new dietary source for decreasing the risk of developing breast cancer. Studies also suggested that immature plum extracts possessed chemopreventive efficacy and may counteract toxic effects of carcinogens, such as B(α)P (Kim et al. 2008). Male ICR mice were pretreated with immature plum extracts (2.5 or 5 g/kg bw/day, for 5 days, i.p.) before treatment with B(α)P(0.5 mg/kg bw, i.p., single dose). The activities of serum aminotransferase, cytochrome P450 (CYPs) and the hepatic content of lipid peroxide were increased on B(α) P-treatment group and control, but those levels were significantly decreased by the pretreatment of immature plum extracts. The pretreatment of immature plum extracts inhibited the induction of CYP1A1 expression. The activities of glutathione peroxidase, superoxide dismutase and catalase were decreased by the pretreatment of immature plum extracts more than with B(α)P alone. Whereas, the hepatic content of glutathione and glutathione S-transferase activity depleted by B(α)P was significantly increased.

In subsequent studies, the scientists reported that in comparison with other cancer cells, the growth inhibition exerted by immature plum extracts was greatest in HepG2 (Yu et al. 2009a). Apoptosis in HepG2 cells mediated by immature plums was associated with "death receptor signaling." The total yield of identified polyphenols in immature plum extract was 10 g/kg dry weight. The major components, (–)-epicatechin and (–)-gallocatechin gallate, were 34.7% and 28.6% of total polyphenols, respectively. (+)-Catechin, (–)-epicatechin gallate, and (–)-catechin gallate were also found. On the basis of these results, the immature plum (P. salicina Lindl. cv. Soldam) and its active compound, (–)-epicatechin, were suggested to be a natural resource for developing novel therapeutic agents for cancer prevention and treatment.

Immature plum extract (IPE) appeared to have a strong inhibitory effect on the PMA (phorbol 12-myristate 13-acetate)-induced matrix metallopeptidase 9 (MMP-9) secretion through suppression of the transcriptional activity of the MMP-9 gene independently of the TIMP gene in HepG2 cells (Yu et al. 2009b). The results indicated that IPE suppressed both AP-1- and NF-κB-mediated MMP-9 gene transcriptional activity through inhibiting the nuclear translocations of AP-1 and NF-κB. IPE elicited a decrease in the migration potential of HepG2 cells in-vitro, and this suggested that the migration inhibition correlated well with its inhibition of MMP-9 expression. MMP-9 is known to play a role in tumour-associated tissue remodeling.

Immunostimulatory Activity

Methanol extract of plum fruit exhibited immunostimulatory effects (Lee et al. 2009). The crude methanol extract stimulated spleen lymphocyte proliferation and NO production by cultured macrophages, and inhibited the viability of tumour cells, significantly greater than media controls.

Glucosyltransferase Inhibition Activity

Among the extracts from 420 kinds of herbs tested in Korea, *Prunus salicina*, showed the highest glucosyltransferase inhibition activity (Won et al. 2007) The active principle was purified and designated GTI-0163. GTI-0163 was revealed to be an oleic acid-based unsaturated fatty acid. Among the unsaturated fatty acids, oleic acid showed a significantly higher GTase inhibitory activity than the saturated fatty acids or the ester form of oleic acid. Glucosyltransferase is an enzyme which transfers residues of glucose to acceptor molecules.

Traditional Medicinal Uses

The fruit is stomachic. It is said to be good for allaying thirst and is given in the treatment of arthritis.

Other Uses

The plant is often grown as an ornamental. Green and dark grey-green dyes can be obtained from the leaves and fruit respectively.

Comments

The species has been divided into two botanically recognised lower taxa:
Prunus salicina var. *pubipes* (Koehne) L. H. Bailey and *Prunus salicina* var. *salicina*.

Selected References

Bown D (1995) Encyclopaedia of herbs and their uses. Dorling Kindersley, London, 424 pp

Chopra RN, Nayar SL, Chopra IC (1986) Glossary of Indian medicinal plants. (Including the supplement). Council Scientific Industrial Research, New Delhi, 330 pp

Facciola S (1990) Cornucopia. A source book of edible plants. Kampong Publ, Vista, 677 pp

Grae I (1974) Nature's colors – dyes from plants. MacMillan Publishing Co, New York

Gu C, Bartholomew B (2003) *Prunus* Linnaeus. In: Wu ZY, Raven PH, Hong DY (eds) Flora of China, vol 9, Pittosporaceae through Connaraceae. Science Press/Missouri Botanical Garden Press, Beijing/St. Louis

Kim HJ, Yu MH, Lee IS (2008) Inhibitory effects of methanol extract of plum (*Prunus salicina* L., cv. 'Soldam') fruits against benzo(alpha)pyrene-induced toxicity in mice. Food Chem Toxicol 46(11):3407–3413

Lee SH, Lillehoj HS, Cho SM, Chun HK, Park HJ, Lim CI, Lillehoj EP (2009) Immunostimulatory effects of oriental plum (*Prunus salicina* Lindl.). Comp Immunol Microbiol Infect Dis 32(5):407–417

Lozano M, Vidal-Aragón MC, Hernández MT, Ayuso MC, Bernalte MJ, García J, Velardo B (2009) Physicochemical and nutritional properties and volatile constituents of six Japanese plum (*Prunus salicina* Lindl.) cultivars. Eur Food Res Technol 228(3):403–410

Robertson JA, Meredith FI, Senter SD (1992) Physical, chemical and sensory characteristics of Japanese-type plums grown in Georgia and Alabama. J Sci Food Agric 60:339–347

Rouse RE, Sherman WB (1997) Plums for Southwest Florida. Proc Fla State Hort Soc 110:184–185

Rumayor-Rodriguez A (1995) Multiple regression models for the analysis of potential cultivation areas for Japanese plums. HortScience 30(3):605–610

Singh SP, Singh Z (2008) Major flavor components in some commercial cultivars of Japanese plum. J Am Pomol Soc 62(4):185–190

Won SR, Hong MJ, Kim YM, Li CY, Kim JW, Rhee HI (2007) Oleic acid: an efficient inhibitor of glucosyltransferase. FEBS Lett 581(25):4999–5002

Yu MH, Im HG, Kim HI, Lee IS (2009a) Induction of apoptosis by immature plum in human hepatocellular carcinoma. J Med Food 12(3):518–527

Yu MH, Im HG, Lee SG, Kim DI, Seo HJ, Lee IS (2009b) Inhibitory effect of immature plum on PMA-induced MMP-9 expression in human hepatocellular carcinoma. Nat Prod Res 23(8):704–718

Yu MH, Im HG, Lee SO, Sung C, Park DC, Lee IS (2007) Induction of apoptosis by immature fruits of *Prunus salicina* Lindl. cv. Soldam in MDA-MB-231 human breast cancer cells. Int J Food Sci Nutr 58(1):42–53

Pseudocydonia sinensis

Scientific Name

Pseudocydonia sinensis (Thouin) C. K. Schneid.

Synonyms

Chaenomeles sinensis (Thouin) Koehne; *Cydonia sinensis* Thouin, *Pyrus sinensis* (Thouin) Spreng., nom. illeg., *Cydonia sinensis* Thouin, *Pseudocydonia sinensis* (Thouin) C. K. Schneider, *Pyrus cathayensis* Hemsley, *Pyrus chinensis* Sprengel, *Pyrus sinensis* (Thouin) Poiret.

Family

Rosaceae

Common/English Name

Chinese-Quince.

Vernacular Names

Chinese: Hai Tang, Mingzha*, Mu Gua*, Mu Li;
French: Cognassier De Chine;
Italian: Cotogno Cinese;
Japanese: Karin;
Korean: Mo-Gua-Na-Moo, Mogwa;
Portuguese: Marmeleiro Da China;
*See comments

Origin/Distribution

The species is native to eastern China. It is widely planted in China and Korea.

Agroecology

The species is adapted to cool temperate climate. It does best in a sunny position in well-drained, fertile soil.

Edible Plant Parts and Uses

The ripe fruit needs to be bletted to soften and become less astringent before being eaten. The fruit is eaten as sweetmeat, candied, preserved in syrup or made into a liqueur. Like quince, it can be made into jams. The juice is mixed with ginger and processed into a beverage.

Botany

A small, unarmed, deciduous tree 5–10 m high with a twiggy crown and reddish branchlets, pubescent when young becoming glabrescent with age. Stipules ovate-oblong, rhomboidal, or

Plate 1 Flowers and leaves of Chinese quince

lanceolate. Leaves alternate, entire, elliptic-ovate or elliptic-oblong 5–8 × 3.5–5.5 cm, tomentose beneath when young becoming glabrescent, base broadly cuneate or rounded, apex acute margin aristate and sharply serrate (teeth glandular at apices) (Plate 1). Flowers solitary, pink, 2.5–3 cm across (Plate 1), hypanthium campanulate; sepals 5, reflexed, deltoid-lanceolate, abaxially glabrous, adaxially brown tomentose; petals 5, obovate, pinkish, base shortly clawed, apex rounded; stamens 20 or more; Styles 3–5, ca. as long as stamens, connate at base. Fruit fragrant subglobose to ellipsoid, pome, yellow, 10–15 cm, hard, on short pedicel.

Nutritive/Medicinal Properties

The fruits were found to contain two major tritepene acids oleanolic acid, ursolic acid (Im and Roh 1994), chaenoside A and B (Im and Roh 1994), maslinic acid, tormentic acid and euscaphic acid (Roh et al. 1995), a-terpineol, isozedoarondiol, isofuranogermacrene, quercetin, rutin quecitrin (NPRI-SNU 1998) and tannins (Matsuo and Ito 1981). A minor flavanone glycoside, 2-hydroxy-naringenin-7-O-β-glucoside was isolated from Chinese quince fruit (Kim et al. 2000). Seven compounds were isolated from the ethanol extract of Chinese quince fruits and identified as: palmitic acid, ursolic acid-3-O-behenate, ursolic acid, 3-acetyl ursolic acid, 3-acetyl pomolic acid, daucosterol and betulinic acid (Sun et al. 1999). From the dried fruits of *Chaenomeles sinensis*, a new pentacyclic triterpenoid, ursolic acid-3-O-behenate, along with 18 known compounds was isolated (Sun and Hong 2000). Seven compounds were isolated from the ethyl acetate fraction of the methanol extract of *C. sinensis* and identified as erythodiol (1), masilinic acid (2), betulinic acid (3), 2α-hydroxy betulinic acid (4), betulin (5), 3-(E) p coumaroylbetulin(6), 3-(Z)-p-coumaroylbetulin (7) (Gao et al. 2003c). Using ultrasonic-assisted extraction, Teng et al. (2010) determined in Chinese quince fruit, total phenolic content 2367.16 mg gallic acid equivalents/100 g, total flavonoid content 544.12 mg rutin equivalents/100 g, and total flavan-3-ols content 709.07 mg catechin equivalents/100 g. Fang et al. (2010) developed an optimized microwave-assisted extraction (MAE) method and an efficient HPLC analysis method for fast extraction and simultaneous determination of oleanolic acid and ursolic acid in the fruit of *Chaenomeles sinensis*.

A total of 111 compounds were identified from Chinese quince peel oil (Mihara et al. 1987). Eighty-four compounds consisted of 4 hydrocarbons, 50 esters, 6 alcohols, 6 aldehydes, 3 ketones, 3 acids, 3 lactones, 6 acetals, and 3 miscellaneous compounds. Of these compounds, the alkyl and alkenyl esters of ω-alkenoic acids and the

5-hexenyl esters of the aliphatic acids, and ethyl 9-decenoate, were found to be the important contributor to the typical Chinese quince flavour. From the flesh, 42 compounds were identified. The flavour concentrate from peel was thought to be more important to Chinese quince aroma than that from flesh. In another study, 145 compounds from the steam volatile concentrate of Chinese quince fruits were identified, consisting of 3 aliphatic hydrocarbons, 1 cyclic hydrocarbon, 4 aromatic hydrocarbons, 9 terpene hydrocarbons, 17 alcohols, 3 terpene alcohols, 6 phenols, 21 aldehydes, 7 ketones, 28 esters, 27 acids, 3 furans, 2 thiazoles, 2 acetals, 3 lactones and 9 miscellaneous ones (Chung et al. 1988b). The greater part of the components except for carboxylic acids were identified from the neutral faction. The neutral fraction gave a much higher yield than others and was assumed to be indispensable for the reproduction of the aroma of the Chinese quince fruits in sensory evaluation. According to the results of the GC-sniff evaluation, 1-hexanal, cis-3-hexenal, trans-2-hexenal, 2-methyl-2-hepten-6-one, 1-hexanol, cis-3-hexenol, trans, trans-2, 4-hexadienal and trans-2-hexenol were considered to be the key compounds of grassy odour and esters appeared to be the major constituents of a fruity aroma in the Chinese quince fruits. Non-volatile flavour components of Chinese quince fruit comprised about 72% of the free amino acids namely valine, asparagine, γ-aminobutyric acid, aspartic acid and serine (Chung et al. 1988a). Arginine, tryosine, methionine and tryptophan were not present. Glutamic acid and glutamine as amino acid for peptides were the major components, whereas cysteic acid, methionine sulfone and tryptophan were not detected. Small amount of nucleotides namely cytosine, uridine-5′-monophosphate and cytidine-5′-monophosphate were detected.

GC-MS analysis identified 62, 60 and 53 different aroma compounds from *Chaenomeles* fruits of the varieties Changjun, Shi Zitou and Yulan separately, among which 21 components were identical (Meng et al. 2007). The most abundant compound in the *Chaenomeles* fruits was 4-methyl-5-penta-1, 3-dienyltetrahydrofuran-2-one, (Changjun 11.38%, Shi Zitou 25.71% and Yulan 17.72%). 2-hexenal, (E)-; 2-hexen-1-ol, (E)-; 3-hexen-1-ol, (Z); were important components that contributed to the fruit aroma, and ionones such as β-ionone were also important. Aroma components of *Chaenomeles* fruits mainly included alcohols, ketones, aldehydes, esters and hydrocarbons and among them alcohols, ketones, aldehydes and esters were critical compounds that contributed to the particular flavor of *Chaenomeles* fruits. Components such as 2-heptenal, (Z)-, hexadecanal, octadecanal, megastigma-4, 6(E), 8(E)-triene and megastigmatrienone were identified for the first time in *Chaenomeles* fruits. Forty-three aromatic compounds comprising aldehydes, alcohols, ketones, furans, alkenes, acids and esters were identified in Chinese quince (Zhou et al. 2008). The major compounds were 2-hexenal; cyclopentanol,2-methyl-,trans-; 2,4-hexadienal,(E,E)-; 2-butanone; 3-hexenal,(Z)-; acetic acid ethenyl ester; 3-hexen-1-ol,(E)-; and theaspirane.

A new dihydrochromone derivative, named chaenomone, was isolated from the twigs of *Chaenomeles sinensis* (Gao et al. 2003a). Seven compounds were isolated from the ethyl acetate fraction of the methanol extract of *C. sinensis* twigs and identified as lup-20(29)-en-3β,24, 28-triol (1), 2α-hydroxyursolic acid (2), euscaphic acid(3), tormentic acid(4), lyoniresinol-9′-O-β-D-glucopyranoside(5), avicularin (6), and (−) epicatechin (7) (Gao et al. 2004).

Chaenomeles sinensis had been reported to contain various chemical constituents, primarily including organic acids, steroid saponins, triterpenoids, lignans flavonoids, (Zhang et al. 2007).

Studies have reported the plant to exhibit many bioactivities such as antioxidant, antiplatelet, antidiabetic, antitumour, antiviral, antibacterial, antiinflammatory, antipruritic and antiulcerogenic activities.

Antioxidant Activity

In comparison to quince and apple, Chinese quince was found to have the largest amount of phenolics consisting mainly of high polymeric procyanidins (Hamauzu et al. 2005). Quince had considerable amounts of hydroxycinnamic derivatives mainly composed of 3-caffeoylquinic acid and 5-caffeoylquinic acid and polymeric procyanidins while Apple (cv. Fuji) had the lowest amount of phenolics, mainly 5-caffeoylquinic acid and monomeric and oligomeric procyanidins. The antioxidant activity of Chinese quince and quince phenolic extracts were higher than chlorogenic acid standard or ascorbic acid evaluated in both the linoleic acid peroxidation system and the DPPH radical scavenging system. However, those extracts were less superior than apple phenolics or (−)-epicatechin in linoleic acid peroxidation system. The strength of antioxidant activity and DPPH radical scavenging activity of Chinese quince, apple and quince fruit phenolics varied according to different in-vitro evaluation systems, whereas the antioxidative property of rat blood increased in all rats orally administered phenolics (Hamauzu et al. 2006). Chinese quince phenolic (mostly procyanidins) solution subjected to heat treatment changed from almost colorless, pale yellow to a reddish color (Hamauzu et al. 2007). Thioacidolysis of denatured reddish phenolics showed that (−)-epicatechin subunits decreased during heat treatment and, in contrast, cyanidin increased. In addition, novel substances that could not be degraded by thioacidolysis were formed. Meanwhile, antioxidant activities, assessed by linoleic acid peroxidation, 1,1-diphenyl-2-picrylhydrazyl (DPPH), Folin-Ciocalteu, and FRAP methods, increased during heat treatment. Administration of heat-treated Chinese quince polyphenols increased the rat plasma levels of protocatechuic and vanillic acids leading to an elevation of plasma antioxidant activity and levels of aromatic health-benefiting compounds derived from procyanidins acid (Hamauzu et al. 2010).

Antiplatelet Activity

Five flavonoid tissue factor inhibitors were isolated from *Chaenomeles sinensis* fruits, namely hovetrichoside C (IC_{50} = 14.0 µg), luteolin-7-O-β-D-glucuronide (IC_{50} = 31.9 µg), hyperin (IC_{50} = 20.8 µg), avicularin (IC_{50} = 54.8 µg) and quercitrin (IC_{50} = 135.7 µg) (Lee et al. 2002). They were isolated along with other inactive compounds such as (±)-(2E,4E)-O-β-D-glucopyranosyl-4′-hydroxy-β-ionylideneacetic acid ester, genistein-7-O-β-D-glucopyranoside, luteolin-3′-methoxy-4′-O-β-D-glucopyranoside, luteolin-7-O-β-D-glucuronide methyl ester, tricetin-3′-methoxy-4′-O-β-D-glucopyranoside (selagin-4′-O-β-D-glucopyranoside), (−)-epicatechin, luteolin-4′-O-β-D-glucopyranoside and apigenin-7-O-β-D-glucuronide methyl ester. Four triterpenoid compounds were isolated from *C. sinensis* fruit and elucidated as 28-O-β-D-glucopyranosyl-2α,3β-dihydroxyolean-12-ene-24,28-dioic acid (2), named chaenomeloside A, trachelosperoside A-1 (1), oleanolic acid (3) and ursolic acid (4) (Lee and Han 2003). Compound 2 and its aglycone 2a, named chaenomelogenin A, inhibited by 50% tissue factor (TF) activity at concentrations of 0.036 and 0.028 mM/unit of TF, respectively. Compounds 3 and 4 were inactive.

Antidiabetic Activity

All *Chaenomeles sinensis* extracts, especially 80% methanol extract and its fractions, n-haxane, methylene chloride, ethyl acetate, n-butanol and aqueous showed remarkable α-glucosidase and β-glucosidase inhbitorty activities ranging from 82–99% to 5–85% respectively (Sancheti et al. 2009). The n-butanol fraction exerted highest (99%) α-glucosidase inhibitory activity whereas minor α-galactosidase (18–35%) and β-galactosidase (10–34%) inhibitions were obtained with all the fractions. *C. sinensis* fruit represented a potent natural antidiabetic source with notable α-glucosidase and β-glucosidase inhibitions and would have potential for management of type 1 diabetes by controlling glucose absorption. Oral treatment of streptozotocin-induced diabetic rats with lipase inhibitory ethyl acetate fraction from the fruits of *Chaenomeles sinensis* at the doses of 50 and 100 mg/kg bw for 14 days significantly reduced blood glucose, triglyceride, total cholesterol, high density lipoprotein cholesterol, alanine aminotransferase, aspartate aminotransaminase, acetylcholinesterase levels (Sancheti et al. 2011). The antioxidant levels were significantly increased in the ethyl acetate treated groups. The observations suggested protective effects of ethyl acetate fruit extract against streptozotocin-induced diabetic dementia model.

Antiviral Activity

In comparison to quince and apple, Chinese quince phenolics showed the strongest anti-influenza viral activity on the hemagglutination inhibition test (Hamauzu et al. 2005). The anti-influenza viral activity of heat denatured reddish phenolics of Chinese quince fruit was inferior to that of intact fruit phenolics; however, they retained moderate activity (Hamauzu et al. 2007). These results indicate that red coloration of fruit products of Chinese quince was mainly due to the spectral (i.e. structural) changes of procyanidins accompanied with formation of cyanidin. Treatment with 50% ethanol extract of the fruit of *Chaenomeles sinensis* at concentrations greater than 5 mg/mL reduced the plaque titers of Type A and B influenza viruses to less than 10% of those of untreated viruses (Sawai et al. 2008). The treatment neutralized influenza virus by inhibiting hemagglutination activity and by suppressing NS2 protein synthesis. Partial purification showed that the neutralizing component consisted of high molecular weight polyphenols.

Antipruritic Activity

A 35% ethanol extract of *Chaenomeles sinensis* fruits, significantly inhibited pruritogenic agent compound 48/80 (COM)-induced scratching behavior in mice (Oku et al. 2003). Antipruritic activity-guided fractionation and purification yielded active quercetin, apigenin, and catechin derivatives, which exhibited significant inhibitory effects on COM-induced scratching behavior. Apigenin, apigenin 7-glucronide, and apigenin 4′-methoxy-7-glucronide (acacetin 7-glucronide) were isolated from the fruits of *C. sinensis* for the first time. The active fraction and these compounds also inhibited serotonin-, platelet activating factor-, and prostaglandin E(2)-induced scratching behavior, but did not inhibit histamine-induced scratching behavior or locomotive behavior. This study also showed that the fruits of *C. sinensis* could be used to treat allergic itching sensation.

Antitumour Activity

Twenty-three compounds from *Chaenomeles sinensis* were isolated in a primary screening of antitumour-promoting activity using soft agar colony assays with JB6 mouse epidermal cells (Gao et al. 2003b). These compounds were lyoniresinol-2a-O-α-L-rhamnopyranoside (1), lyoniresinol-2a-O-β-D-glucopyranoside (2), aviculin (3), betulinic acid (4), betulin (5), 3-O-(E)-p-coumaroylbetulin (6), 3-O-(E)-caffeoylbetulin (7), 3-O-(Z)-p-coumaroylbetulin (8), 3-O-(E)-caffeoyllupeol (9), alphitolic acid (10), sorbikortal II (11), tormentic acid (12), euscaphic acid (13), corosolic acid (14), maslinic acid (15), erythrodiol (16), 1-β-D-glucopyranosyloxy-3,4,5-trimethoxybenzene (17), avicularin (18), 7-O-β-D-glucopyranosylkaempferol (19), 5-O-ββ-D-glucopyranosylgenistein (20), 7-O-β-D-glucopyranosylgenistein (21), epicatechin (22), and β-sitosterol (23). Compound 1, having a rhamnosyl group, showed greater activity than 2, having a glucosyl group, and 3, which was a bis-demethoxy derivative of 1. Betulinic acid (4), having a C-28 carboxyl group, 3-O-(E)-caffeoylbetulin (7), and tormentic acid (12) showed more potent activity than betulin (5), which had a C-28 hydroxymethyl group.

Antiulcerogenic Activity

In the ethanol-induced gastric ulcer, pre-administration of Chinese quince and quince phenolics suppressed the occurrence of gastric lesions in rats, whereas apple phenolics appeared to promote ulceration (Hamauzu et al. 2006). The trend of myeloperoxidase activity was similar to that of the ulcer index. The results showed that Chinese quince and quince phenolics might have health benefits by acting both in blood vessels and on the gastrointestinal tract.

Antiinflammatory Activity

In-vitro studies showed that the ethanol extract of Chinese quince fruit inhibited histamine release from rat mast cells induced by compound 48/80, and on activation of hyaluronidase (Osawa et al. 1999). Chromatographic separation of the extract gave five phenolic compounds and higher molecular weight polyphenols. Some of these compounds showed a potent inhibitory effect on the histamine release and on the hyaluronidase. Five quinic acid derivatives, including one new compound established as 5-O-p-coumaroylquinic acid butyl ester along with caffeic acid and protocatechic acid were isolated from the ethanol Chinese quince fruit extract (Osawa et al. 2001). In the in-vitro assay of inhibitory effect on the histamine release from rat mast cells induced by compound 48/80, this new ester type compound showed high level of inhibition compared with its carboxylic acid form.

Glucosides of *Chaenomeles speciosa* (GCS) was found to have antiinflammatory effect on the collagen induced arthritis (CIA) mice which may be related to the modification of the abnormal immunological function of CIA mice (Zhang et al. 2004). GCS significantly reduced paw-swelling and arthritis scores, reduced the increase of spleen indices of CIA mice, suppressed the ConA or LPS-induced thymocyte or spleen cell proliferation, and the production of IL-1 and IL-2 in CIA mice. GCS also reduced the level of anti-CII antibody and PGE. Histological pathology analysis demonstrated that the synovium of CIA mice was hyperplastic, pannus was formed, and inflammatory cells infiltrated into the synovium. These pathological changes were significantly reduced by GCS.

Antibacterial Activity

Chlorogenic acid isolated from the methanol fruit extract of *C. sinensis* exhibited significant antibacterial activity against *Shigella dysenteriae* and *Staphylococcus aureus* (Hu et al. 2010).

Traditional Medicinal Uses

The fruit is regarded as antitussive and has long been utilized as a folk medicine for cough (Oku et al. 2003). It is used in Korea to treat asthma, the common cold, sore throats, mastitis and tuberculosis. The fruit extract is widely used as a traditional Chinese medicine to treat throat diseases and in Japan, its fruit liquor has been commonly added to lozenges to treat coughs and sore throats (Sawai et al. 2008).

Other Uses

Chinese quince fruits are very aromatic and are placed in a bowl to impart a delightful spicy scent to a room. Its dark red wood is hard and is used for picture frames.

Comments

There is controversy over the accepted scientific name. Some authors maintain that the scientific name should be *Pseudocydonia sinensis* (Thouin) C. K. Schneid and not *Chaenomeles sinensis* (Thouin) Koehne. Tropicos and Flora of China and Korea literature consider the former as a synonym of the latter, while GRIN Taxonomy considers the later as a synonym of the former.

Textual research conducted by Peng and Wang (2009) found that the Chinese name 'Mugua' had been erroneously referred to *Chaenomeles sinensis* (Thouin) Kochne and Tiegenghaitang as the Chinese name for *C. speciosa* (Sweet) Nakai in current literature. They proposed to change the Chinese name of *C. speciosa* (Thouin) Koehne to 'Mugua', *C. sinensis* (Thouin) Koehne to 'Mingzha' and *C. japonica* (Thunb.) Lindl. to 'Tiegenghaitan'g. In a subsequent paper (Han et al. 2010), using nuclear ribosomal internal transcribed spacer (ITS) analysis, they showed that the germplasms of 'Mugua' originated from *C. speciosa* (Sweet) Nakai, not *C. sinensis* (Thouin) Kochne and *C. cathayensis* (Hemsl.) Schneid. The results also showed that 'Yao Mugua' and 'Ornamental Mugua' were the most distantly related species in germplasms.

Selected References

Chung TY, Cho DS, Song JC (1988a) Nonvolatile flavor components in Chinese quince fruits, *Chaenomeles sinensis* Koehne. Korean J Food Sci Technol 20(3): 293–302

Chung TY, Cho DS, Song JC (1988b) Volatile flavor components in Chinese quince fruits, *Chaenomeles sinensis* Koehne. Korean J Food Sci Technol 20(2): 176–187

Fang X, Wang J, Yu X, Zhang G, Zhao J (2010) Optimization of microwave-assisted extraction followed by RP-HPLC for the simultaneous determination of oleanolic acid and ursolic acid in the fruits of *Chaenomeles sinensis*. J Sep Sci 33(8):1147–1155

Gao HZ, Wu B, Li W, Chen DH, Wu LJ (2004) Chemical constituents of *Chaenomeles sinensis* (Thouin) Koehne. Chin J Nat Med 2(6):351–353

Gao HZ, Wu LJ, Hei L (2003c) Chemical constituents of *Chaenomeles sinensis* (Thouin) Koehne. Chin J Nat Med 1(2):82–84

Gao HY, Wu LJ, Kuroyanagi M (2003a) A new compound from *Chaenomeles sinensis* (Thouin) Koehne. Chin Chem Lett 14(3):274–275

Gao HY, Wu LJ, Kuroyanagi M, Harada K, Kawahara N, Nakane T, Umehara K, Hirasawa A, Nakamura Y (2003b) Antitumor-promoting constituents from *Chaenomeles sinensis* Koehne and their activities in JB6 mouse epidermal cells. Chem Pharm Bull(Tokyo) 51(11):1318–1321

Hamauzu Y, Inno T, Kume C, Irie M, Hiramatsu K (2006) Antioxidant and antiulcerative properties of phenolics from Chinese quince, quince, and apple fruits. J Agric Food Chem 54(3):765–772

Hamauzu Y, Kume C, Yasui H, Fujita T (2007) Reddish coloration of Chinese quince (*Pseudocydonia sinensis*) procyanidins during heat treatment and effect on antioxidant and antiinfluenza viral activities. J Agric Food Chem 55(4):1221–1226

Hamauzu Y, Takedachi N, Miyasaka R, Makabe H (2010) Heat treatment of Chinese quince polyphenols increases rat plasma levels of protocatechuic and vanillic acids. Food Chem 118(3):757–763

Hamauzu Y, Yasui H, Inno T, Kume C, Omanyuda M (2005) Phenolic profile, antioxidant property, and anti-influenza viral activity of Chinese quince (*Pseudocydonia sinensis* Schneid.), quince (*Cydonia oblonga* Mill.), and apple (*Malus domestica* Mill.) fruits. J Agric Food Chem 53(4):928–934

Han B, Peng H, Yao Q, Zhou Y, Cheng M, Wang D (2010) Analysis of genetic relationships in germplasms of Mugua in China revealed by internal transcribed spacer and its taxonomic significance. Z Naturforsch C 65(7–8):495–500

Hu ZQ, Hong XD, Yue TL (2010) Extraction of chlorogenic acid from fruit of *Chaenomeles sinensis* Koehne and evaluation of its antibacterial activity. Food Sci 31(24):8–13

Im KS, Roh SB (1994) Structures of chaenoside A and B isolated from the fruits of *Chaenomeles sinensis* Koehne. Pusan Bull Pharm Sci 28:35–41

Kim HK, Jeon WK, Ko BS (2000) Flavanone glycoside from the fruits of *Chaenomeles sinensis*. Nat Prod Sci 6(2):79–81

Ku TC, Spongberg SA (2003) *Chaenomeles* lindley. In: Wu ZY, Raven PH, Hong DY (eds) Flora of China, vol 9, Pittosporaceae through Connaraceae. Science Press/ Missouri Botanical Garden Press, Beijing/St. Louis

Lee MH, Han YN (2003) A new in vitro tissue factor inhibitory triterpene from the fruits of *Chaenomeles sinensis*. Planta Med 69(4):327–331

Lee MH, Son YK, Han YN (2002) Tissue factor inhibitory flavonoids from the fruits of *Chaenomeles sinensis*. Arch Pharm Res 25(6):842–850

Matsuo T, Ito S (1981) Comparative studies of condensed tannins from several young fruits [persimmon, banana, carob bean, Chinese quince]. J Jpn Soc Hortic Sci 50(2):262–269

Meng XM, Liu LQ, Xu HD (2007) GC-MS analysis of aroma components of different *Chaenomeles* fruits. J Northwest Sci Tech Univ Agric For 35(8): 125–130

Mihara S, Tateba H, Nishimura O, Machii Y, Kishino K (1987) Volatile components of Chinese quince (*Pseudocydonia sinensis* Schneid). J Agric Food Chem 35(4):532–537

Natural Products Research Institute, Seoul National University (NPRI-SNU) (1998) Medicinal plants in the Republic of Korea, vol 21, WHO Regional Publications, Western Pacific Series. World Health Organization, Regional Office for the Western Pacific, Manila, 316 pp

Oku H, Ueda Y, Ishiguro K (2003) Antipruritic effects of the fruits of *Chaenomeles sinensis*. Biol Pharm Bull 26(7):1031–1034

Osawa K, Arakawa T, Shimura S, Takeya K (2001) New quinic acid derivatives from the fruits of *Chaenomeles sinensis* (Chinese quince). Nat Med 55(5):255–257

Osawa K, Miyazaki K, Imai H, Arakawa T, Yasuda H, Takeya K (1999) Inhibitory effects of Chinese quince (*Chaenomeles sinensis*) on hyaluronidase and histamine release from rat mast cells. Nat Med 53(4):188–193

Peng HS, Wang DQ (2009) Textual research on the Chinese names of three medicinal plants from *Chaenomeles* Lindl. Zhonghua Yi Shi Za Zhi 39(4):209–213 (In Chinese)

Roh SB, Chang EH, Im KS (1995) Isolation and characterization of acidic triterpenes from the fruits of *Chaenomeles sinensis*. Yakhak Hoechi 39(2):610–615

Sancheti SS, Sancheti S, Seo S (2009) *Chaemnomeles sinensis*: a potent α- and β-glucosidase inhibitor. Am J Pharmacol Toxicol 4:8–11

Sancheti S, Sancheti S, Seo SY (2011) Antidiabetic and antiacetylcholinesterase effects of ethyl acetate fraction of *Chaenomeles sinensis* (Thouin) Koehne fruits in streptozotocin-induced diabetic rats. Exp Toxicol Pathol [Epub ahead of print]

Sawai R, Kuroda K, Shibata T, Gomyou R, Osawa K, Shimizu K (2008) Anti-influenza virus activity of *Chaenomeles sinensis*. J Ethnopharmacol 118(1):108–112

Sun LN, Hong YF (2000) Chemical constituents of *Chaenomeles sinensis* (Thouin.) Koehne. J Chin Pharm Sci 9:6–8

Sun LN, Hong YF, Guo XM, Yang GJ, Zahng GM (1999) Studies on the chemical constituents of *Chaenomeles sinensis* (Thouin) Koehne. Acad J Sec Mil Med Univ 20(10):752–754 (In Chinese)

Tanaka T (1976) Tanaka's Cyclopaedia of edible plants of the world. Keigaku Publishing, Tokyo, 924 pp

Teng H, Jo IH, Choi YH (2010) Optimization of ultrasonic-assisted extraction of phenolic compounds from Chinese quince (*Chaenomeles sinensis*) by response surface methodology. J Korean Soc Appl Biol Chem 53(5):618–625

Zhang LL, Wei W, Yan SX, Hu XY, Sun WY (2004) Therapeutic effects of glucosides of *Chaenomeles speciosa* on collagen-induced arthritis in mice. Acta Pharmacol Sin 25(11):1495–1501

Zhang DS, Gao HY, Wu LJ (2007) Advances in the study of *Chaenomeles sinensis*. J Shenyang Pharm Univ 11:721–726

Zhou GF, Zhao F, Sun Y, Shen GN, Yang XH (2008) Analysis of aromatic compounds from *Chaenomeles sinensis* by headspace solid phase microextraction coupled with gas chromatography mass spectrometry. Chin J Anal Lab 27(8):25–28

Pyrus bretschneideri

Scientific Name

Pyrus x bretschneideri **Rehd.**

Synonyms

Pyrus serotina sensu Hedrick, non Rehder, *Pyrus ussuriensis* var. *chinensis* Kikuchi.

Family

Rosaceae

Common/English Names

Chinese White Pear, Crisp Chinese Pear, Duck Pear, White Pear, Ya Pear, Ya Li Pear.

Vernacular Names

Chinese: Ba-Li, Pai, Li, Guan-Li, Kuan-Li, Lai Yang Zu Li, Ya-Li;
German: Weiße Birne.

Origin/Distribution

This species is native to Northern and north-western China. It is cultivated from central to Eastern China, Japan, Central Asia and USA.

Agroecology

Ya pear is a cold temperate species, preferring sunny, loamy soils and is found from 100 to 2,000 m altitudes. It is often cultivated on slopes.

Edible Plant Parts and Uses

Ya pear is juicy, sweet and crispy and is best eaten fresh or in fruit salads. It also can be blended into drinks.

Botany

A small, deciduous tree 5–8 m tall, with stout hairy branchlets. Leaves, alternate on 2.5–3 cm long petioles, lamina ovate or elliptic-ovate, 5–11 × 3.5–6 cm, both surfaces tomentose when

Plate 1 Harvested Ya pears

Plate 2 Close-view of ya pear

young, soon glabrescent, base obtuse to broadly cuneate, margin spinulose-serrate, apex acuminate. Raceme umbel-like, 7–10-flowered. Flower white, floccose, glabrescent, 2–3.5 cm in diameter, pedicels 1.5–3 cm long, hypanthium cupular, slightly pubescent, sepals caducous, glabrous, petals white, ovate, 1.2–1.4 × 1–1.2 cm, base shortly clawed, apex rounded. Fruit a pome, ovoid to subglobose 2.5–4 cm diameter, yellow, with fine dots (Plates 1 and 2), 4- or 5-loculed, flesh white, gritty, juicy and crisp.

Nutritive/Medicinal Properties

Total sugar content in 13 varieties of *Pyrus bretschneideri* was very variable ranging from 90 to 156 mg/ml juice, fructose 35.2–84.6 mg/ml juice, glucose 25.5–39.2 mg/mL juice, sucrose 3.4–31.6 mg/mL juice and sorbitol 7.4–22.3 mg/mL juice (Pan et al. 2002). Ya pear was found to have in g FW β-cryptoxanthin 0.03 μg/g, δ-tocopherol 2.68 μg/g, γ-tocopherol 1.08 μg/g, α-tocopherol 4.26 μg/g, δ-tocotrienol 0.04 μg/g and α-tocotrienol 0.03 μg/g (Isabelle et al. 2010).

The total organic acid contents in fruits of *P. bretschneideri* cultivars ranged from 1.74 to 4.88 mg/g FW (fresh weight) (Sha et al. 2011). Malic, citric, quinic, oxalic, and shikimic acids were found in the fruit of all 10 cultivars. Malic and citric acids were the major constituents. The malic acid content ranged from 0.71 to 1.98 mg/g FW, accounting for 22–83% of the total organic acid content. The citric acid content ranged from 0.05 to 2.35 mg/g FW, accounting for 2–58% of the total organic acid content. The minor organic acids in the fruit were quinic and oxalic acids with contents of 0.08–0.57 mg/g FW (3–20% of the total) and 0.002–0.19 mg/g FW (0.1–9% of the total), respectively. The content of acetic, shikimic, succinic, fumaric, tartaric, and lactic acids was relatively low. Again, the organic acid composition did not differ significantly between cultivars, but the contents of the individual organic acids varied greatly.

Arbutin and chlorogenic acid were found as the main phenolic constituents in Ya pear (Cui et al. 2005). The two compounds existed in different organs of the Yali pear, one of the major cultivars of *Pyrus bretschnrideri*. The contents of arbutin in the leaf bud, floral bud, flower, and young fruit were 11.9, 12.4, 8.29, and 9.92 mg/g fresh weight (FW), respectively. Chlorogenic acid amounts in the same organs were 2.26, 3.22, 5.32, and 3.72 mg/g FW, respectively. During development, the concentration of the two compounds in Yali pears was the greatest in young fruit (9.92 mg/g FW of arbutin and 3.72 mg/g FW of chlorogenic acid), and then declined swiftly with fruit growth to less than 0.400 and 0.226 mg/g FW, respectively, in mature fruit. Arbutin and chlorogenic acid were also the main phenolic constituents in ya pear fruit skin (Lin and Harnly 2008). Yali pear (group 2) was found to contain significant amounts of dicaffeoylquinic acids compared to other pear groups.

Antioxidant Activity

Pyrus bretschdneideri fruit was found to have the following antioxidant profile H-ORAC (Hydrophilic- Oxygen Absorbance Capacity) 8.45 μmol TE/g FW, total phenolic content 0.6 mg GAE/g FW and ascorbic acid 6.7 μg ascorbic acid/g FW (Isabelle et al. 2010). The N-butanol extract of *Pyrus bretschdneideri* showed the best antioxidant potential as evaluated by the DPPH assay, B-carotene bleaching and FRAP methods (Li et al. 2011a). The main phenolic compounds in five commercial pear cultivars were found to have gallic acid ranging from 5.23 to 10.72 μg/g, catechin from 0.41 to 28.83 μg/g, chlorogenic acid from 485.11 to 837.03 μg/g, caffeic acid from 0 to 1.16 μg/g, epicatechin from 6.73 to 131.49 μg/g, and rutin from 0.92 to 104.64 μg/g (Li et al. 2011b). The total antioxidant capacity was found in descending order: *Shuijing* (*P. pyrifolia*) > *Fengshui* (*P. pyrifolia*) > *Xuehua* (*Pyrus bretschneideri*) > *Ya* (*Pyrus bretschneideri*) > *Xiang* (*Pyrus* sp. nr. *communis*) pears, which was consistent with the total phenol and flavonoid contents. The antioxidant capacity of pears may be attributed to their high contents of phenolics and flavonoids.

Antiinflammatory Activity

The ethanol extract of *P. bretschneideri* was found to have significant antiinflammatory activity (Huang et al. 2010). Its ethyl acetate fraction exhibited the strongest antiinflammatory effect. two sterols β-sitosterol, daucosterol, and two triterpenes, oleanolic acid, and ursolic acid were identified in the fraction. All of the isolated compounds were found to significantly inhibit the ear oedema induced by xylene. Ethyl acetate extract of *Pyrus bretschdneideri* exhibited high antiinflammatory followed by the n-butanol and ethanol extracts (Li et al. 2011a). The antiinflammation activity of five commercial pear cultivars was found in decreasing order: *Xuehua* (*Pyrus bretschneideri*) > *Xiang* (*Pyrus* sp. nr. *communis*) > *Ya* (*Pyrus bretschneideri*) > *Fengshui* (*P. pyrifolia*) > *Shuijing* (*P. pyrifolia*) pear, which indicated that compounds other than antioxidants may be responsible for the antiinflammation effect.

Miscellanous Activity

Pyrus bretschneideri, as a pharmaceutical supplement, is widely used in northern China to treat respiratory diseases (Huang et al. 2010).

Other Uses

The tree is also planted as shelter-belt and its wood can be used for making cabinets and instruments.

Comments

Pyrus × *bretschneideri* is considered an interspecific hybrid of *P. ussuriensis* and *P. pyrifolia*. It is also considered as one of the parents of the hybridogenic *Pyrus singkiangensis*. There are about 200 cultivars of the *Pyrus x bretschneideri*.

Selected References

Bao L, Chen K, Zhang D, Cao Y, Yamamoto T, Teng Y-W (2007) Genetic diversity and similarity of pear (*Pyrus* L.) cultivars native to East Asia revealed by SSR (simple sequence repeat) markers. Genet Resour Crop Evol 54(5):959–971

Cui T, Nakamura K, Ma L, Li JZ, Kayahara H (2005) Analyses of arbutin and chlorogenic acid, the major phenolic constituents in Oriental pear. J Agric Food Chem 53(10):3882–3887

Hu SY (2005) Food plants of China. The Chinese University Press, Hong Kong, 844 pp

Huang LJ, Gao WY, Li X, Zhao WS, Huang LQ, Liu CX (2010) Evaluation of the in vivo anti-inflammatory effects of extracts from *Pyrus bretschneideri* Rehd. J Agric Food Chem 58(16):8983–8987

Isabelle M, Lee BL, Lim MT, Koh WP, Huang D, Ong CN (2010) Antioxidant activity and profiles of common fruit in Singapore. Food Chem 123:77–84

Ku TC, Spongberg SA (1994) *Pyrus* Linnaeus. In: Wu ZY, Raven PH (eds) Flora of China, vol 17, Verbenaceae through Solanaceae. Science Press/Missouri Botanical Garden Press, Beijing/St. Louis, 342 pp

Li X, Gao WY, Huang LJ (2011a) In vivo anti-inflammatory and in-vitro antioxidant potential of crude fraction from *Pyrus bretschdneideri* Rehd. Lat Am J Pharm 30(3):440–445

Li X, Gao WY, Huang LJ, Zhang JY, Guo XH (2011b) Antioxidant and antiinflammation capacities of some pear cultivars. J Food Sci 76:C985–C990

Lin LZ, Harnly JM (2008) Phenolic compounds and chromatographic profiles of pear skins. (*Pyrus* spp.). J Agric Food Chem 56(19):9094–9101

Pan Z, Kawabata S, Sugiyama N, Sakiyama R, Cao Y (2002) Genetic diversity of cultivated resources of pear in north China. Acta Hortic (ISHS) 587:187–194

Sha SF, Li JC, Jun Wu J, Zhang SL (2011) Characteristics of organic acids in the fruit of different pear species. Afr J Agric Res 6(10):2403–2410

Teng Y-W, Tanabe K, Tamura F, Itai A (2001) Genetic relationships of pear cultivars in Xinjiang, China, as measured by RAPD markers. JHortic Sci Biotechnol 76:771–779

Teng Y-W, Tanabe K, Tamura F, Itai A (2002) Genetic relationships of *Pyrus* species and cultivars native to East Asia revealed by randomly amplified polymorphic DNA markers. J Am Soc Hortic Sci 127:262–270

Pyrus communis

Scientific Name

Pyrus communis L.

Synonyms

Pirus communis β *hortensis* Beck, *Pyrus asiae-mediae* Popov, *Pyrus balansae* Decne., *Pyrus bourgaeana* Decne., *Pyrus caucasica* Fed., *Pyrus communis* subsp. *bourgaeana* (Decne.) Nyman, *Pyrus communis* subsp. *caucasica (Fed.) Browicz*, *Pyrus communis* subsp. *pyraster (L.) Ehrh*, *Pyrus communis* var. *sativa* (DC.) DC., *Pyrus communis* var. *mariana* Willk., *Pyrus communis* var. *pyraster* L., *Pyrus domestica* Medik., *Pyrus elata* Rubtzov, *Pyrus medvedevii* Rubtzov, *Pyrus pyraster* (L.) Burgsd., *Pyrus sativa* DC. ex Lam. & DC., *Pyrus sylvestris* Moench, *Sorbus Pyrus* Crantz.

Family

Rosaceae

Common/English Names

Common Pear, European Pear, Dessert Pear, Pear, Pear Tree, Perry Pears, Soft Pear.

Vernacular Names

Brazil: Pêra, Pereira;
Chinese: Xi Yang Li, Yang Li;
Czech: Hrušeň Obecná;
Danish: Almindelig Pære, Pære, Vild Pære;
Dutch: Eetpeer, Gewone Peer, Peer;
Eastonian: Harilik Pirnipuu;
Finnish: Päärynä, Päärynäpuu;
French: Poire, Poirier;
German: Birnbaum, Birne, Birnenbaum, Garten-Birne, Gemeine Birne, Holz-Birne, Holzbirnbaum, Holzbirne Kulturbirne, Mostbirnen, Wild-Birne, Wilder Birnbaum;
Hungarian: Körte(Fa), Nemes Körte;
India: Nashpati (Hindu), Salvag (Malayalam), Kishtabahira (Kashmiri), Naspati (Manipuri), Berikaya, Veripandu (Telugu);
Indonesia: Buah Pir,
Italian: Pera, Perastro, Pero, Pero Comune, Pero Domestic, Pero Selvatico;
Japanese: Seiyou Nashi, Seiyo Nashi;
Korean: Pyongbaenamu;
Malaysia: Buah Pir;
Nepali: Nasapati;
Norwegian: Pære, Pæretre;
Polish: Grusza Domowa, Grusza Pospolita;
Portuguese: Pereira;
Russian: Gruša;
Serbian: Kruska;

Slovaščina: Hruška, Navadna Hruška, Žlahtna Hruška;
Slovencina: Hruška Obyčajná;
Spanish: Pera, Peral, Perello, Piruétano;
Swedish: Päron;
Thai: Sali Thuean;
Vietnamese: Lê.

Origin/Distribution

Pyrus communis is native to central and eastern Europe and southwest Asia. The European Pear is one of the most important fruits of temperate regions, being the species from which most domestic orchard pear cultivars grown in Europe, North America and Australia are developed.

Agroecology

Pyrus communis is a temperate species and thrives in the warmer areas of the temperate zones. It prefers a mild summer and a cool to cold winter. It prefer full sun and rich, well drained, moist sandy loam to clayey loam soils rich in organic matter, with pH of 5.5–8.5.

Edible Plant Parts and Uses

Most pears are dessert pears and eaten raw by themselves or with a robust cheese like Parmesan, Pecorino, Gorgonzola, Stilton or Roquefort. Pears are also a welcome addition to winter salads. Pears are poached in port or red wine spiced with cinnamon, cloves and pared lemon rind or in a vanilla flavoured syrup. A well-known dish is "Pears in Syrup" or "*Perys in Syrip*" using the fifteenth century spelling. For sautéed or grilled pear, pears are peeled, quartered or halved and the cores are scooped out with a melon baller. Dessert and cooking type pears can also be cooked in compotes, tarts, pies, terrines, trifles, pastries, cakes and the famed *Poires Belle Helene* (poached pears with vanilla ice cream and hot chocolate sauce). Pears are excellent for making a pear sherbert where the peeled pear is blended with sugar and a little lemon juice and freeze. Pears also make marvelous fritters and go well with ingredients such as nuts, spices, port and masala. Cooked pears are relished in savoury dishes with game and duck – made into chutney or casseroled with game birds or venison. Pears can be preserved or candied with sugar and vinegar or pickled with mustard seeds and horse radish. Pears are excellent for canning, they are peeled, halved, quartered or diced into pieces and cooked in syrup before canning. Peeled pears are also dried, crystallized and distilled into spirits like *eau-de-vie de poires, Poire William* and *Perry*.

Botany

Small to medium sized, deciduous tree, 10–15 m high with an upright, conical crown and slender branches. Bark gray-brown to reddish brown bark, with shallow furrows and flat-topped scaly ridges. Branchlets tomentose when young, glabrous and brownish red when old. Leaves alternate, simple, ovate to elliptic, 2–7 cm by 2–3.5 cm, acute or short acuminate, base rounded or subcordate, margin serrate-crenate, shiny green above, paler and dull below (Plates 1 and 2); petioles 1.5–5 cm, slender; stipules caducous, linear-lanceolate, membranous. Raceme umbel-like, 6–9-flowered with caducous, linear-lanceolate bracts at the apex of a spur. Flowers 2.5–3 cm across, white, bisexual (Plate 1). Hypanthium

Plate 1 Packham Pear leaves and flowers after anthesis

Plate 2 Developing Packham pear fruits and leaves

Plate 4 Harvested ripe Packham pears

Plate 3 Developing Beurre Bosc pear fruits and leaves

Plate 5 Harvested ripe Beurre Bosc pears

campanulate, abaxially pubescent. Sepals deltoid-lanceolate, 5–9 mm, both surfaces pubescent, persistent. Petals white, obovate, 1.3–1.5 × 1–1.3 cm, base shortly clawed. Stamens 20, half as long as petals. Ovary 5-loculed, with 2 ovules per locule; styles 5, pubescent basally. Fruit a pome obovoid, turbinate or subglobose, 4–8 cm wide × 4.5–9.5 cm long, 5-loculed, pale green, yellow, brown or reddish tinged, dotted (Plates 3, 4, 5, 6, 7, 8a, 8b, 9a, 9b, 10a, 10b). Flesh juicy, soft, pale yellowish-white and sweet when ripe.

Plate 6 Red Williams pears

Nutritive/Medicinal Properties

Pears are rich in nutrients and phytonutrients such as antioxidants. Proximate nutrient composition of fresh pear fruit (*Pyrus communis*) per 100 g edible portion excluding refuse 10% of the core, seeds and stem was reported as follows (UDSDA 2011): Water 83.71 g, energy 58 kcal (242kJ), protein 0.38 g, total lipid (fat) 0.12 g, ash 0.33 g, carbohydrate 15.46 g, fibre (total dietary)

Plate 7 Bartlett pears

Plate 9a Harvested Corella pears

Plate 8a Williams pears (side view)

Plate 9b Close-view Corella pears

Plate 8b Williams pears (top view)

Plate 10a Miniature paradise pears 4–4.5 cm by 3.6–4.5, pale yellow

Plate 10b Miniature paradise pears, sweet, crunchy, white fleshed

3.1 g, sugars (total) 9.8 g, sucrose 0.78 g, glucose 2.76 g, fructose 6.23 g, lactose 0.01 g, maltose 0.01 g; minerals Ca 9 mg, Fe 0.17 mg, Mg 7 mg, P 11 mg, K 119 g, Na 1 mg, Zn 0.10 mg, Cu 0.082 mg, Mn 0.049 mg, F 2.2 µg, Se 0.1 µg; vitamins – vitamin C (ascorbic acid) 4.2 mg, thiamine 0.012 mg, riboflavin 0.025 mg, niacin 0.157 mg, pantothenic acid 0.048 mg, vitamin-6 0.028 mg, folate (total) 7 µg, total choline 5.1 mg, betaine 0.2 mg, vitamin A 23 IU, vitamin E (α-tocopherol) 0.12 mg, γ-tocopherol 0.03 mg, vitamin K (phylloquinone) 4.5 µg; lipids – total saturated fatty acids 0.006 g, 16:0 (palmitic acid) 0.005 g, 18:0 (stearic acid) 0.001 g; total monounsaturated fatty acids 0.029 g, 16:1 undifferentiated (palmitoleic acid) 0.001 g, 18:1 undifferentiated (oleic acid) 0.025 g; total polyunsaturated fatty acids 0.055 g, 18:2 undifferentiated (linoleic acid) 0.029 g; phytosterols 8 mg; amino acids – tryptophan 0.002 g, threonine 0.011 g, isoleucine 0.011 g, leucine 0.019 g, lysine 0.017 g, methionine 0.002 g, cystine 0.002 g, phenylalanine 0.011 g, tyrosine 0.002 g, valine 0.017 g, arginine 0.010 g, histidine 0.002 g, alanine 0.014 g, aspartic acid 0.105 g, glutamic acid 0.030 g, glycine 0.013 g, proline 0.021 g, serine 0.015 g; β-carotene 13 µg, β-cryptoxanthin 2 µg, and lutein+zeaxanthin 45 µg.

Studies showed that different European pear (*P. communis*) cultivars differed in their contents of different sugars and organic acids (Hudina and Štampar 2000). The fructose content in 18 European pear cultivars ranged from 23.7 to 66 g/kg and sorbitol varied from 12.5 to 24.9 g/kg. The early cultivars of pears contained more than 1 g/kg of citric acid and the late ones less than 1 g/kg. The Asian pear cultivars contained more total sugars than the European ones.

The total organic acid content in fruit of *P. communis* cultivars ranged from 0.86 to 3.51 mg/g FW (fresh weight) (Sha et al. 2011). Malic, citric, oxalic, shikimic, and fumaric acids were detected in all 10 cultivars, whereas tartaric and lactic acids were detected in only 4 cultivars. Malic and citric acids were the major constituents. The citric acid content was higher than the malic acid content in 1 cultivar. The malic acid content ranged from 0.69 to 2.61 mg/g FW accounting for 32–82% of the total, followed by citric acid with a content of 0.01–1.35 mg/g FW accounting for 1–52% of the total. The minor organic acids in the fruit were oxalic acid (0.01–0.24 mg/g FW) and acetic acid (0–0.30 mg/g FW), accounting for 0.4–11% and 0–10% of the total organic acid content, respectively. The content of quinic, shikimic, succinic, fumaric, tartaric, and lactic acids were relatively low. The organic acid composition among the cultivars did not differ significantly.

Similar dynamic patterns were found in the glucose, fructose, sucrose and sorbitol contents in leaves and fruits of the genetically related pear cultivars 'Conference' and 'Concorde' (Hudina et al. 2007). Leaf sugar was low at the beginning of the growing season when the leaves were not completely developed. Generally when sucrose increased in leaves it decreased in fruits. At the end of June the total sugar content in leaves reached its peak then rapidly decreased. At the same time, total sugar in fruits increased. From the beginning of August, total sugars in fruits increased regardless of the sugar content in leaves and likely due to decomposition of starch. After harvest, the contents of individual sugars (glucose, fructose, sucrose, and sorbitol) in the leaves decreased until the beginning of October when, just prior to leaf drop, they increased in all cultivars.

The average concentration of phenolic compounds in the Portuguese pear cultivar (*Pyrus* communis L. var. S. Bartolomeu) harvested at commercial maturity stage was 3.7 g/kg of fresh pulp (Ferreira et al. 2002). Procyanidins were the predominant phenolics (96%), hydroxycinnamic acids (2%), arbutin (0.8%), and catechins (0.7%) were also present. The most abundant monomer in the procyanidin structures was (−)-epicatechin (99%). Sun-drying of these pears caused a decrease of 64% (on a dry pulp basis) in the total amount of native phenolic compounds. Hydroxycinnamic acids and procyanidins showed the largest decrease; the B2 procyanidin was not found at all in the sun-dried pear. Less affected were arbutin and catechins. Arbutin and chlorogeinc acids were found to be important phenolic compounds in pears (Cui et al. 2005). The mean concentration of arbutin in 3 pear cultivars was 0.083 mg/g FW and the mean concentration of chlorogenic acid was 0.309 mg/g FW as reported by analysis conducted in China. The main phenolic compounds in the fruit skin were arbutin and chlorogenic acid (Lin and Harnley 2008). Common pear was grouped into Group 4 that included Bartlett, Beurre, Bosc, Comice, D'Anjou, Forelle, Peckham, Red, Red D'Anjou, and Seckel. All were found to contain significant quantities of isorhamnetin glycosides and their malonates and lesser quantities of quercetin glycosides. Red D'Anjou, D'Anjou, and Seckel pears also contained cyanidin 3-O-glucoside.

Arbutin (hydroquinone-β-D-glucoside) was found to be synthesized in young pear leaves from shikimic acid, and more readily from phenylpropanoid compounds (Grisdale and Towers 1960). Flavonoids found in pear leaves included phloretin, phloridizin, (+)-catechin, (−)-catechin, apigenin, cosmosiin, luteolin and cinaroside (Kislichenko and Novosel 2007).

High molecular weight material recovered from the culture filtrate of cell suspension cultured *Pyrus communis* was found to compose of 81% carbohydrate, 13% protein and 5% inorganic material (Webster et al. 2008). The high molecular weight extracellular material consisted of three major and two minor polysaccharides: a (fucogalacto)xyloglucan (36%) in the unbound neutral Fraction A; a type II arabinogalactan (as an arabinogalactan-protein, 29%) and an acidic (glucurono)arabinoxylan (2%) in Fraction B; and a galacturonan (33%) and a trace of heteromannan in Fraction C. The main amino acids in the proteins were Glx, Thr, Ser, Hyp/Pro and Gly. The major proteins detected were two chitanases, two thaumatin-like proteins, a β-1,3-glucanase, an extracellular dermal glycoprotein and a pathogenesis-related protein.

Hydroquinone derivatives (arbutin and pyroside) were found in the flowers of four Polish pear cultivars (Rychlińska and Gudej 2003). Three triterpenoids were isolated from the stem bark of *Pyrus communis* and identified as lup-20(29)-ene-3α, 27-diol, lup-20(29)-ene-3α-ol and lup-20(29)-ene-3α, 28-diol (Mehta et al. 2003).

Antioxidant Activity

When compared to the studied varieties Comice, Abate, General Leclerc and Passe Crassane, Rocha pear (peel and flesh) presented the highest content of total phenolics (Salta et al. 2010). Among them, chlorogenic, syringic, ferulic and coumaric acids, arbutin and (−)-epicatechin were detected as major components. In addition, among the tested varieties, Rocha pear presented the best antioxidant activities in the DPPH radical scavenging and ferric reducing power assays.

Antiulcer Activity

Highly polymerized procyanidins extracted from 'Winter Nélis' pear fruit, orally administered (20 mg/rat) before 60% ethanol treatment, exhibited a high level of antiulcer capacity whereas chlorogenic acid alone seemed to have a negative effect (Hamauzu et al. 2007). The percentage of lesion area to total gastric surface area (ulcer index) increased with increases in ethanol concentration (40–80%) and the length of time after ethanol treatment (60–120 minutes). The trend of myeloperoxidase activity was similar to the trend of the ulcer index. A mixture of those polyphenols had a significant protective effect. The results suggested that the antiulcer effect of pear procyanidins may be due to their strong antioxidant activity.

Anticancer Activity

An aqueous extract of *Pyrus communis* twigs showed inhibitory effect against S-180 sacrcoma cells (Liu and Zuo 1987). Hydroquinone isolated from the twig exhibited inhibitory effect against S-180 sarcoma cells (47.5%).

Antibacterial Activity

Eight compound isolated from *Pyrus communis* twigs were identified as nonacosane (1), lupeol (2), β-sitosterol (3), betulin (4), betulinic acid (5), daucosterol (6), hydroquinone (7) and arbutin (8) (Liu and Zuo 1987). It was shown that the compounds 2, 3, 4, 5, 6 and 8 possessed some bacteriostatic activity against *Escherichia coli*, *Salmonella typhi*, *Shigella flexneri* and *Staphylococus aureus*. Ethyl acetate extract of *Pyrus communis* fruit was reported to exhibit antimicrobial activity against selected bacteria but not against fungi (Guven et al. 2006).

Allergenic Activity

Pear is known as an allergenic food involved in the 'oral allergy syndrome' which affects a high percentage of patients allergic to birch pollen. Karamloo et al. (2001) isolated Pyr c 1, the major allergen from pear (*Pyrus communis*), and characterised as a new member of the Bet v 1 allergen family. The IgE binding characteristics of rPyr c 1 appeared to be similar to the natural pear protein. The biological activity of rPyr c 1 was equal to that of pear extract, as indicated by basophil histamine release in two patients allergic to pears. The related major allergens Bet v 1 from birch pollen and Mal d 1 from apple inhibited to a high degree the binding of IgE to Pyr c 1, whereas Api g 1 from celery, also belonging to this family, had little inhibitory effects, indicating epitope differences between Bet v 1-related food allergens.

Traditional Medicinal Uses

The fruit is regarded as astringent, febrifuge and sedative in folk medicine.

Other Uses

Pear trees are sometimes used as part of a shelter-belt planting. A yellow-tan dye is extracted from the leaves. It provides a heavy, tough, durable, fine grained, hard-wood that is used by cabinet and instrument makers. When covered with black varnish it is an excellent ebony substitute.

Comments

Cultivars of *Pyrus communis* are quite distinct from Asian pear cultivars. *Pyrus communis* exhibits a closer genetic relationship with Xinjiang pear (*P. sinkiangensis*) among the Asian pears cluster.

Selected References

Bailey LH (1976) Hortus third. A concise dictionary of plants cultivated in the United States and Canada. Liberty Hyde Bailey Hortorium/Cornell University, Wiley, 1312 pp

Chopra RN, Nayar SL, Chopra IC (1986) Glossary of Indian medicinal plants. (*Including the supplement*). Council Scientific Industrial Research, New Delhi, 330 pp

Cui T, Nakamura K, Ma L, Li JZ, Kayahara H (2005) Analyses of arbutin and chlorogenic acid, the major phenolic constituents in Oriental pear. J Agric Food Chem 53(10):3882–3887

Ferreira D, Guyot S, Marnet N, Delgadillo I, Renard CMGC, Coimbra MA (2002) Composition of phenolic compounds in a Portuguese pear (*Pyrus communis* L. Var. S. Bartolomeu) and changes after sun-drying. J Agric Food Chem 50(16):4537–4544

Grisdale SK, Towers GHN (1960) Biosynthesis of arbutin from some phenylpropanoid compounds in *Pyrus communis*. Nature 188(4756):1130–1131

Guven K, Yucel E, Cetinta F (2006) Antimicrobial activities of fruits of *Crataegus* and *Pyrus* species. Pharm Biol 44(2):79–83

Hamauzu Y, Forest F, Kohzy Hiramatsu K, Mitsukimi Sugimoto M (2007) Effect of pear (*Pyrus communis* L.) procyanidins on gastric lesions induced by HCl/ethanol in rats. Food Chem 100(1):255–263

Hu SY (2005) Food plants of China. The Chinese University Press, Hong Kong, 844 pp

Hudina M, Colaric M, Štampar F (2007) Primary metabolites in the leaves and fruits of three pear cultivars during the growing season. Can J Plant Sci 87:327–332

Hudina M, Štampar F (2000) Sugars and organic acids contents of European (*Pyrus communis* L.) and Asian (*Pyrus serotina* Rehd.) pear cultivars. Acta Aliment 29(3):217–230

Karamloo F, Scheurer S, Wangorsch A, May S, Haustein D, Vieths S (2001) Pyr c 1, the major allergen from pear (*Pyrus communis*), is a new member of the Bet v 1 allergen family. J Chromatogr B Biomed Sci Appl 756(1–2):281–293

Kislichenko VS, Novosel EN (2007) Flavonoids from leaves of *Pyrus communis, Malus slyvestris and Malus domestica*. Chem Nat Compd 43(6): 704–705

Ku TC, Spongberg SA (1994) Pyrus Linnaeus. In: Wu ZY, Raven PH (eds) Flora of China, vol 17, *Verbenaceae* through *Solanaceae*. Science Press/Missouri Botanical Garden Press, St. Louis/Beijing, 342 pp

Lin LZ, Harnly JM (2008) Phenolic compounds and chromatographic profiles of pear skins. (*Pyrus* spp.). J Agric Food Chem 56(19):9094–9101

Liu JK, Zuo CX (1987) Studies on the chemical constituents of *Pyrus communis*. Acta Bot Sin 29(1):84–87, In Chinese

Mehta BK, Verma M, Jafri M, Neogi R, Desiraju S (2003) Triterpenoids from the stem bark of *Pyrus communis*. Nat Prod Res 17(5):459–463

Pan Z, Kawabata S, Sugiyama N, Sakiyama R, Cao Y (2002) Genetic diversity of cultivated resources of pear in north China. Acta Hort (ISHS) 587:187–194

Rychlińska I, Gudej J (2003) Qualitative and quantitative chromatographic investigation of hydroquinone derivatives in *Pyrus communis* L. flowers. Acta Pol Pharm 60(4):309–312

Salta J, Martins A, Santos RG, Neng NR, Nogueira JMF, Justino J, Rauter AP (2010) Phenolic composition and antioxidant activity of Rocha pear and other pear cultivars – A comparative study. J Funct Food 2(2):153–157

Sha SF, Li JC, Jun Wu J, Zhang SL (2011) Characteristics of organic acids in the fruit of different pear species. Afr J Agric Res 6(10):2403–2410

U.S. Department of Agriculture, Agricultural Research Service (2011) USDA National nutrient database for standard reference, Release 24. Nutrient data laboratory home page. http://www.ars.usda.gov/ba/bhnrc/ndl

van der Zwet T, Childers NF (eds) (1982) The pear. Hort. Publ, Gainesville, 502 pp

Webster JM, Oxley D, Pettolino FA, Bacic A (2008) Characterisation of secreted polysaccharides and (glyco)proteins from suspension cultures of *Pyrus communis*. Phytochemistry 69(4):873–881

Whiteman K (1998) The new guide to fruit. Anness Publishing Limited, London, 128 pp

Pyrus pyrifolia

Scientific Name

Pyrus pyrifolia (Burm.) Nak.

Synonyms

Ficus pyrifolia Burm. f., *Pyrus montana* Nakai, *Pyrus pyrifolia* (Burm. f.) Nakai f. *stapfiana* (Rehder) Rehder, *Pyrus serotina* Rehder, *Pyrus serotina* Rehder var. *stapfiana* Rehder, *Pyrus sinensis* auct. jap., non Poiret.

Family

Rosaceae

Common/English Names

Asian Pear, Apple Pear, Chinese Pear, Korean Pear, Japanese Pear, Nashi, Nashi Pear, Oriental Pear, Sand Pear.

Vernacular Names

Chinese: Sha Li;
Czech: Jabloň Hruškolistá;
Danish: Japanpære, Japansk Pære, Kinesisk Pære, Sandpære;
Dutch: Japanse Peer;
Eastonian: Liiv-Pirnipuu;
French: Poirier Chinois, Poire Nashi, Poirier Des Sables, Poirier Japonais;
German: Nashi-Birne, Sandbirnbaum;
India: Nashpati (Hindu);
Indonesia: Appel Jepang;
Japanese: Nashi, Nihon Nashi, Tyô-Sen, Perusu Serotina, Yama Nashi;
Korean: Bae, Paenamu, Tolpaenamu;
Malaysia: Lai;
Nepal: Naxhpati;
Philippines: Peras;
Russian: Gruša Pesčanaja;
Spanish: Pera;
Thai: Sa Li;
Vietnam: Lê.

Origin/Distribution

The species is native to Eastern Asia – China, Japan and Korea. It is cultivated throughout central and South China, in the Far East of Russia, Korea, Southern Japan, in the northern mountainous parts of Vietnam, Thailand and India and to a lesser extent in Middle Asia, Indonesia and Philippines. Recently, it is also cultivated in Australia, New Zealand and USA (California) and in the warmer regions of Europe (Italy, France).

Agroecology

Nashi pear is adapted to a temperate climate regime with warm (24–32°C), humid (relative humidity 80–90%), rainy summers and cold winters (−5°C to 10°C) when it remains dormant. The variation of annual mean temperature ranges from 10°C to 16°C and mean annual precipitation hovers around 1,500 mm. It is cultivated from 100 m to 1,400 m altitude in full sun and in fertile, well-drained soils.

Edible Plant Parts and Uses

Nashi pear fruit is large with firm, crispy, sweet, juicy flesh and is best eaten raw or fresh out of hand or in fruit salads. It is also eaten cooked. In cooking, ground pears are used in vinegar or soy sauce-based sauces as a sweetener, instead of sugar or as a garnish or ground and added to mixtures. They are also used for marinating meat, especially beef. Nashi pears generally are not baked in pies or made into jams because they have a high water content and a crisp, grainy texture, very different from the buttery European varieties (*Pyrus communis*). Because of the large size and the unique and refreshing taste of Nashi, the fruit commands a premium price and tends to be served to guests or given as gifts, or eaten together in a family context.

Botany

A deciduous, small to medium sized tree, 7–15 m high with a dense broadly pyramidal to rounded canopy. Branchlets terete, purplish brown and tawny tomentose when young becoming dark brown and glabrescent with age and sparsely lenticelled. Stipules caducous, linear-lanceolate; petiole 3–4.5 cm, initially tomentose, glabrescent. Leaves alternate, ovate-elliptic or ovate, 7–12 × 4–6.5 cm, glabrous or brown lanate when young, base rounded or subcordate, margin spinulose-serrate, apex acute, mature leaf shiny, dark green in summer (Plates 1, 2 and 3), yellow, orange, red in autumn. Raceme umbel-like, 6–9-flowered. Flowers 2.5–3.5 cm across, sepals triangular-ovate, 5 mm; petals white, ovate, 1.5–1.7 cm, base shortly clawed, apex rounded; stamens 20; ovary 5- or 4-loculed, with 2 ovules per locule; styles 5, rarely 4, nearly as long as stamens, glabrous (Plate 1). Fruit a pome yellowish-brown or greenish- yellow, with pale dots, subglobose, 5–8 cm in diameter and subglabrous (Plates 3, 4, 5, 6 and 7).

Nutritive/Medicinal Properties

Proximate nutrient composition of fresh nashi fruit (*Pyrus pyrifolia*) per 100 g edible portion excluding refuse 9% of the core and stem was reported as follows (UDSDA 2011): Water 88.25 g, energy 42 kcal (176 kJ), protein 0.50 g, total lipid (fat) 0.23 g, ash 0.37 g, carbohydrate 10.65 g, fibre (total dietary) 3.6 g, sugars (total) 7.05 g; minerals Ca 4 mg, Fe 0.0 mg, Mg 8 mg, P 11 mg, K 121 g, Na 0 mg, Zn 0.02 mg, Cu 0.050 mg, Mn 0.060 mg, Se 0.1 μg; vitamins – vitamin C (ascorbic acid) 3.8 mg, thiamine 0.009 mg, riboflavin 0.010 mg, niacin 0.219 mg, pantothenic acid 0.070 mg, vitamin-6 0.022 mg, folate (total) 8 μg, total choline 5.1 mg, vitamin E (α-tocopherol) 0.12 mg, vitamin K (phylloquinone) 4.5 μg; lipids – total saturated fatty acids 0.012 g, 16:0 (palmitic acid) 0.010 g, 18:0 (stearic acid) 0.002 g; total monounsaturated fatty acids 0.049 g, 16:1 undifferentiated (palmitoliec acid) 0.001 g, 18:1 undifferentiated (oleic acid) 0.047 g, 20:1 0.001 g; total polyunsaturated fatty acids 0.055 g, 18:2 undifferentiated (linoleic acid) 0.055 g, 18:3 undifferentiated (linolenic acid) 0.001 g; amino acids – tryptophan 0.005 g, threonine 0.013 g, isoleucine 0.014 g, leucine 0.025 g, lysine 0.017 g, methionine 0.006 g, cystine 0.005 g, phenylalanine 0.013 g, tyrosine 0.004 g, valine 0.018 g, arginine 0.009 g, histidine 0.005 g, alanine 0.017 g, aspartic acid 0.098 g, glutamic acid 0.036 g, glycine 0.014 g, proline 0.016 g, serine 0.018 g, lutein + zeaxanthin 50 μg.

Studies showed that different Asian pear (*P. serotina*) cultivars differed in their contents of different sugars and organic acids (Hudina and

Plate 1 Nashi flowers and young foliage

Plate 2 Young nashi fruits and leaves

Plate 3 Mature nashi fruit

Štampar 2000). The fructose content in 4 European pear cultivars ranged from 27.9 to 45.7 g/kg and sorbitol varied from 5 to 19.0 g/kg. The Asian pear cultivars contained more total sugars than the European ones.

The total organic acid content in the fruits of *P. pyrifolia* cultivars ranged from 1.84 to 3.46 mg/g FW (fresh weight) (Sha et al. 2011) Malic, citric, quinic, oxalic, shikimic, and fumaric acids were detected in all of the 10 cultivars, tartaric acid was detected in 2 cultivars, and acetic acid was detected in only 1 cultivar. Malic and citric acids were the major constituents. The citric acid content was higher than the malic acid content in 3 cultivars. The malic acid content ranged from 0.61 to 2.11 mg/g FW (= 32–73% of the total) followed by citric acid with a content of 0.36–1.48 mg/g FW (= 13–43% of the total). The minor organic acids in the fruit were quinic and oxalic acids with contents of 0.12–0.44 mg/g FW (= 6–19% of the total) and 0.01–0.17 mg/g FW

Plate 4 Greenish-yellow nashi cultivar

Plate 6 Brown nashi cultivar

Plate 5 Yellow nashi cultivar

Plate 7 Close-up of brown nashi cultivar

(= 0.4–8% of the total), respectively. The content of acetic, shikimic, succinic, fumaric, tartaric, and lactic acids were relatively low. Although all of the cultivars showed similar organic acid composition, the total organic acid content varied significantly.

Nashi pear was found to contain pectin (0.9%) and arbutin and chlorogenic acid as the main phenolic constituents in nashi fruit (Cui et al. 2005). Arbutin and chlorogenic acid were also the main phenolic constituents in the fruit skin (Lin and Harnley 2008). Asian pear (group 2) contained only trace quantities of the remaining phenolics such as dicaffeoylquinic acids, quercetin glycosides, isorhamnetin glycosides and the glycosides of luteolin, apigenin, and chrysoeriol; and cyanidin 3-O-glucoside compared to other pear groups.

Two caffeoylmalic acid methyl esters, 2-O-(trans-caffeoyl)malic acid 1-methyl ester and 2-O-(trans-caffeoyl)malic acid 4-methyl ester were isolated from pear (*Pyrus pyrifolia* Nakai cv. Chuhwangbae) fruit peels (Lee et al. 2011). Further, 5 known hydroxycinnamoylmalic acids and their methyl esters were identified: 2-O-(trans-coumaroyl)malic acid, 2-O-(cis-coumaroyl)malic acid, 2-O-(cis-coumaroyl)malic acid 1-methyl ester, 2-O-(trans-coumaroyl)malic acid 1-methyl ester and 2-O-(trans-caffeoyl)malic acid (phaselic acid).

The aqueous ethanolic extract of *Pyrus pyrifolia* bark was found to exhibit protein glycation inhibitory activity and the extract also showed antioxidative activity (Kim and Kim 2003). The glycation inhibitory activity was significantly correlated with the antioxidative potency of the extract. The positive glycation inhibitory and antioxidative activities of nashi bark extract might suggest a possible role in targeting aging and diabetic complications.

Other Uses

In Japan, nashi seedlings are used as rootstock for various pear cultivars.

Comments

Fruits of the Akanashi cultivar group are brownish (Plates 3, 6 and 7) and those of the Aonashi cultivar group are greenish-yellow (Plates 4 and 5).

Selected References

Cui T, Nakamura K, Ma L, Li JZ, Kayahara H (2005) Analyses of arbutin and chlorogenic acid, the major phenolic constituents in Oriental pea. J Agric Food Chem 53(10):3882–3887

Hudina M, Štampar F (2000) Sugars and organic acids contents of European (*Pyrus communis* L.) and Asian (*Pyrus serotina* Rehd.) pear cultivars. Acta Aliment 29(3):217–230

Huxley AJ, Griffiths M, Levy M (eds) (1992) The new RHS dictionary of gardening, vol 4. Macmillan, New York

Kim HY, Kim K (2003) Protein glycation inhibitory and antioxidative activities of some plant extracts in vitro. J Agric Food Chem 51(6):1586–1591

Ku TC, Spongberg SA (1994) *Pyrus* Linnaeus. In: Wu ZY, Raven PH (eds) Flora of China, vol 17, *Verbenaceae* through *Solanaceae*. Science Press/Missouri Botanical Garden Press, St. Louis/Beijing, 342 pp

Lee KH, Cho JY, Lee HJ, Ma YK, Kwon J, Park SH, Lee SH, Cho JA, Kim WS, Park KH, Moon JH (2011) Hydroxycinnamoylmalic acids and their methyl esters from pear (*Pyrus pyrifolia* Nakai) fruit peel. J Agric Food Chem 59(18):10124–10128

Lin LZ, Harnly JM (2008) Phenolic compounds and chromatographic profiles of pear skins. (*Pyrus spp.*). J Agric Food Chem 56(19):9094–9101

Sha SF, Li JC, Jun Wu J, Zhang SL (2011) Characteristics of organic acids in the fruit of different pear species. Afr J Agric Res 6(10):2403–2410

U.S. Department of Agriculture, Agricultural Research Service (2011) USDA National nutrient database for standard reference, Release 24. Nutrient data laboratory home page. http://www.ars.usda.gov/ba/bhnrc/ndl

Wang YL (1990) Pear breeding in China. Pl Breed Abstr 60:877–879

Pyrus ussuriensis

Scientific Name

Pyrus ussuriensis Maximowicz

Synonyms

Pyrus asiae-mediae Maleev, *Pyrus communis* auct. non. L., *Pyrus communis* sensu Bunge non L., *Pyrus lindleyi* Rehder, *Pyrus ovoidea* Rehder, *Pyrus simonii* Carrière, *Pyrus sinensis* Jard., *Pyrus sinensis* Lindl., *Pyrus sinensis* sensu Decne. Non Poir., *Pyrus sinensis* α *ussuriensis* Makino, *Pyrus sinensis* var. *asiae-mediae* Popov, *Pyrus sinensis* var. *ussuriensis* Makino, *Pyrus sogdiana* Kudr., *Pyrus ussuriensis* Maxim. var. *ovoidea* (Rehder) Rehder.

Family

Rosaceae

Common/English Names

Chinese Pear, Fragrant Pear, Harbin Pear, Manchurian Pear, Mongolian Pear, Siberian Pear, Snow Pear, Ussurian Pear.

Vernacular Names

Chinese: Qiu Zi Li, Chiu-Tzu-Li, Shan-Li, Suan-Li;
Danish: Kinesisk Pære, Sandpære;
Finnish: Paeaerynaepuu;
German: Ussuri-Birne;
Japanese: Chuugoku Nashi, Iwate Yama Nashi, Hokushi Yama Nashi, Michinoku Nashi, Tjosen-Yama-Nashi;
Korean: Santolbaenamu;
Russian: Gruša Ussurijskaja;
Swedish: Manchuriskt Päron;
Thai: Sali Chin.

Origin/Distribution

Fragrant pear is indigenous to North eastern China, Far East of Russia (Amur and Ussuri regions) and Korea.

Agroecology

Fragrant pear is a cold temperate species and the most frost resistant of the *Pyrus* species. It occurs singly or in small groups within cedar-broadleaved, broadleaved and mixed forests along river valleys and on river terraces, in the lower mountain belt in its native range and can be found as high as 1,100–1,300 m above sea level. It grows on a wide range of soil types and in full sun.

Edible Plant Parts and Uses

Fruits are best consumed fresh, they can be used to prepare compotes, drinks and beverages. Fruits can be dried or fermented after addition of salt.

Botany

Deciduous, small to medium tree, growing to 15 m with a broad pyramidal or oval crown, dark grey bark and yellow-gray glabrous branchlets. Leaves orbicular-ovate or ovate, 5–10 cm by 4–6 cm, acuminate, base rounded or subcordate, margin long spinulose-serrate, tomentose when young to glabrescent on 2–5 cm long petioles, green turning (Plate 1) to bright red or bronze in autumn. Corymb densely 5–7-flowered. Flowers white, 3–3.5 cm in diamond 2–5 cm long pedicels. Hypanthium campanulate, slightly tomentose, sepals broadly triangular, glabrous, and persistent, petals white, obovate or broadly ovate, ca. 1.8 × 1.2 cm, glabrous, styles 5 distinct. Fruit ovoid to subglobose, greenish–yellow, with tiny white specks and tinge of pink, sometimes with rusty spots, 3–5 cm diameter, 5-loculed (Plates 2 and 3). Pulp is juicy, with pleasant flavour, sweet to sub-acid.

Plate 1 Fragrant pear leaves

Plate 2 Manchurian (Fragrant) pears

Plate 3 Close up of Fragrant pear cv Hong-hua-guan (red flower jug)

Nutritive/Medicinal Properties

Ussurian pear (*Pyrus ussuriensis*) cultivars and Xinjiang pear (*Pyrus sinkiangensis*) cultivars tend to have smaller fruit weight and higher level of sugars. Total sugar content in 29 varieties of *Pyrus ussuriensis* was very variable ranging from 81.2 to 138 mg/ml juice, fructose 32.2–88.6 mg/ml juice, glucose 7.2–51.1 mg/ml juice, sucrose 2.1–56.5 mg/ml juice and sorbitol 8.2–23.9 mg/ml juice (Pan et al. 2002).

The total organic acid content of fruits of *P. ussuriensis* cultivars ranged from 3.04 to 9.13 mg/g fresh weight (FW) (Sha et al. 2011). The dominant organic acids were malic and citric acid. Malic acid content ranged from 1.51 to 4.78 mg/g FW, accounting for 33–62% of the total organic acid content. Citric acid content ranged from 0.77 to 5.51 mg/g FW, accounting for 20–60% of the total organic acid content. The minor organic acids in the fruit were quinic (0.35–0.95 mg/g-FW) and oxalic acid (0.002–0.18 mg/g-FW), which accounted for 4–15% and 0.3–5.6% of the total organic acid, respectively. The content of acetic, shikimic, succinic, fumaric, tartaric, and lactic acids were relatively low.

Arbutin and chlorogenic acid were found to be the main phenolic constituents in *Pyrus ussuriensis* fruit (Cui et al. 2005). The mean concentration of arbutin and chlorogenic acid was 0.164 mg/g fresh weight (FW) and 0.163 mg/g FW respectively. Arbutin and chlorogenic acid were also the dominant phenolic compounds in the skin (Lin and Harnley 2008). Fragrant pear group was found to contain significant quantities of quercetin glycosides and lesser quantities of isorhamnetin glycosides and the glycosides of luteolin, apigenin, and chrysoeriol compared to other pear groups.

Three antioxidative compounds, 1, 4-dibenzenediol, chlorogenic acid and quercitrin were isolated from the ethyl acetate extract of the fruits of *Pyrus ussuriensis* (Kim et al. 1999). The DPPH (diphenylpicrylhydrazyl) free radical scavenging activities of 1, 4-dibenzenediol (RC_{50} : 0.4 μg) and chlorogenic acid (RC_{50} : 4 μg) were more effective than those of BHA (butylated hydroxyanisole) (RC_{50} : 14 μg) and α-tocopherol (RC_{50}: 12 μg).

Water, ethanol and acetone extracts of *P. ussuriensis* leaf exhibited more than 70%, 80% and 85% DPPH (1,1-diphenyl-2-picryl-hydrazyl) scavenging radical activity at 50 ppm concentration, respectively (Lee et al. 2010). Xanthine oxidase inhibition activity and superoxide dismutase (SOD)-like activity by *P. ussuriensis* extract were higher than 30% and increased with increasing concentration. In the antiinflammatory test, *P. ussuriensis* leaf extract inhibited generation of nitric oxide (NO) stimulated by LPS in the macrophage cell line (raw 264.7) after 12–24 hours. The results suggested *P. ussuriensis* to have great potential as a cosmeceutical raw material as well as antioxidant and antiinflammatory agent.

Other Uses

Pyrus ussuriensis is often used as rootstock for other *Pyrus* species and in pear breeding because of its resistances to frost, fire-blight, Asiatic pear scab and other diseases. Its wood is used to make souvenir craftwork objects.

Comments

Pyrus ussuriensis is cultivated in China since ancient times, domesticated from wild populations in the north. There exist about 150 cultivars with high variability, possibly enhanced by introgression from other pear species. *Pyrus ussuriensis* var. *ovoidea* Rehder is, in fact, a cultivar of *Pyrus ussuriensis*. It is characterized by its ovoid, subglobose, or ellipsoid fruit, longer fruiting pedicels (2–4 cm) and tomentose leaves and corymb.

Pyrus ussuriensis var. *viridis* T. Lee, the fragrant pear (Plates 2 and 3), cultivated in the Shandong province of China, is allowed to be imported into Australia subject to compliance with phytosanitary equirements.

Selected References

Anonymous (2003) Import of Asian ('Shandong') pear (*Pyrus pyrifolia* (Burm.) Nakai and *P. ussuriensis* var. *viridis* T. Lee) fruit from Shandong Province in the People's Republic of China. Agriculture, Fisheries and Forestry Australia. http://www.daff.gov.au/__data/assets/pdf_file/0003/24681/dft_pear_china.pdf

Bailey LH (1976) Hortus third. A concise dictionary of plants cultivated in the United States and Canada. Liberty Hyde Bailey Hortorium/Cornell University, Wiley, 1312 pp

Cui T, Nakamura K, Ma L, Li JZ, Kayahara H (2005) Analyses of arbutin and chlorogenic acid, the major phenolic constituents in Oriental pear. J Agric Food Chem 53(10):3882–3887

Hu SY (2005) Food plants of China. The Chinese University Press, Hong Kong, 844 pp

Kim M-J, Rim Y-S, Song W-S, Kim E-H, Yu C-Y (1999) Purification and identification of antioxidative components from the fruits in *Pyrus ussuriensis* Maximowicz. Korean J Med Crop Sci 7(4):303–307

Ku TC, Spongberg SA (1994) *Pyrus* Linnaeus. In: Wu ZY, Raven PH (eds) Flora of China, vol 17, *Verbenaceae* through *Solanaceae*. Missouri Botanical Garden Press/Science Press, Beijing/St. Louis, 342 pp

Lee CE, Kim YH, Lee BG, Lee DH (2010) Study on anti-oxidant effect of extracts from *Pyrus ussuriensis* leaves. J Korean Forest Soc 99(4):546–552

Lin LZ, Harnly JM (2008) Phenolic compounds and chromatographic profiles of pear skins. (*Pyrus* spp.). J Agric Food Chem 56(19):9094–9101

Pan Z, Kawabata S, Sugiyama N, Sakiyama R, Cao Y (2002) Genetic diversity of cultivated resources of pear in north China. Acta Hort (ISHS) 587:187–194

Sha SF, Li JC, Jun Wu J, Zhang SL (2011) Characteristics of organic acids in the fruit of different pear species. Afr J Agric Res 6(10):2403–2410

Teng Y-W, Tanabe K, Tamura F, Itai A (2001) Genetic relationships of pear cultivars in Xinjiang, China, as measured by RAPD markers. J Hort Sci Biotechnol 76:771–779

Wang YL (1990) Pear breeding in China. Pl Breed Abstr 60:877–879

Rubus fruticosus aggr.

Scientific Name

Rubus fruticosus L. aggr.

Synonyms

There are no synonyms as *Rubus fruticosus* refers not to a single species, but is used in the aggregate sense on a world basis comprising some 2,000 named species, subspecies, hybrids and varieties collectively referred to as taxa, in section *Rubus*, subgenus *Rubus* of the genus *Rubus* (Scher 2010). The name is based on a mixture of *Rubus plicatus* Weihe & Nees and *Rubus ulmifolius* Schott. Many of the species arose as a result of hybridization and apomixis. Some of the *Rubus* taxa in this aggregate include: *Rubus anglocandicans* A. Newton, *Rubus cissburiensis* W.C. Barton & Ridd., *Rubus echinatus* Lindl., *Rubus erythrops* Edees & A. Newton, *Rubus laciniatus* Willd., *Rubus leightonii* Lees ex Leight., *Rubus leucostachys* Schleich. ex Sm., *Rubus phaeocarpus* W.C.R. Watson, *Rubus polyanthemus* Lindeb., *Rubus riddelsdellii* Rilstone, *Rubus rubritinctus* W.C.R. Watson, *Rubus ulmifolius* Schott (including *Rubus ulmifolius* var. *ulmifolius* and *Rubus ulmifolius* var. *anoplothyrsus* Sudre), and *Rubus vestitus* Weihe (Evans et al. 2007).

Family

Rosaceae

Common/English Names

Blackberries, Blackberry, Bramble, Bramble Berry, Cultivated Blackberries Common Blackberry, Dewberry, European Blackberry, Noxious Blackberry, Shrubby Blackberry, Thornless Blackberries, Wild Blackberry.

Vernacular Names

Arabic: Tût Shawkî, 'Ullayq;
Chinese: Ou Zhou Hei Mei;
Czech: Ostružiník Křovitý, Ostružiníky; Ostružiník Křovitý/Ostružiník Mnoholistý/Ostružiník Oasnatý/Ostružiník Řasnatý
Danish: Almindelig Brombær, Brombær, Klynger;
Dutch: Gewone Braam, Braam, Braam Sort, Braambes;
Esperanto: Rubuso;
Estonian: Aedmurakas, Kitsemari, Pampel, Põõsasmurakas;
Finnish: Mustavatukka, Karhunvattu, Karhunvatukka, Oimuvatukka, Poimuvatukka;
French: Catimuron, Aronce, Mûre, Mûre De Ronce, Mûre Sauvage, Mûrier Sauvage, Mûron, Ronce, Ronce Commune, Ronce Des Bois, Ronce Des Haies, Ronce Frutescente, Ronce Sauvage, Ronces;
Gaelic: Dris;
German: Brombeere, Brombeeren, Brombeerestrauch, Echte Brombeere, Echte Brombeer, Falten-Brombeere, Fuchsbeere, Gemeine Brombeere, Hirschbeere, Kratzbeere;

Greek: Vatomuro;
Hungarian: Feketeszeder, Földiszeder, Szeder, Vad Szeder;
Icelandic: Brómber
India: Kaalii Anchhi, Kaalaa Jaamun (Hindu);
Hebrew: Petel Shachor;
Italian: Moro Delle Siepi, Mora Di Bosco, Moro Di Macchia, Mora Di Rovo, Mora Selvatica, Rogo, Rogo Di Macchia, Rovo, Rovo Di Macchia;
Japanese: Seiyou Abu Ichigo;
Norwegian: Bjørnebær, Bjønnbær, Brandbær, Kølabær, Søtbjønnbær, Søtbjørnebær;
Polish: Jeżyna Fałdowana, Jerzyna, Jeżyna, Jezyna Krzewiasta, Jeżyna Pospolita;
Portuguese: Amora , Amora Silvestre;
Slovaščina: Robida Nagubana;
Slovencina: Ostružina Riasnatá;
Spanish: Zarza, Zarza Común, Zarzamora;
Swedish: Blomsterbjörnbär, Sötbjörnbär, Svarthallon
Turkish: Alik, Böyürtlen, Böyürtlen Çalısı, Büğürtlen;
Vietnamese: Quả Mâm Xôi (Fruit).

Origin/Distribution

Most cultivated blackberries are native to temperate Europe in the Northern Hemisphere. Cultivated *Rubus* fruit species belong to two subgenera: the *Idaeobatus* (raspberries) and the *Rubus* formerly *Eubatus* (blackberries). Commercial *Rubus* fruit crops include the red (*R. idaeus* L.), black (*R. occidentalis* L.) and purple (red and black raspberry hybrid) raspberries, blackberries (*Rubus* species and hybrids), Andean blackberries and cloudberries (*R. chamaemorus* L.) (Thompson 1997). Most of the cultivated blackberries are related to or entirely derived from *R. ulmifolius* (=R. *rusticanus*), *R. nitidioides* and *R. thrysiger* (Hall 1990). The cultivated blackberries grown in Australia are all complex hybrids, mostly derived from successive generations of controlled crosses and may be grouped according to derivation as (i) mainly *Rubus ursinus* germplasm; (ii) derivatives of *R. ulmifolius* crossed with *R. argutus* and *R. alleghaniensis*, and (iii) derivatives of *R. alleghaniensis* (McGregor 1998). Cultivated blackberries do not manifest weedy behaviour or represent a threat to Australian ecosystems.

Agroecology

Blackberry is confined to temperate climates with an annual rainfall of at least 700 mm and occur at any altitude up to 1,950 m. They will grown in areas with lower rainfall along waterways, streambanks. They grown an many soil types but thrives best in acidic to slightly acidic, well drained soil rich in organic matter. They do poorly on chalky clayey soils. They fruit well in full sun but will also grown in the shade.

Edible Plant Parts and Uses

Blackberries are consumed fresh. Their culinary uses in prepared foods include desserts, jams, seedless jellies, yogurt, pie fillings, crumbles and sometimes wine. Blackberry flowers are good nectar producers, and afford a medium to dark, fruity honey.

Botany

Blackberry is a perennial, semi-deciduous, prickly (Plate 1) or thornless (Plates 2, 3, 4 and 5), scrambling, semi-prostrate to almost erect shrub to 2 m high and with canes to about 7 m long. Stems mostly arching, green, reddish or purple, ribbed, angled or concave, with or without hairs. Prickles when present are straight or curved rearwards. Leaves compound with 3–5 shortly-stalked oval leaves with obtuse to acute to acuminate apices, serrated margins, adaxial surface darker green, abaxial surface pale green, pubescent or glabrous, with or without short prickles on the mid-vein and petioles (Plates 2, 3 and 4). Stipules narrow, free from petiole or fused at the base. Flowers 2–3 cm across, produce in short racemes or cymes on floricanes with 5 reflexed sepals, 5 mostly white or pinkish, spreading petals, numerous stamens and few to many

Plate 1 Thorny blackberry fruits and leaves

Plate 3 Thornless blackberry blossoms and leaves

Plate 4 Immature thornless blackberries and leaves

Plate 2 Thornless blackberry inflorescence

carpels on an elongated receptacle (Plates 2 and 3). Fruit an aggregate of numerous succulent drupelets, each containing a single seed. Drupelets narrowly to broadly D-shaped or rounded triangular, about 2–3.2 mm long, 1.5–2.8 mm wide, 1–1.8 mm thick. Colour green changing to straw-yellow, amber, orange-red, red, reddish black, to glossy black (Plates 1, 4, 5, 6 and 7) as they ripen.

Nutritive/Medicinal Properties

Nutrient composition of raw blackberries (*Rubus* spp.) per 100 g edible portion was reported by (USDA 2011) as: water 88.15 g, energy 43 kcal (181 kJ), protein 1.39 g, total lipid 0.49 g, as 0.37 g, carbohydrate 9.61 g, total dietary fibre 5.3 g, total sugars 4.88 g, sucrose 0.07 g, glucose 2.31 g, fructose 2.40 g, maltose 0.07 g, galactose 0.03 g, Ca 29 mg, Fe 0.62 mg, Mg 20 mg, P 22 mg, K 162 mg, Na 1 mg, Zn 0.52 mg, Cu 0.165 mg, Mn 0.646 mg, Se 0.4 μg, vitamin C 21 mg, thiamin 0.020 mg, niacin 0.646 mg, pantothenic acid 0.276 mg, vitamin B-6 0.030 mg,

Plate 5 Close-view of thornless immature blackberries

Plate 6 Ripening thornless blackberry fruit

Plate 7 Harvested thornless blackberries

total folate 35 μg, total chlorine 8.5 mg, betaine 0.3 mg, vitamin A 11 μg RAE, vitamin A 214 IU, β-carotene 128 μg, lutein + zeaxanthin 118ug, vitamin E (α-tocopherol) 1.17 mg, β-tocopherol 0.04 mg, γ-tocopherol 1.34 mg, δ-tocopherol 0.90 mg, vitamin K (phylloquinone) 19.8 μg, total saturated fatty acids 0.014 g, 16:0 (palmitic) 0.012 g, 18:0 (stearic) 0.03 g, total monounsaturated fatty acids 0.047 g, 18:1 undifferentiated (oleic acid) 0.044 g, 20:1 0.004 g, total polyunsaturated fatty acids 0.280 g, 18:2 undifferentiated (linoleic acid) 0.186 g, and 18:3 undifferentiated (linolenic acid) 0.094 g.

Another analysis conducted by FSANZ in Australia reported the following nutrient value per 100 g edible portion (FSANZ 2010): energy 211 kJ, water 84.2 g, protein 1.4 g, N 0.22 g, fat 0.3 g, ash 0.3 g, dietary fibre 6.1 g, fructose 3.9 g, glucose 3.6 g, total sugars 7.5 g, lactic acid 0.1 g, malic acid 0.2 g, citric acid 0.4 g, CA, 30 mg, Cu 0.16 mg, Fe 0.42 g, Mg 30 mg, Mn 0.55 mg, P 29 mg, K 114 mg, Se 2 μg, S 16 mg, Zn 0.24 mg, thiamin 0.02 mg, riboflavin 0.03 mg, niain 0.3 mg, niacin equivalents 0.53 mg, pantothenic acid (B5) 0.35 mg, biotin (B7) 1.4 μg, total folate 36 μg, β-carotene 150 μg, cryptoxanthin 340 μg, β-carotene equivalents 320 μg, retinol equivalents 53 μg, vitamin C 38 mg and α-tocopherol (vitamin E) 1.4 mg.

Analysis conducted by the Department of Nutrition, National Food Institute, Technical University of Denmark reported the following nutrient value per 100 g edible portion (Saxholt et al. 2008): moisture 88.2 g, energy 175 kJ, protein 1.4 g, total N 0.2 g, fat 1.0 g, total carbohydrate 9 g, available carbohydrate 4.7 g, total sugars 6.40 g, fructose 2.5 g, glucose 2.7 g, maltose 0.7 g, saccharose 0.5 g, dietary fibre 4.3 g, ash 0.4 g, vitamin A 16.7 RE, β-carotene equivalent 200 µg, vitamin E (α-tocopherol) 5.5 mg, thiamine (vitamin B1) 0.017 mg, riboflavin (vitamin B2) 0.05 mg, niacin 0.5 mg, niacin equivalents 0.7NE, vitamin B6 0.05 mg, pantothenic acid 0.25 mg, biotin 0.4 µg, vitamin C 15 mg, tryptophan 0.2 mg, Na 2 mg, K 266 mg, Ca 27 mg, Mg 23 mg, P 37 mg, Fe 0.12 mg, Cu 0.12 mg, Zn 0.53 mg, I 0.4 µg, Mn 0.646, Cr 1.0 µg and Se 0.1 µg.

Flavonoids (kaempferol, quercetin, myricetin) and phenolic acids (p-coumaric, caffeic, ferulic, p-hydroxybenzoic, gallic and ellagic acids) were detected in the fruits of 19 berries including blackberries (Hákkinen et al. 1999). Ellagic acid was the main phenolic compound in the berries of the genus *Rubus* (red raspberry, Arctic bramble and cloudberry) and genus *Fragaria* (strawberry). The data suggested berries to have potential as good dietary sources of quercetin or ellagic acid. Vrhovsek et al. (2006) reported ellagitannins, a major class of phenolics, to be largely responsible for the astringent and antioxidant properties of raspberries and blackberries. The *Rubus* ellagitannins comprised a complex mixture of monomeric and oligomeric tannins. *Rubus* oligomeric ellagitannins contained, beside the well-known ellagic acid and gallic acid moieties, the sanguisorboyl linking ester group. Phenolic acids (gallic acid, p-hydroxybenzoic acid, caffeic acid, ferulic acid, and ellagic acid) in blueberries and blackberries ranged from 0.19 to 258.90 mg/100 g fresh weight (FW), and flavonoids (catechin, epicatechin, myricetin, quercetin, and kaempferol) ranged from 2.50 to 387.48 mg/100 g FW (Sellapan et al. 2002).

Fresh blackberries were found to contain 24 mg/100 g of the anthocyanin cyanidin-3-rutinoside (Seeram et al. 2001). The fruits of wild Norwegian blackberries and three blackberry (*Rubus fruticosus*) cultivars were found to contain the following anthocyannins: 3-glucoside cyanidin, 3-rutinoside cyanidin, 3-xyloside cyaniding, 3-O-β-(6″-malonylglucoside) cyanidin and cyanidin 3-O-β-(6″-(3-hydroxy-3-methylglutaroyl) glucopyranoside) (Jordheim et al. 2011).

Antioxidant Activity

The amount of total phenolics varied between 617 and 4,350 mg/kg in fresh berries (blackberries, red raspberries, blueberries, sweet cherries and strawberries), as gallic acid equivalent (GAE) (Heinonen et al. 1998). In the copper-catalyzed in-vitro human low-density lipoprotein oxidation assay at 10 µM gallic acid equivalents (GAE), berry extracts inhibited hexanal formation in the order: blackberries>red raspberries>sweet cherries>blueberries>strawberries. In the copper-catalyzed in-vitro lecithin liposome oxidation assay, the extracts inhibited hexanal formation in the order: sweet cherries>blueberries>redraspberries>blackberries>strawberries. HPLC analyses showed high anthocyanin content in blackberries, hydroxycinnamic acid in blueberries and sweet cherries, flavonol in blueberries, and flavan-3-ol in red raspberries. The antioxidant activity for LDL was associated directly with anthocyanins and indirectly with flavonols, and for liposome it correlated with the hydroxycinnamate content. Berries thus contribute a significant source of phenolic antioxidants that may have potential health effects. Studies by Wang and Lin (2000) showed that blackberries and strawberries had the highest ORAC (oxygen radical absorbance capacity) values during the green stages, whereas red raspberries had the highest ORAC activity at the ripe stage. Total anthocyanin content increased with maturity for all three species of fruits. Leaves were found to have higher ORAC values than fruits. ORAC values in fruits varied from 7.8 to 33.7 µmol of Trolox equivalents (TE)/g of fresh berries (35. 0–162.1 µmol of TE/g of dry matter). In leaves, ORAC values ranged from 69.7 to 182.2 µmol of TE/g of fresh leaves (205.0–728.8 µmol of TE/g of dry matter). As the leaves aged, ORAC values and total phenolic contents decreased.

Siriwoharn et al. (2004) found that total anthocyanin pigments increased from 74.7 to 317 mg/100 g fresh weight (FW) from underripe to overripe for Marion blackberries and from 69.9 to 164 mg/100 g FW for Evergreen blackberries. Total phenolics did not show a marked change with maturity with values slightly decreasing from underripe to ripe. Antioxidant activities, while increasing with ripening, also did not show the marked change that total anthocyanins exhibited. They also found that total anthocyanins for 11 blackberry cultivars ranged from 131 to 256 mg/100 g FW (mean=198), total phenolics ranged from 682 to 1,056 mg GAE/100 g FW (mean=900), oxygen radical absorbance capacity ranged from 37.6 to 75.5 µmol TE/g FW (mean=50.2), and ferric reducing antioxidant power ranged from 63.5 to 91.5 µmol TE/g FW (mean=77.5).

The antioxidant activity in blackberry juice of six different thornless blackberry (*Rubus* sp.) cultivars was positively correlated to the activities of most antioxidant enzymes namely superoxide dismutase (SOD) with $R^2=0.902$; glutathione-peroxidase (GSH-POD) with $R^2=0.858$; ascorbate peroxidase (ASA-POD) with $R^2=0.896$; and glutathione reductase (GR) with $R^2=0.862$ (Jiao and Wang 2000). ORAC values were the highest in 'Hull Thornless' and lowest in 'Black Satin'. The highest levels of AsA (ascorbate) and DHAsA (dehydroascorbate) were in the juice of 'Hull Thornless' blackberries with 1.09 and 0.15 µmol/g fresh wt, respectively. 'Hull Thornless' also had the highest ratio of AsA/DHAsA among the six blackberry cultivars studied. The 'Smoothstem' cultivar contained the lowest amounts of AsA and DHAsA. 'Hull Thornless' had the highest glutathione (GSH) content with 78.7 nmol/g fresh wt, while 'Chester Thornless' contained the largest amount of oxidized glutathione (GSSG). The highest GSH/GSSG ratio was 4.90 which was seen in the 'Hull Thornless' cultivar.

Total polyphenols in blueberries and blackberries ranged from 261.95 to 929.62 mg/100 g FW, and total anthocyanins ranged from 12.70 to 197.34 mg/100 g FW. TEAC values varied from 8.11 to a maximum of 38.29 µM/g FW (Sellappan et al. 2002). A linear relationship was observed between TEAC values and total polyphenols or total anthocyanins. The data indicate blueberries and blackberries to be rich sources of antioxidants.

Blackberries (*Rubus* sp.) and strawberries (*Fragaria*×*ananassa*) had the highest ORAC (oxygen radical absorbance capacity) values during the green stages, whereas red raspberries (*Rubus idaeus*) had the highest ORAC activity at the ripe stage (Wang and Lin 2000). Total anthocyanin content increased with maturity for all species of fruits. Compared with fruits, leaves were found to have higher ORAC values. In fruits, ORAC values ranged from 7.8 to 33.7 µmol of Trolox equivalents (TE)/g of fresh berries (35.0–162.1 µmol of TE/g of dry matter), whereas in leaves, ORAC values ranged from 69.7 to 182.2 µmol of TE/g of fresh leaves (205.0–728.8 µmol of TE/g of dry matter). As the leaves become older, the ORAC values and total phenolic contents decreased. The results showed a linear correlation between total phenolic content and ORAC activity for fruits and leaves. For ripe berries, a linear relationship existed between ORAC values and anthocyanin content. Of the ripe fruits tested, on the basis of wet weight of fruit, cv. Jewel black raspberry and blackberries were the richest source for antioxidants. On the basis of the dry weight of fruit, strawberries had the highest ORAC activity followed by black raspberries (cv. Jewel), blackberries, and red raspberries. Ferric reducing antioxidant power (FRAP) values of raspberry (*Rubus idaeus*), blackberry (*Rubus fructicosus*), raspberry × blackberry hybrids, red currant (*Ribes sativum*), gooseberry (*Ribes glossularia*) and Cornelian cherry (*Cormus mas*) cultivars ranged from 41 to 149 µmol ascorbic acid/g dry weight and protection of deoxyribose ranged from 16.1% up to 98.9% (Pantelidis et al. 2007).

Of the fruit juice from different cultivars of thornless blackberries (*Rubus* sp.), blueberries (*Vaccinium* spp.), cranberries (*Vaccinium macrocarpon*), raspberries (*Rubus idaeus* and *Rubus occidentalis*), and strawberries (*Fragaria x ananassa*), blackberries generally exhibited the highest antioxidant capacity against superoxide radicals (O_2^-), hydrogen peroxide (H_2O_2), hydroxyl radicals (·OH) (Wang and Jiao 2000). Of six traditionally used Croatian medicinal plants (*Melissa officinalis, Thymus serpyllum., Lavandula officinalis, Rubus fruticosus, Urtica dioica* and *Olea europea*) the overall highest content of phenolic compounds was found in hydrolyzed extract of blackberry leaves (2,160 mg GAE/L), followed by the non-hydrolyzed extract of lemon balm (Komes et al. 2011). Both extracts also exhibited the highest antioxidant capacity. Fruits of three wild Jamaica-grown blackberry species: *Rubus jamaicensis, Rubus rosifolius* and *Rubus racemosus*, and of the Michigan-grown *Rubus acuminatus, Rubus idaeus* cv. Heritage and *Rubus idaeus* cv. Golden (both raspberries) contained superior levels of anthocyanins (146–2,199 mg/100 g fresh weight) (Bowen-Forbes et al. 2010). Their ethyl acetate and methanol extracts showed good antioxidant activity, the majority of the extracts exhibiting over 50% lipid peroxidation inhibitory activity at 50 μg/ml.

Intake of blackberry juices prepared with water (BJW) and defatted milk (BJM) significantly increased the ascorbic acid content in the plasma (Hassimotto et al. 2008). No changes were observed in the plasma urate and α-tocopherol levels. An increase on the plasma antioxidant capacity, by ORAC assay, was observed only after consumption of BJW. Further, it was noted that plasma catalase increased following intake of blackberry juices. No change was observed on the plasma and erythrocyte catalase and glutathione peroxidase activities. A significant decrease in the urinary antioxidant capacity between 1 and 4 hours after intake of both blackberry juices was observed. A good correlation was observed between total antioxidant capacity and urate and total cyanidin levels. These results suggested association between anthocyanin levels and catalase and a good correlation between antioxidant capacity and ascorbic acid in the human plasma after intake of blackberry juices.

Seeds from five caneberry species: red raspberry, black raspberry, boysenberry, Marion blackberry, and evergreen blackberry were found to have 6–7% protein and 11–18% oil (Bushman et al. 2004). The oils contained 53–63% linoleic acid, 15–31% linolenic acid, and 3–8% saturated fatty acids. Antioxidant capacities were detected both for whole seeds and for cold-pressed oils but did not correlate to total phenolics or tocopherols. Ellagitannins and free ellagic acid were the main phenolics detected in all five caneberry species and were approximately threefold more abundant in the blackberries and the boysenberry than in the raspberries.

Of 11 Sardinian plant species used as beverage tea or medicinal decoction, *R. ulmifolius* exerted the highest in-vitro antioxidant activity as evaluated by Briggs Rauscher (BR), TEAC, DPPH and Folin-Ciocalteu (FC) assays (Dall'Acqua et al. 2008). The bioactive phenolic compounds isolated included caffeic acid, ferulic acid, quercetin-3-O-glucuronide, kaempferol-3-O-glucuronide, kaempferol-3-O-(6″-p-coumaroyl)-β-d-glucopyranoside, kaempferol-3-O-(6″-caffeoyl)-β-d-glucopyranoside, chlorogenic acid, 4-caffeoylquinic acid and 5-caffeoylquinic acid. *Rubus ulmifolius* leaf extract exhibited an antioxidant activity (TEAC value) of 0.12 whilst the antioxidant activity (TEAC values) of the isolated polyphenols compounds ranged from 4.88 (gallic acid) to 1.60 (kaempferol) (Martini et al. 2009).

The anthocyanin-rich fraction from WB-10, a wild blackberry (WB) genotype and the proanthocyanidin-rich fraction from UM-601, a domesticated noncommercial breeding line exhibited the highest NO inhibitory activities ($IC_{50} = 16.1$ and 15.1 μM, respectively) (Cuevas-Rodríguez et al. 2010). Proanthocyanidin-rich fractions from the wild WB-10 showed the highest inhibition of iNOS expression ($IC_{50} = 8.3$ μM). Polyphenolic-rich fractions from WB-7 and UM-601 were potent inhibitors of COX-2 expression ($IC_{50} = 19.1$ and 19.3 μM C3G equivalent, respectively). For most of the extracts, antioxidant capacity was significantly correlated with NO inhibition.

Anticancer Activity

Evidence from studies suggested that edible small and soft-fleshed berry fruits namely, blackberries, black raspberries, blueberries, cranberries, red raspberries, and strawberries may have beneficial effects against several types of human cancers (Seeram et al. 2006; Seeram 2008). The anticancer potential of berries had been related, at least in part, to a multitude of bioactive phytochemicals that these colourful fruits contained, including polyphenols (flavonoids, proanthocyanidins, ellagitannins, gallotannins, phenolic acids), stilbenoids, lignans, and triterpenoids. Studies showed that the anticancer effects of berry bioactives were partially mediated through their abilities to counteract, reduce, and also repair damage resulting from oxidative stress and inflammation. In addition, berry bioactives were also reported to regulate carcinogen and xenobiotic metabolizing enzymes, various transcription and growth factors, inflammatory cytokines, and subcellular signaling pathways of cancer cell proliferation, apoptosis, and tumour angiogenesis.

Extracts of six popularly consumed berries–blackberry, black raspberry, blueberry, cranberry, red raspberry and strawberry were found to inhibit the growth of human oral (KB, CAL-27), breast (MCF-7), colon (HT-29, HCT116), and prostate (LNCaP) tumour cell lines at concentrations ranging from 25 to 200 μg/mL (Seeram et al. 2006). With increasing concentration of berry extract, increasing inhibition of cell proliferation in all of the cell lines were observed, with different degrees of potency between cell lines.

Ellagitannins were also found in blackberry fruit in particular in the seeds; these included pedunculagin, casuarictin/potentillin, castalagin/vescalagin, lambertianin A/sanguiin H-6, lambertianin C, and lambertianin D (Hager et al. 2008). Ellagitannins are known chemopreventive agents. Ellagic acid, an abundant component in these berries, had been shown to inhibit carcinogenesis.

Blackberries (cv Chester Thornless, Hull Thornless and Triple Crown) treated with preharvest methyl jasmonate (0.01 and 0.1 mM) had higher soluble solids content, and lower titratable acids than untreated fruit as well as enhanced content of flavonoids and increased antioxidant capacity (Wang et al. 2008). Extracts of treated fruit showed enhanced inhibition of human lung A549 cancer cell and HL-60 leukemia cell proliferation and induced the apoptosis of HL-60 cells. Cultivar Hull Thornless had higher soluble solids and lower titratable acids compared to cv. Chester Thornless and Triple Crown. Hull Thornless also had significantly higher anthocyanin, total phenolic content, antioxidant and antiproliferation activity than other two cultivars. The hexane extracts of the Jamaican blackberry *Rubus* spp. *Rubus jamaicensis*, *Rubus rosifolius* and *Rubus racemosus*, demonstrated moderate COX (cyclooxygenase) inhibitory activity (27.5–33.1%) at 100 μg/mL, and exhibited the greatest potential to inhibit cancer cell growth, inhibiting colon, breast, lung, and gastric human tumour cells by 50%, 24%, 54% and 37%, respectively (Bowen-Forbes et al. 2010).

Antimicrobial Activity

From the aerial parts of *Rubus ulmifolius* three new anthrones, rubanthrone A, B and C, were isolated (Flamini et al. 2002). Rubanthrone A showed antimicrobial activity against *Staphylococcus aureus* at 4.5 mg/mL. Isolated constituents of *Rubus ulmifolius* namely quercetin-3-O-β-D-glucuronide; kaempferol-3-O-β-D-glucuronide, gallic acid, ferulic acid and tiliroside exhibited high antimicrobial activity (Panizzi et al. 2002).

Plant extracts of *R. ulmifolius* was one of ten plants that demonstrated biofilm inhibition in methicillin-resistant *Staphylococcus aureus* (MRSA) with IC_{50}<=32ug/mL (Quave et al. 2008). *Rubus ulmifolius* leaf extract and the isolated polyphenols showed antibacterial activity against both of the *Helicobacter pylori* strains (Martini et al. 2009). The minimum bactericidal concentrations (MBCs) of the extract for *H. pylori* strains G21 and 10 K, respectively, were 1,200 and 1,500 μg/ml after 24 hours of exposure and 134 and 270 μg/ml after 48 hours exposure. Ellagic acid showed very low MBC values towards both of the *H. pylori* strains after 48 hours (2 and 10 μg/ml for strains G21 and 10 K, respectively) and kaempferol toward G21 strain (MBC=6 μg/ml).

Hypoglycemic Activity

Preliminary study showed *Rubus fruticosus* to have hypoglycaemic activity (Alonso et al. 1980). Daily administration of 5 g/kg of the *R. fruticosus* leaf infusion to alloxan-diabetic rats resulted in a 15% decrease in the blood glucose levels. When the treatment was halted, blood glucose levels reverted back to the initial values. *R. ulmifolius* leaf infusions elicited notable hypoglycaemic effects in both alloxan and streptozotocin induced hyperglycaemic rats (Lemus et al. 1999). *R. ulmifolius* showed decrease of glycaemia in both alloxan and streptozotocin diabetic rats (28% and 29%). Activity-guided fractionation of *R. ulmifolius* showed that petroleum ether extracts elicited a marked hypoglycaemic effect (35%) in the streptozotocin induced model.

Hyaluronidase Inhibitory Activity

Two fractions F3 and F7 obtained from the aqueous extract of blackberry fruits were found to inhibit activity of hyaluronidase enzyme (Marquina et al. 2002).

Traditional Medicinal Uses

Since ancient times, the flowers and fruit have been used as a remedy for venomous bites (Grieve 1971). The ancient Greeks used blackberries as a remedy for gout. The fruits have been used to stop looseness of the bowel and was deemed good for stone. Leaves and root bark contain much tannin and have long been revered as a capital astringent and tonic and valued as a remedy for dysentery and diarrhea. The leaves are useful for piles and are still being used externally for scalds and burns. *Rubus ulmifolius* has been used for burns in traditional Italian medicine (Guarrera 2005).

Other Uses

Blackberry stems have been employed by American Indians to construct a strong rope, and parts of the plant (berries, leaves, roots) have been used to dye hair and fabrics (Anderberg).

Comments

Blackberry (except the domestically cultivated species) is considered a noxious weed of national significance in Australia, because of its invasiveness, potential for spread, and economic and environmental impacts. It is declared noxious in Queensland, New South Wales, Victoria, South Australia, Western Australia and Tasmania but not in the Northern Territory and the Australian Capital Territory. *R. fruticosus* aggr. species can reproduce both vegetatively and by seed. They can also produce sucker plants and reproduce from root fragments and other plant parts.

Selected References

Alonso R, Cadavid I, Calleja JM (1980) A preliminary study of hypoglycemic activity of *Rubus fruticosus*. Planta Med 40:102–106

Bowen-Forbes CS, Zhang YJ, Nair MG (2010) Anthocyanin content, antioxidant, anti-inflammatory and anticancer properties of blackberry and raspberry fruits. J Food Compos Anal 23(6):554–560

Bruzzese E, Mahr F, Fiathful I (2000) Best practice management guide for environmental weeds: no.5 blackberry, *Rubus fruticosus* aggregate. Factsheet produced by the Cooperative Research Centre for Weed Management Systems, University of Adelaide, South Australia

Bushman BS, Phillips B, Isbell T, Ou B, Crane JM, Knapp SJ (2004) Chemical composition of caneberry (*Rubus* spp.) seeds and oils and their antioxidant potential. J Agric Food Chem 52(26):7982–7987

Cuevas-Rodríguez EO, Dia VP, Yousef GG, García-Saucedo PA, López-Medina J, Paredes-López O, de Gonzalez ME, Lila MA (2010) Inhibition of pro-inflammatory responses and antioxidant capacity of Mexican blackberry (*Rubus* spp.) extracts. J Agric Food Chem 58(17):9542–9548

Dall'Acqua S, Cervellati R, Loi MC, Innocenti G (2008) Evaluation of in vitro antioxidant properties of some traditional Sardinian medicinal plants: investigation of the high antioxidant capacity of *Rubus ulmifolius*. Food Chem 106(2):745–749

Daubeny HA (1996) Brambles. In: Janick J, Moore JN (eds) Fruit breeding. Wiley, New York

DPI.VIC (2002) Agricultural notes: Bramblefruit. Victoria Department of Primary Industries

Evans KJ, Symon DE, Whalen MA, Hosking JR, Barker RM, Oliver JA (2007) Systematics of the *Rubus fruticosus* aggregate (Rosaceae) and other exotic *Rubus* taxa in Australia. Aust Syst Bot 20:187–251

Flamini G, Catalano S, Caponi C, Panizzi L, Morelli I (2002) Three anthrones from *Rubus ulmifolius*. Phytochemistry 59(8):873–876

Food Standards Australia New Zealand (FSANZ) (2010) NUTTAB 2006. Australian food composition tables. http://www.foodstandards.gov.au/consumerinformation/nuttab2010/nuttab2010onlinesearchabledatabase/onlineversion.cfm?&action=getFood&foodID=06A10093

Grieve M (1971) A modern herbal. Penguin. 2 Vols. Dover publications, New York, 919 pp

Guarrera PM (2005) Traditional phytotherapy in Central Italy (Marche, Abruzzo, and Latium). Fitoterapia 76(1):1–25

Hager TJ, Howard LR, Liyanage R, Lay JO, Prior RL (2008) Ellagitannin composition of blackberry as determined by HPLC-ESI-MS and MALDI-TOF-MS. J Agric Food Chem 56(3):661–669

Hákkinen S, Heinonen M, Kárenlampi S, Mykkánen H, Ruuskanen J, Törrönen R (1999) Screening of selected flavonoids and phenolic acids in 19 berries. Food Res Int 32:345–353

Hall HK (1990) Blackberry breeding. Plant Breed Rev 8:249–312

Hassimotto NM, Pinto Mda S, Lajolo FM (2008) Antioxidant status in humans after consumption of blackberry (*Rubus fruticosus* L.) juices with and without defatted milk. J Agric Food Chem 56(24):11727–11733

Heinonen IM, Meyer AS, Frankel EN (1998) Antioxidant activity of berry phenolics on human low-density lipoprotein and liposome oxidation. J Agric Food Chem 46(10):4107–4112

Huxley AJ, Griffiths M, Levy M (eds) (1992) The new RHS dictionary of gardening, vol 4. Macmillan, London

Jiao H, Wang SY (2000) Correlation of antioxidant capacities to oxygen radical scavenging enzyme activities in blackberry. J Agric Food Chem 48(11):5672–5676

Jordheim M, Enerstvedt KH, Andersen OM (2011) Identification of cyanidin 3-O-β-(6″-(3-hydroxy-3-methylglutaroyl) glucoside) and other anthocyanins from wild and cultivated blackberries. J Agric Food Chem 59(13):7436–7440

Komes D, Belščak-Cvitanović A, Horžić D, Rusak G, Likić S, Berendika M (2011) Phenolic composition and antioxidant properties of some traditionally used medicinal plants affected by the extraction time and hydrolysis. Phytochem Anal 22(2):172–180

Lemus I, García R, Delvillar E, Knop G (1999) Hypoglycaemic activity of four plants used in Chilean popular medicine. Phytother Res 13(2):91–94

Marquina MA, Corao GM, Araujo L, Buitrago D, Sosa M (2002) Hyaluronidase inhibitory activity from the polyphenols in the fruit of blackberry (*Rubus fruticosus* B.). Fitoterapia 73(7–8):727–729

Martini S, D'Addario C, Colacevich A, Focardi S, Borghini F, Santucci A, Figura N, Rossi C (2009) Antimicrobial activity against *Helicobacter pylori* strains and antioxidant properties of blackberry leaves (*Rubus ulmifolius*) and isolated compounds. Int J Antimicrob Agents 34(1):50–59

McGregor C (1998) Relationships between weedy and commercially grown *Rubus* species. Plant Prot Q 13:157–159

Moore JN, Skirvin RM (1990) Blackberry management. In: Galletta GJ, Himelrick DG (eds) Small fruit crop management. Prentice Hall, Englewood Cliffs, pp 214–244

Panizzi L, Caponi C, Catalano S, Cioni PL, Morelli I (2002) In vitro antimicrobial activity of extracts and isolated constituents of *Rubus ulmifolius*. J Ethnopharmacol 79(2):165–168

Pantelidis GE, Vasilakakis M, Manganaris GA, Diamantidis G (2007) Antioxidant capacity, phenol, anthocyanin and ascorbic acid contents in raspberries, blackberries, red currants, gooseberries and Cornelian cherries. Food Chem 102(3):777–783

Parsons WT, Cuthbertson EG (2001) Noxious weeds of Australia, 2nd edn. CSIRO, Collingwood, p 712

Porcher MH et al (1995–2020) Searchable World Wide Web Multilingual Multiscript Plant Name Database. Published by The University of Melbourne, Australia. http://www.plantnames.unimelb.edu.au/Sorting/Frontpage.html

Quave CL, Plano LR, Pantuso T, Bennett BC (2008) Effects of extracts from Italian medicinal plants on planktonic growth, biofilm formation and adherence of methicillin-resistant *Staphylococcus aureus*. J Ethnopharmacol 118(3):418–428

Saxholt E, Christensen AT, Møller A, Hartkopp HB, Hess Ygil K, Hels OH (2008) Danish food composition databank, revision 7. Department of Nutrition, National Food Institute, Technical University of Denmark. http://www.foodcomp.dk/

Scher J (2010) Federal Noxius Weed Disseminules of the U.S. http://keys.lucidcentral.org/keys/FNW/FNW%20seeds/html/fact%20sheets/Rubus%20fruticosus.htm. Accessed Nov 2010

Seeram NP (2008) Berry fruits for cancer prevention: current status and future prospects. J Agric Food Chem 56:630–635

Seeram NP, Adams LS, Zhang Y, Sand D, Heber D (2006) Blackberry, black raspberry, blueberry, cranberry, red raspberry and strawberry extracts inhibit growth and stimulate apoptosis of human cancer cells in vitro. J Agric Food Chem 2006(54):9329–9339

Seeram NP, Momin RA, Nair MG, Bourquin LD (2001) Cyclooxygenase inhibitory and antioxidant cyanidin glycosides in cherries and berries. Phytomedicine 8(5):362–369

Sellappan S, Akoh CC, Krewer G (2002) Phenolic compounds and antioxidant capacity of Georgia-grown blueberries and blackberries. J Agric Food Chem 50(8):2432–2438

Siriwoharn T, Wrolstad RE, Finn CE, Pereira CB (2004) Influence of cultivar, maturity, and sampling on blackberry (*Rubus* L. Hybrids) anthocyanins, polyphenolics, and antioxidant properties. J Agric Food Chem 52(26):8021–8030

Thompson MM (1997) Survey of chromosome number in *Rubus* (Rosaceae: Rosoideae). Ann Missouri Bot Gard 84:128–164

U.S. Department of Agriculture, Agricultural Research Service (USDA) (2011) USDA National Nutrient Database for Standard Reference, Release 24. Nutrient Data Laboratory Home Page, http://www.ars.usda.gov/ba/bhnrc/ndl

Vrhovsek U, Palchetti A, Reniero F, Guillou C, Masuero D, Mattivi F (2006) Concentration and mean degree of polymerization of *Rubus* ellagitannins evaluated by optimized acid methanolysis. J Agric Food Chem 54(12):4469–4475

Wang SY, Bowman L, Ding M (2008) Methyl jasmonate enhances antioxidant activity and flavonoid content in blackberries (*Rubus* sp.) and promotes antiproliferation of human cancer cells. Food Chem 107(3): 1261–1269

Wang SY, Jiao H (2000) Scavenging capacity of berry crops on superoxide radicals, hydrogen peroxide, hydroxyl radicals, and singlet oxygen. J Agric Food Chem 48(11):5677–5684

Wang SY, Lin HS (2000) Antioxidant activity in fruits and leaves of blackberry, raspberry, and strawberry varies with cultivar and developmental stage. J Agric Food Chem 48:140–146

Rubus idaeus

Scientific Name

Rubus idaeus L.

Synonyms

Batidea idaea (L.) Nieuwl., *Rubus idaeus* subsp. *vulgatus* Arrh.

Family

Rosaceae

Common/English Names

European Raspberry, European Red Raspberry, Nectar Raspberry, Raspberry, Raspberry Bush, Red Raspberry.

Vernacular Names

Aragonese: Chordón;
Arabic: Tût El 'Ullayq;
Brasil: Framboesa (Portuguese);
Bulgarian: Malína;
Catalan: Gerd;
Chinese: Fu Pen Zi, Shan Mei;
Czech: Malina, Maliník Obecný, Ostružiník Maliník;
Danish: Almindelig Hindbær, Hindbær, Hindbærbusk;
Dutch: Framboos, Frambozenstruik;
Eastonian: Harilik Vaarikas;
Esperanto: Frambo;
Finnish: Vaapukka, Vadelma, Vattu;
French: Framboise (Fruit), Framboisier (Plant), Ronce De L'ida;
German: Himbeere, Himbeerstrauch (Plant);
Greek: Smeura; Zmeouriá (Plant), Zméouro, Vatomura Vatomouriá (Plant), Vatómouro;
Haitian: Franbwaz;
Hebrew: Petel Adom;
Hungarian: Málna;
India: Rasabharii (Hindu), Gauriphal (Sanskrit);
Italian: Amponello, Lampone, Lampone Rosso, Malina, Rovo Ideo;
Japanese: Ki Ichigo, Razuberii, Reddo Razuberii, Yezo Ichigo;
Korean: Na Mu Ddal Gi;
Lithuanian: Paprastoji Avietė;
Mexico: Frutilla De La India (Spanish);
Norwegian: Bringebær;
Portuguese: Framboesa Comum (Fruit), Framboeseiro Comum (Plant);
Romanian: Zmeur (Fruit), Zmeură (Plant);
Russian: Malina Obyknovennaia;
Samogitian: Paprastoji Avėitė;
Serbian: Malína;
Slovaščina: Malinjak;
Spanish: Chordón, Frambuesa (Fruit), Frambueso (Plant);

Swedish: Hallon (Fruit), Hallonbuske (Plant), Hallonsnår (Plant);
Taiwan: Fu Pen Zi;
Turkish: Ağaç Çileği, Ahududu;
Ukrainian: Malína;
Upper Sorbian: Čerwjeny Malenowc;
Vietnamese: Quả Mâm Xôi (Fruit), Cây Mâm Xôi (Canes);
Walloon: Amponî;

Origin/Distribution

Red raspberry is native to Europe and northern Asia, Russia (eastward to Eastern Siberia) and Central Asia (Tienschan).

Plate 1 Leaves of red raspberry plant

Agroecology

Red raspberry is a temperate species with optimal growth in the moderate temperature range of 16–25°C and it is frost tolerant down to −40°C. It occurs in sparse forests, forest margins, thickets, valleys, slopes, meadows, roadsides, stream-sides, waste places from 500 to 2,500 m elevation in its native range. Raspberry prefers well drained, fertile loamy, subacid-neutral soil in humid sites that are protected from strong winds.

Plate 2 Red raspberry fruits and leaves (*upper surface*)

Edible Plant Parts and Uses

Red raspberries are grown for the fresh fruit market and for commercial processing into individually quick frozen (IQF) fruit, purée, juice, canned fruits, jams, marmalade, jellies, liqueur (wine and brandy), vinegar or as dried fruit used in a variety of grocery products. Fruits are relished fresh or with ice-cream. Dried leaves are used for tea.

Botany

A shrub, 1–2 m high with reddish-brown, terete, sparsely tomentose and prickled branchlets. Leaves are imparipinnate, petiolate (3–6 cm) with 5–7 leaflets and pubescent, linear stipules. Leaflets narrowly ovate or elliptic, 3–8 cm by 1.5–4.5 cm, tomentose abaxially and glabrous adaxially, base rounded to subcordate, apex acuminate, margin coarsely serrated (Plates 1, 2 and 3). Inflorescences terminal and racemose with several flowers in axillary clusters Flowers 1–1.5 cm across. Calyx with 5 erect, ovate-lanceolate sepals. Petals five, white, spatulate, puberulous or glabrous, base broadly clawed. Stamens many, shorter than petals with filaments broadened and flattened. Pistils shorter than stamens; ovary and base of style densely gray tomentose. Fruit red, subglobose, 1.5–2.5 cm across made up of an aggregate of numerous drupelets around a central core (Plates 2, 3 and 4), the drupelets separate from the core when picked, leaving a hollow fruit, pyrenes prominently pitted (Plate 4).

Plate 3 Red raspberry fruit and leaves (*lower surface*)

Plate 4 Harvested red raspberries

Nutritive/Medicinal Properties

The nutrient composition of raw red raspberries per 100 g edible portion was reported as (Saxholt et al. 2008): energy 228 kJ, total protein 1.4 g, total fat 1.4 g, saturated fatty acids 0.1 g, C 16:0 fatty acids 0.089 g, C 18:0 0.015 g; monounsaturated fatty acids 0.1 g, C 18:1 n-9 0.103 g; polyunsaturated fatty acids 0.9 g, C 18:2 n-6, 0.473 g, C18:3 n-3 0.0401 g; total omega 3 fatty acids 0.401 g, total omega 6 fatty acids 0.473 g, total carbohydrate 11.3 g, available carbohydrate 6.9 g, fructose 1.22 g, glucose 0.75 g, saccharose 0.08 g, total sugars 2.05 g, dietary fibre 4.4 g, moisture 85.9 g, vitamin A 3.50 RE, β-carotene equivalent 42 μg, vitamin E α-tocopherol 1.4 mg, thiamine 0.03 mg, riboflavin 0.05 mg, niacin 0.05 mg, vitamin B6 0.09 mg, pantothenic acid 0.24 mg, biotin 1.9 μg, folate 44 μg, vitamin C 24 mg, ash 0.5 g, Na 2 mg, K 228 mg, Ca 19.7 mg, Mg 17 mg, P 38 mg, Fe 0.55 mg, Cu 0.105 mg, Zn 0.34 mg, Iodine 0.4ug, Mn 1.2 mg, Cr 0.8 ug, Se 0.189 ug, Ni 17.9 μg and tryptophan 15 mg.

Flavonoids (kaempferol, quercetin, myricetin) and phenolic acids (p-coumaric, caffeic, ferulic, p-hydroxybenzoic, gallic and ellagic acids) were detected in the fruits of 19 berries (Hákkinen et al. 1999). Ellagic acid was the main phenolic compound in the berries of the genus *Rubus* (red raspberry, Arctic bramble and cloudberry) and genus *Fragaria* (strawberry). The data suggested berries to have potential as good dietary sources of quercetin or ellagic acid. Eleven anthocyanins, including cyanidin-3-sophoroside, cyanidin-3-(2(G)-glucosylrutinoside), cyanidin-3-glucoside, cyanidin-3-rutinoside, pelargonidin-3-sophoroside, pelargonidin-3-(2(G)-glucosylrutinoside), and pelargonidin-3-glucoside were identified in red raspberries (*R. idaeus*) (Mullen et al. 2002). Significant quantities of an ellagitannin, sanguiin H-6, were detected along with lower amounts of a second ellagitannin, lambertianin C, were detected. Other phenolic compounds that were detected included trace amounts of ellagic acid and its sugar conjugates along with one kaempferol- and four quercetin-based flavonol conjugates. The content of phenolic compounds varied widely and significantly between red raspberry (*Rubus idaeus*) cultivars (Anttonen and Karjalainen 2005). The quercetin content ranged from 0.32 (yellow cultivar) to 1.55 mg/100 g fresh weight (FW) (cv. Balder). The ellagic acid content varied from 38 (cv. Gatineau and cv. Nova) to 118 mg/100 g FW (cv. Ville). The total anthocyanin content varied from close to 0 (yellow cultivars) to 51 mg/100 g FW (cv. Gatineau). The content of total phenolics varied from 192 (cv. Gatineau) to 359 mg/100 g FW (cv. Ville). Among berry fruits, raspberries (*Rubus idaeus*) contained most xylitol, a non-sugar, a sugar alcohol, low calorie sweetener, about 400 μg/g (Makinen and Soderling 1980).

Eight anthocyanins were detected in raspberries, with cyanidin-3-O-sophoroside (375 nmol/g) the major anthocyanin followed by cyanidin-3-O-(2″-O-glucosyl) rutinoside, cyanidin-3-O-sambubioside, and cyanidin-3-O-glucoside (all three totalled 307 nmol/g). Other phenolics included pelargonidin-3-O-sophoroside (44 nmol/g), cyanidin-3-O-rutinoside (85 nmol/g), pelargonidin-3-O-glucoside and pelargonidin-3-O-("-O-glucosyl) rutinoside (both totalled 74 nmol/g) (Borges et al. 2010). The extract also contained the ellagitannins lambertianin C (322 nmol/g) and sanguin H-6 (1,030 nmol/g) along with trace amounts of ellagic acid derivatives and quercetin conjugates namely ellagic acid (11 nmol/g), ellagic acid-O-pentoside (10 nmol/g), ellagic acid-4-O-acetylxyloside (5.1 nmol/g) and quercetin-O-galactosylrhamnoside (7.5 nmol/g), quercetin-3-O-(2″-O-glucosyl) rutinoside (6.7 nmol/g), quercetin-3-O-galactoside (25 nmol/g) and quercetin-3-O-glucoside (28 nmol/g).

Studies found minimal changes in the volatile aroma composition of raspberry produced by the freezing process and frozen storage at −20°C for a 1 year (de Ancos et al. 2000b). Only a significant increase in extraction capacity was obtained for α-ionone (27%) and for caryophyllene (67%) in cv Heritage at 12 months of storage. The stability of anthocyanins to freezing and frozen storage depended on the seasonal period of harvest. Heritage and Autumn Bliss (early cultivars) were less affected by processing and long-term frozen storage (1 year), and the total pigment extracted showed the tendency to increase 17% and 5%, respectively. Rubi and Zeva (late cultivars) suffered a decreased trend on the total anthocyanin content of 4% for Rubi and 17.5% for Zeva. Cyanidin 3-glucoside was degraded most during processing and the storage period.

Cyanidin-3-sophoroside, cyanidin-3-(2(G)-glucosylrutinoside), cyanidin-3-sambubioside, cyanidin-3-rutinoside, cyanidin-3-xylosylrutinoside, cyanidin-3-(2(G)-glucosylrutinoside), and cyanidin-3-rutinoside were the main anthocyanin components in red raspberry (*Rubus idaeu*s L. var. Heritage) extracts (Chen et al. 2007). In addition, in comparison with the conventional solvent extraction, ultrasound-assisted process was found to be more efficient and rapid to extract anthocyanins from red raspberry. Eight anthocyanin cyanidin and pelargonidin glycosides: -3-sophoroside, -3-glucoside, -3-rutinoside and -3-glucosylrutinoside were quantified across two seasons and two environments in progeny from a cross between two *Rubus* subspecies, *Rubus idaeus* (cv. Glen Moy) x *Rubus strigosus* (cv. Latham) (Kassim et al. 2009). Significant seasonal variation was detected across pigments less for different growing environments within seasons.

Anthocyanin contents in 11 red raspberry varieties ranged from 76.22 to 277.06 mg per 100 g of dry weight (dw) and the content of the ellagitannin varied from 135.04 to 547.48 mg per 100 g of dw (Sparzak et al. 2010). The predominant anthocyanin in varieties Heritage and Willamette were cyanidin-3-O-sophoroside and cyanidin-3-O-glucoside, while in the other varieties the predominant compounds were cyanidin-3-O-rutinoside and cyanidin-3-O-(2G-O-glucosylrutinoside). Raspberry ketone (4-(4-hydroxyphenyl) butan-2-one; RK) was found to be a major aromatic compound of red raspberry (*Rubus idaeus*) (Morimoto et al. 2005).

The yield of lipid ratio was between 0.40% (cv ERZ9) and 0.63% (cv Heritage) indicating cultivated raspberry had higher lipid ratio than all wild materials (Celik and Ercisli 2009). The 11 red raspberry genotypes and one cultivar studied contained 10 major compounds, and statistically important differences was observed among genotypes on C16:0 (palmitic acid), C18:1 (oleic), C18:2 (linoleic) and C18:3 (linolenic). Linoleic acid (42.18–52.61%) and linolenic acid (17.83–24.10%) was the main fatty acids for all genotypes studied.

Cold-pressed marionberry, boysenberry, red raspberry, and blueberry seed oils were evaluated for their fatty acid composition, carotenoid content, tocopherol profile, total phenolic content (TPC), oxidative stability index (OSI), peroxide value, and antioxidant properties. Cold-pressed marionberry, boysenberry, red raspberry, and blueberry seed oils were found to contain significant levels of α-linolenic acid ranging from 19.6 to 32.4 g per 100 g of oil, along with a low ratio of n-6/n-3 fatty acids (1.64–3.99) (Parry et al. 2005). The total carotenoid content ranged from 12.5 to 30.0 μmol per kg oil. Zeaxanthin was the major carotenoid compound in all tested berry seed oils, along with β-carotene, lutein, and cryptoxanthin. Total tocopherol was 260.6–2276.9 μmol/kg oil, including α-tocopherol, γ-tocopherol and δ-tocopherol. OSI values were 20.07, 20.30, and 44.76 h for the marionberry, red raspberry, and boysenberry seed oils, respectively. The highest total phenol content of 2.0 mg gallic acid equivalents per gram of oil was observed in the red raspberry seed oil, while the strongest oxygen radical absorbance capacity was in boysenberry seed oil extract (77.9 μmol trolox equivalents per g oil). All tested berry seed oils directly reacted with and quenched DPPH radicals in a dose- and time-dependent manner. The data suggested that the cold-pressed berry seed oils may serve as potential dietary sources of tocopherols, carotenoids, and natural antioxidants. Seeds from red raspberry, black raspberry, boysenberry, Marion blackberry, and evergreen blackberry were found to have 6–7% protein and 11–18% oil (Bushman et al. 2004). The oils contained 53–63% linoleic acid, 15–31% linolenic acid, and 3–8% saturated fatty acids. The two smaller seeded raspberry species had higher percentages of oil, the lowest amounts of saturated fatty acid, and the highest amounts of linolenic acid. Antioxidant capacities were detected both for whole seeds and for cold-pressed oils but did not correlate to total phenolics or tocopherols. Ellagitannins and free ellagic acid were the main phenolics detected in all five caneberry species and were approximately threefold more abundant in the blackberries and the boysenberry than in the raspberries.

Oil yield from *Rubus idaeus* seed was 10.7% and its saponification number was 191; diene value 0.837; p-anisidine value 14.3; peroxide value 8.25 meq/kg; carotenoid content 23 mg/100 g; and viscosity of 26 mPa-s at 25°C (Oomah et al. 2000). Raspberry seed oil showed absorbance in the UV-B and UV-C ranges with potential for use as a broad spectrum UV protectant. The seed oil was rich in tocopherols with the following composition (mg/100 g): α-tocopherol 71, γ-tocopherol 272, δ-tocopherol 17.4, and total vitamin E equivalent of 97 mg. The oil had good oxidation resistance and storage stability. Lipid fractionation of crude raspberry seed oil yielded 93.7% neutral lipids, 3.5% phospholipids, and 2.7% free fatty acids. The main fatty acids of crude oil were C18:2 n-6 (54.5%), C18:3 n-3 (29.1%), C18:1 n-9 (12.0%), and C16:0 (2.7%).

Fruits of red raspberry were found to be rich in elagic acid (Daniel et al. 1989; Zafrilla et al. 2001; Juranic et al. 2005; Salinas-Moreno et al. 2009). The concentration of elagic acid in various fruits was reported as follows: strawberries (630 μg), raspberries (1,500 μg), blackberries (1,500 μg), walnuts (590 μg), pecans (330 μg), and cranberries (120 μg ellagic acid/g dry wt) (Daniel et al. 1989). In strawberries, 95.7% of the ellagic acid was found in the pulp while 4.3% was contained in the seeds. The seeds of raspberries contained 87.8% of the ellagic acid, and 12.2% was present in the pulp. The juice of both fruits contained negligible amounts of ellagic acid. The fruits of "Autumn Bliss" variety red raspberry fruits (*Rubus idaeus*) with deep red colour (degree 3 ripening) was found to have the highest free ellagic acid content, with a value of 5.38 mg/kg of fresh fruit, the anthocyanin profile was more complex as fruits advanced in maturity (Salinas-Moreno et al. 2009). In the immature fruits only four anthocyanins were observed, while in the completely mature fruits there were eight. In the completely mature fruits the anthocyanins with the highest relative percentages were: cyanidin 3-soforoside (46.2%) and cyianidin 3-(2-glucosyl rutinoside) (25.9%). The maximum levels of free ellagic acid and the highest number of anthocyanins were present in the completely mature raspberry fruits. From red raspberries, ellagic acid, its 4-arabinoside, its 4′ (4″-acetyl) arabinoside, and its 4′ (4″-acetyl) xyloside, as well as quercetin and kaempferol 3-glucosides, were identified (Zafrilla et al. 2001). All of the isolated compounds showed antioxidant activity. The flavonol content decreased slightly with processing and more markedly during storage of raspberry jams. The ellagic acid derivatives, with the exception of ellagic acid itself, remained quite stable with processing and during 6 months of jam storage. The content of free ellagic acid increased threefold during the storage period. The initial content (10 mg/kg of fresh weight of raspberries) increased twofold with processing, and it continued increasing up to

35 mg/kg after 1 month of storage of the jam. The increase observed in ellagic acid could be explained by a release of ellagic acid from ellagitannins with the thermal treatment.

The total amount of phenolic compounds in *Rubus idaeus* leaves varied from 0.3 to 2.2 mg of gallic acid equivalents (GAE) in 1 g of dry leaves (Dvaranauskaite et al. 2008). Quercetin glucuronide, quercetin-3-glucoside and quercetin glucosylrhamnoside (rutin) were identified in the leaf. The concentration of ellagic acid in the leaves of *Rubus* species: raspberry (2 wild and 13 cultivars) and blackberry (3 wild and 3 cultivars) after acid hydrolysis ranged from 2.06% to 6.89% (Gudej and Tomczyk 2004). The flavonoid content varied between 0.27% and 1.06%; quercetin and kaempferol were predominant in all samples. The leaves of raspberries were characterized by greater amounts of tannins (varying between 2.62% and 6.87%) than the leaves of other species.

Antioxidant Activity

In raspberry fruit, the antioxidant activity of vitamin C was 681 nmol Trolox/g, sanguine H-6 2,905 nmol Trolox/g, lambertianin C 886 nmol Trolox/g (Borges et al. 2010). Aanthocyanins present at a total concentration of 885 nmol/g contributed 16.5% to the overall AOC (antioxidant capacity), whereas vitamin C (1,014 nmol/g) contributed 10.5%. In contrast to the black currant and blueberry extracts, the ellagitannins lambertianin C and sanguiin H-6 were the main contributors to the AOC, being responsible for >58% of the total.

The amount of total phenolics varied between 617 and 4,350 mg/kg in fresh berries (blackberries, red raspberries, blueberries, sweet cherries and strawberries), as gallic acid equivalents (GAE) (Heinonen et al. 1998). In the copper-catalyzed in-vitro human low-density lipoprotein oxidation assay at 10 μM gallic acid equivalents (GAE), berry extracts inhibited hexanal formation in the order: blackberries > red raspberries > sweet cherries > blueberries > strawberries. In the copper-catalyzed in-vitro lecithin liposome oxidation assay, the extracts inhibited hexanal formation in the order: sweet cherries > blueberries > red raspberries > blackberries > strawberries. Red raspberries were more efficient than blueberries in inhibiting hydroperoxide formation in lecithin liposomes. HPLC analyses showed high anthocyanin content in blackberries, hydroxycinnamic acid in blueberries and sweet cherries, flavonol in blueberries, and flavan-3-ol in red raspberries. The antioxidant activity for LDL was associated directly with anthocyanins and indirectly with flavonols, and for liposome it correlated with the hydroxycinnamate content. Berries thus may contribute a significant source of phenolic antioxidants with potential health effects.

Fruit and vegetables rich in anthocyanins (e.g. strawberry, raspberry and red plum) demonstrated the highest antioxidant activities, followed by those rich in flavanones (e.g. orange and grapefruit) and flavonols (e.g. onion, leek, spinach and green cabbage), while the hydroxycinnamate-rich fruit (e.g. apple, tomato, pear and peach) consistently elicited the lower antioxidant activities (Proteggente et al. 2002) . The TEAC (Trolox Equivalent Antioxidant Capacity), the FRAP (Ferric Reducing Ability of Plasma) and ORAC (Oxygen Radical Absorbance Capacity) values for each extract were relatively similar and well-correlated with the total phenolic and vitamin C contents. The antioxidant activities TEAC in terms of 100 g FW uncooked portion size were in the order: strawberry >> raspberry = red plum >> red cabbage >>> grapefruit = orange > spinach > broccoli > green grape approximately/= onion > green cabbage > pea > apple > cauliflower approximately/= tomato approximately/= peach = leek > banana approximately/= lettuce. Blackberries (*Rubus* sp.) and strawberries (*Fragaria × ananassa*) had the highest ORAC values during the green stages, whereas red raspberries (*Rubus idaeus*) had the highest ORAC activity at the ripe stage (Wang and Lin 2000). All of five types of caneberries [evergreen blackberries (*Rubus laciniatus*), marionberries (*Rubus ursinus*), boysenberries (*Rubus ursinus x idaeus*), red raspberries (*Rubus idaeus*), and black raspberries (*Rubus occidentalis*)] were found to have high oxygen radical absorbance capacity (ORAC) activity ranging from 24 to 77.2 μmol of Trolox equiv/g of fresh berries (Wada and Ou 2002).

Anthocyanin content ranged from 0.65 to 5.89 mg/g, and phenolics ranged from 4.95 to 9.8 mg/g. Black raspberries had the highest ORAC, anthocyanin and phenolic contents. Only red raspberries had detectable amounts of procyanidin oligomers (monomer, dimers, and trimers). All berries had high levels of ellagic acid (47–90 mg/g), but boysenberries had the highest level prior to hydrolysis. The data indicated that these caneberries were high in antioxidant activity and were rich sources of anthocyanins and phenolics.

Among the berry fruits analysed, red raspberry (*Rubus idaeus*), black raspberry (*Rubus occidentalis*), and strawberry (*Fragaria x ananassa*), black raspberries and strawberries had the highest ORAC (oxygen radical absorbance capacity) values during the green stages, whereas red raspberries had the highest ORAC activity at the ripe stage (Wang and Lin 2000). Total anthocyanin content increased with maturity for all species of fruits. Compared with fruits, leaves were found to have higher ORAC values. In fruits, ORAC values ranged from 7.8 to 33.7 μmol of Trolox equivalents (TE)/g of fresh berries (35.0–162.1 μmol of TE/g of dry matter), whereas in leaves, ORAC values ranged from 69.7 to 182.2 μmol of TE/g of fresh leaves (205.0–728.8 μmol of TE/g of dry matter). As the leaves become older, the ORAC values and total phenolic contents decreased. The results showed a linear correlation between total phenolic content and ORAC activity for fruits and leaves. For ripe berries, a linear relationship existed between ORAC values and anthocyanin content. Of the ripe fruits tested, on the basis of wet weight of fruit, cv. Jewel black raspberry and blackberries were the richest source for antioxidants. On the basis of the dry weight of fruit, strawberries had the highest ORAC activity followed by black raspberries (cv. Jewel), blackberries, and red raspberries.

Raspberry cultivar 'Heritage' had the highest total phenolic content (512.70/100 g fruit) followed by 'Kiwigold' (451.06 mg/100 g fruit), 'Goldie' (427.51 mg/100 g fruit) and 'Anne' (359.19 mg/100 g fruit) (Liu et al. 2002; Weber et al. 2002). Similarly, 'Heritage' contained the highest total flavonoids (103.41 mg/100 g fruit) followed by 'Kiwigold' (87.33 mg/100 g fruit), 'Goldie' (84.16 mg/100 g fruit) and 'Anne' (63.53 mg/100 g fruit). 'Heritage' had the highest a/b colorimeter ratio and the darkest colored juice with the highest phenolic/flavonoid content, and 'Anne' had the lowest phytochemical content, the palest color, and lowest a/b ratio. 'Heritage' had the highest total antioxidant activity, followed by 'Kiwigold' and 'Goldie'. 'Anne' had the lowest antioxidant activity of the cultivars tested. The antioxidant activity of each of the cultivars was directly related to the total amount of phenolics and flavonoids. In another study, the highest phenolic compounds were found in wild Yayla raspberry ecotype (26.66 gallic acid equivalents GAE /mg extract). Whilst, the highest flavonoids were determined in wild Yedigöl ecotype (6.09 quercetin equivalents QE/mg extract) (Gülçin et al. 2011). The compounds found in lyophilized aqueous extracts of domesticated and wild ecotypes of raspberry fruits caffeic acid, ferulic acid, syringic acid, ellagic acid, quercetin, α-tocopherol, pyrogallol, p-hydroxybenzoic acid, vanillin, p-coumaric acid, gallic acid, and ascorbic acid. Antioxidant assays using different in vitro assays including DPPH˙, ABTS˙+, DMPD˙+, and O^{-2} radical scavenging activities, H_2O_2 scavenging activity, ferric (Fe^{3+}) and cupric ions (Cu^{2+}) reducing abilities, ferrous ions (Fe^{2+}) chelating activity, showed that p-coumaric acid was the main phenolic acid responsible for the antioxidant and radical scavenging activity of lyophilized aqueous extracts of domesticated and wild ecotypes of raspberry fruits.

Studies by de Ancos et al. (2000a) found that cultivars, freezing time and storage impacted on the ellagic acid, total phenolic, and vitamin C contents of raspberry. Ellagic acid [207–244 mg/kg fresh weight (FW)], total phenolic (137–1,776 mg/kg FW), and vitamin C (221–312 mg/kg FW) contents in raw material were higher in the late cultivars Zeva and Rubi than in the early cultivars Autumn Bliss and Heritage. At the end of long-term −20°C storage (12 months), no significant change of total phenolic content extracted was observed, but significant decreases of 14–21% in ellagic acid and of 33–55% in vitamin C were quantified. Free radical scavenging capacity measured as antiradical efficiency (AE) was found to depend on the seasonal period of harvest. Late cultivars, Rubi (6.1×10^{-4}) and Zeva (10.17×10^{-4}), showed higher AE than early cultivars, Heritage (4.02×10^{-4}) and Autumn Bliss (4.36×10^{-4}). The freezing process produced a decrease of AE values in the four cultivars ranging between 4% and 26%.

The ellagitannin, sanguiin H-6 was a major contributor to the antioxidant capacity of raspberries together with vitamin C and the anthocyanins (Mullen et al. 2002). Berries such as raspberry (*Rubus idaeus*), bilberry (*Vaccinium myrtillus*), lingonberry (*Vaccinium vitis-idaea*), and black currant (*Ribes nigrum*) were found to be rich in monomeric and polymeric phenolic compounds providing protection toward both lipid and protein oxidation as assessed by a lactalbumin-liposome system (Viljanen et al. 2004). In raspberries, ellagitannins were responsible for the antioxidant activity. While the antioxidant effect of berry proanthocyanidins and anthocyanins was dose-dependent, ellagitannins appeared to be equally active at all concentrations. Ferric reducing antioxidant power (FRAP) values of raspberry (*Rubus idaeus*), blackberry (*Rubus fructicosus*), raspberry × blackberry hybrids, red currant (*Ribes sativum*), gooseberry (*Ribes glossularia*) and Cornelian cherry (*Cormus mas*) cultivars ranged from 41 to 149 μmol ascorbic acid/g dry weight and protection of deoxyribose ranged from 16.1% up to 98.9% (Pantelidis et al. 2007). Anthocyanin content ranged from 1.3 mg in yellow-coloured fruit, up to 223 mg cyanidin-3-glucoside equivalents 100/g fresh weight in Cornelian cherry, whereas phenol content ranged from 657 up to 2,611 mg gallic acid equivalents/100gdry weight. Ascorbic acid content ranged from 14 up to 103 mg/100 g fresh weight.

The total amount of phenolic compounds in *R. idaeus* fruits varied from 5.6 to 13.7 mg of gallic acid equivalents per 1 g of plant extract (Dvaranauskaite et al. 2006). All tested raspberry fruit extracts were antioxidatively active; their radical scavenging capacity at the applied concentrations varied from 52.9% to 92.6% in DPPH·reaction system and from 52.5% to 97.8% in ABTS·+ system. The dominant antioxidants in red raspberry fruit could be classified as anthocyanins, ellagitannins, and proanthocyanidin-like tannins (Beekwilder et al. 2005). During fruit ripening, some anthocyanins were freshly produced, while others, like cyanidin-3-glucoside, were already present early in fruit development. The level of tannins, both ellagitannins and proanthocyanidin-like tannins, was reduced strongly during fruit ripening. Among the 14 cultivars, major differences (>20-fold) were observed in the levels of pelagonidin type anthocyanins and some proanthocyanidin type tannins. The content of ellagitannins varied approximately threefold. The antioxidant capacity among wild and cultivated red raspberry samples averaged 14.6 and 14.1 μmol TE/g FW using FRAP and TEAC methods, respectively (Çekiç and Özgen 2010). Significant variability was found for antioxidant capacity, total phenolics, total monomeric anthocyanins, organic acids and sugars of wild raspberries.

Thirty-seven compounds comprising flavanol monomers and oligomers, as well as varieties of ellagitannin components were identified in *R. idaeus* seed methanol extracts (Godevac et al. 2009). Treatment of human lymphocytes with the seed extracts induced a significant decrease in the frequency of micronuclei by 80%. The results demonstrated that the constituents of the seed extracts may be important in the prevention of oxidative lymphocyte damage by reactive oxygen species and may also reduce the level of DNA damage. Further the findings support the potential benefits of polyphenolic compounds from raspberry seeds as efficient antioxidants.

All raspberry (*Rubus idaeus*) leaf ethanol extracts were found to be bioactive, with radical scavenging capacity at the used concentrations from 20.5% to 82.5% in DPPH* reaction system and from 8.0% to 42.7% in ABTS⁺ reaction (Venskutonis et al. 2007). The total amount of phenolic compounds in the leaves varied from 4.8 to 12.0 mg of gallic acid equivalents (GAE) in 1 g of plant extract. Quercetin glucuronide, quercetin-3-O-glucoside and rutin were identified in the extracts

Anticancer Activity

The anticancer potential of berries fruits such as blackberries, black raspberries, blueberries, cranberries, red raspberries, and strawberries had been associated partially, to a multitude of bioactive phytochemicals such as polyphenols (flavonoids, proanthocyanidins, ellagitannins, gallotannins, phenolic acids), stilbenoids, lignans, and triterpenoids (Seeram 2008). Studies showed that the anticancer effects of berry bioactives were partially mediated through their abilities to counteract, reduce, and also repair damage resulting from oxidative stress and inflammation. In addition, berry bioactives also regulated carcinogen and xenobiotic metabolizing enzymes, various transcription and growth factors, inflammatory cytokines, and subcellular signaling pathways of cancer cell proliferation, apoptosis, and tumour angiogenesis. Berry phytochemicals may also potentially sensitize tumour cells to chemotherapeutic agents by inhibiting pathways that lead to treatment resistance, and berry fruit consumption may provide protection from therapy-associated toxicities.

The major classes of phenolics found in blackberry, black raspberry, blueberry, cranberry, red raspberry and strawberry were anthocyanins, flavonols, flavanols, ellagitannins, gallotannins, proanthocyanidins, and phenolic acids (Seeram et al. 2006). With increasing concentration of berry extract from 25 to 200 µg/ml, increasing inhibition of cell proliferation in the human oral (KB, CAL-27), breast (MCF-7), colon (HT-29, HCT116), and prostate (LNCaP) tumour cell lines were observed, with different degrees of potency between cell lines.

A tiliroside, kaempferol-3-O-β-D-(6″-E-p-coumaroyl)-glucopyranoside was isolated from the methanolic extract *Rubus idaeus* plant (Nowak 2003). The compound exhibited cytotoxic activity for human leukaemic cell lines and anti-complement activity. Raspberry fruit extracts significantly inhibited in a dose-dependent manner the proliferation of HepG2 human liver cancer cells (Liu et al. 2002; Weber et al. 2002). The antiproliferative activity of the extract equivalent to 50 mg for cv 'Anne', 'Goldie', 'Heritage', and 'Kiwigold' fruit inhibited the proliferation of those cells by 70.33%, 89.43%, 87.96% and 87.55%, respectively. No significant relationship was found between antiproliferative activity and the total amount of phenolics/flavonoids. Studies showed that water extracts of seeds or pulp of five different raspberry cultivars: K81-6, Latham, Meeker, Tulameen and Willamette rich in ellagic acid possessed the potential for antiproliferative action against malignant human colon carcinoma LS174 cells in-vitro (Juranic et al. 2005). The antiproliferative action of seeds extract was correlated with its content of ellagic acid. The cytotoxic activity of seeds extracts was not pronounced on normal human PBMC.

Studies showed that dietary ellagic acid significantly inhibited the metabolism of [3H] benzo[a]pyrene (BP) by cultured primary keratinocytes prepared from BALB/C mouse epidermis (Mukhtar et al. 1984). Varying concentrations of ellagic acid added to the keratinocyte cultures resulted in a dose-dependent inhibition of the cytochrome P-450-dependent monooxygenases aryl hydrocarbon hydroxylase (AHH) and 7-ethoxycoumarin-O-deethylase (ECD). The results indicated that cultured primary mouse keratinocytes offered a useful model system for studying factors affecting the metabolic activation and detoxification of polycyclic aromatic hydrocarbon carcinogens in the epidermis, and that polyphenolic compounds such as ellagic acid may prove useful in modulating the risk of cutaneous cancer arising from exposure to these environmental chemicals.

Mandal and Stoner (1990) reported that ellagic acid inhibited N-nitrosobenzylmethylamine (NBMA) tumorigenesis in the rat esophagus. Administration of ellagic acid in a semi-purified diet at concentrations of 0.4 and 4 g/kg significantly reduced the average number of NBMA-induced esophageal tumours after 20 and 27 weeks of the bioassay. Ellagic acid exhibited inhibitory effects toward preneoplastic lesions as well as neoplastic lesions. Dietary ellagic acid (EA), had been shown to reduce the incidence of N-2-fluorenyl acetamide-induced hepatocarcinogenesis in rats and N-nitrosomethylbenzylamine (NMBA)-induced rat esophageal tumours (Ahn et al. 1996). Further studies demonstrated that EA caused a fall in total hepatic P450 with a significant effect on hepatic P450 2E1, enhanced some hepatic phase II enzyme activities (GST, NAD(P)H: QR and UDPGT) and decreased hepatic mEH expression. It also inhibited the catalytic activity of some P450 isozymes in-vitro.

Thus, the chemoprotective effect of EA against various chemically induced cancers may involve decreases in the rates of metabolism of these carcinogens by phase I enzymes, in addition to effects on the expression of phase II enzymes, thereby enhancing the ability of the target tissues to detoxify the reactive intermediates.

Results of studies by Narayanan and Re (2001) suggested that growth inhibition of colon cancer cells (SW 480) by dietary ellagic acid was mediated by signaling pathways that mediated DNA damage, triggered p53, which in turn activated p21 and at the same time altered the growth factor expression, resulting in the down regulation of mitogenic insulin like growth factor IGF-II and caused apoptotic cell deaths. The study of Falsaperla et al. (2005) suggested that the use of ellagic acid as support therapy reduced chemotherapy induced toxicity, in particular neutropenia, in hormone refractory prostate cancer (HRPC) patients. Ellagitannins were reported to be hydrolyzed to ellagic acid under physiological conditions in-vivo and then ellagic acid in turn was gradually metabolized by the intestinal microbiota to produce different types of urolithins (Landete 2011). Urolithinshad been reported to be not potent antioxidants as ellagitannins. In contrast urolithins could display estrogenic and/or anti-estrogenic activity and tissue disposition studies revealed urolithins to be rich in prostate, intestinal, and colon tissues in mouse, which could explain why urolithins inhibit prostate and colon cancer cell growth.

In HeLa cells co-transfected with an estrogen response element (ERE)-driven luciferase (Luc) reporter gene and an ERα- or ERβ- expression vector, ellagic acid at low concentrations (10^{-7} to 10^{-9} M) displayed a small but significant estrogenic activity via ERα, whereas it was a complete estrogen antagonist via ERβ (Papoutsi et al. 2005). Further evaluation revealed that ellagic acid was a potent antiestrogen in MCF-7 breast cancer-derived cells. Moreover, ellagic acid induced nodule mineralization in an osteoblastic cell line (KS483), an effect that was abolished by the estrogen antagonist. These findings suggested that ellagic acid may be a natural selective estrogen receptor modulator (SERM).

Antinflammatory Activity

Anthocyanins from raspberries *Rubus idaeus* and sweet cherries *Prunus avium* demonstrated 45% and 47% cyclooxygenase-I and cyclooxygenase-II inhibitory activities, respectively, when assayed at 125 μg/mL (Seeram et al. 2001). The cyclooxygenase inhibitory activities of anthocyanins from these fruits were comparable to those of ibuprofen and naproxen (antiinflammatory agents) at 10 μM concentrations. Anthocyanins cyanidin-3-glucosylrutinoside and cyanidin-3-rutinoside were present in both cherries and raspberry. The yield of the two pure anthocyanins in 100 g raspberries was 13.5 mg, respectively.

Antiurolithiasis Activity

Administration of a herbal decoction of *R. idaeus* to nephrolithic mice for 12 days caused a significant reduction in urinary oxalate, calcium and phosphorus values compared to untreated mice, while creatinine excretion increased (Ghalayini et al. 2011). Serum oxalate, calcium and creatinine were significantly reduced, while phosphorus was not significantly altered. Kidney content of calcium was higher in the untreated group. Mice in treated groups at 12 days had significantly more superoxide dismutase, catalase, glutathione reductase (GSH) and G6PD activities than the untreated group. Hyperoxaluria-induced generation of malondialdehyde (MDA) and protein carbonyls was significantly prevented in the treated groups. A significantly high content of vitamin E was found in the herbal treated groups. The histology showed more calcium oxalate deposition in the kidneys of untreated animals.

Anticariogenic Activity

Raspberry contain xylitol (Makinen and Soderling 1980), an anticariogenic polyol, non-sugar sweetener that has been approved for use in foods, chew gum, toothpaste and other items in many countries (Maguire and Rugg-Gunn 2003). Xylitol was reported to exhibit dental health benefits which were superior to other polyols in all areas where polyols had been shown to have an effect. Xylitol's specific effects on oral flora and especially on certain strains of mutans Streptococci augmented its caries-preventive profile and gave it a unique role in preventive strategies for dental health. The inhibition of mother/child transmission of cariogenic oral flora by xylitol leading to reduced caries development in young children was found to be caries-preventive.

Diuretic Activity

Studies showed that the methanol extract of raspberry (*R. idaeus*) fruits exerted diuretic effect on rats (Zhang et al. 2011). Compared to the control group, significant increase in urine volume was observed from rats treated with wild raspberry methanol extract. There was no increase in potassium excretion following administration of the methanol extract suggesting that this extract had potassium-sparing properties. The results partially elucidated the use of raspberry as a cure for renal diseases in Chinese traditional medical practice.

Antiobesity Activity

Raspberry ketone (4-(4-hydroxyphenyl) butan-2-one; RK) prevented the high-fat-diet-induced increases in body weight and the weights of the liver and visceral adipose tissues (epididymal, retroperitoneal, and mesenteric) (Morimoto et al. 2005). RK also decreased these weights and hepatic triacylglycerol content after they had been increased by a high-fat diet. RK significantly increased norepinephrine-induced lipolysis associated with the translocation of hormone-sensitive lipase from the cytosol to lipid droplets in rat epididymal fat cells. The findings suggested that raspberry ketone prevented and improved obesity and fatty liver by altering the lipid metabolism, or more specifically, in increasing norepinephrine-induced lipolysis in white adipocytes.

Relaxant/Spasmolytic Activity

As a pregnancy tonic, an infusion of dried raspberry leaves have been used by pregnant women in England, China, Europe, and North America to allay the pains of labour (Whitehouse 1941; Lieberman 1995; McFarlin et al. 1999). In a national survey of herbal preparation use by nurse-midwives for labour stimulation conducted by the American College of Nurse-Midwives 90 certified nurse-midwives who used herbal preparations to stimulate labour, reported that 64% used blue cohosh, 45% used black cohosh, 63% used red raspberry leaf, 93% used castor oil, and 60% used evening primrose oil. Burn and Withell (1941) first demonstrated that the raspberry leaf infusion caused a relaxation of the uterus in the non-pregnant cat and a fall in blood pressure. In subsequent experiments the extract produced a contraction of the isolated uteri of both the cat and the guinea-pig. The active principle was identified as the alkaloid, fragarine (Whitehouse 1941). Studies in human puerperal uterus confirmed that fragarine inhibited uterine contraction. Contractions were diminished in force and frequency, secondary contractions were eliminated, and such contractions as occurred were evenly spaced. Whitehouse added that in the absence of more elegant preparation, crude raspberry-leaf tea was being used in one of the Worcestershire maternity hospitals, and the nursing staff report favourably upon its effect in "making things easier".

Beckett et al. (1954) reported that raspberry leaf extracts contained a number of active constituents including a smooth muscle stimulant, an anticholinesterase, and a "spasmolytic". Bamford et al. (1970) demonstrated that dried raspberry leaf extract inhibited uteri of pregnant rat but had no effect on non-pregnant rats. Patel et al. (1995) found that raspberry leaf extract caused a relaxation of intestinal smooth muscles in-vitro. The methanol extract of raspberry leaves exhibited the largest in-vitro relaxant activity on transmurally stimulated guinea-pig ileum (Rojas-Vera et al. 2001). The fractions eluted with chloroform lacked relaxant activity. Samples eluted with chloroform: methanol (95:5) had moderate relaxant activity, while a more polar solvent mixture (chloroform: methanol 50:50) provided strong dose dependent responses. Vasorelaxation activity was restricted to raspberry fruit fractions containing the ellagitannins, lambertianin C and sanguiin H-6 (Mullen et al. 2002). Zheng et al. (2010), found that in pregnant rats red raspberry leaf tea had variable effects on pre-existing oxytocin-induced contractions, sometimes augmenting oxytocin's effect and sometimes causing augmentation followed by inhibition. Their results did not support the hypothesis that red raspberry leaf augmented labour by a direct effect on uterine contractility. Johnson et al. (2009) found that raspberry leaf use during pregnancy was associated with increased gestation length and accelerated reproductive development in the F1 offspring of Wistar rats, raising concerns on the long-term consequences for the health of the offspring and the safety of this herbal preparation for use during pregnancy.

A retrospective observational design study of 108 mothers suggested that women who ingested raspberry leaf might be less likely to receive an artificial rupture of their membranes, or require a caesarean section, forceps or vacuum birth than the women in the control group (Parsons et al. 1999). The findings suggested also that the raspberry leaf herb consumed by women during their pregnancy for the purpose of shortening labour had no identified side effects for the women or their babies. In a double-blind, randomized, placebo-controlled trial, Simpson et al. (2001) found that raspberry leaf, consumed in tablet form, caused no adverse effects for mother or baby, but contrary to popular belief, did not shorten the first stage of labour. The only clinically significant findings were a shortening of the second stage of labour (mean difference=9.59 min) and a lower rate of forceps deliveries between the treatment group and the control group (19.3% vs. 30.4%). No significant relationship was found between tablet consumption and birth outcomes.

After reviewing 12 original publications on the efficacy – safety of the use of raspberry leaves in pregnancy, Holst et al. (2009) concluded that evidence was not convincingly documented and recommendations made of its use was still questionable and proposed suggestions for future studies.

Mutagenic/Antimutagenic Activity

Raspberry (*Rubus idaeus*) extract and ellagitannin and anthocyanin fractions did not show any mutagenic effects in the miniaturized Ames test and were not cytotoxic to Caco-2 cells at the tested concentrations (Kreander et al. 2006). However, the anti-mutagenic properties were changed (i.e. decreased mutagenicity of 2-nitrofluorene in strain TA98, and slightly increased mutagenicity of 2-aminoanthracene in strain TA100) with metabolic activation.

Drug Interaction Activity

Raspberry extract and fractions were found to affect the permeability of some drugs depending on the components (Kreander et al. 2006). The apical-to-basolateral permeability across Caco-2 monolayers of highly permeable verapamil was mostly affected (decreased) during co-administration of the raspberry extract or the ellagitannin fraction. Ketoprofen permeability was decreased by the ellagitannin fraction. Consumption of food rich in phytochemicals, as demonstrated here with chemically characterized raspberry extract and fractions, with well-absorbing drugs would seem to affect the permeability of some of these drugs depending on the components.

Traditional Medicinal Uses

The fruit is antiscorbutic and diuretic, fresh raspberry juice with a little honey, makes an excellent refrigerant beverage for fever. A decongestant face-mask made from the fruit is used cosmetically to alleviate reddened skin. Tea made from the leaves of red raspberry has been used for centuries as a folk medicine to treat wounds, diarrhoea, colic pain and as a uterine. An infusion of dried raspberry leaves has been used as a pregnancy tonic by women to ease labour pains and relieve dysmenorrhea. Leaf and root extracts are regarded to be anti-inflammatory, astringent, decongestant, ophthalmic, oxytocic and stimulant. Leaf and root extracts are used as a gargle to treat tonsillitis and mouth inflammations, as a poultice and wash to treat sores, conjunctivitis, minor wounds, burns, scalds and varicose ulcers.

Other Uses

The plant (especially the leaves) is rich in tannin. A purple to dull blue dye is obtained from the fruit. A fibre obtained from the stems is used in making a kind of brown-coloured paper.

Comments

Red raspberry is commonly propagated from suckers, root cuttings, tip layering and divisions of the stocks.

Selected References

Ahn D, Putt D, Kresty L, Stoner GD, Fromm D, Hollenberg PF (1996) The effects of dietary ellagic acid on rat hepatic and esophageal mucosal cytochromes P450 and phase II enzymes. Carcinogenesis 17(4):821–828

Anttonen MJ, Karjalainen RO (2005) Environmental and genetic variation of phenolic compounds in red raspberry. J Food Compos Anal 18(8):759–769

Bamford DS, Percival RC, Tothill AU (1970) Raspberry leaf tea: a new aspect to an old problem. Br J Pharmacol 40(1):161P–162P

Beckett AH, Belthle FW, Fell KR, Lockett MF (1954) The active constituents of raspberry leaves: a preliminary investigation. J Pharm Pharmacol 6(1):785–796

Beekwilder J, Jonker H, Meesters P, Hall RD, Van Der Meer IM, De Vos CHR (2005) Antioxidants in raspberry: on-line analysis links antioxidant activity to a diversity of individual metabolites. J Agric Food Chem 53:3313–3320

Bell LA (1988) Plant fibres for papermaking. Liliaceae Press, McMinnville, 60 pp

Borges G, Degeneve A, Mullen W, Crozier A (2010) Identification of flavonoid and phenolic antioxidants in black currants, blueberries, raspberries, red currants, and cranberries. J Agric Food Chem 58:3901–3909

Bown D (1995) Encyclopaedia of herbs and their uses. Dorling Kindersley, London, 424 pp

Burn JH, Withell ER (1941) A principle in raspberry leaves which relaxes uterine muscle. Lancet 2:1–3

Bushman BS, Phillips B, Isbell T, Ou B, Crane JM, Knapp SJ (2004) Chemical composition of caneberry (*Rubus* spp.) seeds and oils and their antioxidant potential. J Agric Food Chem 52(26):7982–7987

Çekiç C, Özgen M (2010) Comparison of antioxidant capacity and phytochemical properties of wild and cultivated red raspberries (*Rubus idaeus* L.). J Food Compos Anal 213(6):540–544

Celik F, Ercisli S (2009) Lipid and fatty acid composition of wild and cultivated red raspberry fruits (*Rubus idaeus* L.). J Med Plant Res 3(8):583–585

Chen F, Sun Y, Zhao G, Liao X, Hu X, Wu J, Wang Z (2007) Optimization of ultrasound-assisted extraction of anthocyanins in red raspberries and identification of anthocyanins in extract using high-performance liquid chromatography-mass spectrometry. Ultrason Sonochem 14(6):767–778

Chevallier A (1996) The encyclopedia of medicinal plants. Dorling Kindersley, London, 336 pp

Daniel EM, Krupnick AS, Heur YH, Blinzler JA, Nims RW, Stoner GD (1989) Extraction, stability, and quantitation of ellagic acid in various fruits and nuts. J Food Compos Anal 21(4):338–349

de Ancos B, González EM, Cano MP (2000a) Ellagic acid, vitamin c, and total phenolic contents and radical scavenging capacity affected by freezing and frozen storage in raspberry fruit. J Agric Food Chem 48(10): 4565–4570

de Ancos B, Ibañez E, Reglero G, Cano MP (2000b) Frozen storage effects on anthocyanins and volatile compounds of raspberry fruit. J Agric Food Chem 48(3):873–879

Dvaranauskaite A, Venskutonis PR, Labokas J (2006) Radical scavenging activity of raspberry (*Rubus idaeus* L.) fruit extracts. Acta Aliment 35(1):73–83

Dvaranauskaite A, Venskutonis PR, Labokas J (2008) Comparison of quercetin derivatives in ethanolic extracts of red raspberry (*Rubus idaeus* L.) leaves. Acta Aliment 37(4):449–461

Falsaperla M, Morgia G, Tartarone A, Ardito R, Romano G (2005) Support ellagic acid therapy in patients with hormone refractory prostate cancer (HRPC) on standard chemotherapy using vinorelbine and estramustine phosphate. Eur Urol 47(4):449–454

Foster S, Duke JA (1998) A field guide to medicinal plants in Eastern and Central N. America. Houghton Mifflin Co, Boston, 366 pp

Ghalayini IF, Al-Ghazo MA, Harfeil MNA (2011) Prophylaxis and therapeutic effects of raspberry (*Rubus idaeus*) on renal stone formation in Balb/c mice. Int Braz J Urol 37(2):259–267

Godevac D, Tesević V, Vajs V, Milosavljević S, Stanković M (2009) Antioxidant properties of raspberry seed extracts on micronucleus distribution in peripheral blood lymphocytes. Food Chem Toxicol 47(11): 2853–2859

Grieve M (1971) A modern herbal. Penguin. 2 Vols. Dover publications, New York, 919 pp

Gudej J, Tomczyk M (2004) Determination of flavonoids, tannins and ellagic acid in leaves from *Rubus* L. species. Arch Pharm Res 27(11):1114–1119

Gülçin I, Topal F, Çakmakçı R, Bilsel M, Gören AC, Erdogan U (2011) Pomological features, nutritional quality, polyphenol content analysis, and antioxidant properties of domesticated and 3 wild ecotype forms of raspberries (*Rubus idaeus* L.). J Food Sci 76(4): C585–C593

Hákkinen S, Heinonen M, Kárenlampi S, Mykkánen H, Ruuskanen J, Törrönen R (1999) Screening of selected flavonoids and phenolic acids in 19 berries. Food Res Int 32:345–353

Heinonen IM, Meyer AS, Frankel EN (1998) Antioxidant activity of berry phenolics on human low-density lipoprotein and liposome oxidation. J Agric Food Chem 46(10):4107–4112

Holst L, Haavik S, Nordeng H (2009) Raspberry leaf– should it be recommended to pregnant women? Complement Ther Clin Pract 15(4):204–208

Huxley AJ, Griffiths M, Levy M (eds) (1992) The new RHS dictionary of gardening (4 vols). MacMillan, London

Janick J, Moore JN (eds) (1975) Advances in fruit breeding. Purdue Univ. Press, West Lafayette, 623 pp

Jennings DL (1988) Raspberries and blackberries: their breeding, disease and growth. Academic Press, London, 230 pp

Johnson JR, Makaji E, Ho S, Xiong B, Crankshaw DJ, Holloway AC (2009) Effect of maternal raspberry leaf consumption in rats on pregnancy outcome and the fertility of the female offspring. Reprod Sci 16(6): 605–609

Juranic Z, Zizak Z, Tasic S, Petrovic S, Nidzovic S, Leposavic A, Stanojkovic T (2005) Antiproliferative action of water extracts of seeds or pulp of five different raspberry cultivars. Food Chem 93(1):39–45

Kassim A, Poette J, Paterson A, Zait D, McCallum S, Woodhead M, Smith K, Hackett C, Graham J (2009) Environmental and seasonal influences on red raspberry anthocyanin antioxidant contents and identification of quantitative traits loci (QTL). Mol Nutr Food Res 53(5):625–634

Kreander K, Galkin A, Vuorela S, Tammela P, Laitinen L, Heinonen M, Vuorela P (2006) In-vitro mutagenic potential and effect on permeability of co-administered drugs across Caco-2 cell monolayers of *Rubus idaeus* and its fortified fractions. J Pharm Pharmacol 58(11): 1545–1552

Landete JM (2011) Ellagitannins, ellagic acid and their derived metabolites: a review about source, metabolism, functions and health. Food Res Int 44(5): 1150–1160

Lieberman L (1995) Remedies to file for future reference. Birthkit 5(Spring):1–8

Liu M, Li XQ, Weber C, Lee CY, Brown J, Liu RH (2002) Antioxidant and anti proliferative activities of raspberries. J Agric Food Chem 5:2926–2930

Lu LT, Boufford DE (2003) *Rubus* Linnaeus. In: Wu ZY, Raven PH, Hong DY (eds) Flora of China, vol 9, Pittosporaceae through Connaraceae. Science Press/ Missouri Botanical Garden Press, Beijing/St. Louis

Maguire A, Rugg-Gunn AJ (2003) Xylitol and caries prevention – is it a magic bullet? Br Dent J 194(8): 429–436

Makinen KK, Soderling E (1980) A quantitative study of mannitol, sorbitol, xylitol, and xylose in wild berries and commercial fruits. J Food Sci 45(2):367–371

Mandal S, Stoner GD (1990) Inhibition of N-nitrosobenzylmethylamine-induced esophageal tumorigenesis in rats by ellagic acid. Carcinogenesis 11(1):55–61

McFarlin BL, Gibson MH, O'Rear J, Harman P (1999) A national survey of herbal preparation use by nursemidwives for labor stimulation. J Nurse Midwifery 44(3):205–216

Mills SY (1985) The Dictionary of Modern Herbalism. Thorsons, Wellingborough

Morimoto C, Satoh Y, Hara M, Inoue S, Tsujita T, Okuda H (2005) Anti-obese action of raspberry ketone. Life Sci 77(2):194–204

Mukhtar H, Del Tito BJ, Marcelo CL, Das M, Bickers DR (1984) Ellagic acid: a potent naturally occurring inhibitor of benzo[a]pyrene metabolism and its subsequent glucuronidation, sulfation and covalent binding to DNA in cultured BALB/C mouse keratinocytes. Carcinogenesis 5(12):1565–1571

Mullen W, McGinn J, Lean ME, MacLean MR, Gardner P, Duthie GG, Yokota T, Crozier A (2002) Ellagitannins, flavonoids, and other phenolics in red raspberries and their contribution to antioxidant capacity and vasorelaxation properties. J Agric Food Chem 50(18): 5191–5196

Narayanan BA, Re GG (2001) IGF-II down regulation associated cell cycle arrest in colon cancer cells exposed to phenolic antioxidant ellagic acid. Anticancer Res 21:359–364

Nowak R (2003) Separation and quantification of tiliroside from plant extracts by SPE/RP-HPLC. Pharm Biol 41(8):627–630

Oomah BD, Ladet S, Godfrey DV, Liang J, Girard B (2000) Characteristics of raspberry (*Rubus idaeus* L.) seed oil. Food Chem 69(2):187–193

Pantelidis GE, Vasilakakis M, Manganaris GA, Diamantidis G (2007) Antioxidant capacity, phenol, anthocyanin and ascorbic acid contents in raspberries, blackberries, red currants, gooseberries and Cornelian cherries. Food Chem 102(3):777–783

Papoutsi Z, Kassi E, Tsiapara A, Fokialakis N, Chrousos GP, Moutsatsou P (2005) Evaluation of estrogenic/ antiestrogenic activity of ellagic acid via the estrogen receptor subtypes ERalpha and ERbeta. J Agric Food Chem 53(20):7715–7720

Parry J, Su L, Luther M, Zhou K, Yurawecz MP, Whittaker P, Yu L (2005) Fatty acid composition and antioxidant properties of cold-pressed marionberry, boysenberry, red raspberry, and blueberry seed oils. J Agric Food Chem 53(3):566–573

Parsons M, Simpson M, Ponton T (1999) Raspberry leaf and its effect on labour: safety and efficacy. Aust Coll Midwives Inc J 12(3):20–25

Patel AV, Obiyan J, Patel N, Dacke CG (1995) Raspberry leaf extract relaxes intestinal smooth muscle in-vitro. J Pharm Pharmacol 47(12):1129

Porcher MH et al (1995–2020) Searchable World Wide Web Multilingual Multiscript Plant Name Database.

Published by The University of Melbourne, Australia. http://www.plantnames.unimelb.edu.au/Sorting/Frontpage.html

Proteggente AR, Pannala AS, Paganga G, Van Buren L, Wagner E, Wiseman S, Van De Put F, Dacombe C, Rice-Evans CA (2002) The antioxidant activity of regularly consumed fruit and vegetables reflects their phenolic and vitamin C composition. Free Radic Res 36(2):217–233

Rojas-Vera J, Patel AV, Christopher G, Dacke CG (2001) Relaxant activity of raspberry (*Rubus idaeus*) leaf extract in guinea-pig ileum in-vitro. Phytother Res 16(7):665–668

Salinas-Moreno Y, Almaguer-Vargas G, Peña-Varela G, Ríos-Sánchez R (2009) Ellagic acid and anthocyanin profiles in fruits of raspberry (*Rubus idaeus* L.) in different ripening stages. Rev Chapingo Ser Hortic 15(1):97–101

Saxholt E, Christensen AT, Møller A, Hartkopp HB, Hess Ygil, K, Hels, OH (2008) Danish Food Composition Databank, revision 7. Department of Nutrition, National Food Institute, Technical University of Denmark. http://www.foodcomp.dk/

Seeram NP (2008) Berry fruits for cancer prevention: current status and future prospects. J Agric Food Chem 56(3):630–635

Seeram NP, Adams LS, Zhang Y, Lee R, Sand D, Scheuller HS, Heber D (2006) Blackberry, black raspberry, blueberry, cranberry, red raspberry, and strawberry extracts inhibit growth and stimulate apoptosis of human cancer cells in vitro. J Agric Food Chem 54(25):9329–9339

Seeram NP, Momin RA, Nair MG, Bourquin LD (2001) Cyclooxygenase inhibitory and antioxidant cyanidin glycosides in cherries and berries. Phytomedicine 8(5):362–369

Simpson M, Parsons M, Greenwood J, Wade K (2001) Raspberry leaf in pregnancy: its safety and efficacy in labor. J Midwifery Womens Health 46(2):51–59

Sparzak B, Merino-Arevalo M, Vander Heyden Y, Krauze-Baranowska M, Majdan M, Fecka I, Głód D, Bączek T (2010) HPLC analysis of polyphenols in the fruits of *Rubus idaeus* L. (Rosaceae). Nat Prod Res 24(19): 1811–1822

Venskutonis PR, Dvaranauskaite A, Labokas J (2007) Radical scavenging activity and composition of raspberry (*Rubus idaeus*) leaves from different locations in Lithuania. Fitoterapia 78(2):162–165

Viljanen K, Kylli P, Kivikari R, Heinonen M (2004) Inhibition of protein and lipid oxidation in liposomes by berry phenolics. J Agric Food Chem 52(24): 7419–7424

Wada L, Ou B (2002) Antioxidant activity and phenolic content of Oregon caneberries. J Agric Food Chem 50(12):3495–3500

Wang SY, Lin HS (2000) Antioxidant activity in fruits and leaves of blackberry, raspberry, and strawberry varies with cultivar and developmental stage. J Agric Food Chem 48:140–146

Weber C, Liu RH, Brennan RM, Gordon SL, Williamson B (2002) Antioxidant capacity and anticancer properties of red raspberry. Acta Hortic 585:451–457

Whitehouse B (1941) Fragarine: an inhibitor of uterine action. Br Med J 2(4210):370–371

Zafrilla P, Ferreres F, Tomás-Barberán FA (2001) Effect of processing and storage on the antioxidant ellagic acid derivatives and flavonoids of red raspberry (*Rubus idaeus*) jams. J Agric Food Chem 49(8): 3651–3655

Zhang Y, Zhang Z, Yang Y, Zu X, Guan D, Wang YP (2011) Diuretic activity of *Rubus idaeus* L (Rosaceae) in rats. Trop J Pharm Res 10(3):243–248

Zheng J, Pistilli M, Holloway A, Crankshaw D (2010) The effects of commercial preparations of red raspberry leaf on the contractility of the rat's uterus in vitro. Reprod Sci 17(5):494–501

Rubus occidentalis

Scientific Name

Rubus occidentalis L.

Synonyms

Melanobatus neglectus (Peck) Greene, *Melanobatus occidentalis* (L.) Greene, *Rubus idaeus* L. var. *americanus* Torr., *Rubus idaeus* × *occidentalis* Focke, *Rubus michiganus* Greene, *Rubus neglectus* Peck, *Rubus occidentalis* × *strigosus* Rydb.

Family

Rosaceae

Common/English Names

American Black Raspberry, Blackcap, Black Raspberry, Purple Raspberry, Scotch Cap, Thimble Berry.

Vernacular Names

Chinese: Cao Mei;
Czech: Ostružiník Ojíněný, Ostružiník Západní;
Danish: Sort Hindbær, Ontario-Brombær;
Eastonian: Läänevaarikas

French: Framboisier De Virginie, Framboise Noire (Quebec), Framboisier À Fruits Noirs, Framboisier Noir;
German: Schwarze Himbeere, Schwarzfrüchtige Himbeere;
Greek: Rouvos O Dytikos;
Italian: Lampone Americano;
Japanese: Kuro Mirasu Berii, Kuro Miki Ichigo;
Portuguese: Framboesa Negra;
Russian: Malina Zagadochnaia, Malina Zagadočnaja, Malina Zapadnaia;
Slovaščina: Črna Malina;
Spanish: Frambueso Negro;
Swedish: Svarthallon.

Origin/Distribution

Black Raspberry is indigenous to East and central North America: New Brunswick to Minnesota, south to Georgia, west to Nebraska and Colorado.

Agroecology

A cool climate temperate species. Its natural habitats include openings in deciduous woodlands, woodland borders, savannas, thickets, fence rows, overgrown vacant lots, power-line clearances in wooded areas, and partially shaded areas along buildings. It prefers partial sun, moist to mesic conditions, and thrives best on deep, rich, well-drained, sandy-loam soils well supplied with organic matter and with high moisture holding capacity. a range of 5.8–6.5 is considered optimum. The canes also fail to set fruit if there is too much shade. In areas that are too sunny and dry, the fruit may not develop properly without adequate rain or supplemental irrigation.

Edible Plant Parts and Uses

Black raspberries are eaten raw, cooked, dried, frozen or made into purées and juices or processed as colorants. They are very versatile and can be made into a wide array of food preparations and recipes. They are excellent for jam, jellies, sauces and preserves. They can be used in bread, cakes, cookies, pastries, pudding, pies, tarts, flan, soufflés, pancakes, waffles, soups, salads, salsa, raspberry vinaigrette, and eaten in ice-cream, yoghurt, sorbets, sherberts and slushes. Some popular common recipes using raspberry are fondue-cheesecake, chocolate raspberry streusel squares, chocolate raspberry cheesecake. Pancakes and waffles goes well with raspberry syrup. Black raspberries can be processed into refreshing beverages, syrup, mead, wine and liquor. Raspberry sac-mead is a sweet alcoholic liquor popular in Poland and is made of fermented honey, water and often with spices and black raspberries. Two well known liqueurs predominantly based on black raspberry fruit include France's Chambord Liqueur Royale de France and South Korea's various manufacturers of *Bokbunja*. *Bokbunja* is made from Korean black raspberries and contain 15% alcohol and is considered by many to be especially good for sexual stamina.

Young black raspberry shoots are eaten raw or cooked like rhubarb. They are harvested as they emerge through the soil in the spring, and whilst they are still tender, and then peeled. A tea is made from the leaves and another from the bark of the root.

Botany

An arching, erect, deciduous, armed shrub reaching 1–2 m with stems rooting at the tips. Twigs are reddish-purple, terete, glaucous; eglan-

Plate 1 Black raspberries

dular, sparsely bristly with straight or hooked, stout prickles that is wider at the base. Leaves are alternate, palmately compound, 3–5 foliate. Leaflets are ovate to ovate-lanceolate, 5–19 cm long by 3.8–8.9 cm wide; glabrous above, densely white tomentose beneath; rounded to subcordate at base; margins doubly serrate with occasional shallow lobes; petiole and rachis have many stout prickles. Inflorescence is an umbel of 3–7 monoecious flowers borne on pedicels with stout prickles; sepals are lanceolate, green, tomentose, 6–8 mm long; petals 5, elliptic-oblong, white, less than half the length of the sepals; carpels many, inserted on hypanthium; stamens numerous with filiform filaments and didymous anthers surrounding the carpels; each carpel becomes a drupelet with 1 locule 1 with 1 developed ovule. Fruit is an aggregation of drupelets, 10–15 mm in diameter, hemispherical, yellowish-white to red and purplish-black when ripe (Plate 1), glaucous. Seed is small and pendulous.

Nutritive/Medicinal Properties

Food value of raw, raspberries, *Rubus* spp. (exclude 4% refuse consisting of caps and spoiled berries) per 100 g edible portion was reported as follows (USDA 2011): water 85.75 g, energy 52 kcal (220 kJ), protein 1.20 g, total lipid (fat) 0.65 g, ash 0.46 g, carbohydrate 11.94 g; total dietary fibre 6.5 g, sugars (total) 4.42 g, sucrose 0.20 g, glucose (dextrose) 1.86 g, fructose 2.35 g; minerals – calcium 25 mg, iron, 0.69 mg, magnesium 22 mg, phosphorus 29 mg, potassium 151 mg, sodium 1 mg, Zn 0.42 g, Cu 0.090 g, Mn 0.670 g, Se 0.2 µg; vitamins – vitamin C (total ascorbic acid) 26.2 mg, thiamin 0.032 mg, riboflavin 0.038 mg, niacin 0.598 mg, pantothenic acid 0.329 mg, vitamin-6 0.055 mg, folate (total) 21 µg, choline (total) 12.3 mg, betaine 0.8 mg, β carotene 12 µg, α carotene 16 µg, vitamin A 33 IU, lutein + zeaxanthin 136 µg, vitamin E (α-tocopherol) 0.87 mg, β-tocopherol 0.06 mg, γ-tocopherol 1.42 mg, δ-tocopherol 1.04 mg, vitamin K (phylloquinone) 7.8 µg; total saturated fatty acids 0.014 g, 16:0 (palmitic acid) 0.016 g, 18:0 (stearic acid) 0.004 g; total monounsaturated fatty acids 0.375 g, 18:1 undifferentiated (oleic acid) 0.059 g, 20:1 (gadoleic acid) 0.005 g; total polyunsaturated fatty acids 0.375 g, 18:2 undifferentiated (linoleic acid) 0.249 g and 18:3 undifferentiated (linolenic acid) 0.126 g.

Flavonoids (kaempferol, quercetin, myricetin) and phenolic acids (p-coumaric, caffeic, ferulic, p-hydroxybenzoic, gallic and ellagic acids) were detected in the fruits of 19 berries (Hákkinen et al. 1999). Ellagic acid was the main phenolic compound in the berries of the genus *Rubus* (red raspberry, Arctic bramble and cloudberry) and genus *Fragaria* (strawberry). The data suggested berries to have potential as good dietary sources of quercetin or ellagic acid.

Berry fruits including black raspberries are widely consumed in our diet and have attracted much attention due to their potential human health benefits. Black raspberries are loaded with nutrients such as vitamins, minerals diverse range of phytochemicals including anthocyanins and phenolic compounds with biological properties such as antioxidant, anticancer, anti-neurodegerative, and antiinflammatory activities. Extensive studies have been conducted on their phytochemical contents, pharmacological activities and anticancerous potential in the United States particularly in Ohio.

Antioxidant Activity

Among the berry fruits analysed, black raspberries and strawberries had the highest ORAC (oxygen radical absorbance capacity) values during the green stages, whereas red raspberries had the highest ORAC activity at the ripe stage (Wang and Lin 2000). Total anthocyanin content increased with maturity for all three species of fruits. Compared with fruits, leaves were found to have higher ORAC values. In fruits, ORAC values ranged from 7.8 to 33.7 µmol of Trolox equivalents (TE)/g of fresh berries (35.0–162.1 µmol of TE/g of dry matter), whereas in leaves, ORAC values ranged from 69.7 to 182.2 µmol of TE/g of fresh leaves (205.0–728.8 µmol of TE/g of dry matter). As the leaves become older, the ORAC values and total phenolic contents decreased. Of the ripe fruits tested, on the basis of wet weight of fruit, cv. Jewel black raspberry and blackberries had the richest source for antioxidants. On the basis of the dry weight of fruit, strawberries had the highest ORAC activity followed by black raspberries (cv. Jewel), blackberries, and red raspberries.

Separate studies showed caneberries including black raspberries (*Rubus occidentalis*) to be excellent and rich sources of anthocyanins and phenolics and to be high in antioxidant activity (Wada and Ou 2002). Anthocyanin content ranged from 0.65 to 5.89 mg/g, and phenolics ranged from 4.95 to 9.8 mg/g. All berries had high levels of ellagic acid (47–90 mg/g). All of the berries had high oxygen radical absorbance capacity (ORAC) activity ranging from 24 to 77.2 µmol of Trolox equivalent/g of fresh berries. Black raspberries had the highest ORAC and anthocyanin and phenolic contents.

The black raspberry seed oil contained about 35% α-linolenic acid (18:30–3) and 55% to 50% linoleic acid (Parry and Yu 2004). The meal exhibited strong free radical scavenging activities against 2,2-diphenyl-1-plcrylhydrazyl (DPPH) and 2,2'-azinobi(3-ethylbenzothlazoline-6-sulfonic acid) diammonium salt (ABTS.+) radicals and had a total phenolic content (IPC) of 46 mg gallic acid equivalent/g meal. The ABTS scavenging capacity and TPC of the meal were 300 and 290 times greater than that of the oil. The results from this study suggest the possible food application of black raspberry seed and its fractions in improving human nutrition and potential value-adding opportunities in black raspberry production and processing.

Anticancer Activity

In-Vitro Studies

Due to their content of phenolic and flavonoid compounds, berries including black raspberries were found to exhibit high antioxidant potential, exceeding that of many other foodstuffs (Stoner et al. 2008b). Through their ability to scavenge ROS and reduce oxidative DNA damage, stimulate antioxidant enzymes, inhibit carcinogen-induced DNA adduct formation and enhance DNA repair, berry compounds had been demonstrated to inhibit mutagenesis and cancer initiation. Berry constituents also influenced cellular processes associated with cancer progression including signalling pathways associated with cell proliferation, differentiation, apoptosis and angiogenesis. The bioactive phytochemicals in berries namely blackberry, black raspberry, blueberry, cranberry, red raspberry and strawberry reported included phenolic acids (hydroxycinnamic and hydroxybenzoic acids), flavonoids (anthocyanins, flavanols, flavonols), condensed tannins (proanthocyanins), hydrolyzable tannins (ellagitannins and gallotannins), stilbenoids, lignans, triterpenes and sterols (Seeram 2006; Seeram et al. 2006). Ellagitannins were reported to be hydrolysed to ellagic acid which was gradually metabolised by the intestinal mircobiota to produce different types of urolithins (Landete 2011). Urolithins could exert estrogenic and/or anti-estrogenic activity and tissue disposition studies revealed that urolithins were enriched in prostate, intestinal, and colon tissues in mouse, which explained their inhibitory effect on prostate and colon cancer cell growth. Additionally, anti-proliferative and apoptosis-inducing activities of ellagic acid and urolithins had been demonstrated by the inhibition of cancer cell growth. Studies by Seeram et al. (2006) showed that with concentration of berry extract from 25 to 200 μg/mL, increasing inhibition of cell proliferation in all of the human oral (KB, CAL-27), breast (MCF-7), colon (HT-29, HCT116), and prostate (LNCaP) tumour cell lines were observed, with different degrees of potency between cell lines. Black raspberry and strawberry extracts showed the most significant pro-apoptotic effects against colon cancer cell line, HT-29.

Black raspberry seed flour significantly inhibited HT-29 colon cancer cell line proliferation (Parry et al. 2006). The ORAC value was significantly correlated to the total phenolic content under the experimental condition, and also differed in its total anthocyanin values and Fe(2+)-chelating capacities. The results suggest the potential of developing the value-added use of fruit seed flour as dietary sources of natural antioxidants and antiproliferative agents for optimal human health. Studies showed that the lack of correlation between growth inhibition of HT-29 colon cancer cells and extract total phenolic and total monomeric anthocyanin assays suggested horticultural parameters namely cultivar, production site, and fruit maturity stage influence bioactivity in a complex manner (Johnson et al. 2011).

Five percent and 10% dietary freeze-dried black raspberries (BRBs) inhibited N-nitrosomethylbenzylamine (NMBA) metabolism in the rat oesophageal explants (26% and 20%) and in liver microsomes (22% and 28%), but the inhibition was not dose dependent (Reen et al. 2006). NMBA metabolism in oesophageal explants was inhibited by individual components of BRBs, maximally by cyanidin-3-rutinoside (47%) followed by EA (33%), cyanidin-3-glucoside (23%), and the extract (11%). Similarly, in liver microsomes, the inhibition was maximal by cyanidin-3-rutinoside (47%) followed by EA (33%) and cyanidin-3-glucoside (32%).

In separate studies, black raspberry (*Rubus occidentalis*) extract was found to be antiangiogenic (0.1% w/v) in the human tissue-based in-vitro fibrin clot angiogenesis assay (Liu et al. 2005). At 0.075% (w/v), the active fraction completely inhibited angiogenic initiation and angiogenic vessel growth. The studies suggested that an active black raspberry fraction containing multiple antiangiogenic compounds, one of which had been identified as gallic acid, may be a promising complementary cancer therapy. The multiple active ingredients in the extracts may be additive or synergistic in their antiangiogenic effects. Angiogenesis-inhibiting agents have the potential for inhibiting tumour growth and limiting the dissemination of metastasis, thus keeping cancers in a static growth state for prolonged periods. Han et al. (2005) found that two major chemopreventive components of black raspberries, ferulic acid and β-sitosterol, and a fraction eluted with ethanol (RO-ET) inhibited the growth of premalignant and malignant but not normal human oral epithelial cell lines. Another fraction eluted with CH2Cl2/ethanol (DM:ET) and ellagic acid inhibited the growth of normal as well as premalignant and malignant human oral cell lines. Their results showed that the growth inhibitory effects of black raspberries on premalignant and malignant human oral cells may reside in specific components that target aberrant signalling pathways regulating cell cycle progression.

Rodrigo et el. (2006) demonstrated that a freeze-dried black raspberry ethanol extract suppressed human oral squamous cell carcinoma without affecting viability, inhibited translation of the complete angiogenic cytokine vascular endothelial growth factor, suppressed nitric oxide synthase activity, and induced both apoptosis and terminal differentiation. The data suggested that a freeze-dried black raspberry ethanol extract may have potential to be used as a chemopreventive agent in persons with oral epithelial dysplasia. Studies by Zikri et al. (2009) concluded that the selective effects of the ethanol extract of freeze-dried black raspberries on growth and apoptosis of highly tumorigenic rat esophageal epithelial cells in-vitro may be due to preferential uptake and retention of its component anthocyanins, and this may also be responsible for the greater inhibitory effects of freeze-dried whole berries on tumour cells in-vivo.

Xue et al. (2001) found that freeze-dried strawberries or black raspberries extracts, fractions and ellagic acid exhibited anti-transformation activity in Syrian hamster embryo (SHE) cell transformation model. Ellagic acid and methanol extract from strawberries and black raspberries displayed chemopreventive activity against benzo[a]pyrene (B[a]P)-induced transformation in SHE cells. The possible mechanism by which these methanol fractions inhibited cell transformation was postulated to involve interference of uptake, activation, detoxification of B[a]P and/or intervention of DNA binding and DNA repair. One molecular mechanism through which black raspberries inhibited carcinogenesis may be mediated by impairing signal transduction pathways leading to activation of activator protein 1 and nuclear factor kappaβ (NFκB). Another mechanism for the chemopreventive effect of black raspberry may involve inhibition of the PI-phosphotidylinositol 3-kinase (PI-3K)/Akt pathway and endothelial growth factor (VEGF) pathway (Huang et al. 2006). Li et al. (2008) found that black raspberry fractions inhibited the activation of AP-1, NF-kappaB, and nuclear factor of activated T cells (NFAT) benzoapyrene diol-epoxide (BaPDE) as well as their upstream PI-3K/Akt-p70(S6K) and mitogen-activated protein kinase pathways. In contrast, strawberry fractions inhibited NFAT activation, but did not inhibit the activation of AP-1, NF-kappaB or the PI-3K/Akt-p70(S6K) and mitogen-activated protein kinase pathways. Consistent with the effects on NFAT activation, tumour necrosis factor-α (TNF-α) induction by BaPDE was blocked by extract fractions of both black raspberries and strawberries, whereas vascular endothelial growth factor (VEGF) expression, which depended on AP-1 activation, was suppressed by black raspberry fractions but not strawberry fractions. The results suggested that black raspberry and strawberry components may target different signalling pathways in exerting their anti-carcinogenic effects. Zhang et al. (2011) found that non-toxic levels of a lyophilized black raspberry ethanol extract significantly inhibited the growth of human cervical cancer cell lines HeLa (HPV16-/HPV18+, adenocarcinoma), SiHa (HPV16+/HPV18-, squamous cell carcinoma) and C-33A (HPV16-/HPV18-, squamous cell carcinoma) in a dose-dependent and time-dependent manner to a maximum of 54%, 52% and 67%, respectively.

In-Vivo Studies

Dietary lyophilized black raspberries (LBRs) suppressed N-nitrosomethylbenzylamine (NMBA)-induced oesophageal tumorigenesis in the F344 rat at both the initiation and promotion/progression stages of carcinogenesis (Kresty et al. 2001). LBR at 5% significantly reduced tumour incidence and multiplicity, proliferation indices and preneoplastic lesion formation. Chen et al. (2006a) found dietary black raspberries (BRB) inhibited tumour multiplicity in F344 rats treated with the oesophageal carcinogen, N-nitrosomethylbenzylamine (NMBA). The results suggested a novel tumour suppressive role of BRB through inhibition of COX-2, iNOS, and c-Jun. BRB inhibited mRNA expression of iNOS and c-*Jun*, but not COX-2, in papillomatous lesions of the esophagus. Prostaglandin E_2 and total nitrite levels were also decreased by BRB in papillomas. The scientists (Chen et al. 2006b) also reported that dietary black raspberry powder (BRB) inhibited N-nitrosomethylbenzylamine (NMBA)-induced tumour development in the rat oesophagus by inhibiting the formation of DNA adducts and reducing the proliferation rate of preneoplastic cells. On a molecular level, BRB downregulated the expression of c-Jun, cyclooxygenase-2 (COX-2) and inducible nitric oxide synthase (iNOS). The data also suggested that down-regulation of vascular endothelial growth factor (VEGF) wais correlated with suppression of COX-2 and iNOS. As high vascularity is a risk factor for metastasis and tumour recurrence, BRB may have cancer therapeutic effects in human oesophageal cancer.

Over the years, extensive studies have been conducted on the anticancerous properties of berry fruit especially black raspberry in Ohio state, USA (Stoner et al. 2006). Berry fruits like black raspberries (*Rubus occidentalis*), blackberries (*R. fructicosus*), and strawberries (*Fragaria ananas*) contain known chemopreventive agents which include vitamins A, C, and E and folic acid; calcium and selenium; β-carotene, α-carotene, and lutein; polyphenols such as ellagic acid, ferulic acid, p-coumaric acid, quercetin, and several anthocyanins; and phytosterols such as β-sitosterol, stigmasterol, and kaempferol. Studies had examined and demonstrated the potential cancer chemopreventive activity of freeze-dried diets containing freeze-dried black raspberries, blackberries and strawberries suppressed the development of *N*-nitrosomethylbenzylamine (NMBA)–induced tumours in the rat esophagus. Studies showed that feeding rats with diets containing 5% of either black or red raspberries, strawberries, blueberries, noni, açaí or wolfberry were about equally capable of inhibiting N-nitrosomethylbenzylamine – induced tumour progression in the rat esophagus in spite of known differences in levels of anthocyanins and ellagitannins (Stoner et al. 2010). They also reduced the levels of the serum cytokines, interleukin 5 (IL-5) and GRO/KC (growth related oncogene), the rat homologue for human interleukin-8 (IL-8), and this was associated with increased serum antioxidant capacity.

Studies in male Syrian Golden hamsters showed that ingestion of lyophilized black raspberries (LBR) for 2 weeks prior to treatment with 0.2% 7,12-dimethylbenz(a) anthracene in dimethylsulfoxide inhibited tumour formation in the oral cavity (Casto et al. 2002). Lyophilized (freeze-dried) black raspberries, blackberries, and strawberries had been reported to inhibit carcinogen-induced cancer in the rodent esophagus (Stoner et al. 2006). Some of the known chemopreventive agents in berries include vitamins A, C, and E and folic acid; calcium and selenium; β-carotene, α-carotene, and lutein; polyphenols such as ellagic acid, ferulic acid, p-coumaric acid, quercetin, and several anthocyanins; and phytosterols such as β-sitosterol, stigmasterol, and kaempferol. All three berry types were found to inhibit the number of esophageal tumours (papillomas) in N-nitrosomethylbenzylamine (NMBA) treated Fischer 344 rats. In-vivo mechanistic studies with BRBs indicated that they reduced the growth rate of premalignant oesophageal cells, in part, through down-regulation of cyclooxygenase-2 leading to reduced prostaglandin production and of inducible nitric oxide synthase leading to reduced nitrate/nitrite levels in the oesophagus. Dietary freeze-dried berries were shown to inhibit chemically-induced cancer of the rodent esophagus by 30–60% and of the colon by up- to 80%. The berries were effective at both the initiation and promotion/progression stages of tumour development (Stoner et al. 2007). On a molecular level, berries modulated the expression of genes involved with proliferation, apoptosis, inflammation and angiogenesis. Rats were administered diets containing an alcohol/water-insoluble (residue) fraction of fractions of three berry types, that is, black raspberries (BRBs), strawberries (STRWs), and blueberries (BBs), that differed in their content of ellagitannins in the order BRB>STRW>BB prior to treatment with the esophageal carcinogen N-nitrosomethylbenzylamine (NMBA) (Wang et al. 2010). The residue fractions from all three berry types were about equally effective in reducing NMBA tumorigenesis in the rat esophagus irrespective of their ellagitannin content (0.01–0.62 g/kg of diet). The results suggested that ellagitannins may not be responsible for the chemopreventive effects of the alcohol/water-insoluble fraction of berries. In another study, F344 rats were administered diets containing either (a) 5% whole freeze-dried black raspberries (BRB) powder, (b) an anthocyanin-rich fraction, (c) an organic solvent-soluble, (d) an organic-insoluble (residue) fraction containing (0.02 μmol anthocyanins/g diet), (e) a hexane extract, and (f) a sugar fraction, all derived from BRB prior to treatment with NMBA (Wang et al. 2009). Diets a-c each contained approximately 3.8 μmol anthocyanins/g diet and diets e and f had only trace quantities of anthocyanins. The anthocyanin treatments (diet groups a-c) were about equally effective in reducing NMBA tumorigenesis in the esophagus, indicating that the anthocyanins in BRB had chemopreventive potential. The organic-insoluble (residue) fraction (d) was also effective, suggesting that components other than berry anthocyanins may be chemopreventive. The hexane and sugar diets were inactive. Diet groups a, b, and d all inhibited cell proliferation, inflammation, and angiogenesis and induced apoptosis in both preneoplastic and papillomatous esophageal tissues, suggesting similar mechanisms of action by the different berry components.

Wang et al. (2011b) reported on the effects of black raspberries on the expression of genes associated with the late stages of rat oesophageal carcinogenesis. Treatment with 5% black raspberries (BRBs) reduced the number of dysplastic lesions and the number and size of oesophageal papillomas in N-nitrosomethylbenzylamine (NMBA)-treated rats. When compared to oesophagi from control rats, NMBA treatment led to the differential expression of 4,807 genes in preneoplastic oesophagus and 17,846 genes in oesophageal papillomas. Dietary BRBs modulated 626 of the 4,807 differentially expressed genes in preneoplastic esophagus and 625 of the 17,846 differentially expressed genes in oesophageal papillomas. In both preneoplastic oesophagus and in papillomas, BRBs modulated the mRNA expression of genes associated with carbohydrate and lipid metabolism, cell proliferation and death, and inflammation. In these same tissues, BRBs modulated the expression of proteins associated with proliferation, apoptosis, inflammation, angiogenesis, and both cyclooxygenase and lipoxygenase pathways of arachidonic acid metabolism. Additionally, matrix metalloproteinases involved in tissue invasion and metastasis, and proteins associated with cell-cell adhesion, were also modulated by BRBs. Stoner et al. (2008a) found that 462 of the 2,261 NMBA-dysregulated genes in rat esophagus were restored to near-normal levels of expression by BRB. Further, they identified 53 NMBA-dysregulated genes that were positively regulated by both phenylethyl isothiocyanate (PEITC), a constituent of cruciferous vegetables and BRB. These 53 common genes included genes involved in phase I and II metabolism, oxidative damage, and oncogenes and tumour suppressor genes that regulate apoptosis, cell cycling, and angiogenesis.

Dietary administration of lyophilized black raspberries (LBRs) significantly inhibited chemically induced oral, esophageal, and colon carcinogenesis in animal models (Kresty et al. 2006). Interim results from 6-months chemopreventive pilot study showed that daily consumption of LBRs modulated markers of oxidative stress 8-epi-prostaglandin F2α (8-Iso-PGF2) and, to a lesser more-variable extent, 8-hydroxy-2′-deoxyguanosine (8-OHdG), in Barrett's esophagus patients.

Mallery et al. (2007) formulated and characterised a novel mucoadhesive gel containing freeze dried black raspberries for local delivery to human oral mucosal tissues for oral cancer chemoprevention. The study showed anthocyanin stability was dependent upon gel pH and storage temperature and also demonstrated that the gel composition was well-suited for absorption and penetration into the target oral mucosal tissue site. They also showed show that black raspberry gel application modulated oral intraepithelial neoplasia gene expression profiles, ultimately reducing epithelial COX-2 protein (Mallery et al. 2008). In a patient subset, raspberry gel application also reduced microvascular densities in the superficial connective tissues and induced genes associated with keratinocyte terminal differentiation. A highly sensitive and specific LC-MS/MS assay was developed and validated to simultaneously quantify cyanidin 3-glucoside, cyanidin 3-rutinoside, cyanidin 3-sambubioside and cyanidin 3-(2(G)-xylosyl) rutinoside, the four bioactive constituent black raspberry (FBR) anthocyanins in human saliva, plasma and oral tissue homogenates (Ling et al. 2009). This assay was subsequently used in a pilot pharmacology study to evaluate the effects of topical application of a 10% (w/w) FBR bioadhesive gel to selected mucosal sites in the posterior mandibular gingiva. Measurable saliva and tissue levels of the FBR anthocyanins confirmed that gel-delivered anthocyanins were readily distributed to saliva and easily penetrated human oral mucosa. Studies by Mallery et al. (2011) indicated that interpatient differences in oral mucosal tissue, saliva, and oral microflora could impact on intraoral metabolism, bioactivation and capacities for enteric recycling of black raspberry anthocyanins.

Freeze-dried black raspberries (BRB) were found to be highly effective in inhibiting intestinal tumorigenesis in apc1638$^{+/-}$ and Muc2$^{-/-}$ mouse models of colorectal cancer through targeting multiple signalling pathways (Bi et al. 2010). BRB reduce tumour and tumour multiplicity in both models by suppressing β-catenin signalling in the former and by reducing chronic inflammation in the latter. Montrose et al. (2011) found that dietary BRB markedly reduced dextran sodium sulfate DSS-induced acute injury to the colonic epithelium in C57BL/6 J mice. Body mass were maintained and colonic shortening and ulceration were reduced. However, BRB treatment did not affect the levels of either plasma nitric oxide or colon malondialdehyde, biomarkers of oxidative stress that are otherwise increased by DSS-induced colonic injury. BRB treatment suppressed tissue levels of several key pro-inflammatory cytokines, including tumour necrosis factor α and interleukin 1β. The results demonstrated a potent anti-inflammatory effect of BRB during DSS-induced colonic injury, supporting its possible therapeutic or preventive role in the pathogenesis of ulcerative colitis and related neoplastic events. In a phase I pilot study, Wang et al. (2011a) found that BRBs protectively modulated expression of genes and other epigenetic biomarkers associated with Wnt pathway, proliferation, apoptosis, and angiogenesis in the human colon and rectum of some patients with colorectal adenocarcinomas.

In a clinical trial, 45 g of freeze-dried dietary BRB daily for 7 days was found to be well tolerated by healthy volunteers and resulted in quantifiable anthocyanins and ellagic acid in plasma and urine (Stoner et al. 2005). Overall, less than 1% of these compounds were absorbed and excreted in urine.

Traditional Medicinal Uses

Black raspberry root tea was also part of the traditional pharmacopeia for treatment of haemorrhaging and haemophilia. In south USA, blackberry tea was mixed with whiskey to expel gas. The juice of raspberry fruits was used to flavour medicines. The Kiowa and Apache Indians made a tea from the roots to treat stomach ache. The roots are cathartic. A decoction of the roots has been used as a remedy for gonorrhoea, diarrhoea and dysentery. The root has been chewed in the treatment of coughs and toothache. An infusion of the roots has been employed as a wash for sore eyes. The root has been used, combined with *Hypericum* spp, to treat the early stages of consumption. A decoction of the roots, stems and leaves has been used to treat whooping cough. The leaves are highly astringent. A decoction is employed in the treatment of bowel complaints. A tea made from the leaves is administered as a wash for old and foul sores, ulcers and boils.

Other Uses

A purple to dull blue dye is obtained from the fruit. The plant is used as a disease resistance source against the spur-blight disease of the leaves. Crossing of *R. idaeus* × *R. occidentalis* establish very fruitful hybrids, which usually are called *R.* × *neglectus* Peck.

Comments

Occasional mutations in the genes controlling anthocyanin production has resulted in yellow-fruited variants that still retain the species' distinctive flavour.

Selected References

Bailey LH (1976) Hortus Third. A concise dictionary of plants cultivated in the United States and Canada. Liberty Hyde Bailey Hortorium/Cornell University/Wiley, New York, 1312 pp

Bi X, Fang W, Wang LS, Stoner GD, Yang W (2010) Black raspberries inhibit intestinal tumorigenesis in apc1638$^{+/-}$ and Muc2$^{-/-}$ mouse models of colorectal cancer. Cancer Prev Res (Phila) 3(11):1443–1450

Casto BC, Kresty LA, Kraly CL, Pearl DK, Knobloch TJ, Schut HA, Stoner GD, Mallery SR, Weghorst CM (2002) Chemoprevention of oral cancer by black raspberries. Anticancer Res 22(6C):4005–4015

Chen T, Hwang H, Rose ME, Nines RG, Stoner GD (2006a) Chemopreventive properties of black raspberries in N-nitrosomethylbenzylamine-induced rat esophageal tumorigenesis: down-regulation of cyclooxygenase-2, inducible nitric oxide synthase, and c-Jun. Cancer Res 66(5):2853–2859

Chen T, Rose ME, Hwang H, Nines RG, Stoner GD (2006b) Black raspberries inhibit N-nitrosomethylbenzylamine (NMBA)-induced angiogenesis in rat esophagus parallel to the suppression of COX-2 and iNOS. Carcinogenesis 27(11):2301–2307

Facciola S (1990) Cornucopia. A source book of edible plants. Kampong Publ, Vista, 677 pp

Grae I (1974) Nature's colors – dyes from plants. MacMillan Publishing Co, New York

Hákkinen S, Heinonen M, Kárenlampi S, Mykkánen H, Ruuskanen J, Törrönen R (1999) Screening of selected flavonoids and phenolic acids in 19 berries. Food Res Int 32:345–353

Han C, Ding H, Casto B, Stoner GD, D'Ambrosio SM (2005) Inhibition of the growth of premalignant and malignant human oral cell lines by extracts and components of black raspberries. Nutr Cancer 51(2):207–217

Hecht SS, Huang C, Stoner GD, Li J, Kenney PM, Sturla SJ, Carmella SG (2006) Identification of cyanidin glycosides as constituents of freeze-dried black raspberries which inhibit anti-benzo[a]pyrene-7,8-diol-9,10-epoxide induced NFkappaB and AP-1 activity. Carcinogenesis 27(8):1617–1626

Huang C, HuangY Li J, Hu W, Aziz R, Tang MS, Sun N, Cassady J, Stoner GD (2002) Inhibition of benzo(a) pyrene diol-epoxide-induced transactivation of activated protein 1 and nuclear factor kappaB by black raspberry extracts. Cancer Res 62(23):6857–6863

Huang C, Li J, Song L, Zhang D, Tong Q, Ding M, Bowman L, Aziz R, Stoner GD (2006) Black raspberry extracts inhibit benzo(a)pyrene diol-epoxide-induced activator protein 1 activation and VEGF transcription by targeting the phosphotidylinositol 3-kinase/Akt pathway. Cancer Res 66(1):581–587

Jennings DL (1988) Raspberries and blackberries: their breeding, disease and growth. Academic, London, 230 pp

Johnson JL, Bomser JA, Scheerens JC, Giusti MM (2011) Effect of black raspberry (*Rubus occidentalis* L.) extract variation conditioned by on colon cancer cell proliferation. J Agric Food Chem 59(5):1638–1645

Kresty LA, Frankel WL, Hammond CD, Baird ME, Mele JM, Stoner GD, Fromkes JJ (2006) Transitioning from preclinical to clinical chemopreventive assessments of lyophilized black raspberries: interim results show berries modulate markers of oxidative stress in Barrett's esophagus patients. Nutr Cancer 54:148–156

Kresty LA, Morse MA, Morgan C, Carlton PS, Lu J, Gupta A, Blackwood M, Stoner GD (2001) Chemoprevention of esophageal tumorigenesis by dietary administration of lyophilized black raspberries. Cancer Res 61:6112–6119

Kunkel G (1984) Plants for human consumption. An annotated checklist of the edible phanerogams and ferns. Koeltz Scientific Books, Koenigstein

Landete JM (2011) Ellagitannins, ellagic acid and their derived metabolites: a review about source, metabolism, functions and health. Food Res Int 44(5):1150–1160

Li J, Zhang D, Stoner GD, Huang C (2008) Differential effects of black raspberry and strawberry extracts on BaPDE-induced activation of transcription factors and their target genes. Mol Carcinog 47(4):286–294

Ling Y, Ren C, Mallery SR, Ugalde CM, Pei P, Saradhi UV, Stoner GD, Chan KK, Liu Z (2009) A rapid and sensitive LC-MS/MS method for quantification of four anthocyanins and its application in a clinical pharmacology study of a bioadhesive black raspberry gel. J Chromatogr B Analyt Technol Biomed Life Sci 877(31):4027–4034

Liu Z, Schwimer J, Liu D, Greenway FL, Anthony CT, Woltering EA (2005) Black raspberry extract and fractions contain angiogenesis inhibitors. J Agric Food Chem 53(10):3909–3915

Mallery SR, Budendorf DE, Larsen MP, Pei P, Tong M, Holpuch AS, Larsen PE, Stoner GD, Fields HW, Chan KK, Ling Y, Liu Z (2011) Effects of human oral mucosal tissue, saliva, and oral microflora on intraoral metabolism and bioactivation of black raspberry anthocyanins. Cancer Prev Res (Phila) 4(8):1209–1221

Mallery SR, Stoner GD, Larsen PE, Fields HW, Rodrigo KA, Schwartz SJ, Tian Q, Dai J, Mumper RJ (2007) Formulation and in-vitro and in-vivo evaluation of a mucoadhesive gel containing freeze dried black raspberries: implications for oral cancer chemoprevention. Pharm Res 24(4):728–737

Mallery SR, Zwick JC, Pei P, Tong M, Larsen PF, Shumway BS, Lu B, Fileds HW, Mumper RJ, Stoner GD (2008) Application of a bioadhesive black raspberry gel modulates gene expression and reduces cyclooxygenase 2 protein in human premalignant oral lesions. Cancer Res 68(12):4945–4957

Moerman D (1998) Native American ethnobotany. Timber Press, Oregon, 927 pp

Montrose DC, Horelik NA, Madigan JP, Stoner GD, Wang LS, Bruno RS, Park HJ, Giardina C, Rosenberg DW (2011) Anti-inflammatory effects of freeze-dried black raspberry powder in ulcerative colitis. Carcinogenesis 32(3):343–350

Parry J, Su L, Moore J, Cheng Z, Luther M, Rao JN, Wang JY, Yu LL (2006) Chemical compositions, antioxidant capacities, and antiproliferative activities of selected fruit seed flours. J Agric Food Chem 54(11):3773–3778

Parry J, Yu L (2004) Fatty acid content and antioxidant properties of cold-pressed black raspberry seed oil and meal. J Food Sci 69(3):FCT189–FCT193

Reen RK, Nines R, Stoner GD (2006) Modulation of N-nitrosomethylbenzylamine metabolism by black raspberries in the esophagus and liver of Fischer 344 rats. Nutr Cancer 54(1):47–57

Rodrigo KA, Rawal Y, Renner RJ, Schwartz SJ, Tian Q, Larsen PE, Mallery SR (2006) Suppression of the tumorigenic phenotype in human oral squamous cell carcinoma cells by an ethanol extract derived from freeze-dried black raspberries. Nutr Cancer 54(1):58–68

Seeram NP (2006) Berries. In: Heber D, Blackburn G, Go V, Milner J (eds) Nutritional oncology. Elsevier, Inc, Amsterdam, pp 615–628, Chapter 37

Seeram NP, Adams LS, Zhang Y, Lee R, Sand D, Scheuller HS, Heber D (2006) Blackberry, black raspberry, blueberry, cranberry, red raspberry, and strawberry extracts inhibit growth and stimulate apoptosis of human cancer cells in vitro. J Agric Food Chem 54(25):9329–9339

Stoner GD, Chen T, Kresty LA, Aziz RM, Reinemann T, Nines R (2006) Protection against esophageal cancer in rodents with lyophilized berries: potential mechanisms. Nutr Cancer 54(1):33–46

Stoner GD, Dombkowski AA, Reen RK, Cukovic D, Salagrama S, Wang LS, Lechner JF (2008a) Carcinogen-altered genes in rat esophagus positively modulated to normal levels of expression by both black raspberries and phenylethyl isothiocyanate. Cancer Res 68(15):6460–6467

Stoner GD, Sardo C, Apseloff G, Mullet D, Wargo W, Pound V, Singh A, Sanders J, Aziz R, Casto B, Sun XL (2005) Pharmacokinetics of anthocyanins and ellagic acid in healthy volunteers fed freeze-dried black raspberries daily for 7 days. J Clin Pharmacol 45:1153–1164

Stoner GD, Wang LS, Casto BC (2008b) Laboratory and clinical studies of cancer chemoprevention by antioxidants in berries. Carcinogenesis 29(9):1665–1674

Stoner GD, Wang LS, Seguin C, Rocha C, Stoner K, Chiu S, Kinghorn AD (2010) Multiple berry types prevent N-nitrosomethylbenzylamine-induced esophageal cancer in rats. Pharm Res 27(6):1138–1145

Stoner GD, Wang LS, Zikri N, Chen T, Hecht SS, Huang C, Sardo C, Lechner JF (2007) Cancer prevention with freeze-dried berries and berry components. Semin Cancer Biol 17(5):403–410

Tulio AZ Jr, Reese RN, Wyzgoski FJ, Rinaldi PL, Fu R, Scheerens JC, Miller AR (2008) Cyanidin 3-rutinoside and cyanidin 3-xylosylrutinoside as primary phenolic antioxidants in black raspberry. J Agric Food Chem 56(6):1880–1888

U.S. Department of Agriculture, Agricultural Research Service (USDA) (2011) USDA National Nutrient Database for Standard Reference, Release 24. Nutrient Data Laboratory Home Page, http://www.ars.usda.gov/ba/bhnrc/ndl

Wada L, Ou B (2002) Antioxidant activity and phenolic content of Oregon caneberries. J Agric Food Chem 50(12):3495–3500

Wang LS, Arnold M, Huang YW, Sardo C, Seguin C, Martin E, Huang TH, Riedl K, Schwartz S, Frankel W, Pearl D, Xu Y, Winston J 3rd, Yang GY, Stoner G (2011a) Modulation of genetic and epigenetic biomarkers of colorectal cancer in humans by black raspberries: a phase I pilot study. Clin Cancer Res 17(3):598–610

Wang LS, Dombkowski AA, Seguin C, Rocha C, Cukovic D, Mukundan A, Henry C, Stoner GD (2011b) Mechanistic basis for the chemopreventive effects of black raspberries at a late stage of rat esophageal carcinogenesis. Mol Carcinog 50(4):291–300

Wang LS, Hecht S, Carmella S, Seguin C, Rocha C, Yu N, Stoner K, Chiu S, Stoner G (2010) Berry ellagitannins may not be sufficient for prevention of tumors in the rodent esophagus. J Agric Food Chem 58(7):3992–3995

Wang LS, Hecht SS, Carmella SG, Yu N, Larue B, Henry C, McIntyre C, Rocha C, Lechner JF, Stoner GD (2009) Anthocyanins in black raspberries prevent esophageal tumors in rats. Cancer Prev Res (Phila) 2(1):84–93

Wang SY, Lin HS (2000) Antioxidant activity in fruits and leaves of blackberry, raspberry, and strawberry varies with cultivar and developmental stage. J Agric Food Chem 48(2):140–146

Xue H, Aziz RM, Sun N, Cassady JM, Kamendulis LM, Xu Y, Stoner GD, Klaunig JE (2001) Inhibition of cellular transformation by berry extracts. Carcinogenesis 22(2):351–356

Zhang Z, Knobloch TJ, Seamon LG, Stoner GD, Cohn DE, Paskett ED, Fowler JM, Weghorst CM (2011) A black raspberry extract inhibits proliferation and regulates apoptosis in cervical cancer cells. Gynecol Oncol 123(2):401–406

Zikri NN, Riedl KM, Wang LS, Lechner J, Schwartz SJ, Stoner GD (2009) Black raspberry components inhibit proliferation, induce apoptosis, and modulate gene expression in rat esophageal epithelial cells. Nutr Cancer 61(6):816–826

Rubus ursinus × idaeus 'Boysenberry'

Scientific Name

Rubus ursinus × *idaeus* 'Boysenberry' sensu Wada and Ou (2002).

Synonyms

Rubus loganobaccus × *baileyanus* Britt sensu McGhie et al. (2006), Barnett et al. (2007), *Rubus loganobaccus* cv boysenberry sensu Kiyoko (2005), *Rubus ursinus* var *loganobaccus* cv Boysenberry.

Family

Rosaceae

Common/English Name

Boysenberry

Vernacular Names

Dutch: Boysenbes;
Finnish: Boysenbeere, Boysenbes;
French: Baie De Boysen, Mûre De Boysen, Ronce De Boysen;
German: Boysenbeer, Brombeere;
Slovaščina: Medvedja Robida;
Slovenia: Medvedja Robida;
Swedish: Boysenbär, Boysenbeere, Boysenbes.

Origin/Distribution

Boysenberry is named after by the originator, Rudolf Boysen, a Swedish immigrant and horticulturist who developed the crop during the Great Depression in the Napa Valley region of California. Boysenberry enjoyed commercial success under the growing guidance and development of farmer and berry "expert" Walter Knott of Knott's Berry Farm. The Boysenberry's popularity is the single most reason for making Knott's Berry Farm so famous. Today, Boysenberry is grown as trailing vines throughout the Western Coast of the United States and they have been naturalized in Northern New Zealand, where the fruit is grown for commercial export. Over 60% of the world's Boysenberry production comes from New Zealand. Boysenberry is also grown in Australia.

The exact parentage of boysenberry is obscure. It is thought to be a cross between a European raspberry (*Rubus idaeus*), a common Blackberry (*Rubus fruticosus*), and loganberry (*Rubus* × *loganobaccus*). Scientists surmised, based on analyses of genes, plants and fruits, that Boysenberry, resulted from a cross of Loganberry with an Eastern dewberry.

Agroecology

Boysenberry is a cool climate crop. It grows in full sun to partial shade and does best in a well-drained, moist, loamy soil rich in organic matter from pH 6–7.5. It is usually grown on a support or trellis and needs to be protected from cold weather and strong winds. The vines need to be untied from the trellis and laid flat before a freeze. The canes also should be covered with at least a foot of loose hay or straw.

Plate 1 Prolific boysenberry vine

Edible Plant Parts and Uses

Boysenberry may be eaten fresh, in fruit salads, with ice-cream, in drinks, milk shakes or use alone with creams, in trifles, and to top fruit tarts. It can be frozen and similarly used. It can be used in jams, preserves, pies, cheesecake, muffins, cobblers, bars, pancakes and syrups, or even made into wine. Boysenberry is also used to infuse light spirits and muddle into cocktails.

Botany

A long trailing or climbing shrub usually grown on trellis to 6–7 m long. Primo canes are terete with soft, short prickles. Leaves imparinately compound with 3 shortly-stalked broadly ovate leaves with obtuse to acute to acuminate apices, serrated margins, adaxial surface darker green, abaxial surface pale green, pubescent or glabrous, with or without short prickles on the mid-vein and petioles (Plates 1, 2 and 3). Stipules narrow, free from petiole or fused at the base. Flowers 2–3 cm across, produce in short racemes or cymes on floricanes with 5 reflexed sepals, 5 mostly white spreading petals, numerous short stamens and few to many carpels on an elongated receptacle. Fruit about 3 cm long and from 2 to 3 cm in width, an aggregate of numerous succulent drupelets, each containing a single seed. Colour green changing to straw-yellow, amber, orange-red, red, reddish black, to glossy purplish-black (Plates 1, 2 and 3) as they ripen.

Plate 2 Close-up of boysenberry and leaves

Nutritive/Medicinal Properties

The nutrient value of frozen Boysenberries per 100 g edible portion was reported as follows (USDA 2011): water 85.90 g, energy 50 kcal (209 kJ), protein 1.10 g, total lipid (fat) 0.26 g,

Plate 3 Ripening boysenberries

ash 0.54 g, carbohydrate 12.19 g, total dietary fibre 5.3 g, total sugars 6.89 g, Ca 27 mg, Fe 0.85 mg, Mg 16 mg, P 27 mg, K 139 mg, Na 1 mg, Zn 0.22 mg, Cu 0.080 mg, Mn 0.547 mg, Se 0.2 μg, vitamin C 3.1 mg, thiamin 0.053 mg, riboflavin 0.037 mg, niacin 0.767 mg, pantothenic acid 0.250 mg, vitamin B-6 0.056 mg, total folate 63 μg, total choline 10.2 mg, vitamin A 3 μg RAE, vitamin A 67 IU, β-carotene 40 μg, lutein + zeaxanthin 118 μg, vitamin E (α-tocopherol) 0.87 mg, vitamin K (phylloquinone) 7.8 μg, total saturated fatty acids 0.009 g, 16:0 (palmitic) 0.006 g, 18:0 (stearic) 0.001 g, total monounsaturated fatty acids 0.025 g, 18:1 undifferentiated (oleic) 0.023 g, 20:1 (gadoleic) 0.002 g, total polyunsaturated fatty acids 0.148 g, 18:2 undifferentiated (linoleic) 0.098 g and 18:3 undifferentiated (linolenic) 0.050 g. Studies using labelled glucose-1-C14 showed that glucose was converted to the galacturonic acid and the arabinose moieties of pectin without cleavage of the carbon chain in boysenberry fruit (Seegmiller et al. 1955).

The major anthocyanins in Boysenberries were shown to be cyanidin 3-mono-glucoside and cyanidin 3-diglucoside (Luh et al. 1965). Also present were smaller amounts of cyanidin 3-rhamnoglucoside and cyanidin 3-rhamnoglucosido-5-glucoside. The major anthocyanins of Boysenberry fruit, (*Rubus loganobaccus* x *Rubus baileyanus* Britt), were isolated and elucidated as cyanidin-3-[2-(glucosyl)glucoside] and cyanidin-3-[2-(glucosyl)-6-(rhamnosyl)glucoside] (McGhie et al. 2006).

Compositional analysis revealed that polyphenolic extracts prepared from the waste seeds and commercial juice of Boysenberry contained six polyphenolic classes: flavanol monomers, proanthocyanidins, anthocyanins, ellagic acid, ellagitannins, and flavonol glycosides (Furuuchi et al. 2011). Ellagitannins were the most abundant polyphenols in both extracts. Proanthocyanidins were present as short oligomers consisting of dimeric and trimeric procyanidins and propelargonidins, with the most abundant component being procyanidin B4 in both extracts. The seeds contained a 72-fold higher amount of proanthocyanidins than the juice. These results indicated Boysenberry fruits contained short oligomeric proanthocyanidins along with flavanol monomers and the seeds represented a good source of short oligomeric proanthocyanidins.

Cold-pressed marionberry, boysenberry, red raspberry, and blueberry seed oils were found to contain significant levels of α-linolenic acid ranging from 19.6 to 32.4 g per 100 g of oil, along with a low ratio of n-6/n-3 fatty acids (1.64–3.99) (Parry et al. 2005). The total carotenoid content ranged from 12.5 to 30.0 μmol per kg oil. Zeaxanthin was the major carotenoid compound in all tested berry seed oils, along with β-carotene, lutein, and cryptoxanthin. Total tocopherol was 260.6–2276.9 μmol per kg oil, including α-, γ-, and δ-tocopherols. Seeds from all five caneberry species: red raspberry, black raspberry, Boysenberry, Marion blackberry, and evergreen blackberry were found to have 6–7% protein and 11–18% oil (Bushman et al. 2004). The oils contained 53–63% linoleic acid, 15–31% linolenic acid, and 3–8% saturated fatty acids. Antioxidant capacities were detected both for whole seeds and for cold-pressed oils but did not correlate to total phenolics or tocopherols. Ellagitannins and free ellagic acid were the main phenolics detected in all five caneberry species and were approximately threefold more abundant in the blackberries and the boysenberry than in the raspberries.

Antioxidant Activity

Results of studies by Wada and Ou (2002) indicated that caneberries [evergreen blackberries (*Rubus laciniatus*), marionberries (*Rubus ursinus*),

Boysenberries (*Rubus ursinus x idaeus*), red raspberries (*Rubus idaeus*), and black raspberries (*Rubus occidentalis*)] were high in antioxidant activity and were rich sources of anthocyanins and phenolics. All cane berries had high ORAC (oxygen radical absorbance capacity) activity ranging from 24 to 77.2 μmol of Trolox equivalent/g of fresh berries. Anthocyanin content ranged from 0.65 to 5.89 mg/g, and phenolics ranged from 4.95 to 9.8 mg/g. All berries had high levels of ellagic acid (47–90 mg/g). The oxygen radical absorbance capacity (ORAC) for Boysenberries (42.2 μmol of TE/g) was higher than that seen in either red raspberries or blackberries. Total phenolics in Boysenberry was 5.99 mg/g fresh weight based on gallic acid standard and total anthocyanin content based upon cyanidin 3-glucoside was 1.31 mg/g. Percent contribution of individual anthocyanins to total anthocyanins in boysenberry comprised cyanidin 3-(6′-p-coumaryl) glucoside-5-glucoside 56.27% which coeluted with cyanidin 3,5-diglucoside (minor amount), and cyanidin 3-glucoside 43.73%. Ellagic acid 70 mg/100 g was present primarily as the free form in Boysenberries, thus, there was little change in concentration after hydrolysis. Boysenberry (*Rubus loganobaccus* cv boysenberry), a hybrid cross between a loganberry and a blackberry had been reported to have potential to be an innovative functional ingredient (Kiyoko 2005). Its antioxidant activity was attributed to its anthocyanins with the major components identified as cyanidin-3-O-sophoroside, cyanidin-3-O-glucosylrutinoside and cyanidin-3-O-glucoside. Boysenberry also contained ellagic acid and folate.

Results of studies showed that Boysenberry (*Rubus loganobaccus* x *baileyanus* Britt) fruit extract functioned as an in-vivo antioxidant and raised the antioxidant status of plasma while decreasing some biomarkers of oxidative damage, but the effect was highly modified by basal diet (Barnett et al. 2007). When Boysenberry extract was added to the rat's basal diet containing fish and soybean oils, there was little change in DNA (8-oxo-2′-deoxyguanosine) excretion in urine, oxidative damage to proteins decreased, and plasma malondialdehyde either increased or decreased depending on the basal diet. The results provided further evidence of complex interactions among dietary antioxidants, background nutritional status as determined by diet, and the biochemical nature of the compartments in which antioxidants function.

Oxidative stability index (OSI) values were 20.07, 20.30, and 44.76 h for the marionberry, red raspberry, and Boysenberry seed oils, respectively (Parry et al. 2005). The highest TPC (total phenolic content) of 2.0 mg gallic acid equivalents per gram of oil was observed in the red raspberry seed oil, while the strongest oxygen radical absorbance capacity was in boysenberry seed oil extract (77.9 μmol Trolox equivalents per g oil). All tested berry seed oils directly reacted with and quenched DPPH radicals in a dose- and time-dependent manner. The data suggested that the cold-pressed berry seed oils may serve as potential dietary sources of tocopherols, carotenoids, and natural antioxidants.

Hepatoprotective Activity

Boysenberry anthocyanin was found to have heptoprotective activity. Increases in plasma aspartate aminotrasferase and alanine aminotrasferase, which were all induced by the injection of GalN in rats, were relieved by the intake of boysenberry anthocyanins, suggesting boysenberry anthocyanins to be a food component available to prevent liver injury (Igarashi et al. 2004). Seven polyphenols were isolated from leaves of New Zealand Boysenberry and elucidated as quercetin 3-O-glucuronide, quercetin 3-O-glucoside, quercetin 3-O-arabinoside, kaempferol 3-O-glucuronide, kaempferol 3-O-arabinoside, kaempferol 3-O-(6″-O-p-coumaroyl)-glucoside, and ellagic acid (Kubomura et al. 2006). Increases in plasma aspartate aminotrasferase and alanine aminotrasferase activities in mice, induced with liver injury by the injection of carbon tetrachloride, were suppressed by oral administration of the polyphenol fraction prepared from the leaves, and ellagic acid was found to be its effective component. Thus polyphenol fraction contained in boysenberry leaves may be effective in suppressing liver injury.

Antidiabetic Activity

Increases in the concentration of plasma thiobarbituric acid reactive substances (TBARS), and in the liver 8-hydroxy deoxyguanosine (8-OH dG)/deoxyguanosine (dG) ratio and also in the liver reduced glutathione/oxidized glutathione ratio (GSH/GSSG) ratio, which were all observed in streptozotocin-induced diabetic rats, were restored or tended to be restored to the level of the control rats when a diet with Boysenberry anthocyanins was given to the diabetic animals (Sugimoto et al. 2003). The susceptibility of the liver homogenate of the diabetic rats to the oxidation by AAPH (2,2'-azobis(2-amidinopropane) dihydrochloride) was relieved when boysenberry anthocyanins was fed to them. These results suggested that Boysenberry anthocyanins was effective in protecting the development of in-vivo oxidation involved with diabetes.

Neuroprotective Activity

Results of studies suggested that the putative toxic effects of Abeta or DA (dopamine) might be reduced by high antioxidant fruit extracts such as Boysenberry, cranberry, black currant, strawberry, dried plums, and grapes (Joseph et al. 2004). The fruit extracts antagonized Abeta- or DA-induced deficits in Ca2+ flux in M1-transfected COS-7 cells. The extracts showed differential levels of recovery protection in comparisons to the non-supplemented controls that was dependent upon whether DA or Abeta was used as the pretreatment. Interestingly, assessments of DA-induced decrements in viability revealed that all of the extracts had some protective effects. In-vitro studies showed that SH-SY5Y human neuroblastoma cells were protected against H2O2-induced toxicity by the anthocyanins and phenolic fractions of boysenberries and blackcurrants (Ghosh et al. 2006). The concurrent addition of either fractions of these berries with H2O2 significantly inhibited the increase in intracellular reactive oxygen species production. Studies demonstrated that extracts of Boysenberry and blackcurrant showed significant protective effect and restored the calcium buffering ability of transfected COS-7 cells cells that had been subjected to oxidative stress induced by dopamine and the amyloid β25–35 (Ghosh et al. 2007). Blackcurrant polyphenolics showed slightly higher protective effect against dopamine, whereas Boysenberry polyphenolics had a higher effect against the amyloid β25–35. In viability studies, all extracts showed significant protective effects against dopamine and amyloid β25–35-induced cytotoxicity. Four major anthocyanins, cyanidin glucoside, cyanidin rutinoside, cyanidin sophoroside and cyaniding glucorutinoside were detected in the Boysenberry anthocyanin. The total anthocyanin and phenolic concentrations in the boysenberry extracts 261 and 241 mg/g respectively. The results provided evidence to suggest that the deleterious neurotoxic effect of compounds such as dopamine or Aβ on neurons can be reduced by polyphenolic fractions of Boysenberry and blackcurrant. In further studies, long-term consumption of both the Boysenberry and blackcurrant drinks was found to elevate the plasma total antioxidant capacity of the study of elderly participants suggesting that Boysenberry and blackcurrant may help protect against oxidative stress-related health conditions and degenerative diseases (McGhie et al. 2007). Plasma malondialdehyde decreased in both the Boysenberry and blackcurrant treatments although the decrease was not statistically significant. Measures of oxidative stress for protein oxidation and lipid peroxidation improved for the berryfruit treatments during the study but were not statistically different from the placebo.

Bioavailability Studies

McGhie et al. (2003) investigated the bioabsorption of 15 anthocyanins with structures containing different aglycons and conjugated sugars extracted from blueberry, Boysenberry, black raspberry, and blackcurrant in both humans and rats. They detected intact and unmetabolized anthocyanins in urine of rats and humans following dosing for all molecular structures investigated, thus demonstrating anthocyanins with diverse molecular structure and from different dietary sources were bioavailable at diet relevant dosage rates. In addition, the relative concentrations of anthocyanins detected in urine following dosing varied, indicating that differences in bioavailability were due to variations in chemical structure. Their results suggested that the nature of the sugar conjugate and the phenolic aglycon to be both important determinants of anthocyanin absorption and excretion in rats and humans.

Other Uses

Oxi-fend® Boysenberry Extract is an antioxidant rich extract of Boysenberries from New Zealand where over 60% of the world's Boysenberries are grown (New Zealand Extract Ltd 2011). Oxifend® Plus is a powerful antioxidant formula blended from the extracts of four New Zealand superfruits: Grapes, Boysenberries, Blackcurrants, Kiwifruit.

Comments

Boysenberry is propagated from stem cuttings, or from seeds, stooling, mound layering and tip layering.

Selected References

Barnett LE, Broomfield AM, Hendriks WH, Hunt MB, McGhie TK (2007) The in vivo antioxidant action and the reduction of oxidative stress by boysenberry extract is dependent on base diet constituents in rats. J Med Food 10(2):281–289

Bushman BS, Phillips B, Isbell T, Ou B, Crane JM, Knapp SJ (2004) Chemical composition of caneberry (Rubus spp.) seeds and oils and their antioxidant potential. J Agric Food Chem 52(26):7982–7987

Furuuchi R, Yokoyama T, Watanabe Y, Hirayama M (2011) Identification and quantification of short oligomeric proanthocyanidins and other polyphenols in boysenberry seeds and juice. J Agric Food Chem 59(8):3738–3746

Ghosh D, McGhie TK, Fisher DR, Joseph JA (2007) Cytoprotective effects of anthocyanins and other phenolic fractions of Boysenberry and blackcurrant on dopamine and amyloid β-induced oxidative stress in transfected COS-7 cells. J Sci Food Agric 87: 2061–2067

Ghosh D, McGhie TK, Zhang J, Adaim A, Skinner M (2006) Effects of anthocyanins and other phenolics of boysenberry and blackcurrant as inhibitors of oxidative stress and damage to cellular DNA in SH-SY5Y and HL-60 cells. J Sci Food Agric 86:678–686

Igarashi K, Sugimoto E, Hatakeyama A, Molyneux J, Kubomura K (2004) Preventive effects of dietary boysenberry anthocyanins on galactosamine-induced liver injury in rats. Biofactors 21(1–4):259–261

Joseph JA, Fisher DR, Carey AN (2004) Fruit extracts antagonize Abeta- or DA-induced deficits in Ca2+ flux in M1-transfected COS-7 cells. J Alzheimers Dis 6(4):403–411

Kiyoko K (2005) Boysenberry as a functional food ingredient. J Integr Study Diet Habits 16(1):44–49

Kubomura K, Kurakane S, Molyneux J, Omori M, Igarashi K (2006) Identification of the major polyphenols in boysenberry leaves and their suppressive effect on carbon tetrachloride-induced liver injury in mice. Food Sci Technol Res 12(1):31–37

Luh BS, Stachowicz K, Hsia CL (1965) The anthocyanin pigments of boysenberries. J Food Sci 30:300–306

McGhie TK, Ainge GD, Barnett LE, Cooney JM, Jensen DJ (2003) Anthocyanin glycosides from berry fruit are absorbed and excreted unmetabolized by both humans and rats. J Agric Food Chem 51(16):4539–4548

McGhie TK, Rowan DR, Edwards PJ (2006) Structural identification of two major anthocyanin components of boysenberry by NMR spectroscopy. J Agric Food Chem 54(23):8756–8761

McGhie TK, Walton MC, Barnett LE, Vather R, Martin H, Au J, Alspach PA, Booth CL, Kruger MC (2007) Boysenberry and blackcurrant drinks increased the plasma antioxidant capacity in an elderly population but had little effect on other markers of oxidative stress. J Sci Food Agric 87:2519–2527

New Zealand Extract Ltd (2011) Oxifend Boysenberry extract http://www.nzextracts.co.nz/products.php

Parry J, Su L, Luther M, Zhou K, Yurawecz MP, Whittaker P, Yu L (2005) Fatty acid composition and antioxidant properties of cold-pressed marionberry, boysenberry, red raspberry, and blueberry seed oils. J Agric Food Chem 53(3):566–573

Seegmiller CG, Axelrod B, Mccready RM (1955) Conversion of glucose-1-C14 to pectin in the boysenberry. J Biol Chem 217(2):765–775

Sugimoto E, Igarashi K, Kubo K, Molyneux J, Kubomura K (2003) Protective effects of boysenberry anthocyanins on oxidative stress in diabetic rats. Food Sci Technol Res 9(4):345–349

U.S. Department of Agriculture, Agricultural Research Service (USDA) (2011) USDA National Nutrient Database for Standard Reference, release 24. Nutrient Data Laboratory Home Page. http://www.ars.usda.gov/ba/bhnrc/ndl

Vaughan JC, Geissler CA (2009) The new Oxford book of food plants. Oxford University Press, Oxford, p 88

Wada L, Ou B (2002) Antioxidant activity and phenolic content of Oregon caneberries. J Agric Food Chem 50:3495–3500

Rubus × loganobaccus

Scientific Name

Rubus × loganobaccus L. H. Bailey

Synonyms

Rubus loganobaccus L. H. Bailey, *Rubus ursinus* Cham. & Schltdl. var. *loganobaccus* (L. H. Bailey) L. H. Bailey, *Rubus ursinus* × *Rubus idaeus*.

Family

Rosaceae

Common/English Names

Loganberry, Phenomenal-Berry, Tayberry.

Vernacular Names

Chinese: Luo Gan Mei;
Danish: Loganbær;
Dutch: Loganbes;
Finnish: Boysenmarja, Jättivatukka, Loganinmarja;
French: Ronce De Logan, Ronce-Framboise;
German: Loganbeere;
Italian: Mora Di Rovo;
Japanese: Roogan Berii;
Portuguese: Amora-Framboesa;
Spanish: Mora Logan, Zarza De Logan, Zarza-Frambuesa.

Origin/Distribution

The loganberry, a hexaploid, is generally thought to be derived from a cross between the European red raspberry *Rubus idaeus* 'Red Antwerp' (tetraploid) and the American blackberry *Rubus ursinus* 'Aughinburgh' (octaploid) (Crane 1940). It was accidentally created in 1880 or 1881 in Santa Cruz, California, by the American lawyer and horticulturist James Harvey Logan (1841–1928). It is largely grown in Oregon and Washington in USA. It was introduced to the United Kingdom around 1899. It is cultivated in England and in Tasmania but has escaped and become a weed in Western Australia.

Agroecology

Loganberry is relatively cold-tender and thrives best in areas with mild winters and warm summers. It does best in full sun but requires partially shading during the hot summer. It prefers well drained, humus-rich, loamy soil and is best trained on a wire trellis or other structure.

Edible Plant Parts and Uses

Loganberries produce very large, maroon coloured berries thatare edible raw but may be too acidic as fresh dessert. They make delicious syrup, preserves, jams, crumbles, pie fillings and wines. They can be processed into a nutritious health drink and also mixes well with Sherry wine.

Botany

A spreading or climbing shrub to 4.5 m high when growing on a natural support, through supporting vegetation. Primocane stems are terete, glabrous with soft prickles 3–6 mm long. Flowering stems arise from the leaf axils of the floricane. Primocane leaves are imparipinnately compound, with 3, 5 or rarely (7) leaflets; the terminal leaflet is larger than the rest. Leaflets are sparsely pilose below, broadly-ovate, acute to acuminate tip, rounded bases and coarsely serrated margins and on 3–8.5 cm long petiole (Plate 1). Inflorescence are subcorymbose with 6–12 flowers. Sepals rarely with prickles. Petals 12–18 mm long by 7–9 mm wide, elliptic, white. Stamens shorter than styles. Fruit ovoid to oblong, aggregate fruit, initially green, ripening dark red (Plates 1 and 2) to dark maroon to purplish-black.

Plate 1 Leaves of loganberry seedling

Plate 2 Developing loganberry fruits

Nutritive/Medicinal Properties

Loganberry is a nutritious fruit and the nutrient composition of frozen loganberry per 100 g edible portion had been reported as: water 84.61 g, energy 55 kcal (230 kJ), protein 1.52 g, total lipid (fat) 0.31 g, ash 0.54 g, carbohydrate 13.02 g, total dietary fibre 5.3 g, total sugars 7.70 g, Ca 26 mg, Fe 0.64 mg, Mg 21 mg, P 26 mg, K 145 mg, Na 1 mg, Zn 0.34 mg, Cu 0.117 mg, Mn 1.247 mg, Se 0.2 µg, vitamin C 15.3 mg, thiamin 0.050 mg, riboflavin 0.034 mg, niacin 0.840 mg, pantothenic acid 0.244 mg, vitamin B-6 0.065 mg, total folate 26 µg, total choline 8.5 mg, vitamin A 2 µgRAE, vitamin A 35 IU, β-carotene, 21 µg, lutein + zeaxanthin 118 µg, vitamin E (α-tocopherol) 0.87 mg, vitamin K (phylloquinone) 7.8 µg, total saturated fatty acids 0.011 g, 16:0 (palmitic) 0.007 g, 18:0 (stearic) 0.002 g, total monounsaturated fatty acids 0.030 g, 18:1 undifferentiated (oleic) 0.028 g, 20:1 (gadoleic) 0.002 g, total polyunsaturated fatty acids 0.176 g, 18:2 undifferentiated (linoleic) 0.117 g and 18:3 undifferentiated (linolenic) 0.059 g (USDA 2011).

See also notes under *Rubus ursinus* × *idaeus* 'Boysenberry'.

Other Uses

Loganberry is used as a source of breeding material for the development of *Rubus* hybrids.

Comments

Refer also to notes on blackberries, *Rubus fruticosus* aggr. and *Rubus ursinus* × *idaeus* 'Boysenberry'.

Selected References

Bailey LH (1976) Hortus third. A concise dictionary of plants cultivated in the United States and Canada. Liberty Hyde Bailey Hortorium/Cornell University/Wiley, New York, 1312 pp

Crane MB (1940) The origin of new forms in *Rubus*. II. The loganberry, *R. loganobaccus* Bailey. J Genet 40:129–140

Huxley AJ, Griffiths M, Levy M (eds) (1992) The new RHS dictionary of gardening (4 vols). MacMillan, New York

Kunkel G (1984) Plants for human consumption. An annotated checklist of the edible phanerogams and ferns. Koeltz Scientific Books, Koenigstein

U.S. Department of Agriculture, Agricultural Research Service (USDA) (2011) USDA National Nutrient Database for Standard Reference, release 24. Nutrient Data Laboratory Home Page. http://www.ars.usda.gov/ba/bhnrc/ndl

Sorbus domestica

Scientific Name

Sorbus domestica L.

Synonyms

Cormus domestica Spach, *Malus sorbus* Borkh, *Mespilus domestica* All., *Pirus sorbus* Gaertn., *Pyrus sorbus* Gaertn., *Pyrus domestica* (L.) Ehrh., *Pyrus domestica* (L.) Sm., *Pyrenia sorbus* Clairv., *Sorbus edulis* C. Koch ex Kirch.

Family

Rosaceae

Common/English Names

Beam Tree, Checker Tree, Chess Tree, Service Tree, Service Tree Mountain Ash, Sorb Tree, True Service Tree, Whitty Pear.

Vernacular Names

Croatian: Oskoruša;
Czech: Jeoáb Oskeruše, Jeřáb Oskeruše, Oskeruše Domácí;
Danish: Storfrugtetrøn;
Dutch: Peervormige Lijsterbes;
Eastonian: Aedpihlakas;
Finnish: Vaelimeren Pihlaja;
French: Corbier, Cormier, Sorbier, Sorbier Domestique;
German: Hauseberesche, Speierling, Speierling Sperb, Sperbe, Sperberbaum, Zahme, Zahmer Eberesche;
Hungarian: Kerti Berkenye;
Iceland: Berjareynir;
Italian: Sorbo, Sorbo Comune, Sorbo Domestico;
Japanese: Furanso Nanakamado, Nanakamado;
Polish: Jarzab Domowy;
Portuguese: Sorveira;
Russian: Rjabina Domašnaja, Rjabina Krymskaja;
Slovaščina: Mokovec – Skorž, Navadni Skorš, Oskoruš;
Slovencina: Jarabina Oskorušová, Oskoruša, Oskorušina Mukinja;
Spanish: Acafresna, Azarollo Serbal Común, Sorbo;
Turkish: Üvez, Üvez Ağacı.

Origin/Distribution

Sorb tree is native to western, central and southern Europe, northwest Africa (Atlas Mountains) and southwest Asia (east to the Caucasus).

Agroecology

Sorb tree thrives in most reasonably good soils in an open sunny position. It tolerates light shade, though it fruits better in a sunny position.

Edible Plant Parts and Uses

Ripe fruit is eaten raw or cooked. The fruit is usually bletted if it is going to be eaten raw. This involves storing the fruit in a cool dry place until it is almost but not quite going rotten. At this stage the fruit has a delicious taste, somewhat like a luscious tropical fruit. They are processed into pies, liqueurs, cider-like beverage and marmalade. The fruit can also be dried and used like prunes and after drying and grinding is used as ingredient for baking cake. The fruit is also is used for making traditional fruit wine: addition of about 5% of sorb tree fruits promotes the clearing and improves the taste of the wine. Because of the high sugar content (up to 12%) fermented fruits can be distilled to a fine brandy called "Sorbette" in France, "Sperbel" in SW Germany.

Botany

It is a deciduous tree growing to 15–20 m tall with a trunk up to 1 m diameter and brown bark which is smooth on young trees, becoming fissured and flaky on old trees. The winter buds are green, with a sticky resinous coating. Leaves are imparipinnate, 15–25 cm long, with 6–9 pairs of leaflets; leaflets oblong to oblanceolate, 3–5 cm long, 1–2 cm wide, margins serrated, with a bluntly acute apex, lower surface pubescent and becoming glabrescent with age; petiole 3–5 cm long (Plates 2 and 3). Flowers are produced in corymbs, about 15 mm diameter, hermaphrodite, with five white petals, triangular sepals, 15–25 creamy-white stamens and 5–6 styles. Fruit is a pome 2–3 cm long, 2 cm diameter obovoid or pyriform, greenish (Plate 1) or brownish, often tinged red on the side exposed to sunlight, with numerous stone cells.

Nutritive/Medicinal Properties

Chemical composition of service tree fruit pulp was reported as follows: protein 0.44-0.65g/kg, lipids (not analysed), reducing sugars 14–16%, fructose 9.40–10.20% total FW, glucose 4.10–5.40% total FW, vitamin C 0.89–0.98mg/kg, vitamin E 1.00–2.35mg/kg (Brindza et al. 2009). Chemical composition of the seeds was reported as: protein 32.9g/kg, lipids 205.3g/kg, reducing sugars 2%, fructose 1.83% total FW, glucose 0.57% total FW, vitamin C < 0.1mg/kg, vitamin E 18.95mg/kg (Brindza et al. 2009). The seeds were found to contain 15.6% cis-10-pentadecenoic acid (15:1 ω5, cis), 7.2% palmitic acid (16:0), 3.1% stearic acid (18:0), 28.3% oleic acid (18:1 ω9, cis), 33.1% linoleic acid (18:2 ω6, cis), and 12.7%

Plate 1 Immature fruits

Plate 2 Upper surface of compound leaf

Plate 3 Leaflets with serrated margins

linolenic acid (18:3 ω3, cis). They found that the cation content of service tree fruits was higher than in apples or pears, e.g., there were 3–4 times more potassium and calcium in true service tree fruits.

Antioxidant Activity

Research reported sorb fruit to be rich in antioxidants. Dichloromethane, diethyl ether and ethyl acetate fractions of *Sorbus domestica* fruits possessed significant radical-scavenging activities (DPPH˙ and luminol-induced chemiluminescence methods) which were greater than the activity of Trolox (Termentzi et al. 2006). This seemed to be correlated with their total phenolic content as determined by the Folin-Ciocalteau assayt. Unripe yellow fruits, together with the fruit pulp, were the strongest antioxidants, while the well-matured brown fruits were the weakest ones. Results showed that the fractions of diethyl ether, ethyl acetate and dichloromethane, can be used as antioxidants in food and medicinal preparations. In a subsequent research, 62 different phenolics were identified in *Sorbus domestica* fruits (Termentzi et al. 2008b). There were significant qualitative and quantitative differentiations in the phenolic content among the different maturity stages of the fruits. All stages were rich in benzoic, phenylpropanoic and cinnamoylquinic acids and derivatives. Unripe fruit categories were also rich in flavonoids, while well matured fruit categories had a low content of flavonoids. Fruit pulp, which was proved to be a strong antioxidant, contained very low amounts of both acids and flavonoids, but its phenolic content was highly qualitatively differentiated from the other categories. Egea et al. (2010) found that the hydrogen peroxide scavenging capacity and the Trolox equivalent antioxidant capacity (TEAC) value varied widely for *Sorbus domestica* and *Rosa canina*, ranging between 3.63% and 87.26% inhibition of hydrogen peroxide and between 0.47 and 416.64 mM Trolox/g FW, respectively.

Antidiabetic Activity

The diethyl ether and ethyl acetate fractions of *S. domestica* fruits possessed high aldose reductase inhibitory activity in-vitro (Termentzi et al. 2008a). Detailed phytochemical LC-DAD-MS (ESI+) analysis showed that this aldose reductase inhibitory activity could be attributed to the high content of flavonoids and hydroxycinnamoyl esters. Aldose reductase (ALR2) is a rate-limiting enzyme in the polyol pathway associated with the conversion of glucose to sorbitol and whose activity is implicated in the development of the long-term diabetic complications.

Traditional Medicinal Uses

Sorbus domestica is one of 126 commonly used traditional medicinal plants in Kirklareli Province, Turkey (Kültür 2007). The traditional medicinal plants have been mostly used for the treatment of wounds (25.3%), cold and influenza (24.6%), stomach (20%), cough (19%), kidney ailments (18.2%), diabetes (13.4%).

After drying and grinding the fruit is also reported to be used in the folk medicine against diarrhoea.

Other Uses

The bark is a source of tannin. The wood, the heaviest one of the European deciduous trees is very hard, heavy, hard to split, wonderfully coloured, fine grained can be well polished and is in demand under the trade name "Swiss pear tree" and is used for furniture, screws, wine presses etc.

Comments

The tree is propagated from seeds or cuttings.

Selected References

Aldasoro JJ, Aedo C, Navarro C, Garmendia FM (1998) The genus *Sorbus* (Maloideae, Rosaceae) in Europe and in North Africa: morphological analysis and systematics. Syst Bot 23(2):189–212

Brindza J, Červeňáková J, Tóth D, Bíro D, Sajbidor J (2009) Unutilized potential of true service tree (*Sorbus domestica* L.). Acta Hort (ISHS) 806:717–726

Egea I, Sánchez-Bel P, Romojaro F, Pretel MT (2010) Six edible wild fruits as potential antioxidant additives or nutritional supplements. Plant Foods Hum Nutr 65(2):121–129

Hampton M, Kay QON (1995) *Sorbus domestica* L., new to Wales and the British Isles. Watsonia 20(4):379–384

Huxley AJ, Griffiths M, Levy M (eds) (1992) The new RHS dictionary of gardening, 4 vols. MacMillan, London

Kültür S (2007) Medicinal plants used in Kirklareli Province (Turkey). J Ethnopharmacol 111(2):341–364

Kunkel G (1984) Plants for human consumption. An annotated checklist of the edible phanerogams and ferns. Koeltz Scientific Books, Koenigstein

Rushforth KD (1999) Trees of Britain and Europe. Collins, London, 1336 pp

Termentzi A, Alexiou P, Demopoulos VJ, Kokkalou E (2008a) The aldose reductase inhibitory capacity of Sorbus domestica fruit extracts depends on their phenolic content and may be useful for the control of diabetic complications. Pharmazie 63(9):693–696

Termentzi A, Kefalas P, Kokkalou E (2006) Antioxidant activities of various extracts and fractions of *Sorbus domestica* fruits at different maturity stages. Food Chem 98(4):599–608

Termentzi A, Kefalas P, Kokkalou E (2008b) LC–DAD–MS (ESI+) analysis of the phenolic content of *Sorbus domestica* fruits in relation to their maturity stage. Food Chem 106(3):1234–1245

Usher G (1974) A dictionary of plants used by man. Constable, London, 619 pp

Aegle marmelos

Scientific Name

Aegle marmelos (L.) Correa.

Synonyms

Belou marmelos (L.) A. Lyons, *Belou marmelos* (L.) W.F. Wight, *Bilacus marmelos* Kuntze, *Crataeva marmelos* L. basionym, *Crateva religiosa* Ainslie, *Cydonia indica* Spach., *Feronia pellucida* Roth.

Family

Rutaceae

Common/English Names

Bael Fruit, Bael Fruit Tree, Bael Tree, Ball Tree, Bel Fruit, Bela Tree, Bengal Quince, Elephant Apple, Golden Apple, Holy Fruit, Indian Bael, Indian Quince, Maredoo, Quince-Apple Of India, Stone Apple, Wood Apple.

Vernacular Names

Arabic: Safargal Hindî, Safarjal E Hindî, Shul;
Burmese: Opesheet, Okshit, Mak-Pyin;
Chinese: Ma Bi Za, Ma Di Hang, Mu Ju, Yin Du Gou Qi;
Dutch: Slijm Appelboom;
French: Egle Marmel, Bel Indien, Cognassier Du Bengal, Coing De L'inde, Oranger De Malabar, Oranger Du Malabar;
German: Bälbaum, Belbaum, Bengalische Quitte, Indische Quitte;
Hungarian: Bengálibirs;
India: Bel (Assamese), Maredu (Andhra Pradesh), Bel, Bael (Bengali), Bel, Belethi (Garo), Bel, Beli, Belgiri, Bili (Gujerati), Bael, Bil (Himachal Pradesh), Baelada Mara, Bel, Bael Sripal, Bello, Belpatra, Bilva, Si-Phal, Sirphal, Sri-Phal, Siri-Phal, Bilva, Bel Patri, Bil, Beel, Beelgiri, Beeley, Bael, Bel-Patra, Vilva, Willaw, Willau (Hindi), Belapatre, Bilapatri-Hannu, Bilva, Bilvapatre, Kumbala, Malura, Baelada Mara, Byaalada Hannu, Bilva Patre, Belavina, Bellapatre, Bilapatri, Biliptari, Billadu, Bilpatre, Bilpatri, Bilpattiri, Belavina Mara, Bilpathre, Bilva Pathre, Maaluraa (Kannada), Vilwam, Vilyam (Kerala), Bello (Konkani), Covalam, Koovalam, Kulakam, Kuvalam, Kuvalap-Pazham, Kuvvalam, Mavilavu, Maaredy Vilvam, Covalum, Kovalam (Malayalam), Heirikhagok (Manipuri), Bel (Bael), Bili, Baela, Vel, Bael, Bel Maredu, Vel (Marathi), Belthei, Bel-Thei (Mizoram), Belo, Belthei, Bel (Oriya), Adhararuha, Aritaki, Asholam, Atimangaliya, Balva, Bilva, Bilvah, Bilvam, Bilvaphalam, Bivalva, Duraruha, Gandhagarbha, Gandhapatra, Gohki, Granthila, Hridyagandha, Kantakadhya, Kantaki, Kapitana, Karkatavha, Lakshmiphala, Mahakapitha, Mahakapithakhya, Mahaphala, Mahaphalah, Malura, Malurah, Mangalya, Nilamallika,

Patrashreshtha, Pitaphala, Putivata, Sadaphala, Sailusa, Sailusah, Salatuh, Samirasara, Sandilya, Sandilyah, Sangrahi, Satyadharma, Satyaphala, Shailapatra, Shailusha, Shalatu, Shalya, Shandilya, Shivadruma, Shiveshtha, Shriphala, Sitanuna, Sivadrumah, Sriphal, Sriphala, Sriphalah, Sunitika, Tripatra, Trishakhapatra, Trishikha, Vilva, Vilvah, Vilvaka, Vilvapesika, (Sanskrit), Aluvigam, Vilva-Pazham, Bilvam, Vilvam, Kuvilam Palam, Vilvappu, Kuvilam, Villai, Civatturumam, Kucapi, Kuvilai, Kuvinam, Maluram, Ninmalli, Vilwam, Villvam, Iyalbudi, Mavilangai, Koovilam, Villuvam, Bilva, Vilva, Akuvakananmeccumilai, Alluram, Alukam, Alukam, Aluvikam, Anincil, Aranpucaikkerramaram, Arcanaiyati, Capalukam, Catapalam, Catippattiram, Cattal, Cirettamaruntati, Ciripalam, Ciripalam, Ciripalamaram, Cirivirutcam, Civankam, Civattirumaram, Civattiruvam, Iyalpu, Iyalpupati, Iyalputi, Kantapalai, Kantapattiram, Karkatam, Karuvila, Karuvilakikamaram, Karuvilakitam, Karuvilam, Kentakakarpam, Kentapattiram, Kovaritaki, Kuvalum, Makakapittam, Makapalai, Makapalam, Makapittam, Makavali, Makavalimaram, Makavalli, Maluramaram, Maruntirkati, Mavilamaram, Mikuttikam, Mikuttikamaram, Mikuttiyal, Nirmatalam, Pacunakam, Pacunakamaram, Pattiracirettam, Pattiri, Piracinapanacam, Piracinapanacam, Pitapalam, Pukku, Pukkuli, Pukkulimaram, Putiiratam, Putimarutam, Putimarutamaram, Putivakam, Putivatam, Puttiru, Tiricakam, Tiricakamaram, Tiricam, Tiricikam, Tirucam, Mirutiyal, Mirutiyam, Mirutiyamaram, Mulamukkanai, Munkalantavirutcatti, Mutantamuli, Muvilaicci, Navacikaram, Nilamalikkam, Nimilaiccattumuli, Viccanniyam, Viccanniyamaram, Vil, Villankam, Vilvam, Virinikamaram, Virinka, Virutiyakentam, Viyalputi, Vailavam, Vanamuli, Vatacaram, Vatam, Vatam, Vatamaram (Tamil), Bilva-Pandu, Bilvamu, Bilvapandu, Maluramu, Maredu, Sailushamu, Sandiliyamu, Sriphalamu, Bilva, Maredi, Bilvachettu, Maalaaramu, Maraedu, Shandiliyamu, Shreephalamu, Valaga (Telugu), Bel, Baelphal, Biligiris, Belgiri (Bael), Bel Kham, Guda Belgiri Taza, Chalbel, Belgiri (Urdu);

Indonesia: Kawista, Maja, Maja Batu, Maja Batuh, Maja Pahit, Maja Ingus, Maja Galepung, Maja Gedang, Maja Lumut, Maja Pait, Modjo (Javanese);
Italian: Cotogno Del Bengala, Cotogno D'india;
Japanese: Berunoki,Igure Marumerozu;
Khmer: Bnau, Phneou, Pnoi;
Laos: Toam, Toum;
Malaysia: Bilak, Bila, Bel, Maja Pahit;
Nepalese: Bel,Belapatra, Belpatra;
Pakistan: Bel, Bel Kham, Belgiri;
Persian: Bah Hindi Shull, Shul, Safarjale-Hindi, Shul, Safarjalehindi;
Philippines: Bael;
Polish: Klejowiec Jadalny;
Portuguese: Marmelos, Marmelos De Bengala, Marmeleiro De India;
Spanish: Bela, Milva;
Sri Lanka: Be Li (Sinhalese);
Taiwan: Ying Pi Ju;
Thai: Matoom, Tom, Ma Pin;
Tibetan: Bil Ba, Bi Lva, Bil Ba (D), Ka-Bed;
Turkish: Hind Ayva Agh;
Vietnamese: Cây Bao Nao, Mbau Nau, Trai Mam, Trai Man.

Origin/Distribution

Bael is indigenous to Peninsular India, Pakistan, Bangladesh and Myanmar.

It is also grown elsewhere throughout tropical/subtropical Asia and the Pacific.

Agroecology

Bael fruit grows wild in dry forests on hills and plains of central and southern India and Burma, Pakistan and Bangladesh. It is also found in mixed deciduous and dry dipterocarp forests. It grows from sea level up to 1,200 m altitude in the tropics and sub-tropics. It can grow under harsh conditions, including extremes of temperature in India. The tree can tolerate alkaline soil and is drought tolerant. In Thailand, it only flowers and fruits well after a prominent dry season. It does best on rich, well-drained soil, but it has grown well and fruited on the oolitic limestone of southern Florida. It also grows well in swampy, alkaline or stony soils having pH range from 5 to 8.

Edible Plant Parts and Uses

The sweet fragrant pulp is eaten as fruit or vegetable. Ripe fruit is eaten fresh after removal of mucilaginous seed and is also prepared as sherbet, syrup, jelly and fruit nectar. Beating the seeded pulp together with milk and sugar makes a popular drink called sherbet in India. A beverage is also made by combining bael fruit pulp with that of tamarind. The fruit can also be made into pickles and preserves. The best preparation of bael-fruit is a marmalade made from the full-grown but still tender fruit, cut into thin slices. Young leaves and shoots are also eaten raw as vegetables in Thailand and used as condiment to season food in Indonesia. They are said to reduce the appetite. An infusion of the flower is also used as a cooling drink.

Botany

A small to medium-sized, aromatic deciduous tree up to 8.5 m high with yellowish brown shallowly furrowed corky bark and irregular canopy. The stem and branches are light brown to green with sharp axillary spines present on the branches. Leaves are alternate, pale green comprising 3–5 leaflets, the lateral leaflets are ovate-lanceolate, 4.1 cm long by 2.2 cm wide, with crenate margin, acuminate apex, sessile and are glabrous and densely minutely glandular-punctuate on both surfaces (Plates 1 and 2). The terminal leaflets are 5.7 cm long by 2.8 cm wide, on a long (3.2 cm) petiole. Flowers are axillary, greenish white, large, 3–5 cm across, sweetly scented, bisexual, regular, ebracteate, stalked and borne in lateral panicles of about 10 flowers. The calyx is gamosepalous, five-lobed, pubescent, light green and small. The corolla is polypetalous, with five imbricate, leathery petals, 4 mm long, pale yellow from above and green from beneath. The androecium is polyandrous with numerous basifixed, 4 mm long filaments.

Plate 1 Trifoliate bael leaves

Plate 2 Immature bael fruit and leaves

Plate 3 Bael fruits

The gynoecium is light green, 7 mm long, with a capitate stigma and terminal style. The fruit is green turning yellowish-green to yellow when ripe, with small blotches on the outer surface, oblong, oval to globose, 7–14 cm in diameter (Plates 2 and 3). The fruit rind is tough, woody, 4–5 mm thick and pellucid dotted, and the pulp is yellow-orange, aromatic and mucilaginous.

Seeds are numerous and embedded in the pulp. The seed is oblong, compressed, white with cotton-like hairs on the testa.

Nutritive/Medicinal Properties

The food composition of fresh, ripe bael fruit per 100 g edible portion (Gopalan et al. 2002) was reported as: moisture 61.5 g, energy 137 kcal, protein 1.8 g, fat 0.3 g, carbohydrate 31.8 g, fibre 2.9 g, ash 1.7 g, vitamin A 2.3 µg RE (4.6 µg RAE), total carotene 55 µg, vitamin C 8 mg, thiamin 0.13 mg, riboflavin 0.03 mg, niacin 1.1 mg, iron 0.6 mg, calcium 85 mg and phosphorus 50 mg. The soft, very fragrant and pleasantly flavoured yellowish-orange pulp was found to be rich in protein, carotene, thiamine, riboflavin niacin and vitamin C.

Compared with orange and grapefruit taken as references, the bael fruit contained about three times more total soluble solids and at least 1.5 times more energy (Barthakur and Arnold 1989). Aspartic acid constituted over 32% of the amino acids analyzed. The vitamin C of the fruit was higher than grapefruit but considerably less than that of orange. Of the 11 minerals studied, Fe was found to be 21 times more abundant in bael than in either of the reference fruits. Fully ripe Thai bael fruit pulp was found to have moisture content 67.74%, total soluble solids 39.50°Brix, pH 5.37, reducing sugar 39.60 mg glucose/g fresh weight (fw), total acidity 0.94%, total dietary fibre 19.84 g/100 g dw, soluble dietary fibre 11.22 g, insoluble dietary fibre 8.62 g (Charoensiddhi and Anprungt 2008).

Various phytochemical and biological evaluations have been reported in literature for the medicinal and pharmacological importance and properties of the *Aegle marmelos*.

Other Phytochemicals

From Fruit

Bael fruit contained coumarins like alloimperatorin, and β-sitosteroal (Saha and Chatterjee 1957). An alkaloid named marmeline was isolated and identified as N-2-hydroxy-2-[4-(3′,3′-dimethylallyloxy)phenyl]ethyl cinnamide from unripe fruit (Sharma et al. 1981). Aegline, imperatorin, alloimperatorin and xanthotoxol were also found. Alloimperatorin methyl ether, a coumarin and alkaloids O-isopentenylhalfordinol and O-methylhalfordinol were isolated from the fruit (Sharma and Sharma 1981). Phenolic compounds umbelliferone, psoralen, and eugenol were quantified in the dried fruit pulp of *Aegle marmelos* and in the fruit of *Trachyspermum ammi* and *Foeniculam vulgare* (Dhalwal et al. 2007). The average percentage recovery was found to be 98.88% for umbelliferone, 100.104% for psoralen, and 99.33% for eugenol. A novel lectin isolated from bael fruit pulp was found to be a dimeric protein with N-acetylgalactosamine, mannose and sialic acid binding specificity (Raja et al. 2011a, b).

Twenty-eight volatile compounds were characterized in fully ripe bael fruit pulp (Charoensiddhi and Anprung 2008). Among these components, monoterpenes and sesquiterpenes such as limonene (32.48%), p-cymene (27.19%), dihydro-β-Ionone (3.95%), and β-phellandrene (3.13%) were recognized as the main components and important contributors to fruit aroma. Limonene was one of the major constituents that produces the characteristic bael fruit flavor. The other volatile compounds identified B-ionone 2.95%, caryophyllene oxide 2.57%, (E)-6,10-dimethyl-5,9-Undecadien-2-one 2.16%, isomyl acetate 2.13%, β-caryophyllene 1.88%, hexadecane 1.71%, pulegone 1.71%, α-cubebene 1.75%, dehydro-p-cymene 1.47%, 3,5-octadiene-2-one 1.44%, citronellal 1.42%, verbenone 1.25%, β-cubebene 1.18%, α-humulene 1.08%, (E,E)-2,4-heptadienal 1.08%, carvone 0.98%, hexanal 0.93%, hexadecanoic acid 0.92%, linalool oxide 0.88%, carvyl acetate 0.82%, humulene oxide 0.78%, trans-p-mentha-2,8-dienol 0.71%, acetoin 0.7% and (E)-2-octenal 0.6%.

The homogeneous, neutral polysaccharide isolated from the crude polysaccharide of bael fruit pulp contained arabinose, galactose, and glucose in the molar ratios of 2:3:14 (Basak et al. 1981).

Purified bael gum was found to comprise three neutral and two acidic oligosaccharides, together with monosaccharides (Roy et al. 1975). These sugars were identified as 3-0-β-D-galactopyranosyl-L-arabinose, 5-0-β-D-galactopyranosyl-L-arabinose, and 3-0-β-D-galactopyranosyl-D-galactose, and the acidic oligosaccharides as 3-0-(β-D-galactopyranosyluronic acid)-D-galactose and 3-0-(β-D-galactopyranosyluronic acid)-3-0-βa-D-galactopyranosyl-D-galactose.

From Seeds

The homogeneous fraction from the crude carbohydrate material from bael seed contained 38.5% of carbohydrate and 60.6% of protein, and its carbohydrate moiety consisted of glucose, galactose, rhamnose, and arabinose in the molar ratios of 40:3:1:2 (Mandal and Mukherjee 1981). Luvangetin, a pyranocoumarin was isolated from the seeds (Goel et al. 1997). Antifungal constituents, 2-isopropenyl-4-methyl-1-oxa-cyclopenta[b]anthracene-5,10-dione and (+)-4-(2′-hydroxy-3′-methylbut-3′-enyloxy)-8H-[1,3]dioxolo[4,5-h]chromen-8-one in addition to known compounds imperatorin, β-sitosterol, plumbagin, 1-methyl-2-(3′-methyl-but-2′-enyloxy)-anthraquinone, β-sitosterol glucoside, stigmasterol, vanillin and salicin were isolated from the seeds (Mishra et al. 2010b). A new anthraquinone, 1-methyl-2-(3′-methyl-but-2′-enyloxy)-anthraquinone was also isolated from bael seeds (Mishra et al. 2010a).

From Leaves

Aegelenine, an alkaloid, was isolated from the leaves (Chatterjee and Roy 1957). Alkaloids isolated from the leaves of *A. marmelos* were reported as O-3,3-(dimethylallyl)halfordinol, N-2-ethoxy-2-(4-methoxyphenyl) ethylcinnamamide, N-2-methoxy-2-[4-(3′,3′-dimethylallyloxy) phenyl] ethylcinnamamide, N-2-methoxy-2-(4-methoxyphenyl) ethylcinnamamide and marmeline (Manandhar et al. 1978). Four alkaloids, *N*-2-[4-(3′, 3′-dimethylallyloxy) phenyl]ethyl cinnamide, *N*-2-hydroxy-2-[4-(3′, 3′-dimethylallyloxy) phenyl]ethyl cinnamide, *N*-4-methoxystyryl cinnamide and *N*-2-hydroxy-2-(4-hydroxyphenyl) ethyl cinnamide were isolated from dried bael leaves (Govindachari and Premila 1983). Also isolated were aegeline and an unidentified purple compound. An alkaloid, N-2-hydroxy-2-(4-methoxyphenyl)-ethylcinnamamide (aegeline) was isolated from the leaves (Riyanto et al. 2001). Marmenol, a new 7-geranyloxycoumarin [7-(2,6-dihydroxy-7-methoxy-7-methyl-3-octaenyloxy)coumarin] was isolated from the methanol leaf extract (Ali and Pervez 2004). In addition to marmenol, several known compounds were obtained: praealtin D, trans-cinnamic acid, valencic acid, 4-methoxy benzoic acid, betulinic acid, N-p-cis-coumaroyltyramine, trans-coumaroyltyramine, montanine, and rutaretin. A series of phenylethyl cinnamides, which included new compounds named anhydromarmeline, aegelinosides A and B, were isolated from *Aegle marmelos* leaves (Phuwapraisirisan et al. 2008). Shahidine, a oxazoline alkaloid was isolated as a major constituent from the leaves and found to be the parent compound of aegeline and other amides (Faizi et al. 2009). Periplogenin-3-O-D-glucopyranosyl-(1→6)-D-glucopyaranosyl-(1→4)-D-cymaropyranoside, a cardenolide was isolated from *Aegle marmelos* leaves (Panda and Kar 2009).

Aegle marmelos leaf oil contained eight monoterpene hydrocarbons (71.85%), ten oxygenated monoterpenes (8.54%), four sesquiterpene hydrocarbons (14.02%) and one oxygenated sesquiterpene (0.78%) (Garg et al. 1995). Of the more than 20 compounds identified, myrcene (54.03%) was found to predominate and p-menth-1-en-3 β,5 β-diol (1.04%) was characterized as a new constituent. The leaf essential oil of *Aegle marmelos* was found to contain 15 compounds, including seven monotorpene hydrocarbons (90.7%), three oxygenated monoterpenes (2.9%), four sesquiterpene hydrocarbons (3.1%) and one phenolic compound (0.2%) (Kaur et al. 2006). Limonene (82.4%) was the main constituent.

From Twig/Stem/Wood/Bark

γ-fagarine, marmesin a coumarin and umbellieferone were isolated from the mature bark (Chatterjee and Mitra 1949). Umbelliferone, skimmianine, marmmin and γ-sitosterol were isolated from immature bark (Chatterjee and Bhattacharya 1959). Baslas and Deshpandey (1951) reported the presence of 1,8-cineole and α-phellandrene in the twig oil. Purified hemicellulose isolated from the trunk of a young bael tree contained D-xylose and 4-*O*-methyl-D-glucoronic acid in the molar ratio of 7.43:1; traces of glucose, galactose, rhamnose, and arabinose were also present (Basak et al. 1982). Some constituents isolated from bael heartwood included: acetic acid 2-(2-hydroxy- 4-methoxyphenyl) vinyl ester, lupeol, and sitosterol (Jain et al. 1991). From the bark two new lignan-glucosides, (−)-lyoniresinol 2α-O-β-D-glucopyranoside and (−)-4-epi-lyoniresinol 3α-O-β-D-glucopyranoside, were isolated together with two known lignan-glucosides, (+)-lyoniresinol 3α-O-β-D-glucopyranoside and (−)lyoniresinol 3α-O-β-D-glucopyranoside (Ohashi et al. 1994). Anthraquinones 6-hydroxy-1-methoxy-3-methyl anthraquinone and 7,8- dimethoxy-1- hydroxyl-2-methyl anthraquinone were isolated from bael heartwood (Srivastava et al. 1996).

Mamim, a coumarin, was isolated from the stem bark (Chatterjee and Choudhury 1955). Sixteen compounds were identified in *A. marmelos* twig oil: eight monoterpene hydrocarbons (92.8%), two oxygenated monoterpenes (2.5%), four sesquiterpene hydrocarbons (2.2%), one oxygenated sesquiterpene (0.2%) and one phenolic compound (Kaur et al. 2006). Limonene (51.7%) and (Z)-β-ocimene (39.8%) were the major components found in the twig oil. Bioassay-directed fractionation of the ethyl acetate extracts of bael stem bark afforded a new compound, named skimmiarepin C, along with skimmiarepin A (Samarasekera et al. 2004). The new compound was a senecioate ester analogue of the latter. Marmisin, xanthotoxin and scopoletin were isolated from bael twigs (Laphookhieo et al. 2011).

From Roots

Psoralen, xanthotoxin, o-methylscopoletin, 6,7-dimethoxycoumarin, scopoletin, tembamide, skimmin, marmesin, marmin and skimmianine were found in bael roots (Shoeb et al. 1973). A coumarin decursinol and alkaloid haplopine was isolated from the root bark (Basu and Sen 1974). Auraptene, umbelliferone, marmin, lupeol and skimminanine were isolated from the roots (Chatterjee and Chaudhury 1960). A coumarin 6′,7′-epoxyauraptene was isolated from the root (Bhattacharyya et al. 1989). Two coumarins, named marminal and 7′-O-methylmarmin, along with 19 known compounds: β-sitosteryl pentadecanoate, 4-methoxy-1-methyl-2-quinolone, aurapten, 4-sitosten-3-one, lupeol, imperatorin, xanthotoxin, dictamnine, (+)-epoxyaurapten, β-sitosterol, γ-fagarine, skimmianine, scoparone, umbelliferone, scopoletin, decursinol, marmesin, marmin, and integriquinolone were isolated from the methanolic extract of the roots (Yang et al. 1996b). An alkaloid 4,7,8-trimethoxyfuroquinoline (skimmianine) were isolated from roots (Riyanto et al. 2001). Two coumarins, auraptene and marmin were isolated from roots (Riyanto et al. 2002).

A oxazoline derivative, aeglemarmelosin, skimmianine, imperatorin, aurapten, epoxyaurapten, marmin were isolated from bael roots (Laphookhieo et al. 2011).

Phytochemicals isolated from bael have been proven to be biologically active against several major diseases including cancer, diabetes and cardiovascular diseases and preclinical studies indicated the therapeutic potential of crude extracts of *A. marmelos* in the treatment of many microbial diseases, diabetes and gastric ulcer (Maity et al. 2009). The many pharmacological properties of various parts of the plant including antidiabetic, antiulcer, antioxidant, antimalarial, anti-inflammatory, anticancer, cardiovascular protective, hepatoprotective, gastroprotective, radioprotective, antihyperlipidaemic, antifungal, antibacterial, antiviral and others are elaborated below.

Antioxidant Activity

The potency of scavenging activity of 17 Indian medicinal plants was in the following order: *Alstonia scholaris* > *Cynodon dactylon* > *Morinda citrifolia* > *Tylophora indica* > *Tectona grandis* > *Aegle marmelos* (leaf) > *Momordica charantia* > *Phyllanthus niruri* > *Ocimum sanctum* > *Tinospora cordifolia* (hexane extract) = *Coleus ambonicus* > *Vitex negundo* (alcoholic) > *T. cordifolia* (dichloromethane extract) > *T. cordifolia* (methanol extract) > *Ipomoea digitata* > *V. negundo* (aqueous) > *Boerhaavia diffusa* > *Eugenia jambolana* (seed) > *T. cordifolia* (aqueous extract) > *V. negundo* (dichloromethane/methanol extract) > *Gingko biloba* > *Picrorrhiza kurroa* > *A. marmelos* (fruit) > *Santalum album* > *E. jambolana* (leaf) (Jagetia and Baliga 2004). All the extracts evaluated exhibited a dose-dependent NO scavenging activity.

Bael fruit pulp was found to have high antioxidant property comparable to fruits like mangosteen and persimmon, herbs Asian pennywort and Thai basil, and vegetables beet and sweet potato (Charoensiddhi and Anprung 2008). Determination of antioxidant activities of bael fruit pulp by 2-diphenyl-1-picryhydrazyl (DPPH) free radical scavenging and ferric reducing antioxidant power (FRAP) assays resulted in 6.21 µg dw/µg DPPH and 102.74 µM trolox equivalent (TE)/g dw, respectively Bael fruit pulp was also found to have 87.34 mg gallic acid equivalent (GAE)/g dw total phenolic, 15.20 mg catechin equivalent (CE)/g dw total flavonoid, 32.98 µg/g dw total carotenoid, and 26.17 mg/100 g dw ascorbic acid.

The aqueous alcoholic extracts of bael bark exhibited antioxidant activity (Kumari and Kakkar 2008). It showed superoxide dismutase (SOD) mimetic activity greater than 100 units/g extract and displayed highest NO quenching potential (47.3% per µg extract) and also lipid peroxidation inhibitory potential.

Antilipid Peroxidative Activity

The methanolic root extract of *Aegle marmelos* increased the activities of superoxide dismutase (SOD) and glutathione-peroxidase in the liver cytosol of mice, but showed no significant effect on the activity of catalase, and one of its major constituents, 4-methoxy-1-methyl-2-quinolone (MMQ) increased the activity of SOD in liver tissue of mice intoxicated with $FeCl_2$-ascorbic acid (AA)-ADP in-vivo (Yang et al. 1996a). Other isolated constituents also inhibited in-vitro lipid peroxidation in rat liver homogenate, which was induced by $FeCl_2$-ascorbic acid, CCl_4-NADPH, or ADP-NADPH. Of the test compounds, MMQ and its derivatives integriquinolone were similar to α-tocopherol in inhibiting malondialdehyde production in rat liver microsomes.

Antihyperlipidaemic Activity

Studies found that oral administration of *A. marmelos* fruit extract at doses of 125 and 250 mg/kg to diabetic rats twice daily for 1 month led to a significant lowering of tissue lipids such as total cholesterol, triglycerides, free fatty acids and phospholipids in diabetic rats (Kamalakkannan and Prince 2004b). The effect exerted by the fruit extract at a dose of 250 mg/kg was greater than that of the dose of 125 mg/kg or of glibenclamide (300 µg/kg). The results of this study demonstrated that an aqueous *A. marmelos* fruit extract exhibited an antihyperlipidaemic effect in streptozotocin-induced diabetic rats. The scientist also reported that glucose level and glycosylated hemoglobin were increased and plasma insulin and liver glycogen were decreased in diabetic rats, and that treatment with bael fruit extract reversed the effects of diabetes on these biochemical parameters to near-normal levels. In another paper, they reported that oral administration of aqueous *A. marmelos* fruit extract (125 and 250 mg/kg) twice a day for 4 weeks resulted in significant reductions in blood glucose, plasma thiobarbituric acid reactive substances, hydroperoxides, ceruloplasmin and α-tocopherol and a significant elevation in plasma reduced glutathione and vitamin C in diabetic rats (Kamalakkannan et al. 2003; Kamalakkannan and Prince 2003). The effect of the extract at a dose of 250 mg/kg was more effective than at 125 mg/kg or glibenclamide (300 µg/kg). The results of this study clearly shows antihyperlipidaemic effect of the fruit extract in streptozotocin-induced diabetic rats. The scientist also reported that besides causing a significant decrease in peroxidation products, viz., thiobarbituric acid reactive substances and hydroperoxides in diabetic rats, the activity of antioxidant enzymes such as superoxide dismutase, catalase and glutathione peroxidase was found to be increased in the hepatic and renal tissues of diabetic animals treated with the extract further indicating that the extract also exhibited antidiabetic and anti-oxidative activity in streptozotocin-induced-diabetic rats (Kamalakkannan and Prince 2004a). They also reported that the fruit extract treated groups showed improved functional state of the pancreatic ß-cells and partially reversed the damage caused by streptozotocin to the pancreatic islets (Kamalakkannan and Prince 2005). The findings indicated that *Aegle marmelos* fruit extract also exhibited protective effects on the pancreas. The effects observed in the fruit extract treated animals were better than those in animals treated with glibenclamide (300 μg/kg).

A separate research conducted demonstrated that treatment of severely fasting blood glucose (FBG) diabetic rats for 14 days with a dose of 250 mg/kg of an aqueous bael seed extract reduced the fasting blood glucose by 60.84% and urine sugar by 75% than their pre-treatment levels (Kesari et al. 2006). It decreased the level of total cholesterol (TC) by 25.49% and produced an increase of 33.43% in high density lipoprotein (HDL) and a decrease of 53.97 and 45.77% in low density lipoprotein (LDL) and triglyceride (TG), respectively. These results indicated that aqueous seed extract of *Aegle marmelos* possessed antidiabetic and hypolipidemic effects in diabetic rats. Aegeline 2, an alkaloidal-amide, isolated from bael leaves was found to significantly decrease the plasma triglyceride (Tg) levels by 55%, total cholesterol (TC) by 24%, and free fatty acids by 24%, accompanied with increase in HDL-C by 28% and HDL-C/TC ratio by 66% in dyslipidemic hamster model at the dose of 50 mg/kg body weight (Narender et al. 2007).

Ethanolic bael leaf extract at 125 and 250 mg/kg dose levels inhibited the elevation in serum cholesterol and triglycerides levels on Triton WR 1339 administration in rats (Vijaya et al. 2009). The extract at the same dose levels significantly attenuated the elevated serum total cholesterol and triglycerides with an increase in the high-density lipoprotein cholesterol in high-fat diet-induced hyperlipidaemic rats.

Antidiabetic/Hypoglycaemic Activity

Pretreatment of bael leaf extract decreased the elevated blood urea and serum cholesterol in alloxan diabetic rats. Also the leaf treatment significant decreased liver glycogen to near normal level and improved glucose tolerance (Ponnachan et al. 1993). Similar effect was seen with insulin treatment. The results indicated that the active principle in *A. marmeloe* leaf extract had similar hypoglycaemic activity to insulin treatment. Treatment of bael leaf extract to streptozotocin-induced diabetic rats ameliorated the altered histological changes in the liver such as dilation of veins, liver fibrosis, loss of concentric arrangement of hepatocytes and decrease in glycogen caused by streptozotocin (Das et al. 1996). Bael leaf treatment also improved the functional state of pancreatic β-cells and acinar cells and helped in regeneration of damaged pancreas. Rats orally administered bael leaf extract for 7 consecutive days showed significant improvements in their ability to utilize the external glucose load (Sachdewa et al. 2001). The extract caused 67% average blood glucose lowering and improved the glucose tolerance curve with time. The mechanism of action could be speculated partly to increased utilization of glucose, either by direct stimulation of glucose uptake or via the mediation of enhanced insulin secretion. Studies showed that methanolic extracts of *A. marmelos* and *Syzygium cumini* activated glucose uptake and transport by elevation of glucose transporter (Glut-4), peroxisome proliferator activator receptor γ (PPARγ) and phosphatidylinositol 3' kinase (PI3 kinase) (Anandharajan et al. 2006).

Aegle marmelos methanolic leaf extract was reported to effectively reduce the oxidative stress induced by alloxan and produced a reduction in blood sugar (Sabu and Kuttan 2004). This was evident from a significant decrease in lipid peroxidation, conjugated diene and hydroperoxide levels in serum as well as in liver induced by alloxan. Catalase and glutathione peroxidase activity in blood and liver were found to be increased from 9th day onwards after drug administration. Superoxide dismutase and glutathione levels were found to be increased only on 12th day. Another research (Upadhya et al. 2004) reported that alloxan induced diabetic rats administered bael leaf extract showed a decrease in blood glucose, an increase in erythrocyte glutathione (GSH) and a decrease malondialdehyde and glutathione-S-transferase. This was attributed to hypoglycemic and antioxidant properties, further indicating that bael leaf extract may be useful in the long-term management of diabetes. In another study, *A. marmelos* leaf extract protected against oxidative stress damage in streptozotocin-induced diabetic rats (Narendhirakannan and Subramanian 2010). The extract modulated the activity of primary scavenger enzymes (superoxide dismutase, catalase, and glutathione peroxidase) and nonenzymic antioxidants (vitamin E, C) and enhanced the defence against reactive oxygen species-generated damage in diabetic rats. Histopathological studies also revealed the protective effect of *A. marmelos* on pancreatic β-cells.

Studies in streptozotocin (STZ)-induced diabetic rats showed treatment with umbelliferone, a natural antioxidant, and a benzopyrone found in *Aegle marmelos* fruit, restored plasma insulin and glucose levels and reversed insulin, glucose and lipid peroxidation markers, and diabetes-induced alterations in the activities of membrane-bound ATPases (Ramesh and Pugalendi 2007c). The normalization of membrane-bound ATPases in various tissues, was attributed to improved glycemic control and antioxidant activity by umbelliferone. Treatment with umbellefirone also decreased the elevated levels of glucose, sialic acid, total hoxoses, fucose, and hexosamines to near-normal level in STZ-diabetic rats (Ramesh and Pugalendi 2007b) thereby protecting the risk of diabetic complications. The authors also showed that treatment with umbelliferone normalised the significant decrease in prothrombin, clotting and bleeding time in streptozotocin-diabetic rats (Ramesh and Pugalendi 2007a). The results also indicated that umbelliferone controlled glycaemia and had a beneficial effect on collagen content and its properties, i.e. collagen related parameters, in the tail tendon, which indicated recovery from the risk of complications of collagen-mediated diabetic polyneuropathy and diabetic nephropathy. Aegeline 2, an alkaloidal-amide isolated from bael leaves was found to have antihyperglycemic activity as evidenced by lowering the blood glucose levels by 12.9% and 16.9% at 5 and 24 hours, respectively, in sucrose challenged streptozotocin induced diabetic rats model at the dose of 100 mg/kg body weight (Narender et al. 2007). Oral administration of alcoholic extract of *Momordica charantia, Aegle marmelos* and *Eugenia jambolana* in daily doses of 250 mg and 500 mg/kg for a period of 1 month produced dose- and duration-dependent decrease in serum glutamic oxaloacetate transminase and serum glutamic pyruvate transminase activities as well as decrease in serum urea concentration and restored the serum total protein and albumin concentration and albumin/globulin ratio to a great extent in streptozotocin diabetic rats (Sundaram et al. 2009). The beneficial effects of these plants in 500 mg/kg dose in streptozotocin diabetic rats were comparable to that of glibenclamide (300 μg/kg), a standard oral hypoglycaemic drug used in clinical practice.

In another study, oral administration of bael leaf extract to streptozotocin induced diabetic rats upregulated the decrease in total muscarinic and muscarinic M1 receptors in the cerebral cortex during diabetes indicating its clinical significance in therapeutic management of diabetes (Gireesh et al. 2008). Recent studies by Abraham et al. (2010a, b) suggested that pyridoxine treated alone and in combination with insulin and *A. marmelos* leaf extract had a role in the regulation of insulin synthesis and release, normalizing diabetic related oxidative stress and neurodegeneration affecting the motor ability of an individual by serotonergic receptors through 5-HT(2A) function. This may have clinical significance in the management of diabetes. Administration of a combined aqueous extract of *Eugenia jambolana* seed and *Aegle marmelos* leaf significantly increased serum insulin levels and caused higher reduction in hyperglycemia and hyperlipidemia when compared to the alloxan-diabetic control rats (Gohil et al. 2010). The combined plant extracts treated animals revealed restoration of the shrunken β-cells of islets of langerhans. Treatment with *Aegle marmelos* fruit aqueous extract for 21 days in high fat diet fed-streptozotocin-induced diabetic rats positively modulated the altered parameters (elevated of glucose, insulin, homeostasis model assessment of insulin resistance (HOMA-IR), TNF-α, IL-6, dyslipidemia) in a dose-dependent manner (Sharma et al. 2011). Further, bael extract prevented inflammatory changes and β-cell damage along with a reduction in mitochondrial and endoplasmic reticulum swelling. The findings suggested that the protective effect of bael extract in type 2 diabetic rats was due to the preservation of β-cell function and insulin-sensitivity through increased peroxisome proliferator-activated receptor-γ (PPARγ) expression.

Thai scientists isolated a series of phenylethyl cinnamides, which included new compounds named anhydromarmeline aegelinosides A and B isolated from *Aegle marmelos* leaves as α-glucosidase inhibitors (Phuwapraisirisan et al. 2008). Of the compounds isolated, anhydroaegeline revealed the most potent inhibitory effect against α-glucosidase with IC_{50} value of 35.8 μM. The results further supported the ethnopharmacological use of *A. marmelos* as a remedy for diabetes mellitus.

Cardiovascular Protective Activity

The methanolic extract of bael root bark at a concentration of 100 μg/ml inhibited heart beat rate by approximately 50% (Kakiuchi et al. 1991). Among the isolated constituents, aurapten was the most potent inhibitor; the IC_{50} of aurapten is 0.6 μg/ml, which was comparable with that of verapamil, a calcium antagonist. Addition of aurapten at concentrations higher than 1 μg/ml significantly reduced the ratio of morphologically changed myocardial cells which originated from calcium overload caused by successive treatment with calcium-free and calcium-containing solutions.

Pre-treatment with bael leaf extract at doses of 100 mg/kg and 200 mg/kg bodyweight for 35 days showed a significant effect on the activities of marker enzymes, lipid peroxides, lipids, lipoproteins and antioxidant enzymes in isoprenaline (isoproterenol)-treated rats. The effect of the extract at 200 mg/kg was found to be equal to the effect of α-tocopherol 60 mg/kg. Findings confirmed that *Aegle marmelos* leaves possessed antihyperlipidaemic effect in rats with isoproterenol-induced myocardial infarction (Prince and Rajadurai 2005; Rajadurai and Prince 2005). Linear furanocoumarin, marmesinin isolated from *Aegle marmelos* demonstrated protective effect against the damage caused by experimental myocardial injury in rats (Vimal and Devaki 2004). Marmesinin at a dose of 200 mg/kg, when administered orally, produced a decrease in serum enzyme levels and restored the electrocardiographic changes towards normalcy. Marmesinin oral treatment for 2 days before and during isoproterenol administration was found to decrease the effect of lipidperoxidation. It was also shown to have a membrane stabilizing action by inhibiting the release of β-glucuronidase from the subcellular fractions.

Periplogenin, a cardenolide isolated from bael leaves protected against doxorubicin induced cardiotoxicity and lipid peroxidation (LPO) in rats (Panda and Kar 2009). Cotherapy of the test cardenolide and doxorubicin for 4 weeks reversed all the doxorubicin adverse effects. Periplogenin lowered the elevated levels of doxorubicin induced increase in serum creatine kinase-MB (CK-MB), glutamate-pyruvate transaminase (SGPT), and tissue LPO, with a decrease in superoxide dismutase (SOD), catalase (CAT), and glutathione (GSH). Periplogenin lowered the elevated levels of different serum lipids, and increased the doxorubicin decreased the amount of high-density lipoprotein (HDL). Out of three different concentrations (12.5, 25, and 50 mg/kg p.o.) of the test periplogenin, 25 mg/kg appeared to be most effective. When its efficacy was compared with that of vitamin E (α-tocopherol) periplogenin exhibited a better therapeutic potential.

Hepatoprotective Activity

Treatment with hydroalcoholic (80% ethanol, 20% water) extracts of bael leaves was found to significantly increase the basal levels of acid-soluble sulphydryl (–SH) content, cytochrome P450, NADPH-cytochrome P450 reductase, cytochrome b5, NADH-cytochrome b5 reductase, glutathione S-transferase, DT-diaphorase, superoxide dismutase, catalase, glutathione peroxidase and glutathione reductase in the liver (Singh et al. 2000). Bael acted as a bifunctional inducer since it induced both phase-I and phase-II enzyme systems. The extracts significantly decreased the activity of lactate dehydrogenase and formation of malondialdehyde (MDA) in the liver, suggesting a role in cytoprotection as well as protection against pro-oxidant-induced membrane damage. The extract exhibited modulatory effect in inducing glutathione S-transferase, DT-diaphorase, superoxide dismutase and catalase in the lung, glutathione S-transferase, DT-diaphorase and superoxide dismutase in the fore-stomach. The scientists concluded that these significant changes in the levels of drug-metabolizing enzymes and antioxidative profiles were strongly indicative of the chemopreventive potential of this plant, especially against chemical carcinogenesis. Research showed that the observed values of TBARS (Thiobarbituric acid reactive substances) in healthy, alcohol intoxicated and herbal drug treated animals were 123.35, 235.68 and 141.85 μg/g tissue respectively (Singanan et al. 2007). The results were compared with the standard herbal drug silymarin (133.04 μg/g tissue). The experimental results indicated that bael leaves had excellent hepatoprotective effect.

Pre-treatment of male Wistar rats with *A. marmelos* prior to CCl4 suppressed lipid peroxidation (LPO), xanthine oxidase (XO) and release of serum toxicity marker enzymes viz, SGOT, LDH, SGPT dose-dependently and significantly (Khan and Sultana 2009). Hepatic antioxidant status viz, reduced glutathione (GSH), glutathione reductase (GR), glutathione peroxidase (GPx), quinone reductase (QR), catalase (CAT) were concomitantly restored in *A. marmelos*-treated groups. In addition, *A. marmelos* pretreatment also prevented the CCl4-enhanced ornithine decarboxylase (ODC) and hepatic DNA synthesis significantly. The authors concluded that *A. marmelos* markedly attenuated carbon tetrachloride-induced hepatic oxidative stress, toxicity, tumour promotion and subsequent cell proliferation response in Wistar rats.

Anticancer Activity

Research reported that extracts from *Aegle marmelos* were able to inhibit the in-vitro proliferation of human tumour cell lines, including the leukemic K562, T-lymphoid Jurkat, B-lymphoid Raji, erythroleukemic HEL, melanoma Colo38, and breast cancer MCF7 and MDA-MB-231 cell lines (Lampronti et al. 2003). Three derivatives (butyl p-tolyl sulfide, 6-methyl-4-chromanone and butylated hydroxyanisole) were found to exhibit strong activity in inhibiting in vitro cell growth of human K562 cells. The antiproliferative activity of these compounds was found to be comparable to that of known antitumour agents, including cisplatin, chromomycin, cytosine arabinoside and 5-fluorouracil. In addition, the antiproliferative activity of butyl-p-tolyl sulfide, 6-methyl-4-chromanone and 5-methoxypsoralen was found to be associated with activation of the differentiation pattern of K562 cells. Bael extract exhibited antiproliferative effect on MCF7 and MDA-MB-231 human breast cancer cell lines (Lambertini et al. 2004). *Aegle marmelos* plant extracts inhibited cell proliferation of ERα-negative MDA-MB-231 breast cancer cells and showed no effect on ERα gene expression, but when bael extract were added in combination with the decoy molecule, a strong modulatory effect was observed (Lambertini et al. 2005). Lupeol, a known triterpenoid, was identified as the major bioactive component of *A. marmelos* plant extracts. Similar to the *Aegle marmelos* extracts, lupeol was found to stimulate the decoy effect of RA4 DNA sequence, increasing at a high level ERa gene expression in MDA-MB-231 ERα-negative breast cancer cells, and also inhibited cell proliferation.

Aegle marmelos extract exhibited cytotoxicity in the brine shrimp lethality assay, sea urchin eggs assay, hemolysis assay and MTT assay on leukemia CEM, Human promyelocytic leukemia HL-60 and intestinal adenocarcinoma HCT-8 tumour cell lines (Costa-Lotufo et al. 2005). Administration of an hydroalcoholic extract of *Aegle marmelos* inhibited the proliferation of transplanted Ehrlich ascites carcinoma in mice (Jagetia et al. 2005). Bael treatment resulted in a dose dependent elevation in the median survival time and average survival time up to 400 mg/kg AME and decline thereafter. The acute toxicity study of bael extract showed that it was non-toxic up to a dose of 1,750 mg/kg body weight. The LD10 and LD50 was found to be 2,000 and 2,250 mg/kg respectively.

Marmelin (1-hydroxy-5,7-dimethoxy-2-naphthalene-carboxaldehyde) isolated from ethyl acetate fraction of extracts of *Aegle marmelos*, inhibited the growth of HCT-116 colon cancer tumour xenografts in vivo (Subramaniam et al. 2008). Data revealed marmelin to be a potent anticancer agent that induced apoptosis during G(1) phase of the cell cycle through activation of tumour necrosis factor-α (TNF-α), TNF receptor (TNFR)-associated death domain (TRADD), and caspases. Oral administration of *A. marmelos* fruit extract at 100 mg/kg body weight/day during peri-initiational, postinitiational, and peri- and postinitiational phases of papillomagenesis showed significant reduction in tumour incidence, tumour yield, tumour burden, and cumulative number of papillomas when compared with 7, 12-dimethylbenz (a) anthracene (DMBA)-treated control animals (Agrawal et al. 2010, 2011). Enzyme analysis of skin and liver showed a significant elevation in antioxidant parameters such as superoxide dismutase, catalase, glutathione, and vitamin C in bael extract-treated groups when compared with the carcinogen-treated control. The elevated level of lipid peroxidation in the positive control was significantly inhibited by bael extract administration. These results indicated that bael extract possessed the potential to reduce chemical-induced skin papillomas by enhancing the antioxidant defense system.

Aegle marmelos methanolic extract was found to suppress diethylnitrosamine (DEN) initiated and 2-acetyl aminofluorene (2-AAF) promoted liver carcinogenesis in male Wistar rats (Khan and Sultana 2011). *A. marmelos* (25 and 50 mg/kg body weight) resulted in a marked reduction of the incidence of liver tumours. Pretreatment of *A. marmelos* extract (25 and 50 mg/kg body weight) prevented oxidative stress and hepatic toxicity by restoring the levels of antioxidant enzymes at both the doses. The promotion parameters (ornithine decarboxylase activity and DNA synthesis) induced by 2-AAF administration in the diet with partial hepatectomy (PH) were also significantly suppressed dose-dependently by *A. marmelos*. The authors concluded that ultimately the protection against liver carcinogenesis by *A. marmelos* methanolic extract might be mediated by multiple actions, which include restoration of cellular antioxidant enzymes, detoxifying enzymes, ornithine decarboxylase activity and DNA synthesis.

Antiviral Activity

A series of compounds in extracts from various parts of the bael tree exhibited varying in-vitro antiviral efficacy against human coxsackieviruses B1-B6 (Badam et al. 2002). The inhibitory concentrations (IC_{50}) for leaves (L1 and L2), stem and stem bark (S1, S2, S3 and S4), fruit (F1 and F2μ), root and root bark (R1 and R2) and pure compound, marmelide were 1,000 μg/ml (for L1 and L2), 1,000 μg/ml (for S1, S2, S3 and S4), 1,000 μg/ml (for F1) and 500 μg/ml (for F2), 250 μg/ml (for R1) and 500 μg/ml (for R2) and 62.5 μg/ml for marmelide respectively by plaque inhibition assay for 96 hours. The corresponding value for Ribavirin, a standard antiviral drug, was 2,000 μg/ml for the same viruses at the same time period. These concentrations did not exhibit any toxicity to Vero cells, the host subtoxic concentrations were 5,000 μg/ml for leaf and stem fractions, 2,000 μg/ml for fruit fractions, 500 and 1,000 μg/ml for root fractions, 250 μg/ml for marmelide and 2,000 μg/ml for Ribavirin. The cytotoxic concentrations were 8,000 μg/ml for leaf and stem compounds, 4,000 mg/ml for fruit, 1,000 μg/ml and 2,000 μg/ml for root, 500 μg/ml for marmelide and 4,000 μg/ml for ribavirin at 96 hours. Additionally, pretreatment of host cells, virus inactivation, yield reduction and effect of time of addition assays against coxsackievirus B3 suggested that marmelide was most effective as a virucidal agent besides interfering at early events of its replicative cycle like adsorption, penetration, at various steps in single cycle growth curve and effect of time of addition. Nine extracts of eight different plants including *Aegle marmelos* significantly reduced viral production in human CD4+ T-cell line CEM-GFP cells infected with HIV-1NL4.3 (Sabde et al. 2011).

Gastroprotective Activity

Luvangetin, a pyranocoumarin isolated from bael seeds showed significant protection against pylorus-ligated and aspirin-induced gastric ulcers in rats and cold restraint stress-induced gastric ulcers in rats and guinea pigs (Goel et al. 1997). Studies demonstrated that pre-treatment of animals with unripe bael fruit extract (50 and 100 mg/kg, i.p.) produced a significant inhibition of gastric lesion induced by ethanol but not those induced by restraint stress or indomethacin (Dhuley 2003–2004). A probable involvement of a prostaglandin-independent mechanism of gastro-protection was implicated. At similar doses, both the intestinal transit as well as the accumulation of intestinal fluids induced by castor oil in mice were significantly inhibited by raw fruit extract. Further, bael fruit extract antagonized the contractile responses evoked by different agonists on guinea-pig ileum in-vitro and its inhibitory potential for the drugs are in the order of acetylcholine > histamine > serotonin > barium chloride. These results indicated a possible antidiarrhoeal effect of unripe fruit extract of A. marmelos, since inhibition of intestinal motility and secretion could control clinical diarrhoea.

Antidiarrhoeal Activity

Methanolic plant extract *Aegle marmelos* unripe fruit displayed anti-diarrhoeal potential against castor-oil induced diarrhoea in mice (Shoba and Thomas 2001). The methanolic plant extract was more effective than aqueous plant extract against castor-oil induced diarrhoea. The methanolic plant extract significantly reduced induction time of diarrhoea and total weight of the faeces thus establishing the efficacy of the plant extract as anti-diarrhoeal agent. *Aegle marmelos* root extract treated rats showed significant inhibitory activity against castor oil-induced diarrhoea (Mazumder et al. 2006). Of the 35 tested pathogenic diarrhoea causing strains, the extract was found to be mostly active against the strains of *Vibrio cholerae*, followed by *Escherichia coli* and *Shigella* spp. The in-vitro activity was found to be comparable to that of ciprofloxacin. The results confirmed the efficacy of the extract as an effective antidiarrhoeal agent.

Antibacterial Activity

Methanolic extracts of *A. marmelos* was one of a handful of 54 plants tested for antimicrobial activity that displayed strong antibacterial activity against multi-drug resistant, enteric *Salmonella typhi* (Rani and Khullar 2004). Shahidine, a xazoline alkoid from the leaves showed activity against a few Gram-positive bacteria (Faizi et al. 2009). Ethanolic extract of *Aegle marmelos* was found to have antibacterial activity (Venkatesan et al. 2009). AT 2.5 μg/ml concentration the extract was strongly inhibitory to growth of *Escherichia coli, Pseudomonas aeruginosa* and *Bacillus subtilis* comparable to the control penicillin; however it was not active against *Staphylococcus aureus*. The crude extract was found to contain alkaloids, cardiac glycosides, terpenoids, saponins, tannis, flavonoids, and steroids. When grown in the presence of aqueous bael extract the susceptibility of β-lactam-resistant *Shigella dysenteriae* and *Shigella flexneri* toward β-lactam antibiotics was enhanced (Raja et al. 2008). This was effected by alteration in the outer membrane porin channels. Thus bael extract along with β-lactam could be used for treatment of multidrug-resistant *Shigella* spp. The hot aqueous decoction of unripe bael fruit pulp showed cidal activity against diarrhoel microorganisms Giardia and rotavirus but had limited antibacterial activity against enteropathogenic *Escherichia coli* and enteroinvasive *Shigella flexneri* strains (Brijesh et al. 2009). The decoction significantly reduced bacterial adherence to and invasion of HEp-2 cells. The decoction also affected production of cholera toxin and binding of both labile toxin and cholera toxin to ganglioside monosialic acid receptor but had no effect on stable toxin. The lectin isolated from bael fruit pulp was found to be a dimeric protein with N-acetylgalactosamine, mannose and sialic acid binding specificity (Raja et al. 2011a). It significantly inhibited hemagglutination activity of *Shigella dysenteriae* and its minimum inhibitory concentration was 0.625 μg/well. At this concentration the bael fruit lectin inhibited *Shigella dysenteriae* adherence and invasion of HT29 cells and protected the HT29 cells from *Shigella dysenteriae* induced apoptosis. Studies demonstrated that *Shigella dysenteriae* succumbed to oxidative stress (host defence) due to inhibition of periplasmic copper, zinc super oxide dismutase (pathogen's defence) by imperatorin, an active furocoumarin in the aqueous extract of *A. marmelos* (Raja et al. 2011b).

Antifungal Activity

The oil from bael leaves exhibited variable efficacy against different fungal isolates and 100% inhibition of spore germination of all the fungi tested was observed at 500 ppm (Rana et al. 1997). However, the most resistant fungus, *Fusarium udum* was inhibited 80% at 400 ppm. Kinetic studies showed concentration as well as time dependent complex inhibition of spore germination by the essential oil. A new anthraquinone, 1-methyl-2-(3'-methyl-but-2'-enyloxy)-anthraquinone (1) isolated from bael seeds exhibited significant antifungal activity against pathogenic strains of *Aspergillus* species and *Candida albicans* in disc diffusion assay (MIC value of 6.25 μg/disc), microbroth dilution and percent spore germination inhibition assays (MIC value of 31.25–62.5 μg/ml) (Mishra et al. 2010a). *Aegle marmelos* ethanol leaf extracts exhibited antifungal and antiaflatoxigenic activity at a concentration range of 0.5–2 mg/ml (Patil et al. 2011). The ethanolic extracts were found to contain phytochemicals like phenols, tannins, flavonoids and alkaloids as major constituents.

Bael essential oil showed strong fungitoxicity against some storage fungi-causing contamination of foodstuffs and suppressed aflatoxin B1 production by toxigenic strains of *Aspergillus flavus* at 500 μl/l (Singh et al. 2009). DL-Limonene was found to be major component. It showed no mammalian toxicity, the LD_{50} was found to be 23659.93 mg/kg body weight in mice when administered for acute oral toxicity.

Radioprotective Activity

In studies on radioprotective compounds spanning nearly a decade, it had been observed that bael prevented radiation-induced ill-effects, and the results of these studies indicated that bael possessed the potential to be an effective, non-toxic radioprotective agent (Baliga et al. 2010). Several studies reported that hydroalcoholic fruit extract of *Aegle marmelos* (AME) at 5 μg/ml protected human peripheral blood lymphocytes against radiation-induced DNA damage and genomic instability (Jagetia et al. 2003). It was suggested that its radioprotective activity might be by scavenging of radiation-induced free radicals and increased oxidant status. (AME) was found to inhibit free radicals in a dose-dependent manner up to a dose of 200 μg/ml for the majority of radicals and plateaued thereafter. The administration of hydroalcoholic extract after irradiation was not effective. The acute toxicity study of AME showed that it was nontoxic up to a dose of 6 g/kg body weight, the highest drug dose that could be administered. Irradiation of animals resulted in a dose-dependent elevation in lipid peroxidation in liver, kidney, stomach, and intestine of mice (Jagetia et al. 2004b). Conversely, glutathione (GSH) concentration declined in a dose-dependent manner. Treatment of animals with AME before irradiation caused a significant decrease in the lipid peroxidation accompanied by a significant elevation in the GSH concentration in liver, kidney, stomach, and intestine of mice determined at 31 days post-irradiation. AME pre-treatment protected mice against the gastrointestinal as well as bone marrow deaths, as evidenced by the greater number of survivors on day 10 or 30, respectively. Oral administration of bael extract to mice resulted in an increase in radiation tolerance by 1.6 Gy, and the dose reduction factor was found to be 1.2 (Jagetia and Venkatesh 2005). Preradiation treatment of mice with bael extract caused a significant depletion in lipid peroxidation followed by a significant elevation in glutathione concentration in the liver of mice 31 days after irradiation. The extract was non-toxic up to a dose of 6,000 mg/kg body weight, the highest drug dose that could be tested for acute toxicity. In another study, AME protected mice against the radiation-induced decline in hemoglobin, total leukocyte, and lymphocytes counts and the clonogenicity of hemopoietic progenitor cells assessed by the exogenous spleen colony-forming assay (Jagetia et al. 2006). Treatment of mice with AME before irradiation elevated the peripheral cell count as well as villus height and the crypt number accompanied by a decline in goblet and dead cells when compared with the irradiation control. The recovery and regeneration were faster in AME pretreated animals than the irradiation alone. AME pretreatment significantly decreased lipid peroxidation accompanied by a significant elevation in the GSH concentration in the mouse intestine. The data clearly indicated that the AME significantly reduced the deleterious effect of radiation in the intestine and bone marrow of mouse and could be a useful agent in reducing the side effects of therapeutic radiation

Bael leaf extract treatment of Swiss albino male mice reduced the symptoms of radiation-induced sickness and increased survival (Jagetia et al. 2004a). The radioprotective action was postulated to be due to free-radical scavenging and arrest of lipid peroxidation accompanied by an elevation in glutathione. Treatment of mice with bael leaf extract (AME), once daily for 5 consecutive days, before exposure to 2 Gy resulted in a significant decline in the frequency of micronucleated polychromatic erythrocytes (MPCE) when compared to the non-drug-treated irradiated control (Jagetia and Venkatesh 2007). The greatest reduction in MPCE was observed for 250 mg/kg body weight (AME), accompanied by the highest polychromatic erythrocyte (PCN) to normochromatic erythrocyte (NCE) ratio in comparison with the non-drug-treated irradiated control. Bael leaf extract significantly reduced the frequency of MNCE at all post-irradiation times, when compared to the non-drug-treated irradiated group. Treatment of mice with AME before exposure to different doses of γ-radiation resulted in the inhibition of a radiation-induced decline in the PCE/NCE ratio, when compared with the concurrent irradiated controls. Findings again demonstrated that one of the mechanism of reduction in the radiation-induced DNA damage in mice bone marrow by AME might be due to scavenging of free radicals and elevation in the antioxidant status.

Antiinflammatory Activity

Bael leaf extract caused a significant inhibition of the carrageenan-induced paw oedema and cotton-pellet granuloma in rats (Arul et al. 2005). The percent inhibition obtained with indomethacin and bael root extract in carrageenan induced paw edema were 52.7% and 46% and in cotton pellet induced granuloma were 24.7% and 9.2% respectively (Benni et al. 2011). Indomethacin showed highly significant antiinflammatory activity in both the models. However, Bilwa showed highly significant activity in acute model.

Analgesic/Antipyretic Activities

Bael leaf extract also produced marked analgesic activity by reduction at the early and late phases of paw licking in mice (Arul et al. 2005). A significant reduction in hyperpyrexia in rats was also produced by the extract. Another research showed that methanol extract of leaves of *Aegle marmelos* at a dose level of 200 and 300 mg/kg showed significant analgesic activity on acetic acid-induced writhing and tail flick test in mice (Shankarananth et al. 2007).

Anxiolytic/Antidepressant Activities

Methanol bael leaf extract was found to exhibit anxiolytic and antidepressant activities in mice using elevated plus maze and tail suspension test (Kothari et al. 2010). The extract significantly and dose dependently increased proportionate time spent on and number of entries into open arms while decreased number of stretch attend postures and head dips in closed arms. Dose dependent and significant anti-immobility effect was found in mice treated with the methanol extract. Combination of bael leaf extract (75 mg/kg, po) with imipramine (5 mg/kg, po) or fluoxetine (5 mg/kg, po) also produced significant anxiolytic and antidepressant activity.

Antiallergic/Antiasthmatic Activity

The alcoholic leaf extract was found to produce a complete relaxation of the guinea pig ileum and tracheal chain (Arul et al. 2004). This was attributed to the presence of one or more anti-histaminic constituents present in the alcoholic extract of this plant, therefore supporting to the traditional use of *A. marmelos* in asthmatic complaints and related afflictions. Aegeline, a main alkaloid of bael leaves was found to inhibit histamine release from mast cells (Nugroho et al. 2011). The inhibitory effects of aegeline on the histamine release from mast cells depended on the type of mast cell and also involved some mechanisms related to intracellular Ca(2+) signalling events via the same target of the action of thapsigargin or downstream process of intracellular Ca(2+) signalling in mast cells. In another study, skimmianine isolated from the fresh bael roots extract was found to inhibit histamine release from rats mast cells (Nugroho et al. 2010). Its inhibitory effect may involve some mechanism related to intracellular Ca2+ signaling and protein kinase C signaling having a main role in granule exocytotic processes.

Antigenotoxic Activity

Treatment of mice with 200, 250, 300, 350, and 400 mg/kg body weight of bael extract orally once daily for 5 consecutive days before doxorubicin treatment, significantly reduced the frequency of doxorubicin-induced micronuclei accompanied by a significant elevation in the micronucleated polychromatic/micronucleated normochromatic erythrocytes ratio at all evaluation times (Venkatesh et al. 2007). The greatest protection against doxorubicin-induced genotoxicity was observed at 350 mg/kg of the extract. The protection against doxorubicin – induced genotoxicity by bael extract was suggested to be due to inhibition of free radicals and increased antioxidant status. A brassinosteroid, 24-epibrassinolide, was isolated from *A. marmelos* leaves (Sondhi et al. 2008). It exhibited antigenotixic activity against maleic hydrazide induced genotoxicity in *Allium cepa* chromosomal aberration assay. The percentage of chromosomal aberrations induced by maleic hydrazide (0.01%) declined significantly with 24-epibrassinolide treatment and 10^{-7} M concentration was found to be the most effective concentration with 91.8% inhibition. The methanol and acetone extract of bael fruit was found to be quite effective in decreasing the SOS response induced by hydrogen peroxide and aflatoxin B1 in the SOS chromo test (Kaur et al. 2009). The methanol extract inhibited gemotoxicity of hydrogen peroxide by 70.84% and that of aflatoxin B1 by 84.65%. The antigenotoxic activity may be attributed to the various polyphenolic constituents present in the extracts.

Adoptogenic Activity

Research demonstrated that ethanolic extract of unripe fruits of *Aegle marmelos* exhibited significant antistress, adoptogenic activity in mice using swim endurance test and cold restrain stress (Sumanth and Mustafa 2005). The extract exhibited antistress, adoptogenic activity by improving the swim duration and reducing the elevated WBC, blood glucose and plasma cortisol level. Acute toxicity studies revealed that LD_{50} was more than a dose of 3 g/kg body weight.

Antispermatogenic, Antifertility Effects

The ethanolic extract of *Aegle marmelos* leaf was found to possess antispermatogenic activity (Sur et al. 1999, 2002). Oral administration of the leaf extract for 30 days produced significant decrease in the weight of sex organs in alibino rats, reduced sperm count, increased sperm motility and both protein and RNA contents of testis dose dependently. The scientists also reported that the aqueous extract of bael leaf had anti-motility action on spermatozoa in rats. Increases in concentration of water extract decreased the complete immobility time of sperms accordingly.

Another study reported that aqueous extract of *Aegle marmelos* leaves at the dose 50 mg/100 g body weight, resulted in a significant diminution in the activities of key testicular steroidogenic enzymes along with low levels of plasma testosterone and relative wet weights of sex organs in respect to control without any significant alteration in general body growth (Das et al. 2006). Germ cells numbers in different generation at stage VII of seminiferous epithelial cell cycle were diminished significantly after the treatment of the above extract. The above mentioned dose did not exhibit any toxicity in liver and kidney. The findings indicated that the aqueous leaf extract of *Aegle marmelos* had a potent antitesticular effect at a specific dose. Studies showed that ethanolic bael leaf extract suppressed fertility in male rats (Chauhan et al. 2007; Chauhan and Agarwal 2008, 2009). The weight of the reproductive organs was reduced significantly in all the treatment groups. The extract reduced fertility of male rats by 100% at the 300-mg dose level. There was a marked reduction in motility and density of the sperm derived from cauda epididymis of the treated animals; serum testosterone levels also decreased significantly. Spermatogenesis was impaired. The cross sectional surface area of Sertoli cells and mature Leydig cells was reduced along with a dose dependent reduction of preleptotene and pachytene spermatocytes. The protein, glycogen, fructose, ascorbic acid, acid, alkaline phosphatase contents and lipid peroxidation of the testes were significantly reduced at the highest dose level; a highly significant increase in testicular cholesterol was observed along with a highly significant reduction in the sialic acid contents of testes, epididymis and seminal vesicles. Complete recovery of fertility was observed after a 120 day withdrawal of drug. No clinical signs of side effects on general metabolism were detected throughout the treatment, and after withdrawal, body weight gain was similar in all groups. Absence of any deleterious effect on the vital organs points to the safe use of bael extract.

Antimicrofilarial Activity

Bael leaf methanol extract at 100 ng/ml concentration caused complete loss of motility of *Brugia malayi* microfilariae after 48 hours of incubation (Sahare et al. 2008b). The bioactivity was attributed to the presence of coumarins in the leaf extract. The IC_{50} inhibitory concentration for the extract was 70 ng/ml (Sahare et al. 2008a).

Insecticidal Activity

The ethyl acetate extract of *A. marmelos* and extracts of several other indigenous plants showed the highest larval mortality against fourth-instar larvae of malaria vector *Anopheles subpictus* and Japanese encephalitis vector, *Culex tritaeniorhynchus* (Elango et al. 2009b). *Aegle mamrmelos* leaf extracts exhibited oviposition-deterrent, ovicidal, and repellent activities of indigenous plant extracts against the malarial vector, *Anopheles subpictus* (Elango et al. 2009a). The percentage of effective oviposition repellency of acetone, ethyl acetate, and methanol extracts of *Aegle marmelos* at 500 ppm were 92.60, 93.04, 95.20% and at 31.25 ppm were 47.14, 58.00, 56.52% respectively. The oviposition activity index (OAI) value of acetone, ethyl acetate, and methanol extracts of *Aegle marmelos* at 500 ppm were −0.86, −0.87, −0.90 respectively. Mortality of 100% with ethyl acetate extract of *Aegle marmelos* was found at 1,000 ppm. The maximum repellent activity was observed at 500 ppm in methanol extract of *Aegle marmelos*. Bael leaf extracts showed moderate adulticidal activity and adult emergence inhibition (EI) effects against the malarial vector, *Anopheles subpictus* after 24 hours of exposure at 1,000 ppm (Elango et al. 2011). The effective EI was found in bael leaf acetone extract with LD_{50} of 128.14 ppm and LD_{90} of 713.53 ppm. Studies showed that the hexane extract of *A. marmelos* and *Andrographis paniculata* could serve a potential repellent, ovicidal, and oviposition deterrent against Japanese encephalitis mosquito vector, *Culex tritaeniorhynchus* (Elango et al. 2010a). The maximum repellent activity against *Culex tritaeniorhynchus* was observed at 500 ppm in methanol extracts of *A. marmelos* and acetone extracts of *A. marmelos* showed complete protection in 90 minutes at 250 ppm (Elango et al. 2010b). Methanol bael leaf extract showed moderate larvicidal activity against the early fourth instar larvae of *Aedes aegypti* and *Anopheles stephensi* mosquito (Patil et al. 2010).

Antihyperthyroidism Effect

Research in male mice showed that *Bacopa monnieri* (200 mg/kg), *Aegle marmelos* (1.00 g/kg) and *Aloe vera* (125 mg/kg) leaf extracts could play a role in the regulation of thyroid hormone concentrations (Kar et al. 2002). Findings suggested that *A. marmelos* and *A. vera* may be used in the regulation of hyperthyroidism, while *B. monnieri* in hypothyroidism. Studies showed that daily administration of scopoletin (7-hydroxy-6-methoxy coumarin), isolated from bael leaves, to levo-thyroxine-treated animals for a week decreased the levels of serum thyroid hormones and glucose as well as hepatic glucose-6-phosphatase activity, demonstrating its potential to regulate hyperthyroidism and hyperglycemia (Panda and Kar 2006). Scopoletin also inhibited hepatic lipid peroxidation and increased the activity of antioxidants, superoxide dismutase and catalase. Compared with the standard antithyroid drug, propylthiouracil, scopoletin exhibited a superior therapeutic activity, since unlike propylthiouracil, it also inhibited hepatic lipid peroxidation.

Anti-ocular Hypertension Activity

Aegle marmelos fruit extract showed significant intraocular pressure (IOP)-lowering activity in rabbits (Agarwal et al. 2009). In rabbits with normal IOP, the fruit extract at a concentration of 1% showed the maximum IOP-lowering effect with 22.81% reduction from baseline IOP. The efficacy was comparable to that of timolol (0.25%).

Wound Healing Activity

Both the injection and the ointment of the methanol extract of *Aegle marmelos* produced a significant response in both the excision and incision wound models tested in rats (Jaswanth et al. 1994). In the excision model the extract-treated wounds were found to epithelialise faster and the rate of wound contraction was higher, as compared to control wounds. The extract facilitated the healing process as evidenced by increase in the tensile strength in the incision model. The results were also comparable to those of a standard drug nitrofurazone.

Immunomodulatory Activity

Methanolic bael fruit extract exhibited immunomodulatory effect in mice (Patel and Asdaq 2010). The extract at 100 and 500 mg/kg produced significant increases in adhesion of neutrophils and an increase in phagocytic index in carbon clearance assay. Both high and low doses of the extract significantly prevented the mortality induced by bovine *Pasteurella multocida* in mice. Treatment of animals with the extract significantly increased the circulating antibody titre in indirect haemagglunation test. Among the different doses, the low dose was more effective in augmenting cellular immunity models than the high. However, all the doses exhibited similar protection in humoral immunity procedures.

Toxicological Studies

Safety evaluation of aqueous extracts from *Aegle marmelos* showed no notable abnormalities in any of the pregnant rats (Saenphet et al. 2006). The number of corpus lutea, implanted and dead fetuses, as well as the sizes of the fetuses in the treated rats were not significantly different from those of the controls. Based on these results, the authors concluded that aqueous extracts of *A. marmelos* at the concentrations used in the study did not alter the reproduction of female rats. Intraperitoneal administration of the total alcoholic, total aqueous, whole aqueous and methanolic extracts of the leaves of *A. marmelos* at doses of 50, 70, 90 and 100 mg/kg body wt for 14 consecutive days to male and female Wistar rats did not induce any short-term toxicity (Veerappan et al. 2007). Pathologically, neither gross abnormalities nor histopathological changes were observed. Collectively, the data demonstrated that the extracts of the leaves of *A. marmelos* had a high margin of drug safety.

Traditional Medicinal Uses

Bael is an important medicinal plant of India; its leaves, fruits, seeds, stem and roots have been used in ethno-traditional medicine to exploit its' medicinal properties including astringent, antidiarrheal antidysenteric, demulcent, antipyretic and antiinflammatory activities for treating one human ailment or another (Maity et al. 2009; Dhankhar et al. 2011). All the parts of the tree viz, root, leaf, trunk, fruit, are used in traditional ayurvedic system of medicine: root parts are used in dysentery (pravahika), dyspepsia (agnimandya), chronic diarrhea with mal absorption (graham roga) and diabetes mellitus. The dried roots are used in the disorder of nervous system (vatavadhjy), oedema (uotha), vomiting (chardi), and rheumatism (amavata). The root bark is use in the form of a decoction as remedy for hypochondriasis, melancholia, intermittent fever, and palpitation of the heart. The leaves of *Bael* are astringent, a laxative, and an expectorant and are useful in treatment of ophthalmia, deafness, inflammations, cataract, diabetes, diarrhoea, dysentery, heart palpitation, and asthmatic complications. The leaves are made into a poultice and applied to inflamed parts. The fresh bitter juice is used as a remedy for catarrh and feverishness or mixed with honey as a laxative and febrifuge, and for asthmatic complaints. A sweet-scented extract from the flowers is used as lotion for the eyes.

Bael-marmalade or aromatized confection is useful for breakfast during convalescence from chronic dysentery or diarrhoea and for daily use as a preventive during cholera epidemics. It is also given to prevent the growth of piles. The fruit is prescribed in tuberculosis. The unripe or half-ripe fruit is regarded as an astringent, a digestive, and stomachic, and is said to be an excellent remedy for chronic diarrhoea. Ripe fruit extract is used against rectum inflammation.

Several parts of this plant have been used by the local Thai people in folk medicines. For example, the infusion of dried unripe fruits has been used as antidiarrhea and antidysentery agents, the juice from crushed leaves has been used for the treatment of bronchitis, and the decoction of root barks has also been used as anti-malarial drug (Laphookhieo et al. 2011).

Other Uses

The mucilage (gum) around unripe seeds is used as an adhesive by jewellers and as household glue. It is mixed with lime plaster for waterproofing wells, and is added to cement when building walls. The rind contains up to 20% tannin, which is also present in the leaves. The rind of unripe fruit being rich in tannin can be used as a yellow dye which is utilized for textile prints, calico and for dyeing silk fabrics and as a tanning agent. The mucous fluid of the fruit is rubbed on the hair in place of oil by the poorer classes or is employed as soap in washing garments in India. The Dutch in Ceylon used formerly to prepare an essential oil (or attar) from the rind, known as Marmelle oil. A perfume is also distilled from the flowers. Bael fruit is employed in the treatment of scum in vinegar manufacture.

The leaves and twigs are lopped for fodder. The wood is used for carts and construction, though it is inclined to warp and crack during curing. It is best utilized for carving, small-scale turnery, tool and knife handles, pestles and combs. Leaf extract from *A. marmelos* has been found to have insecticidal activity against the brown plant hopper (*Nilaparvata lugens* Stål), an important pest of rice plant in Asia (Hiremarh et al. 1996). Skimmiarepins A and C fractionated from ethyl acetate extracts of bael stem bark exhibited moderate insecticidal activity against *Phaedon cochleariae* and *Musca domestica* in comparison with natural pyrethrum extract (Samarasekera et al. 2004). The two epimeric acetates of skimmiarepin C were both less active. Methanol bael leaf extract exhibited strong antiparasitic activity against the adult cattle tick *Haemaphysalis bispinosa* with LC_{50} of 395.27 ppm, against the larvae of *Rhipicephalus (Boophilus) microplus* with LC_{50} of 207.7 ppm and sheep fluke *Paramphistomum cervi* with LC_{50} of 254.23 ppm (Elango and Rahuman 2011).

Bael fruit shell activated carbon was found to be an effective and economically viable adsorbent for Cr(VI) toxic metal removal from aqueous system (Anandkumar and Mandal 2009). Studies demonstrated that bael leaves selectively remove Pb(II) in the presence of other metal ions from the effluent of exhausted batteries (Chakravarty et al. 2010).

The bark is also employed as fish poison. Laboratory studies confirmed that dietary *Nelumbo nucifera* and *Aegle marmelos* supplementation of common carp exposed to sublethal concentrations of heavy metals provided a detoxification mechanism for heavy metals in common carp (Vinodhini 2010). The level of red blood cells (RBC), packed cell volume (PCV), haemoglobin concentration were restored to normal levels and glucose and cholesterol level in blood of common carp showed significant reduction compared with heavy-metal-exposed groups.

Comments

Aegle marmelos is a sacred tree, dedicated to the deity, Lord Shiva in India. The offering of bael leaves is a compulsory ritual of the worship of Lord Shiva in the hills in India.

Selected References

Abraham PM, Kuruvilla KP, Mathew J, Malat A, Joy S, Paulose CS (2010a) Alterations in hippocampal serotonergic and INSR function in streptozotocin induced diabetic rats exposed to stress: neuroprotective role of pyridoxine and *Aegle marmelos*. J Biomed Sci 17(1):78

Abraham PM, Paul J, Paulose CS (2010b) Down regulation of cerebellar serotonergic receptors in streptozotocin induced diabetic rats: effect of pyridoxine and *Aegle marmelos*. Brain Res Bull 82(1–2):87–94

Agarwal R, Gupta SK, Srivastava S, Saxena R, Agrawal SS (2009) Intraocular pressure-lowering activity of topical application of *Aegle marmelos* fruit extract in experimental animal models. Ophthalmic Res 42(2):112–116

Agrawal A, Jahan S, Soyal D, Goyal E, Goyal PK (2011) Amelioration of chemical-induced skin carcinogenesis by *Aegle marmelos*, an Indian medicinal plant, fruit extract. Integr Cancer Ther doi: 10.1177/1534735411417127

Agrawal A, Verma P, Goyal PK (2010) Chemomodulatory effects of *Aegle marmelos* against DMBA-induced skin tumorigenesis in Swiss albino mice. Asian Pac J Cancer Prev 11(5):1311–1314

Ali MS, Pervez MK (2004) Marmenol: a 7-geraniloxy coumarin from the leaves of *Aegle marmelos* Corr. Nat Prod Res 18:141–146

Anandharajan R, Jaiganesh S, Shankernarayanan NP, Viswakarma RA, Balakrishnan A (2006) In vitro glucose uptake activity of *Aegles marmelos* and *Syzygium cumini* by activation of Glut-4, PI3 kinase and PPARgamma in L6 myotubes. Phytomedicine 13(6):434–441

Anandkumar J, Mandal B (2009) Removal of Cr(VI) from aqueous solution using Bael fruit (*Aegle marmelos* Correa) shell as an adsorbent. J Hazard Mater 68(2–3):633–640

Arul V, Miyazaki S, Dhananjayan R (2004) Mechanisms of the contractile effect of the alcoholic extract of *Aegle marmelos* Corr. on isolated guinea pig ileum and tracheal chain. Phytomedicine 11(7–8):679–683

Arul V, Miyazaki S, Dhananjayan R (2005) Studies on the anti-inflammatory, antipyretic and analgesic properties of the leaves of *Aegle marmelos* Corr. J Ethnopharmacol 96(1–2):159–163

Backer CA, Bakhuizen van den Brink RC Jr (1965) Flora of Java (spermatophytes only), vol 2. Wolters-Noordhoff, Groningen, 641 pp

Badam L, Bedekar SS, Sonawane KB, Joshi SP (2002) In vitro antiviral activity of Bael (*Aegle marmelos* Corr.) upon human coxsackieviruses B1-B6. J Commun Dis 34(2):88–99

Baliga MS, Bhat HP, Pereira MM, Mathias N, Venkatesh P (2010) Radioprotective effects of *Aegle marmelos* (L.) Correa (Bael): a concise review. J Altern Complement Med 16(10):1109–1116

Barthakur NN, Arnold NP (1989) Certain organic and inorganic constituents in bael (*Aegle marmelos* Correa) fruit. Trop Agric 66(1):65–68

Basak RK, Mandal PK, Mukherjee AK (1981) Studies on a neutral polysaccharide isolated from bael (*Aegle marmelos*) fruit pulp. Carbohydr Res 97(2):315–321

Basak RK, Mandal PK, Mukherjee AK (1982) Investigations on the structure of a hemicelluloses fraction isolated from the trunk of a young bael (*Aegle marmelos*) tree. Carbohydr Res 104(2):309–317

Baslas KK, Deshpandey SS (1951) Essential oil from leaves of Bael. J Indian Chem Soc 28:19–22

Basu D, Sen R (1974) Alkaloids and coumarins from root-bark of *Aegle marmelos*. Phytochemical 13(10): 2329–2330

Benni JM, Jayanthi MK, Suresha RN (2011) Evaluation of the anti-inflammatory activity of *Aegle marmelos* (Bilwa) root. Indian J Pharmacol 43(4):393–397

Bhattacharyya P, Jash SS, Dey AK (1989) 6',7'-Epoxyauraptene. A coumarin from *Aegle marmelos*. J Indian Chem Soc 66(6):424–425

Brijesh S, Daswani P, Tetali P, Antia N, Birdi T (2009) Studies on the antidiarrhoeal activity of *Aegle marmelos* unripe fruit: validating its traditional usage. BMC Complement Altern Med 9:47

Burkill IH (1966) A dictionary of the economic products of the Malay Peninsula. Revised reprint. 2 vols. Ministry of Agriculture and Co-operatives, Kuala Lumpur, vol 1 (A–H), pp 1–1240, vol 2 (I–Z), pp 1241–2444

Chakravarty S, Mohanty A, Sudha TN, Upadhyay AK, Konar J, Sircar JK, Madhukar A, Gupta KK (2010) Removal of Pb(II) ions from aqueous solution by adsorption using bael leaves (*Aegle marmelos*). J Hazard Mater 173(1–3):502–509

Charoensiddhi S, Anprung P (2008) Bioactive compounds and volatile compounds of Thai bael fruit (*Aegle marmelos* (L.) Correa) as a valuable source for functional food ingredients. Int Food Res J 15(3): 287–295

Chatterjee A, Bhattacharya A (1959) The isolation and constitution of marmin, a new coumarin from *Aegle marmelos*, Corres. J Chem Soc Pap 385:1922–1924

Chatterjee A, Chaudhury B (1960) Occurrence of auraptene, umbelliferone, marmin, lupeol and skimminanine in the root of *Aegle marmelos* Correa. J Indian Chem Soc 37:334–336

Chatterjee A, Choudhury A (1955) The structure of marmin, a new coumarin of *Aegle marmelos* Correa. Naturwissenschaften 42(18):512

Chatterjee A, Mitra SS (1949) On the constitution of the active principles isolated from the mature bark of *Aegle marmelos* Correâ. J Am Chem Soc 71(2):606–609

Chatterjee A, Roy SK (1957) Aegelenine, a new alkaloid of the leaves of *Aegle marmelos* Correa. Sci Cult 23:106–107

Chauhan A, Agarwal M (2008) Reversible changes in the antifertility induced by *Aegle marmelos* in male albino rats. Syst Biol Reprod Med 54(6):240–246

Chauhan A, Agarwal M (2009) Assessment of the contraceptive efficacy of the aqueous extract of *Aegle marmelos* Corr. leaves in male albino rats. Hum Fertil (Camb) 12(2):107–118

Chauhan A, Agarwal M, Kushwaha S, Mutreja A (2007) Suppression of fertility in male albino rats following the administration of 50% ethanolic extract of *Aegle marmelos*. Contraception 76(6):474–481

Costa-Lotufo LV, Khan MT, Ather A, Wilke DV, Jimenez PC, Pessoa C, de Moraes ME, de Moraes MO (2005) Studies of the anticancer potential of plants used in Bangladeshi folk medicine. J Ethnopharmacol 99(1):21–30

Council of Scientific and Industrial Research (CSIR) (1948) The wealth of India. A dictionary of Indian raw materials and industrial products. (Raw materials 1). Publications and Information Directorate, New Delhi

Das AV, Padayatti PS, Paulose CS (1996) Effect of leaf extract of *Aegle marmelos* (L.) Correa ex Roxb. on histological and ultrastructural changes in tissues of streptozotocin induced diabetic rats. Indian J Exp Biol 34(4):341–345

Das UK, Maiti R, Jana D, Ghosh D (2006) Effect of aqueous extract of leaf of *Aegle marmelos* on testicular activities in rats. Iranian J Pharmacol Ther 5(1):21–25

Dhalwal K, Shinde VM, Mahadik KR, Namdeo AG (2007) Rapid densitometric method for simultaneous analysis of umbelliferone, psoralen, and eugenol in herbal raw materials using HPTLC. J Sep Sci 30(13): 2053–2058

Dhankhar S, Ruhil S, Balhara M, Dhankhar S, Chhillar AK (2011) *Aegle marmelos* (Linn.) Correa: a potential source of phytomedicine. J Med Plant Res 5(9):1497–1507

Dhuley JN (2003–2004) Investigation on the gastroprotective and antidiarrhoeal properties of *Aegle marmelos* unripe fruit extract. Hindustan Antibiot Bull 45–46:41–46

Elango G, Bagavan A, Kamaraj C, Abduz Zahir A, Abdul Rahuman A (2009a) Oviposition-deterrent, ovicidal, and repellent activities of indigenous plant extracts against *Anopheles subpictus* Grassi (Diptera: Culicidae). Parasitol Res 105(6):1567–1576

Elango G, Rahuman AA (2011) Evaluation of medicinal plant extracts against ticks and fluke. Parasitol Res 108(3):513–519

Elango G, Rahuman AA, Bagavan A, Kamaraj C, Zahir AA, Rajakumar G, Marimuthu S, Santhoshkumar T (2010a) Efficacy of botanical extracts against Japanese encephalitis vector, *Culex tritaeniorhynchus*. Parasitol Res 106(2):481–492

Elango G, Rahuman AA, Bagavan A, Kamaraj C, Zahir AA, Venkatesan C (2009b) Laboratory study on larvicidal activity of indigenous plant extracts against *Anopheles subpictus* and *Culex tritaeniorhynchus*. Parasitol Res 10(6):1381–1388

Elango G, Rahuman AA, Kamaraj C, Bagavan A, Zahir AA (2011) Efficacy of medicinal plant extracts against malarial vector, *Anopheles subpictus* Grassi. Parasitol Res 108(6):1437–1445

Elango G, Rahuman AA, Zahir AA, Kamaraj C, Bagavan A, Rajakumar G, Jayaseelan C, Santhoshkumar T, Marimuthu S (2010b) Evaluation of repellent properties of botanical extracts against *Culex tritaeniorhynchus* Giles (Diptera: Culicidae). Parasitol Res 107(3):577–584

Faizi S, Farooqi F, Zikr-Ur-Rehman S, Naz A, Noor F, Ansari F, Ahmad A, Khan SA (2009) Shahidine, a novel and highly labile oxazoline from *Aegle marmelos*: the parent compound of aegeline and related amides. Tetrahedron 65(5):998–1004

Foundation for Revitalisation of Local Health Traditions (2008) FRLHT database. http://envis.frlht.org

Garg SN, Siddiqui MS, Agarwal SK (1995) p-Menth-1-en-3-beta,5-beta-diol, a new constituent of *Aegle marmelos* leaf oil. J Essent Oil Res 7(3):283–286

Gireesh G, Reas SK, Jobin M, Paulose CS (2008) Decreased muscarinic M1 receptor gene expression in the cerebral cortex of streptozotocin-induced diabetic rats and *Aegle marmelos* leaf extract's therapeutic function. J Ethnopharmacol 116(2):296–304

Goel RK, Maiti RN, Manickam M, Ray AB (1997) Antiulcer activity of naturally occurring pyrano-coumarin and isocoumarins and their effect on prostanoid synthesis using human colonic mucosa. Indian J Exp Biol 35(10):1080–1083

Gohil T, Pathak N, Jivani N, Devmurari V, Patel J (2010) Treatment with extracts of *Eugenia jambolana* seed and *Aegle marmelos* leaf extracts prevents hyperglycemia and hyperlipidemia in alloxan induced diabetic rats. Afr J Pharm Pharmacol 4(5):270–275

Gopalan G, Rama Sastri BV, Balasubramanian SC (2002) Nutritive value of Indian foods. National Institute of Nutrition, Indian Council of Medical Research, Hyderabad

Govindachari TR, Premila MS (1983) Some alkaloids from *Aegle marmelos*. Phytochemistry 22(3):755–757

Hiremarh IG, Ahn YJ, Soon-II K (1996) Insecticidal activity of Indian plant extracts against *Nilaparvata lugens* (Homoptera Delphacidae). Appl Entomol Zool 32(1):159–166

Jagetia GC, Baliga MS (2004) The evaluation of nitric oxide scavenging activity of certain Indian medicinal plants in vitro: a preliminary study. J Med Food 7(3):343–348

Jagetia GC, Venkatesh P (2005) Radioprotection by oral administration of *Aegle marmelos* (L.) Correa in vivo. J Environ Pathol Toxicol Oncol 24(4):315–332

Jagetia GC, Venkatesh P (2007) Inhibition of radiation-induced clastogenicity by *Aegle marmelos* (L.) Correa in mice bone marrow exposed to different doses of γ-radiation. Hum Exp Toxicol 26(2):111–124

Jagetia GC, Venkatesh P, Archana P, Krishnanand BR, Baliga MS (2006) Effects of *Aegle marmelos* (L.) Correa on the peripheral blood and small intestine of mice exposed to gamma radiation. J Environ Pathol Toxicol Oncol 25:611–624

Jagetia GC, Venkatesh P, Baliga MS (2003) Evaluation of the radioprotective effect of *Aegle marmelos* (L.) Correa in cultured human peripheral blood lymphocytes exposed to different doses of γ-radiation: a micronucleus study. Mutagenesis 18(4):387–393

Jagetia GC, Venkatesh P, Baliga MS (2004a) Evaluation of the radioprotective effect of bael leaf (*Aegle marmelos*) extract in mice. Int J Radiat Biol 80(4):281–290

Jagetia GC, Venkatesh P, Baliga MS (2004b) Fruit extract of *Aegle marmelos* protects mice against radiation-induced lethality. Integr Cancer Ther 3(4):323–332

Jagetia GC, Venkatesh P, Baliga MS (2005) *Aegle marmelos* (L.) Correa inhibits the proliferation of transplanted Ehrlich ascites carcinoma in mice. Biol Pharm Bull 28(1):58–64

Jain AK, Srivastava SK, Srivastava SD (1991) Some new constituents from heartwood of *Aegle marmelos* Corr. J Indian Chem Soc 68(8):452–454

Jaswanth A, Akilan D, Loganathan V, Manimaran S, Ruckmani (1994) Wound healing activity of *Aegle marmelos*. Indian J Pharm Sci 56(1):41–44

Kakiuchi N, Senaratne LR, Huang SL, Yang XW, Hattori M, Pilapitiya U, Namba T (1991) Effects of constituents of Beli (*Aegle marmelos*) on spontaneous beating and calcium-paradox of myocardial cells. Planta Med 57(1):43–46

Kamalakkannan N, Prince PS (2003) Hypoglycaemic effect of water extracts of *Aegle marmelos* fruits in streptozotocin diabetic rats. J Ethnopharmacol 87(2–3):207–210

Kamalakkannan N, Prince PSM (2004a) Antidiabetic and anti-oxidant Activity of *Aegle marmelos* extract in Streptozotocin-induced diabetic rats. Pharm Biol 42(2):125–130

Kamalakkannan N, Prince PSM (2004b) Antihyperlipidaemic effect of *Aegle marmelos* fruit extract in streptozotocin-induced diabetes in rats. J Sci Food Agric 85(4):569–573

Kamalakkannan N, Prince PSM (2005) The effect of *Aegle marmelos* fruit extract in streptozotocin diabetes a histopathological study. J Herb Pharmacother 5(3):87–96

Kamalakkannan N, Rajadurai M, Prince PSM (2003) Effect of *Aegle marmelos* fruits on normal and Streptozotocin-diabetic Wistar rats. J Med Food 6(2):93–98

Kar A, Panda S, Bharti S (2002) Relative efficacy of three medicinal plant extracts in the alteration of thyroid hormone concentrations in male mice. J Ethnopharmacol 81(2):281–285

Kaur HP, Garg SN, Sashidhara KV, Yadav A, Naqvi AA, Khanuja SPS (2006) Chemical composition of the essential oil of the twigs and leaves of *Aegle marmelos* (L.) Correa. J Essent Oil Res 18(3):288–289

Kaur P, Walia A, Kumar S, Kaur S (2009) Antigenotixicty activity of polyphenolic rich extracts from *Aegle marmelos* (L.) Correa in human blood lymphocytes and E. coli PQ 37. Rec Nat Prod 3(1):68–75

Kesari AN, Gupta RK, Singh SK, Diwakar S, Watal G (2006) Hypoglycemic and antihyperglycemic activity of *Aegle marmelos* seed extract in normal and diabetic rats. J Ethnopharmacol 107(3):374–379

Khan TH, Sultana S (2009) Antioxidant and hepatoprotective potential of *Aegle marmelos* Correa against CCl4-induced oxidative stress and early tumor events. J Enzyme Inhib Med Chem 24(2):320–327

Khan TH, Sultana S (2011) Effect of *Aegle marmelos* on DEN initiated and 2-AAF promoted hepatocarcinogenesis: a chemopreventive study. Toxicol Mech Methods 21(6):453–462

Kothari S, Minda M, Tonpay SD (2010) Anxiolytic and antidepressant activities of methanol extract of *Aegle marmelos* leaves in mice. Indian J Physiol Pharmacol 54(4):318–328

Kumari A, Kakkar P (2008) Screening of antioxidant potential of selected barks of Indian medicinal plants by multiple in vitro assays. Biomed Environ Sci 21(1):24–29

Lambertini E, Lampronti I, Penolazzi L, Khan MT, Ather A, Giorgi G, Gambari R, Piva R (2005) Expression of estrogen receptor alpha gene in breast cancer cells treated with transcription factor decoy is modulated by Bangladeshi natural plant extracts. Oncol Res 15(2):69–79

Lambertini E, Piva R, Khan MT, Lampronti I, Bianchi N, Borgatti M, Gambari R (2004) Effects of extracts from Bangladeshi medicinal plants on in vitro proliferation of human breast cancer cell lines and expression of estrogen receptor alpha gene. Int J Oncol 24(2):419–423

Lambole VB, Murti K, Kumar U, Sandipkumar PB, Gajera V (2010) Phytopharmacological properties of *Aegle marmelos* as a potential medicinal tree: an overview. Int J Pharm Sci Rev Res 5(2):67–72

Lampronti I, Martello D, Bianchi N, Borgatti M, Lambertini E, Piva R, Jabbar S, Choudhuri MSK, Khan MTH, Gambari R (2003) In vitro antiproliferative effects on human tumor cell lines of extracts from the Bangladeshi medicinal plant *Aegle marmelos* Correa. Phytomedicine 10(4):300–308

Laphookhieo S, Phungpanya C, Tantapakul C, Techa S, Tha-in S, Narmdorkmai W (2011) Chemical constituents from *Aegle marmelos*. J Braz Chem Soc 22(1):176–178

Maity P, Hansda D, Bandyopadhyay U, Mishra DK (2009) Biological activities of crude extracts and chemical constituents of Bael, *Aegle marmelos* (L.) Corr. Indian J Exp Biol 47(11):849–861

Manandhar MD, Shoeb A, Kapil RS, Popli SP (1978) New alkaloids from *Aegle marmelos*. Phytochemistry 17:1814–1815

Mandal PK, Mukherjee AK (1981) Investigations on the partial structure of a glycoprotein from bael (*Aegle marmelos*) seed. Carbohydr Res 98(1):85–91

Mazumder R, Bhattacharya S, Mazumder A, Pattnaik AK, Tiwary PM, Chaudhary S (2006) Antidiarrhoeal evaluation of *Aegle marmelos* (Correa) Linn. root extract. Phytother Res 20(1):82–84

Mishra BB, Kishore N, Tiwari VK, Singh DD, Tripathi V (2010a) A novel antifungal anthraquinone from seeds of *Aegle marmelos* Correa (family Rutaceae). Fitoterapia 81(2):104–107

Mishra BB, Singh DD, Kishore N, Tiwari VK, Tripathi V (2010b) Antifungal constituents isolated from the seeds of *Aegle marmelos*. Phytochemistry 71(2–3):230–234

Molesworth Allen B (1967) Malayan fruits. An introduction to the cultivated species. Moore, Singapore, 245 pp

Narender T, Shweta S, Tiwari P, Papi Reddy K, Khaliq T, Prathipati P, Puri A, Srivastava AK, Chander R, Agarwal SC, Raj K (2007) Antihyperglycemic and antidyslipidemic agent from *Aegle marmelos*. Bioorg Med Chem Lett 17(6):1808–1811

Narendhirakannan RT, Subramanian S (2010) Biochemical evaluation of the protective effect of *Aegle marmelos* (L.), Corr. leaf extract on tissue antioxidant defense system and histological changes of pancreatic beta-cells in streptozotocin-induced diabetic rats. Drug Chem Toxicol 33(2):120–130

Nugroho AE, Riyanto S, Sukari MA, Maeyama K (2010) Effects of skimmianine, a quinoline alkaloid of *Aegle marmelos* Correa roots, on the histamine release from rat mast cells. J Basic Appl Sci 6(2):141–148

Nugroho AE, Riyanto S, Sukari MA, Maeyama K (2011) Effects of aegeline, a main alkaloid of *Aegle marmelos* Correa leaves, on the histamine release from mast cells. Pak J Pharm Sci 24(3):359–367

Ohashi K, Watanabe H, Okumura Y, Uji T, Kitagawa I (1994) Indonesian medicinal plants. XII. Four isomeric lignan-glucosides from the bark of *Aegle marmelos* (Rutaceae). Chem Pharmaceut Bull 42(9):1924–1926

Panda S, Kar A (2006) Evaluation of the antithyroid, antioxidative and antihyperglycemic activity of scopoletin from *Aegle marmelos* leaves in hyperthyroid rats. Phytother Res 20(12):1103–1105

Panda S, Kar A (2009) Periplogenin-3-O-D-glucopyranosyl-(1→6)-D-glucopyaranosyl-(1→4)-D-cymaropyranoside, isolated from *Aegle marmelos* protects doxorubicin induced cardiovascular problems and hepatotoxicity in rats. Cardiovasc Ther 27(2):108–116

Patel P, Asdaq SMB (2010) Immunomodulatory activity of methanolic fruit extract of *Aegle marmelos* in experimental animals. Saudi Pharm J 18(3):161–165

Patil RH, Bhushan C, Sailaxmi S (2011) Antifungal and antiaflatoxigenic activity of *Aegle marmelos* Linn. Pharmacogn J 1(4):Article 12

Patil SV, Patil CD, Salunkhe RB, Salunke BK (2010) Larvicidal activities of six plants extracts against two mosquito species, *Aedes aegypti* and *Anopheles stephensi*. Trop Biomed 27(3):360–365

Phuwapraisirisan P, Puksasook T, Jong-aramruang J, Udom Kokpol U (2008) Phenylethyl cinnamides: a new series of α-glucosidase inhibitors from the leaves of *Aegle marmelos*. Bioorg Med Chem Lett 18(18):4956–4958

Ponnachan PT, Paulose CS, Panikkar KR (1993) Effect of leaf extract of *Aegle marmelos* in diabetic rats. Indian J Exp Biol 31(4):345–347

Prince PS, Rajadurai M (2005) Preventive effect of *Aegle marmelos* leaf extract on isoprenaline-induced myocardial infarction in rats: biochemical evidence. J Pharm Pharmacol 57(10):1353–1357

Purseglove JW (1968) Tropical crops: dicotyledons 1 & 2. Longman, London, 719 pp

Raja SB, Murali MR, Devaraj SN (2008) Differential expression of ompC and ompF in multidrug-resistant *Shigella dysenteriae* and *Shigella flexneri* by aqueous extract of *Aegle marmelos*, altering its susceptibility toward beta-lactam antibiotics. Diagn Microbiol Infect Dis 61(3):321–328

Raja SB, Murali MR, Kumar NK, Devaraj SN (2011a) Isolation and partial characterisation of a novel lectin from *Aegle marmelos* fruit and its effect on adherence and invasion of *Shigellae* to HT29 cells. PLoS One 6(1):e16231

Raja SB, Murali MR, Roopa K, Devaraj SN (2011b) Imperatorin a furocoumarin inhibits periplasmic Cu-Zn SOD of *Shigella dysenteriae* their by modulates its resistance towards phagocytosis during host pathogen interaction. Biomed Pharmacother 65(8):560–568

Rajadurai M, Prince PS (2005) Comparative effects of *Aegle marmelos* extract and alpha-tocopherol on serum lipids, lipid peroxides and cardiac enzyme levels in rats with isoproterenol-induced myocardial infarction. Singapore Med J 46(2):78–81

Ramesh B, Pugalendi KV (2007a) Effect of umbelliferone on tail tendon collagen and haemostatic function in streptozotocin-diabetic rats. Basic Clin Pharmacol Toxicol 101(2):73–77

Ramesh B, Pugalendi KV (2007b) Influence of umbelliferone on glycoprotein components in diabetic rats. Toxicol Mech Methods 17(3):153–159

Ramesh B, Pugalendi KV (2007c) Influence of umbelliferone on membrane-bound ATPases in streptozotocin-induced diabetic rats. Pharmacol Rep 59(3):339–348

Rana BK, Singh UP, Taneja V (1997) Antifungal activity and kinetics of inhibition by essential oil isolated from leaves of *Aegle marmelos*. J Ethnopharmacol 57(1):29–34

Rani P, Khullar N (2004) Antimicrobial evaluation of some medicinal plants for their anti-enteric potential against multi-drug resistant *Salmonella typhi*. Phytother Res 18(8):670–673

Riyanto S, Sukari MA, Rahmani M, Ee GCL, Taufiq-Yap YH, Aimi N, Kitajima M (2001) Alkaloids from *Aegle marmelos* (Rutaceae). Malays J Anal Sci 7(2):463–465

Riyanto S, Sukari MA, Rahmani M, Manas AR, Ali AM, Yusuf UK, Muse R (2002) Coumarins from roots of *Aegle marmelos*. J Trop Med Plants 3(2):159–162

Roy A, Mukherjee AK, Rao CV (1975) Graded-hydrolysis studies on bael (*Aegle marmelos*) gum. Carbohydr Res 41:219–226

Sabde S, Bodiwala HS, Karmase A, Deshpande PJ, Kaur A, Ahmed N, Chauthe SK, Brahmbhatt KG, Phadke RU, Mitra D, Bhutani KK, Singh IP (2011) Anti-HIV activity of Indian medicinal plants. J Nat Med 65(3–4):662–669

Sabu MC, Kuttan R (2004) Antidiabetic activity of *Aegle marmelos* and its relationship with its antioxidant properties. Indian J Physiol Pharmacol 48(1):81–88

Sachdewa A, Raina D, Srivastava AK, Khemani LD (2001) Effect of *Aegle marmelos* and *Hibiscus rosa sinensis* leaf extract on glucose tolerance in glucose induced hyperglycemic rats (Charles foster). J Environ Biol 22(1):53–57

Saenphet K, Aritajat S, Saenphet S, Manosroi J, Manosroi A (2006) Safety evaluation of aqueous extracts from *Aegle marmelos* and *Stevia rebaudiana* on reproduction of female rats. Southeast Asian J Trop Med Public Health 37(Suppl 3):203–205

Saha SK, Chatterjee A (1957) Isolation of allo-imperatorin and beta-sitosterol from the fruits of *Aegle marmelos* Correa. J Indian Chem Soc 34:228–230

Sahare KN, Anandharaman V, Meshram VG, Meshram SU, Gajalakshmi D, Goswami K, Reddy MV (2008a) In vitro effect of four herbal plants on the motility of *Brugia malayi* microfilariae. Indian J Med Res 127(5):467–471

Sahare KN, Anandhraman V, Meshram VG, Meshram SU, Reddy MV, Tumane PM, Goswami K (2008b) Anti-microfilarial activity of methanolic extract of *Vitex negundo* and *Aegle marmelos* and their phytochemical analysis. Indian J Exp Biol 46(2):128–131

Samarasekera JK, Khambay BP, Hemalal KP (2004) A new insecticidal protolimonoid from *Aegle marmelos*. Nat Prod Res 18(2):117–122

Shankarananth V, Balakrishnan N, Suresh D, Sureshpandian G, Edwin E, Sheeja E (2007) Analgesic activity of methanol extract of *Aegle marmelos* leaves. Fitoterapia 78(3):258–259

Sharma BR, Rattan RK, Sharma P (1981) Marmelene, an alkaloid, and other components of unripe fruits of *Aegle marmelos*. Phytochemistry 20:2606–2607

Sharma BR, Sharma P (1981) Constituents of *Aegle marmelos*. II. Alkaloids and coumarin from fruits. Planta Med 43(1):102–103

Sharma AK, Bharti S, Goyal S, Arora S, Nepal S, Kishore K, Joshi S, Kumari S, Arya DS (2011) Upregulation of PPARγ by *Aegle marmelos* ameliorates insulin resistance and β-cell dysfunction in high fat diet fed-streptozotocin induced type 2 diabetic rats. Phytother Res 25(10):1457–1465

Shoba FG, Thomas M (2001) Study of antidiarrhoeal activity of four medicinal plants in castor-oil induced diarrhoea. J Ethnopharmacol 76(1):73–76

Shoeb A, Kapil RS, Popli SP (1973) Coumarins and alkaloids of *Aegle marmelos*. Phytochemistry 12(8):2071–2072

Singanan V, Singanan M, Begum H (2007) The hepatoprotective effect of bael leaves (*Aegle marmelos*) in alcohol induced liver injury in albino rats. Int J Sci Technol 2(2):83–92

Singh P, Kumar A, Dubey NK, Gupta R (2009) Essential oil of *Aegle marmelos* as a safe plant-based antimicrobial against postharvest microbial infestations and aflatoxin contamination of food commodities. J Food Sci 74(6):M302–M307

Singh RP, Banerjee S, Rao AR (2000) Effect of *Aegle marmelos* on biotransformation enzyme systems and protection against free-radical-mediated damage in mice. J Pharm Pharmacol 52(8):991–1000

Sondhi N, Bhardwaj R, Kaur S, Kumar N, Singh B (2008) Isolation of 24-epibrassinolide from leaves of *Aegle marmelos* and evaluation of its antigenotoxicity employing Allium cepa chromosomal aberration assay. Plant Growth Regul 54(3):217–224

Srivastava SD, Srivastava S, Srivastava SK (1996) New anthraquinones from the heartwood of *Aegle marmelos*. Fitoterapia 67(1):83–84

Subhadrabanhdu S (2001) Under utilized tropical fruits of Thailand, FAO RAP Publication 2001/26. FAO, Bangkok, 70 pp

Subramaniam D, Giridharan P, Murmu N, Shankaranarayanan NP, May R, Houchen CW, Ramanujam RP, Balakrishnan A, Vishwakarma RA, Anant S (2008) Activation of apoptosis by 1-hydroxy-5,7-dimethoxy-2-naphthalene-carboxaldehyde, a novel compound from *Aegle marmelos*. Cancer Res 68(20):8573–8581

Sumanth M, Mustafa SS (2005) Antistress, adoptogenic activity of *Aegle marmelos* fruit in mice. Pharm Rev 3(6):Article 5

Sundaram EN, Reddy PU, Singh KP (2009) Effect of alcoholic extracts of Indian medicinal plants on the altered enzymatic activities of diabetic rats. Indian J Pharm Sci 71(5):594–598

Sur TK, Pandit S, Pramanik T (1999) Antispermatogenic activity of leaves of *Aegle marmelos*. Corr. in albino rats: a preliminary report. Biomedicine 19(3):199–202

Sur TK, Pandit S, Paramanik T, Bhattacharyya D (2002) Effect of *Aegle marmelos* leaf on rat sperm motility: an in vitro study. Indian J Pharmacol 34:276–277

Upadhya S, Shanbhag KK, Suneetha G, Naidu BM, Upadhya S (2004) A study of hypoglycemic and antioxidant activity of *Aegle marmelos* in alloxan induced diabetic rats. Indian J Physiol Pharmacol 48:476–480

Veerappan A, Miyazaki S, Kadarkaraisamy M, Ranganathan D (2007) Acute and subacute toxicity studies of *Aegle marmelos* Corr., an Indian medicinal plant. Phytomedicine 14(2–3):209–215

Venkatesan D, Karunakaran M, Kumar SS, Palaniswamy PT, Ramesh G (2009) Antimicrobial activity of *Aegle marmelos* against pathogenic organism compared with control drug. Ethnobot Leaflet 8: Article 1

Venkatesh P, Shantala B, Jagetia GC, Rao KK, Baliga MS (2007) Modulation of doxorubicin-induced genotoxicity by *Aegle marmelos* in mouse bone marrow: a micronucleus study. Integr Cancer Ther 6(1):42–53

Vijaya C, Ramanathan M, Suresh B (2009) Lipid lowering activity of ethanolic extract of leaves of *Aegle marmelos* (Linn.) in hyperlipidaemic models of Wistar albino rats. Indian J Exp Biol 47(3):182–185

Vimal V, Devaki T (2004) Linear furanocoumarin protects rat myocardium against lipidperoxidation and membrane damage during experimental myocardial injury. Biomed Pharmacother 58(6–7):393–400

Vinodhini R (2010) Detoxifying effect of *Nelumbo nucifera* and *Aegle marmelos* on hematological parameters of common ca rp (*Cyprinus carpio* L.). Interdiscip Toxicol 3(4):127–131

Watt G (1972) A dictionary of the economic products of India, vol 1–6. Cosmo Publications, New Delhi

Yang XW, Hattori M, Namba T (1996a) Studies on the anti-lipid peroxidative actions of the methanolic extract of the root of *Aegle marmelos* and its constituents in vivo and in vitro. J Chin Pharm Sci 5(3):132–140

Yang XW, Hattori M, Namba T (1996b) Two new coumarins from the roots of *Aegle marmelos*. J Chin Pharm 5(2):68–73

Citrus 'Meyer'

Scientific Name

Citrus **'Meyer' (sensu Mabberley)**

Synonyms

Citrus meyeri Yu (sensu Hodgson; sensu Tanaka sec. Cottin). *Citrus* × *meyerii* Yu. Tanaka.

Family

Rutaceae

Common/English Names

Chinese Dwarf Lemon, Dwarf Lemon, Grant Lemon, Meyer Lemon, Meyer Lemon Tree, Meyer's Lemon Tree, Valley Lemon.

Vernacular Names

Chinese: Bei Jing Ning Meng, Xiang Ning Meng;
French: Citronneir De Meyer;
German: Meyer's Zitrone;
Japanese: Guranto Remon, Maiyaa Remon, Soorongu.

Origin/Distribution

Meyer lemon is named after USDA plant explorer Frank Meyer who discovered the plant growing as a dooryard tree in Beijing, China in 1908. Its history prior to that is unknown. Its parentage is unknown, but it is considered either an orange-lemon or a mandarin-lemon hybrid or lemon-orange-mandarin hybrid.

Agroecology

Meyer lemon is robust, vigorous and cold and heat hardy. It grows well in warm climates. It grows well in standard citrus producing climates, but also grows in cooler areas, and areas that receive brief freezes. It is fairly cold hardy, surviving temperatures down to −3°C. Its optimum temperature for growth is 25–30°C and growth is halted at >40°C. It needs protection against severe frosts, strong and cold winds.

Edible Plant Parts and Uses

The edible uses in food, juices, lemonade and in confectionary are as described for lemons.

Some culinary recipes with Meyer lemon include: Cornish game hens roasted with slices of Meyer lemon, olives and fennel; *Piri-piri* shrimp with black rice and Meyer lemons; risotto with mascarpone, Parmesan, and grated Meyer lemon peel; thin Meyer lemon slices on a pizza crust topped with goat cheese, rosemary and Picholine olives; grated Meyer lemon peel on Chantilly cream; and smoked salmon sandwich with Meyer lemon, sour cream; Meyer lemon-almond cake, Meyer lemon-cardamom custard, Meyer lemon-cardamom ice-cream, Meyer lemon gimlet comprising Meyer lemon juice, zest, soda water and Meyer lemon syrup; candied Meyer lemon peels.

Botany

Citrus × meyeri trees are unarmed (thornless), short (2–3 m) high at maturity, though they can be pruned smaller, with a compact to weeping habit. Their leaves are large, elliptic-ovate, glossy green, with acute apex and obtuse base and crenulate margin and borne on short, glabrous, wingless petioles (Plate 3). The flowers are hermaphrodite, white, fragrant and borne in terminal fascicles. Flower buds are ovoid, purplish or white with pale purplish tint (Plates 1 and 2). Calyx is cup shaped with 4–5 very short teeth; petals 4–5 white on the upper side and purple or pale purplish-white on the underside; stamens 20–25, connate in several fascicles, filaments white with yellow anthers; ovary superior, subglobose to ovoid with a style and capitate yellow stigma. Fruit large, subglobose to globose-ovoid, 6.5–7 cm with a small nipple, green (Plate 4) turning to an attractive orangey-yellow or deep yellow on maturity, with a smooth, thin rind (Plates 5 and 6). Segments seven with yellow pulp (Plate 7), juicy subacid (low acid and high sugar levels). Seeds few to ten.

Nutritive/Medicinal Properties

The nutritive composition of Meyer lemons has been analysed by USDA (2011) and is elaborated under *Citrus × limon.*

Plate 1 Blossoms in terminal clusters

Plate 2 Open flower and buds

Plate 3 Deep green leaves with crenulate margins

Plate 6 Close-up of harvested fruit

Plate 4 Juvenile fruits

Plate 7 Sliced Meyer lemon with yellowish pulp

Plate 5 Ripe Meyer lemon fruit

Studies in Pakistan reported that *C. limon* juice from Lisbon variety was found to be more acidic (pH 2.50) followed by Eureka (pH 2.56), Meyer (pH 2.62) and Bush (pH 2.69) (Khosa et al. 2011).

Density and viscosity of all the four citrus juices were in the range of 1.026–1.029 g/cm3 and 1.044–1.127 cP, respectively. Among all the citrus varieties, Bush variety had the highest values of total soluble solids (112.38 g/l) while, Meyer contained the lowest (103.15 g/l). The total phenolic contents (TPC) varied widely among varieties ranging from 690.62 to 998.29 mg/l as gallic acid equivalent (GAE). The maximum TPC was found in the Bush variety while the minimum in the Eureka variety. Lisbon and Meyer varieties had 730.46 and 825.37 mg/TPC, respectively. Total flavonoid content (TFC) also varied among varieties in the range of 211.36–220.34 mg/l as catechin equivalents. The maximum TFC was found in the Bush variety while the minimum in the Lisbon variety. Eureka and Meyer varieties of citrus had 219.27 and

216.61 mg/l TFC respectively. Total carotenoids in the juice were very low, 0.05–0.08 mg/l measured as β-carotene equivalent. The vitamin C content was found in the following order: Lisbon (21.13 mg/l) > Eureka > Meyer > Bush (18.87 mg/l).

Other Uses

Meyer lemons are popular as ornamental house plants due to their short stature, compact size, hardiness and productivity. They are highly decorative and suitable for container growing.

Comments

According to Mabberley (1997, 2001) where there is uncertainty around the hybrid parentage of *Citrus* as in the case of Meyer lemon, it is most sensible to refer to it as *Citrus* 'Meyer'.

Selected References

Cottin R (2002) Citrus of the world: a citrus directory. Version 2.0. SRA INRA-CIRAD, France

Hodgson RW (1967) Horticultural varieties of citrus. In: Reuther W, Webber HJ, Batchelor LD (eds) The citrus industry, vol 1: History, world distribution, botany, and varieties. Revised edition, University of California Press, Berkeley, pp 431–591

Khosa MK, Chatha SHS, Hussain AI, Zia KM, Riaz H, Aslam K (2011) Spectrophotometric quantification of antioxidant phytochemicals in juices from four different varieties of *Citrus limon*, indigenous to Pakistan. J Chem Soc Pak 33(2):188–190

Koskinen J (2011) Citrus pages. http://users.kymp.net/citruspages/home.html

Mabberley DJ (1997) A classification for edible *Citrus*. Telopea 7(2):167–172

Mabberley DJ (2001) Citrus reunited. Aust Plant 21(166): 52–55

U.S. Department of Agriculture, Agricultural Research Service (USDA) (2011) USDA National Nutrient Database for Standard Reference, Release 24. Nutrient Data Laboratory Home Page. http://www.ars.usda.gov/ba/bhnrc/ndl

Citrus amblycarpa

Scientific Name

Citrus amblycarpa (Hassk.) Ochse

Synonyms

Citrus limonellus Hassk. var. *amblycarpa* Hassk., *Citrus nobilis* Lour. var. *amblycarpa* (Hassk.) Ochse & De Vries.

Family

Rutaceae

Common/English Names

Jeruk Limo, Leprous Lime, Nasnaran, Nasnaran Mandarin.

Vernacular Names

Chinese: Qiu Luo Ke Li Meng;
Indonesia: Jeruk Limau, Jeruk Limo, Jeruk Sambal (Java);
Japanese: Sherouku Rimoo;
Vietnamese: Quít Ta.

Origin/Distribution

It is indigenous to West Java, Indonesia and cultivated in Java.

Agroecology

The species is found in the hot and humid tropical lowland from near sea level to 350 m elevation in its native range.

Edible Plant Parts and Uses

The fragrant, sour juice of half ripe and young fruit is used as condiment to enhance the flavour of several dishes like *sambal olek* (chilli based condiment with sesame oil, salt), *soto* (spicy soup) and *bahmie* (noodles). Likewise the leaves are used in a variety of meat dishes and curries.

Botany

An erect, branched, arborescent shrub or small tree 2.5–7 m high, with almost thornless, glabrous, terete branches (Plate 1). Leaves are alternate, unifoliolate, shining dark green, ovate oblong or lanceolate, 4–8 cm × 2–4 cm, with

Plate 1 Fruiting Jeruk Limo unarmed shrub

pinnate venation, cuneate base, obtuse to subemarginate apex, shallowly crenate margin, densely pellucid dotted, on short, petioles, fragrant when bruised. Flowers are white, fragrant, 2 cm diameter, solitary and axillary or in short 3–5 flowered, terminal racemes. Calyx is glabrous with 4–5 ovate segments; petals 4–5, oval-oblong, white; stamens 16–20 with filaments forming a tube at the base; ovary obovoid. Fruit is a depressed globose berry, 1.5–3.5 cm in diameter, apex impressed and sunken in the middle, shiny dark green when immature, and turning yellow-green to pale yellow when ripe. Fruit skin is rough and bullate (puckered and blistered) (Plate 1); pulp yellow-green, sour and savoury. Seeds are oblong to pear shaped 0.8–1 cm by 0.4–0.5 cm.

Nutritive/Medicinal Properties

No published information is available on its nutritive and medicinal values.

Other Uses

The species is occasionally used as rootstock for other *Citrus* species. The fruit juice is used for cleaning hands and for shampooing hair.

Comments

This species has nucellar polyembroynic seeds that will grow true to type.

Selected References

Backer CA, Bakhuizen van den Brink RC Jr (1963) Flora of Java, vol 1. Noordhoff, Groningen, 648 pp

Facciola S (1990) Cornucopia. A source book of edible plants. Kampong Publications, Vista, 677 pp

Jansen PCM (1999) *Citrus amblycarpa* (Hassk.) Ochse. In: de Guzman CC, Siemonsma JS (eds) Plant resources of South-East Asia no. 13: spices. Backhuys Publisher, Leiden, p 250

Ochse JJ, Bakhuizen van den Brink RC (1980) Vegetables of the Dutch Indies, 3rd edn. Ascher & Co., Amsterdam, 1016 pp

Porcher MH et al (1995–2020) Searchable World Wide Web multilingual multiscript plant name database. The University of Melbourne, Australia. http://www.plantnames.unimelb.edu.au/Sorting/Frontpage.html

Citrus australasica

Scientific Name

Citrus australasica F. Muell.

Synonyms

Microcitrus australasica var. *australasica* (F. Muell.) Swingle.

Family

Rutaceae

Common/English Names

Australian Finger Lime, Finger Lime, Native Finger Lime, Queensland Finger, Sauvage Lime.

Vernacular Names

French: Lime Digitée d'Australie;
German: Australische Limette.

Origin/Distribution

Citrus australasica is native to Australia, occurring in northern New South Wales and southern Queensland.

Agroecology

In its native range, it grows as an understorey shrub or small tree in dry and subtropical lowland and is rainforest, especially common in regrowth. In cultivation, *C. australasica* is hardy in tropical to temperate climates in well drained conditions.

Edible Plant Parts and Uses

The finger lime is enjoying popularity as a gourmet bush food and is being sold and served in restaurants and exported as fresh fruit. Its juicy vesicles resemble pearls of caviar which are often used as garnish or added into various food recipes. Boutique marmalades and pickles are also made from finger lime. Because of its acidic nature, finger lime is best used for cooking, jams, garnished sauces and drinks. Finger lime peel can be dried and used as a flavouring spice.

Botany

A medium to large armed, shrub or small tree, 2–6 m high with solitary axillary straight spines and compact crown. Leaves are small, glabrous, obovate to elliptic or more or less rhombic, 1–5 cm long by 3–25 mm wide, with notched apex, cuneate base, margins often crenate towards apex, oil glands numerous, aromatic when crushed and borne on 1–3 mm

Plate 1 Leaves and flower buds of finger lime

Plate 2 Mature finger lime (A. Beattie)

wingless petioles (Plate 1). Flowers buds are small and pink in colour on short peduncles 1–3 mm long (Plate 1). The flowers are bisexual and have 6–9 mm long white oblong petals, short 1.5 mm long, free concave sepals and numerous stamens (20–25) with white filaments and yellow anthers, stout ovary with 5–7 locules with 8–16 ovules in each locule. Fruit is cylindrical- fusiform, finger-shaped, 4–8 cm long, sometimes slightly curved, coming in different colours, green to pinkish red to reddish-black (Plate 2); rind slightly rough, pulp green, yellow to pinkish red; seeds 5–6 mm long.

Nutritive/Medicinal Properties

Nutrient composition of the raw fruit (*Microcitrus australasica* var. *australasica*) was reported per 100 g edible portion as: energy including dietary fibre 336 kJ, moisture 65.5 g, protein 2.5 g, Nitrogen 0.40-g, fat 4.9 g, ash 0.7 g, dietary fibre 14 g, carbohydrate 12.4 g, Ca 50 mg, Cu 0.4 mg, Fe 0.8 mg, Mg 31 mg, K 290 mg, Na 9 mg, Zn 0.3 mg, and niacin equivalents 0.42 mg (Brand Miller et al. 1993).

The percentage and concentration of total phenolics in *Microcitrus australasica* var. *sanguinea* leaf, peel and juice were reported by Berhow et al. (1998) as: flavedo (outer pigmented layer of the peel): 40.9% (2.38 mg/g) flavone/flavonol, 0% (0 mg/g) flavanone, 2.5%

(0.4 mg/g) psoralen, 22.1%(0.4 mg/g) coumarin; juice: 11.2% (0.02 mg/g) flavone/flavonol, 0% (0 mg/g) flavanone, 0% psoralen, 74.1% (0.04 mg/g) coumarin.

Fifty-eight out of a total of 65 components were identified from peel oil of *Microcitrus australasica* var. *sanguinea*, bicyclogermacrene (25.9%), α-pinene (10.2%) and spathulenol (9.8%) were the main compounds (Ruberto et al. 2000). Peel oil of *Microcitrus australasica* fruit was found to contain: limonene 51.1%, sabinene 19.6%, β-pinene 7.9%, γ-terpinene 4.9%, geranial 2.0%, myrcene 1.5%, neral 1.2%, β-bisabolene 1.6%, α-pinene 1%, neryl acetate 0.1%, β-phellandrene 0.6%, (Z)-β-ocimene 0.7%, α-terpineol 0.4%, germacrene D 0.2%, *trans*-α-bergamotene 0.4%, (*E,E*)-α-farnesene 0.4%, (*E*)-β-ocimene 0.3%, bicyclogermarcrene 0.3%, 3-carene 0.3%, α-thujuene 0.2%, β-emelene 0.2%, (*E*)-caryophyllene 0.2%, terpinene-4-ol 0.5%, terpinolene 0.2%, citronellal 0.2%, linalool 0.2%, geranyl acetate 0.2%, α-humulene 0.2%, *trans*-sabinene hydrate, 0.1%, globulol 0.1%, undecanal traces, citronellol traces, bornyl acetate traces, *cis*-limonene-1,2-oxide traces, *trans*-limonene-1,2-oxide traces, nonal traces, *allo*-ocimene traces, α-terpinene traces, α-phellandrene traces, camphene traces (Lota et al. 2002).

Limonene and isomenthone (7.5%) were found as the major volatile constituents of *Citrus australasica* peel extract (Delort and Jaquier 2009). Six new terpenyl esters were also identified: citronellyl 2-methylbutanoate; 1,8(10)-*p*-menthadien-9-yl propanoate; 1,8(10)-*p*-menthadien-9-yl 2-methylbutanoate; 1,8(10)-*p*-menthadien-9-yl 3-methylbutanoate; 1-*p*-menthen-9-yl 2-methylbutanoate; and 1-*p*-menthen-9-yl 3-methylbutanoate. Other components included 6-methyloctanal, 4-methylnonanal and 8-methyldecananal.

Australian finger lime *Citrus australasica* was found to be a rich source of antioxidant compounds (Netzel et al. 2007). The radical scavenging activity and total phenolic contents were significantly higher than that of blueberry cv. Biloxi.

Brophy et al. (2001) indentified bicyclogermacrene (19–28%), germacrene-D (2–8%), δ-elemene (0.5–11%) and limonene (12–24%) as the main components of the leaf essential oil.

Other Uses

Finger lime is receiving considerable attention as a commercial food plant, and research is being carried out to develop selected, superior forms of the species and also to develop hybrids with exotic *Citrus* species.

Comments

Finger lime is commercially propagated by grafting or budding onto other *Citrus* rootstocks. It can also be grown from cutting but the strike rate is slow and also seeds but germination is erratic and seedlings may take from 5 to 15 years to reach maturity.

Selected References

Altech Group and Total Earth Care (1999) Improving access to bushfood production and marketing information. RIRDC publication no. 99/158. RIRDC, Canberra, 104 pp

Bayer RJ, Mabberley DJ, Morton C, Miller CH, Sharma IK, Pfeil BE, Rich S, Hitchcock R, Sykes S (2009) A molecular phylogeny of the orange subfamily (Rutaceae: Aurantioideae) using nine cpDNA sequences. Am J Bot 96:668–685

Berhow M, Tisserat B, Kanes K, Vandercook C (1998) Survey of phenolic compounds produced in *Citrus*. USDA ARS Tech Bull 1856:1–154

Brand Miller J, James KW, Maggiore P (1993) Tables of composition of Australian aboriginal foods. Aboriginal Studies Press, Canberra

Brophy JJ, Goldsack RJ, Forster PI (2001) The leaf oils of the Australian species of *Citrus* (Rutaceae). J Essent Oil Res 13:264–268

Cooper W, Cooper WT (2004) Fruits of the Australian tropical rainforest. Nokomis Editions, Melbourne, 616 pp

Cribb AB, Cribb JW (1976) Wild food in Australia. Fontana/Collins, Sydney, 240 pp

Delort E, Jaquier A (2009) Novel terpenyl esters from Australian finger lime (*Citrus australasica*) peel extract. Flavour Fragr J 24(3):123–132

Lota ML, de Rocca SD, Tomi F, Jacquemond C, Casanova J (2002) Volatile components of peel and leaf oils of lemon and lime species. J Agric Food Chem 50(4):796–805

Low T (1991) Wild food plants of Australia. Angus & Robertson, Ryde, 240 pp

Mabberley DJ (2001) Citrus reunited. Aust Plant 21(166):52–55

Mabberley DJ (2002) Limau hantu and limau purut: the story of lime leaves (*Citrus hystrix* DC., Rutaceae)? Gard Bull Singapore 54:185–197

Mabberley DJ (2004) *Citrus* (Rutaceae): a review of recent advances in etymology, systematics and medical applications. Blumea 49:481–488

Macintosh H (2004) Native *Citrus*. In: Salvin S, Bourke MaM, Byrne T (eds) The new crop industries handbook. RIRDC publication no. 04/125. RIRDC, Canberra, pp 358–367

Netzel M, Netzel G, Tian Q, Schwartz S, Konczak I (2007) Native Australian fruits – a novel source of antioxidants for food. Innov Food Sci Emerg Technol 8(3):339–346

Ruberto G, Rocco C, Rapisarda P (2000) Chemical composition of the peel essential oil of *Microcitrus australasica* var. *sanguinea* (F.M. Bail) Swing. J Essent Oil Res 12(3):379–382

Citrus australis

Scientific Name

Citrus australis (Mudie) Planch.

Synonyms

Citrus australis Planch., *Microcitrus australis* (A. Cunn. ex Mudie) Swingle, *Microcitrus australis* (Planch.) Swingle.

Family

Rutaceae

Common/English Names

Australian Round Lime, Australia Sweet, Dooja, Gympie Lime, Native Lime, Native Orange, Round Lime.

Vernacular Name

Australia: Dooja (Aboriginal)

Origin/Distribution

The species is endemic to southeast Queensland.

Agroecology

The species occurs naturally on the fringe of lowland sub-tropical rainforests of southeast Queensland, from Brisbane northwards. It is a hardy plant but slow-growing, needing some protection when young. It thrives in well-drained, organic rich soil in a sheltered position. It values mulching and extra watering during dry periods. It is moderately frost tolerant.

Edible Plant Parts and Uses

The ripe acidic fruit can be eaten raw but is more suitable for making drinks cordials, sauces, jams, marmalades and also as lime flavouring. The thick rind has potential for culinary use, such as grating into spice pastes, or for candied peel and may also have potential for essential oil extraction.

Botany

A compact, dense armed tree growing to 12–20 m high with slender, multiple trunks and trunk diameter of 6–8 mm and with angled, glabrous twigs with slender, 5–10 mm long, axillary thorns. Juvenile leaves linear. Older leaves simple, entire, glabrous, elliptic to obovate or almost rhomboid, 3–4 × 2–3 cm, emarginate or bluntly pointed at tip, cuneate at the base, gland-dotted; petioles short 6 mm, articulated with the leaf

Plate 1 Leaves and fruit of Australian round lime (M. Smith)

Plate 2 Rugose fruit with thick rind and yellowish-white pulp (M. Smith)

blades (Plates 1 and 2). Flowers white or pinkish-white, fragrant about 1–1.5 cm across; solitary in leaf axils, 4- or 5-merous, with 16–20 stamens with free filaments; fruit globose to subglobose, 3.5–5 cm diam., with dark green to green to yellowish-green, rugose 7 mm thick rind and containing six segments with pale yellowish-white vesicles (Plates 1 and 2). Seeds flattened, monoembryonic.

Nutritive/Medicinal Properties

The proximate nutrient composition of the fruit was reported to be: energy 91 g, moisture 74.8 g, protein 2.2 g, nitrogen 0.35 g, , ash 0.8 g, dietary fibre 6.7 g, carbohydrates 15.5 g, Ca 46 mg, Cu 0.2 mg, Fe 0.5 mg, Mg 24 mg, K 270 mg, Na 4 mg, Zn 0.1 mg, niacin 0.37 mg (Brand Miller et al. 1993).

C. australis leaf oil was dominated by α-pinene (68–79%).

Other Uses

The species can be used for *Citrus* hybridisation work.

Comments

Round lime is propagated from fresh seed or cuttings, which are slow to develop roots. It can also be budded onto exotic *Citrus* rootstock.

References

Bayer RJ, Mabberley DJ, Morton C, Miller CH, Sharma IK, Pfeil BE, Rich S, Hitchcock R, Sykes S (2009) A molecular phylogeny of the orange subfamily (Rutaceae: Aurantioideae) using nine cpDNA sequences. Am J Bot 96:668–685

Brand Miller J, James KW, Maggiore P (1993) Tables of composition of Australian aboriginal foods. Aboriginal Studies Press, Canberra

Brophy JJ, Goldsack RJ, Forster PI (2001) The leaf oils of the Australian species of *Citrus* (Rutaceae). J Essent Oil Res 13:264–268

Cottin R (2002) Citrus of the world: a citrus directory. Version 2.0. SRA INRA-CIRAD, France

Mabberley DJ (2004) *Citrus* (Rutaceae): a review of recent advances in etymology, systematics and medical applications. Blumea 49:481–498

Swingle WT, Reece PC (1967) The botany of *Citrus* and its wild relatives. In: Reuther W, Webber HJ, Batchelor LD (eds) The *Citrus* industry vol 1: history, world distribution, botany, and varieties, Revised edn. University of California Press, Berkeley, pp 190–430

Citrus garrawayi

Scientific Name

Citrus garrawayi F.M. Bailey

Synonyms

Microcitrus garrawayae orth. var. (F.M. Bailey) Swingle, *Citrus garrawayae* orth. var. F.M. Bailey.

Family

Rutaceae

Common/English Names

Garraway's Australian Wild Lime, Garraway's Australian Wild Lime, Mount White Lime, Mt White Lime, Thick-Skinned Finger Lime.

Vernacular Names

None

Origin/Distribution

The species is indigenous to Australia – Cape York Peninsula and north east Queensland, and Papua New Guinea.

Agroecology

It is found in monsoonal forest and on rain forest margins on the foothills and upland foothills and upland rainforest of the Cook District, Mt White on Cape York Peninsula in Australia and Goodenough Island in Papua New Guinea from 50 to 450 m altitude. It grows in deciduous vine thickets as an under-storey shrub.

Edible Plant Parts and Uses

The fruit has potential for culinary use, such as grating into spice pastes, or for candied peel.

Botany

A woody armed shrub or small tree, 1–15 m high with a lenticelled stem of 30 cm dbh and leafy green twigs with axillary spines (Plates 1 and 2). Leaves unifoliate, ovate, 2.5–5 × 1–3.5 cm, with subacute to sub-emarginate apex, sub-cuneate base and crenulated margin, pellucid dotted (Plates 2 and 4). Flowers solitary in leaf axils, with 5 white glabrous 7–8 × 3 mm petals; stamens about 15 with yellow anthers, staminal filaments about 4–5 mm long. Disk small, at the base of the ovary, inside the whorl of staminal filaments, stigma knob-shaped and yellow. Fruit finger-shaped, cylindrical 7–10 × 2.5–3.5 cm with 2 mm thick, green, tuberculose rind because of large oil glands beneath the epidermis (Plates 3 and 4),

Plate 1 Mount White lime shrub

Plate 2 Leaves and fruits

Plate 4 Close-up of finger-liked fruits

Plate 3 Leaves and finger-like cylindrical fruit (M. Smith)

Plate 5 Whole and halved fruit showing the greenish-white vesicles

4–5 segments with greenish-white juice vesicles (Plate 4) and 0–3 angular, white seeds.

Nutritive/Medicinal Properties

No information has been published on its nutritive composition.

The leaf oil from *C. garrawayi* appeared to be of two forms, (1) in which the principal components were the monoterpene α-pinene (18–40%), together with β-caryophyllene (7–12%), α-humulene (2–17%), globulol (4–10%) and viridiflorol (4–10%), and (2) which lacked α-pinene and had β-caryophyllene (17–30%), globulol (7–10%) and viridiflorol (7–10%) as principal components (Brophy et al. 2001).

Other Uses

The species is used in *Citrus* breeding and selection studies.

Comments

This species is classified as rare and is protected under the Queensland Nature Conservation Act 1992.

Selected References

Alexander DMcE (1983) Some citrus species and varieties in Australia. CSIRO Publishing, Melbourne, 64 pp

Bailey FM (1904) Contributions to the flora of Queensland. Qld Agric J 15:491–495

Brophy JJ, Goldsack RJ, Forster PI (2001) The leaf oils of the Australian species of *Citrus* (Rutaceae). J Essent Oil Res 13:264–268

Cooper W, Cooper WT (2004) Fruits of the Australian tropical rainforest. Nokomis Editions, Melbourne, 616 pp

Cribb AB, Cribb JW (1980) Wild food in Australia. Fontana/Collins Publication, Sydney, 240 pp

Forster PI (1991) *Microcitrus garrawayae* (Rutaceae) and its distribution in New Guinea and Australia. Telopea 4(2):357–358

Forster PI, Smith MW (2010) *Citrus wakonai* P.I.Forst. & W.M.Sm. (Rutaceae), a new species from Goodenough Island, Papua New Guinea. Austrobaileya 8(2):134–137

Koskinen J (2011) Citrus pages. http://users.kymp.net/citruspages/home.html

Mabberley DJ (1998) Australian Citreae with notes on other Aurantioideae (Rutaceae). Telopea 7(4):338

Swingle WT (1915) *Microcitrus*, a new genus of Australian citrus fruits. J Wash Acad Sci 5:569–578

Swingle WT, Reece PC (1967) The botany of *Citrus* and its wild relatives. In: Reuther W, Webber HJ, Batchelor LD (eds) The *Citrus* industry vol 1: history, world distribution, botany, and varieties. Revised edn. University of California Press, Berkeley, pp 190–430

Citrus hystrix

Scientific Name

Citrus hystrix DC.

Synonyms

Citrus auraria Michel, *Citrus echinata* Saint-Lager, *Citrus hyalopulpa* Tanaka, *Citrus kerrii* (Swingle) Tanaka, *Citrus macroptera* Montrouzier var. *kerrii* Swingle, *Citrus papeda* Miquel, *Fortunella sagittifolia* F.M. Feng & P. I Mao, *Papeda rumphii* Hasskarl.

Family

Rutaceae

Common Names

Caffir Lime, Ichang Lime, Kaffir Lime, Leech Lime, Mauritius Papeda, Porcupine Orange, Porcupine Orange Lime, Rough Lemon, Thai Bai Makrut, Wart Lime, Wild Lime.

Vernacular Names

Burmese: Shauk Cho, Shouk-Pote, Shauk Nu, Shauk Waing;
Chinese: Ma Feng Gan, Mao Li Qiu Si Ku Cheng, Ma Feng Cheng, Ma Feng Mao Gan Fatt-Fung-Kam (Cantonese), Thai-Ko-Kam (Hokkien/Minnan);
Czech: Kaffir Citrus;
Danish: Kaffir Lime;
Dutch: Indonesische Citroenboom, Kaffir Limoen, Djeroek Poeroet;
Eastonian: Kaffir Laimilehed;
Finnish: Kaffir Limetti;
French: Combava, Limettier Hérissé, Citron Combara;
German: Indische Zitrone, Indische Zitronenblätter, Indonesische Zitronenblätter, Kaffir Limette, Makrut Limette, Langdorniger Orangenbaum;
Guam: Limon Admelo;
Hebrew: Aley Kafir Laim;
Hungarian: Kaffercitrom, Kaffir Citrom És Levél;
India: Kolumichai (Tamil);
Indonesia: Juuk Purut (Bali), Jeruk Purut (Java), Jeruk Obat, Jerut Sambal, Limo Purut;
Japanese: Bai Makkuruu, Kobu Mikan, Moorishasu Papeda, Kafaa Raimu, Purutto, Kafiiru Raimu, Kobu Mikan;
Khmer: Kraunch Soeuth, Slirk-Krote Sirk;
Kiribati: Te Remen;
Laos: Khi-Hout, Makgeehoot;
Malaysia: Limau Purut, Limau Hantu, Limau Suwangi;
Palauan: Debechel;
Philippines: Kabog, Kamuntai (Bikol), Kolison, Kolobot (Bisaya), Amongpong, Amontau, Kolo-Oi, Kopalian, Mayagarin (Cebu Bisaya), Kapitan (Ibanag), Kamugau, Kamukau, Kamulan, Kapitan (Iloko), Pinukpuk (Kalinga), Piris

(Pangasingan) Malatbas (Sambali), Muntai (Subanum), Daruga, Duroga (Sulu), Buyak, Buyog, Kabuan, Kabugau, Kobot, Kolobot, Kolong-Kolong (Tagalog);
Reunion Island: Combava (French);
Russian: Kaffir Laim, Kafrskii Laim;
Samoan: Tipolo Patupatu;
Sri Lanka: Kahpiri Dehi, Kudala Dehi, Odu Dehi (Sinhalese);
Spanish: Naranja Puerco-Espín, Hojas De Lima Cafre, Hoja De Lima Kaffir, Lima Kaffir;
Swedish: Kafirlime;
Thailand: Bai Makrut, Luuk Makrut, Luuk Makruut, Ma-Kruut, Magood, Magrood, Makrut, Makut, Som Makrud;
Tongan: Lemani, Moli Lemani;
Tuvaluan: Laim;
Vietnam: Chanh Kaffir, Chanh Sác, Trúc;
Yapese: Gurgur Gurgumimarech.

Origin/Distribution

The species is native to tropical southeast Asia, southern China and Malaysia. It has been introduced and cultivated elsewhere in the tropics and sub-tropics including in northern Australia.

Agroecology

The species thrives in a warm and wet climate. Trees are mildly frost hardy and grow best in areas that receive only short, mild frosts. It thrives in organic rich, fertile and well-drained acid loam soil and requires adequate watering during the dry seasons for good growth and fruiting.

Edible Plant Parts and Uses

The fruit and leaves are popular spice ingredient in Asian cooking. The leaves impart an aromatic, strong, unique and spicy flavour to many food dishes. The leaves can be used fresh or dried, and can be stored frozen. The leaves are widely used in Thai cuisine like *Tom yum gai* (Tom-yum chicken) and *tom yum kung (*Tom–yum prawns), Lao cuisine and Cambodian cuisine for the savoury paste known as *"Krueng"*. In Javanese and Balinese cuisine the leaves are used to prepare *sayur assam* (salty or sour vegetables) and other spicy curry dishes of fish and chicken and also in *sate*. They are also used in Burmese and Malaysian cuisine like beef and chicken *Rendang* (Minangkabau dish comprising coconut milk and a range of spices that include ginger, galangal, turmeric leaf, lemon grass and chillies). The fruit is also used in making *sambal, acar* (vegetable salad pickled in vinegar dried chilies and ground peanuts) and *dalca* (mutton curry). The juice of the fruit is often used for souring dishes. It is also added to fish or meat in order to make them more tender and fragrant or to flavour grilled fish and grilled beef. The jus is also made into refreshing drinks.

The fruit juice is added to *sambal petis* (chilli sauce with shrimp) and also to *rojak* or *petjel* (savoury spicy sauce) in Indonesia. In Thailand, the fruit is used for seasoning and to prepare drinks teas such as *Citrus hystrix* flavonoid-rich sachet, which has been promoted to have great potential as a natural antioxidant health product. The peels are dried, preserved and candied. The zest of the fruit is commonly used in Creole cuisine and also used to impart flavour to rums in the French Reunion Island and Madagascar.

Botany

A small tree 3–6 m high, often not straight, crooked with glabrous and spiny branches. Leaves are alternate, unifoliolate, broadly ovate to ovate-oblong, 7.5–10 cm long, dark green on top, lighter on the bottom, very fragrant with long petiole expanded into prominent wings, 15 cm long by 5 cm wide (Plates 1, 2 and 3). Leaf and expanded petiole appear to be a single "pinched" leaf (Plates 1 and 2). Leaf base is cuneate, or rounded, apex obtuse or slightly acuminate or notched. Inflorescence is both axillary and terminal and is 3–5 flowered. Flowers are small, fragrant, white; calyx cuspidate 4-lobed, white with violet fringe; petals 4–5, ovate-oblong, yellowish white tinged with pink; stamens 24–30

Plate 1 Kaffir lime flowers

Plate 3 Puckered, warty fruit and winged petioles

Plate 2 Pendulous fruiting limb of Kaffir lime

Plate 4 Harvested kaffir lime and leaves on sale in local market

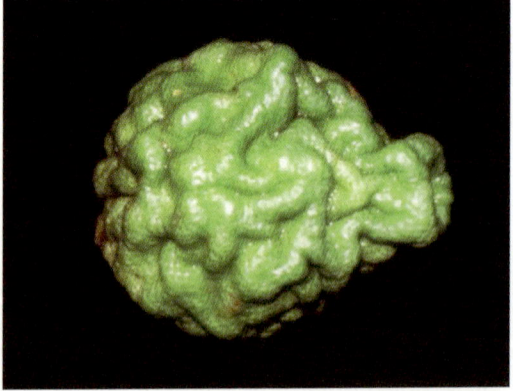

Plate 5 Close up of verrucose, warty fruit rind of Kaffir lime

free (Plate 1). Fruit is large, verrucose, warty or bumpy, globose, ovoid to elliptic, green (Plates 2, 3, 4 and 5) turning yellowish-green when ripe, 5–7 cm diameter, rind thick, pulp yellowish, very acid and bitter. Seeds are numerous, ridged, ovoid-oblong. 1.5–1.8 by 1–1.2 cm, monoembryonic with white cotyledons.

Nutritive/Medicinal Properties

The nutrient composition of the fruit contains per 100 g edible portion was reported as: water 88.6 g, protein 0.8 g, fat 0.6 g, carbohydrate 8.5 g, fibre 0.8 g, Ca 57 mg, P 2 mg, Fe 0.1 mg, K 172 mg, carotene 16 μg, vitamin A 3 μg, vitamin B 1 0.02 mg, vitamin B2 0.07 mg, and vitamin C 37 mg. (Tee et al. 1997).

The fruit is rich in calcium, vitamin C and potassium and also contain vitamin B1 and 2.

The percentage and concentration of total phenolics in *Citrus hystrix* cv Davao lemon leaf, peel and juice were reported by Berhow et al. (1998) as: leaf:- 25.8% (1.22 mg/g) flavone/flavonol, 27% (0.81 mg/g) flavanone, 8.7% psoralen, 20.2% coumarin (0.29 mg/g); juice: 6.3% (0.01 mg/g) flavone/flavonol, 63.5% (0.10 mg/g) flavanone, 25.2% (0.02 mg/g) coumarin.

The percentage and concentration of flavanones reported in *Citrus hystrix* cv Davao lemon by Berhow et al. (1998) were as follows: leaf:- didymin 50% (0.4 mg/g), hesperidin 50% (0.4 mg/g); juice: didymin 28% (traces), hesperidin 72% (0.1 mg/g). The percentage and concentration of flavone reported in *Citrus hystrix* cv Davao lemon by Berhow et al. (1998) were as follows: leaf: rutin 65% (0.8 mg/g).

Citrus microcarpa, Citrus hystrix, Citrus medica var. 1 and 2, and *Citrus suhuiensis* were found to contain high amounts of flavones, flavanones, and dihydrochalcone C- and/or O-glycosides (Roowi and Crozier 2011). Among the major compounds detected were apigenin-6,8-di-C-glucoside, apigenin-8-C-glucosyl-2″-O-rhamnoside, phloretin-3′,5′-di-C-glucoside, diosmetin-7-O-rutinoside, hesperetin-7-O-neohesperidoside, and hesperetin-7-O-rutinoside. Most of the tropical citrus flavanones were neohesperidoside conjugates, which are responsible for imparting a bitter taste to the fruit.

In the peel oil and the juice of kaffir lime, 102 and 90 components were identified, respectively (Motoki et al. 2005). The main components in the peel oil were: β-pinene (22.7%), limonene (17.3%), sabinene (11.9%), citronellal (7.8%), terpinen-4-ol (7.2%), citronellol (3.6%), and linalool (2.6%), while in the juice were β-pinene (35.6%), sabinene (7.0%), limonene (5.9%), terpinen-4-ol (19.7%), γ -terpinene (4.4%), and linalool (2.8%). In another analysis, twenty six (26) compounds were identified from the peel and thirty seven (37) compounds identified from the fruit oils (Safian et al. 2005). The major compounds contained in the fruit peel oil were β-pinene (27%), limonene (24.7%) and sabinene (13.8%) while α-terpeneol (15.8%), β-pinene (15.1%) and limonene (9.1%). Ibrahim et al. (1995) reported the fruit peel oil to have the following major volatile constituents: β-pinene (39.25%), limonene (14.16%), citronellal (11.67%), terpinene-4-ol (8.89%), α-terpineole (3.03%), citronellol (2.96%), γ-terpinene (2.36%), α-pinene (1.99%), *cis*-linalool oxide (1.85%), δ-3-carene (1.44%), terpinolene (1.60%) and geraniol (0.68%).

Citrus hystrix leaf is rich in vitamin E. Among 62 edible tropical plants analysed for α-tocopherol content *Citrus hystrix* leaves (398.3 mg/kg edible portion) ranked second behind *Sauropus androgynus* leaves (426.8 mg/kg) (Ching and Mohamed 2001). Thirty-eight constituents were identified in the leaf essential oil of Kaffir lime representing 89% of the essential oil (Waikedre et al. 2010). The oil was rich in monoterpenes (87%) with β-pinene as the major component (10%) and low in limonene (4.7%). The essential oil of *C. hystrix* was characterized by high contents of terpinen-4-ol (13.0%), α-terpineol (7.6%), 1,8-cineole (6.4%), and citronellol (6.0%). The oil was found inactive against bacteria. Twenty nine compounds were found in the essential oil of kaffir lime leaves (Loh et al. 2011). β-citronellal was the major compound amounting to 66.85% of total oil followed by β-citronellol (6.59%), 5,9-dimethyl-1-decanol (4.96%), unknown (4.75%), linalool (3.90%), methyl citronellate (1.90%), geranyl acetate (1.80%) and citronellol (1.76%). Other compounds included 3-undecanol (1.04%), 2-(2-hydroxy-2-propyl)-5-methyl-cyclohexanol (0.96%), 1,8-terpin (0.95%), isopregol (0.70%), 4,8-dimethyl-1,7,nonadien-4-ol (0.60%), geraniol (0.42%), terpinene-4-ol (0.34%), cis-2,6,-dimethyl-2,6-octadiene (0.33%), (E)-furanoid linalool oxide (0.27%), 4-methyl-6-hepten-3-ol (0.26%), 2,6,-dimethyl-5-heptenal (0.24%), cis-linalool oxide (0.24%), sabinene (0.20%), isopulegol (0.18%), 2-methyl-7-oxabicyclo-heptane (0.13%), α-terpineol (0.11%), β-myrcene (0.08%), (E)-2,5-dimethyl-1,6-octadiene (0.08%), tetrahydro-4-methyl-2-(2-methyl-1-propenyl)-2H-pyran (0.05%), nerolidol (0.04%) and 3-hexen-1ol (0.03%).

Antioxidant Activity

Kaffir lime contains antioxidants and has very good potential to be explored as sources of natural antioxidants. Recent studies showed that ethanol concentration was the most significant factor affecting the TPC (total phenolic content) (Chan et al. 2009). The optimum extraction conditions of TPC from fruit peels were found to be ethanol concentration of 52.9%, extraction temperature of 48.3°C, and extraction time of 126.5 minutes. Under the optimised conditions, the experimental value for TPC was 1291.8 mg GAE/100 g DW, which reasonably close to the predicted value (1268.8 mg GAE/100 g DW). Studies reported that method of processing of kaffir lime can significantly affect the content of flavonoids and their TAC (total antioxidant capacity) values and thereby affect the antioxidative capacity (Butryee et al. 2009). Using three different assays: oxygen radical absorption capacity, ferric reducing/antioxidant power, and scavenging effect on the 2, 2-diphenyl-1-picrylhydrazyl (DPPH) free radical, it was found that boiling decreased TAC values. The amount of total flavonoids calculated as aglycone equivalents of eight identified flavonoids (cyanidin, myricetin, peonidin, quercetin, luteolin, hesperetin, apigenin and isorhamnetin) was 1,129 (deep frying), 1,104 (fresh without processing) and 549 (boiling) mg/100 g freeze-dried weight (dry matter exclude fat). Hesperetin was the predominant flavonoid. The total phenolic content expressed as grams of gallic acid equivalents/100 g fresh weight (excluding fat) was 2.0, 1.9 and 1.8 in fresh, deep fried and boiled samples, respectively. In another study it was found that the antioxidative activities of kaffir lime in descending order were leaves > peels > stems and all were much higher than that of the control but lower than pegaga (*Centella asiatica*) (Jaswir et al. 2004).

The yield of phenolics from *C. hystrix* leaves, total phenolic content (TPC) and DPPH-IC$_{50}$ obtained were 5.06%, 116.53 mg GAE/g extract and IC$_{50}$ of 0.063 mg/ml, respectively (Jamilah et al. 2011). Better inhibition and TPC were obtained using supercritical carbon dioxide extraction method whereas higher yield and phenolic acids were obtained in the ethanol extracts. Phenolic compounds identified were vanillic acid, *p*-coumaric acid, sinapic acid, *m*-coumaric acid, benzoic acid and cinnamic acid.

The antioxidative activity of the extracts in both DPPH radical scavenging activity assay and linoleic acid model system assays followed the decreasing order of: *Cucurma longa* > *Murraya koenigii* > *Citrus hystrix* > *Pandanus amaryllifolius* (Idris et al. 2008). 2,000 ppm was chosen as the optimum concentration to be used in deep frying experiment. Extracts of *Pandanus amaryllifolius* and *Citrus hystrix* exhibited protective activity towards RBD (refined, leached and deodorised) palm olein that was comparable to BHT (butylated hydroxytoluene) during frying. The extracts were useful in improving and also maintaining the sensory characteristics of French fries. The French fries treated with herb extracts were acceptable by panelists until day 5 of frying. The natural antioxidants significantly lowered the rate of oil oxidation during deep-fat frying and maintaining the quality of French fries. They exhibited excellent heat-stable antioxidant properties and presented good natural alternative to existing synthetic antioxidants for the food industry. Studies indicated that the ethanol extract of the *Citrus hystrix* peel exhibited antioxidant properties in frying oil as determined by measuring peroxide value (PV), p-anisidine value (AnV), totox value, iodine value (IV), percent free fatty acids, color and viscosity of the oil and therefore displayed potential as one of the new sources of natural antioxidants.

Anticancer and Antiviral Activities

Citrus hystrix has anticancer and antiviral activities. Three known coumarins isolated from *Citrus hystrix* were found to be inhibitors of both lipopolysaccharide (LPS) and interferon-γ (IFN-γ)-induced nitric oxide (NO) generation in RAW 264.7 cells (Murakami et al. 1999). The inhibitory activity of bergamottin ($IC_{50} = 14$ μM) was comparable to that of N-(iminoethyl)-L-ornithine (L-NIO) ($IC_{50} = 7.9$ μM), whereas oxypeucedanin and 5-[(6′,7′-dihydroxy-3′,7′-dimethyl-2-octenyl)oxy]psoralen, structurally different from bergamottin only in their side-chain moieties, were notably less active. It was found that coumarins in Group (A) bearing an isoprenyl (IP) or a geranyl (GR) group, were highly active; Group (B) bearing an IP group cyclized to a coumarin ring, moderately active; group (C) bearing an IP group modified with hydroxyl group(s) and/or having other functional groups except for the IP were completely inactive. Cellular uptake studies suggested that coumarins in group C were inactive because of poor permeability to the cell membrane.

Ethyl acetate fraction of kaffir lime leaves exhibited the highest cytotoxicity against 4 leukemic cell lines with IC_{50} values of 19.0, 35.3, 21.8 and 19.8 μg/ml against HL60, K562, Molt4 and U937 cell lines respectively (Chueahongthong et al. 2011). These were higher than those fractions from hexane, ethanol and butanol. The methanol fraction exerted no activity (IC_{50} value >100 μg/ml). None of the fractions were cytotoxic on peripheral blood mononuclear cells.

Two glyceroglycolipids isolated from the leaves, namely 1,2-di-O-α-linolenoyl-3-O-β-galactopyranosyl-sn-glycerol (DLGG, 1) and a mixture of two compounds, 1-O-α-linolenoyl-2-O-palmitoyl-3-O-β-galactopyranosyl-sn-glycerol (2a) and its counterpart (2b) (LPGG, 2) were found to be potent inhibitors of tumour promoter-induced Epstein-Barr virus (EBV) activation (Murakami et al. 1995). The IC_{50} values of 1 and 2 were strikingly lower than those of representative cancer preventive agents such as α-linolenic acid, β-carotene, or (−)-epigallocatechin gallate. Compound 1 exhibited anti-tumour-promoting activity even at a dose ten times lower than that of α-linolenic acid in a two-stage carcinogenesis experiment on ICR mouse skin with dimethylbenz[α]anthracene (DMBA) and 12-O-tetradecanoylphorbol 13-acetate (TPA). The inhibition of the arachidonic acid cascade may be involved in antitumour promotion since 1 inhibited TPA-induced edema formation in the antiinflammation test using ICR mouse ears. In another study, *C. hystrix* significantly enhanced 2-amino-3, 8-dimethylimidazo (4,5-f)quinoxaline-associated preneoplastic liver cell focus development while *Boesenbergia pandurata* and *Languas galanga* had borderline effects in F344 male rats (Tiwawech et al. 2000). The results suggested that *C. hystrix* as well as *Boesenbergia pandurata* and *Languas galanga* may contain agents augmenting the hepatocarcinogenicity of 2-amino-3,8-dimethylimidazo(4,5-f)quinoxaline. In a separate study, the methanolic extract of *C. hystrix* leaves (flavonoids and tannins) inhibited the herpes virus (HSV-1) with an ED_{50} (concentration which inhibited viruses by 50%) of 91 μg/ml and a CC_{50} (concentration which reduced cell proliferation by 50%) of 210 μg/ml (Fortin et al. 2002). However the extract was not effective against the poliovirus type 2.

Antifertility Activity

Citrus hystrix has antifertility effects. Thai scientists found that alcohol and chloroform extract of *Citrus hystrix* peels effectively inhibited implantation, produce abortion and slightly hasten labor time when it was given from day 2 to 5, day 8 to 12 and day 15 until labour, respectively (Piyachaturawat et al. 1985). At the same dose level which interrupted pregnancy, the extract did not affect the estrous cycle. Neither uterotrophic effects nor induction of vaginal cornification was observed when it was given to spayed rats. However, the extract enhanced the uterotrophic effect of estradiol when both were simultaneously given. Additionally, the extract stimulated uterine contractions observed in an in situ study. It was suggested that these two effects might had been responsible for the interruption of pregnancy associated with the extract.

Antimicrobial Activity

Citrus hystrix also has antibacterial activity. The volatile oil obtained from leaf, fruit peel and juice inhibited bacterial growth, which could be ascribed to the presence of citronellal (Suri et al. 2002). All Gram negative bacteria were weakly inhibited by the volatile oil. The antibacterial property of methanolic extract of stem, seed, peel, leaf and callus treated with 2.0 mM phenylalanine was also studied. Most of the extracts showed stronger inhibition effect against Gram positive bacteria. In a preliminary study, kaffir oil extract showed higher antibacterial properties compared to the fresh extract (Chaisawadi et al., 2005). The extract was inhibitory to *Bacillus cereus* and *Staphylococcus aureus*. In another study, Thai scientists reported that citrus peels showed stronger antimicrobial activities than their essential oils obtained from hydrodistillation (Chanthaphon et al. 2008). The ethyl acetate extract of kaffir lime (*Citrus hystrix*) peel showed broad spectrum of inhibition against all Gram-positive bacteria, yeast and moulds including *Staphylococcus aureus, Bacillus cereus, Listeria monocytogenes, Saccharomyces cerevisiae* var. *sake* and *Aspergillus fumigatus* TISTR 3180. It exhibited minimum inhibitory concentration (MIC) values of 0.28 and 0.56 mg/ml against *Saccharomyces cerevisiae* var. *sake* and *B. cereus*, respectively while the minimum bactericidal concentration (MBC) values against both microbes were 0.56 mg/ml. The MIC values of the extract against *L. monocytogenes*, *A. fumigatus* TISTR 3180 and *S. aureus* were 1.13 mg/ml while the MBC values against *L. monocytogenes* as well as *A. fumigatus* TISTR 3180 and *S. aureus* were 2.25 and 1.13 mg/ml, respectively. The major components of the ethyl acetate extract from kaffir lime were limonene (31.64%), citronellal (25.96%) and β-pinene (6.83%) whereas β-pinene (30.48%), sabinene (22.75%) and citronellal (15.66%) appeared to be major compounds of the essential oil obtained from hydrodistillation.

In a separate study, Kaffir lime oil was found to be inhibitory to *Propionibacterium acnes* which had a role in the inflammation of acne leading to scar formation (Lertsatitthanakorn et al. 2006). The minimum inhibitory concentration (MIC) and minimum bactericidal concentration (MBC) values of kaffir lime oil were 5 μl/ml. Antiinflammatory activity of the oils as determined using the 5-lipoxygenase inhibition assay found that IC_{50} value kaffir lime oil (0.05 μl/ml) was less than that of nordihydroquaretic acid (1.7 μg/ml).

Cholinesterase Inhibitory Activity

Four new compounds, citrusosides A-D (1–4), and 15 known compounds were isolated from the hexanes and CH_2Cl_2 extracts of *Citrus hystrix* fruit peels (Youkwan et al. 2010). Compound 1 was identified as the 1-O-isopropyl-6-O-β-D-glucopyranosyl ester of 5″,9″-dimethyl-2″,8″-decadienoic acid. Compounds 2–4 possessed a 1-O-isopropyl-β-D-glucopyranosyl and a dihydroxyprenyl furanocoumarin moiety conjugated to the 3-hydroxy-3-methylglutaric acid as diesters. Several furanocoumarins exhibited cholinesterase inhibitory activity. (R)-(+)-6′-hydroxy-7′-methoxybergamottin, (R)-(+)-6′,7′-dihydroxybergamottin, and (+)-isoimparatorin showed IC_{50} values of 11.2, 15.4, and 23 μM, respectively. Bioassay results indicated that the presence of a dioxygenated geranyl chain in the test compounds was crucial for the inhibitory activity.

Skin Conditioning Activity

Studies in Thailand reported that using kaffir lime oil for body massage had a stimulating and activating effect (Hongratanaworakit and Buchbauer 2007). The oil caused a significant increase in blood pressure and a significant decrease in skin temperature. Regarding the behavioural parameters, subjects in the kaffir lime oil group rated themselves more alert, attentive, cheerful and vigorous than subjects in the control group. The findings lend support to its use in aromatherapy such as causing relief from depression and stress in humans. However, Kaffir lime had been reported to cause extensive phytophotodermatitis in a hiker (Koh and Ong 1999). This was caused by the application of the juice as a folk remedy to ward off biting insects.

Insecticidal Activity

Kaffir lime oil exhibited low mosquito repellency against *Aedes aegypti, Anopheles dirus* and *Culex quinquefasciatus,* compared to the volatile oils of turmeric, citronella and hairy basil (Tawatsin et al. 2001). Without vanillin, *Citrus hystrix* volatile oil provided repellency against *Aedes aegypti* for 1 h only. However, in the presence of vanillin, the volatile oil of *Citrus hystrix* had an extended repellency to *Aedes aegypti* for up to 3 h. *Citrus hystrix* essential oil showed good potential for being used as a cockroach repellent (Thavara et al. 2007). *Citrus hystrix* essential oil showed the best repellency over other candidate essential oils (*Boesenbergia rotunda, Curcuma longa, Litsea cubeba, Piper nigrum, Psidium guajava* and *Zingiber officinale*) and naphthalene. The essential oil exhibited complete repellency (100%) against *Periplaneta americana* and *Blattella germanica* and also showed the highest repellency (among the essential oils tested) of about 87.5% against *Neostylopyga rhombifolia* under laboratory conditions. In the field, *Citrus hystrix* essential oil formulated as a 20% active ingredient in ethanol and some additives provided satisfactory repellency of up to 86% reduction in cockroaches, mostly *P. americana* and *N. rhombifolia* with a residual effect lasting a week after treatment.

Traditional Medicinal Uses

The juice and rinds of the kaffir lime are widely used in traditional Indonesian medicine like jamu and the fruit is sometimes referred to in Indonesia as *jeruk obat* (medicine citrus). The leaves can be used to treat stomach ache caused by dyspepsia and insect bites. In Peninsular Malaysia, the fruit and leaves have been used for washing hair; the fruit is halved and the grated rind is rubbed on the head or the whole fruit is boiled and used as shampoo. The peel has also been used in a general tonic medicine "ubat jamu" and the fruit juice in ointment. The rind has also been prescribed for treatment for worms and headache.

Other Uses

Oil is also extracted from the rind for use in cosmetics and beauty products.

Extracts from the fruit skin as well as juice are used as an insecticide for washing the head and treating the feet to kill land leeches and as a bleach to remove tough stains. The timber is used for tool handles and the juice may be used as an adjunct in dyeing with annatto (*Bixa orellana*).

The essential oil from kaffir lime leaves was found effective in killing larvae of tobacco army worm, *Spodoptera litura* with LD50 value of 26.748 ul/g (Loh et al. 2011). The essential oil also exhibited antifeedant properties resulting in severe growth inhibition of the insect.

Comments

The plant is propagated from seeds and grafting.

Selected References

Backer CA, Bakhuizen van den Brink RC Jr (1965) Flora of Java, vol 2. Wolters-Noordhoff, Groningen, 630 pp

Berhow M, Tisserat B, Kanes K, Vandercook C (1998) Survey of phenolic compounds produced in *Citrus*. USDA ARS Tech Bull 1856:1–154

Brown WH (1951–1958) Useful plants of the Philippines 1–3. Technical Bulletin 10, Department Agriculture and National Resources Rep. Manila: 1610 pp

Burkill IH (1966) A dictionary of the economic products of the Malay Peninsula. Revised reprint. 2 vols. Ministry of Agriculture and Co-operatives, Kuala Lumpur, vol 1 (A–H), pp 1–1240, vol 2 (I–Z), pp 1241–2444

Butryee C, Sungpuag P, Chitchumroonchokchai C (2009) Effect of processing on the flavonoid content and antioxidant capacity of *Citrus hystrix* leaf. Int J Food Sci Nutr 1:1–13

Chaisawadi S, Thongbute D, Methawiriyasilp W, Pitakworarat N, Chaisawadi A, Jaturonrasamee K, Khemkhaw J, Tanuthumchareon W (2005) Preliminary study of antimicrobial activities on medicinal herbs of Thai food ingredients. Acta Hort (ISHS) 675:111–114

Chan SW, Lee CY, Yap CF, Wan Aida WM, Ho CW (2009) Optimisation of extraction conditions for phenolic compounds from limau purut (*Citrus hystrix*) peels. Int Food Res J 16:203–213

Chanthaphon S, Chanthachum S, Hongpattarakere T (2008) Antimicrobial activities of essential oils and crude extracts from tropical *Citrus* spp. against food-related microorganisms. Songklanakarin J Sci Technol 30(Suppl 1):125–131

Ching LS, Mohamed S (2001) Alpha-tocopherol content in 62 edible tropical plants. J Agric Food Chem 49(6):3101–3105

Chueahongthong F, Ampasavate C, Okongi S, Tima S, Anuchapeeda S (2011) Cytotoxic effects of crude kaffir lime (*Citrus hystrix* DC) leaf fractional extracts on leukemic cell lines. J Med Plant Res 5(14): 3097–3105

Fortin H, Vigora C, Lohézic-Le Dévéhat F, Robina V, Le Bossé B, Boustiea J, Amoros M (2002) In vitro antiviral activity of thirty-six plants from La Réunion Island. Fitoterapia 73:346–350

Hongratanaworakit T, Buchbauer G (2007) Chemical composition and stimulating effect of *Citrus hystrix* oil on humans. Flavour Frag J 22(5):443–449

Ibrahim J, Abu Said A, Abdul Rashih A, Nor Azah MA, Norsiha A (1995) Chemical composition of some *Citrus* oils from Malaysia. J Essent Oil Res 8: 627–632

Idris NA, Nor MF, Ismail R, Mohamed S, Hassan ZC (2008) Antioxidative activity of Malaysian herb extracts in refined, bleached and deodorized palm olein. J Oil Palm Res 20:517–526

Jamilah B, Abdulkadir Gedi M, Suhaila M, Md Zaidul IS (2011) Phenolics in *Citrus hystrix* leaves obtained using supercritical carbon dioxide extraction. Int Food Res J 18(3):941–948

Jamilah B, Che Man YB, Ching TL (1998) Antioxidant activity of *Citrus hystrix* peel extract in RBD palm olein during frying of fish crackers. J Food Lipids 5(2):149–157

Jaswir I, Hassan JH, Said MZM (2004) Efficacy of Malaysian plant extracts in preventing peroxidation reactions in model and food oil systems. J Oleo Sci 53:525–529

Koh D, Ong CN (1999) Phytophotodermatitis due to the application of *Citrus hystrix* as a folk remedy. Br J Dermatol 140(4):737–738

Lertsatitthanakorn P, Taweechaisupapong S, Aromdee C, Khunkitti W (2006) In vitro bioactivities of essential oils used for acne control. Int J Aromather 16: 43–49

Loh FS, Awang RM, Omar D, Rahmani M (2011) Insecticidal properties of *Citrus hystrix* DC leaves essential oil against *Spodoptera litura* Facricius. J Med Plant Res 5(16):3739–3744

Mabberley DJ (1997) A classification for edible *Citrus*. Telopea 7(2):167–172

Manner HI, Buker RS, Smith VE, Elevitch CR (2006) *Citrus* species. Version 2.1. In: Elevitch CR (ed) Species profiles for Pacific Island agroforestry. Permanent Agriculture Resources (PAR), Holualoa, Hawaii. http://www.agroforestry.net

Molesworth Allen B (1967) Malayan fruits. An introduction to the cultivated species. Donald Moore Press, Moore, 245 pp

Motoki I, Shuichi H, Seiji H (2005) Essential oil components of combava (*Citrus hystrix* D.C.). Koryo, Terupen oyobi Seiyu Kagaku ni kansuru Toronkai Koen Yoshishu 49:27–29 (in Japanese)

Murakami A, Gao GX, Kim OK, Omura M, Yano M, Ito C, Furukawa H, Jiwajinda S, Koshimizu K, Ohigashi H (1999) Identification of coumarins from the fruit of *Citrus hystrix* DC as inhibitors of nitric oxide generation in mouse macrophage RAW 264.7 cells. J Agric Food Chem 47(1):333–339

Murakami A, Nakamura Y, Koshimizu K, Ohigashi H (1995) Glyceroglycolipids from *Citrus hystrix*, a traditional herb in Thailand, potently inhibit the tumor-promoting activity of 12-O-tetradecanoylphorbol 13-acetate in mouse skin. J Agric Food Chem 43(10):2779–2783

Ochse JJ, Bakhuizen van den Brink RC (1980) Vegetables of the Dutch Indies, 3rd edn. Ascher and Co., Amsterdam, 1016 pp

Piyachaturawat P, Glinsukon T, Chanjarunee A (1985) Antifertility effect of *Citrus hystrix* DC. J Ethnopharmacol 13(1):105–110

Roowi S, Crozier A (2011) Flavonoids in tropical citrus species. J Agric Food Chem 59(22): 12217–12225

Safian MF, Mohamad Ali NA, Yury N, Zainal Ariffin Z (2005) Identification of essential oil composition of peel and fruits of *Citrus hystrix* DC. Malaysian J Sci 24(1):109–111

Saidin I (2000) Sayuran Tradisional Ulam dan Penyedap Rasa. Penerbit Universiti Kebangsaan Malaysia, Bangi, pp 228 (in Malay)

Suri R, Radzali M, Aspollah M, Marziah S, Arif ZJ, Samsumaharto RA (2002) Antibacterial assay of *Citrus hystrix* (limau purut) extracts. J Trop Med Plants 3(1): 35–42

Swingle WT, Reece PC (1967) The botany of *Citrus* and its wild relatives. In: Reuther W, Webber HJ, Batchelor LD (eds) The citrus industry, vol 1, History, world distribution, botany, and varieties. University of California, Riverside, pp 190–430

Tawatsin A, Wratten SDR, Scott R, Thavara U, Techadamrongsin Y (2001) Repellency of volatile oils from plants against three mosquito vectors. J Vector Ecol 26(1):76–82

Tee ES, Noor MI, Azudin MN, Idris K (1997) Nutrient composition of Malaysian foods, 4th edn. Institute for Medical Research, Kuala Lumpur

Thavara U, Tawatsin A, Bhakdeenuan P, Wongsinkongman P, Boonruad T, Bansiddhi J, Chavalittumrong P, Komalamisra N, Siriyasatien P, Mulla MS (2007) Repellent activity of essential oils against cockroaches (Dictyoptera: Blattidae, Blattellidae, and Blaberidae) in Thailand. Southeast Asian J Trop Med Public Health 38(4):663–673

Tiwawech D, Hirose M, Futakuchi M, Lin C, Thamavit W, Ito N, Shirai T (2000) Enhancing effects of Thai edible plants on 2-amino-3, 8-dimethylimidazo (4,5-f) quinoxaline hepatocarcinogenesis in a rat medium-term bioassay. Cancer Lett 158(2):195–201

Waikedre J, Dugay A, Barrachina I, Herrenknecht C, Cabalion P, Fournet A (2010) Chemical composition and antimicrobial activity of the essential oils from New Caledonian *Citrus macroptera* and *Citrus hystrix*. Chem Biodivers 7(4):871–877

Youkwan J, Sutthivaiyakit S, Sutthivaiyakit P (2010) Citrusosides A-D and furanocoumarins with cholinesterase inhibitory activity from the fruit peels of *Citrus hystrix*. J Nat Prod 73(11): 1879–1883

Zhang DX, Mabberley DJ (2008) Flora of China. In: Wu ZY, Raven PH, Hong DY (eds) Flora of China vol 11 (Oxalidaceae through Aceraceae). Science Press/Missouri Botanical Garden Press, Beijing/St. Louis

Citrus inodora

Scientific Name

Citrus inodora F.M. Bailey (sensu Mabberley).

Synonyms

Citrus maideniana Domin, *Microcitrus inodora* (Bailey) Swingle, *Microcitrus maideniana* (Domin) Swingle, *Pleurocitrus inodora* (Bailey) T. Tanaka

Family

Rutaceae

Common/English Names

Queensland Wild Lime, Russell River Lime, Lime, Russell River Lime, Queensland Wild, Large Leaf Australian Wild Lime.

Vernacular Names

None

Origin/Distribution

The species is endemic to north-eastern Queensland and is known only from the eastern foothills of the Bellenden Ker Range – Mount Bartle Frere area and also from the Cape Tribulation area.

Agroecology

It occurs as an understory plant in undisturbed well developed lowland tropical rain forest from near sea level to 120 m elevation. It is a fairly rare species from near coastal areas. The plant thrives in shady condition on organically rich, loamy soil and requires plenty of water.

Edible Plant Parts and Uses

The fruit can be used to make drinks or marmalade.

Botany

A bushy, evergreen, armed shrub to 1–3 m high with angular twigs and paired, sharp, 4–10 mm long axillary spines at the leaf axils. Leaf blades alternate, simple, entire, leathery, about 7–17 × 2.5–8.5 cm, diamond-shaped with acute to subacute apex tapering base with widely serrated margins, glossy deep green on a grooved petiole (Plates 1 and 2). Flowers odourless, white to pink flowers, 1 cm across, all parts gland-dotted, calyx tube short with short 1 mm lobes, petals 6–7 mm long. Fruit ellipsoid 3.5–4 cm by 2.5 cm wide, green ripening yellow (Plates 1, 2 and 3). Seeds numerous embedded in vesicular pulp.

Plate 1 Immature fruit and leaves (M. Smith)

Plate 2 Diamond-shaped, large simple leaves (M. Smith)

Plate 3 Near ripe wild lime fruit (M. Smith)

Nutritive/Medicinal Properties

Nutrient values of the fruit have not been published.

Fifty three volatiles were found in the juice of *M. inodora* extracted by dichloromethane (Shaw et al. 2000). The major components were limonene 68.5%, ethanol 14.6%, acetaldehyde 9.4%, myrcene 1.4%, hexanal 0.6%, (*3Z*)-3-hexanol 0.2%, linalool 0.1%. Components of the peel oil were not quantified and included besides the usal monoterpenes, linalool, nerol, geranil, carvone and perillaldehyde. The leaf essential oil was found to be rich in germacrene D (23.7%) and bicyclogermacrene (17.3%) (Brophy et al. 2001).

The percentage and concentration of total phenolics in *Microcitrus inodora* leaf, peel and juice were reported by Berhow et al. (1998) as: leaf:- 12.5% (1.16 mg/g) flavone/flavonol, 67.2% (3.96 mg/g) flavanone, 11.9% 11.9% psoralen, 0% coumarin; flavedo (outer pigmented layer of the peel): 81% (8.75 mg/g) flavone/flavonol, 5.7% (0.39 mg/g) flavanone, 0% psoralen, 0.7% (0.02 mg/g) coumarin; juice: 66% (3.28 mg/g), flavone/flavonol, 5.7% (0.18 mg/g) flavanone, 0% psoralen, 3.5% (0.05) coumarin.

The percentage and concentration of flavanones reported in *Microcitrus inodora* by Berhow et al. (1998) were as follows: leaf:- hesperidin 3% (0.1 mg/g), narirutin 16% (0.6 mg/g), naringin 3% (0.1 mg/g), neohesperidin 16% (0.7 mg/g); flavedo: naringin–6″–malonate (closed form) 100% (0.4 mg/g); juice naringin–6″–malonate (closed form) 100% (0.2 mg/g).

Other Uses

The species has potential to be used in citrus breeding and as rootstock for other *Citrus* species.

Comments

The species is propagated from seeds, by cuttings which are slow to take roots or by budding onto *Citrus* root-stock.

Selected References

Bayer RJ, Mabberley DJ, Morton C, Miller CH, Sharma IK, Pfeil BE, Rich S, Hitchcock R, Sykes S (2009) A molecular phylogeny of the orange subfamily (Rutaceae: Aurantioideae) using nine cpDNA sequences. Am J Bot 96: 668–685

Berhow M, Tisserat B, Kanes K, Vandercook C (1998) Survey of phenolic compounds produced in *Citrus*. USDA ARS Tech Bull 1856:1–154

Brophy JJ, Goldsack RJ, Forster PI (2001) The leaf oils of the Australian species of *Citrus* (Rutaceae). J Essent Oil Res 13:264–268

Mabberley DJ (1998) Australian Citreae with notes on other Aurantioideae (Rutaceae). Telopea 7:338

Shaw PE, Moshonas MG, Bowman KD (2000) Volatile constituents in juice and oil of Australian wild lime (*Microcitrus inodora*). Phytochemistry 53:1083–1086

Swingle WT (1938) A new taxonomic arrangement of the orange subfamily Aurantioideae. J Washington Acad Sci 28:530–533

Tanaka T (1936) The taxonomy and nomenclature of Rutaceae-Aurantioideae. Blumea 2:101–110

Citrus japonica 'Marumi'

Scientific Name

Citrus japonica Thunb. 'Marumi'.

Synonyms

Fortunella japonica (Thunb.) Swingle.

Family

Rutaceae

Common/English Names

Golden Orange, Marumi Kumquat, Morgani Kumquat, Round Kumquat, Sweet-Peeled Kumquat.

Vernacular Names

Chinese: Chin Chü, Jin Gan, Jin Ju, Shan Ju, Yuan Jin Gan (Medicinal Name);
French: Kumquat À Fruits Ronds, Kumquat Marumi, Kumquat Rond, Kumquat Du Japon;
German: Marumi-Kumquat, Runde Kumquat, Rundkumquat;
Hungarian: Japán Kumkvat;
Italian:.Kumquat Rotondo, Kumquat A Frutto Rondo;
Japanese: Maru Kinkan, Marumi Kinkan;
Korean: Dong Gul Gyul, Dung Geun Geum Gam, Geum Gyul, Geum Gyul La Mu;
Portuguese: Cunquato-Marumi;
Russian: Кумкват;
Spanish: Kumquat Redondo, Naranjita Japonesa, Kumquat Redondo;
Vietnamese: Cây Quất Cảnh.

Origin/Distribution

Citrus japonica is native to southern China. It is cultivated in China, Japan, Korea, Taiwan, Europe, southern United States (notably Florida and California), Australia and elsewhere.

Agroecology

Marumi kumquat is more cold hardy than other kumquats like the Nagami and the Meiwa- and even more cold hardy than the Satsuma Orange Tree. Marumi Kumquat is not damaged by temperatures of −20°C.

Edible Plant Parts and Uses

Ripe Marumi kumquat is eaten whole as its rind is extremely sweet, fragrant and pleasant. The fruit is popularly relished as fresh fruit in Korea and Japan. The fruits are easily preserved whole

Plate 1 Marumi kumquat fruits and leaves

Plate 3 Harvested Marumi kumquat

Plate 2 Close-up of ripening fruit

Plate 4 Close view Marumi kumquat fruits

in sugar syrup and bottled or canned. The fruits can also be pickled in jars of water, vinegar, and salt sealed and allowed to stand for 2–3 months or made into sweet pickled by boiling in syrup, vinegar and sugar. The kumquats can also be made into marmalade or jelly (Plates 1, 2 and 3).

Botany

Small evergreen, sparingly armed tree to 2–3 m high with a compact crown and angular green branches when young. Leaves are oval to broadly elliptic, 6–7.5 cm by 3–4.5 cm, medium green, simple, with acute to subacute apex, tapering obtuse base, entire to subcrenulate margin and borne on inconspicuously winged 6–11 mm petioles (Plates 1 and 2). Flowers usually solitary, or paired and axillary, white, bisexual; sepals green glabrous, 5 toothed; petals 5 white, oblong; stamens 20 in 5 or more coherent fascicles; ovary superior subglobose to globose with a simple globose stigma. Fruit is globose to slightly oblate, or subglobose, 2.0–2.7 cm diameter, glossy golden orange to orangey-yellow (Plates 1–4), thin, sweet, fragrant rind and 4–6 pulpy, juicy, orange segments with 1–3 small pointed seeds or seedless.

Nutritive/Medicinal Properties

The nutrient composition of raw kumquat (*Fortunella* spp.) (exclude 7% seeds) per 100 g edible portion was reported as: water 80.85 g, energy 71 kcal (296 kJ), protein 1.88 g, total lipid 0.86 g, ash 0.52 g, carbohydrate 15.90 g, total dietary fibre 6.5 g, total sugars 9.36 g, Ca 62 mg, Fe 0.86 mg, Mg 20 mg, P 19 mg, K 186 mg,

Na 10 mg, Zn 0.17 mg, Cu 0.095 mg, Mn 0.135 mg, vitamin C 43.9 mg, thiamine 0.037 mg, riboflavin 0.090 mg, niacin 0.429 mg, pantothenic acid 0.208 mg, vitamin B-6 0.036 mg, total folate 17 μg, total choline 8.4 mg, vitamin A 290 IU (15 μg RAE), vitamin E (α-tocopherol) 0.15 mg, total saturated fatty acids 0.103 g, 14:0 (myristic) 0.004 g, 16:0 (palmitic) 0.090 g, 18:0 (stearic) 0.004 g; total monounsaturated fatty acids 0.154 g, 16:1 undifferentiated (palmitoleic) 0.021 g, 18:1 undifferentiated (oleic) 0.137 g; total polyunsaturated fatty acids 0.171 g, 18:2 undifferentiated (linoleic) 0.124 g, 18:3 undifferentiated (linolenic) 0.047 g; α-carotene 155 μg, β- cryptoxanthine 193 μg, and luetin + zeaxanthin 129 μg (USDA 2011).

Among a total of 91 volatile constituents identified in round kumquat fruit, 47 were identified for the first time in kumquat fruit (Umano et al. 1994). d-limonene was the most abundant compound, comprising 87% of the sample from steam distillation and 97% of the sample from simultaneous purging/extraction (SPE). In addition to d-limonene, linalool, myrcene, and geranyl acetate were found in the sample from steam distillation as major constituents; myrcene, α-pinene, and β-phellandrene were identified in the sample from SPE as major components.

Ten compounds were isolated from fruit peels of *Fortunella japonica* and identified as α-tocopherol (1), lupenone (2), β-amyrin (3), α-amyrin (4), β-sitosterol (5), β-sitosteryl 3-O-glucopyranoside (6), kaempferide 3-O-rhamnopyranoside (7), 3′,5′-di-C-β-glucopyranosylphloretin (8), acacetin 7-O-neohesperidoside (9), and acacetin 8-C-neohesperidoside (10) (Cho et al. 2005). Eighty-two compounds were identified in kumquat (*Fortunella japonica* Swingle) cold-pressed peel oil (Choi 2005). The major compounds were limonene (93.73%), myrcene (1.84%), and ethyl acetate (1.13%). Camphene, terpinen-4-ol, citronellyl formate, and citronellyl acetate showed high flavour dilution (FD) factors (≥5) and relative flavour activities (RFA) (>20). Citronellyl formate and citronellyl acetate were regarded as the characteristic odour components of the kumquat peel oil. 106 volatile compounds, were identified in round kumquat (*Fortunella japonica*) peel essential oil (Quijano and Pino 2009). Limonene was the most abundant compound, comprising 76.7% of peel oil In addition to limonene, myrcene, germacrene D and linalool and were found as major constituents.

Traditional Medicinal Uses

The plant is regarded as antiphlogistic, antivinous, carminative, deodorant and stimulant. The fresh fruit is antitussive and expectorant; in Vietnam, it is steamed with sugar candy and used in the treatment of sore throats. It is said to be good for infants.

Other Uses

Marumi kumquat is grown as an ornamental plant in the garden, parks and as ornamental house plant in patios and terraces and can be used in bonsai. This plant symbolizes good luck in China, Japan and Korea and other southeast Asian countries, where it is sometimes given as a gift during the Lunar New Year.

Comments

Marumi kumquat is smaller than Meiwa or Nagami kumquats and has the sweetest, golden orange peel.

Selected References

Cho JY, Kawazoe K, Moon JH, Park KH, Murakami K, Takaishi Y (2005) Chemical constituents from the fruit peels of *Fortunella japonica*. Food Sci Biotechnol 14(5):599–603

Choi SC (2005) Characteristic odor components of kumquat (*Fortunella japonica* Swingle) peel oil. J Agric Food Chem 53(5):1642–1647

Cottin R (2002) *Citrus* of the world: a *Citrus* directory, version 2.0, France, SRA INRA-CIRAD

Facciola S (1990) Cornucopia. A source book of edible plants. Kampong Publ, Vista, 677 pp

Hodgson RW (1967) Horticultural varieties of *Citrus* In: Reuther W, Webber HJ, Batchelor LD (eds) The *Citrus* industry vol 1, History, world distribution, botany, and varieties, revised edn. University of California Press, Berkely pp 431–591

Mabberley DJ (1997) A classification for edible *Citrus*. Telopea 7(2):167–172

Mabberley DJ (1998) Australian Citreae with notes on other Aurantioideae (Rutaceae). Telopea 7(4):333–344

Morton J (1987) Kumquat. In: Fruits of warm climates. Julia F. Morton, Miami, pp 182–185

Nguyen VD, Doan TN (1989) Medicinal plants in Vietnam. World Health Organization (WHO), Regional Publications, Western Pacific Series no 3. WHO, Regional Office for the Western Pacific, Manila, the Philippines and Institute of Materia Medica, Hanoi, Vietnam

Porcher MH et al (1995–2020) Searchable world wide web multilingual multiscript plant name database. Published by The University of Melbourne. Australia. http://www.plantnames.unimelb.edu.au/Sorting/Frontpage.html

Quijano CE, Pino JA (2009) Volatile compounds of round kumquat (*Fortunella japonica* Swingle) peel oil from Colombia. J Essent Oil Res 21:483–485

Swingle WT (1946) The botany of *Citrus* and its wild relatives of the orange subfamily (family Rutaceae, subfamily Aurantioideae). In: Webber HJ, Batchelor LD (eds) The *Citrus* industry, vol 1, History, botany and breeding. University of California Press, Berkeley, pp 129–474, 1028 pp

Swingle WT, Reece PC (1967) The botany of *Citrus* and its wild relatives. In: Reuther W, Webber HJ, Batchelor LD (eds) The *Citrus* industry, vol 1, History, world distribution, botany, and varieties. University of California Press, Riverside, pp 190–430

Tanaka T (1922) *Citrus* fruits of Japan; with notes on their history and the origin of varieties through bud variation. J Hered 13:243–253

U.S. Department of Agriculture, Agricultural Research Service (USDA) (2011) USDA National nutrient database for standard reference, Release 24. Nutrient data laboratory home page. http://www.ars.usda.gov/ba/bhnrc/ndl

Umano K, Hagi Y, Tamura T, Shoji A, Shibamoto T (1994) Identification of volatile compounds isolated from round kumquat (*Fortunella japonica* Swingle). J Agric Food Chem 42(9):1888–1890

Zhang DX, Mabberley DJ (2008) *Citrus* Linnaeus. In: Wu ZY, Raven PH, Hong DY (eds) Flora of China, vol 11, Oxalidaceae through Aceraceae. Science Press/Missouri Botanical Garden Press, St. Louis/Beijing

Citrus japonica 'Meiwa'

Scientific Name

Citrus japonica Thunb. 'Meiwa'.

Synonyms

Citrus japonica (Thunb.) 'Nagami' × *Citrus japonica* (Thunb.) 'Marumi', *Fortunella x crassifolia* Swingle pro sp. (sensu Swingle and Reece), *Fortunella crassifolia* Swingle.

Family

Rutaceae

Common/English Names

Meiwa Kumquat, Large Round Kumquat, Sweet Kumquat.

Vernacular Names

Brazil: Kumquat, Kunquat, Laranja De Ouro, Laranja Dos Orientais, Laranja De Ouro Dos Orientais (Portuguese);
Chinese: Chang An Jin Gan, Hou Ye Jin Ju, Jin Dan, Jin Gan, Jin Ju (Fruit), Jin Dan Ju, Ju He (Medicinal Name), Ning Bo Jin Gan, Ning Bo Jin Gan; gīm-gam (Hokkien);
French: Kumquat Doux, Kumquat Meïwa;
Hungarian: Nagy Kumkvat;
Italian: Meiwa Kumquat;
Japanese: Meiwa Kinkan, Nippon Kinkan, Ninpo, Neiha Kinkan;
Korean: Geumgyul;
Nepali: Muntala;
Taiwan: Ning Po Chin Kan;
Thai: Somchíd;
Vietnamese: Cam Quất, Kim Quất.

Origin/Distribution

Citrus japonica is a native of southern China. *Citrus japonica* 'Meiwa' is thought to be a natural hybrid between the oval *Citrus japonica* (Thunb.) 'Nagami' and the round *Citrus japonica* (Thunb.) 'Marumi'. Meiwa kumquat is extensively grown in China and in cooler subtropical areas elsewhere including USA and Australia.

Agroecology

Kumquat including Meiwa kumquat are much hardier than other citrus plants such as oranges. The Meiwa kumquat requires a hot summer, ranging from 25°C to 38°C, but can withstand frost down to about −10°C without injury. It grows in the tea regions of China where the climate is too cold for other citrus fruits. Although slightly less cold-hardy than Nagami it is increasing in popularity.

Edible Plant Parts and Uses

Its edible uses are as described for Marumi and Nagami kumquats. The fresh fruits are eaten raw or can be preserved, pickled or made into maramalade, candies and jellies. Kumquat fruits are also boiled or dried to make a candied snack called *mứt quất*. Like Nagami kumquats Meiwa kumquats are also used as martini garnish, sliced and added to salads or processed into a liqueur in a clear spirit or vodka.

Plate 1 Meiwa kumquat fruits and leaves

Botany

Meiwa kumquat tree is slow-growing, shrubby, compact, 2.4–4.5 m tall, the branches light-green and angled when young, thornless or sparsely spiny. Leaves are elliptic-lanceolate to ovate-lanceolate, 4–9 cm by 2.5–3.5 cm (Plate 1), apex subacute to obtuse, base broadly cuneate, margin crenulate, glossy green, weakly conduplicate, borne on narrowly winged petioles. Flowers fragrant in 1–3 flowered axillary clusters, petals 5, white, 6–9 mm; stamens 15–25, ovary glabrous, globose to subglobose with one style and globose stigma. Fruit globose, subglobose to short oval-oblong, 2.5–3.2 cm diameter, golden-yellow to orangey-yellow, with a smooth thick sweet rind and 4–7 juicy orange segments containing few 2–4 small ovoid, monoembryonic seeds or seedless (Plates 1 and 2).

Plate 2 Harvested Meiwa kumquats

Nutritive/Medicinal Value

The nutrient composition of raw kumquat (*Fortunella* spp.) (exclude 7% seeds) per 100 g edible portion is reported as: water 80.85 g, energy 71 kcal (296 kJ), protein 1.88 g, total lipid 0.86 g, ash 0.52 g, carbohydrate 15.90 g, total dietary fibre 6.5 g, total sugars 9.36 g, Ca 62 mg, Fe 0.86 mg, Mg 20 mg, P 19 mg, K 186 mg, Na 10 mg, Zn 0.17 mg, Cu 0.095 mg, Mn 0.135 mg, vitamin C 43.9 mg, thiamine 0.037 mg, riboflavin 0.090 mg, niacin 0.429 mg, pantothenic acid 0.208 mg, vitamin B-6 0.036 mg, total folate 17 μg, total choline 8.4 mg, vitamin A 290 IU (15 μg RAE), vitamin E (α-tocopherol) 0.15 mg, total saturated fatty acids 0.103 g, 14:0 0.004 g, 16:0 0.090 g, 18:0 0.004 g;total monounsaturated fatty acids 0.154 g, 16:1 undifferentiated 0.021 g, 18:1 undifferentiated 0.137 g; total polyunsaturated fatty acids 0.171 g, 18:2 undifferentiated 0.124 g, 18:3 undifferentiated 0.047 g; α-carotene 155 mcg, β- cryptoxanthine 193 μg, and luetin + zeaxanthin 129 μg (USDA 2011).

The essential oil of kumquat peel contains much of the aroma of the fruit, and is composed of 71 volatile compound of which the monoterpene (8) limonene, makes up around 93% of the total (Koyasako and Bernhard 1983). Besides limonene and α-pinene (0.34%), both monoterpenes, the oil is unusually rich (0.38% total) in sesquiterpenes (13) such as α-bergamotene (0.021%), caryophyllene (0.18%), α-humulene

(0.07%) and α-muurolene (0.06%), and these contribute to the spicy and woody flavour of the fruit. Carbonyl compounds make up much of the remainder, and these are responsible for much of the distinctive flavour; these compounds include esters (13) such as isopropyl propanoate (1.8%) and terpinyl acetate (1.26%); one ketone, carvone (0.175%); and a range of aldehydes (8) such as citronellal (0.6%) and 2-methylundecanal. Other oxygenated compounds include alcohols (11) such as nerol (0.22%) and the furanoid, trans-linalool oxide (0.15%).

The IC_{50} values of superoxide (O_2^-)- and 1-diphenyl-2-picrylhydrazyl (DPPH)-radical scavenging activity in the skin of Meiwa kumquat decreased in the days after full bloom (DAFB) from 60 DAFB to harvest (Kondo et al. 2005). In contrast, the IC_{50} values of those in the flesh increased with DAFB. The IC_{50} values of DPPH in the skin under high photosynthetic photon flux density (PPFD), and O_2^- in the flesh at a light crop load were higher than those of the control. The IC_{50} of O_2^-- and DPPH-radical scavenging activity in fruit that had been administered ethephon treatment were lower than those in the untreated control. Ethephon-treated fruit also inhibited lipid peroxidation in the red blood cell system. Ascorbic acid or β-cryptoxanthin concentrations in the skin or flesh were higher under high PPFD, at a light crop load, and with ethephon application. These results suggested that environmental and growing factors may regulate the antioxidant compounds in the fruit, influencing antioxidant activity.

Other Uses

In Vietnam, kumquat bonsai trees are used as a decoration for the Tết (Lunar New Year) holidays.

Comments

Meiwa kumquat has larger fruits than Marumi kumquat but the fruits are smaller and more globose than Nagami kumquat.

Selected References

Cottin R (2002) *Citrus* of the world: a *Citrus* directory, version 2.0, France, SRA INRA-CIRAD

Facciola S (1990) Cornucopia. A source book of edible plants. Kampong Publ, Vista, 677 pp

Hodgson RW (1967) Horticultural varieties of *Citrus*. In: Reuther W, Webber HJ, Batchelor LD (eds) The *Citrus* industry, vol 1, History, world distribution, botany, and varieties, Revised Edn. University of California Press, Berkely, pp 431–591

Kondo S, Katayama R, Koji Uchino K (2005) Antioxidant activity in Meiwa kumquat as affected by environmental and growing factors. Environ Exp Bot 54(1): 6–68

Koyasako A, Bernhard RA (1983) Volatile constituents of the essential oil of kumquat. J Food Sci 48(6): 1807–1812

Mabberley DJ (1997) A classification for edible *Citrus*. Telopea 7(2):167–172

Mabberley DJ (1998) Australian Citreae with notes on other Aurantioideae (Rutaceae). Telopea 7(4): 333–344

Morton J (1987) Kumquat. In: Fruits of warm climates. Julia F. Morton, Miami, pp 182–185

Porcher MH et al (1995–2020) Searchable world wide web multilingual multiscript plant name database. Published by The University of Melbourne, Australia. http://www.plantnames.unimelb.edu.au/Sorting/Frontpage.html

Swingle WT (1915) A new genus, *Fortunella*, comprising four species of kumquat oranges. J Wash Acad Sci 5:165–176

Swingle WT (1946) The botany of *Citrus* and its wild relatives of the orange subfamily (family Rutaceae, subfamily Aurantioideae). In: Webber HJ, Batchelor LD (eds) The *Citrus* industry, vol 1, History, botany and breeding. University of California Press, Berkeley, pp 129–474, 1028 pp

Swingle WT, Reece PC (1967) The botany of *Citrus* and its wild relatives. In: Reuther W, Webber HJ, Batchelor LD (eds) The *Citrus* industry, vol 1, History, world distribution, botany, and varieties. University of California Press, Riverside, pp 190–430

Tanaka T (1922) *Citrus* fruits of Japan; with notes on their history and the origin of varieties through bud variation. J Hered 13:243–253

U.S. Department of Agriculture, Agricultural Research Service (2011) USDA National nutrient database for standard reference, Release 24. Nutrient data laboratory home page. http://www.ars.usda.gov/ba/bhnrc/ndl

Zhang DX, Mabberley DJ (2008) linnaeus. In: Wu ZY, Raven PH, Hong DY (eds) Flora of China, vol 11, Oxalidaceae through Aceraceae. Science Press/Missouri Botanical Garden Press, St. Louis/Beijing

Citrus japonica 'Nagami'

Scientific Name

Citrus japonica Thunb. 'Nagami'.

Synonyms

Citrus margarita Lour., *Fortunella japonica* (Thunb.) Swingle var. *margarita* (Swingle) Makino, *Fortunella margarita* (Lour.) Swingle var. *margarita*.

Family

Rutaceae

Common/English Names

Bullet Kumquat, Nagami Kumquat, Oval Kumquat, Pearl Lemon, Spicy-peeled Kumquat.

Vernacular Names

Chinese: Jin Ju, Jin Zao, Luo Fou, Nai Kan;
Danish: Dværgappelsin, Guldorange;
French: Kumquat Nagami, Kumquat Ovale, Kumquat À Chair Acide Kumquat À Fruits Oblongs;
German: Ovale Kumquat, Ovalkumquat, Ovaler Kumquat, Chinesische Kumquat;
Hungarian: Kumkvat;
Indonesia: Jeruk Kumquat Nagami;
Italian: Kumquat Ovale, Kumquat A Frutto Ovale;
Japanese: Naga Kinkan, Nagami Kinkan;
Korean: Geum Gam;
Peru: Naranjilla Tunquas, (Spanish);
Portuguese: Cunquato-Nagami;
Spanish: Naranjita China, Kumquat Oval.

Origin/Distribution

Oval kumquat is native to Southern China. It is cultivated in China, Japan, Korea, Taiwan, southeast Asia, Middle East, Europe, southern United States (notably Florida and California), southern Pakistan, Australia, South Africa.

Agroecology

Oval kumquat thrives in areas with mild warm summer of 25–30°C but is tolerant of frost, sometimes withstanding temperatures as low as −8°C. The trees differ also from other *Citrus* species in that they enter into a period of winter dormancy so profound that they will remain through several weeks of subsequent warm weather without putting out new shoots or blossoms. It does best in full sun but will tolerate light partial shade. It abhors calcareous soils and water-logged soils and does best in well-drained, fertile, mildly acidic soil with pH of 6–6.5.

Edible Plant Parts and Uses

Kumquat fruit eaten is eaten whole fresh or pickled. The rind is sweet and the pulp is acidic. Oval kumquat is also processed into candies, kumquat preserves, marmalade, and jelly. Kumquats appear more commonly in the modern market as a martini garnish, replacing the classic olive. They can also be sliced and added to salads. A liqueur can also be made by macerating kumquats in vodka or other clear spirit. In the Philippines, kumquats are a popular addition to both hot and iced tea.

Botany

Small, slow-growing evergreen, compact, shrubby tree to 3.5 m high with light green branches when young, usually thornless or with few spines. Leaves are alternate borne on narrowly winged, 1.2 cm petioles. Lamina is simple, ovate-lanceolate to elliptic, 5–10 × 3–4.5 cm, base broadly cuneate to nearly rounded, apex subacute to acute, dark-green, glossy above, paler below (Plates 1, 2 and 3). Flowers are sweetly fragrant, pentamerous, on 3–5 mm peduncles, in 1–3 per fascicles in the leaf axils, petals white, 6–8 mm, stamens 20–25, ovary elliptic, style slender with clavate stigma. Fruit ellipsoid to ovoid-ellipsoid, 2.3–2.9 cm across by 3.5–4.3 cm long, orangey–yellow to reddish-orange (Plates 2, 3 and 4), with oil dots, 2–5 seeded, pericarp 2 mm thick, sweet, edible; sarcocarp in 4–5 segments, edible, subacid. Seeds small, ovoid, apex pointed, monoembryonic.

Plate 2 Ripe Nagami kumquats and leaves (top surface)

Plate 1 Leaves and immature Nagami kumquats

Plate 3 Nagami kumquats and leaves (bottom surface)

Plate 4 Harvested Nagami kumquats

Nutritive/Medicinal Properties

The nutrient composition of raw kumquat (*Fortunella* spp.) (exclude 7% seeds) per 100 g edible portion was reported as: water 80.85 g, energy 71 kcal (296 kJ), protein 1.88 g, total lipid 0.86 g, ash 0.52 g, carbohydrate 15.90 g, total dietary fibre 6.5 g, total sugars 9.36 g, Ca 62 mg, Fe 0.86 mg, Mg 20 mg, P 19 mg, K 186 mg, Na 10 mg, Zn 0.17 mg, Cu 0.095 mg, Mn 0.135 mg, vitamin C 43.9 mg, thiamine 0.037 mg, riboflavin 0.090 mg, niacin 0.429 mg, pantothenic acid 0.208 mg, vitamin B-6 0.036 mg, total folate 17 μg, total choline 8.4 mg, vitamin A 290 IU (15 μg RAE), vitamin E (α-tocopherol) 0.15 mg, total saturated fatty acids 0.103 g, 14:0 (myristic) 0.004 g, 16:0 (palmitic) 0.090 g, 18:0 (stearic) 0.004 g; total monounsaturated fatty acids 0.154 g, 16:1 undifferentiated (palmitoleic) 0.021 g, 18:1 undifferentiated (oleic) 0.137 g; total polyunsaturated fatty acids 0.171 g, 18:2 undifferentiated (linoleic) 0.124 g, 18:3 undifferentiated (linolenic) 0.047 g; α-carotene 155 μg, β- cryptoxanthine 193 μg, luetin+zeaxanthin 129 μg (USDA 2011).

The percentage and concentration of total phenolics in *Citrus japonica* unidentified cultivar leaf, peel and juice were reported by Berhow et al. (1998) as: leaf:- 18.2% (0.87 mg/g) flavone/flavonol, 45.6% (1.39 mg/g) flavanone, 0% psoralen, 13.6% (0.2 mg/g) coumarin; flavedo (outer pigmented layer of the peel): 4.7% (0.13 mg/g) flavone/flavonol, 72% (1.25 mg/g) flavanone, 0% psoralen, 4.0% (0.03 mg/g) coumarin; albedo (inner layer of the rind): 0% flavone/flavonol, 95.8% (0.92 mg/g) flavanone, 0% psoralen, 0% coumarin; juice: 4.1% (0.6 mg/g), flavone/flavonol, 91.8% (0.86 mg/g) flavanone, 0% psoralen, 2% (0.01 mg/g) coumarin.

The percentage and concentration of flavanones reported in *Citrus japonica* unidentified cultivar by Berhow et al. (1998) were as follows: 12% eriocitrin (0.4 mg/g), hesperidin 75% (2.2 mg/g), naringin 13% (0.4 mg/g); flavedo: 15% eriocitrin (0.1 mg/g), hesperidin 58% (0.4 mg/g), naringin 25% (0.2 mg/g); albedo:- 3% eriocitrin (traces), hesperidin 92% (1.8 mg/g), narirutin 2% (traces), naringin 4% (0.1 mg/g); juice:- eriocitrin 6% (traces), hesperidin 94% (0.6 mg/g). The percentage and concentration of flavone reported *Citrus japonica* unidentified cultivar by Berhow et al. (1998) were as follows: albideo: diosmin 45% (0.1 mg.g); juice: diosmin 35% (traces).

3′,5′-di-C-β-glucopyranosulphoretin was isolated as a major flavonoid from *F. margarita* (Ogawa et al. 2001). Also 2 flavone glycosides were isolated and found to be acacetin 7-O-neohesperidoside (fortunellin) and acacetin 6 C-glucoside (Harborne and Baxter 1999). dl-limonene (61.58%) and carvone (6.36%) were identified as the major components of the essential peel oil (Yang et al. 2010).

Koyasako and Bernhard (1983) found 120 compounds and identified 71 volatile compounds in the essential oil of kumquat. Some 13 sesquiterpenes, 8 terpenes, 11 alcohols, 1 ketone, 8 aldehydes and 13 esters were identified in kumquat oil. Limonene was the most abundant compound comprising 93% of the whole oil.

One hundred and six (106) volatile compounds, representing 99% of the total composition of kumquat (*Fortunella margarita*) leaf oil were isolated and identified (Quijano and Pino 2009). The leaf oil was characterized by being richer in sesquiterpenes than monoterpene compounds. γ-Eudesmol (19.0%), elemol (18.8%) and β-eudesmol (12.4%) were the most abundant compounds of the leaf oil. In addition, germacrene D (8.9%), β-pinene (8.3%) and linalool (8.2%) were also found as secondary constituents.

Antioxidant Activity

F. margarita peel was found to be a rich source of potentially bioactive polyphenols such as C-glycosylated flavones, O-glycosylated flavones, C-glycosylated flavanones, O-glycosylated flavanones, flavonols, chalcones, phenolic acids and derivatives thereof (Sadek et al. 2009). Polyphenols found in the ethyl acetate fraction of samples of Greek and Egyptian cultivars included: dicaffeoyl ester, apigenin 8-C-rutinoside, naringenin 6-C-rhamnoside 8-C-glucoside, 3′,5′-di-C-β-glycopyranoside (a dihydrochalcone derivative) phloretin, apigenin 6-C-rutinoside, quercetin 3-O-glucosyl benzoate, acacetin 8-C-rutinoside, acacetin 6-C rutinoside, poncirin (a flavanone), sinapic acid derivative. Polyphenols found in the dichloromethane fraction included sinapic and p-courmaric acids, chlorogenic acid, (esters) and their derivatives, acacetin 6-C rutinoside, poncirin, acacetin 7-C rutinoside, and two acacetin flavones 7-O-rhamnosyl glucoside (fortunellin) and acacetin 6-C-rhamnosyl glucoside (isomargaritene). Polyphenols found in the n-butanol fraction included: apigenin 6-C-rutinoside, apigenin 8-C-rutinoside, acacetin 6-C- rutinoside, acacetin 7-O- rutinoside. The ethyl acetate fraction of F. margarita peel showed the best antioxidant activity due to their rich and most diverse phenolic profile. Antiradical activity expressed in mg extract/mg DPPH* was as follows: ethyl acetate fraction (EC_{50} = 1.50, (Greek samples), 0.491 (Egyptian sample); dichloromethane (EC_{50} = 2.40, (Greek), 2.062 (Egyptian); n-butanol fraction (EC_{50} = 2.72, (Greek), 2.95 (Egyptian). Hydroxyl free radical scavenging activity expressed in µg/ml was: ethyl acetate fraction IC_{50} = 1.48 µg/ml (Greek), 1.61ug/ml (Egyptian); dichlormethane fraction IC_{50} = 5.10 µg/ml (Greek), 5.10 µg/ml (Egyptian); n-butanol fraction IC_{50} = 4.29 µg/ml (Greek), 4.83 µg/ml (Egyptian). The results indicated that F. margarita peels may be regarded as a rich source of potentially bioactive polyphenols.

Flavones from Fortunella margarita peel exhibited better reducing ability than the antioxidants 2,6- ditert-butyl-4-methyl phenol (BHT) or propyl gallate (PG) at equivalent concentration and also strong scavenging activity for O2-· and ·OH radicals (Tang et al. 2008). Heating, but not oven drying, enhanced the ability of kumquat peels to suppress nitric oxide (NO) and intercept peroxynitrite, as compared with freeze drying production in lipopolysaccharide-activated RAW 264.7 macrophages (Lin et al. 2008). However, heat treatment and oven drying of kumquat flesh attenuated peroxynitrite-mediated nitrotyrosine formation in albumin; these effects were at least partially attributed to heat-susceptible ascorbate. The essential oil from Fortunella japonica var. margarita peel was found to reduce lipopolysaccharide (LPS)-induced secretion of nitric oxide (NO) in RAW 264.7 cells, indicating that they had antiinflammatory effects (Yang et al. 2010). dl-limonene (61.58%) and carvone (6.36%) were identified as the major components of the essential peel oil.

Antimicrobial Activity

The essential oil from Fortunella japonica var. margarita peel exhibited potent antimicrobial activity against Four human skin pathogenic microorganisms, Staphylococcus epidermidis CCARM 3709, Propionibacterium acnes CCARM 0081, Malassezia furfur KCCM 12679, and Candida albicans including the antibiotic-resistant S. epidermidis (Yang et al. 2010).

Other Pharmacological Activities

Eight flavonoid glycosides were isolated from F. margarita peels and some of them were regarded as important constituents, having hypotensive effect on blood pressure. Their structures were assigned to 6,8-di-C-glucosyl apigenin, 3,6-di-C-glucosyl acacetin, 2″-O-α-L-rhamnosyl-4'-O-methyl-vitexin, 2″-O-α-L-rhamnosyl-4'-O-methyl-isovitexin, 2″-O-α-L-rhamnosyl vitexin, 2″-O-α-L-rhamnosoyl orietin, 2″-O-α-L-rhamnosyl-4'O-methyl orietin and poncirin (Kumamoto et al. 1985).

Two bioactive pyranocoumarins 1 (sesselin) and 3 (xanthyletin) and one prenylated coumarin 2 (suberosin), beside three rare kaurene diterpenes 5–7 were isolated from the roots of *Fortunella margarita* (El-Shafae and Ibrahim 2003). Diterpene 5 was found to be a potent stimulator of uterine contraction; it also caused stimulation of brain activity.

Traditional Medicinal Uses

In Taiwan, folk remedies containing dried kumquats (*Fortunella margarita*) are used to cure inflammatory respiratory disorders.

Other Uses

Nagami kumquat is planted as an ornamental in gardens, parks and as ornamental house plant in patios and terraces.

Comments

Israel is a leading producer of Nagami kumquats for the European market.

Selected References

Berhow M, Tisserat B, Kanes K, Vandercook C (1998) Survey of phenolic compounds produced in *Citrus*. USDA ARS Tech Bull 1856:1–154

El-Shafae AM, Ibrahim MA (2003) Bioactive kaurane diterpenes and coumarins from *Fortunella margarita*. Pharmazie 58(2):143–147

Facciola S (1990) Cornucopia. A source book of edible plants. Kampong Publ, Vista, 677 pp

Harborne JB, Baxter H (1999) The handbook of natural flavonoids, vol 1–2. Wiley, England

Koyasako AA, Bernhard RA (1983) Volatile constituents of the essential oil of kumquat. J Food Sci 48(6):1807–1812

Kumamoto H, Matsubara Y, Irzuka Y, Okamoto K, Yokoi K (1985) Structure and hypotensive effect of flavonoid glycosides in kinkan (*Fortunella japonica*) peelings. Agric Biol Chem 49:2613–2618

Lin CC, Hung PF, Ho SC (2008) Heat treatment enhances the NO-suppressing and peroxynitrite-intercepting activities of kumquat (*Fortunella margarita* Swingle) peel. Food Chem 109(1):95–103

Mabberley DJ (1997) A classification for edible *Citrus*. Telopea 7(2):167–172

Morton J (1987) Kumquat. In: Fruits of warm climates. Julia F. Morton, Miami, pp 182–185

Nicolosi E, Deng ZN, Gentile A, Malfa SL, Continella G, Tribulato E (2000) *Citrus* phylogeny and genetic origin of important species as investigated by molecular markers. Theor Appl Genet 100(8):1155–1166

Ogawa K, Kawasaki A, Omura M, Yoshida T, Ikoma Y, YanoM(2001)3′,5′-Di-C-beta-glucopyranosylphloretin, a flavonoid characteristic of the genus *Fortunella*. Phytochemistry 57(5):737–742

Porcher MH et al (1995–2020) Searchable world wide web multilingual multiscript plant name database. Published by The University of Melbourne, Australia. http://www.plantnames.unimelb.edu.au/Sorting/Frontpage.html

Purseglove JW (1968) Tropical crops: dicotyledons, vol 1 & 2. Longman, London, 719 pp

Quijano CE, Pino JA (2009) Volatile compounds of kumquat (*Fortunella margarita* (Lour.) Swingle) leaf oil. J Essent Oil Res 21(3):194–196

Sadek ES, Makris DP, Kefalas P (2009) Polyphenolic composition and antioxidant characteristics of kumquat (*Fortunella margarita*) peel fractions. Plant Foods Hum Nutr 64(4):297–302

Swingle WT (1946) The botany of *Citrus* and its wild relatives of the orange subfamily (family Rutaceae, subfamily Aurantioideae). In: Webber HJ, Batchelor LD (eds) The *Citrus* industry, vol 1, History, botany and breeding. University of California Press, Berkeley, pp 129–474, 1028 pp

Swingle WT, Reece PC (1967) The botany of *Citrus* and its wild relatives. In: Reuther W, Webber HJ, Batchelor LD (eds) The *Citrus* industry, vol 1, History, world distribution, botany, and varieties. University of California Press, Riverside, pp 190–430

Tanaka T (1922) *Citrus* fruits of Japan; with notes on their history and the origin of varieties through bud variation. J Hered 13:243–253

Tang QY, Zhou YF, Zhu YC, Yun X (2008) Extraction of flavone compounds from *Fortunella margarita* peel and its antioxidation in vitro. Nongye Gongcheng Xuebao 24(6):258–261 (in Chinese)

U.S. Department of Agriculture, Agricultural Research Service (2011) USDA National nutrient database for standard reference, Release 24. Nutrient data laboratory home page. http://www.ars.usda.gov/ba/bhnrc/ndl

Yang EJ, Kim SS, Moon JY, Oh TH, Baik JS, Lee NH, Hyun CG (2010) Inhibitory effects of *Fortunella japonica* var. *margarita* and *Citrus sunki* essential oils on nitric oxide production and skin pathogens. Acta Microbiol Immunol Hung 57(1):15–27

Zhang DX, Mabberley DJ (2008) *Citrus* linnaeus. In: Wu ZY, Raven PH, Hong DY (eds) Flora of China, vol 11, Oxalidaceae through Aceraceae. Science Press/Missouri Botanical Garden Press, Beijing/St. Louis

Citrus japonica 'Polyandra'

Scientific Name

Citrus japonica Thunb. 'Polyandra'.

Synonyms

Atalantia polyandra Ridley, *Citrus polyandra* (Ridley) Burkill, non Tanaka, *Citrus swinglei* Burkill ex Harms, *Clymenia polyandra* (Ridley) Swingle, *Fortunella polyandra* (Ridley) Tanaka, *Fortunella swinglei* Tanaka.

Family

Rutaceae.

Common/English Names

Hedge Lime, Malayan Kumquat, Long-Leaved Kumquat, Swingle's Kumquat.

Vernacular Names

Chinese: Chang ye jin gan, Chang ye jin ju, Ma la ya jin gan;
Japanese: Marai kinkan, Nagaba kinkan;
Malaysia: Limau pagar;
Spanish: Kumquat Malayo;
Thai: Somchit.

Origin/Distribution

Malayan kumquat is indigenous to Peninsular Malaya. It is also found in Hainan, south China. In Peninsular Malaysia, it is more abundant in Perak and Malacca.

Agroecology

Malayan kumquat grows well in the warm tropical and subtropical areas. However, it is cold sensitive and cannot be planted in frost prone areas. It thrives in full sun.

It is kept indoor as houseplants in colder areas.

Edible Plant Parts and Uses

The fruits are edible and are eaten along with the skin. The fruits are used for the preparation of drinks and preserves or as flavouring and are sold as a specialty item in local markets of Malaysia.

Botany

A thornless, evergreen large shrub 1–2 m high or small tree 3–5 m high (Plate 3). Branches are angular when young, rounded when older. Leaves are simple, alternate, slender, lanceolate, 3–10 cm long, dark green, pellucid-dotted on the underside,

Plate 1 Flower and immature fruit

Plate 2 Leaves narrowly winged, simple, lanceolate and fruits

Plate 3 Prolific bearing tree

Plate 4 Close-view of ripening fruits and leaves

with acute tip and tapering cuneate base and borne on narrowly winged green petioles (Plates 1, 2, 3 and 4). Flowers borne singly or in clusters of 2–3 in axils of leaves, hermaphrodite, 5-merous, small, white, sweet scented; stamens 20–24, cohering irregularly in bundles; disc cylindrical, ovary 4–6 loculed, stigma capitates (Plate 1). Fruit ovoid to globose, 2.5–3 cm across, glabrous, thin-skinned, green ripening to glossy golden yellow to orange (Plates 1, 2, 3 and 4), 4–6 segmented, seed glabrous and ovoid with green embryo.

Nutritive/Medicinal Properties

No information has been published on the nutritive value of the fruit.

The fruit was found to contain flavonoids such as auraptene (Ogawa et al. 2000) and 3′,5′-di-C-β-glucopyranosylphloretin in the fruit peel and leaves (Ogawa et al. 2001).

Other Uses

Malayan kumquat is popularly planted as an ornamental plant in gardens, potted plant, hedge and as a roadside shrub.

Comments

This kumquat is propagated from seeds or air-layerings.

Selected References

Burkill IH (1966) A dictionary of the economic products of the Malay Peninsula. Revised reprint. 2 vols, Ministry of Agriculture and Co-operatives, Kuala Lumpur, vol 1 (A–H), pp. 1–1240, vol 2 (I–Z), pp 1241–2444

Mabberley DJ (1997) A classification for edible *Citrus*. Telopea 7(2):167–172

Mabberley DJ (2004) *Citrus* (Rutaceae): a review of recent advances in etymology, systematics and medical applications. Blumea 49:481–488

Molesworth Allen B (1967) Malayan fruits. An introduction to the cultivated species. Moore, Singapore, 245 pp

Nicolosi E, Deng ZN, Gentile A, La Malfa S, Continella G, Tribulato E (2000) *Citrus* phylogeny and genetic origin of important species as investigated by molecular markers. Theor Appl Genet 100(8):1155–1166

Ogawa K, Kawasaki A, Yoshida T, Nesumi H, Nakano M, Ikoma Y, Yano M (2000) Evaluation of auraptene content in *Citrus* fruits and their products. J Agric Food Chem 48(5):1763–1769

Ogawa K, Kawasaki A, Omura M, Yoshida T, Ikoma Y, Yano M (2001) 3',5'-Di-C-beta-glucopyranosylphloretin, a flavonoid characteristic of the genus *Fortunella*. Phytochemistry 57(5):737–742

Stone BC (1972) Rutaceae. In: Whitmore TC (ed) Tree flora of Malaya, vol 1, Longman Malaya. Kuala Lumpur, Malaysia, pp 367–387

Swingle WT, Reece PC (1967) The botany of *Citrus* and its wild relatives. In: Reuther W, Webber HJ, Batchelor LD (eds) The *Citrus* industry, vol 1, History, world distribution, botany, and varieties. University of California, Riverside, pp 190–430

Tanaka T (1954) Species problem in *Citrus*. A critical study of wild and cultivated units of *Citrus*, Based upon field studies in their native homes. Japanese Society for the Promotion of Science, Ueno, 152 pp

Verheij EWM, Coronel RE (1992) Plant resources of South-East Asia No. 2 Edible fruits and nuts. Prosea, Bogor, p. 170, 377

Citrus latifolia

Scientific Name

Citrus x latifolia (Yu. Tanaka) Tanaka.

Synonyms

Citrus × aurantiifolia subsp. *latifolia* (Tanaka ex Yu. Tanaka) S. Ríos et al., *Citrus × aurantiifolia* (Christmann) Swingle var. *latifolia* Tanaka ex Yu. Tanaka (basionym), *Citrus aurantiifolia* (Christm. et Panz.) Swingle var. *Tahiti, Citrus latifolia* (Yu. Tanaka) Tanaka.

Family

Rutaceae.

Common/English Names

Bearss Lime, Persian Lime, Seedless Lime, Tahiti Lime, Tahitian Lime.

Vernacular Names

French: Limettier, Limettier De Perse, Limettier Tahiti;
German: Persische Limette, Tahitilimette;
Spanish: Lima De Persia, Lima Tahiti.

Origin/Distribution

The origin of the Tahiti lime is unknown. It is known only in cultivation. It is cultivated commercially in Tahiti, New Caledonia, Florida, California, Mexico, Brazil, Venezuela, Portugal and Australia. It is believed that the lime was introduced into the Mediterranean region by way of Iran (formerly called Persia) from Tahiti.

Agroecology

Tahitian lime thrives in tropical or subtropical environments. It will tolerate cold temperatures but not frost and needs cold protection as seen in Florida. Even in southern Florida, drastic drops in temperature have made it necessary to protect Tahiti lime groves with wind machines or overhead sprinkling. It grows best in full sun on well-drained, sandy or calcareous soils but will not withstand water-logged soils or heavy clays.

Edible Plant Parts and Uses

Fresh fruit is used as garnish for meats and drinks. The fruit is also processed into marmalade and candied peels. Fresh juice is used in beverages especially limeade, cordials, marinating fish and meats and seasoning many foods. The juice is frequently used as an alternative to vinegar in

Plate 1 Leaves and flowers

dressings and sauces. Frozen and canned juice is used in similar ways. Tahiti lime is a great accompaniment with avocado served in the form of wedges. Since the late 1960s, oil from Tahiti lime peel has been accepted by the trade and produced in quantity as a by-product of the juice-extraction process. It is utilized for enhancing lime juice and for most of the other purposes for which Mexican lime peel oil is employed. Flowers petals are edible.

Plate 2 Close view of flowers

Botany

A vigorous, unarmed (thornless) shrub or small tree, 2–5 m high, with widespread, drooping branches. Young shoots are purplish. Leaves are unifoliolate, alternate, broad- lanceolate, medium green, glabrous, pellucid dotted, with acute apex and acute base, and slightly crenulate margin (Plate 1). Flowers are axillary, hermaphrodite, white, fragrant, solitary or in a few-flowered clusters (Plates 1 and 2). Fruit is a berry (hesperidium), oval, obovate, oblong or short-elliptical, usually rounded at the base and apex with papilla, 5–7.5 cm by 4–6.25 cm wide, vivid green to pale yellowish green and seedless (Plates 3, 4, and 5). Pulp is pale greenish in 10

Plate 3 Mature Tahitian lime

Plate 4 Greenish and seedless pulp of Tahitian lime

Plate 5 Tahitian lime fruit on sale in the market

segments (Plate 5), tender and acid. The pollens are not viable.

Nutritive/Medicinal Properties

The proximate nutrient composition of *Citrus latifolia* fruit per 100 g edible portion (16% refuse – peels and seeds) was reported as: water 88.26 g, energy 30 kcal (126 kJ), protein 0.70 g, total lipid 0.2 g, ash 0.3 g, carbohydrates 10.54 g, total dietary fibre 2.8 g, total sugars 1.69 g, minerals – Ca 33 mg, Fe 0.6 mg, Mg 6 mg, P 18 mg, K 102 mg, Na 2 mg, Zn 0.11 g, Cu 0.065 mg, Mn 0.008 mg, Se 0.4 μg, vitamins – vitamin C (ascorbic acid) 29.1 mg, thiamin 0.03 mg, riboflavin 0.02 mg, niacin 0.2 mg, pantothenic acid 0.217 mg, vitamin B-6 0.043 mg, total folate 8 μg, total choline 5.1 mg, vitamin A 50 IU, vitamin E (α-tocopherol) 0.22 mg, vitamin K (phylloquinone) 0.6 μg, lipids – total saturated acids 0.022 g, 14:0 (myristic acid) 0.001 g, 16:0 (palmitic acid) 0.020 g, 18:0 (stearic acid) 0.001 g; total monounsaturated fatty acids 0.019 g, 16:1 undifferentiated (palmitoleic acid) 0.003 g, 18:1 undifferentiated (oleic acid) 0.016 g; total polysaturated fatty acids 0.055 g, 18:2 undifferentiated (linoleic acid) 0.036 g, 18:3 undifferentiated (linolenic acid) 0.019 g; amino acids – tryptophan 0.003 g, lysine 0.014 g, methionine 0.002 g; beta carotene 30 μg (USDA, 2010). The fruit also contained organic acids – citric acid 4.3 g, malic acid 0.6 g, sugars 1.2 g – fructose 0.5 g, sucrose 0.2 g, and glucose 0.5 g (Wills et al. 1985).

In the Tahitian lime oil, 76 components were identified, among which isopiperitenone, trans-dodec-2-enol, trans-α-bisabolene, α-cubebene, and β-cubebene had not been previously reported in the lime oil (Njoroge et al. 1996). Monoterpene hydrocarbons (89.9%) formed the main chemical group of the oil, followed by carbonyls (3.8%), sesquiterpene hydrocarbons (2.7%), esters (2.5%), and alcohols (1.0%). The abundant constituents of the lime oil were limonene (52.2%), γ-terpinene (17.0%), β-pinene (13.0%), citral (3.5%), α-pinene (3.2%), sabinene (2.0%), neryl acetate (1.4%), geranyl acetate (1.0%), α-bergamotene (0.8%), and β-bisabolene (0.8%). The volatile components of lime oil from *Citrus latifolia* by hydrodistillation and supercritical extraction (in brackets) were as follows: δ-limonene 47.5% (48.9%), δ-terpinene 12.3% (17%), β-pinene 12.4%, (14.5%), β-bisabolene 1.8% (3.3), α-pinene 1.9% (2.7%), sabinene 1.6% (2.2%), citral 11.1% (1.8%), myrcene 1.2% (1.4%), cis-α-bergamotene 0.3% (1.2%), geranial 6.4% (1.1%), α-thujune 0.5% (0.8%), α-terpinolene 0.6% (0.7%), terpinene-4-ol 1.2% (0.5%), α-terpineol 2.2% (0.4%), geranil acetate 0.3% (0.4%), α-terpinene 0.3% (0.4%), linalool 1.3% (0.2%). In addition, para-cimene (1.0%), nerol (0.7%) and geraniol (0.3%) were obtained by hydrodistillation but not by supercritical extraction (Atti-Santos et al. 2005). The major components found in petitgrain essential oils of *Citrus latifolia* Tanaka (two varieties, Lime Tahiti and Lime de Perse) and *Citrus limon* (L.) Burm.f. (three varieties, Meyer, Eureka, Doux), were β-pinene, sabinene, limonene, citronellal, linalool, neral, geranial and neryl acetate, in varying amounts according to variety (Smadja et al. 2005).

Twenty-six compounds were identified in the essential oil of *C. latifolia* leaves comprising hydrocarbons, alcohols, aldehydes, ketones, esters and others (Jazet Dongmo et al. 2008). The monoterpenes (51.64%) were dominant represented by mainly by limonene (45.76%) and included Z-β-ocimene (1.66%), camphene (1.32%), myrcene (1.25%), sabinene (0.93%), isocamphene (0.39%) and α-pinene 0.33%). Oxygenated monterpenes (43.94%) were mainly represented by geranial (13.25) and neral (10.35%); others included geraniol (3.91%), neryl acetate (3.64%), nerol (3.34%), *E*-pinocarveol (1.69%), linalol (1.61%), citronellal (0.96%), borneol (0.52%), terpene n-4-ol (0.46%), mrytenol (0.28%), citronellyl acetate (0.25%). Sesquiterpenes (2.49%) were mainly represented by β-caryophyllene (1.49%) and others such as β-elemene (0.48%), β-bisabolene (0.27%), germacrene D (0.25%). Other compounds included caryophyllene oxide 0.29% and farnesyl acetate 0.59%.

The total dietary fibre (TDF) contents of Mexican lime (*Citrus aurantiifolia*) and Persian lime (*Citrus latifolia*) were high; 70.4% and 66.7%, respectively (Ubando-Rivera et al, 2005). Both lime peel varieties had an appropriate ratio of soluble/insoluble fractions. The water-holding capacities (WHC) of DF (dietary fibre) concentrates are high (6.96–12.8 g/g). The WHC was related to the soluble dietary fibre (SDF) which was higher in the DF concentrate of Mexican lime. DF concentrates of Persian lime peel had greater polyphenol contents than those of Mexican lime peel. The polyphenols associated with the DF in both lime peel varieties showed a good antioxidant activity as evaluated by azino-bis (3-ethylbenzothiazoline-6-sulfonic acid) ABTS radical-scavenging activity, α,α-diphenyl–picrylhydrazyl (DPPH) and β-carotene-linoleic acid antioxidant assay. From a nutritional standpoint, DF lime concentrates were suggested to be suitable as food additives.

An earlier study reported that fibres from Valencia orange and Persian lime peels had a high total dietary fibre content (61–69%) with an appreciable amount of soluble fibre (19–22%) (Larrauri et al. 1996). The concentration of antioxidant [AA_{50}] required to achieve a 50% inhibition of oxidation of linoleic acid at 40°C was measured using the ferric-thiocyanate method. Lime peel fibre [AA_{50}] was half the value of DL-α-tocopherol and 23 times lower than orange peel fibre; the [AA_{50}] of commercial butylated hydroxyanisole (BHA) was half the value of lime fibre. The higher the [AA_{50}], the lower the antioxidant capacity. HPLC analyses of the polyphenols extracted from orange and lime peels fibres showed the presence of caffeic and ferulic acids, as well as naringin, hesperidin and myricetin in both fruit fibres. The different antioxidant power of these fibres could be in part explained by the presence in lime peel fibre of ellagic acid, quercetin and kaempferol which are strong antioxidant.

C. latifolia leaf essential oil (SC_{50} = 9930 mg/l) was 1414 times less active in DPPH scavenging capacity than the commercial antioxidant butylated hydroxyl toulene (BHT) (SC_{50} = 7.02 mg/l) (Jazet Dongmo et al. 2008).

The leaves or an infusion of the crushed leaves may be applied to relieve headache. Lime juice, administered quickly, was found to be an effective antidote for the painful oral irritation and inflammation that result from biting into aroids such as *Dieffenbachia spp., Xanthosoma spp., Philodendron spp.,* and their allies. Lime juice has also been applied to relieve the effects of stinging corals. However, excessive exposure to the peel oil of the Tahiti lime may cause dermatitis.

Other Uses

Tahitian lime juice can be used a rinse after shampooing the hair or used as a fresh face lotion. The juice can be used for cleaning the inside of coffee pots and tea kettles. Diluted lime juice will dissolve calcium deposits overnight. A whole lime can be grind in the electric garbage-disposal to eliminate unpleasant odor.

Jazet Dongmo et al. (2008) found that after 40 days of incubation on oil-supplemented medium, the growth of the postharvest pathogen, *Phaeoramularia angolensis* was totally inhibited by 1,600 mg/l of *C. latifolia* leaf essential oil.

Comments

Citrus x latifolia is believed to be a hybrid of the Mexican lime, *Citrus x aurantiifolia* and lemon, *Citrus x limon*, or, the Mexican lime, *Citrus x aurantiifolia* and the citron, *Citrus medica* and is genetically a triploid (Mabberley 2004).

Selected References

Atti-Santos AC, Rossato M, Serafini LA, Cassel E, Moyna P (2005) Extraction of essential oils from lime (*Citrus latifolia* Tanaka) by hydrodistillation and supercritical carbon dioxide. Braz Arch Biol Technol 48(1):155–160

Jazet Dongmo PM, Tatsadjieu NL, Tchinda Sonwa E, Kuate J, Amvam Zollo PH, Menut C (2008) Antiradical potential and antifungal activities of essential oils of the leaves of *Citrus latifolia* against *Phaeoramularia angolensis*. Afr J Biotechnol 7(22):4045–4050

Larrauri JA, Rupérez P, Bravo L, Saura-Calixto F (1996) High dietary fibre powders from orange and lime peels: associated polyphenols and antioxidant capacity. Food Res Int 29(8):757–762

Mabberley DJ (1997) A classification for edible *Citrus*. Telopea 7(2):167–172

Mabberley DJ (2004) *Citrus* (Rutaceae): a review of recent advances in etymology, systematic and medical applications. Blumea 49:481–489

Morton JF (1987) Tahiti lime. In: Fruits of warm climates. Julia F. Morton, Miami, pp 172–175

Njoroge SM, Ukeda H, Kusunose H, Sawamura M (1996) Japanese sour *Citrus* fruits. Part IV. Volatile compounds of naoshichi and Tahiti lime essential oils. Flav Frag J 11(1):25–29

Rouse RE (1988) Major *Citrus* cultivars of the world as reported from selected countries. HortScience 23:680–684

Smadja J, Rondeau P, Sing ASC (2005) Volatile constituents of five *Citrus* Petitgrain essential oils from Reunion. Flav Fragr J 20(4):399–402

Tanaka T (1954) Species problem in *Citrus*. A critical study of wild and cultivated units of *Citrus*, based upon field studies in their native homes. Japanese Society for the Promotion of Science, Ueno, 152 pp

Ubando-Rivera A, Navarro-Ocaña A, Valdivia-López MA (2005) Mexican lime peel: Comparative study on contents of dietary fibre and associated antioxidant activity. Food Chem 89(1):57–61

U.S. Department of Agriculture, Agricultural Research Service (USDA) (2011) USDA National nutrient database for standard reference, Release 24. Nutrient data laboratory home page. http://www.ars.usda.gov/ba/bhnrc/ndl

Webber HJ (1946) Cultivated varieties of *Citrus*. In: Webber HJ, Batchelor LD (eds) The *Citrus* industry, vol 1, History, botany and breeding. University of California Press, Berkeley, pp 475–668, 1028 pp

Wills RBH, Lim JSK, Greenfield H (1985) Composition of Australian foods. 28. *Citrus* fruit. Food Technol Aust 37:308–310

Citrus maxima

Scientific Name

Citrus maxima (Burman) Merrill.

Synonyms

Aurantium maximum Burman, *Aurantium decumanum* (Linnaeus) Miller; *Citrus × aurantium* Linnaeus subsp. *decumana* (Linnaeus) Tanaka, *Citrus × aurantium* var. *decumana* Linnaeus, *Citrus × aurantium* f. *grandis* (Linnaeus) Hiroe, *Citrus × aurantium* var. *grandis* Linnaeus, *Citrus costata* Rafinesque, *Citrus decumana* (Linnaeus) Linnaeus, *Citrus grandis* (Linnaeus) Osbeck; *Citrus grandis* var. *pyriformis* (Hasskarl) Karaya, *Citrus grandis* var. *sabon* (Siebold ex Hayata) Hayata, *Citrus kwangsiensis* Hu, *Citrus medica* Linnaeus subf. *pyriformis* (Hasskarl) Hiroe, *Citrus obovoidea* Yu. Tanaka, *Citrus pampelmos* Risso, *Citrus pompelmos* Risso, *Citrus pyriformis* Hasskarl, *Citrus sabon* Siebold ex Hayata.

Family

Rutaceae

Common/English Names

Forbidden fruit, Paradise apple, Pomelo, Pummelo, Shaddock.

Vernacular Names

Burmese: Shaukpan, Shouk-Ton-Oh;
Chinese: You, Zhu Luan;
Czech: Citroník Největší;
Danish: Pompelmus;
Dutch: Pompelmoes;
Eastonian: Pomelipuu, Vili: Pommel;
Fijian: Moli Kana, Soco Vi Kana;
Finnish: Pummelo;
French: Pamplemousse, Pamplemoussier, Pomme D'Adam, Shadek;
German: Adamsapfel, Pampelmuse, Paradiesapfel, Pomelo, Pompelmus, Pumelo, Riesenorange;
Guam: Kahet Magas, Lalangha;
Hawaiian: Jabon;
Hungarian: Óriás Citrancs, Óriás Narancs;
India: Chakotra (Bengali), Batawi Nimbu, Cakotaraa, Sadaphal (Hindu); Toranji (Konkani), Pamparamasan (Malayalam), Panis, Nobab (Manipuri), Madhukarkati, Mahanimbu (Sanskrit), Gadarangai, Pambalimasu (Tamil), Pampara (Telugu);
Indonesia: Boh Giri, Munter, Nagiri (Aceh), Limau Kesumbe (Alas), Limu Kasumba, Limu Sumba, Limu Wako, Munte Bangko, Munte Kapes, Munte Kasumba, Munte Langhow, Munte Prangi, Munte Wangko, Muntoi Kasumba (Alfoersch, North Sulawesi), Muda Tapo (Alor), Jeruk Muntis, Jeruti Muntis (Bali), Lemo Gola (Baree), Unte Bolon, Unte Godang, Unte Susu (Batak), Jodi (Beak), Limau Serdadu (Besemah), Lemo Walanda (Boengkoe), Limu

Kasumba (Boeol), Fusi Sambute, Pohit Bagut, Puhat Wolanda (Boeroe), Lemo Kaluku, Lemo Pakasumba, Lemon Cilla (Bugis), Muda Kokor (East Ceram), Ahusi A'ate, Umusi Ane-Ane, Susi Ma'a (West Ceram), Musi Walanta, Nusi, Usi Anan (South Ceram), Kelewuku, Kerowohu, Limau Gulong (Dyak), Mudeh (Flores), Limu Banga, Limu Bongo (Gorontalo), Bitie O Haangkari, Sangkari, Wama Hangkari, Wama Lalgo (N. Halmaheira), Fafo Kastela (S. Halmaheira), Jeruk Adas, Jeruk Bali, Jeruk Dalima, Jeruk Gulung, Jeruk Karag, Jeruk Rawa, (Java), Limau Basar (Kambang), Roin Lai (Kai), Paka Pokor (Kisar), Munde Kina (Lalaki), Limau Balak, Limau Kibau (Lampang), Sapruake (Leti), Jeruk Machan, Jeruk Bhali (Madurese), Lemo Kaluku Lemo Pakasumba (Makassar), Jeruk Besar (Malay), (Lemo Kaiyang, Lemo Kasale, Lemo Niande (Mandar), Muntei (Mentawai), Limau Gadang (Minangkabau), Lemo Benu (Mori), Dima Kasumba, Dima Sebua (Nias), Angra Jodi (Noefor), Usil Awalo (Oelias), Yodi (Papua), Munde, Mune (Roti), Munte Kanreang (N. Salajar), Lemo Kanr'eang (S. Salajar), Lemu Gaguwa, Limu, Kakanengang, Limu Kapala, Limu Ralabo (Sangir), Jeruti (Sasak), Jeru Kabau, Jeru Worena (Sawoe), Limu Sorodadu (Serawaj), Alimau Mames, Bolafu, Giri (Simaloer), Muda Belin, Muda Karabaw (Solor), Limau Balak, Liamu Kibak, Limau Gadang (Sumatra), Jeruk Bali, Jeruk Dalima, Jeuk Gede (Sundanese), Plumpa (Talaud), Lemo Lolamo (Ternate), Jodi Lamo (Tidore), Dambua, Dambuak, Lelo Boko (Timur);
Italian: Pompelmo;
Japanese: Buntan, Zabon;
Khmer: Kooch Thlong;
Korean: Dangyuja;
Laos: Ph'uk;
Malaysia: Jambua, Limau Bali, Limau Besar, Limau Betawi, Limau Kedangsa, Liamau Masam, Limau Gading, Limau Tebu (Semang), Pomelo;
Nepal: Sangkatra;
Papua New Guinea: Muli;
Peru: Toronja De La Selva;
Philippines: Lukban (Bikol), Panubang, Taboyog (Bontok), Lubban (Ibanag), Luban (Ifugao), Lukban, Sua (Iloko), Cabugao (Panay Bisaya), Lukban, Suhâ (Tagalog), Gunal (Tinggian);
Polish: Pompela;
Portuguese: Jamboa;
Rotuman: Kurkura;
Samoan: Moli 'Ai Suka, Moli Meleke, Moli Suka, Moli Tonga;
Spanish: Pampelmusa; Toronja;
Swedish: Pompelmus;
Thai: Ma-O, Som-Oh;
Tongan: Moli Tonga;
Vietnamese: Bưởi (General), Biròi, Bòng, Mác Pue (Tày), Ma Pôc Diang (Dao), Plài Plînh (K'ho).

Origin/Distribution

The pummelo is indigenous to southeast Asia (Thailand, Malaysia). It was reported to grow wild on river banks in the Fiji and Friendly Islands. It was reportedly introduced into China around 100 B.C. It is widely cultivated in southern China (Kwang-tung, Kwangsi and Fukien Provinces) and especially in southern Thailand; also in Taiwan and southernmost Japan, southern India, Malaysia, Indonesia, New Guinea and Tahiti. Pummelo was believed to have been brought to the New World in late seventeenth Century by a Captain Shaddock who stopped at Barbados on his way to England. By the end of the seventeenth century, the fruit was being cultivated in Barbados and Jamaica.

Agroecology

The pomelo thrives in the tropical lowlands up to 400 m altitude. It thrives in full sun in tropical climates with mean annual monthly temperatures of 25–32°C and rainfall of 1,500–2,500 mm with 3–4 months of dry period as found in the growing regions of southern Thailand and in Malaysia. Pomelo is adaptable to a wide range soils from coarse sand to heavy clay. However, the tree prefers deep, medium-textured, fertile soils. In southern Thailand and Vietnam, pomelo does well in the rich silt and sand overlying the organically enriched clay loam of the flood plain, and that it is highly tolerant of brackish water pushed inland by high tides and is usually grown one elevated bunds. In Malaysia, it flourishes in the flat, inland calcareous soils around limestone hills and tin tailings (pure sands) as found in Perak and Selangor. It grows well on the sandy tin tailings with frequent supplementation of organic manure.

Edible Plant Parts and Uses

Ripe, sweet pomelo segments is excellent when eaten fresh or made into marmalade, juice or beverage. The skinned segments can be broken apart and used in salads and desserts or made into preserves. The fruit is also used as a source of pectin. There are sweet white-flesh and sweet pink flesh varieties. During the Chinese autumn (mooncake festival) festival, pomelos are an important fruit feature used in offerings to the deity and is much consumed especially the sweet red flesh varieties. The thick rind can be candied, preserved by drying and salted. The fresh or dried rind can be used as an ingredient in cooking meat stews or soup. Segments of acid varieties are used for rujak in Indonesia.

Pomelo essential oil is used as the major flavour ingredients; the leaves are used as a food flavouring in meat and fish dishes, and dried leaves are brewed in water as a drink in Taiwan (Wang et al. 2008). The essential oil is also used for pharmaceutical medicine and in the cosmetic industry.

Plate 1 Pomelo tree

Plate 2 Pomelo leaves with broadly winged petioles

Plate 3 Large subglobose pomelo fruit

Botany

A small evergreen tree, 5–11 m high with spreading, low branches (Plate 1) and pubescent, ridged young branchlets. Spines are usually blunt when present. Leaves are broadly ovate or elliptic, 9–16 × 4–8 cm or larger, thick, dark green, pellucid-dotted, pinnatinerved, base rounded, apex rounded to obtuse and sometimes mucronate, margin crenulated or undulating and borne on broadly winged, green, 2–4 cm long petioles (Plate 2). Flowers, bisexual, pentamerous, large 2–3 cm across, solitary or in few flowered axillary racemes; flower buds purplish or milky white. Calyx 5-lobed; petals 1.5–2 cm, creamy-white; stamens 20–35; ovary with 11–16 loculi and a long thick style. Fruit a globose, oblate, pyriform, or broadly obconic berry, 10–20(–30) cm in diameter (Plates 1, 3, 4, and 5); peel (pericarp) consists of the outermost pigmented layer, flavedo which is densely glandular dotted, greenish-yellow or yellowish, and inner spongy layer, albedo, white

Plate 4 Large, pyriform pomelo fruits

Plate 5 Harvested pomelo fruits

Plate 6 Fresh pomelo segments on sale in a local market

or pinkish, 2–4 cm thick; sarcocarp with 10–15(–19) segments, with large, pale yellow, reddish or pink pulp-vesicles (Plate 6), filled with sweet, bland or acid sweet juice. Seeds few to numerous (>100) in seedy varieties or seedless, large, plump, irregularly shaped, ridged, yellowish, monoembryonic with white cotyledons.

Nutritive/Medicinal Properties

Nutrient composition of fresh, raw pummelo per 100 g edible portion was reported as follows (refuse 44% skin, membrane and seeds): water 89.10 g, energy 38 kcal (159 kJ), protein 0.76 g, total fat 0.04 g, ash 0.48 g, carbohydrate 9.62 g, total dietary fibre 1.0 g, Ca 4 mg, Fe 0.11 mg, Mg 6 mg, P 17 mg, K 216 mg, Na 1 mg, Zn 0.08 mg, Cu 0.048 mg, Mn 0.017 mg, vitamin C 61 mg, thiamine 0.034 mg, riboflavin 0.027 mg, niacin 0.220 mg, vitamin B-6 0.036 mg, β-cryptoxanthin 10 μg, and vitamin A 8 IU (USDA 2011).

Other Fruit Phytochemicals

Pummelo fruit was found to contain per mg/100 g dry weight basis 0.64 mg alkaloids, 0.26 mg flavonoids, 0.03 mg tannins, 0.08 mg phenols and 0.21 mg saponins (Okwu and Emenike 2006). Juice contains per 100 g basis 36.078 mg ascorbic acid, 0.09 mg thiamine, 0.06 mg riboflavin, and 3.6 mg niacin.

A loading-plot analysis revealed that sucrose, glucose, oxaloacetic acid and citric acid were dominant in mature flesh *Citrus grandis* fruit, while naringin, tyramine, proline and alanine were dominant in immature fruit samples (Cho et al. 2009). Three biologically active coumarins imperatorin, meranzin and meranzin hydrate were isolated from the fruit peel of *Citrus maxima* (Feng and Pei 2000; Teng et al. 2005a).

The percentage and concentration of total phenolics in *Citrus maxima* cv Kao Panne leaf, peel and juice were reported by Berhow et al. (1998) as: leaf:- 11.3% (1.08 mg/g) flavone/flavonol, 83.2% (5.06 mg/g) flavanone, 0% psoralen, 3.3% coumarin (0.10 mg/g); flavedo (outer pigmented layer of the peel): 1.4% (0.08 mg/g) flavone/flavonol, 47.6% (1.66 mg/g) flavanone, 19% psoralen, 26.4% (0.45 mg/g) coumarin; albedo (inner layer of the rind): 0% flavone/flavonol, 94.4% (5.67 mg/g) flavanone, 1.6% psoralen, 1.7% (0.05 mg/g) coumarin; juice: 0% flavone/flavonol, 95.5% (3.03 mg/g) flavanone, 0.6% psoralen and 3.5% (0.05 mg/g) coumarin.

The percentage and concentration of flavanones reported in *Citrus maxima* cv Kao Panne by Berhow et al. (1998) were as follows: leaf:- hesperidin 3% (0.2 mg/g), naringin 65% (3.3 mg/g), naringin–4′–glucoside 1% (traces), neohesperidin 2% (0.1 mg/g), naringin–6″–malonate (open form) 29% (1.4 mg/g); flavedo: hesperidin 7% (0.1 mg/g), naringin 51% (0.8 mg/g), naringin–4′–glucoside 12% (0.2 mg/g), neohesperidin 1% (traces), naringin–6″–malonate (open form) 28% (0.5 mg/g); albedo: hesperidin 2% (0.1 mg/g), naringin 76% (4.3 mg/g), naringin–4′–glucoside 4% (0.2 mg/g), neohesperidin 1% (0.1 mg/g), naringin–6″–malonate (open form) 17% (1.0 mg/g); juice: 1% (0.1 mg/g) hesperidin, 1% (trace) narirutin, neoeriocitrin 1% (traces), naringin 77% (2.3 mg/g), naringin–4′–glucoside 2% (traces), neohesperidin 1% (traces), naringin–6″–malonate (open form) 16% (0.5 mg/g). The percentage and concentration of flavone reported in *Citrus maxima* cv Kao Panne leaves by Berhow et al. (1998) were as follows: rhoifolin 56% (0.6 mg/g).

In a study of the distribution of limonin and naringin in various parts of the pummelo fruit, the highest amount of limonin was detected in the seeds of all the Thai cultivars studied (Pichaiyongvongdee and Haruenkit 2009). Lesser amounts were detected in albedo followed by flavedo, segment membranes and juice. The limonin concentration was highest in seeds in all cultivars (1,375.31–2,615.30 ppm), while for the other parts, the ranking in decreasing order was albedo (133.58–352.72 ppm), flavedo (130.16–295.49 ppm), segment membranes (85.81–293.14 ppm), and juice (10.07–29.62 ppm), respectively. Naringin was found in a greater amount than limonin in all fruit parts of the cultivars studied. The order of fruit parts that contained naringin in a decreasing amount was: albedo, flavedo, segment membranes, seeds and juice. Naringin content in albedo ranged from (28,508.01–10,065.06 ppm), in flavedo (8,964.24–2,483.96 ppm), in the segment membranes (4,369.50–1,799.48 ppm), in the seeds. (426.66–257.87 ppm) and in the juice (386.45–242.63 ppm). The chemical composition of the pummelo juice samples showed a high content of ascorbic acid in the range of 37.03–57.59 mg/100 ml. The total soluble solids range was 7.14–9.45 °Brix, the titratable acidity range was 0.38–0.98(g/100 ml) as citric acid, and the pH range was 3.69–4.05. Flavedo or zest is the outer most pigmented layer of the peel and the white or colourless inner layer is the albedo.

Pummelo juice contained an average of 18 ppm limonin and 29 ppm total limonoid glucosides (Ohta and Hasegawa 1995). Compared to other juices, pummelo contained very high concentrations of limonin and very low concentrations of limonoid glucosides.

The total flavonoid content exceeded the total carotenoid content in the peel of 8 Citrus varieties including pomelo (Wang et al. 2008). Two pomelo varieties *C. grandis* Wendun and Peiyou peels had the lowest total carotenoid content (0.036 and 0.021 mg/g dry basis, respectively). Naringin was most abundant in Peiyou and Wendun (peels (29.8 and 23.9 mg/g db, respectively). The peel also contained hesperidin, diosmin and quercetin. Levels of caffeic acid (3.06–80.8 μg/g db) were much lower than that of chlorgenic acid, ferulic acid, sinapic acid and p-coumaric acid in all the varieties. The total pectin content ranged from 36.0 to 86.4 mg/g db.

Pectins from the fruit peel of *Citrus maxima* were composed of two main groups: (1) water-soluble pectin (WSP) [high-methoxyl pectin, degree of methoxylation (DM)=69.32–78.68%], and (2) oxalate-soluble pectin (OSP) (low-methoxyl pectin, DM=21.01–55.41%) Chaidedgumjorn et al. 2009). The contents of WSP and OSP were 8.12–10.87% and 4.89–8.62%, respectively, whereas percentage galacturonic acid of WSP and OSP were 68.31–79.29% and 48.99–74.02%, respectively. Comparison between albedo (inner layer of the peel) and flavedo (outer layer of the peel), the content and percentage galacturonic acid of pectins in albedo were higher. After storing the fruits for 30 days, the molecular weight of WSP in albedo increased from 22 to 25 KDa to 41–50 KDa.

Of the 62 compounds identified, limonene was present in the highest proportion (80.5–83.6%) in the essential oil extracted from the peel of pummelo (cv. Tosa Buntan) fruits, followed by γ-terpinene, myrcene and α-pinene (1.1–6.4%) (Sawamura et al. 1989). The sum of these major components increased gradually during storage. Aldehydes such as n-octanal, n-nonanal, citronellal, n-decanal, neral and geranial did not increase during storage. Nootkatone content was the only component to increase steadily during storage. It was suggested that nootkatone levels could be used as an aid to determining the best time to harvest or ship to market. Ibrahim et al. (1995) isolated the following volatile components from the peel oil: limonene (95.10%), terpineole (1.8%), myrcene (1.60%), terpinene-4-ol (1%), *cis*-linalool oxide (furanoid) (1%), β-pinene (0.63%), α-pinene (0.3%), geranyl acetate (0.22%), linalool (0.21%), α-neral (0.13%) and geranial (0.13%). In another analysis, the main component of the essential oil of the pummelo fruit pericarp was found to be linalool (50.31%), followed by limonene (34.95%) and δ-terpinene (5.70%) (Gonzalez et al. 2002). Also present were δ-3-carene (2.37%), α-terpineole (2.10%), β-pinene (2.00%), ρ-cimene 1.14%, β-myrcene 0.86%) and α-pinene (0.76%). In another study, a total of 52 compounds, amounting to of the volatile oils of pummelo were identified (Njoroge et al. 2005). Monoterpene hydrocarbons constituted 97.5% in the oil, with limonene (94.8%), α-terpinene (1.8%), and α-pinene (0.5%) as the main compounds. Sesquiterpene hydrocarbons constituted 0.4% in the oil. The notable compounds were β-caryophyllene, α-cubebene, and (E,E)-α-farnesene. Oxygenated compounds constituted 2.0% of the pummelo oils, out of which carbonyl compounds (1.3%), alcohols (0.3%), and esters (0.4%) were the major groups. Heptyl acetate, octanal, decanal, citronellal, and (Z)-carvone were the main constituents (0.1–0.5%). Nootkatone, α-sinensal and β-sinensal, methyl-N-methylanthranilate, and (Z,E)-farnesol were prominent in the oils.

In a recent study, 50, 53 and 60 components of the oils and extract from peels of pomelo fruit were obtained by cold-pressing (CP), vacuum steam distillation (VSD) and supercritical carbon dioxide extraction (SC-CO2), respectively (Thavanapong et al. 2010). The main component was limonene (93.74–95.4%), followed by myrcene (1.71–1.93%), β-(+)-pinene (0.37–1.08%), α-(-)-pinene (0.39–0.51%), α-phellandrene (trace – 0.55%), sabinene 0.24–0.35%), linalool (0.1–0.17%), geranial (0.01–0.23%), β-caryophyllene (0.04–0.16%), germacrene D (0.02–1.04%), neral (0.02–0.14%), geraniol (0.02–0.14%), geranyl acetate (0.03–0.12%), α-(-)-selinene (0.04–0.11%) and others (traces to <0.1%).

In another study in China, the main components in the essential oils of shaddock peel were terpene compounds, which accounted for 93.93% (87.1% α-limonene, 7.11% β-myrcene) in Shatian shaddock and 85.843% (65.08% α-limonene 87.1%, 14.60% β-myrcene) in sweet shaddock (Zhou et al. 2006). Nootkatone was the major contributor of shaddock characteristic scent, and its content was 1.069% and 1.749% of essential oils from Sweet shaddock peel and Shatian shaddock peel, respectively. Other characteristic chemicals in Shatian and sweet shaddock were 1S-α-pinene 2.44%, 1.46%; β-pinene 1.02%, 2.17%; β-phellandrene 0.52%, 0.55%; neral 0.87%, 0.28%; geraniol propionate 0.49%, 0.36%; geraniol acetate 0.60%, 0.61%; citronellal 0.27%, 0.22%; terpene-4-ol 0.29%, 1.045; linalool 0.26%, 0.29%; caryophyllene 0.21%, 0.10%; α-citral 0.25%, 0.11% and α-phellandrene 0.14% and 0.17% respectively.

Phytochemicals in the Seeds

Four known limonoids, nomilin, obacunone, limonin and deacetylnomilin, were isolated as bitter components from the seeds of *Citrus grandis* Banhakuyu (Nakatani et al. 1989). Limonin, nomilin, obacunone and trace amounts of deacetyhromilin were found in pummelo seeds (Ohta and Hasegawa 1995). The 17–β-D-glucopyranoside derivatives (glucosides) of nomilin, nomilinic acid and obacunone were also present. Total limonoid aglycone concentration in the seeds ranged from 773 ppm to 9,900 ppm and total limonoid glucosides ranged from 130 ppm to 1,912 ppm.

Phytochemicals in the Flowers

Wang (1979) found the flower oil to contain dimethy anthranilate. Caffeine was detected in the flowers (Stewart 1985). Thirty-five compounds were identified in the flower oil of *C. maxima* and the major components were limonene (18.2%), linalool (16.4%), nerolidol (29.3%) and farnesol (15.7%) (Nguyen et al. 1991). Fifty-one components of the flower extract were identified and the main component was also limonene (86.2–86.8%) (Thavanapong et al. 2010). Other compounds included (E)-β-ocimene (2.85%), myrcene (1.77%), α-phellandrene (1.35%), geranial (1.17%), β-(+)-pinene (0.88%), (E,E)-farnesol (0.79%), neral (0.73%), (E)-nerolidol (0.59%), α-(-)-pinene (0.57%), sabinene (0.56%), germacrene D (0.45%), linalool (0.38%), bicycolgermacrene (0.30%), β-caryophyllene (0.23%) terpinolene (0.22%), nerol (0.19%), geranyl acetate (0.12%), neryl acetate (0.10%) and others (<0.10%).

Phytochemicals In the Roots and Stem

The root and stem bark of pummelo contained many acridone alkaloids and coumarins (Wu et al. 1983; Wu et al. 1986; Wu 1987; Wu 1988; Huang et al. 1989; Wu et al. 1994; Takemura et al. 1995a; Takemura et al. 1995b; Wu et al. 1996; Teng et al. 2005b): acridone alkaloids: grandisinine, grandisine-I, and grandisine-II, baiyumine-A and –B, buntanine, buntanmine-A, buntanbismine (bisacridone alkaloid), honyumine (pyranoacridone alkaloid), citpressine-I and II, citracridone- I and II, glycocitrine-I, grandisine-I and II, grandisinine, 5-hydroxynoracronycine, honyumine and natsucitrine-II, 5-hydroxynoracronycine alcohol, glycocitrine-I, 5-hydroxynoracronycine, citrusinine-I, natsucitrine-II, citracridone-III, acrimarine-F, dimeric acridone alkaloids named citbismine-A, -B and –C; and 2′,2′-dimethyl-(pyrano 5′,6′:3:4)-1,5-dihydroxy, 6-methoxy, 10-methyl acridone, and 2′,2′-dimethyl-(pyrano5′,6′:3:4)-1-hydroxy-5,6-dimethoxy, 10-methyl (Basa and Tripathi 1984); courmarins: 5-methoxyseselin, citrubuntin, buntansin, buntansin A buntansin B and C, buntanin, honyudisin, limonin, thamnosmonin, 6-hydroxymethylherniarin, geibalansine, pubesinol acetonide, cis-isokhellactone, and ulopterol, nordentatin, scopoletin, glycocitrine, imperatorin, meranzin, meranzin hydrate, xanthyletin, and xanthoxyletin; and the flavone: honyucitrin (Wu et al. 1988); an acridonolignoid, acrignineA (Takemura et al. 1993); citbismine E (Ju-Icji et al. 1996).

From the root bark: acridone alkaloids:- grandisinine, grandisine-I, and grandisine-II, citpressine I, citracridone I, citracridone II, citrusinine I, glycocitrine I, preskimmiane, prenylcitpressine and coumarins clausarin, 5-methoxyseselin, xanthyletin, xanthoxyletin and a benzenoid p-hydroquinone (Wu et al. 1983); coumarins xanthyletin and xanthoxyletin and acridone alkaloids 2′,2′-dimethyl-(pyrano 5′,6′:3:4)-1,5-dihydroxy, 6-methoxy, 10-methyl acridone, and 2′,2′-dimethyl-(pyrano5′,6′:3:4)-1-hydroxy-5,6-dimethoxy, 10-methyl acridone (Basa and Tripathi 1984); citropone A, citropone B (McPhail et al. 1985); honyumine (Wu et al. 1986); baiyumine-A and baiyumine -B (Wu 1987); alkaloids citpressine I, citracridone I, citracridone II, glycocitrine I, grandisine I, hoyumine, grandisinine, natsucitrine and coumarins – 5-methoxy seselin, cedrelopsin, honyudisin, clausarin, nordentatin, scopoletin, thamnosin, umbellierone, xanthoxyletin, xanthyletin flavone- honyucitrin (Wu et al. 1988); acridone alkaloid buntanine, citpressine I, citracridone I, citracridone II, citropone A, citropone B, glycocitrine I, grandisine I, grandisine II, grandisinine, prenylcitpressine and coumarins-citrubuntin, clausain, nordentatin, osthole, suberenone, umbelliferone, xanthoxyletin, xanthyletin and benzenoids crenulatin and p-hydroquinone (Wu 1988); grandisine II, gransisine III, pumiline (Takemura et al. 1992); dimeric acridone alkaloids named citbismine-A, citbismine-B and citbismine-C (Takemura et al. 1995b).

From the stem bark: alkaloids – buntanine, buntanmine, citpressine I, citracridone I, coumarin umbelliferone, xanthoxyletin, xanthyletin and a benzenoid crenulatin (Wu 1988); acridone alkaloid:- buntanmine A, 5-hydroxyacronycine, atalafoline, and coumarins buntansin A, 5-methyltoddanol, citributin, suberenone (Huang et al. 1989); two coumarins, buntansin B and buntansin C, together with 10 known compounds, 6-hydroxymethylherniarin, honyumine, limonin, buntanin, thamnosmonin, geibalansine, pubesinol acetonide, cis-isokhellactone, buntansin A and ulopterol (Wu et al. 1994); A C-C linked bisacridone alkaloid, buntanbismine (Wu et al. 1996); acridone alkaloids, 5-hydroxynoracronycine alcohol, glycocitrine-I, 5-hydroxynoracronycine, citrusinine-I, grandisine-I, natsucitrine-II and citracridone-III, (Teng et al. 2005b).

From the plant the acridone alkaloid 5-hydroxynoracronycine and acrimarine-F were isolated (Takemura et al. 1995a).

The acridone alkaloids had been reported to show inhibitory effects on EBV-EA for antitumour promoting activity, and cytotoxic, antiproliferative activities on several leukemia cell lines, antiviral, and antimalarial activities (See sections below). Recently, acridones had attracted broad attention as components of photosensitizers used in photodynamic therapy, a newly introduced cancer treatment. Coumarins also showed antimicrobial activity.

Antioxidant Activity

The antioxidants in the juice and freeze-dried flesh and peel of red pummelo and their ability to scavenge free radicals were higher than those in white pummelo juice (Tsai et al. 2007). The total phenolic content of red pummelo juice extracted by methanol (8.3 mg/mL) was found to be significantly higher than that of white pummelo juice (5.6 mg/mL). The carotenoid content of red pummelo juice was also significantly higher than that in white pummelo juice. The contents of vitamin C and δ-tocopherol in red pummelo juice were 472 and 0.35 μg/mL, respectively. The ability of the antioxidants found in red pummelo juice to scavenge radicals were found by methanol extraction to approximate that of BHA and vitamin C with a rapid rate in a kinetic model. The ability of methanol extracts of freeze-dried peel and flesh from red pummelo to scavenge these radicals was 20–40% that of BHA and vitamin C effects. Fresh red pummelo juice is an excellent source of antioxidant compounds and exhibited great efficiency in scavenging different forms of free radicals including DPPH, superoxide anion, and hydrogen peroxide radicals.

Buntan (*Citrus grandis*) fruit flavedo tissues extracted with n-butanol gave the highest antioxidant activity as evaluated using 2, 2,-diphenyl-l-picrylhydrazyl radical (DPPH) (Mokbel 2005). Of three fractions (Fr. A, Fr. B and Fr. C) of the crude extract Fr. B and Fr. C showed better antioxidant activity by using DPPH antioxidant assay. Those fractions were found to be glycoside, monosaccharide and total soluble sugar. The monosaccharide and glycosides exhibited exceptional DPPH free radical scavenging and β-carotene methods, whereas the identified sucrose, glucose and fructose possessed less antioxidant activity. The findings suggested that the brown materials present in Fr. B had a significant contribution to the antioxidant activity of the crude extract of glycoside.

The neutral methanol extract of pummelo fruit albedo tissues was found to possess maximum antioxidant and antibacterial activity (Mokbel and Suganuma 2006). The oil buntan compound, linoleic acid methyl ester, β-sitosterol, sigmasterol, limonin, nomilin and meranzin hydrate were isolated from the neutral methanol extract of pummelo fruit albedo tissues. Additionally isomeranzin hydrate, p-coumaric acid and caffeic acid compound were isolated. The extract concentration providing 50% inhibition (IC_{50}) was as follows; oil buntan compound 95 μg/ml, caffeic acid 45 μg/mL, -coumaric acid 105 μg/ml, limonin + nomilin (mixture) 135 μg/ml was lower than that of synthetic antioxidant butylated hydroxyanisole (BHA) 40 μg/ml. Ethyl acetate extracts of buntan falvedo exhibited high antioxidative activities, followed by albedo and segment membrane extracts (Mokbel and Hashinaga 2006). Among the six fractions, two fractions (C and D) showed strong antioxidant activities using the free radical scavenging activity (DPPH) antioxidant assay.

The essential oil from buntan pummelo peel had higher total phenolic (214 mM) and flavonoid (134 mg QE/g of dried material) contents than those of different solvent extracts from fruit pulp (Jang et al. 2010). In addition, DPPH free radical-scavenging activity and ferric-reducing antioxidant power values determined for the essential oil were 26.1% and 2.3 mM, respectively, which were significantly higher than those of various fruit pulp extracts. The ethanol-fermented products of pummelo juice had higher total phenolic and flavonoid contents than those of fresh juice. For maintenance of the substantial antioxidant properties of pumelo products, ethanol-fermented juice rather than fresh or acetate-fermented juice is recommended. Through correlation analysis, the phenolic compounds in the fermented pummelo products were found to be the major contributors to the free radical-scavenging activity and ferric-reducing power.

The highest phenolic content in freeze-dried edible parts of pummelo and navel oranges was obtained from ethyl acetate extract and the minimum phenolic content was found in methanol extract (Jayaprakasha et al. 2008). Ethyl acetate extract from navel orange and pummelo was found to be most active radical scavenging activity, whereas hexane extract from pummelo and methanol extract from navel orange was found to be lowest activity. The order of antioxidant capacity of pummelo and navel orange was found to be ethyl acetate>acetone>methanol:water>methanol>hexane and ethyl acetate>methanol:water>acetone>methanol>hexane, respectively. Acetone and methanol extracts from pummelo and navel oranges showed highest reducing power than other extracts at 1,000 µg/ml. Significant differences in antioxidant capacity were found between the values obtained by the same method in different solvents and as well as each extract antioxidant capacity obtained by the different method. The extent of antioxidant capacity of each extract correlated with the amount of carotenoids, phenolics and vitamin C present in the extracts.

Ethyl acetate peel extract of pomelo showed significant antioxidant activity in a dose-dependent manner when compared with ascorbic acid as evaluated by 1,1-diphenyl-2-picrylhydrazyl (DPPH) and hydrogen peroxide (H_2O_2) radical scavenging methods (Sood et al. 2009).

Antiinflammatory Activity

The ethyl acetate pomelo peel extract at the dose of 300 mg/kg produced significant decrease in carrageenan-induced paw volume and pain when compared with reference drug diclofenac and morphine, respectively (Sood et al. 2009). The results indicated that pomelo peel extract may be useful as a natural antioxidant in the treatment of inflammation and pain.

Antimicrobial Activity

Among 7 coumarins and 8 acridone alkaloids investigated for antimicrobial activity, the coumarines nordentatin (in particular) and xanthyletin completely inhibited the growth of several test bacteria pathogenic to man, and scopoletin and glycocitrine were inhibitory to one bacterium (Wu et al. 1988).

Among all the extracts, n-hexane, ethyl acetate, butanol and methanol, the ethyl acetate extracts of buntan flavedo and albedo exhibited the most significant antibacterial activity, no appreciable activity was recorded for the extracts of segment membrane and juice sacs (Mokbel and Hashinaga 2005). Three antioxidant/antimicrobial chemicals were isolated from the ethyl acetate extract of the albedo: β-sitosterol, limonin and oleic acid. The ethyl acetate extracts of flavedo had higher antifungal activity against *Botrytis cinerea*, *Rhizopus stolonifer* and *Penicillium expansum*, the albedo extract at 200–400 ppm. Oleic acid exhibited stronger antibacterial activity than β-sitosterol, limonin had no antibacterial activity. The MIC for β-sitosterol against *Bacillus cereus*, *Bacillus subtilis*, *Staphycoccous aureus*, *Escherichia coli* and *Salmonella enteritidis* were 300, 300, 270, 350, and 300 µg/ml and for oleic acid the MICs were 250, 350, 230, 270 and 150 µg/ml respectively. Isolated compounds from pommel fruit albedo exhibited antibacterial activity (Mokbel and Suganuma 2006). The inhibitory zone (mm) of bacteria tested *Staphylococcus aureus*, *Bacillus subtilis*, *Bacillus cereus*, *Salmonella enteritidis*, *Escherichia coli* were 2.9–4.1 mm for caffeic acid, 7.9–10.3 mm for meranzen hydrate, 3.1–5.3 mm for isomerazin and 11.6–15.1 mm for ρ-coumaric acid.

The essential oil from pummelo pericarp showed potent inhibition against *Staphylococcus aureus* with an inhibition zone of 60 mm diameter (Gonzalez et al. 2002). The oils and peel extracts of pomelo showed fairly good activity against *Staphylococcus aureus* and *Candida albicans* (Thavanapong et al. 2010).

Anticancer Activity

A new acridone alkaloid, 5-hydroxynoracronycine alcohol (4), along with six known acridone alkaloids, glycocitrine-I (1), 5-hydroxynoracronycine (2), citrusinine-I (3), grandisine-I (5), natsucitrine-II (6), and citracridone-III (7), were isolated from the stem bark of *Citrus maxima* f. *buntan* (Teng et al. 2005b). Compounds 1–4 and 7 were found to be cytotoxic on two tumour cell lines. These five alkaloids showed cytotoxic effects (IC_{50} < 50 μM) on HepG2 and KB cell lines, while they did not exhibit cytotoxic activity (IC_{50} > 100 μM) on H2058, COLO 201, HA22T, and Chang liver cell lines. Among these alkaloids, 5-hydroxynoracronycine was most cytotoxic to the KB epidermoid carcinoma cell line (IC_{50} = 19.5 μM), and citracridone-III was most cytotoxic to HepG2 hepatoma cell line (IC_{50} = 17.0 μM). In another study, of 50 acridone alkaloids evaluated for cell growth inhibition human promyelocytic leukemic cells (HL-60), 12 compounds showed IC_{50} < 10 μM; 12 compounds, 10–20 μM; 8 compounds, 20–100 μM; and 18 compounds, >100 μM (Chou et al. 1989). In separate studies, of 15 acridone alkaloids, atalaphyllidine, 5-hydroxy-N-methylseverifoline, atalaphyllinine, and des-N-methylnoracronycine showed potent antiproliferative activity against tumour cell lines, whereas they have weak cytotoxicity on normal human cell lines (Kawaii et al. 1999). In another study, all of the acronycine derivatives (esters and diesters of 1,2-dihydroxy-1,2-dihydroacronycine) were more potent than acronycine itself when tested against L1210 (mouse lymphocytic leukemia) cells in-vitro but the 6-demethoxyacronycine derivatives (esters and diesters of 1,2-dihydroxy-1,2-dihydro-6-demethoxyacronycine) were found to be inactive. (Elomri et al. 1996). Four selected acronycine derivatives were markedly active against murine P388 leukemia at doses 4–16-fold lower than acronycine itself. Against the colon 38 adenocarcinoma, three compounds were highly efficient. 1,2-diacetoxy-1,2-dihydroacronycine was the most active, all the treated mice being tumour-free on day 23. Among a series of 5-hydroxy-4-quinolone and 5-methoxy-4-quinolone derivatives evaluated for antitumour and antiherpes activities, 5-hydroxy-8-methoxy-quinolone showed potent antitumour activity (IC_{50} = 17.7 μM for HL60 (Human promyelocytic leukemia) which was greater than that of acronycine (Chun et al. 1997).

An immature *Citrus grandis* fruit extract exhibited antiproliferative effect against U937 human leukaemia cells (Lim et al. 2009). Maximum cytotoxicity was observed using the hexane fraction (HF) of the extract. Cell death was dose-dependent (IC_{50} = ca. 60 μg/ml) and was characterised by chromatin condensation, apoptotic body formation, and DNA fragmentation. The molecular mechanism underlying HF-induced apoptosis in U937 cells may involve a mitochondria-mediated signalling pathway, as demonstrated by an increase in the Bax/Bcl-2 expression ratio. Nineteen compounds were identified in the hexane fraction including γ-sitosterol (17.5%), 7-methoxy-8-(2-oxo-3-methylbutyl) coumarin (6.8%), stigmasterol (3.8%), and campesterol (3.4%). Together, the results indicated that the hexane fraction of an immature pomelo fruit extract induced apoptosis in U937 cells. Oral administration of pomelo leaf methanol extract at the doses of 200 and 400 mg/kg significantly decreased tumour parameters such as tumour volume, viable tumour cell count and increased body weight, hematological parameters and life span in respect of the Ehrlich's ascites carcinoma control mice (Kundusen et al. 2011).

Since the late 1990s, acridones had attracted much attention as components of photosensitizers used in photodynamic therapy, a newly introduced cancer treatment (Viola et al. 1997). A comparison of the dark and light human leukemic cell line TF-1 survival factor values suggested that irradiation had a significant effect on the toxicity at low concentrations for the porphyrin-chlorambucil diad and to a lesser extent at high concentrations for the porphyrin-acridone diad, the porphyrin-acridine diad and the porphyrin-cholic acid-chlorambucil triad. While the intrinsic antileukemic (via DNA cross-linking) activity of the chlorambucil moiety and the structural details may be responsible for the photoenhancement of the toxicity, the presence of acridine or acridone which were avid intercalators of DNA, was responsible for a similar effect seen for diads.

Antiviral Activity

Intraperitoneal administration of 10-carboxymethyl-9-acridanone sodium salt (CMA) protected at least 50% of mice tested from otherwise lethal infections with Semliki forest, coxsackie B1, Columbia SK, Western equine encephalitis, herpes simplex, and pseudorabies viruses (Kramer et al. 1976). The protective effect against influenza A2/Asian/J305 and coxsackie A21 viruses was less but was statistically significant. When administered either subcutaneously or orally, CMA protected at least 50% of mice against Semliki forest and pseudorabies viruses; the effect against coxsackie B1 and herpes simplex viruses was less but was statistically significant.

Citrusinine-I, an acridone alkaloid isolated from the root bark exhibited potent activity against herpes simplex virus (HSV) type 1 and type 2 at low concentrations relative to their cytotoxicity; with ED_{50} of 0.56 μg/ml and 0.74 μg/ml against HSV-1 and HSV-2, respectively (Yamamoto et al. 1989). Inhibitory action was also demonstrated against cytomegalovirus (CMV) and thymidine kinase-deficient or DNA polymerase mutants of HSV-2. The compound markedly suppressed HSV-2 and CMV DNA synthesis at concentrations which did not inhibit the synthesis of virus-induced early polypeptides. However, citrusinine-I had no inhibitory activity against HSV and CMV DNA polymerases in cell-free extracts. Citrusinine-1, when combined with acyclovir or ganciclovir, synergistically potentiated the antiherpetic activity of these agents In another study, two acridone alkaloids, 5-hydroxynoracronycine and acrimarine-F showed remarkable inhibitory effects against Epstein-Barr virus activation (Takemura et al. 1995b).

Antimalarial Activity

At a concentration of 10 μg/ml in-vitro, seven of the acridone alkaloids from Rutaceous plants namely glycocitrine-I, des-N-methylnoracronycine, atalaphillinine, 5-hydroxy-N-methylseverifoline, atalaphillidine, N-methylatalaphilline, and glycobismine-A, suppressed 90% or more of *Plasmodium yoelii*, causal agent of malaria in rodents (Fujioka et al. 1989). Atalaphillinine, when injected intraperitoneally in a daily dose of 50 mg/kg for 3 days into mice infected with 10^7 erythrocytes parasitized with *Plasmodium berghei* or *Plasmodium vinckei*, completely suppressed the development of malaria parasites, with there being no obvious acute toxic effects from the tested dose. 5-Hydroxynoracronycine and acronycine suppressed almost 90% of *Plasmodium yoelii* in-vitro.

Antidiabetic Activity

Studies in Korea showed that Dangyuja (*Citrus grandis*) extract prepared with enzymatic and microbial treatments could be used for the development of pharmaceutical foods to control the blood glucose level of diabetic patients by inhibiting α-amylase and α-glucosidase in the intestinal tract (Kim et al. 2009). Dangyuja is a *Citrus* fruit of Cheju Island in Korea, which is known to have a high content of flavanone glycosides, such as naringin and neohesperidin. Flavanone glycosides of Dangyuja extract were converted into their aglycones by naringinase and hesperidinase, and into their hydroxylated forms by *Aspergillus saitoi*. Dangyuja extract treated with *A. saitoi* had significantly higher antioxidant and antidiabetic activity than both its glycosides and aglycones in oxygen radical absorption capacity (ORAC) and supercoiled DNA strand scission assay, rat intestinal α-glucosidase and porcine pancreatic α-amylase inhibition, suggesting that the fermentation of flavanone aglycones of Dangyuja extract with *A. saitoi* could increase antioxidant and antidiabetic activity, compared to their glycosides and aglycones.

Two flavone glycosides, rhoifolin and cosmosiin were isolated from pomelo red wendun leaf which was found to be a rich source for rhoifolin (1.1%, w/w) (Rao et al. 2011). In differentiated 3T3-L1 adipocytes, rhoifolin and cosmosiin showed dose-dependent response in concentration range of 0.001–5 μM and 1–20 μM, respectively, in biological studies beneficial to diabetes. Particularly, rhoifolin and cosmosiin at 0.5 and 20 μM, respectively showed nearly similar response to that 10 nM of insulin, on adiponectin secretion level. Furthermore, 5 μM of rhoifolin and 20 μM of cosmosiin showed equal potential with 10 nM of insulin to increase the phosphorylation of insulin receptor-β, in addition to their positive effect on GLUT4 translocation. These findings indicated that rhoifolin and cosmosiin from red wendun leaves may be beneficial for diabetic complications through their enhanced adiponectin secretion, tyrosine phosphorylation of insulin receptor-β and GLUT4 translocation.

Anthelmintic Activity

Alcoholic extract of pomelo rind showed good in-vitro anthelmintic activity against human *Ascaris lumbricoides* (Kalesaraj 1975).

Traditional Medicinal Uses

In the Philippines and Southeast Asia, decoctions of the leaves, flowers, and rind are given for their sedative effect in cases of epilepsy, chorea and convulsive coughing. The hot leaf decoction is applied on swellings and ulcers. The fruit juice is taken as a febrifuge. The seeds are employed against coughs, dyspepsia and lumbago. Gum that exudes from declining trees is collected and taken as a cough remedy in Brazil.

In the Philippines, the leaves are locally used as aromatic baths. The peel dried or in decoction is used in cordials for dyspepsia and coughs. The seeds or pips have similar properties, and are sometimes given for lumbago. In the Himalayas, fruit juice recommended for ulcers; used in diabetes; and mixed with black pepper and a little rock salt, used for malaria. Fruit juice with its pulp, with honey, is given to improve urinary flow. In Nigeria, the fruit has been used against stomach upset, to cure gout, arthritis and kidney stones and ulcer (Okwu and Emenike 2006) Pomelo (Red Wendun) leaves have long been used as herbal remedies in traditional Chinese medicine to promote blood circulation and remove blood stasis in diseases caused by blood stagnation in Taiwan (Li 1980). For these reasons, red wendun is a traditional Chinese antidiabetic medicine.

Other Uses

The bark, leaves, flowers and the peel and juice of the fruits are used medicinally. The flowers are highly aromatic and are gathered in North Vietnam for making perfume. The leaves are used as bath essence. The essential oil is used as food, medicine and in cosmetics for soaps and perfumes. The seed essential oil is of inferior quality and was used to light opium pipes in Indo-China. The wood is heavy, hard, tough, fine-grained and suitable for making tool handles.

Citrus maxima essential oil inhibited the development of *Erwinia amylovora* biofilms on leaves, pomelo oil being more active on *Cydonia* (Aromate) leaves when the leaves were treated for 5 minutes (Măruțescu et al. 2009). The results obtained may contribute to the development of new bio-control agents as alternative strategies to protect fruit trees from fire blight disease.

Comments

Citrus maxima is a parent with *Citrus reticulata* of *Citrus* × *aurantium*.

Selected References

Bailey LH (1976) Hortus third. A concise dictionary of plants cultivated in the United States and Canada. Liberty Hyde Bailey Hortorium/ Cornell University, Wiley, 1312 pp

Basa SC, Tripathy RN (1984) A new acridone alkaloid from *Citrus decumana*. J Nat Prod 47(2):325–330

Berhow M, Tisserat B, Kanes K, Vandercook C (1998) Survey of phenolic compounds produced in *Citrus*. USDA ARS Tech Bull 1856:1–154

Burkill IH (1966) A dictionary of the economic products of the Malay Peninsula. Revised reprint. 2 vols, Ministry of Agriculture and Co-operatives, Kuala Lumpur, vol 1 (A–H), pp. 1–1240, vol 2 (I–Z), pp 1241–2444

Chaidedgumjorn A, Sotanaphun U, Kitcharoen N, Asavapichayont P, Satiraphan M, Sriamornsak P (2009) Pectins from *Citrus maxima*. Pharm Biol 47(6):521–526

Cho SK, Yang S-O, Kim S-H, Kim H, Ko JS, Riu KZ, Lee H-Y, Choi H-K (2009) Classification and prediction of free-radical scavenging activities of dangyuja (*Citrus grandis* Osbeck) fruit extracts using 1H NMR spectroscopy and multivariate statistical analysis. J Pharm Biomed Anal 49(2):567–571

Chou TS, Tzeng CC, Wu TS, Watanabe KA, Su T (1989) Inhibition of cell growth and macromolecule biosynthesis of human promyelocytic leukemic cells by acridone alkaloids. Phytother Res 3(6):237–242

Chun MW, Olmstead KK, Choi YS, Kee CO, Lee CK, Kim JH, Lee J (1997) Synthesis and biological activities of truncated acridone: Structure-activity relationship studies of cytotoxic 5-hydroxy-4-quinolone. Bioorg Med Chem Lett 7(7):789–792

Elomri A, Mitaku S, Michel S, Skaltsounis AL, Tillequin F, Koch M, Pierré A, Guilbaud N, Léonce S, Kraus-Berthier L, Rolland Y, Atassi G (1996) Synthesis and cytotoxic and antitumor activity of esters in the 1,2-Dihydroxy-1,2-dihydroacronycine series. J Med Chem 39(24):4762–4766

Feng B, Pei Y (2000) Study on chemical constituents of coumarins from *Citrus grandis*. Shenyang Yaoke Daxue Xuebao 17(4):253–255 (In Chinese)

Fujioka H, Nishiyama Y, Furukawa H, Kumada N (1989) In vitro and in vivo activities of atalaphillinine and related acridone alkaloids against rodent malaria. Antimicrob Agents Chemother 33:6–9

González CN, Sánchez F, Quintero A, Usubillaga A (2002) Chemotaxonomic value of essential oil compounds in *Citrus* species. Acta Hort (ISHS) 576:49–51

Hanelt P (ed) (2001) Mansfeld's encyclopedia of agricultural and horticultural crops, vol 1-6. Springer, Berlin, 3700 pp

Huang SC, Chen MT, Wu TS (1989) Alkaloids and coumarins from stem bark of *Citrus grandis*. Phytochemistry 28(12):3574–3576

Ibrahim J, Abu Said A, Abdul Rashih A, Nor Azah MA, Norsiha A (1995) Chemical composition of some *Citrus* oils from Malaysia. J Essent Oil Res 8:627–632

Jang H-D, Chang K-S, Chang T-C, Hsu C-L (2010) Antioxidant potentials of buntan pumelo (*Citrus grandis* Osbeck) and its ethanolic and acetified fermentation products. Food Chem 118(3):554–558

Jayaprakasha GK, Girennavar B, Patil BS (2008) Antioxidant capacity of pummelo and navel oranges: extraction efficiency of solvents in sequence. LWT-Food Sci Technol 41(3):376–384

Ju-Ichi M, Takemura Y, Nagareya N, Omura M, Ito C, Furukawa H (1996) Two new bisacridone alkaloids from Citrus plants. Heterocycles 42(1):237–240

Kalesaraj R (1975) Screening of some indigenous plants for anthelmintic action against human *Ascaris lumbricoides*: Part-II. Indian J Physiol Pharmacol 19:47–49

Kawaii S, Tomono Y, Katase E, Ogawa K, Yano M, Takemura Y, Ju-ichi M, Ito C, Furukawa H (1999) The antiproliferative effect of acridone alkaloids on several cancer cell lines. J Nat Prod 62(4):587–589

Kim G-N, Shin J-G, Jang H-D (2009) Antioxidant and antidiabetic activity of Dangyuja (*Citrus grandis* Osbeck) extract treated with *Aspergillus saitoi*. Food Chem 117(1):35–41

Kramer MJ, Cleeland R, Grunberg E (1976) Antiviral Activity of 10-Carboxymethyl-9-Acridanone. Antimicrob Agents Chemother 9(2):233–238

Kundusen S, Gupta M, Mazumder UK, Haldar PK, Saha P, Bala A (2011) Antitumor activity of *Citrus maxima* (Burm.) Merr. leaves in Ehrlich's ascites carcinoma cell-treated mice. ISRN Pharmacol 2011: Article ID 138737

Li SJ (1980) Self-incompatibility in Matou Wendun (*Citrus grandis* (L.) Osb.). Hort Sci 15:298–300

Lim HK, Moon JY, Kim H, Cho M, Cho SK (2009) Induction of apoptosis in U937 human leukaemia cells by the hexane fraction of an extract of immature *Citrus grandis* Osbeck fruits. Food Chem 114(4):1245–1250

Mabberley DJ (1997) A classification for edible *Citrus*. Telopea 7(2):167–172

Manner HI, Buker RS, Smith VE, Elevitch CR (2006). *Citrus* species. Ver. 2.1 In: Elevitch CR (ed) Species profiles for Pacific Island agroforestry. Permanent Agriculture Resources (PAR), Holualoa, Hawaii. http://www.agroforestry.net

Măruțescu L, Saviuc C, Oprea E, Savu B, Bucur M, Stanciu G, Chifiriuc MC, Lazăr V (2009) In vitro susceptibility of *Erwinia amylovora* (Burrill) Winslow et. al. to *Citrus maxima* essential oil. Roum Arch Microbiol Immunol 68(4):223–227

McPhail AT, Ju-Ichi M, Fujitani Y, Inoue M, Wu TS, Furukawa H (1985) Isolation and structures of citropone-A and -B from *Citrus* plants, first examples of naturally-occurring homoacridone alkaloids containing a tropone ring system. Tetrahedon Lett 26(27):3271–3272

Mokbel MS (2005) Antioxidant activities of water-soluble polysaccharides from buntan (*Citrus grandis* Osbeck) fruit flavedo tissues. Pak J Biol Sci 8(10):1472–1477

Mokbel MS, Hashinaga F (2005) Evaluation of the antimicrobial activity of extract from buntan (*Citrus grandis* Osbeck) fruit peel. Pakistan J Biol Sci 8(8): 1090–1095

Mokbel MS, Hashinaga F (2006) Evaluation of the antioxidant activity of extracts from buntan (*Citrus grandis* Osbeck) fruit tissues. Food Chem 94(4):529–534

Mokbel MS, Suganuma T (2006) Antioxidant and antimicrobial activities of the methanol extracts from pummelo (*Citrus grandis* Osbeck) fruit albedo tissues. Eur Food Res Technol 24(1):39–47

Morton JF (1987) Pummelo. In: Fruits of warm climates. Julia F. Morton, Miami, pp 147–151

Nakatani M, Nakama S, Hase T (1989) Bitter limonoids from the seeds of *Citrus grandis* Banhakuyu. Rep Fac Sci Kagoshima Univ (Math Phys Chem) 22:145–151

Nguyen XD, Nguyen MP, Vu NL, An NTK, Leclercq PA (1991) The essential oil from the flowers of *Citrus maxima* (J. Burman) Merrill from Vietnam. J Essent Oil Res 3(5):359–360

Niyomdham C (1992) *Citrus maxima* (Burm.) Merr. In: Coronel RE, Verheij EWM (eds) Plant resources of South-East Asia. No. 2: Edible fruits and nuts. Prosea Foundation, Bogor, pp 128–131

Njoroge SM, Koaze H, Karanja PN, Sawamura M (2005) Volatile constituents of redblush grapefruit (*Citrus paradisi*) and pummelo (*Citrus grandis*) peel essential oils from Kenya. J Agric Food Chem 53(25): 9790–9794

Ochse JJ, van den Brink RCB (1931) Fruits and fruitculture in the Dutch East Indies. G. Kolff & Co, Batavia-C, 180 pp

Ohta H, Hasegawa S (1995) Limonoids in Pummelos [*Citrus grandis* (L.) Osbeck]. J Food Sci 60(6):1284–1285

Okwu DE, Emenike IN (2006) Evaluation of the phytonutrients and vitamin content of *Citrus* fruits. Intl J Mol Med Adv Sci 2(1):1–6

Pichaiyongvongdee S, Haruenkit R (2009) Comparative studies of limonin and naringin distribution in different parts of pummelo [*Citrus grandis* (L.) Osbeck] cultivars grown in Thailand. Kasetsart J (Nat Sci) 43:28–36

Porcher MH et al. (1995–2020) Searchable world wide web multilingual multiscript plant name database. Published by The University of Melbourne. Australia. http://www.plantnames.unimelb.edu.au/Sorting/Frontpage.html

Purseglove JW (1968) Tropical crops: dicotyledons. 1 & 2. Longman, London, 719 pp

Rao YK, Lee MJ, Chen K, Lee YC, Wu WS, Tzeng YM (2011) Insulin-mimetic action of rhoifolin and cosmosiin isolated from *Citrus grandis* (L.) Osbeck leaves: enhanced adiponectin secretion and insulin receptor phosphorylation in 3T3-L1 cells. Evid Based Complem Alternat Med 2011, 624375

Sawamura M, Tsuji T, Kuwahara S (1989) Changes in the volatile constituents of pummelo (*Citrus grandis* Osbeck forma Tosa-buntan) during storage. Agric Biol Chem 53(1):243–246

Scora RW, Nicolson DH (1986) The correct name for the shaddock, *Citrus maxima*, not *C. grandis* (Rutaceae). Taxon 35(3):592–595

Sood S, Aror B, Bansal S, Muthuraman A, Gill NS, Arora R, Bali M, Sharma PD (2009) Antioxidant, anti-inflammatory and analgesic potential of the *Citrus decumana* L. peel extract. Inflammopharmacol 17(5):267–274

Stewart I (1985) Identification of caffeine in *Citrus* flowers and leaves. J Agric Food Chem 33(6):1163–1165

Swingle WT, Reece PC (1967) The botany of *Citrus* and its wild relatives. In: Reuther W, Webber HJ, Batchelor LD (eds) The *Citrus* industry. History, world distribution, botany, and varieties, vol 1. University of California Press, Riverside, pp 190–430

Takemura Y, Abe M, Ju-Ichi M, Ito C, Hatano K, Omura M, Furukawa H (1993) Structure of acrignine-a, the first naturally occurring acridonolignoid from *Citrus* plants. Chem Pharmaceut Bull 41(2):406–407

Takemura Y, Ju-Ichi M, Ito C, Furukawa H, Tokuda H (1995a) Studies on the inhibitory effects of some acridone alkaloids on Epstein-Barr virus activation. Planta Med 61(4):366–368

Takemura Y, Kuwahara J, Nagareya N, Ju-Ichi M, Omura M, Kajiura I, Ito C, Furukawa H (1992) New acridone alkaloids from *Citrus* plant. Heterocycles 34(11):2123–2130

Takemura Y, Matsushita Y, Nagareya N, Abe M, Takaya J, Ju-Ichi M, Hashimmoto T, Kan Y, Takaoka S, Askawa Y, Omura M, Ito C, Furukawa H (1995b) Citbismine-A, -B and -C, new binary acridone alkaloids from *Citrus* plants. Chem Pharm Bull 43(8):1340–1345

Teng W-Y, Chen C-C, Chung R-S (2005a) HPLC comparison of supercritical fluid extraction and solvent extraction of coumarins from the peel of *Citrus maxima* fruit. Phytochem Anals 16(5):459–462

Teng W-Y, Huang Y-L, Shen C-C, Huang R-L, Chung R-S, Chen C-C (2005b) Cytotoxic acridone alkaloids from the stem bark of *Citrus maxima*. J Chin Chem Soc 52:1253–1255

Thavanapong N, Wetwitayaklung P, Charoenteeraboon J (2010) Comparison of essential oils compositions of *Citrus maxima* Merr. peel obtained by cold press and vacuum stream distillation methods and of its peel and flower extract obtained by supercritical carbon dioxide extraction method and their antimicrobial activity. J Essent Oil Res 22(1):71–77

Tsai H-L, Chang SKC, Chang S-J (2007) Antioxidant content and free radical scavenging ability of fresh red pummelo [*Citrus grandis* (L.) Osbeck] juice and freeze-dried products. J Agric Food Chem 55(8): 2867–2872

U.S. Department of Agriculture, Agricultural Research Service (USDA) (2011) USDA National nutrient database for standard reference, Release 24. Nutrient Data Laboratory Home Page, http://www.ars.usda.gov/ba/bhnrc/ndl

Viola A, Mannoni P, Chanon M, Julliard M, Mehta G, Maiya BG, Muthusamy S, Sambaiah TJ (1997) Phototoxicity of some novel porphyrin hybrids against the human leukemic cell line TF-1. J Photochem Photobiol B Biol 40(3):263–272

Wang D (1979) Studies on the constituents of the essential oils of four aromatic flowers. Kexue Fazhan Yuekan 7:1036–1048 (In Chinese)

Wang YC, Chuang YC, Hsu HW (2008) The flavonoid, carotenoid and pectin content in peels of *Citrus* cultivated in Taiwan. Food Chem 106:277–284

Webber HJ (1946) Cultivated varieties of *Citrus*. In: Webber HJ, Batchelor LD (eds) The *Citrus* industry; 1. History, botany and breeding. University of California Press, Berkeley, pp 475–668

Wu TS (1987) Baiyumine-A and -B, two acridone alkaloids from *Citrus grandis*. Phytochemistry 26(3):871–872

Wu TS (1988) Alkaloids and coumarins of *Citrus grandis*. Phytochemistry 27(11):3717–3718

Wu TS, Huang SC, Jong TT, Lai JS (1986) Honyumine, a new linear pyranoacridone alkaloid from *Citrus grandis*. Heterocycles 24(1):41–44

Wu TS, Huang SC, Jong TT, Lai JS, Kuoh CS (1988) Coumarins, acridone alkaloids and a flavone from *Citrus grandis*. Phytochemistry 27(2):585–587

Wu TS, Huang SC, Lai JS (1994) Stem bark coumarins of *Citrus grandis*. Phytochemistry 36(1):217–219

Wu TS, Huang SC, Wu PL (1996) Buntanbismine, a bisacridone alkaloid from *Citrus grandis f. buntan*. Phytochemistry 42(1):221–223

Wu TS, Kuoh CS, Furukawa H (1983) Acridone alkaloids and a coumarin from *Citrus grandis*. Phytochemistry 22(6):1493–1497

Yamamoto N, Furukawa H, Ito Y, Yoshida S, Maeno K, Nishiyama Y (1989) Anti-herpes virus activity of citrusinine-I, a new acridone alkaloid, and related compounds. Antiviral Res 12(1):21–36

Zhang DX, Mabberley DJ (2008) *Citrus* Linnaeus. In: Wu ZY, Raven PH, Hong DY (eds) Flora of China, vol 11, Oxalidaceae through Aceraceae. Science Press/Missouri Botanical Garden Press, St. Louis/Beijing

Zhou J-H, Zhou C-S, Jiang X-Y, Xie L-W (2006) Extraction of essential oil from shaddock peel and analysis of its components by gas chromatography-mass spectrometry. J Central South Univ Technol 13(1):44–48

Citrus medica

Scientific Name

Citrus medica L.

Synonyms

Aurantium medicum (L.) M. Gómez, *Citreum vulgare* Tournefort ex Miller, *Citrus alata* (Yu. Tanaka) Tanaka, *Citrus × aurantium* L. subvar. *amilbed* Engler, *Citrus × aurantium* subvar. *chakotra* Engler, *Citrus aurantium* var. *medica* Wight & Arn., *Citrus cedra* Link, *Citrus cedrata* Raf., *Citrus crassa* Hassk., *Citrus fragrans* Salisb., *Citrus kwangsiensis* Hu, *Citrus medica* subsp. *genuina* Engl., *Citrus medica* var. *medica* proper Hook.f., *Citrus medica vulgaris* Risso & Poit., *Citrus odorata* Roussel, *Citrus tuberosa* Mill.

Family

Rutaceae

Common/English Names

Citron

Vernacular Names

Arabic: Limu, Limun, Limue-Hamiz, Nimu, Qalambak, Utaraj, Utraj, Utrej, Utroj, Uturinji;
Brazil: Cidra;
Burmese: Rhauk Ping;
Chamorro: Limon Real, Setlas, Tronkon Setlas;
Chinese: Gou Yuan, Ju Yuan, Xiang Yuan;
Cook Islands: Rēneme Māori (Maori);
Czech: Citroník Cedrát;
Danish: Cedrat;
Dutch: Cedraat, Sukadeboom;
Eastonian: Näsaviljaline Sidrunipuu, Vili: Näsasidrun;
Fijian: Moli, Moli Karokaro;
Finnish: Sukaattisitruuna;
French: Cédrat, Cédratier, Limettier De Perse, Limettier Tahiti;
German: Cedrat-Zitrone, Zitronat-Zitrone;
Guam: Setlas;
Hungarian: Cedrátcitrom, Citronát, Keseru Citrom;
India: Jora Tenga (Assamese), Bara Nimbu (Bengali), Turanj (Gujarati), Bara Nimbu, Baranimbu, Begpura, Bijaura, Jhamirdi, Kutla, Maphal, Turanj (Hindu), Imbe, Maadala, Mada-Lada-Hannu, Madalada-Hannu, Madavala, Rusakam, Rusaki (Kannada), Mauling (Konkani), Ganapati-Naranna, Ganapatina-rakam, Kaipanaragam, Madalanarakam, Matala-narakam (Malayalam), Heijang (Manipuri), Bijora, Kagi Limbu, Limbu, Mahalung, Mahalunga, Mahalingi, Mavalung (Marathi), Amlakeshara, Antughna, Begapura, Bijahva, Bijaka, Bijaphalaka, Bijapur, Bijapura, Bijapurah, Bijapuraka, Bijapurna, Danturachhada, Jambhhira, Karuna, Madhulunga, Madhuramphala, Mahalunga, Mahaphala, Matulang, Matulunga, Matulungah, Nimbu, Nimbuka, Phalapura, Phalapuraka, Rochanaphala, Rucaka,

Ruchaka, Sukeshar, Supura, Vijapura (Sanskrit), Durunci, Elumiccai, Elumicha, Inippelumiccai, Inippunarattai, Kadara Naraththai, Kadaranarattai, Kommatti Matulai, Komattimadalali, Kotinarattai, Kotiyelumiccai, Marucahagam, Maturacampiram, Narttam-Pazham, Tenrotam, Tentirimiccai, Tentirimiccaimaram, Tiirivatikam, Tirivatikamaram, Tittippelumiccai, Tittippuelumiccai (Tamil), Bija-Pura, Dabba, Lungamu, Madeephalamu, Madhipala-Pandu, Nara-Dabba, Nimma, Pulla-Dabba (Telugu), Utaraj, Turanj (Urdu);
Indonesia: Djĕrook Bodong (Java), Djĕrook Honjè (Sunda Islands);
Italian: Cedrato;
Laos: Manao Ripon;
Malaysia: Limau Susu;
Nepal: Bimbiri;
Persian: Kalinbak, Limu, Limue-Tursh, Turanj;
Polish: Cytron;
Portuguese: Cidra, Cidreira, Cidra, Cidro;
Samoan: Moli Patupatu, Moli Patupatu, Tipolo, Tipolo;
Spanish: Cidra, Cidro, Citrón, Lima De Persia, Lima Tahiti, Toronja;
Swains Island: Tipolo;
Swedish: Suckatcitron;
Tahitian: Tapolo;
Thai: Som-Mu';
Tibetan: Ba Lun Kha Lun, La La Kha Lun, Skyur Rtsi Chen Po;
Turkish: Aaç Kavunu;
Vietnamese: Thank-Yen.

Origin/Distribution

The citron is considered to be native to north east India bordering Myanmar where it is found in valleys at the foot of the Himalaya Mountains, and in the Western Ghats. It has become naturalised in southern China – Guangxi, southwest Guizhou, Hainan, Sichuan, eastern Xizang and Yunnan.

Agroecology

The citron tree is highly sensitive to frost; does not enter winter dormancy as early as other *Citrus* species. Foliage and fruit easily damaged by very intense heat and drought. Best citron locations are those where there are no extremes of temperature. The soils where the citron is grown vary considerably, but the tree requires good aeration.

Edible Plant Parts and Uses

The most important part of the citron is the peel which is a fairly important article in international trade. The fruits are halved, depulped, immersed in seawater or ordinary salt water to ferment for about 40 days, the brine being changed every 2 weeks; rinsed, put in denser brine in wooden barrels for storage and for export. After partial de-salting and boiling to soften the peel, it is candied in a strong sucrose/glucose solution. The candied peel is sun-dried or put up in jars for future use. Candying is done mainly in England, France and the United States. The candied peel is widely employed in the food industry as succade, especially as an ingredient in fruit cake, plum pudding, buns, sweet rolls and candy. In Puerto Rica, the fruits are made into marmalade, jelly, and fruit bars that are crusty on the outside, soft within. In Iran, the rind is made into jams. In Indonesia, citron peel is eaten raw with rice. In Tamil Nadu unripe citron is widely used in pickles and preserves; the young leaves are often used in conjunction with chilli powder and other spices to make a powder, called *'narthellai podi'*. In Guatemala, it is used as flavoring for carbonated soft-drinks. In Malaya, citron juice is used as a substitute for the juice of imported, expensive lemons. A product called "citron water" is made in Barbados and shipped to France for flavoring wine and vermouth. In Spain, a syrup made from the peel is used to flavour unpalatable medical preparations. In Korea, citron fruit is thinly sliced (peel, pith and pulp) and cooked in honey or sugar to create a chunky syrup which is used to make a herbal health tea called Yujacha. Yujacha is popularly consumed during the cold months.

Pectin extracted from citron fruit peel can be used as a gelling agent in pineapple jam (Tangwongchai et al. 2006).

Botany

An armed shrub or small tree, 2–4.5 m high with short, stout axillary spines, about 4 cm long on the branches. Branches, leaf buds, and flower buds are purplish when young. Leaves are simple or rarely 1-foliolate and borne on short, wingless petiole, green, lemon-scented. Leaf lamina is elliptic to ovate-elliptic, 6–12 × 3–6 cm or larger, with serrate margin, rounded obtuse, or rarely mucronate apex. Flowers fragrant, pale purplish or pinkish, pentamerous, pedicellate, bisexual or sometimes male by complete abortion of pistil, solitary or borne in short 12-flowered inflorescences (Plate 1). Calyx is synsepalous, 5-partite, cupular, the lobes are ovate, deciduous. Corolla is apopetalous, the petals 5, broadly lanceolate, white, tinged pink without, fragrant. Stamens are polyadelphous 30–50 inserted round an annular disc and exserted. Ovary one, 5–8 lobed, cylindric, style long and thick, stigma clavate. Fruit is a hesperidium, obovoid, oblong or ellipsoid, obscurely lobed, the apex broadly mammillate, the bases collared, rind leathery and nearly pellucid, greenish-purple, green turning yellow when ripe (Plates 1, 2 and 3), the pulp juicy, acid; seeds numerous, small, smooth, ovoid, endospermic.

Nutritive/Medicinal Properties

Food value of citron fruit per 100 g edible portion was reported as: energy 30 kcal (127 kJ), moisture 91.1 g, protein 1.5 g, total lipid 0.1 g, total carbohydrates 6.9 g, fibre 0.7 g, ash 0.4 g, Ca 25 mg, P 11 mg, Fe 0.3 mg, Na 12 mg, K 192 mg, thiamin 0.06 mg, riboflavin 0.03 mg, niacin 0.2 mg, β-carotene equivalent 12 μg and ascorbic acid 17 mg (Sabry and Rizek 1982). Flavone-di-C-glycosides namely 6,8-Di-C-glucopyranosylapigenin and

Plate 2 Leaves and mature citron fruit

Plate 1 Flowers and young citron fruit

Plate 3 Ripe citron fruit

6,8-di-C-glucopyranosyldiosmetin were detected in a variety of Southern Italian Citrus juices (orange, lemon, bergamot, citron, mandarin, clementine) (Caristi et al. 2006). 8-di-C-glucopyranosyldiosmetin was the most important C-glycoside in lemon and citron juice.

The percentage and concentration of total phenolics in *Citrus medica* cv Citron of Commerce leaf, peel and juice were reported by Berhow et al. (1998) as: leaf:- 24.8% (0.26 mg/g) flavone/flavonol, 58.3% coumarin (0.19 mg/g); flavedo (outer pigmented layer of the peel): 5.7% (0.03 mg/g) flavone/flavonol, 16.8% (0.06 mg/g) flavanone, 0% psoralen, 77.5% (0.13 mg/g) coumarin; albedo (inner layer of the rind): 0% flavone/flavonol, 38.9% (0.03 mg/g) flavanone, 0% psoralen, 61.1% (0.02 mg/g) coumarin; juice: 0% flavone/flavonol, 40.1% (0.02 mg/g) flavanone, 0% psoralen, 0% coumarin; medulla: 27.1% (0.02 mg/g) flavone/flavonol, 70.1% (0.06 mg/g) flavanone, 0% psoralen, 0% coumarin; septa membrane: 12% (0.02 mg/g) flavone/flavonol, 76% (0.07 mg/g) flavanone, 0% psoralen, and 0% coumarin.

The percentage and concentration of flavanones reported in *Citrus medica* cv Citron of Commerce by Berhow et al. (1998) were as follows flavedo: eriocitrin 33% (traces), hesperidin 67% (traces); albedo: eriocitrin 32% (traces), hesperidin 68% (traces); juice: hesperidin 100% (traces); medulla: eriocitrin 76% (traces), hesperidin 25% (traces); septa membrane: eriocitrin 64% (traces), hesperidin 36% (traces). The percentage and concentration of flavone reported in *Citrus medica* cv Citron of Commerce by Berhow et al. (1998) were as follows: leaf: rutin 100% (0.1 mg/g); flavedo: diosmin 99% (traces); and medulla: diosmin 49% (traces).

Citrus microcarpa, Citrus hystrix, Citrus medica var. 1 and 2, and *Citrus suhuiensis* were found to contain high amounts of flavones, flavanones, and dihydrochalcone C- and/or O-glycosides (Roowi and Crozier 2011). Among the major compounds detected were apigenin-6,8-di-C-glucoside, apigenin-8-C-glucosyl-2″-O-rhamnoside, phloretin-3′,5′-di-C-glucoside, diosmetin-7-O-rutinoside, hesperetin-7-O-neohesperidoside, and hesperetin-7-O-rutinoside. Most of the tropical citrus flavanones were neohesperidoside conjugates responsible for imparting a bitter taste to the fruit.

The volatile fraction of samples of fruit oil from small green citron, large, green citron and yellow harvested citron were characterized by a high content of limonene, γ-terpinene, and monoterpene aldehydes and a lower content of α- and β-pinene and myrcene, sesquiterpenes, and aliphatic aldehydes (Gabriele, et al. 2009). Oxypeucedanin was the main component of the oxygenated heterocyclic fraction in the extracts of green fruits, while citropten was the major oxygenated compound in the oil obtained from yellow citron

The essential oils of four varieties of *Citrus medica* (*Citrus medica* var. *sarcodactylis*, *C. medica* var. *ethrog*, *C. medica* var. 1 and *C. medica* var. 2) were found to be rich in monoterpene hydrocarbons with limonene being the principal component (Mohd Ali et al. 2005). The essential oil of the whole fruit of *C. medica* var. *sarcodactylis* contained limonene (48.0%), γ-terpinene (26.3%), (Z)-citral (5.7%) and (E)-citral (6.3%). Limonene and γ-terpinene were also the major components found in the peel oils of *C. medica* var. *ethrog* (72.7% and 13.1%) and *C. medica* var. 2 (70.6% and 15.6%). The peel oils of *C. medica* var. 1 however, was made up entirely of limonene (84.5%) and a-terpineol (4.2%). Limonene and γ-terpinene were the major components of the oils obtained from *Citrus medica* cv. Diamante peel by hydrodistillation (HD) and cold-pressing (CP), while citropten was the major constituent in the oil obtained by supercritical carbon dioxide extraction (SFE) (Menichini et al. 2011).

The main component in both citron leaf and peel oils was limonene (Vekiari et al. 2004). Oil from peel also contained a high content of neral and geranial (80.8 and 133.7 mg/kg of the oil, respectively), with β-pinene and myrcene being the most abundant monoterpene hydrocarbons following limonene. The peel oil was also characterized by appreciable proportions of the monoterpene alcohols citronellol nerol and geraniol (20.5 and 22.3 mg/kg of the oil, respectively). The content not only of monoterpene but also of aliphatic aldehydes was particularly higher in leaf oil than in the peel oil. Concerning the monoterpinic hydrocarbons in leaf oil, myrcene was the second most abundant after limonene. Three chemotypes: limonene, limonene/γ-terpinene and limonene/geranial/neral were observed for peel oils while leaf oils exhibited the limonene/geranial/neral composition (Lota et al. 1999).

Another research reported that the chemical constituents of leaf essential oil of *Citrus medica* consisted of 19 components accounting for 99.9% of the oil (Bhuiyan et al. 2009). The major constituents were dominated by erucylamide (28.43%), limonene (18.36%), citral (12.95%), mehp (8.96%),2,6-octadien-1-ol, 3,7-dimethyl-, acetate, (Z)- (5.23%), 6-octenal, 3,7-dimethyl- (4.39%), 1,2-cyclohexanediol, 1-methyl-4-(1-methylethenyl)- (3.98%) and methoprene (3.51%). The peel oil contains 43 components accounting for 99.8% of the total oil and the major components were isolimonene (39.37%), citral (23.12%), limonene (21.78%), β-myrcene (2.70%), neryl acetate (2.51%), neryl alcohol (2.25%), linalool (0.94%), β-bisabolene (0.71%), neryl acetate (0.6%) and caryophyllene (0.59%), β-terpinyl acetate (0.52%) and α-bergamotene (0.48%) (Bhuiyan et al. 2009). In addition, *C. medica* peel essential oil was found to contain d-limonene, other terpenes found in the flavedo oil fraction were linalool, geraniol, citronellol, α-terpineol, valencene, myrcene and α-pinene (Verzera et al. 2005). Another research (Habashi et al. 2009) conducted in Iran found 25 components in the hydrodistilled oil. The main components were limonene (58.3%), γ-terpinene (16.8%), geranial (6%) neral (4.8%), geranyl acetate (1.4%) and geraniol (1.36%). Twenty-three components were characterized in cold-pressed oil with limonene (63.7%), γ-terpinene (21.7%) and geranial (1.3%) as the main components.

Antioxidant Activity

The n-hexane citron extract showed significant antioxidant activity that was carried out using different assays (DPPH test, β-carotene bleaching test and bovine brain peroxidation assay) (Conforti et al. 2007). The n-hexane extract of *Citrus medica* cv. Diamante peel was characterized by the presence of monoterpenes and sesquiterpenes. The most abundant constituents were two monoterpenes: limonene and γ-terpinene. Oxidative damage, caused by the action of free radicals, may initiate and promote the progression of a number of chronic diseases, including diabetes and Alzheimer's disease. Diamante citron peel extract also showed hypoglycaemic activity and an anticholinesterase effect.

Acetylcholinesterase Inhibition and Antiinflammatory Activities

The peel was found to contain citroflavonoids comprising a mixture of hesperidoside (rhamnoglucoside of hesperetol), naringoside and ecryodietyoside (flavanones). The citroflavonoids were reported to have an antiinflammatory, antihistamine and diuretic activities and could cause dilatation of the coronaries (Bhuiyan et al. 2009; Nicolosi et al. 2005).

The essential oil obtained from *Citrus medica* cv. Diamante peel by hydrodistillation exerted the highest inhibitory activity against butyrylcholinesterase (BChE) (IC_{50} value of 154.6 μgml) and acetylcholinesterase (AChE) (IC_{50} value of 171.3 μg/ml) (Menichini et al. 2011). The oil obtained by cold-pressing exhibited a selective inhibitory activity against AChE. The oil obtained by hydrodistillation exerted a significant inhibition of NO production in LPS induced RAW 264.7 macrophages with an IC_{50} value of 17 μg/ml (IC_{50} of positive control 53 μg/ml).

Estrogenic Activity

In one study, the petroleum ether extract of *Citrus medica* seeds -treated rats was shown to exhibit estrogenic effects, which included increase in uterine weight and vaginal epithelial cell cornification (Sharangouda and Patil 2008). The micrometric measurements of the uterus and its components were increased and glands showed high secretory activity. When the above extract was tested in 30-day-old immature rats, they exhibited opening of vagina on the fifth day and cornification of vaginal epithelial cells, which was about 10 days earlier compared to controls, further supporting the estrogenic activity of the extract. The results indicated the potent estrogenic nature of petroleum ether extract of *Citrus medica* seeds, which may be used as an antifertility agent.

Anticancer/Antimutagenic Activity

Studies showed that *Citrus medica* fruit juice had antimutagenic and anticancer activities; activities were higher from half-ripe fruit than from ripe fruit (Entezari et al. 2009). In the MTT assay, human astrocytoma cancer cell line sustained significant mortality when compared with control. In the Ames test the fruit juice prevented the reverted mutations and the inhibition percent of half-ripe fruit was 71.7% and ripe fruit was 34.4% in antimutagenicity test, and this value in the anticancer test was 83.3% and 50% for half-ripe fruit and ripe fruit respectively.

Antimicrobial Activity

All ethanolic extracts of root, leaf, bark, peel and pulp of citron and fruit juice showed varied level of antibacterial activity against one or more test bacteria (*Bacillus subtilis, Staphylococcus aureus, Enterococcus faecalis, Escherichia coli, Klebsiella pneumoniae, Pseudomonas aeruginosa* and *Proteus vulgaris*). (Sah et al. 2011) Antifungal activity (*Aspergillus flavus, A. niger*) was shown by only root extract and fruit juice while *Candida albicans* was resistant to all tested plant samples. Broad spectrum antimicrobial activity was shown by fruit juice (MIC <1–3.5% and MBC 1–7% v/v) and fruit pulp (MIC 25 mg/ml and MBC 30–75 mg/ml). Root extract was found highly potent with MIC as small as 0.5 mg/ml and MBC 1 mg/ml against *Staphylococcus aureus*. Among all tested plant samples, leaf and peel extracts displayed less antimicrobial activity. It was concluded that fruit juice and juiceless fruit pulp extract exhibited broad antimicrobial activity while root extract was very effective against some tested microorganisms.

Anthelmintic Activity

Studies demonstrated that the petroleum ether citron leaf extract had anthelmintic activity (Bairagi et al. 2011). The extract paralyzed earthworm and caused its death at the higher dose (80 mg/ml) after 30.86 minutes.

Traditional Medicinal Uses

In ancient times and in the Middle Ages, the 'Etrog' was employed as a remedy for seasickness, pulmonary troubles, intestinal ailments and other disorders. Citron juice with wine was considered an effective purgative to rid the system of poison. In India, the peel is a remedy for dysentery and is eaten to overcome halitosis. The distilled juice is given as a sedative. The candied peel is sold in China as a stomachic, stimulant, expectorant and tonic. In West Tropical Africa, the citron is used against rheumatism. The flowers are used medicinally by the Chinese. In Malaya, a decoction of the fruit is taken to drive off evil spirits. A decoction of the shoots of wild plants is administered to improve appetite, relieve stomachache and expel intestinal worms. The leaf juice, combined with that of *Polygonum* and *Indigofera* is taken after childbirth. A leaf infusion is given as an antispasmodic. In Southeast Asia, citron seeds are given as a vermifuge. In Panama, they are ground up and combined with other ingredients and given as an antidote for poison. The essential oil of the peel is regarded as an antibiotic. In Myanmar, the young fruit has been reported to be used as carminative; expectorant and for biliousness; the ripe fruit for heartburn; appetizer, haematemesis, dyspepsia, hiccough, whooping cough, oedema, cough, laryngitis, leprosy and as antiseptic, appetizer; analgesic, carminative and mucolytic (Tun et al. 2008).

The buds and flowers are listed as antiemetic; appetizer; and for energy, asthma, whooping cough, giddiness and to promote peristalsis of the intestine.

Other Uses

The citron is much esteemed in China and Japan for its fragrance and the fruit is commonly placed on tables to freshen up and perfume rooms. The dried fruits are placed with stored clothing to repel moths. In southern China, the juice is used to wash fine linen. In some islands in Oceania, "Cedrat Petitgrain Oil" is distilled from the leaves and twigs of citron trees for the French perfume industry. Essential oil can also be distilled from the flowers. Citron essential oil can be used as a fungicide. Studies reported that ctron essential oil exhibited a wide spectrum of fungitoxicity, inhibiting 14 storage fungi tested of peanuts (Essien et al. 2008). The wood is heavy and hard and employed for agricultural implements in India, while the branches are used as walking sticks.

Comments

Fingered citron, *Citrus medica* L. var. sarcodactylis Swingle is covered in a separate chapter.

Selected References

Bailey LH (1976) Hortus third. A concise dictionary of plants cultivated in the United States and Canada. Liberty Hyde Bailey Hortorium, Cornell University, Wiley, New York, 1312 pp

Bairagi GB, Kabra AO, Mandade RJ (2011) Anthelmintic activity of *Citrus medica* L. leaves in Indian adult earthworm. Int J Pharm Technol Res 3(2):664–667

Berhow M, Tisserat B, Kanes K, Vandercook C (1998) Survey of phenolic compounds produced in *Citrus*. USDA ARS Technol Bull 1856:1–154

Bhuiyan MNI, Begum J, Sardar PK, Rahman MS (2009) Constituents of peel and leaf essential oils of *Citrus medica* L. J Sci Res 1(2):387–392

Caristi C, Bellocco E, Gargiulli C, Toscano G, Leuzzi U (2006) Flavone-di-C-glycosides in *citrus* juices from Southern Italy. Food Chem 5(3):431–437

Conforti F, Statti GA, Tundis R, Loizzo MR, Menichini F (2007) In vitro activities of *Citrus medica* L. cv. Diamante (Diamante citron) relevant to treatment of diabetes and Alzheimer's disease. Phytother Res 21(5):427–433

Entezari M, Majd A, Falahian F, Mehrabian S, Hashemi M, Lajimi AA (2009) Antimutagenicity and anticancer effects of *Citrus medica* fruit juice. Acta Med Iran 47(5):373–377

Essien EP, Essien JP, Ita BN, Ebong GA (2008) Physicochemical properties and fungitoxicity of the essential oil of *Citrus medica* L. against groundnut storage fungi. Turk J Bot 32:161–164

Foundation for Revitalisation of Local Health Traditions (2008) FRLHT Database. http://envis.frlht.org

Gabriele B, Fazio A, Dugo P, Costa R, Mondello L (2009) Essential oil composition of *Citrus medica* L. Cv. Diamante (Diamante citron) determined after using different extraction methods. J Sep Sci 32(1):99–108

Habashi M, Mirza M, Mostofi Y, Jaimand K (2009) Identification and comparison of the essential oil components from the peel of Citron (*Citrus medica* L.) by using two extraction methods (hydrodistillation and cold press). Iran J Med Aromc Plants 24(4):428–436

Hanelt P (ed) (2001) Mansfeld's encyclopedia of agricultural and horticultural crops, vol 1–6. Springer, Berlin, 3700 pp

Lota ML, Serra DR, Tomi P, Bessiere J, Casanova J (1999) Chemical composition of peel and leaf essential oils of *Citrus medica* L. and *C. limonimedica* Lush. Flav Fragr J 14(3):161–166

Mabberley DJ (1997) A classification for edible *Citrus*. Telopea 7(2):167–172

Manner HI, Buker RS, Smith VE, Elevitch CR (2006) *Citrus* species. Ver. 2.1. In: Elevitch CR (ed) Species profiles for Pacific Island Agroforestry. Permanent Agriculture Resources (PAR), Holualoa, Hawaii. http://www.agroforestry.net

Menichini F, Tundis R, Bonesi M, de Cindio B, Loizzo MR, Conforti F, Statti GA, Menabeni R, Bettini R, Menichini F (2011) Chemical composition and bioactivity of *Citrus medica* L. cv. Diamante essential oil obtained by hydrodistillation, cold-pressing and supercritical carbon dioxide extraction. Nat Prod Res 25(8):789–799

Mohd Ali NA, Ayub N, Mohd Zaki Z, Jalil AM, Ahmad AS, Zolpatah MF (2005) Essential oil constituents of four varieties of *Citrus medica* (Rutaceae). Malays J Sci 24(1):85–88

Morton JF (1987) Citron. In: Fruits of warm climates. Julia F. Morton, Miami, FL, pp 179–182

Nicolosi E, La-Malfa S, El-Otmani M, Negbi M, Goldschmidt EE (2005) The search for the authentic citron (*Citrus medica* L.): historic and genetic analysis. Hortic Sci 40(7):1963–1968

Pacific Island Ecosystems at Risk (PIER) (2004) *Citrus medica* L., Rutaceae. http://www.hear.org/Pier/species/citrus_medica.htm

Perry LM (1980) Medicinal plants of East and Southeast Asia. Attributed properties and uses. MIT Press, Cambridge/London, 620 pp

Porcher MH et al (1995–2020) Searchable world wide web multilingual multiscript plant name database. Published by The University of Melbourne. Australia. http://www.plantnames.unimelb.edu.au/Sorting/Frontpage.html

Roowi S, Crozier A (2011) Flavonoids in tropical *Citrus* species. J Agric Food Chem 59(22):12217–12225

Sabry ZI, Rizek RL (1982) Food composition tables for the near East, FAO food and nutrition paper. FAO, Rome, 275 pp

Sah AN, Juyal V, Melkani AB (2011) Antimicrobial activity of six different parts of the plant *Citrus medica* Linn. Pharmacogn J 3(21):80–83

Samson JA (1992) *Citrus* L. In: Verheij EWM, Coronel RE (eds) Plant resources of South-East Asia no 2. Edible fruits and nuts. PROSEA, Bogor, pp 119–138

Sharangouda SB, Patil SB (2008) Estrogenic activity of petroleum ether extract of seeds of *Citrus medica* on immature albino rats. Int J Green Pharm 2(2):91–94

Swingle WT (1946) The botany of *Citrus* and its wild relatives of the orange subfamily (family Rutaceae, subfamily Aurantioideae). In: Webber HJ, Batchelor LD (eds) The *citrus* industry. vol. 1. History, botany and breeding. Univ. of California Press, Berkeley, pp 129–474, 1028 pp

Tangwongchai R, Lerkchaiyaphum K, Nantachai K, Rojanakorn T (2006) Pectin extraction from citron peel (*Citrus medica* Linn.) and its use in food system. Songklanakarin J Sci Technol 28(6):1351–1363

Tun UK, Than UP, staff of Tun Institute of Learning (2008) Myanmar Medicinal Plant Database. http://www.tuninst.net/MyanMedPlants/indx-DB.htm#Top

Vekiari SA, Protopapadakis EE, Gianovits AN (2004) Composition of the leaf and peel oils of *Citrus medica* L. 'Diamante' from Crete. J Essent Oil Res 16(6):528–530

Verzera A, Trozzi A, Zappala M, Condurso C, Cotroneo A (2005) Essential oil composition of *Citrus meyerii* Y. Tan. and *Citrus medica* L. cv. Diamante and their lemon hybrids. J Agric Food Chem 53(12):4890–4894

Zhang DX, Mabberley DJ (2008) *Citrus* Linnaeus. In: Wu ZY, Raven PH, Hong DY (eds) Flora of China, vol 11, Oxalidaceae through Aceraceae. Science Press/Missouri Botanical Garden Press, Beijing/St. Louis

Citrus medica var. sarcodactylis

Scientific Name

Citrus medica L. var. *sarcodactylis* (Hoola van Nooten) Swingle.

Synonyms

Citrus medica f. *monstrosa* Guillaumin, *Citrus medica* L. var. *digitata* (Lour.) Risso, *Citrus sarcodactylis* Hoola van Nooten, *Sarcodactilis helicteroides* C. F. Gaertner.

Family

Rutaceae

Common/English Names

Buddha-Hand Citron, Buddha's-Hand, Buddha's Fingers, Fingered Citron, Flesh-Finger Citron.

Vernacular Names

Chinese: Fo Shou, Fo Shou Gan, Fo Shou Kan, Fo Shou Pian, Fo Shou P'ien, Fu Shou;
Danish: Buddhafinger;
Dutch: Hand Van Boeddha;
French: Cédrat Digité, Cédrat Main De Bouddha, Sarcodactyle;
Finnish: Sormisukaattisitruuna;
German: Buddhafinger, Gefingerte Zitrone;
Indonesia: Jeruk Tangan;
Italian: Cedro A Mano Di Budda;
Japanese: Bushukan;
Malaysia: Jeruk Tangan, Limau Jari, Limau Kerat Lintang;
Thai: Som Mue;
Vietnamese: Cây Phât Thu, Phật Thủ.

Origin/Distribution

Wild distribution is found in south and southeast Asia. The plant is cultivated in the tropics and sub-tropics like India, Sri Lanka, Thailand, Vietnam, China and Japan.

Agroecology

In the tropics and subtropics, Buddha finger citron grows very well in fukll sun at elevations below 1,300 m. The soil should be moist, well-drained, deep, rich in organic matter and fertile. It is very sensitive to frost and to intense heat and drought.

Edible Plant Parts and Uses

The fruit is edible and has been eaten in China for centuries. In Japan and the southern provinces of China and Taiwan, it is commonly used as functional vegetables and preserved as sweetmeats. The fruit often has no pulp but consist entirely of the edible fleshy peel that can be steamed and candied fresh or pickled and made into sweetmeats and jam. The fruit is first pickled in salt for a month to remove the bitter taste of the tannin, drained, washed and then steamed to soften the fruit. The fruit can then be cut into pieces and candied as you would lemon zest. The resulting candy is an aromatic and tasty snack that is perfect for indigestion and sore throat. The peels also add a citron tang to seafood dishes. A twinge or slice is used in fish marinade, Italian desserts, and other Chinese specialities such as "*chicken rolls steamed in lotus leaf*", "*stewed fruit with white fungus*" or in pork dishes. In Indonesia, the peel is eaten as *lalab* (salad) with rice and the fruit is also eaten whole even though it has no or scanty pulp.

Botany

An erect perennial shrub or small tree, up to 3 m tall, with light grey bark and relatively soft wood. The twigs are angular and purplish when young, turning terete, glabrous, with single axillary spines. Leaves are elliptic-ovate to ovate-lanceolate, 5–20 × 3–9 cm, cuneate or rounded at base, margins serrate, apex bluntly pointed or rounded and borne on short, wingless or nearly so petioles (Plate 1). Flower buds are large, pinkish. Flowers are 3–4 cm in diameter, perfect or staminate, in axillary, few-flowered racemes. Flower has 5 petals pinkish externally, 30–40 (-60) stamens, 10–13-locular ovary and thick style. Fruit is an ovoid to oblongoid berry, 10–20 cm long, slightly to considerably rough-tuberculate with very thick, fragrant and aromatic peel, green turning to yellow or golden yellow when ripe (Plates 1, 2 and 3). Fruit is split into a number of finger-like sections (Plates 1–4), without or with very scanty acid-sweet pulp and can be seedless or seedy. Seeds are numerous when present, ovoid, hanging loosely in the locules.

Plate 1 Leaves and fingered citron fruit

Plate 2 Unripe, green Buddha-hand citron fruit

Nutritive/Medicinal Properties

More than 40 compounds were identified, accounting for almost 99% of the oils of the fruit peel of *Citrus medica* var. *sarcodactylis* (Dung

Plate 3 Ripe yellow and golden yellow fruit with characteristic finger-segments

Plate 4 Harvested ripe Fingered Citron

et al. 1996). The major oil constituents of one chemotype were found to be limonene (48.4%) and p-cymene (33.7%), while the oil of the other chemotype contained predominantly limonene (55.5%) and γ-terpinene (22.5%).

The essential oil of the whole fruit of *C. medica* var. *sarcodactylis* contained limonene (48.0%), γ-terpinene (26.3%), (Z)-citral (5.7%) and (E)-citral (6.3%) (Mohd Ali et al. 2005). The main volatile components in the peel oil were limonene (47.8%), γ-terpinene (32.1%), α-pinene (2.9%), α-pinene (2.7%), geranial (2.5%), myrcene (1.7%), neral (1.6%), terpinolene (1.4%) and γ-thujene (1.33%). The peel oil was characterized by the smaller content of limonene and the larger content of β-terpinene, geranial and neral. β-ionone, one of the aroma volatile components, was responsible for the characteristic osmanthus-like odour of the fruit.

Two flavonoid compounds, 3, 5, 6-trihydroxy-4′, 7-dimethoxyflavone and 3, 5, 6-trihydroxy-3′, 4′, 7-trimethoxyflavone, were isolated from dried fingered citron fruit (He and Ling 1985). Two photochemical products, cis-head-to-tail-lemittin dimer and cis-head-to-head-lemittin dimer were isolated from the fruit of fingered citron, a traditional Chinese medicinal plant (He et al. 1987). The trans-head-to-head-lemittin dimer were synthesized from lemittin upon UV irradiation. A crystalline substance which separated from the peel oil was identified as 5,7-dimethoxycoumarin (limettin) (Shiota 1990). The peel also contains cyclic peptides (Matsumoto et al. 2002). Nine compounds were isolated from the fruit and identified as 5-methoxyfurfural; 5-hydroxy-2-hydroxymethyl-4H-pyran-4-one; diosmetin; diosmin; obacunone; aviprin; 3-(3-methoxy-4-hydroxyphenyl)-acrylic acid; vanillic acid; and 3,4-dihydroxy-benzoic acid. *C. medica* var. *sarcodactylis* also contained limettin, stigmasta-5, 22-dien-3-ol and palmitic acid (Gao et al. 2002).

Ten compounds were isolated *Citrus medica* var. *Sarcodactylis* and identified as 5,7-dimethoxycoumarin (limettin); 7-hydroxy-6-methoxycoumarin (scopoletin); 7-hydroxycoumarin (umbelliferon); 7-hydroxy-5-methoxycoumarin; *p*-coumaric acid; 6,7-dimethoxycoumarin; limonin; nomilin; stigmasterol and β-*D*-glucoside (Yin and Lou 2004).

Antiinflammatory Activity

The antiinflammatory components isolated from the stem and root barks of *Citrus medica* var. *sarcodactylis* were xanthyletin (2), nordentatin (3), atalantoflavon (4) and lonchocarpol A (5) which displayed potent nitric oxide (NO)-reducing activity in microglial cells (Chan et al. 2010). Fingered citron is used in traditional Chinese medicine for the treatment of allergic response and inflammatory conditions.

Antidyspepsia Activity

C. medica var. *sacordactylis* is 1 of 15 herbal components of the Xiayou decoction (Jiang et al.

2001). Studies showed that after treatment with Xiaoyu decoction plus psychotherapy, the mental symptoms such as depression and anxiety in the FD (functional dyspepsia) patients were markedly improved, as compared with those before treatment, the difference was significant.

Anticancer Activity

Lonchocarpol A, a flavanol 1 of 23 constituents isolated from the stem bark of *Citrus medica* L. var. *sarcodactylis* showed marginally cytotoxic activity against four cancer cell lines: Daoy, Hep2, MCF-7 and Hela cell lines (Chan et al. 2009). The constituents included a new biscoumarin, citrumedin-A [6,6′,7,7′-tetramethoxy-3,3′-biscoumarin].

Hypoglycaemic Activity

By kinetic analysis on the hypoglycemic patterns of the intraperitoneal glucose tolerance (IPGTT) and the insulin–glucose tolerance tests (IGTT), the insulin secretagogue effect of fingered citron was confirmed in Sprague-Dawley–SPF rats and Wistar DIO rats by Taiwanese scientists (Peng et al. 2009). Additionally, the major compounds in essential oils (% in EO) of the fruits were determined to be d-limonene (51.24), γ-terpenene (33.71), α-pinene (3.40), and β-pinene (2.88) were identified. The study showed finger citron fruits concomitantly possessed insulin secretagogue and slimming effects and would be very beneficial to type 2 diabetes mellitus patients.

Traditional Medicinal Uses

Buddha's Hand citron is prescribed as a stimulant, expectorant, and tonic in traditional medicine. In China, it has been used in combination with other medicinal herbs for a range of ailments. In Chinese traditional herbal medicine, the fruit of *C. medica* var. *sacordactylis* is used to treat: (a) distension and pain the chest and hypondriac regions in combination with Rhizoma Cyperi (rhizome of *Cyperus rotundus*) and Radix Curcumae (rhizome of *Curcuma longa*); (b) bloated stomach, anorexia, belching and vomiting used in combination with Radix Aucklandiae (root of *Aucklandia lappa*) and Fructus Aurantii (fruit of *Citrus aurantium*); (c) cough and copious sputum and chest pain in combination with Retinervus Luffae Fructus (*Luffa cylindrica* fruit sponge) and folium Eriobotryae (leaf of *Eriobtory japonica*) (Wu 2005).

Other Uses

The fragrant fruits are used in Asia for perfuming clothing and rooms. In earlier times it was much used as an insect repellent and an air-freshener. The Fingered citron is also used for religious purposes as an offering in temples. When grown as a dwarf plant it is also a valued ornamental throughout the Far East.

Comments

Some botanists assert that the Buddha-hand Citron with separated segments more or less surrounded by pericarp, is best treated as a cultivar, correctly *Citrus medica* 'Fingered.'

Selected References

Backer CA, van den Bakhuizen Brink RC Jr (1965) Flora of Java, vol 2, Spermatophytes only. Wolters-Noordhoff, Groningen, p 641
Chan YY, Li CH, Shen YC, Wu TS (2010) Anti-inflammatory principles from the stem and root barks of *Citrus medica*. Chem Pharm Bull(Tokyo) 58(1): 61–65
Chan YY, Wu TS, Kuo YH (2009) Chemical constituents and cytotoxicity from the stem bark of *Citrus medica*. Heterocycles 78(5):1309–1316
Dung NX, Pha NM, Lo VN, Thien NH, Leclercq PA (1996) Chemical investigation of the fruit peel oil of *Citrus medica* L. var. *sarcodactylis* (Noot.) swingle from Vietnam. J Essent Oil Res 8(1):15–18
Facciola S (1990) Cornucopia. A source book of edible plants. Kampong Publications, Vista, 677 pp
Gao Y, Huang H, Xu H, Diao Y, Dong Z (2002) Studies on the chemical constituents of *Citrus medica* var. *sarcodactylis*. J Chin Med Mater (Zhong Yao Cai) 25(9): 639–640 (in Chinese)

He HY, Ling LQ (1985) Chemical studies of a Chinese traditional drug fingered citron (*Citrus medica* L. var. *sarcodactylis* (Noot.) Swingle. Yao Xue Xue Bao 20(6):433–435

He HY, Ling L, Zhou M (1987) Isolation and structure elucidation of two dimeric limettins from fingered citron. Youji Huaxue 3:193–196

Hu S-Y (2005) Food plants of China. The Chinese University Press, Hong Kong, 844 pp

Jiang B, Lin J, Zhang Y (2001) Clinical observation on Xiaoyu decoction plus psychotherapy in treating functional dyspepsia. Chin J Integr Med 7(1):19–21 (in Chinese)

Jones DT (1992) *Citrus medica* L. In: Verheij EWM, Coronel RE (eds) Plant resources of South-East Asia. No. 2: edible fruits and nuts. Prosea Foundation, Bogor, pp 131–133

Mabberley DJ (2004) *Citrus* (Rutaceae): a review of recent advances in etymology, systematics and medical applications. Blumea 49:481–498

Matsumoto T, Nishimura K, Takeya K (2002) New cyclic peptides from *Citrus medica* var. *sarcodactylis* Swingle. Chem Pharm Bull(Tokyo) 50(6):857–860

Mohd Ali NA, Ayub N, Mohd Zaki Z, Jalil AM, Ahmad AS, Zolpatah MF (2005) Essential oil constituents of four varieties of *Citrus medica* (Rutaceae). Malays J Sci 24(1):85–88

Molesworth Allen B (1967) Malayan fruits. An introduction to the cultivated species. Moore, Singapore, 245 pp

Ochse JJ, Bakhuizen van den Brink RC (1980) Vegetables of the Dutch Indies, 3rd edn. Ascher & Co., Amsterdam, 1016 pp

Peng C-H, Ker Y-B, Weng C-F, Peng C-C, Huang C-N, Lin L-Y, Peng R-Y (2009) Insulin secretagogue bioactivity of finger citron fruit (*Citrus medica* L. var. *Sarcodactylis* Hort, Rutaceae). J Agric Food Chem 57(19):8812–8819

Shiota H (1990) Volatile components in the peel oil from fingered citron (*Citrus medica* L. var. *sarcodactylis* Swingle). Flav Fragr J 5(1):33–37

Swingle WT, Reece PC (1967) The botany of *Citrus* and its wild relatives. In: Reuther W, Webber HJ, Batchelor LD (eds) The *Citrus* industry volume 1: history, world distribution, botany, and varieties, Revised edn. University of California Press, Berkeley, pp 190–430

Wu J-N (2005) An illustrated Chinese materia medica. Oxford University Press, New York, 706 pp

Yin F, Lou FC (2004) Studies on the constituents of *Citrus medica* L.var. *Sarcodactylis*. Chin Pharm J 40(1):20–21

Zhang DX, Mabberley DJ (2008) *Citrus* Linnaeus. In: Wu ZY, Raven PH, Hong DY (eds) Flora of China, vol 11, Oxalidaceae through Aceraceae. Science Press/Missouri Botanical Garden Press, Beijing/St. Louis

Citrus reticulata

Scientific Name

Citrus reticulata Blanco.

Synonyms

Citrus × *aurantium* Linnaeus f. *deliciosa* (Tenore) Hiroe, *Citrus aurantium* subsp. *suntra* Engl., *Citrus aurantium* var. *1* Parker, *Citrus* × *aurantium* var. *tachibana* Makino, *Citrus chrysocarpa* Lush., *Citrus daoxianensis* S. W. He & G. F. Liu, *Citrus deliciosa* Tenore, *Citrus depressa* Hayata, *Citrus erythrosa* Yu. Tanaka, *Citrus khasia* Marc., *Citrus madurensis* Loureiro var. *deliciosa* (Tenore) Sagot, *Citrus mangshanensis* S. W. He & G. F. Liu, *Citrus nobilis* Andrews, non Lour., nom. illeg., *Citrus nobilis* var. *poonensis* Hayata, *Citrus* × *nobilis* Loureiro subf. *deliciosa* (Tenore) Hiroe, *Citrus* × *nobilis* var. *deliciosa* (Tenore) Guillaumin, *Citrus* × *nobilis* subf. *erythrosa* (Yu. Tanaka) Hiroe, *Citrus* × *nobilis* var. *major* Ker Gawler, *Citrus* × *nobilis* var. *ponki* Hayata, *Citrus* × *nobilis* subf. *reticulata* (Blanco) Hiroe, *Citrus* × *nobilis* var. *spontanea* Ito, *Citrus* × *nobilis* subf. *succosa* (Tanaka) Hiroe, *Citrus* × *nobilis* var. *sunki* Hayata, *Citrus* × *nobilis* subf. *tachibana* (Makino) Hiroe, *Citrus* × *nobilis* var. *tachibana* (Makino) Ito, *Citrus* × *nobilis* subf. *unshiu* (Marcowicz) Hiroe, *Citrus* × *nobilis* var. *unshiu* (Marcowicz) Tanaka ex Swingle, *Citrus* × *nobilis* var. *vangasy* (Bojer) Guillaumin, *Citrus ponki* Yu. Tanaka, *Citrus poonensis* Yu. Tanaka, *Citrus reticulata* var. *austera* Swingle, *Citrus reticulata* subsp. *deliciosa* (Tenore) Rivera et al., *Citrus reticulata* subsp. *tachibana* (Tanaka) Rivera et al., *Citrus reticulata* subsp. *unshiu* (Marcowicz) Rivera et al., *Citrus reticulata* var. *chrysocarpa* Tanaka, *Citrus succosa* Tanaka, *Citrus suhuiensis* Hayata, *Citrus sunki* Tanaka, *Citrus tachibana* (Makino) Yu. Tanaka, *Citrus tachibana* subf. *depressa* (Hayata) Hiroe, *Citrus tachibana* subf. *ponki* (Hayata) Hiroe, *Citrus tachibana* subf. *suhuiensis* (Hayata) Hiroe, *Citrus tachibana* subf. *sunki* (Hayata) Hiroe, *Citrus tangerina* Yu. Tanaka, *Citrus tankan* Hayata, *Citrus unshiu* Marcowicz, *Citrus vangasy* Bojer., *Sinocitrus reticulata* (Blanco) Tseng.

Family

Rutaceae

Common/English Names

Culate Mandarin, European Mandarin, Loose-Skinned Orange, Mandarin, Mandarin Orange, Maltese Orange, Sasutma, Sasutma Orange, Suntara Orange, Swatow Tangerine, Tangerine, Tangerine Orange, True Mandarin.

Vernacular Names

Brazil: Bergamota, Tangerine;
Chinese: Chu, Gan Ju, Jie, Ju, Mi Gan, Kuan Pi Gan, Kuan Pi Jie;

Banaban: Te Mantarin;
Czech: Mandarinka Obecná;
Danish: Mandarin;
Dutch: Mandarijn;
Eastonian: Harilik Mandariinipuu;
Esperanto: Mandarino;
Fijian: Moli Madarini, Soco Madarini;
Finnish: Klementiin, I Mandariini, Satsuma, Tangeriini;
French: Mandarine, Mandarinier, Mandarinier Commun;
German: Mandarinen, Mandarinenbaum, Tangerine;
Greek: Mantapinia I Koini;
Guam: Kahe Na Kikiki, Lalanghita;
Hawaiian: Alani-Pake, Tacibana;
Hungarian: Mandarin, Mandarin-Narancs;
India: Kamala, Sumothira (Assamese), Kamala, Kamala Leboo (Bengali), Naramgi, Santara (Hindu), Kanchi Kaayi, Kiththale, Kodagina Kithaale, Naagapuri Kitthale, Naarangi (Kannada), Madhuranaranna (Malayalam), Komola (Manipuri), Naarangi, Nowrangi, Santhara (Marathi), Kamala (Oriya), Aravata, Naranga, Narangah, Svadunarangah (Sanskrit), Kamalappalam, Narangam (Tamil), Naarinja (Telugu), Bahar Naranj, Gul-E-Bahar, Sangtara (Urdu);
Indonesia: Juruk Semaga (Bali), Jeruk Chempaga, Jeruk Jepun, Jeruk Keprok, Jeruk Paseh, (Sundanese); Jeruk Garut, Jeruk Kaprok, Jeruk Keprok, Jeruk Jepun, Jeruk Mandarin, Jeruk Ragi, Jeruk Siem;
Italian: Mandarina (Fruit), Mandarino (Tree);
Japanese: Mikan, Ponkan;
Korean: Gyur Na Mu, Geul Na Mu;
Kosraean: Muhsrisrik;
Lithuanian: Mandarininis Citrinmedis;
Malaysia: Limau Cina, Liamu Jepun, Limau Kupas Masak Hijau, Limau Mandarin, Limau Manis;
Nepalese: Kamala, Suntala;
New Calidonia: Le Mandarinnier;
Norway: Clementin, Mandarin, Tangerin
Pakistan: Santra;
Palauan: Kerekur;
Philippines: Ransas (Bikol), Ukban (Bisaya), Alsem (Bontok), Darangita (Iloko), Naranjita (Spanish), Daladan, Dalanghita, Darangita, Sintonis, Sinturis, Tison (Tagalog);
Polish: Mandarynka;
Portuguese: Mandarina;
Rotuman: Mor Jaene;
Samoan: Moli Saina;
Slovak: Mandarínkovník;
Spanish: Mandarina, Mandarino;
Sri Lanka: Jamanaran;
Swedish: Mandarin, Småcitrus;
Taiwan: Peng Gan, Tu Gan;
Thailand: Som Cheen, Som Khiao Waan, Som Keo Wahn;
Tibetan: Skyur Rtsi Chun Na;
Turkish: Mandalina;
Vietnamese: Cam Ngọt, Khoan Bì Cam, Khoan Bì Quất, Quít Ngọt, Quýt;
Yapese: Goligao.

Origin/Distribution

Mandarin orange is possibly a native of south east China and or south Japan. It is extensively cultivated in China, south of the Qin Ling, Taiwan, India, Sri Lanka, Myanmar, Japan, Java, Thailand, Vietnam, Australia and the Philippines. It was introduced to Europe only at the beginning of the nineteenth century. The commercial cultivation of mandarin oranges in the United States has developed mostly in Alabama, Florida and Mississippi and, to a lesser extent, in Texas, Georgia and California. Mexico and Brazil also has extensive plantings.

Agroecology

Mandarin oranges are subtropical in requirement but can also grow in the tropics. In it native range they are found in low elevations in hillside forests. They are much more cold-hardy than the sweet orange, and the tree is more tolerant of drought. The fruits are tender and readily damaged by cold. They perform and fruit best in full sun but will tolerate a partially shaded area. Like all citrus, they prefer fertile, free draining soils that are well-mulched, though they will grow in other soil types such as sandy soils and even clayey soils. It's best to plant them in mounded soil to ensure adequate drainage, as they don't tolerate water-logging of the roots. Regular watering will ensure bountiful harvest.

Edible Plant Parts and Uses

Mandarin oranges of all kinds are primarily eaten fresh, out-of-hand, or the segments are utilized in fruit salads, gelatins, puddings, confectionery, or on cakes. Very small types are canned in syrup. The fruit is also made into juice, drinks and beverages. The rind is also eaten dried or preserved in salt. Dried Mandarin rind (Plate 9) has a sweet, spicy flavour and is used in cakes, some Chinese dishes and green gram or red-bean porridge dessert.

The essential oil expressed from the peel is employed commercially in flavoring hard candy, gelatins, ice cream, chewing gum, and bakery goods. Mandarin essential oil paste is a standard flavoring for carbonated beverages. The essential oil, with terpenes and sesquiterpenes removed, is utilized in liqueurs. Petitgrain mandarin oil, distilled from the leaves, twigs and unripe fruits, has the same food applications. Tangerine oil is not suitable for flavoring purposes. Satsuma mandarin peel components may be valuable for the flavouring of foods, where floral-fresh-fruity aromas are required, such as chewing gums, sweets, teas, soft and energy drinks and milk products (Tao et al. 2008).

Plate 1 Short mandarin tree in fruit

Plate 2 Harvested local mandarins in Vietnam

Plate 3 Harvested local Mandarin in Jogjarkata, Indonesia

Botany

A small, evergreen tree growing to 3–5 m in height and width with numerous spiny branchlets (Plate 1). Leaves are unifoliolate, 6–8 cm long, lanceolate, ovate-lanceolate to elliptic-lanceolate, margin crenulate, base broadly cuneate, apex emarginate and petiole narrowly winged (Plates 2 and 8). Flowers are white, bisexual, 1–3 in axillary fascicles. Calyx is irregularly 3–5 lobed, petals white, stamens 20–25 more or less united into a tube, style long and slender with a clavate stigma. Fruit is subglobose, oblate or pyriform, 5–9 cm across, rind green turning glossy bright yellow, orange to reddish-orange with sunken oil glands, smooth, rough or warty; rind thin or thick, easily separable from sarcocarp which has 7–14 segments (Plates 1, 2, 3, 4, 5, 6, 7 and 8), sweet to

Plate 4 Mandarin fruit and leaves (close-view)

Plate 6 Emperor mandarin whole and peeled

Plate 5 Emperor Mandarin

acidic and sometimes bitter, with few to many seeds or rarely seedless; pulp vesicles plump, short, rarely slender and long. Seeds usually ovoid, smooth, base rounded, apex narrow and acute; embryos numerous, rarely solitary; cotyledons green.

Nutritive/Medicinal Properties

Nutrient composition of fresh, raw tangerines (mandarin oranges) per 100 g edible portion was reported as follows: water 85.17 g, energy 53 kcal

Plate 7 Honey green mandarin (Sarawak)

(223 kJ), protein 0.81 g, total fat 0.31 g, ash 0.38 g, carbohydrate 13.34 g, total dietary fibre 1.8 g, total sugars 10.58 g, sucrose 6.05 g, glucose

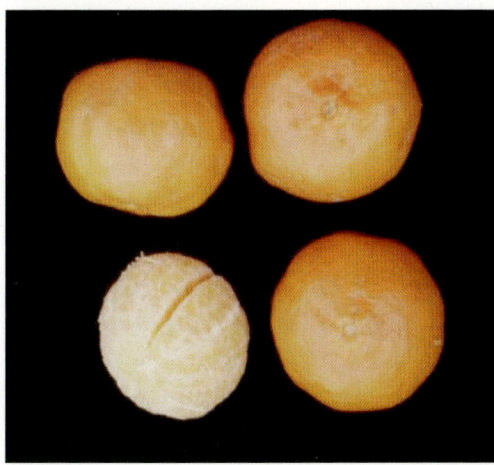

Plate 8 Imperial mandarin whole and peeled

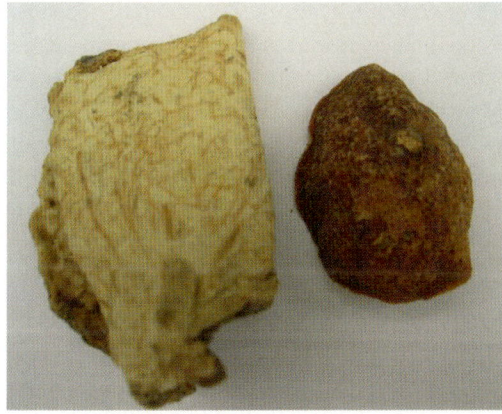

Plate 9 Dried Mandarin peel

2.13 g, fructose 2.40 g, Ca 37 mg, Fe 0.15 mg, Mg 12 mg, P 20 mg, K 166 mg, Na 2 mg, Zn 0.07 mg, Cu 0.042 mg, Mn 0.039 mg, Se 0.1 µg, vitamin C 26.7 mg, thiamine 0.058 mg, riboflavin 0.036 mg, niacin 0.376 mg, pantothenic acid 0.216 mg, vitamin B-6 0.078 mg, total folate 16 µg, total choline 10.21 mg, vitamin A 681 IU (34 µg RAE), vitamin E (α-tocopherol) 0.20 mg, total saturated fatty acids 0.039 g, 14:0 (myristic acid) 0.002 g, 16:0 (palmitic acid) 0.035 g, 18:0 (stearic acid) 0.002 g; total monounsaturated fatty acids 0.060 g, 16:1 undifferentiated (palmitoleic acid) 0.007 g, 18:1 undifferentiated (oleic acid) 0.053 g; total polyunsaturated fatty acids 0.065 g, 18:2 undifferentiated (linoleic acid) 0.048 g, 18:3 undifferentiated (linolenic acid) 0.018 g, β-carotene 155 µg, α-carotene 101 µg, β-cryptoxanthin 407 µg, lutein+zeaxanthin 138 µg; amino acids tryptophan 0.002 g, threonine 0.016 g, isoleucine 0.017 g, leucine 0.028 g, lysine 0.032 g, methionine 0.002 g, cystine 0.002 g, phenylalanine 0.018 g, tyrosine 0.015 g, valine 0.021 g, arginine 0.068 g, histidine 0.011 g, alanine 0.028 g, aspartic acid 0.129 g, glutamic acid 0.061 g, glycine 0.019 g, proline 0.074 g and serine 0.033 g (USDA 2011).

Nutrient composition of Imperial mandarin per 100 g value was reported as: energy 191 kJ, moisture 88.2 g, protein 0.9 g, Nitrogen 0.15 g, fat 0.2 g, ash 0.2 g, dietary fibre 1.4 g, available carbohydrates 9 g, total sugars 9 g, fructose 2.8 g, glucose 2 g, sucrose 4.3 g, malic acid 0.2 g, citric acid 0.9 g, Ca 31 mg, Mg 15 mg, P 16 mg, K 150 mg, Mn 0.022 mg, Cu 0.032 mg, Fe 0.46 mg, Na 3 mg, Zn 0.12 mg, thiamine (B1) 0.058 mg, riboflavin (B2) 0.023 mg, niacin (B3) 0.23 mg, pyridosine (B6) 0.05 mg, α-carotene 12 µg, β-carotene 46 µg, β-carotene equivalent 52 µg, xanthophyl 601.6 µg, retinol equivalent 9 µg, vitamin C 58 mg (Wills et al. 1985) and tryptophan 4 mg (Fox et al. 1988).

Citrus fruits and juices are an important source of bioactive phyotchemicals including antioxidants such as ascorbic acid, flavonoids, phenolic compounds, limonids and pectins that are important to human nutrition and health. *Citrus* fruits including mandarins are rich sources of flavonoids such as flavanones, flavones and flavonols (Gattuso et al. 2007). Mandarin juice was found to be quite similar to sweet orange juice, since they were characterized by the same distinctive flavanones (Gattuso et al. 2007). Hesperidin (24.3 mg/100 ml) was found to be the main component, followed by narirutin (3.92 mg/100 ml), didymin (1.44 mg/100 ml), eriocitrin (0.31 mg/100 ml) and neoeriocitrin (0.05 mg/100 ml), with neohesperidin and naringin being virtually absent, confirming *C. sinensis* and *C. reticulata* to be closely related. Mandarin juice also contained polymethoxyflavones that include sinensetin (1.05 mg/100 ml), tangeretin (0.26 mg/100 ml), nobiletin (0.23 mg/100 ml), heptamethoxyflavon (0.07 mg/100 ml) and quercetogetin (0.06 mg/100 ml). It also contained the aglycone, acacetin (0.02 mg/100 ml). Sweet oranges, tangerines and tangors were found to have similar amounts and characteristic flavones – hesperidin and narirutin with virtually no naringin or neohesperidin (Peterson et al. 2006). The following flavanones (mg aglycone/100 g juice or edible fruit) were found in sweet orange: didymin 1.11 mg, eriocitrin 0.02 mg, hesperidin 19.26 mg, narirutin 2.70 mg, naringin 0 mg, neoeriocitrin 0 mg, neohesperidine 0 mg. Tangerines contained 23 mg total aglycones from rutinose glycosides and none from neohesperidose glycosides. Five bioactive flavonoids, hesperidin, nobiletin, 3,5,6,7,8,3′,4′-heptamethoxyflavone, tangeretin, and 5-hydroxy-6,7,8,3′,4′-pentamethoxyflavone were isolated from *C. reticulata* pericarp (Zheng et al. 2009). Many osmolytes l-proline, N-methyl-l-proline (hygric acid), N,N-dimethyl-l-proline (stachydrine), 4-hydroxy-l-prolinebetaine (betonicine), 4-hydroxy-l-proline, γ-aminobutyric acid (Gaba), 3-carboxypropyltrimethylammonium (GabaBet), N-methylnicotinic acid (trigonelline), and choline were found in the fruit juices of yellow orange, blood orange, lemon, mandarin, bitter orange (*Citrus aurantium*), chinotto (*Citrus myrtifolia*), and grapefruit (Servillo et al. 2011).

The total flavonoid content exceeded the total carotenoid content in eight Taiwanese *Citrus* varieties (Wang et al. 2008). Ponkan (*C. reticulata*) peel had the highest total carotenoid content (2.04 mg/g dwb) and Wendun (*C. grandis*) and Peiyou (*C. grandis* Osbeck) peels, the lowest (0.036 and 0.021 mg/g dwb, respectively). Naringin was abundant in Peiyou and Wendun peels (29.8 and 23.9 mg/g dwb, respectively) and hesperidin was abundant in ponkan, tonkan (*C. tankan*), and Liucheng (*C. sinensis*) peels (29.5, 23.4, 20.7 mg/g db, respectively). Kumquat (*C. fortunella*) peel contained the most diosmin (1.12 mg/g dwb) and quercetin (0.78 mg/g dwb). Levels of caffeic acid (3.06–80.8 μg/g dwb) were much lower than that of chlorgenic acid, ferulic acid, sinapic acid and ρ-coumaric acid. Ponkan, kumquat and Liucheng peels contained the most total amounts of lutein, zeaxanthin, β-cryptoxanthin, and β-carotene (114, 113, and 108 mg/g dwb, respectively). The total pectin content ranged from 36.0 to 86.4 mg/g dwb. Satsuma mandarin (Owari) group contained the highest amount of flavanone glycosides hesperidin and narirutin in the rind, as well as the highest amounts of the carotenoid and β-cryptoxanthin, compared with Clementine mandarin (Fino, Loretina and Marisol), navel orange (Navelate and Navelina) and common orange (Valencia Late) verities (Bermejo et al. 2011). Both mandarin and orange varieties studied showed similar tendencies in phenolic compounds and total ascorbic acid concentrations. Limonene was the most abundant peel essential oil in all cultivars studied, followed by myrcene. Calcium and potassium were the dominant macronutrients for each cultivar studied.

Forty-one compounds were found in the volatile components of Philippine Dalandan (*Citrus reticulata*) juice (Dharmawan et al. 2007). They comprised terpenes, carbonyls, alcohols, esters and hydrocarbons, with limonene as the main compound including isopiperitenone. Mandarin peel oil contained decylaldehyde, γ-phellandrene, *p*-cymene, linalool, terpineol, nerol, linalyl, terpenyl acetate, aldehydes, citral, citronellal, and *d*-limonene. Petitgrain mandarin oil contained α-pinene, dipentene, limonene, *p*-cymene, methyl anthranilate, geraniol, and methyl methylanthranilate. Two major chemotypes, limonene and limonene/γ-terpinene, were distinguished for peel oils while three major chemotypes, sabinene/linalool, linalool/γ-terpinene and methyl N-methylanthranilate, were observed for leaf oils (Lota et al. 2000). The following components were found in the essential oil of *C. reticulata* fruit pericarp: limonene 80.81%, linalool 11.33%, δ-terpinene 4.50%, β-myrcene 1.37%, α-terpineole 1.06% and α-pinene 0.92% (González et al. 2002).

Twenty-eight components were isolated in the essential oil from Satsuma mandarin peel (Tao et al. 2008). Among the 25 components identified, the monoterpenes hydrocarbons ($C_{10}H_{16}$) group was predominant, accounting for 86.62% (w/w) of the total oil. Of these, limonene was the most abundant (67.44%), followed by β-myrcene (7.15%), 3-carene (4.4%), α-pinene (2.52%), p-cymene (2.43%), β-pinene (1.46%), sabinene (0.77%), terpinolene (0.47%) and α-thujene (0.45%). Oxygenated monoterpenes ($C_{10}H_{16}O$)

represented 6.33% of the total oil, with *cis* limonene oxide (1.45%) and *trans*-limonene oxide (0.81%) being the most abundant constituents of this group and 3-cyclohexene-1-acetaldehyde, α, 4-dimethyl (0.28%). The 1-pentanol (0.75%), *trans*-carveol (0.56%), *cis*-carveol (0.41%), carvone (0.7%) and limonene dioxide ($C_{10}H_{16}O_2$) (0.41%) were also detected. Oxygenated monoterpenes ($C_{10}H_{18}O$) were also found: neraniol (0.35%), β-terpineol (0.42%) (1S, %R)-2-cyclohexen-1-ol-2-methyl-5-(1-methylethenyl) (0.94%). The sesquiterpene ($C_{15}H_{24}$) contents were found to be only 3.43% of the total oil, mainly α-farnesene (1.11%) and β-elemene (1.04%) and also copaene (0.23%), germacrene-D(0.27%) and decahydro-4a-methyl-1-methylene-7-(1-methylethylidene) naphthalene (0.78%).

Several polymethoxylated flavones including nobiletin, 3,5,6,7,8,3′,4′-heptamethoxyflavone, tangeretin and 5-hydroxy-6,7,8,3′,4′-pentamethoxyflavone were separated from tangerine peel (Juhong in Chinese) by high-speed countercurrent chromatography (Wang et al. 2005). 26 mg of nobiletin, 6 mg of 3,5,6,7,8,3′,4′-hepta methoxyflavone, 35 mg of tangeretin and 11 mg of 5-hydroxy-6,7,8,3′,4′-pentamethoxyflavone were obtained from 150 mg crude extracts and their purities were 98.6%, 95.9%, 99.8% and 96.8%, respectively.

In addition to the main flavanone glycosides (i.e., hesperidin and naringin) in *Citrus* peel, polymethoxylated flavones and numerous hydroxycinnamates were also found and constituted the major phenolic constituents of the molasses byproduct generated from fruit processing (Manthey and Grohmann 2001). Although a small number of the hydroxycinnamates in citrus occur as amides, most occur as esters. The highest concentrations of hydroxycinnamates occurred in molasses of orange [*C. sinensis*] and tangerine (*C. reticulata*) compared to grapefruit (*C. paradisi*) and lemon [*C. limon*]. Concentrations of two phenolic glucosides, phlorin (phloroglucinol-β-O-glucoside) and coniferin (coniferyl alcohol-4-β-O-glucoside), were also determined. Of the polymethoxylated flavones in molasses from several tangerine and orange varieties, the highest amounts occurred in Dancy tangerine, whereas samples from two other tangerine molasses contained significantly lower levels, similar to those in the molasses samples from late- and early/mid-season oranges. Polymethoxylated flavones (PMFs) found in tangerine peel included isosinensetin (0.2%) (1), sinensetin (1.7%) (4), tetramethyl-o-isoscutellarein (0.3%) (5), nobiletin (40.5%) (6), tetramethyl-o-scutellarein (1.2%) (7), tangeretin (45.6%) (10), 5-demethylnobiletin (8.7%) (12), 5-demethyl tangeretin (0.8%) (14) and other flavonoids including heptamethoxyflavone (1.0%) (9) (Wang et al. 2007). In addition, compound 14 was isolated and identified for the first time from Citrus. Four compounds were isolated from the pericarp of *Citrus reticulata* and three of them were identified as natsudaidai, nobiletin and 3,5,6,7,8,3′,4′-hepta methoxyflavone (Qian and Chen 1998). The major polymethoxyflavones in the fruit peels of *Citrus reticulata* cv. ponkan were identified as isosinensetin, sinensetin, nobiletin and tetramethyl-o-scutellarein (Du and Chen 2010).

Fifty-eight constituents, amounting to 97.2% of the total volatiles were identified in mandarin peel essential oil (Muhoho Njoroge et al. 2006). Monoterpene hydrocarbons accounted for the most abundant chemical group (94.7%). Limonene was the most prominent constituent (84.8%), followed by γ-terpinene (5.4%), myrcene (2.2%) and α-pinene (1.1%). Sesquiterpene hydrocarbons accounted for a minor quantity (0.2%), where germacrene D and valencene were the main constituents. Oxygenated compounds of various chemical groups constituted 2.3%. Aliphatic aldehydes (0.7%) and terpene alcohols (0.7%) were the major chemical groups. The main constituents were linalool (0.7%), octanal (0.5%) and decanal (0.2%). Octyl acetate, α-sinensal, decanol and perillaldehyde occurred at 0.1% levels. Thymol, α-sinensal, methyl thymol, and the acetate esters, bornyl, α-terpinyl, geranyl, citronellyl and decyl acetates were detected. Another analysis identified 37 different components constituting approximately ≥99% of the mandarin peel essential oil (Chutia et al. 2009). The major components were limonene (46.7%), geranial (19.0%), neral

(14.5%), geranyl acetate (3.9%), geraniol (3.5%), β-caryophyllene (2.6%), nerol (2.3%), neryl acetate (1.1%) etc. Forty-four volatile components were identified in the fruit juice of composition of 65 cross pollinated hybrid fruits and their parents: mandarin (*Citrus reticulata* var. Willow Leaf) and clementine (*Citrus reticulata* x *Citrus sinensis* var. Commune) (Barboni et al. 2009). They accounted for 90.2–99.8% of the volatile fraction and limonene (56.8–93.3%) and γ-terpinene (0.1–36.4%) were the major components in all samples. The clementine juice was characterised by the dominance of limonene (90.0%) and a minor amount of γ-terpinene (1.2%) while the mandarin juice exhibited high amount of limonene (66.3%) and γ-terpinene (21.1%). All hybrid juices showed qualitatively similar composition but differing in the quantitative profile of the couple limonene/γ-terpinene.

Citrus reticulata (ponkan) cold-pressed oil and its oxygenated fraction consisted predominantly of the monoterpene group which accounted for more than 89.6% (w/w), of which limonene was the most abundant (80.3%) (Sawamura et al. 2004). Among the oxygenated compounds, octanal and decanal were the major ones among 12 aldehydes accounting for >1.5%; six alcohols were identified with a total concentration of >0.7%, while oxides, ketones and esters did not quantitatively or qualitatively contribute to the oil. Sniffing the ponkan cold-pressed oil and its oxygenated fraction demonstrated that octanal and decanal were the characteristic odor components of ponkan. A total of 45 components were identified in the Mandarin orange peel oil (Kırba lar et al. 2009). The major monoterpenes of the mandarin oils were found limonene (90.7%), γ-terpinene (3.9%), myrcene (2.1%), α-pinene (0.5%), sabinene (0.3%). The major sesquiterpene component was (E)-β-farnesene (0.1%). The following major oxygenated components of the oils found were: aldehyde components – octanal (0.2%), decanal (0.1%); alcohol components – linalool (0.4%), α-terpineol (0.1%) and ester components -geranyl acetate (0.2%) and neryl acetate (0.1%).

The percentage and concentration of total phenolics in *Citrus reticulata* cv Sunburst leaf, peel and juice were reported by Berhow et al. (1998) as: leaf: 23.5% (1.04 mg/g) flavone/flavonol, 54.6% (1.54 mg/g) flavanone, 0% psoralen, 0% coumarin; flavedo (outer pigmented layer of the peel): 24% (1.30 mg/g) flavone/flavonol, 41.7% (1.44 mg/g) flavanone, 0% psoralen, 10.9% (0.18 mg/g) coumarin; albedo (inner layer of the rind): 16.2% (0.98 mg/g), flavone/flavonol, 52.2% (2.01 mg/g) flavanone, 0% psoralen, 14.8% (0.27 mg/g) coumarin; juice: 8.2% (0.03 mg/g), flavone/flavonol, 68.3% (0.18 mg/g) flavanone, 0% psoralen, 0% coumarin.

The percentage and concentration of flavanones reported in *Citrus reticulata* cv Sunburst by Berhow et al. (1998) were as follows: leaf:- hesperidin 100% (1.5 mg/g); flavedo: hesperidin 97% (1.4 mg/g), narirutin 3% (traces); albedo: didymin 3% (0.1 mg/g), hesperidin 93% (1.9 mg/g), narirutin 4% (0.1 mg/g); juice hesperidin 100% (0.2 mg/g).

The percentage and concentration of flavone reported in *Citrus reticulata* cv Sunburst by Berhow et al. (1998) were as follows: leaf: diosmin 37% (0.4 mg/g), flavedo: rutin 8% (0.1 mg/g).

Five flavonoids in the parts of fruits of *C. reticulata* cv "Chachi" had been quantified (in μg/g) by Sun et al. (2010) as: Peel – naringin 811.5 μg, hesperidin 55260.4 μg, didymin 1232.7 μg, tangeretin 7702.1 μg, nobiletin 1520.4 μg; Pith – naringin 7083.1 μg, hesperidin 8538.2 μg, didymin 2228.6 μg, tangeretin 5.1 μg, nobiletin 194.2 μg; Endocarp – 3180.2 μg, hesperidin 1810.8 μg, didymin 1668.6 μg, tangeretin 1.6 μg, nobiletin 65.7 μg; Pulp – 584.0 μg, hesperidin 8369.4 μg, didymin 291.3 μg, tangeretin 10.7 μg, nobiletin 17.8 μg; Seed – 79.7 μg, hesperidin 241.6 μg, didymin 24.9 μg, tangeretin 3.0 μg and nobiletin 4.2 μg.

A new limonoid derivative, isolimonexic acid methyl ether was isolated from *C. reticulata* seeds besides limonin, deacetylnomilin, obacunone and ichangin (Khalil et al. 2003). A new limonoid, named citriolide-A was isolated from the liposoluble extract of *Citrus reticulata* seeds (Liao et al. 2011).

Pharmacological Activities of Citrus Flavonoids

Flavonoids had been reported to show a strong antioxidant and radical scavenging activity (Pietta 2000; Rice-Evans 2001) and to possess a host of biological activities that include prevention of cardiovascular disorders, anti cancerous, antiviral, antimicrobial, antiinflammatory, antiplatelet aggregation, antiulcerogenic, and antiallergenic activities.

A recent review of *Citrus* flavonoids discussed their properties and uses in anticancer, cardiovascular protection, and antiinflammatory and their role in degenerative diseases (Benavente-García and Castillo 2008). Epidemiological and animal studies pointed to a possible protective effect of flavonoids against cardiovascular diseases and some types of cancer. Many of the pharmacological properties of *Citrus* flavonoids could be attributed to the abilities of these compounds to inhibit enzymes involved in cell activation. In coronary heart disease, the protective effects of flavonoids include mainly antithrombotic, anti-ischemic, antioxidant, and vasorelaxant activities. Several key studies had shown that the antiinflammatory properties of *Citrus* flavonoids were due to its inhibition of the synthesis and biological activities of different pro-inflammatory mediators, mainly the arachidonic acid derivatives, prostaglandins E 2, F 2, and thromboxane A 2. The antioxidant and antiinflammatory properties of *Citrus* flavonoids could play a key role in their activity against several degenerative diseases and particularly brain diseases. The most abundant Citrus flavonoids reported were flavanones, such as hesperidin, naringin, or neohesperidin. However, generally, the flavones, such as diosmin, apigenin, or luteolin, possessed higher biological activity, even though they occurred in much lower concentrations. Diosmin and rutin had demonstrated activity as a venotonic agent and are present in several pharmaceutical products. Apigenin and their glucosides had exhibited good antiinflammatory activity without the side effects of other antiinflammatory products.

Currently *Citrus* limonoids are under investigation for a wide variety of therapeutic effects such as antiviral or viricide, antifungal, antibacterial, antineoplastic and antimalarial.

Antioxidant Activity

Of 20 *Citrus* fruit types screened based on rat liver microsomal lipid peroxidation induced by dihydronicotinamide adenine dinucleotide phosphate (NADPH) and adenosine diphosphate (ADP), the strongest antioxidative activity was found in ponkan (*Citrus reticulata*) (Tanizawa et al. 1992). The activities of the exocarp were stronger than those of the sarcocarp and the activities from immature fruits were greater than those from mature fruits.

Inhibitory effects of citrus peels on free radical generation were be higher than those of the corresponding juice sac parts, in particular, the peel portion of Dancy tangerine showed marked anti-oxidative activities in (1) O^{2-}generation by the xanthine (XA)-xanthine oxidase (XOD) system; (2) O^{2-}generation induced by 12-O-tetradecanoylphorbol-13-acetate (TPA) in differentiated human promyelocytic HL-60 cells; and (3) NO generation in murine macrophage RAW264.7 cells stimulated with lipopolysaccharide (LPS) and interferon (IFN)-γ (Murakami et al. 2000a).

The antioxidant activities of *Citrus unshiu* peels extracts increased as heating temperature increased (Jeong et al. 2004). Heat treatment of the ethanol or water extracts of citrus peel at 150°C for 60 minutes increased the total phenol contents (TPC), radical scavenging activity (RSA), and reducing power. Several low molecular weight phenolic compounds such as 2,3-diacetyl-1-phenylnaphthalene, ferulic acid, p-hydroxybenzaldoxime, 5-hydroxyvaleric acid, 2,3-diacetyl-1-phenylnaphthalene, and vanillic acid were newly formed in the citrus peel heated at 150°C for 30 minutes.

Moisture, fat, protein and ash content for all dried peel samples showed statistical differences among orange (*Citrus sinensis*), tangerine

(*Citrus reticulata*) and white grapefruit (*Citrus paradisi*) (Rincón et al. 2005). Tangerine's peel showed the highest magnesium and carotenoid content, while highest ascorbic acid and carotenoid content was found in the grapefruit's peel. Dietary fiber content was highest in tangerine peel. All samples presented high content of extractable polyphenols (4.33; 7.6 and 5.1 g/100 g respectively). The highest antiradical efficiency (DPPH) was shown by the tangerine's peel, which correlated with the polyphenol content. The results suggested that tangerine peel should be the most suitable, to reduce risk of some diseases such as cardiovascular and some associated to lipid oxidation and presented good sources of dietary fiber, phenolic compounds and antioxidants in the formulation of functional foods.

Two flavanones (naringin and hesperidin) were identified by HPLC; hesperidin accounted for 18.5–38.5% of the total phenolics in the species *Citrus unshiu, Citrus reticulata*, and *Citrus sinensis*, while naringin was only found in *Citrus changshanensis* and it accounted for 53.7% of the total phenolics in the segment membrane of this species (Abeysinghe et al. 2007). In the segment membrane of all selected species, the contents of phenolic compounds and total antioxidant capacities (TAC) were significantly higher than those in juice sacs and segment. Highest total phenolics, total flavonoids, naringin, and TAC were found in segment membrane of *C. changshanensis*, while the highest carotenoid content was found in juice sacs of *C. reticulata*. The contribution of vitamin C to TAC ranged from 26.9% to 45.9% in juice sacs and segments of all selected species. In segment membrane, however, a high contribution from hesperidin was observed in *C. unshiu* (54.0%), *C. sinensis* (46.7%) and *C. reticulata* (30.0%). The results indicated that segment membrane of citrus fruit were high in contents of bioactive compounds and TAC; the authors recommended to consume citrus fruit with all edible tissues rather than juice or juice sacs alone.

IC_{50} for antioxidant activity of the methanolic extracts of 13 commercially available *Citrus* species including *C. reticulata* in Iran ranged from 0.6 to 3.8 mg/ml (Ghasemi et al. 2009). Total phenolic content of the *Citrus spp.* samples (based on Folin Ciocalteu method) varied from 66.5 to 396.8 mg gallic acid equivalent/g of extract and flavonoids content (based on colorimetric AlCl3 method) varied from 0.3 to 31.1 mg quercetin equivalent/g of extract. The total phenolic and flavonoid contents were usually higher in peels than in tissues. *C. reticulata* var. Clementine and *C. reticulata* var. Page tissues showed the highest values (396.8 and 226.2 mg gallic acid equivalent/g of extract powder, respectively). There were no correlation between the total phenolic and/or flavonoids contents and antioxidant activity in tissues and/or peels. *C. reticulata* var. Ponkan peels showed highest antioxidant activity (IC_{50}=0.6 mg/ml) and *C. aurantium* tissues showed the weakest one (IC_{50}=3.9/ml). The IC_{50} values for ascorbic acid, quercetin and BHA were 5.05, 5.28 and 53.96 μg/ml, respectively. Studies in Brazil found the cravo tangerine to have the highest content of citric acid, while the pera orange was richest in ascorbic acid (Duzzioni et al. 2009). The lima orange had the highest total phenolic contents, and the ponkan tangerine the highest total carotenoids. The antioxidant activities, expressed as DPPH EC_{50}, ranged from 139.1 to 182.2 g extract/l for orange juice varieties and 186.3 to 275.5 g extract/l for tangerine juice varieties. In methanolic extracts the EC_{50} ranged from 192.5 to 267.4 g extract/l for orange varieties and from 225.2 to 336.3 g extract/l for tangerine varieties. For all the *Citrus* varieties studied, the radical scavenging capacity was higher in the aqueous than in the methanolic or acetone fractions.

Anticancer Activity

Citrus flavonoids act as suppressing agents preventing the formation of new cancers from procarcinogens, and blocking agents preventing carcinogenic compounds from reaching critical initiation sites, and transformation agents acting to facilitate the metabolism of carcinogenic components into less toxic materials or prevent their biological actions (Benavente-García and Castillo 2008).

The extract of immature tangerine peel (50 µg/ml) was found to increase apoptosis of human gastric cancer cells with typical apoptotic characteristics, including morphological changes of chromatin condensation and apoptotic body formation (Kim et al. 2005). The extract (50 µg/ml) reduced the expression of BCL-2, whereas the expression of BAX and CASP-3 was increased compared with the control group. Additionally, caspase-3 activity and caspase-3 protein expression in the CR-treated group was significantly increased compared with that in control group. These results suggested that the extract may induce apoptosis through the caspase-3 pathway in human gastric cancer cells.

Polymethoxylated Flavonoids

Citrus fruits including mandarins were found to contain high concentrations of several classes of phenols, including numerous hydroxycinnamates, flavonoid glycosides, and polymethoxylated flavones (Manthey and Guthrie 2009). The polymethoxylated flavones occurring without glycosidic linkages had been shown to inhibit the proliferation of a number of human cancer cell lines. In many cases the IC_{50} values occurred below 10 µm. Other hydroxylated flavone and flavanone aglycons also exhibited antiproliferative activities against the cancer cell lines, with the flavones showing greater activities than the flavanones. Glycosylation of these compounds removed their activity. The strong antiproliferative activities of the polymethoxylated flavones suggested that they may have use as anticancer agents in humans. Studies had shown polymethoxyflavones such as tangeretin and nobiletin to have cancer chemopreventive properties that may make them particularly useful as cancer chemopreventive agents (Walle 2007). At the cancer initiation stage, bioactivation of polyaromatic hydrocarbon carcinogens and binding to DNA are markedly diminished through effects on CYP1A1/1B1 transcription but also through direct interactions with the proteins. At the cancer promotion stage, the proliferation of cancer cells, but not normal cells, is inhibited with greater potency than with the unmethylated flavones. Flavonoid compounds isolated from tangerine oil solids, nobiletin, tangeretin, and tetra-O-methylisoscutellarein had been reported with significant activity against various strains of carcinoma cells (Chen et al. 1997).

In-vitro studies showed that of four plant flavonoids (quercetin, taxifolin, nobiletin and tangeretin), nobiletin and tangeretin, markedly inhibited cell growth a human squamous cell carcinoma cell line (HTB43) at all concentrations (2–8 µg/ml) tested (Kandaswami et al. 1991) Quercetin and taxifolin exhibited no significant inhibition at any of the concentrations tested. In another study, tangeretin and nobiletin, markedly inhibited the proliferation of squamous cell carcinoma (HTB 43) and a gliosarcoma (9L) cell line at 2–8 µg/ml concentrations (Kandaswami et al. 1992). Hirano et al. (1995) demonstrated that tangeretin (5,6,7,8,4′-pentamethoxyflavone) induced apoptosis in human promyelocytic leukaemia HL-60 cells, whereas the flavone showed no cytotoxicity against immune human peripheral blood mononuclear cells (PBMCs). Tangeretin showed no cytotoxicity against either HL-60 cells or mitogen-activated PBMCs even at high concentration (27 µM). Cycloheximide significantly decreased the tangeretin effect on HL-60 cell growth, suggesting that protein synthesis was required for flavonoid-induced apoptosis. The IC_{50} values range between 0.062 and 0.173 µM. The tangeretin effect was significantly attenuated in the presence of Zn^{2+}, which was known to inhibit Ca^{2+}-dependent endonuclease activity. Ca^{2+} and Mg^{2+}, in contrast, promoted the effect of tangeretin.

Citrus reticulata pericarp extract inhibited murine myeloid leukemic cell clone WEHI 3B (JCS) in a dose-dependent manner (Mak et al. 1996). The extract also induced differentiation of JCS cells into macrophages and granulocytes concomitant with an increase in phagocytic activity of the cells. In addition, both in-vitro clonogenicity and in-vivo growth of extract-treated JCS leukemic cells in syngeneic BALB/c mice were significantly reduced. The survival rate of mice receiving PCRJ treated JCS tumour cells was also increased. Two active components isolated from the extract were identified as nobiletin and tangeretin. Tangeretin and nobiletin showed potent inhibitory effects on human T lymphoblastoid leukemia MOLT-4 and its daunorubicin-resistant cells with the IC50 values of 7.1–14.0 μmol/L (Ishii et al. 2010). Both polymethoxyflavonoids inhibited the P-glycoprotein function and significantly influenced the cell cycle but they did not induce apoptosis. Nobiletin inhibited the tumour growth and angiogenesis by reducing vascular endothelial growth factor (VEGF) expression of human leukemia K562 cells xenograft in nude mice (Wang et al. 2009).

Studies indicated nobiletin to be a functionally novel and possible chemopreventive agent in inflammation-associated tumorigenesis (Murakami et al. 2000b). Nobiletin significantly inhibited two distinct stages of skin inflammation induced by double 12-O-tetradecanoylphorbol-13-acetate (TPA) application [first stage priming (leukocyte infiltration) and second stage activation (oxidative insult by leukocytes)] by decreasing the inflammatory parameters. Nobiletin also inhibited the expression of cyclooxygenase-2 and inducible NO synthase proteins and prostaglandin E2 release. Nobiletin inhibited dimethylbenz[a]anthracene/TPA-induced skin tumour formation by reducing the number of tumours per mouse. In-vitro studies showed that among putative anti-invasive agents, swainsonine (a locoweed alkaloid), captopril (an anti-hypertensive drug), tangeretin and nobiletin (both citrus flavonoids), the *Citrus* flavonoids, particularly nobiletin, showed the greatest downregulation of secretion of matrix metalloproteinase MMP-2 and MMP-9 parameters of brain tumour invasion (Rooprai et al. 2001). In addition, captopril and nobiletin were most efficient at inhibiting invasion, migration and adhesion in four representative cell lines (an ependymoma, a grade II oligoastrocytoma, an anaplastic astrocytoma and a glioblastoma multiforme). The results suggested that *Citrus* flavonoids, could be of future therapeutic value to combat brain tumours.

Nobiletin, inhibited proliferation of the cancer cell line, TMK-1, and its production of matrix metalloproteinase MMPs (Minagawa et al. 2001). In the SCID mouse model, we found that nobiletin inhibited the formation of peritoneal dissemination nodules from TMK-1 in the SCID mouse model, we found that nobiletin inhibited the formation of peritoneal dissemination nodules from TMK-1. The results suggested that nobiletin may be a candidate anti-metastatic drug for prevention of peritoneal dissemination of gastric cancer. Yoshimizu et al. (2004) found that nobiletin exerted antitumour effects on human gastric cancer cell lines TMK-1, MKN-45, MKN-74 and KATO-III by direct cytotoxicity, induction of apoptosis and modulation of cell cycle. Treatment with nobiletin 24 h prior to CDDP, a conventional anticancer drug, administration showed a synergistic effect compared to the control. Kawabata et al. (2005) reported that nobiletin attenuated proMMP-7 protein and its mRNA expression both concentration- and time-dependently in HT-29 human colorectal cancer cells via a reduction of activator protein-1 (AP-1) DNA binding activity, suggesting it to be a promising agent for suppression of cancer cell invasion and metastasis through MMP-7 gene repression. Nobiletin was found to exhibit antitumour metastasis in human fibrosarcoma HT-1080 cells (Miyata et al. 2008). Nobiletin directly inhibited mitogen-activated protein/extracellular signal-regulated kinase (MEK) activity and decreased the sequential phosphorylation of extracellular regulated kinases ERK, exhibiting the antitumour metastatic activity by suppressing matrix metalloproteinase expression in HT-1080 cells. Studies showed that 3 hydroxylated polymethoxyflavones (PMFs) from *Citrus*, namely: 5-hydroxy-3,6,7,8,3′,4′-hexamethoxyflavone, and 5-hydroxy-6,7,8,4′-tetramethoxyflavone showed much stronger inhibitory effects on the growth of human colon cancer HCT116 and HT29 cells in comparison with their permethoxylated counterparts, namely: nobiletin, 3,5,6,7,8,3′,4′-heptamethoxylflavone, and tangeretin suggesting the pivotal role of hydroxyl group at 5-position in the enhanced inhibitory activity by 5-hydroxy PMFs (Qiu et al. 2010). The results demonstrated that the inhibitory effects of 5-hydroxy PMFs were associated with their ability in modulating key signaling proteins related to cell proliferation and apoptosis, such as p21(Cip1/Waf1), CDK-2, CDK-4, phosphor-Rb, Mcl-1, caspases 3 and 8, and poly ADP ribose polymerase (PARP). Studies suggested that the three dietary flavonoids (chrysin, quercetin and nobiletin) were able to suppress the early phase of colon carcinogenesis in obese mice, partly through inhibition of proliferation activity caused by serum growth factors (Miyamoto et al. 2010). Each flavonoid (100 ppm), given in the diet throughout the experimental period, significantly reduced the numbers of aberrant crypt foci by 68–91% and β-catenin-accumulated crypts (BCACs) by 64–71%, as well as proliferation activity in the colon cancer lesions. Lee et al. (2011a, b) demonstrated that nobiletin suppressed invasion and migration involving FAK/PI3K/Akt and small GTPase signals in human gastric adenocarcinoma AGS cells. Feeding nobiletin to mice with azoxymethane (AOM)- and dextran sulfate sodium (DSS)-induced colon carcinogenesis, abolished colonic malignancy and notably decreased the serum leptin level by 75% (Miyamoto et al. 2008). Further, nobiletin dose-dependently suppressed the leptin-dependent proliferation of HT-29 colon cancer cells and decreased leptin secretion through inactivation of mitogen-activated protein kinase/extracellular signaling-regulated protein kinase, in differentiated 3T3-L1 mouse adipocytes. The results suggested that higher levels of leptin in serum promoted colon carcinogenesis in mice, whereas nobiletin had chemopreventive effects against colon carcinogenesis, partly through regulation of leptin levels.

Zhang et al. (2006) found that nobiletin significantly inhibited Heps tumour growth in mice inducing necrosis and apoptosis by stimulating T lymphocyte transformation and the production of TNF α and IL-2 (Zhang et al. 2006). Nobiletin exhibited inhibitory effect on the proliferation of human lung adenocarcinoma cell line A549 cells, while having a minimal effect on human umbilical vein endothelial cell line ECV-304 cells (Luo et al. 2008). It exerted marked notable inhibitory effect on the tumour growth in nude mice model which may be related to up-regulation of Bax and Caspase-9 and down-regulation of Bcl-2 proteins (Luo et al. 2008, 2009).

Diet supplementation of nobiletin and auraptene to had a protective effect in transgenic rats with prostate carcinogenesis (Tang et al. 2007). Nobiletin caused significant reduction in the ventral, lateral and dorsal prostate lobes, while decreasing high grade lesions in the ventral and lateral lobes. Feeding of auraptene also effectively reduced the epithelial component and high grade lesions in the lateral prostate. Growth of androgen sensitive LNCaP and androgen insensitive DU145 and PC3 human prostate cancer cells, was suppressed by both nobiletin and to a lesser extent auraptene in a dose-dependent manner, with significant increase in apoptosis. The authors further reported that feeding of nobiletin to rats with 2-amino-1-methyl-6-phenylimidazo[4,5-b] pyridine (PhIP)-induced prostate and colon carcinogenesis significantly reduced the relative prostate and testes weights as well as the Ki67 labeling index in the normal epithelium in the ventral prostate (Tang et al. 2011). However, the reduction in incidence and multiplicity of adenocarcinomas in nobiletin-treated ventral prostate were not statistically significant from controls. Nevertheless, nobiletin did significantly reduce the total number of colonic aberrant crypt foci (ACF) compared to the control value. The results suggested that nobiletin may have potential for chemoprevention of early changes associated with carcinogenesis in both the prostate and colon.

Tangeretin was found to inhibit the invasion of MO4 cells (Kirsten murine sarcoma virus transformed fetal mouse cells) into embryonic chick heart fragments in-vitro (Bracke et al. 1989). Tangeretin inhibited a number of intracellular processes, which led to an inhibition of cell motility and hence of invasion. Of six flavonoids tested, tangeretin was the most effective of the flavonoids in inhibiting B16F10 and SK-MEL-1 melanoma cell growth, showing a clear dose-response curve after 72 hours exposure (Rodriguez et al. 2002).

When cultures were treated for 24 hours, only slight inhibition at the highest concentrations (25 and 50 µM) of tangeretin and luteolin were observed. Quercetin, hesperetin, 7,3′-dimethylhesperetin and eriodictyol did not produce any effect at 24 hours on B16F10 or SK-MEL-1 cells. After 72 hours of exposure culture growth was inhibited by 7,3′-dimethylhesperetin at 50 µM, but lower concentrations had no effect. In-vitro studies showed that 24 hours tangeretin (5,6,7,8,4′-pentamethoxyflavone) exposure inhibited growth of colorectal carcinoma COLO 205 cells through modulation of the activities of several key G1 regulatory proteins, such as Cdk2 and Cdk4 or mediating the increase of Cdk inhibitors p21 and p27 (Pan et al. 2002). Tangeretin suppressed growth of human T47D mammary cancer cells and cytolysis by natural killer cells via inhibition of extracellular-signal-regulated kinases 1/2 (ERK1/2) phosphorylation in a dose- and time-dependent manner (Van Slambrouck et al. 2005). Tangeretin and nobiletin inhibited proliferation human breast cancer cell lines MDA-MB-435 and MCF-7 and human colon cancer line HT-29 in a dose- and time-dependent manner, and blocked cell cycle progression at G1 in all three cell lines (Morley et al. 2007). Tangeretin and nobiletin were found to be cytostatic and significantly suppressed proliferation by cell cycle arrest without apoptosis. In-vitro studies showed that tangeretin and 5-demethyl nobiletin induced apoptosis in human neuroblastoma SH-SY5Y cells by reducing the mitochondrial membrane potential suggesting that an intrinsic pathway of apoptosis was synergistically activated by the combination treatment with tangeretin and 5-demethyl nobiletin (Akao et al. 2008). On the other hand, in the combined treatment including nobiletin, the growth inhibitory activity of tangeretin was reduced. The results indicated the relevance of the combination of phytochemicals for the enhancement of the anticancer effect.

In-vitro studies revealed that the tangeretin inhibited IL-1beta-induced COX-2 expression in human lung epithelial carcinoma cells, A549 (Chen et al. 2007) This was mediated partly through suppression of NF-kappaB transcription factor as well as through suppression of the signaling proteins of p38 MAPK, JNK, and PI3K, but not of ERK. Separate in-vitro studies indicated that tangeretin exposure preconditioned A2780/CP70 and 2008/C13 cisplatin-resistant human ovarian cancer cells for a conventional response to low-dose cisplatin-induced cell death occurring through downregulation of PI3K/Akt signaling pathway (Arafael et al. 2009). Thus, effectiveness of tangeretin-cisplatin combinations, may have a promising modality in the treatment of resistant cancers. Results of separate in-vitro studies by Yoon et al. (2011) suggested that the antiinflammatory effects of tangeretin was due to its modulation of cell signaling and suppression of intracellular ROS generation reactive oxygen species (ROS) thereby protecting JB6 P+ mouse skin epidermal cells against oxidative stress. Tangeretin may have a potent chemopreventive effect in skin cancer.

Studies on the antiproliferative activity of four polymethoxyflavones from *C. reticulata* var ponkan fruit peels, namely isosinensetin, sinensetin, nobiletin and tetramethyl-o-scutellarein against four cancer cell lines (A549, HL-60, MCF-7 and HO8910) showed that isosinensetin had a lower IC_{50} value for MCF-7 and HO8910 cancer cell lines (Du and Chen 2010). The results showed ponkan peels to be excellent sources of functional polymethoxyflavones that may help prevent female cancers, such as ovarian cancer and breast cancer.

Limonoids

Topical applications of limonin to dimethylbenz[a]anthracene (DMBA)-induced buccal pouch epidermoid carcinomas in Syrain hamsters elicited a 60% reduction in tumour burden due to a 20% decrease in tumour number and a 50% decrease in tumour mass (Miller et al. 1989). The results for nomilin was considerably less effective as an inhibitor of DMBA-induced neoplasia. Limonin and nomilin, two bitter principles found in lemon, lime, orange and grapefruit were found to induce increased activity of the detoxifying enzyme glutathione S-transferase and to inhibit chemically induced carcinogenesis in laboratory animals (Lam et al. 1994). Administration of nomilin by gavage to ICR/Ha mice reduced the incidence and number of forestomach tumours per mouse induced by benzo[α]-pyrene (BP). Addition of nomilin and limonin to the diet at various concentrations inhibited BP-induced lung tumour formation in A/J mice and inhibited the formation of BP-DNA adducts in the lung. Topical application of the limonoids was found to inhibit both the initiation and the promotion phases of carcinogenesis in the skin of SENCAR mice. Nomilin appeared to be more effective during the initiation stage while limonin was more potent as an inhibitor during the promotion phase of carcinogenesis. These findings suggested *Citrus* limonoids may be useful as cancer chemopreventive agents.

In a pilot study, dietary feeding of male F344 rats with *Citrus* limonoids obacunone and limonin at dose levels of 200 and 500 p.p.m. during azoxymethane exposure for 4 weeks ('initiation' feeding) or after azoxymethane treatment for 4 weeks ('post-initiation' feeding) significantly inhibited (55–65%) aberrant crypt foci formation (Tanaka et al. 2000). In a long-term study dietary exposure to obacunone or limonin during the initiation phase and feeding during the post-initiation phase reduced the frequency of colonic adenocarcinoma (Tanaka et al. 2001).

Limonoids (obacunone 17 β-D-glucopyranoside, nomilinic acid 17 β-D-glucopyranoside, limonin, nomilin, and a limonoid glucoside mixture), found in high concentrations in mandarin (*Citrus reticulata*) was tested against a series of human cancer cell lines that included leukemia (HL-60), ovary (SKOV-3), cervix (HeLa), stomach (NCI-SNU-1), liver (Hep G2), and breast (MCF-7) (Tian et al. 2001). The growth-inhibitory effects of the four limonoids and the limonoid glucoside mixture against MCF-7 cells were significant. No significant effects were observed on growth of the other cancer cell lines treated with the four individual limonoids at 100 μg/ml. Obacunone exhibited dose and time dependant inhibition of pancreatic cancer (MDA Panc-28) cells proliferation (Chidambara Murthy et al. 2011) Obacunone upregulated expression of tumour suppressor protein p53, pro-apoptotic protein Bax and downregulated anti-apoptotic protein Bcl2. It also resulted in down regulation of vital inflammatory mediators such as NFκB and Cox-2 suggesting the antiinflammatory potential. Results provided evidence on activation of caspase dependant, cytochrome-c mediated intrinsic apoptosis and antiinflammatory activity in Panc-28 cells by citrus obacunone.

Citrus limonoids, obacunone and deoxylimonin were found to inhibit the development of 7,12-dimethylbenz[a]anthracene -induced oral tumours in hamster cheek pouch model (Miller et al. 2004). Obacunone reduced tumour number and burden by 25% and 40% respectively, whereas deoxylimonin reduced tumour number and burden by 30% and 50%, respectively. Citriolide-A a new limonoid isolated from the liposoluble extract of *Citrus reticulata* seeds exhibited medium cytotoxicity against murine leukemia P-388 and A-549 (human lung adenocarcinoma epithelial) cell lines (Liao et al. 2011).

Nomilin a triterpenoid present in common edible *Citrus* fruits including *C. reticulata* showed an inhibition of tumorous B16F-10 melanoma cell invasion and activation of matrix metalloproteinases (Pratheeshkumar et al. 2011). Treatment with nomilin induced apoptotic response. Nomilin treatment also exhibited a downregulated Bcl-2 and cyclin-D1 expression and upregulated p53, Bax, caspase-9, caspase-3, p21, and p27 gene expression in B16F-10 cells. Proinflammatory cytokine production and gene expression were found to be downregulated in nomilin-treated cells. The study also revealed that nomilin could inhibit the activation and nuclear translocation of antiapoptotic transcription factors such as nuclear factor (NF)-κB, CREB, and ATF-2 in B16F-10 cells.

Flavanones and Flavones

In-vivo studies in mice suggested that diet supplementation of the flavonoids diosmin (a flavone) and hesperidin (a flavanone), individually and in combination, were effective in significantly reducing inhibiting the frequency of bladder carcinoma and preneoplasia (Yang et al. 1997). The inhibition was attributed partly related to suppression of cell proliferation.

Iwase et al. (2001) found that 3,5,6,7,8,3′,4′-heptamethoxy (HPT) flavone from *Citrus* peel exhibited significant anti-tumour-initiating effect on mouse skin induced by a nitric oxide donor, (+/−)-(E)-methyl-2-[(E)-hydroxyimino]-5-nitro-6-methoxy-3-hexenamide indicating the possibility of HPT being a chemopreventive agent against nitric oxide (NO) carcinogenesis. Hesperidin and other crude peel extracts showed very low cytotoxicity to the B16 cell. Martínez Conesa et al. (2005) analysed the antimetastatic effects of the flavonoids tangeretin, rutin, and diosmin on pulmonary metastasis and the melanoma B16F10 cell subline highly metastatic in the lung. The greatest reduction in the number of metastatic nodules (52%) was obtained with diosmin; similarly, the percentages of implantation, growth index, and invasion index (79.40%, 67.44%, and 45.23%, respectively), were all compared with those of the ethanol group, considered to be an effective control group. Rutin- and tangeretin-treated groups also showed reductions of the same index compared with the ethanol group. Ghorbani et al. (2011) demonstrated that hesperidin-mediated proapoptotic and antiproliferative actions were regulated via both PPARγ-dependent and PPARγ-independent pathways in human pre-B NALM-6 cells suggesting hesperidin, a flavanone, could be developed as an agent against hematopoietic malignancies. Naringin (a flavanone) treatment resulted in significant dose-dependent growth inhibition of urinary bladder cancer 5637 cells together with G1-phase cell-cycle arrest at a dose of 100 mM (the half maximal inhibitory concentration) (Kim et al. 2008). Results suggested that the naringin-induced inhibition of cell growth appeared to be linked to the activation of Ras/Raf/ERK through p21WAF1-mediated G1-phase cell-cycle arrest.

Antiviral Activity

The limonoids, limonin and nomilin were found to possess antiretroviral activity (Battinelli et al. 2003) Limonin and nomilin were found to inhibit the human immunodeficiency virus-1 (HIV-1) replication in a culture of human peripheral blood mononuclear cells (PBMC) and on monocytes/macrophages. Both compounds were found to inhibit viral replication (EC_{50} 52.2 μM for nomilin and 60 μM for limonin). At all concentrations studied, the two terpenoids inhibited the production of HIV-p24 antigen even when PBMCs employed were chronically infected (EC_{50} values of 61.0 μM for limonin and 76.2 μM for nomilin). The limonoids were effective at inhibiting HIV-1 replication in infected monocytes/macrophages at concentrations ranging from 20 to 80 μM. The mechanism of action was attributed to in-vitro inhibition of HIV-1 protease activity.

Suzuki et al. (2005) showed that both the ethyl acetate layer of *Citrus unshiu* peel extract and fraction 7 decreased hepatitis C virus absorption in MOLT-4 cells (a human lymphoblastoid leukemia cell line). Further they demonstrated that 3′,4′,5,6,7,8-hexamethoxyflavone (nobiletin) the active ingredient markedly inhibited hepatitis C virus infection in MOLT-4 cells.

Antimicrobial Activity

The ethanol soluble fraction of *C. reticulata* peel was found to be most effective against different gram positive and gram negative bacteria compared to the hexane, chloroform and acetone (Jayaprakasha et al. 2000). Fractionation of ethanol-soluble fraction yielded three polymethoxylated flavones, namely desmethylnobiletin, nobiletin and tangeretin.

The essential oils of lemon (*Citrus lemon*), mandarin (*Citrus reticulata*), grapefruit (*Citrus paradisi*) and orange (*Citrus sinensis*) showed antifungal activity against all the moulds commonly associated with food spoilage: *Aspergillus niger, Aspergillus flavus, Penicillium chrysogenum* and *Penicillium verrucosum* (Viuda-Martos et al. 2008). Orange essential oil was the most effective against *A. niger*, mandarin essential oil was most effective at reducing the growth of *Aspergillus flavus* while grapefruit was the best inhibitor of the moulds *P. chrysogenum* and

P. verrucosum. Mandarin orange peel oil exhibited high antibacterial activity against Gram positive bacteria (*Staphylococcus aureus*, *Bacillus cereus*, *Mycobacterium smegmatis*, *Listeria monocytogenes*, *Micrococcus luteus*) and Gram negative bacteria (*Escherichia coli*, *Klebsiella pneumoniae*, *Pseudomonas aeruginosa*, *Proteus vulgaris*) and antifungal activity against *Kluyveromyces fragilis*, *Rhodotorula rubra*, *Candida albicans*, *Hanseniaspora guilliermondii* and *Debaryomyces hansenii* yeasts (Kırbaşlar et al. 2009).

Antiinflammatory Activity

Pretreatment of rats with hesperidin (50 and 100 mg/kg, s.c.) reduced the paw oedema induced by carrageenan by 47% and 63%, respectively (Emim et al. 1994). The effect was equivalent to that produced by indomethacin (10 mg/kg, p.o.). Hesperidin also inhibited pleurisy induced by carrageenan, reducing the volume of exudate and the number of migrating leucocytes by 48% and 34%, respectively, of control values. Equal doses of duartin and claussequinone were ineffective in all the above tests. Pretreatment of mice with hesperidin (100 mg/kg, s.c.) reduced acetic acid-induced abdominal constriction by 50%, but did not affect the tail flick response. Hyperthermia induced by yeast in rats was slightly reduced by hesperidin. No lesions of the gastric mucosae were detected in rats pretreated with hesperidin. The results indicated that hesperidin may present a potential therapeutical use as a mild antiinflammatory agent.

Nobiletin showed a higher antiinflammatory activity than indomethacin in a 12-O-tetradeca-noylphorbol-13-acetate (TPA)-induced edema formation test in mouse ears (Murakami et al. 2000a). Nobiletin was found to suppress the interleukin (IL)-1-induced production of prostaglandin PGE(2) in human synovial cells in a dose-dependent manner (<64 μM) (Lin et al. 2003). Further, it selectively decreased COX-2, but not COX-1 mRNA expression. Nobiletin also interfered with the lipopolysaccharide-induced production of PGE(2) and the gene expression of proinflammatory cytokines including IL-1α, IL-1beta, TNF-α and IL-6 in mouse J774A.1 macrophages. In addition, nobiletin decreased the IL-1-induced gene expression and production of proMMP-1/procollagenase-1 and proMMP-3/prostromelysin-1 in human synovial fibroblasts but augmented production of the endogenous MMP inhibitor, TIMP-1. These antiinflammatory actions of nobiletin were very similar to those of antiinflammatory steroids such as dexamethasone, and the enhancement of TIMP-1 production was a unique action of nobiletin. The results further supported nobiletin to be a candidate as a novel immunomodulatory and antiinflammatory drug.

Citrus reticulata immature peel extract significantly inhibited the lipopolysaccharide (LPS)-induced nitric oxide (NO) production, inducible nitric oxide synthase (iNOS) protein and mRNA expression in RAW 264.7 macrophage cells (Jung et al. 2007). The extract also reduced the iNOS promoter activity in piNOS-LUC-transfected cells and significantly inhibited the activity of NF-κB DNA binding activity in LPS-induced macrophage cells. The results suggested that the extract may suppress LPS-stimulated NO production by inhibiting nuclear factor kappa B.

Among the seven tested *Citrus* peels, ponkan (*Citrus reticulata*) and tonkan (*Citrus tankan*) exerted outstanding inhibitory effect on prostaglandin E2 (PGE2) and nitric oxide secretion in lipopolysaccharide (LPS) activated RAW 264.7 cells (Huang and Ho 2010). The polymethoxy flavone content, especially nobiletin, appeared to correlate well with the antiinflammatory activities of the citrus peel extracts. Release of nitric oxide (NO), the major inflammatory mediator in microglia, was markedly suppressed in a dose-dependent manner following nobiletin treatment (1–50 μM) in LPS-stimulated BV-2 microglia cells (Cui et al. 2010). The inhibitory effect of nobiletin from *Citrus* peel was similar to that of minocycline, a well-known microglial inactivator. Nobiletin significantly inhibited the release of the pro-inflammatory cytokine tumour necrosis factor (TNF-α) and interleukin-1β (IL-1β). LPS-induced phosphorylations of extracellular signal-regulated kinase (ERK), c-Jun NH(2)-terminal kinase

(JNK), and p38 mitogen-activated protein kinases (MAPKs) were also significantly inhibited by nobiletin treatment. Additionally, nobiletin markedly inhibited the LPS-induced pro-inflammatory transcription factor nuclear factor κB (NF-κB) signaling pathway by suppressing nuclear NF-κB translocation from the cytoplasm and subsequent expression of NF-κB in the nucleus. These results indicated the therapeutic potential and molecular mechanism of nobiletin in relation to neuroinflammation and neurodegenerative diseases. Results of separate studies in mice suggested that naringin exhibited anti-inflammatory effects through inhibiting lung edema, myeloperoxidase (MPO: a marker enzyme of neutrophil granule) and inducible nitric oxide synthase (iNOS) activities, tumour neurosis factor-α (TNF-α) and pulmonary neutrophil infiltration by blockade of NF-κB in lipopolysaccharide (LPS)-induced acute lung injury (Liu et al. 2011).

In-vitro studies by Nazari et al. (2011) demonstrated that hesperidin inhibited proliferation of Ramos Burkitt's lymphoma cells and sensitized them to doxorubicin-induced apoptosis through inhibition of both constitutive and doxorubicin-mediated NF-κB activation in a PPARγ (Peroxisome Proliferator-Activated Receptor-γ) -independent manner. Supplementation of irradiated mice with hesperidin at 50 mg/kg of body weight increased splenocyte proliferation, CD4(+) and CD8(+) lymphocytes on Day 10 after irradiation compared with that of the control group without irradiation on Day 30 after irradiation (Lee et al. 2011a, b). The concentration of serum cytokines (interleukin-1β, interleukin-6, and tumour necrosis factor-α) decreased in the radiation group treated with hesperidin at 50 and 200 mg/kg of body weight compared with the control group on Day 10 after irradiation. Irradiated mice fed 50 mg/kg of body weight hesperidin had significantly higher levels of total protein and albumin compared with the other groups 30 days after irradiation. The findings suggested that hesperidin may enhance immunocompetence with beneficial effects on nutritional status, and decrease irradiation-induced inflammation in mice. Using the rat air pouch–carrageenan model of inflammation, naringin and hesperidin were found to have antiinflammatory and antioxidative properties (Jain and Parmar 2011). Hesperidin proved to be better than indomethacin and naringin because of more pronounced pharmacological actions without tissue toxicity.

Antitussive Activity

Naringin exhibited antitussive effect on experimentally induced cough in guinea pigs (Gao et al. 2011). Results suggested that naringin was not a central antitussive drug and did not exert its peripheral antitussive effect through either the sensory neuropeptides system or the modulation of ATP-sensitive K (+) channels.

Gastroprotective Activity

Oral pretreatment of rats with the highest dose of naringin (400 mg/kg), was found to have a cytoprotective effect against ethanol-induced gastric ulcers (Martín et al. 1994). This activity appeared to be mediated by non-prostaglandin-dependent mechanisms

Antihypercholesterolemic/Antiobesity Activity

In-vitro studies found that tangeretin modulated apoB-containing lipoprotein metabolism through multiple mechanisms in the human hepatoma cell-line HepG2 (Kurowska et al. 2004). The marked reduction in apoB secretion observed in cells incubated with tangeretin was rapid, apoB-specific, and partly reversible. A 24-h exposure of cells to tangeretin decreased intracellular synthesis of cholesteryl esters, free cholesterol, and TAG (triacylglycerol) and mass of cellular TAG. This suppression was associated with decreased activities of DAG (diacylglycerol) acyltransferase and microsomal triglyceride transfer protein. Tangeretin was also found to activate the peroxisome proliferator-activated receptor, a transcription factor with a positive regulatory impact on fatty acid oxidation and TAG availability.

Male mice fed a high fat diet supplemented with 0.2%w/w nomilin for 77 days had lower body weight, serum glucose, serum insulin, and enhanced glucose tolerance compared to mice fed a HFD alone (Ono et al. 2011). The results suggested a novel biological function of nomilin as an agent having anti-obesity and anti-hyperglycemic effects that are likely to be mediated through the activation of TGR5, a member of the G protein-coupled receptor family. A dose-response analysis showed that 200 μmol/L hesperetin, a flavanone present as a disaccharide in oranges, increased LDL receptor (LDLR) mRNA levels 3.6- to 4.7-fold of the untreated control, whereas nobiletin, a PMF found at the highest concentration in oranges and tangerines, achieved maximal stimulation of 1.5- to 1.6-fold of control at only 5 μmol/L (Morin et al. 2008). Hesperetin sustained induction, whereas nobiletin was inhibitory at high doses, resulting in an inverted-U dose response. Thus, citrus flavonoids were likely to act through the sterol regulatory element -binding proteins, with PMF initially activating these mechanisms at considerably lower concentrations than flavanones. This was in accord with earlier studies that showed polymethoxylated flavones (PMF) reduced plasma cholesterol levels at lower doses than required for flavanones.

In-vitro studies showed that naringin, a flavanone, protected vascular smooth muscle cells (VSMCs) from the mitogenic effect of lysophosphatidylcholine, an atherogenic lysophospholipid of oxidized low density lipoprotein (LDL), known to impair endothelial release of nitric oxide, to up-regulate the expression of adhesion molecules, and to promote the proliferation of VSMCs (Kim et al. 2003).

Radioprotective Activity

Treatment of mice with 2 mg/kg body weight naringin before exposure to various doses of γ-radiation resulted in a significant reduction in the frequencies of aberrant cells and chromosomal aberrations like acentric fragments, chromatid and chromosome breaks, centric rings, dicentrics and exchanges (Jagetia et al. 2003). Naringin, was also found to scavenge free radicals of hydroxyl, superoxide and 2,2-diphenyl-1-picryl hydrazyl, in a dose-dependent manner. Naringin at 5 μM scavenged the 2,2-azino-bis-3-ethyl benzothiazoline-6-sulphonic acid cation radical very efficiently, where a 90% scavenging was observed. The study demonstrated that naringin could protect mouse bone marrow cells against radiation-induced chromosomal damage.

Antiplatelet Activity

Flavonoid compounds sinensetin and nobiletin, isolated from tangerine oil solids, decreased erythrocyte aggregation and sedimentation in-vitro (Chen et al. 1997).

Melanogenesis Activity

Nobiletin was found to increase melanin content and tyrosinase activity in murine B16/F10 melanoma cells via the extracellular signal-regulated kinase (ERK) pathway (Yoon et al. 2007).

Antimalarial and Mosquitocidal Activity

Three limonoids, namely limonin, nomilin and obacunone, isolated from the seeds of *Citrus reticulata* exhibited moult inhibiting activity in mosquito *Culex quinquefasciatus* larvae (Jayaprakasha et al. 1997). With fourth instar larvae of mosquito *Culex quinquefasciatus*, the EC_{50} for inhibition of adult emergence was 6.31, 26.61 and 59.57 ppm for obacunone, nomilin and limonin, respectively. The efficacy of a limonoid enriched fraction was similar to that of limonin, probably due to the high content (70%) of limonin. Isolimonexic acid methyl ether from *C. reticulata* seeds showed marginal activity against *Plasmodium falciparum* with $IC_{50}>$ 4.76 μg/ml, with selectivity index >1 and without any cytotoxicity to Vero cells (Vero African Green Monkey kidney cells) (Khalil et al. 2003).

Antipulmonary Fibrosis Activity

A Chinese herbal formula, Hu-qi-yin possessed an antipulmonary fibrosis effect (Zhou et al. 2009). Pericarp of *Citrus reticulata*, one of the herbal drugs contained in this formula showed the most potent inhibitory activity on the proliferation of human embryonic lung fibroblasts (HELF). The ethanol extract showed the strongest inhibitory activity on HELF proliferation. Further research using BLM (bleomycin)-induced rat model revealed that the ethanol extract at the doses of 100 and 200 mg/(kg day) caused a marked increase of body weight at first 7 days, significantly lowered the hydroxyproline levels in lung, greatly improved the pathologic scores, as well as inhibited the overexpressions of TGF-β1 protein and mRNA The results suggest that the ethanol extract of *Citrus reticulata* had antipulmonary fibrosis effects and might have a great potential for the treatment of fibrosis of lung.

Antiosteoporotic Activity

Nobiletin significantly suppressed bone loss in ovariectomized ddY mice and type II collagen-induced arthritis in DBA/1J mice (Murakami et al. 2007). Nobiletin attenuated receptor activator of nuclear factor kappaB ligand (RANKL)-induced osteoclastogenesis of RAW264.7 cells. Nobiletin also down-regulated RANKL-activated extracellular signal-regulated kinase1/2, c-Jun N-terminal kinase1/2, and p38 mitogen-activated protein kinase activities, and thereby modulated the promoter activation of nuclear factor kappaB (NFkappaB) and activator protein-1, key transcription factors. The results suggested nobiletin to be a promising phytochemical for the prevention or treatment of osteoclastogenesis-related disorders, including osteoporosis and rheumatoid arthritis. In-vitro, nobiletin suppressed osteoclast formation and bone resorption induced by interleukin (IL)-1 (Harada et al. 2011). Nobiletin suppressed the expression of cyclooxygenase-2, NFκB-dependent transcription, and prostaglandin E (PGE) production induced by IL-1 in osteoblasts. In-vivo, in ovariectomized mice showing severe bone loss in the femur by increased bone resorption due to estrogen deficiency, nobiletin significantly restored the bone mass. Thus, nobiletin could be beneficial to bone health in postmenopausal women.

Wound Healing Activity

In-vitro studies showed that *Citrus reticulata* extract inhibited proliferation of fibroblasts in human hypertrophic scar taken from burnt victims by promoting apoptosis and collagen degradation (Qi et al. 2006). The authors concluded that *Citrus reticulata* extract might be beneficial in the management of hypertrophic scars.

Antiangiogenic Activity

Citrus reticulata peel extract was found to have antiangiogenic activity (Chrisnanto et al. 2008). The extract inhibited new blood vessels formation in chorio allantoic membrane induced by bFGF (basic fibroblast growth factor). Also macrophage cells in the extract treated group was less than the bFGF+vehicle group. Recent studies showed nobiletin to be a novel antiangiogenic compound that exhibited its activity through combined inhibition of multiple angiogenic endothelial cells functions (Kunimasa et al. 2010) Nobiletin suppressed the proliferation, migration and tube formation on matrigel of human umbilical vein EC (HUVEC) stimulated with endothelial cell growth supplement (ECGS), a mixture of acidic and basic fibroblast growth factors. nobiletin suppressed pro-matrix metalloproteinase-2 (proMMP-2) production and MMP-2 mRNA expression in ECGS-stimulated HUVEC. Nobiletin also down-regulated cell-associated plasminogen activator (PA) activity and urokinase-type PA mRNA expression. It also down-regulated angiogenesis-related signalling molecules, such as extracellular signal-regulated kinase 1/2 and c-Jun N-terminal kinase, and transcriptional factors (c-Jun and signal transducer and activator of transcription 3), and activation of the caspase pathway.

Caries Control Activity

In-vitro studies showed that lesion depth and mineral loss were reduced in root dentine collagen incubated with *Citrus* hesperidin or chlorhexidine (Hiraishi et al. 2011). Remineralization in deep lesions was found when the matrix was incubated in hesperidin, whilst no mineral uptake occurred in deep lesion when incubated in chlorhexidine. The results indicated that hesperidin preserved collagen and inhibited demineralization, and enhanced remineralization even under the fluoride-free condition.

Antischistosomal Activity

Schistosoma mansoni infection of mice livers caused a marked reduction in succinate dehydrogenase (SDH); lactate dehydrogenase (LDH), aspartate aminotransferase (AST); alanine aminotransferase (ALT) enzyme activities and a significant increase in G-6-Pase, AP, 5′-nucleotidase, and alkaline phosphatase (ALP) (Hamed and Hetta 2005). Treatment with the ethanolic extract of *Citrus reticulata* root or the oleo-resin extract from myrrh, *Commiphora molmol* tree (Mirazid), a new antishistosomal drug, ameliorated all these enzyme activities with a noticeable reduction in ova count and worm burden.

Traditional Medicinal Uses

The dried and mature peel of *Citrus reticulata* (Guang Chenpi) is an important Chinese herbal medicine and has been recorded in the Chinese Pharmacopoeia as appropriate for medical use (Sun et al. 2010). Various parts of the fruit are used for various ailments (Yeung 1985; Chopra et al. 1986; Lu 2005; Stuart 2010). The pericarp is deemed as analgesic, antiasthmatic, anticholesterolemic, anti-inflammatory, antiscorbutic, antiseptic, antitussive, carminative, expectorant and stomachic; the peel as warm, pungent, anodyne, aperitif, digestive, diuretic, expectorant, febrifuge, sedative, refrigerant, stomachic; the endocarp as carminative and expectorant, unripened green exocarp as carminative and stomachic; and seed as analgesic and carminative. The pericarp, endocarp, peel is used in the treatment of dyspepsia, gastro-intestinal discomfort, cough with profuse phlegm, hiccup, vomiting, and the peel also for lack of appetite, revitalising tonic, fish and crab poisoning and acute mastitis. The unripened green exocarp is employed for the treatment of chest pains, hypochondrium, gastro-intestinal distension, swelling of the liver and spleen and cirrhosis of the liver and the is used in the treatment of hernia, lumbago, mastitis and pain or swellings of the testes. In Ayurvedic medicine, tangerine has been used for general debility, urinary retention, indigestion, anorexia, hepatitis, and sexual weakness.

Other Uses

Mandarin essential oil, Petitgrain oil and tangerine oil, and their various tinctures and essences, are employed in perfume-manufacturing, particularly in the formulation of floral compounds and colognes. They are produced mostly in Italy, Sicily and Algiers.

Mandarin peel essential oil displayed inhibitory effect against several plant pathogenic fungi (Chutia et al. 2009).

Comments

Mandarins are smaller than Satsuma mandarin or King mandarins.

Selected References

Abeysinghe DC, Li X, Sun CD, Zhang WS, Zhou CH, Chen KS (2007) Bioactive compounds and antioxidant capacities in different edible tissues of *Citrus* fruit of four species. Food Chem 104(4):1338–1344

Akao Y, Itoh T, Ohguchi K, Iinuma M, Nozawa Y (2008) Interactive effects of polymethoxy flavones from citrus on cell growth inhibition in human neuroblastoma SH-SY5Y cells. Bioorg Med Chem 16(6):2803–2810

Arafael SA, Zhu Q, Barakat BM, Wani G, Zhao Q, El-Mahdy MA, Wani AA (2009) Tangeretin sensitizes cisplatin-resistant human ovarian cancer cells through downregulation of phosphoinositide 3-kinase/ Akt signaling pathway. Cancer Res 69(23): 8910–8917

Barboni T, Luro F, Chiaramonti N, Desjobert JM, Muselli A, Costa J (2009) Volatile composition of hybrids *Citrus* juices by headspace solid-phase micro extraction/gas chromatography/mass spectrometry. Food Chem 116(1):382–390

Battinelli L, Mengoni F, Lichtner M, Mazzanti G, Saija A, Mastroianni CM, Vullo V (2003) Effect of limonin and nomilin on HIV-1 replication on infected human mononuclear cells. Planta Med 69:910–913

Bayer RJ, Mabberley DJ, Morton C, Miller CH, Sharma IK, Pfeil BE, Rich S, Hitchcock R, Sykes S (2009) A molecular phylogeny of the orange subfamily (Rutaceae: Aurantioideae) using nine cpDNA sequences. Am J Bot 96:668–685

Benavente-García O, Castillo J (2008) Update on uses and properties of *Citrus* flavonoids: new findings in anticancer, cardiovascular, and anti-inflammatory activity. J Agric Food Chem 56(15):6185–6205

Berhow M, Tisserat B, Kanes K, Vandercook C (1998) Survey of phenolic compounds produced in *Citrus*. USDA ARS Tech Bull 1856:1–154

Bermejo A, Llosá MJ, Cano A (2011) Analysis of bioactive compounds in seven *Citrus* cultivars. Food Sci Technol Int 17(1):55–62

Bown D (1995) Encyclopaedia of herbs and their uses. Dorling Kindersley, London, 424 pp

Bracke ME, Vyncke BM, Van Larebeke NA, Bruyneel EA, De Bruyne GK, De Pestel GH, De Coster WJ, Espeel MF, Mareel MM (1989) The flavonoid tangeretin inhibits invasion of MO4 mouse cells into embryonic chick heart in vitro. Clin Exp Metastasis 7(3):283–300

Burkill IH (1966) A dictionary of the economic products of the Malay Peninsula. Revised reprint, 2 vols. Ministry of Agriculture and Co-operatives, Kuala Lumpur, vol 1 (A–H), pp 1–1240, vol 2 (I–Z), pp 1241–2444

Chen J, Montanari AM, Widmer WW (1997) Two new polymethoxylated flavones, a class of compounds with potential anticancer activity, isolated from cold pressed dancy tangerine peel oil solids. J Agric Food Chem 45(2):364–368

Chen KH, Weng MS, Lin JK (2007) Tangeretin suppresses IL-1beta-induced cyclooxygenase (COX)-2 expression through inhibition of p38 MAPK, JNK, and AKT activation in human lung carcinoma cells. Biochem Pharmacol 73(2):215–227

Chidambara Murthy KN, Jayaprakasha GK, Patil BS (2011) Apoptosis mediated cytotoxicity of *Citrus* obacunone in human pancreatic cancer cells. Toxicol In Vitro 25(4):859–867

Chopra RN, Nayar SL, Chopra IC (1986) Glossary of Indian medicinal plants (including the supplement). Council Scientific Industrial Research, New Delhi, 330 pp

Chrisnanto E, Adelina R, Dyaningtyas Dewi PP, Abdi Sahid MN, Setyaningtias D, Jenie RI, Edy Meiyanto E (2008) Antiangiogenic effect of ethanol extract of *Citrus reticulata* peel in the chorio allantoic membrane (CAM) induced by bFGF. In: Proceedings of the international symposium on molecular targeted therapy, Faculty of Pharmacy UGM, pp 57–66

Chutia M, Bhuyan PD, Pathak MG, Sarma TC, Boruah P (2009) Antifungal activity and chemical composition of *Citrus reticulata* Blanco essential oil against phytopathogens from North East India. LWT- Food Sci Technol 42(3):777–780

Cottin R (2002) *Citrus* of the world: a *Citrus* directory, version 2.0. SRA INRA-CIRAD, France

Cui Y, Wu J, Jung SC, Park DB, Maeng YH, Hong JY, Kim SJ, Lee SR, Kim SJ, Kim SJ, Eun SY (2010) Anti-neuroinflammatory activity of nobiletin on suppression of microglial activation. Biol Pharm Bull 33(11):1814–1821

Dharmawan J, Kasapis S, Curran P, Johnson JR (2007) Characterization of volatile compounds in selected citrus fruits from Asia. Part I: freshly-squeezed juice. Flavor Fragr J 22(3):228–232

Du QZ, Chen H (2010) The methoxyflavones in *Citrus reticulata* Blanco cv. ponkan and their antiproliferative activity against cancer cells. Food Chem 119:567–572

Duzzioni AG, Franco AG, De Sylos CM (2009) Radical scavenging activity of orange and tangerine varieties cultivated in Brazil. Int J Food Sci Nutr 20:1–9

Emim JA, Oliveira AB, Lapa AJ (1994) Pharmacological evaluation of the anti-inflammatory activity of a *Citrus* bioflavonoid, hesperidin, and the isoflavonoids, duartin and claussequinone, in rats and mice. J Pharm Pharmacol 46(2):118–122

Facciola S (1990) Cornucopia. A source book of edible plants. Kampong, Vista, 677 pp

Foundation for Revitalisation of Local Health Traditions (2008) FRLHT Database. http://envis.frlht.org.

Fox M, Rayner C, Wu P (1988) Amino acid composition of Australian foods. Food Technol Aust 40:320–323

Gao S, Li P, Yang H, Fang S, Su W (2011) Antitussive effect of naringin on experimentally induced cough in Guinea pigs. Planta Med 77(1):16–21

Gattuso G, Barreca D, Gargiulli C, Leuzzi U, Caristi C (2007) Flavonoid composition of *Citrus* juices. Molecules 12:1641–1673

Ghasemi K, Ghasemi Y, Ebrahimzadeh MA (2009) Antioxidant activity, phenol and flavonoid contents of 13 *Citrus* species peels and tissues. Pak J Pharm Sci 22(3):277–281

Ghorbani A, Nazari M, Jeddi-Tehrani M, Zand H (2011) The citrus flavonoid hesperidin induces p53 and inhibits NF-κB activation in order to trigger apoptosis in NALM-6 cells: involvement of PPARγ-dependent mechanism. Eur J Nutr (in press)

González CN, Sánchez F, Quintero A, Usubillaga A (2002) Chemotaxonomic value of essential oil compounds in *Citrus* species. Acta Hortic (ISHS) 576:49–51

Hamed MA, Hetta MH (2005) Efficacy of *Citrus reticulata* and Mirazid in treatment of *Schistosoma mansoni*. Mem Inst Oswaldo Cruz 100(7):771–778

Hanelt P (ed) (2001) Mansfeld's encyclopedia of agricultural and horticultural crops, vol 1–6. Springer, Berlin, 3700 pp

Harada S, Tominari T, Matsumoto C, Hirata M, Takita M, Inada M, Miyaura C (2011) Nobiletin, a polymethoxy flavonoid, suppresses bone resorption by inhibiting NFκB-dependent prostaglandin E synthesis in osteoblasts and prevents bone loss due to estrogen deficiency. J Pharmacol Sci 115(1):89–93

Hiraishi N, Sono R, Islam MS, Otsuki M, Tagami J, Takatsuka T (2011) Effect of hesperidin in vitro on root dentine collagen and demineralization. J Dent 39(5):391–396

Hirano T, Abe K, Gotoh M, Oka K (1995) Citrus flavone tangeretin inhibits leukaemic HL-60 cell growth partially through induction of apoptosis with less cytotoxicity on normal lymphocytes. Br J Cancer 72:1380–1388

Huang YS, Ho SC (2010) Polymethoxy flavones are responsible for the anti-inflammatory activity of citrus fruit peel. Food Chem 119(3):868–873

Ishii K, Tanaka S, Kagami K, Henmi K, Toyoda H, Kaise T, Hirano T (2010) Effects of naturally occurring polymethyoxyflavonoids on cell growth, p-glycoprotein function, cell cycle, and apoptosis of daunorubicin-resistant T lymphoblastoid leukemia cells. Cancer Invest 28(3):220–229

Iwase Y, Takemura Y, Ju-ichi M, Yano M, Ito C, Furukawa H, Mukainaka T, Kuchide M, Tokuda H, Nishino H (2001) Cancer chemopreventive activity of 3,5,6,7,8,3',4'-heptamethoxyflavone from the peel of citrus plants. Cancer Lett 163(1):7–9

Jagetia GC, Venkatesha VA, Reddy TK (2003) Naringin, a citrus flavonone, protects against radiation-induced chromosome damage in mouse bone marrow. Mutagenesis 18(4):337–343

Jain M, Parmar HS (2011) Evaluation of antioxidative and anti-inflammatory potential of hesperidin and naringin on the rat air pouch model of inflammation. Inflamm Res 60(5):483–491

Jayaprakasha GK, Singh RP, Pereira J, Sakariah KK (1997) Limonoids from *Citrus reticulata* and their moult inhibiting activity in mosquito *Culex quinquefasciatus* larvae. Phytochemistry 44(5):843–846

Jayaprakasha GK, Negi PS, Sikder S, Rao LJ, Sakariah KK (2000) Antibacterial activity of *Citrus reticulata* peel extracts. Z Naturforsch C 55(11–12):1030–1034

Jeong SM, Kim SY, Kim DR, Jo SC, Nam KC, Ahn DU, Lee SC (2004) Effect of heat treatment on the antioxidant activity of extracts from *Citrus* peels. J Agric Food Chem 52(11):3389–3393

Jung KH, Ha E, Kim MJ, Won HJ, Zheng LT, Kim HK, Hong SJ, Chung JH, Yim SV (2007) Suppressive effects of nitric oxide (NO) production and inducible nitric oxide synthase (iNOS) expression by *Citrus reticulata* extract in RAW 264.7 macrophage cells. Food Chem Toxicol 45(8):1545–1550

Kandaswami C, Perkins E, Drzewiecki G, Soloniuk DS, Middleton E Jr (1992) Differential inhibition of proliferation of human squamous cell carcinoma, gliosarcoma and embryonic fibroblast-like lung cells in culture by plant flavonoids. Anticancer Drugs 3(5):525–530

Kandaswami C, Perkins E, Soloniuk DS, Drzewiecki G, Middleton E Jr (1991) Antiproliferative effects of *Citrus* flavonoids on a human squamous cell carcinoma in vitro. Cancer Lett 56(2):147–152

Kawabata K, Murakami A, Ohigashi H (2005) Nobiletin, a citrus flavonoid, down-regulates matrix metalloproteinase-7 (matrilysin) expression in HT-29 human colorectal cancer cells. Biosci Biotechnol Biochem 69(2):307–314

Khalil AT, Maatooq GT, El Sayed KA (2003) Limonoids from *Citrus reticulata*. Z Naturforsch 58c:165–170

Kim DI, Lee SJ, Lee SB, Park KR, Kim WJ, Moon SK (2008) Requirement for Ras/Raf/ERK pathway in naringin-induced G1-cell-cycle arrest via p21WAF1 expression. Carcinogenesis 29(9):1701–1709

Kim MJ, Park HJ, Hong MS, Park HJ, Kim MS, Leem KH, Kim JB, Kim YJ, Kim HK (2005) *Citrus reticulata* Blanco induces apoptosis in human gastric cancer cells SNU-668. Nutr Cancer 51(1):78–82

Kim SH, Zo JH, Kim MA, Hwang KK, Chae IH, Kim HS, Kim CH, Sohn DW, Oh BH, Lee MM, Young-Bae Park YB (2003) Naringin suppresses the mitogenic effect of lysophosphatidylcholine on vascular smooth muscle cells. Nutr Res 23(12):1671–1683

Kırbaşlar FG, Tavman A, Dülger B, Türker G (2009) Antimicrobial activity of Turkish *Citrus* peel oils. Pak J Bot 41(6):3207–3212

Kunimasa K, Ikekita M, Sato M, Ohta T, Yamori Y, Ikeda M, Kuranuki S, Oikawa T (2010) Nobiletin, a *Citrus* polymethoxyflavonoid, suppresses multiple angiogenesis-related endothelial cell functions and angiogenesis in vivo. Cancer Sci 101(11):2462–2469

Kurowska EM, Manthey JA, Casaschi A, Theriault AG (2004) Modulation of HepG2 cell net apolipoprotein B secretion by the *Citrus* polymethoxyflavone, tangeretin. Lipids 9(2):143–151

Lam LKT, Zhang J, Hasegawa S, Schut HAJ (1994) Inhibition of chemically induced carcinogenesis by *Citrus* limonoids. In: Schut HAJ, Huang MT, Osawa CTH, Rosen RT (eds) Phytochemicals for cancer prevention, vol 546. American Chemical Society, Washington, DC, pp 209–219

Lee YC, Cheng TH, Lee JS, Chen JH, Liao YC, Fong Y, Wu CH, Shih YW (2011a) Nobiletin, a *Citrus* flavonoid, suppresses invasion and migration involving FAK/PI3K/Akt and small GTPase signals in human gastric adenocarcinoma AGS cells. Mol Cell Biochem 347(1–2):103–115

Lee YR, Jung JH, Kim HS (2011b) S Hesperidin partially restores impaired immune and nutritional function in irradiated mice. J Med Food 14(5):475–482

Liao J, Xu T, Liu YH, Wang SZ (2011) A new limonoid from the seeds of *Citrus reticulata* Blanco. Nat Prod Res (in press)

Lin N, Sato T, Takayama Y, Mimaki Y, Sashida Y, Yano M, Ito A (2003) Novel anti-inflammatory actions of nobiletin, a *Citrus* polymethoxy flavonoid, on human synovial fibroblasts and mouse macrophages. Biochem Pharmacol 65(12):2065–2071

Liu Y, Wu H, Nie YC, Chen JL, Su WW, Li PB (2011) Naringin attenuates acute lung injury in LPS-treated mice by inhibiting NF-κB pathway. Int Immunopharmacol 11:1606–1612

Lota ML, Serra DR, Tomi F, Casanova J (2000) Chemical variability of peel and leaf essential oils of mandarins from *Citrus reticulata* Blanco. Biochem Syst Ecol 28(1):61–78

Lu HC (2005) Chinese natural cures. Black Dog & Leventhal Publishers, New York, p 512

Luo G, Guan X, Zhou L (2008) Apoptotic effect of *Citrus* fruit extract nobiletin on lung cancer cell line A549 in vitro and in vivo. Cancer Biol Ther 7(6):966–973

Luo G, Zeng Y, Zhu L, Zhang YX, Zhou LM (2009) Inhibition effect and its mechanism of nobiletin on proliferation of lung cancer cells. Sichuan Da Xue Xue Bào Yi Xue Ban 40(3):449–453 (in Chinese)

Mabberley DJ (1997) A classification for edible *Citrus* . Telopea 7(2):167–172

Mak NK, Wong-Leung YL, Chan SC, Wen JM, Leung KN, Fung MC (1996) Isolation of anti-leukemia compounds from *Citrus reticulata*. Life Sci 58(15):1269–1276

Manner HI, Buker RS, Smith VE, Elevitch CR (2006) *Citrus* species. Ver 2.1. In: Elevitch CR (ed) Species profiles for Pacific island agroforestry. Permanent Agriculture Resources (PAR), Holualoa, http://www.agroforestry.net

Manners GD (2007) Citrus limonoids: analysis, bioactivity, and biomedical prospects. J Agric Food Chem 55:8285–8294

Manthey JA, Grohmann K (2001) Phenols in *Citrus* peel byproducts. Concentrations of hydroxycinnamates and polymethoxylated flavones in citrus peel molasses. J Agric Food Chem 49(7):3268–3273

Manthey JA, Guthrie N (2009) Antiproliferative activities of *Citrus* flavonoids against six human cancer cell lines. J Agric Food Chem 50(21):5837–5843

Martín MJ, Marhuenda E, Pérez-Guerrero C, Franco JM (1994) Antiulcer effect of naringin on gastric lesions induced by ethanol in rats. Pharmacology 49(3):144–150

Martínez Conesa C, Vicente Ortega V, Yáñez Gascón MJ, Alcaraz Baños M, Canteras Jordana M, Benavente-García O, Castillo J (2005) Treatment of metastatic melanoma B16F10 by the flavonoids tangeretin, rutin, and diosmin. J Agric Food Chem 53(17):6791–6797

Miller EG, Fanous R, Rivera-Hidalgo F, Binnie WH, Hasegawa S, Lam LK (1989) The effect of *Citrus* limonoids on hamster buccal pouch carcinogenesis. Carcinogenesis 10:1535–1537

Miller EG, Porter JL, Binnie WH, Guo IY, Hasegawa S (2004) Further studies on the anticancer activity of *Citrus* limonoids. J Agric Food Chem 52(15):4908–4912

Minagawa A, Otani Y, Kubota T, Wada N, Furukawa T, Kumai K, Kameyama K, Okada Y, Fujii M, Yano M, Sato T, Ito A, Kitajima M (2001) The *Citrus* flavonoid, nobiletin, inhibits peritoneal dissemination of human gastric carcinoma in SCID mice. Jpn J Cancer Res 92(12):1322–1328

Miyamoto S, Yasui Y, Ohigashi H, Tanaka T, Murakami A (2010) Dietary flavonoids suppress azoxymethane-induced colonic preneoplastic lesions in male C57BL/KsJ-db/db mice. Chem Biol Interact 183(2):276–283

Miyamoto S, Yasui Y, Tanaka T, Ohigashi H, Murakami A (2008) Suppressive effects of nobiletin on hyperleptinemia and colitis-related colon carcinogenesis in male ICR mice. Carcinogenesis 29(5):1057–1063

Miyata Y, Sato T, Imada K, Dobashi A, Yano M, Ito A (2008) A citrus polymethoxyflavonoid, nobiletin, is a novel MEK inhibitor that exhibits antitumor metastasis in human fibrosarcoma HT-1080 cells. Biochem Biophys Res Commun 366(1):168–173

Morin B, Nichols LA, Zalasky KM, Davis JW, Manthey JA, Holland LJ (2008) The *Citrus* flavonoids hesperetin and nobiletin differentially regulate low density lipoprotein receptor gene transcription in HepG2 liver cells. J Nutr 138(7):1274–1281

Morley KL, Ferguson PJ, Koropatnick J (2007) Tangeretin and nobiletin induce G1 cell cycle arrest but not apoptosis in human breast and colon cancer cells. Cancer Lett 251(1):168–178

Morton JF (1987) Mandarin orange. In: Fruits of warm climates. Julia F. Morton, Miami, pp 142–145

Muhoho Njoroge S, Njoki Mungai H, Koaze H, Nguyen TLP, Sawamura M (2006) Volatile constituents of mandarin (*Citrus reticulata* Blanco) peel oil from Burundi. J Essent Oil Res 18(6):659–662

Murakami A, Nakamura Y, Ohto Y, Yano M, Koshiba T, Koshimizu K, Tokuda H, Nishino H, Ohigashi H (2000a) Suppressive effects of *Citrus* fruits on free radical generation and nobiletin, an anti-inflammatory polymethoxyflavonoid. Biofactors 12(1–4):187–192

Murakami A, Nakamura Y, Torikai K, Tanaka T, Koshiba T, Koshimizu K, Kuwahara S, Takahashi Y, Ogawa K, Yano M, Tokuda H, Nishino H, Mimaki Y, Sashida Y, Kitanaka S, Ohigashi H (2000b) Inhibitory effect of *Citrus* nobiletin on phorbol ester-induced skin inflammation, oxidative stress, and tumor promotion in mice. Cancer Res 60(18):5059–5066

Murakami A, Song M, Katsumata S, Uehara M, Suzuki K, Ohigashi H (2007) *Citrus* nobiletin suppresses bone loss in ovariectomized ddY mice and collagen-induced arthritis in DBA/1 J mice: possible involvement of receptor activator of NF-kappaB ligand (RANKL)-induced osteoclastogenesis regulation. Biofactors 30(3):179–192

Nazari M, Ghorbani A, Hekmat-Doost A, Jeddi-Tehrani M, Zand H (2011) Inactivation of nuclear factor-κB by *Citrus* flavanone hesperidin contributes to apoptosis and chemo-sensitizing effect in Ramos cells. Eur J Pharmacol 650(2–3):526–533

Ono E, Inoue J, Hashidume T, Shimizu M, Sato R (2011) Anti-obesity and anti-hyperglycemic effects of the dietary *Citrus* limonoid nomilin in mice fed a high-fat diet. Biochem Biophys Res Commun 410(3):677–681

Pan MH, Chen WJ, Lin-Shiau SY, Ho CT, Lin JK (2002) Tangeretin induces cell-cycle G1 arrest through inhibiting cyclin-dependent kinases 2 and 4 activities as well as elevating Cdk inhibitors p21 and p27 in human colorectal carcinoma cells. Carcinogenesis 23(10): 1677–1684

Peterson JJ, Dwyer JT, Beecher GR, Bhagwat SA, Gebhardt SE, Haytowitz DB, Holden JM (2006) Flavanones in oranges, tangerines (mandarins), tangors, and tangelos: a compilation and review of the data from the analytical literature. J Food Compos Anal 19:S66–S73

Pietta PG (2000) Flavonoids as antioxidants. J Nat Prod 63:1035–1042

Porcher MH et al (1995–2020) Searchable world wide web multilingual multiscript plant name database. Published by The University of Melbourne, Australia. http://www.plantnames.unimelb.edu.au/Sorting/Frontpage.html

Poulose SM, Harris ED, Patil BS (2005) Citrus limonoids induce apoptosis in human neuroblastoma cells and have radical scavenging activity. J Nutr 135:870–877

Poulose SM, Harris ED, Patil BS (2006) Antiproliferative effects of *Citrus* limonoids against human neuroblastoma and colonic adenocarcinoma cells. Nutr Cancer 56:103–112

Pratheeshkumar P, Raphael TJ, Kuttan G (2011) Nomilin inhibits metastasis via induction of apoptosis and regulates the activation of transcription factors and the cytokine profile in B16F-10 cells. Integr Cancer Ther (in press)

Qi SH, Xu YB, Bian HN, Liu P, Xie JL, He JH, Shu B, Li TZ (2006) Effects of *Citrus reticulata* Blanco extract on fibroblasts from human hypertrophic scar in vitro. Zhonghua Shao Shang Za Zhi 22(4):269–272 (in Chinese)

Qian S, Chen L (1998) Studies on the chemical constituents of *Citrus reticulata*. Zhong Yao Cai 21(6):301–302 (in Chinese)

Qiu P, Dong P, Guan H, Li S, Ho CT, Pan MH, McClements DJ, Xiao H (2010) Inhibitory effects of 5-hydroxy polymethoxyflavones on colon cancer cells. Mol Nutr Food Res 54(Suppl 2):S244–S252

Rice-Evans C (2001) Flavonoid antioxidants. Curr Med Chem 8:797–807

Rincón AM, Vásquez AM, Padilla FC (2005) Chemical composition and bioactive compounds of flour of orange (*Citrus sinensis*), tangerine (*Citrus reticulata*) and grapefruit (*Citrus paradisi*) peels cultivated in Venezuela. Arch Latinoam Nutr 55(3):305–310 (in Spanish)

Rodriguez J, Yáñez J, Vicente V, Alcaraz M, Benavente-García O, Castillo J, Lorente J, Lozano JA (2002) Effects of several flavonoids on the growth of B16F10 and SK-MEL-1 melanoma cell lines: relationship between structure and activity. Melanoma Res 12(2):99–107

Rooprai HK, Kandanearatchi A, Maidment SL, Christidou M, Trillo-Pazos G, Dexter DT, Rucklidge GJ, Widmer W, Pilkington GJ (2001) Evaluation of the effects of swainsonine, captopril, tangeretin and nobiletin on the biological behaviour of brain tumour cells in vitro. Neuropathol Appl Neurobiol 27(1):29–39

Sawamura M, Tu NTM, Onishi Y, Ogawa E, Choi H-S (2004) Characteristic odor components of *Citrus reticulata* Blanco (ponkan) cold-pressed oil. Biosci Biotechnol Biochem 68(8):1690–1697

Servillo L, Giovane A, Balestrieri ML, Bata-Csere A, Cautela D, Castaldo D (2011) Betaines in fruits of *Citrus* genus plants. J Agric Food Chem 59(17):9410–9416

Stuart GU (2010) Philippine alternative medicine. Manual of some Philippine medicinal plants. http://www.stuartxchange.org/OtherHerbals.html

Sun Y, Wang J, Gu S, Liu Z, Zhang Y ZX (2010) Simultaneous determination of flavonoids in different parts of *Citrus reticulata* 'Chachi' fruit by high performance liquid chromatography – photodiode array detection. Molecules 15:5378–5388

Suzuki M, Sasaki K, Yoshizaki F, Oguchi K, Fujisawa M, Cyong JC (2005) Anti-hepatitis C virus effect of *Citrus unshiu* peel and its active ingredient nobiletin. Am J Chin Med 33(1):87–94

Swingle WT, Reece PC (1967) The botany of *Citrus* and its wild relatives. In: Reuther W, Webber HJ, Batchelor LD (eds) The *Citrus* industry, vol 1, History, world distribution, botany, and varieties. University of California, Riverside, pp 190–430

Tanaka T (1954) Species problem in *Citrus*. A critical study of wild and cultivated units of *Citrus*, based upon field studies in their native homes. Japanese Society for the Promotion of Science, Tokyo, 152 pp

Tanaka T, Kohno H, Tsukio Y, Honjo S, Tanino M, Miyake M, Wada K (2000) *Citrus* limonoids obacunone and limonin inhibit azomethane-induced carcinogenesis in rats. Biofactors 13:213–218

Tanaka T, Maeda M, Kohno H, Murakami M, Kagami S, Miyake M, Wada K (2001) Inhibition of azoxymethane-induced colon carcinogenesis in male F-344 rats by the *Citrus* limonoids obacunone and limonin. Carcinogenesis 22:193–198

Tang M, Ogawa K, Asamoto M, Chewonarin T, Suzuki S, Tanaka T, Shirai T (2011) Effects of nobiletin on PhIP-induced prostate and colon carcinogenesis in F344 rats. Nutr Cancer 63(2):227–233

Tang M, Ogawa K, Asamoto M, Hokaiwado N, Seeni A, Suzuki S, Takahashi S, Tanaka T, Ichikawa K, Shirai T (2007) Protective effects of *Citrus* nobiletin and auraptene in transgenic rats developing adenocarcinoma of the prostate (TRAP) and human prostate carcinoma cells. Cancer Sci 98(4):471–477

Tanizawa H, Ohkawa Y, Takino Y, Miyase T, Ueno A, Kageyama T, Hara S (1992) Studies on natural antioxidants in Citrus species. I. Determination of antioxidative activities of citrus fruits. Chem Pharm Bull(Tokyo) 40(7):1940–1942

Tao NG, Liu YJ, Zhang JH, Zeng HY, Tang YF, Zhang ML (2008) Chemical composition of essential oil from the peel of Satsuma mandarin. Afr J Biotechnol 7(9):1261–1264

Tian Q, Miller EG, Ahmad H, Tang L, Patil BS (2001) Differential inhibition of human cancer cell proliferation by citrus limonoids. Nutr Cancer 40:180–184

U.S. Department of Agriculture, Agricultural Research Service (USDA) (2011) USDA National Nutrient Database for Standard Reference, Release 24. Nutrient Data Laboratory Home Page, http://www.ars.usda.gov/ba/bhnrc/ndl

Van Slambrouck S, Parmar VS, Sharma SK, De Bondt B, Foré F, Coopman P, Vanhoecke BW, Boterberg T, Depypere HT, Leclercq G, Bracke ME (2005) Tangeretin inhibits extracellular-signal-regulated kinase (ERK) phosphorylation. FEBS Lett 579(7):1665–1669

Viuda-Martos M, Fernández-López J, Ruiz-Navajas Y, Pérez-Álvarez J (2008) Antifungal activity of lemon (*Citrus lemon* L.), mandarin (*Citrus reticulata* L.), grapefruit (*Citrus paradisi* L.) and orange (*Citrus sinensis* L.) essential oils. Food Control 19(12):1130–1138

Walle T (2007) Methoxylated flavones, a superior cancer chemopreventive flavonoid subclass? Semin Cancer Biol 17(5):354–362

Wang D, Wang J, Huang X, Tu Y, Ni K (2007) Identification of polymethoxylated flavones from green tangerine peel (*Pericarpium Citri Reticulatae Viride*) by chromatographic and spectroscopic techniques. J Pharm Biomed Anal 44(1):63–69

Wang X, Li F, Zhang H, Geng Y, Yuan J, Jiang T (2005) Preparative isolation and purification of polymethoxylated flavones from Tangerine peel using high-speed counter-current chromatography. J Chromatogr A 1090(1–2):188–192

Wang Y, Su M, Yin J, Zhang H (2009) Effect of nobiletin on K562 cells xenograft in nude mice. Zhongguo Zhong Yao Za Zhi 34(11):1410–1414 (in Chinese)

Wang YC, Chuang YC, Hsu HW (2008) The flavonoid, carotenoid and pectin content in peels of *Citrus* cultivated in Taiwan. Food Chem 106(1):277–284

Wills RBH, Lim JSK, Greenfield H (1985) Composition of Australian foods. 28. *Citrus* fruit. Food Technol Aust 37:308–310

Yang M, Tanaka T, Hirose Y, Deguchi T, Mori H, Kawada Y (1997) Chemopreventive effects of diosmin and hesperidin on N-butyl-N-(4-hydroxybutyl)nitrosamine-induced urinary-bladder carcinogenesis in male ICR mice. Int J Cancer 73(5):719–724

Yeung HC (1985) Handbook of Chinese herbs and formulas. Institute of Chinese Medicine, Los Angeles

Yoon HS, Lee SR, Ko HC, Choi SY, Park JG, Kim JK, Kim SJ (2007) Involvement of extracellular signal-regulated kinase in nobiletin-induced melanogenesis in murine B16/F10 melanoma cells. Biosci Biotechnol Biochem 71(7):1781–1784

Yoon JH, Lim TG, Lee KM, Jeon AJ, Kim SY, Lee KW (2011) Tangeretin reduces ultraviolet B (UVB)-induced cyclooxygenase-2 expression in mouse epidermal cells by blocking mitogen-activated protein kinase (MAPK) activation and reactive oxygen species (ROS) generation. J Agric Food Chem 59(1): 222–228

Yoshimizu N, Otani Y, Saikawa Y, Kubota T, Yoshida M, Furukawa T, Kumai K, Kameyama K, Fujii M, Yano M, Sato T, Ito A, Kitajima M (2004) Anti-tumour effects of nobiletin, a citrus flavonoid, on gastric cancer include: antiproliferative effects, induction of apoptosis and cell cycle deregulation. Aliment Pharmacol Ther Suppl 1:95–101

Zhang C, Lu Y, Tao L, Tao X, Su X, Wei D (2007) Tyrosinase inhibitory effects and inhibition mechanisms of nobiletin and hesperidin from *Citrus* peel crude extracts. J Enzyme Inhib Med Chem 22(1):91–98

Zhang DX, Mabberley DJ (2008) *Citrus* Linnaeus. In: Wu ZY, Raven PH, Hong DY (eds) Flora of China, vol 11, Oxalidaceae through Aceraceae. Science Press/Missouri Botanical Garden Press, Beijing/St. Louis

Zhang HQ, Ge H, Cheng MY (2006) Antitumor effects of nobiletin on Heps and its mechanism. Yao Xue Xue Bao 41(8):797–800 (in Chinese)

Zheng GD, Yang DP, Wang DM, Zhou F, Yang X, Jiang L (2009) Simultaneous determination of five bioactive flavonoids in pericarpium Citri reticulatae from China by high-performance liquid chromatography with dual wavelength detection. J Agric Food Chem 57(15):6552–6557

Zhou XM, Huang MM, He CC, Li JX (2009) Inhibitory effects of *Citrus* extracts on the experimental pulmonary fibrosis. J Ethnopharmacol 126(1):143–148

Citrus reticulata Satsuma Group

Scientific Name

Citrus reticulata Blanco Satsuma Group (sensu Swingle and Reece 1967; Mabberley 1997).

Synonyms

Citrus nobilis Lour. var. *unshiu* Swingle, *Citrus reticulata* Blanco cv. '*Wenzhou*', *Citrus reticulata* Blanco cv. '*Wenzhou Migan*', *Citrus reticulata* Blanco cv. '*Unshu*', *Citrus reticulata* Blanco var. *unshiu* (Marcov.) H. H. Hu, *Citrus unshiu* (Makino) Marcov. (sensu Hodgson 1967, sensu Tanaka sec. Cottin 2002).

Family

Rutaceae

Common/English Names

Cold Hardy Mandarin, Japanese Mandarin, Japanese Sweet Orange, Mikan, Satsuma Mandarin, Satsuma Orange, Satsuma tangerine, Satsuma, Seedless Mandarin, Unshiu Mandarin, Unshiu Orange.

Vernacular Names

Afrikaans: Naartjie;
Chinese: Wenzhou Migan, Wēnzhōu Mìgān, Wúhé Jú;
Danish: Satsumamandarin;
Finnish: Satsuma;
French: Mandarinier Satsuma;
German: Satsuma;
Greek: Satsuma;
Italian: Mandarino Satsuma;
Japanese: Mikan, Unshiu Mikan, Unshyuu Mikan;
Korean: Guel, Geul Na Mu, Gyur Na Mu;
Taiwan: Wenzhou Migan;
Vietnamese: Cam Ngọt Ôn Châu, Cam Nhật Bản, Cam Satsuma.

Origin/Distribution

An easy-peel *Citrus* mutant of Chinese origin but introduced to the West via Japan. Mikan is held to have originated from Wenzhou, a city in Zhejiang province in China some 2,400 years. It was listed as a tribute item for Imperial consumption in the Tang Dynasty. The mikan was introduced to Japan by the Buddhist monk Chie, who passed through Wenzhou on his way back from Wutai Mountain. This was further developed into new cultivars, with one mutation recorded as early as 1429.

Plate 1 Leaves and rough-skinned Satsuma mandarins

Now Mikan is grown in the cool subtropical regions of Japan, Spain, central China, Korea, Turkey, along the Black Sea in Russia, southern South Africa, South America, and on a small scale in central California and northern Florida. The world's largest Satsuma industry is located in southern Japan. Controlled pollinations over many years have resulted in a collection of over 100 cultivars that differ in date of maturity, fruit shape, colour and quality.

Agroecology

The species grows best in the cool subtropics with cool winters and hot summers and is relatively frost hardy as it withstands brief frost periods at temperatures down to −7°C. Satsuma is the most cold tolerant of mandarins. It thrives best in full sun, well-drained fertile soil.

Edible Plant Parts and Uses

The attractive, bright orange fruit is usually seedless and the flesh is sweet and juicy and is widely eaten. The fruit is bigger than the normal mandarin but smaller than an orange. The dried peel is also used as condiment in cooking. The fruit is also canned as segments or juice in China, Japan and Spain.

Botany

A small to medium sized tree to 5 m tall. Leaves are alternate, lanceolate or broad-lanceolate, obtuse at base, 5–7 cm long, entire or serrulate, dark green, borne on narrowly winged or wingless and glandular-dotted, slender petioles. Both main and primary lateral veins are prominent above as well as below. Leaves are generally wider than other mandarins and tangerine (Plates 1 and 2). Flowers are white, hermaphrodite, fragrant; calyx 5-lobed; petals 5; stamens about 20, united in bundles with 1 carpel. Fruit an aromatic, thin, leathery-skinned, glabrous, smooth or rough, oblate berry (a hesperidium), 3–4 cm in diameter, glandular-dotted, green to bright orange when ripened (Plates 3 and 4), with 8–12 cells or internal segments with juicy orange pulp.

Nutritive/Medicinal Properties

Four cyclic peptides were isolated from young unripe *C. unshiu* fruit peelings (Matsubara et al. 1991). They were each found to consist of seven or eight amino acids. In six cultivars of satsuma mandarin, Kawano Wase, Owari, Silverhill, Foley, Sugiyama and Nobilis, linear hydrocarbons accounted for more than 53% of the saturates in the juice sacs and more than 87% of the monoenes (Nordby and Nagy 1975). In the

Plate 2 Leaves and smooth-skinned Satsuma mandarins

Plate 3 Rough-skinned Satsuma mandarins

Plate 4 Box of smooth-skinned Satsuma mandarin

saturated fraction the major linear hydrocarbon was C25 while in the monoene fraction C25 and C29 predominated. Limonene, linalol, p-cymene, β-elemene are found to accumulate intensively in *C unshui* fruits prior to ripening (Kekelidze et al. 1989). From thence till the physiological fruit maturity, the biosynthesis of γ-terpinene, β-elemene and especially that of limonene increased, while the content of linalol, β-caryophyllene, p-cymene and α-terpineol decreased. From the start of blossoming until the commencement of fruit ripening, the content of γ-terpinene and p-cymene varied greatly in the leaves. From October the content of mono and sesquiterpenoids decreased sharply and remained practically unchanged. Synephrine (protoalkaloid) content of Satsuma mandarin juice from ten different groves located in a major growing region in California were found to range from 73.3 to 158.1 mg/l and was grove and maturity dependent (Dragull et al. 2008).

Recoveries of pure eriocitrin, naringin, hesperidin and tangeretin from Unshiu (*Citrus unshiu*) and Hirado-buntan (*Citrus grandis*) and subsequent extraction were 97.47–103.13% from the mesocarp and 96.87–104.93% from the juice (Nogata et al. 1994).

Six flavonoids, including naringin, naringenin, hesperidin, hesperetin, neohesperidin, and luteolin, were tentatively identified from three Korean citrus varieties Yuza (*Citrus junos*), Kjool (*Citrus unshiu*), and Dangyooja (*Citrus grandis*) (Yoo et al. 2009). Naringin, hesperidin, and neohesperidin were the predominant flavonoids in three *Citrus* species.

Six flavonoid glycosides namely limocitrin 3-β-D-glucose (1), limocitrin-3-α-L-rhamnose (2), 3, 6-di-C-glucosylapigenin (3), narirutin (4), rutin (5) and narcissin (6) were isolated from the water extract of *C. unshui* peelings (Matsubara et al. 1985).

Citrus unshui fruit peelings were found to contain flavanonol glycosides: limocitrin 3-*O*-{[3-hydroxy-3-methylglutaryl(1→6)]-β-D-glucopyranoside}; 3,7,4'-trihydroxy-5,6,8,3'-tetramethoxyflavone 3-*O*-{[3-hydroxy-3-methylglutaryl (1→6)-β-D-glucopyranoside}; limocitrin 3-*O*-{[5-α-glucopyrannosyl-3-hydroxy-3-methylglutaryl(1→2)]-β-D-glocopyranoside}, narcissin; flavanone glycosides: naringenin 4'-*O*-glucopyranosyl-7-*O*-rutinose, narirutin, prunin, hesperidin, phenylpropanoid glycosides: coniferin, syringing, citrusin A, citrusin B, citrusin D, terpenoid glycosides: *trans*-carveol 6-β-D-glucopyranoside, (2*E*, 6*R*)-2,6-dimethy-2,7-octadien-6-ol-1-*O*-β-D-glucopyranoside, (1*S*,4*R*,6*S*)-1,3,3-trimethyl-2-oxabicyclo[2.2.2]octane-6-*O*- β-D-glucopyranoside, vomifoliol 9-*O*-β-D-glucopyranoside, (6*R*, 7*E*, 9*R*)-9-hydroxymegastigma-4,7-dien-3-one-9-*O*-β-D- glucopyranoside; limonoid glycosides: ichangin 4-β-D-glucopyranoside, nomilinic acid 4-β-D-glucopyranoside (Sawabe and Matsubara 1999). The biological activities of the compounds were studied for the utilization as hypotensive and hypertensive drugs. Three limonoid glycosides were isolated from *Citrus unshiu* peels, and their structures were elucidated as nomilinic acid 17-O-β-*D*-glucopyranoside (1), methyl nomilinate 17-O-β-*D*-glucopyranoside (2), and obacunone 17-O-β-*D*-glucopy-ranoside (3) (Sawabe et al. 1999). Five volatiles were identified as the most common odorants in Satsuma mandarin peel: (Z)-hex-3-enal, decanal, linalool, yuzuol, and (2*E*)-trans-4,5-epoxydec-2-ena (Miyazawa et al. 2010). The (R)-isomer of linalool was dominant in Satsuma mandarin peel oil (90%). Seven phenolic compounds of two families including cinnamic acids (caffeic, p-coumaric, ferulic, sinapic acid), and benzoic acids (protocatechuic, p-hydroxybenzoic, vanillic acid) from citrus (*Citrus unshiu* Marc) were evaluated by ultrasound-assisted extraction (Ma et al. 2009).

The contents of seven phenolic acids (caffeic acid, p-coumaric acid, ferulic acid, sinapic acid, protocatechuic acid, p-hydroxybenzoic acid, and vanillic acid) and two flavanone glycosides (narirutin and hesperidin) in Satsuma mandarin peel extracts obtained by ultrasonic treatment were significantly higher than in extracts obtained by the maceration method (Ma et al. 2008). The contents of extracts increased as both treatment time and temperature increased. However, the phenolic acids may be degraded by ultrasound at higher temperature for a long time, for example, after ultrasonic treatment at 40°C for 20 minutes, the contents of caffeic acid, p-coumaric acid, ferulic acid, and p-hydroxybenzoic acid decreased by 48.90%, 44.20%, 48.23%, and 35.33%, respectively.

Two flavanonol glycosides were isolated from *C. unshui* molasses, along with several flavonoids such as hesperidine, narirutin, eriodictyol, 3',4',5,6,7,8-hexamethoxy-3-O-β-D-glucopyranosyloxyflavone, and 3',4',5,6,7,8-hexamethoxy-3-β-D-[4-O-(3-hydroxy-3-methylglutaloyl)]-glucopyranosyloxyflavone, and limonoids such as limonin, nomilin, and cyclic peptide, citrusin III (Kuroyanagi et al. 2008). The structures of the new flavanonol glycosides were determined as (2R, 3R)-7-O-(6-O-α-L-rahmnopyranosyl-β-D-glucopyranosyl)-aromadendrin and 7-O-(6-O-α-L-rahmnopyranosyl-β-D-glucopyranosyl)-3,3',5,7-tetrahydroxy-4'-methoxyflavanone. Of these compounds, flavanone glycoside, hesperidin and narirutin were isolated as the main constituents, indicating *Citrus* molasses to be a promising source of flavonoid glycosides.

There were 73, 71, and 66 aroma components in the three Satsuma mandarin varieties Guoqing 1, Miyagawa Wase, and Owari, and the total contents were 584.67, 505.29, and 494.63 µg/g, respectively (Qiao et al. 2007). A total 29 constituents were common amongst the three varieties. It was also found that Guoqing 1, Miyagawa Wase, and Owari had 12, 5, and 2 unique components, respectively. The key aroma components were limonene, linalool, γ-terpinene, β-myrcene, α-pinene, and octanal in the three fruits. Guoqing 1 contained more key aroma compounds than cvs Miyagawa Wase and Owari.

In Satsuma mandarin, 1,8 cineole and (E)-β-ocimene were comparatively more abundant in flowers and then decreased toward fruit development (Shimada et al. 2005).

A total of 96 volatile constituents were found in *C. unshui* blossoms (Choi 2003). Monoterpene hydrocarbons were prominent in *C. unshiu* blossom accounting for 84.1% in fresh blossoms, 60.0% in shade-dried, 88.4% in microwave-dried and 29.9% in freeze-dried blossoms. , p-Cymene (23.3%) was the most abundant component in fresh *C. unshiu* blossom; γ-terpinene was the most abundant in shade-dried and microwave-dried samples (26.8% and 31.2%, respectively) and β-caryophyllene (10.5%) in freeze-dried sample.

Two new glycosides citroside A (3) and B (4), together with 2-phenylethyl β-D-glucopyranoside (1) and 2-phenlethyl D-rutinoside (2), were isolated from the methanol extract of leaves of *Citurs unshiu*, in addition to two known terpenoids, limonin (5) and friedelin (6) (Umehara et al. 1988). The structures of the new glycosides were elucidated as (5-dehydroxy-grasshopper ketone-5-yl) 5-β-D-glucopyraoside (3) and a (5-dehydroxy-allenic ketodiol-5-yl) 5-β-D-glucopyranoside (4).

Some of the many pharmacological properties of *Citrus unshui* and its phytochemicals are elaborated below.

Antioxidant Activity

The cold-and hot-water extracts of fresh Satsuma mandarin peels showed significant suppressive activity against hydroperoxide generation in a dose-dependent manner but not the methanol or acetone extracts of fresh peels (Higashi-okai et al. 2002). At equivalent concentrations the commercially available ascorbic acids showed roughly equal antioxidative activities compared with those of the water extracts of fresh peels. Although the cold- and hot- water extracts of dried peels indicated a considerable reduction of ascorbic acid concentration, they exhibited much higher antioxidative activities than those of the fresh peels. The methanol extract of dried peels also showed significant antioxidative activities, but did not contain significant ascorbic acid. These results suggested that the fresh peels of Satsuma mandarin had potential antioxidant activities, and the drying treatment of fresh peels caused an enhancement of the antioxidant activity. A linear relationship was observed between Trolox equivalent antioxidant capacity (TEAC) values and total phenolic contents (TPC) of Satsuma mandarin peel; the correlation coefficient, R^2, was 0.8288 at 15°C, 0.7706 at 30°C, and 0.8626 at 40°C, respectively (Ma et al. 2008). The data indicated that Satsuma mandarin peels were rich sources of antioxidants. Three Korean citrus varieties Yuza (*Citrus junos*), Kjool (*Citrus unshiu*), and Dangyooja (*Citrus grandis*) studied exhibited antioxidant capacity of 731–1,221 µmol of Trolox equivalent/g, total phenolic contents of 334–411 mg of chlorogenic acid equivalent/100 g, total flavonoid contents of 214–281 mg of catechin equivalent/100 g and total carotenoid contents of 63–84 mg/100 g (Yoo et al. 2009). Six flavonoids, including naringin, naringenin, hesperidin, hesperetin, neohesperidin, and luteolin, were identified. Naringin, hesperidin, and neohesperidin were the predominant flavonoids in the three citrus varieties. Cold and hot water extracts of Chenpi (dried peels of Satsuma mandarin) exhibited a strong inhibitory activity against linoleic acid peroxidation and 1,1-diphenyl-2-picryl-hydrazyl (DPPH) radical-scavenging activity compared with o-methanol extract (Higashi-Okai et al. 2009). The results indicated that the antioxidant activity against lipid peroxidation in the water extracts was dominantly associated with citric acid, and the DPPH radical-scavenging activity of the water extracts was primarily responsible for ascorbic acid, suggesting a compensatory action of ascorbic acid and citric acid in expression of the antioxidant and radical-scavenging activities of Chenpi.

The methanolic extract of *Citrus* peel powder had been shown to possess strong antioxidant activity and the major flavonoid isolated from *C. unshiu* peel was identified as quercetagetin (Yang et al. 2011). Quercetagetin showed strong DPPH radical-scavenging activity (IC_{50} 7.89 µmol/l) but much lower hydroxyl radical-scavenging activity (IC_{50} 203.82 µmol/l).

Additionally, it significantly reduced reactive oxygen species (ROS) in Vero cells and showed a strong protective effect against hydrogen peroxide-induced DNA damage. The results of this study suggested that quercetagetin could be used in the functional food, cosmetic and pharmaceutical industries. Satsuma mandarin (*Citrus unshiu*) peel extracts were utilized to develop a drink using a mixture design and optimization process with statistical modeling using 1,1-diphenyl-2-picrylhydrazyl (DPPH) radical scavenging activity, total flavonoid content and taste tests, all important target constraints in a drink (Kim et al. 2011). The optimal formulation of the drink was set at 1.974% *Citrus* peel extract, 27.543% fructo-oligosaccharide syrup and 70.364% water. The data showed that *Citrus* peels extract can be utilized for functional drinks having narirutin and hesperidin.

Antiviral Activity

Both the ethyl acetate fraction of *Citrus unshiu* peel extract and fraction 7 decreased hepatitis C virus (HCV) absorption in MOLT-4 cells (a human lymphoblastoid leukemia cell line) (Suzuki et al. 2005). In addition it was demonstrated that 3′,4′,5,6,7,8-hexamethoxyflavone (nobiletin), the active ingredient also markedly inhibited HCV infection in MOLT-4 cells.

Anticancer Activity

Dietary feeding of a polymethoxyflavonoid nobiletin isolated from *Citrus unshiu* inhibited azoxymethane-induced colonic aberrant crypt foci in male rats (Kohno et al. 2001). Dietary administration of nobiletin caused significant reduction in the frequency of colonic aberrant crypt foci. It significantly lowered antibody MIB-5 in aberrant crypt foci and significantly reduced prostaglandin E2 content in the colonic mucosa. The findings suggested possible chemopreventive ability of nobiletin, through suppression of cell proliferating activity of aberrant crypt foci in the development of colonic aberrant crypt foci. Studies showed that supplementation of *Citrus unshiu* segment membrane in the diet had a chemopreventive effect against the early phase of azoxymethane-induced colon carcinogenesis in the db/db as well as db/+ and +/+ mice (Suzuki et al. 2007). Feeding with segment membrane caused reduction in the frequency of aberrant crypt foci in all genotypes of mice and the potency was high in order of the db/db mice, db/+ mice and +/+ mice indicating potential use of *Citrus unshiu* segment membrane in cancer chemoprevention in obese people. Administration with *Citrus unshiu* segment membrane (CUSM) at three doses in diets significantly inhibited development of aberrant crypts foci induced by 5 weekly subcutaneous injections of azoxymetahene (15 mg/kg bw) in male db/db mice: 53% inhibition by 0.02% CUSM, 54% inhibition by 0.1% CUSM, and 59% inhibition by 0.5% CUSM. CUSM treatment also decreased serum level of triglycerides (Tanaka et al. 2008). The findings suggested that certain citrus materials were capable of inhibiting clitis-related and obesity-related colon carcinogenesis. In separate studies, nobiletin and petroleum ether extract of *C. unshui* peels exerted efficacious antiproliferation effects on the B16 mouse melanoma cell with IC_{50} values 88.6 μM and 62.96 μg/ml, respectively, using the MTT (methylthiazole trazolium) assay (Zhang et al. 2007).

Satsuma mandarin content and peel extracts exhibited anti-tumour properties in tumour-bearing mice with renal carcinoma cell, Renca (Lee et al. 2011). The mechanism underlying the anti-tumour effects of Satsuma mandarin extracts was strongly suggested to be via boosting cytokines such as IFN-γ and TNF-α, enhancing immune-mediated anti-tumour properties.

Hypotensive Activity

Six flavonoid glycosides namely limocitrin 3-β-D-glucose (1), limocitrin-3-α-L-rhamnose (2), 3, 6-di-C-glucosylapigenin (3), narirutin (4), rutin (5) and narcissin (6) were isolated from the water extract of *C. unshui* peelings (Matsubara et al. 1985). Compounds 3 and 5 were found to be hypotensive, lowering blood pressure of SHR-SP rats when intravenously injected.

Antiallergic/Antiinflammatory Activities

The 50% ethanolic extract of *Citrus unshiu* green fruit exhibited anti-allergic actions against the type I, II and IV reactions (Kubo et al. 1989). Forty eight-hour homologous passive cutaneous anaphylaxis (PCA) in rats as a typical model of the type I reaction was significantly inhibited by the oral administration of the extract. Complement-dependent cytolysis of sheep red blood cell as a typical model of the type II reaction was significantly inhibited by incubation with the extract. Contact dermatitis in mice as a model of the type IV reaction caused by picryl chloride or sheep red blood cell was significantly inhibited by the oral application of the extract.

The flavonoid component, hesperidin, isolated from the methanolic extract of *Citrus unshiu* green fruit, was found to inhibit histamine release from peritoneal mast cells of rats induced by compound 48/80 (Matsuda et al. 1991). Forty eight-hour homologous passive cutaneous anaphylaxis (PCA) in intact rats was significantly inhibited by the oral administration of hesperidin. However, the anti-allergic action on PCA was not observed in adrenalectomized rats. The results suggested hesperidin to be an effective component of the fruit of *Citrus unshiu* with the antiallergic action against the type I reaction. Of 3 polymethoxyflavones, 3',4',5,6,7,8-hexamethoxyflavone (1), 5-hydroxy-3',4',6,7,8-pentamethoxyflavone (II) and 3',4',5,7,8,-pentamethoxyflavone (III) isolated from the immature peels of *Citrus unshiu*, compounds I and II inhibited dose-dependently histamine release from the rat peritoneal mast cells activated by compound 48/80 or anti-DNP IgE (Kim et al. 1999).

The flavanones naringenin, hesperetin isolated from *Citrus unshiu* pericarp potently inhibited IgE-induced β-hexosaminidase release from RBL-2H3 cells and the PCA reaction (Park et al. 2005). Among the flavanones examined, naringenin was the most potent with an IC_{50} value for β-hexosaminidase release from RBL-2H3 cells of 0.029 mM. The inhibitory activity of naringenin was found to be comparable to that of azelastine, a commercially available antiallergic drug. When the flavanone glycosides were administered to rats, the aglycones, but not the flavanone glycosides, were excreted in urine. This suggested that the flavanone glycosides could be activated by intestinal bacteria, and may be effective toward IgE-induced atopic allergies.

The 50% methanol extract of *Citrus unshiu* powder exhibited potent inhibitory activity against histamine release from basophils of patients suffering from seasonal allergic rhinitis to cedar pollen (Kobayashi and Tanabe 2006). It was demonstrated that the extract significantly inhibited IgE-induced histamine and β-hexosaminidase release from rat basophlilic leukemia RBL-2H3 cells. Among the extract flavonoids (nobiletin, hesperetin, and hesperidin) tested, hesperetin was the most potent, while hesperidin had far less, if any, inhibitory activity. In addition it was found that hesperetin and nobiletin suppressed the phosphorylation of Akt-1, direct downstream effector of phosphatidylinositol 3-kinase (PI3-K). Thus, it was suggested that proper intake of *Citrus unshiu* would be favorable for managing seasonal allergic rhinitis to cedar pollen. Separate studies showed that a 50% methanol extract of *Citrus unshiu* powder suppressed pollen-induced TNF-α release and increased IFN-γ release from peripheral blood mononuclear cells (PBMCs) obtained from patients with seasonal allergic rhinitis to cedar pollen (Tanabe et al. 2007). The results suggested that *Citrus unshiu* powder had an immunomodulatory effect in-vitro and that its use could improve seasonal allergic rhinitis symptoms.

Oral administration of ethanolic extract (50%) of unripe *C. unshiu* fruit and subcutaneous administration of prednisolone to mice elicited inhibition of ear swelling during both induction and effector phases of picryl chloride-induced contact dermatitis (Fujita et al. 2008). The inhibitory activities of combinations of the extract (per os) and prednisolone (sub-cutaneous) during induction phase of picryl chloride-induced contact dermatitis were more potent than those of the extract alone and prednisolone alone. Similarly, successive oral administration of hesperidin, a major flavanone glycoside of the extract, inhibited ear swelling during induction phase of picryl chloride-induced contact dermatitis. The inhibitory activities of combinations of hesperidin (p.o.) and prednisolone (s.c.) were more potent than those of hesperidin alone and prednisolone alone. These results indicated that the combinations of prednisolone and CU-ext or hesperidin exerted a synergistic effect.

Naringin, a bioflavonoid derivative found in *Citrus* species including *C. unshui*, was found to significantly inhibit tumour necrosis factor α (TNF-α)/interferon γ (IFN-γ)-induced RANTES (regulated upon activation normal T-cell expressed and secreted) production in HaCaT (Human keratinocyte) cell partially via NF-κB-dependent signal pathway (Wang et al. 2011).

Antidiabetic Activity

A 10-week administration of *Citrus unshiu* fruit extract to streptozotocin (STZ)-induced diabetic rats preserved acetylcholine (ACh)-induced endothelium-dependent relaxation, but not sodium nitroprusside (SNP)-induced endothelium-independent relaxation, in the diabetic aorta (Kamata et al. 2005). In age-matched control rats, chronic administration of the fruit extract had no influence on the ACh- or SNP-induced aortic relaxation. The increased total cholesterol, low-density lipoprotein (LDL) cholesterol, and triglyceride levels seen in STZ-induced diabetic rats were not normalized by fruit-extract treatment. The results suggested that *Citrus unshiu* extract preserved endothelial function in the aorta in STZ-induced diabetic rats without lowering plasma cholesterol. This beneficial effect may be due to this extract protecting of nitric oxide against inactivation by oxygen free radicals. After a 10-week administration of Satsuma mandarin (1% or 3%), superoxide dismutase (SOD), catalase, and glutathione-peroxidase (GPx) activities, and glutathione levels in streptozotocin-diabetic rat livers were significantly higher than those in the normal diet-fed STZ-diabetic rat livers. Further, although the serum alanine aminotransferase and γ-glutamyl-aminotransferase concentrations of normal diet-fed STZ-diabetic rats were significantly higher than those of the age-matched normal rats, these increments of serum liver enzymes were diminished by the chronic administration of Satsuma mandarin. These results suggested that Satsuma mandarin protected against liver cell damage and may inhibit the progression of liver dysfunction induced by chronic hyperglycemia. They also reported that after 10 weeks administration of *Citrus unshiu* fruit extract to Type 2 diabetic GK rats, intraperitoneal glucose tolerance tests revealed significant decrements of blood glucose levels after glucose loading (Sugiura et al. 2006a, b). The findings further support an advantageous association of citrus fruit consumption with diabetes.

Antiosteoporotic Activity

Oral administration of β-cryptoxanthin isolated from Satsuma mandarin induced anabolic effects on bone components in the femoral tissues of rats in vivo (Uchiyama et al. 2004). Administration of β-cryptoxanthin (10, 25, or 50 μg/100 g body weight) caused a significant increase in calcium content and alkaline phosphatase activity in the femoral-diaphyseal and femoral-metaphyseal tissues. Femoral-diaphyseal and femoral-metaphyseal DNA contents were significantly increased by the dose of 25 or 50 μg/100 g body weight. A significant increase in metaphyseal DNA content was also seen with the dose of 10 μg/100 g body weight of β-cryptoxanthin. The scientists also found that β-cryptoxanthin had a direct stimulatory effect on bone formation and an inhibitory effect on bone resorption in tissue culture in-vitro (Yamaguchi and Uchiyama 2004) and exhibited a synergistic effect of β-cryptoxanthin and zinc sulfate on the bone component in rat femoral tissues in-vitro (Uchiyama et al. 2005). Yamaguchi et al. (2004, 2006) demonstrated that the prolonged intake of Satsuma mandarin juice fortified with β-cryptoxanthin had stimulatory effects on bone formation and inhibitory effects on bone resorption in humans, and that the intake had an effect in menopausal women. In menopausal women, bone-specific alkaline phosphatase activity and γ-carboxylated osteocalcin concentration were significantly increased after the intake of juice containing β-cryptoxanthin (3.0 or 6.0 mg/day) for 56 days as compared with the value obtained after placebo intake. Also, this intake caused a significant decrease in bone tartrate-resistant acid phosphatase activity and type I collagen N-telopeptide concentration.

Antihyperlipidemic Activity

Hesperidin, a major flavonoid isolated from *C. unshui* peels inhibited lipase activity from porcine pancreas and that from *Pseudomonas*, and their IC_{50} was 32 and 132 µg/ml, respectively but narirutin and naringinn, other main flavonoids in *Citrus unshiu* did not show these effects (Kawaguchi et al. 1997). Neohesperidin another flavonid also inhibited the lipase from procine pancreas, but did not have any effect on *Pseudomonas*. In animal experiments, the concentration of plasma triglyceride in rats fed a diet containing 10% hesperidin were significantly lower than that fed the control diet.

Tyrosinase Inhibitory Activity

Nobiletin from *Citrus* peels was found to be a tyrosinase inhibitor of the production of melanin (Sasaki and Yoshizaki 2002). Nobiletin (IC_{50} of 46.2 µM) exhibited more potency than Kojic acid (IC_{50} of 77.4 µM) used as a positive control. Nobiletin and hesperidin from *Citrus unshiu* peel crude extracts exhibited inhibitory effects on tyrosinase diphenolase activity; IC_{50} of nobiletin and hesperidin were 1.49 and 16.08 mM, respectively (Zhang et al. 2007). The crude peel extract ethanol fraction and nobiletin gave efficacious antiproliferation effects on the B16 mouse melanoma cell with IC_{50} values 62.96 µg/ml and 88.6 µM, respectively. Hesperidin and other crude extracts showed very low cytotoxicity to the B16 cell.

Traditional Medicinal Uses

The pericarp has medicinal attributes. It is reported to exhibit the following actions: stimulation of gastric secretion and movement, serotonin antagonism, tyrosinase inhibition, antibacterial and smooth muscle relaxant activity. Traditionally it is used for treating dyspepsia, cough, phlegm, common cold and liver disease. *Citrus unshiu* (Satsuma mandarin,) fruit peel has been used as a traditional Chinese medicine to treat common cold, relieve exhaustion, and cancer (Lee et al. 2011).

Other Uses

In cosmetics, essential oil from Satsuma mandarin peel was found to have a characteristic floral-fresh-fruity odour impressions and may be used in shampoos, soaps, shower gels, body lotions and tooth pastes, while an application of the oils in fine perfumery seems to be interesting as top-notes in perfumes (Tao et al. 2008).

In East Asia and southeast Asia, Satsuma mandarins are highly prized as gift items and religious offerings during special and festive occasions such as the Lunar New Year.

Comments

Owari and Mikan are well-known seedless Satsuma mandarins.

Selected References

Bayer RJ, Mabberley DJ, Morton C, Miller CH, Sharma IK, Pfeil BE, Rich S, Hitchcock R, Sykes S (2009) A molecular phylogeny of the orange subfamily (Rutaceae: Aurantioideae) using nine cpDNA sequences. Am J Bot 96:668–685

Choi HS (2003) Characterization of *Citrus unshiu* (*C. unshiu* Marcov. forma Miyagawa-wase) blossom aroma by solid-phase microextraction in conjunction with an electronic nose. J Agric Food Chem 51(2): 418–423

Cottin R (2002) *Citrus* of the World: a *Citrus* directory. Version 2.0. SRA INRA-CIRAD, Tokyo

Dragull K, Breksa AP 3rd, Cain B (2008) Synephrine content of juice from Satsuma mandarins (*Citrus unshiu* Marcovitch). J Agric Food Chem 56(19): 8874–8878

Fujita T, Shiura T, Masuda M, Tokunaga M, Kawase A, Iwaki M, Gato T, Fumuro M, Sasaki K, Utsunomiya N, Matsuda H (2008) Anti-allergic effect of a combination of *Citrus unshiu* unripe fruits extract and prednisolone on picryl chloride-induced contact dermatitis in mice. J Nat Med 62(2):202–206

Higashi-Okai K, Ishikawa A, Yasumoto S, Okai Y (2009) Potent antioxidant and radical-scavenging activity of Chenpi–compensatory and cooperative actions of ascorbic acid and citric acid. J UOEH 31(4):311–324

Higashi-Okai K, Kamimoto K, Yoshioka A, Okai Y (2002) Potent suppressive activity of fresh and dried peels from satsuma mandarin *Citrus unshiu* (Marcorv.) on hydroperoxide generation from oxidized linoleic acid. Phytother Res 16(8):781–784

Hodgson RW (1967) Horticultural varieties of *Citrus*. In: Reuther W, Webber HJ, Batchelor LD (eds) The *Citrus* industry, vol 1, History, world distribution, botany, and varieties. University of California Press, Berkeley, pp 431–591, Revised edition

Institute Natural Products Research (1998) Medicinal plants in the Republic of Korea. WHO Regional Publications, Western Pacific Series, Manila, 316 pp

Kamata K, Kobayashi T, Matsumoto T, Kanie N, Oda S, Kaneda A, Sugiura M (2005) Effects of chronic administration of fruit extract (*Citrus unshiu* Marc) on endothelial dysfunction in streptozotocin-induced diabetic rats. Biol Pharm Bull 28(2):267–270

Kawaguchi K, Mizuno T, Aida K, Uchino K (1997) Hesperidin as an inhibitor of lipases from porcine pancreas and *Pseudomonas*. Biosci Biotechnol Biochem 61(1):102–104

Kekelidze NA, Lomidze EP, Janikashvili MI (1989) Analysis of terpene variation in leaves and fruits of *Citrus unshiu* Marc. during ontogenesis. Flavor Fragr J 4(1):37–41

Kim DK, Lee KT, Eun JS, Zee OP, Lim JP, Eum SS, Kim SH, Shin TY (1999) Anti-allergic components from the peels of *Citrus unshiu*. Arch Pharm Res 22(6): 642–645

Kim JK, Baik MY, Hahm YT, Kim BY (2011) Development and optimization of a drink utilizing *Citrus* (*Citrus unshiu*) peel extract. J Food Process Eng. doi:10.1111/j.1745-4530.2010.00608.x

Kobayashi S, Tanabe S (2006) Evaluation of the anti-allergic activity of *Citrus unshiu* using rat basophilic leukemia RBL-2 H3 cells as well as basophils of patients with seasonal allergic rhinitis to pollen. Int J Mol Med 17(3):511–515

Kohno H, Yoshitani S, Tsukio Y, Murakami A, Koshimizu K, Yano M, Tokuda H, Nishino H, Ohigashi H, Tanaka T (2001) Dietary administration of *Citrus* nobiletin inhibits azoxymethane-induced colonic aberrant crypt foci in rats. Life Sci 69(8):901–913

Kubo M, Yano M, Matsuda H (1989) Pharmacological study on citrus fruits. I. Anti-allergic effect of fruit of *Citrus unshiu* Markovich. Yakugaku Zasshi 109(11):835–842 (in Japanese)

Kuroyanagi M, Ishii H, Kawahara N, Sugimoto H, Yamada H, Okihara K, Shirota O (2008) Flavonoid glycosides and limonoids from *Citrus* molasses. Nat Med (Tokyo) 62(1):107–111

Lee S, Ra J, Song JY, Gwak C, Kwon HJ, Yim SV, Hong SP, Kim J, Lee KH, Cho JJ, Park YS, Park CS, Ahn HJ (2011) Extracts from *Citrus unshiu* promote immune-mediated inhibition of tumor growth in a murine renal cell carcinoma model. J Ethnopharmacol 133(3): 973–979

Ma YQ, Ye XQ, Fang ZX, Chen JC, Xu GH, Liu DH (2008) Phenolic compounds and antioxidant activity of extracts from ultrasonic treatment of Satsuma Mandarin (*Citrus unshiu* Marc.) peels. J Agric Food Chem 56(14):5682–5690

Ma YQ, Chen JC, Liu DH, Ye XQ (2009) Simultaneous extraction of phenolic compounds of *Citrus* peel extracts: effect of ultrasound. Ultrason Sonochem 16(1):57–62

Mabberley DJ (1997) A classification for edible *Citrus*. Telopea 7(2):167–172

Matsubara Y, Kumamoto H, Iizuka Y, Murakami T, Okamoto K, Miyake H, Yokoi K (1985) Structure and hypotensive effect of flavonoid glycosides in *Citrus unshiu* peelings. Agric Biol Chem 49(4):909–914

Matsubara Y, Yusa T, Sawabe A, Iizuka Y, Takekuma S, Yoshida Y (1991) Structures of new cyclic peptides in young unshiu (*Citrus unshiu* Marcov.), orange (*Citrus sinensis* Osbeck.) and amanatsu (*Citrus natsudaidai*) peelings. Agric Biol Chem 55(12):2923–2929

Matsuda H, Yano M, Kubo M, Iinuma M, Oyama M, Mizuno M (1991) Pharmacological study on citrus fruits. II. Anti-allergic effect of fruit of *Citrus unshiu* Markovich (2). On flavonoid components. Yakugaku Zasshi 111(3):193–198 (in Japanese)

Miyazawa N, Fujita A, Kubota K (2010) Aroma character impact compounds in Kinokuni mandarin orange (*Citrus kinokuni*) compared with Satsuma mandarin orange (*Citrus unshiu*). Biosci Biotechnol Biochem 74(4):835–842

Nogata Y, Ohta H, Yoza K, Berhow M, Hasegawa S (1994) High-performance liquid chromatographic determination of naturally occurring flavonoids in *Citrus* with a photodiode-array detector. J Chromatogr A 667(1–2):59–66

Nordby HE, Nagy S (1975) Saturated and mono-unsaturated long-chain hydrocarbon profiles from *Citrus unshiu* juice sacs. Phytochemistry 14(1):183–187

Park SH, Park EK, Kim DH (2005) Passive cutaneous anaphylaxis-inhibitory activity of flavanones from *Citrus unshiu* and *Poncirus trifoliata*. Planta Med 71(1):24–27

Qiao Y, Xie BJ, Zhang Y, Zhou HY, Pan SY (2007) Study on aroma components in fruit from three different Satsuma mandarin varieties. Agric Sci China 6(12): 1487–1493

Sasaki K, Yoshizaki F (2002) Nobiletin as a tyrosinase inhibitor from the peel of *Citrus* fruit. Biol Pharm Bull 25(6):806–808

Sawabe A, Matsubara Y (1999) Bioactive glycosides in *Citrus* fruit peels. Stud Plant Sci 6:261–274

Sawabe A, Morita M, Kiso T, Kishine H, Ohtsubo Y, Minematsu T, Matsubara Y, Okamoto T (1999) Isolation and characterization of new limonoid glycosides from *Citrus unshiu* peels. Carbohydr Res 315(1–2):142–147

Shimada T, Endo T, Fujii H, Hara M, Omura M (2005) Isolation and characterization of (E)-beta-ocimene and 1,8 cineole synthases in *Citrus unshiu* Marc. Plant Sci 168(4):987–995

Sugiura M, Ogawa K, Yano M (2006a) Effect of chronic administration of fruit extract (*Citrus unshiu* Marc.) on glucose tolerance in GK rats, a model of type 2 diabetes. Biosci Biotechnol Biochem 70(1):293–295

Sugiura M, Ohshima M, Ogawa K, Yano M (2006b) Chronic administration of Satsuma mandarin fruit (*Citrus unshiu* Marc.) improves oxidative stress in streptozotocin-induced diabetic rat liver. Biol Pharm Bull 29(3):588–591

Suzuki M, Sasaki K, Yoshizaki F, Oguchi K, Fujisawa M, Yong JC (2005) Anti-hepatitis C virus effect of *Citrus unshiu* peel and its active ingredient nobiletin. Am J Chin Med 33(1):87–94

Suzuki R, Kohno H, Yasui Y, Hata K, Sugie S, Miyamoto S, Sugawara K, Sumida T, Hirose Y, Tanaka T (2007) Diet supplemented with *Citrus unshiu* segment membrane suppresses chemically induced colonic preneoplastic lesions and fatty liver in male db/db mice. Int J Cancer 120(2):252–258

Swingle WT, Reece PC (1967) The botany of *Citrus* and its wild relatives. In: Reuther W, Webber HJ, Batchelor LD (eds) The *Citrus* industry, vol 1, History, world distribution, botany, and varieties. University of California Press, Berkeley, pp 190–430, Revised edition

Tanabe S, Kinuta Y, Yasumatsu H, Takayanagi M, Kobayashi S, Takido N, Sugiyama M (2007) Effects of *Citrus unshiu* powder on the cytokine balance in peripheral blood mononuclear cells of patients with seasonal allergic rhinitis to pollen. Biosci Biotechnol Biochem 71(11):2852–2855

Tanaka T (1922) *Citrus* fruits of Japan; with notes on their history and the origin of varieties through bud variation. J Hered 13:243–253

Tanaka T (1954) Species problem in *Citrus*. A critical study of wild and cultivated units of *Citrus*, based upon field studies in their native homes. Japanese Society for the Promotion of Science, Tokyo, 152 pp

Tanaka T (1976) Tanaka's cyclopedia of edible plants of the world. Keigaku Publishing, Tokyo

Tanaka T, Yasui Y, Ishigamori-Suzuki R, Oyama T (2008) Citrus compounds inhibit inflammation- and obesity-related colon carcinogenesis in mice. Nutr Cancer 60(Suppl 1):70–80

Tao NG, Liu YJ, Zhang JH, Zeng HY, Tang YF, Zhang ML (2008) Chemical composition of essential oil from the peel of Satsuma mandarin. Afr J Biotechnol 7(9):1261–1264

Uchiyama S, Ishiyama K, Hashimoto K, Yamaguchi M (2005) Synergistic effect of beta-cryptoxanthin and zinc sulfate on the bone component in rat femoral tissues in vitro: the unique anabolic effect with zinc. Biol Pharm Bull 28(11):2142–2145

Uchiyama S, Sumida T, Yamaguchi M (2004) Oral administration of beta-cryptoxanthin induces anabolic effects on bone components in the femoral tissues of rats in vivo. Biol Pharm Bull 27(2):232–235

Umehara K, Hattori I, Miyase T, Ueno A, Hara S, Kageyama C (1988) Studies on the constituents of leaves of *Citrus unshiu* Marcov. Chem Pharm Bull 36(12):5004–5008

Wang SS, Liu L, Zhu L, Yang YX (2011) Inhibition of TNF-α/IFN-γ induced RANTES expression in HaCaT cell by naringin. Pharm Biol 49(8):810–814

Webber HJ (1946) Cultivated varieties of *Citrus*. In: Webber HJ, Batchelor LD (eds) The *Citrus* industry, vol 1, History, botany and breeding. University of California Press, Berkeley, pp 475–668, 1028 pp

Yamaguchi M, Uchiyama S (2004) Beta-Cryptoxanthin stimulates bone formation and inhibits bone resorption in tissue culture in vitro. Mol Cell Biochem 258(1–2):137–144

Yamaguchi M, Igarashi A, Uchiyama S, Morita S, Sugawara K, Sumida T (2004) Prolonged intake of juice (*Citrus Unshiu*) reinforced with β-crypthoxanthin has an effect on circulating bone biochemical markers in normal individuals. J Health Sci 50(6):619–624

Yamaguchi M, Igarashi A, Uchiyama S, Sugawara K, Sumida T, Morita S, Ogawa H, Nishitani M, Kajimoto Y (2006) Effect of β-crytoxanthin on circulating bone metabolic markers: intake of juice (*Citrus unshiu*) supplemented with β-cryptoxanthin has an effect in menopausal women. J Health Sci 52(6):758–768

Yang X, Kang SM, Jeon BT, Kim YD, Ha JH, Kim YT, Jeon YJ (2011) Isolation and identification of an antioxidant flavonoid compound from *Citrus*-processing by-product. J Sci Food Agric 91(10):1925–1927

Ye Y (1988) General conditions of *Citrus* germplasm research in China. In: Suzuki S (ed) Crop genetic resources of East Asia. Proceedings of the international workshop of crop genetic resources of East Asia, 10–13 Nov 1987, Tsukuba, Japan. IBPGR, Kuala Lumpur, Malaysia, pp 91–94

Yoo KM, Hwang IK, Park JH, Moon B (2009) Major phytochemical composition of 3 native Korean *Citrus* varieties and bioactive activity on V79-4 cells induced by oxidative stress. J Food Sci 74(6):C462–C468

Zhang C, Lu Y, Tao L, Tao X, Su X, Wei D (2007) Tyrosinase inhibitory effects and inhibition mechanisms of nobiletin and hesperidin from *Citrus* peel crude extracts. J Enzym Inhib Med Chem 22(1):91–98

Citrus reticulata 'Shiranui'

Scientific Name

Citrus reticulata Blanco 'Shiranui'.

Synonyms

[*Citrus unshiu* Marcov × *Citrus sinensis* Osbeck] × *Citrus reticulata* Blanco.

Family

Rutaceae

Common Names

Dekopon, Dekopon Mandarin, Hallabong, Hallabong Mandarin, Shiranui, Shiranui Dekopon.

Vernacular Names

Japanese: Dekopon, Kiyopon, Shiranui;
Korean: Hallabong, Bujihwa.

Origin/Distribution

Shiranui or Dekopon or Hallabong is a tangerine mandarin, a three-way hybrid and between Cheonggyeon (Kiyomi tangor) [*Citrus unshiu* Marcov × *Citrus sinensis* Osbeck] and Ponggang or Ponkan, (*Citrus reticulata* Blanco) developed in Japan by the Japanese Department of Agriculture in 1972 and introduced to Jeju Island in the early 1990s. Hallabong was named so because it resembles the peak of Mt. Halla in Jeuju Island, Korea. Hallabong ia also widely cultivated in Njau, Korea.

Dekopon is a brand name but has become a generic name for the hybrid in Japan. Dekopon is most likely derived from a Portmanteau between the kanji, 'deko' as a meaning uneven protruding bump, and the 'Pon' in Ponkan to create 'Dekopon. The tangor is also called Shiranui in Japan. There are many market names for Dekopon. For instance, Dekopon is the market name for the fruits originating from Kumamoto. The ones grown in Hiroshima are marketed as *Kiyopon*. It is also produced in most areas of Oita but particularly in Taketa and Usuki. Brazilian farmers of Japanese origin have succeeded in adapting the variety to tropical to temperate climate in the highlands of São Paulo state. In Brazil, dekopon is marketed under the brand name of *Kinsei* which derived from the Japanese word for Venus.

Agroecology

Shiranui is a cool climate citrus, thriving in areas with a cool winter and warm summer with the annual mean temperature of 16.5°C. It is cold hardy and abhors waterlogged soil. It grows in full sun on well drained, friable, fertile soil.

Edible Plant Parts and Uses

Shiranui is very sweet, tender, and juicy and a highly priced fruit in Korea, Japan and Brazil. It is an excellent dessert fresh fruit and is also used for jams, marmalades teas, other beverages and vinegar. Its peel improves the taste of many dishes especially baked fish, sashimi and hot pot dishes. In Japan, a dekapon flavoured candy is very popular.

Botany

A small tree to 3–4 m, low spreading growth habit and usually unarmed. Leaves are dark green, glabrous, lanceolate to elliptic tapering at the base and apex and are typically broader than other tangerines, shallowly crenate margin, with prominent lateral venation on narrowly winged petioles. Flowers are white, bisexual, with calyx of 3–5 irregular lobes, and white petals, 20–25 stamens and a slender style with clavate stigma. Fruit is a hesperidium, rounded with a prominent protruding neck (bump), 6.5–9.7 cm in diameter, green turning to golden yellow or deep, bright orange when ripe (Plate 1). Pericarp is rough, moderately thick but loose enclosing a sweet (15% Brix), juicy and orangey-yellow, seedless flesh.

Plate 1 Harvested Hallbaong fruits

Nutritive/Medicinal Properties

β-Cryptoxanthin contents in both peel and flesh of Shiranuhi mandarin were found to increase gradually as the citrus fruits ripened fully until harvesting season (February) in Korea (Heo et al. 2005). The peels had much higher concentrations of β-cryptoxanthin 1.71 mg% than the flesh 0.65 mg% and thus have potential for use as a functional food ingredient.

Analysis in Japan identified a total of 127 volatile chemicals in the peel and flesh of a citrus fruit, dekopon, Shiranuhi mandarin (Umano et al. 2002). They included 11 monoterpenes, 32 monoterpenoids, 9 sesquiterpenes, 5 sesquiterpenoids, 20 aliphatic alcohols, 14 aliphatic esters, 15 aliphatic aldehydes and ketones, 7 aliphatic acids, and 10 miscellaneous compounds. The major volatile constituents of the extract from the peel were d-limonene (2380.33 mg/kg), myrcene (36.54 mg/kg), bisabolene (30.03 mg/kg), sabinene (21.12 mg/kg), trans-β-ocimene (16.96 mg/kg), valencene (12.84 mg/kg), decanal (8.14 mg/kg), β-phellandrene (4.53 mg/kg), citronellol (4.51 mg/kg), 4-terpineol (4.50 mg/kg), linalool (4.13 mg/kg), and citronellyl acetate (3.63 mg/kg). The major volatile constituents of the extract from the flesh were ethyl acetate (21.54 mg/kg), acetoin (7.23 mg/kg), 3-methylbutanol (2.79 mg/kg), p-mentha-cis-2,8-dien-1-ol (1.01 mg/kg), 3-methylbutanoic acid (0.95 mg/kg), isobutanol (0.59 mg/kg), trans-isopiperitenol (0.58 mg/kg), p-mentha-trans-2,8-dien-1-ol, and trans-carveol (0.44 mg/kg). Compositions of volatile chemicals in peel and flesh extract were considerably different: the peel extract was richer in terpenes, whereas the flesh extract was richer in aliphatic compounds.

Forty-four flavour components were identified in Hallabong peel oil grown in open field and 45 flavour components in Hallabong peel oil grown in green house (Song et al. 2005). (E)-Limonene-1,2-epoxide and neral were identified only in Hallabong oil grown in open field, while β-cubebene, β-elemene and decyl acetate were detected only in green house oil. Limonene was the most abundant component in both oils as more than 86% of peak weight, followed by sabinene (1.8 approx 3.6%) and myrcene (2.4 approx 2.6%). The difference of the volatile profile between open field and green house oils were significantly characterized by identification and quantity of alcohol group. The total alcohols in open field and green house oils accounted for 1.8% and 0.8%, respectively. Among alcohols, the level of linalool was relatively high in open field oil (1.2%), however, it accounted for 0.5% in green house oil. Flavour properties of fresh Hallabong peel and flesh were also examined by sensory evaluation. Flavor properties of fresh Hallabong grown in open field were relatively stronger on both peel and flesh by sensory analysis. Sweetness was strong in Hallabong flesh from open field, and sourness in that from green house. The sensory evaluation of the preference in consideration of taste and aroma was significantly high in Hallabong grown in open field. From the present study, the stronger flavor properties and the preference of Hallabong from open field by sensory evaluation seem to be associated with the high level of linalool in its peel oil, and the composition of monoterpene hydrocarbons such as sabinene and (E)-β-ocimene.

Other analyses reported that d-Limonene (96.98%) in Hallabong was the main constituent of the aroma components emitted from Hallabong peel, and relatively higher peaks of cis-β-ocimene (0.60%), valencene (0.51%) and α-farnesene isomer(0.92%) were observed (Yoo et al. 2004). Other volatile aromas, such as sabinene (0.13%), santrolina triene (0.14%), linalool (0.13%) α-terpineol (0.12%), isopulegol (0.12%), geranial (0.06%), α-farnesene and δ-elemene (0.05%) were observed as smaller peaks. Separate analyses reported that the volatile components of Hallabong, was predominated by limonene (90.68%) followed by sabinene (2.15%), myrcene (1.86%), and γ-terpinene (0.88%) (Choi 2003). The highest flavor dilution factors were found for citronellal and citronellyl acetate, and δ-murollene showed a higher relative flavour activity. Sniff testing of the original oil and its oxygenated fraction revealed that citronellal, cis-β-farnesene, and citronellyl acetate were the character impact odorants of Hallabong peel oil, and citronellal gave the most odour-active character of Hallabong aroma. Yang et al. 2009 reported that the Hallabong peel waste essential oil had a total of six compounds which made up 94.5% of the total oil (Yang et al. 2009). The oil contained limonene (80.51%), γ-terpinene (6.80%), cymene (4.02%), β-myrcene (1.59%), α-pinene (1.02%) and α-terpinolene (0.56%).

Differences in the volatile profiles between Korean and Japanese Shiranui cold-pressed peel oil were reported by Song et al. (2006). Limonene was the most abundant in the Japanese (91.8%) and Korean (86.4%) peel oil. Alcohols accounted for 1.8% in the Korean oil, and 0.2% in the Japanese oil, in which the respective linalool levels were 1.2% and 0.1% respectively. The level of aldehydes was also higher in the Korean oil (1.6%) than in the Japanese oil (0.7%).

The main constituents in the Hallabong citrus peel waste essential oil were limonene (80.51%), γ-terpinene (6.80%), cymene (4.02%), β-myrcene (1.59%), α-pinene (1.02%) and α-terpinolene (0.56%) (Yang et al. 2009).

Seven polymethoxyflavones (PMFs) were isolated from the dried peels of Hallabong, the hybrid *Citrus* and identified as 5,6,7,3′,4′-pentamethoxyflavone (1), 6,7,8,3′,4′-pentamethoxyflavone (2), 3-hydroxy-5,6,7,4′-tetramethoxyflavone (3), 5,6,7,8,3′,4′-hexamethoxyflavone (4), 3,6,7,4′-tetramethoxyflavone (5), 3,5,6,7,8,3′,4′-heptamethoxyflavone (6), and 5,6,7,8,4′-pentamethoxyflavone (7) (Han et al. 2010) The major PMFs of Hallabong were 5 in the dried peels (15.4 mg/g) and 7 in the dried leaves (12.2 mg/g).

Several compounds in the *Citrus* oil extracted from the Hallabong's peel have various therapeutic biological activities including, antiinflammatory, anticancerous, antioxidant and antimicrobial activities. A large body of evidence implicated flavonoids extracted from citrus fruit peels as being the most bioactive as supported by the high biological activities of peel components compared to fruit juice sac components.

Antioxidant Activity

The antioxidant activities of marmalade with varying combinations of peel and flesh were investigated in three types of citrus fruit: satsuma mandarin (*Citrus unshiu* Marc.), shiranui [(*Citrus unshiu* Marc. × *Citrus unshiu* Osbeck) × *Citrus reticulata* Blanco] and navel orange (*Citrus sinensis* Osbeck) (Yoshikawa et al. 2006). The antioxidant activity was estimated in three ways: according to the superoxide (O^{2-}) -radical scavenging activity, according to the l-diphenyl-2-picrylhydrazyl (DPPH)-radical scavenging activity and by an assay of the haemolysis of red blood cells. Although the levels of O^{2-} and DPPH IC_{50} in the peel were lower than those in the flesh, the levels of IC_{50} for both O^{2-} and DPPH in the marmalade that was composed of only peel (P) were higher than those in the marmalade made with a combination (PFJ) of peel and flesh juice or a combination (PFB) of peel and flesh juice, and the water in which the peel had been boiled (BW). The scientists were of the opinion that the decrease of total ascorbic acid and total phenolics in the peel that was boiled in water were caused by the higher IC_{50}. In contrast, O^{2-} and DPPH IC_{50} in the PFB and BW marmalade were low. The haemolysis of red blood cells was slowest in BW. They asserted that adding BW to marmalade increased the antioxidant activity in marmalade.

Antiinflammatory Activity

Studies showed that *Citrus* nobiletin had potential as a novel immunomodulatory and antiinflammatory drug (Lin et al. 2003) Nobiletin suppressed the interleukin (IL)-1-induced prtoduction of PGE(2) in human synovial cells in a dose-dependent manner (<64 µM). Additionally, it selectively downregulated COX-2, but not COX-1 mRNA expression. Nobiletin also interfered with the lipopolysaccharide-induced production of PGE(2) and the gene expression of proinflammatory cytokines including IL-1α, IL-1β, TNF-α and IL-6 in mouse J774A.1 macrophages. In addition, nobiletin downregulated the IL-1-induced gene expression and production of proMMP-1/procollagenase-1 and proMMP-3/prostromelysin-1 in human synovial fibroblasts. In contrast, production of the endogenous MMP inhibitor, TIMP-1, was augmented by nobiletin. These anti-inflammatory actions of nobiletin were very similar to those of antiinflammatory steroids such as dexamethasone, and the upregulation of TIMP-1 production, a unique action of nobiletin.

Studies reported that nobiletin could inhibit the eosinophilic airway inflammation in asthmatic rats (Wu et al. 2006) Lowering the levels of Eotaxin, relieving airway infiltration of eosinophils and promoting apoptosis of eosinophils by enhancing expression of Fas mRNA may be important mechanisms for nobiletin to antagonize eosinophilic airway inflammation of asthmatic rats. *Citrus* flavonoids showed potent inhibitory activity against histamine release from basophils of patients suffering from seasonal allergic rhinitis to ceder pollen (Kobayashi and Tanabe 2006). Among the flavonoids tested, hesperetin was the most potent, while hesperidin had far less, if any, inhibitory activity. The mechanism by which flavonoids like hesperetin and nobiletin s inhibited the degranulation process was by suppressing the suppressed IgE mediated stimulation of basophils through phosphatidylinositol 3-kinase (PI3-K) pathway. The results indicated ingestion of these flavonoids and proper intake of *Citrus unshiu* would be favourable for managing seasonal allergic rhinitis to cedar pollen.

Citrus varieties tested had varying peel contents of seven flavonoids (naringin, naringenin, hesperidin, hesperetin, rutin, nobiletin, and tangeretin) (Choi 2005). The contents of nobiletin and tangeretin, which were contained in all 20 fruit peels including hallabong orange, showed a positive and significant correlation with each other ($R^2 = 0.879$ for immature fruit peels; $R^2 = 0.858$ for mature fruit peels). Flavonoid content in hallabong peel extracts consisted of the flavanones: narigengin 9.55 mg/g and hesperidin 26.60 mg/g and the polymethoxylated flavones: nobitelin 8.26 mg/g and tangeretin 1.27 mg/g; while in mature peels hesperidin

21.07 mg/g, nobiletin 4.06 mg/g and tangeretin 0.78 mg/g were found. All citrus peel extracts dose-dependently inhibited LPS-induced NO production in RAW 264.7 cells. This inhibitory effect was significantly and positively correlated with the content of nobiletin and tangeretin. Nobiletin showed a more potent NO production inhibitory activity ($IC_{50}=26.5$ μM) compared to tangeretin ($IC_{50}=136.6$ μM). Inhibitory activity of hallabong fruit peel extracts on NO production in LPS-Activated RAW 264.7 cells were IC_{50} value of 353.7 μg/ml for immature hallabong peel and 815.9 μg/ml for mature peel. This result supported the premise that nobiletin-rich citrus may provide protection against disease resulting from excessive NO production. Biological screening of citrus nobiletin and its metabolites 3′- and 4′-demethylnobiletin and 3′,4′-didemethylnobiletin revealed that the metabolites possessed more potent antiinflammatory activity against LPS-induced NO production and iNOS, COX-2 protein expression in RAW264.7 macrophage than their parent compound (Li et al. 2007).

Anticancer Activity

Studies reported that that *Citrus* nobiletin might be a possible chemopreventive agent against colon cancer development (Suzuki et al. 2004). *Citrus* nobiletin inhibited azoxymethane-induced large bowel carcinogenesis in rats. Starting 1 week after the initiation, rats received the diet mixed with 100 or 500 ppm nobiletin for 34 weeks. The inhibition rates of colonic adenocarcinoma in rats that received 100 or 500 ppm nobiletin after azoxymethane (AOM) exposure were 18% and 48%, respectively. Also, nobiletin feeding suppressed the prostaglandin E2 levels and cell proliferation activity, and enhanced apoptosis in colonic adenocarcinomas. *Citrus* nobiletin also suppressed HT-29 human colorectal cancer cell invasion and metastasis by down-regulating matrix m metalloproteinase-7 expression in HT-29 human colorectal cancer cells (Kawabata et al. 2005).

Tangeretin and nobiletin are *Citrus* flavonoids that are among the most effective at inhibiting cancer cell growth in-vitro and in-vivo (Morley et al. 2007). Both flavonoids inhibited proliferation of human breast cancer cell lines MDA-MB-435 and MCF-7 and human colon cancer line HT-29 in a dose- and time-dependent manner, and blocked cell cycle progression at G1 in all three cell lines. At concentrations that resulted in significant inhibition of proliferation and cell cycle arrest, neither flavonoid induced apoptosis or cell death in any of the tumour cell lines. These data indicated that, in these cell lines at concentrations that inhibited proliferation up to 80% over 4 days, tangeretin and nobiletin were cytostatic and significantly suppressed proliferation by cell cycle arrest without apoptosis. Such an agent could be expected to spare normal tissues from toxic side effects. Thus, the scientists asserted that tangeretin and nobiletin could be effective cytostatic anticancer agents. Inhibition of proliferation of human cancers without inducing cell death may be advantageous in treating tumours as it would restrict proliferation in a manner less likely to induce cytotoxicity and death in normal, non-tumour tissues.

Citrus nobiletin exerted anti-tumour effect against human gastric cancer cell lines TMK-1, MKN-45, MKN-74 and KATO-III in several ways, namely by direct cytotoxicity, induction of apoptosis and modulation of cell cycle (Yoshimizu et al. 2004). Treatment with nobiletin 24 hour prior to a conventional anticancer drug, CDDP (cisplatin) administration showed a synergistic effect compared to the control.

Antimicrobial Activity

Hallabong peel oil exhibited promising antibacterial effects against drug-susceptible and drug resistant skin pathogens as evidenced by the diameter of zones of inhibition and MIC values (Yang et al. 2009). *Citrus* peel waste essential oil (CPWE) including that of hallabong showed excellent antibacterial activities against acne-causing skin pathogens – *Propionobacterium acnes, Staphylococcus epidermidis* and *Candida albicans*. The antibacterial activity present in the CPWE were attributed to the presence of limonene, γ-terpinene, cymene, β-myrcene, α-pinene and α-terpinolene. The effects of *Citrus* peel waste essential oil on nitric oxide (NO) production in lipopolysaccharide (LPS)-activated RAW 264.7 macrophages were also examined. The addition of CPWE (100 μg/ml) to the medium with LPS largely inhibited the production of NO by 42.5%. The number of viable activated macrophages was not altered by CPWE, as determined by MTT assays, indicating that the inhibition of NO synthesis by CPWE was not simply due to cytotoxic effects. In addition CPWE (6.125–100 μg/ml) inhibited NO production in a dose-dependent manner. Based on these results, the scientists suggested that CPWE could be considered as an attractive antibacterial and antiinflammatory material for topical application.

Skin Whitening Activity

Among *Citrus* peel waste (CWP) by-products such CW- citrus waste ethanol extract, CWE – *Citrus* waste ethyl acetate extract, CWER – *Citrus* waste acid lysate, and CWEA – *Citrus* waste acid lysate by autoclave, CWER was the most potent tyrosinase inhibitor (IC_{50} = 109 μg/ml), and CWEA (IC_{50} = 167 μg/ml) showed good antioxidative effects (Kim et al. 2008). CWE and CWEA samples had dose-dependent inhibitory effects on melanin production. The cytotoxicity of the 4 CPW by-product extract s exhibited low cytotoxicity at 100ug/ml indicating they have good potential for topical application on the human skin.

Other Uses

Hallabong is also planted as an ornamental. Peel of *Citrus* fruits including hallabong peel has potential to be used as:

(a) raw material in the production of a high quality traditional hanji – a paper made from mulberry fibres in Korea (Kim et al. 2007),
(b) fodder for the growth of juvenile flounder, *Paralichthys olivaceus* in Fisheries (Song et al. 2002). Groups of juvenile flounder fed on diets containing 0.1% and 0.2% EM (effective microbial) fermented orange grew significantly faster; their feed coefficient and daily feeding rate were also higher.
(c) cosmeticeuticals (Kim et al. 2008).

Comments

Some scientists have erroneously referred hallabong as *Citrus sphaerocarpa* Tanaka nom nud. which is *Kabosu* a juicy, green citrus fruit closely related to the *yuzu* and is believed to be an ichang papeda – sour orange hybrid. The *Kabosu* was brought over from China in the Edo Period and became a popular fruit in Japan.

Selected References

Choi HS (2003) Character impact odorants of *Citrus* Hallabong [(*C. unshiu* Marcov × *C. sinensis* Osbeck) × *C. reticulata* Blanco] cold-pressed peel oil. J Agric Food Chem 51:2687–2692

Choi HS (2005) Character impact odorants of *Citrus* Hallabong ([*C. unshiu* Marcov × *C. Sinensis* Osbeck] × *C. reticulata* Blanco) cold-pressed peel oil. Acta Hortic (ISHS) 678:119–126

Han S, Kim HM, Lee JM, Mok SY, Lee S (2010) Isolation and identification of polymethoxyflavones from the hybrid *Citrus*, hallabong. J Agric Food Chem 58(17): 9488–9491

Heo JM, Kim DH, Kim IJ, Lee SP, Kim CS (2005) Effect of harvesting season on the β – cryptoxanthin in Shiranuhi mandarin fruit cultivated in Jeju Island. J Food Sci Nutr 10(3):219–223

Kawabata K, Murakami A, Ohigashi H (2005) Nobiletin, a citrus flavonoid, down-regulates matrix metalloproteinase-7 (matrilysin) expression in HT-29 human colorectal cancer cells. Biosci Biotechnol Biochem 69:307–314

Kim HG, Lim HA, Kim SY, Kang SS, Lee HY, Yun PY (2007) Development of functional Hanji added *Citrus* peel. J Korea TAPPI 39:38–47

Kim SS, Lee JA, Kim JY, Lee NH, Hyun CG (2008) *Citrus* peel wastes as functional materials for cosmeceuticals. J Appl Biol Chem 51:7–12

Kobayashi S, Tanabe S (2006) Evaluation of the anti-allergic activity of *Citrus unshiu* using rat basophilic leukemia RBL-2 H3 cells as well as basophils of patients with seasonal allergic rhinitis to pollen. Int J Mol Med 17:511–515

Li S, Sang S, Pan MH, Lai CS, Lo CY, Ang CS, Ho CT (2007) Anti-inflammatory property of the urinary metabolites of nobiletin in mouse. Bioorg Med Chem Lett 17(18):5177–5181

Lin N, Sato T, Takayama Y, Mimaki Y, Sashida Y, Yano M, Ito A (2003) Novel anti-inflammatory actions of nobiletin, a *Citrus* polymethoxy flavonoid, on human synovial fibroblasts and mouse macrophages. Biochem Pharmacol 65(12):2065–2071

Morley K, Ferguson P, Koropatnick J (2007) Tangeretin and nobiletin induce G1 cell cycle arrest but not apoptosis in human breast and colon cancer cells. Cancer Lett 251(1):168–178

Song BS, Moon SW, Kim SJ, Lee YD (2002) Effect of EM fermented orange in commercial diet on growth of juvenile flounder, *Paralichthys olivaceus*. J Aquac 15:103–110

Song HS (2004) Antioxidant activity of juices and peel extracts from Hallabong (*Citrus kiyomi*×*C. ponkan*) and Yuza (*Citrus junos* Tanaka). J Kwangju Health Coll 29:129–138

Song HS, Nguyen TLP, Park YH, Sawamura M (2006) Volatile profiles in cold-pressed peel oil from Korean and Japanese Shiranui (*Citrus unshiu* Marcov. × *C. sinensis* Osbeck×*C. reticulata* Blanco). Biosci Biotechnol Biochem 70:737–739

Song HS, Park YH, Moon DG (2005) Volatile flavor properties of Hallabong grown in open field and green house by gc/gc-ms and sensory evaluation. J Korean Soc Food Sci Nutr 34(8):1239–1245

Suzuki R, Kohno H, Murakami A, Koshimizu K, Ohigashi H, Yano M, Tokuda H, Nishino H, Tanaka T (2004) *Citrus* flavonoid nobiletin suppresses azoxymethane-induced rat colon tumorigenesis. Biofactors 22:111–114

Umano K, Hagi Y, Shibamoto T (2002) Volatile chemicals identified in extracts from newly hybrid *Citrus*, dekopon (Shiranuhi mandarin Suppl. J.). J Agric Food Chem 50(19):5355–5359

Wu YQ, Zhou CH, Tao J, Li SN (2006) Antagonistic effects of nobiletin, a polymethoxyflavonoid, on eosinophilic airway inflammation of asthmatic rats and relevant mechanisms. Life Sci 78:2689–2696

Yang EJ, Kim SS, Oh TH, Baik JS, Lee NH, Hyun CG (2009) Essential oil of *Citrus* fruit waste attenuates LPS-induced nitric oxide production and inhibits the growth of skin pathogens. Int J Agric Biol 11:791–794

Yoo ZW, Kim NS, Lee DS (2004) Comparative analyses of the flavors from hallabong (*Citrus sphaerocarpa*) with lemon, orange and grapefruit by SPTE and HS-SPME combined with GC-MS. Bull Korean Chem Soc 25(2):271–278

Yoshikawa H, Ogawa A, Fukuhara K, Kondo S (2006) Antioxidant activity of tropical fruit jam and marmalade processed with different combinations of peel and flesh in *Citrus* fruit. Int J Food Agric Environ 4(2):78–84

Yoshimizu N, Otani Y, Saikawa Y, Kubota T, Yoshida M, Furukawa T, Kumai K, Kameyama K, Fujii M, Yano M, Sato T, Ito A, Kitajima M (2004) Anti-tumour effects of nobiletin, a citrus flavonoid, on gastric cancer include: antiproliferative effects, induction of apoptosis and cell cycle deregulation. Aliment Pharmacol Ther 20(Suppl 1):95–101

Citrus wintersii

Scientific Name

Citrus wintersii Mabb.

Synonyms

Microcitrus papuana H. F. Winters.

Family

Rutaceae

Common/English Names

Brown River Finger Lime, Papuan Wild Lime.

Vernacular Names

French: Lime Digité De Brown River;
German: Fingerlimette.

Origin/Distribution

The species occurs in Papua New Guinea.

Agroecology

The species occurs naturally in the transition area between eucalyptus savannah and tropical rain forest at the Forestry Station, Brown River, Central District, Papua New Guinea at an elevation of 37 m.

Edible Plant Parts and Uses

The fruit is edible but very acid.

Botany

Shrub usually 1–1.5 m, spiny, with 3–4 erect stems arising from a common base close to the ground (Plate 1). Twigs minutely puberulous when juvenile becoming glabrous. Leaves are small, slender linear to narrowly elliptic when mature, lime-green with blunt apex and borne on short, glabrous, wingless petioles (Plate 2). Crushed leaves with a distinct citric smell. Flowers solitary, axillary, small, cream at bud stage. Flowers with 5 pale green sepals, 5 white separate petals, 20 free stamens with white filaments and yellow anthers. Pistil yellowish-green short, ovary 3–5 loculed. Fruit elongate cylindric, round in cross-section, slightly curved, abruptly beaked at apex and abruptly

Plate 1 Small shrub with small slender leaves

Plate 2 Slender leaves on short, wingless petioles on spiny twigs

narrowed at base, 5–8 cm long by 1.0–1.3 cm at the widest part, green turning to yellow at maturity, 3–5-celled, each cell usually containing 5 seeds. Fruit with strong odour of lime when crushed.

Nutritive/Medicinal Properties

No information has been published on its nutritive value or medicinal uses.

Other Uses

The species is used in breeding and hybridisation work with other *Citrus* species; and also used as *Citrus* rootstock.

Comments

An uncommon *Citrus* species closely related to *C. warburgiana* F. M. Bailey (*Microcitrus warburgiana* F. M. Bailey) Tanaka which is also native to Papua New Guinea and the genus *Eremocitrus*.

Selected References

Bayer RJ, Mabberley DJ, Morton C, Miller CH, Sharma IK, Pfeil BE, Rich S, Hitchcock R, Sykes S (2009) A molecular phylogeny of the orange subfamily (Rutaceae: Aurantioideae) using nine cpDNA sequences. Am J Bot 96:668–685

Cottin R (2002) *Citrus* of the world: a *Citrus* directory, version 2.0. SRA INRA-CIRAD, France

Mabberley DJ (1998) Australian Citreae with notes on other Aurantioideae (Rutaceae). Telopea 7: 333–344

Winters HF (1976) *Microcitrus papuana*, a new species from Papua New Guinea (Rutaceae). Baileya 20(1): 19–24

Citrus x aurantiifolia

Scientific Name

Citrus x aurantiifolia (Christm. & Panzer) Swingle.

Synonyms

Citrus acida Roxb., *Citrus × acida* Persoon, *Citrus × aurantiifolia* subsp. *murgetana* Garcia Lidón et al., *Citrus aurantium* L. subsp. *aurantiifolia* Guillaumin, *Citrus × aurantium* Linnaeus subsp. *aurantiifolia* (Christmann) Guillaumin, *Citrus aurantium* L. var. *aurantifolia*, *Citrus × aurantium* var. *proper* Guillaumin, *Citrus excelsa* Wester, *Citrus hystrix* Candolle ssp. *acida* (Roxb.) Engl., *Citrus javanica* Blume, *Citrus × javanica* Blume, *Citrus lima* Lunan, *Citrus × lima* Macfadyen, *Citrus limetta* auct., non Risso, *Citrus limetta* var. *aromatica* Wester, *Citrus limonellus* Hassk., *Citrus maxima × Citrus* sp 1., *Citrus medica* subsp. *acida* (Roxb.) Engl., *Citrus medica* Linnaeus subf. *aurantiifolia* (Christmann) Hiroe, *Citrus medica* var. *acida* (Roxb.) Hook.F., *Citrus notissima* Blanco, *Limonia acidissima* Houtt., *Limonia aurantiifolia* Christm. & Panz., *Limonia x aurantiifolia* Christmann.

Family

Rutaceae

Common/English Names

Acid Lime, Bartender's Lime, Common Lime, Egyptian Lime, Indian Lime, Key Lime, Lime, Mexican Lime, Omani Lime, Sour Lime, West Indian Lime.

Vernacular Names

Amharic: Limeti;
Armenian: Limoni Desag, Limoni Tesak;
Banaban: Te Raim;
Basque: Lima, Limondo;
Benin: Donuti, Mumoe (Ewe);
Brazil: Limao Galego, Lintao Miudo, Limão-Taiti (Portuguese);
Chamorro: Kahel, Lemon, Limon;
Chinese: Cheng Nihng, Loih Mung, Loih Mou (Cantonese), Qing Ning, Lai Meng, Lai Mu, Suan Ning Meng, Suan Ning Meng Ma Zhuan (Mandarin);
Chuukese: Laimis, Layimes, Limes, Naimis, Náymis;
Cook Islands: Tīpolo, Tīporo, Tiporo (Maori);
Cote D'ivoire: Ankama (Fante), Domunli (Nzema);
Croatian: Limeta;
Czech: Citroník Kyselý Lajm, Limeta;
Danish: Lime;
Dutch: Limoen, Lemmetje, Limmetje;
Egypt: Limûn Baladi, Baladi;
Esperanto: Limeo;

Eastonian: Hapu Laimipuu, Vili: Hapulaim;
Eritea: Lemin (Tigrinya);
Ethiopia: Lemin (Tigrinya);
Farsi: Limoo, Limou Torsh; Omani, Amani (Dried);
Fijian: Laimi, Laini, Moli Kara, Moli Laimi;
Finnish: Limetti;
French: Citronnier, Citron Vert, Citronnier Gallet, Limette (Acide), Limettier, Limettier Des Antilles, Limettier Mexicain, Limon;
Gaelic: Liomaid Uaine;
German: Limette, Limettenbaum, Limone, Loomi, Saure Limette;
Ghana: Donuti, Mumoe (Ewe), Nyamsa (Dagbani), Ankama (Fante), Abonua, Kpete (Ga-Dangme), Domunli (Nzema), Akenkaa, Twaree, Ankaatwaree (Twi);
Greek: Laim;
Guam: Limon;
Hebrew: Laim, Laym;
Hungarian: Apró, Zöld Citromfajta, Lime, Zöldcitrom, Savanyú Citrom;
Icelandic: Límóna;
I-Kiribati: Te Raim;
India: Jombir, Nemu, Nemu-tenga, Sokola Tenga (Assamese), Kaghzinimbu, Kagji lebu, Kagji nebu (Bengali), Limbu (Gujerati), Kaghzi-Nimbu, Kagzinimbu, Khatta, Nibu, Nimbu (Hindu), Cherunaarangi, Hulinimbe, Ilimichai, Imbe, Limbe (Kannada), Nembo (Maithili), Cheru-naragam, Cheru-naranga, Churnnam, Vatukappulinarakam (Malayalam), Kagadilimbu, Limbu (Marathi), Khata kamala, Nembu (Oriya), Nimbu karaji (Punjabi), Brhatjambirah, Jambira, Limpaka, Nagranga, Nimbuka, Swadunaringa, Vijapura (Sanskrit), Elumichai, Elumichchan Chaaru, Elumiccai, Elumiccam Palam, Kich Chilip-Pazham, Kiccilippalam, Narattai, Palam (Tamil), Nimma, Nimma Pandu, Nimmakaya (Telugu), Leemu Kaghzi, Nibu (Urdu);
Indonesia: Kedangsa, Kelangsa, Kruet Kedangsa, Krut Kelangsa (Aceh), Inta, Limo Inta, Limu Manipis, Limu Nipis, Limu onta, Lemo Nipis, Mute Inta, Munte Koyawas, Mutoi Onta (Alfurese, North, Sulawesi), Mausi Nipise, Usi Nipis, Usi Nipise (Amboinese), Jeruk Alit, Kaputungan, Limo (Bali), Unte Bunga, Unte Hapas, Ute Hapas (Batak), Dungga 'Nceta (Bima), Puhat Emnipi (Boeroe, Amboina), Makolona Nipi (Boetoeng, Sulawesi), Lemo Ape, Lemon Kapasa (Bugis), Ahusi Binsi, Ahusi Binci, Amusi Binsi, Ausi Pipis (South Ceram, Amboina), Limau Nipis (Dyak), Mude Telo, Mude Telong (Flores), Asam Kedengsa (Gajo), Makanilu (Goram), Wana Bi Udo, Wana Dabu-Dabu, Lemo Jawa (North Halmaheira), Jeruk Pěcěl (Javanese), Jeruk Tipis, Limau Tallui, Limau Kunci, Limau Tipis, Lemau Telo (Lampong), Jeruk Dhurga, Jeruk Pěcěl (Madurese), Lemo Kadasa (Makassar), Jeruk Nipis, Jeruk Tipis (Malay), Kedangsa, Limau Kapeh (Malay, Jambi), Lemon Nipis (Malay, Manado), Lemon Nifis (Malay, Moluccas), Limau Hapas (Malay, Singkep), Dima Adulo (Nias), Ahusi Binsi, Ahusi Bincil, Usi Nipis, Usi Nipise, Usil Binsi, Usil Bincil (Oelias), Delo Makii (Roti), Jeroe Kii (Sawoe), Muda Kenelu (Solor), Jeruk Nipis, Jeruk Tipis (Sundanese);
Irish: Líoma;
Italian: Lima, Limetta;
Japanese: Raimu;
Kampuchea: Krooch Chhmaa Muul;
Kashmiri: Limb;
Korean: Raim;
Kosraean: Laim;
Laotian: Kok Mak Nao, Naaw;
Lithuanian: Swing Laimas, Rūgščiavaisis, Rūgščiavaisis Citrinmedis, Limai;
Macedonian: Limata;
Malaysia: Limau Amkian, Limau Bo (Semang), Limau Assam, Limau Masam, Limau Nipis;
Maldives: Lun'boa (Dhivehi);
Marquesas: Hitoto, Ihitoro;
Marshall Islands: Laim;
Morocco: Doc;
Nauruan: Deraim, Derem;
Nepal: Kagati (Newari), Nibuwa, Kagati (Nepali);
New Caledonia: Le Limier;
Niger: Olomankilisi, Lemu (Huasa);
Nigeria: Olomankilisi, Lemu (Huasa);
Niuean: Sipolo Fau Ikiiki;
Norway: Lime;
Pakistan: Kaghzi Nimboo;
Palauan: Malchianged;

Papua New Guinea: Muli (Pidgin);
Papiamento: Lamoentsji, Lamunchi;
Philippines: Sua (Bikol), Muyong (Bontok), Lidalaya (Ibanag), Gugulo (Ifugao), Dalayap, Gorong-Gorong (Iloko), Dulugot (Negrito), Bilolo, Dayap (Tagalog);
Pohnpeian: Karer, Karrer, Kuruhkur;
Polish: Lima, Limetka
Portuguese: Limão Gelego, Lima Âcida, Limeira;
Pukapukan: Tīpolo;
Romanian: Lămâi Mexican, Lămâi Mici, Lămâie Verde, Limă, Limetă;
Rotuman: Lemen Riri'i;
Russian: Lajm, Lajm Nastoyaschi;
Samoan: Moli Tipolo, Tipolo, Tipolo;
Serbian: Limeta, Lajm;
Spanish: Lima, Lima Ácida, Lima Chica, Lima Boba, Lima Mejicana, Limero, Limón Agria, Limón Chiquito, Limón Corriente, Limón Criollo, Limón Su'til;
Slovak: Limeta, Kyselý Lajm;
Slovenian: Citronovka, Limeta;
Sri Lanka: Dehi (Sinhalese);
Sudan: Baladi, Limûn Baladi;
Suriname: Lemty (Sranan);
Swahili: Ndimu;
Swedish: Lime, Cedratträd;
Tahitian: Taporo, Te Tumu Taporo;
Thai: Ma Nao, Somma Nao;
Togo: Donuti, Mumoe (Ewe);
Tokelau Islands: Tipolo;
Tongarevan: Tīpolo, Tīporo;
Turkish: Tatlı Limon;
Vietnamese: Chấp, Chanh, Chanh Ta;
Yapese: Remong;
Yiddish: Grine Limene, Laym.

Origin/Distribution

Common lime is believed to have originated in the Indo-Malayan region. It is now cultivated and naturalised throughout the tropics and in warm subtropical areas. Important production areas are Mexico, Florida, the West Indies, and Egypt.

Agroecology

Common lime being sensitive to cold temperatures as such is restricted to the tropics and subtropics. It thrives in full sun in a warm, humid climate with moderate annual rainfall though it is drought tolerant. Diseases can be a liming factor in the wet tropics. In subtropical or sub-temperate areas it needs protection from cold winds. It can be cultivated from near sea level to 1,000 m elevation but is more commonly grown in the lowlands. Lime is adaptable to a wide variety of soils including poor, infertile sandy soils, calcareous soils and clayey soils provided that good drainage prevents water-logging. Heavy clays and saline soils should be avoided. Lime thrives best on fertile, well-drained sandy-loam and is also cultivated in rich, well-drained, gravelly and volcanic soils.

Edible Plant Parts and Uses

The sour, juicy limes are used fresh in drinks, as garnish and for flavouring in food dishes. The fruit is also made into pickles, salted preserves, dried limes, processed into beverages, jams, jellies and marmalade or distilled into an edible oil. With terpenes and sesquiterpenes removed, the essential oil is extensively used in flavouring soft drinks, confectionery, chewing gum sweets, ice cream, sherbet, and other food products. The settled juice is marketed for beverage manufacturing. The residue can be processed to recover citric acid.

Lime juice has a special bouquet and a rich unique flavour that renders it ideally suitable for flavouring food. The juice is used for flavouring sauces, meat, fish, beverages, teas, alcoholic drinks, papaya, and confectionary. In south Asian and southeast Asian homes, lime is commonly used to flavour and acidy food including hot and spicy fish and meat dishes, curries, laksa and for making sauces. Lime juice is an important ingredient in Vietnamese everyday sauce, *nuớc chấm* which also comprises sugar, the ubiquitous fish sauce *nuớc chấm,* garlic and fresh chillies. *Nuoc cham* is served as a table condiment to almost every South Vietnamese cuisine. In Kampuchea, a similar, but less pungent sauce, *tik marij* is made from ground pepper, salt and lime juice, but without fish flavouring. In Indonesia, lime juice is used in sambal, rujak or pecel (peanut sambal). In Thailand, lime is used as an important accompaniment in Pad Thai noodles and other savoury dishes In Malaysia, the fruit is first incised, preserved in brine and pickled in vinegar; it is then enjoyed as an appetizer when fried in oil with added sugar or other flavouring. Lime fruits are also preserved in syrup. In the Philippines, lime peel is open chopped in small pieces and mixed with milk to make popular desserts like *leche plan,* sometimes *macapuno* coconut can be added. In India, the fruit is quartered, layered with salt and sundried. Green chilli peppers, turmeric, ginger or other spices may be included at the beginning and coconut or other edible oil may be added last to enhance the keeping quality. Another method of pickling involves scraping the fruits, steeping them in lime juice, then salting and drying in the sun. Limes are also an essential element in many Tamil cuisine.

In the Middle East countries, ripe limes are boiled in salt water and sun-dried until their interior turns dark (black lime). The resultant spice is called *loomi* or *lumi* in the countries of the Arab peninsula; in Iran, it is known as *amani* or *omani*. Black limes are often used to impart a distinct citrus odour and a sour tang to legumes and meat dishes. The limes are either crushed or pierced with a skewer before usage, and then added to slow-simmering foods. Examples are *majboos,* an aromatic rice dish prepared in the Gulf States and the Irani herb sauce *ghorme.* In Iran and Northern India, powdered *loomi* is also used to flavour rice especially the long-grain Indian basmati rice. Hard, dried limes are exported from India to Iraq for making a special beverage. Lime (or lemon) juice is often used in the Yemeni spice paste *zhoug.* Throughout Malaysia, lime is grown mainly to flavour prepared foods and beverages. Commercially bottled lime juice is prized the world over for use in mixed alcoholic drinks

Lime is a very common ingredient in authentic Mexican, South-western United States, and Caribbean cuisines. Lime juice is made into limeade, syrup, sauce and pies similar to lemon pie. "Key Lime Pie" is a famous dish of the Florida Keys and southern Florida, but today is largely made from the frozen concentrate of the 'Tahiti' lime. In the Bahamas, fishermen and others who spend days in their sailboats, always have with them their bottles of homemade "old sour"– lime juice and salt. Lime is also employed for its pickling properties in *ceviche*, a common method to prepare fresh fish in Polynesia and Latin America. Raw fish is marinated with ample lime juice overnight and, on the next day, seasoned with fresh chillis and coriander leaves (or long coriander, culantro), other ingredients include onion and tomatoes. The recipe appears to be of Polynesian origin, but is today often found along the Pacific coast from México to Perú, and on the Caribbean islands.

Additionally, the leaves of lime are used as a herb in southeast Asian cuisine. The leaves are minced and eaten as a condiment with sate and other dishes in Java. In Vietnam, chicken is cooked with lime leaves and a mixture of salt, black pepper and lime juice. In tropical Africa, lime twigs are popular chew-sticks.

Botany

Small, shrubby trees to 5 m high with numerous, irregular branchlets, armed with short stout spines. Leaves alternate, unifoliolate, slightly stiff with a short, conspicuous, narrowly winged petiole; leaf blade broadly ovate to elliptic, $5-8 \times 2-4$ cm, base rounded, margin slightly crenulate, apex obtuse (Plate 1) and sometimes mucronate. Inflorescences short, axillary racemes with 1–7 flowers. Flowers small, fragrant, white in bud; calyx cupular, 4–6-lobed broadly deltoid, white and obtuse; petals 5, white, 8–12 mm long, ovate-oblong or lanceolate; stamens 20–25 (Plate 1), thick filaments connate half-way up, ovary depressed globose, 9–12(–15)-celled, style abruptly distinct with large, globose, yellow stigma. Fruit green turning to greenish-yellow or yellow, globose, ellipsoid, or obovoid, usually 3–5 cm in diameter, smooth, with prominent oil

Plate 1 Lime flower and leaves

Plate 3 Ripe yellow limes

Plate 2 Mature limes

Plate 4 Lime fruits on sale in a local market

glands, apex with a papilla (Plates 2, 3 and 4); pericarp thin, densely glandular; sarcocarp with 9–12, yellowish-green segments, very acidic, juicy and fragrant. Seeds few, small, ovoid-oblong; seed coat smooth; with white to greenish-white embryos (polyembryonic).

Nutritive/Medicinal Properties

Analyses conducted in Denmark (Saxholt et al. 2008) reported the nutrient composition of raw lime per 100 g/edible portion as: energy 173 kJ, water 88.3 g, total protein 0.7 g, total fat 0.2 g, total carbohydrate 10.5 g, available carbohydrate 7.7 g, dietary fibre 2.8 g, ash, 0.3 g, vitamin A 0.5RE 6 μg β-carotene equivalent, thiamin 0.030 mg, riboflavin 0.02 mg, niacin 0.2 mg, vitamin B-6 0.043 mg, pantothenic acid 0.217 mg, folates 8 μg, vitamin C (L-ascorbic acid) 29.1 mg, Na 2 mg, K 102 mg, Ca 33 mg, Mg 6 mg, P 18 mg, Fe 0.6 mg, Cu 0.065 mg, Zn 0.11 mg, Mn 0.08 mg, lysine 13 mg, methionine 2 mg, tryptophan 3 mg, saturated fatty acids 0.024 g, C14:0 (myristic acid) 0.001 g, C16:0 (palmitic acid) 0.022 g, C:18 (stearic acid) 0.001 g; monounsaturated fatty acids 0.021 g, C16:1 (palmitoleic acid) 0.003 g, C18:1 (n-9) (oleic acid, omega -9 fatty acids) 0.018 g; polyunsaturated fatty acids

0.061 g, C18:2 (n-6) (linoleic acid, omega-6 fatty acid) 0.04 g, and C 18:3 (n-3) (γ-linolenic, omega-3 fatty acid) 0.021 g.

Lemon juice and lime juice were found to be rich sources of citric acid, containing 1.44 and 1.38 g/oz, respectively (Penniston et al. 2008). Lemon and lime juice concentrates contained 1.10 and 1.06 g/oz, respectively. The citric acid content of commercially available lemonade and other juice products varied widely, ranging from 0.03 to 0.22 g/oz. Lemon and lime juice, both from the fresh fruit and from juice concentrates, were found to provide more citric acid per liter than ready-to-consume grapefruit juice, ready-to-consume orange juice, and orange juice squeezed from the fruit.

A total of 31 compounds (tentatively) representing around 85% of total volatiles were identified in the peel of 'Kagzi' lime (*Citrus aurantifolia*) (Shivashankara et al. 2002). Total hydrocarbon content was increased from 66.18% in green fruits to 77.10% in the full yellow, ripe fruits. Total alcohols, aldehydes and esters decreased as the fruit ripened from dark green to full yellow stage. Some of the major compounds identified include limonene (33.74–39.95%), β-pinene (17.82–22.16%), γ-terpinene (6.77–9.11%), geraniol (3.01–7.37%) and neral (1.13–4.31%). sesquiterpene hydrocarbons such as β-elemene, β-farnesene, β-caryophyllene, δ-cadinene and α-humulene were present in very less quantity. It was also observed that the total organoleptic oxygenated compounds decreased from dark green stage (19.24%) to full yellow stage (8.91%) in which neral, geranial and geraniol were high at dark green stage and declined during ripening.

The main volatile components identified from lime in Malaysia were limonene (39.28%), β-pinene (28.44%), geraniol (7.5%), neral (5.3%), α-terpineol (2.39%), geranial (2.05%), terpinene-4-ol (2.01%), α-pinene (1.45%), linalool (1.40%), γ-terpinene (0.79%), geranyl acetate (0.59%) and δ-elemene (0.44%) (Ibrahim et al. 1995).

In another study, over 50 odour-active volatiles were detected in the extracted oil from the peels of key lime, and over 60 in the distilled oil (Chisholm et al. 2003). Geranial, neral and linalool were found to dominate the lime oil aroma in both oils, which accounted for their fresh, floral citrus-like character. Many simple aldehydes contributed to the aroma of the extracted oil, giving it a characteristic lime tone. The distilled oil contained many new odorants, resulting from the acid-catalysed distillation, which imparted woody, musty, spicy and balsamic notes to its aroma. These are C_{10} alcohols and ethers and many sesquiterpenes, most of which possessed lower odour activity and contributed to the more intense piney aroma of the distilled oil. 7-methoxycoumarin was found to be one of the more intense odorants in the extracted oil. Caryophyllene oxide and humulene oxide II were found to be major odorants in the distilled oil; their intensities varied with the method of analysis of the oil. They contributed a sawdust-like and skunky odour, respectively, to the distilled lime oil. Many sesquiterpenes were found to possess some odour activity.

Four coumarins separated, purified and identified from *Citrus aurantifolia*, were limettin, bergapten, imperatorin and isopimpinellin (Khatoon 1995).

The percentage and concentration of total phenolics in *Citrus aurantiifolia* cv Indian leaf, peel and juice were reported by Berhow et al. (1998) as: leaf:- 44.3% (3.35 mg/g) flavone/flavonol, 39.7% (1.92 mg/g) flavanone, 0% psoralen, 9.7% coumarin (0.23 mg/g); flavedo (outer pigmented layer of the peel): 34% (0.83 mg/g) flavone/flavonol, 60.7% (0.95 mg/g) flavanone, 1.9% psoralen, 0% coumarin; albedo (inner layer of the rind): 31.5% (0.71 mg/g) flavone/flavonol, 59.3% (0.85 mg/g) flavanone, 0% psoralen, 3.3% (0.02 mg/g) coumarin; juice: 3.1% (0.061 mg/g) flavone/flavonol, 84.5% (0.23 mg/g) flavanone, 0% psoralen, and 0% coumarin.

The percentage and concentration of flavanones reported in the *Citrus aurantiifolia* cv Indian by Berhow et al. (1998) were as follows: leaf:- narirutin–4'–glucoside 4% (0.1 mg/g), eriocitrin 22% (0.4 mg/g), hesperidin 74% (1.4 mg/g); flavedo: eriocitrin 16% (0.2 mg/g), hesperidin 84% (0.8 mg/g); albedo: eriocitrin 15% (0.1 mg/g), hesperidin 85% (0.7 mg/g); juice: eriocitrin 11% (traces), and hesperidin 89% (0.2 mg/g). The percentage and concentration of flavone reported in the same lime cultivar by

Berhow et al. (1998) were as follows: leaf: diosmin 12% (0.4 mg/g); flavedo: rutin 15% (0.1 mg/g), and diosmin 8% (0.1 mg/g).

Lime fruit, leaves and roots contains many bioactive compounds which impart many pharmacological prosperities that include antioxidant, anticancer, immunomodulatory, antimicrobial, antiviral and other activities.

Antioxidant Activity

A comparative study between the antioxidant properties of peel (flavedo and albedo) and juice of some commercially grown citrus fruit (Rutaceae), grapefruit (*Citrus paradisi*), lemon (*Citrus limon*), lime (*Citrus x aurantiifolia*) and sweet orange (*Citrus sinensis*) revealed that Reducing sugars and phenolics were the main antioxidant compounds found in all the extracts (Guimarães et al. 2009). Peel polar fractions revealed the highest contents in phenolics, flavonoids, ascorbic acid, carotenoids and reducing sugars, which certainly contribute to the highest antioxidant potential found in these fractions.

All of the 34 *Citrus* essential oils were found to have scavenging effects on 1,1-diphenyl-2-picrylhydrazyl (DPPH) in the range of 17.7–64.0% (Choi et al. 2000). The radical-scavenging activities of 31 kinds of *Citrus* essential oils were comparable with or stronger than that of Trolox, a standard antioxidant. The oils of Ichang lemon (64.0%, 172.2 mg of Trolox equiv/ml), Tahiti lime (63.2%, 170.2 mg of Trolox equiv/ml), and Eureka lemon (*Citrus limon*) (61.8%, 166.2 mg of Trolox equiv/ml) were stronger radical scavengers than other citrus oils. *Citrus* volatile components such as geraniol (87.7%, 235.9 mg of Trolox equiv/ml), terpinolene (87.4%, 235.2 mg of Trolox equiv/ml), and γ-terpinene (84.7%, 227.9 mg of Trolox equiv/ml) showed marked scavenging activities on DPPH.

Among the different solvents, namely, chloroform, acetone, MeOH, and MeOH/water (8:2) used for extracting freeze-dried lime juice, the chloroform extract showed the highest (85.4% and 90%) radical-scavenging activity by DPPH and 2,2′-azino-bis(3-ethylbenzthiazoline-6-sulfonic acid) (ABTS) methods at 624 µg/ml, whereas the MeOH/water extract showed the lowest (<20%) activity (Patil et al. 2009b). The bioactive components were identified to be rutin, neohesperidin, hesperidin, and hesperitin and the limonoids, limonexic acid, isolimonexic acid and limonin.

Anticancer Activity

All of the extracts (chloroform, acetone, MeOH, and MeOH/water) of lime juice inhibited Panc-28 cancer cell growth (Patil et al. 2009b). The MeOH extract exhibited the maximum activity, with an IC_{50} value of 81.20 µg/ml after 72 hours. The inhibition of Panc-28 cells was in the range of 73–89%, at 100 µg/ml at 96 hours. The involvement of apoptosis in induction of cytotoxicity was confirmed by expression of Bax, Bcl-2, casapase-3, and p53. The results of the present study clearly indicated that antioxidant activity was proportionate to the content of flavonoids and proliferation inhibition ability was proportionate to the content of both flavonoids and limonoids.

Recent studies showed that lime volatile oil had potential benefits in colon cancer prevention (Patil et al. 2009a). Twenty-two compounds representing more than 89.5% of the volatile oil were identified. d-limonene (30.13%) and d-dihydrocarvone (30.47%) were found to be the major compounds in the lime volatile oil. This oil showed 78% inhibition of human colon cancer cells (SW-480) with 100 µg/ml concentration at 48 h. Lime volatile oil elicited DNA fragmentation and induction of caspase-3 up to 1.8 and 2-folds after 24 and 48 hours, respectively, which may be due to the involvement of apoptosis. Analysis of apoptosis-related protein expression further confirmed apoptosis induction by lime volatile oil. Separate studies reported that concentrated lime juice (CLJ) extract had no significant effect on human breast carcinoma cell line MDA-MB-453, however, using the concentrations of 125, 250, and 500 µg/ml of CLJ extract, a significant inhibition of the spontaneous proliferation of human lymphoblastoid B cell line RPMI-8866 was detected (Gharagozloo et al. 2002). Due to the protein nature of the biologically active macromolecules of the CLJ extract, the researchers asserted that the protein components of concentrated lime juice extract may have antiproliferative effects on tumour cell lines.

Antiviral Activity

Nigerian women reportedly apply lime juice intravaginally to protect themselves against HIV. In an interview survey of 300 sexually active women comprising 200 female sex workers (FSWs) and 100 family planning clients (FPCs), lemon and lime juice were found to be widely used for douches among women at high risk of HIV transmission (Imade et al. 2005). The juice was used either neat 44/167 respondees (26%) or diluted in water 75/167 (45%) either before or after sex. Nineteen per cent (32/167) found the juice painful. Over half of the women believed that it protected them from pregnancy and/or sexually transmitted infections; they did not know their HIV status. In another survey of 398 female sex workers, no associations were found between the use of citrus douching and other sexually transmissible infections (Imade et al. 2008). There were no significant associations between the prevalence of STI and HIV and lime or lemon juice usage. HIV prevalence was high for both users of lemon or lime (UL) or non-users (NUL) 48.8% and 48.2%, respectively. The rates of bacterial vaginosis were not significantly higher in UL.

In-vitro data indicated lime juice to be virucidal, but only at cytotoxic concentrations. Preclinical studies showed that pre-treatment of HIV with lime juice demonstrated direct virucidal activity, with 10% juice inactivating the virus within 5 min (Fletcher et al. 2008). However, the virucidal activity of lime juice was severely compromised in the presence of seminal plasma. Potentially, to be effective against HIV in-vivo, women would need to apply a volume of neat lime juice equal to that of an ejaculate, and maintain this ratio vaginally for 5–30 minutes after ejaculation. Data presented in studies suggested that this would have significant adverse effects on the genital mucosa (cervicovaginal epithelial cells). These data raised serious questions about the plausibility and safety of such a prevention approach. This was further confirmed in a randomised trial with 47 women applying water or lime juice (25%, 50%, or undiluted) intravaginally twice daily for two 6-day intervals, separated by a 3-week washout period (Mauck et al. 2008). Data showed that the brief reduction in pH after vaginal lime juice application in women was unlikely to be virucidal in the presence of semen. Lime juice was unlikely to protect against HIV and may actually be harmful. Separate studies reported that lime juice up to 20% concentration had an acceptable safety profile for vaginal use (Hemmerling et al. 2007). However, as new in-vitro research showed that the effectiveness of lime juice to prevent HIV transmission in concentrations lower than > or =50% was unlikely and concentrations of 50% had been shown to be toxic, the authors advised that women should be discouraged from commencing or continuing the vaginal use of lime juice.

Renal Protective Activity

Hypocitraturia, a low amount of citrate in the urine, is an important risk factor for kidney stone formation. The administration of potassium citrate 2×20 mEq/day in 6 months was found to improve patient's complaints and occurrence of renal colic by producing an increase of urine volume, pH level and total citrate value, and total potassium level, and the decrease of calcium ratio to citrate urine (Sja'bani et al. 2007). Similar results were obtained (except for calcium level, which was not decreased) from the administration of *Citrus aurantifolia* juice in patients with idiopathic hypocitraturic calcium nephrolithiasis.

Separate studies in Thailand reported that consumption of in-house lime powder exerted citraturic and alkalinizing actions as efficient as consumption of potassium citrate (Tosukhowong et al. 2008). In addition, it provided an antioxidative effect and was able to attenuate renal tubular damage. These pharmacological properties may be clinically useful to diminish the stone-forming potential in kidney stone patients and hence for preventing recurrent calculi. Potassium citrate has long been used as a prophylactic remedy for nephrolithiasis recurrence. Patients with kidney stone were enrolled and randomly assigned to three treatment programs for 3 month period consisting of consumption of solution containing lime powder (Group 1, n=13), potassium citrate (Group 2, n=11) and lactose as placebo regimen (Group 3, n=7). Lime powder and potassium citrate contained equal amounts of potassium (21 mEq) and citrate (63 mEq). After treatment, there was an increase in urinary pH, potassium and citrate in Group 1 and 2. Increased plasma potassium and red blood cell glutathione (R-GSH) and decreased urinary malondialdehyde were found in Group 1, but not observed in Group 2. R-GSH was decreased in Group 2. Urinary N-acetyl-β-glucosaminidase activity and fractional excretion of magnesium, as renal tubular damage indicators, were decreased only in Group 1. In Group 3, all measured parameters were unaltered except for an increased urinary chloride.

Immunomodulatory Activity

Results of studies indicated that proliferation of phytohemagglutinin (PHA) activated mononuclear cells were significantly inhibited by 250 and 500 µg/ml of concentrated lime juice (CLJ) extract, whereas only 500 µg/ml of the extract could inhibit proliferation of staphylococcal protein A (SPA) activated mononuclear cells (Gharagozloo and Ghaderi 2001) . The abrogation of this inhibitory effect of the CLJ extract was noted by adding anti-CLJ antibody to the lymphocyte culture. The researchers concluded that the CLJ extract possessed immunomodulatory principles, which may mainly be due to the protein components of the extract.

Antiosteoporotic Activity

Studies showed that *Citrus aurantifolia* and *Citrus sinensis* extracts reduced bone loss in ovariectomized rats (Shalaby et al. 2011). Administration of *Citrus* extracts increased trabecular bone mineral content and bone mineral density of tibia, improved the levels of phosphorus and calcium. Eighteen compounds were obtained from both plants, eight coumarins (1–8), eight flavonoids (9–16) and two sterols analogues (17 and 18). Leaves and peels of *C. aurantifolia* afforded 11 components, while those of *C. sinensis* afforded 14 compounds. Isobergapten (6), marmesin (7), myricetin (11), 4′,5,7-trihydroxy-3,6-dimethoxy flavone (12), and quercetin-3-O-robinobioside (14), were reported to first time from the two *Citrus* species. The authors attributed the antiosteoporotic activity of the extracts to coumarins, flavonoids and sterols.

Antimicrobial Activity

Among 12 medicinal plants tested, *Thymus vulgaris* and *Citrus aurantifolia* essential oils were found to inhibit both *Aspergillus parasiticus* and aflatoxin production (Razzaghi-Abyaneh et al. 2009). *Carum carvi* effectively inhibited aflatoxin production without any obvious effect on fungal growth. The IC_{50} values of *T. vulgaris*, *C. aurantifolia* and *C. carvi* for aflatoxin inhibition were reported as 93.5, 285.6, and 621.9 µg/ml for aflatoxin B_1, while they were calculated as 11.7, 50.1, and 56.0 µg/ml for aflatoxin G_1. The authors results indicated that the essential oils of the medicinal plants may be considered as potential candidates to protect foods and feeds from toxigenic fungal growth and subsequent aflatoxin contamination.

Citrus aurantifolia crude aqueous extract showed promising broad spectrum in-vitro antibacterial effects when tested on 53 fresh human bacterial pathogens isolated in Ladoke Akintola University Teaching Hospital, Osogbo, Nigeria (Taiwo et al. 2007). This included isolates of 27 *Staphylococcus* spp., 15 *Escherichia coli*, 3 *Klebsiella* spp., 4 *Proteus* spp. and 4 *Pseudomonas* spp. In another study, undiluted lime-juice exhibited in-vitro inhibitory activity against *Staphylococcus aureus*, *Bacillus* spp., *Escherichia coli* and *Salmonella* spp. (Onyeagba et al. 2004).

The aqueous and ethanolic extracts of garlic and ginger singly did not inhibit any of the test organisms. The highest inhibition was observed with a combination of extracts on *Staphylococcus aureus*. *Salmonella* spp. were resistant to almost all the extracts except lime.

Studies reported that crude extracts of lime juice and burnt lime peel in the following solvents: palm-wine (a local alcoholic drink tapped from palm trees), Seaman's Schnapps 40% alcoholic drink, water, ethanol and fermented water from 3 days soaked milled maize, varied in degree of in-vitro microbial inhibition (Aibinu et al. 2007). The anaerobes (*Bacteroides* spp., *Porphyromonas* spp. and *Clostridium* spp.) and the Gram-positive bacteria (*Staphylococcus aureus, Streptococcus faecalis*) were susceptible to all the extracts (juice in different solvents and distilled fruit oil) with minimum inhibitory concentration (MIC) ranging from 32 mg/ml to 128 g/ml. The activity against the fungi showed only the oil extract potent for *Aspergillus niger*, while *Candida albicans* was susceptible to all the extracts with MIC ranging from 256 to 512 mg/ml. The Gram-negative bacteria (*Salmonella paratyphi, Shigella flexnerii, Citrobacter spp, Serratia spp, Klebsiella pneumoniae, Pseudomonas aeruginosa, Escherichia coli*) had MIC ranging from 64 to 512 mg/ml. Minimum bactericidal concentration (MBC) ranged between 32 and 512 mg/ml depending on isolates and extracting solvent. The oil and palm-wine extract of "epa-ijebu" (burnt lime peel) showed greater activity than the other extracts. The killing rate of the Schnapps extract on *S. aureus* and *E. coli* was 1 and 3.5 hour respectively. Separate studies showed that five bacteria (*Pseudomonas aeruginosa, Staphylococcus aureus, Escherichia coli, Klebseilla sp.* and *Proteus mirabilis*) were susceptible to the crude extract of bitter leaf (*Vernonia amygdalina*) and lime juice (Ojiezeh et al. 2011).

Studies show that during epidemics of cholera in areas without safe sources of drinking water, juice from *Citrus* fruits added to water and food in palatable concentrations may be appropriate measures in reducing the transmission of cholera (Dalsgaard et al. 1997). A 5-log reduction in *Vibrio cholerae* counts to <100 CFU/ml of well water was shown for two test strains after exposure to 0.5% lime juice (pH 3.3) for 12 min. In tap water, no significant reduction in CFU of *V. cholerae* was found after exposure to 0% (pH 8.3) and 0.5% (pH 5.6) lime juice whereas exposure to 1.0% lime juice (pH 4.4) for 120 min caused a 5-log reduction to <100 CFU/ml of tap water for the test strains. A 3-log reduction of *V. cholerae* was found in food samples containing 3.5% and 5.0% lime juice after 120 min exposure. However, local characteristics of the water, in particular its alkalinity, should be considered before applying this measure.

Lime roots were found to contain tannins, phlobatannins, polyphenols, hydroxymethyl anthraquinones, glucides, saponins, alkaloids, cardiac glycosides, flavanoids and reducing compounds (Ebana et al. 1991). The aqueous and alcoholic extracts as well as alkaloids and cardiac glycosides were found to inhibit such organisms as *Staphylococcus aureus, Klebsiella pneumoniae, Proteus mirabilis, Pseudomonas aeruginosa*, β-haemolytic Streptococci, *Escherichia coli* and *Neisseria gonorrhoeae*.

Antihypertensive Activity

Studies showed that the aqueous extract of *Citrus aurantifolia* produced a dose-dependent and significant decrease in rabbit blood pressure (Souza et al. 2011). The extract (4–16 mg/kg b.w.) also dose dependently reduced hypertension evoked by adrenalin (30 µg/kg b.w.). The extract (10-8 mg/ml–10-2 mg/ml) induced both negative inotropic and chronotropic effects on the heart contractile activity and induced a dose-dependent relaxation of contractions produced by adrenalin (3.10–3 mM) or by KCl (80 mM). The extract-evoked vasorelaxant effects were totally abolished by removal of the endothelium layer or by a pre-treatment with L-NAME (mg/ml). The researchers concluded that the extract possessed an antihypertensive activity which could be related to both cardiodepression and the vasorelaxation and that endothelium-dependent mechanisms might be involved. The results supported the use of lime used in African folk medicine for the management of hypertension.

Antiobesity Activity

Studies showed that mice administered with lime essential oil displayed a reduction in body weight and food consumption in mice, possibly through promoting anorexia which might have played a role in weight loss (Asnaashari et al. 2010). Co-administration of the lime essential oil and ketotifen caused significant suppression in gaining weight, as well as decreased body weights of mice. While mice that received ketotifen demonstrated an enhancement both in the amount of food intake and body weight compared with the control group. The data obtained suggested that lime essential oil played an important role in weight loss and could be useful in the treatment of drug-induced obesity and related diseases.

Traditional Medicinal Uses

The juice of the Mexican lime has been regarded as an antiseptic, tonic, an antiscorbutic, an astringent, and as a diuretic in liver ailments, a digestive stimulant, a remedy for intestinal haemorrhage and haemorrhoids, heart palpitations, headache, convulsive cough, rheumatism, arthritis, falling hair, bad breath, and as a disinfectant for all kinds of ulcers when applied in a poultice. Lime juice was used to dispel various cutaneous irritations and swelling of mosquito bites. In Malaysia, lime juice was taken as a tonic 'ubat jamu', remedy for coughs and to relieve stomach-ache. Lime juice was used with arsenic for yaws. Lime pounded in vinegar has been employed as poultice on the head to allay neuralgia. Lime was mixed with other ingredients in a past and smeared over the body of women after confinement. Gonorrhoea was treated by poulticing *Phyllanthus* pulp in lime juice. In Senegal and Sierra Leone, lime juice was sometimes given as a vermifuge, mitigated by being mixed with oil. Externally the fresh juice has been used as a cleanser or stimulant of wound surfaces, or cuts limes, roasted, are applied to chronic sores, yaws, etc.

The leaf decoction has been used as eye drops and to bathe a feverish patient; also as a mouth wash and gargle in cases of sore throat and thrush. In Malaysia, lime leaves were used for poulticing against evil spirits and for skin complaints, and a lotion of the pounded leaves used for headache. Leaves pounded with betel-nut, *Areca catechu* maybe drunk for stomach-ache. Pounded leaves were used in mixture like a cosmetic on the abdomen of women after confinement. In Africa, an infusion of the leaves was given for fever with slight jaundice. In fevers it was applied externally by friction, and internally as a refrigerant and febrifuge, sometimes being combined with tamarind to give laxative effect. It was given in small doses for the vomiting of bilious fever. For sore throat, thrush, etc., it was used as a mouthwash and gargle, and is painted on the throat. In Africa, gonorrhoea was also treated by fumigation, using a steaming brew of the leaves under a blanket.

Roots were used as a decoction for dysentery in Malaysia. In Yucatan, Mexico, a decoction of the root was used against gonorrhoea. In Africa, that the root-bark has been regarded a good febrifuge, and a decoction was given for gonorrhoea and accompanying troubles.

Other Uses

Lime is also used occasionally by many Asian martial artists to enhance vision. In India, the lime is used in Tantra for removing evil spirits. It is also combined with Indian chillies to make a protective charm to repel the evil eye. In addition, there are many purely superstitious uses of the lime. In Malaysia, lime juice was used in ceremonial bathing of the bride groom in marriage, to invoke spirits and the name of girl written on the lime fruit in love charms. Fallen ripe fruit was given to a demented man possessed by evil elephant spirits. Evil spirits were believed to be kept away by chewing lime leaves and breathing over a sleeping patient. In India, the lime was used in Tantra for removing evil spirits. It is also combined with Indian chillies to make a protective charm to repel the evil eye.

Lime extracts and essential oils from the peel and leaves are frequently used in perfumes, soap, detergents, cleaning products, and aromatherapy. Lime juice has been used in the process of dyeing leather. Lime juice and arsenic were used together in cleaning the blades of kris in Malaysia. In India, the powdered dried peel and the sludge remaining after clarifying lime juice have been employed for cleaning metal. Lime peels has been used as cattle feed.

Comments

As a common name, lime covers a number of different *Citrus* species and hybrid species. Those referred to as *C.* ×*aurantiifolia* are hybrids involving *C. medica* and possibly *C. hystrix*

From intergeneric hybridization between the lime and species of the genus *Fortunella* Swingle the limequat hybrids resulted.

Selected References

Aibinu I, Adenipekun T, Adelowotan T, Ogunsanya T, Odugbemi T (2007) Evaluation of the antimicrobial properties of different parts of *Citrus aurantifolia* (lime fruit) as used locally. Afr J Tradit Complement Altern Med 4(2):185–190

Asnaashari S, Delazar A, Habibi B, Vasfi R, Nahar L, Hamedeyazdan S, Sarker SD (2010) Essential oil from *Citrus aurantifolia* prevents ketotifen-induced weight-gain in mice. Phytother Res 24(12):1893–1897

Berhow M, Tisserat B, Kanes K, Vandercook C (1998) Survey of phenolic compounds produced in *Citrus*. USDA ARS Tech Bull 1856:1–154

Burkill IH (1966) A dictionary of the economic products of the Malay Peninsula. Revised reprint, 2 vols. Ministry of Agriculture and Co-operatives, Kuala Lumpur, Malaysia, vol 1 (A–H), pp 1–1240, vol 2 (I–Z), pp 1241–2444

Chevallier A (1996) The encyclopedia of medicinal plants. Dorling Kindersley, London, 336 pp

Chisholm MG, Wilson MA, Gaskey GM (2003) Characterization of aroma volatiles in key lime essential oils (*Citrus aurantifolia* Swingle). Flavor Fragr J 18(2):106–115

Choi HS, Song HS, Ukeda H, Sawamura M (2000) Radical-scavenging activities of citrus essential oils and their components: detection using 1,1-Diphenyl-2-picrylhydrazyl. J Agric Food Chem 48(9):4156–4161

Chopra RN, Nayar SL, Chopra IC (1986) Glossary of Indian medicinal plants (including the supplement). Council Scientific Industrial Research, New Delhi, 330 pp

Dalsgaard A, Reichert P, Mortensen HF, Sandström A, Kofoed P-E, Larsen JL, Mølbak K (1997) Application of lime (*Citrus aurantifolia*) juice to drinking water and food as a cholera-preventive measure. J Food Prot 60(11):1329–1333

Ebana RU, Madunagu BE, Ekpe ED, Otung IN (1991) Microbiological exploitation of cardiac glycosides and alkaloids from *Garcinia kola, Borreria ocymoides, Kola nitida* and *Citrus aurantifolia*. J Appl Bacteriol 71(5):398–401

Facciola S (1990) Cornucopia. A source book of edible plants. Kampong, Vista, 677 pp

Fletcher PS, Harman SJ, Boothe AR, Doncel GF, Shattock RJ (2008) Preclinical evaluation of lime juice as a topical microbicide candidate. Retrovirology 5:3

Foundation for Revitalisation of Local Health Traditions (2008) FRLHT database. http://envis.frlht.org

Gharagozloo M, Ghaderi A (2001) Immunomodulatory effect of concentrated lime juice extract on activated human mononuclear cells. J Ethnopharmacol 77(1): 85–90

Gharagozloo M, Doroudchi M, Ghaderi A (2002) Effects of *Citrus aurantifolia* concentrated extract on the spontaneous proliferation of MDA-MB-453 and RPMI-8866 tumor cell lines. Phytomedicine 9(5):475–477

Grieve M (1971) A modern herbal. Penguin, 2 vols. Dover publications, New York, 919 pp

Guimarães R, Barros L, Barreira JC, Sousa MJ, Carvalho AM, Ferreira IC (2009) Targeting excessive free radicals with peels and juices of *Citrus* fruits: grapefruit, lemon, lime and orange. Food Chem Toxicol 48(1): 99–106

Hemmerling A, Potts M, Walsh J, Young-Holt B, Whaley K, Stefanski DA (2007) Lime juice as a candidate microbicide? An open-label safety trial of 10% and 20% lime juice used vaginally. J Womens Health (Larchmt) 16(7):1041–1051

Ibrahim J, Abu Said A, Abdul Rashih A, Nor Azah MA, Norsiha A (1995) Chemical composition of some *Citrus* oils from Malaysia. J Essent Oil Res 8:627–632

Imade GE, Sagay A, Egah D, Onwuliri V, Grigg M, Egbodo C, Thacher T, Potts M, Short R (2008) Prevalence of HIV and other sexually transmissible infections in relation to lemon or lime juice douching among female sex workers in Jos, Nigeria. Sex Health 5(1):55–60

Imade GE, Sagay AS, Onwuliri VA, Egah DZ, Potts M, Short RV (2005) Use of lemon or lime juice douches in women in Jos, Nigeria. Sex Health 2(4):237–239

Khatoon T (1995) Phytochemical investigations of *Citrus* species of Pakistan. PhD thesis, Islamia University, Bahawalpur, Pakistan

Mabberley DJ (1997) A classification for edible *Citrus*. Telopea 7(2):167–172

Manner HI, Buker RS, Smith VE, Elevitch CR (2006). *Citrus* species, ver. 2.1. In: Elevitch CR (ed) Species profiles for Pacific island agroforestry. Permanent Agriculture Resources (PAR), Holualoa. http://www.agroforestry.net

Mauck CK, Ballagh SA, Creinin MD, Weiner DH, Doncel GF, Fichorova RN, Schwartz JL, Chandra N, Callahan MM (2008) Six-day randomized safety trial of intravaginal lime juice. J Acquir Immune Defic Syndr 49(3):243–250

Molesworth Allen B (1967) Malayan fruits. An introduction to the cultivated species. Moore, Singapore, 245 pp

Morton JF (1987) Mexican lime. In: Fruits of warm climates. Julia F. Morton, Miami, pp 168–172

Ochse JJ, Bakhuizen van den Brink RC (1931) Fruits and fruitculture in the Dutch East Indies. G. Kolff & Co, Batavia-C, 180 pp

Ochse JJ, Bakhuizen van den Brink RC (1980) Vegetables of the Dutch Indies, 3rd edn. Ascher & Co., Amsterdam, 1016 pp

Ojiezeh TI, Nwachukwu SE, Udoh SJ (2011) Antimicrobial effect of *Citrus aurantifolia* juice and *Veronica amygdalina* on common bacteria isolates. Pharm Chem 3(1):1–7

Onyeagba RA, Ugbogu OC, Okeke CU, Iroakasi O (2004) Studies on the antimicrobial effects of garlic (*Allium sativum* Linn), ginger (*Zingiber officinale* Roscoe) and lime (*Citrus aurantifolia* Linn). Afr J Biotechnol 3(10):552–554

Pacific Island Ecosystems at Risk (PIER) (2004) *Citrus aurantiifolia* (Christm.) Swingle, Rutaceae. http://www.hear.org/Pier/species/citrus_aurantiifolia.htm

Patil JR, Jayaprakasha GK, Murthy KNC, Tichy SE, Chetti MB, Patil BS (2009a) Apoptosis-mediated proliferation inhibition of human colon cancer cells by volatile principles of *Citrus aurantifolia*. Food Chem 114(4):1351–1358

Patil JR, Murthy KNC, Jayaprakasha GK, Chetti MB, Patil BS (2009b) Bioactive compounds from Mexican lime (*Citrus aurantifolia*) juice induce apoptosis in human pancreatic cells. J Agric Food Chem 57(22):10933–10942

Penniston KL, Nakada SY, Holmes RP, Assimos DG (2008) Quantitative assessment of citric acid in lemon juice, lime juice, and commercially-available fruit juice products. J Endourol 22(3):567–570

Peterson JJ, Beecher GR, Bhagwat SA, Dwyer JT, Gebhardt SE, Haytowitz DB, Holden JM (2006) Flavanones in grapefruit, lemons, and limes: a compilation and review of the data from the analytical literature. J Food Compos Anal 19(Suppl 1):S74–S80

Porcher MH et al (1995–2020) Searchable world wide web multilingual multiscript plant name database. Published by The University of Melbourne, Australia. http://www.plantnames.unimelb.edu.au/Sorting/Frontpage.html

Razzaghi-Abyaneh M, Shams-Ghahfarokhi M, Rezaee M-B, Jaimand K, Alinezhad S, Saberi R, Yoshinari T (2009) Chemical composition and antiaflatoxigenic activity of *Carum carvi L., Thymus vulgaris* and *Citrus aurantifolia* essential oils. Food Control 20(11):1018–1024

Saxholt E, Christensen AT, Møller A, Hartkopp HB, Hess Ygil K, Hels OH (2008) Danish food composition databank, revision 7. Department of Nutrition, National Food Institute, Technical University of Denmark. http://www.foodcomp.dk/

Sethpakdee R (1992) *Citrus aurantifolia* (Christm. & Panzer) Swingle. In: Coronel RE, Verheij EWM (eds) Plant resources of South-East Asia. no. 2: edible fruits and nuts. Prosea Foundation, Bogor, pp 126–128

Shalaby NMM, Abd-Alla HI, Ahmed HH, Basoudan N (2011) Protective effect of *Citrus sinensis* and *Citrus aurantifolia* against osteoporosis and their phytochemical constituents. J Med Plant Res 5(4):579–588

Shivashankara KS, Roy TK, Rao VK (2002) A study of volatile composition of Indian 'Kagzi' lime (*Citrus aurantifolia* Swingle) peel by SPME, during ripening. Indian Perfum 46(4):315–319

Sja'bani M, Ismadi M, Ismiati SD, Sidabutar RP, Rahardjo D (2007) The therapeutic effect of *Citrus aurantifolia* Swingle in idiopathic hypocitraturic calcium nephrolithiasis. Berk Ilmun Kedokt 39(4):199–208

Souza A, Lamidi M, Ibrahim B, Aworet Samseny RRR, Boukandou Mounanga M, M'Batchi B (2011) Antihypertensive effect of an aqueous extract of *Citrus aurantifolia* (Rutaceae) (Christm.) Swingle, on the arterial blood pressure of mammal. Int Res J Pharm Pharmacol 1(7):142–148

Stuart GU (2010) Philippine alternative medicine. Manual of some Philippine medicinal plants. http://www.stuartxchange.org/OtherHerbals.html

Swingle WT (1946) The botany of *Citrus* and its wild relatives of the orange subfamily (family Rutaceae, subfamily Aurantioideae). In: Webber HJ, Batchelor LD (eds) The *Citrus* industry, vol 1, History, botany and breeding. University of California Press, Berkeley, pp 129–474, 1028 pp

Taiwo SS, Oyekanmi BA, Adesiji YO, Opaleye OO, Adeyeba OA (2007) In vitro antimicrobial activity of crude extracts of *Citrus aurantifolia* and *Tithonia diversifolia* on clinical bacterial isolates. Int J Trop Med 2(4):113–117

Tanaka T (1954) Species problem in *Citrus*. A critical study of wild and cultivated units of *Citrus*, based upon field studies in their native homes. Japanese Society for the Promotion of Science, Tokyo, 152 pp

Tosukhowong P, Yachantha C, Sasivongsbhakdi T, Ratchanon S, Chaisawasdi S, Boonla C, Tungsanga K (2008) Citraturic, alkalinizing and antioxidative effects of limeade-based regimen in nephrolithiasis patients. Urol Res 36(3–4):149–155

Webber HJ (1946) Cultivated varieties of *Citrus*. In: Webber HJ, Batchelor LD (eds) The *Citrus* industry, vol 1, History, botany and breeding. University of California Press, Berkeley, pp 475–668, 1028 pp

Zhang DX, Mabberley DJ (2008) *Citrus* Linnaeus. In: Wu ZY, Raven PH, Hong DY (eds) Flora of China, vol 11, Oxalidaceae through Aceraceae. Science Press/Missouri Botanical Garden Press, Beijing/St. Louis

Citrus x aurantium Grapefruit Group

Scientific Name

Citrus x aurantium L., pro sp. [Grapefruit Group] (sensu Mabberley 1997, Bayer et al. 2009).

Synonyms

Citrus grandis (L.) Osbeck var. *racemosa* (Roem.) B. C. Stone, *Citrus decumana* L., *Citrus decumana* var. *paradisi* Nichols, *Citrus decumana* var. *racemosa* Roem., *Citrus maxima* var. *racemosa* Osbeck, *Citrus maxima* (Burman f.) Merrill, var. *uvacarpa* Merrill & H. A. Lee, *Citrus paradisi* Macfad. (sensu Swingle and Reece 1967; sensu Tanaka sec. Cottin 2002), *Citrus pompelmos racemosus* Risso & Poit., *Citrus racemosa* Marcovitch ex Tanaka, *Citrus x paradisi* Macfad.

Family

Rutaceae.

Common/English Names

Common Grapefruit, Grapefruit.

Vernacular Names

Brazil: Toranja;
Chinese: Yuan You, Pu Tao You, Xi You (Hong Kong);
Czech: Citroník Grapefruit;
Danish: Grapefrugt, Pompelmus;
Dutch: Pompelmoes;
Eastonian: Kreibipuu, Vili: Kreip;
Finnish: Greippi;
French: Pomelo, Pamplemousse, Pamplemoussier;
German: Grapefruit, Paradisapfel, Pampelmuse, Pampelmusenbaum;
Hungarian: Citrancs, Grépfrút;
Indonesia: Buah Jeruk Kecut, Limau Gedang;
Italian: Pompelmo, Pampelino, Arancio Maggiore;
Japanese: Gureepufuruushu, Gureepufuruutsu;
Khmer: Krôôch Thlông;
Laotian: Kièngz S'aangz;
Norwegian: Grapefrukt;
Philippines: Toronja (Tagalog);
Portuguese: Pomelo;
Russian: Greipfrut, Pompel'mus;
Spanish: Citrumelo, Pomelo, Toronja;
Swedish: Grapefrukt, Pompelmus;
Thai: Grapefruit;
Turkish: Greyfurt;
Vietnamese: Bưởi Đắng, Bưởi Chùm, Bưởi Vỏ Dính, Cây Bưởi Chùm, Quả Bưởi Chum.

Origin/Distribution

The grapefruit was originally thought to be a spontaneous sport of the pummelo and was separated from the pummelo by James MacFayden, in his *Flora of Jamaica,* in 1837 and given the botanical name *Citrus paradisi*. About 1948, citrus researchers found that grapefruit was not a sport of the pummelo but an accidental hybrid between the pummelo, *Citrus maxima* and the orange, *Citrus sinensis* and altered to the botanical name to *Citrus x paradisi* to reflect the hybrid nature. The grapefruit is widely cultivated in subtropical and tropical countries. The most important grapefruit growing countries are the southern USA, Philippines, the Caribbean Island, Israel, South Africa, Morocco, Brazil, Argentina and Australia.

Agroecology

The grapefruit thrives in a warm subtropical climate in areas with mean annual rainfall of 900–1,200 mm evenly distributed throughout the year, cool winters and warm or hot summers. Grapefruit are moderately susceptible to frosts. Young trees, flowers and fruit will not tolerate sub-zero temperatures of −7°C. Grapefruit tolerates a wide range of soil from acidic to alkaline soils but does best in well-drained, deep sandy loams with soil pH of 6–7. Successful grapefruit culture depends mainly on the choice of rootstock best adapted to each type of soil. Maximum amount of sunlight is desirable for the growth, setting and maturity of fruit. Locations that are low and frost-prone, or exposed to strong winds, particularly sea winds should be avoided.

Edible Plant Parts and Uses

Grapefruit is commonly consumed as a breakfast fruit, chilled, cut in half and the pulp spooned from the "half-shell". The pulp may be sweetened with white or brown sugar, or a dash of honey. Grapefruit is also eaten as pre-dinner appetizer whence the halves may be similarly sweetened, lightly broiled, and served hot, and topped with a maraschino cherry. The fruit segments are commonly used in fruit cups, fruit salads, in gelatins, puddings, tarts and processed into jellies. They are also commercially canned in syrup. In Australia, grapefruit is commercially processed as marmalade. In Costa Rica, grapefruit is eaten cooked to remove its acidity, rendering them as sweets; the fruit is also stuffed with *dulce de leche* resulting in the *toronja rellana* dessert.

Grapefruit juice is marketed as a fresh beverage, canned juice, frozen concentrated and as dehydrated powder. It can be made into an excellent vinegar or carefully fermented into wine. Grapefruit juice has increased in popularity especially after its promotion as a diet drink. Many weight-loss diets incorporate grapefruit juice. The pulp left over after commercial juice extraction provides an important source of grapefruit oil, which is used as a flavouring in many soft drinks, confectionery and food.

Grapefruit peel is utilized candied and provides an important source of pectin for preservation of other fruits. The peel essential oil, expressed or distilled, is commonly used as a food flavouring for soft drinks, chewing gum, sweets, gelatine desserts, candies, baked goods, ice crème etc. An ingredient of the outer peel (flavedo) oil, nookatone is extracted and added to grapefruit juice powder, enhancing the flavour of the reconstituted juice. Citric acid and naringin can be extracted from the inner peel (albedo). Naringin is used as a bitter agent in "tonic" beverages, bitter chocolate, ice cream and ices. Naringin can be chemically converted into a sweetener about 1,500 times sweeter than sugar. After the extraction of naringin, the albedo can be further processed to recover pectin.

Grapefruit seeds afford a dark and exceedingly bitter oil, when bleached and refined, it becomes pale-yellow, bland, much like olive oil in flavour, and can be used similarly for cooking. Being high in unsaturated fats, its production has increased appreciably since 1960.

Botany

An evergreen armed tree, usually 5–6 m high but can reach 13–15 m high with a trunk of 15 cm diameter and glabrous branches and twigs with short axillary spines. Grafted trees are shorter and fruit earlier (Plate 1). Leaves alternate, similar to those of *Citrus maxima* but leaf blade elliptic-ovate smaller and narrower with ciliate mid-vein and shallowly crenulate margin, sub-emarginate to sub-acute apex and rounded base, on narrowly winged petioles (Plates 1, 2 and 3). Flowers in 2–6 flowered axillary clusters, calyx 4–5 glabrous lobes, petals 4 linear oblong rounded white petals, stamens 25–26 with narrow yellow anthers (Plate 2). Fruit pale green flushed with pink, or yellow or flushed with pink or red, oblate to globose, 8–14 cm, smaller than those of *C. maxima*; pericarp thin; sarcocarp with 12–15 segments, yellowish-white, pink, red to deep red, tender, juicy, slightly fragrant and acidic. Seeds few with numerous embryos or absent (Plates 1, 4, 5 and 6).

Nutritive/Medicinal Properties

Nutrient composition of fresh, raw pink-, red-fleshed grapefruit per 100 g edible portion was reported as follows: water 88.06 g, energy 42 kcal (176 kJ), protein 0.77 g, total fat 0.14 g, ash 0.36 g, carbohydrate 10.66 g, total dietary fibre 1.6 g, total sugars 6.89 g, sucrose 3.51 g, glucose 1.61 g, Ca 22 mg, Fe 0.08 mg, Mg 9 mg, P 18 mg, K 135 mg, Na 0 mg, Zn 0.07 mg, Cu 0.032 mg, Mn 0.022 mg, Se 0.1 μg, vitamin C 31.2 mg, thiamine 0.043 mg, riboflavin 0.031 mg, niacin 0.204 mg, pantothenic acid 0.262 mg, vitamin B-6 0.053 mg, total folate 13 μg, total choline 7.7 mg, betaine 0.1 mg, vitamin A 1150 IU (58 μg RAE), total saturated fatty acids 0.021 g, 16:0 (palmitic acid) 0.018 g, 18:0 (stearic acid) 0.002 g; total monounsaturated fatty acids 0.020 g, 16:1 undifferentiated (palmitoleic acid) 0.002 g, 18:1 undifferentiated (oleic acid) 0.018 g; total polyunsaturated fatty acids 0.036 g, 18:2 undifferentiated (linoleic acid) 0.029 g, 18:3 undifferentiated (linolenic acid) 0.008 g, β-carotene 686 μg, α-carotene 3 μg, β-cryptoxanthin 6 μg, lycopene 1,419 μg, lutein + zeaxanthin 5 μg, tryptophan 0.008 g, threonine 0.013 g, isoleucine 0.008 g,

Plate 1 Grafted fruiting Marsh grapefruit tree

Plate 2 Grapefruit flowers

Plate 3 Juvenile grapefruit and leaves

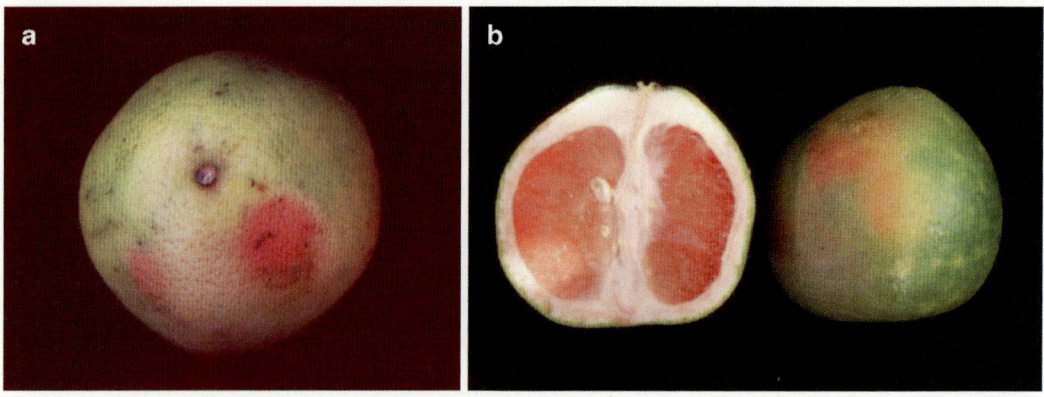

Plate 4 Ruby red grapefruit: (**a**) whole; (**b**) halved showing reddish flesh

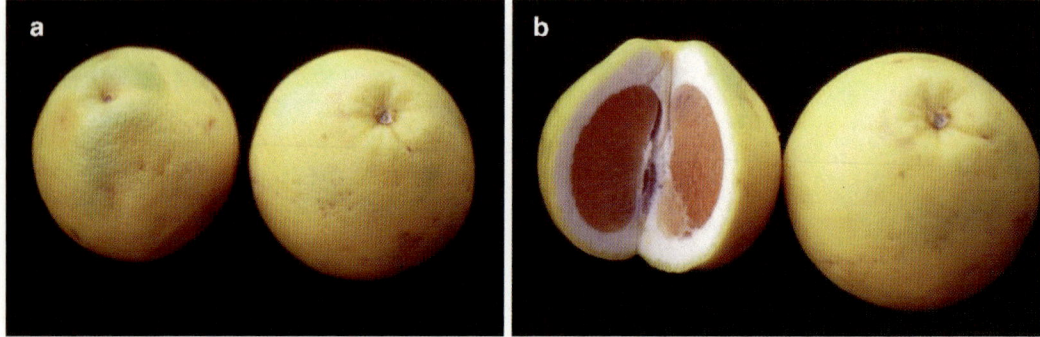

Plate 5 Pink-fleshed seedless grapefruit: (**a**) whole; (**b**) halved

Plate 6 Star Ruby grapefruit with deep red flesh

Nutrient composition of fresh, raw white-fleshed grapefruit per 100 g edible portion was reported as follows: water 90.48 g, energy 33 kcal (138 kJ), protein 0.69 g, total fat 0.10 g, ash 0.33 g, carbohydrate 8.41 g, total dietary fibre 1.1 g, total sugars 7.31 g, Ca 12 mg, Fe 0.06 mg, Mg 9 mg, P 8 mg, K 148 mg, Na 0 mg, Zn 0.07 mg, Cu 0.050 mg, Mn 0.013 mg, Se 1.4 μg, vitamin C 33.3 mg, thiamine 0.037 mg, riboflavin 0.020 mg, niacin 0.269 mg, pantothenic acid 0.283 mg, vitamin B-6 0.043 mg, total folate 10 μg, total choline 7.7 mg, vitamin A 33 IU (2 μg RAE), total saturated fatty acids 0.014 g, 16:0 (palmitic acid) 0.012 g, 18:0 (stearic acid) 0.001 g; total monounsaturated fatty acids 0.013 g, 16:1 undifferentiated (palmitoleic acid) 0.001 g, 18:1 undifferentiated (oleic acid) 0.012 g; total polyunsaturated fatty acids 0.024 g, 18:2 undifferentiated (linoleic acid) 0.019 g, 18:3 undifferentiated (linolenic acid) 0.005 g, physterols leucine 0.015 g, lysine 0.019 g, methionine 0.007 g, cystine 0.008 g, phenylalanine 0.046 g, tyrosine 0.008 g, valine 0.015 g, arginine 0.087 g, histidine 0.008 g, alanine 0.024 g, aspartic acid 0.138 g, glutamic acid 0.197 g, glycine 0.015 g, proline 0.063 g and serine 0.028 g (USDA 2011).

17 mg, β-carotene 14 μg, α-carotene 8 μg, β-cryptoxanthin 3 μg, lycopene 0 μg, lutein + zeaxanthin 10 μg, tryptophan 0.007 g, threonine 0.012 g, isoleucine 0.007 g, leucine 0.013 g, lysine 0.017 g, methionine 0.007 g, cystine 0.007 g, phenylalanine 0.041 g, tyrosine 0.007 g, valine 0.0154 g, arginine 0.078 g, histidine 0.007 g, alanine 0.022 g, aspartic acid 0.123 g, glutamic acid 0.176 g, glycine 0.013 g, proline 0.056 g and serine 0.025 g (USDA 2011).

Lycopene was found to be the major coloured pigment in the juice of six red grapefruit cultivars with lesser amounts of β-carotene (Lee 2000). Two new lycopene β-cyclases (LCYBs) were cloned and characterized from *Citrus paradisi* 'Flame' (red flesh) and 'Marsh' (white flesh) fruits (Mendes et al. 2011).

Other Fruit Phytochemicals

1-p-menthene-8-thiol was identified as a potent flavor impact constituent of grapefruit juice (Demole et al. 1982). Fourteen sesquiterpene ketones D–Q pertaining to the valencane (D–J) and eudesmane (K–Q) groups were identified in grapefruit juice flavour (Demole and Enggist 1983). In particular, (+)-8,9-didehydronootkatone (E) was found to have a powerful flavor with good grapefruit juice character. The glycoside 7 β-neohesperidosyl-4-(β-D-glucopyranosyl) naringenin was found in grapefruit pulp segments (Aurnhammer et al. 1971).

The percentage and concentration of total phenolics in *Citrus paradisi* cv Star Ruby peel and juice were reported by Berhow et al. (1998) as: flavedo (outer pigmented layer of the peel): 6.9% (0.22 mg/g) flavone/flavonol, 59.6% (1.21 mg/g) flavanone, 5.9% psoralen, 19.3% (0.08 mg/g) coumarin; albedo (inner layer of the rind): 0.4% (0.10 mg/g), flavone/flavonol, 93.9% (15.03 mg/g) flavanone, 0% psoralen, 3.4% (0.26 mg/g) coumarin; juice: 2% (0.05 mg/g), flavone/flavonol, 83.8% (1.39 mg/g) flavanone, 1.3% psoralen, 9.5% (0.08 mg/g) coumarin.

The percentage and concentration of flavanones reported in *Citrus paradisi* cv Star Ruby by Berhow et al. (1998) were as follows: flavedo: narirutin 6% (0.1 mg/g), naringin 82% (1.0 mg/g), neohesperidin 5% (traces), naringin–6″–malonate (open form) 7% (0.1 mg/g); albedo: didymin 1% (0.1 mg/g), narirutin 9% (1.4 mg/g), neoerocitrin 1% (0.1 mg/g), naringin 75% (11.3 mg/g), naringin–4′–glucoside 2% (0.3 mg/g), neohesperidin 1% (0.2 mg/g), naringin–6″–malonate (open form) 4% (0.7 mg/g), naringin–6″–malonate (closed form) 5% (0.7 mg/g); juice: narirutin–4′–glucoside 2% (traces), narirutin 15% (0.2 mg/g), naringin 75% (1.0 mg/g), neohesperidin 2% (traces), naringin–6″–malonate (open form) 2% (traces), naringin–6″–malonate (closed form) 3% (traces).The percentage and concentration of flavone reported in *Citrus paradisi* cv Star Ruby by Berhow et al. (1998) were as follows: flavedo: rhoifolin 44% (0.1 mg/g); albedo: rhoifolin 100% (0.1 mg/g).

The following flavonoid compounds were isolated and identified from a commercial naringin product prepared from grapefruit: neohesperidin, isosakuranetin-7-rhamnoglucoside, rhoifolin, and kaempferol (Dunlap and Wender 1962). Segments of Texas Ruby Red grapefruit were found to contain 7-β-rutinosides of the flavanones naringenin, hesperetin, and isosakuranetin (Mizelle et al. 1965) . These tasteless compounds, were found to be isomers of three bitter neohesperidosides (naringin, neohesperidin, and ponoirin) previously known to be present in grapefruit. Two isomeric rhamnodiglucosides of naringenin have been isolated from the segments of Texas Ruby Red grapefruit and identified as 4′β-D-glucosyl-7β-neohesperidosylnaringenin and 4′β-D-glucosyl-7β-rutinosylnaringenin (Mizelle et al. 1967). Grapefruit juice and juice sacs were found to contain the bitter flavanone glycosides naringin, neohesperidin, and poncirin, and their tasteless isomers, naringenin-7β-rutinoside, hesperidin, and isosakuranetin-7β-rutinoside (Hagen et al. 1965). Grapefruit had a total flavanone content (summed means) of 27 mg/100 g as aglycones and a distinct flavanone profile, dominated by naringin (Peterson et al. 2006). The sugar neohesperidose (2-O-α-Lrhamnosyl-β-D-glucose), was found to be high in grapefruits, imparting the tangy or bitter taste to the glycoside naringin. White grapefruit varieties tended to be slightly but not significantly higher in total flavanones than pink and red varieties. The flavanone contents (means in mg aglycone/100 g fresh fruit or juice) in white grapefruit were as follows: didymin 0.09 mg, eriocitrin 0.16 mg, hesperidin 3.95 mg, naringin 16.90 mg, narirutin 5.36 mg, neoeriocitrin 0.05 mg, neohesperidin 0.25 mg, poncirin 0.20 mg, total flavanone 26.96 mg. The flavanone contents (means in mg aglycone/100 g fresh fruit or juice) in red and pink grapefruit were as follows: didymin 0 mg, eriocitrin 0 mg, hesperidin 0.27 mg, naringin 13.87 mg, narirutin 3.34 mg, neoeriocitrin 0 mg, neohesperidin 0.42 mg, poncirin 0 mg, and total flavaonone 17.9 mg.

Saturated and mono-unsaturated long-chain hydrocarbons were determined in juice sacs of three grapefruit varieties—Marsh Seedless, Redblush and Foster (Nagy and Norby 1972). In the saturated fraction, the dominant linear hydrocarbon was C25 while C29 predominated in the monoene fraction. Iso- and anteiso-branched hydrocarbons comprised between 52% and 53% of the saturated fraction but only 9–12% of the monoene. The saturated and mono-unsaturated hydrocarbon profiles of Marsh Seedless, Redblush and Foster were similar. Trans-hydroxycinnamic acids (caffeic, p-coumaric, ferulic, and sinapic) were more abundant in the fruits of white grapefruit than in its hybrid (Jaffa Sweeties) (Gorinstein et al. 2004c). Nine new furanocoumarins were detected in different fractions of grapefruit juice retentate and tentatively designated as FC 338, FC 420, FC 524, FC 530, FC 540, FC 546, FC 552, FC 570 and FC 614 (Yu et al. 2009).

The flavonoids, naringin and naringenin and the furanocoumarin, bergapten (5-methoxypsoralen), were detected in some fresh grapefruit and commercial grapefruit juices but were not detected in other fruit juices tested (orange; orange with apple base; dark grape; orange and mango with apple base; orange, peach, passion fruit juice) (Ho et al. 2000). Flavonoid glycosides narirutin, naringin, hesperidin, neohesperidin, didymin, and poncirin have been identified in all the grapefruit juices examined (Ross et al. 2000). The aglycone quercetin was detected in only two brands. All the juices were free from methoxylated flavonoid aglycones. The concentration of total flavonoids ranged between 19.44 and 84.28 mg/100 ml juice. Naringin (14.56–63.8 mg/100 ml juice) was found to be the major flavonoid followed by narirutin 2.25–12.20 mg/100 ml juice and hesperidin 0.24–3.12 mg/100 ml juice.

Five compounds including furanocoumarin monomers (bergamottin and 6′,7′-dihydroxybergamottin (DHB)), furanocoumarin dimers (4-[[6-hydroxy-7[[1-[(1-hydroxy-1-methyl) ethyl]-4-methyl-6-(7-oxo-7H-furo[3,2-g][1]benzopyran-4-yl)-4-hexenyl]oxy]-3,7-dimethyl- 2-octenyl]oxy]-7H-furo[3,2-g][1]benzopyran-7-one (GF-I-1) and 4-[[6-hydroxy-7[[4-methyl-1-(1-methylethenyl)-6-(7-oxo-7H-furo[3,2-g][1] benzopyran-4-yl)-4-hexenyl]oxy]-3,7-dimethyl-2-octenyl]oxy]-7H-furo[3,2-g][1]benzopyran-7-one (GF-I-4), and a sesquiterpene nootkatone were isolated from grapefruit juice (Tassaneeyakul et al. 2000). Content variation of furanocoumarin 6,7-dihydroxybergamottin, bergamottin, and 6 furanocoumarin dimers were determined for 58 commercial grapefruit juices collected over two seasons (Widmer and Haun 2005). The content of 6,7-dihydroxybergamottin ranged from 0.2 to 7.7 ppm in all juices tested, and averaged 1.8 ppm in the red compared with 2.9 ppm for white grapefruit juices. Bergamottin content ranged from 1.6 to 7.3 ppm for all juices and averaged 3.4 and 4.2 ppm and in red and white grapefruit juices, respectively. Only 1 dimer varied significantly between the red and white juice types. Dihydroxybergamottin and 2 dimer compounds were significantly lower in shelf-stable or non-refrigerated products compared with refrigerated products whereas bergamottin and 1 dimer compound occurred at higher levels in the shelf-stable products. Individual measured dimer compounds varied up to 60-fold in all juices, but the sum of all 6 dimers varied only 14-fold. Manthey and Buslig (2005) separated freshly extracted grapefruit juice into four fractions, namely raw finished juice (approximately 5% fine pulp), centrifugal retentate (approximately 35% fine pulp), centrifuged supernatant (<1% pulp), and coarse finisher pulp. They found that the centrifugal retentate had the highest furanocoumarin content, containing 892 ppm of bergamottin, 628 ppm of 6′,7′-dihydroxybergamottin, 116 ppm of 6′,7′-epoxybergamottin, 105 ppm of 7-geranyloxycoumarin, and approximately 467 ppm of furanocoumarin dimers. Recent studies reported that furanocoumarins related to bergamottin [5-[(3′,7′-dimethyl-2′,6′-octadienyl)oxy]psoralen] were primarily responsible for the grapefruit inhibitory effect on cytochrome P450 3A4 (CYP3A4) which mediates grapefruit drug interactions. The most important compounds of grapefruit juice considered to be involved in the pharmacokinetic interaction comprised flavonoids (naringin, naringenin, narirutin, quercetin, kaempferol, hesperidin, neohesperidin, didymin, and poncirin), furanocoumarins (6′,7′-dihydroxybergamottin, bergamottin, bergamottin-6′,7′-epoxide, bergapten, epoxy-bergamottin) and sesquiterpens (noot-katone) (Harapu et al. 2010). The concentrations of flavonoid compounds in total grapefruit juice were 143.86 mg/100 ml in red, 131.47 mg/100 ml in pink and 84.21 mg/100 ml in white grapefruit juice. In juice prepared from fruit pulp, the concentrations were 81.92, 108.23 and 65.76 mg/100 ml, respectively. The content of naringin, the most important flavanone in citrus fruit, varied between 1.98 and 51.2 mg/100 ml juice.

Thirty-seven odor-active compounds were detected in extract prepared from fresh grapefruit juice in the flavor dilution (FD) factor range of 4–256 and subsequently identified (Buettner and Schieberle 1999). Among them the highest odor activities (FD factors) were determined for ethyl butanoate, p-1-menthene-8-thiol, (Z)-3-hexenal, 4,5-epoxy-(E)-2-decenal, 4-mercapto-4-methylpentane-2-one, 1-heptene-3-one, and wine lactone. Besides the five last mentioned compounds, a total of 13 further odorants were identified for the first time as flavor constituents of grapefruit. The data confirmed results of the literature on the significant contribution of 1-p-menthene-8-thiol in grapefruit aroma but clearly showed that a certain number of further odorants were necessary to elicit the typical grapefruit flavor.

Grape fruit peel oil comprised 90% limonene; the volatile fraction (2–3%) consisted mainly of oxygen compounds and sesquiterpenes; the waxy fraction (7–8%) comprised C_8 and C_{10} aldehydes, plus geraniol, cadinene and small amounts of citral and dimethyl arthranilate, plus acid (Morton 1987). Also present were nine coumarins and 0.88% 22-dihydrostigmasterol. The dried pulp and seeds contain β-sitosteryl-D-glucoside and β-sitosterol. A total of 67 volatile compounds amounting to 97.9% were found in cold-pressed essential oil of red-blush grapefruit (*Citrus paradisi*) peel (Njoroge et al. 2005). Monoterpene hydrocarbons constituted 93.3% in the oil with limonene (91.1%), α-terpinene (1.3%) and α-pinene (0.5%) as the main components. Sesquiterpene hydrocarbons constituted 0.4% in the oil. The notable compounds were β-caryophyllene, α-cubebene, and (E,E)-α-farnesene. Oxygenated compounds constituted 4.2%, out of which carbonyl compounds (2.0%), alcohols (1.4%), and esters (0.7%) were the major groups. Heptyl acetate, octanal, decanal, citronellal, and (Z)-carvone were the main constituents (0.1–0.5%). Perillene, (E)-carveol, and perillyl acetate were also found. Nootkatone, α-sinensal, β-sinensal, methyl-N-methylanthranilate, and (Z,E)-farnesol were also prominent in the oil. A total of 27 components were identified in Turkish grapefruit peel oil (Kırbaşlar et al. 2006, 2009): monoterpene hydrocarbons (96.4%) of which limonene (92.5%) and myrcene (2.6%) were the first two major components; the major sesquiterpene components (0.8%) were β-caryophyllene (0.4%) and δ-cadinene (0.2%) and nootkatone (0.2%). The major oxygenated components (1.2%) comprised aldehydes (0.7%) octanal (0.2%), decanal (0.2%) geranial (0.1%) and neral (0.1%); alcohol components (0.3%) linalool (0.2%), α-terpineol (0.1%) and ester components (0.2%) neryl acetate (0.1%) and geranyl acetate (0.1%).

The content of two major bitter components, naringin and limonin, in grapefruit peel albedo decreased with fruit maturity and were affected by some treatments with gibberellic acid (Shaw et al. 1991). In Marsh grapefruit albedo, naringin was the major flavanone glycoside, representing more than 80% of the total flavanones. In a non-bitter grapefruit hybrid, only traces of naringin were present in the albedo, and the non-bitter isomer naringenin 7β-rutinoside was the major flavonoid present. Debittering of red grapefruit juice was found to remove more than 78% of bitterness in grapefruit juice, based on naringin content (Lee and Kim 2003). Also, some of the non-bitter flavonoids, such as narirutin and hesperidin, were nearly completely removed. Other values, such as vitamin C, and total phenolics were also reduced after debittering. Star Ruby grapefruit peel was found to have high content of naringin (Ortuño et al. 2006). The presence of the polymethoxyflavones nobiletin, heptamethoxyflavone and tangeretin, were found in all the grapefruit varieties analysed. Five compounds namely friedelin, β-sitosterol, 7(3′,7′,11′,14′-tetramethy)pentadec-2′,6′,10′-trienyloxycoumarin, limonin and cordialin B were isolated from grapefruit peel (Meera and Kalidhar 2008). A new cyclic acetal of marmin (6′,7′-dihydroxy-7-geranyloxycoumarin), two new cyclic acetals of 6′,7′-dihydroxybergamottin, and the known compounds marmin, 7-geranyloxycoumarin, bergamottin and 6′,7′-dihydroxybergamottin were isolated from grapefruit peel oil (César et al. 2009).

Phytochemicals in Seeds

Citrus limonoids such as limonin and limonin-17-β-d-glucopyranoside (LG) and flavonoid naringin were extracted from grapefruit seeds using a supercritical carbon dioxide (SC-CO2) extraction technique (Yu et al. 2007). Highest yield of limonin (6.3 mg/g seeds) was achieved at 48.3 MPa pressure, 50°C temperature and 60 minutes of extraction time while highest yield of LG (0.62 mg/g seeds) was achieved at 41.4 MPa pressure, 60°C temperature and 30% ethanol concentration in 40 minutes. Highest yield of naringin (0.2 mg/g seeds) was achieved at 41.4 MPa pressure, 50°C temperature and 20% ethanol concentration in 40 minutes.

Phytochemicals in Leaves

Feruloylputrescine was found in grapefruit juice and leaves (Wheaton and Stewart 1965). Mature grapefruit leaves contain the flavonoid, apigenin 7 β-rutinoside. Young leaves contain the 7 β-neohesperidoside and 7 β-rutinoside of naringenin (Fisher 1968). Apigenin 7β-rutinoside, an isomer of apigenin 7β-neohesperidoside (rhoifolin) was isolated from mature leaves of *Citrus paradisi* (Nordby et al. 1968). The percentage and concentration of total phenolics in *Citrus paradisi* cv Star Ruby leaf, were reported by Berhow et al. (1998) as: 34.2% (1.95 mg/g) flavone/flavonol, 57.3% (2.09 mg/g) flavanone, 0% psoralen, 0% coumarin; percentage and concentration of flavanones in the leaf as narirutin 15% (0.3 mg/g), naringin 43% (0.9 mg/g), neohesperidin 15% (0.3 mg/g), naringin–6″–malonate (open form) 26% (0.5 mg/g); and percentage and concentration of flavones in the leaf as isorhoifolin 22% (0.4 mg/g), rhoifolin 51% (1.0 mg/g).

The many pharmacological properties of grapefruit are elaborated below.

Antioxidant Activity

Freeze-dried orange peel showed the highest, lemon peel somewhat less and grapefruit peel the lowest but still appreciable antioxidant activity (Kroyer 1986). This could be significantly improved by preparing methanolic extracts of the peels. Autoxidation studies with the flavanon glycosides hesperidin and naringin as well as with their aglycones hesperetin and naringenin showed that the former were mainly responsible for the antioxidative activity of the citrus peel and extracts.

Fruit and vegetables rich in anthocyanins (e.g. strawberry, raspberry and red plum) demonstrated the highest antioxidant activities, followed by those rich in flavanones (e.g. orange and grapefruit) and flavonols (e.g. onion, leek, spinach and green cabbage), while the hydroxycinnamate-rich fruit (e.g. apple, tomato, pear and peach) consistently elicited lower antioxidant activities (Proteggente et al. 2002). The TEAC (Trolox Equivalent Antioxidant Capacity), the FRAP (Ferric Reducing Ability of Plasma) and ORAC (Oxygen Radical Absorbance Capacity) values for each extract were relatively similar and well-correlated with the total phenolic and vitamin C contents. The antioxidant activities TEAC in terms of 100 g fresh weight uncooked portion size were in the order: strawberry >> raspberry = red plum >> red cabbage >>> grapefruit = orange > spinach > broccoli > green grape approximately/= onion > green cabbage > pea > apple > cauliflower approximately/= tomato approximately/= peach = leek > banana approximately/= lettuce. On a fresh weight basis, grapefruit's hybrid Jaffa Sweeties was found to have a higher total phenol content as well as a higher antioxidant capacity in comparison with white grapefruit (Gorinstein et al. 2004e). A linear relationship existed between total antioxidant activity (TAA) and anthocyanins ($R2 = 0.8068$), TAA and flavonoids ($R2 = 0.9320$) and TAA and total phenols ($R2 = 0.9446$). Trans-hydroxycinnamic acids (caffeic, p-coumaric, ferulic, and sinapic) were more abundant in the fruits of white grapefruit than in its hybrid (Jaffa Sweeties).

To ascertain the antioxidant properties of the peel and juice of grapefruit, lemon, lime and sweet orange, different in-vitro assays were applied to the volatile and polar fractions of peels and to crude and polar fraction of juices: 2,2-diphenyl-1-picrylhydrazyl (DPPH) radical scavenging capacity, reducing power and inhibition of lipid peroxidation using β-carotene-linoleate model system in liposomes and thiobarbituric acid reactive substances (TBARS) assay in brain homogenates (Guimarães et al. 2010). Reducing sugars and phenolics were the main antioxidant compounds found in all the extracts. Peels polar fractions exhibited the highest contents in phenolics, flavonoids, ascorbic acid, carotenoids and reducing sugars, which evidently contributed to the highest antioxidant potential found in these fractions. Peels volatile fractions were clearly separated using discriminant analysis, which was in accord with their lowest antioxidant potential.

The methanol:water (80:20) fraction of Rio Red grapefruit showed the highest radical scavenging activity 42.5%, 77.8% and 92.1% at 250, 500 and 1,000 ppm, respectively (Jayaprakasha et al. 2008). Ethyl acetate extract of Rio Red grapefruit was found to contain maximum phenolics. All the fractions showed good antioxidant capacity by the formation of phosphomolybdenum complex at 200 ppm. All the extracts showed variable superoxide radical scavenging activity using non-enzymatic (NADH/phenaxine methosulfate) superoxide generating system. In addition, methanol:water (80:20) extract of Rio Red exhibited marked reducing power in potassium ferricyanide reduction method. Acetone and methanol extracts of defatted seed powder of red Mexican grapefruit, respectively, showed the highest (85.7%) and lowest (53.3%) radical scavenging activity, at 500 ppm (Mandadi et al. 2007). The total phenolic contents were found to be highest in the acetone extract (15.94%) followed by the methanol extract (5.92%), ethyl acetate extract (5.54%) and water extract (5.26%). Antioxidant capacity of the extracts as equivalents to ascorbic acid (μmol/g of the extract) was in the order, ethyl acetate extract > acetone extract > water extract > methanol extract. Five limonoids were identified in the ethyl acetate and acetone extracts: obacunone, nomilin, limonin, deacetylnomilin (DAN) and limonin-17-β-D-glucopyranoside (LG). Citrus limonoids had been shown to inhibit the growth of cancer in colon, lung, mouth, stomach and breast in animal and cell culture studies. Bergaptol and geranylcoumarin from grapefruit juice showed very good radical scavenging activity using 2,2′-azobis (3-ethylbenz-thiazoline-6-sulfonic acid) (ABTS) and 2,2-diphenyl-1-picrylhydrazil (DPPH) methods (Girennavar et al. 2007a, b).

Aqueous extract of grapefruit seed glyceric exhibited higher antioxidant activity than the ethanol extract as evaluated by DPPH, 5-lipoxygenase and luminol/xanthine/xanthine oxidase chemiluminescence assays (Giamperi et al. 2004).

Antimicrobial Activity

Of five of the six extracts commercially available, grapefruit seed showed potent growth inhibiting activity against the test microbes *Bacillus subtilis, Micrococcus flavus, Staphylococcus aureus, Serratia marcescens, Escherichia coli, Proteus mirabilis* and *Candida maltosa* (von Woedtke et al. 1999). In all of the antimicrobial active grapefruit seed extracts, the preservative benzethonium chloride was detected by thin layer chromatography. In only one of the grapefruit seed extracts tested no preservative agent was found. However, with this extract as well as with several self-made extracts from seed and juiceless pulp of grapefruits (*Citrus paradisi*) no antimicrobial activity could be detected (standard serial broth dilution assay, agar diffusion test). Thus, it was concluded that the potent as well as nearly universal antimicrobial activity being attributed to grapefruit seed extract was in fact due to the synthetic preservative agents contained within.

Commercial grapefruit seed extracts (GSE) were extracted with chloroform and found to have three major constituents benzyldimethyldodecylammonium chloride, benzyldimethyltetradecylammonium chloride, and benzyldimethylhexadecylammonium chloride (Takeoka et al. 2005). This mixture of homologues is commonly known as benzalkonium chloride, a widely used synthetic antimicrobial ingredient used in cleaning and disinfectant agents. An in-vitro study revealed that polymethoxyflavones and flavanone glycosides from *C. sinensis* and *C. paradisi* acted as antifungal agents against *Penicillium digitatum*, the polymethoxyflavones being more active than the flavanones in this respect (Ortuño et al. 2006). Grapefruit essential oil was found to inhibit growth of moulds commonly associated with food spoilage: *Aspergillus niger*, *Aspergillus flavus*, *Penicillium chrysogenum* and *Penicillium verrucosum* (Viuda-Martos et al. 2008b). Grapefruit essential oil showed the highest inhibition effect upon *Penicillium chrysogenum* and *Penicillium verrucosum*.

Preliminary studies showed that the grapefruit oil component 4-[[(E)-5-(3,3-dimethyl-2-oxiranyl)-3-methyl-2-pentenyl]oxy]-7H-furo[3,2-g]chromen-7-one (2) enhanced the susceptibility of test of methicillin-resistant *Staphylococcus aureus* (MRSA) to antibacterial agents, e.g., ethidium bromide and norfloxacin, to which these micro-organisms were normally resistant (Abulrob et al. 2004). Ethanolic extracts of grapefruit (*Citrus paradisi*) seed and pulp exhibited the strongest antimicrobial effect against *Salmonella enteritidis* (MIC 2.06%, m/V) (Cvetnić and Vladimir-Knezević 2004). Other tested bacteria (19) and yeasts (10) were sensitive to extract concentrations ranging from 4.13% to 16.50% (m/V). The presence of flavanones naringin and hesperidin in the extracts was confirmed by TLC analysis. Preliminary studies thus suggested that dried or fresh grapefruit seeds (*C. paradisi*) when taken at a dosage of 5–6 seeds every 8 hours, was comparable to that of proven antibacterial drugs (Oyelami et al. 2005). Of four patients with urinary tract infections, only the female was asymptomatic, the three males had *Pseudomonas aeruginosa*, *Klebsiella* species, and *Staphylococcus aureus*, respectively, in their urine samples, while the female had *Escherichia coli*. All four patients treated with grapefruit seeds orally for 2 weeks responded satisfactorily to the treatment except the man with *P. aeruginosa* infection. Grapefruit peel oil exhibited antibacterial activity against Gram positive bacteria (*Staphylococcus aureus*, *Bacillus cereus*, *Mycobacterium smegmatis*, *Listeria monocytogenes*, *Micrococcus luteus*) and Gram negative bacteria (*Escherichia coli*, *Klebsiella pneumoniae*, *Pseudomonas aeruginosa*, *Proteus vulgaris*) and antifungal activity against *Kluyveromyces fragilis*, *Rhodotorula rubra*, *Candida albicans*, *Hanseniaspora guilliermondii* and *Debaryomyces hansenii* yeasts (Kırbaşlar et al. 2009).

Using the *Vibrio harveyi* based autoinducer bioassay, naturally occurring furocoumarins from grapefruit exhibited >95% inhibition of autoinducer (AI) and boronated-diester molecules (AI-2) signalling activities (Girennavar et al. 2008). Autoinducers (AI) are small diffusible molecules that mediate cell-to-cell communications in bacteria and boronated-diester molecules (AI-2) are involved in inter-specific communication among both Gram-positive and Gram-negative bacteria. Grapefruit juice and furocoumarins also inhibited biofilm formation by *Escherichia coli* O157:H7, *Salmonella typhimurium* and *Pseudomonas aeruginosa*. The results suggested that grape fruit juice and furocoumarins could serve as a source to develop bacterial intervention strategies targeting microbial cell signaling processes. Laboratory studies demonstrated that grapefruit juice afforded a harsh environment for microbial survival (Nicolò et al. 2011). The results showed that *Listeria monocytogenes* lost both culturability and viability and did not resuscitate within 24 hours independently on inoculum size, whereas *Escherichia coli* O157:H7 was able to resuscitate after 24 hours but did not after 48 hours. *Salmonella enterica* serovar Typhimurium and *Shigella flexneri*, depending on inoculum size, lost culturability but maintained viability and were able to resuscitate. *Shigella flexneri* was still able to form colonies after 48 hours at high inoculum size. In-vitro studies showed that naringenin a flavonone, present predominantly in grapefruit, specifically repressed 24 genes in the *Salmonella typhimurium* pathogenicity island 1 and down-regulated 17 genes involved in flagellar and motility (Vikram et al. 2011). Additionally, naringenin appeared to repress SPI-1 in pstS/hilD-dependent manner. The data suggested that naringenin attenuated *S. typhimurium* virulence and cell motility. Grapefruit essential oil was found to inhibit growth of some bacteria commonly used in the food industry, (*Lactobacillus curvatus, Lactobacillus sakei, Staphylococcus carnosus* and *Staphylococcus xylosus*) or related to food spoilage (*Enterobacter gergoviae* and *Enterobacter amnigenus*) (Viuda-Martos et al. 2008a). Grapefruit essential oil showed the highest inhibition effect upon *Staphylococcus xylosus*, *Lactobacillus curvatus* and *Lactobacillus sakei*.

Antiviral Activity

Grapefruit seed extract is listed as a natural food additive in Japan and products containing the seed extract are used as disinfectants (Sugimoto et al. 2008). Japanese researchers found that grapefruit seed extract could prevent the growth of norovirus responsible for acute gasteroenteritis in humans. Nahmias et al. (2008) demonstrated that Hepatitis C virus was actively secreted by infected liver cells through a Golgi-dependent mechanism while bound to very low density lipoprotein (vLDL). Silencing apolipoprotein B (ApoB) messenger RNA in infected cells caused a 70% reduction in the secretion of both ApoB-100 and Hepatitis C virus. The grapefruit flavonoid naringenin inhibited the microsomal triglyceride transfer protein activity as well as the transcription of 3-hydroxy-3-methyl-glutaryl-coenzyme A reductase and acyl-coenzyme A: cholesterol acyltransferase 2 in infected cells. Stimulation with naringenin reduced Hepatitis C virus secretion in infected cells by 80%. Naringenin was effective at concentrations that were an order of magnitude below the toxic threshold in primary human hepatocytes and in mice. These results suggested a novel therapeutic approach for the treatment of Hepatitis C virus infection.

Goldwasser et al. (2011) showed naringenin to be a non-toxic assembly inhibitor of Hepatitis C virus (HCV) and that other PPARα agonists played a similar role in blocking viral production. Naringenin dose-dependently inhibited HCV production without affecting intracellular levels of the viral RNA or protein. Naringenin blocked the assembly of intracellular infectious viral particles. This antiviral effect was mediated in part by the activation of PPARα, leading to a decrease in VLDL production without causing hepatic lipid accumulation in Huh7.5.1 cells and primary human hepatocytes. The researchers asserted that the combination of naringenin with STAT-C agents could potentially bring a rapid reduction in HCV levels during the early treatment phase, an outcome associated with sustained virological response. In another paper the researchers demonstrated that naringenin regulated the activity of nuclear receptors PPARα, PPARgamma, and LXRα by activating the ligand-binding domain of both PPARα and PPARgamma, while inhibiting LXRα in GAL4-fusion reporters (Goldwasser et al. 2010).

Anticancer Activity

Two *Citrus* flavonoids, hesperetin and naringenin, found in oranges and grapefruit, respectively, and four non-citrus flavonoids, baicalein, galangin, genistein, and quercetin, were found to inhibit proliferation and growth of a human breast carcinoma cell line, MDA-MB-435, singly and in one-to-one combinations (So et al. 1996). IC$_{50}$ varied from 5.9 to 140 µg/ml for the single flavonoids, with the most potent being baicalein and IC_{50} values for the one-to-one combinations ranged from 4.7 µg/ml (quercetin+hesperetin, quercetin+naringenin) to 22.5 µg/ml (naringenin+hesperetin). All the flavonoids showed low cytotoxicity (>500 µg/ml for 50% cell death). Naringenin, present in grapefruit mainly as its glycosylated form, naringin, as well as grapefruit and orange juice concentrates, were tested for their ability to inhibit development of mammary tumours induced by 7,12-dimethylbenz[a]anthracene (DMBA) in female Sprague-Dawley rats. As expected, rats fed the high-fat diet developed more tumours than rats fed the low-fat diet, but in both experiments tumour development was delayed in the groups given orange juice or fed the naringin-supplemented diet compared with the other three groups. Although tumour incidence and tumour burden (grams of tumour/rat) were somewhat variable in the different groups, rats given orange juice had a smaller tumour burden than controls, although they grew better than any of the other groups. The results indicated Citrus flavonoids to be effective inhibitors of human breast cancer cell proliferation in-vitro, especially when paired with quercetin.

Rats allowed free access to grapefruit juice for 5 days prior to aflatoxin B1 (AFB1) administration resulted in clearly reduced DNA damage in liver, to 65% of the level in rats that did not receive grapefruit juice (Miyata et al. 2004). Further grapefruit juice extract also reduced the DNA damage to 74% of the level in rats that did not receive grapefruit juice. In an Ames assay with AFB1 using *Salmonella typhimurium* TA98, lower numbers of revertant colonies were detected with hepatic microsomes prepared from rats administered grapefruit juice, compared with those from control rats. Microsomal testosterone 6 β-hydroxylation was also lower with rats given grapefruit juice than with control rats. In microsomal systems, grapefruit juice extract inhibited AFB1-induced mutagenesis in the presence of a microsomal activation system from livers of humans as well as rats. The results suggested that grapefruit juice intake suppressed AFB1-induced liver DNA damage through inactivation of the metabolic activation potency for AFB1 in rat liver.

Citrus aurantium var. *dulcis* (sweet orange), *Citrus paradisi* (grapefruit) and *Citrus limon* (lemon) essential oils were found to induce apoptosis in human leukemic HL-60 cells and the apoptosis activities were related to the limonene content of the essential oils (Hata et al. 2003). Moreover, sweet orange essential oil and grapefruit essential oil may contain components besides limonene that have apoptotic activity. Analysis found that the n-hexane fraction contained limonene, and the dichloromethane fraction (DF) contained aldehyde compounds and nootkatone. Decanal, octanal and citral in the dichloromethane fraction DF showed strong apoptotic activity, suggesting that the aldehyde compounds induced apoptosis strongly in HL-60 cells. Studies by Vanamala et al. (2006) found that in azoxymethane-injected rats, the experimental diets (untreated or irradiated (300 Gy, 137Cs)) grapefruit pulp powder (13.7 g/kg), naringin (200 mg/kg) or limonin (200 mg/kg) suppressed aberrant crypt foci (ACF) and high multiplicity ACF (HMACF) formation and the proliferative index (compared with the control diet). Only untreated grapefruit and limonin suppressed HMACF/cm and expansion of the proliferative zone that occurred in the azoxymethane-injected rats consuming the control diet. All diets elevated the apoptotic index in azoxymethane -injected rats, compared with the control diet; however, the greatest enhancement was seen with untreated grapefruit and limonin. Untreated grapefruit and limonin diets suppressed elevation of both iNOS and COX-2 levels observed in azoxymethane-injected rats consuming the control diet. The results suggested that consumption of grapefruit or limonin may help to suppress colon cancer development.

Miller et al. (2008) demonstrated that the citrus flavonoids, naringin and naringenin significantly lowered tumour number in hamster cheek pouches. The other four flavonoids (hesperetin, neohesperidin, tangeretin, and nobiletin) were inactive. Naringin also significantly reduced tumour burden. The data suggested that naringin and naringenin, two flavonoids found in high concentrations in grapefruit, may be able to inhibit the development of cancer. Oral administration of naringenin significantly decreased the number of metastatic tumour cells in the lung and extended the life span of tumour resected mice (Qin et al. 2011). Flow cytometry analysis revealed that T cells displayed enhanced antitumour activity in naringenin treated mice, with an increased proportion of IFN-γ and IL-2 expressing T cells. The results indicated that orally administered naringenin could inhibit the outgrowth of metastases after surgery via regulating host immunity. Recent studies showed that *Citrus* limonids, obacunone and obacunone glucoside (OG) from seeds of Marsh White grapefruit inhibited human colon cancer (SW480) cells by the induction of apoptosis via activation of intrinsic apoptosis pathway and activation of p21 leading to arresting cells at G2/M phase of the cell cycle (Chidambara Murthy et al. 2011).

In-vitro studies showed that exposure of melanoma cells to naringenin resulted in morphological changes accompanied by the induction of melanocyte differentiation-related markers, such as melanin synthesis, tyrosinase activity, and the expression of tyrosinase and microphthalmia-associated transcription factor (MITF) (Huang et al. 2011). There was an increase in the intracellular accumulation of β-catenin as well as the phosphorylation of glycogen synthase kinase-3β (GSK3β) protein after treatment with naringenin. Naringenin also augmented the activity of phosphatidylinositol 3-kinase (PI3K). Based on these results, the researcher concluded that naringenin induced melanogenesis through the Wnt-β-catenin-signalling pathway in mouse melanoma cells.

Antigenotoxic Activity

Naringin a major flavonone of grapefruit was found to be neither genotoxic nor cytotoxic in mice (Alvarez-González et al. 2001). It was found to be antigenotoxic decreasing micronucleated polychromatic erythrocytes (MNPE) produced by ifosfamide. This effect was dose-dependent, showing the highest reduction in MNPE frequency (54.2%) at 48 hours with 500 mg/kg naringin. However, no protection on the cytotoxicity produced by ifosfamide was observed. Alvarez-González et al. (2004) showed that grapefruit juice inhibited micronucleated polychromatic erythrocytes (MNPE) produced by daunorubicin (Dau) in an acute assay in mice. It generated a significant reduction of the MNPE formed by daunorubicin and exhibited no toxic or genotoxic damage. It also caused a 50% decrease in liver microsomal lipid peroxidation produced by daunorubicin. The results established an efficient anticlastogenic potential of grapefruit, probably related to its antioxidant capacity, or to alterations of daunorubicin metabolism. Separate studies showed that naringin exhibited antigenotoxic effect against the chromosome damage induced by daunorubicin in mouse hepatocytes and cardiocytes (Cariño-Cortés et al. 2010). It reduced damage caused by daunorubicin by 71.3% and 51.1% in hepatocytes and cardiocytes, respectively. Further, 20 mM of naringin produced a free radical scavenging activity as high as 95%. The protective effect by naringin was probably related to its capacity to trap free radicals. Subsequent studies suggested that the protective effect of grapefruit juice against the genotoxicity of benzo(a)pyrene (BaP), an environmental contaminant and strong genotixc agent, may be related to the inhibition of Cyp1a1 enzyme activity (Alvarez-Gonzalez et al. 2011). A reduction in microsomal hepatic and intestinal ethoxyresorufin-O-deethylase (EROD) activity of 20% and 44%, respectively, was found in mice treated with BaP and grapefruit juice compared to BaP-only treated animals. Further, when EROD inhibition was tested in-vitro, a concentration-dependent EROD inhibition by grapefruit, which reached 85% of the maximum level was found. Animal studies showed that consumption of grapefruit juice protected against chromosome damage induced by the antineoplastic alkylating agent ifosfamide (Alvarez-González et al. 2010). Grapefruit juice consumption at a dose of 1,000 mg/kg resulted in a reduction of 72% in the rate of sister chromatid exchanges and a mean reduction of 65.3% in the rate of micronucleated polychromatic erythrocytes induced ifosfamide. No modification was caused by grapefruit juice either on the cellular proliferation kinetics or in the mitotic index and no induction of bone marrow cytotoxicity.

Antiobesity Activity

In a randomised placebo controlled study of 91 obese patients, after 12 weeks of diet supplementation with grapefruit resulted in the following weight loss regime: the fresh grapefruit group had lost 1.6 kg, the grapefruit juice group had lost 1.5 kg, the grapefruit capsule group had lost 1.1 kg, and the placebo group had lost 0.3 kg (Fujioka et al. 2006). The fresh grapefruit group lost significantly more weight than the placebo group. A secondary analysis of those with the metabolic syndrome in the four treatment groups demonstrated a significantly greater weight loss in the grapefruit, grapefruit capsule, and grapefruit juice groups compared with placebo. There was also a significant reduction in 2-hour post-glucose insulin level in the grapefruit group compared with placebo. Insulin resistance was improved with fresh grapefruit. The authors asserted that although the mechanism of this weight loss was unknown it would appear reasonable to include grapefruit in a weight reduction diet.

Studies demonstrated that grapefruit oil efficiently inhibited adipogenesis in cultured subcutaneous preadipocytes and adipocytes (Haze et al. 2010). Grapefruit oil inhibited the accumulation of triglycerides in a dose-dependent manner at concentrations of 50–400 μg/ml. Further, it suppressed the expression of GPDH and caused a 70% decrease in the enzymatic activity of GPDH at a concentration of 50 μg/ml. Grapefruit oil also caused a nearly two-fold increase in the intracellular concentration of Ca2+ and suppressed the expression of PPAR γ genes. Studies indicated incorporating consumption of a low energy dense dietary preload such as grapefruit, grapefruit juice or water in a caloric restricted diet to be a highly effective weight loss strategy (Silver et al. 2011). Eighty-five obese adults (BMI 30–39.9) were randomly assigned to (127 g) grapefruit (GF), grapefruit juice (GFJ) or water preload for 12 weeks after completing a 2-week caloric restriction phase. After preloads were combined with caloric restriction, average dietary energy density and total energy intakes decreased by 20–29% from baseline values. Subjects experienced 7.1% weight loss overall, with significant decreases in percentage body, trunk, android and gynoid fat, as well as waist circumferences (−4.5 cm). However, differences were not statistically significant among groups. The data showed that the form of the preload did not have differential effects on energy balance, weight loss or body composition. Nevertheless, the amount and direction of change in serum HDL-cholesterol levels in GF (+6.2%) and GFJ (+8.2%) preload groups was significantly greater than water preload group (−3.7%) indicating that subjects in GF and GFJ preload groups experienced significantly greater benefits in lipid profiles. In hypelipidemic patients on extended stable atorvastatin treatment, addition of daily grapefruit juice in typical quantities slightly elevated serum atorvastatin levels, but had no meaningful effect on the serum lipid profile, and caused no detectable adverse liver or muscle effects (Reddy et al. 2011).

Antiinflammatory Activity

Studies showed that rats pre-treated with a solution of naringin, rats pre-treated with a solution of naringenin (in concentrations equal as in grapefruit juice) and rats treated with a solution of naringenin and naringin, revealed a significant reduction on edema formation, 6 hours after λ-carrageenan injection (Ribeiro et al. 2008). Naringenin demonstrated a high in-vivo antiinflammatory activity, only 8% of paw edema was observed in rats pre-treated with a solution of naringenin. Comparative studies, in rats administered orally with grapefruit juice (before and after processing), showed that enzymatic processing did not affect the antiinflammatory properties of the juice.

Naringin, was found to exert antiinflammatory effects (Wang et al. 2011). Studies showed that it inhibited tumour necrosis factor α (TNF-α)/interferon γ (IFN-γ)-induced elevated production of RANTES (regulated upon activation normal T-cell expressed and secreted) in human HaCaT cells. The results showed that this occurred partially through the nuclear factor kappa B (NF-κB) P65 protein- dependent signal pathway. Naringin decreased the expression of NF-κB P65 protein in the nuclei.

Antihypercholesterolemic/ Antilipidemic Activity

In a 16-week double-blind, crossover (placebo or pectin) study involving 27 human volunteers Grapefruit pectin supplementation decreased plasma cholesterol 7.6%, low-density lipoprotein cholesterol 10.8%, and the low-density lipoprotein:high-density lipoprotein cholesterol ratio 9.8% (Cerda et al. 1988). The other plasma lipid fractions studied showed no significant differences. The study showed that a grapefruit pectin-supplemented diet, without change in lifestyle, can significantly reduce plasma cholesterol.

In-vitro studies showed that citrus flavonoids, naringenin and hesperetin lowered plasma cholesterol (Wilcox et al. 2001). Both flavonoids decreased the availability of lipids for assembly of apoB-containing lipoproteins in human hepatoma cell line, HepG2. This hypocholesterolemic effect was postulated to be mediated by (1) reduced activities of acyl CoA:cholesterol acyltransferase ACAT1 and ACAT2, (2) a selective decrease in ACAT2 expression, and (3) microsomal triglyceride transfer protein. In-vitro studies suggested that hesperetin and naringenin may, in part, reduce net apoB secretion by HepG2 cells by inhibiting cholesteryl ester synthesis (Borradaile et al. 1999). In subsequent studies, naringenin was found to reduce plasma lipids in-vivo and inhibited apoB secretion, cholesterol esterification, and microsomal triglyceride transfer protein (MTP) activity in HepG2 human hepatoma cells (Borradaile et al. 2003). Further, the researchers found that naringenin inhibited apoB secretion in oleate-stimulated HepG2 cells and selectively increased intracellular degradation via a largely proteasomal, rapid kinetic pathway (Borradaile et al. 2002). Although naringenin inhibited ACAT, cholesteryl ester availability in the endoplasmic reticulum (ER) lumen did not appear to regulate apoB secretion in HepG2 cells. Rather, inhibition of triglyceride accumulation in the ER lumen via inhibition of MTP was the primary mechanism blocking apoB secretion.

The contents of total polyphenols, flavonoids, and anthocyanins in fresh orange and grapefruit juices were determined as 962.1 and 906.9; 50 and 44.8; and 69.9 and 68.7 µg/ml, respectively (Gorinstein et al. 2004d). The antioxidant potential measured by the scavenging activity against nitric oxide, the β-carotene-linoleate model system (β-carotene), and the 1,1-diphenyl-2-picrylhydrazyl and 2,2′-azino-bis(3-ethylbenzothiazoline-6-sulfonic acid) diammonium salt assays was higher in orange juice but not significantly. A high level of correlation between contents of total polyphenols and flavonoids and antioxidant potential values of both juices was found. Diets supplemented with orange and to a lesser degree with grapefruit juices improved plasma lipid metabolism only in rats fed added cholesterol. However, an increase in the plasma antioxidant activity was observed in both groups. The findings indicated fresh orange and grapefruit juices to contain high quantities of bioactive compounds, which imparted high antioxidant potential, and positive influence on plasma lipid metabolism and plasma antioxidant activity making fresh orange and grapefruit juices to be valuable supplement for disease-preventing diets. Gorinstein et al. (2005b) found that the antioxidant activity of a correlated quantity of red grapefruit juice was higher than that of naringin. After 30 days of feeding, it was found that diets supplemented with red grapefruit juice and to a lesser degree with naringin improved the plasma lipid levels mainly in rats fed cholesterol and increased the plasma antioxidant activity. Fresh red grapefruit was found more preferable than naringin as it more effectively influenced plasma lipid levels and plasma antioxidant activity and, therefore, could be used as a valuable supplement for disease-preventing diets. In another study, they reported that diets supplemented with peeled red and blond qualities of Jaffa grapefruits and their peels increased plasma antioxidant capacity and improved plasma lipid levels, especially in rats fed with cholesterol added diet (Gorinstein et al. 2005a). The antioxidant potentials of red peeled grapefruits and their peels were significantly higher than of blond peeled grapefruits and their peels. Both types of Jaffa grapefruits contained high quantities of bioactive compounds, but the antioxidant potential of red grapefruits was significantly higher. In a study of hyperlipidemic patients, ages 39–72 years, diet supplementation with red and blond grapefruit decreased serum lipid levels of all fractions total cholesterol, low-density lipoprotein cholesterol, triglycerides (Gorinstein et al. 2006). The decrease was greater with red grapefruit which exhibited significantly higher antioxidant potential than blond grapefruit. No changes in the serum lipid levels in patients of the control group were found. The addition of fresh red grapefruit to generally accepted diets could be beneficial for hyperlipidemic, especially hypertriglyceridemic, patients suffering from coronary atherosclerosis. In another study with hypercholesterolemic patients, ages 43–71 years, daily supplementation for 30 days with sweetie (pummelo-grapefruit hybrid) juice decreased serum lipid levels of total cholesterol, low-density lipoprotein cholesterol and total glycerides (Gorinstein et al. 2004a). A significant increase in the serum, albumin, and fibrinogen antioxidant capacities were observed especially in the higher dose 200 ml sweetie juice group and to a lesser extent in the 100 ml sweetie group. Diet supplemented with sweetie juice positively influences serum lipid, albumin, and fibrinogen levels and their antioxidant capacities. Similarly, the addition of peeled sweeties, rich in dietary fibre and antioxidant compounds, to a generally accepted antiatherosclerotic diet may be beneficial in prevention of atherosclerosis, mainly in hypercholesterolemic patients (Gorinstein et al. 2004b). Peeled sweeties-supplemented diets decreased plasma lipids levels of total cholesterol, low-density lipoprotein cholesterol and total glycerides in hypercholesterolemic volunteers and increased plasma antioxidant capacity. No changes in the studied indices in the patients of the control group were observed.

Antidiabetic Activity

Administration of grapefruit methanol seed extract to Wistar rats induced a significant dose related lowering effects on fasting plasma glucose (FPG), total cholesterol (TC), low density lipoprotein-cholesterol (LDL-c) and very low density lipoprotein-cholesterol (VLDL-c) (Adeneye 2008a, b). High density lipoprotein cholesterol (HDL-c) level was elevated. The extract also induced significant dose related weight loss in the treated rats in the latter 15 days of treatment. Phytochemical result showed the presence of alkaloids, flavonoids, cardiac glycosides, tannins and saponin in varying concentrations. The results supported the traditional use of grapefruit seeds in the management of type 1 diabetic patients.

Antiosteoporotic Activity

Administration of grapefruit pulp improved the antioxidant status and reduced osteoporosis in orchidectomized old male Sprague-Dawley rats (Deyhim et al. 2008b). Independent of dosage, the antioxidant status, bone density, and delayed time-induced femoral fracture were higher in the grapefruit pulp groups, whereas fecal calcium excretion and urinary deoxypyridinoline excretion were lowered. Grapefruit dose-dependently slowed down bone turnover, elevated bone calcium and magnesium contents, tended to lower urinary excretion of magnesium, and numerically improved bone strength. Similarly feeding orchidectomized rats with grapefruit juice increased antioxidant status, bone density, and bone mineral contents, delayed femoral fracture, and slowed down bone turnover rate and tended to have a decrease in urinary deoxypridinoline (Deyhim et al. 2008a). In sham-treated animals, drinking grapefruit juice increased bone density and tended to increase the femoral strength. The concentration of IGF-I in the plasma was not affected across treatments. The results suggested that the beneficial effects of eating red grapefruit on bone quality of orchidectomized rats was due to bone mineral deposition, improved bone density and slowed-down bone turnover.

Cardiovascular Activity

In the Langendorff isolated and perfused heart model and in the heart and lung dog preparation, *Citrus paradisi* peel extract decreased coronary vascular resistance and mean arterial pressure when compared with control values (Díaz-Juárez et al. 2009). In humans, grapefruit juice decreased diastolic arterial pressure and systolic arterial pressure both in normotensive and hypertensive subjects. Grapefruit juice elicited a greater decrease in mean arterial pressure when compared with *Citrus sinensis* juice, cow milk and a vitamin C-supplemented beverage.

Scholz et al. (2005) demonstrated that grapefruit flavonoid naringenin inhibited cardiac human ether-à-go-go-related gene (HERG) channels. Also they observed that grapefruit juice induced mild QTc prolongation in healthy subjects most likely via HERG blockade by naringenin. Naringenin blocked HERG potassium channels but did not affect HERG current activation. Naringenin inhibited HERG channels with pharmacological characteristics similar to those of well-known HERG antagonists. From a clinical point of view, the authors maintained that this effect could have both proarrhythmic and antiarrhythmic consequences and may have important implications for phytotherapy and for dietary recommendations for cardiologic patients. In a study of patients with dilated or hypertensive cardiomyopathy and in healthy subjects, administration of fresh pink grapefruit juice increased significantly indices of corrected QT (QT_c) and QT variability index (QTVI) from values observed after placebo (Piccirillo et al. 2008). Presumably because of its high naringenin glycoside content, pink grapefruit juice prolonged cardiac repolarization and concurrently increased temporal cardiac repolarization dispersion. The findings suggested that the potential proarrhythmic actions of pink grapefruit juice might be of concern in patients with major myocardial structural disorders. Grapefruit juice causes significant QT prolongation in healthy volunteers and naringenin had been identified as the most potent human ether-a-go-go-related gene (HERG) channel blocker among several dietary flavonoids (Lin et al. 2008). Naringenin exhibited an additive inhibitory effect on HERG current when combined with I(Kr)-blocking antiarrhythmic drugs azimilide, amiodarone, dofetilide and quinidine. This additive HERG inhibition could pose an increased risk of arrhythmias by increasing repolarization delay and possible repolarization heterogeneity.

Neuroprotective Activity

Naringin, a grapefruit flavonoid, inhibited rotenone-induced cell death in human neuroblastoma SH-SY5Y cells. (Kim et al. 2009). Naringin also prevented rotenone-induced phosphorylation of Jun NH2-terminal protein kinase (JNK) and P38, and blocked changes in B-cell CLL/lymphoma 2 (BCL2) and BCL2-associated X protein (BAX) expression levels. Further, naringin reduced the enzyme activity of caspase 3 and cleavages of caspase 9, poly (ADP-ribose) polymerase (PARP), and caspase 3.

Hepatoprotective Activity

Studies in Male Sprague-Dawley rats showed that grapefruit and oroblanco consumption enhanced hepatic detoxification enzymes, protecting against the carcinogenesis induced by the procarcinogen 1,2-dimethylhydrazine (Hahn-Obercyger et al. 2005). Grapefruit juice significantly increased activity and expression of the hepatic phase I enzyme, cytochrome P450 CYP1A1, with a marked trend toward enhanced NAD(P)H:quinone reductase (QR) activity. Oroblanco juice significantly increased glutathione S-transferase phase II enzyme activity along with CYP1A1 expression and a notable trend toward increased activity of both CYP1A1 and QR. The results indicated these citrus fruits to be bifunctional inducers, modulating both phase I and phase II drug-metabolizing enzymes to enhance hepatic detoxification.

Gastroprotective Activity

Animal studies showed that pretreatment of grapefruit seed extract rich in flavonoids exerted a potent gastroprotective activity against ethanol and water immersion and restraint stress-induced gastric lesions in rats via an increase in endogenous prostaglandin generation, suppression of lipid peroxidation and hyperemia possibly mediated by nitric oxide and calcitonine gene-related peptide released from sensory nerves (Brzozowski et al. 2005).

Nootropic and CNS Activity

Nootkatone and auraptene were isolated from *C. paradisi* essential oil and showed 17–24% inhibition of acetylcholinesterase activity at the concentration of 1.62 μg/ml (Miyazawa et al. 2001).

Anti-periodontitis Activity

In a study of 58 patients with chronic periodontitis (smokers and non-smokers) and 22 healthy subjects, periodontitis patients were found to have significantly reduced plasma vitamin C levels in the test group and diseased controls in comparison with the healthy controls (Staudte et al. 2005). Smokers showed lower levels of vitamin C than non-smokers. After grapefruit consumption for 2 weeks, the mean plasma vitamin C levels rose significantly in the test group compared to the diseased controls (non-smokers and smokers). Further, sulcus bleeding index was reduced in the test group whereas plaque index and probing pocket depths were unaffected. The present results showed that periodontitis patients had plasma vitamin C levels below the normal range, especially in smokers and that intake of grapefruit led to an increase in plasma vitamin C levels and improved sulcus bleeding scores.

Testicular Protective Activity

In-vivo studies with Wistar rats showed that doxorubicin-induced reduction in sperm motility and epididymal sperm concentrations as well as increase in total abnormal sperm rates were all normalized in the group pretreated with an antioxidant-rich ethanolic seed extract of *Citrus paradisi* (Saalu et al. 2010). Pretreatment with grapefruit extract ameliorated the testicular content of glutathione (GSH) and superoxide dismutase (SOD), catalase (CAT) and glutathione peroxidase (GPx) activities. Similarly, the extract attenuated the doxorubicin-induced increase in testicular lipid peroxidation reflected by malondialdehyde (MDA) levels. The data indicated that grapefruit protected the rat testis against doxorubicin-induced oxidative stress and deranged sperm characteristics.

Anti-hypocitraturia/ Anti-nephrolithiasis Activity

In a study of seven healthy subjects, with no history of kidney stones, intake of grapefruit juice significantly increased urinary excretion of citrate, calcium and magnesium compared to mineral water intake (Trinchieri et al. 2002). Citrus fruit juices could represent a natural alternative to potassium citrate in the management of nephrolithiasis, because they could be better tolerated and cost-effective than pharmacological calcium treatment. In patients with mild to moderate hypocitraturia, dietary supplementation with citrus-based juices may be an effective alternative to increasing urinary citrate excretion and correcting hypocitraturia to medical management while not requiring large serving sizes (Haleblian et al. 2008). Quantitative analysis revealed the highest concentration of citrate was in grapefruit juice (64.7 mmol/L), followed in decreasing concentrations by lemon juice (47.66 mmol/L), orange juice (47.36 mmol/L), pineapple juice (41.57 mmol/L), reconstituted lemonade (38.65 mmol/L), lemonade flavored Crystal Light (38.39 mmol/L), ready to consume not from concentrate lemonade (38.24 mmol/L), cranberry juice (19.87 mmol/L), lemon-flavored Gatorade (19.82 mmol/L), homemade lemonade (17.42 mmol/L), Mountain Dew (8.84 mmol/L), and Diet 7Up (7.98 mmol/L), respectively.

Anti-pancreatitis Activity

Pretreatment of rats with grapefruit seed extract exerted protective activity against ischemia/reperfusion-induced pancreatitis with maximal protective effect at the dose 250 μl (Dembinski et al.

2004). The was probably due to the activation of antioxidative mechanisms in the pancreas and the improvement of pancreatic blood flow. The extract administered alone increased significantly pancreatic tissue content of lipid peroxidation products, malondialdehyde and 4-hydroxyalkens, and when administered before ischemia/reperfusion, the extract reduced the pancreatitis-induced lipid peroxidation.

Grapefruit Intake and Breast Cancer

There are mixed reports on the association of grapefruit intake and the risk of breast cancer in women. Kim et al. (2008) found that in women with oestrogen and progesterone receptor negative cancers, there was a significant decrease in breast cancer risk with increased consumption of grapefruit. In a cross-sectional examination on the relationship between consumption of grapefruit and grapefruit juice and plasma levels of oestrogens among 701 postmenopausal women not using hormone replacement, no significant correlation was observed (grapefruit, grapefruit juice) for plasma oestradiol, oestrone, or oestrone sulphate. Their findings did not support an adverse effect of consumption of grapefruit or grapefruit juice on risk of breast cancer or on endogenous hormone levels. Spencer et al. (2009) found no evidence of an association between grapefruit intake and risk of breast cancer in a prospective study in the European Prospective Investigation into Cancer and Nutrition (EPIC). They examined and followed up 114,504 women with information on dietary intake of grapefruit and on reproductive and lifestyle risk factors for a median 9.5 years and identified 3,747 incident breast cancers. They found no relationship between grapefruit intake and breast cancer risk among premenopausal women, all postmenopausal women, or postmenopausal women categorized by hormone replacement therapy use. There was no association between grapefruit intake and estradiol or estrone among postmenopausal women. In contrast, in the Hawaii-Los Angeles Multiethnic Cohort Study, a prospective cohort that included over 50,000 postmenopausal women from five racial/ethnic groups with a total of 1,657 incident breast cancer Monroe et al. (2007) found that grapefruit intake was significantly associated with an increased risk of breast cancer for subjects in the highest category of intake, that is, one-quarter grapefruit or more per day, compared to non-consumers. An increased risk of similar magnitude was seen in users of oestrogen therapy, users of oestrogen + progestin therapy, and among never users of hormone therapy. They concluded that grapefruit intake may increase the risk of breast cancer among postmenopausal women.

Anti-lice Activity

Licatack(R), an anti-louse medicinal product containing extracts of grapefruit besides high quality shampoo components was found to be efficacious against head lice (*Pediculus humanus capitis*) in children 2–9 years old (Abdel-Ghaffar et al. 2010). It was effective against lice larvae and larval stages inside nits. This new anti-louse medicinal product possessed a very quick and efficient activity besides being non-inflammable, skin safe, and nice smelling.

Effects on Drug Pharmacokinetics and Pharmacodynamics Activity

Grapefruit juice and grapefruit products are consumed for nutritious and health benefits in certain cancers and cardiovascular diseases. However, people on medication should be wary that consumption of grapefruit can interact with many drugs and grapefruit juice can alter drug bioavailability, and alter their pharmacokinetic and pharmacodynamic parameters since its initial discovery in 1989 (Bailey et al. 1998; Dresser and Bailey 2003; Bailey and Dresser 2004; Dahan and Altman 2004; Sica 2006; Kiani and Imam 2007; Cuciureanu et al. 2010; Seden et al. 2010; Hanley et al. 2011) Most vulnerable populations were reported to be elderly, cirrhotics, subjects with genetic polymorphisms and individuals taking other CYP3A4 inhibitors. (Cuciureanu et al. 2010). Grapefruit-drug interactions had been reported for drugs like calcium channel blockers, immunosuppressants, antihistamines, antiallergics, antibiotics, antimalaria drugs, anxiolytics, HIV protease inhibitors, HMG-CoA reductase inhibitors. Clinically relevant interactions may appear likely for most dihydropyridine calcium channel antagonists, felodipine, terfenadine, saquinavir, cyclosporin, midazolam, triazolam and verapamil, cisapride and astemizole and may also occur with HMG-CoA reductase inhibitors atorvastatin, lovastatin, or simvastatin, (Bailey et al. 1998; Bailey and Dresser 2004). Ueda et al. (2009) found that grapefruit juice affected the pharmacokinetics of sertraline in healthy volunteers. Grapefruit juice increased the mean peak concentrations in plasma (C(max)) of sertraline from and the mean area under the plasma sertraline concentration-time curve (AUC) of sertraline. In the setting of a controlled clinical study involving 30 liver transplant patients, the co-administration of grapefruit juice with tacrolimus an immunosuppressant drug increased the bioavailability of tacrolimus (Liu et al. 2009).

Numerous medications used in the prevention or treatment of coronary artery disease and its complications had been observed or are predicted to interact with grapefruit juice (Bailey and Dresser 2004). Such interactions with the HMG-CoA reductase inhibitors atorvastatin, lovastatin, or simvastatin may increase the risk of rhabdomyolysis when treating dyslipidemia. Such interactions with the dihydropyridines felodipine, nicardipine, nifedipine, nisoldipine, or nitrendipine for treating hypertension may also cause excessive vasodilatation. Grapefruit juice may reduce the therapeutic effect of the angiotensin II type 1 receptor antagonist losartan. Grapefruit juice interacting with the antidiabetic agent repaglinide may cause hypoglycaemia. Grapefruit juice interaction with the appetite suppressant sibutramine may cause elevated blood pressure and heart rate. Administration of grapefruit juice could result in atrioventricular conduction disorders with verapamil in angina pectoris or attenuate antiplatelet activity with clopidrogel. Grapefruit juice may enhance drug toxicity for antiarrhythmic agents such as amiodarone, quinidine, disopyramide, or propafenone, and for the congestive heart failure drug, carvediol. Grapefruit juice may also interact with some drugs used for the treatment of peripheral or central vascular disease. Grapefruit juice interaction with sildenafil, tadalafil, or vardenafil for erectile dysfunction, may cause serious systemic vasodilatation especially when combined with a nitrate. Interaction between ergotamine for migraine and grapefruit juice may cause gangrene or stroke. In stroke, interaction with nimodipine may cause systemic hypotension. Taniguchi et al. (2007) reported that an elderly patient taking cilostazol and aspirin atherosclerosis obliterans and ingesting grapefruit juice for a month developed purpura which disappeared upon cessation of grapefruit juice, although his medication was not altered. The most probable cause of his purpura was an increase in the blood level of cilostazol because of the inhibition of cilostazol metabolism by components of grapefruit juice. Chronic treatment with simvastatin alone, co-administration of simvastatin with single and double strength grapefruit juice or single and double strength grapefruit alone significantly decreased plasma cholesterol levels in rats over a 4 week period (Butterweck et al. 2009). The results suggested that toxic effects in rats of concomitant intake of simvastatin and grapefruit juice were not more pronounced than those of simvastatin alone and that dose relationships between the administration of the juice and the drug may be important in determining the magnitude of the interaction.

Grapefruit juice can alter oral drug pharmacokinetics by different mechanisms (Dresser and Bailey 2003; Bailey and Dresser 2004; Dahan and Altman 2004; Kiani and Imam 2007; Seden et al. 2010; Hanley et al. 2011). The primary mechanism through which interactions are mediated is mechanism-based intestinal cytochrome P450 3A4 inhibition as a result, presystemic (first pass) metabolism is reduced and oral drug bioavailability increased. Four furocoumarin derivatives were isolated from grapefruit juice that inhibited human CYP3A-mediated drug oxidation (Fukuda et al. 1997). They included two new furocoumarins, 4-[[6-hydroxy-7-[[1-[(1-hydroxy-

1-methyl)ethyl]-4-methyl-6-(7-oxo-7H-furo[3,2-g][1]benzopyran-4-yl)-4-hexenyl]oxy]-3,7-dimethyl- 2-octenyl] oxy]-7H-furo[3,2-g][1]benzopyran-7-one (GF-I-1) and 4-[[6-hydroxy-7-[[4-methyl-I-(1-methylethenyl)-6-(7-oxo-7H-furo[3,2-g][1]benzopyran-4-yl)-4- hexenyl] oxy]-3,7-dimethyl-2-octenyl]oxy]-7H-furo[3,2-g][1]benzopyran-7-one (GF-I-4). These furocoumarins exhibited an inhibition potential that was equal to or stronger than the prototypical CYP3A4 inhibitor, ketoconazole, on liver microsomal testosterone 6 β-hydroxylation. Tassaneeyakul et al. (2000) found that four coumarins in grapefruit juice namely furanocoumarin monomers bergamottin and 6′,7′-dihydroxybergamottin (DHB) and two furanocoumarin dimers, GF-I-1 and GF-I-4, clearly inhibited CYP3A4-catalyzed nifedipine oxidation in concentration- and time-dependent manners, suggesting that these compounds were mechanism-based inhibitors of CYP3A4. The two furanocoumarin dimers were the most potent inhibitors of CYP3A4. The furanocoumarin, 6′,7′-dihydroxybergamottin isolated from grapefruit juice was found to be a potent inhibitor of CYP3A activity (Edwards et al. 1996). Grapefruit juice reduced CYP3A activity to a significantly greater extent than did orange juice, which contained no measurable 6′,7′-dihydroxybergamottin suggesting that the furnocoumarin may be primarily responsible for the effects of grapefruit juice on cytochrome P450 activity in humans. Studics by Guo ct al. (2000) and Guo and Yamazoe (2004) found that furanocoumarins derivatives (bergamottin and 6′,7′-dihydroxybergamottin) in grapefruit juice contributed to the CYP3A inhibitory properties of grapefruit juice. Although the inhibition appeared to be stronger in the dimers than that in the monomers, all contribute comprehensively to the grapefruit juice-drug interaction.

Studies with healthy volunteers showed that a usual single exposure to grapefruit juice appeared to impair the enteric, but not the hepatic, component of presystemic extraction of oral midazolam (Greenblatt et al. 2003). Recovery was largely complete within 3 days, consistent with enzyme regeneration after mechanism-based inhibition. 6′7′-Dihydroxybergamottin was conformed as a potent mechanism-based inhibitor of midazolam α-hydroxylation by CYP3A in-vitro than bergamottin. Of the furanocoumarins investigated, furanocoumarin dimers, GF-I-1 and GF-I-4, were the most potent inhibitors of CYP3A4. Apparent selectivity toward CYP3A4 did occur with the furanocoumarin dimers. In contrast, bergamottin showed rather stronger inhibitory effect on CYP1A2, CYP2C9, CYP2C19, and CYP2D6 than on CYP3A4. DHB inhibited CYP3A4 and CYP1A2 activities at nearly equivalent potencies. Among P450 forms investigated, CYP2E1 was the least sensitive to the inhibitory effect of furanocoumarin components. A sesquiterpene nootkatone had no significant effect on P450 activities investigated except for CYP2A6 and CYP2C19.

Three bioactive furocoumarins, bergamottin, 6′,7′-dihydroxybergamottin (DHB), and paradisin-A from grapefruit juice exhibited inhibitory effects on hydroxylase and O-dealkylase activities of human CYP 3A4 and CYP 1B1 isoenzymes (Girennavar et al. 2006). Paradisin-A was found to be a potent CYP 3A4 inhibitor with an IC_{50} of 1.2 μM followed by DHB and bergamottin. All three compounds showed a substantial inhibitory effect on CYP 3A4 below 10 μM. Inhibitory effects on CYP 1B1 exhibited a greater variation due to the specificity of substrates. The researchers also found grapefruit and pummelo juices to be potent inhibitors of human cytochrome CYP3A4 and CYP2C9 isoenzymes at 25% concentration, while CYP2D6 was inhibited significantly lower at all the tested concentration of juices (Girennavar et al. 2007b). Among the 5 furocoumarins isolated from grapefruit juice, the inhibitory potency was in the order of paradisin A > dihydroxybergamottin > bergamottin > bergaptol > geranylcoumarin at 0.1 μM to 0.1 mM concentrations. The IC_{50} value was lowest for paradisin A for CYP3A4 with 0.11 μM followed by DHB for CYP2C9 with 1.58 μM. Bergaptol and geranylcoumarin from grapefruit juice were found to be potent inhibitors of debenzylation activity of CYP3A4 enzyme with an IC_{50} value of 24.92 and 42.93 μM, respectively (Girennavar et al. 2007a).

A new cyclic acetal (1) of marmin (6′,7′-dihydroxy-7-geranyloxycoumarin), two new cyclic acetals (5, 6) of 6′,7′-dihydroxybergamottin, and the known compounds marmin (2), 7-geranyloxycoumarin (3), bergamottin (4), and 6′,7′-dihydroxybergamottin (7) were isolated from grapefruit peel oil (César et al. 2009). Coumarins (1–3) exhibited negligible inhibitory activity against intestinal cytochrome P450 3A4, an enzyme involved in the "grapefruit/drug" interactions in humans, while the furanocoumarins (4–7) showed potent in-vitro inhibitory activity with IC_{50} values of 2.42, 0.13, 0.27, and 1.58 μM, respectively. Furanocoumarins bergamottin, bergaptol and 6′,7′-dihydroxybergamottin (DHB) were also isolated from grapefruit juice, citrus fruit of 20 species and health food (Sakamaki et al. 2008). The contents of bergamottin were 0–16 μg/g, 0–16 μg/g and 0–5.6 μg/g, bergaptol were 0–39 μg/g, 0–13 μg/g and 0–28 μg/g, DHB were 0–10 μg/g, 0–35 μg/g and 0–6.2 μg/g, respectively. Bergapten was not detected. The results suggested that patients prescribed calcium antagonists or antiallergic agents should be cautions about their intake of furanocoumarins from grapefruit juice, citrus and health foods. Studies suggested that grapefruit juice extract containing no peel extract may have a lower potential for interactions with CYP3A4 or P-glycoprotein (Brill et al. 2009). Grapefruit juice containing peels (with higher content of the furanocoumarin bergamottin) increased expression of multidrug resistance of MDR1 and cytochrome CYP3A4 in LS180 cells, whereas grapefruit juice without peel (containing more the flavonoid naringin and its aglycone naringenin) showed no significant effect on MDR1 and CYP3A4 mRNA expression. Grapefruit juice was found to inhibit the CYP3A4-mediated first-pass metabolism of oxycodone, decreased the formation of noroxycodone and noroxymorphone and increased that of oxymorphone (Nieminen et al. 2010). It was concluded that dietary consumption of grapefruit products may increase the concentrations and effects of oxycodone in clinical use.

In an open, randomized, cross-over study in 8 ovariectomized women, it was demonstrates that grapefruit juice may alter the metabolic degradation of estrogens, and increase the bioavailable amounts of 17 β-estradiol and its metabolite estrone, presumably by affecting the oxidative degradation of estrogens (Schubert et al. 1994). This food interaction may be one factor behind the inter-individual variability in 17 β-estradiol, estrone and estriol serum concentrations after exogenous administration of 17 β-estradiol to patients. The researchers found that naringenin, quercetin and kaempferol, which may be found in glycoside form in grapefruit inhibited the in-vitro hepatic metabolism of 17 β-estradiol (Schubert et al. 1995). At the highest concentrations tested of the respective flavonoid, there was approximately 75–85% inhibition of estriol formation. However, naringenin was a less potent inhibitor of 17 β-estradiol metabolism as compared to quercetin and kaempferol. The most likely mechanism of action of the flavonoids on 17 β-estradiol metabolism was inhibition of the cytochrome P-450 IIIA4 enzyme, which catalysed the reversible hydroxylation of 17 β-estradiol into estrone and further into estriol.

Results of a two-phase, randomized, placebo-controlled crossover study with male volunteers suggested that consumption of grapefruit juice prior to administration of probe drugs midazolam and IV [14GN-methyl] erythromycin, inhibited intestinal and hepatic CYP3A4 in an exposure-dependent manner (Veronese et al. 2003). Alterations of midazolam AUC and Cmax induced by nine glasses of double-strength grapefruit juice were significantly greater than those produced by one glass of single strength or double-strength grapefruit juice. The results also suggested that patients taking medication based on CYP3A4 substrates are at risk for developing drug-related adverse events if they consume large amounts of grapefruit juice. In four separate trials with an open, randomized, cross-over design of healthy subjects, administration of grapefruit juice increased the plasma concentrations of two benzodiazepine hypnotic drugs triazolam and quazepam and of the active metabolite of quazepam, 2-oxoquazepam (Sugimoto et al. 2006). Triazolam and quazepam produced similar sedative-like effects, none of which were enhanced by grapefruit juice. The results suggested that the effects of grapefruit juice on the pharmacodynamics of triazolam were greater than those on quazepam. These GFJ-related different effects were partly elucidated by the fact that triazolam is presystemically metabolized by CYP3A4, while quazepam is presystemically metabolized by CYP3A4 and CYP2C9.

In a study of ten volunteers, grapefruit juice inhibited the metabolism of nicotine to cotinine, a pathway mediated by CYP2A6, and increased the renal clearance of nicotine and cotinine (Hukkanen et al. 2006). Nicotine oral clearance was not affected by grapefruit juice because the inhibition of hepatic metabolism was offset by the increase in the renal clearance of nicotine. Studies in healthy subjects showed that stereoselective disposition of manidipine enantiomers was altered by grapefruit juice, as an inhibitor of CYP3A4 (Uno et al. 2006). Grapefruit juice appeared to affect this metabolic disposal of (R)-manidipine to a greater extent than that of (S)-manidipine. In a randomized, four-phase crossover study of healthy subjects, grapefruit juice increased the mean AUC (area under the curve) of atorvastatin acid by 83% and that of pitavastatin acid by 13% indicating the tpitavastatin, unlike atorvastatin, appeared to be scarcely affected by the CYP3A4-mediated metabolism (Ando et al. 2005).

Edwards and Bernier (1996) found that neither naringin nor naringenin flavonone components of grapefruit juice were primarily responsible for the inhibitory effect on CYP3A activity. The flavonoids, naringin and naringenin and the furanocoumarin, bergapten (5-methoxypsoralen), known inhibitors of CYP3A4 were detected in varying concentrations in some fresh grapefruit and commercial grapefruit juices (Ho et al. 2000). Differences in the concentrations of these three constituents may have potential for drug interaction and may contribute to the variability in pharmacokinetics of CYP3A4 drugs and some contradictory results of drug interaction studies with grapefruit juice. Fuhr et al. (1993) found that grapefruit juice and naringenin inhibit CYP1A2 activity in man based on caffeine as a probe substrate. In-vivo grapefruit juice (1.2 l per day containing 0.5 g/l naringin, the glycone form of naringenin) decreased the oral clearance of caffeine by 23% and prolonged its half-life by 31%. The small effect on caffeine clearance in-vivo suggested that in general the ingestion of grapefruit juice should not cause clinically significant inhibition of the metabolism of other drugs that are substrates of CYP1A2.

Another possible mechanism of grapefruit juice-drug interaction involved inhibition of P-glycoprotein (P-gp) increasing oral drug bioavailability by reducing intestinal and/or hepatic efflux transport (Dresser and Bailey 2003; Bailey and Dresser 2004; Dahan and Altman 2004; Kiani and Imam 2007; Seden et al. 2010; Hanley et al. 2011). Grapefruit juice may also inhibit intestinal P-glycoprotein-mediated efflux transport of drugs such as cyclosporine to increase its oral bioavailability (Dresser and Bailey 2003). Schwarz et al. (2005) found that grapefruit juice ingestion significantly reduced P-glycoprotein-transported drug talinolol bioavailability in humans. Panchagnula et al. (2005) found that grapefruit juice extract inhibited P-gp-mediated efflux in co-treatment with paclitaxel, whereas chronic administration with indinavir to rats led to increased levels of P-gp expression, thus having a profound effect on intestinal absorption and grapefruit juice-drug interactions in-vivo.

Studies showed that talinolol permeability across Caco-2 cells monolayers was selectively inhibited by grapefruit juice and its components, bergamottin, 6′,7′-dihydroxybergamottin, 6′,7′-epoxybergamottin, naringin, and naringenin (de Castro et al. 2007). The furanocoumarin, 6′,7′-epoxybergamottin, was the most potent inhibitor ($IC_{50}=0.7$ μM), followed by 6′,7′-dihydroxybergamottin ($IC_{50}=34$ μM) and bergamottin that did not show any inhibition at concentrations up to 10 μM. The flavonoid aglycone naringenin was around ten-fold more potent than its glycoside naringin with IC_{50} values of 236 and 2,409 μM, respectively. The in-vitro data suggested that compounds present in grapefruit juice were able to inhibit the P-glycoprotein activity modifying the disposition of drugs that were P-glycoprotein substrates such as talinolol. Ali et al. (2009) found that pre-treatment by i.p. administration of naringin for 3 consecutive days prior to doxorubicin (the most common used anticancer drug which induces multidrug resistance) administration was able to significantly lower P-glycoprotein expression reaching nearly the level of animals treated with verapamil. Pretreatment with naringin prior to doxorubicin increased the sensitivity to the drug. Naringin inhibited the doxorubicin-stimulated ATPase activity demonstrating that naringin may interact directly with the transporter. Induction of both glutathione (GSH) and glutathione-S-transferase (GST) by doxorubicin was consistent with an increased ATP-dependent doxorubicin transport. The study indicated the dual modulation of P-gp expression and function by the flavonoid naringin. Contrary to previous in-vitro results that showed naringin and naringenin known as P-glycoprotein (P-gp) inhibitors to interact with doxorubicin, in-vivo studies in rats showed the plasma concentration, biliary and urinary clearance, and tissue distribution of doxorubicin were not altered by pre-treatment with naringin and naringenin (Park et al. 2011) Biliary clearance and urinary clearance were slightly decreased by quercetin, but there was no statistical difference. It was concluded that naringin, naringenin and quercetin did not affect the in-vivo pharmacokinetics of intravenously administered doxorubicin.

Another mechanism of grapefruit juice-drug interaction involved the inhibition of organic anion-transporting polypeptides by grape fruit juice, inhibiting intestinal uptake and decreasing drug bioavailability (Dresser and Bailey 2003; Dresser et al. 2005; Seden et al. 2010; Hanley et al. 2011). Dresser et al. (2005) found that grapefruit juice at a commonly consumed volume diminished the oral bioavailability of fexofenadine possibly by direct inhibition of uptake by intestinal organic anion transporting polypeptide A (OATP-A; new nomenclature, OATP1A2). Bailey et al. (2007) found naringin, a major flavonoid component of grapefruit juice to be a major selective clinical inhibitor of organic anion-transporting polypeptide 1A2 (OATP1A2) and decreased oral fexofenadine bioavailability clinically. Animal studies showed that oral single dose exposure to grapefruit juice showed no effect on P-glycoprotein (P-gp), whereas multiple dose administration of grapefruit juice resulted in increased levels of P-gp expression and decreased levels of organic anion transporting polypeptide (OATP), thus showing a varied effect on intestinal absorption, and therefore overcoming the inhibition of diltiazem metabolism in rats (Boddu et al. 2009). Shirasaka et al. (2010) found that bioavailability of talinolol, a $\beta(1)$-adrenergic receptor antagonist, was enhanced by coadministration with grapefruit juice in rats, whereas grapefruit juice ingestion markedly reduced the absorption of talinolol in humans. They found that the species difference in the effect of grapefruit juice on intestinal absorption of talinolol between humans and rats may be due to differences in the affinity of naringin for OATP/Oatp and P-gp multidrug resistance 1 (MDR1/Mdr1) transporters between the two species. Results of studies in rats suggested that the decrease of pravastatin absorption in the presence of grapefruit juice was due to the inhibitory effect of naringin on organic anion transporting polypeptide (Oatp), whereas the increase of pitavastatin was due to the inhibition of P-glycoprotein (P-gp) (Shirasaka et al. 2011). Oatp and/or P-gp contributed to the intestinal absorption of statins, and the differential effect of grapefruit juice on pravastatin and pitavastatin absorption was at least partly accounted for by the different inhibitory effects of naringin on these transporters.

Another mechanism of grapefruit juice-drug interaction could be attributable to inhibition of esterase activity (Li et al. 2007; Hanley et al. 2011). Study demonstrated that grapefruit juice inhibited purified porcine esterase activity toward p-nitrophenyl acetate and the prodrugs lovastatin and enalapril. In rat and human hepatic or gut S9 fractions and rat gut lumen, grapefruit juice inhibited the hydrolysis of enalapril and lovastatin, known to be metabolized principally by esterases, lovastatin being metabolized also by CYP3A. Overall, along with the CYP3A inactivation by grapefruit juice, the decreased esterase activity also played a significant role in increasing the metabolic stability and permeability of ester prodrugs leading to enhancement of exposure to the active drug.

Another mechanism involved the inhibition of sulfotransferase activities by grapefruit juice

(Tamura and Matsui 2000; Nishimuta et al. 2005, 2007; Hanley et al. 2011).

Tamura and Matsui (2000) showed that grape juice exhibited the most potent inhibitory action on the phenol sulfotransferases activity of mouse intestines and human colon carcinoma cells. The inhibitory activity of grape juice was located mainly in the skin and seeds. Flavonols, such as quercetin and kaempferol, inhibited the P-ST activity at low concentrations. In-vitro studies suggested that ritodrine sulfation activities of sulfotransferase (SULT) 1A1 and SULT1A3 were significantly inhibited by all beverages (grapefruit juice, orange juice, green tea, and black tea) at a concentration of 10% (Nishimuta et al. 2005). The grapefruit constituent, quercetin, completely inhibited SULT1A1, while quercetin and naringin both partially inhibited SULT1A3 (Nishimuta et al. 2007). The results of the studies suggested that concomitant ingestion of such beverages may increase the bioavailability of orally administered ritodrine, and perhaps other beta2-agonists, and may lead to an increase in the clinical effects or adverse reactions.

Animal studies showed that administration of grapefruit juice to rats significantly enhanced the antinociception of morphine by increasing the intestinal absorption of this agent (Okura et al. 2008).

Traditional Medicinal Uses

An essence prepared from the flowers is as a stomachic, and cardiac tonic and to overcome insomnia. The pulp is regarded as an effective aid in the treatment of urinary disorders.

Other Uses

Grapefruit peel oil is used in aromatherapy. Pulp and molasses waste are after fruit juice extraction can be used as cattle feed. After oil extraction, the seed hulls can be used for soil conditioning, or, combined with the dried pulp, as cattlefeed. A detoxification process must precede the feeding of this product to pigs or poultry.

Old grapefruit trees can be used as timber. The sapwood is pale-yellow or nearly white, the heartwood yellow to brownish, hard, fine-grained, and useful for domestic purposes. Mainly, pruned branches and felled trees are cut up for firewood.

Comments

Grapefruit comes in many varieties, determinable by color, which is caused by the pigmentation of the fruit in respect of both its state of ripeness and genetic bent. The most popular varieties cultivated today are red, white, and pink hues, referring to the internal pulp color of the fruit. The family of flavors range from highly acidic and somewhat bitter to sweet and tart.

Selected References

Abdel-Ghaffar F, Semmler M, Al-Rasheid K, Klimpel S, Mehlhorn H (2010) Efficacy of a grapefruit extract on head lice: a clinical trial. Parasitol Res 106(2):445–449

Abulrob AN, Suller MT, Gumbleton M, Simons C, Russell AD (2004) Identification and biological evaluation of grapefruit oil components as potential novel efflux pump modulators in methicillin-resistant *Staphylococcus aureus* bacterial strains. Phytochemistry 65(22):3021–3027

Adeneye AA (2008a) Hypoglycemic and hypolipidemic effects of methanol seed extract of *Citrus paradisi* Macfad (Rutaceae) in alloxan-induced diabetic Wistar rats. Nig Q J Hosp Med 18(4):211–215

Adeneye AA (2008b) Methanol seed extract of *Citrus paradisi* Macfad lowers blood glucose, lipids and cardiovascular disease risk indices in normal Wistar rats. Nig Q J Hosp Med 18(1):16–20

Ali MM, Agha FG, El-Sammad NM, Hassan SK (2009) Modulation of anticancer drug-induced P-glycoprotein expression by naringin. Z Naturforsch C 64(1–2):109–116

Alvarez-González I, Madrigal-Bujaidar E, Dorado V, Espinosa-Aguirre JJ (2001) Inhibitory effect of naringin on the micronuclei induced by ifosfamide in mouse, and evaluation of its modulatory effect on the Cyp3a subfamily. Mutat Res 480–481:171–178

Alvarez-González I, Madrigal-Bujaidar E, Martino-Roaro L, Espinosa-Aguirre JJ (2004) Antigenotoxic and antioxidant effect of grapefruit juice in mice treated with daunorubicin. Toxicol Lett 152(3):203–211

Alvarez-González I, Madrigal-Bujaidar E, Sánchez-García VY (2010) Inhibitory effect of grapefruit juice on the genotoxic damage induced by ifosfamide in mouse. Plant Foods Hum Nutr 65(4):369–373

Alvarez-Gonzalez I, Mojica R, Madrigal-Bujaidar E, Camacho-Carranza R, Escobar-García D, Espinosa-Aguirre JJ (2011) The antigenotoxic effects of grapefruit juice on the damage induced by benzo(a)pyrene and evaluation of its interaction with hepatic and intestinal cytochrome P450 (Cyp) 1a1. Food Chem Toxicol 49(4):807–811

Ando H, Tsuruoka S, Yanagihara H, Sugimoto K, Miyata M, Yamazoe Y, Takamura T, Kaneko S, Fujimura A (2005) Effects of grapefruit juice on the pharmacokinetics of pitavastatin and atorvastatin. Br J Clin Pharmacol 60(5):494–497

Aurnhammer G, Wagner H, Hörhammer L, Farkas L (1971) Synthese des 7-β-neohesperidosyl-4′-β-D-glucopyranosylnaringenins, eines flavanontriglykosids aus Citrusfrüchten. Chem Ber 104:473–478 (In German)

Bailey DG, Dresser GK (2004) Interactions between grapefruit juice and cardiovascular drugs. Am J Cardiovasc Drugs 4(5):281–297

Bailey DG, Dresser GK, Leake BF, Kim RB (2007) Naringin is a major and selective clinical inhibitor of organic anion-transporting polypeptide 1A2 (OATP1A2) in grapefruit juice. Clin Pharmacol Ther 81(4):495–502

Bailey DG, Malcolm J, Arnold O, Spence JD (1998) Grapefruit juice-drug interactions. Br J Clin Pharmacol 46(2):101–110

Bailey DG, Spence JD, Munoz C, Arnold JMO (1991) Interaction of Citrus juices with felodipine and nifedipine. Lancet 337:268–269

Bayer RJ, Mabberley DJ, Morton C, Miller CH, Sharma IK, Pfeil BE, Rich S, Hitchcock R, Sykes S (2009) A molecular phylogeny of the orange subfamily (Rutaceae: Aurantioideae) using nine cpDNA sequences. Am J Bot 96:668–685

Berhow M, Tisserat B, Kanes K, Vandercook C (1998) Survey of phenolic compounds produced in Citrus. USDA ARS Tech Bull 1856:1–154

Boddu SP, Yamsani MR, Potharaju S, Veeraraghavan S, Rajak S, Kuma SV, Avery BA, Repka MA, Varanasi VS (2009) Influence of grapefruit juice on the pharmacokinetics of diltiazem in Wistar rats upon single and multiple dosage regimens. Pharmazie 64(8): 525–531

Borradaile NM, Carroll KK, Kurowska EM (1999) Regulation of HepG2 cell apolipoprotein B metabolism by the citrus flavanones hesperetin and naringenin. Lipids 34(6):591–598

Borradaile NM, de Dreu LE, Barrett PH, Behrsin CD, Huff MW (2003) Hepatocyte apoB-containing lipoprotein secretion is decreased by the grapefruit flavonoid, naringenin, via inhibition of MTP-mediated microsomal triglyceride accumulation. Biochemistry 42(5):1283–1291

Borradaile NM, de Dreu LE, Barrett PH, Huff MW (2002) Inhibition of hepatocyte apoB secretion by naringenin: enhanced rapid intracellular degradation independent of reduced microsomal cholesteryl esters. J Lipid Res 43(9):1544–1554

Brill S, Zimmermann C, Berger K, Drewe J, Gutmann H (2009) In vitro interactions with repeated grapefruit juice administration–to peel or not to peel? Planta Med 75(4):332–335

Brzozowski T, Konturek PC, Drozdowicz D, Konturek SJ, Zayachivska O, Pajdo R, Kwiecien S, Pawlik WW, Hahn EG (2005) Grapefruit-seed extract attenuates ethanol-and stress-induced gastric lesions via activation of prostaglandin, nitric oxide and sensory nerve pathways. World J Gastroenterol 11(41):6450–6458

Buettner A, Schieberle P (1999) Characterization of the most odor-active volatiles in fresh, hand-squeezed juice of grapefruit (Citrus paradisi Macfayden). J Agric Food Chem 47(12):5189–5193

Butterweck V, Zdrojewski I, Galloway C, Frye R, Derendorf H (2009) Toxicological and pharmacokinetic evaluation of concomitant intake of grapefruit juice and simvastatin in rats after repeated treatment over 28 days. Planta Med 75(11):1196–1202

Cariño-Cortés R, Alvarez-González I, Martino-Roaro L, Madrigal-Bujaidar E (2010) Effect of naringin on the DNA damage induced by daunorubicin in mouse hepatocytes and cardiocytes. Biol Pharm Bull 33(4): 697–701

Cerda JJ, Robbins FL, Burgin CW, Baumgartner TG, Rice RW (1988) The effects of grapefruit pectin on patients at risk for coronary heart disease without altering diet or lifestyle. Clin Cardiol 11(9):589–594

César TB, Manthey JA, Myung K (2009) Minor furanocoumarins and coumarins in grapefruit peel oil as inhibitors of human cytochrome P450 3A4. J Nat Prod 72(9):1702–1704

Chidambara Murthy KN, Jayaprakasha GK, Patil BS (2011) Obacunone and obacunone glucoside inhibit human colon cancer (SW480) cells by the induction of apoptosis. Food Chem Toxicol 49(7):1616–1625

Cuciureanu M, Vlase L, Muntean D, Varlan I, Cuciureanu R (2010) Grapefruit juice–drug interactions: importance for pharmacotherapy. Rev Med Chir Soc Med Nat Iasi 114(3):885–891

Cvetnić Z, Vladimir-Knezević S (2004) Antimicrobial activity of grapefruit seed and pulp ethanolic extract. Acta Pharm 54(3):243–250

Dahan A, Altman H (2004) Food-drug interaction: grapefruit juice augments drug bioavailability–mechanism, extent and relevance. Eur J Clin Nutr 58(1):1–9

de Castro WV, Mertens-Talcott S, Derendorf H, Butterweck V (2007) Grapefruit juice-drug interactions: grapefruit juice and its components inhibit P-glycoprotein (ABCB1) mediated transport of talinolol in Caco-2 cells. J Pharm Sci 96(10):2808–2817

Dembinski A, Warzecha Z, Konturek SJ, Ceranowicz P, Dembinski M, Pawlik WW, Kusnierz-Cabala B, Naskalski JW (2004) Extract of grapefruit-seed reduces acute pancreatitis induced by ischemia/reperfusion in rats: possible implication of tissue antioxidants. J Physiol Pharmacol 55(4):811–821

Demole E, Enggist P (1983) Further investigation of grapefruit juice flavor components (Citrus paradisi Macfayden). Valencane- and eudesmane-type sesquiterpene ketones. Helv Chim Acta 66:1381–1391

Demole E, Enggist P, Ohloff G (1982) 1-p-Menthene-8-thiol: a powerful flavor impact constituent of grapefruit juice (*Citrus paradisi* Macfayden). Helv Chim Acta 65(6):1785–1794

Deyhim F, Mandadi K, Faraji B, Patil BS (2008a) Grapefruit juice modulates bone quality in rats. J Med Food 11(1):99–104

Deyhim F, Mandadi K, Patil BS, Faraji B (2008b) Grapefruit pulp increases antioxidant status and improves bone quality in orchidectomized rats. Nutrition 24(10):1039–1044

Díaz-Juárez JA, Tenorio-López FA, Zarco-Olvera G, Valle-Mondragón LD, Torres-Narváez JC, Pastelín-Hernández G (2009) Effect of *Citrus paradisi* extract and juice on arterial pressure both in vitro and in vivo. Phytother Res 23(7):948–954

Dresser GK, Bailey DG (2003) The effects of fruit juices on drug disposition: a new model for drug interactions. Eur J Clin Invest 33(Suppl 2):10–16

Dresser GK, Kim RB, Bailey DG (2005) Effect of grapefruit juice volume on the reduction of fexofenadine bioavailability: possible role of organic anion transporting polypeptides. Clin Pharmacol Ther 77(3):170–177

Dunlap WJ, Wender SH (1962) Identification studies on some minor flavonoid constituents of the grapefruit. Anal Biochem 4(2):110–115

Edwards DJ, Bellevue FH 3rd, Woster PM (1996) Identification of 6′,7′-dihydroxybergamottin, a cytochrome P450 inhibitor, in grapefruit juice. Drug Metab Dispos 24(12):1287–1290

Edwards DJ, Bernier SM (1996) Naringin and naringenin are not the primary CYP3A inhibitors in grapefruit juice. Life Sci 59(13):1025–1030

Facciola S (1990) Cornucopia. A source book of edible plants. Kampong, Vista, 677 pp

Fisher JF (1968) A procedure for obtaining radioactive naringin from grapefruit leaves fed L-phenylalanine-14C. Phytochemistry 7(5):769–771

Fuhr U, Klittich K, Staib AH (1993) Inhibitory effect of grapefruit juice and its bitter principal, naringenin, on CYP1A2 dependent metabolism of caffeine in man. Br J Clin Pharmacol 35(4):431–436

Fujioka K, Greenway F, Sheard J, Ying Y (2006) The effects of grapefruit on weight and insulin resistance: relationship to the metabolic syndrome. J Med Food 9(1):49–54

Fukuda K, Ohta T, Oshima Y, Ohashi N, Yoshikawa M, Yamazoe Y (1997) Specific CYP3A4 inhibitors in grapefruit juice: furocoumarin dimers as components of drug interaction. Pharmacogenetics 7(5):391–396

Garvan C, Kane GC, Lipsky JJ (2000) Drug-grapefruit juice interactions. Mayo Clin Proc 75:933–942

Giamperi L, Fraternale D, Bucchini A, Ricci D (2004) Antioxidant activity of *Citrus paradisi* seeds glyceric extract. Fitoterapia 75(2):221–224

Girennavar B, Cepeda ML, Soni KA, Vikram A, Jesudhasan P, Jayaprakasha GK, Pillai SD, Patil BS (2008) Grapefruit juice and its furocoumarins inhibits autoinducer signaling and biofilm formation in bacteria. Int J Food Microbiol 125(2):204–208

Girennavar B, Jayaprakasha GK, Jadegoud Y, Nagana Gowda GA, Patil BS (2007a) Radical scavenging and cytochrome P450 3A4 inhibitory activity of bergaptol and geranylcoumarin from grapefruit. Bioorg Med Chem 15(11):3684–3691

Girennavar B, Jayaprakasha GK, Patil BS (2007b) Potent inhibition of human cytochrome P450 3A4, 2D6, and 2C9 isoenzymes by grapefruit juice and its furocoumarins. J Food Sci 72(8):C417–C421

Girennavar B, Poulose SM, Jayaprakasha GK, Bhat NG, Patil BS (2006) Furocoumarins from grapefruit juice and their effect on human CYP 3A4 and CYP 1B1 isoenzymes. Bioorg Med Chem 14(8):2606–2612

Goldwasser J, Cohen PY, Lin W, Kitsberg D, Balaguer P, Polyak SJ, Chung RT, Yarmush ML, Nahmias Y (2011) Naringenin inhibits the assembly and long-term production of infectious hepatitis c virus particles through a PPAR-mediated mechanism. J Hepatol 55(5):963–971

Goldwasser J, Cohen PY, Yang E, Balaguer P, Yarmush ML, Nahmias Y (2010) Transcriptional regulation of human and rat hepatic lipid metabolism by the grapefruit flavonoid naringenin: role of PPARalpha, PPARgamma and LXRalpha. PLoS One 5(8):e12399

Gorinstein S, Caspi A, Libman I, Katrich E, Lerner HT, Trakhtenberg S (2004a) Fresh israeli jaffa sweetie juice consumption improves lipid metabolism and increases antioxidant capacity in hypercholesterolemic patients suffering from coronary artery disease: studies in vitro and in humans and positive changes in albumin and fibrinogen fractions. J Agric Food Chem 52(16):5215–5222

Gorinstein S, Caspi A, Libman I, Katrich E, Lerner HT, Trakhtenberg S (2004b) Preventive effects of diets supplemented with sweetie fruits in hypercholesterolemic patients suffering from coronary artery disease. Prev Med 38(6):841–847

Gorinstein S, Caspi A, Libman I, Lerner HT, Huang D, Leontowicz H, Leontowicz M, Tashma Z, Katrich E, Feng S, Trakhtenberg S (2006) Red grapefruit positively influences serum triglyceride level in patients suffering from coronary atherosclerosis: studies in vitro and in humans. J Agric Food Chem 54(5): 1887–1892

Gorinstein S, Cvikrova M, Machackova I, Haruenkit R, Park YS, Jung ST, Yamamota K, Ayala ALM, Katrich E, Trakhtenberg S (2004c) Characterization of antioxidant compounds in Jaffa sweeties and white grapefruits. Food Chem 84:503–510

Gorinstein S, Leontowicz H, Leontowicz M, Drzewiecki J, Jastrzebski Z, Tapia MS, Katrich E, Trakhtenberg S (2005a) Red Star Ruby (Sunrise) and blond qualities of Jaffa grapefruits and their influence on plasma lipid levels and plasma antioxidant activity in rats fed with cholesterol-containing and cholesterol-free diets. Life Sci 77(19):2384–2397

Gorinstein S, Leontowicz H, Leontowicz M, Krzeminski R, Gralak M, Delgado-Licon E, Martinez Ayala AL, Katrich E, Trakhtenberg S (2005b) Changes in plasma lipid and antioxidant activity in rats as a result of naringin and red grapefruit supplementation. J Agric Food Chem 53(8):3223–3228

Gorinstein S, Leontowicz H, Leontowicz M, Krzeminski R, Gralak M, Martin-Belloso O, Delgado-Licon E, Haruenkit R, Katrich E, Park YS, Jung ST, Trakhtenberg S (2004d) Fresh Israeli Jaffa blond (Shamouti) orange and Israeli Jaffa red Star Ruby (Sunrise) grapefruit juices affect plasma lipid metabolism and antioxidant capacity in rats fed added cholesterol. J Agric Food Chem 52(15):4853–4859

Gorinstein S, Zachwieja Z, Katrich E, Pawelzik E, Haruenkit R, Trakhtenberg S, Martin-Belloso O (2004e) Comparison of the contents of the main antioxidant compounds and the antioxidant activity of white grapefruit and his new hybrid. Lebensm Wissensch Technol 37(3):337–343

Greenblatt DJ, von Moltke LL, Harmatz JS, Chen G, Weemhoff JL, Jen C, Kelley CJ, LeDuc BW, Zinny MA (2003) Time course of recovery of cytochrome p450 3A function after single doses of grapefruit juice. Clin Pharmacol Ther 74(2):121–129

Guimarães R, Barros L, Barreira JC, Sousa MJ, Carvalho AM, Ferreira IC (2010) Targeting excessive free radicals with peels and juices of citrus fruits: grapefruit, lemon, lime and orange. Food Chem Toxicol 48(1):99–106

Guo LQ, Fukuda K, Ohta T, Yamazoe Y (2000) Role of furanocoumarin derivatives on grapefruit juice-mediated inhibition of human CYP3A activity. Drug Metab Dispos 28(7):766–771

Guo LQ, Yamazoe Y (2004) Inhibition of cytochrome P450 by furanocoumarins in grapefruit juice and herbal medicines. Acta Pharmacol Sin 25(2):129–136

Hagen RE, Dunlap WJ, Mizelle JW, Wender SH, Lime BJ, Albach RF, Griffiths FP (1965) A chromatographic-fluorometric method for determination of naringin, naringenin rutinoside, and related flavanone blycosides in grapefruit juice and juice sacs. Anal Biochem 12(3):472–482

Hahn-Obercyger M, Stark AH, Madar Z (2005) Grapefruit and oroblanco enhance hepatic detoxification enzymes in rats: possible role in protection against chemical carcinogenesis. J Agric Food Chem 53(5):1828–1832

Haleblian GE, Leitao VA, Pierre SA, Robinson MR, Albala DM, Ribeiro AA, Preminger GM (2008) Assessment of citrate concentrations in *Citrus* fruit-based juices and beverages: implications for management of hypocitraturic nephrolithiasis. J Endourol 22(6):1359–1366

Hanley MJ, Cancalon P, Widmer WW, Greenblatt DJ (2011) The effect of grapefruit juice on drug disposition. Expert Opin Drug Metab Toxicol 7(3):267–286

Harapu CD, Miron A, Cuciureanu M, Cuciureanu R (2010) Flavonoids–bioactive compounds in fruits juice. Rev Med Chir Soc Med Nat Iasi 114(4):1209–1214 (in Romanian)

Hata T, Sakaguchi I, Mori M, Ikeda N, Kato Y, Minamino M, Watabe K (2003) Induction of apoptosis by *Citrus paradisi* essential oil in human leukemic (HL-60) cells. In Vivo 17(6):553–559

Haze S, Sakai K, Gozu Y, Moriyama M (2010) Grapefruit oil attenuates adipogenesis in cultured subcutaneous adipocytes. Planta Med 76(10):950–955

Ho PC, Saville DJ, Coville PF, Wanwimolruk S (2000) Content of CYP3A4 inhibitors, naringin, naringenin and bergapten in grapefruit and grapefruit juice products. Pharm Acta Helv 74(4):379–385

Huang YC, Yang CH, Chiou YL (2011) *Citrus* flavanone naringenin enhances melanogenesis through the activation of Wnt/β-catenin signalling in mouse melanoma cells. Phytomedicine 18(14):1244–1249

Hukkanen J, Jacob P 3rd, Benowitz NL (2006) Effect of grapefruit juice on cytochrome P450 2A6 and nicotine renal clearance. Clin Pharmacol Ther 80(5):522–530

Jayaprakasha GK, Girennavar B, Patil BS (2008) Radical scavenging activities of Rio Red grapefruits and Sour orange fruit extracts in different in vitro model systems. Bioresour Technol 99(10):4484–4494

Kiani J, Imam SZ (2007) Medicinal importance of grapefruit juice and its interaction with various drugs. Nutr J 6:33

Kim EH, Hankinson SE, Eliassen AH, Willett WC (2008) A prospective study of grapefruit and grapefruit juice intake and breast cancer risk. Br J Cancer 98(1):240–241

Kim HJ, Song JY, Park HJ, Park HK, Yun DH, Chung JH (2009) Naringin protects against rotenone-induced apoptosis in human neuroblastoma SH-SY5Y cells. Korean J Physiol Pharmacol 13(4):281–285

Kırbaşlar FG, Tavman A, Dülger B, Türker G (2009) Antimicrobial activity of Turkish *Citrus* peel oils. Pak J Bot 41(6):3207–3212

Kırbaşlar Şİ, Boz İ, Kırbaşlar FG (2006) Composition of Turkish lemon and grapefruit peel oils. J Essent Oil Res 18(5):525–543

Kroyer G (1986) The antioxidant activity of *Citrus* fruit peels. Z Ernahrungs Wiss 25(1):63–69 (In German)

Lee HS (2000) Objective measurement of red grapefruit juice color. J Agric Food Chem 48(5):1507–1511

Lee HS, Kim JG (2003) Effects of debittering on red grapefruit juice concentrate. Food Chem 82(2):177–180

Li P, Callery PS, Gan LS, Balani SK (2007) Esterase inhibition attribute of grapefruit juice leading to a new drug interaction. Drug Metab Dispos 35(7):1023–1031

Lin C, Ke X, Ranade V, Somberg J (2008) The additive effects of the active component of grapefruit juice (naringenin) and antiarrhythmic drugs on HERG inhibition. Cardiology 110(3):145–152

Liu C, Shang YF, Zhang XF, Zhang XG, Wang B, Wu Z, Liu XM, Yu L, Ma F, Lv Y (2009) Co-administration of grapefruit juice increases bioavailability of tacrolimus in liver transplant patients: a prospective study. Eur J Clin Pharmacol 65(9):881–885

Lundahl J, Regardh CG, Edgar B, Johnsson G (1995) Relationship between time of intake of grapefruit juice and its effect on pharmacokinetics and pharmacodynamics of felodipine in healthy subjects. Eur J Clin Pharmacol 49:61–67

Mabberley DJ (1997) A classification for edible *Citrus*. Telopea 7(2):167–172

Mandadi KK, Jayaprakasha GK, Bhat NG, Patil BS (2007) Red Mexican grapefruit: a novel source for bioactive limonoids and their antioxidant activity. Z Naturforsch C 62(3–4):179–188

Manthey JA, Buslig BS (2005) Distribution of furanocoumarins in grapefruit juice fractions. J Agric Food Chem 53(13):5158–5163

Meera, Kalidhar SB (2008) A new coumarin from *Citrus paradisi* Macf. Indian J Pharm Sci 70(4):517–519

Mendes AF, Chen C, Gmitter FG Jr, Moore GA, Costa MG (2011) Expression and phylogenetic analysis of two new lycopene β-cyclases from *Citrus paradisi*. Physiol Plant 141(1):1–10

Miller EG, Peacock JJ, Bourland TC, Taylor SE, Wright JM (2008) Inhibition of oral carcinogenesis by *Citrus* flavonoids. Nutr Cancer 60(1):69–74

Miyata M, Takano H, Guo LQ, Nagata K, Yamazoe Y (2004) Grapefruit juice intake does not enhance but rather protects against aflatoxin B1-induced liver DNA damage through a reduction in hepatic CYP3A activity. Carcinogenesis 25(2):203–209

Miyazawa M, Tougo H, Ishihara M (2001) Inhibition of acetylcholinesterase activity by essential oil from *Citrus paradisi*. Nat Prod Lett 15(3):205–210

Mizelle JW, Dunlap WJ, Hagen RE, Wender SH, Lime BJ, Albach RF, Griffiths FP (1965) Isolation and identification of some flavanone rutinosides of the grapefruit. Anal Biochem 12(2):316–324

Mizelle JW, Dunlap WJ, Wender SH (1967) Isolation and identification of two isomeric naringenin rhamnodiglucosides from grapefruit. Phytochemistry 6(9): 1305–1307

Monroe KR, Murphy SP, Kolonel LN, Pike MC (2007) Prospective study of grapefruit intake and risk of breast cancer in postmenopausal women: the multiethnic cohort study. Br J Cancer 97(3):440–445

Morton JF (1987) Grapefruit. In: Fruits of warm climates. Julia F. Morton, Miami, pp 152–158

Nagy S, Norby HE (1972) Long-chain hydrocarbon profiles of grapefruit juice sacs. Phytochemistry 11(9): 2789–2794

Nahmias Y, Goldwasser J, Casali M, van Poll D, Wakita T, Chung RT, Yarmush ML (2008) Apolipoprotein B-dependent hepatitis C virus secretion is inhibited by the grapefruit flavonoid naringenin. Hepatology 47(5): 1437–1445

Nicolò MS, Gioffrè A, Carnazza S, Platania G, Silvestro ID, Guglielmino SP (2011) Viable but nonculturable state of foodborne pathogens in grapefruit juice: a study of laboratory. Foodborne Pathog Dis 8(1):11–17

Nieminen TH, Hagelberg NM, Saari TI, Neuvonen M, Neuvonen PJ, Laine K, Olkkola KT (2010) Grapefruit juice enhances the exposure to oral oxycodone. Basic Clin Pharmacol Toxicol 107(4):782–788

Nishimuta H, Ohtani H, Tsujimoto M, Ogura K, Hiratsuka A, Sawada Y (2007) Inhibitory effects of various beverages on human recombinant sulfotransferase isoforms SULT1A1 and SULT1A3. Biopharm Drug Dispos 28(9):491–500

Nishimuta H, Tsujimoto M, Ogura K, Hiratsuka A, Ohtani H, Sawada Y (2005) Inhibitory effects of various beverages on ritodrine sulfation by recombinant human sulfotransferase isoforms SULT1A1 and SULT1A3. Pharm Res 22(8):1406–1410

Njoroge SM, Koaze H, Karanja PN, Sawamura M (2005) Volatile constituents of redblush grapefruit (*Citrus paradisi*) and pummelo (*Citrus grandis*) peel essential oils from Kenya. J Agric Food Chem 53(25):9790–9794

Nordby HE, Fisher JF, Kew TJ (1968) Apigenin 7β-rutinoside, a new flavonoid from the leaves of *Citrus paradisi*. Phytochemistry 7(9):1653–1657

Ochse JJ, Soule MJ Jr, Dijkman MJ, Wehlburg C (1961) Tropical and subtropical agriculture, 2 vols. Macmillan, New York, 1446 pp

Okura T, Ozawa T, Ito Y, Kimura M, Kagawa Y, Yamada S (2008) Enhancement by grapefruit juice of morphine antinociception. Biol Pharm Bull 31(12): 2338–2341

Ortuño A, Báidez A, Gómez P, Arcas MC, Porras I, García-Lidón A, Del Río JA (2006) *Citrus paradisi* and *Citrus sinensis* flavonoids: their influence in the defence mechanism against *Penicillium digitatum*. Food Chem 98(2):351–358

Oyelami OA, Agbakwuru EA, Adeyemi LA, Adedeji GB (2005) The effectiveness of grapefruit (*Citrus paradisi*) seeds in treating urinary tract infections. J Altern Complement Med 11(2):369–371

Panchagnula R, Bansal T, Varma MV, Kaul CL (2005) Co-treatment with grapefruit juice inhibits while chronic administration activates intestinal P-glycoprotein-mediated drug efflux. Pharmazie 60(12):922–927

Park HS, Oh JH, Lee J, Lee YJ (2011) Minor effects of the citrus flavonoids naringin, naringenin and quercetin, on the pharmacokinetics of doxorubicin in rats. Pharmazie 66(6):424–429

Peterson JJ, Beecher GR, Bhagwat SA, Dwyer JT, Gebhardt SE, Haytowitz DB, Holden JM (2006) Flavanones in grapefruit, lemons, and limes: a compilation and review of the data from the analytical literature. J Food Compos Anal 19(Suppl 1):S74–S80

Piccirillo G, Magrì D, Matera S, Magnanti M, Pasquazzi E, Schifano E, Velitti S, Mitra M, Marigliano V, Paroli M, Ghiselli A (2008) Effects of pink grapefruit juice on QT variability in patients with dilated or hypertensive cardiomyopathy and in healthy subjects. Transl Res 151(5):267–272

Porcher MH et al. (1995–2020) Searchable world wide web multilingual multiscript plant name database. Published by The University of Melbourne, Australia. http://www.plantnames.unimelb.edu.au/Sorting/Frontpage.html

Proteggente AR, Pannala AS, Paganga G, Van Buren L, Wagner E, Wiseman S, Van De Put F, Dacombe C, Rice-Evans CA (2002) The antioxidant activity of regularly consumed fruit and vegetables reflects their phenolic and vitamin C composition. Free Radic Res 36(2):217–233

Purseglove JW (1968) Tropical crops: dicotyledons, vol 1 and 2. Longman, London, 719 pp

Qin L, Jin L, Lu L, Lu X, Zhang C, Zhang F, Liang W (2011) Naringenin reduces lung metastasis in a breast cancer resection model. Protein Cell 2(6):507–516

Reddy P, Ellington D, Zhu Y, Zdrojewski I, Parent SJ, Harmatz JS, Derendorf H, Greenblatt DJ, Browne K Jr (2011) Serum concentrations and clinical effects of atorvastatin in patients taking grapefruit juice daily. Br J Clin Pharmacol 72(3):434–441

Reuther W (1988) Major commercial *Citrus* varieties of the United States. HortScience 23:693–697

Ribeiro IA, Rocha J, Sepodes B, Mota-Filipe H, Ribeiro MH (2008) Effect of naringin enzymatic hydrolysis towards naringenin on the anti-inflammatory activity of both compounds. J Mol Catal B Enzym 52–53:13–18

Ross SA, Ziska DS, Zhao K, ElSohly MA (2000) Variance of common flavonoids by brand of grapefruit juice. Fitoterapia 71(2):154–161

Saalu LC, Osinubi AA, Jewo PI, Oyewopo AO, Ajayi GO (2010) An evaluation of influence of *Citrus paradisi* seed extract on doxorubicin-induced testicular oxidative stress and impaired spermatogenesis. Asian J Sci Res 3:51–61

Sakamaki N, Nakazato M, Matsumoto H, Hagino K, Hirata K, Ushiyama H (2008) Contents of furanocoumarins in grapefruit juice and health foods. Shokuhin Eiseigaku Zasshi 49(4):326–331 (in Japanese)

Samson JA (1992) *Citrus x paradisi* Macf. In: Verheij EWM, Coronel RE (eds) Plant resources of South-East Asia, no. 2. Edible fruits and nuts. Prosea Foundation, Bogor, pp 133–135

Scholz EP, Zitron E, Kiesecker C, Lück S, Thomas D, Kathöfer S, Kreye VA, Katus HA, Kiehn J, Schoels W, Karle CA (2005) Inhibition of cardiac HERG channels by grapefruit flavonoid naringenin: implications for the influence of dietary compounds on cardiac repolarisation. Naunyn Schmiedebergs Arch Pharmacol 371(6):516–525

Schubert W, Cullberg G, Edgar B, Hedner T (1994) Inhibition of 17 beta-estradiol metabolism by grapefruit juice in ovariectomized women. Maturitas 20(2–3):155–163

Schubert W, Eriksson U, Edgar B, Cullberg G, Hedner T (1995) Flavonoids in grapefruit juice inhibit the in vitro hepatic metabolism of 17 beta-estradiol. Eur J Drug Metab Pharmacokinet 20(3):219–224

Schwarz UI, Seemann D, Oertel R, Miehlke S, Kuhlisch E, Fromm MF, Kim RB, Bailey DG, Kirch W (2005) Grapefruit juice ingestion significantly reduces talinolol bioavailability. Clin Pharmacol Ther 77(4):291–301

Seden K, Dickinson L, Khoo S, Back D (2010) Grapefruit-drug interactions. Drugs 70(18):2373–2407

Shaw PE, Calkins CO, McDonald RE, Greany PD, Webb JC, Nisperos-Carriedo MO, Barros SM (1991) Changes in limonin and naringin levels in grapefruit albedo with maturity and the effects of gibberellic acid on these changes. Phytochemistry 30(10):3215–3219

Shirasaka Y, Kuraoka E, Spahn-Langguth H, Nakanishi T, Langguth P, Tamai I (2010) Species difference in the effect of grapefruit juice on intestinal absorption of talinolol between human and rat. J Pharmacol Exp Ther 332(1):181–189

Shirasaka Y, Suzuki K, Nakanishi T, Tamai I (2011) Differential effect of grapefruit juice on intestinal absorption of statins due to inhibition of organic anion transporting polypeptide and/or P-glycoprotein. J Pharm Sci 100(9):3843–3853

Sica DA (2006) Interaction of grapefruit juice and calcium channel blockers. Am J Hypertens 19(7):768–773

Silver HJ, Dietrich MS, Niswender KD (2011) Effects of grapefruit, grapefruit juice and water preloads on energy balance, weight loss, body composition, and cardiometabolic risk in free-living obese adults. Nutr Metab (Lond) 8(1):8

So FV, Guthrie N, Chambers AF, Moussa M, Carroll KK (1996) Inhibition of human breast cancer cell proliferation and delay of mammary tumorigenesis by flavonoids and citrus juices. Nutr Cancer 26(2):167–181

Spencer EA, Key TJ, Appleby PN, van Gils CH, Olsen A, Tjønneland A, Clavel-Chapelon F, Boutron-Ruault MC, Touillaud M, Sánchez MJ, Bingham S, Khaw KT, Slimani N, Kaaks R, Riboli E (2009) Prospective study of the association between grapefruit intake and risk of breast cancer in the European Prospective Investigation into Cancer and Nutrition (EPIC). Cancer Causes Control 20(6):803–809

Staudte H, Sigusch BW, Glockmann E (2005) Grapefruit consumption improves vitamin C status in periodontitis patients. Br Dent J 199(4):213–217

Sugimoto K, Araki N, Ohmori M, Harada K, Cui Y, Tsuruoka S, Kawaguchi A, Fujimura A (2006) Interaction between grapefruit juice and hypnotic drugs: comparison of triazolam and quazepam. Eur J Clin Pharmacol 62(3):209–215

Sugimoto N, Tada A, Kuroyanagi M, Yoneda Y, Yun YS, Kunugi A, Sato K, Yamazaki T, Tanamoto K (2008) Survey of synthetic disinfectants in grapefruit seed extract and its compounded products. Shokuhin Eiseigaku Zasshi 49(1):56–62

Swingle WT, Reece PC (1967) The botany of *Citrus* and its wild relatives. In: Reuther W, Webber HJ, Batchelor LD (eds) The *Citrus* industry, vol 1, History, world distribution, botany, and varieties. University of California, Riverside, pp 190–430

Takeoka GR, Dao LT, Wong RY, Harden LA (2005) Identification of benzalkonium chloride in commercial grapefruit seed extracts. J Agric Food Chem 53(19):7630–7636

Tamura H, Matsui M (2000) Inhibitory effects of green tea and grape juice on the phenol sulfotransferase activity of mouse intestines and human colon carcinoma cell line, Caco-2. Biol Pharm Bull 23(6):695–699

Tanaka T (1954) Species problem in *Citrus*. A critical study of wild and cultivated units of *Citrus*, based upon field studies in their native homes. Japanese Society for the Promotion of Science, Tokyo, 152 pp

Taniguchi K, Ohtani H, Ikemoto T, Miki A, Hori S, Sawada Y (2007) Possible case of potentiation of the antiplatelet effect of cilostazol by grapefruit juice. J Clin Pharm Ther 32(5):457–459

Tassaneeyakul W, Guo LQ, Fukuda K, Ohta T, Yamazoe Y (2000) Inhibition selectivity of grapefruit juice components on human cytochrome P450. Arch Biochem Biophys 378:356–363

Trinchieri A, Lizzano R, Bernardini P, Nicola M, Pozzoni F, Romano AL, Serrago MP, Confalanieri S (2002) Effect of acute load of grapefruit juice on urinary excretion of citrate and urinary risk factors for renal stone formation. Dig Liver Dis 34(Suppl 2):S160–S163

U.S. Department of Agriculture, Agricultural Research Service (USDA) (2011) USDA national nutrient database for standard reference, release 24. Nutrient Data Laboratory Home Page. http://www.ars.usda.gov/ba/bhnrc/ndl

Ueda N, Yoshimura R, Umene-Nakano W, Ikenouchi-Sugita A, Hori H, Hayashi K, Kodama Y, Nakamura J (2009) Grapefruit juice alters plasma sertraline levels after single ingestion of sertraline in healthy volunteers. World J Biol Psychiatry 10(4 Pt 3):832–835

Uno T, Ohkubo T, Motomura S, Sugawara K (2006) Effect of grapefruit juice on the disposition of manidipine enantiomers in healthy subjects. Br J Clin Pharmacol 61(5):533–537

Vanamala J, Leonardi T, Patil BS, Taddeo SS, Murphy ME, Pike LM, Chapkin RS, Lupton JR, Turner ND (2006) Suppression of colon carcinogenesis by bioactive compounds in grapefruit. Carcinogenesis 27(6):1257–1265

Veronese ML, Gillen LP, Burke JP, Dorval EP, Hauck WW, Pequignot E, Waldman SA, Greenberg HE (2003) Exposure-dependent inhibition of intestinal and hepatic CYP3A4 in vivo by grapefruit juice. J Clin Pharmacol 43(8):831–839

Vikram A, Jesudhasan PR, Jayaprakasha GK, Pillai SD, Jayaraman A, Patil BS (2011) *Citrus* flavonoid represses *Salmonella* pathogenicity island 1 and motility in *S. typhimurium* LT2. Int J Food Microbiol 145(1):28–36

Viuda-Martos M, Ruiz-Navajas Y, Fernández-López J, Perez-ÁLvarez J (2008a) Antibacterial activity of lemon (*Citrus lemon* L.), mandarin (*Citrus reticulata* L.), grapefruit (*Citrus paradisi* L.) and orange (*Citrus sinensis* L.) essential oils. J Food Saf 28:567–576

Viuda-Martos M, Ruiz-Navajas Y Fernández-López J, Pérez-Álvarez J (2008b) Antifungal activity of lemon (*Citrus lemon* L.), mandarin (*Citrus reticulata* L.), grapefruit (*Citrus paradisi* L.) and orange (*Citrus sinensis* L.) essential oils. Food Control 19(12):1130–1138

von Woedtke T, Schlüter B, Pflegel P, Lindequist U, Jülich WD (1999) Aspects of the antimicrobial efficacy of grapefruit seed extract and its relation to preservative substances contained. Pharmazie 54(6):452–456

Wang SS, Liu L, Zhu L, Yang YX (2011) Inhibition of TNF-α/IFN-γ induced RANTES expression in HaCaT cell by naringin. Pharm Biol 49(8):810–814

Webber HJ (1946) Cultivated varieties of *Citrus*. In: Webber HJ, Batchelor LD (eds) The *Citrus* industry, vol 1, History, botany and breeding. University of California Press, Berkeley, pp 475–668

Wheaton TA, Stewart I (1965) Feruloylputrescine: isolation and identification from citrus leaves and fruit. Nature 206:620–621

Widmer W, Haun C (2005) Variation in furanocoumarin content and new furanocoumarin dimers in commercial grapefruit (*Citrus paradisi* Macf.) juices. J Food Sci 70(4):C307–C312

Wilcox LJ, Borradaile NM, de Dreu LE, Huff MW (2001) Secretion of hepatocyte apoB is inhibited by the flavonoids, naringenin and hesperetin, via reduced activity and expression of ACAT2 and MTP. J Lipid Res 42(5):725–734

Yu J, Dandekar DV, Toledo RT, Singh RK, Patil BS (2007) Supercritical fluid extraction of limonoids and naringin from grapefruit (*Citrus paradisi* Macf.) seeds. Food Chem 105(3):1026–1031

Yu J, Buslig BS, Haun C, Cancalon P (2009) New furanocoumarins detected from grapefruit juice retentate. Nat Prod Res 23(5):498–506

Zhang DX, Mabberley DJ (2008) *Citrus* Linnaeus. In: Wu ZY, Raven PH, Hong DY (eds) Flora of China, vol 11, Oxalidaceae through Aceraceae. Science Press/Missouri Botanical Garden Press, Beijing/St. Louis

Citrus x aurantium Sour Orange Group

Scientific Name

Citrus x aurantium L. Sour Orange Group.

Synonyms

Aurantium acre Mill., *Aurantium* var. *citrus* L., *Citrus amara* Link, *Citrus aurantium* L. (sensu Swingle and Reece 1967; sec. Hodgson 1967; sensu Tanaka sec. Cottin 2002), *Citrus aurantium* ssp. *amara* (L.) Engl., *Citrus aurantium* L. subsp. *aurantium*, *Citrus aurantium* var. *bigaradia* Hook f., *Citrus bigaradia* Loisel., *Citrus bigaradia* Risso et Poit., *Citrus florida* Salisb., *Citrus fusca* Lour., *Citrus vulgaris* Risso.

Family

Rutaceae

Common/English Names

Bigarade Orange, Bigrade, Bitter Orange, Chinese Bitter Orange, Clementine, Marmalade Orange, Neroli, Rough Seville, Seville, Seville Orange, Sour Orange.

Vernacular Names

Arabic: Burtuqân, Burtuqâl, Burtuqual, Kabbâd, Limu, Limue-Hamiz, Naffâsh, Naranj, Nâring, Zahr (Flower);
Basque: Larando;
Brazil: Bergamoteira, Laranja-Amarga, Laranja-Azeda, Laranja Da Terra, Laranjeira;
Burmese: Kabala, Leinmaw;
Catalan: Taronger Agre;
Chinese: Cheng, Daidai Hua, Jin Qiu, Shangzhou Zhiqiao, Suan Che'ng, Zhi Ke, Zhi Shi, Zhĭ Shí, Zhi Qiao;
Croatian: Gorka Naranča;
Czech: Citroník Birgádie, Pomerančovník Citrus, Pomerančovník Hořký, Pomerančovník Kyselý;
Danish: Pomerans;
Dutch: Bittere Sinaasappel, Citroenboom, Oranjeappel, Oranjeappel Sort, Pomerans, Sinaasappel, Sinasappelboom, Zure Sinaasappel;
Eastonian: Pomerantsipuu;
Esperanto: Bigarado;
Fijian: Moli Kula, Moli Jamu;
Finnish: Hapanappelsiini, Hunaja-Appelsiini, Makea Appelsiini, Pomeranssi;
French: Bigarade, Bigaradier, Bigardier, Orange Amer, Orange Amère, Oranger Amer;
Futuna: Moli kai;
German: Bittere Orange, Bitterorange, Bitterorangen, Pomeranze;

Greek: Neratzi (Fruit), Neratzia (Tree);
Guam: Kahet;
Haiti: Zorange Si;
Hawaiian: 'Alani;
Hungarian: Keserű Narancs, Narancs, Savanyú Narancs, Sevillai Narancs;
Icelandic: Beiskjuappelsína;
India: Khatta, Naarangii, Naaringii (Hindu), Jambha, Kanchikai, Kanci, Limbe, Nimbe (Kannada), Cerunarakam, Cherunarakam, Conakanarakam, Jambhalam, Jambham, Jonakanarakam, Karna (Malayalam), Nimbuka (Oriya), Dantaharshana, Dantakarshana, Dantashatha, Jadyari, Jambha, Jambhaka, Jambhala, Jambhara, Jambir, Jambira-Phalam, Jantujita, Mukashodhi, Nagarunga, Nimbuka, Rewatavakrashodhi, Rochanaka (Sanskrit), Camiranam, Elumiccai, Narandam, Narandai, Narattai (Tamil), Jambhalamu, Jambhamu, Kiccili, Kittali, Narangamu, Naranji, Narija, Narinja (Telugu);
Indonesia: Lemon Itam (Moluccas);
Israel: Chushchash, Hushhash, Khushkhash, Tapuz Marir (Hebrew);
Italian: Arancio, Arancia Amara, Arancia Forte, Arancio Amaro, Cedrangola (Fruit), Cedrangolo (Tree), Melangolo, Melarancia (Fruit), Melarancio (Tree);
Japanese: Bitaa Orenji, Daidai, Kaisei-To, Kikoku, Kijitu, Sawa-Orenji;
Korean: Biteo Orenji, Pito Orenji; Gwang-Gyul, Kwang-Kyul, Kwanggyulnamu;
Kampuchea: Leang Sat;
Lithuanian: Karčiavaisis Citrinmedis;
Macedonian: Turunka;
Malaysia: Limau Samar;
Nepalese: Kali Jyamir, Suntala;
New Caledonia: L'Oranger;
Pakistan: Khatta;
Papiamento: Laraha;
Persian: Limeh, Limu, Limuetursh, Naranj;
Philippines: Cajel (Bikol), Cajel, Kahil (Bisaya), Talamisan, Tamamisan, Tamisan (Cebu-Bisaya), Cajel (Ibanag), Valachinuk, Volatino (Ivatan), Daladan, Kahil (Tagalog), Cabiso;
Polish: Pomarańcza Gorzka;
Portuguese: Laranja-Azeda, Laranjeira Azêda;
Russian: Bigarad, Pomeranets;
Samoan: Moli 'Aina;
Serbian: Naranča Gorka;
Slovaščina: Grenka Pomaranča;
Spanish: Amarga, Narnja Acida, Naranja Agria, Naranja Amargo, Naranjo De Fruta Agria, Naranja Mateca;
Swains Island: Moli;
Swedish: Pomerans;
Taiwan: Lai Mu;
Thai: Som, Som Kliang;
Tongan: Kola;
Tubuai: Anani, Bigarade;
Turkish: Turunç;
Vietnamese: Bồng, Cam Chua, Cam Đắng, Chanh Đắng, Dại Dại Hoa, Toan Đắng;
Yiddish: Bitere Marants, Khushkhosh.

Origin/Distribution

The sour orange is believed to have originated in north-eastern India and adjoining areas of northern Myanmar and southeastern China. It spread eastward to Japan and westward through India to the Middle East and from there to Europe, where it rapidly became established in the Mediterranean about 1,000 years ago. It became especially common in Spain, hence its vernacular name Seville orange. It has been cultivated in France since the early 1400s, initially mainly as an ornamental, then as special perfumery cultivars grown for their fragrant flowers in the French Riviera region and became known as 'Bouquetiers'. It was one of the first *Citrus* taken to South America in the sixteenth century, where it soon escaped from cultivation and naturalized in many areas. It is now cultivated in many tropical and subtropical countries and important areas of cultivation are Paraguay, Morocco, Spain. *C. x aurantium* is rarely grown in South-East Asia.

Agroecology

The sour orange thrives in the warm subtropical, near-tropical climates with mean annual temperature: 22–24°C and mean annual precipitation of 1,000 mm in Spring and summer is ideal. It has considerable tolerance of adverse conditions such as frost, temperatures up to 45°C, excess soil moisture, and neglect and is resistant to gummosis (*mal di gomma*) disease. It can withstand frost for brief periods. A few consecutive days of frost with minimum temperatures of −4°C will devastate the tree and severe frost is fatal to the tree. Very strong winds may cause tree damage, while dry hot winds in spring may reduce leaf size and number and cause extensive withering during flowering.

Sour orange does best in full sun in altitudes below 300 m but will also grown up to an elevation of 500 m. It prefers rich, fertile, well-drained soils with a high water table and is adapted to a wide range of soil conditions but does poorly on compacted, poorly drained soils that contain excessive amounts of clay, lime or silica.

Edible Plant Parts and Uses

Sour orange is too sour to be eaten raw but in Mexico the fruit is halved, consumed raw with salt and chilli pepper. The immature fruits are eaten pickled in salt or vinegar as well as eaten fried in oil. In southern India, he fresh fruit is also used frequently in *pachadis*, a yogurt-based side dish; the immature fruit is used in Tamil cuisine. Immature fruit is pickled by cutting it into spirals and stuffing it with salt. The pickle is usually consumed with *thayir sadam* (yogurt rice). The juice of the fruit is added to meat and fish for flavouring in Spain and Cuba and is used as vinegar substitute as in Yucatan. Ripe fruits are processed for juice and drinks. The juice is valued for orangeade, compote and syrup. In Egypt and elsewhere, it has been fermented to make wine. Sour orange is most prized for making marmalade and without equal being higher in pectin than the sweet orange, and therefore giving a better set and a higher yield. The fruits are largely exported to England and Scotland for making marmalade. Sour oranges are used primarily for marmalade in South Africa.

Bitter orange peel is used as an ingredient in bitters. The peel is used to spiced Belgian beer made from wheat called *Witbier* (white beer). Bitter orange oil (also called 'bitter orange peel oil'), expressed from the peel, with terpenes and sesquiterpenes removed, is in high demand for flavouring candy, ice cream, baked products, confectionaries, gelatins, puddings, chewing gum, soft drinks, liqueurs (orange sec and triple sec liquer flavours), sauces for meats and poultry, and pharmaceutical products. The essential oil derived from the dried peel of immature fruit, particularly from the selected types -'Jacmel' in Jamaica and the much more aromatic 'Curacao orange' (var. *curassaviensis*) imparts a distinctive flavour to certain liqueurs such as Curaçao and Cointreau. Bitter orange oil is produced primarily in Sicily, Spain, West Africa, the West Indies, Brazil, Mexico and Taiwan. The ripe peel of the sour orange has been reported to contain 2.4–2.8%, and the green peel up to 14%, neohesperidin dihydrochalcone which is 20 times sweeter than saccharin and 200 times sweeter than cyclamate.

The petitgrain oil, derived from the leaves, young shoots and immature fruits, and with terpenes removed, is used like the bitter orange oil, for flavouring food products, drinks etc. It is used to enhance the fruit flavours (peach, apricot, gooseberry, black currant, etc.) in food products, candy, ginger ale, and various condiments.

"Neroli oil", or "Neroli Bigarade Oil", distilled from bitter orange flowers has limited use in flavoring candy, soft-drinks and liqueurs, ice cream, baked goods and chewing gum like the petitgrain oil, neroli oil is widely used in the perfume and soap industry. Orange flower-water is also used for flavoring cakes.

Botany

An evergreen, armed, branched, erect tree 3–10-m high (Plate 1). Juvenile twigs trigonous and bearing slender short spines, older branches with stout spines up to 8 cm long. Leaves simple, alternate, subcoriaceous, dotted with glands, aromatic when bruised; obovate-winged petiole 2–3 cm long (Plates 2 and 4). Leaves alternate, broadly ovate to elliptical, 7–12 cm×4–7 cm, base cuneate or obtuse, margin subentire to slightly crenulate, apex obtuse to bluntly pointed. Flowers axillary, single or in 2–3 flowered cymes, very

Plate 1 Sour orange tree habit

Plate 2 Leaves and sour orange fruit

fragrant, white, usually bisexual but 5–12% male flowers may occur. Flower calyx cupular, 4–5 mm long, 3-5-lobed, lobes ovate-triangular, glabrous to pubescent; petals 4–5, oblong, 1.5 cm × 4 mm, fleshy and glandular; stamens 20–25, often in 4–5 groups, filaments 6–10 mm long connate at the base, anthers oblong; pistil with glabrous ovary, stout style and capitate stigma. Fruit oblate to globose hesperidium, 5–8 cm in diameter, with 8–12 segments, central axis hollow, peel thick, glandular, smooth to rugose, yellow to orangey-yellow, strongly aromatic, pulp pale-orange, very acid and slightly bitter (Plates 3 and 4). Seeds few to numerous, polyembryonic.

Plate 3 Ripe sour oranges

Nutritive/Medicinal Properties

Analyses of the raw sour orange fruits conducted in Guatemala and El Salvador reported the following nutrient composition per 100 g edible portion (Morton 1987): energy 37–66 cal., moisture 83–89.2 g, protein 0.6–1.0 g, fat trace-0.1 g, carbohydrates 9.7–15.2 g, fibre 0.4 g, ash 0.5 g, calcium 18–50 mg, iron 0.2 mg, phosphorus 12 mg, vitamin A 290 μg (200 I.U.), thiamine 100 μg, riboflavin 40 μg, niacin 0.3 mg and ascorbic acid 45–90 mg.

The percentage and concentration of total phenolics in sour orange cv Keen peel and juice

Plate 4 Close-view of leaf, fruit and flesh

were reported by Berhow et al. (1998) as follows: flavedo (outer pigmented layer of the peel): 8.4% (0.78 mg/g) flavone/flavonol, 83% (4.90 mg/g) flavanone, 1.9% psoralen, 2.5% (0.07 mg/g) coumarin; albedo (inner layer of the rind): 8% (0.08 mg/g) flavone/flavonol, 80.1% (5.43 mg/g) flavanone, 0.4% psoralen, 7.8% (0.26 mg/g) coumarin; juice: 4.2% (0.06 mg/g) flavone/flavonol, 86.0% (1.01 mg/g) flavanone, 0% psoralen and 3.3% (0.02 mg/g) coumarin.

The percentage and concentration of flavanones reported in the same sour orange cultivar by Berhow et al. (1998) were as follows: flavedo: didymin 9% (0.5 mg/g), eriocitrin 1% (0.1 mg/g), narirutin 1% (traces), neoeriocitrin 17% (0.8 mg/g), naringin 35% (1.7 mg/g), neohesperidin 30% (1.5 mg/g), naringin–6″–malonate (open form) 5% (0.2 mg/g); albedo: eriocitrin 1% (0.1 mg/g), narirutin 1% (traces), neoeriocitrin 17% (0.9 mg/g), naringin 42% (2.3 mg/g), neohesperidin 33% (1.8 mg/g), naringin–6″–malonate (open form) 5% (0.3 mg/g); juice: didymin 4% (traces), eriocitrin 2% (traces), neoeriocitrin 18% (0.2 mg/g), naringin 35% (0.3 mg/g), neohesperidin 3% (traces) and naringin–6″–malonate (open form) 5% (traces).

Peterson et al. (2006) listed the following flavones in sour orange in mg/100 g juice or edible fruit: didymin 2.89 mg, eriocitrin 0.53 mg, naringin 18.83 mg, narirutin 0.08 mg, neoeriocitrin 14.01 mg, neohesperidin 11.09 mg. Sour orange was distinct from sweet oranges, tangerines and tangor, they were devoid of hesperidin and higher in naringin, neoeriocitrin and neohesperidin. It also had the highest overall mean total flavanones content of 48 mg aglycone/100 g juice or edible fruit. It had 44 mg of total aglycones from neohesperidose glycosides and comparatively less of total aglycones, 4 mg from rutinose glycosides.

Two flavonoid glycosides, naringin and rhoifolin were isolated from the ripe peel of Citrus aurantium f. kabusu and Citrus aurantium f. cyathifera and hesperidin from the flower petals of Citrus aurantium f. kabusu (Hattori et al. 1952). A flavonoid 5-hydroxyauranetin was isolated from C. aurantium peel (Sarin and Seshadri 1960). The main constituents of the C. aurantium peel were the volatile oil and an amorphous, bitter glucoside called aurantiamarin (Schneider et al. 1968). Other constituents included hesperidin occurring mainly in the white zest of the peel, isohesperidin, hesperic acid, aurantiamaric acid, and a bitter acrid resin. In the peel of immature fruits, the chief constituents were naringin and hesperidin, while in the fruit flesh it was umbelliferone (Wu and Sheu 1992). Ripe bitter orange peels were found to contain higher concentrations of aliphatic aldehydes and oxygen-containing monoterpenes and sesquiterpenes than the peels of fully developed unripe fruits (Boelens and Jimenez 1989). Changes were found in the concentrations of linalol and linalyl acetate (together 0.3–3.2%) and in those of limonene (92–95%) in the peel oils from bitter oranges. The sesquiterpenes nootkatone and α-selinenone could not be detected in the peel oils from fully developed unripe bitter oranges, whereas such oils from ripe fruits contained up to 0.15% of them. Thus, some oxygen-containing sesquiterpenes appeared to be formed during ripening.

Eight flavonoids: isonaringin, naringin, hesperidin, neohesperidin, naringenin, hesperitin, nobiletin and tangeritin were identified from sour orange (He et al. 1997). Several polymethoxylated flavones such as nobiletin, heptamethoxyflavone, sinensetin, quercetogetin and tangeretin were found in C. aurantium fruits (Del-Rio et al. 1998). The highest concentration of polymethoxylated flavones were associated with the early stages of fruit growth and were located in the peel. Polymethoxyflavonoids isolated from the dichloromethane fraction of sour orange were identified as tetra-O-methylscutellarein, sinensetin and nobiletin (Miyazawa et al. 1999). Five compounds, auraptene, marmin, tangeretin, nobiretin and 5-[(6′,7′-dihydroxy-3′,7′-dimethyl-2-octenyl)oxy]psoralen were isolated from *Citrus aurantium* (Satoh et al. 1996).

Synephrine was the main component in *Citrus aurantium* fruits (0.10–0.35%) and in dry extracts (3.00–3.08%) and was present in the range 0.25–0.99% in herbal medicines (Pellati et al. 2004). Flavanones identified and quantified in the same samples included neoeriocitrin, narirutin, naringin, hesperidin, neohesperidin, naringenin and hesperetin. *C. aurantium* fruits and derivatives contained mainly glycosylated flavanones: in particular, naringin and neohesperidin were found to be the major flavonoids and their concentrations ranged from 1.80 to 26.30 and from 3.90 to 14.71 mg/g, respectively. The levels of aglycones were very low in all samples tested.

Volatile compounds and methyl esters in bitter orange were identified using GC; limonene was the main compound in the monoterpenes, and the mean concentration of 18 fatty acids identified was 678 mg/l (Moufida and Marzouk 2003). The main constituents found in the essential oil of sour orange fruit were: monoterpenes (limonene, 77.90%; β-pinene, 3.40%; myrcene, 1.81%; and *trans*-ocimene, 1.16%), sesquiterpenes (valencene, 0.52%), aldehydes (decanal, 3.51%; dodecanal, 0.36% and geranial, 0.29%), alcohols (β-nerolidol, 0.85% and linalool, 0.89%), and nootkatone as the only ketone (Quintero et al. 2003). Terpinene-4-ol, a product of limonene degradation, was found in traces and thus no unpleasant odor was present. Major changes occurred in the cold-pressed sour orange peel oil stored at 20°C and 5°C, but no changes were found at −21°C (Njoroge et al. 2003). Monoterpene hydrocarbons decreased from 98.0% to 66.4% upon 12 months at 20°C, while sesquiterpene hydrocarbons and alcohols increased from 0.1% to 2.4% and from 0.3% to 7.9%, respectively. Notable decreases of germacrene D, myrcene, linalyl acetate, and limonene occurred. Marked increases of *cis*-carveol, *trans*-β-farnesene, *trans*-p-2,8-menthadien-1-ol, linalool, and β-caryophyllene were found. Thirty-four artifact compounds constituting 17.0% of the total volatile compounds were formed upon 12 months at 20°C. The artifacts consisted of 13 alcohols (6.0%), five carbonyl compounds (5.3%), seven esters (4.9%), three epoxides (0.4%), four hydrocarbons (0.3%), and two unidentified. The prominent artifact compounds were (+)-carvone, *trans*-farnesyl acetate, sabinene hydrate, 1-octen-3-ol, *cis, cis*-farnesyl acetate, and dihydrocarveol acetate. Twenty-nine components were identified in the Turkish bitter orange peel oil (Kırbaşlar and Kırbaşlar 2003; Kırbaşlar et al. 2009): monoterpenes (97.3%) with limonene (94.1%), myrcene (1.8%), β-pinene (0.5%) as the major components; sesquiterpenes (0.1%) with β-caryophyllene (0.1%) predominating. The oxygenated components (2.5%) comprised aldehyde components (0.5%): decanal (0.2%), geranial (0.1%); alcohol components (0.5%): linalool (0.4%) and esters (1.4%): linalyl acetate (1.2%), geranyl acetate (0.08%).

The essential oil of bitter orange was characterized by 19 major odour compounds with different aromatic notes among which were α-pinene (floral), octanal (green), limonene (citrus), linalool (floral), α-terpineol (indeterminate: 'ND', green), linalyl acetate (floral), (E,E) or (E,Z)-2,4-decadienal (frying), dodecanal (ND), caryophyllene (green) and an unknown compound (plastic note) (Deterre et al. 2011). The heart cut of the distillate presented a complex mixture of several components in low concentrations, which characterized the bitter orange flavour. The heart cut of the distillate was characterized by seven high FD (flavour dilution factor) odour compounds, three of them (myrcene, α-phellandrene, and limonene) with citrus and mint aromatic notes, and the other four ((Z)-linalool oxide, α-terpinolene, linalool and neral) with floral notes.

Four coumarins (osthol, meranzin, isomeranzin, and meranzin hydrate), three psoralens (bergapten, epoxybergamottin, and epoxybergamottin hydrate), and four polymethoxyflavones (tangeretin; 3,3′,4′,5,6,7,8-heptamethoxyflavone; nobiletin; and tetra-O-methylscutellarein) were identified from genuine, industrial, cold-pressed Italian and Spanish bitter orange essential oils, commercial bitter orange oils, laboratory hand-extracted bitter orange oils and mixtures of bitter orange oil with sweet orange, lemon, lime, and grapefruit oils (Dugo et al. 1996). In addition, three unknown coumarins were found. Meranzin was the main component isolated. Meranzin hydrate, the formation of which was probably due to the hydration of meranzin during the industrial extraction, was not found in the laboratory hand-extracted samples. Meranzin was not present in some industrial samples. Italian essential oils usually exhibited a higher content of oxygen heterocyclic compounds than the Spanish oils.

Adrenergic protoalkaloids found in bitter orange included predominantly synephrine and minor amounts of tyramine, N-methyltyramine, octopamine, and hordenine (Nelson et al. 2007). Synephrine was reputed to have thermogenic properties and used as a dietary supplement to enhance energy and promote weight loss. However, there were some concerns that the consumption of dietary supplements containing synephrine or similar protoalkaloids may contribute to adverse cardiovascular events. Three bioactive compounds viz. osthol (7-methoxy-8-(3′-methyl-2′-butenyl)-2H-1-benzopyran-2-one); bergapten (4-methoxy-7H-furo[3,2- g]benzopyran-7-one); and 6′,7′-epoxybergamottin (4-((E)-3′-methyl-5′-(3″,3″-dimethyloxiran-2″-yl)pent-2′-enyloxy)-7H-furo[3,2- g][1]benzopyran-7-one) were isolated from bitter orange fruit peel petroleum ether extract (Siskos et al. 2008).

The content of the adrenergic amines dl-octopamine, dl-synephrine, and tyramine in fruits, extracts, and herbal products of *C. aurantium* var. *amara* were determined using HPLC with UV detection; the direct separation of synephrine enantiomers was done with HPLC on a β-cyclodextrin stationary phase (Pellati et al. 2002). Enantioselective LC analysis of *C. auratium* samples showed that (−)-synephrine was the main component (Pellati et al. 2005). (+)-Synephrine was not detected in C. *aurantium* fruits and was present in low concentration in the phytoproducts. Biogenic amines found in bitter orange raw material, extracts, and dietary supplements included for p-synephrine and 5 other biogenic amines: octopamine, phenylephrine (m-synephrine), tyramine, N-methyltyramine, and hordenine (Roman et al. 2007). p-Synephrine was found to be the primary biogenic amine present in all materials tested, accounting for >80% of the total biogenic amine content in all samples. The levels of synephrine, octopamine, tyramine, N-methyltyramine, hordenine, total alkaloids, and caffeine in a suite of three dietary supplement standard reference materials (SRMs) containing bitter orange were determined by as many as six analytical methods (Sander et al. 2008). Excellent agreement was obtained among the measurements, with data reproducibility for most methods and analytes better than 5% relative standard deviation. Chizzali et al. (2011) reported the use of capillary electrochromatography (CEC) for the determination of tyramine, (±) synephrine, and (±) octopamine, the major alkaloids in bitter orange peel. The application of the CEC assay on *C. aurantium* var. *amara* plant material and dietary supplements found synephrine (0.17–0.82%) to be the dominant alkaloid.

Flower Phytochemicals

The following 33 chemical components were identified in the essential oils in the flowers, the peels, the leaves of sour orange: α-thujene, α-pinene, camphene, β-pinene, myrcene, limonene, β-ocimene, trans-linalooloxide (furanoid), cis-linalooloxide (furanoid), linalool, 1,4-p-methadien-7-ol, trans-pinocarveol, camphor, terpinen-4-ol α-terpineol, nerol, citral-b, geraniol, linalylacetate, citrala, trans-linalooloxide (pyranoid), methyl anthranilate, terpinyl acetate, cis-linalooloxide (pyranoid), neryl acetate, geranyl acetate, nonanal, β-caryophyllene, α-humulene, γ-muurolene, β-nerolidol, farnesol, α-nerolidol (Lin et al. 1986). Dried flowers and leaves of *Citrus aurantium* were found to have a similar flavonoid pattern, but the flavonoid levels of flowers were higher than those of leaves (Carnat et al. 1999). The mean levels of the principal flavonoid compounds were respectively: total flavonoids 12.35 and 1.06%, neohesperidin 5.44 and 0.08%, naringin 1.93 and 0.06%, eriocitrin 0.38 and 0.25%. Eleven compounds were isolated from sour orange flowers and identified as neohesperidin, synephrin, 5,8-epidioxyergosta-6,22-dien-3-β-ol, adenosine, asparagine, tyrosine, valine, isoleucine, alanine, β-sitosterol and β-daucosterol (Huang et al. 2001a).

Qualitative and quantitative components of Neroli (*C. aurantium* flower) essential oil comprised: linalyl acetate 31.47%, linalool 22.12%, limonene 14.99%, geranyl acetate 4.81%, methyl anthranilate 4.55%, α-terpineol 3.14%, nerol 3.19%, neryl acetate 2.96%, α-(−)-pinene 2.91%, phenyl ethyl alcohol 1.60%, myrcene 1.40%, (*E*)-β-ocimene 1.37%, β-(+)-pinene 0.96%, geranial 0.09%, α-terpinyl acetate 0.09%, (*Z*)-β-ocimene 0.86%, α-phellandrene 0.57%, β-caryophyllene 0.34% *n*-decanal 0.32%, sabinene 0.18%, camphene 0.12%, bicycolgermacrene 0.08%, geranyl formate 0.06%, (*E*)-(+)-linalool oxide 0.05%, (*Z*)-limonene oxide 0.05%, α-humulene 0.04%, neral 0.04%, (*E*)-nerolidol 0.03%, citronellal 0.03%, geraniol 0.02%, α-thujene 0.02%, γ-terpinene 0.02%, (*E*)-limonene oxide 0.02%, indole 0.02%, β-(−)-elemene 0.02%, dihydrolinayl acetate 0.01%, α-copaene 0.01%, dodecanal 0.01%, germacrene D 0.01%, α-(−)-selinene 0.01%, and β-copaene traces (British Pharmacopoeia Commission 2002).

Seed Phytochemicals

The following bioactive limonoid compounds were isolated from sour orange seed: deacetyl nomilinic acid glucoside, deacetyl nomilin and limonin (Dandekar et al. 2008).

Leaf Phytochemicals

The mature leaf was reported to contain 1-stachyhydrine (Yoshimura and Trier 1912). The percentage and concentration of total phenolics in sour orange cv Keen leaf were reported by Berhow et al. (1998) as follows: 40.2% (5.51 mg/g) flavone/flavonol, 49.5% (4.33 mg/g) flavanone, 0% psoralen, 0% coumarin. The percentage and concentration of flavanones reported in the same sour orange cultivar were as follows: didymin 24% (1.0 mg/g), neoeriocitrin 25% (1.1 mg/g), naringin 20% (0.9 mg/g), neohesperidin 17% (0.7 mg/g); and the percentage and concentration of flavone reported were rhoifolin 9% (0.5 mg/g).

Pharmacological Properties

C. aurantium fruits are rich in vitamin C, flavonoids and volatile oil and contains bioactive phenethylamine alkaloids octopamine, synephrine, tyramine, N-methyltyramine and hordenine that impart a diverse range of phamacological activities. Its fruit extracts have been employed for the treatment of various diseases such as gastrointestinal disorders, insomnia, headaches, cardiovascular diseases, cancer, antiseptic, antioxidant, antispasmodic, aromatic, astringent, carminative, digestive, sedative, stimulant, stomachic and tonic and obesity problems (Suryawanshi 2011). Owing to the publicity concerning adverse reactions of Ephedra, many companies had withdrawn Ephedra from the market and had been substituting *Citrus aurantium* as a thermogenic, weight-reduction replacement for ephedra in their formulations (Preuss et al. 2002). The various pharmacological properties of sour orange are elaborated as follows.

Antimicrobial Activity

Sour orange fruit essential oil was moderately inhibitory against *Staphylococcus aureus* but was inactive against *Escherichia coli* and *Pseudomonas* (Quintero et al. 2003). In a clinical study of patients with tinea corporis, tinea cruris, and tinea pedis, a 25% emulsion of bitter orange oil treatment (thrice/daily) cured 80% of patients in 1–2 weeks and 20% in 2–3 weeks (Ramadan et al. 1996). In patients treated with 20% bitter orange oil in alcohol (thrice daily) 50% were cured in 1–2 weeks, 30% in 2–3 weeks and 20% in 3–4 weeks. In patients treated with pure bitter orange oil once daily, 33.3% were cured in 1 week, 60% in 1–2 weeks, and 6.7% in 2–3 weeks. Bitter orange oil was found to exert fungistatic and fungicidal activity against a variety of pathogenic dermatophyte species in-vitro. Bitter orange peel oil exhibited high antibacterial activity against Gram positive bacteria (*Staphylococcus aureus, Bacillus cereus, Mycobacterium smegmatis, Listeria monocytogenes, Micrococcus luteus*) and Gram negative bacteria (*Escherichia coli, Klebsiella pneumoniae, Pseudomonas aeruginosa, Proteus vulgaris*) and antifungal activity against *Kluyveromyces fragilis, Rhodotorula rubra, Candida albicans, Hanseniaspora guilliermondii* and *Debaryomyces hansenii* yeasts (Kırbaşlar et al. 2009).

Anticancer Activity

Five compounds, auraptene, marmin, tangeretin, nobiretin and 5-[(6′,7′-dihydroxy-3′,7′-dimethyl-2-octenyl)oxy]psoralen isolated from *Citrus aurantium,* all showed a cell-growth inhibitory effect against mouse leukemia L1210 and K562 in-vitro (Satoh et al. 1996). Two bioactive triterpenoids, isolimonoic acid and ichanexic acid isolated from the ethyl acetate extract of sour orange exhibited significant differential inhibition of human colon cancer cell (HT-29) (Jayaprakasha et al. 2008). Significant arrest of cell growth by isolimonoic acid was observed within 24 h of treatment on the HT-29 colon cancer cells at a concentration as low as 5.0 μM and by ichanexic acid at 10.0 μM. None of the compounds exerted any apparent cytostatic effects on the non-cancerous COS-1 fibroblast cells. Both the compounds exerted nearly 4- to 5-fold increase in the counts of G2/M stage cells at 5.0 μM indicating a potential role in the cell cycle arrest. In subsequent studies, two other bioactive compounds, limonexic acid and β-sitosterol glucoside isolated from the ethyl acetate extract of sour orange exhibited similar significant inhibition of human colon cancer cell (HT-29) (Jayaprakasha et al. 2010). Both compounds caused 4- to 5-fold increases in the counts of G2/M stage cells at 50 μM, indicating a potential role in cell cycle arrest and were not toxic to non-cancerous cells. The findings indicated limonoids and phytosterols to be effective apoptosis-promoting agents and incorporation of enriched fractions of these compounds in the diet may serve to prevent colon cancer.

Nobiletin, a polymethoxylated flavonoid from citrus fruits including sour orange, showed anti-angiogenic activity in-vivo in zebrafish embryos and in-vitro in human umbilical vein endothelial cells (HUVECs) (Lam et al. 2011). In HUVECs, nobiletin inhibited endothelial cell proliferation and, to a greater extent, tube formation in a dose-dependent manner. As in the in vivo study, nobiletin induced G0/G1 cell cycle arrest in HUVECs. However, this arrest was not accompanied by an increase in apoptosis, indicating a cytostatic effect of nobiletin. The results suggested the potential of nobiletin as a cytostatic anti-proliferative agent. Nobiletin, a citrus polymethoxyflavonoid had been reported to suppress tumour growth and metastasis, both of which depend on angiogenesis (Kunimasa et al. 2010). Nobiletin exerted concentration-dependent inhibitory effects on multiple functions of angiogenesis-related endothelial cells (EC); it suppressed the proliferation, migration and tube formation on matrigel of human umbilical vein EC (HUVEC) stimulated with endothelial cell growth supplement (ECGS), a mixture of acidic and basic fibroblast growth factors (FGFs). Nobiletin also suppressed pro-matrix metalloproteinase-2 (proMMP-2) production and MMP-2 mRNA expression in ECGS-stimulated HUVEC. Nobiletin also downregulated cell-associated plasminogen activator (PA) activity and urokinase-type PA mRNA expression. Further, nobiletin inhibited angiogenic differentiation induced by vascular endothelial growth factor and FGF, an in-vitro angiogenesis model. In a chick embryo chorioallantoic membrane assay, nobiletin showed an antiangiogenic activity, the ID_{50} value being 10 μg (24.9 nmol) per egg. These results indicated nobiletin to be a novel antiangiogenic compound that exhibited its activity through combined inhibition of multiple angiogenic endothelial cell functions.

Flavonoids isolated from *Citrus aurantium* exerted antiproliferative effect on human gastric cancer AGS cells (Lee et al. 2012). The flavonoids inhibited cell cycle progression in the G2/M phase, decreased expression level of cell cycle related proteins cyclin B1, cdc 2, cdc 25c and induced apoptosis through activation of caspase and inactivation of PARP. The findings suggested *Citrus aurantium* flavonoids may be useful agent for the chemoprevention of gastric cancer.

Antimutagenic Activity

Citrus aurantium methanol extract exhibited a suppressive effect on umu gene expression of SOS response in *Salmonella typhimurium* TA1535/pSK1002 against the mutagen 2-(2-furyl)-3-(5-nitro-2-furyl)acrylamide (furylfuramide) (Miyazawa et al. 1999). Its dichloromethane fraction also showed a suppressive effect. The suppressive compounds in the dichloromethane fraction were identified as tetra-O-methylscutellarein (1), sinensetin (2), and nobiletin (3). These compounds suppressed the furylfuramide-induced SOS response in the umu test and gene expression was suppressed 67, 45, and 25% at a concentration of 0.6 μmol/mL, respectively. The ID_{50} value of compound 1 was 0.19 μmol/ml. Similar results were obtained with other mutagens in the umu test. In addition, compounds 1–3 exhibited antimutagenic activity in the *S. typhimurium* TA100 Ames test.

Antiinflammatory Activity

Results of studies suggested that the methanol extract of sour orange had antiinflammatory properties by reducing the expression of COX-2, iNOS, and proinflammatory cytokines, such as TNF-α and IL-6, in lipopolysaccharide–stimulated macrophage RAW 264.7 cells via the NF-κB pathway (Kang et al. 2011a). Flavonoids (nobiletin, naringin and hesperidin) isolated from *Citrus aurantium* exhibited antiinflammatory effects by suppressing expression of cyclooxygenase-2 (COX-2), inducible nitric oxide synthase (iNOS) and cytokines by blocking the nuclear factor kappa B (NF-κB) and mitogen activated protein kinase (MAPK) signalling in RAW 264.7 macrophages (Kang et al. 2011b). The flavonoids reduced lipid polysaccharide-induced COX-2, iNOS, TNF-α and IL-6 via NF-κB and MAPK signal pathway. The flavonoids could be used as therapeutic agent for inflammatory diseases.

Hepatoprotective Activity

Administration of sour orange extract to diabetic mice significantly reduced blood glucose levels and activities of glutathione peroxidase (GSHPx), the contents of malon dialdehyde (MDA) and nitric oxide (NO) (Jiao et al. 2007). The levels of glutathione (GSH) were markedly increased and of superoxide dismutase (SOD) in the liver was also increased. The liver histological damages of experimental diabetic mice treated with *C. aurantium* extract were abated in comparison with the experimental diabetic mice. The results suggested that the extract of *C. aurantium* could effectively enhance the liver antioxidant function and decrease hepatocyte damages.

Anxiolytic and Antidepressant Activities

Norepinephrine, octopamine and phenethylamine when injected intraperitoneally elicited an "antidepressant" effect that accorded with adrenergic and dopaminergic stimulations (Bulach et al. 1984). Results of studies suggested that p-synephrine elicited an antidepressant-like activity in mouse models of immobility tests using in the tail suspension and the forced swimming tests, through the stimulation of α1 adrenoceptors (Song et al. 1996). Further studies suggested that S-(+)-p-synephrine had more effective antidepressant-like activity than R-(−)-p-synephrine (Kim et al. 2001).

In another study, sour orange peel essential oil (0.5 g/kg) increased the latency period of tonic seizures in the convulsing experimental models (elevated plus maze – EPM) and anticonvulsant activity (induced by pentylenetetrazole – PTZ or by maximal electroshock) (Carvalho-Freitas and Costa 2002). Treatment with 1.0 g/kg increased the sleeping time induced by barbiturates and the time spent in the open arms of the elevated plus maze. Both doses used, did not promote deficits in general activity or motor coordination.

The hexane and dichloromethane fractions (1.0 g/kg) did not interfere in the epileptic seizures, but were able to enhance the sleeping time induced by barbiturates. The results obtained with the essential oil in the anxiety model, and with essential oil, hexane and dichloromethane fractions in the sedation model, accorded with the ethnopharmacological use of *Citrus aurantium*. Pultrini Ade et al. (2006) used two other experimental models: the light–dark box and the marble-burying test, respectively related to generalized anxiety disorder and to obsessive compulsive disorder to evaluate the anxiolytic activity of *C. aurantium* essential oil in mice. In the light–dark box test, single treatment with the essential oil increased the time spent by mice in the light chamber and the number of transitions between the two compartments. Single and repeated treatments with essential oil were able to suppress marble-burying behavior. At effective doses in the behavioral tests, mice showed no impairment on rotarod procedure after both single and repeated treatments with essential oil, denoting absence of motor deficit. Results observed in marble-burying test, related to obsessive compulsive disorder, appeared more consistent than those observed in light–dark box. In another study, *Citrus aurantium* essential oil at the concentration of 2.5% increased both the time of the animals in the open arms of the elevated plus maze and the time of active social interaction in the open-field being longer than that of the diazepam group (1.5 mg/kg i.p) (Leite et al. 2008). The decrease in the level of emotionality of the animals observed in the two experimental models suggested a possible central action, which was in accord with the phytochemical profile of the oil under study, since it showed the presence of limonene (96.24%) and myrcene (2.24%), components with a well-known depressant activity on the central nervous system.

Central Nervous System Activity

Octopamine (50–250 μg ivc) was found to activate both noradrenergic and dopaminergic system of the rat (Jagiello-Wojtowicz 1979). The mechanism of octopamine, phenylethylamine and adrenaline central nervous system activity was found to be related to stimulation of noradrenergic and dopaminergic systems (Jagiełło-Wójtowicz 1983). The central effect of octopamine and phenylethylamine seems to be related to the interaction with specific binding sites. The behavioral studies in rats showed that octopamine (50, 100 and 250 μg/rat ivc) significantly increased locomotor activity at all doses tested (Jagiello-Wojtowicz and Chodkowska 1984). In rats pretreated with GABA (γ-aminobutyric acid)-ergic drugs such muscimol (0.5 mg/kg sc) or baclofen (2.5 mg/kg ip), octopamine did not produce hyperactivity. Biochemical studies showed that octopamine decreased the cerebral concentration of GABA and reduced activity of generalized anxiety disorder in rat's brain. Octopamine depressed the noradrenaline level in the rat brain and increased utilization of the amine, but did not affect the level and utilization of dopamine. Aging studies by David et al. (1989) established that that p-octopamine and catecholamine metabolisms may have some independent steps and, that p-octopamine may have a role in the normal activity of the brain. Chance et al. (1985) demonstrated that infusing large amounts of the trace amine octopamine depressed behavior in the rat and that the depression was closely associated with depletion of stores of norepinephrine in the brain.

Ortho-, meta-, and para-octopamine failed to show significant β-adrenergic activity in rats as assessed by initiation of thirst and by increase in tail skin temperature (Fregly et al. 1979). All three isomers increased mean blood pressure in pentolinium-blocked rats. Of the three isomers, m-octopamine possessed the greatest α-adrenergic activity. When the responses were compared with those induced by L-norepinephrine, the order of activities was: 1:0.01:0.0005:0.0007 for norepinephrine, m-, p- and o-octopamine, respectively. Thus, DL-m-octopamine had about 1/100th the α-adrenergic activity of L-norepinephrine. The vascular reactivity potencies of the three meta-, para- and ortho-octopamine isomers were: 0.7500, 0.0075 and 0.0038 times that of phenylephrine respectively when tested in-vitro in aortic smooth muscle of the rat (Ress et al. 1980). Thus, of the three isomers, DL-m-octopamine had the greatest α-adrenergic activity in-vitro as it did in-vivo in earlier studies.

Cardiovascular and Hemodynamics Activity

Similar to the well-known cardiac stimulants dopamine (DA), dobutamine (DB), crude extract of *Citrus aurantium* and its active components were found to have inotropic action by increasing myocardial contractility in dogs (Chen et al. 1980). Their anti-shock effect was attributed to their pump-stimulant action and elevation of peripheral vascular resistance. As to the relative efficiency on peripheral vascular resistance, N-methyltyramine was weaker than crude extract of *Citrus aurantium* and synephrine. As to the heart rate stimulant action, the reverse was true. Synephrine and N-methyltyrosamine (chemicals found in immature *C. aurantium* fruit) were shown to be effective antishock (i.e., primarily cardiotonic and vasoconstrictive) agents in dogs (Zhao et al. 1989). In one study, 48 of 50 children with infective shock were cured when treated with synthetic synephrine and N-methyltyrosamine (1.66–24 mg/kg) (Zhao et al. 1989).

Varma and Chemtob (1993) reported that tyramine caused concentration-dependent relaxation of rat aorta, which was endothelium independent and was not exerted via α 1-adrenoceptors (AR), α 2AR, β 1AR, β 2AR, or receptors for 5-hydroxytryptamine, histamine, and adenosine. Further, the scientists concluded that tyramine and several other phenylethylamines produced relaxation of rat aorta, which did not involve any of the known adrenoceptors but may be exerted via novel tyramine receptors (Varma et al. 1995).

Both synephrine and *C. aurantium* extract were found to raise blood pressure in animal studies (Huang et al. 1995, 2001b; Calapai et al. 1999). Unripe fruit infusion of *C. aurantium* when injected into portal hypertensive rats dose-dependently reduced portal pressure in portal vein ligation and sham rats, with the percentage change in portal pressure more pronounced in portal vein ligation rats and dose-dependently elevated mean arterial pressure (Huang et al. 1995). Synephrine also dose-dependently reduced portal pressure and elevated mean arterial pressure in PVL and Sham rats. *Citrus aurantium* (2.8–280 µg/ml) induced dose-dependent contractile responses mainly in aorta and mesenteric artery, but little response in portal vein. The results showed that *Citrus aurantium* infusion reduced portal pressure, possibly by way of arterial vasoconstriction. In a subsequent study, synephrine, a sympathomimetic α1-adrenoceptor agonist, significantly ameliorated the hyperdynamic state in both portal vein ligation (PVL) or bile duct ligation (BDL) rats (Huang et al. 2001b). The portal venous pressure in PVL and BDL rats (−13.5% and −10.1%, respectively), portal tributary blood flow (−19.5% and −20.4%) and cardiac index (−12.1% and −18.8%) were significantly reduced, while mean arterial pressure (10.4% and 23.4%) and systemic (26.3% and 51.0%) as well as portal territory (47.1% and 67.7%) vascular resistance were enhanced by treatment of synephrine as compared with vehicle treatment. The results showed that 8-day administration of synephrine exerted beneficial hemodynamic effects in two models of portal hypertensive rats. In a prospective, randomized, double-blind, placebo-controlled, crossover study involving 15 young, healthy, adult subjects, results showed that the systolic blood pressure, diastolic blood pressure and heart rate were higher for up to 5 hours after a single dose of bitter orange (a 900 mg dietary supplement extract standardized to 6% synephrine) versus placebo in young, healthy adults (Bui et al. 2006).

Antiobesity Activity

An extensive search of MEDLINE, EMBASE, BIOSIS, and the Cochrane Collaboration Database identified only 1 eligible randomized placebo controlled trial, which followed 20 patients for 6 weeks, demonstrated no statistically significant benefit for weight loss, and provided limited information about the safety of *Citrus aurantium* (Bent et al. 2004). In their review, Fugh-Berman and Myers (2004) asserted that there was little evidence that products containing *C. aurantium* and its synephrine alkaloids were effective in weight loss. Synephrine had lipolytic effects in human adipocytes and only at high doses, but not octopamine. They added that although no adverse events had been associated with intake of *C. aurantium* products thus far, synephrine increased blood pressure in humans and animals, and had the potential to increase cardiovascular events. Additionally, *C. aurantium* contained 6′,7′-dihydroxybergamottin and bergapten, inhibitors of cytochrome P450-3A, and thus would be expected to increase serum levels of many drugs. In 2006, Haaz et al. asserted in their review that although some evidence appeared promising, there was a need for larger and more rigorous clinical trials to draw adequate conclusions regarding the safety and efficacy of *C. aurantium* and synephrine alkaloids for promoting weight loss. In a recent review, Stohs et al. (2011b) asserted that based on current knowledge on human, animal and in vitro assessments as well as receptor binding and mechanistic studies, the use of bitter orange extract and p-synephrine appeared to be exceedingly safe with no serious adverse effects being directly attributable to these ingredients. They stated that millions of doses of dietary supplements containing *C. aurantium* extract have been consumed by possibly millions of individuals in recent years without any reported adverse effects. Further, they added that millions of people consume on a daily basis various juices and food products from *Citrus* species that contain p-synephrine.

Of 42 *Citrus* species, the extract of *Citrus aurantium* exhibited the highest inhibitory activity against lipoxygenase (IC_{50} = 56 µg/ml) (Nogata et al. 1996). Studies found that octopamine stimulated lipolysis through β(3)-rather than β(1)-or β(2)-AR activation in white adipocytes from different mammalian species (Galitzky et al. 1993; Carpéné et al. 1999). Further studies found that octopamine was fully lipolytic in garden dormouse and Siberian hamster while tyramine was ineffective (Fontana et al. 2000). Octopamine reduced insulin-dependent glucose transport in rat fat cells, a response also observed with noradrenaline and selective β(3)-AR agonists but not with β(1)-or β(2)-agonists. Human adipocytes, which endogenously express a high level of α(2)-ARs, exhibited a clear α(2)-adrenergic antilipolytic response to adrenaline but not to octopamine. Thus, octopamine could be considered as an endogenous selective β(3)-AR agonist. Studies found that besides stimulating fat cell lipolysis in mammals via activation of β3-adrenoceptors, octopamine exerted, at millimolar dose, dual effect on glucose transport in adipocytes: counteracting insulin action via β3-AR activation and stimulating basal transport via its oxidation by monoamine oxidase or semicarbazide-sensitive amine oxidase (Visentin et al. 2001). Separate studies showed that octopamine could reduce body weight gain in obese Zucker rats, without apparent adverse effects, but with less efficacy than β3-AR agonists (Bour et al. 2003). The lipolytic responses to isoprenaline or octopamine and the stimulation of glucose transport by insulin or by the amine oxidase substrate tyramine were unmodified by the treatments. Additionally, the elevated plasma insulin of obese rats was lowered by octopamine.

Recent studies reported that in rat, octopamine was slightly more lipolytic than synephrine, while tyramine and N-methyl tyramine did not stimulate-and even suppressed lipolysis (Mercader et al. 2011). In human adipocytes, none of these amines were lypolytic when tested up to 10 µg/ml. At higher doses (≥100 µg/ml), tyramine and N-methyl tyramine induced only 20% of the maximal lipolysis and displayed antilipolytic properties. Synephrine and octopamine were partially stimulatory at high doses. Since synephrine is more abundant than octopamine in *C. aurantium*, it should be the main principle responsible for the putative lipolytic action of the extracts claimed to mitigate obesity. Also, the common isopropyl derivative, isopropylnorsynephrine (also named isopropyloctopamine or betaphrine), was clearly lipolytic: active at 1 µg/ml and exerting more than 60% of isoprenaline maximal effect in human adipocytes. The authors asserted that this compound, not detected in *C. aurantium*, and which had few reported adverse effects to date, might be useful for in-vivo triglyceride breakdown.

Both isolated synephrine and *C. aurantium* extract have been shown to raise blood pressure in animal studies (Huang et al. 1995, 2001b; Calapai et al. 1999). *C. aurantium* administration to rats significantly reduced food intake and body weight gain (Calapai et al. 1999). Arterial blood pressure was not modified, but electrocardiogram (ECG) alterations (ventricular arrhythmias) were evident in animals treated with both extracts. The data indicated that, in the rat, antiobesity effects of *C. aurantium* were accompanied by toxic effects probably due to cardiovascular toxicity. These data suggested that compounds containing *C. aurantium* may be harmful to individuals with cardiovascular conditions such as hypertension or dysrhythmias. Studies in humans, however, are necessary to confirm or refute such speculation (Penzak et al. 2001). In a cross-over study of normotensive adults, hemodynamics (heart rate; systolic, diastolic, and mean arterial pressure) did not differ significantly between groups that ingested Seville orange juice extract containing significant amount of synephrine, or water (Penzak et al. 2001).

Colker et al. (1999) reported that in a double-masked, randomized, placebo-controlled study, receiving a combination of *Citrus aurantium*, caffeine and St John's Wort, lost significant amounts of total body weight while on a strict diet and exercise. Those in the placebo and control groups who also were on the same restricted diet did not. However, intergroup analysis showed no statistical significance among the weight changes in the three groups. In contrast, the loss of fat mass in the test group was significantly greater compared to the placebo and control groups. No significant changes in blood pressure, heart rate, electrocardiographic findings, serum chemistries, or urinalysis findings were observed in any of the groups. Based on these results, it was concluded that the combination of *C aurantium* extract, caffeine, and St. John's Wort was safe and effective when combined with mild caloric restriction and exercise for promoting both body weight and fat loss in healthy overweight adults. Gougeon et al. (2005) found that thermic effect of food (TEF) was significantly lower in women than men independent of age and magnitude of adiposity. The thermic response to adrenergic amines extracted from *Citrus aurantium* (CA) alone was higher in men, but, when added to the meal, CA increased TEF only in women and to values no longer different from men. CA had no effect on blood pressure and pulse rate but increased epinephrine excretion by 2.4-fold. However, this acute response may not translate into a chronic effect or a clinically significant weight loss over time.

Stohs et al. (2011a) in a double-blinded, randomized, placebo-controlled protocol with 10 subjects per treatment group found that oral ingestion of bitter orange p-synephrine alone or in combination with flavonoids naringin and hesperidin elicited a thermogenic effect, increasing metabolic rates but without corresponding elevations in blood pressure and heart-rates. None of the treatment groups exhibited changes in heart rate or blood pressure relative to the control group, nor there were no differences in self-reported ratings of 10 symptoms between the treatment groups and the control group. They suggested that longer term studies were warranted to assess its value as a weight control agent.

Antiatherogenic Activity

Treatment with naringenin and hesperetin was found to enhance adiponectin transcription in differentiated 3 T3-L1 cells (Liu et al. 2008). Both naringenin and hesperetin induced peroxisome proliferator-activated receptor (PPAR) γ-controlled luciferase expression in a dose-dependent manner (20–160 μM), whereas only naringenin possessed significant activity to stimulate PPARα. The results suggested the two flavonoids might exert antiatherogenic effects partly through activating PPAR and up-regulating adiponectin expression in adipocytes.

Antiestrogenic Activity

Oral gavage of immature female rats with *Ephedra sinica* (85.5 and 855.0 mg/kg/day), *C. aurantium* (25.0 and 50.0 mg/kg/day), ephedrine (5.0 mg/kg/day) and p-synephrine (50.0 mg/kg/day) for three consecutive days revealed that all analyzed substances showed an antiestrogenic potential, but only ephedrine at 0.5 mg/kg/day presented a significant antiestrogenic effect (Arbo et al. 2009a). Adrenals relative mass were reduced in all tested compounds when compared to the control, which appeared to be related to the α-1-adrenoceptor agonist activity, which promoted a vasoconstriction and reduction of the liquid in the organ.

Effects on Cytochrome Enzymes (Drug-Drug Interaction)

Sour orange juice and *Citrus aurantium* crude drugs were found to inhibit human cytochrome P450 3A (CYP3A) as determined by microsomal testosterone 6beta-hydroxylation, whereas sweet orange juices did not (Guo et al. 2001). Thus drugs containing *C. aruantium* may impact on xenobiotics metabolism. Octopamine, but not synephrine, was found to inhibit cytochrome P450c11 in-vitro (Louw et al. 2000).

Toxicity Studies

Analysis of samples of unripe fruits and leaves from *Citrus aurantium, C. sinensis, C. deliciosa, C. limon* and *C. limonia* revealed the presence of p-synephrine in amounts that range from 0.012% to 0.099% in the unripe fruits and 0.029–0.438% in the leaves (Arbo et al. 2008). Acute oral administration of *C. aurantium* extracts (2.5% p-synephrine, 300–5,000 mg/kg) in mice decreased locomotor activity while p-synephrine (150–2,000 mg/kg) produced piloerection, gasping, salivation, exophtalmia and reduction in locomotor activity. The toxic effects observed seem to be related with adrenergic stimulation and should alert for possible side effects of p-synephrine and *C. aurantium*. Oral gavage of mice with a commercial *C. aurantium* dried extract (containing 7.5% p-synephrine) 400, 2,000 or 4,000 mg/kg and p-synephrine 30 or 300 mg/kg revealed a reduction in body weight gain of animals treated with both doses of p-synephrine (Arbo et al. 2009b). Organs relative weight, biochemical and hematological parameters were not altered in all treated mice. There was an increase in reduced glutathione (GSH) concentration in groups treated with *C. aurantium* 4,000 mg/kg and p-synephrine 30 and 300 mg/kg. In glutathione peroxidase (GPx), there was an inhibition of the activity in *C. aurantium* 400 and 2,000 mg/kg and p-synephrine 30 and 300 mg/kg treated animals, respectively, and with no alteration in malondialdehyde (MDA) levels. The results indicated a low subchronic toxicity of the tested materials in mice and a possible alteration in the oxidative metabolism.

Recent studies in Sprague–Dawley rats showed that at doses up to 100 mg synephrine/kg body weight, there were no adverse effects on embryo lethality, fetal weight, or incidences of gross, visceral, or skeletal abnormalities (Hansen et al. 2011). Doses of up to 100 mg synephrine/kg body weight did not produce developmental toxicity in the animals.

Traditional Medicinal Uses

The leaves, flowers, fruit, peel, bark and volatile oil of bitter orange are official in many pharmacopoeias. *Citrus aurantium* is reported to be an expectorant, laxative, hypertensive, nervine, tonic, and diuretic. Bitter orange was used medicinally to treat colds, fevers, hepatic disorders, gall bladder problems, rheumatism, epilepsy, emotional shock, bruising internally and externally, skin blemishes and digestive problems in Haiti (Paul and Cox 1995). The dried, immature fruit of *Citrus aurantium* 'Zhiqiao' in Chinese, has been used to treat cardiovascular diseases in traditional Chinese medicine for centuries (Liu et al. 2008). In Haiti, *C. aurantium* was used medicinally to treat colds, fevers, hepatic disorders, gall bladder problems, rheumatism, epilepsy, emotional shock, internal and external bruising, skin blemishes and digestive problems (Paul and Cox 1995). Bitter orange extract has been added to many dietary supplements and herbal weight loss formulas (as an alternative to ephedra).

Bitter orange peel is used in many pharmacopoeial preparations as a flavoring agent, stomachic, carminative, tonic, astringent and emmenagogue. The dried rind is useful in a tonic for dyspepsia and general debility and to check vomiting. Fresh peel is rubbed on the face to treat acne and body to treat eczema. Bitter orange juice is antiseptic, antibilious and hemostatic. Africans apply the cut-open orange on ulcers and yaws and areas of the body afflicted with rheumatism. In Italy, Mexico and Central America, leaf decoctions are employed as sudorific, antispasmodic, stimulant, tonic and stomachic. The leaves are applied to reduce swollen legs and prescribed as pectorals and for bronchitis. The peels, leaves and flowers are used as a stomachic and antiscorbutic in the Philippines (Stuart 2010). The flowers, prepared as a syrup, is used as a sedative in nervous disorders, hysterical cases and to induce sleep and also used as an antispasmodic. The bark infusion is used as a tonic, stimulant, febrifuge and vermifuge.

Other Uses

Sour orange has been used extensively as a rootstock for other citrus species like the lemon, grapefruit and especially sweet orange as it produces a well-developed root system and is highly resistant to many important diseases (e.g. gummosis, root rot) and to cold but its use as rootstock had declined because of its susceptibility to tristeza virus disease and verrucosis (scab). The flowers provide exude nectar for honeybees. The wood is hard and strong, fine-grained and is valued for cabinetwork, furniture and turnery.

The flower essential oil of *C. aurantium* is called 'neroli bigarade oil'. The best quality neroli oil -is obtained from Bouquetier cultivars, formerly from southern France and Italy and nowadays from Morocco and Tunisia. Bouquetier cultivars are small, subspineless trees, flowering profusely with very large single flowers. Neroli oil is highly prized in the perfume industry. It is used for high quality *eau de cologne* since the seventeenth century. It is also widely used in blends with other floral oils and absolutes – jasmine oil, lavender oil, rose oil and its absolute, ylang-ylang, vanilla as well as petigrain oil, lemon oil, lime oil, grapefruit oil etc. Significant amounts (up to 25%) of aroma compounds remain in solution in the water left in the still after distillation of *C. aurantium* flowers. The oil obtained by fat and solvent extraction of these compounds is traded as 'orange flower water absolute' and is mainly used in the reconstitution of other essential oils and of the formerly popular 'orange flower water'.

Petitgrain oil is distilled from the leaves, twigs and immature fruits and is called 'petitgrain bigarade oil'. Both Petitgrain and the oil of the ripe peel are of great importance in formulating scents for perfumes and cosmetics. Petitgrain oil is indispensable in fancy *eau-de-cologne*. Petitgrain oils are often used as a substitute for the much more costly neroli oils. Petitgrain water absolute or 'eau-de-brouts' is the equivalent of orange flower water absolute and is obtained as a by-product from petitgrain bigarade oil. It enhances the 'naturalness' of several other fragrances, e.g. jasmine, neroli, ylang-ylang and gardenia. The seed oil is employed in soaps.

In Haiti, the sour orange was also found to be valuable in food preparation, agriculture, construction and voodoo (Paul and Cox 1995). In Guam, the leaves and pulp of sour orange are used as soap for washing clothes and shamppoin the hair. In Zanzibar and Pemba, the fruits are used for scouring floors and brass. The petroleum ether extract from *C. aurantium* fruit showed insecticidal activity against olive fruit fly, *Bactrocera oleae* (Siskos et al. 2007). The bioactivity was limited to the flavedo, and this activity was significantly higher than that of the whole fruit extract. Extracts of ripe fruit were more effective than those of unripe fruit. Silica gel fractionation of bitter orange (*Citrus aurantium*) fruit peel petroleum ether extract yielded a fraction that inflicted up to 96% mortality to adults of the olive fruit fly *Bactrocera oleae*

(Gmelin) 3 days post-treatment (Siskos et al. 2008). Subsequent HPLC purification of the active fraction resulted in the isolation of three components, osthol, bergapten and 6′,7′-epoxybergamottin. 6′,7′-Epoxybergamottin was toxic when tested individually, while bergapten and osthol were found to act synergistically to 6′,7′-epoxybergamottin.

Comments

Citrus aurantium is believed to have originated as a hybrid between the mandarin (*C. reticulata* Blanco) and the pummelo (*C. maxima* (Burm.) Merrill). Its taxonomy at species and at subspecific level is still very confused and may need revision, as is the taxonomy of the entire genus *Citrus* L.

Selected References

Arbo MD, Franco MT, Larentis ER, Garcia SC, Sebben VC, Leal MB, Dallegrave E, Limberger RP (2009a) Screening for in vivo (anti)estrogenic activity of ephedrine and p-synephrine and their natural sources *Ephedra sinica* Stapf. (Ephedraceae) and *Citrus aurantium* L. (Rutaceae) in rats. Arch Toxicol 83(1):95–99

Arbo MD, Larentis ER, Linck VM, Aboy AL, Pimentel AL, Henriques AT, Dallegrave E, Garcia SC, Leal MB, Limberger RP (2008) Concentrations of p-synephrine in fruits and leaves of *Citrus* species (Rutaceae) and the acute toxicity testing of *Citrus aurantium* extract and p-synephrine. Food Chem Toxicol 46(8):2770–2775

Arbo MD, Schmitt GC, Limberger MF, Charão MF, Moro AM, Ribeiro GL, Dallegrave E, Garcia SC, Leal MB, Limberger RP (2009b) Subchronic toxicity of *Citrus aurantium* L. (Rutaceae) extract and p-synephrine in mice. Regul Toxicol Pharmacol 54(2):114–117

Bayer RJ, Mabberley DJ, Morton C, Miller CH, Sharma IK, Pfeil BE, Rich S, Hitchcock R, Sykes S (2009) A molecular phylogeny of the orange subfamily (Rutaceae: Aurantioideae) using nine cpDNA sequences. Am J Bot 96:668–685

Bent S, Padula A, Neuhaus JO (2004) Safety and efficacy of *Citrus aurantium* for weight loss. Am J Cardiol 94(10):1359–1361

Berhow M, Tisserat B, Kanes K, Vandercook C (1998) Survey of phenolic compounds produced in *Citrus*. USDA ARS Tech Bull 1856:1–154

Boelens MH, Jimenez R (1989) The chemical composition of the peel oils from unripe and ripe fruits of bitter orange, *Citrus aurantium* L. ssp. *amara* Engl. Flav Fragr J 4(3):139–142

Bour S, Visentin V, Prevot D, Carpene C (2003) Moderate weight-lowering effect of octopamine treatment in obese Zucker rats. J Physiol Biochem 59:175–182

British Pharmacopoeia Commission (2002) Bitter-orange flower oil. British Pharmacopoeia, vol 1. Stationery Office, London, pp 1267–1269

Bui LT, Nguyen DT, Ambrose PJ (2006) Blood pressure and heart rate effects following a single dose of bitter orange. Ann Pharmacother 40:53–57

Bulach C, Doare L, Massari B, Simon P (1984) Antidepressive effects of 3 endogenous monoamines: psychopharmacologic profiles of noradrenaline, octopamine, and phenethylamine. J Pharmacol 15(1):1–15 (in French)

Calapai G, Firenzuoli F, Saitta A, Squadrito F, Arlotta MR, Costantino G, Inferrera G (1999) Antiobesity and cardiovascular toxic effects of *Citrus aurantium* extracts in the rat: a preliminary report. Fitoterapia 70:586–592

Carnat A, Carnat AP, Fraisse D, Lamaison JL (1999) Standardization of the sour orange flower and leaf. Ann Pharm Fr 57(5):410–414 (In French)

Carpéné C, Galitzky J, Fontana E, Atgié C, Lafontan M, Berlan M (1999) Selective activation of beta3-adrenoceptors by octopamine: comparative studies in mammalian fat cells. Naunyn Schmiedebergs Arch Pharmacol 359(4):310–321

Carvalho-Freitas MI, Costa M (2002) Anxiolytic and sedative effects of extracts and essential oil from *Citrus aurantium* L. Biol Pharm Bull 25(12):1629–1633

Chance WT, Bernardini AP, James JH, Edwards LL, Minnema K, Fischer JE (1985) Behavioral depression after intraventricular infusion of octopamine in rats. Am J Surg 150(5):577–584

Chen X, Huang QX, Zhou TJ, Da HY (1980) Studies of *Citrus autantium* and its hypertensive ingredients on the cardiac functions and hemodynamics in comparison with dopamine and dobutamine. Yao Hsueh Hsueh Pao 15:71–77 (in Chinese)

Chizzali E, Nischang I, Ganzera M (2011) Separation of adrenergic amines in *Citrus aurantium* L. var. *amara* by capillary electrochromatography using a novel monolithic stationary phase. J Sep Sci 34(16–17):2301–2304

Colker CM, Kalman DS, Torina GC, Perlis T, Street C (1999) Effects of *Citrus aurantium* extract, caffeine, and St. John's Wort on body fat loss, lipid levels, and mood states in overweight healthy adults. Curr Ther Res Clin Exp 60(3):145–153

Cottin R (2002) *Citrus* of the world: a *Citrus* directory. Version 2.0. SRA INRA-CIRAD, France

Dandekar DV, Jayaprakasha GK, Patil BS (2008) Simultaneous extraction of bioactive limonoid aglycones and glucoside from *Citrus aurantium* L. using hydrotropy. Z Naturforsch C 63(3–4):176–180

David JC, Coulon JF, Cavoy A, Delacour J (1989) Effects of aging on p- and m-octopamine, catecholamines, and their metabolizing enzymes in the rat. J Neurochem 53(1):149–154

Del-Rio JA, Cruz Arcas M, Benavente O, Sabater F, Ortuno A (1998) Changes of polymethoxylated flavones levels during development of *Citrus aurantium* (cv. Sevillano) fruits. Planta Med 64(6):575–576

Deterre S, Rega B, Delarue J, Decloux M, Lebrun M, Giampaoli P (2011) Identification of key aroma compounds from bitter orange (*Citrus aurantium* L.) products: essential oil and macerate–distillate extract. Flav Fragr J 27(1):77–88

Dugo P, Mondello L, Cogliandro E, Verzera A, Dugo G (1996) On the genuineness of *Citrus* essential oils. 51. Oxygen heterocyclic compounds on bitter orange oil (*Citrus aurantium* L.). J Agric Food Chem 44(2):544–549

Facciola S (1990) Cornucopia. A source book of edible plants. Kampong Publication, Vista, 677 pp

Fontana E, Morin N, Prévot D, Carpéné C (2000) Effects of octopamine on lipolysis, glucose transport and amine oxidation in mammalian fat cells. Comp Biochem Physiol C Pharmacol Toxicol Endocrinol 125(1):33–44

Foundation for Revitalisation of Local Health Traditions (2008) FRLHT database. http://envis.frlht.org

Fregly MJ, Kelleher DL, Williams CM (1979) Adrenergic activity of ortho-, meta-, and para-octopamine. Pharmacology 18(4):180–187

Fugh-Berman A, Myers A (2004) *Citrus aurantium*, an ingredient of dietary supplements marketed for weight loss: current status of clinical and basic research. Exp Biol Med (Maywood) 229(8):698–704

Galitzky J, Carpene C, Lafontan M, Berlan M (1993) Specific stimulation of adipose tissue adrenergic beta 3 receptors by octopamine. C R Acad Sci III 316(5):519–523

Gougeon R, Harrigan K, Tremblay JF, Hedrei P, Lamarche M, Morais JA (2005) Increase in the thermic effect of food in women by adrenergic amines extracted from *Citrus aurantium*. Obes Res 13(7):1187–1194

Guo LQ, Taniguchi M, Chen QY, Baba K, Yamazoe Y (2001) Inhibitory potential of herbal medicines on human cytochrome P450-mediated oxidation: Properties of Umbelliferous or *Citrus* crude drugs and their relative prescriptions. Jpn J Pharmacol 85(4): 399–408

Haaz S, Fontaine KR, Cutter G, Limdi N, Perumean-Chaney S, Allison DB (2006) *Citrus aurantium* and synephrine alkaloids in the treatment of overweight and obesity: an update. Obes Rev 7(1):79–88

Hansen DK, Juliar BE, White GE, Pellicore LS (2011) Developmental toxicity of *Citrus aurantium* in rats. Birth Defects Res B Dev Reprod Toxicol 92(3): 216–223

Hattori S, Shimokoriyama M, Kanao M (1952) Studies on flavanone glycosides. IV. The glycosides of ripe fruit peel and flower petals of *Citrus aurantium* L. J Am Chem Soc 74:3614–3615

He XG, Lian LZ, Lin LZ, Bernart MW (1997) High-performance liquid chromatography-electrospray mass spectrometry in phytochemical analysis of sour orange (*Citrus aurantium* L.). J Chromatogr A 791 (1–2):127–134

Hodgson RW (1967) Horticultural varieties of *Citrus*. In: Reuther W, Webber HJ, Batchelor LD (eds) The *Citrus* industry, vol 1, History, world distribution, botany, and varieties. Revised edition. University of California Press, Berkeley, pp 431–591

Huang YT, Wang GF, Chen CF, Chen CC, Hong CY, Yang MC (1995) Fructus aurantii reduced portal pressure in portal hypertensive rats. Life Sci 57(22):2011–2020

Huang S, Hu S, Shi J, Yang Y (2001a) Studies on chemical constituents from the flower of *Citrus aurantium*. Zhong Yao Cai 24(12):865–867 (in Chinese)

Huang YT, Lin HC, Chang Y-Y, Yan YY, Lee SD, Hong CY (2001b) Hemodynamic effects of synephrine treatment in portal hypertensive rats. Jpn J Pharmacol 85(2):183–188

Jagiello-Wojtowicz E (1979) Mechanism of central action of octopamine. Pol J Pharmacol Pharm 31(5): 509–516

Jagiełło-Wójtowicz E (1983) Comparison of central actions of intraventricularly administered octopamine, phenyl-ethylamine and adrenaline in rats. Pol J Pharmacol Pharm 35(1):59–62

Jagiello-Wojtowicz E, Chodkowska A (1984) Effects of octopamine on GABA-ergic transmission in rats. Pol J Pharmacol Pharm 36(6):595–601

Jayaprakasha GK, Mandadi KK, Poulose SM, Jadegoud Y, Nagana Gowda GA, Patil BS (2008) Novel triterpenoid from *Citrus aurantium* L. possesses chemopreventive properties against human colon cancer cells. Bioorg Med Chem 16(11):5939–5951

Jayaprakasha GK, Jadegoud Y, Nagana Gowda GA, Patil BS (2010) Bioactive compounds from sour orange inhibit colon cancer cell proliferation and induce cell cycle arrest. J Agric Food Chem 58(1):180–186

Jiao S, Huang C, Wang H, Yu S (2007) Effects of *Citrus aurantium* extract on liver antioxidant defense function in experimental diabetic mouse. Wei Sheng Yan Jiu 36(6):689–692 (in Chinese)

Kang SR, Han DY, Park KI, Park HS, Cho YB, Lee HJ, Lee WS, Ryu CH, Ha YL, Lee do H, Kim JA, Kim GS (2011a) Suppressive effect on lipopolysaccharide-induced proinflammatory mediators by *Citrus aurantium* L. in macrophage raw 264.7 cells via NF-κB signal pathway. Evid Based Complement Alternat Med 2011: 248592

Kang SR, Park KI, Park HS, Lee DH, Kim JA, Nagappan A, Kim EH, Lee WS, Shin SC, Park MK, Han DY, Kim GS (2011b) Anti-inflammatory effect of flavonoids isolated from Korea *Citrus aurantium* L. on lipopolysaccharide-induced mouse macrophage RAW 264.7 cells by blocking of nuclear factor-kappa B (NF-κB) and mitogen-activated protein kinase (MAPK) signalling pathways. Food Chem 129(4):11721–1728

Kim KW, Kim HD, Jung JS, Woo RS, Kim HS, Suh HW, Kim YH, Song DK (2001) Characterization of antidepressant-like effects of p-synephrine stereoisomers.

Naunyn Schmiedebergs Arch Pharmacol 364(1): 21–26

Kırbaşlar FG, Kırbaşlar Sİ (2003) Composition of cold-pressed bitter orange peel oil from Turkey. J Essent Oil Res 15:6–9

Kırbaşlar FG, Tavman A, Dülger B, Türker G (2009) Antimicrobial activity of Turkish *Citrus* peel oils. Pak J Bot 41(6):3207–3212

Kunimasa K, Ikekita M, Sato M, Ohta T, Yamori Y, Ikeda M, Kuranuki S, Oikawa T (2010) Nobiletin, a *Citrus* polymethoxyflavonoid, suppresses multiple angiogenesis-related endothelial cell functions and angiogenesis in vivo. Cancer Sci 101(11):2462–2469

Lam KH, Alex D, Lam IK, Tsui SK, Yang ZF, Lee SM (2011) Nobiletin, a polymethoxylated flavonoid from *Citrus*, shows anti-angiogenic activity in a zebrafish in vivo model and HUVEC in vitro model. J Cell Biochem 112(11):3313–3321

Lee DH, Park KI, Park HS, Kang SR, Nagappan A, Kim JA, Kim EH, Lee WS, Hah YS, Chung HJ, An SJ, Kim GS (2012) Flavonoids Isolated from Korea *Citrus aurantium* L. induce G2/M phase arrest and apoptosis in human gastric cancer AGS cells. Evid Based Complement Alternat Med 2012:515901

Leite MP, Fassin J Jr, Baziloni EMF, Almeida RN, Mattei R, Leite JR (2008) Behavioral effects of essential oil of *Citrus aurantium* L. inhalation in rats. Rev Bras Farmacogn 18:661–666

Lin ZK, Hua YF, Gu YH (1986) Chemical constituents of the essential oil from the flowers, leaves and peel of *Citrus aurantium* (Chin.). Acta Botanica Sinica 28(6):635–640

Liu L, Shan S, Zhang K, Ning ZQ, Lu XP, Cheng YY (2008) Naringenin and hesperetin, two flavonoids derived from *Citrus aurantium* up-regulate transcription of adiponectin. Phytother Res 22(10): 1400–1403

Louw A, Allie F, Swart AC, Swart P (2000) Inhibition of cytochrome P450c11 by biogenic amines and an aziridine precursor, 2-(4-acetoxyphenyl)-2-chloro-N-methyl-ethylammonium chloride. Endocr Res 26(4): 729–736

Mabberley DJ (1997) A classification for edible *Citrus*. Telopea 7(2):167–172

Manner HI, Buker RS, Smith VE, Elevitch CR (2006). *Citrus* species. Ver. 2.1. In: Elevitch CR (ed) Species profiles for Pacific Island Agroforestry. Permanent Agriculture Resources (PAR), Holualoa, Hawaii. http://www.agroforestry.net

Mercader J, Wanecq E, Chen J, Carpéné C (2011) Isopropylnorsynephrine is a stronger lipolytic agent in human adipocytes than synephrine and other amines present in *Citrus aurantium*. J Physiol Biochem 67(3):443–452

Miyazawa M, Okuno Y, Fukuyama M, Nakamura S, Kosaka H (1999) Antimutagenic activity of polymethoxyflavonoids from *Citrus aurantium*. J Agric Food Chem 47(12):5239–5244

Morton JF (1987) Sour orange. In: Fruits of warm climates. Julia F. Morton, Miami, pp 130–133

Moufida S, Marzouk B (2003) Biochemical characterization of blood orange, sweet orange, lemon, bergamot and bitter orange. Phytochemistry 62(8):1283–1289

Nelson BC, Putzbach K, Sharpless KE, Sander LC (2007) Mass spectrometric determination of the predominant adrenergic protoalkaloids in bitter orange (*Citrus aurantium*). J Agric Food Chem 55(24):9769–9775

Njoroge SM, Ukeda H, Sawamura M (2003) Changes of the volatile profile and artifact formation in Daidai (*Citrus aurantium*) cold-pressed peel oil on storage. J Agric Food Chem 51(14):4029–4035

Nogata Y, Yoza KI, Kusumoto KI, Kohyama N, Sekiya K, Ohta H (1996) Screening for inhibitory activity of *Citrus* fruit extracts against platelet cyclooxygenase and lipoxygenase. J Agric Food Chem 44(3):725–729

Ochse JJ, Soule MJ Jr, Dijkman MJ, Wehlburg C (1961) Tropical and subtropical agriculture. 2 vols. Macmillan, New York, 1446 pp

Oyen LPA, Jansen PCM (1999) *Citrus aurantium* L. cv. group Bouquetier. In: Oyen LPA, Nguyen XD (eds) Plant resources of South-East Asia. No. 19: essential-oils plants. Prosea Foundation, Bogor, pp 78–83

Paul A, Cox PA (1995) An Ethnobotanical survey of the uses for *Citrus aurantium* (Rutaceae) in Haiti. Econ Bot 49(3):249–256

Pellati F, Benvenuti S, Melegari M, Firenzuoli F (2002) Determination of adrenergic agonists from extracts and herbal products of *Citrus aurantium* L. var. *amara* by LC. J Pharm Biomed Anal 29(6):1113–1119

Pellati F, Benvenuti S, Melegari M (2004) High-performance liquid chromatography methods for the analysis of adrenergic amines and flavanones in *Citrus aurantium* L. var. *amara*. Phytochem Anal 15(4):220–225

Pellati F, Benvenuti S, Melegari M (2005) Enantioselective LC analysis of synephrine in natural products on a protein-based chiral stationary phase. J Pharm Biomed Anal 37:839–849

Penzak SR, Jann MW, Cold JA, Hon YY, Desai HD, Gurley BJ (2001) Seville (sour) orange juice: synephrine content and cardiovascular effects in normotensive adults. J Clin Pharmacol 41:1059–1063

Peterson JJ, Dwyer JT, Beecher GR, Bhagwat SA, Gebhardt SE, Haytowitz DB, Holden JM (2006) Flavanones in oranges, tangerines (mandarins), tangors, and tangelos: a compilation and review of the data from the analytical literature. J Food Comp Anal 19:S66–S73

Porcher MH et al (1995–2020) Searchable World Wide Web multilingual multiscript plant name database. Published by The University of Melbourne, Australia. http://www.plantnames.unimelb.edu.au/Sorting/Frontpage.html

Preuss HG, DiFerdinando D, Bagchi M, Bagchi D (2002) *Citrus aurantium* as a thermogenic, weight-reduction replacement for ephedra: an overview. J Med 33(1–4):247–264

Pultrini Ade M, Galindo LA, Costa M (2006) Effects of the essential oil from *Citrus aurantium* L. in experimental anxiety models in mice. Life Sci 78(15):1720–1725

Purseglove JW (1968) Tropical crops: Dicotyledons. *1 & 2*. Longman, London, 719 pp

Quintero A, de Gonzalez CN, Sanchez F, Usubillaga A, Rojas L, Szoke E, Mathe I, Blunden G, Kery A (2003) Constituents and biological activity of *Citrus aurantium amara* L. essential oil. Acta Horticult (597):115–117

Ramadan W, Mourad B, Ibrahim S, Sonbol F (1996) Oil of bitter orange: new topical antifungal agent. Int J Dermatol 35(6):448–449

Ress RJ, Rahmani MA, Fregly MJ, Field FP, Williams CM (1980) Effect of isomers of octopamine on in vitro reactivity of vascular smooth muscle of rats. Pharmacology 21(5):342–347

Roman MC, Betz JM, Hildreth J (2007) Determination of synephrine in bitter orange raw materials, extracts, and dietary supplements by liquid chromatography with ultraviolet detection: single-laboratory validation. J AOAC 90(1):68–81

Sander LC, Putzbach K, Nelson BC, Rimmer CA, Bedner M, Thomas JB, Porter BJ, Wood LJ, Schantz MM, Murphy KE, Sharpless KE, Wise SA, Yen JH, Siitonen PH, Evans RL, Nguyen Pho A, Roman MC, Betz JM (2008) Certification of standard reference materials containing bitter orange. Anal Bioanal Chem 391(6): 2023–2034

Sarin PS, Seshadri TR (1960) New components of *Citrus aurantium*. Tetrahedron 8(1–2):64–66

Satoh Y, Tashiro S, Satoh M, Fujimoto Y, Xu JY, Ikekawa T (1996) Studies on the bioactive constituents of Aurantii Fructus Immaturus. Yakugaku Zasshi 116(3):244–250 (in Japanese)

Schneider G, Unkrich G, Pfaender P (1968) Methoxylated flavones of *Citrus aurantium* L. subspecies *amara* L. Arch Pharm Ber Dtsch Pharm Ges 301(10):785–792 (in German)

Siskos EP, Konstantopoulou MA, Mazomenos BE, Jervis M (2007) Insecticidal activity of *Citrus aurantium* fruit, leaf, and shoot extracts against adult olive fruit flies (Diptera: Tephritidae). J Econ Entomol 100(4):1215–1220

Siskos EP, Mazomenos BE, Konstantopoulou MA (2008) Isolation and identification of insecticidal components from *Citrus aurantium* fruit peel extract. J Agric Food Chem 56(14):5577–5581

Song DK, Suh HW, Jung JS, Wie MB, Son KH, Kim YH (1996) Antidepressant-like effects of p-synephrine in mouse models of immobility tests. Neurosci Lett 23(214):107–110

Stohs SJ, Preuss HG, Keith SC, Keith PL, Miller H, Kaats GR (2011a) Effects of p-synephrine alone and in combination with selected bioflavonoids on resting metabolism, blood pressure, heart rate and self-reported mood changes. Int J Med Sci 8(4):295–301

Stohs SJ, Preuss HG, Shara M (2011b) The safety of *Citrus aurantium* (bitter orange) and its primary protoalkaloid p-synephrine. Phytother Res 25(10):1421–1428

Stuart GU (2010) Philippine alternative medicine. Manual of some Philippine medicinal plants. http://www.stuartxchange.org/OtherHerbals.html

Suryawanshi JAS (2011) An overview of *Citrus aurantium* used in treatment of various diseases. Afr J Plant Sci 5(7):390–395

Swingle WT (1946) The botany of *Citrus* and its wild relatives of the orange subfamily (family Rutaceae, subfamily Aurantioideae). In: Webber HJ, Batchelor LD (eds) The *Citrus* industry, vol 1. History, botany and breeding. University of California Press, Berkeley, pp 129–474, 1028 pp

Swingle WT, Reece PC (1967) The botany of *Citrus* and its wild relatives. In: Reuther W, Webber HJ, Batchelor LD (eds) The *Citrus* industry, vol 1, History, world distribution, botany, and varieties. Revised edition. University of California Press, Berkeley, pp 190–430

Tanaka T (1954) Species problem in *Citrus*. A critical study of wild and cultivated units of *Citrus*, based upon field studies in their native homes. Japanese Society for the Promotion of Science, Ueno, 152 pp

Varma DR, Chemtob S (1993) Endothelium- and beta-2 adrenoceptor-independent relaxation of rat aorta by tyramine and certain other phenylethylamines. J Pharmacol Exp Ther 265(3):1096–1104

Varma DR, Deng XF, Chemtob S, Nantel F, Bouvier M (1995) Characterization of the vasorelaxant activity of tyramine and other phenylethylamines in rat aorta. Can J Physiol Pharmacol 73(6):742–746

Visentin V, Morin N, Fontana E, Prévot D, Boucher J, Castan I, Valet P, Grujic D, Carpéné C (2001) Dual action of octopamine on glucose transport into adipocytes: inhibition via beta3-adrenoceptor activation and stimulation via oxidation by amine oxidases. J Pharmacol Exp Ther 299(1):96–104

Webber HJ (1946) Cultivated varieties of *Citrus*. In: Webber HJ, Batchelor LD (eds) The *Citrus* industry, 1. History, botany and breeding. University of California Press, Berkeley, pp 475–668, 1028 pp

Wu FJ, Sheu SJ (1992) Analysis and processing of Chinese herbal drugs (XII) The study of Fructus Aurantii Immaturus. Chin Pharm J 44:257–263

Yoshimura K, Trier G (1912) Weitere beiträge über das vorkommen von betainen im pflanzenreich. Z Physiol Chem 77:290–302 (In German)

Zhang DX, Mabberley DJ (2008) *Citrus* Linnaeus. In: Wu ZY, Raven PH, Hong DY (eds) Flora of China, vol 11 (Oxalidaceae through Aceraceae). Science Press/Missouri Botanical Garden Press, Beijing/St. Louis

Zhao XW, Li JX, Zhu ZR, Sun DQ, Liu SC (1989) Anti-shock effects of synthetic effective compositions of Fructus aurantii immaturus. Experimental study and clinical observation. Chin Med J 102(2): 91–93

Citrus x aurantium Sweet Orange Group

Scientific Name

Citrus x *aurantium* L. pro sp. [Sweet Orange Group] (sensu Mabberley 1997; Bayer et al. 2009).

Synonyms

Aurantium sinense (L.) Mill., *Citrus aurantium* Lour. non L., *Citrus aurantium* β *sinensis* L., *Citrus aurantium* subsp. *sinensis* Engl., *Citrus aurantium* L. subsp. *sinensis* (L.) P. Fourn., *Citrus aurantium* Hook.f. var. *aurantium*, *Citrus aurantium* var. *sinensis* Thell., *Citrus macracantha* Hassk, *Citrus sinensis* (L.) Osbeck (sensu Swingle and Reece 1967; sensu Tanaka sec. Cottin 2002), *Citrus* × *aurantium* L. var. *sinensis* L.

Family

Rutaceae

Common/English Names

Blood Orange, China Orange, Citrange, Coolie Orange, Navel Orange, Orange, Portugal Orange, Shamuti Orange, Sweet Orange, Tight-Skinned Orange, Valencia Orange.

Vernacular Names

Albanian: Nerënxë, Portokall;
Amharic: Bertukan, Birtukan, Oranje;
Armenian: Narinch, Narinjh, Narenjhatsaghik;
Arabic: Al-Burtuqal, Burtuqal, Narenj, Naranjah, Naranshi;
Azeri: Narınc, Portağal;
Basque: Laranja, Limoi, Limonondo;
Belarusian: Apelsin;
Brazil: Laranja-Amarga, Laranja-Azeda, Laranja-Da-Terra, Laranja-De-Sevilha, Laranjeira (Portuguese);
Bulgarian: Portokal;
Burmese: Shonsi, Thanbaya, Tung Chin Thi;
Catalan: Taronger Doç;
Chinese: Chaang (Cantonese), Cheng, Guang Gan, Huang Guo, Tian Cheng, Zhi Shi;
Cook Islands: 'Anani;
Creole: Zoran'y;
Croatian: Slatka Naranča;
Czech: Citroník Čínský, Pomerančovník Čínský;
Danish: Appelsin, Appelsintrae, Orange;
Dutch: Sinaasappel, Zoete Djeroek;
Eastonian: Apelsinipuu;
Esperanto: Oranĝo;
Fijian: Mitha Nimbu, Molidawa, Molilecau, Molitaiti, Moli Ni Tati, Moli Unumi;
Finnish: Appelsiini, Hapanappelsiini, Hunaja-Appelsiini, Makea, Pomeranssi;
French: Orange De Malte, Orange Douce, Oranger, Oranger De Portugal, Oranger Doux, Sanguine;

Futuna: Moil;
German: Apfelsine, Apfelsinenbaum, Blutapfelsine, Orange, Orangenbaum, Süssorangenbaum;
Georgian: Narinjis, Phortokhali;
Greek: Portakali, Portokalia;
Guam: Kahet;
Hawaiian: Alani Hawaii, Ka'u Orange, Waialua Orange;
Hebrew: Tapuz, Tappuah Zahav;
Hungarian: Édes Narancs, Narancs;
Icelandic: Appelsína, Glóaldin;
India: Kamala Tenga, Komola, Xumthira, Xumthira-tenga (Assamese), Kamala, Kamala Nembu (Bengali), Foniliboa (Dhivehi), Narangi, Naringi, Santara, Santru (Gujarati), Musambi, Naarangii, Narangy, Santara (Hindu), Kittale, Naranga (Kannada), Madhura-Naranga, Naragam, Narakam, Oranchu (Malayalam), Komla (Manipuri), Mosambi, Santra (Marathi), Sangtra (Punjabi), Nagaruka, Naranga (Sanskrit), Aranchu, Nagarugam, Nagarukam, Nariyagam, Sathagudi (Tamil), Battavinarinja, Kamalakaya, Kicchilipandu, Naranji (Telugu), Naarangii, Narangi, Naranj, Santra (Urdu);
Indonesia: Jeruk Manis, Limau Manis;
Irish: Oráiste;
Italian: Arancia, Arancio, Arancio Della Cina, Arancio Dolce, Portogallo, Purtualle;
Japanese: Ama Daidai, Orenji, Shina Mikan, Suiito Orenji;
Kashmiri: Santara;
Kazakh: Apelsin;
Khasi: Soh Ñiamtra;
Khmer: Krôôch Pôôsat';
Kikuyu: Njugwa;
Kiribati: Te Aoranti;
Korean: Kamkyulnamu, Tanggyulnamu;
Korean: Tungja-Namu, Deungja-Namu, Kyul-Lamu, Orenji, Suwitu Orenji;
Kosrae: Muhluhlahp;
Laotian: Kièngz;
Latin: Citrangulum;
Latvian: Apelsīns;
Lithuanian: Apelsinai, Apelsininis Citrinmedis;
Macedonian: Portokal;
Maltese: Lariṅ, Laringa Helwa;
Malaysia: Limau Manis, Limau Wangkang, Oren;
Mizo: Serthlum;
Mongolian: Zhürzh;
Nepal: Deshi Suntala, Mausam Suntala, Sunttala (Nepali), Santarasi (Nepalbhasa);
New Caledonia: L'Oranger;
Niue: Moli;
Norwegian: Appelsin;
Palauan: Meradel;
Papiamento: Apusina;
Persian: Narang, Madani, Porteqal;
Philippines: Sankis (Bisaya), Kahel (Tagalog);
Polish: Pomarańcza, Pomarańcza Słodka;
Portuguese: Laranja, Laranja Doce, Laranjeira, Laranjeira Doce;
Romanina: Portocală (Fruit), Portocal (Tree);
Rotuman: Mori;
Russian: Apel'sin;
Samoan: Moli 'Aiga, Moli'Aina;
Serbian: Naranča, Nerandža, Pomoranča, Pomorandža;
Slovaščina: Pomaranča;
Slovenia: Pomaranč Sladký;
Society Islands:'Arani;
Spanish: China Dulce, Naranja, Naranja De China, Naranja Dulce, Naranjo Dulce;
Sri Lanka: Dodan, Peni-Dodan (Sinhala);
Swahili: Mchunhwa;
Swedish: Apelsin, Apelsinträd;
Taiwan: Tian Cheng, Xue Gan;
Tajik: Aflesin;
Thai: Som, Som Kliang (Central Thailand), Som Tra (Bangkok), Makhun;
Tibetan: Tsha lu ma, Tsaluma;
Tigrinya: Bertuan;
Tongan: Moli Inu, Moli Kai;
Turkish: Portakal;
Turkmen: Apelsin, Narynç;
Ukrainian: Apelsyn;
Uzbek: Po'rtahol;
Vietnamese: Cam, Cam Ngọt, Cáy Cam;
Yiddish: Marants.

Origin/Distribution

The orange is only known in the cultivated state. Its origin is assumed to be in southern China, northeastern India, Myanmar, and perhaps in

Indochina. At the present time, orange is widely cultivated in the subtropics and to a lesser extent in the tropics. The main orange producing countries are USA (California, Florida and Arizona), Mexico, Brazil, Spain, Italy, Israel, Australia, South Africa, Japan and China.

Agroecology

The orange is adapted to a semi-arid and subtropical climatic regime, with temperature for growth ranging from 12.78°C to 37.78°C (optimal of 16–20°C) and a winter dormancy temperature of 1.67–10°C. The tree is frost sensitive, young trees are killed by brief periods of frost and mature trees are damaged by freezing temperatures of −1°C to −3°C. Orange requires good amount of water and do well in areas with well-distributed annual precipitation of 500–800 mm. Where rainfall is uneven, supplementary irrigation is necessary.

The orange can be grown in a wide range of soils provided they are deep, well-drained and with a soil pH of 5.5–6.5 such as alluvial, sandy loam to loams, red sand soils to black clayey soils. A well-drained, fertile sandy loam to clay loam is preferred by sweet orange.

Edible Plant Parts and Uses

Ripe fruits are peeled and eaten fresh or made into orange juice, marmalade or wine. Frozen orange juice concentrate is made from freshly squeezed and filtered orange juice. The fruit segments or slices are also consumed in salads and desserts as well as on cakes and meats or candied. The dried and pulverized fruits are utilized for flavouring baked products. Orange peel is also used candied as confection. The grated outermost layer of the rind, zest, is used in cooking and cake making. Finisher pulp, comprising mainly the juice sacs after the extraction of orange juice, has become a major by-product; dried to a moisture content of <10%, it has many uses as an emulsifier and binder in the food and beverage industries. Pectin for use in fruit preserves and otherwise, is derived from the white inner layer (albedo) of the peel. The essential oils extracted from the peel are used commercially for flavouring drinks, ice-cream, pudding, desserts, chewing gum, and sweetmeats. Orange wine and brandy are made in Brazil from fruits which have been processed for peel oil and then crushed. Marmalade a conserve is usually made with Seville orange.

Orange blossom petals are used to make a delicate citrus scented rose-water which is commonly used in French and Middle Eastern cuisines as an ingredient in desserts and baked products. In USA, orange blossom water is used to make orange blossom scones and marshmallows. In Spain, fallen orange blossoms are dried and used to make herbal teas. Likewise orange leaves are dried to make tea. Orange blossom honey is highly prized with a unique orangey citrus taste is derived from honey bees with hives placed in the orange grove.

Studies showed that incorporation of dietary fibre-rich orange bagasse product (DFROBP) (10%) to a bakery product like muffins could be an alternative for people requiring low glyceamic response (Romero-Lopez et al. 2011). DFROBP showed low fat and high dietary fibre contents. The soluble and insoluble dietary fibre fractions were balanced, which is of importance for the health beneficial effects of fibre sources. DFROBP-containing muffins showed the same rapidly digestible starch content as the reference muffin, whilst the slowly digestible starch level increased with the addition of DFROBP. The addition of DFROBP to muffin decreased the predicted glyceamic index.

Botany

Small, armed, branched tree to 10 m tall with large, axillary spines on younger branches. Leaves alternate on narrowly winged-petioles (3–5 mm wide), lamina the blade ovate to ovate-elliptic, 6–10 cm long by 3–5 cm wide, bluntly toothed to crenulate and oil-gland dotted (Plates 1, 2 and 4). Flowers axillary, borne singly or in a small cluster, white and very fragrant; calyx 5 pointed lobes; corolla 5 white obovate petals; 20–25 stamens with distinct yellow anthers; style

Plate 1 Orange blossoms and leaves

Plate 3 Harvested Valencia oranges

Plate 2 Ripening Valencia oranges

Plate 4 Harvested Washington Navels and leaves

Plate 5 Washington Navel orange top and bottom view

stout with a large stigma (Plate 1). Fruit is globose, subglobose, oblate or somewhat oval, 6–7.5 cm across by 7–8 cm long, epicarp pellucid with minute glands, yellow, orangey yellow, bright orange to orangey red with or without navel (Plates 2, 3, 4, 5, 6, 7 and 8); mesocarp or albedo (inner rind) yellowish-white, sarcocarp with 9–12 segments, yellow, orange, crimson or purplish, sweet or slightly acidic (Plates 8 and 9). Seeds few or absent; seed coat slightly ridged, with numerous embryos containing white cotyledons.

There is also a Valencia cultivar with variegated fruits (Plates 10 and 11).

Plate 6 Harvested blood oranges

Plate 9 Blood orange with ten blood red segments

Plate 7 Blood orange side and bottom views

Plate 10 Variegated oranges

Plate 8 Blood orange whole and halved

Plate 11 Close-up of unripe variegated orange

Nutritive/Medicinal Properties

The nutrient value of raw navel orange per 100 g edible portion (exclude 32% peel and navel) was reported as: water 85.97 g, energy 49 kcal (207 kJ), protein 0.91 g, total lipid (fat) 0.15 g, ash 0.43 g, carbohydrate 12.54 g, total dietary fibre 2.2 g, total sugars 8.50 g, sucrose 4.28 g, glucose 1.97 g, Ca 43 mg, Fe 0.13 mg, Mg 11 mg, P 23 mg, K 166 mg, Na 1 mg, Zn 0.08 mg, Cu

0.039 mg, Mn 0.029 mg, vitamin C 59.1 mg, thiamin 0.068 mg, riboflavin 0.051 mg, niacin 0.425 mg, pantothenic acid 0.261 mg, vitamin B-6 0.079 mg, total folate 34 µg, total choline 8.4 mg, betaine 0.1 mg, vitamin A 247 IU, vitamin A 12 µg RAE, vitamin E (α-tocopherol) 0.15 mg, total saturated fatty acids 0.017 g, 16:0 (palmitic acid) 0.017 g, total monounsaturated fatty acids 0.030 g, 16:1 undifferentiated (palmitoleic acid) 0.003 g, 18:1 undifferentiated (oleic acid) 0.026 g, total polyunsaturated fatty acids 0.031 g, 18:2 undifferentiated (linoleic acid) 0.023 g, 18:3 undifferentiated (linolenic acid) 0.009 g, phytosterols 24 mg, tryptophan 0.009 g, threonine 0.018 g, isoleucine 0.017 g, leucine 0.029 g, lysine 0.038 g, methionine 0.009 g, cystine 0.010 g, phenylalanine 0.053 g, tyrosine 0.013 g, valine 0.026 g, arginine 0.115 g, histidine 0.013 g, alanine 0.032 g, aspartic acid 0.139 g, glutamic acid 0.247 g, glycine 0.023 g, proline 0.181 g, serine 0.037 g, β-carotene, 87 µg, α-carotene 7 µg, β-cryptoxanthin 116 µg, lutein+zeaxanthin 129 µg (USDA 2011).

A new apocarotenal carotenoid found in Valencia orange juice was shown to be 3-hydroxy-5,8-epoxy-5,8-dihydro-8′-apo-β-caroten-8′-al (Gross et al. 1975). Two UV-fluorescent apocarotenols designated trollixanthin and trollichrome, were assigned the following structures 5,6-dihydro-β,β-carotene-3-3′,5,6-tetrol and 5,8-epoxy-5,8,5′,6′-tetrahydro-β,β-carotene-3,3′,5′,6′-tetrol, both containing a trihydroxylated ring as in heteroxanthin. More than 25 carotenoid pigments were found in a new sweet orange, Earlygold (Lee et al. 2001). Major carotenoids identified included violaxanthin, lutein, β-cryptoxanthin, antheraxanthin, luteoxanthin, zeaxanthin, β-carotene, and α-carotene. Several geometrical isomers of the carotenoid violaxanthin were detected in orange juice, of which (all-E)-violaxanthin and, above all, (9Z)-violaxanthin, was the most important in quantitative terms (Meléndez-Martínez et al. 2007). Other isomers detected included (15Z)-violaxanthin, (13Z)-violaxanthin and (Di-Z)-violaxanthin.

A concentration range of 0.11–1.21 mg/l was determined for total carotenoids with β-carotene, the most important source of Vitamin A, being found in the highest concentration in the Pera orange variety followed by Valencia, Natal, Lima and Baía varieties (Pupin et al. 1999). Lutein, zeaxanthin, β-cryptoxanthin, α-carotene and β-carotene were found in the orange juices. The total carotenoids present in samples of frozen concentrated orange juice (FCOJ) obtained from factories ranged from 0.26 to 0.48 mg/l, while retail samples of this product contained slightly more (0.46–0.81 mg/l). Frozen concentrated orange pulp-wash presented much lower concentrations, ranging from 0.04 to 0.08 mg/l of total carotenoids. Fourteen samples of retail freshly-squeezed orange juice contained carotenoids ranging from 0.04 to 0.55 mg/l. Juice from hand-squeezed orange fruit presented narirutin and hesperidin concentrations of 16–142 mg/l and 104–537 mg/l, respectively (Pupin et al. 1998). Frozen concentrated orange juice contained higher quantities of flavanone glycosides with narirutin ranging from 62 to 84 mg/l and hesperidin from 531 to 690 mg/l. In frozen concentrated orange juice pulp-wash, the narirutin level ranged from 155 to 239 mg/l and hesperidin from 1,089 to 1,200 mg/l. Thirty-four components, including seven esters, two aldehydes, five alcohols, five terpenes, twelve terpenols, and three ketones were identified and quantified in the orange juice obtained from the cv. Kozan (Selli et al. 2004). The major flavour components were linalool, limonene, β-phellandrene, terpinene-4-ol and ethyl 3-hydroxy hexanoate. Nine amines were detected in orange juice at mean total levels of 53.5 mg/l (Vieira et al. 2007). There were significant differences among orange juice brands, in the levels of bioactive amines spermidine, synephrine, spermine, octopamine, pH and total acidity. Five amines were detected in soft drinks with mean total levels of 3.85 mg/l. There were significant differences, among orange soft drink brands, in the levels of most amines and the physicochemical characteristics. The predominant amine was putrescine, followed by synephrine and spermidine, in both orange juices and soft drinks. The levels of these amines in the soft drink varied from 5.0% to 7.6% of the mean levels in orange juice, suggesting that less than 10% of orange juice could have been used in the soft drink. Samples of orange juices were found to be characterized by the presence of relatively high levels of the biogenic amine putrescine (range, 550–2,210 μg/l) but tryptamine and phenylethylamine were not detected, histamine and spermidine were detected (Basheer et al. 2011).

Flavonoid composition of *C. sinensis* juice (mg/100 ml) was reported by Gattuso et al. (2007) as flavanones: didymin 1.89 mg, eriocitrin 0.31 mg, hesperidin 28.6 mg, narirutin 5.2 mg; flavones: neoeriocitrin 0.59 mg, poncirin 1.04 mg, 6,8-di-*C*-Glu-apigenin 5.72 mg, 6,8-di-*C*-Glu-diosmetin 0.35 mg, rhoifolin 0.05 mg, isorhoifolin 0.07 mg, diosmin 0.09 mg, neodiosmin 0.08 mg; polymethoxyflavones: heptamethoxyflavone 0.08 mg, nobiletin 0.33 mg, sinensetin 0.37 mg, tangeretin 0.04 mg; aglycones: taxifolin 0.03 mg, and acacetin 0.03 mg. Flavonoid composition of commercial sweet orange juice (mg/100 ml) was reported as flavanones: didymin 1.89 mg, hesperidin 37.5 mg, naringin 2.13 mg, narirutin 5.9 mg, neohesperidin 0.95 mg; flavones: 6,8-di-*C*-Glu-apigenin 4.16 mg, diosmin 3.46 mg; polymethoxyflavones: quercetogetin 0.04 mg, heptamethoxyflavone 0.05 mg, nobiletin 0.26 mg, sinensetin 0.24 mg, tangeretin 0.04 mg; aglycones: naringenin 0.8 mg, and isoscutellarein 0.05 mg (Gattuso et al. 2007). Limonene was the most abundant volatile compound of monoterpene hydrocarbons for blood orange, sweet orange, lemon, bergamot and bitter orange juices (Moufida and Marzouk 2003). Eighteen fatty acids were identified in the studied *Citrus* juices, with unsaturated acids predominating over the saturated ones. Mean concentration of fatty acids varied from 311.8 mg/l in blood orange juice to 678 mg/l in bitter orange juice. *Citrus sinensis* (blood orange) juice was found to have high quantity of ascorbic acid (36.90 mg/ml) and also limonene, oleic acid and phenolic acids (Tounsi et al. 2011).

Flavone-di-C-glycosides namely 6,8-Di-C-glucopyranosylapigenin and 6,8-di-C-glucopyranosyldiosmetin were detected in a variety of Southern Italian *Citrus* juices (orange, lemon, bergamot, citron, mandarin, Clementine) (Caristi et al. 2006). 6,8-Di-C-glucopyranosylapigenin was found to be characteristic of orange juice. Cao et al. (2010) showed that cyanidin-3-glucoside (35.2%) and cyaniding-3-(6″-malonyl) glucoside (42.9%) were the major anthocyanins of blood orange. In addition, cyanidin-3-(3″-malonyl) glucoside, cyanidin 3-(6″-dioxalyl) glucoside and cyanidin-3-glucoside adduct: 4-vinylcatechol were identified. Cyanidin-3-glucoside, cyanidin-3-(6″-malonyl)-glucoside, and cyanidin-3-glucoside-derived pyranoanthocyanins were three major anthocyanins found in blood orange (Liu et al. 2011). A total of 18 phenolic compounds were identified and quantified in *Citrus sinensis* cvs Moro and Sanguinello blood orange juices, including hydroxybenzoic acids (2), hydroxycinnamic acids (5), flavanones (5), and anthocyanins (6) (Kelebek et al. 2008). Ferulic acid was the most dominant hydroxycinnamic acid and cyanidin 3-(6″-malonyl glucoside) and cyanidin 3-glucoside were the most dominant anthocyanins in both cultivars. A total of 58 volatile components, including esters (nine), terpenes (19), terpenols (13), aldehydes (two), ketones (three), alcohols (four) and acids (eight) were identified and quantified in Dortyol yerli orange juice (Kelebek and Selli 2011). Organic acids, sugars and phenolic compositions were also found. The major organic acid and sugar found were citric acid and sucrose, respectively. With regard to phenolics, 14 compounds were identified and quantified in the orange juice. In terms of aroma contribution to orange juice, 12 compounds were prominent based on the odour activity values (OAVs). The highest OAV values were recorded for ethyl butanoate, nootkatone, linalool and DL-limonene. The composition of Dortyol orange juice was similar to Valencia and Navel orange juices.

A total of 83 and 78 aroma compounds, including alcohols, esters, terpenes, terpenols, aldehydes, acids, ketones, volatile phenols, and lactones were identified in the blood orange cultivars Moro and Sanguinello juices, respectively (Selli and Kelebek 2011). Of these, 15 volatile components presented odour activity values (OAVs) greater than 1, with dl-limonene, nootkatone and linalool being those with the highest OAVs in both cultivars. Three organic acids (citric, malic and ascorbic acids) and three sugars (sucrose, glucose and fructose) were determined in orange juice and orange wine made from a

Turkish cv. Kozan (Kelebek et al. 2009). Citric acid was the major organic acid found. Sucrose was present in the largest amounts in orange juice and wine. A total of 13 phenolic compounds were identified and quantified in orange juice and wine, including hydroxybenzoic acids (2), hydroxycinnamic acids (5), and flavanones (6). Hesperidin, narirutin and ferulic acid were the most abundant phenolic compounds in orange juice and wine. Antioxidant activities of orange juice and wine were measured using the DPPH• (2,2-diphenyl-1-picrylhydrazyl) assay, and the antioxidant capacity of orange juice was found to be higher than that of orange wine.

From sweet orange *Citrus sinensis* peel, the flavanone glycosides naringin and isosakuranetin 7-rhamnoglucoside were isolated in small quantity and purified (Dunlap and Wender 1966). Hesperidin, the major flavanone glycoside was also isolated. The main groups of flavonoids found in Greek navel sweet orange peel were polymethoxylated flavones, C-glycosylated flavones, O-glycosylated flavones, O-glycosylated flavanones, flavonols and phenolic acids and their derivatives (Anagnostopoulou et al. 2005). The C-glycosylated flavones found were 6-C-β-glucosyldiosmin, 6,8-di-C-glucopyranosylapigenin, 6,8-di-C-β-glucosyldiosmin and two unknown. In addition to the main flavanone glycosides (i.e., hesperidin and naringin) in *Citrus* peel, polymethoxylated flavones and numerous hydroxycinnamates were also found and constituted the major phenolic constituents of the molasses byproduct generated from fruit processing (Manthey and Grohmann 2001). Although a small number of the hydroxycinnamates in *Citrus* occur as amides, most occur as esters. The highest concentrations of hydroxycinnamates occurred in molasses of orange (*C. sinensis*) and tangerine (*C. reticulata*) compared to grapefruit (*C. paradisi*) and lemon (*C. limon*). Concentrations of two phenolic glucosides, phlorin (phloroglucinol-β-O-glucoside) and coniferin (coniferyl alcohol-4-β-O-glucoside), were also determined. Of the polymethoxylated flavones in molasses from several tangerine and orange varieties, the highest amounts occurred in Dancy tangerine, whereas samples from two other tangerine molasses contained significantly lower levels, similar to those in the molasses samples from late- and early/mid-season oranges.

Eight hydroxylated polymethoxyflavones (PMFs): 5-hydroxy-6,7,4'-trimethoxyflavone; 5-demethyltangeretin (5-hydroxy-6,7,8,4'-tetramethoxyflavone); 3-hydroxy-5,6,7,4'-tetramethoxyflavone; 3-hydroxytangeretin (3-hydroxy-5,6,7,8,4'-pentamethoxyflavone); 5-hydroxy-3,6,7,8,3',4'-hexamethoxyflavone; 5-hydroxy-3,7,3',4'-tetramethoxyflavone; 5-hydroxy-3,7,8,3',4'-pentamethoxyflavone; 5-demethylnobiletin (5-hydroxy-6,7,8,3',4'-pentamethoxyflavone); six polymethoxyflavones (PMFs): sinensetin (5,6,7,3',4'-pentamethoxyflavone); 3-methoxysinensetin (3,5,6,7,3',4'-hexamethoxyflavone); nobiletin (5,6,7,8,3',4'-hexamethoxyflavone); 5,6,7,4'-tetramethoxyflavone; 3-methoxynobiletin (3,5,6,7,8,3',4'-heptamethoxyflavone); tangeretin (5,6,7,8,4'-pentamethoxyflavone); one polymethoxyflavanone: 5,6,7,4'-tetramethoxyflavanone; one hydroxylated polymethoxyflavanone: 5-hydroxy-6,7,8,3',4'-pentamethoxyflavanone and two hydroxylated polymethoxychalcones: 2'-hydroxy-3,4,4',5',6'-pentamethoxychalcone and 2'-hydroxy-3,4,3',4',5',6'-hexamethoxychalcone were isolated from sweet orange peel (Li et al. 2006a). The main flavonoid groups found within the methanolic extract fractions of Navel orange peel were polymethoxylated flavones, O-glycosylated flavones, C-glycosylated flavones, O-glycosylated flavonols, O-glycosylated flavanones and phenolic acids along with their ester derivatives (Kanaze et al. 2009). Hesperidin was the major flavonoid glycoside found in the orange peel and its quantity at 48 mg/g of dry peel permits the commercial use of orange peel as a source for the production of hesperidin.

"Sanguinelli" orange peel was found to have high content of hesperidin; higher polymethoxyflavone levels were recorded in orange than in grape fruit peel, with "Valencia Late" showing the greatest nobiletin, sinensetin and tangeretin contents and "Navelate" the highest heptamethoxyflavone levels (Ortuño et al. 2006). The polymethoxyflavone (PMF) fraction of orange (*C. sinensis*) peels contained mainly nonhydroxylated PMFs (75.1%) and a small amount (5.44%)

of hydroxylated polymethoxyflavones (PMF-OH), while total content of PM-FOHs in the PMF-OH fraction was 97.2% as PMF-OH (Sergeev et al. 2007). The following polymethoxyflavones were found in the PMF fraction of: heptamethoxyflavone 10.43%, tangeretin 17.60%, 5,6,7,49-tetramethoxyflavone 12.02%, nobiletin 29.63%, sinesetin 5.41%. The following hydroxylated polymethoxyflavones were found in the PMF fraction: 5-demethyltangeretin 0%, 5-hydroxy-6,7,49-trimethoxyflavone 1.11%, 5-hydroxy-3,6,7,8,3′,49-hexamethoxyflavone <0.1%, 5-demethylnobiletin 2.82%, 5-hydroxy-3,6,7,39,49-pentamethoxyflavone 1.25% and 5-hydroxy-6,7,39,49-tetramethoxyflavone 0.26%. The following PMF-0Hs were found in the PMF-OH fraction: 5-demethyltangeretin 0.68%, 5-hydroxy-6,7,49-trimethoxyflavone 19.00%, 5-hydroxy-3,6,7,8,3′,49-hexamethoxyflavone 28.56%, 5-demethylnobiletin 38.4%, 5-hydroxy-3,6,7,39,49-pentamethoxyflavone 5.96% and 5-hydroxy-6,7,39,49-tetramethoxyflavone 4.64%. Ultrasound-assisted extraction of polyphenols (flavanone glycosides) from orange peel yielded an extraction yield of 10.9%, high total phenolic content (275.8 mg of gallic acid equivalent/100 g FW) and flavanone concentrations (70.3 mg of naringin and 205.2 mg of hesperidin/100 g FW) (Khan et al. 2010).

Li et al. (2007a) employed supercritical chromatography for a relatively large-scale isolation of four PMFs nobiletin, tangeretin, 3,5,6,7,8,3′,4′-heptamethoxyflavone and 5,6,7,4′-tetramethoxyflavone from sweet orange (*Citrus sinensis*) peel. Earlier, they (Li et al. 2006b) reported the an efficient isolation method for the large quantity isolation of nobiletin from orange peel using a silica gel flash column and eluted by a mixed solvent system and eluting the fractions under pressure and loading the resultant residue a Regis chiral column. 5,6,7,4′-tetramethoxyflavone was also eluted.

Eighteen compounds were obtained from *Citrus aurantifolia* cv. Swingle and *Citrus sinensis*. cv. Liucheng plant extracts, namely eight coumarins (1–8), eight flavonoids (9–16) and two sterols analogues (17 and 18) (Shalaby et al. 2011). Leaves and peels of *C. aurantifolia* afforded eleven components, while those of *C. sinensis* afforded fourteen compounds. isobergapten (6), marmesin (7), myricetin (11), 4′,5,7-trihydroxy-3,6-dimethoxy flavone (12), and quercetin-3-*O*-robinobioside (14), were reported for the first time from both *Citrus* species.

The percentage and concentration of total phenolics in *Citrus sinensis* Navel orange cv Thompson leaf, peel and juice were reported by Berhow et al. (1998) as: leaf:- 21.9% (0.62 mg/g) flavone/flavonol, 70.8% (1.29 mg/g) flavanone, 0% psoralen, 0% coumarin; flavedo (outer pigmented layer of the peel): 35.7% (1.40 mg/g) flavone/flavonol, 39.9% (1.00 mg/g) flavanone, 0% psoralen, 10.2% (0.04 mg/g) coumarin; albedo (inner layer of the rind): 8.3% (0.12 mg/g), flavone/flavonol, 68.3% (0.62 mg/g) flavanone, 0% psoralen, 10.2% (0.04 mg/g) coumarin; juice: 5.4% (0.05 mg/g), flavone/flavonol, 87.3% (0.47 mg/g) flavanone, 0% psoralen, and 1.4% coumarin.

The percentage and concentration of flavanones reported in *Citrus sinensis* Navel orange cv Thompson by Berhow et al. (1998) were as follows: leaf:- eriocitrin 4% (traces), hesperidin 88% (1.5 mg/g), narirutin 8% (0.1 mg/g); flavedo: eriocitrin 8% (traces), hesperidin 88% (1.0 mg/g); albedo: narirutin–4′–glucoside 1% (traces), didymin 2% (traces), eriocitrin 2% (trces), hesperidin 90% (0.4 mg/g), narirutin 5% (0.1 mg/g); juice: narirutin–4′–glucoside 5% (traces), eriocitrin 1% (traces), hesperidin 80% (0.3 mg/g), narirutin 14% (traces). The percentage and concentration of flavone reported in *Citrus sinensis* Navel orange cv Thompson by Berhow et al. (1998) were as follows: leaf: isorhoifolin 16% (0.2 mg/g), diosmin 14% (trace), flavedo: rutin 5% (traces), isorhoifolin 16% (0.1 mg/g), and diosmin 5% (trace).

The percentage and concentration of total phenolics in *Citrus sinensis* Blood orange cv Rotuma Island leaf, peel and juice were reported by Berhow et al. (1998) as: leaf:- 46.8% (3.73 mg/g) flavone/flavonol, 43.9% (2.23 mg/g) flavanone, 0% psoralen, 1.5% coumarin (0.04 mg/g); juice: 5.6% (0.04 mg/g), flavone/flavonol, 80.9% (0.41 mg/g) flavanone, 0% psoralen, and 2.4% (0.01 mg/g) coumarin.

The percentage and concentration of flavanones reported in *Citrus sinensis* Blood orange cv Rotuma Island by Berhow et al. (1998) were as follows: leaf:- hesperidin 100% (2.2 mg/g), juice:- hesperidin 80% (0.3 mg/g), narirutin 11% (traces). The percentage and concentration of flavone reported in *Citrus sinensis* Blood orange cv Rotuma Island by Berhow et al. (1998) were as follows: leaf: rutin 3% (0.1 mg/g), isorhoifolin 8% (0.3 mg/g), and diosmin 14% (0.5 mg/g).

Sweet oranges, tangerines and tangors were similar in both amounts and the characteristic flavones – hesperidin and narirutin with virtually no naringin or neohesperidin (Peterson et al. 2006). The following flavanones in mg aglycone/100 g juice or edible fruit were found in sweet orange: didymin 0.45 mg, eriocitrin 0.28 mg, hesperidin 15.25 mg, narirutin 2.33 mg, naringin 0.17 mg, neoeriocitrin 0.04 mg. Sweet oranges contained 17 mg total aglycones from rutinose glycosides and none from neohesperidose glycosides. Four polymethoxyflavones namely tangeretin, nobiletin, tetramethoxyflavone and sinensitin were isolated from peels of Cleopatra mandarin (*Citrus reshni*) and Marrs sweet orange (*Citrus sinensis*) (Uckoo et al. 2011).

Several polymethoxy flavones such as sinensetin, nobiletin, tangeretin, quercetogetin, heptamethoxyflavone, and other derivatives were found in cold-pressed peel oils of *Citrus sinensis* (Weber et al. 2006). In addition, proceranone, an acetylated tetranortriterpenoid with limonoid structure, was identified. Fifty-four components were identified in orange (*C. sinensis*) peel oil (Kirbaslar et al. 2009a, b). The major monoterpenes were limonene (91.16%), myrcene (1.30%), sabinene (1%) and α-pinene (0.9%). The major sesquiterpene components were α-copaene (0.1%) and β-caryophyllene (0.1%). The major oxygenated components of the oils were aldehyde components, octanal (1.4%), decanal (0.2%) and geranial (0.2%); alcohol components, linalool (0.4%), α-terpineol (0.1%), geraniol (0.1%) and nerol (0.05–0.06%); and ester components, geranyl acetate (0.1%) and neryl acetate (0.1%). The flavanone glycoside hesperidin was the main flavonoid in orange peel, precipitation of this compound during processing resulted in dramatic losses in hesperidin in filtered peel juice and filtered molasses byproducts of orange processing (Manthey and Grohmann 1996). Hesperidin occurred at very high levels in dimethyl sulfoxide extracts of unfiltered molasses (5,718 ppm) and in the centrifuged insoluble solids of orange peel molasses (65,642 ppm). The polymethoxylated flavone aglycons were the only flavonoids in cold-pressed orange peel oil. These compounds also occurred in high concentrations in light-density oil solids and in commercial wax isolated from the cold-pressed peel oil.

A new, non-bitter limonoid, 17-dehydrolimonoate A-ring lactone (III), was isolated from orange peel and juice (Hsu et al. 1973). Eighteen compounds were obtained from *Citrus aurantifolia cv.* Swingle and *Citrus sinensis. cv.* Liucheng plant extracts, namely eight coumarins (1–8), eight flavonoids (9–16) and two sterols analogues (17 and 18) (Shalaby et al. 2011). Leaves and peels of *C. aurantifolia* afforded eleven components, while those of *C. sinensis* afforded fourteen compounds. Isobergapten, marmesin, myricetin, 4′,5,7-trihydroxy-3,6-dimethoxy flavone and quercetin-3-*O*-robinobioside were reported for the first time from both *Citrus* species. Six compounds characterized as tetracosane, ethyl pentacosanoate, tetratriacontanoic acid, tangertin, β-sitosteryl-β-D-glucoside and 3,5, 4′-trihydroxy-7,3′-dimethoxy flavanone 3-O-β-glucoside were isolated from *Citrus sinensis* flavedo var. Pineapple (Rani et al. 2009).

Obacunone-17-β-D-glucopyranoside was found to be the predominant limonoid glucoside in *Citrus sinensis* seeds followed by nomilin glucoside (Tian et al. 2000).

Phytochemicals in Other Plant Parts

Caffeine was isolated and identified in extracts from flower buds of several *Citrus* cultivars and from leaves of Valencia oranges (*Citrus sinensis*) (Stewart 1985).

Fifty-four constituents accounting for 82.3–98.2% were identified in the volatile oils extracted from *C. sinensis* leaves (Kasali et al. 2011). Sabinene (20.9–49.1%), δ-3-carene (0.3–14.3%), (E)-β-ocimene (4.4–12.6%), linalool (3.7–11.1%) and terpinen-4-ol (1.7–12.5%) were the major constituents.

Three new acridone alkaloids namely citrusinie-I, -II and citbrasine and a new coumarin, ethylsuberenol were isolated from the acetone root bark extract of *C. sinensis* var. *Brasiliensis* (Wu and Furukawa 1983). A new chalcone, citrunobin together with citflavanone and lonchocarpol-A (senegalensein) were isolated from the root bark of *Citrus sinensis* var. *brasiliensis* (Wu 1989). The structure of citrunobin was elucidated as (E)-1-[7'-hydroxy-5'-methoxy-2',2'-dimeth-yl-2'H-chromen-8'-yl]-3-(4-hydroxyphenyl)prop-2-enone.

Citracridone-I, suberenol, suberosin, crenulatin, xanthoxyletin, xanthyletin, nordentatin, elemol, and p-hydroquinone were also isolated. A new flavonoid, 5, 8-dihydroxy-6, 7, 4'-trimethoxyflavone was isolated from the ethyl acetate extract of *C. sinensis* roots (Intekhab and Aslam 2009).

Of the *Citrus* flavonoids, flavones and flavonols although found in low concentrations in *Citrus* tissues, compared to flavanones, had been shown to be powerful antioxidants and free radical scavengers (Benavente-García et al. 1997). The antioxidant capacity of any flavonoid was reported to be determined by a combination of the O-dihydroxy structure in the B-ring, the 2,3-double bond in conjugation with a 4-oxo function and the presence of both hydroxyl groups in positions 3 and 5. Polymethoxyflavones (PMFs) from *Citrus* genus had been of particular interest because of their broad spectrum of biological activities, including antiinflammatory, anticarcinogenic, neuroprotective, antiviral, antiinflammatory activities, effects on capillary fragility, antiplatelet aggregation and antiatherogenic properties (Benavente-García et al. 1997; Li et al. 2006a; Lee et al. 2010).

Antioxidant Activity

Fruit and vegetables rich in anthocyanins (e.g. strawberry, raspberry and red plum) demonstrated the highest antioxidant activities, followed by those rich in flavanones (e.g. orange and grapefruit) and flavonols (e.g. onion, leek, spinach and green cabbage), while the hydroxycinnamate-rich fruit (e.g. apple, tomato, pear and peach) consistently elicited lower antioxidant activities (Proteggente et al. 2002). The TEAC (Trolox Equivalent Antioxidant Capacity), the FRAP (Ferric Reducing Ability of Plasma) and ORAC (Oxygen Radical Absorbance Capacity) values for each extract were relatively similar and well-correlated with the total phenolic and vitamin C contents.

Ethyl acetate extract of the freeze-dried edible part of navel orange exhibited most active radical scavenging activity while the hexane extract the lowest as evaluated by α,α-diphenyl-β-picrylhydrazyl (DPPH), 2,2'-azino-bis (3-ethylbenzthiazoline-6-sulfonic acid) (ABTS) radical assay and ORAC methods (Jayaprakasha et al. 2008). The order of antioxidant capacity of was ethyl acetate > methanol:water > acetone > methanol > hexane. Acetone and methanol extracts from navel oranges showed highest reducing power (evaluated by potassium ferricyanide reduction assay) than other extracts at 1,000 μg/ml. The results accorded with the amount of carotenoids, phenolics and vitamin C present in the extracts. In another study, Jayaprakasha and Patil (2007) found that methanol:water (80:20) fraction of blood orange showed lowest DPPH radical scavenging activity at various tested concentrations (250, 500 and 1,000 ppm) but showed remarkable antioxidant capacity by the formation of phosphomolybdenum complex. The acetone extract of blood orange was found to contain maximum phenolics. IC_{50} for antioxidant activity (DPPH assay) of the methanolic extracts (peels and tissues) of 13 commercially available Iranian citrus were found to range from 0.6 to 3.8 mg/ml (Ghasemi et al. 2009). Total phenolic content of the *Citrus* spp. samples varied from 66.5 to 396.8 mg gallic acid equivalent/g of extract and flavonoids content varied from 0.3 to 31.1 mg quercetin equivalent/g of extract. The total flavonoid contents were usually higher in peels with the range of 0.3–31.1 respect to tissues with the range of 0.3–17.1 mg quercetin equivalent/g of extract powder. *C. unshiu* var. Mahalli and *C. sinensis* var. Washington Navel peels showed the highest values (31.1 and 23.2 mg quercetin equivalent/g of extract powder, respectively).

Using in-vitro antioxidant assays, namely bleaching of the stable 1,1-diphenyl-2-picrylhydrazyl radical; peroxidation, induced by the water-soluble radical initiator 2,2'-azobis(2-amidinopropane) hydrochloride, on mixed dipalmitoylphosphatidylcholine/linoleic acid unilamellar vesicles; scavenging activity against nitric oxide; total antioxidant status, the antioxidant profile of fresh orange juices from three pigmented varieties, Moro, Sanguinello, and Tarocco, and two blond varieties, Valencia late and Washington navel were determined (Rapisarda et al. 1999). The findings indicated that the antioxidant efficiency of orange juices may be attributed, largely to their content of total phenols while ascorbic acid played a minor role. The antioxidant efficiency of the pigmented juices appeared to be strongly influenced by the anthocyanin content. The antioxidant activity of orange juices was related not only to structural features of phytochemicals contained in them, but also to their capability to interact with biomembranes. Differences in phenolic compositions, the ascorbic acid contents and the antioxidant activities of fresh Sicilian orange juices from pigmented (Moro, Tarocco and Sanguinello) and non-pigmented (Ovale, Valencia and Navel) varieties of orange (*Citrus sinensis*), was found (Proteggente et al. 2003). Differences in flavanone glycoside content, particularly hesperidin were observed, with the Tarocco juices presenting the highest content. The anthocyanins, cyanidin-3-glucoside and cyanidin-3-(6″-malonyl)-glucoside were predominant in all the pigmented varieties, but their concentration was higher in the Moro variety juice. Quantitatively, the major antioxidant component of all juices was ascorbic acid and its concentration was significantly correlated with the total antioxidant activity of the juices, determined in-vitro using the ABTS radical cation decolorization assay. Similarly, hydroxycinnamates and anthocyanins content showed a good correlation with the ascertained antioxidant capacity.

The contents of total polyphenols, flavonoids, and anthocyanins in fresh Jaffa blond orange and Star Ruby grapefruit juices were 962.1 and 906.9; 50.1 and 44.8; and 69.9 and 68.7 µg/ml, respectively (Gorinstein et al. 2004). The antioxidant potential measured by the scavenging activity against nitric oxide, the β-carotene-linoleate model system, DPPH* and 2,2'-azino-bis(3-ethyl-benzothiazoline-6-sulfonic acid) diammonium salt assays was higher in orange juice but not significantly. A high level of correlation between contents of total polyphenols and flavonoids and antioxidant potential values of both juices was found. Diets supplemented with Jaffa blond orange and to a lesser degree with Star Ruby grapefruit juices improved plasma lipid metabolism only in rats fed added cholesterol. However, an increase in the plasma antioxidant activity was observed in both groups. The results indicated that fresh orange and grapefruit juices contained high quantities of bioactive compounds, which guaranteed their high antioxidant potential, and the positive influence on plasma lipid metabolism and plasma antioxidant activity could make fresh orange and grapefruit juices a valuable supplement for disease-preventing diets.

Thermal treatment induced an increase in the main phenolic substances of orange juice, such as anthocyanins and total cinnamates, while ascorbic acid was decreased (Scalzo et al. 2004). Antioxidant properties, evaluated in a lipoxygenase–linoleic acid system, were higher in thermally-treated samples, while free radical scavenging activity, evaluated by ESR spin trapping of hydroxyl radical and DPPH* quenching, were enhanced in untreated juices. The ascorbic acid content decreased significantly after storage in orange juice reconstituted from frozen concentrate, but did not change in chilled juice (Johnston and Hale 2005). The data indicated that the loss of ascorbic acid in refrigerated juice may impact postprandial oxidative stress. In a separate study the total phenolic and flavonoid contents of drying treated orange (*C. sinensis*) peels were decreased by lower drying temperature (50°C and 60°C) and increased by higher drying temperature (70°C, 80°C, 90°C and 100°C) (Chen et al. 2011). Contents of phenolic compounds in the 100°C treated sample extract were significantly higher than the amounts in samples heated at other temperatures. EC_{50} values of orange peel extracts by DPPH radical scavenging effects and ABTS·+scavenging effects were higher with lower drying temperature and decreased with higher drying temperature, and the values of 100°C treated sample extract were significantly lower than the samples heated at other temperatures. However, the chelating Fe^{2+} activities of samples showed the opposite trend. Separate studies found that vitamin C and free and conjugated hydroxycinnamic acids in orange juice were the most affected by both duration 2, 4 and 6 months and temperature of storage at 18°C, 28°C and 38°C (Klimczak et al. 2007). The decrease in the content of polyphenols and vitamin C upon storage was reflected by the decrease in the antioxidant capacity of orange juices. Small changes in flavanone content were observed, indicating high stability of these compounds upon storage.

Valencia orange and Persian lime peels were found to have high total dietary fibre content (61–69%) with an appreciable amount of soluble fibre (19–22%) (Larrauri et al. 1996). The antioxidant capacity of orange peel fibre was 23 times lower than Persian lime peel fibre, which was half the value of DL-α-tocopherol. Both fruit peel fibres were found to contain polyphenols caffeic and ferulic acids, as well as naringin, hesperidin and myricetin. The ethyl acetate fraction of sweet orange peel which exhibited the best radical scavenging activity was found to contain C-glycosylated flavones, polymethoxylated flavones, O-glycosylated flavones, O-glycosylated flavanones, two phenolic acid derivatives and two unknown compounds, all in low concentrations (Anagnostopoulou et al. 2005). The C-glycosylated flavones found were 6-C-β-glucosyldiosmin; 6,8-di-C-glucopyranosylapigenin; 6,8-di-C-β-glucosyldiosmin and two unknown. The results suggested that the ethyl acetate fraction of navel *Citrus sinensis* peel contained significant antioxidant compounds and could be used as a food additive of natural origin or a pharmaceutical supplement using as a source of peel the byproducts of the orange juice industry. High phenolic content and radical scavenging activities (assessed by DPPH* and luminol induced chemiluminescence) were found for the ethyl acetate fraction of Navel sweet orange peel (Anagnostopoulou et al. 2006). The radical scavenging activity of the ethyl acetate fraction was comparable to the activity of the standards, Trolox, ascorbic acid and quercetin. Of the three different doses of *Citrus sinensis* peel extract (12.5 mg/kg, 25 mg/kg and 50.0 mg/kg), 25 mg/kg was found to be the most effective and antiperoxidative, while 50 mg/kg was proved to be hepatotoxic (Parmar and Kar 2008). The results of the DPPH* assay and the Co(II)/EDTA-induced luminol chemiluminescence method showed orange peel methanolic extracts possessed moderate antioxidant activity as compared with the activity seen in tests where the corresponding aglycones, diosmetin and hesperetin were assessed in different ratios (Kanaze et al. 2009). Reducing sugars and phenolics were the main antioxidant compounds found in all the peel and juice extracts of grapefruit, lemon, lime and sweet orange (Guimarães et al. 2010). Peels polar fractions revealed the highest contents in phenolics, flavonoids, ascorbic acid, carotenoids and reducing sugars, which contributed to the highest antioxidant potential found in these fractions as determined by DPPH* radical scavenging capacity, reducing power and inhibition of lipid peroxidation using β-carotene-linoleate model system in liposomes and thiobarbituric acid reactive substances (TBARS) assay in brain homogenates. Hesperidin from *C. sinensis* peel was found to be moderately active as an antioxidant agent against free radical diphenylpicrylhydrazyl (DPPH*) with a capacity of 36% (Al-Ashaal and El-Sheltawy 2011).

Freeman et al. (2010) analysed the interactions of individual phenolic compounds (chlorogenic acid, hesperidin, luteolin, myricetin, naringenin, p-coumaric acid, and quercetin) at the concentrations found in navel oranges (*Citrus sinensis*) for their antioxidant capacity using the Oxygen Radical Absorbance Capacity (ORAC) assay. They found that three different combinations of 2 compounds and 5 combinations of 3 compounds were synergistic. One antagonistic combination of 2 was also found. No additional synergism occurred with the addition of a 4th compound. They asserted that understanding how combinations of fruit antioxidants work together will support their future use in preservation of foods and/or beverages.

Anticancer Activity

Epidemiological and animal studies suggest that flavonoids have a protective effect against cardiovascular diseases and some types of cancer. Over the past two decades, studies had provided growing evidence supporting the beneficial action of flavonoids on multiple cancer-related biological pathways (carcinogen bio-activation, cell-signaling, cell cycle regulation, angiogenesis and inflammation) (Benavente-García et al. 2007; Manthey et al. 2001). Scientific reports on flavonoid activity had largely been associated with enzyme inhibition and anti-proliferative activity. Studies by Benavente-García and coworkers (2008) had shown that structural factors would explain the antioxidant, antiproliferative and antimetastatic properties of some citrus flavonoids. These antiinflammatory and anticancer properties were consistent with their effects on the microvascular endothelial tissue (Manthey et al. 2001). The citrus flavonoids showed little effect on normal, healthy cells, and thus typically exhibited remarkably low toxicity in animals.

The polymethoxylated flavonoids, nobiletin and tangeretin, markedly inhibited cell growth of a human squamous cell carcinoma cell line (HTB43) in-vitro while quercetin and taxifolin exhibited no significant inhibition at any of the concentrations tested (Kandaswami et al. 1991). Kandaswami et al. (1992) also reported that tangeretin and nobiletin markedly inhibited the proliferation of a squamous cell carcinoma (HTB 43) and a gliosarcoma (9 L) cell line at 2–8 μg/ml concentrations. Quercetin displayed no effect on 9 L cell growth at these concentrations, while at 8 μg/ml it inhibited HTB 43 cell growth. Taxifolin slightly inhibited HTB 43 cell growth at 8 μg/ml, while moderately inhibiting HTB 43 cell growth at 2–8 μg/ml. The proliferation of a human lung fibroblast-like cell line (CCL 135) was relatively insensitive to low concentrations of the above flavonoids. Miller et al. (1992) tested limonin 17-β-D-glucopyranoside, nomilin 17-β-D-glucopyranoside, and nomilinic acid 17-β-D-glucopyranoside, three limonoid glucosides isolated from oranges, for cancer chemopreventive activity as a pretreament in buccal pouches of female Syrian hamsters before assault with the carcinogen 7,12-di-methylbenz[a]anthracene (DMBA). Multiple tumours were common in the animals treated with DMBA; however, the animals treated with limonin 17-β-D-glucopyranoside exhibited a 55% decrease in average tumour burden. Of 27 Citrus flavonoids, 7 flavonoids were judged to be active against the tumour cell lines, while they had weak antiproliferative activity against the normal human cell lines (Kawaii et al. 1999a). The rank order of potency was luteolin, natsudaidain, quercetin, tangeretin, eriodictyol, nobiletin, and 3,3′,4′,5,6,7,8-heptamethoxyflavone. The flavones and flavanones with the orthocatechol moiety in ring B and a C2-C3 double bond were found to be important for the antiproliferative activity. As to polymethoxylated flavones, C-3 hydroxyl and C-8 methoxyl groups were essential for high activity. Of the 27 Citrus flavonoids, 10 flavonoids were judged to be active in induction of terminal differentiation of human promyelocytic leukemia cells (HL-60) and the rank order of potency was natsudaidain, luteolin, tangeretin, quercetin, apigenin, 3, 3, ′4, ′5, 6, 7, 8-heptamethoxyflavone, nobiletin, acacetin, eriodictyol, and taxifolin (Kawaii et al. 1999b). These flavonoids exerted their activity in a dose-dependent manner. HL-60 cells treated with these flavonoids differentiated into mature monocyte/macrophage.

A standardized extract of red orange juice was shown to inhibit proliferation of fibroblast and epithelial prostate cells (Vitali et al. 2006). The antiproliferative properties of the juice extract could not be ascribed to cytotoxic effect, nevertheless highlighted its potential usefulness in the management of benign prostatic hyperplasia. The fresh fruit juices extracted from Citrus sinensis (cv. Washington Navel and cv. Sanguinello) showed antiproliferative activity against K562 (human chronic myelogenous leukemia), HL-60 (human leukemia) and MCF-7 (human breast adenocarcinoma) cell lines (Camarda et al. 2007).

Fifteen polymethoxyflavones (PMFs) and hydroxylated PMFs isolated from sweet orange (*Citrus sinensis*) peel extract were tested in HL-60 cancer cell proliferation and apoptosis induction assays (Li et al. 2007b). While some PMFs and hydroxylated PMFs had moderate anti-carcinogenic activities, 5-hydroxy-6,7,8,3′,4′-pentamethoxyflavone and 5-hydroxy-3,6,7,8,3′,4′-hexamethoxyflavone showed strong inhibitory activities against the proliferation and induced apoptosis of HL-60 cell lines. Sergeev et al. (2006) demonstrated that polymethoxyflavones (PMFs) derived from sweet orange (*Citrus sinensis*) inhibited growth of human breast cancer cells via Ca^{2+}-dependent apoptotic mechanism. The treatment of MCF-7 breast cancer cells with 5-hydroxy-3,6,7,8,3′,4′-hexamethoxyflavone (5-OH-HxMF) and 3′-hydroxy-5,6,7,4′-tetramethoxyflavone (3′-OH-TtMF) induced a sustained increase in concentration of intracellular Ca^{2+} ($[Ca^{2+}](i)$) resulting from both depletion of the endoplasmic reticulum Ca^{2+} stores and Ca^{2+} influx from the extracellular space. Corresponding non-hydroxylated PMFs, 3,5,6,7,8,3′,4′-heptamethoxyflavone (HpMF) and 5,6,7,3′,4′-pentamethoxyflavone (PtMF), were dramatically less active in inducing Ca^{2+}-mediated apoptosis. They also found that the fraction of orange peel extract containing a mixture of non-hydroxylated PMFs (75.1%) and hydroxylated PMFs (5.44%) and the fraction containing only hydroxylated PMFs (97.2%) induced apoptosis in human MCF-7 breast cancer cells in a concentration- and time-dependent manner (Sergeev et al. 2007). The results strongly implied that bioactive PMFs from orange peel exerted proapoptotic activity in human breast cancer cells, which depended on their ability to induce an increase in intracellular Ca^{2+} and thus, activate Ca^{2+}-dependent apoptotic proteases.

In-vitro studies showed that tangeretin and 5-demethyl nobiletin induced apoptosis in human neuroblastoma SH-SY5Y cells by reducing the mitochondrial membrane potential suggesting that an intrinsic pathway of apoptosis was synergistically activated by the combination treatment with tangeretin and 5-demethyl nobiletin (Akao et al. 2008). On the other hand, in the combined treatment including nobiletin, the growth inhibitory activity of tangeretin was reduced. The results indicated the relevance of the combination of phytochemicals for the enhancement of the anticancer effect.

Two major polymethoxyflavones (PMFs), namely, nobiletin and 3,5,6,7,8,3′,4′-heptamethoxyflavone (HMF), and two major monodemethylated polymethoxyflavones, namely 5-hydroxy-3,7,8,3′,4′-pentamethoxyflavone (5HPMF), and 5-hydroxy-3,6,7,8,3′,4′-hexamethoxyflavone (5HHMF) from orange peels, inhibited growth of human lung cancer H1299, H441, and H460 cells (Xiao et al. 2009). Monodemethylated PMFs were much more potent in growth inhibition of lung cancer cells than their permethoxylated counterpart PMFs. In H1299 cells, monodemethylated PMFs caused significant increase in sub-G0/G1 phase, suggesting possible role of apoptosis in the growth inhibition observed, whereas the permethoxylated counterpart PMFs did not affect cell cycle distribution at same concentrations tested. The results strongly suggested that the phenolic group was essential for the growth inhibitory activity of monodemethylated PMFs. Further studies in H1299 cells demonstrated that monodemethylated PMFs downregulated oncogenic proteins, such as iNOS, COX-2, Mcl-1, and K-ras, as well as induced apoptosis evidenced by activation of caspase-3 and cleavage of PARP. The results provided rationale to develop orange peel extract enriched with monodemethylated PMFs into value-added nutraceutical products for cancer prevention. Hesperidin from *C. sinensis* peel exhibited pronounced anticancer activity against larynx, cervix, breast and liver carcinoma cell lines (Al-Ashaal and El-Sheltawy 2011). IC_{50} values were 1.67, 3.33, 4.17, 4.58 μg/ml, respectively. Tangeretin and nobiletin inhibited proliferation of human breast cancer cell lines MDA-MB-435 and MCF-7 and human colon cancer line HT-29 cell lines (Morley et al. 2007). Both flavonoids significantly inhibited proliferation in a dose- and time-dependent manner, and blocked cell cycle progression at G1 in all three cell lines but did not induce apoptosis or cell death. In another study, the polymethoxyflavonoids, tangeretin and nobiletin exhibited potent inhibitory effects against daunorubicin-resistant T lymphoblastoid leukemia cells with IC_{50} values of 7.1–14.0 μmol/l (Ishii et al. 2010). Both polymethoxyflavonoids inhibited the P-glycoprotein function and significantly influenced the cell cycle but did not induce apoptosis. Results of animal studies suggested that 3 dietary flavonoids (chrysin, quercetin and nobiletin) suppressed the early phase of colon carcinogenesis in obese mice, partly through inhibition of proliferation activity caused by serum growth factors and may be useful for prevention of colon carcinogenesis in obese humans (Miyamoto et al. 2010).

Antiinflammatory Activity

Flavonoids have antiinflammatory property. Several mechanisms of action had been proposed to explain in-vivo antiinflammatory actions of flavonoids, such as antioxidant activity, inhibition of eicosanoid generating enzymes or the inhibition of the synthesis and biological activities of different pro-inflammatory mediators, mainly the arachidonic acid derivatives, prostaglandins E 2, F 2, and thromboxane Λ 2 (Benavente-García and Castillo 2008; García-Lafuente et al. 2009). Recent studies had also shown that some flavonoids acted as modulators of proinflammatory gene expression, thus leading to the attenuation of the inflammatory response. The antiinflammatory activity of flavonoids have implications on the protection against cancer, cardiovascular and brain diseases. Apigenin and their glucosides had been shown a good antiinflammatory activity without the side effects of other antiinflammatory products (Benavente-García and Castillo 2008). Ramachandran et al. (2002) found that *Citrus sinensis* peel extract at two doses of 150 and 300 mg/kg exhibited significant antiinflammatory activity in the carrageenan-induced acute pedal paw edema model and Freund's complete adjuvant-induced chronic inflammatory model acute. A red orange (*Citrus sinensis* varieties: Moro, Tarocco, Sanguinello) complex extract, characterized by high levels of anthocyanins, flavanones, hydroxycinnamic acids and ascorbic acid displayed notable antiinflammatory properties in human keratinocyte cells NCTC 2544 (Cardile et al. 2010). Addition of the complex extract at different concentrations together with IFN-γ and histamine induced a dose-dependent inhibition of inter-cellular adhesion molecule-1 (ICAM-1) expression and monocyte chemoattractant protein-1 (MCP-1) and interleukin-8 (IL-8) release. The findings suggested that the complex extract could have a topical employment and mitigate the consequences of some skin pathologies.

Cardiovascular Activity

Flavonoid glycosides from orange peels were found to have vasodilatory and hypotensive effects in animal studies (Kumamoto et al. 1986). Epidemiological and animal studies pointed to a possible protective effect of citrus flavonoids against cardiovascular diseases and some types of cancer (Benavente-García and Castillo 2008). The most abundant Citrus flavonoids are flavanones, such as hesperidin, naringin, or neohesperidin. However, generally, the flavones, such as diosmin, apigenin, or luteolin, exhibit higher biological activity, even though they occur in much lower concentrations. Many epidemiological studies had shown regular flavonoid intake to be associated with a reduced risk of cardiovascular diseases. In coronary heart disease, the protective effects of flavonoids included mainly antithrombotic, anti-ischemic, anti-oxidant, and vasorelaxant. Research findings suggested that flavonoids decreased the risk of coronary heart disease by three major actions: improving coronary vasodilatation, decreasing the ability of platelets in the blood to clot, and preventing low-density lipoproteins (LDLs) from oxidizing. Red orange (*Citrus sinensis*) extract and its active components rutin, its aglycon quercetin, cyanidin 3-O-β-d-glucoside, its aglycon cyanidin, hydroxycinnamic acids were found to inhibit low-density lipoprotein oxidation involved in the development of cardiovascular diseases (Sorrenti et al. 2004). Zhou et al. (2009) found that nobiletin inhibited angiotensin II-induced vascular smooth muscle cells (VSMCs) proliferation. This was attributed, in part, to its inhibitory effect on Ca^{2+}-dependent c-Jun N-terminal kinases (JNK) activation in VSMCs. Their results suggested the usefulness of nobiletin for the treatment of cardiovascular diseases relevant to VSMCs growth.

The consumption of 500 ml/daily of orange juice associated with aerobic training in overweight women decreased cardiovascular disease risk by reducing LDL-C levels and increasing HDL-C levels (Aptekmann and Cesar 2010). This association also decreased blood lactate concentration and increased anaerobic threshold, showing some improvement in the physical performance with less muscle fatigue and better response to training. Consumption of orange juice by the experimental group was associated with increased dietary intake of vitamin C and folate by 126% and 61% respectively.

In a randomized, controlled, crossover study of healthy, middle-aged, moderately overweight men, orange juice was found to significantly decrease diastolic blood pressure (DBP) when regularly consumed for 4 weeks and postprandially increased endothelium-dependent microvascular reactivity. A similar trend was observed with consumption of a control drink with hesperidin (Morand et al. 2011). The data suggested that hesperidin could be causally linked to the beneficial effect of orange juice.

Antiobesity/Antihypercholesterolemic Activity

Dietary supplementation of Moro (a blood orange) juice, but not Navelina (a blond orange) juice significantly reduced body weight gain and fat accumulation regardless of the increased energy intake because of sugar content in mice (Titta et al. 2010). Furthermore, mice drinking Moro juice were resistant to high-fat diet -induced obesity with no alterations in food intake. Only the anthocyanin extract, but not the purified cyanidin-3-glucoside (C3G), slightly affected fat accumulation. Moro juice antiobesity effect on fat accumulation could not be explained only by its anthocyanin content. The findings suggested that multiple components present in the Moro orange juice might act synergistically to inhibit fat accumulation.

Studies showed that orange juice consumption for 60 days decreased low-density lipoprotein cholesterol (160–141) in the hypercholesterolemic subjects but not in the normolipidemic group (Cesar et al. 2010). HDL-cholesterol and triglycerides remained unchanged in both groups. Free-cholesterol transfer to HDL increased whereas triglyceride and phospholipid transfers decreased in both groups. Cholesteryl-ester transfer decreased only in hypercholesterolemic group (3.6 to 3.1), but not in the normolipidemic group. Thus, by decreasing atherogenic low-density lipoprotein cholesterol in hypercholesterolemic subjects and increasing HDL ability to take up free cholesterol in hypercholesterolemic and normolipidemic subjects, orange juice may be beneficial to both groups as free-cholesterol transfer to HDL was crucial for cholesterol esterification and reverse cholesterol transport.

Antithyroidal Activity

Administration of *Citrus sinensis* extract (25 mg/kg/day) in hyperthyroid L-thyroxine-induced mice reversed most of the adverse effects of increased levels of thyroxine (T4) and triiodothyronine (T3), serum glucose concentration, α-amylase activity, heart/body weight ratio (HW/BW), kidney/body weight ratio (KW/BW) and cardiac as well as hepatic lipid peroxidation (Parmar and Kar 2008). The peel extract also ameliorated the decreased concentration of different serum lipids such as total cholesterol (TC), triglycerides (TG), high-density lipoprotein cholesterol (HDL-C), low-density lipoprotein cholesterol (LDL-C) and very low-density lipoprotein cholesterol (VLDL-C). The findings suggested that the ameliorating effect of orange peel extract may be attributed to its antioxidative/free radical-scavenging, antithyroidal and HDL-C stimulating properties.

Antiosteoporotic Activity

In-vivo studies in male rats showed that orchidectomy plus orange juice and orchidectomy plus grapefruit juice reversed orchidectomy-induced antioxidant suppression, decreased alkaline phosphatase and acid phosphatase activities, moderately restored femoral density, increased femoral strength, significantly delayed time-induced femoral fracture, and decreased urinary excretion of hydroxyproline (Deyhim et al. 2006). Studies in ovariectomized showed that nobiletin significantly suppressed the reduction of whole bone mineral density by 61%, and markedly decreased type II collagen-induced arthritis by 45% (Murakami et al. 2007). Nobiletin attenuated receptor activator of nuclear factor kappaB ligand (RANKL)-induced osteoclastogenesis of RAW264.7 cells, and suppressed RANKL-activated extracellular signal-regulated kinase1/2, c-Jun N-terminal kinase1/2, and p38 mitogen-activated protein kinase activities, and thereby regulated the promoter activation of nuclear factor kappaB (NFkappaB) and activator protein-1, key transcription factors. The results suggested nobiletin to be a promising phytochemical for the prevention or treatment of osteoclastogenesis-related disorders, including osteoporosis and rheumatoid arthritis. In another study, feeding orchidectomized male rats with the crude citrus extract or the purified bioactive compounds (limonin or naringin) increased the plasma antioxidant status, plasma IGF-I (insulin-like growth factor –I), and bone density, preserved the concentration of calcium in the femur and in the 5th lumbar, and numerically improved bone strength (Mandadi et al. 2009). The crude extract and the bioactive compounds decreased faecal excretion of calcium, numerically lowered the urinary excretion of calcium, and suppressed the plasma TRAP (total radical-trapping potential) activity without affecting urinary excretion of deoxypyridinoline in comparison to the orchidectomized group. The results indicated the potential benefit of the citrus crude extract and its bioactive compounds on bone quality in preserving bone calcium concentration and increased antioxidant status. In separate study, feeding orchidectomized male rats with orange pulp significantly increased bone volume fraction and trabecular number and decreased trabecular separation in the fourth lumbar trabecular cores of orchidectomized rats and improved microarchitectural properties of vertebral bones and in cortical thickness of long bones (Morrow et al. 2009). Orchidectomy decreased antioxidant status, while orange pulp as low as 2.5% maintained the antioxidant capacity of orchidectomized rats comparable to that of the SHAM group. Cortical thickness at the tibial mid-shaft was significantly decreased by orchidectomy and increased by orange pulp, and urinary deoxypyridinoline was significantly increased by orchidectomy and decreased by orange pulp.

Studies demonstrated that *Citrus aurantifolia* cv. Swingle and *Citrus sinensis. cv.* Liucheng extracts reduced bone loss in ovariectomized rats (Shalaby et al. 2011). Administration of *Citrus* extracts increased trabecular bone mineral content and bone mineral density of tibia, and improved the levels of phosphorus and calcium.

Antidiabetic Activity

Administration of 25 mg/kg of *C. sinensis* peel extract to alloxan-diabetic rats was found to normalize all the adverse diabetic changes namely elevated serum levels of glucose and α-amylase activity, rate of water consumption and lipid peroxidation (LPO) in hepatic, cardiac and renal tissues with a parallel decrease in serum insulin level induced by alloxan (Parmar and Kar 2007). The antidiabetic and antiperoxidative effects of orange peels may be related to its high content of total polyphenols. Studies by Lee et al. (2010) suggested that nobiletin improved hyperglycemia and insulin resistance in obese diabetic ob/ob mice. This was achieved by regulating expression of glucose transporter Glut1 and Glut4 in white adipose tissue and muscle, and expression of inflammatory adipokines such as interleukin (IL)-6 and monocyte chemoattractant protein (MCP)-1 and increasing the mRNA expression levels of adiponectin, peroxisome proliferator-activated receptor (PPAR)-γ and its target genes in white adipose tissues.

Ionotropoic Activity

Two *Citrus* flavonoids, 3,5,6,7,8,3′,4′-heptamethoxyflavone (HEPTA) and natsudaidain were found to have cardiotonic effect (Itoigawa et al. 1994). Both showed a positive inotropic effect on the guinea pig right ventricle. Natsudaidain was more potent than HEPTA, but the maximum positive inotropic effect of HEPTA was greater than that of natsudaidain and the effect of HEPTA was related to catecholamine release from cardiac tissue. Some flavonoids isolated from *C. sinensis* presented a positive inotropic effect on myocardium, however the crude leaf extract presented an inotropic depression instead of an increase in force on the atria of guinea pigs of both sexes (Oliveira et al. 2005). The EC_{50} for crude, ethanol, acetic, aqueous, and acetone extracts was 300, 300, 600, 1,000, and 140 μg/ml, respectively. The study showed that the effect was concentration-dependent and indicated the existence of another inhibitory contractile mechanism such as the simultaneous activation of some of the membrane potassium channels reducing the myocardial action potential duration and further decreasing the cellular calcium entry.

Venotonic Activity

Citrus flavones, diosmin and rutin had a demonstrated activity as a venotonic agent and are present in several pharmaceutical products (Benavente-García and Castillo 2008). Both have a long tradition in polish herbal medicine for their venotonic and anti-oedematous properties (Bylka and Kornobis 2005). Diosmin and other flavane derivatives had been shown beneficial in heavy leg syndrome characterized by venous type symptoms that included painful sensation of heavy, swollen or restless legs (Carpentier and Mathieu 1998).

Anxiolytic Activity

Citrus sinenis essential oil demonstrated anxiolytic activity in male Wistar rats and, at the highest dose of 400 microl it displayed significant anxiolytic effects, as indicated by increased exploration of the open arms of the elevated plus-maze and of the lit chamber of the light/dark paradigm (Brito Faturi et al. 2010). The results lend scientific support to its use as a tranquilizer by aroma therapists.

Premenstrual Syndrome Management

In a randomized, double-blind, placebo-controlled trial done on 80 students suffering from premenstrual syndrome, students administered *C. sinensis* essence during the luteal phase for two cycles witnessed a significant reduction of 46.08% premenstrual symptoms compared to the group on placebo 14.21% (Ozgoli et al. 2011). After the intervention, there were also significant decreases in the severity of physical and psychological symptoms. *Citrus* essence was found to contain compounds such as limonene, flounder and citral with proposed sedative and antispasmodic effects in addition to its antidepressant properties. These effects were similar to those of fluoxetine an effective medication in the treatment of premenstrual syndrome.

Sedative Activity

The methanol and dichloromethane extracts of *Citrus sinensis* flowers exhibited a dose-dependent sedative effect in the exploratory cylinder model in mice, with ED_{50} (ip) values of 47.04 mg/kg and 129.15 mg/kg, respectively (Guzmán-Gutiérrez et al.). Hesperidin (ED_{50} = 11.34 mg/

kg) was identified in the methanol extract as the sedative active principle and the results suggested that adenosine receptors might be involved in the sedative action of hesperidin.

Antimicrobial Activity

The hydrodistilled essential oils of orange (*Citrus sinensis* cvv. "Washington navel", "Sanguinello", "Tarocco", "Moro", "Valencia late", and "Ovale") exhibited antifungal action on *Penicillium digitatum* and *Penicillium italicum* (Caccioni et al. 1998). Findings showed a positive correlation between monoterpenes other than limonene and sesquiterpene content of the oils and the pathogen fungi inhibition. An in-vitro study revealed that flavanone glycosides and polymethoxyflavones from *C. sinensis* and *C. paradisi* acted as antifungal agents against *Penicillium digitatum*, the polymethoxyflavones being more active than the flavanones in this respect (Ortuño et al. 2006). *C. sinensis* extract was found to have significant antimicrobial effect at 10% dosage against Gram positive microorganism, *Staphylococcus aureus, Bacillus subtilis*; Gram negative *Enterococcus feacalis, Escherichia coli, Klebsiella pneumonia, Pseudomonas aeruginosa* and fungi *Candida albicans* and *Aspergillus niger* (Osarumwense et al. 2011). Orange terpenes, single-folded d-limonene, and orange oil essence terpenes all exhibited inhibitory activity against 11 serotypes/strains of *Salmonella* spp. in the disc diffusion assay (O'Bryan et al. 2008). Orange terpenes and d-limonene both had MICs of 1%. The most active compound, terpenes from orange essence, produced an MIC that ranged from 0.125% to 0.5% against the 11 *Salmonella* tested. The orange essence oil was found to compose principally of d-limonene, 94%, and myrcene at about 3%. Of the citrus essential oils/blends tested, a blend of 1:1 (v/v) orange (*Citrus sinensis*) and bergamot (*Citrus bergamia*) essential oil was the most effective against vancomycin-susceptible and vancomycin-resistant *Enterococcus faecium* and *E. faecalis* strains with a minimum inhibitory concentration (MIC), at 25°C and pH 5.5, of 0.25–0.5% (v/v) and a minimum inhibitory dose (MID) of 50 mg/l, at 50°C at pH 7.5, when viable counts were reduced by 5.5–10 log10 colony forming units (cfu)/ml (Fisher and Phillips 2009). The Turkish Citrus (grapefruit, lemon, bergamot, bitter orange, sweet orange and mandarin) peel oils showed strong antimicrobial activity against the Gram (+) and Gram (−) bacteria and the fungi cultures *Kluyveromyces fragilis, Rhodotorula rubra, Candida albicans, Hanseniaspora guilliermondii* and *Debaryomyces hansenii* yeasts studied (Kirbaşlar et al. 2009b). All of the Citrus peel oils were more effective towards *Proteus vulgaris* than the other microorganisms. Lemon and bergamot peel oils showed higher antimicrobial activity.

Insecticidal Activity

The mosquito, *Culex quinquefasciatus* larvae were found to be susceptible to orange peel essential oil (Mwaiko 1992). Volatile extracts of orange peel – *Citrus sinensis* (sweet orange) and *Citrus aurantifolia* (lime) exhibited insecticidal activity against mosquito, cockroach and housefly (Ezeonu et al. 2001). Volatile extracts of orange showed greater insecticidal potency, while the cockroach was the most susceptible to the orange peels among the three insects studied. *C. sinensis* extract was found to have significant larvicidal effect at 10% dosage against mosquito (Osarumwense et al. 2011).

Antitrypanosomal Activity

Essential oil from *Citrus sinensis* was found to have antitrypanosomal activity in-vitro against *Trypanosoma brucei brucei* and *Trypanosoma evansi* in a dose-dependent pattern in a short period of time (Habila et al. 2010). The drop in number of parasite over time was achieved doses of 0.4 g/ml, 0.2 g/ml, and 0.1 g/ml. GC-MS analysis revealed the presence of cyclobutane (96.09%) in the essential oil.

Drug Pharmacokinetic Activity

Oral administration of orange juice or a 0.079% hesperidin (a component of orange juice) suspension to male Sprague–Dawley rats 2 days prior to administration of pravastatin significantly increased the AUC, C(max), and t(1/2) values of pravastin and significantly decreased multidrug resistance-associated protein 2 (Mrp2) and mRNA levels in the small intestine and liver (Watanabe et al. 2011). The same results were obtained with hesperidin. These results suggested that the changes in pravastatin pharmacokinetic parameters and the decrease in Mrp2 expression caused by orange juice were due to hesperidin in the juice.

Iron Absorption Enhancement Activity

Studies demonstrated an overall benefit to iron absorption from ferrous fumarate provided with orange juice (Balay et al. 2010). Ferrous fumarate is a common, inexpensive iron form increasingly used instead of ferrous sulfate as a food iron supplement. The absorption of ferrous fumarate given with orange juice and enhancement of absorption by the presence of juice were significantly positively related to height, weight, and age. The effect was age related such that in children older than 6 years of age, there was a nearly twofold increase in iron absorption from ferrous fumarate given with orange juice.

Toxicological Study

A mineral analysis of *Citrus sinensis* fruit peel found that the levels of Cu, Zn, Cd, Mg and Ca in the fruit peel were within the acceptable limits for human consumption (Dhiman et al. 2011). The order of concentration of elements in the sample showed the following trend: $Mg > Ca > Al > Zn > Cu > Cd > Hg = As = Se$. The content of Hg, As and Se in *C. sinensis* fruit peel was significantly low and below detection limit. The content of toxic metals in tested fruit peel was found to be low when compared with the limits prescribed by various authorities (World Health Organization, WHO; International Centre for Materials Research, ICMR; American Public Health Association, APHA).

Traditional Medicinal Uses

The fruit, leaves and bark have reported uses in traditional folk medicine (Kirtikar and Basu 1984; Holdsworth 1992; WHO 1998). The fruit is used as a cardiotonic, laxative, anthelmintic and to remove fatigue and for strengthening in India. The fruit juice is used together with coconut oil or castor oil as a purgative and a whole peeled orange is swallowed to remove a fishbone stuck in the throat by Cook Islanders. In Tonga, an infusion of the leaves is used to treat relapse sickness in women who return to strenuous work too soon after giving birth. Crushed leaves are ingested to ease abdominal pains in Samoa. In Papua New Guinea leaf extracts have been used to treat neurological disorders and to facilitate the digestion of food. In Tahiti, the leaves are used as remedies for internal ailments and fractures as well as other sicknesses. In Samoa, a bark infusion is used to treat an illness similar to relapse sickness, as well as to treat postpartum sickness.

Other Uses

Sweet orange oil contains about 90% d-limonene, a solvent used in various household chemicals, such as a furniture wood conditioner and along with other citrus oils in grease removal agents and in hand-cleansing agent. It is an efficient cleaning agent which is promoted as being environmentally friendly and preferable to petroleum distillates. Orange peels are used as a slug repellant. Orange peel essential oils are also utilized in perfume manufacturing like the essential oils, obtained from the flowers and leaves by distillation and in aromatherapy. The oil of the seeds is used in the soap industry or as cooking oil.

Orange blossoms are popularly used in bridal bouquets and head wreaths for weddings as it is traditionally associated with good fortune and for its strong fragrance. Orange wood sticks are used as cuticle pushers in manicures and pedicures. They are also used as spudgers for manipulating slender electronic wires.

Comments

Valencia orange is a sweet orange first hybridized by California pioneer agronomist and land developer William Wolfskil. This variety is primarily grown for processing and juice production.

Navel orange is a variety of sweet orange characterised by the growth of a second rudimentary fruit at the apex, which protrudes slightly and resembles a human navel. They are primarily used for eating. It is the most commonly grown eating orange.

Blood orange is a variety of orange with crimson-blood coloured segments caused by the presence of anthocyanins and has a thicker skin than the navel or Valencia orange.

Selected References

Akao Y, Itoh T, Ohguchi K, Iinuma M, Nozawa Y (2008) Interactive effects of polymethoxy flavones from *Citrus* on cell growth inhibition in human neuroblastoma SH-SY5Y cells. Bioorg Med Chem 16(6):2803–2810

Al-Ashaal HA, El-Sheltawy ST (2011) Antioxidant capacity of hesperidin from *Citrus* peel using electron spin resonance and cytotoxic activity against human carcinoma cell lines. Pharm Biol 49(3):276–282

Anagnostopoulou MA, Kefalas P, Kokkalou E, Assimopoulou AN, Papageorgiou VP (2005) Analysis of antioxidant compounds in sweet orange peel by HPLC-diode array detection-electrospray ionization mass spectrometry. Biomed Chromatogr 19(2): 138–148

Anagnostopoulou MA, Kefalas P, Papageorgiou VP, Assimopoulou AN, Boskou D (2006) Radical scavenging activity of various extracts and fractions of sweet orange peel (*Citrus sinensis*). Food Chem 94(1):19–25

Aptekmann NP, Cesar TB (2010) Orange juice improved lipid profile and blood lactate of overweight middle-aged women subjected to aerobic training. Maturitas 67(4):343–347

Bailey LH (1976) Hortus third. A concise dictionary of plants cultivated in the United States and Canada. Liberty Hyde Bailey Hortorium, Cornell University, Wiley, 1312 pp

Balay KS, Hawthorne KM, Hicks PD, Griffin IJ, Chen Z, Westerman M, Abrams SA (2010) Orange but not apple juice enhances ferrous fumarate absorption in small children. J Pediatr Gastroenterol Nutr 50(5): 545–550

Basheer C, Wong W, Makahleh A, Tameem AA, Salhin A, Saad B, Lee HK (2011) Hydrazone-based ligands for micro-solid phase extraction-high performance liquid chromatographic determination of biogenic amines in orange juice. J Chromatogr A 1218(28):4332–4339

Bayer RJ, Mabberley DJ, Morton C, Miller CH, Sharma IK, Pfeil BE, Rich S, Hitchcock R, Sykes S (2009) A molecular phylogeny of the orange subfamily (Rutaceae: Aurantioideae) using nine cpDNA sequences. Am J Bot 96:668–685

Benavente-García O, Castillo J (2008) Update on uses and properties of *Citrus* flavonoids: new findings in anti-cancer, cardiovascular, and anti-inflammatory activity. J Agric Food Chem 56(15):6185–6205

Benavente-García O, Castillo J, Marin FR, Ortuño A, Del Río JA (1997) Uses and properties of *Citrus* flavonoids. J Agric Food Chem 45(12):4505–4515

Benavente-García O, Castillo J, Alcaraz M, Vicente V, Del Río JA, Ortuño A (2007) Beneficial action of *Citrus* flavonoids on multiple cancer-related biological pathways. Curr Cancer Drug Targets 7(8): 795–809

Berhow M, Tisserat B, Kanes K, Vandercook C (1998) Survey of phenolic compounds produced in *Citrus*. USDA ARS Tech Bull 1856:1–154

Brito Faturi C, Leite JR, Barreto Alves P, Canton AC, Teixeira-Silva F (2010) Anxiolytic-like effect of sweet orange aroma in Wistar rats. Prog Neuropsychopharmacol Biol Psychiatry 34(4):605–609

Bylka W, Kornobis J (2005) Butcher's Broom, in the treatment of venous insufficiency. Pol Merkur Lekarski 19(110):234–236 (in Polish)

Caccioni DR, Guizzardi M, Biondi DM, Renda A, Ruberto G (1998) Relationship between volatile components of *Citrus* fruit essential oils and antimicrobial action on *Penicillium digitatum* and *Penicillium italicum*. Int J Food Microbiol 43:73–79

Camarda L, Stefano VD, Bosco SFD, Schillaci D (2007) Antiproliferative activity of *Citrus* juices and HPLC evaluation of their flavonoid composition. Fitoterapia 78(6):426–429

Cao SQ, Pan SY, Yao XL, Fu HF (2010) Isolation and purification of anthocyanins from blood oranges by column chromatography. Agric Sci China 9(2):207–215

Cardile V, Frasca G, Rizza L, Rapisarda P, Bonina F (2010) Antiinflammatory effects of a red orange extract in human keratinocytes treated with interferon-gamma and histamine. Phytother Res 24(3):414–418

Caristi C, Bellocco E, Gargiulli C, Toscano G, Leuzzi U (2006) Flavone-di-C-glycosides in *Citrus* juices from Southern Italy. Food Chem 95(3):431–437

Carpentier PH, Mathieu M (1998) Evaluation of clinical efficacy of a venotonic drug: lessons of a therapeutic trial with hemisynthesis diosmin in "heavy legs syndrome". J Mal Vasc 23(2):106–112 (in French)

Cesar TB, Aptekmann NP, Araujo MP, Vinagre CC, Maranhão RC (2010) Orange juice decreases low-density lipoprotein cholesterol in hypercholesterolemic subjects and improves lipid transfer to high-density lipoprotein in normal and hypercholesterolemic subjects. Nutr Res 30(10):689–694

Chen ML, Yang DJ, Liu SC (2011) Effects of drying temperature on the flavonoid, phenolic acid and antioxidative capacities of the methanol extract of *Citrus* fruit (*Citrus sinensis* (L.) Osbeck) peels. Int J Food Sci Technol 46:1179–1185

Cottin R (2002) *Citrus* of the world: a *Citrus* directory. Version 2.0. SRA INRA-CIRAD, France

Deyhim F, Garica K, Lopez E, Gonzalez J, Ino S, Garcia M, Patil BS (2006) *Citrus* juice modulates bone strength in male senescent rat model of osteoporosis. Nutrition 22(5):559–563

Dhiman A, Nanda A, Ahmad S (2011) Metal analysis in *Citrus sinensis* fruit peel and *Psidium guajava* leaf. Toxicol Int 18(2):163–167

Dunlap WJ, Wender SH (1966) Purification and identification of flavanone glycosides in the peel of the sweet orange. Arch Biochem Biophys 87(2):228–231

Ezeonu FC, Chidume GI, Udedi SC (2001) Insecticidal properties of volatile extracts of orange peels. Bioresour Technol 76(3):273–274

Fisher K, Phillips C (2009) In vitro inhibition of vancomycin-susceptible and vancomycin-resistant *Enterococcus faecium and E. faecalis* in the presence of citrus essential oils. Br J Biomed Sci 66(4):180–185

Freeman BL, Eggett DL, Parker TL (2010) Synergistic and antagonistic interactions of phenolic compounds found in navel oranges. J Food Sci 75(6):C570–C576

García-Lafuente A, Guillamón E, Villares A, Rostagno MA, Martínez JA (2009) Flavonoids as anti-inflammatory agents: implications in cancer and cardiovascular disease. Inflamm Res 58(9):537–552

Gattuso G, Barreca D, Gargiulli C, Leuzzi U, Caristi C (2007) Flavonoid composition of *Citrus* juices. Molecules 12:1641–1673

Ghasemi K, Ghasemi Y, Ebrahimzadeh MA (2009) Antioxidant activity, phenol and flavonoid contents of 13 *Citrus* species peels and tissues. Pak J Pharm Sci 22(3):277–281

Gorinstein S, Leontowicz H, Leontowicz M, Krzeminski R, Gralak M, Martin-Belloso O, Delgado-Licon E, Haruenkit R, Katrich E, Park YS, Jung ST, Trakhtenberg S (2004) Fresh Israeli Jaffa blond (Shamouti) orange and Israeli Jaffa red Star Ruby (Sunrise) grapefruit juices affect plasma lipid metabolism and antioxidant capacity in rats fed added cholesterol. J Agric Food Chem 52(15):4853–4859

Gross J, Carmon M, Lifshitz A, Sklarz B (1975) Structural elucidation of some orange juice carotenoids. Phytochemistry 14(1):249–252

Guimarães R, Barros L, Barreira JC, Sousa MJ, Carvalho AM, Ferreira IC (2010) Targeting excessive free radicals with peels and juices of *Citrus* fruits: grapefruit, lemon, lime and orange. Food Chem Toxicol 48(1):99–106

Guzmán-Gutiérrez SL, Navarrete A (2009) Pharmacological exploration of the sedative mechanism of hesperidin identified as the active principle of *Citrus sinensis* flowers. Planta Med 75(4):295–301

Habila N, Agbaji AS, Ladan Z, Bello IA, Haruna E, Dakare MA, Atolagbe TO (2010) Evaluation of in vitro activity of essential Oils against *Trypanosoma brucei brucei* and *Trypanosoma evansi*. J Parasitol Res 2010:Article ID 534601

Holdsworth DK (1992) Medicinal plants of the East and West Sepik Provinces, Papua, New Guinea. Int J Pharm 30(2):18–22

Hsu AC, Hasegawa S, Maie VP, Bennett RD (1973) 17-dehydrolimonoate A-ring lactone: a possible metabolite of limonoate A-ring lactone in *Citrus* fruits. Phytochemistry 12(3):563–567

Intekhab J, Aslam M (2009) Isolation of a flavonoid from the roots of *Citrus sinensis*. Malays J Pharm Sci 7(1):1–8

Ishii K, Tanaka S, Kagami K, Henmi K, Toyoda H, Kaise T, Hirano T (2010) Effects of naturally occurring polymethyoxyflavonoids on cell growth, p-glycoprotein function, cell cycle, and apoptosis of daunorubicin-resistant T lymphoblastoid leukemia cells. Cancer Invest 28(3):220–229

Itoigawa M, Takeya K, Furukawa H (1994) Cardiotonic flavonoids from *Citrus* plants (Rutaceae). Biol Pharm Bull 17(11):1519–1521

Jayaprakasha GK, Girennavar B, Patil BS (2008) Antioxidant capacity of pummelo and navel oranges: extraction efficiency of solvents in sequence. LWT-Food Sci Technol 41(3):376–384

Jayaprakasha GK, Patil BS (2007) In vitro evaluation of the antioxidant activities in fruit extracts from citron and blood orange. Food Chem 101(1):410–418

Johnston CS, Hale JC (2005) Oxidation of ascorbic acid in stored orange juice is associated with reduced plasma vitamin C concentrations and elevated lipid peroxides. J A Diet Assoc 105(1):106–109

Kanaze FI, Termentzi A, Gabrieli C, Niopas I, Georgarakis M, Kokkalou E (2009) The phytochemical analysis and antioxidant activity assessment of orange peel (*Citrus sinensis*) cultivated in Greece-Crete indicates a new commercial source of hesperidin. Biomed Chromatogr 23(3):239–249

Kandaswami C, Perkins E, Drzewiecki G, Soloniuk DS, Middleton E Jr (1992) Differential inhibition of proliferation of human squamous cell carcinoma, gliosarcoma and embryonic fibroblast-like lung cells in culture by plant flavonoids. Anticancer Drugs 3(5):525–530

Kandaswami C, Perkins E, Soloniuk DS, Drzewiecki G, Middleton E Jr (1991) Antiproliferative effects of *Citrus* flavonoids on a human squamous cell carcinoma in vitro. Cancer Lett 56(2):147–152

Kasali AA, Lawal OA, Eshilokun AO, Olaniyan AA, Opoku AR, Setzer WN (2011) *Citrus* essential oil of Nigeria. Part V: volatile constituents of sweet orange leaf oil (*Citrus sinensis*). Nat Prod Commun 6(6):875–878

Kawaii S, Tomono Y, Katase E, Ogawa K, Yano M (1999a) Antiproliferative activity of flavonoids on several cancer cell lines. Biosci Biotechnol Biochem 63(5):896–899

Kawaii S, Tomono Y, Katase E, Ogawa K, Yano M (1999b) Effect of *Citrus* flavonoids on HL-60 cell differentiation. Anticancer Res 19(2A):1261–1269, 19

Kelebek H, Canbas A, Selli S (2008) Determination of phenolic composition and antioxidant capacity of blood orange juices obtained from cvs. Moro and Sanguinello (*Citrus sinensis* (L.) Osbeck) grown in Turkey. Food Chem 107(4):1710–1716

Kelebek H, Selli S (2011) Determination of volatile, phenolic, organic acid and sugar components in a Turkish cv. Dortyol (*Citrus sinensis* L. Osbeck) orange juice. J Sci Food Agric 91(10):1855–1862

Kelebek H, Selli S, Canbas A, Cabaroglu T (2009) HPLC determination of organic acids, sugars, phenolic compositions and antioxidant capacity of orange juice and orange wine made from a Turkish cv. Kozan. Microchem J 91(2):187–192

Khan MK, Abert-Vian M, Fabiano-Tixier AS, Dangles O, Chemat F (2010) Ultrasound-assisted extraction of polyphenols (flavanone glycosides) from orange (*Citrus sinensis* L.) peel. Food Chem 119(2):851–858

Kirbaşlar FG, Kirbaşlar SI, Pozan G, Boz I (2009a) Volatile constituents of Turkish orange (*Citrus sinensis* (L.) Osbeck) peel oils. J Essent Oil Bear Plant 12(5):586–604

Kirbaşlar FG, Tavman A, Dülger B, Türker G (2009b) Antimicrobial activity of Turkish *Citrus* peel oils. Pak J Bot 41(6):3207–3212

Kirtikar KR, Basu BD (1984) Indian medicinal plants, vol 2. Singh and Singh, Dehradun

Klimczak I, Małecka M, Szlachta M, Gliszczyńska-Świgło A (2007) Effect of storage on the content of polyphenols, vitamin C and the antioxidant activity of orange juices. J Food Comp Anal 20(3–4):313–322

Kumamoto H, Matsubara Y, Iizuka Y, Okamoto K, Yokoi K (1986) Structure and hypotensive effect of flavonoid glycosides in orange (*Citrus sinensis* Osbeck) peelings. Agric Biol Chem 50:781–783

Larrauri JA, Rupérez P, Bravo L, Fulgencio aura-Calixto F (1996) High dietary fibre powders from orange and lime peels: associated polyphenols and antioxidant capacity. Food Res 29(8):757–762

Lee HS, Castle WS, Coates GA (2001) High-performance liquid chromatography for the characterization of carotenoids in the new sweet orange (Earlygold) grown in Florida, USA. J Chromatogr 913(1–2):371–377

Lee YS, Cha BY, Saito K, Yamakawa H, Choi SS, Yamaguchi K, Yonezawa T, Teruya T, Nagai K, Woo JT (2010) Nobiletin improves hyperglycemia and insulin resistance in obese diabetic ob/ob mice. Biochem Pharmacol 79(11):1674–1683

Li S, Lambros T, Wang Z, Goodnow R, Ho CT (2007a) Efficient and scalable method in isolation of polymethoxyflavones from orange peel extract by supercritical fluid chromatography. J Chromatogr B Analyt Technol Biomed Life Sci 846(1–2):291–297

Li S, Lo CY, Ho CT (2006a) Hydroxylated polymethoxyflavones and methylated flavonoids in sweet orange (*Citrus sinensis*) peel. J Agric Food Chem 54(12): 4176–4185

Li S, Pan MH, Lai CS, Lo CY, Dushenkov S, Ho CT (2007b) Isolation and syntheses of polymethoxyflavones and hydroxylated polymethoxyflavones as inhibitors of HL-60 cell lines. Bioorg Med Chem 15(10):3381–3389

Li S, Yu H, Ho CT (2006b) Nobiletin: efficient and large quantity isolation from orange peel extract. Biomed Chromatogr 20(1):133–138

Liu L, Cao SQ, Pan SY (2011) Thermal degradation kinetics of three kinds of representative anthocyanins obtained from blood orange. Agric Sci China 10(4):642–649

Mabberley DJ (1997) A classification for edible *Citrus*. Telopea 7(2):167–172

Mabberley DJ (2001) *Citrus* reunited. Aust Plant 21(166): 52–55

Mandadi K, Ramirez M, Jayaprakasha GK, Faraji B, Lihono M, Deyhim F, Patil BS (2009) *Citrus* bioactive compounds improve bone quality and plasma antioxidant activity in orchidectomized rats. Phytomedicine 16(6–7):513–520

Manner HI, Buker RS, Smith VE, Elevitch CR (2006) *Citrus* species. Ver. 2.1. In: Elevitch CR (ed) Species profiles for Pacific Island Agroforestry, Permanent Agriculture Resources (PAR), Holualoa, Hawaii. http://www.agroforestry.net

Manthey JA, Grohmann K (1996) Concentrations of hesperidin and other orange peel flavonoids in *Citrus* processing byproducts. J Agric Food Chem 44(3): 811–814

Manthey JA, Grohmann K (2001) Phenols in *Citrus* peel byproducts. Concentrations of hydroxycinnamates and polymethoxylated flavones in *Citrus* peel molasses. J Agric Food Chem 49(7):3268–3273

Manthey JA, Grohmann K, Guthrie N (2001) Biological properties of *Citrus* flavonoids pertaining to cancer and inflammation. Curr Med Chem 8(2):135–153

Meléndez-Martínez AJ, Vicario IM, Heredia FJ (2007) Geometrical isomers of violaxanthin in orange juice. Food Chem 104(1):169–175

Miller EG, Gonzales-Sanders AP, Couvillon AM, Wright JM, Hasegawa S, Lam LKT (1992) Inhibition of hamster buccal pouch carcinogenesis by limonin 17-B-D-glucopyranoside. Nutr Cancer 17:1–7

Miyamoto S, Yasui Y, Ohigashi H, Tanaka T, Murakami A (2010) Dietary flavonoids suppress azoxymethane-induced colonic preneoplastic lesions in male C57BL/KsJ-db/db mice. Chem Biol Interact 183(2):276–283

Morand C, Dubray C, Milenkovic D, Lioger D, Martin JF, Scalbert A, Mazur A (2011) Hesperidin contributes to the vascular protective effects of orange juice: a randomized crossover study in healthy volunteers. Am J Clin Nutr 93(1):73–80

Morley KL, Ferguson PJ, Koropatnick J (2007) Tangeretin and nobiletin induce G1 cell cycle arrest but not apoptosis in human breast and colon cancer cells. Cancer Lett 251(1):168–178

Morrow R, Deyhim F, Patil BS, Stoecker BJ (2009) Feeding orange pulp improved bone quality in a rat model of male osteoporosis. J Med Food 12(2):298–303

Morton JF (1987) Orange. In: Fruits of warm climates. Julia F. Morton, Miami, pp 134–142

Moufida S, Marzouk B (2003) Biochemical characterization of blood orange, sweet orange, lemon, bergamot and bitter orange. Phytochemistry 62(8):1283–1289

Murakami A, Song M, Katsumata S, Uehara M, Suzuki K, Ohigashi H (2007) *Citrus* nobiletin suppresses bone loss in ovariectomized ddY mice and collagen-induced arthritis in DBA/1 J mice: possible involvement of receptor activator of NF-kappaB ligand (RANKL)-induced osteoclastogenesis regulation. Biofactors 30(3):179–192

Mwaiko GL (1992) *Citrus* peel oil extracts as mosquito larvae insecticides. East Afr Med J 69:223–226

O'Bryan CA, Crandall PG, Chalova VI, Ricke SC (2008) Orange essential oils antimicrobial activities against *Salmonella* spp. J Food Sci 73(6):M264–M267

Oliveira ED, Leite TS, Silva BA, Conde-Garcia EA (2005) Inotropic effect of *Citrus sinensis* (L.) Osbeck leaf extracts on the guinea pig atrium. Braz J Med Biol Res 38(1):111–118

Ortuño A, Báidez A, Gómez P, Arcas MC, Porras I, García-Lidón A, Del Río JA (2006) *Citrus paradisi* and *Citrus sinensis* flavonoids: their influence in the defence mechanism against *Penicillium digitatum*. Food Chem 98(2):351–358

Osarumwense PO, Okunrobo LO, Imafidon KE (2011) Phytochemical composition of *Citrus sinensis* (l.) Osbeck and its larvicidal and antimicrobial activities. Contin J Pharm Sci 5:15–19

Ozgoli G, Shahveh M, Esmaielli S, Nassiri N (2011) Essential oil of *Citrus sinensis* for the treatment of premenstrual syndrome; a randomized double-blind placebo-controlled trial. J Reprod Infertility 12(2):123–129

Parmar HS, Kar A (2007) Antidiabetic potential of *Citrus sinensis* and *Punica granatum* peel extracts in alloxan treated male mice. Biofactors 31(1):17–24

Parmar HS, Kar A (2008) Antiperoxidative, antithyroidal, antihyperglycemic and cardioprotective role of *Citrus sinensis* peel extract in male mice. Phytother Res 22(6):791–795

Peterson JJ, Dwyer JT, Beecher GR, Bhagwat SA, Gebhardt SE, Haytowitz DB, Holden JM (2006) Flavanones in oranges, tangerines (mandarins), tangors, and tangelos: a compilation and review of the data from the analytical literature. J Food Comp Anal 19:S66–S73

Porcher MH et al (1995–2020) Searchable World Wide Web multilingual multiscript plant name database. Published by The University of Melbourne, Australia. http://www.plantnames.unimelb.edu.au/Sorting/Frontpage.html

Proteggente AR, Pannala AS, Paganga G, Van Buren L, Wagner E, Wiseman S, Van De Put F, Dacombe C, Rice-Evans CA (2002) The antioxidant activity of regularly consumed fruit and vegetables reflects their phenolic and vitamin C composition. Free Radic Res 36(2):217–233

Proteggente AR, Saija A, De Pasquale A, Rice-Evans CA (2003) The compositional characterisation and antioxidant activity of fresh juices from sicilian sweet orange (*Citrus sinensis* L. Osbeck) varieties. Free Radic Res 37(6):681–687

Pupin AM, Dennis MJ, Toledo MCF (1998) Flavanone glycosides in Brazilian orange juice. Food Chem 61(3):275–280

Pupin AM, Dennis MJ, Toledo MCF (1999) HPLC analysis of carotenoids in orange juice. Food Chem 64(2):269–275

Ramachandran S, Anbu J, Saravanan M, Gnanasam SSK (2002) Antioxidant and antiinflammatory properties of *Citrus sinensis* peel extract. Indian J Pharm Sci 64:66–68

Rani G, Yadav L, Kalidhar SB (2009) Chemical examination of *Citrus sinensis* flavedo variety pineapple. Indian J Pharm Sci 71:677–679

Rapisarda P, Tomaino A, Cascio RO, Bonina F, De Pasquale A, Saija A (1999) Antioxidant effectiveness as influenced by phenolic content of fresh orange juices. J Agric Food Chem 47(11):4718–4723

Romero-Lopez MR, Osorio-Diaz P, Bello-Perez LA, Tovar J, Bernardino-Nicanor A (2011) Fiber concentrate from orange (*Citrus sinensis* L.) bagase: characterization and application as bakery product ingredient. Int J Mol Sci 12(4):2174–2186

Samson JA (1992) *Citrus sinensis* (L.) Osbeck. In: Coronel RE, Verheij EWM (eds) Plant resources of South-East Asia. No. 2: edible fruits and nuts. Prosea Foundation, Bogor, pp 139–141

Scalzo RL, Iannoccari T, Summa C, Morelli R, Rapisarda P (2004) Blood orange juice is noted for its antioxidant properties, due to its rich phenolic profile. Effect of thermal treatments on antioxidant and antiradical activity of blood orange juice. Food Chem 85(1):41–47

Scora RW (1975) IX. On the history and origin of *Citrus*. Bull Torrey Bot Club 102:369–375

Selli S, Kelebek H (2011) Aromatic profile and odour-activity value of blood orange juices obtained from Moro and Sanguinello (*Citrus sinensis* L. Osbeck). Ind Crop Prod 33(3):727–733

Selli S, Cabaroglu T, Canbas A (2004) Volatile flavour components of orange juice obtained from the cv. Kozan of Turkey. J Food Comp Anal 17(6):789–796

Sergeev IN, Ho CT, Li S, Colby J, Dushenkov S (2007) Apoptosis-inducing activity of hydroxylated polymethoxyflavones and polymethoxyflavones from orange peel in human breast cancer cells. Mol Nutr Food Res 51(12):1478–1484

Sergeev IN, Li S, Colby J, Ho CT, Dushenkov S (2006) Polymethoxylated flavones induce Ca(2+)-mediated apoptosis in breast cancer cells. Life Sci 80(3):245–253

Shalaby NMM, Abd-Alla HI, Ahmed HH, Basoudan N (2011) Protective effect of *Citrus sinensis* and *Citrus aurantifolia* against osteoporosis and their phytochemical constituents. J Med Plant Res 5(4):579–588

Sorrenti V, Giacomo CD, Russo A, Acquaviva R, Barcellona ML, Vanella A (2004) Inhibition of LDL oxidation by red orange (*Citrus sinensis*) extract and its active components. J Food Sci 69(6):C480–C484

Stewart I (1985) Identification of caffeine in *Citrus* flowers and leaves. J Agric Food Chem 33:1163–1165

Swingle WT, Reece PC (1967) The botany of *Citrus* and its wild relatives. In: Reuther W, Webber HJ, Batchelor LD (eds) The *Citrus* industry, vol 1. History, world distribution, botany, and varieties. University of California Press, Riverside, pp 190–430

Tian QG, Dai J, Ding XL (2000) Research on the separation of limonoid glucosides by reversed-phase preparative high performance liquid chromatography. Se Pu 18(2):109–111 (In Chinese)

Titta L, Trinei M, Stendardo M, Berniakovich I, Petroni K, Tonelli C, Riso P, Porrini M, Minucci S, Pelicci PG, Rapisarda P, Recupero GR, Giorgio M (2010) Blood orange juice inhibits fat accumulation in mice. Int J Obes (Lond) 34(3):578–588

Tounsi MS, Wannes WA, Ouerghemmi I, Jegham S, Ben Njima Y, Hamdaoui G, Zemni H, Marzouk B (2011) Juice components and antioxidant capacity of four Tunisian *Citrus* varieties. J Sci Food Agric 91(1):142–151

Uckoo RM, Jayaprakasha GK, Patil BS (2011) Rapid separation method of polymethoxyflavones from *Citrus* using flash chromatography. Sep Purif Technol 81(2):151–158

U.S. Department of Agriculture, Agricultural Research Service (USDA) (2011) USDA National Nutrient Database for Standard Reference, Release 24. Nutrient Data Laboratory Home Page, http://www.ars.usda.gov/ba/bhnrc/ndl

Vieira SM, Theodoro KH, Glória MBA (2007) Profile and levels of bioactive amines in orange juice and orange soft drink. Food Chem 100(3):895–903

Vitali F, Pennisi C, Tomaino A, Bonina F, Pasquale AD, Saija A, Beatrice Tita B (2006) Effect of a standardized extract of red orange juice on proliferation of human prostate cells in vitro. Fitoterapia 77(3):151–155

Watanabe M, Matsumoto N, Takeba Y, Kumai T, Tanaka M, Tatsunami S, Takenoshita-Nakaya S, Harimoto Y, Kinoshita Y, Kobayashi S (2011) Orange juice and its component, hesperidin, decrease the expression of multidrug resistance-associated protein 2 in rat small intestine and liver. J Biomed Biotechnol 2011:Article 502057

Webber HJ (1946) Cultivated varieties of *Citrus*. In: Webber HJ, BatchelorL D (eds) The *Citrus* industry, 1. History, botany and breeding. University of California Press, Berkeley, pp 475–668, 1028 pp

Weber B, Hartmann B, Stöckigt D, Schreiber K, Roloff M, Bertram HJ, Schmidt CO (2006) Liquid chromatography/mass spectrometry and liquid chromatography/nuclear magnetic resonance as complementary analytical techniques for unambiguous identification of polymethoxylated flavones in residues from molecular distillation of orange peel oils (*Citrus sinensis*). J Agric Food Chem 54(2):274–278

World Health Organisation (WHO) (1998) Medicinal plants in the South Pacific, Western Pacific series no 19. WHO Regional Publications, Manila, 151 pp

Wu TS (1989) Flavonoids from root bark of *Citrus sinensis* and *C. nobilis*. Phytochemistry 28(12):3558–3560

Wu TS, Furukawa F (1983) Acridone alkaloids. VII. Constituents of *Citrus sinensis* Osbeck var. brasiliensis Tanaka. Isolation and characterization of three new acridone alkaloids, and a new coumarin. Chem Pharm Bull 31(3):901–906

Xiao H, Yang CS, Li S, Jin H, Ho CT, Patel T (2009) Monodemethylated polymethoxyflavones from sweet orange (*Citrus sinensis*) peel inhibit growth of human lung cancer cells by apoptosis. Mol Nutr Food Res 53(3):398–406

Zhang DX, Mabberley DJ (2008) *Citrus* Linnaeus. In: Wu ZY, Raven PH, Hong DY (eds) Flora of China, vol 11 (Oxalidaceae through Aceraceae). Science Press/Missouri Botanical Garden Press, Beijing/St. Louis

Zhou CH, Wu XH, Wu YQ (2009) Nobiletin, a dietary phytochemical, inhibits vascular smooth muscle cells proliferation via calcium-mediated c-Jun N-terminal kinases pathway. Eur J Pharmacol 615(1–3):55–60

Citrus x aurantium Tangelo Group

Scientific Name

Citrus x aurantium pro sp. Tangelo Group (sensu Mabberley 1997, 2004).

Synonyms

Citrus reticulata Blanco x *Citrus paradisi* Macfad. (sensu Swingle and Reece 1967), *Citrus tangerina* Yu. Tanaka x *Citrus paradisi* Macfad. (sec. NPGS/GRIN 2010), *Citrus x paradisi x Citrus reticulata* (Tangelo group) (sensu Mabberley 1997, 2004), *Citrus x reticulata* Blanco 'Tangelo' Swingle, *Citrus x tangelo* J.W. Ingram & H.E. Moore.

Family

Rutaceae

Common/English Names

Tangelo, Ugli fruit.
 Common Cultivars: Minneola (Honeybell), Saccaton, Orlando, Pearl, Sunshine, Allspice, Mandalo, Wekiwa.

Vernacular Names

Finnish: Rumeliini, tangel, ugli.

Origin/Distribution

Many of the tangelo cultivars were developed in USA. Minneola tangelo (Plates 1 and 2) is a hybrid of Duncan grapefruit and Dancy tangerine (mandarin) produced in Florida by the U.S. Department of Agriculture and named and released in 1931. Orlando is of the same parentage as Minneola and Seminole—a hybrid of Duncan grapefruit and Dancy tangerine. Pearl is a cross of Imperial grapefruit and Willowleaf. The cross was made by Howard. B Frost at the California Citrus Experiment Station in 1929. Sunshine is of the same parentage as Minneola and Orlando. Allspice redulted from a cross of Imperial grapefruit and Willowleaf mandarin. Mandalo has King Mandarin, Dancy tangerine and Siamese sweet pommel genotypes.

Agroecology

Like all *Citrus* species, tangelos require hot summers and mild winters to thrive. They tend to be fairly cold hardy, tolerating sub-zero temperatures down to −5°C with only minor damages to the shoots. Tangelos require full sun and are adaptable to most soils, even poor soils, with good drainage. They are moderately drought tolerant once established, but require regular, adequate watering for good fruit production. Regular fertilization is also required as they are heavy feeders. Tangelo trees are large and need plenty of space for their crowns to grow.

Plate 1 Box of Mineola tangelo

Plate 2 Close-up of Mineola tangelo

Edible Plant Parts and Uses

Fruit is eaten raw as dessert fruit. The juice and zest from the peels are used in pies, puddings, confectionery and food dishes. Juice is also made into drinks.

Botany

Evergreen vigorous tree 10–15 m high with grayish trunk, and branches and a rounded crown. Leaves are alternate, glossy green, ellipsoid to ellipsoid-linear, acuminate tip with entire margins. Flowers are solitary or in small corymbs, fragrant, 2.5–3.5 cm across, hermaphrodite, with 5 green sepals, 5 white petals, numerous stamens in whorl, partially connate at the base surrounding a long style and superior ovary of about 8–15 fused carpels. The fruit is a hesperidium, globose to oblate, or necked and bell shaped depending on cultivar, 6–10 cm diameter, bright orange to reddish-orange (Plates 1 and 2) when ripe with a thick, loose rind and juicy segments.

Nutritive/Medicinal Properties

Nutrient values per 100 g edible portion of tangelo was reported as: energy 172 kJ, moisture 85.9 g, protein 0.6 g, nitrogen 0.1 g, fat 0.1 g, ash 0.4 g, dietary fibre 2 g, total sugars 7.8 g, fructose 1 g, glucose 1.7 g, sucrose 5.1 g, available carbohydrate without sugar alcohols 7.8 g, available carbohydrate with sugar alcohols 7.8 g, malic acid 0.3 g, citric acid 1.4 g, Ca 22 mg, Fe 0.3 mg, Mg 10 mg, K 140 mg, Na 4 mg, Zn 0.3 mg, thiamine 0.05 mg, riboflavin 0.03 mg, niacin 0.3 mg, niacin equivalents 0.4 mg, vitamin C 28 mg, total folate 10 µg, α-carotene 10 µg, β-carotene 290 µg, and xanthophylls 1,050 µg (Wills et al. 1985).

Some nutrient values reported for tangelo fruit were: water 87.73–88.09%, protein 0.66–0.74%, ash 0.37–0.42%, citric acid 1.41–1.83%, sucrose 3.95%, reducing sugars 5.05%, total sugar 9–9.75%, total solid 11.91–12.27% (Jaffa and Goss 1921–22). Tangelo is devoid of the bitter taste, so pronounced in the grapefruit.

The percentage and concentration of total phenolics in tangelo cv Sacaton leaf, peel and juice were reported by Berhow et al., (1998) as leaf: 7% (0.41 mg/g) flavone/flavonol, 81.9% (3.06 mg/g) flavanone, 0% psoralen, 2.7% (0.05 mg/g) coumarin; flavedo outer pigmented layer of the peel: 21.4% (0.0.96 mg/g) flavone/flavonol, 65.2% (1.87 mg/g) flavanone, 0% psoralen, 5.9% (0.08 mg/g) coumarin; albedo (inner layer of the rind): 2.5% (0.23 mg/g) flavone/flavonol, 94.0% (5.52 mg/g) flavanone, 0% psoralen, 1.5% (0.05 mg/g) coumarin; juice: 1.7% (0.07 mg/g) flavone/flavonol, 92.5% (2.48 mg/g) flavanone, 0% psoralen, and 3% (0.04 mg/g) coumarin.

The percentage and concentration of flavanones reported in the same cultivar by Berhow et al., (1998) were as follows: leaf:- hesperidin 8%

(0.2 mg/g), narirutin 6% (0.2 mg/g), neoeriocitrin 2% (0.1 mg/g), naringin 13% (0.5 mg/g), neohesperidin 61% (1.9 mg/g); flavedo: hesperidin 13% (0.2 mg/g), narirutin 4% (0.21 mg/g), neoeriocitrin 2% (trace mg/g), naringin 17% (0.3 mg/g), naringin–4′–glucoside 2% (trace mg/g), neohesperidin 61% (1.1 mg/g), naringin–6″–malonate (close form) 1% (trace mg/g); albedo: hesperidin 5% (0.3 mg/g), narirutin 3% (0.20 mg/g), neoeriocitrin 1% (0.1 mg/g), naringin 35% (2.2 mg/g), naringin–4′–glucoside 2%(0.1 mg/g), neohesperidin 52% (2.7 mg/g), naringin–6″–malonate (open form) 1%, (trace mg/g), naringin–6″–malonate (close form) 1% (0.1 mg/g); juice: didymin 1% (trace mg/g), hesperidin 8% (0.1 mg/g), narirutin 5% (0.1 mg/g), neoeriocitrin 1% (trace mg/g), naringin 29% (0.2 mg/g), naringin–4′–glucoside 1% (trace mg/g), neohesperidin 53% (1.2 mg/g), naringin–6″–malonate (open form) 1%, (trace mg/g), and naringin–6″–malonate (close form) 1% (trace mg/g).

The percentage and concentration of flavones reported in the same cultivar by Berhow et al., (1998) were as follows: leaf:- diosmin 11% (0.1 mg/g); flavedo: rutin 3% (trace mg/g).

Tangor Elendale and tangelo Minneola flavedo and albedo extracts contained 6-7-fold total phenolics compared to the pulp extracts. The amount ranged from 6,343 to 7,667 μg/g of FW in the following increasing order: tangor albedo<tangelo flavedo<tangelo albedo<tangor flavedo (Ramful et al. 2010b). Highest levels of total flavonoids were obtained in tangelo albedo extracts (4,207 μg/g of FW), followed by its flavedo extracts (3,171 μg/g of FW). Tangor flavedo and albedo contained lower levels of flavonoids. The pulps were relatively poor in flavonoids. Flavonoid glycosides were higher in flavedo and albedo extracts compared to pulp extracts. Naringin was not detected in any of the Citrus extracts, while poncirin, didymin, isorhoifolin, and hesperidin were present in all of them. Hesperidin was present at highest concentrations in both Citrus varieties, and its content ranged from 301 mg/g of FW (tangor albedo) to 17 mg/g of FW (tangelo pulp). Narirutin, which ranked second after hesperidin, was present in highest concentrations in albedo extracts of both tangelo and tangor. Total phenolic content in tangelo fruit was (μg of gallic acid/g of FW) 6,339 μg flavedo, 7,203 μg albedo, pulp 1,062 μg and total flavonoid content was (μg of quercetin/g of FW) 3,171 μg flavedo, 4,206 μg albedo, 625 μg pulp. The flavanone, flavones and flavonol glycosides in tangelo flavedo, albido and pulp were: Flavedo 5.61 mg/g poncirin, rhoifolin not detected (nd), didymin 5.26 mg/g, naringin nd, rutin 11.71 mg/g, diosmin 5.66 mg/g, isorhoifolin 5.13 mg/g, neohespeidin 10.33 mg/g, hesperidin 163.21 mg/g, neoeriocitrin 15.75 mg/g, and narirutin 13.99 mg/g.

Albedo: 8.26 mg/g poncirin, rhoifolin nd, didymin 16.67 mg/g, naringin nd, rutin ndetected, diosmin nd, isorhoifolin 3.42 mg/g, neohespeidin 10.18 mg/g, hesperidin 217.67 mg/g, neoeriocitrin nd, and narirutin 60.72 mg/g.

Pulp: 0.71 mg/g poncirin, rhoifolin 0.56, didymin 3.60 mg/g, naringin nd, rutin 0.73 mg/g, diosmin nd, isorhoifolin 0.41 mg/g, neohespeidin nd, hesperidin 17.04 mg/g, neoeriocitrin 0.56 mg/g, and narirutin 20.91 mg/g.

Antioxidant Activity

Studies conducted on 21 Mauritius varieties of citrus fruits (oranges, satsumah, clementine, mandarins, tangor, bergamot, lemon, tangelos, kumquat, calamondin and pamplemousses) found that that polyphenolic-rich extracts exhibited important antioxidant propensities in trolox equivalent antioxidant capacity (TEAC), ferric reducing antioxidant capacity (FRAP) and hypochlorous acid (HOCl) scavenging activity assays (Ramful et al., 2010a). Nine flavedo extracts exhibited good DNA protecting ability in the cuphen assay with IC_{50} values ranging from 6.3 to 23.0 mg FW/m. The flavedos were able to chelate metal ions however, tangor was most effective with an IC_{50} value of 9.1 mg FW/ml.

Antioxidant activities of tangelo Mineola flavedo, albedo, and pulp extracts as measured by the Trolox equivalent antioxidant capacity (TEAC), ferric-reducing antioxidant power

(FRAP), hypochlorous acid (HOCL) scavenging, copper-phenanthroline, and iron(II) chelation assays were as follows: TEAC (μmol of trolox/g-of FW) flavedo 43.1, albedo 61.1, pulp 7.1; FRAP (μmol of Fe^{2+}/g of FW) flavedo 55, albedo 48.9, pulp 7.8; HOCl (IC_{50}, mg of FW/ml) flavedo 5.2, albedo 5.6, pulp 53.6; copper-phenanthroline (IC_{50}, mg of FW/ml) flavedo 6.3, albedo 71.7; iron(II) chelating activity (IC_{50}, mg of FW/ml) flavedo10.8, albedo132.1. Flavedo extracts exhibited higher antioxidant activities than the albedo and pulp. The ability of the tangelo extracts to retard free-radical-induced hemolysis of human erythrocytes was also investigated (Ramful et al. 2010a). Tangelo flavedo was found to increase the hemolysis time of red blood cells, compared to the control, in a dose-dependent manner. In the assessment of antioxidant activity of *Citrus* fruit extracts in the SW872 human liposarcoma cell line system, the flavedo extracts decreased cell proliferation in a dose-response manner, with significant reductions at 0.75% and 1%. Albedo and pulp extracts had no adverse effects on cell viability, even at the highest concentration tested (1%). Non-cytotoxic concentrations of tangelo and tangor flavedo extracts significantly decreased the levels of protein carbonyls in response to advanced glycation end products (AGEs) generated by albumin glycation in SW872 cells. Flavedo extracts lowered carbonyl accumulation in H_2O_2-treated adipocytes, while tangelo and tangor flavedo, albedo, and pulp extracts suppressed ROS (reactive oxygen species) production in SW872 cells with or without the addition of H_2O_2. The results clearly indicated Mauritian *Citrus* fruit extracts to be an important source of antioxidants, with a novel antioxidative role at the adipose tissue level.

Drug Interaction Activity

None of tangerine or tangelo cultivars tested were found to contain any furanocoumarins except for the K-Early variety, which contained trace amounts of 6,7-dihydroxybergamottin (0.028 ppm), bergapten (0.011 ppm), and bergamottin (0.025 ppm) (Widmer 2005). These amounts were low and insignificant in that they were not at levels that would significantly inhibit CYP3A4 enzyme and affect absorption of drugs metabolized by this enzyme.

Other Uses

Tangelos are also used as pollinizer for other citrus.

Comments

Minneola tangelo is less cold resistant than Orlando tangelo.

Selected References

Berhow M, Tisserat B, Kanes K, Vandercook C (1998) Survey of phenolic compounds produced in *Citrus*. USDA ARS Technol Bull 1856:1–154

Cottin R (2002) *Citrus* of the world: a *Citrus* directory. Version 2.0. SRA INRA-CIRAD, France

Hodgson RW (1967) Horticultural varieties of *Citrus*. In: Reuther W, Webber HJ, Batchelor LD (eds) The *Citrus* industry volume 1. History, world distribution, botany, and varieties. Revised edition. University of California Press, Berkeley, pp 431–591

Jaffa ME, Goss H (1922) Nutrition studies of some new varieties of *Citrus* fruits and avocados. Calif Avocado Soc Yearbook 1921–22 8:58–64

Koskinen I (2011) *Citrus* pages. http://users.kymp.net/citruspages/home.html

Mabberley DJ (1997) A classification for edible *Citrus*. Telopea 7(2):167–172

Mabberley DJ (2004) *Citrus* (Rutaceae): a review of recent advances in etymology, systematics and medical applications. Blumea 49:481–488

Morton JF (1987) Tangelo. In: Fruits of warm climates. Julia F. Morton, Miami, pp 158–160

Peterson JJ, Dwyer JT, Beecher GR, Bhagwat SA, Gebhardt SE, Haytowitz DB, Holden JM (2006) Flavanones in oranges, tangerines (mandarins), tangors, and tangelos: a compilation and review of the data from the analytical literature. J Food Compos Anal 19:S66–S73

Ramful D, Bahorun T, Bourdon E, Tarnus E, Aruoma OI (2010a) Bioactive phenolics and antioxidant propensity of flavedo extracts of Mauritian *Citrus* fruits: potential prophylactic ingredients for functional foods application. Toxicology 278(1):75–87

Ramful D, Tarnus E, Rondeau P, Da Silva CR, Bahorun T, Bourdon E (2010b) *Citrus* fruit extracts reduce advanced glycation end products (AGEs)- and H2O2-induced oxidative stress in human adipocytes. J Agric Food Chem 58:11119–11129

Rouseff RL, Martin SF, Youtsey CO (1987) Quantitative survey of narirutin, naringin, hesperidin, neohesperidin in citrus. J Agric Food Chem 35:1027–1030

Swingle WT, Reece PC (1967) The botany of *Citrus* and its wild relatives. In: Reuther W, Webber HJ, Batchelor LD (eds) The *Citrus* industry volume 1: history, world distribution, botany, and varieties. Revised Edition. University of California Press, Berkeley, pp 190–430

Tucker DPH, Futch SH, Gmitter FG, Kesinger MC (1998) Florida *Citrus* varieties. SP-102. University of Florida, Institute of Food and Agric. Sciences, Cooperative Extension Service, Gainesville, p 30

Widmer W (2005) One tangerine/grapefruit hybrid (tangelo) contains trace amounts of furanocoumarins at a level too low to be associated with grapefruit/drug interactions. J Food Sci 70(6):c419–c422

Wills RBH, Lim JSK, Greenfield H (1985) Composition of Australian foods. 28. Citrus fruit. Food Technol Aust 37:308–310

Citrus x aurantium Tangor Group

Scientific Name

Citrus x aurantium L. pro sp. Tangor Group sensu Mabberley 1997, 2004.

Synonyms

Citrus reticulata Blanco 'Tangor' Swingle, *Citrus reticulata* Blanco x *Citrus* x *aurantium* L., pro sp. [Sweet Orange Group] sensu Mabberley, *Citrus* × *nobilis* Lour., *Citrus reticulata* Blanco x *Citrus sinensis* (L.) Osbeck (sensu Swingle and Reece 1967; sensu Tanaka sec. Cottin 2002), *Citrus* × *tangor* Tanaka.

Family

Rutaceae

Common/English Names

Tangor, King of Siam, King Orange, Temple Orange.

Vernacular Names

French: Mandarinier King, Roi De Siam;
Japanese: Kunembo.

Common Varieties: Afourer (Nardorcott, W. Murcott), Ambersweet, Sweet, Ellendale, King, Murcott (Honey Murcott, Smith, Honey), Ortanique, Temple.

Origin/Distribution

Tangor is a hybrid of tangerine (mandarin) and sweet orange. Ellendale tangor originated in Queensland, Australia where it was found growing on the Ellendale property of E. A. Burridge as early as 1878. Today it is an important variety in parts of Australia as well as other citrus-growing regions of the world (Plate 1).

Murcott also known as Honey Murcott, Smith, Honey is a tangor of unknown origin resulting from the breeding program of the U.S. Department of Agriculture in Florida. The oldest known tree is from 1922 in Charles Murcott Z Smith in Bayview, Florida.

Agroecology

Tangors are subtropical in requirement. Tangors like tangerines prefers a warm sunny site with free draining fertile, sandy or loamy soils with pH 6–6.5. Waterlogged areas should be avoided. Although cold hardy, they need protection from frosts. Mulching and regular watering especially during the dry months will ensure good yield.

Plate 1 Ellendale tangor fruits

Plate 3 Afourer Murcott top and bottom views

Plate 2 Murcott tangor

Plate 4 Afourer Murcott deep orange segments (10) and few seeds

Edible Plant Parts and Uses

Ripe fruits are peeled and eaten fresh or made into juice, marmalade or wine.

Botany

Tree erect and branched, medium vigour, 3–6 m high, unusually thornless to very sparsely thorny, spreading with a rounded open crown. Leaves are large, glossy dark-green, broadly-lanceolate, broadly elliptic, 7.8–8.5 cm long by 3.5–4.3 cm, acute apex, base obtuse, margin entire to crenulate margin, the petioles narrowly winged to wingless. Flowers white, fragrant, hermaphrodite in a few-flowered axillary clusters, calyx 4–5 lobed, green; petals 4–5, elliptic; stamens 16–25; ovary subglobose, superior with a single style and stigma. Fruit medium large to large, oblate to subglobose hesperidium, 7–8 cm cross and 4.7–5.2 cm high; base rounded; apex flat or slightly depressed. Rind medium-thin 2–4 mm thick, smooth to faintly pebbled, colour yellow-orange, orange or red-orange orange-red. Segments 10–12, readily separable; axis solid to semi-hollow. Flesh bright orange-coloured; very juicy; flavour rich and pleasantly sweet to subacid. Seeds few (3–4) to many (>10) (Plate 4).

Nutritive/Medicinal Properties

Nutrient value of Elendale tangor per 100 g edible portion was reported by Wills et al. (1985) as follows: energy 196 kJ, moisture 85.6 g, protein 1.1 g, N 0.18 g, fat 0.2 g, ash 0.3 g, dietary fibre 1.6 g, total sugars 8.8 g, fructose 2 g, sucrose 5.2 g, available carbohydrate without sugar alcohols 8.8 g, available carbohydrate with sugar alcohols 8.8 g, Ca 28 mg, P 20 mg, K 164 mg, Na 2 mg, Zn 0.11 mg, Fe 0.22 mg, Mn 0.034 mg, Cu 0.052 mg, vitamin C 48 mg, thiamine 0.076 mg, riboflavin 0.033 mg, niacin equivalents 0.51 mg, vitamin B-6 0.04 mg, total folate 24 µg, α-carotene 22 µg, β-carotene 98 µg, β-carotene equivalents 109 µg, xanthophylls 2258.1 µg, retinol equivalent 18 µg, α-tocopherol 0.4 mg, vitamin E 0.36 mg, malic acid 0.3 g and citric acid 1.3 g.

The percentage and concentration of total phenolics in *Citrus reticulata x Citrus sinensis* tangor cv. Sue Linda Temple (a Floridian cultivar) leaf, peel and juice were reported by Berhow et al. (1998) as: leaf:- 22.9% (1.41 mg/g) flavone/flavonol, 62.8% (2.47 mg/g) flavanone, 0% psoralen, 2% (0.04 mg/g) coumarin; flavedo (outer pigmented layer of the peel): 2.7% (0.13 mg/g) flavone/flavonol, 61.1% (1.93 mg/g) flavanone, 0% psoralen, 25.9% (0.39 mg/g) coumarin; albedo (inner layer of the rind): 9.8% (0.47 mg/g), flavone/flavonol, 70.6% (2.16 mg/g) flavanone, 0% psoralen, 13.4% (0.20 mg/g) coumarin; juice: 2.7% (0.02 mg/g), flavone/flavonol, 84.6% (0.42 mg/g) flavanone, 0% psoralen, and 7% (0.02 mg/g) coumarin.

The percentage and concentration of flavanones reported in *Citrus reticulata x Citrus sinensis* tangor cv. Sue Linda Temple by Berhow et al. (1998) were as follows: leaf:- eriocitrin 6% (0.1 mg/g), hesperidin 85% (2.1 mg/g), narirutin 9% (0.2 mg/g); flavedo: narirutin–4′–glucoside 4% (0.1 mg/g), didymin 3% (traces), hesperidin 75% (1.4 mg/g), narirutin 17% (0.3 mg/g); albedo: narirutin–4′–glucoside 4% (0.1 mg/g), didymin 2% (traces), eriocitrin 5% (0.1 mg/g), hesperidin 70% (1.6 mg/g), narirutin 12% (0.3 mg/g); juice: narirutin–4′–glucoside 4% (traces), eriocitrin 5% (trces), hesperidin 75% (0.3 mg/g), and narirutin 16% (0.1 mg/g).

The percentage and concentration of flavone reported in *Citrus reticulata x Citrus sinensis* tangor cv. Sue Linda Temple by Berhow et al. (1998) were as follows: leaf: rutin 8% (0.1 mg/g), diosmin 11% (0.1 mg/g). Citrus juice containing tangor juice could be differentiated by a simultaneous liquid chromatographic method that allow the quantitation of six FGs (narirutin, naringin, hesperidin, neohesperidin, didymin, poncirin) and six PMFs (sinensetin, hexamethoxyflavone, nobiletin, scutellarein, heptamethoxyflavone and tangeretin) (Mouly et al. 1998).

Sweet oranges, tangerines and tangors were similar in both amounts and the characteristic flavones – hesperidin and narirutin with virtually no naringin or neohesperidin (Peterson et al. 2006). The following flavanones in mg aglycone/100 g juice or edible fruit were found in tangor: didymin 1.17 mg, eriocitrin 1.01 mg, hesperidin 15.42 mg, narirutin 7.10 mg, naringin 0 mg, neoeriocitrin 0 mg, neohesperidin 0 mg. Tangors contained 25 mg total aglycones from rutinose glycosides and none from neohesperidose glycosides. Tangors were significantly higher in narirutin than sweet oranges or tangerines.

Ortanique peel contained the highest quantity of polymethoxylated flavones (34,393 ppm), followed by tangerine (28,389 ppm) and Mexican sweet orange (sample 1; 21,627 ppm) (Green et al. 2007). The major polymethoxylated flavones, i.e. sinensetin, nobiletin, tangeretin, heptamethoxyflavone, tetramethylscutellarein and hexamethyl-o-quercetagetin, present in the peels of 20 citrus cultivars, was quantified. From 1,000 kg of Ortanique peel extract for example, approximately 33 kg of total polymethoxylated flavones was recoverable, comprising 9.4 kg of tangeretin, 7.9 kg of nobiletin, 3.5 kg of sinensitin, 7.5 kg of tetramethylscutellarein, 1.2 kg of heptamethoxyflavone and 3.6 kg hexamethyloquercetagetin (Green et al. 2009). In tangor Murcott juice, narirutin was the main flavanone followed by hesperidin (Moura and Sylos 2008).

Tangor and tangelo flavedo and albedo extracts contained 6–7-fold more total phenolics compared to the pulp extracts (Ramful et al.

2010b). Total phenolics ranged from 6,343 to 7,667 μg/g of fresh weight in the following increasing order: tangor albedo<tangelo flavedo<tangelo albedo <tangor flavedo. Highest concentrations of total flavonoids were obtained in tangelo albedo extracts (4,207 μg/g of FW), followed by its flavedo extracts (3,171 μg/g of FW). Tangor flavedo and albedo contained lower levels of flavonoids. The pulps were relatively poor in flavonoids. In both tangor and tangelo. Flavonoid glycosides were higher in flavedo and albedo extracts compared to pulp extracts. poncirin, didymin, isorhoifolin, and hesperidin were present in all of the extracts which were devoid of naringin. Hesperidin was present at highest concentrations in both *Citrus* varieties, and its content ranged from301mg/g of FW(tangor albedo) to 17 mg/g of FW (tangelo pulp). Narirutin, which ranked second after hesperidin, was present in highest concentrations in albedo extracts of both tangelo and tangor. Total phenolic and total flavonoid contents in tangor fruit were: total phenolic (μg of gallic acid/g of FW) 7,667 μg flavedo, 6,343 μg albedo, pulp 1,009 μg; total flavonoids (μg of quercetin/g of FW) 2,863 μg flavedo, 2,282 μg albedo, 446 μg pulp. The flavanone, flavones and flavonol glycosides in tangor flavedo, albido and pulp were as follows: flavedo:- 473 mg/g poncirin, rhoifolin 78gmg, didymin 13.94 mg/g, naringin nd, rutin 10.62 mg/g, diosmin 5.91 mg/g, isorhoifolin 8.90 mg/g, neohespeidin 5.67 mg/g, hesperidin 234.08 mg/g, neoeriocitrin 8.43 mg/g, narirutin 20.03 mg/g; albedo:- 2.85 mg/g poncirin, rhoifolin nd, didymin 44.74 mg/g, naringin nd, rutin 5.72 mg/g, diosmin nd, isorhoifolin 2.62 mg/g, neohesperidin 4.87 mg/g, hesperidin 301.18 mg/g, neoeriocitrin 47.22 mg/g, narirutin 177.80 mg/g; pulp:- 0.35 mg/g poncirin, rhoifolin nd, didymin 4.86 mg/g, naringin nd, rutin 0.74 mg/g, diosmin nd, isorhoifolin 0.69 mg/g, neohespeidin 0.73 mg/g, hesperidin 26.98 mg/g, neoeriocitrin 1.34 mg/g, narirutin not detected (Plate 2).

Fruits of hybrids of murcott tangor were found to contain S-methylmethionine sulfonium, an off-flavour which increased during fruit ripening (Plate 3) (Sakamoto et al. 1996).

Antioxidant Activity

Studies conducted on twenty-one Mauritius varieties of *Citrus* fruits (oranges, satsumah, clementine, mandarins, tangor, bergamot, lemon, tangelos, kumquat, calamondin and pamplemousses) found that that polyphenolic-rich extracts exhibited important antioxidant propensities in trolox equivalent antioxidant capacity (TEAC), ferric reducing antioxidant capacity (FRAP) and hypochlorous acid (HOCl) scavenging activity assays (Ramful et al. 2010a). Nine flavedo extracts exhibited good DNA protecting ability in the cuphen assay with IC_{50} values ranging from 6.3 to 23.0 mg FW/m. The flavedos were able to chelate metal ions however, tangor was most effective with an IC_{50} value of 9.1 mg FW/ml. The flavedo extracts of *Citrus* fruits represent a significant source of phenolic antioxidants with potential prophylactic properties for the development of functional foods.

Antioxidant activities of tangor, Elendale flavedo, albedo, and pulp extracts as measured by the Trolox equivalent antioxidant capacity (TEAC), ferric-reducing antioxidant power (FRAP), hypochlorous acid (HOCL) scavenging, copper-phenanthroline, and iron(II) chelation assays were as follows: TEAC (μmol of trolox/g of FW) flavedo 46.1, albedo 36.6, pulp 62; FRAP (μmol of Fe^{2+}/g of FW) flavedo 57.8, albedo 49.5, pulp 92; HOCl (IC_{50}, mg of FW/ml) flavedo 4.4, albedo 7.5, pulp 99.8; copper-phenanthroline (IC_{50}, mg of FW/ml) flavedo 7.5, albedo 67.7; iron(II) chelating activity (IC_{50}, mg of FW/ml) flavedo 9.1, albedo167.1 (Ramful et al. 2010b). Tangor flavedo extracts exhibited higher antioxidant activities than the albedo and pulp.

The ability of the tangor extracts to retard free-radical-induced hemolysis of human erythrocytes was also investigated. Tangor flavedo was found to increase the hemolysis time of red blood cells, compared to the control, in a dose-dependent manner (Ramful et al. 2010b). These increases in HT_{50} reached 100% and 50% when red blood cells were incubated with flavedo and albedo extracts, respectively. In the assessment of antioxidant activity of *Citrus* fruit extracts in the

SW872 human liposarcoma cell line system, the flavedo extracts decreased cell proliferation in a dose-response manner, with significant reductions at 0.75% and 1%. Albedo and pulp extracts had no adverse effects on cell viability, even at the highest concentration tested (1%). In subsequent experiments, cells were thus incubated in the presence of low concentrations of flavedo, albedo, and pulp extracts (0.05% and 0.1%), for which no cytotoxic effect was noted. Non-cytotoxic concentrations of tangelo and tangor flavedo extracts significantly decreased the levels of protein carbonyls in response to advanced glycation end products (AGEs) generated by albumin glycation in SW872 cells. Flavedo extracts lowered carbonyl accumulation in H_2O_2-treated adipocytes, while tangelo and tangor flavedo, albedo, and pulp extracts suppressed ROS (reactive oxygen species) production in SW872 cells with or without the addition of H_2O_2. The results clearly indicated Mauritian Citrus fruit extracts to be an important source of antioxidants, with a novel antioxidative role at the adipose tissue level.

Hypocholesterolemic Activity

The Citrus flavonoids, naringenin and hesperetin, were found to lower plasma cholesterol in-vivo (Wilcox et al. 2001). They found that both naringenin and hesperetin decrease the availability of lipids for assembly of apoB-containing lipoproteins in the human hepatoma cell line, HepG2. The effect was mediated by reduced activities of acyl CoA:cholesterol acyltransferase ACAT1 and ACAT2, a selective decrease in ACAT2 expression, and reduced microsomal triglyceride transfer protein activity. Together with an enhanced expression of the LDL receptor, these mechanisms may explain the hypocholesterolemic properties of the citrus flavonoids. Diets containing 1% citrus polymethoxylated flavones (PMF) such as tangeretin significantly reduced serum total and very low-density lipoprotein (VLDL)+LDL cholesterol (by 19–27 and 32–40%, respectively) and either reduced or tended to reduce serum triacylglycerols in hamsters with diet-induced hypercholesterolemia (Kurowska and Manthey 2004). Comparable reductions were achieved by feeding a 3% mixture of hesperidin and naringin (1:1, w/w). High serum, liver, and urine concentrations of tangeretin metabolites including monohydroxytetramethoxyflavone glucuronides and aglycones and dihydroxytrimethoxyflavone were detected. The data suggested PMFs to benovel flavonoids with cholesterol- and triacylglycerol-lowering potential and that elevated levels of PMF metabolites in the liver might be directly responsible for their hypolipidemic effects in-vivo. Tangeretin was also found to modulate apolipoprotein B (apoB) and lipid metabolism in the human hepatoma cell-line HepG2 (Kurowska et al. 2004). Tangeretin caused a marked reduction in apoB secretion in HepG2 cells. A 24-h exposure of cells to 72.8 μM tangeretin decreased intracellular synthesis of cholesteryl esters, free cholesterol, and triacylglyceride by 82%, 45%, and 64%, respectively; tangeretin also reduced the mass of cellular triacylglyceride by 37%. The tangeretin-induced suppression of triacylglyceride synthesis and mass were associated with decreased activities of DAG acyltransferase and microsomal triglyceride transfer protein. Tangeretin was also found to activate the peroxisome proliferator-activated receptor, a transcription factor with a positive regulatory impact on fatty acid oxidation and triacylglyceride availability. The data suggested that tangeretin modulated apoB-containing lipoprotein metabolism through multiple mechanisms.

Findings by Li et al. (2006) suggested that PMF could ameliorate hypertriglyceridemia and its anti-diabetic effects may occur as a consequence of adipocytokine regulation and PPARalpha and PPARgamma activation. PMF-treated hamsters showed a statistically significant decrease in serum triglyceride (TG) and cholesterol levels compared to the fructose-fed control group. PMF-supplemented animals experienced a reversal of the elevated serum insulin and impaired insulin sensitivity including a decrease in insulin level and an improvement in glucose tolerance. PMF-supplementation significantly increased PPARalpha and PPARgamma protein expression in the liver.

Other Uses

Tangor mandarin is used as a pollinizer for other *Citrus* and in *Citrus* breeding research.

Comments

Tangor mandarin is propagated by growing cuttings and by grafting buds or cuttings onto disease-resistant rootstocks.

Selected References

Berhow M, Tisserat B, Kanes K, Vandercook C (1998) Survey of phenolic compounds produced in *Citrus*. USDA ARS Tech Bull 1856:1–154

Bown D (1995) Encyclopaedia of herbs and their uses. Dorling Kindersley, London, 424 pp

Cottin R (2002) *Citrus* of the world: a *Citrus* directory. Version 2.0. SRA INRA-CIRAD, France

Green CO, Wheatley AO, Osagie AU, St A Morrison EY, Asemota HN (2007) Determination of polymethoxylated flavones in peels of selected Jamaican and Mexican citrus (*Citrus* spp.) cultivars by high-performance liquid chromatography. Biomed Chromatogr 21(1):48–54

Green CO, Wheatley AO, Dilworth LL, Asemota HN (2009) Ortanique peel biomolecules: characterization and nanoparticularization for potential biomedical application. USEACANI workshop 21–26 June 2009. World J Eng (Supplement):P297

Hodgson RW (1967) Horticultural varieties of *Citrus*. In: Reuther W, Webber HJ, Batchelor LD (eds) The *Citrus* industry, vol 1: history, world distribution, botany, and varieties. Revised edition, University of California Press, Berkeley, pp 431–591

Koskinen J (2011) Citrus pages. http://users.kymp.net/citruspages/home.html

Kurowska EM, Manthey JA (2004) Hypolipidemic effects and absorption of citrus polymethoxylated flavones in hamsters with diet-induced hypercholesterolemia. J Agric Food Chem 52(10):2879–2886

Kurowska EM, Manthey JA, Casaschi A, Theriault AG (2004) Modulation of HepG2 cell net apolipoprotein B secretion by the *Citrus* polymethoxyflavone, tangeretin. Lipids 39(2):143–151

Li RW, Theriauet AG, Au K, Douglas TD, Casaschi A, Kurowska EM, Mukherjee R (2006) *Citrus* polymethoxylated flavones improve lipid and glucose homeostasis and modulate adipocytokines in fructose induced insulin resistant hamsters. Life Sci 79(4):365–373

Mabberley DJ (1997) A classification for edible *Citrus*. Telopea 7(2):167–172

Mabberley DJ (2004) *Citrus* (Rutaceae): a review of recent advances in etymology, systematics and medical applications. Blumea 49:481–488

Morton JF (1987) Tangor. In: Fruits of warm climates. Julia F. Morton, Miami, pp 145–146

Mouly P, Gaydou EM, Auffray A (1998) Simultaneous separation of flavanone glycosides and polymethoxylated flavones in citrus juices using liquid chromatography. J Chromatogr A 800(2):171–179

Moura LM, Sylos CM (2008) The effect of the manufacturing process on the *Citrus* juice on the concentrations of flavanones. Alim Nutr Araquara 19(4):379–384

Peterson JJ, Dwyer JT, Beecher GR, Bhagwat SA, Gebhardt SE, Haytowitz DB, Holden JM (2006) Flavanones in oranges, tangerines (mandarins), tangors, and tangelos: a compilation and review of the data from the analytical literature. J Food Comp Anal 19:S66–S73

Ramful D, Bahorun T, Bourdon E, Tarnus E, Aruoma OI (2010a) Bioactive phenolics and antioxidant propensity of flavedo extracts of Mauritian *Citrus* fruits: potential prophylactic ingredients for functional foods application. Toxicol 278(1):75–87

Ramful D, Tarnus E, Rondeau P, Da Silva CR, Bahorun T, Bourdon E (2010b) *Citrus* fruit extracts reduce advanced glycation end products (AGEs)- and H2O2-induced oxidative stress in human adipocytes. J Agric Food Chem 58:11119–11129

Sakamoto K, Inoue A, Nakatani M, Kozuka H, Ohta H, Osajima Y (1996) S-methylmethionine sulfonium in fruits of *Citrus* hybrids. Biosci Biotechnol Biochem 60(9):1486–1487

Swingle WT, Reece PC (1967) The botany of *Citrus* and its wild relatives. In: Reuther W, Webber HJ, Batchelor LD (eds) The *Citrus* industry, vol 1, History, world distribution, botany, and varieties. University of California, Riverside, pp 190–430

Wilcox LJ, Borradalie NM, de Dreu LE, Huff MW (2001) Secretion of hepatocyte apoB is inhibited by flavonoids, naringenin and hesperetin, via reduced activity and expression of ACTAT2 and MTP. J Lipid Res 42:725–734

Wills RBH, Lim JSK, Greenfield H (1985) Composition of Australian foods. 28. *Citrus* fruit. Food Technol Aust 37:308–310

Citrus x floridana

Scientific Name

Citrus x floridana (**J. Ingram & H. Moore**) **Mabb.**

Synonyms

X *Citrofortunella floridana* J. W. Ingram & H. E. Moore, X*Citrofortunella swinglei* J. W. Ingram & H. E. Moore, *Citrus* x *aurantiifolia* (Chrism.) Swingle x *Citrus japonica* Thunb., *Citrus aurantiifolia* (Christm.) Swingle, X *Fortunella* sp.

Family

Rutaceae

Common/English Name

Limequat.

Vernacular Name

Finnish: Limekvatti

Origin/Distribution

Limequat is a hybrid citrus resulting from a crossing of key lime (*Citrus aurantiifolia*) and the round kumquat (*Citrus japonica*) developed by Dr Walter Swingle in 1909. The plant is now grown in Australia, Japan, Israel, Spain, Malaysia, South Africa, Armenia, the United Kingdom and the United States in California, Florida, and Texas.

Agroecology

Limequat does best in full sun in well-drained, well-composted, fertile soil in areas with temperatures between 10°C and 30°C. It is more cold tolerant than limes but less cold-hardy than kumquats. It prefers a moderately heavy loam with a generous amount of compost and sand added. They need frequent sufficient watering to ensure that the soil does not dry out but not excessive watering and abhors manures. It can also be gown in pots indoors.

Edible Plant Parts and Uses

Limequat can be eaten whole or the juice and rind can be used to flavour drinks and dishes. The fruit can be made into jams, jellies and preserves.

Botany

A small, compact, evergreen tree with bushy canopy and glabrous, nearly spineless or sparsely thorny twigs. Petioles are short, glabrous and narrowly-winged to wingless. Leaves simple, oval to ovate-elliptic, 3–4.5 cm by 2–3.5 cm,

Plate 1 Flower cluster and buds

Plate 3 Glossy, simple dark green leaves with crenulate margin

Plate 2 Fertilised flower with persistent calyx

Plate 4 Developing fruits with persistent calyx and stigma remnant

entire with crenulated margin, subacute apex and cuneate base, darker green adaxially and paler green abaxially (Plates 2 and 3). Flowers are borne singly or in clusters in leaf axils, white with tinge of pink in the bud stage, calyx with five persistent, acute lobes (Plates 1, 2 and 3), five white-purplish tinged petals, numerous connate stamens with yellow anthers, superior ovary with one long style and one stigma. Fruit oval to obovate, about 3.2–4 cm long by 2.5–2.8 cm wide at the broadest diameter, pale green to glossy (Plates 4 and 5) , yellow or orangey yellow when fully ripe, rind smooth or glabrous, gland-dotted, containing a few (5–10) small seeds.

Nutritive/Medicinal Properties

No published information is available on its nutritive composition or medicinal uses. Refer to kumquat and other *Citrus* species.

Other Uses

Also planted as indoor ornamental.

Comments

Limequat is a hybrid from a cross of key lime and kumquat:

Plate 5 Mature, unripe fruit

Citrus x aurantiifolia (Chrism.) Swingle X *Citrus japonica* Thunb.] (sensu Mabberley 2004); *Citrus aurantiifolia* (Christm.) Swingle X *Fortunella* sp. (sensu Swingle and Reece 1967; sensu Tanaka sec. Cottin 2002).

Most of the named varieties of limequats, Eustis, Lakeland, etc., are hybrids of *Fortunella japonica* with *Citrus aurantifolia* 'Mexican,' the round kumquat being the male parent. Another named variety of limequat, the Tavares, has the oval kumquat as the male parent (*Citrus aurantifolia* 'Mexican' X *Fortunella margarita*).

Selected References

Cottin R (2002) *Citrus* of the world: a *Citrus* directory. Version 2.0. SRA INRA-CIRAD, France

Hodgson RW (1967) Horticultural varieties of *Citrus*. In: Reuther W, Webber HJ, Batchelor LD (eds) The *Citrus* industry, vol 1: history, world distribution, botany, and varieties. Revised edition, University of California Press, Berkeley, pp 431–591

Huxley AJ, Griffiths M, Levy M (eds) (1992) The new RHS dictionary of gardening, 4 vols, MacMillan, London

Koskinen J (2011) *Citrus* pages. http://users.kymp.net/citruspages/home.html

Mabberley DJ (2004) *Citrus* (Rutaceae): a review of recent advances in etymology, systematics and medical applications. Blumea 49:481–488

Swingle WT, Reece PC (1967) The botany of *Citrus* and its wild relatives. In: Reuther W, Webber HJ, Batchelor LD (eds) The *Citrus* industry, vol 1: history, world distribution, botany, and varieties. Revised edition, University of California Press, Berkeley pp 190–430

Swingle WT, Robinson TR (1923) Two important new types of *Citrus* hybrids for the home garden, citrangequats and limequats. J Agric Res 23:229–238

Webber HJ (1946) Cultivated varieties of *Citrus*. In: Webber HJ, Batchelor LD (eds) The *Citrus* industry, vol 1, History, botany and breeding. University of California Press, Berkeley/Los Angeles, pp 475–668

Citrus x taitensis

Scientific Name

Citrus x taitensis **Risso (sensu Mabberley 2004; Bayer et al. 2009).**

Synonyms

Citrus aurantium subsp. *jambhiri* Engl., *Citrus* x *jambhiri* Lush. (sensu Hodgson 1967; sensu Tanaka sec. Cottin 2002), *Citrus* x *limon* (L.) Osbeck, pro sp. x *Citrus reticulata* Blanco (sensu Federici et al. 2000), *Citrus* x *limonia, Citrus limonia* s. lat. sensu Richards, *Citrus verrucosa* Hort. ex Tanaka.

Family

Rutaceae

Common/English Names

Bush Lemon, Citronelle, Jamberi, Jambhiri Orange, Mazoe Lemon, Rough Lemon, Wild Lemon.

Vernacular Names

Chinese: Cu Pi Ning Meng, Cu Ning Meng;
Finnish: Jambhiri, Karheakuorinen Sitruuna;
French: Citron Verruqueux;
German: Rauhschalig Zitrone;
India: Jamberi, Jatti Khatti, Jhatti Khatti (Hindu);
Italian: Rugoso;
Japanese: Rafu Remon;
Spanish: Limón Rugoso, Rugoso;
Vietnamese: Chanh Vỏ Thô.

Origin/Distribution

Rough lemon is native to the Himalayan foothills in India and has naturalised in many parts in Asia, Australia and the Pacific Islands. In Assam, India semi-wild populations are found. The rough lemon is one of the most important rootstocks of *Citrus* species, cultivated for this purpose in India, USA, South Africa, and Australia. At the end of the sixteenth century it had been introduced to southeast Africa by the Portuguese, and later it was distributed to Europe and the New World. In Australia, it has naturalized along creek banks and on rainforest margins; often persists around old habitations in New South Wales and southern Queensland.

Agroecology

Rough lemon has been used as a rootstock for *Citrus* in Australia for more than 100 years. It was most widely used in Australia in the first half of the twentieth Century. A wide range of Rough lemon selections exist and the most commonly used selections in Australia are McKillop and Lockyer. It has good tolerance to environmental and soil conditions. It is adapted to a wide range of soil types from well drained and aerated sandy soils to clayey soil and alkaline and saline soils. It has moderate tolerance to alkaline and saline soils and is highly drought tolerant. It produces an extensive root system enables trees on to forage effectively for soil nutrients and water.

Edible Plant Parts and Uses

Rough lemon fruit is edible fresh but is more used more for cooking e.g. in Fiji. Rough lemons are used just like the common lemon. Slices of rough lemon are served as a garnish on fish or meat or with iced or hot tea, to be squeezed for the flavoursome juice. In India, the fruits are used as vegetables. The fruit is utilized occasionally as a substitute for lemon.

Plate 1 Rough lemon fruit sold in a Fijian market

Plate 2 Rough lemon fruits sold in a Solok market, Sumatra

Botany

Rough lemon is a shrub or small tree up to 5–6 m high, branches with stiff spines to about 2–3 cm. Leaves are alternate, unifoliolate, oval to elliptic, 5–11 cm by 3.5–6 cm, with broadly acute base and crenate-dentate, blunt and retuse margin, dark green above paler green below, lemon–scented on 3–10 mm long channelled, un-winged petiole. Inflorescence axillary with small, solitary or few-flowered, fragrant flowers. Sepals 4, 1–1.5 mm long; petals 4, 1.5 cm long, white with pinkish tinge, stamens 20–40; ovary 8–12-locular. Fruit is globose to subglobose; pericarp is thick, rough and knobbly or bullate, gland-dotted, with a short, somewhat depressed, conical protuberance (mammilla) at apex, green turning yellow to golden-yellow (Plates 1 and 2), 4–10 cm by 3.5–9 cm, skin thick 2–5 mm, sarcocarp (pulp) contains 10–12 fleshy, pale-yellow, strongly acidic segments.

Nutritive/Medicinal Properties

Its nutritive value has not been published.

The percentage and concentration of total phenolics in *Citrus jambhiri* cv Limoneira leaf, peel and juice were reported by Berhow et al. (1998) as: leaf:- 64% (5.43 mg/g) flavone/flavonol, 30.1% (1.63 mg/g) flavanone, 0% psoralen, 2.7% coumarin (0.07 mg/g); flavedo (outer pigmented layer of the peel): 45% (1.54 mg/g) flavone/flavonol, 38% (0.83 mg/g) flavanone, 0% psoralen, 4.8% (0.05 mg/g) coumarin; albedo (inner layer of the rind): 30.6% (1.00 mg/g) flavone/flavonol, 45.4% (0.95 mg/g) flavanone, 0% psoralen, 3.37.9% (0.08 mg/g) coumarin; juice: 24.5% (0.21 mg/g) flavone/flavonol, 62.4% (0.34 mg/g) flavanone, 0% psoralen, and 3.7% (0.01 mg/g) coumarin.

The percentage and concentration of flavanones reported in *Citrus jambhiri* cv Limoneira by Berhow et al. (1998) were as follows: leaf:- eriocitrin 40% (0.6 mg/g), hesperidin 60% (1.0 mg/g); flavedo: eriocitrin 20% (0.2 mg/g), hesperidin 74% (0.6 mg/g); albedo: eriocitrin 14% (0.1 mg/g), hesperidin 84% (0.8 mg/g); juice:

eriocitrin 12% (traces), hesperidin 88% (0.3 mg/g). The percentage and concentration of flavone reported in *Citrus jambhiri* cv Limoneira cultivar by Berhow et al. (1998) were as follows: leaf: isorhoifolin 3% (0.2 mg/g), diosmin 10% (0.5 mg/g); flavedo: rutin 6% (0.1 mg/g), diosmin 5% (0.1 mg/g); and albedo diosmin 4% (trace).

Twenty-five identified compounds were identified of 33 isolated from the volatiles of *Citrus taitensis* oil (Saleh et al. 1998). The main components were linalool, indole, linalyl acetate, methyl anthranilate, α-terpineol and benzyl cyanide (60.2, 9.0, 6.25, 5.67, 4.26 and 2.18% respectively). Among the flavonoids two major components were dihydrorobinetin and genistein. Both compounds exhibited a high percentage inhibition of the chemiluminescence in polymorphonuclear cells stimulated by N-formylmethionyl-leucylphenylalanine, but a much lower inhibition with stimulation by opsonized zymosan.

Other Uses

Citrus jambhiri is widely used as a rootstock for cultivated lemons against Tristeza virus disease. One of the dwarf potted cultivar *Citrus x jambhiri* 'Otaheite' is used as an ornamental by florists.

Comments

RFLP (restriction fragment length polymorphism) molecular studies by Federici et al. (2000) indicated that Citrus limon could be one progenitor of *Citrus jambhiri* in combination with *C. aurantium, C. bergamia, C. natsudaidai* or *C. taiwanica*. However, except for *C. limonia* the other species were not around in the geographic area in India where *C. jambhiri* arose. They maintained that *C. reticulata* could be another progenitor of *C. jambhiri* on the basis that they share the following common attributes: i) hollow fruit axis, rough and bumpy rind and loosely adhering albedo, ii) high content of monoterpene sabine and presence of polyphenol oxidase, iii) high content of the flavonid hesperid and iv) tolerance to citrus tristeza virus disease. The presence of eriocitrin points to lemon as the other parent. Analysis of published isozyme data would also support *C. limon* x *C. reticulata* (mandarin) as progenitors of C. jambhiri. They concluded that *C. jambhiri* was created by hybridisation of a mandarin by *C. limon*. Earlier Mabberley (1997) reported rough lemon to be a hybrid of *Citrus medica* x *Citrus reticulata* or *Citrus medica* x *Citrus x limon* x *Citrus reticulata*.

Selected References

Bayer RJ, Mabberley DJ, Morton C, Miller CH, Sharma IK, Pfeil BE, Rich S, Hitchcock R, Sykes S (2009) A molecular phylogeny of the orange subfamily (Rutaceae: Aurantioideae) using nine cpDNA sequences. Am J Bot 96:668–685

Berhow M, Tisserat B, Kanes K, Vandercook C (1998) Survey of phenolic compounds produced in *Citrus*. USDA ARS Tech Bull 1856:1–154

Cottin R (2002) *Citrus* of the world: a *Citrus* directory. Version 2.0. SRA INRA-CIRAD, France

Facciola S (1990) Cornucopia. A source book of edible plants. Kampong Publ., Vista, 677 pp

Federici CT, Roose ML, Scora RW (2000) RFLP analysis of the origin of *Citrus bergamia, Citrus jambhiri,* and *Citrus limonia*. Acta Hort (ISHS) 535:55–64

Green P (1994) Norfolk Island species list. In: Flora of Australia, vol 49 Oceanic Islands 1, AGPS. Australian Biological Resources Study

Hodgson RW (1967) Horticultural varieties of *Citrus*. In: Reuther W, Webber HJ, Batchelor LD (eds) The *Citrus* industry, vol 1: History, world distribution, botany, and varieties. Revised edition, University of California Press, Berkeley, pp 431–591

Mabberley DJ (1997) A classification for edible *Citrus* (Rutaceae). Telopea 7(2):167–172

Mabberley DJ (2004) *Citrus* (Rutaceae): a review of recent advances in etymology, systematic and medical applications. Blumea 49:481–489

Morton JF (1987) Lemon. In: Fruits of warm climates. Julia F. Morton, Miami, pp 160–168

Saleh MM, Hashem FAE-M, Glombitza KW (1998) Study of *Citrus taitensis* and radical scavenger activity of the flavonoids isolated. Food Chem 63(3):397–400

Swingle WT (1946) The botany of *Citrus* and its wild relatives of the orange subfamily (family Rutaceae, subfamily Aurantioideae). In: Webber HJ, Batchelor LD (eds) The *Citrus* industry, vol 1, History, botany and breeding. University of California Press, Berkeley, pp 129–474, 1028 pp

Webber HJ (1946) Cultivated varieties of *Citrus*. In: Webber HJ, Batchelor LD (eds) The *Citrus* industry, vol 1, History, botany and breeding. University of California Press, Berkeley, pp 475–668, 1028 pp

Citrus x limon

Scientific Name

Citrus x limon (Linnaeus) Osbeck.

Synonyms

Citrus abyssinica (Engl.) Engl., *Citrus inaequalis* (DC.) Benth., *Citrus limon* (Linn.) Burm. f., *Citrus limon* (L.) Osbeck, *Citrus limonum* Risso, *Citrus medica* subsp. *limonia* Hook. f. ex Engl., *Citrus medica* Linnaeus var. *limon* Linnaeus, *Citrus x aurantium* Linnaeus subsp. *bergamia* (Risso) Engler, *Citrus x aurantium* var. *bergamia* (Risso) Brandis; *Citrus x aurantium* var. *mellarosa* (Risso) Engler; *Citrus x bergamia* Risso, *Citrus x bergamia* subsp. *mellarosa* (Risso) Rivera et al., *Citrus x bergamota* Rafinesque, *Citrus x limodulcis* Rivera et al., *Citrus x limonum* Risso; *Citrus medica* Linnaeus f. *limon* (Linnaeus) Hiroe, *Citrus medica* subsp. *limonum* (Risso) J. D. Hooker, *Citrus medica* var. *limonum* (Risso) Brandis, *Citrus x mellarosa* Risso, *Citrus x meyeri* Yu. Tanaka, *Limon x vulgaris* Ferrarius ex Miller, not *Citrus x vulgaris* Risso.

Family

Rutaceae

Common/English Names

Canton Lemon, Chinese Lemon, Lemon.

Vernacular Names

Albanian: Limoni;
Amharic: Li-Mo-Ne, Lo-Mi;
Arabic: Laymūn, Līmūn, Qalambak, Limoon, Laymûn Adâlyâ Mâlehh;
Aragonese: Limón;
Armenian: Gidron, Gidronakhod, Kitron, Kitrona ot, Kitronaxot;
Azerbaijan: Limon;
Basque: Limoi;
Brazil: Limão-Siciliano (Portuguese);
Bulgaria: Limon;
Burmese: Shauktakera;
Catalan: Llimonera;
Chamorro: Lemon Reat, Limon Real;
Chinese: Ning Meng, Yang Ning Meng;
Cook Islands: Rēmene, Rēmene PapaʻA, Tiporo (Maori);
Croatia: Limun;
Czech: Citrón, Citroník Kantonský, Citroník Kyselý, Citroník Limetta, Citroník Limonový, Citroník Obecný;
Danish: Citron;
Dutch: Citroenboom, Citroen;
Eastonia: Harilik Sidrunipuu;
Esperanto: Citron;
Farsi: Limoo, Limou, Limou Khagi, Lýmw, Lýmw agý;
Fijian: Moli Karokaro, Moli Ni Vavalangi, Moli Sosoriatia;
Finnish: Sitruuna, Sitruunapuu;
French: Citronnier, Citron, Lemonier, Limon;
Frisian: Sitroen;

Galician: Follas De Lima Cafre;
Georgian: Limoni;
German: Citron, Citrone, Limone, Sauer-Zitrone, Zitrone, Zitronenbaum;
Greek: Lemóni;
Guam: Limon Real;
Haiti: Sitron;
Hawaiian: Kukane, Lemi;
Hebrew: Limon;
Hungarian: Citrom, Európai Citrom, Közönséges Citrom, Valódi Citrom;
Icelandic: Sítróna;
I-Kiribati: Remen, Te Remen, Te Remon;
India: Gol Nemu (Assamese), Baranebu, Nebu (Bengali), Limbu (Gujerati), Bara-Nimbu, Jamiri Nimbu, Khatta, Nimbu, Pahari-Nimbu (Hindu), Brihat Nimbe, Dodda Nimbe (Kannada), Cerunarakam, Gilam (Malayalam) Champra, Nobab (Manipuri), Limbu, (Marathi), Lembu (Oriya), Bijauri, Galgal (Punjabi), Amla, Dantasathah, Jambira, Jambirah, Limpaka, Mahajambiraphalam, Mahanimbu Tvak, Naranga, Nimbu, Nimbaka, Nimbuka, Ruchaka, Vijapura (Sanskrit), Cīta ai, Elumiccai, Elumichai, Elumicham, Sidalai (Tamil), Dabba, Nimmapandu (Telugu), Lemu, Lembu, Nembu, Nenbu, Bara Nebu, Utraj (Urdu);
Indonesia: Jerul Nipis, Sitrun;
Italian: Limone;
Japanese: Kanton Remon, Remon, Re-Mo-N;
Kazakhsistan: Limon, Lymon;
Korean: Re-Mon;
Latvia: Citrons;
Lingala: Ndímo;
Lithuanian: Citrinos, Tikrasis Citrinmedis;
Malaysia: Limau;
Maltese: Lumi;
Nepalese: Kaagati, Nibuwa;
New Caledonia: Le Citronnier;
Niuean: Sipolo, Sipolo Fua Lalahi, Tipolu;
Norway: Sitron;
Pakistan: Jatti Khatti, Gulgul, Jhambheri;
Palauan: Debechel, Malchianged;
Persian: Kalanbak, Lemu-E-Tursh, Limoo Khagi;
Philippines: Lemon (Tagalog);
Polish: Cytryna, Cytryna Zwyczajna;
Portuguese: Limão, Limão Cravo, Limoeiro Azedo;
Pukapukan: Lēmene Papā;
Romanian: Lămâi;
Russian: Limon;
Samoan: Moli, Moli Tipolo;
Scotland: Crann Limoin (Gaelic);
Slovaščina: Limona;
Slovencina: Citrón, Citrónovník, Citróny;
Spanish: Arbol De Limón, Lemonero, Limón, Limón Agria, Limón Amarillo, Limón De Canton, Limón Frances, Limón Grande, Limón Real, Limón Sútil, Limonero;
Sri Lanka: Sedaran;
Swahili: Limau;
Swedish: Citron, Citronträd, Suckat;
Thai: Ma Nao Leung, Manao Farang, Manao Thet, Som Saa;
Tibetan: Dzam Bi Ri (D), Dzi Ma Bi Ra;
Tongan: Lemani, Moli Lemani;
Tongarevan: Lēmene, Rēmene;
Turkish: Limon;
Ukrainian: Limon, Lymon;
Vietnamese: Chanh Tây, Nịnh Mong;
Yiddish: Limene, Tsitrin, Zitstrin.

Common Lemon Varieties: Eureka, Lisbon, Fino, Verna.

Origin/Distribution

The exact origin of the lemon is unclear but is thought to have originated in the foothills of the Himalayas, Assam (India) and northern Burma to China. Lemon has been introduced all over the world and has naturalised in many areas.

Agroecology

Lemon has a wide climatic range; it can be cultivated in a tropical, subtropical and temperate regions, in humid, semi-arid or arid conditions. Where water is not limiting, temperature and wind are the two most important constraints to growth and development. Lemons are mostly grown in the semi arid and arid subtropical regions of the world where temperatures are above −4°C. Lemons are sensitive to freezing temperatures and frosts and should not be grown in areas with severe winter frosts and where temperatures fall below −4°C. Lemon trees are defoliated at −4.4°C to −5.6°C, wood is damaged at −6.7°C, young fruits and flowers are destroyed at −1.7°C and mature fruits are badly damaged at −2°C (Morton 1987).The optimum temperature for growth is 25–30°C with the maximum rate of photosynthesis occurring at 30°C. There is little or no growth at temperatures above 40°C. Good fruit set and yields can be obtained in warm regions if trees have adequate water and nutrients. High temperatures combined with low humidity can cause problems of sunburnt fruit.

Lemon trees are susceptible to strong winds or cold winds that reduce yields and scar fruit. Trees should be planted in protected sites from cold winds. Lemon tree is adaptable to many soil types excepting those that are too acidic < pH 4.5 or too alkaline above pH 8. A preferable pH range is 6.0–7.0. The lemon tree has the reputation of tolerating very infertile, very poor soil.

Eureka cultivar is less cold tolerant than the Lisbon cultivar. Lisbon is hardier than Eureka with greater cold and heat tolerance.

Edible Plant Parts and Uses

Lemons are mostly used for their juice, peel and oil. The oil is present in the peel, juice sacs and seeds and is used as flavouring agents in drinks, food, pharmaceutical and other produces. Lemons are highly acid and therefore used in refreshing drinks, cordials, and in syrups and to flavour and garnish foods. Lemons alone or combined with orange are used to make marmalade. In Morocco, fresh ripe lemons are incised with large amount of salt to make a unique pickled lemons delicacy called *al-hamid al-marqad* which for flavouring. The peel is used in the production of brined rind, candied peels, pectin, citric acid and flavonoids. Pickled lemon peel is an indispensable spice of Moroccan cuisine and frequently employed in meat or fish stews known as *tagine* or *tajine* which are slowly braised in a conical-shaped clay pot carrying the same name. *Ritschert*, a traditional South Austrian stew made from white beans, smoked meat and pearl barley, is usually prepared with a dash of lemon peel. Lemon peel goes well for types of food that are prepared with lemon juice as well, for example fish soups or fish stews. In South Italy, whole chopped lemons, or lemon juice plus lemon peel are made into pasta sauces and a liqueur called *limoncello* is made from lemon peel. Many mixed drinks, soft drinks, iced tea, and water are often served with a wedge or slice of lemon in the glass or on the rim. Lemon zest, the yellow part of the peel on the outside of a lemon has many uses. Lemon zest imparts a strong citrus flavour to foods. Lemon zest, depending on its use, comes in many forms: long wide strips, julienne, or finely grated. Lemon zest (or juice) is used in muffins, stir-fries, cakes, cookies, sherbert, cheesecakes, walnut rolls, stews, casseroles, tarts, pies, preserves and rice. A few drops of lemon juice added to cream before whipping gives stability to the whipped cream. Lemon juice is also used to help set jams.

More importantly, lemons are valued for its lemon juice in cuisines all over the world. Fish are marinated in lemon juice to neutralize the odor and neutralise the amines, converting them into ammonium salts. Fried or grilled fish is nearly always served with a few splashes of lemon juice which mitigates the typical 'fishy' smell and makes it more pleasant. Lemon juice, alone or in combination with other ingredients, is used to marinate and tenderise meat before cooking. Lemon juice sprinkled over peeled apples, avocados and bananas retards phenol oxidase browning and deterioration. Lemon juice intensifies the flavour of many fruits, and a few drops of lemon juice plus a dash of sugar creates a slightly sweet-sour tang that can make many vegetables more interesting. Outside of the tropics, lemon juice is often used as a substitute for lime juice. Lemon juice is especially popular in the East Mediterranean, e.g., in Lebanese *tabbouleh* and is the key ingredient famed Greek yolk-lemon sauce *avgolemono* the ingredients of which include fish or meat broth, lemon juice, egg yolks, and a pinch of black pepper with or without corn starch as thickener. This sauce is usually served to boiled meats or vegetables, but it can also be made into a more hearty soup by adding rice or noodles. Lemon juice is frequently employed to prepare refreshing salads, especially in the Mediterranean countries. Lemon juice, fresh, canned, concentrated or frozen or in the form of dehydrated powder is primarily used for lemonade, carbonated beverages or other mixed drinks.

Studies showed that a polyphenol rich drink made of 75% of pomegranate juice and 25% of lemon juice (v:v), has potential for development of new healthy beverages or food products, emphasised by its high antioxidant capacity determined by its phenolic composition – punicalagin isomers, anthocyanins and vitamin C – and improved colour properties (González-Molina et al. 2009). A high dietary fibre powder with good functional, microbial quality and favourable physicochemical characteristics for use in food formulations was obtained from lemon juice industry by-products (Lario et al. 2004).

Dried lemon leaves are sometimes mixed with tea leaves for use as a flavouring. Lemon flowers are eaten in ice creams, fritters, jams, etc.

Botany

Lemon is a small, thornless to thorny, medium vigour tree, reaching 3–6 m high. Leaves are alternate, reddish when young turning green, ovate to elliptic, 8–14×4–6 cm, with conspicuously crenulate margin and mucronate, acuminate to shallowly emarginate apex on narrowly winged petioles (Plates 1, 2 and 3). Flowers solitary or several in axillary or terminal fascicles (Plate 1). Flowers bisexual or male by complete abortion of pistil. Calyx is cup shaped with 4–5 lobes; petals, 4 or 5, are white inside and purplish outside, stamens 20–40 more or less united stamens with yellow anthers, ovary subcylindric to barrel-shaped with a clavate stigma. Fruit green turning pale green and yellow when ripe, ellipsoid to ovoid, narrowed at both ends, with a nipple-like protuberance (mammilla) at the apex, up to 12 × 6 cm, with a fairly thick, coarse, pellucid-dotted rind (Plates 2, 3, 4, 5 and 6). Sarcocarp (pulp) in 8–11 segments, pale yellow, acidic. Seeds ovoid, elliptic or ovate, pointed, smooth, small up to 10 mm long with one to numerous embryo(s) and milky-white cotyledons.

Plate 1 Eureka lemon flowers and leaves

Plate 2 Immature Eureka lemons

Plate 3 Ripe Eureka lemons

Eureka (Plates 1, 2, 3 and 4) has a spreading habit with sparse foliage and is virtually thornless. Fruits are borne in terminal clusters, ellipsoid to oblong with a short neck and nipple, seedless to a few (5) seeds. Rind is yellow, rough with ridges on the surface.

Lisbon (Plate 5) is thorny, large and upright, vigorous growth habit with dense foliage. Fruit is produced inside the tree. The fruit is medium in size, ellipsoid to oblong with an inconspicuous neck and prominent nipple. The fruit is smoother and less ridged than Eureka, yellow when mature with medium thick rind and pale greenish yellow flesh, very juicy and acid, seedless to a few seeds per fruit.

There is an Indonesian lemon with an elongated nipple (Plate 6).

Plate 4 Eureka lemon fruits and leaves

Nutritive/Medicinal Properties

Nutrient composition of fresh, raw lemons per 100 g edible portion (USDA 2011) has been reported as follows: water 88.98 g, energy 29 kcal (121 kJ), protein 1.10 g, total fat 0.30 g, ash 0.30 g, carbohydrate 9.32 g, total dietary fibre 2.8 g, total sugars 2.50 g, Ca 26 mg, Fe 0.60 mg,

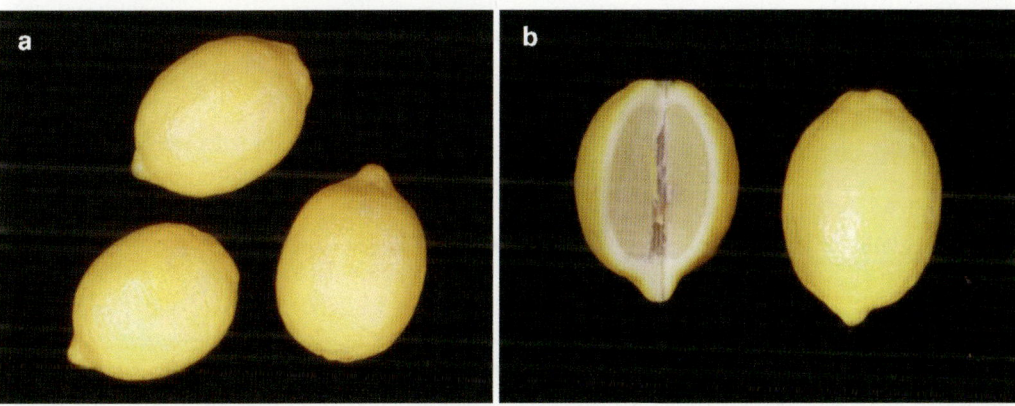

Plate 5 Harvested Lisbon lemons: (**a**) whole and (**b**) sliced

Plate 6 (**a**, **b**) Indonesian lemon cultivar with elongated nipple

Mg 8 mg, P 16 mg, K 138 mg, Na 2 mg, Zn 0.06 mg, Cu 0.037 mg, Mn 0.030 mg, Se 0.4 μg, vitamin C 53 mg, thiamine 0.040 mg, riboflavin 0.020 mg, niacin 0.10 mg, pantothenic acid 0.190 mg, vitamin B-6 0.080 mg, total folate 11 μg, total choline 5.1 mg, vitamin A 22 IU (1 μg RAE), vitamin E (α-tocopherol) 0.15 mg, total saturated fatty acids 0.039 g, 14:0 (myristic acid) 0.001 g, 16:0 (palmitic acid) 0.035 g, 18:0 (stearic acid) 0.002 g; total monounsaturated fatty acids 0.011 g, 16:1 undifferentiated (palmitoleic acid) 0.001 g, 18:1 undifferentiated (oleic acid) 0,010 g; total polyunsaturated fatty acids 0.089 g, 18:2 undifferentiated (linoleic acid) 0.063 g, 18:3 undifferentiated (linolenic acid) 0.026 g, β-carotene 3 μg, α-carotene 1 μg, β cryptoxanthin 20 μg, lutein + zeaxanthin 11 μg.

Lemons, both mature-green and yellow, yielded a complex mixture of carotenoids (Yokoyama and Vandercook 1967). The lighter color of the pulp and flavedo of yellow lemons may be attributed to much lower concentrations of carotenoids than in oranges. As the lemons ripened, chlorophyll and α-carotene disappeared and ζ-carotene and an η-carotene-like compound appeared in both the pulp and peel. Further, small but significant amounts of β-carotene 5,6-monoepoxide and its isomeric 5,8-epoxide, mutatochrome, were detected in the pulp of yellow lemons. Studies reported that grapefruit, lemons, and limes contained the following flavanones: didymin, eriocitrin, hesperidin, naringin, narirutin, neoeriocitrin, neohesperidin, poncirin (Peterson et al. 2006). Grapefruit had a total flavanone content (summed means) of 27 mg/100 g as aglycones and a distinct flavanone profile, dominated by naringin. For lemons, total flavanones (summed means) were 26 mg/100 g and for limes 17 mg/100 g. The flavanone profiles of both lemons and limes were dominated by hesperidin and eriocitrin.

The average total carotenoid contents for reconstituted lemon juice concentrate and for fresh juice from coastal, desert, and Italian lemons were respectively 49, 50, 40, and 45 μg β-carotene/l00 ml juice (Vandercook and Yokoyama 1965). The average respective sterol values were 8.28, 8.99, 9.10, and 8.46 mg β-sitosterol/100 ml juice. The major lemon juice sterol was β-sitosterol. Lemon juice and lime juice were found to be rich sources of citric acid, containing 1.44 and 1.38 g/oz, respectively (Penniston et al. 2008). Lemon and lime juice concentrates contained 1.10 and 1.06 g/oz, respectively. The citric acid content of commercially available lemonade and other juice products varied widely, ranging from 0.03 to 0.22 g/oz. Lemon and lime juice, both from the fresh fruit and from juice concentrates, provided more citric acid per liter than ready-to-consume grapefruit juice, ready-to-consume orange juice, and orange juice squeezed from the fruit. Flavone-di-C-glycosides namely 6,8-Di-C-glucopyranosylapigenin and 6,8-di-C-glucopyranosyldiosmetin were detected in a variety of Southern Italian Citrus juices (orange, lemon, bergamot, citron, mandarin, clementine) (Caristi et al. 2006). 8-di-C-glucopyranosyldiosmetin was the most important C-glycoside in lemon and citron juice.

A total of 51 constituents were identified and quantified in the essential oil composition from fruit peels of *C. limon* grown in Zulia State, Venezuela (de Rodríguez et al. 1998): 28 mono and sesquiterpene hydrocarbons, 8 aldehydes, 10 alcohols, 3 esters, 1 ketone, and 1 oxide were identified. Limonene was the major component (65.65%). Among the other monoterpene hydrocarbons, there was a high proportion of β-pinene (11.0%) and γ-terpinene (9.01%). Oxygenated compounds were found in amounts of 3.79%. Aldehydes were the most abundant constituents of this oxygenated fraction (2.70%) specially the quality indicators: geranial (1.43%) and neral (0.87%). Alcohols were represented in 0.53%, esters in 0.56% and just a little percentage of artifacts was found (0.01%). Monoterpenes (0.94%) included: limonene 65.65%, β-pinene 11.00%, γ -terpinene 9.01%, α-pinene 1.88%, sabinene 1.05%, myrcene 1.01%, α-thujene 0.42%, terpinolene 0.39%, α-terpinene 0.22%, ρ-cymene 0.10%, *trans*-β-ocymene 0.09%, camphene 0.06%, α-phellandrene 0.05%, δ-3-carene 0.01%. Aldehydes (2.71%) included: geranial 1.44%, neral 0.87%, citronellal 0.14%, nonanal

0.12%, octanal 0.07%, decanal 0.04%, undecanal 0.02%, dodecanal/decyl acetate 0.01%. Alcohols (0.52%) included: α-terpineol 0.17%, linalool/*cis*-sabinene hydrate 0.16%, terpinen-4-ol 0.06%, *trans*-sabinene hydrate/octanol 0.05%, citronellol/nerol 0.04%, geraniol 0.03%, borneol 0.01%. Ketone and oxide (0.02%) included: camphor 0.01%, and piperitone 0.01%. Esters (0.57%) included: neryl acetate 0.35%, geranyl acetate 0.22%. Sesquiterpenes (1.57%) included: *trans*-α-bergamotene 0.41%, β-bisabolene 0.40%, β-caryophyllene 0.25%, germacrene B 0.11%, α-bisabolol 0.09%, β-santalene/cis-β-farnesene 0.08%, cis-α-bergamotene 0.05%, *trans*-β-farnesene 0.04%, campherenol 0.03%, 2,3-dimethyl-3-(4-methyl-3-pentenyl)-2-norbornanol 0.03%, γ-elemene 0.03%, valencene 0.03%, and α-humulene 0.02%.

González et al. (2002) reported that the major component of the essential oil of *Citrus x limon* fruit pericarp was limonene 75.73% followed in descending order by δ-terpinene 9.19%, sabinene 4.77%, 1-isopropyl-2-methoxy-4-methyl-bencene 2.28%, b-myrcene 2.17%, α-pinene 2.04%, p-cimene 1.46%, β-pinene 1.16%, α-terpinene 0.64% and α-thujene 0.56%.

A total 42 components were identified in Turkish lemon peel oil (Kırbaşlar et al. 2006, 2009): monoterpene hydrocarbons (89.9%) with limonene (61.8%), γ-terpinene (10.6%) and β-pinene (8.1%) as the major components; sesquiterpene hydrocarbons (3.3%) with β-bisabolene (1.6%), trans-α-bergamotene (1.0%) and β-caryophyllene (0.7%) as major components. The major oxygenated components (5.1%) comprised aldehydes components (2.4%) geranial (1.3%), neral (0.7%), octanal (0.1%), decanal (0.1%); alcohol components (0.9%); linalool (0.2%), nerol (0.1%), geraniol (0.1%) and ester components (1.8%) neryl acetate (1.2%) and geranyl acetate (0.6%).

Hydrodistilled essential oils from the fresh leaves of *Citrus limon* cv. Pusa, was found to contain approximately 95% monoterpenes and 2% sesquiterpenes (Goel et al. 2004). Limonene (40.9%) was the predominant monoterpenic component followed by linalool 12.2%, geranial 12.1%, neral 9.1% and (Z)-β-ocimene 7.3%.

Citrus limon is a rich source of flavonoids. The flavonoids identified in lemon were hesperidin, glycoside of eriodictyol, two flavones glycosides – diosmin and luteolin (Horowitz 1956). Some of the medicinal properties of lemons are due to the flavonoids they contain since they are involved in many biological activities and have many health-related functions. Studies showed that the immature fruits from cultivars Lisbon and Fino-49 were ideal for obtaining the flavanone hesperidin, while the mature fruits of cultivar Fino-49 and the leaves of cultivar Eureka were the most interesting for obtaining the flavone diosmin and the flavanone eriocitrin (Del Rio et al. 2004).

Four flavonoids: 5,7,8,3′,4′-pentamethoxy flavone; 5,7,8,4′-tetramethoxy flavone; 5, 6,7, 8, 3′, 4′-hexamethoxy flavone; 5, 6, 7, 8, 4′-pentamethoxy flavone; and four coumarins xanthoxyletin, limettin, umbelliferone and scopoletin were purified and structurally confirmed from *Citrus limon* (Linn) Burm f. (var. Jatti Khatti) (Khatoon 1995). The percentage and concentration of total phenolics in *Citrus limon* cv Bergamot leaf, peel and juice were reported by Berhow et al. (1998) as: leaf:- 50.2% (3.70 mg/g) flavone/flavonol, 38.8% (3.92 mg/g) flavanone, 0.6% psoralen, 5.1% coumarin (1.92 mg/g); flavedo (outer pigmented layer of the peel): 51% (0.8 mg/g) flavone/flavonol, 16.4% (5.17 mg/g) flavanone, 8.8% psoralen, 10.2% (1.06 mg/g) coumarin; albedo (inner layer of the rind): 34.8% (1.0 mg/g) flavone/flavonol, 46% (1.83 mg/g) flavanone, 7.3% psoralen, 3.1% (1.55 mg/g) coumarin; juice: 19.9% (1.2 mg/g) flavone/flavonol, 67.7% (0.27 mg/g) flavanone, 3% psoralen, and 0% coumarin.

The percentage and concentration of flavanones reported in *Citrus limon* cv Bergamot by Berhow et al. (1998) were as follows: leaf:- hesperidin 13% (0.2 mg/g), neoeriocitrin 27% (0.05 mg/g), naringin 6% (0.1 mg/g), naringin–6″–malonate (closed form) 17% (0.3 mg/g); flavedo: neoeriocitrin 46% (0.5 mg/g), naringin 25% (0.3 mg/g); albedo: hesperidin 6% (0.1 mg/g), neoeriocitrin 57% (0.9 mg/g), naringin 20% (0.3 mg/g), naringin–6″–malonate (closed form) 10% (0.2 mg/g); juice: didymin 16% (0.1 mg/g), eriocitrin 37% (0.2 mg/g), naringin 3% (traces), neohesperidin 9% (traces), naringin–6″–malonate (open form) 10% (0.1 mg/g). The percentage and concentration of flavone reported in *Citrus limon* cv Bergamot by Berhow et al. (1998) were as follows: leaf: rhoifolin 9% (0.4 mg/g), diosmin 16% (0.6 mg/g); flavedo: diosmin 16% (0.8 mg/g); albedo: rhoifolin 11% (0.2 mg/g); juice: rhoifolin 23% (0.1 mg/g).

Citrus genus is the most important fruit tree crop in the world and lemon is the third most important *Citrus* species (González-Molina et al. 2010). Several studies highlighted lemon as an important health-promoting fruit rich in phenolic compounds as well as vitamins, minerals, dietary fibre, essential oils and carotenoids. Lemon fruit has a strong commercial value for the fresh products market and food industry. Moreover, lemon productive networks generate high amounts of wastes and by-products that constitute an important source of bioactive compounds with potential for animal feed, manufactured foods, and health care. Lemon fruit and leaf extracts and essential oil have antioxidant, anticancer, hepatoprotective, antimicrobial and insecticidal bioactivities.

Antioxidant Activity

All of the 34 *Citrus* essential oils were found to have scavenging effects on 1,1-diphenyl-2-picrylhydrazyl (DPPH) in the range of 17.7–64.0% (Choi et al. 2000). The radical-scavenging activities of 31 kinds of *Citrus* essential oils were comparable with or stronger than that of Trolox, a standard antioxidant. The oils of Ichang lemon (64.0%, 172.2 mg of Trolox equiv/ml), Tahiti lime (63.2%, 170.2 mg of Trolox equiv/ml), and Eureka lemon (*Citrus limon*) (61.8%, 166.2 mg of Trolox equiv/ml) were stronger radical scavengers than other *Citrus* oils. *Citrus* volatile components such as geraniol (87.7%, 235.9 mg of Trolox equiv/ml), terpinolene (87.4%, 235.2 mg of Trolox equiv/ml), and γ-terpinene (84.7%, 227.9 mg of Trolox equiv/ml) showed marked scavenging activities on DPPH.

An antioxidant was isolated from the peel and juice of lemon fruit and identified as eriocitrin (eriodictyol 7-rutinoside) a flavanone glycoside (Miyake et al. 1997b). The distribution of eriocitrin in *Citrus* fruits was found to be especially abundant in lemons and limes, however, it was scarcely found in other citrus fruits. In the case of lemon fruit, eriocitrin was primarily distributed in the peel (about 1,500 ppm) composed of the albedo (mesocarp), flavedo (epicarp), and pulp vesicles. It was also significantly present in the juice (about 200 ppm) but was not detected in the seed. Two varieties of lemon fruits, eureka and lisbon, almost had the same eriocitrin content. The antioxidative activity of eriocitrin in the linoleic acid autoxidation system was equal to that of α-tocopherol, and it was enhanced when used together with citric acid. The eriocitrin had a synergistic effect on α-tocopherol. Studies showed that eriocitrin was effective in the prevention of oxidative damages caused by acute exercise-induced oxidative stress (Minato et al. 2003). Without eriocitrin administration the contents of Nε- (hexanoyl)lysine (HEL); o,o-dityrosine (DT); and nitrotyrosine, NT, oxidative stress markers in the rat's liver were markedly increased by exercise. These increases were significantly suppressed by eriocitrin administration before exercise. The level of reduced glutathione after exercise was maintained by administration of eriocitrin. The increase in the concentration of oxidized glutathione caused by exercise was significantly suppressed by eriocitrin. This result suggested that eriocitrin might play an important role in the control of the change in glutathione redox status in the rat liver during exercise.

In a another study, eriocitrin (eriodictyol 7-rutinoside), an antioxidant in lemon fruit was hydrolyzed to eriodictyol, its aglycon, by intestinal bacteria, *Bacteroides distasonis* and *Bacteroides uniformis* (Miyake et al. 1997c). Eriodictyol was converted to 3,4-dihydroxyhydrocinnamic acid and phloroglucinol by *Clostridium butyricum*. Using linoleic acid autoxidation and a rabbit erythrocyte membrane oxidation system, it was shown that eriodictyol exhibited stronger antioxidant activity than α-tocopherol. The antioxidant activities of

3,4-dihydroxyhydrocinnamic acid and phloroglucinol were weaker. Eriocitrin, a flavonoid glycoside present in lemon fruit, was metabolized in-vivo by intestinal bacteria to a series of eriodictyol, methylated eriodictyol, 3,4-dihydroxyhydrocinnamic acid, and their conjugates (Miyake et al. 2000). Plasma antioxidant activity increased following oral administration of aqueous eriocitrin solutions to rats. Eriocitrin was not detected in plasma and renal excreted urine, but eriodictyol, homoeriodictyol, and hesperetin in their conjugated forms were detected in plasma of 4 hours following administration of eriocitrin. In urine for 24 hours, both nonconjugates and conjugates of these metabolites were detected. 3,4-Dihydroxyhydrocinnamic acid, which was metabolized from eriodictyol by intestinal bacteria, was detected in slight amounts with each form in 4-hours plasma and 24-hours urine. Following administration of eriocitrin, plasma exhibited an elevated resistance effect to lipid peroxidation.

Recent studies revealed that the alcoholic extract of lemon peel had the maximal content of both phenolics and flavanoids compared to the hexane and aqueous extracts (Akhila et al. 2009). Antioxidant activity was evaluated by using the β-carotene The alcoholic extract was found to have good free radical inhibitory property as well as nitric oxide radical scavenging activity. The hexane extract, however, showed only good hydrogen peroxide scavenging activity. Thus, the higher antioxidant activity of the alcoholic and hexane extracts of *Citrus limon* peel could have wide therapeutic utility against various diseases. Antioxidants play an important role in the treatment of a number of human neurodegenerative disorders like diabetes, inflammation, Alzheimer's disease, autoimmune pathologies, and digestive system disorders.

Six flavanone glycosides: eriocitrin, neoeriocitrin, narirutin, naringin, hesperidin, and neohesperidin, and three flavone glycosides: diosmin, 6,8-di-*C*-β-glucosyldiosmin (DGD), and 6-*C*-β-glucosyldiosmin (GD) were identified in lemon peel (Miyake et al. 1998a). Lemon fruit contained abundant amounts of eriocitrin and hesperidin and also contained narirutin, diosmin, and DGD, but GD, neoeriocitrin, naringin, and neohesperidin were present only in trace amounts. The antioxidative activity of eriocitrin, neoeriocitrin and DGD was stronger than that of the others as assessed using a linoleic acid autoxidation system. Two antioxidative C-glucosylflavones were isolated from the peel of lemon fruit and were identified as 6,8-di-C-β-glucosyldiosmin (LE-B) and 6-C-β-glucosyldiosmin (LE-C) (Miyake et al. 1997a). LE-B and LE-C showed antioxidative activity in linoleic acid autoxidation and the liposome oxidation systems but exhibited weaker activity than eriocitrin, its eriodictyol of its aglycon. Eriocitrin and its metabolites (eriodictyol, 3,4-dihydroxyhydrocinnamic acid, and phloroglucinol) produced by intestinal bacteria exhibited stronger antioxidative activity than α-tocopherol in the LDL oxidation system and had approximately the same activity as (−)-epigallocatechin gallate. Eriocitrin and its metabolites are powerful antioxidants using an in vitro oxidation model for heart disease. A comparative study between the antioxidant properties of peel (flavedo and albedo) and juice of some commercially grown citrus fruit (Rutaceae), grapefruit (*Citrus paradisi*), lemon (*Citrus limon*), lime (*Citrus x aurantiifolia*) and sweet orange (*Citrus sinensis*) revealed that reducing sugars and phenolics were the main antioxidant compounds found in all the extracts (Guimarães et al. 2010). Peels polar fractions revealed the highest contents in phenolics, flavonoids, ascorbic acid, carotenoids and reducing sugars, which certainly contribute to the highest antioxidant potential found in these fractions.

Fermented lemon peel prepared by fermentation of the lemon peel with *Aspergillus saitoi*, exhibited higher radical-scavenging activity against DPPH than lemon peel (Miyake et al. 2004). The administration of lemon peel and fermented lemon peel prior to exercise significantly suppressed the increases in thiobarbituric acid-reactive substance caused by lipid peroxidation during exhaustive exercise, with fermented lemon peel having a tendency of higher activity than lemon peel. Further, fermented lemon peel showed a significant suppressive effect of Ne-(hexanonyl)lysine content, a primary oxidative stress marker, which increased due to exhaustive exercise, whereas lemon peel did not. The high antioxidative activity of fermented lemon peel was assumed to be related to the production of antioxidative flavonoids, as a hydroxyflavanone and aglycones, by the fermentation of lemon peel with *Aspergillus saitoi*.

A compound isolated from lemon oil, called Lem 1 was found to be endowed with a strong antioxidant activity and that it was capable of inhibiting free radical-mediated reactions, evaluated by both in-vitro and in-vivo biochemical systems (Calabrese et al. 1999). It was effective in controlling free radical-induced lipid peroxidation and tissue damage in the skin. Lem 1 was deemed to have potential for biotechnological application in cosmetic dermatology.

The antioxidant activity and the contents of antioxidants of lemon leaf methanol extracts were studied in different lipid peroxidation systems (Chang et al. 2004). Antioxidant capacity of methanol extracts of both freeze-dried and heat-dried, old age lemon leaves were higher than those from middle-age or young leaves. Compared with methanol extracts from hot air-dried lemon leaves (MEHDL), methanol extracts from freeze-dried lemon leaves (MEFDL) exhibited higher antioxidant activity. MEFDL has 63.1%, 67.9%, and 75.6% scavenging effect on α-diphenyl-β-picrylhydrazyl (DPPH), superoxide anion, and hydrogen peroxide, respectively. MEFDL also showed high reducing power and chelating ability when combined with copper. MEFDL exhibited the highest antioxidative activity in the autooxidation system compared with the other induced systems, and had higher inhibited peroxidation effect at higher dosage than any peroxidation systems. The primary antioxidant contents of ascorbic acid, α-tocopherol, β-carotene, and total phenol in MEFDL were 0.93, 0.8, 1.69, and 2.8 mg/g, respectively. Essential oil of *C. limon* leaves exhibited strong antioxidant activity according to the scavenging assays (Campêlo et al. 2011b). The essential oil exerted a significant antioxidant effect against peroxyl radicals generated by AAPH in the TBARS assay, protecting lipids from oxidation, especially at the highest dose. It demonstrated significant scavenging effect against NO, but lower concentrations reversed this NO-inhibiting effect and led to a small increase in NO production. GC-MS analysis showed a mixture of monoterpenes, with *limon*ene (52.77%), geranyl acetate (9.92%) and trans-*limon*ene-oxide (7.13%) as the main components in the essential oil.

Antidiabetic Activity

After the 28-day feeding period, the concentration of the thiobarbituric acid- reactive substance in the serum, liver, and kidney of on streptozotocin-induced diabetic rats administered crude flavonoids, eriocitrin, and hesperidin (both lemon flavonoids) significantly decreased as compared with that of the untreated diabetic group (Miyake et al. 1998a). The levels of 8-hydroxydeoxyguanosine, in the urine of diabetic rats administered eriocitrin and hesperidin significantly decreased as compared with that of the diabetic rat group. Crude flavonoids, eriocitrin, and hesperidin. The results demonstrated that dietary lemon flavonoids of eriocitrin and hesperidin suppressed the oxidative stress in the diabetic rats.

Anticancer Activity

The flavonoid from lemon fruit and its metabolites, particularly eriodictyol, 3,4-dihydroxyhydrocinnamic acid and phloroglucinol were found to exhibit the function of DNA fragmentation in HL-60 cells in a dose and time dependent manner (Ogata et al. 2000). An apoptotic DNA ladder and chromatin condensation were observed in HL-60 cells when treated with these compounds. The order of potency as an inducer of apoptosis was eriodictyol, 3, 4-dihydroxyhydrocinnamic acid, phloroglucinol and eriocitrin. These compounds were deemed to be useful for medical purposes.

Three coumarins were isolated as significant inhibitors of tumour promoter 12-O-tetradecanoylphorbol-13-acetate (TPA)-induced Epstein–Barr virus (EBV) activation in Raji cells from the peel of lemon fruit (Miyake et al. 1999). They were identified as 8-geranyloxypsolaren (LE−1), 5-geranyloxypsolaren (bergamottin, LE−2), and

5-geranyloxy-7-methoxycoumarin (LE–3). They had no potential O^{2-}-scavenging but markedly suppressed TPA-induced superoxide (O^{2-}) generation in differentiated human promyelocytic HL-60 cells. Further, LE–1 and LE–3 reduced both lypopolysaccharide (LPS) and interferon-γ (IFN-γ)-induced nitric oxide (NO) generation in mouse macrophage RAW 264.7 cells. Similarly, they were found to be Nitric oxide (NO) generation inhibitors rather than scavengers by measuring the extracellular l-citrulline levels. The occurrence of these coumarins in a lemon fruit was abundant in the flavedo of the peel. The present study suggested the coumarins in lemon fruit to be promising chemopreventive agents by inhibiting radical generation.

Hepatoprotective Activity

The ethanol extract of *Citrus limon* fruit was evaluated for its effects on experimental liver damage induced by carbon tetrachloride, and the ethyl acetate soluble fraction of the extract was evaluated on HepG2 cell line. The ethanol extract of lemon fruits was found to normalize the levels of aspartate aminotransferase (ASAT), alanine aminotransferase (ALAT), alkaline phosphatase (ALP), and total and direct bilirubin, which were altered due to carbon tetrachloride intoxication in rats (Bhavsar et al. 2007). In the liver tissue, treatment significantly reduced the levels of malondialdehyde (MDA), hence the lipid peroxidation, and raised the levels of antioxidant enzymes superoxide dismutase (SOD) and catalase. It improved the reduced glutathione (GSH) levels in treated rats in comparison with CCl4-intoxicated rats. In the histopathologic studies, treated animals exhibited restoration of the liver architecture toward normal. The effect seen was dose dependent, and the effect of the highest dose (500 mg/kg) was almost equal to the standard silymarin. In the investigation carried out on human liver–derived HepG2 cell line, significant reduction in cell viability was observed in cells exposed to CCl4. A dose-dependent increase in the cell viability was observed when CCl4-exposed HepG2 cells were treated with different concentrations of ethyl acetate soluble fraction of the ethanol extract. The highest percentage viability of HepG2 cells was observed at a concentration of 100 μg/ml. The results from the current investigation also indicated good correlation between the in-vivo and in-vitro studies.

Antimicrobial Activity

The essential oil of *Citrus limon* peel exhibited fungitoxicitiy against dermatophytic fungi *Epidermophyton floccosum* and *Microsporum gypseum* at 1,000 ppm and *Trichophyton mentagrophytes* at 900 ppm (Misra et al. 1988). The fungitoxicity of the essential oil was not affected by autoclaving and storage. Citral was found to be the fungitoxic factor. In another study, the essential oils of lemon (*Citrus limon*), mandarin (*Citrus reticulata*), grapefruit (*Citrus paradisi*) and orange (*Citrus sinensis*) showed antifungal activity against all the moulds commonly associated with food spoilage: *Aspergillus niger, Aspergillus flavus, Penicillium chrysogenum* and *Penicillium verrucosum* (Viuda-Martos et al. 2008b). Lemon peel oil exhibited high antibacterial activity against Gram positive bacteria (*Staphylococcus aureus, Bacillus cereus, Mycobacterium smegmatis, Listeria monocytogenes, Micrococcus luteus*) and Gram negative bacteria (*Escherichia coli, Klebsiella pneumoniae, Pseudomonas aeruginosa, Proteus vulgaris*) and antifungal activity against *Kluyveromyces fragilis, Rhodotorula rubra, Candida albicans, Hanseniaspora guilliermondii* and *Debaryomyces hansenii* yeasts (Kırbaşlar et al. 2009).

Studies showed that the essential oils of lemon (*Citrus limon*), mandarin (*C. reticulata*), grapefruit (*C. paradisi*) and orange (*C. sinensis* L.) inhibited the growth of some bacteria commonly used in the food industry, (*Lactobacillus curvatus, L. sakei, Staphylococcus carnosus* and *S. xylosus*) or related to food spoilage (*Enterobacter gergoviae* and *E. amnigenus*) (Viuda-Martos et al. 2008a). Lemon essential oil showed the highest inhibition effect upon *S. carnosus, E. gergoviae* and *E. amnigenus*. In another study, the highest antibacterial activity (*Staphylococcus aureus, Bacillus subtilis, Escherichia coli, Klebsiella pneumonia* and *Salmonella typhi*) was exhibited by the acetone peel extract of *Citrus sinensis* followed by the ethyl acetate peel extract of *Citrus limon* (Ashok Kumar et al. 2011). The peel extract of *Citrus sinensis* and *Citrus limon* could be considered to be as equally potent as the antibiotics, such as metacillin and penicillin. MICs were tested at concentrations ranging from 50–6.25 mg/ml as wells as their MBCs. The phytochemical analysis of the citrus peel extracts showed the presence of flavonoids, saponins, steroids, terpenoids, tannins and alkaloids.

Antihyperlipidemic Activity

Hesperidin, the most important flavanone of *Citrus* sp., significantly increased HDL and lowered cholesterol, LDL, total lipid and triglyceride plasma levels in normolipidemic rats and in rats with diet-induced and triton-induced hyperlipidemia (Monforte et al. 1995).

Antihypertension Activity

In a survey of hypertensive patients in a local region of northern Turkey, 156 (72.5%) of hypertensive patients were using alternative therapy that include lemon juice and 86 patients (40%) were drinking lemon juice (Adibelli et al. 2009).

Antinociceptive Activity

Citrus limon leaf essential oil exhibited antinociceptive in mice (Campêlo et al. 2011a). Oral administration of the oil (50, 100, and 150 mg/kg) significantly reduced the number of writhes in the acetic acid-induced writhing test, and, at highest doses, it reduced the number of paw licks in the formalin test. Naloxone antagonized the antinociceptive action of the essential oil, suggesting at least, the participation of the opioid system in the central analgesic action of essential oil.

Neuroprotective Activity

Lemon essential oil treatment significantly reduced the lipid peroxidation level and nitrite content but increased the glutathione reduced (GSH) levels and the superoxide dismutase, catalase, and glutathione peroxidase activities in mice hippocampus (Campêlo et al. 2011b). The findings strongly supported the hypothesis that oxidative stress in hippocampus could occur during neurodegenerative diseases, proving that hippocampal damage induced by the oxidative process played a crucial role in brain disorders, and also implied that a strong protective effect could be achieved using lemon essential oil as an antioxidant.

Insecticidal Activity

Lemon essential oil extracted from lemon leaves (by steam distillation) exhibited repellancy against sandfly, *Lutzomyia youngi* both in the laboratory and in the field (in a coffee plantation in Venezuela) by topical application to the skin of volunteers (Rojas and Scorza 1991). The lemon essential oil gave 70% protection against bites of the sandflies (20–30-min exposure) compared with 63% for Deet and 33% for Citronellal. The lemon essential oil was stated to be easy to extract and cheap.

Studies in Iran reported that extracts and essential oils of *Citrus limon* (lemon) and *Melissa officinalis*, had repellent effect against *Anopheles stephensi* in laboratory on animal and humans (Oshaghi et al. 2003). Results of statistical analysis revealed significant differences between oils and extracts ($P < 0.05$) against the tested species, thus oils were more effective than extracts. There was no significant difference between synthetic repellent, N,Ndiethyl-3-methylbenzamide (Deet) and lemon oil, whereas the difference between lemon and Melissa oils was significant. Relative efficacy of lemon oil to Deet was 0.88 whereas it was 0.71 for melissa oil. The results were found marginally superior in repellency for animals than human. Application of lemon *Citrus* repellent compounds gave acceptable percentage biting protection against *A. stephensi* which was less than but not significant with the percentage protection seen in Deet under similar conditions. No adverse effects on the skin or other parts of the body of the human volunteers were observed during the study period and through 1 month after application. This study showed that oils provided better protection than extracts. The researchers concluded that because of the simplicity of production, cheapness, and protective activity, lemon essential oil was recommended as an effective alternative to Deet with potential as a means of personal protection against mosquito vectors of disease. Studies showed that the fourth instar larvae of *Culex pipiens* was highly susceptible to *Bacillus thuringiensis israelensis* followed by *Citrus limon* and *Allium sativum* (Zayed et al. 2009). In all case, a positive relationship was found between the time of exposure and mortality percentage of larvae. The tissues of gut, muscle, fat body and cuticle were the most severely damaged by the treatment. Also, the destructive damages described in the investigation depend mainly up on the time of exposure and the dosage.

Traditional Medicinal Uses

Lemons are rich in vitamin C which helps the body to fight off infections and also to prevent or treat scurvy. The lemon has an alkaline effect in the human body, despite the high acid content, once it has been sufficiently digested in the stomach it tends to alkalinize. This property of the plant makes it very effective and useful in the treatment of rheumatic conditions involving acidity as one of the contributory factors in the genesis of the disorder. Lemon peel is carminative and stomachic. The essential oil from the rind is strongly rubefacient and when taken internally in small doses has stimulating and carminative properties. The antiseptic property of the volatile oil and its bactericidal effects are also useful in the treatment of many disease states affecting the body such as athlete's foot, stings and bites of insects, as well as ringworm, sunburn, and warts on the skin. An essential oil from the fruit rind is used in aromatherapy as massage and bath oil.

In traditional medicine, lemon juice is wieldy regarded as a diuretic, antiscorbutic, astringent and febrifuge. Lemon juice have been used to relive gingivitis, stomatitis and inflammation of the tongue, sore throat gargle and common colds and flu and as a laxative. However, prolonged daily used has been reported to erode the enamel of the teeth. Lemon juice with honey or salt or ginger is taken as a cold remedy. Lemon juice is also a good antiperiodic and has been used as a substitute for quinine in treating malaria and other fevers.

Other Uses

Prior to the development of fermentation-based processes, lemons were used as the primary source for the production of citric acid. Lemon juice may be used to lighten hair and for bleaching freckles and has been used in some facial creams. When mixed with baking soda lemon juice can remove stains from plastic food containers. A halved lemon dipped in salt or baking powder can be used to brighten copper cookware. Lemon is often employed, as a kitchen sanitary deodoriser – to remove grease, bleach stain and disinfect. Fruit pulp has also been used as animal fodder. Lemon oil has been used on the unsealed rosewood fingerboards of guitars and other stringed instruments. Lemon oil is also used in furniture polishes, detergents, soaps and shampoos, in body toning lotions. It is also important in perfume blending and cologne. Petitgrain oils (containing up to 50% citral) have been distilled from lemon leaves, twigs and immature fruits. With monoterpenes and sesquiterpenes removed, it is prized in colognes and floral perfumes. Lemonade when added to potted plants has been reported to keep their flowers fresh longer than normal. Lemon tree wood is fine-grained and easy to work and is used for toys, chessman, small spoons and other articles. D-limonene in lemon oil is used as a non-toxic insecticide treatment.

Comments

The parents of the lemon are *Citrus* × *aurantium* and *C. medica* (Zhang and Mabberley 2008). Backcrosses with either parent give a range of sour to sweet lemons which go under various names and perhaps would best be considered as forming cultivar groups, e.g., Bergamot Group. The rough lemon, *C.* ×*taitensis* Risso (*C.* ×*aurantium* subsp. *jambhiri* Engler; *C.* ×*jambhiri* Lushington; *C.* ×*sinensis* subsp. *jambhiri* (Lushington) Engler), sometimes included here, is perhaps *C. medica* × *C. reticulata*. The name "*Citrus limonia*" has been misapplied to other *Citrus* taxa.

Selected References

Adibelli Z, Dilek M, Akpolat T (2009) Lemon juice as an alternative therapy in hypertension in Turkey. Int J Cardiol 135(2):e58–e59

Akhila S, Bindu AR, Bindu KA, Aleykutty NA (2009) Comparative evaluation of extracts of *Citrus limon* Burm. peel for antioxidant activity. J Young Pharm 1:136–140

Ashok Kumar K, Narayani M, Subanthini A, Jayakumar M (2011) Antimicrobial activity and phytochemical analysis of *Citrus* fruit peels – utilization of fruit waste. Int J Eng Sci Technol 3(6):15414–15421

Berhow M, Tisserat B, Kanes K, Vandercook C (1998) Survey of phenolic compounds produced in *Citrus*. USDA ARS Tech Bull 1856:1–154

Bhavsar SK, Joshi P, Shah MB, Santani DD (2007) Investigation into hepatoprotective activity of *Citrus limon*. Pharm Biol 45(4):303–311

Calabrese V, Randazzo SD, Catalano C, Rizza V (1999) Biochemical studies on a novel antioxidant from lemon oil and its biotechnological application in cosmetic dermatology. Drugs Exp Clin Res 25(5):219–225

Campêlo LML, de Almeida AAC, de Freitas RLM, Cerqueira GS, de Sousa GF, Saldanha GB, Feitosa CM, de Freitas RM (2011a) Antioxidant and antinociceptive effects of *Citrus limon* essential oil in mice. J Biomed Biotechnol 2011:678673

Campêlo LML, Gonçalves M, Custódio F, Feitosa CM, de Freitas RM (2011b) Antioxidant activity of *Citrus limon* essential oil in mouse hippocampus. Pharm Biol 49(7):709–715

Caristi C, Bellocco E, Gargiulli C, Toscano G, Leuzzi U (2006) Flavone-di-C-glycosides in *Citrus* juices from Southern Italy. Food Chem 5(3):431–437

Chang MC, Wang TC, Liu PC, Wu MC, Chang HM, Liai JW (2004) Studies on antioxidative activity of methanol extracts from lemon (*Citrus limon* Burm.) leaves. J Chin Soc Hort Sci 50(1):85–98 (in Chinese)

Chevallier A (1996) The encyclopedia of medicinal plants. Dorling Kindersley, London, 336 pp

Choi HS, Song HS, Ukeda H, Sawamura M (2000) Radical-scavenging activities of *Citrus* essential oils and their components: detection using 1,1-Diphenyl-2-picrylhydrazyl. J Agric Food Chem 48(9):4156–4161

Chopra RN, Nayar SL, Chopra IC (1986) Glossary of Indian medicinal plants. (Including the supplement). Council Scientific Industrial Research, New Delhi, 330 pp

de Rodríguez GO, de Godoy VM, de Colmenares NG, Salas LC, de Ferrer BS (1998) Composition of Venezuelan lemon essential oil *Citrus limon* (L.) Burm.f. Rev Fac Agron (LUZ) 15:343–349

Del Rio JA, Fuster MD, Gomez P, Porras I, Garcia-Lidon A, Ortuno A (2004) *Citrus limon*: a source of flavonoids of pharmaceutical interest. Food Chem 84(3):457–461

Facciola S (1990) *Cornucopia*. A source book of edible plants. Kampong Publications, Vista, 677 pp

Foundation for Revitalisation of Local Health Traditions (2008) FRLHT database. http://envis.frlht.org

Goel D, Ali M, Mir SR (2004) Chemical composition of the leaf essential oil of *Citrus limon* grown in Delhi. J Med Arom Plant Sci 26(3):500–502

González CN, Sánchez F, Quintero A, Usubillaga A (2002) Chemotaxonomic value of essential oil compounds in *Citrus* species. Acta Hortic (ISHS) 576:49–51

González-Molina E, Domínguez-Perles R, Moreno DA, García-Viguera C (2010) Natural bioactive compounds of *Citrus limon* for food and health. J Pharm Biomed Anal 51(2):327–345

González-Molina E, Moreno DA, García-Viguera C (2009) A new drink rich in healthy bioactives combining lemon and pomegranate juices. Food Chem 115(4):1364–1372

Grieve M (1971) A modern herbal, 2 vols. Penguin/Dover publications, New York, 919 pp

Guimarães R, Barros L, Barreira JC, Sousa MJ, Carvalho AM, Ferreira IC (2010) Targeting excessive free radicals with peels and juices of *Citrus* fruits: grapefruit, lemon, lime and orange. Food Chem Toxicol 48(1):99–106

Hardy S (2004) Growing lemons in Australia – a production manual. NSW Department of Primary Industries. http://www.dpi.nsw.gov.au/agriculture/horticulture/citrus/lemon-manual

Horowitz RM (1956) Notes – flavonoids of Citrus. I. Isolation of diosmin from lemons (*Citrus limon*). J Org Chem 21(10):1184–1185

Khatoon T (1995) Phytochemical investigations of *Citrus* species of Pakistan. PhD thesis, Islamia University, Bahawalpur, Pakistan

Kırbaşlar FG, Tavman A, Dülger B, Türker G (2009) Antimicrobial activity of Turkish *Citrus* peel oils. Pak J Bot 41(6):3207–3212

Kırbaşlar Şİ, Boz İ, Kırbaşlar FG (2006) Composition of Turkish lemon and grapefruit peel oils. J Essent Oil Res 18(5):525–543

Lario Y, Sendra E, García-Pérez J, Fuentes C, Sayas-Barberá E, Fernández-López J, Pérez-Alvarez JA (2004) Preparation of high dietary fiber powder from lemon juice by-products. Innov Food Sci Emerg Technol 5(1):113–117

Mabberley DJ (1997) A classification for edible *Citrus*. Telopea 7(2):167–172

Manner HI, Buker RS, Smith VE, Elevitch CR (2006) *Citrus* species. Ver. 2.1 In: Elevitch CR (ed) Species profiles for Pacific Island Agroforestry. Permanent Agriculture Resources (PAR), Holualoa, Hawaii. http://www.agroforestry.net

Minato K, Miyake Y, Fukumoto S, Yamamoto K, Kato Y, Shimomura Y, Osawa T (2003) Lemon flavonoid, eriocitrin, suppresses exercise-induced oxidative damage in rat liver. Life Sci 72(14):1609–1616

Misra N, Batra S, Mishra D (1988) Fungitoxic properties of the essential oil of *Citrus limon* (L.) Burm. against a few dermatophytes. Mycoses 31(7):380–382

Miyake Y, Minato K, Fukumoto S, Shimomura Y, Osawa T (2004) Radical-scavenging activity in vitro of lemon peel fermented with *Aspergillus saitoi* and its suppressive effect against exercise-induced oxidative damage in rat liver. Food Sci Technol Res 10(1):7–74

Miyake Y, Murakami A, Sugiyama Y, Isobe M, Koshimizu K, Ohigashi H (1999) Identification of coumarins from lemon fruit (*Citrus limon*) as inhibitors of in vitro tumor promotion and superoxide and nitric oxide generation. J Agric Food Chem 47(8):3151–3157

Miyake Y, Shimoi K, Kumazawa S, Yamamoto K, Kinae N, Osawa T (2000) Identification and antioxidant activity of flavonoid metabolites in plasma and urine of eriocitrin-treated rats. J Agric Food Chem 48(8):3217–3224

Miyake Y, Yamamoto K, Morimitsu Y, Osawa T (1997a) Isolation of C-glucosylflavone from lemon peel and antioxidative activity of flavonoid compounds in lemon fruit. J Agric Food Chem 45(12):4619–4623

Miyake Y, Yamamoto K, Morimitsu Y, Osawa T (1998a) Characterizatics of antioxidative flavonoids glycosides in lemon fruit. Food Sci Technol Int 4(1):48–53

Miyake Y, Yamamoto K, Osawa T (1997b) Isolation of eriocitrin (eriodictyol 7-rutinoside) from lemon fruit (*Citrus limon* Burm. f.) and its antioxidative activity. Food Sci Technol Int Tokyo 3(1):84–8

Miyake Y, Yamamoto K, Osawa T (1997c) Metabolism of antioxidant in lemon fruit (*Citrus limon* Burm. f.) by human intestinal bacteria. J Agric Food Chem 45(10):3738–3742

Miyake Y, Yamamoto K, Tsujihara N, Osawa T (1998b) Protective effects of lemon flavonoids on oxidative stress in diabetic rat. Lipids 33(7):689–695

Monforte MT, Trovato A, Kirjavainen S, Forestieri AM, Galati EM, Lo Curto RB (1995) Biological effects of hesperidin a *Citrus* flavonoid. (Note II): hypolipidemic activity on experimental hypercholesterolemia in rat. Farmaco 50(9):595–599

Morton JF (1987) Lemon. In: Fruits of warm climates. Julia F. Morton, Miami, pp 160–168

Ogata S, Miyake Y, Yamamoto K, Okumura K, Taguchi H (2000) Apoptosis induced by the flavonoid from lemon fruit (*Citrus limon* Burm. f.) and its metabolites in HL-60 cells. Biosci Biotechnol Biochem 64(5):1075–1078

Oshaghi MA, Ghalandari R, Vatandoost H, Shayeghi M, Kamali-nejad M, Tourabi-Khaledi H, Abolhassani M, Hashemzadeh M (2003) Repellent effect of extracts and essential oils of *Citrus limon* (Rutaceae) and *Melissa officinalis* (Labiatae) against main malaria vector, *Anopheles stephensi* (Diptera: Culicidae). Iran J Public Health 32(4):47–52

Pacific Island Ecosystems at Risk (PIER) (2004) *Citrus limon* (L.) Burm.f., Rutaceae http://www.hear.org/Pier/scientificnames/..%5Cspecies%5Ccitrus_limon.htm

Penniston KL, Nakada SY, Holmes RP, Assimos DG (2008) Quantitative assessment of citric acid in lemon juice, lime juice, and commercially-available fruit juice products. J Endourol 22(3):567–570

Peterson JJ, Beecher GR, Bhagwat SA, Dwyer JT, Gebhardt SE, Haytowitz DB, Holden JM (2006) Flavanones in grapefruit, lemons, and limes: a compilation and review of the data from the analytical literature. J Food Compos Anal 19(Supplement 1):S74–S80

Porcher M H et al (1995–2020) Searchable World Wide Web multilingual multiscript plant name database. The University of Melbourne, Melbourne, Australia. http://www.plantnames.unimelb.edu.au/Sorting/Frontpage.html

Rojas E, Scorza JV (1991) The use of lemon essential oil as a sandfly repellent. Trans Roy Soc Trop Med Hyg 85(6):803

Swingle WT (1946) The botany of *Citrus* and its wild relatives of the orange subfamily (family Rutaceae, subfamily Aurantioideae). In: Webber HJ, Batchelor LD (eds) The Citrus industry, vol 1, History, botany and breeding. University of California Press, Berkeley, pp 129–474, 1028

Swingle WT, Reece PC (1967) The botany of *Citrus* and its wild relatives. In: Reuther W, Webber HJ, Batchelor LD (eds) The *Citrus* industry, vol 1: History, world distribution, botany, and varieties, Revised Edition. University of California Press, Berkeley, pp 190–430

U.S. Department of Agriculture, Agricultural Research Service (2011) USDA national nutrient database for standard reference, Release 24. Nutrient Data Laboratory Home Page, http://www.ars.usda.gov/ba/bhnrc/ndl

Vandercook CE, Yokoyama H (1965) Lemon juice composition. IV. Carotenoid and sterol content. J Food Sci 30(5):865–868

Viuda-Martos M, Ruiz-Navajas Y, Fernández-López J, Perez-Álvarez J (2008a) Antibacterial activity of lemon (*Citrus lemon* L.), mandarin (*Citrus reticulata* L.), grapefruit (*Citrus paradisi* L.) and orange (*Citrus sinensis* L.) essential oils. J Food Safety 28(4):567–576

Viuda-Martos M, Ruiz-Navajas Y, Fernández-López J, Pérez-Álvarez J (2008b) Antifungal activity of lemon (*Citrus lemon* L.), mandarin (*Citrus reticulata* L.), grapefruit (*Citrus paradisi* L.) and orange (*Citrus sinensis* L.) essential oils. Food Control 19(12):1130–1138

Yokoyama H, Vandercook CE (1967) *Citrus* carotenoids. I. Comparison of carotenoids of mature-green and yellow lemons. J Food Sci 32(1):42–48

Zayed AA, Saeed RMA, El-Namaky AH, Ismail HM, Mady HY (2009) Influence of *Allium sativum* and *Citrus limon* oil extracts and *Bacillus thuringiensis israelensis* on some biological aspects of *Culex pipiens* larvae (Diptera: Culicidae). World J Zool 4(2):109–121

Zhang DX, Mabberley DJ (2008) *Citrus* Linnaeus. In: Wu ZY, Raven PH, Hong DY (eds) Flora of China, vol 11 (Oxalidaceae through Aceraceae). Science Press/Missouri Botanical Garden Press, Beijing/St. Louis

Citrus x microcarpa

Scientific Name

Citrus x microcarpa Bunge.

Synonyms

× *Citrofortunella microcarpa* (Bunge) Wijnands, × *Citrofortunella mitis* (Blanco) J.Ingram & H. E. Moore, *Citrus madurensis* Lour., *Citrus (Fortunella) margarita* (= *Citrus japonica*, Nagami kumquat) × *Citrus reticulata* Blanco, x *Citrus microcarpa* (Bunge), Wijnands, *Citrus* x *mitis* Blanco, *Citrus medica.* var. *limau kasturi* Ridley.

Family

Rutaceae

Common/English Names

Calamandarin, Calamansi, Calmondin, Calamondin Orange, China Orange, Chinese Orange, Golden Lime, Musk Lime, Philippine Lime, Panama Orange, Scarlet Lime.

Vernacular Names

Chinese: Jin Ju, Si Ji Ju, Szu Kai Kat, Yue Ju;
Danish: Stueappelsin;
German: Zwergapfelsine;
Finnish: Kalamondiini;
Hawaiian: 'Alani 'Aawa 'Awa;
India: Hazara;
Indonesia: Jeruk Kasturi, Jeruk Peres, Jeruk Potong;
Japanese: Karamonjin, Shikikikat, Shiki Kitsu, Shiki Kitsu, Tokinkan;
Malaysia: Limau Chuit, Limau Kesturi;
Palauan: Kingkang;
Philippines: Agridolsi, Limonsito (Bisayan), Aldonisis, Calamansi, Kalamansi, Kalamondin, Kalamunding (Tagalog);
Polish: Kalamondina;
Portuguese: Limoeiro Do Japão;
Samoan: Tipolo Iapani;
Spanish: Lemonsito, Naranjita De San José;
Thailand: Ma-Nao-Wan, Som Chit (Bangkok), Som Mapit;
Vietnamese: Tâc, Hanh.

Origin/Distribution

Calamondin originated in China as a natural hybrid, however, the exact hybrid nature of remains to be established. It is commonly accepted to be a hybrid of a loose-skinned, sour mandarin type probably *Citrus reticulata* var. *austera* Swingle and a kumquat the Oval or 'Nagami' kumquat *Citrus japonica* Thunb. 'Nagami'. It is believed to have been introduced in early times to Indonesia and the Philippines where it became the most important Citrus juice source in the Philippine Islands. Today, it is widely grown in India and throughout southern Asia and southeast Asia especially in Malaysia. It is a common ornamental dooryard tree in Hawaii, the Bahamas, some islands of the West Indies, and parts of Central America.

Agroecology

Calamondin thrives in a warm climate but is cold tolerant and frost sensitive. In fact, it is as cold-hardy as the Satsuma orange and can be grown all along the Gulf Coast of the southern United States in frost free areas. Areas with uniformly distributed rainfall of 1,500–2,000 mm per year is ideal. It is also adaptable to areas with long dry periods provided irrigation is available. Calamondin is mainly grown in the lowlands and is adaptable to a wide range of soils from clay-loams to calcareous soils to sandy soils. It does best in well-drained, sandy loams or clay loams rich in organic matter at pH of 5.5–7. It is moderately drought-tolerant and intolerant of strong winds.

Edible Plant Parts and Uses

Calamondin can be eaten fresh but it usually make into juice, drinks, dried and used as preserves, pickles or to flavour food. The fruit is used for marmalade and chutneys, preserved whole in sugar syrup, and used as flavouring in seafood and meat dishes. It is frequently used to impart acidity to noodles like Pad Thai, laksa, grilled fish or smoked fish and in a variety of sambal sauces in southeast Asia. In Malaysia, the calamondin is an ingredient in chutney. Whole fruits, fried in coconut oil with various seasonings, are eaten with curry. The preserved peel is added as flavoring to other fruits stewed or preserved. Calamondin halves or quarters may be served with iced tea, seafood and meats, or noodles to be squeezed for the acid juice.

Calamondin fruits can be sliced and boiled with cranberries to make a tart sauce. Calamondins are also preserved whole in sugar syrup, or made into sweet pickles, or marmalade. A superior marmalade is made by using equal quantities of calamondins and kumquats. In Hawaii, a calamondin-papaya marmalade is popular.

The acid fruit juice is processed into bottled concentrate and juice. The juice is primarily valued for making acid beverages. The juice is also concentrated with syrup. It is often employed like lime or lemon juice to make gelatin salads or desserts, custard pie or chiffon pie. In the Philippines, the extracted juice, with the addition of gum tragacanth as an emulsifier, is pasteurized and bottled commercially. This product must be stored at low temperature to keep well. Pectin is recovered from the peel as a by-product of juice production.

Botany

Evergreen shrub or small tree to 4 m high, densely branched beginning close to the ground sparsely spiny with 3 mm long spines or none. Branchlets strongly angled, glabrous, tap root deep. Leaves elliptic to broad-oval, crenulate, glabrous; lamina 3–7 cm long, dark-green, glossy on the upper surface, paler green to yellowish-green beneath; petiole 8–12 mm long, very narrowly winged, 1–2 mm wide (Plates 1, 2 and 3). Flowers axillary or terminal, usually solitary, sometimes 2–3 flowered, bisexual, fragrant; rachis 5 mm long, glabrous; pedicels 6–8 mm long, glabrous. Calyx 1 mm long, deeply 5-lobed; lobes acute, minutely pubescent. Petals 5, elliptic-oblong, pure-white c. 12 mm long. Filaments 7 mm long; anthers

Plate 1 Calamondin plant with mature and immature fruits

Plate 2 Immature fruits and leaves

Plate 4 Calamondin fruits on sale in a local market

Plate 3 Ripe calamondin fruits, thin skin and juicy orange segments

Plate 5 Bundles of calamondin fruits

ellipsoidal, c. 1 mm long. Style c. 3 mm long. Fruit globose or oblate, 2–4.5 cm diameter, green turning to yellow, orange yellow or deep orange; rind thin dotted with numerous small oil glands, easily peeled; skin thin; pulp, fleshy orange in 6–10 segments, juicy, highly acid (Plates 2, 3, 4 and 5). Seeds large, smooth 1–5 small, obovoid seeds, green within, polyembryonic.

Nutritive/Medicinal Properties

The nutrient value of calamondin fruit per 100 g edible portion is reported as: water 93.6 g, protein 0.4 g, fat 0.3 g, carbohydrate 5.1 g, fibre 0.1 g, Ka 79 mg, carotene 470 μg, vitamin A 78 ug, thiamine 0.03 mg, riboflavin 0.04 mg, niacin 0.2 mg, and vitamin C 41.6 mg (Tee et al. 1997).

The percentage and concentration of total phenolics in *Citrus microcarpa* cv Calasnu leaf, peel and juice were reported by Berhow et al. (1998) as: leaf:- 21.2% (1.10 mg/g) flavone/flavonol, 71.3% (2.35 mg/g) flavanone, 0% psoralen, 0% coumarin; flavedo (outer pigmented layer of the peel): 36.1% (1.99 mg/g) flavone/flavonol, 36.4% (1.28 mg/g) flavanone, 0% psoralen, 16.8% (0.28 mg/g) coumarin; albedo (inner layer of the rind): 2.5% (0.28 mg/g) flavone/flavonol, 87% (6.15 mg/g) flavanone, 0% psoralen, 2.2% (0.08 mg/g) coumarin; juice: 9.5% (0.23 mg/g) flavone/flavonol, 77.4% (1.19 mg/g) flavanone, 0% psoralen, and 3.7% (0.03 mg/g) coumarin.

The percentage and concentration of flavanones reported in *Citrus microcarpa* cv Calasnu by Berhow et al. (1998) were as follows: leaf:- eriocitrin 3% (0.1 mg/g), hesperidin 32% (0.7 mg/g), narirutin 7% (0.2 mg/g), naringin 48% (1.1 mg/g), naringin–6″–malonate (open form) 5% (0.1 mg/g), naringin–6″–malonate (closed form) 5% (0.1 mg/g); flavedo: hesperidin 89% (1.1 mg/g); albedo: didymin 2% (1.0 mg/g), eriocitrin 3% (0.2 mg/g), hesperidin 25% (1.6 mg/g), narirutin 1% (0.1 mg/g), neoeriocitrin 1% (0.1 mg/g), naringin 55% (3 mg/g), neohesperidin 1% (0.1 mg/g), naringin–6″–malonate (open form) 1% (0.1 mg/g); juice: didymin 3% (traces), eriocitrin 3% (traces), hesperidin 40% (0.5 mg/g), naringin 50% (0.6 mg/g), naringin–6″–malonate (open form) 4% (traces mg/g). The percentage and concentration of flavone reported in *Citrus microcarpa* cv Calasnu by Berhow et al. (1998) were as follows: flavedo: diosmin 6% (0.1 mg/g); juice diosmin 14% (traces).

Citrus microcarpa, *Citrus hystrix*, *Citrus medica* var. 1 and 2, and *Citrus suhuiensis* were found to contain high amounts of flavones, flavanones, and dihydrochalcone C- and/or O-glycosides (Roowi and Crozier 2011). Among the major compounds detected were apigenin-6,8-di-C-glucoside, apigenin-8-C-glucosyl-2″-O-rhamnoside, phloretin-3′,5′-di-C-glucoside, diosmetin-7-O-rutinoside, hesperetin-7-O-neohesperidoside, and hesperetin-7-O-rutinoside. Most of the tropical citrus flavanones were neohesperidoside conjugates, which are responsible for imparting a bitter taste to the fruit. *C. microcarpa* contained a high amount of phloretin-3′,5′-di-C-glucoside.

The volatile components of calamondin fruit was determined to consist of 5 aldehydes (acetaldehyde. decanal, nonal, octanal and perillaldehyde), 2 esters (geranyl acetate and neryl acetate), 5 alcohols (ethanol, linalool, methanol, terpinene-4-ol and a-terpineol) and 8 hydrocarbons (3-carene, limonene, myrcene, a-pinene, b-pinene, g-terpene, terpinolene and valencene) (Nisperos-Carriedo et al. 1992) It also contained glucose, fructose, sucrose and ascorbic, dehydroascorbic, citric and malic acids.

Fifty-eight compounds, including monoterpene hydrocarbons such as limonene (58.2%), β-phellandrene (6.0%) and γ-terpinene (4.8%), and sesquiterpene hydrocarbons such as β-elemene (4.6%) and (E,E)-α-farnesene (3.4%) were identified in the volatile concentrate of calamondin peel (Takeuchi et al. 2005). In contrast, 98 compounds, including alcohols such as linalool (11.8%), α-terpineol (11.0%) and terpinen-4-ol (10.0%) were identified in calamondin juice extract. Limonene, cis-linalool oxide, linalool, α-terpineol, (E,E)-2,4-decadienal, and methyl N-methyl anthranilate had high flavour dilution factors. It was found that methyl N-methyl anthranilate (a characteristic component of mandarin), thymol (a characteristic component of yuzu), citronellal, can/one and citronellol (characteristic components of sudachi) were included in the extract of calamondin. In contrast, neral and geranial (characteristic components of lemon) were hardly detected in it. The flavour of calamondin peel extract was found to be characterized by the citrus-like odor of limonene, the petit-grain-like odour of linalool, the citrus and orange-like odour of (E,E)-2,4-decadienal and the sweet, floral and fruity odour of methyl N-methyl anthranilate. The flavour of calamondin juice extract was found to be characterized by the sweet, woody and floral odour of cis-linalool oxide, the petit-grain-like odour of linalool and the lilac-like odour of a-terpineol.

In Malaysia, calamondin essential peel oil was found to have the following main volatile components: limonene (94.03%), myrcene (1.81%), α-pinene (0.45%), linalool (0.37%), geranyl acetate (0.19%), decanal (0.12%), γ-terpinene (0.05%), α-phellandrene (0.05%), terpinene-4-ol (0.10%) and β-pinene (0.05%) (Ibrahim et al. 1995).

The seeds of musk lime (*Citrus microcarpa*) represented a substantial waste product of small-scale citrus-processing factories, as they constitute about 100.0 g/kg of the whole fruit and contain a considerable amount of crude fat (338.0 g/kg) (Manaf et al. 2008). The iodine and saponification values and unsaponifiable matter and free fatty acid contents of the freshly extracted seed oil were 118.0 g I2 per 100 g oil, 192.6 mg KOH g-1 oil, 22 mg/g oil and 18 mg oleic acid g/1 oil respectively. Its fatty acid profile indicated that 73.6% of the fatty acids present were unsaturated. Linoleic (L, 31.8%), oleic (O, 29.6%) and palmitic (P, 21.4%) acids were the predominant fatty acids, existing mainly as the triacylglycerols

POL (18.9%), PLL (13.7%) and OLL (11.9%). The melting and cooling points of the oil were 10.7°C and −45.2°C respectively. Musk lime seeds were found to have a rich and unusual source of oil, having linoleic, oleic and palmitic acids dominating the fatty acid composition. This property should make the oil both relatively stable to thermal oxidation owing to the combined presence of oleic and palmitic acids (61.0%) and highly nutritive owing to its high concentration of unsaturated fatty acids (73.6%).

Antimicrobial Activity

A study showed that the bioactive compound isolated from *C. microcarpa*, 2-hydroxypropane-1,2,3-tricarboxylic acid monohydrate possessed antimicrobial property against fish pathogenic bacteria, namely 18 isolates of *Edwardsiella tarda* and 7 bacterial reference strains, namely, *Escherichia coli, Citrobacter freundii, Aeromonas hydrophila, Pseudomonas aeruginosa, Streptococcus agalatiae,* and *Yersinia enterocolitica* (Lee and Najlah 2009). The MIC values for both the natural 2-hydroxypropane-1,2,3-tricarboxylic acid and the synthetic one were ranging from 15.6 to 62.5 mg/ml, whereas that of the crude extract was ranging from 7.8 to 31.3 mg/ml. The findings showed that both the crude extract of *C. microcarpa* and its bioactive component might have potential as antimicrobial agent for aquaculture use.

Citrus microcarpa was one of the most active of 9 tropical plants in exhibiting antibacterial activity against 12 clinical and pathogenic bacterial strains isolated from aquatic animals including *Vibrio cholera, Escherichia coli, Vibro parahemolytics, Salmonella* sp. and *Streptococcus* sp. (Lee et al. 2008).

Traditional Medicinal Uses

Kalamansi fruits may be crushed with the saponaceous bark of *Entada Phaseoloides* for shampooing the hair, or the fruit juice applied to the scalp after shampooing. It eliminates itching and promotes hair growth. Rubbing calamondin juice on insect bites banishes the itching and irritation. It bleaches freckles and helps to clear up *acne vulgaris* and *pruritus vulvae*. Kalamansi juice is considered as a refrigerant (being a rich source of Vitamin C) and is taken orally as a cough remedy and antiphlogistic. Slightly diluted and drunk warm, it serves as a laxative. Combined with pepper, it is prescribed in Peninsular Malaysia to expel phlegm. Fruit juice also serves as a body deodorant. The root enters into a treatment given at childbirth. The distilled oil of the leaves serves as a carminative with more potency than peppermint oil. It is also used in shampoos, bath lotion and skin cleanser cosmetics.

Plate 6 Variegated calamondin plant

Plate 7 Variegated calamondin plant with ripe fruits

Other Uses

Calamondin is also a popular ornamental in many homes in southeast Asia. The fruit juice is used in the Philippines to bleach ink stains from fabrics and clothes. Calamondin is also used as rootstock for lemons and the oval kumquat.

Comments

Variegated calamondin plants (Plates 6 and 7) are often grown as ornamental potted plants.

Selected References

Berhow M, Tisserat B, Kanes K, Vandercook C (1998) Survey of phenolic compounds produced in *Citrus*. USDA ARS Technol Bull 1856:1–154

Burkill IH (1966) A dictionary of the economic products of the Malay Peninsula. Revised reprint. 2 volumes. Ministry of Agriculture and Co-operatives, Kuala Lumpur, Malaysia, vol. 1 (A–H) pp 1–1240, vol 2 (I–Z) pp 1241–2444

Ibrahim J, Abu Said A, Abdul Rashih A, Nor Azah MA, Norsiha A (1995) Chemical composition of some *Citrus* oils from Malaysia. J Essent Oil Res 8:627–632

Lee SW, Najlah M (2009) Antimicrobial property of 2-hydroxypropane-1,2,3-tricarboxylic acid isolated from *Citrus microcarpa* extract. Agric Sci China 8(7):880–886

Lee SW, Najlah M, Chuah TS, Wee W, Noor Azhar MS (2008) Antimicrobial properties of tropical plants against 12 pathogenic bacteria isolated from aquatic organisms. Afr J Biotechnol 7(13):2275–2278

Mabberley DJ (1998) Australian Citreae with notes on other Aurantioideae (Rutaceae). Telopea 7(4):333–344

Manaf YNA, Osman A, Lai OM, Long K, Ghazali HM (2008) Characterisation of musk lime (*Citrus microcarpa*) seed oil. J Sci Food Agric 88(4):676–683

Manner HI, Buker RS, Smith VE, Elevitch CR (2006) *Citrus* species. Ver. 2.1. In: Elevitch CR (ed) Species profiles for Pacific Island Agroforestry. Permanent Agriculture Resources (PAR), Holualoa, Hawaii. http://www.agroforestry.net

Morton JF (1987) Calamondin. In: Fruits of warm climates. Julia F. Morton, Miami, pp 176–178

Nisperos-Carriedo MO, Baldwin EA, Moshonas MG, Shaw PE (1992) Determination of volatile flavor components, sugars, and ascorbic, dehydroascorbic, and other organic acids in calamondin (*Citrus mitis* Blanco). J Agric Food Chem 40:2464–2466

Roowi S, Crozier A (2011) Flavonoids in tropical *Citrus* species. J Agric Food Chem 59(22):12217–12225

Saidin I (2000) Sayuran Tradisional Ulam dan Penyedap Rasa. Penerbit Universiti Kebangsaan Malaysia, Bangi, Malaysia. p 228 (In Malay)

Sotto RC (1992) *X Citrofortunella macrocarpa* (Bunge) Wijnands. In: Coronel RE, Verheij EWM (eds) Plant resources of South-East Asia. No. 2: edible fruits and nuts. Prosea Foundation, Bogor, pp 117–119

Stuart GU (2010) Philippine alternative medicine. Manual of some Philippine medicinal plants. http://www.stuartxchange.org/OtherHerbals.html

Swingle WT, Reece PC (1967) The botany of *Citrus* and its wild relatives. In: Reuther W, Webber HJ, Batchelor LD (eds) The *Citrus* industry volume 1: history, world distribution, botany, and varieties. Revised edition. University of California Press, Berkeley, pp 190–430

Takeuchi H, Ubukata Y, Hanafusa M, Hayashi S, Hashimoto S (2005) Volatile constituents of calamondin peel and juice (*Citrus madurensis* Lour.) cultivated in the Philippines. J Essent Oil Res 17(1):23–26

Tee ES, Mohd Ismail N, Mohd Nasir A, Khatijah I (1997) Nutrient composition of Malaysian foods, 4th edn, Malaysian food composition database programme. Institute for Medical Research, Kuala Lumpur

Clausena lansium

Scientific Name

Clausena lansium (Lour.) Skeels.

Synonyms

Aulacia punctata (Sonn.) Raeusch., *Clausena punctata* (Sonn.) Rehder & E.H. Wilson, *Clausena wampi* (Blanco) Oliver, *Cookia punctata* Sonn., *Cookia wampi* Blanco, *Quinaria lansium* Loureiro, *Sonneratia punctata* (Sonn.) J.F. Gmel.

Family

Rutaceae

Common/English Names

Fool's Curry Leaf, Wampee, Wampi.

Vernacular Names

Chinese: Huang Pi, Huang-P'i-Kuo, Huang P'i Ho, Huang P'i Kan, Huang-P'i-Tzu;
Dutch: Vampi, Wampi;
French: Vampi;
Khmer: Kantrop;
Korean: Hwangphinamu;
Laotian: Somz Mafai;
Malaysia: Wampi, Wang-Pei (Cantonese);
Philippines: Huampit, Galumpi Uampit, Wampi, Wampit (Tagalog);
Singapore: Wampoi, Wong-Pei;
Thailand: Mafia-Jean, Som-Mafai;
Vietnamese: Giôr, Hoàng Bi, Hong Bi.

Origin/Distribution

C. lansium is indigenous to and commonly cultivated in Southern China – Fujian, Guangdong, Guangxi, South Guizhou, Hainan, Sichuan, South East Yunnan and North to Central Vietnam. The fruit has been introduced to southeast Asia, i.e. Cambodia, Indonesia, Laos, Malaysia, the Philippines, Singapore and Thailand. Outside this region it is occasionally grown in India, Sri Lanka, Australia (north Queensland, Northern Territory), the United States (Hawaii and Florida) and in Central America.

Agroecology

Clausena lansium occurs in brushwood and open forest, at low and medium altitudes in its native range. Its climatic requirement is subtropical to tropical, it survives light, brief frost (−2°C) and is killed at temperatures of −6°C and lower. The tree is quite tolerant of a range of soils, including the deep sand and the oolitic limestone of southern Florida but thrives best in rich, well-drained, loamy soils. It requires watering in dry periods and is intolerant of water-logged conditions and drought.

Edible Plant Parts and Uses

Wampee is eaten fresh after peeling the skin or added to fruit cups, gelatins, other desserts or made into jellies, jams, pies. In China, Wampee fruit is eaten fresh, served with meat dishes or in preserves. In Viet Nam, fermenting the fruit with sugar and straining off the juice makes a bottled, carbonated beverage reminiscent of champagne. Dried Wampee is currently a popular product in Thailand.

Botany

An evergreen shrub or small tree, 3–12 m tall, trunk up to 40 cm in diameter with rough, grey-brown bark and usually low-branched (Plate 1). Leaves are dark green, spirally arranged resinous, 10–30 cm long, pinnate, with 7–15 alternate, elliptic or elliptic-ovate leaflets on 2–6 mm petiolules (Plate 2). Leaflets are 7–10 cm long, oblique at the base, margin repand to crenulated, thin, with minute hairs on the veins of the upper side and with a prominent yellow, warty midrib on the underside. The petiole is also warty and hairy. Inflorescences are terminal and paniculate (Plate 3). Flowers globose in bud, pentamerous, 14 mm across and sweet-scented. Calyx 5-lobed, tiny, pubescent, broadly ovate, 1 mm long; petals whitish to pale yellowish-green, oblong, 5 mm; stamens ten with linear filaments, basal portion slightly expanded; disk s short; ovary 5-loculed, rounded and hirsute; style thick, short and caduceus (Plate 4). Fruit pale green turning to pale yellow or brownish-yellow when ripe, globose, ellipsoid, or broadly ovoid, berry 1.5–3 × 1–2 cm, sparsely puberulent, glandular dotted (Plates 5, 6 and 7), 1–5-seeded. Seed is 10–15 mm long, bright-green with one brown tip. Pulp is mucilaginous, juicy, semi-translucent, sweet to acid-sweet.

Plate 2 Glossy, green, pinnate leaves of wampee

Plate 1 Rough, grey brown bark of wampee

Plate 3 Inflorescence and fruits of wampee

Plate 4 Wampee flowers

Plate 5 Cluster of ripe wampee fruits

Plate 6 Close-up of wampee fruits

Plate 7 Relative size of wampee fruits

fibre 0.8 g, ash 0.9 g, P 19 mg, K 281 mg, Ca 15 mg, Fe traces, β-carotene equivalent 0, thiamin 0.02 mg, riboflavin 0.11 mg, niacin 3.3 mg and vitamin C 148 mg. (Leung et al. 1972). Proximate nutrient composition of fruit pulp of *Clausena lansium* percent by weight was found to be water 71.97%, carbohydrate 18.64%, fat 0.33%, fibre 4.58%, protein 1.88% and ash 2.6% (Chokeprasert et al. 2006).

Various phytochemicals were identified in various plant parts:

Phytochemicals in Aerial Parts (Unspecified)

Compounds isolated from the plant included an oxirane carboxamide from the hexane extract (Milner et al. 1996); wampetin, a furocoumarin (Khan et al. 1983), and a tetracyclic triterpene alcohol 3β-hydroxy-23,24,24-trimethyllanosta-9(11)-25-diene (Lakshmi et al. 1989).

Fruit

In the fresh wampee fruit, 27 volatile components were identified in the essential oil, amounting to 99.01% of the relative area (Chokeprasert et al.

Nutritive/Medicinal Properties

Proximate nutrient composition of fruit pulp of *Clausena lansium* per 100 g edible portion was reported as: energy 55 kcal, moisture 84.0%, protein 0.9 g, fat 0.1 g, carbohydrate 14.1, dietary

2006). Ten monoterpene hydrocarbons, 2 sesquiterpene hydrocarbons, 7 alcohols, 1 aldehyde, 3 ketones, 1 carboxylic acid, and 3 terpene oxides were found with the monoterpene hydrocarbons fractions (94.48%) dominating. Sabinene (66.1%), α-pinene (11.7%), l-phellandrene (7.2%), and myrcene (43%) were the major components. The oxygenated components represented 4.3% with α-terpineol (1.18%) and 2- cyclohexen-1-one (1.02%) as the main components. The sesquiterpene hydrocarbons content was smaller with isosativene (0.13%) and curcumene (0.03%). Three components curcumene (0.03%), α–campholene aldehyde (0.06%), and α-pinene oxide (0.06%) were lost or changed to other compounds after drying. Thirteen components remained after different drying methods included: α-pinene, camphene, myrcene, l-phellandrene, 4 – carene, sabinene, γ-terpinene, isosativene, fenchol, 3-cyclohexen-1-ol, α-terpineol, 2-cyclohexen -1-one, and acetic acid. After vacuum drying, 7 new compounds; tricyclene (0.04%), 1H-3a,7-methanoazulene (0.04%), bergamotene (0.03%), β-famesene (0.04%), valencene 2 (0.03%), aromadendrene (0 .04%), and santalol (0.21%) were found. After sun drying, 6 new compounds; d-limonene (0.03%), β-pinene (0.1%), trans-ocimene (0.17%), trans-β-farnesene (0.05%), γ-curcumene (0.04%), and carvota acetone (0.05%) were produced. After hot air drying at 60°C, 5 new compounds: δ-3-carene (0.22%), β-biabolene (0.04%), δ-cadinene (0.02%), l-octanol (0.03%), and benzaldehyde (0.05%) were produced. After hot air drying at 45°C, 3 new compounds: α-muurolene (0.03%), cis-calamenene (0.04%), and 1-pentanol (0.05%) were produced.

In another study, Chokeprasert et al. (2007) reported the volatile components of fruits, seeds and leaves of *Clausena lansium* comprised mainly monoterpenes with 76–98% in flesh, skin and seed (Chokeprasert et al. 2007). Sabinene was the main component in leaf (14.9%), flesh (50.6%), skin (69.1%) and seed (83.6%). Other major compounds in the fruit flesh were, 3-cyclohexen-1-ol (15%), cyclohexene (6.5%), 1,4 cyclohexadiene (6.2%) and α-phellandrere (5%); in the fruit skin, α-phellandrene (10.6%) and α-pinene (9.4%) and isosativene (1.4%); and in the seed, α-pinene (4.3%), α-phellandrene (3.0%), and myrcene (2.9%). All the components identified in the fruit flesh and skin were as follows:

Fruit flesh:- sabinene (50.64%), 3-cyclohexen-1-ol (15.17%), cyclohexene (6.5%), 1,4 cyclohexadiene (6.19%), α-phellandrene (5.03%), (+)4-carene (3.98%), α-pinene (2.08%), acetic acid (2.65%), ethanol (2.46%), and the rest <1%: hexanal, β-pinene, myrcene, limonene, 3-methyl-4-brendene, β-fenchyl alcohol, and *ar*-curcumene. Zhao et al. (2004) found the main components of the fruit flesh essential oil to be β-santalol (52.0%), α-santalol (15.5%), farnesol (5.2%) and sinensal (4.0%) (Zhao et al. 2004). The following components were found in the fruit skin:- sabinene (69.07%), acetic acid, %, α-phellandrene (10.63%), α-pinene (9.41%), myrcene (3.15%), and the rest (<1%) hexanal, tricyclene, α-thujene, camphene, β-pinene, 3-carene, (+)4-carene, 1,3,6-octatriene, 1,4 cyclohexadiene,, γ-terpinene, cyclohexene, linalool, 3-cyclohexen-1-ol, 2-cyclohexen-1-one, 3-cyclohexen-1-methanol, cis-3-hexenyl 2-methylbutanoate, geranyl acetate, α-bergamotene, isosativene, β-santalene, α-humulene, *ar*-curcumene, allaromadendrene, α-farnesene, β-sesquiphellandrene. Prasad et al. (2010) found 8-hydroxypsoralen in wampee fruit peel.

Seed

Chokeprasert et al. (2007) reported the following volatile components in wampee seeds: sabinene (83.56%), α-pinene (4.26%), α-phellandrene (3.08%), myrcene (2.94%), γ terpinene (1.95%), and the rest <1%: acetic acid, α-thujene, camphene, benzaldehyde, (+)4-carene, trans-β-ocimene, cyclohexene, 2-nonanone, 3-cyclohexen-1-ol, 2-cyclohexen-1-one, β-fenchyl alcohol, bornyl acetate, β-caryophllene, α-bergamotene, isosativene, α-humulene, *ar*-curcumene, α-zingiberene, bicyclogermacrene, and β-bisabolene. Phellandrene (54.8%), limonene (23.6%), and *p*-menth-1-en-4-ol (7.5%) were the main constituents in the essential oil of the seeds (Zhao et al. 2004). The seed was also reported to contain amides: lansiumamide A (N-cis-styryl-cinnamamide), lansiumamideB (N-methyl-N-cis-styryl-cinnamamide) and lansiumamide C (N-methyl-N-phenethyl-cinnamamide) (Lin 1989), lansamide-I (Lin 1989); Lansiumamide A and SB-204900 from seeds (Tha-in et al. 2009); and a sporamin-type trypsin inhibitor (Ng et al. 2003). N-methyl-N-styrylcinnamamide (a lansamide) was also isolated from seeds (Luger et al. 2009).

Leaf

Seventy compounds of the oil of *C. lansium* leaf were identified of which caryophyllene oxide (16.8%) and (Z)-α-santalol (11.7%) were the major constituents (Pino et al. 2006). In another study, sabinene was the main component in wampee leaf (14.9%), and other major components were β-bisabolene (9.9%), β-caryophyllene (7.7%) and α-zingiberene (6.5%) Chokeprasert et al. (2007). The components in the leaf were: – sabinene (14.92%), β-bisabolene (9.88%), butanal (8.61%), β-caryophllene (7.72%), α-zingiberene (6.52%), 2-ethylfuran (4.61%), nonanone (3.42%), 2-propanone (3.02%), benzaldehyde (2.56%), 6-methyl-5-hepten-2-one, (2.26%), linalool,(2.25%) and the following compounds <2%: bicyclogermacrene, δ-cadinene, β-sesquiphellandrene, propanal, 2-methylfuran, 1-pentene, ethanone, acetic acid, cis-2-pentenol, hexanal, 2-hexanal, 3-hexen-1-ol, styrene, α-pinene, camphene, myrcene, α-phellandrene, benzeneacetaldehyde, 1,3,6-octatriene, 2-(E)-4,8-dimethyl-1,3,7-nonatriene, benzoic acid, cis-3-hexenyl 2-methylbutanoate, copaene, α-bergamotene, (+)-aromadendrene, isosativene, α-humulene, and *ar*-curcumene.

Wampee leaf was also reported to contain amides: neoclausenamide and dehydrocyclo-clausenamide (Yang et al. 1987), clausenamide, neoclausenamide, and cycloclausenamide (Yang et al. 1988), clausenamide I, clausenamide II, phenylethylcinnamamide, N-methylcinnamamide (Li et al. 1996); secoclausenamide, secodem-ethyl-clausenamide (Yang and Huang, 199), homoclausenamide, ζ-clausenamide (Yang et al. 1991), lansamides-1 (Prakash et al. 1980), lansamide-2, lansamide-3, lansamide-4 (Lakshmi et al. 1998), lansimide-2, lansimide-3 (Lakshmi et al. 2005), N-2-lansimide (Li et al. 1996), zetaclausenamide (Ma et al. 2008); carbazole alkaloids: heptaphylline, lansine (Prakash et al. 1980), coumarins: clausenacoumarin (Shen and Chen 1989), 3-benzylcoumarin (Li et al. 1996); triterpenoids: lansiol (Lakshmi et al. 1989), essential oils: caryophyllene oxide, (Z)-α-santalol (Pino et al. 2006), cis-β-farnesene (Luo et al. 1998); β-santalol (35.2%), bisabolol (13.7%), methyl santalol (6.9%), ledol (6.5%) and sinensal (5.6%) β-were found to be the main components of the leaf essential oil (Zhao et al. 2004). Seven compounds were isolated from the leaves and identified as corchoionoside C (1), 1′-O-β-D-glucopyranosyl (2R,3S)-3-hydroxynodakenetin (2), quercetin-3-O-robinobioside (3), rutin (4), quercetin-3-O-scillabioside (5), kaempferol-3-O-α-L-rhamnopyranosyl(1→2)[α-L-rhamnopyranosyl(1→6)]-β-D-glucopyranoside (6) and mauritianin (7) (Zhao et al. 2010). A new megastigmane glucoside (6S,7E,9S)-6,9,10-trihydroxy-4,7-megastigmadien-3-one 9-O-β-d-glucopyranoside and a new amide alkaloid (E)-N-(4-methoxyphenethyl)-2-methylbut-2-enamide were isolated from the leaves (Zhao et al. 2011).

Flowers

The main constituents of the essential oils extracted from wampee flowers were β-santalol (50.6%), 9-octade-cenamide (17.2%) and sinensal (4.1%) (Zhao et al. 2004).

Stem/Branches

The stem was reported to contain carbazole alkaloids: 3-formylcarbazole (Adebajo et al. 2009); 3-methylcarbazole, 3-formylcarbazole, methyl carbazole-3-carboxylate (Wu and Li 1999); coumarins: imperatorin, xanthotoxol, 8-geranoxypsoralen, wampetin (Wu and Li 1999), chalepin, phellopterin (Adebajo et al. 1998), imperatorin, chalepin and phellopterin

(Adebajo et al. 2009); sesquiterpenoids: oplopanone (Wu and Li 1999). Coumarins were isolated from the branch: lansiumarins A, B, C (Ito et al. 1998). Alkaloids: phenethyl cinnamide, daurine and inidazaline from the twigs (Tha-in et al. 2009). A carbazole, indizoline ([systematic name: 1-meth-oxy-2-(3-methyl-but-2-en-yl)-9H-carbazole-3-carbaldehyde], was isolated from *C. lansium* twigs (Fun et al. 2009).

Root

The root was reported to contain: carbazole alkaloids:- 3-formyl-6-methoxycarbazole, methyl 6-methoxycarbazole-3-carboxylate and 3-formyl-1,6-dimethoxycarbazole, murrayanine, glycozoline and indizoline and two carbazole derivatives, 3-formyl carbazole and methyl carbazole-3-carboxylate (Li et al. 1991), 2,7-dihydroxy-3-formyl-1-(3′-methyl-2′-butenyl)carbazole (Kumar et al. 1995); amides:- angustifoline; and coumarins:- chalepensin, chalepin, and gravelliferone (Kumar et al. 1995). A carbazole, glycozolidal (systematic name: 2,7-dimeth-oxy-9H-carbazole-3-carbaldehyde) was isolated from *C. lansium* roots (Fun et al. 2011).

Many of the isolated bioactive compounds such as cyclic amides, coumarins and sesquiterpene alcohols from wampee were found to exhibit antioxidant, antinociceptive, hepaprotective, anticancerous, antimicrobial, cerebral protective and nootropic properties.

Antioxidant Activity

Clausenamide isolated from *Clausena lansium* (p.o.) inhibited ethanol-induced lipid peroxidation in the mouse, causing the decrease of effect level of thiobarbituric acid. It also significantly activated the activity of glutathione peroxidase (GSH-Px) in the serous fluid of brain and liver tissues (Liu et al. 1991). From Fenton reaction studies, clausenamide had clearing effects on hydroxyl free radicals and superoxide anions. In addition, clausenamide could also clear the free radicals produced by the stimulation of human polymorphonuclear leukocyte (PML) by phorbol myristate acetate (PMA) (Lin et al. 1992). Fruits of wampee also contain a significant amount of coumarins with many health benefits. In one study, the ethyl acetate fraction (EAF) was found to exhibit the highest antioxidant activity compared to other fractions, even higher than synthetic antioxidant butylated hydroxyl toluene (BHT) (Prasad et al. 2010). The total phenolic content of wampee fraction was positively correlated with the antioxidant activity. The study indicated that the ethyl acetate fraction (EAF) possessed the highest phenolic content than other fractions. One coumarin compound, 8-hydroxypsoralen isolated from the ethyl acetate fraction of the fruit exhibited good scavenging activities against DPPH radical and superoxide anion as well as significant reducing power. Also, the EAF exhibited strong antioxidant and anticancer activities, which were comparable to the commercial antioxidant BHT and the anticancer drug cisplatin. The findings indicated that the wampee peel extract could be used as natural antioxidant and anticancer agent.

Anticancer Activity

Both α- and β-santalol were also reported to affect some tumours (Zhao et al. 2004). In particular, studies in CD mice and SENCAR mice (sensitive to carcinogenesis) showed that α-santalol had chemopreventive action and might prevent skin tumour development (Johnson et al. 2001; Dwivedi et al. 2003).

The ethyl acetate fraction of wampee fruit exhibited strong anticancer activities against human gastric carcinoma (SGC-7901), human hepatocellular liver carcinoma (HepG-2) and human lung adenocarcinoma (A-549) cancer cell lines, higher than cisplatin, a conventional anticancer drug (Prasad et al. 2010). 8-hydroxypsoralen isolated from the ethyl acetate fraction of the fruit also showed potent proliferation inhibitory activity against human hepatocellular liver carcinoma cell line (HepG2), human lung adenocarcinoma epithelial cell line (A549) and human cervical carcinoma cell line (HELA). Three new carbazole alkaloids, mafaicheenamine A-C (1–3), along with five know compounds (4–8) were isolated from the twigs of *Clausena lansium* (Maneerat and Laphookhieo 2010). Compounds 1, 2 and 4–8 exhibited antitumoral activity against three human cancer cell lines, KB, MCF-7 and NCI-H187. Two new coumarins namely clausenalansimin A (5) and B (9) together with seven known coumarins (1–4 and 6–8), xanthotoxol (1), imperatorin (2), heraclenin (3), heraclenol (4), wampetin (6), indicolactonediol (7), and isoscopoletin (8)were isolated from twigs of *Clausena lansium* (Maneerat et al. 2010). Of these compounds, (1–3, 5 and 6) showed weak activity with cytotoxicity against KB (oral human epidermal carcinoma), MCF7 (breast cancer) and NCI-H187(human, small cell lung cancer) cell lines, except coumarins 3 and 6 which were found to be inactive against MCF7 cancer cell line.

Hepatoprotective Activity

Huangpi leaf chloroform extract, clausenamide, or secoclausenamide (i.g.) showed significant protective effects against liver damage of the mouse induced by CCl_4, paracetamol, and thioacetamide. They reduced serum glutamic pyruvic transaminase (sGPT) level and liver pathological damage, and enhanced the antitoxic functions of the liver (Liu et al. 1996). Clausenamide, neoclausenamide, and cycloclausenamide were also found to reduce aminotransferase (Yang et al. 1988). In-vitro, clausenamide compounds protected the rat from liver cell unscheduled DNA synthesis (UDS) induced by aflatoxin B_1 (AFB_1) (Wu and Liu 2006). They also inhibited the release of glutamic pyruvic transaminase. In another study, nine cyclic amides excluding demethylsecoclausenamide, isolated from leaves were found to have significant hepatoprotective activity in test mice (Liu et al. 1996). The compounds at the dose of 250 mg/kg, significantly depressed the elevated serum transaminase in mice intoxicated with carbon tetrachloride (CCl_4). Two compounds seco-clausenamide and clausenamide, further decreased the hepatotoxicity of thioacetamide and acetaminophen in test mice. Clausenamide was shown to significantly inhibit CCl4-induced lipid peroxidation of liver microsomes and 14C-CCl_4 covalent binding to microsomal lipids. Another cyclic amide, zeta-clausenamide, isolated from the leaves also displayed hepatoprotective activity (Ma et al. 2008). Clausenamide, neoclausenamide and cyclo-clausenamide, isolated front the leaves were also found to lower elevated blood serum glutamate pyruvate transaminase (SGPT) levels in test animals. In another study, 100 and 200 mg/kg of methanolic stem bark extract administered i.p., reduced CCl_4-induced hepatotoxicity firstly by 5.3% and 8.4% reduction in phenobarbitone-sleeping time respectively, secondly by reversing the reduction in serum liver proteins by 7.0–8.8%, serum AST (aspartate aminotransferase), ALT (Alanine transaminase) and ALP (alkaline phosphatase) activities by 27.7–107.9% and thirdly by diminishing increased values of plasma AST, ALT and ALP activities by 13.2–83.8% (Adebajo et al. 2009). The extract exhibited antioxidant activities. The hepatoprotective activity of *C. lansium* was partly due to its anti-oxidant and antiinflammatory properties and confirmed its folkloric use in the treatment of gastro-intestinal inflammation, bronchitis and hepatitis.

Antinociceptive Activity

Recent studies had indicated that β-santalol, which was the main constituent of the essential oils from leaves, flowers and sarcocarps of *C. lansium*, had an effect on the central nervous system in mice, i.e. antinociceptive, thus it could be considered as a neuroleptic (Okugawa and Ueda 1995).

Nootropic and Cerebral Protective Activity

Clausenamide, isolated from aqueous extract of dry leaves of *Clausena lansium*, a Chinese folk medicine, was found to have potent activity in enhancing long-term potentiation (LTP) and show nootropic activity in animal tests (Feng et al. 2009). The results of LTP assay showed that the nootropic activity of the stereoisomers of clausenamide is closely related to the configuration of stereoisomers. Earlier studies showed that effects of clausenamide on the synaptic transmission in the dentate gyrus depended on its chirality (Liu and Zhang 1998). The potentiating effects of (−) clausenamide on synaptic transmission in hippocampus strongly supported the nootropic action of (−) clausenamide. They (Liu and Zhang 1999) found that the activation of opioid receptors contributed to the induction of l-clausenamide-induced long-term potentiation of synaptic transmission in dentate gyrus of anaesthetized rats. Studies revealed that (−) clausenamide could protect against neuron loss in the cortex and striatum regions (Tang and Zhang 2002). The results indicated a beneficial effect of (−)clausenamide for synaptic plasticity and cognitive function impaired by transient focal cerebral ischemia in rats. Tang and Zhang (2003) further demonstrated that (−) clausenamide could improve long-term potentiation (LTP) impairment in the ipsilateral dentate gyrus and enhanced cell survival in the striatum compared to the vehicle-treated rats 4 days following ischemic damage and its protective effect on mitochondria may partially underlie its action. Separate studies on clausenamide induced Ca^{2+} signaling in primary cultures of rat cortical neurons revealed that (−) clausenamide induced calcium transient in primary culture of rat cortical neurons through the IP3 receptor pathway (Tang and Zhang 2004). Studies by Liu et al. (1999), Xu et al. (2005) on the effect of clausenamide on synaptic transmission in the dentate gyrus of rats in vivo found that: (1) (−)-clausenamide potentiated synaptic transmission in both anesthetized and freely moving animals while (+)-clausenamide showed no or little effect; (2) (−)-clausenamide increased the magnitude of long-term potentiation (LTP) induced by high-frequency stimulation (HFS) in anaesthetized animals whereas (+)-clausenamide had no effect; (3) voltage-dependent calcium channels (VDCCs) calcineurin and calpain were involved in (−)-clausenamide-induced potentiation of synaptic transmission. Thier findings may provide the pharmacological basis for understanding the nootropic mechanisms of (−)-clausenamide, which was the first chiral nootropic agent developed in China. Another paper reported (−) clausenamide effects included the promotion of intelligence development to be resisting oxidation, scavenging free radicals, suppressing Aβ neurotoxicity and nerve cell apoptosis, inhibiting over phosphorylation of tau protein, and improving central cholinergic system (Liu and Zhu 2007). (−) Clausenamide was found to stimulate central cholinergic neuron development in-vitro (Duan and Zhang 1998b). It increased the activity of choline acetyltransferase (ChAT) and protein content in cultured neurons, as well as stimulated proliferation of neuronal cells, support survival and neurite outgrowth of neurons. The neurotrophic action of (−) clausenamide (0.001–10 umol/l) was similar to that of nerve growth factor. The (+)clausenamide had no neurotrophic action, even at high concentrations (0.1–10 umol/l), but neurons were damaged.

Studies showed that (−) clausenamide (i.g.) improved anisondine-induced amnesia in mice (Duan and Zhang 1998a) and β-amyloid peptide-induced spacious learning disabilities of the rat (Zhang et al. 2001). (−) clausenamide was reported to be a cognition enhancer and may affect regional acetylcholine (ACh) levels (Duan and Zhang 1998a). (+) clausenamide had no effect on ACh decrease in all examined brain regions. The protective action of (−) clausenamide was postulated to be due to the acetylcholine level reduction caused by anisondine inhibition in the cortex, hippocampus, and corpus striatum (Duan and Zhang 1998a; Zhang et al. 2001), the increased concentration of brain NMDA (N-methyl-D-asparate) receptors (Duan and Zhang 1997), the increased activity of cortex choline acetyltransferase (ChAT), the increased activities of brain protein phosphatase neurocalcin and calpain (Zhang et al. 2001), as well as the enhanced transmission of postsynaptic hypocampal dentate gyrus (Liu et al. 1999).

Clausenamide (p.o.) significantly inhibited 5-hydroxytryptamine (5-HT)-, prostaglandin F2α (PGF2α)-, and arachidonic acid-induced contraction of brain basilar artery of the rat, relieving blood vessel spasm and increasing cerebral blood flow (Liu et al. 1991). Clausenamide p.o. 100–200 mg/kg prolonged both the duration of gasping after decapitation and the survival time after subcutaneous administration of 225 mg/kg. In addition, multiple doses of clausenamide were shown to inhibit the liver lipid peroxidation caused by 50% alcohol and increase the glutathione-peroxidase (GSH-peroxidase) activity significantly in rat liver and brain cytosol indicating clausenamide to be a cerebral protective agent. (−) Clausenamide (p.o., consecutive days) increased the density of hippocampal synapses and the number of mossy nerve fibre buddings in ablactated mice and adult rats (Zhang et al. 2001). In-vitro (−) clausenamide inhibited PC12 cell apoptosis (Wang and Zhang 2000), and antagonized the neurotoxic effects of sodium nitroprusside and apoptosis of hippocamal neurons. The mechanism may due to the increased expression of anti-apoptosis gene bcl-2, the decreased expression of pro-apoptotic bax gene, and the increased ratio of bcl-2/bax (Liu and Zhu 2007). Studies showed that clausenamide could enhance the expression of Bcl-2 protein and inhibit considerably apoptosis and may coordinate with Bcl-2 in inhibiting apoptosis after focal cerebral ischemia/reperfusion in renovascular hypertensive rats (Jiang et al. 2005). This may be the mechanism of clausenamide protection of brain cells from ischemic damage.

Studies showed that (−) clausenamide exerted protective effects against neurotoxicity induced by okadaic acid and Abeta25-35 on the neurons of hippocampus and cerebral cortex (Zhang et al. 2007). Pre-treatment of (−) clausenamide and LiCl decreased the rate of cell death from MTT [3-(4,5-dimethylthiazol-2-yl)-2,5-diphenyltetrazolium bromide], LDH (lactic dehydrogenase) release, and apoptosis from Hoechst 33258 staining in SH-SY5Y cell line. The step-through tests showed (−) clausenamide could improve the ability of learning and memory.

Studies showed that (−) clausenamide significantly elevated cell viability of differentiated PC12 cells (Hu et al. 2010). Further, (−) clausenamide arrested the apoptotic cascade by reversing overload of calcium, prevented ROS generation, moderated the dissipation of mitochondrial transmembrane potential and the imbalance of Bcl-2 and Bax, inhibited the activation of p38 MAPK and the expression of P53 and cleaved caspase 3. The results suggested that (−) clausenamide may be a therapeutic agent for Alzheimer's disease.

Bu-7, a flavonoid isolated from *Clausena lansium* leaves, protected PC 12 cells against rotenone-induced injury (Li et al. 2011). Pretreatment with Bu-7 (0.1 and 10 μmol/L) decreased rotenone-induced apoptosis, attenuated rotenone-induced mitochondrial potential reduction and suppressed rotenone-induced protein phosphorylation and expression. The protective effect thus could be attributed to MAP kinase cascade (JNK and p38) signalling pathway and suggested that Bu-7 may be a potential bioactive compound for the treatment of Parkinson's disease.

Antidiabetic Activity

A coumarine compound, clausenacoumarine isolated from leaves was found to lower blood glucose level in normal mice and alloxan diabetic mice at 200 mg/kg/day for 3 days orally, and antagonize the elevation of blood glucose caused by injecting adrenaline in normal mice (Shen and Chen 1989; Shen et al. 1989). No effect on blood lactic acid was observed.

The methanolic stem bark extract (100 mg/kg) induced maximum and significant anti-hyperglycaemic activity of and increase in plasma insulin at 60 min, compared to control (Adebajo et al. 2009). The significant 174.6% increase of insulin release from insulinoma cell line, INS-1 cells (in-vitro) at 0.1 mg/ml indicated that it mediated its antidiabetic action mainly by stimulating insulin release. Imperatorin and chalepin were the major active constituents increasing in-vitro insulin release to 170.3% and 137.9%, respectively.

Anti-Trichomonal Activity

A dichloromethane extract of the stem bark was found to be superior over methanolic extract with respect to an anti-trichomonal activity which was measured after 24 and 48 hours (Adebajo et al. 2009). The isolated compounds imperatorin and 3-formylcarbazole had the main anti-trichomonal activity (LC_{50}s of 6.0, 3.0 and 3.6, 9.7 µg/mL after 24 and 48 h, respectively). 100 mg/kg of the methanolic extract produced an antiinflammatory activity after 4 hours. Thus, the use of *C. lansium* stem bark was suggested to be useful in trichomoniasis and diabetes.

Antimicrobial Activity

In India, four novel lansamides isolated from *Clausena lansium* leaves were found to have antifungal activity (Lakshmi et al. 2006). Lansimide-1 and lansiol exhibited promising in-vitro activity against *Trichophyton mentagrophytes* and *Crytococcus neoformans*. Lansimide-2 showed promise against *Sporothrix schenckii* and also *Crytococcus neoformans*. Lansimide-3 was inactive.

Trypsin Inhibition Activity

A homodimeric trypsin inhibitor isolated from the seeds was found to inhibit trypsin with an IC_{50} of 2.2 nM but was without any inhibitory effect on chymotrypsin and proteinase K (Ng et al. 2003). The uptake of MTT (3-(4,5-dimethylthiazol-2-yl)-2,5-diphenyltetrazolium bromide) by human leukemia HL60 and hepatoma Hep G2 cells was inhibited with an IC_{50} of 100 µM. Translation in the cell-free rabbit reticulocyte lysate system was inhibited with an IC_{50} of 3.6 µM. The activity of HIV-1 reverse transcriptase was reduced in the presence of the trypsin inhibitor. The trypsin inhibitor exerted antifungal activity toward *Physalospora piricola* but not *Mycosphaerella arachidicola*, *Botrytis cinerea*, *Fusarium oxysporum* or *Coprinus comatus*.

Traditional Medicinal Uses

The unripe fruits, the leaves, and roots have been used medicinally in folkloric medicine. In Malaysia and China, the leaves are used in traditional medicine to treat coughs, asthma, hepatitis and dermatological diseases. The dried unripe fruits and dried sliced roots are used as a remedy for bronchitis. Ripe fruits are said to have stomachic and cooling effects and to act as a vermifuge. The leaves, fruits or seeds are widely employed for gastro-intestinal problems, including acute and chronic inflammation and ulcers. In Taiwan the roots are used in traditional medicine to treat bronchitis and malaria. In Vietnam, a decoction of the leaves is used to treat dandruff and preserve the colour of the hair. In the Philippines, the fruit is used as a folkloric medicine for stomach upsets and indigestion, for coughs, influenza, colds and abdominal colic pains. It was reported to have folkloric use in the treatment of gastro-intestinal inflammation, bronchitis and hepatitis in Africa (Adebajo et al. 2009).

Other Uses

Clausena lansium is sometimes planted as a hedge, or as a lane tree.

Comments

Wampee is related to *Citrus* species.

Selected References

Adebajo AC, Iwalewa EO, Obuotor EM, Ibikunle GF, Omisore NO, Adewunmi CO, Obaparusi OO, Klaes M, Adetogun GE, Schmidt TJ, Verspohl EJ (2009) Pharmacological properties of the extract and some isolated compounds of *Clausena lansium* stem bark: anti-trichomonal, antidiabetic, anti-inflammatory, hepatoprotective and antioxidant effects. J Ethnopharmacol 122(1):10–19

Adebajo AC, Kumar V, Reish J (1998) 3-Formylcarbazole and furocoumarins from *Clausena lansium*. Nig J Nat Prod Med 2:57–58

Backer CA, van den Brink RCB Jr (1965) Flora of Java, vol 2. Wolters-Noordhoff, Groningen, 630 pp

Brown WH (1951–1957) Useful plants of the Philippines. Reprint of the 1941-1943 edition. 3 volumes. Technical Bulletin 10. Department of Agriculture and Natural Resources. Bureau of Printing, Manila, the Philippines, vol 1 (1951) 590 p, vol 2 (1954) 513 p, vol 3 (1957) 507 p

Burkill IH (1966) A dictionary of the economic products of the Malay Peninsula. Revised reprint. 2 volumes.

Ministry of Agriculture and Co-operatives, Kuala Lumpur, Malaysia, vol 1 (A–H) pp 1–1240, vol 2 (I–Z) pp 1241–2444

Campbell CW (1975) The wampee, a fruit well adapted to Southern Florida. Proc Fla State Hortic Soc 87: 390–392

Chokeprasert P, Khotavivattana S, Oupadisskoon C, Huang TC, Chen HH (2006) Volatile components of wampee fruits [*Clausena lansium* (Lour.) Skeels] treated by different drying conditions. Thammasat Int J Sci Technol 11(1):66–71

Chokeprasert P, Charles AL, Sue KH, Huang TC (2007) Volatile components of the leaves, fruits and seeds of wampee [*Clausena lansium* (Lour.) Skeels]. J Food Compos Anal 20(1):52–56

De Bruijn J (1992) *Clausena lansium* (Lour.) Skeels. In: Verheij EWM, Coronel RE (eds) Plant resources of South-East Asia no 2. Edible fruits and nuts. Prosea Foundation, Bogor, pp 141–143

Duan WZ, Zhang JT (1997) Effects of (−), (+)clausenamide on central N-methyl-D-asparate receptors in rodents. Yao Xue Xue Bao 32(4):259–263

Duan W, Zhang J (1998a) Effects of clausenamide on anisodine-induced acetylcholine decrease and memory deficits in the mouse brain. Chin Med J (Engl) 111(11): 1035–1038

Duan WZ, Zhang JT (1998b) Stimulation of central cholinergic neurons by (−) clausenamide in vitro. Zhongguo Yao Li Xue Bao 19(4):332–336

Dwivedi C, Guan X-M, Harmsen WL, Voss AL, Goetz-Parten DE, Koopman EM, Johnson KM, Valluri HB, Matthees DP (2003) Chemopreventive effects of α-santalol on skin tumor development in CD-1 and SENCAR mice. Cancer Epidemiol Biomarkers Prev 12:151–156

Feng Z, Li X, Zheng G, Huang L (2009) Synthesis and activity in enhancing long-term potentiation (LTP) of clausenamide stereoisomers. Bioorg Med Chem Lett 19(8):2112–2115

Fun HK, Maneerat W, Laphookhieo S, Chantrapromma S (2009) Indizoline. Acta Crystallogr Sect E Struct Rep Online 65(Pt 10):o2497–o2498

Fun HK, Maneerat W, Laphookhieo S, Chantrapromma S (2011) Glycozolidal. Acta Crystallogr Sect E Struct Rep Online 67(Pt 7):o1811–o1812

Hu SY (2005) Food plants of China. The Chinese University Press, Hong Kong, 844 pp

Hu JF, Chu SF, Ning N, Yuan YH, Xue W, Chen NH, Zhang JT (2010) Protective effect of (−) clausenamide against Abeta-induced neurotoxicity in differentiated PC12 cells. Neurosci Lett 483(1):78–82

Ito C, Katsuno S, Furukawa H (1998) Structures of lansiumarin-A, -B, -C, three new furocoumarins from *Clausena lansium*. Chem Pharm Bull 46(2):341–343

Jiang ZC, Bi GN, Shi SL (2005) Effect of clausenamide on the expression of Bcl-2 protein and apoptosis after focal cerebral ischemia/reperfusion in renovascular hypertensive rats. Zhongguo Wei Zhong Bing Ji Jiu Yi Xue 17(5):289–292 (in Chinese)

Johnson K-M, Koopman E-M, Guan X-M, Dwivedi C (2001) Alpha-santalol prevents skin tumour development in SENCAR mice. FASEB J 15:237

Khan NU, Naqvi SWI, Ishratullah K (1983) Wampetin, a furocoumarin from *Clausena wampi*. Phytochemistry 22(11):2624–2625

Kumar V, Vallipuram K, Adebajo AC, Reisch J (1995) 2,7-dihydroxy-3-formyl-1-(3′-methyl-2′-butenyl) carbazole from *Clausena lansium*. Phytochemistry 40: 1563–1565

Lakshmi V, Kumar R, Agarwal SK (2005) Novel cyclic amide heterodimer lansamide-2 from *Clausena lansium*. Nat Prod Res 19(4):355–358

Lakshmi V, Kumar R, Varshneya V, Chaturvedi A, Shukla PK, Agarwal SK (2006) Antifungal activity of lansimides from *Clausena lansium* (Lour.). Nig J Nat Prod Med 9:61–62

Lakshmi V, Raj K, Kapil RS (1989) A triterpene alcohol, lansiol, from *Clausena lansium*. Phytochemistry 28(3):943–945

Lakshmi V, Raj K, Kapil RS (1998) Chemical constituents of *Clausena lansium*: part III. Structure of lansamide-3 and 4. Indian J Chem 37B(4):422–424

Leung W-TW, Butrum RR, Huang Chang F, Narayana Rao M, Polacchi W (1972) Food composition table for use in East Asia. FAO, Rome, 347 pp

Li BY, Yuan YH, Hu JF, Zhao Q, Zhang DM, Chen NH (2011) Protective effect of Bu-7, a flavonoid extracted from *Clausena lansium*, against rotenone injury in PC12 cells. Acta Pharmacol Sin 32(11):1321–1326

Li SH, Wu SL, Li WS (1996) Amides and coumarin from the leaves of *Clausena lansium*. Chin Pharm J 48(5):367–373

Li WS, McChesney JD, El-Feraly FS (1991) Carbazole alkaloids from *Clausena lansium*. Phytochemistry 30(1):343–346

Lin JH (1989) Cinnamamide derivatives from *Clausena lansium*. Phytochemistry 28(2):621–622

Lin TJ, Liu GT, Li XJ, Zhao BL, Xin WJ (1992) Antilipid peroxidation and oxygen free radical scavenging activity of clausenamide. Zhongguo Yaolixue Yu Dulixue Zazhi 6(2):97–102 (in Chinese)

Liu GT, Li WX, Chen YY, Wei HL (1996) Hepatoprotective action of nine constituents isolated from the leaves of *Clausena lansium* in mice. Drug Dev Res 39(2): 174–178

Liu Y, Shi CZ, Zhang JT (1991) Anti-lipidperoxidation and cerebral protective effects of clausenamide. Yao Xue Xue Bao 26(3):166–170 (in Chinese)

Liu SL, Zhang JT (1998) Difference between the effects of (−) clausenamide and (+) clausenamide on the synaptic transmission in the dentate gyrus of anesthetized rats. Yao Xue Xue Bao 33(4):254–258 (in Chinese)

Liu SL, Zhang JT (1999) Effects of naloxone on l-clausenamide-induced long-term potentiation in dentate gyrus of anesthetized rats. Zhongguo Yao Li Xue Bao 20(2):112–116 (in Chinese)

Liu SL, Zhao MR, Zhang JT (1999) Effects of clausenamide on synaptic transmission of the dentate gyrus in freely-moving rats. Yao Xue Xue Bao 34(5):325–328 (in Chinese)

Liu Y-J, Zhu Q-F (2007) Effect of clausenamide on hippocampal neuron apoptosis induced by sodium nitroprusside. Neural Regen Res 2(1):33–37

Luger P, Weber M, Thang TD, Luu HV, Dung NX (2009) N-Methyl-N-styrylcinnamamide (lansamide) from *Clausena lansium* in Vietnam. Acta Crystallogr 65(4):809

Luo H, Cai C, Zhang JH, Mo LE (1998) Study on the chemical constituents of essential oil from *Clausena lansium* leaves. J Chin Med Mat 21(8):405–406

Ma NC, Wu K, Huang L (2008) An elegant synthesis of Zetaclausenamide. Eur J Med Chem 43(4):893–896

Maneerat W, Laphookhieo S (2010) Antitumoral alkaloids from *Clausena lansium*. Heterocycles 81:1261–1269

Maneerat W, Prawat U, Seawan N, Laphookhieo S (2010) New coumarins from *Clausena lansium* twigs. J Braz Chem Soc 21:665–668

Milner PH, Coates NJ, Gilpin ML, Spear SR (1996) SB-204900, a novel oxirane carboxamide from *Clausena lansium*. J Nat Prod 59(4):400–402

Molesworth Allen B (1967) Malayan fruits. An introduction to the cultivated species. Moore, Singapore, 245 pp

Morton J (1987) Wampee. Fruits of warm climates. Julia F. Morton, Miami, pp 197–198

Ng TB, Lam SK, Fong WP (2003) A homodimeric sporamin-type trypsin inhibitor with antiproliferative, HIV reverse transcriptase inhibitory and antifungal activities from wampee (*Clausena lansium*) seeds. Biol Chem 384:289–293

Nguyen VD (1993) Medicinal plants of Vietnam, Cambodia and Laos. Mekong Printing, Santa Ana, 528 pp

Ochse JJ, Soule MJ Jr, Dijkman MJ, Wehlburg C (1961) Tropical and subtropical agriculture. 2 vols. Macmillan, New York, 1446 p

Okugawa H, Ueda R (1995) Effect of alpha-santalol and beta-santalol from sandalwood on the central nervous system in mice. Phytomedicine 2:119–126

Pino JA, Marbot R, Fuentes V (2006) Aromatic plants from western Cuba IV. Composition of the leaf oils of *Clausena lansium* (Lour.) Skeels and *Swinglea glutinosa* (Blanco) Merr. J Essent Oil Res 18(2):139–141

Prakash D, Raj K, Kapil RS, Popli SP (1980) Chemical constituents of *Clausena lansium*: part I. Structure of lansamide-I and lansine. Indian J Chem 19B(12): 1075–1076

Prasad KN, Xie HH, Hao J, Yang B, Qiu SX, Wei XY, Chen F, Jiang YM (2010) Antioxidant and anticancer activities of 8-hydroxypsoralen isolated from wampee [*Clausena lansium* (Lour.) Skeels] peel. Food Chem 118(1):62–66

Purseglove JW (1968) Tropical crops: dicotyledons, vol 1 & 2. Longman, London, 719 pp

Schmelzer GH (2001) *Clausena lansium* (Lour.) Skeels. In: van Valkenburg JLCH, Bunyapraphatsara N (eds) Plant resources of South-East Asia no. 12(2): medicinal and poisonous plants 2. Prosea Foundation, Bogor, pp 166–167

Shen ZF, Chen QM (1989) The hypoglycemic effect of clausenacoumarine. Acta Pharm Sin 24(5):391–392

Shen ZF, Chen QM, Liu HF, Xie MZ (1989) The hypoglycemic effect of clausenacoumarine. Yao Xue Xue Bao 24(5):391–392 (in Chinese)

Stuart GU (2010) Philippine alternative medicine. Manual of some Philippine medicinal plants. http://www.stuartxchange.org/OtherHerbals.html

Tang K, Zhang JT (2002) The effects of (−)clausenamide on functional recovery in transient focal cerebral ischemia. Neurol Res 24(5):473–478

Tang K, Zhang JT (2003) (−)Clausenamide improves long-term potentiation impairment and attenuates apoptosis after transient middle cerebral artery occlusion in rats. Neurol Res 25(7):713–717

Tang K, Zhang J-T (2004) Mechanism of (−)clausenamide induced calcium transient in primary culture of rat cortical neurons. Life Sci 74(11):1427–1434

Tha-in S, Maneerat W, Machan T, Dachathai S, Laphookhieo, S (2009) Alkaloids from *Clausena lansium*. In: Proceedings of the 35th congress of science and technology of Thailand, Chonburi, Thailand, 15–17 Oct 2009

Wang RS, Zhang JT (2000) Construction of Bax α high expressing PC-12 cell line and the mechanisms of (−) clausenamide in inhibiting apoptosis. Yao Xue Xue Bao 35(6):404–407 (In Chinese)

Wong KC, Wong SN, Sam TW, Chee SG (1998) Volatile constituents of *Clausena lansium* (Lour.) Skeels fruit. J Essent Oil Res 10:700–702

Wu SL, Li WS (1999) Chemical constituents from the stems of *Clausena lansium*. Chin Pharm J 51(3):227–240

Wu YQ, Liu GT (2006) Protective effect of enantiomers of clausenamides on aflatoxin B1-induced damage of unscheduled DNA synthesis of isolated rat hepatocytes. Zhongguo Yaolixue Yu Dulixue Zazhi 20(5):393–398 (in Chinese)

Xu L, Liu SL, Zhang JT (2005) (−)-Clausenamide potentiates synaptic transmission in the dentate gyrus of rats. Chirality 17(5):239–244

Yang MH, Chen YY, Huang L (1987) Studies on the chemical constituents of (*Clausena Lansium*) (Lour.) Skeels II: the isolation and structural elucidation of neoclausenamide and dehydrocycloclausenamide. Acta Chim Sin 5(3):267–272

Yang MH, Chen Y-Y, Huang L (1988) Three novel cyclic amides from *Clausena lansium*. Phytochemistry 27(2):445–450

Yang MH, Chen YY, Huang L (1991) Studies on the chemical constituents of *Clausena lansium* (Lour.) Skeels. III. The structural elucidation of homo- and ζ-clausenamide. Chin Chem Lett 2(4):291

Yang MH, Huang L (1991) Studies on the chemical constituents of *Clausena lansium* (Lour.) Skeels. IV. The structural elucidation of seco- and secodemethylclausenamide. Chin Chem Lett 2(10):775–776

Zhang DX, Hartley TG (2008) *Clausena* N. L. Burman. In: Wu ZY, Raven PH, Hong DY (eds) Flora of China, vol 11, Oxalidaceae through Aceraceae. Science Press/Missouri Botanical Garden Press, Beijing/St. Louis

Zhang J, Cheng Y, Zhang JT (2007) Protective effect of (−) clausenamide against neurotoxicity induced by okadaic acid and beta-amyloid peptide25-35. Yao Xue Xue Bao 42(9):935–942 (in Chinese)

Zhang JT, Duan WZ, Liu SL, Wang RS (2001) Antidementia effects of (−)clausenamide. Yiyao Daobao 20(7):403–404

Zhao J, Nan P, Zhong Y (2004) Chemical composition of the essential oils of *Clausena lansium* from Hainan Island, China. Z Naturforsch C 59(3–4):153–156

Zhao Q, Li C, Yang J, Zhang D (2010) Chemical constituents of *Clausena lansium*. Zhongguo Zhong Yao Za Zhi 35(8):997–1000 (in Chinese)

Zhao Q, Yang JZ, Li CJ, Chen NH, Zhang DM (2011) A new megastigmanle glucoside and a new amide alkaloid from the leaves of *Clausena lansium* (Lour.) Skeels. J Asian Nat Prod Res 13(4):361–366

Limonia acidissima

Scientific Name

Limonia acidissima L.

Synonyms

Feronia elephantum Correa, *Feronia limonia* (L.) Swingle, *Schinus limonia* L.

Family

Rutaceae

Common/English Names

Curd Fruit, Elephant Apple, Indian Wood-Apple, Monkey Fruit, Wood Apple.

Vernacular Names

Arabic: Tuffâhh El Fîl;
Bangladesh: Kath Bel;
Burmese: Thibin, Thanaka, Tha Nap-Hka;
Chinese: Mu Ping Guo, Mu Ping Kuo;
Danish: Elefantæble;
Dutch Olifants Appel;
Fiji: Kabeet, Kabut, Vellam Pelam, Vakandra (Hindu);
French: Citron Des Mois, Citronnier Des Éléphants, Féronie De L'inde, Pomme De Bois, Pomme Des Éléphants, Pommier D'éléphant;
German: Elefantenapfel;
India: Kath Baei (Assamese), Bela, Kait, Kainta, Kapittha, Kath Bel, Kathbel, Kayet Bel, Kayetabela (Bengali), Kotha (Gujarati), Barnah, Barnahi Billan, Barnasi, Beli, Bernahi, Bilan, Bilin, Dadhiphal, Dantasath, Kabut, Kaith, Kaitha, Katbel, Kath Bel Kabeet, Katori, Kavita, Khatha, Manamath, Pushpaphal, (Hindu), Aranamullu, Aruna Mullu, Arunamullu, Baela, Baelada, Baeladahannu, Baelada Hannina Mara, Baelada Hannu, Baelada Mara, Baeladakaayi, Baluvali, Baloola Kaayi, Bela, Belada Mara, Belakudi, Belaruhi, Belavu, Byaalada Mara, Byala, Dadhiphala, Danthashata, Graaha, Kaadu Baela, Kadbela, Kadinimbi, Kadubela, Kadunimbe, Kaduvelada, Kapithha, Kotamiria, Malura Mara, Manmatha Mara, Naibel, Naibela, Naibyalada, Nayibel, Nayibela, Nimbai, (Kannada), Belpatre (Konkani), Blanka, Cerukattunarakam, Cherrukatnarragam, Cherukattunarakam, Katnaragam, Kattunarakam, Mlanka, Naivelam, Tsjeroukatounarigam, Tsjerucaatnaregam, Vilankai, Vilarmaram, Vilavu (Malayalam), Kapith, Kauth, Kavant, Kavanti, Kavat, Kaveet, Kawath Kovit, Sit-Ranlimbi (Marathi), Kaintha, Koyito (Oriya), Katha (Punjabi), Dadhinama, Dadhiphala, Dadhittha, Danthashata, Kapipriya, Kapita, Kapitha, Kapithama, Kapittha, Kapitya, Kapityama, Pushpaphala

(Sanskrit), Kapittam, Kavittam, Narivila, Nilavila, Norivila, Tantacatam, Vila, Vilaa, Vilamaram, Vilampazam, Vilanga, Vilankay Maram (Tamil), Kapithhamu, Parupuvelaga, Pulivelaga, Pushpaphalamu, Taruvelaga, Tholuvelaga, Thorelaga, Thorrivelaga, Tolielaga, Tollivelam, Torelega, Toriallega, Toriyelaka, Torravelaga, Torravelagu, Torrayelaka, Torriyelaka, Velaga, Velagachettu, Velagapandu, Volaga (Telugu);
Indonesia: Kawista (Java), Kusta (Bali);
Italian: Pomo D'elefante;
Japanese: Feronia Rimonia, Rimonia Akidisshima, Tanaka, Zou No Ringo;
Khmer: Kramsang;
Laotian: Ma Fit;
Malaysia: Belinggai, Gelinggai;
Pakistan: Kaith Bel;
Philippines: Ponoan;
Reunion Island: Tanaka;
Russian: Feroniia Limonnaia, Slonovaia Iabloni, Slonove Iabloko;
Thai: Makhwit (Central), Mafit (Northern);
Tibetan: Ka Bi Ta, Ka Pi Ta, Ka-Pi-Rtha;
Vietnamese: Cân Thâng.

Origin/Distribution

This species is indigenous to the Indian subcontinent and Sri Lanka. In India, it is reported from the states of Punjab, Delhi, Rajasthan, Madhya Pradesh, West Bengal, Arunachal Pradesh, Maharashtra, Goa, Karnataka, Tamil Nadu and Andhra Pradesh occurring throughout the plains especially in the drier regions. It was introduced, cultivated and naturalised throughout southeast Asia – Myanmar, Thailand, Malaysia, Vietnam, Kampuchea, Laos and Indonesia.

Agroecology

Wood-apple thrives in a monsoonal or seasonally dry tropical climate. It is found on a wide diversity of soils from sea level to 450 m altitude. It is best adapted to light soils and is tolerant to drought and mild water-logging.

Edible Plant Parts and Uses

The hard rind of the ripe fruit must be cracked open with a hammer and the pulp scooped out. The pulp is eaten raw with or without sugar, or is blended with coconut milk and palm-sugar syrup. In Indonesia, the pulp is beaten with pal suagra and eaten at breakfast. The pulp can also be processed and drunk as a beverage or sherbet, or frozen as an ice cream. Ripe fruit juice is a good health drink and provides a rich source of antioxidants. The pulp is also made into chutney, preserves, jelly, marmalade, jams or processed into treacle or toffee. In Sri Lanka, wood apple cream is processed from the fruit pulp and is canned and exported. Young leaves, tender shoots and ripe fruits are also eaten fresh in Thailand. The bark also produces an edible gum. Seeds contain a bland, non-bitter, oil high in unsaturated fatty acids.

Botany

A small to medium sized, thorny, deciduous tree growing up to 6–15 m high with shallowly furrowed, grey bark. The spines are axillary, short, straight and the young branchlets and foliage are covered with minute, short hairs becoming glabrous with age. Leaves are alternate, imparipinnate with narrowly winged rachis. Leaflets number 5–9, are subsessile, opposite in arrangement with obovate, 25–35 mm long by 10–20 mm wide lamina with tip often notched, entire to crenulated margin (Plates 1 and 2) and dotted with fragrant filled glands. Flowers are small, fragrant, reddish, bisexual and formed in axillary and terminal panicles. Flower are 4–6-merous, calyx small, deltoid, flat, dentate and caducous; petals are spreading, oblong to ovate-lanceolate, imbricate in bud with a short, finely pilose disk; stamens are usually twice as many as petals bearing large, linear-oblong, basifixed anthers; ovary is globose, incompletely 4–6 (–7)-locular (empty in male flowers), becoming 1-locular with parietal placentae. Fruits are large, 7–10 cm diameter, globose, berry with a hard, woody shell, borne on

Plate 1 Fruits and leaves of wood apple

Plate 2 Close-up of compound leaves and spines of wood apple

Plate 3 Large, woody fruit with a stout woody stalk

stout, woody stalks (Plates 1 and 3) and are unilocular, the parietal placentae bearing numerous oblong, compressed, white seeds embedded in sticky brown, edible pulp.

Nutritive/Medicinal Properties

Nutrient composition of the fresh, ripe fruit per 100 g edible portion was reported (Gopalan et al. 2002) as: energy 87 kcal, moisture 77.5 g, protein 2.6 g, fat 0.2 g , carbohydrate 18.8 g, fibre 2.9 g, ash 0.9 g, β carotene 55 ug, vitamin C 9 mg, Fe 0.6 mg, Ca 38 mg. The fruit also contained polysaccharides - pectic polymers with galactosyl and arabinofuranosyl groups, hemicellulosic polymers and insoluble cellulose-rich material (Mondal et al. 2002). The unripe fruits contained 0.015% stigmasterol. Leaves contained stigmasterol (0.012%) and bergapten (0.01%) (Morton 1987). The bark also contained marmesin and the root bark contained aurapten, bergapten, isopimpinellin and other coumarins (Morton 1987).

Chemical investigations on different parts of *Limonia acidissima* plant have afforded various constituents, including coumarins, steroids, triterpenoids, benzoquinones, and tyramine derivatives with various pharmacological activities.

Anticancer Activity

An acidic heteropolysaccharide with a partially carboxymethylated α-(1–4) polygalacturonan backbone structure with 2- and 2,4-O-α-L-rhamnopyranosyl, 2- and 2,3-O-α-L-arabinofuranosyl and 3-, 2,4-and terminal α-D-galactopyranosyl bearing side chains was isolated from *F. limonia* and showed significant in-vivo Ehrlich ascites carcinoma cell growth inhibition (Saima et al. 2000).

Diuretic Activity

Methanolic leaf extracts exhibited diuretic effect (Parial et al. 2009). At a dose of 200 mg/kg a significant increase in urine output was produced. The extract increased the urinary excretion of sodium, potassium and chloride ions. The findings supported the traditional uses of *Limonia acidissima* leaves as diuretic agents.

Antimicrobial Activity

The leaf on distillation yielded 0.45% light green essential oil containing estragol (72.68%) and anethol (26.24%) and the oil exhibited antimicrobial activity against several fungi and bacteria (Mehta et al. 1983). Feronia leaf extract was found ineffective on *Bacillus pumilus* and *Xanthomonas campestris*, but *Vibrio cholerae* was quite sensitive to the extract.

The heartwood was found to contain ursolic acid and a flavanone glycoside, 7-methylporiol-D-xylopyranosyl-D-glucopyranoside. The stem bark yielded a pyranoflavanone (−)-(2S)-5,3′-dihydroxy-4′-methoxy-6″,6″-dimethylchromeno-(7,8,2″,3″)-flavanone along with several known compounds including an alkaloid, five coumarins, a flavanone, a lignan, three sterols and a triterpene (Rahman and Gray 2002). The antimicrobial screening of these compounds by a microdilution technique resulted in MICs in the range 25–100 µg/ml.

Antiinflammatory Activity

Three benzamide derivatives, N-{[P-(3,7-dimethyl-6 R,7-dihydroxy-4 R-octadecanoyloxy-2-octenyloxy)phenyl]ethyl} benzamide (1); N-{[P-(3,7-dimethyl-6 R,7-dihydroxy-4 R-9‴(E)-octadecenoyloxy-2-octenyloxy)phenyl]ethyl} benzamide (2) and N-{[P-(3,7-dimethyl-6 R,7-epoxy-4 R-9‴(E)-octadecenoyloxy-2-octenyoxy)phenyl] ethyl} benzamide (3) together with ten known compounds (4–13), were isolated from *Limonia acidissima* bark (Kim et al. 2009a). Among the isolates, 13 α,14 β,17 α-lanosta-7,9,24-triene-3 β,16 α-diol (8); 4-methoxy-1-methyl-2(1H)-quinolinone (10); and 13 α,14 β,17 α-lanosta-7,24-diene-3 β,11 β,16 α-triol (13) potently inhibited nitric oxide (NO) production in microglia cells. In additional studies, the scientists (Kim et al. 2009b) isolated a new dimeric coumarin, limodissimin A (1), together with four known coumarins: osthenol (2), 17 (2′R)-7-hydroxy-8-(2′,3′-dihydroxy-3′-methylbutyl)-2H-1-benzopyran-2-one (3), columbianetin (4), and seselin (5). Limodissimin A (1) was isolated as a yellow gum and showed a fluorescence at UV-365 nm. Compounds 2, 3, 4 and 5 significantly inhibited the LPS-induced NO production, with IC_{50} values of 22.3, 21.6, 33.5 and 23.1 µM, respectively. Compound 1 did not show the significant inhibitory effect on NO production in ranges from 5 to 20 µM. The study demonstrated that coumarins (compounds 2, 3, 4 and 5) isolated from *L. acidissima* exerted antiinflammatory effects in LPS- stimulated microglia cells and might be good lead compounds to modulate neurological diseases associated with inflammatory processes.

Insecticidal Activity

In Myanmar, wood apple root bark paste is used as a cosmetic locally, leaves are used to treat epilepsy, the roots are used as a purgative, and the fruit as a tonic. Pregnant Karen women in the camps at the Thai–Myanmar border used insect repellents which were mixed with 'thanaka', a root paste made from the wood apple tree against malarial parasites transmitted by *Anopheles minimus* and *Anopheles maculatus* (Lindsay et al. 1998). Bioassays using a laboratory strain of *Aedes aegypti* demonstrated that thanaka to be slightly repellent at high dosages and the mixture with deet provided protection for over 10 h. The treatment would therefore also provide some personal protection against dengue transmitted by *Aedes aegypti* and *Aedes albopictus* biting during the daytime. Another study reported that dried leaves afforded a potent mosquito larvicide, identified as n-hexadecanoic acid that was found to be effective against fourth instar larvae of *Culex quinquefasciatus*, *Anopheles stephensi* and *Aedes aegypti*, with LC_{50} of 129.24, 79.58 and 57.23 ppm, respectively (Rahuman et al. 2000).

Traditional Medicinal Uses

Various parts of *Limonia acidissima* have been used in folkloric medicine. Ripe fruit exhibit cooling, astringent and tonic properties and is used as a stomachic. Ripe fruit juice is a good health drink and it is a rich source of antioxidants. In Indo-China, the spines and bark are used in medicinal preparations for the treatment of excessive menstruation (menorrhagia), liver disorders, bites and stings and nausea. In India, the fruit is more popular as medicine than as food. The tannin in the fruit has an astringent effect that once led to its use as a general tonic and as a traditional cure for dysentery, diarrhoea, liver ailments, chronic cough and indigestion. In India, after the rainy season, a reddish brown gum like substance called *Feronia* gum is obtained from the trunk, which is of medicinal value and a substitute for gum Arabic. The powdered gum, mixed with honey, is given to overcome dysentery and diarrhoea in children. The bark, leaves, fruits and gum is used to treat snake-bites, diarrhoea, anorexia, vomiting, cough, bronchitis, hiccough, gingivitis and cardiac debility. Unripe fruit is astringent and useful in diarrhoea, dysentery and provides an effective treatment for hiccough, sore throat and diseases of the gums. The pulp and the powdered rind are used as poultices on bites and stings of venomous insects. The seed oil is a purgative, and the leaf juice mixed with honey is a folk remedy for fever. The tannin-rich and alkaloid-rich bark decoction is a folk cure for malaria. Sap of young leaves is mixed with milk and sugar candy and given as a remedy for biliousness and intestinal disorders of children.

Other Uses

The rind of the fruit is so thick and hard it can be carved and used as a utensil such as a bowl or ashtray. The tree has hard wood which can be used for woodworking. The timber is employed in the construction of houses, posts, rollers for mills, hubs and agricultural implements. The wood also serves as fuel-wood. A gum (*Feronia* gum), obtained from the trunk and branches, is utilized like gum Arabic and used for medicinal purposes and in making water colour paints, ink, dyes and varnish. The leaves are lopped for fodder. In Thailand the plant also has been used as root-stock for *Citrus* because of its water tolerance.

Comments

The wood-apple is usually propagated from seeds. It is also propagated by root cuttings, air-layers, or by budding.

Selected References

Backer CA, van den Brink RCB Jr (1965) Flora of Java, vol 2. Wolters-Noordhoff, Groningen, 630 pp

Burkill IH (1966) A dictionary of the economic products of the Malay Peninsula. Revised reprint. 2 volumes. Ministry of agriculture and Co-operatives, Kuala Lumpur, Malaysia, vol 1 (A–H) pp 1–1240, vol 2 (I–Z) pp 1241–2444

Chopra RN, Nayar SL, Chopra IC (1986) Glossary of Indian medicinal plants. (Including the supplement). Council Scientific Industrial Research, New Delhi, 330 pp

Council of Scientific and Industrial Research (CSIR) (1956) The wealth of India. A dictionary of Indian raw materials and industrial products, vol 4, Raw materials. Publications and Information Directorate, New Delhi

Facciola S (1990) Cornucopia. A source book of edible plants. Kampong Publ, Vista, 677 pp

Foundation for Revitalisation of Local Health Traditions (2008) FRLHT database. htttp://envis.frlht.org.

Gangrade SK, Jain NK, Mishra PK, Moghe MN (2002) *Limonia acidissima* L.: a multipurpose tree species having essential oil. Indian Perfum 46(2):109–113

Gopalan G, Rama Sastri BV, Balasubramanian SC (2002) Nutritive value of Indian foods. National Institute of Nutrition/Indian Council of Medical Research, Hydrabad

Jones DT (1992) *Limonia acidissima* L. In: Verheij EWM, Coronel RE (eds) Plant resources of south-east Asia no 2. Edible fruits and nuts. PROSEA, Bogor, pp 190–191

Kim KH, Ha SK, Kim SY, Kim SH, Lee KR (2009a) Limodissimin A: a new dimeric coumarin from *Limonia acidissima*. Bull Korean Chem Soc 30(9):2135–2137

Kim KH, Lee IK, Kim KR, Ha SK, Kim SY, Lee KR (2009b) New benzamide derivatives and NO production inhibitory compounds from *Limonia acidissima*. Planta Med 75(10):1146–1151

Lindsay SW, Ewald JA, Samung Y, Apiwathnasorn C, Nosten F (1998) Thanaka (*Limonia acidissima*) and deet (di-methyl benzamide) mixture as a mosquito repellent for use by Karen women. Med Vet Entomol 12(3):295–301

Mehta P, Chopra S, Mehta A (1983) Antimicrobial properties of some plant extracts against bacteria. Folia Microbiol (Praha) 28(6):467–469

Molesworth AB (1967) Malayan fruits. An introduction to the cultivated species. Moore, Singapore, 245 pp

Mondal SK, Ray B, Thibault JF, Ghosal PK (2002) Cell-wall polysaccharides from the fruits of *Limonia acidissima*: isolation, purification and chemical investigation. Carbohydr Polym 48(2):209–212

Morton JF (1987) Wood apple. In: Fruits of warm climates. Julia F. Morton, Miami, FL, pp 190–191

Parial S, Jain DC, Joshi SB (2009) Diuretic activity of the extracts of *Limonia acidissima* in rats. Rasāyan J Chem 2(1):53–56

Perry LM (1980) Medicinal plants of east and southeast Asia. Attributed properties and uses. MIT Press, Cambridge/London, 620 pp

Pongpangan S, Poobrasert S (1985) Edible and poisonous plants in Thai forests. Science Society of Thailand/Science Teachers Section, Bangkok, 206 pp

Purseglove JW (1968) Tropical crops: dicotyledons, vol 1 & 2. Longman, London, 719 pp

Rahman MM, Gray AI (2002) Antimicrobial constituents from the stem bark of *Feronia limonia*. Phytochem 59(1):73–77

Rahuman A, Gopalakrishnan G, Ghouse BS, Arumugam S, Himalayan B (2000) Effect of *Feronia limonia* on mosquito larvae. Fitoterapia 71(5):553–555

Saima Y, Das AK, Sarkar KK, Sen AK, Sur P (2000) An antitumor pectic polysaccharide from *Feronia limonia*. Int J Biol Macromol 27(5):333–335

Merrillia caloxylon

Scientific Name

Merrillia caloxylon (Ridley) Swingle.

Synonyms

Murraya caloxylon Ridley.

Family

Rutaceae

Common/English Names

Flowering Merrillia, Katinga, Malay lemon.

Vernacular Names

Malaysia: Katinga, Ketenggah, Kemuning gajah.

Origin/Distribution

The species is indigenous to Malaysia (Peninsular, Sabah), Thailand (Southern Thailand) and Indonesia (Sumatra).

Agroecology

In its native range, the tree is found scattered in lowland moist primary and secondary tropical forest on stream banks and hill sides and ridges up to 400 m altitude. It is able to grow in on various well-drained soils.

Edible Plant Parts and Uses

The rind has a bitter turpentine flavour and the scanty pulp is quite tasteless (Swingle 1918). The potential of the thick rind, fresh or dried for culinary purposes has not been investigated.

Botany

An evergreen upright, unarmed small to medium sized tree up to 20 (–30) m high and a short bole of 50 cm diameter, bushy multi-branched crown and pale greyish-brownish, flaky bark (Plate 1). Leaves imparipinnate, with 5–13 leaflets, rachis narrowly-winged, leaflets sub-opposite, entire, unequal sized, deep green, basal pairs smaller (Plates 2, 3 and 5) , much reduced suborbicular, 5–10 mm across, and resembling stipules, lateral leaflets gradually increase in size to 7.6–10 cm long, oblanceolate, sub-acuminate and cuneate

Plate 1 Small to medium sized tree with a thin multi-branched crown

Plate 4 Large, oval-oblong immature fruit

Plate 2 Leaves and immature fruit

Plate 5 Large subglobse near ripe fruit

Plate 3 Imparipinnate leaf

Plate 6 Fruit with thick rind and scanty pulp

base, margin wavy and obscurely serrated, petiolule subsessile. Flowers axillary single or in pair, bisexual, pentamerous; calyx cupulate, petals, 1.8 cm long free, trumpet shaped but with free petals, greenish-white to white; stamens 10 unequal, slender; ovary superior on a gynophore, 5(-6) fused carpels, each with 8–10 ovules, style long slender, hairy and stigma capitate. Fruit subglobose, ovoid to oblong, green ripening to dull yellow, up to 11 cm long and 7.5 cm wide, pericarp gland dotted, rough to warty, 1.25 cm thick and full of long resin cells (Plates 3, 4, 5 and 6).

Seeds numerous, ovate, flattened, olive-grey covered with fimbriate scales and embedded in scanty olive coloured, mucilaginous pulp (Plate 6).

Nutritive/Medicinal Properties

The fruit juice of *Merrillia caloxylon* was found to contain at least 0.1% of the cytotoxic flavone eupatorin (Adams and Lewis 1976). One pentaoxygenated chalcone, 2-hydroxy-3,4,4′,5,6′-pentamethoxychalcone and two oxygenated chalcones, 2′- hydroxy-3,4,4′,6′-tetramethoxychalcone and 2′,3-dihydroxy-4,4′,6′- trimethoxychalcone together with 3 flavones: 3′,4′,5,7-tetramethoxyflavone, 3′,4′,5,5′,7-pentamethoxyflavone and 5-hydroxy-3′,4′,6,7-tetramethoxyflavone were isolated from fruit of *Merrillia caloxylon* (Zakaria et al. 1989).

The root and stem bark of *Merrillia caloxylon* contained the anti-implantation indole alkaloid, yuehchukene, and the 8-prenylated coumarins, sibiricin and phebalosin, as well as 3-(3-methylbuta-1,3-diene) indole and eupatorin (But et al. 1988; Kong et al. 1988); Zakaria et al. (1989) isolated five isoprenylcoumarins (−)-sibiricin, (−)-phebalosin, (−)-murrangatin, (−)-mexoticin and merillin from the roots of *Merrillia caloxylon*.

An infusion of the wood is applied medicinally for stomach-ache, whereas powder wood is rubbed on the skin against aches and pain.

Other Uses

The tree is valued for its handsome, light yellow coloured wood with dark brown streaks and stains, fairly hard wood that takes a good polish. The wood is highly prized for making walking sticks, kris handles and sheaths, furniture, boxes, and other small objects like pipes, amulets and rings.

Comments

The species is categorised as vulnerable by the World Conservation Monitoring Centre.

Selected References

Adams JH, Lewis JR (1976) Eupatorin, a constituent of *Merrillia caloxylon*. Planta Med 32(1):86–87

But PPH, Kong YC, Li Q, Chang HT, Chang KL, Wong KM, Gray AI, Waterman PG (1988) Chemotaxonomic relationship between *Murraya* and *Merrillia* (Rutaceae). Acta Phytotaxon Sin 26:205–210

Jones DT (1987) Rare plant profile no. 1: *Merrillia caloxylon* (Rutaceae). Bot Gard Conserv News 1(1): 38–42

Kong YC, But PPH, Nguyen KH, Cheng KF, Chang KL, Wong KM, Gray AI, Waterman PG (1988) The biochemical systematics of *Merrillia*; in relationship to *Murraya*, the Clauseneae and the Aurantioideae. Biochem Syst Ecol 16:47–50

Ong HC (1998) *Merrillia* Swingle. In: Sosef MSM, Hong LT, Prawirohatmodjo S (eds) Plant resources of South-East Asia no. 5(3) timber trees: lesser-known timbers. Prosea Foundation, Bogor, pp 371–373

Ridley HN (1908) New or rare Malayan plants. Series 1V. J Straits Branch R Asiat Soc 50:111–114

Soepadmo E, Wong KM (1995) Tree flora of Sabah and Sarawak. Ampang Press Sdn. Bhd, Kuala Lumpur

Stone BC (1972) Rutaceae. In: Whitmore TC (ed) Tree flora of Malaya, vol 1. Longman Malaya, Kuala Lumpur, pp 367–387

Stone BC, Jones DT (1988) New and noteworthy Rutaceae–Aurantioideae from Northern Borneo. Studies in Malesian Rutaceae, V. Proc Acad Nat Sci Phila 140:267–274

Swingle WT (1918) *Merrillia*, a new Rutaceous genus of the tribe Citreae from the Malay Peninsula. Philipp J Sci 13(6):335–343

Swingle WT, Reece RC (1967) The botany of *Citrus* and its wild relatives. In: Reuther W, Webber HJ, Batchelor LD (eds) The *Citrus* industry, volume I, history, world distribution botany, and varieties. Division of Agricultural Sciences/University of California, Berkeley, pp 190–430

World Conservation Monitoring Centre (1998) *Merrillia caloxylon*. In: IUCN 2011. IUCN red list of threatened species. Version 2011.1. www.iucnredlist.org

Zakaria MB, Saito I, Matsuura T (1989) Coumarins of *Merrillia caloxylon*. Phytochem 28(2):657–659

Poncirus trifoliata

Scientific Name

Poncirus trifoliata (L.) Rafin.

Synonyms

Aegle sepiaria DC., *Citrus trifolia* Thunb., *Citrus trifoliata* L., *Citrus trifoliata* var. *monstrosa* T. Itô, *Citrus triptera* Carr., *Limonia trichocarpa* Hance, *Poncirus trifoliata* var. *monstrosa* (T. Itô) Swingle, *Pseudaegle sepiaria* Miq., *Pseudaegle trifoliata* Makino.

Family

Rutaceae

Common/English Names

Bitter Orange, Citrange, Citrangequat, Flying Dragon, Hardy Orange, Japanese Bitter Orange, Japanese Hardy Orange, Mock Orange, Trifoliata, Trifoliata Orange, Trifoliate-Orange.

Vernacular Names

Chinese: Ch'ou Chü, Ju Ju, Ju Jie, Zhi Ke, Zhi Qiao, Zhi Shi;
Czech: Citronečník Trojlistý, Trifoliata;
Danish: Dværg-Citrontræ, Trebladet Dværgcitron;
Dutch: Driebladige Citroenboom;
Eastonian: Pontsirus;
French: Citronnier Trifolié, Oranger Trifolié, Oranger Trifoliolé, Poncir, Poncire, Poncirus, Poncire Commun;
German: Bitterorange, Dreiblättrige Citrus, Dreiblättrige Zitrone;
Hungarian: Tövisescitrom, Vadcitrom;
Italian: Egle;
Japanese: Karatachi, Shi, Kikoku;
Korean: Taengdzanamu, Tang-Ja-Na-Moo;
Portuguese: Limoeiro-Trifoliado;
Spanish: Limonero Trifoliado, Naranjo Trébol.

Origin/Distribution

The species is native to North and Central China, Korea and Japan.

Agroecology

In its native subtemperate-temperate areas, it occurs in the forests in mountains and hills. It is a hardy and robust plant. It grows on lime-free soil, is drought intolerant and can withstand cold (frost) and wet conditions. However impeded drainage can result in 'sudden death' of trees. It prefers fertile well-drained, clays to loams and is intolerant of highly acid or highly alkaline soils and grows in full sun. It also reacts adversely to saline conditions, it readily take up chloride through the leaves, a problem most often observed in trees watered by overhead irrigation.

Edible Plant Parts and Uses

Fruit is bitter and not commonly consumed raw and is made into marmalade, jams or juice. The fruits are used as a spice or for medicinal purposes. When dried and powdered, the fruits can be used as a condiment. Occasionally the boiled young leaves are eaten.

Botany

A deciduous, sparsely -leaved shrub, reaching 3.5 m tall with characteristic crooked, twisted and tangled green shoots and branches, armed with large, sharp, 3–5 cm long spines (Plate 2). Leaves alternate, trifoliolate, leaflets sessile, 22–40 × 10–20 mm, obovate to ovate, margin crenate, base cuneate, apex often emarginate, shiny dark green

Plate 1 Trifoliata orange fruit

Plate 2 Crooked, entangled pricky shoots

Plate 3 Trifoliate leaves with crenate margins

borne on slightly winged, 2.5 cm long green petioles (Plate 3). Floral buds scaly. Flowers almost sessile, fragrant, solitary, terminal or axillary, 1–3 cm across. Calyx 5-segmented. Petals 5, white, obovate, 2 mm, shortly clawed. Stamens pink, 20–23, free. Ovary 6–7-locular. Fruit green turning to yellow or yellowish-orange when ripe, globose, 3–5 cm diameter, rough, finely pubescent, fragrant (Plate 1). Pulp greenish, acidic, juicy and oily. Seeds numerous, glabrous.

Nutritive/Medicinal Properties

The fruits of *Poncirus trifoliata* are widely used in oriental traditional medicine as a remedy for allergic inflammation and in a traditional Korean folk medicine for the treatment of digestive dysfunction like gastritis. Other traditional uses include hernia, sputum remedy and sinusitis. It is sedative and angiotonic.

Numerous studies have reported that *Poncirus trifoliata* extracts possess anti-allergic, anti-inflammatory, anti-cancerous, anti-platelet, anti-allergic, mosquitocidal, bactericidal and antiviral activities. The fruit contains the following main chemical components: flavonoids, coumarins, monoterpenes and alkaloids.

Anticancer and Antiviral Activities

Coumarin compounds (poncimarin, heraclenol 3′-methyl ester and oxypeucedanin methanolate) from *Poncirus trifoliata* selectively increased glutathione S-transferase (GST)α-protein expression in the H4IIE cell-line (a rat hepatocyte cell line) (Pokharel et al. 2006b). Poncimarin most potently increased GST enzyme and most actively induced GSTα. The findings suggested that these three coumarin compounds possessed phase II enzyme inducible functions, and that poncimarin had chemopreventive potential. Chemopreventive agents induce a battery of genes whose protein products can protect cells from chemical-induced carcinogenesis. Another coumarin compound, isoimperatorin from *Poncirus trifoliata* was shown to possess potent hepatoprotective effect against aflatoxin B1 (AFB1)-induced cytotoxicity in H4IIE cells, presumably through the induction of GSTα and the direct inhibition of cytochrome P450 enzymes (CYP1A) (Pokharel et al. 2006a). Two other bioactive compounds from fruits, β-sitosterol and 2-hydroxy-1, 2, 3-propanetricarboxylic acid 2-methyl ester (HPCME) exhibited differential inhibition of cancer cell proliferation and apoptosis using human colon cancer cell line (HT-29) at various concentrations (Jayaprakasha et al. 2007). Significant arrest of cell growth was observed with β-sitosterol even at low concentration of 0.63 μM and none of the compounds exerted any apparent cytostatic effects on the non-cancerous COS-1 fibroblast cells. Growth inhibition assay suggested the potential use of bioactive compounds as cancer chemopreventive and therapeutic agents.

Separate studies showed that the cytotoxic activity of the fruit extract in HL-60 cells was increased in a concentration- and time-dependent manner. The fruit extract caused cell shrinkage, cell membrane blebbing, apoptotic body and DNA fragmentation (Yi et al. 2004). Fruit extract-induced apoptosis was accompanied by the activation of caspase-3 and the specific proteolytic cleavage of poly-ADP-ribose polymerase. The findings provided evidence that the fruit extract could be a candidate as an anti-leukemic agent through apoptosis of cancer cells and supported the use of *Poncirus trifoliata* fruits for the treatment of various cancers among Korean Oriental medical doctors. A triterpenoid, 25-methoxyhispidol A, isolated from the fruit of *P. trifoliata* exhibited antiproliferative effects; it arrested the cell cycle in the G1 phase and subsequently induced apoptosis of the SK-HEP-1 human hepatocellular carcinoma cells (Hong et al. 2008). These findings suggested the potential of 25-methoxyhispidol A as an antitumour agent against human hepatocarcinoma cells. Using an in-vitro cell culture system *Poncirus trifoliata* was also reported to display anti-HCV (Hepatitis C virus) activities (Ho et al. 2003).

Antiinflammatory Activity

Poncirin, a flavanone glycoside isolated from the EtOAc extract of the dried immature fruits of *Poncirus trifoliata*, was found to be an anti-inflammatory compound that inhibited PGE(2) and IL-6 production (Kim et al. 2007). Poncirin reduced lipopolysaccharide (LPS)-induced protein levels of inducible nitric oxide synthase (iNOS) and cyclooxygenase-2 (COX-2) and the mRNA expressions of iNOS, COX-2, tumour necrosis factor-α (TNF-α) and interleukin-6 (IL-6) in a concentration-dependent manner. Further, poncirin inhibited the LPS-induced DNA binding activity of nuclear factor-kappaB (NF-kappaB). Additionally, this effect was accompanied by a parallel reduction in IkappaB-α degradation and phosphorylation. The data indicated that anti-inflammatory properties of poncirin might be the result from the inhibition iNOS, COX-2, TNF-α and IL-6 expression via the down-regulation of NF-kappaB binding activity.

The fruits of *Poncirus trifoliata* are widely used in Oriental medicine as a remedy for allergic inflammation (Lee et al. 2004). Two isomers of 21-methylmelianodiol (MMD) isolated from the fruits of *P. trifoliata* were found to inhibit nitric oxide (NO) production in lipopolysaccharide (LPS)-stimulated RAW 264.7 macrophages. 21α-MMD and 21β-MMD attenuated LPS-induced inducible nitric oxide synthase (iNOS) and cyclooxygenase (COX)-2 protein expressions as well as the mRNA levels of iNOS, COX-2, tumour necrosis factor-α (TNF-α) and interleukin-1beta (IL-1β). Both 21α-MMD and 21β-MMD significantly suppressed LPS-induced NF-kappaB transcriptional activity in RAW 264.7

macrophages. In the carrageenan-induced paw edema model, administration of 21α-MMD (20 and 100 mg/kg, i.p.) dose-dependently reduced paw swelling (Zhou et al. 2007). The data indicated 21-methylmelianodiol to be an important constituent of the fruit of *P. trifoliata*, and that the inhibition of iNOS and COX-2 expression by 21α-MMD and 21β-MMD might be one of the mechanisms responsible for their anti-inflammatory effects. Additional studies reported that 21α-methylmelianodiol and 21β-methylmelianodiol inhibited IL-5-dependent growth of Y16 pro-B cells in a dose-dependent manner with IC_{50} values of 17 μM and 15 μM, respectively (Lee et al. 2008). 21α-methylmelianodiol and 21β-methylmelianodiol caused G1 arrest of IL-5-induced cell cycle progression of Y16 cells, and also reduced IL-5-dependent survival of the cells by apoptosis. This study indicated a pharmacological potential for *P. trifoliata* in treatment of Interleukin IL-5-associated inflammatory disorders. Interleukin (IL)-5 plays an important role in the progression of allergic inflammation.

In-vitro studies provided evidence that the fruit extract of *P. trifoliata* might contribute to the treatment of mast cell-derived allergic inflammatory diseases (Shin et al. 2006). It was reported that the fruit extract dose dependently decreased the gene expression and production of TNF-α and IL-6 on phorbol 12-myristate 13-acetate (PMA) and calcium ionophore A23187-stimulated human mast cell HMC-1 cells. In addition, the fruit extract attenuated PMA and A23187-induced activation of NF-kappaB indicated by inhibition of degradation of IkappaBalpha, nuclear translocation of NF-kappaB, NF-kappaB/DNA binding, and NF-kappaB-dependent gene reporter assay.

Another research showed that hesperidin, hesperidin methyl chalone and phellopterin derived from *P. trifoliata* reduced tumour necrosis factor-α (TNF α)-induced vascular cell adhesion molecule-1 expression through regulation of the Akt and PKC pathway, which contributed to inhibit the adhesion of monocytes to endothelium (Nizamutdinova et al. 2008).

Betaine and hesperidin derived from *P. trifoliata* was found to increase mucin release by directly acting on airway mucin-secreting cells suggesting the possible use of these chemicals as mild expectorants during the treatment of chronic airway diseases (Lee et al. 2004).

Antiallergic Activity

Poncirus trifoliata was found to have anti-anaphylactic activity (Lee et al. 1996). An aqueous extract of *Poncirus trifoliata* inhibited compound 48/80-induced anaphylaxis almost 100% with doses above 0.4 mg/g body weight intraperitoneally administered. The extract (1–1,000 μg/ml) also dose-dependently inhibited the histamine release induced by compound 48/80 (5 μg/ml) in rat peritoneal mast cells. The level of cAMP in peritoneal mast cells was also significantly increased by the extract. Further, the extract inhibited intracellular calcium release induced by compound 48/80. These results suggested that the fruit extract had anti-anaphylactic activity by stabilizing the peritoneal mast cell membrane. Additionally research showed that oral administration of an aqueous extract from immature fruit of *Poncirus trifoliata* (200 mg/kg) significantly inhibited passive cutaneous anaphylaxis in rats (Lee et al. 1997). It also inhibited histamine release from rat peritoneal mast cells (RPMC) induced by mouse anti-dinitrophenyl (DNP)-IgE and dinitrophenyl-human serum albumin (DNP-HSA). These results suggested that the fruit extract had anti-allergic action against the type I hypersensitivity reaction. In another study, aqueous fruit extract of *Poncirus* was found to dose-dependently inhibit active systemic anaphylaxis and serum IgE production induced by immunization with ovalbumin, *Bordetella pertussis* toxin and aluminum hydroxide gel (Kim et al. 1999b). The extract potently suppressed interleukin 4 (IL-4)-dependent IgE production by lipopolysaccharide-stimulated murine whole spleen cells. In the case of U266 human IgE-bearing B cells, the extract also showed an inhibitory effect on the IgE production. These results suggested that *Poncirus* fruit extract had an anti-allergic activity by inhibition of IgE production from B cells.

Park et al. (2011) revealed that the extracts of trifoliate orange fruit suppressed IgE production in U266B1 human myeloma cells in a dose dependent manner with or without lipopolysaccharide (LPS) stimulation of the cells. Further, chemically elevated blood IgE concentrations were dramatically decreased by oral administration, ventral injection, and topical application of the trifoliate orange extracts. The results indicated that the fraction of the extract contained an important compound of the fruit of trifoliate orange, and had the ability to reduce IgE concentrations in cultured human myeloma cells and 1-chloro-2,4-dinitrobenzene (DNCB) sensitized in-vivo mouse model.

Antiplatelet Aggregation Activity

Poncitrin a coumarin from *Poncirus trifoliata* was also reported to inhibit the aggregation and ATP release of rabbit platelets induced by arachidonic acid, collagen, ADP, platelet-activating factor (PAF) or U46619 (a thromboxane A2 analog) (Teng et al. 1992). Thrombin-induced ATP release, but not the aggregation, was also inhibited. The findings confirmed that the antiplatelet actions of coumarin compounds were due to the inhibition on thromboxane A2 formation and phosphoinositides breakdown.

Prokinetic Activity

Aqueous extracts from dried immature fruit of *Poncirus trifoliata* are used as a traditional Korean folk medicine for the treatment of digestive dysfunction (Lee et al. 2005a, b). Aqueous fruit extract had a unique prokinetic activity, which accelerated the transit of intestinal contents, but had no effect on the gastric emptying rate. Research reported that fruit extract of *P. trifoliata* had the potential for development as a prokinetic agent that may prevent or alleviate gastrointestinal motility dysfunctions (GMD) in human patients. Among the compounds isolated from the ethyl acetate extract of dried immature fruit of *Poncirus trifoliata*, a new flavanone glycoside, (2R)-5-hydroxy-4′-methoxyflavanone-7-O-{β-glucopyranosyl-(1→2)-β-glucopyranoside} (1),

and three known compounds, (2S)-poncirin, (2S)-naringin, and (2S)-poncirenin, (2S)-poncirin exhibited considerable inhibitory activity against lipopolysaccharide (LPS)-induced prostaglandin E(2) (PGE(2)) and interleukin-6 (IL-6) production, and mRNA expression in RAW 264.7 murine macrophage cells (Han et al. 2007). The findings supported the traditional use of the dried immature fruit of *Poncirus trifoliata for* uterine contraction.

Antiosteoporosis Activity

The hexane extract of *Poncirus trifoliata*, a Korean medicinal plant, was found to inhibit apoptotic cell death in dexamethasone-induced osteoblastic cell lines, C3H10T1/2 and MC3T3-E1 (Kim et al. 2011). In vivo mouse, the extract not only inhibited bone loss caused by glucocorticoid, but also promoted bone formation. The extract also significantly decreased expression level of AnxA6 in Dex-induced osteoblastic cells and prednisolone (PD)-treated GIO-model mice. The findings suggested that the extract had a strong in-vitro and in- vivo inhibitory effect on glucocorticoid-induced osteoporosis, and decreased expression of AnxA6 that may play a key role in this inhibition.

Antimicrobial Activity

Ponciretin (5,7-dihydroxy-4′-methoxyflavanone), the main metabolite of poncirin from aqueous fruit extract of *P. trifoliata* most potently inhibited the growth of *Helicobacter pylori*, with a minimum inhibitory concentration (MIC) of 10–20 μg/ml (Kim et al. 1999a).

Mosquitocidal Activity

Four flavonoid compounds, namely poncirin, rhoifolin, naringin and marmesin, from *Poncirus trifoliata* exhibited oviposition-deterrent activity against gravid *Aedes aegypti* female mosquitoes (Rajkumar and Jebanesan 2008). Oviposition decreased with an increase in concentration of flavonoid compounds. Rhoifolin provided maximum protection (365 minutes) and also 100% repellency against mosquito bite followed by poncirin, marmesin and naringin. None of the 25 volunteers of either sex exposed to 10% (w/v) flavonoid compounds (4-hour patch test) showed a positive skin irritant reaction. All of the tested compounds proved to have various activities against different life stages of *A. aegypti*. These flavonoid compounds from *P. trifoliata* could be potential candidates for use in the development of commercial mosquitocidal products that may be an alternative to conventional synthetic chemicals, particularly in integrated vector control applications.

Traditional Medicinal Uses

The fruit of trifoliate orange is widely used for treating allergies including allergenic dermatitis and inflammation in traditional oriental medicine (Park et al. 2011).

Other Uses

Ponicrus trifoliata is planted mainly as a medicinal plant, ornamental or as rootstock for *Citrus* and *Fortunella* species for its frost tolerance and viroid disease (e.g. citrus exocortis viroid) resistance. It is also commonly planted as a fence/hedge.

Comments

Crosses between *Poncirus trifoliata* and *Citrus sinensis* has yielded the citrange. The hybrids between citrange and species of *Fortunella* are called citrangequat

Selected References

Bailey LH (1976) Hortusthird. A concise dictionary of plants cultivated in the United States and Canada. Liberty Hyde Bailey Hortorium/Cornell University/Wiley, New York, 1312pp

Facciola S (1990) Cornucopia. A source book of edible plants. Kampong Publ, Vista, 677 pp

Han AR, Kim JB, Lee J, Nam JW, Lee IS, Shim CK, Lee KT, Seo EK (2007) A new flavanone glycoside from the dried immature fruits of *Poncirus trifoliata*. Chem Pharm Bull (Tokyo) 55(8):1270–1273

Ho TY, Wu SL, Lai IL, Cheng KS, Kao ST, Hsiang CY (2003) An in vitro system combined with an in-house quantitation assay for screening hepatitis C virus inhibitors. Antiviral Res 58(3):199–208

Hong J, Min HY, Xu GH, Lee JG, Lee SH, Kim YS, Kang SS, Lee SK (2008) Growth inhibition and G1 cell cycle arrest mediated by 25-methoxyhispidol A, a novel triterpenoid, isolated from the fruit of *Poncirus trifoliata* in human hepatocellular carcinoma cells. Planta Med 74(2):151–155

Jayaprakasha GK, Mandadi KK, Poulose SM, Jadegoud Y, Nagana Gowda GA, Patil BS (2007) Inhibition of colon cancer cell growth and antioxidant activity of bioactive compounds from *Poncirus trifoliata* (L.) Raf. Bioorg Med Chem 15(14):4923–4932

Kim BY, Yoon HY, Yun SI, Woo ER, Song NK, Kim HG, Jeong SY, Chung YS (2011) In vitro and in vivo inhibition of glucocorticoid-induced osteoporosis by the hexane extract of *Poncirus trifoliata*. Phytother Res 25(7):1000–1010

Kim DH, Bae EA, Han MJ (1999a) Anti-*Helicobacter pylori* activity of the metabolites of poncirin from *Poncirus trifoliata* by human intestinal bacteria. Biol Pharm Bull 22:422–424

Kim HM, Kim HJ, Park ST (1999b) Inhibition of immunoglobulin E production by *Poncirus trifoliata* fruit extract. J Ethnopharmacol 66(3):283–288

Kim JB, Han AR, Park EY, Kim JY, Cho W, Lee J, Seo EK, Lee KT (2007) Inhibition of LPS-induced iNOS, COX-2 and cytokines expression by poncirin through the NF-kappaB inactivation in RAW 264.7 macrophage cells. Biol Pharm Bull 30(12):2345–2351

Lee CJ, Lee JH, Seok JH, Hur GM, Park JJ, Bae S, Lim JH, Park YC (2004) Effects of betaine, coumarin and flavonoids on mucin release from cultured hamster tracheal surface epithelial cells. Phytother Res 18:301–305

Lee HT, Seo EK, Chung SJ, Shim CKJ (2005a) Effect of an aqueous extract of dried immature fruit of *Poncirus trifoliata* (L.) Raf. on intestinal transit in rodents with experimental gastrointestinal motility dysfunctions. J Ethnopharmacol 102(2):302–306

Lee HT, Seo EK, Chung SJ, Shim CKJ (2005b) Prokinetic activity of an aqueous extract from dried immature fruit of *Poncirus trifoliata* (L.) Raf. J Ethnopharmacol 102(2):131–136

Lee IJ, Xu GH, Ju JH, Kim JA, Kwon SW, Lee SH, Han SB, Kim Y (2008) 21-Methylmelianodiols from *Poncirus trifoliata* as inhibitors of interleukin-5 bioactivity in Pro-B cells. Planta Med 74(4):396–400

Lee YM, Kim CY, Kim YC, Kim HM (1997) Effects of *Poncirus trifoliata* on type I hypersensitivity reaction. Am J Chin Med 25(1):51–56

Lee YM, Kim DK, Kim SH, Shin TY, Kim HM (1996) Antianaphylactic activity of *Poncirus trifoliata* fruit extract. J Ethnopharmacol 54(2–3):77–84

Natural Products Research Institute (1998) Medicinal plants in the Republic of Korea, vol 21, WHO Regional Publications, Western Pacific Series. Seoul National University, Manila, 316 pp

Nizamutdinova IT, Jeong JJ, Xu GH, Lee SH, Kang SS, Kim YS, Chang KC, Kim HJ (2008) Hesperidin, hesperidin methyl chalone and phellopterin from *Poncirus trifoliata* (Rutaceae) differentially regulate the expression of adhesion molecules in tumor necrosis factor-alpha-stimulated human umbilical vein endothelial cells. Int Immunopharmacol 8(5):670–678

Park HJ, Kim JB, Han IS, Kim JH, Kim BY, Kang SH, Jung HS, Moon SH, Sung SH, Song H (2011) Extract of *Poncirus trifoliata* fruit reduces elevated serum IgE concentrations in chemically induced allergic model mice. Hortic Environ Biotechnol 25(2):224–231

Pokharel YR, Han EH, Kim JY, Oh SJ, Kim SK, Woo ER, Jeong JE, Kang KW (2006b) Potent protective effect of isoimperatorin against aflatoxin B1-inducible cytotoxicity in H4IIE cells: bifunctional effects on glutathione S-transferase and CYP1A. Carcinogenesis 27(12):2483–2490

Pokharel YR, Jeong JE, Oh SJ, Kim SK, Woo ER, Kang KW (2006a) Screening of potential chemopreventive compounds from *Poncirus trifoliata* Raf. Pharmazie 61(9):796–798

Purseglove JW (1968) Tropical crops: dicotyledons, vol 1 & 2. Longman, London, 719 pp

Rajkumar S, Jebanesan A (2008) Bioactivity of flavonoid compounds from *Poncirus trifoliata* L. (Family: Rutaceae) against the dengue vector, *Aedes aegypti* L. (Diptera: Culicidae). Parasitol Res 104(1):19–25

Shin T-Y, Oh JM, Choi B-J, Park W-H, Kim C-H, Jun C-D, Kim S-H (2006) Anti-inflammatory effect of *Poncirus trifoliata* fruit through inhibition of NF-kappaB activation in mast cells. Toxicol In Vitro 20(7):1071–1076

Swingle WT (1946) The botany of *Citrus* and its wild relatives of the orange subfamily (family Rutaceae, subfamily Aurantioideae). In: Webber HJ, Batchelor LD (eds) The *Citrus* industry. Vol. 1 History, botany and breeding. Univ. of California Press, Berkeley, pp 129–474, 1028 pp

Teng CM, Li HL, Wu TS, Huang SC, Huang TF (1992) Antiplatelet actions of some coumarin compounds isolated from plant sources. Thromb Res 66(5):549–557

Yi J-M, Kim M-S, Koo H-N, Song B-K, Yoo Y-H, Kim H-M (2004) *Poncirus trifoliata* fruit induces apoptosis in human promyelocytic leukemia cells. Clin Chim Acta 340(1–2):179–185

Zhou HY, Shin EM, Guo LY, Zou LB, Xu GH, Lee SH, Ze KR, Kim EK, Kang SS, Kim YS (2007) Anti-inflammatory activity of 21(alpha, beta)-methylmelianodiols, novel compounds from *Poncirus trifoliata* Rafinesque. Eur J Pharmacol 572(2–3):239–248

Triphasia trifolia

Scientific Name

Triphasia trifolia (**Burm. f.**) **P. Wilson.**

Synonyms

Limonia diacantha DC., *Limonia trifolia* Burm.f., (basionym), *Limonia trifoliata* L., *Triphasia aurantiola* Lour., *Triphasia trifoliata* DC.

Family

Rutaceae

Common/English Names

Lemon China, Limeberry, Lime Chinese, Lime Orange Berry, Trifoliate Limeberry, Trifoliate Limeberry, Triphasia, Triphasia Limeberry.

Vernacular Names

Chamorro: Lemon China, Lemon De China, Lemoncito, Lemondichina, Limon-China, Limoncito;
French: Orangine;
India: Chini Naranghi (Hindu);
Indonesia: Jeruk Kingkip, Jeruk Kingkit (Java), Kingkip, Kaliyage (Sundanese);
Malaysia: Limau Kiah, Limau kaya, Limau Kikir, Limau Kingkip, Limau Kingkit, Limau Kelinket, Limau Kerinket, Limau Kerisek (Peninsular);
Philippines: Sua-Sua, Limonsitong-Kastila, Suaang-Kastila (Bikol), Kalamansito (Ibanag), Kalamansito (Iloko), Tagimunau (Negrito), Limonsito (Spanish), Limonsito, Kamalitos (Tagalog);
Spanish: Lemon De China, Lemoncito, Lemondichina, Limon-China, Limoncito;
Thailand: Manao Tet.

Origin/Distribution

Its origin is uncertain, it is believed to have come from China or elsewhere in southeast Asia. It is an ancient introduction to Peninsular Malaysia, now naturalised. It was also introduced and naturalised in Thailand and the Philippines and elsewhere in tropical Asia. Now it has been widely introduced and cultivated in other subtropical to tropical countries and has also naturalized on a number of islands in the Pacific.

Agroecology

A tropical species but will grow well also in the subtropics. Commonly found in the lowlands from near sea level to 300 m altitude. A common naturalized shrub of limestone, in the undergrowth, sometimes forming dense, spiny thickets or in dry waste places. It is also cultivated as ornamental hedge plants.

Edible Plant Parts and Uses

The ripe, red fruits (Plates 1 and 2) are eaten raw, candied or are made into preserves, jams, marmalade, beverages and pickles.

Botany

An evergreen, erect, glabrous, perennial shrub or small tree, 1–3 (–7) m high with terete twigs bearing paired spines in the axils of the leaves. Leaves are trifoliate, glossy dark green, at the

Plate 1 Ripe and unripe fruit

Plate 2 Ripe crimson subglobose fruit

Plate 3 Ovoid fruit, trifoliate leaves with sharp axillary spines

base of each leaf there are two, slender, sharp, and straight spines (Plate 3). Leaflets are ovate to oblong-ovate, the terminal one being 2–4 cm long by 1.2-cm wide, and the lateral ones smaller, margin crenate; petiolules and petioles short (5 mm or less) and wingless. Flowers are axillary, 1–3 in axils on peduncles 3–4 mm long, fragrant, bisexual, trimerous, calyx cupular with 3 ovate triangular small, green, persistent lobes, petals rounded and pellicid near apex, white; stamens 6 with slender filaments and oblong anthers; ovary ovoid or fusiform, 3-locular; locules 1-seeded, style filiform-clavate, with a capitate 3-lobed stigma. The fruit is ovoid or subglobose berry, fleshy, dull reddish-orange or crimson when ripe (Plates 1, 2 and 3), somewhat resinous, 1.2–1.5 cm across, flesh mucilaginous, pulpy acid-sweet. Seeds 1–3, flattened, 1–3 mm long embedded in mucilaginous pulpy flesh.

Nutritive/Medicinal Properties

Phytochemicals

Two naturally occurring carotenoid ketones semi-β-carotenone and β-carotenone were isolated from *T. trifolia* (Yokoyama and White 1968); α- and β-carotene, and cryptoxanthin were found in immature fruit of *Triphasia trifolia*, but not in fully ripe fruits being pigmented by semi-α-carotene and semi-β-carotenone, triphasiaxanthin (I) and β-carotenone (Yokoyama and White 1970). The essential oils obtained from leaves, stems and fruits of *Triphasia trifolia* were characterized by a high amount of sabinene (leaf: 31.1%, stem: 21.1%, fruit: 23.9%) and β-pinene (leaf: 40.8%, stem: 36.2%, fruit: 32.4%) (Zoghbi and Andrade 2009). Sabinene, β-pinene and γ-terpinene also were the major compounds identified in the pentane extract.

The main components of the leaf oil of *T. trifolia* of Mauritius were the monoterpenes terpinen-4-ol (29.0%) and carvacrol (14.2%) (Gurib-Fakin et al. 1995). The major constituents of oil obtained from *Triphasia trifolia* leaves from Cuba were sabinene (35.4%) and myrcene (34.1%) while the prevalent compounds in oil from fruits were sabinene (37.2%), β-pinene (23.95) and γ-terpinene (16.3%) Another study reported the 81 compounds were identified in the oil of *T. trifolia* leaves and the major constituent was germacrene B (16.3%) (Pino et al. 2006). *Triphasia trifolia* also had two other coumarins, heraclenol and isomeranzin (Rastogi et al. 1998), and a bicoumarin derivative from mexoticin and meranzin hydrate which were known constituents of the plant (Dondon et al. 2006).

A high percentage of the chemical composition of the oils was determined (99.9% for leaves and 98.4% for fruit) (Santos et al. 2008). The essential oil of the leaves turned out to be composed mainly of monoterpenes (90.0%), of which sabinene (35.4%) and myrcene (34.1%) were the main components. Other components were identified in significant levels in the leaves include sesquisabine hydrate (6.3%), γ-terpinene (5.7%), limonene (4.5%), terpinen-4-ol (4.1%) and α-pinene (2.5%). Similarly, in the essential oil of fruits monoterpenes dominated, accounting for 92.0% of the oil. The major constituents were sabinene (37.2%), β-pinene (23.9%) and γ-terpinene (16.3%), followed by limonene (5.3%), α-pinene (3%), terpinen-4-ol (3.3%). In both oils, a small fraction of sesquiterpenes, representing 6.0–8.2% was identified. Except for the presence of sesquisabine hydrate, which contributed 6.3% of the oil from the leaves and 2.8% of the oil of the fruit, no other sesquiterpene was identified in high amount. The presence of low levels of aliphatic aldehydes n-dodecanal (1.7%) and n-decanal (1.6%) were detected in the leaf oil but not in the fruit oil.

The fruit decoction also yielded the coumarins isopimpinelin, (R)-byakangelicin and (S)-mexoticin. From leaves were isolated the coumarins (R)-byakangelicin, aurapten, (S)-mexoticin, isosibiricin, isomerazin and coumurrayin and the flavonoid vitexin (Santos et al. 2008).

Antimicrobial Activity

Both *Triphasia trifolia* fruit and leaf oils showed moderate antimicrobial activity (Santos et al. 2008). The fruit oil was active against only two types of bacteria, *Bacillus subtilis* and *Chromobacterium violaceum*. In contrast, the leaf oil was moderately effective against all strains tested (*Bacillus subtilis* and *Chromobacterium violaceum, Enterobacter aerogenesis, Klebsiiella pneumoniae, Pseudomonas aeruginosa, Salmonella choleraeasuis, Staphylococcus aureus*).

Cholinesterase Inhibitory Activity

The fruit decoction yielded the coumarins isopimpinelin, (R)-byakangelicin and (S)-mexoticin. From leaves were isolated the coumarins (R)-byakangelicin, aurapten, (S)-mexoticin, isosibiricin, isomerazin and coumurrayin and the flavonoid vitexin (Santos et al. 2008). All coumarins showed cholinesterase inhibition on TLC tests of the courmarins, Aurapten was the most active followed by isopimpinelin, (R)-byakangelicin, isomerazin, mexoticin and isosibiricin while coumurrayin was less active. The data suggested the plant and it coumarins to be promising anticholinesterase candidates for the development of natural drugs with potential to treat Alzheimer's disease.

Antiviral Activity

A total of 25 phenolic compounds were studied for their inhibitory effects against herpes simplex virus (HSV)-1, HSV-2, and human immunodeficiency virus (HIV)-1 (Likhitwitayawuid et al. 2005). These include five flavonoids (1–5) and two dimeric stilbenes (6,7) from *Artocarpus gomezianus*, five phloroglucinol derivatives (8–12) from *Mallotus pallidus* and 13 courmarins (13–25) from *Triphasia trifolia*. The results suggested the bis-hydroxyphenyl structure as a potential target for anti-HSV and HIV drugs development.

Traditional Medicinal Uses

The plant has been used in local traditional medicine. In the Philippines, aromatic bath salts are made from the leaves of this plant. A medicine for diseases of the chest is made from the sweetened fruit. In Indonesia, the leaves are utilised for various complaints, such as diarrhoea, colic, and skin diseases. They also use them in cosmetics.

Other Uses

It makes a good bonsai and good hedge plant and is cultivated for its ornamental features – small, red fruit, deep glossy green leaves and fragrant white flowers. The plant is used as rootstock for *Citrus* species. Wood is useful for small objects such as tool-handles; also provides satisfactory charcoal. Young fruits produce a good glue.

Comments

The species is deemed a weed in other introduced locations.

Selected References

Backer CA, van den Brink RCB Jr (1965) Flora of Java, vol 2, Spermatophytes only. Wolters-Noordhoff, Groningen, 630 pp

Burkill IH (1966) A dictionary of the economic products of the Malay Peninsula. Revised reprint. 2 volumes. Ministry of agriculture and Co-operatives, Kuala Lumpur, Malaysia, vol 1 (A–H) pp 1–1240, vol 2 (I–Z) pp 1241–2444

Chopra RN, Nayar SL, Chopra IC (1986) Glossary of Indian medicinal plants. (Including the supplement). Council Scientific Industrial Research, New Delhi, 330 pp

Dondon R, Bourgeois P, Fery-Forgues S (2006) A new bicoumarin from the leaves and stems of *Triphasia trifolia*. Fitoterapia 77(2):129–133

Gurib-Fakin A, Sewraj MD, Narod F, Menut C (1995) The composition of the essential oils of *Coleus aromaticus, Triphasia trifolia,* and *Eucalyptus kirtoniana*. J Essent Oil Res 8:1–4

Huxley AJ, Griffiths M, Levy M (eds) (1992) The new RHS dictionary of gardening. (4 Vols), MacMillan

Jansen PCM, Jukema J, Oyen LPA, van Lingen TG (1992) Minor edible fruits and nuts. In: Verheij EWM, Coronel RE (eds) Plant resources of south-east Asia 2. Edible fruits and nuts. Prosea Foundation, Bogor, p 363

Likhitwitayawuid K, Supudompol B, Sritularak B, Lipipun V, Rapp K, Schinazi RF (2005) Phenolics with anti-HSV and anti-HIV activities from *Artocarpus gomezianus, Mallotus pallidus,* and *Triphasia trifolia*. Pharm Biol 43(8):651–657

Pacific Island Ecosystems at Risk (PIER) (1999) *Triphasia trifolia* (Burm.f.) Paul G.Wilson, Rutaceae http://www.hear.org/Pier/species/triphasia_trifolia.htm

Pino JA, Marbot R, Fuentes V (2006) Aromatic plants from western Cuba. VI. composition of the leaf oils of *Murraya exotica* L., *Amyris balsamifera* L., *Severinia buxifolia* (Poir.) Ten. and *Triphasia trifolia* (Burm. F.) P. Wilson. J Essent Oil Res 18(1):24–28

Rastogi N, Abaul J, Goh KS, Devallois A, Philogène E, Bourgeois P (1998) Antimycobacterial activity of chemically defined natural substances from the Caribbean flora in Guadeloupe. FEMS Immunol Med Microbiol 20(4):267–273

Santos RPD, Maria Teresa Salles Trevisan MTS, Edilberto R, Silveira ER, Otilia Deusdênia L, Pessoa ODL, Melo VMM (2008) Chemical composition and biological activity of leaves and fruits of *Triphasia trifolia*. Quím Nova 31(1):53–58

Swingle WT (1946) The botany of *Citrus* and its wild relatives of the orange subfamily (family Rutaceae, subfamily Aurantioideae). In: Webber HJ, Batchelor LD (eds) The *Citrus* industry. Vol. 1. History, botany and breeding. Univ. of California Press, Berkeley, pp 129–474, 1028 pp

Yokoyama H, White MJ (1968) *Citrus* carotenoids-VIII: the isolation of semi-β-carotenone and β-carotenone from *Citrus* relatives. Phytochem 7(6):1031–1034

Yokoyama H, White MJ (1970) Carotenone formation in *Triphasia trifolia*. Phytochem 9(8):1795–1797

Zoghbi MDGB, Andrade EHA (2009) Chemical composition of the leaf, stem and fruit essential oils from *Triphasia trifolia* (Burm. f.) P. Wilson cultivated in north of Brazil. J Essent Oil Bear Plants 12(1):81–86

Zanthoxylum simulans

Scientific Name

Zanthoxylum simulans Hance.

Synonyms

Fagara podocarpa (Hemsl.) Engl., *Fagara setosa* (Hemsl.) Engl., *Zanthoxylum acanthophyllum* Hayata, *Zanthoxylum argyi* H. Lév., *Zanthoxylum bungei* Planch., *Zanthoxylum bungei* var. *inermis* Franch., *Zanthoxylum bungeanum* Maxim., *Zanthoxylum fraxinoides* Hemsl., *Zanthoxylum podocarpum* Hemsl., *Zanthoxylum setosum* Hemsl., *Zanthoxylum simulans* var. *podocarpum* (Hemsl.) C.C. Huang.

Family

Rutaceae

Common Names

Anise Pepper, Chinese Pepper, Chinese Prickly Ash, Fagara, Flatspine Prickly Ash, Flower pepper, Sichuan Pepper, Sichuan Peppercorn, Szechwan Pepper, Toothache Tree.

Vernacular Names

Chinese: Ci Hua Jiao, Chuan-Jiao, Ch'uan-Chiao, Da Hongpao Huajiao, Hu Jiao Mu, Hua Chiao, Hua Jiao, Hua Jiao, Qin Jiao, Qing Dao Hua, Ya Jiao, Ye Hua Jiao;
Eastonian: Sarnas-Koldpuu;
French: Poivre Chinois, Clavalier, Clavalier De Bunge;
German: Täuschende Stachelesche;
Indonesia: Andaliman (North Sumatra, Batak Toba), Tuba (North Sumatra, Batak Karo);
India: Tepal, Tirphal (Konkani);
Japanese: Sanshō;
Polish: Pieprz Chiński;
Taiwan: Ci Hua Jiao';
Tibetan: Yer Ma.

Origin/Distribution

The species is native to China (Western Szechuan) and Taiwan. It is cultivated in China and occasionally planted in the former Soviet Union.

Agroecology

A temperate species, it grows wild in the semi-desert, plains, arid hillside and upland forests of Western Szechuan. It grows in well-drained,

T.K. Lim, *Edible Medicinal And Non-Medicinal Plants: Volume 4, Fruits*,
DOI 10.1007/978-94-007-4053-2_105, © Springer Science+Business Media B.V. 2012

loamy soils in full sun or partial shade and is drought tolerant and cold hardy.

Edible Plant Parts and Uses

The dried fruits (Sichuan peppercorns) are used ground or whole as a spice in cooking, applied like pepper but more popularly and more widely used is the ground, powdered pericarp of the fruit. Deep frying the fruit and pericarp enhances the peppery flavour. Sichuan pepper is widely used in Chinese (especially Szechwan), Japanese, Korean, Tibetan, Nepalese, Bhutanese, Batak Toba (North Sumatra) and Konkani (India) cuisines. In Chinese cuisines, Sichuan pepper is widely used in poultry and meat dishes through out the world. *Hua jiao yen* is a mixture of salt and Sichuan pepper, roasted and browned in a wok and served as a condiment to accompany chicken, duck and pork dishes. *Ma la* is spicy combination of Sichuan pepper and chilli pepper commonly used in Sichuan cooking.

Sichuan pepper, star anise and ginger constitute the prime important ingredients in Sichuan cuisine. Sichuan pepper (ground and dry-roasted fruit pericarp) is an ingredient in Chinese five-spice powder and also *shichimi togarashi*, a Japanese seven-flavour spice mixture. Sichuan peer oil known as *Hwajiaw oil* is best used in stir fry noodles with ginger oil, brown sugar, rice vinegar, chopped spring onions and other vegetables. In Sumatran Batak cuisine, Sichuan peppercorn is ground into a green *sambal Tinombur* or chili paste, by mixing with chillis and other seasonings to accompany grilled pork, carp and other regional dishes. The national dish of Tibet is *momo*, a dumpling stuffed with vegetables, cottage cheese or minced yak meat, beef or pork and seasoned with Sichuan pepper, garlic, ginger and onion.

Botany

A small, branched, armed, aromatic, deciduous shrub or treelet up to 3–7 m tall. The stem, branches and branchlets are armed with spiny

Plate 1 Szechuan pepper – fruit, husk and seeds

thorns and the bark is aromatic. Leaves are alternate, odd-pinnate, on winged rachis with curved spines. Leaflets, sessile, 5–15, ovate-orbicular, oblong-ovate or rhombic-ovate, 2.5–7 cm long by 1.8–4.0 cm wide, green, crenulate, punctate, apex acute to mucronate or with a retuse tip, strigose adaxially. Flowers small, inconspicuous, greenish yellow, unisexual in axillary or terminal corymbose panicles 1–5 cm long, perianth in 1 or 2 series with 5–8 undifferentiated tepals. Male flowers with 5–8 stamens. Female flowers with 2–3 carples and recurved styles. Fruit red, globose, 4–5 mm diameter, dehiscent follicles, with verrucose pericarp punctuate with oil-glands and white and smooth inside surface (Plate 1). Seed globose to sub-globose, glossy black, 3–4 mm in diameter (Plate 1).

Nutritive/Medicinal Properties

According to Yang (2008), the two most commercially popular Szechuan pepper species are *Zanthoxylum bungeanum* (red huajiao) and *Z. schinifolium* (green huajiao). Fresh huajiao was found to have a very high content of essential oil, up to 11%, which was described as having fresh, spicy, floral, cooling, and green aroma notes. A total of 120 aroma compounds for each species had been found. In the essential oils, linalyl acetate (15%), linalool (13%), and limonene (12%) were the major components of red huajiao, whereas linalool (29%), limonene (14%), and sabinene (13%) were the main components of green huajiao Despite the differences in major components, both species had six common compounds of top aroma character impact: linalool, α-terpineol, myrcene, 1,8-cineole, limonene, and geraniol. The tingling sensation of huajiao was caused mainly by the alkylamide hydroxy-α-sanshool. The tingling compound decomposed easily under hydrolytic conditions or under UV light.

The tender buds of Z. bungeanum were found to rich (g/kg dw) in nutrients: protein 87.3 g, carbohydrate 21.1 g, cellulosic 15.8 g, fat 8.41 g, carotene 179.6 mg, vitamin B11.23 mg, vitamin D 34.67 μg, amino-acids 244.78 g accounting for 24.48% (Deng et al. 2005). Among the amino acids glutamic acid, proline and serine predominated, accounting for amino 13.8%, 10.8% and 8.8% of the total respectively. Among the minerals, calcium, iron, phosphorus were high with levels of 933, 73, 1 700 mg/kg respectively.

From the pericarps of Zanthoxylum bungeanum six unsaturated aliphatic acid amides (1–6) were isolated (Mizutani et al. 1988). Of these, three were identical with the known amides, hydroxy-α-sanshool, hydroxy-β-sanshool and hydroxy-γ-sanshool (1, 2 and 3). The other three are new compounds, and their structures were established as (2E, 4E, 8E, 10E, 12E)-2′-hydroxy-N-isobutyl-2, 4, 8, 10, 12-tetradecapentaenamide (4), (2E, 4E, 8Z, 11Z)- tetradecatetraenamide (5) and (2E, 4E, 8Z, 11E)-2′-hydroxy-N-isobutyl-2, 4, 8, 11-tetradecatetraenamide (6). Two new flavonol glucosides, namely quarcetin 3′,4′-dimethyl ether 7-glucoside and tamarixetin 3,7-bis-glucoside, together with hyperin, quercetin, quercitrin, foeniculin, isorhamnetin 7-glucoside, rutin, 3,5,6-trihydroxy-7,4′-dimethoxyflavone, arbutin, sitosterol β-glucoside, L-sesamin and palmitic acid were isolated from the pericarps of Zanthoxylum bungeanum (Xiong et al. (1988). Ten unsaturated alkylamides were isolated from the pericarps of Zanthoxylum bungeanum. Three were novel and identified as (2E,4E)-2′-hydroxy-N-isobutyl-2,4-tetradecadienamide (named tetrahydrobungeanool), (2E,4E,8Z)-2′-hydroxy-N-isobutyl-2,4,8-tetradecatrienamide (named dihydrobungeanool) and (2E,4E,8Z,10E,12E)-1′-isopropenyl-N-(2′-bisobutenyl)-2,4,8,10,12-tetradecapentaenamide (named dehydro-γ-sanshool). The main volatile compounds of the secretory mixture in the secretory glands of Zanthoxylum bungeanum fruits were β-phellandrene (36.68%), hydroxy-α-sanshool (19.51%), piperitone (9.29%) and β-pinene (9.23%) (Tirillini and Stoppini 1994). Forty-five compounds which account for 86.76% were identified in the pericarp of Zanthoxylum bungeanun (Li et al. 2001). The main components were β-pinene (10.33%), 1,8-terpinene (7.6%), cis-piperitol acetate (7.07%), oleic acid (5.46%), palmitic acid (5.41%) and 4-terpineol, (4.37%).

The essential oils of the pericarp of Z. bungeanum were found to compose of alkenes (80.96%), alcohols (12.45%), ketones (3.63%), epoxides (1.51%), and esters (1.43%) (Tirillini et al. 1991). Limonene (27%) was the major component and other significant constituents included β-phellandrene (6.1%), β-myrcene (16.6%), and β-ocimene (9.7%). In another study, Trillini et al. (1991) reported that the major components in the fruit oil were β-pinene (25.3%), limonene (20.5%), β-phellandrene (14.1%) and (Z)-β-ocimene (12.2%). In another study, the essential oil obtained from Zanthoxylum bungeanum fruits was found to contain 35 compounds made up of mainly oxygenated monoterpenes (59.7%) and monoterpene hydrocarbons (34.4%) (Gong et al. 2009). Four major oxygenated monoterpenes were terpinen-4-ol (19.7%), 1,8-cineole (16.0%), α-terpineol (7.2%) and α-terpinyl acetate (5.4%). Also notable were the monoterpenes p-cymene (7.9%), γ-terpinene (7.3%) and, δ-3-carene(4.6%) Other significant components included linalool (3.7%), piperitone (3.3%), o-cymene (3%), myrcene (2.8%) and α-pinene (2.5%). Other minor components included α-thujene (0.9%), camphene (0.1%), sabinene (1.6%), α-phellandrene (0.8%), (Z)-β-ocimene (1.15), (E)-β-ocimene (0.4%), cis-p-menth-2-en-1-ol (0.6%), trans- p-menth-2-en-1-ol (0.5%), borneol (0.3%), phellandral (0.6%), bornyl acetate (0.3%), carvacrol (1.0%), 2-endo-acetoxy-1,8-cineole (0.4%), β-elemene (1.3%), β-caryophyllene (0.6%), α-humulene (0.2%), γ-cadinene (0.3%), spathulenol (0.3%), caryophyllene oxide (0.4%), α-muurolol (0.6%), α-cadinol (0.5%) and sesquiterpene hydrocarbon (0.7%).

Forty-three compounds were identified in the volatiles of *Zanthoxylum simulans* fruits, including 20 terpenes, 10 alcohols, 8 esters, and 5 other components (Chyau et al. 1996). The yield in steam distillation was 1.69% (w/w), while that in liquid carbon dioxide extraction was 6.38% (w/w). The major volatile components (>10%) found in both volatiles were β-myrcene, limonene, 1,8-cineole, and (Z)-β-ocimene. Some minor components, including isobutyl acetate, isoamyl acetate, and α-terpinene, were not found in the extract of liquid carbon dioxide. In another study, 108 compounds of the volatile oil of *Zanthoxylum simulans* pericarps and 46 compounds of the seeds were identified separately (Chang et al. 1981). The main constituents were 1, 8-cineole (17.91%), limonene (12.66%), β-elemene (9.81%), (−)-α-terpineol (7.61%), β-selinene (4.81%), α-selinene(3.79%) in the pericarp oil, and 9-hexadecenoic acid (35.89%), hexadecanoic acid (17.70%), oleic acid (11.68%), linoleic acid (4.72%) and 9, 12, 15-octadecatrienoic acid, methyl ester (4.72%) in the seed oil.

Six compounds were isolated from the fruit of *Zanthoxylum simulans* and two were identified as trielaidin and arbutin; 11 compounds were isolated from the root and 7 were identified as glyceryl trilinoleate, β-eudesmol, β-sitosterol, β-amyrin, des-N-methyc-helerythrine, heptacosane and nodakenetin (Chang et al. 1981).

The major alkaloid in *Zanthoxylum simulans*, was chelerythrine with smaller quantities of dihydro-andoxy-chelerythrine,N-acetylanomine, skimmianine, fagarine, sitosterol and sesamine (Gray and O'Sullivan 1980). Other alkaloids isolated from the stem, and root bark (Chen et al. 1994a; 1994b; Chen et al. 1997; Yang et al. 2002) included pyrrole alkaloid, pyrrolezanthine [5-hydroxymethyl-1-[2-(4-hydroxyphenyl)-ethyl]-1H-pyrrole-2-carbaldehyde]; a lignan, (−)-simulanol [4- [3-hydroxymethyl-5-((E)-3-hydroxypropenyl)-7-methoxy-2,3-dihydrobenzofuran-2-yl]-2,6-dimethoxy-phenol] and a monocyclic γ-pyrone, zanthopyranone [3,5-dimethoxy-2-methyl-pyran-4-one]; benzo[c]phenanthridine alkaloids, 6-methyldihydrochelerythrine and 6-methylnorchelerythrine; pyranoquinoline alkaloids, zanthosimuline, and huajiaosimuline, simulenoline, peroxysimulenoline, benzosimuline and zanthodioline; dimeric 2-quinolone alkaloid, zanthobisquinolone, bis-(4-hydroxy-2-keto-1-methyl-3-quinolinyl) methylene; furoquinoline alkaloids and N-acetylanonaine. Four alkaloids skimmianine, edulinine, (±) ribalinine and (±) araliopsine were isolated from the root and root bark of *Zanthoxylum simulans* (Chang et al. 1981).

Antioxidant Activity

The seed oil of *Z. bungeanum* demonstrated marked antioxidant activity in the DPPH radical-scavenging assay (Xia et al. 2011). Both the extraction yield and the antioxidant activity were strongly dependent on the pressure and the amount of modifier. The saponification and unsaponifiable matter of the *Z. bungeanum* seed oil obtained by supercritical CO_2 fluid extraction were 193.4 mg/g and 0.58%, respectively. The main fatty acids were C18:3 linolenic acid, C22:6 docosahexenoic acid, C20:4 arachidonic acid, C18:2 linoleic acid, C16:0 palmitic acid, C18:1 oleic acid, C18:0 stearic acid and C20:1 eicosenoic acid with corresponding mass percentages (%) of 27.01, 0.46, 5.54, 27.81, 12.37, 23.81, 1.81 and 1.19, respectively. The seed oil was rich in unsaturated fatty acids, including C18:3, C22:6, C20:4, C18:2, C18:1 and C20:1, which accounted for 84.0% (mass percentage) of the total amount, which was one of the main reasons for its strong antioxidant activity. Total tocopherols amounted to 27.7 mg/100 g, α-tocopherol 27.3 mg/100 g, γ-tocopherol 0.4 mg/100 g.

Antimicrobial Activity

The essential oil of *Z. simulans* exhibited antibacterial activity in-vitro against *Staphylococcus aureus, Escherichia coli* and *Bacillius subtilis* (Chang et al. 1981).

Anticancer/Cytotoxic Activity

Two new benzo[c]phenanthridine alkaloids, 6-methyldihydrochelerythrine and 6-methylnorchelerythrine, together with 23 known compounds, were isolated from the root bark of *Zanthoxylum simulans*. Among the compounds, the pyranoquinoline alkaloids, zanthosimuline, and huajiaosimuline, exhibited cytotoxic activity. In addition, compound 4 induced terminal differentiation with cultured HL-60 (Human promyelocytic leukemia) cells (Chen et al. 1994b).

Chelerythrine, alkaloid from *Z. simulans*, had been found to act as a specific blocker of protein kinase C (PKC) phosphorylation (Chao et al. 1998). Protein kinase C (PKC) is regarded as an important signal in cellular responses. Chelerythrine inhibited the translocation of PKC from cytosol to membrane. An increase of PKC translocation from cytosol to the membrane was observed in isolated ileal synaptosomes incubated with phorbol 12-myristate 13-acetate (PMA) to reach the plateau. Pretreatment with chelerythrine dose-dependently attenuated this action of PMA. The inhibition of PKC α translocation was similar to that of PKC β. An inhibitory effect of chelerythrine on the translocation of PKC was considered in addition to the inhibition of PKC phosphorylation. Chelerythrine a protein kinase C inhibitor, had also been shown to inhibit the anti-apoptotic Bcl-2 family proteins (Funakoshi et al. 2011). Recent studies showed that chelerythrine induced the rapid mitochondrial apoptotic death of H9c2 cardiomyoblastoma cells in a manner that was independent of the generation of ROS from mitochondria.

Antiinflammatory Activity

Methanol extract of *Z. bungeanum* exhibited significant inhibition of nitric oxide production in lipopolysaccharide-stimulated J774.1 macrophages and iNOS mRNA expression (Tezuka et al. 2001). The inhibitory compound in the methanol extract was identified as 4-O-β-D-glucopyranosyldihydroferulic acid.

Antiplatelet Activity

Chelerythrine chloride an alkaloid from *Zanthoxylum simulans* was found to be a potent antiplatelet agent (Ko et al. 1990) Aggregation and ATP release of washed rabbit platelets caused by ADP (adenosine 5′-diphosphate), arachidonic acid, PAF (platelet activating factor), collagen, ionophore A23187 and thrombin were inhibited by chelerythrine chloride. Less inhibition was observed in platelet-rich plasma. Data indicated that the inhibitory effect of chelerythrine chloride on rabbit platelet aggregation and release reaction was due to the inhibition on thromboxane formation and phosphoinositides breakdown. The pyranoquinoline alkaloid, huajiaosimuline from the root bark, exhibited significant antiplatelet aggregation activity (Chen et al. 1994b).

A pyrrole alkaloid, pyrrolezanthine [5-hydroxymethyl-1-[2-(4-hydroxyphenyl)-ethyl]-1H-pyrrole-2-carbaldehyde]; a lignan, (−)-simulanol [4- [3-hydroxymethyl-5-((E)-3-hydroxypropenyl)-7-methoxy-2,3-dihydro-benzofuran-2-yl]-2,6-dimethoxy-phenol] and a monocyclic γ-pyrone, zanthopyranone [3,5-dimethoxy-2-methyl-pyran-4-one], together with 28 known compounds were isolated from the stem wood of Formosan *Zanthoxylum simulans* (Yang et al.2002). Among the isolates, 11 compounds showed antiplatelet aggregation activity in-vitro.

Cardiovascular Activity

Water and methanol extracts of the *Z. bungeanum* pericarp significantly increased the spontaneous beating rate (BR) of cultured embryonic mouse myocardial cell sheets (Huang et al. 1993). Through bioassay directed fractionation of the extracts, hydroxy-β-sanshool (1b), xanthoxylin (2) and two quercetin glycosides, hyperin (4) and quercitrin (6), were found to increase the BR in a standard medium (2.1 mM Ca2+), but in a low Ca2+ medium (0.5 mM Ca2+), these compounds suppressed the decrease of BR. Of 16 flavonoids related in structure with hyperin (4) and quercitrin (6), quercetin, isoquercitrin, rutin, myricetin and myricitrin also increased the BR in the standard medium, while kaempferol and luteorin decreased the BR in the standard medium. When compared with control, hydroxy-β-sanshool (1b) and xanthoxylin (2) stimulated 13–15 fold calcium uptake of the cultured myocardial cells, which might have caused the positive chronotropic effect. Hyperin (4) and quercitrin (6) did not affect calcium uptake of the myocardial cells, $Na^{+}-K^{+}$ ATPase activity or Ca^{2+}-ATPase activity of sarcoplasmic reticulum.

Cosmeticeutical Activity

The lipophilic hydroxyalkamides hydroxy α-sanshool and β-sanshool-rich extract from fruit husks of Z. bungeanum were found to be efficacious, immediate-action lifting agents for wrinkles (Artaria et al. 2011). Previous studies had validated a sanshool-rich lipophilic extract from the fruit husks of Z. bungeanum (Zanthalene®) as an anti-itching cosmetic ingredient. The lipophilic hydroxyalkamides hydroxy α- and β-sanshools had been identified as the tingling principles for this spice.

Traditional Medicinal Uses

Zanthoxylum simulans is one the of most common Chinese medicinal herbal plant, whose fruit, leaf and root can be used as medicine (Chang et al. 1981; Duke and Ayensu 1985; Yeung 1985). The pericarp is regarded to be pungent, warm, and used to relieve pain, expel intestinal worms, invigorate stomach, cure pathogenic dampness, skin itch and dental necrosis pain (Chang et al. 1981). The seed is bitter, pungent, cool, and it can induce diuresis to reduce edema. The root is pungent, a little warm, and it can cure the symptoms of fall injury, chest and abdomen pains, snake bite, stomach and enteralgia pains.

The fruit is reported to be carminative, diuretic, stimulant, stomachic, tonic, astringent, diaphoretic, emmenagogue anaesthetic, diuretic parasiticide and vasodilator. It is used in the treatment of gastralgia and dyspepsia due to cold with vomiting, diarrhoea, abdominal pain, ascariasis and dermal diseases. It has a local anaesthetic action and is parasiticidal against the pork tapeworm. The leaves are considered carminative, stimulant and sudorific; the seed is reported to be antiphlogistic and diuretic. A decoction of the root is digestive and used in treating snakebites. The resin contained in the bark, and especially in that of the roots, is reported to be a strong stimulant and tonic.

Other Uses

Yang et al. (2008) produced a methyl ester biodiesel from Zanthoxylum bungeanum seed oil (ZBSO) using methanol, sulfuric acid, and potassium hydroxide in a two-stage process. A maximum yield of 96% of methyl esters in ZBSO biodiesel was achieved using a 6.5:1 M ratio of methanol to oil, 0.9% KOH (percent oil), and reaction time of 0.5 h at 55°C. The fuel properties of the ZBSO biodiesel obtained were similar to those of no. 0 petroleum diesel fuel, and most of the parameters complied with the limits established by specifications for biodiesel.

Ethanol extract of Zanthoxylum bungeanum was found to have insecticidal activity against Plutella xylostella (Wei et al. 2008). The extract had a oviposition repellency rate over 80%.

The essential oil obtained from Zanthoxylum bungeanum fruits exhibited antifungal activity against 15 plant pathogenic fungi assayed by the mycelial growth inhibition method and the values of median inhibition concentration (IC_{50}) were below 1.0 mg/ml (Gong et al. 2009). The results showed that the oil had a broad spectrum of inhibitory activity against the tested fungi. The volatile components of the oil also markedly inhibited the mycelial growth of Rhizoctonia solani and Rhizoctonia cerealis.

Comments

Sichuan pepper is produced from several major Zanthoxylum species:

China: *Zanthoxylum ailanthoides* Sieb. & Zucc., *Zanthoxylum armatum* DC., *Zanthoxylum piperitum* (L.) DC., *Zanthoxylum simulans* Hance (*Z. bungeanum* Max.), *Zanthoxylum schinifolium* Sieb. & Zucc.;

Japan: *Zanthoxylum piperitum* (L.) DC.;

Korea: *Zanthoxylum piperitum* and *Zanthoxylum schinifolium* Sieb. & Zucc.

Selected References

Artaria C, Maramaldi G, Bonfigli A, Rigano L, Appendino G (2011) Lifting properties of the alkamide fraction from the fruit husks of *Zanthoxylum bungeanum*. Int J Cosmet Sci 33(4):328–333

Chang ZQ, Liu F, Wang SL, Zhao TZ, Wang MT (1981) Studies on the chemical constituents of *Zanthoxylum simulans* Hance. Yao Xue Xue Bao 16(5):394–396 (In Chinese)

Chao MD, Chen IS, Cheng JT (1998) Inhibition of protein kinase C translocation from cytosol to membrane by chelerythrine. Planta Med 64(7):662–663

Chen IS, Wu SJ, Lin YC, Tsai IL, Seki H, Ko FN, Teng CM (1994a) Dimeric 2-quinolone alkaloid and antiplatelet aggregation constituents of *Zanthoxylum simulans*. Phytochemistry 36(1):237–239

Chen IS, Wu SJ, Tsai IL (1994b) Chemical and bioactive constituents from *Zanthoxylum simulans*. J Nat Prod 57(9):1206–1211

Chen IS, Tsai IW, Teng CH, Chen JJ, Chang YL, Ko FK, Lu MC, Pezzuto JM (1997) Pyranoquinoline alkaloids from *Zanthoxylum simulans*. Phytochemistry 46(3):525–529

Chyau CC, Mau JL, Wu CM (1996) Characteristics of the steam-distilled oil and carbon dioxide extract of *Zanthoxylum simulans*. Fruits J Agric Food Chem 44(4):1096–1099

Deng ZY, Sun BY, Kang KG, Dong YG (2005) Analysis of the main nutritional labeling in the tender bud of *Zanthoxylum bungeanum*. J Northwest Forest Univ 20(1):179–180, 185. (In Chinese)

Duke JA, Ayensu ES (1985) Medicinal plants of China. Reference Publications, Inc, Algonac, 705 pp

Facciola S (1990) Cornucopia. A source book of edible plants. Kampong Publ., Vista, 677 pp

Funakoshi T, Aki T, Nakayama H, Watanuki Y, Imori S, Uemura K (2011) Reactive oxygen species-independent rapid initiation of mitochondrial apoptotic pathway by chelerythrine. Toxicol In Vitro 25(8):1581–1587

Gong Y, HuangY ZL, Shi X, Guo Z, Wang M, Jiang W (2009) Chemical composition and antifungal activity of the fruit oil of *Zanthoxylum bungeanum* Maxim. (Rutaceae) from China. J Essent Oil Res 21:174–178

Gray AI, O'Sullivan JJ (1980) Alkaloid, lignan and sterol constituents of *Zanthoxylum Simulans*. Planta Med 39(3):209

Hu SY (2005) Food plants of China. The Chinese University Press, Hong Kong, 844 pp

Huang XL, Kakiuchi N, Che QM, Huang SL, Sheng L, Hattori M, Namba T (1993) Effects of extracts of *Zanthoxylu*m fruit and their constituents on spontaneous beating rate of myocardial cell sheets in culture. Phytother Res 7(1):41–48

Ko FN, Chen IS, Wu SJ, Lee LG, Haung TF, Teng CM (1990) Antiplatelet effects of chelerythrine chloride isolated from *Zanthoxylum simulans*. Biochim Biophys Acta 1052(3):360–365

Li Y, Zeng J, Liu L, Jin X (2001) GC-MS analysis of supercritical carbon dioxide extraction products from pericarp of *Zanthoxylum bungeanum*. Zhong Yao Cai 24(8):572–573 (In Chinese)

Mizutani K, Fukunaga Y, Tanaka O, Takasugi N, Saruwatari Y, Fuwa T, Yamauchi T, Wang J, Jia MR (1988) Amides from Huajiao, pericarps of *Zanthoxylum bungeanum* Maxim. Chem Pharm Bull 36(7):2362–2365

Stuart RGA (1979) Chinese materia medica: vegetable kingdom. Southern Materials Centre Inc, Taipei

Tezuka Y, Irikawa S, Kaneko T, Banskota AH, Nagaoka T, Xiong Q, Hase K, Kadota S (2001) Screening of Chinese herbal drug extracts for inhibitory activity on nitric oxide production and identification of an active compound of *Zanthoxylum bungeanum*. J Ethnopharmacol 77(2–3):209–217

Tirillini B, Stoppini AM (1994) Volatile constituents of the fruit secretory glands of *Zanthoxylum bungeanum* Maxim. J Essent Oil Res 6(3):249–252

Tirillini B, Manunta A, Stoppini AM (1991) Constituents of the essential oil of the fruits of *Zanthoxylum bungeanum*. Planta Med 57(1):90–91

Uphof JC Th (1968) Dictionary of economic plants, 2nd edn. (1st edn. 1959). Cramer, Lehre, 591 pp

Wei H, Hou Y, Yang G, You M (2008) Repellent and antifeedant effect of secondary metabolites of non-host plants on *Plutella xylostella*. Ying Yong Sheng Tai Xue Baa 15(3):473–476 (In Chinese)

Wikipedia (2010) http://en.wikipedia.org/wiki/Sichuan_pepper

Xia L, You J, Li G, Sun Z, Suo Y (2011) Compositional and antioxidant activity analysis of *Zanthoxylum bungeanum* seed oil obtained by supercritical CO_2 fluid extraction. J Am Oil Chem Soc 88(1):23–32

Xiong QB, Shi DW, Mizuno M (1988) Flavonol glucosides in pericarps of *Zanthoxylum bungeanum*. Phytochemistry 39(3):723–725

Xiong QB, Shi DW, Yamamoto H, Mizuno M (1997) Alkylamides from pericarps of *Zanthoxylum bungeanum*. Phytochemistry 46(6):1123–1126

Yang FX, Su YQ, Li XH, Zhang Q, Sun RC (2008) Studies on the preparation of biodiesel from *Zanthoxylum bungeanum* Maxim seed oil. J Agric Food Chem 56(17):7891–7896

Yang X (2008) Aroma constituents and alkylamides of red and green huajiao (*Zanthoxylum bungeanum* and

Zanthoxylum schinifolium). J Agric Food Chem 56(5):1689–1696

Yang YP, Cheng MJ, Teng CM, Chang YL, Tsai IL, Chen IS (2002) Chemical and anti-platelet constituents from Formosan *Zanthoxylum simulans*. Phytochemistry 61(5):567–572

Yeung HC (1985) Handbook of Chinese herbs and formulas. Institute of Chinese Medicine, Los Angeles

Zhang DX, Hartley TG (2008) *Zanthoxylum* Linnaeus. In: Wu ZY, Raven PH, Hong DY (eds) Flora of China. Vol. 11 (Oxalidaceae through Aceraceae). Science Press/Missouri Botanical Garden Press, Beijing/St. Louis

Zhang J, Jiang L (2008) Acid-catalyzed esterification of *Zanthoxylum bungeanum* seed oil with high free fatty acids for biodiesel production. Bioresour Technol 99(18):8995–8998

Medical Glossary

AAD allergic airway disease, an inflammatory disorder of the airways caused by allergens.

AAPH 2,2′-azobis(2-amidinopropane) dihydrochloride, a water-soluble azo compound used extensively as a free radical generator, often in the study of lipid peroxidation and the characterization of antioxidants.

Abdominal distension referring to generalised distension of most or all of the abdomen. Also referred to as stomach bloating often caused by a sudden increase in fibre from consumption of vegetables, fruits and beans.

Abeta aggregation amyloid beta protein (Abeta) aggregation is associated with Alzheimer's disease (AD); it is a major component of the extracellular plaque found in AD brains.

Ablation therapy the destruction of small areas of myocardial tissue, usually by application of electrical or chemical energy, in the treatment of some tachyarrhythmias.

Abortifacient a substance that causes or induces abortion.

Abortivum a substance inducing abortion.

Abscess a swollen infected, inflamed area filled with pus in body tissues.

ABTS 2.2 azinobis-3-ethylhenthiazoline-6-sulfonic acid, a type of mediator in chemical reaction kinetics of specific enzymes.

ACAT acyl CoA: cholesterol acyltransferase.

ACE see angiotensin-converting enzyme.

Acetogenins natural products from the plants of the family Annonaceae, are very potent inhibitors of the NADH-ubiquinone reductase (Complex I) activity of mammalian mitochondria.

Acetylcholinesterase (AChE) is an enzyme that degrades (through its hydrolytic activity) the neurotransmitter acetylcholine, producing choline.

Acidosis increased acidity.

Acne vulgaris also known as chronic acne, usually occurring in adolescence, with comedones (blackheads), papules (red pimples), nodules (inflamed acne spots), and pustules (small inflamed pus-filled lesions) on the face, neck, and upper part of the trunk.

Acquired immunodeficiency syndrome (AIDS) an epidemic disease caused by an infection by human immunodeficiency virus (HIV-1, HIV-2), retrovirus that causes immune system failure and debilitation and is often accompanied by infections such as tuberculosis.

Acridone an organic compound based on the acridine skeleton, with a carbonyl group at the 9 position.

ACTH adrenocorticotropic hormone (or corticotropin), a polypeptide tropic hormone produced and secreted by the anterior pituitary gland. It plays a role in the synthesis and secretion of gluco- and mineralo-corticosteroids and androgenic steroids.

Activating transcription factor (ATF) a protein (gene) that binds to specific DNA sequences regulating the transfer or transcription of information from DNA to mRNA.

Activator protein-1 (AP-1) a heterodimeric protein transcription factor that regulates gene expression in response to a variety of stimuli, including cytokines, growth factors, stress, and bacterial and viral infections. AP-1 in turn regulates a number of cellular processes including differentiation, proliferation, and apoptosis.

Acyl-CoA dehydrogenases A group of enzymes that catalyzes the initial step in each cycle of fatty acid β-oxidation in the mitochondria of cells.

Adaptogen a term used by herbalists to refer to a natural herb product that increases the body's resistance to stresses such as trauma, stress and fatigue.

Adaptogenic increasing the resistance of the body to stress.

Addison's disease is a rare endocrine disorder. It occurs when the adrenal glands cannot produce sufficient hormones (corticosteroids). It is also known as chronic adrenal insufficiency, hypocortisolism or hypocorticism.

Adenocarcinoma a cancer originating in glandular tissue.

Adenoma a benign tumour from a glandular origin.

Adenopathy abnormal enlargement or swelling of the lymph node.

Adenosine receptors a class of purinergic, G-protein coupled receptors with adenosine as endogenous ligand. In humans, there are four adenosine receptors. A1 receptors and A2A play roles in the heart, regulating myocardial oxygen consumption and coronary blood flow, while the A2A receptor also has broader antiinflammatory effects throughout the body. These two receptors also have important roles in the brain, regulating the release of other neurotransmitters such as dopamine and glutamate, while the A2B and A3 receptors are located mainly peripherally and are involved in inflammation and immune responses.

ADH see alcohol dehydrogenase.

Adipocyte a fat cell involved in the synthesis and storage of fats.

Adipocytokine bioactive cytokines produced by adipose tissues.

Adiponectin a protein in humans that modulates several physiological processes, such as metabolism of glucose and fatty acids, and immune responses.

Adipose tissues body fat, loose connective tissue composed of adipocytes (fat cells).

Adoptogen containing smooth pro-stressors which reduce reactivity of host defense systems and decrease damaging effects of various stressors due to increased basal level of mediators involved in the stress response.

Adrenal glands star-shaped endocrine glands that sit on top of the kidneys.

Adrenalectomized having had the adrenal glands surgically removed.

Adrenergic having to do with adrenaline (epinephrine) and/or noradrenaline (norepinephrine).

Adrenergic receptors a class of G protein-coupled receptors that are targets of the noradrenaline (norepinephrine) and adrenaline (epinephrine).

Adulterant an impure ingredient added into a preparation.

Advanced Glycation End products (AGEs) resultant products of a chain of chemical reactions after an initial glycation reaction. AGEs may play an important adverse role in process of atherosclerosis, diabetes, aging and chronic renal failure.

Aegilops an ulcer or fistula in the inner corner of the eye.

Afferent something that so conducts or carries towards, such as a blood vessel, fibre, or nerve.

Agalactia lack of milk after parturition (birth).

Agammaglobulinaemia an inherited disorder in which there are very low levels of protective immune proteins called immunoglobulins. Cf. x-linked agammaglobulinaemia.

Agglutination clumping of particles.

Agglutinin a protein substance, such as an antibody, that is capable of causing agglutination (clumping) of a particular antigen.

Agonist a drug that binds to a receptor of a cell and triggers a response by the cell.

Ague a fever (such as from malaria) that is marked by paroxysms of chills, fever, and sweating that recurs with regular intervals.

AHR AhR, aryl hydrocarbon receptor, a cytosolic protein transcription factor.

AIDS see Acquired Immunodeficiency Syndrome.

Akathisia a movement disorder in which there is an urge or need to move the legs to stop unpleasant sensations. Also called restless leg syndrome, the disorder is often caused by long-term use of antipsychotic medications.

Akt/FoxO pathway Cellular processes involving Akt and FoxO transcription factors that play a role in angiogenesis and vasculogenesis.

Akt signaling pathway Akt are protein kinases involved in mammalian cellular signaling, inhibits apoptotic processes.

Alanine transaminase (ALT) also called Serum Glutamic Pyruvate Transaminase (SGPT) or Alanine aminotransferase (ALAT), an enzyme present in hepatocytes (liver cells). When a cell is damaged, it leaks this enzyme into the blood.

ALAT, (Alanine aminotransferase) see Alanine transaminase.

Albumin water soluble proteins found in egg white, blood serum, milk, various animal tissues and plant juices and tissues.

Albuminaria excessive amount of albumin in the urine, a symptom of severe kidney disease.

Alcohol dehydrogenase (ADH) an enzyme involved in the break-down of alcohol.

Aldose reductase, aldehyde reductase an enzyme in carbohydrate metabolism that converts glucose to sorbitol.

Alexipharmic an antidote, remedy for poison.

Alexiteric a preservative against contagious and infectious diseases, and the effects of poisons.

Algesic endogenous substances involved in the production of pain that is associated with inflammation, e.g. serotonin, bradykinin and prostaglandins.

Alkaline phosphatase (ALP) an enzyme in the cells lining the biliary ducts of the liver. ALP levels in plasma will rise with large bile duct obstruction, intrahepatic cholestasis or infiltrative diseases of the liver. ALP is also present in bone and placental tissues.

Allergenic having the properties of an antigen (allergen), immunogenic.

Allergic pertaining to, caused, affected with, or the nature of the allergy.

Allergic conjunctivitis inflammation of the tissue lining the eyelids (conjunctiva) due to allergy.

Allergy a hypersensitivity state induced by exposure to a particular antigen (allergen) resulting in harmful immunologic reactions on subsequent exposures. The term is usually used to refer to hypersensitivity to an environmental antigen (atopic allergy or contact dermatitis) or to drug allergy.

Allogeneic cells or tissues which are genetically different because they are derived from separate individuals of the same species. Also refers to a type of immunological reaction that occurs when cells are transplanted into a genetically different recipient.

Allografts or homografts, a graft between individuals of the same species, but of different genotypes.

Alloknesis itch produced by innocuous mechanical stimulation.

Allostasis the process of achieving stability, or homeostasis, through physiological or behavioral change.

Alopecia is the loss of hair on the body.

Alopecia areata is a particular disorder affecting hair growth (loss of hair) in the scalp and elsewhere.

ALP see Alkaline phosphatase.

Alpha-adrenoceptor receptors postulated to exist on nerve cell membranes of the sympathetic nervous system in order to explain the specificity of certain agents that affect only some sympathetic activities (such as vasoconstriction and relaxation of intestinal muscles and contraction of smooth muscles).

Alpha amylase α-amylase a major form of amylase found in humans and other mammals that cleaves alpha-bonds of large sugar molecules.

ALT see Alanine transaminase.

Alterative a medication or treatment which gradually induces a change, and restores healthy functions without sensible evacuations.

Alveolar macrophage a vigorously phagocytic macrophage on the epithelial surface of lung alveoli that ingests carbon and other inhaled particulate matter. Also called coniophage or dust cell.

Alzheimer's disease a degenerative, organic, mental disease characterized by progressive brain deterioration and dementia, usually occurring after the age of 50.

Amastigote refers to a cell that does not have any flagella, used mainly to describe a certain phase in the life-cycle of trypanosome protozoans.

Amenorrhea the condition when a woman fails to have menstrual periods.

Amidolytic cleavage of the amide structure.

Amoebiasis state of being infected by amoeba such as *Entamoeba histolytica*.

Amoebicidal lethal to amoeba.

AMPK (5′ AMP-activated protein kinase) or 5′ adenosine monophosphate-activated protein kinase, enzyme that plays a role in cellular energy homeostasis.

Amyloid beta (Aβ or Abeta) a peptide of 39–43 amino acids that appear to be the main constituent of amyloid plaques in the brains of Alzheimer's disease patients.

Amyotrophic lateral sclerosis or ALS, is a disease of the motor neurons in the brain and spinal cord that control voluntary muscle movement.

Amyotrophy progressive wasting of muscle tissues. *adj.* amyotrophic.

Anaemia a blood disorder in which the blood is deficient in red blood cells and in haemoglobin.

Anaesthesia condition of having sansation temporarily suppressed.

Anaesthetic a substance that decreases partially or totally nerve the sense of pain.

Analeptic a central nervous system (CNS) stimulant medication.

Analgesia term describing relief, reduction or suppression of pain. *adj.* analgetic.

Analgesic a substance that relieves or reduces pain.

Anaphoretic an antiperspirant.

Anaphylactic *adj.* see anaphylaxis.

Anaphylaxis a severe, life-threatening allergic response that may be characterized by symptoms such as reduced blood pressure, wheezing, vomiting or diarrhea.

Anaphylotoxins are fragments (C3a, C4a or C5a) that are produced during the pathways of the complement system. They can trigger release of substances of endothelial cells, mast cells or phagocytes, which produce a local inflammatory response.

Anaplasia a reversion of differentiation in cells and is characteristic of malignant neoplasms (tumours).

Anaplastic *adj.* see anaplasia.

Anasarca accumulation of great quantity of fluid in body tissues.

Androgen male sex hormone in vertebrates. Androgens may be used in patients with breast cancer to treat recurrence of the disease.

Android adiposity centric fat distribution patterns with increased disposition towards the abdominal area, visceral fat – apple shaped cf gynoid adiposity.

Angina pectoris, Angina chest pain or chest discomfort that occurs when the heart muscle does not get enough blood.

Angiogenic *adj.* see angiogenesis.

Angiogenesis a physiological process involving the growth of new blood vessels from pre-existing vessels.

Angiotensin an oligopeptide hormone in the blood that causes blood vessels to constrict, and drives blood pressure up. It is part of the renin-angiotensin system.

Angiotensin-converting enzyme (ACE) an exopeptidase, a circulating enzyme that participates in the body's renin-angiotensin system (RAS) which mediates extracellular volume (i.e. that of the blood plasma, lymph and interstitial fluid), and arterial vasoconstriction.

Anglioplasty medical procedure used to open obstructed or narrowed blood vessel resulting usually from atherosclerosis.

Anisonucleosis a morphological manifestation of nuclear injury characterized by variation in the size of the cell nuclei.

Ankylosing spondylitis (AS) is a type of inflammatory arthritis that targets the joints of the spine.

Annexitis also called adnexitis, a pelvic inflammatory disease involving the inflammation of the ovaries or fallopian tubes.

Anodyne a substance that relieves or soothes pain by lessening the sensitivity of the brain or nervous system. Also called an analgesic.

Anoikis apoptosis that is induced by inadequate or inappropriate cell-matrix interactions.

Anorectal relating to the rectum and anus.

Anorectics appetite suppressants, substances which reduce the desire to eat. Used on a short term basis clinically to treat obesity. Also called anorexigenics.

Anorexia lack or loss of desire to eat.

Anorexic having no appetite to eat.

Anorexigenics see anorectics.

Antagonist a substance that acts against and blocks an action.

Antalgic a substance used to relive a painful condition.

Antecubital vein This vein is located in the antecubital fossa -the area of the arm in front of the elbow.

Anterior uveitis is the most common form of ocular inflammation that often causes a painful red eye.

Anthelmintic an agent or substance that is destructive to worms and used for expulsion of internal parasitic worms in animals and humans.

Anthocyanins a subgroup of antioxidant flavonoids, are glucosides of anthocyanidins. Which are beneficial to health. They occur as water-soluble vacuolar pigments that may appear red, purple, or blue according to pH in plants.

Anthrax a bacterial disease of cattle and ship that can be transmitted to man though unprocessed wool.

Anthropometric pertaining to the study of human body measurements.

Antiamoebic a substance that destroys or suppresses parasitic amoebae.

Antiamyloidogenic compounds that inhibit the formation of Alzheimer's β-amyloid fibrils (fAβ) from amyloid β-peptide (Aβ) and destabilize fAβ.

Antianaphylactic agent that can prevent the occurrence of anaphylaxis (life threatening allergic response).

Antiangiogenic a drug or substance used to stop the growth of tumours and progression of cancers by limiting the pathologic formation of new blood vessels (angiogenesis).

Antiarrhythmic a substance to correct irregular heartbeats and restore the normal rhythm.

Antiasmathic drug that treats or ameliorates asthma.

Antiatherogenic that protects against atherogenesis, the formation of atheromas (plaques) in arteries.

Antibacterial substance that kills or inhibits bacteria.

Antibilious an agent or substance which helps remove excess bile from the body.

Antibiotic a chemical substance produced by a microorganism which has the capacity to inhibit the growth of or to kill other microorganisms.

Antiblennorrhagic a substance that treats blennorrhagia a conjunctival inflammation resulting in mucus discharge.

Antibody a gamma globulin protein produced by a kind of white blood cell called the plasma cell in the blood used by the immune system to identify and neutralize foreign objects (antigen).

Anticarcinomic a substance that kills or inhibits carcinomas (any cancer that arises in epithelium/tissue cells).

Anticephalalgic headache-relieving or preventing.

Anticestodal a chemical destructive to tapeworms.

Anticholesterolemic a substance that can prevent the build up of cholesterol.

Anticlastogenic having a suppressing effect of chromosomal aberrations.

Anticoagulant a substance that thins the blood and acts to inhibit blood platelets from sticking together.

Antidepressant a substance that suppresses depression or sadness.

Antidiabetic a substance that prevents or alleviates diabetes. Also called antidiabetogenic.

Antidiarrhoeal having the property of stopping or correcting diarrhoea, an agent having such action.

Antidopaminergic a term for a chemical that prevents or counteracts the effects of dopamine.

Antidote a remedy for counteracting a poison.

Antidrepanocytary anti-sickle cell anaemia.

Antidysenteric an agent used to reduce or treat dysentery and diarrhea.

Antidyslipidemic agent that will reduce the abnormal amount of lipids and lipoproteins in the blood.

Anti-edematous reduces or suppresses edema.

Anti-emetic an agent that stops vomiting.

Anti-epileptic a drug used to treat or prevent convulsions, anticonvulsant.

Antifebrile a substance that reduces fever, also called antipyretic.

Antifeedant preventing something from being eaten.

Antifertility agent that inhibits formation of ova and sperm and disrupts the process of fertilization (antizygotic).

Antifilarial effective against human filarial worms.

Antifungal an agent that kills or inhibits the growth of fungi.

Antiganacratia anti- menstruation.

Antigen a substance that prompts the production of antibodies and can cause an immune response. *adj.* antigenic.

Antigastralgic preventing or alleviating gastric colic.

Antigenotoxic an agent that inhibits DNA adduct formation, stimulates DNA repair mechanisms, and possesses antioxidant functions.

Antihematic agent that stops vomiting.

Antihemorrhagic an agent which stops or prevents bleeding.

Antihepatotoxic counteracting injuries to the liver.

Antiherpetic having activity against Herpes Simplex Virus (HSV).

Antihistamine an agent used to counteract the effects of histamine production in allergic reactions.

Antihyperalgesia the ability to block enhanced sensitivity to pain, usually produced by nerve injury or inflammation, to nociceptive stimuli. *adj.* antihyperalgesic.

Antihypercholesterolemia term to describe lowering of cholesterol level in the blood or blood serum.

Antihypercholesterolemic agent that lowers cholesterol level in the blood or blood serum.

Antihyperlidemic promoting a reduction of lipid levels in the blood, or an agent that has this action.

Antihypersensitive a substance used to treat excessive reactivity to any stimuli.

Antihypertensive a drug used in medicine and pharmacology to treat hypertension (high blood pressure).

Antiinflammatory a substance used to reduce or prevent inflammation.

Antileishmanial inhibiting the growth and proliferation of *Leishmania* a genus of flagellate protozoans that are parasitic in the tissues of vertebrates.

Antileprotic therapeutically effective against leprosy.

Antilithiatic an agent that reduces or suppresses urinary calculi (stones) and acts to dissolve those already present.

Antileukaemic anticancer drugs that are used to treat leukemia.

Antilithogenic inhibiting the formation of calculi (stones).

Antimalarial an agent used to treat malaria and/or kill the malaria-causing organism, *Plasmodium* spp.

Antimelanogenesis obstruct production of melanin.

Antimicrobial a substance that destroys or inhibits growth of disease-causing bacteria, viruses, fungi and other microorganisms.

Antimitotic inhibiting or preventing mitosis.

Antimutagenic an agent that inhibits mutations.

Antimycotic antifungal.

Antineoplastic said of a drug intended to inhibit or prevent the maturation and proliferation of neoplasms that may become malignant, by targeting the DNA.

Antineuralgic a substance that stops intense intermittent pain, usually of the head or face, caused by neuralgia.

Antinociception reduction in pain: reduction in pain sensitivity produced within neurons when an endorphin or similar opium-containing substance opioid combines with a receptor.

Antinociceptive having an analgesic effect.

Antinutrient are natural or synthetic compounds that interfere with the absorption of nutrients and are commonly found in food sources and beverages.

Antioestrogen a substance that inhibits the biological effects of female sex hormones.

Antiophidian anti venoms of snake.

Antiosteoporotic substance that can prevent osteoporosis.

Antiovulatory substance suppressing ovulation.

Antioxidant a chemical compound or substance that inhibits oxidation and protects against free radical activity and lipid oxidation such as vitamin E, vitamin C, or beta-carotene (converted to vitamin B), carotenoids and flavonoids which are thought to protect body cells from the damaging effects of oxidation. Many foods including fruit and vegetables contain compounds with antioxidant properties. Antioxidants may also reduce the risks of cancer and age-related macular degeneration(AMD).

Antipaludic antimalarial.

Antiperiodic substance that prevents the recurrence of symptoms of a disease e.g. malaria.

Antiperspirant a substance that inhibits sweating. Also called antisudorific, anaphoretic.

Antiphlogistic a traditional term for a substance used against inflammation, an anti-inflammatory.

Antiplatelet agent drug that decreases platelet aggregation and inhibits thrombus formation.

Antiplasmodial suppressing or destroying plasmodia.

Antiproliferative preventing or inhibiting the reproduction of similar cells.

Antiprostatic drug to treat the prostate.

Antiprotozoal suppressing the growth or reproduction of protozoa.

Antipruritic alleviating or preventing itching.

Antipyretic a substance that reduces fever or quells it. Also known as antithermic.

Antirheumatic relieving or preventing rheumatism.

Antiscorbutic a substance or plant rich in vitamin C that is used to counteract scurvy.

Antisecretory inhibiting or diminishing secretion.

Antisense refers to antisense RNA strand because its sequence of nucleotides is the complement of message sense. When mRNA forms a duplex with a complementary antisense RNA sequence, translation of the mRNA into the protein is blocked. This may slow or halt the growth of cancer cells.

Antiseptic preventing decay or putrefaction, a substance inhibiting the growth and development of microorganisms.

Anti-sickling agent an agent used to prevent or reverse the pathological events leading to sickling of erythrocytes in sickle cell conditions.

Antispasmodic a substance that relieves spasms or inhibits the contraction of smooth muscles; smooth muscle relaxant, muscle-relaxer.

Antispermatogenic preventing or suppressing the production of semen or spermatozoa.

Antisudorific see antiperspirant.

Antisyphilitic a drug (or other chemical agent) that is effective against syphilis.

Antithermic a substance that reduces fever and temperature. Also known as antipyretic.

Antithrombotic preventing or interfering with the formation of thrombi.

Antitoxin an antibody with the ability to neutralize a specific toxin.

Antitumoral substance that acts against the growth, development or spread of a tumour.

Antitussive a substance that depresses coughing.

Antiulcerogenic an agent used to protect against the formation of ulcers, or is used for the treatment of ulcers.

Antivenin an agent used against the venom of a snake, spider, or other venomous animal or insect.

Antivinous an agent or substance that treats addiction to alcohol.

Antiviral substance that destroys or inhibits the growth and viability of infectious viruses.

Antivomitive a substance that reduces or suppresses vomiting.

Antizygotic see antifertility.

Anuria absence of urine production and excretion. *adj.* anuric.

Anxiolytic a drug prescribed for the treatment of symptoms of anxiety.

APAF-1 apoptotic protease activating factor 1.

Apelin also known as APLN, a peptide which in humans is encoded by the APLN gene.

Aperient a substance that acts as a mild laxative by increasing fluids in the bowel.

Aperitif an appetite stimulant.

Aphonia loss of the voice resulting from disease, injury to the vocal cords, or various psychological causes, such as hysteria.

Aphrodisiac an agent that increases sexual activity and libido and/or improves sexual performance.

Aphthae white, painful oral ulcer of unknown cause.

Aphthous ulcerr also known as a canker sore, is a type of oral ulcer, which presents as a painful open sore inside the mouth or upper throat.

Aphthous stomatitis a canker sore, a type of painful oral ulcer or sore inside the mouth orupper throat, caused by a break in the mucous membrane. Also called aphthous ulcer.

Apnoea suspension of external breathing.

Apoliprotein B (APOB) primary apolipoprotein of low-density lipoproteins which is responsible for carrying cholesterol to tissues.

Apoplexy a condition in which the brain's function stops with loss of voluntary motion and sense.

Apoprotein the protein moiety of a molecule or complex, as of a lipoprotein.

Appendicitis is a condition characterized by inflammation of the appendix. Also called epityphlitis.

Appetite stimulant a substance to increase or stimulate the appetite. Also called aperitif.

Apolipoprotein A-I (APOA1) a major protein component of high density lipoprotein (HDL) in plasma. The protein promotes cholesterol efflux from tissues to the liver for excretion.

Apolipoprotein B (APOB) is the primary apolipoprotein of low-density lipoproteins (LDL or "bad cholesterol"), which is responsible for carrying cholesterol to tissues.

Apolipoprotein E (APOE) the apolipoprotein found on intermediate density lipoprotein and chylomicron that binds to a specific receptor on liver and peripheral cells.

Apoptogenic ability to cause death of cells.

Apoptosis death of cells.

Apurinic lyase a DNA enzyme that catalyses a chemical reaction.

Arachidonate cascade includes the cyclooxygenase (COX) pathway to form prostanoids and the lipoxygenase (LOX) pathway to generate several oxygenated fatty acids, collectively called eicosanoids.

Ariboflavinosis a condition caused by the dietary deficiency of riboflavin that is characterized by mouth lesions, seborrhea, and vascularization.

Aromatase an enzyme involved in the production of estrogen that acts by catalyzing the conversion of testosterone (an androgen) to estradiol (an estrogen). Aromatase is located in estrogen-producing cells in the adrenal glands, ovaries, placenta, testicles, adipose (fat) tissue, and brain.

Aromatherapy a form of alternative medicine that uses volatile liquid plant materials, such as essential oils and other scented compounds from plants for the purpose of affecting a person's mood or health.

Aromatic having a pleasant, fragrant odour.

Arrhythmias abnormal heart rhythms that can cause the heart to pump less effectively. Also called dysrhythmias.

Arsenicosis see arsenism.

Arsenism an incommunicable disease resulting from the ingestion of ground water containing unsafe levels of arsenic, also known as arsenicosis.

Arteriosclerosis imprecise term for various disorders of arteries, particularly hardening due to fibrosis or calcium deposition, often used as a synonym for atherosclerosis.

Arthralgia is pain in the joints from many possible causes.

Arthritis inflammation of the joints of the body.

Aryl hydrocarbon receptor (AhR) a ligand-activated transcription factor best known for mediating the toxicity of dioxin and other exogenous contaminants and is responsible for their toxic effects, including immunosuppression.

ASAT or AST aspartate aminotransferase, see aspartate transaminase,

Ascaris a genus of parasitic intestinal round worms.

Ascites abnormal accumulation of fluid within the abdominal or peritoneal cavity.

Ascorbic acid See vitamin C.

Aspartate transaminase (AST) also called Serum Glutamic Oxaloacetic Transaminase (SGOT) or aspartate aminotransferase (ASAT) is similar to ALT in that it is another enzyme associated with liver parenchymal cells. It is increased in acute liver damage, but is also present in red blood cells, and cardiac and skeletal muscle and is therefore not specific to the liver.

Asphyxia failure or suppression of the respiratory process due to obstruction of air flow to the lungs or to the lack of oxygen in inspired air.

Asphyxiation the process of undergoing asphyxia.

Asthenia a nonspecific symptom characterized by loss of energy, strength and feeling of weakness.

Asthenopia weakness or fatigue of the eyes, usually accompanied by headache and dimming of vision. *adj.* asthenopic.

Asthma a chronic illness involving the respiratory system in which the airway occasionally constricts, becomes inflamed, and is lined with excessive amounts of mucus, often in response to one or more triggers.

Astringent a substance that contracts blood vessels and certain body tissues (such as mucous membranes) with the effect of reducing secretion and excretion of fluids and/or has a drying effect.

Astrocytes collectively called astroglia, are characteristic star-shaped glial cells in the brain and spinal cord.

Ataxia (loss of co-ordination) results from the degeneration of nerve tissue in the spinal cord and of nerves that control muscle movement in the arms and legs.

Ataxia telangiectasia and Rad3-related protein (ATR) also known as serine/threonine-protein kinase ATR, FRAP-related protein 1 (FRP1), is an enzyme encoded by the ATR gene. It is involved in sensing DNA damage and activating the DNA damage checkpoint, leading to cell cycle arrest.

ATF-2 activating transcription factor 2.

Atherogenic having the capacity to start or accelerate the process of atherogenesis.

Atherogenesis the formation of lipid deposits in the arteries.

Atheroma a deposit or degenerative accumulation of lipid-containing plaques on the innermost layer of the wall of an artery.

Atherosclerosis the condition in which an artery wall thickens as the result of a build-up of fatty materials such as cholesterol.

Atherothrombosis medical condition characterized by an unpredictable, sudden disruption (rupture or erosion/fissure) of an atherosclerotic plaque, which leads to platelet activation and thrombus formation.

Athlete's foot a contagious skin disease caused by parasitic fungi affecting the foot, hands, causing itching, blisters and cracking. Also called dermatophytosis.

Athymic mice laboratory mice lacking a thymus gland.

Atonic lacking normal tone or strength.

Atony insufficient muscular tone.

Atopic dermatitis an inflammatory, non-contagious, pruritic skin disorder of unknown etiology; often called eczema.

Atresia a congenital medical condition in which a body orifice or passage in the body is abnormally closed or absent.

Atretic ovarian follicles an involuted or closed ovarian follicle.

Atrial fibrillation is the most common cardiac arrhythmia (abnormal heart rhythm) and involves the two upper chambers (atria) of the heart.

Attention-deficit hyperactivity disorder (ADHD, ADD or AD/HD) is a neurobehavioral developmental disorder, primarily characterized by the co-existence of attentional problems and hyperactivity.

Auditory brainstem response (ABR) also called brainstem evoked response (BSER) is an electrical signal evoked from the brainstem of a human by the presentation of a sound such as a click.

Augmerosen a drug that may kill cancer cells by blocking the production of a protein that makes cancer cells live longer. Also called bcl-2 antisense oligonucleotide.

Auricular of or relating to the auricle or the ear in general.

Aurones [2-benzylidenebenzofuran-3(2H)-ones] are the secondary plant metabolites and is a subgroup of flavonoids. See flavonoids.

Autoantibodies antibodies manufactured by the immune system that mistakenly target and damage specific tissues and organs of the body.

Autolysin an enzyme that hydrolyzes and destroys the components of a biological cell or a tissue in which it is produced.

Autophagy digestion of the cell contents by enzymes in the same cell.

Autopsy examination of a cadaver to determine or confirm the cause of death.

Avidity Index describes the collective interactions between antibodies and a multivalent antigen.

Avulsed teeth is tooth that has been knocked out.

Ayurvedic traditional Hindu system of medicine based largely on homeopathy and naturopathy.

Azoospermia is the medical condition of a male not having any measurable level of sperm in his semen.

Azotaemia a higher than normal blood level of urea or other nitrogen containing compounds in the blood.

Babesia a protozoan parasite (malaria–like) of the blood that causes a hemolytic disease known as Babesiosis.

Babesiosis malaria-like parasitic disease caused by Babesia, a genus of protozoal piroplasms.

Bactericidal lethal to bacteria.

Balanitis is an inflammation of the glans (head) of the penis.

BALB/c mice Balb/c mouse was developed in 1923 by McDowell. It is a popular strain and is used in many different research disciplines, but most often in the production of monoclonal antibodies.

Balm aromatic oily resin from certain trees and shrubs used in medicine.

Baroreceptor a type of interoceptor that is stimulated by pressure changes, as those in blood vessel wall.

Barrett's esophagus (Barrett esophagitis) a disorder in which the lining of the esophagus is damaged by stomach acid.

Basophil a type of white blood cell with coarse granules within the cytoplasm and a bilobate (two-lobed) nucleus.

BCL-2 a family of apoptosis regulator proteins in humans encoded by the B-cell lymphoma 2 (BCL-2) gene.

BCL-2 antisense oligonucleotide see augmereson.

BCR/ABL a chimeric oncogene, from fusion of BCR and ABL cancer genes associated with chronic myelogenous leukemia.

Bechic a remedy or treatment of cough.

Bed nucleus of the stria terminalis (BNST) act as a relay site within the hypothalamic-pituitary-adrenal axis and regulate its activity in response to acute stress.

Belching, or burping refers to the noisy release of air or gas from the stomach through the mouth.

Beri-beri is a disease caused by a deficiency of thiamine (vitamin B1) that affects many systems of the body, including the muscles, heart, nerves, and digestive system.

Beta-carotene naturally-occurring retinol (vitamin A) precursor obtained from certain fruits and vegetables with potential antineoplastic and chemopreventive activities. As an antioxidant, beta carotene inhibits free-radical damage to DNA. This agent also induces cell differentiation and apoptosis of some tumour cell types, particularly in early stages of tumorigenesis, and enhances immune system activity by stimulating the release of natural killer cells, lymphocytes, and monocytes.

Beta-catenin is a multifunctional oncogenic protein that contributes fundamentally to cell development and biology, it has been implicated as an integral component in the Wnt signaling pathway.

Beta cells a type of cell in the pancreas in areas called the islets of Langerhans.

Beta-lactamase enzymes produced by some bacteria that are responsible for their resistance to beta-lactam antibiotics like penicillins.

Beta-thalassemia an inherited blood disorder that reduces the production of hemoglobin.

BHT butylated hydroxytoluene (phenolic compound), an antioxidant used in foods, cosmetics, pharmaceuticals, and petroleum products.

Bifidobacterium is a genus of Gram-positive, non-motile, often branched anaerobic bacteria. Bifidobacteria are one of the major genera of bacteria that make up the gut flora. Bifidobacteria aid in digestion, are associated with a lower incidence of allergies and also prevent some forms of tumour growth. Some bifidobacteria are being used as probiotics.

Bifidogenic promoting the growth of (beneficial) bifidobacteria in the intestinal tract.

Bile fluid secreted by the liver and discharged into the duodenum where it is integral in the digestion and absorption of fats.

Bilharzia, bilharziosis see Schistosomiasis.

Biliary relating to the bile or the organs in which the bile is contained or transported.

Biliary infections infection of organ(s) associated with bile, comprise: (a) acute cholecystitis: an acute inflammation of the gallbladder wall; (b) cholangitis: inflammation of the bile ducts.

Biliousness old term used in the eighteenth and nineteenth centuries pertaining to bad digestion, stomach pains, constipation, and excessive flatulence.

Bilirubin a breakdown product of heme (a part of haemoglobin in red blood cells) produced by the liver that is excreted in bile which causes a yellow discoloration of the skin and eyes when it accumulates in those organs.

Biotin also known as vitamin B7. See vitamin B7.

Bitter a medicinal agent with a bitter taste and used as a tonic, alterative or appetizer.

Blackhead see comedone.

Blackwater fever dangerous complication of malarial whereby the red blood cells burst in the blood stream (haemolysis) releasing haemoglobin directly into the blood.

Blain see chilblain.

Blastocyst blastocyst is an embryonic structure formed in the early embryogenesis of mammals, after the formation of the morula, but before implantation.

Blastocystotoxic agent that suppresses further development of the blastocyst through to the ovum stage.

Blebbing Bulging e.g. membrane blebbing also called membrane bulging or ballooning.

Bleeding diathesis is an unusual susceptibility to bleeding (hemorrhage) due to a defect in the system of coagulation.

Blennorrhagia gonorrhea.

Blennorrhea inordinate discharge of mucus, especially a gonorrheal discharge from the urethra or vagina.

Blepharitis inflammation of the eyelids.

Blister thin vesicle on the skin containing serum and caused by rubbing, friction or burn.

Blood brain barrier (BBB) is a separation of circulating blood and cerebrospinal fluid (CSF) in the central nervous system (CNS). It allows essential metabolites, such as oxygen and glucose, to pass from the blood to the brain and central nervous system (CNS) but blocks most molecules that are more massive than about 500 Da.

Boil localized pyrogenic, painful infection, originating in a hair follicle.

Borborygmus rumbling noise caused by the muscular contractions of peristalsis, the process that moves the contents of the stomach and intestines downward.

Bouillon a broth in French cuisine.

Bowman Birk inhibitors type of serine proteinase inhibitor.

Bradicardia as applied to adult medicine, is defined as a resting heart rate of under 60 beats per minute.

Bradyphrenia referring to the slowness of thought common to many disorders of the brain.

Brain derived neutrophic factor (BDNF) a protein member of the neutrophin family that plays an important role in the growth, maintenance, function and survival of neurons. The protein molecule is involved in the modulation of cognitive and emotional functions and in the treatment of a variety of mental disorders.

Bright's disease chronic nephritis.

Bronchial inflammation see bronchitis.

Bronchiectasis a condition in which the airways within the lungs (bronchial tubes) become damaged and widened.

Bronchitis is an inflammation of the main air passages (bronchi) to your lungs.

Bronchoalveolar lavage (BAL) a medical procedure in which a bronchoscope is passed through the mouth or nose into the lungs and fluid is squirted into a small part of the lung and then recollected for examination.

Bronchopneumonia or bronchial pneumonia; inflammation of the lungs beginning in the terminal bronchioles.

Broncho-pulmonary relating to the bronchi and lungs.

Bronchospasm is a difficulty in breathing caused by a sudden constriction of the muscles in the walls of the bronchioles as occurs in asthma.

Brown fat brown adipose tissue (BAT) in mammals, its primary function is to generate body heat in animals or newborns that do not shiver.

Bubo inflamed, swollen lymph node in the neck or groin.

Buccal of or relating to the cheeks or the mouth cavity.

Bullae blisters; circumscribed, fluid-containing, elevated lesions of the skin, usually more than 5 mm in diameter.

Bursitis condition characterized by inflammation of one or more bursae (small sacs) of synovial fluid in the body.

C-jun NH(2)-terminal kinase enzymes that belong to the family of the MAPK superfamily of protein kinases. These kinases mediate a plethora of cellular responses to such stressful stimuli, including apoptosis and production of inflammatory and immunoregulatory cytokines in diverse cell systems. *cf:* MAPK.

c-FOS a cellular proto-oncogene belonging to the immediate early gene family of transcription factors.

C-reactive protein a protein found in the blood the levels of which rise in response to inflammation.

c-Src a cellular non-receptor tyrosine kinase.

CAAT element-binding proteins-alpha (c/EBP-alpha) regulates gene expression in adipocytes in the liver.

Cachexia physical wasting with loss of weight, muscle atrophy, fatigue, weakness caused by disease.

Caco-2 cell line a continuous line of heterogeneous human epithelial colorectal adenocarcinoma cells.

Cacogeusia a bad taste not due to ingestion of specific substances.

Cadaver a dead body, corpse.

Ca2+ ATPase (PMCA) is a transport protein in the plasma membrane of cells that serves to remove calcium (Ca2+) from the cell.

Calcium (Ca) is the most abundant mineral in the body found mainly in bones and teeth. It is required for muscle contraction, blood vessel expansion and contraction, secretion of hormones and enzymes, and transmitting impulses throughout the nervous system. Dietary sources include milk, yoghurt, cheese, Chinese cabbage, kale, broccoli, some green leafy vegetables, fortified cereals, beverages and soybean products.

Calcium ATPase is a form of P-ATPase which transfers calcium after a muscle has contracted.

Calcium channel blockers (CCBs) a class of drugs and natural substances that disrupt the calcium (Ca2+) conduction of calcium channels.

Calculi infection most calculi arise in the kidney when urine becomes supersaturated with a salt that is capable of forming solid crystals. Symptoms arise as these calculi become impacted within the ureter as they pass toward the urinary bladder.

Calculus (calculi) hardened, mineral deposits that can form a blockage in the urinary system.

Caligo dimness or obscurity of sight, dependent upon a speck on the cornea.

Calmodulin is a calcium modulated protein that can bind to and regulate a multitude of different protein targets, thereby affecting many different cellular functions.

cAMP dependent pathway cyclic adenosine monophosphate is a G protein-coupled receptor triggered signaling cascade used in cell communication in living organisms.

CAMP factor diffusible, heat-stable, extracellular protein produced by Group B *Streptococcus* that enhances the hemolysis of sheep erythrocytes by *Staphylococcus aureus.* It is named after Christie, Atkins, and Munch-Peterson, who described it in 1944.

Cancer a malignant neoplasm or tumour in nay part of the body.

Candidiasis infections caused by members of the fungus genus *Candida* that range from superficial, such as oral thrush and vaginitis, to systemic and potentially life-threatening diseases.

Canker see chancre.

Carboxypeptidase an enzyme that hydrolyzes the carboxy-terminal (C-terminal) end of a peptide bond. It is synthesized in the pancreas and secreted into the small intestine.

Carbuncle is an abscess larger than a boil, usually with one or more openings draining pus onto the skin.

Carcinogenesis production of carcinomas. *adj.* carcinogenic.

Carcinoma any malignant cancer that arises from epithelial cells.

Carcinosarcoma a rare tumour containing carcinomatous and sarcomatous components.

Cardiac relating to, situated near or affecting the heart.

Cardiac asthma acute attack of dyspnoea with wheezing resulting from a cardiac disorder.

Cardialgia heartburn.

Cardinolides cardiac glycosides with a 5-membered lactone ring in the side chain of the steroid aglycone.

Cardinolide glycoside cardenolides that contain structural groups derived from sugars.

Cardioactive having an effect on the heart.

Cardiogenic shock is characterized by a decreased pumping ability of the heart that causes a shock like state associated with an inadequate circulation of blood due to primary failure of the ventricles of the heart to function effectively.

Cardiomyocytes cardiac muscle cells.

Cardiomyopathy heart muscle disease.

Cardiopathy disease or disorder of the heart.

Cardioplegia stopping the heart so that surgical procedures can proceed in a still and bloodless field.

Cardiotonic something which strengthens, tones, or regulates heart functions without overt stimulation or depression.

Cardiovascular pertaining to the heart and blood vessels.

Caries tooth decay, commonly called cavities.

Cariogenic leading to the production of caries.

Carminative substance that stops the formation of intestinal gas and helps expel gas that has already formed, relieving flatulence: relieving flatulence or colic by expelling gas.

Carnitine palmitoyltransferase I (CPT1) also known as carnitine acyltransferase I or CAT1 is a mitochondrial enzyme, involved in converting long chain fatty acid into energy.

Carotenes are a large group of intense red and yellow pigments found in all plants ; these are hydrocarbon carotenoids (subclass of tetraterpenes) and the principal carotene is beta-carotene which is a precursor of vitamin A.

Carotenodermia yellow skin discoloration caused by excess blood carotene.

Carotenoids a class of natural fat-soluble pigments found principally in plants, belonging to a subgroup of terpenoids containing eight isoprene units forming a C40 polyene chain. Carotenoids play an important potential role in human health by acting as biological antioxidants. See also carotenes.

Carpopedal spasm spasm of the hand or foot, or of the thumbs and great toes.

Capases cysteine-aspartic acid proteases, are a family of cysteine proteases, which play essential roles in apoptosis (programmed cell death).

Catalase (CAT) enzyme in living organism that catalyses the decomposition of hydrogen peroxide to water and oxygen.

Catalepsy indefinitely prolonged maintenance of a fixed body posture; seen in severe cases of catatonic schizophrenia.

Catamenia menstruation.

Cataplasia Degenerative reversion of cells or tissue to a less differentiated form.

Cataplasm a medicated poultice or plaster. A soft moist mass, often warm and medicated, that is spread over the skin to treat an inflamed, aching or painful area, to improve the circulation.

Cataractogenesis formation of cataracts.

Catarrh, Catarrhal inflammation of the mucous membranes especially of the nose and throat.

Catechins are polyphenolic antioxidant plant metabolites. They belong to the family of flavonoids; tea is a rich source of catechins. See flavonoids.

Catecholamines hormones that are released by the adrenal glands in response to stress.

Cathartic is a substance which accelerates defecation.

Cathepsin K A cysteine protease that plays an essential role in osteoclast function in bone remodelling and resorption in diseases such as osteoporosis, osteolytic bone metastasis and rheumatoid arthritis.

Caustic having a corrosive or burning effect.

Cauterization a medical term describing the burning of the body to remove or close a part of it.

cdc2 Kinase a member of the cyclin-dependent protein kinases (CDKs).

CDKs cyclin-dependent protein kinases, a family of serine/threonine kinases that mediate many stages in mitosis.

CD 28 is one of the molecules expressed on T cells that provide co-stimulatory signals, which are required for T cell (lymphocytes) activation.

CD31 also known as PECAM-1 (Platelet Endothelial Cell Adhesion Molecule-1), a member of the immunoglobulin superfamily, that mediates cell-to-cell adhesion.

CD36 an integral membrane protein found on the surface of many cell types in vertebrate animals.

CD40 an integral membrane protein found on the surface of B lymphocytes, dendritic cells, follicular dendritic cells, hematopoietic progenitor cells, epithelial cells, and carcinomas.

CD68 a glycoprotein expressed on monocytes/macrophages which binds to low density lipoprotein.

Cecal ligation tying up the cecam.

Cell adhesion molecules (CAM) glycoproteins located on the surface of cell membranes involved with binding of other cells or with the extra-cellular matrix.

Cellular respiration is the set of the metabolic reactions and processes that take place in organisms' cells to convert biochemical energy from nutrients into adenosine triphosphate (ATP), and then release waste products. The reactions involved in respiration are catabolic reactions that involve the oxidation of one molecule and the reduction of another.

Cellulitis a bacterial infection of the skin that tends to occur in areas that have been damaged or inflamed.

Central nervous system part of the vertebrate nervous system comprising the brain and spinal cord.

Central venous catheter a catheter placed into the large vein in the neck, chest or groin.

Cephalagia pain in the head, a headache.

Cephalic relating to the head.

Ceramide oligosides oligosides with an N-acetyl-sphingosine moiety.

Cerebral embolism a blockage of blood flow through a vessel in the brain by a blood clot that formed elsewhere in the body and traveled to the brain.

Cerebral infarction is the ischemic kind of stroke due to a disturbance in the blood vessels supplying blood to the brain.

Cerebral ischemia is the localized reduction of blood flow to the brain or parts of the brain due to arterial obstruction or systematic hyperfusion.

Cerebral tonic substance that can alleviate poor concentration and memory, restlessness, uneasiness, and insomnia.

Cerebrosides are glycosphingolipids which are important components in animal muscle and nerve cell membranes.

Cerebrovascular disease is a group of brain dysfunctions related to disease of the blood vessels supplying the brain.

Cerumen ear wax, a yellowish waxy substance secreted in the ear canal of humans and other mammals.

cGMP cyclic guanosine monophosphate is a cyclic nucleotide derived from guanosine triphosphate (GTP). cGMP is a common regulator of ion channel conductance, glycogenolysis, and cellular apoptosis. It also relaxes smooth muscle tissues.

Chalcones a subgroup of flavonoids.

Chancre a painless lesion formed during the primary stage of syphilis.

Chemoembolization a procedure in which the blood supply to the tumour is blocked surgically or mechanically and anticancer drugs are administered directly into the tumour.

Chemokines are chemotactic cytokines, which stimulate migration of inflammatory cells towards tissue sites of inflammation.

Chemosensitizer a drug that makes tumour cells more sensitive to the effects of chemotherapy.

Chemosis edema of the conjunctiva of the eye.

Chickenpox is also known as varicella, is a highly contagious illness caused by primary infection with varicella zoster virus (VZV). The virus causes red, itchy bumps on the body.

Chilblains small, itchy, painful lumps that develop on the skin. They develop as an abnormal response to cold. Also called perniosis or blain.

Chlorosis iron deficiency anemia characterized by greenish yellow colour.

Cholagogue is a medicinal agent which promotes the discharge of bile from the system.

Cholecalcifereol a form of vitamin D, also called vitamin D3. See vitamin D.

Cholecyst gall bladder.

Cholecystitis inflammation of the gall bladder.

Cholecystokinin a peptide hormone that plays a key role in facilitating digestion in the small intestine.

Cholera an infectious gastroenteritis caused by enterotoxin-producing strains of the bacterium *Vibrio cholera* and characterized by severe, watery diarrhea.

Choleretic stimulation of the production of bile by the liver.

Cholestasis a condition caused by rapidly developing (acute) or long-term (chronic) interruption in the excretion of bile.

Cholesterol a soft, waxy, steroid substance found among the lipids (fats) in the bloodstream and in all our body's cells.

Cholethiasis presence of gall stones (calculi) in the gall bladder.

Choline a water soluble, organic compound, usually grouped within the Vitamin B complex. It is an essential nutrient and is needed for physiological functions such as structural integrity and signaling roles for cell membranes, cholinergic neuro-transmission (acetylcholine synthesis).

Cholinergic activated by or capable of liberating acetylcholine, especially in the parasympathetic nervous system.

Cholinergic system a system of nerve cells that uses acetylcholine in transmitting nerve impulses.

Cholinomimetic having an action similar to that of acetylcholine; called also parasympathomimetic.

Chonotropic affecting the time or rate, as the rate of contraction of the heart.

Choriocarcinoma a quick-growing malignant, trophoblastic, aggressive cancer that occurs in a woman's uterus (womb).

Chromium (Cr) is required in trace amounts in humans for sugar and lipid metabolism. Its deficiency may cause a disease called chromium deficiency. It is found in cereals, legumes, nuts and animal sources.

Chromosome long pieces of DNA found in the center (nucleus) of cells.

Chronic persisting over extended periods.

Chyle a milky bodily fluid consisting of lymph and emulsified fats, or free fatty acids.

Chylomicrons are large lipoprotein particles that transport dietary lipids from the intestines to other locations in the body. Chylomicrons are one of the five major groups of lipoproteins (chylomicrons, VLDL, IDL, LDL, HDL) that enable fats and cholesterol to move within the water-based solution of the bloodstream.

Chylorus milky (having fat emulsion).

Chyluria also called chylous urine, is a medical condition involving the presence of chyle (emulsified fat) in the urine stream, which results in urine appearing milky.

Chymase member of the family of serine proteases found primarily in mast cell.

Chymopapain an enzyme derived from papaya, used in medicine and to tenderize meat.

Cicatrizant the term used to describe a product that promotes healing through the formation of scar tissue.

Cirrhosis chronic liver disease characterized by replacement of liver tissue by fibrous scar tissue and regenerative nodules/lumps leading progressively to loss of liver function.

C-Kit Receptor a protein-tyrosine kinase receptor that is specific for stem cell factor. this interaction is crucial for the development of hematopoietic, gonadal, and pigment stem cells.

Clastogen is an agent that can cause one of two types of structural changes, breaks in chromosomes that result in the gain, loss, or rearrangements of chromosomal segments. *adj.* clastogenic.

Claudication limping, impairment in walking.

Climacterium refers to menopause and the bodily and mental changes associated with it.

Clonic seizures consist of rhythmic jerking movements of the arms and legs, sometimes on both sides of the body.

Clyster enema.

C-myc codes for a protein that binds to the DNA of other genes and is therefore a transcription factor.

CNS Depressant anything that depresses, or slows, the sympathetic impulses of the central nervous system (i.e., respiratory rate, heart rate).

Coagulopathy a defect in the body's mechanism for blood clotting, causing susceptibility to bleeding.

Cobalamin vitamin B12. See vitamin B12.

Co-carcinogen a chemical that promotes the effects of a carcinogen in the production of cancer.

Cold an acute inflammation of the mucous membrane of the respiratory tract especially of the nose and throat caused by a virus and accompanied by sneezing and coughing.

Collagen protein that is the major constituent of cartilage and other connective tissue; comprises the amino acids hydroxyproline, proline, glycine, and hydroxylysine.

Collagenases enzymes that break the peptide bonds in collagen.

Colic a broad term which refers to episodes of uncontrollable, extended crying in a baby who is otherwise healthy and well fed.

Colitis inflammatory bowel disease affecting the tissue that lines the gastrointestinal system.

Collyrium a lotion or liquid wash used as a cleanser for the eyes, particularly in diseases of the eye.

Colorectal relating to the colon or rectum.

Coma a state of unconsciousness from which a patient cannot be aroused.

Comedone a blocked, open sebaceous gland where the secretions oxidize, turning black. Also called blackhead.

Comitogen agent that is considered not to induce cell growth alone but to promote the effect of the mitogen.

Concoction a combination of crude ingredients that is prepared or cooked together.

Condyloma, Condylomata acuminata genital warts, venereal warts, anal wart or anogenital wart, a highly contagious sexually transmitted infection caused by epidermotropic human papillomavirus (HPV).

Conglutination becoming stuck together.

Conjunctival hyperemia enlarged blood vessels in the eyes.

Conjunctivitis sore, red and sticky eyes caused by eye infection.

Constipation a very common gastrointestinal disorder characterised by the passing of hard, dry bowel motions (stools) and difficulty of bowel motion.

Constitutive androstane receptor (CAR, NR1I3) is a nuclear receptor transcription factor that regulates drug metabolism and homoeostasis.

Consumption term used to describe wasting of tissues including but not limited to tuberculosis.

Consumptive afflicted with or associated with pulmonary tuberculosis.

Contraceptive an agent that reduces the likelihood of or prevents conception.

Contraindication a condition which makes a particular treatment or procedure inadvisable.

Contralateral muscle muscle of opposite limb (leg or arm).

Contralateral rotation rotation occurring or originating in a corresponding part on an opposite side.

Contusion another term for a bruise. A bruise, or contusion, is caused when blood vessels are damaged or broken as the result of a blow to the skin.

Convulsant a drug or physical disturbance that induces convulsion.

Convulsion rapid and uncontrollable shaking of the body.

Coolant that which reduces body temperature.

Copper (Cu) is essential in all plants and animals. It is found in a variety of enzymes, including the copper centers of cytochrome C oxidase and the enzyme superoxide dismutase (containing copper and zinc). In addition to its enzymatic roles, copper is used for biological electron transport. Because of its role in facilitating iron uptake, copper deficiency can often produce anemia-like symptoms. Dietary sources include curry powder, mushroom, nuts, seeds, wheat germ, whole grains and animal meat.

Copulation to engage in coitus or sexual intercourse. *adj.* copulatory.

Cordial a preparation that is stimulating to the heart.

Corn or callus is a patch of hard, thickened skin on the foot that is formed in response to pressure or friction.

Corticosteroids a class of steroid hormones that are produced in the adrenal cortex, used clinically for hormone replacement therapy, for suppressing ACTH secretion, for suppression of immune response and as antineoplastic, anti-allergic and anti-inflammatory agents.

Corticosterone a 21-carbon steroid hormone of the corticosteroid type produced in the cortex of the adrenal glands.

Cortisol is a corticosteroid hormone made by the adrenal glands.

Cornification is the process of forming an epidermal barrier in stratified squamous epithelial tissue.

Coryza a word describing the symptoms of a head cold. It describes the inflammation of the mucus membranes lining the nasal cavity which usually gives rise to the symptoms of nasal congestion and loss of smell, among other symptoms.

COX-1 see cyclooxygenase -1.

COX-2 see cyclooxygenase-2.

CpG islands genomic regions that contain a high frequency of CpG sites.

CpG sites the cytosine-phosphate-guanine nucleotide that links two nucleosides together in DNA.

cPLA(2) cytosolic phospholipases A2, these phospholipases are involved in cell signaling processes, such as inflammatory response.

CPY1B1, CPY1A1 a member of the cytochrome P450 superfamily of heme-thiolate monooxygenase enzymes.

Corticosterone a 21-carbon corticosteroid hormone produced in the cortex of the adrenal glands that functions in the metabolism of carbohydrates and proteins.

Creatin a nitrogenous organic acid that occurs naturally in vertebrates and helps to supply energy to muscle.

Creatine phosphokinase (CPK, CK) enzyme that catalyses the conversion of creatine and consumes adenosine triphosphate (ATP) to create phosphocreatine and adenosine diphosphate (ADP).

CREB cAMP response element-binding, a protein that is a transcription factor that binds to certain DNA sequences called cAMP response elements.

Crohn Disease an inflammatory disease of the intestines that affect any part of the gastrointestinal tract.

Crossover study a longitudinal, balance study in which participants receive a sequence of different treatments or exposures.

Croup is an infection of the throat (larynx) and windpipe (trachea) that is caused by a virus (Also called laryngotracheobronchitis).

Crytochidism (cryptochism) a developmental defect characterized by the failure of one or both testes to move into the scrotum as the male fetus develops.

Curettage surgical procedure in which a body cavity or tissue is scraped with a sharp instrument or aspirated with a cannula.

Cutaneous pertaining to the skin.

CXC8 also known as interleukin 8, IL-8.

Cyanogenesis generation of cyanide. *adj.* cyanogenetic.

Cyclooxygenase (COX) an enzyme that is responsible for the formation of prostanoids – prostaglandins, prostacyclins, and thromboxanes that are each involved in the inflammatory response. Two different COX enzymes existed, now known as COX-1 and COX-2.

Cyclooxygenase-1 (COX-1) is known to be present in most tissues. In the gastrointestinal tract, COX-1 maintains the normal lining of the stomach. The enzyme is also involved in kidney and platelet function.

Cyclooxygenase-2 (COX-2) is primarily present at sites of inflammation.

Cysteine proteases are enzymes that degrade polypeptides possessing a common catalytic mechanism that involves a nucleophilic cysteine thiol in a catalytic triad. They are found in fruits like papaya, pineapple, and kiwifruit.

Cystitis a common urinary tract infection that occurs when bacteria travel up the urethra, infect the urine and inflame the bladder lining.

Cystorrhea discharge of mucus from the bladder.

Cytochrome bc-1 complex ubihydroquinone: cytochrome c oxidoreductase.

Cytochrome P450 3A CYP3A a very large and diverse superfamily of heme-thiolate proteins found in all domains of life. This group of enzymes catalyzes many reactions involved in drug metabolism and synthesis of cholesterol, steroids and other lipids.

Cytokine non-antibody proteins secreted by certain cells of the immune system which carry signals locally between cells. They are a category of signaling molecules that are used extensively in cellular communication.

Cytopathic any detectable, degenerative changes in the host cell due to infection.

Cytoprotective protecting cells from noxious chemicals or other stimuli.

Cytosolic relates to the fluid of the cytoplasm in cells.

Cytostatic preventing the growth and proliferation of cells.

Cytotoxic of or relating to substances that are toxic to cells; cell-killing.

D-galactosamine an amino sugar with unique hepatotoxic properties in animals.

Dandruff scurf, dead, scaly skin among the hair.

Dartre condition of dry, scaly skin

Debility weakness, relaxation of muscular fibre.

Debridement is the process of removing non-living tissue from pressure ulcers, burns, and other wounds.

Debriding agent substance that cleans and treats certain types of wounds, burns, ulcers.

Decidual stromal cells like endometrial glands and endothelium, express integrins that bind basement components.

Deciduogenic relating to the uterus lining that is shed off at childbirth.

Decoction a medical preparation made by boiling the ingredients.

Decongestant a substance that relieves or reduces nasal or bronchial congestion.

Defibrinated plasma blood whose plasma component has had fibrinogen and fibrin removed.

Degranulation cellular process that releases antimicrobial cytotoxic molecules from secretory vesicles called granules found inside some cells.

Delayed after depolarizations (DADs) abnormal depolarization that begins during phase 4 – after repolarization is completed, but before another action potential would normally occur.

Delirium is common, sudden severe confusion and rapid changes in brain function that occur with physical or mental illness; it is reversible and temporary.

Demulcent an agent that soothes internal membranes. Also called emollient.

Dendritic cells are immune cells and form part of the mammalian immune system, functioning as antigen presenting cells.

Dentition a term that describes all of the upper and lower teeth collectively.

Deobstruent a medicine which removes obstructions; also called an aperient.

Deoxypyridinoline (Dpd) a crosslink product of collagen molecules found in bone and excreted in urine during bone degradation.

Depilatory an agent for removing or destroying hair.

Depressant a substance that diminish functional activity, usually by depressing the nervous system.

Depurative an agent used to cleanse or purify the blood, it eliminates toxins and purifies the system.

Dermatitis inflammation of the skin causing discomfort such as eczema.

Dermatophyte a fungus parasitic on the skin.

Dermatosis is a broad term that refers to any disease of the skin, especially one that is not accompanied by inflammation.

Dermonecrotic pertaining to or causing necrosis of the skin.

Desquamation the shedding of the outer layers of the skin.

Detoxifier a substance that promotes the removal of toxins from a system or organ.

Diabetes a metabolic disorder associated with inadequate secretion or utilization of insulin and characterized by frequent urination and persistent thirst. See diabetes mellitus.

Diabetes mellitus (DM) (sometimes called "sugar diabetes") is a set of chronic, metabolic disease conditions characterized by high blood sugar (glucose) levels that result from defects in insulin secretion, or action, or both. Diabetes mellitus appears in two forms.

Diabetes mellitus type I (formerly known as juvenile onset diabetes), caused by deficiency of the pancreatic hormone insulin as a result of destruction of insulin-producing beta cells of the pancreas. Lack of insulin causes an increase of fasting blood glucose that begins to appear in the urine above the renal threshold.

Diabetes mellitus type II (formerly called non-insulin-dependent diabetes mellitus or adult-onset diabetes), the disorder is characterized by high blood glucose in the context of insulin resistance and relative insulin deficiency in which insulin is available but cannot be properly utilized.

Diads two adjacent structural units in a polymer molecule.

Dialysis is a method of removing toxic substances (impurities or wastes) from the blood when the kidneys are unable to do so.

Diaphoresis is profuse sweating commonly associated with shock and other medical emergency conditions.

Diaphoretic a substance that induces perspiration. Also called sudorific.

Diaphyseal pertaining to or affecting the shaft of a long bone (diaphysis).

Diaphysis the main or mid section (shaft) of a long bone.

Diarrhoea a profuse, frequent and loose discharge from the bowels.

Diastolic referring to the time when the heart is in a period of relaxation and dilatation (expansion). *cf.* systolic.

Dieresis surgical separation of parts.

Dietary fibre is a term that refers to a group of food components that pass through the stomach and small intestine undigested and reach the large intestine virtually unchanged. Scientific evidence suggest that a diet high in dietary fibre can be of value for treating or preventing such disorders as constipation, irritable bowel syndrome, diverticular disease, hiatus hernia and haemorrhoids. Some components of dietary fibre may also be of value in reducing the level of cholesterol in blood and thereby decreasing a risk factor for coronary heart disease and the development of gallstones. Dietary fibre is beneficial in the treatment of some diabetics.

Digalactosyl diglycerides are the major lipid components of chloroplasts.

Diosgenin a steroid-like substance that is involved in the production of the hormone progesterone, extracted from roots of *Dioscorea* yam.

Dipsomania pathological use of alcohol.

Discutient an agent (as a medicinal application) which serves to disperse morbid matter.

Disinfectant an agent that prevents the spread of infection, bacteria or communicable disease.

Diuresis increased urination.

Diuretic a substance that increases urination (diuresis).

Diverticular disease is a condition affecting the large bowel or colon and is thought to be caused by eating too little fibre.

DMBA 7,12-Dimethylbenzanthracene. A polycyclic aromatic hydrocarbon found in tobacco smoke that is a potent carcinogen.

DNA deoxyribonucleic acid, a nucleic acid that contains the genetic instructions used in the development and functioning of all known living organisms.

DOCA desoxycorticosterone acetate – a steroid chemical used as replacement therapy in Addison's disease.

Dopamine a catecholamine neurotransmitter that occurs in a wide variety of animals, including both vertebrates and invertebrates.

Dopaminergic relating to, or activated by the neurotransmitter, dopamine.

Double blind refer to a clinical trial or experiment in which neither the subject nor the researcher knows which treatment any particular subject is receiving.

Douche a localised spray of liquid directed into a body cavity or onto a part.

DPPH 2,2 diphenyl -1- picryl-hydrazyl – a crystalline, stable free radical used as an inhibitor of free radical reactions.

Dracunculiasis also called guinea worm disease (GWD), is a parasitic infection caused by the nematode, *Dracunculus medinensis*.

Dropsy an old term for the swelling of soft tissues due to the accumulation of excess water. *adj.* dropsical.

Dysentery (formerly known as flux or the bloody flux) is a disorder of the digestive system that results in severe diarrhea containing mucus and blood in the feces. It is caused usually by a bacterium called *Shigella*.

Dysesthesia an unpleasant abnormal sensation produced by normal stimuli.

Dysgeusia distortion of the sense of taste.

Dyskinesia the impairment of the power of voluntary movement, resulting in fragmentary or incomplete movements. *adj.* dyskinetic.

Dyslipidemia abnormality in or abnormal amount of lipids and lipoproteins in the blood.

Dysmenorrhea is a menstrual condition characterized by severe and frequent menstrual cramps and pain associated with menstruation.

Dysmotility syndrome a vague, descriptive term used to describe diseases of the muscles of the gastrointestinal tract (esophagus, stomach, small and large intestines).

Dyspedia indigestion followed by nausea.

Dyspepsia refers to a symptom complex of epigastric pain or discomfort. It is often defined as chronic or recurrent discomfort centered in the upper abdomen and can be caused by a variety of conditions.

Dysphagia swallowing disorder.

Dysphonia a voice disorder, an impairment in the ability to produce voice sounds using the vocal organs.

Dysplasia refers to abnormality in development.

Dyspnoea shortness of breath, difficulty in breathing.

Dysrhythmias see arrhythmias.

Dystocia abnormal or difficult child birth or labour.

Dystonia a neurological movement disorder characterized by prolonged, repetitive muscle contractions that may cause twisting or jerking movements of muscles.

Dysuria refers to difficult and painful urination.

E- Selectin also known as endothelial leukocyte adhesion molecule-1 (ELAM-1), CD62E, a member of the selectin family. It is transiently expressed on vascular endothelial cells in response to IL 1 beta and TNF-alpha

EC 50 median effective concentration that produces desired effects in 50% of the test population.

Ecbolic a drug (as an ergot alkaloid) that tends to increase uterine contractions and that is used especially to facilitate delivery.

Ecchymosis skin discoloration caused by the escape of blood into the tissues from ruptured blood vessels.

ECG see electrocardiography.

EC–SOD extracellular superoxide dismutase, a tissue enzyme mainly found in the extracellular matrix of tissues. It participates in the detoxification of reactive oxygen species by catalyzing the dismutation of superoxide radicals.

Eczema is broadly applied to a range of persistent skin conditions. These include dryness and recurring skin rashes which are characterized by one or more of these symptoms: redness, skin edema, itching and dryness, crusting, flaking, blistering, cracking, oozing, or bleeding.

Eczematous rash dry, scaly, itchy rash.

ED 50 is defined as the dose producing a response that is 50% of the maximum obtainable.

Edema formerly known as dropsy or hydropsy, is characterized swelling caused by abnormal accumulation of fluid beneath the skin, or in one or more cavities of the body. It usually occurs in the feet, ankles and legs, but it can involve the entire body.

Edematogenic producing or causing edema.

EGFR proteins epidermal growth factor receptor (EGFR) proteins – Protein kinases are enzymes that transfer a phosphate group from a phosphate donor onto an acceptor amino acid in a substrate protein.

EGR-1 early growth response 1, a human gene.

Eicosanoids are signaling molecules made by oxygenation of arachidonic acid, a 20-carbon essential fatty acid, includes prostaglandins and related compounds.

Elastase a serine protease that also hydrolyses amides and esters.

Electrocardiography or ECG, is a transthoracic interpretation of the electrical activity of the heart over time captured and externally recorded by skin electrodes.

Electromyogram (EMG) a test used to record the electrical activity of muscles. An electromyogram (EMG) is also called a myogram.

Electuary a medicinal paste composed of powders, or other medical ingredients, incorporated with sweeteners to hide the taste, suitable for oral administration.

Elephantiasis a disorder characterized by chronic thickened and edematous tissue on the genitals and legs due to various causes.

Embolism Obstruction or occlusion of a blood vessel by a blood clot, air bubble or other foreign matter.

Embrocation lotion or liniment that relieves muscle or joint pains.

Embryotoxic term that describes any chemical which is harmful to an embryo.

Emesis vomiting, throwing up.

Emetic an agent that induces vomiting, *cf*: antiemetic.

Emetocathartic causing vomiting and purging.

Emmenagogue a substance that stimulates, initiates, and/or promotes menstrual flow. Emmenagogues are used in herbal medicine to balance and restore the normal function of the female reproductive system.

Emollient an agent that has a protective and soothing action on the surfaces of the skin and membranes.

Emulsion a preparation formed by the suspension of very finely divided oily or resinous liquid in another liquid.

Encephalitis inflammation of the brain.

Encephalopathy a disorder or disease of the brain.

Endocrine *adj.* of or relating to endocrine glands or the hormones secreted by them.

Endocytosis is the process by which cells absorb material (molecules such as proteins) from outside the cell by engulfing it with their cell membrane.

Endometriosis is a common and often painful disorder of the female reproductive system. The two most common symptoms of endometriosis are pain and infertility.

Endometritis refers to inflammation of the endometrium, the inner lining of the uterus.

Endometrium the inner lining of the uterus.

Endoplasmic reticulum is a network of tubules, vesicles and sacs around the nucleus that are interconnected.

Endostatin a naturally-occurring 20-kDa C-terminal protein fragment derived from type XVIII collagen. It is reported to serve as an anti-angiogenic agent that inhibits the formation of the blood vessels that feed cancer tumours.

Endosteum the thin layer of cells lining the medullary cavity of a bone.

Endosteul pertaining to the endosteum.

Endothelial progenitor cells population of rare cells that circulate in the blood with the ability to differentiate into endothelial cells, the cells that make up the lining of blood vessels.

Endothelin any of a group of vasoconstrictive peptides produced by endothelial cells.

Endotoxemia the presence of endotoxins in the blood, which may result in shock. *adj.* endotoxemic.

Endotoxin toxins associated with certain bacteria, unlike an 'exotoxin' that is not secreted in soluble form by live bacteria, but is a structural component in the bacteria which is released mainly when bacteria are lysed.

Enema liquid injected into the rectum either as a purgative or medicine, Also called clyster.

Enophthalmos a condition in which the eye falls back into the socket and inhibits proper eyelid function.

Enteral term used to describe the intestines or other parts of the digestive tract.

Enteral administration involves the esophagus, stomach, and small and large intestines (i.e., the gastrointestinal tract).

Enteritis refers to inflammation of the small intestine.

Enterocolic disorder inflamed bowel disease.

Enterocytes tall columnar cells in the small intestinal mucosa that are responsible for the final digestion and absorption of nutrients.

Enterohemorrhagic causing bloody diarrhea and colitis, said of pathogenic microorganisms.

Enterolactone a lignin formed by the action of intestinal bacteria on lignan precursors found in plants; acts as a phytoestrogen.

Enteropooling increased fluids and electrolytes within the lumen of the intestines due to increased levels of prostaglandins.

Enterotoxigenic of or being an organism containing or producing an enterotoxin.

Enterotoxin is a protein toxin released by a microorganism in the intestine.

Entheogen a substance taken to induce a spiritual experience.

Enuresis bed-wetting, a disorder of elimination that involves the voluntary or involuntary release of urine into bedding, clothing, or other inappropriate places.

Envenomation is the entry of venom into a person's body, and it may cause localised or systemic poisoning.

Eosinophilia the state of having a high concentration of eosinophils (eosinophil granulocytes) in the blood.

Eosinophils (or, less commonly, acidophils), are white blood cells that are one of the immune system components.

Epididymis a structure within the scrotum attached to the backside of the testis and whose coiled duct provides storage, transit and maturation of spermatozoa.

Epididymitis a medical condition in which there is inflammation of the epididymis.

Epigastralgia pain in the epigastric region.

Epigastric discomfort bloated abdomen, swelling of abdomen, abdominal ditension.

Epilepsy a common chronic neurological disorder that is characterized by recurrent unprovoked seizures.

Epileptiform resembling epilepsy or its manifestations. *adj.* epileptiformic.

Epileptogenesis a process by which a normal brain develops epilepsy, a chronic condition in which seizures occur. *adj.* epileptogenic.

Episiotomy a surgical incision through the perineum made to enlarge the vagina and assist childbirth.

Epithelioma a usually benign skin disease most commonly occurring on the face, around the eyelids and on the scalp.

Epitrochlearis the superficial-most muscle of the arm anterior surface.

Epistaxis acute hemorrhage from the nostril, nasal cavity, or nasopharynx (nose-bleed).

Epstein Barr Virus herpes virus that is the causative agent of infectious mononucleosis. It is also associated with various types of human cancers.

ERbeta estrogen receptor beta, a nuclear receptor which is activated by the sex hormone, estrogen.

Ergocalciferol a form of vitamin D, also called vitamin D2. See vitamin D.

Ergonic increasing capacity for bodily or mental labor especially by eliminating fatigue symptoms.

ERK (extracellular signal regulated kinases) widely expressed protein kinase intracellular signaling molecules which are involved in functions including the regulation of meiosis, mitosis, and post mitotic functions in differentiated cells.

Eructation the act of belching or of casting up wind from the stomach through the mouth.

Eruption a visible rash or cutaneous disruption.

Erysipelas is an intensely red *Streptococcus* bacterial infection that occurs on the face and lower extremities.

Erythema abnormal redness and inflammation of the skin, due to vasodilation.

Erythematous characterized by erythema.

Erythroleukoplakia an abnormal patch of red and white tissue that forms on mucous membranes in the mouth and may become cancer. Tobacco (smoking and chewing) and alcohol may increase the risk of erythroleukoplakia.

Erythropoietin (EPO) a hormone produced by the kidney that promotes the formation of red blood cells (erythrocytes) in the bone marrow.

Eschar a slough or piece of dead tissue that is cast off from the surface of the skin.

Escharotic capable of producing an eschar; a caustic or corrosive agent.

Estradiol is the predominant sex hormone present in females, also called oestradiol.

Estrogen female hormone produced by the ovaries that play an important role in the estrous cycle in women.

Estrogen receptor (ER) is a protein found in high concentrations in the cytoplasm of breast, uterus, hypothalamus, and anterior hypophysis cells; ER levels are measured to determine a breast CA's potential for response to hormonal manipulation.

Estrogen receptor negative (ER−) tumour is not driven by estrogen and need another test to determine the most effective treatment.

Estrogen receptor positive (ER+) means that estrogen is causing the tumour to grow, and that the breast cancer should respond well to hormone suppression treatments.

Estrogenic relating to estrogen or producing estrus.

Estrus sexual excitement or heat of female; or period of this characterized by changes in the sex organs.

Euglycaemia normal blood glucose concentration.

Exanthematous characterized by or of the nature of an eruption or rash.

Excitotoxicity is the pathological process by which neurons are damaged and killed by glutamate and similar substances.

Excipient a pharmacologically inert substance used as a diluent or vehicle for the active ingredients of a medication.

Exocytosis the cellular process by which cells excrete waste products or chemical transmitters.

Exophthalmos or exophthalmia or proptosis is a bulging of the eye anteriorly out of the orbit. *adj.* exophthalmic.

Exotoxin a toxin secreted by a microorganism and released into the medium in which it grows.

Expectorant an agent that increases bronchial mucous secretion by promoting liquefaction of the sticky mucous and expelling it from the body.

Exteroceptive responsiveness to stimuli that are external to an organism.

Extrapyramidal side effects are a group of symptoms (tremor, slurred speech, akathisia, dystonia, anxiety, paranoia and bradyphrenia) that can occur in persons taking antipsychotic medications.

Extravasation discharge or escape, as of blood from the vein into the surrounding tissues.

FADD Fas-associated protein with death domain, the protein encoded by this gene is an adaptor molecule which interacts with other death cell surface receptors and mediates apoptotic signals.

Familial amyloid polyneuropathy (FAP) also called Corino de Andrade's disease, a neurodegenerative autosomal dominant genetically transmitted, fatal, incurable disease.

Familial adenomatous polyposis (FAP) is an inherited condition in which numerous polyps form mainly in the epithelium of the large intestine.

Familial dysautonomia a genetic disorder that affects the development and survival of autonomic and sensory nerve cells.

FasL or CD95L Fas ligand is a type-II transmembrane protein that belongs to the tumour necrosis factor (TNF) family.

FAS: fatty acid synthase (FAS) a multi-enzyme that plays a key role in fatty acid synthesis.

Fas molecule a member of the Tumour Necrosis Factor Receptors, that mediates apoptotic signal in many cell types.

Fauces the passage leading from the back of the mouth into the pharynx.

Favus a chronic skin infection, usually of the scalp, caused by the fungus, *Trichophyton schoenleinii* and characterized by the development of thick, yellow crusts over the hair follicles. Also termed tinea favosa.

Fc epsilon RI or FcεRI is a high-affinity IgE receptor for the Fc region of immunoglobulin E (IgE), an antibody isotype in allergy and resistance to parasites.

Febrifuge an agent that reduces fever. Also called an antipyretic.

Febrile pertaining to or characterized by fever.

Fetotoxic toxic to the fetus.

Fibrates hypolipidemic agents primarily used for decreasing serum triglycerides, while increasing High density lipoprotein (HDL).

Fibril a small slender fibre or filament.

Fibrin insoluble protein that forms the essential portion of the blood clot.

Fibrinolysis a normal ongoing process that dissolves fibrin and results in the removal of small blood clots.

Fibrinolytic causing the dissolution of fibrin by enzymatic action.

Fibroblast type of cell that synthesizes the extracellular matrix and collagen, the structural framework (stroma) for animal tissues, and play a critical role in wound healing.

Fibrogenic promoting the development of fibres.

Fibromyalgia a common and complex chronic pain disorder that affects people physically, mentally and socially. Symptoms include debilitating fatigue, sleep disturbance, and joint stiffness. Also referred to as FM or FMS.

Fibrosarcoma a malignant tumour derived from fibrous connective tissue and characterized by immature proliferating fibroblasts or undifferentiated anaplastic spindle cells.

Fibrosis the formation of fibrous tissue as a reparative or reactive process.

Filarial pertaining to a thread-like nematode worm.

Filariasis a parasitic and infectious tropical disease that is caused by thread-like filarial nematode worms in the superfamily Filarioidea.

Fistula an abnormal connection between two parts inside of the body.

Fistula-in-ano a track connecting the internal anal canal to the skin surrounding the anal orifice.

5′-Nucleotidase (5′-ribonucleotide phosphohydrolase), an intrinsic membrane glycoprotein present as an ectoenzyme in a wide variety of mammalian cells, hydrolyzes 5′-nucleotides to their corresponding nucleosides.

Flatulence is the presence of a mixture of gases known as flatus in the digestive tract of mammals expelled from the rectum. Excessive flatulence can be caused by lactose intolerance, certain foods or a sudden switch to a high fibre.

Flavans a subgroup of flavonoids. See flavonoids.

Flavanols a subgroup of flavonoids, are a class of flavonoids that use the 2-phenyl-3,4-dihydro-2H-chromen-3-ol skeleton. These compounds include the catechins and the catechin gallates. They are found in chocolate, fruits and vegetables. See flavonoids.

Flavanones a subgroup of flavonoids, constitute >90% of total flavonoids in citrus. The major dietary flavanones are hesperetin, naringenin and eriodictyol.

Flavivirus A family of viruses transmitted by mosquitoes and ticks that cause some important diseases, including dengue, yellow fever, tick-borne encephalitis and West Nile fever.

Flavones a subgroup of flavonoids based on the backbone of 2-phenylchromen-4-one (2-phenyl-1-benzopyran-4-one). Flavones are mainly found in cereals and herbs.

Flavonoids (or bioflavonoids) are a group of polyphenolic antioxidant compounds in that are occur in plant as secondary metabolites. They are responsible for the colour of fruit and vegetables. Twelve basic classes (chemical types) of flavonoids have been recognized: flavones, isoflavones, flavans, flavanones, flavanols, flavanolols, anthocyanidins, catechins (including proanthocyanidins), leukoanthocyanidins, chalcones, dihydrochalcones, and aurones. Apart from their antioxidant activity, flavonoids are known for their ability to strengthen capillary walls, thus assisting circulation and helping to prevent and treat bruising, varicose veins, bleeding gums and nosebleeds, heavy menstrual bleeding and are also anti-inflammatory.

Flourine F is an essential chemical element that is required for maintenance of healthy bones and teeth and to reduce tooth decay. It is found in sea weeds, tea, water, seafood and dairy products.

Fluorosis a dental health condition caused by a child receiving too much fluoride during tooth development.

Flux an excessive discharge of fluid.

FMD (Flow Mediated Dilation) a measure of endothelial dysfunction which is used to evaluate cardiovascular risk.

Folin Ciocalteu (Total Phenolics) assay used as a measure of total phenolic compounds in natural products.

Follicle stimulating hormone (FSH) a hormone produced by the pituitary gland. In women, it helps control the menstrual cycle and the production of eggs by the ovaries.

Follicular atresia the break-down of the ovarian follicles.

Fomentation treatment by the application of war, moist substance.

Fontanelle soft spot on an infant's skull.

Framboesia see yaws.

FRAP (Ferric Reducing Antioxidant Power) assay measure the total concentration of redox-active compounds. It measure not only antioxidants that are able to scavenge free oxygen radicals but also other redox-active compounds.

Friedreich's ataxia is a genetic inherited disorder that causes progressive damage to the nervous system resulting in symptoms ranging from muscle weakness and speech problems to heart disease. *cf.* ataxia.

Fulminant hepatitis acute liver failure.

Functional food is any fresh or processed food claimed to have a health-promoting or disease-preventing property beyond the basic function of supplying nutrients. Also called medicinal food.

Furuncle is a skin disease caused by the infection of hair follicles usually caused by *Staphylococcus aureus,* resulting in the localized accumulation of pus and dead tissue.

Furunculosis skin condition characterized by persistent, recurring boils.

GABA gamma aminobutyric acid, required as an inhibitory neurotransmitter to block the transmission of an impulse from one cell to another in the central nervous system, which prevents over-firing of the nerve cells. It is used to treat both epilepsy and hypertension.

GADD 152 a pro-apoptotic gene.

Galctifuge or lactifuge, causing the arrest of milk secretion.

Galactogogue a substance that promotes the flow of milk.

Galactophoritis inflammation of the milk ducts.

Galactopoietic increasing the flow of milk; milk-producing.

Gall bladder a small, pear-shaped muscular sac, located under the right lobe of the liver, in which bile secreted by the liver is stored until needed by the body for digestion. Also called cholecyst, cholecystis.

Gallic Acid Equivalent (GAE) measures the total phenol content in terms of the standard Gallic acid by the Folin-Ciocalteau assay.

Gamma GT (GGT) Gamma-glutamyl transpeptidase, a liver enzyme.

Gastralgia (heart burn) - pain in the stomach or abdominal region. It is caused by excess of acid, or an accumulation of gas, in the stomach.

Gastric pertaining to or affecting the stomach.

Gastric emptying refers to the speed at which food and drink leave the stomach.

Gastritis inflammation of the stomach.

Gastrocnemius muscle the big calf muscle at the rear of the lower leg.

Gastrotonic (Gastroprotective) substance that strengthens, tones, or regulates gastric functions (or protects from injury) without overt stimulation or depression.

Gavage forced feeding.

Gene silencing suppression of the expression of a gene.

Genotoxic describes a poisonous substance which harms an organism by damaging its DNA thereby capable of causing mutations or cancer.

Genotoxin a chemical or other agent that damages cellular DNA, resulting in mutations or cancer.

Geriatrics is a sub-specialty of internal medicine that focuses on health care of elderly people.

Gestational hypertension development of arterial hypertension in a pregnant woman after 20 weeks gestation.

Ghrelin a gastrointestinal peptide hormone secreted by epithelial cells in the stomach lining, it stimulates appetite, gastric emptying, and increases cardiac output.

Gingival Index an index describing the clinical severity of gingival inflammation as well as its location.

Gingivitis refers to gingival inflammation induced by bacterial biofilms (also called plaque) adherent to tooth surfaces.

Gin-nan sitotoxism toxicity caused by ingestion of ginkgotoxin and characterised mainly by epileptic convulsions, paralysis of the legs and loss of consciousness.

Glaucoma a group of eye diseases in which the optic nerve at the back of the eye is slowly destroyed, leading to impaired vision and blindness.

Gleet a chronic inflammation (as gonorrhea) of a bodily orifice usually accompanied by an abnormal discharge.

Glial cells support, non-neuronal cells in the central nervous system that maintain homeostasis, form myelin and provide protection for the brain's neurons.

Glioma is a type of tumour that starts in the brain or spine. It is called a glioma because it arises from glial cells.

Glioblastoma multiforme most common and most aggressive type of primary brain tumour in humans, involving glial cells.

Glomerulonephritis (GN) a renal disease characterized by inflammation of the glomeruli, or small blood vessels in the kidneys. Also known as glomerular nephritis. *adj.* glomerulonephritic.

Glomerulosclerosis a hardening of the glomerulus in the kidney.

Glossal pertaining to the tongue.

GLP-1 glucagon-like peptide-1 is derived from the transcription product of the proglucagon gene, associate with type 2-diabetes therapy.

Gluconeogenesis a metabolic pathway that results in the generation of glucose from non-carbohydrate carbon substrates such as lactate. *adj.* gluconeogenic.

Glucose transporters (GLUT or SLC2A family) are a family of membrane proteins found in most mammalian cells.

Glucosyltranferase an enzyme that enable the transfer of glucose.

Glucuronidation a phase II detoxification pathway occurring in the liver in which glucuronic acid is conjugated with toxins.

Glutamic Oxaloacetate Transaminase (GOT) catalyzes the transfer of an amino group from an amino acid (Glu) to a 2-keto-acid to generate a new amino acid and the residual 2-keto-acid of the donor amino acid.

Glutamic pyruvate transaminase (GPT) see Alanine aminotransferase.

Glutathione (GSH) a tripeptide produced in the human liver and plays a key role in intermediary metabolism, immune response and health. It plays an important role in scavenging free radicals and protects cells against several toxic oxygen-derived chemical species.

Glutathione peroxidase (GPX) the general name of an enzyme family with peroxidase activity whose main biological role is to protect the organism from oxidative damage.

Glutathione S-transferase (GST) a major group of detoxification enzymes that participate in the detoxification of reactive electrophilic compounds by catalysing their conjugation to glutathione.

Glycaemic index (GI) measures carbohydrates according to how quickly they are absorbed and raise the glucose level of the blood.

Glycaemic load (GL) is a ranking system for carbohydrate content in food portions based on their glycaemic index and the amount of available carbohydrate, i.e. GI x available carbohydrate divided by 100. Glycemic load combines both the quality and quantity of carbohydrate in one 'number'. It's the best way to predict blood glucose values of different types and amounts of food.

Glycation or glycosylation a chemical reaction in which glycosyl groups are added to a protein to produce a glycoprotein.

Glycogenolysis is the catabolism of glycogen by removal of a glucose monomer through cleavage with inorganic phosphate to produce glucose-1-phosphate.

Glycometabolism metabolism (oxidation) of glucose to produce energy.

Glycosuria or glucosuria is an abnormal condition of osmotic diuresis due to excretion of glucose by the kidneys into the urine.

Glycosylases a family of enzymes involved in base excision repair.

Goitre an enlargement of the thyroid gland leading to swelling of the neck or larynx.

Goitrogen substance that suppresses the function of the thyroid gland by interfering with iodine uptake, causing enlargement of the thyroid, i.e. goiter.

Goitrogenic *adj.* causing goiter.

Gonadotroph a basophilic cell of the anterior pituitary specialized to secrete follicle-stimulating hormone or luteinizing hormone.

Gonatropins protein hormones secreted by gonadotrope cells of the pituitary gland of vertebrates.

Gonorrhoea a common sexually transmitted bacterial infection caused by the bacterium *Neisseria gonorrhoeae*.

Gout a disorder caused by a build-up of a waste product, uric acid, in the bloodstream. Excess uric acid settles in joints causing inflammation, pain and swelling.

G-protein-coupled receptors (GPCRs) comprise a large and diverse family of proteins whose primary function is to transduce extracellular stimuli into cells.

Granulation the condition or appearance of being granulated (becoming grain-like).

Gravel sand-like concretions of uric acid, calcium oxalate, and mineral salts formed in the passages of the biliary and urinary tracts.

Gripe water is a home remedy for babies with colic, gas, teething pain or other stomach ailments. Its ingredients vary, and may include alcohol, bicarbonate, ginger, dill, fennel and chamomile.

Grippe an epidemic catarrh; older term for influenza.

GSH see Glutathione.

GSH-Px Glutathione peroxidase, general name of an enzyme family with peroxidase activity whose main biological role is to protect the organism from oxidative damage.

GSSG glutathione disulfides are biologically important intracellular thiols, and alterations in the GSH/GSSG ratio are often used to assess exposure of cells to oxidative stress.

GSTM glutathione S transferase M1, a major group of detoxification enzymes.

GSTM 2 glutathione S transferase M2, a major group of detoxification enzymes.

G2-M cell cycle the phase where the cell prepare for mitosis and where chromatids and daughter cells separate.

Gynecopathy any or various diseases specific to women.

Gynoid adiposity fat distribution mainly to the hips and thighs, pear shaped.

Haemagogic promoting a flow of blood.

Haematemesis, Hematemesis is the vomiting of blood.

Haematinic improving the quality of the blood, its haemoglobin level and the number of erythrocytes.

Haematochezia passage of stools containing blood.

Haematochyluria, hematochyluria the discharge of blood and chyle (emulsified fat) in the urine, see also chyluria.

Haematoma, hematoma a localized accumulation of blood in a tissue or space composed of clotted blood.

Haematometra, hematometra a medical condition involving bleeding of or near the uterus.

Haematopoiesis, hematopoiesis formation of blood cellular components from the haematopoietic stem cells.

Haematopoietic *adj.* relating to the formation and development of blood cells.

Haematuria, Hematuria is the presence of blood in the urine. Hematuria is a sign that something is causing abnormal bleeding in a person's genitourinary tract.

Haeme oxygenase (HO-1, encoded by Hmox1) is an inducible protein activated in systemic inflammatory conditions by oxidant stress, an enzyme that catalyzes degradation of heme.

Haemochromatosis is a condition in which the body takes in too much iron.

Haemodialysis, Hemodialysis a method for removing waste products such as potassium and urea, as well as free water from the blood when the kidneys are in renal failure.

Haemolyis lysis of red blood cells and the release of haemoglobin into the surrounding fluid (plasma). *adj.* haemolytic.

Haemoptysis, hemoptysis is the coughing up of blood from the respiratory tract. The blood can come from the nose, mouth, throat, and the airway passages leading to the lungs.

Haemorrhage, hemaorrhage bleeding, discharge of blood from blood vessels.

Haemorrhoids, Hemorrhoids a painful condition in which the veins around the anus or lower rectum are enlarged, swollen and inflamed. Also called piles.

Haemostasis, hemostasis a complex process which causes the bleeding process to stop.

Haemostatic, hemostatic something that stops bleeding.

Halitosis (bad breath) a common condition caused by sulfur-producing bacteria that live within the surface of the tongue and in the throat.

Hallucinogen drug that produces hallucinogen.

Hallucinogenic inducing hallucinations.

Haplotype a set of alleles of closely linked loci on a chromosome that tend to be inherited together.

Hapten a small molecule that can elicit an immune response only when attached to a large carrier such as a protein.

HBeAg hepatitis B e antigen.

HBsAg hepatitis B s antigen.

Heartburn burning sensation in the stomach and esophagus caused by excessive acidity of the stomach fluids.

Heat rash any condition aggravated by heat or hot weather such as intertrigo.

Heat Shock Chaperones (HSC) ubiquitous molecules involved in the modulation of protein conformational and complexation states, associated with heat stress or other cellular stress response.

Heat Shock Proteins (HSP) a group of functionally related proteins the expression of which is increased when the cells are exposed to elevated temperatures or other cellular stresses.

Helminthiasis a disease in which a part of the body is infested with worms such as pinworm, roundworm or tapeworm.

Hemagglutination a specific form of agglutination that involves red blood cells.

Hemagglutination–inhibition test measures of the ability of soluble antigen to inhibit the agglutination of antigen-coated red blood cells by antibodies.

Hemagglutinin refers to a substance that causes red blood cells to agglutinate.

Hemangioma blood vessel.

Hematocrit is a blood test that measures the percentage of the volume of whole blood that is made up of red blood cells.

Hematopoietic pertaining to the formation of blood or blood cells.

Hematopoietic stem cell is a cell isolated from the blood or bone marrow that can renew itself, and can differentiate to a variety of specialized cells.

Heme oxygenase-1 (HO-1) an enzyme that catalyses the degradation of heme; an inducible stress protein, confers cytoprotection against oxidative stress in-vitro and in-vivo.

Hemoglobinopathies genetic defects that produce abnormal hemoglobins and anemia.

Hemolytic anemia anemia due to hemolysis, the breakdown of red blood cells in the blood vessels or elsewhere in the body.

Hemorheology study of blood flow and its elements in the circulatory system. *adj.* hemorheological.

Hemorrhagic colitis an acute gasteroenteritis characterized by overtly bloody diarrhea that is caused by *Escherichia coli* infection.

Hemolytic-uremic syndrome is a disease characterized by hemolytic anemia, acute renal failure (uremia) and a low platelet count.

Hepa-1c1c7 a type of hepatoma cells.

Hepatalgia pain or discomfort in the liver area.

Hepatectomy the surgical removal of part or all of the liver.

Hepatic relating to the liver.

Hepatic cirrhosis affecting the liver, characterize by hepatic fibrosis and regenerative nodules.

Hepatitis inflammation of the liver.

Hepatitis A (formerly known as infectious hepatitis) is an acute infectious disease of the liver caused by the hepatovirus hepatitis A virus.

Hepatocarcinogenesis represents a linear and progressive cancerous process in the liver in which successively more aberrant monoclonal populations of hepatocytes evolve.

Hepatocellular carcinoma (HCC) also called malignant hepatoma, is a primary malignancy (cancer) of the liver.

Hepatocytolysis cytotoxicity (dissolution) of liver cells.

Hepatoma cancer of the liver.

Hepatopathy a disease or disorder of the liver.

Hepatoprotective (liver protector) a substance that helps protect the liver from damage by toxins, chemicals or other disease processes.

Hepatoregenerative a compound that promotes hepatocellular regeneration, repairs and restores liver function to optimum performance.

Hepatotonic (liver tonic) a substance that is tonic to the liver - usually employed to normalize liver enzymes and function.

Hernia occurs when part of an internal organ bulges through a weak area of muscle.

HER- 2 human epidermal growth factor receptor 2, a protein giving higher aggressiveness in breast cancer, also known as ErbB-2, ERBB2.

Herpes a chronic inflammation of the skin or mucous membrane characterized by the development of vesicles on an inflammatory base.

Herpes simplex virus 1 and 2 – (HSV-1 and HSV-2) are two species of the herpes virus family which cause a variety of illnesses/infections in humans such cold sores, chickenpox or varicella, shingles or herpes zoster (VZV), cytomegalovirus (CMV), and various cancers, and can cause brain inflammation (encephalitis). HSV-1 is commonly associated with herpes outbreaks of the face known as cold sores or fever blisters, whereas HSV-2 is more often associated with genital herpes. They are also called Human Herpes Virus 1 and 2 (HHV-1 and HHV-2) and are neurotropic and neuroinvasive viruses; they enter and hide in the human nervous system, accounting for their durability in the human body.

Herpes zoster or simply zoster, commonly known as shingles and also known as zona, is a viral disease characterized by a painful skin rash with blisters.

Heterophobia term used to describe irrational fear of, aversion to, or discrimination against heterosexuals.

Heterophoria An eye condition where the motion of the eyes are not parallel to each other.

HDL-C (HDL Cholesterol) high density lipoprotein-cholesterol, also called "good cholesterol". See also high-density lipoprotein.

Hiatus hernia occurs when the upper part of the stomach pushes its way through a tear in the diaphragm.

High-density lipoprotein (HDL) is one of the five major groups of lipoproteins which enable cholesterol and triglycerides to be transported within the water based blood stream. HDL can remove cholesterol from atheroma within arteries and transport it back to the liver for excretion or re-utilization—which is the main reason why HDL-bound cholesterol is sometimes called "good cholesterol", or HDL-C. A high level of HDL-C seems to protect against cardiovascular diseases. cf. LDL.

HGPRT, HPRT (hypoxanthine-guanine phosphoribosyl transferase) an enzyme that catalyzes the conversion of 5-phosphoribosyl-1-pyrophosphate and hypoxanthine, guanine, or 6-mercaptopurine to the corresponding

5′-mononucleotides and pyrophosphate. The enzyme is important in purine biosynthesis as well as central nervous system functions.

Hippocampus a ridge in the floor of each lateral ventricle of the brain that consists mainly of gray matter.

Hippocampal pertaining to the hippocampus.

Histaminergic liberated or activated by histamine, relating to the effects of histamine at histamine receptors of target tissues.

Histaminergic receptors are types of G-protein coupled receptors with histamine as their endogenous ligand.

HIV see Human immunodeficiency virus.

Hives (urticaria) is a skin rash characterised by circular wheals of reddened and itching skin.

HMG-CoAr 3-hydroxy-3-methyl-glutaryl-CoA reductase or (HMGCR) is the rate-controlling enzyme (EC 1.1.1.88) of the mevalonate pathway.

HMG-CoA 3-hydroxy-3-methylglutaryl-coenzyme A, an intermediate in the mevalonate pathway.

Hodgkin's disease disease characterized by enlargement of the lymph glands, spleen and anemia.

Homeodomain transcription factor a protein domain encoded by a homeobox. Homeobox genes encode transcription factors which typically switch on cascades of other genes.

Homeostasis the maintenance of a constant internal environment of a cell or an organism, despite fluctuations in the external.

Homeotherapy treatment or prevention of disease with a substance similar but not identical to the causative agent of the disease.

Homocysteine an amino acid in the blood.

Homograft see allograft.

Hormonal (female) substance that has a hormone-like effect similar to that of estrogen and/or a substance used to normalize female hormone levels.

Hormonal (male) substance that has a hormone-like effect similar to that of testosterone and/or a substance used to normalize male hormone levels.

HRT hormone replacement therapy, the administration of the female hormones, oestrogen and progesterone, and sometimes testosterone.

HSP27 is an ATP-independent, 27 kDa heat shock protein chaperone that confers protection against apoptosis.

HSP90 a 90 kDa heat shock protein chaperone that has the ability to regulate a specific subset of cellular signaling proteins that have been implicated in disease processes.

hTERT – **(TERT) t**elomerase reverse transcriptase is a catalytic subunit of the enzyme telomerase in humans. It exerts a novel protective function by binding to mitochondrial DNA, increasing respiratory chain activity and protecting against oxidative stress–induced damage.

HT29 cells are human intestinal epithelial cells which produce the secretory component of Immunoglobulin A (IgA), and carcinoembryonic antigen (CEA).

Human cytomegalovirus (HCMV) a DNA herpes virus which is the leading cause of congenital viral infection and mental retardation.

Human factor X a coagulation factor also known by the eponym Stuart-Prower factor or as thrombokinase, is an enzyme involved in blood coagulation. It synthesized in the liver and requires vitamin K for its synthesis.

Human immunodeficiency virus (HIV) a retrovirus that can lead to acquired immunodeficiency syndrome (AIDS), a condition in humans in which the immune system begins to fail, leading to life-threatening opportunistic infections.

Humoral immune response (HIR) is the aspect of immunity that is mediated by secreted antibodies (as opposed to cell-mediated immunity, which involves T lymphocytes) produced in the cells of the B lymphocyte lineage (B cell).

HUVEC human umbilical vein endothelial cells.

Hyaluronidase enzymes that catalyse the hydrolysis of certain complex carbohydrates like hyaluronic acid and chondroitin sulfates.

Hydatidiform a rare mass or growth that forms inside the uterus at the beginning of a pregnancy.

Hydrocholagogue see cholagogue.
Hydrocholeretic an agent that stimulates an increased output of bile of low specific gravity.
Hydrogogue a purgative that causes an abundant watery discharge from the bowel.
Hydronephrosis is distension and dilation of the renal pelvis and calyces, usually caused by obstruction of the free flow of urine from the kidney.
Hydrophobia a viral neuroinvasive disease that causes acute encephalitis (inflammation of the brain) in warm-blooded animals. Also called rabies.
Hydropsy see dropsy.
Hyperaemia the increase of blood flow to different tissues in the body.
Hyperalgesia an increased sensitivity to pain (enhanced pricking pain), which may be caused by damage to nociceptors or peripheral nerves.
Hyperammonemia, hyperammonaemia a metabolic disturbance characterised by an excess of ammonia in the blood.
Hypercholesterolemia high levels of cholesterol in the blood that increase a person's risk for cardiovascular disease leading to stroke or heart attack.
Hyperemia is the increased blood flow that occurs when tissue is active.
Hyperemesis severe and persistent nausea and vomiting (morning sickness) during pregnancy.
Hyperglycaemia high blood sugar; is a condition in which an excessive amount of glucose circulates in the blood plasma.
Hyperglycemic a substance that raises blood sugar levels.
Hyperhomocysteinemia is a medical condition characterized by an abnormally large level of homocysteine in the blood.
Hyperinsulinemia a condition in which there are excess levels of circulating insulin in the blood; also known as pre-diabetes.
Hyperkalemia is an elevated blood level of the electrolyte potassium.
Hyperknesis enhanced itch to pricking.
Hyperleptinemia increased serum leptin level.
Hypermethylation an increase in the inherited methylation of cytosine and adenosine residues in DNA.
Hyperpiesia persistent and pathological high blood pressure for which no specific cause can be found.
Hyperplasia increased cell production in a normal tissue or organ.
Hyperprolactinaemia the presence of abnormally high levels of prolactin in the blood.
Hyperpropulsion using water pressure as a force to move objects; used to dislodge calculi in the urethra.
Hyperpyrexia is an abnormally high fever.
Hypertension commonly referred to as "high blood pressure" or HTN, is a medical condition in which the arterial blood pressure is chronically elevated.
Hypertensive characterized or caused by increased tension or pressure as abnormally high blood pressure.
Hypertriglyceridaemia or hypertriglycemia a disorder that causes high triglycerides in the blood.
Hypertrophy enlargement or overgrowth of an organ.
Hyperuricemia is a condition characterized by abnormally high level of uric acid in the blood.
Hypoadiponectinemia low plasma adiponectin concentrations associated with obesity and type 2 diabetes; that is closely related to the degree of insulin resistance and hyperinsulinemia than to the degree of adiposity and glucose tolerance.
Hypoalbuminemia a medical condition where levels of albumin in blood serum are abnormally low.
Hypocalcemic tetany a disease caused by an abnormally low level of calcium in the blood and characterized by hyperexcitability of the neuromuscular system and results in carpopedal spasms.
Hypochlorhydria refer to states where the production of gastric acid in the stomach is absent or low.
Hypocholesterolemic (cholesterol-reducer), a substance that lowers blood cholesterol levels.
Hypocorticism see Addison's disease.
Hypocortisolism see Addison's disease.

Hypoglycemic an agent that lowers the concentration of glucose (sugar) in the blood.

Hypoperfusion decreased blood flow through an organ, characterized by an imbalance of oxygen demand and oxygen delivery to tissues.

Hypophagic under-eating.

Hypospadias an abnormal birth defect in males in which the urethra opens on the under surface of the penis.

Hypotensive characterised by or causing diminished tension or pressure, as abnormally low blood pressure.

Hypothermia a condition in which an organism's temperature drops below that required for normal metabolism and body functions.

Hypothermic relating to hypothermia, with subnormal body temperature.

Hypoxaemia is the reduction of oxygen specifically in the blood.

Hypoxia a shortage of oxygen in the body. *adj.* hypoxic.

ICAM-1 (Inter-Cellular Adhesion Molecule 1) also known as CD54 (Cluster of Differentiation 54), is a protein that in humans is encoded by the ICAM1 gene.

IC 50 the median maximal inhibitory concentration; a measure of the effectiveness of a compound in inhibiting biological or biochemical function.

I.C.V. (intra-cerebroventricular) injection of chemical into the right lateral ventricle of the brain.

Iceterus jaundice, yellowish pigmentation of the skin.

Ichthyotoxic a substance which is poisonous to fish.

Icteric hepatitis an infectious syndrome of hepatitis characterized by jaundice, nausea, fever, right-upper quadrant pain, enlarged liver and transaminitis (increase in alanine aminotransferase (ALT) and/or aspartate aminotransferase (AST)).

Icterus neonatorum jaundice in newborn infants.

Idiopathic of no apparent physical cause.

Idiopathic sudden sensorineural hearing loss (ISSHL) is sudden hearing loss where clinical assessment fails to reveal a cause.

IgE Immunoglobin E – a class of antibody that plays a role in allergy.

IGFs insulin-like growth factors, polypeptides with high sequence similarity to insulin.

IgG Immunoglobin G – the most abundant immunoglobin (antibody) and is one of the major activators of the complement pathway.

IgM Immunoglobin M – primary antibody against A and B antigens on red blood cells.

IKAP is a scaffold protein of the IvarKappaBeta kinase complex and a regulator for kinases involved in pro-inflammatory cytokine signaling.

IKappa B or IkB-beta, a protein of the NF-Kappa-B inhibitor family.

Ileus a temporary disruption of intestinal peristalsis due to non-mechanical causes.

Immune modulator a substance that affects or modulates the functioning of the immune system.

Immunodeficiency a state in which the immune system's ability to fight infectious disease is compromised or entirely absent.

Immunogenicity the property enabling a substance to provoke an immune response.

Immunomodulatory capable of modifying or regulating one or more immune functions.

Immunoreactive reacting to particular antigens or haptens.

Immunostimulant agent that stimulates an immune response.

Immunosuppression involves a process that reduces the activation or efficacy of the immune system.

Immunotoxin a man-made protein that consists of a targeting portion linked to a toxin.

Impetigo a contagious, bacterial skin infection characterized by blisters that may itch, caused by a *Streptoccocus* bacterium or *Staphylococcus aureus* and mostly seen in children.

Impotence a sexual dysfunction characterized by the inability to develop or maintain an erection of the penis.

Incontinence (fecal) the inability to control bowel's movement.

Incontinence (Urine) the inability to control urine excretion.

Index of structural atypia (ISA) index of structural abnormality.

Induration hardened, as a soft tissue that becomes extremely firm.

Infarct an area of living tissue that undergoes necrosis as a result of obstruction of local blood supply.

Infarction is the process of tissue death (necrosis) caused by blockage of the tissue's blood supply.

Inflammation a protective response of the body to infection, irritation or other injury, aimed at destroying or isolating the injuries and characterized by redness, pain, warmth and swelling.

Influenza a viral infection that affects mainly the nose, throat, bronchi and occasionally, lungs.

Infusion a liquid extract obtained by steeping something (e.g. herbs) that are more volatile or dissolve readily in water, to release their active ingredients without boiling.

Inguinal hernia a hernia into the inguinal canal of the groin.

Inhalant a medicinal substance that is administered as a vapor into the upper respiratory passages.

iNOS, inducible nitric oxide synthases through its product, nitric oxide (NO), may contribute to the induction of germ cell apoptosis. It plays a crucial role in early sepsis-related microcirculatory dysfunction.

Inotropic affecting the force of muscle contraction.

Insecticide an agent that destroys insects. *adj.* insecticidal.

Insomnia a sleeping disorder characterized by the inability to fall asleep and/or the inability to remain asleep for a reasonable amount of time.

Insulin a peptide hormone composed of 51 amino acids produced in the islets of Langerhans in the pancreas causes cells in the liver, muscle, and fat tissue to take up glucose from the blood, storing it as glycogen in the liver and muscle. Insulin deficiency is often the cause of diabetes and exogenous insulin is used to control diabetes.

Insulin-like growth factors (IGFs) polypeptides with high sequence similarity to insulin. They are part of a complex system that cells employ to communicate with their physiologic environment.

Insulin-mimetic to act like insulin.

Insulinogenic associated with or stimulating the production of insulin.

Insulinotropic changing the action of insulin.

Integrase an enzyme produced by a retrovirus (such as HIV) that enables its genetic material to be integrated into the DNA of the infected cell.

Interferons (IFNs) are natural cell-signaling glycoproteins known as cytokines produced by the cells of the immune system of most vertebrates in response to challenges such as viruses, parasites and tumour cells.

Interleukins a group of naturally occurring proteins and is a subset of a larger group of cellular messenger molecules called cytokines, which are modulators of cellular behavior.

Interleukin-1 (IL-1) a cytokine that could induce fever, control lymphocytes, increase the number of bone marrow cells and cause degeneration of bone joints. Also called endogenous pyrogen, lymphocyte activating factor, haemopoetin-1 and mononuclear cell factor, amongst others that IL-1 is composed of two distinct proteins, now called IL-1α and IL-1β.

Interleukin 1 Beta (IL-1β) a cytokine protein produced by activated macrophages. cytokine is an important mediator of the inflammatory response, and is involved in a variety of cellular activities, including cell proliferation, differentiation, and apoptosis.

Interleukin 2 (IL-2) a type of cytokine immune system signaling molecule that is instrumental in the body's natural response to microbial infection.

Interleukin-2 receptor (IL-2R) a heterotrimeric protein expressed on the surface of certain immune cells, such as lymphocytes, that binds and responds to a cytokine called IL-2.

Interleukin-6 (IL-6) an interleukin that acts as both a pro-inflammatory and anti-inflammatory cytokine.

Interleukin 8 (IL-8) a cytokine produced by macrophages and other cell types such as epithelial cells and is one of the major mediators of the inflammatory response.

Intermediate-density lipoproteins (IDL) is one of the five major groups of lipoproteins (chylomicrons, VLDL, IDL, LDL, and HDL) that enable fats and cholesterol to move within the water-based solution of the bloodstream. IDL is further degraded to form LDL particles and, like LDL, can also promote the growth of atheroma and increase cardiovascular diseases.

Intermittent claudication an aching, crampy, tired, and sometimes burning pain in the legs that comes and goes, caused by peripheral vascular disease. I t usually occurs with walking and disappears after rest.

Interoceptive relating to stimuli arising from within the body.

Interstitium the space between cells in a tissue.

Interstitial pertaining to the interstitium.

Intertrigo an inflammation (rash) caused by microbial infection in skin folds.

Intima innermost layer of an artery or vein.

Intoxicant substance that produce drunkenness or intoxication.

Intraperitoneal (i.p.) the term used when a chemical is contained within or administered through the peritoneum (the thin, transparent membrane that lines the walls of the abdomen).

Intrathecal (i.t.) through the theca of the spinal cord into the subarachnoid space.

Intromission the act of putting one thing into another.

Intubation refers to the placement of a tube into an external or internal orifice of the body.

Iodine (I) is an essential chemical element that is important for hormone development in the human body. Lack of iodine can lead to an enlarged thyroid gland (goitre) or other iodine deficiency disorders including mental retardation and stunted growth in babies and children. Iodine is found in dairy products, seafood, kelp, seaweeds, eggs, some vegetables and iodized salt.

IP see Intraperitoneal.

Iron (Fe) is essential to most life forms and to normal human physiology. In humans, iron is an essential component of proteins involved in oxygen transport and for haemoglobin. It is also essential for the regulation of cell growth and differentiation. A deficiency of iron limits oxygen delivery to cells, resulting in fatigue, poor work performance, and decreased immunity. Conversely, excess amounts of iron can result in toxicity and even death. Dietary sources include, certain cereals, dark green leafy vegetables, dried fruit, legumes, seafood, poultry and meat.

Ischemia an insufficient supply of blood to an organ, usually due to a blocked artery.

Ischuria retention or suppression of urine.

Isoflavones a subgroup of flavonoids in which the basic structure is a 3-phenyl chromane skeleton. They act as phytoestrogens in mammals. See flavonoids.

Isomers substances that are composed of the same elements in the same proportions and hence have the same molecular formula but differ in properties because of differences in the arrangement of atoms.

Isoprostanes unique prostaglandin-like compounds generated in vivo from the free radical-catalysed peroxidation of essential fatty acids.

Jamu traditional Indonesian herbal medicine.

Jaundice refers to the yellow color of the skin and whites of the eyes caused by excess bilirubin in the blood.

JNK (Jun N-terminal Kinase), also known as Stress Activated Protein Kinase (SAPK), belongs to the family of MAP kinases.

Jurkat cells a line of T lymphocyte cells that are used to study acute T cell leukemia.

KB cell a cell line derived from a human carcinoma of the nasopharynx, used as an assay for antineoplastic (anti-tumour) agents.

Kallikreins peptidases (enzymes that cleave peptide bonds in proteins), a subgroup of the serine protease family; they liberate kinins from kininogens. Kallikreins are targets of active investigation by drug researchers as possible biomarkers for cancer.

Kaposi sarcoma a cancerous tumour of the connective tissues caused by the huma herpesvirus 8 and is often associated with AIDS.

Kaposi sarcoma herpes virus (KSHV) also known as human herpesvirus-8, is a gamma 2 herpesvirus or rhadinovirus. It plays an important role in the pathogenesis of Kaposi sarcoma (KS), multicentric Castleman disease (MCD) of the plasma cell type, and primary effusion lymphoma and occurs in HIV patients.

Karyolysis dissolution and disintegration of the nucles when a cell dies.

Karyorrhexis destructive fragmentation of the nucleus of a dying cell whereby its chromatin disintegrates into formless granules.

Keratin a sulphur-containing protein which is a major component in skin, hair, nails, hooves, horns, and teeth.

Keratinocyte is the major constituent of the epidermis, constituting 95% of the cells found there.

Keratinophilic having an affinity for keratin.

Keratitis inflammation of the cornea.

Keratomalacia an eye disorder that leads to a dry cornea.

Kidney stones (calculi) are hardened mineral deposits that form in the kidney.

Kinin is any of various structurally related polypeptides, such as bradykinin, that act locally to induce vasodilation and contraction of smooth muscle.

Kininogen either of two plasma α2-globulins that are kinin precursors.

Knockout gene knockout is a genetic technique in which an organism is engineered to carry genes that have been made inoperative.

Kunitz protease inhibitors a type of protein contained in legume seeds which functions as a protease inhibitor.

Kupffer cells are resident macrophages of the liver and play an important role in its normal physiology and homeostasis as well as participating in the acute and chronic responses of the liver to toxic compounds.

L-Dopa (L-3,4-dihydroxyphenylalanine) is an amino acid that is formed in the liver and converted into dopamine in the brain.

Labour process of childbirth involving muscular contractions.

Lacrimation secretion and discharge of tears.

Lactagogue an agent that increases or stimulates milk flow or production. Also called a galactagogue.

Lactate dehydrogenase (LDH) enzyme that catalyzes the conversion of lactate to pyruvate.

Lactation secretion and production of milk.

Lactic acidosis is a condition caused by the buildup of lactic acid in the body. It leads to acidification of the blood (acidosis), and is considered a distinct form of metabolic acidosis.

LAK cell a lymphokine-activated killer cell i.e. a white blood cell that has been stimulated to kill tumour cells.

Laminin a glycoprotein component of connective tissue basement membrane that promotes cell adhesion.

Laparotomy a surgical procedure involving an incision through the abdominal wall to gain access into the abdominal cavity. *adj.* laparotomized.

Larvacidal an agent which kills insect or parasite larva.

Laryngitis is an inflammation of the larynx.

Laxation bowel movement.

Laxatives substances that are used to promote bowel movement.

LC 50 median lethal concentration, see LD 50.

LD 50 median lethal dose – the dose required to kill half the members of a tested population. Also called LC 50 (median lethal concentration).

LDL see low-density lipoprotein.

LDL Cholesterol see low-density lipoprotein.

LDL receptor (LDLr) a low-density lipoprotein receptor gene.

Lectins are sugar-binding proteins that are highly specific for their sugar moieties, that agglutinate cells and/or precipitate glycoconjugates. They play a role in biological recognition phenomena involving cells and proteins.

Leishmaniasis a disease caused by protozoan parasites that belong to the genus *Leishmania* and is transmitted by the bite of certain species of sand fly.

Lenticular opacity also known as or related to cataract.

Leprosy a chronic bacterial disease of the skin and nerves in the hands and feet and, in some cases, the lining of the nose. It is caused by the *Mycobacterium leprae*. Also called Hansen's disease.

Leptin is a 16 kDa protein hormone with important effects in regulating body weight, metabolism and reproductive function.

Lequesne Algofunctional Index is a widespread international instrument (10 questions survey) and recommended by the World Health Organization (WHO) for outcome measurement in hip and knee diseases such as osteoarthritis.

Leucocyte white blood corpuscles, colourless, without haemoglobin that help to combat infection.

Leucoderma a skin abnormality characterized by white spots, bands and patches on the skin; they can also be caused by fungus and tinea. Also see vitiligo.

Leucorrhoea commonly known as whites, refers to a whitish discharge from the female genitals.

Leukemia, leukaemia a cancer of the blood or bone marrow and is characterized by an abnormal proliferation (production by multiplication) of blood cells, usually white blood cells (leukocytes).

Leukemogenic relating to leukemia, causing leukemia.

Leukocytopenia abnormal decrease in the number of leukocytes (white blood cells) in the blood.

Leukomyelopathy any diseases involving the white matter of the spinal cord.

Leukopenia a decrease in the number of circulating white blood cells.

Leukoplakia condition characterized by white spots or patches on mucous membranes, especially of the mouth and vulva.

Leukotriene a group of hormones that cause the inflammatory symptoms of hay-fever and asthma.

Luteolysis degeneration of the corpus luteum and ovarian luteinized tissues. adj. luteolytic.

Levarterenol see Norepinephrine.

LexA repressor or Repressor LexA is repressor enzyme that represses SOS response genes coding for DNA polymerases required for repairing DNA damage.

Libido sexual urge.

Lichen planus a chronic mucocutaneous disease that affects the skin, tongue, and oral mucosa.

Ligroin a volatile,, inflammable fraction of petroleum, obtained by distillation and used as a solvent.

Liniment liquid preparation rubbed on skin, used to relieve muscular aches and pains.

Linterized starch starch that has undergone prolonged acid treatment.

Lipodiatic having lipid and lipoprotein lowering property.

Lipodystrophy a medical condition characterized by abnormal or degenerative conditions of the body's adipose tissue.

Lipogenesis is the process by which acetyl-CoA is converted to fats.

Lipolysis is the breakdown of fat stored in fat cells in the body.

Liposomes artificially prepared vesicles made of lipid bilayer.

Lipotoxicity refers to tissues diseases that may occur when fatty acids spillover in excess of the oxidative needs of those tissues and enhances metabolic flux into harmful pathways of nonoxidative metabolism.

Lipotropic refers to compounds that help catalyse the breakdown of fat during metabolism in the body. e.g. chlorine and lecithin.

Lipoxygenase a family of iron-containing enzymes that catalyse the dioxygenation of polyunsaturated fatty acids in lipids containing a cis,cis-1,4- pentadiene structure.

Lithiasis formation of urinary calculi (stones) in the renal system (kidneys, ureters, urinary bladder, urethra) can be of any one of several compositions.

Lithogenic promoting the formation of calculi (stones).

Lithontripic removes stones from kidney, gall bladder.

Liver X receptors nuclear hormones that function as central transcriptional regulators for lipid homeostasis.

Lotion a liquids suspension or dispersion of chemicals for external application to the body.

Lovo cells colon cancer cells.

Low-density lipoprotein (LDL) is a type of lipoprotein that transports cholesterol and triglycerides from the liver to peripheral tissues. High levels of LDL cholesterol can signal medical problems like cardiovascular disease, and it is sometimes called "bad cholesterol".

LRP1 low-density lipoprotein receptor-related protein-1, plays a role in intracellular signaling functions as well as in lipid metabolism.

LTB4 a type of leukotriene, a major metabolite in neutrophil polymorphonuclear leukocytes. It stimulates polymorphonuclear cell function (degranulation, formation of oxygen-centered free radicals, arachidonic acid release, and metabolism). It induces skin inflammation.

Luciferase is a generic name for enzymes commonly used in nature for bioluminescence.

Lumbago is the term used to describe general lower back pain.

Lung abscess necrosis of the pulmonary tissue and formation of cavities containing necrotic debris or fluid caused by microbial infections.

Lusitropic an agent that affects diastolic relaxation.

Lutein a carotenoid, occurs naturally as yellow or orange pigment in some fruits and leafy vegetables. It is one of the two carotenoids contained within the retina of the eye. Within the central macula, zeaxanthin predominates, whereas in the peripheral retina, lutein predominates. Lutein is necessary for good vision and may also help prevent or slow down atherosclerosis, the thickening of arteries, which is a major risk for cardiovascular disease.

Luteinising hormone (LH) a hormone produced by the anterior pituitary gland. In females, it triggers ovulation. In males, it stimulates the production of testosterone to aid sperm maturation.

Luteolysis is the structural and functional degradation of the corpus luteum (CL) that occurs at the end of the luteal phase of both the estrous and menstrual cycles in the absence of pregnancy.

Lymphadenitis-cervical inflammation of the lymph nodes in the neck, usually caused by an infection.

Lymphatitis inflammation of lymph vessels and nodes.

Lymphadenopathy a term meaning disease of the lymph nodes – lymph node enlargement.

Lymphoblastic pertaining to the production of lymphocytes.

Lymphocyte a small white blood cell (leucocyte) that plays a large role in defending the body against disease. Lymphocytes are responsible for immune responses. There are two main types of lymphocytes: B cells and T cells. Lymphocytes secrete products (lymphokines) that modulate the functional activities of many other types of cells and are often present at sites of chronic inflammation.

Lymphocyte B cells the B cells make antibodies that attack bacteria and toxins.

Lymphocyte T cells T cells attack body cells themselves when they have been taken over by viruses or have become cancerous.

Lymphoma a type of cancer involving cells of the immune system, called lymphocytes.

Lymphopenia abnormally low number of lymphocytes in the blood.

Lysosomes are small, spherical organelles containing digestive enzymes (acid hydrolases) and other proteases (cathepsins).

Maceration softening or separating of parts by soaking in a liquid.

Macrophage a type of large leukocyte that travels in the blood but can leave the bloodstream and enter tissue; like other leukocytes it protects the body by digesting debris and foreign cells.

Macular degeneration a disease that gradually destroys the macula, the central portion of the retina, reducing central vision.

Macules small circumscribed changes in the color of skin that are neither raised (elevated) nor depressed.

Maculopapular describes a rash characterized by raised, spotted lesions.

Magnesium (Mg) is the fourth most abundant mineral in the body and is essential to good health. It is important for normal muscle and nerve function, steady heart rhythm, immune system, and strong bones. Magnesium also helps regulate blood sugar levels, promotes normal blood pressure, and is known to be involved in energy metabolism and protein synthesis and plays a role in preventing and managing disorders such as hypertension, cardiovascular disease, and diabetes. Dietary sources include legumes (e.g. soya bean and by-products), nuts, whole unrefined grains, fruit (e.g. banana, apricots), okra and green leafy vegetables.

MAK cell macrophage-activated killer cell, activated nacrophage that is much more phagocytic than monocytes.

Malaise a feeling of weakness, lethargy or discomfort as of impending illness.

Malaria is an infection of the blood by *Plasmodium* parasite that is carried from person to person by mosquitoes. There are four species of malaria parasites that infect man: *Plasmodium falciparum*, so called 'malignant tertian fever', is the most serious disease, *Plasmodium vivax*, causing a relapsing form of the disease, *Plasmodium malariae*, and *Plasmodium ovale*.

Malassezia a fungal genus (previously known as *Pityrosporum*) classified as yeasts, naturally found on the skin surfaces of many animals including humans. It can cause hypopigmentation on the chest or back if it becomes an opportunistic infection.

Mammalian target of rapamycin (mTOR) pathway that regulates mitochondrial oxygen consumption and oxidative capacity.

Mammogram an x-ray of the breast to detect tumours.

Mandibular relating to the mandible, the human jaw bone.

Manganese is an essential element for heath. It is an important constituent of some enzymes and an activator of other enzymes in physiologic processes. Manganese superoxide dismutase (MnSOD) is the principal antioxidant enzyme in the mitochondria. Manganese-activated enzymes play important roles in the metabolism of carbohydrates, amino acids, and cholesterol. Manganese is the preferred cofactor of enzymes called glycosyltransferases which are required for the synthesis of proteoglycans that are needed for the formation of healthy cartilage and bone. Dietary source include whole grains, fruit, legumes (soybean and by-products), green leafy vegetables, beetroot and tea.

MAO activity monoamine oxidase activity.

MAPK (Mitogen-activated protein kinase) these kinases are strongly activated in cells subjected to osmotic stress, UV radiation, disregulated K+ currents, RNA-damaging agents, and a multitude of other stresses, as well as inflammatory cytokines, endotoxin, and withdrawal of a trophic factor. The stress-responsive MAPKs mediate a plethora of cellular responses to such stressful stimuli, including apoptosis and production of inflammatory and immunoregulatory cytokines in diverse cell systems.

Marasmus is one of the three forms of serious protein-energy malnutrition.

Mastectomy surgery to remove a breast.

Masticatory a substance chewed to increase salivation. Also called sialogue.

Mastitis a bacterial infection of the breast which usually occurs in breastfeeding mothers.

Matrix metalloproteinases (MMP) a member of a group of enzymes that can break down proteins, such as collagen, that are normally found in the spaces between cells in tissues (i.e., extracellular matrix proteins). Matrix metalloproteinases are involved in wound healing, angiogenesis, and tumour cell metastasis. See also metalloproteinase.

MBC minimum bacterial concentration – the lowest concentration of antibiotic required to kill an organism.

MCP-1 monocyte chemotactic protein-1, plays a role in the recruitment of monocytes to sites of infection and injury. It is a member of small inducible gene (SIG) family.

MDA malondialdehyde is one of the most frequently used indicators of lipid peroxidation.

Measles an acute, highly communicable rash illness due to a virus transmitted by direct contact with infectious droplets or, less commonly, by airborne spread.

Medial Preoptic Area is located at the rostral end of the hypothalamus, it is important for the regulation of male sexual behavior.

Megaloblastic anemia an anemia that results from inhibition of DNA synthesis in red blood cell production, often due to a deficiency of vitamin B12 or folate and is characterized by many large immature and dysfunctional red blood cells (megaloblasts) in the bone marrow.

Melaene (melena) refers to the black, "tarry" feces that are associated with gastrointestinal hemorrhage.

Melanogenesis production of melanin by living cells.

Melanoma malignant tumour of melanocytes which are found predominantly in skin but also in the bowel and the eye and appear as pigmented lesions.

Melatonin a hormone produced in the brain by the pineal gland, it is important in the regulation of the circadian rhythms of several biological functions.

Menarche the first menstrual cycle, or first menstrual bleeding, in female human beings.

Menorrhagia heavy or prolonged menstruation, too-frequent menstrual periods.

Menopausal refer to permanent cessation of menstruation.

Menses see menstruation.

Menstruation the approximately monthly discharge of blood from the womb in women of childbearing age who are not pregnant. Also called menses. *adj.* menstrual.

Mesangial cells are specialized cells around blood vessels in the kidneys, at the mesangium.

Metabonome complete set of metabolically regulated elements in cells.

Metallogeusia metallic taste in the mouth.

Metalloproteinase enzymes that breakdown proteins and requiring zinc or calcium atoms for proper function.

Meta-analysis a statistical procedure that combines the results of several studies that address a set of related research hypotheses.

Metaphysis is the portion of a long bone between the epiphyses and the diaphysis of the femur.

Metaphyseal pertaining to the metaphysis.

Metaplasia transformation of one type of one mature differentiated cell type into another mature differentiated cell type.

Metastasis is the movement or spreading of cancer cells from one organ or tissue to another.

Metetrus the quiescent period of sexual inactivity between oestrus cycles.

Metroptosis the slipping or falling out of place of an organ (as the uterus).

Metrorrhagia uterine bleeding at irregular intervals, particularly between the expected menstrual periods.

Mevinolin a potent inhibitor of 3-hydroxy-3-methylglutaryl coenzyme A reductase (HMG-CoA reductase).

MHC acronym for major histocompatibility complex, a large cluster of genes found on the short arm of chromosome 6 in most vertebrates that encodes MHC molecules. MHC molecules play an important role in the immune system and autoimmunity.

MIC minimum inhibitory concentration – lowest concentration of an antimicrobial that will inhibit the visible growth of a microorganism.

Micelle a submicroscopic aggregation of molecules.

Micellization formation process of micelles.

Microangiopathy (or microvascular disease) is an angiopathy affecting small blood vessels in the body

Microfilaria a pre-larval parasitic worm of the family Onchocercidae, found in the vector and in the blood or tissue fluid of human host.

Micronuclei small particles consisting of acentric fragments of chromosomes or entire chromosomes, which lag behind at anaphase of cell division.

Microsomal PGE2 synthase is the enzyme that catalyses the final step in prostaglandin E2 (PGE2) biosynthesis.

Microvasculature the finer vessels of the body, as the arterioles, capillaries, and venules.

Micturition urination, act of urinating.

Migraine a neurological syndrome characterized by altered bodily perceptions, severe, painful headaches, and nausea.

Mimosine is an alkaloid, β-3-hydroxy-4 pyridone amino acid, it is a toxic non-protein free amino acid and is an antinutrient.

Mineral apposition rate MAR, rate of addition of new layers of mineral on the trabecular surfaces of bones.

Miscarriage spontaneous abortion.

Mitochondrial complex I the largest enzyme in the mitochondrial respiratory oxidative phosphorylation system.

Mitochondrial permeability transition (MPT) is an increase in the permeability of the mitochondrial membranes to molecules of less than 1,500 Da in molecular weight. MPT is one of the major causes of cell death in a variety of conditions.

Mitogen an agent that triggers mitosis, elicit all the signals necessary to induce cell proliferation.

Mitogenic able to induce mitosis or transformation.

Mitogenicity process of induction of mitosis.

Mitomycin a chemotherapy drug that is given as a treatment for several different types of cancer, including breast, stomach, oesophagus and bladder cancers.

Mitosis cell division in which the nucleus divides into nuclei containing the same number of chromosomes.

MMP matrix metalloproteinases, a group of peptidases involved in degradation of the extracellular matrix (ECM).

Mnestic pertaining to memory.

Molecular docking is a key tool in structural molecular biology and computer-assisted drug design.

Molluscidal destroying molluscs like snails.

Molt 4 cells MOLT4 cells are lymphoblast-like in morphology and are used for studies of apoptosis, tumour cytotoxicity, tumorigenicity, as well as for antitumour testing.

Molybdenum (Mo) is an essential element that forms part of several enzymes such as xanthine oxidase involved in the oxidation of xanthine to uric acid and use of iron. Molybdenum concentrations also affect protein synthesis, metabolism, and growth. Dietary sources include meat, green beans, eggs, sunflower seeds, wheat flour, lentils, and cereal grain.

Monoamine oxidase A (MAOA) is an isozyme of monoamine oxidase. It preferentially deaminates norepinephrine (noradrenaline), epinephrine (adrenaline), serotonin, and dopamine.

Monoaminergic of or pertaining to neurons that secrete monoamine neurotransmitters (e.g., dopamine, serotonin).

Monoclonal antibodies are produced by fusing single antibody-forming cells to tumour cells grown in culture.

Monocyte large white blood cell that ingest microbes, other cells and foreign matter.

Monogalactosyl diglyceride are the major lipid components of chloroplasts.

Monorrhagia is heavy bleeding and that's usually defined as periods lasting longer than 7 days or excessive bleeding.

Morbidity a diseased state or symptom or can refer either to the incidence rate or to the prevalence rate of a disease.

Morelloflavone a biflavonoid extracted from *Garcinia dulcis*, has shown antioxidative, antiviral, and anti-inflammatory properties.

Morphine the major alkaloid of opium and a potent narcotic analgesic.

MTTP microsomal triglyceride transfer protein that is required for the assembly and secretion of triglyceride -rich lipoproteins from both enterocytes and hepatocytes.

MUC 5AC mucin 5AC, a secreted gel-forming protein mucin with a high molecular weight of about 641 kDa.

Mucositis painful inflammation and ulceration of the mucous membranes lining the digestive tract.

Mucous relating to mucus.

Mucolytic capable of reducing the viscosity of mucus, or an agent that so acts.

Mucus viscid secretion of the mucous membrane.

Multidrug resistance (MDR) ability of a living cell to show resistance to a wide variety of structurally and functionally unrelated compounds.

Muscarinic receptors are G protein-coupled acetylcholine receptors found in the plasma membranes of certain neurons and other cells.

Mutagen an agent that induces genetic mutation by causing changes in the DNA.

Mutagenic capable of inducing mutation (used mainly for extracellular factors such as X-rays or chemical pollution).

Myc codes for a protein that binds to the DNA of other genes and is therefore a transcription factor, found on chromosome 8 in human.

Mycosis an infection or disease caused by a fungus.

Myelocyte is a young cell of the granulocytic series, occurring normally in bone marrow, but not in circulating blood.

Myeloid leukaemia (Chronic) a type of cancer that affects the blood and bone marrow, characterized by excessive number of white blood cells.

Myeloma cancer that arise in the plasma cells a type of white blood cells.

Myeloperoxidase (MPO) is a peroxidase enzyme most abundantly present in neutrophil granulocytes (a subtype of white blood cells). It is an inflammatory enzyme produced by activated leukocytes that predicts risk of coronary heart disease.

Myeloproliferative disorder disease of the bone marrow in which excess cells are produced.

Myocardial relating to heart muscles tissues.

Myocardial infarction (MI) is the rapid development of myocardial necrosis caused by a critical imbalance between oxygen supply and demand of the myocardium.

Myocardial ischemia an intermediate condition in coronary artery disease during which the heart tissue is slowly or suddenly starved of oxygen and other nutrients.

Myogenesis the formation of muscular tissue, especially during embryonic development.

Myopia near – or short-sightedness.

Myosarcoma a malignant muscle tumour.

Myotonia dystrophica an inherited disorder of the muscles and other body systems characterized by progressive muscle weakness, prolonged muscle contractions (myotonia), clouding of the lens of the eye (cataracts), cardiac abnormalities, balding, and infertility.

Myringosclerosis also known as tympanosclerosis or intratympanic tympanosclerosis, is a condition caused by calcification of collagen tissues in the tympanic membrane of the middle ear.

Mytonia a symptom of certain neuromuscular disorders characterized by the slow relaxation of the muscles after voluntary contraction or electrical stimulation.

Myotube a developing skeletal muscle fibre with a tubular appearance.

N-nitrosmorpholine a human carcinogen.

N-nitrosoproline an indicator for N-nitrosation of amines.

NADPH The reduced form of nicotinamide adenine dinucleotide phosphate that serves as an electron carrier.

NAFLD Non-alcoholic fatty liver disease.

Narcotic an agent that produces narcosis, in moderate doses it dulls the senses, relieves pain and induces sleep; in excessive dose it cause stupor, coma, convulsions and death.

Nasopharynx upper part of the alimentary continuous with the nasal passages.

Natriorexia excessive intake of sodium evoked by sodium depletion. *adj.* natriorexic, natriorexigenic.

Natriuresis the discharge of excessive large amount of sodium through urine. *adj.* natriuretic.

Natural killer cells (NK cells) a type of cytotoxic lymphocyte that constitute a major component of the innate immune system.

Natural killer T (NKT) cells a heterogeneous group of T cells that share properties of both T cells and natural killer (NK) cells.

Nausea sensation of unease and discomfort in the stomach with an urge to vomit.

Necropsy see autopsy.

Necrosis morphological changes that follow cell death, usually involving nuclear and cytoplasmic changes.

Neointima a new or thickened layer of arterial intima formed especially on a prosthesis or in atherosclerosis by migration and proliferation of cells from the media.

Neonatal *adj.* of or relating to newborn infants or an infant.

Neoplasia abnormal growth of cells, which may lead to a neoplasm, or tumour.

Neoplasm tumour; any new and abnormal growth, specifically one in which cell multiplication is uncontrolled and progressive. Neoplasms may be benign or malignant.

Neoplastic transformation conversion of a tissue with a normal growth pattern into a malignant tumour.

Neovasculature formation of new blood vessels.

Nephrectomised kidneys surgically removed.

Nephrectomy surgical removal of the kidney.

Nephric relating to or connected with a kidney.

Nephrin is a protein necessary for the proper functioning of the renal filtration barrier.

Nephritic syndrome is a collection of signs (known as a syndrome) associated with disorders affecting the kidneys, more specifically glomerular disorders.

Nephritis is inflammation of the kidney.

Nephrolithiasis process of forming a kidney stone in the kidney or lower urinary tract.

Nephropathy a disorder of the kidney.

Nephrotic syndrome nonspecific disorder in which the kidneys are damaged, causing them to leak large amounts of protein from the blood into the urine.

Nephrotoxicity poisonous effect of some substances, both toxic chemicals and medication, on the kidney.

Nerve growth factor (NGF) a small protein that induces the differentiation and survival of particular target neurons (nerve cells).

Nervine a nerve tonic that acts therapeutically upon the nerves, particularly in the sense of a sedative that serves to calm ruffled nerves.

Neuralgia is a sudden, severe painful disorder of the nerves.

Neuraminidase glycoside hydrolase enzymes that cleaves the glycosidic linkages of neuraminic acids.

Neuraminidase inhibitors a class of antiviral drugs targeted at the influenza viruses whose mode of action consists of blocking the function of the viral neuraminidase protein, thus preventing the virus from reproducing.

Neurasthenia a condition with symptoms of fatigue, anxiety, headache, impotence, neuralgia and impotence.

Neurasthenic a substance used to treat nerve pain and/or weakness (i.e. neuralgia, sciatica, etc).

Neurite refers to any projection from the cell body of a neuron.

Neuritis an inflammation of the nerve characterized by pain, sensory disturbances and impairment of reflexes. *adj.* neuritic.

Neuritogenesis the first step of neuronal differentiation, takes place as nascent neurites bud from the immediate postmitotic neuronal soma.

Neuroblastoma a common extracranial cancer that forms in nerve tissues, common in infancy.

Neuroendocrine *adj.* of, relating to, or involving the interaction between the nervous system and the hormones of the endocrine glands.

Neuroleptic refers to the effects on cognition and behavior of antipsychotic drugs that reduce confusion, delusions, hallucinations, and psychomotor agitation in patients with psychoses.

Neuropharmacological relating the effects of drugs on the neurosystem.

Neuroradiology is a subspecialty of radiology focusing on the diagnosis and characterization of abnormalities of the central and peripheral nervous system. *adj.* neuroradiologic.

Neurotrophic relating to neutrophy i.e. the nutrition and maintenance of nervous tissue.

Neutropenia a disorder of the blood, characterized by abnormally low levels of neutrophils.

Neutrophil a type of white blood cell, specifically a form of granulocyte.

Neutrophin protein that induce the survival, development and function of neurons.

NF-kappa B (NF-kB) nuclear factor kappa B, is an ubiquitous rapid response transcription factor in cells involved in immune and inflammatory reactions.

Niacin vitamin B3. See vitamin B3.

Niacinamide an amide of niacin, also known as nicotinamide. See vitamin B3.

NIH3T3 cells a mouse embryonic fibroblast cell line used in the cultivation of keratinocytes.

Nitrogen (N) is an essential building block of amino and nucleic acids and proteins and is essential to all living organisms. Protein rich vegetables like legumes are rich food sources of nitrogen.

NK cells natural killer cells, a type of cytotoxic lymphocyte that constitute a major component of the innate immune system.

NMDA receptor N-methyl-D-aspartate receptor, the predominant molecular device for controlling synaptic plasticity and memory function. A brain receptor activated by the amino acid glutamate, which when excessively stimulated may cause cognitive defects in Alzheimer's disease.

Nociceptive causing pain, responding to a painful stimulus.

Non-osteogenic fibroma of bone a benign tumour of bone which show no evidence of ossification.

Nootropics are substances which are claimed to boost human cognitive abilities (the functions and capacities of the brain). Also popularly referred to as "smart drugs", "smart nutrients", "cognitive enhancers" and "brain enhancers".

Noradrenalin see Norepinephrine.

Norepinephrine a substance, both a hormone and neurotransmitter, secreted by the adrenal medulla and the nerve endings of the sympathetic nervous system to cause vasoconstriction and increases in heart rate, blood pressure, and the sugar level of the blood. Also called levarterenol, noradrenalin.

Normoglycaemic having the normal amount of glucose in the blood.

Normotensive having normal blood pressure.

Nosocomial infections infections which are a result of treatment in a hospital or a healthcare service unit, but secondary to the patient's original condition.

NK1.1+ T (NKT) cells a type of natural killer T (NKT) cells. See natural killer T cells.

Nuclear factor erythroid 2-related factor 2 (Nrf2) a transcription factor that plays a major role in response to oxidative stress by binding to antioxidant-responsive elements that regulate many hepatic phase I and II enzymes as well as hepatic efflux transporters.

Nucleosomes fundamental repeating subunits of all eukaryotic chromatin, consisting of a DNA chain coiled around a core of histones.

Nulliparous term used to describe a woman who has never given birth.

Nyctalopia night blindness, impaired vision in dim light and in the dark, due to impaired function of certain specialized vision cells.

Nycturia excessive urination at night; especially common in older men.

Occlusion closure or blockage (as of a blood vessel).

Occlusive peripheral arterial disease (PAOD) also known as peripheral vascular disease (PVD), or peripheral arterial disease (PAD) refers to the obstruction of large arteries not within the coronary, aortic arch vasculature, or brain. PVD can result from atherosclerosis, inflammatory processes leading to stenosis, an embolism, or thrombus formation.

Oculomotor nerve the third of twelve paired cranial nerves.

Odds ratio a statistical measure of effect size, describing the strength of association or non-independence between two binary data values.

Odontalgia toothache. *adj.* odontalgic.

Odontopathy any disease of the teeth.

Oedema see edema.

Oligoanuria insufficient urine volume to allow for administration of necessary fluids, etc.

Oligoarthritis an inflammation of two, three or four joints.

Oligonucleosome a series of nucleosomes.

Oligospermia or oligozoospermia refers to semen with a low concentration of sperm, commonly associated with male infertility.

Oliguria decreased production of urine.

Omega 3 fatty acids are essential polyunsaturated fatty acids that have in common a final carbon–carbon double bond in the n−3 position. Dietary sources of omega-3 fatty acids include fish oil and certain plant/nut oils. The three most nutritionally important omega 3 fatty acids are alpha-linolenic acid, eicosapentaenoic acid (EPA) and docosahexaenoic acid (DHA). Research indicates that omega 3 fatty acids are important in health promotion and disease and can help prevent a wide range of medical problems, including cardiovascular disease, depression, asthma, and rheumatoid arthritis.

Omega 6 fatty acids are essential polyunsaturated fatty acids that have in common a final carbon–carbon double bond in the n–6 position. Omega-6 fatty acids are considered essential fatty acids (EFAs) found in vegetable oils, nuts and seeds. They are essential to human health but cannot be made in the body. Omega-6 fatty acids – found in vegetable oils, nuts and seeds – are a beneficial part of a heart-healthy eating. Omega-6 and omega-3 PUFA play a crucial role in heart and brain function and in normal growth and development. Linoleic acid (LA) is the main omega-6 fatty acid in foods, accounting for 85% to 90% of the dietary omega-6 PUFA. Other omega 6 acids include gamma-linolenic acid or GLA, sometimes called gamoleic acid, eicosadienoic acid, arachidonic acid and docosadienoic acid.

Omega 9 fatty acids are not essential polyunsaturated fatty acids that have in common a final carbon–carbon double bond in the n–9 position. Some n–9s are common components of animal fat and vegetable oil. Two n–9 fatty acids important in industry are: oleic acid (18:1, n–9), which is a main component of olive oil and erucic acid (22:1, n–9), which is found in rapeseed, wallflower seed, and mustard seed.

Oncogenes genes carried by tumour viruses that are directly and solely responsible for the neoplastic (tumorous) transformation of host cells.

Ophthalmia severe inflammation of eye, or the conjunctiva or deeper structures of the eye. Also called ophthalmitis.

Ophthalmia (Sympathetic) inflammation of both eyes following trauma to one eye.

Opiate drug derived from the opium plant.

Opioid receptors a group of G-protein coupled receptors located in the brain and various organs that bind opiates or opioid substances.

Optic placode an ectodermal placode from which the lens of the embryonic eye develops; also called lens placode.

ORAC (Oxygen radical absorbance capacity) a method of measuring antioxidant capacities in biological samples. It evaluates the ability of sample to scavenge reactive oxygen species.

ORAC-H (Oxygen radical absorbance capacity_hydrophilic) reflects the watersoluble (hydrophilic) antioxidant capacity.

ORAC-L (Oxygen radical absorbance capacity –lipohilic) reflects the lipid soluble (lipholic) antioxidant capacity. It measures predominantly the activity of tocopherols and doses not measure the antioxidant of carotenoids.

ORAC-T (Oxygen radical absorbance capacity-total) represents the sum of ORAC-H and ORAC-L.

Oral submucous fibrosis a chronic debilitating disease of the oral cavity characterized by inflammation and progressive fibrosis of the submucosa tissues.

Oral thrush an infection of yeast fungus, *Candida albicans*, in the mucous membranes of the mouth.

Orchidectomy surgery to remove one or both testicles.

Orchidectomised with testis removed.

Orchitis an acute painful inflammatory reaction of the testis secondary to infection by different bacteria and viruses.

Orexigenic increasing or stimulating the appetite.

Orofacial dyskinesia abnormal involuntary movements involving muscles of the face, mouth, tongue, eyes, and occasionally, the neck—may be unilateral or bilateral, and constant or intermittent.

Oropharyngeal relating to the oropharynx.

Oropharynx part of the pharynx between the soft palate and the epiglottis.

Ostalgia, Ostealgia pain in the bones. Also called osteodynia.

Osteoarthritis is the deterioration of the joints that becomes more common with age.

Osteoarthrosis chronic noninflammatory bone disease.

Osteoblast a mononucleate cell that is responsible for bone formation.

Osteoblastic relating to osteoblasts.

Osteocalcin a noncollagenous protein found in bone and dentin, also refer to as bone gamma-carboxyglutamic acid-containing protein.

Osteoclasts a kind of bone cell that removes bone tissue by removing its mineralized matrix.
Osteoclastogenesis the production of osteoclasts.
Osteodynia pain in the bone.
Osteogenic derived from or composed of any tissue concerned in bone growth or repair.
Osteomalacia refers to the softening of the bones due to defective bone mineralization.
Osteomyelofibrosis a myeloproliferative disorder in which fibrosis and sclerosis finally lead to bone marrow obliteration.
Osteopenia reduction in bone mass, usually caused by a lowered rate of formation of new bone that is insufficient to keep up with the rate of bone destruction.
Osteoporosis a disease of bone that leads to an increased risk of fracture.
Osteoprotegerin also called osteoclastogenesis inhibitory factor (OCIF), a cytokine, which can inhibit the production of osteoclasts.
Osteosacrcoma a malignant bone tumour. Also called osteogenic sarcoma.
Otalgia earache, pain in the ear.
Otic placode a thickening of the ectoderm on the outer surface of a developing embryo from which the ear develops.
Otitis inflammation of the inner or outer parts of the ear.
Otorrhea running drainage (discharge) exiting the ear.
Ovariectomised with one or two ovaries removed.
Ovariectomy surgical removal of one or both ovaries.
Oxidation the process of adding oxygen to a compound, dehydrogenation or increasing the electro-negative charge.
Oxidoreductase activity catalysis of an oxidation-reduction (redox) reaction, a reversible chemical reaction. One substrate acts as a hydrogen or electron donor and becomes oxidized, while the other acts as hydrogen or electron acceptor and becomes reduced.
Oxygen radical absorbance capacity (ORAC) a method of measuring antioxidant capacities in biological samples.
Oxytocic *adj.* hastening or facilitating childbirth, especially by stimulating contractions of the uterus.
Oxytocin is a mammalian hormone that also acts as a neurotransmitter in the brain. It is best known for its roles in female reproduction: it is released in large amounts after distension of the cervix and vagina during labor, and after stimulation of the nipples, facilitating birth and breastfeeding, respectively.
Oxyuriasis infestation by pinworms.
Ozoena discharge of the nostrils caused by chronic inflammation of the nostrils.
p.o. per os, oral administration.
P-glycoprotein (P-gp, ABCB1, MDR1) a cell membrane-associated drug-exporting protein that transports a variety of drug substrates from cancer cells.
P- Selectin also known as CD62P, GMP-140, LLECAM-3, PADGEM, a member of the selectin family. It is expressed by activated platelets and endothelial cells.
Palpebral ptosis the abnormal drooping of the upper lid, caused by partial or total reduction in levator muscle function.
Palpitation rapid pulsation or throbbing of the heart.
Paludism state of having symptoms of malaria characterized by high fever and chills.
Pancreatectomized having undergone a pancreatectomy.
Pancreatectomy surgical removal of all or part of the pancreas.
Pancreatitis inflammation of the pancreas.
Pantothenic acid vitamin B5. See vitamin B5.
Papain a protein degrading enzyme used medicinally and to tenderize meat.
Papilloma a benign epithelial tumour growing outwardly like in finger-like fronds.
Papule a small, solid, usually inflammatory elevation of the skin that does not contain pus.
Paradontosis is the inflammation of gums and other deeper structures, including the bone.
Paralytic person affected with paralysis, pertaining to paralysis.
Parasitemia presence of parasites in blood. *adj.* parasitemic.
Parasympathetic nervous system subsystem of the nervous systems that slows the heart rate and increases intestinal and gland activity and relaxes the sphincter muscles.

Parasympathomimetic having an action resembling that caused by stimulation of the parasympathetic nervous system.

Parenteral administration administration by intravenous, subcutaneous or intramuscular routes.

Paresis a condition characterised by partial loss of movement, or impaired movement.

Paresthesia is an abnormal sensation of the skin, such as burning, numbness, itching, hyperesthesia (increased sensitivity) or tingling, with no apparent physical cause.

Parotitis inflammation of salivary glands.

Paroxysm a sudden outburst of emotion or action, a sudden attack, recurrence or intensification of a disease.

Paroxystic relating to an abnormal event of the body with an abrupt onset and an equally sudden return to normal.

PARP see poly (ADP-ribose) polymerase.

Parturition act of child birth.

PCE/PCN ratio polychromatic erythrocyte/normochromatic erythrocyte ratio use as a measure of cytotoxic effects.

pCREB phosphorylated cAMP (adenosine 3'5' cyclic monophosphate)-response element binding protein.

PDEF acronym for prostate-derived ETS factor, an ETS (epithelial-specific E26 transforming sequence) family member that has been identified as a potential tumour suppressor.

Pectoral pertaining to or used for the chest and respiratory tract.

pERK phosphorylated extracellular signal-regulated kinase, protein kinases involved in many cell functions.

p53 also known as protein 53 or tumour protein 53, is a tumour suppressor protein that in humans is encoded by the TP53 gene.

Peliosis see purpura.

Pellagra is a systemic nutritional wasting disease caused by a deficiency of vitamin B3 (niacin).

Pemphigus neonatorum Staphylococcal scalded skin syndrome, a bacterial disease of infants, characterized by elevated vesicles or blebs on a normal or reddened skin .

Peptic ulcer a sore in the lining of the stomach or duodenum, the first part of the small intestine.

Percutanous pertains to a medical procedure where access to inner organs or tissues is done via needle puncture of the skin.

Perfusion to force fluid through the lymphatic system or blood vessels to an organ or tissue.

Periapical periodontitis is the inflammation of the tissue adjacent to the tip of the tooth's root.

Perifuse to flush a fresh supply of bathing fluid around all of the outside surfaces of a small piece of tissue immersed in it.

Perilipins highly phosphorylated adipocyte proteins that are localized at the surface of the lipid droplet.

Perimenopause is the phase before menopause actually takes place, when ovarian hormone production is declining and fluctuating. *adj.* perimenopausal.

Periodontal ligament (PDL) is a group of specialized connective tissue fibres that essentially attach a tooth to the bony socket.

Periodontitis is a severe form of gingivitis in which the inflammation of the gums extends to the supporting structures of the tooth. Also called pyorrhea.

Peripheral arterial disease (PAD) see peripheral artery occlusive disease.

Peripheral neuropathy refers to damage to nerves of the peripheral nervous system.

Peripheral vascular disease (PVD) see peripheral artery occlusive disease .

Peristalsis a series of organized, wave-like muscle contractions that occur throughout the digestive tract.

Perlingual through or by way of the tongue.

Perniosis an abnormal reaction to cold that occurs most frequently in women, children, and the elderly. Also called chilblains.

Per os (P.O.) oral administration.

Peroxisome proliferator-activated receptors (PPARs) a family of nuclear receptors that are involved in lipid metabolism, differentiation, proliferation, cell death, and inflammation.

Peroxisome proliferator-activated receptor alpha (PPAR-alpha) a nuclear receptor protein, transcription factor and a major regulator of lipid metabolism in the liver.

Peroxisome proliferator-activated receptor gamma (PPAR-γ) a type II nuclear receptor protein that regulates fatty acid storage and glucose metabolism.

Pertussis whooping cough, sever cough.

Peyers Patches patches of lymphoid tissue or lymphoid nodules on the walls of the ileal-small intestine.

P53 also known as protein 53 or tumour protein 53, is a tumour suppressor protein that in humans is encoded by the TP53 gene.

PGE-2 Prostaglandin E2, a hormone-like substance that is released by blood vessel walls in response to infection or inflammation that acts on the brain to induce fever.

Phagocytes are the white blood cells that protect the body by ingesting (phagocytosing) harmful foreign particles, bacteria and dead or dying cells. *adj.* phagocytic.

Phagocytosis is process the human body uses to destroy dead or foreign cells.

Pharmacognosis the branch of pharmacology that studies the composition, use, and history of drugs.

Pharmacodynamics branch of pharmacology dealing with the effects of drugs and the mechanism of their action.

Pharmacokinetics branch of pharmacology concerned with the movement of drugs within the body including processes of absorption, distribution, metabolism and excretion in the body.

Pharmacopoeia authoritative treatise containing directions for the identification of drug samples and the preparation of compound medicines, and published by the authority of a government or a medical or pharmaceutical society and in a broader sense is a general reference work for pharmaceutical drug specifications.

Pharyngitis, Pharyngolaryngitis inflammation of the pharynx and the larynx.

Pharyngolaryngeal pertaining to the pharynx and larynx.

Phenolics class of chemical compounds consisting of a hydroxyl group (–OH) bonded directly to an aromatic hydrocarbon group.

Pheochromocytoma is a rare neuroendocrine tumour that usually originates from the adrenal glands' chromaffin cells, causing overproduction of catecholamines, powerful hormones that induce high blood pressure and other symptoms.

Phlebitis is an inflammation of a vein, usually in the legs.

Phlegm abnormally viscid mucus secreted by the mucosa of the respiratory passages during certain infectious processes.

Phlegmon a spreading, diffuse inflammation of the soft or connective tissue due to infection by Streptococci bacteria.

Phoroglucinol a white, crystalline compound used as an antispasmodic, analytical reagent, and decalcifier of bone specimens for microscopic examination.

Phosphatidylglycerol is a glycerophospholipid found in pulmonary active surface lipoprotein and consists of a L-glycerol 3-phosphate backbone ester-bonded to either saturated or unsaturated fatty acids on carbons 1 and 2.

Phosphatidylinositol 3-kinases (PI 3-kinases or PI3Ks) a group of enzymes involved in cellular functions such as cell growth, proliferation, differentiation, motility, survival and intracellular trafficking, which in turn are involved in cancer.

Phosphatidylserine a phosphoglyceride phospholipid that is one of the key building blocks of cellular membranes, particularly in the nervous system. It is derived from soy lecithin.

Phosphodiesterases a diverse family of enzymes that hydrolyse cyclic nucleotides and thus play a key role in regulating intracellular levels of the second messengers cAMP and cGMP, and hence cell function.

Phospholipase an enzyme that hydrolyzes phospholipids into fatty acids and other lipophilic substances.

Phospholipase A2 (PLA2) a small lipolytic enzyme that releases fatty acids from the second carbon group of glycerol. Plays an essential role in the synthesis of prostaglandins and leukotrienes.

Phospholipase C enzymes that cleaves phospholipase.

Phospholipase C gamma (PLCgamma) enzymes that cleaves phospholipase in cellular proliferation and differentiation, and its enzymatic activity is upregulated by a variety of growth factors and hormones.

Phosphorus (P) is an essential mineral that makes up 1% of a person's total body weight and is found in the bones and teeth. It plays an important role in the body's utilization of carbohydrates and fats; in the synthesis of protein for the growth, maintenance, and repair of cells and tissues. It is also crucial for the production of ATP, a molecule the body uses to store energy. Main sources are meat and milk; fruits and vegetables provides small amounts.

Photoaging is the term that describes damage to the skin caused by intense and chronic exposure to sunlight resulting in premature aging of the skin.

Photocarcinogenesis represents the sum of a complex of simultaneous and sequential biochemical events that ultimately lead to the occurrence of skin cancer.

Photophobia abnormal visual intolerance to light.

Photopsia an affection of the eye, in which the patient perceives luminous rays, flashes, coruscations, etc.

Photosensitivity sensitivity toward light.

Phthisis an archaic name for tuberculosis.

Phytohemagglutinin a lectin found in plant that is involved in the stimulation of lymphocyte proliferation.

Phytonutrients certain organic components of plants, that are thought to promote human health. Fruits, vegetables, grains, legumes, nuts and teas are rich sources of phytonutrients. Phytonutrients are not 'essential' for life. Also called phytochemicals.

Phytosterols a group of steroid alcohols, cholesterol-like phytochemicals naturally occurring in plants like vegetable oils, nuts and legumes.

Piebaldism rare autosomal dominant disorder of melanocyte development characterized by distinct patches of skin and hair that contain no pigment.

Piles see haemorrhoids.

Pityriasis lichenoides is a rare skin disorder of unknown aetiology characterised by multiple papules and plaques.

PKC protein kinase C, a membrane bound enzyme that phosphorylates different intracellular proteins and raised intracellular Ca levels.

PKC Delta inhibitors Protein Kinase C delta inhibitors that induce apoptosis of haematopoietic cell lines.

Placebo a sham or simulated medical intervention.

Placode a platelike epithelial thickening in the embryo where some organ or structure later develops.

Plasma the yellow-colored liquid component of blood, in which blood cells are suspended.

Plasma kallikrien a serine protease, synthesized in the liver and circulates in the plasma.

Plasmalemma plasma membrane.

Plasmin a proteinase enzyme that is responsible for digesting fibrin in blood clots.

Plasminogen the proenzyme of plasmin, whose primary role is the degradation of fibrin in the vasculature.

Plaster poultice.

Platelet activating factor (PAF) is an acetylated derivative of glycerophosphorylcholine, released by basophils and mast cells in immediate hypersensitive reactions and macrophages and neutrophils in other inflammatory reactions. One of its main effects is to induce platelet aggregation.

PLC gamma phospholipase C gamma plays a central role in signal transduction.

Pleurisy is an inflammation of the pleura, the lining of the pleural cavity surrounding the lungs, which can cause painful respiration and other symptoms. Also known as pleuritis.

Pneumonia an inflammatory illness of the lung caused by bacteria or viruses.

Pneumotoxicity damage to lung tissues.

Poliomyelitis is a highly infectious viral disease that may attack the central nervous system and is characterized by symptoms that range from a mild non-paralytic infection to total paralysis in a matter of hours; also called polio or infantile paralysis.

Poly (ADP-ribose) polymerase (PARP) a protein involved in a number of cellular processes especially DNA repair and programmed cell death.

Polyarthritis is any type of arthritis which involves five or more joints.

Polychromatic erythrocyte (PCE) an immature red blood cell containing RNA, that can be differentiated by appropriate staining techniques from a normochromatic erythrocyte (NCE), which lacks RNA.

Polycystic kidney disease is a kidney disorder passed down through families in which multiple cysts form on the kidneys, causing them to become enlarged.

Polycythaemia a type of blood disorder characterised by the production of too many red blood cells.

Polymorphnuclear having a lobed nucleus. Used especially of neutrophilic white blood cells.

Polyneuritis widespread inflammation of the nerves.

Polyneuritis gallinarum a nervous disorder in birds and poultry.

Polyp a growth that protrudes from a mucous membrane.

Polyphagia medical term for excessive hunger or eating.

Polyuria a condition characterized by the passage of large volumes of urine with an increase in urinary frequency.

Pomade a thick oily dressing.

Porphyrin any of a class of water-soluble, nitrogenous biological pigments.

Postpartum Depression depression after pregnancy; also called postnatal depression.

Postprandial after mealtime.

Potassium (K) is an element that's essential for the body's growth and maintenance. It's necessary to keep a normal water balance between the cells and body fluids, for cellular enzyme activities and plays an essential role in the response of nerves to stimulation and in the contraction of muscles. Potassium is found in many plant foods and fish (tuna, halibut): chard, mushrooms, spinach, fennel, kale, mustard greens, Brussels sprouts, broccoli, cauliflower, cabbage winter squash, eggplant, cantaloupe, tomatoes, parsley, cucumber, bell pepper, turmeric, ginger root, apricots, strawberries, avocado and banana.

Poultice is a soft moist mass, often heated and medicated, that is spread on cloth over the skin to treat an aching, inflamed, or painful part of the body. Also called cataplasm.

PPARs peroxisome proliferator-activated receptors – a group of nuclear receptor proteins that function as transcription factors regulating the expression of genes.

Prebiotics a category of functional food, defined as non-digestible food ingredients that beneficially affect the host by selectively stimulating the growth and/or activity of one or a limited number of bacteria in the colon, and thus improve host health. *cf.* probiotics.

Pre-ecamplasia toxic condition of pregnancy characterized by high blood pressure, abnormal weight gain, proteinuria and edema.

Prepubertal before puberty; pertaining to the period of accelerated growth preceding gonadal maturity.

Pregnane X receptor (PXR; NR1I2) is a ligand-activated transcription factor that plays a role not only in drug metabolism and transport but also in various other biological processes.

Pregnenolone a steroid hormone produced by the adrenal glands, involved in the steroidogenesis of other steroid hormones like progesterone, mineralocorticoids, glucocorticoids, androgens, and estrogens.

Prenidatory referring to the time period between fertilization and implantation.

Prenylated flavones flavones with an isoprenyl group in the 8-position, has been reported to have good anti-inflammatory properties.

Proangiogenic promote angiogensis (formation and development of new blood vessels).

Probiotics are dietary supplements and live microorganisms containing potentially beneficial bacteria or yeasts that are taken into the alimentary system for healthy intestinal functions. *cf.* prebiotics.

Procyanidin also known as proathocyanidin, oligomeric proathocyanidin, leukocyanidin, leucoanthocyanin, is a class of flavanols found in many plants. It has antioxidant activity and plays a role in the stabilization of collagen and maintenance of elastin.

Progestational of or relating to the phase of the menstrual cycle immediately following ovulation, characterized by secretion of progesterone.

Proglottid one of the segments of a tapeworm.

Prognosis medical term to describe the likely outcome of an illness.

Prolactin a hormone produced by the pituitary gland, it stimulates the breasts to produce milk in pregnant women. It is also present in males but its role is not well understood.

Prolapsus to fall or slip out of place.

Prolapus ani eversion of the lower portion of the rectum, and protruding through the anus, common in infancy and old age.

Proliferating cell nuclear antigen (PCNA) a new marker to study human colonic cell proliferation.

Proliferative vitreoretinopathy (PVR) a most common cause of failure in retinal reattachment surgery, characterised by the formation of cellular membrane on both surfaces of the retina and in the vitreous.

Promastigote the flagellate stage in the development of trypanosomatid protozoa, characterized by a free anterior flagellum.

Promyelocytic leukemia a subtype of acute myelogenous leukemia (AML), a cancer of the blood and bone marrow.

Pro-oxidants chemicals that induce oxidative stress, either through creating reactive oxygen species or inhibiting antioxidant systems.

Prophylaxis prevention or protection against disease.

Proptosis see exophthalmos.

Prostacyclin a prostaglandin that is a metabolite of arachidonic acid, inhibits platelet aggregation, and dilates blood vessels.

Prostaglandins a family of C 20 lipid compounds found in various tissues, associated with muscular contraction and the inflammation response such as swelling, pain, stiffness, redness and warmth.

Prostaglandin E2 (PEG -2) one of the prostaglandins, a group of hormone-like substances that participate in a wide range of body functions such as the contraction and relaxation of smooth muscle, the dilation and constriction of blood vessels, control of blood pressure, and modulation of inflammation.

Prostaglandin E synthase an enzyme that in humans is encoded by the glutathione-dependent PTGES gene.

Prostanoids term used to describe a subclass of eicosanoids (products of COX pathway) consisting of: the prostaglandins (mediators of inflammatory and anaphylactic reactions), the thromboxanes (mediators of vasoconstriction) and the prostacyclins (active in the resolution phase of inflammation).

Prostate a gland that surround the urethra at the bladder in the male.

Prostate cancer a disease in which cancer develops in the prostate, a gland in the male reproductive system. Symptoms include pain, difficulty in urinating, erectile dysfunction and other symptoms.

Prostate –specific antigen (PSA) a protein produced by the cells of the prostate gland.

Protein kinase C (PKC) a family of enzymes involved in controlling the function of other proteins through the phosphorylation of hydroxyl groups of serine and threonine amino acid residues on these proteins. PKC enzymes play important roles in several signal transduction cascades.

Protein tyrosine phosphatase (PTP) a group of enzymes that remove phosphate groups from phosphorylated tyrosine residues on proteins.

Proteinase a protease (enzyme) involved in the hydrolytic breakdown of proteins, usually by splitting them into polypeptide chains.

Proteinuria means the presence of an excess of serum proteins in the urine.

Proteolysis cleavage of the peptide bonds in protein forming smaller polypeptides. *adj.* proteolytic.

Proteomics the large-scale study of proteins, particularly their structures and functions.

Prothrombin blood-clotting protein that is converted to the active form, factor IIa, or thrombin, by cleavage.

Prothyroid good for thyroid function.

Protheolithic proteolytic see proteolysis.

Proto-oncogene A normal gene which, when altered by mutation, becomes an oncogene that can contribute to cancer.

Prurigo a general term used to describe itchy eruptions of the skin.

Pruritis defined as an unpleasant sensation on the skin that provokes the desire to rub or scratch the area to obtain relief; itch, itching. *adj.* pruritic.

PSA Prostate Specific Antigen, a protein which is secreted into ejaculate fluid by the healthy prostate. One of its functions is to aid sperm movement.

Psoriasis a common chronic, non-contagious autoimmune dermatosis that affects the skin and joints.

Psychoactive having effects on the mind or behavior.

Psychonautics exploration of the psyche by means of approaches such as meditation, prayer, lucid dreaming, brain wave entrainment etc.

Psychotomimetic hallucinogenic.

Psychotropic capable of affecting the mind, emotions, and behavior.

Ptosis also known as drooping eyelid; caused by weakness of the eyelid muscle and damage to the nerves that control the muscles or looseness of the skin of the upper eyelid..

P13-K is a lipid kinase enzyme involved in the regulation of a number of cellular functions such as cell growth, proliferation, differentiation, motility, survival and intracellular trafficking, which in turn are involved in cancer.

P13-K/AKT signaling pathway shown to be important for an extremely diverse array of cellular activities – most notably cellular proliferation and survival.

Pthysis silicosis with tuberculosis.

Ptosis drooping of the upper eye lid.

PTP protein tyrosine phosphatase.

PTPIB protein tyrosine phosphatase 1B.

P21 also known as cyclin-dependent kinase inhibitor 1 or CDK-interacting protein 1, is a potent cyclin-dependent kinase inhibitor.

Puerperal pertaining to child birth.

Pulmonary embolism a blockage (blood clot) of the main artery of the lung.

Purgative a substance used to cleanse or purge, especially causing the immediate evacuation of the bowel.

Purpura is the appearance of red or purple discolorations on the skin that do not blanch on applying pressure. Also called peliosis.

Purulent containing pus discharge.

Purulent sputum sputum containing, or consisting of, pus.

Pustule small, inflamed, pus-filled lesions.

Pyelonephritis an ascending urinary tract infection that has reached the pyelum (pelvis) of the kidney.

Pyodermatitis refers to inflammation of the skin.

Pyorrhea see periodontitis.

Pyretic referring to fever.

Pyrexia fever of unknown origin.

Pyridoxal a chemical form of vitamin B6. See vitamin B6.

Pyridoxamine a chemical form of vitamin B6. See vitamin B6.

Pyridoxine a chemical form of vitamin B6. See vitamin B6.

Pyrolysis decomposition or transformation of a compound caused by heat. *adj.* pyrolytic.

PYY Peptide a 36 amino acid peptide secreted by L cells of the distal small intestine and colon that inhibits gastric and pancreatic secretion.

QT interval is a measure of the time between the start of the Q wave and the end of the T wave in the heart's electrical cycle. A prolonged QT interval is a biomarker for ventricular tachyarrhythmias and a risk factor for sudden death.

Quorum sensing (QS) the control of gene expression in response to cell density, is used by both gram-negative and gram-positive bacteria to regulate a variety of physiological functions.

Radiolysis the dissociation of molecules by radiation.

Radioprotective serving to protect or aiding in protecting against the injurious effect of radiations.

RAGE is the receptor for advanced glycation end products, a multiligand receptor that propagates cellular dysfunction in several inflammatory disorders, in tumours and in diabetes.

RAS see renin-angiotensin system or recurrent aphthous stomatitis.

Rash a temporary eruption on the skin, see uticaria.

Reactive oxygen species species such as superoxide, hydrogen peroxide, and hydroxyl radical. At low levels, these species may function in cell signaling processes. At higher levels, these species may damage cellular macromolecules (such as DNA and RNA) and participate in apoptosis (programmed cell death).

Rec A is a 38 kDa *Escherichia coli* protein essential for the repair and maintenance of DNA.

Receptor for advanced glycation end products (RAGE) is a member of the immunoglobulin superfamily of cell surface molecules; mediates neurite outgrowth and cell migration upon stimulation with its ligand, amphoterin.

Recticulocyte non-nucleated stage in the development of the red blood cell.

Recticulocyte lysate cell lysate produced from reticulocytes, used as an in-vitro translation system.

Recticuloendothelial system part of the immune system, consists of the phagocytic cells located in reticular connective tissue, primarily monocytes and macrophages.

Recurrent aphthous stomatitis, or RAS is a common, painful condition in which recurring ovoid or round ulcers affect the oral mucosa.

Redox homeostasis is considered as the cumulative action of all free radical reactions and antioxidant defenses in different tissues.

Refrigerant a medicine or an application for allaying heat, fever or its symptoms.

Renal calculi kidney stones.

Renal interstitial fibrosis damage sustained by the kidneys' renal tubules and interstitial capillaries due to accumulation of extracellular waste in the wall of the small arteries and arterioles.

Renin also known as an angiotensinogenase, is an enzyme that participates in the body's renin-angiotensin system (RAS).

Renin-angiotensin system (RAS) also called the renin-angiotensin-aldosterone system (RAAS) is a hormone system that regulates blood pressure and water (fluid) balance.

Reperfusion the restoration of blood flow to an organ or tissue that has had its blood supply cut off, as after a heart attack.

Reporter gene a transfected gene that produces a signal, such as green fluorescence, when it is expressed.

Resistin a cysteine-rich protein secreted by adipose tissue of mice and rats.

Resolutive a substance that induces subsidence of inflammation.

Resolvent reduce inflammation or swelling.

Resorb to absorb or assimilate a product of the body such as an exudates or cellular growth.

Restenosis is the reoccurrence of stenosis, a narrowing of a blood vessel, leading to restricted blood flow.

Resveratrol is a phytoalexin produced naturally by several plants when under attack by pathogens such as bacteria or fungi. It is a potent antioxidant found in red grapes and other plants.

Retinol a form of vitamin A, see vitamin A.

Retinopathy a general term that refers to some form of non-inflammatory damage to the retina of the eye.

Revulsive counterirritant, used for swellings.

Rheumatic pertaining to rheumatism or to abnormalities of the musculoskeletal system.

Rheumatism, Rheumatic disorder, Rheumatic diseases refers to various painful medical conditions which affect bones, joints, muscles, tendons. Rheumatic diseases are characterized by the signs of inflammation – redness, heat, swelling, and pain.

Rheumatoid arthritis (RA) is a chronic, systemic autoimmune disorder that most commonly causes inflammation and tissue damage in joints (arthritis) and tendon sheaths, together with anemia.

Rhinitis irritation and inflammation of some internal areas of the nose and the primary symptom of rhinitis is a runny nose.

Rhinoplasty is surgery to repair or reshape the nose.

Rhinorrhea commonly known as a runny nose, characterized by an unusually significant amount of nasal discharge.

Rhinosinusitis inflammation of the nasal cavity and sinuses.

Rho GTPases Rho-guanosine triphosphate hydrolase enzymes are molecular switches that regulate many essential cellular processes, including actin dynamics, gene transcription, cell-cycle progression and cell adhesion.

Ribosome inactivating proteins protein that are capable of inactivating ribosomes.

Rickets is a softening of the bones in children potentially leading to fractures and deformity.

Ringworm dermatophytosis, a skin infection caused by fungus.

Roborant restoring strength or vigour, a tonic.

Rotavirus the most common cause of infectious diarrhea (gastroenteritis) in young children and infants, one of several viruses that causes infections called stomach flu.

Rubefacient a substance for external application that produces redness of the skin e.g. by causing dilation of the capillaries and an increase in blood.

Ryanodine receptor intracellular Ca^{++} channels in animal tissues like muscles and neurons.

S.C. abbreviation for sub-cutaneous, beneath the layer of skin.

S-T segment the portion of an electrocardiogram between the end of the QRS complex and the beginning of the T wave. Elevation or depression of the S-T segment is the characteristics of myocardial ischemia or injury and coronary artery disease.

Sapraemia see septicaemia.

Sarcoma cancer of the connective or supportive tissue (bone, cartilage, fat, muscle, blood vessels) and soft tissues.

Sarcopenia degenerative loss of skeletal muscle mass and strength associated with aging.

Sarcoplasmic reticulum a special type of smooth endoplasmic reticulum found in smooth and striated muscle.

SARS Severe acute respiratory syndrome, the name of a potentially fatal new respiratory disease in humans which is caused by the SARS coronavirus (SARS-CoV).

Satiety state of feeling satiated, fully satisfied (appetite or desire).

Scabies a transmissible ectoparasite skin infection characterized by superficial burrows, intense pruritus (itching) and secondary infection.

Scarlatina scarlet fever, an acute, contagious disease caused by infection with group A streptococcal bacteria.

Schwann cells or neurolemmocytes, are the principal supporting cells of the peripheral nervous system, they form the myelin sheath of a nerve fibre.

Schistosomiasis is a parasitic disease caused by several species of fluke of the genus *Schistosoma*. Also known as bilharzia, bilharziosis or snail fever.

Schizophrenia a psychotic disorder (or a group of disorders) marked by severely impaired thinking, emotions, and behaviors.

Sciatica a condition characterised by pain deep in the buttock often radiating down the back of the leg along the sciatic nerve.

Scleroderma a disease of the body's connective tissue. The most common symptom is a thickening and hardening of the skin, particularly of the hands and face.

Scrofula a tuberculous infection of the skin on the neck caused by the bacterium *Mycobacterium tuberculosis*.

Scrophulosis see scrofula.

Scurf abnormal skin condition in which small flakes or sales become detached.

Scurvy a state of dietary deficiency of vitamin C (ascorbic acid) which is required for the synthesis of collagen in humans.

Secretagogue a substance that causes another substance to be secreted.

Sedative having a soothing, calming, or tranquilizing effect; reducing or relieving stress, irritability, or excitement.

Seizure the physical findings or changes in behavior that occur after an episode of abnormal electrical activity in the brain.

Selectins are a family of cell adhesion molecules; e.g. selectin-E, selectin –L, selectin P.

Selenium (Se) a trace mineral that is essential to good health but required only in tiny amounts; it is incorporated into proteins to make selenoproteins, which are important antioxidant enzymes. It is found in avocado, brazil nut, lentils, sunflower seeds, tomato, whole grain cereals, seaweed, seafood and meat.

Sensorineural bradyacuasia hearing impairment of the inner ear resulting from damage to the sensory hair cells or to the nerves that supply the inner ear.

Sepsis a condition in which the body is fighting a severe infection that has spread via the bloodstream.

Sequela an abnormal pathological condition resulting from a disease, injury or trauma.

Serine proteinase peptide hydrolases which have an active centre histidine and serine involved in the catalytic process.

Serotonergic liberating, activated by, or involving serotonin in the transmission of nerve impulses.

Serotonin a monoamine neurotransmitter synthesized in serotonergic neurons in the central nervous system.

Septicaemia a systemic disease associated with the presence and persistence of pathogenic microorganisms or their toxins in the blood.

Sequelae a pathological condition resulting from a prior disease, injury, or attack.

Sexual potentiator increases sexual activity and potency, enhances sexual performance due to increased blood flow and efficient metabolism.

Sexually transmitted diseases (STD) infections that are transmitted through sexual activity.

SGOT, Serum glutamic oxaloacetic transaminase an enzyme that is normally present in liver and heart cells. SGOT is released into blood when the liver or heart is damaged. Also called aspartate transaminase (AST).

SGPT, Serum glutamic pyruvic transaminase an enzyme normally present in serum and body tissues, especially in the liver; it is released into the serum as a result of tissue injury, also called Alanine transaminase (ALT),

Shiga–like toxin a toxin produced by the bacterium *Escherichia coli* which disrupts the function of ribosomes, also known as verotoxin.

Shiga toxigenic *Escherichia coli* (STEC) comprises a diverse group of organisms capable of causing severe gastrointestinal disease in humans.

Shiga toxin a toxin produced by the bacterium *Shigella dysenteriae*, which disrupts the function of ribosomes.

Shingles skin rash caused by the Zoster virus (same virus that causes chicken pox) and is medically termed Herpes zoster.

Sialogogue salivation-promoter, a substance used to increase or promote the excretion of saliva.

Sialoproteins glycoproteins that contain sialic acid as one of their carbohydrates.

Sialyation reaction with sialic acid or its derivatives; used especially with oligosaccharides.

Sialyltransferases enzymes that transfer sialic acid to nascent oligosaccharide.

Sickle cell disease is an inherited blood disorder that affects red blood cells. People with sickle cell disease have red blood cells that contain mostly hemoglobin S, an abnormal type of hemoglobin. Sometimes these red blood cells become sickle-shaped (crescent shaped) and have difficulty passing through small blood vessels.

Side stitch is an intense stabbing pain under the lower edge of the ribcage that occurs while exercising.

Signal transduction cascade refers to a series of sequential events that transfer a signal through a series of intermediate molecules until final regulatory molecules, such as transcription factors, are modified in response to the signal.

Silicon (Si) is required in minute amounts by the body and is important for the development of healthy hair and the prevention of nervous disorders. Lettuce is the best natural source of Silicon.

Sinapism signifies an external application, in the form of a soft plaster, or poultice.

Sinusitis inflammation of the nasal sinuses.

SIRC cells Statens Seruminstitut Rabbit Cornea (SIRC) cell line.

SIRT 1 stands for sirtuin (silent mating type information regulation 2 homolog) 1. It is an enzyme that deacetylates proteins that contribute to cellular regulation.

6-Keto-PGF1 alpha a physiologically active and stable hydrolysis product of Epoprostenol, found in nearly all mammalian tissues.

Skp1 (S-phase kinase-associated protein 1) is a core component of SCF ubiquitin ligases and mediates protein degradation.

Smads a family of intracellular proteins that mediate signaling by members of the TGF-beta (transforming growth factor beta) superfamily.

Smad2/3 a key signaling molecule for TGF-beta.

Smad7 a TGFβ type 1 receptor antagonist.

Smallpox is an acute, contagious and devastating disease in humans caused by *Variola* virus and have resulted in high mortality over the centuries.

Snuff powder inhaled through the nose.

SOD superoxide dismutase, is an enzyme that repairs cells and reduces the damage done to them by superoxide, the most common free radical in the body.

Sodium (Na) is an essential nutrient required for health. Sodium cations are important in neuron (brain and nerve) function, and in influencing osmotic balance between cells and the interstitial fluid and in maintenance of total body fluid homeostasis. Extra intake may cause a harmful effect on health. Sodium is naturally supplied by salt intake with food.

Soleus muscle smaller calf muscle lower down the leg and under the gastrocnemius muscle.

Somites mesodermal structures formed during embryonic development that give rise to segmented body parts such as the muscles of the body wall.

Somites Mesodermal structures formed during embryonic development that give rise to segmented body parts such as the muscles of the body wall.

Somnolence Sleepiness or drowsiness.

Soporific a sleep inducing drug.

SOS response a global response to DNA damage in which the cell cycle is arrested and DNA repair and mutagenesis are induced.

Soyasapogenins triterpenoid products obtained from the acid hydrolysis of soyasaponins, designated soyasapogenols A, B, C, D and E.

Soyasaponins bioactive saponin compounds found in many legumes.

Spasmolytic checking spasms, see antispasmodic.

Spermatorrhoea medically an involuntary ejaculation/drooling of semen usually nocturnal emissions.

Spermidine an important polyamine in DNA synthesis and gene expression.

Sphingolipid a member of a class of lipids derived from the aliphatic amino alcohol, sphingosine.

Spleen organ that filters blood and prevents infection.

Spleen tyrosine kinase (SYK) is an enigmatic protein tyrosine kinase functional in a number of diverse cellular processes such as the regulation of immune and inflammatory responses.

Splenitis inflammation of the spleen.

Splenocyte is a monocyte, one of the five major types of white blood cell, and is characteristically found in the splenic tissue.

Splenomegaly is an enlargement of the spleen.

Sprain to twist a ligament or muscle of a joint without dislocating the bone.

Sprue is a chronic disorder of the small intestine caused by sensitivity to gluten, a protein found in wheat and rye and to a lesser extent oats and barley. It causes poor absorption by the intestine of fat, protein, carbohydrates, iron, water, and vitamins A, D, E, and K.

Sputum matter coughed up and usually ejected from the mouth, including saliva, foreign material, and substances such as mucus or phlegm, from the respiratory tract.

SREBP-1 see sterol regulatory element-binding protein-1.

Stanch to stop or check the flow of a bodily fluid like blood from a wound.

Statin a type of lipid-lowering drug.

Status epilepticus refers to a life-threatening condition in which the brain is in a state of persistent seizure.

STD sexually transmitted disease.

Steatorrhea is the presence of excess fat in feces which appear frothy, foul smelling and floats because of the high fat content.

Steatohepatitis liver disease, characterized by inflammation of the liver with fat accumulation in the liver.

Steatosis refer to the deposition of fat in the interstitial spaces of an organ like the liver, fatty liver disease.

Sterility inability to produce offspring, also called asepsis.

Steroidogenic relating to steroidogenisis.

Steroidogenisis the production of steroids.

Sterol regulatory element-binding protein-1 (SREBP1) is a key regulator of the transcription of numerous genes that function in the metabolism of cholesterol and fatty acids.

Stimulant a substance that promotes the activity of a body system or function.

Stomachic (digestive stimulant), an agent that stimulates or strengthens the activity of the stomach; used as a tonic to improve the appetite and digestive processes.

Stomatitis oral inflammation and ulcers, may be mild and localized or severe, widespread, and painful.

Stomatology medical study of the mouth and its diseases.

Stool faeces.

Strangury is the painful passage of small quantities of urine which are expelled slowly by straining with severe urgency; it is usually accompanied with the unsatisfying feeling of a remaining volume inside and a desire to pass something that will not pass.

Straub tail condition in which an animal carries its tail in an erect (vertical or nearly vertical) position.

STREPs sterol regulatory element binding proteins, a family of transcription factors that regulate lipid homeostasis by controlling the expression of a range of enzymes required for endogenous cholesterol, fatty acid, triacylglycerol and phospholipid synthesis.

Stria terminalis a structure in the brain consisting of a band of fibres running along the lateral margin of the ventricular surface of the thalamus.

Striae gravidarum a cutaneous condition characterized by stretch marks on the abdomen during and following pregnancy.

Stricture an abnormal constriction of the internal passageway within a tubular structure such as a vessel or duct

Strongyloidiasis an intestinal parasitic infection in humans caused by two species of the parasitic nematode *Strongyloides*. The nematode or round worms are also called thread worms.

Styptic a short stick of medication, usually anhydrous aluminum sulfate (a type of alum) or titanium dioxide, which is used for stanching blood by causing blood vessels to contract at the site of the wound. Also called hemostatic pencil. see antihaemorrhagic.

Subarachnoid hemorrhage is bleeding in the area between the brain and the thin tissues that cover the brain.

Sudatory medicine that causes or increases sweating. Also see sudorific.

Sudorific a substance that causes sweating.

Sulfur Sulfur is an essential component of all living cells. Sulfur is important for the synthesis of sulfur-containing amino acids, all polypeptides, proteins, and enzymes such as glutathione an important sulfur-containing tripeptide which plays a role in cells as a source of chemical reduction potential. Sulfur is also important for hair formation. Good plant sources are garlic, onion, leeks and other Alliaceous vegetables, Brassicaceous vegetables like cauliflower, cabbages, Brussels sprout, Kale; legumes – beans, green and red gram, soybeans; horse radish, water cress, wheat germ.

Superior mesenteric artery (SMA) arises from the anterior surface of the abdominal aorta, just inferior to the origin of the celiac trunk, and supplies the intestine from the lower part of the duodenum to the left colic flexure and the pancreas.

Superoxidae mutase (SOD) antioxidant enzyme.

Suppuration the formation of pus, the act of becoming converted into and discharging pus.

Supraorbital located above the orbit of the eye.

SYK , Spleen tyrosine kinase is a human protein and gene. Syk plays a similar role in transmitting signals from a variety of cell surface receptors including CD74, Fc Receptor, and integrins.

Sympathetic nervous system the part of the autonomic nervous system originating in the thoracic and lumbar regions of the spinal cord that in general inhibits or opposes the physiological effects of the parasympathetic nervous system, as in tending to reduce digestive secretions or speed up the heart.

Synaptic plasticity the ability of neurons to change the number and strength of their synapses.

Synaptogenesis the formation of synapses.

Synaptoneurosomes purified synapses containing the pre- and postsynaptic termini.

Synaptosomes isolated terminal of a neuron.

Syncope fainting, sudden loss of consciousness followed by the return of wakefulness.

Syndactyly webbed toes, a condition where two or more digits are fused together.

Syneresis expulsion of liquid from a gel, as contraction of a blood clot and expulsion of liquid.

Syngeneic genetically identical or closely related, so as to allow tissue transplant; immunologically compatible.

Synovial lubricating fluid secreted by synovial membranes, as those of the joints.

Synoviocyte located in the synovial membrane, there are two types. Type A cells are more numerous, have phagocytic characteristics and produce degradative enzymes. Type B cells produce synovial fluid, which lubricates the joint and nurtures nourishes the articular cartilage.

Syphilis is perhaps the best known of all the STD's. Syphilis is transmitted by direct contact with infection sores, called chancres, syphitic skin rashes, or mucous patches on the tongue and mouth during kissing, necking, petting, or sexual intercourse. It can also be transmitted from a pregnant woman to a fetus after the 4th month of pregnancy.

Systolic the blood pressure when the heart is contracting. It is specifically the maximum arterial pressure during contraction of the left ventricle of the heart.

T cells or T lymphocytes, a type of white blood cell that play a key role in the immune system.

Tachyarrhythmia any disturbance of the heart rhythm in which the heart rate is abnormally increased.

Tachycardia a false heart rate applied to adults to rates over 100 beats per minute.

Tachyphylaxia a decreased response to a medicine given over a period of time so that larger doses are required to produce the same response.

Tachypnea abnormally fast breathing.

Taenia a parasitic apeworm or flatworm of the genus, *Taenia*.

Taeniacide an agent that kills tapeworms.

TBARS see thiobarbituric acid reactive substances.

T-cell a type of white blood cell that attacks virus-infected cells, foreign cells and cancer cells.

TCA cycle see Tricarboxylic acid cycle.

TCID50 median tissue culture infective dose; that amount of a pathogenic agent that will produce pathological change in 50% of cell cultures.

Telencephalon the cerebral hemispheres, the largest divisions of the human brain.

Telomerase enzyme that acts on parts of chromosomes known as telomeres.

Temporomandibular joint disorder (TMJD or TMD syndrome) a disorder characterized by acute or chronic inflammation of the temporomandibular joint, that connects the mandible to the skull.

Tendonitis is inflammation of a tendon.

Tenesmus a strong desire to defaecate.

Teratogen is an agent that can cause malformations of an embryo or fetus. *adj.* teratogenic.

Testicular torsion twisting of the spermatic cord, which cuts off the blood supply to the testicle and surrounding structures within the scrotum.

Tetanus an acute, potentially fatal disease caused by tetanus bacilli multiplying at the site of an injury and producing an exotoxin that reaches the central nervous system producing prolonged contraction of skeletal muscle fibres. Also called lockjaw.

Tete acute dermatitis caused by both bacterial and fungal infection

Tetter any of a number of skin diseases.

TGF-beta transforming growth factor beta is a protein that controls proliferation, cellular differentiation, and other functions in most cells.

Th cells or T helper cells a subgroup of lymphocytes that helps other white blood cells in immunologic processes.

Thermogenic tending to produce heat, applied to drugs or food (fat burning food)

Thiobarbituric acid reactive substances (TBARS) a well-established method for screening and monitoring lipid peroxidation.

Thixotropy the property exhibited by certain gels of becoming fluid when stirred or shaken and returning to the semisolid state upon standing.

Thrombocythaemia a blood condition characterize by a high number of platelets in the blood.

Thrombocytopenia a condition when the bone marrow does not produce enough platelets (thrombocytes) like in leukaemia.

Thromboembolism formation in a blood vessel of a clot (thrombus) that breaks loose and is carried by the blood stream to plug another vessel.

Thrombogenesis formation of a thrombus or blood clot.

Thrombophlebitis occurs when there is inflammation and clot in a surface vein.

Thromboplastin an enzyme liberated from blood platelets that converts prothrombin into thrombin as blood starts to clot, also called thrombokinase.

Thrombosis the formation or presence of a thrombus (clot).

Thromboxanes any of several compounds, originally derived from prostaglandin precursors in platelets that stimulate aggregation of platelets and constriction of blood vessels.

Thromboxane B2 the inactive product of thromboxane.

Thrombus a fibrinous clot formed in a blood vessel or in a chamber of the heart.

Thrush a common mycotic infection caused by yeast, *Candida albicans*, in the digestive tract or vagina. In children it is characterized by white spots on the tongue.

Thymocytes are T cell precursors which develop in the thymus.

Thyrotoxicosis or hyperthyroidism – an overactive thyroid gland, producing excessive circulating free thyroxine and free triiodothyronine, or both.

TIMP-3 a human gene belongs to the tissue inhibitor of matrix metalloproteinases (MMP) gene family. see MMP.

Tincture solution of a drug in alcohol.

Tinea ringworm, fungal infection on the skin.

Tinea cruris ringworm of the groin.

Tinea favosa See favus.

Tinea pedis fungal infection of the foot, also called athletes's foot.

Tinnitus a noise in the ears, as ringing, buzzing, roaring, clicking, etc.

Tisane a herbal infusion used as tea or for medicinal purposes.

Tissue plasminogen activator a serine protease involved in the breakdown of blood clots.

TNF alpha cachexin or cachectin and formally known as tumour necrosis factor-alpha, a cytokine involved in systemic inflammation. primary role of TNF is in the regulation of immune cells. TNF is also able to induce apoptotic cell death, to induce inflammation, and to inhibit tumorigenesis and viral replication.

Tocolytics medications used to suppress premature labor.

Tocopherol fat soluble organic compounds belonging to vitamin E group. See vitamin E.

Tocotrienol fat soluble organic compounds belonging to vitamin E group. See vitamin E.

Toll-like receptors (TLRs) a class of proteins that play a key role in the innate immune system.

Tonic substance that acts to restore, balance, tone, strengthen, or invigorate a body system without overt stimulation or depression

Tonic clonic seizure a type of generalized seizure that affects the entire brain.

Tonsillitis an inflammatory condition of the tonsils due to bacteria, allergies or respiratory problems.

Topoisomerases a class of enzymes involved in the regulation of DNA supercoiling.

Topoisomerase inhibitors a new class of anticancer agents with a mechanism of action aimed at interrupting DNA replication in cancer cells.

Total parenteral nutrition (TPN) is a method of feeding that bypasses the gastrointestinal tract.

Toxemia is the presence of abnormal substances in the blood, but the term is also used for a serious condition in pregnancy that involves hypertension and proteinuria. Also called preeclampsia.

Tracheitis is a bacterial infection of the trachea; also known as bacterial tracheitis or acute bacterial tracheitis.

Trachoma a contagious disease of the conjunctiva and cornea of the eye, producing painful sensitivity to strong light and excessive tearing.

TRAIL acronym for tumour necrosis factor-related apoptosis-inducing ligand, is a cytokine that preferentially induces apoptosis in tumour cells.

Tranquilizer a substance drug used in calming person suffering from nervous tension or anxiety.

Transaminase also called aminotransferase is an enzyme that catalyzes a type of reaction between an amino acid and an α-keto acid.

Transaminitis increase in alanine aminotransferase (ALT) and/or aspartate aminotransferase (AST) to >5 times the upper limit of normal.

Transcatheter arterial chemoembolization (TACE) is an interventional radiology procedure involving percutaneous access of to the hepatic artery and passing a catheter through the abdominal artery aorta followed by radiology. It is used extensively in the palliative treatment of unresectable hepatocellular carcinoma (HCC)

Transcriptional activators are proteins that bind to DNA and stimulate transcription of nearby genes.

Transcriptional coactivator PGC-1 a potent transcriptional coactivator that regulates oxidative metabolism in a variety of tissues.

Transcriptome profiling to identify genes involved in peroxisome assembly and function.

Transforming growth factor beta (TGF-β) a protein that controls proliferation, cellular differentiation, and other functions in most cells.

TRAP 6 thrombin receptor activating peptide with 6 amino acids.

Tremorine a chemical that produces a tremor resembling Parkinsonian tremor.

Tremulous marked by trembling, quivering or shaking.

Triacylglycerols or triacylglyceride, is a glyceride in which the glycerol is esterified with three fatty acids.

Tricarboxylic acid cycle (TCA cycle) a series of enzymatic reactions in aerobic organisms involving oxidative metabolism of acetyl units and producing high-energy phosphate compounds, which serve as the main source of cellular energy. Also called citric acid cycle, Krebs cycle.

Trichophytosis infection by fungi of the genus *Trichophyton*.

Trigeminal neuralgia (TN) is a neuropathic disorder of one or both of the facial trigeminal nerves, also known as prosopalgia.

Triglycerides a type of fat (lipids) found in the blood stream.

Trismus continuous contraction of the muscles of the jaw, specifically as a symptom of tetanus, or lockjaw; inability to open mouth fully.

TrKB receptor also known as TrKB tyrosine kinase, a protein in humans that acts as a catalytic receptor for several neutrophins.

Trolox Equivalent measures the antioxidant capacity of a given substance, as compared to the standard, Trolox also referred to as TEAC (Trolox equivalent antioxidant capacity).

Trypanocidal destructive to trypanosomes.

Trypanosomes protozoan of the genus *Trypanosoma*.

Trypanosomiasis human disease or an infection caused by a trypanosome.

Trypsin an enzyme of pancreatic juice that hydrolyzes proteins into smaller polypeptide units.

Trypsin inhibitor small protein synthesized in the exocrine pancreas which prevents conversion of trypsinogen to trypsin, so protecting itself against trypsin digestion.

Tuberculosis (TB) is a bacterial infection of the lungs caused by a bacterium called *Mycobacterium tuberculosis,* characterized by the formation of lesions (tubercles) and necrosis in the lung tissues and other organs.

Tumorigenesis formation or production of tumours.

Tumour an abnormal swelling of the body other than those caused by direct injury.

Tussis a cough.

Tympanic membrane ear drum.

Tympanitis infection or inflammation of the inner ear.

Tympanophonia increased resonance of one's own voice, breath sounds, arterial murmurs, etc., noted especially in disease of the middle ear.

Tympanosclerosis see myringoslcerosis.

Tyrosinase a copper containing enzyme found in animals and plants that catalyses the oxidation of phenols (such as tyrosine) and the production of melanin and other pigments from tyrosine by oxidation.

UCP1 an uncoupling protein found in the mitochondria of brown adipose tissue used to generate heat by non-shivering thermogenesis.

UCP – 2 enzyme uncoupling protein 2 enzyme, a mitochondrial protein expressed in adipocytes.

Ulcer an open sore on an external or internal body surface usually accompanied by disintegration of tissue and pus.

Ulcerative colitis is one of 2 types of inflammatory bowel disease – a condition that causes the bowel to become inflamed and red.

Ulemorrhagia bleeding of the gums.

Ulitis inflammation of the gums.

Unguent ointment.

Unilateral ureteral obstruction unilateral blockage of urine flow through the ureter of one kidney, resulting in a backup of urine, distension of the renal pelvis and calyces, and hydronephrosis.

Uraemia an excess in the blood of urea, creatinine and other nitrogenous end products of protein and amino acids metabolism, more correctly referred to as azotaemia.

Urethra tube conveying urine from the bladder to the external urethral orifice.

Urethritis is an inflammation of the urethra caused by infection.

Uricemia an excess of uric acid or urates in the blood.

Uricosuric promoting the excretion of uric acid in the urine.

Urinary pertaining to the passage of urine.

Urinogenital relating to the genital and urinary organs or functions.

Urodynia pain on urination.

Urokinase a serine protease enzyme in human urine that catalyzes the conversion of plasminogen to plasmin.

Urolithiasis formation of stone in the urinary tract (kidney bladder or urethra).

Urticant a substance that causes wheals to form.

Urticaria (or hives) is a skin condition, commonly caused by an allergic reaction, that is characterized by raised red skin welts.

Uterine relating to the uterus.

Uterine relaxant an agent that relaxes the muscles in the uterus.

Uterine stimulant an agent that stimulates the uterus (and often employed during active childbirth).

Uterotonic giving muscular tone to the uterus.

Uterotrophic causing an effect on the uterus.

Uterus womb.

Vagotomy the surgical cutting of the vagus nerve to reduce acid secretion in the stomach.

Vagus nerve a cranial nerve, that is, a nerve connected to the brain. The vagus nerve has branches to most of the major organs in the body, including the larynx, throat, windpipe, lungs, heart, and most of the digestive system

Variola or smallpox, a contagious disease unique to humans, caused by either of two virus variants, *Variola major* and *Variola minor*. The disease is characterised by fever, weakness and skin eruption with pustules that form scabs that leave scars.

Varicose veins are veins that have become enlarged and twisted.

Vasa vasorum is a network of small blood vessels that supply large blood vessels. *plur.* vasa vasori.

Vascular endothelial growth factor (VEGF) a polypeptide chemical produced by cells that stimulates the growth of new blood vessels.

Vasoconstrictor drug that causes constriction of blood vessels.

Vasodilator drug that causes dilation or relaxation of blood vessels.

Vasodilatory causing the widening of the lumen of blood vessels.

Vasomotor symptoms menopausal symptoms characterised by hot flushes and night sweats.

Vasospasm refers to a condition in which blood vessels spasm, leading to vasoconstriction and subsequently to tissue ischemia and death (necrosis).

Vasculogenesis process of blood vessel formation occurring by a de novo production of endothelial cells.

VCAM-1 (vascular cell adhesion molecule-1) also known as CD106, contains six or seven immunoglobulin domains and is expressed on both large and small vessels only after the endothelial cells are stimulated by cytokines.

VEGF Vascular endothelial growth factor.

Venereal disease (VD) term given to the diseases syphilis and gonorrhoea.

Venule a small vein, especially one joining capillaries to larger veins.

Vermifuge a substance used to expel worms from the intestines.

Verotoxin a Shiga-like toxin produced by *Escherichia coli,* which disrupts the function of ribosomes, causing acute renal failure.

Verruca plana is a reddish-brown or flesh-colored, slightly raised, flat-surfaced, well-demarcated papule on the hand and face, also called flat wart.

Vertigo an illusory, sensory perception that the surroundings or one's own body are revolving; dizziness.

Very-low-density lipoprotein (VLDL) a type of lipoprotein made by the liver. VLDL is one of the five major groups of lipoproteins (chylomicrons, VLDL, intermediate-density lipoprotein, low-density lipoprotein, high-density lipoprotein (HDL)) that enable fats and cholesterol to move within the water-based solution of the bloodstream. VLDL is converted in the bloodstream to low-density lipoprotein (LDL).

Vesical calculus calculi (stones) in the urinary bladder

Vesicant a substance that causes tissue blistering.

Vestibular relating to the sense of balance.

Vestibular disorders includes symptoms of dizziness, vertigo, and imbalance; it can be result from or worsened by genetic or environmental conditions.

Vestibular system includes parts of the inner ear and brain that process sensory information involved with controlling balance and eye movement.

Vibrissa stiff hairs that are located especially about the nostrils.

Viremia a medical condition where viruses enter the bloodstream and hence have access to the rest of the body.

Visceral fat intra-abdominal fat, is located inside the peritoneal cavity, packed in between internal organs and torso.

Vitamin any complex, organic compound, found in various food or sometimes synthesized in the body, required in tiny amounts and are essential for the regulation of metabolism, normal growth and function of the body.

Vitamin A retinol, fat-soluble vitamins that play an important role in vision, bone growth, reproduction, cell division, and cell differentiation, helps regulate the immune system in preventing or fighting off infections. Vitamin A that is found in colorful fruits and vegetables is called provitamin A carotenoid. They can be made into retinol in the body. Deficiency of vitamin A results in night blindness and keratomalacia.

Vitamin B1 also called thiamine, water-soluble vitamins, dissolve easily in water, and in general, are readily excreted from the body they are not readily stored, consistent daily intake is important. It functions as coenzyme in the metabolism of carbohydrates and branched chain amino acids, and other cellular processes. Deficiency results in beri-beri disease.

Vitamin B2 also called riboflavin, an essential water-soluble vitamin that functions as coenzyme in redox reactions. Deficiency causes ariboflavinosis.

Vitamin B3 comprises niacin and niacinamide, water-soluble vitamin that function as coenzyme or co-substrate for many redox reactions and is required for energy metabolism. Deficiency causes pellagra.

Vitamin B5 also called pantothenic acid, a water-soluble vitamin that function as coenzyme in fatty acid metabolism. Deficiency causes paresthesia.

Vitamin B6 water-soluble vitamin, exists in three major chemical forms: pyridoxine, pyridoxal, and pyridoxamine. Vitamin B6 is needed in enzymes involved in protein metabolism, red blood cell metabolism, efficient functioning of nervous and immune systems and hemoglobin formation. Deficiency causes anaemia and peripheral neuropathy.

Vitamin B7 also called biotin or vitamin H, an essential water-soluble vitamin, is involved in the synthesis of fatty acids amino acids and glucose, in energy metabolism. Biotin promotes normal health of sweat glands, bone marrow, male gonads, blood cells, nerve tissue, skin and hair, Deficiency causes dermatitis and enteritis.

Vitamin B9 also called folic acid, an essential water-soluble vitamin. Folate is especially important during periods of rapid cell division and growth such as infancy and pregnancy. Deficiency during pregnancy is associated with birth defects such as neural tube defects. Folate is also important for production of red blood cells and prevent anemia. Folate is needed to make DNA and RNA, the building blocks of cells. It also helps prevent changes to DNA that may lead to cancer.

Vitamin B12 a water-soluble vitamin, also called cobalamin as it contains the metal cobalt. It helps maintain healthy nerve cells and red blood cells, and DNA production. Vitamin B12 is bound to the protein in food. Deficiency causes megaloblastic anaemia.

Vitamin C also known as ascorbic acid is an essential water-soluble vitamin. It functions as cofactor for reactions requiring reduced copper or iron metallonzyme and as a protective antioxidant. Deficiency of vitamin C causes scurvy.

Vitamin D a group of fat-soluble, prohormone vitamin, the two major forms of which are vitamin D2 (or ergocalciferol) and vitamin D3 (or cholecalciferol). Vitamin D obtained from sun exposure, food, and supplements is biologically inert and must undergo two hydroxylations in the body for activation. Vitamin D is essential for promoting calcium absorption in the gut and maintaining adequate serum calcium and phosphate concentrations to enable normal growth and mineralization of bone and prevent hypocalcemic tetany. Deficiency causes rickets and osteomalacia. Vitamin D has other roles in human health, including modulation of neuromuscular and immune function, reduction of inflammation and modulation of many genes encoding proteins that regulate cell proliferation, differentiation, and apoptosis.

Vitamin E is the collective name for a group of fat-soluble compounds and exists in eight chemical forms (alpha-, beta-, gamma-, and delta-tocopherol and alpha-, beta-, gamma-, and delta-tocotrienol). It has pronounced antioxidant activities stopping the formation of Reactive Oxygen Species when fat undergoes oxidation and help prevent or delay the chronic diseases associated with free radicals. Besides its antioxidant activities, vitamin E is involved in immune function, cell signaling, regulation of gene expression, and other metabolic processes. Deficiency is very rare but can cause mild hemolytic anemia in newborn infants.

Vitamin K a group of fat soluble vitamin and consist of vitamin K1 which is also known as phylloquinone or phytomenadione (also called phytonadione) and vitamin K2 (menaquinone, menatetrenone). Vitamin K plays an important role in blood clotting. Deficiency is very rare but can cause bleeding diathesis.

Vitamin P a substance or mixture of substances obtained from various plant sources, identified as citrin or a mixture of bioflavonoids, thought to but not proven to be useful in reducing the extent of hemorrhage.

Vitiligo a chronic skin disease that causes loss of pigment, resulting in irregular pale patches of skin. It occurs when the melanocytes, cells responsible for skin pigmentation, die or are unable to function. Also called leucoderma.

Vitreoretinopathy see proliferative vitreoretinopathy.

VLA-4 very late antigen-4, expressed by most leucocytes but it is observed on neutrophils under special conditions.

VLDL see very low density lipoproteins.

Vomitive substance that causes vomiting.

Vulnerary (wound healer), a substance used to heal wounds and promote tissue formation.

Wart an infectious skin tumour caused by a viral infection.

Welt see wheal.

Wheal a firm, elevated swelling of the skin. Also called a weal or welt.

White fat white adipose tissue (WAT) in mammals, store of energy. cf. brown fat.

Whitlow painful infection of the hand involving one or more fingers that typically affects the terminal phalanx.

Whooping cough acute infectious disease usually in children caused by a *Bacillus* bacterium and accompanied by catarrh of the respiratory passages and repeated bouts of coughing.

Wnt signaling pathway is a network of proteins involved in embryogenesis and cancer, and also in normal physiological processes.

X-linked agammaglobulinemia also known as X-linked hypogammaglobulinemia, XLA, Bruton type agammaglobulinemia, Bruton syndrome, or sex-linked agammaglobulinemia; a rare x-linked genetic disorder that affects the body's ability to fight infection.

Xanthine oxidase a flavoprotein enzyme containing a molybdenum cofactor (Moco) and (Fe2S2) clusters, involved in purine metabolism. In humans, inhibition of xanthine oxidase reduces the production of uric acid, and prevent hyperuricemia and gout.

Xanthones unique class of biologically active phenol compounds with the molecular formula C13H8O2 possessing antioxidant properties, discovered in the mangosteen fruit.

Xenobiotics a chemical (as a drug, pesticide, or carcinogen) that is foreign to a living organism.

Xenograft a surgical graft of tissue from one species to an unlike species.

Xerophthalmia a medical condition in which the eye fails to produce tears.

Yaws an infectious tropical infection of the skin, bones and joints caused by the spirochete bacterium *Treponema pertenue,* characterized by papules and papilloma with subsequent deformation of the skins, bone and joints; also called framboesia.

Yellow fever is a viral disease that is transmitted to humans through the bite of infected mosquitoes. Illness ranges in severity from an influenza-like syndrome to severe hepatitis and hemorrhagic fever. Yellow fever virus (YFV) is maintained in nature by mosquito-borne transmission between nonhuman primates.

Zeaxanthin a common carotenoid, found naturally as coloured pigments in many fruit vegetables and leafy vegetables. It is important for good vision and is one of the two carotenoids contained within the retina of the eye. Within the central macula, zeaxanthin predominates, whereas in the peripheral retina, lutein predominates.

Zinc (Zn) is an essential mineral for health. It is involved in numerous aspects of cellular metabolism: catalytic activity of enzymes, immune function, protein synthesis, wound healing, DNA synthesis, and cell division. It also supports normal growth and development during pregnancy, childhood, and adolescence and is required for proper sense of taste and smell. Dietary sources include beans, nuts, pumpkin seeds, sunflower seeds, whole wheat bread and animal sources.

Scientific Glossary

Abaxial facing away from the axis, as of the surface of an organ.

Abscission shedding of leaves, flowers, or fruits following the formation of the abscission zone.

Acaulescent lacking a stem, or stem very much reduced.

Accrescent increasing in size after flowering or with age.

Achene a dry, small, one-seeded, indehiscent one-seeded fruit formed from a superior ovary of one carpel as in sunflower.

Acid soil soil that maintains a pH of less than 7.0.

Acidulous acid or sour in taste.

Actinomorphic having radial symmetry, capable of being divided into symmetrical halves by any plane, refers to a flower, calyx or corolla.

Aculeate having sharp prickles.

Acuminate tapering gradually to a sharp point.

Acute (Botany) tapering at an angle of less than 90° before terminating in a point as of leaf apex and base.

Adaxial side closest to the stem axis.

Adherent touching without organic fusion as of floral parts of different whorls.

Adnate united with another unlike part as of stamens attached to petals.

Adpressed lying close to another organ but not fused to it.

Adventitious arising in abnormal positions, e.g. roots arising from the stem, branches or leaves, buds arising elsewhere than in the axils of leaves.

Adventive Not native to and not fully established in a new habitat or environment; locally or temporarily naturalized. e.g. an adventive weed.

Aestivation refers to positional arrangement of the floral parts in the bud before it opens.

Akinete a thick-walled dormant cell derived from the enlargement of a vegetative cell. It serves as a survival structure.

Albedo inner layer of the peel of citrus fruit.

Aldephous having stamens united together by their filaments.

Alfisols soil with a clay-enriched subsoil and relatively high native fertility, having undergone only moderate leaching, containing aluminium, iron and with at least 35% base saturation, meaning that calcium, magnesium, and potassium are relatively abundant.

Alkaline soil soil that maintains a pH above 7.0, usually containing large amounts of calcium, sodium, and magnesium, and is less soluble than acidic soils.

Alkaloids naturally occurring bitter, complex organic-chemical compounds containing basic nitrogen and oxygen atoms and having various pharmacological effects on humans and other animals.

Allomorphic with a shape or form different from the typical.

Alluvial soil a fine-grained fertile soil deposited by water flowing over flood plains or in river beds.

Alluvium soil or sediments deposited by a river or other running water.

Alternate leaves or buds that are spaced along opposite sides of stem at different levels.

Amplexicaul clasping the stem as base of certain leaves.

Anatomizing interconnecting network as applied to leaf veins.

Anatropous With the ovule completely inverted.

Andisols are soils formed in volcanic ash and containing high proportions of glass and amorphous colloidal materials.

Androdioecious with male flowers and bisexual flowers on separate plants.

Androecium male parts of a flower; comprising the stamens of one flower.

Androgynophore a stalk bearing both the androecium and gynoecium above the perianth of the flower.

Androgynous with male and female flowers in distinct parts of the same inflorescence.

Andromonoecious having male flowers and bisexual flowers on the same plant.

Angiosperm a division of seed plants with the ovules borne in an ovary.

Annual a plant which completes its life cycle within a year.

Annular shaped like or forming a ring.

Annulus circle or ring-like structure or marking; the portion of the corolla which forms a fleshy, raised ring.

Anthelate an open, paniculate cyme.

Anther the part of the stamen containing pollen sac which produces the pollen.

Antheriferous containing anthers.

Anthesis the period between the opening of the bud and the onset of flower withering.

Anthocarp a false fruit consisting of the true fruit and the base of the perianth.

Anthocyanidins are common plant pigments. They are the sugar-free counterparts of anthocyanins.

Anthocyanins a subgroup of antioxidant flavonoids, are glucosides of anthocyanidins. They occur as water-soluble vacuolar pigments that may appear red, purple, or blue according to pH in plants.

Antipetala situated opposite petals.

Antisepala situated opposite sepals.

Antrorse directed forward upwards.

Apetalous lacking petals as of flowers with no corolla.

Apical meristem active growing point. A zone of cell division at the tip of the stem or the root.

Apically towards the apex or tip of a structure.

Apiculate ending abruptly in a short, sharp, small point.

Apiculum a short, pointed, flexible tip.

Apocarpous carpels separate in single individual pistils.

Apopetalous with separate petals, not united to other petals.

Aposepalous with separate sepals, not united to other sepals.

Appressed pressed closely to another structure but not fused or united.

Aquatic a plant living in or on water for all or a considerable part of its life span.

Arachnoid (Botany) formed of or covered with long, delicate hairs or fibers.

Arborescent resembling a tree; applied to non-woody plants attaining tree height and to shrubs tending to become tree-like in size.

Arbuscular mycorrhiza (AM) a type of mycorrhiza in which the fungus (of the phylum Glomeromycota) penetrates the cortical cells of the roots of a vascular plant and form unique structures such as arbuscules and vesicles. These fungi help plants to capture nutrients such as phosphorus and micronutrients from the soil.

Archegonium a flask-shaped female reproductive organ in mosses, ferns, and other related plants.

Arenasols sandy soils with very weak or no soil development.

Areolate with areolea.

Areole (Botany) a small, specialized, cushion-like area on a cactus from which hairs, glochids, spines, branches, or flowers may arise; an irregular angular spaces marked out on a surface e.g. fruit surface. *pl.* areolea.

Aril specialized outgrowth from the funiculus (attachment point of the seed) (or hilum) that encloses or is attached to the seed. *adj.* arillate.

Arillode a false aril; an aril originating from the micropyle instead of from the funicle or chalaza of the ovule, e.g. mace of nutmeg.

Aristate bristle-like part or appendage, e.g. awns of grains and grasses.

Aristulate having a small, stiff, bristle-like part or appendage; a diminutive of aristate

Articulate jointed; usually breaking easily at the nodes or point of articulation into segments.

Ascending arched upwards in the lower part and becoming erect in the upper part.

Ascospore spore produced in the ascus in Ascomycete fungi.

Ascus is the sexual spore-bearing cell produced in Ascomycete fungi. *pl.* asci.

Asperulous refers to a rough surface with short, hard projections.

Attenuate tapered or tapering gradually to a point.

Auricle an ear-like appendage that occurs at the base of some leaves or corolla.

Auriculate having auricles.

Awn a hair-like or bristle-like appendage on a larger structure.

Axil upper angle between a lateral organ, such as a leaf petiole and the stem that bears it.

Axile situated along the central axis of an ovary having two or more locules, as in axile placentation.

Axillary arising or growing in an axil.

Baccate beery-like, pulpy or fleshy.

Barbate bearded, having tufts of hairs.

Barbellae short, stiff, hair-like bristles. *adj.* barbellate.

Bark is the outermost layers of stems and roots of woody plants.

Basal relating to, situated at, arising from or forming the base.

Basaltic soil soil derived from basalt, a common extrusive volcanic rock.

Basidiospore a reproductive spore produced by Basidiomycete fungi.

Basidium a microscopic, spore-producing structure found on the hymenophore of fruiting bodies of Basidiomycete fungi.

Basifixed attached by the base, as certain anthers are to their filaments.

Basionym the synonym of a scientific name that supplies the epithet for the correct name.

Beak a prominent apical projection, especially of a carpel or fruit. *adj.* beaked.

Bearded having a tuft of hairs.

Berry a fleshy or pulpy indehiscent fruit from a single ovary with the seed(s) embedded in the fleshy tissue of the pericarp.

Biconvex convex on both sides.

Biennial completing the full cycle from germination to fruiting in more than one, but not more than 2 years.

Bifid forked, divided into two parts.

Bifoliolate having two leaflets.

Bilabiate having two lips as of a corolla or calyx with segments fused into an upper and lower lip.

Bipinnate twice pinnate; the primary leaflets being again divided into secondary leaflets.

Bipinnatisect refers to a pinnately compound leaf, in which each leaflet is again divided into pinnae.

Biserrate doubly serrate; with smaller regular, asymmetric teeth on the margins of larger teeth.

Bisexual having both sexes, as in a flower bearing both stamens and pistil, hermaphrodite or perfect.

Biternate Twice ternate; with three pinnae each divided into three pinnules.

Blade lamina; part of the leaf above the sheath or petiole.

Blotched see variegated.

Bole main trunk of tree from the base to the first branch.

Brachyblast a short, axillary, densely crowded branchlet or shoot of limited growth, in which the internodes elongate little or not at all.

Bracket fungus shelf fungus.

Bract a leaf-like structure, different in form from the foliage leaves, associated with an inflorescence or flower. *adj.* bracteate.

Bracteate possessing bracts.

Bracteolate having bracteoles.

Bracteole a small, secondary, bract-like structure borne singly or in a pair on the pedicel or calyx of a flower. *adj.* bracteolate.

Bristle a stiff hair.

Bulb a modified underground axis that is short and crowned by a mass of usually fleshy, imbricate scales. *adj.* bulbous.

Bulbil A small bulb or bulb-shaped body, especially one borne in the leaf axil or an inflorescence, and usually produced for asexual reproduction.

Bullate puckered, blistered.

Burr type of seed or fruit with short, stiff bristles or hooks or may refer to a deformed type of wood in which the grain has been misformed.

Bush low, dense shrub without a pronounced trunk.

Buttress supporting, projecting outgrowth from base of a tree trunk as in some Rhizophoraceae and Moraceae.

Caducous shedding or falling early before maturity refers to sepals and petals.

Caespitose growing densely in tufts or clumps; having short, closely packed stems.

Calcareous composed of or containing lime or limestone.

Calcrete a hardpan consisting gravel and sand cemented by calcium.

Callus a condition of thickened raised mass of hardened tissue on leaves or other plant parts often formed after an injury but sometimes a normal feature. A callus also can refer to an undifferentiated plant cell mass grown on a culture medium. *n.* callosity. *pl.* calli, callosities. *adj.* callose.

Calyptra the protective cap or hood covering the spore case of a moss or related plant.

Calyptrate operculate, having a calyptra.

Calyx outer floral whorl usually consisting of free sepals or fused sepals (calyx tube) and calyx lobes. It encloses the flower while it is still a bud. *adj.* calycine.

Calyx lobe one of the free upper parts of the calyx which may be present when the lower part is united into a tube.

Calyx tube the tubular fused part of the calyx, often cup shaped or bell shaped, when it is free from the corolla.

Cambisol soil with incipient soil formation, characterized by weak horizon differentiation.

Campanulate shaped like a bell refers to calyx or corolla.

Campylotropous With the ovule partially inverted and curved.

Canaliculate having groove or grooves.

Candelabriform having the shape of a tall branched candle-stick.

Canescent covered with short, fine whitish or grayish hairs or down.

Canopy uppermost leafy stratum of a tree.

Cap see pileus.

Capitate growing together in a head. Also means enlarged and globular at the tip.

Capitulum a flower head or inflorescence having a dense cluster of sessile, or almost sessile, flowers or florets.

Capsule a dry, dehiscent fruit formed from two or more united carpels and dehiscing at maturity by sections called valves to release the seeds. *adj.* capsular.

Carinate keeled.

Carpel a simple pistil consisting of ovary, ovules, style and stigma. *adj.* carpellary.

Carpogonium female reproductive organ in red algae. *pl.* carpogonia.

Carpophore part of the receptacle which is lengthened between the carpels as a central axis; any fruiting body or fruiting structure of a fungus.

Cartilaginous sinewy, having a firm, tough, flexible texture (in respect of leaf margins).

Caryopsis a simple dry, indehiscent fruit formed from a single ovary with the seed coat united with the ovary wall as in grasses and cereals.

Cataphyll a reduced or scarcely developed leaf at the start of a plant's life (i.e., cotyledons) or in the early stages of leaf development.

Catkin a slim, cylindrical, pendulous flower spike usually with unisexual flowers.

Caudate having a narrow, tail-like appendage.

Caudex thickened, usually underground base of the stem.

Caulescent having a well developed aerial stem.

Cauliflory botanical term referring to plants which flower and fruit from their main stems or woody trunks. *adj.* cauliflorus.

Cauline borne on the aerial part of a stem.

Chaffy having thin, membranous scales in the inflorescence as in the flower heads of the sunflower family.

Chalaza the basal region of the ovule where the stalk is attached.

Chartaceous papery, of paper-like texture.

Chasmogamous describing flowers in which pollination takes place while the flower is open.

Chloroplast a chlorophyll-containing organelle (plastid) that gives the green colour to leaves and stems. Plastids harness light energy that is used to fix carbon dioxide in the process called photosynthesis.

Chromoplast plastid containing colored pigments apart from chlorophyll.

Chromosomes thread-shaped structures that occur in pairs in the nucleus of a cell, containing the genetic information of living organisms.

Cilia hairs along the margin of a leaf or corolla lobe.

Ciliate with a fringe of hairs on the margin as of the corolla lobes or leaf.

Ciliolate minutely ciliate.

Cilium a straight, usually erect hair on a margin or ridge. *pl.* cilia.

Cincinnus a monochasial cyme in which the lateral branches arise alternately on opposite sides of the false axis.

Circinnate spirally coiled, with the tip innermost.

Circumscissile opening by a transverse line around the circumference as of a fruit.

Cladode the modified photosynthetic stem of a plant whose foliage leaves are much reduced or absent. *cf.* cladophyll, phyllode.

Cladophyll A photosynthetic branch or portion of a stem that resembles and functions as a leaf, like in asparagus. *cf.* cladode, phyllode.

Clamp connection In the Basidiomycetes fungi, a lateral connection or outgrowth formed between two adjoining cells of a hypha and arching over the septum between them.

Clavate club shaped thickened at one end refer to fruit or other organs.

Claw the conspicuously narrowed basal part of a flat structure.

Clay a naturally occurring material composed primarily of fine-grained minerals like kaolinite, montmorrillonite-smectite or illite which exhibit plasticity through a variable range of water content, and which can be hardened when dried and/or fired.

Clayey resembling or containing a large proportion of clay.

Cleft incised halfway down.

Cleistogamous refers to a flower in which fertilization occurs within the bud i.e. without the flower opening. *cf.* chasmogamous.

Climber growing more or less upwards by leaning or twining around another structure.

Clone all the plants reproduced, vegetatively, from a single parent thus having the same gentic make-up as the parent.

Coccus one of the sections of a distinctly lobed fruit which becomes separate at maturity; sometimes called a mericarp. *pl.* cocci.

Coenocarpium a fleshy, multiple pseudocarp formed from an inflorescence rather than a single flower.

Coherent touching without organic fusion, referring to parts normally together, e.g. floral parts of the same whorl. *cf.* adherent, adnate, connate.

Collar boundary between the above- and below ground parts of the plant axis.

Colliculate having small elevations.

Column a structure formed by the united style, stigma and stamen(s) as in Asclepiadaceae and Orchidaceae.

Commisure the place along which two structures, such as carpels, are joined.

Comose tufted with hairs at the ends as of seeds.

Composite having two types of florets as of the flowers in the sunflower family, Asteraceae.

Compost organic matter (like leaves, mulch, manure, etc) that breaks down in soil releasing its nutrients.

Compound describe a leaf that is further divided into leaflets or pinnae or flower with more than a single floret.

Compressed flattened in one plane.

Conceptacles specialised cavities of marine algae that contain the reproductive organs.

Concolorous uniformly coloured, as in upper and lower surfaces. *cf.* discolorous

Conduplicate folded together lengthwise.

Cone a reproductive structure composed of an axis (branch) bearing sterile bract-like organs and seed or pollen bearing structures.

Applied to Gymnospermae, Lycopodiaceae, Casuarinaceae and also in some members of Proteaceae.

Conic cone shaped, attached at the broader end.

Conic-capitate a cone-shaped head of flowers.

Connate fused to another structure of the same kind. *cf.* adherent, adnate, coherent.

Connective the tissue separating two lobes of an anther.

Connivent converging.

Conspecific within or belonging to the same species.

Contorted twisted.

Convolute refers to an arrangement of petals in a bud where each has one side overlapping the adjacent petal.

Cooperage making of barrels and casks.

Cordate heart-shaped as of leaves.

Core central part.

Coriaceous leathery texture as of leaves.

Corm a short, swollen, fleshy, underground plant stem that serves as a food storage organ used by some plants to survive winter or other adverse conditions

Cormel a miniature, new corm produced on a mature corm.

Corolla the inner floral whorl of a flower, usually consisting of free petals or a petals fused forming a corolla tube and corolla lobes. *adj.* corolline.

Corona a crown-like section of the staminal column, usually with the inner and outer lobes as in the **Stapelieae**.

Coroniform crown shaped, as in the pappus of Asteraceae.

Cortex the outer of the stem or root of a plant, bounded on the outside by the epidermis and on the inside by the endodermis containing undifferentiated cells.

Corymb a flat-topped, short, broad inflorescence, in which the flowers, through unequal pedicels, are in one horizontal plane and the youngest in the centre. *adj.* corymbose

Costa a thickened, linear ridge or the midrib of the pinna in ferns. *adj.* costate.

Costapalmate having definite costa (midrib) unlike the typical palmate leaf, but the leaflets are arranged radially like in a palmate leaf.

Cotyledon the primary seed leaf within the embryo of a seed.

Cover crop crop grown in between trees or in fields primarily to protect the soil from erosion, to improve soil fertility and to keep off weeds.

Crenate round-toothed or scalloped as of leaf margins.

Crenulate minutely crenate, very strongly scalloped.

Crisped with a curled or twisted edge.

Cristate having or forming a crest or crista.

Crozier shaped like a shepherd's crook.

Crustaceous like a crust; having a hard crust or shell.

Cucullate having the shape of a cowl or hood, hooded.

Culm the main aerial stem of the Graminae (grasses, sedges, rushes and other monocots).

Culm sheath the plant casing (similar to a leaf) that protects the young bamboo shoot during growth, attached at each node of culm.

Cultigen plant species or race known only in cultivation.

Cultivar cultivated variety; an assemblage of cultivated individuals distinguished by any characters significant for the purposes of agriculture, forestry or horticulture, and which, when reproduced, retains its distinguishing features.

Cuneate wedge-shaped, obtriangular.

Cupular cup-shaped, having a cupule.

Cupule a small cup-shaped structure or organ, like the cup at the base of an acorn.

Cusp an elongated, usually rigid, acute point. *cf.* mucro.

Cuspidate terminating in or tipped with a sharp firm point or cusp. *cf.* mucronate.

Cuspidulate constricted into a minute cusp. *cf.* cuspidate.

Cyathiform in the form of a cup, a little widened at the top.

Cyathium a specialised type of inflorescence of plants in the genus Euphorbia and Chamaesyce in which the unisexual flowers are clustered together within a bract-like envelope. *pl.* cyathia.

Cylindric tubular or rod shaped.

Cylindric-acuminate elongated and tapering to a point.
Cymbiform boat shaped, elongated and having the upper surface decidedly concave.
Cyme an inflorescence in which the lateral axis grows more strongly than the main axis with the oldest flower in the centre or at the ends. *adj.* cymose.
Cymule a small cyme or one or a few flowers.
Cystidium a relatively large cell found on the hymenium of a Basidiomycete, for example, on the surface of a mushroom.
Cystocarp fruitlike structure (sporocarp) developed after fertilization in the red algae.
Deciduous falling off or shedding at maturity or a specific season or stage of growth.
Decompound as of a compound leaf; consisting of divisions that are themselves compound.
Decorticate to remove the bark, rind or husk from an organ; to strip of its bark; to come off as a skin.
Decumbent prostrate, laying or growing on the ground but with ascending tips. *cf.* ascending, procumbent.
Decurrent having the leaf base tapering down to a narrow wing that extends to the stem.
Decussate having paired organs with successive pairs at right angles to give four rows as of leaves.
Deflexed bent downwards.
Dehisce to split open at maturity, as in a capsule.
Dehiscent splitting open at maturity to release the contents. *cf.* indehiscent.
Deltate triangular shape.
Deltoid shaped like an equilateral triangle.
Dendritic branching from a main stem or axis like the branches of a tree.
Dentate with sharp, rather coarse teeth perpendicular to the margin.
Denticulate finely toothed.
Diadelphous having stamens in two bundles as in Papilionaceae flowers.
Diageotropic the tendency of growing parts, such as roots, to grow at right angle to the line of gravity.
Dichasium a cymose inflorescence in which the branches are opposite and approximately equal. *pl.* dichasia. *adj.* dichasial.
Dichotomous divided into two parts.
Dicotyledon angiosperm with two cotyledons.
Didymous arranged or occurring in pairs as of anthers, having two lobes.
Digitate having digits or fingerlike projections.
Dikaryophyses or dendrophydia, irregularly, strongly branched terminal hyphae in the Hymenomycetes (class of Basidiomyctes) fungi.
Dimorphic having or occurring in two forms, as of stamens of two different lengths or a plant having two kinds of leaves.
Dioecious with male and female unisexual flowers on separate plants. *cf.* monoecious.
Diploid a condition in which the chromosomes in the nucleus of a cell exist as pairs, one set being derived from the female parent and the other from the male.
Diplobiontic life cycle life cycle that exhibits alternation of generations, which features of spore-producing multicellular sporophytes and gamete-producing multicellular gametophytes. mitoses occur in both the diploid and haploid phases.
Diplontic life cycle or gametic meiosis, wherein instead of immediately dividing meiotically to produce haploid cells, the zygote divides mitotically to produce a multicellular diploid individual or a group of more diploid cells.
Dipterocarpous trees of the family Dipterocarpaceae, with two-winged fruit found mainly in tropical lowland rainforest.
Disc (Botany) refers to the usually disc shaped receptacle of the flower head in Asteraceae; also the fleshy nectariferous organ usually between the stamens and ovary; also used for the enlarged style-end in Proteaceae.
Disc floret the central, tubular 4 or 5-toothed or lobed floret on the disc of an inflorescence, as of flower head of Asteraceae.
Disciform flat and rounded in shaped. *cf.* discoid, radiate.
Discoid resembling a disc; having a flat, circular form; disk-shaped *cf.* disciform, radiate.
Discolorous having two colours, as of a leaf which has different colors on the two surfaces. *cf.* concolorous.
Dispersal dissemination of seeds.

Distal site of any structure farthest from the point of attachment. *cf.* proximal.

Distichous referring to two rows of upright leaves in the same plane.

Dithecous having two thecae.

Divaricate diverging at a wide angle.

Domatium a part of a plant (e.g., a leaf) that has been modified to provide protection for other organisms. *pl.* domatia.

Dormancy a resting period in the life of a plant during which growth slows or appears to stop.

Dorsal referring to the back surface.

Dorsifixed attached to the back as of anthers.

Drupaceous resembling a drupe.

Drupe a fleshy fruit with a single seed enclosed in a hard shell (endocarp) which is tissue embedded in succulent tissue (mesocarp) surrounded by a thin outer skin (epicarp). *adj.* drupaceous.

Drupelet a small drupe.

Ebracteate without bracts.

Echinate bearing stiff, stout, bristly, prickly hairs.

Edaphic refers to plant communities that are distinguished by soil conditions rather than by the climate.

Eglandular without glands. *cf.* glandular.

Ellipsoid a 3-dimensional shape; elliptic in outline.

Elliptic having a 2-dimensional shape of an ellipse or flattened circle.

Emarginate refers to leaf with a broad, shallow notch at the apex. *cf.* retuse.

Embryo (Botany) a minute rudimentary plant contained within a seed or an archegonium, composed of the embryonic axis (shoot end and root end).

Endemic prevalent in or peculiar to a particular geographical locality or region.

Endocarp The hard innermost layer of the pericarp of many fruits.

Endosperm tissue that surrounds and nourishes the embryo in the angiosperm seed.

Endospermous refers to seeds having an endosperm.

Endotrophic as of mycorrhiza obtaining nutrients from inside.

Ensilage the process of preserving green food for livestock in an undried condition in airtight conditions. Also called silaging.

Entire having a smooth, continuous margin without any incisions or teeth as of a leaf.

Entisols soils that do not show any profile development other than an A horizon.

Eongate extended, stretched out.

Ephemeral transitory, short-lived.

Epicalyx a whorl of bracts, subtending and resembling a calyx.

Epicarp outermost layer of the pericarp of a fruit.

Epicormic attached to the corm.

Epicotyl the upper portion of the embryonic axis, above the cotyledons and below the first true leaves.

Epigeal above grounds with cotyledons raised above ground.

Epiparasite an organism parasitic on another that parasitizes a third.

Epipetalous borne on the petals, as of stamens.

Epiphyte a plant growing on, but not parasitic on, another plant, deriving its moisture and nutrients from the air and rain e.g. some Orchidaceae. *adj.* epiphytic.

Erect upright, vertical.

Essential oils volatile products obtained from a natural source; refers to volatile products obtained by steam or water distillation in a strict sense.

Etiolation to cause (a plant) to develop without chlorophyll by preventing exposure to sunlight.

Eutrophic having waters rich in mineral and organic nutrients that promote a proliferation of plant life, especially algae, which reduces the dissolved oxygen content and often causes the extinction of other organisms.

Excentric off the true centre.

Excrescence abnormal outgrowth.

Excurrent projecting beyond the tip, as the midrib of a leaf or bract.

Exserted sticking out, protruding beyond some enclosing organ, as of stamens which project beyond the corolla or perianth.

Exstipulate without stipules. *cf.* stipulate.

Extra-floral outside the flower.

Extrose turned outwards or away from the axis as of anthers. *cf.* introrse, latrorse.

Falcate sickle shaped, crescent-shaped.

Faronaceous mealy, resembling flour.

Fascicle a cluster or bundle of stems, flowers, stamens. *adj.* fasciculate.

Fasciclode staminode bundles.

Fastigiate a tree in which the branches grow almost vertically.

Ferrosols soils with an iron oxide content of greater than 5%.

Ferruginous rust coloured, reddish-brown.

Fertile having functional sexual parts which are capable of fertilisation and seed production. *cf.* sterile.

Filament the stalk of a stamen supporting and subtending the anther.

Filiform Having the form of or resembling a thread or filament.

Fimbriate fringed.

Fixed oils non volatile oils, triglycerides of fatty acids.

Flaccid limp and weak.

Flag leaf the uppermost leaf on the stem.

Flaky in the shape of flakes or scales.

Flavedo the outermost pigmented layer of the rind of citrus fruits containing essential oils.

Flexuous zig-zagging, sinuous, bending, as of a stem.

Floccose covered with tufts of soft woolly hairs.

Floral tube a flower tube usually formed by the basal fusion of the perianth and stamens.

Floret one of the small individual flowers of sunflower family or the reduced flower of the grasses, including the lemma and palea.

Flower the sexual reproductive organ of flowering plants, typically consisting of gynoecium, androecium and perianth or calyx and/or corolla and the axis bearing these parts.

Fluted as of a trunk with grooves and folds.

Fodder plant material, fresh or dried fed to animals.

Foliaceous leaf-like.

Foliar pertaining to a leaf.

Foliolate pertaining to leaflets, used with a number prefix to denote the number of leaflets.

Foliose leaf-like.

Follicle (Botany) a dry fruit, derived from a single carpel and dehiscing along one suture.

Forb any herb that is not grass or grass-like.

Free central placentation The arrangement of ovules on a central column that is not connected to the ovary wall by partitions, as in the ovaries of the carnation and primrose.

Frond the leaf of a fern or cycad.

Fruit ripened ovary with adnate parts.

Fugacious shedding off early.

Fulvous yellow, tawny.

Funiculus (Botany) short stalk which attaches the ovule to the ovary wall.

Fusiform a 3-dimensional shape; spindle shaped, i.e. broad in the centre and tapering at both ends thick, but tapering at both ends.

Gamete a reproductive cell that fuses with another gamete to form a zygote. Gametes are haploid, (they contain half the normal (diploid) number of chromosomes); thus when two fuse, the diploid number is restored.

Gametophyte The gamete-producing phase in a plant characterized by alternation of generations.

Gamosepalous with sepals united or partially united.

Gall-flower short styled flower that do not develop into a fruit but are adapted for the development of a specific wasp within the fruit e.g. in the fig.

Geniculate bent like a knee, refer to awns and filaments.

Geocarpic where the fruit are pushed into the soil by the gynophore and mature.

Geophyte a plant that stores food in an underground storage organ e.g. a tuber, bulb or rhizome and has subterranean buds which form aerial growth.

Geotextile are permeable fabrics which, when used in association with soil, have the ability to separate, filter, reinforce, protect, or drain.

Glabrescent becoming glabrous.

Glabrous smooth, hairless without pubescence.

Gland a secretory organ, e.g. a nectary, extrafloral nectary or a gland tipped, hair-like or wart-like organ. *adj.* glandular. *cf.* eglandular.

Glaucous pale blue-green in colour, covered with a whitish bloom that rubs off readily.

Gley soils a hydric soil which exhibits a greenish-blue-grey soil color due to wetland conditions.

Globose spherical in shape.
Globular a three-dimensional shape; spherical or orbicular; circular in outline.
Glochidiate having glochids.
Glochidote plant having gkochids.
Glochids tiny, finely barbed hair-like spines found on the areoles of some cacti and other plants.
Glume one of the two small, sterile bracts at the base of the grass spikelet, called the lower and upper glumes, due to their position on the rachilla. Also used in Apiaceae, Cyperaceae for the very small bracts on the spikelet in which each flower is subtended by one floral glume. *adj.* glumaceous.
Guttation the appearance of drops of xylem sap on the tips or edges of leaves of some vascular plants, such as grasses and bamboos.
Guttule small droplet.
Gymnosperm a group of spermatophyte seed-bearing plants with ovules on scales, which are usually arranged in cone-like structures and not borne in an ovary. *cf.* angiosperm.
Gynoecium the female organ of a flower; a collective term for the pistil, carpel or carpels.
Gynomonoecious having female flowers and bisexual flowers on the same plant. *cf.* andromonoecious.
Gynophore stalk that bears the pistil/carpel.
Habit the general growth form of a plant, comprising its size, shape, texture and stem orientation, the locality in which the plant grows..
Halophyte a plant adapted to living in highly saline habitats. Also a plant that accumulates high concentrations of salt in its tissues. *adj.* halophytic.
Hapaxanthic refer to palms which flowers only once and then dies. c.f. pleonanthic.
Haploid condition where nucleus or cell has a single set of unpaired chromosomes, the haploid number is designated as n.
Haplontic life cycle or zygotic meiosis wherein meiosis of a zygote immediately after karyogamy, produces haploid cells which produces more or larger haploid cells ending its diploid phase.
Hastate having the shape of an arrowhead but with the basal lobes pointing outward at right angles as of a leaf.

Hastula a piece of plant material at the junction of the petiole and the leaf blade; the hastula can be found on the top of the leaf, adaxial or the bottom, abaxial or both sides.
Heartwood wood from the inner portion of a tree.
Heliophilous sun-loving, tolerates high level of sunlight.
Heliotropic growing towards sunlight.
Herb a plant which is non-woody or woody at the base only, the above ground stems usually being ephemeral. *adj.* herbaceous.
Herbaceous resembling a herb, having a habit of a herb.
Hermaphrodite bisexual, bearing flowers with both androecium and gynoecium in the same flower. *adj.* hermaphroditic.
Hesperidium modified fruit with a tough leathery rind as found in Citrus fruits.
Heterocyst a differentiated cyanobacterial cell that carries out nitrogen fixation.
Heterogamous bearing separate male and female flowers, or bisexual and female flowers, or florets in an inflorescence or flower head, e.g. some Asteraceae in which the ray florets may be neuter or unisexual and the disk florets may be bisexual. *cf.* homogamous.
Heteromorphous having two or more distinct forms. *cf.* homomorphous.
Heterophyllous having leaves of different form.
Heterosporous producing spores of two sizes, the larger giving rise to megagametophytes (female), the smaller giving rise to microgametophytes (male). Refer to the ferns and fern allies. *cf.* homosporous.
Heterostylous having styles of two different lengths or forms.
Heterostyly the condition in which flowers on polymorphous plants have styles of different lengths, thereby facilitating cross-pollination.
Hilar of or relating to a hilum.
Hilum The scar on a seed, indicating the point of attachment to the funiculus.
Hirsute bearing long coarse hairs.
Hispid bearing stiff, short, rough hairs or bristles.
Hispidulous minutely hispid.

Histosol soil comprising primarily of organic materials, having 40 cm or more of organic soil material in the upper 80 cm.

Hoary covered with a greyish layer of very short, closely interwoven hairs.

Holdfast an organ or structure of attachment, especially the basal, root-like formation by which certain seaweeds or other algae are attached to a substrate.

Holocarpic having the entire thallus developed into a fruiting body or sporangium.

Homochromous having all the florets of the same colour in the same flower head *cf.* heterochromous.

Homogamous bearing flowers or florets that do not differ sexually *cf.* heterogamous.

Homogenous endosperm endosperm with even surface that lacks invaginations or infoldings of the surrounding tissue.

Homogonium a part of a filament of a cyanobacterium that detaches and grows by cell division into a new filament. *pl.* homogonia.

Homomorphous uniform, with only one form. *cf.* heteromorphous.

Homosporous producing one kind of spores. Refer to the ferns and fern allies. *cf.* heterosporous.

Hurd fibre long pith fibre of the stem.

Hyaline colourless, almost transparent.

Hybrid the first generation progeny of the sexual union of plants belonging to different taxa.

Hybridisation the crossing of individuals from different species or taxa.

Hydathode a type of secretory tissue in leaves, usually of Angiosperms, that secretes water through pores in the epidermis or margin of the leaf.

Hydrophilous water loving; requiring water in order to be fertilized, referring to many aquatic plants.

Hygrochastic applied to plants in which the opening of the fruits is caused by the absorption of water.

Hygrophilous living in water or moist places.

Hymenial cystidia the cells of the hymenium develop into basidia or asci, while in others some cells develop into sterile cells called cystidia.

Hymenium spore-bearing layer of cells in certain fungi containing asci (Ascomycetes) or basidia (Basidiomycetes).

Hypanthium cup-like receptacles of some dicotyledonous flowers formed by the fusion of the calyx, corolla, and androecium that surrounds the ovary which bears the sepals, petals and stamens.

Hypha is a long, branching filamentous cell of a fungus, and also of unrelated Actinobacteria. *pl.* hyphae.

Hypocotyl the portion of the stem below the cotyledons.

Hypodermis the cell layer beneath the epidermis of the pericarp.

Hypogeal below ground as of germination of seed.

Hysteresis refers to systems that may exhibit path dependence.

Imbricate closely packed and overlapping. *cf.* valvate.

Imparipinnate pinnately compound with a single terminal leaflet and hence with an odd number of leaflets. *cf.* paripinnate.

Inceptisols old soils that have no accumulation of clays, iron, aluminium or organic matter.

Incised cut jaggedly with very deep teeth.

Included referring to stamens which do not project beyond the corolla or to valves which do not extend beyond the rim of a capsular fruit. *cf.* exserted.

Incurved curved inwards; curved towards the base or apex.

Indefinite numerous and variable in number.

Indehiscent not opening or splitting to release the contents at maturity as of fruit. *cf.* dehiscent.

Indumentum covering of fine hairs or bristles commonly found on external parts of plants.

Indurate to become hard, often the hardening developed only at maturity.

Indusium an enclosing membrane, covering the sorus of a fern. Also used for the modified style end or pollen-cup of some Goodeniaceae (including Brunoniaceae). *adj.* indusiate.

Inferior said of an ovary or fruit that has sepals, petals and stamens above the ovary. *cf.* superior.

Inflated enlarged and hollow except in the case of a fruit which may contain a seed. *cf.* swollen.

Inflexed Bent or curved inward or downward, as petals or sepals.

Inflorescence a flower cluster or the arrangement of flowers in relation to the axis and to each other on a plant.

Infrafoliar located below the leaves.

Infraspecific referring to any taxon below the species rank.

Infructescence the fruiting stage of an inflorescence.

Inrolled curved inwards.

Integuments two distinct tissue layers that surround the nucellus of the ovule, forming the testa or seed coat when mature.

Intercalary of growth, between the apex and the base; of cells, spores, etc., between two cells.

Interfoliar inter leaf.

Internode portion of the stem, culm, branch, or rhizome between two nodes or points of attachment of the leaves.

Interpetiolar as of stipules positioned between petioles of opposite leaves.

Intrastaminal within the stamens.

Intricate entangled, complex.

Introduced not indigenous; not native to the area in which it now occurs.

Introrse turned inwards or towards the axis or pistil as of anthers. *cf.* extrorse, latrorse.

Involucre a whorl of bracts or leaves that surround one to many flowers or an entire inflorescence.

Involute having the margins rolled inwards, referring to a leaf or other flat organ.

Jugate of a pinnate leaf; having leaflets in pairs.

Juvenile young or immature, used here for leaves formed on a young plant which are different in morphology from those formed on an older plant.

Keel a longitudinal ridge, at the back of the leaf. Also the two lower fused petals of a 'pea' flower in the Papilionaceae, which form a boat-like structure around the stamens and styles, also called carina. *adj.* keeled. *cf.* standard, wing.

Labellum the modified lowest of the three petals forming the corolla of an orchid, usually larger than the other two petals, and often spurred.

Laciniate fringed; having a fringe of slender, narrow, pointed lobes cut into narrow lobes.

Lamella a gill-shaped structure: fine sheets of material held adjacent to one another.

Lamina the blade of the leaf or frond.

Lanate wooly, covered with long hairs which are loosely curled together like wool.

Lanceolate lance-shaped in outline, tapering from a broad base to the apex.

Landrace plants adapted to the natural environment in which they grow, developing naturally with minimal assistance or guidance from humans and usually possess more diverse phenotypes and genotypes. They have not been improved by formal breeding programs.

Laterite reddish–coloured soils rich in iron oxide, formed by weathering of rocks under oxidizing and leaching conditions, commonly found in tropical and subtropical regions. *adj.* lateritic.

Latex a milky, clear or sometimes coloured sap of diverse composition exuded by some plants.

Latrorse turned sideways, i.e. not towards or away from the axis as of anthers dehiscing longitudinally on the side. *cf.* extrorse, introse.

Lax loose or limp, not densely arranged or crowded.

Leaflet one of the ultimate segments of a compound leaf.

Lectotype a specimen chosen after the original description to be the type.

Lemma the lower of two bracts (scales) of a grass floret, usually enclosing the palea, lodicules, stamens and ovary.

Lenticel is a lens shaped opening that allows gases to be exchanged between air and the inner tissues of a plant, commonly found on young bark, or the surface of the fruit.

Lenticellate dotted with lenticels.

Lenticular shaped like a biconvex lens. *cf.* lentiform.

Lentiform shaped like a biconvex lens, *cf.* lenticular.

Leptomorphic temperate, running bamboo rhizome; usually thinner then the culms they support and the internodes are long and hollow.

Liane a woody climbing or twining plant.

Lignescent tending towards woodiness.

Lignotuber a woody, usually underground, tuberous rootstock often giving rise to numerous aerial stems.

Ligulate small and tongue shaped or with a little tongue shaped appendage or ligule, star shaped as of florets of Asteraceae.

Ligule a strap-shaped corolla in the flowers of Asteraceae; also a thin membranous outgrowth from the inner junction of the grass leaf sheath and blade. *cf.* ligulate.

Limb the expanded portion of the calyx tube or the corolla tube, or the large branch of a tree.

Linear a 2-dimensional shape, narrow with nearly parallel sides.

Linguiform tongue shaped *cf.* ligulate.

Lithosol a kind of shallow soils lacking well-defined horizons and composed of imperfectly weathered fragments of rock.

Littoral of or on a shore, especially seashore.

Loam a type of soil made up of sand, silt, and clay in relative concentration of 40-40-20% respectively.

Lobed divided but not to the base.

Loculicidal opening into the cells, when a ripe capsule splits along the back.

Loculus cavity or chamber of an ovary. *pl.* loculi.

Lodicules two small structures below the ovary which, at flowering, swell up and force open the enclosing bracts, exposing the stamens and carpel.

Luvisol a characteristic soil of forested region found in temperate and Mediterranean regions with an argic horizon (a subsurface horizon with a distinct higher clay content than the overlying horizon).

Lyrate pinnately lobed, with a large terminal lobe and smaller laterals ones which become progressively smaller towards the base.

Macronutrients chemical elements which are needed in large quantities for growth and development by plants and include nitrogen, phosphorus, potassium, and magnesium.

Maculate spotted.

Mallee a growth habit in which several to many woody stems arise separately from a lignotuber; usually applied to certain low-growing species of *Eucalyptus*.

Mangrove a distinctive vegetation type of trees and shrubs with modified roots, often viviparous, occupying the saline coastal habitats that are subject to periodic tidal inundation.

Marcescent withering or to decay without falling off.

Margin the edge of the leaf blade.

Medulla the pith in the stems or roots of certain plants; or the central portion of a thallus in certain lichens; or pith of fruit.

Megasporangium the sporangium containing megaspores in fern and fern allies. *cf.* microsporangium.

Megaspore the large spore which may develop into the female gametophyte in heterosporous ferns and fern allies. *cf.* microspore.

Megasporophyll a leaflike structure that bears megasporangia.

Megastrobilus female cone, seed cone, or ovulate cone, contains ovules within which, when fertilized by pollen, become seeds. The female cone structure varies more markedly between the different conifer families.

Meiosis the process of cell division that results in the formation of haploid cells from diploid cells to produce gametes.

Mericarp a 1-seeded portion of an initially syncarpous fruit (schizocarp) which splits apart at maturity. *Cf.* coccus.

Meristem the region of active cell division in plants, from which permanent tissue is derived. *adj.* meristematic

-merous used with a number prefix to denote the basic number of the 3 outer floral whorls, e.g. a 5-merous flower may have 5 sepals, 10 petals and 15 stamens.

Mesic moderately wet.

Mesocarp the middle layer of the fruit wall derived from the middle layer of the carpel wall. *cf.* endocarp, exocarp, pericarp.

Mesophytes terrestrial plants which are adapted to neither a particularly dry nor particularly wet environment.

Micropyle the small opening in a plant ovule through which the pollen tube passes in order to effect fertilisation.

Microsporangium the sporangium containing microspores in petridophyes. *cf.* megasporangium.

Microspore a small spore which gives rise to the male gametophyte in heterosporous pteridophytes. Also for a pollen grain. *cf.* megaspore.

Midvein the main vascular supply of a simple leaf blade or lamina. Also called mid-rib.

Mitosis is a process of cell division which results in the production of two daughter cells from a single parent cell.

Mollisols soils with deep, high organic matter, nutrient-enriched surface soil (A horizon), typically between 60–80 cm thick.

Monadelphous applied to stamens united by their filaments into a single bundle.

Monocarpic refer to plants that flower, set seeds and then die.

Monochasial a cyme having a single flower on each axis.

Monocotyledon angiopsrem having one cotyledon.

Monoecious having both male and female unisexual flowers on the same individual plant. *cf.* dioecious.

Monoembryonic seed the seed contains only one embryo, a true sexual (zygotic) embryo. polyembryonic seed.

Monolete a spore that has a simple linear scar.

Monopodial with a main terminal growing point producing many lateral branches progressively. *cf.* sympodial.

Monotypic of a genus with one species or a family with one genus; in general, applied to any taxon with only one immediately subordinate taxon.

Montane refers to highland areas located below the subalpine zone.

Mucilage a soft, moist, viscous, sticky secretion. *adj.* mucilaginous.

Mucous (Botany) slimy.

Mucro a sharp, pointed part or organ, especially a sharp terminal point, as of a leaf.

Mucronate ending with a short, sharp tip or mucro, resembling a spine. cf. cuspidate, muticous.

Mucronulate with a very small mucro; a diminutive of mucronate.

Mulch protective cover of plant (organic) or non-plant material placed over the soil, primarily to modify and improve the effects of the local microclimate and to control weeds.

Multiple fruit a fruit that is formed from a cluster of flowers.

Muricate covered with numerous short hard outgrowths. *cf.* papillose.

Muriculate with numerous minute hard outgrowths; a diminutive of muricate.

Muticous blunt, lacking a sharp point. *cf.* mucronate.

MYB proteins are a superfamily of transcription factors that play regulatory roles in developmental processes and defense responses in plants.

Mycorrhiza the mutualistic symbiosis (non-pathogenic association) between soil-borne fungi with the roots of higher plants.

Mycorrhiza (vesicular arbuscular) endomycorrhiza living in the roots of higher plants producing inter-and intracellular fungal growth in root cortex and forming specific fungal structures, referred to as vesicles and arbuscles. *abbrev.* VAM.

Native a plant indigenous to the locality or region.

Naviculate boat-shaped.

Necrotic applied to dead tissue.

Nectariferous having one or more nectaries.

Nectary a nectar secretory gland; commonly in a flower, sometimes on leaves, fronds or stems.

Nervation venation, a pattern of veins or nerves as of leaf.

Node the joint between segments of a culm, stem, branch, or rhizome; the point of the stem that gives rise to the leaf and bud.

Nodule a small knoblike outgrowth, as those found on the roots of many leguminous, that containing *Rhizobium* bacteria which fixes nitrogen in the soil.

Nomen Dubium an invalid proposed taxonomic name because it is not accompanied by a definition or description of the taxon to which it applies. *abbrev.* nom. dub.

Nomen Illegitimum illegitimate taxon deemed as superfluous at its time of publication either because the taxon to which it was applied already has a name, or because the name has already been applied to another plant. *abbrev.* nom. illeg.

Nomen Nudum the name of a taxon which has never been validated by a description. *abbrev.* nom. nud.

Nucellar embryony a form of seed reproduction in which the nucellar tissue which surrounds the embryo sac can produce additional embryos (polyembryony) which are genetically identical to the parent plant. This is found in many citrus species and in mango.

Nucellus central portion of an ovule in which the embryo sac develops.

Nut a dry indehiscent 1-celled fruit with a hard pericarp.

Nutlet a small. 1-seeded, indehiscent lobe of a divided fruit.

Ob- prefix meaning inversely or opposite to.

Obconic a 3-dimensional shape; inversely conic; cone shaped, conic with the vertex pointing downward.

Obcordate inversely cordate, broad and notched at the tip; heart shaped but attached at the pointed end.

Obdeltate inversely deltate; deltate with the broadest part at the apex.

Oblanceolate inversely lanceolate, lance-shaped but broadest above the middle and tapering toward the base as of leaf.

Oblate having the shape of a spheroid with the equatorial diameter greater than the polar diameter; being flattened at the poles.

Oblong longer than broad with sides nearly parallel to each other.

Obovate inversely ovate, broadest above the middle.

Obpyramidal resembling a 4-sided pyramid attached at the apex with the square base facing away from the attachment.

Obpyriform inversely pyriform, resembling a pear which is attached at the narrower end. *cf.* pyriform.

Obspathulate inversely spathulate; resembling a spoon but attached at the broadest end. *cf.* spathulate.

Obtriangular inversely triangular; triangular but attached at the apex. *cf.* triangular.

Obtrullate inversely trullate; resembling a trowel blade with the broadest axis above the middle. *cf.* trullate.

Obtuse with a blunt or rounded tip, the converging edges separated by an angle greater than 90°.

-oid suffix denoting a 3-dimensional shape, e.g. spheroid.

Ochraceous a dull yellow color.

Ocreate having a tube-like covering around some stems, formed of the united stipules; sheathed.

Oleaginous oily.

Oligotrophic lacking in plant nutrients and having a large amount of dissolved oxygen throughout.

Operculum a lid or cover that becomes detached at maturity by abscission, e.g. in *Eucalyptus*, also a cap or lid covering the bud and formed by fusion or cohesion of sepals and/or petals. *adj.* operculate.

Opposite describing leaves or other organs which are borne at the same level but on opposite sides of the stem. *cf.* alternate.

Orbicular of circular outline, disc-like.

Order a taxonomic rank between class and family used in the classification of organisms, i.e. a group of families believed to be closely related.

Orifice an opening or aperture.

Organosols soils not regularly inundated by marine waters and containing a specific thickness of organic materials within the upper part of the profile.

Ovary the female part of the pistil of a flower which contains the ovules (immature seeds).

Ovate egg-shaped, usually with reference to two dimensions.

Ovoid egg-shaped, usually with reference to three dimensions.

Ovule the young, immature seed in the ovary which becomes a seed after fertilisation. *adj.* ovular.

Ovulode a sterile reduced ovule borne on the placenta, commonly occurring in Myrtaceae.

Oxisols refer to ferralsols.

Pachymorphic describes the short, thick, rhizomes of clumping bamboos with short, thick and solid internode (except the bud-bearing internodes, which are more elongated). *cf.* sympodial.

Palate (Botany) a raised appendage on the lower lip of a corolla which partially or completely closes the throat.

Palea the upper of the two membraneous bracts of a grass floret, usually enclosing the lodicules, stamens and ovary. *pl.* paleae. *adj.* paleal. *cf.* lemma.

Paleate having glumes.

Palm heart refers to soft, tender inner core and growing bud of certain palm trees which are eaten as vegetables. Also called heart of palm, palmito, burglar's thigh, chonta or swamp cabbage.

Palmate describing a leaf which is divided into several lobes or leaflets which arise from the same point. *adj.* palmately.

Palmito see palm heart.

Palustrial paludal, swampy, marshy.

Palustrine marshy, swampy.

Palustrine herb vegetation that is rooted below water but grows above the surface in wetland system.

Panduriform fiddle shaped, usually with reference to two dimensions.

Panicle a compound, indeterminate, racemose inflorescence in which the main axis bears lateral racemes or spikes. *adj.* paniculate.

Pantropical distributed through-out the tropics.

Papilionaceous butterfly-like, said of the pea flower or flowers of Papilionaceae, flowers which are zygomorphic with imbricate petals, one broad upper one, two narrower lateral ones and two narrower lower ones.

Papilla a small, superficial protuberance on the surface of an organ being an outgrowth of one epidermal cell. *pl.* papillae. *adj.* papillose.

Papillate having papillae.

Papillose covered with papillae.

Pappus a tuft (or ring) of hairs, bristles or scales borne above the ovary and outside the corolla as in Asteraceae often persisting as a tuft of hairs on a fruit. *adj.* pappose.

Papyraceous resembling parchment of paper.

Parenchyma undifferentiated plant tissue composed of more or less uniform cells.

Parietal describes the attachment of ovules to the outer walls of the ovaries.

Paripinnate pinnate with an even number of leaflets and without a terminal leaflet. *cf.* imparipinnate.

-partite divided almost to the base into segments, the number of segments written as a prefix.

Patelliform shaped like a limpet shell; cap-shaped and without whorls.

Patent diverging from the axis almost at right angles.

Peat is an accumulation of partially decayed vegetation matter.

Pectin a group of water-soluble colloidal carbohydrates of high molecular weight found in certain ripe fruits.

Pectinate pinnatifid with narrow segments resembling the teeth of a comb.

Pedicel the stalk of the flower or stalk of a spikelet in Poaceae. *adj.* pedicellate.

Pedicellate having pedicel.

Peduncle a stalk supporting an inflorescence. *adj.* pedunculate

Pellucid allowing the passage of light; transparent or translucent.

Pellucid-dotted copiously dotted with immersed, pellucid, resinous glands.

Peltate with the petiole attached to the lower surface of the leaf blade.

Pendant hanging down.

Pendulous drooping, as of ovules.

Penniveined or **penni-nerved** pinnately veined.

Pentamerous in five parts.

Perennial a plant that completes it life cycle or lives for more than 2 years. *cf.* annual, biennial.

Perfoliate a leaf with the basal lobes united around–and apparently pierced by–the stem.

Pergamentaceous parchment-like.
Perianth the two outer floral whorls of the Angiosperm flower; commonly used when the calyx and the corolla are not readily distinguishable (as in monocotyledons).
Pericarp (Botany). The wall of a ripened ovary; fruit wall composed of the exocarp, mesocarp and endocarp.
Persistent remaining attached; not falling off. *cf.* caduceus.
Petal free segment of the corolla. *adj.* petaline. *cf.* lobe.
Petiolar relating to the petiole.
Petiolate having petiole.
Petiole leaf stalk. *adj.* petiolate.
Petiolulate supported by its own petiolule.
Petiolule the stalk of a leaflet in a compound leaf. adj. petiolulate.
pH is a measure of the acidity or basicity of a solution. It is defined as the cologarithm of the activity of dissolved hydrogen ions (H+).
Phenology the study of periodic plant life cycle events as influenced by seasonal and interannual variations in climate.
Phyllary a bract of the involucre of a composite plant, term for one of the scale-like bracts beneath the flower-head in Asteraceae.
Phylloclade a flattened, photosynthetic branch or stem that resembles or performs the function of a leaf, with the true leaves represented by scales.
Phyllode a petiole that function as a leaf. *adj.* phyllodineous. *cf.* cladode.
Phyllopodia refer to the reduced, scale-like leaves found on the outermost portion of the corm where they seem to persist longer than typical sporophylls as in the fern Isoetes.
Phytoremediation describes the treatment of environmental problems (bioremediation) through the use of plants which mitigate the environmental problem without the need to excavate the contaminant material and dispose of it elsewhere.
Pileus (Botany) cap of mushroom.
Piliferous (Botany) bearing or producing hairs, as of an organ with the apex having long, hair-like extensions.
Pilose covered with fine soft hairs.

Pinna a primary division of the blade of a compound leaf or frond. *pl.* pinnae.
Pinnate bearing leaflets on each side of a central axis of a compound leaf; divided into pinnae.
Pinnatifid, pinnatilobed a pinnate leaf parted approximately halfway to midrib; when divided to almost to the mid rib described as deeply pinnatifid or pinnatisect.
Pinnatisect lobed or divided almost to the midrib.
Pinnule a leaflet of a bipinnate compound leaf.
Pistil female part of the flower comprising the ovary, style, and stigma.
Pistillate having one or more pistils; having pistils but no stamens.
Placenta the region within the ovary to which ovules are attached. *pl.* placentae.
Placentation the arrangement of the placentae and ovules in the ovary.
Plano- a prefix meaning level or flat.
Pleonanthic refer to palms in which the stem does not die after flowering.
Plicate folded like a fan.
Plumose feather-like, with fine hairs arising laterally from a central axis; feathery.
Pneumatophore modified root which allows gaseous exchange in mud-dwelling shrubs, e.g. mangroves.
Pod a dry 1 to many-seeded dehiscent fruit, as applied to the fruit of Fabaceae ie. Caesalpiniaceae, Mimosaceae and Papilionaceae.
Podzol, Podsolic soil any of a group of acidic, zonal soils having a leached, light-coloured, gray and ashy appearance. Also called spodosol.
Pollen cone male cone or microstrobilus or pollen cone is structurally similar across all conifers, extending out from a central axis are microsporophylls (modified leaves). Under each microsporophyll is one or several microsporangia (pollen sacs).
Pollinia the paired, waxy pollen masses of flowers of orchids and milkweeds.
Polyandrous (Botany) having an indefinite number of stamens.
Polyembryonic seed seeds contain many embryos, most of which are asexual (nucellar) in origin and genetically identical to the maternal parent.

Polygamous with unisexual and bisexual flowers on the same or on different individuals of the same species.

Polymorphic with different morphological variants.

Polypetalous (Botany) having a corolla composed of distinct, separable petals.

Pome a fleshy fruit where the succulent tissues are developed from the receptacle.

Pore a tiny opening.

Porrect extended forward.

Premorse Abruptly truncated, as though bitten or broken off as of a leaf.

Procumbent trailing or spreading along the ground but not rooting at the nodes, referring to stems. *cf.* ascending, decumbent, erect.

Prophyll a plant structure that resembles a leaf.

Prostrate lying flat on the ground.

Protandous relating to a flower in which the anthers release their pollen before the stigma of the same flower becomes receptive.

Proximal end of any structure closest to the point of attachment. *cf.* distal.

Pruinose having a thick, waxy, powdery coating or bloom.

Pseudocarp a false fruit, largely made up of tissue that is not derived from the ovary but from floral parts such as the receptacle and calyx.

Pseudostem The false, herbaceous stem of a banana plant composed of overlapping leaf bases.

Pteridophyte a vascular plant which reproduces by spores; the ferns and fern allies.

Puberulent covered with minute hairs or very fine down; finely pubescent.

Puberulous covered with a minute down.

Pubescent covered with short, soft hairs.

Pulvinate having a swelling, pulvinus at the base as a leaf stalk.

Pulvinus swelling at the base of leaf stalk.

Pulviniform swelling or bulging.

Punctate marked with translucent dots or glands.

Punctiform marked by or composed of points or dots.

Punctulate marked with minute dots; a diminutive of punctate.

Pusticulate characterized by small pustules.

Pyrene the stone or pit of a drupe, consisting of the hardened endocarp and seed.

Pyriform pear-shaped, a 3-dimensional shape; attached at the broader end. *cf.* obpyriform.

Pyxidium seed capsule having a circular lid (operculum) which falls off to release the seed.

Raceme an indeterminate inflorescence with a simple, elongated axis and pedicellate flowers, youngest at the top. *adj.* racemose.

Rachilla the main axis of a grass spikelet.

Rachis the main axis of the spike or other inflorescence of grasses or a compound leaf.

Radiate arranged around a common centre; as of an inflorescence of Asteraceae with marginal, female or neuter, ligulate ray-florets and central, perfect or functionally male, tubular, disc florets. *cf.* disciform, discoid.

Radical arising from the root or its crown, or the part of a plant embryo that develops into a root.

Ray the marginal portion of the inflorescence of Asteraceae and Apiaceae when distinct from the disc. Also, the spreading branches of a compound umbel.

Receptacle the region at the end of a pedicel or on an axis which bears one or more flowers. *adj.* receptacular.

Recurved curved downwards or backwards.

Reflexed bent or turned downward.

Regosol soil that is young and undeveloped, characterized by medium to fine-textured unconsolidated parent material that maybe alluvial in origin and lacks a significant horizon layer formation.

Reniform kidney shaped in outline.

Repand with slightly undulate margin.

Replicate folded back, as in some corolla lobes.

Resinous producing sticky resin.

Resupinate twisted through 180°.

Reticulate having the appearance of a network.

Retrorse bent or directed downwards or backwards. *cf.* antrorse.

Retuse with a very blunt and slightly notched apex. *cf.* emarginated.

Revolute with the margins inrolled on the lower (abaxial) surface.

Rhizine a root-like filament or hair growing from the stems of mosses or on lichens.

Rhizoid root-like filaments in a moss, fern, fungus, etc. that attach the plant to the substratum.

Rhizome a prostrate or underground stem consisting of a series of nodes and internodes with adventitious roots and which generally grows horizontally.

Rhizophore a stilt-like outgrowth of the stem which branches into roots on contact with the substrate.

Rhombic shaped like a rhombus.

Rhomboid shaped like a rhombus.

Rib a distinct vein or linear marking, often raised as a linear ridge.

Riparian along the river margins, interface between land and a stream.

Rosette a tuft of leaves or other organs arranged spirally like petals in a rose, ranging in form from a hemispherical tuft to a flat whorl. *adj.* rosetted, rosulate.

Rostrate beaked; the apex tapered into a slender, usually obtuse point.

Rostrum a beak-like extension.

Rosulate having a rosette.

Rotate wheel shaped; refers to a corolla with a very short tube and a broad upper part which is flared at right angles to the tube. *cf.* salverform.

Rotundate rounded; especially at the end or ends.

Rugae refers to a series of ridges produced by folding of the wall of an organ.

Rugose deeply wrinkled.

Rugulose finely wrinkled.

Ruminate (Animal) chew repeatedly over an extended period.

Ruminate endosperm uneven endosperm surface that is often highly enlarged by ingrowths or infoldings of the surrounding tissue. cf. homogenous endosperm.

Rz value is a numerical reference to the mesh/emulsion equalization on the screen.

Saccate pouched.

Sagittate shaped like an arrow head.

Saline soils soils that contain excessive levels of salts that reduce plant growth and vigor by altering water uptake and causing ion-specific toxicities or imbalances.

Salinity is characterised by high electrical conductivities and low sodium ion concentrations compared to calcium and magnesium

Salverform applies to a gamopetalous corolla having a slender tube and an abruptly expanded limb.

Samara an indehiscent, winged, dry fruit.

Sand a naturally occurring granular material composed of finely divided rock and mineral particles range in diameter from 0.0625 μm to 2 mm. *adj.* sandy

Saponins are plant glycosides with a distinctive foaming characteristic. They are found in many plants, but get their name from the soapwort plant (*Saponaria*).

Saprophytic living on and deriving nourishment from dead organic matter.

Sapwood outer woody layer of the tree just adjacent to and below the bark.

Sarcotesta outermost fleshy covering of Cycad seeds below which is the sclerotesta.

Scabrid scurfy, covered with surface abrasions, irregular projections or delicate scales.

Scabrous rough to the touch.

Scale dry bract or leaf.

Scandent refer to plants, climbing.

Scape erect flowering stem, usually leafless, rising from the crown or roots of a plant. *adj.* scapose.

Scapigerous with a scape.

Scarious dry, thin and membranous.

Schizocarp a dry fruit which splits into longitudinally multiple parts called mericarps or cocci. *adj.* schizocarpous.

Sclerotesta the innermost fleshy coating of cycad seeds, usually located directly below the sarcotesta.

Scorpioid resembling a scorpion's tail, circinnate.

Scutellum (Botany) any of various parts shaped like a shield.

Secondary venation arrangement of the lateral veins arising from the midrib in the leaf lamina.

Secund with the flowers all turned in the same direction.

Sedge a plant of the family Apiaceae, Cyperaceae.

Segmented constricted into divisions.

Seminal root or seed root originate from the scutellar node located within the seed embryo and are composed of the radicle and lateral seminal roots.

Senescence refers to the biological changes which take place in plants as they age.

Sensu lato In a broad or wide sense.

Sensu stricto In a narrow or strict sense.

Sepal free segment of the calyx. *adj.* sepaline.

Septum a partition or cross wall. *pl.* septa. *adj.* septate.

Seriate arranged in rows.

Sericeous silky; covered with close-pressed, fine, straight silky hairs.

Serrate toothed like a saw; with regular, asymmetric teeth pointing forward.

Serrated toothed margin.

Serratures serrated margin.

Serrulate with minute teeth on the margin.

Sessile without a stalk.

Seta a bristle or stiff hair. *pl.* setae. *adj.* setose, setaceous.

Setaceous bristle-like.

Setate with bristles.

Setiform bristle shaped.

Setulose with minute bristles.

Sheathing clasping or enveloping the stem.

Shrub a woody plant usually less than 5 m high and many-branched without a distinct main stem except at ground level.

Silicula a broad, dry, usually dehiscent fruit derived from two or more carpels which usually dehisce along two sutures. *cf.* siliqua.

Siliqua a silicula which is at least twice as long as broad.

Silt is soil or rock derived granular material of a grain size between sand and clay, grain particles ranging from 0.004 to 0.06 mm in diameter. *adj.* silty.

Simple refer to a leaf or other structure that is not divided into parts. *cf.* compound.

Sinuate with deep wavy margin.

Sinuous wavy.

Sinus an opening or groove, as occurs between the bases of two petals.

Sodicity is characterised by low electrical conductivities and high sodium ion concentrations compared to calcium and magnesium.

Sodic soils contains high levels of sodium salts that affects soil structure, inhibits water movement and causes poor germination and crop establishment and plant toxicity.

Soil pH is a measure of the acidity or basicity of the soil. See pH.

Solitary usually refer to flowers which are borne singly, and not grouped into an inflorescence or clustered.

Sorocarp fruiting body formed by some cellular slime moulds, has both stalk and spore mass.

Sorophore stalk bearing the sorocarp.

Sorosis fleshy multiple fruit formed from flowers that are crowded together on a fleshy stem e.g. pineapple and mulberry.

Sorus a discrete aggregate of sporangia in ferns. *pl.* sori

Spadix fleshy spike-like inflorescence with an unbranched, usually thickened axis and small embedded flowers often surrounded by a spathe. *pl.* spadices.

Spathe a large bract ensheathing an inflorescence or its peduncle. *adj.* spathaceous.

Spatheate like or with a spathe.

Spathulate spatula or spoon shaped; broad at the tip and narrowed towards the base.

Spicate borne in or forming a spike.

Spiculate spikelet-bearing.

Spike an unbranched, indeterminate inflorescence with sessile flowers or spiklets. *adj.* spicate, spiciform.

Spikelet a small or secondary spike characteristics of the grasses and sedges and, generally composed of 2 glumes and one or more florets. Also applied to the small spike-like inflorescence or inflorescence units commonly found in Apiaceae.

Spine a stiff, sharp, pointed structure, formed by modification of a plant organ. *adj.* spinose.

Spinescent ending in a spine; modified to form a spine

Spinulate covered with small spines.

Spinulose with small spines over the surface.

Spodosol see podsol.

Sporidia asexual spores of smut fungi.

Sporangium a spore bearing structure found in ferns, fern allies and gymnosperms. *pl.* sporangia. *adj.* sporangial.

Sporocarp a stalked specialized fruiting structure formed from modified sporophylls, containing sporangia or spores as found in ferns and fern allies.

Sporophore a spore-bearing structure, especially in fungi.

Sporophyll a leaf or bract which bears or subtends sporangia in the fern allies, ferns and gymnosperms.

Sporophyte the spore-producing phase in the life cycle of a plant that exhibits alternation of generations.

Spreading bending or spreading outwards and horizontally.

Spur a tubular or saclike extension of the corolla or calyx of a flower.

Squama structure shaped like a fish scale. *pl.* squamae.

Squamous covered in scales.

Squarrose having rough or spreading scale-like processes.

Stamen the male part of a flower, consisting typically of a stalk (filament) and a pollen-bearing portion (anther). *adj.* staminal, staminate.

Staminate unisexual flower bearing stamens but no functional pistils.

Staminode a sterile or abortive stamen, often reduced in size and lacking anther. *adj.* staminodial.

Standard refers to the adaxial petal in the flower of Papilionaceae. cf. keel, wing.

Starch a polysaccharide carbohydrate consisting of a large number of glucose units joined together by glycosidic bonds α-1-4 linkages.

Stellate star shaped, applies to hairs.

Stem the main axis of a plant, developed from the plumule of the embryo and typically bearing leaves.

Sterile lacking any functional sexual parts which are capable of fertilisation and seed production.

Stigma the sticky receptive tip of an ovary with or without a style which is receptive to pollen.

Stilt root a supporting root arising from the stem some distance above the ground as in some mangroves, sometimes also known as a prop root.

Stipe a stalk that support some other structure like the frond, ovary or fruit.

Stipel secondary stipule at the base of a leaflet. *pl.* stipellae. *adj.* stipellate.

Stipitate having a stalk or stipe, usually of an ovary or fruit.

Stipulated having stipules.

Stipule small leaf-like, scale-like or bristle-like appendages at the base of the leaf or on the petiole. *adj.* stipulate.

Stolon a horizontal, creeping stem rooting at the nodes and giving rise to another plant at its tip.

Stoloniferous bearing stolon or stolons.

Stoma a pore in the epidermis of the leaf or stem for gaseous exchange. *pl.* stomata.

Stone the hard endocarp of a drupe, containing the seed or seeds.

Stramineous chaffy; straw-liked.

Striae parallel longitudinal lines or ridges. *adj.* striate.

Striate marked with fine longitudinal parallel lines or ridges.

Strigose bearing stiff, straight, closely appressed hair; often the hairs have swollen bases.

Strobilus a cone-like structure formed from sporophylls or sporangiophores. *pl.* strobili

Style the part of the pistil between the stigma and ovary.

Stylopodium enlargement of the base of the style in some members of the Apiaceae family.

Sub- a prefix meaning nearly or almost, as in subglobose or subequal.

Subcarnose nearly fleshy.

Sub-family taxonomic rank between the family and tribe.

Subglobose nearly spherical in shape.

Subretuse faintly notched at the apex.

Subsessile nearly stalkless or sessile.

Subshrub intermediate between a herb and shrub.

Subspecies a taxonomic rank subordinate to species.

Substrate surface on which a plant or organism grows or attached to.

Subtend attached below something.

Subulate narrow and tapering gradually to a fine point, awl-shaped.

Succulent fleshy, juicy, soft in texture and usually thickened.

Suckers young plants sprouting from the underground roots of a parent plant and appearing around the base of the parent plant.

Sulcate grooved longitudinally with deep furrows.

Sulcus a groove or depression running along the internodes of culms or branches.

Superior refers to the ovary is free and mostly above the level of insertion of the sepals, and petals. *cf.* inferior.

Suture line of dehiscence.

Swidden slash-and-burn or shifting cultivation.

Symbiosis describes close and often long-term mutualistic and beneficial interactions between different organisms.

Sympetalous having petals united.

Sympodial refers to a specialized lateral growth pattern in which the apical meristem. *cf* monopodial.

Synangium an organ composed of united sporangia, divided internally into cells, each containing spores. *pl.* synangia.

Syncarp an aggregate or multiple fruit formed from two or more united carpels with a single style. *adj.* syncarpous.

Syncarpous carpels fused forming a compound pistil.

Syconium a type of pseudocarp formed from a hollow receptacle with small flowers attached to the inner wall. After fertilization the ovaries of the female flowers develop into one-seeded achenes, e.g. fig.

Tannins group of plant-derived phenolic compounds.

Taxon the taxonomic group of plants of any rank. e.g. a family, genus, species or any infraspecific category. *pl.* taxa.

Tendril a slender, threadlike organ formed from a modified stem, leaf or leaflet which, by coiling around objects, supports a climbing plant.

Tepal a segment of the perianth in a flower in which all the perianth segments are similar in appearance, and are not differentiated into calyx and corolla; a sepal or petal.

Tetrasporangium a sporangium containing four haploid spores as found in some algae.

Terete having a circular shape when cross-sectioned or a cylindrical shape that tapers at each end.

Terminal at the apex or distal end.

Ternate in threes as of leaf with three leaflets.

Testa a seed coat, outer integument of a seed.

Thallus plant body of algae, fungi, and other lower organisms.

Thyrse a dense, panicle-like inflorescence, as of the lilac, in which the lateral branches terminate in cymes.

Tomentose refers to plant hairs that are bent and matted forming a wooly coating.

Tomentellose mildly tomentose.

Torus receptacle of a flower.

Transpiration evaporation of water from the plant through leaf and stem pores.

Tree that has many secondary branches supported clear of the ground on a single main stem or trunk.

Triangular shaped like a triangle, 3-angled and 3-sided.

Tribe a category intermediate in rank between subfamily and genus.

Trichome a hair-like outgrowth of the epidermis.

Trichotomous divided almost equally into three parts or elements.

Tridentate three toothed or three pronged.

Trifid divided or cleft into three parts or lobes.

Trifoliate having three leaves.

Trifoliolate a leaf having three leaflets.

Trifurcate having three forks or branches.

Trigonous obtusely three-angled; triangular in cross-section with plane faces.

Tripartite consisting of three parts.

Tripinnate relating to leaves, pinnately divided three times with pinnate pinnules.

Tripliveined main laterals arising above base of lamina.

Triveined main laterals arising at the base of lamina.

Triploid describing a nucleus or cell that has three times (3n) the haploid number (n) of chromosomes.

Triquetrous three-edged; acutely 3-angled.

Trullate with the widest axis below the middle and with straight margins; ovate but margins straight and angled below middle, trowel-shaped.

Truncate with an abruptly transverse end as if cut off.

Tuber a stem, usually underground, enlarged as a storage organ and with minute scale-like leaves and buds. *adj.* tuberous.

Tubercle a wart-like protuberance. *adj.* tuberculate.

Tuberculate bearing tubercles; covered with warty lumps.

Tuberization formation of tubers in the soil.

Tuft a densely packed cluster arising from an axis. *adj.* tufted.

Turbinate having the shape of a top; cone-shaped, with the apex downward, inversely conic.

Turgid distended by water or other liquid.

Turion the tender young, scaly shoot such as asparagus, developed from an underground bud without branches or leaves.

Turnery articles made by the process of turning.

Twining winding spirally.

Ultisols mineral soils with no calcareous material, have less than 10% weatherable minerals in the extreme top layer of soil, and with less the 35% base saturation throughout the soil.

Umbel an inflorescence of pedicellate flowers of almost equal length arising from one point on top of the peduncle. *adj.* umbellate.

Umbellet a secondary umbel of a compound umbel. *cf.* umbellule.

Umbellule an, a secondary umbel of a compound umbel. *cf.* umbellet.

Umbo projection at the scale tip of a seed bearing cone.

Uncinate bent at the end like a hook; unciform.

Undershrub subshrub; a small, usually sparsely branched woody shrub less than 1 m high. *cf.* shrub.

Undulate with an edge/margin or edges wavy in a vertical plane; may vary from weakly to strongly undulate or crisped. *cf.* crisped.

Unifoliolate a compound leaf which has been reduced to a single, usually terminal leaflet.

Uniform with one form, e.g. having stamens of a similar length or having one kind of leaf. *cf.* dimorphic.

Uniseriate arranged in one row or at one level.

Unisexual with one sex only, either bearing the anthers with pollen, or an ovary with ovules, referring to a flower, inflorescence or individual plant. *cf.* bisexual.

Urceolate shaped like a jug, urn or pitcher.

Utricle a small bladdery pericarp.

Valvate meeting without overlapping, as of sepals or petals in bud. *cf.* imbricate.

Valve one of the sections or portions into which a capsule separates when ripe.

Variant any definable individual or group of individuals which may or may not be regarded as representing a formal taxon after examination.

Variegate, variegated diverse in colour or marked with irregular patches of different colours, blotched.

Variety a taxonomic rank below that of subspecies.

Vein (Botany) a strand of vascular bundle tissue.

Velum a flap of tissue covering the sporangium in the fern, Isoetes.

Velutinous having the surface covered with a fine and dense silky pubescence of short fine hairs; velvety. *cf.* sericeous

Venation distribution or arrangement of veins in a leaf.

Veneer thin sheet of wood.

Ventral (Botany) facing the central axis, opposed to dorsal.

Vernation the arrangement of young leaves or fronds in a bud or at a stem apex. *cf.* circinnate

Verrucose warty

Verticil a circular arrangement, as of flowers, leaves, or hairs, growing about a central point; a whorl.

Verticillaster false whorl composed of a pair of opposite cymes as in Lamiaceae.

Verticillate whorled, arranged in one or more whorls.

Vertisol a soil with a high content of expansive montmorillonite clay that forms deep cracks in drier seasons or years.

Vertosols soils that both contain more than 35% clay and possess deep cracks wider than 5 mm during most years.

Vesicle a small bladdery sac or cavity filled with air or fluid. *adj.* vesicular.

Vestigial the remaining trace or remnant of an organ which seemingly lost all or most of its original function in a species through evolution.

Vestiture covering; the type of hairiness, scaliness or other covering commonly found on the external parts of plants. *cf.* indumentums.

Vibratile capable of to and for motion.

Villose covered with long, fine, soft hairs, finer than in pilose.

Villous covered with soft, shaggy unmatted hairs.

Vine a climbing or trailing plant.

Violaxanthin is a natural xanthophyll pigment with an orange color found in a variety of plants like pansies.

Viscid sticky, being of a consistency that resists flow.

Vittae oil tubes in some plants e.g. in the Apiaceae.

Viviparous describes seeds or fruit which sprout before they fall from the parent plant.

Whorl a ring-like arrangement of leaves, sepals, stamens or other organs around an axis.

Winged having a flat, often membranous expansion or flange, e.g. on a seed, stem or one of the two lateral petals of a Papilionaceous flower or one of the petal-like sepals of Polygalaceae. *cf.* keel, standard.

Xanthophylls are yellow, carotenoid pigments found in plants. They are oxidized derivatives of carotenes.

Xeromorphic plant with special modified structure to help the plant to adapt to dry conditions.

Xerophyte a plant which naturally grows in dry regions and is often structurally modified to withstand dry conditions.

Zest scraping of the pigmented outermost layer of the rind of citrus fruit used as a culinary ingredient.

Zygomorphic having only one plane of symmetry, usually the vertical plane, referring to a flower, calyx or corolla. *cf.* actinomorphic.

Zygote the fist cell formed by the union of two gametes in sexual reproduction. *adj.* zygotic.

Common Name Index

A
A549. *See* Lung carcinoma cell lines (A549)
Acid lime, 742
A2780/CP70, 708
A431 human epidermoid carcinoma, 12
Alder Leaf Shadbush, 358
Aleppo Oak, 19
Almond nut, 3, 483
Almonds, 12, 84, 298, 305, 312, 442, 443, 452, 480–490, 507, 620
Amalki, 263
American Black Raspberry, 570
Amla, 258, 259, 262, 265–267, 274–276, 278–282, 285, 286, 288, 290
Andean blackberries, 545
Androgen insensitive DU145, 707
Androgen sensitive LNCaP, 707
Anise, 111, 112, 120, 905
Anise Pepper, 904
Annatto, 641
Apple pear, 423, 428, 535
Apples, 4, 132, 359, 371, 372, 375, 376, 378, 400, 401, 410, 412–432, 459, 468, 469, 472, 517–519, 533, 535, 560, 594, 667, 760, 762, 763, 816, 884–888
Apple-shaped Quince, 371
Apple tree, 413, 887
Apricot, 3, 442–450, 464, 465, 788
Aprium, 449
Arctic bramble, 398, 403, 548, 557, 572
Ashes, 2
Asian Holly-Oak, 16
Asian pear, 531, 533, 535–538
Asian pennywort, 600
Asian plum, 509
Asiatic apple, 410
Astrocytoma, 687, 706
Australian Finger Lime, 625
Australian Round Lime, 629, 630
Australia sweet, 629

B
Bacang, 238
Baccaurea, 3, 225–250
Baden, 181
Bael, 4, 594–613
Bael fruit, 594–597, 600, 606, 607, 609, 611–613

Bael fruit tree, 594
Bael tree, 594, 598, 605
Bakong, 124
Balinese Pepper, 351
Bali Vanilla, 106
Ball tree, 594
Banana, 118, 122, 144, 182, 382, 400, 401, 406, 419, 423, 452, 469, 560, 763, 851
Barbadine, 174, 181, 184
Bartender's Lime, 742
Basil, 600, 641
Bawang Hutan, 77
Beach Pandan, 138
Beam tree, 590
Bearss Lime, 662
Beech, 1
Beet, 600
Bela tree, 594
Bel fruit, 594
Bencoi, 243, 246, 247
Bengal Quince, 594
Benign prostatic hyperplasia (BPH), 316, 819
Beniseed, 187
Benneseed, 187
Bergamot, 685, 762, 775, 777, 812, 825, 834, 840, 854–856, 862
B16F10. *See* Murine melanoma cells (B16F10)
Bigarade Orange, 786
Bignay, 220–224
Bigrade, 786
Bilberry, 34, 403, 404, 562
Bing cherry, 451, 452, 458
Bird cherry, 451
Bitter almond, 480, 482, 490
Bitter leaf, 751
Bitter melon, 386, 405
Bitter orange, 700, 786, 788, 790–793, 797–800, 812, 825, 893
Blackberry, 3, 48, 49, 54, 400–403, 456, 459, 544–552, 559–563, 572, 573, 575, 577, 581, 583, 584, 587, 589
Blackcap, 570
Black currants, 1, 27–37, 39, 40, 44, 47–49, 54, 57, 58, 197, 220, 459, 560, 562, 585, 586, 788
Black currant tree, 220
Black olive, 82, 84
Black Passionfruit, 147, 148

Black pepper, 3, 322–333, 335, 336, 338, 339, 341, 343–345, 356, 678, 745, 851
Black raspberry, 32, 49, 401, 403, 545, 549–551, 559–563, 570–577, 583–585
Blond orange, 817, 822
Blood orange, 700, 806, 810, 812, 814–816, 822, 827
Blueberry, 32, 33, 49, 358, 360, 400, 402–404, 456, 459, 548–551, 558, 560, 562, 563, 573, 575, 576, 583, 585, 627
B-lymphoid Raji, 269, 604
B16 melanoma, 170, 336
Bourbon Vanilla, 106
Boysenberry, 3, 550, 558–561, 581–586, 588, 589
BPH. *See* Benign prostatic hyperplasia (BPH)
Bramble, 398, 403, 548, 557, 572
Bramble Berry, 544
Breast cancer cell lines, 49, 74, 88, 91, 92, 94, 158, 159, 170, 205, 212, 270, 271, 389, 424, 489, 513, 551, 604, 708, 763, 766, 820
  MCF-7, 69, 92, 158, 160, 270, 271, 316, 335, 341, 403, 404, 424, 425, 487, 551, 563, 564, 573, 604, 693, 708, 709, 736, 820, 877
  MDA-MB-231, 270, 425, 513, 604
  T47-D, 121, 403
Breast carcinoma Mcf-7 and Mcf-7:Her18 cells, 424
Broccoli, 401, 469, 560, 763, 960
Brown plant hopper, 613
Brown River Finger Lime, 739
Buah Merah, 117, 120, 121
Buccal pouch epidermoid carcinomas, 708
Buddha's fingers, 690
Buddha's-hand, 690
Buddha's-hand citron, 690, 691, 693
Bullace, 474
Bullace Plum Damask Plum, 476
Bullet kumquat, 654
Burmese Grape, 248
Bush Lemon, 846
Bush Passion fruit, 166

C

C-33A (HPV16-/HPV18-, squamous cell carcinoma), 574
Cabbage, 135, 143, 401, 469, 560, 763
Caco-2 colon cancer cells, 35, 62, 93, 341, 342, 403, 421, 423, 424, 429, 472, 566, 777
Caffir Lime, 634
Calamandarin, 865
Calamansi, 865
Calamondin, 834, 840, 865–870
Calamondin Orange, 865
Californian Plum, 509
Calmondin, 865
Cancer cell line, 49, 69, 257, 271, 425, 573, 574, 687, 706, 865
  Cancer cell line A549, 120, 121, 269, 270, 341, 389, 403, 551, 707–709, 876, 877
  Cancer cell line HL-60, 68, 92, 93, 335, 389, 424, 551, 604, 639,676, 703, 705, 708, 709, 766, 819, 820, 858, 859, 880, 895
  Cancer cell line HO8910, 708
  Cancer cell line MCF-7, 69, 92, 158, 160, 270, 271, 316, 341, 403, 404, 424, 425, 487, 551, 563, 564, 604, 693, 708, 709
    ODS (Osteogenic Disorder Shionogi), 270
Caneberries, 560, 561, 572, 583
Canton Lemon, 849
Capsicum, 84, 312, 335
Carberry, 51
Cardamom, 343, 344, 620
CaSki cervical cancer cell, 403
Castor oil, 286, 336, 355, 565, 606, 826
Cattle tick, 288, 613
Cauliflower, 401, 469, 560, 763
Cedar Pine, 297
Cedars, 3
CeHa cell lines, 698
CEM-SS leukemia cell line, 80
Cervix (HeLa), 389, 425, 709
Cervix epithelioid carcinoma (HeLa), 425
Cervix epitheloid (HeLa) cell lines, 389
Chagas' disease, 317
Checker tree, 590
Cherry, 3, 48, 360, 452, 453, 455–460, 464, 564, 756
Chess tree, 590
Chestnuts, 1, 6–14, 249
Chili, 84, 190, 905
Chili pepper, 745, 788, 905
China Orange, 806, 865
Chinese almond, 442
Chinese Bitter Orange, 786
Chinese Crabapple, 410
Chinese Dwarf Lemon, 619
Chinese Flowering Apple, 410
Chinese Flowering Quince, 364
Chinese Laurel, 220
Chinese Lemon, 849
Chinese Loquat, 381, 392
Chinese Orange, 865
Chinese Pear, 535, 540
Chinese Pepper, 904
Chinese Pinenut, 297
Chinese Plum, 509
Chinese Prickly Ash, 904
Chinese Quince, 364, 376, 378, 515–520
Chinese star anise, 120
Chinese White Pear, 523
Chinotto, 700
Chokeberry, 34
Chokecherry, 360
Cholera, 213, 368, 431, 607, 612, 751, 869
Chupak, 228
Cinnamon, 111, 112, 382, 443, 528
Cinquefoils, 3
Cisplatin-resistant human ovarian cancer cells (2008/C13), 708
Citrange, 806, 893, 898
Citrangequat, 893, 898
Citronelle, 846
Citrons, 634, 666, 682–688, 691, 812, 849, 850, 854

Clementine, 685, 702, 704, 786, 812
Clementine mandarin, 700
Cleopatra mandarin, 815
Cloudberry, 403, 545, 548, 557, 572
Clymenia, 3, 4, 659
Coconut, 107, 121, 125, 129, 140, 144, 382, 443, 635, 745, 885
Coconut oil, 121, 132, 145, 256, 338, 826, 886
Cold Hardy Mandarin, 721
Colitis-associated colon cancer, 426
COLO 201, 676
Colon cells. *See* Human colon cancer cell lines
Colonic adenocarcinoma, 709, 736
Colorectal cancers, 121, 425, 426, 577, 706, 736
Colorectal carcinoma COLO 205, 708
Columbia SK, 677
Commercial Vanilla, 106
Common Apricot, 442
Common Blackberry, 544
Common carp, 613
Common Currant, 43
Common Grapefruit, 755
Common Lime, 742, 744
Common Olive, 82, 97
Common Orange, 700
Common Pear, 527, 532
Common Plums, 463, 474
Common Quince, 371
Common Rambai, 239
Common Vanilla, 106
Coolie Orange, 806
Coral Ardisia, 65
Coralberry, 65
Coralberry tree, 65
Coral Bush, 65
Cork oak, 1
Cornelian cherry, 48, 54, 549, 562
Country Gooseberry, 252
Coxsackie A21 viruses, 677
Coxsackie B1, 677
Coxsackie virus B3 (CVB3), 205, 206, 277
Crabapples, 410–412
Crab Cherry, 451
Cranberry, 49, 402–404, 419, 549–551, 559, 563, 573, 585, 772, 866
Crisp Chinese Pear, 523
CT-26 colon carcinoma, 504
Cubeb, 311, 312, 319, 320
Cubeb Pepper, 311, 313, 314, 319
Culate Mandarin, 695
Cultivated blackberry, 544
Cultivated Currant, 43
Cultivated Olive, 82
Curculigo, 1, 59, 61, 63
Curd fruit, 884
Currant tree, 220
Currentwood, 220
Curry leaf, 4, 325, 635, 866
CVB3. *See* Coxsackie virus B3 (CVB3)
Cyprus Oak, 16

D

Dalton's lymphoma ascites (DLA), 268, 269, 274
Damson plums, 3, 469, 474, 476–478
Danshen, 392
Deberries, 51
Dekopon, 732, 733
Dekopon mandarin, 732
Dessert pear, 528, 572
Dewberry, 544, 581
DLA. *See* Dalton's lymphoma ascites (DLA)
Dooja, 629
Doubleflower Chinese Crabapple, 410
Downy Oak, 16
Dried plums, 465, 471–474, 585
DU-145. *See* Prostate cancer cell lines
Duck Pear, 523
Duck's Eye, 72
Durian daun, 238
Durian nyekak, 238
Dwarf lemon, 619
Dyer's oak, 16

E

EBV. *See* Epstein-Barr virus (EBV)
Egyptian Lime, 742
Ehrlich ascites carcinoma cell, 37, 604, 676, 886
Ehrlich ascites cells, 37
Elderberry, 404
Elephant apple, 4, 594, 884
Elliptical-Leaf Ardisia, 72
Emblic, 258, 260, 261, 264, 267, 269, 272, 274, 275, 278, 279, 281, 283–286, 289, 290
Emblica, 258–290
Emblic Myrobalans, 260, 289
English Gooseberry, 51
Ependymoma, 706
Epstein-Barr virus (EBV), 639, 674, 677, 858
ER-negative breast cancer (MDA-MB-231), 270, 389, 425, 513, 604
Erythroleukemic HEL, 269, 604
Erythroleukemic HEL cell lines, 269
Eurasian Wild Plum, 476
Eureka lemon, 748, 852, 853
European Blackberry, 544
European Blackcurrant, 27
European chestnuts, 6, 9
European Gooseberry, 51
European Mandarin, 695
European Olive, 82
European Pear, 527, 528, 531, 537
European Plum, 463, 464
European Raspberry, 555, 581
European Red Raspberry, 555, 581
European Stone Pine, 304
Evening primrose, 300, 301, 565
Evergreen blackberry, 549, 550, 559, 560, 583, 584

F
Fagara, 904
False Pepper, 311
False sandalwood, 2
Feaberry, 51
Feabes, 51
Feronia, 887
Feverberry, 51
Fingered Citron, 688, 690, 692, 693
Finger Lime, 625, 627
Firs, 3
Flat-Leaved Vanilla, 106
Flatspine Prickly Ash, 904
Flaxseed, 204, 206, 211
Flesh-Finger Citron, 690
Flounder, 737
Flowering Merrillia, 890
Flowering-Quince, 364
Flower pepper, 904
Flying Dragon, 893
Foetid Passion Flower, 166
Fool's Curry Leaf, 871
Forbidden fruit, 667
Forest Onion, 2, 77
Forsythia, 2
Fragrant Pear, 540–542
Fresh Nectarine, 492
Fringetrees, 2

G
Gage, 463, 468, 470, 471
Gall nuts, 1, 16–20, 23, 24
Gall Oak, 16
Galls, 17, 18, 21, 22, 24
Garden Blackcurrant, 27
Garden Currant, 43
Garden Plum, 463, 466
Garden Strawberry, 395
Garlic, 2, 77, 78, 81, 84, 197, 312, 744, 751, 903
Garlic Nut, 77
Garraway's Australian Wild Lime, 631
Gastric human tumour cells, 551
Gean, 451
Giant Granadilla, 181–183, 185, 186
Giardia and rotavirus, 606
Gingelly Sesame, 187
Ginger, 111, 112, 144, 189, 515, 635, 745, 751, 788, 861, 905
Glioblastoma multiforme, 706
Gliosarcoma (9 L) cell line, 705, 819
Golden Apple, 417, 594
Golden Lime, 865
Golden Orange, 154, 647–649
Gooseberry, 1, 30, 33, 47, 48, 51–55, 58, 549, 562, 788
Goosegogs, 51
Granadilla, 147–149, 174, 175, 181–184
Granadilla Cimarrona, 166
Granadille True, 181
Grant Lemon, 619

Grapefruits, 4, 400, 419, 459, 468, 560, 597, 700, 747, 755–779, 792, 813, 832, 854
Grape marc, 11
Greater Tampoi, 236
Green and white currants, 32, 47
Green cabbage, 400, 401, 468, 469, 560, 763, 816
Greengage, 468, 470, 474
Green gooseberries, 33, 48, 54
Green grape, 401, 469, 560, 763
Green huajiao, 905, 906
Green Olives, 82, 86
Green Pepper, 322, 324, 325, 329, 330, 332
Green peppercorns, 325, 332
Green tea, 376, 377, 779
Grenadia, 174
Grenadine, 181
Grenedilla, 181
Groser, 51
Grozet, 51
Gympie Lime, 629

H
H441 Human lung cancer, 820
H2058 tumour, 676
Hala tree, 138, 139
Hallabong, 732–737
Hallabong Mandarin, 732
Harbin Pear, 540
Hardy Orange, 893
HA22T tumour, 676
Hawthorns, 3
Hazelnut, 486
H9c2 cardiomyoblastoma, 908
H460 Human lung cancer cells, 820
HCT-116 colon cancer cells, 425, 605
Head lice, 773
Hedge Lime, 659
HeLa. *See* Human cervica cancer (HeLa)
Hemlocks, 3
Hen's-Eyes, 65
Hepatitis C virus (HCV), 23, 317, 710, 726, 765, 895
Hepatocellular carcinoma BEL-7404, 270, 279
Hepatoma cells. *See* Human hepatocellular carcinoma (HepG2)
HepG2. *See* Human hepatocellular carcinoma (HepG2)
Heps tumour, 707
Herpes simplex virus type 1 (HSV-1), 36, 74, 277, 427, 639, 677, 903, 940
Herpes simplex virus type 2 (HSV-2), 36, 74, 277, 427, 677, 903, 940
High-bush lueberry, 49
Hilo Holly, 65
Hinggan Red Pine, 297
Hispid granadilla, 166
HIV, 69, 121, 122, 276, 377, 390, 710, 749, 774, 880
HIV-1NL4.3, 276, 606
HIV-1(IIIB) virus, 276
HL-60. *See* Human promyelocytic leukaemia (HL-60) cells

Hog-plum, 2
Holy Fruit, 594
Honeyblobs, 51
Honeydew melon, 400
HPV. *See* Human papillomavirus (HPV)
HSC-3 squamous cell carcinoma, 424
HT-29 human colon cancer cells, 93, 403, 425, 426,
 458, 551, 563, 573, 607, 706–708, 736, 764,
 794, 820, 895
Huangpi, 877
Human breast cancer cell lines
 MDA-MB-435, 708, 736, 820
 MDA-MB-453, 748, 765
Human cancer cells derived from colon (HT29) and
 stomach (MKN45), 458
Human cervica cancer (HeLa), 403, 574, 877
 cervical, 270
 human uterine carcinoma, 269
Human cervical cancer cell lines HeLa (HPV16-/
 HPV18+, adenocarcinoma), 574
Human cholangiocarcinoma CL-6 cell, 354
Human chronic myelogenous leukemia (K562), 269,
 604, 639, 706, 794, 820
Human colon adenocarcinoma, 93
Human colon adenoma cells (LT97), 426
Human colon cancer cell lines, 93, 377, 403, 404, 426,
 458, 551, 563, 573
 Caco-2, 403
 HCT116, 706
 HT-29, 425, 706, 708, 736, 794, 820, 895
 SW-480, 748, 767
Human colorectal adenocarcinoma HT-29, 93
Human coxsackieviruses B1-B6, 605
Human fibrosarcoma HT-1080 cells, 706
Human gastric adenocarcinoma
 AGS cells, 707
 MK-1, 269, 707
Human gastric cancer cell lines
 MKN-45, MKN-74 and KATO-III, 706, 736
 TMK-1, 706, 736
Human gastric cancer cells, 705
Human gastric carcinoma (SGC-7901), 876
Human hepatocellular carcinoma (HepG2), 37, 69, 269,
 277, 335, 423–425, 513, 563, 709, 712, 769, 841,
 876, 880
 cells, 69, 269, 277, 390, 425, 513, 769, 841, 859
 hepatoma cell line, 676, 712, 769, 841
 liver cancer, 37, 69, 270, 401, 423, 424
Human hepatoma Bel-7402 cells, 69
Human immunodeficiency virus HIV-1NL4.3, 276
Human immunodeficiency virus type 1 (HIV-1), 276,
 390, 710
Human laryngeal (Hep-2), 354
Human leukaemia cells (U937), 639, 676
Human lung adenocarcinoma cancer cell (A-549), 551,
 707–709, 876
Human lung cancer (H1299), 820
Human lung fibroblast-like cell line (CCL 135), 819
Human lymphoblastoid B cell line RPMI-8866,
 748–749

Human lymphoblastoid T leukemia cell line (MOLT-4),
 639, 706, 710, 726, 821
Human lymphoid leukemia Molt 4B cells, 205
Human monocytic leukaemia (THP-1) cell, 96, 316, 343
Human neuroblastoma (IMR32), 354
Human neuroblastoma SH-SY5Y cells, 585, 708,
 771, 820
Human oesophageal cancer, 575
Human oral (KB, CAL-27), 403, 551, 563, 573
Human pancreatic cancer cells, 271
Human papillomavirus (HPV), 276
Human papillomavirus type 16 (HPV-16), 276
Human promyelocytic leukaemia (HL-60) cells, 68, 80,
 92, 93, 205, 335, 389, 406, 424, 551, 586, 604,
 674, 676, 703, 705, 706, 708–710, 726, 727, 766,
 794, 819–821, 858, 859, 880, 895, 908
Human prostate cancer cells (LNCaP), 316, 403, 404,
 551, 563, 573, 707
Human prostate carcinoma CWR22Rnu1
 and DU145 cells, 424
Human prostate epithelial cell line (P21), 404
Human salivary gland tumor cell lines, 388, 389
Human squamous cell carcinoma cell line (HTB43), 388,
 389, 705, 819
Human T47D mammary cancer cells, 708
Hybrid Strawberry, 395

I
Iberian Olive, 82
Ichang Lime, 634
Immunodeficiency virus (HIV)-1, 903
Indian Bael, 594
Indian gooseberry, 252, 258, 259
Indian Lime, 742
Indian Ocean Vanilla, 106
Indian Pepper, 322
Indian Quince, 594
Indian Wood-Apple, 884
Indonesian Vanilla, 106
Influenza A2/Asian/J305, 677
Influenza viral, 519
Inkberry, 72
Italian Chestnut, 6
Italian Stone Pine, 304

J
Jaborandi Pepper, 351
Jamberi, 846
Jambhiri Orange, 846
Jambu mawar, 238
Jambu susu, 238
Japanese Bitter Orange, 893
Japanese Blood Plum, 509, 511, 512
Japanese Box Plum, 509
Japanese Hardy Orange, 893
Japanese Holy, 65
Japanese Loquat, 381
Japanese Mandarin, 721

Japanese Medlar, 381
Japanese Pear, 535
Japanese Plums, 381, 510–513
Japanese Sweet Orange, 721
Jasmines, 2, 801
Java Long Pepper, 351–353, 355, 356
Javanese Long Pepper, 351
Javanese Pepper, 311
Javanese Peppercorn, 311
Java Pepper, 311, 351
Java Vanilla, 106
JCS leukemic cells, 706
Jentik-jentik, 238
Jeruk Limo, 623, 624
Jostaberry, 54, 56–58
Juneberry, 3, 358
Jungle Belimbing, 115
Juniper berries, 84

K

K562. *See* Human chronic myelogenous leukemia (K562)
Kaffir Lime, 634, 636–641
Kalamansi, 865, 869
Kapundung, 243
Karuka, 128–130
Karuka Nut, 128, 130
Katinga, 890
KATO III, 472, 706, 736
KB (oral human epidermal carcinoma), 676, 877
Ketupa, 228
Key Lime, 742, 747, 843
Kidney bean, 62
King mandarins, 715
King of Siam, 837
King Orange, 837
Kirsten murine sarcoma virus, 707
Kiwi fruit, 400, 459, 486
Knob-Fruited Screwpine, 124
Korean Cedar, 297
Korean Pear, 535
Korean Pine, 297, 299–302
Korean Pinenut, 297, 299
KU812F chronic myelogenous leukemia cell, 80
Kuini, 238
Kumquats, 4, 647–649, 651–660, 700, 834, 840, 843–845, 865, 866, 870

L

L1210. *See* Mouse lymphocytic leukemia (L1210)
Lampaong, 232
LanternTree, 248
Larches, 3
Large-Fruited Strawberry, 395
Large leaf Australian wild lime, 644
Large Round Kumquat, 651
Leech Lime, 634
Leek, 400, 401, 468, 469, 560, 762, 763, 816

Lemba, 59, 63, 181
Lemon balm, 550
Lemon China, 900
Lemons, 4, 60, 84, 147, 152, 167, 174, 182, 325, 365, 377, 378, 382, 419, 423, 528, 550, 619–622, 634, 635, 637, 654, 666, 668, 683–685, 691, 700, 701, 708, 710, 742, 743, 745, 747–749, 762, 763, 766, 772, 787, 792, 801, 812, 813, 818, 825, 834, 840, 846–862, 865, 866, 868, 870, 890, 900
Leprous Lime, 623
Lettuce, 161, 401, 469, 560, 763
Lewis lung cancer and human nonsmall lung cancer, 121
Licorice, 111, 112, 392
Lilacs, 2
Lima orange, 704
Lime, 708, 792
Limeberry, 4, 900
Lime Chinese, 900
Lime Orange Berry, 900
Limequat, 753, 843–845
Limes, 4, 182, 213, 612, 623, 625–627, 629–632, 634–641, 644, 645, 659, 662–666, 708, 739, 740, 742–753, 763, 787, 788, 792, 801, 818, 843, 844, 851, 854–857, 865, 866, 868, 869, 893, 900
Lingonberry, 34, 403, 562
Liver cancer cells. *See* Human hepatocellular carcinoma (HepG2)
LNCaP. *See* Human prostate cancer cells (LNCaP)
Loganberry, 3, 581, 584, 587, 588
Long-Leaved Kumquat, 659
Long Pepper, 345, 351–354, 356
Loose-Skinned Orange, 695
Loquat, 3, 378, 381–392, 405
Love-In-A-Mist Passionflower, 166
Low-bush blueberry, 49
Lumbah, 59
Lung carcinoma cell lines (A549), 120, 121, 269, 270, 341, 389, 403, 551, 707–709, 876, 877

M

Mafai, 239, 248–250, 871, 877
Maize, 301, 751
Malacca Tree, 258
Malarial parasites, 677, 887
Malayan Kumquat, 659, 660
Malay Gooseberry, 252
Malay lemon, 890
Maltese Orange, 695
Mammary carcinoma, 271, 272
Manchurian Pear, 540
Mandarin oranges, 4, 480, 619, 623, 685, 695–702, 705, 709–711, 715, 721–726, 728, 729, 732, 733, 735, 742, 802, 812, 815, 825, 832, 834, 837, 840, 842, 848, 854, 859, 865, 868
Mango, 760, 989
Mangosteen, 600
Maracuya, 147, 149, 181
Maredoo Quince-Apple Of India, 594

Marionberry, 558–560, 583, 584
Marion blackberry, 550, 559, 583
Marita, 117, 118, 122
Marmalade Orange, 786
Marron, 6, 11, 166, 167
Marumi kumquats, 647–649, 652, 653
Mauritius Papeda, 634
Mauritius Vanilla, 106
Mazoe Lemon, 846
Mazzard, 451
MCF-7. *See* Breast cancer cell lines, MCF-7
MDA-HB-435 human breast cancer cell, 708, 736, 765, 820
MDA-MB-231 breast cancer cell lines, 270, 425, 513, 604
Meadowsweets, 3
Mediterranean Olive, 82
Mediterranean Stone Pine, 304
Medlar, 3, 381, 437–440
Medlar Tree, 437
Meiwa kumquat, 651–653
Melanoma cells. *See* Murine melanoma cells (B16F10)
Melanoma Colo38, 604
Menteng, 239, 243–247
Meth-A-fibrosarcoma, 389
Mexican lime, 663, 665, 666, 742, 752
Mexican Vanilla, 106
Meyer lemons, 619–622
Meyer Lemon Tree, 619
Mikan, 634, 696, 721, 722, 729, 807
Mint, 111, 112, 791, 869
Mirabelle, 470, 471, 474
MK-1 human gastric cancer cell line, 269
MKN-45 human gastric cancer cell line, 706, 736
MKN-74 human gastric cancer cell line, 706, 736
Mock Orange, 893
MOLT-4 cells. *See* Human lymphoblastoid T leukemia cell line
Mongolian Pear, 540
Monkey Fruit, 884
Monkey-Guzzle, 178
Mosquito, 23, 113, 288, 345, 355, 611, 641, 713, 752, 825, 861, 887, 894, 897, 898
Mossy Passionflower, 166
Mountain ash, 3, 590
Mountain Juneberry, 358
Mount White Lime, 631, 632
Mouse embryonal carcinoma (PCC4), 354
Mouse lymphocytic leukemia (L1210), 676, 794
Mouse Sarcoma-180, 121
Mouse thymic lymphoma (barcl-95), 271
Mulberry, 386, 405, 737
Murine B16/F10 melanoma cells, 713
Murine leukemia P-388, 709
Murine melanoma cells (B16F10), 68, 211, 269, 334, 707–710
Murine myeloid leukemic cell clone WEHI 3B, 705
Musk Lime, 865, 868–870
Myrobalan, 258, 260, 289
Myrrh, 715

N

Nagami kumquats, 649, 652–658, 865
Nai Plum, 509
Nashi, 527, 535–540
Nashi Pear, 535, 536, 538
Nasnaran, 623
Nasnaran Mandarin, 623
Naspli, 381
Native Finger Lime, 625
Native Lime, 629
Native Orange, 629
Navel oranges, 675, 700, 735, 806, 809, 810, 812–814, 816–819, 822, 825, 827
Navel sweet orange, 813, 818
NCI-H460 Liver cancer cell lines, 69
NCI-H187 (human, small cell lung cancer) cell lines, 877
Nectarines, 3, 465, 492–496, 500, 503, 504
Nectar Raspberry, 555
Neroli, 786, 788, 793
Nicobar-Breadfruit, 131
Nigger's Cord, 220
Non-melanoma skin cancer, 505
Norovirus, 765
Noxious Blackberry, 544
Nut-Galls, 16–18, 20, 22
Nutmeg, 111, 112, 182, 346
Nutwood, 77

O

Oak, 1, 16, 72
Oil Pandan, 117
Oligoastrocytoma, 706
Olives, 83–85, 89, 91, 94, 96, 101
Olive Tree, 82–84, 88, 89, 101
Omani Lime, 742
Onion, 77, 371, 401, 405, 468, 469, 560, 745, 762, 763, 813, 905
Oral (KB, CAL-27), 403, 551, 563, 573
Oral epithelial dysplasia, 574
Orangeberry, 4
Oranges, 4, 402, 406, 651, 675, 696–699, 701, 708, 713, 765, 788–790, 792, 809, 810, 813, 815, 816, 818, 819, 834, 839, 840, 842, 854
Orchid, 2
Oriental Pear, 535
Ornamental Quince Japanese Quince, 364
Oroblanco, 771
Osmanthuses, 2
Otaheite Gooseberry, 252, 254
Oval Kumquat, 654, 655, 845, 870
Ovarian cancer, 708

P

Pacific Serviceberry, 358
Pamplemousses, 667, 755, 834, 840
Panama Orange, 865
Panc-28 cancer cell, 748

Pancreatic cancer (MDA Panc-28), 271, 404, 709
Pandan, 138, 139
Pandanas, 138, 139
Pandanas Palm, 138
Pandanus Nut, 117, 124, 128
Pandanus Palm, 134
Pandan Wong, 131
Papaya, 122, 182, 744, 866
Papuan Wild Lime, 739
Paradise apple, 667
Parasol Pine, 304
Passion flower, 166, 181
Passion fruit, 2, 147–149, 151–158, 160, 161, 178, 760
PC-3. *See* Prostate cancer cell lines
Pea, 17, 401, 469, 560, 763
Peach, 3, 400, 401, 419, 422, 423, 449, 465, 468, 469, 481, 492, 493, 495, 496, 498–507, 560, 760, 762, 763, 788, 816
Peachcot, 449
Peach Tree, 496, 498, 499, 503
Peanut, 134, 201, 482, 486, 635, 688, 745
Pear, 3, 372, 378, 422, 524, 525, 527–533, 536, 541
Pearl lemon, 654
Pear Tree, 527, 533, 593
Pegaga, 638
Pepper 3, 81, 182, 189, 256, 323, 745, 869, 905
 black pepper, 3, 322–333, 335, 336, 338, 339, 341, 343–345, 356, 678, 745, 851
 green pepper, 322, 324, 325, 329, 330, 332
 white pepper, 322, 324–328, 330, 331, 333, 341, 342, 345
Pepper tree
Perry Pears, 527
Persian Gum, 476
Persian lime, 662, 665, 818
Persimmon, 405, 600
Petitgrain, 664, 688, 697, 700, 715, 788, 801, 861
petitgrain mandarin oil, 697, 700
Phenomenal-Berry, 587
Philippine Lime, 865
Pickling Olive, 82
Pignolia-Nut Pine, 304
Pimentos, 84
Pineapple, 142, 182, 419, 452, 459, 683, 772
Pineapple strawberry, 395
Pine nuts, 3, 298, 299, 301, 302, 305, 307–309
Pines, 3, 305, 308, 309
Pine seed, 300–302, 307
Pine strawberry, 395
Pink grapefruit, 400, 760, 771
P388 leukemia cells?, 676
Plumcot, 449
Plums, 3, 449, 464–466, 468–470, 474, 477, 478, 513
Pluot, 449
Poliovirus type 2, 639
Pomelo, 667–672, 675, 676, 678, 755
Ponkan, 696, 700–704, 708, 711, 732
Porcupine orange, 634
Porcupine orange lime, 634
Portugal Orange, 806

Portuguese Chestnut, 6, 9, 10
Privets, 2
Prostate cancer cell lines, 270, 316, 403, 404, 424, 564, 707
Prostate tumor cell. *See* Human prostate cancer cells (LNCaP)
Prune Plum, 463, 467
Prunes, 465, 468, 470–474, 591
Pseudorabies viruses, 677
Pumello, 4
Pummelo, 667, 668, 670–675, 756, 770, 775, 802
Purple and yellow passion fruit, 153, 154
Purple Granadilla, 147, 148
Purple Passion Fruit, 147–151, 153, 154, 159, 160
Purple Raspberry, 545, 570
Purple Water Lemon, 147
Pyrene Oil, 82

Q
Queensland Finger, 625
Queensland Wild, 644
Queensland Wild Lime, 644
Quince, 3, 372–378, 515, 517–519
Quince Seeds, 371, 375, 378
Quince Tree, 371

R
Rambai, 221, 228, 232, 236, 239, 241, 243, 248
Rambi, 239, 243
Raspberry, 3, 34, 48, 49, 54, 360, 400–404, 459, 468, 469, 545, 548–550, 555–566, 571, 583, 762, 816
Raspberry Bush, 555
Raspberry seeds, 403, 404, 559, 562
Raspberry x blackberry hybrids, 48, 54, 549, 562
Rat histiocytoma (BC-8), 354
Red Angled Tampoi, 225
Red currants, 1, 32, 33, 43–49, 54, 459, 549, 562
Red delicious, 415–417, 421–423, 425, 459
Red fruit, 66, 117, 118, 120–122, 556, 903, 905
Red Granadilla, 178
Red grape, 400, 419, 423, 758, 759, 761, 763, 765, 769, 770
Red huajiao, 905
Red orange, 819, 821, 838
Red Pandanus, 117
Red Passion Flower, 178
Red passionfruit, 147, 150
Red Passion Vine, 178
Red peppercorns, 324
Red plum, 400, 401, 468, 469, 560, 762, 763, 816
Red raspberry, 398, 400, 401, 403, 456, 548–551, 555–563, 565–567, 572, 573, 575, 583, 584, 587
Renal adenocarcinoma, 377
Rhinovirus, 390
Riversii Chinese Crabapple, 410
Rocky Mountain Blueberry, 358
Roses, 3, 265, 275, 276, 384, 772, 801, 993
Rough lemon, 634, 846–848

Rough Seville, 786
Round kumquat, 647, 649, 651, 843, 845
Round Lime, 629, 630
Rowan, 3
Running Pop, 167
Russell River Lime, 644

S

Safflower, 202, 300, 301
Salamander Tree, 220
Sandfly, 860
Sand Pear, 535
Sarcoma 180 cells, 334
Sarvisberry, 358
Saskatoon, 358, 360, 362
Saskatoon berry, 358–362
Saskatoon Serviceberry, 358
Satsuma, 695, 721–729, 834, 840
Satsuma mandarins, 697, 700, 715, 721–726, 728, 729, 735
Satsuma Orange, 647, 695, 721, 866
Satsuma tangerine, 721
Sauvage Lime, 625
Scarlet Lime, 865
Scarlet Passion Flower, 178
Scotch Cap, 570
Screw Palm, 134
Screw Pine, 124, 134, 138
Sea buckthorn, 49, 360
Seashore Ardisia, 72
Seedless Lime, 662
Seedless Mandarin, 721
Semliki forest, 677
Semsem, 187
Serviceberry, 3, 49, 358
Service tree, 590
Service Tree Mountain Ash, 590
Sesame, 3, 187–213, 288, 623
Seville, 786
Seville Orange, 786, 787, 808
Seville orange juice, 799
Seychelles Vanilla, 106
Shadbush, 358
Shaddock, 667, 668, 672
Shamuti Orange, 806
Sheep fluke, 288, 613
Shiranui, 732–737
Shiranui Dekopon, 732
Shoebutton, 72
Shoebutton Ardisia, 72
Shrubby Blackberry, 544
Siberian Apricot, 442
Siberian Pear, 540
Sichuan pepper, 189, 904, 905, 909
Sichuan peppercorn(s), 904, 905
SiHa (HPV16+/HPV18-, squamous cell carcinoma), 574
SiHa cervical cancer cell, 403
Sindbis virus, 390
Sindu, 77

SKBR3 human breast adenocarcinoma, 74, 92, 170
SK-HEP-1 human hepatocellular carcinoma cells, 895
SK-MEL-1 melanoma cell, 707, 708
SK-OV3 (ovarian) human cancer cell lines, 270
Sloe, 3, 464
Smoky berries, 360
Smooth-Skinned Peach, 492
Snow Pear, 540
Soft Pear, 527
Sorb tree, 590, 591
Sour cherry, 48, 402, 452, 455, 457, 459
Sour Lime, 742, 745
Spanish chestnut, 6, 13
Spear Flower, 65
Spiceberry, 65, 66
Spicy-peeled kumquat, 654
Spinach, 400–402, 468, 469, 560, 762, 763, 816
Spruces, 3
Squamous cell carcinoma HSC-2, 424
Square Stalked Passion Flower, 181
Square-Stem Passion Flower, 181
S-180 sacrcoma cells, 533
Starfruit, 2
Star Gooseberry, 252–254, 256
Stinking Granadilla, 167
Stinking Passionflower, 167
St. John's Wort, 799
Stomach (NCI-SNU-1), 709
Stone Apple, 594
Stone pine, 305, 307, 309
Stone pine nut, 305
Strawberry, 3, 48, 49, 360, 386, 395–407, 411, 419, 423, 456, 459, 468, 469, 548–551, 557, 559–561, 563, 572–576, 585, 762, 816
Submandibular gland carcinoma HSG, 424
SunAmla, 265, 279, 280
Suntara Orange, 695
Swatow Tangerine, 695
Sweet Almond, 480, 490
Sweet cherry, 400, 451–460, 548, 560, 564
Sweet chestnut, 6, 7, 11, 13
Sweet Cup, 147
Sweet granadilla, 174–177
Sweet kumquat, 651
Sweet orange(s), 696, 699, 700, 721, 748, 763, 766, 788, 790, 792, 800, 801, 806–827, 837, 839, 857
Sweet Passion Fruit, 174
Sweet-Peeled kumquat, 647
Sweet potato, 118, 600
SW620 (colorectal) human cancer cell lines, 270
Swingle's kumquat, 659
Szechuan pepper, 4, 189, 904, 905

T

Table Nectarine, 492
Tagua Passion Flower, 167
Tahitian Gooseberry, 252
Tahitian Lime, 662–665
Tahitian Screwpine, 138

Tahitian Vanilla, 106
Tahiti Lime, 662, 663, 665, 745, 748, 856
Tailed Pepper, 311
Tampoi, 225, 226, 228, 230, 231, 236, 238, 243, 248
Tampoi kuning, 236, 238, 248
Tampoi Merah, 230, 248
Tampoi putih, 236, 238
Tampui, 236, 243
Tangelo(s), 832–835, 839–841
Tangerines, 695, 698, 700, 713, 733, 790, 815, 837, 839
Tangor mandarin, 842
Tangors, 700, 815, 837, 839
Tart cherries, 457
Tayberry, 587
Temple Orange, 837
Textile Screw-Pine, 138
Thai Bai Makrut, 634
Thai cobra, 24
Thatch Screw-Pine, 138
Thick-Skinned Finger Lime, 631
Thimble Berry, 570
Thornless blackberry, 1, 546, 547, 549
THP-1 cells. *See* Human monocytic leukaemia (THP-1) cell
Tight-skinned orange, 806
T-lymphoid Jurkat, 269, 604
TMK-1. *See* Human gastric cancer cell lines, TMK-1
Tobacco army worm, 641
Tomato, 400, 401, 468, 469, 560, 745, 762, 763, 816
Tonkan, 700, 711
Toothache tree, 904
Trifoliata, 893–898, 900
Trifoliata Orange, 893, 894
Trifoliate Limeberry, 900
Triphasia, 4, 900–903
Triphasia Limeberry, 4, 900
Tristeza virus disease, 801, 848
True Mandarin, 695
True Service tree, 590
Type A and B influenza viruses, 36, 519

U

U937. *See* Human leukaemia cells (U937)
U14 cervical cancer, 300
Ugli fruit, 832
Umbrella Pine, 304
Unshiu, 695, 703, 704, 710, 721–729, 732, 735, 737, 817
Urinary bladder cancer 5637 cells, 710
Ussurain Plum, 509
Ussurian Pear, 540, 541

V

Valencia, 665, 700, 806, 809, 811–813, 815, 817, 818, 825, 827
Valencia oranges, 665, 809, 811, 815, 818, 827

Valley Lemon, 619
Vanilla, 2, 106–113, 149, 456, 528, 801
Vanilla vine, 106, 108
Varicella-zoster virus, 36
Veitch Screw-Pine, 138
Velvet leaf blueberry, 49
Village Ardisia, 65

W

Wampee, 4, 871–876, 880
Wampi, 871
Wart Lime, 634
Washington Navel, 809, 817, 819, 825
Washington Navel orange, 809
Water lemon, 147, 167, 174
Weevil Lily, 59
Western equine encephalitis, 677
Western Serviceberry, 358
Western Shadbush, 358
West Indian Lime, 742
West Indian Vanilla, 106
White currants, 32, 43, 44, 47–49
White grape, 400, 704, 759–761, 763, 767
White Pear, 523
White Pepper, 322, 324–328, 330, 331, 333, 341, 342, 345
White peppercorns, 325, 327
White Pine, 56, 297
White sapote, 4
Whitty Pear, 590
Wild bilberry, 403, 404
Wild Blackberry, 544, 550
Wild blueberry, 360, 403
Wild Carambola, 225
Wild Cherry, 220, 451
Wild chestnut, 9
Wild Gooseberry, 51
Wild Lemon, 846
Wild Lime, 631, 634, 644, 645, 739
Wild Maracuja, 167
Wild Passionfruit, 167
Wild Water Lemon, 167
Willow-leaf cherry, 509
Wood apple, 594, 884–888
Wood Garlic, 2, 77–81
Woodland Onion, 77
Wood sorrel, 2

X

Xinjiang pear, 533, 541

Y

Ya Li Pear, 523, 524
Ya Pear, 523, 524
Yellow passion fruit, 147–154, 156

Scientific Name Index

A
Acacia catechu, 19, 20
Acinetobacter baumannii, 20, 37
Adenocrepis lanceolatus, 232
Aedes
  aegypti, 344, 345, 355, 611, 641, 887, 897, 898
  albopictus, 23, 887
  togoi, 344
Aegiceras, 2
Aegle
  marmelos, 4, 269, 594–613
  sepiaria, 893
Aeromonas hydrophila, 869
Aframomum melegueta, 312
Albizia lebbeck, 264, 282
Alchemilla, 3
Allium
  cepa, 609
  sativum, 861
Aloe vera, 275, 276, 611
Alstonia scholaris, 599
Amelanchier, 441
  alnifolia, 358–362
  canadensis var. *alnifolia*, 358
  carrii, 358
  leptodendron, 358
  macrocarpa, 358
  sanguinea var. *alnifolia*, 358
Amygdalus
  amara, 480
  armeniaca, 442
  communis, 480
  communis var. *amara*, 480
  communis var. *dulcis*, 480
  communis var. *fragilis*, 480
  dulcis, 480
  fragilis, 480
  persica, 498
  persica [unranked] *aganonucipersica*, 498
  persica [unranked] *aganopersica*, 498
  persica [unranked] *scleronucipersica*, 498
  persica [unranked] *scleropersica*, 498
  persica var. *aganonucipersica*, 498
  persica var. *compressa*, 498
  persica var. *nectarina*, 492
  persica var. *nucipersica*, 492–496
  persica var. *scleronucipersica*, 498
  persica var. *scleropersica*, 498
  sativa, 480
Andrographis paniculata, 611
Anguillaria solanacea, 72
Anopheles
  dirus, 641
  maculates, 887
  minimus, 887
  stephensi, 23, 611, 861, 887
  stephensi mosquito, 611
  subpictus, 610, 611
Anthadenia sesamoides, 187
Antiaris toxicaria, 80
Antidesma, 3
  andamanicum, 220
  bunius, 220–224
  bunius var. *cordifolium*, 220
  bunius var. *floribundum*, 220
  bunius var. *genuinum*, 220
  bunius var. *pubescens*, 220
  bunius var. *sylvestre*, 220
  bunius var. *wallichii*, 220
  ciliatum, 220
  colletii, 220
  cordifolium, 220
  crassifolium, 220
  floribundum, 220
  glabellum, 220
  glabrum, 220
  retusum, 220
  rumphii, 220
  stilago, 220
  sylvstre, 220
  thorelianum, 220
Aphanamixis polystachya, 269
Apinus
  koraiensis, 297
  pinea, 304
Ardisia
  bicolor, 65
  compressa, 69
  crenata, 2, 65–70, 74
  crenata var. *bicolor*, 65
  crenulata, 65
  crispa, 65, 68, 69
  crispa var. *taquetii*, 65
  elliptica, 2, 72–75

Ardisia (cont.)
  hainanensis, 72
  henryii, 65
  humilis, 72, 74, 75
  ketoensis, 72
  konishii, 65
  kusukusensis, 65
  labordei, 65
  lentiginosa, 65
  linangensis, 65
  littoralis, 72
  miaoliensis, 65
  pyrgus, 72
  solanacea, 72
  squamulosa, 72, 74
Areca, 107
Areca catechu, 24, 63, 752
Armeniaca
  armeniaca, 442
  bericoccia, 442
  communis, 442
  epirotica, 442
  macrocarpa, 442
  vulgaris, 442
Aronia alnifolia, 358, 362
Artemia salina, 223
Arthrobacter luteus, 21
Artocarpus gomezianus, 903
Ascaris lumbricoides, 678
Aspergillus
  flavus, 317, 607, 687, 710, 764, 859
  fumigates, 161, 640
  niger, 342, 343, 687, 710, 751, 764, 825, 859
  ochraceous, 80
  ochraceus, 342
  parasiticus, 750
  saitoi, 677, 678, 857, 858
Aspergillus spp., 607
Atalantia polyandra, 659
Aucklandia lappa, 693
Aulacia punctata, 871
Aurantiifolia, 742
Aurantium
  acre, 786
  decumanum, 667
  maximum, 667
  medicum, 682
  sinensis, 806
  var. *citrus*, 786
Aurota latifolia, 59
Australasica var. *sanguinea*, 626, 627
Averrhoa, 2
Averrhoa acida, 252

B

Baccaurea, 3, 226, 229
  angulata, 225–227
  bhaswatii, 243
  borneensis, 236
  cauliflora, 248
  dulcis, 228–229
  edulis, 230–231
  flaccida, 248
  glabriflora, 232
  griffithii, 236
  lanceolata, 232–234
  macrocarpa, 236–238
  motleyana, 239–242
  oxycarpa, 248
  pierardi, 248
  polyneura, 238
  propinqua, 248
  pubescens, 239
  pyrrhodasya, 232
  racemosa, 243–247
  ramiflora, 248–250
  sapida, 248
  suvrae, 228
  wallichii, 243
  wrayi, 248
Bacillius
  subtilis, 21, 37, 58, 100, 160, 317, 342, 343, 448, 606, 675, 687, 763, 825, 860, 902, 907
  cereus, 20, 37, 58, 100, 112, 241, 317, 427, 640, 675, 764, 794, 859
  megaterium, 161
  pumilus, 317, 887
  sphaericus, 342, 355
  thuringiensis israelensis, 861
Bacillus spp., 750
Bacopa monnieri, 611
Bacteroides
  distasonis, 856
  fragilis, 431
  uniformis, 856
Bacteroides spp., 751
Bactrocera oleae, 801
Ballota nigra, 13
Barrotia
  gaudichaudii, 124
  macrocarpa, 124
  tetrodon, 124
Batidea idaea, 555
Belou marmelos, 594
Bergera, 3
Bergera kongii, 4
Bifidobacteria, 489, 922
Bilacus marmelos, 594
Bixa orellana, 641
Bladhia
  crenata, 65
  crispa, 65
  crispa var. *taquetii*, 65
  elliptica, 72
  kotoensis, 72
  lentiginosa var. *lanceolata*, 65
  solanacea, 72
Blastocystis hominis, 23
Blattella germanica, 641

Boerhaavia diffusa, 599
Boesenbergia
    pandurata, 639
    rotunda, 641
Boronia, 4
Bothrops atrox, 185
Botrycarpum nigrum, 27
Botrytis cinerea, 675, 880
Brevibacterium fermentans, 21
Brucea javanica, 23
Brugia malayi, 610
Bryantia butyrophora, 117
Bulinus truncates, 23
Bunius sativus, 220
Bursaphelenchus xylophilus, 256

C
Calloselasma rhodostoma, 24
Camellia sinensis, 376
Campylobacter jejuni, 100
Candida
    albicans, 54, 80, 100, 256, 275, 317, 342, 343, 449, 459, 607, 657, 676, 687, 711, 736, 751, 764, 794, 825, 859, 969
    glabrata, 37–38, 58, 275, 302
    krusei, 54
    lipolytica, 38, 58, 80
    lusitaniae, 54
    maltosa, 763
    norwegica, 38, 58
    parapsilosis, 38, 58
    pulcherrima, 54
    tropicalis, 38, 58, 275
    zeylanoides, 38, 58
Candida spp., 37
Capraria integerrima, 187
Carum carvi, 750
Casimiroa edulis, 4
Castanea, 1
    castanea, 6
    prolifera, 6
    sativa, 1, 6–14
    sativa f. *discolor*, 6
    sativa var. *hamulata*, 6
    sativa var. *microcarpa*, 6
    sativa var. *prolifera*, 6
    sativa var. *spicata*, 6
    sativa var. *typica*, 6
    vesca, 6
    vulgaris, 6
Cattleya, 2
Cellulosimicrobium cellulans, 21
Centella asiatica, 638
Cerapu 2 *(Garcinia prainiana)*, 238
Cerasus
    avium, 451
    avium var. *aspleniifolia*, 451
    dulcis, 451
    nigra, 451

Chaenomeles, 3, 378, 517
    lagenaria, 364
    sinensis, 515–520, 699–701, 704, 710
    speciosa, 364–369, 520, 521
Chavica
    maritima, 351
    officinarum, 351
    peepuloides, 351
    retrofracta, 351
Chromobacterium violaceum, 342, 427, 902
Chrysosporium tropicum, 160
Cicca
    acida, 252
    acidissima, 252
    disticha, 252
    emblica, 258
    nodiflora, 252
    racemosa, 252
Cinnamomum zeylanicium, 21
Citreum vulgare, 682
Citrobacter freundii, 869
Citrobacter spp., 751
Citrofortunella
    microcarpa, 865
    mitis, 865
X *Citrofortunella floridana*, 843
X *Citrofortunella swinglei*, 843
Citrus, 3, 627, 645, 798, 840, 989
    abyssinica, 849
    acida, 742
    alata, 682
    albicans, 54, 459
    amara, 786
    amblycarpa, 623–624
    aurantifolia, 747, 750, 814, 815, 823, 825
    aurantiifolia var. *latifolia*, 662
    aurantiifolia var. *tahiti*, 662
    aurantium, 682, 693, 695, 700, 742, 786, 791–802, 806, 846
    aurantium β *sinensis*, 806
    aurantium f. *cyathifera*, 790
    aurantium f. *deliciosa*, 695
    aurantium f. *kabusu*, 790
    aurantium f. var. *aurantium*, 806
    aurantium L. subsp. *sinensis*, 806
    aurantium ssp. *amara*, 786
    aurantium subsp. *aurantium*, 806
    aurantium subsp. *jambhiri*, 846, 862
    aurantium subsp. *sinensis*, 806
    aurantium subsp. *suntra*, 695
    aurantium var. *1*, 695
    aurantium var. *amara*, 792
    aurantium var. *aurantifolia*, 742
    aurantium var. *bigaradia*, 786
    aurantium var. *dulcis*, 766
    aurantium var. *medica*, 682
    aurantium var. *sinensis*, 806
    auraria, 634
    australasica, 625–627
    australis, 629–630

Citrus (cont.)
  bergamia, 825
  bigaradia, 786
  cedrata, 682
  changshanensis, 704
  chrysocarpa, 695
  costata, 667
  crassa, 682
  daoxianensis, 695
  decumana, 667, 755
  decumana var. *paradisi*, 755
  decumana var. *racemosa*, 755
  deliciosa, 695, 800
  depressa, 695
  echinata, 634
  erythrosa, 695
  fortunella, 700
  fragrans, 682
  fusca, 786
  garrawayae orth. var, 631
  garrawayi, 631–633
  grandis, 667–672, 670–672, 674, 676, 677, 700, 723, 725, 755
  grandis var. *pyriformis*, 667
  grandis var. *racemosa*, 755
  grandis var. *sabon*, 667
  guilliermondii, 38
  hyalopulpa, 634
  hystrix, 634–641, 685, 742, 753, 868
  inaequalis, 849
  inconspicua, 38, 58
  inodora, 644–645
  jambhiri, 847, 848
  japonica, 364, 520, 647, 651, 656, 843
  javanica, 742
  junos, 723, 725
  kerrii, 634
  khasia, 695
  kwangsiensis, 667, 682
  latifolia, 662–666
  lemon, 710
  lima, 742
  limetta, 742
  limetta var. *aromatic*, 742
  limon, 621, 701, 748, 766, 800, 813, 848, 854, 856, 858–861
  limonellus, 742
  limonellus var. *amblycarpa*, 623
  limonia, 800, 848, 862
  limonia s. lat., 846
  limonum, 849
  macracantha, 806
  macroptera var. *kerrii*, 634
  madurensis, 695, 865
  madurensis var. *deliciosa*, 695
  maideniana, 644
  mangshanensis, 695
  margarita, 654
  maxima, 667–678
  maxima var. *racemosa*, 755
  maxima var. *uvacarpa*, 755
  medica, 682–688
  medica f. *limon*, 849
  medica f. *monstrosa*, 690
  medica L. var. *digitata*, 690
  medica subf. *aurantiifolia*, 742
  medica subf. *pyriformis*, 667
  medica subsp. *acida*, 742
  medica subsp. *genuina*, 682
  medica subsp. *limonia*, 849
  medica subsp. *limonum*, 849
  medica var. *acida*, 742
  medica var. *ethrog*, 685
  medica var. *limau kasturi*, 865
  medica var. *limon*, 848
  medica var. *limonum*, 849
  medica var. *medica*, 682
  medica var. *sacordactylis*, 692, 693
  medica var. *sarcodactylis*, 685, 690–693
  medica vulgaris, 682
  'Meyer,' 619–622
  meyeri, 619
  microcarpa, 143, 637, 685, 868, 869
  myrtifolia, 700
  nobilis, 623, 695
  nobilis var. *amblycarpa*, 623
  nobilis var. *poonensis*, 695
  nobilis var. *unshiu*, 695
  notissima, 742
  obovoidea, 667
  odorata, 682
  pampelmos, 667
  papeda, 634
  paradisi, 701, 704, 748, 755, 756, 759, 762, 764, 766, 770, 825, 857, 859
  polyandra, 659
  pompelmos, 667
  pompelmos racemosus, 755
  ponki, 695
  poonensis, 695
  pyriformis, 667
  racemosa, 755
  reticulata, 695–715
  reticulata Blanco, 721, 732, 837
  reticulata cv. 'Unshu', 721
  reticulata cv. 'Wenzhou,' 721
  reticulata cv. 'Wenzhou Migan', 721
  reticulata 'Shiranui,' 732–737
  reticulata subsp. *deliciosa*, 695
  reticulata subsp. *tachibana*, 695
  reticulata subsp. *unshiu*, 695
  reticulata Tangor, 837
  reticulata var. *austera*, 695, 865
  reticulata var. *austeres*, 695, 865
  reticulata var. *chrysocarpa*, 695
  reticulate Satsuma Group, 721–729
  sabon, 667
  sarcodactylis Hoola, 690
  sinenis, 703, 704, 710, 732, 735, 748, 750, 756, 764, 771, 800, 806, 812–826, 837, 857, 859, 860, 862, 898
  sphaerocarpa, 737

Scientific Name Index

succosa, 695
suhuiensis, 637, 685, 695, 868
sunki, 695
swinglei, 659
tachibana, 695
tachibana subf. *depressa*, 695
tachibana subf. *ponki*, 695
tachibana subf. *suhuiensis*, 695
tachibana subf. *sunki*, 695
taitensis, 848
tangerina, 695
tankan, 695, 700, 711
trifolia, 893
trifoliata var. *Monstrosa*, 893
triptera, 893
tuberosa, 682
unshiu, 695, 703, 704, 710, 721–729, 735, 817
vangasy, 695
verrucosa, 846
vulgaris, 786
warburgiana, 740
wintersii, 739–741
Citrus *hystrix* ssp. *acida*, 742
Citrus x *acida*, 742
Citrus x *aurantiifolia*, 662, 666, 742–753, 843, 845, 857
Citrus x *aurantiifolia* subsp. *latifolia*, 662
Citrus x *aurantiifolia* subsp. *murgetana*, 742
Citrus x *aurantium*, 667, 678, 695, 742, 755–779, 786–802, 806–827, 832–835, 837–842
Citrus x *aurantium* L. pro parte [Sweet Orange group], 755–779, 806–827
Citrus x *aurantium* f. *grandis*, 667
Citrus x *aurantium* pro sp. [Grapefruit Group] (chapter 91), 755–779
Citrus x *aurantium* pro sp. Tangelo Group(ch 94), 832–835
Citrus x *aurantium* Sour orange group (ch 92), 786–802
Citrus x *aurantium* subsp. *aurantiifolia*, 742
Citrus x *aurantium* subsp. *bergamia*, 849
Citrus x *aurantium* subvar. *amilbed*, 682
Citrus x *aurantium* subvar. *chakotra*, 682
Citrus x *aurantium* Tangor Group (ch 95), 837–842
Citrus x *aurantium* var. *bergamia*, 849
Citrus x *aurantium* var. *decumana*, 667
Citrus x *aurantium* var. *mellarosa*, 849
Citrus x *aurantium* var. *proper*, 742
Citrus x *aurantium* var. *sinensis*, 806
Citrus x *aurantium* var. *tachibana*, 695
Citrus x *bergamia*, 849
Citrus x *bergamia* subsp. *mellarosa*, 849
Citrus x *bergamota*, 849
Citrus x *floridana*, 843–845
Citrus x *jambhiri*, 846, 848
Citrus x *javanica*, 742
Citrus x *latifolia*, 662–666
Citrus x *lima*, 742
Citrus x *limodulcis*, 849
Citrus x *limon*, 620, 849–862
Citrus x *limon*, 666, 846, 848–862

Citrus x *limonia*, 846
Citrus x *limonum*, 849
Citrus x *meyeri*, 620
Citrus x *meyeri*, 849
Citrus x *meyerii*, 619
Citrus x *microcarpa*, 865–870
Citrus x *mitis*, 865
Citrus x *nobilis*, 837
Citrus x *nobilis* subf. *deliciosa*, 695
Citrus x *nobilis* subf. *erythrosa*, 695
Citrus x *nobilis* subf. *reticulate*, 695
Citrus x *nobilis* subf. *succosa*, 695
Citrus x *nobilis* subf. *tachibana*, 695
Citrus x *nobilis* subf. *unshiu*, 695
Citrus x *nobilis* var. *deliciosa*, 695
Citrus x *nobilis* var. *major*, 695
Citrus x *nobilis* var. *ponki*, 695
Citrus x *nobilis* var. *spontanea*, 695
Citrus x *nobilis* var. *sunki*, 695
Citrus x *nobilis* var. *tachibana*, 695
Citrus x *nobilis* var. *unshiu*, 695, 721
Citrus x *nobilis* var. *vangasy*, 695
Citrus x *paradisi*, 755, 756
Citrus x *paradisi* x Citrus *reticulata* (Tangelo group), 832
Citrus x *reticulata* Tangelo, 832
Citrus x *sinensis* subsp. *jambhiri*, 862
Citrus x *taitensis*, 846–848, 862
Citrus x *tangelo*, 832
Citrus x *tangor*, 837
Citrus *maxima* x Citrus sp 1., 742
Citrus *medica* x Citrus *reticulata*, 848, 862
Citrus *medica* x Citrus *reticulate*, 848
Citrus *medica* x Citrus x *limon* x Citrus *reticulate*, 848
Citrus *reticulata* x Citrus *paradise*, 832
Citrus *reticulata* x Citrus *sinensis* tangor, 839
Citrus *reticulata* Citrus *sinensis* var. *Commune*, 702
Citrus *tangerina* x Citrus *paradısi*, 832
Citrus *unshiu* x Citrus *sinensis* × Citrus *reticulata*, 732
Citrus *unshiu* Marc. x Citrus *unshiu* Osbeck × Citrus *reticulata*, 735
Cladosporium fulvum, 213
Clausena
    lansium, 4, 871–880
    punctata, 871
    wampi, 871
Clostridium
    butyricum, 856
    perfringens, 431
Clostridium spp., 751
Clymenia, 3
Clymenia *polyandra*, 4, 659
Coccomelia *racemosa*, 243
Coleus *ambonicus*, 599
Collinsela *aerofaciense*, 431
Colubrina *asiatica*, 317
Commiphora *molmol*, 715
Connaropsis
    acuminata, 115
    diversifolia, 115
    grandiflora, 115

Cookia
  punctata, 871
  wampi, 871
Coprinus comatus, 880
Coptis chinensis, 19
Cormus
  domestica, 590
  mas, 48, 54, 549, 562
Corynebacterium, 449
Corynebacterium diptheriae, 448
Corypha laevis, 136
Cotoneaster, 3
Crataegus, 3
Crataegus bibas, 381
Crataeva marmelos, 594
Crategus, 441
Crateva religiosa, 594
Cryptococcus neoformans, 100, 302
Cubeba officinalis, 311
Culex
  pipiens, 23, 861
  pipiens pallens, 344, 345
  quinquefasciatus, 355, 641, 713, 825, 887
  tritaeniorhynchus, 288, 610, 611
Curculigo, 1
  latifolia, 1, 59, 61, 63
  latifolia var. *latifolia*, 63
Curcuma longa, 641, 693
Cuscuta reflexa, 269
Cydonia, 3, 378, 678
  communis, 371
  cydonia, 371
  europaea, 371
  indica, 594
  japonica var. *lagenaria*, 364
  lagenaria, 364, 366
  oblonga, 371–378
  sinensis, 515
  speciosa, 364
  vulgaris, 371
Cynips quercufolii, 17
Cynodon dactylon, 599
Cyperus rotundus, 693

D
Debaryomyces hansenii, 711, 764, 794, 825, 859
Dendrobium, 2
Diasperus emblica, 258
Dichelactina nodicaulis, 258
Dieffenbachia spp., 665
Dione juno, 171
Diplolepis gallae tinctoriae, 17
Drosophila melanogaster, 335
Druparia avium, 451
Durio kutejensis, 238
Dysosmia
  ciliata, 166
  fluminensis, 166
  foetida, 166
  gossypifolia, 166
  polyadena, 166
Dysosmon amoenum, 187

E
Echinochloa crusgalli, 160, 161
Edwardsiella tarda, 869
Eleutherococcus senticosis, 275
Embelia, 2
  arborea, 258
  officinalis, 263, 265, 267, 269, 273, 277, 281–285, 287, 288
Entada phaseoloides, 869
Entamoeba histolytica, 23, 915
Enterobacter
  aerogenes, 448
  aerogenesis, 902
  amnigenus, 765, 859
  faecalis, 20
  gergoviae, 765, 859
Enterococcus, 431
  faecalis, 275, 317, 687, 825
  faecium, 825
Ephedra sinica, 800
Epidendrum, 2
Epidendrum vanilla, 106
Epidermophyton floccosum, 859
Eremocitrus, 3, 741
Eriobotrya, 3
Eriobotrya japonica, 381–392
Erwinia amylovora, 678
Escherichia coli, 19–21, 37, 58, 100, 112, 113, 160, 161, 171, 241, 256, 275, 276, 317, 427, 431, 448, 533, 606, 675, 687, 711, 750, 751, 763–765, 793, 794, 825, 859, 860, 869, 907, 940, 963, 965, 972
Eubacterium
  limosum, 431
  rectale, 489
Eugenia jambolana, 600, 602

F
Fagara
  podocarpa, 904
  setosa, 904
Fagus
  castanea, 6
  castanea var. *variegate*, 6
  procera, 6
  f. *albiplena*, 412
  f. *edulis*, 148
Feronia, 888
  elephantum, 884
  limonia, 884, 886
  pellucid, 594
Ficus, 535
Ficus pyrifolia, 535
Filipendula, 3

f. *nodiflorus*, 252
Foeniculam vulgare, 597
Fortunella, 3, 753, 898
   crassifolia, 651
   japonica, 647, 649
   japonica var. *margarita*, 654, 657
   margarita, 654, 656–658, 845
   margarita var. *margarita*, 654
   polyandra, 659
   sagittifolia, 634
   swinglei, 659
Fortunella spp., 648, 652, 656, 843, 845, 898
X *Fortunella* spp., 843
Fortunella x *crassifolia*, 651
Fragaria, 3, 399
   ananas, 575
   ananassa, 48
   bonariensis, 395
   calyculata, 395
   caroliniensis, 395
   chilensis β *ananassa*, 395
   chilensis δ *tincta*, 395
   chilensis γ *calyculata*, 395
   chiloensis, 395, 396, 398
   chiloensis var. *ananassa*, 395
   grandiflora, 395
   hybrida, 395
   moschata, 398
   tincta, 395
   vesca, 395, 398
   vesca ananassa, 395
   vesca ε *ananas*, 395
   vesca var. *sativa*, 395
   virginiana, 396
Fragaria x *ananassa*, 395–407, 549, 560, 561
Fragaria x *cultorum*, 395
Fragaria x *magna*, 395
Fraxinus, 2
f. *stapfiana*, 535
Fusarium
   graminearum, 342
   oxysporum, 161, 432, 880
   udum, 607

G

Garcinia mangostana, 20
Gatnaia annamica, 248
Gingko biloba, 600
Glycosmis pentaphylla, 4
Glycyrrhiza
   glabra, 281
   uralensis, 19, 392
Grossularia
   nigra, 27
   reclinata, 51
   uva, 51
   uva-crispa, 51
   vulgaris, 51

H

Haemaphysalis bispinosa, 288, 613
β-Haemolytic Streptococci, 751
Hanseniaspora guilliermondii, 711, 764, 794, 825, 859
Hedycarpus lanceolatus, 232
Helicobacter pylori, 20, 37, 100, 403, 429, 551, 897
Hibiscus rosa-sinensis, 63
Holarrhena antidysenterica, 19
Hombronia edulis, 124
Houttuynia cordata, 268
Hypoxis, 1

I

Icacorea solanacea, 72
Indigofera, 687
Indoplanorbis exustus, 319
Ipomoea digitata, 599

K

Kaempferia
   galanga, 19
   parviflora, 268
Kerria, 3
Klebseilla
   aerogenes, 342
   pneumonia, 20, 100, 112, 113, 160, 171, 275, 276, 687, 711, 751, 764, 794, 825, 859, 860
Klebseilla spp., 751
Kluyveromyces fragilis, 711, 764, 794, 825, 859

L

Lactobacillus, 431
   acidophilus, 19
   curvatus, 765, 859
   sakei, 765, 859
Lagerstroemia speciosa, 269
Languas galanga, 639
Lavandula officinalis, 550
Leishmania donovani, 318, 355
Limon x *vulgaris*, 849
Limonia
   acidissima, 4, 742, 884–888
   aurantiifolia, 742
   diacantha, 900
   trichocarpa, 893
   trifolia, 900
Limonia x *aurantiifolia*, 742
Linusm usitatissium, 21
Listeria monocytogenes, 20, 459, 640, 711, 764, 794, 859
Litsea
   chinensis, 317
   cubeba, 641
Luffa cylindrica, 693
Lutzomyia youngi, 860

M

Macaranga nicobarica, 132
Maesa, 2
Mallotus pallidus, 903
Malus, 3, 414
   communis, 413
   dasyphylla, 413
   dasyphylla var. *domestica*, 413
   domestica, 413–432
   domestica subsp. *pumila*, 413
   domestica var. *spectabilis*, 410
   malus, 413
   microcarpa var. *spectabilis*, 410
   niedzwetzkyana, 413
   pumila, 261, 413
   pumila auct. var. *domestica*, 413
   sorbus, 590
   spectabilis, 410–412
   sylvestris, 413, 414
   sylvestris subsp. *mitis*, 413
   sylvestris var. *domestica*, 413
Mangifera
   foetida, 238
   indica, 263, 266
   odorata, 238
Mappa borneensis, 236
Melanobatus, 570
   neglectus, 570
   occidentalis, 570
Melissa officinalis, 550, 860
Merrillia, 3
Merrillia caloxylon, 890–892
Mespilus, 3, 441
   domestica, 590
   germanica, 437–441
   japonica, 381
   sylvestris, 437
Microcitrus, 3
   australasica, 627
   australasica var. *astralasica*, 625, 626
   australis, 629
   garrawayae orth. var, 631
   inodora, 644, 645
   maideniana, 644
   papuana, 739
   warburgiana, 740
Micrococcus
   flavus, 763
   luteus, 448, 711, 764, 794, 859
Microsporum
   canis, 473
   gypseum, 160, 859
Mirobalanus embilica, 258
Molineria
   latifolia, 59–63
   latifolia (Curculigo latifolia), 1, 61
Momordica charantia, 599, 602
Monochoria vaginalis, 161
Morinda citrifolia, 268, 599
Mortierella alpine, 204

Muldera
   multinervis, 322
   wightiana, 322
Murraya, 3, 638
Murraya caloxylon, 890
Musca domestica, 613
Mus musculus, 272, 273
Mycobacterium
   fortuitum, 448
   smegmatis, 343, 448, 711, 764, 794, 859
Mycosphaerella arachidicola, 880
Myrobroma fragrans, 106
Myrsine, 1, 2
Myrtus communis, 22

N

'Nagami' x *Citrus japonica*, 651
Naja
   kaouthia, 23, 24, 287
   naja kaouthia, 24
Neisseria gonorrhoeae, 275, 276, 751, 938
Nelumbo nucifera, 613
Neostylopyga rhombifolia, 641
Nilaparvata lugens, 613
Notylia planifolia, 106

O

Ocimum sanctum, 599
Oerskovia xanthineolytica, 21
Olea. europaea, 2, 82–101
Olea
   ferruginea, 101
   laperrinii, 101
   officinarum, 82
   pallida, 82
Orchis, 2
Oroxylum indicum, 269
Oxalis, 2
Oxanthera, 3
Oxyacanthus uva-crispa, 51

P

Pandanas tectorius, 145
Pandanus, 124, 134, 139, 140, 142–144
   absonus, 136
   adscendens, 136
   aequor, 136
   aitutakiensis, 136
   alloios, 136
   andamanensium, 124
   bagea, 124
   bergmanii, 136
   bicurvatus, 136
   bidoer, 124
   bidur, 124
   brosimos, 130
   butyrophorus, 117

ceramicus, 117
chamissonis, 136
citraceus, 136
collatus, 136
cominsii, 117
cominsii var. *augustus*, 117
cominsii var. *micronesicus*, 117
complanatus, 136
compressus, 124
conoideous, 117–122
convexus, 134, 137
cooperi, 136
coronatus, 136
crassiaculeatus, 136
crassus, 136
cylindricus, 136
cylindricus var. *sinnau*, 136
cymatilis, 136
darwinensis, 134
darwinensis var. *darwinensis*, 134
darwinensis var. *latifructus*, 134
dubius, 124–126
dubius var. *compressus*, 124
englerianus, 117
erythros, 117
eyesyes, 136
fahina, 136
fischerianus f. *bergmanii*, 136
fischerianus f. *bryanii*, 136
fischerianus var. *bryanii*, 136
fischerianus var. *rockii*, 136
grantii, 136
guamensis, 136
haapaiensis, 136
heronensis, 136
hivaoaensis, 136
hollrungii, 117
hollrungii f. *caroliniana*, 117
hombronia, 117
horneinsularum, 137
indicus, 131
integer, 134
julianettii, 128–130
kafu var. *confluentus*, 124
latericius, 117
latifolius, 124
latifructus, 134
latissimus, 124
leram, 124, 131–133
leram var. *macrocarpus*, 131
macrocarpus, 124
magnificus, 117
mellori, 131
minusculus, 117
odoratissimus var. *suvaensis*, 137
odoratus, 124
oronatus f. *minor*, 136
pacificus, 124
plicatus, 117
rubber, 113
semiarmatus, 134
spiralis, 134–135
spiralis var. *convexus*, 134
spiralis var. *flammeus*, 134
spiralis var. *multimammillatus*, 134
spiralis var. *septemlocularis*, 134
spiralis var. *spiralis*, 134
spiralis var. *thermalis*, 134
subumbellatus, 117
sylvestris, 117
tahitensis var. *exiguous*, 138
tahitensis var. *niueana*, 138
tectorius, 136–145
tectorius var. *zollingeri*, 138
temehaniensis, 138
tetrodon, 124
thermalis, 134
yamagutii, 124
yirrkalaensis, 134
yorkensis, 138
yunckeri, 138
Papeda rumphii, 634
Paracoccidioides brasiliensis, 161
Paralichthys olivaceus, 737
Paramphistomum cervi, 288, 613
Passiflora
   alata, 155, 156, 158
   ciliata var. *polyadena*, 166
   ciliata var. *quinqueloba*, 166
   ciliata var. *riparia*, 166
   ciliate, 166
   coccinea, 178–180
   diaden, 147
   edulis, 147–162
   edulis 'edulis', 159
   edulis f. *edulis*, 149, 151
   edulis f. *flavicarpa*, 149, 151, 153, 155, 158, 159
   edulis var. *edulis*, 154
   edulis var. *flavicarpa*, 154, 155, 157
   edulis var. *pomifera*, 147
   edulis var. *rubricaulis*, 147
   edulis var. *verrucifera*, 147
   foetida, 157, 166–172
   foetida forma *latifolia*, 166
   foetida forma *longifolia*, 166
   foetida forma *suberecta*, 166
   foetida var. *albiflora*, 170
   foetida var. *balansae*, 166
   foetida var. *galapagensis*, 166
   foetida var. *gardneri*, 166
   foetida var. *glaziovii*, 166
   foetida var. *gossypifolia*, 166
   foetida var. *hastate*, 166
   foetida var. *hibiscifolia*, 166
   foetida var. *hirsute*, 166
   foetida var. *hirsutissima*, 166
   foetida var. *hisbiscifolia*, 169
   foetida var. *hispida*, 166, 169
   foetida var. *subpalmata*, 166
   gossypiifolia, 166

Passiflora (cont.)
  grandiflora, 181
  gratissima, 147
  incarnate, 162
  iodocarpa, 147
  ligularis, 148, 174–177, 174–180
  lowei, 174
  macrocarpa, 181
  middletoniana, 147
  miniata, 178–180
  minima, 147
  pallidiflora, 147
  picroderma, 147
  polyadena, 166
  pomifera, 147
  quadrangularis, 181–186
  quadrangularis var. *variegata*, 181
  rigidula, 147
  rubricaulis, 147
  serratistipula, 174
  sulcata, 181
  tetragona, 181
  tiliaefolia, 174
  variegate, 166
  vernicosa, 147
  verrucifera, 147
  vesicaria, 166
Pediculus humanus capitis, 773
Peltophorum pterocarpum, 19, 20
Penicillium
  chrysogenum, 710, 764, 859
  digitatum, 764, 825
  expansum, 432, 675
  italicum, 825
  verrucosum, 710, 711, 764, 859
  viridicatum, 317, 342
Penicillium spp, 342, 343
Pentace burmanica, 24
Peperomia, 3
Periplaneta Americana, 641
Persica
  ispahanensis, 498
  platycarpa, 498
  vulgaris, 492, 498
  vulgaris var. *compressa*, 498
Phaedon cochleariae, 613
Phaeoramularia angolensis, 666
Phalaenopsis, 2
Phaseolus vulgaris, 62
Philodendron spp, 665
Photinia, 3
Photinia japonica, 381
Phyllanthus, 752
  acidus, 252–256
  cicca, 252
  cicca var. *bracteosa*, 252
  cochinchinensis, 252
  distichus, 252
  emblica, 258–290
  fluitans, 3

  glomeratus, 258
  longifolius, 252
  mairei, 258
  mimosifolius, 258
  niruri, 599
Physalis angulata, 316
Physalospora piricola, 880
Picrorrhiza kurroa, 600
Pierandia
  racemosa, 243
  dulcis, 228
  flaccid, 248
  macrocarpa, 236
  motleyana, 239
  pyrrhodasya, 232
  sapida, 248
Pine nut (*Pinus pinea*), 307, 308
Pinus, 3
  armandii, 302, 309
  cembra β-*excelsa*, 297
  cembra var. *manchurica*, 297
  cembra var. *mandschurica*, 297
  domestica, 304
  esculenta, 304
  fastuosa, 304
  gerardiana, 302, 309
  halepensis, 308
  koraiensis, 297–302, 309
  maderiensis, 304
  mandschurica, 297
  maritima, 11
  pinaster, 308
  pinea, 298, 302, 304–309
  pinea var. *maderiensis*, 304
  prokoraiensis, 297
  sativa, 304
  Sibirica, 302, 309
  strobes, 297
  umbraculifera, 304
Piper
  aromaticum, 322
  betel, 280
  brachystachyum, 332
  chaba, 351, 353–355
  cubeba, 311–320, 332, 333, 344, 355
  guineense, 333, 338
  longum, 264, 282, 319, 328, 332, 334, 338, 343, 345, 351
  nigrum, 3, 264, 282, 315, 322–346, 356, 641, 3315
  nigum, 338, 343
  officinarum, 351
  retrofractum, 318, 351–356
  trioicum, 322
  umbellatum, 333, 338
Pirus
  communis β *hortensis*, 527
  sorbus, 590
Pithecellobium dulce, 24
Plasmodium

Scientific Name Index

berghei, 677
falciparum, 69, 75, 288, 713
vinckei, 677
yoelii, 677
Pleurocitrus inodora, 644
Plutella xylostella, 909
Polyandra, 659–660
Polygonum, 687
Poncirus, 3, 893, 896
   trifoliata var. *monstrosa*, 893
   trifoliate, 893–898
   trifoliata, 893–898
Porphyromonas spp, 751
Potentilla, 3, 395
Potentilla x ananassa, 395
Propionibacterium acnes, 640, 657, 736
Proteus
   mirabilis, 276, 448, 751, 763
   vulgaricus, 241
Proteus spp., 750
Prunus, 3
   amygdalus, 480
   amygdalus var. *amara*, 480
   amygdalus var. *dulcis*, 480
   amygdalus var. *fragilis*, 480
   amygdalus var. *sativa*, 480
   armeniaca, 442–450
   armeniaca L.var. *vulgaris*, 442
   avium, 48, 451–461, 564
   avium var. *aspleniifolia*, 451
   avium var. *sylvestris*, 451
   campanulata, 452
   cerasifera, 464
   cerasus, 48
   cerasus var. *avium*, 451
   communis, 463, 480
   communis var. *dulcis*, 480
   communis var. *fragilis*, 480
   communis var. *sativa*, 480
   domestica, 463–474, 477, 478
   domestica L. subsp. *insititia*, 476
   domestica L. var. *insititia*, 476
   domestica ssp. *domestica*, 474
   domestica ssp. *institia*, 476–478
   domestica ssp. *italica*, 474
   domestica subsp. *insititia*, 476–478
   domestica subsp. *oeconomica*, 463
   domestica var. *damascene*, 463
   dulcis, 3, 480–490
   dulcis var. *amara*, 580
   insititia, 478
   insititia L., 476
   macrophylla, 451
   nucipersica, 492
   persica, 474, 498, 504, 505
   persica (L) Batsch var. *persica*, 498
   persica nucipersica, 492
   persica subsp. *platycarpa*, 498
   persica var. *compressa*, 498
   persica var. *nectarine*, 492
   persica var. *nucipersica*, 492–496
   persica var. *platycarpa*, 498
   pleioceracus, 452
   salicina, 509–514
   salicina var. *pubipes*, 514
   salicina var. *salicina*, 514
   sativa L. subsp. *domestica*, 463
   species, 460
   spinosa, 464
   thibetica, 509
   tiliaefolia, 442
   triflora, 509
Pseudaegle
   sepiaria, 893
   trifoliate, 893
Pseudocydonia sinensis, 515–521
Pseudomonas, 431, 729, 750, 793
   aeruginosa, 20, 21, 37, 75, 100, 112, 113, 160, 161, 171, 241, 276, 281, 317, 342, 343, 459, 606, 687, 711, 751, 764, 794, 825, 859, 869, 902
   putida, 171
   solanacearum, 317
Pseudomonas spp., 750
Psidium guajava, 19, 641
Punica granatum, 19, 20, 263
Pyracantha, 3
Pyrenia sorbus, 590
Pyrus
   asiae-mediae, 527, 540
   balansae, 527
   bourgaeana, 527
   bretschneideri, 523–525
   cathayensis, 515
   chinensis, 515
   communis, 463, 480, 527–533, 536, 540
   communis subsp. *bourgaeana*, 527
   communis subsp. *caucasica*, 527
   communis subsp. *pyraste*, 527
   communis var. *mariana*, 527
   communis var. *pyraster*, 527
   communis var. *sativa*, 527
   cydonia, 371
   domestica, 527, 590
   elata, 527
   lindleyi, 540
   malus, 413
   malus var. *pumila*, 413
   medvedevii, 527
   montana, 535
   ovoidea, 540
   pyraster, 527
   pyrifolia, 525, 535–539, 537
   sativa, 527
   serotina, 523, 535, 536
   serotina var. *stapfiana*, 535
   simonii, 540
   sinensis, 515, 535, 540
   sinensis α *ussuriensis*, 540
   sinensis var. *asiae-mediae*, 540

Pyrus (cont.)
   sinensis var. *ussuriensis*, 540
   singkiangensis, 525
   sinkiangensis, 533, 541
   sogdiana, 540
   sorbus, 590
   spectabilis, 410
   sylvestris, 527
   ussuriensis, 540–542
   ussuriensis var. *chinesis*, 523
   ussuriensis var. *ovoidea*, 542
   ussuriensis var. *viridis*, 542
Pyrus sp. nr. *communis*, 525
Pyrus x *bretschneideri*, 523, 525

Q

Quercus, 1, 22
   infectoria, 1, 16–24
   infectoria ssp. *euinfectoria*, 16
   lusitania var. *infectoria*, 23
   lusitanica ssp. *infectoria*, 16
   lusitanica var. *infectoria*, 16
   suber, 1
Quinaria lansium, 871

R

Reshni, 815
Rhipicephalus (Boophilus) microplus, 613
Rhizoctonia
   cerealis, 909
   solani, 161, 909
Rhizopus stolonifer, 675
Rhizotonia solani, 160
Rhodiola rosea, 275
Rhodohypoxis, 1
Rhodotorula rubra, 711, 764, 794, 825, 859
Rhodotypos, 3
Ribes, 1, 30, 32, 33, 34, 45, 46, 48, 54, 55, 58
   divaricatum, 56, 57
   domesticum, 43
   glossularia, 48, 54, 549, 562
   grossularia, 30, 45, 54, 56
   nigrum, 27–40, 48, 54, 56, 562
   nigrum, 30, 31, 33, 36, 38, 39, 45, 46, 54, 56, 57
   nigrum forma *chlorocarpum*, 27
   nigrum var. *chlorocarpum*, 27
   nigrum var. *europaeum*, 27
   nigrum var. *pauciflorum*, 27
   nigrum var. *sibiricum*, 27
   olidum, 27
   pauciflorum, 27
   pubescens, 51
   reclinatum, 51
   rubrum, 30, 31, 34, 43–49, 54
   rubrum L. var. *sativum*, 43
   rubrum subsp. *vulgare*, 43
   sativum, 43, 48, 54, 549, 562
   uva-crispa, 30, 51–57
   vulgare, 43
   vulgare var. *macrocarpum*, 43
   vulgare var. *sylvestre*, 43
Ribes x *nidigrolaria*, 54, 56–58
Ribes x *pallidum*, 32
Rosa, 3
   canina, 592
   damascene, 22
   hybrid, 316
Rourea diversifolia, 115
Roussinia indica, 131
Rubus, 3, 32, 34, 48, 544–546, 548, 549, 551, 557, 571, 572, 588
   acuminatus, 550
   alleghaniensis, 545
   anglocandicans, 544
   argutus, 545
   armeniacus, 3
   chamaemorus, 545
   cissburiensis, 544
   echinatus, 544
   erythrops, 544
   fructicosus, 48, 54, 549, 562, 575
   fruticosus, 544, 548, 550, 552
   fruticosus aggr, 544–552, 589
   fruticosus L. aggr, 544
   idaeus, 3, 34, 48, 54, 401, 459, 545, 549, 550, 555–567, 578, 581, 584, 587
   idaeus subsp. *vulgates*, 555
   idaeus var.*americanus*, 570
   jamaicensis, 550
   laciniatus, 3, 544, 560, 583
   leightonii, 544
   leucostachys, 544
   loganobaccus, 581, 587–589
   loganobaccus cv. *boysenberry*, 581
   michiganus, 570
   neglectus, 570, 578
   nitidioides, 545
   occidentalis, 3, 545, 560, 561, 570–578, 584
   phaeocarpus, 544
   plicatus, 117, 544
   polyanthemus, 544
   racemosus, 550, 551
   riddelsdellii, 544
   rosifolius, 550, 551
   rubritinctus, 544
   strigosus, 558
   thrysiger, 545
   ulmifolius (= *Rubus* rusticanus), 545
   ulmifolius, 544, 545, 550–552
   ulmifolius var. *anoplothyrsus*, 544
   ulmifolius var. *ulmifolius*, 544
   ursinus, 545, 560, 583, 587
   ursinus var. *loganobaccus*, 581, 587
   ursinus var *loganobaccus* cv *Boysenberry*, 581
   vestitus, 544
Rubus x *loganobaccus*, 581, 587–589
Rubus idaeus x *occidentalis*, 570
Rubus loganobaccus x *baileyanus*, 581, 583, 584
Rubus occidentalis x *strigosus*, 558, 570

Rubus spp., 3, 32, 34, 48, 544, 545, 549, 551, 571
Rubus ursinus x *idaeus*, 560
Rubus ursinus x *idaeus* Boysenberry, 581–586, 588
Rubus ursinus x *Rubus idaeus*, 587
Rume x *maritimus*, 269
Ruta, 4

S

Saccharomyces
    boulardii, 13
    cerevesiae, 254
    cerevisiae var. *sake*, 640
    lipolytica, 80
Salmonella, 20, 75, 473, 750, 751, 825, 869
    choleraeasuis, 902
    enterica serovar Typhimurium, 765
    enteritidis, 675, 764
    paratyphi, 160, 751
    paratyphi A, 276, 448
    paratyphi B, 275, 276, 448
    typhi, 112, 113, 160, 161, 276, 448, 533, 606, 860
    typhimurium, 21, 273, 459, 764–766, 795
Salmonella spp., 20, 750, 751
Salvia miltiorrhiza, 392
Sambucus ebulus, 13
Santalodes diversifolium, 115
Santalum album, 600
Sapindus mukorossi, 275, 276
Sapium crassifolium, 220
Saraca asoka, 269
Sarcina lutea, 161
Sarcodactilis helicteroides, 690
Sarcotheca
    acuminata, 115
    diversifolia, 2, 115–116
Sarcotheca
    glauca, 116
    griffithii, 116
    macrophylla, 116
    rubrinervis, 116
    subtriplinervis, 115
Sauropus androgynus, 637
Schinifolium, 905, 910
Schinus limonia, 884
Schistosoma mansoni, 318, 715
Scorodocarpus borneensis, 2, 77–81
Semecarpus anacardium, 271, 272, 282, 283, 289
Serratia marcescens, 37, 58, 112, 276, 763
Serratia spp., 751
Serviceberry, 3, 49, 358
Sesamum
    africanum, 187
    alatum, 193
    angustifolium, 193
    brasiliense, 187
    indicum, 2, 187–213
    luteum, 187
    malabaricum, 187
    mulayanum, 187

    oleiferum, 187
    orientale, 187
    radiatum, 193
    trifoliatum, 187
Shigella
    boydii, 161
    dysenteriae, 161, 448, 520, 606, 607
    flexneri, 20, 171, 275, 533, 606, 607, 751, 765
    flexneri, 171
    sonnei, 275
Shigella spp., 606
Sinocitrus reticulata, 695
Sarcotheca ochracea, 116
Sonneratia punctata, 871
Sorbus, 3
    cydonia, 371
    domestica, 590–593
    edulis, 590
    pyrus, 527
Spiraea, 3
Spodoptera litura, 345
Staphycoccous
    aureus, 13, 19–21, 37, 54, 58, 69, 80, 100, 160, 161, 241, 256, 275, 371, 342, 427, 431, 448, 449, 459, 520, 533, 551, 606, 640, 675–676, 687, 711, 750, 752, 763, 764, 793, 794, 825, 859–860, 902, 907
    carnosus, 765, 859
    epidermidis, 21, 54, 58, 657, 736
    epidermis, 448
    xylosus, 765, 859
Staphylococcus spp., 750
Stilago bunius, 220
Streptococcus, 171, 431
    agalatiae, 869
    faecalis, 112, 317, 448, 751
    feacalis, 113
    mitis, 317
    mutans, 19, 20
    pyogenes, 20, 171, 448
    salivarius, 19, 317
    sanguis, 19
α-*Streptococcus hemolyticus*, 69
Streptococcus spp. 869
Strobus koraiensis, 297
Syzygium
    aromaticum, 23, 263, 288
    cumini, 601
    jambos, 238
    malaccense, 238

T

Tectona grandis, 599
Terminalia
    arjuna, 269
    belerica, 264, 266, 270, 275, 287, 288
    bellerica, 264, 273, 280, 282, 288
    bellirica, 289
    chebula, 264, 266, 268, 270, 273, 275, 280, 282, 287, 288, 289

Thymus
  serpyllum., 550
  vulgaris, 750
Tinospora cordifolia, 278, 279, 281, 599, 600
Tinus
  humilis, 72
  squamulosa, 72
Trachyspermum ammi, 597
Tribulus terrestris, 319
Tricarium cochinchinense, 252
Trichoderma harzianum, 161
Trichoderma spp., 343
Trichophyton
  mentagrophytes, 859, 880
  simmi, 473
  terrestre, 160
Triphasia
  aurantiola, 900
  trifoliate, 900
  trifolia, 4, 900–903
Tripsilina foetida, 166
Trypanosoma
  brucei brucei, 825
  cruzi, 317
  evansi, 825
Tylophora indica, 599
Typhimurium, 765

U
Uncaria gambir, 19
Upaca, 3
Urtica dioica, 550

V
Vaccinium, 32, 34, 48
  macrocarpon, 549
  myrtillus, 34, 562
  vitis-idaea, 34, 562
Vaccinium spp, 549
Vanilla, 2
  aromatica, 106
  bampsiana, 106
  carinata, 106
  domestica, 106
  duckei, 106
  epidendrum, 106
  fragrans, 106, 110, 113
  majaijensis, 106
  mexicana, 106
  planifolia, 2, 106–113
  rubra, 106
  sativa, 106
  sylvestris, 106
  tahitensis, 111
  viridflora, 106
Vernonia amygdalina, 751
Vibrio
  cholerae, 171, 275, 606, 751, 869, 887
  harveyi, 764
  mimicus, 161
  parahemolyticus, 161
  parahemolytics, 869
Vinca rosea, 337
Vipera russellii, 287
Vitex negundo, 599, 600
Volkameria
  orientalis, 187
  sesamoides, 187

W
Walsura robusta, 19

X
Xanthomonas campestris, 887
Xanthosoma spp, 665
Ximenia, 2
Ximenia borneensis, 77

Y
Yersinia enterocolitica, 869

Z
Zanthoxylum, 4
  acanthophyllum, 904
  ailanthoides, 910
  argyi, 904
  armatum, 910
  bungeanum, 904–910
  bungei, 904
  bungei var. *inermis*, 904
  fraxinoides, 904
  piperitum, 910
  podocarpum, 904
  schinifolium, 905, 910
  setosum, 904
  simulans, 4, 904–910
  simulans var. *podocarpum*, 904
Zingiber officinale, 264, 282, 641
Zingiber spp.,356